TROPICAL RAIN FOREST IN SOUTHEAST ASIA

At one time, throughout the world, tropical forests cloaked regions that were, collectively, twice the size of Europe. For ten thousand years or more, the forests endured in rich complexity, as the homes of an estimated fifty to ninety percent of the world's land-dwelling species.

In less than four decades, we destroyed more than half of these ancient forests, and most of their spectacular arrays of species may be lost forever. With each passing year, we log over an additional 38 million acres for fuel, lumber, and other products. That is the equivalent of leveling thirty-four city blocks every minute. At this writing, the human population has exceeded 5.6 billion. It is still growing, exponentially, and so is the demand for forest products.

Rampant deforestation extends beyond the tropics. Today, highly mechanized logging operations are proceeding in temperate forests throughout the United States, Canada, Europe, Siberia, and elsewhere.

Biologists rightly decry the mass extinction, the assault on species diversity, the depletion of a major portion of the world's genetic reservoir. But something else is going on here. Too many of us experience a vague sense of unease when we hike or drive through a logged-over part of our own country, when we see before-and-after photographs of wholesale destruction. Is it because we are losing the comfort of our evolutionary heritage, a connection with our own past?

Many millions of years ago, our primate ancestors moved into the trees of tropical forests. Through countless generations, novel connections formed among increasing numbers of cells in their brains; and vision, hearing, and other senses became highly responsive to information-rich, arboreal worlds.

Does our neural wiring still resonate with rustling leaves, with shafts of light and mosaic shadows? Are we innately attuned to the forests of Eden—or have time and change buried recognition of home?

General Advisors / Contributors

JOHN ALCOCK
Arizona State University

AARON BAUER
Villanova University

ROBERT COLWELL
University of Connecticut

GEORGE COX
San Diego State University

PAUL HERTZ
Barnard College

DANIEL FAIRBANKS
Brigham Young University

JOHN JACKSON
North Hennepin Community College

EUGENE KOZLOFF
University of Washington

ROBERT LAPEN
Central Washington University

WILLIAM PARSON
University of Washington

CLEON ROSS
Colorado State University

SAMUEL SWEET
University of California, Santa Barbara

STEPHEN WOLFE
University of California, Davis

Developmental Editor

BEVERLY MCMILLAN

Primary Biological Illustrator

RAYCHEL CIEMMA

BIOLOGY

THE UNITY AND DIVERSITY OF LIFE

SEVENTH EDITION

CECIE STARR

BELMONT, CALIFORNIA

RALPH TAGGART

MICHIGAN STATE UNIVERSITY

WADSWORTH PUBLISHING COMPANY

I(T)P™ AN INTERNATIONAL THOMSON PUBLISHING COMPANY

Belmont • Albany • Bonn • Boston
Cincinnati • Detroit • London • Madrid
Melbourne • Mexico City • New York
Paris • San Francisco • Singapore
Tokyo • Toronto • Washington

BIOLOGY PUBLISHER: Jack C. Carey

ASSISTANT EDITOR: Kristin Milotich

EDITORIAL ASSISTANT: Kerri Abdinoor

PRINT BUYER: Randy Hurst

PRODUCTION SERVICES COORDINATOR: Sandra Craig

PRODUCTION: Mary Douglas, Rogue Valley Publications

TEXT AND COVER DESIGN, ART DIRECTION: Gary Head, Gary Head Design

EDITORIAL PRODUCTION: Myrna Engler-Forkner, Melissa Andrews, Rosaleen Bertolino, Marilyn Evenson, Susan Gall, Ed Serdziak, Karen Stough

ARTISTS: Raychel Ciemma, Robert Demarest, Hans & Cassady, Inc. (Hans Neuhart), Darwen Hennings, Vally Hennings, Betsy Palay, Precision Graphics (Jan Flessner), Nadine Sokol, Kevin Somerville, Lloyd Townsend

PHOTO RESEARCH AND PERMISSIONS: Marion Hansen

COVER PHOTOGRAPH: © Frans Lanting/Minden Pictures

COMPOSITION: American Composition & Graphics, Inc. (Jim Jeschke, Jody Ward, and Valerie Norris)

COLOR PROCESSING: H & S Graphics, Inc. (Tom Anderson, Nancy Dean, and John Deady)

PRINTING AND BINDING: R. R. Donnelley & Sons Company/Willard

BOOKS IN THE WADSWORTH BIOLOGY SERIES

Biology: Concepts and Applications, 2nd, Starr

Biology: The Unity and Diversity of Life, 7th, Starr and Taggart

Human Biology, Starr and McMillan

Laboratory Manual for Biology, Perry and Morton

Introduction to Microbiology, Ingraham and Ingraham

Living in the Environment, 8th, Miller

Environmental Science, 5th, Miller

Sustaining the Earth, Miller

Environment: Problems and Solutions, Miller

Introduction to Cell and Molecular Biology, Wolfe

Molecular and Cellular Biology, Wolfe

Cell Ultrastructure, Wolfe

Marine Life and the Sea, Milne

Essentials of Oceanography, Garrison

Oceanography: An Invitation to Marine Science, Garrison

Oceanography: An Introduction, 5th, Ingmanson and Wallace

Plant Physiology, 4th, Salisbury and Ross

Plant Physiology Laboratory Manual, Ross

Plant Physiology, 4th, Devlin and Witham

Exercises in Plant Physiology, Witham et al.

Plants: An Evolutionary Survey, 2nd, Scagel et al.

Psychobiology: The Neuron and Behavior, Hoyenga and Hoyenga

Sex, Evolution, and Behavior, 2nd, Daly and Wilson

Dimensions of Cancer, Kupchella

Evolution: Process and Product, 3rd, Dodson and Dodson

The ITP logo is a trademark under license.

Printed in the United States of America
2 3 4 5 6 7 8 9 10—01 00 99 98 97 96 95

Library of Congress Cataloging-in-Publication Data

Starr, Cecie.
    Biology: the unity and diversity of life/Cecie Starr. Ralph Taggart.—7th ed.
      p. cm.
    Includes bibliographical references and index.
    ISBN 0-534-21060-0
    1. Biology. I. Taggart, Ralph. II. Title.
QH308.2.S72 1995
574—dc20
                              94-44432
                              CIP

For more information, contact Wadsworth Publishing Company:

Wadsworth Publishing Company
10 Davis Drive, Belmont, California 94002, USA

International Thomson Publishing Europe
Berkshire House 168-173, High Holborn
London, WC1V 7AA, England

Thomas Nelson Australia
102 Dodds Street
South Melbourne 3205, Victoria, Australia

Nelson Canada
1120 Birchmount Road
Scarborough, Ontario, Canada M1K 5G4

International Thomson Editores
Campos Eliseos 385, Piso 7
Col. Polanco, 11560 México D.F. México

International Thomson Publishing GmbH
Königswinterer Strasse 418
53227 Bonn, Germany

International Thomson Publishing Asia
221 Henderson Road, #05-10 Henderson Building
Singapore 0315

International Thomson Publishing Japan
Hirakawacho Kyowa Building, 3F
2-2-1 Hirakawacho, Chiyoda-ku, Tokyo 102, Japan

CONTENTS IN BRIEF

INTRODUCTION

1   Methods and Concepts in Biology 2

I   THE CELLULAR BASIS OF LIFE

2   Chemical Foundations for Cells 16
3   Carbon Compounds in Cells 32
4   Cell Structure and Function 50
5   A Closer Look at Cell Membranes 76
6   Ground Rules of Metabolism 92
7   Energy-Acquiring Pathways 106
8   Energy-Releasing Pathways 122

II   PRINCIPLES OF INHERITANCE

9   Cell Division and Mitosis 140
10   Meiosis 154
11   Observable Patterns of Inheritance 168
12   Chromosomes and Human Genetics 186
13   DNA Structure and Function 208
14   From DNA to Proteins 218
15   Control of Gene Expression 234
16   Recombinant DNA and Genetic Engineering 246

III   PRINCIPLES OF EVOLUTION

17   Emergence of Evolutionary Thought 260
18   Microevolution 270
19   Speciation 286
20   The Macroevolutionary Puzzle 300

IV   EVOLUTION AND DIVERSITY

21   The Origin and Evolution of Life 322
22   Bacteria and Viruses 346
23   Protistans 362
24   Fungi 378
25   Plants 390
26   Animals: The Invertebrates 408
27   Animals: The Vertebrates 444
28   Human Evolution: A Case Study 470

V   PLANT STRUCTURE AND FUNCTION

29   Plant Tissues 482
30   Plant Nutrition and Transport 500
31   Plant Reproduction 514
32   Plant Growth and Development 530

VI   ANIMAL STRUCTURE AND FUNCTION

33   Tissues, Organ Systems, and Homeostasis 546
34   Information Flow and the Neuron 560
35   Integration and Control: Nervous Systems 574
36   Sensory Reception 592
37   Integration and Control: Endocrine Systems 612
38   Protection, Support, and Movement 630
39   Circulation 652
40   Immunity 674
41   Respiration 694
42   Digestion and Human Nutrition 714
43   Water-Solute Balance and Temperature Control 734
44   Principles of Reproduction and Development 752
45   Human Reproduction and Development 772

VII   ECOLOGY AND BEHAVIOR

46   Population Ecology 802
47   Community Interactions 822
48   Ecosystems 844
49   The Biosphere 864
50   Human Impact on the Biosphere 892
51   An Evolutionary View of Behavior 910
52   Adaptive Value of Social Behavior 920

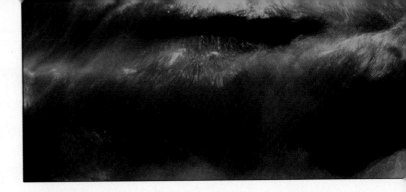

DETAILED CONTENTS

INTRODUCTION

1 METHODS AND CONCEPTS IN BIOLOGY

■ BIOLOGY REVISITED 2

Key Concepts 3

1.1 Shared Characteristics of Life 3

Energy, DNA, and Life 3
Levels of Organization in Nature 4
Metabolism: Life's Energy Transfers 4
Interdependency Among Organisms 5
Sensing and Responding to Change 6
Reproduction 6
Mutation: Source of Variations in Heritable Traits 7

1.2 Life's Diversity 8

So Much Unity, Yet So Many Species 8
An Evolutionary View of Diversity 10

1.3 The Nature of Biological Inquiry 11

On Scientific Methods 11
About the Word "Theory" 11
■ FOCUS ON SCIENCE:
DARWIN'S THEORY AND DOING SCIENCE 12

1.4 The Limits of Science 12

Summary 13

I THE CELLULAR BASIS OF LIFE

2 CHEMICAL FOUNDATIONS FOR CELLS

■ THE CHEMISTRY IN AND AROUND YOU 16

Key Concepts 17

2.1 Organization of Matter 18

The Structure of Atoms 19
Isotopes: Variant Forms of Atoms 19
■ FOCUS ON SCIENCE:
USING RADIOISOTOPES TO DATE FOSSILS, TRACK CHEMICALS,
AND SAVE LIVES 20

2.2 The Nature of Chemical Bonds 22

What Is a Chemical Bond? 22
Electrons and Energy Levels 22

2.3 Important Bonds in Biological Molecules 24

Ionic Bonding 24
Covalent Bonding 24

Hydrogen Bonding 25

2.4 Properties of Water 26

2.5 Water, Dissolved Ions, and pH Values 28

The pH Scale 28
Acids, Bases, and Salts 28
Buffers and the pH of Body Fluids 29

2.6 Chemical Interactions and the World of Cells 30

Summary 30

3 CARBON COMPOUNDS IN CELLS

■ MOM, DAD, AND CLOGGED ARTERIES 32

Key Concepts 33

3.1 Properties of Organic Compounds 34

Carbon-to-Carbon Bonds and the Stability of
Organic Compounds 34
Carbon-to-Carbon Bonds and the Shape of
Organic Compounds 34
Hydrocarbons and Functional Groups 35

3.2 How Cells Use Organic Compounds 36

Five Classes of Reactions 36
The Molecules of Life 36

3.3 The Small Carbohydrates 37

Monosaccharides—The Simple Sugars 37
Oligosaccharides 37

3.4 Complex Carbohydrates: The Polysaccharides 38

3.5 Lipids 40

Fatty Acids 40
Neutral Fats (Triglycerides) 40
Phospholipids 41
Waxes 41
Sterols and Their Derivatives 41

3.6 Proteins 42

Primary Structure of Proteins 42
Three-Dimensional Structure of Proteins 44
Lipoproteins and Glycoproteins 46
Protein Denaturation 46

3.7 Nucleotides and Nucleic Acids 46

Nucleotides With Key Roles in Metabolism 46
Arrangement of Nucleotides in Nucleic Acids:
DNA and RNA 47
Summary 48

4   CELL STRUCTURE AND FUNCTION

■ ANIMALCULES AND CELLS FILL'D WITH JUICES   50

Key Concepts   51

4.1   The Cell Theory   51

4.2   The Nature of Cells   52

Basic Aspects of Cell Structure and Function   52
Structure and Functions of Cell Membranes   52
Surface-to-Volume Constraints on the
Size and Shape of Cells   53
■ FOCUS ON SCIENCE:
MICROSCOPES—GATEWAYS TO THE CELL   54

4.3   Prokaryotic Cells—The Bacteria   56

4.4   Eukaryotic Cells   58

Functions of Organelles   58
Organelles Characteristic of Plants   58
Organelles Characteristic of Animals   60

4.5   The Nucleus   62

Nucleolus   62
Nuclear Envelope   63
Chromosomes   63

4.6   Cytomembrane System   64

Endoplasmic Reticulum   65
Peroxisomes   65
Golgi Bodies   66
Lysosomes   66

4.7   Mitochondria   67

4.8   Specialized Plant Organelles   68

Chloroplasts and Other Plastids   68
Central Vacuole   68

4.9   The Cytoskeleton   69

4.10   The Structural Basis of Cell Movements   70

The Internal Structure of Flagella and Cilia   70
Microtubule Organizing Centers   71

4.11   Cell Surface Specializations   72

Cell Walls and Cell Junctions in Plants   72
Intercellular Material in Animals   72
Cell Junctions in Animals   72
Summary   73

5   A CLOSER LOOK AT CELL MEMBRANES

■ IT ISN'T EASY BEING SINGLE   76

Key Concepts   77

5.1   Membrane Structure and Function   78

The Lipid Bilayer   78
The Fluid Mosaic Model of Membrane Structure   79

5.2   Functions of Membrane Proteins   80

■ FOCUS ON SCIENCE:
DISCOVERING DETAILS ABOUT MEMBRANE STRUCTURE   80

5.3   Diffusion   82

Concentration Gradients and Diffusion   82
Factors Influencing the Rate and Direction
of Diffusion   82

5.4   Osmosis   82

■ FOCUS ON THE ENVIRONMENT:
WILTING PLANTS AND SQUIRTING CELLS   84

5.5   Routes Across Cell Membranes   85

5.6   Protein-Mediated Transport   86

Characteristics of Transport Proteins   86
Passive Transport   86
Active Transport   87

5.7   Exocytosis and Endocytosis   88

Summary   90

6   GROUND RULES OF METABOLISM

■ GROWING OLD WITH MOLECULAR MAYHEM   92

Key Concepts   93

6.1   Energy and Life   94

How Much Energy Is Available?   94
The One-Way Flow of Energy   94

**6.2   Energy and the Direction of
Metabolic Reactions**   96

Energy Losses and Energy Gains   96
Reversible Reactions   96

6.3   Metabolic Pathways   98

6.4   Enzymes   98

Characteristics of Enzymes   98
Enzyme-Substrate Interactions   98
Effects of Temperature and pH on Enzyme Activity   100
Control of Enzyme Activity   100

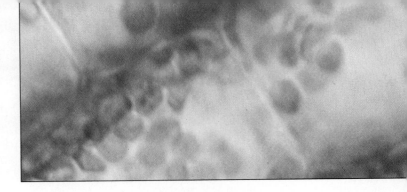

6.5 **Enzyme Helpers** 101

Coenzymes 101

Metal Ions 101

6.6 **Electron Transfers in Metabolic Pathways** 102

■ FOCUS ON HEALTH:
YOU LIGHT UP MY LIFE—VISIBLE EFFECTS OF
EXCITED ELECTRONS 103

6.7 **ATP—The Main Energy Carrier** 104

Structure and Function of ATP 104

The ATP/ADP Cycle 104

Summary 105

7 ENERGY-ACQUIRING PATHWAYS

■ SUN, RAIN, AND SURVIVAL 106

Key Concepts 107

7.1 **Photosynthesis: An Overview** 108

Energy and Materials for the Reactions 108

Where the Reactions Take Place 109

7.2 **Light-Trapping Pigments** 110

7.3 **Light-Dependent Reactions** 112

Photosystems 112

ATP and NADPH: Loading Up Energy,
Hydrogen, and Electrons 112

The Legacy—A New Atmosphere 113

A Closer Look at ATP Formation in Chloroplasts 114

7.4 **Light-Independent Reactions** 115

Capturing Carbon 115

Building the Glucose Subunits 115

7.5 **The Reactions, Start to Finish** 116

■ FOCUS ON THE ENVIRONMENT:
PASTURES OF THE SEAS 117

7.6 **Fixing Carbon—So Near, Yet So Far** 118

Fixing Carbon Twice, in Two Cell Types 118

Fixing and Storing Carbon by Night, Using It by Day 119

7.7 **Chemosynthesis** 119

Summary 120

8 ENERGY-RELEASING PATHWAYS

■ THE KILLERS ARE COMING! 122

Key Concepts 123

8.1 **How Cells Make ATP** 124

Comparison of Three Types of Energy-Releasing
Pathways 124

Overview of Aerobic Respiration 124

8.2 **Glycolysis: First Stage of the
Energy-Releasing Pathways** 126

8.3 **Completing the Aerobic Pathway** 128

Preparatory Steps and the Krebs Cycle 128

Functions of the Second Stage 129

Third Stage of the Aerobic Pathway—
Electron Transport Phosphorylation 130

Summary of the Energy Harvest 130

8.4 **Anaerobic Routes** 132

Fermentation Pathways 132

Anaerobic Electron Transport 133

8.5 **Alternative Energy Sources in the Human Body** 134

Carbohydrate Breakdown in Perspective 134

Energy From Fats 134

Energy From Proteins 134

■ COMMENTARY:
PERSPECTIVE ON LIFE 136

Summary 136

II PRINCIPLES OF INHERITANCE

9 CELL DIVISION AND MITOSIS

■ SILVER IN THE STREAM OF TIME 140

Key Concepts 141

9.1 **Dividing Cells: The Bridge Between Generations**
142

Overview of Division Mechanisms 142

Some Key Points About Chromosomes 142

Mitosis and the Chromosome Number 143

9.2 **Mitosis and the Cell Cycle** 144

■ FOCUS ON HEALTH:
HENRIETTA'S IMMORTAL CELLS 145

9.3 **Stages of Mitosis: An Overview** 146

9.4 **A Closer Look at Mitosis** 148

Prophase: Mitosis Begins 148

Transition to Metaphase 148

Anaphase 149

Telophase 149

9.5 **Division of the Cytoplasm** 150

Summary 152

10 MEIOSIS

■ OCTOPUS SEX AND OTHER STORIES 154

Key Concepts 155

10.1 **On Asexual and Sexual Reproduction** 156

10.2 **Overview of Meiosis** 156

Think "Homologues" 156

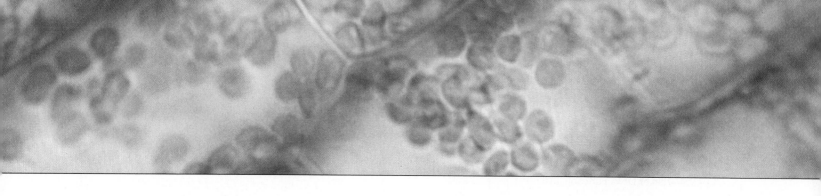

Overview of the Two Divisions 157

10.3 Key Events During Meiosis I 160

Prophase I Activities 160

Metaphase I Alignments 161

10.4 Formation of Gametes 162

Gamete Formation in Plants 162

Gamete Formation in Animals 162

10.5 More Gene Shufflings at Fertilization 162

10.6 Meiosis and Mitosis Compared 164

Summary 166

11 OBSERVABLE PATTERNS OF INHERITANCE

■ A SMORGASBORD OF EARS AND OTHER TRAITS 168

Key Concepts 169

11.1 Mendel's Insight Into Patterns of Inheritance 170

Mendel's Experimental Approach 170

Some Terms Used in Genetics 171

11.2 Mendel's Theory of Segregation 172

Predicting the Outcome of Monohybrid Crosses 172

Testcrosses 173

11.3 Independent Assortment 174

Predicting the Outcome of Dihybrid Crosses 174

The Theory in Modern Form 175

11.4 Dominance Relations 176

11.5 Multiple Effects of Single Genes 176

■ FOCUS ON HEALTH:
ABO BLOOD TYPING 177

11.6 Interactions Between Gene Pairs 178

Hair Color in Mammals 178

Comb Shape in Poultry 179

11.7 Less Predictable Variations in Traits 180

Continuous Variation in Populations 180

11.8 Examples of Environmental Effects on Phenotype 182

Summary 183

12 CHROMOSOMES AND HUMAN GENETICS

■ TOO YOUNG TO BE OLD 186

Key Concepts 187

12.1 Return of the Pea Plant 187

12.2 The Chromosomal Basis of Inheritance—An Overview 188

■ FOCUS ON SCIENCE:
PREPARING A KARYOTYPE DIAGRAM 188

Autosomes and Sex Chromosomes 188

Karyotype Analysis 189

12.3 Sex Determination in Humans 190

12.4 Early Questions About Gene Locations 193

X-Linked Genes: Clues to Patterns of Inheritance 193

Linkage Groups and Crossing Over 193

12.5 Recombination Patterns and Chromosome Mapping 194

12.6 Human Genetic Analysis 195

12.7 Regarding Human Genetic Disorders 196

12.8 Patterns of Autosomal Inheritance 196

Autosomal Recessive Inheritance 196

Autosomal Dominant Inheritance 197

12.9 Patterns of X-Linked Inheritance 198

X-Linked Recessive Inheritance 198

X-Linked Dominant Inheritance 198

A Few Qualifications 198

12.10 Changes in Chromosome Structure 200

Deletions 200

Inversions and Translocations 200

Duplications 200

12.11 Changes in Chromosome Number 201

Categories of Change 201

Mechanisms of Change 201

12.12 When the Number of Autosomes Changes 202

12.13 When the Number of Sex Chromosomes Changes 203

Turner Syndrome 203

Klinefelter Syndrome 203

XYY Condition 203

■ FOCUS ON BIOETHICS:
PROSPECTS AND PROBLEMS IN HUMAN GENETICS 204

Summary 206

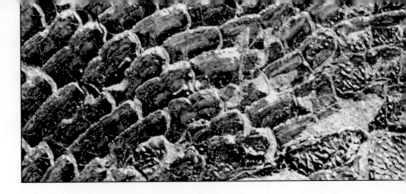

13   DNA STRUCTURE AND FUNCTION

■ CARDBOARD ATOMS AND BENT-WIRE BONDS 208

Key Concepts 209

13.1   Discovery of DNA Function 210

Early Clues 210
Confirmation of DNA Function 210

13.2   DNA Structure 212

Components of DNA 212
Patterns of Base Pairing 213

13.3   DNA Replication and DNA Repair 214

13.4   Organization of DNA in Chromosomes 216

Summary 216

14   FROM DNA TO PROTEINS

■ BEYOND BYSSUS 218

Key Concepts 219

■ FOCUS ON SCIENCE:
DISCOVERING THE CONNECTION BETWEEN GENES
AND PROTEINS 220

14.1   Transcription and Translation: An Overview 221

14.2   Transcription of DNA into RNA 222

How RNA Is Assembled 222
Finishing Touches on mRNA Transcripts 222

14.3   From mRNA to Proteins 224

The Genetic Code 224
Roles of tRNA and rRNA 225

14.4   Stages of Translation 226

14.5   Mutation and Protein Synthesis 228

■ COMMENTARY:
MUTATION, GENE PRODUCTS, AND EVOLUTION 229

Summary 230

15   CONTROL OF GENE EXPRESSION

■ GENES, PROTEINS, AND CANCER 234

Key Concepts 235

15.1   The Nature of Gene Control 236

15.2   Examples of Gene Control in Prokaryotic Cells 236

Negative Control of Transcription 236
Positive Control of Transcription 236

15.3   Gene Control in Eukaryotic Cells 238

15.4   Evidence of Gene Control 240

Transcription in Lampbrush Chromosomes 240
X Chromosome Inactivation 240

15.5   Examples of Signaling Mechanisms 242

Hormonal Signals 242
Sunlight as a Signal 242

15.6   Genes Implicated in Cancer 244

Summary 244

16   RECOMBINANT DNA AND GENETIC ENGINEERING

■ MAKE WAY FOR DESIGNER GENES 246

Key Concepts 247

16.1   Recombination in Nature— And in the Laboratory 248

16.2   Producing Restriction Fragments 249

16.3   Working with DNA Fragments 250

Amplification Procedures 250
Sorting Out and Sequencing Specific DNA Fragments 250

16.4   RFLP Analysis 252

■ FOCUS ON SCIENCE:
RIFF-LIPS AND DNA FINGERPRINTS 252

16.5   Modified Host Cells 252

Use of DNA Probes 252
Use of cDNA 253

16.6   Genetic Engineering of Bacteria 254

16.7   Genetic Engineering of Plants 254

16.8   Genetic Engineering of Animals 256

Supermice and Biotech Barnyards 256
Applying the New Technology to Humans 256

■ FOCUS ON BIOETHICS:
SOME IMPLICATIONS OF HUMAN GENE THERAPY 257

Summary 257

III   PRINCIPLES OF EVOLUTION

17   EMERGENCE OF EVOLUTIONARY THOUGHT

■ FIRE, BRIMSTONE, AND HUMAN HISTORY 260

Key Concepts 261

17.1   Early Beliefs and Confounding Discoveries 262

The Great Chain of Being 262
Questions From Biogeography 262
Questions From Comparative Anatomy 262
Questions About Fossils 263

17.2   A Flurry of New Theories 264

17.3   Voyage of the *Beagle* 264

17.4   Darwin's Theory Takes Form 266

A Clue from Geology 266
Old Bones and Armadillos 266

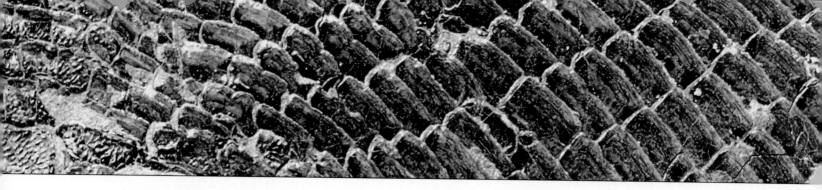

Darwin's Insight—Natural Selection of
Adaptive Versions of Heritable Traits 266

17.5   Regarding Wallace 268

17.6   Regarding the "Missing Links" 268

Summary 268

18   MICROEVOLUTION

■ DESIGNER DOGS   270

Key Concepts 271

18.1   Individuals Don't Evolve; Populations Do 272

Variation Within Populations 272
Sources of Variation 272
Environmental Influences on Variation 273

18.2   Stability Versus Change in Allele Frequencies 274

18.3   Mutation 275

18.4   Genetic Drift 276

Founder Effect 276
Bottlenecks 276
■ FOCUS ON THE ENVIRONMENT:
GENES OF ENDANGERED SPECIES   277

18.5   Gene Flow 277

18.6   Natural Selection—Darwin Revisited 278

**18.7   Directional Change in the
Range of Variation** 278

**18.8   Selection Against or In Favor of
Extreme Phenotypes** 280

Stabilizing Selection 280
Disruptive Selection 280

18.9   Special Cases of Selection 282

Balanced Polymorphism 282
Sexual Selection 282
■ FOCUS ON HEALTH:
SICKLE-CELL ANEMIA—LESSER OF TWO EVILS?   282
Summary 284

19   SPECIATION

■ THE CASE OF THE ROAD-KILLED SNAILS   286

Key Concepts 287

19.1   What Is a Species? 288

If It Looks Like a Duck, . . . 288

19.2   Isolating Mechanisms 289

Regarding Genetic Divergence 289

Categories of Isolating Mechanisms 289

19.3   On the Road to Speciation 290

Prezygotic Isolation 290
Postzygotic Isolation 291
Speciation Defined 291

19.4   Models of Speciation—An Overview 292

19.5   Allopatric Speciation 292

■ FOCUS ON SCIENCE:
SPECIATION AND THE ISTHMUS OF PANAMA   292

19.6   Less Prevalent Speciation Routes 294

Sympatric Speciation and the Crater Lake Cichlids 294
Polyploidy 294
Parapatric Speciation 295

19.7   Patterns of Speciation 296

Branching and Unbranched Speciation 296
Evolutionary Trees and Rates of Change 296
Adaptive Radiations 297

19.8   Extinctions—End of the Line 297

Summary 298

20   THE MACROEVOLUTIONARY PUZZLE

■ OF FLOODS AND FOSSILS   300

Key Concepts 301

20.1   Fossils—Evidence of Ancient Life 302

Fossilization 302
Interpreting the Geologic Tombs 303
Interpreting the Fossil Record 303

20.2   Evidence From Comparative Embryology 304

Developmental Programs Among the Vertebrates 304
Developmental Programs Among the Larkspurs 304

20.3   Evidence of Morphological Divergence 306

Homologous Structures 306
Potential Confusion From Analogous Structures 306

20.4   Evidence From Comparative Biochemistry 308

Molecular Clocks 308
Protein Comparisons 308
Nucleic Acid Comparisons 308

**20.5   Organizing the Evidence—
Classification Schemes** 310

20.6   Identifying Species 312

Pitfalls Awaiting the Taxonomist 312
If It's So Difficult, Why Bother? 313

20.7 **Reconstructing Evolutionary History** 314

Systematics 314

Portraying Relative Relationships
Among Organisms 314

■ FOCUS ON SCIENCE:
CONSTRUCTING A CLADOGRAM 316

20.8 **A Five-Kingdom Scheme** 318

Summary 319

IV   EVOLUTION AND DIVERSITY

21   THE ORIGIN AND EVOLUTION OF LIFE

■ IN THE BEGINNING 322

Key Concepts 323

21.1 **Conditions on the Early Earth** 324

Origin of the Earth 324
The First Atmosphere 324
The Synthesis of Organic Compounds 324

21.2 **Emergence of the First Living Cells** 326

Origin of Agents of Metabolism 326
Origin of Self-Replicating Systems 326
Origin of the First Plasma Membranes 327

21.3 **Life on a Changing Geologic Stage** 328

Drifting Continents and Changing Seas 328
The Geologic Time Scale 329

21.4 **Life in the Archean and Proterozoic Eras** 330

■ FOCUS ON SCIENCE:
THE RISE OF EUKARYOTIC CELLS 331

21.5 **Life in the Paleozoic Era** 334

21.6 **Life in the Mesozoic Era** 336

■ FOCUS ON THE ENVIRONMENT:
PLUMES, GLOBAL IMPACTS, AND THE DINOSAURS 337

21.7 **Life in the Cenozoic Era** 341

Summary 344

22   BACTERIA AND VIRUSES

■ THE UNSEEN MULTITUDES 346

Key Concepts 347

22.1 **Characteristics of Bacteria** 348

Splendid Metabolic Diversity 348
Bacterial Sizes and Shapes 348
Structural Features 349

22.2 **Bacterial Reproduction** 350

22.3 **Bacterial Classification** 351

22.4 **Archaebacteria** 352

22.5 **Eubacteria** 352

Photoautotrophic Eubacteria 352
Chemoautotrophic Eubacteria 353
Chemoheterotrophic Eubacteria—Wonderfully
Beneficial Types 353

■ FOCUS ON HEALTH:
ANTIBIOTICS 354

22.6 **Regarding the Pathogenic Eubacteria** 354

22.7 **Regarding the "Simple" Bacteria** 355

22.8 **The Viruses** 356

Defining Characteristics 356
Examples of Viruses 356

22.9 **Viral Multiplication Cycles** 358

■ FOCUS ON HEALTH:
HUMAN DISEASE—MODES AND PATTERNS OF INFECTION 359

Summary 360

23   PROTISTANS

■ KINGDOM AT THE CROSSROADS 362

Key Concepts 363

23.1 **What Is a Protistan?** 363

23.2 **Chytrids and Water Molds** 364

23.3 **Slime Molds** 364

23.4 **Concerning the Animal-Like Protistans** 366

Protozoan Classification 366
Protozoan Life Cycles 366

23.5 **Amoeboid Protozoans** 366

23.6 **Ciliated Protozoans** 367

23.7 **Flagellated Protozoans** 368

23.8 **Sporozoans** 368

■ FOCUS ON HEALTH:
MALARIA AND THE NIGHT-FEEDING MOSQUITOES 369

23.9 **Euglenoids** 370

23.10 **Concerning "The Algae"** 370

23.11 **Chrysophytes and Dinoflagellates** 371

23.12 **Red Algae** 372

23.13 **Brown Algae** 372

23.14 **Green Algae** 374

23.15 **On the Road to Multicellularity** 376

Summary 376

24   FUNGI

■ DRAGON RUN   378

Key Concepts   379

24.1   Major Groups of Fungi   379

24.2   Characteristics of Fungi   380

Nutritional Modes   380

Fungal Body Plans   380

Reproductive Modes   380

24.3   Zygomycetes   380

24.4   Sac Fungi   382

■ COMMENTARY:
A FEW FUNGI WE WOULD RATHER DO WITHOUT   383

24.5   Club Fungi   384

A Sampling of Spectacular Diversity   384

Generalized Life Cycle   385

24.6   Beneficial Associations Between Fungi and Plants   386

Lichens   386

Mycorrhizae   386

■ FOCUS ON THE ENVIRONMENT:
WHERE HAVE ALL THE FUNGI GONE?   387

24.7   "Imperfect" Fungi   388

Summary   388

25   PLANTS

■ PIONEERS IN A NEW WORLD   390

Key Concepts   391

25.1   Classification of Plants   391

25.2   Evolutionary Trends Among Plants   392

Evolution of Roots, Stems, and Leaves   392

From Haploid to Diploid Dominance   392

Evolution of Pollen and Seeds   393

25.3   Bryophytes   394

25.4   Seedless Vascular Plants   396

Whisk Ferns   396

Lycophytes   396

■ FOCUS ON THE ENVIRONMENT:
ANCIENT CARBON TREASURES   396

Horsetails   398

Ferns   398

25.5   The Seed-Bearing Plants   400

25.6   Gymnosperms   401

Conifers   401

Lesser Known Gymnosperms   402

25.7   Angiosperms   403

Flowering Plant Diversity   403

Classes of Flowering Plants   405

Key Aspects of the Life Cycles   405

Summary   406

26   ANIMALS: THE INVERTEBRATES

■ ANIMALS OF THE BURGESS SHALE   408

Key Concepts   409

26.1   Overview of the Animal Kingdom   410

General Characteristics of Animals   410

Diversity in Body Plans   410

26.2   An Evolutionary Road Map   412

26.3   The One Placozoan   412

26.4   Sponges   412

26.5   Cnidarians   414

Cnidarian Body Plans   414

Some Colonial Cnidarians   417

Cnidarian Life Cycles   417

26.6   Comb Jellies   417

26.7   Flatworms   418

Turbellarians   418

Flukes   418

Tapeworms   419

26.8   Roundworms   419

■ FOCUS ON HEALTH:
A ROGUES' GALLERY OF PARASITIC WORMS   420

26.9   Ribbon Worms   422

26.10   Rotifers   422

26.11   A Major Divergence   423

26.12   Mollusks   424

Gastropods   424

Chitons   425

Bivalves   426

Cephalopods   426

26.13 Annelids 428

Earthworms: A Case Study of Annelid Adaptations 428

Polychaetes 430

Leeches 430

Regarding the Segmented, Coelomic Body Plan 431

26.14 Arthropods 432

Adaptations of Insects and Other Arthropods 432

Chelicerates 433

Crustaceans 434

Insects and Their Relatives 435

Insect Diversity 436

26.15 Echinoderms 438

Summary 440

27 ANIMALS: THE VERTEBRATES

■ MAKING DO (RATHER WELL) WITH WHAT YOU'VE GOT 444

Key Concepts 445

27.1 The Chordate Heritage 446

Characteristics of Chordates 446

Chordate Classification 446

27.2 Invertebrate Chordates 447

Tunicates 447

Lancelets 448

27.3 Origin of Vertebrates 448

27.4 Evolutionary Trends Among the Vertebrates 450

27.5 Fishes 452

The First Vertebrates 452

Existing Jawless Fishes 453

Cartilaginous Fishes 454

Bony Fishes 454

About Those Lobed-Finned, Air-Gulping Fishes . . . 456

27.6 Amphibians 456

Challenges of Life on Land 456

Salamanders 456

Frogs and Toads 457

Caecilians 457

27.7 Reptiles 458

The Rise of Reptiles 458

Turtles 460

Lizards and Snakes 460

Tuataras 461

Crocodilians 461

27.8 Birds 462

27.9 Mammals 464

Egg-Laying Mammals 465

Pouched Mammals 466

Placental Mammals 466

Summary 468

28 HUMAN EVOLUTION: A CASE STUDY

■ THE CAVE AT LASCAUX AND THE HANDS OF GARGAS 470

Key Concepts 471

28.1 Primate Classification 472

28.2 From Primate to Human: Evolutionary Trends 473

Upright Walking 473

Precision Grips and Power Grips 474

Enhanced Daytime Vision 474

Teeth for All Occasions 474

Better Brains, Bodacious Behavior 474

28.3 Primate Origins 475

28.4 The First Hominids 476

28.5 On The Road to Modern Humans 478

Early *Homo* 478

From *Homo erectus* to *H. sapiens* 478

Summary 480

V PLANT STRUCTURE AND FUNCTION

29 PLANT TISSUES

■ THE GREENING OF THE VOLCANO 482

Key Concepts 483

29.1 Overview of the Plant Body 484

Shoots and Roots 484

Three Plant Tissue Systems 485

Meristems 485

29.2 Simple Tissues 486

29.3 Complex Tissues 486

Vascular Tissues 486

Dermal Tissues 487

29.4 Monocots and Dicots Compared 488

29.5 Shoot Primary Structure 488

Internal Structure of Stems 488

Characteristics of Leaves 490

Leaf Internal Structure 491

Origin of Leaves 492

■ COMMENTARY:
USES AND ABUSES OF LEAVES 493

29.6 Root Primary Structure 494

Taproot and Fibrous Root Systems 494

Internal Structure of Roots 494

29.7   **Woody Plants** 496

Nonwoody and Woody Plants Compared 496

Growth at Lateral Meristems 496

Early and Late Wood 497

Summary 498

30   PLANT NUTRITION AND TRANSPORT

■ FLIES FOR DINNER 500

Key Concepts 501

30.1   **Uptake of Water and Nutrients** 502

Nutritional Requirements 502

Specialized Absorptive Structures 502

30.2   **Control of Nutrient Uptake** 504

■ FOCUS ON THE ENVIRONMENT:
PUTTING DOWN ROOTS 504

30.3   **Water Transport** 506

30.4   **Control of Water Loss** 508

The Water-Conserving Cuticle 508

Controlled Water Loss at Stomata 509

30.5   **Transport of Organic Substances** 510

Storage and Transport Forms of Organic Compounds 510

Translocation 510

Pressure Flow Theory 510

Summary 512

31   PLANT REPRODUCTION

■ CHOCOLATE FROM THE TREE'S POINT OF VIEW 514

Key Concepts 515

31.1   **Reproductive Modes** 516

31.2   **Floral Structure** 516

Components of Flowers 516

Where Pollen and Eggs Develop 517

31.3   **A New Generation Begins** 518

From Microspores to Pollen Grains 518

From Megaspores to Eggs 518

From Pollination to Fertilization 518

■ COMMENTARY:
FLOWERING PLANTS AND POLLINATORS—
A COEVOLUTIONARY TALE 520

31.4   **From Zygote to Seed** 522

Formation of the Embryo Sporophyte 522

Seed and Fruit Formation 522

31.5   **Fruit and Seed Dispersal** 524

■ FOCUS ON SCIENCE:
WHY SO MANY FLOWERS AND SO FEW FRUITS? 524

31.6   **Asexual Reproduction of Flowering Plants** 526

Asexual Reproduction in Nature 526

Induced Propagation 526

Summary 528

32   PLANT GROWTH AND DEVELOPMENT

■ FOOLISH SEEDLINGS AND GORGEOUS GRAPES 530

Key Concepts 531

32.1   **Patterns of Growth and Development** 532

Seed Germination 532

Prescribed Growth Patterns 532

32.2   **Plant Hormones** 534

Auxins 534

Gibberellins 534

■ FOCUS ON HEALTH:
ABOUT THOSE HERBICIDES 535

Cytokinins 535

Abscisic Acid 535

Ethylene 535

Other Hormones and Hormone-Like Substances 535

32.3   **The Tropisms** 536

Phototropism 536

Gravitropism 536

Thigmotropism 537

32.4   **Responses to Mechanical Stress** 537

32.5   **Biological Clocks and Their Effects** 538

Circadian Rhythms 538

Photoperiodism 538

32.6   **The Flowering Process** 539

32.7   **Vernalization** 540

32.8   **Senescence** 540

32.9   **Dormancy** 541

Entering Dormancy 541

Breaking Dormancy 541

■ FOCUS ON THE ENVIRONMENT:
THE RISE AND FALL OF A GIANT 542

Summary 543

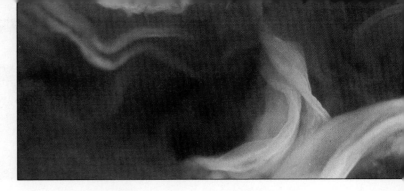

VI ANIMAL STRUCTURE AND FUNCTION

33 TISSUES, ORGAN SYSTEMS, AND HOMEOSTASIS

■ MEERKATS, HUMANS, IT'S ALL THE SAME 546

Key Concepts 547

33.1 **Animal Structure and Function: An Overview** 547

33.2 **Epithelial Tissue** 548

General Characteristics 548
Cell-to-Cell Contacts 549
Glandular Epithelium 549

33.3 **Connective Tissue** 550

Connective Tissue Proper 550
Specialized Connective Tissue 550

33.4 **Muscle Tissue** 552

33.5 **Nervous Tissue** 553

■ FOCUS ON SCIENCE:
FRONTIERS IN TISSUE RESEARCH 553

33.6 **Organ Systems** 554

Tissue and Organ Formation 554
Overview of Major Organ Systems 554

33.7 **Homeostasis and Systems Control** 556

The Internal Environment 556
Mechanisms of Homeostasis 556
Summary 558

34 INFORMATION FLOW AND THE NEURON

■ TORNADO! 560

Key Concepts 561

34.1 **Cells of the Nervous System** 561

34.2 **How the Neuron Responds to Stimulation** 562

Functional Zones of a Neuron 562
A Neuron at Rest, Then Moved to Action 562
Restoring and Maintaining Readiness 562

34.3 **Action Potentials** 564

Approaching Threshold 564
An All-or-Nothing Spike 564
Propagation of Action Potentials 565

34.4 **Node-to-Node-Hopping Along Sheathed Axons** 566

34.5 **Chemical Synapses** 566

34.6 **Synaptic Integration** 568

A Smorgasbord of Signals 568
Summation of Incoming Signals 568
Removing Neurotransmitter From the Synaptic Cleft 568

■ FOCUS ON HEALTH:
DEADLY IMBALANCES AT SYNAPSES 569

34.7 **Paths of Information Flow** 570

Blocks and Cables of Neurons 570

Reflex Arcs 570
Summary 572

35 INTEGRATION AND CONTROL: NERVOUS SYSTEMS

■ WHY CRACK THE SYSTEM? 574

Key Concepts 575

35.1 **Invertebrate Nervous Systems** 576

Life-Styles and Nervous Systems 576
Regarding the Nerve Net 576
On the Importance of Having a Head 576

35.2 **Vertebrate Nervous Systems** 578

Evolutionary High Points 578
Functional Divisions of the Human Nervous System 578

35.3 **Peripheral Nervous System** 580

Somatic and Autonomic Subdivisions 580
The Sympathetic and Parasympathetic Nerves 580

35.4 **Central Nervous System: The Spinal Cord** 581

35.5 **Central Nervous System: The Vertebrate Brain** 582

Hindbrain 582
Midbrain 582
Forebrain 583
The Reticular Formation 583
Brain Cavities and Canals 583

35.6 **A Closer Look at the Human Cerebrum** 584

Organization of the Cerebral Hemispheres 584
The Cerebral Cortex 584

■ FOCUS ON SCIENCE:
SPERRY'S SPLIT-BRAIN EXPERIMENTS 586

35.7 **Memory** 587

35.8 **States of Consciousness** 587

35.9 **The Brain and Behavior** 588

■ FOCUS ON HEALTH:
DRUGS, THE BRAIN, AND BEHAVIOR 588

Summary 590

36 SENSORY RECEPTION

■ DIFFERENT STROKES FOR DIFFERENT FOLKS 592

Key Concepts 593

36.1 **Sensory Systems** 594

Types of Sensory Receptors 594
Sensory Pathways 594

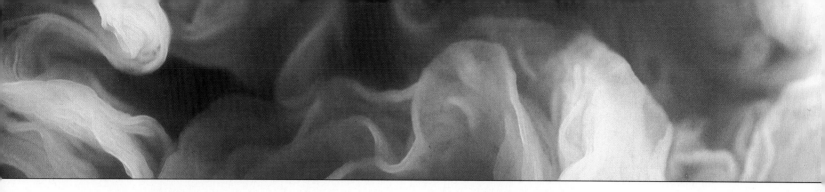

36.2   **Somatic Sensations** 596

Touch and Pressure 596

Temperature 596

Pain 596

Muscle Sense 597

36.3   **Taste and Smell** 598

36.4   **Balance** 599

36.5   **Hearing** 600

36.6   **Vision** 602

36.7   **Invertebrate Eyes** 602

36.8   **Vertebrate Eyes** 604

Eye Structure 604

Focusing Mechanisms 604

■ FOCUS ON HEALTH:
DISORDERS OF THE EYE 606

36.9   **From Signaling to Visual Perception** 608

Organization of the Retina 608

Neuronal Responses to Light 609

Summary 610

37   **INTEGRATION AND CONTROL:
ENDOCRINE SYSTEMS**

■ HORMONE JAMBOREE 612

Key Concepts 613

37.1   **The Endocrine System** 614

Hormones and Other Signaling Molecules 614

Discovery of Hormones and Their Sources 614

37.2   **Signaling Mechanisms** 616

Steroid Hormone Action 616

Nonsteroid Hormone Action 617

37.3   **The Hypothalamus and the Pituitary Gland** 618

Posterior Lobe Secretions 618

Anterior Lobe Secretions 618

37.4   **Examples of Abnormal Pituitary Output** 620

37.5   **Selected Examples of Hormonal Control** 621

37.6   **Pancreatic Islets** 622

37.7   **Adrenal Glands** 623

Adrenal Cortex 623

Adrenal Medulla 623

37.8   **The Thyroid and Parathyroid Glands** 624

Thyroid Functioning 624

Parathyroid Functioning 624

37.9   **Pineal Gland** 625

■ FOCUS ON THE ENVIRONMENT:
RHYTHMS AND BLUES 625

37.10   **Some Final Examples of Integration
and Control** 626

Hormones of the Thymus 626

Hormonal Control of the Gonads 626

Effects of Some Local Signaling Molecules 626

Comparative Look at a Few Invertebrate Hormones 626

Summary 628

38   **PROTECTION, SUPPORT, AND MOVEMENT**

■ "ALL RIGHT—GO!" 630

Key Concepts 631

38.1   **Integumentary System** 632

Vertebrate Skin and Its Derivatives 632

Functions of Skin 632

38.2   **A Closer Look at the Structure of Skin** 634

Epidermis 634

Dermis 634

■ FOCUS ON HEALTH:
SUNLIGHT AND SKIN 635

38.3   **Invertebrate and Vertebrate Skeletons** 636

Operating Principle for Skeletons 636

Examples From the Invertebrates 636

Examples From the Vertebrates 637

38.4   **Characteristics of Bone** 638

Functions of Bone 638

Bone Structure 638

38.5   **Human Skeletal System** 640

Divisions of the Skeleton 640

Skeletal Joints 640

38.6   **Muscle Systems** 642

How Muscles and Bones Interact 642

Human Muscle System 643

38.7   **Muscle Function** 644

Functional Organization of a Skeletal Muscle 644

Sliding-Filament Model of Muscle Contraction 645

38.8   **Control of Contraction** 646

The Control Pathway 646

The Control Mechanism 646

Energy for Contraction 647

38.9   Muscle Tension, Strength, and Fatigue 648

Muscle Tension 648

Muscle Twitches and Tetanic Contraction 648

Decline in Tension—Muscle Fatigue 648

■ FOCUS ON HEALTH:
MUSCLE MANIA 649

Summary 650

39   CIRCULATION

■ HEARTWORKS 652

Key Concepts 653

39.1   Circulatory Systems—An Overview 654

General Characteristics 654

Functional Links with the Lymphatic System 654

39.2   Characteristics of Blood 656

Functions of Blood 656

Blood Volume and Composition 656

39.3   Human Cardiovascular System 658

General Organization 658

Blood Circulation Routes 658

39.4   The Human Heart 660

Heart Structure 660

Cardiac Cycle 661

Mechanisms of Contraction 661

39.5   Blood Pressure in the Cardiovascular System 662

Arterial Blood Pressure 662

Resistance at Arterioles 662

Controls Over Blood Pressure and Blood Distribution 663

39.6   From Capillary Beds Back to the Heart 664

Capillary Function 664

Venous Pressure 664

■ FOCUS ON HEALTH:
CARDIOVASCULAR DISORDERS 666

39.7   Hemostasis 668

39.8   Blood Typing 668

ABO Blood Typing 668

Rh Blood Typing 668

39.9   Lymphatic System 670

Lymph Vascular System 671

Lymphoid Organs and Tissues 671

Summary 672

40   IMMUNITY

■ RUSSIAN ROULETTE, IMMUNOLOGICAL STYLE 674

Key Concepts 675

40.1   Three Lines of Defense 676

Surface Barriers to Invasion 676

Nonspecific and Specific Responses 676

40.2   Complement Proteins 677

40.3   Inflammation 678

The Roles of Macrophages and Their Kin 678

The Inflammatory Response 678

40.4   The Immune System 680

Defining Features 680

Antigen-Presenting Cells—The Triggers for
Immune Responses 680

Key Players in Immune Responses 680

Control of Immune Responses 681

40.5   Cell-Mediated Responses 682

Helper T Cells and Cytotoxic T Cells 682

■ FOCUS ON HEALTH:
CANCER AND IMMUNOTHERAPY 683

Regarding the Natural Killer Cells 683

40.6   Antibody-Mediated Responses 684

B Cells and the Targets of Antibodies 684

The Immunoglobulins 685

40.7   Lymphocyte Battlegrounds 685

40.8   Specificity and Memory 686

How Lymphocytes Produce Antigen-Specific
Receptors 686

Immunological Memory 687

40.9   Immunization 688

40.10   Abnormal or Deficient Immune Responses 689

Allergies 689

Autoimmune Disorders 689

Deficient Immune Responses 689

■ FOCUS ON HEALTH:
AIDS—THE IMMUNE SYSTEM COMPROMISED 690

Summary 692

41   RESPIRATION

■ CONQUERING CHOMOLUNGMA 694

Key Concepts 695

41.1   The Nature of Respiration 696

Why Oxygen and Carbon Dioxide Move
Into and Out of the Body 696

Factors That Influence Gas Exchange 696

41.2   Modes of Respiration 698

Integuments as Respiratory Surfaces 698

Gills 698

Lungs 699

41.3   Human Respiratory System 700

Airways to the Lungs 700

Sites of Gas Exchange in the Lungs 701

41.4   Ventilation 702

How Breathing Changes Air Pressure in the Lungs 702

Lung Volumes 702

Breathing and Sound Production 703

41.5   Gas Exchange and Transport 704

Gas Exchange in Alveoli 704

Gas Transport To and From Metabolically Active Tissues 704

Controls Over Gas Exchange 705

◼ FOCUS ON HEALTH:
WHEN THE LUNGS BREAK DOWN   706

41.6   Respiration in Unusual Environments 708

Breathing at High Altitudes 708

Carbon Monoxide Poisoning 708

Respiration in the Deep Ocean 709

◼ FOCUS ON SCIENCE:
THE LEATHERBACK SEA TURTLE   710

Summary 712

42   DIGESTION AND HUMAN NUTRITION

◼ SORRY, HAVE TO EAT AND RUN   714

Key Concepts 715

42.1   Digestive Systems 716

General Features 716

The Human Digestive System 716

42.2   Food Preparation and Storage 718

Into the Mouth, Down the Tube 718

The Stomach 718

42.3   Digestion in the Small Intestine 720

Pancreatic Secretions 721

Bile and the Digestion of Fats 721

42.4   Absorption From the Small Intestine 722

42.5   Functions of the Large Intestine 723

42.6   Controls Over the System 723

42.7   Human Nutritional Requirements 724

New, Improved Food Pyramids 724

Carbohydrates 724

Lipids 725

Proteins 725

42.8   Vitamins and Minerals 726

42.9   Energy and Body Weight 728

◼ FOCUS ON HEALTH:
WHEN DOES DIETING BECOME A DISORDER?   729

42.10   Nutrition and Organic Metabolism 730

◼ FOCUS ON HEALTH:
CASE STUDY: FEASTING, FASTING, AND SYSTEMS INTEGRATION 731

Summary 732

43   WATER-SOLUTE BALANCE AND TEMPERATURE CONTROL

◼ TALE OF THE DESERT RAT   734

Key Concepts 735

43.1   Maintaining the Extracellular Fluid 736

Water Gains and Losses 736

Solute Gains and Losses 736

Urinary System of Mammals 737

43.2   Kidney Structure and Function 738

Functional Units of Kidneys—The Nephrons 738

Urine-Forming Processes—An Overview 739

43.3   Urine Formation 740

Factors Influencing Filtration 740

Reabsorption of Water and Sodium 740

43.4   Acid-Base Balance 742

43.5   Kidney Malfunctions 742

43.6   On Fish, Frogs, and Kangaroo Rats 743

43.7   Maintaining the Body's Core Temperature 744

Temperatures Suitable for Life 744

Heat Gains and Heat Losses 744

43.8   Classification of Animals Based on Temperature 746

Ectotherms 746

Endotherms 746

Heterotherms 747

Thermal Strategies Compared 747

43.9   Temperature Regulation in Mammals 748

Responses to Cold Stress 748

Responses to Heat Stress 749

Fever 749

Summary 750

44 PRINCIPLES OF REPRODUCTION AND DEVELOPMENT

■ FROM FROG TO FROG AND OTHER MYSTERIES 752

Key Concepts 753

44.1 The Beginning: Reproductive Modes 754

44.2 Stages of Development 756

44.3 A Visual Tour of Frog and Chick Development 758

44.4 Early Marching Orders 760

Developmental Information in the Egg Cytoplasm 760
Reorganization of the Egg Cytoplasm at Fertilization 760

44.5 Emergence of the Early Embryo 762

Cleavage 762
Gastrulation 763

44.6 Formation of Specialized Tissues and Organs 764

Cell Differentiation 764
Morphogenesis 764

44.7 Pattern Formation 766

Inducer Signals in the Embryo 766
Homeobox Genes 767

44.8 From the Embryo Onward 768

Post-Embryonic Development 768
Aging and Death 768
■ COMMENTARY:
DEATH IN THE OPEN 769
Summary 770

45 HUMAN REPRODUCTION AND DEVELOPMENT

■ THE JOURNEY BEGINS 772

Key Concepts 773

45.1 Male Reproductive System 774

Where Sperm Form 774
Where Semen Forms 774
Cancer of the Testis 774

45.2 Male Reproductive Function 776

Sperm Formation 776
Hormonal Controls 776

45.3 Female Reproductive System 778

The Reproductive Organs 778
Overview of the Menstrual Cycle 778

45.4 Female Reproductive Function 780

Cyclic Changes in the Ovary 780
Cyclic Changes in the Uterus 781

45.5 Summing Up—Key Events of the Menstrual Cycle 782

45.6 Pregnancy Happens 783

Sexual Intercourse 783
Fertilization 783

45.7 Formation of the Early Embryo 784

Early Cleavage and Implantation 784
Extraembryonic Membranes 785

45.8 Emergence of the Vertebrate Body Plan 786

45.9 On the Importance of the Placenta 787

45.10 Emergence of Distinctly Human Features 788

■ FOCUS ON HEALTH:
MOTHER AS PROTECTOR, PROVIDER, POTENTIAL THREAT 790

45.11 From Birth Onward 792

The Process of Birth 792
Lactation 792
Postnatal Development and Aging 793

45.12 Control of Human Fertility 794

Some Ethical Considerations 794
Birth Control Options 794
■ FOCUS ON HEALTH:
SEXUALLY TRANSMITTED DISEASES 796

45.13 Two Issues of Clinical and Ethical Importance 798

Termination of Pregnancy 798
In Vitro Fertilization 798
Summary 798

VII ECOLOGY AND BEHAVIOR

46 POPULATION ECOLOGY

■ A TALE OF NIGHTMARE NUMBERS 802

Key Concepts 803

46.1 Characteristics of Populations 804

Regarding Population Density 804
Spatial Distribution 804
Some Qualifiers 805

46.2 Population Size and Exponential Growth 806

How the Number of Individuals Can Change 806
The Exponential Growth Pattern 806
Regarding the Biotic Potential 807

46.3 Limits on the Growth of Populations 808

Limiting Factors 808
Carrying Capacity and Logistic Growth 808
Density-Dependent Controls 809
Density-Independent Controls 809

46.4 Life History Patterns 810

Life Tables 810

Survivorship Curves and Reproductive Patterns 811

■ FOCUS ON SCIENCE:
NATURAL SELECTION AMONG THE GUPPIES OF TRINIDAD 812

46.5 Human Population Growth 814

How We Began Sidestepping Controls 814

Present and Future Growth 815

46.6 Control Through Family Planning 816

46.7 Population Growth and Economic Development 818

■ FOCUS ON THE ENVIRONMENT:
POPULATIONS AND RESOURCE CONSUMPTION 819

46.8 Questions Concerning Zero Population Growth 820

Summary 820

47 COMMUNITY INTERACTIONS

■ NO PIGEON IS AN ISLAND 822

Key Concepts 823

47.1 Community Characteristics 824

Factors That Shape Community Structure 824

The Niche 824

Types of Species Interactions 824

47.2 Mutualism 825

47.3 Competitive Interactions 826

Categories of Competition 826

Competitive Exclusion 826

Resource Partitioning 827

47.4 Predation 828

"Predator" Versus "Parasite" 828

Dynamics of Predator-Prey Interactions 828

47.5 Prey Defenses 830

Camouflage 830

Warning Coloration and Mimicry 831

Moment-of-Truth Defenses 831

47.6 Adaptive Responses to Prey 832

47.7 Parasitic Interactions 833

Parasites 833

Parasitoids 833

Parasites as Biological Control Agents 833

47.8 Succession—Directional Change in Communities 834

Primary and Secondary Succession 834

The Climax-Pattern Model 835

47.9 Regarding Community Stability 836

Cyclic, Nondirectional Change 836

So Many Communities, So Few All-Encompassing Models 836

47.10 Effects of Dispersal 838

Modes of Dispersal 838

Examples of Species Introductions 838

■ FOCUS ON THE ENVIRONMENT:
HELLO LAKE VICTORIA, GOODBYE CICHLIDS 839

47.11 Patterns of Biodiversity 840

Mainland and Marine Patterns 840

Island Patterns 840

Summary 842

48 ECOSYSTEMS

■ CRÊPES FOR BREAKFAST, PANCAKE ICE FOR DESSERT 844

Key Concepts 845

48.1 The Nature of Ecosystems 846

Overview of the Participants 846

Structure of Ecosystems 846

48.2 Energy Flow Through Ecosystems 848

Primary Productivity 848

Major Pathways of Energy Flow 848

Ecological Pyramids 848

48.3 Case Study: Energy Flow at Silver Springs, Florida 850

48.4 Biogeochemical Cycles—An Overview 851

48.5 Hydrologic Cycle 852

48.6 Carbon Cycle 854

■ FOCUS ON THE ENVIRONMENT:
FROM GREENHOUSE GASES TO A WARMER PLANET? 856

48.7 Nitrogen Cycle 858

Cycling Processes 858

Nitrogen Scarcity 858

Human Intervention in the Nitrogen Cycle 858

48.8 Phosphorus Cycle 860

48.9 Ecosystem Modeling 861

■ FOCUS ON THE ENVIRONMENT:
TRANSFER OF HARMFUL COMPOUNDS THROUGH ECOSYSTEMS 861

Summary 862

49 THE BIOSPHERE

■ DOES A CACTUS GROW IN BROOKLYN? 864

Key Concepts 865

49.1 Air Circulation Patterns and Regional Climates 866

49.2 The Ocean's Effect on Regional Climates 868

49.3 Effects of Topography on Regional Climates 869

49.4 The World's Biomes 870

49.5 Soils of Major Biomes 872

49.6 Deserts 873

49.7 Dry Shrublands and Woodlands 874

49.8 Grasslands and Savannas 874

49.9 Broadleaf Forests 876

Evergreen Broadleaf Forests 876
Deciduous Broadleaf Forests 877

49.10 Coniferous Forests 878

49.11 Tundra 879

49.12 Regarding the Water Provinces 880

49.13 Lake Ecosystems 880

Lake Zonation 880
Seasonal Changes in Lakes 880
Trophic Nature of Lakes 881

49.14 Stream Ecosystems 882

49.15 Coastal Ecosystems 882

Estuaries 882
Life Along the Coasts 883

49.16 The Open Ocean 884

49.17 Coral Reefs and Banks 886

49.18 Upwelling 888

■ FOCUS ON THE ENVIRONMENT:
EL NIÑO AND SEESAWS IN THE WORLD'S CLIMATES 888

Summary 890

50 HUMAN IMPACT ON THE BIOSPHERE

■ TROPICAL FORESTS—DISAPPEARING BIOMES? 892

Key Concepts 893

50.1 Human Population Growth and the Environment 894

50.2 Local Effects of Air Pollution 895

50.3 Global Effects of Air Pollution 896

Acid Deposition 896
Damage to the Ozone Layer 897

50.4 Water Scarcity and Water Pollution 898

Consequences of Large-Scale Irrigation 898
Maintaining Water Quality 899

50.5 Changes in the Land 900

What To Do With Solid Wastes? 900
Conversion of Marginal Lands for Agriculture 900

50.6 Desertification 901

50.7 Deforestation 902

50.8 A Question of Energy Inputs 904

Fossil Fuels 904
Nuclear Energy 904

■ FOCUS ON THE ENVIRONMENT:
INCIDENT AT CHERNOBYL 905

50.9 Alternative Sources of Energy 906

■ FOCUS ON BIOETHICS:
BIOLOGICAL PRINCIPLES AND THE HUMAN IMPERATIVE 906

Summary 908

51 AN EVOLUTIONARY VIEW OF BEHAVIOR

■ DECK THE NEST WITH SPRIGS OF GREEN STUFF 910

Key Concepts 911

51.1 The Heritable Basis of Behavior 912

Genes and Behavior 912
Hormones and Behavior 912

51.2 Instinctive and Learned Behavior 914

51.3 The Adaptive Value of Behavior 916

Selection Theory and Feeding Behavior 916
Selection Theory and Mating Behavior 916
Summary 918

52 ADAPTIVE VALUE OF SOCIAL BEHAVIOR

■ CONSIDER THE TERMITE 920

Key Concepts 921

52.1 Signaling Mechanisms 922

Functions of Communication Signals 922
Types of Communication Signals 922

52.2 Costs and Benefits of Social Life 924

Disadvantages to Sociality 924
Advantages of Sociality 924

52.3 Social Life and Self-Sacrifice 926

Self-Sacrificing Behavior and Dominance Hierarchies 926
Costs and Benefits of Parenting 927

52.4 The Evolution of Altruism 928

■ FOCUS ON SCIENCE:
ABOUT THOSE SELF-SACRIFICING NAKED MOLE-RATS 930

52.5 An Evolutionary View of Human Social Behavior 931

Summary 932

Appendix I   Brief Classification Scheme
Appendix II   Units of Measure
Appendix III Answers to Genetics Problems
Appendix IV   Answers to Self-Quizzes
Appendix V   A Closer Look at Some Major Metabolic Pathways
Appendix VI   Periodic Table of the Elements
A Glossary of Biological Terms

Learning Guide
for Students

Before you start reading, spend a little time looking over the next few pages. They provide a quick guide to the learning tools throughout the text that have been designed to enhance your understanding of biology.

Take note, in particular, of the topic spreads found throughout this new edition of BIOLOGY: THE UNITY AND DIVERSITY OF LIFE. The entire text for each topic and its reinforcing illustrations are positioned on the same page or a facing page, all in one place. No more flipping from page to page, losing sight of the development of major concepts.

Self-contained topic spreads are just one of several learning aids you'll find in this text. Equally important are visual summaries that are incorporated within key illustrations. Reading descriptions within illustrations will help you create a mental image of concepts before you read the text proper.

Self-Contained Topic Spreads

In this book, all of the text and illustrations for a particular topic are organized together on one page or on two pages that face each other. These easy-to-identify "topic spreads" start with a boxed number and heading at the upper left

---

5.1 MEMBRANE STRUCTURE AND FUNCTION

The Lipid Bilayer

A cell's organization as well as its activities depends on membranes. Fluid bathes both surfaces of each cell membrane. How, then, does the membrane remain structurally intact? Is it like a solid wall separating two fluid regions? Not at all. Push a fine needle into and out of a cell, and cytoplasm will not ooze out. Instead, the cell surface will appear to flow over the puncture and seal it. This observation suggests that the plasma membrane shows fluid behavior!

How does a "fluid" cell membrane remain distinct from fluid surroundings? For the answer, start with phospholipids—the most abundant component of cell membranes. A **phospholipid**, recall, consists of a hydrophilic (water-loving) head and two fatty acid tails that are mainly hydrophobic (water-dreading):

} polar (hydrophilic) head

} nonpolar (hydrophobic) tails

Like other lipids, phospholipid molecules cluster together spontaneously when immersed in water. Hydrophobic interactions may even force them into two layers, with all the fatty acid tails sandwiched between the hydrophilic heads. This arrangement, a **lipid bilayer**, is the structural basis of cell membranes:

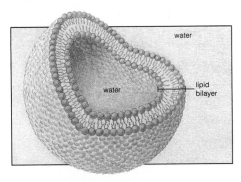

water

water

lipid bilayer

Because lipid bilayers minimize the number of hydrophobic groups exposed to water, the fatty acid tails do not have to spend a lot of energy fighting the water molecules, so to speak. In fact, that plasma membrane you "punctured" earlier exhibited sealing behavior precisely because a puncture is energetically

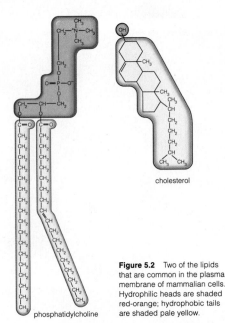

cholesterol

Figure 5.2 Two of the lipids that are common in the plasma membrane of mammalian cells. Hydrophilic heads are shaded red-orange; hydrophobic tails are shaded pale yellow.

phosphatidylcholine

unfavorable—it leaves too many hydrophobic groups exposed to the surrounding water.

Ordinarily, few cells are ever punctured with fine needles. But the self-sealing behavior of membrane lipids is useful for more than damage control. It has roles in the formation of those vesicles described in the preceding chapter. For example, as vesicles bud off ER or Golgi membranes, hydrophobic interactions with cytoplasmic water push lipid molecules together and close off the ruptured site. The same thing happens during exocytosis and endocytosis, as you will see later in the chapter.

Lipid bilayers of cell membranes incorporate a variety of phospholipids, glycolipids, and sterols. The phospholipids differ in their hydrophilic heads and in the length and saturation of their fatty acid tails. (Unsaturated fatty acids have one or more double bonds in the carbon backbone; fully saturated ones have none.) Figure 5.2 shows the structure of phosphatidylcholine, one of the most abundant phospholipids. Glycolipids of the bilayer are structurally similar to this, and their head region also incorporates one or more sugar monomers. Cholesterol is an abundant sterol of animal cell membranes; phytosterols are common to plant cell membranes.

corner of a page. They end with concise statements that summarize the key concepts within that spread. These self-contained page layouts allow you to focus easily on one subject at a time.

This topic spread, on the structure and function of cell membranes, provides clear, simple illustrations and concise, yet well-developed, text—all in the space of two pages.

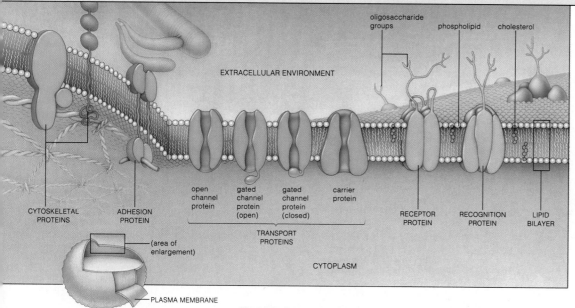

Figure 5.3 Cutaway view of a plasma membrane, based on the fluid mosaic model of membrane structure.

The Fluid Mosaic Model of Membrane Structure

Figure 5.3 shows the **fluid mosaic model** of membrane structure. By this model, a membrane bilayer shows "fluid" behavior as a result of the constant motion of lipid molecules and their interactions with one another. Membrane lipids spin about their long axis, they move sideways, and their tails flex back and forth—all of which helps keep neighboring lipids from packing into a solid layer. Lipids with short tails and unsaturated (kinked) tails further disrupt the packing. Being a composite of lipids and proteins, the membrane is said to have a "mosaic" quality. The proteins carry out most membrane functions, such as the transport of substances into and out of the cell.

The mosaic is not symmetrical. The two sides of a given membrane are not the same; they differ in terms of the number, kind, and arrangement of their lipids and proteins. In later chapters, you will see how the asymmetry relates to differences in the tasks carried out on the two sides of the membrane.

Bear in mind, the fluid mosaic model is only a starting point for discussing membranes—which differ in fluidity as well as in composition and molecular arrangements. Why bother to think about this fluidity? Cell survival depends on it, as a simple example will make clear. When outside temperatures fall,

membranes tend to stiffen and disrupt the functions of membrane proteins. Bacterial and yeast cells rapidly synthesize unsaturated fatty acids at such times. Infusion of these kinky lipids into a membrane helps keep it from stiffening up.

The first fluid mosaic model of membrane structure was put together in 1972 by S. J. Singer and G. Nicolson. Evidence favoring this model comes from many sources, including freeze-fracture microscopy, as described in the *Focus* essay on the next page. The features of this model can be summarized this way:

---

1. Cell membranes are composed mostly of lipids (especially phospholipids) and proteins.

2. The lipid molecules have their hydrophilic heads at the two outer faces of a bilayer and their fatty acid tails sandwiched in between.

3. A lipid bilayer imparts structure to cell membranes and serves as a hydrophobic barrier between two solutions. (A plasma membrane separates the fluids inside and outside the cell. Internal cell membranes separate different fluids in the cytoplasm.)

4. Proteins embedded in the bilayer or positioned at its surfaces carry out most membrane functions.

Visual Summaries

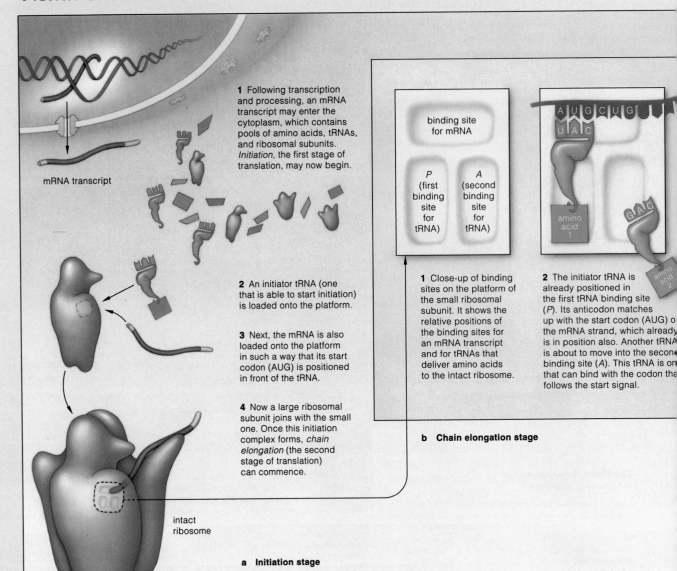

1 Following transcription and processing, an mRNA transcript may enter the cytoplasm, which contains pools of amino acids, tRNAs, and ribosomal subunits. *Initiation*, the first stage of translation, may now begin.

mRNA transcript

2 An initiator tRNA (one that is able to start initiation) is loaded onto the platform.

3 Next, the mRNA is also loaded onto the platform in such a way that its start codon (AUG) is positioned in front of the tRNA.

4 Now a large ribosomal subunit joins with the small one. Once this initiation complex forms, *chain elongation* (the second stage of translation) can commence.

intact ribosome

a   Initiation stage

binding site for mRNA

P
(first binding site for tRNA)

A
(second binding site for tRNA)

1 Close-up of binding sites on the platform of the small ribosomal subunit. It shows the relative positions of the binding sites for an mRNA transcript and for tRNAs that deliver amino acids to the intact ribosome.

amino acid 1

2 The initiator tRNA is already positioned in the first tRNA binding site (*P*). Its anticodon matches up with the start codon (AUG) o[f] the mRNA strand, which already is in position also. Another tRNA is about to move into the secon[d] binding site (*A*). This tRNA is on[e] that can bind with the codon tha[t] follows the start signal.

b   Chain elongation stage

Who hasn't stared at the text on a complex topic, stared at the illustrations, then stared back at the text, wondering how it all fits together? Visual summaries can give you perspective on the big picture. These illustrations have detailed, often step-by-step descriptions integrated within the diagrams. The visual summary shown here on the topic of translation provides an example.

A visual summary can work for you in two ways. First, you can use it as a preview, to organize your thoughts about a concept before you read a detailed discussion about it. Alternatively, you can read the text first, then study the illustration, step by step, to reinforce your understanding of the details.

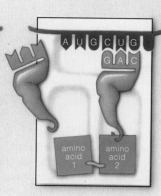

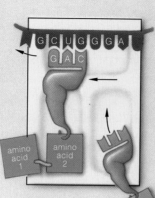

3 Through enzyme action, the bond between the initiator tRNA and the amino acid hooked to it is broken. At the same time, an enzyme catalyzes the formation of a peptide bond between the two amino acids. After these bonding events are over, the initiator tRNA will be released from the ribosome.

4 Now the first amino acid is attached only to the second one—which is still hooked to the second tRNA. This tRNA will move into the *P* site on the ribosomal platform, sliding the mRNA with it by one codon. When it does so, the third codon will become aligned above the *A* site.

5 A third tRNA is about to move into the *A* site. Its anticodon is capable of base-pairing with the third codon of the mRNA transcript. Next, through enzyme action, a peptide bond will form between amino acids 2 and 3.

6 Steps 3 through 5 are repeated again and again. The polypeptide chain continues to grow this way until enzymes reach a stop codon in the mRNA transcript. Now *termination*, the last stage of transcription, can begin, as shown in (**c**).

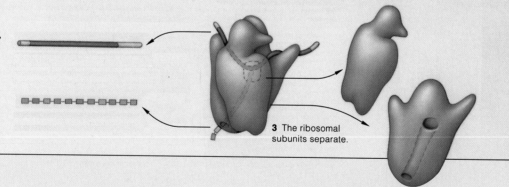

Once a stop codon is reached, the mRNA transcript is released from the ribosome.

The newly formed polypeptide chain also is released.

Chain termination stage

3 The ribosomal subunits separate.

Focus Essays

We've devised two tracks throughout the book. First, the text proper deals with theoretical aspects of biology—the concepts and the connections between them. Second, throughout the book, numerous examples of applications show how those concepts can be used to explain something that's going on in the world around you. Focus essays are the most prominent of these applications, all of which are indexed on the end papers at the back of the book.

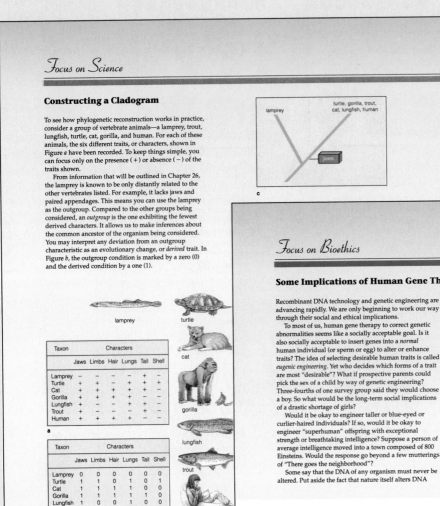

Focus on Science

Constructing a Cladogram

To see how phylogenetic reconstruction works in practice, consider a group of vertebrate animals—a lamprey, trout, lungfish, turtle, cat, gorilla, and human. For each of these animals, the six different traits, or characters, shown in Figure *a* have been recorded. To keep things simple, you can focus only on the presence (+) or absence (−) of the traits shown.

From information that will be outlined in Chapter 26, the lamprey is known to be only distantly related to the other vertebrates listed. For example, it lacks jaws and paired appendages. This means you can use the lamprey as the outgroup. Compared to the other groups being considered, an *outgroup* is the one exhibiting the fewest derived characters. It allows us to make inferences about the common ancestor of the organism being considered. You may interpret any deviation from an outgroup characteristic as an evolutionary change, or *derived* trait. In Figure *b*, the outgroup condition is marked by a zero (0) and the derived condition by a one (1).

lamprey turtle

Taxon	Characters					
	Jaws	Limbs	Hair	Lungs	Tail	Shell
Lamprey	−	−	−	−	+	−
Turtle	+	+	−	+	+	+
Cat	+	+	+	+	+	−
Gorilla	+	+	+	+	−	−
Lungfish	+	−	−	+	+	−
Trout	+	−	−	−	+	−
Human	+	+	+	+	−	−

a

cat

gorilla

Taxon	Characters					
	Jaws	Limbs	Hair	Lungs	Tail	Shell
Lamprey	0	0	0	0	0	0
Turtle	1	1	0	1	0	1
Cat	1	1	1	1	0	0
Gorilla	1	1	1	1	1	0
Lungfish	1	0	0	1	0	0
Trout	1	0	0	0	0	0
Human	1	1	1	1	1	0

b

lungfish

trout

human

316

Just about every day of your life, television, newspapers, and magazines bombard you with "facts" and opinions of "experts" on everything from your personal diet and hygiene to planetwide changes in the atmosphere. The *Focus on Science* essays can help you understand the advantages of critical thinking—of evaluating ideas and observations with an objective eye.

Abortion. Human gene therapy. Pollution versus economic growth. These are a few of the most explosive issues of our time. In the *Focus on Bioethics* essays, we address these bioethical issues in an objective way and invite *you* to think about their implications.

Focus on Bioethics

Some Implications of Human Gene Therapy

Recombinant DNA technology and genetic engineering are advancing rapidly. We are only beginning to work our way through their social and ethical implications.

To most of us, human gene therapy to correct genetic abnormalities seems like a socially acceptable goal. Is it also socially acceptable to insert genes into a *normal* human individual (or sperm or egg) to alter or enhance traits? The idea of selecting desirable human traits is called *eugenic engineering*. Yet who decides which forms of a trait are most "desirable"? What if prospective parents could pick the sex of a child by way of genetic engineering? Three-fourths of one survey group said they would choose a boy. So what would be the long-term social implications of a drastic shortage of girls?

Would it be okay to engineer taller or blue-eyed or curlier-haired individuals? If so, would it be okay to engineer "superhuman" offspring with exceptional strength or breathtaking intelligence? Suppose a person of average intelligence moved into a town composed of 800 Einsteins. Would the response go beyond a few mutterings of "There goes the neighborhood"?

Some say that the DNA of any organism must never be altered. Put aside the fact that nature itself alters DNA

much of the time. The concern is that we don't have the wisdom to bring about beneficial changes without causing harm to ourselves or to the environment.

When it comes to manipulating human genes, one is reminded of our human tendency to leap before we look. When it comes to restricting genetic modifications of any sort, one also is reminded of an old saying: "If God had wanted us to fly, he would have given us wings." And yet, something about the human experience gave us the *capacity* to imagine wings of our own making—and that capacity has carried us to the frontiers of space.

Where are we going from here with recombinant DNA technology, this new product of our imagination? To gain perspective on the question, spend some time reading the history of our species. It is a history of survival in the face of all manner of new challenges, threats, bumblings, and sometimes disasters on a grand scale. It is also a story of our connectedness with the environment and with one another.

The questions confronting you today are these: Should we be more cautious, believing that one day the risk takers may go too far? And what do we as a species stand to lose if the risks are *not* taken?

This chapter opened with the first human application of this technology, meant to save the life of a four-year-old girl. It closes with a *Focus* essay that invites you to consider some prospects and problems associated with the application of recombinant DNA technology to our rapidly advancing knowledge of the human genome.

As with any new technology, the potential benefits of recombinant DNA technology and genetic engineering should be weighed carefully against the potential risks.

SUMMARY

1. Gene mutations and other genetic "experiments" have been going on in nature for billions of years. Through artificial selection, humans have been manipulating the genetic character of different species for many thousands of years. Today, recombinant DNA technol-

ogy has enormously expanded our capacity to modify organisms genetically.

2. With recombinant DNA technology, researchers isolate, cut, and splice together gene regions from different species, then greatly amplify the number of copies of those regions into usefully large quantities. Enzymes make the cuts and do the splicing.

a. Researchers use restriction enzymes and DNA ligase to cut and insert DNA into plasmids from bacterial cells. (The plasmids are small, circular DNA molecules that contain extra genes, besides those of the bacterial chromosome.) A DNA library is a collection of DNA fragments, produced by restriction enzymes, and incorporated into plasmids.

b. Restriction fragments may be amplified by a population of rapidly dividing cells (such as bacteria or yeasts). Short DNA fragments may be amplified more rapidly in a test tube by the polymerase chain reaction (PCR).

c. Following amplification, the DNA fragments can be sorted out (according to length), and their nucleotide sequences can be determined.

The *Focus on Health* essays help dispel much of the misleading or downright inaccurate information that exists today on health topics. We worked closely with dozens of specialists to give you information you can use with a high degree of confidence. These current essays highlight behavioral, genetic, and environmental factors that can affect your health, from malaria (the essay excerpted here) to eating disorders.

Focus on the Environment

Pastures of the Seas

Drifting through the surface waters of the world ocean are uncountable numbers of single cells. You can't see them without a microscope—a row of 7 million cells of one species would be less than a quarter-inch long. Yet are they abundant! In some parts of the world, a cup of seawater may hold 24 million cells of one species, and that doesn't even include all the cells of *other* aquatic species.

Many of the drifters are photosynthetic bacteria, protistans, and plants. Together they are the pastures of the seas, the food base for diverse consumers. The pastures "bloom" in spring, when waters become warmer and

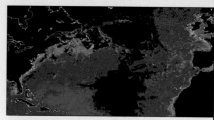

a Photosynthetic activity in winter. (In this color-enhanced image, red-orange shows where chlorophyll concentrations are greatest.)

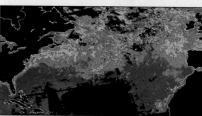

b Photosynthetic activity in spring.

Focus on Health

Malaria and the Night-Feeding Mosquitoes

Each year, about 150 million people come down with *malaria*, and 1 to 2 million die from it. At one time, this long-lasting disease was prevalent mostly in tropical and subtropical parts of Africa. However, the number of cases in North America and elsewhere is increasing dramatically, owing to globe-hopping travelers and unprecedented levels of immigration.

A sporozoan (*Plasmodium*) causes malaria. The mosquito *Anopheles* transmits its sporozoites to human (or bird) hosts. The sporozoites repeatedly engage in multiple fission, producing thousands of progeny that infect and multiply inside of red blood cells. The cells rupture and release the parasite's metabolic wastes, which cause fever and chills.

The *Plasmodium* life cycle (Figure *a*) is attuned to the host's body temperature, which normally rises and falls rhythmically every 24 hours. Gametocytes, the infective stage that is transmitted to new hosts, are ready for travel at night—when mosquitoes feed.

Travelers in countries with high rates of malaria are advised to use antimalarial drugs such as chloroquine. Drug resistance is common, however. And a vaccine has been difficult to develop. Vaccines induce the body to build up its resistance to a specific pathogen. Experimental vaccines for malaria have not been equally effective against all the different stages that develop during sporozoan life cycles. This is true of most parasites with complex life cycles.

When studying Figure 7.10, remind yourself of how photosynthesis "fits" in the world of living things. Think of the mind-boggling numbers of single-celled and multicelled photosynthetic organisms in which the reactions are proceeding at this very moment, on land and in the sunlit waters of the earth. The sheer volume of the reactant and product molecules that the photosynthesizers deal with may surprise you. The *Focus* essay above is a case in point.

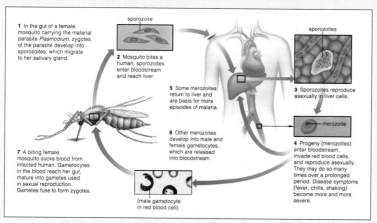

1 In the gut of a female mosquito carrying the malarial parasite *Plasmodium*, zygotes of the parasite develop into sporozoites, which migrate to her salivary gland.

sporozoite

2 Mosquito bites a human, sporozoites enter bloodstream and reach liver.

5 Some merozoites return to liver and are basis for more episodes of malaria.

sporozoites

3 Sporozoites reproduce asexually in liver cells.

6 Other merozoites develop into male and female gametocytes, which are released into bloodstream.

merozoite

4 Progeny (merozoites) enter bloodstream, invade red blood cells, and reproduce asexually. They may do so many times over a prolonged period. Disease symptoms (fever, chills, shaking) become more and more severe.

7 A biting female mosquito sucks blood from infected human. Gametocytes in the blood reach her gut, mature into gametes used in sexual reproduction. Gametes fuse to form zygotes.

(male gametocyte in red blood cell)

a *Plasmodium*, a sporozoan that causes malaria. Its life cycle requires a human host and an insect intermediate host.

Similarly, a great deal of conflicting information exists on many environmental issues. The *Focus on the Environment* essays present the thinking of respected environmentalists and ecologists to give you a balanced picture of the major issues in this critical area, such as the health of our oceans, noise pollution, and many others.

Vignettes

You can view biology as just another course that you have to take in order to graduate. Or you can open your mind to a fantastically rewarding way to gain perspective on the world around you—and on your place in it. No matter what comes to mind—be it stings from bees or jellyfishes, or firestorms through the canyons of Malibu—you will be armed with knowledge that can help you make sense of it all.

Beautifully illustrated short stories—vignettes—start each chapter. Many, including the one shown here, convey *how* biologists think critically and conduct their research. Each vignette leads you, in a nontechnical way, to the chapter's key concepts. Right after the vignette, the concepts are listed as simple summary statements. At each chapter's end, those same concepts are reinforced in a "Summary" section (see Built-In Study Aids on the next page of this Guide).

---

19 SPECIATION

The Case of the Road-Killed Snails

If you happen to be a snail living in a garden in Bryan, Texas, it doesn't take much to keep your genes away from snails in a backyard across the street. By day, the sunbaked asphalt would be about as inviting to a snail as a desert would be to a catfish. Besides, day or night, a street-traversing snail is vulnerable to cars, trucks, skateboards, and bicycles (Figure 19.1). Whatever else it might be, that strip of asphalt is a formidable barrier to gene flow between populations. For snails, that is.

Whether or not a physical barrier deters gene flow depends partly on an organism's mode of locomotion or dispersal. It also depends on how fast and how long an organism *can* move in response to environmental factors or its own hormones.

A snail is not swift, and it does not roam far from its home population. Compare it to the wild black duck, banded in Virginia in 1969, that turned up eight years later in Korea. Compare it to the wandering albatross, one of the supreme barrier busters. After lifting off from Kerguelen Island in the Indian Ocean, one of these birds soared westward past Africa, across the Atlantic, and around Cape Horn. Thirteen thousand kilometers later, it landed in Chile. With a lightweight body and a wingspan of 3.65 meters (12 feet across), the albatross

was able to exploit the great prevailing winds of the Southern Hemisphere.

And yet, in 1859, snails did cross an ocean. Humans transported garden-variety snails (*Helix aspersa*) from France to California, then turned them loose. The idea was that the snails would multiply and so meet the demand for that French delicacy, *escargot aux fines herbes*. The snails not only exceeded expectations, they became an absolute nuisance in gardens throughout the southwestern United States.

By the 1930s, *H. aspersa* had hitched rides, possibly as eggs in the soil of plant containers, to Bryan, Texas, and founded small colonies in the local vegetation. Forty years later, Robert Selander, now of Pennsylvania State University, was down on his hands and knees with a few graduate students, scouring patches of vegetation on two adjacent city blocks. Why? Selander wanted to determine the effect of genetic drift on the introduced populations. He and his students collected every single snail—2,218 of them—from fourteen local populations. Each population was analyzed to determine the allele frequencies at five different gene locations. In all five cases, the results pointed to some genetic variation between colonies on the same block—

b

and to major differences in allele frequencies *between* blocks. Figure 19.1*a* shows the results for one of the genes studied.

Assume that genetic differences between the populations continue to increase, as through genetic drift and, later, natural selection. Will the time come when snails from opposite sides of the street can no longer interbreed successfully, even if they do manage to get together? In other words, *will they become members of separate species*? Or will selection work against increases in genetic differences by eliminating extreme phenotypes from the populations? After all, how much can a workable package of *H. aspersa* genes evolve in adjacent patches of vegetation—and under similar environmental pressures—in a small town in Texas?

With this chapter, we turn to **speciation**—to changes in allele frequencies that are significant enough to mark the formation of daughter species from a parental species. Obviously, no one was around to watch the formation of species during the past. No one lives long enough to know whether many existing populations are at some intermediate stage leading to speciation. Evolutionary biologists are still working out their theories about speciation processes, often quite contentiously. What you are about to read may change in the near or distant future.

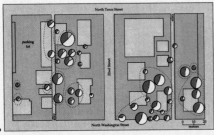

Figure 19.1 (a) Results from a study of neighboring populations, all descended from snails (*Helix aspersa*) that arrived in a town in Texas in the 1930s. For each population, a circle represents relative abundances of three alleles (coded *yellow*, *gold*, or *brown*) for leucine aminopeptidase (an enzyme). Greater genetic variation exists between the populations living on opposite sides of the street. (b) A snail about to encounter a barrier to gene flow.

KEY CONCEPTS

1. A species consists of one or more populations of individuals that can interbreed under natural conditions and produce fertile offspring, and that are reproductively isolated from other such populations. This definition is restricted to sexually reproducing species.

2. The populations of a species have a shared genetic history, they are maintaining genetic contact over time, and they are evolving independently of other species.

3. Species are distinct, well-adapted entities that resist change unless the environmental conditions to which they are adapted change.

4. Speciation is the formation of daughter species from a parent species.

5. Three models of speciation guide current research. In allopatric speciation, species form in separated populations or subpopulations. In parapatric speciation, they form in a region where populations share a border. In sympatric speciation, they form within the home range of the parent species.

6. Speciation requires irreversible genetic divergence of one population from others. Isolating mechanisms maintain genetic divergence.

7. The timing, rate, and direction of speciation vary within and between lineages. The extinction of some number of species is inevitable for all lineages.

287

Built-In Study Aids

Besides the visual summaries, in-text summary statements, and other features presented on previous pages of this guide, each chapter concludes with the following useful student aids:

Summaries

presented in a list format reinforce the main chapter concepts.

Review Questions

are keyed to italicized and boldfaced sentences to further reinforce main concepts. These questions are cross-referenced to page numbers in the text. Detailed answers to all review questions are available in a separate booklet.

Key Terms

consist of a list of all the **boldfaced** terms defined in the chapter. Check to see if you understand their meaning. The *italicized* page numbers provide fast entry back into the chapter for the ones you aren't sure of. All terms are also defined in the book's glossary.

SUMMARY

1. Parent cells provide each daughter cell with the hereditary instructions (DNA) and cytoplasmic machinery necessary to start up its own operation.

2. Mitosis divides the hereditary instructions in the nucleus of a parent cell into two equivalent parcels. It is the basis for growth and tissue repair in multicelled eukaryotes. It also is the basis of asexual reproduction among many single-celled and multicelled eukaryotes. (Meiosis, another nuclear division mechanism, occurs only in germ cells set aside for sexual reproduction.)

3. The cytoplasm divides by different mechanisms near the end of nuclear division or at some point thereafter. (Animal cells are cleaved in two. A cell plate forms and divides plant cells.)

4. A chromosome is a single DNA molecule. In eukaryotes, many proteins are attached to it. A cell's chromosomes differ from one another in length, shape, and which portion of the hereditary instructions they carry. Members of the same species have the same total number of chromosomes in their somatic cells.

5. "Chromosome number" is not the same thing as the total number of chromosomes. It is the number of *each type* of chromosome characteristic of a species. Mitosis keeps the chromosome number constant from one cell generation to the next. Thus, if a parent cell is diploid (with two chromosomes of each type), the daughter cells resulting from mitosis will be diploid also.

6. The cell cycle starts when a new cell forms, proceeds through interphase, and ends when the cell reproduces by mitosis and cytoplasmic division. At interphase, a cell increases in mass, doubles its number of cytoplasmic components, and duplicates all chromosomes.

7. After duplication, each chromosome consists of two DNA molecules attached at the centromere. For as long as the two stay attached, they are called sister chromatids of the chromosome.

8. Mitosis proceeds through four continuous stages:
a. Prophase. Duplicated, threadlike chromosomes start to condense. New microtubules assemble in organized arrays near the nucleus; they will form a spindle. The nuclear envelope starts to break up.
b. Metaphase. During the *transition* to metaphase, the nuclear envelope breaks up completely. Microtubules of a forming spindle attach to the sister chromatids of each chromosome and orient the two toward opposite spindle poles. *At* metaphase, all chromosomes are aligned at the spindle equator.
c. Anaphase. Microtubules pull the sister chromatids of each chromosome away from each other, to opposite

spindle poles. Every chromosome that was present in the parent cell is now represented by a daughter chromosome at both poles.
d. Telophase. The chromosomes decondense to the threadlike form. A new nuclear envelope forms around them. Each nucleus has the same chromosome number as the parent cell. Mitosis is completed.

Review Questions

1. Define the two types of nuclear division mechanisms that occur in eukaryotes. Does either one divide the cytoplasm? *142*

2. Define somatic cell and germ cell. *142*

3. What is a chromosome? What is a chromosome called in its unduplicated state? In its duplicated state (that is, with two sister chromatids)? *142–143*

4. Describe the spindle apparatus and its general function in nuclear division. *148–149*

5. Name and describe the main features of the four stages of mitosis. At what stages were the following micrographs of a plant cell taken? *148–150*

6. Correctly label the component parts of this human chromosome. *148*

7. Describe cytoplasmic division as it occurs in plant cells and animal cells. *150*

Self-Quiz (Answers in Appendix IV)

1. A eukaryotic chromosome consists of _____ .
a. DNA only
b. DNA plus proteins
c. DNA plus membrane
d. DNA plus lipids

2. A cell with pairs of the chromosomes characteristic of the species is a(an) _____ cell.
a. diploid
b. mitotic
c. abnormal
d. a and c

3. A duplicated chromosome has _____ chromatid(s).
a. one
b. two
c. three
d. four

4. The _____ is the constricted region of a duplicated chromosome, with attachment sites for microtubules.
a. chromatid
b. centromere
c. cell plate
d. cleavage furrow

5. In the cell cycle, the interphase stage immediately following DNA replication is _____ .
a. G₂
b. G₁
c. S
d. none of the above

6. Interphase is the stage when _____ .
a. nothing occurs
b. a germ cell forms its spindle apparatus
c. a cell grows and duplicates its DNA
d. mitosis occurs

7. After mitosis, the chromosome number of a daughter cell is _____ the parent cell's.
a. the same as
b. one-half
c. rearranged compared to
d. doubled compared to

8. The function of mitosis and cytoplasmic division is _____ .
a. asexual reproduction of single-celled eukaryotes
b. growth, repair, and asexual reproduction in many eukaryotes
c. gamete formation in eukaryotes
d. both a and b are correct

9. Only _____ is not a stage of mitosis.
a. prophase
b. interphase
c. metaphase
d. anaphase

10. Match each stage with its key events.
____ metaphase
____ prophase
____ telophase
____ anaphase
a. sister chromatids of each chromosome separate
b. threadlike chromosomes start to condense
c. chromosomes decondense, daughter nuclei form
d. all chromosomes are aligned at the equator

Selected Key Terms

anaphase *146*	kinetochore *148*
cell cycle *144*	meiosis *142*
cell plate formation *150*	metaphase *146*
centriole *148*	mitosis *142*
centromere *142*	prophase *146*
chromosome *142*	reproduction *142*
chromosome number *143*	sister chromatid *142*
cleavage *150*	somatic cell *142*
diploid cell *143*	spindle apparatus *146*
germ cell *142*	telophase *146*
interphase *144*	

Readings

Gallagher, G. March 1990. "Evolution: The Mitotic Spindle." *The Journal of NIH Research.*

Murray, A., and M. Kirschner. March 1991. "What Controls the Cell Cycle?" *Scientific American* 264(3): 56–63.

Science. November 3, 1989. "Frontiers in Biology: The Cell Cycle."

Wolfe, S. 1993. *Molecular and Cellular Biology.* Belmont, California: Wadsworth.

Genetics Problems

are available in the unit on inheritance. They give you a better sense of Mendelian patterns of inheritance. Appendix III gives detailed solutions to assist understanding.

Self-Quizzes

are a quick way to test your recall of important definitions and key concepts with the help of objective questions. Appendix IV lists the answers.

Recommended Readings

provide you with a starting point for exploring a topic in more depth.

Consistent Color Coding

Carbohydrates

Lipid heads

Lipid tails

Lipid bilayer of cell membranes

Proteins

DNA, chromosomes

RNAs

ATP

Coenzymes (NADP$^+$, NAD$^+$, FAD)

$\Longrightarrow$ Energy flow

} polar (hydrophilic) head

} nonpolar (hydrophobic) tails

From page 78

Color coding helps remind you of the different classes of molecules. It can also help you remember the structural and functional uses of these molecules, as when they must be tracked through metabolic pathways. Shown here is a guide to the color coding, plus several examples from the text that illustrate the consistency of the color coding throughout.

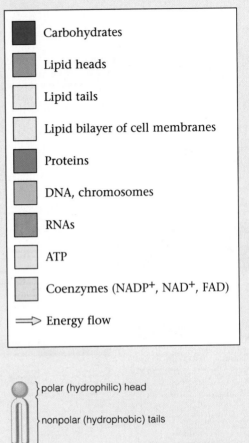

From page 41

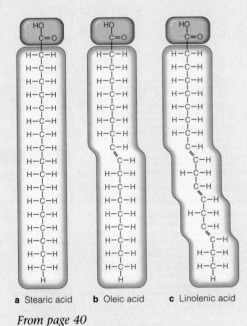

a Stearic acid **b** Oleic acid **c** Linolenic acid

From page 40

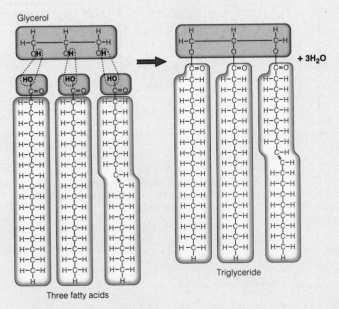

Glycerol

Three fatty acids

Triglyceride

+ 3H$_2$O

From page 40

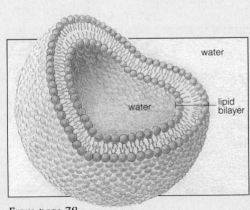

water

water

lipid bilayer

From page 78

THIRTY-TWO SUPPLEMENTS

Full-color transparencies and *35mm slides* are available for nearly all of the book's illustrations. All illustrations have vivid colors and large, bold-lettered labels. All diagrams are on CD-ROM. For easy access, these images are delivered in Kudo run time catalog format. The catalog has an image drag-and-drop feature for presentation tools.

Collaborating with Carolina Biological Supply, Bill Surver of Clemson University developed a new *text-specific videodisc*. Its material corresponds to each chapter of *Biology: The Unity and Diversity of Life*. All the animations and films are new. It includes nearly all of the line art, clearly labeled in large, boldface type. Many book illustrations either are broken down into step-by-step format or are animated in full motion. They are also available with fill-in-the-blank labels for testing purposes. There are 3,500 photographic stills. A directory and bar code guide come with the disk. A separate videodisc *Correlation Directory and Barcode Guide* correlates the text with three generic general biology videodiscs. *HyperCard Stacks for the Mac* and *Toolbook for the IBM* correlate the text with all four videodiscs.

Two interactive disks, entitled *Liquid Assets: The Ecology of Water* explore endangered species, environmental restoration, and other topics. A study guide accompanies the disks. The *Test Items* booklet has more than 5,000 questions by outstanding test writers. Questions exist in electronic form for IBM and Macintosh in a test-generating data manager. Questions are also available in Word-Perfect for DOS and Microsoft Word for the Mac.

The *Instructor's Resource Manual* has the following items for each chapter: an outline, objectives, a list of boldface or italic terms, and a detailed lecture outline. The manual includes hundreds of suggestions for lecture presentations, classroom and laboratory demonstrations, suggested discussion questions, research paper topics, and annotations for filmstrips and videos. *Lecture outlines* in the *Instructor's Resource Manual* exist on a Mac or IBM disk for those who wish to modify the material.

The new, active-oriented *Study Guide and Workbook* asks students to respond in writing to almost all of the questions given. The guide has questions arranged by chapter section for ease of assignment. Each of its chapters includes a set of critical thinking questions. For those who wish to modify or select portions of the material, *chapter objectives* of the guide are available on disk as part of the testing file. An *electronic study guide* has more than 50 multiple-choice questions per chapter; these differ from those in the test-item booklet. After students respond to each question, an onscreen prompt lets them review their answers and learn why they are correct or incorrect.

The 100-page *Answer Booklet* has answers to the end-of-chapter, critical thinking, and review questions in the textbook. (The self-quiz questions are answered in the textbook itself.)

A special version of *STELLA II*, a software tool for developing critical thinking skills, together with a work-book, is available to adopters. In addition to the critical thinking questions in the textbook itself, 150 questions, arranged by chapter, exist in a booklet, *Critical Thinking Questions*.

One thousand glossary items exist on *flashcards*. Also available is a booklet, *Building Your Life Science Vocabulary*, that helps students learn biological terms by explaining root words and their applications.

Eight additional-readings supplements are available:

- *Contemporary Readings in Biology* has articles on applications of interest to students.
- *Science and the Human Spirit: Contexts for Writing and Learning* helps students learn to write effectively about biology.
- *A Beginner's Guide to Scientific Method* is a useful supplement for those who wish to cover this topic in detail.
- *The Game of Science* gives students a realistic view of what science is and what scientists do.
- *Environment: Problems and Solutions* provides a brief, 120-page introduction to the environment.
- *Green Lives, Green Campuses* is a hands-on workbook to help students evaluate the environmental impact of their own lifestyles.
- *Watersheds: Classic Cases in Environmental Ethics* provides major case studies.
- *Environmental Ethics: An Introduction to Environmental Philosophy* offers a survey of environmental ethics examining current philosophical positions.

The new *Laboratory Manual* by Jim Perry and David Morton has 38 experiments and exercises, with more than 600 full-color labeled photographs and diagrams. The authors have divided many experiments into parts that can be assigned individually, depending on time available. For each experiment, the manual includes objectives, discussion (introduction, background, and relevance), a list of materials for each portion of an experiment, procedural steps, pre-lab questions, and post-lab questions. An *Instructor's Manual* accompanies the laboratory manual. It covers quantities, procedures for preparing reagents, time requirements for each portion of an exercise, hints to make the lab a success, and vendors of materials with item numbers. It offers additional investigative exercises that can be copied for laboratory use.

Another laboratory manual by Phillip Shelp and its accompanying instructor's manual can be customized for your course. The new *Photo Atlas* contains 800 labeled full-color photographs of organisms and micrographs that students view in the lab.

PREFACE

Think of what you might pay for a night out on the town or a single top-of-the-line athletic shoe. The book you are about to read probably costs about the same. Will it be worth it? That depends on what you expect to get out of a course in general biology. More than this, it depends on what you expect to get out of life.

Reading *Biology: The Unity and Diversity of Life* is like having private audiences with speakers from colleges, universities, and research laboratories the world over. With it, you gain insights from 2,000 biologists who, over the past two decades, contributed time and energy to shaping each new edition. You benefit from classroom feedback from a great number of the several million students who used earlier editions and helped us dispel patches of foggy writing.

Together with your biology instructor, this book can sharpen your view of life. The biological perspective happens to be one of the most useful and rewarding tools for interpreting the world. With it, you can cut your own intellectual paths through social, environmental, and medical thickets. With it, you can come to understand the past and to predict possible futures for ourselves and all other organisms.

OBJECTIVES FOR THIS EDITION

People who write biology textbooks have the same basic objectives. Their writing should be engaging, logical, clear, and never patronizing. The book as a whole should be framed in light of the most inclusive principles of biochemistry, inheritance, and evolution. The power of evolutionary theory to explain life's unity and diversity should be conveyed through examples of problem solving and experiments. For each major field of inquiry, key concepts must be identified, and the topics selected should give students a sense of current understandings and research trends. The structure and functioning of a good sampling of organisms should be explained in enough detail to help students build a working vocabulary about life's parts and processes.

Four more objectives guided this seventh revision of *Biology*. First, we decided to organize the entire book as self-contained topic spreads. Second, we refined and updated our selection of applications. Third, at the start of each chapter, we wrote a lively or sobering short story that leads to key concepts in a comfortable, non-threatening way. Fourth, we created many more visual summaries. With near-universal enthusiasm, students

report back that the summary illustrations made it much easier for them to work their way through difficult concepts.

REVISION HIGHLIGHTS

Major Refinements in Content

Biology has no room for minimalist authorship; it simply advances too rapidly on too many fronts. Are we obsessed with the idea of presenting each field of inquiry in updated, unbiased fashion? Maybe. Each time around, we have evaluated every chapter's content, line by line, for currency and evenhandedness in presentation, and this revision was no exception.

The treatment of evolution and diversity (Units III and IV) differs in significant ways from the preceding edition. We now treat microevolutionary processes, speciation, and macroevolution in separate chapters. We worked with leading evolutionists to reorganize and clarify the existing material on these topics. We incorporated new text and illustrations on recent field work, experiments, and working models.

The diversity unit now opens with a chapter on the origin and evolution of life. This overview chapter presents ideas about the chemical origins of the first prokaryotic cells and includes a major essay on the possible endosymbiotic origins of eukaryotic cells. To provide for greater clarity and more flexibility in classroom assignments, we accorded the bacteria and viruses their own chapter. After consulting at great length with numerous specialists, we accorded the protistans their own chapter and now group the red, brown, and green algae in this kingdom. The classification scheme in Appendix I reflects the reorganization. The chapter on human evolution now follows the chapter on vertebrate evolution and concludes the diversity unit.

We reordered some topics in Unit VI, which deals with animal anatomy and physiology. Now sensory reception immediately follows the chapters on neurophysiology. Circulation and respiration now precede the topics of digestion and human nutrition. Users tell us that the new chapter on immunity is notable for its clarity, given the fog-shrouded lexicon of its practitioners. The case study of human reproduction and development (Chapter 45), whether assigned or not, draws intense student interest. Besides updating the chapter, this time around we created spectacular, informative illustrations for it.

Self-Contained Topic Spreads

Skim through this new edition and a major feature becomes apparent. *All of the text and illustrations for a particular topic appear together, on the same page or on two pages that face each other*. Students can work their way through one topic at a time, without page-flipping searches for information. These topic spreads are easy to identify. They start with a boxed number and heading at the upper left corner of a page. They typically end with concise statements or lists, identified by box rules, that summarize key concepts within that spread.

Consider aerobic respiration, a fear-inspiring topic for most students. Now look at Chapter 8, with its self-contained page layouts on glycolysis, the Krebs cycle, and electron transport phosphorylation. The text on each of these three stages of reactions has a clearly identified beginning and end. It is not cluttered with details that could obscure the main points. We do present details, in nonthreatening illustrations. Simple captions within illustrations walk students through the reactions, one step at a time. Icons next to the illustrations serve as reminders of how the stages fit together and where the reactions proceed in cells.

This is not to say that format dictated content, like the tail wagging the proverbial dog. First we developed text, illustrations, and captions for each chapter as part of an integrated, continuous story. We blocked out headings and subheadings to highlight the relative importance of sections to one another. Any good story has this hierarchy of information, with major and minor characters, background development, and high points where everything comes together. Without such a hierarchy, information has all the excitement, flow, and drama of an encyclopedia.

Only after we were satisfied with the content did we assemble topic spreads. This entailed paragraph tinkerings and major creative adjustments in the size of illustrations and photographs. We introduced subtle variations in page layouts to avoid the monotonous, encyclopedic look that makes students wonder whether they are brain dead. Students get the best of both worlds: appropriate, interesting content—*and* cleanly served-up content, one topic at a time.

Illustrations

Instructors often ask us how we come up with such unique, pedagogically sound illustrations. All it takes, really, is a twenty-year obsession with writing and creating illustrations simultaneously, as inseparable parts of the same story. Researching, developing, and integrating illustrations with the text takes almost as much time as the writing.

Visual Summaries One of the most helpful things we do for students is to create summary illustrations. Each contains step-by-step, written descriptions of a topic *within* the illustration itself. When we break down information into a series of illustrated steps, students find this far less threatening than a complicated, "wordless" diagram. *In this edition, all major concepts now have their own visual summaries.*

New to this edition are visual summaries for anatomical drawings—they integrate structure and function. Students need not jump back and forth from text, to tables, to illustrations in order to get an overview of how an organ system is put together and what its assorted parts do. Even the individual descriptions are arranged to reflect the organ system's structural and functional organization.

Zoom-Sequence Illustrations Many illustrations progress from macroscopic to microscopic views. Figure 7.2 is a fine example; this zoom sequence shows where the reactions of photosynthesis proceed. Another example, in Chapter 38, proceeds down through levels of skeletal muscle contraction.

Color Coding Careful use of color helps students track information on hard-to-visualize molecular topics. Throughout the book, for instance, proteins are color-coded *green*, and DNA is color-coded *blue*. The *Guide to Learning* on page xxiii lists all the color codes.

Icons These small pictorial representations relate the topic of an illustration to the big picture. For instance, small but detailed representations of a cell remind students that a particular event proceeds in the nucleus or the cytoplasm. Some icons in the text serve as gentle reminders of the evolutionary relationships among animal phyla.

Improvements in the Writing

The illustrations, vignettes, and *Focus* essays should not distract you from the *line-by-line* judgment calls made with respect to the text. We tossed out verbosity and cleaned up the logic, it is as simple as that.

Actually, it is not as simple as that. We worked just as hard to make this edition *interesting* to read. Years ago, we had an idea that students could be enticed into the core science of biology with lively writing, memorable analogies, and engaging bits of natural history. The prevailing view was that such an idea was inappropriate for something as serious as science. So for fifteen years, we focused primarily on making the writing clear and the science accurate.

Today, students often pick up biology textbooks with apprehension. If the words do not engage them, they sometimes end up hating the book *and* the subject. Instructors still ask for a scientifically accurate book—but now they also ask for one that puts the life back in.

We could not be more pleased. Because we devoted years to writing with precision and objectivity, we have a strong sense of when to leave core material alone and when to loosen it up. Interrupting a description of, say, the mechanisms of mitosis with a distracting anecdote does a struggling student no good. Plunking humorous asides into a chapter that ties together organismic and geologic evolution trivializes a magnificent story. By contrast, it certainly is appropriate to liven up a paragraph on, say, the structure of the nuclear envelope, the role of mitosis in growth and development, and the functions of skin.

Balancing Concepts and Applications

At strategic points, examples of applications parallel the flow of concepts—not so many as to be distracting, but enough to keep minds perking along with the core material. We index the applications separately, on the back end papers, for easy reference. Many are in *Focus* essays on scientific methods, human health, bioethics, and the environment. They provide in-depth information for interested students but do not interrupt the text flow. For example, one essay asks students to look beyond clinical trials for human gene therapy, as described in the chapter, to the question of whether we as a species are ready to engage in such gene tinkerings.

Vignettes

Taking up valuable reading time with interesting stories is pointless—unless they lead students to the big concepts. Take a look at how the vignettes on growing old (Chapter 6) and on "killer bees" (Chapter 8) reinforce the concept of life's biochemical unity. Consider how another describes the effects of crack cocaine on the nervous system (Chapter 35) and puts this in evolutionary perspective. Or consider how the vignette on the trashing of Antarctica (Chapter 48) drives home the concept of life's interconnectedness. Classroom responses indicate that the accompanying photographs are arresting enough to make students sit up and take notice.

Doing Science

We continue to use examples of biologists at work to help students develop an understanding of critical thinking. Students get ample descriptions of experimental evidence for the concepts being discussed, as in the sections on plant development (Chapter 32). Several vignettes, including those on pages 286 and 910, provide lively glimpses into how experiments are set up. The index lists these and other selections (under the entry, *Experiments*).

We even selected and wrote certain chapters to give students a sense of how science proceeds. Examples are Chapters 11 (Mendelian genetics), 13 (DNA structure and function), and 51 (an evolutionary approach to animal behavior). Also, *Focus on Science* essays in each unit provide detail on biologists in action. For example, John Alcock contributed two *Focus on Science* essays. The first, an entertaining account of scientific methods, concludes with a real-life example (Chapter 1). The other describes how DNA fingerprinting was used to help explain self-sacrificing behavior among a fascinating group of mammals (Chapter 52).

As another example, the *Focus on Science* essay on page 331 helps students perceive that biology is not a closed book, so to speak. Even when research suddenly brings a sweeping story into sharper focus—in this case, the origins of the great prokaryotic and eukaryotic kingdoms of life—questions abound about the details. Today's students may well be among the new scientific detectives who eventually find answers to such fascinating questions.

STUDY AIDS

Following each chapter introduction is a list of *key concepts*. Topic spreads end with *summary statements* or *summary lists* that help keep readers on track. End-of-chapter study aids reinforce the key concepts. They include a *summary* in list form, *review questions*, a *self-quiz*, *selected key terms*, and *recommended readings*. Page numbers tie review questions and key terms to relevant text pages. *Genetics problems* help students grasp the principles of inheritance.

A *glossary* includes the boldfaced terms in the text, as well as pronunciation guides and origins of words when such information will make formidable words less so. Our *index*, with its fine divisions of topics, helps students find a door to the text more quickly.

Students can use the detailed *classification scheme* in Appendix I for reference purposes. Appendix II includes *metric-English conversion charts*. The third appendix has *detailed answers* to genetics problems, and the fourth has *answers* to self-quizzes. For interested students and professors who prefer the added detail, Appendix V gives structural formulas for the *major metabolic pathways*, and Appendix VI shows the periodic table of the elements.

Appendixes and the *Glossary* are printed on paper of different tints to preclude frustrating searches for where one ends and the next starts.

One, two, or a smattering of authors can write accurately and often well about their field of interest, but it takes more than this to deal with the full breadth of the biological sciences. For us, it takes an educational network of more than 2,000 teachers, researchers, and photographers in the United States, Canada, England, Germany, France, Australia, Sweden, and elsewhere.

The two pages that follow list the reviewers whose recent contributions continue to shape our thinking. There simply is no way to describe the thoughtful effort that these individuals and other reviewers before them gave to our books. We can only salute their commitment to quality in education.

Through their long-term commitment, kindness, and good humor, certain biologists helped chisel major portions of *Biology: The Unity and Diversity of Life*. John Alcock, Aaron Bauer, Rob Colwell, George Cox, Daniel Fairbanks, Eugene Kozloff, Paul Hertz, Bill Parson, Cleon Ross, Sam Sweet, Steve Wolfe—these are not only exceptional advisors, they are contributors of resource manuscripts. In large part, *Biology* has become widely respected because it reflects their thorough understanding of their respective fields of interest.

New to this group is Robert Lapen, who wrote the definitive manuscript on immunology and lived to tell about it.

This time around, Jerry Coyne and John Endler independently pummeled us into doing a better job of representing the divergent schools of thought on evolutionary theory. Lynn Margulis and others persuaded us finally to give those underappreciated protistans their due.

Kate Denniston, Tom Garrison, Ron Hoham, Bruce Holmes, Tyler Miller, David Morton, and Frank Salisbury have been fine guardians of accuracy and teachability. Tom and Tyler especially have the good grace and curious minds to discuss fine points with me even at home, even in the dead of night. John Jackson, another ongoing, cheerful scrutinizer, also wrote many of the self-quiz questions for this edition.

The current edition benefits from the fine editorial analyses of Beverly McMillan, Elmarie Hutchinson, and Mary Arbogast. Bev, who also drafted several of the vignettes and essays, is now busily coauthoring other biology textbooks for Wadsworth.

Mary Douglas is the only one between us and extinction as publishing lurches into the electronics age. We simply cannot imagine these complex revisions moving through computerized design and production without the talented, flexible, even-tempered Mary and her calm yet tenacious art coordinator, Myrna Engler-Forkner.

Raychel Ciemma, a friend of long standing, works directly with us to turn our rough sketches truly into works of art. Her paintings in the chapter on human reproduction and development especially are stunners.

As if designing our remarkable new format weren't enough, Gary Head went out on photographic assignment and came back with such treats as Fred-and-Ginger, the streetwise snail in Figure 19.1*b*. Amy Head served as apprentice in these novel enterprises.

Bless Marion Hansen's amnesia; this splendid photo researcher and permissions editor keeps on forgetting that she vows never to get sucked into another edition.

Thank goodness for Kristin Milotich, Kerri Abdinoor, and Sandra Craig, each a true editorial professional with endurance, a big heart, and no attitude problem. Kristin is now in charge of the book's supplements.

And thank goodness for Kathie Head, Pat Brewer, Stephen Rapley, Randy Hurst, Bill Ralph, and Peggy Meehan for their steady guidance over the years. We are glad that Gary Carlson and Dave Garrison contribute to the team effort. We are grateful for the contributions of Rosaleen Bertolino, Marilyn Evenson, Bob Demarest, Natalie Hill, Carol Lawson, Jill Turney, Ed Serdziak, John Douglas, and the other individuals listed on the copyright page, and for the many kindnesses of Verbal Clark and Laurie Campbell.

Jan Flessner and her colleagues at Precision Graphics, Tom Anderson and Nancy Dean at H&S Graphics, and Jim Jeschke and Jody Ward at American Composition & Graphics, Inc., accommodated our pursuit of excellence, which often sends people up the walls and clinging to the ceiling.

Twenty years ago, Jack Carey convinced us to do the first edition of this book. Ever since, he has remained our closest counselor, our abiding friend. And nothing, in all that time, has shaken our shared belief in the intrinsic capacity of biology to enrich the lives of each new generation of students.

REVIEWERS

ABBAS, ABDUL, *Brigham Womens Hospital*
ADAMS, THOMAS, *Michigan State University*
ALDRIDGE, DAVID, *North Carolina State University*
AMOS, BONNIE, *Angelo State University*
ANDERSON, DEBRA, *Oregon Health Sciences University*
ARMSTRONG, PETER, *University of California, Davis*
ARNOLD, STEVAN, *University of Chicago*
ATCHISON, GARY, *Iowa State University*
BAJER, ANDREW, *University of Oregon*
BAKKEN, AIMEE, *University of Washington*
BAPTISTA, LUIS, *California Academy of Sciences*
BARKWORTH, MARY, *Utah State University*
BATZING, J. L., *State University of New York, Cortland*
BECK, CHARLES, *University of Michigan*
BEECHER, MICHAEL, *University of Washington*
BEEVERS, HARRY, *University of California, Santa Cruz*
BELL, MICHAEL, *Collin County Community College*
BELL, ROBERT, *University of Wisconsin, Stevens Point*
BENACERRAF, BARUJ, *Harvard School of Medicine*
BENDER, KRISTEN, *California State University, Long Beach*
BENES, ELINOR, *California State University, Sacramento*
BERGSTROM, BRADLEY, *Valdosta State College*
BIDLACK, JAMES, *University of Central Oklahoma*
BIRKEY, C. WILLIAM, *Ohio State University*
BISHOP, VERNON, *University of Texas, San Antonio*
BLYSTONE, ROBERT, *Trinity University*
BOHR, D. F., *University of Michigan Medical School*
BONNER, JAMES, *California Institute of Technology*
BOOHAR, RICHARD, *University of Nebraska, Lincoln*
BOTTRELL, CLYDE, *Tarrant County Junior College South*
BRADBURY, JACK, *University of California, San Diego*
BRENGELMANN, GEORGE, *University of Washington*
BRINKLEY, B., *University of Alabama, Birmingham*
BRINSON, MARK, *East Carolina University*
BROMAGE, TIM, *City University of New York*
BROOKS, VIRGINIA, *Oregon Health Sciences University*
BROWN, ARTHUR, *University of Arkansas*
BROWNING, MARK, *Purdue University*
BUCKNER, VIRGINIA, *Johnson County Community College*
BURNES, E. R., *University of Arkansas, Little Rock*
BURTON, HAROLD, *State University of New York, Buffalo*
CABOT, JOHN, *State University of New York, Stony Brook*
CALVIN, CLYDE, *Portland State University*
CARLSON, ALBERT, *State University of New York, Stony Brook*
CASE, CHRISTINE, *Skyline College*
CASE, TED, *University of California, San Diego*
CASEY, KENNETH, *Veterans Administration Medical Center*
CENTANNI, RUSSELL, *Boise State University*
CHAMPION, REBECCA, *Kennesaw State College*
CHAPMAN, DAVID, *University of California, Los Angeles*
CHARLESWORTH, BRIAN, *University of Chicago*
CHERNOFF, BARRY, *Field Museum of Natural History*
CHISZAR, DAVID, *University of Colorado*
CHRISTENSEN, A. KENT, *University of Michigan Medical School*
CHRISTIAN, DONALD, *University of Minnesota*
CLAIRBORNE, JAMES, *Georgia Southern University*
CLARK, NANCY, *University of Connecticut*
COLAVITO, MARY, *Santa Monica College*
CONKEY, JIM, *Truckee Meadows Community College*
CONNELL, MARY, *Appalachian State University*
COQUELIN, ARTHUR, *Texas Tech University*
COTTER, DAVID, *Georgia College*
COYNE, JERRY, *University of Chicago*
CROWCROFT, PETER, *University of Texas at Austin*
DANIELS, JUDY, *Washtenau Community College*
DAVIS, DAVID GALE, *University of Alabama*
DAVIS, JERRY, *University of Wisconsin*
DAVIS, ROGER, *University of Maryland, Baltimore County*
DELCOMYN, FRED, *University of Illinois, Urbana*
DEMMANS, DANA, *Finger Lakes Community College*
DEMPSEY, JEROME, *University of Wisconsin*
DENGLER, NANCY, *University of Toronto*
DENISON, WILLIAM, *Oregon State University*
DENNISTON, KATHERINE, *Towson State University*
DESAIX, JEAN, *University of North Carolina*
DEWALT, RICHARD, *Louisiana State University*
DIBARTOLOMEIS, SUSAN, *Millersville University of Pennsylvania*
DICKSON, KATHRYN, *California State University, Fullerton*
DIEHL, FRED, *University of Virginia*
DLUZEN, DEAN, *University of Illinois, Urbana*
DOYLE, PATRICK, *Middle Tennessee State University*

DRUGER, MARVIN, *Syracuse University*
DULING, BRIAN, *University of Virginia, Charlottesville*
DUOBINIS-GRAY, LEON, *Murray State University*
ECK, GERALD, *University of Washington*
EDLIN, GORDON, *University of Hawaii, Manoa*
EICHINGER, DAVID, *Purdue University Main Campus*
ELMORE, HAROLD, *Marshall University*
ENDLER, JOHN, *University of California, Santa Barbara*
ENGLISH, DARREL, *Northern Arizona University*
ERICKSON, GINA, *Highline Community College*
ERWIN, CINDY, *City College of San Francisco*
ESCH, GERALD, *Wake Forest University*
ESTEP, DAN, *University of Georgia*
EWALD, PAUL, *Amherst College*
FALK, RICHARD, *University of California, Davis*
FISHER, DAVID, *University of Hawaii, Manoa*
FISHER, DONALD, *Washington State University*
FLESCH, DAVID, *Mansfield University*
FLESSA, KARL, *University of Arizona*
FONDACARO, JOSEPH, *Marion Merrell Dow*
FORTNEY, SUSAN, *Johnson Space Center*
FOYER, CHRISTINE, *Laboratoire du Metabolisme*
FRAILEY, CARL, *Johnson County Community College*
FROEHLICH, JEFFREY, *University of New Mexico*
FROST-MASON, SALLY, *University of Kansas Main Campus*
FRYE, BERNARD, *University of Texas, Arlington*
FULCHER, THERESA, *Pellissippi State Technical Community College*
FULLER, STEPHEN, *Mary Washington College*
FUNK, FRED, *Northern Arizona University*
GAGLIARDI, GRACE, *Bucks County Community College*
GAINES, MICHAEL, *University of Miami*
GENUTH, SAUL, *Case Western Reserve University*
GHOLZ, HENRY, *University of Florida*
GIBSON, THOMAS, *San Diego State University*
GIDDAY, JEFFREY, *University of Virginia School of Medicine*
GOFF, CHRISTOPHER, *Haverford College*
GOODMAN, MAURICE, *University of Massachusetts Medical School*
GORDON, ALBERT, *University of Washington*
GOSZ, JAMES, *University of New Mexico*
GRAHAM, LINDA, *University of Wisconsin*
GRANT, BRUCE, *College of William and Mary*
GREEN, GARETH, *Johns Hopkins University*
GREENE, HARRY, *University of California, Berkeley*
GREGG, KATHERINE, *West Virginia Wesleyan College*
HALL, RICHARD, *Capitol University*
HAMM, MICHAEL, *Rutgers, the State University of New Jersey*
HANKEN, JAMES, *University of Colorado*
HANRATTY, PAMELA, *University of Southern Mississippi*
HARDIN, JOYCE, *Hendrix College*
HARLEY, JOHN, *Eastern Kentucky University*
HARRIS, JAMES, *Utah Valley Community College*
HARTNEY, KRISTINE BEHRENTS, *California State University, Fullerton*
HASSAN, ASLAM, *University of Illinois, Urbana*
HELGESON, JEAN, *Collin County Community College*
HELLER, LOIS JANE, *University of Minnesota*
HEPFER, CAROL, *Millersville University of Pennsylvania*
HESS, WILLIAM, *Brigham Young University*
HEWITSON, WALTER, *Bridgewater State College*
HILDEBRAND, MILTON, *University of California, Davis*
HILFER, S. ROBERT, *Temple University*
HINCK, LARRY, *Arkansas State University*
HODGSON, RONALD, *Central Michigan University*
HOHAM, RONALD, *Colgate University*
HOLLINGER, TOM, *University of Florida, Gainesville*
HOLMES, BRUCE, *Western Illinois University*
HOLMES, KENNETH, *University of Illinois, Urbana*
HOOKE, ANNE, *Miami University of Ohio*
HOSICK, HOWARD, *Washington State University*
HUERTA, ALFREDO, *Miami University of Ohio*
JAKOBSON, ERIC, *University of Illinois, Urbana*
JENSEN, STEVEN, *Southwest Missouri State University*
JOHNSON, LEONARD, *University of Tennessee School of Medicine*
JOHNSON, TED, *St. Olaf College*
JONES, PATRICIA, *Stanford University*
JUILLERAT, FLORENCE, *Indiana University/Purdue University*
KAYE, GORDON, *Albany Medical School*
KAYNE, MARLENE, *Trenton State College*
KEIM, MARY, *Seminole College*
KELLY, DOUGLAS, *University of Southern California*
KELSEN, STEVEN, *Temple University Hospital*
KENDRICK, BRYCE, *University of Waterloo*

KIGER, JOHN, *University of California, Davis*
KILLIAN, JOELLA, *Mary Washington College*
KIMBALL, JOHN, *Tufts University*
KIRK, HELEN, *University of Western Ontario*
KIRKPATRICK, LEE, *Glendale Community College*
KNUTTGEN, HAROLD, *Boston University*
KREBS, CHARLES, *University of British Columbia*
KREBS, JULIA, *Francis Marion University*
KURIS, ARMAND, *University of California, Santa Barbara*
KUTCHAI, HOWARD, *University of Virginia Medical School*
LASSITER, WILLIAM, *University of North Carolina, Chapel Hill*
LATIES, GEORGE, *University of California, Los Angeles*
LATTA, VIRGINIA, *Jefferson State Junior College*
LEVY, MATTHEW, *School of Medicine, City University of New York*
LEWIS, LARRY, *Bradford University*
LINDSEY, JERRI, *Tarrant County Junior College*
LITTLE, ROBERT, *Medical College of Georgia*
LOCKE, MICHAEL, *University of Western Ontario*
LOHMEIER, LYNNE, *Mississippi Gulf Coast Community College*
MACLEAN, STEPHEN, *University of Alaska Fairbanks*
MACKLIN, MONICA, *Northeastern State University*
MADIGAN, MICHAEL, *Southern Illinois University, Carbondale*
MALLOCH, DAVID, *University of Toronto*
MANN, ALAN, *University of Pennsylvania*
MARGULIES, MAURICE, *Rockville, Maryland*
MARR, ELEANOR, *Dutchess Community College*
MARTIN, JAMES, *Reynolds Community College*
MATHEWS, ROBERT, *University of Georgia*
MATSON, RONALD, *Kennesaw State College*
MATTHAI, WILLIAM, *Tarrant County Junior College*
MAXSON, LINDA, *Pennsylvania State University*
MAXWELL, JOYCE, *California State University, Northridge*
MCCLINTIC, J. ROBERT, *California State University, Fresno*
MCCULLOCH, DAVID, *Collin County Community College*
MCEDWARD, LARRY, *University of Florida*
MCKEE, DOROTHY, *Auburn University, Montgomery*
MCNABB, ANNE, *Virginia Polytechnic Institute and State University*
MCREYNOLDS, JOHN, *University of Michigan Medical School*
MERTENS, THOMAS, *Ball State University*
MILLER, G. TYLER, *Pittsboro, North Carolina*
MIMMS, CHARLES, *University of Georgia*
MITZNER, WAYNE, *Johns Hopkins University*
MOCK, DOUG, *University of Washington*
MOHRMAN, DAVID, *University of Minnesota*
MOISES, HYLAND, *University of Michigan Medical School*
MONTVILO, JEROME, *Bryant College*
MOORE-LANDECKER, ELIZABETH, *Glassboro State University*
MORBECK, MARY ELLEN, *University of Arizona*
MORRISON, WILLIAM, *Shippensburg University*
MORTON, DAVID, *Frostburg State University*
MOUNT, DAVID, *University of Arizona*
MUCH, DAVID, *Muhlenberg College*
MURPHY, RICHARD, *University of Virginia Medical School*
MURRISH, DAVID, *State University of New York, Binghamton*
MYERS, NORMAN, *Oxford University, England*
MYRES, BRIAN, *Cypress College*
NAGLE, JAMES, *Drew University*
NELSON, RILEY, *University of Texas at Austin*
NEMEROFSKY, ARNOLD, *State University of New York, New Paltz*
NICHOLS-KIRK, HELEN, *University of Western Ontario*
NORRIS, DAVID, *University of Colorado*
O'BRIEN, ELINOR, *Boston College*
OJANLATVA, ANSA, *Sacramento, California*
OLSON, MERLE, *University of Texas Health Science Center*
ORR, CLIFTON, *University of Arkansas*
ORR, ROBERT, *California Academy of Sciences*
PAI, ANNA, *Montclair State College*
PALMBLAD, IVAN, *Utah State University*
PAPPENFUSS, HERBERT, *Boise State University*
PARSONS, THOMAS, *University of Toronto*
PAULY, JOHN, *University of Arkansas for Medical Sciences*
PECHENIK, JAN, *Tufts University*
PECK, JAMES, *University of Arkansas, Little Rock*
PERRY, JAMES, *Frostburg State University*
PETERSON, GARY, *South Dakota State University*
PIERCE, CARL, *Washington University*
PIKE, CARL, *Franklin and Marshall College*
PIPERBERG, JOEL, *Millersville University*
PLEASANTS, BARBARA, *Iowa State University*
PLETT, HAROLD, *Fullerton College*
POWELL, FRANK, *University of California, San Diego*
RALPH, CHARLES, *Colorado State University*

RAWN, CARROLL, *Seton Hall University*
RICKETT, JOHN, *University of Arkansas, Little Rock*
RIEDER, CONLY, *Wadsworth Center for Laboratories and Research*
ROMANO, FRANK, *Jacksonville State University*
ROSE, GREIG, *West Valley College*
ROSEN, FRED, *Harvard University School of Medicine*
ROSS, GERALDINE, *Highline Community College*
ROSS, GORDON, *University of California Medical Center, Los Angeles*
ROSS, IAN, *University of California, Santa Barbara*
ROST, THOMAS, *University of California, Davis*
RUIBAL, RUDOLFO, *University of California, Riverside*
SACHS, GEORGE, *University of California, Los Angeles*
SACKETT, JAMES, *University of California, Los Angeles*
SALISBURY, FRANK, *Utah State University*
SCHAPIRO, HARRIET, *San Diego State University*
SCHECKLER, STEPHEN, *Virginia Polytechnic Institute and State University*
SCHIMEL, DAVID, *NASA Ames Research Center*
SCHLESINGER, WILLIAM, *Duke University*
SCHMID, RUDI, *University of California, Berkeley*
SCHMOYER, IRVIN, *Muhlenberg College*
SCHNERMANN, JURGEN, *University of Michigan School of Medicine*
SCHREIBER, DAN, *Mesa Community College*
SEARLES, RICHARD, *Duke University*
SEHGAL, PREM, *East Carolina University*
SHARP, ROGER, *University of Nebraska, Omaha*
SHEPHERD, JOHN, *Mayo Medical University*
SHERMAN, PAUL, *Cornell University*
SHOEMAKER, DAVID, *Emory University School of Medicine*
SHONTZ, NANCY, *Grand Valley State University*
SHOPPER, MARILYN, *Johnson County Community College*
SILK, WENDY, *University of California, Davis*
SLATKIN, MONTGOMERY, *University of California, Berkeley*
SLOBODA, ROGER, *Dartmouth College*
SLONE, J. HENRY, *Francis Marion College*
SMILES, MICHAEL, *State University of New York, Farmingdale*
SMITH, MICHAEL, *Valdosta State University*
SOLOMON, NANCY, *Miami University*
SOLOMON, TRAVIS, *Kansas City V.A. Medical Center*
STARR, LISA, *Scripps Clinic and Research Foundation*
STEELE, KELLY, *Appalachian State University*
STEIN-TAYLOR, JANET, *University of Illinois, Chicago*
STEINERT, KATHLEEN, *Bellevue Community College*
STITT, JOHN, *John B. Pierce Foundation Laboratory*
SULLIVAN, LAWRENCE, *University of Kansas*
SUMMERS, GERALD, *University of Missouri*
SUNDBERG, MARSHALL, *Louisiana State University*
SWAIN, SARAH, *Middle Tennessee State University*
TERHUNE, JERRY, *Jefferson Community College/University of Kentucky*
THAMES, MARC, *Medical College of Virginia*
TIFFANY, LOIS, *Iowa State University*
TIZARD, IAN, *Texas A & M University*
TRAMMELL, JAMES, JR., *Arapahoe Community College*
TROUT, RICHARD, *Oklahoma City Community College*
TUTTLE, JEREMY, *University of Virginia, Charlottesville*
TYSER, ROBIN, *University of Wisconsin, La Crosse*
VALENTINE, JAMES, *University of California, Santa Barbara*
VALTIN, HEINZ, *Dartmouth Medical School*
WAALAND, ROBERT, *University of Washington*
WADE, MICHAEL, *University of Chicago*
WAHLERT, JOHN, *City University of New York, Baruch College*
WALDVOGEL, JERRY, *Clemson University*
WALSH, BRUCE, *University of Arizona*
WARING, RICHARD, *Oregon State University*
WARMBRODT, ROBERT, *University of Maryland*
WARNER, MARGARET, *Indiana University, Krennert Institute*
WEBB, JACQUELINE, *Villanova University*
WEIGL, ANN, *Winston-Salem State University*
WEISBRODT, NORMAN, *University of Texas Medical School, Houston*
WEISS, MARK, *Wayne State University*
WELKIE, GEORGE, *Utah State University*
WENDEROTH, MARY PAT, *University of Washington*
WHEELIS, MARK, *University of California, Davis*
WHIPP, BRIAN, *University of California, Los Angeles*
WHITENBERG, DAVID, *Southwest Texas State University*
WHITTOW, G. CAUSEY, *University of Hawaii School of Medicine*
WILSON, THOMAS, *Miami University*
WINICUR, SANDRA, *Indiana University, South Bend*
WISE, ROBERT, *Francis Scott Key Medical Center*
WOMBLE, MARK, *University of Michigan Medical School*
YONENAKA, SHANNA, *San Francisco State University*
ZIHLMAN, ADRIENNE, *University of California, Santa Cruz*

Current configurations of the earth's oceans and land masses—the geologic stage upon which life's drama continues to unfold. Thousands of separate images were pieced together to create this remarkable cloud-free view of our planet.

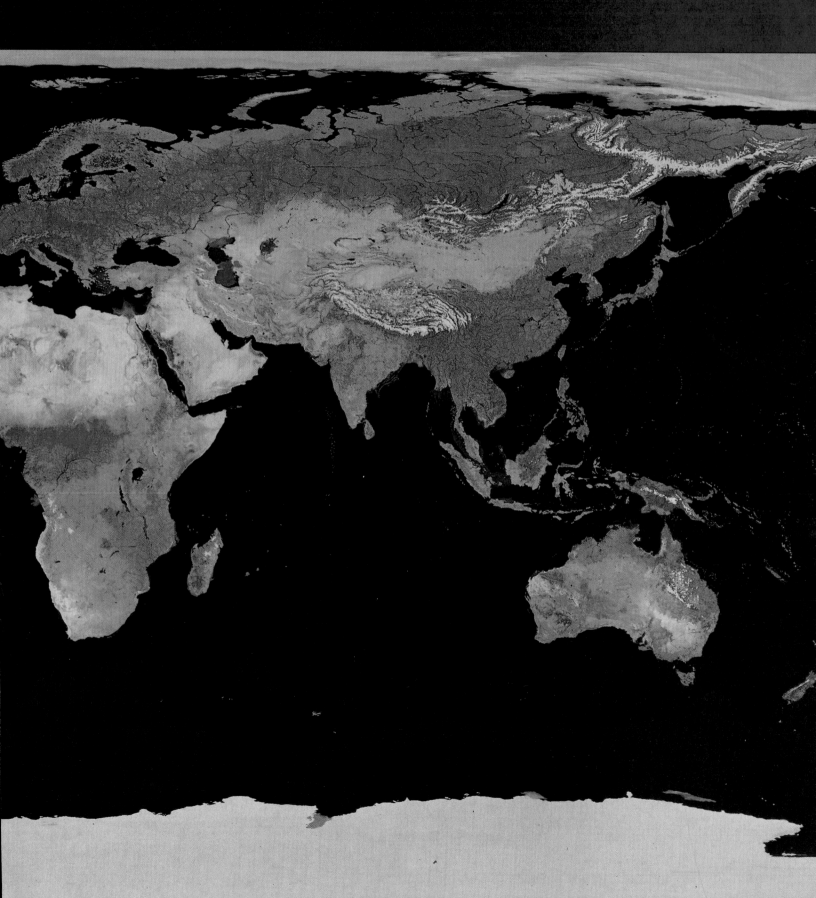

1 METHODS AND CONCEPTS IN BIOLOGY

Biology Revisited

Buried somewhere in that mass of tissue just above and behind your eyes are memories of your first encounters with the living world. Still in residence are memories of discovering your hands and feet, your family, friends, the change of seasons, the smell of rain-drenched earth and grass. In that brain are memories of early introductions to a great disorganized parade of insects, spiders, flowers, frogs, and furred things—mostly living, sometimes dead. There are memories of questions—*"What is life?"* and, inevitably, *"What is death?"* There are memories of answers, some satisfying, others less so.

Figure 1.1 Think back on all you have known and seen. This is a foundation for your deeper probes into the world of life.

By observing, asking questions, and accumulating answers, you have built up a store of knowledge about the world of life. Experience and education have been refining your questions, and no doubt some answers are difficult to come by. Think of a young man whose brain is functionally dead as a result of a motorcycle accident. If his breathing and other basic functions proceed only as long as he remains hooked up to mechanical support systems, is he no longer "alive"? Think of a recently fertilized egg growing inside a pregnant woman, but currently no more than a cluster of a few dozen tiny cells. At what point in its development is it a definably "human" life? If questions like this have crossed your mind, your thoughts about life obviously run deep.

The point is, this book isn't your introduction to biology—"the study of life"—for you have been studying life ever since information began penetrating your brain. This book simply is biology *revisited*, in ways that may help carry your thoughts to more organized levels of understanding.

Return to the question, *What is life?* Offhandedly, you might reply that you know it when you see it. To biologists, however, the question opens up a story that has been unfolding in countless directions for several billion years! "Life" is an outcome of ancient events by which nonliving materials became assembled into the first living cells. "Life" is a way of capturing and using energy and raw materials. "Life" is a way of sensing and responding to specific changes in the environment. "Life" is a capacity to reproduce, grow, and develop. And "life" evolves, meaning that details in the body plan and functions of organisms can change through successive generations.

Yet this short description only hints at the meaning of life. Deeper insight requires wide-ranging study of life's characteristics.

Throughout this book you will come across many examples of how organisms are constructed, how they function, where they live, and what they do. The examples support certain concepts which, taken together, will give you a sense of what "life" is. This chapter provides an overview of the basic concepts. As you continue reading the book, you may find it useful to return to this overview to reinforce your grasp of details.

1. There is unity in the living world, for all organisms are alike in key respects. Their structural organization and functions depend on properties of matter and energy. They obtain and use energy and materials from the environment. They make controlled responses to changing conditions. They grow and reproduce, based on instructions contained in DNA.

2. There is diversity in the living world, for organisms vary immensely in body plans, body functions, and behavior. Evolutionary theories explain this diversity.

3. Biology, like other branches of science, is based on systematic observations, hypotheses, predictions, and relentless tests. The external world, not internal conviction, is the testing ground for scientific theories.

1.1 SHARED CHARACTERISTICS OF LIFE

Energy, DNA, and Life

Picture a frog on a rock, busily croaking. Without even thinking about it, you know the frog is alive and the rock is not. At a much deeper level, however, the difference between them blurs. They and all other things are composed of the same particles (protons, electrons, and neutrons). The particles are organized as atoms, according to the same physical laws. At the heart of those laws is something called **energy**—a capacity to make things happen, to do work. Energetic interactions bind atom to atom in predictable patterns, giving rise to the structured bits of matter we call molecules. Energetic interactions among molecules hold a rock together—and they hold a frog together.

It takes a special molecule called deoxyribonucleic acid, or **DNA**, to set living things apart from the nonliving world. No chunk of granite or quartz has it. DNA molecules contain instructions for assembling new organisms from "lifeless" molecules that contain carbon and a few other kinds of atoms. By analogy, with proper instructions and a little effort, you can turn a disordered heap of ceramic tiles—even just two kinds of tiles—into ordered patterns such as these:

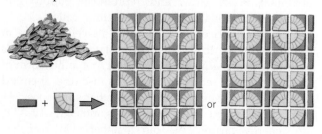

Similarly, life emerges from lifeless matter with DNA "directions," raw materials, and energy inputs.

Levels of Organization in Nature

Look carefully at Figure 1.2, which outlines the levels of organization in nature. The properties of life emerge at the level of cells. A **cell** is an organized unit that can survive and reproduce on its own, given DNA instructions and sources of energy and raw materials. In other words, the cell is the basic *living* unit. This definition obviously fits a free-living, single-celled organism such as an amoeba. Does it fit a **multicelled organism**, that has specialized cells organized into tissues and organs? Yes. You may find this a strange answer. After all, your own cells could never live all by themselves in nature. They must be bathed by fluids inside your body. Yet even human cells can be isolated and kept alive under controlled laboratory conditions. Researchers around the world routinely maintain human cells for use in important experiments, including cancer studies.

Referring to Figure 1.2, we find a more inclusive level of organization—the **population**. This is a group of single-celled or multicelled organisms of the same kind, such as a breeding colony of Emperor penguins in Antarctica. Next is the **community**, which includes all populations of all species (penguins, whales, seals, fishes, and so on) living in the same area. The next level, the **ecosystem**, includes the community and its physical and chemical environment. The most inclusive level of organization in nature is the **biosphere**. The word refers to all regions of the earth's waters, crust, and atmosphere in which organisms live.

Within the hierarchy of organization in nature, the properties of life emerge at the level of cells. Cells emerge through a convergence of raw materials, sources of energy, and instructions contained in DNA molecules.

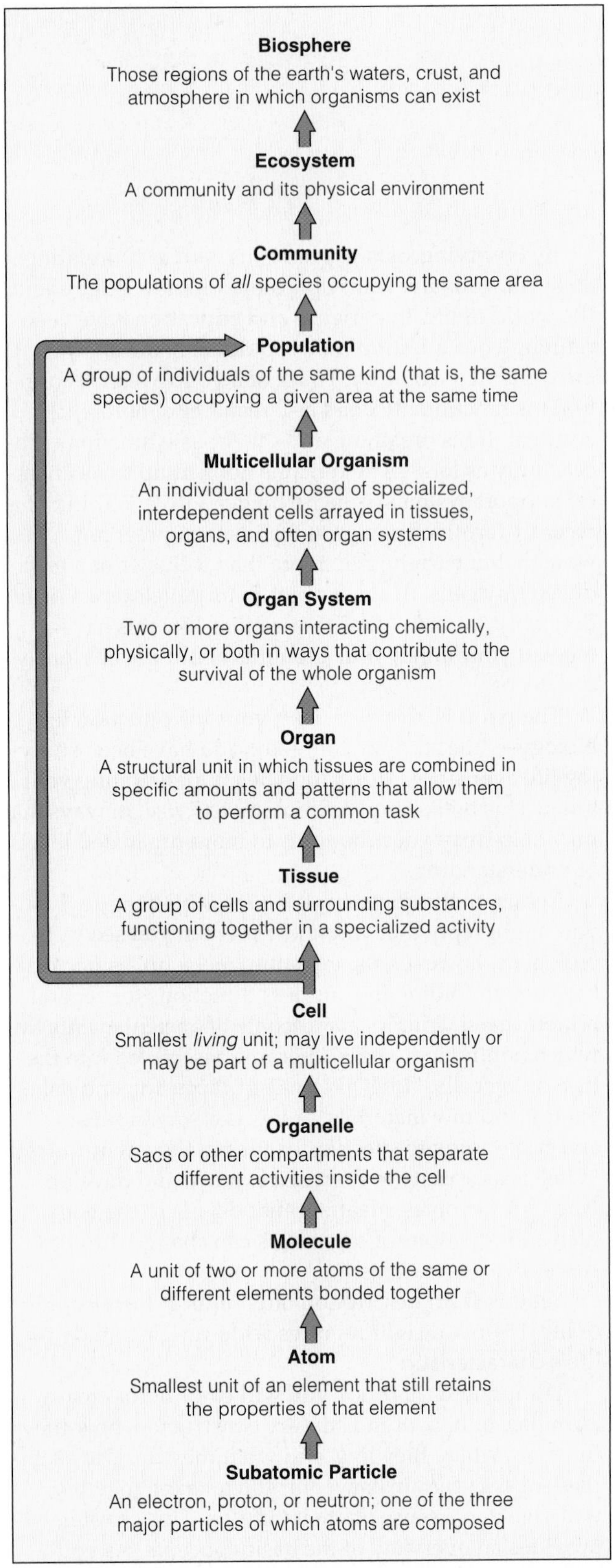

Metabolism: Life's Energy Transfers

You never, ever will find a rock engaged in metabolic activities. Only living cells can do this. **Metabolism** refers to the cell's capacity to (1) extract and convert energy from its surroundings and (2) use energy and so maintain itself, grow, and reproduce. Simply put, metabolism means *energy transfers* within cells.

Think of a rice plant. Many of its cells engage in **photosynthesis**. In the first stage of this process, cells trap sunlight energy, then convert it to another form of energy. In the second stage, cells use the chemical energy to build sugars, starch, and other substances. As part of the process of photosynthesis, molecules of **ATP**, an "energy carrier," are put together. ATP transfers energy to other molecules that function as metabolic workers (enzymes), building blocks, or energy reserves.

Biosphere
Those regions of the earth's waters, crust, and atmosphere in which organisms can exist

Ecosystem
A community and its physical environment

Community
The populations of *all* species occupying the same area

Population
A group of individuals of the same kind (that is, the same species) occupying a given area at the same time

Multicellular Organism
An individual composed of specialized, interdependent cells arrayed in tissues, organs, and often organ systems

Organ System
Two or more organs interacting chemically, physically, or both in ways that contribute to the survival of the whole organism

Organ
A structural unit in which tissues are combined in specific amounts and patterns that allow them to perform a common task

Tissue
A group of cells and surrounding substances, functioning together in a specialized activity

Cell
Smallest *living* unit; may live independently or may be part of a multicellular organism

Organelle
Sacs or other compartments that separate different activities inside the cell

Molecule
A unit of two or more atoms of the same or different elements bonded together

Atom
Smallest unit of an element that still retains the properties of that element

Subatomic Particle
An electron, proton, or neutron; one of the three major particles of which atoms are composed

Figure 1.2 Levels of organization in nature. Cells represent the first level at which the properties of life emerge.

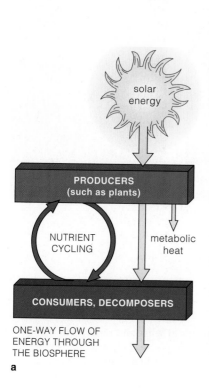

solar
energy

PRODUCERS
(such as plants)

NUTRIENT
CYCLING

metabolic
heat

CONSUMERS, DECOMPOSERS

ONE-WAY FLOW OF
ENERGY THROUGH
THE BIOSPHERE

a

b

c

d

Figure 1.3 (**a**) Direction of energy flow and the cycling of materials through the biosphere.

(**b**) Example of interdependency through nutrient cycling, although this cast of characters may seem a bit improbable at first. In the warm, dry grassland called the African savanna, we come across an adult male elephant. It eats huge quantities of plants to maintain its eight-ton self, and it produces huge piles of solid wastes—dung—that still contain some unused nutrients. Thus, although most organisms would not recognize it as such, elephant dung *is* an exploitable food source.

(**c**) And so we next have little dung beetles rushing to the scene almost simultaneously with the uplifting of an elephant tail. Working rapidly, they carve fragments of moist dung into round balls, which they roll off and bury in burrows. In these balls the beetles lay eggs, a reproductive behavior that assures forthcoming offspring (**d**) of a compact food supply.

Thanks to beetles, dung does not pile up and dry out into rock-hard mounds in the intense heat of the day. Instead, the surface of the land is tidied up, beetle offspring are fed, and leftover dung accumulates in burrows—there to enrich the soil that nourishes the plants that sustain (among others) the elephants.

In rice plants, some of the stored energy becomes concentrated in starchy seeds—rice grains. Energy reserves in countless trillions of rice grains provide energy for billions of rice-eating humans around the world. How? In humans, as in most animals and plants, stored energy is released and transferred to ATP by way of **aerobic respiration**, another metabolic process.

Living things show metabolic activity. Their cells acquire and use energy to stockpile, tear down, build, and eliminate materials in ways that promote survival and reproduction.

Interdependency Among Organisms

With few exceptions, a flow of energy from the sun maintains the great pattern of organization in nature. Plants and some other photosynthetic organisms are the entry point for this flow. They are the food **producers**. Animals are **consumers**. Directly or indirectly, they feed on energy stored in plant parts. Thus zebras tap directly into the stored energy when they nibble on grass, and lions tap into it indirectly when they chomp on zebras. Many kinds of bacteria and fungi are **decomposers**. When they feed on tissues or remains of other organisms, they break down sugars and other biological molecules to simple raw materials—which can be cycled back to producers.

And so we have interdependency among organisms, based on a one-way flow of energy *through* them and a cycling of materials *among* them (Figure 1.3).

Such interactions among organisms influence populations, communities, and ecosystems. They even influence the global environment. Understand the extent of the interactions and you will gain insight into amplification of the greenhouse effect, acid rain, and many other modern-day problems.

Webs of organization connect all organisms in nature, in that organisms depend directly or indirectly on one another for energy and raw materials.

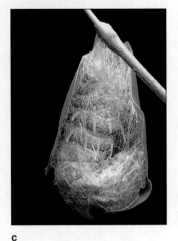

a   b   c   d

Figure 1.4 "The insect"—a continuous series of stages in development. Different adaptive properties emerge at each stage. Shown here, a silkworm moth, from egg (**a**) to larval stage (**b**), to pupal form (**c**), to the splendid adult form (**d**,**e**).

Sensing and Responding to Change

It is often said that only organisms "respond" to the environment. Yet a rock also "responds" to the environment, as when it yields to gravity and tumbles downhill or changes shape slowly under the battering of wind, rain, or tides. The real difference is this: *Organisms have the cellular means to sense changes in the environment and make controlled responses to them.* They do so with the help of **receptors**, which are molecules and structures that can detect specific information about the environment. When cells receive signals from receptors, they adjust their activities in ways that bring about an appropriate response.

Your body, for example, can withstand only so much heat or cold. It must rid itself of harmful substances. Certain foods must be available, in certain amounts. Yet temperatures shift, harmful substances may be encountered, and food is sometimes plentiful or scarce.

Think about what happens after you eat and simple sugar molecules enter your bloodstream. Blood is part of the body's "internal environment" (the other part is the tissue fluid bathing your cells). When the sugar level in blood rises, cells of the pancreas step up their secretion of insulin. Most cells in your body have receptors for insulin, a hormone that prods the cells into taking up sugar molecules. With so many cells taking up sugar, the blood sugar level returns to normal.

Suppose you skip breakfast, then lunch, and the blood sugar level falls. Now a different hormone prods liver cells to dig into their stores of energy-rich molecules. Those molecules are broken down to simple sugars, which are released into the bloodstream—and again the blood sugar level returns to normal.

Usually, the internal environment of a multicelled organism is kept fairly constant. When conditions in the internal environment are being maintained within tolerable limits, we call this a state of **homeostasis**.

Organisms have the means to sense and respond to changes in their environment. The responses help maintain favorable operating conditions inside the cell or multicelled body.

Reproduction

We humans tend to think we enter the world rather abruptly and leave it the same way. Yet we and all other organisms are more than this. *We are part of an immense, ongoing journey that began billions of years ago.* Think of the first cell produced when a human sperm penetrates an egg. The cell would not even exist if the sperm and egg had not formed earlier, according to DNA instructions passed down through countless generations. With those time-tested instructions, a new human body develops in ways that will prepare it, ultimately, for helping to produce individuals of the next generation. With **reproduction**—that is, the production of offspring—life's journey continues.

Or think of a moth. Do you simply picture a winged insect? What of the tiny fertilized egg deposited on a branch by a female moth (Figure 1.4)? The egg contains the instructions necessary to become an adult. By those instructions, the egg develops into a caterpillar, a larval stage adapted for rapid feeding and growth. The caterpillar eats and grows until an internal "alarm clock" goes off. Then its body enters a so-called pupal stage of development, which involves wholesale remodeling. Some cells die, and others multiply and become organized in different patterns. In time an adult moth emerges. It has organs that contain eggs or sperm. Its wings are brightly colored and flutter at a frequency appropriate for attracting a mate. In short, the adult stage is adapted for reproduction.

None of these stages is "the insect." The insect is a series of organized stages from one fertilized egg to the

6 Introduction

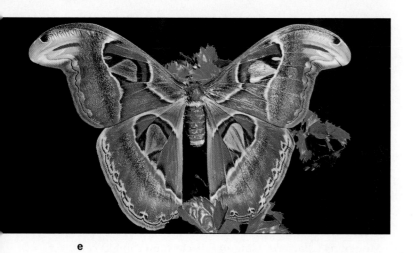

e

a

b

Figure 1.5 An example of how two different forms of the same trait (coloration of moths) are each adaptive under different environmental conditions.

next. Each stage is vital for the ultimate production of new moths. The instructions for each stage were written into moth DNA long before each moment of reproduction—and so the ancient moth story continues.

Each organism arises through reproduction.

Each organism is part of a reproductive continuum that extends back through countless generations.

Mutation: Source of Variations in Heritable Traits

Reproduction involves **inheritance**. The word means that parents transmit DNA instructions for duplicating their traits, such as body form, to offspring.

DNA has two striking qualities. Its instructions assure that offspring will resemble parents—and they also permit *variations* in the details of traits. For example, having five fingers on each hand is a human trait. Yet some humans are born with six fingers on each hand instead of five! Variations in traits arise through **mutations**, which are abnormal, heritable changes in the structure of DNA molecules.

Many mutations are harmful. A change in even a bit of DNA may be enough to sabotage the steps necessary to produce a vital trait. In *hemophilia A*, for example, a tiny mutation leads to an impaired ability to clot blood. Bleeding continues for an abnormally long time after even a small cut or bruise.

Yet some mutations are harmless, even beneficial, under prevailing conditions. A classic example is a mutation in light-colored moths that leads to dark-colored offspring. Moths fly by night and rest during the day, when birds that eat them are active. What happens when a light moth rests on a light-colored tree

trunk (Figure 1.5)? Birds simply don't see it. Suppose, as a result of heavy industry, light trunks in a forested region become soot covered—and dark. The dark moths are less conspicuous, so they have a better chance of living long enough to reproduce. Under sooty conditions, the dark form of the trait is more adaptive.

An **adaptive trait** simply is one that helps an organism survive and reproduce under a given set of environmental conditions.

DNA is the molecule of inheritance in organisms. Its instructions for reproducing traits are passed on from parents to offspring.

Mutations introduce variations in heritable traits.

Although many mutations are harmful, some give rise to variations in form, function, or behavior that turn out to be adaptive under prevailing conditions.

LIFE'S DIVERSITY

So Much Unity, Yet So Many Species

Until now, we have focused on the *unity* of life—on characteristics shared by all organisms. Superimposed on the shared heritage is immense *diversity*. Many millions of different kinds of organisms, or **species**, inhabit the earth. Many millions more lived in the past and became extinct. Attempts to make sense of diversity led to a classification scheme in which each species is assigned a two-part name. The first part designates the **genus** (plural, genera). A genus encompasses all species related by descent from a common ancestor. The second part designates a particular species within that genus. For instance, *Quercus alba* is the scientific name of the white oak. *Q. rubra* is the name of the red oak. (Once the genus name is spelled out in a document, subsequent uses of it can be abbreviated.)

Life's diversity is further classified by assigning species to groups at more encompassing levels. Genera that share a common ancestor are placed in the same *family*, related families are placed in the same *order*, then related orders are placed in the same *class*. Related classes are placed in a *phylum* (plural, phyla) or *division*, which is assigned to one of five *kingdoms*:

Monerans   Bacteria (singular, bacterium). Single cells, all prokaryotic (their DNA is not enclosed in a membrane-bound compartment called a nucleus). Producers, consumers, decomposers. Kingdom of greatest metabolic diversity.

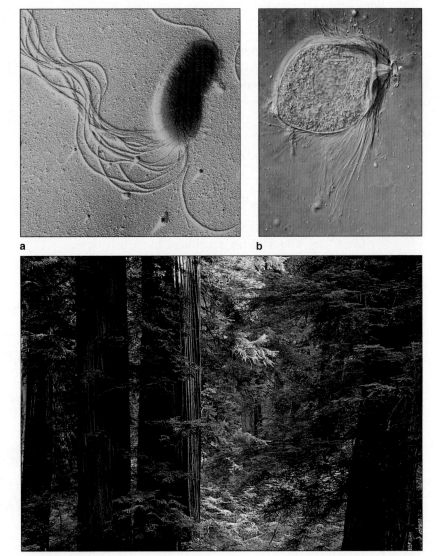

Figure 1.6   Representatives of life's diversity.

Kingdom Monera.   (**a**) A bacterium, a microscopically small single cell. Bacteria live nearly everywhere, including in or on other organisms. The ones in your gut and on your skin outnumber the cells of your body.

Kingdom Protista.   (**b**) A trichomonad, living as a parasite in a termite's gut. Most protistans are single celled, but they generally are much larger and have much greater internal complexity than bacteria.

Kingdom Plantae.   (**c**) A grove of California coast redwoods. Like nearly all members of the plant kingdom, they produce their own food through photosynthesis. (**e**) From a plant called a composite, a flower having a pattern that guides bees to nectar. The bees get food, the plants get help in reproducing. Many organisms are locked in mutually helpful interactions.

Kingdom Fungi.   (**d**) A stinkhorn fungus. The kingdom of fungi includes many major decomposers, which break down the remains and wastes of organisms. Without decomposers, communities would gradually become buried in their own garbage.

Kingdom Animalia.   (**f**) Male bighorn sheep competing for females. Like all members of the animal kingdom, they cannot produce their own food; they depend on other organisms for it. They generally move about far more than other kinds of organisms.

a   b   c

Protistans Mostly single cells, larger than bacteria. Eukaryotic (DNA enclosed in a nucleus). Producers, consumers. Diverse life-styles.

Fungi Mostly multicelled. Eukaryotic. Decomposers, consumers. Their secretions digest food outside the fungal body; their cells absorb the breakdown products.

Plants Mostly multicelled. Eukaryotic. Nearly all producers that rely on photosynthesis.

Animals Multicelled. Eukaryotic. Consumers, typically motile, with diverse life-styles.

Figure 1.6 shows a few members of these kingdoms. Table 1.1 summarizes the characteristics of life that have been described so far. With very few exceptions, the living organisms in all five kingdoms display these characteristics.

Table 1.1 Characteristics of Living Organisms

1. Complex structural organization based on instructions contained in DNA molecules.

2. Directly or indirectly, dependence on other organisms for energy and material resources.

3. Metabolic activity by the single cell or multiple cells composing the body.

4. Use of homeostatic controls that maintain favorable operating conditions in the body within tolerable limits.

5. Reproductive capacity, by which the instructions for heritable traits are passed from parents to offspring.

6. Diversity in body form, in the functions of various body parts, and in behavior. Such traits are adaptations to prevailing conditions in the environment.

7. The capacity to evolve, based ultimately on variations in traits that arise through mutations in DNA.

d

e

f

a   b   c

Figure 1.7 Representatives of the more than 300 varieties of domesticated pigeons resulting from artificial selection practices. Pigeon breeders started with variant forms of traits among captive populations of wild rock doves (**a**).

d   e

f   g

An Evolutionary View of Diversity

Given that organisms are so much alike, what could account for their diversity? One key explanation is called evolution by means of natural selection. A few simple examples will be enough to introduce you to the main premises of this explanation.

Evolution Defined Suppose a DNA mutation gives rise to a different form of a trait in a few members of a population. We can use those dark moths in a sooty forest as an example. Birds see and eat many of their light-colored relatives, but dark moths escape detection and live to reproduce. So do their dark offspring—and so do *their* offspring. The dark version of the trait is popping up with greater frequency. In time it may become the more common form—and we might end up referring to "the dark moth population." **Evolution** is taking place—meaning that the features that character-ize a population are changing through successive generations. Because heritable changes in the DNA are the source of variation in those features, they are the starting point for evolution.

Natural Selection Defined Long ago, Charles Darwin used pigeons to explain the link between varia-tion in traits and evolution. Domesticated pigeons vary greatly in size, feather color, and other traits (Figure 1.7). As Darwin knew, pigeon breeders "select" certain traits. Suppose the desired traits are black tail feathers with curly edges. Only those pigeons having the most black and the most curl in their tail feathers are allowed to mate. Over time, those forms of the traits become most common as others are eliminated from the pigeon population.

Because pigeon breeders do their selecting in an artificial environment rather than in nature, the practice is an example of *artificial* selection. Even so, Darwin rec-ognized it as a way to explain evolution by *natural* selection—that is, selection of adaptive traits in nature. Later in the book, we will consider the mechanisms by which organisms evolve. In the meantime, keep the fol-lowing key points of Darwin's explanation in mind. They are absolutely central to biological inquiry, and we will be using them to explain topics throughout the book. We express them here in modern terms:

1. Members of a population vary in form, function, and behavior. Much of this variation is heritable.

2. Some forms of heritable traits are more adaptive than others—they improve chances of surviving and reproducing. As a result, those forms become more common in the population.

3. **Natural selection** is simply the result of differences in survival and reproduction that have occurred among individuals that differ in one or more traits.

4. Any population *evolves* when some forms of traits become more or less common, or even disappear, over the generations. In this manner, variations have accumulated in different lines of organisms. Life's diversity is the sum total of those variations.

1.3 THE NATURE OF BIOLOGICAL INQUIRY

On Scientific Methods

Biology, like science generally, is an ongoing, methodical search for information that helps reveal the secrets of the natural world. Since Darwin's time, the searchers have branched out and established a great number of specialized subdivisions. Biologists now pursue topics that range from the molecular structure of the virus that causes AIDS to the ozone hole in the stratosphere. There is still so much to be learned about each topic that few now claim the whole of nature as their research interest. And no one claims that one method alone can be used to study its complexity.

Despite the specialization, scientists everywhere still have methods in common. *They ask questions, make educated guesses about possible answers, then devise ways to test predictions that will hold true if their guesses are good ones.* The following list is a more formal description of how scientists generally proceed with an investigation:

---

1. Ask a question (or identify a problem) about some aspect of nature, then develop one or more **hypotheses**, or educated guesses, about what the answer (or solution) might be. This might involve sorting through existing information about related phenomena.

2. Using hypotheses as a guide, make a **prediction**—that is, a statement of what you should be able to observe in nature, if you were to go looking for it. This is often called the "if-then" process. (*If* gravity pulls objects toward the earth, *then* it should be possible to observe apples falling down, not up, from a tree.)

3. Devise ways to test the accuracy of predictions. You might do so by making observations, developing models, and doing experiments.

4. If the tests do not turn out as expected, check to see what might have gone wrong. (Maybe you overlooked something that influenced the results. Or maybe the hypothesis isn't a good one.)

5. Repeat the tests or devise new ones—the more the better. (Hypotheses supported by many different tests are more likely to be correct.) Then objectively report the test results and conclusions drawn from them.

In broad outline, a scientific approach to studying nature is that simple. You yourself can use this approach to advantage. You can use it to satisfy curiosity about mammoths or moth wings. You can use it to pick your way logically through environmental, medical, and social land mines of the sort described later in the book. And you can use it to understand the past and predict possible futures for life on this planet.

About the Word "Theory"

How does a hypothesis differ from a theory? By way of example, go back to Darwin's ideas about evolution. When Darwin proposed these ideas more than a century ago, he ushered in one of the most dramatic of all scientific revolutions. The core of his thinking became popularly known as "the theory of evolution."

In science, a **theory** is a related set of hypotheses that, taken together, form a broad-ranging, testable explanation about some fundamental aspect of the natural world. A scientific theory differs from a scientific hypothesis in its *breadth of application*. Darwin's theory fits this description—it is a big, encompassing "Aha!" explanation that, in a few intellectual strokes, makes sense of a huge number of observable phenomena. Think of it. Darwin's theory explains how most of the diversity among many millions of different living things came about! The *Focus* essay on the next page is but one example of how biologists use the theory to formulate and test ideas about the natural world.

Yet is any theory an "absolute truth" in science? Ultimately, no. Why? It would be impossible to perform the infinite number of tests required to show that a theory holds true under all possible conditions! Objective scientists say only that they are *relatively* certain that a theory is (or is not) correct.

Even so, "relative certainty" can be impressive. Especially after exhaustive tests by many scientists, a theory may be as close to the truth as we can get with the evidence at hand. After more than a century's worth of thousands of different tests, Darwin's theory still stands, with only minor modification. Most biologists accept the modified theory—although they still keep their eyes open for contradictory evidence.

Scientists must keep asking themselves: "Will some other evidence show my idea to be incorrect?" They are expected to put aside pride or bias by testing ideas, even in ways that might prove them wrong. If an individual doesn't (or won't) do this, *others will*—for science proceeds as a community that is both cooperative and competitive. Ideas are shared, with the understanding that it is just as important to expose errors as it is to applaud insights. Individuals can change their mind when presented with new evidence—and this is a strength of science, not a weakness.

A scientific theory is a testable explanation about the cause or causes of a broad range of related phenomena. As is true of hypotheses, theories are open to tests, revision, and tentative acceptance or rejection.

Darwin's Theory and Doing Science

A time-tested theory serves as a general frame of reference for studying nature. Consider Charles Darwin's theory of evolution by natural selection. How might you use it to explain the patterned wings of the moth shown in Figure 1.4? According to this theory, the traits of moths and all other organisms exist because they have contributed to reproductive success. So your question might be this: "I wonder how the wing pattern helps the moth leave descendants." Then you hypothesize about the answer.

"Maybe a wing pattern is a mating flag that helps males and females of the same species identify each other." This may be correct, but you don't limit yourself to one hypothesis. Why? Nearly always, there's more than one possible answer to a question about some aspect of nature. So you also come up with an alternative: "Maybe the pattern camouflages moths during the day." In science, *alternative hypotheses are the rule, not the exception*.

Testing Hypotheses   Of any number of alternative hypotheses, how do you identify the most plausible one? The trick is to let each one guide you in making testable predictions. If wing patterns help moths identify mates (the hypothesis), then it follows that moths should mate only when the patterns are visible (the prediction). Moths mate at night. If wing pattern is a mating flag, then on moonless nights, moths shouldn't be able to see it and you won't see moths mating. To test the prediction, you watch moths on a moonlit and then on a moonless night. You notice they mate with or without help from the light of the moon. Here is evidence that the prediction—and, by extension, the hypothesis—might be wrong. Then again, maybe you overlooked something important. For example, maybe moths (like cats) see better than you do in the dark.

The Role of Experiments   Now you decide to test the same hypothesis by experimentation. An **experiment** is a test in which nature is manipulated to reveal one of its secrets. It requires careful design of a set of controls to evaluate possible side effects of the manipulation.

If wing pattern is a mating flag, then moths with altered color patterns might have a tough time attracting a mate. To test this new prediction, you capture new moths, paint an altered pattern on their wings (Figure *a*), then put them in a cage with unaltered moths to see what happens.

You also set up a **control group** to evaluate possible side effects of a test involving an experimental group. Ideally, members of a control group should be identical to those of an experimental group in all respects—except for the key **variable** (the factor being investigated). The number of individuals in both groups also must be large enough so results won't be due to chance alone.

Besides wing pattern, what other variables between the two groups might affect the outcome of your experiment? Maybe paint fumes are as repulsive as a painted-on pattern to a potential mate. Maybe when you paint the moths you somehow rough them up, making them less desirable than those in the control group. Maybe the paint weighs enough to change the flutter frequency of the wings.

So you decide the control group also must be painted with the same kind of paint, using the same brushes, and they must be handled the same way. But for this group, you *duplicate* the natural wing color pattern as you paint. Now your experimental and control groups are identical except for the variable under study. If only those moths with altered wing patterns turn out to be unlucky in love, then the greater reproductive success of your control group will help substantiate your hypothesis.

1.4   THE LIMITS OF SCIENCE

The call for objective testing strengthens the theories that emerge from scientific studies. Yet it also puts limits on the kinds of studies that can be carried out. Beyond the realm of scientific analysis, some events remain unexplained. Why do we exist, for what purpose? Why does any one of us have to die at a particular moment and not another? Answers to such questions are *subjective*. This means they come from within us, as an outcome of all the experiences and mental connections that shape our consciousness. Because people differ so enormously in this regard, subjective answers do not readily lend themselves to scientific analysis.

This is not to say that subjective answers are without value. No human society can function without a shared commitment to standards for making judgments, even if the judgments are subjective. Moral, aesthetic, economic, and philosophical standards vary from one society to the next. But all guide their mem-

Generally, experiments are devised to disprove a hypothesis. Why? It would be impossible to prove beyond a shadow of a doubt that a hypothesis is correct. It would take an infinite number of experiments to demonstrate that it holds under all possible conditions.

Have you been thinking that painting moths is a rather fanciful example of a scientific approach? As reported in *Nature* in 1993, Karen Marchetti, a graduate student of the University of California at Davis, wielded a paintbrush on birds in a forest in Kashmir, India. She found evidence for her hypothesis that bright feather color, not patterning, gives male yellow-browed leaf warblers an edge in mating. In early tests, Marchetti put a patternless dab of yellow paint on head feathers. In later tests, she painted larger-than-normal yellow bands on wings of one group and painted out part of the bands of another group. She used transparent paint for a third group. (Can you guess why?) As Marchetti discovered, color-enhanced birds secured larger territories and produced more offspring than toned-down birds.

bers in deciding what is important and good, and what is not. All attempt to give meaning to what we do.

Every so often, scientists stir up controversy when they explain part of the world that was considered beyond natural explanation—that is, belonging to the "supernatural." This is sometimes true when moral codes are interwoven with religious narratives. Exploring some longstanding view of the world from a scientific perspective may be misinterpreted as questioning morality, even though the two are not at all the same thing.

For example, centuries ago Nicolaus Copernicus studied the movements of planets and stated that the earth circles the sun. Today the statement seems obvious. Back then, it was heresy. The prevailing belief was that the Creator had made the earth (and, by extension, humanity) the immovable center of the universe! Not long afterward a respected professor, Galileo Galilei, studied the Copernican model of the solar system. He thought it was a good one and said so. He was forced to retract his statement publicly, on his knees, and to put the earth back as the fixed center of things. (Word has it that when he stood up he muttered, "Even so, it does move.")

Today, as then, society has its sets of standards. Today, as then, those standards may be called into question when a new, natural explanation runs counter to supernatural belief. This doesn't mean the scientists who raise the questions are less moral, less lawful, less sensitive, or less caring than anyone else. It simply means one more standard guides their work: *The external world, not internal conviction, must be the testing ground for scientific beliefs.*

Systematic observations, hypotheses, predictions, tests—in all these ways, science differs from systems of belief that are based on faith, force, or simple consensus.

SUMMARY

1. All organisms share the following characteristics:

a. Their structure, organization, and interactions arise from basic properties of matter and energy.

b. They use metabolic and homeostatic processes.

c. They have the capacity for growth, development, and reproduction, based on instructions contained in their DNA molecules.

2. There are many millions of different organisms. Each kind of organism is a species. In classification schemes, species are placed in increasingly inclusive groupings, from genus on up through family, order, class, phylum (or division), and kingdom.

3. Diversity among organisms arises through mutation. Mutations introduce changes in the DNA. The changes may lead to variation in heritable traits (traits that parents transmit to offspring, including most details of the body's form and functioning).

4. Different versions of the same heritable trait occur among individuals of a population. These influence the ability to survive and reproduce. Under prevailing conditions, some may be more adaptive than others and

will become more common ("selected") in subsequent generations. Others will become less common and may disappear. Thus the population changes over time; it evolves. These points are central to the theory of evolution by natural selection.

5. There are many specialized scientific methods, corresponding to many different fields of inquiry. The following terms are important to all of them:

a. Theory: An explanation of a broad range of phenomena that has successfully withstood intensive testing.

b. Hypothesis: A possible explanation of a specific phenomenon. Sometimes called an educated guess.

c. Prediction: A claim about what you can expect to see in nature if a theory or hypothesis is correct.

d. Test: An attempt to produce actual observations that match predicted or expected observations.

e. Conclusion: A statement about whether a theory or hypothesis should be accepted, rejected, or modified, based on tests of the predictions derived from it.

6. Scientific theories are based on systematic observations, hypothesizing, predictions, and tests. The external world, not internal conviction, is the testing ground for those theories.

Review Questions

1. Why is it difficult to give a simple definition of life? *3* (For this and subsequent chapters, *italic numbers* following review questions indicate the pages on which the answers may be found.)

2. What characteristics do all organisms have in common? *3*

3. What is energy? What is DNA? *3*

4. Study Figure 1.2. Then, on your own, arrange and define the levels of biological organization. *4*

5. Define metabolic activity. *4–5*

6. Make a sketch of the one-way flow of energy and the cycling of materials through the biosphere. *5*

7. What is mutation? How is it related to the diversity of life? *7*

8. Witnesses in a court of law are asked to "swear to tell the truth, the whole truth, and nothing but the truth." What are some problems inherent in the question? Can you think of a better alternative?

9. Design a test to support or refute the following hypothesis: Body fat appears yellow in certain rabbits—but only when those rabbits also eat leafy plants that contain a yellow pigment molecule called xanthophyll.

Self-Quiz *(Answers in Appendix IV)*

1. The _____ is the smallest unit of life.

2. _____ is the ability of cells to extract and transform energy from the environment and use it to maintain themselves, grow, and reproduce.

3. _____ is a state in which the body's internal environment is being maintained within tolerable limits.

4. If a form of a trait improves chances for surviving and reproducing in a particular environment, it is a(n) _____ trait.

5. The capacity to evolve is based on variations in heritable traits, which originally arise through _____ .

6. You have some number of traits that also were present in your great-great-great-great-grandmothers and -grandfathers. This is an example of _____ .
   a. metabolism   c. a control group
   b. homeostasis   d. inheritance

7. DNA molecules _____ .
   a. contain instructions for traits
   b. undergo mutation
   c. are transmitted from parents to offspring
   d. all of the above

8. For many years in a row, a dairy farmer allowed his best milk-producing cows but not the poor producers to mate. Over the generations, milk production increased. This outcome is an example of

   _____ .
   a. natural selection   c. evolution
   b. artificial selection d. both b and c

9. A related set of hypotheses that explains some aspect of the natural world is a scientific _____ .
   a. prediction   c. theory
   b. test   d. observation

Selected Key Terms

For this and subsequent chapters, these are the **boldface** terms that occur in the text on the pages indicated by *italic* numbers. Make a list of these terms, write a definition next to each, then check it against the one in the text. (You will be using these terms later on.)

adaptive trait *7*	energy *3*	natural selection *10*
aerobic respiration *5*	evolution *10*	photosynthesis *4*
animal *9*	experiment *12*	plant *9*
ATP *4*	fungus *9*	population *4*
biosphere *4*	genus *8*	prediction *11*
cell *4*	homeostasis *6*	producer *5*
community *4*	hypothesis *11*	protistan *9*
consumer *5*	inheritance *7*	receptor *6*
control group *12*	metabolism *4*	reproduction *6*
decomposer *5*	moneran *8*	species *8*
DNA *3*	multicelled organism *4*	theory *11*
ecosystem *4*	mutation *7*	variable *12*

Readings

Committee on the Conduct of Science. 1989. *On Being a Scientist.* Washington, D.C.: National Academy of Sciences. Paperback.

Larkin, T. June 1985. "Evidence vs. Nonsense: A Guide to the Scientific Method." *FDA Consumer* 19:26–29.

FACING PAGE: *Living cells of a green plant* (Elodea), *as seen with the aid of a microscope. Each rectangular cell contains efficient chemical factories called chloroplasts (the green spheres).*

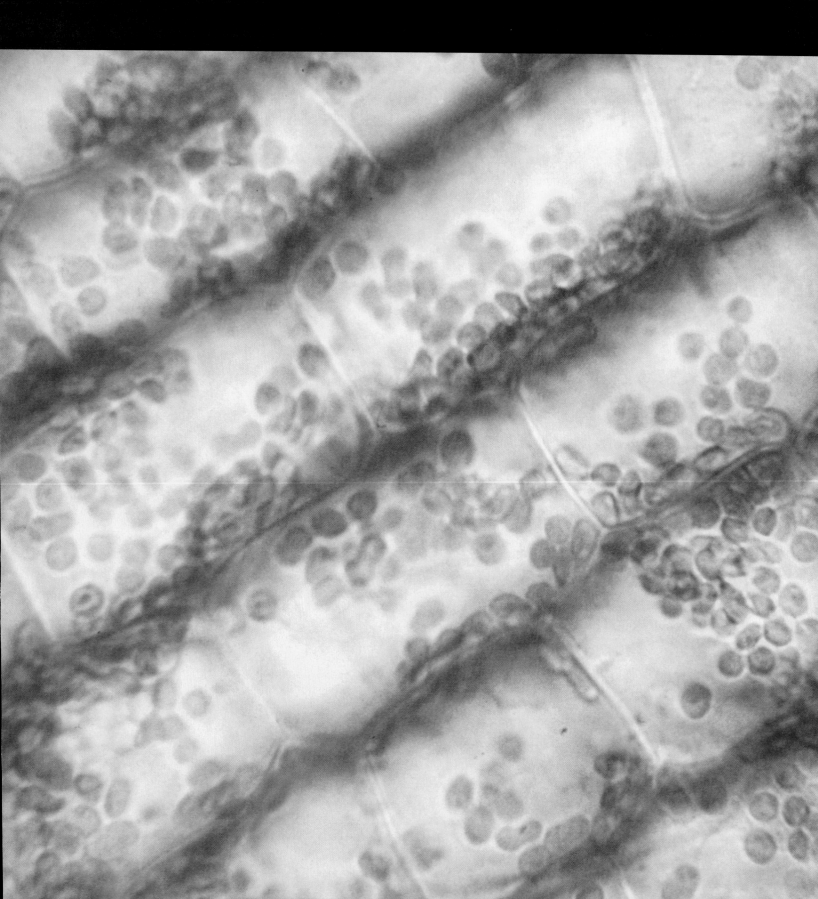

2 CHEMICAL FOUNDATIONS FOR CELLS

The Chemistry In and Around You

Right now you are breathing in oxygen. You would die without it. But where did the oxygen come from? Whether it's noon or midnight, countless plants on the sunlit side of our planet are busily converting energy from the sun to forms of energy they can use—and they release oxygen during the conversions. Aquatic algae have been doing this for more than 900 million years. Some types of bacteria have been doing the same thing for more than 3 *billion* years! All of that released oxygen adds up.

In the past two centuries (which is a mere blip of evolutionary time), we managed to discover what substances are made of and how they can be converted into different forms. With this amazing knowledge of chemistry, we developed such products as fertilizers, nylons, vaccines, lipsticks, antibiotics, and plastic

parts of refrigerators, computers, television sets, jet planes, and cars.

Our chemical "magic" brings benefits *and* problems. Without synthetic fertilizers to help grow crops, more humans than you might imagine would starve to death. But weeds don't know that fertilizers are for crop plants, and animal pests don't know that crops aren't being raised for them. Each year pests ruin or gobble up nearly half of what we grow. In 1945 we began using synthetic pesticides. Among other things, pesticides kill weeds, moths, worms, and rats that threaten our food supplies, health, pets, and ornamental plants. In 1988 alone, Americans spread more than a billion pounds of pesticides through homes, gardens, offices, industries, and farmlands (Figure 2.1).

Among the pesticides, we find the carbamates, organophosphates (including malathion), and halogenated compounds (including chlordane). Most are toxins that block vital communication signals between cells of the nervous system. Some remain active for days, others for weeks or years. Besides pests, they kill great numbers of pest-eaters, including dragonflies and birds. Besides this, pest populations build up resistance to pesticides, for reasons that will become apparent in later chapters.

We, too, inhale pesticides, ingest them with food, or absorb them through skin. Many pesticides cause headaches, rashes, and asthma in susceptible people. Some trigger hives, joint pain, even life-threatening allergic reactions in millions of people in the United States alone.

Maintaining our crops, industries, and health depends on chemistry. So does our chance of reducing harmful side effects of the application of chemistry. You owe it to yourself and others to gain understanding of chemical substances. By demystifying chemistry's "magic," you will be better equipped to assess its benefits and risks.

KEY CONCEPTS

1. All atoms of each element have the same number of protons and electrons, but they may differ slightly in their number of neutrons. These variant forms of atoms are called isotopes.

2. Atoms give up, acquire, or share electrons with other atoms in specific ways. These interactions are the basis for the structural organization and activities of all organisms.

3. Chemical bonds are unions between the electron structures of different atoms. The number and arrangement of electrons in the atoms of different elements give rise to differences in their bonding behavior. Some atoms tend to give up, gain, or share one or more electrons with another atom. Carbon, hydrogen, nitrogen, and oxygen—the main building blocks of life—are like this.

4. In biological molecules, the most common bonds are ionic bonds, hydrogen bonds, and covalent bonds.

5. An ion is an atom or molecule that has gained or lost one or more electrons, and so has acquired an overall positive or negative charge.

6. Life depends on the properties of water, including its temperature-stabilizing effects, cohesiveness, and capacity to dissolve many substances. Hydrophilic substances dissolve in water; hydrophobic ones are repelled by it.

7. Life depends on the controlled formation, use, and disposal of hydrogen ions (H^+). The pH scale is a measure of the concentration of these ions in different solutions.

Figure 2.1 A low-flying cropduster, with its rain of pesticides.

2.1 ORGANIZATION OF MATTER

"Matter" is anything that occupies space and has mass. All solids, liquids, and gases within and around you are forms of matter, and each consists of one or more kinds of elements. An **element** is a fundamental substance that cannot be broken down to a different substance, at least by ordinary means. About ninety-two elements occur naturally on earth. It takes only four kinds—oxygen, carbon, hydrogen, and nitrogen—to make up most of the human body.

You and all other organisms also require seemingly insignificant amounts of other elements. Collectively, these so-called *trace* elements make up less than 0.01 percent of any organism, yet normal body functioning depends on them. Copper is an example. Carefully dry out and analyze tissues of a maple tree and you will find they are about 0.006 percent copper. If that tree has a copper deficiency, its leaf buds will die and its growth will suffer.

Take a look at Figure 2.2. It shows the proportions of elements that make up a human and the fruit of pumpkin plants, and it compares them to the proportions of elements in the "nonliving" materials of the earth's crust. In what respects are they similar? In what respects do they differ?

By international agreement, a one- or two-letter chemical symbol stands for each element, regardless of the element's name in different languages. What we call nitrogen is *azoto* in Italian and *stickstoff* in German—but the symbol remains N. Similarly, the symbol for sodium is Na (from the Latin *natrium*). Table 2.1 lists the chemical symbols of elements that are common in living things.

Figure 2.2 Proportions of different elements in the earth's crust, the human body, and a pumpkin (the fruit of a pumpkin plant).

EARTH'S CRUST		HUMAN		PUMPKIN	
Oxygen	46.6	Oxygen	65	Oxygen	85
Silicon	27.7	Carbon	18	Hydrogen	10.7
Aluminum	8.1	Hydrogen	10	Carbon	3.3
Iron	5.0	Nitrogen	3	Potassium	0.34
Calcium	3.6	Calcium	2	Nitrogen	0.16
Sodium	2.8	Phosphorus	1.1	Phosphorus	0.05
Potassium	2.6	Potassium	0.35	Calcium	0.02
Magnesium	2.1	Sulfur	0.25	Magnesium	0.01
Other		Sodium	0.15	Iron	0.008
elements:	1.5	Chlorine	0.15	Sodium	0.001
		Magnesium	0.05	Zinc	0.0002
		Iron	0.004	Copper	0.0001
		Iodine	0.0004	Other:	0.00005

Table 2.1	Atomic Number and Mass Number of Elements Common in Living Things		
Element	Symbol	Atomic Number	Most Common Mass Number
Hydrogen	H	1	1
Carbon	C	6	12
Nitrogen	N	7	14
Oxygen	O	8	16
Sodium	Na	11	23
Magnesium	Mg	12	24
Phosphorus	P	15	31
Sulfur	S	16	32
Chlorine	Cl	17	35
Potassium	K	19	39
Calcium	Ca	20	40
Iron	Fe	26	56
Iodine	I	53	127

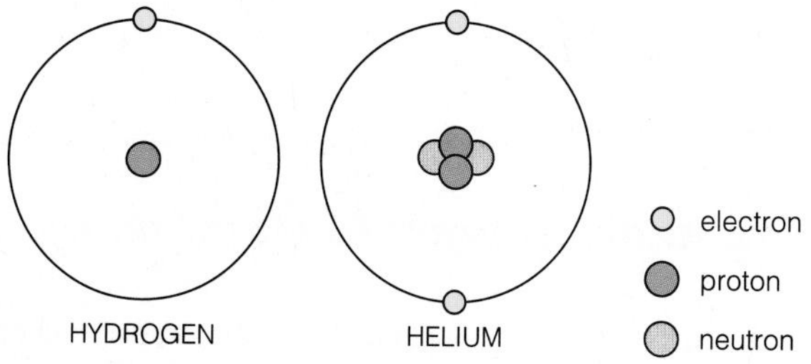

Figure 2.3 Simplified model of atomic structure, using hydrogen and helium as examples. At this scale, the nucleus actually would be an invisible speck at the atom's center.

The Structure of Atoms

Each kind of **atom** is the smallest unit of matter that is unique to a particular element. A **molecule** is two or more joined-together atoms of the same or different elements. Pure water (if there still is such a thing) consists of a stupendous number of water molecules, each composed of one oxygen and two hydrogen atoms.

Water, incidentally, is a compound. A **compound** is a substance in which the relative proportions of two or more elements never vary. Water in rainclouds or a Siberian lake or your bathtub always has twice as many hydrogen atoms as oxygen atoms, period.

Atomic Building Blocks The organization and activities of living things start with charged particles called **protons** and **electrons**. Together with **neutrons**, which have no charge, these are building blocks of atoms (Figure 2.3). An atom's core region, or nucleus, consists of some number of protons and (except for hydrogen) neutrons. Protons carry a positive charge (p^+). Electrons move rapidly around the nucleus, they occupy most of the atom's volume, and they carry a negative charge (e^-). Because an atom has just as many electrons as protons, it has no *net* charge, overall.

Atomic Number and Mass Number The number of protons in the nucleus is called the **atomic number**. It differs for each element. For example, that number is 1 for the hydrogen atom (with one proton) and 6 for the carbon atom (with six protons). Table 2.1 lists other examples. The **mass number** is the number of protons *and* neutrons in the nucleus. A carbon atom with six protons and six neutrons has a mass number of 12.

The relative masses of atoms are also called atomic weights. This term is not precise—mass is not quite the same thing as weight—but its use continues.

As you will see, knowing the atomic numbers and mass numbers gives us an idea of whether certain atoms can lose, gain, or share electrons. *Such electron activity is the basis for the organization of materials and the flow of energy through the living world.*

Isotopes: Variant Forms of Atoms

All the atoms of an element have the same number of protons and electrons, but they can differ in the number of neutrons. Such atoms are **isotopes**. For example, "a carbon atom" might be carbon 12 (six protons, six neutrons), carbon 13 (six protons, seven neutrons), or carbon 14 (six protons, eight neutrons). You can write these as ^{12}C, ^{13}C, and ^{14}C. All isotopes of an element interact with other atoms the same way. Thus cells can use any carbon isotope for a metabolic reaction.

You have probably heard of radioactive isotopes, or **radioisotopes**. These are unstable and tend to break apart (decay) into more stable atoms. The *Focus* essay on the pages that follow describes some uses of radioisotopes in research, medicine, and studies of the history of life.

Each kind of atom is the smallest unit of matter that is unique to a given element.

All atoms of an element have the same number of protons and electrons, but they can vary slightly in the number of neutrons.

Variant atoms of the same element are isotopes.

Using Radioisotopes to Date Fossils, Track Chemicals, and Save Lives

In the winter of 1896, the physicist Henri Becquerel tucked a heavily wrapped rock of uranium into a desk drawer, on top of an unexposed photographic plate. A few days later, he opened the drawer and discovered a faint image of the rock on the plate—apparently caused by energy emitted from the rock. His coworker, Marie Curie, named the phenomenon "radioactivity."

As we now know, radioisotopes are unstable atoms with dissimilar numbers of protons and neutrons. They emit electrons and energy. In this spontaneous process (radioactive decay), an isotope changes to a new, stable one that is not radioactive.

Radioactive Dating Each type of radioisotope has a certain number of protons and neutrons, and it decays spontaneously at a certain rate into a particular isotope of a different element. "Half-life" is the time it takes for half the nuclei in any given amount of a radioactive element to decay into another element. Half-life can't be modified by temperature, pressure, chemical reactions, or any other environmental factor. That's why radioactive dating is a reliable way to discern the age of different rock layers in the earth—and of fossils contained in the layers. To discover a rock's age, we can compare the amount of one of its radioisotopes with the amount of the decay product for that isotope.

For example, ^{40}potassium has a half-life of 1.3 billion years and decays to ^{40}argon, a stable isotope. The age of anything that contains ^{40}potassium can be determined by measuring the ratio of ^{40}argon to ^{40}potassium. In such ways, researchers have dated ancient fossils. Figure *a* shows two examples. Similarly, researchers who used ^{238}uranium (with a half-life of 4.5 billion years) discovered that the earth formed more than 4.6 billion years ago. The following table lists useful ranges of some radioisotopes that are employed in dating methods:

Main Radioisotopes Used in Dating

Radioisotope (unstable)	Stable Product	Half-Life (years)	Useful Range (years)
^{87}rubidium ⟶	^{87}strontium	49 billion	100 million
^{232}thorium ⟶	^{208}lead	14 billion	200 million
^{238}uranium ⟶	^{206}lead	4.5 billion	100 million
^{40}potassium ⟶	^{40}argon	1.3 billion	100 million
^{235}uranium ⟶	^{207}lead	704 million	100,000
^{14}carbon ⟶	^{14}nitrogen	5,730	0–60,000

a (*Left*) A fossilized frond of a tree fern, one of many species that lived more than 250 million years ago. (*Right*) A fossilized sycamore leaf that dropped 50 million years ago.

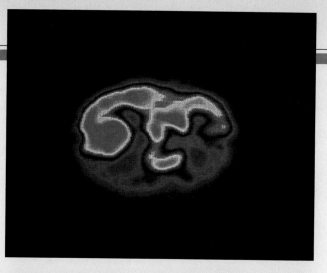

Tracking Chemicals Scintillation counters and other devices can detect emissions from radioisotopes. Thus radioisotopes can be used as **tracers**. Tracers reveal a pathway or destination of a substance that has entered a cell, the human body, an ecosystem, or some other "system."

Carbon provides an example. All isotopes of an element have the same number of electrons, so they all interact with other atoms the same way. This means cells can use any isotope of carbon in reactions that require carbon atoms. For example, such reactions are part of the photosynthetic pathway. By putting plant cells in a medium enriched in ^{14}carbon, researchers identified the steps by which plants take up carbon and incorporate it into carbohydrates during photosynthesis. Researchers also are using tracers to identify how plants use naturally occurring nutrients and synthetic fertilizers. This knowledge may lead to improvements in crop production.

What about medical applications? Consider the human thyroid, the only gland of ours that takes up iodine. If a tiny amount of the radioisotope ^{123}iodine is injected into a patient's blood, the thyroid can be scanned with a photographic imaging device. Figure *b* shows examples of the resulting images.

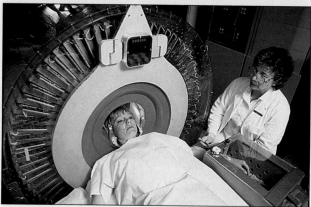

c A patient being moved into a PET scanner. The top photo shows a vivid image of a brain scan of a child who has a severe neurological disorder. Normally, different colors in a brain scan signify differences in metabolic activity in the brain. Notice how one half of this child's brain shows no activity.

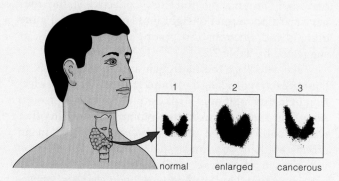

b Scans of human thyroid glands, taken after ^{123}iodine was injected into the bloodstream of three different patients. A normal thyroid takes up iodine (including radioisotopes of iodine) and uses it in hormone production. The scans show (1) a normal gland, (2) the enlarged gland of a patient with a thyroid disorder, and (3) a cancerous thyroid gland.

Saving Lives Radioisotopes are used in nuclear medicine to diagnose and treat diseases. For example, patients with irregular heartbeats are given artificial pacemakers, powered by energy from ^{238}plutonium. This dangerous radioisotope is sealed in a case to prevent its emissions from damaging body tissues.

As another example, PET (short for positron-emission tomography) yields images of metabolically active and inactive tissues. Radioisotopes are attached to glucose or some other biological molecule. Then they are injected into a patient, who is moved into a PET scanner. When cells in certain tissues absorb glucose, radioisotope emissions are used to produce a vivid image of variations or abnormalities in metabolic activity (Figure *c*).

Finally, in some cancer treatments, radioisotopes are used to destroy or impair the function of living cells. In radiation therapy, localized cancers are deliberately bombarded with energy from a ^{226}radium or ^{60}cobalt source.

2.2 THE NATURE OF CHEMICAL BONDS

We turn now to interactions between atoms. Take a moment to review Figure 2.4, which summarizes a few conventions used to describe these events.

What Is a Chemical Bond?

A **chemical bond** is a union between electron structures of atoms. In other words, *it is an energy relationship.* Most chemical bonds form when an atom gives up, gains, or shares one or more electrons with another atom. Some atoms enter into such relationships rather

---

Figure 2.4 Chemical bookkeeping.

We use symbols for elements when writing *formulas*, which identify the composition of compounds. For example, water has the formula H_2O. The subscript indicates that two hydrogen (H) atoms are present for every oxygen (O) atom. We use symbols and formulas in *chemical equations*—representations of reactions among atoms and molecules. An arrow in such an equation means "yields." Substances entering a reaction (reactants) are to the left of the arrow, and products are to the right, as shown by this equation for photosynthesis:

$$6CO_2 \ + \ 6H_2O \ \longrightarrow \ C_6H_{12}O_6 \ + \ 6O_2$$

6 carbons 12 hydrogens	6 carbons 12 oxygens
12 oxygens 6 oxygens	12 hydrogens
	6 oxygens

Notice there are as many atoms of each element to the right of the arrow as there are to the left, even though they are combined in different forms. Atoms taking part in chemical reactions may be rearranged, but they are never destroyed. By the *law of conservation of mass*, the total mass of all materials entering a reaction equals the total mass of all the products. Keep in mind, the equations that you use to represent cellular reactions must be balanced this way, because no atoms are lost.

Reactants and products of reactions can be expressed in moles. A "mole" is a certain number of atoms or molecules of any substance, just as "a dozen" can refer to any twelve cats, roses, and so forth. Its weight, in grams, equals the total atomic weight of the atoms that compose the substance.

For example, the atomic weight of carbon is 12, so one mole of carbon weighs 12 grams. A mole of oxygen (atomic weight 16) weighs 16 grams. Can you show why a mole of water (H_2O) weighs 18 grams, and why a mole of glucose ($C_6H_{12}O_6$) weighs 180 grams?

---

easily, but others do not. Whether one atom will bond with another depends on the *number* and *arrangement* of its electrons.

Electrons and Energy Levels

Picture three actresses arriving at the Academy Awards ceremony wearing the same bright red designer dress. Each seeks recognition but dreads being caught next to the others. Two might maneuver themselves *near* the center of attention while avoiding each other. But by unspoken agreement, all three never, ever are in the same place at the same time.

Electrons behave roughly the same way. They are attracted to an atom's protons but repelled by other electrons. They spend as much time as possible near the nucleus and far away from each other by moving in different orbitals. Think of **orbitals** as regions of space around an atom's nucleus in which electrons are likely to be at any instant. Each orbital has enough room for two electrons, at most.

In all atoms, the orbital closest to the nucleus is shaped like a ball (Figure 2.5a). Each hydrogen atom has a lone electron in that orbital; each helium atom has two. Any electron in the orbital closest to the nucleus is said to be at the *lowest energy level*.

Atoms larger than helium have two electrons in the first orbital. They also have other electrons that occupy different orbitals. On the average, those other electrons are farther away from the nucleus, and they are said to be at *higher energy levels*.

There is a simple although not quite accurate way to think about this. Imagine that all the different electron orbitals are arranged within a series of *shells* around the nucleus. As shown in Figure 2.5b, one of the orbitals, with space for two electrons, fits inside a shell closest to the nucleus. Four additional orbitals fit inside the second shell around the nucleus. Collectively, the four have space for a total of eight more electrons. Still more orbitals and electrons can occupy successive shells, as suggested by Table 2.2.

Take a look at the hydrogen and sodium atoms in Figure 2.6. Hydrogen, the simplest atom, has one electron in its first and only shell. Sodium, with eleven electrons, has a lone electron in its outermost shell. In other words, both kinds of atoms have a "vacancy" in an orbital in their outermost shell. Atoms with electron vacancies tend to react with other atoms. Hydrogen, oxygen, carbon, and nitrogen—*the most abundant components of organisms*—are like this.

An atom tends to react with other atoms when its outermost shell is only partly filled with electrons.

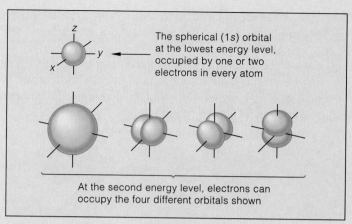

The spherical (1s) orbital at the lowest energy level, occupied by one or two electrons in every atom

At the second energy level, electrons can occupy the four different orbitals shown

Figure 2.5 Arrangement of electrons in atoms. (**a**) One or at most two electrons occupy a ball-shaped volume of space (an orbital) close to the nucleus. Electrons occupying this orbital are at the lowest energy level. At the next (higher) energy level, there can be as many as eight more electrons (four more orbitals, two electrons each). Orbital shapes get tricky, but for our purposes we can ignore them. Simply think of the total number of orbitals at a given energy level as being somewhere inside a "shell" around the nucleus. As suggested in (**b**), higher energy levels correspond to shells farther from the nucleus.

a Shapes of electron orbitals that occur at the first and second energy levels, which correspond to the first and second "shells" in (**b**).

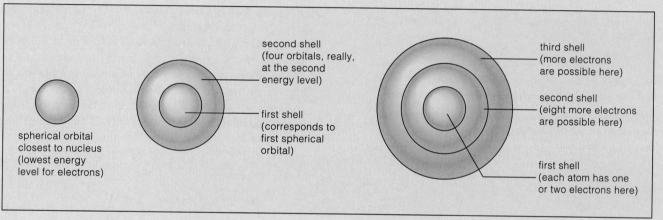

second shell (four orbitals, really, at the second energy level)

first shell (corresponds to first spherical orbital)

spherical orbital closest to nucleus (lowest energy level for electrons)

third shell (more electrons are possible here)

second shell (eight more electrons are possible here)

first shell (each atom has one or two electrons here)

b The shell model of electron distribution in atoms.

Table 2.2	Electron Distribution for a Few Elements				
			Electron Distribution		
Element	Chemical Symbol	Atomic Number	First Shell	Second Shell	Third Shell
Hydrogen	H	1	1	—	—
Helium	He	2	2	—	—
Carbon	C	6	2	4	—
Nitrogen	N	7	2	5	—
Oxygen	O	8	2	6	—
Neon	Ne	10	2	8	—
Sodium	Na	11	2	8	1
Magnesium	Mg	12	2	8	2
Phosphorus	P	15	2	8	5
Sulfur	S	16	2	8	6
Chlorine	Cl	17	2	8	7

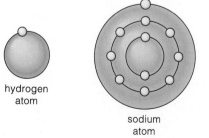

hydrogen atom

sodium atom

Figure 2.6 Distribution of electrons (*yellow dots*) in hydrogen and sodium atoms. Each atom has a lone electron (and room for more) in its outermost shell. Atoms having such partly filled shells tend to enter into reactions with other atoms.

2.3 IMPORTANT BONDS IN BIOLOGICAL MOLECULES

Ionic Bonding

Electrons can be knocked out of atoms, pulled away from them, or added to them. When an atom loses or gains one or more electrons, it becomes positively or negatively charged. In this state, it is an **ion**.

When do atoms lose or gain electrons? They do so when another atom of the right kind is nearby to accept or donate those electrons. Because one atom loses and one gains, *both* become ionized. Depending on the surroundings, two ions can go their separate ways or stay together through the mutual attraction of their opposite charges.

An association of two oppositely charged ions is known as an **ionic bond**. Table salt, or NaCl, has ions of sodium (Na^+) and chloride (Cl^-) linked together this way. Figure 2.7 shows the arrangement of the two kinds of ions in this substance.

When an atom gains or loses one or more electrons, it becomes an ion with an overall positive or negative charge.

In an ionic bond, a positive and a negative ion stay together by the mutual attraction of opposite charges.

Covalent Bonding

In a **covalent bond**, one atom cannot pull electrons completely away from another atom, and the two end up sharing electrons.

We can use a single line to represent a covalent bond when writing out the structural formula for a molecule, as in H—H. Or we can use dots to represent the shared electrons, as shown here:

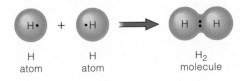

In a double covalent bond, two atoms share two pairs of electrons. This happens in the O_2 molecule, or O=O. In a triple covalent bond, such as N≡N, two atoms share three pairs of electrons.

Covalent bonds are nonpolar or polar. In a *nonpolar* covalent bond, both atoms exert the same pull on shared electrons. The term "nonpolar" implies no difference at the two ends (the two poles) of the bond. An example is the H—H molecule. The two hydrogen atoms, each with one proton, attract the shared electrons equally.

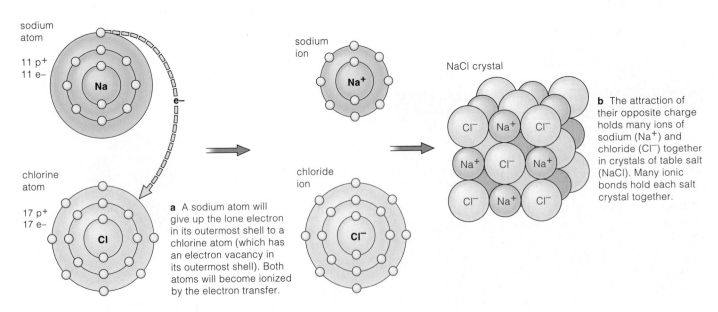

b The attraction of their opposite charge holds many ions of sodium (Na^+) and chloride (Cl^-) together in crystals of table salt (NaCl). Many ionic bonds hold each salt crystal together.

a A sodium atom will give up the lone electron in its outermost shell to a chlorine atom (which has an electron vacancy in its outermost shell). Both atoms will become ionized by the electron transfer.

Figure 2.7 Ionic bonding in sodium chloride (NaCl).

In a *polar* covalent bond, atoms of different elements (which have different numbers of protons) do not exert the same pull on shared electrons. The more attractive atom ends up with a slight negative charge (that atom is "electronegative"). It is balanced by the other atom, which ends up with a slight positive charge.

In other words, taken together, two atoms that are interacting in a polar covalent bond have no *net* charge, but the charge is distributed unevenly between the two ends of that bond.

There are two polar covalent bonds in a water molecule (H—O—H). In this case, the electrons are less attracted to the hydrogens than to the oxygen, which has more protons. Thus, even though a water molecule carries no *net* charge, it can weakly attract other atoms in the vicinity (because of its polarity).

In short, patterns of electron sharing in covalent bonds hold atoms together in specific arrangements in molecules. Additionally, some of these patterns are responsible for weak attractions and repulsions *between* molecules. As you will see in later chapters, all of these bonding forces are absolutely crucial to the structure and functioning of biological molecules.

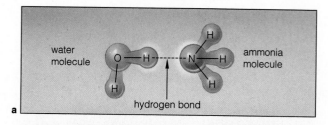

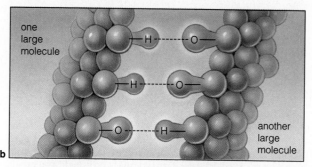

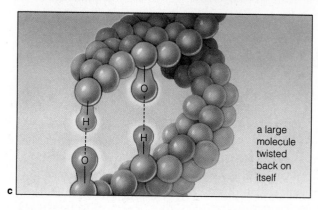

Figure 2.8 Examples of hydrogen bonds. Their collective action gives water and other substances some of their properties.

In a covalent bond, atoms share electrons.

If electrons are shared equally, the bond is nonpolar. If they are not shared equally, the bond is polar (slightly positive at one end and slightly negative at the other).

Hydrogen Bonding

In a **hydrogen bond**, an atom of a molecule weakly interacts with a neighboring hydrogen atom that is already taking part in a polar covalent bond. (The hydrogen, with its slight positive charge, is attracted to the other atom's slight negative charge.) As Figure 2.8 shows, hydrogen bonds can form between two different molecules. Such bonds also can form between two different regions of the same molecule where it twists back on itself.

Hydrogen bonds are common in large biological molecules. For example, many occur between the two strands of a DNA molecule. Individually, the hydrogen bonds are easily broken, but collectively they help stabilize DNA's structure. In later chapters, we will look at the energy it takes to break these and other bonds. Here, we will turn next to the manner in which hydrogen bonds give water some of its life-sustaining properties.

In a hydrogen bond, an atom or molecule interacts weakly with a neighboring hydrogen atom that is already taking part in a polar covalent bond.

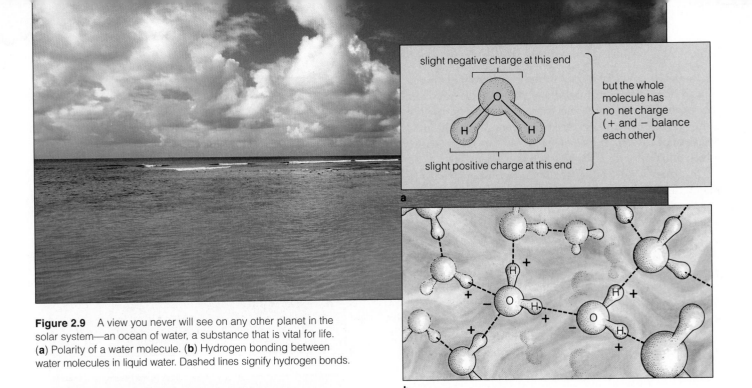

Figure 2.9 A view you never will see on any other planet in the solar system—an ocean of water, a substance that is vital for life. (**a**) Polarity of a water molecule. (**b**) Hydrogen bonding between water molecules in liquid water. Dashed lines signify hydrogen bonds.

2.4 PROPERTIES OF WATER

Life originated in water. Many organisms still live in it, and the ones that don't carry water around with them, in cells and tissue spaces. Many metabolic reactions require water as a reactant. The very shape and internal structure of cells depends on it. We will consider its effects on organisms in many chapters, so you may find it useful to study the following points concerning its major properties.

1. *The polarity of the water molecule influences the behavior of other substances.* A water molecule carries no *net* charge, but remember the charges it does carry are unevenly distributed. As a result of its electron arrangements and bond angles, the water molecule's oxygen "end" is a bit negative and its other end is a bit positive (Figure 2.9*a*). The polarity allows water molecules to form hydrogen bonds with one another and with other polar substances, such as sugars. All polar molecules are attracted to water; they are **hydrophilic** (water loving).

By contrast, water's polarity repels oil and other nonpolar substances, which are **hydrophobic** (water dreading). Shake a bottle containing water and salad oil, then put it on a counter. In time, hydrogen bonds reunite the water molecules. (They replace bonds that were broken when you shook the bottle.) As they do, they push oil molecules aside and force them to cluster in droplets or in a film at the water's surface. As you will see, such hydrophobic interactions help organize the rather oily, sheetlike layers of cell membranes.

2. *Water has temperature-stabilizing effects.* The first forms of life on earth originated in *liquid* water—not water in a frozen or gaseous form. And most of the earth's water tends to stay liquid because of hydrogen bonds between water molecules.

Consider that molecules of water or any other substance are in constant motion, and that energy inputs make them move faster. **Temperature** is a measure of this molecular motion. Compared to most other fluids, water requires a greater input of heat energy before its temperature increases measurably. Why? Hydrogen bonds in water absorb much of the incoming energy, so the motion of individual molecules does not increase as fast. This property helps stabilize temperatures in cells, which are mostly water. It helps cells resist temperature changes that could disrupt vital activities, such as enzyme action.

When water is liquid, its hydrogen bonds are constantly breaking and forming again. During **evaporation**, an energy input converts liquid water to the gaseous state. The energy input increases the molecular motion so much that hydrogen bonds stay broken, and molecules at the water's surface escape into the air. As molecules break free and depart in large numbers, they carry away energy and lower the water's surface temperature. That is why you may cool off when you work up a sweat on hot, dry days. Under such conditions, sweat—which is 99 percent water—evaporates from your skin.

Below 0°C, hydrogen bonds resist breaking and they lock water molecules in the bonding pattern of ice (Figure 2.10). During winter freezes, ice sheets form on ponds, lakes, and streams. They hold in water's heat and help protect aquatic organisms against freezing.

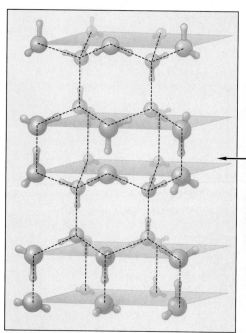

Figure 2.10 Hydrogen bonding between water molecules in ice. Below 0°C, each molecule becomes locked by four hydrogen bonds into a crystal lattice. In this bonding pattern, molecules are spaced farther apart than in liquid water at room temperature. When water is liquid, constant molecular motion usually prevents the maximum number of hydrogen bonds from forming.

3. *Water has cohesive properties.* Swimming on a hot summer night can be refreshing, if you don't mind the night-flying bugs that hit the water's surface and float about on it. Because of hydrogen bonds, the water shows cohesion. This means it resists rupturing when placed under tension—that is, stretched—as by weighty bugs (Figure 2.11). At the surface, hydrogen bonds pull the uppermost water molecules inward. A high surface tension results from the ongoing, collective bonding. In most land plants, cohesion helps move water through pipelines that extend from roots to leaves. When water evaporates from leaves, hydrogen bonds "pull" more water molecules up into leaf cells.

4. *Water has outstanding solvent properties.* Water is a great solvent, in that ions and polar molecules readily dissolve in it. Dissolved substances are called **solutes**. But what does "dissolved" mean? By way of example, pour some table salt into a glass of water. In time the salt crystals disappear, for they separate into Na^+ and Cl^- ions. Each Na^+ attracts the negative end of water molecules. Each Cl^- attracts the positive end of other water molecules. Many water molecules cluster as "spheres of hydration" around ions and keep them dispersed in fluid (Figure 2.12). A substance is "dissolved" when spheres of hydration form around its individual ions or molecules. This happens to solutes in cells, body fluids, maple tree sap, and so on.

Water's internal cohesion, temperature-stabilizing effects, and capacity to dissolve many substances influence the structure and functioning of organisms.

Figure 2.11 Water's cohesion, as demonstrated by a water strider "walking" on water molecules held together by hydrogen bonds.

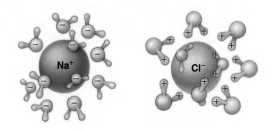

Figure 2.12 Spheres of hydration around charged ions.

WATER, DISSOLVED IONS, AND PH VALUES

The pH Scale

A variety of ions are dissolved in the fluids inside and outside of cells, and these profoundly influence the structure and activities of biological molecules. Among the most influential kinds are **hydrogen ions** (H^+), which are the same thing as free (unbound) protons.

At any given time in liquid water, a few molecules are breaking apart into hydrogen ions and **hydroxide ions** (OH^-). This ionization of water is the basis of the **pH scale**, shown in Figure 2.13. The pH scale is used to measure the concentration of H^+ in blood, tree sap, or any other aqueous solution.

At 25°C, pure water always has just as many H^+ as OH^- ions. Whether for water or any other fluid, this condition represents *neutrality* on the pH scale, and it is assigned a value of 7. It is the midpoint of the pH scale, which ranges from 0 (highest H^+ concentration) to 14 (lowest concentration). *The greater the H^+ concentration, the lower the pH value.*

Starting at neutrality, each change by one unit of the pH scale corresponds to a *tenfold* increase or decrease in H^+ concentration. To get a sense of this logarithmic difference between units, dissolve some baking soda (pH 9) on your tongue or taste some egg white (pH 8.0). Next, sip pure water (pH 7), then vinegar (3) or lemon juice (2.3).

The fluid in most of your cells hovers around pH 7. The pH of your blood and most tissue fluids ranges between 7.35 and 7.45. The pH of an aquatic habitat may be higher or lower than it is inside organisms dwelling there. Certain airborne industrial wastes are acidic (Figure 2.14), and they affect the pH of rain. We will consider the effects of acid rain in Chapter 50.

The pH value of any solution represents its concentration of hydrogen ions.

Acids, Bases, and Salts

Acids are any substances that release H^+ ions when they dissolve in water, and **bases** are substances that combine with those ions. *Acidic* solutions such as lemon juice have more H^+ than OH^- ions, and their pH is below 7. *Basic* (or alkaline) solutions, such as egg whites, have fewer H^+ than OH^- ions, and their pH is above 7.

Think of what happens when you sniff, chew, then swallow fried chicken and so send it on its way to the gastric fluid in your saclike stomach. As an outcome of

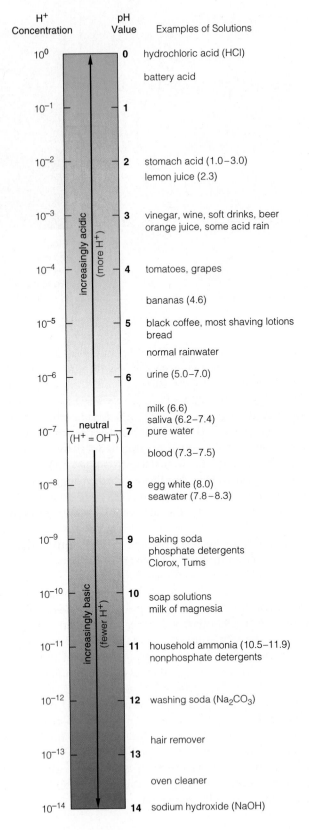

Figure 2.13 The pH scale, in which a liter of fluid is assigned a number according to the number of hydrogen ions in it. The scale ranges from 0 (most acidic) to 14 (most basic). A change of only 1 on the scale means a tenfold change in the H^+ concentration.

Figure 2.14 Sulfur dioxide emissions from a coal-burning power plant. Special camera filters revealed these otherwise invisible emissions. Together with other airborne pollutants, sulfur dioxides dissolve in atmospheric water to form acidic solutions. They are a major component of acid rain.

your feeding activities, cells in the stomach's lining are stimulated to secrete hydrochloric acid (HCl), which separates into H^+ and Cl^-. These ions make the gastric fluid more acidic, and a good thing, too. Increased acidity switches on enzymes that can digest the chicken proteins. It also helps kill bacteria that may be lurking in or on the chicken.

If you eat too much of the fried chicken, you may end up with an "acid stomach." And you might reach for an antacid tablet. Milk of magnesia is one kind of antacid. When dissolved, it releases magnesium ions and hydroxide ions. The hydroxide ions may then *combine with* some of the excess hydrogen ions in your gastric fluid and help settle things down.

Acids commonly combine with bases. The results are ionic compounds called **salts**. Salts often dissolve and form again, depending on the pH. Consider how sodium chloride forms, then dissolves:

$$HCl + NaOH \longrightarrow NaCl + H_2O$$

hydrochloric acid   sodium hydroxide (a base)   sodium chloride (a salt)

$$\swarrow \searrow$$

$$Na^+ \quad Cl^-$$

Many other salts also dissolve into ions. Such ions serve key functions in cells. For example, many help maintain the body's acid-base balance, described next.

Buffers and the pH of Body Fluids

Chemical reactions in cells are sensitive to even slight shifts in pH. Yet hydrogen ions are continually being added to the cellular environment and withdrawn from

it. Control mechanisms counter the potentially disruptive shifts and help maintain cellular pH.

In multicelled organisms, controls also maintain the pH of blood and tissue fluids, which together constitute the body's internal fluid environment. Later in the book, you will see how lungs and kidneys help maintain the acid-base balance of this internal environment, at levels suitable for cellular life. For now, simply keep in mind that many of the control mechanisms involve buffer molecules.

A **buffer** is any molecule that can combine with hydrogen ions, release them, or both, and so help stabilize pH. Some weak acids and bases work as a buffer system. Consider how bicarbonate (HCO_3^-) helps restore pH when blood becomes too acidic. It combines with excess H^+ to form carbonic acid:

$$HCO_3^- + H^+ \longrightarrow H_2CO_3$$

bicarbonate   carbonic acid

Conversely, bicarbonate releases H^+ when blood is not acidic enough. Its buffering action is crucial. For instance, some lung diseases interfere with carbon dioxide elimination, so the blood level of carbonic acid (hence of H^+) increases. This abnormal condition, a form of *acidosis*, makes breathing difficult and weakens the body.

In multicelled organisms, buffering mechanisms help counter slight shifts in the pH of the internal fluid environment that sustains the body's living cells.

Electrons and protons. Atoms, ions, and molecules. With this chapter, we have started thinking about interactions among these bits of matter. These interactions will occupy our attention in many other parts of the book, for they are the foundation for the organization and activities characteristic of life.

For now, merely reflect on the manner in which water influences the functions of a biological molecule—a protein, for instance. The surface of a protein that is dissolved in some cellular fluid may be positively or negatively charged, overall. The charged regions and polar groups attract water molecules. They also attract ions—which in turn attract more water molecules. In this way, an electrically charged "cushion" of ions and water forms around the protein. Figure 2.15 is a simple way of showing this cushioning effect.

Through such interactions with water and ions, the protein remains dispersed in the cellular fluid rather than randomly settling against some cell structure. Why is it important to prevent settling? Many chemical reactions proceed on specific molecular regions of proteins. *Cells must have access to those surfaces, for they are stages for life-sustaining tasks.*

Many more examples could be given of interactions between the polar water molecule and other substances characteristic of the cellular world. For now, the point to keep in mind is this:

Interactions among atoms, ions, and molecules are the start of the organization and behavior of substances that make up cells and the cellular environment.

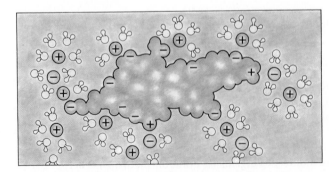

Figure 2.15 Simple diagram of a dissolved protein, surrounded by an electrically charged cushion of ions and water molecules.

SUMMARY

1. Protons, neutrons, and electrons are building blocks of atoms. All atoms of an element have the same number of protons and electrons. Interaction between atoms depends on the number and arrangement of their electrons, which carry a negative charge.

2. All atoms of an element have the same number of protons and electrons. The ones that differ slightly in their number of neutrons are isotopes. Radioisotopes decay spontaneously into atoms of different types.

3. An orbital is a region of space around an atom's nucleus that can be occupied by one or at most two electrons. There are many different orbitals. We can think of them as being arranged inside a series of imaginary shells around the nucleus. Electrons in the shell closest to the nucleus are at the lowest energy level. Electrons in shells farther away are at higher energy levels.

4. Atoms of hydrogen, carbon, nitrogen, and oxygen are the main elements of biological molecules. They have one or more unfilled orbitals in their outermost shell and tend to form bonds with other elements.

5. Atoms have no net charge. If an atom gains or loses one or more electrons, it becomes an ion, with an overall positive or negative charge.

6. In ionic bonds, a positive ion and a negative ion stay together by mutual attraction of their opposite charges. In covalent bonds, atoms share one or more electrons. In hydrogen bonds, an atom or molecule weakly interacts with a neighboring hydrogen atom that is already taking part in a polar covalent bond.

7. The pH scale is a measure of the concentration of hydrogen ions (H^+) that are dissolved in some fluid. Acids are any substances that release H^+, and bases are any substances that combine with them. At pH7, the H^+ and OH^- concentrations in a solution are equal.

8. Buffers and other mechanisms maintain pH values of blood, tissue fluids, and the fluid inside cells.

9. A water molecule shows polarity. Due to its electron arrangements, one end of the molecule carries a partial negative charge, and the other end carries a partial positive charge.
   a. Other polar molecules are attracted to water; they are hydrophilic.
   b. Nonpolar molecules are repelled by water; they are hydrophobic.

10. Water takes part in many reactions and helps give cells shape and internal organization. Because of hydrogen bonds between its molecules, water resists temperature changes. It shows internal cohesion and a notable capacity to dissolve other substances.

Review Questions

1. Define element, atom, molecule, and compound. What are the six main elements (and their symbols) in most organisms? *18–19*

2. Define an atom, an ion, and an isotope. *19, 24*

3. Explain the differences among covalent, ionic, and hydrogen bonds. *24–25*

4. What is the difference between a hydrophilic and a hydrophobic interaction? Is a film of oil on water an outcome of bonding between the molecules making up the oil? *26*

5. What type of bond is associated with the temperature-stabilizing, cohesive, and solvent properties of water? Is that bond also important in hydrophobic interactions? *26–27*

6. Define an acid, a base, and a salt. On a pH scale from 0 to 14, what is the acid range? Why are buffers important in cells? *28–29*

Self-Quiz *(Answers in Appendix IV)*

1. Atoms are constructed of protons, neutrons, and _____ .

2. An _____ has a net charge of zero; an _____ has gained or lost one or more electrons, and so has become negatively or positively charged.
   a. ion; ion
   b. ion; atom
   c. atom; atom
   d. atom; ion

3. Interactions between atoms as they give up, acquire, or share _____ help determine the organization and activities of living things.

4. _____ are atoms of the same element that vary only in the number of neutrons they possess.

5. The main chemical elements found in biological molecules are:
   a. hydrogen, sulfur, nitrogen, oxygen
   b. phosphorus, hydrogen, carbon, oxygen
   c. carbon, oxygen, hydrogen, nitrogen
   d. carbon, oxygen, nitrogen, sulfur

6. Orbitals within shells around the nucleus of an atom can each hold no more than _____ electrons.
   a. one
   b. two
   c. three
   d. four

7. Electrons are shared unequally in a(n) _____ bond.
   a. nonpolar covalent
   b. ionic
   c. hydrogen
   d. polar covalent

8. Polar substances are _____; nonpolar substances are _____ .
   a. hydrophilic; also hydrophilic
   b. hydrophilic; hydrophobic
   c. hydrophobic; also hydrophobic
   d. hydrophobic; hydrophilic

9. Which characterizes the internal pH of most cells?
   a. high concentration of H^+
   b. nearly equal concentration of H^+ and OH^-
   c. high concentration of OH^-
   d. both b and c are correct

10. A(n) _____ can combine with hydrogen ions or release them in response to changes in cellular pH.
   a. acid
   b. salt
   c. base
   d. buffer

11. Match these chemistry concepts appropriately:
   ____ water molecule's polarity
   ____ common bonds in biological molecules
   ____ cellular pH
   ____ hydrogen bonds between water molecules
   ____ salt

   a. close to neutral
   b. temperature-stabilizing and cohesive properties
   c. permits ions and polar molecules to dissolve more easily
   d. produced by reaction between acid and base
   e. ionic, covalent, and hydrogen

Selected Key Terms

acid *28*	hydroxide ion *28*
atom *19*	ion *24*
atomic number *19*	ionic bond *24*
base *28*	isotope *19*
buffer *29*	mass number *19*
chemical bond *22*	molecule *19*
compound *19*	neutron *19*
covalent bond *24*	orbital *22*
electron *19*	pH scale *28*
element *18*	proton *19*
evaporation *26*	radioisotope *19*
hydrogen bond *25*	salt *29*
hydrogen ion *28*	solute *27*
hydrophilic substance *26*	temperature *26*
hydrophobic substance *26*	tracer *21*

Readings

Lehninger, A., D. Nelson, and M. Cox. 1993. *Principles of Biochemistry.* Second edition. New York: Worth. Many excellent illustrations in this new edition of a highly respected textbook.

Science. 26 June 1992. "A New Blueprint for Water's Architecture." 256:1,764.

3 CARBON COMPOUNDS IN CELLS

Mom, Dad, and Clogged Arteries

Butter! Bacon and eggs! Ice cream! Cheesecake! Possibly you think of such foods as enticing, off-limits, or both. After all, who in America doesn't know about animal fats and the dreaded cholesterol?

Soon after you feast on these fatty foods, cholesterol from them enters your bloodstream. Cholesterol has useful roles. It is a structural component of animal cell membranes, and without membranes, there would be no cells. Cells also remodel cholesterol into various molecules, including the vitamin D that is necessary for good bones and teeth. Normally, however, your liver synthesizes enough cholesterol for your cells.

Also circulating in the blood are Apos (short for apolipoproteins). These protein molecules combine

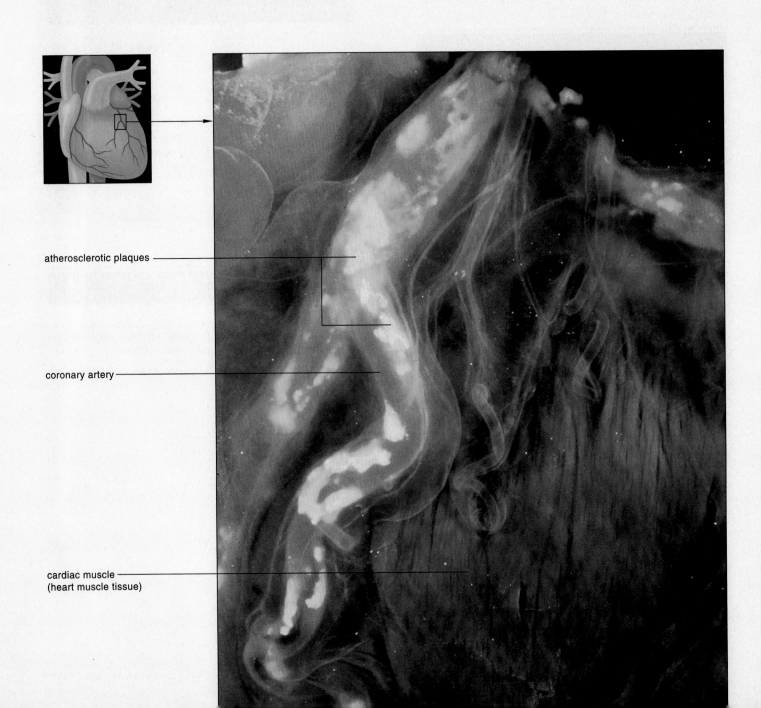

atherosclerotic plaques

coronary artery

cardiac muscle
(heart muscle tissue)

with cholesterol and other substances, forming particles called lipoproteins. Some particles are HDLs (high-density lipoproteins). They collect cholesterol, then transport it to the liver for metabolism. Other particles are VDLs and LDLs (very-low and low-density lipoproteins). If all goes well, these end up in cells that store or use cholesterol.

When too many LDLs form, the excess infiltrates the walls of blood vessels called arteries. There, calcium accumulates and a fibrous net forms. The resulting abnormal masses are atherosclerotic plaques, and they may interfere with blood flow (Figure 3.1). Plaques can narrow arteries, a condition called *atherosclerosis*. When they clog a small-diameter artery that delivers blood to the heart, symptoms can range from mild chest pains to a heart attack.

Do high-cholesterol diets put you at risk? That largely depends on what you inherited from your parents. For instance, each parent provided you with DNA instructions (a gene) for building a certain Apo molecule. However, there are three slightly different versions of this gene in the human population. They specify three different forms of the molecule—Apo 2, 3, or 4—which behave in different ways. If both genes specify Apo E2, your blood cholesterol level will remain so low that your arteries won't get clogged no matter what you eat. If one specifies Apo E2 and the other, Apo E3, the level will be higher. If they both specify Apo E4 , you definitely should forgo the butter and bacon.

This example takes us into the world of large biological molecules—the complex carbohydrates, proteins, lipids, and nucleic acids. These molecules are the foundation for the structure and function of every cell, every organism. Your own body uses them as building materials, as "worker" molecules such as enzymes and transporters, and as storehouses of energy that drive all of your activities—from chewing on an apple to thinking about the hazards of too much cholesterol.

KEY CONCEPTS

1. In organic compounds, carbon atoms covalently bonded to each other serve as a backbone to which hydrogen, oxygen, nitrogen, and other atoms are attached.

2. Cells assemble simple sugars, fatty acids, amino acids, and nucleotides. All of the large biological molecules—the complex carbohydrates, lipids, proteins, and nucleic acids—can be assembled from these four families of small organic compounds.

3. The simple sugars, including glucose, are carbohydrates. So are organic compounds composed of two or more sugar units, of the same or different kind, that are covalently bonded together. The most complex carbohydrates are the polysaccharides, many of which consist of hundreds or thousands of sugar units.

4. Lipids are greasy or oily compounds that dissolve in one another but show little tendency to dissolve in water. They include neutral fats, phospholipids, waxes, and sterols.

5. Cells use carbohydrates and lipids as sources of energy and as building blocks.

6. Proteins have truly diverse roles. Many are structural materials. Many are enzymes that enhance the rate of specific metabolic reactions. Other kinds transport cell substances, contribute to cell movements, trigger changes in cell activities, and defend the body against disease.

7. The nucleic acids called DNA and RNA are the basis of inheritance and cell reproduction.

Figure 3.1 Potentially life-threatening plaques (*bright yellow*) inside one of the arteries that deliver blood to the heart. Such plaques are a legacy of abnormally high levels of an otherwise vital organic compound—cholesterol.

3.1 PROPERTIES OF ORGANIC COMPOUNDS

An **organic compound** consists of carbon and one or more additional elements, covalently bonded to one another. The term is a holdover from a time when chemists thought "organic" substances were the ones they obtained from animals and vegetables, as opposed to the "inorganic" substances obtained from minerals. The term persists even though researchers now synthesize organic compounds in laboratories. And it persists even though there are reasons to believe that organic compounds were present on the earth *before* organisms were.

Carbon-to-Carbon Bonds and the Stability of Organic Compounds

By far, you and all other organisms are composed of oxygen, hydrogen, and carbon. Much of the oxygen and hydrogen is in the form of water. Remove the water, and carbon makes up more than half of what's left.

Carbon's importance in life arises from its versatile bonding behavior. *Each carbon atom can share pairs of elec-trons with as many as four other carbon atoms. Each covalent bond formed this way is quite stable.* Such bonds link carbon atoms in chains and rings, and these serve as a backbone to which hydrogen, oxygen, and other elements become attached (Figure 3.2).

Carbon-to-Carbon Bonds and the Shape of Organic Compounds

The flattened structural formulas in Figure 3.2 might lead you to believe that the molecules they represent are flat, like paper dolls. They actually show stunning diversity in their three-dimensional shapes, which begin with bonding arrangements in carbon backbones.

A carbon atom taking part in four single covalent bonds has this arrangement:

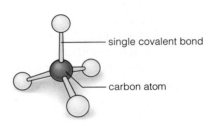

single covalent bond

carbon atom

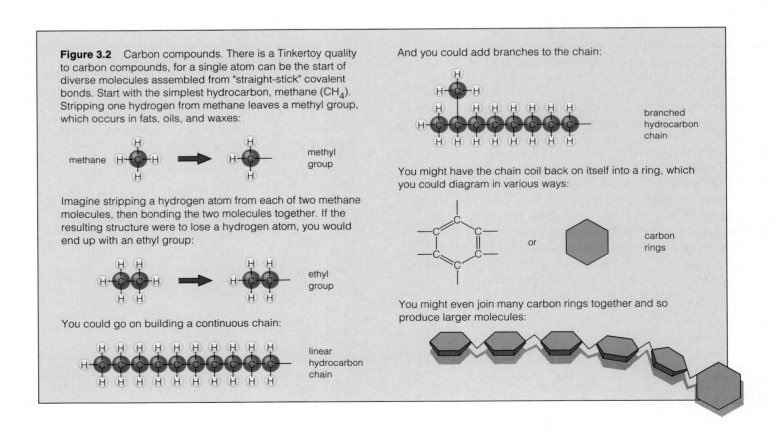

Figure 3.2 Carbon compounds. There is a Tinkertoy quality to carbon compounds, for a single atom can be the start of diverse molecules assembled from "straight-stick" covalent bonds. Start with the simplest hydrocarbon, methane (CH_4). Stripping one hydrogen from methane leaves a methyl group, which occurs in fats, oils, and waxes:

methane

methyl group

Imagine stripping a hydrogen atom from each of two methane molecules, then bonding the two molecules together. If the resulting structure were to lose a hydrogen atom, you would end up with an ethyl group:

ethyl group

You could go on building a continuous chain:

linear hydrocarbon chain

And you could add branches to the chain:

branched hydrocarbon chain

You might have the chain coil back on itself into a ring, which you could diagram in various ways:

or

carbon rings

You might even join many carbon rings together and so produce larger molecules:

Often, the carbon atoms can rotate freely such bonds:

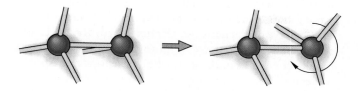

By contrast, a *double* covalent bond uniting a carbon atom with another atom restricts rotation. Where such bonds exist in a carbon backbone, the two atoms rigidly hold their position in space:

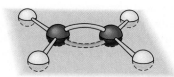

Now imagine a long carbon backbone, with some atoms locked in rigid positions and others free to rotate at different angles in space. The resulting orientations of atoms attached to the backbone may enhance or discourage interactions with neighboring atoms or molecules in the vicinity. As you will see shortly, such interactions help give rise to the three-dimensional shapes and functions of biological molecules.

Flexible and rigid bonding arrangements in carbon backbones are a starting point for the three-dimensional shapes and functions of organic compounds.

Hydrocarbons and Functional Groups

In **hydrocarbons**, only hydrogen atoms are attached to the carbon backbone. Hydrocarbons don't break apart easily, and they form stable portions of most biological molecules. These molecules also have other kinds of atoms attached to the backbone, either alone or in clusters. Atoms covalently bonded to a carbon backbone are **functional groups**. Many influence the electron arrangements of neighboring atoms and so affect the structure and behavior of the whole molecule. Table 3.1 lists some of the major functional groups.

Consider sugars and other organic compounds that have one or more hydroxyl groups (—OH) attached to carbon atoms. These compounds, classified as **alcohols**, readily dissolve in water, because water molecules can form hydrogen bonds with —OH groups. Or consider proteins. Their backbone forms by repeated reactions

Table 3.1	Functional Groups Common to Biological Molecules	
Group	Structural Formula	Common Locations
Methyl	$-\overset{\displaystyle H}{\underset{\displaystyle H}{C}}-H$	Fats, oils, waxes
Hydroxyl	—OH	Sugars, other alcohols
Aldehyde	$-C\overset{H}{\underset{O}{\diagdown}}$	Sugars
Ketone	$-C=O$	Sugars
Carboxyl	$-C\overset{O}{\underset{OH}{\diagup}}$	Sugars, fats, amino acids
Amino	$-N\overset{H}{\underset{H}{\diagdown}}$ or $-\overset{H}{\underset{H}{N}}-H$	Amino acids, proteins
Phosphate	$-O-\overset{O^-}{\underset{O}{P}}-O^-$	Symbolized DNA, RNA, ATP as —Ⓟ
Sulfhydryl	—S—H	Proteins

between many amino groups and carboxyl groups. And the particular bonding patterns associated with this backbone contribute to the three-dimensional structure of proteins. Amino groups also can combine with H^+ and so act as buffers against decreases in pH.

The structure and behavior of an organic compound are influenced by the functional groups covalently bonded to its carbon backbone.

3.2 HOW CELLS USE ORGANIC COMPOUNDS

Five Classes of Reactions

Enzymes, a special class of proteins, speed up specific metabolic reactions. We will study these remarkable proteins in later chapters. For now, it is enough to know that they mediate five categories of reactions by which most biological molecules are assembled, rearranged, and broken apart:

1. **Functional-group transfer**. One molecule gives up a functional group, which another molecule accepts.

2. **Electron transfer**. One or more electrons stripped from one molecule are donated to another molecule.

3. **Rearrangement**. A juggling of internal bonds converts one type of organic compound into another.

4. **Condensation**. Through covalent bonding, two molecules combine to form a larger molecule.

5. **Cleavage**. A molecule splits into two smaller ones.

To get a sense of what goes on, consider just two examples of these reactions. In many condensation reactions, enzymes remove a hydroxyl group from one molecule and an H atom from another, then a covalent bond forms between the two molecules at the exposed sites (Figure 3.3a). The discarded atoms (now H^+ and OH^- ions) combine to form a water molecule. Cells assemble starch and other polymers by repeated condensation reactions. A "polymer" is a large molecule composed of three to millions of subunits, which may or may not be identical. The individual subunits are often called "monomers," as in the sugar monomers of starch.

One type of cleavage reaction, **hydrolysis**, is like condensation in reverse (Figure 3.3b). Enzyme action breaks covalent bonds at functional groups and splits a molecule into two or more parts. At the same time, —H and —OH derived from a water molecule are attached to the exposed sites. Cells commonly hydrolyze starch and other polymers, then use the released subunits as building blocks or energy sources.

The Molecules of Life

Under the physical conditions that now exist on earth, only living cells can synthesize the organic compounds known as carbohydrates, lipids, proteins, and nucleic acids. Together, *these are the molecules characteristic of life*. Different kinds function as energy packets, energy stores, structural materials, metabolic workers, and libraries of hereditary information.

Some biological molecules are rather small organic compounds. The simple sugars, fatty acids, amino acids, and nucleotides are like this. As you will see, cells use these four families of small organic compounds as subunits for the synthesis of larger carbohydrates, lipids, proteins, and nucleic acids.

The large molecules of life are composed of one or more units called simple sugars, fatty acids, amino acids, and nucleotides.

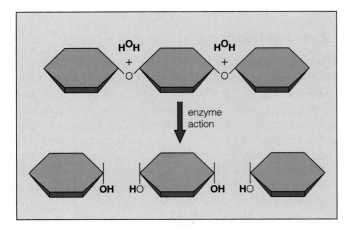

a By two condensation reactions, these three molecules covalently bond into a larger molecule; two water molecules form as by-products.

b Hydrolysis, a water-requiring cleavage reaction. Covalent bonds in a molecule are broken, and H^+ and OH^- derived from a water molecule bond to the molecular fragments.

Figure 3.3 Examples from two categories of enzyme-mediated reactions: condensation and hydrolysis.

3.3 THE SMALL CARBOHYDRATES

A **carbohydrate** is a simple sugar or a molecule composed of two or more sugar units. Carbohydrates are the most abundant biological molecules. All cells use them as structural materials, transportable packets of energy, and stored forms of energy.

We recognize three classes of carbohydrates. These are **monosaccharides** (simple sugars), **oligosaccharides**, and **polysaccharides** (complex carbohydrates).

Monosaccharides—The Simple Sugars

"Saccharide" comes from a Greek word meaning sugar. A *mono*saccharide, or one sugar unit, is the simplest carbohydrate of all. It has at least two —OH groups and an aldehyde or ketone group. Most simple sugars are sweet tasting, and they dissolve readily in water.

The most common monosaccharides have a backbone of five or six carbon atoms that tends to form a ring structure when dissolved in body fluids or cells. Ribose and deoxyribose (sugar components of RNA and DNA, respectively) have five carbon atoms. Glucose has six (Figure 3.4a). You will encounter glucose repeatedly in this book. It is the main energy source for most organisms. It is a precursor (parent molecule) of many compounds and a building block for larger carbohydrates.

You also will encounter three other compounds derived from sugar monomers. Glycerol (a sugar) is a component of fats. Vitamin C (a sugar acid) has roles in nutrition. Glucose-6-phosphate (a sugar phosphate) is a premier entrant into major reaction pathways, including aerobic respiration.

Oligosaccharides

An *oligo*saccharide is a short chain of two or more covalently bonded sugar units. Among those with two sugars—the *di*saccharides—are lactose, sucrose, and maltose. You probably know that lactose (a glucose and a galactose unit) is present in milk.

Sucrose, the most plentiful sugar in nature, consists of a glucose and a fructose unit (Figure 3.4). Leafy plants continually convert carbohydrates to sucrose, which is easily transported through the fluid-filled pipelines that service all living cells in leaves, stems, and roots. Crystallizing sucrose extracts from sugar cane and other plants gives us table sugar.

When seeds of barley and other plants first start to sprout, the forthcoming seedlings produce an abundance of hydrolytic enzymes. These enzymes break down starch, stored in tissues inside the seed coat, into

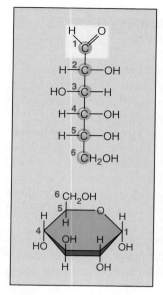
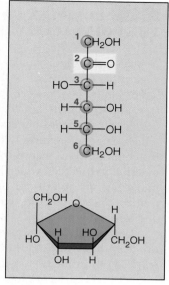

a Glucose   **b** Fructose

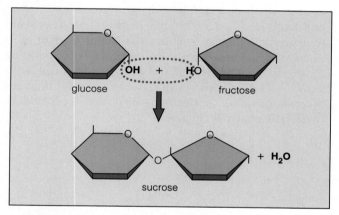

c Formation of sucrose

Figure 3.4 Straight-chain and ring forms of glucose (**a**) and fructose (**b**). For reference purposes, the carbon atoms of sugars are often numbered in sequence, starting at the end of the molecule closest to the aldehyde or ketone group.

(**c**) Condensation of two monosaccharides (glucose and fructose) into a disaccharide (sucrose).

maltose (two glucose units). Seedlings take up maltose as a source of energy for rapid growth. In malt breweries, the hydrolytic enzymes in large batches of carefully tended sprouting seeds hasten the conversion of starch to many sugar molecules that can be fermented.

Proteins and other large molecules often have oligosaccharides attached to them as side chains. These chains are composed of three or more sugar monomers, but they are all short. Some have roles in cell membrane function and in immunity, which are topics of later chapters.

3.4 COMPLEX CARBOHYDRATES: THE POLYSACCHARIDES

A *poly*saccharide is a straight or branched chain of sugar units—hundreds or thousands of the same or different kinds. The most common of these—starch, cellulose, and glycogen—consist only of glucose.

Plant cells store sugars as large starch molecules, which enzymes readily hydrolyze back to sugar units. By contrast, they incorporate cellulose as a structural material in their cell walls. Fibers composed of cellulose molecules are tough and insoluble. Like steel rods in reinforced concrete, they withstand considerable weight and stress (Figure 3.5).

If both cellulose and starch consist of glucose, why do they have such different properties? The answer starts with differences in the pattern of covalent bonding between adjacent glucose units. Long chains of glucose form in both substances. In cellulose, however, the chains stretch out, side by side, and hydrogen-bond to one another at —OH groups (Figure 3.6). The many hydrogen bonds help stabilize the chains in tight bundles that resist breakdown. In starch, a flexible bonding arrangement allows the chains to twist into a coil, with many —OH groups facing outward. The linkages become oriented in positions that are accessible to hydrolytic enzymes (Figure 3.7).

Figure 3.5 A few uses of polysaccharides. Stems gain strength from cellulose, a structural material in plant cell walls. Cells in flowers store starch, which is readily hydrolyzed into sugars that enrich bird-attracting nectar. Birds can convert nectar sugars into glycogen, the animal's equivalent of starch.

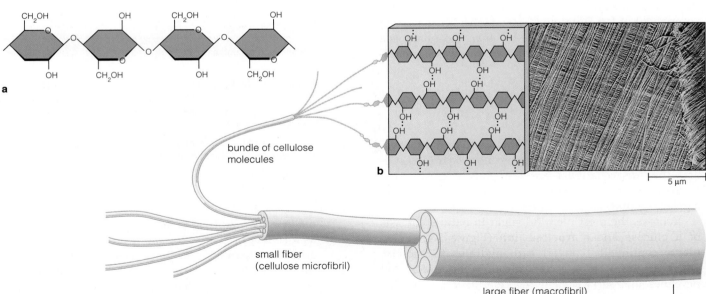

Figure 3.6 Structure of cellulose, an insoluble structural material composed of glucose units. Very few organisms have enzymes that can digest it. (**a**) Bonding pattern between the adjacent glucose units. (**b**) In large fibers of cellulose, chains of glucose units are stretched out, side by side. The chains hydrogen-bond to one another at —OH groups. The bonds stabilize the chains in tight bundles that are arranged into increasingly larger fibers. The micrograph shows large cellulose fibers from the cell wall of *Cladophora*, a green alga.

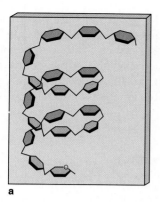

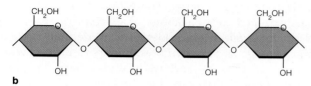

Figure 3.7 (**a**) Structure of amylose, a readily soluble form of starch composed of glucose units. (**b**) Bonding pattern between adjacent glucose units. These particular linkages cause chains of glucose to coil. Coiling orients the linkages in such a way that they are accessible to enzymes of hydrolysis.

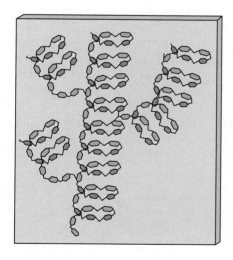

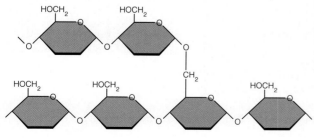

Figure 3.8 Branching structure of glycogen. Like cellulose and starch, glycogen also is composed of glucose units.

a

b

Figure 3.9 (**a**) Scanning electron micrograph of a tick. Its body covering is a protective, chitin-reinforced cuticle. (**b**) Big laughing mushroom (*Gymnophilus*), the cell walls of which are structurally reinforced with chitin.

Like starch in plants, glycogen is a sugar storage form in animals—notably so in liver and muscle tissues. When blood sugar levels fall, liver cells break down glycogen and release glucose units to the blood. During exercise, muscle cells tap into their glycogen stores for quick access to energy. Figure 3.8 shows a few of glycogen's numerous branchings.

Cells of many animals and most fungi secrete chitin. This polysaccharide has nitrogen atoms attached to the backbone. Chitin is the main structural material in external skeletons and other hard body parts of many animals, including crabs and insects (Figure 3.9*a*). It also is the main structural material in the cell walls of many fungal species (Figure 3.9*b*).

3.5 LIPIDS

The greasy or oily compounds called **lipids** dissolve readily enough in one another but show very little tendency to dissolve in water. In nearly all organisms, certain lipids function as the main reservoirs of stored energy. Other lipids function as structural materials in cellular components (such as membranes) and products (such as surface coatings). Here, we consider neutral fats, phospholipids, and waxes, all of which have fatty acid components. We also consider sterols, each with a backbone of four carbon rings.

Fatty Acids

The hydrocarbons called **fatty acids** have a backbone of up to thirty-six carbon atoms, a carboxyl (—COOH) group at one end, and hydrogen atoms occupying most or all of the remaining bonding sites. When combined with other molecules, fatty acids typically stretch out, like flexible tails. Tails with one or more double bonds in their backbone are said to be *unsaturated*. Those with single bonds only are said to be *saturated*. Figure 3.10 shows examples.

When many saturated fatty acid tails are present in a substance, weak attractions cause them to snuggle in parallel, and this gives the substance a rather solid consistency. By contrast, the double and triple bonds in unsaturated fatty acids put rigid kinks in the tails. The packing arrays are less stable, and this imparts fluidity to substances.

Neutral Fats (Triglycerides)

Butter, lard, and oils are examples of **triglycerides**, or neutral fats—the body's most abundant lipids and its richest source of energy. These lipids have fatty acid tails attached to a backbone of glycerol (Figure 3.11). Gram for gram, triglycerides yield more than twice as much energy as carbohydrates. Energy is released when bonds are broken—and triglycerides have far more covalent bonds than carbohydrates do.

In vertebrates, cells of adipose tissue store great quantities of triglycerides as fat droplets. In penguins, seals, and some other animals, a very thick layer of triglycerides also functions as insulation against near-freezing environmental temperatures (Figure 3.12).

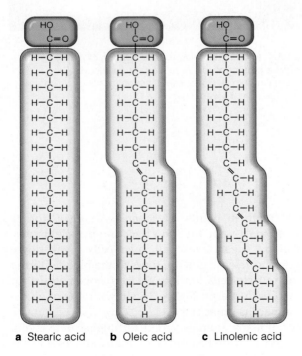

a Stearic acid **b** Oleic acid **c** Linolenic acid

Figure 3.10 Structural formulas for three fatty acids. (**a**) Stearic acid's carbon backbone is fully saturated with hydrogen atoms. (**b**) Oleic acid, with its double bond in the carbon backbone, is unsaturated. (**c**) Linolenic acid, with three double bonds, is a "polyunsaturated" fatty acid.

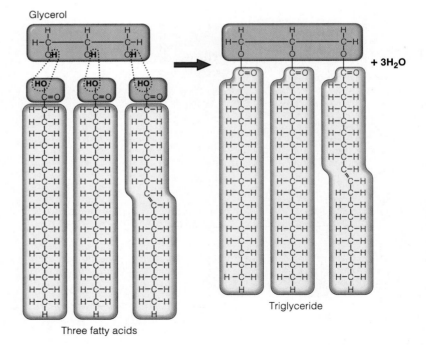

Figure 3.11 Condensation of fatty acids into a triglyceride.

Figure 3.12 Penguins taking the plunge. They can swim in icy waters for long periods, thanks to a thick, insulative layer of triglycerides under their skin.

Figure 3.13 Structural formula of a typical phospholipid of animal cell membranes. The hydrophilic head is shaded orange. Are the hydrophobic tails (*yellow parts*) saturated or unsaturated?

Figure 3.14 (**a**) Honeycomb, constructed from a firm, water-repellant, waxy secretion called beeswax. (**b**) Demonstration of the water-repelling attribute of the cherry cuticle.

Phospholipids

A **phospholipid** has a backbone of glycerol, two fatty acid tails, and a hydrophilic "head" that includes a phosphate group (Figure 3.13). It is the main component of the two lipid layers of cell membranes. The phospholipid heads of one layer are dissolved in cellular fluids, those of the other layer are dissolved in the surroundings, and all fatty acid tails are sandwiched between the two.

Waxes

The lipids called **waxes** have long-chain fatty acids, tightly packed and linked to long-chain alcohols or to carbon rings. Waxes have a firm consistency and repel water. In many animals, cells secrete waxes for coatings that keep skin or hair protected, lubricated, and pliable. In waterfowl and other birds, wax secretions help keep feathers dry. Honeycomb, shown in Figure 3.14*a*, consists of beeswax (after *weax*, an archaic English word meaning "the honeycomb material"). In many plants, waxes and another lipid (cutin) form a cuticle, a surface covering on aboveground parts that helps restrict water loss and hinder some parasites. The waxy cherry cuticle is an example (Figure 3.14*b*).

Sterols and Their Derivatives

Sterols are among the lipids that have no fatty acid tails. Sterols differ in the number, position, and type of their functional groups, but they all have a rigid backbone of four fused-together carbon rings:

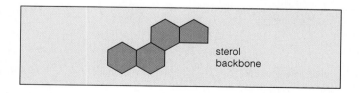

sterol backbone

Sterols are structural components of eukaryotic cell membranes. Cholesterol, described at the start of this chapter, is the major kind in animal tissues. It is the precursor of steroid hormones such as testosterone and estrogen, which influence gamete formation and sexual traits in humans. Some athletes and bodybuilders use hormonelike steroids to increase muscle mass, but these have bad side effects (page 649). Bile salts, which assist fat digestion in the small intestine, also are derived from cholesterol. So is vitamin D, which has roles in calcium and phosphate metabolism.

Of all the large biological molecules, **proteins** are by far the most diverse. The ones called enzymes make reactions proceed much faster than they otherwise would. The structural types are the stuff of bone and cartilage, webs and feathers, and a dizzying variety of other biological parts and products. Transport proteins move cargo across cell membranes or through body fluids. Nutritious proteins abound in milk, eggs, and assorted seeds. Regulatory types, including protein hormones, function as signals for change in cellular activities. Many proteins even function in the body's defense against invasion or predation. Amazingly, cells build all of these diverse molecules from their pools of only twenty or so kinds of amino acids.

An **amino acid** is a small organic compound having an amino group, an acid group, a hydrogen atom, and one or more atoms called its R group. All of these parts are covalently bonded to the same carbon atom:

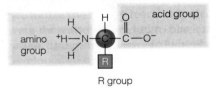

R group

Figure 3.15 shows some amino acids that we will consider later in the book.

Primary Structure of Proteins

When cells synthesize proteins, amino acids become linked, one after the other, by *peptide* bonds. As Figure 3.16 shows, this type of covalent bond forms between the amino group of one amino acid and the acid (carboxyl) group of another. Three or more amino acids joined this way form a **polypeptide chain**. The backbone of such chains incorporates nitrogen atoms in this regular pattern: —N—C—C—N—C—C—.

Amino acid units are not chosen at random for a given chain. They are specifically selected, one at a time, from the twenty kinds available. *And the resulting sequence of amino acids is unique for each particular kind of protein.*

Figure 3.17 shows the sequence for two chains that make up insulin molecules in cattle. (Insulin, a protein hormone, prods cells to take up glucose.) Although you and other vertebrates also produce insulin molecules, the sequences in your versions are not quite the same as the ones in cattle. The overall sequence of amino acids is unique for each kind of protein and represents its *primary* structure.

A protein consists of one or more chains of amino acids.

The particular amino acids that follow one another in sequence in such chains are unique for each kind of protein and represent its primary structure.

---

valine (val)

cysteine (cys)

glycine (gly)

leucine (leu)

isoleucine (ile)

phenylalanine (phe)

methionine (met)

tyrosine (tyr)

glutamate (glu)

lysine (lys)

Figure 3.15   Structural formulas for ten of the twenty common amino acids. Green boxes highlight their particular R groups.

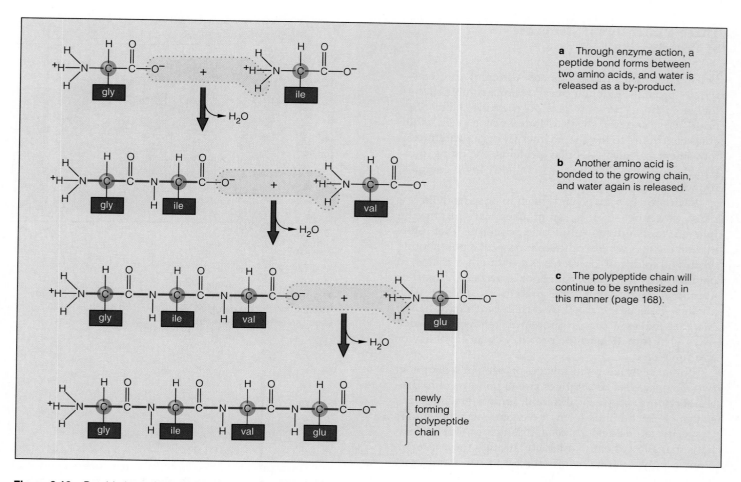

a Through enzyme action, a peptide bond forms between two amino acids, and water is released as a by-product.

b Another amino acid is bonded to the growing chain, and water again is released.

c The polypeptide chain will continue to be synthesized in this manner (page 168).

newly forming polypeptide chain

Figure 3.16 Peptide bond formation during protein synthesis. The amino acids in this example are the first four in the sequence for one of the two polypeptide chains that make up the protein insulin in cattle (Figure 3.17).

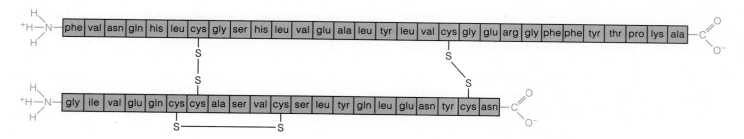

Figure 3.17 Linear sequence of amino acids in cattle insulin, as deduced by Frederick Sanger in 1953. This protein is composed of two polypeptide chains. Disulfide bridges (—S—S—), formed by condensation at two sulfhydryl groups, link the two chains together.

Three-Dimensional Structure of Proteins

Cells synthesize thousands of different proteins—fibrous proteins, globular ones, or some combination of the two. Fibrous proteins have polypeptide chains organized as strands or sheets. Many such protein molecules contribute to the shape, internal organization, and movement of cells. Globular proteins, including most enzymes, have their chains folded into compact, rounded shapes.

Each protein's shape and function arise from its primary structure—that is, from chemical information inherent in its unique amino acid sequence. That information dictates whether different parts of a polypeptide chain will fold, coil, or stretch out with other chains, in parallel array. A folded, coiled, or stretched-out part of a protein might allow the protein to interact with another molecule or a cell. A certain part might be the site that makes a certain protein an enzyme. In such ways, structure dictates the protein's chemical behavior and functions.

Primary structure influences a protein's shape in two major ways. *First*, it gives rise to patterns of hydrogen bonding between different amino acids in the sequence. Such bonds form at oxygen and other atoms of specific amino acids. *Second*, it puts R groups in positions that allow them to interact. Through these interactions, a polypeptide chain is forced to bend and twist into its three-dimensional shape.

Think of a polypeptide chain as a set of dominoes held together by links that can swivel a bit. Each domino corresponds to a peptide group in the chain, as shown here:

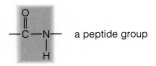

a peptide group

Such groups are rigidly oriented in the same plane. They cannot rotate around covalent bonds, owing to the way that electrons are being shared. But atoms on either side of these planes can rotate somewhat. And some can form bonds with neighboring atoms.

In many cases, hydrogen bonds form between every third amino acid, so that the "dominoes" become helically coiled (Figure 3.18a). The polypeptide chains of hemoglobin, an oxygen-carrying protein, have such coils. In other cases, the chain is extended, and hydrogen bonds hold two or more chains side by side in a sheetlike structure (Figure 3.18b). The proteins of silk are like this.

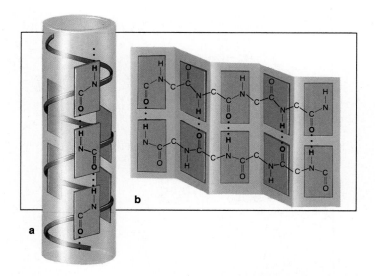

Figure 3.18 Hydrogen bonds (dotted lines) in a polypeptide chain. Such bonds can give rise to a coiled chain (**a**) or to a sheetlike array of chains (**b**).

Thus, proteins have *secondary* structure—a coiled or extended pattern, brought about by hydrogen bonds at regular intervals along polypeptide chains.

Most coiled chains continue to fold into distinctive shapes when one R group interacts with another R group some distance away, with the backbone of the chain itself, or with substances present in the cell. The folding that arises through interactions among the R groups of a polypeptide chain represents the *tertiary* structure of protein molecules. Figure 3.19a shows an example of this.

Finally, some proteins have *quaternary* structure, meaning that they incorporate two or more polypeptide chains. Hemoglobin is a good example of a globular protein. Keratin, the structural material of hair and fur, is an example of a fibrous one (Figure 3.20). Collagen, the most common animal protein, also is fibrous. Skin, bone, tendons, cartilage, blood vessels, heart valves, corneas—these and other structural components of the animal body depend on the strength inherent in collagen.

The amino acid sequence in polypeptide chains gives rise to the three-dimensional structure of proteins. That structure influences a protein's functions.

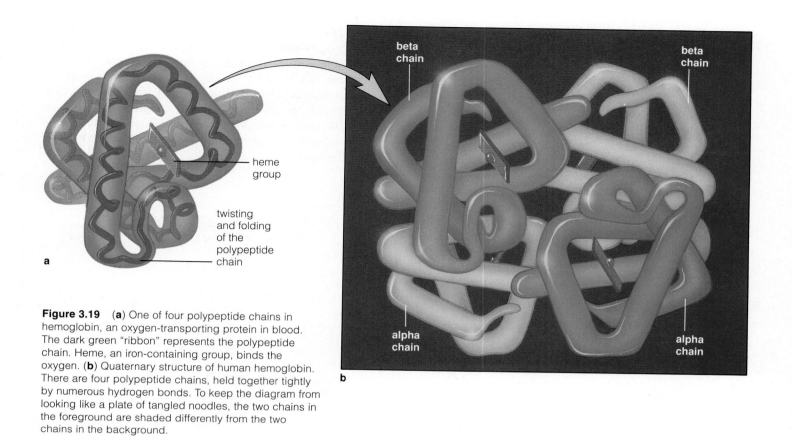

beta chain

beta chain

heme group

twisting and folding of the polypeptide chain

alpha chain

alpha chain

a

b

Figure 3.19 (a) One of four polypeptide chains in hemoglobin, an oxygen-transporting protein in blood. The dark green "ribbon" represents the polypeptide chain. Heme, an iron-containing group, binds the oxygen. (b) Quaternary structure of human hemoglobin. There are four polypeptide chains, held together tightly by numerous hydrogen bonds. To keep the diagram from looking like a plate of tangled noodles, the two chains in the foreground are shaded differently from the two chains in the background.

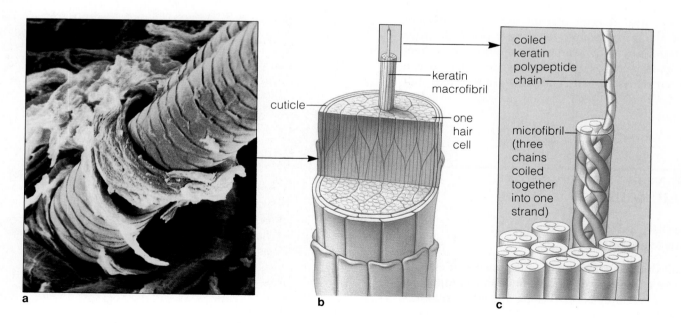

keratin macrofibril

cuticle

one hair cell

coiled keratin polypeptide chain

microfibril (three chains coiled together into one strand)

a

b

c

Figure 3.20 Structure of hair. Polypeptide chains of the protein keratin are synthesized inside hair cells, which are derived from cells of the skin. The chains become organized into fine fibers (microfibrils), which become bundled together into larger, cablelike fibers (macrofibrils). These practically fill the cells, which eventually die. Dead, flattened cells form a tubelike cuticle around the developing hair shaft.

Lipoproteins and Glycoproteins

Many proteins combine with other types of molecules. Think back on the **lipoproteins**, described at the start of this chapter. They form when apolipoproteins that circulate freely in the bloodstream encounter and combine with cholesterol, triglycerides, and phospholipids that have been absorbed from the gut. Lipoproteins differ in their combinations of protein and lipid components.

By contrast, most **glycoproteins** form when newly synthesized polypeptide chains are undergoing modification into mature proteins. These are proteins to which oligosaccharides are covalently bonded. Some of the attached oligosaccharides are linear chains; others are branched. Nearly all proteins on the outer surface of animal cells are glycoproteins. So are most protein secretions from cells and most of the proteins in blood.

Protein Denaturation

If a protein, nucleic acid, or any other molecule loses its three-dimensional shape following the disruption of weak bonds, we call this event **denaturation**. For example, being individually weak, sensitive to heat, and sensitive to pH, hydrogen bonds that help hold a protein in its normal, three-dimensional shape can be disrupted. Then, its polypeptide chains unwind or change shape, and the protein can no longer function.

Consider the protein albumin, concentrated in the "egg white" of uncooked chicken eggs. When you cook an egg, the heat doesn't disrupt the strong covalent bonds of albumin's primary structure. But it destroys the weaker bonds contributing to albumin's three-dimensional shape. For some proteins, denaturation can be reversed when normal conditions are restored—but albumin isn't one of them. There is no way to uncook a cooked egg.

3.7 NUCLEOTIDES AND NUCLEIC ACIDS

Nucleotides with Key Roles in Metabolism

The small organic compounds called **nucleotides** have three parts: a five-carbon sugar, a phosphate group, and a nitrogen-containing base. The sugar is either ribose or deoxyribose. The base has either a single- or double-carbon ring structure (Figure 3.21).

One nucleotide is absolutely central to metabolism; we call it **ATP** (adenosine triphosphate). ATP can deliver energy from one reaction site to virtually any other reaction site in cells. Other kinds of nucleotides

function as **coenzymes**, or enzyme helpers. They accept hydrogen atoms and electrons that are being stripped from molecules and transfer them elsewhere. The coenzymes NAD^+ (nicotinamide adenine dinucleotide) and FAD (flavin adenine dinucleotide) are examples. Still other nucleotides function as **chemical messengers** within and between cells. You will encounter one of these (cyclic adenosine monophosphate, or cAMP) later in the book.

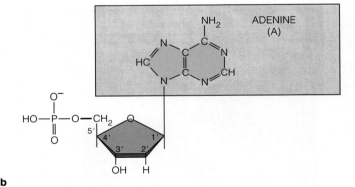

Figure 3.21 Examples of a nucleotide having (**a**) a single ring structure and (**b**) a double ring structure. (**c**) Pattern of covalent bonding between the nucleotides *within* a strand of RNA or DNA.

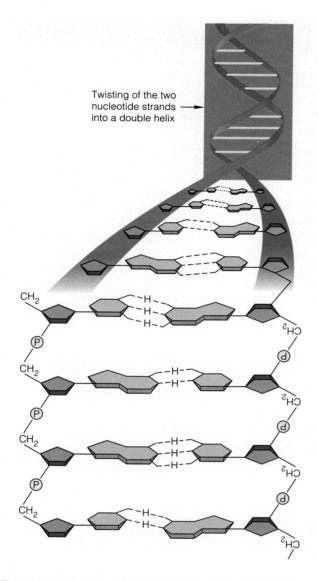

Figure 3.22 Pattern of hydrogen bonding *between* the two nucleotide strands of a DNA molecule. Hydrogen bonds connect bases of one strand with the bases of the other strand.

Twisting of the two nucleotide strands into a double helix

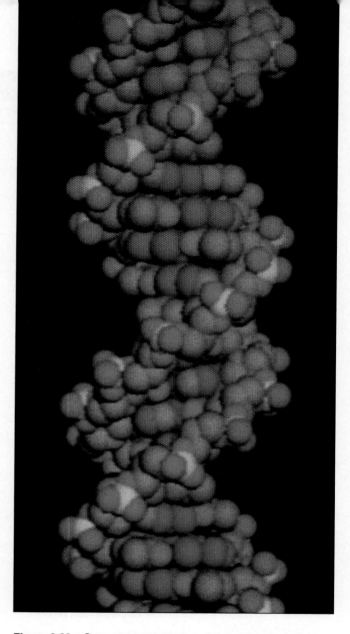

Figure 3.23 Computer-generated model showing the helical coiling of DNA, the double-stranded molecule that is central to maintaining and reproducing the cell.

Arrangement of Nucleotides in Nucleic Acids: DNA and RNA

In **nucleic acids**, four different kinds of nucleotides are strung together, forming large single- or double-stranded molecules. Each strand's backbone consists of joined-together sugars and phosphate groups of adjacent nucleotides. The nucleotide bases stick out to the side (Figure 3.21*b*). As is true of the amino acid sequence of proteins, the sequence of particular nucleotides is unique to each kind of nucleic acid.

You have probably heard of **RNA** (ribonucleic acid) as well as **DNA** (deoxyribonucleic acid). RNA is a sin-gle nucleotide strand, most often. DNA is usually a double-stranded molecule that twists helically, like a spiral staircase, with hydrogen bonds holding the two strands together. Figures 3.22 and 3.23 show this helical configuration.

You will read more about RNA and DNA in later chapters. For now, it is enough to know this:

Genetic instructions are encoded in the sequence of bases in DNA.

RNA molecules function in the processes by which genetic instructions are used to build proteins.

Table 3.2 Summary of the Main Carbon Compounds in Living Things

Category	Main Subcategories	Some Examples and Their Functions	
CARBOHYDRATES *contain an aldehyde or a ketone group, and one or more hydroxyl groups*	**Monosaccharides** (simple sugars)	Glucose	Energy source
	Oligosaccharides	Sucrose (a disaccharide)	Form of sugar transported in plants
	Polysaccharides (complex carbohydrates)	Starch Cellulose	Energy storage Structural roles
LIPIDS *are largely hydrocarbon; generally do not dissolve in water but dissolve in nonpolar substances*	**Lipids with fatty acids:** *Glycerides:* one, two, or three fatty acid tails attached to glycerol backbone	Fats (e.g., butter) Oils (e.g., corn oil)	Energy storage
	Phospholipids: phosphate group, another polar group, and (often) two fatty acids attached to glycerol backbone	Phosphatidylcholine	Key component of cell membranes
	Waxes: long-chain fatty acid tails attached to alcohol	Waxes in cutin	Water retention by plants
	Lipids with no fatty acids: *Sterols:* four carbon rings; the number, position, and type of functional groups vary	Cholesterol	Component of animal cell membranes; can be converted into a variety of steroid molecules; precursor of vitamin D
PROTEINS *are polypeptides (up to several thousand amino acids, covalently linked)*	**Fibrous proteins:** Individual polypeptide chains, often linked into tough, water-insoluble molecules	Keratin Collagen	Structural element of hair, nails Structural element of bones and cartilage
	Globular proteins: One or more polypeptide chains folded and linked into globular shapes; many roles in cell activities	Enzymes Hemoglobin Insulin Antibodies	Increase in rates of reactions Oxygen transport Control of glucose metabolism Tissue defense
NUCLEIC ACIDS (AND NUCLEOTIDES) *are chains of units (or individual units) that each consist of a five-carbon sugar, phosphate, and a nitrogen-containing base*	**Adenosine phosphates**	ATP	Energy carrier
	Nucleotide coenzymes	NAD^+, $NADP^+$	Transport of protons (H^+) and electrons from one reaction site to another
	Nucleic acids: Chains of thousands to millions of nucleotides	DNA, RNAs	Storage, transmission, translation of genetic information

SUMMARY

1. Organic compounds consist of carbon and one or more additional elements. Each carbon atom can form up to four covalent bonds with other atoms. Often they are bonded together in linear or ring structures that serve as the backbone of organic compounds.

2. A hydrocarbon has only hydrogen atoms bonded to the carbon backbone. Most organic compounds in cells are hydrocarbon derivatives, in that they also have functional groups covalently bonded to the backbone. Such groups are reactive arrangements of atoms that impart specific chemical properties.

3. Cells assemble, rearrange, and break apart most organic compounds by four kinds of enzyme-mediated reactions: functional-group transfers, electron transfers, internal rearrangements, condensation reactions, and cleavages (including hydrolysis).

4. Cells have pools of simple sugars, fatty acids, amino acids, and nucleotides, all of which are small organic compounds that include no more than twenty or so carbon atoms. They are building blocks for the large biological molecules—the polysaccharides, lipids, proteins, and nucleic acids.

5. The structure and function of proteins and nucleic acids arise from chemical information contained in the precise sequences of their subunits.

6. Table 3.2 summarizes the main categories of biological molecules described in this chapter. We will have occasion to return to the nature and roles of these molecules in diverse life processes.

Review Questions

1. Pantothenic acid is a type of vitamin. Identify the functional groups attached to its backbone: 35

2. Identify the carbohydrate, fatty acid, amino acid, and polypeptide in the following list: 42, 37, 40
   a. $^+NH_3$—CHR—COO$^-$ c. (glycine)$_{20}$
   b. $C_6H_{12}O_6$ d. $CH_3(CH_2)_{16}COOH$

3. Explain how the amino acid sequence of one or more polypeptides gives rise to the three-dimensional structure and function of proteins. 44–45

4. Distinguish between the following:
   a. monosaccharide, polysaccharide 37–39
   b. peptide bond, polypeptide 42
   c. glycerol, fatty acid 40
   d. nucleotide, nucleic acid 46–47

5. A clerk in a health-food store tells you that certain "natural" vitamin C tablets extracted from rose hips are better for you than synthetic vitamin C tablets. Given your understanding of the structure of organic compounds, what would be your response?

Self-Quiz (Answers in Appendix IV)

1. The backbone of organic compounds is formed by the chemical bonding of _____ atoms into chains and rings.

2. A carbon atom can form up to _____ bonds with other atoms.
   a. four   c. eight
   b. six   d. sixteen

3. Four categories of large biological molecules are the _____ , _____ , _____ , and _____ .

4. All of the following *except* _____ are small organic compounds that serve as building blocks for large biological molecules or as energy sources.
   a. fatty acids   d. nucleotides
   b. simple sugars   e. amino acids
   c. triglycerides

5. Which of the following would *not* be included in the family of carbohydrates?
   a. glucose molecules   c. waxes
   b. simple sugars   d. polysaccharides

6. _____ molecules enhance the rate of specific reactions.
   a. DNA   c. Steroid
   b. Amino acid   d. Enzyme

7. Chemical information that is the basis of inheritance and cell reproduction resides in the structure of _____ .
   a. polysaccharides   c. proteins
   b. DNA and RNA   d. simple sugars

8. Match each molecule with the correct description.
   _____ long sequence of amino acids   a. carbohydrate
   _____ the main energy carrier   b. phospholipid
   _____ glycerol, fatty acids, phosphate c. protein
   _____ two strands of nucleotides   d. DNA
   _____ one or more sugar monomers   e. ATP

Selected Key Terms

alcohol 35	hydrolysis 36
amino acid 42	lipid 40
ATP 46	lipoprotein 46
carbohydrate 37	monosaccharide 37
chemical messenger 46	nucleic acid 47
cleavage 36	nucleotide 46
coenzyme 46	oligosaccharide 37
condensation 36	organic compound 34
denaturation 46	phospholipid 41
DNA 47	polypeptide chain 42
electron transfer 36	polysaccharide 37
enzyme 36	protein 42
fatty acid 40	rearrangement 36
functional group 35	RNA 47
functional-group transfer 36	sterol 41
glycoprotein 46	triglyceride 40
hydrocarbon 35	wax 41

Readings

Goodsell, D. September–October 1992. "A Look Inside the Living Cell." *American Scientist.* 80:457–465. Current models of biological molecules.

Scientific American. October 1985. "The Molecules of Life." Entire issue devoted to biological molecules.

4 CELL STRUCTURE AND FUNCTION

Animalcules and Cells Fill'd With Juices

Early in the seventeenth century, Galileo Galilei arranged two glass lenses in a cylinder. With this instrument he happened to look at an insect, and afterward he described the stunning geometric patterns of its tiny eyes. Thus Galileo, who was not a biologist, was the first to record a biological observation made through a microscope. The study of the cellular basis of life was about to begin. First in Italy, then in France and England, biologists set out to explore a world whose existence had not even been suspected.

At midcentury Robert Hooke, Curator of Instruments for the Royal Society of England, was at the forefront of these studies. When Hooke first turned one of his microscopes to thinly sliced cork from a mature tree, he observed tiny compartments (Figure 4.1a). He gave them the Latin name *cellulae* (meaning small rooms); hence the origin of the biological term "cell." They were actually the walls of dead cells, which is what cork is made of, but Hooke did not think of them as being dead because he did not know cells could be alive. In other plant tissues, he discovered cells "fill'd with juices" but could not imagine what they represented.

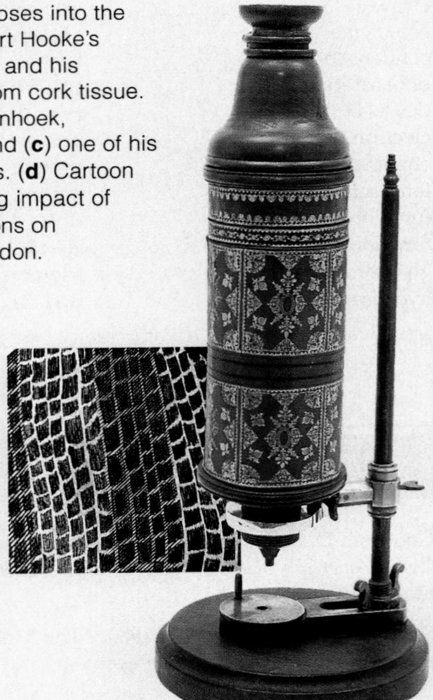

Figure 4.1 Early glimpses into the world of cells. (**a**) Robert Hooke's compound microscope and his drawing of cell walls from cork tissue. (**b**) Antony van Leeuwenhoek, microscope in hand, and (**c**) one of his sketches of sperm cells. (**d**) Cartoon evidence of the startling impact of microscopic observations on nineteenth-century London.

a

Given the simplicity of their instruments, it is amazing that the pioneers in microscopy saw as much as they did. Antony van Leeuwenhoek, a Dutch shopkeeper, had great skill in constructing lenses and possibly the keenest vision (Figure 4.1b). By the late 1600s he was turning up wonders everywhere, including "many very small animalcules, the motions of which were very pleasing to behold," in scrapings of tartar from his teeth. He observed diverse protistans, sperm, even a bacterium—an organism so small it would not be seen again for another two centuries!

In the 1820s, improvements in lens design brought cells into sharper focus. Robert Brown, a botanist, noticed the constant presence of an opaque spot in egg cells, pollen cells, then cells of the growing tissues of orchid plants. He called the spot a "nucleus." In 1838 Matthias Schleiden, another botanist, suggested that the nucleus and cell development are closely related. On the basis of his own studies, Schleiden decided further that each plant cell leads a double life—one independent, pertaining to its own development, the other as an integral part of the plant.

By 1839, after years of studying the structure and growth of animal tissues, the zoologist Theodor Schwann had this to say: Animals as well as plants consist of cells and cell products—and even though the cells are part of a whole organism, they have, to some extent, an individual life of their own.

Yet a question remained: Where do cells come from? A decade later Rudolf Virchow, a physiologist, completed studies of cell growth and reproduction—that is, their division into two cells. Every cell, he concluded, comes from an already existing cell.

And so, by the middle of the nineteenth century, microscopic analysis yielded these insights: *the cell is the smallest unit of life—and the very continuity of life arises directly from the growth and division of single cells.* The insights still hold true.

This chapter provides an overview of our current understanding of cell structure and function. It introduces some of the modern microscopes that transport us ever more deeply into the spectacular worlds of juice-fill'd cells and animalcules. The chapter provides background information for answering a question that you might have asked of Virchow—*If every cell comes from a cell, then where did the very first cell come from?*—but that's another story.

b

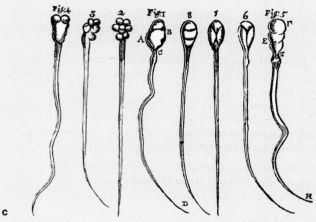

c

d

4.1 THE CELL THEORY

Within your body and at its moist surfaces, trillions of cells live together in interdependency. In scummy pondwater, a single-celled amoeba thrives on its own. For humans, amoebas, and all other organisms, the **cell** is the smallest biological entity that still retains the characteristics of life. A cell either can survive on its own or has the potential to do so. It is structurally organized in specific ways, and it shows metabolic behavior. It senses and responds to specific changes in the surrounding environment. And its inherited DNA instructions give it the potential to reproduce.

Under the physical and chemical conditions that have prevailed on earth for about the past 2 billion years, living cells have not been able to arise spontaneously from nonliving matter. Only cells that already exist have been able to divide and give rise to new cells. The early microscopists sensed this when they observed cells growing and dividing in plant and animal tissues. They came up with the following generalizations, which constitute the **cell theory**:

All organisms are composed of one or more cells.

The cell is the basic unit of life.

New cells arise only from cells that already exist.

4.2 THE NATURE OF CELLS

Basic Aspects of Cell Structure and Function

Cells differ greatly in size, shape, and activities, as you might gather by comparing a bacterium with one of your liver cells. Yet they are alike in three respects. All cells start out life with a plasma membrane, a region of DNA, and a region of cytoplasm:

1. **Plasma membrane.** This outermost membrane maintains the cell as a distinct entity, apart from the environment, and allows metabolic events to proceed in organized, controlled ways. The plasma membrane does not *isolate* the cell interior; substances and signals continually move across it.

2. **DNA-containing region.** DNA occupies part of the cell interior, along with molecules that can copy or read its hereditary instructions.

3. **Cytoplasm.** The cytoplasm is everything enclosed by the plasma membrane, *except* for the region of DNA. It is a semifluid substance in which particles, filaments, and often membranous parts are organized.

This chapter introduces two fundamentally different kinds of cells. **Eukaryotic cells** contain distinctive arrays of organelles (sacs and other compartments formed by internal membranes). One organelle houses the DNA. It is the **nucleus**, and it is the key defining feature of eukaryotic cells:

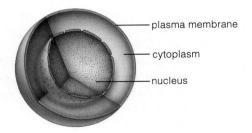

plasma membrane

cytoplasm

nucleus

By contrast, **prokaryotic cells** have no nucleus; no membranes intervene between the region of DNA and the surrounding cytoplasm. Bacteria are the only prokaryotic cells. Outside the realm of bacteria, all other organisms—from amoebas to peach trees and puffball mushrooms to zebras—are eukaryotes.

Structure and Functions of Cell Membranes

Two thin sheets of lipid molecules serve as the structural framework for cell membranes. Figure 4.2 shows this "lipid bilayer" arrangement. The bilayer of a plasma membrane is a continuous boundary that bars the free passage of water-soluble substances into and out of the cell. Within many cell types, membrane bilayers also subdivide the cytoplasm into compartments in which specific substances accumulate.

Proteins, positioned in the lipid bilayer or at its surfaces, carry out most membrane functions. Some of the proteins are passive channels for water-soluble substances. Others carry electrons or pump substances across the bilayer. The receptor types latch onto hormones and other substances that trigger alterations in cell activities. In multicelled organisms, certain proteins even help cells recognize each other and stick together in a tissue.

The next chapter provides a closer look at the structure and functions of cell membranes. For now, keep these points in mind:

A lipid bilayer imparts structure to cell membranes and serves as a barrier to water-soluble substances.

Proteins embedded in the bilayer or positioned at its surfaces carry out most membrane functions.

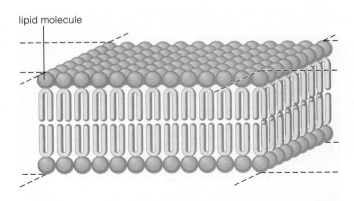

lipid molecule

Figure 4.2 The arrangement of lipid molecules in a layer two molecules in thickness. The double layer provides the framework of all biological membranes.

Surface-to-Volume Constraints on the Size and Shape of Cells

Can any cell be seen with the unaided human eye? There are a few, including the "yolks" of bird eggs, cells in the red part of a watermelon, and the fish eggs we call caviar. Generally, however, cells are too small to be observed without microscopes (see Figure 4.3 and the *Focus* essay on the next page). To give you a sense of cell sizes, one of your red blood cells is only about 8 *millionths* of a meter in diameter. You could fit a string of about 2,000 of them across your thumbnail!

Why are most cells so small? A physical relationship, the **surface-to-volume ratio**, constrains increases in cell size. By this relationship, an object's volume increases with the cube of the diameter, but its surface area increases with the square (Figure 4.4). Simply put, *when a cell expands in diameter, its volume increases more rapidly than its surface area does.*

Suppose we figure out a way to make a round cell grow four times wider. Its volume increases sixty-four times (4^3) and its surface area increases sixteen times (4^2). Unlike fat cells and chicken eggs (which are chockful of fat, food, and so on), our expanded cell is chockful of cytoplasmic machinery. Unfortunately, each unit of plasma membrane must serve four times as much cytoplasm as before! Past a certain point, the inward flow of nutrients and the outward flow of wastes will not be fast enough, and the cell will die.

A very large, round cell also would have trouble moving nutrients and wastes *through* the cytoplasm. By contrast, the random, tiny motions of molecules easily distribute substances through small or skinny cells. If a cell isn't small, it probably is long and thin, or has outfoldings and infoldings that increase its surface relative to its volume. *The smaller or more stretched out or frilly-surfaced the cell, the more efficiently materials can cross its surface and become distributed through the interior.*

We also see the influence of surface-to-volume constraints in the body plans of multicelled organisms. For example, in some algae, cells are attached end to end, forming delicate strands. The arrangement allows each cell to interact directly with the environment. Other algae and a few protistans are sheetlike, with all cells at or near the body surface.

Complex plants and animals also provide evidence of surface-to-volume constraints. For example, they have transport systems that move materials to and from millions, billions, or trillions of cells packed together in tissues. That is the point of having an incessantly pumping heart and an elaborate network of blood vessels inside your own body. This efficient circulatory system quickly delivers materials from the environment to all tissues and sweeps away wastes. Its "highways" cut through the volume of tissue and so shrink the distance to and from individual cells.

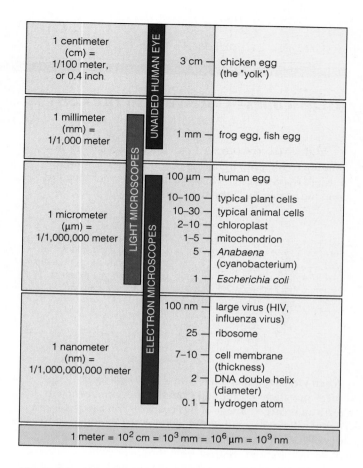

Figure 4.3 Units of measure used in microscopy. Biologists use the micrometer when describing whole cells or large cell structures. They use the nanometer when describing smaller cell structures and large organic molecules.

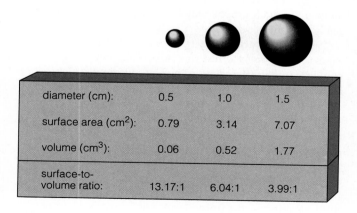

Figure 4.4 Relationship between surface area and volume when a sphere is enlarged. As the diameter increases, the volume increases more rapidly than the surface area does.

Microscopes—Gateways to the Cell

Light Microscopes   Light microscopes work by bending (refracting) light rays. Light rays pass straight through the center of a curved lens. The farther they are from the center, the more they bend:

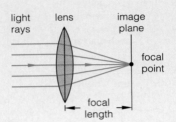

light rays · lens · image plane · focal point · focal length

The angle at which a ray of light enters a glass lens and the molecular structure of the glass dictate the extent to which the ray will bend

When viewing a specimen with a *compound light microscope*, light emanating from the specimen bends to form an enlarged image of it (Figure *a*). You can observe a living cell if it is small or thin enough for light to pass through. Its internal structures will be visible only if they differ in color and contrast from the surroundings. Because most cell structures are nearly colorless and optically uniform in density, you may wish to stain the specimen (expose it to dyes that react with some cell structures but not others). Bear in mind, staining usually alters the structures and kills the cells. Also, dead cells begin to break down at once, so they must be preserved (fixed) before staining. Most micrographs show dead, fixed, or stained cells.

A **micrograph** is simply a photograph of an image formed with a microscope. Fine cell structures are the first to fall apart when cells die, so some of the structures that show up in micrographs may be artifacts—they are not really present in cells.

If living cells are mostly transparent, you can observe them with a *phase-contrast microscope*. This microscope converts small differences in the way different structures bend light to larger variations in brightness.

Suppose you keep magnifying a cell with a light microscope until its image diameter is 2,000 times larger. Cell structures are no longer clear. Why? Think of what happens when you hold a magnifying glass close to a newspaper photograph. You see only black dots. You cannot see a detail as small as or smaller than a dot; the dot covers it up. In microscopy, something like dot size intervenes to limit resolution—the property that determines whether small objects close together can be discerned as separate entities. That limiting factor is the physical size of wavelengths of visible light.

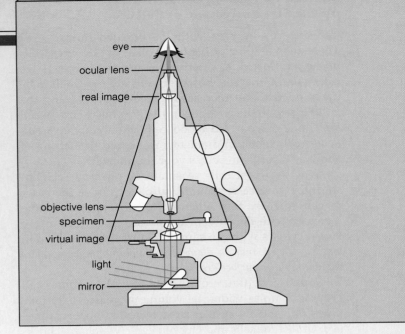

eye · ocular lens · real image · objective lens · specimen · virtual image · light · mirror

a Light microscope

Light travels as different **wavelengths**, which correspond to different colors. Imagine a train of waves moving across an ocean. Each "wavelength" is the distance from one wave's peak to the peak of the wave behind it. For a given wavelength of light, that distance remains constant. It is about 750 nanometers for red wavelengths and 400 nanometers for violet; all other colors fall in between. Suppose a cell structure is less than one-half of a wavelength long. Light rays passing by it will overlap so much that the object won't be visible. The best light microscopes resolve detail only to about 200 nanometers.

Transmission Electron Microscopes   Electrons are particles that also behave like waves. Electron microscopes employ electrons with wavelengths of about 0.005 nanometer—about 100,000 times shorter than those of visible light! Ordinary glass lenses would scatter rather than focus such accelerated streams of electrons. But the electric charge of each electron responds to the force of a magnetic field. In *transmission electron microscopes*, a magnetic field acts as a lens; it diverts electrons along defined paths and channels them to a focal point (Figure *b*).

Electrons must travel in a vacuum (molecules in the air would randomly scatter them). Cells can't live in a vacuum, so they don't stay alive in a transmission electron microscope. To observe fine details, you must slice cells extremely thin. Then, electrons will be scattered in patterns that correspond to the density of different structures. Dense cell parts will be the darkest areas in the final image formed. Most cell structures are somewhat transparent to

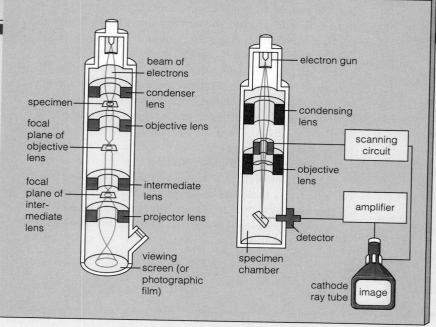

beam of electrons

specimen

focal plane of objective lens

focal plane of intermediate lens

condenser lens

objective lens

intermediate lens

projector lens

viewing screen (or photographic film)

electron gun

condensing lens

scanning circuit

objective lens

amplifier

specimen chamber

detector

cathode ray tube

image

b Transmission electron microscope   **c** Scanning electron microscope

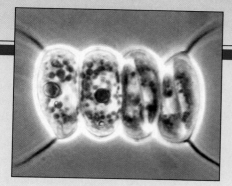

d Light micrograph (phase-contrast)

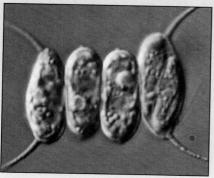

e Light micrograph (Nomarski process)

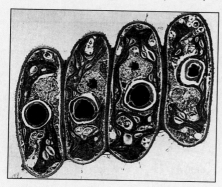

f Transmission electron micrograph, thin section

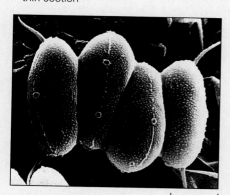

10 µm

g Scanning electron micrograph

electrons, so they must first be stained with heavy metal "dyes."

Compared to standard electron microscopes, *high-voltage* electron microscopes excite electrons until they are ten times more energetic. The energy boost allows electrons to penetrate intact cells several micrometers thick. Like an x-ray plate, the resulting image shows some of the internal organization of cells.

Scanning Electron Microscopes   A narrow beam of electrons is used in *scanning electron microscopes* (Figure *c*). The beam moves back and forth across a specimen's surface, which has been coated with a thin metal layer. The metal responds by emitting some of its own electrons. Equipment similar to a television camera detects the emission patterns, and an image is formed. Scanning electron microscopy does not approach the high resolution of transmission instruments. However, its images have fantastic depth.

We conclude this essay with a comparison of how different types of microscopes reveal different aspects of the same organism—*Scenedesmus*, a green alga. All four specimens in Figures *d* through *g* are at the same magnification. The usefulness of light micrographs has been enhanced by phase-contrast and Nomarski processes, which create optical contrasts without staining the cells.

As for other micrographs in the book, the short bar below the micrograph in Figure *g* provides a reference for size. A micrometer (µm) is 1/1,000,000 of a meter.

4.3 PROKARYOTIC CELLS—THE BACTERIA

We turn now to specific cell types, starting with bacteria. All bacterial cells are prokaryotic; their DNA is not enclosed in a nucleus. *Prokaryotic* means "before the nucleus." The word implies that bacteria existed on earth before the nucleus evolved in the forerunners of all other cells.

Bacteria are the smallest cells. Usually they are not much more than a micrometer wide, and even the rod-shaped ones aren't much more than a few micrometers long. Many have one or more long, threadlike motile structures that extend from the cell surface, as shown in Figure 4.5*a*. These structures are called **bacterial flagella** (singular, flagellum). They permit rapid movements through fluid environments.

In structural terms, bacteria are the simplest cells to think about. Most types have a continuous, semirigid or rigid **cell wall** around the plasma membrane. The wall supports the cell and imparts shape to it (Figure 4.5*b*). Often, polysaccharides cover the wall and help its owner attach to rocks, teeth, and other surfaces. In many disease-causing bacteria, polysaccharides form a jellylike capsule that deters counterattacks.

The plasma membrane controls the movement of substances into and out of the cytoplasm. It has receptors for molecules that activate built-in machinery for key metabolic reactions, including the breakdown of energy-rich molecules. Photosynthetic types have special clusters of membrane proteins that harness light energy and convert it to the chemical energy of ATP.

Bacterial cells have a small volume of cytoplasm. Suspended within the cytoplasm is an irregularly

a

0.5 μm

plasma membrane   nucleoid (region of DNA)

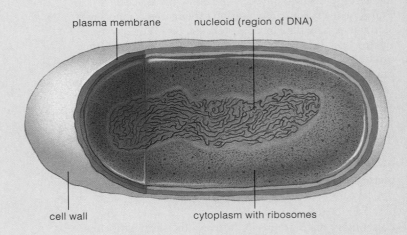

cell wall   cytoplasm with ribosomes

Figure 4.5 Prokaryotic body plans. (**a**) Micrograph and sketch of a common bacterium, *Escherichia coli*. (**b**) This *E. coli* cell has been shocked into releasing its circular molecule of DNA.

Your own gut is home to a large population of a normally harmless strain of *E. coli*. In 1993, a dangerous strain contaminated meat that was sold to some fast-food restaurants. The same strain also contaminated hard apple cider sold at a few roadside stands. Cooking the meat thoroughly (or boiling the cider) would have killed the bacterial cells. Where this was not done, people who ate the meat or drank the cider became quite sick, and some died.

shaped region of DNA. No cell membranes surround this region, which sometimes is called a nucleoid. It contains a single, circular DNA molecule. In the best-studied bacterium, *Escherichia coli*, the circle of DNA is about 1,500 μm long.

The cytoplasm often is densely stained, owing to the presence of large numbers of ribosomes. A **ribosome** consists of two subunits, each composed of RNA and protein molecules. Each is about 20 to 30 nanometers in diameter. *In all living cells, not just bacteria, proteins are synthesized at ribosomes located in the cytoplasm.* At the ribosomal surface, enzymes speed the assembly of polypeptide chains. Each new protein consists of one or more of those chains (page 44).

Figure 4.6 shows only a few representative bacteria. As a group, bacteria are the most metabolically diverse of all organisms. They have managed to exploit energy and raw materials in just about every kind of environment. Besides this, ancient members of their kingdom gave rise to all the protistans, plants, fungi, and animals ever to appear on earth.

The evolution, structure, and functioning of bacteria are topics that will be addressed later in the book. In the pages that follow, our focus will be on nucleated cells, the eukaryotes.

Bacteria alone are prokaryotic cells; their DNA is not confined in a membranous envelope. Their cytoplasm contains many ribosomes and is enclosed in a plasma membrane, which in most species is enclosed within a cell wall.

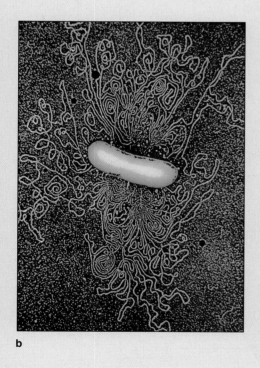

b

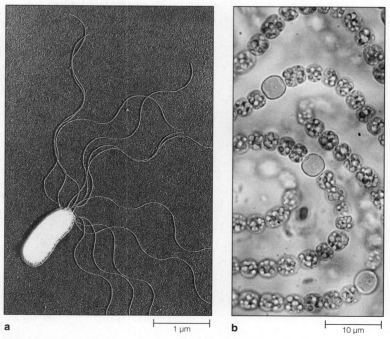

a 1 μm   b 10 μm

Figure 4.6 Interesting variations on the bacterial body plan. (**a**) Like the *Pseudomonas marginalis* cell shown here, many species have surface appendages such as bacterial flagella, which propel the cell through fluid environments. (**b**) Cells of assorted species are shaped like rods, corkscrews, or balls. The ball-shaped cells of *Nostoc*, a photosynthetic bacterium, stick together inside a thick, gelatin-like sheath. Chapter 22 gives other splendid examples.

4.4 EUKARYOTIC CELLS

Functions of Organelles

By definition, an **organelle** is an internal, membrane-bounded sac or some other kind of compartment that has a specific metabolic function within a cell. As Robert Brown perceived when he tracked that opaque spot through generations of orchid cells, the nucleus is a most conspicuous organelle. The cells that have it are *eukaryotic* (which means "true nucleus").

No human-built apparatus matches the eukaryotic cell for the sheer number of chemical activities that can proceed simultaneously in so small a space. For example, many metabolic reactions are incompatible with others. Think about a plant cell putting together a starch molecule by some reactions and breaking it down by others. The cell would gain nothing if the synthesis and breakdown reactions proceeded at the same time on the same starch molecule! Such reactions can proceed smoothly at the same time, largely because organelle membranes keep them physically separated.

Organelle membranes also permit compatible, interconnected reactions to proceed at different times. Thus starch molecules are produced and stored by reactions in one organelle—then later released for use in other reactions in the same plant cell.

Organelles physically separate chemical reactions, many of which are incompatible, inside the cell.

Organelles separate different reactions in time, as when molecules are produced in one organelle, then used later in other reaction sequences.

Organelles Characteristic of Plants

Figure 4.7 can start you thinking about the location of organelles in a typical plant cell. Keep in mind that calling this cell "typical" is like calling a cactus or a crocus a "typical" plant. As is true of animal cells, variations on the basic plan are mind-boggling. With this qualification in mind, also take a look at Figure 4.8. The sketches accompanying the micrograph outline the functions of organelles and other structures that you are likely to find in the plant kingdom.

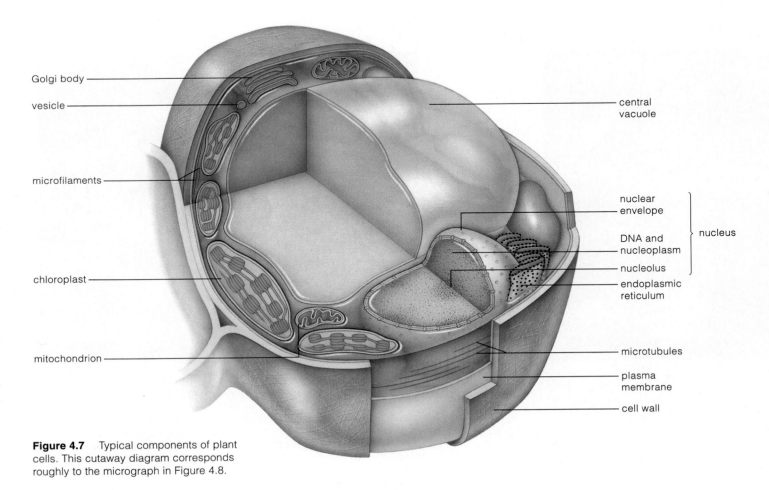

Figure 4.7 Typical components of plant cells. This cutaway diagram corresponds roughly to the micrograph in Figure 4.8.

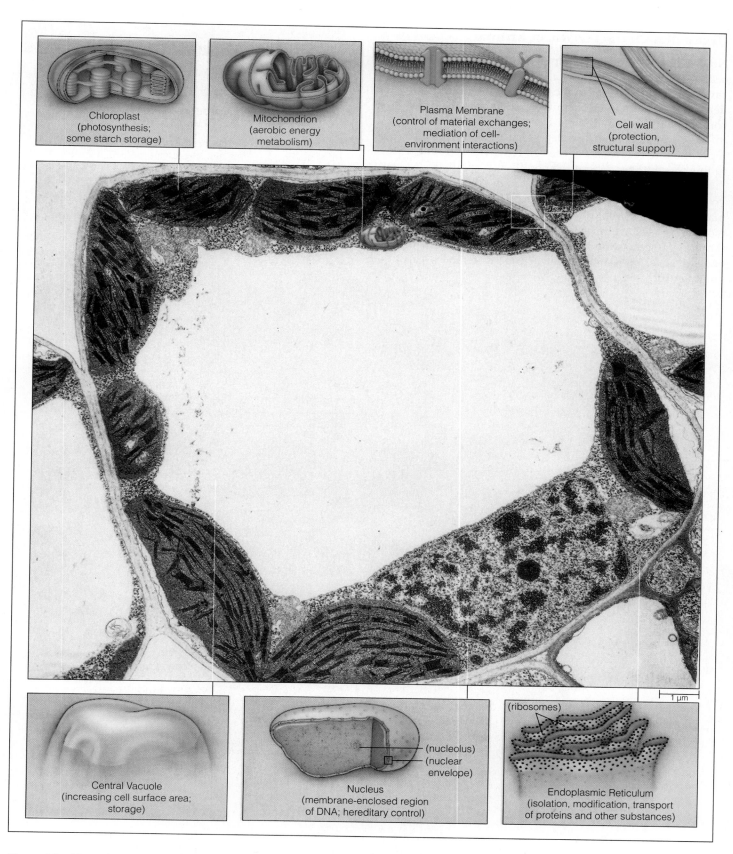

Chloroplast
(photosynthesis;
some starch storage)

Mitochondrion
(aerobic energy
metabolism)

Plasma Membrane
(control of material exchanges;
mediation of cell-
environment interactions)

Cell wall
(protection,
structural support)

1 µm

Central Vacuole
(increasing cell surface area;
storage)

Nucleus
(membrane-enclosed region
of DNA; hereditary control)

(nucleolus)
(nuclear
envelope)

Endoplasmic Reticulum
(isolation, modification, transport
of proteins and other substances)

(ribosomes)

Figure 4.8 Transmission electron micrograph of a plant cell, cross-section, from a blade of Timothy grass.

Organelles Characteristic of Animals

Now take a look at the organelles of a typical animal cell, as shown in Figures 4.9 and 4.10. Right away, you can see that it is similar to the plant cell illustrated earlier, in that it has a nucleus, mitochondria, and the other components listed in Table 4.1. Such similarities point to basic functions that are necessary for survival and reproduction, regardless of the cell type. We will return to this concept throughout the book.

Comparisons of Figures 4.7 through 4.10 also give you an initial idea of the ways in which plant and animal cells are structurally different. For example, you won't ever see an animal cell surrounded by a cell wall. (You might see assorted fungal and protistan cells with one, however.) What other structural differences can you identify?

Table 4.1	Features Typical of Most Eukaryotic Cells
Nucleus	Physical isolation and organization of DNA
Ribosomes	Synthesis of polypeptide chains
Endoplasmic reticulum (ER)	Initial modification of new polypeptide chains; lipid synthesis
Golgi bodies	Further modification of polypeptide chains into mature proteins; sorting, shipping of proteins and lipids for secretion or use in cell
Diverse vesicles	Transport or storage of substances; digestion inside cell; other functions
Mitochondria	Efficient ATP formation
Cytoskeleton	Cell movement, shape, internal organization

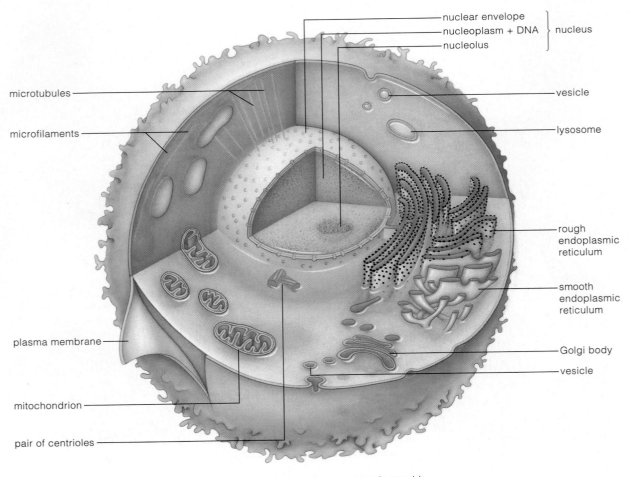

Figure 4.9 Typical components of animal cells. This cutaway diagram corresponds roughly to the micrograph in Figure 4.10.

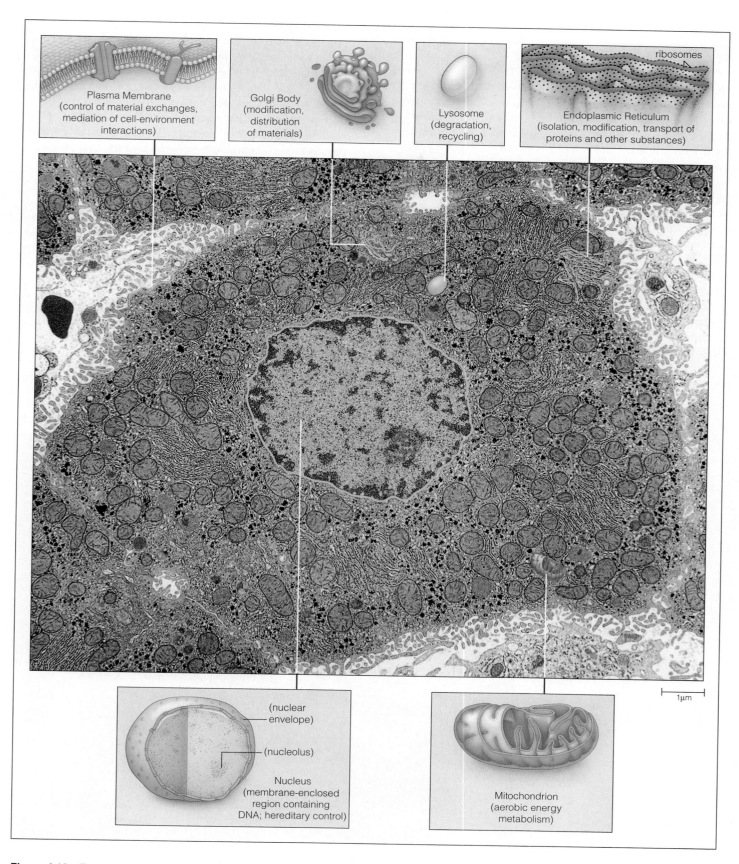

Plasma Membrane
(control of material exchanges,
mediation of cell-environment
interactions)

Golgi Body
(modification,
distribution
of materials)

Lysosome
(degradation,
recycling)

ribosomes

Endoplasmic Reticulum
(isolation, modification, transport of
proteins and other substances)

1μm

(nuclear
envelope)

(nucleolus)

Nucleus
(membrane-enclosed
region containing
DNA; hereditary control)

Mitochondrion
(aerobic energy
metabolism)

Figure 4.10 Transmission electron micrograph of an animal cell, cross-section.
This cell is from a rat liver.

4.5 THE NUCLEUS

There would be no cells whatsoever without complex carbohydrates, lipids, proteins, and nucleic acids. It takes a special class of proteins—enzymes—to build and use those molecules. Thus, *cell structure and function begin with proteins—and instructions for building the proteins themselves are contained in DNA.*

Unlike bacteria, eukaryotic cells have their hereditary instructions distributed in several to many DNA molecules of various lengths. Your own body cells, for example, contain forty-six DNA molecules. Stretched out end to end, they would be about 1 meter long. The DNA in frog cells would be 10 meters, end to end. Compared to the single molecule in bacteria, that's a lot of DNA!

Eukaryotic DNA resides in the nucleus. This type of organelle has a characteristic structure, shown by the example in Figure 4.11, and it serves two functions. First, the nucleus sequesters all of the DNA molecules, out of the way of the complex metabolic machinery in the cytoplasm. This makes it easier to sort out hereditary instructions when the time comes for a cell to divide. The DNA molecules can be assorted into parcels—one parcel for each new cell that forms. Second, membranes that form the outermost portion of the nucleus help control the signals and substances that pass between the nuclear material and the cytoplasm.

Nucleolus

You may be wondering about the dense, globular mass of material within the nucleus shown in Figure 4.11. As eukaryotic cells grow, one or more of these appear in the nucleus. Each mass is a **nucleolus** (plural, nucleoli). Nucleoli are sites where the proteins and RNA subunits of ribosomes are assembled. These subunits are shipped out of the nucleus, into the cytoplasm. When proteins are about to be synthesized, they join together into intact, functional ribosomes.

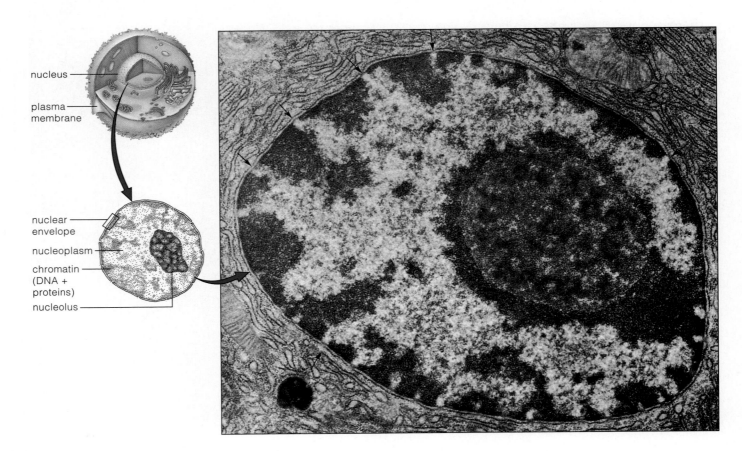

nucleus

plasma membrane

nuclear envelope

nucleoplasm

chromatin (DNA + proteins)

nucleolus

Figure 4.11 Transmission electron micrograph of the nucleus from a pancreatic cell, thin section. Arrows point to pores in its nuclear envelope.

Nuclear Envelope

Notice, in Figures 4.11 and 4.12, that the outermost part of the nucleus has *two* lipid bilayers, one wrapped around the other. They completely surround the nucleoplasm, which is the fluid portion of the nucleus. This double-membrane system is the **nuclear envelope**. As is true of all cell membranes, its lipid bilayers act as a barrier to water-soluble substances. In this case, clusters of membrane proteins span both bilayers, forming pore complexes (Figure 4.12). The pore complexes allow ions and small molecules to enter and leave the nucleus—but they selectively control the passage of larger molecules and particles. For example, they permit ribosomal subunits to pass into the cytoplasm, but they restrict enzymes of the nucleus and the cytoplasm to their respective "compartments" in the cell.

The innermost surface of the nuclear envelope has attachment sites for tiny filaments that help organize the DNA. Attached to the outermost surface are ribosomes, which take part in protein synthesis.

Chromosomes

Between cell divisions, eukaryotic DNA is threadlike, with many enzymes and other proteins attached to it like beads on a string. Except at extreme magnification, the beaded threads look grainy, as they do in Figure 4.11. Prior to division, however, DNA molecules are duplicated (so each new cell will get a set of hereditary instructions). Before the molecules are sorted into two sets, they fold and twist into condensed structures, proteins and all.

Early microscopists named the grainy material *chromatin* and the condensed structures *chromosomes*. Today, we define **chromatin** as the total collection of DNA molecules and associated proteins in the nucleus. We define a **chromosome** as an individual DNA molecule and its associated proteins, regardless of whether it is in threadlike or condensed form.

The point is, the "chromosome" does not always look the same during the life of a eukaryotic cell. We will consider different aspects of chromosomes in chapters to come. It will help to keep this point, and the following sketch, in mind:

unduplicated, uncondensed chromosome (a DNA double helix + proteins)

duplicated, uncondensed chromosome (two DNA double helices + proteins)

duplicated, condensed chromosome

Table 4.2 summarizes the components of the nucleus.

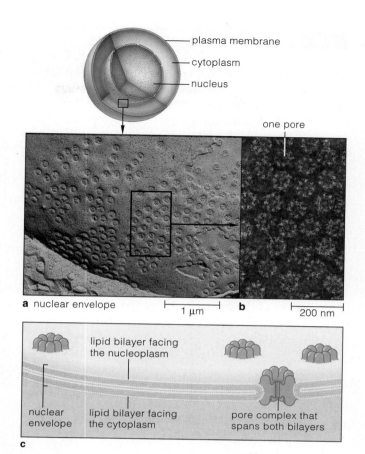

Figure 4.12 (**a**) Surface view of a nuclear envelope, which consists of two pore-studded lipid bilayers. This specimen was deliberately fractured for microscopy, by the method described on page 80. (**b**) Closer look at the pores, each a protein complex that permits the selective transport of larger molecules into and out of the nucleus. (**c**) Diagram of the nuclear envelope's structure.

Table 4.2	Components of the Nucleus
Nuclear envelope	Pore-riddled double membrane system that selectively controls passage of substances into and out of the nucleus
Nucleolus	Dense cluster of RNA and proteins used in the assembly of ribosomal subunits
Nucleoplasm	Fluid portion of interior of nucleus
Chromosome	One DNA molecule and the proteins that are associated with it
Chromatin	Total collection of all DNA molecules and their associated proteins in the nucleus

4.6 CYTOMEMBRANE SYSTEM

In eukaryotic cells, enzymes assemble polypeptide chains on the many thousands of ribosomes that are present in the cytoplasm. What happens to the new chains? Many become stockpiled in the cytoplasm. Others pass through a series of organelles called the **cytomembrane system**. These include the endoplasmic reticulum, Golgi bodies, and certain vesicles (Figure 4.13). In this system, lipids as well as proteins take on final form, then become packaged in vesicles.

Some vesicles deliver proteins and lipids to regions where new membrane must be built. Others accumulate and store proteins or lipids for specific uses. Still other vesicles are part of a "secretory pathway." They move to the plasma membrane, fuse with it, and release their contents to the surroundings. Figure 4.13 shows where the secretory pathway fits in the system.

Figure 4.13 (**a**) Destinations of the proteins and lipids that are modified or assembled in this system, then packaged and shipped out of it. (**b**) The cytomembrane system.

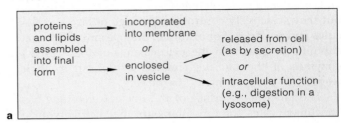

a

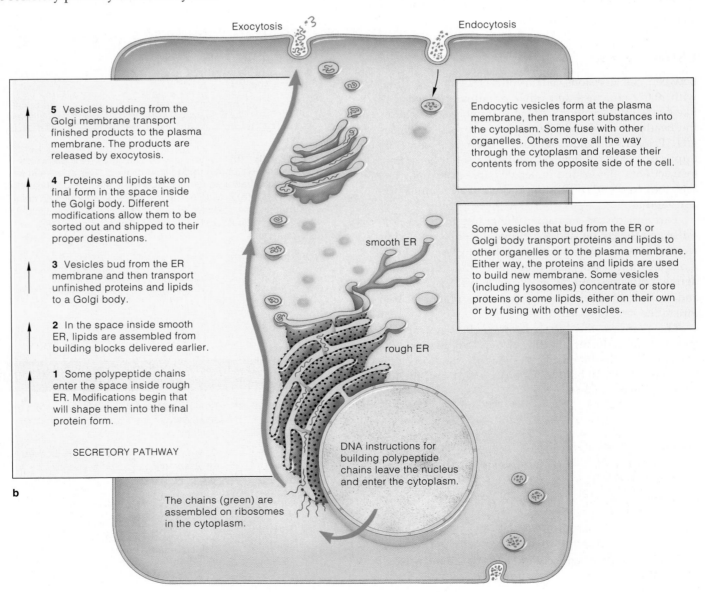

Exocytosis   Endocytosis

5 Vesicles budding from the Golgi membrane transport finished products to the plasma membrane. The products are released by exocytosis.

4 Proteins and lipids take on final form in the space inside the Golgi body. Different modifications allow them to be sorted out and shipped to their proper destinations.

3 Vesicles bud from the ER membrane and then transport unfinished proteins and lipids to a Golgi body.

2 In the space inside smooth ER, lipids are assembled from building blocks delivered earlier.

1 Some polypeptide chains enter the space inside rough ER. Modifications begin that will shape them into the final protein form.

SECRETORY PATHWAY

Endocytic vesicles form at the plasma membrane, then transport substances into the cytoplasm. Some fuse with other organelles. Others move all the way through the cytoplasm and release their contents from the opposite side of the cell.

Some vesicles that bud from the ER or Golgi body transport proteins and lipids to other organelles or to the plasma membrane. Either way, the proteins and lipids are used to build new membrane. Some vesicles (including lysosomes) concentrate or store proteins or some lipids, either on their own or by fusing with other vesicles.

smooth ER

rough ER

DNA instructions for building polypeptide chains leave the nucleus and enter the cytoplasm.

The chains (green) are assembled on ribosomes in the cytoplasm.

b

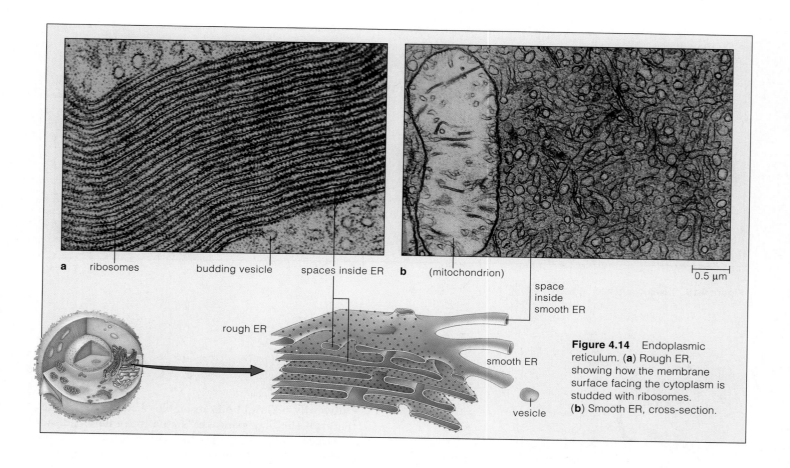

a ribosomes   budding vesicle   spaces inside ER   **b** (mitochondrion)   0.5 µm

space inside smooth ER

rough ER

smooth ER

vesicle

Figure 4.14 Endoplasmic reticulum. (**a**) Rough ER, showing how the membrane surface facing the cytoplasm is studded with ribosomes. (**b**) Smooth ER, cross-section.

Endoplasmic Reticulum

In animal cells, **endoplasmic reticulum**, or **ER**, is a membrane that begins at the nucleus and curves through the cytoplasm. It has rough and smooth regions, due largely to the presence or absence of ribosomes on the side facing the cytoplasm.

Rough ER often is arranged as stacked, flattened sacs and has many ribosomes attached (Figure 4.14*a*). Polypeptide chains are assembled on the ribosomes. Only the newly forming chains that incorporate a "signal" of about fifteen to twenty specific amino acids can enter spaces inside rough ER. Once inside, some acquire side chains (oligosaccharides, mostly).

Many kinds of cells specialize in secreting proteins, and rough ER is notably abundant in them. In your own body, for example, some ER-rich cells of the pancreas produce and secrete specific enzymes that end up in your small intestine, where they function in digestion.

Smooth ER is free of ribosomes and curves through the cytoplasm like connecting pipes (Figure 4.14*b*). It is the main site of lipid synthesis in many cells. Smooth ER is highly developed in seeds. The smooth ER of liver cells inactivates certain drugs and harmful by-products of metabolism. Sarcoplasmic reticulum, a type of smooth ER in skeletal muscle cells, stores and releases calcium ions for muscle contraction.

Peroxisomes

Among the vesicles that form in cells are **peroxisomes**. These sacs of enzymes have a specialized metabolic role: they break down fatty acids and amino acids. The reactions produce hydrogen peroxide, a potentially harmful substance. Before hydrogen peroxide can do harm, another enzyme converts it to water and oxygen or uses it to break down alcohol. If you drink alcohol, nearly half of it is degraded in peroxisomes of your liver and kidney cells.

Similar vesicles, called glyoxysomes, form in plant cells. Their enzymes help convert stored fats and oils to the sugars necessary for rapid, early growth. Glyoxysomes are abundant in lipid-rich seeds, such as those of peanut plants.

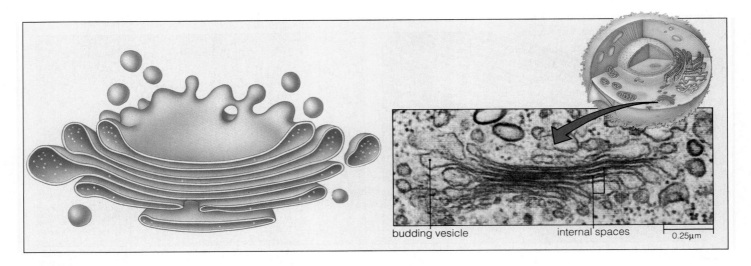

Figure 4.15 Sketch and micrograph of a Golgi body.

budding vesicle   internal spaces   0.25μm

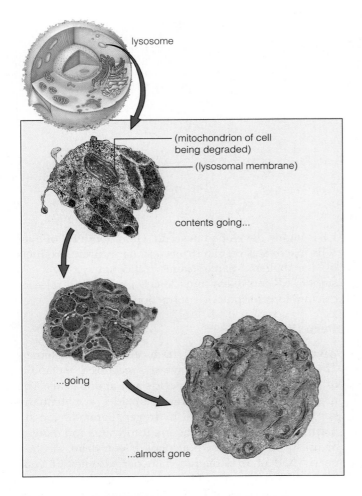

lysosome

(mitochondrion of cell being degraded)

(lysosomal membrane)

contents going...

...going

...almost gone

Figure 4.16 Digestion of "foreign" organelles inside one of the lysosomes of a phagocytic cell. Some of your own white blood cells are phagocytic (which means "cell-eater"). They engulf and digest bacteria and other disease-causing cells that enter your body.

Golgi Bodies

Outwardly, a **Golgi body** resembles a stack of pancakes (Figure 4.15). The "pancakes" actually are flattened sacs in which lipids and proteins are modified. Specific modifications allow them to be sorted out and packaged in vesicles for transport to specific locations. For example, an enzyme in one of the sacs may attach a phosphate group to a particular protein, and this serves as a mailing tag to the protein's proper destination. At the topmost Golgi membrane, vesicles form when a portion of the membrane bulges out, then breaks away.

Lysosomes

Among the vesicles that bud from the Golgi membranes of animal cells and certain fungal cells are **lysosomes**. Lysosomes are organelles of intracellular digestion. They contain different enzymes that can break down virtually all polysaccharides, proteins, and nucleic acids, and some lipids.

In many types of cells, lysosomes fuse with endocytic vesicles, which contain substances or foreign cells that docked against the plasma membrane. As described in the next chapter, such vesicles form when the plasma membrane dimples inward and forms a membrane sac around the trapped particles or cells. Figure 4.16 shows what can happen from that point on.

Often, lysosomes take part in the digestion of some or all of the cell's own parts. For example, when a tadpole turns into an adult frog, its tail disappears as lysosomal enzymes destroy tail cells in a controlled, genetically programmed way.

4.7 MITOCHONDRIA

Energy that ATP molecules carry from one reaction site to another drives nearly all cell activities. In **mitochondria** (singular, mitochondrion), energy stored in breakdown products of glucose and other organic compounds is used to form *many* ATP molecules. These organelles use oxygen to extract far more energy than can be done by any other means. When you breathe in, you are taking in oxygen primarily for mitochondria.

As Figure 4.17 shows, a mitochondrion has two membranes. The outer one faces the cytoplasm. The inner membrane usually folds inward repeatedly. Each fold is a crista (plural, cristae). This double-membrane system creates two compartments in the mitochondrion. As described in Chapter 8, enzymes and other proteins associated with the inner membrane serve as the machinery for ATP formation, and oxygen helps keep the machinery running.

All eukaryotic cells have one or more mitochondria. You might find only one in a single-celled yeast. You might find a thousand or more in energy-demanding cells, such as those in many muscles. Take another look at all the mitochondria in Figure 4.10, which is a micrograph of merely one thin slice from a liver cell. This alone tells you that your liver must be an exceptionally active organ.

In their size and biochemistry, mitochondria resemble bacteria. They have their own DNA and some ribosomes. They even divide on their own. Possibly they evolved from ancient bacteria that were engulfed by a predatory, amoebalike cell, yet managed to escape digestion (page 331). Perhaps they were able to reproduce inside the cell and its descendants. If they became permanent, protected residents, they may have lost many structures and functions required for independent life while becoming mitochondria.

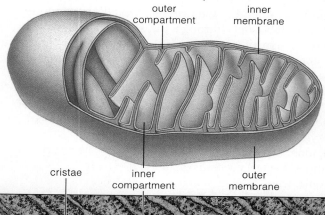

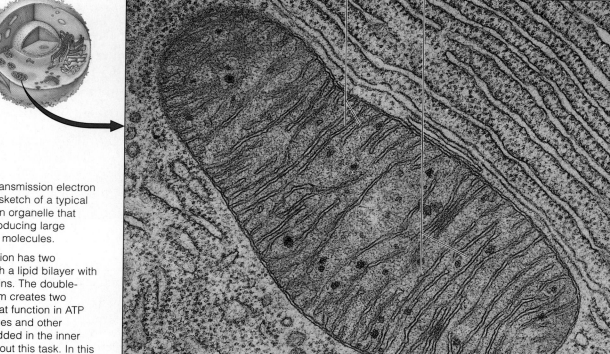

Figure 4.17 Transmission electron micrograph and sketch of a typical mitochondrion, an organelle that specializes in producing large quantities of ATP molecules.

Each mitochondrion has two membranes, each a lipid bilayer with associated proteins. The double-membrane system creates two compartments that function in ATP formation. Enzymes and other molecules embedded in the inner membrane carry out this task. In this micrograph, the dark spots between the membrane folds may be lipid deposits.

0.5 μm

4.8 SPECIALIZED PLANT ORGANELLES

Chloroplasts and Other Plastids

Many types of plant cells contain one or more plastids, which are organelles of photosynthesis or storage. Three kinds are common: chloroplasts, chromoplasts, and amyloplasts.

Only plant cells that specialize in photosynthesis have **chloroplasts**. Within these organelles, sunlight energy is absorbed, ATP forms, and organic compounds are synthesized from simple raw materials (water and carbon dioxide).

Chloroplasts often are oval or shaped like disks. Their outermost portion consists of two membrane layers, one wrapped around the other. These surround a semifluid interior, the stroma. *Another* membrane weaves elaborately through the stroma and forms an inner system of compartments that connect with one another (Figure 4.18). Often you will observe disk-shaped compartments stacked together. Each stack is called a granum (plural, grana).

The energy-acquiring and ATP-forming stage of photosynthesis proceeds at light-trapping pigments, enzymes, and other proteins of the inner membrane system. The synthesis stage proceeds in the stroma, where sugars, starch, and other organic compounds are put together. Here also, starch grains (clusters of new starch molecules) may be stored temporarily.

Chloroplast pigments impart green, yellow-green, or golden-brown coloration to plant parts. Chlorophyll molecules (green) are the most abundant of these, followed by the carotenoids (yellow, orange, and red).

In many ways, chloroplasts resemble certain photosynthetic bacteria. Like mitochondria, they may have evolved from ancient bacterial ancestors that became engulfed but not digested by predatory cells. We return to this idea in Chapter 21.

Unlike chlorophylls, the **chromoplasts** have an abundance of carotenoids but no chlorophylls. They often are the source of the yellow-to-red colors of many flowers, autumn leaves, ripening fruits, and carrots and other roots. Their pigments also may visually attract animals that pollinate plants and disperse seeds.

The **amyloplasts** have no pigments whatsoever. They often store starch grains and are abundant in cells of stems, potato tubers, and many seeds.

Central Vacuole

Mature, living plant cells often have a fluid-filled **central vacuole** (Figure 4.7). It usually occupies 50 to 90 percent of the cell interior, leaving room for only a narrow zone of cytoplasm. A central vacuole stores amino acids, sugars, ions, and toxic wastes. It also increases cell size and surface area. During growth, fluid pressure builds up in the vacuole and forces the still-pliable cell wall to enlarge. The cell itself enlarges under this force, and its increased surface area enhances the rate at which substances can move into and out of it.

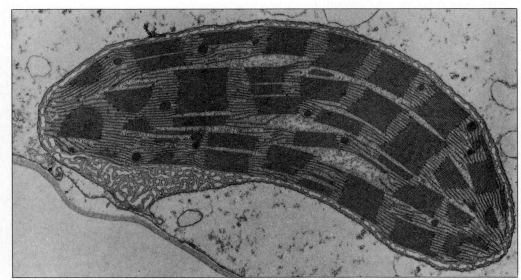

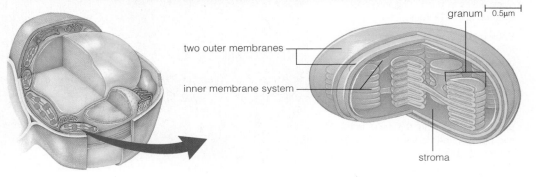

granum　0.5μm

two outer membranes

inner membrane system

stroma

Figure 4.18 Transmission electron micrograph and generalized structure of a chloroplast, an organelle that specializes in trapping sunlight energy, converting it to the chemical energy of ATP—then using the ATP to form sugars and other compounds.

THE CYTOSKELETON

Extending from the plasma membrane and nucleus of eukaryotic cells is an interconnected system of bundled fibers, slender threads, and lattices. This system, the **cytoskeleton**, gives eukaryotic cells their internal organization, overall shape, and capacity to move. Some parts of the cytoskeleton reinforce the plasma membrane and hold its proteins in place; other parts do the same for the nuclear envelope. Still other parts are scaffolds for specific cytoplasmic regions. They may even stabilize the machinery—enzyme systems—of protein synthesis and other metabolic activities.

In all eukaryotic cells, the cytoskeleton consists mainly of **microtubules** and **microfilaments**. The cytoskeletons of animal cells also include various **intermediate filaments**. All three types of components are assembled from protein subunits (Figure 4.19). Microtubules consist of tubulin subunits, arranged in parallel rows. The subunits of microfilaments differ, depending on cell type, but they all contain actin and myosin sub-

units. Keratin subunits derived from intermediate filaments occur in hair cells.

Although many parts of the cytoskeleton are permanent, other parts appear only at certain times in a cell's life. Before a cell divides, for instance, new microtubules are assembled to form a "spindle" structure that moves chromosomes about, and then they disassemble when the task is done.

Figure 4.19 (**a**, **b**) Examples of cytoskeletons, which give eukaryotic cells their internal organization, shape, and capacity for movements. (**c**) Their main components are microtubules and microfilaments. Cytoskeletons of animal cells also contain intermediate filaments.

The cytoskeleton in (**a**) is from a cell of the African blood lily (*Haemanthus*). The cell took up molecules previously labeled with fluorescent dyes. One type of labeled molecule bound only to tubulins, the protein subunits of microtubules. These glowing, green-stained microtubules extend from the cell's nucleus (with purple-stained chromosomes) to its plasma membrane. The cytoskeleton in (**b**) is from a fibroblast, a type of cell that gives rise to certain animal tissues. In this composite of three images, green identifies microtubules, and blue and red identify two different kinds of microfilaments.

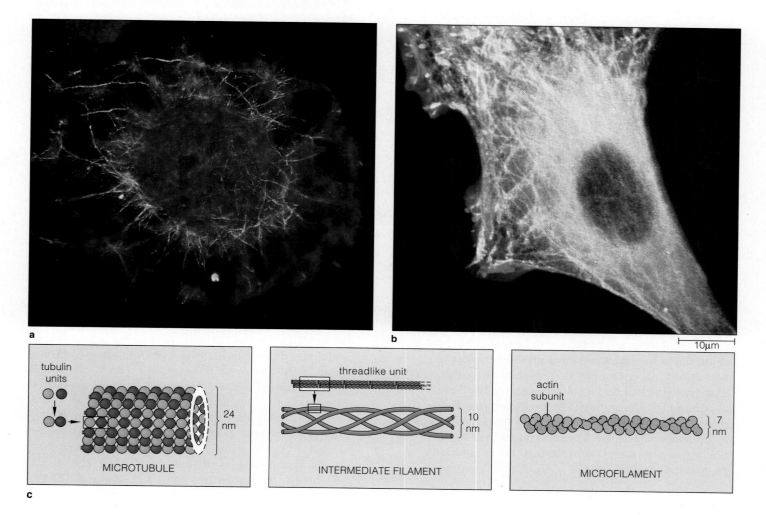

a

b

10μm

c

tubulin units

24 nm

MICROTUBULE

threadlike unit

10 nm

INTERMEDIATE FILAMENT

actin subunit

7 nm

MICROFILAMENT

4.10 THE STRUCTURAL BASIS OF CELL MOVEMENTS

If a cell lives, it moves. This is true even of cells embedded in the tissues of multicelled organisms. The simplest movements are slow, limited rearrangements of internal structures. With more intricate movements, a cell might propel itself through a fluid environment, rapidly shunt organelles or chromosomes to different locations, bulge forward, or produce skinny, temporary surface lobes called **pseudopods** ("false feet").

Microfilaments, microtubules, or both produce most movements. And they do so by three mechanisms:

1. *Through the controlled assembly and disassembly of their subunits, microtubules and microfilaments grow or shrink in length, and structures attached to them are thereby pushed or dragged through the cytoplasm.* For example, this is how spindle microtubules separate chromosomes before a cell divides.

2. *Microfilaments or microtubules actively slide past one another.* As an example, muscle cells contain permanent arrays of actin and myosin microfilaments, arranged in parallel, that bring about **contraction**. Briefly, tiny molecular heads projecting from the myosin microfilaments repeatedly bind and release their actin neighbors. When bound to actin, each head bends in a short, ATP-driven power stroke that slides the actin past it in a given direction. The cumulative, directional sliding causes the arrays of microfilaments to shorten (contract).

A similar kind of sliding mechanism produces the **amoeboid motion** of certain free-living cells and cancer cells. In this case, contractions in one part of a cell squeeze some cytoplasm into other parts, causing the cell body to bulge forward or pseudopods to form. Still another kind of sliding mechanism underlies the beating of flagella and cilia, as described shortly.

3. *Microtubules or microfilaments shunt organelles from one location to another.* For example, between dawn and dusk, the sun's overhead position changes. In response to the changing angle of incoming rays of light, chloroplasts in photosynthetic cells seem to flow to new light-intercepting positions! Actually, myosin filaments attached to the chloroplasts are "walking" over bundles of actin microfilaments, carrying the chloroplasts with them. This mechanism is the basis of **cytoplasmic streaming**. The term refers to an active flowing of cytoplasmic components that is especially pronounced in plant cells.

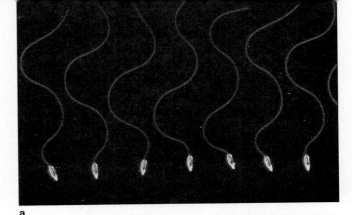

a

b   ├─── 20 µm

Figure 4.20 (**a**) Composite micrographs of the beating pattern of a sperm flagellum. The waves start at the base of the flagellum and move toward the tip. (**b**) Scanning electron micrograph of *Paramecium*, a ciliated protistan.

The Internal Structure of Flagella and Cilia

Later chapters provide closer looks at the mechanisms just described. For now, let's limit our attention to the structure and operation of the **flagellum** (plural, flagella) and **cilium** (plural, cilia). Both structures project from the cell surface and beat in distinctive patterns. Flagella are whiplike tails that propel many free-living cells, such as sperm, through fluid environments (Figure 4.20*a*). Cilia are shorter and usually more profuse than flagella (Figure 4.20*b*). In multicelled organisms, some ciliated epithelial cells stir their surroundings. For example, thousands of these cells line airways to your respiratory tract, and the beating of their cilia directs mucus-trapped bacteria and other particles away from the lungs.

In both flagella and cilia, nine pairs of microtubules ring two central microtubules. This arrangement is called a "9 + 2 array." A system of spokes and links holds the arrangement together. When the sliding mechanism described in Figure 4.21 operates, the system forces the flagellum or cilium to bend.

Figure 4.21 Internal organization of flagella and cilia. Both kinds of motile structures have a system of microtubules and a connective system of spokes and linking elements that serve as the basis of beating movements. Nine pairs (doublets) of the microtubules are arranged as an outer ring around two central microtubules. This is called a 9 + 2 array.

In an unbent flagellum (or cilium), all microtubule doublets extend the same distance into the tip. When the flagellum bends, the doublets on the side that is bending the most are displaced farthest from the tip, as shown here:

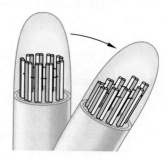

When a flagellum bends, it is not contracting. Rather, a sliding mechanism is operating between microtubule doublets of the outer ring. A series of short arms, composed of the protein dynein, extend from each doublet toward the next doublet in the ring. Inputs of ATP energy cause the dynein arms to attach to the doublet in front of it, tilt in a short power stroke that pulls the attached doublet along with it, then release its hold. Through repeated power strokes, doublets in front of the arms are forced to slide in a prescribed direction. As one doublet slides down toward the base of the flagellum, its cross-bridging arms force the next doublet in line to slide down farther, and so on in sequence.

Because the system of connecting spokes and links extends the length of the flagellum, it prevents the doublets from sliding completely out of their 9 + 2 array. This restriction forces the entire flagellum to bend to accommodate the internal displacement of sliding doublets. Thus, energized dynein arms cause doublet sliding, and restrictions imposed by the spokes and links convert doublet sliding into bending movements. That, in essence, is the model proposed by cell biologists K. Summers and R. Gibbons.

Microtubule Organizing Centers

Microtubules of a flagellum or cilium arise from **centrioles**, which remain at the base of the completed structure as a basal body (Figure 4.21). Centrioles often are present in some kinds of **MTOCs** (short for microtubule organizing centers). These are sites of dense material that generate large numbers of microtubules. (For example, centrosomes, a type of MTOC near the cell nucleus, give rise to the spindle microtubules.)

In short, this is the point to remember about the cytoskeleton of eukaryotic cells:

Microtubules and other cytoskeletal elements are the basis of a cell's shape, internal structure, and capacity for movements.

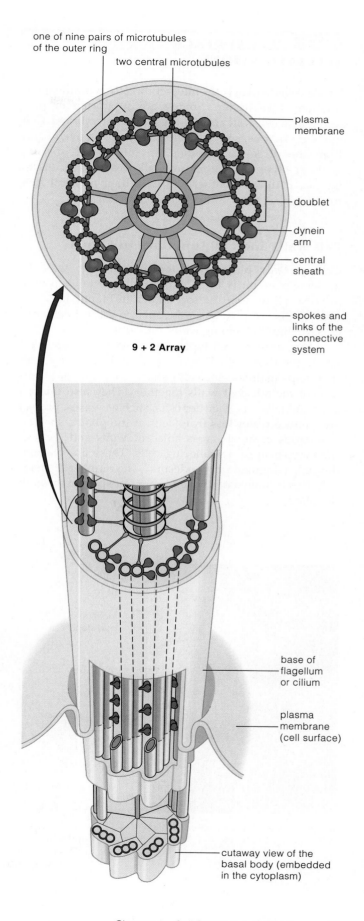

one of nine pairs of microtubules of the outer ring

two central microtubules

plasma membrane

doublet

dynein arm

central sheath

spokes and links of the connective system

9 + 2 Array

base of flagellum or cilium

plasma membrane (cell surface)

cutaway view of the basal body (embedded in the cytoplasm)

4.11 CELL SURFACE SPECIALIZATIONS

Single-celled eukaryotes interact directly with the environment. Like bacteria, many are supported or protected by a cell wall around the plasma membrane. Cell walls are porous, so water and solutes can move to and from the plasma membrane. In multicelled species, walls or other surface features allow adjacent cells to interact with one another and with their physical surroundings, as a few examples will illustrate.

Cell Walls and Cell Junctions in Plants

Cells of leafy plants stick together, wall to wall. In growing plant parts, bundles of cellulose strands form a *primary* cell wall, of the sort shown in Figure 4.22. Primary walls are pliable, so the surface area of cells can enlarge under incoming water pressure.

Later, in many cell types, more layers are deposited inside the first wall. They form a rigid, *secondary* wall that helps maintain the cell's shape. Deposits of pectin cement the adjacent walls together. (They also thicken jams and jellies.) Other deposits, such as waxes, protect and reduce water loss from cells at the plant's surface. Numerous channels cross adjacent walls and connect the cytoplasm of neighboring cells. These are plasmodesmata (singular, plasmodesma). The number of channels affects how fast substances are transported through the plant.

Intercellular Material in Animals

Cell secretions and other materials often intervene between cells of animal tissues. Think of the cartilage at the knobby ends of your leg bones. Cartilage consists of cells scattered in a "ground substance" (of modified polysaccharides) and protein fibers (of collagen or elastin). Through this material, nutrients and other substances diffuse from cell to cell. In mature bone and other tissues, intercellular material accounts for much of your body weight.

Cell Junctions in Animals

Three cell-to-cell junctions are common in animal tissues. *Tight* junctions link cells of epithelial tissues, which line the body's outer surface, inner cavities, and organs. The junctions form seals that keep molecules from freely crossing the tissue, as when they keep acids from leaking out of the stomach. *Adhering* junctions are like spot welds. They keep cells together in tissues of the skin, heart, and other organs that are subject to stretching. *Gap* junctions link the cytoplasm of adjacent cells; they are open channels for the rapid flow of signals and substances.

We will return to these and other cell-to-cell interactions in later chapters. For now, the point to remember is this: *In multicelled species, coordinated cell activities depend on specialized forms of linkages and communication between cells.*

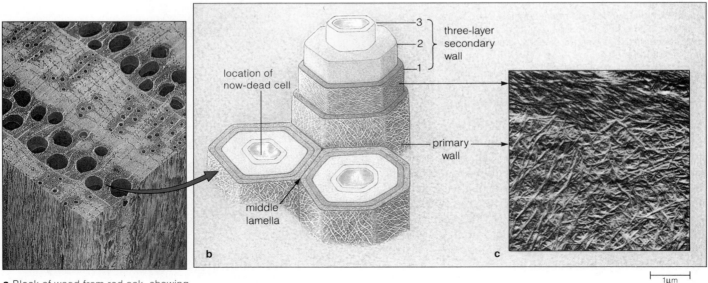

a Block of wood from red oak, showing water-conducting tubes

Figure 4.22 Primary and secondary walls of plant cells. Many cell types have only a primary wall. Other types form a secondary wall after they have stopped growing and their primary wall is completed. The cells in the sketch died; only their walls remain. Walls of many such cells, arranged end to end and side by side, conduct water and strengthen plant parts.

Table 4.3 Summary of Typical Components of Prokaryotic and Eukaryotic Cells

Cell Component	Function	Prokaryotic	Eukaryotic			
		Moneran	Protistan	Fungus	Plant	Animal
Cell wall	Protection, structural support	✓*	✓*	✓	✓	none
Plasma membrane	Control of substances moving into and out of cell	✓	✓	✓	✓	✓
Nucleus	Physical separation and organization of DNA	none	✓	✓	✓	✓
DNA	Encoding of hereditary information	✓	✓	✓	✓	✓
RNA	Transcription, translation of DNA messages into specific proteins	✓	✓	✓	✓	✓
Nucleolus	Assembly of ribosomal subunits	none	✓	✓	✓	✓
Ribosome	Protein synthesis	✓	✓	✓	✓	✓
Endoplasmic reticulum (ER)	Initial modification of many newly forming proteins; lipid synthesis	none	✓	✓	✓	✓
Golgi body	Final modification of proteins, lipids; sorting and packaging them for use inside cell or for export	none	✓	✓	✓	✓
Lysosome	Intracellular digestion	none	✓	✓*	✓*	✓
Mitochondrion	ATP formation	**	✓	✓	✓	✓
Photosynthetic pigment	Light–energy conversion	✓*	✓*	none	✓	none
Chloroplast	Photosynthesis, some starch storage	none	✓*	none	✓	none
Central vacuole	Increasing cell surface area, storage	none	none	✓*	✓	none
Cytoskeleton	Cell shape, internal organization, basis of cell motion	none	✓*	✓*	✓*	✓
9+2 flagellum, cilium	Movement	none	✓*	✓*	✓*	✓

*Known to occur in at least some groups.
**Aerobic reactions do occur in many groups, but mitochondria are not involved.

SUMMARY

1. The cell theory has three key points: All living things are made of one or more cells. Each cell is a basic living unit (it lives independently or has the inherent capacity to do so). A new cell arises only from cells that already exist.

2. At the minimum, each newly formed cell has a plasma membrane, a region of cytoplasm, and a region of DNA.

3. Cell membranes are composed of lipids and proteins. The lipids become arranged in two layers, and these serve as a barrier to water-soluble substances. The proteins carry out most membrane functions, such as transport of large molecules into and out of the cell.

4. In eukaryotic cells, internal membranes divide the cytoplasm into functional compartments, which are called organelles. The nucleus is a prime example. Prokaryotic cells (bacteria) do not have comparable organelles.

5. Organelle membranes separate different chemical reactions in the space of the cytoplasm and so allow them to proceed in orderly fashion.

6. Table 4.3 summarizes the organelles and other structures of both prokaryotic and eukaryotic cells.

1. Label the organelles indicated in this diagram of a typical plant cell. *58*

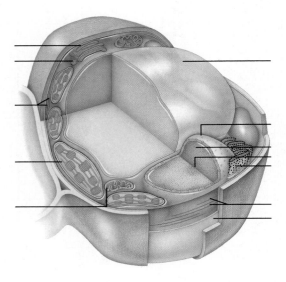

2. Label the organelles indicated in this diagram of a typical animal cell. *60*

3. State the three key points of the cell theory. *51*

4. Why is it likely that you will never encounter a predatory two-ton living cell on the sidewalk? *53*

5. Suppose you want to observe the three-dimensional surface of an insect's eye. Would you benefit most from a compound light microscope, transmission electron microscope, or scanning electron microscope? *54–55*

6. Describe the three features that all cells have in common. Then, after reviewing pp. 56–57 and Table 4.1, write a paragraph describing the key differences between prokaryotic and eukaryotic cells. *52, 56–58, 60*

7. Which organelles are part of the cytomembrane system? Sketch their arrangement in an animal cell, from the nuclear envelope to the plasma membrane. *64*

8. Is the following statement true or false? Plant cells contain chloroplasts, but not mitochondria. *67–68*

9. What are the functions of the central vacuole in mature, living plant cells? *68*

10. What is a cytoskeleton? How does it aid in cell functioning? *69*

11. Are all components of the cytoskeleton permanent? *69*

12. What gives rise to the 9 + 2 microtubular array of cilia and flagella? *71*

13. Cell walls are features of which organisms: bacteria, protistans, plants, fungi, or animals? Are cell walls solid or porous? *72–73*

14. In plants, is a secondary wall deposited inside or outside the surface of the primary cell wall? Do all plant cells have secondary walls? *72*

15. What are some functions of the intercellular matrix in animal tissues? *72*

16. In multicelled organisms, coordinated interactions depend on linkages and communications between adjacent cells. What types of junctions occur between adjacent animal cells? Plant cells? *72*

1. The plasma membrane _____ .
   a. surrounds cytoplasm
   b. separates nucleus from cytoplasm
   c. acts as a nucleus in prokaryotic cells
   d. both a and b

2. Unlike eukaryotic cells, prokaryotic cells _____ .
   a. do not have a plasma membrane
   b. have RNA, not DNA
   c. do not have a nucleus
   d. all of the above

3. The _____ is responsible for cell shape, internal structural organization, and cell movement.
   a. flagellum   c. cytoskeleton
   b. cilium   d. both a and b

4. Cell membranes consist mainly of a _____ .
   a. carbohydrate bilayer and proteins
   b. protein bilayer and phospholipids
   c. phosopholipid bilayer and proteins
   d. none of the above

5. Organelles _____ .
   a. are membrane-bound compartments
   b. are typical of eukaryotic cells, not prokaryotic cells
   c. separate chemical reactions in time and space
   d. all of the above are functions of organelles

6. Plant cells but not animal cells have _____ .
   a. mitochondria   c. ribosomes
   b. a plasma membrane   d. a cell wall

7. Eukaryotic DNA is contained within the _____ .
   a. central vacuole   d. Golgi body
   b. nucleus   e. b and d are correct
   c. lysosome

8. The cytomembrane system does *not* include:
   a. ER   d. plastids
   b. transport vesicles e. all of the above are
   c. Golgi bodies   parts of the system

9. Match each cell component with its function.
   _____ mitochondrion   a. synthesis of polypeptide chains
   _____ chloroplast   b. movement
   _____ ribosome   c. digestion in cell
   _____ flagellum   d. initial modification of new
   _____ rough ER   polypeptide chains
   _____ Golgi body   e. modify, sort, ship proteins and
   _____ nucleolus   lipids
   _____ cytoskeleton   f. cell shape, organization,
   _____ lysosome   movement
                           g. photosynthesis
                           h. ATP formation
                           i. ribosome subunit assembly

Selected Key Terms

amoeboid motion *70*
amyloplast *68*
bacterial flagella *56*
cell *51*
cell theory *51*
cell wall *56*
central vacuole *68*
centriole *71*
chloroplast *68*
chromatin *63*
chromoplast *68*
chromosome *63*

cilium *70*
contraction *70*
cytomembrane system *64*
cytoplasm *52*
cytoplasmic streaming *70*
cytoskeleton *69*
ER (endoplasmic reticulum) *65*
eukaryotic cell *52*
flagellum *70*
Golgi body *66*
intermediate filament *69*
lysosome *66*
microfilament *69*
micrograph *54*
microtubule *69*
mitochondrion (mitochondria)
MTOC (microtubule organizing center) *71*
nuclear envelope *63*
nucleolus *62*
nucleus *52*
organelle *58*
peroxisome *65*
plasma membrane *52*
prokaryotic cell *52*
pseudopod *70*
ribosome *57*
surface-to-volume ratio *53*
wavelength *54*

Readings

Bloom, W., and D. Fawcett. 1986. *A Textbook of Histology*. Eleventh edition. Philadelphia: Saunders.

Burgess, J., M. Marten, and R. Taylor. 1987. *Under the Microscope—A Hidden World Revealed*. Cambridge: Cambridge University Press. Paperback.

deDuve, C. 1985. *A Guided Tour of the Living Cell*. New York: Freeman. Beautifully illustrated introduction to the cell; two short volumes.

Weibe, H. 1978. "The Significance of Plant Vacuoles." *Bioscience* 28: 327–331.

Wolfe, S. 1993. *Molecular and Cellular Biology*. Belmont, California: Wadsworth. Outstanding reference text.

5 A CLOSER LOOK AT CELL MEMBRANES

It Isn't Easy Being Single

As small as it may be, a cell is a *living* thing engaged in the risky business of survival. Consider how something as ordinary as water can challenge its very existence. Water bathes cells inside and out, donates its molecules to metabolic reactions, and dissolves vital ions. If all goes well, each cell holds on to enough water and dissolved ions—not too little, not too much—to survive. But who is to say that life consistently goes well?

Think of a goose barnacle attached to the submerged side of a log that is drifting offshore, at the mercy of ocean currents. At feeding time, the barnacle opens its hinged shell and extends featherlike appendages, which trap bacteria and other bits of food suspended in the water (Figure 5.1*a*). The fluid bathing each living cell in the barnacle's body is salty, rather like the salt composition of seawater. And seawater usually is in balance with the salty fluid inside barnacle cells.

But suppose the log drifts into dilute waters from a melting glacier and stays there (Figure 5.1*b*). For reasons explored in this chapter, salts inevitably move out of the barnacle's body—and out of its cells. The salt-water balance is gradually destroyed, and the cells die. And so, in time, does the barnacle.

The same thing can happen to burrowing worms and other soft-bodied organisms that live between the high and low tide marks along a rocky shore. During an unpredictably heavy storm, when the seawater becomes dilute with runoff from the land, the balance between the surrounding water and the body fluids that bathe the organisms' cells is similarly upset. At such times, the resulting death toll in the intertidal zone can be catastrophic.

With these examples we begin to see the cell for what it is: a tiny, organized bit of life in a world that is, by

a

Figure 5.1 (**a**) Goose barnacles, which live attached to logs and other floating objects in the seas. As is true of every organism, drastic changes in salt concentration can threaten the cells of these marine animals. Such changes result when the barnacles accidentally end up in glacial meltwaters. (**b**) The dark-blue seawater in this photograph is quite salty. The lighter blue water flowing out from the glacier in the background is much, much lower in salts.

comparison, unorganized and sometimes harsh. How exquisitely adapted a living cell must be to its environment! *The cell must be built in such a way that it can bring in certain substances, release or keep out other substances, and conduct its internal activities with great precision.*

For this bit of life, precision begins at the plasma membrane—a flimsy bilayer of lipids, dotted with diverse proteins, that surrounds the cytoplasm. Across this membrane, every cell exchanges substances with its surroundings in highly selective ways. For eukaryotic cells, precision continues at the internal membranes of those specialized compartments called organelles. Photosynthesis, respiration, and many other metabolic reactions depend absolutely on the selective movement of substances across organelle membranes. Understand the structure and function of a cell's membranes, and you will gain insight into survival at life's most fundamental level.

b

KEY CONCEPTS

1. Cell membranes are mostly phospholipids and proteins. The phospholipid molecules are organized as a double layer. This "bilayer" gives the membrane its basic structure and serves as a barrier to water-soluble substances. Proteins in the bilayer or at one of its surfaces carry out most membrane functions.

2. Transport proteins span the bilayer of all cell membranes. Different kinds serve as open channels, gated channels, carriers, or pumps for specific water-soluble substances.

3. Receptor proteins at the outer surface of all plasma membranes receive chemical signals that can trigger alterations in cell activities. Recognition proteins at the outer surface are like molecular fingerprints that identify a cell as being of a certain type.

4. Molecules or ions of a substance tend to move from one region to an adjacent region where they are less concentrated. This behavior is called diffusion.

5. Osmosis is the movement of water across a membrane in response to concentration gradients, a pressure gradient, or both.

6. With passive transport, a solute crosses a membrane by diffusing through the interior of a channel protein or carrier protein that spans the bilayer.

7. With active transport, an energized carrier protein actively pumps a solute across the membrane, against the concentration gradient.

5.1 MEMBRANE STRUCTURE AND FUNCTION

The Lipid Bilayer

A cell's organization as well as its activities depends on membranes. Fluid bathes both surfaces of each cell membrane. How, then, does the membrane remain structurally intact? Is it like a solid wall separating two fluid regions? Not at all. Push a fine needle into and out of a cell, and cytoplasm will not ooze out. Instead, the cell surface will appear to flow over the puncture and seal it. This observation suggests that the plasma membrane shows fluid behavior!

How does a "fluid" cell membrane remain distinct from fluid surroundings? For the answer, start with phospholipids—the most abundant component of cell membranes. A **phospholipid**, recall, consists of a hydrophilic (water-loving) head and two fatty acid tails that are mainly hydrophobic (water-dreading):

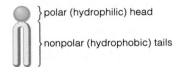

} polar (hydrophilic) head

} nonpolar (hydrophobic) tails

Like other lipids, phospholipid molecules cluster together spontaneously when immersed in water. Hydrophobic interactions may even force them into two layers, with all the fatty acid tails sandwiched between the hydrophilic heads. This arrangement, a **lipid bilayer**, is the structural basis of cell membranes:

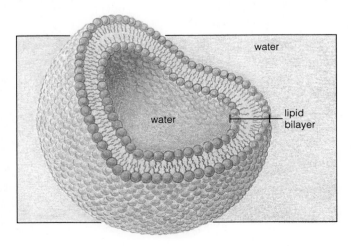

water

water

lipid bilayer

Because lipid bilayers minimize the number of hydrophobic groups exposed to water, the fatty acid tails do not have to spend a lot of energy fighting the water molecules, so to speak. In fact, that plasma membrane you "punctured" earlier exhibited sealing behavior precisely because a puncture is energetically

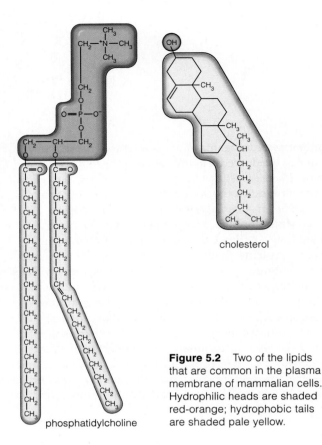

cholesterol

phosphatidylcholine

Figure 5.2 Two of the lipids that are common in the plasma membrane of mammalian cells. Hydrophilic heads are shaded red-orange; hydrophobic tails are shaded pale yellow.

unfavorable—it leaves too many hydrophobic groups exposed to the surrounding water.

Ordinarily, few cells are ever punctured with fine needles. But the self-sealing behavior of membrane lipids is useful for more than damage control. It has roles in the formation of those vesicles described in the preceding chapter. For example, as vesicles bud off ER or Golgi membranes, hydrophobic interactions with cytoplasmic water push lipid molecules together and close off the ruptured site. The same thing happens during exocytosis and endocytosis, as you will see later in the chapter.

Lipid bilayers of cell membranes incorporate a variety of phospholipids, glycolipids, and sterols. The phospholipids differ in their hydrophilic heads and in the length and saturation of their fatty acid tails. (Unsaturated fatty acids have one or more double bonds in the carbon backbone; fully saturated ones have none.) Figure 5.2 shows the structure of phosphatidylcholine, one of the most abundant phospholipids. Glycolipids of the bilayer are structurally similar to this, and their head region also incorporates one or more sugar monomers. Cholesterol is an abundant sterol of animal cell membranes; phytosterols are common to plant cell membranes.

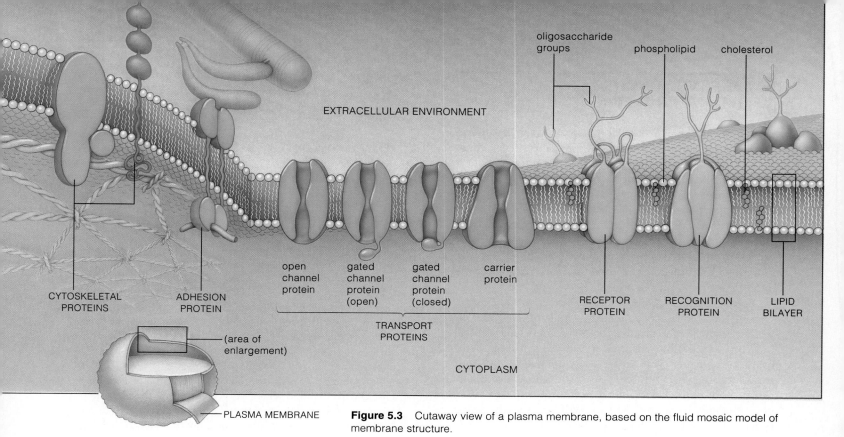

EXTRACELLULAR ENVIRONMENT

oligosaccharide groups   phospholipid   cholesterol

CYTOSKELETAL PROTEINS   ADHESION PROTEIN   open channel protein   gated channel protein (open)   gated channel protein (closed)   carrier protein   RECEPTOR PROTEIN   RECOGNITION PROTEIN   LIPID BILAYER

TRANSPORT PROTEINS

(area of enlargement)

CYTOPLASM

PLASMA MEMBRANE

Figure 5.3 Cutaway view of a plasma membrane, based on the fluid mosaic model of membrane structure.

The Fluid Mosaic Model of Membrane Structure

Figure 5.3 shows the **fluid mosaic model** of membrane structure. By this model, a membrane bilayer shows "fluid" behavior as a result of the constant motion of lipid molecules and their interactions with one another. Membrane lipids spin about their long axis, they move sideways, and their tails flex back and forth—all of which helps keep neighboring lipids from packing into a solid layer. Lipids with short tails and unsaturated (kinked) tails further disrupt the packing. Being a composite of lipids and proteins, the membrane is said to have a "mosaic" quality. The proteins carry out most membrane functions, such as the transport of substances into and out of the cell.

The mosaic is not symmetrical. The two sides of a given membrane are not the same; they differ in terms of the number, kind, and arrangement of their lipids and proteins. In later chapters, you will see how the asymmetry relates to differences in the tasks carried out on the two sides of the membrane.

Bear in mind, the fluid mosaic model is only a starting point for discussing membranes—which differ in fluidity as well as in composition and molecular arrangements. Why bother to think about this fluidity? Cell survival depends on it, as a simple example will make clear. When outside temperatures fall,

membranes tend to stiffen and disrupt the functions of membrane proteins. Bacterial and yeast cells rapidly synthesize unsaturated fatty acids at such times. Infusion of these kinky lipids into a membrane helps keep it from stiffening up.

The first fluid mosaic model of membrane structure was put together in 1972 by S. J. Singer and G. Nicolson. Evidence favoring this model comes from many sources, including freeze-fracture microscopy, as described in the *Focus* essay on the next page. The features of this model can be summarized this way:

---

1. Cell membranes are composed mostly of lipids (especially phospholipids) and proteins.

2. The lipid molecules have their hydrophilic heads at the two outer faces of a bilayer and their fatty acid tails sandwiched in between.

3. A lipid bilayer imparts structure to cell membranes and serves as a hydrophobic barrier between two solutions. (A plasma membrane separates the fluids inside and outside the cell. Internal cell membranes separate different fluids in the cytoplasm.)

4. Proteins embedded in the bilayer or positioned at its surfaces carry out most membrane functions.

5.2 FUNCTIONS OF MEMBRANE PROTEINS

Spanning the bilayer of all plasma membranes and organelle membranes are proteins with transport functions. Plasma membranes also incorporate proteins that function in signal reception, adhesion, and cell-to-cell recognition.

Transport proteins allow water-soluble substances to move through their interior, which opens on both sides of the bilayer. They are either *channel* proteins or *carrier* proteins. As you will see, most types weakly bind ions or molecules on one side of the membrane and release them on the other.

Some channel proteins remain open at all times. Others have molecular gates that open and close in controlled ways. (For example, gated channels help control the directional flow of ions across the plasma membrane of neurons. This is the basis of "messages" that travel through nervous systems.) The interior of a carrier protein remains closed off to solutes until it undergoes changes in shape. As you will see shortly, some types of carrier proteins passively transport their weakly bound cargo. Other types require an energy input. Once energized, they actively pump their cargo across the membrane.

Receptor proteins have binding sites for hormones and other extracellular substances that can trigger alterations in cell activities. For example, enzymes that crank up machinery for cell growth and division are switched on when the hormone somatotropin binds to cell receptors. Different cell types have different combinations of receptors. Among vertebrates, some kinds of receptors are present at the surface of most cells. Others are restricted to only a few cell types.

Recognition proteins are like molecular fingerprints at the cell surface; they identify a cell as being of a specific type. For example, your body cells bear "self" recognition proteins. Infection-fighting white blood cells recognize these and leave the cells alone. They also recognize certain proteins on bacterial cells and other invaders as being "nonself"—and destroy them.

Adhesion proteins help cells stay connected to one another in a given tissue. They are glycoproteins (with oligosaccharides attached). While the tissues are forming, glycoproteins connect neighboring cells. Some of these attachment sites become the cell-to-cell junctions mentioned earlier, on page 72.

Different classes of membrane proteins function in the transport of substances across the lipid bilayer and in signal reception, cell-to-cell recognition, and cell-to-cell adhesion.

Focus on Science

Discovering Details About Membrane Structure

Suppose you'd like to explore for yourself some of the methods that researchers used to deduce the structure of the plasma membrane. You could start by attempting to identify its molecular components. Your first challenge is to secure a large sample of plasma membrane that isn't contaminated with organelle membranes. Using red blood cells will simplify the task. These are abundant, easy to collect—and structurally simple. While red blood cells are maturing, they discard their nucleus. They have only a few components, such as hemoglobin and ribosomes, but these are enough to keep them functioning for their four-month life span.

By placing a sample of red blood cells in a test tube filled with distilled water, you can separate the plasma membranes from the rest of the cell components. Red blood cells have more solutes and fewer water molecules than distilled water. The difference in solute concentrations between the two regions causes water to move across a plasma membrane, into the cells. These particular cells have no mechanisms for actively expelling the excess water, so they swell up. In time they burst and their contents spill out.

Now you have a mixture of membranes, hemoglobin, and other cell parts in the water. How do you separate the bits of membrane? You can place a tube containing a special solution of cell parts in a **centrifuge**, a device that spins test tubes at high speed. Each component of a cell has its own molecular composition that gives it a characteristic density. As the centrifuge spins at the appropriate speed, components of the greatest density move toward the bottom of the tube. Other components take up layered positions above them, according to their relative densities.

With proper centrifugation, one layer in the solution will be shreds of cell membrane only. You can carefully draw off this layer and examine it with a microscope to verify that your membrane sample is not contaminated. Standard chemical analysis tells you that the plasma membranes consist of lipids and proteins.

In the past, there were two competing models of membrane structure. According to the "protein coat" model, the membrane was composed of a bilayer of lipids, coated on both surfaces with a layer of proteins. According to the "fluid mosaic" model, proteins were largely embedded within the bilayer.

Suppose you decide to test the protein coat model. First you calculate how much protein actually would be required to coat the inner and outer surface of a known number of red blood cells. Then you separate and measure a membrane sample into its lipid and protein fractions. In

1. With freeze-fracture techniques, specimens being prepared for electron microscopy are rapidly frozen, then fractured by a sharp blow from the edge of a fine blade.

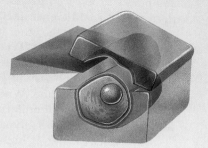

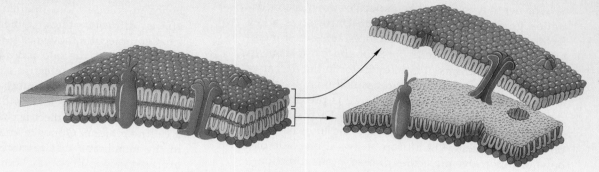

outer membrane layer exposed by etching

fracture edge

2. Fractured membranes commonly split down the middle of the lipid bilayer. Typically, one inner surface is studded with particles and depressions, and the other is a complementary pattern of depressions and particles. The particles are membrane proteins.

this way, you can compare the *observed* ratio of proteins to lipids against the ratio *predicted* on the basis of the model. Such tests were done with membrane samples. They reveal that there is far too little protein to cover both surfaces of a lipid bilayer. They provide evidence against the protein coat model, at least for red blood cells.

Suppose you now want to do a direct test of the fluid mosaic model. You could employ the *freeze-fracturing* and *freeze-etching* methods of preparing cells for electron microscopy. You first immerse a sample of cells in liquid nitrogen, an extremely cold fluid. The cells freeze instantly. You can now strike them with a very small chisel. As shown in Figure *a*, a properly directed blow will fracture the cell in such a way that one layer of the lipid bilayer separates from the other. These preparations can then be inspected under extremely high magnifications.

If the *protein coat* model were correct, what do you suppose examination of one such layer would reveal?

What you actually see is not a perfectly smooth, pure lipid layer. Freeze-fractured cell membranes reveal many bumps and other irregularities in lipid layers. The bumps are proteins, incorporated directly in the bilayer—just as the fluid mosaic model suggests.

exposed by etching

3. Specimens are often freeze-etched. More ice is evaporated from the fracture face to expose the outer membrane surface.

(deposition of carbon and metal in thin layer on specimen surface)

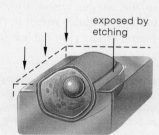

4. With metal shadowing techniques, the fractured surface is coated with a layer of carbon and heavy metal, such as platinum. The coating is thin enough to replicate details of the exposed specimen surface. The metal replica, not the specimen itself, is used for micrographs.

a Freeze-fracturing and freeze-etching. The micrograph shows part of a replica of a red blood cell that was prepared by the techniques described here.

5.3 DIFFUSION

Concentration Gradients and Diffusion

Cellular life depends on the energy inherent in molecules (or ions), which keeps them in constant motion. It depends also on **concentration gradients**. "Concentration" refers to the number of molecules of a substance in a specified volume of fluid. Add the word "gradient," and this means one region of the fluid contains more molecules than a neighboring region.

In the absence of other forces, molecules move down their concentration gradient. They do so because they constantly collide with one another, millions of times a second. Random collisions send the molecules back and forth, but the *net* movement is away from the place of greater concentration.

The net movement of like molecules down a concentration gradient is called **diffusion**. Diffusion is a key factor in the movement of substances across cell membranes and through fluid parts of the cytoplasm. In multicelled organisms, it is a key factor in the exchange of substances between cells and tissue fluids.

Suppose more than one substance is present in the same fluid. This makes no difference. Each substance still diffuses in some direction according to its *own* concentration gradient. Put a drop of dye at one side of a bowl of water. Dye molecules diffuse in one direction—to the region where they are less concentrated. And water molecules move in the opposite direction, to the region where *they* are less concentrated (Figure 5.4).

Diffusion is the net movement of like molecules or ions down their concentration gradient. They show a net outward movement from a region where they are most concentrated to a neighboring region where they are less concentrated.

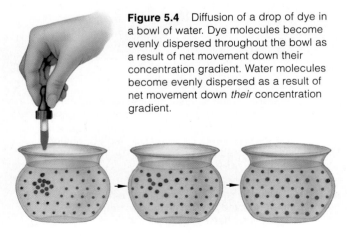

Figure 5.4 Diffusion of a drop of dye in a bowl of water. Dye molecules become evenly dispersed throughout the bowl as a result of net movement down their concentration gradient. Water molecules become evenly dispersed as a result of net movement down *their* concentration gradient.

Factors Influencing the Rate and Direction of Diffusion

The rate at which ions or molecules move down their concentration gradient depends on several factors. When the concentration gradient is steep, diffusion is faster. Then, far more molecules are moving outward, compared to the number moving in. As the gradient decreases, the difference in the number of molecules moving one way or the other becomes less pronounced, and diffusion is slower. When the gradient is gone, individual molecules are still in motion. But the total number moving one way or the other is just about the same at any given time. When the net distribution of molecules is just about uniform through the two adjoining regions, we call this "dynamic equilibrium."

Diffusion is faster at higher temperatures, because heat energy causes molecules to move more rapidly—hence to collide more frequently. And molecular size affects diffusion rates. Other factors being equal, smaller molecules move faster than large ones do.

Besides this, the rate and direction of diffusion may be modified by an **electric gradient**—a difference in charge between two adjoining regions. For example, a variety of dissolved ions and charged particles contribute to the overall electric charge near each surface of a cell membrane. Opposite charges attract. So the side of the membrane that is more *negatively* charged will exert the most pull on, say, sodium ions—which are *positively* charged. The pull can enhance the flow of sodium across the membrane. Many processes, such as the flow of information through your nervous system, depend on the combined force of electric and concentration gradients to attract specific substances across cell membranes.

As you will see shortly, a **pressure gradient**—a difference in pressure between two adjoining regions—also can influence the rate and direction of diffusion.

5.4 OSMOSIS

Some small molecules readily diffuse through the lipid bilayer of a cell membrane or through protein channels across it. Glucose and other molecules cannot do this; membrane proteins must pump them across. Because the membrane shows this selective permeability, concentrations of dissolved substances (that is, solutes) can increase on one side of the membrane and not the other. The resulting solute concentration gradients affect the movement of water into and out of cells.

Water by itself cannot get more or less concentrated. (Hydrogen bonds keep water molecules from crowding closer together or drifting apart.) Water becomes "less concentrated" only when solutes are dissolved in it.

Said another way, *a water concentration gradient is influenced by the number of molecules of all the solutes that are present on both sides of the membrane.*

Thus, the direction in which water moves across a membrane is influenced by **tonicity**—that is, the relative concentrations of solutes in two fluids. As Figures 5.5 and 5.6 show, water tends to move where solute concentrations are greater. When a membrane separates two *iso*tonic fluids, solute concentrations are equal—so there is no net movement of water in either direction. When solute concentrations are not equal, water tends to move from the *hypo*tonic solution (less solutes) to the *hyper*tonic solution (more solutes).

Unless cells have built-in mechanisms for adjusting to differences in tonicity between cytoplasmic fluid and the surroundings, they can shrivel or burst. As the *Focus* essay on the next page makes clear, some cells are better than others at doing this.

Bear in mind, other factors can modify the direction of water movement. For example, water also tends to move from regions of higher to lower pressure. Think of the fluid pressure generated by your beating heart. In your kidneys, the pressure forces some water and small solutes to leave blood capillaries and enter the surrounding tissue fluid. From there, water and solutes cross the membranes of kidney cells. Normally, tissue fluid is about three times more dilute than plasma (the fluid portion of blood), yet the pressure forces water to move against the gradient. The movement of molecules of water and solutes in the same direction in response to a pressure gradient is called **bulk flow**.

There is a special name for the movement of water across membranes in response to solute concentration gradients, fluid pressure, or both. It is called **osmosis**.

Osmosis is the movement of water across a selectively permeable membrane in response to concentration gradients, fluid pressure, or both.

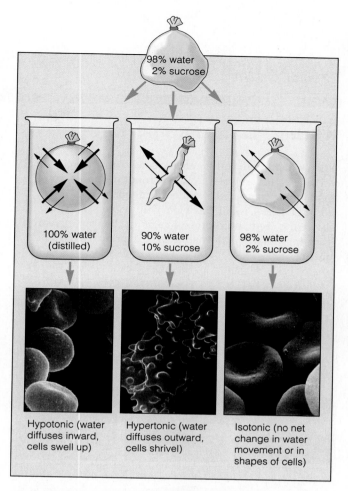

Figure 5.5 Tonicity and the diffusion of water. "Tonicity" refers to the relative concentrations of solutes in two fluids. In the sketches, membrane-like bags through which water but not sucrose can move are placed in three different solutions having different solute levels. Arrow widths indicate the relative amounts of water movement in each container.

The micrographs correspond to the sketches. They show the kinds of shapes that might be seen in human red blood cells placed in solutions that are hypotonic (water will move into cells), hypertonic (water will move out), or isotonic (no net movement). Solute levels of the fluids inside and outside red blood cells normally are in balance. These cells cannot make large adjustments in the concentration of cytoplasmic water. Drastic decreases in solutes in their fluid surroundings would make them burst; increases would make them shrivel.

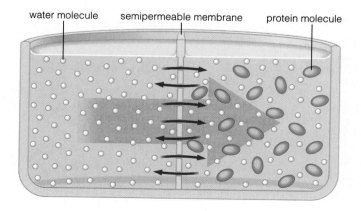

Figure 5.6 Osmosis. In this diagram, a membrane-like barrier divides a container into two compartments. Pure water was poured into the left compartment. A protein-rich solution of equal volume was poured into the one on the right. Water molecules can move across this membrane. The larger protein molecules cannot, so they occupy some of the available space in their compartment on an ongoing basis. Thus, the right compartment has fewer water molecules than the left. Water molecules follow this water concentration gradient. There is a net osmotic movement from left to right (*large blue arrow*).

Wilting Plants and Squirting Cells

When a plant with soft green leaves is growing well, you can safely bet that water in the surrounding soil is dilute (hypotonic), compared to water inside the plant's living cells. Plant cells, recall, have fairly rigid walls. As water moves into them, fluid pressure against the wall increases. "Turgor pressure" is the name for this internal fluid pressure on a cell wall.

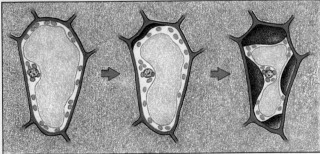

a

Fluid pressure builds up against any walled cell when water moves inward by osmosis. But water will also be squeezed back out when the pressure is great enough to counter the attractive force of cytoplasmic fluid (which has more solutes than water from the soil). Both forces have the potential to cause the directional movement of water. The sum of these two opposing forces is called the "water potential."

When the soft parts of a plant remain erect, the movements of water into and out of its cells are equal. The constant turgor pressure keeps the cell walls plumped. But suppose the soil dries out. Suppose the inward movement of water dwindles or stops. Then, water will move out of cells and the soft parts of the plant body wilt.

Plants also wilt when the soil has too many solutes. A simple experiment will allow you to observe the wilting effect for yourself. Put 10 grams of table salt (NaCl) into 60 milliliters of water, then pour this salty solution into the soil around a potted tomato plant. As indicated by the clocks in the Figure *a* photographs, the plant will start to collapse after about five minutes. In less than thirty minutes, wilting will be severe. The sketches corresponding to the photographs show progressive plasmolysis—a shrinking of cytoplasm away from the cell wall.

Paramecium is not as vulnerable. This single-celled organism lives in freshwater. Because its cytoplasm is hypertonic relative to the surroundings, water tends to move into the cell by osmosis. If unchecked, the influx would bloat the cell and rupture its plasma membrane. *Paramecium* expels the excess through an energy-requiring mechanism involving contractile vacuoles (Figure *b*). Tube-like extensions of this type of organelle extend through the cytoplasm. Water enters the extensions and collects in a central vacuolar space. When filled, the vacuole contracts—squirting the excess water out through a small pore that opens to the environment.

contractile vacuoles

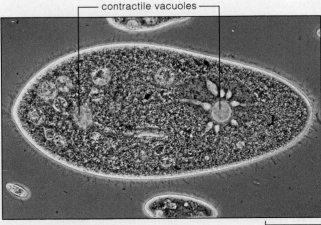

b   10 μm

5.5 ROUTES ACROSS CELL MEMBRANES

We turn now to the mechanisms available to move solutes across cell membranes. As Figure 5.7 shows, the mechanisms range from unassisted diffusion, facilitated diffusion, and active transport to endocytosis and exocytosis. Through a combination of these mechanisms, cells or organelles are supplied with numerous raw materials and they are rid of wastes, at controlled rates. These mechanisms help maintain pH and volume inside the cell or organelle within functional ranges.

Oxygen, carbon dioxide, and other small molecules with no net charge readily diffuse across the membrane's lipid bilayer:

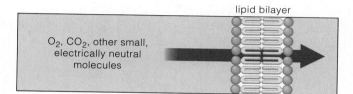

Glucose and other large, water-soluble molecules with no net charge almost never diffuse freely across the bilayer. Neither do positive or negative ions:

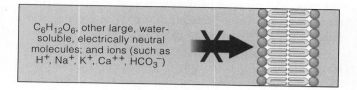

These substances must cross the membrane through the interior of transport proteins that span the bilayer.

With **passive transport**, water-soluble substances diffuse through the interior of channel proteins and many types of carrier proteins. With **active transport**, the membrane crossing occurs only when some type of carrier protein receives an energy boost, usually from an ATP molecule. Let's now take a closer look at how these mechanisms operate.

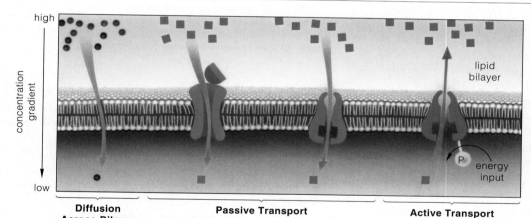

Diffusion Across Bilayer

Lipid-soluble substances diffuse across lipid bilayer.

Passive Transport

Water-soluble substances diffuse through interior of channel or carrier protein; no energy input required. Also called facilitated diffusion.

Active Transport

Solute pumped through interior of carrier protein; requires energy input.

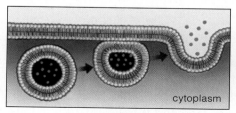

Exocytosis

Vesicle moves to plasma membrane and fuses with it; contents released outside.

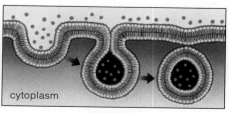

Endocytosis

Vesicle forms at surface of plasma membrane, then moves into cytoplasm.

Figure 5.7 Overview of mechanisms by which solutes move across cell membranes. Exocytosis and endocytosis proceed at the plasma membrane only.

5.6 PROTEIN-MEDIATED TRANSPORT

Characteristics of Transport Proteins

The type and number of transport proteins that span a membrane influence whether a given substance will move into or out of a given cell type, and how fast.

The transport proteins that serve as channels differ from one kind of organism to the next. For example, channel proteins in the plasma membrane of bacteria are like large pores; they are rather nonselective about which solutes travel through them. By contrast, channel proteins in the plasma membranes of animal and plant cells are chemically picky; they permit some solutes but not others to cross.

Cells also differ in their types and numbers of carrier proteins. Each type of carrier protein is highly selective, with a binding site that attracts a particular substance or a number of related substances. The carrier that transports amino acids, for instance, will not transport glucose.

To understand how the selective transport proteins work, you have to know that they are not rigid blobs of atoms. When they interact with molecules or ions, the proteins change from one shape into another shape, then back again.

Before a transport protein will allow passage of a solute, the solute must become weakly bound to it at a specific site somewhere on the protein surface. Binding leads to changes in the protein shape. Part of the protein closes in behind the bound solute—and part opens up to the opposite side of the membrane. There, the solute dissociates from the site. Think of the solute as hopping onto the protein on one side of the membrane, then hopping off on the other side.

Passive Transport

All channel proteins and many types of carrier proteins permit the passive diffusion of solutes across the membrane, down their concentration gradient. That is why the mechanism of passive transport is sometimes called "facilitated diffusion."

Consider the manner in which a carrier protein functions. It can move molecules (or ions) of a solute *both ways* across the membrane, depending on which way the carrier's binding site faces (Figure 5.8). At any given time, the *net* direction of movement depends on how many solute molecules are actually making random contact with vacant binding sites. Where the solute is more concentrated, binding and transport proceed more often. Where solute molecules are less concentrated, they still bind to the sites—but they do so at a lower rate. (With fewer molecules around, the random encounters with binding sites are not as frequent.)

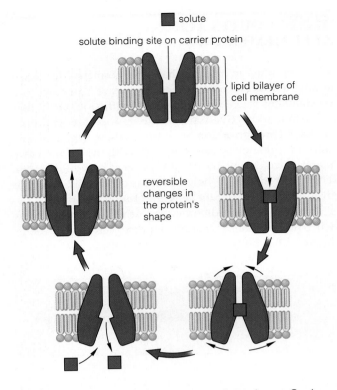

Figure 5.8 Passive transport across a cell membrane. Carrier proteins can move the ions or molecules of a solute in both directions. The *net* movement will be down the gradient (from higher to lower solute concentration) until concentrations become the same on both sides of the membrane.

By itself, the passive two-way transport would continue until solute concentrations became equal on both sides of the membrane. In most cases, however, other processes influence the outcome.

For example, the bloodstream delivers glucose to tissues throughout the body. Nearly all cells use glucose for energy and as building blocks. When the concentration of glucose in blood (and tissues) is high, the cells take up glucose. However, as fast as glucose is passively transported into the cells, glucose usually is being withdrawn from the cytoplasm for metabolic reactions. Thus, glucose metabolism helps maintain the concentration gradient that favors the movement of glucose into cells.

---

1. In passive transport, a transport protein moves molecules (or ions) of a specific solute across a cell membrane by undergoing a reversible change in shape.

2. Passive transport is based on diffusion. Barring other influences, it will continue until solute concentrations are the same on both sides of the membrane.

Active Transport

Carrier proteins that function in active transport pump a solute across the cell membrane, *against* its concentration gradient. Unlike passive transport (which will continue until concentrations are the same on both sides of the membrane), active transport continues until the solute becomes *more* concentrated on the side of the membrane where it is being pumped.

Active transport does not proceed spontaneously; it is an uphill process, so to speak. As Figure 5.9 shows, the carrier requires an input of energy, most often from ATP. When ATP donates energy to the carrier protein, the solute binding site becomes altered. Molecules (or ions) of the solute bind to the site more easily when it is in its altered configuration. After a molecule has been moved to the opposite side of the membrane, it dissociates from the binding site—which thereupon reverts to its less attractive configuration. At any given time, fewer molecules can make the return trip. So the net movement is to the side of the membrane where the solute is more concentrated.

By analogy, imagine skiers rapidly hopping on a ski lift that moves them up a mountain. At the top, the lift chairs tilt in a way that is not very inviting for skiers who want to hop on for a downhill ride. More skiers will be transported uphill than downhill.

One active transport system, the **calcium pump**, helps keep the calcium concentration in cells at least a thousand times lower than the concentration outside.

The **sodium-potassium pump** is one of the major cotransport systems. In this particular case, the binding of sodium ions to an energized carrier protein alters a binding site so that it becomes more easily occupied by potassium—the solute. Operation of many sodium-potassium pumps sets up concentration and electric gradients across a plasma membrane. And energy that is inherent in these specific gradients can be used to drive many cell activities.

We will return to active transport mechanisms in chapters that describe muscle contraction, information flow through nervous systems, and other physiological processes. For now, keep these points in mind:

---

1. In active transport, carrier proteins pump molecules or ions of a specific solute across a cell membrane, against their concentration gradient.

2. Active transport is not spontaneous; it requires an energy input, most often from ATP.

3. The energy input changes the solute binding site in a way that favors the net movement of the solute in the direction where it is most concentrated.

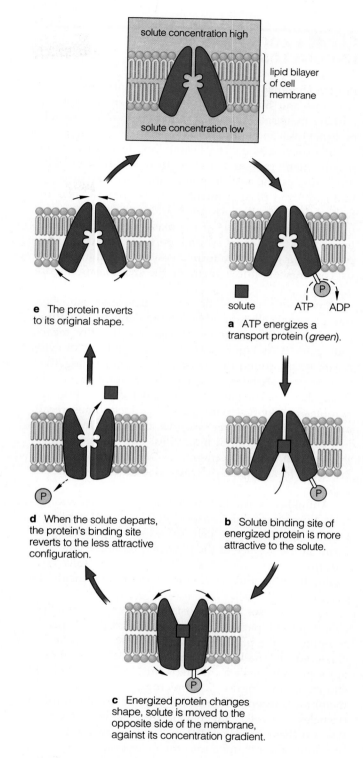

solute concentration high

lipid bilayer of cell membrane

solute concentration low

e The protein reverts to its original shape.

solute   ATP   ADP

a ATP energizes a transport protein (*green*).

d When the solute departs, the protein's binding site reverts to the less attractive configuration.

b Solute binding site of energized protein is more attractive to the solute.

c Energized protein changes shape, solute is moved to the opposite side of the membrane, against its concentration gradient.

Figure 5.9 Active transport across a cell membrane. This carrier protein moves molecules of a specific solute in both directions. However, with an energy input from ATP, the chemical fit between the molecules and their binding site improves. Once a bound molecule is released, the site reverts to its unaltered, less attractive state. Even at low concentrations, molecules occupy the altered sites and move across the membrane faster. That is why there can be a greater *net* movement of molecules against the concentration gradient.

5.7 EXOCYTOSIS AND ENDOCYTOSIS

Protein channels and carriers can deal with one to several ions and molecules at a time. When it comes to moving materials in bulk across membranes, cells rely on exocytosis and endocytosis.

In **exocytosis**, a cytoplasmic vesicle moves to the plasma membrane and fuses with it. As shown in Figure 5.10a, as the vesicle membrane becomes incorporated into the plasma membrane, its contents end up fully exposed to the surroundings.

In **endocytosis**, part of the plasma membrane sinks inward and balloons around particles, fluid, or tiny prey. It seals on itself to form a vesicle, which transports its contents or stores them in the cytoplasm (Figure 5.10b).

Free-living phagocytic cells, including amoebas, trap bacterial cells and other foreign items by extending pseudopods around them. (Phagocyte literally means "cell eater.") As Figure 5.11 shows, these lobes of cytoplasm wrap around the trapped item and seal together, so that part of the plasma membrane forms a vesicle. After these particular vesicles move deeper into the cytoplasm, they fuse with lysosomes. Lysosomes, recall, are filled with digestive enzymes that can break down just about all biological molecules. The enzymes digest the trapped item into fragments and molecules. These are used in a variety of ways, depending on the cell type.

Animal cells also take up liquid droplets by endocytosis. Sometimes this transport process is called pinocytosis (which means "cell drinking"). A depression forms at the surface of the plasma membrane and dimples inward around solute-rich extracellular fluid. Here again, an endocytic vesicle forms and moves inside the cytoplasm.

In many cases, endocytosis is *receptor-mediated*. Specific types of particles or molecules bind to specific receptors at the plasma membrane. These receptors are clustered in shallow depressions in the membrane, appropriately called "coated pits." The coated pit shown in the Figure 5.12 micrographs is lined with membrane receptors that are specific for lipoprotein particles. All cells that metabolize or store cholesterol take up these particles. When lipoproteins bind to the receptors, the pit sinks into the cytoplasm, forming an endocytic vesicle.

Figure 5.13 summarizes the possible destinations of the vesicles formed by exocytosis and endocytosis.

Exocytosis and endocytosis are transport processes that move materials in bulk across the plasma membrane.

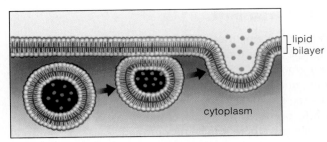

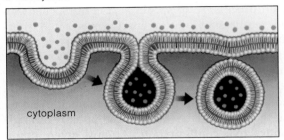

a Exocytosis

b Endocytosis

Figure 5.10 (**a**) Exocytosis. The membrane of a vesicle fuses with the plasma membrane, and its contents are released outside the cell. (**b**) Endocytosis. Part of the plasma membrane balloons inward and seals back on itself, forming a separate vesicle.

Figure 5.11 Phagocytosis, the means by which certain cells, including predatory amoebas and macrophages (infection-fighting white blood cells), engulf their targets. A vesicle forms around the target and moves into the cytoplasm, where it fuses with lysosomes—those sacs of digestive enzymes described on page 66.

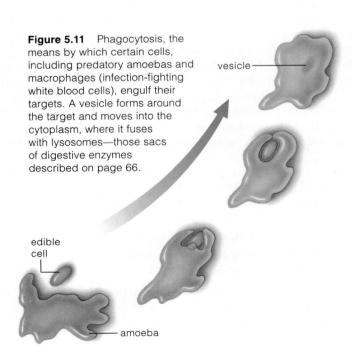

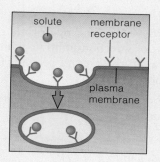

Figure 5.12 Example of receptor-mediated endocytosis, a transport process that begins when specific molecules bind with receptors at the surface of a plasma membrane. The electron micrographs show part of the plasma membrane from an immature chicken egg. Before it was plucked from the chicken, the egg was taking up lipoproteins and other substances necessary for its growth and development.

The shallow indentation in (**a**) is an example of a "coated pit." On the cytoplasmic side of the pit, molecules of clathrin, a type of protein, form a dense lattice. (**b**) Receptors at the membrane's outer surface bind lipoprotein particles. (**c**) The pit deepens and rounds out. (**d**) The fully formed coated vesicle encloses lipoproteins, which will be used or stored by the cell. Receptor-laden membrane that was appropriated by the vesicle will be recycled.

indentation on extracellular
side of plasma membrane

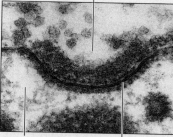 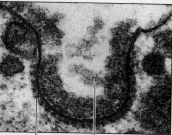 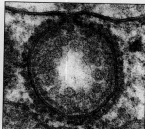 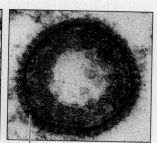

a cytoplasm plasma membrane **b** proteins of cytoskeleton lipoprotein particles bound to membrane receptors **c** 0.1μm **d** vesicle completely formed, moving into cytoplasm

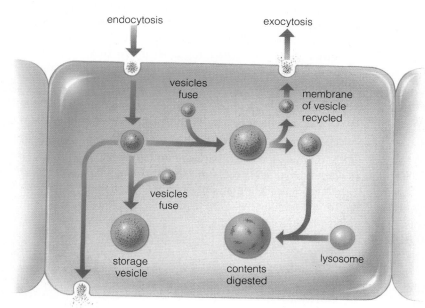

Figure 5.13 Examples of the use and cycling of membrane lipids and proteins. Besides incorporating new membrane from some of the vesicles that budded earlier from ER membranes and Golgi bodies (page 64), the plasma membrane gives up bits of itself to endocytic vesicles and gets membrane bits back from exocytic vesicles.

SUMMARY

Membrane Structure and Function

1. Fluid of one sort or another bathes living cells, the cytoplasm of which is largely fluid, also. The cell's plasma membrane is the structural and functional boundary between the two fluids. In eukaryotic cells, internal cell membranes divide the fluid portion of cytoplasm into functionally diverse organelles.

2. Each cell membrane is two layers thick. Lipid molecules—phospholipids especially—make up this bilayer, which has many proteins embedded in it or positioned at one of its surfaces.

   a. The lipids are arranged with all hydrophobic tails sandwiched in between the hydrophilic heads.

   b. The lipid bilayer gives the membrane its water-impermeable structure. The proteins carry out most membrane functions.

3. These are the main features of the fluid mosaic model of membrane structure:

   a. Cell membranes are fluid structures, mainly because their lipids twist, move laterally, and flex their tails. Also, lipids with short tails, kinked tails, or ring structures disrupt what might otherwise be tight packing among the fatty acid tails.

   b. Cell membranes have a mosaic quality, in that they are a composite of diverse lipids, integrated proteins, and surface-bound proteins.

   c. The "mosaic" is asymmetrical, in that the two sides of a given membrane differ in the number, kind, and arrangement of lipids and proteins.

Functions of Membrane Proteins

1. Cells differ in the number and type of membrane proteins. The differences affect a cell's metabolism, volume, and pH, its responsiveness to extracellular substances, and its interactions with other cells.

2. All cell membranes incorporate transport proteins. Plasma membranes also incorporate receptor proteins, recognition proteins, and adhesion proteins.

3. Transport proteins function as channels and carriers.

   a. The interior of transport proteins opens to both sides of the membrane, and solutes move through them. Some channel proteins have molecular gates that can be closed and opened in controlled ways.

   b. Many carrier proteins function in active transport; they move specific solutes across the membrane only when they receive an energy boost, as from ATP.

   c. Solutes do not diffuse freely through most transport proteins. They bind to an exposed site on the protein, then the protein's shape changes, exposing the bound solute to the opposite side of the membrane.

4. Receptor proteins bind hormones and other extracellular substances that can trigger alterations in cell behavior or metabolism.

5. Recognition proteins are molecular fingerprints at the surface of each cell type. Infection-fighting white blood cells distinguish between recognition proteins of the body's own cells and those of invaders.

6. Adhesion proteins help cells adhere to one another and form cell-to-cell junctions.

Movement of Substances Into and Out of Cells

1. Specific solutes move across cell membranes by the following mechanisms:

   a. Diffusion is the unassisted movement of solutes from a region of higher to lower concentration.

   b. Osmosis is the movement of water across a selectively permeable membrane in response to solute concentration gradients, fluid pressure, or both.

   c. Passive transport is the movement of a solute through the interior of a channel protein or a carrier protein that does not require an energy input to operate.

   d. Active transport is an energy-requiring pumping of a solute through a carrier protein, against its concentration gradient.

2. Oxygen, carbon dioxide, and other small molecules with no net charge diffuse across the lipid bilayer. Ions and large, water-soluble molecules such as glucose are actively or passively transported across.

3. Tonicity refers to the relative concentrations of solutes of two fluids. When conditions are isotonic (equal concentrations), there is no osmotic movement in either direction. Water tends to move from hypotonic fluids (lower concentration) to hypertonic fluids (higher concentration).

4. Through exocytosis, cells eject larger volumes of dissolved substances across the plasma membrane, compared to membrane transport processes.

5. Through mechanisms of endocytosis, cells take in large particles or fluid droplets. Amoebas and other phagocytic cells use endocytosis when they trap their targets.

Review Questions

1. Describe the fluid mosaic model of plasma membranes. What makes the membrane fluid? What parts constitute the mosaic? *78–79*

2. List the structural features that all cell membranes have in common. *78–79*

3. Describe some features of membrane proteins. *80*

4. How does diffusion work? *82*

5. What is osmosis and what causes it? *82–83*

6. Explain the difference between active and passive transport mechanisms. *85*

7. What types of substances can readily diffuse across the lipid bilayer of a cell membrane? What types of substances must be actively or passively transported across? *85–87*

8. Can you explain the difference between exocytosis and endocytosis? What is receptor-mediated endocytosis? *88*

Self-Quiz *(Answers in Appendix IV)*

1. Cell membranes consist mainly of a _____ .
   a. carbohydrate bilayer and proteins
   b. protein bilayer and phospholipids
   c. phospholipid bilayer and proteins
   d. none of the above are correct

2. The most abundant components of cell membranes are _____ .
   a. phospholipids   c. proteins
   b. cholesterols   d. a and c are correct

3. In a lipid bilayer, the _____ of lipid molecules are sandwiched between their _____ .
   a. hydrophilic tails, hydrophobic heads
   b. hydrophilic heads, hydrophilic tails
   c. hydrophobic tails, hydrophilic heads
   d. hydrophobic heads, hydrophilic tails

4. Most membrane functions are carried out by _____ .
   a. proteins   c. nucleic acids
   b. phospholipids   d. hormones

5. Internal cell membranes as well as the plasma membrane incorporate _____ .
   a. transport proteins   d. recognition proteins
   b. receptor proteins   e. all are correct
   c. adhesion proteins

6. When a cell is placed in a hypotonic solution, water tends to _____ .
   a. move into the cell   c. show no net movement
   b. move out of the cell   d. move by endocytosis

7. A _____ is a device that spins test tubes and separates cell components according to their relative densities.

8. The direction in which a solute diffuses from one region to another will *not* be influenced by _____ .
   a. how steep the concentration gradient is
   b. an electric gradient
   c. a pressure gradient
   d. the presence of other solutes
   e. all of these factors influence diffusion

9. _____ can diffuse across a lipid bilayer.
   a. glucose   c. carbon dioxide
   b. oxygen   d. b and c are correct

10. Sodium ions move across a membrane, through the interior of a protein that has received an energy boost. This is an example of _____ .
    a. passive transport   c. facilitated diffusion
    b. active transport   d. a and c are correct

11. Match each event or condition with the appropriate description.
    ____ movement of water across a membrane in response to solute concentration or pressure gradients
    ____ the energy-requiring pumping of a solute against its concentration gradient
    ____ exocytosis
    ____ isotonic, hypotonic, hypertonic
    ____ unassisted movement of solutes down a concentration gradient
    ____ endocytosis

    a. refers to tonicity
    b. diffusion
    c. cells take in large particles or fluid droplets
    d. active transport
    e. vesicle fuses with plasma membrane, contents released to outside
    f. osmosis

Selected Key Terms

active transport *85*   lipid bilayer *78*
adhesion protein *80*   osmosis *83*
bulk flow *83*   passive transport *85*
calcium pump *87*   phospholipid *78*
centrifuge *80*   pressure gradient *82*
concentration gradient *82*   receptor protein *80*
diffusion *82*   recognition protein *80*
electric gradient *82*   sodium-potassium pump *87*
endocytosis *88*   tonicity *83*
exocytosis *88*   transport protein *80*
fluid mosaic model *79*

Readings

Bretscher, M. October 1985. "The Molecules of the Cell Membrane." *Scientific American* 253(4):100–108.

Dautry-Varsat, A., and H. Lodish. May 1984. "How Receptors Bring Proteins and Particles Into Cells." *Scientific American* 250(5):52–58. Describes receptor-mediated endocytosis.

Singer, S., and G. Nicolson. 1972. "The Fluid Mosaic Model of the Structure of Cell Membranes." *Science* 175:720–731.

Unwin, N., and R. Henderson. February 1984. "The Structure of Proteins in Biological Membranes." *Scientific American* 250(2): 78–94.

6 GROUND RULES OF METABOLISM

Growing Old With Molecular Mayhem

Somewhere in those slender strands of DNA in your cells are snippets of instructions for constructing two wonderful proteins. Those proteins go by the names superoxide dismutase and catalase. Both are enzymes—they make metabolic reactions proceed much, much faster than they would on their own. And both help keep you from growing old before your time.

The two enzymes help your cells clean house, so to speak. Together, they produce and then hack up hydrogen peroxide (H_2O_2), a normal but toxic outcome of certain oxygen-requiring reactions. Oxygen (O_2) picks up electrons from those reactions, and sometimes it does not get quite as many as it is supposed to. So it takes on a negative charge (O_2^-).

Like other unbound, molecular fragments with the wrong number of electrons, O_2^- is a "free radical." Free radicals are *so* reactive, they even attach to molecules that usually do not take part in random reactions—molecules like DNA.

Enter superoxide dismutase. Under its prodding, two of the rogue oxygen molecules combine with hydrogen ions, forming H_2O_2 and O_2.

Enter catalase. Under *its* prodding, two molecules of hydrogen peroxide react and split into ordinary water and ordinary oxygen: $2H_2O_2 \longrightarrow 2H_2O + O_2$.

As people age, their capacity to produce functional proteins—including enzymes—begins to falter. Among those enzymes are superoxide dismutase and catalase.

a

b

Figure 6.1 (**a**) Owner of a good supply of functional enzymes that keep free radicals in check. (**b**) Owner of skin with age spots—visible evidence of free radicals on the loose.

They are produced in diminishing numbers, crippled form, or both. Now free radicals and hydrogen peroxide can accumulate. Like loose cannons, they career through cells with tiny blasts at the structural integrity of proteins, DNA, membranes, and other vital parts. Cells suffer or die outright. Those brown "age spots" on an older person's skin are evidence of assaults by free radicals (Figure 6.1). The spots are masses of brown pigment molecules that build up in cells when free radicals take over—all for the want of two enzymes.

With this chapter, we start to examine activities that keep cells alive and functioning smoothly. Sometimes the topics may seem remote from the world of your interests. But they help define who *you* are and who you will become, age spots and all.

KEY CONCEPTS

1. Cells trap and use energy for building, stockpiling, breaking apart, and eliminating substances in ways that help them survive and reproduce. These activities are called metabolism.

2. Metabolism proceeds as long as cells acquire and use energy. To stay alive, then, cells must replace the energy they inevitably lose during each metabolic reaction. The sun is the original source of energy replacements through most of the biosphere.

3. Different metabolic pathways, operating in coordinated ways, help maintain, increase, or decrease the relative amounts of substances inside cells.

4. Enzymes increase the rate of specific reactions. They take part in nearly all metabolic pathways. So does ATP, an organic compound that transfers energy from one reaction site to another in cells.

Find a light microscope somewhere and use it to peer down on a living cell suspended in a water droplet. The image is eerie—something that small is practically pulsating with movements. Even as you watch it, the cell takes in energy-rich solutes from the water. It busily builds membranes, stores things, checks out the DNA, and replenishes pools of enzymes. It is alive; it is growing; it may divide in two. Multiply these proceedings by 65 trillion cells and you have an idea of what goes on in your own body as you sit quietly, observing that energetic single cell!

Metabolism is the somewhat dreary name for this dynamic activity. It refers to the cell's capacity to acquire energy and use it to build, store, break apart, and eliminate substances in controlled ways. Metabolism is the basis for survival and reproduction—and it all begins with energy.

Energy is a capacity to make things happen, to do work. You use energy to wax a car, even to watch a movie. In both cases, cells of your muscles, brain, and other body parts are being put to work. Energy in muscle cells drives the contractions that move your body parts and hold them in various positions. Energy allows your brain cells to chatter among themselves and guide what you are doing. Gain insight into how cells secure, use, and lose energy, and you gain deeper understanding into the nature of life.

How Much Energy Is Available?

You, like cells, cannot create your own energy from scratch. You must get it from someplace else. That is the message of the **first law of thermodynamics:**

The total amount of energy in the universe remains constant. More energy cannot be created, and existing energy cannot be destroyed. It can only be converted from one form to another.

Consider what this law means. The universe has only so much energy, distributed in a variety of forms. One form can be converted to another, as when corn plants absorb sunlight energy and convert it to the chemical energy of starch. By eating corn, you can extract and convert some of the energy in starch to other forms, such as mechanical energy for your movements. With each conversion, a little energy escapes to the surroundings as heat. Even when you are "doing nothing," your body steadily gives off about as much heat as a 100-watt light bulb because of ongoing conversions in your cells. When released energy "heats up" your surroundings, it really is being transferred to the rather disorderly array of molecules that make up the air (Figure 6.2). It increases the number of ongoing, ran-

dom collisions among those molecules, and with each encounter, a bit more energy spreads outward. However, *none of the energy has vanished*. It simply has become dissipated, in a random, disorderly way, among molecules in the air.

The One-Way Flow of Energy

In the world of life, most of the energy available for energy conversions is stored in covalent bonds. Glucose, starch, glycogen, fatty acids, and other biological molecules with a number of covalent bonds have a high energy content. When reactions break a bond or rearrange atoms around it, atoms move—and add to the molecular commotion (heat energy) of the surroundings. Cells generally cannot recapture energy that has been lost as heat. They can only direct atoms into different bonds.

For example, most cells can break apart glucose until all that remains is six molecules of carbon dioxide and six of water. The breakdown reactions are energetically favored; the carbon and hydrogen atoms of glucose form more stable arrangements when bonded to oxygen, which the cells readily secure from the atmosphere. The total bond energy of glucose is many times greater than the total bond energy of carbon dioxide molecules formed when glucose is broken down. The differences reflect the amount of energy lost as heat during many steps of the breakdown reactions. (We

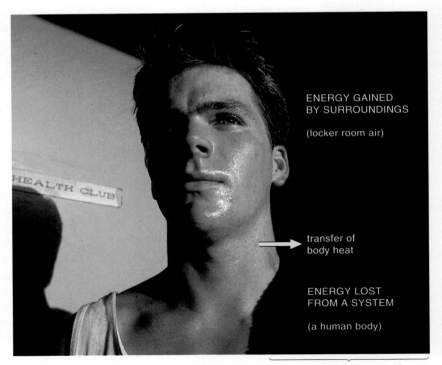

Figure 6.2 Example of how the total energy content of a system *together with its surroundings* remains constant.

"System" means all matter in a specific region, such as a human body, a plant, a DNA molecule, or a galaxy.

"Surroundings" can be a small region in contact with the system or as vast as the entire universe. The system shown here is giving off heat to the surroundings by evaporative water loss from sweat. What one loses, the other gains, so the total energy content of both doesn't change.

ENERGY GAINED BY SURROUNDINGS

(locker room air)

transfer of body heat

ENERGY LOST FROM A SYSTEM

(a human body)

net energy change = 0

measure such differences in terms of kilocalories per mole. A *kilocalorie* is the same thing as a thousand calories—the amount of energy that will heat 1,000 grams of water from 14.5°C to 15.5°C at standard pressure.)

With their high energy content, glucose and other molecules of life are "high-quality" forms of energy. By contrast, a small amount of heat energy spread out in the air is "low quality," because it doesn't lend itself to conversions in cells.

Bad news for cells of the remote future: The amount of low-quality energy in the universe is increasing. Because no energy conversion can ever be 100 percent efficient, the total amount of energy in the universe is spontaneously moving from forms of higher to lower energy content. That, basically, is the point to remember about the **second law of thermodynamics:**

Taken as a whole, the amount of energy in the universe is spontaneously flowing from forms of higher to lower energy content.

Without energy to maintain it, any organized system tends to become disorganized over time. **Entropy** is a measure of the degree of a system's disorder. Think about the Egyptian pyramids—originally organized, presently crumbling, and many thousands of years from now, dust. The ultimate destination of the pyramids and everything else in the universe is a state of maximum entropy. Billions of years from now, all of the energy available for conversions will be dissipated, and nothing we know of will pull it together again.

Can it be that life is one glorious pocket of resistance to the depressing flow toward maximum entropy? After all, every time a new organism grows, atoms become linked in precise arrays, and energy becomes more concentrated and organized, not less so! Yet a simple example will show that the second law does indeed apply to life on earth.

The primary energy source for life on earth is the sun, which is steadily losing energy. Plants capture some sunlight energy, convert it in various ways, then lose energy to other organisms that feed, directly or indirectly, on plants. At each energy transfer along the way, some energy is lost, usually as heat that joins the universal pool. Overall, energy still flows in one direction. The world of life maintains a high degree of organization only because it is being resupplied with energy lost from someplace else (Figure 6.3).

A steady flow of sunlight energy into the interconnected web of life compensates for the steady flow of energy leaving it.

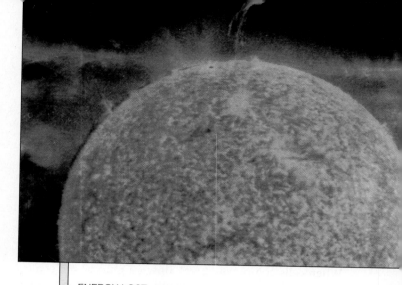

ENERGY LOST
one-way flow of energy away from the sun

ENERGY GAINED
one-way flow of energy from the sun into organisms

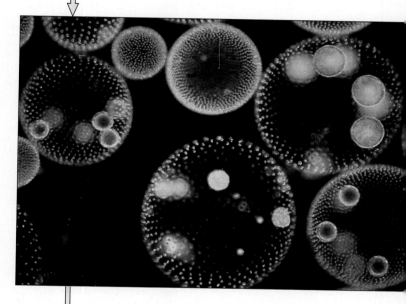

ENERGY LOST
one-way flow of energy from organisms to the surroundings

Figure 6.3 The one-way flow of energy into the world of life that compensates for the one-way flow out of it.

Living cells capture some of the energy being lost from the sun and convert it to useful forms. Energy is released during each conversion and lost to the surroundings, mostly as heat.

Green "dots" in the lower photograph are photosynthetic cells (*Volvox*), organized in tiny, spherical colonies. Orange "dots" are cells set aside for reproduction. They use energy to form new colonies inside the parent sphere.

6.2 ENERGY AND THE DIRECTION OF METABOLIC REACTIONS

Most of the one-way flow of energy through the world of life proceeds in cells that engage in photosynthesis and aerobic respiration. The following sections will help you make sense of these activities, both of which proceed through a step-by-step series of reactions.

Energy Losses and Energy Gains

When cells completely break down a glucose molecule, the covalent bonds of the resulting carbon dioxide and water molecules are more stable, in that it takes more energy to break them. But the total bond energies are lower than they were in glucose. This is an example of a reaction that ends with a net *loss* in energy (Figure 6.4a). It is an *exergonic* reaction, a word that means "energy out." Every day in an adult human of average size, the breakdown of glucose and other molecules by such reactions provides the body with an average of 1,200 to 2,800 kilocalories of energy.

How can a cell build starch, fatty acids, and other energy-rich molecules from smaller, energy-poor ones? Doesn't the overall synthesis of such molecules run counter to the spontaneous direction of energy flow? It does indeed. However, *inputs of energy*—as from ATP molecules—can drive reactions in an energetically unfa-

vorable direction. Such reactions typically end with a *net gain* in energy (Figure 6.4b). They are said to be *endergonic*, a word that means "energy in."

Reversible Reactions

Most reactions are reversible. They proceed in the forward direction, from starting substances (reactants) to products. They also proceed in reverse—with products being converted back to reactants. Figure 6.5 is an example of a reversible reaction, as the two opposing arrows signify.

Whether a reversible reaction runs more strongly in the forward or reverse direction depends partly on the ratio of reactant to product molecules. Remember, molecules are in constant random motion that puts them on collision courses with one another. The more concentrated they are, the more often they will collide. And energy released during a collision may be enough to make molecules combine, change shape, or split into smaller parts.

With a high concentration of reactant molecules, the reaction proceeds strongly in the forward direction. When the product concentration increases enough, a greater number of product molecules will be available to revert to reactants.

A reversible reaction tends spontaneously toward **chemical equilibrium**, at which time it proceeds at about the same pace in the forward and reverse direc-

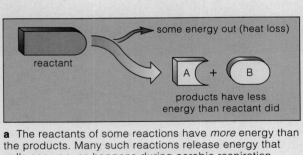

a The reactants of some reactions have *more* energy than the products. Many such reactions release energy that cells can use, as happens during aerobic respiration.

Figure 6.4 Energy changes in metabolic reactions.

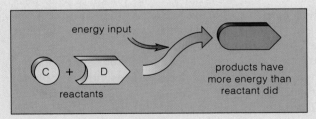

b The reactants of other reactions have *less* energy than the products. Energy inputs drive such reactions, as when an energy input from the sun drives photosynthesis.

c Energy inputs drive reactions by which cells build energy-rich molecules. When such molecules are split apart, some energy is released and can be harnessed to do cellular work.

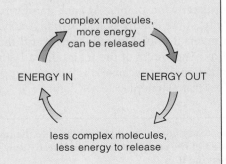

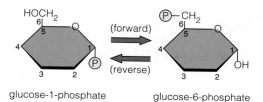

glucose-1-phosphate   glucose-6-phosphate

Figure 6.5 A reversible reaction. Glucose is primed to enter reactions when a phosphate group becomes attached to it. With high concentrations of glucose-1-phosphate, the reaction tends to run in the forward direction. With high concentrations of glucose-6-phosphate, it runs in reverse. (The 1 and 6 of these names simply identify which particular carbon atom of the glucose ring has a phosphate group attached to it.)

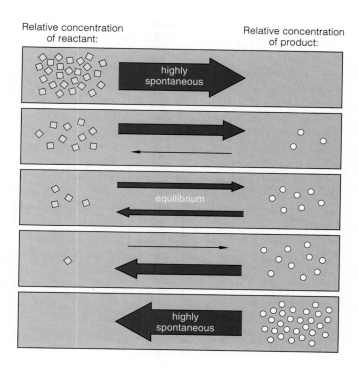

Relative concentration of reactant: Relative concentration of product:

Figure 6.6 Chemical equilibrium. With a high concentration of reactant molecules, a reaction tends to proceed most strongly in the forward direction (to product molecules). With a high concentration of product molecules, it proceeds most strongly in reverse. At equilibrium, the rates of the forward and reverse reactions are the same.

tions (Figure 6.6). The *amounts* of reactant and product molecules may or may not be the same at equilibrium. Think of a party with as many people wandering in as wandering out of two adjoining rooms. The total number in each room stays the same—say, thirty in one and ten in the other—even though the mix of people in each room continually changes.

Every reaction has its own characteristic ratio of products to reactants at equilibrium. For example, in one reaction, glucose-1-phosphate is rearranged into glucose-6-phosphate. (As Figure 6.5 shows, these are simply glucose molecules with a phosphate group attached to one of their six carbon atoms.) Suppose we allow the reaction to proceed when the concentrations of both substances are the same. At that point, the forward reaction proceeds 19 times faster than the reverse reaction does. Said another way, the forward reaction is producing more molecules in the same amount of time. The forward and reverse reactions eventually proceed at the same rate—but only when there are nineteen molecules of glucose-6-phosphate for every glucose-1-phosphate molecule. For this reaction, the ratio of products to reactants at equilibrium is 19:1.

Cells never stop building and tearing down molecules and moving them about in specific directions until they die. What does their great juggling act accomplish? Think it through. Cells can use only so many molecules at a given time and have only so much internal space to hold any excess. If they produce more of a substance than they can use, put into storage, or secrete, the excess might cause problems.

Consider *phenylketonuria* (PKU). People affected by this genetic (heritable) disorder produce a defective enzyme that prevents cells from using phenylalanine, one of the amino acids. When this amino acid accumu-

lates, the excess enters reactions that produce phenyl-ketones. An accumulation of *these* product molecules can damage the brain in less than a few months. In most developed countries, routine screening programs identify affected newborns, who can grow up symptom-free on a phenylalanine-restricted diet.

1. The substances present at the conclusion of a reaction—the end products—may have less or more energy than the starting substances (the reactants).

2. Most reactions are reversible. Besides proceeding in the forward direction, from reactants to end products, they can proceed in the reverse direction—from end products back to reactants.

3. A reversible reaction moves spontaneously toward equilibrium, a state in which it proceeds at about the same rate in both directions.

6.3 METABOLIC PATHWAYS

All cells maintain, increase, and decrease the concentrations of substances by coordinating different metabolic pathways. A **metabolic pathway** is an orderly sequence of reactions, with specific enzymes acting at each step. The pathways are linear or circular (Figure 6.7). Often they are coupled, with products of one serving as reactants for others.

The main metabolic pathways are biosynthetic or degradative, overall. In *biosynthetic* pathways, small molecules are assembled into proteins, lipids, and other large molecules of higher energy content. In *degradative* pathways, large molecules are broken down to products of lower energy content. Both types have these participants:

Substrates: Substances able to enter a reaction; also called reactants or precursors.

Intermediates: Compounds formed between the start and the end of a metabolic pathway.

Enzymes: Proteins that catalyze (speed up) reactions.

Cofactors: Organic molecules or metal ions that assist enzymes or carry atoms or electrons from one reaction site to another.

Energy Carriers: Mainly ATP, which readily donates energy to diverse reactions.

End Products: Substances present at the end of a metabolic reaction or pathway.

Let's take a quick look at the roles of these substances before turning to the chapters on the main metabolic pathways.

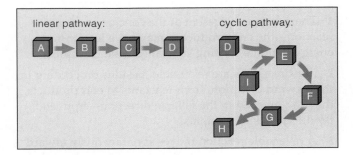

Figure 6.7 Example of a linear metabolic pathway that is coupled to a cyclic (circular) metabolic pathway.

6.4 ENZYMES

Characteristics of Enzymes

Enzymes are catalytic molecules, meaning that they greatly speed up specific reactions. They enhance the rate at which the reactions approach equilibrium. With few exceptions, enzymes are proteins.

Enzymes have four features in common. *First*, they do not make anything happen that could not happen on its own. But they usually make it happen at least a million times faster. *Second*, enzymes are not altered permanently or used up in the reactions they mediate; the same enzyme may be used over and over again. *Third*, the same enzyme works for the forward and reverse directions of a reaction. *Fourth*, each type of enzyme is highly selective about its substrates.

Substrates are molecules that a specific enzyme can chemically recognize, bind, and modify in a specific way. For example, thrombin, an enzyme involved in blood clotting, only recognizes and splits a peptide bond between two amino acids, arginine and glycine:

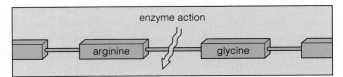

Enzyme-Substrate Interactions

An enzyme has one or more crevices in its surface where substrates interact with it. Such a crevice is an **active site**, the place where a specific reaction is catalyzed (Figure 6.8).

According to Daniel Koshland's **induced-fit model**, each substrate has a surface region that almost *but not quite* matches chemical groups in an active site. When substrates first settle into the site, the contact strains some of their bonds. Strained bonds are easier to break, and this helps pave the way for new bonds (within the products). Also, interaction with charged or polar groups in the site favors a redistribution of electric charge that primes the substrates for conversion to an activated state.

Substrates reach an activated, transition state when they fit most precisely in the active site (Figure 6.9). The enzyme induces the fit in several ways. For instance, weak but extensive bonding at the active site puts substrates in positions that make them collide and so promotes reaction. If those same molecules were colliding on their own, they would do so from random directions. The reaction rate would not be very impressive, because mutually attractive chemical groups would meet up less frequently.

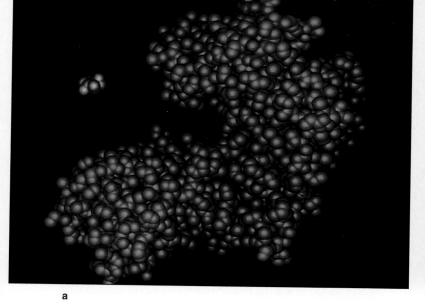

a

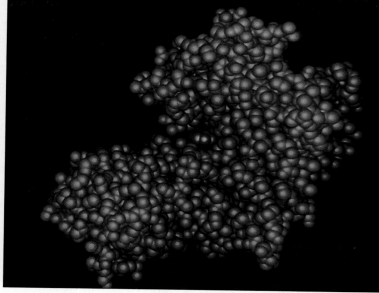

b

Figure 6.8 An enzyme at work. (**a**) Model of the enzyme hexokinase (*green shading*) and its substrate, a glucose molecule (*red*). The glucose is heading toward the active site, a cleft in the enzyme. (**b**) When glucose contacts the site, the upper and lower parts of the enzyme temporarily close in around it and prod the molecule to enter a specific reaction.

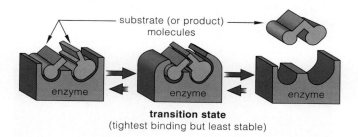

transition state
(tightest binding but least stable)

Figure 6.9 Induced-fit model of enzyme-substrate interactions. Only when the substrate is bound in place is an enzyme's active site complementary to it. The fit is most precise during a transition state of a reaction. The enzyme-substrate complex is short-lived, partly because the bonds holding it together are usually weak.

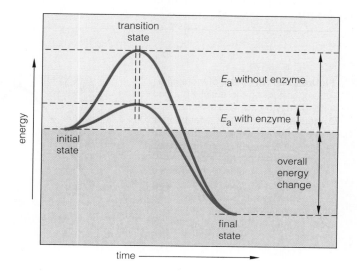

Figure 6.10 Energy hill diagram showing the effect of enzyme action. An enzyme greatly enhances the rate at which a given number of molecules complete a reaction because it lowers the required activation energy (E_a). It takes less energy to boost reactants to the crest (transition state) of a lower energy hill.

Once substrates are in the transition state, they react spontaneously, just as a boulder pushed up and over the crest of a hill rolls down on its own. However, they simply won't reach that state unless they collide with some minimum amount of energy. That amount, the **activation energy**, is like an "energy hill" that must be surmounted (Figure 6.10). By putting its substrates on a precise collision course, an enzyme makes the energy hill smaller, so to speak. And this means the reaction will proceed more rapidly.

An enzyme enhances the rate at which a reaction proceeds. It does this by lowering the amount of activation energy necessary to make its substrates react.

Effects of Temperature and pH on Enzyme Activity

Each type of enzyme functions best within a certain temperature range. As Figure 6.11 indicates, reaction rates decrease sharply when temperatures become too high. The increase in heat energy disrupts weak bonds holding the enzyme in its three-dimensional shape. This alters the active site, and substrates cannot bind to it. Often, exposure to temperatures that are much higher than an organism typically encounters destroys enzymes—and disrupts metabolism. This happens during a dangerously high *fever*. Humans usually die when internal body temperature reaches 44°C (112°F).

Each enzyme also functions best within a certain pH range. Higher or lower pH values generally disrupt its structure and function (Figure 6.12). Most enzymes function best in neutral solutions (pH 7). Pepsin is one of the exceptions. This protein-digesting enzyme performs its task in gastric fluid, which is extremely acidic.

Enzymes function best within limited ranges of temperature and pH.

Control of Enzyme Activity

Earlier you read that cells maintain, increase, and decrease concentrations of substances by coordinating different metabolic pathways. They do this mostly by controlling enzymes. Some controls govern enzyme synthesis and so cut down or beef up the number of enzyme molecules that are operating at any time at a key step in a pathway. Other controls stimulate or inhibit enzymes that are already formed.

Internal controls come into play when concentrations of substances change within the cell itself. Think of a bacterium, busily synthesizing tryptophan and other amino acids necessary to construct its proteins.

After a bit, protein synthesis slows and no more tryptophan is needed. But the tryptophan pathway is still in full swing, so the cellular concentration of its end product rises. Now a control mechanism called **feedback inhibition** operates. By this mechanism, when production of a substance triggers a cellular change, the substance itself shuts down its further production (Figure 6.13). In this case, unused molecules of tryptophan inhibit a key enzyme in the pathway.

This enzyme happens to be governed by *allosteric* control. Besides the active site, allosteric enzymes have

Figure 6.11 (**a**) Effect of increases in temperature on the activity of one kind of enzyme. (**b**) Observable evidence of temperature effects on an enzyme that influences the coat color of Siamese cats. Hairs on the ears and paws of this cat contain more dark-brown pigment (melanin) than hairs on the rest of the body. A heat-sensitive enzyme that controls melanin production is less active in warmer body regions—which have lighter hair as a result.

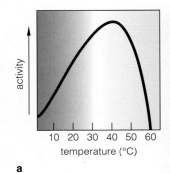

a

b

Figure 6.12 Effect of pH on the activity of three different enzymes. One enzyme (*black*) functions best in neutral solutions. Another kind of enzyme (*orange*) functions best in basic solutions, and another (*blue*), in acidic solutions.

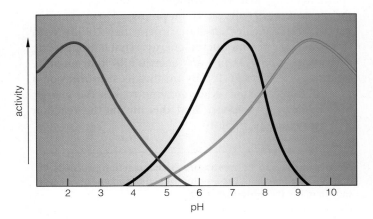

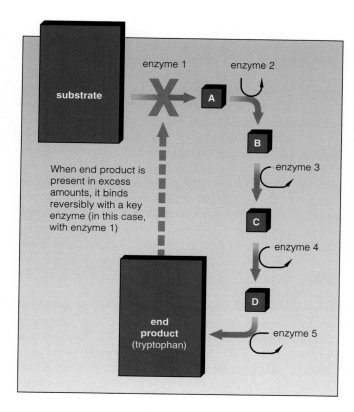

Figure 6.13 Feedback inhibition of a metabolic pathway. In this example, five enzymes act in sequence to convert a substrate into tryptophan, the end product. When an excessive number of end-product molecules accumulate, some of them bind to the first enzyme in the sequence and so block the entire pathway.

When end product is present in excess amounts, it binds reversibly with a key enzyme (in this case, with enzyme 1)

end product (tryptophan)

control sites where specific substances can bind and alter enzyme activity.

When the pathway is blocked, fewer tryptophan molecules are around to inhibit the key enzyme—so production rises. In such ways, feedback inhibition quickly adjusts concentrations of substances in cells.

In humans and other multicelled organisms, control of enzyme activity is just amazing. Cells not only work to keep themselves alive, they work with other cells in coordinated ways that benefit the whole body! Hormones are signaling agents in this vast enterprise. Specialized cells release these chemical signals into the bloodstream. Any cell having receptors for a particular hormone will take it up, and its program for constructing a particular protein or some other internal activity will be altered. The hormone trips internal control agents into action, and the activities of specific enzymes change.

Enzymes act only on specific substrates, and controls over their activity are central to the directed flow of substrates into, through, and out of the cell.

6.5 ENZYME HELPERS

During a metabolic reaction, enzymes speed the transfer of one or more electrons, atoms, or functional groups from one substrate to another. "Cofactors" help catalyze the reaction or serve briefly as transfer agents.

Some of the cofactors are specific types of organic compounds. Others are nothing more than metal ions that associate with the enzyme.

Coenzymes

The enzyme helpers called **coenzymes** are complex organic molecules, many of which are derived from vitamins. The coenzymes abbreviated **NAD**$^+$ (short for "nicotinamide adenine dinucleotide") and **FAD** (flavin adenine dinucleotide) are examples.

Both of these coenzymes can pick up hydrogen atoms that are liberated during glucose breakdown. They transfer these unbound protons (H$^+$) to other reaction sites. Electrons, being attracted to the opposite charge of the protons, go along for the ride. When carrying their cargo, the two coenzymes are abbreviated NADH and FADH$_2$, respectively.

The coenzyme **NADP**$^+$ (for nicotinamide adenine dinucleotide phosphate) functions in photosynthesis. It also transfers protons and electrons from certain degradative pathways to biosynthetic pathways, such as the ones by which fatty acids are assembled. Discerning enzymes recognize its phosphate "tag" (the P$^+$ attached to the molecule) and accept its proton and electron cargo. Like NAD$^+$, this coenzyme is derived from the vitamin niacin. When carrying its cargo, it is abbreviated NADPH.

Metal Ions

The metal ions that serve as cofactors include ferrous iron (Fe^{++}). This metal ion is a component of cytochrome molecules. The cytochromes are carrier proteins that show enzyme activity. They are embedded in cell membranes, such as the membranes of chloroplasts and mitochondria.

a cytochrome

Coenzymes and metal ions assist enzymes in the transfer of electrons, atoms, and functional groups from one substrate to another.

6.6 ELECTRON TRANSFERS IN METABOLIC PATHWAYS

If you were to throw some glucose into a wood fire, its atoms would quickly let go of one another and combine with oxygen in the atmosphere, forming CO_2 and H_2O. Of course, all of the released energy would be lost as heat to the surroundings.

Cells do not "burn" glucose all at once. Doing so would waste its stored energy. Instead, enzymes pluck atoms away from the molecule during controlled steps of a degradative pathway, so that intermediate molecules form along the route. Thus, only some of the bond energy is released at each enzyme-mediated step.

Recall that chemical bonds are interactions between electrons of different atoms. Breaking those bonds puts electrons up for grabs, so to speak. And the liberated electrons can be sent through **electron transport systems**. As microscopists discovered, such systems are bound in the internal membrane systems of chloroplasts and mitochondria. They actually are organized arrays of enzymes and coenzymes that transfer elec-

trons in a highly organized sequence. One molecule donates electrons, and the next in line accepts them.

You may hear of *oxidation-reduction* reactions. This rather obtuse name simply refers to electron transfers. A donor that gives up electrons is said to be oxidized. An acceptor that acquires electrons is reduced.

Electron transport systems operate by accepting electrons at higher energy levels, releasing them at lower energy levels, and making use of the energy they release. Think of a transport system as a staircase (Figure 6.14). Excited electrons at the top step have the most energy. They drop down, one step at a time. At certain steps, energy being released is harnessed to do work—for instance, to move ions in ways that set up ion concentration and electric gradients across a membrane. Such gradients, you will discover, are central to the formation of ATP.

Some transfers proceed after atoms or molecules absorb enough energy to "excite" electrons—that is, to boost the electrons farther from the nucleus. When an excited electron returns to the lowest energy level available, it gives off energy. The *Focus* essay describes some interesting evidence of this energy release.

a

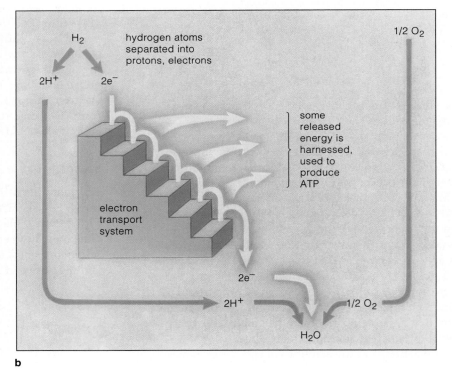

b

Figure 6.14 Controlled release of energy in metabolic reactions. (**a**) Hydrogen and oxygen exposed to an electric spark will react and release energy all at once. (**b**) In cells, the same type of reaction proceeds in many small steps that allow some of the released energy to be harnessed in useful forms. The "steps" are electron transfers, often between molecules that operate together as an electron transport system.

Focus on Health

You Light Up My Life—Visible Effects of Excited Electrons

At night, during certain metabolic reactions, fireflies, various beetles, and some other organisms flash with light. These displays of **bioluminescence** take place when enzymes called luciferases excite the electrons of luciferins, a class of highly fluorescent substances. When the excited electrons quickly return to a lower energy level, they release energy—in the form of light.

In the tropical forests of Jamaica, click beetles, known locally as kittyboos, fly about at night and give startling displays of bioluminescence (Figure *a–c*). These particular beetles belong to the genus *Pyrophorus*, and different varieties emit green, greenish-yellow, yellow, or orange flashes.

Molecular biologists have identified the kittyboo genes responsible for the flashes. More than this, they have managed to insert copies of the genes into other organisms, including bacteria (Figure *d*). How they do these gene transfers is a topic of Chapter 16.

Besides being fun to think about, transfers of the genes for bioluminescence have practical applications. Each year, for example, 3 million people die from a lung disease caused by *Myobacterium tuberculosis*. Different strains of this bacterium are resistant to different antibiotics. A patient cannot be treated effectively until the strain causing her or his particular infection is identified. Before 1993, the tests used for identification took weeks to complete.

Now, bacterial cells from a patient are rapidly cultured and exposed to luciferase genes. Some cells take up the genes, which may become inserted into the bacterial DNA. Clinicians then expose colonies of the genetically modified bacteria to different antibiotics. If an antibiotic has no effect, the cells churn out gene products—*including luciferase*—and the colony glows. If a colony doesn't glow, this is evidence that the antibiotic has stopped the infectious cells in their metabolic tracks.

a

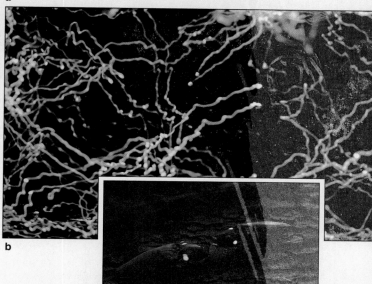

b

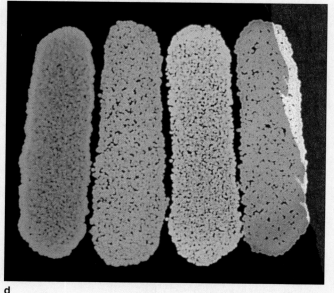

c

(**a**) One of the Jamaican kittyboo beetles (*Pyrophorus noctilucus*). The time-lapse photograph in (**b**) reveals the random paths of kittyboos on the wing, as they light up the night with bioluminescent flashes. The flashes help potential mates find each other in the dark (**c**). The micrograph in (**d**) shows four colonies of bacterial cells that have taken up kittyboo genes responsible for bioluminescence.

d

6.7 ATP—THE MAIN ENERGY CARRIER

Structure and Function of ATP

Photosynthetic cells do not run directly on sunlight. First they convert sunlight energy to the chemical energy of **ATP**. This is the abbreviation for adenosine triphosphate, a type of nucleotide.

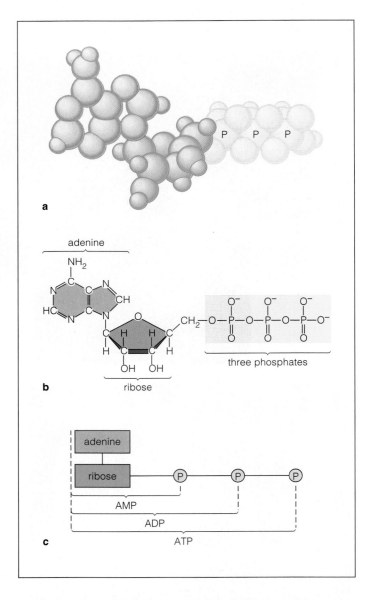

a

b

adenine

ribose

three phosphates

c

AMP

ADP

ATP

adenine
ribose
P — P — P

Figure 6.15 ATP—adenosine triphosphate, the main energy carrier in cells. (**a**) Three-dimensional model depicting its component atoms. (**b**) The structural formula for ATP. (**c**) Adenosine diphosphate (ADP) forms when ATP gives up one phosphate group. Adenosine monophosphate (AMP) forms when another phosphate group is removed.

Besides this, no cell whatsoever *directly* uses energy released during glucose breakdown. First the energy is harnessed and converted to ATP energy.

As Figure 6.15 shows, an ATP molecule consists of adenine (a nitrogen-containing compound), the five-carbon sugar ribose, and a string of three phosphate groups. Covalent bonds hold these components together—and the bonding arrangement is not that stable. Many hundreds of different enzymes can easily split off the outermost phosphate group by hydrolysis reactions. When they do, a great deal of the energy that is released can be tapped for hundreds of different tasks—such as breaking apart molecules, actively transporting substances across cell membranes, and triggering muscle contraction.

Thus ATP molecules are like coins of a nation—they are the cell's common currency of energy.

The ATP/ADP Cycle

Given the central role of ATP in metabolism, it comes as no surprise that cells have mechanisms for renewing the molecule after it gives up phosphate. In the **ATP/ADP cycle**, an energy input drives the binding of adenosine diphosphate (ADP) to a phosphate group or to unbound phosphate (P_i), forming ATP. Then the ATP is ready to donate a phosphate group elsewhere and revert back to ADP (Figure 6.16).

The attachment of phosphate to a molecule is called **phosphorylation**. Keep this molecular event in mind in chapters to follow. Generally speaking, when a molecule becomes phosphorylated, as by ATP, its store of energy increases *and it becomes primed to enter a specific reaction.*

ATP delivers energy to or picks up energy from almost all metabolic pathways.

With the ATP/ADP cycle, cells have a renewable means of conserving energy and transferring it to specific reactions.

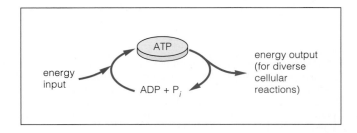

energy input

ATP

ADP + P_i

energy output (for diverse cellular reactions)

Figure 6.16 The ATP/ADP cycle.

SUMMARY

1. Cells acquire and use energy to build, store, break down, and rid themselves of substances. These activities are called metabolism. They underlie growth, maintenance, and reproduction of all organisms.

2. Energy flows in one direction through the world of life. The sun is the primary energy source. It replaces metabolically generated energy that all cells inevitably lose to their surroundings (as heat, mostly).

3. A metabolic pathway is a stepwise sequence of reactions in cells. In biosynthetic pathways, large molecules are assembled, and energy becomes stored in them. In degradative pathways, large molecules are broken down to smaller ones, and energy is released.

4. These substances take part in metabolic reactions:

   a. Substrates (or reactants): the substances that enter a specific reaction.

   b. Enzymes: proteins that serve as catalysts (they speed up reactions).

   c. Cofactors: coenzymes (including NAD^+) and metal ions that help catalyze reactions or carry electrons, hydrogen, or functional groups stripped from substrates.

   d. Energy carriers: mainly ATP, which readily donates energy to other molecules. Most biosynthetic pathways are driven by ATP energy.

   e. End products: the substances formed at the end of a metabolic pathway.

5. Control mechanisms stimulate or inhibit the activity of enzymes at key steps in metabolic pathways. They help coordinate the flow of substances into, through, and out of cells.

6. Electron transport systems are organized sequences of enzymes and coenzymes. They are built into cell membranes, such as those of chloroplasts and mitochondria. Electrons stripped from substrates are transferred through these systems. During certain transfers, energy is released that can be used to do work—for example, to make ATP.

Review Questions

1. State the first and second laws of thermodynamics. Does life violate the second law? *94–95*

2. In metabolic reactions, does equilibrium imply equal amounts of reactants and products? Think of a cellular activity that might keep a reaction from approaching equilibrium. *96–97*

3. Describe an enzyme and its role in metabolic reactions. *98–99*

4. What are the three components of ATP? Why is phosphorylation of a molecule by ATP so important? *104*

Self-Quiz *(Answers in Appendix IV)*

1. A cell's capacity to acquire and use energy for building and breaking apart molecules is called _____ .

2. Two laws of _____ govern how cells acquire, convert, and transfer energy during metabolic reactions.

3. _____ is the primary source of energy for life on earth.
   - a. Food
   - b. Water
   - c. The sun
   - d. ATP

4. Which is *not* true of chemical equilibrium?
   - a. Product and reactant concentrations are always equal.
   - b. The rates of the forward and reverse reactions are the same.
   - c. There is no further net change in product and reactant concentrations.

5. In a biosynthetic pathway, _____ .
   - a. large molecules are broken down to smaller ones
   - b. energy is not required for the reactions
   - c. large molecules are assembled from simpler ones
   - d. both a and b

6. Enzymes _____ .
   - a. enhance reaction rates
   - b. are affected by pH
   - c. act on specific substrates
   - d. all of the above

7. Electron transport systems involve _____ .
   - a. enzymes and cofactors
   - b. electron transfers and released energy
   - c. cell membranes
   - d. all of the above

8. The main energy carriers in cells are _____ .
   - a. NAD^+ molecules
   - b. cofactors
   - c. ATP molecules
   - d. enzymes

9. Match each substance with its correct description.
   - _____ a coenzyme or metal ion
   - _____ mainly ATP
   - _____ substance entering a reaction
   - _____ protein that catalyzes a reaction
     - a. substrate
     - b. enzyme
     - c. cofactor
     - d. energy carrier

Selected Key Terms

activation energy *99*	end product *98*	intermediate *98*
active site *98*	energy *93*	metabolic pathway *98*
ATP (adenosine triphosphate) *104*	energy carrier *98*	metabolism *93*
	entropy *95*	NAD^+ *101*
ATP/ADP cycle *104*	enzyme *98*	$NADP^+$ *101*
bioluminescence *103*	FAD (flavin adenine dinucleotide) *101*	phosphorylation *104*
chemical equilibrium *96*	feedback inhibition *100*	second law of thermodynamics *95*
coenzyme *101*	first law of thermodynamics *94*	substrate *98*
cofactor *98*		
electron transport system *102*	induced-fit model *98*	

Readings

Fenn, J. 1982. *Engines, Energy, and Entropy.* New York: Freeman. Deceptively simple paperback.

Rusting, R. December 1992. "Why Do We Age?" *Scientific American* 267(6):131–141.

7 ENERGY-ACQUIRING PATHWAYS

Sun, Rain, and Survival

Just before dawn in the Midwest, the air is dry and motionless. The heat that has scorched the land for weeks still rises from the earth and hangs in the air of a new day. There are no clouds in sight. There is no promise of rain. For hundreds of miles, crops stretch out, withered and nearly dead. All the marvels of modern agriculture can't save them. In the absence of one vital substance—water—life in each cell of those many hundreds of thousands of plants has ceased.

In Los Angeles, a student wonders if the Midwest drought will bump up food prices. In Washington, D.C., economists analyze crop failures in terms of tonnage available for domestic consumption and export.

Thousands of kilometers away, in the vast Sahel Desert of Africa, grasses and cattle are dying after a similar unrelenting drought. Children with bloated bellies and spindly legs wait passively for death. Deprived of nourishment for too long, cells of their bodies will never function normally again.

You are about to explore pathways by which cells trap and use energy. At first these pathways may seem to be far removed from your everyday world. *Yet the food that nourishes you and nearly all other organisms cannot be produced or used without them.*

We will return to this point in later chapters, when we address major concerns such as human population

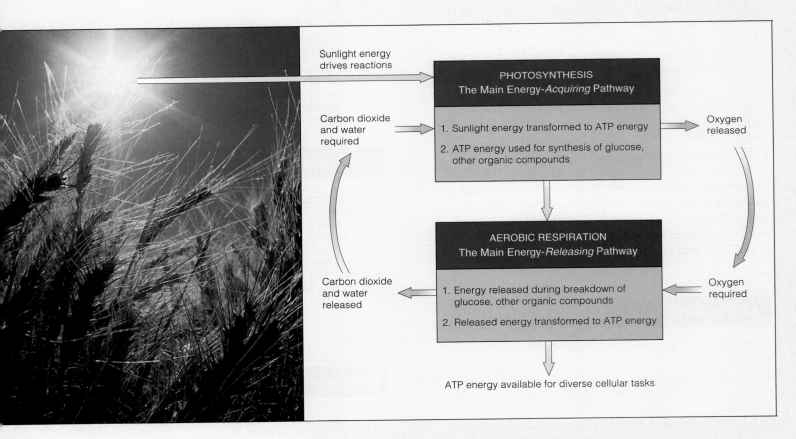

Figure 7.1 Links between photosynthesis and aerobic respiration, the main energy-acquiring and energy-releasing pathways in the world of life.

growth, nutrition, limits on agriculture, genetic engineering of plants, and effects of pollution on crops. Here, our point of departure is the *source* of food—which isn't a farm or a supermarket or a refrigerator. What we call "food" was put together somewhere in the world by living cells from glucose and other organic compounds. Such compounds are built on a framework of carbon atoms, so the questions become these:

1. Where does the carbon come from in the first place?

2. Where does the energy come from to drive the assembly of carbon-based compounds?

Answers to these questions depend on an organism's mode of nutrition.

Organisms classified as **autotrophs** are "self-nourishing," which is what *autotroph* means. Their carbon source is carbon dioxide (CO_2), a gaseous substance all around us in the air and dissolved in water. The world's plants, certain protistans, and certain bacteria are *photo*autotrophs; sunlight is their energy source. A few bacteria are *chemo*autotrophs; they can extract energy from an inorganic substance, such as sulfur.

Most organisms classified as **heterotrophs** get their carbon *and* energy from organic compounds that other organisms have already put together. They eat autotrophs, each other, and organic wastes. (*Hetero*- means other, as in "nourished by other organisms.") That is how animals, fungi, many of the protistans, and most bacteria stay alive.

When you think about this grand pattern of who eats whom, one thing becomes clear. *Survival of nearly all organisms ultimately depends on photosynthesis, the main pathway by which carbon and energy enter the world of life.*

Once glucose and other organic compounds are assembled, cells put them in storage or use them as building blocks. When cells require energy, they can break such compounds apart by several pathways. However, *the predominant energy-releasing pathway is called aerobic respiration.* Figure 7.1 provides you with a preview of the links between photosynthesis and aerobic respiration—the focus of this chapter and the next.

KEY CONCEPTS

1. Carbon-based compounds are the building blocks and energy stores of life. Plants assemble these compounds by photosynthesis. First they trap sunlight energy and convert it to chemical energy (in the form of bonds in ATP molecules). Then ATP delivers energy to reactions in which glucose is put together from carbon dioxide and water. Finally, glucose subunits are combined to form starch and other molecules.

2. Photosynthesis is the biosynthetic pathway by which most carbon and energy enter the web of life.

3. In plant cells, photosynthesis proceeds in organelles called chloroplasts. The pathway starts at a membrane system inside the chloroplast. Its machinery includes light-absorbing pigments, enzymes, and a coenzyme ($NADP^+$), which delivers hydrogen and electrons to the synthesis reactions.

4. Photosynthesis is often summarized this way:

$$12H_2O + 6CO_2 \xrightarrow{\text{sunlight}} 6O_2 + C_6H_{12}O_6 + 6H_2O$$

7.1 PHOTOSYNTHESIS: AN OVERVIEW

Energy and Materials for the Reactions

Photosynthesis is an ancient pathway, and it evolved in distinct ways in different organisms. To keep things simple, let's focus on what goes on in lettuce, weeds, and other leafy plants.

The pathway consists of two stages, each with its own set of reactions. In the *light-dependent* reactions, energy from sunlight is absorbed and converted to ATP energy. Water molecules are split, and the coenzyme $NADP^+$ picks up the liberated hydrogen and electrons, thereby becoming NADPH.

In the *light-independent* reactions, ATP donates energy to sites where glucose ($C_6H_{12}O_6$) is put together from carbon, hydrogen, and oxygen. Carbon dioxide (CO_2) provides the carbon and oxygen. Water provides the hydrogen, as delivered by NADPH.

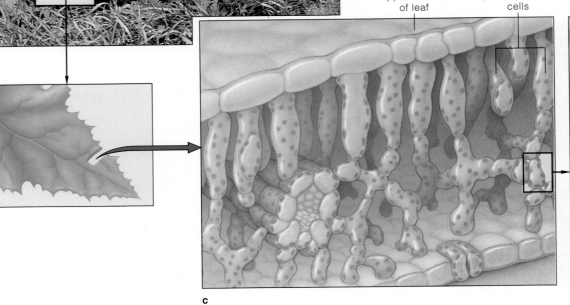

upper surface of leaf   photosynthetic cells   single photosynthetic cell

starch grain

e chloroplast

Figure 7.2 Where the reactions of photosynthesis take place inside the leaves of sow thistle (*Sonchus*), a common plant in many parts of the world.

(**a**) This plant is growing alongside a country road in Germany. The subsequent photographs and diagrams take us inside one of the leaves.

(**b,c**) Here is a close-up of one leaf, then of a small section cut from the leaf. (**d**) Within that leaf section are numerous photosynthetic cells, one of which is illustrated. (**e**) Inside that cell are chloroplasts, the organelle of photosynthesis.

Photosynthesis is often summarized in this manner:

$$12H_2O + 6CO_2 \xrightarrow{\text{sunlight}} 6O_2 + C_6H_{12}O_6 + 6H_2O$$

This summary equation shows glucose as an end product in order to keep the chemical bookkeeping simple. However, the reactions don't really stop with glucose. Glucose and other simple sugars combine at once to form sucrose, starch, and other carbohydrates—the true end products of photosynthesis.

Where the Reactions Take Place

The two stages of photosynthesis proceed at different sites inside the **chloroplast**. Only the photosynthetic cells of plants and a few protistans contain this type of organelle (page 68). Each chloroplast has two outer membranes wrapped around its interior, the **stroma**. An inner membrane weaves through the stroma, which is largely fluid. Often the membrane has the form of flattened channels and disks, which are arranged in stacks called grana (singular, granum). The first stage of photosynthesis proceeds at this inner membrane, which is the **thylakoid membrane system**. The spaces inside the disks and channels connect as a single compartment for hydrogen ions, which are used in ATP production. The second stage of photosynthesis—the reactions by which sugars are assembled—takes place in the stroma.

Figure 7.2 marches down through a chloroplast from sow thistle (*Sonchus*), a common weed. Two thousand of those chloroplasts, lined up single file, would be no wider than a dime. Imagine all the chloroplasts in just one weed or lettuce leaf, each a tiny factory for producing sugars and starch—and you get an idea of the magnitude of metabolic events required to feed you and all other organisms living together on this planet.

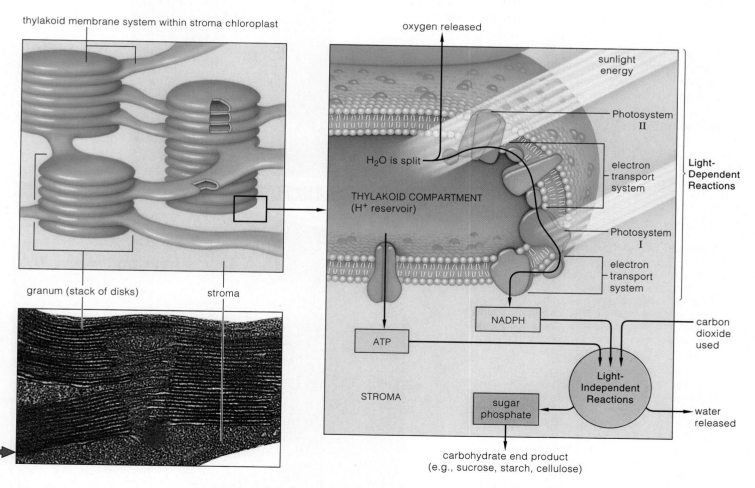

thylakoid membrane system within stroma chloroplast

granum (stack of disks)

stroma

oxygen released

sunlight energy

Photosystem II

H_2O is split

electron transport system

Light-Dependent Reactions

THYLAKOID COMPARTMENT (H^+ reservoir)

Photosystem I

electron transport system

NADPH

carbon dioxide used

ATP

STROMA

Light-Independent Reactions

sugar phosphate

water released

carbohydrate end product (e.g., sucrose, starch, cellulose)

(**f**) Inside the photosynthetic cell, light-dependent reactions proceed at the thylakoid membrane system. Light-independent reactions proceed within the stroma.

(**g**) This diagram provides an overview of the sites where the key steps of both stages of reactions proceed.

7.2 LIGHT-TRAPPING PIGMENTS

Generally speaking, **pigments** are molecules that can absorb light. In animals, melanin and other pigments have roles in vision, coloration of body surfaces, and other functions. In photosynthetic organisms, a variety of pigment molecules trap photons from the sun.

Photons are packets of light energy that travel through space in undulating motion, rather like ocean waves. A photon's wavelength (the distance from one wave peak to the next) is related to its energy. The most energetic photons travel as short wavelengths, and the least energetic as long wavelengths. Humans and some other animals can perceive part of the spectrum of different wavelengths as different colors of light (Figure 7.3a).

In the thylakoid membranes of chloroplasts, clusters of pigments trap light of certain wavelengths. Just as some ocean waves are more exciting than others to surfers, certain wavelengths are more exciting to these pigments. For example, **chlorophylls** absorb violet-to-blue as well as red wavelengths (Figure 7.3b). They are key photosynthetic pigments in green algae and plants (Figure 7.4). They transmit green wavelengths, so that is why plant parts having an abundance of chlorophyll appear green to us. As another example, a variety of **carotenoids** absorb violet and blue wavelengths but transmit red, orange, and yellow.

In all plants, chlorophyll *a* is the main pigment of photosynthesis. Chlorophyll *b*, carotenoids, and other pigments absorb energy of different wavelengths. They don't use the energy themselves; they transfer it to the main pigment and so enhance its effectiveness.

In green leaves, the carotenoids are far less abundant than the chlorophylls. Often they become visible in autumn, when many plants stop producing chlorophyll (Figure 7.5). Several other pigments contribute to the distinctive coloration of various organisms. Among these are the red and blue **phycobilins**, the signature pigments of red algae and cyanobacteria.

Chlorophylls and other pigment molecules absorb different wavelengths of light, which correspond to different colors.

Figure 7.3 Wavelengths of visible light. These fall within a much larger range of wavelengths, called the electromagnetic spectrum (**a**). Various organisms use the wavelengths that range from about 400 to 750 nanometers for photosynthesis, vision, and other light-requiring processes. Shorter wavelengths, including ultraviolet light and x-rays, are energetic enough to break bonds in organic compounds. They can destroy cells.

(**b**) Wavelengths absorbed by photosynthetic pigments. Colors within the graph line for each kind of pigment correspond to the wavelengths that it absorbs. Three pigment classes are represented: the chlorophylls (*green* lines), carotenoids (*yellow-orange* line), and phycobilins (*purple* and *blue* lines). Peaks in the ranges of absorption correspond to the measured amount of energy absorbed and used in photosynthesis. The dashed line shows what would happen if we *combined* the amounts of energy that all the different pigments absorb. Taken together, photosynthetic pigments can absorb most of the wavelengths of visible light.

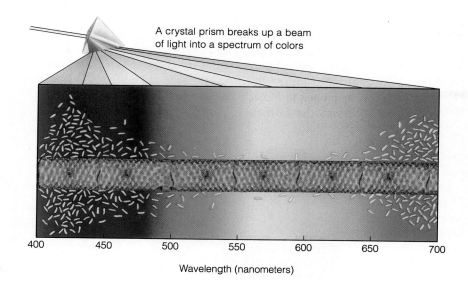

A crystal prism breaks up a beam of light into a spectrum of colors

Wavelength (nanometers)

400   450   500   550   600   650   700

Figure 7.4 T. Englemann's 1882 experiment that correlated photosynthetic activity of a strandlike green alga (*Spirogyra*) with certain wavelengths of light.

Oxygen is a by-product of photosynthesis. Many organisms use oxygen (for aerobic respiration). Among them are certain free-living bacteria that tumble about and so move through their aquatic habitat. Englemann reasoned that oxygen-requiring bacteria living in the same habitats as the alga would congregate where oxygen was being produced. He put a strand of the alga in a drop of water containing such bacteria, then mounted it on a microscope slide. He used a crystal prism to cast a spectrum of colors across the slide. Bacteria congregated mostly where violet and red wavelengths fell across the strand. Those wavelengths were most effective for photosynthesis (and oxygen production).

maple leaf in summer

maple leaf in autumn

Figure 7.5 Changes in leaf color in autumn.

Chloroplasts of mature leaves contain chlorophylls, carotenoids, and other pigments. (The carotenoids include the yellow carotenes and xanthophylls.) Intensely green leaves have an abundance of chlorophyll pigments that are masking the presence of other pigments. In many species, the gradual reduction in daylength in autumn and other factors trigger the breakdown of chlorophyll, and more colors show through.

Also in autumn, water-soluble anthocyanins accumulate in the central vacuoles of leaf cells. These pigments appear red when the plant fluids are slightly acidic, blue when the fluids are basic (alkaline), or colors in between when fluids are at intermediate pH levels. Soil conditions contribute to the differences in pH.

Birch, aspen, and other tree species always have the same characteristic color in autumn. The color of other species, including maple, ash, and sumac, varies around the country. Color even varies from one leaf to the next, depending on the pigment combinations.

7.3 LIGHT-DEPENDENT REACTIONS

Three events unfold during the **light-dependent reactions**, the first stage of photosynthesis. *First*, pigments absorb sunlight energy and give up electrons. *Second*, electron and hydrogen transfers lead to ATP and NADPH formation. *Third*, the pigments that gave up electrons in the first place get electron replacements.

Photosystems

Embedded in thylakoid membranes are light-trapping clusters called **photosystems**. There may be many thousands of them, and each includes 200 to 300 pigment molecules. Most of the pigments merely "harvest" sunlight. When they absorb a photon, one of their electrons gets boosted to a higher energy level:

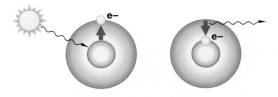

Energy boost from absorbed photon raises an electron in a pigment to a higher energy level (next "shell" away from nucleus)

Electron gives off extra energy, returns to lower energy level

When the electron returns to a lower level, it quickly gives up the added energy. The energy bounces among pigments, and a bit is lost at each bounce (as heat). Soon the energy remaining corresponds to a certain wavelength that only a few special chlorophylls can trap. Only these chlorophylls give up the electrons used in photosynthesis. As you will see, they will transfer them to an electron-accepting molecule poised at the start of a neighboring transport system.

ATP and NADPH: Loading Up Energy, Hydrogen, and Electrons

Recall that **electron transport systems** are organized sequences of enzymes and other proteins bound in a cell membrane. When excited electrons are transferred through these systems, they release their extra energy, some of which is harnessed to drive specific reactions. In thylakoid membranes, two kinds of photosystems give up electrons to different transport systems. They allow plants to make ATP by two different pathways, one cyclic and the other noncyclic.

Cyclic Pathway One pathway starts at an electron-donating chlorophyll of a "type I" photosystem. In the **cyclic pathway of ATP formation**, electrons "cycle" from chlorophyll P700, through a transport system, then back to P700. Energy linked with the electron flow drives the formation of ATP from ADP and unbound phosphate:

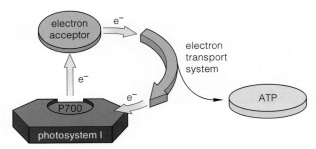

The cyclic pathway is probably the oldest means of ATP production. The first cells to use it were as tiny as existing bacteria, so their body-building programs were scarcely enormous. ATP alone would have provided enough energy to build their organic compounds. Building larger organisms requires far more organic compounds—*and vast amounts of hydrogen atoms and electrons*. Long ago, in the forerunners of multicelled plants, the cyclic pathway's machinery underwent expansion and became the basis of a more efficient ATP-forming pathway. Amazingly, this tiny bit of new machinery changed the course of evolution.

Noncyclic Pathway Today, the cyclic pathway still operates in trees, weeds, and other leafy members of the plant kingdom. But the **noncyclic pathway of ATP formation** dominates. Electrons are not cycled through this pathway. They depart (in NADPH), and electrons from *water molecules* replace them.

The pathway goes into operation when the sun's rays bombard a "type II" photosystem. Photon energy makes this photosystem's special chlorophyll (P680) give up electrons. It also triggers **photolysis**, a reaction sequence in which water molecules split into oxygen, hydrogen ions, and electrons. P680 attracts the rather unexcited electrons as replacements for the excited ones that got away (Figure 7.6).

Meanwhile, the excited electrons are transferred through a transport system—then to chlorophyll P700 of *photosystem I*. Electrons arriving at P700 haven't lost all of their extra energy. Because photons are also bombarding P700, the incoming energy boosts electrons to a higher energy level that allows them to enter a second transport system. One enzyme of this system has a helper—$NADP^+$. This coenzyme picks up two electrons and a hydrogen ion, and so becomes NADPH. It carts hydrogen and electrons to sites where organic compounds are built.

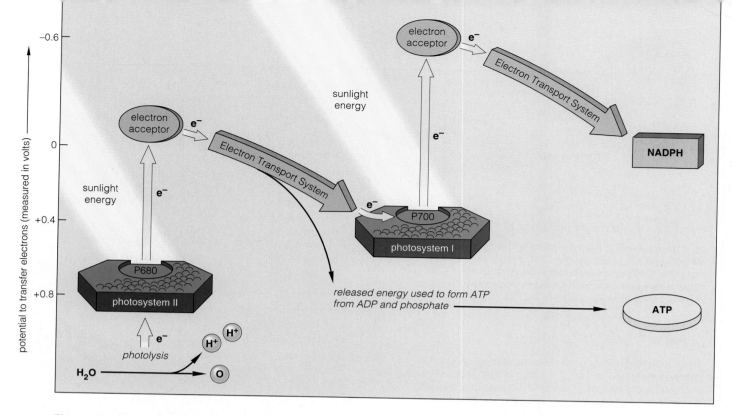

Figure 7.6 Noncyclic pathway of ATP formation, which also yields NADPH. Electrons derived from the splitting of water molecules (photolysis) travel through two photosystems. By working together, these boost electrons to an energy level that is high enough to lead to NADPH formation.

Figure 7.7 Visible evidence of photosynthesis—oxygen emerging from leaves of *Elodea*, an aquatic plant.

The Legacy—A New Atmosphere

On sunny days, on the surfaces of aquatic plants, you can see bubbles of oxygen. Figure 7.7 shows an example of this. The oxygen is a by-product of the noncyclic pathway of photosynthesis. As described in Chapter 21, this pathway may have evolved more than 2 billion years ago. At first, the oxygen simply dissolved in seas, lakes, wet mud, and other bacterial habitats. By about 1.5 billion years ago, however, large amounts of dissolved oxygen were escaping into what had been an oxygen-free atmosphere. Its accumulation changed the atmosphere forever. And it made possible aerobic respiration, the most efficient pathway for extracting energy

from organic compounds. The emergence of the non-cyclic pathway ultimately allowed you and all other animals to be around today, breathing the oxygen that helps keep your cells alive.

In the light-dependent reactions, energy from the sun drives the formation of ATP (which carries energy) and NADPH (which carries hydrogen and electrons).

Oxygen, a by-product of photosynthesis, profoundly changed the early atmosphere and made aerobic respiration possible.

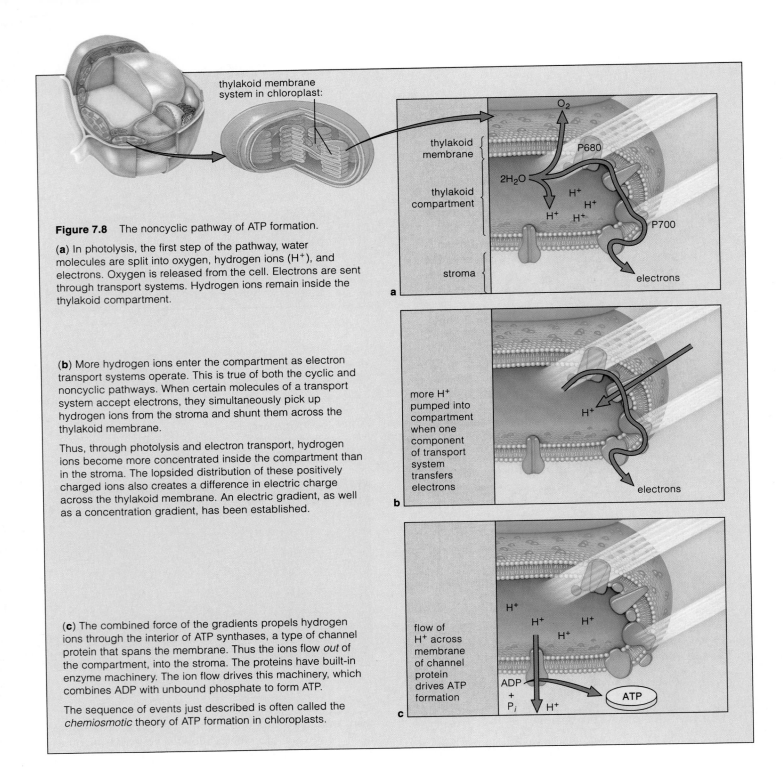

Figure 7.8 The noncyclic pathway of ATP formation.

(a) In photolysis, the first step of the pathway, water molecules are split into oxygen, hydrogen ions (H$^+$), and electrons. Oxygen is released from the cell. Electrons are sent through transport systems. Hydrogen ions remain inside the thylakoid compartment.

(b) More hydrogen ions enter the compartment as electron transport systems operate. This is true of both the cyclic and noncyclic pathways. When certain molecules of a transport system accept electrons, they simultaneously pick up hydrogen ions from the stroma and shunt them across the thylakoid membrane.

Thus, through photolysis and electron transport, hydrogen ions become more concentrated inside the compartment than in the stroma. The lopsided distribution of these positively charged ions also creates a difference in electric charge across the thylakoid membrane. An electric gradient, as well as a concentration gradient, has been established.

(c) The combined force of the gradients propels hydrogen ions through the interior of ATP synthases, a type of channel protein that spans the membrane. Thus the ions flow *out* of the compartment, into the stroma. The proteins have built-in enzyme machinery. The ion flow drives this machinery, which combines ADP with unbound phosphate to form ATP.

The sequence of events just described is often called the *chemiosmotic* theory of ATP formation in chloroplasts.

A Closer Look at ATP Formation in Chloroplasts

As you may have noticed, we've saved the trickiest question for last. *How*, exactly, does ATP form during the noncyclic (and cyclic) pathways?

As Figure 7.8 shows, when electrons flow through the membrane-bound transport systems, they pick up hydrogen ions (H$^+$) outside the membrane and dump them into the thylakoid compartment. This sets up H$^+$ concentration and electric gradients across the membrane. Hydrogen ions that were split away from water molecules increase the gradients. The ions respond by flowing out through the interior of channel proteins that span the membrane. Energy associated with the flow drives the binding of unbound phosphate to ADP, the result being ATP.

7.4 LIGHT-INDEPENDENT REACTIONS

The **light-independent reactions** are the "synthesis" part of photosynthesis. ATP molecules deliver the required energy for the reactions. NADPH molecules deliver the required hydrogen and electrons. Carbon dioxide (CO_2) in the air around photosynthetic cells provides the carbon and oxygen.

We say the reactions are light-independent because they don't depend directly on sunlight. They can proceed even in the dark, as long as ATP and NADPH are available.

Capturing Carbon

Let's track a CO_2 molecule that diffuses into the air spaces inside a sow thistle leaf and ends up next to a photosynthetic cell (Figure 7.2d). From there, it diffuses across the plasma membrane and on into the stroma of a chloroplast. The light-independent reactions start when the carbon atom of the CO_2 becomes attached to **RuBP** (ribulose bisphosphate), a molecule with a backbone of five carbon atoms. This is called **carbon dioxide fixation**.

Building the Glucose Subunits

Attaching carbon to RuBP is the first step of a cyclic pathway that produces a sugar phosphate molecule *and* regenerates the RuBP. The pathway is named the **Calvin-Benson cycle**, in honor of its discoverers. A specific enzyme catalyzes each step. For our purposes, we can simply focus on the carbon atoms of the pathway's substrates, intermediates, and end products. Figure 7.9 shows the carbon atoms as red circles.

The attachment of a carbon atom to RuBP produces an unstable six-carbon intermediate. This splits into two molecules of **PGA** (phosphoglycerate), each with a three-carbon backbone. ATP donates a phosphate group to each PGA. NADPH donates hydrogen and electrons to the resulting intermediate, thereby forming **PGAL** (phosphoglyceraldehyde).

The reaction steps just outlined proceed not once but *six* times. In other words, six CO_2 molecules are fixed, and twelve PGAL molecules are produced. Most of the PGAL becomes rearranged to form new RuBP molecules, which can be used to fix more carbon. But *two* of the PGAL combine, forming a six-carbon sugar phosphate. When a sugar has a phosphate group attached, it is primed for further reaction (refer to Figure 7.9).

The Calvin-Benson cycle produces enough RuBP molecules to replace the ones used in carbon dioxide fixation. The ADP, NADP$^+$, and phosphate leftovers

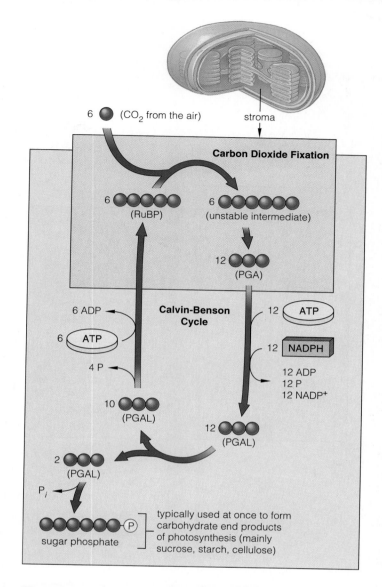

Figure 7.9 Summary of the light-independent reactions of photosynthesis. Carbon atoms of the key molecules are shown in red. All of the intermediates have one or two phosphate groups attached. However, for simplicity, only the phosphate on the resulting sugar phosphate is shown.

diffuse back to sites of the light-dependent reactions and can be converted once more to NADPH and ATP. The sugar phosphate formed in the cycle can serve as a building block for sucrose, starch, or cellulose—the plant's main carbohydrates. Synthesis of these large organic compounds by other pathways marks the conclusion of the light-independent reactions.

During the Calvin-Benson cycle, carbon is "captured" from carbon dioxide, a sugar phosphate forms in reactions that require ATP and NADPH, and RuBP (needed to capture the carbon) is regenerated.

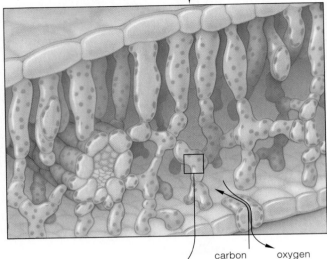

7.5 THE REACTIONS, START TO FINISH

Be sure to look carefully at Figure 7.10. It summarizes the key reactants, intermediates, and products of both the light-dependent and light-independent reactions of photosynthesis.

During daylight hours, photosynthetic cells convert the newly formed sugar phosphates to sucrose or starch. Of all plant carbohydrates, sucrose is the most easily transportable, and starch is the most common storage form. The cells convert excess PGAL to starch

Figure 7.10 Summary of the main reactants, intermediates, and products of photosynthesis, corresponding to the equation:

$$12H_2O + 6CO_2 \xrightarrow{\text{sunlight}} 6O_2 + C_6H_{12}O_6 + 6H_2O$$

Starting with the light-dependent reactions, the splitting of twelve water molecules yields twenty-four electrons and six molecules of O_2. For every four electrons, three ATP and two NADPH form.

During the light-independent reactions, *each turn* of the Calvin-Benson cycle requires one CO_2, three ATP, and two NADPH. Each sugar phosphate that forms has a backbone of six carbon atoms—so its formation requires six turns of the cycle.

carbon dioxide diffuses in

oxygen diffuses out

one chloroplast from one photosynthetic cell within a leaf

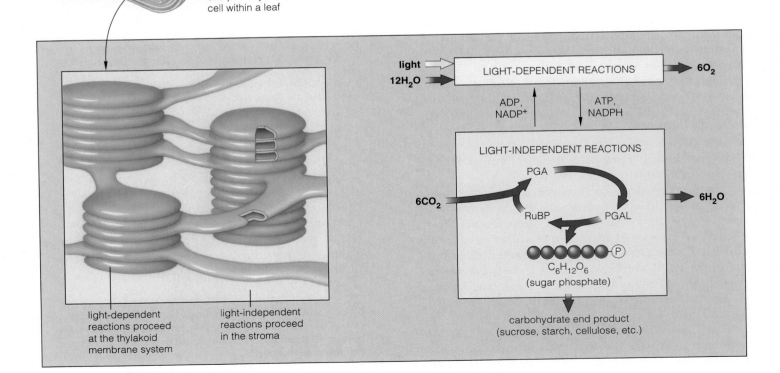

light-dependent reactions proceed at the thylakoid membrane system

light-independent reactions proceed in the stroma

Focus on the Environment

Pastures of the Seas

Drifting through the surface waters of the world ocean are uncountable numbers of single cells. You can't see them without a microscope—a row of 7 million cells of one species would be less than a quarter-inch long. Yet are they abundant! In some parts of the world, a cup of seawater may hold 24 million cells of one species, and that doesn't even include all the cells of *other* aquatic species.

Many of the drifters are photosynthetic bacteria, protistans, and plants. Together they are the pastures of the seas, the food base for diverse consumers. The pastures "bloom" in spring, when waters become warmer and enriched with nutrients churned up from the deep by winter currents. Then, populations burgeon as cells divide again and again. Biologists had no idea that the number of cells and their distribution were so mind-boggling until satellites provided photographs from space. For example, the satellite image in Figure *b* shows a springtime bloom in the North Atlantic stretching from North Carolina all the way to Spain!

Those single cells have enormous impact on the world's climate. As they collectively photosynthesize, they sponge up nearly half the carbon dioxide we humans release each year (as when we burn fossil fuels and forests). Without them, carbon dioxide in the air would accumulate more rapidly and possibly accelerate global warming, as described in Chapter 49. If our planet warms too much, vast coastal regions may become submerged, and the current food-producing nations may be hit hard.

Amazingly, we daily dump industrial wastes, fertilizers, and raw sewage into the ocean and alter the living conditions for those drifting cells. How much of that noxious chemical brew can they tolerate?

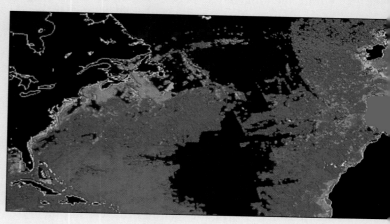

a Photosynthetic activity in winter. (In this color-enhanced image, red-orange shows where chlorophyll concentrations are greatest.)

b Photosynthetic activity in spring.

also. They briefly store the starch as grains in the stroma (you can see one of these in Figure 7.2*e*). After the sun goes down, the cells convert their starch to sucrose, for export to other living cells in leaves, stems, and roots. Ultimately, the products and intermediates of photosynthesis end up as energy sources and building blocks for all of the lipids, amino acids, and other organic compounds required for plant growth, survival, and reproduction.

When studying Figure 7.10, remind yourself of how photosynthesis "fits" in the world of living things. Think of the mind-boggling numbers of single-celled and multicelled photosynthetic organisms in which the reactions are proceeding at this very moment, on land and in the sunlit waters of the earth. The sheer volume of the reactant and product molecules that the photosynthesizers deal with may surprise you. The *Focus* essay above is a case in point.

7.6 FIXING CARBON— SO NEAR, YET SO FAR

If light intensity, air temperature, rainfall, and soil composition were uniform all year long, everywhere in the world, then maybe the photosynthetic reactions would proceed in exactly the same way in all plants. However, environments obviously differ—and so do the details of photosynthesis among plants that have evolved in different places. A brief comparison of two different carbon-fixing adaptations to stressful environmental conditions will be enough to emphasize this point.

Fixing Carbon Twice, In Two Cell Types

Think of a Kentucky bluegrass plant, with its narrow, blade-shaped leaves. As is true of all plants, its growth depends on CO_2 uptake. Athough CO_2 is plentiful in the air, it is not always abundantly available to photosynthetic cells *inside the leaves*. Leaves have a waxy cover that restricts water loss. Water escapes mainly through stomata (singular, stoma), which are tiny openings across the leaf surface. Most CO_2 diffuses into the leaves, and most O_2 diffuses out through these openings.

On hot, dry days, all plants close their stomata and so conserve water. But when they do, CO_2 can't diffuse into their leaves. Meanwhile, photosynthetic cells are busy, so oxygen builds up in each leaf. The stage is set for a wasteful process called "photorespiration." By this process, oxygen instead of CO_2 becomes attached to the RuBP used in the Calvin-Benson cycle, with different results:

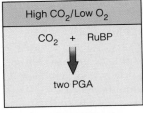
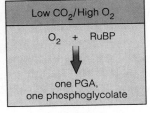

High CO_2/Low O_2	Low CO_2/High O_2
CO_2 + RuBP	O_2 + RuBP
↓	↓
two PGA	one PGA, one phosphoglycolate
Calvin-Benson cycle predominates	photorespiration predominates

Formation of sugar phosphates depends on PGA. When photorespiration wins out, less PGA forms—and the bluegrass plant's capacity for growth suffers.

Kentucky bluegrass and many other species are known as **C3 plants**, because the three-carbon PGA is the first intermediate formed by carbon fixation. By contrast, when corn, crabgrass, sugarcane, and many other plants fix carbon, the resulting intermediate is not PGA. It is the four-carbon oxaloacetate (Figure 7.11). Hence the name, **C4 plants**.

Figure 7.11 C4 pathway. (**a**) Internal structure of a leaf from corn (*Zea mays*), a typical C4 plant.

Photosynthetic mesophyll cells (*light green*) surround bundle-sheath cells (*dark green*). These in turn surround veins, the transport tubes that carry water into leaves and photosynthetic products away from them.

(**b**) A carbon-fixing system precedes the Calvin-Benson cycle in C4 plants.

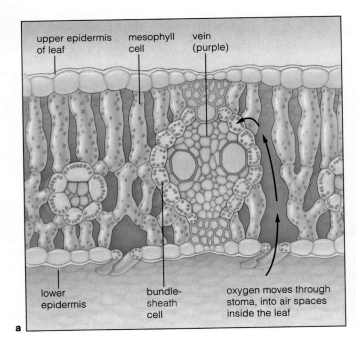
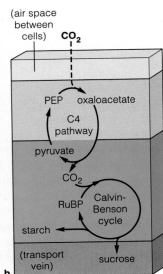

C4 plants maintain adequate amounts of CO_2 inside their leaves even though they, too, close stomata on hot, dry days. They fix CO_2 not once but twice, in two different types of photosynthetic cells. *Mesophyll* cells have first crack at the CO_2 in the leaf and use it to form oxaloacetate. It is a temporary fix. The oxaloacetate is quickly transferred to *bundle-sheath* cells that wrap around every vein in C4 leaves. In those cells, CO_2 is released and fixed again—in the Calvin-Benson cycle (Figure 7.11).

With their more efficient carbon-fixing mechanism, C4 species get by with tinier stomata—and so lose less water—than C3 species. Generally speaking, C4 plants are better adapted where temperatures are highest during the growing season.

For example, of all plant species that evolved in Florida, 80 percent are C4 plants—compared to 0 percent in Manitoba, Canada. Kentucky bluegrass and other C3 species have an advantage where temperatures drop below 25°C; they are less sensitive to cold. Mix C3 and C4 species from different regions in a garden, and one or the other kind will do better at least part of the year. That is why a lawn of Kentucky bluegrass does well during cool spring weather in San Diego, only to be overwhelmed during the hot summer by a C4 plant—crabgrass.

Figure 7.12 Giant saguaro (*Carnegiea gigantea*) and prickly pear (*Opuntia*) growing in the Sonoran desert of Arizona. Both kinds of cacti have reduced leaves, protective spines, and fleshy stems that function in photosynthesis as well as in water storage.

Fixing and Storing Carbon by Night, Using It by Day

We find a different carbon-fixing adaptation in deserts and other severely dry environments. Consider the **succulents**—plants with juicy, water-storing tissues and very thick surface layers that restrict water loss. They include cacti, such as the giant saguaro and prickly pear of the Sonoran desert (Figure 7.12). Like many flowering plants, they cannot open their stomata during the day without losing precious water. They open them and fix carbon dioxide only *at night*. Their cells store the resulting intermediate in the central vacuole, then use it in photosynthesis *the next day*—when stomata close.

Species that show this adaptation are called **CAM plants** (a blessedly shorter version of the name Crassulacean Acid Metabolism). Unlike C4 species, CAM plants do not fix carbon in separate cells. They fix it at different times.

During murderously prolonged droughts, when most plants wither and die, some CAM plants survive by keeping their stomata closed even at night. They get carbon by repeatedly fixing the carbon dioxide that forms during aerobic respiration. Not that much forms, but it is enough to allow these plants to maintain very low rates of metabolism.

As you may have deduced, CAM plants grow slowly. Try growing a cactus in Seattle or some other place with a mild climate and it will compete poorly with C3 and C4 plants.

As is true of other plant characteristics, modifications in the basic photosynthetic machinery are adaptations to specific kinds of environments.

7.7 CHEMOSYNTHESIS

Photosynthesis is so pervasive, it sometimes is easy to overlook other, less common energy-acquiring routes. Bacteria that are classified as chemoautotrophs obtain energy not from sunlight but rather by pulling away hydrogen and electrons from ammonium ions, iron or sulfur compounds, and other inorganic substances. Many influence the global cycling of nitrogen and other vital elements through the biosphere. We will return later to their environmental effects. In this unit, we turn next to pathways by which energy is released from glucose and other biological molecules—the chemical legacy of autotrophs.

SUMMARY

1. Plants and other photoautotrophs use sunlight (as an energy source) and carbon dioxide (as the carbon source) for building organic compounds. Animals and other heterotrophs obtain carbon and energy from organic compounds already synthesized by plants and other autotrophs.

2. Photosynthesis is the main biosynthetic pathway by which carbon and energy enter the web of life. It consists of two sets of reactions that start with the trapping of sunlight energy ("photo") and proceed through the assembly reactions ("synthesis").

   a. In chloroplasts, the light-dependent reactions take place at the thylakoid membrane system. These reactions produce ATP and NADPH.

   b. The light-independent reactions take place in the stroma around the membrane system. They produce sugar phosphates that are used in building sucrose, starch, and other end products of photosynthesis.

3. These are the key points concerning the light-dependent reactions:

   a. Photons are absorbed by photosynthetic pigments of photosystems (light-absorbing clusters of molecules embedded in the thylakoid membrane). Photon energy drives the transfer of electrons from a specific chlorophyll to an acceptor molecule, which donates them to a transport system in the membrane.

   b. In the cyclic pathway of ATP formation, excited electrons leave the P700 chlorophyll of photosystem I, give up energy in the transport system, and return to that photosystem.

   c. In the noncyclic pathway of ATP formation, electrons from the P680 chlorophyll of photosystem II pass through a transport system, enter photosystem I, then pass through another transport system, then end up in NADPH. Also, water molecules split into hydrogen ions, oxygen, and electrons. These electrons replace the ones that P680 gives up.

   d. Operation of electron transport systems moves many hydrogen ions into the inner compartment of the thylakoid membrane. Hydrogen ions split from water molecules accumulate here also. The accumulation sets up concentration and electric gradients that drive ATP formation.

4. These are the key points concerning the light-independent reactions:

   a. ATP delivers energy and NADPH delivers hydrogen and electrons to the stroma of chloroplasts, where sugar phosphates form and are combined into starch, cellulose, and other end products.

   b. Sugar phosphates form during the Calvin-Benson cycle. This cyclic pathway begins when carbon dioxide from the air is affixed to RuBP, making an unstable intermediate that splits into two PGA. ATP donates a phosphate group to each PGA. The resulting molecule receives H^+ and electrons from NADPH to form PGAL.

   c. For every six CO_2 molecules that enter the cycle, twelve PGAL are produced. Two of those are used to produce a six-carbon sugar phosphate. The remainder are used to regenerate RuBP for the cycle.

5. Plants differ in their carbon-fixing machinery. C3 plants can fix carbon by the Calvin-Benson cycle only. C4 plants can fix carbon twice, in two cell types. CAM plants fix and store carbon at night, then use it by day.

Review Questions

1. A caterpillar chewing on a weed is speared and eaten by a bird, which in turn is eaten by a cat. Which of these organisms are autotrophs? heterotrophs? *107*

2. Summarize the photosynthesis reactions as an equation. State the key events of the light-dependent reactions, then of the light-independent reactions. *107–108*

3. Which of the substances listed accumulates in the thylakoid compartment of chloroplasts: glucose, photosynthetic pigments, hydrogen ions, fatty acids? *109*

4. Which substance is *not* required for the light-independent reactions: ATP, NADPH, RuBP, carotenoids, free oxygen, carbon dioxide, enzymes? *115*

5. Suppose a plant busily photosynthesizing is exposed to CO_2 molecules that contain radioactively labeled carbon atoms ($^{14}CO_2$). Identify the compound in which the labeled carbon will first appear: NADPH, PGAL, pyruvate, or PGA. *115*

6. How many CO_2 molecules must enter the Calvin-Benson cycle to produce one sugar phosphate? Why? *115*

7. Fill in the blanks (red lines) in the diagram on the next page. *109*

Self-Quiz (Answers in Appendix IV)

1. Molecules with a backbone of _____ serve as the main building blocks of all organisms.

2. Photosynthetic autotrophs use _____ from the air as their carbon source and _____ as their energy source.

3. In plants, light-*dependent* reactions occur at the _____ .
   a. cytoplasm   c. stroma
   b. plasma membrane   d. thylakoid membrane

4. The light-*independent* reactions occur in the _____ .
   a. cytoplasm   c. stroma
   b. plasma membrane   d. grana

5. In the light-dependent reactions, _____ .
   a. carbon dioxide is fixed
   b. ATP and NADPH form
   c. carbon dioxide accepts electrons
   d. sugar phosphates form

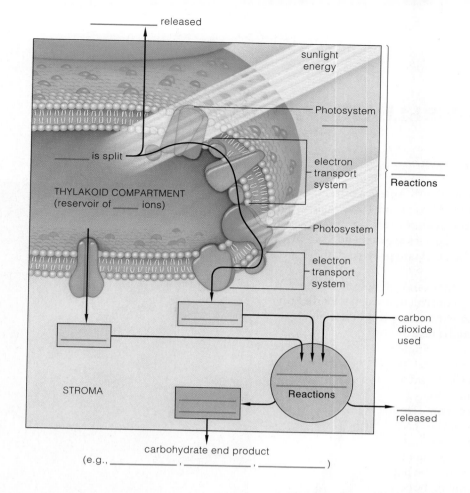

6. When a photosystem absorbs light, _____ .
   a. sugar phosphates are produced
   b. electrons are transferred to an acceptor molecule
   c. RuBP accepts electrons
   d. light-dependent reactions begin
   e. both b and d are correct

7. The Calvin-Benson cycle starts when _____ .
   a. light is available
   b. light is not available
   c. carbon dioxide is attached to RuBP
   d. electrons leave a photosystem

8. In the light-independent reactions, ATP furnishes phosphate groups to _____ .
   a. RuBP   c. PGA
   b. NADP$^+$ d. PGAL

9. Match each event in photosynthesis with its correct description.
   _____ RuBP used; PGA formed   a. cyclic pathway
   _____ ATP and NADPH used   b. noncyclic pathway
   _____ NADPH formed   c. carbon dioxide
   _____ ATP and NADPH formed   fixation
   _____ only ATP formed   d. PGAL formation
                                      e. transfer of H$^+$ and
                                         electrons to NADP$^+$

Selected Key Terms

autotroph 107
C3 plant 118
C4 plant 118
Calvin-Benson cycle 115
CAM plant 119
carbon dioxide fixation 115
carotenoid 110
chlorophyll 110
chloroplast 109
cyclic pathway of ATP
    formation 112
electron transport system 112
heterotroph 107
light-dependent reaction 112
light-independent reaction 115
noncyclic pathway of ATP
    formation 112

PGA (phosphoglycerate) 115
PGAL
    (phosphoglyceraldehyde) 115
photolysis 112
photon 110
photosystem 112
phycobilin 110
pigment 110
RuBP (ribulose
    bisphosphate) 115
stroma 109
succulent 119
thylakoid membrane
    system 109

Readings

Daviss, B. February 1992. "Going for the Green." *Discover* 13:20. Artificial systems for photosynthesis.

Hendry, George. May 1990. "Making, Breaking, and Remaking Chlorophyll." *Natural History*, pp. 36–41.

Youvan, D., and B. Marrs. 1987. "Molecular Mechanisms of Photosynthesis." *Scientific American* 256:42–50.

8 ENERGY-RELEASING PATHWAYS

The Killers Are Coming!

In 1990, "killer" bees from South America buzzed across the border between Mexico and Texas. In less than four years, they spread west to the border between Arizona and California. The bees are descended from African queen bees. When they are provoked, they can be terrifying.

In 1994, for instance, thousands of bees flew into action simply because a construction worker started up a tractor a few hundred yards away from their hive. The agitated bees entered a nearby subway station and started stinging passengers on the platform and in the trains. One person died, and a hundred others were injured.

Where did these bees come from? In the 1950s, some queen bees had been shipped from Africa to Brazil for selective breeding experiments. Why? Honeybees happen to be big business. Besides producing honey, they are rented out to pollinate commercial orchards—and they make a big difference. Put a screened cage around an orchard tree in bloom and less than 1 percent of the flowers will set fruit. Put a hive of honeybees in the same cage and 40 percent of the flowers will set fruit.

Compared with their African relatives, the bees in Brazil are sluggish pollinators and honey producers. By cross-breeding the two varieties, researchers hoped to

Figure 8.1 One of the mild-mannered honeybees buzzing in for a landing on a flower, wings beating with energy provided by ATP. If this were one of its Africanized relatives protecting a hive, possibly you would not stay around to watch the landing. Both kinds of bees look alike. How can we tell them apart? From our own biased perspective, the Africanized bees are the ones with an attitude problem.

produce a strain of mild-mannered but zippier bees. They put the local and imported bees together in artificial hives that were enclosed in nets; then they let nature take its course.

Twenty-six African queen bees escaped. That was bad enough. Then beekeepers got wind of preliminary experimental results. After learning that the first few generations of offspring were more energetic but not overly aggressive, they imported hundreds of African queens and encouraged them to mate with locals. And they set off a genetic time bomb.

Before long, African bees became established in commercial hives—and in wild bee populations. And their traits became dominant. The "Africanized" bees do everything other bees do, but they do more of it faster. Their eggs develop into adults more quickly. Adults fly more rapidly, outcompete other bees for nectar, and even die sooner.

When something disturbs their hives or swarms, Africanized bees become extremely agitated. Whereas a mild-mannered honeybee might chase an intruding animal 50 yards or so, a squadron of Africanized bees will chase it a quarter of a mile. If they catch up to it, they can sting it to death.

Doing things faster means having a nonstop supply of energy. The stomach of an Africanized bee can hold 30 milligrams of sugar-rich nectar—enough fuel to fly 60 kilometers (more than 35 miles). Compared to other kinds of bees, its flight muscle cells have larger mitochondria. And mitochondria are the organelles that specialize in releasing a great deal of energy from sugars and other organic compounds, then converting it to the energy of ATP.

In their ability to tap into the stored energy of organic compounds, Africanized bees are like all other organisms on earth. To be sure, the energy-releasing pathways differ in some details from one organism to the next. But they all require characteristic starting materials, then yield predictable products and by-products. And *all* of the energy-releasing pathways yield ATP.

In fact, throughout the biosphere, there is startling similarity in the uses to which energy and raw materials are put. *At the biochemical level, there is undeniable unity among all forms of life.* We will return to this idea in the *Commentary* that concludes this chapter.

KEY CONCEPTS

1. All organisms produce ATP by releasing energy stored in glucose and other organic compounds. Energy-releasing pathways differ from one another. But they all begin with the initial breakup of glucose by a set of reactions called glycolysis.

2. Three kinds of energy-releasing pathways dominate. Fermentation routes and anaerobic electron transport proceed in the cytoplasm. They yield a small amount of ATP for each glucose molecule metabolized. An oxygen-requiring pathway, aerobic respiration, releases far more energy from glucose. It alone proceeds in mitochondria.

3. Aerobic respiration has three stages. First (in glycolysis), glucose is broken down to pyruvate. Second, the pyruvate is broken down to carbon dioxide and water, and coenzymes pick up electrons liberated by these reactions. Third, coenzymes carry unbound hydrogen (H$^+$ ions) and electrons to a transport system. That system gives up electrons to oxygen, and its operation leads to ATP formation.

4. Over evolutionary time, photosynthesis and aerobic respiration have become linked on a global scale. Oxygen released during photosynthesis is required for the aerobic pathway. And the carbon dioxide and water released during aerobic respiration are raw materials used in building organic compounds during photosynthesis:

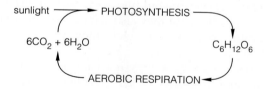

HOW CELLS MAKE ATP

Organisms stay alive by taking in energy. Plants get energy from the sun. Animals get energy secondhand, thirdhand, and so on, by eating plants and one another. Regardless of its source, energy must be put in a form that can drive the life-sustaining metabolic reactions. Energy carried by adenosine triphosphate, ATP, serves that function.

Plants make ATP during photosynthesis. They and all other organisms also make ATP by breaking covalent bonds of carbohydrates (glucose especially), lipids, and proteins. When they do, energy stored in those bonds is released, and some of it is used to drive the formation of ATP. Electron transfers of the sort described on page 102 are at the heart of the energy-releasing pathways.

Comparison of Three Types of Energy-Releasing Pathways

For most types of cells, **aerobic respiration** is the main energy-releasing pathway leading to ATP formation. The "aerobic" part of the name means that the pathway cannot be completed without oxygen. With every breath, you provide your busily respiring cells with oxygen.

Other energy-releasing pathways are "anaerobic," in that they can be completed without using oxygen. The most common are the **fermentation pathways** and **anaerobic electron transport**. Many bacteria and many protistans depend exclusively on anaerobic pathways to make ATP. The cells of your own body run on ATP from aerobic respiration, but they also can use a fermentation pathway for short periods when they are not getting enough oxygen.

All three types of energy-releasing pathways start with the same reactions, called **glycolysis**. These reactions split and rearrange each glucose molecule into two pyruvate molecules. Glycolysis proceeds in the cytoplasm, and oxygen has no role in it.

Once this first stage is over, the energy-releasing pathways differ. Most important, the aerobic pathway continues inside a mitochondrion (Figure 8.2). There, oxygen serves as the final acceptor of electrons that are released during the reactions. Anaerobic pathways end in the cytoplasm, and a substance other than oxygen is the final electron acceptor.

As we examine the three types of pathways, keep in mind that the reaction steps do not proceed all by themselves. Enzymes catalyze each reaction step, and the intermediate produced at a given step serves as a substrate for the next enzyme in the pathway.

Overview of Aerobic Respiration

Of all energy-releasing pathways, aerobic respiration produces the most ATP for each glucose molecule. Whereas fermentation has a net yield of two ATP, the aerobic route may yield thirty-six or more. If you were a bacterium, you would not require much ATP. Being large, complex, and highly active, you depend on the high yield of the aerobic route.

When glucose is the starting material, aerobic respiration can be summarized this way:

$$C_6H_{12}O_6 + 6O_2 \longrightarrow 6CO_2 + 6H_2O$$

glucose   carbon
                                             dioxide

The summary equation tells us only what the substances are at the start and finish of the pathway. In between are three stages of reactions.

Figure 8.3 outlines the reactions. The first stage, again, is glycolysis. In the second stage, which includes the **Krebs cycle**, pyruvate is broken down completely to carbon dioxide and water. Neither stage produces much ATP. But hydrogen and electrons are stripped from intermediates during both stages—and coenzymes deliver these to an electron transport system. Such systems, along with neighboring enzymes, serve as the machinery for **electron transport phosphorylation**. This third and final stage of reactions yields many ATP molecules. As it draws to a close, oxygen accepts the "spent" electrons from the transport system.

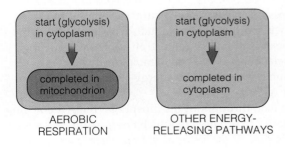

Figure 8.2 Where the main energy-releasing pathways proceed. All three start with glycolysis in the cytoplasm. Aerobic respiration alone is completed in mitochondria.

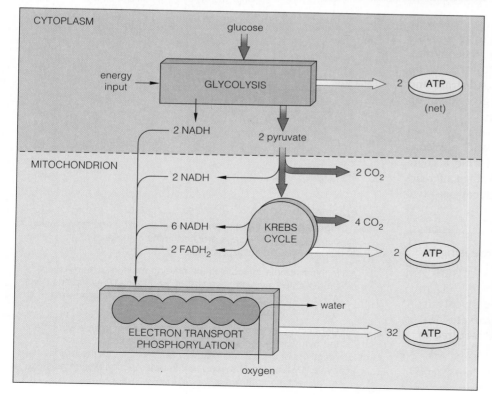

CYTOPLASM

glucose

energy input → GLYCOLYSIS → 2 ATP (net)

2 NADH

2 pyruvate

MITOCHONDRION

2 NADH ← ← 2 CO_2

6 NADH ← KREBS CYCLE → 4 CO_2

2 $FADH_2$ ← → 2 ATP

ELECTRON TRANSPORT PHOSPHORYLATION → water → 32 ATP

oxygen

Typical Energy Yield: 36 ATP

Figure 8.3 Overview of the three stages of aerobic respiration, the main energy-releasing pathway. Only this pathway delivers enough ATP to build and maintain giant redwoods and other large, multicelled organisms. Only ATP delivers enough energy to sustain birds, humans, and other highly active animals.

During the first stage (glycolysis), glucose is partially broken down to pyruvate. During the second stage, which includes the Krebs cycle, pyruvate is broken down to carbon dioxide. In these two stages, hydrogen and electrons stripped from intermediates are loaded onto coenzymes (NAD^+ and FAD).

During the final stage, electron transport phosphorylation, the loaded coenzymes (NADH and $FADH_2$) give up hydrogen and electrons. As the electrons pass through transport systems, energy is released that indirectly drives ATP formation. Oxygen accepts the electrons at the end of the transport system.

From start (glycolysis) to finish, the aerobic pathway typically has a net energy yield of thirty-six ATP for every glucose molecule.

8.2 GLYCOLYSIS: FIRST STAGE OF THE ENERGY-RELEASING PATHWAYS

Recall, from Figure 3.4, that glucose is a simple sugar with a backbone of six carbon atoms. In glycolysis, glucose (or some other carbohydrate) present in the cytoplasm is partially broken down to **pyruvate**, a molecule with a backbone of three carbon atoms.

We can think about the glucose backbone in this simplified way:

As Figure 8.4 indicates, the first steps of glycolysis are energy-requiring. They proceed only when two ATP molecules each donate energy to the glucose backbone by transferring a phosphate group to it. Such transfers are called phosphorylations. The attachments cause the backbone to split apart into two molecules of PGAL (phosphoglyceraldehyde).

PGAL formation marks the start of the energy-releasing steps of glycolysis. Each PGAL is converted to an unstable intermediate that gives up a phosphate group to ADP, forming ATP. The next intermediate in the sequence does the same thing. Thus, a total of four ATP molecules have been formed by **substrate-level phosphorylation**—the direct transfer of a phosphate group from a substrate of the reactions to ADP. Remember, though, two ATP were invested to start the reactions. So the net energy yield is only two ATP.

Meanwhile, hydrogen atoms and electrons that were released from each PGAL are transferred to an enzyme helper—the coenzyme NAD$^+$. Like other coenzymes, NAD$^+$ is reusable. As you read on page 101, it picks up hydrogen and electrons stripped from a substrate and so becomes NADH. When the coenzyme gives them up at a different site, it becomes NAD$^+$ again.

In sum, glycolysis converts a bit of the energy stored in glucose to ATP energy. Hydrogen and electrons stripped from glucose have been loaded onto a coenzyme, and these have roles in the next stage of reactions. So do the end products of glycolysis—two molecules of pyruvate.

---

1. Aerobic respiration and other energy-releasing pathways start with glycolysis, a series of reactions that partially break down a glucose molecule.

2. Two NADH and four ATP form during glycolysis, which ends with the formation of two pyruvate molecules. After subtracting the two ATP required to start the reactions, the net energy yield of glycolysis is two ATP.

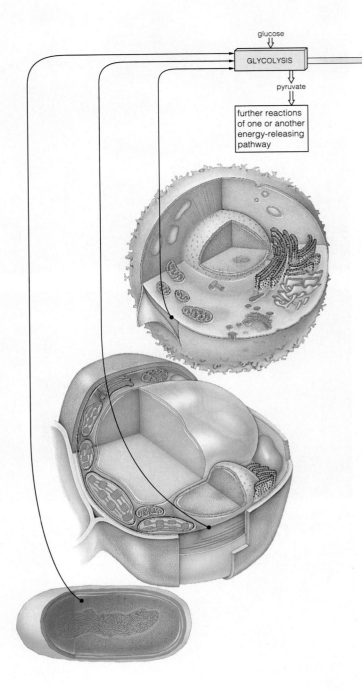

Figure 8.4 Glycolysis, first stage of the main energy-releasing pathways. The reactions proceed in the cytoplasm of all prokaryotic and eukaryotic cells.

In this example, glucose is the starting material. The reactions produce two pyruvate, two NADH, and four ATP molecules. Two ATP are invested to start glycolysis, so the net energy yield of glycolysis is two ATP.

Depending on the cell type and on conditions in its environment, the pyruvate molecules may be used in the second set of reactions of the aerobic pathway, which includes the Krebs cycle. Or they may be used in different reactions, such as those of fermentation pathways.

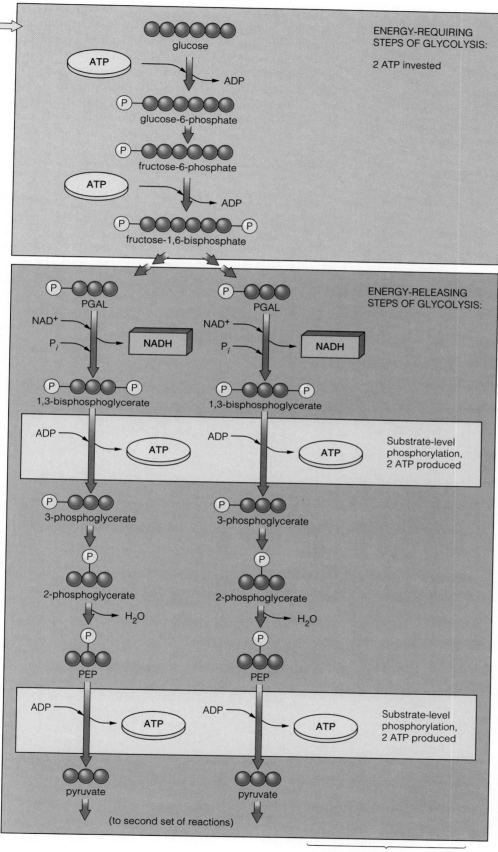

ENERGY-REQUIRING
STEPS OF GLYCOLYSIS:

2 ATP invested

glucose

ATP → ADP

glucose-6-phosphate

fructose-6-phosphate

ATP → ADP

fructose-1,6-bisphosphate

ENERGY-RELEASING
STEPS OF GLYCOLYSIS:

PGAL   PGAL

NAD⁺   NAD⁺
P$_i$ → NADH   P$_i$ → NADH

1,3-bisphosphoglycerate   1,3-bisphosphoglycerate

ADP → ATP   ADP → ATP   Substrate-level
                                       phosphorylation,
                                       2 ATP produced

3-phosphoglycerate   3-phosphoglycerate

2-phosphoglycerate   2-phosphoglycerate

→ H₂O   → H₂O

PEP   PEP

ADP → ATP   ADP → ATP   Substrate-level
                                       phosphorylation,
                                       2 ATP produced

pyruvate   pyruvate

(to second set of reactions)

NET ENERGY YIELD: 2 ATP

a *Glycolysis starts with an energy investment of two ATP.* First, enzyme action promotes the transfer of a phosphate group from ATP to glucose, which has a backbone of six carbon atoms. With this transfer, the glucose molecule becomes slightly rearranged.

b Enzyme action promotes the transfer of a phosphate group from another ATP to the rearranged molecule.

c The resulting fructose-1,6-bisphosphate molecule splits at once into two molecules, each with a three-carbon backbone. We can call these two PGAL.

d During enzyme-mediated reactions, two NADH form after each PGAL gives up two electrons and a hydrogen atom to two NAD⁺. Each PGAL also combines with inorganic phosphate (P$_i$) present in the cytoplasm, then donates a phosphate group to ADP.

e *Thus two ATP have formed by the direct transfer of phosphate from two intermediate molecules that serve as substrates in the reactions.* With this formation of two ATP, the original energy investment of two ATP is paid off.

f In the next two enzyme-mediated reactions, each of the two intermediate molecules releases a hydrogen atom and an —OH group, which combine to form water.

g The resulting intermediates (two molecules of 3–phosphoenolpyruvate, or PEP) are rather unstable. Each gives up a phosphate group to ADP. *Once again, two ATP have formed by substrate-level phosphorylation.*

h Thus the net energy yield from glycolysis is two ATP for each glucose molecule entering the reactions. The end products of glycolysis are two molecules of pyruvate, each with a three-carbon backbone.

8.3 COMPLETING THE AEROBIC PATHWAY

Preparatory Steps and the Krebs Cycle

Suppose pyruvate, the product of glycolysis, leaves the cytoplasm and enters a mitochondrion, as shown in Figure 8.5. Only the aerobic pathway continues inside this organelle, and two more stages are required to complete it. During the *second* stage, a bit more ATP forms. Carbon and oxygen atoms depart from pyruvate, in carbon dioxide and water. And coenzymes accept hydrogen and electrons released during the reactions.

Take a look at Figure 8.6. During a few preparatory steps, a carbon atom is stripped from each pyruvate molecule, leaving an acetyl group that gets picked up by a coenzyme (forming acetyl-CoA). The acetyl group becomes attached to oxaloacetate, the point of entry into the Krebs cycle. The name of this cyclic pathway honors Hans Krebs, who began working out its details in the 1930s. Notice that three carbon atoms enter the second stage of reactions (as a pyruvate backbone) and three leave (in three carbon dioxide molecules) during the preparatory reactions and the cycle proper.

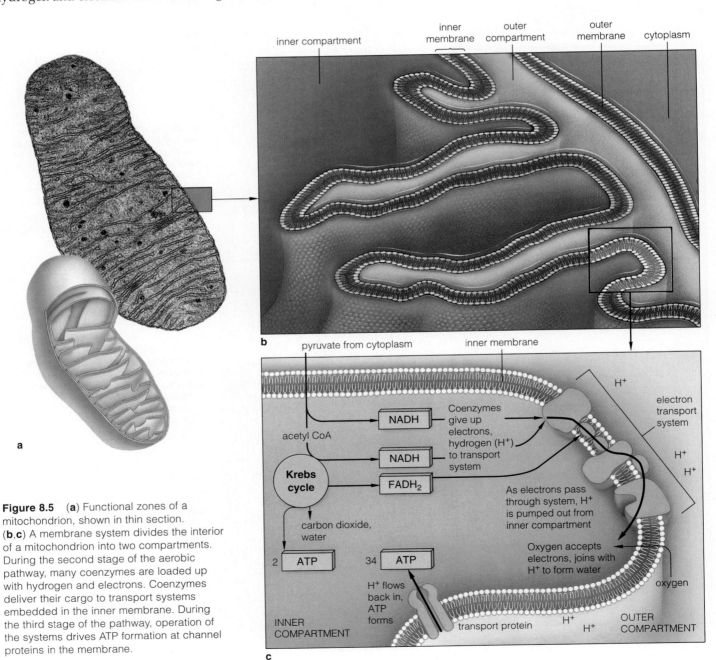

Figure 8.5 (**a**) Functional zones of a mitochondrion, shown in thin section. (**b,c**) A membrane system divides the interior of a mitochondrion into two compartments. During the second stage of the aerobic pathway, many coenzymes are loaded up with hydrogen and electrons. Coenzymes deliver their cargo to transport systems embedded in the inner membrane. During the third stage of the pathway, operation of the systems drives ATP formation at channel proteins in the membrane.

Functions of the Second Stage

The second-stage reactions serve three functions. First, hydrogen and electrons are transferred to NAD^+ and FAD, forming NADH and $FADH_2$. Second, substrate-level phosphorylations produce two ATP. Third, intermediates are rearranged into oxaloacetate. (Cells have only so much oxaloacetate, and it must be regenerated to keep the cyclic reactions going.)

The two ATP do not add much to the small yield from glycolysis. But the reactions also load up *many* coenzymes with hydrogen and electrons that can be used during the third stage of the aerobic pathway:

Glycolysis:	2 NADH
Pyruvate conversion that precedes Krebs cycle:	2 NADH
Krebs cycle:	2 $FADH_2$ + 6 NADH
Total coenzymes sent to third stage:	2 $FADH_2$ + 10 NADH

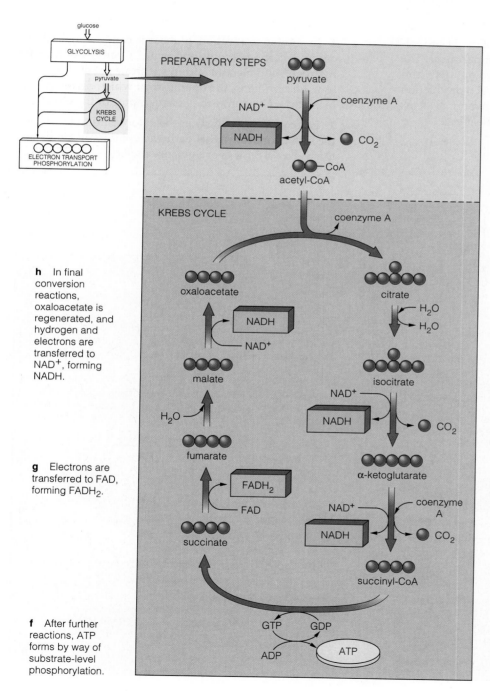

a A pyruvate molecule enters a mitochondrion. It undergoes preparatory conversions before entering cyclic reactions (Krebs cycle).

b First, the pyruvate is stripped of a functional group (COO^-), which departs as CO_2. Next, it gives up hydrogen and electrons to NAD^+, forming NADH. A coenzyme joins with the two-carbon fragment, forming acetyl-CoA.

c The acetyl-CoA is transferred to oxaloacetate, a four-carbon compound that is the point of entry into the Krebs cycle. The result is citrate, with a six-carbon backbone.

d Citrate enters conversion reactions in which a COO^- group departs (as CO_2). Also, hydrogen and electrons are transferred to NAD^+, forming NADH.

e Another COO^- group departs (as CO_2) and another NADH forms. *At this point, three carbon atoms have been released, balancing out the three that entered the mitochondrion (in pyruvate).*

h In final conversion reactions, oxaloacetate is regenerated, and hydrogen and electrons are transferred to NAD^+, forming NADH.

g Electrons are transferred to FAD, forming $FADH_2$.

f After further reactions, ATP forms by way of substrate-level phosphorylation.

Figure 8.6 Second stage of aerobic respiration: the Krebs cycle *and* a few reactions that immediately precede it. For each three-carbon pyruvate molecule that enters the cycle, three CO_2, one ATP, four NADH, and one $FADH_2$ are formed. The steps shown proceed twice (remember, the glucose molecule was broken down initially to two pyruvate molecules).

Third Stage of the Aerobic Pathway— Electron Transport Phosphorylation

ATP production goes into high gear in the third stage of the aerobic pathway. The production machinery runs on electrons and unbound hydrogen (that is, H^+ ions) delivered by coenzymes. Electron transport systems and neighboring channel proteins serve as the machinery. These are embedded in the inner membrane that divides the mitochondrion into two compartments (Figure 8.7).

As electrons are transferred through the transport systems, H^+ ions are tossed into the outer compartment. This sets up H^+ concentration and electric gradients across the membrane. Channel proteins allow H^+ to follow the gradients, back into the inner compartment. The flow drives the formation of ATP from ADP and unbound phosphate. Free oxygen keeps ATP production going. It withdraws electrons from transport systems and combines with H^+ to form water molecules.

Summary of the Energy Harvest

In many cells, the third stage of the aerobic pathway produces thirty-two ATP. Add these to the net yield from the preceding stages, and we have a total energy harvest of thirty-six ATP from one glucose molecule. Think of this as a typical yield only. The amount depends on cellular conditions, as when molecules of some intermediate are required elsewhere and pulled out of a reaction sequence. As Figure 8.8 indicates, the amount also depends on how a given cell uses the electrons from NADH that formed in the cytoplasm.

---

1. In aerobic respiration, glucose is completely broken down to carbon dioxide and water. Coenzymes transfer hydrogen and electrons from substrates to electron transport systems, the operation of which drives ATP formation. Oxygen is the final acceptor of the electrons.

2. From start (glycolysis in the cytoplasm) to finish (in the mitochondrion), this pathway commonly yields thirty-six ATP for every glucose molecule.

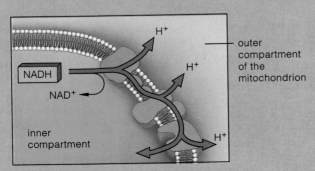

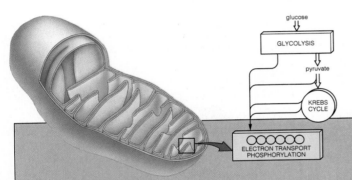

Figure 8.7 Electron transport phosphorylation, the third and final stage of aerobic respiration. The reactions proceed at transport systems and at channel proteins (ATP synthases) that are embedded in the inner mitochondrial membrane. Each transport system consists of enzymes, cytochromes, and other proteins that act in sequence.

The inner membrane divides the mitochondrion into two compartments. The third-stage reactions start in the inner compartment, when NADH and $FADH_2$ give up hydrogen (as H^+ ions) and electrons to the transport system. The electrons are transferred through the system, but the ions are left behind—in the outer compartment:

Soon there is a higher concentration of H^+ in the outer compartment than in the inner one. Concentration and electric gradients now exist across the membrane. The ions follow the gradients and flow across the membrane, through the interior of the channel proteins. Energy associated with the flow drives the formation of ATP from ADP and unbound phosphate (hence the name, electron transport *phosphorylation*):

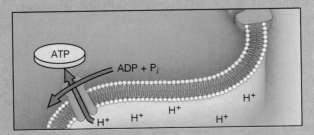

Do these events sound familiar? They should. ATP forms in much the same way in chloroplasts (Figure 7.8). The *chemiosmotic theory* that concentration and electric gradients across a membrane drive ATP formation applies also to mitochondria—even though ions flow in the opposite direction.

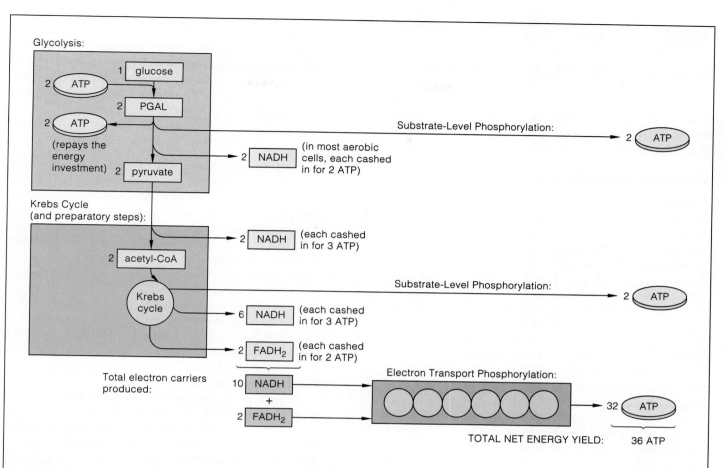

Figure 8.8 Summary of the harvest from one glucose molecule sent through the energy-releasing pathway of aerobic respiration. Commonly, 36 ATP form for each glucose molecule. By contrast, the anaerobic pathways would require 18 glucose molecules to produce as many ATP.

Remember, glucose has more energy (stored in more covalent bonds) than carbon dioxide or water. When it is broken down completely to those more stable end products, about 686 kilocalories of energy are released. Most escapes (as heat energy), but about 7.5 kilocalories are conserved in each ATP molecule. Thus, when 36 ATP form during the breakdown of one glucose molecule, the energy-conserving efficiency of this pathway is (36)(7.5)/(686), or 39 percent.

Bear in mind, the net energy yield is not *always* 36 ATP. The outcome depends not only on cellular conditions, but also on the type of cell and its electron-shuttling mechanisms. Two examples will make the point.

First, think about what happens when NADH forms *inside* a mitochondrion (during the second stage of reactions). It delivers electrons to the highest possible point of entry into a transport system. When it does, enough H^+ can be pumped across the inner mitochondrial membrane to produce *three* ATP. When $FADH_2$ forms inside the mitochondrion, it makes its delivery at a lower point of entry into the transport system. Fewer H^+ can be pumped, and only *two* ATP can be produced.

Now think about NADH formed in the cytoplasm. It can only deliver electrons *to* the mitochondrion, not *into* it. In liver, heart, and kidney cells, a protein shuttle built into the outer membrane accepts the electrons and donates them to NAD^+ in the mitochondrion. NADH formed this way puts electrons at the top of a transport system, so three ATP form, and the overall energy harvest is 38 ATP.

More commonly, as in skeletal muscle and brain cells, *a different shuttle accepts electrons from cytoplasmic NADH.* It donates these to FAD inside the mitochondrion—so only two ATP form. In such cells, the overall energy harvest is 36 ATP.

8.4 ANAEROBIC ROUTES

So far, we have tracked the fate of a glucose molecule through the pathway of aerobic respiration. We turn now to its use as a substrate for fermentation pathways. These are anaerobic pathways; they do *not* use oxygen as the final acceptor of the electrons that ultimately drive the ATP-forming machinery.

Fermentation Pathways

Which kinds of organisms use fermentation pathways? Many are bacteria and protistans that make their homes in marshes, bogs, mud, the animal gut, canned foods, sewage treatment ponds, and other oxygen-free settings. Some kinds actually die if exposed to oxygen. Bacteria responsible for many diseases, including botulism and tetanus, are like this. Other kinds, such as the bacterial "employees" of yogurt manufacturers, are indifferent to the presence of oxygen. Still others can use oxygen, but they switch to fermentation when oxygen levels decline. Even your muscle cells do this.

As is true of aerobic respiration, glycolysis also is the first stage of the fermentation pathways. Here also, a glucose molecule is split and rearranged into two pyruvate molecules, two NADH form, and the net energy yield is two ATP. However, as Figure 8.9 shows, the reactions do not break down glucose completely to carbon dioxide and water, and they produce no more ATP beyond the yield from glycolysis. *The final steps serve only to regenerate NAD$^+$—the coenzyme that is central to the pathway's operation.*

The energy yield from fermentation is enough for many single-celled anaerobic organisms. It even helps carry some aerobic cells through times of stress. It is not enough to sustain large, multicelled organisms (this being one reason why you never will come across an anaerobic elephant).

Lactate Fermentation Figure 8.10 shows the main reaction steps of the energy-releasing pathway called **lactate fermentation**. As you can see, pyruvate itself, formed during the first stage of reactions (glycolysis), accepts hydrogen and electrons from NADH. With this transfer, pyruvate is converted to a different three-carbon compound, lactate. Sometimes lactate is called "lactic acid." However, the ionized form (lactate) is far more common in cells.

A few bacteria, including species of *Lactobacillus*, rely exclusively on this anaerobic pathway. Left to their own devices, their fermentation activities can spoil food. Yet some fermenters also have commercial uses, as when they produce yogurt from milk, and sauerkraut from raw cabbage.

Some types of animal cells also can switch to lactate fermentation for a quick ATP fix. When your demands for energy are intense but brief—say, during a short race—your muscle cells use this anaerobic pathway. They cannot do this for long—too much of glucose's stored energy would be thrown away for too little ATP. When glucose stores are depleted, muscles fatigue and lose their ability to contract.

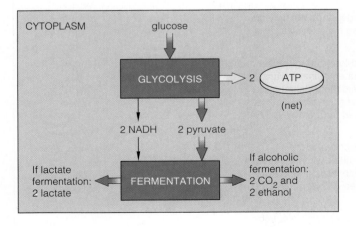

Figure 8.9 Overview of two fermentation routes. In this type of energy-releasing pathway, the net ATP yield is from the initial reactions (glycolysis).

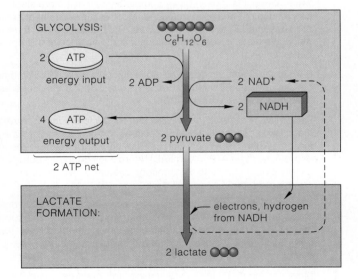

Figure 8.10 Lactate fermentation. In this anaerobic pathway, electrons end up in the reaction product (lactate).

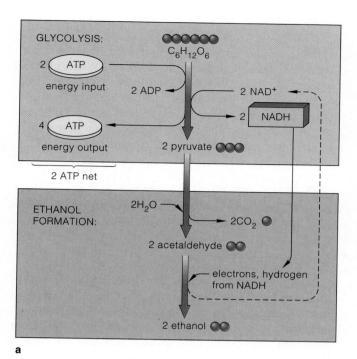

a

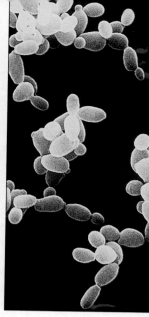

b   c

Figure 8.11 (**a**) Alcoholic fermentation. In this anaerobic pathway, an intermediate of the reactions (acetaldehyde) is the final electron acceptor, and ethanol is the end product. Yeasts, single-celled organisms, use this pathway. (**b**) One species of *Saccharomyces* lives on sugar-rich, ripened grapes. Another (**c**) makes bread dough rise.

Alcoholic Fermentation In this anaerobic pathway, each pyruvate molecule from the first stage of reactions (glycolysis) is rearranged into an intermediate form called acetaldehyde. When this intermediate accepts hydrogen and electrons from NADH, it is converted to ethanol, the end product of this pathway (Figure 8.11).

The single-celled fungi called yeasts use this pathway. And commercial bakers use certain yeasts. *Saccharomyces cerevisiae* can make bread dough rise. Bakers mix this yeast with sugar and blend the mixture into dough. As the yeast cells use the sugar, they release carbon dioxide, a gas that expands the dough (makes it "rise"). Oven heat forces the gas out, leaving a porous product.

Beer and wine manufacturers use yeasts on a large scale. Vintners use wild yeasts that live on grapes. They also use cultivated strains of *S. ellipsoideus*, which remain active until the alcohol concentration in the vats exceeds 14 percent. Wild yeasts die when it exceeds 4 percent—but 4 percent still packs a punch. Robins get drunk on the naturally fermenting berries of pyracantha shrubs. So do wild turkeys when they gobble up fermenting apples in untended orchards.

In fermentation pathways, the net energy yield of two ATP is from glycolysis. The remaining reactions simply regenerate the NAD^+ required for glycolysis.

Anaerobic Electron Transport

Although aerobic respiration and fermentation are the most common, there are other kinds of energy-releasing pathways, especially among the bacteria. As described in Chapter 48, some of these less common pathways influence the global cycling of sulfur, nitrogen, and other vital elements—and so influence the availability of nutrients for organisms everywhere.

Consider anaerobic electron transport, as employed by a variety of bacteria. Electrons stripped from an organic compound are sent through transport systems bound in the plasma membrane. The energy yield varies. An inorganic compound in the environment often serves as the final electron acceptor. While you are reading this, for example, certain anaerobic bacteria living in waterlogged soil are stripping electrons from a variety of compounds. Then they dump the electrons on sulfate (SO_4^-), leaving hydrogen sulfide (H_2S), a very foul-smelling gas. Sulfate-reducing bacteria also live in aquatic habitats that are rich in decomposed organic material. They even live deep in the ocean, around hydrothermal vents. As described in Chapter 49, these bacteria form the food production base for unique communities.

In anaerobic electron transport, an inorganic substance (but not oxygen) usually serves as the final electron acceptor.

8.5 ALTERNATIVE ENERGY SOURCES IN THE HUMAN BODY

Carbohydrate Breakdown in Perspective

So far, you've looked at what happens after a lone glucose molecule enters an energy-releasing pathway. Now you can start thinking about what cells do when they have too many or too few of these molecules.

Consider what happens after you or any other mammal finishes a meal. Your body absorbs a great deal of glucose and other small organic molecules, which your bloodstream delivers to tissues throughout your body. There, cells rapidly take up excess glucose, then "trap" it by converting it to glucose-6-phosphate (which cannot be transported across the plasma membrane). Glucose-6-phosphate, recall, is the first intermediate of glycolysis.

When your food intake exceeds cellular demands for energy, ATP-producing machinery goes into high gear. In time, the increased ATP concentration inhibits glycolysis. Now glucose-6-phosphate is diverted into a biosynthesis pathway that ends with glycogen—a storage polysaccharide composed of glucose units.

Between meals, when the input of free glucose dwindles, cells break down the glycogen to glucose-6-phosphate, and they use this in glycolysis. Liver cells do more. They convert glucose-6-phosphate back to free glucose and release it. The bloodstream delivers this glucose to energy-demanding cells of muscles and other organs that have depleted their own glycogen stores.

Don't let the preceding paragraphs lead you to believe that *all* cells squirrel away large amounts of glycogen. Liver and muscle cells maintain the largest stores. Even then, glycogen represents a mere 1 percent or so of the total stored energy in the body of an adult human. On the average, 78 percent is stored in fats and another 21 percent in proteins.

Energy from Fats

Maybe you avoid butter, ice cream, and other fatty foods, thinking it is better to fill up on carbohydrates and proteins. This is a good idea, as long as you don't stuff yourself with these organic compounds—because your body converts the excess to fats. Fat molecules, recall, have a glycerol head and one, two, or three fatty acid tails. Most of the fats stored in your body are in the form of triglycerides (with three tails). These accumulate inside the fat cells of adipose tissues, at strategic locations beneath the skin.

Between meals or during exercise, triglycerides are tapped as alternatives to glucose. At such times, enzymes in fat cells cleave the bonds holding glycerol and fatty acids together, then release the breakdown products. These enter the bloodstream. When glycerol reaches the liver, it is converted to PGAL—a key intermediate of glycolysis. Most cells can take up the circulating fatty acids. Enzymes cleave the carbon backbone of the tails and convert the fragments to acetyl-CoA, which can enter the Krebs cycle (Figures 8.6 and 8.12).

Each fatty acid tail has many more carbon-bound hydrogen atoms than glucose—so its breakdown yields much more ATP. Between meals or during sustained exercise, fatty acid conversions supply about half of the ATP that your muscle, liver, and kidney cells require.

Energy from Proteins

Eat more protein molecules than your body requires to grow and maintain itself, and your cells won't store them. Enzymes split the proteins into amino acid units, and then they remove the amino group (NH_3) from each unit. Depending on cellular conditions, the carbon backbones that remain may be converted to fats or carbohydrates. Or they may enter the Krebs cycle. There, hydrogen and electrons stripped from the carbon atoms are transferred to the cycle's coenzymes. The amino groups undergo conversions that produce urea, a nitrogen-containing waste product that is excreted from the body in urine.

In humans and other mammals, the entrance of glucose or some other organic compound into an energy-releasing pathway depends on its concentrations inside and outside the cell, and on the type of cell.

This concludes our look at aerobic respiration and other energy-releasing pathways. To gain insight into how these pathways fit into the greater picture of life's evolution and interconnectedness, take a moment to read the *Commentary* that concludes this unit.

Figure 8.12 Points of entry into the aerobic pathway for complex carbohydrates, fats, and proteins after they have been reduced to their simpler components in the human digestive system.

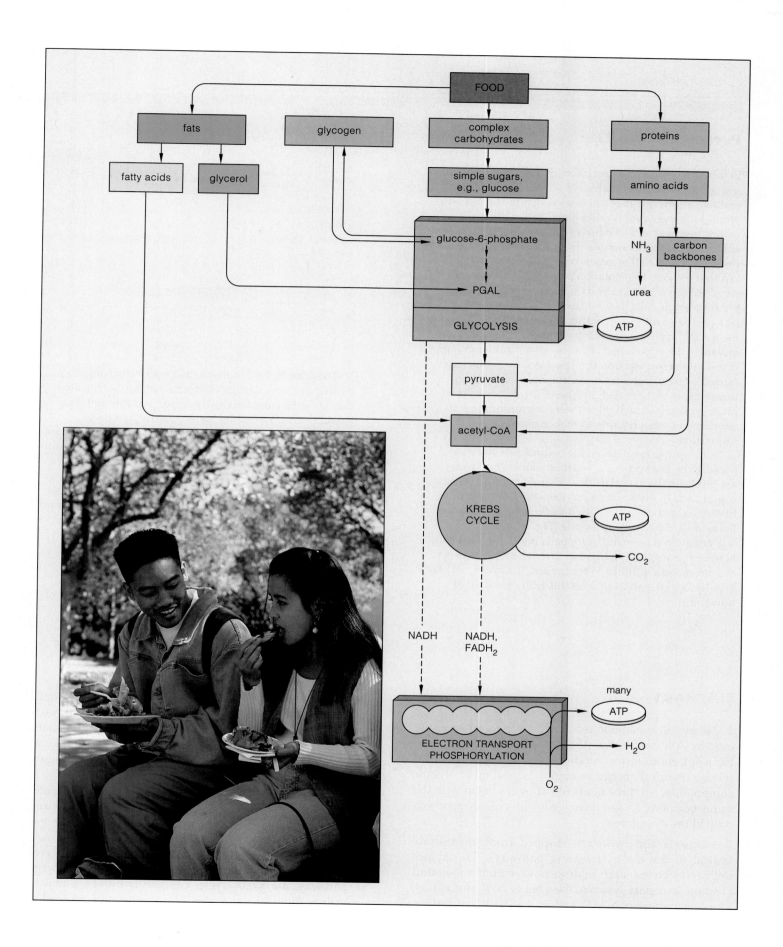

Commentary

Perspective on Life

In this unit, you read about photosynthesis and aerobic respiration—the main pathways by which cells trap, store, and release energy. Those pathways became linked on a grand scale over evolutionary time.

Life began about 3.8 billion years ago, when the atmosphere had little free oxygen. Early single-celled organisms probably made ATP by reactions similar to glycolysis. Without oxygen, fermentation pathways must have dominated. About 1.5 billion years later, oxygen-producing photosynthetic cells had emerged, and they turned out to be a profound force in evolution. Oxygen, a by-product of the noncyclic pathway of photosynthesis, began to accumulate in the atmosphere. Possibly as a result of mutations in the proteins of electron transport systems, some cells started using oxygen as an electron acceptor. In time, descendants of those fledgling aerobic cells abandoned photosynthesis. Among them were the forerunners of animals and other organisms that engage in aerobic respiration.

With aerobic respiration, life became self-sustaining, because the final products—carbon dioxide and water—are precisely the materials that are used to build organic compounds in photosynthesis! Thus the flow of carbon, hydrogen, and oxygen through the metabolic pathways of living organisms came full circle (Figure *a*).

Perhaps you have difficulty perceiving the connection between yourself—a living, intelligent being—and such remote-sounding things as energy and the cycling of carbon, hydrogen, and oxygen. Is this really the stuff of humanity?

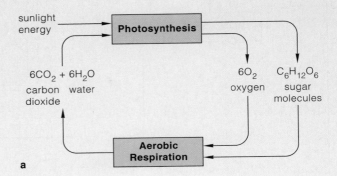

Think back, for a moment, to a water molecule. Two hydrogen atoms sharing electrons with an oxygen atom doesn't seem close to our daily lives. But through that sharing, water molecules show polarity, and they hydrogen-bond with one another. That is a beginning for the organization of lifeless matter that leads to the organization of all living things.

For now you can imagine different molecules in water. The nonpolar ones resist interaction with water; the polar ones dissolve in it. The phospholipids among them spontaneously assemble into a two-layered film. Such lipid bilayers are the basis for all cell membranes, hence all cells. From the beginning, the cell has been the fundamental *living* unit.

The essence of life is not some mysterious force. It is metabolic control. With a cell membrane to contain them,

SUMMARY

1. Nearly all metabolic reactions run on energy delivered by ATP molecules. ATP can be produced by aerobic respiration, fermentation, and other pathways that release chemical energy from glucose and other organic compounds. All three kinds of pathways begin with the same reactions, called glycolysis. Glycolysis proceeds only in the cytoplasm.

2. Electrons and hydrogen stripped from glucose are central to the energy-releasing pathways. Coenzymes deliver electrons and hydrogen to membrane-bound electron transport systems, the sites of ATP production. The coenzymes are NAD^+ and, in one pathway, FAD.

3. In aerobic respiration, oxygen is the final acceptor of electrons stripped from glucose. The pathway proceeds through three stages of reactions: glycolysis, the Krebs cycle (and preparatory steps), and electron transport phosphorylation. Its net energy yield is commonly thirty-six or more ATP molecules.

a. Glycolysis partly breaks down a glucose molecule. Two pyruvate, two NADH, and four ATP form. The net energy yield is only two ATP (two ATP had to be invested up front to get the reactions going).

b. The next stage proceeds in mitochondria. Pyruvate is converted to an intermediate that enters a cyclic pathway, the Krebs cycle. Glucose is broken down to carbon dioxide and water. Ten coenyzmes are loaded

metabolic reactions *can* be controlled. Cells can respond to energy changes and to the kinds of molecules in their environment. Their response mechanisms operate by "telling" protein molecules—enzymes—when and what to build or tear down.

And it is not some mysterious force that creates the proteins themselves. DNA, the slender double strand of heredity, has the chemical structure—*the chemical message*—that allows molecule to reproduce molecule, one generation after the next. Those DNA strands tell trillions of cells in your body how countless molecules must be built and torn apart for their stored energy.

So yes, carbon, hydrogen, oxygen, and other organic molecules represent the stuff of you, and us, and all of life. But it takes more than molecules to complete the picture. Life exists only as long as a constant flow of energy maintains its organization. Molecules are assembled into cells, cells into organisms, organisms into communities, and so on up through the biosphere. It takes energy, primarily from the sun, to maintain all these levels of organization. And energy flows through time in one direction—from organized to less organized forms. Only as long as sunlight flows into the web of life can life continue in all its rich diversity.

In short, life is no more *and no less* than a marvelously complex system of prolonging order. Sustained by energy transfusions, life continues because of a capacity for self-reproduction—a handing down of hereditary instructions in DNA. With DNA, energy and materials can be organized, generation after generation. Even with the death of

b

the individual, life is prolonged. With death, molecules are released and recycled once more, providing raw materials for new generations. In this flow of energy and cycling of material through time, each birth is affirmation of our ongoing capacity for organization, each death a renewal.

with hydrogen and electrons (eight NADH and two FADH$_2$), and two ATP form.

c. The third stage also proceeds in mitochondria. An inner membrane divides the mitochondrion into two compartments. Electron transport systems and channel proteins are embedded in this inner membrane.

d. Coenzymes deliver electrons to a transport system. Operation of the system sets up H$^+$ concentration and electric gradients across the membrane. H$^+$ follows the gradients and leaves the inner compartment, through channel proteins. Energy associated with the flow drives the formation of ATP from ADP and unbound phosphate. Oxygen withdraws electrons from the transport system and combines with H$^+$ to form water.

4. Alcoholic fermentation, lactate fermentation, and anaerobic electron transport use an organic or inorganic compound (not oxygen) as the final acceptor of electrons from the reactions.

a. The fermentation pathways start with glycolysis. Two ATP are invested, glucose is partially broken down to pyruvate, and two NADH and four ATP form.

b. In lactate fermentation, pyruvate accepts hydrogen and electrons from coenzymes and is itself rearranged into the end product, lactate.

c. In alcoholic fermentation, pyruvate is converted to an intermediate (acetaldehyde), and carbon dioxide is released. That intermediate accepts hydrogen and electrons from coenzymes and so becomes ethanol.

1. Which energy-releasing pathway yields the most ATP for each glucose molecule? *124*

2. Think of the various energy-releasing pathways. Which of their reactions occur only in cytoplasm? Which occur only in mitochondria? *124*

3. State the function of coenzymes in the energy-releasing pathways. *124*

4. Briefly describe glycolysis, the first stage of the energy-releasing pathways. *126–127*

5. Describe the two stages of aerobic respiration that follow glycolysis:
    a. the Krebs cycle (and preparatory conversions). *128–129*
    b. electron transport phosphorylation. *130*

6. Describe the functional zones of a mitochondrion. Explain where the electron transport systems and transport proteins required for ATP formation are located. *128*

7. Is the following statement true? Your muscle cells cannot function at all without oxygen. *132*

8. In fermentation, conversions following the net production of two ATP do not yield more energy. What do they accomplish? *132*

9. Cells do not use nucleic acids as an alternative energy source. Speculate why.

10. Summarize the net energy yield from one glucose molecule for aerobic respiration. *131*

1. Stored energy can be released from glucose and used to produce _____ , an energy carrier.

2. In the first stage of the main energy-releasing pathways, glucose is partly broken down to _____ .

3. ATP is _____ .
    a. a phosphate compound
    b. an energy carrier
    c. produced by all organisms
    d. all of the above

4. Which of the following is *not* produced during glycolysis?
    a. NADH   c. FAD
    b. pyruvate   d. ATP

5. Glycolysis occurs in the _____ .
    a. nucleus   c. plasma membrane
    b. mitochondrion   d. cytoplasm

6. The final acceptor of electrons stripped from glucose during aerobic respiration is _____ .
    a. water   c. oxygen
    b. hydrogen   d. NADH

7. For the aerobic pathway, electron transport systems are located in the _____ .
    a. cytoplasm
    b. inner mitochondrial membrane
    c. outer mitochondrial compartment
    d. stroma

8. The flow of _____ through channel proteins drives the formation of ATP from ADP and phosphate.
    a. electrons   c. NADH
    b. hydrogen ions   d. $FADH_2$

9. Match the events with the metabolic reaction.
    ____ glycolysis
    ____ fermentation
    ____ Krebs cycle
    ____ electron transport phosphorylation

    a. ATP, NADH, and CO_2 form
    b. glucose to two pyruvate
    c. NAD^+ regenerated, a net yield of two ATP
    d. H^+ flows through channel proteins, ATP forms

aerobic respiration *124*
anaerobic electron transport *124*
electron transport phosphorylation *124*
fermentation pathway *124*
glycolysis *124*
Krebs cycle *124*
lactate fermentation *132*
pyruvate *126*
substrate-level phosphorylation *126*

Levi, P. October 1984. "Travels with C." *The Sciences.* Journey of a carbon atom through the world of life.

Roberts, L. August 28, 1987. "Discovering Microbes with a Taste for PCBs." *Science* 237:975–977. Bacteria make ATP by breaking down pollutants.

Wolfe, S., 1993. *Molecular and Cellular Biology.* Belmont, California: Wadsworth.

FACING PAGE: *Human sperm, one of which will penetrate this mature egg and so set the stage for the development of a new individual in the image of its parents.*

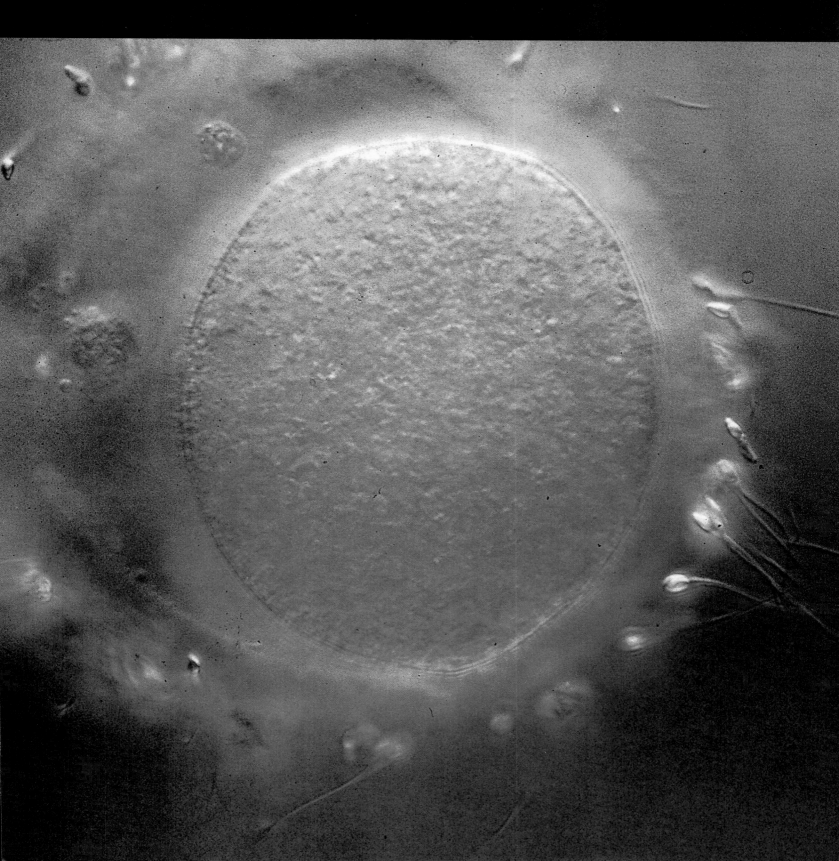

9 CELL DIVISION AND MITOSIS

Silver in the Stream of Time

Five o'clock, and the first rays of the sun dance over the wild Alagnak River of the Alaskan tundra. It is September, and life is both ending and beginning in the clear, frigid waters. By the thousands, mature silver salmon have returned from the open ocean to spawn in their shallow native home. The females are tinged with red, the color of spawners, and they are dying.

On this morning, a female salmon releases translucent pink eggs into a "nest," hollowed out by her fins in the gravel riverbed (Figure 9.1). Within moments, a male salmon sheds a cloud of sperm, and fertilization follows. Trout and other predators eat most of the eggs, but some survive and give rise to a new generation.

Within three years, the pea-size eggs have become streamlined salmon, fashioned from billions of cells. A few of those cells will develop into eggs or sperm. In time, on some September morning, they will take part in an ongoing story of birth, growth, death, and rebirth.

For you, as for salmon and all other multicelled organisms, growth as well as reproduction depends on *cell division*. In your mother's body, a single fertilized egg divided in two, then the two into four, and so on until billions of cells were growing, developing in specialized ways, and dividing at different times to produce your genetically prescribed body parts. Your body now has roughly 65 trillion cells—and many of the cells are still dividing. Every five days, for instance, cell divisions replace the lining of your small intestine.

Understanding cell division—and, ultimately, how new individuals are put together in the image of their parents—begins with answers to three questions. *First*, what instructions are necessary for inheritance? *Second*, how are those instructions duplicated for distribution into daughter cells? *Third*, by what mechanisms are those instructions divided into daughter cells? We will require more than one chapter to consider cell reproduction and other mechanisms of inheritance. However, the points made in the first part of this chapter can help you keep the overall picture in focus.

1. When a cell divides, its two daughter cells must each receive a required number of DNA molecules as well as cytoplasm. In eukaryotes, mitosis sorts out the DNA into two new nuclei. A separate mechanism divides the cytoplasm in two.

2. Mitotic cell division is the basis of growth and tissue repair in multicelled eukaryotes. It also is the means by which single-celled eukaryotes and many multicelled eukaryotes reproduce asexually.

3. DNA molecules are also called chromosomes. Members of the same species normally have the same total number of chromosomes in their body cells. The chromosomes have different lengths and shapes, and they carry different portions of the hereditary instructions.

4. When cells prepare for division, every chromosome is duplicated. Each now consists of two DNA molecules. Mitosis separates the two for distribution to daughter cells. In this way, each cell gets the same total number and the same types of chromosomes as the parent cell.

5. For many species, body cells have two of each type of chromosome, inherited from two parents. "Chromosome number" is the number of *each type* of chromosome for the species. Mitosis keeps the chromosome number constant from one cell generation to the next.

Figure 9.1 The last of one generation and the first of the next in the Alagnak River of Alaska.

9.1 DIVIDING CELLS: THE BRIDGE BETWEEN GENERATIONS

Overview of Division Mechanisms

In biology, the word **reproduction** means producing a new generation of cells or multicelled individuals. Reproduction begins with the division of single cells. And the ground rule for cell division is this:

Parent cells must provide their daughter cells with hereditary instructions (encoded in DNA) and enough cytoplasmic machinery to start up their own operation.

DNA, recall, contains instructions for synthesizing proteins. Some proteins are structural materials. Many are enzymes that put together carbohydrates, lipids, and other building blocks of cells. Unless a daughter cell receives the necessary instructions for making proteins, it cannot grow or function properly.

Also, the parent cell's cytoplasm already has operating machinery—enzymes, organelles, and so on. When a daughter cell inherits what looks like a blob of cytoplasm, it really is getting start-up machinery for its operation, until it has time to use its inherited DNA for growing and developing on its own.

The cells of plants, animals, and other eukaryotic organisms divide the DNA by **mitosis** or **meiosis**. Both mechanisms sort out and package DNA molecules into new nuclei for forthcoming daughter cells. In other words, mitosis and meiosis only divide the nuclear material. Different mechanisms divide the cytoplasm and so split a parent cell into daughter cells.

Multicelled organisms grow by mitosis and cytoplasmic division of body cells, which are called **somatic cells**. They also repair tissues that way. Nick yourself peeling a potato, and mitotic cell divisions will replace the cells the knife sliced away. Besides this, many protistans, fungi, plants, and even some animals reproduce asexually by mitotic cell division.

By contrast, meiosis occurs only in **germ cells**, a cell lineage set aside for sexual reproduction. Meiosis must precede the formation of gametes, such as sperm and eggs. As you will see in the next chapter, it has much in common with mitosis, but the end result is different.

What about prokaryotic cells—the bacteria? They reproduce asexually by way of a different mechanism. We will consider the bacteria later, in Chapter 22. For now, take a moment to study Table 9.1, which lists the major division mechanisms.

Some Key Points About Chromosomes

Before a cell starts preparing for mitosis, its DNA molecules are stretched out like threads, with many proteins attached to them. Each DNA molecule, along with its attached proteins, is a **chromosome**.

Chromosomes undergo duplication while they are in this threadlike form. Once duplicated, each consists of *two* DNA molecules, which will stay together until late in mitosis. For as long as they remain attached, the two are called **sister chromatids** of the chromosome:

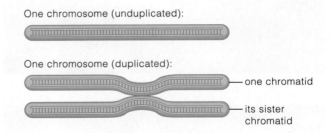

One chromosome (unduplicated):

One chromosome (duplicated):
— one chromatid
— its sister chromatid

"Sister chromatids" is a name that seems to confuse almost everybody. It might help to stretch their dictionary meaning a bit and think of them as the forthcoming daughters of a parent chromosome.

Notice how the duplicated chromosome narrows down in a small region. This is the **centromere**, a region having attachment sites for microtubules that will help move the chromosome during nuclear division:

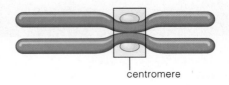

centromere

Table 9.1	Cell Division Mechanisms
Mechanisms	Used by
Mitosis, cytoplasmic division	*Single-celled eukaryotes (for asexual reproduction)*
	Multicelled eukaryotes (for bodily growth; also for asexual reproduction in many species)
Meiosis, cytoplasmic division	*Eukaryotes (basis of gamete formation and sexual reproduction)*
Prokaryotic fission	*Bacterial cells (for asexual reproduction)*

Keep in mind, the preceding sketches are simplified. As Figure 9.2 suggests, the centromere location is not the same for all chromosomes. Also, a DNA molecule's two parallel strands don't look like a ladder—they are twisted together repeatedly like a spiral staircase and are much longer than can be shown here.

Mitosis and the Chromosome Number

All normal members of the same species have the same total number of chromosomes in their somatic cells. Humans have forty-six, gorillas have forty-eight, and pea plants have fourteen.

Figure 9.3 shows all forty-six human chromosomes, lined up as twenty-three pairs. Think of them as two sets of books on how to build a house. Your father gave you one set. Your mother had her own ideas about plumbing, storage, and so forth, so she gave you a revised edition. Her set covers the exact same topics but has different things to say about many of them.

Your chromosomes are like the volumes of two sets of books. Each set is numbered 1 to 23. Thus, for example, you have two "volumes" of chromosome 22—that is, a pair of them. Generally, both members of each pair have the same length and shape, and they carry instructions for the same traits. In these respects, they are not like any of the other pairs of chromosomes.

Any cell having two of each type of chromosome is a **diploid cell**. Such cells exist in gorillas, pea plants, and a great many other organisms besides humans.

With mitosis, a diploid parent cell produces two diploid daughter cells. This doesn't mean each merely gets forty-six or forty-eight or fourteen chromosomes. If only the total mattered, one cell might get, say, two pairs of chromosome 22 and no pairs of chromosome 9. However, neither cell would function properly without *one pair of each type of chromosome*.

Chromosome number tells you how many of each type of chromosome are present in a cell. Diploid cells, with two of each type, are 2n. The n stands for chromosome number.

With mitosis, the chromosome number remains constant, division after division, from one cell generation to the next. Thus, if a parent cell is diploid, its two daughter cells will be diploid also.

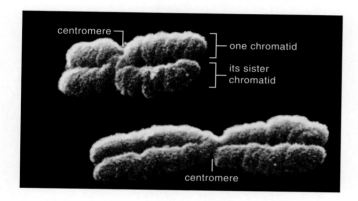

Figure 9.2 Scanning electron micrograph of two human chromosomes, each in the duplicated state.

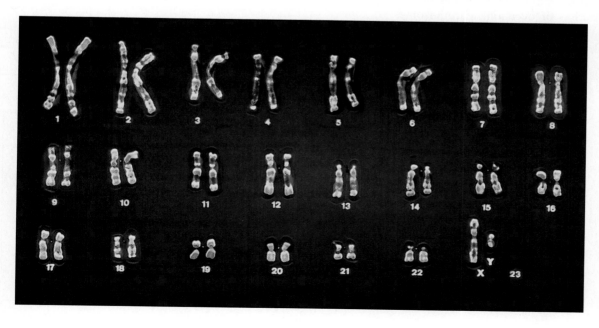

Figure 9.3 Forty-six chromosomes from a human male. Each chromosome is in the duplicated state.

9.2 MITOSIS AND THE CELL CYCLE

Mitosis is only one phase of the **cell cycle**. Such cycles start at the time new cells are produced, and they end when those cells complete their own division. The cycle starts again for each new daughter cell. **Interphase** usually is the longest part of a cell cycle. At this time, a cell increases its mass, roughly doubles the number of its cytoplasmic components, and finally duplicates its chromosomes. The following abbreviations designate the phases of the cell cycle (see also Figure 9.4):

M   *Mitosis*; nuclear division, commonly followed by cytoplasmic division

G_1 Of interphase, a "*Gap*" (interval) before the onset of DNA replication

S   Of interphase, the time of "*Synthesis*" (replication) of DNA and associated proteins

G_2 Of interphase, a second "*Gap*" between the completion of DNA replication and the onset of mitosis

The cell cycle lasts about the same length of time for cells of a given type. Its duration differs among cells of different types. For example, all of the neurons in your brain are arrested at interphase and typically will not divide again. Cells of a new sea urchin may double in number every two hours.

Adverse conditions may disrupt a cell cycle. When deprived of a vital nutrient, for instance, the free-living cells called amoebas do not leave interphase. Even so, when a cell proceeds past a certain point in interphase, the cycle normally continues regardless of outside conditions, owing to built-in controls over its duration.

Good health depends on the proper timing and completion of events in the cell cycle. Mistakes in the duplication or distribution of even one chromosome during interphase may lead to a genetic disorder. As described on page 234, a mature tissue will be destroyed

if controls are lost that otherwise keep its cells from dividing, and cancer may follow. The *Focus* essay describes a landmark case of unchecked cell divisions.

We turn now to mitosis and how it can maintain the chromosome number through turn after turn of the cell cycle. Figure 9.5 only hints at the precise division mechanisms that take over as a cell leaves interphase.

A cell cycle begins at interphase, when a new cell (formed by mitosis and cytoplasmic division) increases its mass, doubles the number of its cytoplasmic components, then duplicates its chromosomes. The cycle ends when the cell divides.

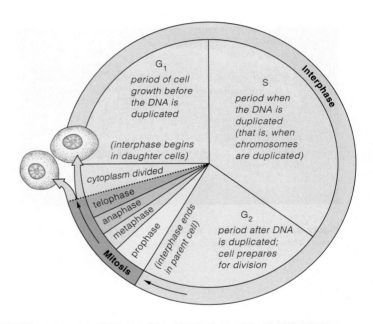

Figure 9.4 Generalized eukaryotic cell cycle. The length of different stages differs among cells.

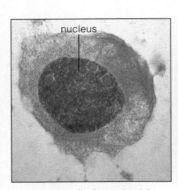

a Interphase (before mitosis)

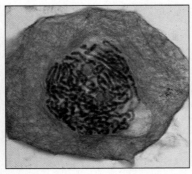

b Early prophase

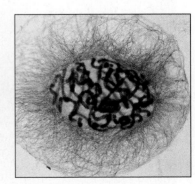

c Prophase

Henrietta's Immortal Cells

Each human starts out as a single fertilized egg. At birth a human body has about a trillion cells. Even in an adult, trillions of cells are still dividing. Cells in the stomach's lining divide every day. Liver cells usually don't divide—but if part of the liver becomes injured or diseased, they will divide repeatedly and produce new cells until the damaged part is replaced.

In 1951, George and Margaret Gey of Johns Hopkins University were trying to develop a way to keep human cells dividing outside the body. (Researchers could study basic life processes with such cells. They also could study cancer and other diseases, without having to experiment directly on humans.) Local physicians had provided them with normal or diseased human cells from patients. But the Geys just couldn't stop the cell lines from dying out within a few weeks.

Mary Kubicek, one of their assistants, was about to give up after dozens of failed attempts. Still, in 1951, she prepared another sample of cancer cells for culture. The sample was code-named HeLa, for the first two letters of the patient's first and last names.

The cells began to divide. And divide. And divide again. By the fourth day there were so many cells that they had to be subdivided into more tubes. As months passed, the culture continued to thrive. Unfortunately, the tumor cells inside the patient's body were just as vigorous. Six months after the patient was first diagnosed as having cancer, tumor cells had spread through her body. Two months later, Henrietta Lacks, a young woman from Baltimore, was dead.

Although Henrietta was gone, some of her cells lived on in the Geys' laboratory as the first successful human cell culture. HeLa cells were soon being shipped to other researchers, who passed cells on to others, and so HeLa cells came to live in laboratories all over the world. Some even traveled into space aboard the *Discoverer XVII* satellite. Every year, research that is described in hundreds of scientific papers is based on work with HeLa cells.

Henrietta was only thirty-one years old when runaway cell divisions killed her. Now, more than forty years later, her legacy is still benefiting humans everywhere, in cells that are still alive and dividing, day after day after day.

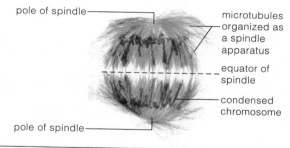

pole of spindle — microtubules organized as a spindle apparatus — equator of spindle — condensed chromosome — pole of spindle

Figure 9.5 (*Below*) Light micrographs showing the progress of mitosis in a cell from the African blood lily (*Haemanthus*). As indicated by the icon to the right, the chromosomes are stained blue, and microtubules that are moving them about are stained red.

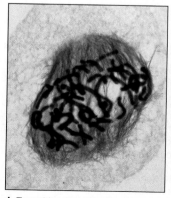

d Transition to metaphase

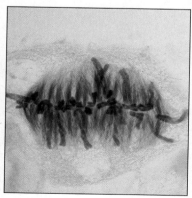

e Metaphase

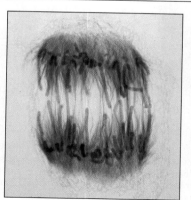

f Anaphase

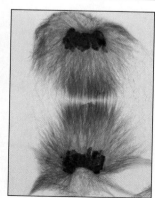

g Telophase

9.3 STAGES OF MITOSIS: AN OVERVIEW

When a cell makes the transition from interphase to mitosis, it has stopped constructing new cell parts, and its DNA has been replicated. Within that cell, profound changes now proceed smoothly, one after the other, through four stages. The sequential stages of mitosis are **prophase**, **metaphase**, **anaphase**, and **telophase**.

Figure 9.6 shows these stages for a dividing animal cell. Compare this illustration with the preceding one of a plant cell. When you look closely, it becomes quite clear that the chromosomes within both cells move about dramatically during mitosis. They do not move about on their own. A **spindle apparatus** moves them.

In all eukaryotic cells, a fully formed spindle consists of two organized sets of microtubules. The microtubules extend from the spindle's two end points (poles) and overlap at its equator, which lies midway

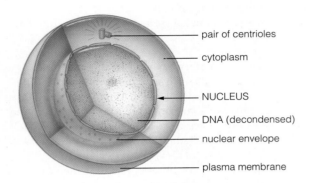

— pair of centrioles

— cytoplasm

— NUCLEUS

— DNA (decondensed)

— nuclear envelope

— plasma membrane

Cell at Interphase
The DNA is duplicated, then the cell prepares for division.

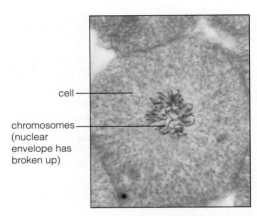

cell —

chromosomes (nuclear envelope has broken up) —

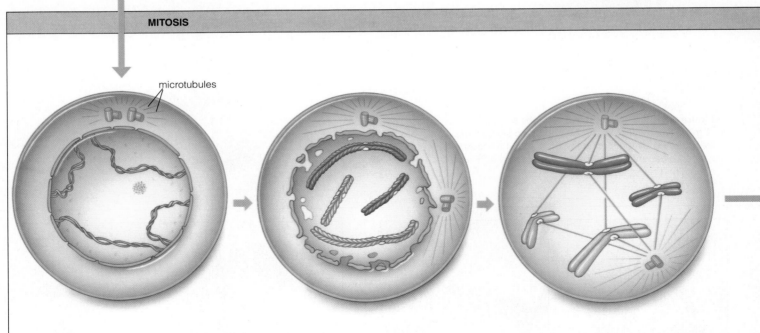

MITOSIS

microtubules

Early Prophase

The DNA and its associated proteins start to condense. The two chromosomes shaded purple were inherited from the male parent. The other two (blue) are their counterparts, inherited from the female parent.

Late Prophase

Chromosomes continue to condense. New microtubules are assembled, and they move one of two centriole pairs toward the opposite end of the cell. The nuclear envelope starts to break up.

Transition to Metaphase

Microtubules penetrate the nuclear region. Together they form a spindle apparatus. They become attached to the sister chromatids of each chromosome.

between the poles. The spindle poles establish the destinations of chromosomes during mitosis.

Mitosis proceeds through four stages, called prophase, metaphase, anaphase, and telophase.

A bipolar spindle, composed of two sets of microtubules, positions the chromosomes and moves them to specific locations during these stages.

Figure 9.6 Mitosis. This nuclear division mechanism assures that daughter cells will have the same chromosome number as the parent cell. For clarity, the diagram shows only two pairs of chromosomes from a diploid (2*n*) animal cell. With rare exceptions, the picture is more involved than this, as indicated by the micrographs of mitosis in a whitefish cell.

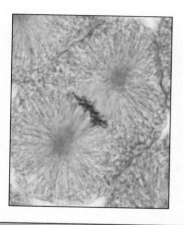
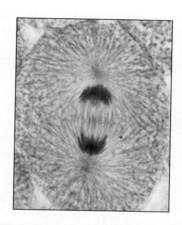
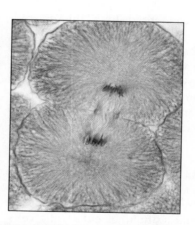

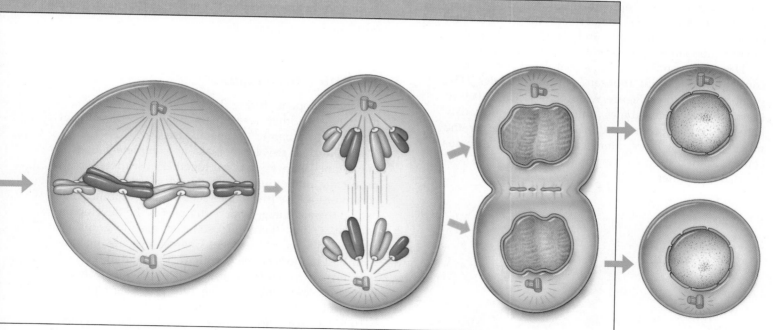

Metaphase
All chromosomes are lined up at the equator of the spindle. They are now in their most condensed form.

Anaphase
The attachment between the two sister chromatids of each chromosome breaks; the two are now chromosomes in their own right. They move to opposite spindle poles.

Telophase
Chromosomes decondense. New patches of membrane join to form nuclear envelopes around them. Most often, cytoplasmic division occurs before telophase is over.

Interphase
Two daughter cells have formed. Each is diploid, with two of each type of chromosome—just like the parent cell.

9.4 A CLOSER LOOK AT MITOSIS

Prophase: Mitosis Begins

Chromosomes Start Condensing Prophase, the first stage of mitosis, is evident when chromosomes become visible in the light microscope as threadlike forms. ("Mitosis" comes from the Greek *mitos*, meaning thread.) Every chromosome was duplicated earlier, during interphase. In other words, each already consists of two sister chromatids joined at the centromere. Early in prophase, both chromatids are twisting and folding into a more compact form. By late prophase, all of the chromosomes will be condensed to a thicker, rod-shaped form.

Figure 9.7 shows what one type of duplicated chromosome will look like when it reaches its most condensed form. Notice the centromere. At this constricted region, each sister chromatid of the chromosome has a disk-shaped structure on its surface. This structure, a **kinetochore**, will serve as an attachment site for microtubules of the spindle.

The Spindle Starts Forming Meanwhile, in the cytoplasm near the nucleus, microtubules are forming. Microtubules, recall, are tube-shaped proteins of the cytoskeleton, the internal scaffold of eukaryotic cells. They are composed of numerous subunits called tubulins (Figure 4.19). When a nucleus is about to divide, most of the cytoskeletal microtubules disassemble, then the unbound subunits reassemble as *new* microtubules.

Depending on the species, a completed spindle may consist of ten to many thousands of microtubules. (The plant cell shown in Figure 9.5 probably has 10,000 of them.) Some microtubules will extend continuously between a kinetochore and one spindle pole or the other. The rest will not interact at all with the chromosomes. Rather, they will extend from both poles and overlap each other.

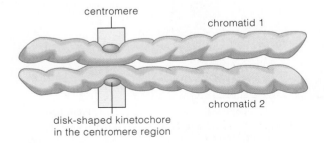

centromere

chromatid 1

disk-shaped kinetochore in the centromere region

chromatid 2

Figure 9.7 Kinetochores of a duplicated chromosome. Spindle microtubules become attached to the kinetochore of each sister chromatid.

In many cells, an MTOC near the nucleus organizes the assembly of microtubules so that they lengthen in specific directions (hence "MTOC," for microtubule organizing center). While the new microtubules are lengthening, the nuclear envelope prevents them from interacting with the chromosomes inside the nucleus. However, as prophase draws to a close, the nuclear envelope starts to break up.

The Spindle Separates the Centrioles In many cells, an MTOC near the nucleus includes two barrel-shaped **centrioles**. Each centriole started duplicating itself during interphase, and by the time prophase is under way, the cell has two pairs of them. Now microtubules start moving one pair to the opposite pole of the newly forming spindle. You can see this carefully orchestrated movement in Figure 9.6.

Centrioles, recall, give rise to flagella or cilia. If you observe them in the cells of a particular organism, you can safely bet that flagellated or ciliated cells develop during at least some part of its life cycle. For example, the protistan responsible for *trichomoniasis*, a sexually transmitted disease, swims through vaginal fluids by beating its flagella. Even "stationary" plants and fungi commonly produce flagellated, motile gametes. And ciliated cells line airways to your lungs. Ultimately, these and other functions depend on centrioles, so it is no wonder that they are parceled out carefully for the forthcoming daughter cells.

Transition to Metaphase

So much happens during the transition from prophase to metaphase that researchers give this transitional period its own name, "prometaphase."

The nuclear envelope breaks up completely. Its two membrane layers have fused together at scattered sites, producing numerous tiny, flattened vesicles. Now the chromosomes are free to interact with microtubules that extend toward them, from both poles of the developing spindle. At first the chromosomes appear to go into a frenzy. This happens when kinetochores are making their first random contacts with the microtubules.

Once a chromosome has both of its kinetochores harnessed, microtubules from *both* poles start pulling on it. The two-way pulling orients the two sister chromatids of that chromosome toward opposite poles. Meanwhile, the overlapping spindle microtubules are ratcheting past each other in such a way that the two poles are pushed farther apart. All of the push–pull forces are balanced when the chromosomes reach the midpoint of the spindle. The formation of the bipolar spindle is now complete.

When all of the duplicated chromosomes are aligned midway between the poles of a completed spindle, we call this metaphase (*meta-* means midway between). The alignment is crucial for the next stage of mitosis.

Anaphase

During anaphase, the two sister chromatids of each chromosome are separated from each other and moved to opposite poles. They do this at about the same time, and they move at the same rate. Once they do separate, they are no longer referred to as chromatids. Each is now a chromosome in its own right:

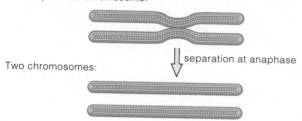

One duplicated chromosome:

Two chromosomes: separation at anaphase

Said another way, every chromosome that was present in the parent cell now has a daughter chromosome at both spindle poles.

Two mechanisms account for the anaphase movements. First, microtubules attached to the kinetochores shorten, pulling the chromosomes toward the spindle poles (Figure 9.8a). Second, the spindle elongates when overlapping microtubules ratchet past each other and push the two spindle poles farther apart (Figure 9.8b).

Telophase

Telophase begins once the two daughter chromosomes arrive at opposite spindle poles. The chromosomes are no longer harnessed to microtubules, and they return to threadlike form. Vesicles of the old nuclear envelope fuse together to form patches of membrane around the chromosomes. Patch joins with patch, and soon a new nuclear envelope separates each cluster of chromosomes from the cytoplasm. If the parent cell was diploid, each cluster will contain a pair of each type of chromosome. With mitosis, each new nucleus has the same chromosome number as the parent nucleus.

Once the two nuclei form, telophase is over—and so is mitosis.

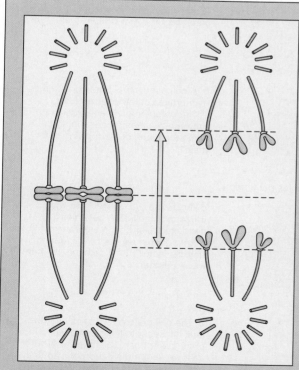

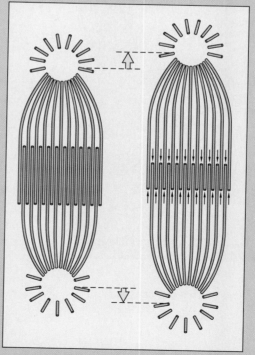

Figure 9.8 Two of the mechanisms responsible for separating the sister chromatids of a chromosome at anaphase.

a Microtubules attached to the chromatids shorten, decreasing the distance between the kinetochore and the spindle poles.

b Overlapping microtubules ratchet past each other, moving the poles farther apart and so increasing the distance between sister chromatids.

9.5 DIVISION OF THE CYTOPLASM

The cytoplasm usually divides at some time between late anaphase and the end of telophase. Cytoplasmic division (or cytokinesis, as it is often called) proceeds by different mechanisms in plant and animal cells.

The cells of most land plants have fairly rigid walls, and they cannot be merely pinched in two. Their cytoplasm divides by a mechanism that involves **cell plate formation**. As Figure 9.9 shows, vesicles filled with wall-building material fuse with remnants from the spindle, forming a disklike structure (the "cell plate"). At this location, cellulose deposits accumulate and form a crosswall that divides the parent cell into two daughter cells.

By contrast, animal cells do not have a cell wall, and most of them can divide by "pinching in two." By a mechanism called **cleavage**, a layer of deposits forms around microtubules at the midsection of most animal cells. A shallow, ringlike depression appears above the layer, at the cell surface (Figure 9.10). At this depression, the cleavage furrow, microfilaments are at work.

Microfilaments, recall, are components of the cytoskeleton. These particular ones are arranged to slide past one another. As they do, they also pull the plasma membrane inward and cut the cell in two.

Most plant cells divide by the formation of a cell plate and then a crosswall between the developing plasma membranes of daughter cells.

Most animal cells divide by cleavage. Microfilaments, arranged in a ring around the cell's midsection, slide past one another in such a way that the cell is cut in two.

This concludes our picture of mitotic cell division. Look now at your hands—and think of all the cells in your palms, thumbs, and fingers. Find a microscope and look at cells in a sample from your skin. Imagine all the divisions of all the cells that preceded them when you were developing early on, inside your mother (Figure 9.11). And be grateful for the astonishing precision that led to their formation. Even from the *Focus* essay presented earlier in this chapter, you have an inkling that the alternatives can be terrible indeed.

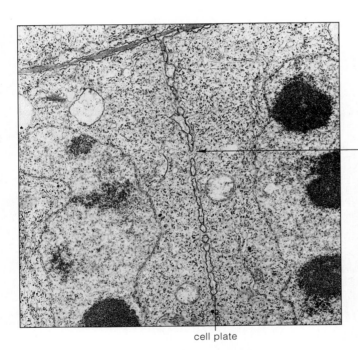

cell plate

Figure 9.9 Cytoplasmic division of a plant cell, as brought about by cell plate formation.

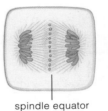

spindle equator

a As mitosis draws to a close, vesicles gather at the equator of the microtubular spindle. They contain cementing materials and starting materials for a new primary wall.

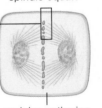

vesicles gathering

b A cell plate starts forming as the membranes of the vesicles fuse. Their contents become sandwiched between two membranes that form along the plane of the growing cell plate.

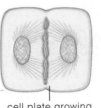

cell plate growing

c Inside the "sandwich," two walls will form as cellulose becomes deposited over both membranes. Other substances inside will form the "middle lamella" that cements the new walls together.

two new primary walls

d The cell plate grows at its margins until it fuses with the plasma membrane of the parent cell. During growth, when cells are expanding and walls are thin, new material also is deposited over the old primary wall.

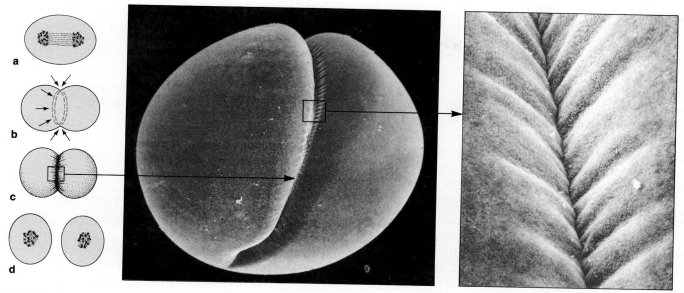

Figure 9.10 Cytoplasmic division of an animal cell. (**a**) Mitosis is completed; the microtubular spindle is disassembling. (**b**) Just beneath the plasma membrane of the parent cell, microfilament rings at the former spindle equator contract, like a purse string. (**c,d**) Contractions continue and in time will divide the cell in two. The micrographs show how the plasma membrane sinks inward, defining the cleavage plane. Cleavages helped form the human hand shown below in Figure 9.11*d*.

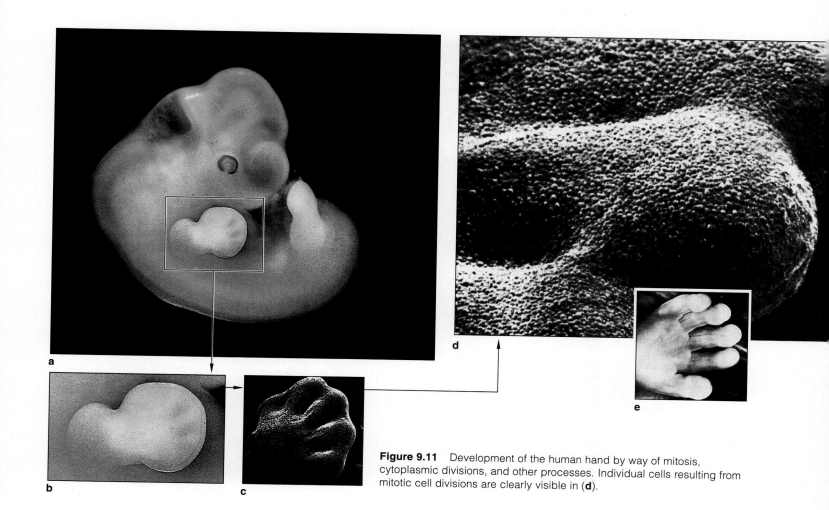

Figure 9.11 Development of the human hand by way of mitosis, cytoplasmic divisions, and other processes. Individual cells resulting from mitotic cell divisions are clearly visible in (**d**).

SUMMARY

1. Parent cells provide each daughter cell with the hereditary instructions (DNA) and cytoplasmic machinery necessary to start up its own operation.

2. Mitosis divides the hereditary instructions in the nucleus of a parent cell into two equivalent parcels. It is the basis for growth and tissue repair in multicelled eukaryotes. It also is the basis of asexual reproduction among many single-celled and multicelled eukaryotes. (Meiosis, another nuclear division mechanism, occurs only in germ cells set aside for sexual reproduction.)

3. The cytoplasm divides by different mechanisms near the end of nuclear division or at some point thereafter. (Animal cells are cleaved in two. A cell plate forms and divides plant cells.)

4. A chromosome is a single DNA molecule. In eukaryotes, many proteins are attached to it. A cell's chromosomes differ from one another in length, shape, and which portion of the hereditary instructions they carry. Members of the same species have the same total number of chromosomes in their somatic cells.

5. "Chromosome number" is not the same thing as the total number of chromosomes. It is the number of *each type* of chromosome characteristic of a species. Mitosis keeps the chromosome number constant from one cell generation to the next. Thus, if a parent cell is diploid (with two chromosomes of each type), the daughter cells resulting from mitosis will be diploid also.

6. The cell cycle starts when a new cell forms, proceeds through interphase, and ends when the cell reproduces by mitosis and cytoplasmic division. At interphase, a cell increases in mass, doubles its number of cytoplasmic components, and duplicates all chromosomes.

7. After duplication, each chromosome consists of two DNA molecules attached at the centromere. For as long as the two stay attached, they are called sister chromatids of the chromosome.

8. Mitosis proceeds through four continuous stages:

   a. Prophase. Duplicated, threadlike chromosomes start to condense. New microtubules assemble in organized arrays near the nucleus; they will form a spindle. The nuclear envelope starts to break up.

   b. Metaphase. During the *transition* to metaphase, the nuclear envelope breaks up completely. Microtubules of a forming spindle attach to the sister chromatids of each chromosome and orient the two toward opposite spindle poles. *At* metaphase, all chromosomes are aligned at the spindle equator.

   c. Anaphase. Microtubules pull the sister chromatids of each chromosome away from each other, to opposite

spindle poles. Every chromosome that was present in the parent cell is now represented by a daughter chromosome at both poles.

   d. Telophase. The chromosomes decondense to the threadlike form. A new nuclear envelope forms around them. Each nucleus has the same chromosome number as the parent cell. Mitosis is completed.

Review Questions

1. Define the two types of nuclear division mechanisms that occur in eukaryotes. Does either one divide the cytoplasm? *142*

2. Define somatic cell and germ cell. *142*

3. What is a chromosome? What is a chromosome called in its unduplicated state? In its duplicated state (that is, with two sister chromatids)? *142–143*

4. Describe the spindle apparatus and its general function in nuclear division. *148–149*

5. Name and describe the main features of the four stages of mitosis. At what stages were the following micrographs of a plant cell taken? *148–150*

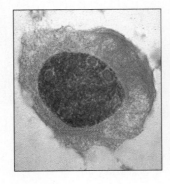
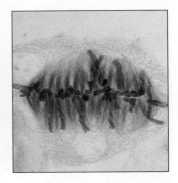
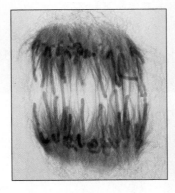
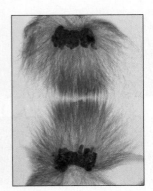

6. Correctly label the component parts of this human chromosome. *148*

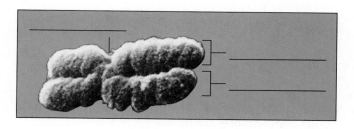

7. Describe cytoplasmic division as it occurs in plant cells and animal cells. *150*

8. The function of mitosis and cytoplasmic division is _____ .
   a. asexual reproduction of single-celled eukaryotes
   b. growth, repair, and asexual reproduction in many eukaryotes
   c. gamete formation in eukaryotes
   d. both a and b are correct

9. Only _____ is not a stage of mitosis.
   a. prophase   c. metaphase
   b. interphase   d. anaphase

10. Match each stage with its key events.
    _____ metaphase   a. sister chromatids of each
    _____ prophase   chromosome separate
    _____ telophase   b. threadlike chromosomes start
    _____ anaphase   to condense
                         c. chromosomes decondense,
                            daughter nuclei form
                         d. all chromosomes are aligned
                            at the equator

Self-Quiz *(Answers in Appendix IV)*

1. A eukaryotic chromosome consists of _____ .
   a. DNA only   c. DNA plus membrane
   b. DNA plus proteins d. DNA plus lipids

2. A cell with pairs of the chromosomes characteristic of the species is a(an) _____ cell.
   a. diploid   c. abnormal
   b. mitotic   d. a and c

3. A duplicated chromosome has _____ chromatid(s).
   a. one   c. three
   b. two   d. four

4. The _____ is the constricted region of a duplicated chromosome, with attachment sites for microtubules.
   a. chromatid   c. cell plate
   b. centromere   d. cleavage furrow

5. In the cell cycle, the interphase stage immediately following DNA replication is _____ .
   a. G_2 c. S
   b. G_1 d. none of the above

6. Interphase is the stage when _____ .
   a. nothing occurs
   b. a germ cell forms its spindle apparatus
   c. a cell grows and duplicates its DNA
   d. mitosis occurs

7. After mitosis, the chromosome number of a daughter cell is _____ the parent cell's.
   a. the same as   c. rearranged compared to
   b. one-half   d. doubled compared to

Selected Key Terms

anaphase *146*
cell cycle *144*
cell plate formation *150*
centriole *148*
centromere *142*
chromosome *142*
chromosome number *143*
cleavage *150*
diploid cell *143*
germ cell *142*
interphase *144*

kinetochore *148*
meiosis *142*
metaphase *146*
mitosis *142*
prophase *146*
reproduction *142*
sister chromatid *142*
somatic cell *142*
spindle apparatus *146*
telophase *146*

Readings

Gallagher, G. March 1990. "Evolution: The Mitotic Spindle." *The Journal of NIH Research.*

Murray, A., and M. Kirschner. March 1991. "What Controls the Cell Cycle?" *Scientific American* 264(3): 56–63.

Science. November 3, 1989. "Frontiers in Biology: The Cell Cycle."

Wolfe, S. 1993. *Molecular and Cellular Biology.* Belmont, California: Wadsworth.

10 MEIOSIS

Octopus Sex and Other Stories

The couple clearly are interested in each other. He caresses her with one tentacle, then another—then another and another. She reciprocates with a hug here, a squeeze there. This goes on for hours. Finally the male reaches under his mantle, a tissue fold that drapes around most of his body. He removes a packet of sperm and inserts it into an egg chamber under the female's mantle. For every sperm that fertilizes an egg, a new octopus may develop.

Unlike the coupling between a male and female octopus, sex for the slipper limpet is a group activity. Slipper limpets are marine animals, relatives of land snails. Before one of them becomes transformed into a sexually mature adult, it passes through a free-living larval stage. When the time comes for the larva to undergo transformation, it settles down on sand or sediments. If it settles down alone, it will become a

female. If a second larva settles down on the first one, *it* will become a male—and so will any subsequent larvae. Adult slipper limpets almost always live in such piles, with the bottom one always being female (Figure 10.1*a*). The males release sperm on a perpetual basis. When the female finally dies, the male at the bottom of the pile becomes transformed into a female, and so it goes.

For many sexually reproducing organisms, the life cycle also has asexual episodes, based on mitotic cell divisions. Many kinds of plants, including orchids, can reproduce without sex. Flatworms can split lengthwise into two roughly equivalent parts that each grow into a new flatworm. In summer, nearly all aphids are females, which busily produce more females from unfertilized egg cells (Figure 10.1*b*). When autumn approaches, male aphids finally develop and do their part in the sexual phase of the life cycle. Still, females

Figure 10.1 Variations in reproductive modes. (**a**) Slipper limpets busily perpetuating the species by group participation in sexual reproduction. (**b**) Live birth of an aphid, a type of insect that reproduces sexually in autumn but switches to an asexual mode in summer.

a

that survive the winter can do without the males, and come summer they begin another round of producing offspring all by themselves.

These examples only hint at the immense variation in reproductive modes. Yet despite the variation, sexual reproduction dominates eukaryotic life cycles, and it always involves certain events. Chromosomes are duplicated in germ cells. The germ cells undergo meiosis and cytoplasmic division, and then gametes develop. When gametes join at fertilization, they form the first cell of a new individual. *Meiosis, gamete formation, and fertilization are the hallmarks of sexual reproduction.* As you will see in this chapter, these processes contribute to life's splendid diversity.

b

10.1 ON ASEXUAL AND SEXUAL REPRODUCTION

What kind of offspring do orchids, flatworms, and aphids get with **asexual reproduction**? By this process, one parent alone passes on a duplicate of all its genes to each new individual. **Genes** are specific stretches of chromosomes—that is, DNA molecules. Genes are all the heritable bits of information that are required to produce new individuals. Rare mutations aside, this means that asexually produced offspring can only be clones—genetically identical copies of the parent.

Inheritance is much more interesting with **sexual reproduction**. In humans and many other species, *two* parents—each with pairs of genes—engage in this process. Both parents pass on one of each gene to offspring by way of meiosis, gamete formation, and fertilization (the union of two gametes). Thus the first cell of a new individual also ends up with pairs of genes.

If instructions in every pair of genes were identical down to the last detail, then sexual reproduction would produce clones, also. Just imagine—you, everyone you know, and the entire human population would be clones and might end up looking exactly alike.

But a pair of genes may *not* be identical. Why not? The molecular structure of genes can change; this is what we mean by mutation. Depending on its structure, one gene of a pair may "say" slightly different things about a trait. Each unique molecular form of the same gene is called an **allele**.

Such tiny differences affect thousands of traits. Figure 10.2 is an example. One allele that may occur at a certain location in a human chromosome says "put a dimple in the chin" and another says "no dimple."

This brings us to a key reason why members of any sexually reproducing species don't all look alike. *Through sexual reproduction, offspring end up with new combinations of alleles, and these lead to variations in their physical and behavioral traits.*

This chapter describes the cellular basis of sexual reproduction. More importantly, it starts us thinking about the far-reaching effects of gene shufflings at different stages of the process. The resulting variation among offspring may be acted upon by agents of natural selection. Thus, *variation in traits is a foundation for evolutionary change.*

Asexual reproduction produces genetically identical copies of the parent. Sexual reproduction introduces variations in the details of traits among offspring.

10.2 OVERVIEW OF MEIOSIS

Think "Homologues"

Think back on the preceding chapter, which focused on mitotic cell division. Unlike mitosis, **meiosis** divides the chromosomes in a nucleus not once but twice prior to cell division. Unlike mitosis, it is the first step leading to gamete formation. **Gametes** are sex cells, such as sperm and eggs. In multicelled organisms, gametes arise only from germ cells, which are produced in specific reproductive structures and organs (Figure 10.3).

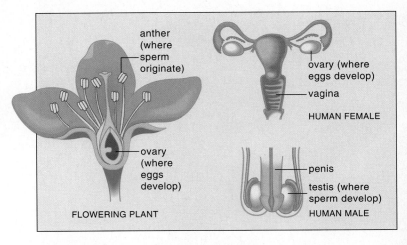

Figure 10.3 Examples of gamete-producing structures.

Figure 10.2 (**a**) The chin fissure, a heritable trait arising from an uncommon form of a gene. Actor Kirk Douglas received a gene that influences this trait from each of his parents. One gene called for a chin fissure and the other didn't, but one is all it takes in this case. (**b**) This photograph shows what his chin might have looked like if he had inherited two ordinary forms of the gene instead.

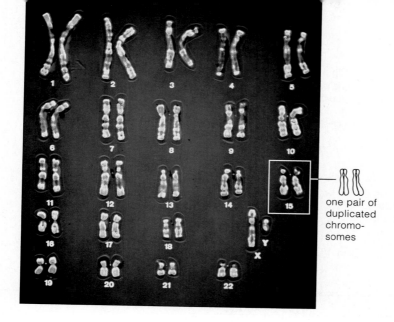

one pair of duplicated chromosomes

Figure 10.4 From a diploid cell of a human male, twenty-three pairs of homologous chromosomes. The sketch corresponding to the two boxed chromosomes indicates that the chromosomes are in the duplicated state (each consists of two sister chromatids).

Germ cells have the same chromosome number as the rest of the body's cells. The ones that are **diploid** (2*n*) have *two* of each type of chromosome, often from two parents. The two chromosomes have the same length and shape. Their genes deal with the same traits. And they line up with each other at meiosis. Think of them as **homologous chromosomes** (*hom*- means alike).

As you can deduce from Figure 10.4, there are 23 + 23 homologous chromosomes in your germ cells. After meiosis, 23 chromosomes—one of each type—end up in gametes. Said another way, meiosis halves the chromosome number, so the gametes are **haploid** (*n*).

Overview of the Two Divisions

Meiosis resembles mitosis in some respects, even though the outcome is different. Before interphase gives way to meiosis, a germ cell duplicates its DNA. Now each duplicated chromosome consists of two DNA molecules. These remain attached at a narrowed-down region (the centromere). For as long as the two remain attached, they are called **sister chromatids** of the chromosome:

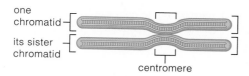

one chromatid —
its sister chromatid —
centromere

As in mitosis, microtubules of a spindle apparatus move the chromosomes in prescribed directions.

With meiosis alone, however, *chromosomes proceed through two consecutive divisions, which end with the formation of four haploid nuclei.* The two divisions are called meiosis I and II:

DNA duplication during interphase	MEIOSIS I	No DNA duplication between divisions	MEIOSIS II
	Prophase I Metaphase I Anaphase I Telophase I		Prophase II Metaphase II Anaphase II Telophase II

During meiosis I, each duplicated chromosome lines up with its partner, *homologue to homologue*; then the partners are separated from each other:

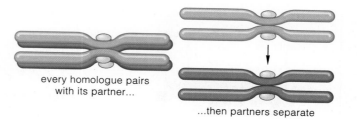

every homologue pairs with its partner...

...then partners separate

The cytoplasm typically divides after the separation of homologues. The two daughter cells are haploid, with only one of each type of chromosome. But remember, those chromosomes are still duplicated.

During meiosis II, *the two sister chromatids of each chromosome are separated from each other:*

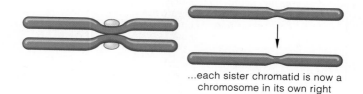

...each sister chromatid is now a chromosome in its own right

After four nuclei form, the cytoplasm often divides once more, the outcome being four haploid cells.

On the next two pages, Figure 10.5 illustrates the key events of meiosis I and II.

Meiosis is a type of nuclear division that reduces the parental chromosome number by half—to the haploid number (*n*).

Meiosis is the first step leading to the formation of gametes, which are required for sexual reproduction.

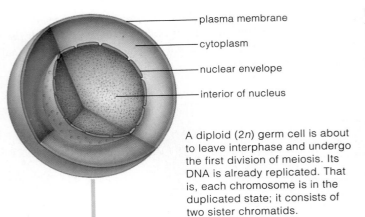

plasma membrane

cytoplasm

nuclear envelope

interior of nucleus

A diploid (2n) germ cell is about to leave interphase and undergo the first division of meiosis. Its DNA is already replicated. That is, each chromosome is in the duplicated state; it consists of two sister chromatids.

Figure 10.5 Meiosis: the nuclear division mechanism by which the parental number of chromosomes is reduced by half (to the haploid number) for forthcoming gametes. Only two pairs of homologous chromosomes are shown. Maternal chromosomes are shaded *purple,* and paternal ones *blue*.

Meiosis I

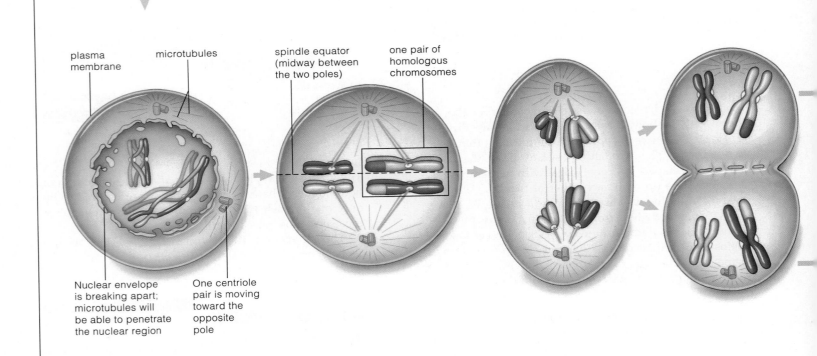

plasma membrane

microtubules

spindle equator (midway between the two poles)

one pair of homologous chromosomes

Nuclear envelope is breaking apart; microtubules will be able to penetrate the nuclear region

One centriole pair is moving toward the opposite pole

Prophase I

Each chromosome starts to twist and fold into more condensed form. It pairs up with its homologue, and the two typically swap segments. This swapping, called crossing over, is indicated by the break in color on the pair of larger chromosomes. As in mitosis, chromosomes become attached to a newly forming microtubular spindle.

Metaphase I

Microtubules have pushed and pulled all chromosomes into position, midway between the two spindle poles. The spindle is now completely formed, owing to the chromosome-microtubule interactions.

Anaphase I

Each chromosome is separated from its homologue. The two are moved to opposite poles of the spindle.

Telophase I

When the cytoplasm divides, two haploid (2n) cells are the result. Each cell has one chromosome of each type, although these are still in the duplicated state.

Of the four haploid cells that may form by way of meiosis, one or all may develop into gametes. (In plants, the cells that form by way of meiosis may develop into spores, which precede the formation of gametes.)

Meiosis II

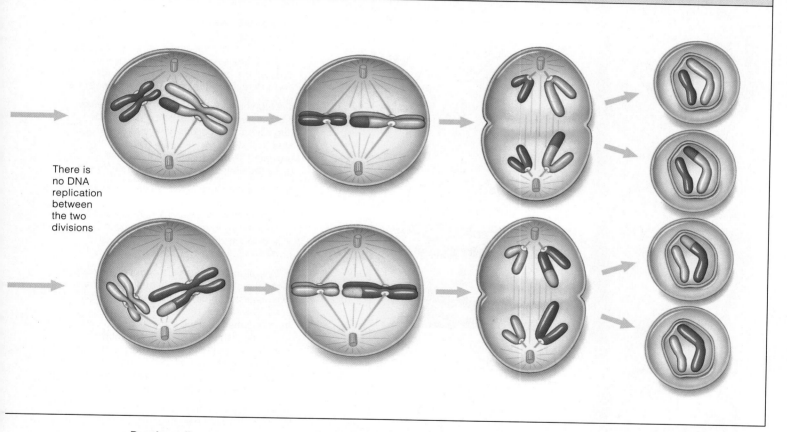

There is no DNA replication between the two divisions

Prophase II

During the transition to prophase II, the two centrioles were moved apart and a new spindle was assembled in each newly formed cell. Now, at prophase II, chromosomes become attached to microtubules and start moving toward the spindle equator.

Metaphase II

All of the duplicated chromosomes are now positioned midway between the two poles of the spindle.

Anaphase II

The attachment between the two chromatids of each chromosome breaks. Now the former "sister chromatids" are chromosomes in their own right and are moved to opposite poles of the spindle.

Telophase II

Four daughter nuclei form. When the cytoplasm divides, each new cell has a haploid number of chromosomes, all in the unduplicated state.

10.3 KEY EVENTS DURING MEIOSIS I

The preceding overview is enough to convey the over-riding function of meiosis—that is, *reduction of the chromosome number by half for forthcoming gametes.* However, as you will now see, two other events that unfold during prophase I and metaphase I contribute enormously to the key adaptive advantage of sexual reproduction—*the production of offspring with new combinations of alleles, hence variations in traits.*

Prophase I Activities

Prophase I of meiosis is a time of major gene shufflings. Consider Figure 10.6a, which shows two chromosomes that have condensed to threadlike form. All chromosomes in a germ cell condense this way. When they do, each is drawn close to its homologue by a process called synapsis. It is as if they become stitched together point by point along their length, with little space between. The intimate, parallel array favors **crossing over**, a type of molecular interaction between the chromatids of a pair of homologous chromosomes. These *nonsister* chromatids break at the same places along their length and exchange corresponding segments—that is, genes—at the break points. Figure 10.6c is a simplified picture of this interaction.

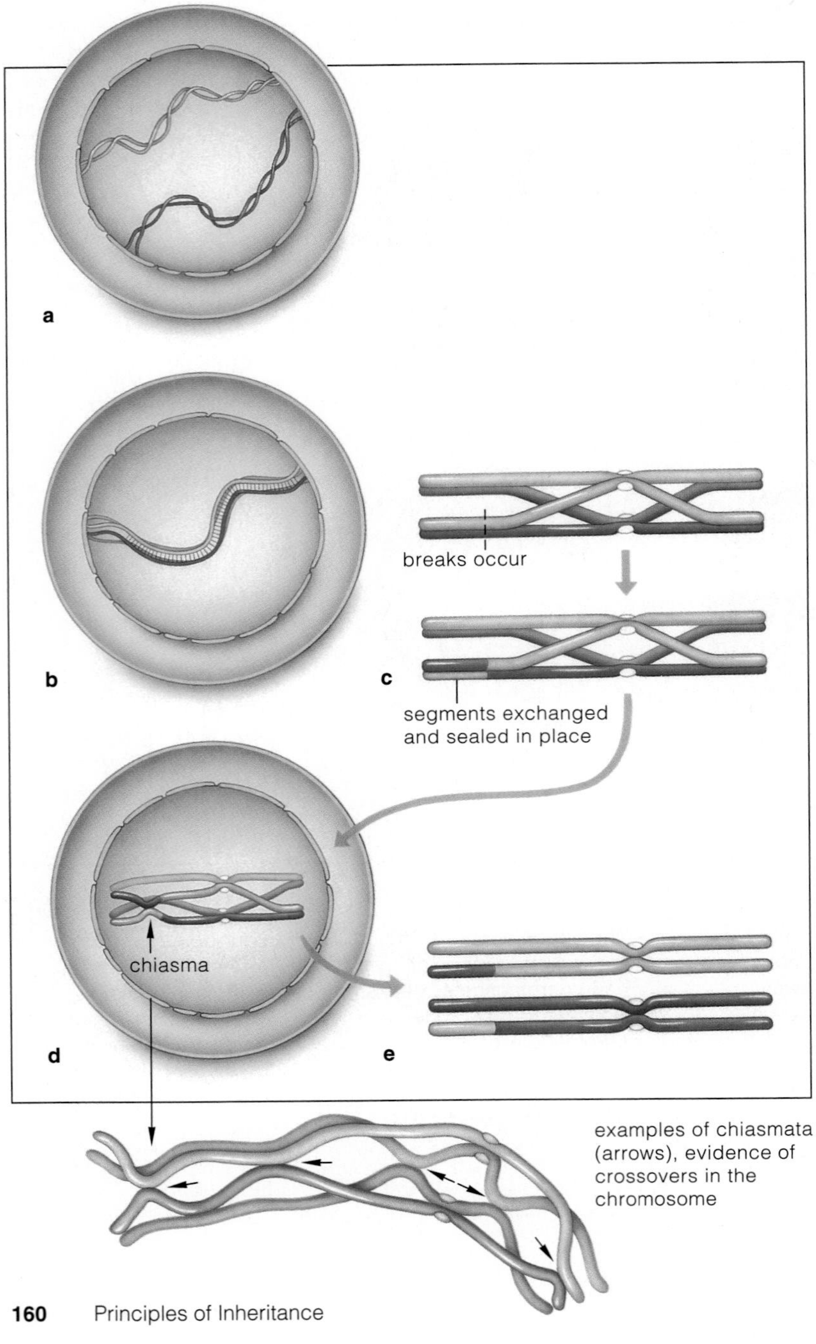

breaks occur

segments exchanged and sealed in place

chiasma

examples of chiasmata (arrows), evidence of crossovers in the chromosome

Figure 10.6 Key events during prophase I, the first stage of meiosis. For clarity, only a single pair of homologous chromosomes and only one crossover event are shown. *Blue* signifies the paternal chromosome, and *purple* signifies its maternal homologue.

a Chromosomes become duplicated before meiosis begins. Early in prophase I, each duplicated chromosome is in threadlike form, attached at both ends to the nuclear envelope. Its two sister chromatids are so close together, they look like a single thread.

b The two chromosomes become zippered together, so that all four chromatids are intimately aligned. (X and Y chromosomes, which are two forms of sex chromosomes, also become zippered together for a small part of their length.)

c One or more crossovers occur at intervals along the chromosomes. In each crossover, two nonsister chromatids break at identical sites. They swap segments at the breaks, then enzymes seal the broken ends. (For clarity, the two chromosomes are shown in condensed form and pulled apart. Crossing over may seem more plausible when you realize that it occurs while chromosomes are extended like threads and tightly aligned.)

d As prophase I ends, the chromosomes continue to condense, becoming thicker, rodlike forms. They detach from the nuclear envelope and from each other—except at "chiasmata." Each chiasma is indirect evidence that a crossover occurred at some point in the chromosomes.

e Crossing over breaks up old combinations of alleles and puts new ones together in pairs of homologous chromosomes.

Gene swapping would be rather pointless if each type of gene never varied from one chromosome to the next. But remember, a gene can have slightly different forms—alleles. You can safely bet that some alleles on one chromosome will *not* be identical to those on the homologue. So each crossover represents a chance to swap a *slightly different version* of the hereditary instructions for a particular trait.

We will look at the mechanism of crossing over in later chapters. For now, it is enough to know that crossing over leads to genetic recombination, which in turn leads to variation in the traits of offspring.

Crossing over is an interaction between a pair of homologous chromosomes. It breaks up old combinations of alleles and puts new ones together.

Metaphase I Alignments

Major shufflings of whole chromosomes begin during the transition from prophase I to metaphase I, the second stage of meiosis. Suppose the shufflings are proceeding right now in one of your germ cells. By now, crossovers have made genetic mosaics out of the chromosomes, but let's put this aside to simplify tracking. Just call the twenty-three chromosomes inherited from your mother the *maternal* chromosomes and call their twenty-three homologues from your father, the *paternal* chromosomes.

Microtubules have harnessed and oriented one chromosome of each pair toward one spindle pole and its homologue toward the other. Now they are moving all the chromosomes, which soon will be positioned midway between the spindle poles.

Are all the maternal chromosomes attached to one pole and all the paternal ones attached to the other? Maybe, but probably not. Remember, the first contacts between spindle microtubules and chromosomes are random. Because of this random harnessing, *the eventual positioning of a maternal or paternal chromosome at the spindle equator at metaphase I follows no particular pattern.* Carrying this one step further, either chromosome of a pair may end up at either pole of the spindle when the two move apart at anaphase I.

Think of the possibilities when merely three pairs of homologues are involved. As Figure 10.7 shows, by metaphase I, these may be arranged in any one of four possible positions. In this case, 2^3 or 8 combinations of maternal and paternal chromosomes are possible for the forthcoming gametes.

Of course, a human germ cell has twenty-three pairs of homologous chromosomes, not just three. So 2^{23} or *8,388,608 combinations* of maternal and paternal chromo-

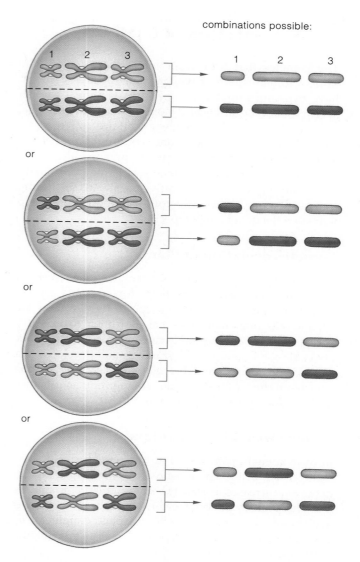

Figure 10.7 The possible outcomes of the random alignment of three pairs of homologous chromosomes at metaphase I of meiosis. The three types of chromosomes are labeled 1, 2, and 3. Maternal chromosomes are *purple;* paternal ones are *blue.* With merely four possible alignments, eight different combinations of maternal and paternal chromosomes are possible in forthcoming gametes.

somes are possible every time a germ cell gives rise to sperm or eggs! And in each sperm or egg, many hundreds of alleles inherited from the mother might differ from their counterparts from the father. Are you beginning to get an idea of why such splendid mixes of traits show up even in the same family?

The random attachment and subsequent positioning of each pair of maternal and paternal chromosomes at metaphase I lead to different combinations of maternal and paternal traits in offspring.

10.4 FORMATION OF GAMETES

The gametes that form after meiosis are not all the same. Human sperm have one tail, opossum sperm have two, and roundworm sperm have none. Crayfish sperm look like pinwheels. Most eggs are microscopic, yet ostrich eggs tucked inside a shell are as large as a softball. From appearance alone, you might not believe a plant gamete is even remotely like an animal's.

Later chapters contain details of gamete formation in the life cycles of representative organisms, including humans. Figure 10.8 and the following points may help you keep the details in perspective.

Gamete Formation in Plants

For pine trees, roses, and other familiar plants, certain events intervene between the time of meiosis and gamete formation. Among other things, spores form. **Spores** are haploid cells, often with walls that allow them to resist dry periods or other adverse environmental conditions. Under favorable conditions, spores germinate and develop into a haploid body or structure that will produce gametes. Thus gamete-producing bodies and spore-producing bodies develop during the life cycle (Figure 10.8*a*).

Gamete Formation in Animals

In male animals, gametes form by a process called spermatogenesis. Inside a male reproductive system, a diploid germ cell increases in size. It becomes a large, immature cell (the primary spermatocyte) that undergoes meiosis and cytoplasmic divisions. Its four haploid daughter cells develop into immature cells called spermatids (Figure 10.9). Spermatids change in form and develop a tail. In this way, each becomes a **sperm**, a male gamete.

In female animals, gametes form by a process called oogenesis. Each diploid germ cell develops into an immature egg (oocyte). Compared to a sperm, an oocyte accumulates more cytoplasmic components. Also, its daughter cells differ in size and function (Figure 10.10). When the oocyte divides after meiosis I, one cell (the secondary oocyte) gets nearly all the cytoplasm. So the other cell (the first polar body) is quite small. Both may undergo meiosis II and cytoplasmic division. Here again, one cell gets most of the cytoplasm. It develops into a gamete. A mature female gamete is called an ovum or, more commonly, an **egg**.

Division of the smaller cell means there are now three polar bodies. These do not function as gametes. In effect, they are dumping grounds for chromosomes, so the egg ends up with the required haploid number. Because polar bodies do not have much cytoplasm, they do not have much in the way of nutrients and metabolic machinery. In time they degenerate.

10.5 MORE GENE SHUFFLINGS AT FERTILIZATION

The parental chromosome number is restored at **fertilization**, when the nuclei of two haploid gametes fuse. Unless meiosis precedes it, fertilization would result in a doubling of the chromosome number in every new generation.

Fertilization contributes to variation in offspring. Reflect on the possibilities for humans alone. During prophase I of meiosis, an average of two or three crossovers takes place in each human chromosome. Even without crossovers, the random positioning of pairs of paternal and maternal chromosomes at metaphase I results in one of 8,388,608 possible chromosome combinations in each gamete. And of all the male and female gametes that are produced, *which two* actually get together is a matter of chance. Thus the sheer number of combinations that can exist at fertilization is staggering!

Cumulatively, crossing over, the distribution of random mixes of homologous chromosomes into gametes, and fertilization contribute to variation in the traits of offspring.

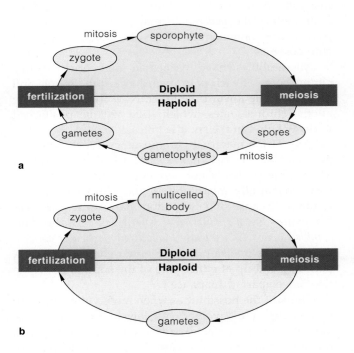

Figure 10.8 Generalized life cycles for (**a**) multicelled plants and (**b**) animals. A zygote is the first cell formed after two gametes fuse at fertilization. For plants, a sporophyte (spore-producing body) forms from the zygote, and gametophytes (gamete-producing bodies) form after meiosis. A pine tree is a sporophyte. Gametophytes develop in its cones (page 401).

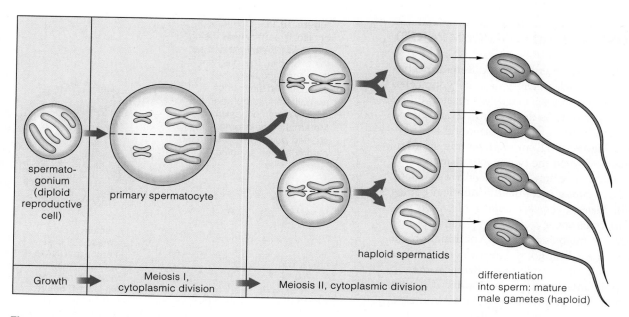

Figure 10.9 Generalized picture of sperm formation in male animals.

spermato-
gonium
(diploid
reproductive
cell)

primary spermatocyte

haploid spermatids

differentiation
into sperm: mature
male gametes (haploid)

Growth

Meiosis I,
cytoplasmic division

Meiosis II, cytoplasmic division

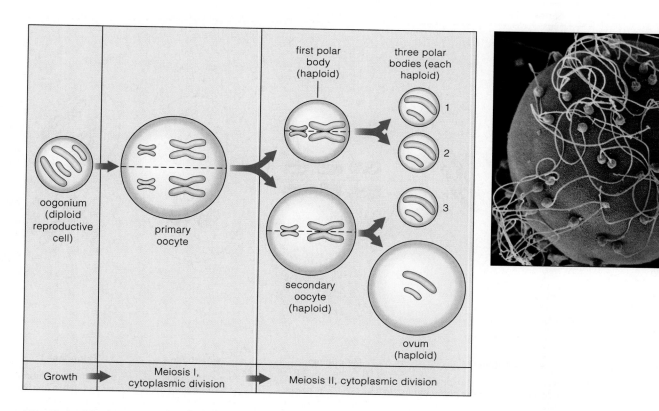

Figure 10.10 Generalized picture of egg formation in female animals. The diagram is not at the same scale as Figure 10.9. An egg is *much* larger than a sperm, as indicated by the micrograph of sea urchin gametes. Also, the three polar bodies are extremely small compared to an egg.

first polar
body
(haploid)

three polar
bodies (each
haploid)

oogonium
(diploid
reproductive
cell)

primary
oocyte

secondary
oocyte
(haploid)

ovum
(haploid)

Growth

Meiosis I,
cytoplasmic division

Meiosis II, cytoplasmic division

10.6 MEIOSIS AND MITOSIS COMPARED

In this unit, our focus has been on two different mechanisms that divide the nuclear DNA of eukaryotic cells. Single-celled species use mitosis in asexual reproduction; multicelled species use mitosis during growth and tissue repair. Meiosis is the basis of gamete formation and sexual reproduction. Figure 10.11 summarizes the similarities and differences between the two mechanisms.

The two mechanisms differ in a crucial way. *Mitotic cell division produces clones* (genetically identical copies of a parent cell). *Meiosis, together with fertilization, promotes variation in traits among offspring.* First, crossing over at prophase I of meiosis puts new combinations of alleles in chromosomes. Second, the movement of either member of each pair of homologous chromosomes to either spindle pole after metaphase I puts different mixes of maternal and paternal alleles into gametes. Third, different combinations of alleles are brought together by chance at fertilization. Later chapters describe how meiosis and fertilization contribute to the evolution of sexually reproducing organisms.

Figure 10.11 Comparison of mitosis and meiosis, using a diploid (2n) animal cell as the example. This diagram is arranged to help you compare the similarities and differences between the two division mechanisms. Maternal chromosomes are shaded *purple*, and paternal chromosomes are *blue*.

A diploid (2n) *somatic* cell is at interphase. DNA is replicated (all chromosomes are duplicated) before nuclear division begins.

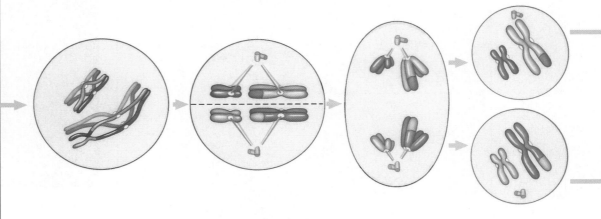

Meiosis I

A diploid (2n) *reproductive* cell is at interphase. DNA is replicated (all chromosomes are duplicated) before nuclear division begins.

Prophase I

Each duplicated chromosome (consisting of two sister chromatids) condenses to threadlike form, then rodlike form. *Crossing over* occurs. Each chromosome unzips from its homologue. Each gets attached to the spindle in transition to metaphase.

Metaphase I

All chromosomes are now positioned at the spindle's equator.

Anaphase I

Each chromosome is separated from its homologue. They are moved to opposite poles of the spindle.

Telophase I

When the cytoplasm divides, there are two cells. Each has a haploid (n) number of chromosomes, but these are still in the duplicated state.

Mitosis

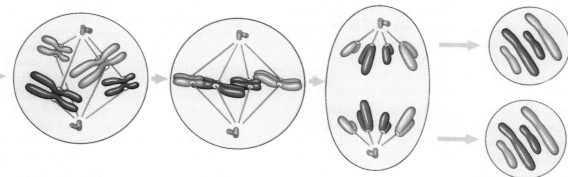

Prophase

Each duplicated chromosome (consisting of two sister chromatids) condenses from threadlike form to rodlike form. Each gets attached to the spindle during the transition to metaphase.

Metaphase

All chromosomes are now positioned at the spindle's equator.

Anaphase

Sister chromatids of each chromosome are separated from each other. These new, daughter chromosomes are moved to opposite poles of the spindle.

Telophase

When the cytoplasm divides, there are two cells. Each is diploid (2*n*)—*it has the same chromosome number as the parent cell.*

Meiosis II

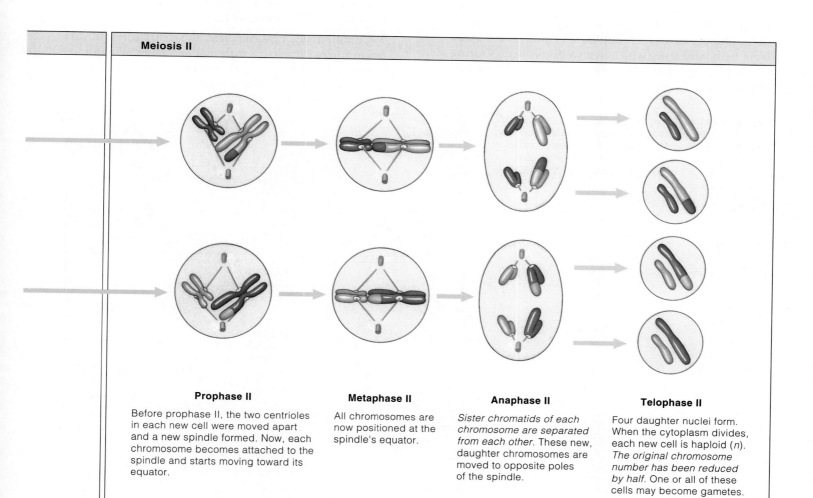

Prophase II

Before prophase II, the two centrioles in each new cell were moved apart and a new spindle formed. Now, each chromosome becomes attached to the spindle and starts moving toward its equator.

Metaphase II

All chromosomes are now positioned at the spindle's equator.

Anaphase II

Sister chromatids of each chromosome are separated from each other. These new, daughter chromosomes are moved to opposite poles of the spindle.

Telophase II

Four daughter nuclei form. When the cytoplasm divides, each new cell is haploid (*n*). *The original chromosome number has been reduced by half.* One or all of these cells may become gametes.

SUMMARY

1. Life cycles of sexually reproducing species proceed through meiosis, gamete formation, and fertilization. Meiosis reduces the chromosome number of a parent cell by half. It precedes the formation of haploid gametes (typically, sperm in males, eggs in females). Fusion of a sperm nucleus and an egg nucleus at fertilization restores the chromosome number (Figure 10.12).

2. If cells of sexually reproducing organisms are diploid (2n), they have two of each type of chromosome characteristic of the species. Commonly, one of the two is maternal (inherited from a female parent), and the other is paternal (from a male parent).

3. Each pair of maternal and paternal chromosomes shows homology (the two are alike). Generally, the two have the same length, same shape, and same sequence of genes. They interact during meiosis.

4. Chromosomes are duplicated during interphase. So before meiosis, each consists of two DNA molecules that remain attached (as sister chromatids).

5. Meiosis consists of two consecutive divisions that require a spindle apparatus. In meiosis I, microtubules of the spindle move each duplicated chromosome away from its homologue (which is also duplicated). During meiosis II, microtubules separate the sister chromatids of each chromosome.

6. The following activities proceed during meiosis I:

   a. At prophase I, nonsister chromatids of homologous chromosomes break at corresponding sites and exchange segments. Crossing over puts new combinations of alleles together. Alleles (slightly different molecular forms of the same gene) code for different forms of the same trait. So new allelic combinations lead to variation in the details of a given trait among offspring.

   b. At metaphase I, all pairs of homologous chromosomes are positioned at the spindle equator. In each case, either the maternal chromosome or its homologue can be oriented toward either pole.

   c. At anaphase I, each chromosome separates from its homologue, and the two move to opposite poles.

7. The following activities proceed during meiosis II:

   a. At metaphase II, chromosomes are still duplicated, and they are positioned at the spindle equator.

   b. During anaphase II, the sister chromatids of each chromosome are moved apart. Each is now a separate, unduplicated chromosome.

   c. During telophase II, four haploid nuclei form.

8. When the cytoplasm divides, there are four haploid cells. One or all of those cells may function as gametes (or as spores, in the case of flowering plants).

9. Cumulatively, crossing over, the distribution of different mixes of pairs of maternal and paternal chromosomes into gametes, and fertilization contribute to immense variation in the traits of offspring.

Figure 10.12 Summary of changes in the chromosome number at different stages of sexual reproduction, using diploid (2n) germ cells as the example. Meiosis reduces the chromosome number by half (n). The union of haploid nuclei of two gametes at fertilization restores the diploid number.

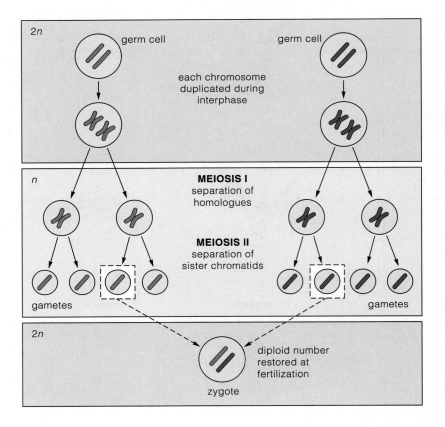

1. Name the three key events of sexual reproduction. *156*

2. What is the key difference between sexual and asexual reproduction? *156*

3. Refer to the following numbers of chromosomes in the diploid body cells of a few organisms. In each case, how many chromosomes would end up in gametes? *157*

Fruit fly, *Drosophila melanogaster*	8
Garden pea, *Pisum sativum*	14
Corn, *Zea mays*	20
Frog, *Rana pipiens*	26
Earthworm, *Lumbricus terrestris*	36
Human, *Homo sapiens*	46
Chimpanzee, *Pan troglodytes*	48
Amoeba, *Amoeba*	50
Horsetail, *Equisetum*	216

4. Suppose a diploid germ cell has four pairs of homologous chromosomes, designated AA, BB, CC, and DD. How would the chromosomes of the gametes be designated? *156–157*

5. Define meiosis and characterize its main stages. In what respects is meiosis *not* like mitosis? *156–159, 164–165*

6. Is this cell at anaphase I or anaphase II?

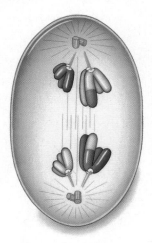

7. Outline the steps involved in the formation of sperm and eggs in animals. *162–163*

Self-Quiz *(Answers in Appendix IV)*

1. Sexual reproduction requires _____ .
   a. meiosis
   b. gamete formation
   c. fertilization
   d. all of the above

2. Meiosis is a division mechanism that produces _____ .
   a. two cells
   b. two nuclei
   c. four cells
   d. four nuclei

3. An animal cell with two of each type of chromosome characteristic of the species is _____ .
   a. diploid
   b. haploid
   c. probably not a normal gamete
   d. both b and c

4. Meiosis _____ the parental chromosome number.
   a. doubles
   b. reduces
   c. maintains
   d. corrupts

5. Generally, a pair of homologous chromosomes _____ .
   a. carry the same genes
   b. are the same length and shape
   c. interact at meiosis
   d. all of the above

6. Before the onset of meiosis, all chromosomes are _____ .
   a. condensed
   b. released from protein
   c. duplicated
   d. b and d

7. Each chromosome moves away from its homologue and ends up at the opposite spindle pole during _____ .
   a. prophase I
   b. prophase II
   c. anaphase I
   d. anaphase II

8. Sister chromatids of each chromosome move apart and end up at opposite spindle poles during _____ .
   a. prophase I
   b. prophase II
   c. anaphase I
   d. anaphase II

9. Match each term and its description.
   ____ chromosome number
   ____ alleles
   ____ metaphase I
   ____ interphase

   a. different molecular forms of the same gene
   b. none between meiosis I, II
   c. pairs of homologues aligned at spindle equator
   d. the number of each type of chromosome present in cell

Selected Key Terms

allele *156*
asexual reproduction *156*
crossing over *160*
diploid *157*
egg *162*
fertilization *162*
gamete *156*
gene *156*

haploid *157*
homologous chromosome *157*
meiosis *156*
sexual reproduction *156*
sister chromatid *157*
sperm *162*
spore *162*

Readings

Klug, W., and M. Cummings. 1994. *Concepts of Genetics*. Fourth edition. New York: Macmillan.

Wolfe, S. 1993. *Molecular and Cellular Biology*. Belmont, California: Wadsworth.

11 OBSERVABLE PATTERNS OF INHERITANCE

A Smorgasbord of Ears and Other Traits

Basketball ace Charles Barkley has them. So does actor Tom Cruise. Actress Joan Chen doesn't, and neither did a monk named Gregor Mendel. To see how *you* fit in with these folks, use a mirror to check your ears. Is the fleshy lobe at the base of each ear attached to the side of your head? If so, you and Barkley and Cruise have something in common. Or is the fleshy lobe unattached, so that you can flap it back and forth? If so, you are like Chen and Mendel (Figure 11.1).

Whether a person is born with detached or attached earlobes depends on a single kind of gene. That gene comes in slightly different molecular forms—alleles. Only one form has information about detached lobes. The information is put to use while a human body is developing inside the mother. It calls for a death signal, which is sent to all the cells positioned between the newly forming lobes and the head. Without the signal, the cells don't die, and earlobes don't detach.

We all have genes for thousands of traits, including earlobes, cheeks, lashes, and eyeballs. Most of the traits vary in their details from one person to the next. Remember, humans inherit pairs of genes, on pairs of chromosomes. In some pairings, one allele has power-ful effects and overwhelms the other's contribution to a trait. The outgunned allele is said to be recessive to the dominant one. If you have *detached* earlobes, *dimpled* cheeks, *long* lashes, or *large* eyeballs, you carry at least one and possibly two dominant alleles that influence the trait in a particular way.

When both alleles of a pair are recessive, nothing masks their effect on a trait. You get *attached* earlobes with one pair of recessive alleles (and *flat* feet with another, a *straight* nose with another, and so on).

How did we discover such remarkable things about our genes? It all started with Gregor Mendel. By ana-lyzing pea plants generation after generation, Mendel found indirect but *observable* evidence of how parents transmit units of hereditary information—genes—to offspring. This chapter focuses on Mendel's experimen-tal methods and results. They remain a classic example of how a scientific approach can pry open important secrets about the natural world. And to this day, they serve as the foundation for modern genetics.

a Tom Cruise

b Charles Barkley

c Joan Chen

d Gregor Mendel

Figure 11.1 The attached and detached earlobes of a few representative humans. This sampling provides observable evidence of a trait that is governed by a single gene. Do you have one or the other version of this trait? It depends on whether you inherited a specific molecular form of that gene from your mother, your father, or both of your parents.

Earlobe attachment, chin dimpling, cheek dimpling, and many other single-gene traits vary from one individual to the next. As Gregor Mendel perceived, such easily observable traits can be used to identify patterns of inheritance that exist from one generation to the next.

KEY CONCEPTS

1. Genes are units of information about heritable traits. Each gene has a specific location in the chromosomes of a species. But its molecular form may differ slightly from one chromosome to the next. Different molecular forms of a gene (alleles) specify different versions of the same trait.

2. The two genes of a pair are segregated from each other during meiosis and end up in different gametes. Gregor Mendel found indirect evidence of this when he crossbred plants having observable differences in the same trait.

3. The sorting of each gene pair into different gametes tends to be independent of how the other gene pairs are sorted out. Mendel found evidence of this when he tracked plants having observable differences in *two* traits, such as flower color and height.

4. If the two genes of a pair specify different versions of a trait (if they are nonidentical alleles), one may have more pronounced effects on the trait. Besides this, two or more gene pairs often influence the same trait, and some single genes influence many traits. Finally, environmental conditions may alter gene expression.

More than a century ago, Charles Darwin explained how natural selection might bring about evolutionary change. According to his key premise, individuals vary in heritable traits. Variations that improve chances of surviving and reproducing are favored more often in each generation. Those that don't become less frequent. Thus the population changes over time—it evolves.

Darwin's theory did not fit with a prevailing idea about inheritance. It was common knowledge that sperm and eggs both transmit information about traits to offspring. But almost no one suspected the information is organized in *units* (genes). Instead, the idea was that a father's blob of information "blended" with a mother's blob at fertilization, like cream into coffee.

Carried to its logical conclusion, blending would eventually dilute hereditary information until there was only one version left of each trait. Yet why did *freckled* children keep turning up among nonfreckled generations? Why weren't all the descendants of a herd of white stallions and black mares *gray*? The blending theory scarcely explained what people could see with their own eyes. But few disputed it. Blending happened, so populations "had to be" uniform—and with uniformity in traits, evolution could not occur.

Even before Darwin presented his theory, however, someone was gathering evidence that eventually would support his key premise. A monk, Gregor Mendel, was about to prove that sperm and eggs do indeed carry "units" that deal with separate heritable traits.

11.1 MENDEL'S INSIGHT INTO PATTERNS OF INHERITANCE

Mendel's Experimental Approach

Mendel spent most of his life in a monastery in Brünn, an Austrian village that has since become part of the Czech Republic. The monastery of St. Thomas was somewhat removed from the European capitals, which were then the centers of scientific inquiry. Yet Mendel was not a man of narrow interests who simply stumbled by chance onto principles of great import. Having been raised on a farm, he was well aware of agricultural principles and their application. He kept abreast of breeding experiments and developments described in the available books. Mendel was a member of the regional agricultural society. He won several awards for developing improved varieties of vegetables and fruits. After entering the monastery, he spent two years studying mathematics at the University of Vienna. Few schol-

ars of his time had similarly combined talents in plant breeding and mathematics.

Shortly after his university training, Mendel began experiments with the garden pea plant, *Pisum sativum* (Figure 11.2). This plant is self-fertilizing. Its flowers produce sperm and eggs, which meet up in the same flower. Some of the plants are **true-breeding**. In other words, successive generations are just like the parents in one or more traits, as when all plants grown from seeds of white-flowered plants have white flowers.

As Mendel knew, pea plants also will cross-fertilize when sperm and eggs from different plants are brought together under controlled conditions. For his experiments, he could open flower buds of a plant that bred true for a trait (say, white flowers) and snip out the stamens. (Stamens bear pollen grains in which sperm develop.) Then he could brush the "castrated" buds with pollen from a plant that bred true for a *different* version of the same trait (purple flowers). Mendel hypothesized that such clearly observable differences

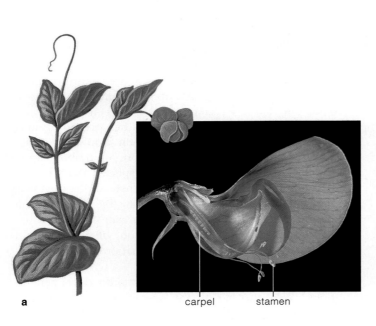

a carpel   stamen

Figure 11.2 The garden pea plant (*Pisum sativum*), the focus of Mendel's experiments. A flower has been sectioned to show the location of its stamens and carpel. Sperm-producing pollen grains form in stamens. Eggs develop, fertilization takes place, and seeds mature inside the carpel.

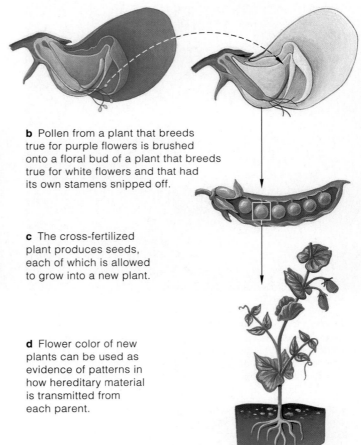

b Pollen from a plant that breeds true for purple flowers is brushed onto a floral bud of a plant that breeds true for white flowers and that had its own stamens snipped off.

c The cross-fertilized plant produces seeds, each of which is allowed to grow into a new plant.

d Flower color of new plants can be used as evidence of patterns in how hereditary material is transmitted from each parent.

could be used to track a trait through many generations. If there were patterns to the trait's inheritance, *those patterns might tell him something about the hereditary material itself.*

Some Terms Used in Genetics

Having read the chapter on meiosis, you already have insight into the mechanisms of sexual reproduction—which is more than Mendel had. He did not know about chromosomes. So he could not have known that the chromosome number is reduced by half in gametes, then restored at fertilization. Yet Mendel sensed what was going on. As we follow his thinking, let's simplify things by substituting a few modern terms used in studies of inheritance (see also Figure 11.3):

---

1. **Genes** are units of information about specific traits, and they are passed from parents to offspring. Each gene has a specific location (locus) on a chromosome.

2. Diploid cells have a pair of genes for each trait, on a pair of homologous chromosomes.

3. Although both genes of a pair deal with the same trait, they may vary in their information about it. This happens when they have slight molecular differences, as when one gene for flower color specifies purple and another specifies white. The different molecular forms of a gene are called **alleles** of that gene.

4. If it turns out that the two alleles of a pair are the same, this is a *homozygous* condition. If different, this is a *heterozygous* condition.

5. An allele is *dominant* when its effect on a trait masks that of any *recessive* allele paired with it. We use capital letters for dominant alleles and lowercase letters for recessive ones (for instance, *A* and *a*).

6. Putting this together, a **homozygous dominant** individual has a pair of dominant alleles (*AA*) for the trait that is being studied. A **homozygous recessive** individual has a pair of recessive alleles (*aa*) for the trait. A **heterozygous** individual has a pair of nonidentical alleles (*Aa*) for the trait.

7. Two terms help keep the distinction clear between genes and the traits they specify. **Genotype** refers to the genes present in an individual. **Phenotype** refers to an individual's observable traits.

8. When tracking the inheritance of traits through generations of offspring, these abbreviations apply:

P parental generation
F_1 first-generation offspring
F_2 second-generation offspring

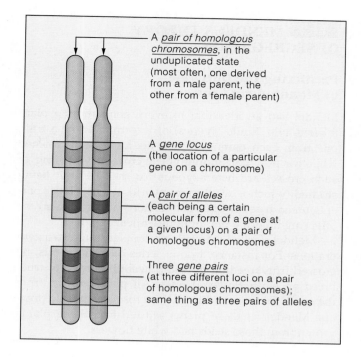

A *pair of homologous chromosomes*, in the unduplicated state (most often, one derived from a male parent, the other from a female parent)

A *gene locus* (the location of a particular gene on a chromosome)

A *pair of alleles* (each being a certain molecular form of a gene at a given locus) on a pair of homologous chromosomes

Three *gene pairs* (at three different loci on a pair of homologous chromosomes); same thing as three pairs of alleles

Figure 11.3 A few genetic terms illustrated. Diploid organisms have pairs of genes, on pairs of homologous chromosomes. (In your case, one chromosome of each pair was inherited from your mother, and the other homologous chromosome was inherited from your father.)

The genes themselves may have different molecular forms, called alleles. Different alleles specify slightly different versions of the same trait. An allele at one location on a chromosome may or may not be identical to its partner on the homologous chromosome.

When the offspring of genetic crosses inherit identical alleles for a given trait, generation after generation, they are said to be a true-breeding lineage. When offspring of a genetic cross inherit nonidentical alleles for a trait under study, they are said to be hybrid offspring.

11.2 MENDEL'S THEORY OF SEGREGATION

Predicting the Outcome of Monohybrid Crosses

Mendel had an idea that in every generation, a plant inherits two "units" (genes) of information for a trait, one from each parent. To test his idea, he performed what we now call **monohybrid crosses**. Offspring of such crosses are heterozygous for the one trait being studied (which is what monohybrid means). Their parents breed true for different versions of the trait, so the offspring inherit a pair of nonidentical alleles.

Mendel tracked many single traits through two generations. For instance, in one series of experiments, he crossed true-breeding purple-flowered plants and true-breeding white-flowered ones. All plants grown from the seeds that resulted from this cross had purple flowers. Mendel let these plants self-fertilize. Some plants grown from those seeds had white flowers!

If Mendel's hypothesis were correct—if each plant had inherited two units of information about flower color—then the "purple" unit would have to be dominant, for it masked the "white" unit in F_1 plants.

Let's rephrase Mendel's thinking. Germ cells of garden pea plants are diploid, with pairs of homologous chromosomes. Assume one parent plant is homozygous dominant (*AA*) for flower color and the other is homozygous recessive (*aa*). After meiosis, their sperm or eggs will carry only one allele for flower color (Figure 11.4). Thus, when a sperm fertilizes an egg, only one outcome is possible: $A + a = Aa$.

Before continuing, you should know that Mendel crossed hundreds of plants and tracked thousands of offspring. He also counted and recorded the plants showing dominance or recessiveness. As Figure 11.5 indicates, an intriguing ratio emerged. On the average, three of every four F_2 plants had the dominant phenotype and one had the recessive phenotype.

To Mendel, this ratio suggested that fertilization is a chance event, with a number of possible outcomes. And

homozygous dominant parent
homozygous recessive parent
(chromosomes duplicated before meiosis)
meiosis I
meiosis II
(gametes)
(gametes)
fertilization produces heterozygous offspring

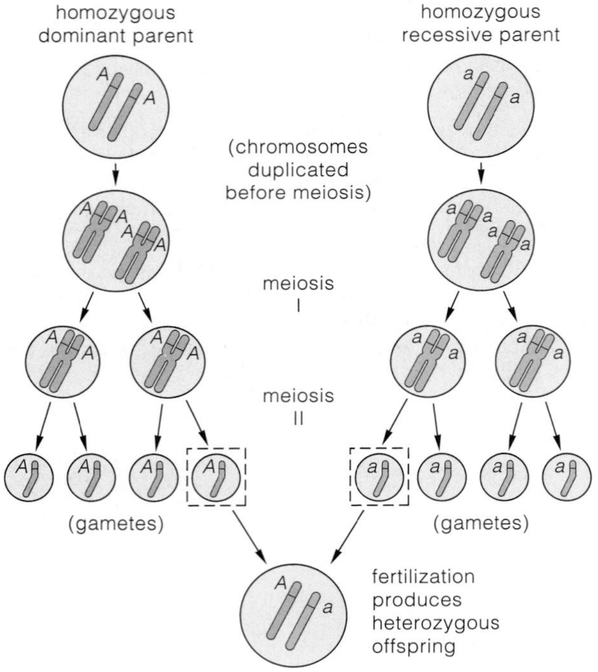

Figure 11.4 Example of a monohybrid cross, showing how one gene of a pair segregates from the other. Two parents that breed true for two different versions of a trait produce heterozygous offspring only.

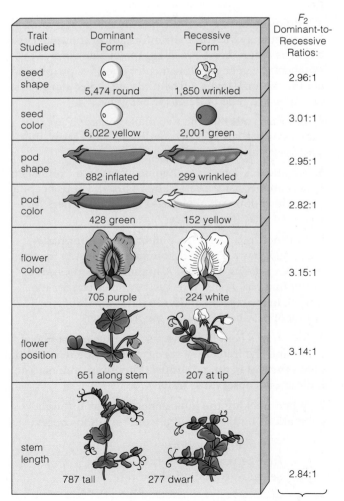

Trait Studied
Dominant Form
Recessive Form
F_2 Dominant-to-Recessive Ratios:
seed shape
5,474 round
1,850 wrinkled
2.96:1
seed color
6,022 yellow
2,001 green
3.01:1
pod shape
882 inflated
299 wrinkled
2.95:1
pod color
428 green
152 yellow
2.82:1
flower color
705 purple
224 white
3.15:1
flower position
651 along stem
207 at tip
3.14:1
stem length
787 tall
277 dwarf
2.84:1
Average ratio for all traits studied:
3:1

Figure 11.5 (*Right*) Results from Mendel's monohybrid cross experiments with the garden pea. The numbers are his counts of F_2 plants that he assumed were carrying dominant or recessive hereditary "units" (alleles) for the trait. On the average, the dominant-to-recessive ratio was 3:1.

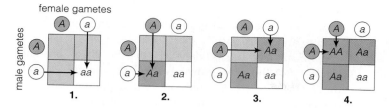

female gametes
male gametes

1.   2.   3.   4.

Figure 11.6 Punnett-square method of predicting the probable outcome of a genetic cross. Circles represent gametes. Letters on gametes represent dominant or recessive alleles. The different squares depict the different genotypes possible among offspring. In this example, gametes are from a self-fertilizing plant that is heterozygous (*Aa*) for a trait.

Figure 11.7 Results from one of Mendel's monohybrid crosses. On the average, the dominant-to-recessive ratio among F_2 plants was 3:1.

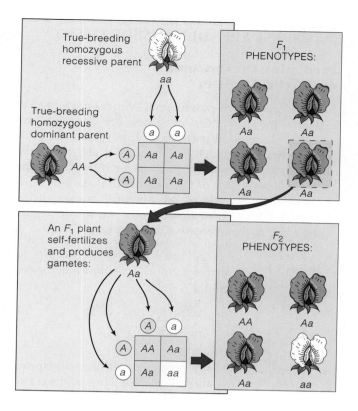

he had an understanding of **probability**, which applies to chance events *and so could help him predict the possible outcomes of crosses.* (Probability simply means that the chance of each outcome occurring is proportional to the number of ways it can be reached.) The **Punnett-square method**, explained in Figure 11.6 and applied in Figure 11.7, can help you visualize the possibilities. As you can see, if half of a plant's sperm (or eggs) were *a* and half were *A*, four outcomes were possible every time a sperm fertilized an egg:

Possible event:	Probable outcome:
sperm *A* meets egg *A*	1/4 *AA* offspring
sperm *A* meets egg *a*	1/4 *Aa*
sperm *a* meets egg *A*	1/4 *Aa*
sperm *a* meets egg *a*	1/4 *aa*

By this prediction, an F_2 plant had three chances in four of getting at least one dominant allele (purple flowers). It had one chance in four of getting two recessive alleles (white flowers). That is a probable phenotypic ratio of three purple to one white, or 3:1.

Mendel's observed ratios weren't *exactly* 3:1. You can see this for yourself by studying the numerical results in Figure 11.5. Why did Mendel put aside the deviations? To understand why, flip a coin a couple of times. As we all know, a coin is just as likely to end up heads as tails. But often it ends up heads, or tails, several times in a row. So if you flip the coin only a few times, the observed ratio may differ greatly from the predicted ratio of 1:1. Only by flipping the coin many, many times will you come close to the predicted ratio. Mendel knew the rules of probability—and he performed a large number of crosses. Almost certainly, this kept him from being confused by minor deviations from the predicted results of the experimental crosses.

Testcrosses

Mendel gained support for his prediction with the **testcross**. In this type of experimental cross, an organism shows dominance for a trait but its genotype is unknown, so it is crossed to a known homozygous recessive individual. Results may reveal whether the organism is homozygous dominant or heterozygous.

With respect to the monohybrid crosses described above, Mendel tested his prediction that the purple-flowered F_1 offspring were heterozygous. He crossed F_1 plants with true-breeding, white-flowered plants. If the F_1 plants were homozygous dominant, all of their offspring would show dominance. If heterozygous, there would be about as many recessive as dominant offspring. Sure enough, about half the offspring had purple flowers (*Aa*) and half had white (*aa*). Can you construct two Punnett squares that show the two possible outcomes of this testcross?

On the basis of results from Mendel's monohybrid crosses and testcrosses, it is possible to formulate a theory, which we state here in modern terms:

MENDEL'S THEORY OF SEGREGATION.   **Diploid cells have pairs of genes (on pairs of homologous chromosomes). During meiosis, the two genes of each pair segregate from each other. As a result, they end up in different gametes.**

11.3 INDEPENDENT ASSORTMENT

Predicting the Outcome of Dihybrid Crosses

By another series of experiments, Mendel attempted to explain how *two* pairs of genes are assorted into gametes. He selected true-breeding pea plants that differed in two traits, such as flower color and height. In such **dihybrid crosses**, the F_1 offspring inherit two gene pairs, each consisting of two nonidentical alleles.

Let's diagram one of his dihybrid crosses, using dominant alleles *A* for flower color and *B* for height, and *a* and *b* as their recessive counterparts:

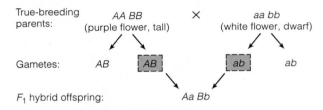

As Mendel would have predicted, all the F_1 offspring are purple-flowered and tall (*Aa Bb*). How will the two

gene pairs be assorted into gametes when those plants mature and reproduce?

The answer partly depends on the chromosomal locations of the gene pairs. Assume the *Aa* alleles and the *Bb* alleles are located on two different pairs of homologous chromosomes. Next, think about how all chromosomes become positioned at the spindle equator during metaphase I of meiosis (page 161 and Figure 11.8). The chromosome with the *A* allele may be positioned to move to either spindle pole (and on into one of four gametes). The same is true for its homologue. And the same is true for the chromosomes with the *B* and *b* alleles. Therefore, after meiosis, four combinations of alleles are possible in sperm or eggs: 1/4 *AB*, 1/4 *Ab*, 1/4 *aB*, and 1/4 *ab*.

This permits several different combinations at fertilization. Simple multiplication (four kinds of sperm times four kinds of eggs) tells us sixteen combinations of gametes are possible in the F_2 offspring of a dihybrid cross. Use the Punnett-square method to diagram the probabilities (Figure 11.9). Add up the possible phenotypes and you get 9/16 tall purple-flowered, 3/16 dwarf purple-flowered, 3/16 tall white-flowered, and 1/16 dwarf white-flowered plants. That is a probable phenotypic ratio of 9:3:3:1. Results from all of Mendel's dihybrid F_2 crosses were close to a 9:3:3:1 ratio.

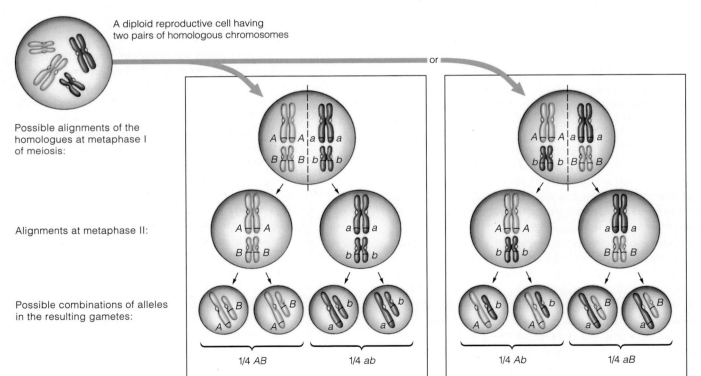

Figure 11.8 Example of independent assortment using just two pairs of homologous chromosomes. An allele at one chromosome locus may or may not be identical to its partner allele on the homologous chromosome. During meiosis, either chromosome of a pair may get attached to either spindle pole. So two different lineups are possible at metaphase I.

Figure 11.9 Results from Mendel's dihybrid cross between parent plants that bred true for different versions of two traits (flower color and height). *A* and *a* represent dominant and recessive alleles for flower color. *B* and *b* represent dominant and recessive alleles for height. The probabilities of certain combinations of phenotypes among F_2 offspring occur in a 9:3:3:1 ratio, on the average.

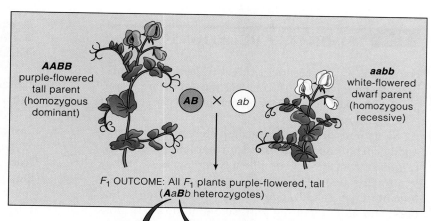

AABB
purple-flowered tall parent (homozygous dominant)

AB ✕ **ab**

aabb
white-flowered dwarf parent (homozygous recessive)

F_1 OUTCOME: All F_1 plants purple-flowered, tall (**AaBb** heterozygotes)

The Theory in Modern Form

Mendel didn't know that a pea plant's "units" of inheritance are divided up among a number of different chromosomes, so he could do no more than analyze the numbers from his dihybrid crosses. It seemed to him that the units for different traits were assorting independently into gametes, and so he formulated a theory of independent assortment. As you will read in the next chapter, we now know there are exceptions to this. So we state Mendel's theory in updated form:

MENDEL'S THEORY OF INDEPENDENT ASSORTMENT. At meiosis, the gene pairs on homologous chromosomes tend to be sorted out for distribution into one gamete or another independently of how gene pairs on other chromosomes are sorted out.

Independent assortment and hybrid crossing can lead to staggering variety among offspring. In a monohybrid cross (involving a single gene pair), three genotypes are possible (*AA*, *Aa*, and *aa*). We can represent this as 3^n, where *n* is the number of gene pairs. With more gene pairs, the number of possible combinations increases dramatically. Even if parents differ in only ten gene pairs, almost 60,000 different genotypes are possible among their offspring. If they differ in twenty gene pairs, the number approaches 3.5 billion!

On the basis of his experimental results, Mendel was convinced that the hereditary material comes in units that retain their physical identity from one generation to the next, despite being segregated and assorted before gametes form. In 1865 he reported this idea before the Brünn Society for the Study of Natural Science. His report made no impact whatsoever. The following year his paper was published. Apparently it was read by few, and its near-universal applicability was understood by no one.

In 1871 Mendel became an abbot of the monastery, and his experiments gave way to administrative tasks. He died in 1884, never to know that his work was the starting point for the development of modern genetics.

AaBb   **AaBb**

meiosis, gamete formation   meiosis, gamete formation

	1/4 **AB**	1/4 **Ab**	1/4 **aB**	1/4 **ab**
1/4 **AB**	1/16 **AABB**	1/16 **AABb**	1/16 **AaBB**	1/16 **AaBb**
1/4 **Ab**	1/16 **AABb**	1/16 **AAbb**	1/16 **AaBb**	1/16 **Aabb**
1/4 **aB**	1/16 **AaBB**	1/16 **AaBb**	1/16 **aaBB**	1/16 **aaBb**
1/4 **ab**	1/16 **AaBb**	1/16 **Aabb**	1/16 **aaBb**	1/16 **aabb**

Possible outcomes of cross-fertilization

ADDING UP THE F_1 COMBINATIONS POSSIBLE:

■ 9/16 or 9 purple-flowered, tall

□ 3/16 or 3 purple-flowered, dwarf

▨ 3/16 or 3 white-flowered, tall

□ 1/16 or 1 white-flowered, dwarf

11.4 DOMINANCE RELATIONS

For the most part, Mendel studied traits having clearly dominant or recessive forms. As the following examples will make clear, the expression of other traits is not as straightforward.

Sometimes one allele is incompletely dominant over another. This condition, called **incomplete dominance**, results in a heterozygous phenotype that is *somewhere between* the two homozygous phenotypes. Cross true-breeding red snapdragons with true-breeding white ones, and all F_1 offspring will have pink flowers. Cross two F_1 plants, and you get red, pink, or white snapdragons in a predictable ratio, as shown in Figure 11.10.

What is the basis of this inheritance pattern? Red snapdragons have two alleles of a gene that specifies a red pigment. White snapdragons have two alleles that specify "no pigment." Pink snapdragons are heterozygous. They have one red allele, but its expression results in only enough pigment molecules to make flowers pink, not red.

Or consider **codominance**, a condition in which two nonidentical alleles of a pair specify two different phenotypes, yet one cannot mask expression of the other. If you have type AB blood, for instance, you have a pair of codominant alleles that are both being expressed in your red blood cells (see the *Focus* essay).

11.5 MULTIPLE EFFECTS OF SINGLE GENES

Expression of a single gene can influence two or more traits, an effect called **pleiotropy**. Consider the genetic disorder *sickle-cell anemia*. It arises from a mutated gene for hemoglobin (Hb^S instead of Hb^A). Hemoglobin is an oxygen-transporting protein in red blood cells. Heterozygotes (Hb^A/Hb^S) typically show few symptoms; they produce enough normal hemoglobin molecules to cover for the abnormal ones. Homozygotes (Hb^S/Hb^S) can only produce abnormal hemoglobin, and drastic changes in phenotype may follow. The changes start with disruptions to the oxygen concentration in blood.

Humans depend on oxygen for aerobic respiration. Oxygen enters the lungs, then is picked up by the bloodstream and distributed to tissues throughout the body. In those tissues, oxygen diffuses out of thin-walled blood vessels called capillaries, then diffuses into cells. The concentration gradient for oxygen is steep across capillary walls. Most of the oxygen dissolved in blood moves out, into the surrounding tissues. In response to the decreased oxygen concentration, abnormal hemoglobin molecules in the red blood cells stick together in long, rodlike arrangements. The rods distort the cells into a shape rather like a sickle, a short-handled farm tool with a crescent-shaped blade. The cells rupture easily, and they clog and rupture capillaries. Body tissues become oxygen-starved, and metabolic wastes build up. Figure 11.11 tracks the resulting phenotypic changes. Other aspects of this disorder are described in later chapters.

homozygous dominant parent × homozygous recessive parent

All F_1 offspring are heterozygous.

cross between two F_1 plants

F_2 offspring show three phenotypes in a 1:2:1 ratio.

Figure 11.10 Example of incomplete dominance. Red-flowering and white-flowering homozygous snapdragons produce pink-flowering plants in the first generation. In heterozygotes, the red allele is only partly dominant over the white allele.

Focus on Health

ABO Blood Typing

Various molecules at the surface of your cells are "self" markers—they identify the cells as being part of your own body. One kind of marker on red blood cells has different molecular forms. **ABO blood typing**, a method of analysis, can reveal which form a person has.

In the human population, there are three alleles of the gene that specifies this marker. They influence its form in different ways. Two alleles, I^A and I^B, are codominant when paired with each other. A third allele, i, is recessive. When paired with either I^A or I^B, its effect is masked. Whenever three or more alleles of a gene exist in a population, we call this a **multiple allele system**.

Think about a cell that is making new molecules of the marker. Before each molecule becomes positioned at the cell surface, it takes on final form in the cytomembrane system (page 64). A carbohydrate chain is attached to a lipid. Then an enzyme attaches a sugar unit to the chain. Alleles I^A and I^B code for two versions of that enzyme. The two attach different sugar units, and this gives the marker a special identity—either A or B.

Which two alleles for this enzyme do you have? With either $I^A I^A$ or $I^A i$, your blood is type A. With $I^B I^B$ or $I^B i$, it is

type B. With codominant alleles $I^A I^B$, *both* versions of the enzyme are produced. And both attach sugar units to the marker molecules. In this case, your blood is type AB. If you are homozygous recessive (*ii*), the markers never did get a sugar unit attached to them. Then your blood type is neither A nor B (that's what type "O" means).

When blood of two people mixes during transfusions, self markers must be compatible. Without the proper markers, red blood cells from a donor will be recognized as foreign. The immune system, which defends the body against "nonself" will act against the foreign cells and may cause death (Chapter 40).

range of genotypes:

$I^A I^A$		$I^B I^B$	
or		or	
$I^A i$	$I^A I^B$	$I^B i$	*ii*
A	AB	B	O

resulting blood types

a Possible combinations of alleles that are associated with ABO blood typing.

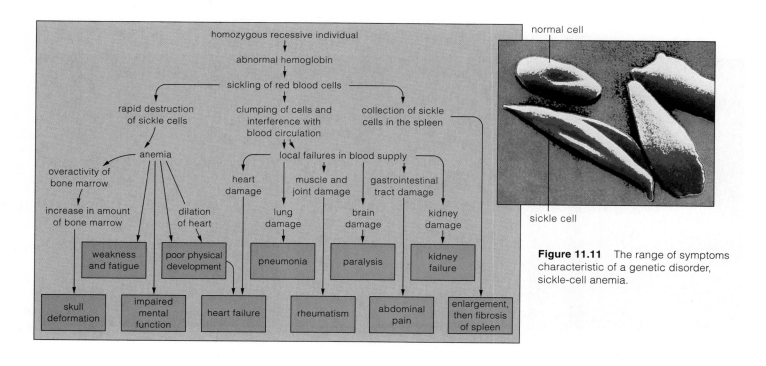

Figure 11.11 The range of symptoms characteristic of a genetic disorder, sickle-cell anemia.

11.6 INTERACTIONS BETWEEN GENE PAIRS

Often a trait results from interactions among two or more gene pairs. For example, two alleles of one gene can mask alleles of another, and some expected phenotypes may not occur. Such interactions are called **epistasis**.

Hair Color in Mammals

Epistasis is common among the gene pairs responsible for the color of fur or skin in mammals. Consider the black, brown, or yellow fur of Labrador retrievers (Figure 11.12). Variations in the amount and distribution of melanin, a brownish black pigment, produce the different colors. Many gene pairs affect different steps in melanin production and its deposition in certain body regions. Alleles of one gene specify an enzyme required to produce melanin. Allele *B* (black) has a more pronounced effect and is dominant to *b* (brown). Alleles of a different gene control whether melanin will be deposited in a retriever's hairs. Allele *E* permits deposition, but a pair of recessive alleles (*ee*) will block it, and the fur will be yellow.

The interactions between the two gene pairs just described may not even be possible if a certain allelic combination exists at still another gene locus. Here, a gene (*C*) codes for tyrosinase, the first of several enzymes in the metabolic pathway that produces melanin. An individual with one or two dominant alleles (*CC* or *Cc*) can produce the functional enzyme. An individual with two recessive alleles (*cc*) cannot. When the biosynthetic pathway leading to melanin production is blocked, the resulting condition is called *albinism*. Figure 11.13 shows an example.

Figure 11.12 Coat color of Labrador retrievers, as dictated by interactions among alleles of two gene pairs. Allele *B* (black) of one kind of gene involved in melanin production is dominant to allele *b* (brown). Allele *E* of a different gene allows melanin pigment to be deposited in individual hairs. Two recessive alleles (*ee*) of this gene block deposition, and a yellow coat results.

F_1 offspring of a dihybrid cross produce F_2 offspring in a 9:3:4 ratio, as the Punnett-square diagram shows. The yellow Labrador in photograph (**b**) probably has genotype *BB ee*, because it can produce melanin but can't deposit pigment in hairs. (Looking at that photograph, can you say why?)

P homozygotes: *BB EE* × *bb ee*

All F_1: *Bb Ee*

F_2 combinations possible:

	BE	*Be*	*bE*	*be*
BE	*BB EE*	*BB Ee*	*Bb EE*	*Bb Ee*
Be	*BB Ee*	*BB ee*	*Bb Ee*	*Bb ee*
bE	*Bb EE*	*Bb Ee*	*bb EE*	*bb Ee*
be	*Bb Ee*	*Bb ee*	*bb Ee*	*bb ee*

RESULTING PHENOTYPES:

- 9/16 or 9 black
- 3/16 or 3 brown
- 4/16 or 4 yellow

a Black Labrador

b Yellow Labrador

c Chocolate Labrador

Figure 11.13 A rare albino rattlesnake. Like other animals that cannot produce melanin, it has pink eyes and white surface coloration. (Eyes look pink because the absence of melanin from a tissue layer in the inner eyeball allows red light to be reflected from blood vessels in the eyes.) In birds and mammals, surface coloration is due to pigments in feathers, fur, or skin. In fishes, amphibians, and reptiles, color-bearing cells give skin its surface coloration. Some of the cells contain melanin pigments or yellow-to-red pigments. Others contain crystals that reflect light and alter the effect of other pigments present.

The mutation affecting melanin production in the snake shown here had no effect on the production of yellow-to-red pigments and light-reflecting crystals. So the snake's skin appears iridescent yellow as well as white.

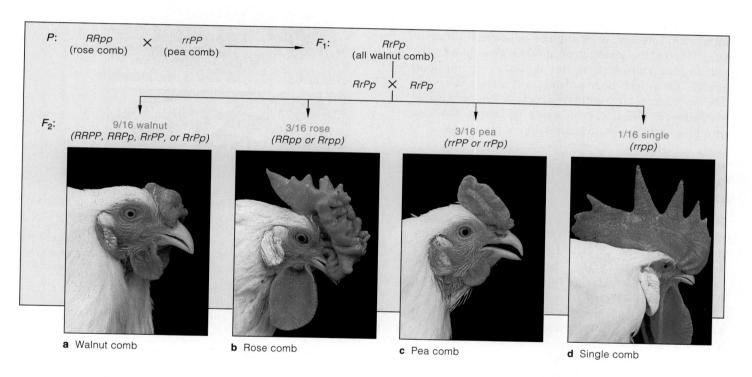

a Walnut comb   b Rose comb   c Pea comb   d Single comb

Comb Shape in Poultry

In some cases of epistasis, interaction between two gene pairs results in a phenotype that neither pair can produce alone. The geneticists W. Bateson and R. Punnett identified two interacting gene pairs (*R* and *P*) that produce comb shape in chickens. The allelic combinations of *rr* at one chromosome locus and *pp* at the other locus result in the most common phenotype, the single comb. The presence of dominant alleles (*R*, *P*, or both) results in different phenotypes. Figure 11.14 shows the allelic combinations that lead to rose, pea, or walnut combs.

Figure 11.14 Interaction between two genes affecting the same trait in domestic breeds of chickens. The initial cross is between a Wyandotte (with a rose comb, **b**, on the crest of its head) and a brahma (with a pea comb, **c**). With complete dominance at the locus for pea comb and at the locus for rose comb, the products of these two gene pairs interact and give a walnut comb (**a**). With complete recessiveness at both loci, the products interact and give rise to a single comb (**d**).

11.7 | LESS PREDICTABLE VARIATIONS IN TRAITS

As Mendel demonstrated, the phenotypic effects of one or two pairs of certain genes show up in predictable ratios from one generation to the next. Certain interactions among gene pairs also may produce phenotypes in predictable ratios, as the example of Labrador coat color demonstrated.

However, even when you are tracking a single gene through the generations, the resulting phenotypes may be not quite as expected. Consider *camptodactyly*. Some people who carry the gene that is responsible for this rare genetic disorder have immobile, bent fingers on both hands. Others have immobile, bent fingers on the left hand only—or on the right hand only. Still others who carry the mutated gene have normal fingers.

Why the confounding variation? Bear in mind, the path from most genes to their end products (proteins) is a series of small metabolic steps. Maybe one gene is mutated in any one of a number of slightly different ways. Maybe the product of another gene blocks the pathway, or causes it to run nonstop, or not long enough. Or maybe poor nutrition or some other variable environmental condition affects a key enzyme in the pathway. These are the kinds of factors that can introduce variation in the expression of phenotype.

Continuous Variation in Populations

Generally, the individuals of a population show a range of small differences in most traits. We call this **continuous variation**. The greater the number of genes and environmental factors that can influence a trait, the more continuous will be the expected distribution of all the versions of that trait.

Think about the color of your eyes. As is true of all humans, the color is the cumulative result of many genes that are involved in the stepwise production and distribution of a light-absorbing pigment, melanin. Black eyes have abundant melanin deposits in the iris and absorb most of the incoming light. Dark brown eyes have less melanin, so some light is not absorbed but rather is reflected from the iris. Pale brown or hazel eyes have even less (Figure 11.15). Green, gray, and blue eyes do not have green, gray, or blue pigments. In these cases, the iris contains different quantities of melanin, but not very much of it, and most of the blue wavelengths of light are reflected.

How can you describe the continuous variation of a trait within a group—say, the students in Figure 11.16? They range from very short to very tall, with average heights more common than either extreme. Start out by

Figure 11.15 Samples from the range of continuous variation in human eye color. Different gene pairs interact to produce and deposit melanin. Among other things, this pigment helps color the eye's iris. Different combinations of alleles result in small differences in eye color. So the frequency distribution for the eye-color trait appears to be continuous over the range from black to light blue.

dividing the full range of different phenotypes into measurable categories. Next, count the individual students in each category. This gives you the relative frequencies of phenotypes that are distributed across the range of measurable values.

Figure 11.16*b* is a bar chart that plots the proportion of students in each category against the range of measured phenotypes. Vertical bars that are shortest represent the categories with the least number of students. The bar that is tallest represents the category with the greatest number of students. Draw a graph line around all of the bars and you get a "bell-shaped" curve. Such curves are typical of populations that show continuous variation in a trait.

| 1 | 4 | 8 | 10 | 16 | 16 | 15 | 15 | 14 | 13 | 13 | 11 | 9 | 8 | 8 | 5 | 1 | 2 |

number of individuals

| 60 (5 feet) | 61 | 62 | 63 | 64 | 65 | 66 | 67 | 68 | 69 | 70 | 71 | 72 | 73 | 74 | 75 | 76 | 77 |

height (inches)

a Students at Brigham Young University in a splendid example of continuous variation in height.

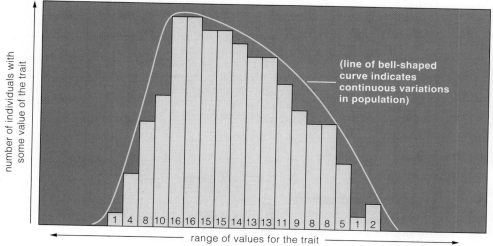

number of individuals with some value of the trait

(line of bell-shaped curve indicates continuous variations in population)

| 1 | 4 | 8 | 10 | 16 | 16 | 15 | 15 | 14 | 13 | 13 | 11 | 9 | 8 | 8 | 5 | 1 | 2 |

range of values for the trait

b

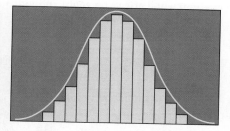

c Idealized bell-shaped curve for populations showing continuous variation in some trait.

Figure 11.16 Continuous variation in a trait that is characteristic of the human population. (**a**) Suppose you want to know the frequency distribution for height in a group of 169 biology students at Brigham Young University. You decide on how finely the range of possible heights should be divided. Then you measure and assign each student to the proper category. Finally, you divide the number in each category by the total number of all students in all categories.

(**b**) Often a bar graph is used to depict continuous variation. Here, the proportion of students in each category is plotted against the range of measured phenotypes. Notice the curved line above the bars. It is a specific example of the kind of "bell-shaped" curve that emerges for populations showing continuous variation in some trait. The bell-shaped curve in (**c**) is a generalized model for this phenomenon.

11.8 EXAMPLES OF ENVIRONMENTAL EFFECTS ON PHENOTYPE

We have mentioned in passing that environmental conditions can contribute to variable gene expression among individuals of a population. Before leaving this chapter, consider just two examples of environmentally induced variations in phenotype.

The environment influences plant growth. You may have observed water buttercups growing half in and half out of a pond. The submerged leaves have a quite different appearance, compared to the leaves above the water's surface (Figure 11.17). In this type of plant, the genes responsible for leaf shape produce different phenotypes under the two environmental conditions.

Or think of Himalayan rabbits and Siamese cats. Both carry the Himalayan allele (*ch*), which specifies a heat-sensitive version of one of the enzymes used in melanin production. At the surface of warm body regions, the enzyme is less active. There, fur grows in lighter than it does at cooler regions, including the ears and other extremities. The experiment described in Figure 11.18 provides observable evidence of the variation.

And so we conclude this chapter, which has dealt with the heritable and environmental aspects of phenotypic variations. What is the take-home lesson? Simply this:

Owing to gene mutations, cumulative gene interactions, and environmental effects on genes, individuals of a population show degrees of variation in their traits.

Figure 11.17 Effect of different environmental conditions on gene expression in plants. Leaves growing from submerged stems of the water buttercup (*Ranunculus aquatilis*) are more finely divided than leaves growing in air. The variation occurs even in the same leaf if it develops half in and half out of water.

Figure 11.18 Effect of different environmental conditions on gene expression in animals. A Himalayan rabbit normally has black hair only on its long ears, nose, tail, and lower leg limbs. In one experiment, a patch of a rabbit's white fur was plucked clean, then an icepack was secured over the hairless patch. While the cold conditions were maintained, the hairs that grew back were black.

Himalayan rabbits are homozygous for the *ch* allele of a gene that codes for tyrosinase (an enzyme needed to produce melanin). The allele specifies a heat-sensitive version of the enzyme that functions only when temperatures are below about 33°C. When hairs grow under warmer conditions, they appear light (no melanin is produced). Light fur normally covers warm body regions. Ears and other slender extremities are cooler; they tend to lose metabolic heat more rapidly.

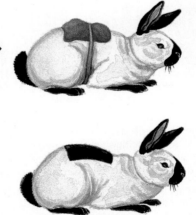

SUMMARY

1. A gene is a unit of information about a heritable trait. The alleles of a gene are slightly different versions of that information. Through experimental crosses with pea plants, Mendel gathered evidence that diploid organisms have two genes for each trait and that genes retain their identity when transmitted to offspring.

2. Mendel performed monohybrid crosses (between two true-breeding plants showing different versions of a single trait). The crosses provided evidence that a gene can have different molecular forms (alleles), some of which are dominant over other, recessive forms.

3. Homozygous dominant individuals have two dominant alleles (AA) for the trait being studied. Homozygous recessives have two recessive alleles (aa). Heterozygotes have two nonidentical alleles (Aa).

4. In Mendel's monohybrid crosses ($AA \times aa$), all F_1 offspring were Aa. Crosses between F_1 plants resulted in these combinations of alleles in F_2 offspring:

	A	a
A	AA	Aa
a	Aa	aa

AA (dominant)
Aa (dominant)
Aa (dominant)
aa (recessive)

This produced the expected phenotypic ratio of 3:1.

5. Results from Mendel's monohybrid crosses led to the formulation of a theory of segregation. In modern terms, diploid organisms have pairs of genes, on pairs of homologous chromosomes. The two genes of each pair segregate from each other during meiosis, such that each gamete formed ends up with one or the other.

6. Mendel also performed dihybrid crosses (between two true-breeding plants showing different versions of two traits). Results from several experiments were close to a 9:3:3:1 phenotypic ratio:

9 dominant for both traits
3 dominant for A, recessive for b
3 dominant for B, recessive for a
1 recessive for both traits

7. Mendel's dihybrid crosses led to the formulation of a theory of independent assortment. In modern terms, the gene pairs of two homologous chromosomes tend to be sorted into one gamete or another independently of how the gene pairs of other chromosomes are sorted out.

8. Four factors can influence gene expression. First, degrees of dominance exist between some gene pairs. Second, gene pairs can interact to influence the same trait. Third, a single gene can influence many seemingly unrelated traits. Fourth, environmental conditions can affect gene expression.

Review Questions

1. State the theory of segregation. Does segregation occur during mitosis or meiosis? *173*

2. Define the difference between these terms. *171*
   a. gene and allele
   b. dominant allele and recessive allele
   c. homozygote and heterozygote
   d. genotype and phenotype

3. Define true-breeding. What is a hybrid? *170, 172*

4. Distinguish between monohybrid and dihybrid crosses. What is a testcross, and why is it useful in genetic analysis? *172–174*

5. State the theory of independent assortment. Does independent assortment occur during mitosis or meiosis? *175*

Self-Quiz *(Answers in Appendix IV)*

1. Alleles are _____ .
   a. different molecular forms of a gene
   b. different molecular forms of a chromosome
   c. self-fertilizing, true-breeding homozygotes
   d. self-fertilizing, true-breeding heterozygotes

2. A heterozygote has _____ for the trait being studied.
   a. a pair of identical alleles
   b. a pair of nonidentical alleles
   c. a haploid condition, in genetic terms
   d. a and c

3. The observable traits of an organism are its _____ .
   a. phenotype   c. genotype
   b. sociobiology   d. pedigree

4. Offspring of a monohybrid cross $AA \times aa$ are _____ .
   a. all AA d. 1/2 AA and 1/2 aa
   b. all aa e. none of the above
   c. all Aa

5. Second-generation offspring from a cross are the _____ .
   a. F_1 generation   c. hybrid generation
   b. F_2 generation   d. none of the above

6. Assuming complete dominance, offspring of the cross $Aa \times Aa$ will show a phenotypic ratio of _____ .
   a. 3:1   c. 9:1
   b. 1:2:1   d. 9:3:3:1

7. Crosses between F_1 individuals resulting from the cross $AABB \times aabb$ lead to F_2 phenotypic ratios close to _____ .
   a. 1:2:1   c. 3:1
   b. 1:1:1:1   d. 9:3:3:1

8. Match each genetic term appropriately.
   _____ dihybrid cross   a. *bb*
   _____ monohybrid cross   b. *AaBb* $\times$ *AaBb*
   _____ homozygous condition   c. *Aa*
   _____ heterozygous condition d. *Aa* $\times$ *Aa*

1. One gene has alleles *A* and *a*. Another has alleles *B* and *b*. For each genotype listed, what type(s) of gametes will be produced? (Assume independent assortment occurs.)

 a. *AA BB*
 b. *Aa BB*
 c. *Aa bb*
 d. *Aa Bb*

2. Still referring to Problem 1, what will be the genotypes of offspring from the following matings? With what frequency will each genotype show up?

 a. *AA BB* × *aa BB*
 b. *Aa BB* × *AA Bb*
 c. *Aa Bb* × *aa bb*
 d. *Aa Bb* × *Aa Bb*

3. In one experiment, Mendel crossed a pea plant that bred true for green pods with one that bred true for yellow pods. All the F_1 plants had green pods. Which form of the trait (green or yellow pods) is recessive? Explain how you arrived at your conclusion.

4. At one gene location on a human chromosome, a dominant allele controls whether you can curl the sides of your tongue upward (*see photo*). People homozygous for the recessive allele cannot roll their tongue. At a different gene location, a dominant allele controls whether earlobes are attached or detached (see Figure 11.1). These two gene pairs assort independently.

   Suppose a tongue-rolling woman with detached earlobes marries a man who has attached earlobes and can't roll his tongue. Their first child has attached earlobes and can't roll its tongue.

 a. What are the genotypes of the mother, father, and child?
 b. What is the probability that a second child will have detached earlobes and won't be a tongue roller?

5. Go back to Problem 1, and assume you now study a third gene having alleles *C* and *c*. For each genotype listed, what type(s) of gametes will be produced?

 a. *AA BB CC*
 b. *Aa BB cc*
 c. *Aa BB Cc*
 d. *Aa Bb Cc*

6. Mendel crossed a true-breeding tall, purple-flowered pea plant with a true-breeding dwarf, white-flowered plant. All F_1 plants were tall and purple-flowered. If an F_1 plant self-fertilizes, what is the probability that a randomly selected F_2 plant will be heterozygous for the genes specifying height and flower color?

7. Assume that a new gene has been identified in mice. One of its alleles specifies yellow fur color. A second allele specifies brown fur color. Suppose you are asked to determine whether the relationship between the two alleles is one of simple dominance, incomplete dominance, or codominance. What types of crosses would give you the answer? On what types of observations would you base your conclusions?

8. The ABO blood-typing system has been used to settle cases of disputed paternity. Suppose, as a geneticist, you must testify during a case in which the mother has type A blood, the child has type O blood, and the alleged father has type B blood. How would you respond to the following statements?

 a. *Attorney of the alleged father:* "The mother has type A blood, so the child's type O blood must have come from the father. Because my client has type B blood, he could not have fathered this child."

 b. *Mother's attorney:* "Further tests prove this man is heterozygous, so he must be the father."

9. As in Labrador retrievers (page 178), fur color in mice is governed by genes concerned with producing and distributing melanin. At one gene location, a dominant allele (B), specifies dark brown and a recessive allele (b) specifies light brown, or tan. At another gene location, a dominant allele (C) permits melanin production and a recessive allele (c) shuts it down and results in albinism.

a. A homozygous bb cc albino mouse mates with a homozygous BB CC brown mouse. State the probable genotypic and phenotypic ratios for the F_1 and F_2 offspring.

b. If an F_1 mouse from Problem 9a is backcrossed with its albino parent, what phenotypic and genotypic ratios would you expect?

10. A dominant allele W confers black fur on guinea pigs. If the guinea pig is homozygous recessive (ww), it has white fur. Fred would like to know if his pet black-furred guinea pig is homozygous WW or heterozygous (Ww). How might he determine his pet's genotype?

11. In snapdragons, red-flowering plants are homozygous for a certain allele (R^1R^1). White-flowering plants are homozygous for a different allele (R^2R^2). Heterozygotes (R^1R^2) bear pink flowers. What phenotypes are likely to appear among the F_1 offspring of the following crosses? And what are the expected proportions for each phenotype?

a. $R^1R^1 \times R^1R^2$
b. $R^1R^1 \times R^2R^2$
c. $R^1R^2 \times R^1R^2$
d. $R^1R^2 \times R^2R^2$

(Note: in cases of incomplete dominance, it is inappropriate to refer to either allele of a pair as dominant or recessive. When the phenotype of a heterozygote is halfway between those of the two homozygotes, then there is no dominance. Such alleles are usually designated with superscript numerals, as shown above, rather than by uppercase letters for dominance and lowercase letters for recessiveness.)

12. An allele codes for a mutant form of hemoglobin (Hb^S instead of Hb^A). Homozygotes (Hb^SHb^S) are affected by sickle cell anemia. Heterozygotes (Hb^AHb^S) show few outward symptoms. Suppose a female whose father was homozygous for the Hb^S allele marries a heterozygous male, and they consider having children. For *each* pregnancy,

a. What is the probability of having a child homozygous for the Hb^S allele?

b. What is the probability of having a child homozygous for the Hb^A allele?

c. What is the probability of having a child heterozygous Hb^AHb^S?

13. An inability to produce melanin results in albinism. A recessive allele, a, is responsible for this phenotype. State the possible genotypes of both parents and offspring of the following crosses.

a. Both parents have normal phenotypes but have both albino and normal children.

b. Both parents are albino and have only albino children.

c. A normal woman and an albino man who have two albino and two normal children.

14. In chickens two pairs of genes affect the comb type (p. 179). When both genes are in the recessive form, the chicken will have a single (or normal) comb. Pea comb is determined by the dominant allele of one of these genes, P. The dominant allele of the other gene, R, causes rose comb. An epistatic interaction occurs if a chicken has at least one of both dominants, $P\_$ $R\_$, producing a walnut comb. Predict the offspring ratios of a cross between two walnut-combed chickens that are heterozygous for both genes ($PpRr$).

15. If their selection of mates were truly random, human populations might show continuous variation for skin pigmentation. Hypothetically speaking, assume that this is the case and that two pairs of genes control the trait. Assume further that alleles A and B contribute equally to the production of the pigment melanin, and that alleles a and b are incapable of contributing to its formation. $AABB$ would be associated with the greatest amount of pigmentation, $aabb$ with the least amount, and all other allelic combinations with intermediate amounts.

a. Predict the amount of pigmentation of the offspring from a mating of two individuals with the genotypes $AABB \times aabb$.

b. Predict the genotypes of offspring if both parents are $AaBb$.

Selected Key Terms

ABO blood typing 177	homozygous recessive 171
allele 171	incomplete dominance 176
codominance 176	monohybrid cross 172
continuous variation 180	multiple allele system 177
dihybrid cross 174	phenotype 171
epistasis 178	pleiotropy 176
gene 171	probability 173
genotype 171	Punnett-square method 173
heterozygous 171	testcross 173
homozygous dominant 171	true-breeding 170

Readings

Cummings, M. 1994. *Human Heredity*. Third edition, St. Paul: West Publishing Company.

Griffiths, A., et al. 1993. *An Introduction to Genetic Analysis*. Fourth edition. New York: Freeman.

Mendel, G. 1966. "Experiments on Plant Hybrids." Translation in C. Stern and E. Sherwood, *The Origin of Genetics: A Mendel Source Book*. New York: Freeman.

Orel, V. 1984. *Mendel*. New York: Oxford University Press.

12 CHROMOSOMES AND HUMAN GENETICS

Too Young To Be Old

Imagine being ten years old, trapped in a body that with each passing day becomes a bit more shriveled, more frail, *old*. You are just tall enough to peer over the top of the kitchen counter, and you weigh less than thirty-five pounds. Already you are bald, and your nose is crinkled and beaklike. Possibly you have only a few more years to live. Yet, like Mickey Hayes and Fransie Geringer (Figure 12.1), you still play, laugh, and hug your friends.

Of every 8 million newborns, one is destined to grow old far too soon. That rare individual possesses a mutated gene on just one of forty-six chromosomes, inherited either from its mother or father. Through hundreds, thousands, then many millions of DNA replications and mitotic cell divisions, terrible information encoded in that gene was methodically distributed to every cell in the embryo, then in the newborn's body. The outcome will be accelerated aging and a greatly

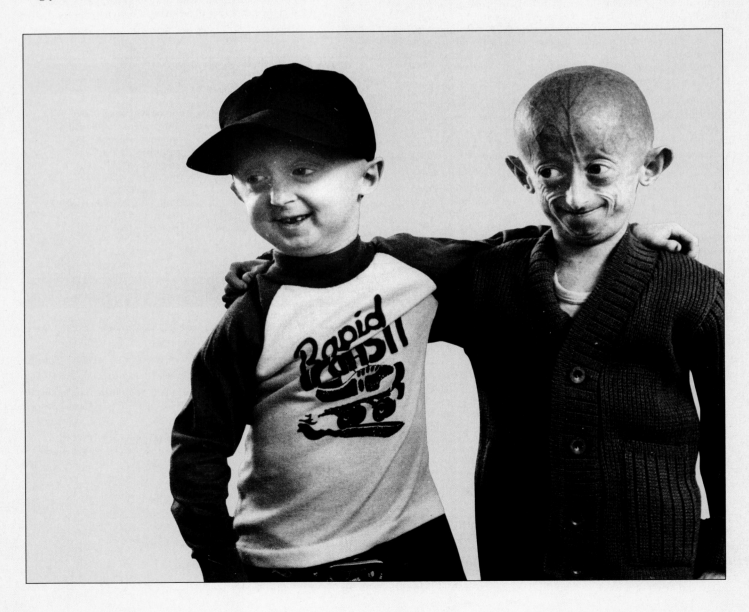

reduced life expectancy. This is the defining feature of *Hutchinson-Gilford progeria syndrome*. There is no cure.

The mutation leads to disruptions in the gene interactions underlying cell division, growth, and development. Outward symptoms start to emerge when the child is less than two years old. The skin becomes thinner. Muscles become flabby. Limb bones that should lengthen and become stronger start to soften instead. Hair loss is pronounced and typically ends in baldness.

Most progeriacs die in their early teens from strokes or heart attacks. These are the final insults, brought on by a hardening of the arteries—a condition that is typical of advanced age.

There are no documented cases of progeria running in families, which suggests that the gene must mutate spontaneously, at random. The disorder can develop with equal frequency in either boys or girls, so the gene cannot be on a sex chromosome. Because the characteristic phenotype always develops, the mutated gene must be dominant over its partner on the homologous chromosome.

We began this unit of the book by looking at cell division, the starting point of inheritance. Then we started thinking about how chromosomes—and the genes they carry—are shuffled at meiosis and fertilization. In this chapter we delve more deeply into the patterns of chromosomal inheritance, with emphasis on humans. At times the methods of analysis might seem abstract. But keep in mind that we are talking about messages of inheritance in yourself and in other human individuals. When Mickey Hayes turned eighteen, he was the oldest living progeriac. Fransie was seventeen when he died.

Figure 12.1 Two boys, both less than ten years old, who met during a gathering of progeriacs at Disneyland, California. Progeria is a heritable (genetic) disorder. It is characterized by accelerated aging and extremely reduced life expectancy.

KEY CONCEPTS

1. One gene follows another along the length of a chromosome. Each has its own position in that sequence.

2. A chromosome's gene sequence does not necessarily remain intact through meiosis and gamete formation. Through crossing over, some genes leave the sequence, and genes from the homologous chromosome replace them.

3. A chromosome's structure may change, as when a segment is deleted, duplicated, inverted, or moved to a new location. Also, improper separation of duplicated chromosomes during meiosis or mitosis may lead to changes in chromosome number.

4. Allele shufflings, changes in chromosome structure, and changes in chromosome number contribute to variation in traits. Often they lead to genetic abnormalities or disorders.

12.1 RETURN OF THE PEA PLANT

The year was 1884. Mendel's paper on pea plants had been gathering dust in a hundred libraries for nearly two decades, and Mendel himself had just passed away. Ironically, the experiments described in that forgotten paper were about to be devised once again.

Improvements in microscopy had rekindled efforts to locate the hereditary material within cells. By 1882, Walther Flemming had observed threadlike bodies—chromosomes—in the nucleus of dividing cells. By 1884, a question was taking shape: Could chromosomes be the hereditary material?

Then researchers realized each gamete has half the number of chromosomes of a fertilized egg. In 1887, August Weismann proposed that a special division process must reduce the chromosome number by half before gametes form. Sure enough, in that same year meiosis was discovered. Weismann began to promote his theory of heredity: The chromosome number is halved during meiosis, then restored at fertilization. Thus a cell's hereditary material is half paternal in origin, and half maternal. His theory was hotly debated, and it prompted a flurry of experimental crosses—just like the ones Mendel had carried out.

Finally, in 1900, researchers came across Mendel's paper while checking literature related to their own genetic crosses. To their surprise, their results merely confirmed what Mendel had already proposed. Diploid cells have two units (genes) for each heritable trait, and the units segregate before gametes form.

Researchers learned a great deal about chromosomes during the decades that followed. Let's start with a few high points of their work, which will serve as background for understanding human inheritance.

12.2 THE CHROMOSOMAL BASIS OF INHERITANCE—AN OVERVIEW

From the preceding chapters in this unit, you already are familiar with some concepts concerning the structure of chromosomes and their behavior during meiosis. Let's now integrate them with a few concepts that will help explain patterns of human inheritance:

1. **Genes** are units of information about heritable traits. Each kind of gene has a particular location in a particular type of chromosome.

2. For each eukaryotic species, the different genes are distributed among different types of chromosomes.

3. Many cells that give rise to gametes are diploid ($2n$), with pairs of **homologous chromosomes**. In humans, one chromosome of each type is inherited from the mother, and its homologue from the father.

4. Each type of chromosome has the same length, shape, and gene sequence as its homologous partner, and interacts with it during meiosis. Different sex chromosomes (X and Y) are homologous in one small region only.

5. A given pair of homologous chromosomes may be carrying identical or nonidentical alleles at corresponding locations in their gene sequence. **Alleles** are slightly different molecular forms of the same gene, and they arise through mutation.

6. There is no pattern to the way that maternal and paternal chromosomes become attached to the microtubular spindle just before metaphase I of meiosis, so different combinations are possible at anaphase I. As a result, gametes and then new individuals receive a mix of alleles of maternal and paternal chromosomes.

7. Genes close together on the same chromosome *tend* to stay together through meiosis, but the chromosome breaks and exchanges corresponding segments with its homologous partner. This event, called **crossing over**, puts new combinations of alleles into chromosomes.

8. A chromosome's structure may change, as when a segment is deleted, duplicated, inverted, or moved to a new location. Also, improper separation of chromosomes that have been duplicated in advance of meiosis or mitosis may lead to changes in chromosome number in gametes and the new individual.

9. For any given trait, the particular combination of alleles inherited by the new individual may have neutral, beneficial, or harmful effects.

Focus on Science

Preparing a Karyotype Diagram

In laboratories throughout the world, karyotype diagrams help answer questions about an individual's chromosomes. You can use the diagram in Figure *a* as a basis for understanding this chapter's examples of karyotypes that point to human genetic disorders.

Chromosomes are most condensed and easiest to identify in dividing cells, particularly those at metaphase of mitosis. Technicians don't count on finding a cell that happens to be dividing in the body when they go looking for it. Instead, they culture cells *in vitro* (literally, "in glass"). They put a sample of blood, skin, bone marrow, amniotic fluid (which bathes embryos), or some other tissue in a glass container. In that container is a solution that stimulates cell growth and mitotic division for many generations.

The dividing cells can be arrested at metaphase by adding colchicine to the culture medium. Colchicine is an extract of the autumn crocus (*Colchicum autumnale*). Technicians and researchers use it to block formation of microtubular spindles—and to prevent duplicated chromosomes from separating during nuclear division. With the proper colchicine concentration and exposure time, many metaphase cells can accumulate, and this increases the chances of finding good candidates for karyotype diagrams.

Next, the culture medium is transferred to tubes of a centrifuge, a motor-driven spinning device. The cells have greater mass and density than the surrounding solution, so the spinning force moves them farthest away from the

Autosomes and Sex Chromosomes

Generally speaking, a pair of homologous chromosomes is exactly alike in length, shape, and gene sequence. In the late 1800s, however, microscopists discovered an exception to this. In humans and many other species, a distinctive chromosome is present in females *or* males, but not in both. The *Focus* essay shows a lineup of the twenty-three pairs of human chromosomes. In males only, the last two in the series are physically different. A long, X-shaped chromosome is paired with a shorter one, shaped like an upside-down Y. These two are called the **X chromosome** and **Y chromosome**. Despite the physical difference, the two are able to synapse in a small region and function as homologues during meiosis. Unlike the males (XY), females are XX; they have two X chromosomes.

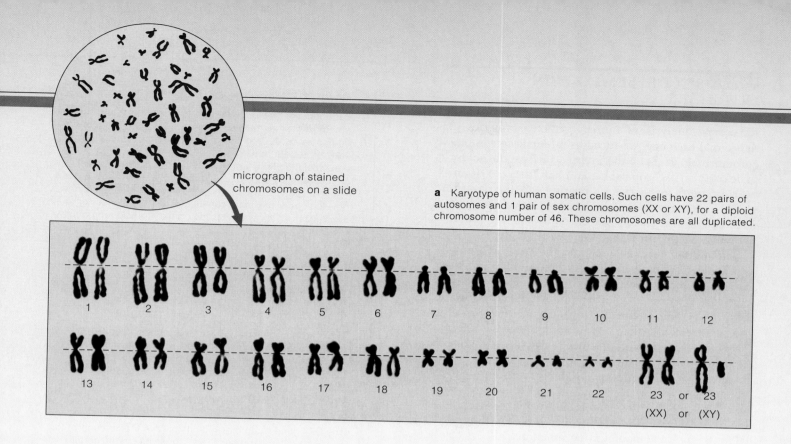

micrograph of stained
chromosomes on a slide

a Karyotype of human somatic cells. Such cells have 22 pairs of autosomes and 1 pair of sex chromosomes (XX or XY), for a diploid chromosome number of 46. These chromosomes are all duplicated.

1 2 3 4 5 6 7 8 9 10 11 12

13 14 15 16 17 18 19 20 21 22 23 or 23
(XX) or (XY)

motor, to the bottom of the attached tubes. (Separation in response to the spinning force is called *centrifugation*.) Afterward, the cells are transferred to a saline solution. They swell (by osmosis), then move apart—and so do their metaphase chromosomes. The cells are ready to be dropped onto a microscope slide, fixed (as by air-drying), and stained.

Chromosomes take up some stains (orcein and Giemsa dyes) uniformly. This is enough to allow identification of

their size and shape, as in Figure *a*. With new staining procedures, horizontal bands appear in human chromosomes that fluoresce when viewed in ultraviolet light. Examples of these appear in several chapters of the book.

Finally, a microscope image of the chromosomes is photographed and enlarged, so that the chromosomes can be cut apart individually. Cutouts can be arranged by size, shape, and length of arms, then the homologous pairs can be aligned horizontally by their centromeres (Figure *a*).

Human X and Y chromosomes are examples of **sex chromosomes**. Inheritance of one type or the other governs gender—that is, whether a new individual will be male or female. All other chromosomes, which are the same in both sexes, are designated **autosomes**.

Karyotype Analysis

Today, microscopists routinely analyze a cell's sex chromosomes and autosomes. Chromosomes, recall, are most highly condensed at metaphase of mitosis. At that time, each has a characteristic size, length, and centromere location. Also, the chromosomes of many species show distinct banding patterns when they are stained in certain ways. Regions that are the most con-

densed often take up more stain and form darker bands. Figure 10.4 shows an example of this.

The number of metaphase chromosomes and their defining characteristics are called the **karyotype** of an individual (or species). The *Focus* essay explains how to construct a karyotype diagram. Such a diagram is a cut-up, rearranged photograph in which all the autosomes are lined up, largest to smallest, and the sex chromosomes are positioned last.

A sex chromosome is one whose presence determines a new individual's gender. An autosome is any chromosome other than a sex chromosome.

12.3 SEX DETERMINATION IN HUMANS

Karyotype analysis of human gametes provides evidence that each egg produced by a female carries one X chromosome. Half of the sperm cells produced by a male carry an X chromosome and half carry a Y. If an X-bearing sperm fertilizes an X-bearing egg, the new individual will develop into a female. Conversely, if the fertilizing sperm carries a Y chromosome, the individual will develop into a male (Figure 12.2).

Among the very few genes on the Y chromosome is a "male-determining gene." As Figure 12.3 indicates, expression of this gene leads to the formation of testes, which are the primary male reproductive organs. In the absence of this gene, ovaries form automatically. Ovaries are the primary female reproductive organs. Testes and ovaries both produce hormones that govern the development of particular sexual traits.

A human X chromosome probably carries more than 300 genes. Like other chromosomes, it carries some genes associated with sexual traits, such as the distribution of body fat. But most of its genes deal with *nonsexual* traits, such as blood-clotting functions.

A certain gene on the human Y chromosome dictates that a new individual will develop into a male. In the absence of the Y chromosome (and the gene), a female develops.

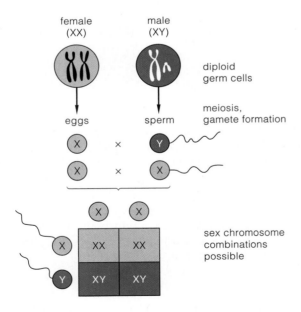

Figure 12.2 Pattern of sex determination in humans. Males transmit their Y chromosome to sons but not to daughters. Males receive their X chromosome only from their mother.

Figure 12.3 Boys, girls, and the Y chromosome.

For about the first four weeks of its existence, a human embryo is neither male nor female, even though it normally carries XY or XX chromosomes. However, ducts and other structures start forming that can go either way. In XY embryos, the primary male reproductive organs (testes) start to form during the next four to six weeks (**a–c**). A gene region on the Y chromosome seems to govern a fork in the developmental road that can lead either to maleness or to femaleness. In XX embryos, the primary female reproductive organs (ovaries) start to form. They form automatically in the absence of a Y chromosome.

The testes start producing testosterone and other sex hormones that influence the development of the male reproductive system. The ovaries also start producing sex hormones that influence the development of the female reproductive system.

The master gene for sex determination is named SRY (short for *Sex-determining Region* of the *Y* chromosome). So far, the same gene has been identified in DNA from male humans, chimpanzees, rabbits, pigs, horses, cattle, and tigers. None of the females tested had the gene. Tests with mice indicate that the gene region becomes active about the time testes start developing.

The SRY gene resembles regions of DNA that are known to specify regulatory proteins. Such proteins bind to certain parts of DNA and so turn genes on and off. Apparently, the SRY gene product regulates a cascade of reactions that are necessary for sex determination.

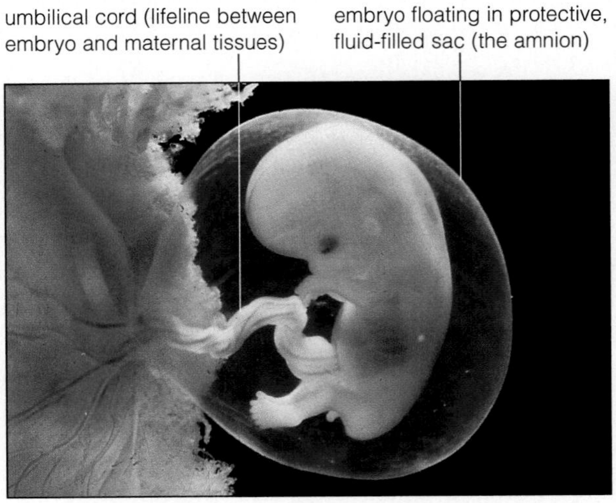

umbilical cord (lifeline between embryo and maternal tissues)

embryo floating in protective, fluid-filled sac (the amnion)

a A human embryo, eight weeks old. Male reproductive organs have already started to develop.

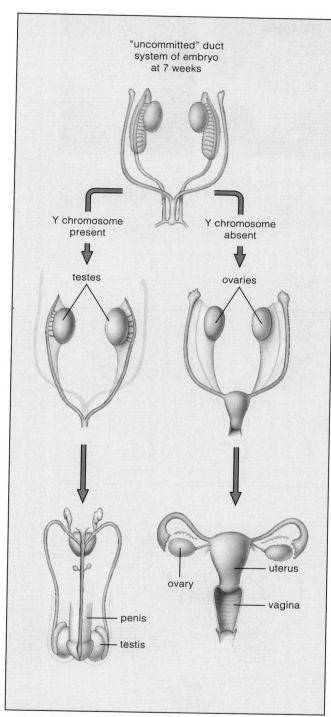

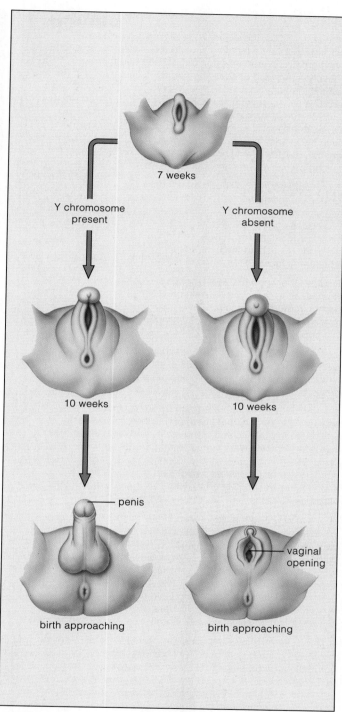

b Duct system in the early embryo that develops into a male or female reproductive system. Compare pages 774 and 778.

c External appearance of the developing reproductive organs in embryos.

Figure 12.4 X-linked genes: clues to inheritance patterns.

In the early 1900s, the embryologist Thomas Morgan was studying inheritance patterns. During those studies, he and his coworkers discovered an apparent genetic basis for the connection between gender and certain nonsexual traits. For example, human males and females both have blood-clotting mechanisms. Yet hemophilia (a blood-clotting disorder) shows up most often in the males, not females, of a family lineage. This gender-specific outcome was not like anything Mendel saw in his hybrid crosses between pea plants. (Either parent plant could carry a recessive allele. It made no difference; the resulting phenotype was the same.)

Morgan studied eye color and other nonsexual traits of *Drosophila melanogaster*. These fruit flies can be grown in bottles on bits of cornmeal, molasses, and agar. A female lays hundreds of eggs in a few days, and new flies that develop from the eggs can themselves reproduce in less than two weeks. Morgan could track hereditary traits through nearly thirty generations of thousands of flies in a year's time.

At first, all the flies were wild-type for eye color; they had brick-red eyes, as in (**a**). (*Wild-type* simply means the normal or most common form of a trait in a population.) Then, through an apparent mutation in a gene controlling eye color, a white-eyed male appeared (**b**).

Morgan established true-breeding strains of white-eyed males and females. Then he did a series of *reciprocal* crosses. These are pairs of crosses. In the first, one parent displays the trait in question. In the second, the other parent displays it.

White-eyed males were mated with homozygous red-eyed females. All the F_1 offspring of the cross had red eyes. But of the F_2 offspring, only some of the males had white eyes. Then white-eyed females were mated with true-breeding red-eyed males. Of the F_1 offspring of that second cross, half were red-eyed females and half were white-eyed males. Of the F_2 offspring, 1/4 were red-eyed females, 1/4 white-eyed females, 1/4 red-eyed males, and 1/4 white-eyed males!

The seemingly odd results implied a relationship between the eye-color gene and gender. Probably the gene was located on a sex chromosome. But which one? Because females (XX) could be white-eyed, the recessive allele would have to be on one of their X chromosomes. Suppose white-eyed males (XY) also carry the recessive allele on their X chromosome—and suppose there is no corresponding eye-color allele on the Y chromosome. The males would have white eyes, for they have no dominant allele to mask the effect of the recessive one.

(**c**) This diagram shows the expected results when the idea of an X-linked gene is combined with Mendel's concept of segregation. By proposing that a specific gene is located on the X chromosome but not on the Y, Morgan was able to explain the outcome of his reciprocal crosses. The results of the experiments matched the predicted outcomes.

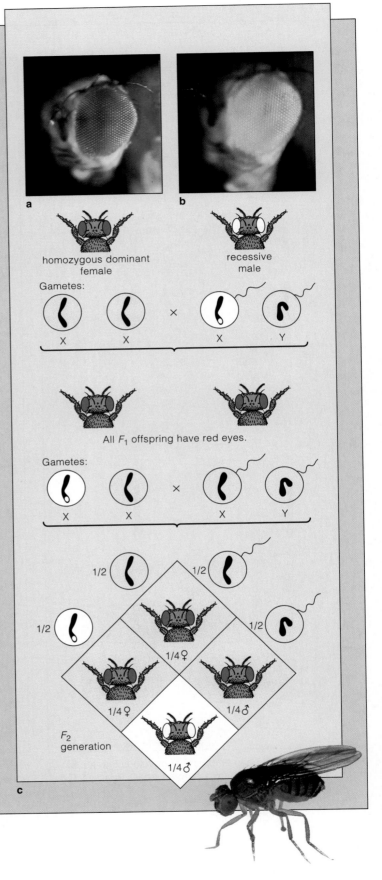

a

b

homozygous dominant female   recessive male

Gametes:

X   X   ×   X   Y

All F_1 offspring have red eyes.

Gametes:

X   X   ×   X   Y

1/2   1/2

1/2   1/4♀   1/2

1/4♀   1/4♂

F_2 generation

1/4♂

c

12.4 EARLY QUESTIONS ABOUT GENE LOCATIONS

X-Linked Genes: Clues to Patterns of Inheritance

Some time ago, researchers only suspected that each gene has a specific location on a chromosome. Through a series of hybridization experiments with fruit flies (*Drosophila melanogaster*), Thomas Hunt Morgan and his coworkers helped confirm this. For example, as described in Figure 12.4, they found strong evidence of a gene for eye color on the *Drosophila* X chromosome. For a time, genes located on sex chromosomes were called "sex-linked genes." Today researchers use the more precise terms, **X-linked** and **Y-linked genes**.

Linkage Groups and Crossing Over

It seemed, during the early *Drosophila* experiments, that the genes "linked" to one chromosome or another travel together and end up in the same gamete. In time, researchers identified a large number of genes, and the genes located on each type of chromosome came to be called **linkage groups**. For example, *D. melanogaster* has four linkage groups, which correspond to its four pairs of homologous chromosomes. Indian corn (*Zea mays*) has ten linkage groups, corresponding to ten pairs of homologous chromosomes.

As we now know, linkage groups are vulnerable to crossing over (Figure 12.5). We also know that crossing over is not a rare event. In most eukaryotic organisms, humans included, meiosis cannot be completed properly unless *every* pair of homologous chromosomes takes part in at least one crossover.

Imagine any two genes at two different locations on the same chromosome. The probability of a crossover disrupting their linkage is proportional to the distance separating them. Suppose genes *A* and *B* are twice as far apart as two other genes, *C* and *D*:

We would expect crossing over to disrupt the linkage between *A* and *B* much more often.

Two genes are very closely linked when the distance between them is small; they nearly always end up in the same gamete. Linkage is more vulnerable to crossing over when the distance between two genes is greater. When two genes are very far apart, crossing over disrupts their linkage so often that the genes assort independently of each other into gametes.

The farther apart two genes are on a chromosome, the greater will be the frequency of crossing over and recombination between them.

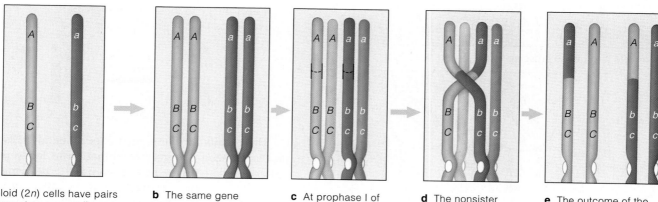

a Diploid (2*n*) cells have pairs of genes, at corresponding locations, on pairs of homologous chromosomes. The two genes of each pair may or may not be identical. In this example, nonidentical alleles are present at all three locations (*A* with *a*, *B* with *b*, and *C* with *c*).

b The same gene regions at interphase, after DNA replication at interphase. Both of these homologous chromosomes are now in the duplicated state.

c At prophase I of meiosis, two of the nonsister chromatids break while aligned very tightly together. (compare Figure 10.6).

d The nonsister chromatids swap segments, then enzymes seal the broken ends. The breakage and exchange represent one crossover event.

e The outcome of the crossover is genetic recombination between two of four chromatids. (They are shown here following meiosis, as separate, unduplicated chromosomes.)

Figure 12.5 Crossing over, an event that influences inheritance. A preceding chapter started us thinking about crossing over by showing the time of its occurrence in meiosis (Figure 10.6). Here we see how it disrupts part of a linkage group—that is, the genes that otherwise would stay together on the same chromosome during meiosis and gamete formation.

12.5 RECOMBINATION PATTERNS AND CHROMOSOME MAPPING

Even without knowing about chromosomes, Mendel almost certainly would have recognized Figure 12.6a as an illustration of the theory that the genes on one chromosome are assorted into gametes independently of the genes on a different chromosome. Had he known about crossing over, he would have been able to make sense of Figure 12.6b, which is an example of the resulting genetic recombination.

For a specific example, consider an experimental cross between two watermelon plants. One plant is true-breeding for green-rind, round melons, the dominant phenotypes. The other is true-breeding for stripe-rind, oblong melons, the recessive phenotypes. The gene for rind color is closely linked to the gene for melon shape. As you might expect, all of the F_1 offspring of the cross produced green, round melons.

In a testcross, F_1 plants were hybridized with the homozygous recessive parent. Knowing that the genes for the two traits are closely linked, you predict that the phenotypic ratio of the F_2 offspring will be 1:1:

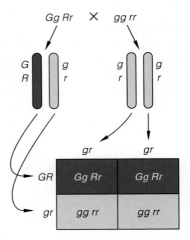

Yet the actual F_2 phenotypes were not as expected:

46 green, round   4 green, oblong
47 striped, oblong   3 striped, round

How can you explain these results? You can assume that 7 percent of the offspring (the 4 green, oblong and the 3 striped, round) inherited a chromosome that had undergone a crossover between the gene locus for rind color and the gene locus for rind shape:

The results of many such experimental crosses tell us that genes undergo recombination in fairly regular patterns. The patterns have been used to determine the positions of genes relative to one another along the length of a chromosome, an activity called linkage mapping. For example, of the several thousand known genes in the four types of *Drosophila* chromosomes, the positions of about a thousand have been mapped.

What about the estimated 50,000 to 100,000 genes on the twenty-three types of human chromosomes? These must be mapped by different methods, of the sort described in Chapter 16. Humans, unlike watermelon plants, do not lend themselves to experimental crosses. And with this jarring thought fresh in your mind, you are ready to begin the next section.

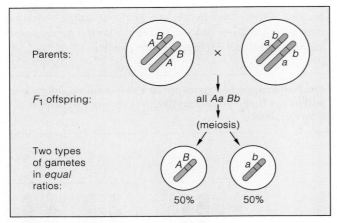

a Complete linkage (no crossing over)

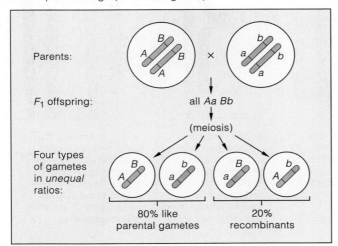

b Incomplete linkage owing to crossing over

Figure 12.6 How crossing over can affect gene linkage, using two genes on the same chromosome as the example.

12.6 HUMAN GENETIC ANALYSIS

Some organisms, including pea plants and fruit flies, lend themselves to genetic analysis. They grow and reproduce rapidly in small spaces, under controlled conditions. It doesn't take long to track a trait through many generations.

Humans are another story. We live under variable conditions in diverse environments. We select our own mates and reproduce if and when we want to. Human subjects live as long as the geneticists who study them, so tracking traits through generations can be rather tedious. Most human families are so small, there aren't enough offspring for easy inferences about inheritance.

To get around some of the problems associated with analyzing human inheritance, geneticists put together **pedigrees**. A pedigree is a chart that is constructed, by standardized methods, to show the genetic connections among individuals. Figure 12.7 includes an example, along with definitions of a few of the standardized symbols used to represent individuals.

When analyzing pedigrees, geneticists rely on their knowledge of probability and Mendelian inheritance patterns, which may yield clues to a trait's genetic basis. For instance, they might determine that the responsible allele is dominant or recessive, or that it is located on an autosome or a sex chromosome. Gathering many family pedigrees increases the numerical base for their analysis. When a trait clearly shows a simple Mendelian inheritance pattern, a geneticist may be justified in predicting the probability of its occurrence among children of prospective parents. We will return to this topic later in the chapter.

For many genes, pedigree analysis may reveal simple Mendelian inheritance patterns that permit inferences about the probability of their transmission to children.

□ male

○ female

□—○ marriage/mating

offspring (in order of birth, from left to right)
1 2 3 4

■ ● individual showing trait being tracked

◇ sex unknown; numerals present indicate number of children

a I, II, III, IV... successive generations

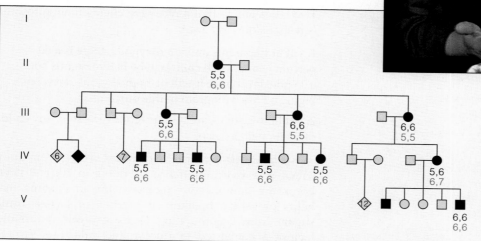

b

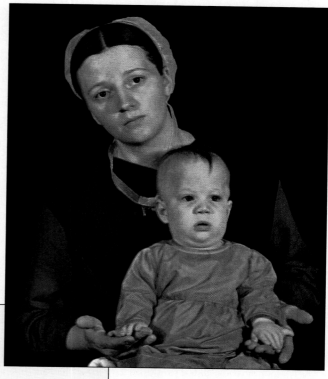

Figure 12.7 (**a**) Some symbols used in constructing pedigree diagrams. (**b**) This is an example of a pedigree for *polydactyly*. An individual with this condition has extra fingers, extra toes, or both. Expression of the gene governing polydactyly can vary from one individual to the next. Here, black numerals designate the number of fingers on each hand. Blue ones designate the number of toes on each foot.

12.7 REGARDING HUMAN GENETIC DISORDERS

Table 12.1 lists some heritable traits that have been studied in detail. A few of these are abnormalities, or deviations from the average condition. Said another way, a **genetic abnormality** is simply a rare or less common version of a trait, as when a person is born with six toes on each foot instead of five. Whether we view an abnormal trait as disfiguring or merely interesting is subjective. There is nothing inherently life-threatening or even ugly about it.

By contrast, a **genetic disorder** is an inherited condition that results in mild to severe medical problems. Alleles underlying severe genetic disorders do not abound in populations, for they put individuals at great risk. They do not disappear entirely for two reasons. First, rare mutations put new copies of the alleles in the population. Second, in heterozygotes, such an allele is paired with a normal one that may cover its functions, so it can be passed to offspring.

You may hear someone call a genetic disorder a disease, but the terms are not always interchangeable. "Diseases" result from infection by bacteria, viruses, or some other environmental agent that invades and multiplies inside the body. Illness follows only if the agent's activities damage tissues and interfere with normal body functions. When a person's genes increase susceptibility or weaken the response to an infection, the resulting illness might be called a genetic disease.

With these qualifications in mind, let's turn to some patterns of inheritance in the human population.

A genetic abnormality simply is a rare or less common version of an inherited trait. A genetic disorder is an inherited condition that results in mild to severe medical problems.

Table 12.1 Examples of Human Genetic Disorders	
Disorder or Abnormality*	**Main Consequences**
Autosomal Recessive Inheritance:	
Albinism *178*	Absence of pigmentation
Sickle-cell anemia *176*	Severe tissue, organ damage
Galactosemia *196*	Brain, liver, eye damage
Phenylketonuria *204*	Mental retardation
Autosomal Dominant Inheritance:	
Achondroplasia *197*	One form of dwarfism
Amyotrophic lateral sclerosis *199*	Motor neurons deteriorate, muscles waste away
Camptodactyly *180*	Rigid, bent little fingers
Familial hypercholesterolemia *33, 197*	High cholesterol levels in blood, clogged arteries
Huntington's disorder *197*	Nervous system degenerates progressively, irreversibly
Polydactyly *195*	Extra fingers, toes, or both
Progeria *187, 197*	Drastic premature aging
X-Linked Dominant Inheritance:	
Faulty enamel trait *198*	Problems with teeth
X-Linked Recessive Inheritance:	
Hemophilia A *198*	Deficient blood-clotting
Duchenne muscular dystrophy *198*	Muscles waste away
Testicular feminization *617*	XY individual but female traits, sterility
Red-green color blindness *606*	Reds, greens appear the same
Changes in Chromosome Structure:	
Cri-du-chat *200*	Retardation, skewed larynx
Fragile X syndrome *200*	Mental retardation
Changes in Chromosome Number:	
Down syndrome *202*	Mental retardation, heart defects
Turner syndrome *203*	Sterility, abnormal ovaries and sexual traits
Klinefelter syndrome *203*	Sterility, retardation
XYY condition *203*	Mild retardation in some; no symptoms in others

*Number indicates the page(s) on which the disorder is described.

12.8 PATTERNS OF AUTOSOMAL INHERITANCE

Autosomal Recessive Inheritance

Two clues tell us that a recessive allele on an autosome is responsible for a trait:

1. If both parents are heterozygous, there is a 50 percent chance each child will be heterozygous and a 25 percent chance it will be homozygous recessive. Figure 12.8 is a diagram of this outcome.

2. If both parents are homozygous recessive, any child of theirs will be, also.

Galactosemia is a genetic disorder arising from autosomal recessive inheritance. About 1 in 100,000 newborns are homozygous recessive for an enzyme that helps prevent a breakdown product of lactose (milk sugar) from accumulating to toxic levels. Normally, lactose is converted to glucose and galactose, then to glucose-1-phosphate, which can be broken down by

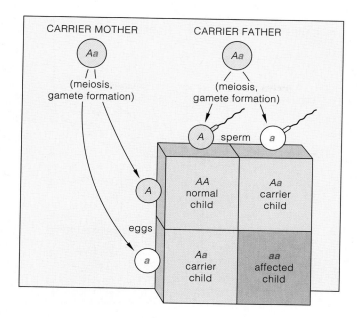

Figure 12.8 One pattern for autosomal recessive inheritance. This example shows the phenotypic outcomes possible when both parents are heterozygous carriers of the recessive allele (shown in *red*).

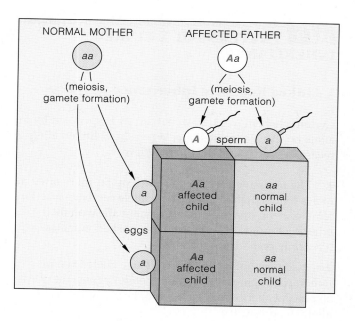

Figure 12.9 One pattern for autosomal dominant inheritance. In this example, assume the dominant allele (shown in *red*) is fully expressed in the carriers.

glycolysis or converted to glycogen (page 134). The defective enzyme blocks the full conversion:

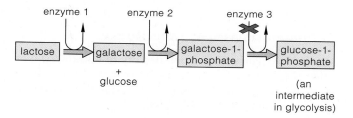

A high blood level of galactose can damage the eyes, liver, and brain. Malnutrition, diarrhea, and vomiting are early symptoms. Untreated galactosemics often die in childhood. However, the telling symptom—a high galactose level—can be detected in urine samples. When affected individuals are placed on a diet that includes milk substitutes and excludes dairy products, they can grow up symptom-free.

Autosomal Dominant Inheritance

Figure 12.9 shows an example of what can happen when a dominant allele on an autosome is responsible for a trait. Two clues point to this condition:

1. Usually the trait appears in each generation, for the allele usually is expressed even in heterozygotes.

2. If one parent is heterozygous and the other homozygous recessive, there is a 50 percent chance that any child of theirs will be heterozygous.

In Chapter 3, you read about a genetic disorder that can lead to cholesterol-clogged arteries. The disorder, called *familial hypercholesterolemia*, arises from an autosomal dominant allele. Affected individuals can be placed on a diet low in cholesterol and saturated fats. Also, certain drugs can lower cholesterol levels.

Even though a few dominant alleles cause severe genetic disorders, they persist in populations. Some are perpetuated by spontaneous mutations. This is the case for progeria, the rare aging disorder described in the introduction to this chapter.

In other cases, a dominant allele may not prevent reproduction. Consider *achondroplasia*, which affects about 1 in 10,000 humans. Usually, the homozygous dominant condition results in a stillbirth. However, heterozygotes are able to reproduce. They cannot form cartilage properly when limb bones are growing, and this leads to abnormally short arms and legs. Adults are less than 4 feet, 4 inches tall. The dominant allele often has no other phenotypic effects than this. In others, severe symptoms may not appear until after a person reproduces and so transmits the allele.

In still other cases, affected persons have children before their symptoms become severe. In *Huntington's disorder*, a progressive deterioration of the nervous system, symptoms may not even begin until about age forty. Most people have already reproduced by then.

12.9 PATTERNS OF X-LINKED INHERITANCE

X-Linked Recessive Inheritance

Certain clues may tell you that a trait is specified by an allele on the X chromosome. When the X-linked allele is recessive, you will detect this pattern:

1. The recessive phenotype shows up far more often in males than females. (A recessive allele can be masked in females, who may inherit a dominant allele on their other X chromosome. It cannot be masked in males, who have only one X chromosome.)

2. A son cannot inherit the recessive allele from his father. A daughter can. If she is heterozygous, there is a 50 percent chance that each son of hers will inherit the allele (Figure 12.10).

Hemophilia A is an example of X-linked recessive inheritance. In most people, a blood-clotting mechanism quickly stops bleeding from minor injuries. The reactions that lead to clot formation require the products of several genes. If any of the genes is mutated, its defective product can cause affected persons to bleed for an abnormal time.

A mutated gene for "clotting factor VIII" causes hemophilia A. It affects about 1 in 7,000 males, who run the risk of dying from untreated bruises, cuts, or internal bleeding. Blood-clotting time is more or less normal in heterozygous females.

The frequency of hemophilia A was unusually high among the males in royal families of nineteenth-century Europe. As Figure 12.11 indicates, Queen Victoria of England was a carrier. At one time, the recessive allele was present in eighteen of her sixty-nine descendants. One hemophilic great-grandchild, Crown Prince Alexis, was a focus of political intrigue that helped usher in the Russian revolution.

Duchenne muscular dystrophy is another example of X-linked recessive inheritance. Although rare, it is the most common of several heritable disorders in which certain groups of muscles slowly degenerate over time. The muscles themselves enlarge as fat and connective tissue is deposited in them, but the muscle cells atrophy (waste away).

At first, children with this form of dystrophy may appear to be normal. But somewhere between their second and tenth birthday, their muscles start to weaken. Such children become progressively clumsier and fall frequently. Usually, after their twelfth birthday, they no longer can walk. In time, muscles of the chest and head degenerate. Affected individuals typically die of respiratory failure in their early twenties. There is no cure.

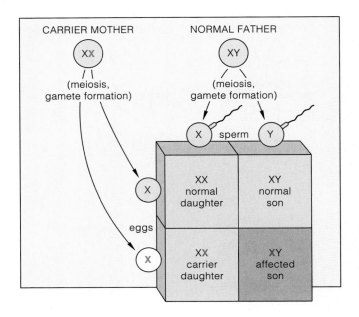

Figure 12.10 One pattern for X-linked inheritance. This example shows the phenotypic outcomes possible when the mother carries a recessive allele on one of her X chromosomes (shown in *red*).

X-Linked Dominant Inheritance

The *faulty enamel trait* is one of very few known examples of X-linked dominant inheritance. With this disorder, the hard, thick enamel coating that normally protects teeth fails to develop properly.

The inheritance pattern is similar to that for X-linked recessive alleles, except that the trait is expressed in heterozygous females. The expression tends to be less pronounced in females than in males. A son cannot inherit the dominant allele responsible for the trait from an affected father, but all of his daughters will. If a woman is heterozygous, she will transmit the allele to half of her offspring, regardless of their sex.

A Few Qualifications

Don't take the preceding examples of autosomal and X-linked traits too seriously. We include them not to turn you into an armchair geneticist, but rather to give you a general idea of the kinds of clues that hold meaning for trained geneticists.

Before diagnosing a case, geneticists often find it necessary to pool together many pedigrees. Typically they make detailed analyses of clinical data and keep

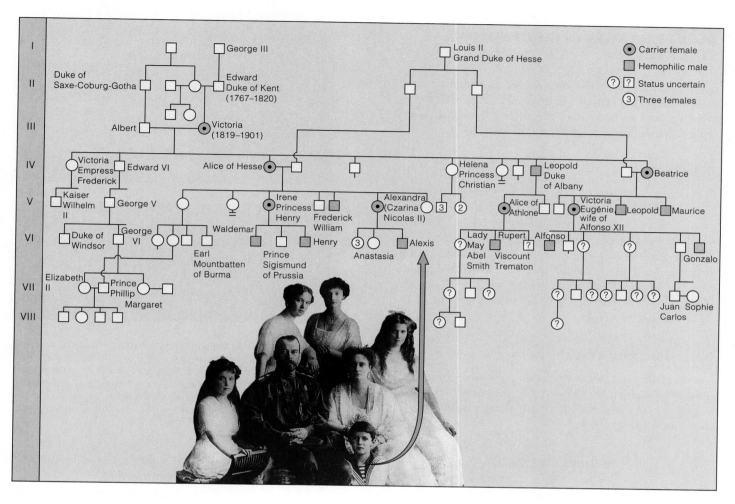

Figure 12.11 Partial pedigree of the descendants of Queen Victoria of England. The chart shows carriers and affected males that possessed the X-linked allele conferring the disorder hemophilia A. Many individuals of later generations are not included. The photograph shows the Russian royal family members. The mother was a carrier of the mutated allele. Crown Prince Alexis was hemophilic.

abreast of current research. Why? Consider that more than one type of gene may be responsible for a given phenotype. Geneticists already know of dozens of conditions that can arise from a mutated gene on an autosome *or* a mutated gene on the X chromosome. They know of genes on autosomes that show dominance in males and recessiveness in females—so they appear to be due to X-linked recessive inheritance, even though they are not.

Besides this, don't assume that errant genes automatically relegate a person to the sidelines of life. More than twenty years ago, Stephen Hawking noticed his muscles were starting to weaken. In time it became difficult for him to speak, to swallow, to use his hands. Motor neurons in his brain and spinal cord were deteriorating, and scar tissue was forming along his spinal

cord. Without normal control signals from his nervous system, his skeletal muscles began to waste away. Such are the symptoms of a rare, ultimately fatal disorder, *amyotrophic lateral sclerosis*. Possibly a virus issues the death warrant, but an autosomal dominant allele serves as the executioner.

And yet, through his acclaimed research in astrophysics and his accessible, eloquent writings, Hawking has changed the way we view time and the universe. His affliction has not immobilized his mind, which dances freely around the most challenging of questions—*When did time begin? Does the universe have boundaries? If we know the past, why can't we know the future?* He lives to move beyond the horizons of our knowledge, believing that when we find the ultimate answers, we will know the mind of God.

12.10 CHANGES IN CHROMOSOME STRUCTURE

So far, we have focused on genetic abnormalities and disorders associated with autosomal or X-linked genes. We turn now to some rare conditions that arise when chromosomes undergo structural alterations.

Deletions

A **deletion** is the loss of a chromosome region by irradiation, viral attack, chemical action, or other environmental factors. One or more genes may be lost, and this nearly always causes problems. A certain deletion from human chromosome 5 causes mental retardation and an abnormally shaped larynx. When an affected infant cries, he or she produces meowing sounds. Hence the name of the disorder, *cri-du-chat* (cat-cry). Figure 12.12*a* shows an affected child.

Inversions and Translocations

An **inversion** is a segment that separated from a chromosome and then was inserted at the same place—but in reverse. The reversal alters the position and order of the chromosome's genes:

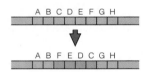

Most often, a **translocation** is part of one chromosome that has exchanged places with a corresponding part of another, *non*homologous chromosome. As an example, chromosome 14 may end up with a segment of chromosome 8 (and chromosome 8 with a segment of 14). Controls over the segment's genes are lost at the new location, and a form of cancer results.

Duplications

Even a normal chromosome contains **duplications**, which are gene sequences that are repeated several to many times. Often the same gene sequence is repeated thousands of times. Some duplications are built into the DNA of the species; others are not.

Years ago, geneticists noticed there were twice as many males as females in mental institutions. They thought X-linked recessive inheritance might be one reason why. (Can you say why?) Later, they found an abnormal constricted region on the X chromosomes of mentally impaired, related males (Figure 12.12*b*). Abnormally expanded repeats cause the constriction.

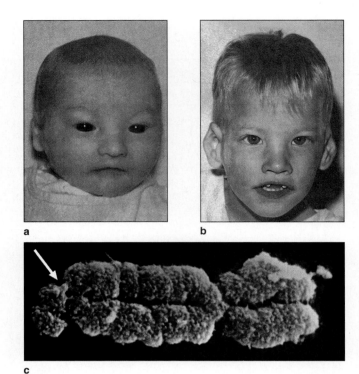

Figure 12.12 (**a,b**) Cri-du-chat syndrome, a result of a deletion on the short arm of chromosome 5. The left photograph shows the patient just after birth. The right photograph, taken four years later, shows how facial features change. By this age, patients no longer make the mewing sounds typical of the syndrome. (**c**) On the long arm of the X chromosome, the constricted region (arrow) responsible for the fragile X chromosome.

They give rise to the *fragile X syndrome*, the second most common form of mental impairment.

Not all duplications are harmful. Many have had roles in evolution. Cells require specific gene products, so mutations that alter most genes are probably selected against. But *duplicates* of some gene sequences that do no harm could be retained and would be free to mutate, for the normal gene would still provide the required product. In time, duplicated and then slightly modified sequences could yield products with related functions or even new ones. This apparently happened in gene regions coding for the polypeptide chains of the hemoglobin molecule. In humans and other primates, those regions have multiple copies of strikingly similar gene sequences, and they produce whole families of slightly different chains.

Together with duplications, some inversions and translocations probably helped put the primate ancestors of humans on a unique evolutionary road. Of the twenty-three pairs of human chromosomes, eighteen are nearly identical to their counterparts in chimpanzees and gorillas. The other five pairs differ at inverted and translocated regions.

On rare occasions, a segment of a chromosome may be lost, inverted, moved to a new location, or duplicated.

12.11 CHANGES IN CHROMOSOME NUMBER

Various abnormal cellular events can put too many or too few chromosomes into gametes. New individuals end up with the wrong chromosome number. The effects range from minor physical changes to lethal disruption of organ systems. More often, affected individuals are miscarried, or spontaneously aborted before birth.

Categories of Change

New individuals may end up with one extra or one less chromosome. This condition, called **aneuploidy**, is a major cause of human reproductive failure, affecting possibly half of all fertilized eggs. Most miscarried and autopsied human embryos were aneuploids.

New individuals also may end up with three or more of each type of chromosome. This condition is called **polyploidy**. About half of all flowering plant species are polyploid (page 294). So are some insects, fishes, and other animals. Polyploidy is lethal for humans. It may disrupt interactions between the genes of autosomes and sex chromosomes at key steps in the complex pathways of development and reproduction. All but 1 percent of human polyploids die before birth, and the rare newborns die within a month.

Mechanisms of Change

A chromosome number can change during mitotic or meiotic cell divisions or during the fertilization process. Suppose a cell cycle proceeds through DNA duplication and mitosis, then is arrested before the cytoplasm divides. This polyploid cell is "tetraploid," with four of each type of chromosome. Or suppose one or more pairs of chromosomes fail to separate properly during mitosis or meiosis. Such events are called **nondisjunction**. As Figure 12.13 shows, some or all of the resulting cells end up with too many or too few chromosomes. Researchers routinely induce nondisjunction by exposing cells to colchicine (compare the *Focus* essay on page 188).

What if a gamete with an extra chromosome ($n + 1$) unites with a normal gamete at fertilization? The new individual will be "trisomic," with three of one type of chromosome ($2n + 1$). If the gamete is missing a chromosome, then the individual will be "monosomic" ($2n - 1$). The following pages provide specific examples of such changes.

Compared to the parental chromosome number, aneuploids have one extra or one less chromosome. Polyploids have three or more of each type of chromosome.

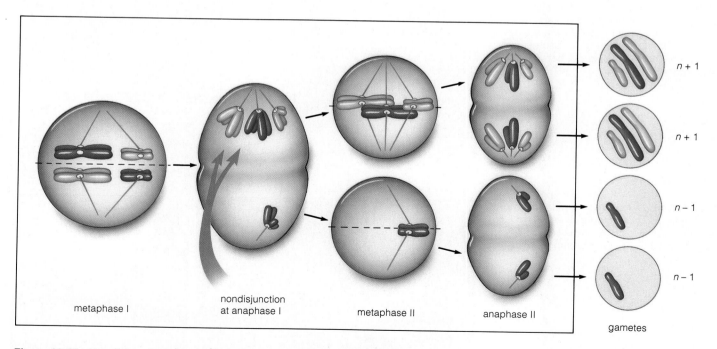

metaphase I   nondisjunction at anaphase I   metaphase II   anaphase II $n + 1$ $n + 1$ $n - 1$ $n - 1$ gametes

Figure 12.13 Nondisjunction. In this example, chromosomes fail to separate during anaphase I of meiosis and so change the chromosome number in the resulting gametes. As another example, make a similar sketch of nondisjunction at anaphase II of meiosis. What will the chromosome numbers be in the resulting gametes?

12.12 WHEN THE NUMBER OF AUTOSOMES CHANGES

Most changes in the number of autosomes arise through nondisjunction during the formation of gametes. Here we consider the most common of the resulting disorders.

Chromosome 21 is one of the smallest chromosomes in human cells. Someone who inherits three of them is categorized as a trisomic 21 and will show the effects of *Down syndrome*. ("Syndrome" simply means a set of symptoms that characterize a disorder.) Figure 12.14*a* shows a karyotype of an affected girl.

Symptoms vary considerably. However, most affected individuals show moderate to severe mental impairment. About 40 percent develop heart defects. As a result of abnormal skeletal development, older children have shortened body parts, loose joints, and poorly aligned bones of the hips, fingers, and toes. Muscles and muscle reflexes are weaker than normal, and development of speech and other motor functions is quite slow.

Nevertheless, with special training, trisomic 21 individuals often take part in normal activities. As a group,

they seem truly to enjoy life (Figure 12.15). These are cheerful, impish, affectionate people who characteristically derive great pleasure from music and dancing.

Down syndrome is one of many genetic disorders that can be detected by prenatal diagnosis, as described in the *Focus* essay that concludes this chapter. Before detection procedures were widespread, about 1 in 700 newborns of all ethnic groups were trisomic 21. Today the number is closer to 1 of every 1,100. The risk is much greater when women are over thirty-five years old (Figure 12.14*b*).

The genes responsible for the disorder seem to reside in a certain band region on the long arm of chromosome 21. One of the genes codes for the precursor of *beta*-amyloid protein. Excessive amounts of this protein are present in plaques and tangles in the brain of trisomic 21 patients. The same is true of *Alzheimer disease*, which is characterized by confusion, memory loss, increasing inability to perform simple tasks and, in time, the loss of all motor functions. Intriguingly, past thirty years of age, trisomic 21 patients show many of the same symptoms. They also show similar forms of deterioration in the same regions of the brain.

Figure 12.14 Down syndrome.
(**a**) Karyotype of an affected girl. The *red* arrows identify the trisomy of chromosome 21. (**b**) Relationship between the frequency of Down syndrome and the mother's age at the time her child was born. Results are from a study of 1,119 children with the disorder who were born in Victoria, Australia, between 1942 and 1957.

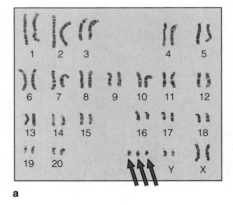

a

b

Figure 12.15 A few children with Down syndrome. The photographs at center and to the right show two of the enthusiastic participants in the Special Olympics.

12.13 WHEN THE NUMBER OF SEX CHROMOSOMES CHANGES

Most sex chromosome abnormalities arise through nondisjunction during gamete formation, as shown in Figure 12.16. Let's look at a few phenotypic outcomes.

Turner Syndrome

Inheritance of one X chromosome without a partner X or Y chromosome gives rise to *Turner syndrome*, which affects about 1 in 2,500 to 10,000 newborn girls. About 75 percent of these cases are results of nondisjunction in the father.

Turner syndrome is not as common as other sex chromosome abnormalities. Probably this is because at least 98 percent of all X0 zygotes are spontaneously aborted early in pregnancy. And X0 embryos represent 20 percent of all spontaneous abortions of embryos with chromosome abnormalities.

Despite the near-lethality, X0 survivors are not as disadvantaged as other aneuploids. They grow up well proportioned, albeit short (4 feet, 8 inches tall, on the average). Generally, their behavior is normal during childhood. But most Turner females are infertile. They do not have functional ovaries and so cannot produce eggs or sex hormones. Without sex hormones, breast enlargement and the development of other secondary sexual traits cannot occur. Possibly as a result of their arrested sexual development and small size, females often become passive and are easily intimidated by peers during their teens. Some patients have benefited from hormone therapy and corrective surgery.

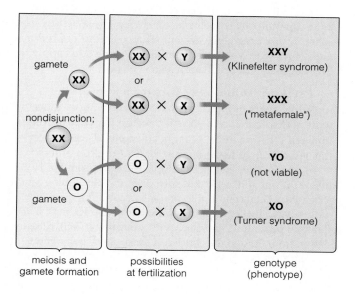

Figure 12.16 Examples of genetic disorders resulting from nondisjunction of X chromosomes followed by fertilization involving normal sperm.

Klinefelter Syndrome

About 1 in 500 to 2,000 liveborn males inherits two X chromosomes as well as one Y chromosome. About 67 percent of the time, the XXY condition results from nondisjunction in the mother, and the other 33 percent in the father.

Symptoms of the resulting *Klinefelter syndrome* do not develop until after the onset of puberty. XXY males are taller than average, and they are sterile or show low fertility. Their testes usually are much smaller than average, although the scrotum and penis are normal in size. Facial hair is often sparse, and there may be some breast enlargement. Injections of the hormone testosterone can reverse the feminized traits but not the low fertility. Some XXY males show mild mental impairment, but many fall within the normal range of intelligence. Except for their low fertility, many show no outward symptoms at all.

XYY Condition

About 1 in every 1,000 males has one X and two Y chromosomes. These XYY males tend to be taller than average. Some may be mildly retarded, but most are phenotypically normal.

At one time, XYY males were thought to be genetically predisposed to become criminals. This erroneous conclusion was based on small numbers of cases in highly selected groups, such as inmates in prisons. The investigators often knew in advance who the XYY males were, and this may have biased their evaluations. There were no double-blind studies, in which data on karyotypes and on personal histories were gathered independently by entirely different investigators, then matched up only after both sets of data were completed. Fanning the stereotype was a sensationalized report in 1968 that Richard Speck, a mass-murderer of young nurses, was an XYY male. He wasn't.

In 1976, a Danish geneticist reported on a large-scale study based on the records of 4,139 tall males, twenty-six years old, who had reported to their draft board. Besides the results of a physical examination and intelligence test, the records also provided clues to socioeconomic status, educational history, and any criminal convictions. Twelve of the males were XYY, which left more than 4,000 for the control group. The only significant finding was that tall, mentally impaired males who engage in criminal activity are more likely to get caught—irrespective of karyotype.

Nondisjunction during gamete formation accounts for most changes in chromosome number.

Prospects and Problems in Human Genetics

With the arrival of a newborn, parents typically want to know whether it is a girl or boy. Then most apprehensively ask, "Is the baby normal?" Quite naturally, they want their baby to be free of genetic disorders, and most of the time it is. Chapter 45 describes the story of human reproduction and development when all goes well. But what are the options when it does not?

We do not approach diseases and heritable disorders the same way. "Diseases" are the outcome of infection by bacteria, viruses, and other agents from the outside. We eliminate or control infectious agents with antibiotics and other weapons. By contrast, how do we attack an "enemy" within our genes?

Do we institute regional, national, or global programs to identify people carrying harmful alleles? Do we tell them they are "defective" and run a risk of bestowing their disorder on their children? Who decides which alleles are "harmful"? Should society bear the cost of treating Down syndrome and other genetic disorders? If so, should society also have a say in whether affected embryos will be born at all, or aborted? These questions are only the tip of an ethical iceberg. And answers have not been worked out in universally acceptable ways.

Phenotypic Treatments   Genetic disorders cannot be cured. But often their symptoms can be suppressed or minimized. For example, this can be done by controlling diets, making environmental adjustments, or intervening surgically or with hormone therapy.

Diet control works for several genetic disorders, including galactosemia (page 196). *Phenylketonuria* (or PKU for short) is another case in point. A certain gene codes for an enzyme that converts one amino acid to another (phenylalanine to tyrosine). In people who are homozygous for a recessive mutated form of the gene, phenylalanine accumulates. If the excess is diverted to other pathways, phenylpyruvate and other compounds may be produced. A high level of phenylpyruvate can lead to mental impairment.

But suppose affected persons restrict the amount of phenylalanine in their diet. Their body does not have to dispose of excess amounts, so they can lead normal lives. Phenylketonurics are usually aware that diet soft drinks and many other products often are artificially sweetened with aspartame, which contains phenylalanine. Such products carry warning labels alerting them to this.

Environmental adjustments help counter symptoms of some disorders. Sickle-cell anemics (page 176) can avoid strenuous activity when oxygen levels are low, as at high altitudes. Albinos (page 178) can avoid direct sunlight.

Surgical reconstructions can correct or minimize many phenotypic problems. In one form of *cleft lip,* a vertical fissure cuts through the lip midsection and often extends into the roof of the mouth. Surgery usually corrects the lip's appearance and function.

Genetic Screening   Some genetic disorders can be detected early enough to start preventive measures before symptoms develop. "Genetic screening" refers to large-scale programs to detect affected persons or carriers in a population. For example, most hospitals in the United States routinely screen all newborns for PKU, so it is now less common to see people with symptoms of this disorder.

Genetic Counseling   Sometimes, prospective parents suspect that they are likely to produce a severely afflicted child. (Their first child or a close relative may suffer from a genetic disorder.) Parents at risk may request information from clinical psychologists, geneticists, and social workers.

Counseling starts with accurate diagnosis of parental genotypes. This may reveal the risk of a specific disorder. Biochemical tests can be used to detect many metabolic disorders. Detailed family pedigrees may be constructed to aid the diagnosis.

For disorders showing simple Mendelian inheritance, geneticists can predict the chances of having an affected child. But as this chapter made clear, not all disorders follow Mendelian patterns. Even ones that do can be influenced by other factors. Even when the risk has been defined with some confidence, prospective parents must know that the risk is the same for each pregnancy. For example, if a pregnancy has one chance in four of producing a child with a genetic disorder, the same odds apply to every subsequent pregnancy, also.

Prenatal Diagnosis   Suppose a woman who is forty-five years old and pregnant wants to know if her fetus is trisomic 21 and so will develop Down syndrome. Prenatal ("before birth") diagnostic methods can detect this genetic disorder and more than a hundred others.

In **amniocentesis**, a fluid sample is drawn from fluid inside the amnion, a sac that surrounds the developing fetus (page 190). Fetal cells in the sample are cultured, then they are analyzed by karyotyping and other tests (Figure *a*). In a different procedure, **chorionic villi sampling** (CVS), cells are drawn from fluid in the chorion, a sac around the amnion.

Like amniocentesis, CVS is risky. Either procedure may accidentally cause infection or puncture the fetus. Besides this, amniocentesis is performed during the fifth month of pregnancy, when a mother already feels kicks and movements inside her. CVS can be performed earlier, between the ninth and twelfth weeks. However, even though the

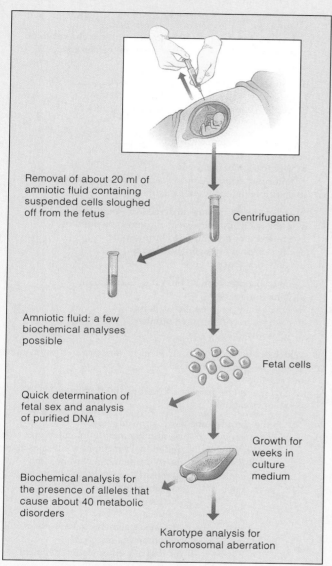

Removal of about 20 ml of amniotic fluid containing suspended cells sloughed off from the fetus

Centrifugation

Amniotic fluid: a few biochemical analyses possible

Fetal cells

Quick determination of fetal sex and analysis of purified DNA

Growth for weeks in culture medium

Biochemical analysis for the presence of alleles that cause about 40 metabolic disorders

Karotype analysis for chromosomal aberration

a Steps in amniocentesis.

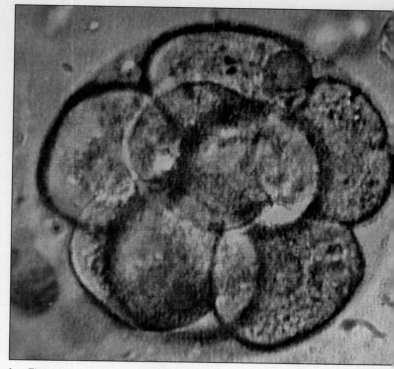

b Eight-cell stage of human development.

fetus is only about half as long as the little finger, its major organs have already started to form.

What choice do prospective parents make if either procedure reveals a devastating genetic disorder? Do they opt for induced **abortion**—an induced expulsion of the embryo from the uterus? This can be an agonizing decision. They must weigh awareness of the crushing severity of the disorder against ethical and religious beliefs. Worse, they must play out their personal tragedy on a larger stage, dominated now by a nationwide battle between fiercely vocal "pro-life" and "pro-choice" factions.

In 1992, clinical trials of preimplantation diagnosis proved successful. This new procedure relies on **in vitro fertilization**. Sperm and eggs donated by prospective parents are put in an enriched medium in a petri dish. There, one or more eggs may become fertilized. Two days later, cell divisions convert each fertilized egg into a ball of eight cells (Figure *b*). The tiny ball might be considered a prepregnancy stage. Like unfertilized eggs flushed monthly from a woman's body, it is free-floating; it is not connected to the uterus.

All cells in that ball have the same genes, and they are not differentiated. They are not yet committed to giving rise to specialized cells of the heart, toes, and other tissues. Researchers pluck one of the cells from each ball and analyze its genes for suspected disorders. Only a ball that is free of genetic defects is inserted back into the uterus.

At this writing, several couples who are at risk of passing on muscular dystrophy, cystic fibrosis, and other severe disorders have opted for the procedure. The procedure is still highly experimental, and it is costly. Yet a number of "test-tube" babies have been born. All are in good health—and free of the harmful genes.

SUMMARY

1. Genes, the units of instruction for heritable traits, are arranged one after the other along chromosomes.

2. Human cells are diploid (2*n*), with twenty-three pairs of homologous chromosomes that interact during meiosis. Except for X and Y chromosomes, each pair has the same length, shape, and gene sequence.

3. Human females have a pair of X chromosomes. Males have X paired with Y. All other pairs of chromosomes are autosomes (the same in both females and males). A gene on the Y chromosome determines gender.

4. Clues to inheritance often show up in pedigrees (charts of genetic connections through lines of descent). Certain patterns are characteristic of dominant or recessive alleles on autosomes or on the X chromosome.

5. Genes on the same chromosome represent a linkage group. Crossing over (the breakage and exchange of segments between homologues) disrupts linkages. The farther apart two genes are, the greater will be the frequency of crossovers between them.

6. A chromosome's structure may change on rare occasions. A segment may be deleted, inverted, moved to a new location, or duplicated.

7. The chromosome number may change on rare occasions. New individuals may end up with one more or one less chromosome than the parents (aneuploidy). Or they may end up with three or more of each type of chromosome (polyploidy). Nondisjunction during meiosis and gamete formation accounts for most of these chromosome abnormalities.

8. Crossing over and changes in chromosome number or in a chromosome's structure may influence the course of evolution. The changes in genotype (genetic makeup) lead to variations in phenotype (observable traits) among members of a population, so that evolution is possible.

---

1. _____ segregate during _____ .
   a. Homologues; mitosis
   b. Genes on one chromosome; meiosis
   c. Homologues; meiosis
   d. Genes on one chromosome; mitosis

2. The genes of one chromosome and the genes of the homologous chromosome end up in separate _____ .
   a. body cells
   b. gametes
   c. nonhomologous chromosomes
   d. offspring
   e. both b and d are possible

3. Genes on the same chromosome tend to remain together during _____ and end up in the same _____ .
   a. mitosis; body cell
   b. mitosis; gamete
   c. meiosis; body cell
   d. meiosis; gamete
   e. both a and d

4. The probability of a crossover occurring between two genes on the same chromosome is _____ .
   a. unrelated to the distance between them
   b. increased if they are closer together on the chromosome
   c. increased if they are farther apart on the chromosome
   d. impossible

5. Chromosome structure can be altered by _____ .
   a. deletions
   b. duplications
   c. inversions
   d. translocations
   e. all of the above

6. Nondisjunction can be caused by _____ .
   a. crossing over in meiosis
   b. segregation in meiosis
   c. failure of chromosomes to separate during meiosis
   d. multiple independent assortments

7. A gamete affected by nondisjunction would have _____ .
   a. a change from the normal chromosome number
   b. one extra or one missing chromosome
   c. the potential for a genetic disorder
   d. all of the above

8. Genetic disorders can be caused by _____ .
   a. gene mutations
   b. changes in chromosome structure
   c. changes in chromosome number
   d. all of the above

9. Which of the following contributes to variation in a population?
   a. independent assortment
   b. crossing over
   c. changes in chromosome structure and number
   d. all of the above

10. Match the chromosome terms appropriately.
   _____ crossing over
   _____ deletion
   _____ nondisjunction
   _____ translocation
   _____ karyotype
   a. number and defining features of individual's metaphase chromosomes
   b. movement of a chromosome segment to a nonhomologous chromosome
   c. disrupts gene linkages during meiosis
   d. causes gametes to have abnormal chromosome numbers
   e. loss of a chromosome segment

1. Human females are XX and males are XY.
   a. Does a male inherit his X chromosome from his mother or father?
   b. With respect to an X-linked gene, how many different types of gametes can a male produce?
   c. If a female is homozygous for an X-linked gene, how many different types of gametes can she produce with respect to that gene?
   d. If a female is heterozygous for an X-linked gene, how many different types of gametes can she produce with respect to that gene?

2. One allele of a presumed Y-linked gene results in nonhairy ears in males (*see photograph*). Another allele of the same gene results in *hairy pinnae* (relatively long hairs at the outer ear).

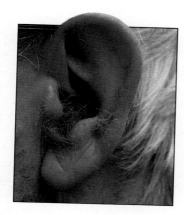

    a. Why would you *not* expect females to have hairy pinnae?

    b. A hairy-eared male's son will have hairy ears, but no daughter will. Explain why.

3. Suppose that you have two linked genes with alleles *A,a* and *B,b* respectively. An individual is heterozygous for both genes, as in the following:

$$A \qquad\qquad B$$
$$A \qquad\qquad B$$

$$a \qquad\qquad b$$
$$a \qquad\qquad b$$

If the crossover frequency between these two genes is 0 percent, what genotypes would be expected among gametes from this individual, and with what frequencies?

4. *Linkage maps* are plots of the relative distance between genes that undergo crossing over and other chromosomal rearrangements. They are based on the patterns in which such genes are distributed into gametes. They don't show *exact* physical distances between genes, because the probability of crossing over isn't equal along a chromosome's length. The distances are measured in map units that are based on the frequency of recombination between genes. One genetic map unit equals 1 percent recombination. Thus, for example, if the crossover frequency for two genes is 10 percent, then ten map units separate them.

Consider two linked genes on a *D. melanogaster* chromosome. One gene, which influences eye color, has a red (dominant) allele and a purple (recessive) allele. Another gene dictates wing length. A dominant allele at this locus codes for long wings; a recessive allele codes for vestigial (short) wings. Suppose a fully homozygous dominant female having red eyes and long wings mates with a male having purple eyes and vestigial wings. Suppose next that the F_1 females mate with males having purple eyes and vestigial wings. The second cross results in offspring with the following traits:

252 red eyes, long wings
276 purple eyes, vestigial wings
 42 red eyes, vestigial wings
 30 purple eyes, long wings
_____
600 offspring total

Based on these data, how many map units separate the two linked genes?

5. Suppose you cross a homozygous dominant long-winged fruit fly with a homozygous recessive vestigial-winged fly. Shortly after mating, the fertilized eggs are exposed to a level of x-rays known to cause mutation and chromosomal deletions. When these fertilized eggs subsequently develop into adults, most of the flies are long-winged and heterozygous. However, a few are vestigial-winged. Provide possible explanations for the unexpected appearance of these vestigial-winged adults.

6. Individuals affected by Down syndrome typically have an extra chromosome 21, so their cells have a total of 47 chromosomes. However, in a few cases of Down syndrome, 46 chromosomes are present. Included in this total are two normal-appearing chromosomes 21 and a longer-than-normal chromosome 14. Interpret this observation and indicate how these few individuals can have a normal chromosome number.

7. The mugwump, a type of tree-dwelling mammal, has a reversed sex-chromosome condition. The male is XX and the female is XY. However, perfectly good sex-linked genes are found to have the same effect as in humans. For example, a recessive, X-linked allele *c* produces red-green color blindness. If a normal female mugwump mates with a phenotypically normal male mugwump whose mother was color blind, what is the probability that a son from that mating will be color blind? A daughter?

8. One type of childhood muscular dystrophy is a recessive, X-linked trait in humans. A slowly progressing loss of muscle function leads to death, usually by age twenty or so. Unlike color blindness, this disorder is restricted to males, not ever having been found in a female. Suggest why.

Selected Key Terms

abortion 205	inversion 200
allele 188	in vitro fertilization 205
amniocentesis 204	karyotype 189
aneuploidy 201	linkage group 193
autosome 189	nondisjunction 201
chorionic villi sampling 204	pedigree 195
crossing over 188	polyploidy 201
deletion 200	sex chromosome 189
duplication 200	translocation 200
gene 188	X chromosome 188
genetic abnormality 196	X-linked gene 193
genetic disorder 196	Y chromosome 188
homologous chromosome 188	Y-linked gene 193

Readings

Cummings, M. 1994. *Human Heredity: Principles and Issues*. Third edition. St. Paul, Minnesota: West.

Edlin, G. 1988. *Genetic Principles: Human and Social Consequences*. Second edition. Portola Valley, California: Jones & Bartlett.

Fuhrmann, W., and F. Vogel. 1986. *Genetic Counseling*. Third edition. New York: Springer-Verlag.

Holden, C. 1987. "The Genetics of Personality." *Science* 237:598–601. For students interested in human behavioral genetics.

13 DNA STRUCTURE AND FUNCTION

Cardboard Atoms and Bent-Wire Bonds

Linus Pauling in 1951 did something no one had done before. Through his training in biochemistry, a talent for model building, and a few educated guesses, he deduced the three-dimensional structure of a protein (collagen). His discovery electrified the scientific community. If the secrets of proteins could be pried open, why not other biological molecules? Further, wouldn't structural details provide clues to molecular functions? And who would go down in history as having discovered the molecule that contains instructions for reproducing parental traits in offspring—*the very secrets of inheritance*? Scientists around the world started scrambling after that ultimate prize.

Maybe hereditary instructions were encoded in the structure of some unknown class of proteins. After all, heritable traits are spectacularly diverse. Surely the molecules encoding information about those traits were structurally diverse also. Proteins are put together from potentially limitless combinations of amino acid subunits, so they almost certainly could function as the sentences (genes) in each cell's book of inheritance.

Yet there was something about another substance—DNA—that excited many researchers. Among them were James Watson, a young postdoctoral student from Indiana University, and Francis Crick, an energetic researcher at Cambridge University. How could DNA, a molecule consisting of only four kinds of subunits, hold genetic information? Watson and Crick spent long hours arguing over everything they had read about the size, shape, and bonding requirements of the subunits of DNA. They fiddled with cardboard cutouts of the subunits. They badgered chemists to identify potential bonds they might have overlooked. They assembled models from bits of metal held together with wire "bonds" bent at appropriate angles.

In 1953, they put together a model that fit all the pertinent biochemical rules and all the facts about DNA they had gleaned from other sources (Figure 13.1). They had discovered the structure of DNA. The breathtaking simplicity of that structure also enabled them to solve a long-standing riddle—*how life can show unity at the molecular level and yet give rise to so much diversity at the level of whole organisms*.

With this chapter, we turn to investigations that led to our current understanding of DNA structure and function. They are revealing of how ideas are generated in science. On the one hand, having a shot at fame and fortune quickens the pulse of men and women in any profession, and scientists are no exception. On the other hand, science proceeds as a community effort, with individuals sharing not only what they can explain but also what they do not understand. Even if an experiment "fails," it may turn up information that others can use or lead to questions that others can answer. Unexpected results, too, might be clues to something important about the natural world.

Figure 13.1 James Watson and Francis Crick posing in 1953 by their newly unveiled model of the structure of DNA. The photograph on the facing page is a more recent model. This computer-generated model is more sophisticated in appearance, yet it is basically the same as the prototype that Watson and Crick put together nearly four decades before.

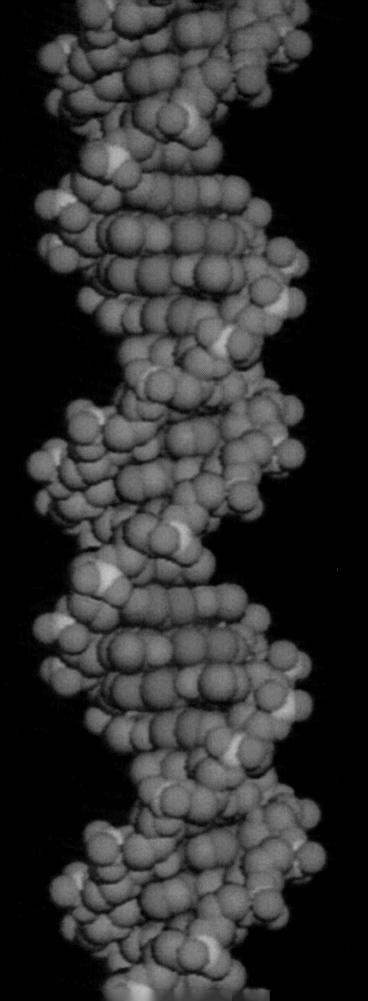

1. In living cells, DNA is the storehouse of information about heritable traits. The hereditary information is encoded in the nucleotide subunits that make up the DNA molecule.

2. DNA consists of four kinds of nucleotides that differ in only one component, a nitrogen-containing base. The four bases are adenine, guanine, thymine, and cytosine. Great numbers of these occur one after another in a DNA molecule, and the order in which one kind follows another is different for each species of organism.

3. In a DNA molecule, two strands of nucleotides twist together like a spiral stairway; they form a double helix. Hydrogen bonds connect the bases of one strand to bases of the other. As a rule, adenine pairs (hydrogen-bonds) only with thymine, and guanine only with cytosine.

4. Before a cell divides, its DNA is replicated with the help of enzymes and other proteins. Each double-stranded DNA molecule starts unwinding. A new, complementary strand is assembled on the exposed bases of each parent strand according to the base-pairing rule.

5. There is only one DNA molecule in an unduplicated chromosome. Except in bacteria, great numbers of proteins are attached to the DNA and function in its structural organization.

One might have wondered, in the spring of 1868, why Johann Friedrich Miescher was collecting cells from the pus of open wounds and, later, from the sperm of a fish. Miescher, a physician, wanted to identify the chemical composition of the nucleus. He was interested in those cells because they are composed mostly of nuclear material, with very little cytoplasm. Miescher succeeded in isolating an acidic substance, one with a notable amount of phosphorus. He called it "nuclein." He had discovered what came to be known as deoxyribonucleic acid, or **DNA**.

The discovery caused scarcely a ripple through the scientific community. At the time, no one knew much about the physical basis of inheritance—that is, *which* chemical substance in cells actually encodes the instructions for reproducing parental traits in offspring. Only a few researchers suspected that the nucleus might hold the answer. In fact, seventy-five years passed before DNA was recognized as having profound biological importance.

13.1 DISCOVERY OF DNA FUNCTION

Early Clues

In 1928, an Army medical officer, Fred Griffith, was attempting to create a vaccine against *Streptococcus pneumoniae*, which causes a type of pneumonia. Many vaccines are preparations of killed or weakened bacterial cells that, when introduced into the body, can mobilize defenses against a real attack. Griffith never did create a vaccine, but his work unexpectedly opened a door to the molecular world of heredity.

Griffith isolated and cultured two strains of the bacterium. He noticed that colonies of one strain had a rough surface appearance and those of the other strain appeared smooth. He designated the strains *R* and *S* and used them in a series of four experiments:

1. Laboratory mice were injected with live R cells. As Figure 13.2 indicates, they did not develop pneumonia. *The R strain was harmless.*

2. Mice were injected with live S cells. The mice died. Blood samples from them teemed with live S cells. *The S strain was pathogenic* (disease-causing).

3. S cells were killed by exposure to high temperature. Mice injected with these cells did not die.

4. Live R cells were mixed with heat-killed S cells and injected into mice. The mice died—and blood samples from them teemed *with* live S cells!

What was going on in the fourth experiment? Maybe heat-killed S cells in the mixture weren't really dead. But if that were true, then mice injected with heat-killed S cells alone (experiment 3) would have died. Maybe harmless R cells in the mixture had mutated into a killer form. But if that were true, then mice injected with the R cells alone (experiment 1) would have died.

The simplest explanation was this: *Heat did kill the S cells but did not destroy their hereditary material, including the part that specified "how to cause infection."* Somehow, that material had been transferred from dead S cells to living R cells—where it was put to use.

Further experiments showed that harmless cells had indeed picked up information about infection and had been permanently transformed into pathogens. Hundreds of generations of bacteria descended from the transformed cells also caused infections!

Griffith's unexpected results intrigued a microbiologist, Oswald Avery, and his colleagues. They were able to transform harmless cells with extracts of killed pathogens. In 1944, they reported that the hereditary substance in the extracts was probably DNA, not pro-

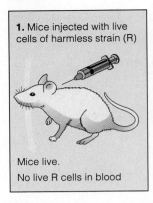
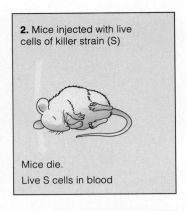
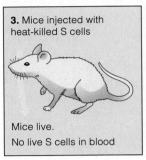
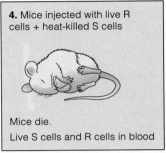

1. Mice injected with live cells of harmless strain (R)

Mice live.
No live R cells in blood

2. Mice injected with live cells of killer strain (S)

Mice die.
Live S cells in blood

3. Mice injected with heat-killed S cells

Mice live.
No live S cells in blood

4. Mice injected with live R cells + heat-killed S cells

Mice die.
Live S cells and R cells in blood

Figure 13.2 Summary of results from Griffith's experiments with a harmless strain and a disease-causing strain of *Streptococcus pneumoniae*, as described in the text.

teins, as was widely believed. They had added protein-digesting enzymes to some extracts, but cells were transformed anyway. Then they added an enzyme that breaks apart DNA but not proteins—and that enzyme blocked hereditary transformation.

Yet how were Avery's impressive findings received? Many (if not most) biochemists refused to give up on the proteins. His experimental results, they said, probably applied only to bacteria.

Confirmation of DNA Function

By the early 1950s, Max Delbrück, Alfred Hershey, Salvador Luria, and other molecular detectives were using certain viruses as their experimental subjects. The viruses, called **bacteriophages**, infect bacterial cells such as *Escherichia coli*. Viruses are about as biochemically simple as you can get. They are not alive, but they do contain hereditary information for building new virus particles. At some point after they infect a host cell, viral enzymes take over a portion of the cell's metabolic machinery—which starts churning out the substances required to construct new virus particles.

By 1952, researchers knew that some bacteriophages consist only of DNA and protein. Also, electron micrographs revealed that the main part of these viruses

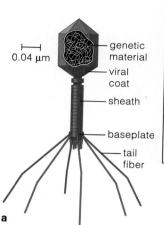

genetic material
viral coat
sheath
baseplate
tail fiber

0.04 μm

a

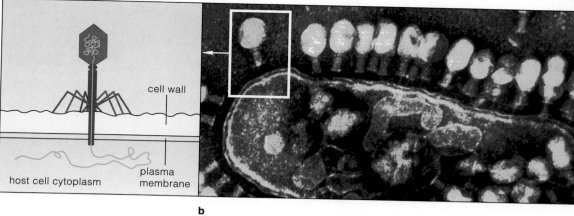

cell wall
host cell cytoplasm
plasma membrane

b

Figure 13.3 (**a**) Structural organization of a T4 bacteriophage. (**b**) Electron micrograph of T4 virus particles infecting a host bacterial cell (*Escherichia coli*). The diagram shows DNA, the genetic material of this bacteriophage, being injected into the cell cytoplasm.

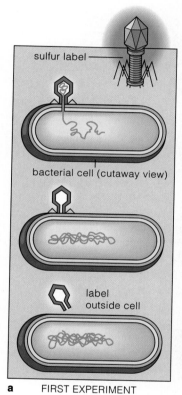

sulfur label

bacterial cell (cutaway view)

label outside cell

a FIRST EXPERIMENT

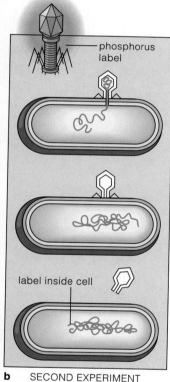

phosphorus label

label inside cell

b SECOND EXPERIMENT

Figure 13.4 Two examples of the landmark experiments pointing to DNA as the substance of heredity. In the 1940s, Alfred Hershey and his colleague, Martha Chase, were studying the biochemical basis of inheritance. They were aware that certain bacteriophages were composed of proteins and DNA. Did the proteins, DNA, or both contain the viral genetic information?

To find out, Hershey and Chase designed two experiments, based on the following biochemical facts: First, proteins incorporate sulfur (S) but not phosphorus (P). Second, DNA incorporates phosphorus but not sulfur.

(**a**) In one experiment, bacterial cells were grown on a culture medium that included the radioisotope ^{35}S and no other form of sulfur. To synthesize proteins, the bacterial cells would have to use the radioisotope—which would serve as a tracer. After the cells became labeled with the tracer, bacteriophages were allowed to infect them. Subsequently, viral proteins were synthesized inside the host cells. These proteins also became labeled with ^{35}S. So did the new generation of virus particles.

Next, the labeled bacteriophages were allowed to infect a new batch of unlabeled bacteria that were suspended in a fluid culture medium. Afterward, Hershey and Chase whirred the fluid in a kitchen blender. Whirring dislodged the infectious particles from the cells. The particles became suspended in the fluid medium. Analysis revealed the presence of labeled protein in the fluid—*not* inside the bacterial cells.

(**b**) In the second experiment, new bacterial cells were cultured. The only phosphorus available for synthesizing DNA was the radioisotope ^{32}P. Bacteriophages were allowed to infect the cells. As predicted, the viral DNA assembled inside the infected cells became labeled, as did the new generation of virus particles. Next, the labeled particles were allowed to infect bacteria in a fluid medium, and then were dislodged from the host cells. Analysis revealed that the labeled DNA was not in the fluid. It remained *inside* the host cells, where its instructions had to be put to use. Here was evidence that DNA is the genetic material of these bacteriophages.

remains *outside* of the cells they are infecting (Figure 13.3). Such viruses probably were injecting genetic material alone *into* host cells. If that were true, was the material DNA, protein, or both?

Through many experiments, researchers came up with convincing evidence that DNA, not proteins, functions as the molecule of heredity. Figure 13.4 describes two of these landmark experiments.

Information for producing the heritable traits of single-celled and multicelled organisms is encoded in DNA.

13.2 DNA STRUCTURE

Components of DNA

Long before the bacteriophage studies were under way, biochemists knew that DNA contains only four types of nucleotides, the building blocks of nucleic acids. A **nucleotide** consists of a five-carbon sugar (which is deoxyribose in DNA), a phosphate group, and one of the following nitrogen-containing bases:

| adenine | guanine | thymine | cytosine |
| (A) | (G) | (T) | (C) |

As you can see from Figure 13.5, all four types of nucleotides in DNA have these components joined together in much the same way. But T and C are pyrimidines, which are single-ring structures. A and G are purines, which are double-ring structures. These are larger, bulkier molecules.

By 1949, the biochemist Erwin Chargaff had shared two crucial insights into the composition of DNA with the scientific community. First, the amount of adenine relative to guanine differs from one species to the next. Second, the amount of adenine always equals the amount of thymine, and the amount of guanine always equals the amount of cytosine. We may show this as:

$$A = T \quad \text{and} \quad G = C$$

The relative proportions of the four different nucleotides were tantalizing clues to their arrangement in DNA. In some way, the four had to follow one after another along the length of the molecule.

The first convincing evidence of that arrangement came from Maurice Wilkins's laboratory in England. One of Wilkins's coworkers, Rosalind Franklin, had obtained good **x-ray diffraction images** of DNA fibers.

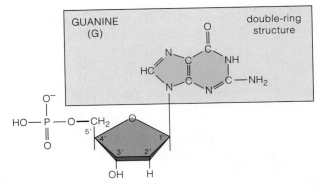

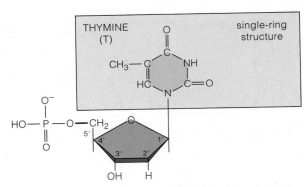

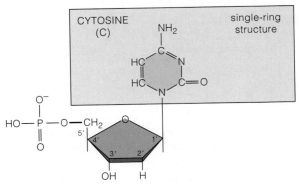

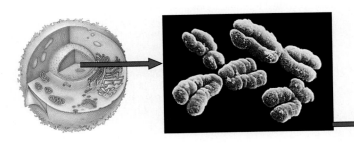

Figure 13.5 The four kinds of nucleotide subunits of DNA. Small numerals on the structural formulas identify the carbon atoms to which other parts of the molecule are attached.

All chromosomes contain DNA. What does DNA contain? Four kinds of nucleotides. Each nucleotide has a five-carbon sugar (shaded red). That sugar has a phosphate group attached to the fifth carbon atom of its ring structure. It also has one of four kinds of nitrogen-containing bases (shaded blue) attached to its first carbon atom. The nucleotides differ only in which base is attached to that atom.

DNA does not readily lend itself to x-ray diffraction. However, the researchers rapidly spun a suspension of DNA molecules, spooled them onto a rod, and gently pulled them into gossamer fibers, like cotton candy. If the atoms in DNA were arranged in a regular order, an x-ray beam directed at a fiber would scatter in a regular pattern that could be captured on film. The pattern would consist only of dots and streaks. It would not, in itself, reveal molecular structure. But it could be used to calculate the positions of atoms in DNA.

According to calculations in Wilkins's laboratory, DNA had to be long and thin, with a 2-nanometer diameter. Some molecular configuration was being repeated every 0.34 nanometer along its length, and another one, every 3.4 nanometers.

Could the sequence of nucleotide bases be twisting, like a circular stairway? Certainly Pauling thought about this possibility. After all, he was the one who discovered that a collagen chain has a helical shape, held in place by hydrogen bonds. He wrote to Wilkins, requesting copies of the x-ray diffraction images. The response was lukewarm, perhaps a delaying tactic. Why should Wilkins help a formidable competitor win the race to glory?

As it turned out, neither won the race. Watson and Crick were about to close in on the answer.

Patterns of Base Pairing

As Watson and Crick perceived, DNA consists of *two* strands of nucleotides, held together at their bases by hydrogen bonds. These bonds form when the two strands run in opposing directions and twist together into a double helix (Figure 13.6). Only two kinds of base pairings form along the entire length of the molecule: A–T and G–C.

This bonding pattern permits variation in the order of bases in any given strand. For example, in even a tiny stretch of DNA from a rose, gorilla, human, or any other organism, the sequence might be:

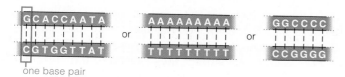

one base pair

And so DNA molecules show constancy and variation from one species to the next. This is the molecular foundation for the unity and diversity of life.

Base pairing between the two nucleotide strands in DNA is *constant* for all species (A with T, and G with C).

The base sequence (that is, which base follows another in a nucleotide strand) is *different* from species to species.

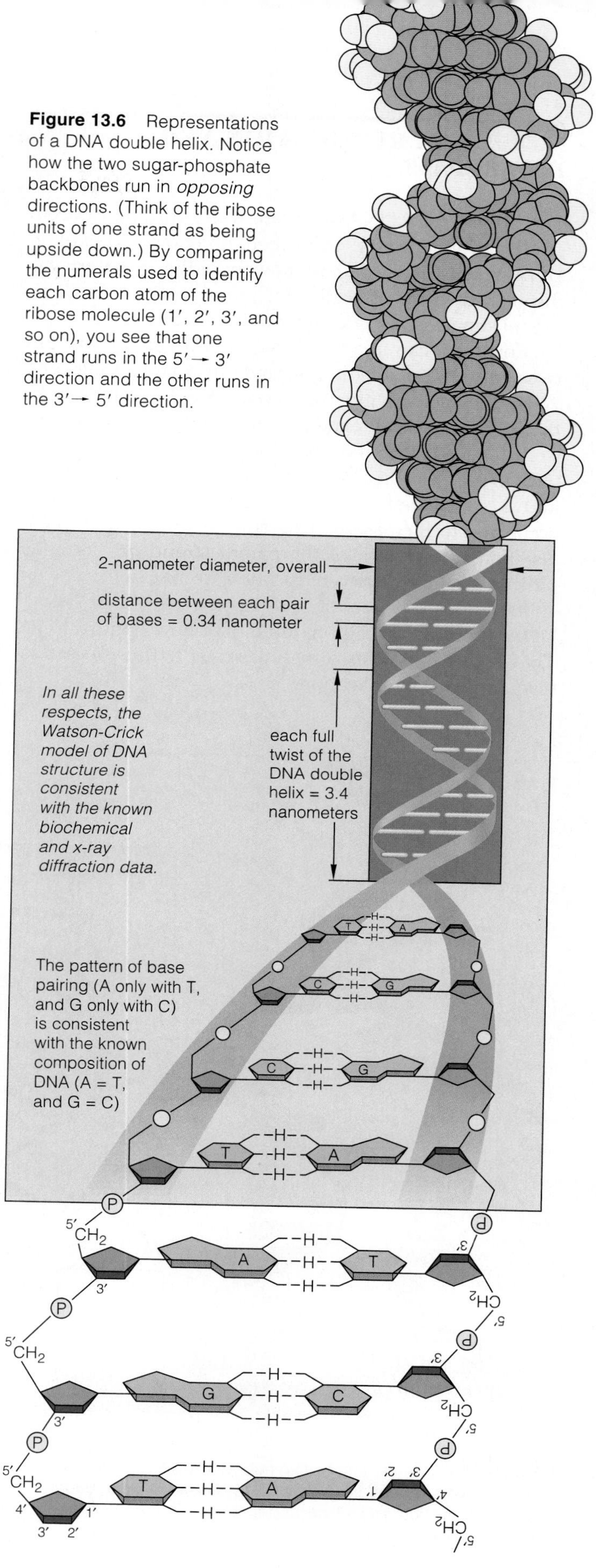

Figure 13.6 Representations of a DNA double helix. Notice how the two sugar-phosphate backbones run in *opposing* directions. (Think of the ribose units of one strand as being upside down.) By comparing the numerals used to identify each carbon atom of the ribose molecule (1′, 2′, 3′, and so on), you see that one strand runs in the 5′ → 3′ direction and the other runs in the 3′ → 5′ direction.

2-nanometer diameter, overall

distance between each pair of bases = 0.34 nanometer

In all these respects, the Watson-Crick model of DNA structure is consistent with the known biochemical and x-ray diffraction data.

each full twist of the DNA double helix = 3.4 nanometers

The pattern of base pairing (A only with T, and G only with C) is consistent with the known composition of DNA (A = T, and G = C)

13.3 DNA REPLICATION AND DNA REPAIR

The discovery of DNA structure was a turning point in studies of inheritance. Until then, no one could explain **DNA replication**—that is, how hereditary material is duplicated prior to cell division. The Watson-Crick model suggested at once how this might be done.

Enzymes can readily break hydrogen bonds between the two nucleotide strands of a DNA molecule. When they act at a given site, one strand unwinds from the other, thereby exposing some nucleotide bases. Cells have stockpiles of free nucleotides, and these pair with exposed bases. Each parent strand remains intact, and a companion strand is assembled on each one, according to the base-pairing rule (A to T, and G to C). As soon as a stretch of the parent strand has its new, partner strand, the two twist into a double helix.

Because the parent strand is conserved, each "new" DNA molecule is really half old, half new (Figure 13.7). That is why biologists sometimes refer to this process as *semiconservative* replication.

unwinding enzyme

binding proteins (stabilize the DNA in single-strand form)

primer-synthesizing enzyme

replication fork

a Replication of eukaryotic DNA begins at many short, specific base sequences called *origins*. Organized complexes of enzymes and other proteins unwind the two strands of the parent DNA molecule, prevent rewinding, and assemble new strands on each one. They do this only in *replication forks*. These are limited, V-shaped regions that advance in both directions away from an origin. Enzymes rewind the half-old, half-new molecules while they are being completed, behind each advancing replication fork.

origin

DNA replication requires a large team of molecular workers. For example, as Figure 13.8 indicates, one kind of enzyme unwinds the two nucleotide strands, and many proteins bind to them and hold them apart. **DNA polymerases** are key players; they attach free nucleotides to a growing strand. **DNA ligases** seal new short stretches of nucleotides into one continuous strand. Chapter 16 will describe how some of these enzymes have uses in recombinant DNA technology.

DNA polymerases, DNA ligases, and other enzymes also engage in a process called **DNA repair**. If the sequence of bases in one strand of a double helix becomes altered, DNA polymerases "read" the complementary sequence on the other strand. With the aid of other repair enzymes, they restore the original sequence. Figure 13.9 gives an example of what can happen when the excision-repair function is impaired.

Prior to cell division, the double-stranded DNA molecule unwinds and is replicated. Each parent strand remains intact—it is conserved—and enzymes assemble a new, complementary strand on each one.

Certain enzymes also correct base-pairing errors in the nucleotide sequence of DNA.

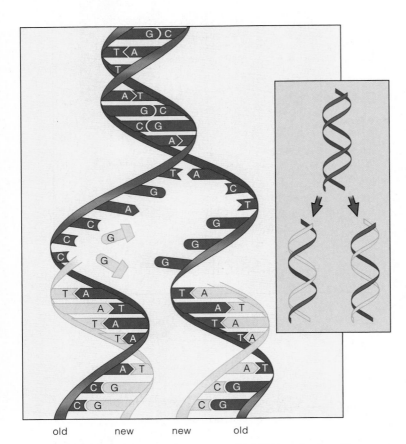

old   new   new   old

Figure 13.7 Semiconservative nature of DNA replication. The original two-stranded DNA molecule is shown in *blue*. Each parent strand remains intact, and a new strand (*yellow*) is assembled on each one.

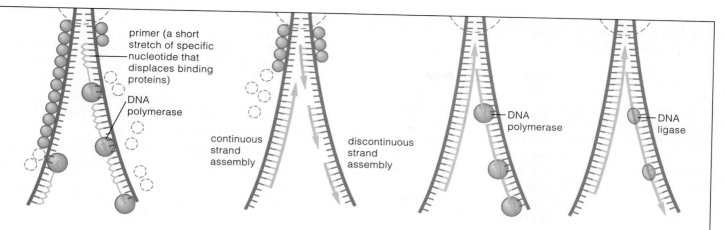

b At each replication fork, some binding proteins are displaced when enzymes synthesize primers at intervals along both of the parent templates. Other enzymes (DNA polymerases) recognize these primers as "start" tags. They join nucleotide units together behind primers. Exposed bases of the parent templates dictate which kind of nucleotide is added in sequence, according to the base-pairing rule.

c As discovered by Reiji Okazaki, strand assembly is *continuous* on one parent template. It is *discontinuous* on the other, where nucleotides must be assembled in short stretches only. Why? They can be joined together in the 5'⟶3' direction only, because this leaves one of the −OH groups of the sugar-phosphate backbone exposed (**f**). This exposed group is the only site where nucleotide units can be joined together.

d Primers are removed. DNA polymerases fill in the gaps but cannot make the final connections between stretches of nucleotides.

e DNA ligases seal the tiny nicks that remain, the result being continuous strands on the parent templates.

Figure 13.8 A closer look at the functions of enzymes and other proteins in DNA replication. The free nucleotides brought up for strand assembly also provide energy to drive the process. Each has three phosphate groups attached. DNA polymerase splits away two of these and uses some of the released energy to attach the nucleotide to a growing strand.

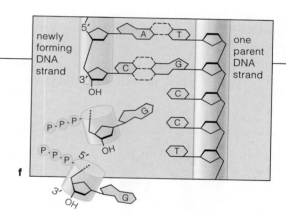

Figure 13.9 When DNA can't be fixed. A newborn destined to develop *xeroderma pigmentosum* faces a dim future, literally. Individuals affected by this genetic disorder cannot be exposed to sunlight, even briefly, without risking disfiguring skin tumors and possible early death from cancer. The disorder is a result of faulty DNA repair in the body's cells. One or more of the genes thought to govern repair processes have become damaged.

The ultraviolet wavelengths in light from the sun, tanning lamps, and other sources can cause molecular changes in DNA (page 635). Among other things, the wavelengths can promote covalent bonding between two adjacent thymine bases in a nucleotide strand. Thus the two nucleotides to which the bases belong are combined into an abnormal, bulky molecule called a "thymine dimer."

Normally, a DNA repair mechanism gets rid of such bulky lesions. The mechanism requires at least seven gene products. Mutation in one or more of the required genes can skew the repair machinery. Thymine dimers can accumulate in skin cells of individuals with such mutations. The accumulation triggers development of lesions, including skin cancer. The photograph shows a common skin cancer, called *basal cell carcinoma*.

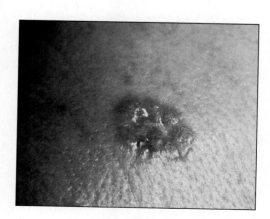

basal cell carcinoma

13.4 ORGANIZATION OF DNA IN CHROMOSOMES

Each chromosome has one DNA molecule. If the DNA of all forty-six chromosomes in one of your somatic cells were stretched out end to end, the string might extend from your shoulder to your fingertips. What keeps all that DNA from becoming a tangled mess? Proteins.

Eukaryotic DNA has many **histones** and other protein molecules bound tightly to it (Figure 13.10). Some histones are like spools for winding up small stretches of DNA. Each histone-DNA spool is a **nucleosome**. Another histone stabilizes the spools.

Histone-DNA interactions can make a chromosome coil back on itself again and again. The coiling greatly increases its diameter. Further folding results in a series of loops. Proteins other than histones serve as a structural "scaffold" for the loops.

Apparently, scaffold regions intervene between the genes, not in regions that contain information for building proteins. Are the scaffold proteins organizing the chromosome into functional "domains"? Would that organization make it easier for DNA replication and protein synthesis to proceed? These are just two of the possibilities being probed by the new generation of molecular detectives.

Figure 13.10 Levels of organization of DNA in a eukaryotic chromosome.

SUMMARY

1. Hereditary information of cells and multicelled organisms is encoded in DNA (deoxyribonucleic acid).

2. DNA is composed of nucleotide subunits. Each of these has a five-carbon sugar (deoxyribose), a phosphate group, and one of four nitrogen-containing bases (adenine, thymine, guanine, or cytosine).

3. A DNA molecule consists of two nucleotide strands twisted together into a double helix. The bases of one strand pair (hydrogen-bond) with bases of the other.

4. The bases of the two strands in a DNA double helix pair in constant fashion. Adenine always pairs with thymine (A to T), and guanine with cytosine (G to C). *Which* base pair follows the next (A–T, T–A, G–C, or C–G) varies along the length of the strands.

5. Overall, the DNA of one species includes a number of unique stretches of base pairs that set it apart from the DNA of all other species.

6. During DNA replication, enzymes unwind the two strands of a double helix and assemble a new strand of complementary sequence on each one. Two double-stranded molecules result. One strand of each molecule is "old" (it is conserved) and the other is "new."

7. Many histones and other proteins are tightly bound to eukaryotic DNA. The structural organization of chromosomes arises from DNA-protein interactions.

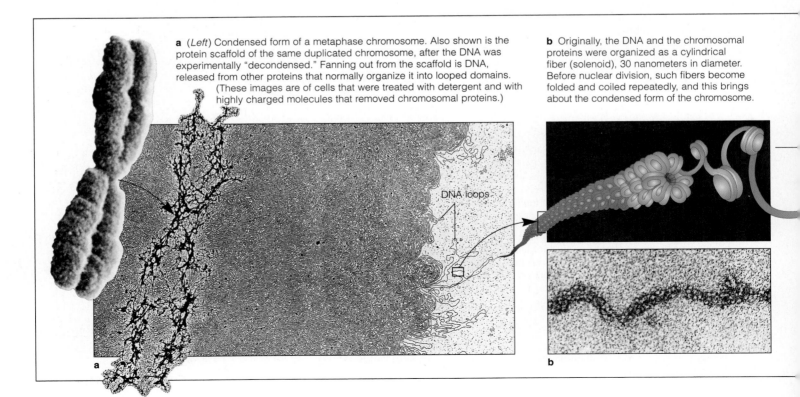

a (*Left*) Condensed form of a metaphase chromosome. Also shown is the protein scaffold of the same duplicated chromosome, after the DNA was experimentally "decondensed." Fanning out from the scaffold is DNA, released from other proteins that normally organize it into looped domains. (These images are of cells that were treated with detergent and with highly charged molecules that removed chromosomal proteins.)

b Originally, the DNA and the chromosomal proteins were organized as a cylindrical fiber (solenoid), 30 nanometers in diameter. Before nuclear division, such fibers become folded and coiled repeatedly, and this brings about the condensed form of the chromosome.

DNA loops

1. Name the three molecular parts of a nucleotide in DNA. Name the four different kinds of nitrogen-containing bases that occur in the nucleotides of DNA. *212*

2. What kind of bond holds two DNA chains together in a double helix? Which nucleotide base-pairs with adenine? Which pairs with guanine? *212, 213*

3. The four bases in DNA may differ greatly in relative amounts from one species to the next—yet the relative amounts are always the same among members of a single species. How does base pairing explain these twin properties—the unity and diversity—of DNA molecules? *213*

4. When regions of a double helix are unwound during DNA replication, do the two unwound strands join back together again after a new DNA molecule has formed? *214, 215*

Self-Quiz *(Answers in Appendix IV)*

1. Which is *not* a nucleotide base in DNA?
   a. adenine   c. uracil   e. guanine
   b. thymine   d. cytosine

2. What are the base-pairing rules for DNA?
   a. A–G, T–C   c. A–U, C–G
   b. A–C, T–G   d. A–T, C–G

3. A DNA strand with the sequence C–G–A–T–T–G would be complementary to the sequence _____ .
   a. C–G–A–T–T–G   c. T–A–G–C–C–T
   b. G–C–T–A–A–G   d. G–C–T–A–A–C

4. One species' DNA differs from others in its _____ .
   a. sugars   c. base sequence
   b. phosphate groups   d. all of the above

5. When DNA replication begins, _____ .
   a. the two DNA strands start to unwind from each other
   b. the two DNA strands condense for base transfers
   c. two DNA molecules bond
   d. old strands move to find new strands

6. DNA replication results in _____ .
   a. two half-old, half-new molecules
   b. two molecules, one with old strands and one with newly assembled strands
   c. three double-stranded molecules, one with new strands and two that are discarded
   d. none of the above

7. DNA replication requires _____ .
   a. a supply of free nucleotides   c. many enzymes and other proteins
   b. new hydrogen bonds   d. all of the above

8. Match the DNA concepts appropriately.
   ____ base sequence variation   a. two nucleotide strands twisted together
   ____ metaphase chromosome   b. A = T, G = C
   ____ constancy in base pairing   c. duplication of hereditary material
   ____ replication   d. accounts for diversity among species
   ____ DNA double helix   e. condensed DNA with protein scaffold

c A chromosome immersed in a salt solution loosens up to a beads-on-a-string organization. The "string" is DNA. Each "bead" is a nucleosome (**d**). Short stretches of the chromosome may look like this when DNA is being replicated or when a gene's instructions are being read. Each nucleosome consists of a double loop of DNA around a core of proteins (eight histone molecules). Another histone (H1) stabilizes the arrangement.

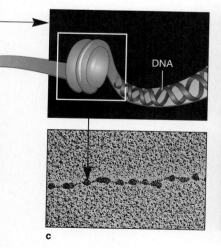

c

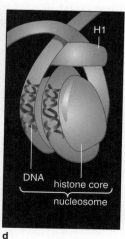

d

Selected Key Terms

bacteriophage *210*
DNA *209*
DNA ligase *214*
DNA polymerase *214*
DNA repair *214*

DNA replication *214*
histone *216*
nucleosome *216*
nucleotide *212*
x-ray diffraction image *212*

Readings

Cairns, J., G. Stent, and J. Watson, eds. 1966. *Phage and the Origins of Molecular Biology.* Cold Spring Harbor, New York: Cold Spring Harbor Laboratories. Gives a sense of the insights, wit, humility, and personalities of the founders of and converts to molecular genetics.

Radman, M., and R. Wagner. August 1988. "The High Fidelity of DNA Duplication." *Scientific American* 259(2): 40–46.

Watson, J. 1978. *The Double Helix.* New York: Atheneum. Highly personal view of scientists and their methods, interwoven into an account of how DNA structure was discovered.

Wolfe, S. 1993. *Molecular and Cellular Biology.* Belmont, California: Wadsworth. Comprehensive, current, and accessible.

14 FROM DNA TO PROTEINS

Beyond Byssus

Picture a mussel, of the sort shown in Figure 14.1. Hard-shelled but soft of body, it is using its muscular foot to probe a wave-scoured rock. At any moment, pounding waves can whack the mussel into the water, hurl it repeatedly against the rock with shell-shattering force, and so offer up a gooey lunch for gulls.

By chance, the mussel's foot comes across a crevice in the rock. The foot moves, broomlike, and sweeps the crevice clean. It presses down, forcing air out from under it, then arches up. The result is a vacuum-sealed chamber, rather like the one that forms when a plumber's rubber plunger is being squished down and up to unclog a drain. Into this vacuum chamber the mussel spews a fluid composed of keratin and other proteins, which bubbles into a sticky foam. And now, by curling its foot into a small tubular shape and pump-ing the foam through it, the mussel forms sticky threads about as wide as a human whisker. As a final touch, it varnishes the threads with another protein and thereby ends up with an adhesive called byssus—which anchors the mussel to the rock.

Byssus is the world's premier underwater adhesive. Nothing that humans have manufactured comes close. (Sooner or later, water chemically degrades or deforms synthetic adhesives.) Byssus fascinates biochemists, dentists, and surgeons looking for better ways to do tissue grafts and to rejoin severed nerves. Genetic engi-neers are inserting mussel DNA into yeast cells, which reproduce in large numbers and serve as "factories" for translating mussel genes into useful amounts of pro-teins. This exciting work, like the mussel's own byssus building, starts with one of life's universal concepts.

Every protein is synthesized according to instructions contained in DNA.

You are about to trace the steps leading from DNA to protein. Many enzymes are players in this pathway, as are molecules of RNA. The same steps produce *all* of the world's proteins, from mussel-inspired adhesives to the keratin in your fingernails to the insect-digesting enzymes of a Venus flytrap.

Figure 14.1 Mussels, busily demonstrating the importance of proteins for survival. When mussels come across a suitable anchoring site, they use their foot like a plumber's plunger to create a vacuum chamber. In this chamber they manufacture the world's best underwater adhesive from a mix of proteins. The adhesive anchors them to substrates in their wave-swept habitat.

KEY CONCEPTS

1. Life cannot exist without enzymes and other proteins. Proteins consist of polypeptide chains, which consist of amino acids. The sequence of amino acids corresponds to a gene—which is a sequence of nucleotide bases in DNA. And the path leading from genes to proteins has two steps, called transcription and translation.

2. In transcription, the double-stranded DNA molecule is unwound at a gene region, then an RNA molecule is assembled on the exposed bases of one of the strands.

3. In translation, RNA directs the linkage of one amino acid after another, in the sequence required to produce a specific kind of polypeptide chain.

4. With few exceptions, the genetic "code words" by which DNA instructions are translated into proteins are the same in all species of organisms.

5. A gene's base sequence is vulnerable to permanent changes, called mutations. Such changes are the original source of genetic variation. Mutations lead to alterations in protein structure, protein function, or both, and these alterations may lead to differences in traits among the individuals of a given population.

Discovering the Connection Between Genes and Proteins

Garrod's Hypothesis   In the early 1900s a physician, Archibald Garrod, was puzzling over certain illnesses. They appeared to be heritable, for they kept recurring in the same families. They also appeared to be metabolic disorders, for blood or urine samples from the affected patients contained abnormally high levels of a substance that was known to be produced at a certain step in a metabolic pathway. Most likely, the enzyme at the *next* step in the pathway was defective in a way that prevented it from using the substance as a substrate. If this were true, then the pathway would be blocked from that step onward, as this diagram indicates:

$$A \longrightarrow B \longrightarrow \overset{\textstyle C\,C\,C}{\underset{\textstyle C}{C}} \;\times\; D$$

pathway is blocked

This hypothesis could explain why molecules of a particular substance were accumulating in excess amounts in affected individuals. Garrod suspected that his patients differed from normal individuals in only one respect: they had inherited one metabolic defect. Thus, Garrod concluded, *specific "units" of inheritance (genes) function through the synthesis of specific enzymes.*

Beadle, Tatum, and a Bread Mold   Thirty-three years later, George Beadle and Edward Tatum were experimenting with a bread mold (*Neurospora crassa*). This fungus is a common spoiler of baked goods (Figure *a*), but it also lends itself to laboratory experiments. *N. crassa* can be grown easily on an inexpensive culture medium that contains only sucrose, mineral salts, and biotin (one of the B vitamins). It synthesizes all other nutrients it requires, including other vitamins.

Suppose an enzyme of a synthesis pathway is defective as a result of a gene mutation. The researchers suspected this had happened in some strains of *N. crassa*. One strain grew only when supplied with vitamin B_6, another with vitamin B_1, and so on. Chemical analysis of cell extracts revealed a different defective enzyme in each mutant strain. In other words, *each inherited mutation corresponded to a defective enzyme.* Here was evidence favoring the "one-gene, one-enzyme" hypothesis.

Gel Electrophoresis Studies   The one-gene, one-enzyme hypothesis was refined through studies of *sickle-cell anemia.* This heritable disorder arises from the presence of a protein, hemoglobin, in red blood cells. The abnormal hemoglobin is designated HbS instead of HbA.

In 1949 Linus Pauling and Harvey Itano subjected molecules of HbS and HbA to **gel electrophoresis**. This laboratory procedure uses an electric field to move molecules through a viscous gel and separate them according to their size, shape, and net surface charge. Often the gel is sandwiched between glass or plastic plates to form a viscous slab. The two ends of the slab are suspended in two salt solutions, which are connected by electrodes to a power source (Figure *b*). When voltage is applied to the appara-

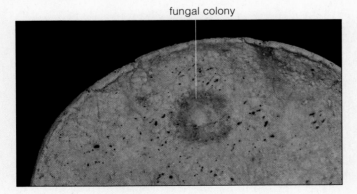

fungal colony

a   Colony of the red bread mold (*Neurospora crassa*) growing on a stale tortilla.

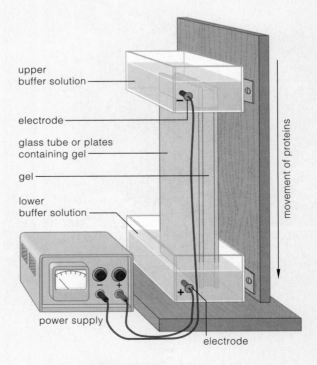

upper buffer solution

electrode

glass tube or plates containing gel

gel

lower buffer solution

power supply

electrode

movement of proteins

b   One type of apparatus used in gel electrophoresis.

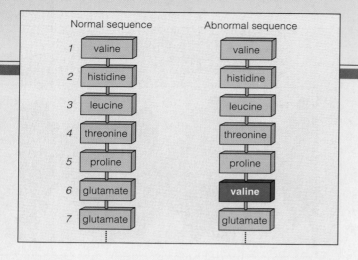

Normal sequence	Abnormal sequence
1 valine	valine
2 histidine	histidine
3 leucine	leucine
4 threonine	threonine
5 proline	proline
6 glutamate	**valine**
7 glutamate	glutamate

c The single amino acid substitution that gives rise to abnormal HbS hemoglobin molecules.

tus, molecules present in the gel will migrate through the electric field according to their individual charge. The molecules move away from one another in the gel and can be identified by staining the gel after a predetermined period of electrophoresis.

Pauling and Itano carefully layered a mixture of HbS and HbA molecules onto the gel at the top of the slab. The molecules gradually migrated down through the gel and separated into distinct bands. The band moving fastest carried the greatest surface charge, and this turned out to be composed of HbA molecules. HbS molecules moved more slowly.

One Gene, One Polypeptide   Later, Vernon Ingram pinpointed the difference between HbS and HbA. Hemoglobin, recall, consists of four polypeptide chains (page 44). Two are designated alpha and the other two, beta. An abnormal HbS chain arises from a single gene mutation that affects protein synthesis. The mutation causes the addition of valine instead of glutamate as the sixth amino acid of the beta chain (Figure c). Whereas glutamate carries an overall negative charge, valine has no net charge—and so HbS behaved differently in the electrophoresis studies.

As a result of this mutation, HbS hemoglobin has a "sticky" (hydrophobic) patch. In blood capillaries, where oxygen concentrations are at their lowest, hemoglobin molecules interact at the sticky patches. They aggregate into rods and distort red blood cells, and this affects organs throughout the body (page 176).

The discovery of the genetic difference between the alpha and beta chains of hemoglobin suggested that *two* genes code for hemoglobin—one for each kind of polypeptide chain. More importantly, it suggested further that genes must code for proteins in general, not just for enzymes.

And so a more precise hypothesis emerged: *The amino acid sequences of polypeptide chains—the structural units of proteins—are encoded in genes.*

DNA is like a book of instructions in each cell. The alphabet used to create the book is simple enough: A, T, G, and C. But how do we get from an alphabet to a protein? The answer starts with the structure of DNA.

DNA, recall, is a double-stranded molecule (Figure 13.6). The strands consist of four types of nucleotides, which differ in one component only. That component, a nitrogen-containing base, may be adenine, thymine, guanine, or cytosine. Which base follows the next along the length of a strand—that is, the **base sequence**—differs from one kind of organism to the next.

Before a cell divides, its DNA is replicated and the two strands unwind entirely from each other. At other times in a cell's life, the two strands unwind only in certain regions to expose certain genes. *And those genes contain instructions for building proteins.*

Researchers discovered this connection between genes and proteins through many studies, of the sort described in the *Focus* essay.

It takes two steps, **transcription** and **translation**, to carry out a gene's protein-building instructions. In eukaryotic cells, transcription proceeds in the nucleus. In this step, the nucleotide bases of DNA serve as a structural pattern (template) for assembling a strand of **ribonucleic acid** (**RNA**) from the cell's pool of free nucleotides. Afterward, the RNA moves to the cytoplasm, where translation proceeds. In this second step, RNA directs the assembly of amino acids into polypeptide chains. Later, the chains become folded into the three-dimensional shapes of proteins.

In short, DNA guides the synthesis of RNA, then RNA guides the synthesis of proteins:

DNA   *transcription*→   RNA   *translation*→   protein

The new proteins will have structural and functional roles in cells. Some even will have roles in building more DNA, RNA, and proteins.

From this overview, you may have the impression that protein synthesis requires only one class of RNA molecules. Actually, it requires three. Transcription of most genes produces **messenger RNA** (**mRNA**), the only class of RNA that carries protein-building instructions. Transcription of some other genes produces **ribosomal RNA** (**rRNA**), which are the main components of ribosomes. Ribosomes, recall, are the sites where polypeptide chains are assembled. Transcription of still other genes produces **transfer RNA** (**tRNA**).

As you will see, tRNA delivers amino acids one by one to a ribosome, in the order specified by mRNA.

14.2 TRANSCRIPTION OF DNA INTO RNA

How RNA Is Assembled

An RNA molecule is almost, but not quite, like a single strand of DNA. It is composed of only four types of nucleotides, each consisting of a five-carbon sugar (ribose), a phosphate group, and a base. The bases are adenine, cytosine, guanine, and uracil (Figure 14.2). **Uracil** is like the thymine in DNA; it pairs with adenine. Thus a new RNA strand can be put together on a DNA region according to the base-pairing rule:

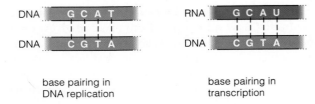

base pairing in
DNA replication

base pairing in
transcription

Transcription resembles DNA replication in another respect. Enzymes add the nucleotide units to a growing RNA strand one at a time, in the 5' → 3' direction. (Here you may wish to refer to Figure 13.8.)

Transcription *differs* from DNA replication in three key respects. First, only a certain stretch of one DNA strand—not the whole molecule—serves as the template. Second, different enzymes, called **RNA polymerases**, catalyze the addition of nucleotides to the 3' end of a growing strand. Third, transcription results in a *single* strand of nucleotides in RNA.

Transcription starts at a **promoter**, a base sequence that signals the start of a gene. Proteins help position an RNA polymerase on the DNA so that it binds with the promoter. The enzyme moves along the DNA, joining

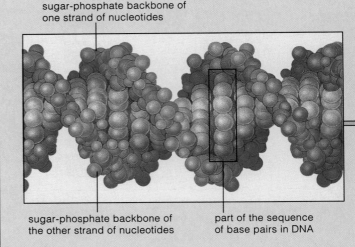

sugar-phosphate backbone of
one strand of nucleotides

sugar-phosphate backbone of
the other strand of nucleotides

part of the sequence
of base pairs in DNA

a This is a gene region of a DNA double helix. The base sequence of one of the nucleotide strands in this region is about to be transcribed into an RNA molecule.

Figure 14.3 Gene transcription: the synthesis of an RNA molecule on a DNA template.

nucleotides together (Figure 14.3). When it reaches a base sequence that serves as a stop signal, the RNA is released as a free transcript.

Finishing Touches on mRNA Transcripts

Newly formed mRNA is an unfinished molecule, one that must be modified before its protein-building instructions can be put to use. Just as a dressmaker might snip off some threads or add bows on a dress before it leaves the shop, so does a eukaryotic cell tailor this pre-mRNA.

Very quickly, enzymes attach a cap to the 5' end of the pre-mRNA molecule. The cap is a nucleotide that has a methyl group and phosphate groups bonded to it. Enzymes also attach about 100 to 200 adenine-containing nucleotides to the 3' end of most pre-mRNA molecules (hence the name, "poly-A tail"). The tail becomes wound up with proteins. Later on, in the cytoplasm, the cap will promote the binding of mRNA to a ribosome. There also, enzymes will gradually destroy the wound-up tail from the tip on up, then the mRNA. Clearly, such tails "pace" enzyme access to mRNA. Perhaps they help keep protein-building messages intact for as long as the cell requires them.

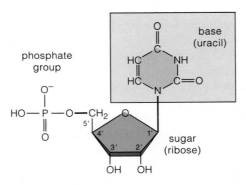

Figure 14.2 Structure of one of the four nucleotides of RNA. The other three have a different base (adenine, guanine, or cytosine instead of the uracil shown here). Compare Figure 13.5, which shows the nucleotides of DNA.

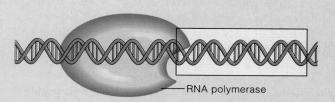

RNA polymerase

b An RNA polymerase molecule binds to a promoter in the DNA. It will use the base sequence positioned downstream from that site as a template for linking nucleotides into a strand of RNA.

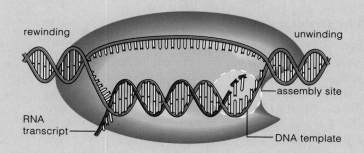

rewinding   unwinding

RNA transcript

assembly site

DNA template

d Throughout transcription, the DNA double helix unwinds just in front of the RNA polymerase. Short lengths of the newly forming RNA strand temporarily wind up with the DNA template strand. Then they unwind from it—and the two strands of DNA wind up together again.

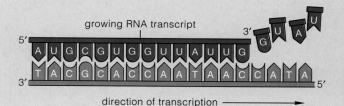

growing RNA transcript

direction of transcription ⟶

c During transcription, RNA nucleotides are based-paired, one after another, with the exposed bases on the DNA template.

new RNA transcript

e At the end of the gene region, the last stretch of the RNA is unwound from the DNA template and so is released.

Besides this, the mRNA message itself is modified. Most eukaryotic genes have one or more **introns**, which are base sequences that do not get translated into an amino acid sequence. Introns intervene between **exons**, the only parts of mRNA that become translated into protein. As Figure 14.4 shows, introns are transcribed right along with exons, but enzymes snip them out before the mRNA leaves the nucleus in mature form.

Some introns may be evolutionary junk, the left-overs of past mutations that led nowhere. However, other introns are sites where instructions for building a particular protein can be snipped apart and spliced back together in different ways. Alternative splicing allows different cells in your body to produce modified versions of an mRNA transcript—and modified versions of the resulting protein—from the same gene. We will return to this topic in the next chapter.

In gene transcription, a sequence of bases in one of the two strands of a DNA molecule serves as the template for assembling an RNA strand. The assembly follows base-pairing rules (adenine with uracil, cytosine with guanine).

Before leaving the nucleus, new RNA transcripts undergo modification into final form.

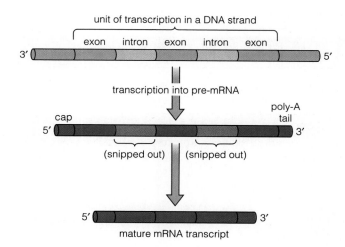

unit of transcription in a DNA strand

exon   intron   exon   intron   exon

transcription into pre-mRNA

cap   poly-A tail

(snipped out)   (snipped out)

mature mRNA transcript

Figure 14.4 Transcription and modification of newly formed mRNA in the nucleus of eukaryotic cells. The cap simply is a nucleotide with functional groups attached. The poly-A tail is a string of adenine nucleotides.

14.3 FROM mRNA TO PROTEINS

The Genetic Code

Like a DNA strand, mRNA is a linear sequence of nucleotides. What are the protein-building "words" encoded in its sequence? Gobind Khorana, Marshall Nirenberg, and others came up with the answer. They deduced that enzymes recognize nucleotide bases *three at a time*, as triplets. In mRNA, the base triplets are called **codons**. Figure 14.5 shows how the order of different codons in an mRNA molecule dictates the order in which particular amino acids are added to a growing polypeptide chain. Count the codons listed in Figure 14.6, and you see there are sixty-four kinds.

Most of the twenty kinds of amino acids correspond to more than one codon. (Glutamate corresponds to the code words GAA *or* GAG, for example.) The codon AUG also establishes the reading frame for translation. That is, enzymes start their "three-bases-at-a-time"

selections at an AUG that serves as the "start" signal in an mRNA strand. Three codons (UAA, UAG, UGA) are stop signals. They stop enzymes from adding any more amino acids to a new polypeptide chain.

The set of sixty-four different codons is the **genetic code**. It the basis of protein synthesis in all organisms.

Protein-building instructions are encoded in the nucleotide sequence of DNA and mRNA. The genetic code is a set of sixty-four base triplets (nucleotide bases, read in blocks of three). A codon is a base triplet in mRNA.

Different combinations of codons specify the amino acid sequence of different polypeptide chains, start to finish.

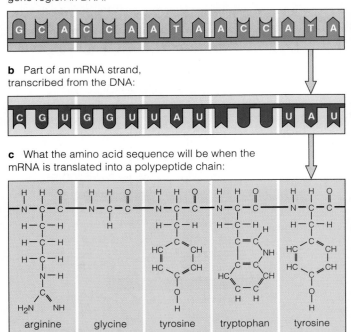

a Base sequence of a gene region in DNA:

b Part of an mRNA strand, transcribed from the DNA:

c What the amino acid sequence will be when the mRNA is translated into a polypeptide chain:

arginine   glycine   tyrosine   tryptophan   tyrosine

Figure 14.5 The steps from genes to proteins. (**a**) This region of a DNA double helix was unwound during transcription. (**b**) Exposed bases on one strand served as a template for assembling an mRNA strand. In the mRNA, every three nucleotide bases equaled one codon. Each codon called for one amino acid in this polypeptide chain. (**c**) Referring to Figure 14.6, can you fill in the blank codon for tryptophan in the chain?

First Letter	Second Letter				Third Letter
	U	C	A	G	
U	phenylalanine	serine	tyrosine	cysteine	U
	phenylalanine	serine	tyrosine	cysteine	C
	leucine	serine	stop	stop	A
	leucine	serine	stop	tryptophan	G
C	leucine	proline	histidine	arginine	U
	leucine	proline	histidine	arginine	C
	leucine	proline	glutamine	arginine	A
	leucine	proline	glutamine	arginine	G
A	isoleucine	threonine	asparagine	serine	U
	isoleucine	threonine	asparagine	serine	C
	isoleucine	threonine	lysine	arginine	A
	(start) methionine	threonine	lysine	arginine	G
G	valine	alanine	aspartate	glycine	U
	valine	alanine	aspartate	glycine	C
	valine	alanine	glutamate	glycine	A
	valine	alanine	glutamate	glycine	G

Figure 14.6 The genetic code. The codons in mRNA are nucleotide bases, read in blocks of three. Sixty-one of these base triplets correspond to specific amino acids. Three others serve as signals that stop translation. The left column of the diagram shows the first of the three nucleotides in each codon in mRNA. The middle columns show the second nucleotide. The right column shows the third. Reading from left to right, for instance, the triplet U G G corresponds to tryptophan. Both U U U and U U C correspond to phenylalanine.

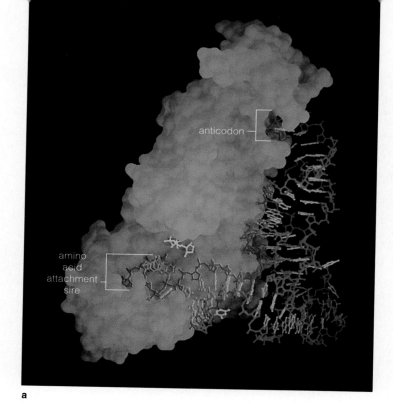

a

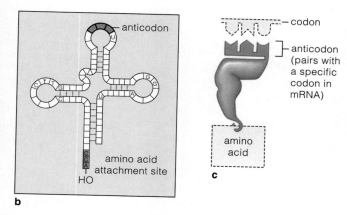

b

c

Figure 14.7 (**a**) Computer-generated, three-dimensional model of one type of tRNA molecule. The tRNA (*reddish brown*) is shown attached to a bacterial enzyme (*green*), along with an ATP molecule (*gold*). This particular enzyme attaches amino acids to tRNAs. (**b**) Structural features common to all tRNAs. (**c**) Simplified model of tRNA that is used in subsequent illustrations. The "hook" at one end is the site to which a specific amino acid can become attached.

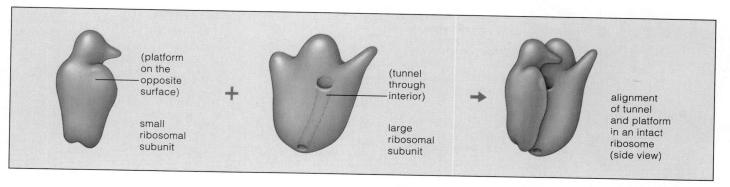

Figure 14.8 Model of eukaryotic ribosomes. Polypeptide chains are assembled on the small subunit's platform. Newly forming chains may move through the large subunit's tunnel.

Roles of tRNA and rRNA

Cells have pools of free amino acids and free tRNA molecules in the cytoplasm. tRNAs have a molecular "hook," an attachment site for amino acids. They also have an **anticodon**, a nucleotide triplet that can base-pair with codons (Figure 14.7). When tRNAs bind to codons, they automatically position their attached amino acids in the order specified by mRNA.

A cell has more than sixty kinds of codons but fewer kinds of tRNAs. How do they match up? By the base-pairing rules, adenine must pair with uracil, and cytosine with guanine. However, for codon-anticodon interactions, the rules loosen up for the third base. For example, CCU, CCC, CCA, and CCG all specify proline. Such freedom in codon-anticodon pairing at the third base is called the "wobble effect."

tRNAs interact with mRNA at binding sites on the surface of **ribosomes**. Each ribosome has two subunits

(Figure 14.8). These are assembled in the nucleus from rRNA and proteins, then are shipped separately to the cytoplasm. There they will combine as functional units only during translation, as described next.

mRNAs are the only molecules that carry protein-building instructions from DNA to the cytoplasm.

tRNAs bind to codons and so position their attached amino acids in the order specified by mRNA. Thus they translate the mRNA into a corresponding sequence of amino acids.

rRNAs are components of ribosomes, the sites where amino acids are assembled into polypeptide chains.

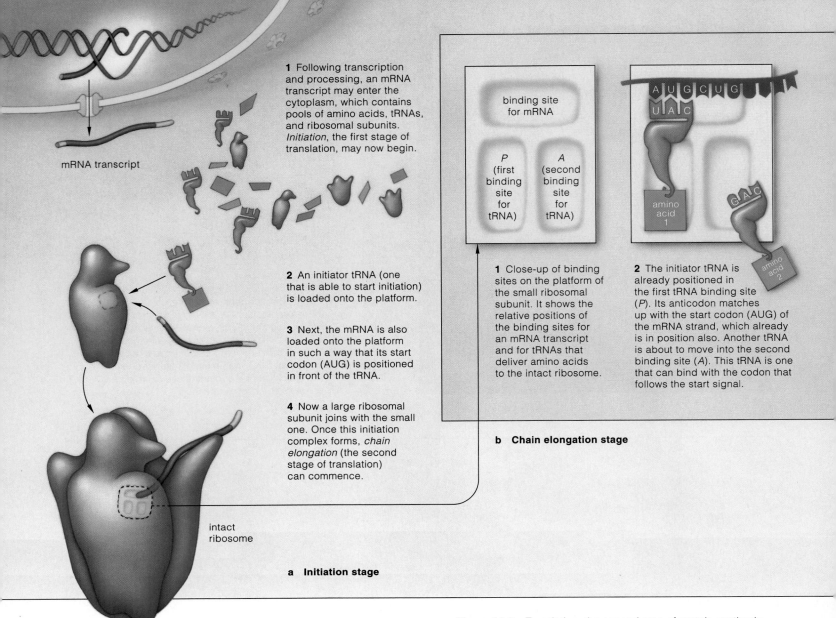

1 Following transcription and processing, an mRNA transcript may enter the cytoplasm, which contains pools of amino acids, tRNAs, and ribosomal subunits. *Initiation*, the first stage of translation, may now begin.

mRNA transcript

2 An initiator tRNA (one that is able to start initiation) is loaded onto the platform.

3 Next, the mRNA is also loaded onto the platform in such a way that its start codon (AUG) is positioned in front of the tRNA.

4 Now a large ribosomal subunit joins with the small one. Once this initiation complex forms, *chain elongation* (the second stage of translation) can commence.

intact ribosome

a Initiation stage

binding site for mRNA

P (first binding site for tRNA)

A (second binding site for tRNA)

1 Close-up of binding sites on the platform of the small ribosomal subunit. It shows the relative positions of the binding sites for an mRNA transcript and for tRNAs that deliver amino acids to the intact ribosome.

2 The initiator tRNA is already positioned in the first tRNA binding site (*P*). Its anticodon matches up with the start codon (AUG) of the mRNA strand, which already is in position also. Another tRNA is about to move into the second binding site (*A*). This tRNA is one that can bind with the codon that follows the start signal.

b Chain elongation stage

Figure 14.9   Translation, the second step of protein synthesis.

14.4 STAGES OF TRANSLATION

Translation occurs in the cytoplasm. It requires three stages, called initiation, elongation, and termination.

In *initiation*, a tRNA that can start transcription and an mRNA transcript become loaded onto an intact ribosome. First, the initiator tRNA binds with the small ribosomal subunit. So does the start codon (AUG) of the mRNA transcript. After this, a large ribosomal subunit binds with the small one (Figure 14.9a). The next stage can begin.

In *elongation*, a polypeptide chain forms as the mRNA strand passes between the ribosomal subunits, like a thread being moved through the eye of a needle. Some proteins in the ribosome that function as enzymes join amino acids together in the sequence dictated by mRNA's codons. As Figure 14.9b indicates, they catalyze the formation of a peptide bond between every

two amino acids delivered to the ribosome. Here you may also wish to refer to an earlier diagram of peptide bond formation (Figure 3.16).

In *termination*, a stop codon is reached and there is no corresponding anticodon. Now the enzymes bind to certain proteins (release factors). The binding causes the ribosome as well as the polypeptide chain to detach from the mRNA (Figure 14.9c). The detached chain may join the pool of free proteins in the cytoplasm. Or it may enter the cytomembrane system, starting with the compartments of rough ER. As indicated on page 64, many newly formed chains take on their final form in that system before being shipped to their destinations.

Often, enzymes and tRNAs repeatedly translate the same mRNA transcript in a given period. The transcript threads through many ribosomes, which are arranged

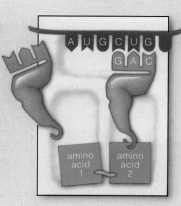
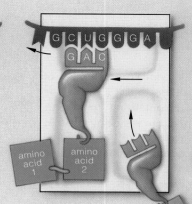

3 Through enzyme action, the bond between the initiator tRNA and the amino acid hooked to it is broken. At the same time, an enzyme catalyzes the formation of a peptide bond between the two amino acids. After these bonding events are over, the initiator tRNA will be released from the ribosome.

4 Now the first amino acid is attached only to the second one—which is still hooked to the second tRNA. This tRNA will move into the *P* site on the ribosomal platform, sliding the mRNA with it by one codon. When it does so, the third codon will become aligned above the *A* site.

5 A third tRNA is about to move into the *A* site. Its anticodon is capable of base-pairing with the third codon of the mRNA transcript. Next, through enzyme action, a peptide bond will form between amino acids 2 and 3.

6 Steps 3 through 5 are repeated again and again. The polypeptide chain continues to grow this way until enzymes reach a stop codon in the mRNA transcript. Now *termination*, the last stage of transcription, can begin, as shown in (**c**).

1 Once a stop codon is reached, the mRNA transcript is released from the ribosome.

2 The newly formed polypeptide chain also is released.

c Chain termination stage

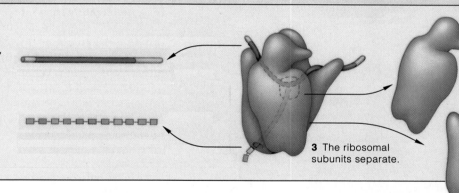
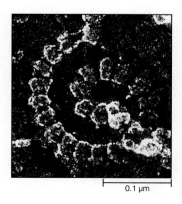
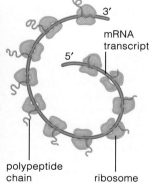

3 The ribosomal subunits separate.

one after another, closely together in assembly-line fashion. Figure 14.10 shows one of these arrangements, which are called **polysomes**. Their presence indicates that a cell is producing many copies of a polypeptide chain from the same transcript.

Translation is initiated by the convergence of a small ribosomal unit, an initiator tRNA, an mRNA transcript, then a large ribosomal subunit.

Next, anticodons of tRNAs base-pair with mRNA codons. A polypeptide chain grows through peptide bond formation between every two amino acids delivered to the ribosome.

Translation ends when a stop codon triggers events that cause the chain and the mRNA to detach from the ribosome.

0.1 µm

Figure 14.10 From a eukaryotic cell, a polysome (many ribosomes simultaneously translating the same mRNA molecule).

3′
mRNA transcript
5′
polypeptide chain
ribosome

14.5 MUTATION AND PROTEIN SYNTHESIS

Whenever a cell puts its genetic code into action, it is synthesizing precisely those proteins that it requires for its structure and functioning. If something changes a gene's code words, we can expect the resulting protein to change. If that protein is central to cell architecture or metabolism, we can expect the outcome to be a dead or damaged cell.

Every so often, genes do change. Maybe one base is substituted for another in the DNA sequence. Or maybe an extra base is inserted or a base is lost. These small-scale changes in the nucleotide sequence of a DNA molecule are **gene mutations**.

Some mutations result from exposure to **mutagens**. These are agents that increase the risk of heritable alterations in the structure of DNA. Ultraviolet light, ionizing radiation, and certain substances in tobacco smoke are common mutagens. So are free radicals, those rogue molecular fragments described on page 92.

Even in the absence of mutagens, however, mutations arise in cells. For example, spontaneous mutations may follow replication errors, as when adenine wrongly pairs with a cytosine unit on a DNA template strand. DNA repair enzymes may simply detect an error, then "fix" the wrong base (Figure 14.11).

Regardless of whether it is spontaneous or induced by a mutagen, "base-pair substitutions" of this sort may lead to the substitution of one amino acid for another during protein synthesis. Earlier in the chapter, you saw how one of these substitutions results in sickle-cell anemia—a genetic disorder that has wide-ranging structural and physiological consequences.

Or consider the "frameshift mutation," in which one to several base pairs are inserted into a DNA molecule or deleted from it. Remember, polymerases read a nucleotide sequence in blocks of three. As Figure 14.12 shows, an insertion or deletion in a gene region can shift this reading frame. Because the gene is not read correctly, an abnormal protein is synthesized.

As Barbara McClintock discovered, spontaneous mutations also result when "jumping genes" are on the move. These DNA regions also are called transposable elements. They can move from one location to another in the same DNA molecule or in a different one. Often they inactivate genes into which they are inserted. As Figure 14.13 suggests, the unpredictability of the jumps can give rise to interesting changes in phenotype.

In terms of individual lifetimes, gene mutations are rather rare events. Whether they prove to be harmful, neutral, or beneficial will depend on how the resulting proteins interact with other genes and with the environment. And the outcome may have evolutionary consequences, as the *Commentary* makes clear.

Gene mutations are heritable, small-scale alterations in the nucleotide sequence of DNA. They may be harmful, neutral, or beneficial, depending on how the proteins they specify interact with other genes and with the environment.

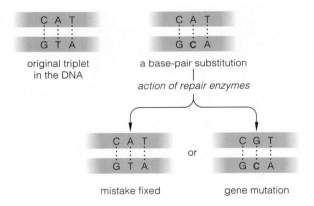

Figure 14.11 Example of a base-pair substitution.

Figure 14.12 Outcomes of three frameshift mutations. In this hypothetical example, three different base insertions occur in the same nucleotide sequence of DNA. The word *cat* represents a code word (base triplet) in mRNA. In the first mutation, an extra nucleotide (arrow) is inserted into a gene. When mRNA is transcribed off the gene, the insertion puts the reading frame out of phase and changes the code words. Thus the wrong amino acids are called up during translation. The second base insertion does not improve matters. The third one restores most of the code words, so only part of the resulting protein will be defective.

Insertions of extra bases into a DNA molecule often give rise to mutant phenotypes. Studies of such mutations in bacteriophages led to the discovery of the genetic code.

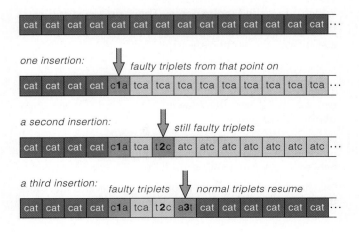

Mutation, Gene Products, and Evolution

In the natural world, gene mutations are rare, chance events. We cannot predict exactly when or in which individual they will appear. Even so, each gene does have a characteristic *mutation rate*, which simply is the probability of its mutating between or during DNA replications. The mutation rate for a typical gene is one in a million (10^6) replications.

Genes also mutate independently of one another. To predict the probability of any two mutations occurring in a given cell, we would have to multiply the individual mutation rates for two of its genes. For example, if the rate for the first gene is one in a million per cell generation and the rate for the second is one in a billion (10^9), then there is only one chance in a million billion (10^{15}) that both genes will mutate in the same cell. Many trillions of replications occur during human growth and development, so chances are that at least one cell will contain a new mutation. If the cell happens to be a reproductive cell, the mutation may be passed on to offspring.

What will be the phenotypic outcome of a given gene mutation? That depends on how the resulting protein functions in the individual. Many mutations turn out to be harmful. No matter what the species, each individual generally inherits a combination of many finely tuned genes that work well together under a certain range of operating conditions. A mutation that has drastic effects on essential traits increases the odds that an individual will not survive or reproduce. Such *lethal* genes obviously do not accumulate in the DNA. Yet many other mutations have little, if any, effect on traits. As you will read in the next unit, these *neutral* mutations have accumulated in the DNA of all species.

Natural selection has played a major role in perpetuating *beneficial* genes as well as the mechanisms that protect them. For example, we can readily understand the adaptive advantage of genes that specify effective repair enzymes, ones that have little tolerance of mismatched base pairs. Such enzymes have helped protect the overall stability of all the DNA molecules that have been replicated countless times for nearly 4 billion years.

Reflect on some of the concepts you have read about so far in this book. You have an idea that all living things on earth share the same chemical heritage. Your DNA has the same kinds of substances, and follows the same base-pairing rules, as the DNA of earthworms in Missouri and grasses on the Mongolian steppes. Your DNA is replicated in much the same way as theirs, and the same genetic code is followed in translating its messages into proteins. In the evolutionary view, the reason you don't look like an earthworm or a grass plant is largely a result of natural selection of different mutations that originated in different lines of descent. Thus the sequence of base pairs in DNA has come to be different in you, the plant, and the worm.

And so we have three concepts of profound importance. *First, DNA is the source of the unity of life. Second, gene mutations are the original source of life's diversity. Finally, the changing environment is the testing ground for the success or failure of the proteins specified by each novel gene product that appears on the evolutionary scene.*

Figure 14.13 Mutation in different kernels of Indian corn (*Zea mays*). All of the kernel cells have the same pigment-coding genes. Yet some kernels are colorless or spottily colored.

In an ancestor of this plant, a transposable element (jumping gene) in one of its cells left its position in a DNA molecule, invaded another, and shut down a pigment gene. This plant inherited the mutations. As cell divisions proceeded in the growing plant, none of the mutated cell's descendants could produce pigment molecules. They gave rise to colorless kernel tissue.

Later, in some cells, the transposable element slipped out of the pigment gene. All of the descendants of *those* cells were able to produce pigment. They gave rise to colored kernel tissue.

SUMMARY

1. A gene is a sequence of nucleotide bases in DNA. Most genes contain protein-building instructions; their nucleotide sequence corresponds to the sequence of amino acids in a polypeptide chain. Proteins are composed of one or more polypeptide chains.

2. Some genes contain instructions for building RNA molecules, which are necessary for protein synthesis. There are three classes of RNA molecules:

a. Messenger RNA (mRNA) carries protein-building instructions from DNA to the cytoplasm. It is the only class of RNA molecules to do this.

b. Transfer RNA (tRNA) is the translator that converts the sequential message of mRNA into a corresponding sequence of amino acids.

c. Ribosomal RNA (rRNA) is a component of ribosomes, the actual sites where amino acids are assembled into polypeptide chains.

3. As summarized in Figure 14.14, the path leading from genes to proteins has two steps. In the first step, called transcription, a DNA template is used to synthesize RNA. In the second step, called translation, three classes of RNA interact in ways that lead to the synthesis of polypeptide chains, which later will become folded into the three-dimensional shape of proteins:

4. Here are the key points concerning transcription:

a. A double-stranded DNA molecule is unwound at a particular gene region, then an RNA molecule is assembled on the exposed bases of one of the strands, which serves as a template.

b. Transcription of DNA into RNA follows the same base-pairing rule that applies to DNA replication. However, uracil takes the place of thymine in an RNA strand. It pairs with adenine:

c. In eukaryotic cells, new RNA transcripts undergo modification into final form before being shipped from the nucleus. Typically, enzymes attach a cap at the 5′ end and a poly-A tail at the 3′ end. For mRNAs, the noncoding portions (introns) are excised and the coding portions (exons) are spliced together. Only mature mRNA transcripts are translated in the cytoplasm.

5. Here are the key points concerning translation:

a. mRNA interacts with tRNAs and ribosomes in such a way that amino acids become linked one after another, in the sequence required to produce a specific kind of polypeptide chain.

b. In all species, translation is based on the genetic code, a set of sixty-four base triplets (nucleotide bases, which enzymes read in blocks of three).

c. In mRNA, a base triplet is called a codon. An anticodon is a complementary triplet in tRNA. Some combination of codons specifies the amino acid sequence of a polypeptide chain, start to finish.

6. Translation proceeds through three stages:

a. Initiation. A small ribosomal subunit binds with an initiator tRNA, then with an mRNA transcript. The small subunit then binds with a large ribosomal subunit to form the initiation complex.

b. Chain elongation. tRNAs deliver amino acids to the ribosome. Each has an anticodon, a base triplet that is complementary to a codon in mRNA and that can base-pair with it. Each amino acid brought to the ribosome by tRNAs is linked (by a peptide bond) to the growing polypeptide chain.

c. Chain termination. A stop codon triggers events that cause the polypeptide chain and mRNA to detach from the ribosome.

7. Gene mutations are heritable, small-scale alterations in the nucleotide sequence of DNA. They may be harmful, neutral, or beneficial, depending on how the proteins they specify interact with other genes and with the environment.

a. Mutations can be induced by exposure to mutagens, which are agents that cause heritable alterations in the structure of DNA. They also may arise spontaneously, as through replication errors and movements of transposable elements.

8. DNA is the source of the unity of life. Gene mutations are the original source of life's diversity. The changing environment is the testing ground for the success or failure of the proteins specified by each novel gene product that appears on the evolutionary scene.

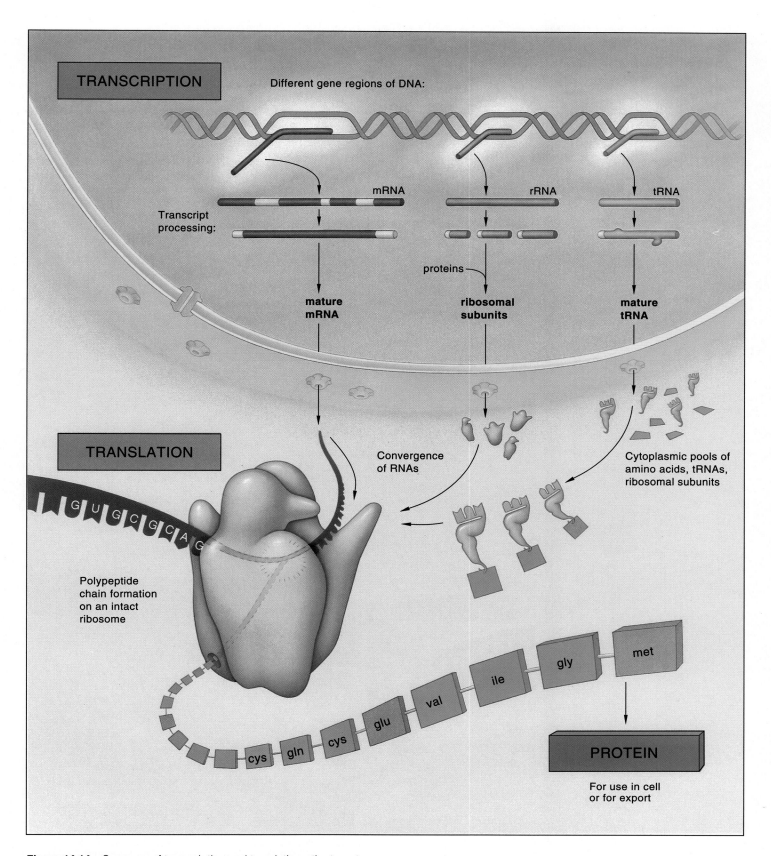

Figure 14.14 Summary of transcription and translation—the two steps leading to protein synthesis—as they proceed in eukaryotic cells.

1. Are the proteins specified by eukaryotic DNA assembled *on* the DNA molecule? If so, state how. If not, tell where they are assembled, and on which molecules. *221*

2. Protein synthesis requires the interaction of three classes of RNA molecules. Name the three classes and define the functions of each. *221, 224, 225*

3. In what key respect does an RNA molecule differ from a DNA molecule? *222*

4. Describe the steps of gene transcription. In what respects does this process resemble DNA replication? In what respects does it differ? *222*

5. Before an mRNA transcript leaves the nucleus, are its introns or exons snipped out? *223*

6. What is the genetic code? Is the code the basis of protein synthesis in all organisms? *224*

7. Distinguish between a codon and an anticodon. *224, 225–227*

8. If sixty-one codons in mRNA actually specify amino acids, and if there are only twenty common amino acids, then more than one codon combination must specify some of the amino acids. How do triplets that code for the same thing usually differ? *225*

9. Describe the events that unfold during the three steps of translation. *226–227*

10. Describe one kind of gene mutation that leads to an altered protein. What determines whether the altered protein will have beneficial, neutral, or harmful effects on the individual? *228, 229*

11. Fill in the blanks on the diagram below. *231*

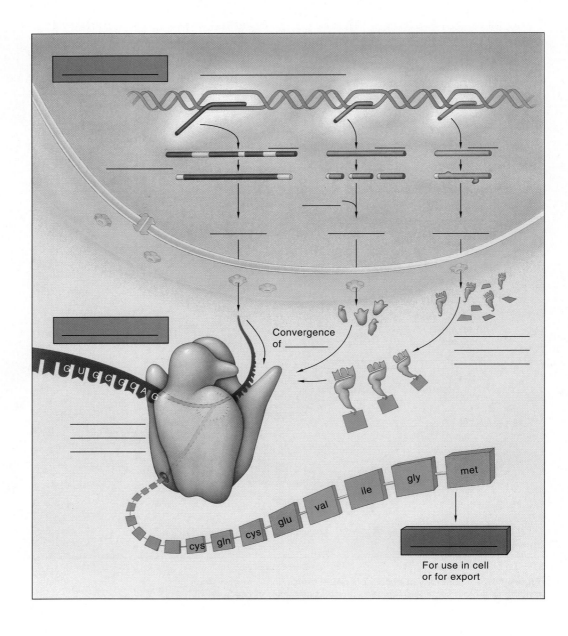

1. Nucleotide bases, read _____ at a time, serve as the "code words" of genes.

2. DNA contains different genes that are transcribed into _____ .
   a. proteins
   b. mRNAs
   c. rRNAs
   d. tRNAs
   e. b, c, and d

3. The RNA molecule is _____ .
   a. a double helix
   b. usually single-stranded
   c. always double-stranded
   d. usually double-stranded

4. mRNA is produced by _____ .
   a. replication
   b. duplication
   c. transcription
   d. translation

5. _____ carries coded instructions for an amino acid sequence to the ribosome.
   a. DNA
   b. rRNA
   c. mRNA
   d. tRNA

6. tRNA _____ .
   a. delivers amino acids to ribosomes
   b. picks up genetic messages from rRNA
   c. synthesizes mRNA
   d. all of the above

7. Each codon calls for a specific _____ .
   a. protein
   b. polypeptide
   c. amino acid
   d. carbohydrate

8. How many amino acids are coded for in this mRNA sequence: CGUUUACACCGUCAC?
   a. three
   b. five
   c. six
   d. seven
   e. more than seven

9. An anticodon pairs with the nitrogen-containing bases of _____ .
   a. mRNA codon
   b. DNA codons
   c. tRNA anticodon
   d. amino acids

10. The loading of mRNA onto an intact ribosome occurs during _____ .
    a. transcription
    b. transcript processing
    c. translation
    d. chain elongation

11. The presence of _____ in a micrograph indicates that a cell is producing many copies of a polypeptide chain from the same transcript.
    a. RNA polymerase
    b. DNA polymerase
    c. polysomes
    d. both a and b

12. Use the genetic code (Figure 14.6) to translate the mRNA sequence UAUCGCACCUCAGGAUGAGAU. Which amino acid sequence is being specified?
    a. tyr-arg-thr-ser-gly-stop-asp . . .
    b. tyr-arg-thr-ser-gly . . .
    c. tyr-arg-tyr-ser-gly-stop-asp . . .
    d. none of the above

13. Match the terms related to protein building.
    _____ disrupts genetic instructions
    _____ genetic code word
    _____ transcription
    _____ translation
    _____ translation stages

    a. initiation, elongation, termination
    b. RNAs convert genetic messages into polypeptide chains
    c. base triplet for an amino acid
    d. one DNA strand is template for the process
    e. gene mutation

Selected Key Terms

anticodon *225*
base sequence *221*
codon *224*
exon *223*
gel electrophoresis *220*
gene mutation *228*
genetic code *224*
intron *223*
messenger RNA (mRNA) *221*
mutagen *228*

polysome *227*
promoter *222*
ribonucleic acid (RNA) *221*
ribosomal RNA (rRNA) *221*
ribosome *225*
RNA polymerase *222*
transcription *221*
transfer RNA (tRNA) *221*
translation *221*
uracil *222*

Readings

Amato, I. January 1991. "Stuck on Mussels." *Science News* 139:8–15.

Grunstein, M. October 1992. "Histones as Regulators of Genes." *American Scientist* 267(4):68–74B.

Liotta, L. February 1992. "Cancer Cell Invasion and Metastasis." *Scientific American* 266(2): 54–63.

Murray, A., and M. Kirschner. March 1991. "What Controls the Cell Cycle?" *Scientific American* 264(3):56–63.

Wolfe, S. 1993. *Molecular and Cellular Biology.* Belmont, California: Wadsworth.

15 CONTROL OF GENE EXPRESSION

Genes, Proteins, and Cancer

Every second, millions of cells in your skin, gut lining, liver, and other body regions divide and replace their worn-out, dead, and dying predecessors. They do not divide willy-nilly. Certain genes specify the enzymes and other proteins that are required for cell growth, DNA replication, spindle formation, chromosome movements, and cytoplasmic division. Certain regulatory proteins control the synthesis and use of these gene products, and they control when the division machinery is put to rest.

On rare occasions, cell division goes out of control. Once controls are lost, divisions cannot stop as long as conditions for growth remain favorable. When cells are not responding to normal controls over growth and division, they form a tissue mass called a **tumor**.

Cells of common skin warts and other *benign* tumors grow rather slowly, in an unprogrammed way, and they still have surface recognition proteins that hold them together in their home tissue. Surgically remove a benign tumor, and you remove its threat to the surrounding tissues.

By contrast, abnormal cells grow and divide more rapidly in a *malignant* tumor, and their physical and metabolic effects on surrounding tissues are destructive. These are grossly disfigured cells (Figure 15.1).

They cannot construct a proper cytoskeleton or plasma membrane, and they cannot synthesize normal versions of recognition proteins. When such cells break loose from their home tissue and enter lymph or blood vessels, they can travel through the body—and become lodged where they don't belong. You may have heard about this process of migration and invasion. It is called **metastasis**.

Basal cell carcinomas, of the sort shown in Figure 13.9, are merely one of more than 200 types of malignant tumors that have been identified to date. All of these malignant tumors are grouped into the general

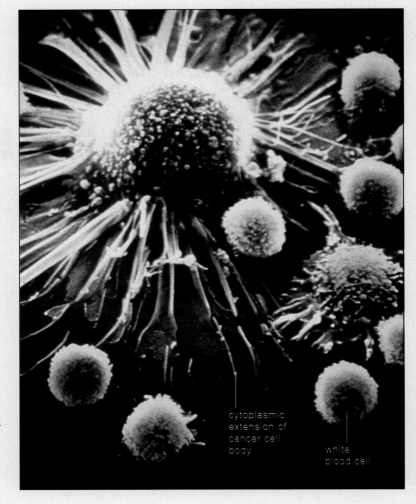

cytoplasmic
extension of
cancer cell
body

white
blood cell

Figure 15.1 Visible evidence of an absence of gene control. This scanning electron micrograph shows a cancer cell surrounded by a number of the body's defenders—white blood cells—that may or may not be able to destroy it. Cancer arises through the loss of controls that govern certain genes with roles in cell growth and division.

category of **cancer**. Each year in the United States and other developed countries, cancer accounts for 15 to 20 percent of all deaths. It is not just a human affliction. It has been observed in most of the animals that have been studied. Similar abnormalities have even been observed in many kinds of plants.

Researchers have traced most cancers to a certain category of genes, called oncogenes. You will read about oncogenes in this chapter. In some cases of cancer, base substitutions or deletions have altered one of these genes or one of the controls over its transcription. In other cases, a gene has been abnormally duplicated over and over again. In still other cases, a whole gene or part of it moves to a new location in the same chromosome or to an entirely different chromosome. With such alterations, transformations may begin that lead, ultimately, to cancer.

Consider the *myc* gene. It is located in human chromosome 8, and it specifies a regulatory protein that helps control the cell cycle. Sometimes a break occurs in this gene region *and* in a region of chromosome 14. The two chromosomes swap segments, so that part of the *myc* gene ends up in a new location. Possibly the translocated portion escapes the normal controls over its transcription and comes under the influence of new ones. Possibly it leaves behind part of the message concerning how the mRNA transcripts are supposed to be processed. Whatever the case, transformations begin that can lead, ultimately, to a form of *Burkitt's lymphoma*.

Burkitt's lymphoma is a malignant tumor of B lymphocytes. B lymphocytes are the only type of cell that produces the antibody molecules, which are protein weapons against specific agents of infectious disease (page 680). In activated B cells, a gene region that specifies antibody is transcribed at an exceptionally high rate. The translocation between chromosomes 8 and 14 puts the *myc* gene next to this gene region and makes it abnormally active.

Perhaps more than any other example, cancerous transformations bring home the extent to which you and all other organisms depend on gene controls. A dizzying variety of controls govern transcription and translation. They influence when and where gene products will be used. And they are the topic of this chapter.

KEY CONCEPTS

1. In cells, controls govern when, how, and to what extent genes are expressed. The control elements operate in response to changing chemical conditions and to signals from the outside environment.

2. Control is exerted through regulatory proteins and other molecules that interact with DNA, with RNA that has been transcribed off the DNA, or with the gene products. Among most multicelled organisms, the regulatory molecules called hormones have major roles in controlling gene expression.

3. Control elements come into play during transcription and translation. They also act on the resulting polypeptide chains and proteins. In eukaryotic cells, additional controls govern the processing of new RNA transcripts and their shipment out of the nucleus for translation in the cytoplasm.

4. All cells in a multicelled organism have the same genes, but they activate or suppress many of those genes in different ways. The controlled, selective use of genes leads to synthesis of the proteins that give each type of cell its distinctive structure, function, and products.

15.1 THE NATURE OF GENE CONTROL

At this very moment, bacteria are feeding on nutrients in your gut. Your own red blood cells are binding, transporting, or giving up oxygen, and many epithelial cells in your skin are busily synthesizing the protein keratin. Like cells everywhere, they are functioning by virtue of the protein products of genes.

Cells don't express all of their genes all of the time. Some genes and their products are used only once. Others are used some or all of the time or not at all. Which genes are expressed depends on the type of cell, on its responses to chemical conditions and signals from the environment, and on its built-in control systems. These systems consist of molecules, including **regulatory proteins**, that interact with DNA, RNA, or the actual gene products.

For example, nutrient availability shifts rapidly and often for enteric bacteria, which reside in animal intestines. These prokaryotic cells rapidly transcribe genes and synthesize many nutrient-digesting enzyme molecules when nutrients are moving past—then restrict synthesis when they are not. By contrast, the composition and solute concentrations of the fluid bathing your cells do not shift drastically. And few of your cells exhibit such rapid shifts in transcription.

Different kinds of control systems operate during transcription, translation, and after translation, on the gene product. For example, cells use two common control systems to block or enhance transcription. With **negative control systems**, a regulatory protein can bind to DNA at a certain site and block transcription. It can be removed from the DNA by the action of an inducer, a type of molecule that promotes transcription. With **positive control systems**, a regulatory protein binds to the DNA and promotes the initiation of transcription.

The binding of a regulatory protein to DNA can be reversed, as when the conditions that called for the synthesis of a particular protein change.

Summing up, these are the points to keep in mind as you read through the remainder of this chapter:

---

1. Cells exert control over when, how, and to what extent each of their genes is expressed.

2. The expression of a given gene depends on the type of cell and its functions, on chemical conditions, and on signals from the outside environment.

3. Regulatory proteins and other molecules govern gene expression through interactions with DNA, RNA, and gene products (polypeptide chains, and then proteins).

15.2 EXAMPLES OF GENE CONTROL IN PROKARYOTIC CELLS

When nutrients are in good supply and the environment favors growth, bacterial cells tend to grow and divide indefinitely. Controls promote rapid synthesis of the enzymes required for nutrient digestion and other growth-related activities. Transcription is rapid, and translation begins even before mRNA transcripts are completed. (Bacteria have no nucleus; nothing separates the DNA from ribosomes in the cytoplasm.)

If a nutrient-degrading pathway requires several enzymes, all genes for those enzymes are transcribed, often into one continuous mRNA molecule. They are not transcribed when conditions turn unfavorable. Let's look at two examples of this all-or-nothing control of transcription, which is called **cooperative regulation**.

Negative Control of Transcription

Escherichia coli is one of the enteric bacteria residing in the mammalian gut. It lives on glucose, lactose, and other ingested nutrients. Lactose is a sugar in milk. Like other adult mammals, you probably don't drink milk around the clock. When you do drink milk—and only then—*E. coli* cells rapidly transcribe three genes for enzymes with roles in certain breakdown reactions that begin with lactose (Figure 15.2).

A promoter precedes the three genes, which are next to one another. A **promoter**, recall, is a base sequence that signals the start of a gene. Another sequence, an **operator**, intervenes between the promoter and the genes. It is a binding site for a **repressor** protein, which can block transcription. Such an arrangement, in which a promoter and operator service more than one bacterial gene, is an **operon**. Elsewhere in *E. coli* DNA, another gene codes for the repressor, which is able to bind with the operator *or* with a lactose molecule.

When the concentration of lactose is low, a repressor binds with the operator. Being a large molecule, it overlaps the promoter and *prevents* transcription (Figure 15.2b). *Thus lactose-degrading enzymes are not produced when they are not required.*

When the lactose concentration is high, the odds are greater that a lactose molecule will bind with the repressor. Binding alters the repressor's shape, so it cannot bind with the operator. RNA polymerase can now transcribe the genes (Figure 15.2c). *Thus lactose-degrading enzymes are produced only when required.*

Positive Control of Transcription

E. coli cells pay far more metabolic attention to glucose than to lactose. (Can you give a few reasons why?) Even when lactose is available, the lactose operon isn't used

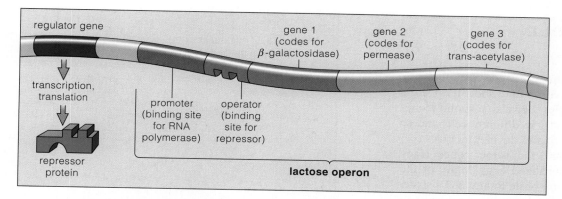

a A repressor protein exerts negative control over three genes of the lactose operon by binding to the operator and inhibiting transcription.

regulator gene

gene 1 (codes for β-galactosidase)

gene 2 (codes for permease)

gene 3 (codes for trans-acetylase)

transcription, translation

promoter (binding site for RNA polymerase)

operator (binding site for repressor)

repressor protein

lactose operon

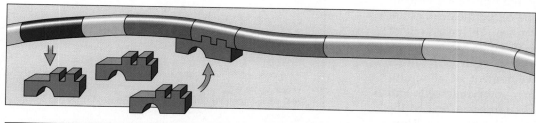

b When the lactose concentration is low, the repressor is free to block transcription. Being a bulky molecule, it overlaps the promoter and prevents binding by RNA polymerase. The enzymes (which are not needed) are not produced.

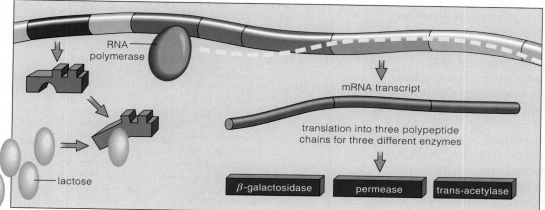

RNA polymerase

mRNA transcript

translation into three polypeptide chains for three different enzymes

lactose

β-galactosidase

permease

trans-acetylase

c At high concentration, lactose is an inducer of transcription. It binds to and distorts the shape of the repressor—which now cannot bind to the operator. The promoter is exposed and the genes can be transcribed.

Figure 15.2 Negative control of the lactose operon. The first gene of the operon codes for an enzyme that splits the disaccharide lactose into its two subunits (glucose and galactose). The second gene codes for an enzyme that transports lactose molecules into the cell. The third enzyme functions in metabolizing certain sugars.

much—*unless glucose is absent*. By contrast, the genes for glucose breakdown are transcribed continuously, at faster rates.

The lactose operon also is subject to *positive* control by an **activator protein**, called CAP. It happens that the operon's promoter is not very efficient at binding RNA polymerase. It does a much better job when CAP adheres to it first. However, CAP will not do this without being activated by a small molecule, called cAMP. Among other things, cAMP is produced from ATP—which *E. coli* produces by breaking down glucose. Not much cAMP is available when glucose is plentiful and

glycolysis is proceeding full bore. At such times, the activator protein is not primed to adhere to the promoter, and transcription of lactose operon genes slows almost to a standstill. When glucose is scarce, cAMP accumulates, the CAP-cAMP complex forms, and the lactose operon genes receive appropriate attention.

Prokaryotic cells, which must respond rapidly to changing conditions, commonly rely on a small number of regulatory proteins that exert rapid, on-off control of transcription.

15.3 GENE CONTROL IN EUKARYOTIC CELLS

In eukaryotic cells, gene control is complicated business. Like prokaryotes, these cells respond to short-term shifts in diet and level of activity. But they also respond to signals that influence gene activity in different cells and so bring about the body's growth and development.

Your own cells inherited the same genes (they descended from the same fertilized egg). Many of the genes specify proteins that are basic to any cell's structure and functions. That is why the protein subunits of ribosomes are the same from one cell to the next, as are many enzymes used in metabolism. Yet nearly all of your cells have become specialized in composition, structure, and function. This process, **cell differentiation**, happens during the development of all multicelled organisms. It arises when embryonic cells and their descendants activate and suppress a few of the total number of genes in unique, selective ways. Thus, only the precursors of red blood cells use the genes for hemoglobin. Only certain white blood cells use the genes for weapons called antibodies.

As outlined in Figure 15.3, selective control of gene expression is exerted at many levels. Some genes become amplified or undergo rearrangement before being transcribed. Many are shut down, temporarily or permanently, through chemical modification and chromosome packaging. On-off transcription is rare. Usually, transcription rates rise or fall by degrees, depending on changing concentrations of molecular signals, substrates, or products. Post-transcriptional controls govern transcript processing and transport of mature RNAs from the nucleus. Controls govern rates of translation. And controls govern the modification of new polypeptides or the activation, inhibition, and degradation of the protein products.

Many genes specify enzymes. Thus, controls over the transcription, translation, and expression of those genes extend to *all* metabolic reactions in the cell. They indirectly govern the type and number of all molecules that will be synthesized and degraded at a given time. *And so they govern all short-term and long-term aspects of cell structure and function.*

In multicelled organisms, all cells inherit the same genes and use most of them in the same way. But each differentiated cell type also activates and suppresses a small number of genes in a selective way.

Gene controls operating at many different levels underlie short-term housekeeping tasks *and* intricate, long-term patterns of growth and development.

Figure 15.3 Examples of the levels of control over gene expression in eukaryotes. Compared with prokaryotes, the cells of multicelled eukaryotes require more complex control mechanisms. Gene activity changes as the developmental program unfolds, as cells physically contact one another in the developing tissues, and as they start interacting by way of hormones and other molecular signals.

a Controls Related to Transcription. With respect to transcription, we see less of the rapid, all-or-nothing control that is common among bacteria. (The intercellular environment provides more stable operating conditions.) Many eukaryotic genes concerned with housekeeping tasks are under positive controls that promote continuous, low levels of transcription. Transcription rates for many other genes are often adjusted by degrees, rather than being abruptly switched on or off.

Also, even before some genes are transcribed, a portion of their base sequences may become amplified, rearranged, or chemically modified in temporary or reversible ways. These are not mutations; they are programmed events that affect the manner in which particular genes will be expressed (if at all).

(1) *Gene amplification.* Some cells that require enormous numbers of certain molecules temporarily increase the number of the required genes. In response to a molecular signal, multiple rounds of DNA replication sometimes produce hundreds or thousands of gene copies prior to transcription. This happens in immature amphibian eggs and in glandular cells of some insect larvae (page 242).

(2) *DNA rearrangements.* In a few cell types, many base sequences are alternative regions for the same gene. Before transcription, they are snipped out of the DNA molecule. Combinations of these snippets are spliced together as the gene's final base sequence.

This happens when B lymphocytes, a special class of white blood cells, are forming. As shown in Figure 40.9, these cells transcribe and translate the rearranged DNA into staggeringly diverse versions of the protein weapons called antibody molecules.

(3) *Chemical modification.* Numerous histones and other proteins interact with eukaryotic DNA in highly organized fashion. The DNA-protein packaging, as well as chemical modifications to particular sequences, influences gene activity. Usually, only a small fraction of genes are available for transcription. A more dramatic shutdown is called X chromosome inactivation in female mammals, as described on page 240.

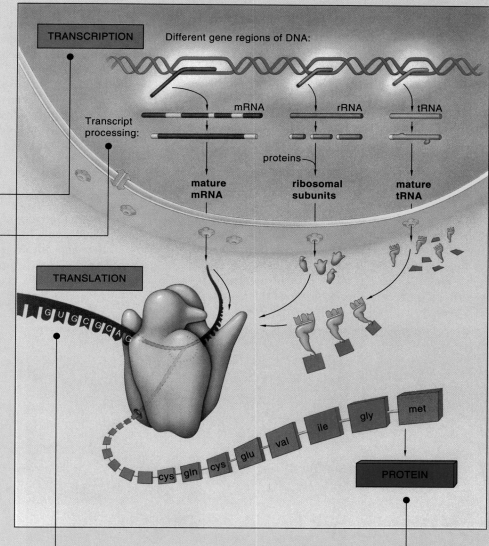

TRANSCRIPTION

Different gene regions of DNA:

mRNA   rRNA   tRNA

Transcript processing:

proteins

mature mRNA   **ribosomal subunits**   **mature tRNA**

TRANSLATION

GUGCGCAG

cys gln cys glu val ile gly met

PROTEIN

b Transcript Processing Controls. As indicated on page 222, pre-mRNA transcripts undergo modifications in the nucleus, as when introns are removed and exons are spliced together in more than one way. Thus, a transcript from a single gene may undergo *alternative splicing*.

Pre-mRNA from a gene that specifies a contractile protein (troponin-T) is like this. Enzymes excise different portions of the mRNA transcript in different cells. Thus, when the exons are spliced together, the final protein-building message is slightly different from one cell to the next. The resulting proteins are all very similar, but each is unique in a certain region of its amino acid sequence. The proteins function in slightly different ways—which may account for subtle variations in the functioning of different types of muscles in the body.

c Controls Over Translation. Controls govern when, how rapidly, and how often an mRNA transcript will be translated. Pages 616, 760, and elsewhere in the book provide elegant examples.

The stability of an mRNA transcript influences the number of protein molecules that can be produced from it. Enzymes destroy the transcripts from the poly-A tail on up (page 222). The tail's length and its attached proteins influence the pace of degradation.

Also, after leaving the nucleus, some kinds of mRNA transcripts are temporarily or permanently inactivated. For example, in unfertilized eggs, many transcripts are chemically inactivated and stored in the cytoplasm. These "masked messengers" will not be available for translation until after fertilization, when many protein molecules will be required for the early cell divisions of the new individual (page 760).

d Controls Following Translation. Before they can become fully functional, many polypeptide chains must pass through the cytomembrane system (page 64). There they undergo modification, as by having specific oligosaccharides or phosphate groups attached to them.

Also, a variety of control mechanisms govern the activation, inhibition, and stability of enzymes and other molecules that control protein synthesis. Allosteric control of tryptophan synthesis, described earlier on page 100, is an example.

15.4 EVIDENCE OF GENE CONTROL

By some estimates, the cells of a multicelled organism rarely use more than 5 to 10 percent of their genes at a given time. One way or another, most of their genes are being repressed. However, *which* genes are being expressed varies, depending on the stage of growth and development that the organism is passing through. The following sections provide you with a few specific examples of the control mechanisms that are operating at different stages.

Transcription in Lampbrush Chromosomes

As they mature, the unfertilized eggs of amphibians grow extremely fast. Growth requires numerous copies of enzymes and other proteins. The germ cells that give rise to these eggs stockpile many RNAs and ribosomes that will be required for rapid protein synthesis. As these germ cells proceed through prophase I of meiosis,

their chromosomes decondense into thousands of loops. The chromosomes look so bristly at this time, they are said to have a "lampbrush" configuration (Figure 15.4). At such times, proteins that structurally organize the DNA loosen their grip, making RNA genes accessible. Among these are the proteins of nucleosomes.

Nucleosomes are the basic unit of organization in eukaryotic chromosomes, as described on page 216. Each consists of a stretch of DNA looped around a core of histone molecules. The diagram in Figure 15.5 is based on micrographs that show the nucleosome packaging being loosened up during transcription.

X Chromosome Inactivation

A mammalian zygote that is destined to develop into a female has inherited two X chromosomes, one from the mother and one from the father. During development, one of the two chromosomes condenses in each cell. In that chromosome, cytosines in the DNA base sequences become heavily methylated, and this may block tran-

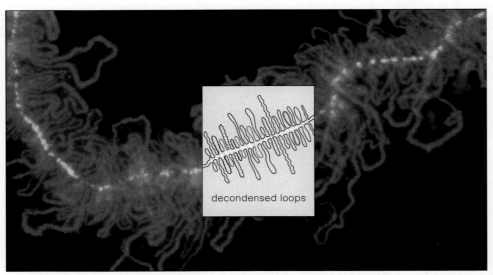

a

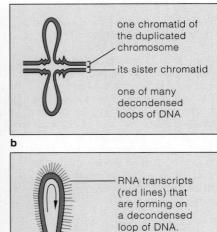

b

RNA transcripts (red lines) that are forming on a decondensed loop of DNA. (The arrow shows the direction of transcription.)

c

Figure 15.4 Lampbrush chromosome from a germ cell of a newt (*Notophthalmus viridiscens*). During prophase I, gene regions of this duplicated chromosome decondensed into thousands of loops. A *red* fluorescent dye labeled one of the protein components of ribosomal subunits. Ribonucleoproteins are evidence of gene activity. (A *white* fluorescent dye labeled proteins making up the duplicated chromosome's framework.)

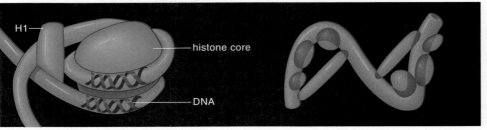

a   b

Figure 15.5 Model of changes in nucleosome structure during gene transcription. (**a**) A nucleosome consists of a portion of the DNA double helix, looped twice around a core of histone proteins. (**b**) During the transcription of specific gene regions, the tight DNA-histone packing loosens up.

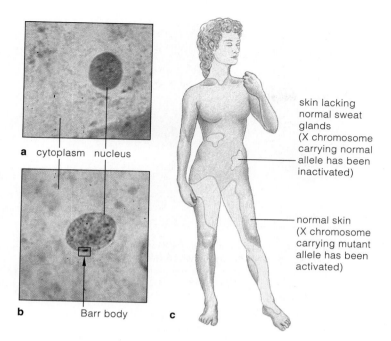

a cytoplasm nucleus

b Barr body **c**

skin lacking normal sweat glands (X chromosome carrying normal allele has been inactivated)

normal skin (X chromosome carrying mutant allele has been activated)

Figure 15.6 Comparison of the interphase nucleus of somatic cells from a human male and female. (**a**) In the male's cell, the X chromosome is not condensed. (**b**) In the female's cell, a Barr body (one inactivated X chromosome) is visible as a dark-staining spot on the inside of the nuclear envelope. (**c**) Anhidrotic ectodermal dysplasia. A mosaic pattern of gene expression, arising from random X chromosome inactivation, leads to patches of skin with and without normal sweat glands.

scription of most of its genes. Condensation is a programmed event, but the outcome is random. *Either chromosome may be inactivated.* The condensed X chromosome is visible as a dark spot in the interphase nucleus (Figure 15.6). It is called a **Barr body** after its discoverer, Murray Barr.

When a maternal X chromosome is inactivated in a cell, the same chromosome becomes inactivated in all of the cell's descendants—and in tissue regions that they produce. The same thing happens following inactivation of the paternal X chromosome in a cell. By adulthood, every adult female is a "mosaic" for inactivated X chromosomes. She has patches of tissues in which maternal genes are being expressed—and patches of tissue in which paternal genes are being expressed. Because any pair of alleles on the two X chromosomes may or may not be identical, the tissue patches may or may not have the same characteristics.

The mosaic tissue effect arising from random X chromosome inactivation is called **Lyonization** (after its discoverer, Mary Lyon). We see its effects in human females who are heterozygous for a recessive allele that prevents sweat glands from forming. The allele resides on the X chromosome. In these females, the X chromosome bearing the normal dominant allele has been condensed—and genes on the one bearing the mutated allele are being transcribed in patches of skin that have no sweat glands (Figure 15.6c). This condition is called *anhidrotic ectodermal dysplasia.*

The mosaic effect is wonderfully apparent in female calico cats, which are heterozygous for black and yellow coat-color alleles on the X chromosome. Coat color in a given body region depends on which X chromosome's genes are being transcribed (Figure 15.7).

Figure 15.7 Why is this female calico cat "calico"? One X chromosome in her cells carries a dominant allele for the brownish-black pigment melanin. The allele on her other X chromosome specifies yellow fur. At an early stage of the cat's embryonic development, one of the two X chromosomes was inactivated at random in each cell that had formed by then. In all descendants of those cells, the same chromosome also became inactivated, leaving only one functional allele for the coat-color trait. We see patches of different colors, depending on which allele was inactivated in cells that formed a given tissue region. (The white patches result from a gene interaction involving the "spotting gene," which blocks melanin synthesis entirely.)

15.5 EXAMPLES OF SIGNALING MECHANISMS

A dizzying variety of signals influence gene activity. The following examples from animals and plants will give you an idea of their action at the molecular level.

Hormonal Signals

Hormones are major signaling molecules that can stimulate or inhibit gene activity in target cells. Any cell with receptors for a given hormone is a target. Animal cells secrete hormones into interstitial fluid. Most hormones are picked up by the bloodstream and distributed to cells some distance away. Plant hormones do not travel far from cells that secrete them.

Certain hormones bind to membrane receptors at the surface of target cells. Others enter target cells, bind with regulatory proteins, and so help initiate transcription. In many cases, a hormone molecule must first touch bases, so to speak, with an enhancer. **Enhancers** are base sequences that serve as binding sites for suitable activator proteins. They may or may not be adjacent to a promoter in the same DNA molecule. When some distance separates the two, a loop forms in the DNA and brings them together. RNA polymerase binds avidly to this looped complex.

Consider the effect of ecdysone, a hormone with key roles in the life cycle of many insects. Larvae of these insects grow rapidly and feed continuously on plant or animal tissues. The larvae require copious amounts of saliva to prepare food for digestion. In their salivary gland cells, the DNA has been replicated repeatedly. The copies of the DNA molecules have remained together in parallel array, forming what is called the "polytene chromosome." In response to ecdysone, multiple copies of genes in the DNA are rapidly transcribed. The gene regions affected by this hormonal signal puff out during transcription (Figure 15.8). Afterward, translation of the mRNA transcripts produces the protein components of saliva.

In vertebrates, some hormones have widespread effects on gene expression in many cell types. For instance, the pituitary gland secretes somatotropin (growth hormone). This hormonal signal stimulates synthesis of the proteins required for cell division and, ultimately, the body's growth. Most cells have receptors for somatotropin.

Other vertebrate hormones signal only certain cells at certain times. Prolactin, a secretion from the pituitary gland, is like this. Beginning a few days after a female mammal gives birth, prolactin can be detected in the bloodstream. Prolactin activates genes in certain cells of mammary glands. Those genes alone have responsibil-

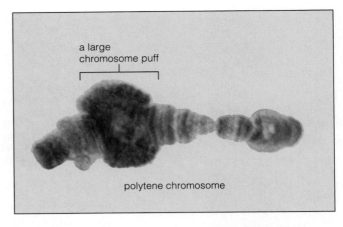

Figure 15.8 Visible evidence of transcription in a polytene chromosome from a midge larva. (A midge is a type of insect.) Staining techniques reveal banding patterns in these chromosomes, in which the DNA has been replicated repeatedly.

Ecdysone, a hormone, serves as a regulatory protein that helps promote transcription of the amplified genes in these chromosomes. Where genes are being transcribed, the chromosome loosens and puffs out. The puffs become large and diffuse when transcription is most intense.

ity for milk production. Liver cells and heart cells have those genes also, but they have no means of responding to signals from prolactin.

Explaining hormonal control of gene activity is like explaining a full symphony orchestra to someone who has never seen one or heard it perform. Many separate parts must be defined before their interactions can be understood! We will return to this topic, starting with Chapter 37 on the endocrine system. As you will see in Chapters 44 and 45, some of the most elegant examples of hormonal controls are drawn from studies of animal reproduction and development.

Sunlight as a Signal

Plant a few seeds from a corn or bean plant in a pot that contains moist, nutrient-rich soil. Let the seeds germinate—but keep them in total darkness. After eight days have passed, they will have developed into seedlings that are spindly and pale, owing to the absence of chlorophyll (Figure 15.9). Next, expose the seedlings to a single burst of dim light from a flashlight. Within ten minutes, they will start converting stockpiled molecules to active forms of chlorophyll, the light-trapping pigment molecules necessary for photosynthesis!

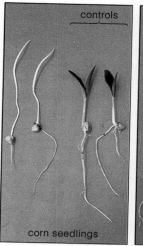
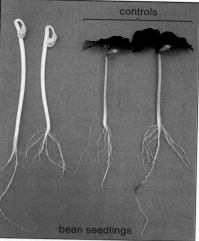

corn seedlings

bean seedlings

Figure 15.9 Effect of the absence of light on corn and bean seedlings. The two seedlings to the right in each photograph served as the control group; they were grown in a greenhouse. The two seedlings positioned next to them were grown in total darkness for eight days. The dark-grown plants could not convert their stockpiled precursors of chlorophyll molecules to active form, and they never did green up.

Phytochrome, a blue-green pigment, is a signaling molecule that helps plants adapt to changes in light conditions. This molecule exists in active and inactive forms. As described on page 538, it becomes inactive at sunset, at night, or even in shade, where far-red wavelengths predominate. The molecule becomes activated at sunrise, when red wavelengths dominate the sky. As days alternate with nights, as days grow shorter and then longer with the changing seasons, the quantity of incoming red or far-red wavelengths varies. By controlling phytochrome activity, these variations control transcription of certain genes at certain times of day and year. The genes specify enzymes and other proteins with roles in germination, stem elongation, branching, leaf expansion, and formation of flowers, fruits, and seeds.

Phytochrome control of gene transcription may work along the lines of the model shown in Figure 15.10. Experiments conducted by Elaine Tobin and her coworkers at the University of California, Los Angeles, provide evidence in favor of such a model. Working with dark-grown seedlings of duckweed (*Lemna*), they discovered a marked increase in the number of certain mRNA transcripts following a one-minute exposure to red light. Exposure enhanced transcription of the genes for proteins that bind the chlorophylls and for the enzyme that mediates carbon fixation. All of these molecules are required for the development and greening of chloroplasts.

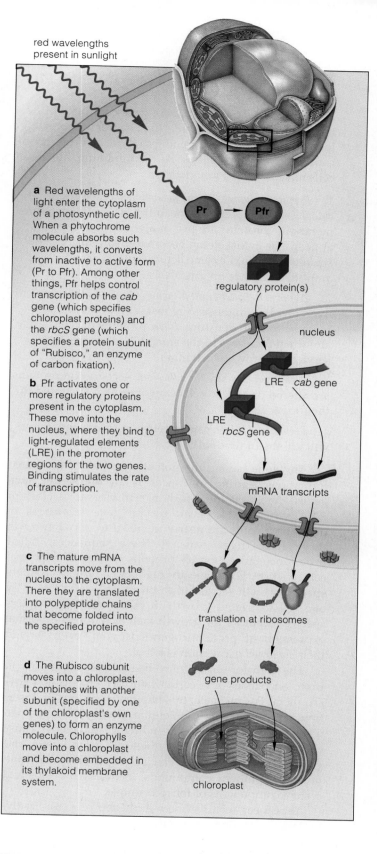

red wavelengths present in sunlight

a Red wavelengths of light enter the cytoplasm of a photosynthetic cell. When a phytochrome molecule absorbs such wavelengths, it converts from inactive to active form (Pr to Pfr). Among other things, Pfr helps control transcription of the *cab* gene (which specifies chloroplast proteins) and the *rbcS* gene (which specifies a protein subunit of "Rubisco," an enzyme of carbon fixation).

b Pfr activates one or more regulatory proteins present in the cytoplasm. These move into the nucleus, where they bind to light-regulated elements (LRE) in the promoter regions for the two genes. Binding stimulates the rate of transcription.

c The mature mRNA transcripts move from the nucleus to the cytoplasm. There they are translated into polypeptide chains that become folded into the specified proteins.

d The Rubisco subunit moves into a chloroplast. It combines with another subunit (specified by one of the chloroplast's own genes) to form an enzyme molecule. Chlorophylls move into a chloroplast and become embedded in its thylakoid membrane system.

Pr → Pfr

regulatory protein(s)

nucleus

LRE   *cab* gene

LRE
rbcS gene

mRNA transcripts

translation at ribosomes

gene products

chloroplast

Figure 15.10 A proposed model for the mechanism by which phytochrome helps control gene transcription in plants. Light of red wavelengths converts the phytochrome molecule from inactive form (here designated Pr) to active form (Pfr). In this form, the phytochrome can serve as a regulator of transcription.

15.6 GENES IMPLICATED IN CANCER

Having touched on the kinds of gene controls that exist in eukaryotes, let's conclude the chapter with a closer look at what may be happening to those controls during cancerous transformations. At the minimum, all cancer cells have these characteristics:

1. *Profound changes in the plasma membrane and cytoplasm.* Membrane permeability is amplified. Some membrane proteins are lost or altered; new ones appear. The cytoskeleton becomes disorganized, shrinks, or both. Enzyme activity shifts, as in an amplified reliance on glycolysis.

2. *Abnormal growth and division.* Controls that prevent overcrowding in tissues are lost. Cell populations increase to high densities. New proteins trigger abnormal increases in the small blood vessels that service the growing cell mass.

3. *Weakened capacity for adhesion.* Recognition proteins are altered or lost, so cells cannot remain anchored in the proper tissue.

4. *Lethality.* Unless cancer cells are eradicated, they will kill the individual.

Any gene having the potential to induce the transformations leading to cancer is an **oncogene**. Such genes were first identified in retroviruses (a class of RNA viruses). However, certain gene sequences in many organisms are nearly identical to oncogenes, yet they rarely trigger cancer! These sequences, called **proto-oncogenes**, specify proteins required for normal cell function. Some specify regulatory proteins. Others specify growth factors (transcriptional signals sent by one cell to trigger growth in other cells). Still others specify receptors for growth factors. Thus, the normal expression of proto-oncogenes is vital, even though their *abnormal* expression is lethal.

The transformation may begin with mutations in proto-oncogenes or in certain control elements that govern them. For instance, this happens with certain insertions of viral DNA into cellular DNA. It can happen when **carcinogens** (cancer-inducing mutagens) cause changes in the DNA. Ultraviolet radiation, x-rays, and gamma rays are common carcinogens. So are many natural and synthetic compounds, including asbestos and certain components of tobacco smoke.

Yet cancer seems to be a multistep process, involving more than one oncogene. Thus, in 1994, researchers identified three such genes that contribute to nearly all colon cancers. About 1 of every 200 individuals carries these oncogenes. As a result of this discovery, it may become possible to diagnose carriers early and observe them closely for cancer development.

SUMMARY

1. Cells exert control over gene expression—that is, which gene products appear, at what times, and in what amounts. At any time, the control mechanisms that are operating depend on the type of cell and its functions, on prevailing chemical conditions, and on outside signals that can change the cell's activities.

2. Regulatory proteins, enzymes, and hormones are common control elements. They interact with one another, with DNA and RNA, and with gene products (polypeptide chains, the building blocks of proteins).

   a. As an example, in prokaryotes, control is exerted at promoters (DNA binding sites that signal the start of a gene), operators (binding sites for regulatory proteins that can inhibit transcription), and enhancers (binding sites for activator proteins that can promote transcription).

   b. Different kinds of controls operate before, during, and after gene transcription and translation.

3. Two common types of control systems operate in cells to block or enhance transcription.

   a. With negative control systems, a regulatory protein binds at a specific DNA sequence and prevents one or more genes from being transcribed; its effect is reversed by the action of a regulatory protein molecule. Repressor proteins in prokaryotes are examples.

   b. With positive control systems, a regulatory protein binds to DNA and promotes the initiation of transcription. Activator proteins, which bind at enhancer sites in eukaryotic DNA, are examples.

4. Most prokaryotes (bacteria) do not live long and do not require great numbers of genes to grow, divide, and reproduce. Commonly, systems involving a few control elements enhance transcription rates only when gene products are required. This is true of operons, which are groupings of related genes and their control elements.

5. Eukaryotes require complex gene controls. Besides changing rapidly in response to short-term shifts in the surroundings, gene activity must change over the long term, during growth and development. Then, cells multiply, physically contact one another in tissues, and chemically interact, as by hormonal signals.

6. All cells in a multicelled eukaryote inherit the same genes, but different cell types activate and suppress some fraction of the genes in different ways.

7. Selective gene control brings about cell differentiation. In other words, cell lineages become specialized in appearance, composition, and function. Later chapters describe the phenotypic consequences of selected gene control.

1. Define these terms: promoter, operator, repressor protein, and activator protein. *236, 237*

2. Cells depend on controls over which gene products are synthesized, at what times, at what rates, and in what amounts. Describe one type of control over transcription in *E. coli*, a type of prokaryote. Then list five general kinds of gene controls involved in eukaryotes. *236–239*

3. Define cell differentiation. How does it arise? *238*

4. A plant, fungus, or animal is composed of diverse cell types. How might this diversity arise, given that the body cells in each organism inherit the same set of genetic instructions? *238*

5. What are the characteristics of cancer cells? Explain the difference between a benign tumor and one that is malignant. *244, 234*

Self-Quiz *(Answers in Appendix IV)*

1. The expression of a given gene depends on _____ .
   a. cell type and functions
   b. changing chemical conditions
   c. environmental signals
   d. all of the above

2. Regulatory proteins are part of control systems that interact with _____ to govern gene expression.
   a. DNA
   b. RNA
   c. gene products
   d. all of the above

3. A negative control system that governs the lactose operon in *E. coli* operates during _____ .
   a. replication
   b. transcription
   c. translation
   d. post-translation

4. A base sequence that signals the start of a gene is called a _____ .
   a. promoter
   b. operator
   c. enhancer
   d. activator protein

5. An operon is a control system that most typically governs _____ .
   a. a bacterial gene
   b. a eukaryotic gene
   c. genes of all types
   d. DNA replication

6. Prokaryotic cells rely heavily on _____ controls over transcription.
   a. slow, continuous
   b. rapid, continuous
   c. slow, on-off
   d. rapid, on-off

7. Cell differentiation _____ .
   a. occurs in all multicelled organisms
   b. requires different genes in different cells
   c. involves selective gene expression
   d. both a and c
   e. all of the above

8. In eukaryotic cells, controls over gene expression include _____ .
   a. amplifying genes
   b. rearranging DNA
   c. chemically modifying DNA
   d. processing RNA transcripts
   e. all of the above

9. Lampbrush chromosomes are evidence of _____ .
   a. mutated genes
   b. heavily methylated DNA
   c. intense transcription
   d. both b and c

10. Inactivation of one of the two X chromosomes in cells of mammalian females appears to involve regions of _____ .
    a. mutated genes
    b. heavily methylated DNA
    c. intense transcription
    d. both a and b

11. In eukaryotic cells, various types of hormones interact with _____ .
    a. membrane receptors
    b. regulatory proteins
    c. enhancers
    d. all of the above

12. Match the terms with their most suitable descriptions.
    ____ phytochrome
    ____ enhancer
    ____ Barr body
    ____ proto-oncogene
    ____ Lyonization
    ____ oncogene

    a. helps plants adapt to changes in light conditions
    b. a gene having the potential to induce cancerous transformation
    c. normal expression is vital, abnormal expression is lethal
    d. condensed X chromosome in an interphase nucleus
    e. mosaic tissue effect arising from random X chromosome inactivation
    f. base sequence that serves as binding site for activator protein

Selected Key Terms

activator protein *237*	oncogene *244*
Barr body *241*	operator *236*
cancer *235*	operon *236*
carcinogen *244*	phytochrome *243*
cell differentation *238*	positive control system *236*
cooperative regulation *236*	promoter *236*
enhancer *242*	proto-oncogene *244*
hormone *242*	regulatory protein *236*
Lyonization *241*	repressor *236*
metastasis *234*	tumor *234*
negative control system *236*	

Readings

Feldman, M., and L. Eisenbach. November 1988. "What Makes a Tumor Cell Metastatic?" *Scientific American* 259(5):60–85.

Kupchella, C. 1987. *Dimensions of Cancer*. Belmont, California: Wadsworth.

Murray, A., and M. Kirschner. March 1991. "What Controls the Cell Cycle." *Scientific American* 264(3):56–63.

Ptashne, M. January 1989. "How Gene Activators Work." *Scientific American* 260(1):41–47.

Weintraub, H. January 1990. "Antisense RNA and DNA." *Scientific American* 262(1):40–46.

16 RECOMBINANT DNA AND GENETIC ENGINEERING

Make Way for Designer Genes

In 1990, when she was four years old, Ashanthi DeSilva received a historic genetic reprieve. Ashanthi was born without defenses against viruses, bacteria, and other agents of disease. She has no immune system. Of her forty-six chromosomes, one bears a defective gene that normally would specify adenosine deaminase (ADA), an enzyme.

Without the enzyme, Ashanthi's cells cannot properly break down excess amounts of a nucleotide (AMP), and a reaction product accumulates that is toxic to lymphoblasts in the bone marrow. Lymphoblasts give rise to white blood cells—the immune system's

army. With too few of those cells (or none at all), the outcome is a *severe combined immune deficiency* (SCID), and it leads to a devastating set of disorders. In this particular case, symptoms included dangerous infections of the ears and lungs, high fevers, severe diarrhea, and an inability to gain weight. Ashanthi was so vulnerable to germs, her parents would not let her attend school.

Bone marrow transplants help some individuals with SCID when the donated lymphoblasts go on to produce functional white blood cells. ADA injections help others. But these are not permanent cures.

Figure 16.1 White blood cells on patrol inside a blood vessel. Ashanthi DeSilva and other individuals have drastically reduced numbers of these infection-fighting cells—or none at all. They are candidates for gene therapy, one of the beneficial applications of recombinant DNA research.

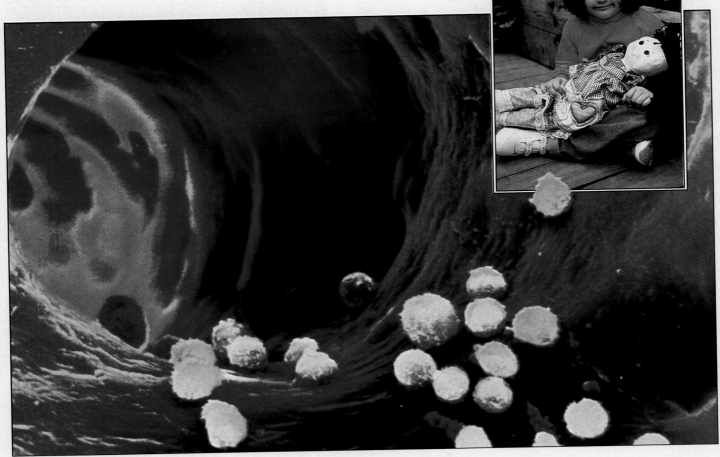

Given the options, the girl's parents consented to the first federally approved gene therapy test for humans. Using recombinant DNA methods of the sort described in this chapter, medical researchers had already identified and isolated the ADA gene, and they were producing quantities of it in the laboratory. They also were able to splice those genes into the genetic material of a harmless type of virus that could serve roughly the same function as a hypodermic needle. That is, by infecting targeted cells, the harmless virus could deliver copies of the "good" gene to them.

The researchers harvested some of Ashanthi's white blood cells, cultured them in petri dishes, then exposed them to the modified virus. The ADA gene became incorporated in some cells—which started to synthesize ADA. Later, the researchers inserted about a billion copies of the genetically modified cells into the girl's bloodstream.

The gene therapy worked. At first Ashanthi received additional infusions of fortified cells every month. Now she receives them once a year. And she is attending school. Four more ADA-deficient patients also have started treatment—and they are doing well at this writing.

Some medical researchers are now employing lymphoblasts. They retrieved small amounts of lymphoblasts from blood in the umbilical cord (which is discarded following childbirth). They exposed these stem cells to viruses that could deliver copies of the ADA gene into them. Afterward, they exposed the cells to factors that stimulated growth and division, then reinserted the cells into the newborn patients. It is too soon to tell whether these patients have permanent stem cells that are producing functional copies of the ADA gene. If this turns out to be the case, gene therapy will have given them a continuous, lifelong supply of the crucial enzyme—and of functional disease fighters (Figure 16.1).

As this example suggests, recombinant DNA technology has staggering potential for medicine. It has equally staggering potential for agriculture and industry. It does not come without risks. With this chapter, we consider some basic aspects of the new technology, and we address ecological, social, and ethical questions related to its application.

KEY CONCEPTS

1. Genetic experiments have been occurring in nature for billions of years as a result of gene mutations, crossing over and recombination, and other events. Humans are now purposefully bringing about genetic changes by way of recombinant DNA technology. Such enterprises are called genetic engineering.

2. With this technology, researchers isolate, cut, and splice together gene regions from different species, then greatly amplify the number of copies of the genes that interest them. The genes, and in some cases their protein products, are produced in quantities that are large enough for research and for practical applications.

3. Three activities are at the heart of recombinant DNA technology. First, procedures based on specific enzymes are used to cut DNA molecules into fragments. Second, the fragments are inserted into cloning tools, such as plasmids. Third, fragments containing the genes of interest are identified, then copied rapidly and repeatedly.

4. Genetic engineering involves isolating, modifying, and inserting particular genes back into the same organism or into a different one. The goal is to modify traits influenced by those genes in beneficial ways. Human gene therapy, which focuses on controlling or curing genetic disorders, is an example.

5. The new technology raises social, legal, ecological, and ethical questions regarding its benefits and risks.

16.1 RECOMBINATION IN NATURE— AND IN THE LABORATORY

For more than 3 billion years, nature has been conducting genetic experiments by way of mutation, then by crossing over between chromosomes and other events. Genetic messages have changed countless times, and this is the source of life's diversity.

Figure 16.2 Plasmids (*blue* arrows) released from a ruptured bacterial cell (*Escherichia coli*).

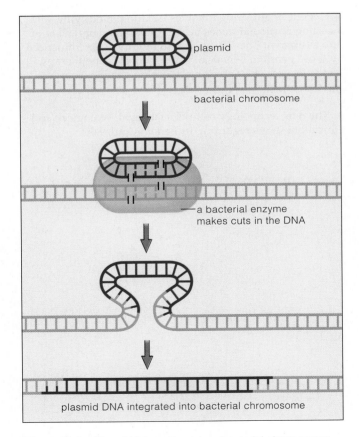

plasmid

bacterial chromosome

a bacterial enzyme makes cuts in the DNA

plasmid DNA integrated into bacterial chromosome

Figure 16.3 Plasmid integration into a bacterial chromosome.

For many thousands of years, we humans have been changing genetically based traits of species. Through artificial selection, we produced modern crop plants and new breeds of cattle, birds, dogs, and cats from wild ancestral stocks. We developed meatier turkeys, sweeter oranges, seedless watermelons, flamboyant ornamental roses, and other wonderfully useful or novel plants. We produced the tangelo (tangerine × grapefruit) and the mule (donkey × horse).

Currently, researchers employ **recombinant DNA technology** to analyze genetic changes. With this technology, they cut and splice together DNA from different species, then insert the modified molecules into bacteria or other types of cells that engage in rapid replications and cell divisions. The cells copy the foreign DNA right along with their own. In short order, huge bacterial populations can produce useful quantities of recombinant DNA molecules. The new technology also is the basis of **genetic engineering**. Genes are being isolated, modified, and inserted back into the same organism or into a different one.

Believe it or not, this astonishing technology had its origins in the innards of bacteria. Bacterial cells have a single chromosome, a circular DNA molecule with all the genes needed for growth and reproduction. Many bacteria also have **plasmids**, which are small, circular molecules of "extra" DNA having a few genes (Figure 16.2). Replication enzymes can copy plasmid DNA, just as they copy the chromosomal DNA.

Many bacteria can transfer plasmid genes to a bacterial neighbor (page 350). Transferred plasmids may even get integrated into a recipient's chromosome, forming a recombinant DNA molecule (Figure 16.3). Specific enzymes mediate the recombination events. A bacterial enzyme recognizes a short nucleotide sequence present in both the plasmid and the chromosome. It cuts the molecules at that sequence, then another enzyme splices the cut ends together. In nature, viruses as well as bacteria dabble in gene transfers and recombinations. So do most eukaryotic organisms.

As we turn now to some key features of the new technology, keep these concepts in mind:

---

1. With recombinant DNA technology, DNA from different species is cut and spliced together, then the recombinant molecules are amplified (by way of copying mechanisms) to produce useful quantities of the genes of interest.

2. The new technology is being used for basic research. It also is being used for genetic engineering—the deliberate modification of genes, followed by their insertion into the same individual or a different one.

16.2 PRODUCING RESTRICTION FRAGMENTS

Bacteria are equipped with many restriction enzymes. These enzymes cut apart foreign DNA that enters the bacterial cell, often during viral infection. Each type of restriction enzyme makes a cut wherever a specific, very short nucleotide sequence occurs in the DNA. Cuts at two identical sequences in the same molecule produce a fragment. Because some types of enzymes make staggered cuts, the fragments may have single-stranded portions at both ends. These are sometimes referred to as sticky ends:

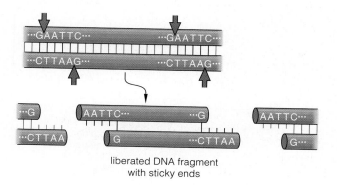

liberated DNA fragment
with sticky ends

By "sticky," we mean that the short, single-stranded ends of a DNA fragment can base-pair with any other DNA molecule that also has been cut by the same restriction enzyme.

Suppose you use the same restriction enzyme to cut plasmids and the DNA from a human cell. When you mix the cut-up molecules together, they base-pair at the cut sites. Then you add **DNA ligase** to the mixture. This enzyme seals the sugar-phosphate backbone of DNA at the cut sites, just as it does during DNA replication. In this way, you create "recombinant plasmids," which have pieces of DNA from another organism inserted into them.

You now have a **DNA library**—a collection of DNA fragments, produced by restriction enzymes, that have been incorporated into plasmids (Figure 16.4).

Recombination is made possible by restriction enzymes that make specific cuts in DNA molecules and by DNA ligases that seal the cut ends.

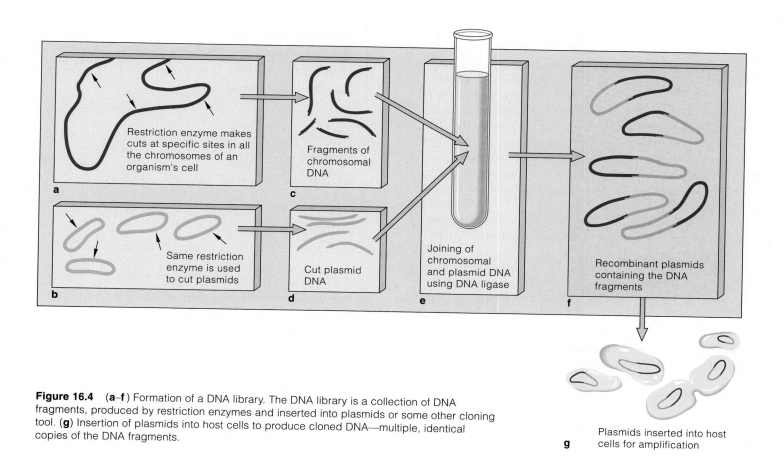

Figure 16.4 (**a–f**) Formation of a DNA library. The DNA library is a collection of DNA fragments, produced by restriction enzymes and inserted into plasmids or some other cloning tool. (**g**) Insertion of plasmids into host cells to produce cloned DNA—multiple, identical copies of the DNA fragments.

16.3 WORKING WITH DNA FRAGMENTS

Amplification Procedures

A DNA library is almost vanishingly small. It must be amplified—copied again and again—into useful amounts. One way to do this is to employ "factories" of bacteria, yeasts, or some other kind of cell that reproduces rapidly and can take up plasmids. In short order, a growing population of such cells can amplify a DNA library. Their repeated replications and divisions yield cloned DNA that has been inserted into plasmids. The "cloned" part of the name simply refers to the multiple, identical copies of DNA fragments.

The **polymerase chain reaction**, or **PCR**, is an alternative method of amplifying fragments of DNA. The reactions proceed in test tubes, not in microbial factories. First, researchers identify short nucleotide sequences located just before and just after a region of DNA from an organism that interests them. Then they synthesize primers (nucleotide sequences that will base-pair with the ones in the DNA). They mix the primers with enzyme molecules, free nucleotides, and all the DNA from one of the organism's cells. Then they subject the mixture to precise temperature cycles.

During the cycles, the two strands of all the DNA molecules unwind from each other. Then, as Figure 16.5 indicates, the short sequences become positioned at the specific target sites, in accordance with the exposed, complementary bases. **DNA polymerase**, a replication enzyme, recognizes the synthesized primers as signals to get busy. This particular DNA polymerase comes from a bacterium that thrives in hot springs and even in hot water heaters. The enzyme is not affected by the elevated temperatures required to unwind the DNA or by the lower temperatures required for base-pairing.

With each round of reactions, the number of DNA molecules doubles. For example, if there are 10 such molecules in the test tube, there soon will be 20, then 40, 80, 160, 320, and so on. Very quickly, a target region from a single DNA molecule can be amplified to many *billions* of molecules.

Thus PCR can amplify samples with very little DNA. As you will see shortly, such tiny samples can be obtained from some fossils—even from a single hair left at the scene of a crime.

Sorting Out and Sequencing Specific DNA Fragments

Gel Electrophoresis of DNA When restriction enzymes cut DNA, the resulting fragments are not all the same length. Researchers can employ gel electrophoresis to separate the fragments from one another

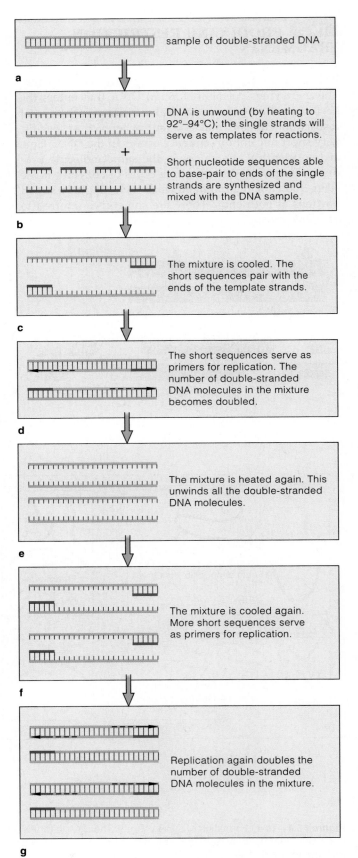

Figure 16.5 The polymerase chain reaction (PCR).

Figure 16.6 One method of sequencing DNA, as initially developed by Frederick Sanger. The method can be used to determine the nucleotide sequence of specific DNA fragments, as this example illustrates.

a Single-stranded DNA fragments are added to a solution in four different test tubes:

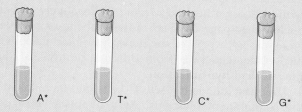

All four tubes contain DNA polymerases, short nucleotide sequences that can serve as primers for replication, and the nucleotide subunits of DNA (that is, A, T, C, and G). One of the four kinds of nucleotides in each tube is modified and radioactively labeled. *Which* one is different in each tube. We can designate these as A*, T*, C*, and G*. Let's follow what happens in the tube with the A* subunits.

b As expected, the DNA polymerase recognizes a primer that has become attached to a fragment, which it uses as a template strand. The enzyme assembles a complementary strand according to base-pairing rules (A only to T, and C only to G). Sooner or later, the enzyme picks up an A* subunit for pairing with a T on the template strand. The modified nucleotide is a chemical roadblock—it prevents the enzyme from adding more nucleotides to the growing complementary strand. In time, the tube contains radioactively labeled strands of different lengths, as dictated by the location of each A* in the sequence:

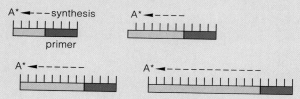

The same thing happens in the other three tubes, with strand lengths dictated by the location of T*, C*, and G*.

c DNA from each of the four tubes is placed in four parallel lanes in the same gel. Then it can be subjected to electrophoresis. The resulting nucleotide sequence can be read off the resulting bands in the gel. Look at the numbers running down the side of this diagram:

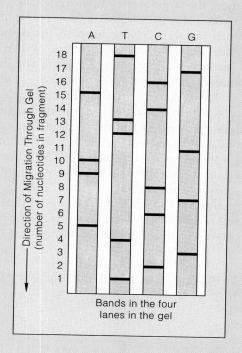

Bands in the four lanes in the gel

Start with "l" and read across the four lanes (A, T, C, G). As you can see, T is closest to the start of the nucleotide sequence; it has migrated farthest through the gel. At "2," the next nucleotide is C, and so on. The entire sequence, read from the first nucleotide to the last, is

T C G T A C G C A A G T T C A C G T

Now, using the rules of base-pairing, you can deduce the sequence of the DNA fragment that served as the template.

according to their length. (As you read on page 220, this procedure also can be used to separate a sample of protein molecules according to their overall electric charge and size.)

A sample of restriction fragments is placed near one end of a slab of gel that is mounted on a glass plate and bathed in an appropriate solution. When an electric current passes through the gel, the fragments respond to it (because of their charged phosphate groups). The extent to which fragments of a given length migrate through the gel depends only on their size—larger ones cannot move as fast through it.

DNA Sequencing Once DNA fragments have been sorted out according to length, researchers can work out the nucleotide sequence of each type. The Sanger method of DNA sequencing will give you a sense of one of the ways this can be done (Figure 16.6).

Restriction fragments can be rapidly amplified by PCR and by large populations of bacterial or yeast cells.

Restriction fragments can be sorted out according to length, and their nucleotide sequences can be determined.

16.4 RFLP ANALYSIS

Imagine that a researcher has used restriction enzymes to break some DNA from one of your cells and now is subjecting the DNA to gel electrophoresis. As the fragments migrate through a slab of gel in response to an electric current, the large ones can't move through it as fast as small ones do. They separate from one another, forming bands in the gel. Now suppose she does the same to DNA samples from other people. When she compares the resulting banding patterns, she finds small variations among them. Variations in the banding patterns of DNA fragments from different individuals have a name. They are called **restriction fragment length polymorphisms**, or **RFLPs** for short. They arise because some base sequences in the DNA vary in molecular form from one person to the next. The molecular differences shift the number and location of sites where restriction enzymes make their cuts.

RFLP analysis is a wonderful new procedure for basic research. As the *Focus* essay suggests, it also has a few rather startling applications in society at large.

16.5 MODIFIED HOST CELLS

Use of DNA Probes

When you mix DNA with living cells, how can you find out which ones take up the DNA and contain a gene of interest? You can use **DNA probes**—short DNA sequences synthesized from radioactively labeled nucleotides (page 21). Part of the probe must be able to base-pair with part of the gene. Base-pairing between nucleotide sequences from different sources is called **nucleic acid hybridization**.

The first step is to "select" cells that have taken up the recombinant plasmids. As it happens, most plasmids contain genes that make their bacterial owners resistant to antibiotics. You put the prospective host cells on a culture medium that has the antibiotic added to it. The antibiotic prevents growth of all cells *except* the ones housing plasmids (because the plasmids carry the antibiotic-resistance genes).

The next step is to locate the particular cells that contain the recombinant plasmids with the gene being

Focus on Science

RIFF-lips and DNA Fingerprints

The DNA molecules of any two people are alike in most respects. Yet they also differ in some base sequences, and this affects the lengths of restriction fragments cut from the DNA. Comparisons of DNA electrophoresis patterns, such as the one in Figure *a*, reveal RFLPs (pronounced RIFF-lips), or "restriction fragment length polymorphisms" among individuals. In fact, each individual of any sexually reproducing species has a **DNA fingerprint**, a unique array of RFLPs, inherited from each parent in a Mendelian pattern.

RFLP analysis has uses in basic research, such as the human genome project. This is an effort to establish the nucleotide sequence of human DNA, with its 3.2 billion base pairs. As another example, evolutionary biologists have analyzed RFLPs to decipher DNA from mummies, from fossilized insects and plants, even from mammoths and humans that were preserved many thousands of years ago in glacial ice. Enterprising investigators used RFLP analysis to confirm that bones exhumed from a shallow pit in Siberia belonged to five members of the Russian imperial family, all shot to death in 1918.

RFLP analysis has medical applications. For example, researchers have identified unique restriction sites within and near several mutated genes, including those responsi-

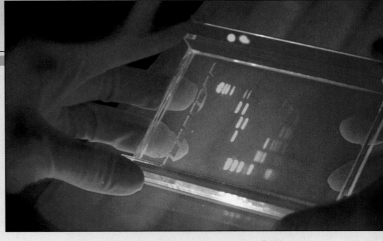

a Running from left to right, columns of separated DNA fragments (labeled with a dye that fluoresces *pink*). The banding pattern reflects differences in their lengths.

ble for sickle-cell anemia, cystic fibrosis, and other genetic disorders. The sites are used to determine whether an individual carries the mutated gene. Thus RFLP analysis can be used for prenatal diagnosis (page 204).

As another example, paternity and maternity cases can be resolved by carefully comparing the child's DNA fingerprint with the disputed parent's. As a final example, murderers can be identified if they lose even a few drops of blood at the scene of the crime or if drops of the victim's blood stain their clothing. A bloodstain may provide enough DNA to identify the perpetrator. Similarly, a rapist can be identified from semen recovered from the victim.

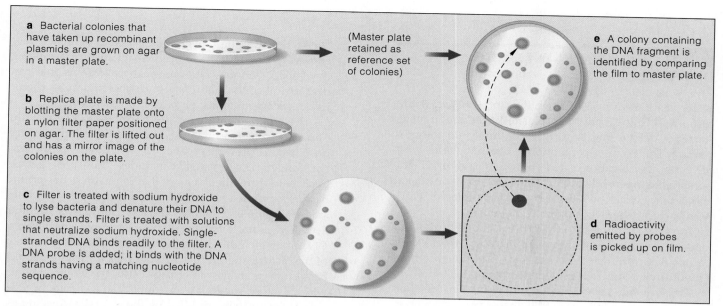

a Bacterial colonies that have taken up recombinant plasmids are grown on agar in a master plate.

b Replica plate is made by blotting the master plate onto a nylon filter paper positioned on agar. The filter is lifted out and has a mirror image of the colonies on the plate.

c Filter is treated with sodium hydroxide to lyse bacteria and denature their DNA to single strands. Filter is treated with solutions that neutralize sodium hydroxide. Single-stranded DNA binds readily to the filter. A DNA probe is added; it binds with the DNA strands having a matching nucleotide sequence.

(Master plate retained as reference set of colonies)

e A colony containing the DNA fragment is identified by comparing the film to master plate.

d Radioactivity emitted by probes is picked up on film.

Figure 16.7 Use of a DNA probe to identify a colony of bacterial cells that have taken up plasmids containing a specific DNA fragment.

studied. As the cells divide, they form colonies. Suppose the colonies are on agar (a gel-like substance) in a petri dish. You blot the agar against a nylon filter. Some cells stick to the filter in locations that mirror the locations of the original colonies (Figure 16.7). You use solutions to rupture the cells, fix the released DNA onto the filter, and make the double-stranded DNA molecules unwind. Then you add the DNA probes. The probes hybridize only with the gene region having the proper base sequence. Probe-hybridized DNA emits radioactivity and allows you to tag the colonies that harbor the gene of interest.

Use of cDNA

We study genes because we want to learn about or use their protein products. Even if a host bacterial cell takes up a gene, however, the protein product may not materialize. For example, recall that human genes contain noncoding sequences (introns). The mRNA transcripts of these genes cannot be used until the introns are snipped out and coding regions (exons) are spliced together into mature form. Bacterial enzymes don't recognize the splice signals, so bacterial host cells cannot always properly translate human DNA.

Sometimes this problem can be circumvented by using **cDNA**, which has been "copied" from a mature mRNA transcript for the desired gene from which the introns have been removed. By a backwards process called **reverse transcription**, a matching DNA strand is assembled on mRNA. The result is a hybrid DNA-RNA molecule (Figure 16.8). Enzymes remove the RNA, then

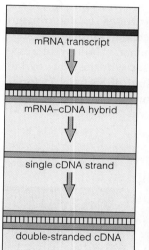

mRNA transcript

mRNA–cDNA hybrid

single cDNA strand

double-stranded cDNA

a An mRNA transcript of a desired gene is used as a template for assembling a DNA strand. An enzyme (reverse transcriptase) does the assembling.

b An mRNA-DNA hybrid molecule results.

c Enzyme action removes the mRNA and assembles a second strand of DNA on the remaining DNA strand.

d The result is double-stranded cDNA, "copied" from an mRNA template.

Figure 16.8 Formation of cDNA from an mRNA transcript.

they synthesize a complementary DNA strand to make double-stranded DNA.

Double-stranded cDNA can be further modified by the addition of a bacterial promoter and other signals that are required for transcription and translation. Then the modified cDNA can be inserted into a plasmid for amplification. The recombinant plasmids can be inserted into bacteria, which may then use the cDNA directions for making a protein.

Host cells that take up modified genes may be identified by procedures involving DNA probes.

Gene expression does not automatically follow the successful uptake of a modified gene by a host cell. Genes must be suitably engineered first.

16.6 GENETIC ENGINEERING OF BACTERIA

Many years have passed since the first transfer of foreign DNA into a bacterial plasmid, yet the transfer started a debate that is sure to continue into the next century. The point of contention is this: Do the benefits of gene modifications and transfers outweigh the potential dangers? Before coming to any conclusion about this, reflect on the following examples of work with bacteria, then with plants and animals.

Imagine a miniaturized factory that churns out insulin or another protein having medical value. This is an apt description for the huge stainless steel vats of genetically engineered, protein-producing bacteria.

Think of the diabetics who require insulin injections for as long as they live (page 623). Insulin is a protein hormone of the pancreas, and until the 1970s, medical supplies of it had to be obtained from cattle and pigs. Some diabetics developed allergic reactions to the foreign insulin. Also, the demand started outstripping the supplies. Then synthetic genes for human insulin were transferred into *E. coli* cells. (Can you say why the genes had to be synthesized?) This was the start of bacterial factories for human insulin and, later, growth hormone, hemoglobin, serum albumin, interferon, and other proteins.

Figure 16.9 Spraying an experimental strawberry patch in California with "ice-minus" bacteria. The sprayer used elaborate protective gear to meet government regulations in effect then.

At this writing, several other lines of genetically engineered bacteria have been established or are being developed. Among them are bacterial strains that can degrade oil spills, manufacture alcohol and other chemicals, process minerals, or leave crop plants alone.

The strains being used are harmless to begin with. They are grown in confined settings, behind barriers designed to prevent escape. As an added precaution, the foreign DNA usually includes "fail-safe" genes. Such genes are silent *unless* the engineered bacterium becomes exposed to conditions characteristic of the outside environment. Upon exposure, the genes become activated, with lethal results. For example, the foreign DNA may include a *hok* gene with an adjacent promoter of the lactose operon (page 236). Thus, if the engineered bacterium manages to escape into the environment, where lactose sugars are common, the *hok* gene is activated. The protein specified by the gene destroys membrane function and so destroys the cell.

Even so, there is concern about possible risks of introducing genetically engineered bacteria into the environment. Consider how Steven Lindow altered a bacterium that can make many crop plants less vulnerable to frost. Proteins at the bacterial surface promote the formation of ice crystals. Lindow excised the ice-forming gene from some cells. He hypothesized that spraying these "ice-minus bacteria" on strawberry plants in an isolated field just before a frost would make the plants more resistant to freezing.

Here was an organism from which a harmful gene had been deleted, yet it triggered a bitter legal debate on the risks of releasing genetically engineered microbes into the environment. The courts finally ruled in favor of allowing the genetically engineered bacteria to be released, and researchers sprayed a small patch of strawberries (Figure 16.9). Nothing bad happened. Since then, rules governing the release of genetically engineered organisms have become less restrictive.

16.7 GENETIC ENGINEERING OF PLANTS

Years ago, Frederick Steward and his coworkers cultured cells of carrot plants and induced the cells to grow into small embryos. Some embryos actually grew into whole plants. Today many plant species, including major crop plants, are regenerated from cultured cells. The culturing methods increase mutation rates, so the cultures are a source of genetic modifications.

Researchers can pinpoint a useful mutation among millions of cells. Suppose a culture medium contains a toxin that is produced by a disease agent. If a few cells have a mutated gene that confers resistance to the toxin, they will end up being the only live cells in the culture.

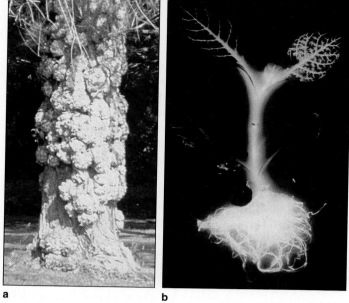

a

b

Figure 16.10 Gene transfers in plants. **(a)** A tumor-inducing plasmid from a common bacterium (*Agrobacterium tumefaciens*) causes crown gall tumors on willow trees and other woody plants. Scientists put this plasmid to work in cultures of plant cells, where it moves desirable genes into the plant chromosomes. **(b)** Evidence of a successful gene transfer: a modified tobacco plant that glows in the dark. A gene from a firefly has become incorporated into its own DNA and is being translated into functional protein, which in this case is the enzyme luciferase.

a

b

c

d

Figure 16.11 Examples of commercially valuable plants that benefit from genetic engineering. **(a)** Buds of cotton plants are vulnerable to worm attack. **(b)** Buds of a modified cotton plant resist attack. **(c)** A virus damages potato plants, but a modified plant **(d)** resists attack.

Now suppose plants are regenerated from the cells, then hybridized with other varieties. The hybrid plants may end up with the new gene that confers disease resistance.

Today, researchers are successfully inserting genes into cultured plant cells. For example, they have inserted DNA fragments into the "Ti" plasmid from *Agrobacterium tumefaciens*, a bacterium that infects many flowering plants. Some plasmid genes become integrated into the DNA of infected plants, and they induce the formation of crown gall tumors (Figure 16.10*a*). Before the plasmid is introduced into a plant, the tumor-inducing genes are removed from it and desired genes are inserted into it. Then the genetically modified bacterial cells are grown with cultured plant cells. Selected plants that are regenerated from infected cultures contain the foreign genes within their DNA. In some cases, the foreign genes are expressed in the plant tissues, with observable effects.

Vivid evidence of a successful gene transfer came from researchers who used *A. tumefaciens* to deliver a firefly gene into cultured tobacco plant cells. The gene codes for luciferase, an enzyme required for bioluminescence (page 103). Plants regenerated from the infected culture cells have the peculiar ability to glow in the dark (Figure 16.10*b*).

A. tumefaciens only infects the plants called dicots. Wheat, corn, rice, oats, and other major food crop plants are monocots. In some cases, genetic engineers use chemicals or electric shocks to deliver DNA directly into protoplasts (plant cells stripped of their walls). For some species, however, regenerating whole plants from protoplast cultures is not yet possible.

Not long ago, someone came up with the idea to deliver genes into cultured plant cells by shooting them with a pistol. Instead of bullets, blanks are used to drive DNA-coated, microscopic tungsten particles into the cells. Although this "gene gun" might seem analogous to using a battleship cannon to light a match, the shooters are reporting some success.

Despite the obstacles, many improved varieties of crop plants have been developed or are on the horizon. For example, certain cotton plants have been genetically engineered for resistance to worm attacks (Figure 16.11*b*). Such gene insertions are ecologically safer than pesticide applications. They kill only the targeted pest and do not interfere with beneficial insects, including the ladybird beetles that prey on aphids. Also on the horizon are genetically engineered plants that may serve as pharmaceutical factories. Two years ago, genetically engineered tobacco plants that produce human hemoglobin, melanin, and other proteins were planted in a field in North Carolina on a trial basis. A year after the trial was completed, ecologists found no trace of the foreign genes or proteins in the soil or in any plants or animals in the vicinity.

16.8 GENETIC ENGINEERING OF ANIMALS

Supermice and Biotech Barnyards

Mice were the first mammals subjected to genetic engineering experiments. Consider how R. Hammer, R. Palmiter, and R. Brinster managed to correct a hormone deficiency that causes dwarfism in mice. Such mice have trouble producing somatotropin (also called growth hormone). The researchers used a microneedle to inject the gene for rat somatotropin into fertilized mouse eggs, then they implanted the eggs into an adult female. The gene was successfully integrated into the mouse DNA. The baby mice in which the foreign gene was expressed were 1-1/2 times larger than their dwarf littermates. In other experiments, the gene for human somatotropin was transferred into a mouse embryo, where it became integrated into the DNA. A "supermouse" resulted (Figure 16.12).

Today, as part of research into the genetic basis of Alzheimer disease and other genetic disorders, several other human genes are being inserted into mouse embryos. Besides microneedles, microscopic laser beams are being used to open temporary holes in the plasma membrane of cultured cells, although such

Figure 16.12 Ten-week-old mouse littermates, the one on the left weighing 29 grams, and the one on the right, 44 grams. The larger mouse grew from a fertilized egg into which the gene for human somatotropin (growth hormone) had been inserted.

methods have varying degrees of success. Retroviruses also are used to insert genes into cultured cells, virtually all of which incorporate the foreign genes into their DNA. However, the genetic material of retroviruses often undergoes rearrangments, deletions, and other alterations that render the introduced genes ineffective. There is also the possibility that viral particles can escape and infect other individuals.

Soon, "biotech barnyards" may be competing with bacterial factories as genetically engineered sources of proteins. Farm animals produce the proteins in far greater quantities, at less cost. Consider Herman, a Holstein bull that received the human gene for lactoferrin, a milk protein, when he was just an embryo. His female offspring may mass-produce the protein, which can be used as a supplement for infant formulas. Similarly, goats are providing CFTR protein (used in the treatment of cystic fibrosis), and TPA (which lessens the severity of heart attacks). Sheep are producing alpha-1 antitrypsin, used in the treatment of emphysema. Cattle may soon be producing human collagen, a key component of skin, cartilage, and bone.

Applying the New Technology to Humans

Researchers in laboratories throughout the world are now working their way through the 3.2 billion base pairs that are present in the twenty-three pairs of human chromosomes. This ambitious effort is called the **human genome project**. Some researchers are working on specific chromosomes, others on specific gene regions only. For example, rather than busying themselves with noncoding sequences (introns), J. Venter and Sidney Brenner are isolating mRNAs from brain cells and using them to make cDNA. By sequencing only the cDNAs, they already have identified hundreds of previously unknown genes.

About 99.9 percent of the nucleotide sequence is the same in every human on earth. Thus, once the sequencing project is completed, we will have the ultimate reference book on human biology and genetic disorders. (Or should we say reference *books*—the complete sequence will fill the equivalent of 200 Manhattan telephone directories.)

What will we do with the information? Certainly we will use it in the search for effective treatments and cures for genetic disorders. Of 2,000 or so genes studied so far, 400 already have been linked to genetic disorders. The knowledge opens doors to **gene therapy**, the transfer of one or more normal or modified genes into body cells of an individual to correct a genetic defect or boost resistance to disease. But what about forms of human gene expression that are neither disabling nor life-threatening? Will we tinker with these, also?

Some Implications of Human Gene Therapy

Recombinant DNA technology and genetic engineering are advancing rapidly. We are only beginning to work our way through their social and ethical implications.

To most of us, human gene therapy to correct genetic abnormalities seems like a socially acceptable goal. Is it also socially acceptable to insert genes into a *normal* human individual (or sperm or egg) to alter or enhance traits? The idea of selecting desirable human traits is called *eugenic engineering*. Yet who decides which forms of a trait are most "desirable"? What if prospective parents could pick the sex of a child by way of genetic engineering? Three-fourths of one survey group said they would choose a boy. So what would be the long-term social implications of a drastic shortage of girls?

Would it be okay to engineer taller or blue-eyed or curlier-haired individuals? If so, would it be okay to engineer "superhuman" offspring with exceptional strength or breathtaking intelligence? Suppose a person of average intelligence moved into a town composed of 800 Einsteins. Would the response go beyond a few mutterings of "There goes the neighborhood"?

Some say that the DNA of any organism must never be altered. Put aside the fact that nature itself alters DNA much of the time. The concern is that we don't have the wisdom to bring about beneficial changes without causing harm to ourselves or to the environment.

When it comes to manipulating human genes, one is reminded of our human tendency to leap before we look. When it comes to restricting genetic modifications of any sort, one also is reminded of an old saying: "If God had wanted us to fly, he would have given us wings." And yet, something about the human experience gave us the *capacity* to imagine wings of our own making—and that capacity has carried us to the frontiers of space.

Where are we going from here with recombinant DNA technology, this new product of our imagination? To gain perspective on the question, spend some time reading the history of our species. It is a history of survival in the face of all manner of new challenges, threats, bumblings, and sometimes disasters on a grand scale. It is also a story of our connectedness with the environment and with one another.

The questions confronting you today are these: Should we be more cautious, believing that one day the risk takers may go too far? And what do we as a species stand to lose if the risks are *not* taken?

This chapter opened with the first human application of this technology, meant to save the life of a four-year-old girl. It closes with a *Focus* essay that invites you to consider some prospects and problems associated with the application of recombinant DNA technology to our rapidly advancing knowledge of the human genome.

As with any new technology, the potential benefits of recombinant DNA technology and genetic engineering should be weighed carefully against the potential risks.

SUMMARY

1. Gene mutations and other genetic "experiments" have been going on in nature for billions of years. Through artificial selection, humans have been manipulating the genetic character of different species for many thousands of years. Today, recombinant DNA technology has enormously expanded our capacity to modify organisms genetically.

2. With recombinant DNA technology, researchers isolate, cut, and splice together gene regions from different species, then greatly amplify the number of copies of those regions into usefully large quantities. Enzymes make the cuts and do the splicing.

a. Researchers use restriction enzymes and DNA ligase to cut and insert DNA into plasmids from bacterial cells. (The plasmids are small, circular DNA molecules that contain extra genes, besides those of the bacterial chromosome.) A DNA library is a collection of DNA fragments, produced by restriction enzymes, and incorporated into plasmids.

b. Restriction fragments may be amplified by a population of rapidly dividing cells (such as bacteria or yeasts). Short DNA fragments may be amplified more rapidly in a test tube by the polymerase chain reaction (PCR).

c. Following amplification, the DNA fragments can be sorted out (according to length), and their nucleotide sequences can be determined.

3. Each individual of a species has a DNA fingerprint: a unique array of RFLPs (restriction fragment length polymorphisms). Molecular variations in the base sequence of their DNA lead to small, identifiable differences in restriction fragments cut from the DNA.

4. Cells that take up foreign genes can be identified by DNA probes, which base-pair with the genes of interest. Base-pairing of nucleotide sequences from different sources is called nucleic acid hybridization.

5. In genetic engineering, genes are isolated, modified, and inserted into the same organism or a different one. In gene therapy, genes are inserted into an individual to correct a genetic defect.

6. Recombinant DNA technology and genetic engineering have enormous potential for research and applications in medicine, agriculture, the home, and industry. As with any new technology, potential benefits must be weighed against potential risks, including ecological and social.

Review Questions

1. In which type of organisms are restriction enzymes produced naturally? What is the function of these enzymes in nature? *249*

2. Recombinant DNA technology involves producing DNA restriction fragments, amplifying DNA, and identifying modified host cells. Briefly describe some examples of how these activities are carried out. *249–250; 252–253*

3. Name two enzymes used in recombinant DNA technology and define their function. *249–250*

4. Besides this chapter's examples, list what you believe might be some potential benefits and risks of genetic engineering. *254–257*

Self-Quiz *(Answers in Appendix IV)*

1. _____ are small circles of bacterial DNA that are separate from the bacterial chromosome.

2. DNA fragments result when _____ cut DNA molecules at specific sites.
   a. DNA polymerases   c. restriction enzymes
   b. DNA probes   d. RFLPs

3. Recombinant DNA technology involves _____ .
   a. producing DNA fragments   d. all of the above
   b. making DNA libraries   e. a and c
   c. amplifying DNA

4. PCR stands for _____ .
   a. polymerase chain reaction
   b. polyploid chromosome restrictions
   c. polygraphed criminal rating
   d. politically correct research

5. A _____ is a collection of DNA fragments, produced by restriction enzymes and incorporated into plasmids.
   a. DNA clone   c. DNA probe
   b. DNA library   d. gene map

6. A _____ is a collection of multiple, identical copies of DNA fragments.
   a. DNA clone   c. DNA probe
   b. DNA library   d. gene map

7. In reverse transcription, _____ is assembled on _____ .
   a. mRNA; DNA   c. DNA; enzymes
   b. DNA; mRNA   d. DNA; agar

8. _____ is the transfer of normal genes into body cells to correct a genetic defect.
   a. Reverse transcription   c. Gene mutation
   b. Nucleic acid hybridization d. Gene therapy

9. Tobacco plant leaves that produce hemoglobin are a result of _____ .
   a. gene therapy   c. pressure on tobacco growers
   b. genetic engineering   d. a and b

10. Match the terms appropriately.
    _____ DNA library   a. mutation, crossing over
    _____ plasmid   b. raises social, legal, and
    _____ nature's genetic   ethical questions
          experiments   c. rapid DNA amplification
    _____ polymerase chain   d. cut DNA fragments
          reaction   incorporated into plasmids
    _____ human gene therapy   e. extra bacterial genes

Selected Key Terms

cDNA *253*
DNA fingerprint *252*
DNA library *249*
DNA ligase *249*
DNA polymerase *250*
DNA probe *252*
gene therapy *256*
genetic engineering *248*

human genome project *256*
nucleic acid hybridization *252*
plasmid *248*
polymerase chain reaction (PCR) *250*
recombinant DNA technology *248*
restriction fragment length
   polymorphism (RFLP) *252*
reverse transcription *253*

Readings

Anderson, W. F. 1992. Human Gene Therapy. *Science* 256:808–813.

Gasser, C. S. and R. T. Fraley. 1989. "Genetically Engineering Plants for Crop Improvement." *Science* 244:1293–1299.

Joyce, G. December 1992. "Directed Molecular Evolution." *Scientific American* 267(6):90–97.

Pursel, V. G. et al. 1989. "Genetic Engineering of Livestock." *Science* 244:1281–1288.

Watson, J. D. 1990. "The Human Genome Project: Past, Present, and Future." *Science* 248:44–49.

White, R., and J. Lalouel. February 1988. "Chromosome Mapping with DNA Markers." *Scientific American* 258(2):40–48.

FACING PAGE: *Millions of years ago, a bony fish died, and sediments gradually buried it. Today its fossilized remains are studied as one more piece of the evolutionary puzzle.*

17 EMERGENCE OF EVOLUTIONARY THOUGHT

Fire, Brimstone, and Human History

With this unit, we turn to evolutionary theories and the ways in which they can be used to interpret the past and present, even to predict possible futures for the natural world. Today the theories are widely accepted, but this wasn't the case when naturalists in Europe first started working out the details. Those early evolutionists rankled more than a few people with their astounding explanations, which tied together the geologic record, the fossil record, and the sweep of existing biological diversity.

It was as if they shook out bits of recorded history from the Bible, the premier book of the Western world, and assembled them into a radically different story. For generation after generation, that book had been helping people cope with the uncertainties of life, with death, and with unpredictable forces of nature.

People did have a lot to worry about. Consider Thera, a crescent-shaped island in the Mediterranean waters between Greece and Crete. We now know that Thera wraps around a submerged crater 400 meters deep. The crescent above the water's surface is a remnant of a volcano that blew apart approximately 3,500 years ago. The eruption was so violent, it generated seismic sea waves that were probably 100 meters high. Within twenty minutes, the giant waves slammed into the island of Crete to the south (Figure 17.1). Most likely, this catastrophe brought about the abrupt collapse of the Minoan civilization on Crete, the earliest civilization in the history of Europe. Knossos, the Minoan capital, was virtually leveled at about the same time as the eruption. Prevailing winds carrying huge volumes of volcanic ash across the island would have darkened the skies for days, further terrifying those who survived the deluge.

Or consider the Jordan Valley, a long depression in the earth's crust that lies between the Red Sea to the south and the Dead Sea to the north. Around 4,000 years ago, the notorious cities of Sodom and Gomorrah apparently flourished in this valley, at the south end of the Dead Sea. By Biblical account, "brimstone and fire rained upon the cities . . . and the smoke of the land went up like the smoke of a furnace." Both cities were destroyed, but the stories of the horrified survivors became part of recorded history.

Today, scientists look at satellite images of the straight walls of the Jordan Valley. They see evidence of hot springs, past and present. They see evidence of ancient, great lava flows and violent earth tremors.

Taken together, the straight valley walls, hot springs, lava flows, and earthquakes are signs of deep cracks (faults) in the earth's crust. Severe earthquakes must have tilted part of the crust along the fault at the southern end of the Dead Sea. In what surely must have been one of the all-time nightmares, the earth heaved, incandescent lava poured out from the depths, hot springs spewed forth a sulfurous brew—and the violently displaced waters of the sea flooded Sodom and Gomorrah.

Catastrophic events of this sort certainly would have impressed people of the early Mediterranean civilizations. After all, they still scare us, even though we have faith that seismographs, satellite imaging, and other technological advances will help us understand why the earth trembles and skies darken and giant waves sweep across oceans.

So imagine yourself living 2,000 years ago, ignorant of the geologic forces responsible for huge floods, lava flows, and other natural disasters. How would you have interpreted what was going on? Most likely then, as now, *your interpretation would have been shaped by the prevailing beliefs of your society*.

Floods, fire, and brimstone were interpreted as punishment for bad human behavior. When some people were particularly obnoxious, their cities were obliterated. When nearly everybody was heading in a bad direction, a great flood submerged the entire world. When the floodwaters receded, countless bones and shells were buried in mud and turned into fossils.

Biological science emerged within the framework of such prevailing beliefs. So did awareness of change not only in the earth, but also in its creatures. Understand this fact, and perhaps you will see why acceptance of the very idea of biological evolution was so long in coming.

Figure 17.1 Map showing the location of Thera, a volcanic island in the eastern Mediterranean Sea, and the area most devastated by its violent eruption in prehistoric times. The photograph shows an incandescent lava flow from a modern-day volcanic eruption.

1. We define biological evolution as heritable changes in lineages (lines of descent). Evidence of evolution comes from investigations that began nearly two centuries ago.

2. Through some investigations, relationships were discerned among major groups of animals, based on comparisons of body structure and patterning. At the same time, explorers discovered differences in the world distribution of species that could not be explained unless those species had evolved in different places. Besides this, geologists discovered apparent sequences of changing fossils in distinct layers of the earth.

3. As Charles Darwin and Alfred Wallace perceived long ago, members of the same population share the same heritable traits—yet they don't look or act exactly the same, and they don't all survive and reproduce. Differences in survival and reproduction among individuals that differ in one or more traits are the basis of natural selection.

4. As we now know, variation in any heritable trait results when members differ in their alleles for the trait. A population changes (evolves) when some alleles increase in frequency and others decrease or disappear over time, as by natural selection.

a

Figure 17.2 Examples of three species that are native to three geographically separate parts of the world. (**a**) The emu of Australia, (**b**) rhea of South America, and (**c**) ostrich of Africa. All three species of birds have features in common.

17.1 EARLY BELIEFS AND CONFOUNDING DISCOVERIES

The Great Chain of Being

More than 2,000 years ago, the seeds of biological inquiry were taking hold among the ancient Greeks. At the time, popular belief held that supernatural beings intervened directly in human affairs. For example, the gods were said to cause a common ailment known as the sacred disease. Yet from a physician of the school of Hippocrates, these thoughts come down to us:

It seems to me that the disease called sacred . . . has a natural cause, as other diseases have. Men think it divine merely because they do not understand it. But if they called everything divine that they did not understand, there would be no end of divine things! . . . If you watch these fellows treating the disease, you see them use all kinds of incantations and magic—but they are also very careful in regulating diet. Now if food makes the disease better or worse, how can they say it is the gods who do this? . . . It does not really matter whether you call such things divine or not. In Nature, all things are alike in this, in that they can be traced to preceding causes.

— On the Sacred Disease (400 B.C.)

Passages such as this reflect the early attempts to find natural explanations for observable events.

Aristotle was foremost among the early naturalists, and he described the world around him in excellent detail. He had no reference books or instruments to guide him, for biological science in the Western world began with the great thinkers of this age. Yet here was a man who was no mere collector of random bits of information. In his descriptions is evidence of a mind perceiving connections between observations and attempting to explain the order of things.

Aristotle believed (as did others) that each kind of organism was distinct from all the rest. Yet he wondered about organisms that seemed to have rather blurred positions in nature. For example, even though some sponges look like plants, they have traits that make them animals. In time, Aristotle came to view nature as a continuum of organization, from lifeless matter through complex forms of plant and animal life.

By the fourteenth century, this idea had become transformed into a rigid view of life. A great Chain of Being was seen to extend from the lowest forms of life to humans and on to spiritual beings. Each kind of being, or *species* as it was called, was a separate link in the chain. All the links were designed and forged at the same time, at the same center of creation, and had not changed since. Scholars thought that once they had discovered, named, and described all the links, the meaning of life would be revealed to them.

Questions from Biogeography

As long as naturalists were unaware that the world is a great deal bigger than Europe, the task of locating and describing all species seemed manageable. Then, with the global explorations of the sixteenth century, "the world" expanded enormously. Naturalists were soon overwhelmed by descriptions of thousands of plants and animals discovered in Asia, Africa, the Pacific Islands, and the New World.

The naturalist Thomas Moufet, in attempting to sort through the bewildering array, simply gave up and recorded such gems as this description of grasshoppers and locusts: "Some are green, some black, some blue. Some fly with one pair of wings; others with more; those that have no wings they leap; those that cannot fly or leap they walk; some have long shanks, some shorter. Some there are that sing, others are silent. . . . " This passage comes from his *Theater of Insects*, written in 1590. It was not exactly a time of subtle distinctions.

Even so, a few scholars began to examine the world distribution of plants and animals, a discipline now called **biogeography**. It became apparent that many species are unique to oceanic islands and other isolated places. It also became apparent that some similar-looking species live in places that are separated by impenetrable mountains and vast oceans (Figure 17.2). *How did so many species get from the center of creation to islands and other isolated places?*

Questions from Comparative Anatomy

In the eighteenth century, scholars were noticing similarities and differences in the body plans of different

b

c

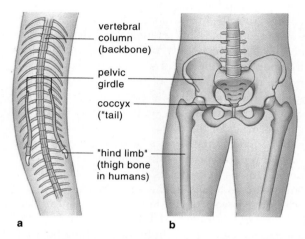

vertebral
column
(backbone)

pelvic
girdle

coccyx
("tail)

"hind limb"
(thigh bone
in humans)

a   b

Figure 17.3 (**a**) Python bones corresponding to the pelvic girdle of other vertebrates, including humans (**b**). Small "hind limbs" protrude through skin on the underside of the snake.

mammals, reptiles, and other major groups, a discipline called **comparative anatomy**. Consider how a human arm, whale flipper, and bat wing differ in size, shape, and function. All three body parts have similar body locations. They are composed of the same tissues, arranged in the same patterns. They develop similarly in embryos. Comparative anatomists who discovered all of this wondered: Why are animals as different as humans, whales, and bats so much alike?

According to one explanation, some body plans were so perfect there was no need to come up with new ones for each organism at the time of creation. Yet if that were true, then how could there be body parts with no apparent function at all? For instance, certain snakes have bones that correspond to a pelvic girdle, a set of bones to which hind legs attach (Figure 17.3). Snakes don't have legs, so why are the bones there? Similarly, humans have bones corresponding to a few of the tail bones of many other mammals. Humans don't have a tail. Why do they have parts of one?

Questions About Fossils

Geologists added to the confusion. By midcentury, they were mapping **sedimentary beds** at sites where erosion or quarrying had cut deep into the earth. They found similar beds throughout the world. Such beds consist of distinct, multiple layers of sedimentary rock. (Figure 19.3 shows an example of this.) Most geologists agreed that sand and other sediments were deposited at these sites at different times and gradually were compacted into rock layers. They soon realized that certain layers contain a characteristic array of **fossils**, which are evidence of organisms that lived in the past. For example, very deep layers contained fossils of very simple marine organisms. Layers above these contained fossils of organisms that were more complex in their structure. In layers above these, fossils closely resembled living

marine organisms. Were these *sequences* of fossils? Were they evidence of organisms that were somehow connected through geologic time?

Taken together, the findings from biogeography, comparative anatomy, and geology did not fit well with prevailing beliefs. George-Louis Leclerc de Buffon and a few others started formulating some rather startling hypotheses. If dispersal of species from a single center of creation were not possible, given the oceans and other barriers, *then perhaps species originated in more than one place.* If species were not created in a perfect state, as suggested by the layering of different fossils and the occurrence of "useless" body parts in various organisms, *then perhaps species became modified over time.* Awareness of **evolution**—of changes in lines of descent—was in the wind.

Awareness of evolution emerged over centuries, through the cumulative observations of naturalists, biogeographers, comparative anatomists, and geologists.

17.2 A FLURRY OF NEW THEORIES

In the nineteenth-century, naturalists tried to reconcile the growing evidence of change in lines of descent with a traditional conceptual framework that did not allow for change. Foremost among them was Georges Cuvier, a respected anatomist.

For many years, Cuvier had compared body plans of fossils and living organisms. He acknowledged that there were abrupt changes in the fossil record, corresponding to discontinuities between certain layers of sedimentary beds. Were these evidence of changes in the populations of organisms that inhabited ancient environments? Cuvier thought so. And he was right, as you will see from the evolutionary story in Chapter 21.

Cuvier incorporated his inference in a theory of **catastrophism**. There was only one time of creation, he said, that populated the world with all species. A global catastrophe destroyed many of them. A few survivors repopulated the world. It was not that the survivors were new species; naturalists simply hadn't yet discovered earlier fossils of them that would date to the time of creation. Later catastrophes wiped out still more species, and repopulation by the survivors followed, as recorded by fossils in the rocks.

Many scholars accepted Cuvier's theory, yet others kept chipping away at the puzzle. Consider Jean-Baptiste Lamarck's **theory of inheritance of acquired characteristics**. During an individual's life, he said, environmental pressures and internal "desires" bring about permanent changes in the body, and offspring inherit the desired changes. Thus life, created long ago in a simple state, has gradually improved. The force for change is a drive for perfection, up the Chain of Being. The drive is centered in nerves that directed an unknown "fluida" to body parts in need of change.

For instance, suppose the ancestor of modern giraffes was a short-necked animal. Pressed by the need to find food, it constantly stretched its neck to browse on leaves beyond the reach of other animals. Stretching directed fluida to its neck, which lengthened permanently. The longer neck was bestowed on offspring, which stretched their necks also. Thus generations of animals desiring to reach higher leaves led to the modern giraffe.

Larmarck had inferred, correctly, that the environment is a factor in evolution. But his overall theory, like others proposed at this time, never was supported by findings from any of the investigations carried out since then. Yet it is useful to reflect on these theories, because they reinforce this point:

Prevailing beliefs can influence how we interpret clues to natural processes and their observable outcomes.

17.3 VOYAGE OF THE *BEAGLE*

In 1831, in the midst of the confusion, Charles Darwin was twenty-two years old and wondering what to do with his life. Ever since he was eight, he had wanted to hunt, fish, collect shells, or just watch insects and birds—anything but sit in school. Later, at his father's insistence, he tried to study medicine in college. The crude, painful procedures used on patients at that time sickened him. Then his father urged him to become a clergyman, so Darwin packed for Cambridge. His grades were good enough to earn him a degree in theology. But he spent most of his time among faculty members with leanings toward natural history.

John Henslow, a botanist, perceived Darwin's real interests. He arranged for Darwin to become ship's naturalist aboard H.M.S. *Beagle*. Shortly thereafter, the *Beagle* set out on a five-year voyage that would take Darwin around the world (Figure 17.4). The young man who hated work, who had no formal training as a naturalist, suddenly began to work enthusiastically. In time he would even formulate a theory—and it would become the centerpiece of modern biology.

a

Figure 17.4 (**a**) Charles Darwin and a blue-footed booby, one of the species he encountered during his five-year voyage around the world on H.M.S. *Beagle*. A replica of this ship is shown in (**b**), sailing off the coast of South America. (**c**) During stops along Argentina's coast, Darwin explored parts of the Andes. He saw fossils of marine organisms embedded in rocks 3.6 kilometers above sea level. (**d–f**) The Galápagos Islands, isolated bits of land in the ocean, far to the west of Ecuador. We now know they arose through volcanic action about 5 million years ago, so organisms could not have originated there; winds or ocean currents must have brought them.

b

c

d

Darwin

Wolf

Pinta

Marchena   Genovesa

EQUATOR

Santiago

Bartolomé

Seymour

Rábida   Baltra

Fernandina   Pinzón

Santa Cruz

Santa Fe

Tortuga   San Cristóbal

Isabela   Española

Floreana

e

EQUATOR

Galápagos
Islands

f

17.4 DARWIN'S THEORY TAKES FORM

A Clue from Geology

The *Beagle* sailed first to South America, where work could be completed on mapping the coastline. During the Atlantic crossing, Darwin collected and examined marine life. He read Henslow's parting gift, the first volume of Charles Lyell's *Principles of Geology*. During stops along the coast and at various islands, he observed species in environments ranging from sandy shores to high mountains. And he started circling the question of evolving life, now on the mind of many respected individuals—including his own grandfather.

This was the time when Darwin started mulling over a theory that Lyell was advancing in his book. According to the **theory of uniformity**, the same geologic processes that are now slowly sculpting the earth's surface also were at work in the past. Over millions of years, mountain ranges rose from ancient seafloors. Seasonal rains and winds eroded mountain flanks, and layers of sediments slowly accumulated in valleys and seafloors. Major earthquakes and other catastrophes were like abrupt punctuation points in this history of gradual geologic change.

The theory bothered many scholars, who believed the earth was less than 6,000 years old. They thought people had recorded everything that happened during those thousands of years, and in all that time, no one had ever mentioned seeing a species evolve. However, by Lyell's calculations, it must have taken millions of years to mold the present landscape. Wasn't that enough time for species to evolve in diverse ways?

Old Bones and Armadillos

After returning to England in 1836, Darwin talked with other naturalists and studied his notes for evidence of evolving life. He identified some possibilities. For example, while in Argentina, he had observed fossils of glyptodonts—peculiar animals, now extinct. Of all animals on earth, only living armadillos bear resemblances to glyptodonts (Figure 17.5). And of all

places on earth, armadillos live only in the same places where glyptodonts once lived. If two kinds of animals had been created at the same time, lived in the same place, and were so much alike, why is only one kind still alive? Wouldn't it be reasonable to assume that armadillos were the descendants of glyptodonts? If so, many of the same traits were retained through the generations. But other traits were modified, over a long span of time. *Descent with modification*—it did seem possible.

Darwin's Insight—Natural Selection of Adaptive Versions of Heritable Traits

While Darwin was assessing his notes, an influential essay by Thomas Malthus, a clergyman and economist, made him take a closer look at a topic of current social interest. The essay correlated famine, disease, and war with population size. According to Malthus, people tend to produce children faster than food supplies, living space, and other resources can be sustained. The larger the population becomes, the more individuals there are to reproduce in each generation. Population size burgeons, resources dwindle—and the struggle for existence heightens. Many people starve, get sick, and engage in war and other forms of competition for the remaining resources.

From his own observations, Darwin suspected that all populations have the capacity to produce more individuals than the environment can support. To give a simple example of this, a single sea star can release 2,500,000 eggs each year, but the seas obviously are not filled with sea stars. In each generation, nearly all of the offspring end up in the belly of predators. Others starve or succumb to some other environmental insult.

Assuming that the environment keeps the number of reproducing individuals in check, *which* individuals are the winners and losers? What might affect the outcome? Darwin thought about the plant and animal populations he had observed during his voyage. He recalled that individuals in those populations varied in size, coloration, and other traits. *And it dawned on him that the variations in traits might affect an individual's ability to secure resources—and therefore to survive and reproduce—in particular environments.*

a

b

Figure 17.5 (**a**) An armadillo. (**b**) Reconstruction of a glyptodont. The shared, unusual features and restricted geographic distribution of these two animals provided Darwin with a clue that helped him develop a theory of evolution by natural selection.

Figure 17.6 Examples of different finch species of the Galápagos Islands.

(**a**) *Geospiza conirostris* and (**b**) *G. scandens*, two species that have a bill adapted for eating cactus flowers and fruits. Other finches have thick, strong bills adapted for crushing cactus seeds. (**c**) *Certhidea olivacea*, a tiny tree-dwelling finch, resembles warblers in song and behavior. It uses its slender beak to probe for insects. (**d**) *Camarhynchus pallidus* feeds on wood-boring insects such as termites. It swings its small body like a woodpecker does, to hammer at bark. It does not have the woodpecker's long, probing tongue. It has learned to break cactus spines and twigs to suitable lengths, then hold the "tools" in its beak and use them as probes.

Would the Galápagos Islands provide evidence of this? About 900 kilometers of open ocean separate these volcanic islands from the South American coast. They provide a variety of habitats along rocky shorelines, in deserts, and on mountain flanks. Most of their inhabitants live nowhere else in the world—although they resemble species on the mainland. Were the islands colonized by species that flew, floated, or were blown over from the mainland? If so, island-hopping descendants of the colonizers that became established in different habitats could have become adapted over time to the local conditions.

Through conversations with other naturalists, Darwin learned that populations of thirteen species of finches were distributed among the Galápagos Islands. He himself had collected specimens of these, and he now attempted to correlate some of their traits with environmental challenges. For example, all the birds of one population have a large, strong bill suitable for cracking seeds (Figure 17.6). Yet some birds have a bill that is a bit stronger—so they can crack open seeds that other birds find too tough to deal with. If most of the seeds being produced in a particular environment have hard coats, a strong-billed bird will be more likely to survive and reproduce. Assuming the trait has a heritable basis, the same will be true of its strong-billed descendants. If factors in the natural environment continue to "select" the most adaptive version of the trait, the population may become composed of mostly strong-billed birds. *And a population is evolving when its heritable traits are changing through successive generations.*

Recall, from Chapter 1, that Darwin used pigeon breeding and other examples of artificial selection as a way to explain selection of adaptive traits in nature. When breeders favor pigeons with black tail feathers, every newly hatched black-tailed pigeon will be allowed to mate, but the white-tailed ones might be destined for a stewpot. It was an easy way to show how their selections could cause an increase in the frequency of one version of a trait in the captive population, and how they might eliminate the other version.

Putting all of these observations and inferences together, Darwin came up with a theory of evolution by natural selection:

Observation: All natural populations have the reproductive capacity to exceed the resources that sustain them, yet they cannot increase in size indefinitely without running out of food, living space, and other resources.

Inference: Because more individuals are produced than can survive to reproductive age, the members of a population must compete for the available resources.

Observation: Individuals of a population differ in the details of their form, function, and behavior. Much of the variation is heritable (parents bestow their version of a given trait on their offspring).

Observation: Some forms of heritable traits are more adaptive than others in the competition for resources.

Inference: Because adaptive traits promote survival and reproduction, they must increase in frequency over the generations, and less adaptive traits must decrease in frequency or disappear.

Inference: A population can evolve by **natural selection**: The traits characterizing the population can change over time when its individuals differ in one or more heritable traits that are responsible for differences in survival and reproduction.

17.5 REGARDING WALLACE

Anticipating that his theory of evolution by natural selection would generate storms of controversy, Darwin waited to announce it and laboriously searched for flaws in his reasoning. He waited too long. In 1858—more than a decade after he wrote up but never published an essay on his theory—Darwin received a paper from the naturalist Alfred Wallace (Figure 17.7). Independently, Wallace had developed the same theory, then quickly wrote up and circulated his ideas. Even so, Darwin's colleagues prevailed upon him to formally present a paper along with Wallace's (who also believed that Darwin deserved most of the credit). The next year Darwin's detailed evidence in support of the theory was published in book form.

Figure 17.7 Alfred Wallace, who studied in Southeast Asia. More than a decade after Darwin wrote up but did not publish his theory of natural selection, Wallace independently developed—and circulated—the same theory.

17.6 REGARDING THE "MISSING LINKS"

Darwin's theory faced a crucial test. If he were correct, then all existing species are related, by way of descent, to ancestral species. Given the rapid outcomes of artificial selection, *natural* selection, operating on populations in specific environments, could explain the evolution of one kind of organism into another. Darwin saw this as happening gradually, in small increments, over hundreds or thousands of generations.

But where were the "missing links"? If the theory were correct, then there would have to be fossils of transitional forms between major groups of organisms.

In 1861, such a fossil was unearthed. *Archaeopteryx* appears to be a transitional form between reptiles and birds (Figure 17.8). Like modern birds, its body was covered with feathers. Like the fossils of small, bipedal reptiles, *Archaeopteryx* had teeth and a long, bony tail. (Bipedal means being able to walk upright, on two legs.) Other fossils turned up during Darwin's lifetime. Even so, nearly seventy years passed before advances in genetics led to widespread acceptance of his theory of natural selection. In the meantime, his name was associated mostly with the idea that life evolves—something others had proposed before him.

With this bit of history behind us, we are ready to consider some of the current views of evolution. The chapter that follows shows how modern genetics gives us a better understanding of the mechanisms by which evolution occurs. Later, in Chapter 19 and in Unit IV, we will turn to the results of those processes—that is, to the large-scale patterns of evolution through time and their relationship with earth history.

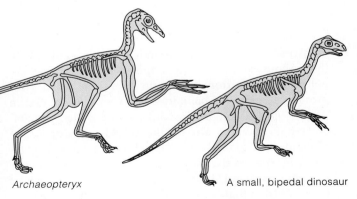

Archaeopteryx   A small, bipedal dinosaur

Figure 17.8 One of six fossils of *Archaeopteryx* from limestone deposits that are more than 140 million years old. *Archaeopteryx* was a transitional form on (or very near) the evolutionary road leading from reptiles to birds. The fossils might have been classified as reptilian, had it not been for the imprints of feathers in the finely grained limestone. *Archaeopteryx* had teeth and hind limbs like its reptilian ancestors, but in other features (such as its feathers and wishbone) it was like modern birds.

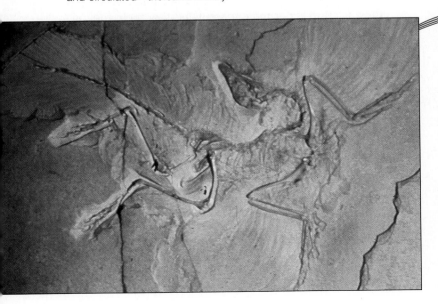

SUMMARY

1. Awareness of evolution—changes in lines of descent over time—emerged through the following:

a. Comparisons of body structure and patterning for major groups of animals (comparative anatomy).

b. Questions about the world distribution of plants and animals (biogeography).

c. Observations of fossils of different types present in different layers of sedimentary rocks.

2. A population evolves when some forms of a trait become more or less common relative to the other forms.

3. Here are the key points of the theory of evolution by natural selection, as proposed by Darwin and Wallace:

a. All populations have the reproductive capacity to produce more individuals than the environment can support, so individuals must compete for resources.

b. Individuals of a population vary in size, form, and other traits. The variant forms of a trait may be more or less adaptive under prevailing conditions.

c. Assuming a form of a trait is adaptive under prevailing conditions, and assuming it has a heritable basis, its bearers tend to survive and reproduce more often than individuals with less adaptive forms of the trait. Over the generations the adaptive version becomes more common in the population.

d. Natural selection is a difference in survival and reproduction among individuals of a population that differ from one another in one or more traits.

e. Natural selection results in modifications within lines of descent. Over time, it may bring about the evolution of new species.

4. One "test" of Darwin's theory would be evidence of one major kind of organism changing into another kind. *Archaeopteryx*, an extinct transitional form between reptiles and birds, provided early evidence. But not until many decades later would studies at many levels of biological organization provide a large body of substantiating evidence for his theory of evolution by natural selection.

Review Questions

1. Define biogeography and comparative anatomy. How did studies in both disciplines contradict the idea that species have remained unchanged since the time of creation? *262–263*

2. Cuvier and Lamarck interpreted the fossil record differently. Briefly state how their interpretations differed. *264*

3. Summarize the observations and inferences made by Darwin that allowed him to formulate the theory of evolution by natural selection. *266, 267*

4. Define evolution. Define evolution by natural selection. Can an individual evolve? *263*

Self-Quiz (Answers in Appendix IV)

1. Given that traits are inherited, individuals in a population who possess the _____ will tend to make up more of the reproductive base for the next generation.
   a. greatest variation in traits
   b. least variation in traits
   c. most adaptive traits
   d. most ancestors in the fossil record
   e. widest world distribution of their species

2. Natural selection may occur when there are _____ .
   a. heritable traits
   b. variation in traits within a population
   c. adaptive traits
   d. differences in survival and reproduction among members of a population
   e. all of the above

3. Most biologists strongly support the theory that species evolve because _____ .
   a. it was first suggested by Darwin and Wallace
   b. the discovery of the *Archaeopteryx* fossil makes it certain
   c. studies at many levels of biological organization provide a large body of evidence that the theory is correct
   d. another, better explanation of the great body of observations and test results has yet to appear
   e. both c and d are correct

4. Match the following individuals and ideas or perceptions.
   ____ Cuvier   a. theory of natural selection
   ____ Lamarck   b. populations outgrow resources
   ____ Lyell   c. catastrophism
   ____ Malthus   d. inheritance of traits acquired
   ____ Darwin   through environmental pressures
      and Wallace   and internal desires for change
                      e. geologic evidence that the earth is
                         extremely ancient

Selected Key Terms

biogeography *262*	natural selection *267*
catastrophism *264*	sedimentary bed *263*
comparative anatomy *263*	theory of inheritance of
evolution *263*	acquired characteristics *264*
fossil *263*	theory of uniformity *266*

Readings

Darwin, C. 1957. *Voyage of the Beagle.* New York: Dutton. In his own words, what Darwin saw and thought about during his voyage.

Singer, C. 1962. *A History of Biology to About the Year 1900.* New York: Abelard-Schuman. Out of date, but contains absorbing portrayals of the men and women who led the way in developing basic biological concepts.

18 MICROEVOLUTION

Designer Dogs

We humans have tinkered rather ruthlessly with the modern descendants of a long and distinguished lineage. That lineage began some 40 million years ago with small, weasel-shaped, tree-dwelling carnivores, the forerunners of bears, raccoons, pandas, badgers—and dogs. About 10,000 years ago, humans began domesticating wild dogs.

No doubt the advantages of doing so were obvious. Times were tough in the days before supermarkets and police protection. Dogs could guard us and our possessions. They could eat and so dispose of rats and other unwelcome vermin in our shelters.

It didn't take us long to develop different varieties (breeds) through artificial selection. Individual dogs

having the desired forms of traits were selected from each new litter and, later, encouraged to breed. Those having undesired forms of traits were passed over.

After favoring the pick of the litter through hundreds or thousands of generations, we ended up with sheep-herding collies, badger-hunting dachshunds, wily retrievers, and snow-traversing, sled-pulling huskies. And at some point we began to delight in the odd, extraordinary dog. Imagine! In no time at all, evolutionarily speaking, we picked our way through the pool of variant dog genes and came up with such extremes as Great Danes and Chihuahuas (Figure 18.1).

Of course, canine designs can exceed the limits of biological common sense. For example, how long would a tiny, finicky-eating, nearly hairless, nearly defenseless Chihuahua last in the wild? Not very long. Or what about the English bulldog, bred for a short snout and compressed face? Long ago, breeders thought these particular traits would allow the dogs to get a better grip on the nose of bulls. (Why they wanted dogs to bite bulls is a story in itself.) The roof of the bulldog mouth is now ridiculously wide and often flabby, so bulldogs have trouble breathing. They sometimes get so short of air, they pass out.

Through our centuries-old fascination with artificial selection, we produced thousands of varieties of crop plants, cats, cattle, and birds as well as dogs. With the currently available technologies of genetic engineering, we are now mixing genes of different species and getting astounding new varieties, such as hemoglobin-producing tobacco plants (page 255).

So, when you hear someone wonder about whether "evolution" takes place, remind yourself that evolution simply means *change through time*. Our selective breeding practices provide abundant, tangible evidence that heritable changes do, indeed, occur. *How* those changes are brought about in a given line of descent is the subject of this chapter.

KEY CONCEPTS

1. All individuals of a population generally have the same number and kinds of genes, which give rise to the same assortment of traits.

2. In the population as a whole, each gene may exist in two or more slightly different molecular forms, called alleles. Individuals may or may not inherit the same alleles, so they may not be exactly alike in the details of their traits.

3. Any allele may become more or less common in the population, relative to the other kinds, or it may disappear. *Microevolution* refers to changes in a population's allele frequencies over time.

4. Allele frequencies change through mutation, gene flow, genetic drift, and natural selection. Mutation alone produces new alleles. Gene flow, genetic drift, and natural selection shuffle existing alleles into, through, or out of populations.

5. Natural selection is not an "agent," purposefully searching for the "best" individuals in a population. It simply is the difference in survival and reproduction among individuals that differ in one or more traits.

Figure 18.1 Two designer dogs. About 10,000 years ago, humans began domesticating wild dogs. From that ancestral stock, artificial selection produced diverse yet rather closely related breeds, such as the Great Dane (*legs, left*) and the Chihuahua (*possibly fearful of being stepped on, right*).

18.1 INDIVIDUALS DON'T EVOLVE; POPULATIONS DO

Variation Within Populations

As Charles Darwin perceived, an individual doesn't evolve; populations do. By definition, a **population** is a group of individuals of the same species occupying a given area. To understand how a population evolves, we must begin with the variation in traits that exists among its individuals.

Certain traits help define a population. Its members have the same overall form and appearance, as when bluejays have cobalt blue feathers, perching feet with three toes forward and one toe back, and so on. These are *morphological* traits (*morpho-* means form). The cells and body parts of all individuals in the population operate in much the same ways during growth, day-to-day tasks, and reproduction. These are *physiological* traits (they relate to body functions). Also, individuals of a population generally respond the same way to certain basic stimuli, as when human babies instinctively imitate facial expressions of adults (Figure 51.3). These are *behavioral* traits.

Yet the details of traits vary from one individual to the next, particularly among sexually reproducing species. Pigeon feathers or snail shells can differ in patterning, coloration, or both from one individual to the next (Figures 1.7 and 18.2). Frogs of the same popula-tion may differ in their sensitivity to winter cold or in their success at courtship. Humans differ in the color, texture, amount, and distribution of their hair. And these examples only hint at the staggering variation that exists in populations.

Sources of Variation

Most variation in traits has a heritable basis. This is apparent simply by observing that children resemble their parents more than they resemble anyone else. Information about heritable traits resides in hundreds to many thousands of genes, which are specific regions of DNA molecules.

Individuals of the same population generally have the same number and kinds of genes. (We say "gener-ally" because males and females differ in some genes on their sex chromosomes and because genes occasion-ally become duplicated or lost, owing to chromosome abnormalities.) Think of all the genes in the entire pop-ulation as a **gene pool**—a pool of genetic resources that, in theory at least, is available to all of its reproducing members.

In this gene pool, each kind of gene usually exists in two or more slightly different molecular forms, called **alleles**. Individuals inherit different combinations of alleles. And this leads to variation in **phenotype**—to differences in morphological, physiological, and behav-ioral traits from one individual to the next.

Figure 18.2 Variation in shell color and banding patterns among populations of one species of snail found on islands of the Caribbean. For most traits, different members can carry different alleles. Alleles are alternative molecular forms of the same gene, and they code for alternative forms of the same trait. Members of a population vary in their traits because they can carry different combinations of alleles at most gene locations along their chromosomes.

For example, whether your hair is black, brown, red, or blond depends on which alleles of certain genes you inherited from your mother and father. As described in earlier chapters, the particular mix of alleles for any sexually reproducing organism depends on five factors:

1. Gene mutation (produces new alleles)

2. Crossing over at meiosis (leads to new combinations of alleles in chromosomes)

3. Independent assortment at meiosis (leads to mixes of maternal and paternal chromosomes in gametes)

4. Fertilization (puts together combinations of alleles from two parents)

5. Changes in chromosome structure or number (lead to the loss, duplication, or alteration of alleles)

Of these factors, mutation alone creates alleles. The others shuffle existing alleles into new combinations. But what a shuffle! For example, a human gamete ends up with one of 10^{600} possible combinations of alleles. Not even 10^{10} humans are alive today. So unless you have an identical twin, it is extremely unlikely that another person with your exact genetic makeup has ever lived—or ever will.

Environmental Influences on Variation

Variation among phenotypes is not entirely attributable to genes. In any number of ways, the environment can mediate how the genes governing different traits are expressed in an individual. Ivy plants grown from cuttings of the same parent all have the same genes, but the amount of sunlight affects the genes governing leaf growth. The plants (or even parts of a single plant) growing in full sun have smaller leaves than those growing in full shade. Figure 18.3 shows another example of how the environment can bring about variations in gene expression.

When studying the genetic basis of evolution, environmental factors that might be influencing gene expression must be identified. (Offspring inherit genes, not phenotypes.) Artificial selection experiments and molecular experiments make it possible to distinguish between the genetic and environmental components.

"Genetic variation" refers to differences in the combinations of alleles carried by the individuals of a population.

For populations of sexually reproducing species, the gene pool is a source of potentially enormous variation in traits.

a

b

Figure 18.3 An example of environmental effects on phenotype. Two plants of the same species (*Hydrangea macrophylla*) may carry the same alleles for flower color. Even so, the color of their floral clusters may vary from pink to blue—depending on the acidity of the soil in which a plant happens to be growing.

18.2 STABILITY VERSUS CHANGE IN ALLELE FREQUENCIES

Imagine yourself in a large flower garden in summer. Gradually you notice that butterflies are flitting about and that they are all the same, except in wing coloration. A few wings are white, but most are blue. Most likely, you muse, the "blue" allele must be more common than the "white" allele. If it were possible to count up all the individuals and perform a suitable genetic analysis, you possibly could identify the **allele frequencies**—the abundance of each kind of allele in the butterfly population as a whole.

Why bother identifying the allele frequencies of a population? Doing so will allow you to track whether the population is evolving. You start with the "Hardy-Weinberg formula," which helps set up a theoretical reference point for measuring the rates of change. That reference point, **genetic equilibrium**, implies stability in allele frequencies through the generations. In other words, a population at genetic equilibrium would *not* be evolving (Figure 18.4). This happens only when the following five conditions are being met.

First, no genes are undergoing mutation. Second, the population is very, very large. Third, the population is isolated from other populations of the species. Fourth, all members survive and reproduce (there is no natural selection). Fifth, mating is random.

Rarely (if ever) do all five conditions exist at the same time in a natural population. Three processes—gene flow, genetic drift, and natural selection—can drive a population away from genetic equilibrium, even in the span of a few generations. Besides this, mutations are rare but inevitable events. The term **microevolution** refers to small-scale changes in allele frequencies as brought about by mutation, genetic drift, gene flow, and natural selection.

The evolutionary story begins with changes in the frequencies of alleles that are distributed among all the individuals of a population.

Hardy-Weinberg Rule:

In a population at genetic equilibrium, the proportions of genotypes for a locus with two alleles are

$$p^2\ AA + 2pq\ Aa + q^2\ aa$$

where p is the frequency of allele A, and q is the frequency of allele a. The frequencies will remain the same through the generations *if* there is no mutation, *if* the population is infinitely large and isolated from other populations, *if* mating is random, and *if* all genotypes are equally viable and fertile.

Figure 18.4 Hardy-Weinberg rule. To prove the rule stated above, let's track two alleles through an imaginary population at genetic equilibrium. The alleles influence wing color in a population of butterflies, which are diploid organisms. A dominant allele A specifies medium-blue wings; allele a is associated with white wings. In the population as a whole, the frequencies of A and a must add up to 1. For example, if A occupies half of all the gene loci of the population's members, then $0.5 + 0.5 = 1$. If A occupies 90 percent of all the gene loci, then a must occupy the remaining 10 percent ($0.9 + 0.1 = 1$). No matter what the proportions of the two kinds of alleles,

$$p + q = 1$$

When diploid organisms reproduce, each pair of alleles segregate and end up in separate gametes. Thus p is also the proportion of gametes with the A allele, and q the proportion with the a allele. To find the expected frequencies of the possible genotypes AA, Aa, and aa in the next generation, let's construct a Punnett square:

	p Ⓐ	q ⓐ
p Ⓐ	AA (p^2)	Aa (pq)
q ⓐ	Aa (pq)	aa (q^2)

The frequencies of genotypes add up to 1:

$$p^2 + 2pq + q^2 = 1$$

To see how the allele frequencies and genotypic frequencies remain the same through the generations, let's work through an example. We have a population of 1,000 diploid individuals that each produce two gametes:

```
490  AA individuals produce 980  A gametes
420  Aa individuals produce 420  A and  420 a gametes
 90  aa individuals produce 180  a gametes
```

Notice that p, the frequency of A among the 2,000 gametes, is $(980 + 420)/2,000 = 0.7$. Also, $q = (420 + 180)/2,000 = 0.3$. These gametes combine at random to produce the next generation as given in the Punnett square, so we will have:

$$
\begin{aligned}
p^2 \quad AA &= 0.7 \times 0.7 = 0.49, \text{ or} & 490\ AA \text{ individuals} \\
2pq \quad Aa &= 2 \times 0.7 \times 0.3 = 0.42, \text{ or} & 420\ Aa \text{ individuals} \\
q^2 \quad aa &= 0.3 \times 0.3 = 0.09, \text{ or} & 90\ aa \text{ individuals}
\end{aligned}
$$

and $p^2 + 2pq + q^2 = 0.49 + 0.42 + 0.09 = 1$.

18.3 MUTATION

A **mutation** is a heritable change in DNA that can alter gene expression. The phenotypic outcome may be neutral, beneficial, harmful, or even lethal, depending on how the altered gene product functions under the prevailing conditions in the environment.

Recall, from page 229, that harmful mutations alter traits in ways that lower the individual's chances of survival and reproduction. Lethal mutations always lead to death. Consider cartilage, a flexible, collagen-containing tissue in your nose, rib cage, and many other parts of the body. Certain mutations affect where, when, and how much collagen is produced. Suppose a developing human embryo carries a mutated gene that results in abnormal collagen deposition in the nose and rib cage. If the embryo's nostrils become blocked and its ribs become too thick, normal breathing will be impossible.

A neutral mutation is neither harmful nor helpful to the individual. Suppose you have the mutated gene that leads to attached earlobes (page 168). You still should be able to make your way in the world. You might also have a mutated gene with unpleasant but not lethal effects. Besides, if environmental conditions change, that not-great but neutral mutation might turn out to be beneficial.

Other mutations enhance prospects of surviving and reproducing under prevailing conditions. For example, a mutated growth-regulating gene might make a plant grow larger or faster, and so give it the best shot at sunlight and nutrients.

By themselves, mutations are too rare to cause significant change in a population's allele frequencies. Their only notable effect over the short term is the replacement of rare alleles that are lost, for one reason or another. But mutations have been accumulating in many millions of different lineages for several billion years. Through all that time, they have been the only source of alleles—the genetic foundation for the staggering range of biological diversity, past and present.

Mutations are the original source of alleles, hence of all variation in heritable traits.

Notice that the allele frequencies have not changed:

$$A = \frac{2 \times 490 + 420}{2,000 \text{ alleles}} = \frac{1,400}{2,000} = 0.7 = p$$

$$a = \frac{2 \times 90 + 420}{2,000 \text{ alleles}} = \frac{600}{2,000} = 0.3 = q$$

Genotype frequencies have not changed either. They will stay the same over the generations as long as the assumptions stated in the green box hold true. To verify this, calculate the allele frequencies in the gametes of the next generation:

F_1 genotypes: 0.49 AA 0.42 Aa 0.09 aa

Gametes: A A A a a a

0.49 + 0.21   0.21 + 0.09

0.7 A 0.3 a

which is back where we started. Because the frequencies of alleles for medium-blue wings and white wings are the same as they were in the original gametes, they will yield the same phenotypic frequencies as in the second generation. You could do similar calculations for other wing colors. You could go on with the calculations until you ran out of paper (or patience). As long as the five assumptions hold true, allele frequencies and the range of values for the wing-color trait will not change, as shown in the diagram to the right. This is an example of the theoretical possibility known as genetic equilibrium.

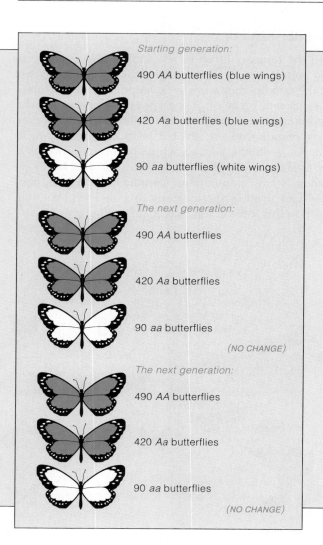

Starting generation:

490 AA butterflies (blue wings)

420 Aa butterflies (blue wings)

90 aa butterflies (white wings)

The next generation:

490 AA butterflies

420 Aa butterflies

90 aa butterflies

(NO CHANGE)

The next generation:

490 AA butterflies

420 Aa butterflies

90 aa butterflies

(NO CHANGE)

18.4 GENETIC DRIFT

Genetic drift is the random change in allele frequencies through the generations owing to chance events alone. Genetic drift often is most rapid in small populations. It may decrease variation within a population. And it may increase variation between populations.

Suppose most members of a small population of butterflies carry the allele for blue wings. Whether you call it chance or bad luck, many of them die during a freak summer storm (Figure 18.5). Next summer, an amateur butterfly collector snags the only blue-winged butterflies remaining. It makes no difference whether the blue-wing allele is "best" (most advantageous) for a given trait. Within a couple of generations, it has disappeared from the population. If the population had been larger, with many hundreds of blue-winged butterflies, the allele might still be around.

Founder effects and bottlenecks are two extreme cases of genetic drift. Both may result in severely limited genetic variation, because the population originates or is rebuilt from very few individuals.

Founder Effect

With the **founder effect**, a few individuals establish a new population. By chance, their allele frequencies are not the same as they were in the population left behind, so a different range of phenotypes is tested in the new environment. The outcome of this chance event is sometimes pronounced on isolated islands. Long ago, for example, seabirds, winds, or ocean currents carried a few seeds from the Pacific Northwest to the Hawaiian Islands. Plant populations became established and rapidly evolved in spectacularly diverse ways. We will return to this topic in the next chapter.

Bottlenecks

With **bottlenecks**, disease, starvation, or some other stressful event nearly wipes out a large population. The population recovers, but frequencies of its various alleles have been altered at random. Before the turn of the century, hunters killed all but twenty of a large population of northern elephant seals. The population recovered. Researchers who studied twenty-four genes found that 30,000 seals alive at the time carried the same alleles. The absence of variation suggests a number of alleles were lost during the bottleneck.

The *Focus* essay describes what such losses may mean for **endangered species**, the populations of which are too small and vulnerable to extinction.

Genetic drift is a random change in allele frequencies over the generations, as brought about by chance events alone.

Figure 18.5 Hypothetical example of genetic drift in a small population. Each symbol designates two butterflies that breed true either for *blue* wings (allele *A*) or *white* wings (allele *a*). In generation I, four of the six butterflies have *blue* wings.

All generation I individuals reproduce. Therefore, the relative abundances of the two alleles is unchanged in the generation II butterflies, all of which go on to reproduce. But 50 percent of the white-winged offspring and all but two of the blue-winged offspring are pummeled to death during a sudden hailstorm. All the survivors reproduce, but allele frequencies shift. But now an ecologically naïve butterfly collector flits amongst the generation IV offspring. She snags 25 percent of the carriers of allele *a*—and all remaining carriers of allele *A*. Through chance alone, allele *a* comes to dominate generation V—and allele *A* disappears from the population.

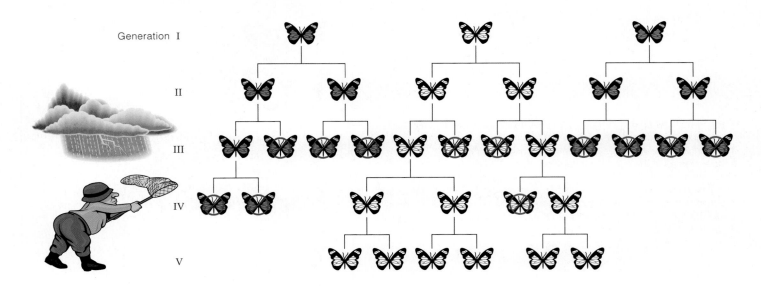

18.5 GENE FLOW

Allele frequencies change as individuals leave a population (emigration) or new individuals enter it (immigration). This physical flow of alleles, or **gene flow**, helps keep neighboring populations genetically similar. Over time, it tends to counter the differences between populations that are brought about through mutation, genetic drift, and natural selection.

Think of blue jays dispersing acorns when they store nuts for the winter. Each fall they may make hundreds of round trips from acorn-bearing oak trees to soil in their home territories, which may be a mile away (Figure 18.6). The alleles flowing in with "immigrant acorns" help neutralize variations that arise among neighboring stands of oaks. Or think of the millions of people from economically bankrupt and politically explosive countries who are immigrating to more stable places. The scale is unprecedented, but not unique. Throughout history, immigrations may have neutralized much of the genetic differences that otherwise would have accumulated among human populations.

Gene flow is the physical movement of alleles into and out of populations.

Figure 18.6 One travel agent for gene flow among oak populations. Blue jays hoard acorns in their home territory, but they might shop at nut-bearing trees up to a mile away. Some acorns contribute to the allele pool of an oak population some distance away from the parent tree.

Focus on the Environment

Genes of Endangered Species

Barbara Durrant and many other biologists at zoos, private conservation centers, and wildlife refuges are scrambling to buy time for the world's most endangered species. Populations of such species are so small, they are poised at the brink of extinction. The biologists face daunting problems, including the absence of genetic diversity.

Consider the cheetah. Only 20,000 of these sleek, swift cats have survived to the present. Apparently, cheetahs went through a severe bottleneck in the nineteenth century. Parents mated with their own offspring when no other mates were available. As a result of the inbreeding among the survivors and their descendants, cheetahs now carry strikingly similar alleles. They are so genetically uniform, a patch of skin from one can be successfully grafted to another. (This seldom works with other mammals.)

Among the shared alleles are previously rare bad ones. Some mutated alleles affect fertility. Typically, 70 percent of a male's sperm is abnormal, and sperm counts are low.

Other alleles are responsible for weakened resistance to disease. Infections that are seldom life-threatening to other cats can reach epidemic proportions among cheetahs. During one outbreak of *feline infectious peritonitis* in a wild animal park, the infectious agent (a coronavirus) had little effect on the captive lions. It killed the majority of cheetahs. The infection triggers an inflammatory response that goes out of control. The body cavity that houses the heart, gut, and other internal organs fills with fluid, and the cat dies in agony. There is no vaccine.

Another big cat, the Florida panther, is endangered as a result of intense hunting and other human activities. With fewer than fifty members, the last remaining population is highly inbred. Researchers at the National Zoo retrieve DNA from unfertilized eggs of "road-killed" female panthers. With genetic engineering, the DNA may find use in captive breeding programs.

18.6 NATURAL SELECTION— DARWIN REVISITED

Natural selection probably accounts for more changes in allele frequencies than any other microevolutionary process. It has been observed during hundreds of field studies of populations from all five kingdoms. Darwin, recall, was able to explain natural selection after correlating his understanding of inheritance with certain features of populations and the environment. Before we consider the modes of natural selection, let's restate his correlations in terms of modern genetics:

1.   Individuals of a population tend to produce more offspring than can survive and reproduce in a given environment. In time, its individuals must compete for resources.

2.   All members of the population have the same genes and are characterized by certain traits. Collectively, their genes represent a pool of heritable information.

3.   Individuals differ from one another in phenotypic details, because the gene pool contains two or more alleles for most (if not all) heritable traits.

4.   Some alleles promote survival and reproduction, because their version of a trait is more adaptive than other versions in the competition for resources.

5.   Successful reproducers contribute more alleles to the gene pool than unsuccessful ones do. Thus the relative abundances of alleles change (the population evolves).

6.   For any natural population, a difference in survival and reproduction among individuals that differ in one or more heritable traits is called **natural selection**.

Natural selection has different outcomes. As you will see from the following examples, it may shift a range of values for a trait in some direction, or it may stabilize or disrupt an existing range of values.

Figure 18.7 Directional selection, using phenotypic variation within a population of butterflies as the example. The bell-shaped curve (*brown*) represents the range of continuous variation in wing color. The most common forms (*powder blue*) are between extreme forms of the trait (*white* at one end of the curve, *deep purple* at the other). *Orange* arrows signify which forms are being selected against over time.

18.7 DIRECTIONAL CHANGE IN THE RANGE OF VARIATION

In **directional selection**, allele frequencies shift in a steady, consistent direction (Figure 18.7). This type of shift is a response to a directional change in the environment or to new environmental conditions. Forms of traits at one end of the range of variation become more common than midrange forms.

Peppered Moths   In England, directional selection has affected about a hundred species of moths, including the peppered moth (*Biston betularia*). Peppered moths feed and mate at night. During daylight, they rest motionless on birches and other trees. Their wings are mottled, in shades ranging from light gray to

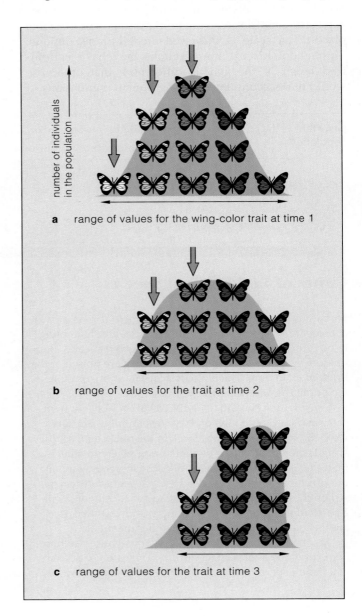

a   range of values for the wing-color trait at time 1

b   range of values for the trait at time 2

c   range of values for the trait at time 3

Table 18.1	Marked *Biston betularia* Moths Recaptured in a Polluted Area and a Nonpolluted Area			
	Light-Gray Moths		Dark-Gray Moths	
Area	Released	Recaptured	Released	Recaptured
Near Birmingham (pollution high)	64	16 (25%)	154	82 (53%)
Near Dorset (pollution low)	393	54 (13.7%)	406	19 (4.7%)

Data after H. B. Kettlewell.

Figure 18.8 Individuals of a population that underwent directional selection in response to changes in the environment. Light-winged and dark-winged peppered moths (*Biston betularia*) are resting on a lichen-covered tree trunk in (**a**) and on a soot-darkened tree trunk in (**b**).

nearly black. Their behavior, coloration, and wing patterns apparently help camouflage them from moth-eating birds, which are active during the day.

When an industrial revolution began in the 1800s, conditions changed in many parts of the country. Before then, light-winged moths were common and a dark-winged form was rare. Also before then, light-gray speckled lichens grew profusely on tree trunks. The lichens camouflaged light-winged moths that rested on them, but not dark-winged ones (Figure 18.8*a*).

Lichens are sensitive to air pollution. Between 1848 and 1898, soot and other pollutants from factories started killing the lichens and darkening tree trunks. Dark-winged moths were better camouflaged against the sooty backgrounds (Figure 18.8*b*). It seemed to many researchers that conditions previously had favored light wings—but now favored dark wings. Dark wings apparently were becoming the more common phenotype as a result of directional selection in this environment.

In the 1950s, H. B. Kettlewell used a **mark-release-recapture method** to test this hypothesis. He bred both forms of the moth in captivity, then marked hundreds of them so that they could be identified. He released moths near the heavily industrialized area around Birmingham and in an unpolluted area (Dorset). After a time, he recaptured as many moths as he could. More dark moths were recaptured in the polluted area—and more light moths in the pollution-free area (Table 18.1). By stationing watchers in blinds near groups of moths tethered to tree trunks, Kettlewell directly observed birds capturing more light moths around Birmingham and more dark moths around Dorset. Directional selection was operating on the population.

In 1952, strict pollution controls went into effect. Lichens made comebacks. Tree trunks became free of soot, for the most part. As you might predict, directional selection is now operating in the reverse direction. Where pollution levels have declined, the frequency of dark-winged moths is declining, also.

Pesticide Resistance Directional selection also is an outcome of the widespread use of chemical pesti-

cides in agriculture. Initial applications kill most of the insects, worms, or other pests, but some usually manage to survive. Some aspect of their structure, physiology, or behavior allows them to resist the chemical effects. If the resistance has a heritable basis, it becomes more common in the next generation, and the next—and the next. The chemicals are agents of selection; they favor the most resistant forms! Today, 450 different species are resistant to one or more pesticides. Worse yet, the pesticides also kill the natural predators of pests. Freed from natural constraints, the populations of resistant pests burgeon—and crop damage is greater than ever. This outcome is called **pest resurgence**.

Genetic engineering can help reduce pesticide use (page 255). Biological control can help, too. Huge populations of parasitic wasps, predatory beetles, and other natural enemies of pests are being raised in insectaries. But farmers who pay for these controls must replace the ones that migrate away from the fields or are destroyed at harvest time. Reality hits home at the market; many consumers are leery of genetically engineered food.

In directional selection, allele frequencies shift in a consistent direction in response to directional change in the environment.

18.8 SELECTION AGAINST OR IN FAVOR OF EXTREME PHENOTYPES

As you have seen, natural selection can result in steady, directional shifts in a population's range of phenotypic variation. Depending on environmental conditions, natural selection also may favor either the most common or the most extreme phenotypes in that range.

Stabilizing Selection

In **stabilizing selection**, the most common forms of a trait in a population are favored (Figure 18.9). Over time, alleles for uncommon forms are eliminated. Therefore, this mode of selection tends to counter the effects of mutation, genetic drift, and gene flow.

Stabilizing selection may account for the persistence of certain phenotypes over time. For example, through studies conducted in diverse societies over the past few centuries, we know stabilizing selection favors human newborns who weigh an average of 7 pounds. Without medical intervention, newborns who weigh significantly more or less than this tend to die soon after birth (Figure 18.10c). Also, you may have seen horsetails (*Equisetum*) along roadsides. These plants still bear resemblances to their ancient relatives (Figure 18.11).

Disruptive Selection

In **disruptive selection**, forms at both ends of the range of variation are favored and intermediate forms are selected against (Figure 18.11). Thomas Smith discovered a clear example of this in a remote rain forest in Cameroon, West Africa. Smith had read about unusual

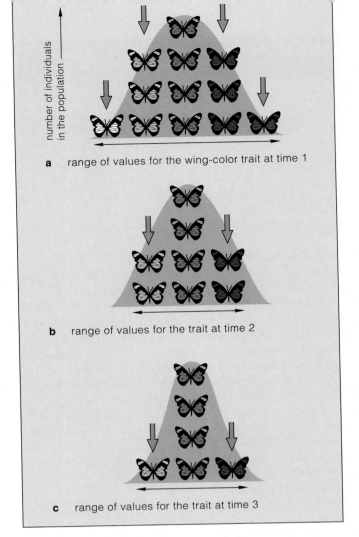

a range of values for the wing-color trait at time 1

b range of values for the trait at time 2

c range of values for the trait at time 3

Figure 18.9 Stabilizing selection, using phenotypic variation within a population of butterflies as the example.

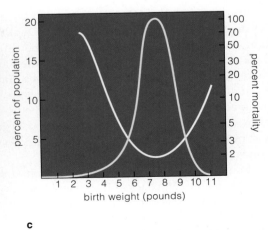

birth weight (pounds)

Figure 18.10 Stabilizing selection. (**a**) Fossilized sphenophyte, a seedless vascular plant that lived 380 million years ago. (**b**) Certain traits persist in horsetails (*Equisetum*), the only living representatives of the sphenophyte lineage. (**c**) Weight distribution for 13,730 human newborns (*yellow* curve) correlated with mortality rate (*white* curve). Stabilizing selection operates against newborns weighing significantly higher or lower than midrange values.

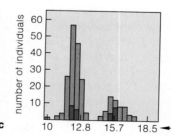

a range of values for the wing-color trait at time 1

b range of values for the trait at time 2

c range of values for the trait at time 3

Figure 18.11 Disruptive selection, using phenotypic variation within a population of butterflies as the example.

variation in bill size in populations of the black-bellied seedcracker (*Pyrenestes ostrinus*). These African finches have large *or* small bills, but no sizes in between. The pattern holds for females and males, throughout their geographic range. This is remarkable. (Imagine every person in Texas being 4 feet *or* 6 feet tall.) If the pattern is unrelated to gender or geography, what brings it about?

Smith hypothesized that if the persistence of only two bill sizes in the populations relates to seed-cracking ability (which directly affects survival), then disruptive selection may be eliminating birds with bills of intermediate size. What factors could affect seed hardness and so act as a selection pressure on feeding performance? In Cameroon, swamp forests flood during a very wet season. Natural fires occur during the dry season. Two species of grasslike, fire-resistant plants called sedges dominate the region. One has hard seeds; the other has soft seeds. When finches reproduce, hard and soft seeds are abundant. All birds prefer soft seeds as long as they can get them. At the height of the dry season, food supplies dwindle. Young birds especially are at a competitive disadvantage; many do not survive (Figure 18.12).

Smith also performed experimental crosses between large- and small-billed birds. All offspring had large *or* small bills. Together with other data, this suggests that bill size (and feeding performance) may be largely controlled by a single autosomal gene with two alleles.

Stabilizing selection favors the most common phenotypes and eliminates extreme phenotypes.

In disruptive selection, allele frequencies shift through selection against intermediate forms in the range of variation.

Figure 18.12 Directional selection among African finches. (**a**) A large-billed bird eating seeds in its natural habitat. In feeding trials conducted by the biologist Thomas Smith, large-billed birds were more efficient at using hard seeds. Small-billed birds were most efficient at using soft seeds, not hard ones. (**b**) Two specimens displaying the only two bill sizes in the finch populations. (**c**) Survival of juvenile birds during the dry season, when competition for resources is most intense. Individuals with very small, very large, or intermediate-sized bills cannot feed efficiently on either type of seed, and they survive poorly. The only survivors are indicated by the black shading. This finding is based on Smith's measurements of 2,700 netted individuals.

18.9 SPECIAL CASES OF SELECTION

Balanced Polymorphism

Smith's work with African finches is a fine example of how selection can maintain two or more alleles for the same trait in fairly steady proportions, generation after generation. When selection has this outcome, a population is said to display **balanced polymorphism** for the trait (after the Greek *polymorphos*, meaning "having many forms"). A population is in this state when nonidentical alleles for a trait are being maintained at frequencies greater than 1 percent. Even if the frequencies shift slightly, they bounce back to the same values.

Balanced polymorphism often results when environmental conditions favor the heterozygotes over homozygotes. Heterozygotes, recall, have a pair of nonidentical alleles for a given trait, and homozygotes have a pair of identical ones. One of the two nonidentical alleles may even be harmful in one environmental situation, yet it may also help the individual through a different situation. As an example of this, the *Focus* essay describes the environmental pressures that favor the combination of one normal allele and a mutated partner—which is responsible for sickle-cell anemia.

Balanced polymorphism also may result when heterozygotes are selected against. For example, Rh markers are among the recognition proteins at the surface of red blood cells. Two alleles for the markers, Rh^+ and Rh^-, apparently have persisted in the human population since prehistoric times. As described on page 669, if a homozygous recessive woman becomes pregnant by a heterozygous man, she may produce antibodies against the Rh^+ markers. If she becomes pregnant again, red blood cells of the new fetus will be damaged. In extreme cases, the fetus dies before birth. For unknown reasons, this unstable polymorphism persists in the human population, despite its potentially lethal effects.

Sexual Selection

As a final example, consider how natural selection works to alter or maintain differences in appearance between the males and females of a species. The occurrence of phenotypic differences between males and females of a species is called **sexual dimorphism** (after *dimorphos* meaning "having two forms").

Striking sexual dimorphism exists among many species of birds and mammals, including those shown in Figures 18.13 and 18.14. As Darwin noticed, in some

Focus on Health

Sickle-Cell Anemia—Lesser of Two Evils?

Sickle-cell anemia is a potentially severe human genetic disorder. It results from a mutation in the allele that codes for hemoglobin (page 176). In tropical and subtropical regions of Africa, the harmful allele (HbS) remains in the population along with the normal one (HbA). The HbS/HbS homozygotes often die in their early teens or early twenties. Yet nearly a third of the population are HbS/HbA *heterozygotes*. Notice that the harmful allele is maintained at high frequency *when it is paired with a nonidentical allele* (HbS with HbA). This balancing act is an outcome of natural selection. It is most pronounced in regions where malaria is most prevalent.

A parasitic protistan causes malaria (page 359). The parasite is transmitted to humans by a mosquito that is most common in the tropics and subtropics. Individuals who don't carry the HbS allele are far more likely to survive and reproduce than individuals who do—as long as they

don't get malaria. The combination of abnormal and normal hemoglobin molecules has an interesting effect. It interferes with the spread of the parasite through the body. As a result, heterozygotes are more apt to survive severe infections. In one study, there were *twice* as many severe or fatal infections among homozygotes than among heterozygotes.

Thus the persistence of a harmful allele becomes a matter of relative evils. It hurts the individual in one situation, yet has adaptive effects in a different situation.

In Central Africa, malaria has been an agent of selection for less than 2,000 years. In tropical and subtropical regions of the Middle East and Asia, it has been around much longer. Even though the sickle-cell trait occurs at high frequencies in these regions also, the symptoms are not as pronounced as they are in Africa. Alleles of other genes must be countering the serious effects of the HbS allele.

cases the males are larger, more splendidly colored and patterned, and more aggressive than the females.

Remember the male bighorn sheep butting heads in Figure 1.6*f*? Fighting wastes time, it uses up energy, and it may cause injuries. Yet reproductive benefits offset the costs. Males fight only to control areas where receptive females gather during the winter rutting season. Winners mate often, with many females. Losers will not challenge a stronger, larger male. But a gang of losers may invade his area and overwhelm his capacity to drive them all away. Losers attempt to mate—literally on the run. (Even though females are running away from them, they sometimes succeed—and so assure their representation in the gene pool of the next generation.)

Sometimes females of the species are the agents of selection for these traits. By choosing mates, they directly affect reproductive success. This is a form of **sexual selection**, which is based on any trait that gives an individual a competitive edge in mating and producing offspring. We will return to this intriguing topic in Chapters 51 and 52.

Balanced polymorphism is a state in which natural selection is maintaining two or more alleles over the generations at frequencies greater than 1 percent.

In sexual selection, a trait gives an individual an advantage in reproductive success. Sexual dimorphism is one outcome.

Figure 18.14 One outcome of sexual selection. This male bird of paradise (*Paradisaea raggiana*) is engaged in a spectacular courtship display. He caught the eye (and, perhaps, sexual interest) of the smaller, less colorful female. Males of this species compete fiercely for females, which serve as selective agents. (Why do you suppose drab-colored females have been favored?)

Figure 18.13 Northern seal with his harem. Only the largest, strongest males secure a territory, mate, and contribute to the gene pool of the next generation. Through sexual selection, males weigh about twice as much as females.

SUMMARY

1. Individuals of a population share the same genes. But the genes come in different allelic forms, and this leads to variations in traits.

2. A population is evolving when some forms of a trait (and the alleles that specify them) are becoming more or less common, relative to the other kinds.

3. Allele frequencies change as a result of four microevolutionary processes: mutation, gene flow, genetic drift, and natural selection (Table 18.2).

4. Mutations are heritable changes in DNA, and they are the only source of new alleles.

5. Gene flow is a change in allele frequencies brought about by the physical movement of alleles into and out of a population (by immigration and emigration).

Table 18.2	Summary of Major Microevolutionary Processes
Mutation	A heritable change in DNA
Gene flow	Change in allele frequencies as individuals leave or enter a population
Genetic drift	Random fluctuation in allele frequencies over time, due to chance occurrences alone
Natural selection	Change or stabilization of allele frequencies due to differences in survival and reproduction among variant members of a population

6. Genetic drift is a change in allele frequencies over the generations due to chance events alone. Two extreme cases are called the founder effect and bottlenecks.

   a. With the founder effect, a new population is founded by a few individuals that by chance may differ in allele frequencies from the original population.

   b. With bottlenecks, disease or some other stressful event destroys all but a few individuals of a large population. By chance, the alleles of the survivors give rise to a different range of phenotypes.

7. Natural selection is a difference in survival and reproduction among members of a population that vary in one or more traits. One form of a trait may be more adaptive than others under prevailing conditions. This may cause the relative abundances of the alleles responsible for the different forms of the trait to shift over the generations.

8. With natural selection, the range of variation for a trait may shift steadily in one direction (directional selection). The range may be disrupted at or near midrange (disruptive selection). Or forms at both ends of the range of variations are favored, and intermediate forms are selected against (stabilizing selection).

9. Natural selection can also bring about balanced polymorphism. A population is in this state when nonidentical alleles for a given trait are being maintained over the generations at frequencies greater than 1 percent, even after slight perturbations in those frequencies.

10. Sexual selection favors the forms of traits that give the individual an advantage in reproductive success. Sexual dimorphism, the persistence of phenotypic differences between males and females of a species, is one outcome of sexual selection.

a   b   c

Review Questions

1. What is the Hardy-Weinberg baseline against which changes in allele frequencies may be measured? *274*

2. Define the following microevolutionary processes, and explain how each one can send allele frequencies out of equilibrium:
   a. mutation *275*   c. gene flow *277*
   b. genetic drift *276*   d. natural selection *278–281*

3. What implications might the effect of genetic drift hold for an earlier concept of "survival of the fittest"? *276*

4. The three diagrams to the left represent stabilizing, directional, and disruptive modes of natural selection. Identify each mode, and give a brief example of each. *278–281*

5. Define balanced polymorphism. Does this mode of selection always favor heterozygotes over homozygotes? *282*

6. Define sexual selection, and give an example of one of its outcomes. *282–283*

time 1

time 2

time 3

7. This photograph shows a brilliantly hued male sugarbird and a subdued-hued female. Explain this difference between the two in terms of selection theory.

8. Disruptive selection _____ .
   a. eliminates uncommon forms of alleles
   b. shifts allele frequencies in a steady, consistent direction
   c. doesn't favor intermediate forms of traits
   d. works against adaptive traits

9. Speciation is _____ .
   a. an extinction of one population that makes way for another
   b. a buildup of environmental factors leading to geographic isolation
   c. a process whereby species originate
   d. the means by which gene frequencies change
   e. both c and d

10. Match the evolution concepts appropriately.
    ____ gene flow
    ____ natural selection
    ____ mutation
    ____ genetic drift

    a. source of new alleles
    b. changes in a population's allele frequencies due to chance alone
    c. immigration, emigration change allele frequencies
    d. differences in survival and reproduction among variant members of population

Self-Quiz (Answers in Appendix IV)

1. Individuals don't evolve; _____ do.

2. The allele responsible for sickle-cell anemia first appeared in tropical and subtropical regions of Asia, the Middle East, and Africa. It entered the United States population when individuals were forcibly brought over from Africa prior to the Civil War. In microevolutionary terms, this is an example of _____ .
   a. mutation   c. gene flow
   b. genetic drift   d. natural selection

3. Allele frequencies change as a result of _____ .
   a. mutation   d. natural selection
   b. gene flow   e. all of the above
   c. genetic drift

4. The only source of new alleles is _____ .
   a. mutation   d. natural selection
   b. gene flow   e. all of the above
   c. genetic drift

5. Existing alleles are shuffled into, through, or out of populations by _____ .
   a. mutation   d. natural selection
   b. gene flow   e. b, c, and d only
   c. genetic drift

6. A bottleneck is _____ .
   a. a new population established by individuals who left an old one
   b. a chance event that wipes out nearly all members of a population
   c. an extreme case of genetic drift
   d. both b and c

7. Directional selection _____ .
   a. eliminates uncommon forms of alleles
   b. shifts allele frequencies in a steady, consistent direction
   c. doesn't favor intermediate forms of a trait
   d. works against adaptive traits

Selected Key Terms

allele 272
allele frequencies 274
balanced polymorphism 282
bottleneck 276
directional selection 278
disruptive selection 280
endangered species 276
founder effect 276
gene flow 277
gene pool 272
genetic drift 276
genetic equilibrium 274

mark-release-recapture method 279
microevolution 274
mutation 275
natural selection 278
pest resurgence 279
phenotype 272
population 272
sexual dimorphism 282
sexual selection 283
stabilizing selection 280

Readings

Ayala, F. J., and J. W. Valentine. 1979. *Evolving.* Menlo Park, California: Benjamin/Cummings. Short introduction to evolutionary theory.

de Blieu, J. *Meant to Be Wild.* 1991. Golden, Colorado: Fulcrum Publishing. Describes captive breeding programs to save endangered species.

Cook, L., G. Mani, and M. Varley. 1986. "Postindustrial Melanism in the Peppered Moth." *Science* 231:611–613.

Futuyma, D. 1987. *Evolutionary Biology.* Second edition. Sunderland, Massachusetts: Sinauer.

Grant, V. 1981. *Plant Speciation.* Second edition. New York: Columbia University Press. Good discussion of speciation in plants.

Smith, T. B. January 1991. "A Double-Billed Dilemma." *Natural History* 14–21.

19 SPECIATION

The Case of the Road-Killed Snails

If you happen to be a snail living in a garden in Bryan, Texas, it doesn't take much to keep your genes away from snails in a backyard across the street. By day, the sunbaked asphalt would be about as inviting to a snail as a desert would be to a catfish. Besides, day or night, a street-traversing snail is vulnerable to cars, trucks, skateboards, and bicycles (Figure 19.1). Whatever else it might be, that strip of asphalt is a formidable barrier to gene flow between populations. For snails, that is.

Whether or not a physical barrier deters gene flow depends partly on an organism's mode of locomotion or dispersal. It also depends on how fast and how long an organism *can* move in response to environmental factors or its own hormones.

A snail is not swift, and it does not roam far from its home population. Compare it to the wild black duck, banded in Virginia in 1969, that turned up eight years later in Korea. Compare it to the wandering albatross, one of the supreme barrier busters. After lifting off from Kerguelen Island in the Indian Ocean, one of these birds soared westward past Africa, across the Atlantic, and around Cape Horn. Thirteen thousand kilometers later, it landed in Chile. With a lightweight body and a wingspan of 3.65 meters (12 feet across), the albatross was able to exploit the great prevailing winds of the Southern Hemisphere.

And yet, in 1859, snails did cross an ocean. Humans transported garden-variety snails (*Helix aspersa*) from France to California, then turned them loose. The idea was that the snails would multiply and so meet the demand for that French delicacy, *escargot aux fines herbes*. The snails not only exceeded expectations, they became an absolute nuisance in gardens throughout the southwestern United States.

By the 1930s, *H. aspersa* had hitched rides, possibly as eggs in the soil of plant containers, to Bryan, Texas, and founded small colonies in the local vegetation. Forty years later, Robert Selander, now of Pennsylvania State University, was down on his hands and knees with a few graduate students, scouring patches of vegetation on two adjacent city blocks. Why? Selander wanted to determine the effect of genetic drift on the introduced populations. He and his students collected every single snail—2,218 of them—from fourteen local populations. Each population was analyzed to determine the allele frequencies at five different gene locations. In all five cases, the results pointed to some genetic variation between colonies on the same block—

Figure 19.1 (**a**) Results from a study of neighboring populations, all descended from snails (*Helix aspersa*) that arrived in a town in Texas in the 1930s. For each population, a circle represents relative abundances of three alleles (coded *yellow*, *gold*, or *brown*) for leucine aminopeptidase (an enzyme). Greater genetic variation exists between the populations living on opposite sides of the street. (**b**) A snail about to encounter a barrier to gene flow.

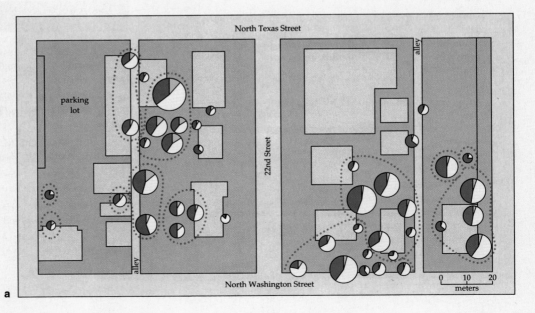

b

and to major differences in allele frequencies *between* blocks. Figure 19.1a shows the results for one of the genes studied.

Assume that genetic differences between the populations continue to increase, as through genetic drift and, later, natural selection. Will the time come when snails from opposite sides of the street can no longer interbreed successfully, even if they do manage to get together? In other words, *will they become members of separate species?* Or will selection work against increases in genetic differences by eliminating extreme phenotypes from the populations? After all, how much can a workable package of *H. aspersa* genes evolve in adjacent patches of vegetation—and under similar environmental pressures—in a small town in Texas?

With this chapter, we turn to **speciation**—to changes in allele frequencies that are significant enough to mark the formation of daughter species from a parental species. Obviously, no one was around to watch the formation of species during the past. No one lives long enough to know whether many existing populations are at some intermediate stage leading to speciation. Evolutionary biologists are still working out their theories about speciation processes, often quite contentiously. What you are about to read may change in the near or distant future.

KEY CONCEPTS

1. A species consists of one or more populations of individuals that can interbreed under natural conditions and produce fertile offspring, and that are reproductively isolated from other such populations. This definition is restricted to sexually reproducing species.

2. The populations of a species have a shared genetic history, they are maintaining genetic contact over time, and they are evolving independently of other species.

3. Species are distinct, well-adapted entities that resist change unless the environmental conditions to which they are adapted change.

4. Speciation is the formation of daughter species from a parent species.

5. Three models of speciation guide current research. In allopatric speciation, species form in separated populations or subpopulations. In parapatric speciation, they form in a region where populations share a border. In sympatric speciation, they form within the home range of the parent species.

6. Speciation requires irreversible genetic divergence of one population from others. Isolating mechanisms maintain genetic divergence.

7. The timing, rate, and direction of speciation vary within and between lineages. The extinction of some number of species is inevitable for all lineages.

19.1 WHAT IS A SPECIES?

If It Looks Like a Duck, . . .

"If it looks like a duck, walks like a duck, and quacks like a duck, probably it's a duck." Let's use this familiar aphorism as a starting point for defining what is and what is not a species. **Species** is a Latin word. Generally speaking, it simply means "kind," as in "a particular kind of duck."

Few of us have trouble identifying a duck. But how do we go about distinguishing one kind from another? By outward appearance alone? Suppose some ducks with vibrant green, white, and black feathers are paddling in a pond. You have never seen anything quite like them. Probably the ducks belong to the same population and, by extension, the same species. Then you see another duck, smaller than the rest, with drab-brown feathers (Figure 19.2). Is it a different species? Or is it a female paddling amongst males of the same species? Outward appearances can be deceiving; they have been known to fool even knowledgeable observers.

If appearances alone do not define a species, then what does? This is not easy to answer. The biologist John Endler put it this way: "'Species' are moving targets; they are changing even as we describe them." Throughout the history of any species, microevolutionary processes have been introducing changes in the gene pools of its populations. Remember, particular alleles give rise to the heritable traits of populations. As an outcome of mutation, genetic drift, gene flow, and natural selection, different alleles can increase or decrease in frequency, or even disappear over time. Besides varying in time, alleles vary within and between populations, so that some phenotypes are more common than others.

Phenotype, recall, refers to the observable aspects of an individual's form, physiology, and behavior. Given the variation in phenotypic details, perhaps we should move past the details, down to a more basic function that unites members of a species and isolates them from all other species.

For example, could reproduction be a defining function? This possibility is central to the **biological species concept**. Ernst Mayr, a prominent evolutionary biologist, phrases the concept this way: "Species are groups of interbreeding natural populations that are reproductively isolated from other such groups." No matter how extensive the phenotypic variation, individuals remain members of the same species as long as their morphology, physiology, and behavior permit them to interbreed and produce fertile offspring. They belong to the same species because they can contribute to a shared gene pool.

The biological species concept avoids the trap of defining the species by appearance alone. However, with its focus on interbreeding, it cannot work for the entire spectrum of life. It cannot be applied easily to the kingdom of bacteria and other organisms that rely most or all of the time on asexual reproduction. Also, applying the concept to the fossil record is a bit awkward, although this would be true of any species concept. Clues to form, function, and behavior do show up among the fossils of extinct organisms. Intact DNA retrieved from some fossils has even yielded clues to genetic histories. But fossils formed long before humans learned to write. We have no written record and very little physical evidence of whether the different populations represented by particular fossils were interbreeding or not.

Even though it is not universally applicable, Mayr's concept has proved useful for research into the factors that define sexually reproducing species.

A species consists of one or more populations of individuals that can interbreed under natural conditions and produce fertile offspring, and that are reproductively isolated from other such populations.

Figure 19.2 What is a species? The possibility of extreme sexual dimorphism is one reason why species identification cannot always be based on appearances alone. This drab female and gaudy male are both wood ducks (*Aix sponsa*).

19.2 ISOLATING MECHANISMS

Regarding Genetic Divergence

Even when the range of phenotypic diversity within a species is stunning, individuals maintain membership in that species as long as they continue to interbreed successfully. They will do so even if they belong to geographically distant populations, provided that *gene flow* persists among them. Gene flow, recall, is the physical movement of alleles into and out of populations, through immigration and emigration. Over time, it tends to counter the differences between populations that arise through mutation, genetic drift, and natural selection, and so maintains the common pool of alleles.

But suppose something prevents gene flow between two populations (or subpopulations). The stage is set for **genetic divergence**, a buildup of differences in the separated pools of alleles. Genetic drift, natural selection, and mutation are now free to operate independently in the isolated populations.

Consider the zebroids in Figure 19.3. These hybrids resulted from unnatural dalliances between wild zebras that were captured and confined in pastures with domesticated horses. (*Hybrid* individuals simply are offspring of parents of different genotypes.) The lineages leading to horses and zebras diverged more than 3 million years ago. The existing members of this family are diverse; they range from true horses (*Equus*) to donkeys and asses (*Asinus*). Intermediate forms include a variety of zebras (*Hippotrigris* and *Dolichohippus*). Yet the production of zebroids, even in captivity, attests to a lingering genetic compatibility.

Categories of Isolating Mechanisms

Normally you will not come across such extreme examples of gene flow—and breaches of species integrity—in nature. So what keeps horses and zebras from interbreeding in the wild? Something about their morphology, physiology, or behavior must be working to prevent gene flow. Most likely, **isolating mechanisms** began operating at some time in the past, and they have been reinforcing genetic divergence ever since.

Table 19.1 lists the main mechanisms by which interbreeding can be prevented between genetically divergent populations. One way or another, these isolating mechanisms discourage gene flow.

An isolating mechanism is any heritable aspect of body form, physiology, or behavior that prevents interbreeding (hence gene flow) between genetically divergent populations.

Figure 19.3 A mixed herd of zebroids and horses. Zebroids are interspecific hybrids, resulting from crosses between horses and zebras.

Table 19.1	Categories of Isolating Mechanisms
Prezygotic Isolation (*Mating or zygote formation is blocked*)	
Temporal isolation	Potential mates occupy overlapping ranges but reproduce at different times
Behavioral isolation	Potential mates meet but cannot figure out what to do about it
Mechanical isolation	Potential mates attempt engagement, but sperm cannot be successfully transferred
Gametic mortality	Sperm is transferred but egg is not fertilized (gametes die or gametes are incompatible)
Ecological isolation	Potential mates occupy different local habitats within the same area
Postzygotic Isolation (*Hybrids don't work*)	
Zygotic mortality	Egg is fertilized, but zygote or embryo dies
Hybrid inviability	First-generation hybrid forms but shows very low fitness
Hybrid offspring	Hybrid is sterile or partially so

After Alan Templeton, Jerry Coyne, and others.

Prezygotic Isolation

Certain isolating mechanisms take effect before or during fertilization. Because they prevent mating or pollination between genetically divergent populations, hybrid zygotes cannot form. The same isolating mechanisms operate to prevent interbreeding between individuals of different species.

Temporal Isolation Temporal isolation results from interpopulational differences in the timing of reproduction. For most animals or plants, mating or pollination is a seasonal event that is over rather quickly, sometimes in less than a day. Even closely related species may be isolated from each other if their reproductive cycles do not coincide.

Consider the cicada, a type of insect that spends nearly all of its life underground in immature form, feeding on juices from plant tissues. One species matures, emerges, and reproduces every 13 years. The other species does this every 17 years. Only once every 221 years do they release gametes at the same time!

Behavioral Isolation This type of isolation is a major barrier to gene flow among related species in the same territory. For example, before male and female birds copulate, they engage in complex courtship rituals (Figure 19.4). A female of the same species is neurologically equipped to recognize a male's singing, wing-spreading, head-bobbing, or prancing as an overture to sex. Females of another species usually do not.

Mechanical Isolation Incompatibility between the body parts of individuals is a form of mechanical isolation. Consider two sage species. Both depend on insects to carry pollen from flower to flower—but the flower petals differ in size and arrangement. One species has small flower petals arranged as a "landing platform" for only one kind of pollinator. Larger pollinators of other flowers tend not to cross-fertilize these plants (Figure 19.5).

Gametic Isolation Reproductive isolation may also result through the evolution of molecular incompatibilities between the gametes of different species. For example, if pollen lands on a plant of a different species, it may not receive the molecular signals that trigger its growth down through the plant tissues, to the egg (compare Figure 25.13).

Or consider sea urchin gametes, which are fertilized externally. Gametes from different species might be released at the same time, in the same place, but they must be molecularly mismatched. They rarely cross-fertilize.

Figure 19.4 A sampling of courtship displays that precede copulation between a male and female albatross. Such mutual displays have visual, acoustical, and tactile components that are recognizable by members of the same species, but usually not by members of different albatross species.

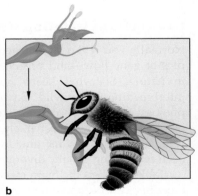

a   b

Figure 19.5 Mechanical isolation between two species of sage. (**a**) Certain petals of the flowers of *Salvia mellifera* are arranged as a small landing platform for small or medium-size pollinators. Other petals are arranged as a pollinator-attracting nectar cup. (**b**) *S. apiana* has a large landing platform and long, pollen-bearing stamens, which extend some distance away from the nectar cup. Small bees that land on the large platform usually don't brush against the stamens. Larger pollinators do. Hence the small pollinators of *S. mellifera* are mostly incapable of spreading pollen to flowers of *S. apiana*. The large pollinators of *S. apiana* cannot land on and cross-pollinate *S. mellifera*.

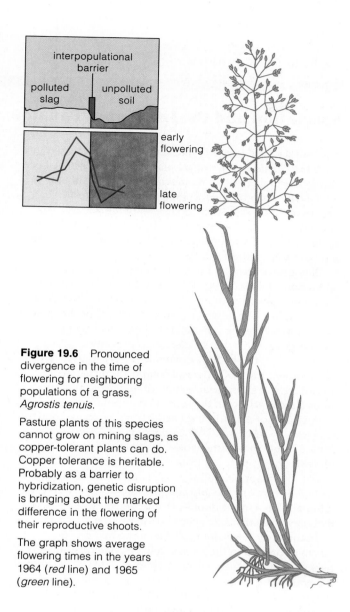

Figure 19.6 Pronounced divergence in the time of flowering for neighboring populations of a grass, *Agrostis tenuis*.

Pasture plants of this species cannot grow on mining slags, as copper-tolerant plants can do. Copper tolerance is heritable. Probably as a barrier to hybridization, genetic disruption is bringing about the marked difference in the flowering of their reproductive shoots.

The graph shows average flowering times in the years 1964 (*red* line) and 1965 (*green* line).

Ecological Isolation   Reproductive isolation also may arise when potential mates occupy and become adapted to different microenvironments that exist in their habitat.

For example, as shown in Figure 19.6, A. Bradshaw and T. McNeilly identified an abrupt divergence in flowering time between adjacent populations of a grass (*Agrostis tenuis*) that are growing near copper mines. Copper-polluted slag heaps (the residues of mining activities) have piled up next to pastures. Grass populations growing within 20 meters of this environmental discontinuity differ greatly in copper tolerance. Transplant a pasture plant to polluted soil just a few meters away from its home range, and it will die. Transplant offspring from a copper-tolerant plant to a pasture, and they show reduced fitness.

Apparently, ecological isolation may be driving the two populations of *Agrostis* in the direction of speciation. A barrier to hybridization already is becoming established, as evidenced by the marked divergence in the time at which their reproductive shoots flower.

Postzygotic Isolation

Interspecific hybridization also may be discouraged by isolating mechanisms that take effect *after* fertilization, while the embryo is developing. These mechanisms lead to early death, sterility, or hybrids of low fitness. Genes, the mechanisms that control gene expression, or the gene products themselves may fail to interact properly. Even when hybrids get through embryonic development, they commonly are weak, and their survival rate is not good. A few types are sturdy but sterile. Mules, which result from a cross between a female horse and a male donkey, are like this. So are most zebroids, owing to chromosomal differences between the parents.

Speciation Defined

Suppose two genetically diverging populations remain isolated from each other. Suppose, further, that enough differences accumulate between them to prevent their individuals from interbreeding. By Mayr's concept, this is the key indicator that speciation—the process by which species form—is completed.

Because genetic divergence is commonly a gradual process, it is impossible to say precisely when daughter species often form. Consider this diagram:

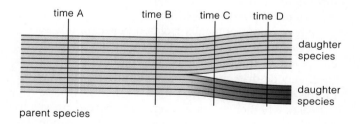

Each horizontal line in the diagram represents a different population of a species. At time A, there is only one species. At D, there are two. At B and C, divergence is under way but is far from complete.

The evolution of reproductive isolating mechanisms paves the way for genetic divergence and speciation.

19.4 MODELS OF SPECIATION—AN OVERVIEW

Speciation may unfold slowly or rapidly, and it varies in its details, depending on the distribution patterns and interactions among the affected populations. Three current models for speciation account for these differences:

1. **Allopatric speciation**. In the absence of gene flow between geographically separate populations, daughter species form gradually, by divergence. (*Allo-* means different, and *patria* can be taken to mean homeland.)

2. **Sympatric speciation**. Daughter species arise, sometimes rapidly, from a small proportion of individuals within an existing population. (*Sym-* means together with, as in "together with others in the homeland.")

3. **Parapatric speciation**. Daughter species form from a small proportion of individuals along a common border between two populations. (*Para-* means near, as in "near another homeland.")

19.5 ALLOPATRIC SPECIATION

This model for speciation emphasizes the disruption of gene flow, and it may represent the most prevalent speciation route. By this model, a physical barrier cuts off gene flow between two or more populations. Through divergence and the evolution of isolating mechanisms, the separated populations become reproductively incompatible and cannot interbreed even if circumstances put them back together.

When a geographically isolated population is small, chances are greater that its allele frequencies will differ, compared to the bulk of the population. If this is the case, genetic drift might have a significant initial effect on the direction of change.

Also, whether or not a geographic barrier proves to be effective at cutting off gene flow between populations depends on a particular organism's built-in means of travel, how fast it can travel, and whether it is inclined or compelled to disperse. You have only to think back on those garden snails and the wandering albatross described at the start of this chapter to know that this is so.

Allopatry is thought to be the main speciation route because the populations of most species are not strung out continuously, with one merging into the others. Most often some distance separates them, and gene flow is more of an intermittent trickle than a steady stream. Sometimes barriers form and shut off even the

Focus on Science

Speciation and the Isthmus of Panama

The "solid earth" beneath your feet is actually cracked into huge fragments, called crustal plates. The plates have been slowly drifting about on molten material, probably for as long as the earth has had a crust. Roughly 3,100,000 years ago, two crustal plates collided beneath the waters of an ancient ocean. As the advancing edge of one plate thrust under the other, a sinuous ridge of land was uplifted. We now call it the Isthmus of Panama.

Biologists interested in the mechanisms of allopatric speciation could scarcely have asked for a better natural laboratory. The formation of the Isthmus divided an ocean basin in two. Gene flow between populations on either side of the newly forming coastlines declined, then stopped.

This geographic barrier disrupted prevailing ocean currents, so that they now flowed in different directions. The temperature and chemical composition changed accordingly. We can assume that different selection pressures started acting on the separated populations.

A variety of Atlantic and Pacific fishes are now known to be closely related. Recent evidence comes from comparisons of mutations in the molecular structure of the same proteins (gene products) from different species. As you will read in the chapter to follow, an accumulation of mutations that are neutral or possibly adaptive might signify an evolutionary "foot in the door"—that is, genetic divergences that may be evidence of speciation in progress.

In the early 1980s, John Graves compared certain enzymes present in the muscle cells of related Isthmus fishes, all of which are strong swimmers. He started with the knowledge that enzyme activity varies with increases in temperature, and that different enzymes differ in their responses to variations in temperature. He knew that seawater on the Pacific side of the Isthmus is cooler by about 2°–3°C than it is on the Atlantic side, and that it varies more with the changing seasons.

trickles. This has happened abruptly, as when a major earthquake changed the course of the Mississippi River in the 1800s and isolated some populations of insects that could not swim or fly.

Geographic isolation also has happened slowly, over great spans of time. The fossil record provides ample evidence of the breakup of formerly continuous environments. For example, many times in the past, populations became geographically cut off from each other by immense glaciers that advanced down through North America and Europe. When the glaciers retreated and

Strenuous muscle activity requires efficient enzyme activity. Have the temperature differences on either side of the Isthmus put the related fishes under different selection pressures? To test this hypothesis, Graves studied four pairs of enzymes from closely related fishes. He compared their performance under a range of temperatures. He also subjected them to gel electophoresis so that he could pinpoint any differences in their molecular structure (page 220).

Graves discovered that all of the "Pacific" enzymes function better at lower temperatures than the "Atlantic" enzymes do. Two of the four pairs of enzymes differ slightly in electric charge. Electrophoresis results revealed a slight difference in their amino acid sequences.

Graves drew three tentative conclusions. First, even small differences in environmental conditions may favor molecular adaptation. Second, mutations in the genes he studied apparently had neutral effects; even with slight molecular differences, they all remained active. Finally, the enzymes may have undergone molecular changes that cannot be detected by electrophoresis; they all showed detectable differences in their activity.

Two kinds of fishes, apparently related by descent from a common ancestral population that became divided when geologic forces created the Isthmus of Panama. (**a**) *Thalassoma bifasciatum*, a blue-headed wrasse from the Atlantic side of the Isthmus, and (**b**) *T. lucasanum*, a Cortez rainbow wrasse from the Pacific side. In this case, differences in body coloration and patterning do not provide much evolutionary insight. Why? Reef fishes commonly show stunning phenotypic variation among individuals of the *same* species. Information more revealing of evolutionary relationship comes from studies such as the one described in this essay.

descendant populations came in contact, sometimes they were no longer reproductively compatible; they had become separate species. In other cases, divergences had not proceeded far enough in separated populations, and they can still interbreed; speciation was not completed.

Also, many millions of years ago, imperceptibly slow but colossal movements in the earth's crust resulted in the breakup and crunching together of huge land masses. Such movements brought about an uplifting of the seafloor in the region we now call the Isthmus of Panama. The uplifting divided an ocean basin, setting the stage for allopatric speciation among populations of fishes and other marine species. The *Focus* essay describes one investigation into the genetic divergence that followed.

In the absence of gene flow between geographically separate populations, genetic divergence may become sufficient to bring about allopatric speciation.

LESS PREVALENT SPECIATION ROUTES

Sympatric Speciation and The Crater Lake Cichlids

The model of sympatric speciation does not invoke geographic isolation. It suggests that species can form *within* the range of an existing species, in the absence of physical or ecological barriers. The model has been controversial, but strong evidence in its favor comes from fishes within the confines of crater lakes.

In 1994, Ulrich Schliewen and his coworkers reported an apparent case of sympatric speciation among cichlids, a type of fish, in two ecologically uniform crater lakes in Cameroon, East Africa. (The "craters" are collapsed volcanic cones.) Both lakes are small, yet eleven cichlid species are native to one of them, and nine are native to the other. The researchers analyzed mitochondrial DNA from all of the species. They did the same for related species in neighboring lake and river systems, as a basis for comparison. They found that all species in each lake are alike in their mitochondrial DNA sequences—and unlike the neighboring "outgroup" species—so there is good reason to believe they descended from the same founder species. For example, only the nine species in one lake carry an unusual base-pair substitution in the gene coding for cytochrome *b*.

The lakes are so small, it is safe to assume that individuals of different species encounter one another often. That is, they must live in sympatry. As Figure 19.7 suggests, the lakes have uniform shorelines, with no small-scale geographic barriers—so allopatric populations could not have formed even on a small scale. Besides, cichlids are not sluggish; they frequently move through different parts of the lake. And even if the lake levels fluctuated over time, isolated basins could not have formed because each crater is uniformly conical. Finally, crater rims isolate the lakes from all but small creeks; gene flow from the outside is restricted. The lakes must have been colonized in the past, before their connection with a source river system was broken.

The species in each lake do show a small degree of ecological separation, with some feeding in the open waters and others on the lake bottom. Possibly this separation may have been enough to promote selective mating and, eventually, the reproductive isolation that can trigger speciation. However, even if there were sorting based on feeding preferences, all of the species breed close to the lake bottom; they still mate in sympatry.

Polyploidy

Sympatric speciation also may have proceeded among flowering plants, about half of which are polyploid species. The word **polyploidy** refers to the inheritance

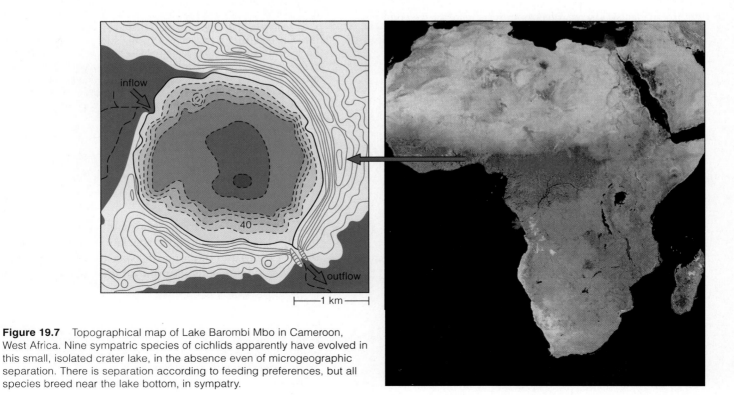

Figure 19.7 Topographical map of Lake Barombi Mbo in Cameroon, West Africa. Nine sympatric species of cichlids apparently have evolved in this small, isolated crater lake, in the absence even of microgeographic separation. There is separation according to feeding preferences, but all species breed near the lake bottom, in sympatry.

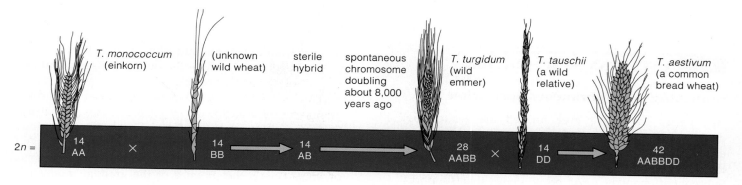

a About 11,000 years ago, humans start cultivating wild wheats. The species *Triticum monococcum* has diploid number 14 (two sets of 7 chromosomes, shown as 14AA). It hybridizes with another species that has the same chromosome number.

b AB hybrid offspring are sterile but self-fertilizing; an interbreeding population of AB plants arises by asexual reproduction. About 8,000 years ago, by unknown events, polyploidy arises in the population. Some plants (*T. turgidum*) are tetraploid (AABB), with a chromosome number of 28 (two sets of 14). They are fertile. (A chromosomes can pair with each other, and so can B chromosomes, during meiosis.)

c Later, an AABB plant hybridizes with *T. tauschii*, a wild relative with a diploid number of 14 (two sets of 7 DD). Today, populations of the hybrid descendants, *T. aestivum*, provide wheat for bread. Their chromosome number is 42 (six sets of 7 AABBDD).

Figure 19.8 Presumed sympatric speciation in wheat through polyploidy and hybridizations. Wheat grains 11,000 years old have been found in the Near East. Diploid wild wheats still grow there.

of three or more of each type of chromosome characteristic of the parental stock. It can result from improper separation of chromosomes during meiosis or mitosis (page 201). It also can result when a germ cell duplicates its DNA but fails to divide, then goes on to function as a gamete.

Speciation may have been instantaneous for many flowering plant species; about half of these species are polyploid. Many can self-fertilize or reproduce asexually, so their offspring can inherit the novel chromosome number. Their extra chromosomes pair successfully at meiosis, and the extra genes do no harm. Polyploidy also may result in speciation when followed by cross-fertilization between plant species. Figure 19.8 provides an example of this. However, most hybrids between species are sterile.

Polyploidy is less common in animals, probably because it upsets complex gene interactions that are crucial for animal development. Usually, embryos that develop from animal gametes with mismatched chromosomes cannot survive. Those that do commonly have severe genetic disorders, shorter lives, or both.

Parapatric Speciation

Some biologists argue that parapatric speciation may occur where populations share a common border, even though the borders are permeable to gene flow. In some cases, most gene exchange may be confined to a **hybrid zone**, where adjoining populations meet, interbreed, and produce hybrid offspring. One hybrid zone extends through Nebraska and adjoining states (Figure 19.9). To the east of this overlapping zone of contact are Balti-

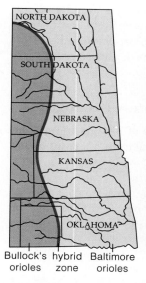

Bullock's hybrid Baltimore
orioles zone orioles

Bullock's oriole

Figure 19.9 A hybrid zone where parapatric speciation among orioles (*Icterus*) may be occurring along the borders.

more orioles (*Icterus*). To the west are Bullock's orioles. Hybridization was so common in the past, Baltimore and Bullock's orioles were said to be variants of a single species. Today, hybrid orioles are becoming less and less frequent. Isolating mechanisms may be reinforcing divergence in the zone of overlap.

Although data are sketchy, it may be that speciation can proceed even in the absence of geographic isolation.

19.7 PATTERNS OF SPECIATION

Branching and Unbranched Speciation

All species that have ever formed are related by descent. They are genetically connected through lineages that extend back in time to the molecular origin of the first prototypic cells, some 3.8 billion years ago. Subsequent chapters focus on evidence that supports this view. In anticipation of those chapters, let's start thinking about ways to interpret the large-scale histories of species.

The fossil record points to two speciation patterns, one branching, the other unbranched. The first of these patterns is called **cladogenesis** (from the Greek *klados*, meaning branch, and *genesis*, meaning origin). It applies to populations that become isolated from one another and subsequently diverge in different directions, as described earlier in this chapter.

Species also may form within a single, unbranched line of descent. This pattern of speciation is called **anagenesis**. (In this context, *ana-* means renewed.) In this case, the changes in allele frequencies accumulate along a single evolutionary road. In time, someone decides that a descendant population is morphologically different enough from the ancestral form and so names it a new species.

Evolutionary Trees and Rates of Change

Information about the continuity of relationship among species is often summarized in the form of **evolutionary tree diagrams**.

Take a look at Figure 19.10, which is a simple way to start thinking about how the diagrams are constructed. Each branch in a tree diagram represents one line of descent from a common ancestor. A branch point is a time of divergence and speciation, as brought about by microevolutionary processes.

Figure 19.10 also shows how evolutionary tree diagrams can be used to convey rates of change—that is, the approximate length of time between speciation events. Branches with slight angles imply that species emerge by many small changes in form over long spans of time. This is the premise of the **gradual model of speciation**, which shows a good fit with many fossil sequences. For example, in layers of sedimentary rock we find sequences of beautifully preserved, beautifully perforated shells of ancient foraminiferans (Figure 23.5). The sequences provide evidence of slow change.

Alternatively, some evolutionary tree diagrams have short, horizontal branches that abruptly make a 90-degree turn (Figure 19.10*b*). These diagrams are consistent with the **punctuation model of speciation**. By this model, most morphological changes are compressed into a brief period when populations first start to diverge—say, within hundreds or thousands of years. The idea is that founder effects, bottlenecks, strong directional selection, or some combination of these bring about rapid speciation. The daughter species quickly recover from the adaptive wrenching and change very little after that—say, for 2 million to 6 million years. Thus, adaptive cohesion usually prevails for about 99 percent of the history of most lineages. This pattern seems to pervade many parts of the fossil record.

Apparently, change can be gradual, abrupt, or both. Species originate at different times and differ in their persistence on the evolutionary stage. Remember, some lineages have endured without much change, producing a species here, losing a species there, over tens of millions of years. Still others have branched bushily and sometimes spectacularly, during times of adaptive radiation.

Figure 19.10 How to read an evolutionary tree diagram.

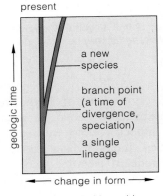

present

geologic time

← change in form →

a new species

branch point (a time of divergence, speciation)

a single lineage

a Softly angled branching means speciation occurred through gradual changes in traits over geologic time.

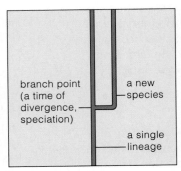

branch point (a time of divergence, speciation)

a new species

a single lineage

b Horizontal branching means traits changed rapidly around the time of speciation. Vertical continuation of a branch means traits of the new species did not change much thereafter.

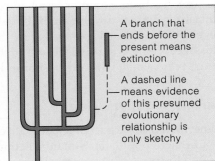

A branch that ends before the present means extinction

A dashed line means evidence of this presumed evolutionary relationship is only sketchy

c Many branchings of the same lineage at or near the same point in geologic time means that an adaptive radiation occurred.

Adaptive Radiations

An **adaptive radiation** is a burst of microevolutionary activity within a lineage, one that results in the formation of new species in a wide range of habitats. Figure 19.11 shows an example. Lineages often spread out this way when their members are presented with unfilled adaptive zones. **Adaptive zones** are most easily defined as ways of life, such as "burrowing in the seafloor" or "catching insects in the air at night."

A lineage may radiate into an adaptive zone when it has physical, evolutionary, or ecological access to it. *Physical* access means a lineage happens to be there when adaptive zones open up. Consider how mammals were once distributed through uniform tropical regions of a single continent. That continent split into several land masses. Habitats and resources changed differently on these land masses and set the stage for independent radiations.

Evolutionary access means that modification of some structure or function will permit a lineage to exploit the environment in improved or novel ways. When the forelimbs of certain five-toed vertebrates evolved into wings, for example, this opened new adaptive zones to the ancestors of birds and bats.

Ecological access means the lineage can enter an unoccupied adaptive zone or outcompete the resident species. We will say more about this in Chapter 47.

19.8 EXTINCTIONS— END OF THE LINE

Some number of species within a lineage inevitably disappear as local conditions change. This expected rate of disappearance over time is called **background extinction**. Table 19.2 lists examples. By contrast, an abrupt rise in extinction rates above the background level is called **mass extinction**. It is a catastrophic, global event in which families and other major groups of species are wiped out simultaneously.

Lineages have tended to survive mass extinctions when their members have been widely dispersed. The hardest hit have been highly specialized for life in tropical regions. Even so, luck has a lot to do with it. For example, the playing field was once leveled when an asteroid hit the earth. As described in Chapter 21, some lucky survivors radiated into adaptive zones previously occupied by dinosaurs. Among them were mammals that had previously remained, wisely, hidden in the shrubbery.

Taken together, the persistence, branchings, and extinctions of species account for the full range of biological diversity at any point in time.

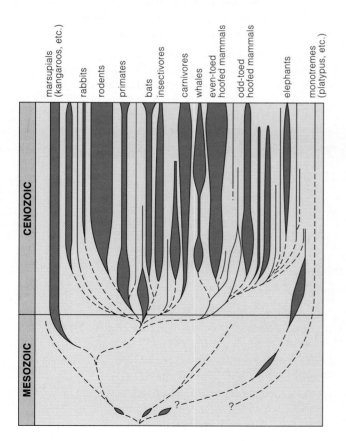

Figure 19.11 Evolutionary tree diagram of the great adaptive radiation of mammals, which began about 65 million years ago at the start of a geologic era called the Cenozoic. Variations in the width of a given branch correspond to the range of diversity within the lineage represented at different points in time. The wider the branch, the greater the diversity.

Table 19.2	Examples of the Estimated Average Durations of Species	
	Group	Duration (millions of years)
Protistans:	Foraminiferans	20–30
	Diatoms	25
Plants:	Bryophytes	20+
	Higher plants	8–20+
Animals:	Bivalves	11–14
	Gastropods	10–13.5
	Graptolites	2–3
	Ammonites	1–2, 6–15
	Trilobites	1+
	Beetles	2+
	Freshwater fishes	3
	Snakes	2+
	Mammals	1–2+

From Stanley, 1985.

SUMMARY

1. For sexually reproducing organisms, a species consists of one or more populations of individuals that can interbreed under natural conditions and produce fertile offspring, and that are reproductively isolated from other such populations.

2. For as long as populations with a shared genetic history are maintaining genetic contact over time (by gene flow), they retain membership in the same species.

a. Microevolutionary processes underlie the history of species and lineages (Table 19.3).

b. Species are distinct, well-adapted, cohesive entities. Gene flow and natural selection tend to maintain the phenotypes that are best adapted to the environment in which the species has evolved.

c. The directions in which a species can change depend on its genetic luggage and on the kinds of environments through which its ancestral members traveled.

3. Species evolve through genetic divergence, a buildup of differences in allele frequencies between two populations (or between two subpopulations).

4. Genetic divergence is promoted by geographic isolation and by reproductive isolating mechanisms.

a. Prezygotic isolating mechanisms prevent mating or pollination. They include differences in the timing of reproduction, differences in reproductive behavior, incompatibilities between reproductive structures and gametes, and occupation of different microhabitats within an area.

b. Postzygotic isolating mechanisms lead to early death, sterility, or unfit hybrid offspring. They take effect after fertilization, during the development of the embryo.

5. There are three current models of speciation:

a. Allopatric speciation. Gene flow is prevented between geographically separated populations, which proceed to diverge genetically into new species.

b. Sympatric speciation. Species form from a small proportion of individuals in the home range of an existing population. Geographic isolation is not required. Speciation by polyploidy is an example.

c. Parapatric speciation. Species form from a small proportion of individuals along the common border between two populations, where gene flow is localized.

6. Lineages vary in the times, rates, and direction of speciation. The speciation process may be gradual, rapid, or both within a lineage. It can proceed in a single, unbranched line of descent or by branching (divergence of one or more populations in different directions). Extensive branching during the same period of geologic time is called an adaptive radiation.

7. Evolutionary tree diagrams show the relationship among groups of species. Each branch of such trees represents a line of descent (lineage). Branch points are speciation events, as brought about by genetic drift, natural selection, and other microevolutionary processes.

8. All lineages inevitably lose some number of species over time. This loss is called background extinction. In a mass extinction, major groups of species perish abruptly in a catastrophic, global event.

9. Taken together, the persistence, branchings, and extinctions of species account for the full range of biological diversity during any period of geologic time.

Table 19.3 Summary of the Processes and Patterns of Evolution

Microevolutionary Processes

Mutation	Original source of alleles	Stability or change in a species is the outcome of balances or imbalances among all of these processes, the effects of which are influenced by population size and by prevailing environmental conditions
Gene flow	Preserves species cohesion	
Genetic drift	Erodes species cohesion	
Natural selection	Preserves or erodes species cohesion, depending on environmental pressures	

Macroevolutionary Patterns

Genetic persistence	The basis of the unity of life. Genetic continuity extends from the molecular origin of the first cells through all subsequent lines of descent
Genetic divergence	The basis of the diversity of life, brought about by adaptive shifts, branchings, and radiations. The rates and times of change have varied within and between lineages
Genetic disconnect	The steady loss (background extinction) or abrupt, catastrophic loss (mass extinction) of lineages

1. Explain the value of the "biological species concept" as proposed by evolutionary biologist Ernst Mayr. *288*

2. Define and cite an example of each of the three speciation models. *292–295*

3. Describe the appearance of the following features of an evolutionary tree diagram and how they are interpreted: a single lineage, softly angled branching, horizontal branching, vertical continuation of a branch, many branchings of the same lineage; a dashed line, a branch that ends before the present. *296*

Self-Quiz *(Answers in Appendix IV)*

1. Sexually reproducing individuals belong to the same species if they _____ .
   a. can interbreed under natural conditions
   b. can produce fertile offspring
   c. are reproductively isolated from other such populations
   d. have a shared genetic history
   e. all of the above

2. The biological species concept _____ .
   a. avoids the trap of defining species by appearance alone
   b. applies to all kingdoms of life
   c. helps us interpret the fossil record
   d. all of the above

3. Genetic divergence is a buildup of differences in allele frequencies between _____ .
   a. species
   b. populations of a species
   c. subpopulations of a species
   d. both b and c

4. Isolating mechanisms _____ .
   a. prevent interbreeding   c. reinforce genetic divergence
   b. prevent gene flow   d. all of the above

5. Speciation occurs through _____ .
   a. the evolution of reproductive isolating mechanisms
   b. genetic divergence
   c. mutation, genetic drift, and natural selection
   d. both b and c
   e. all of the above

6. When potential mates occupy overlapping ranges but reproduce at different times, this is a case of _____ isolation.
   a. postzygotic   c. temporal
   b. mechanical   d. gametic

7. Probably the most prevalent speciation route is _____ .
   a. allopatric speciation   c. parapatric speciation
   b. sympatric speciation   d. both a and c

8. Whether or not a geographic barrier deters gene flow between populations of the same species depends on an organism's _____ .
   a. built-in means of travel, speed, and endurance
   b. environmental proddings to disperse
   c. inclination to disperse
   d. all of the above

9. In evolutionary tree diagrams, each branch represents a _____ , and each branch point represents _____ .
   a. species; genetic divergence
   b. line of descent; genetic divergence
   c. species; speciation event
   d. line of descent; speciation event

10. An evolutionary tree diagram with horizontal branches that abruptly become vertical is consistent with _____ .
   a. the gradual model of speciation
   b. the punctuation model of speciation
   c. the idea of small changes in form over long spans of time
   d. both a and c
   e. both b and c

11. Match the terms with the suitable descriptions.
   ____ cladogenesis   a. unbranched lineage
   ____ anagenesis   b. burst of microevolutioary
   ____ adaptive radiation   activity within a lineage
   ____ background extinction c. catastrophic disappearance
   ____ mass extinction   of major groups of organisms
                              d. branching lineages
                              e. expected rate of species loss
                                 within a lineage

Selected Key Terms

adaptive radiation *297*
adaptive zone *297*
allopatric speciation *292*
anagenesis *296*
background extinction *297*
biological species concept *288*
cladogenesis *296*
evolutionary tree diagram *296*
genetic divergence *289*
gradual model of speciation *296*

hybrid zone *295*
isolating mechanism *289*
mass extinction *297*
parapatric speciation *292*
polyploidy *294*
punctuation model of speciation *296*
speciation *287*
species *288*
sympatric speciation *292*

Readings

Mayr, E. 1976. *Evolution and the Diversity of Life*. Cambridge, Massachusetts: Belknap Press of Harvard University Press. Collection of the author's essays on major issues in evolutionary biology, revised to reflect his present views.

Otte, D., and J. Endler (editors). 1989. *Speciation and Its Consequences*. Sunderland, Massachusetts. Thought-provoking essays by evolutionists who specialize in a broad range of subjects. Paperback.

Ridley, M. 1993. *Evolution*. Boston: Blackwell Scientific Publications.

20 THE MACROEVOLUTIONARY PUZZLE

Of Floods and Fossils

About 500 years ago, Leonardo da Vinci was brooding about seashells entombed in the layered rocks of northern Italy's high mountains, hundreds of kilometers from the sea. How did they get there? If he accepted the traditional explanation, he would have to agree that turbulent floodwaters deposited shells in the mountains during the Great Deluge (Figure 20.1). But many shells were thin, fragile—and intact. Surely they would have been battered to bits if they had been swept across such distances, then up the mountains.

Da Vinci also brooded about the rock layers. They were stratified (stacked like cake layers), and some had shells but others had none. Then he remembered how large rivers deposit silt during spring floods. Had the layers slowly accumulated one atop the other, as a series of silt deposits in the past? If so, the shells in the mountains would be evidence of a series of vanished communities of organisms that had been gradually buried *in the sea!* Da Vinci did not announce his novel idea, perhaps knowing it would have been met with deafening silence, imprisonment, or worse.

By the 1700s, fossils were being accepted as evidence of past life. They were still interpreted in a traditional way, as when a Swiss naturalist excitedly unveiled the remains of a giant salamander and announced they were the skeleton of a man who drowned in the Deluge.

By midcentury, however, scholars began to question such interpretations. Extensive mining, quarrying, and canal excavations were under way. The diggers were finding similar rock layers and similar fossil sequences in distant places, such as the cliffs on both sides of the English Channel. And some scholars started analyzing the discoveries for possible connections between earth history and the history of life.

Ever since, fossils have been analyzed in more and more refined ways. Together with biochemical studies and other modern sources of information, they provide good evidence of evolution over vast spans of time—*of changes in the geologic stage, and changes in the organisms that have marched across it.*

Figure 20.1 (*Left*) From the Sistine Chapel, Michelangelo's painting of the onset of a catastrophic flood, traditionally called the Great Deluge. Events of this magnitude actually punctuated geologic time, although they have been explained in different ways throughout recent human history. (*Below*) A modern-day photographer captures the intricate structural pattern of the fossilized shells of marine animals called ammonites. About 65 million years ago, all ammonites perished, along with many other groups of organisms. That mass extinction is but one intriguing piece of the macroevolutionary puzzle.

KEY CONCEPTS

1. In the evolutionary view, all species that have ever lived are related—some closely, others remotely. This is because each new species had to evolve from variant individuals of existing species, starting with the first living cells to appear on earth. The term "macroevolution" refers to the patterns, trends, and rates of change among lineages over geologic time.

2. The fossil record, geologic record, and radioactive dating of rocks yield evidence of life's macroevolution. Anatomical comparisons between different lineages help us understand and reconstruct patterns of change through time. Among the most revealing aspects of anatomy are homologous structures in separate lineages; these suggest descent from a common ancestor. Biochemical comparisons within and between major groups of organisms also provide insight into their evolution.

3. Organisms show great diversity in their distribution through time and through the environment. Biological systematics attempts to discern patterns in diversity in three ways. With *taxonomy*, organisms are identified and named. With *phylogenetic reconstruction*, their evolutionary connections are worked out. With *classification*, the information is organized into retrieval systems.

The history of life spans nearly 4 billion years. Ultimately, it is a story of *species*—how each kind formed, whether its traits persisted or changed over time, and whether it endured or became extinct.

Biologists have already identified about 2 million species, which they have been grouping into genera, families, and so on up to kingdoms. And they have been hard at work on the macroevolutionary puzzle. **Macroevolution** refers to large-scale patterns, trends, and rates of change among these groups of species.

Evolution, recall, can proceed only within existing populations of species. Therefore, *a continuity of relationship must connect all species that have ever appeared on earth.* This principle of evolution guides efforts to make sense of scattered and sometimes puzzling scraps of evidence of past life. It guides the task of identifying and sorting out lines of descent, or lineages, that connect all species, past and present. As you will now read, the fossil record and earth history yield evidence of evolutionary connectedness. So do anatomical and biochemical comparisons.

20.1 FOSSILS—EVIDENCE OF ANCIENT LIFE

Fossilization

"Fossil" comes from a Latin word for something that has been "dug up." In general, **fossils** are recognizable, physical evidence of ancient life. Figure 20.2 shows examples. The most common fossils are bones, teeth, shells, spore capsules, seeds, and other hard parts. (Soft parts usually are the first to decompose when an organism dies.) Indirect evidence comes from *trace* fossils, which include imprints of leaves and stems, tracks, trails, and burrows. Even fossilized feces (coprolites) provide clues to what ancient organisms ate.

Figure 20.2 Representative fossils—evidence of life in the distant past. (**a**) A fossil hunter's dream—the complete skeleton of a bat that lived 50 million years ago. Erosion, earth movements, and other forces of nature have left few burial sites undisturbed, so intact fossils are extremely rare. Even the jumbled skeletal parts of several ducklike birds in (**b**) are considered a good find. Many hours of careful preparation and study will be required to establish the identity of the species and identify its morphological traits.

(**c**) Fossilized parts of *Cooksonia*, the oldest known type of land plant. Its stems were less than 7 centimeters tall. (**d**) One of the most magnificent, complete fossils ever recovered—skeletal remains of an ichthyosaur, a dolphin-like marine reptile that lived about 200 million years ago.

a
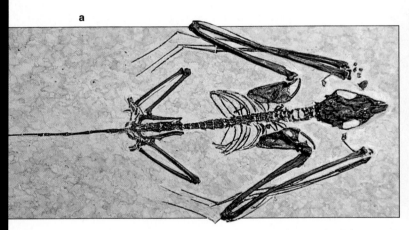

b

c

d (*Below*)

Fossilization begins with burial in sediments or volcanic ash. Sooner or later, water infiltrates the organic remains, infusing them with dissolved metal ions and other inorganic compounds. As more and more sediments accumulate above the burial site, the remains become subjected to increasing pressure. Over great spans of time, the chemical changes and pressure transform them to stony hardness.

Preservation is favored when organisms are buried rapidly in the absence of oxygen. Gentle entombment by volcanic ash or anaerobic mud is best. Preservation also is favored when a burial site is undisturbed. Most often, though, erosion and other geologic insults have crushed, deformed, broken, or scattered the fossils.

Interpreting the Geologic Tombs

You can find similar fossil-containing layers of sedimentary rock over vast areas, even on different continents. Such layers formed long ago, when silt, volcanic ash, and other materials were gradually deposited, one above the other. This layering of sedimentary deposits is called **stratification**. Figure 20.3 shows an example. The deepest layers originally were the first ones formed; layers closest to the surface were formed last. Because particles tend to settle in response to gravity, most sedimentary layers must have formed horizontally. Where they are ruptured or tilted, this generally is evidence of later geologic disturbance.

Understand how rock layers form, and you realize that fossils within a given layer are from a similar age in earth history. Therefore, the older the layer, the older the fossils. Early naturalists had no way of determining the age of either the fossils or the rocks. They simply divided the past into great eras, on the basis of abrupt changes in fossil sequences. Much later, actual dates were assigned to this "geologic time scale," which forms the chronological framework for the next chapter.

Interpreting the Fossil Record

At present, we have fossils for about 250,000 species, and they provide rich insights into evolutionary history. However, judging from the current sweep of diversity, there must have been many millions of ancient, now-extinct species, and we will not be able to recover fossils for most of them. Thus, our "record" of past life is incomplete, with built-in biases.

Most important, large-scale movements in the earth's crust have obliterated evidence from crucial periods in the history of life. Besides this, most members of ancient communities simply have not been preserved. For example, hard-shelled mollusks and bony fishes are well represented in the fossil record. Jelly-fishes and soft-bodied worms are not, even though they

Figure 20.3 A splendid slice through time: the Grand Canyon of the American Southwest, once part of an ocean basin. Its layers of sedimentary rocks formed gradually over hundreds of millions of years. Tectonic forces lifted them above sea level. Later, the erosive force of rivers carved the deep canyon walls.

may have been just as common or more so. Population density and body size further skew the record. A population of ancient plants may have produced millions of spores in a single growing season, whereas the earliest members of the human lineage lived in very small groups and produced very few offspring. What are your chances of finding a fossilized skeleton of an early human, compared to spores of plant species that lived at the same time?

The fossil record also is heavily biased toward certain environments. Most of the species represented lived on land or in shallow seas that, through geologic uplifting, became part of continents. We have few fossils from sediments beneath the ocean—which extends across three-fourths of the earth's surface. Finally, most fossils have been found in the Northern Hemisphere, because that's where most geologists have lived.

Fossils, the stone-hard physical evidence of ancient life, are present in layers of sedimentary rocks. The deeper the layers, the older the fossils.

The completeness of the fossil record varies as a function of the kinds of organisms represented, where they lived, and the stability of their burial sites.

20.2 EVIDENCE FROM COMPARATIVE EMBRYOLOGY

Evidence of macroevolution comes from anatomical comparisons of major lineages. This work is called **comparative morphology**. A few examples will provide you with insight into one of its guiding principles—namely, when it comes to changes that affect the body's development, evolution tends to follow the line of least resistance.

Developmental Programs Among the Vertebrates

Recall, from Chapter 19, that most constraints on evolution are expressed at the time of development. In one sense, these constraints work to preserve existing species. In another sense, they allow evolution to proceed, but only in particular directions.

Consider the vertebrates, which range from fishes to amphibians, reptiles, birds, and mammals. These distinct lineages are spectacularly diverse. Even so, comparisons of the ways in which their embryos develop provide compelling evidence of their evolutionary connection to one another.

In each lineage, individuals start out life as a fertilized egg. They proceed through various stages of embryonic development before becoming an adult. Early in the developmental program, embryos of different lineages proceed through strikingly similar stages. (Take a look at Figure 20.4. Without the labels, would you know which embryo is from a fish, frog, chicken, or human?)

Vertebrate embryos strongly resemble one another because they inherit the same ancient plan for development. According to the plan, tissues form only when cells divide in certain patterns and interact in prescribed ways. Later on, the heart, bones, muscles, and other body parts grow in prescribed ways, provided that each developmental step is properly completed before the next step begins.

During vertebrate evolution, most mutations that disrupted an early stage of development would have had devastating effects on the organized interactions required for later stages. Embryos of different groups remained similar because mutations that altered steps in early developmental stages were selected against.

How, then, did the adults of different vertebrate groups get to be so different? At least some differences must have resulted from mutations that affected the onset, rate, or time of completion of certain developmental steps. Such mutations can bring about changes in shape, due to increases or decreases in the size of certain body parts. (These are "allometric" changes.) They

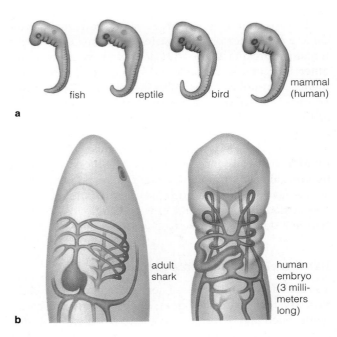

Figure 20.4 From comparative embryology, evidence of evolutionary relationship among vertebrates. (**a**) Even though adult vertebrates show great diversity, the very early embryos retain striking similarities. This is evidence of change in a shared developmental program. (**b**) Fishlike structures still form in early embryos of reptiles, birds, and mammals. For example, a two-chambered heart (*orange*), certain veins (*blue*), and portions of arteries called aortic arches (*red*) develop in fish embryos and persist in adult fishes. The same structures form in an early human embryo.

also can lead to adult forms that retain some juvenile features.

Figure 20.5 illustrates how changes in the growth rate at a key step of development could be responsible for changes in the proportions of chimpanzee and human skull bones. The skull bones of both primates are alike at the time of birth. Thereafter, they change dramatically for chimps but only slightly for humans. Chimps and humans arose from the same ancestral stock. *Their genes are nearly identical.* However, at some point on the separate evolutionary road leading to humans, regulatory genes probably mutated. Since then, instead of promoting the rapid growth required for dramatic changes in skull bones, the mutated genes have blocked it.

Developmental Programs Among the Larkspurs

Flowering plants also provide evidence of evolution through changes in development. Consider the flowers of the most common species of larkspur, *Delphinium*

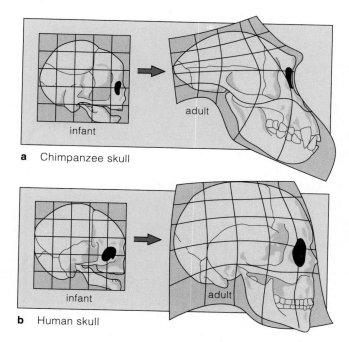

a Chimpanzee skull

a *D. decorum*

b *D. nudicaule*

b Human skull

c Bud and mature flower (side and front views) of *D. decorum*

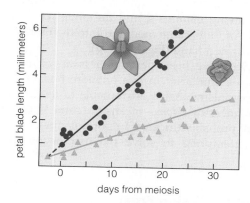

d Mature flower of *D. nudicaule*

Figure 20.5 Morphological differences between two lines of descent, presumably an outcome of changes in the timing of developmental steps. This example compares proportional changes in the skull bones of a chimpanzee and a human. Both skulls are quite similar in infants. Think of the representations of infant skulls as paintings on a blue rubber sheet divided into a grid. Stretching the sheet deforms the grid's squares. For the adult skulls, differences in size and shape within corresponding grid sections reflect differences in growth patterns.

e Lengthening of petals, plotted against development of pollen and eggs in the floral structures, beginning with meiosis

decorum. A ringlike array of petals guides honeybee pollinators to the nectar-storing tube (Figure 20.6). At the flower's center, outward-bulging reproductive structures give the bees something to hang onto while they draw nectar (and pollinate flowers).

D. nudicaule, a larkspur of more recent origin, has tight flowers that discourage bees. Hummingbirds, which hover in front of flowers, are its pollinators. As Figure 20.6b–d shows, its mature flowers strongly resemble the buds of the more common larkspur. The morphological difference is known to be an outcome of a slowed-down rate of floral development in the less common species.

Similar patterns of embryonic development may be a clue to evolutionary relationship among lineages.

Mutations affecting certain steps in the developmental program may be sufficient to bring about major differences in the adult forms of related lineages.

Figure 20.6 From comparative embryology, evidence of evolutionary relationship among two species of larkspurs (*Delphinium*). Although the time required for flowers to develop and mature is much the same in both species, the *rate* of development is much slower for most of the floral structures of *D. nudicaule*, so the changes in petal shape are not nearly as great as they are in the more common species, *D. decorum*.

20.3 EVIDENCE OF MORPHOLOGICAL DIVERGENCE

Homologous Structures

Recall, from the preceding chapter, that when populations of the same species become separated from one another, they start to diverge genetically. Over large spans of time, they diverge in appearance, functions, or both. This macroevolutionary pattern of change from a common ancestor is called **morphological divergence**. (*Morpho-* means body form.)

Even with extensive morphological divergence, however, evolutionarily related species remain alike in many ways. They started out with the same body plan—and their evolution could proceed only through modifications to that shared plan. By looking carefully enough, we can discover their underlying similarities.

For example, reptiles, birds, and mammals are all land-dwelling vertebrates, presumably descended from amphibians that started to venture onto land. Our knowledge of the earliest land vertebrates is based on fossilized, five-toed limb bones of a rather squat reptile (Figure 20.7a). The descendants of this early reptile apparently expanded into new environments on land. Some of the later descendants even retreated from land, through adaptations that allowed them to live in the seas.

And so, in the separate lineages of pterosaurs, birds, and bats, modifications to the five-toed limb gradually evolved into wings (Figure 20.7b–d). In other lineages, the five-toed limb was evolutionary clay that became molded into flippers of porpoises and penguins (Figure 20.7e–f). It became modified into the long, one-toed limbs of modern horses, the stubby limbs of moles and other burrowing mammals, and the pillarlike limbs of elephants. And it became modified into the human arm and five-fingered hand (Figure 20.7g).

What we have been describing is an example of **homology**—a similarity in one or more body parts in different organisms that is attributable to descent from a common ancestor. (*Homo-* means the same.) Such similarities are probably the most important clues for those who are attempting to solve the macroevolutionary puzzle.

Potential Confusion From Analogous Structures

Body parts that have similar form and functions in different lineages aren't *always* homologous. If the last shared connection between two lineages was in the remote past, major differences in form may have developed during their separate evolutionary journeys. Suppose that, by chance, some of their descendants encountered similar environmental pressures. Suppose

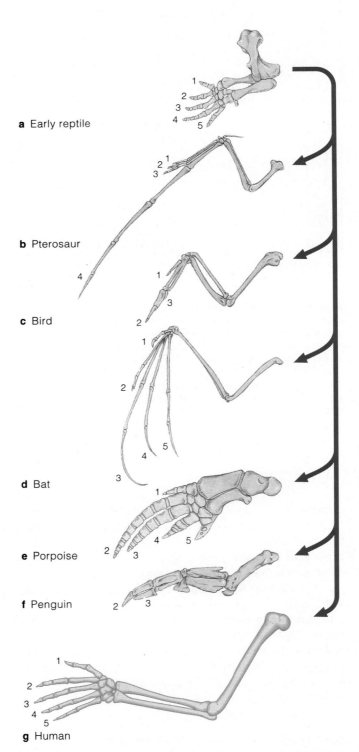

a Early reptile

b Pterosaur

c Bird

d Bat

e Porpoise

f Penguin

g Human

Figure 20.7 Morphological divergence in the vertebrate forelimb, starting with the generalized form of ancestral early reptiles. Diverse forms evolved even while similarities in the number and position of bones were preserved. The drawings are not to the same scale.

Figure 20.8 Morphological convergence among three separate lineages—sharks, penguins, and porpoises. In all three lineages, natural selection favored adaptations that promoted rapid swimming. As a consequence, sharks, penguins, and porpoises show superficial similarities in body form, even though they are only remotely related.

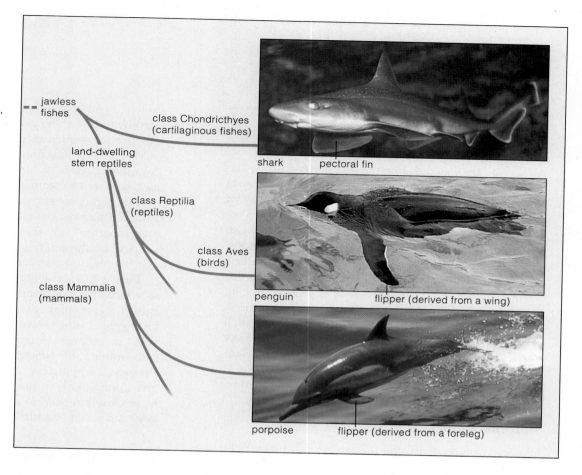

those distantly related organisms started putting comparable body parts to similar uses. Most likely, they would become modified through natural selection and end up resembling one another in structure and function. Such a pattern of long-term change in similar directions among remotely related lineages is called **morphological convergence**.

For example, sharks, penguins, and porpoises are all fast-swimming predators of the seas. As you can see from Figure 20.8, penguin and porpoise flippers are similar in shape to the shark pectoral fin. All three structures stabilize the body in water—but they are not homologous. The shark, penguin, and porpoise lineages diverged very early in vertebrate evolution, so they are only remotely related.

Sharks never left the water, and shark fins haven't changed much from the ancestral form. By contrast, penguins are descended from four-legged vertebrates that evolved into birds, which later became adapted to life in water instead of air. Penguin flippers are modified wings, which are modified forelimbs. Porpoises descended from four-legged vertebrates that evolved

into mammals, which later became adapted to life in the seas—their flippers are modified front legs. Thus, the penguin and porpoise flippers converged on the pectoral fins of sharks.

The structures just described are an example of **analogy**. The word refers to similarity in body parts among distantly related lineages that were once quite different but that converged in structure and function, owing to similar environmental pressures. (*Analogos* means similar to one another.)

Homologous structures are evidence of morphological divergence. They are the same body parts that became modified in different ways in different lines of descent from a common ancestor.

Analogous structures are evidence of morphological convergence. These body parts were once dissimilar in separate lineages but then were put to comparable uses in the same kinds of environments. They became similar in form and function.

+NH₃-gly asp val glu lys gly lys lys ile phe ile met lys cys ser gln cys his thr val glu lys gly gly lys his lys thr gly pro asn leu his gly leu phe gly arg lys thr gly gln ala pro gly tyr ser

+NH₃-ala ser phe ser glu ala pro pro gly asn pro asp ala gly ala gly ile ile phe lys thr lys cys ala gln cys his thr val asp ala gly ala gly his lys gln gly pro asn leu his gly leu phe gly arg gln ser gly thr thr ala gly tyr ser

+NH₃-thr glu phe lys ala gly ser ala lys lys gly ala thr leu phe lys thr arg cys leu gln cys his thr val glu lys gly gly pro his lys val gly pro asn leu his gly ile phe gly arg his ser gly gln ala glu gly tyr ser

EVIDENCE FROM COMPARATIVE BIOCHEMISTRY

All species are a mix of ancestral and derived traits. The kinds and numbers of traits they do or do not share are clues to how closely they are related. This is also true of their biochemical traits.

For example, from morphological studies, you might already have an idea that monkeys, humans, chimpanzees, and other primates are related to one another. You could test your idea by studying small differences in the amino acid sequences of proteins that all three primates produce. Or you could determine whether the nucleotide sequences in their DNA match up closely or not much at all. Logically, the most closely related species share the greatest biochemical similarities. This premise is central to comparative analysis at the molecular level.

Molecular Clocks

Like its close relatives, the human species has unique traits, the outcome of unique gene mutations that accumulated after divergence from the ancestral primate stock. Humans also share many genes (and traits) with other species. For example, the last shared ancestor of yeasts and humans lived many hundreds of millions of years ago. Yet more than 90 percent of the known gene products (proteins) of certain yeasts and of humans have remained the same.

Even though the overall structure of such genes has been highly conserved, it may be that neutral mutations introduced small differences in their structure in different lineages. **Neutral mutations**, recall, have little or no effect on survival or reproduction. They can slip past agents of selection and accumulate in the DNA.

By some calculations, neutral mutations in highly conserved genes accumulated at a regular rate. Think of the accumulation of neutral mutations in any given lineage as a series of predictable ticks of a **molecular clock**. Then turn the clock back—so that the total number of ticks "unwinds" down through the geologic time scale. Where the last tick stops, that is roughly the time of origin for the lineage.

Protein Comparisons

Now suppose two species have the same conserved gene. The amino acid sequences of the gene's product, a protein, are the same or nearly so. The absence of mutation suggests that the two species are closely related. Or suppose the sequences differ. If there are many differences, many neutral mutations must have accumulated. Presumably, then, a long time passed since the two species shared a common ancestor.

One highly conserved gene specifies cytochrome c, a protein component of electron transport chains. Organisms ranging from aerobic bacteria to corn plants to humans synthesize this protein, which has a primary structure of 104 amino acids.

Figure 20.9 shows the strikingly similar amino acid sequences for cytochrome c from a fungus, plant, and animal. Now think about this: The *entire* sequence is identical in humans and chimps. It differs by *one* amino acid in rhesus monkeys. Compared to human cytochrome c, the sequence differs by 18 amino acids in chickens, 19 in turtles, and 56 in yeasts. On the basis of this information, would you assume that humans are more closely related to a chimpanzee or a rhesus monkey? A chicken or a turtle?

Nucleic Acid Comparisons

Typically, structural alterations resulting from gene mutations pepper the nucleotide sequences of DNA and RNA. Therefore, the extent to which a DNA or RNA strand from one species will base-pair with a comparable strand from another species is a rough measure of evolutionary distance between them.

As an example, some nucleic acid comparisons are based on **DNA-DNA hybridization**. In this method, recall, the two strands of a double-stranded DNA molecule are induced to unwind from each other (page 252). A DNA molecule from another species is induced to unwind, also. Then the single strands are allowed to recombine into "hybrid" molecules. Next, the two hybridized strands are subjected to heating, which breaks the hydrogen bonds holding them together. The amount of heat energy required to pull them apart is a measure of their similarity. It takes more heat energy to disrupt the hybrid DNA of closely related species. Figure 20.10 describes one application of this method.

Biochemical similarities and differences among species are clues to their evolutionary relatedness.

ala asn lys asn lys gly ile ile trp gly glu asp thr leu met glu tyr leu glu asn pro lys lys tyr ile pro gly thr lys met ile phe val gly ile lys lys lys glu glu arg ala asp leu ile ala tyr leu lys lys ala thr asn glu-COO⁻

ala asn lys asn lys ala val glu trp glu glu asn thr leu tyr asp tyr leu leu asn pro lys lys tyr ile pro gly thr lys met val phe pro gly leu lys lys pro gln asp arg ala asp leu ile ala tyr leu lys lys ala thr ser ser-COO⁻

ala asn ile lys lys asn val leu trp asp glu asn asn met ser glu tyr leu thr asn pro lys lys tyr ile pro gly thr lys met ala phe gly gly leu lys lys glu lys asp arg asn asp leu ile thr tyr leu lys lys ala cys glu-COO⁻

Figure 20.9 Stretching out across the top of these two pages, the primary structure of three different molecules of cytochrome *c*. The amino acid sequence of this protein has not changed much, even in evolutionarily distant lineages. The three sequences shown are from a representative yeast (*top row*), wheat (*middle row*), and primate (*bottom row*). *Gold* shading marks the amino acids that are identical in all three species. The probability that this striking molecular resemblance resulted from chance alone is extremely low.

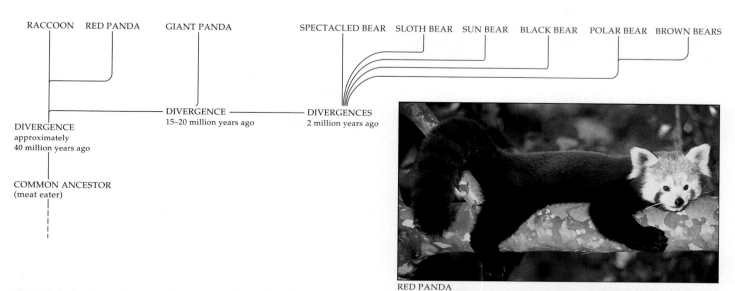

RACCOON   RED PANDA   GIANT PANDA   SPECTACLED BEAR   SLOTH BEAR   SUN BEAR   BLACK BEAR   POLAR BEAR   BROWN BEARS

DIVERGENCE
15–20 million years ago

DIVERGENCES
2 million years ago

DIVERGENCE
approximately
40 million years ago

COMMON ANCESTOR
(meat eater)

RED PANDA

GIANT PANDA

BROWN BEAR

Figure 20.10 How bears, the giant panda, and the red panda are related, according to DNA-DNA hybridization studies.

A giant panda is a plant-eating mammal with a peculiar assortment of traits. Unlike antelope, deer, or cattle, the panda cannot efficiently digest the tough cellulose fibers in plants. Antelope pummel plant material in an elaborately chambered gut, and they benefit from symbiotic gut-dwelling bacteria that produce cellulose-digesting enzymes for them (page 719). By contrast, the panda has the same type of gut as a meat eater. Yet it eats only bamboo—huge amounts of it—to get enough carbohydrates, proteins, and fats. Its rounded, short-toed paws are not much good at holding and stripping leaves from bamboo stalks. But the panda also has a thumblike bony digit on each front paw. When it uses a "thumb" in opposition to the five toes on one of its paws, a bamboo meal is easier to deal with.

Is the giant panda more closely related to bears or to red pandas? It certainly looks like a bear. It also resembles the red panda, which lives in the same general area in China and eats bamboo, although it doesn't have thumbs.

Researchers studied the extent to which single-stranded DNA from the giant panda would hybridize with DNA from the red panda and from bears. The more mismatched nucleotide bases, the greater the evolutionary distance between these animals.

Results from DNA-DNA hybridization studies supported an earlier view that the three kinds of animals are related by descent from a common ancestor, which lived more than 40 million years ago. A divergence occurred among the descendants, with one branch eventually leading to modern raccoons and the red panda, and another to bears. Much later (between 20 million and 15 million years ago), a split from the bear lineage put the ancestors of the giant panda on a unique evolutionary road. Therefore, the giant panda seems to be more closely related to bears than it is to the red panda.

20.5 ORGANIZING THE EVIDENCE—CLASSIFICATION SCHEMES

We have mentioned only a tiny number of the many millions of species that have originated and vanished over the past 3.8 billion years. Yet the sampling is enough to convey the challenge facing taxonomists, who attempt to identify, name, and classify species in ever more inclusive groupings. These groupings of species are the **higher taxa** (singular, taxon). They are the organizing units of **classification schemes**. In es-

Scientific Name: *Bothrops asper*

Sampling of the Local Names:

YELLOW-JAW TOMMYGOFF (Belize)

BARBA AMARILLA (Colombian coast)

BOQUIDORÁ, BOQUIDORADA (Caribbean coast)

CUATRO NARICES, EQUIS, EQUIS NEGRA, EQUIS RABO DE CHUCHA (Isla Gorgona, Cauca)

GATA, MACABREL, MACAUREL, MAPANÁ, MAPANÁ DE UÑA, MAPANÁ EQUIS, MAPANÁ PRIETA, MAPANÁ RABO BLANCO, MAPANÁ TIGRE (Chocó)

EQUIS (Ecuador, coast)

BARBA AMARILLA, CANTIL BOCA DORADA (Guatemala, south coast)

BARBA AMARILLA (Honduras)

AHUEYACTLI, BARBA AMARILLA, COLA BLANCA, CUATRO NARICES, NAUHYACACÓATL, NAUYACA, NAUYACA COLA DE HUESO, NAUYACA REAL, NAUYAQUE, PALANCA, PALANCA LOCA, PALANCA LORA, PALANCACOATE, PALANCACÓATL, PALANCACUATE, RABO DE HUESO, TEPOCHO, TEPOTZO, VÍBORA SORDA, XOCHINAUYAQUE (Mexico)

BARBA AMARILLA, TERCIOPELO (Nicaragua)

FER-DE-LANCE, MAPEPIRE BALSAIN (also spelled: balsin, balcin, valsin, barcin), RABO FRITO (juveniles) (Trinidad)

MACAGUA, MACAUREL (Venezuela, central coast)

MAPANARÉ (Andes)

TERCIOPELO (western Andes)

CACHETE DE PUINCA, CHANGUANGA, CHIGDU, CUAIMA CARBÓN, CUATRONARIZ, DEROYA, DOROYA; JUBA-VITU (Indigenous tribal name), MACAO, MAPANARÉ TERCIOPELO, RABOAMARILLO, SAPAMANARE, TALLA EQUIS, TIGRA, TUKEKA (Venezuela)

Figure 20.11 Common names applied to the same species of snake that is widely distributed through Latin America. To add to the confusion, some of the names, including vibora, are used for as many as fifty other species in the region.

sence, these schemes are ways of retrieving information about particular species.

Whenever a new species is identified, it is given a scientific name. The modern practice of providing scientific names for organisms has its origins in the work of an eighteenth-century Swedish naturalist. We are referring to Carl von Linné, better known by his more prestigious, latinized name, Linnaeus.

Linnaeus was a naturalist whose enthusiasm knew no bounds. He sent ill-prepared students around the world to gather specimens of plants and animals, and is said to have lost a third of his collectors to the rigors of their expeditions. Although perhaps not very commendable as a student adviser, Linnaeus did go on to develop a **binomial system** by which species are assigned a two-part Latin name. The first part was generic; it was descriptive of similar species that were thought to be the same type of organism and so were grouped together. Such a grouping is still called a **genus** (plural, genera). The second part was a specific name which, in combination with the generic name, referred to one kind of organism only.

For example, there is only one *Ursus maritimus*, or polar bear. Other bears are *Ursus arctos* (the brown bear) and *Ursus americanus* (the black bear). Notice that the first letter of the generic name is capitalized and the second names are uncapitalized. The specific name is never used without the full or abbreviated generic name preceding it, for it also can be the second name of a species in an entirely different group. As an example, *U. americanus* does indeed mean black bear, but *Homarus americanus* means Atlantic lobster, and *Bufo americanus* means American toad. Hence one would not order *americanus* for dinner unless one is willing to take what one gets.

Latin or scientific names and the conventions associated with them provide a universal way for people to discuss the same organism unambiguously, everywhere. Figure 20.11 shows the many different common names that are used in Latin America for the same venomous snake (*Bothrops asper*). Many different names are often used for the same organism even within a single country, and the same common name is sometimes used for more than one species. Only by using the standardized Latin name can we avoid confusion about which organism is being discussed and, possibly, be warned in time that we are about to step on a bad snake.

Shortly after it was devised, the binomial system became the heart of a classification scheme that was thought to mirror the patterns of links in the great Chain of Being (page 262). The scheme was based only on perceived similarities or differences in physical features—the number of legs, body size, wings or no wings, coloration, and so forth. In time, more inclusive

Table 20.1 Classification of Four Organisms

Category (Taxon)	Corn	Vanilla Orchid	Housefly	Human
Kingdom	Plantae	Plantae	Animalia	Animalia
Phylum (or Division)*	Anthophyta (flowering plants)	Anthophyta	Arthropoda	Chordata
Class	Monocotyledonae (monocots)	Monocotyledonae	Insecta	Mammalia
Order	Commelinales	Orchidales	Diptera	Primates
Family	Poaceae	Orchidaceae	Muscidae	Hominidae
Genus	*Zea*	*Vanilla*	*Musca*	*Homo*
Species	*Z. mays*	*V. planifolia*	*M. domestica*	*H. sapiens*

* At its recent meeting in Tokyo, the International Botanical Congress formally recommended the use of phylum rather than division in plant classification schemes.

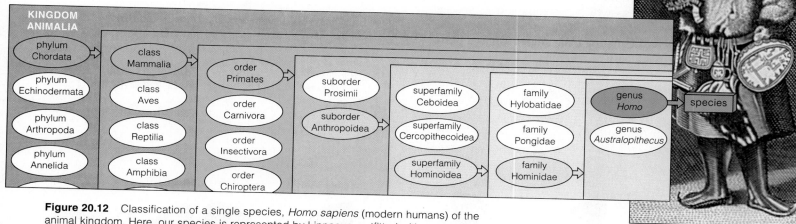

Figure 20.12 Classification of a single species, *Homo sapiens* (modern humans) of the animal kingdom. Here, our species is represented by Linnaeus, outfitted with equipment suitable for exploration. Species are the only real entities in classification schemes. Higher taxa represent our perceptions of relationships among groups of species. Appendix I includes additional taxa.

Homo sapiens (only living species of this genus)

groupings were invented to show broader relationships. Consider the following taxa:

  kingdom
    phylum
      class
        order
          family
            genus
              species

Now look at Table 20.1, which shows how the taxa are used in the classification of four different organisms. As you can see, ranking alone does not provide information about the exact pattern of relationships. The groupings reflect only *relative degrees* of relationship. A corn plant and a vanilla orchid are close relatives in the sense that they are both monocots, but they belong to separate orders. A housefly and a human are both animals, but they are only distantly related—they are not even in

the same phylum. Similarly, Figure 20.12 shows that humans are more closely related to other members of the superfamily Hominoidea than to those of the Ceboidea, but they share a more recent common ancestry with the Ceboidea than with any member of the suborder Prosimii.

In time, traditional classification schemes became modified to reflect **phylogeny**—the evolutionary relationships among species, starting with the most ancestral forms and including all the branches leading to all of their descendants. As you will see next, pitfalls as well as promises still await the new taxonomists.

Classification schemes organize information about species and simplify its retrieval. The newer, phylogenetic schemes attempt to reflect evolutionary relationships among species.

20.6 | IDENTIFYING SPECIES

Pitfalls Awaiting the Taxonomist

The preceding chapter mentioned in passing that identifying species is not always easy. Let's expand on this point with a few more examples.

Individuals of a species generally show a range of phenotypic variation. Besides this, dramatic differences in color, body form, and behavior may unfold as part of the life history of the species, so that the young may not even look like the adults. Also, sexual dimorphism may reach positively jarring levels of expression. For example, consider how adult butterflies are dramatically different from their wriggly larval beginnings. Or consider how a male African swallowtail butterfly (*Papilio dardanus*) scarcely resembles the females—which resemble other species (Figure 20.13). As another example, at first glance, you might dismiss a busily mating male angler fish as a puny extension of the female's body (Figure 20.14).

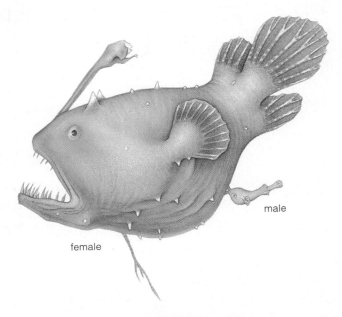

male

female

Figure 20.14 Sexual dimorphism in a deep sea angler fish (*Linophryne*), also known as the net devil. The female is about 10 centimeters long. The tiny male is attached to her.

male phenotype
Papilio dardanus

Figure 20.13 Sexual dimorphism within populations of the African swallowtail butterfly (*Papilio dardanus*). In every population, the males have yellow and black wings with "tails" at the tips. In most of tropical Africa, the females are notably different in appearance from the males—and from one another.

Mimicry accounts for the variation among females. (With mimicry, one species is deceptively similar in coloration, form, or behavior to another species that has a greater chance of survival.) The species that serve as models for the mimicking females have warning coloration that advertises their inedibility to bird predators. The *P. dardanus* females are tasty, but as mimics they have a better chance of being left alone. The wing patterns and coloration of each female variant resemble those of an inedible species present in the same region. In areas where models are absent, there are no mimicking females.

P. dardanus males are not mimics. Being easily recognized by females probably has greater reproductive advantage for the males than does the avoidance of predation by mimicry.

Inedible Models: **Mimics:**

Danaus niavius dominicanus   *P. dardanus* (female)

Amauris jacksoni   *P. dardanus* (female)

Bematistes poggeii   *P. dardanus* (female)

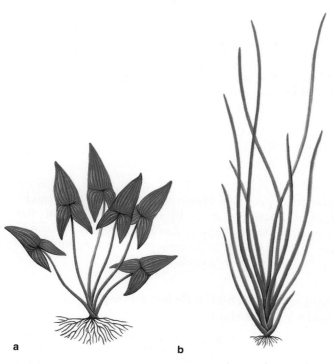

a   b

Figure 20.15 Adaptive responses by members of the same population to different environmental conditions. The shape of mature leaves of *Sagittaria sagittifolia* depends on whether the individual plant grows on land (**a**) or in the water (**b**). The difference is not attributable to genetic differences.

a

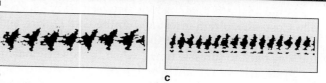

b   c

Figure 20.16 (**a**) Which frog is shown here—*Hyla versicolor* or *H. chrysoscelis*? You cannot tell the two species apart on the basis of external appearance. Identification depends on other clues, such as the female-attracting calls made by the male frogs during the breeding season. The sound spectrograms of (**b**) *H. versicolor* and (**c**) *H. chrysoscelis* reveal a behavioral difference between the two.

As you know, individuals of a species show heritable phenotypic variation. However, some pronounced variations may not be attributable directly to genes. They may be responses to differences in environmental conditions. The arrowhead plants shown in Figure 20.15 are an example.

Alternatively, organisms that outwardly appear to be almost identical may be genetically distinct and have separate evolutionary histories. Consider two species of gray treefrogs (*Hyla versicolor* and *H. chrysoscelis*), both of which live throughout the central and eastern United States. Both have warty skin and prominent sticky pads on their toes. Both have a white belly, yellow or yellow-orange coloration on the inside of the hind legs, and a large white spot below each eye (Figure 20.16). Both live in the same type of forest habitats, eating the same kind of prey.

How were *H. versicolor* and *H. chrysoscelis* identified as separate species? During the breeding season, the males attract only females of the same species with a distinctive call. Besides this behavioral isolating mechanism, the two treefrogs show a chromosomal difference—one is diploid and the other is tetraploid.

If It's So Difficult, Why Bother?

Classifying organisms has its own rewards for anyone who burns with curiosity about life's diversity. But it has its practical applications, as well. For example, how might you go about deciding which type of habitat should be protected to help conserve a rare, poorly understood species? You might start with species that are classified as being closely related to it. If those species are better understood, you could make inferences based on their habitat requirements. How might you go about testing the effectiveness of a new antibiotic against a dangerous bacterium, such as the one responsible for Legionnaires' disease in humans? You might conduct tests on a close primate relative of humans, the assumption being that its physiological responses to the infection and to the antibiotic will be similar, if not the same.

Correct identification of species usually requires knowledge of the morphology, biochemistry, physiology, ecology, and behavior of the organisms of interest.

20.7 RECONSTRUCTING EVOLUTIONARY HISTORY

Systematics

When reconstructing evolutionary history, which assessments will provide useful insights? A simple count of the total number of species wouldn't be very informative. A tally of, say, lizard species doesn't tell us anything about the lizards themselves—how various types came to live where they do, and how their characteristics emerged.

We might learn more by using methods that help us assess *patterns of diversity* among organisms through time and in the space of the environment. For example, we could explore relationships between lizard species in a Costa Rican rain forest and in the Gila Desert, and between living and extinct species of lizards—maybe all the way back to Jurassic times, when lizards originated.

Systematics, a branch of biology, deals with patterns of diversity in an evolutionary context. Systematics approaches the patterns in three ways. It relies on taxonomy, the naming and identification of organisms. It also relies heavily on phylogenetic reconstruction, the identification of evolutionary patterns that unite different organisms. Third, it uses classification, the retrieval systems that consist of many hierarchical levels, or ranks. This work feeds back into taxonomy, given that naming species or higher taxa should reflect their place in the hierarchy.

Reconstructing the history of life should reflect microevolutionary processes and macroevolutionary patterns of diversity. For example, we may hypothesize that the forerunners of reptiles evolved by way of mutation, natural selection, and other microevolutionary processes. We may hypothesize further that dinosaurs, lizards, and other reptilian groups arose from the ancestral stock by way of divergences and speciation events. As a practical matter, though, *reconstruction must be based mostly on analysis of the fossil record and on observations of living organisms.* It must be attempted without knowledge of genotypes or of actual environmental pressures that influenced the direction of evolution.

There are three schools of thought about how to go about reconstructing life's history. In one approach (evolutionary systematics), differences as well as similarities between organisms are compared, in a relatively imprecise, subjective way. This approach has produced many useful results, but researchers sometimes interpret the same data in different ways. In the second approach (phenetics), organisms are grouped according to similarities. This approach may be helpful, provided that perceived similarities are really the outcome of morphological divergences, *not* convergences. In the third approach (cladistics), organisms are grouped according to similarities that are derived from a com-

mon ancestry. (*Clad-* means branching). As you might discern from Figure 20.17, the differences in the approaches to assessing patterns of diversity can translate into some very large differences in how biologists end up grouping organisms together.

Portraying Relative Relationships Among Organisms

At present, cladistics may be the least treacherous way to portray evolutionary relatedness. Why? When we construct a phylogeny, we are making *hypotheses* about how organisms are evolutionarily related. Phylogenetic reconstruction requires thorough understanding of the structure, evolution, and development of the organisms of interest. And it depends on accurate identification of homologous structures. By examining the patterns of distribution of many homologous structures, it is possible to construct a **cladogram**, a branching diagram that represents patterns of relationships of the whole organism.

Cladograms do not convey direct information about ancestors and descendants—"who came from whom." Rather, they portray *relative* relationships among organisms. Figure 20.17b is an example of this. Taxa that are closer together on the cladogram simply share a more recent common ancestor than those that are farther apart.

The *Focus* essay that follows describes how cladograms are constructed and interpreted. Briefly, an organism is analyzed in terms of discrete morphological, physiological, and behavioral traits, or "characters," that vary among the taxa being studied. The analysis requires at least some understanding of likely relationships among organisms in the group under study. This in turn requires selection of an "outgroup," a different organism (but one not too distantly related) to serve as a point of departure for evaluating relative evolutionary distances within the "ingroup."

The methods of taxonomy, phylogenetic reconstruction, and classification currently are employed to assess patterns of diversity in time and space.

Figure 20.17 How a few major groups of vertebrates are defined according to two different schools of systematics. In evolutionary systematics (**a**), the groups are defined by traditionally accepted morphological features, such as scales, feathers, and fur. (**b**) A cladistic (phylogenetic) scheme links together organisms having shared ancestries.

a

time before present

turtles
snakes
lizards
crocodiles
dinosaurs
birds
mammals

reptiles

change in morphology

b

mammals
turtles
lizards
snakes
crocodiles
dinosaurs
birds

evolutionary distance

Constructing a Cladogram

To see how phylogenetic reconstruction works in practice, consider a group of vertebrate animals—a lamprey, trout, lungfish, turtle, cat, gorilla, and human. For each of these animals, the six different traits, or characters, shown in Figure *a* have been recorded. To keep things simple, you can focus only on the presence (+) or absence (−) of the traits shown.

From information that will be outlined in Chapter 26, the lamprey is known to be only distantly related to the other vertebrates listed. For example, it lacks jaws and paired appendages. This means you can use the lamprey as the outgroup. Compared to the other groups being considered, an *outgroup* is the one exhibiting the fewest derived characters. It allows us to make inferences about the common ancestor of the organism being considered. You may interpret any deviation from an outgroup characteristic as an evolutionary change, or *derived* trait. In Figure *b*, the outgroup condition is marked by a zero (0) and the derived condition by a one (1).

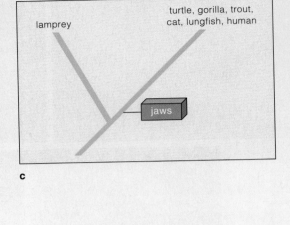

c

The next step is to look for the *shared* derived traits. For example, you see that the human and the gorilla share five derived traits, whereas the human and the cat share only four. For such a small set of traits, you can simply sketch out a cladogram, a type of family tree that shows patterns of relationships of organisms, based on their derived traits. (In practice, systematists often resort to computers. They may use many, many traits to examine relationships between many taxa, and they require a computer to find the pattern that is best supported by so much data.)

Figure *c* shows how you would sketch information based on the jaw trait. All of the vertebrates other than the outgroup have jaws. This is a derived trait. Figure *d* shows what happens when you add information based on lungs. All of the vertebrates except the lamprey and the trout have lungs. The lungfish, human, gorilla, turtle, and cat thus appear to be a *monophyletic* group, meaning they have a common evolutionary heritage. (They share two derived traits with respect to the lamprey and one with respect to the trout.)

Figures *e* through *g* show the resulting changes in the cladogram as information about other derived traits is added to it.

Notice that one character, the presence of a shell, was not used. Why not? Only the turtle has a shell. Whereas a shell is a derived trait relative to the outgroup, it is not shared with any of the other animals under consideration. Therefore, it is not useful in reconstructing this phylogeny.

Of course, every species has many such characters. The traits may be useful in identifying the species but not in determining relationships. In this example, none of the traits provides information that is contradictory to other traits. In reality, such contradictions often show up. Why?

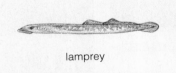

lamprey   turtle

cat

gorilla

lungfish

trout

human

Taxon	Characters					
	Jaws	Limbs	Hair	Lungs	Tail	Shell
Lamprey	−	−	−	−	+	−
Turtle	+	+	−	+	+	+
Cat	+	+	+	+	+	−
Gorilla	+	+	+	+	−	−
Lungfish	+	−	−	+	+	−
Trout	+	−	−	−	+	−
Human	+	+	+	+	−	−

a

Taxon	Characters					
	Jaws	Limbs	Hair	Lungs	Tail	Shell
Lamprey	0	0	0	0	0	0
Turtle	1	1	0	1	0	1
Cat	1	1	1	1	0	0
Gorilla	1	1	1	1	1	0
Lungfish	1	0	0	1	0	0
Trout	1	0	0	0	0	0
Human	1	1	1	1	1	0

b

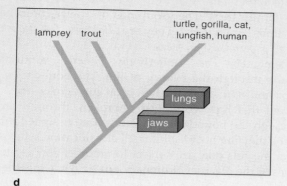

d

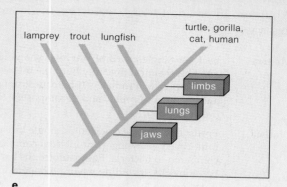

e

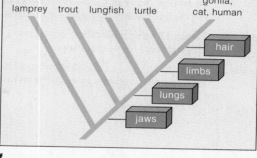

f

All organisms are a mixture of evolutionarily ancient and derived traits. Also, not all traits you initially regarded as homologous may actually be homologous. As you saw on page 307, morphological convergences may have occurred.

How do you read the cladogram in Figure *g*? Keep in mind that cladograms give you information about relative relatedness, not about ancestors. Your cladogram tells you that the human is more closely related to the gorilla than to the cat. The cat, human, and gorilla as a group are more closely related to each other than to the turtle.

These examples are easy to follow. Perhaps less obviously, your cladogram also tells you that the lungfish is more closely related to the human than to the trout. Follow the lines or branches of the cladogram from the human and lungfish back to the intersection, or node, where they meet (Figure *h*, node 1). Next, trace back the branches of the lungfish and trout to their intersection (Figure *h*, node 2). Note that the trout-lungfish intersection is nearer to the base of the cladogram than the lungfish-human intersection. The higher up the cladogram, the more derived traits are shared by the taxa. The lower they are on the cladogram, the fewer are shared.

Another way to think about this is to regard the axis of the cladogram as a time bar, but one without absolute dates. It simply signifies this: the lower the position of a group on the cladogram, the more distant is the most recent common ancestor of the taxa being compared.

Finally, even though the axis of the cladogram can be viewed as a time line, keep in mind that the ordering of taxa from left to right is not important. No matter how the taxa are arranged across the top of the cladogram, following the branches back to the lungfish-human and lungfish-trout intersections will give the same result.

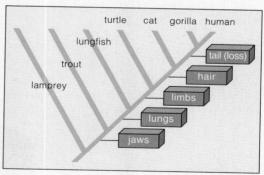

g

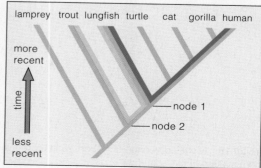

h

20.8 A FIVE-KINGDOM SCHEME

We recognize five kingdoms of organisms in this book, following a classification scheme established by Robert Whittaker. In broadest terms, we define the kingdoms this way:

Monera   Bacteria. Single-celled prokaryotes showing great biochemical diversity but little internal complexity. Producers (photoautotrophs and chemoautotrophs) and decomposers (heterotrophs).

Protista   Protistans. Diverse eukaryotes, single celled or multicelled, showing great internal complexity compared to bacteria. Photoautotrophs and heterotrophs, including major pathogens and parasites.

Fungi   Fungi. Multicelled eukaryotes that feed by extracellular digestion and absorption. Heterotrophs; many major decomposers (and nutrient cyclers). Many pathogens and parasites.

Plantae   Plants. Diverse multicelled eukaryotes. Nearly all producers (photoautotrophs).

Animalia   Animals. Diverse multicelled eukaryotes. Heterotrophs, including predators and parasites.

The existing members of each kingdom are the current representatives of very long, independent lines of evolution. The evolutionary tree of Figure 20.18 only hints at their relationships to one another. It is important for you to understand that only *species* are real entities in such family trees. The groupings of "twigs, branches, and limbs" are not real. They are *categories of relationship that we superimpose on species*.

Remember, an underlying genetic continuity threads through the history of life. The boundaries of kingdoms and other higher taxa cut across lines of descent that have continued unbroken, through time. Although the boundaries have not been assigned arbitrarily, they are not absolute. All classification schemes, including this one, are subject to modification as more pieces of the macroevolutionary puzzle come together.

A widely accepted classification scheme groups organisms into five kingdoms, as determined partly by cell structure and modes of nutrition. No classification scheme is etched in granite, including this one.

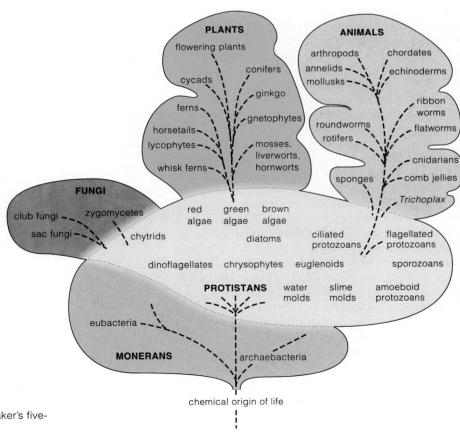

Figure 20.18 Modified version of Robert Whittaker's five-kingdom system of classification.

SUMMARY

1. A continuity of relationship exists among all species, past and present. The patterns, trends, and rates of change among groups of species over long spans of time are called macroevolution.

2. Evidence of macroevolution comes from the fossil record, earth history, comparative morphology, and comparative biochemistry. The evidence reveals similarities and differences in body form, functions, behavior, and biochemistry.

3. The fossil record varies as a function of the kinds of organisms represented, where they lived, and the stability of the region since the time of fossilization.

4. Comparative morphology may reveal similarities in embryonic development that indicate evolutionary relationship. It may reveal homologous structures, shared as a result of descent from a common ancestor.

5. Comparative biochemistry relies on gene mutations that have accumulated in different species. Neutral mutations serve as a molecular clock for dating divergences from a common ancestor. Nucleic acid hybridization and other methods reveal degrees of similarity between gene products of different species.

6. Systematics deals with phylogeny, the evolutionary relationships among species, beginning with the most ancestral forms and including all the branches leading to all of their descendants. This branch of biology takes three approaches to assess patterns of diversity in the distribution of organisms in space and time:
   a. Taxonomy (identifying and naming organisms).
   b. Phylogenetic reconstruction (identifying evolutionary patterns that give evidence of inherited relationships among organisms).
   c. Classification (categorizing information about species into a hierarchal scheme that serves as a useful retrieval system).

7. In the binomial system, developed by Linnaeus, each kind of organism is assigned a two-part Latin name. The first part, the genus, identifies a group of similar species that share a recent common ancestor. The second part is the name of the particular species.

8. Cladistics is a widely used approach to studying patterns of diversity. Organisms are grouped by shared similarities that are perceived to be derived from common ancestry.

9. Current classification systems consist of a series of ever more inclusive taxa, from species to kingdoms. Although the classification does not reveal explicit phylogenetic patterns, it does reflect the general level of relatedness among organisms.

Review Questions

1. Will the fossil record ever be complete? Why or why not? *303*

2. Explain the difference between:
   a. microevolution and macroevolution *301*
   b. homologous and analogous structures *306–307*
   c. morphological divergence and convergence *306–307*

3. What kind of mutation serves as the basis of a molecular clock, and why? *308*

4. Name a protein specified by a gene that has been highly conserved in organisms ranging from bacteria to humans. *308*

5. Why do evolutionary biologists apply heat energy to hybrid molecules of DNA from two species? *308*

6. Give reasons why two organisms that are quite different in outward appearances may belong to the same species. *312*

7. Give reasons why two organisms that seem identical in outward appearance may not belong to the same species. *313*

8. This chapter outlines the major taxa of classification schemes, but many more exist. For example, using Table 20.1, Figure 20.12, and Appendix I as guides, fill in as many groupings as you can for the following organisms:

corn

human

vanilla orchid

housefly

Superkingdom _____
Kingdom _____
Subphylum _____
Superclass _____
Class _____
Subclass _____
Order _____
Suborder _____
Infraorder _____
Superfamily _____
Family _____
Genus _____
Species _____

1. Macroevolution is defined as _____ .
   a. small changes that occur in the gradual transition of an organism through time
   b. the evolution of large organisms
   c. the large-scale patterns, trends, and rates of change among groups of species
   d. large evolutionary patterns that are so confused they do not allow phylogenetic reconstruction

2. Evidence for macroevolution comes from _____ .
   a. the fossil record
   b. earth history
   c. comparative morphology
   d. comparative biochemistry
   e. all of the above

3. The fossil record _____ .
   a. is incomplete and biased
   b. presents a better record of ancient organisms that lived on ocean floors and uplands than those on lowlands
   c. represents poor evidence of evolutionary history
   d. will eventually be complete

4. Morphological divergences may lead to _____ .
   a. analogous structures   c. convergent structures
   b. homologous structures   d. both a and c

5. Morphological convergences may lead to _____ .
   a. analogous structures   c. divergent structures
   b. homologous structures   d. both a and c

6. A classification system based on evolutionary relationship is _____ .
   a. epigenetic   c. credited to Linnaeus
   b. phylogenetic   d. both b and c

7. *Pinus banksiana, Pinus strobus,* and *Pinus radiata* represent _____ .
   a. three families of pine trees
   b. three different names for the same organism
   c. several species belonging to the same genus
   d. both a and c

8. Increasingly inclusive taxa range from _____ to _____ .
   a. kingdom; species   c. genera; kingdom
   b. kingdom; genera   d. species; kingdom

9. Match these terms appropriately.
   ____ phylogeny   a. molecular clock
   ____ fossil   b. recognizable physical evidence of ancient life
   ____ stratification   c. layers of sedimentary rock
   ____ homology   d. penguin flippers and shark fins
   ____ accumulation of neutral mutations
   ____ analogy   e. evolutionary relationships among species, ancestors to descendants
                           f. similarity in body parts in different organisms attributable to common descent

Selected Key Terms

analogy *307*	homology *306*
Animalia *318*	macroevolution *301*
binomial system *310*	molecular clock *308*
cladogram *314*	Monera *318*
classification scheme *310*	morphological convergence *307*
comparative morphology *304*	morphological divergence *306*
DNA-DNA hybridization *308*	neutral mutation *308*
fossil *302*	phylogeny *311*
fossilization *303*	Plantae *318*
Fungi *318*	Protista *318*
genus *310*	stratification *303*
higher taxa *310*	systematics *314*

Readings

Brooks, D. R., and D. A. McLennan. 1991. *Phylogeny, Ecology, and Behavior.* Chicago: University of Chicago Press. An up-to-date examination of the approaches to studying the diversity of life.

Sneath, P., and R. Sokal. 1973. *Numerical Taxonomy.* San Francisco: Freeman. A leading text in the phenetic approach to systematics.

Wiley, E. 1981. *Phylogenetics: The Theory and Practice of Phylogenetic Systematics.* New York: Wiley.

FACING PAGE: *Patterns of diversity in nature, here represented by different species of plants and fungi.*

21 THE ORIGIN AND EVOLUTION OF LIFE

In the Beginning . . .

Some evening, watch the moon as it rises from the horizon and think about the 380,000 kilometers between it and you. *Five billion trillion* times the distance between the moon and you are galaxies—systems of stars—at the boundary of the known universe. Wavelengths of light travel faster than anything else, 300 million meters per second, yet long wavelengths from those faraway galaxies are only now reaching the earth.

Near and distant galaxies are suspended in the vast space of the universe. By every known measure, they are all moving away from one another—which means the universe is expanding. And the prevailing theory of that expansion accounts for every bit of matter in the universe, in every living thing.

Think about how you rewind a video tape on a VCR, then imagine "rewinding" the universe. As you do, the galaxies start moving back together. After 10 to 20 billion years of rewinding, all galaxies, all matter, and all of space have become compressed into a hot, dense volume about the size of the sun. You have arrived at time zero.

That enormously hot, dense state lasted only for an instant. What happened next is known as the "big bang"—a truly inadequate name for the stupendous, nearly instantaneous distribution of all matter and energy everywhere, through all of the known universe. About a minute later, temperatures dropped to a billion degrees and fusion reactions produced most of the light elements, including helium, which are still the most abundant elements throughout the universe. Radio telescopes have detected the relics of the big bang—cooled and diluted background radiation, left over from the beginning of time.

Over the next billion years, stars started forming by the gravitational contraction of gaseous material. When the stars became massive enough, nuclear reactions were ignited in their central region, and they gave off tremendous light and heat. As contraction continued, many stars may have become dense enough to promote the formation of heavier elements.

All stars have a life history, from birth to a sometimes spectacularly explosive death. In what might be

a

b

called the original stardust memories, the heavier elements released during the explosions became swept up during the gravitational contraction of new stars—and they became raw materials for the formation of even heavier elements. Today, the Hubble space telescope is providing astounding glimpses of such star-forming activity in the dust clouds of Orion and other constellations (Figure 21.1).

Long ago, explosions of dying stars ripped through our own galaxy and left behind a dense cloud of dust and gas that extended trillions of kilometers in space. As the cloud cooled, countless bits of matter gravitated toward one another. By 4.6 billion years ago, the cloud had flattened out into a slowly rotating disk. At the dense, hot center of that disk, the star of our solar system—the sun—was born.

The remainder of this chapter is a sweeping slice through time, one that cuts back to the formation of the earth and the chemical origins of life. It is the starting point for the next four chapters, which will take us along major lines of descent that lead to the present range of species diversity. The story is not complete. Even so, all of the available evidence points to a principle that can be used to organize separate bits of information about the past: *Life is a magnificent continuation of the physical and chemical evolution of the universe, of galaxies and stars, and of the planet earth.*

KEY CONCEPTS

1. The available evidence suggests that life originated more than 3.8 billion years ago. Its origin and subsequent evolution have been linked to the physical and chemical evolution of the universe, the stars, and the planet earth.

2. All of the inorganic and organic compounds necessary for self-replication, membrane assembly, and metabolism —that is, for the structure and functioning of living cells— could have formed spontaneously under conditions that existed on the early earth.

3. Not long after the time of life's origin, divergences led to three great prokaryotic lineages—the archaebacteria, the eubacteria, and the ancestors of eukaryotes. A theory of endosymbiosis helps explain the profusion of specialized organelles that characterize eukaryotes.

4. Earth history is divided into five great eras, the boundaries of which reflect times of the largest mass extinctions. Prokaryotes dominated the first two eras, the Archean and Proterozoic. Eukaryotes originated during the late Proterozoic and became spectacularly diverse. All five kingdoms of organisms are characterized by the persistences, extinctions, and radiations of different lineages.

5. Throughout the history of life, asteroid impacts, drifting and colliding continents, and other environmental insults have had profound impact on the direction of evolution.

Figure 21.1 (**a**) In winter in the Northern Hemisphere, brilliant stars of the constellation Orion dominate the southern sky. (**b**) Shown here, the great nebula in Orion, a cloud of dust, silicates, ice, and water. (The Latin *nebula* means mist.) The Hubble space telescope provided this remarkable detailed view of part of the nebula—a hotbed of star formation.

Origin of the Earth

Cloudlike remnants of stars, including those shown in Figure 21.1, are mostly hydrogen gas. But they also contain water, iron, silicates, hydrogen cyanide, methane, ammonia, formaldehyde, and many other simple inorganic and organic substances. Most likely, the contracting cloud from which our solar system evolved was similar in composition. Between 4.6 and 4.5 billion years ago, the outer regions of the cloud cooled. Swarms of mineral grains and ice orbited in parallel around the sun. By electrostatic attraction, by gravitational pull, they started clumping together (Figure 21.2). In time, the larger, faster clumps started colliding and shattering. Some of the larger ones became more massive by sweeping up asteroids, meteorites, and other rocky remnants of collisions—and gradually they evolved into planets.

While the early earth was forming, much of its inner rocky material melted. Most likely, stupendous asteroid impacts generated the required heat, although heat released during the radioactive decay of minerals contributed to it. As rocks melted, nickel, iron, and other heavy materials moved toward the interior, and lighter minerals floated toward the surface. This process of differentiation resulted in the formation of a crust of basalt, granite, and other types of low-density rock, a rocky region of intermediate density (the mantle), and a high-density, partially molten core of nickel and iron.

Four billion years ago, the earth was a thin-crusted inferno (Figure 21.3a). Yet within 200 million years, life had originated on its surface! We have no record of the event. As far as we know, movements in the mantle and crust, volcanic activity, and erosion obliterated all traces of it. Still, we can put together a plausible explanation of how life originated by considering three questions:

1. What were the prevailing physical and chemical conditions on earth at the time of life's origin?

2. Based on physical, chemical, and evolutionary principles, could large organic molecules have formed spontaneously and then evolved into molecular systems displaying the characteristics of life?

3. Can we devise experiments to test whether living systems could have emerged by chemical evolution?

The First Atmosphere

When patches of crust were forming, heat and gases blanketed the earth. This first atmosphere probably consisted of gaseous hydrogen (H_2), nitrogen (N_2), carbon monoxide (CO), and carbon dioxide (CO_2). Were gaseous oxygen (O_2) and water also present? Probably not. Rocks release very little oxygen during volcanic eruptions. Even if oxygen were released, those small amounts would have reacted at once with other elements, and any water would have evaporated because of the intense heat.

When the crust finally cooled and solidified, water condensed into clouds and the rains began. For millions of years, runoff from rains stripped mineral salts and other compounds from the earth's parched rocks. Salt-laden waters collected in depressions in the crust and formed the early seas.

Without an oxygen-free atmosphere, the organic compounds that started the story of life never would have formed on their own. (As described on page 92, free oxygen would have attacked them.) Without liquid water, cell membranes never would have formed. Cells, recall, are the basic units of life. Each has a capacity for independent existence.

The Synthesis of Organic Compounds

When we reduce cells to their lowest common denominator, we are left with proteins, complex carbohydrates, lipids, and nucleic acids. Today, cells assemble these molecules from small organic compounds—simple sugars, fatty acids, amino acids, and nucleotides. Energy from the environment drives the synthesis reactions. Were small organic compounds also present on the early earth? Were there sources of energy that drove their spontaneous assembly into the large molecules of life?

Mars, meteorites, the earth's moon, and the earth formed at the same time, from the same cosmic cloud. Rocks collected from Mars, meteorites, and the moon contain precursors of biological molecules—so the same precursors must have been present on the early earth.

Figure 21.2 Representation of the cloud of dust, gases, and clumps of rock and ice around the early sun.

a

Figure 21.3 (**a**) Representation of the earth during its formation, when the moon's orbit was much closer. If the earth had condensed into a smaller planet, its gravitational mass would not have been great enough to hold onto an atmosphere. If it had settled into an orbit closer to the sun, water would have evaporated from its surface. If the orbit had been more distant, the surface would have been too cold, locking up water as ice. Without liquid water, life never would have originated on earth. (**b**) Stanley Miller's apparatus, used to study the synthesis of organic compounds under conditions that presumably existed on the early earth. The condenser cools circulating steam so that water droplets form.

b

Sunlight, lightning, or heat escaping from the earth's crust could have supplied the energy to drive their condensation into complex organic molecules.

In the first of many tests of that hypothesis, Stanley Miller mixed hydrogen, methane, ammonia, and water in a reaction chamber (Figure 21.3b). He recirculated the mixture and bombarded it with a spark discharge to simulate lightning. Within a week, amino acids and other small organic compounds had formed. In other experiments that simulated conditions on the early earth, glucose, ribose, deoxyribose, and other sugars were produced from formaldehyde. Adenine was produced from hydrogen cyanide. Adenine plus ribose are present in ATP, NAD, and other nucleotides.

Even if amino acids did form in the early seas, they wouldn't have lasted long. In water, the favored direction of most spontaneous reactions is toward hydrolysis, not condensation (page 36).

Maybe more lasting bonds formed at the margins of seas. By one scenario, clay in the rhythmically drained muck of tidal flats and estuaries served as templates (structural patterns) for the spontaneous assembly of proteins and other organic compounds. Clay consists of thin, stacked layers of aluminosilicates with metal ions at its surface. Clay and metal ions attract amino acids. When clay is first warmed by sunlight, then alternately dried out and moistened, it actually promotes condensation reactions that yield complex organic compounds. We know this from experiments.

Suppose proteins that formed on some clay templates had the shape and chemical behavior necessary to function as weak enzymes in hastening bonds between amino acids. If certain templates promoted such bonds, they would have selective advantage over other templates in the chemical competition for available amino acids. Perhaps selection was at work before the origin of cells, favoring the chemical evolution of enzymes.

Many experiments provide indirect evidence that the complex organic molecules characteristic of life could have formed under conditions that existed on the early earth.

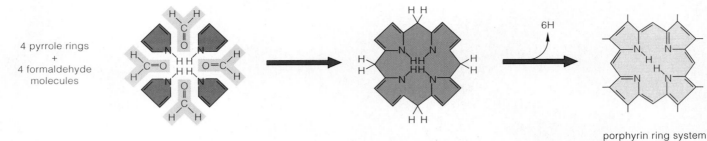

4 pyrrole rings
+
4 formaldehyde
molecules

porphyrin ring system

Figure 21.4 How formaldehyde, a substance present on the early earth, might have undergone chemical evolution into porphyrin. Today, porphyrin is the light-trapping and electron-donating component of chlorophyll molecules. It also is a component of cytochrome, which has roles in electron transport systems of a variety of metabolic pathways.

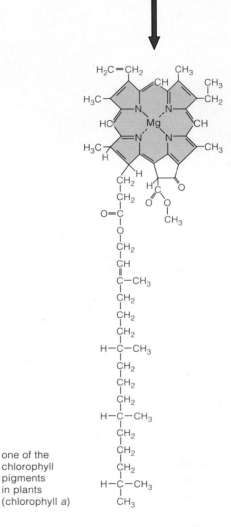

one of the chlorophyll pigments in plants (chlorophyll *a*)

21.2 EMERGENCE OF THE FIRST LIVING CELLS

Origin of Agents of Metabolism

A defining characteristic of life is metabolism. The word refers to all the reactions by which cells harness energy and use it to drive their activities, such as biosynthesis. During the first 600 million years of earth history, enzymes, ATP, and other molecules could have assembled spontaneously at the same locations. If so, their close association would have promoted chemical interactions—and the beginning of metabolic pathways.

Imagine an ancient sunlit estuary, rich in clay deposits. Countless aggregations of organic molecules stick to the clay. At first there are quantities of an amino acid called *D*. All over the estuary, *D* is incorporated into protein molecules—until the supply dwindles. Suppose a certain enzyme can promote the formation of *D* from a simpler substance (*C*)—which is still abundant. Aggregations having that enzyme will be favored in the chemical competition for starting materials. Now suppose *C* becomes scarce also. At that point, the advantage tilts to aggregations that can promote formation of *C* from the even simpler organic substances *B* and *A*—say, carbon dioxide and water. These substances are present in essentially unlimited amounts in the atmosphere and the sea. Selection has favored a synthetic pathway:

$$A + B \longrightarrow C \longrightarrow D$$

Suppose, finally, that some aggregations in the estuary are better than others at harnessing and using energy. What kind of molecules could give them this advantage? Think of photosynthesis, the energy-trapping pathway that now dominates the world of life. This metabolic route starts at light-trapping pigment molecules called chlorophylls. The part of chlorophyll that absorbs sunlight energy and gives up electrons is a porphyrin ring structure. Porphyrins also are part of cytochromes. And cytochromes are part of electron transport systems in every photosynthetic and aerobically respiring cell. As

Figure 21.4 shows, porphyrins can spontaneously assemble from formaldehyde—one of the legacies of cosmic clouds. Was porphyrin an electron transporter of the first metabolic pathways? Perhaps.

Origin of Self-Replicating Systems

Another defining characteristic of life is a capacity for reproduction, and reproduction starts with DNA's protein-building instructions. DNA is a fairly stable molecule, and it is easily replicated before each cell division. Enzymes and RNA carry out its encoded instructions.

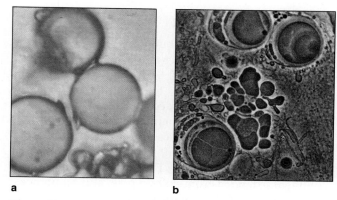

Figure 21.5 Microscopic spheres of (**a**) proteins and (**b**) lipids that self-assembled under abiotic conditions.

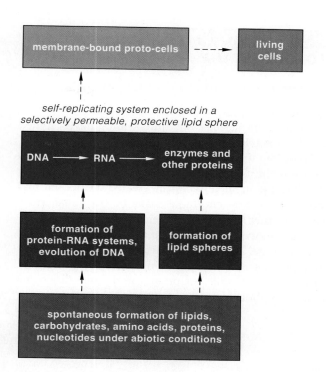

Figure 21.6 One possible sequence of events that led to the first self-replicating systems, then to the first living cells.

Small molecules called coenzymes assist most of the existing enzymes. Intriguingly, the structure of some coenzymes is the same as that of RNA nucleotides. Another clue: If you heat nucleotide precursors and short chains of phosphate groups together, they will assemble into RNA strands. On the early earth, sunlight energy alone could have driven the spontaneous formation of RNA molecules.

Simple, self-replicating systems of RNA, enzymes, and coenzymes have been created in the laboratory. Experiments with these systems suggest that RNA could have formed on clay templates. Did RNA later replace clay as information-storing templates for protein synthesis? Perhaps, although existing RNA is too chemically inept to serve this role. And we still don't know how DNA entered the picture.

Until we identify the likely chemical ancestors of RNA and DNA, the story of life's origin will be incomplete. Filling in the details will require imaginative sleuthing. For example, researchers ran a simple program on one of the most rapid, advanced computers. The program contained information on simple compounds, of the sort that are believed to have been present on the early earth. It asked the computer to subject the compounds to random chemical competition and natural selection. The program ran repeatedly, and the outcome was always the same. Simple precursors inevitably evolved into interacting systems of large, complex molecules.

Origin of the First Plasma Membranes

Experiments are more revealing of the origin of the plasma membrane, which surrounds every living cell. The membrane is a lipid bilayer, studded with proteins that carry out diverse functions. The membrane's primary function is to control which substances move into and out of the cell. Without this control, cells cannot exist.

Before cells, there must have been "proto-cells." These may have been little more than membrane sacs, protecting information-storing templates and various metabolic agents from the environment. We know that simple membrane sacs can form spontaneously.

In one experiment, Sidney Fox heated amino acids until they formed protein chains, which he put in hot water. After cooling, the chains assembled into small, stable spheres (Figure 21.5*a*). Like cell membranes, the spheres were selectively permeable to different substances. They picked up free lipids, and a lipid-protein film formed at their surface.

In other experiments, fatty acids and glycerol combined to form long-tail lipids under conditions that simulated evaporating tidepools. The lipids self-assembled into small, water-filled sacs, which in many were like cell membranes (Figure 21.5*b*).

In short, there are major gaps in the story of life's origins. But there also is strong experimental evidence that chemical evolution could have led to the molecules and structures characteristic of life. Figure 21.6 summarizes the key events in that chemical evolution, which preceded the first cells.

Although the story is far from complete, laboratory experiments and computer simulations indirectly suggest that chemical and molecular evolution gave rise to proto-cells.

21.3 LIFE ON A CHANGING GEOLOGIC STAGE

Drifting Continents and Changing Seas

By about 3.8 billion years ago, the earth's crust was becoming more stable—but things have never settled down completely.

According to **plate tectonic theory**, plumes of molten rock from the mantle spread out beneath the crust or broke through it. Such fracturing created the oceanic and continental plates—enormous slabs of crust that move apart and crunch together at their margins (Figure 21.7). The plates move no more than a few centimeters a year, on the average. Yet that small displacement causes pressure to decrease or increase along vast crustal margins. Earthquakes, volcanic eruptions, and lava flows are among the short-term consequences. Crustal uplifting and subsidence, mountain building, seafloor spreading, and continental drift are long-term consequences.

Take a look at Figure 21.8. In the remote past, land masses collided together to form supercontinents, which later split open at deep rifts that became new ocean basins. One early continent, Gondwana, drifted southward from the tropics, across the south polar region, then northward until it crunched together with other land masses to form Pangea. This single world continent extended from pole to pole, and an immense ocean spanned the rest of the globe.

Besides this, the erosive forces of wind and water have continually resculpted the earth's surface. Asteroids and meteorites have continued to bombard the crust, and huge impacts have had long-term effects on global temperature and climate.

All such changes in land masses, the oceans, and the atmosphere profoundly influenced the evolution of life.

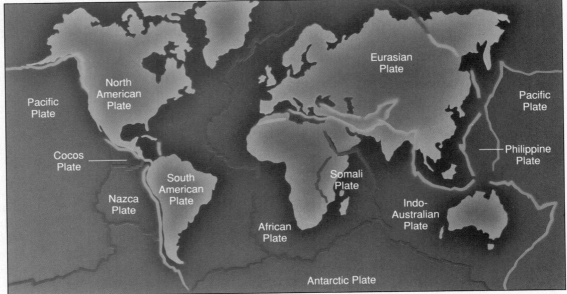

a

Figure 21.7 Some forces of geologic change. (**a**) Plate tectonics. The crust is fractured in rigid plates that split, drift, and collide. (**b**) Plumes of molten material drive the crustal movements. They well up from the earth's interior and spread out beneath the crust. Such plumes have ruptured the crust at vast ridges on the ocean floor. Here, molten material escapes, cools, and forces the seafloor away from the ridge. Seafloor spreading displaces plates away from the ridge. Often the opposite plate margin is thrust under another plate and contributes to mountain building. Plumes also cause deep rifts and splitting in the interior of continents. Such rifting is taking place in Missouri, at Lake Baikal in Russia, and in eastern Africa. In the past, superplumes violently ruptured the Earth's crust. Volcanoes still form at the ruptured "hot spots." The Hawaiian Islands formed this way.

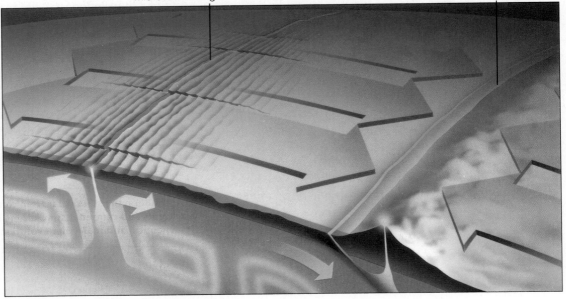

b

mid-oceanic ridge

zone where one crustal plate is being thrust under another

For example, life flourished in warm, shallow waters along the vast shorelines of the early continents. When continents collided, the shorelines vanished. These and other changes had dire consequences for many lineages. Yet even as old habitats were lost, new ones opened up for the survivors—and evolution took off in new directions.

The Geologic Time Scale

As you read in the preceding chapter, early naturalists had no way of dating the earth's rocks or the fossils that were contained in sedimentary beds. But there were obviously abrupt transitions in the kinds of fossils represented. They interpreted these as boundaries between four great eras in time. The most ancient fossils were from the first geologic era, the **Proterozoic**. They were followed by fossils from the **Paleozoic**, the **Mesozoic**, and finally the "modern" era, the **Cenozoic**. The boundaries between eras still serve as the basis of a **geologic time scale**, which is shown in Figure 21.8. Today we know the boundaries mark the times of mass extinctions. Thanks to radioisotope dating work (page 20), actual dates have been assigned to the time scale. The work also revealed that the Proterozoic lasted for an astoundingly immense time. The Proterozoic has since been subdivided. The first era of the geologic time scale is now called the **Archean** ("the beginning").

Over the past 3.8 billion years, changes in the earth's crust, the atmosphere, and the oceans profoundly affected the evolution of life. The boundaries of the geologic time scale mark abrupt shifts in the direction of evolution.

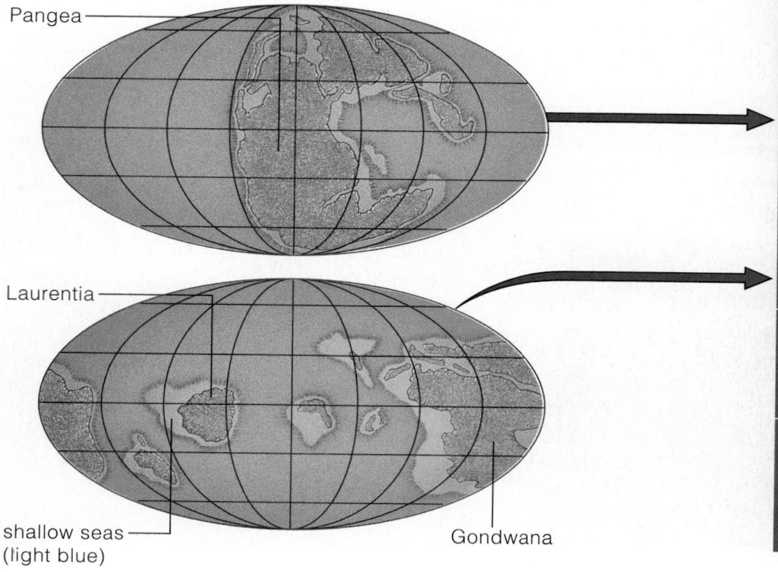

Pangea

Laurentia

shallow seas (light blue)

Gondwana

Figure 21.8 The geologic time scale. The time spans of the different eras are not to scale. If they were, the Archean and Paleozoic portions of the chart would run off the page.

Think of the spans of time as minutes on a clock. Life originated at midnight. The Paleozoic began at 10:04 A.M., the Mesozoic at 11:09 A.M., and the Cenozoic at 11:47 A.M. The recent epoch of the Cenozoic started during the last 0.1 second before noon.

Era	Period	Epoch	Millions of Years Ago (mya)
CENOZOIC	Quaternary	Recent	0.01–
		Pleistocene	1.65
	Tertiary	Pliocene	5
		Miocene	25
		Oligocene	38
		Eocene	54
		Paleocene	65
MEZOZOIC	Cretaceous	Late	100
		Early	138
	Jurassic		205
	Triassic		240
PALEOZOIC	Permian		290
	Carboniferous		360
	Devonian		410
	Silurian		435
	Ordovician		505
	Cambrian		550
PROTEROZOIC			2,500
ARCHEAN			4,600

21.4 LIFE IN THE ARCHEAN AND PROTEROZOIC ERAS

The first cells emerged as molecular extensions of the evolving universe, the solar system, and the planet earth. They may have originated in tidal flats or on the seafloor, in muddy sediments warmed by heated water from volcanic vents. (Such *hydrothermal* vents are described on page 884.) Like existing bacteria, the first were **prokaryotic cells**, without a nucleus. Possibly they were little more than self-replicating, membranous sacs of DNA and other complex organic molecules. Given the absence of free oxygen, they must have secured energy through anaerobic routes—fermentation pathways, most likely. Energy and carbon sources were plentiful; dissolved organic compounds had accumulated in the seas by natural geologic processes. Thus, "food" was available, predators were absent, and biological molecules were free from oxygen attacks.

Early in the Archean era, the original prokaryotic lineage diverged in three evolutionary directions. Two of the branchings led to the domains of **archaebacteria** and **eubacteria**. You will read about these bacteria in the chapter to follow. As described in the *Focus* essay, the third branching led to the prokaryotic ancestors of **eukaryotic cells**.

Between 3.5 and 3.2 billion years ago, the cyclic pathway of photosynthesis evolved in some species of eubacteria. This pathway is described on page 112. Those eubacteria were anaerobic, but components of their electron transport systems had become modified in ways that allowed them to tap an unlimited energy source—sunlight. For nearly 2 billion years, their photosynthetic descendants dominated the world of life. Their populations formed large mats in which sediments collected. Mats accumulated one atop the other, becoming structures called **stromatolites** (Figure 21.9).

By the dawn of the Proterozoic, 2.5 billion years ago, the noncyclic pathway of photosynthesis had evolved in some eubacterial species. Oxygen, one of its by-products, started to accumulate—imperceptibly at first, but in time it had two irreversible effects. First, *an oxygen-rich atmosphere stopped the further chemical origin of living cells.* Except in very few anaerobic habitats, spontaneously formed organic compounds would not survive attacks by free oxygen. Second, *aerobic respiration became the dominant energy-releasing pathway.* In many prokaryotic lineages, selection now favored metabolic equipment that could "neutralize" oxygen by using it as an electron acceptor. This innovation foreshadowed the evolution of multicelled eukaryotes and their invasion of far-flung environments.

Eukaryotes originated during the Proterozoic, by 1.2 billion years ago and possibly earlier. We have fossils, 900 million years old, of well-developed red algae, green algae, plant spores, and fungi. Diverse internal membranous compartments—organelles—are the hallmark of eukaryotic cells. Where did they come from? The *Focus* essay describes a few of the most plausible hypotheses.

About 800 million years ago, stromatolites began a dramatic decline. They had become an untapped food source for newly evolved, tiny, bacteria-eating animals. About the same time, small, soft-bodied animals were leaving tracks and burrows on the seafloor. They lived near the shores of a supercontinent, Laurentia. About 570 million years ago, in "Precambrian" times, their descendants may have started the first adaptive radiation of animals.

The first living cells had evolved by about 3.8 billion years ago. About 600 million years later, oxygen released by some of their photosynthetic descendants was starting to change the atmosphere.

In time, an oxygen-rich atmosphere precluded further spontaneous chemical evolution of life—and it was a key selection pressure in the evolution of eukaryotic cells.

Figure 21.9 (**a**) One of the oldest known fossils: a strand of walled cells, 3.5 billion years old. The strand resembles certain modern-day filamentous bacteria. (**b**) From Western Australia, stromatolites that formed between 2,000 and 1,000 years ago in shallow seawater. Calcium deposits preserved their structure. They are identical to stromatolites more than 3 billion years old.

10 µm

a   b

The Rise of Eukaryotic Cells

The key defining feature of eukaryotic cells is the profusion of membrane-bound organelles in the cytoplasm. Where did they come from? Speculations abound. Some organelles probably evolved gradually, through gene mutations and natural selection. For others, researchers make a good case for evolution by way of endosymbiosis.

Origin of the Nucleus and ER   Prokaryotic cells do not have a profusion of organelles. Yet some species do have infoldings of the plasma membrane (Figure *a*). Enzymes and other metabolic agents are embedded in that membrane. In ancient forerunners of eukaryotic cells, similar infoldings may have extended far enough into the cell interior to serve as a channel to the surface. Perhaps they evolved into ER channels and into an envelope around the DNA.

What would be the advantage of such membranous enclosures? Maybe they protected genes and protein products from "foreigners." Remember how bacterial species can transfer plasmid DNA among themselves? Yeasts, which are simple eukaryotic cells, also make such transfers. Some yeast species contain up to fifty plasmids. Initially, a nuclear envelope may have been favored because it got the cell's genes, replication enzymes, and transcription enzymes out of the cytoplasm. It would have allowed vital genetic messages to be produced free of metabolic competition from an unmanageable hodgepodge of foreign genes. Similarly, ER channels might have kept vital proteins and other organic compounds away from metabolically hungry "guests"—foreign cells that became permanent residents inside the host cell.

A Theory of Endosymbiosis   Accidental partnerships between different prokaryotic species must have

formed countless times on the evolutionary road to eukaryotes. Some partnerships resulted in the origin of mitochondria, chloroplasts, and other organelles. This is a **theory of endosymbiosis**, developed in greatest detail by Lynn Margulis. *Endo-* means within, and *symbiosis* means living together. In endosymbiosis, one cell (a guest species) lives permanently inside another kind of cell (the host species), and the interaction benefits both. By this theory, eukaryotes arose after the noncyclic pathway of photosynthesis emerged and oxygen accumulated to significant levels in the atmosphere.

In some bacterial groups, electron transport systems had already become expanded to include "extra" cytochromes. As it happened, those cytochromes were able to donate electrons to oxygen. In other words, the bacteria could extract energy from organic compounds by way of aerobic respiration. By 1.2 billion years ago, and possibly much earlier, the forerunners of eukaryotes were engulfing aerobic bacteria. They may have been like existing amoebalike cells that weakly tolerate free oxygen. If so, they trapped food by sending out cytoplasmic extensions of the cell body. Endocytic vesicles formed around the food and delivered it to the cytoplasm for digestion.

Some aerobic bacteria resisted digestion. They actually thrived in the new, protected, nutrient-rich environment. In time they were releasing extra ATP—which the hosts came to depend on for growth, greater activity, and assembly of more structures, such as hard body parts. The guests were no longer duplicating metabolic functions that the hosts were performing for them. The anaerobic and aerobic cells were now incapable of independent existence. The guests had become mitochondria, supreme suppliers of ATP.

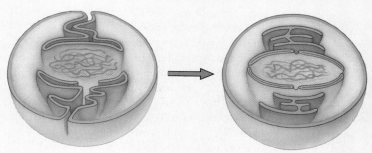

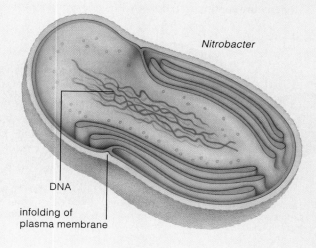

Nitrobacter

DNA

infolding of
plasma membrane

a   Over evolutionary time, infoldings of the plasma membrane may have given rise to the nuclear envelope and endoplasmic reticulum now present in eukaryotic cells. Such infoldings are present in the cytoplasm of many existing bacteria, including *Nitrobacter*, sketched here in cutaway view.

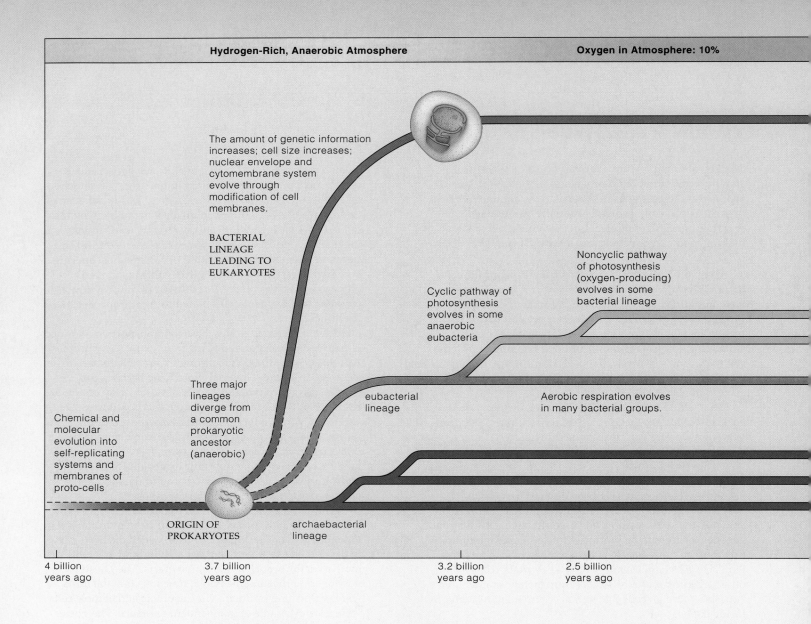

The amount of genetic information increases; cell size increases; nuclear envelope and cytomembrane system evolve through modification of cell membranes.

BACTERIAL LINEAGE LEADING TO EUKARYOTES

Noncyclic pathway of photosynthesis (oxygen-producing) evolves in some bacterial lineage

Cyclic pathway of photosynthesis evolves in some anaerobic eubacteria

Chemical and molecular evolution into self-replicating systems and membranes of proto-cells

Three major lineages diverge from a common prokaryotic ancestor (anaerobic)

eubacterial lineage

Aerobic respiration evolves in many bacterial groups.

ORIGIN OF PROKARYOTES

archaebacterial lineage

4 billion years ago

3.7 billion years ago

3.2 billion years ago

2.5 billion years ago

Evidence of Endosymbiosis   Strong evidence supports Margulis's theory. There are plenty of examples that nature continues to tinker with endosymbionts, includ-

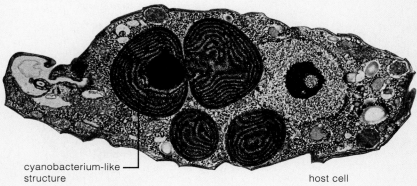

cyanobacterium-like structure

host cell

ing the one shown in the micrograph below. In such eukaryotic cells, mitochondria are like bacteria in size and structure. The inner mitochondrial membrane is like a bacterial plasma membrane. A mitochondrion replicates its own DNA and divides independently of the host cell's division. Its DNA contains instructions for building a few proteins required for specialized mitochondrial tasks.

A few of the genetic code words have uniquely mitochondrial meanings. That is, the "mitochondrial code" has a few slight variations, compared to the genetic code of cells.

As another example, consider the food-producing factories called chloroplasts. Chloroplasts may be descended from aerobic eubacteria that engaged in oxygen-producing photosynthesis. Such cells may have been engulfed, and they may have resisted digestion. By providing their respiring host cell with

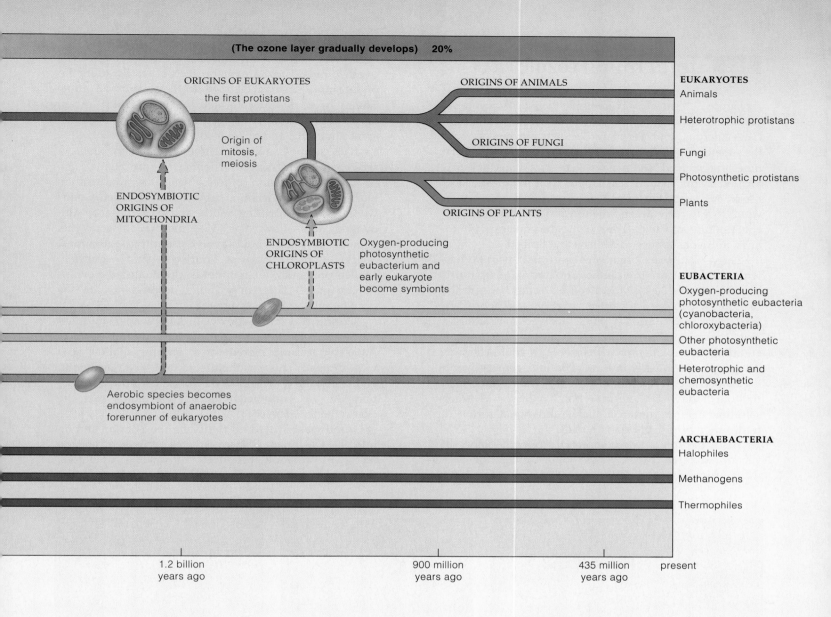

(The ozone layer gradually develops)   20%

ORIGINS OF EUKARYOTES
the first protistans

ORIGINS OF ANIMALS

EUKARYOTES
Animals

Heterotrophic protistans

Origin of
mitosis,
meiosis

ORIGINS OF FUNGI

Fungi

Photosynthetic protistans

ENDOSYMBIOTIC
ORIGINS OF
MITOCHONDRIA

Plants

ENDOSYMBIOTIC
ORIGINS OF
CHLOROPLASTS

Oxygen-producing
photosynthetic
eubacterium and
early eukaryote
become symbionts

ORIGINS OF PLANTS

EUBACTERIA
Oxygen-producing
photosynthetic eubacteria
(cyanobacteria,
chloroxybacteria)

Other photosynthetic
eubacteria

Aerobic species becomes
endosymbiont of anaerobic
forerunner of eukaryotes

Heterotrophic and
chemosynthetic
eubacteria

ARCHAEBACTERIA
Halophiles

Methanogens

Thermophiles

1.2 billion
years ago

900 million
years ago

435 million
years ago

present

oxygen, their endosymbiotic existence would have been favored. And they could have evolved into chloroplasts.

In metabolism and overall nucleic acid sequence, chloroplasts resemble some eubacteria. Their DNA is self-replicating, and they divide independently of the cell's division. Chloroplasts vary in shape and in their array of light-absorbing pigments, just as different photosynthetic eubacteria do. They may have originated a number of times, in a number of different lineages.

An Evolutionary Tree for Eukaryotes   Whatever the routes, new kinds of cells appeared on the evolutionary stage. They were equipped with a nucleus, ER membranes, and mitochondria or chloroplasts (or both). They were the first eukaryotes, the first **protistans**.

b   An evolutionary tree of life that reflects mainstream thinking about the connections among major lineages. The diagram incorporates a few ideas about the origins of some of the key eukaryotic organelles: the nucleus, ER membrane, mitochondria, and chloroplasts.

With their efficient metabolic strategies, the early protistans underwent rapid divergences and adaptive radiations. In no time at all, evolutionarily speaking, some of their descendants gave rise to the kingdoms of animals, fungi, and plants. Pulling this all together, we can put together a plausible evolutionary tree for eukaryotes. Take a moment to study Figure *b*. You can use it as a road map for chapters to follow.

21.5 LIFE IN THE PALEOZOIC ERA

As Figure 21.8 shows, we subdivide the Paleozoic into six periods: the Cambrian, Ordovician, Silurian, Devonian, Carboniferous, and Permian. Tectonic movements split the supercontinent Laurentia before this era. By Cambrian times, fragmented land masses were straddling the equator, and warm, shallow seas lapped their margins. Global conditions restricted pronounced seasonal changes in winds, ocean currents, and the upward churning of nutrients from deep waters. Thus, along the equatorial shorelines, nutrient supplies were stable but limited.

Early Cambrian organisms had flattened bodies, with a good surface-to-volume ratio for taking up nutrients (Figure 21.10a–c). Most lived on or just beneath the seafloor, where dead organisms and organic debris settled. Most major animal phyla evolved in Cambrian seas. They ranged from sponges to simple vertebrates— and they were exuberantly diverse. How could so much diversity arise so early in time? Possibly, genes governing embryonic development were less intertwined than they are today, so there may not have been as much selection against mutated alleles and novel forms of traits among the Cambrian animals.

Also, the long, warm shorelines offered plenty of vacant adaptive zones, with opportunities for new ways of securing food. Now we start seeing fossilized organisms with missing chunks, punctures, and healed-over wounds. These are not artifacts of fossilization; the organisms were damaged while they were still alive. Things were starting to get lively! In short order, diverse predators and prey with armorlike shells, spines, mouths, and novel feeding structures evolved.

Late in the Cambrian, temperatures in the shallow seas changed drastically. Trilobites (Figure 21.10d), one of the most common animals, nearly became extinct. During the Ordovician, a supercontinent called Gondwana had been drifting south, and parts became submerged in shallow seas. Vast marine environments opened up, and they favored adaptive radiations. Species of reef organisms flourished. Fast-swimming, shelled predators, the nautiloids, dominated the evolutionary stage. Their only living descendants are the chambered nautiluses (page 887).

During the late Ordovician, Gondwana straddled the South Pole. Immense glaciers formed on its surface. As water became locked up as ice, the shallow seas were drained. This was the first ice age we know about, and it may have triggered the first global mass extinction. At the Ordovician-Silurian boundary, reef life everywhere collapsed.

Gondwana drifted north during the Silurian and on into the Devonian. Reef organisms recovered, and an adaptive radiation produced formidable predators: armor-plated, massive-jawed fishes. Plants, fungi, and a

Figure 21.10 Representative Cambrian animals. (**a,b**) From Australia's Ediacara Hills, fossils of two animals, about 600 million years old. (**c**) From the Burgess Shale of British Columbia, a fossilized marine worm. (**d**) A beautifully preserved fossil of one of the earliest trilobites.

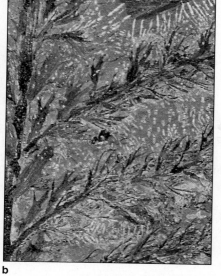

a　　　　　　　　　　　b

Figure 21.11 (**a**) Fossils of a Devonian plant (*Psilophyton*), possibly one of the earliest ancestors of seed-bearing plants. Compare this to Figure 20.2c, which shows some fossilized stems of the oldest known genus of land plants, *Cooksonia*. (**b**) Fossilized leaves of *Archaeopteris*, a land plant of the upper Devonian that may have belonged to the lineage leading to gymnosperms (such as pine trees) and angiosperms (the flowering plants). Some *Archaeopteris* trees were over three stories tall.

Figure 21.12 Lobe-finned fishes venturing onto a Devonian shoreline. They probably gulped air at the surface of shallow, stagnant water. In some lineages, lobed fins evolved into the limbs of amphibians, reptiles, birds, and mammals. Do fish out of water seem farfetched? A catfish thrashes sideways and uses its fins to pull its body forward across land (*see the photograph*).

variety of invertebrates, including segmented worms, became established in the wet lowlands. The small stalked plants foreshadowed a major radiation of terrestrial plants (Figure 21.11). Later in the Devonian, lobe-finned fishes that were ancestral to amphibians invaded the land (Figure 21.12).

At the Devonian-Carboniferous boundary, an unknown catastrophic event caused sea levels to swing dramatically, and another mass extinction occurred. Afterward, land plants and insects embarked on major adaptive radiations. Throughout the Carboniferous, land masses were gradually submerged and drained many times. Organic debris accumulated, became compacted, and was converted to coal, in the manner described on page 396.

In Permian times, insects, amphibians, and early reptiles flourished in vast swamp forests, composed in part of the ancestors of modern-day cycads, ginkgos, and conifers. As the Permian drew to a close, the greatest of all mass extinctions occurred. Nearly all known species on land and in the seas perished. At that time, all land masses were colliding to form Pangea. This vast supercontinent extended from pole to pole. A single world ocean lapped its margins:

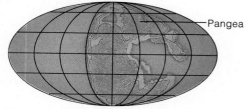

As you will see, the new distribution of oceans, land masses, and land elevations had catastrophic effects on global climates and on the course of evolution.

Early in the Paleozoic era, organisms of all five kingdoms were flourishing in the seas. By the end of the era, many lineages had successfully invaded land environments, including the wet lowlands of the supercontinent Pangea.

21.6 LIFE DURING THE MESOZOIC ERA

The Mesozoic lasted about 175 million years. It is divided into the Triassic, Jurassic, and Cretaceous periods. Adaptive radiations that began during this era greatly expanded the range of global diversity. For example, early in the Mesozoic, divergences in a few lineages gave rise to mammals and wonderful reptilian "monsters," including the dinosaurs. At the Triassic-Jurassic boundary, many marine organisms perished during a mass extinction.

Early in the Cretaceous, the supercontinent Pangea started to break up. As its huge fragments gradually drifted apart, the resulting geographic isolation favored divergences and speciation on a grand scale:

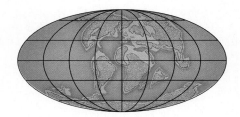

The Mesozoic was a time when dinosaurs and angiosperms (flowering plants) were the most visibly dominant lineages. The flowering plants emerged about 127 million years ago, and they began a spectacular radiation. Within 30 to 40 million years, they would displace the already declining conifers and related plants in nearly all environments (Figure 21.13). Marine invertebrates, fishes, and insects underwent equally spectacular radiations.

About 120 million years ago, global temperatures shot up by 25 degrees. By one theory, a superplume (or a rash of them) originated deep in the earth, then broke through the crust. Huge plumes spreading out beneath the crust "greased" the earth's plates into moving twice as fast. Around the globe, volcanoes spewed nutrient-rich ashes. Basalt and lava poured from huge fissures. The South Atlantic crust opened up like a zipper. Simultaneously, the plumes released huge amounts of carbon dioxide. This is one of the "greenhouse" gases that absorb heat radiating from the earth before it can escape into space. The nutrient-enriched planet warmed up—and stayed warm for 20 million years. On land and in shallow seas, photosynthetic organisms flourished. Their remains were slowly buried and converted into the world's oil reserves.

About 65 million years ago, the last dinosaurs and many marine organisms disappeared in a mass extinction. As the *Focus* essay indicates, their disappearance coincided with an asteroid impact, at a site that eventually drifted into a position that we now call the northern Yucatán peninsula.

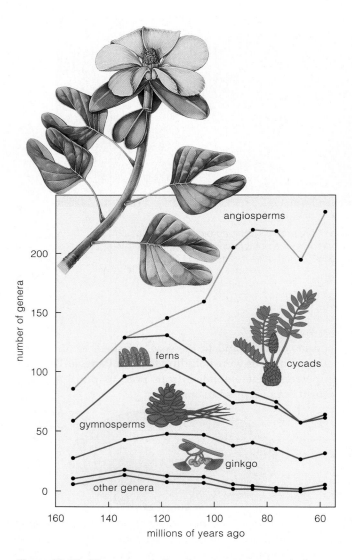

Figure 21.13 The range of diversity among vascular plants during the Mesozoic era. Accompanying the spectacular adaptive radiation of flowering plants were significant declines in the other groups shown. The painting depicts a floral shoot of *Archaeanthus linnenbergeri*, a flowering plant that lived during the Cretaceous. This now-extinct species resembled living magnolias in many respects.

The Mesozoic was a time of major adaptive radiations and of a mass extinction in which the last dinosaurs and many marine organisms disappeared.

Focus on the Environment

Plumes, Global Impacts, and the Dinosaurs

Rise of the Ruling Reptiles   Early in the Mesozoic era, a reptilian lineage evolved into the first dinosaurs. They weren't much larger than a wild turkey. Many sprinted about on two legs. Most of the early dinosaurs had high metabolic rates, and they may have been warm-blooded. They were not the dominant land animals during the Triassic period. Center stage belonged to *Lystrosaurus* and other plant-eating, mammal-like reptiles that were too large to be bothered by most predators.

Then *Lystrosaurus* ran out of luck, and adaptive zones opened up for the dinosaurs, possibly when an asteroid struck the earth. There is a crater about the size of Rhode Island in central Quebec (Figure *a*). Even if it is not *the* impact crater, it certainly shows how huge such an impact could have been. The blast wave, firestorm, earthquakes, and lava flows would have been stupendous. The atmospheric distribution of rocks and water must have vaporized upon impact, which could have darkened the skies. If so, months of acid rain followed. What kinds of animals survived this time of mass extinction (and later ones)? Most of them were small, metabolically active, and less vulnerable than others to drastic swings in climate.

The surviving dinosaurs underwent a major adaptive radiation. For 140 million years, their descendants were the ruling reptiles. Some, including the ultrasaurs, reached monstrous proportions. These were 15 meters tall.

Many of the dinosaur lineages perished in another mass extinction at the end of the Jurassic, then in a pulse of extinctions during the Cretaceous. Perhaps plumes of molten material from deep in the earth ruptured the crust, releasing enough carbon dioxide in the atmosphere to change the global temperature and climate. Or perhaps major environmental disturbances followed a swarm of asteroid bombardments (Figure *b*). Whatever the causes, conditions changed for the dinosaurs, and not all of them lived through it.

Yet some lineages recovered and new forms replaced them. By the late Cretaceous, there were perhaps a hundred different genera of dinosaurs. Duckbilled dinosaurs appeared in forests and swamps. Tanklike *Triceratops* and other plant eaters flourished in more open regions. They were prey for the swift, agile, and fearsomely toothed *Velociraptor* of motion picture fame.

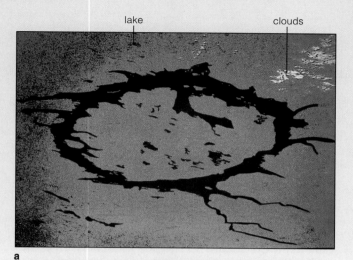

lake   clouds

a

b

a   What an impact crater looks like. Aerial view of Manicougan Crater, Quebec, where an asteroid struck the earth approximately 210 million years ago. Asteroids are rocky, metallic bodies with diameters ranging from 1,000 kilometers (600 miles) to a few meters. Most were swept from the sky when the planets were forming. About 6,000 known asteroids are still orbiting the sun in a belt between Mars and Jupiter. Unfortunately, the orbits of many dozens of others take them across the earth's orbit. Think of it as Russian roulette on a cosmic scale.

b   What one asteroid looks like, courtesy of the Galileo spacecraft that flew past on its way to Jupiter. This is one of the smaller asteroids; it would only extend from Washington, D.C., halfway to Baltimore.

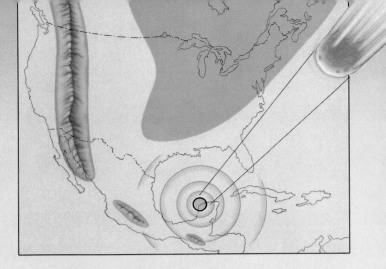

c Artist's interpretation of what might have happened during the last few minutes of the Cretaceous.

Horrendous End to Dominance The final blow came at the Cretaceous-Tertiary (K-T) boundary. Then, all (or nearly all) of the remaining dinosaurs perished when an asteroid apparently hit the earth. A thin layer of iridium-rich rock distributed around the world dates precisely to the K-T boundary. Iridium is rare on the earth's surface but common in asteroids.

By analyzing gravity maps, iridium levels in soils, and other evidence, researchers identified what appears to be the impact site. Massive movements in the crust transported the site to what is now the northern Yucatán peninsula of Mexico (Figure c). The crater itself is 9.6 kilometers deep and 300 kilometers across—wider than the state of Connecticut. This means the asteroid hit the earth at 160,000+ kilometers (100,000 miles) per hour. At least 200,000 cubic kilometers of debris and dense gases were blasted skyward—enough to shut out sunlight for months. Monstrous, 120-meter waves raced across the oceans; the entire crust heaved with earthquakes.

Things haven't settled down much since. About 2.3 million years ago, for example, a huge object from space hit the Pacific Ocean. About the same time, vast ice sheets started forming abruptly in the Northern Hemisphere. Long-term shifts in climate may have been ushering in this most recent ice age, but a global impact might have accelerated the process. Water vaporized during the impact would have formed a cloud cover that prevented sunlight from reaching the earth's surface. As you will see from the next chapter, the early ancestors of humans were around when all of this happened. The extreme shift in climate surely put their adaptability to the test.

In short, the formation of ice sheets following the global impact is one more bit of information that compels us to look skyward, also, in our attempts to piece together the environments in which life evolved.

d If dinosaurs of this sort had not disappeared by the dawn of the Cenozoic, would the then-tiny mammals ever have ventured out from under the shrubbery? Would you even be here today?

Figure 21.14 A few representatives of the Cenozoic. (**a**) From the Eocene, small, four-toed horses (*Orohippus*) and a much larger herbivore (*Uintatherium*). (**b**) From the Pleistocene, the saber-tooth cat (*Smilodon*), a large bird (*Teratornis*), and an extinct horse. This reconstruction is based on fossils recovered from a pitch pool at Rancho LaBrea, California.

a

b

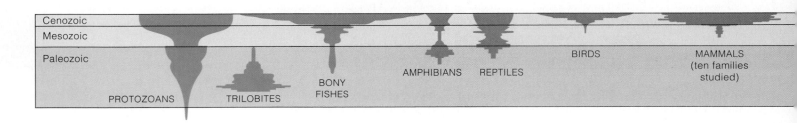

Cenozoic							
Mesozoic							
Paleozoic						BIRDS	MAMMALS (ten families studied)
	PROTOZOANS	TRILOBITES	BONY FISHES	AMPHIBIANS	REPTILES		

21.7 LIFE DURING THE CENOZOIC ERA

The breakup of Pangea set events in motion that have continued to the present. At the dawn of the Cenozoic, land masses were on collision courses:

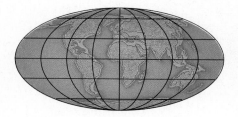

Coastlines fractured. Intense volcanic activity and uplifting produced mountains along the margins of massive rifts and plate boundaries. The Alps, Andes, Himalayas, and Cascades were born through these upheavals. The geologic changes caused major shifts in climate that affected the further evolution of life. As you will read in later chapters, vast, semiarid, cooler grasslands emerged. Diverse plant-eating mammals and their predators radiated into these new adaptive zones (Figure 21.14).

Today the distribution of land masses has favored unparalleled richness in species diversity. The tropical forests of South America, Madagascar, and Southeast Asia may well be the richest ecosystems ever to appear on earth. Marine ecosystems of the island chains of the tropical Pacific are probably not far behind. Yet we are in the middle of what may turn out to be one of the greatest of all mass extinctions. About 50,000 years ago, early humans started following migrating herds of wild animals around the Northern Hemisphere. Within a few thousand years, major groups of large mammals disappeared. The pace of extinction has been accelerating ever since, as humans hunt animals for food, fur, feath-ers, or fun, and as they destroy habitats to clear land for farm animals or crops. Chapter 49 focuses on the repercussions, which are global in scope.

The Cenozoic era has been a time of unparalleled richness in species diversity. It also is a time when the human species may be causing one of the greatest of all mass extinctions.

On the next two pages, we will be concluding our overview of life's history with a summary illustration that correlates milestones in the evolution of the earth and life. As you study this figure, keep in mind that it is only a generalized summary. It shows the five greatest mass extinctions—but there were many others in between. It shows the shrinking and expanding range of species diversity—but the range is for all the major groups combined. As you can see from the sampling of Figure 21.15, each major group has its own distinctive history of persistences, radiations, and extinctions.

With these qualifications in mind, we are ready to turn to the next chapters in this unit. They will provide you with richer detail of the history and current range of diversity for all five kingdoms of organisms.

Figure 21.15 A few representative evolutionary histories. Notice the differences in the patterns of adaptive radiations and mass extinction for this limited sampling. Notice also the spectacular current success of the insects. (For plants and animals, the widths of the lineages, shown in blue, represent the approximate numbers of families.)

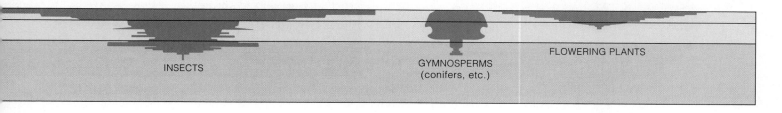

INSECTS

GYMNOSPERMS
(conifers, etc.)

FLOWERING PLANTS

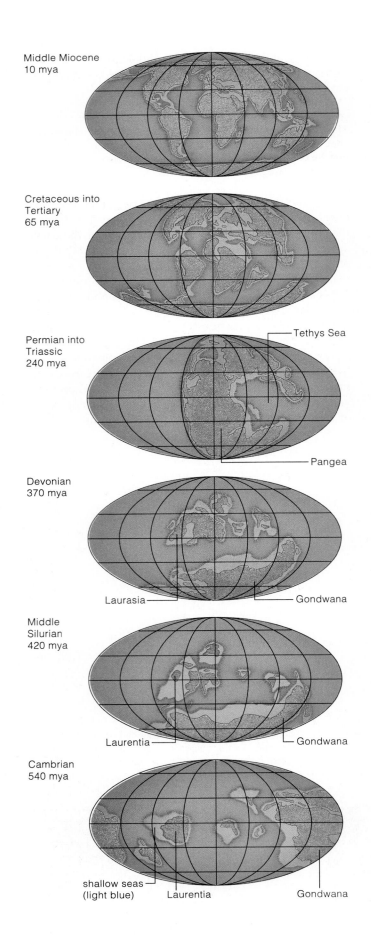

Middle Miocene
10 mya

Cretaceous into
Tertiary
65 mya

Permian into
Triassic
240 mya

Tethys Sea

Pangea

Devonian
370 mya

Laurasia — Gondwana

Middle
Silurian
420 mya

Laurentia — Gondwana

Cambrian
540 mya

shallow seas
(light blue) — Laurentia — Gondwana

Era	Period	Epoch	Millions of Years Ago (mya)	
CENOZOIC	Quaternary	Recent	0.01-	
		Pleistocene	1.65	
	Tertiary	Pliocene	5	
		Miocene	25	
		Oligocene	38	
		Eocene	54	
		Paleocene	65	
MESOZOIC	Cretaceous	Late	100	
		Early	138	
	Jurassic		205	
	Triassic		240	
PALEOZOIC	Permian		290	
	Carboniferous		360	
	Devonian		410	
	Silurian		435	
	Ordovician		505	
	Cambrian		550	
PROTEROZOIC			2,500	
ARCHEAN				

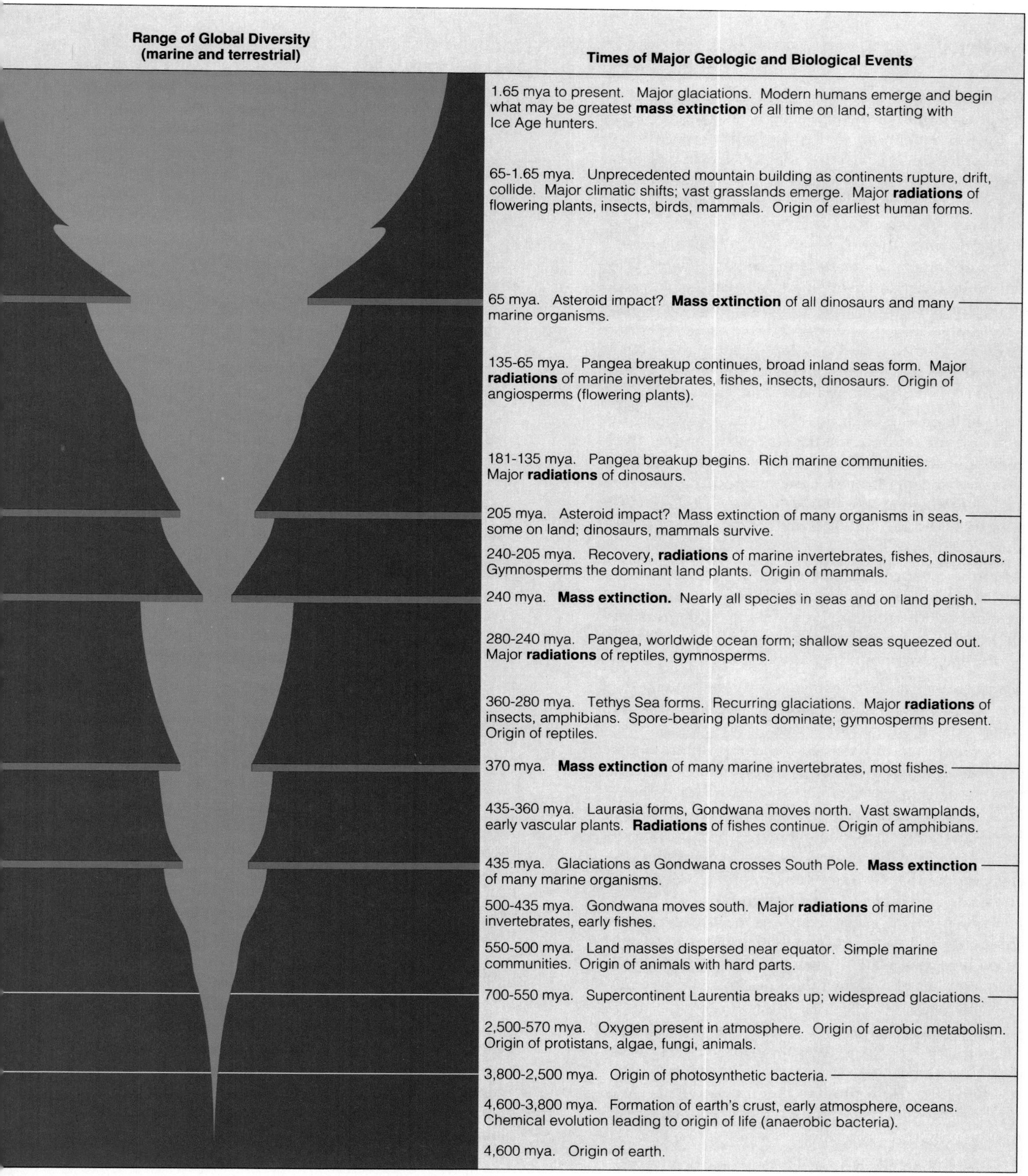

Range of Global Diversity
(marine and terrestrial)

Times of Major Geologic and Biological Events

1.65 mya to present. Major glaciations. Modern humans emerge and begin what may be greatest **mass extinction** of all time on land, starting with Ice Age hunters.

65-1.65 mya. Unprecedented mountain building as continents rupture, drift, collide. Major climatic shifts; vast grasslands emerge. Major **radiations** of flowering plants, insects, birds, mammals. Origin of earliest human forms.

65 mya. Asteroid impact? **Mass extinction** of all dinosaurs and many marine organisms.

135-65 mya. Pangea breakup continues, broad inland seas form. Major **radiations** of marine invertebrates, fishes, insects, dinosaurs. Origin of angiosperms (flowering plants).

181-135 mya. Pangea breakup begins. Rich marine communities. Major **radiations** of dinosaurs.

205 mya. Asteroid impact? Mass extinction of many organisms in seas, some on land; dinosaurs, mammals survive.

240-205 mya. Recovery, **radiations** of marine invertebrates, fishes, dinosaurs. Gymnosperms the dominant land plants. Origin of mammals.

240 mya. **Mass extinction.** Nearly all species in seas and on land perish.

280-240 mya. Pangea, worldwide ocean form; shallow seas squeezed out. Major **radiations** of reptiles, gymnosperms.

360-280 mya. Tethys Sea forms. Recurring glaciations. Major **radiations** of insects, amphibians. Spore-bearing plants dominate; gymnosperms present. Origin of reptiles.

370 mya. **Mass extinction** of many marine invertebrates, most fishes.

435-360 mya. Laurasia forms, Gondwana moves north. Vast swamplands, early vascular plants. **Radiations** of fishes continue. Origin of amphibians.

435 mya. Glaciations as Gondwana crosses South Pole. **Mass extinction** of many marine organisms.

500-435 mya. Gondwana moves south. Major **radiations** of marine invertebrates, early fishes.

550-500 mya. Land masses dispersed near equator. Simple marine communities. Origin of animals with hard parts.

700-550 mya. Supercontinent Laurentia breaks up; widespread glaciations.

2,500-570 mya. Oxygen present in atmosphere. Origin of aerobic metabolism. Origin of protistans, algae, fungi, animals.

3,800-2,500 mya. Origin of photosynthetic bacteria.

4,600-3,800 mya. Formation of earth's crust, early atmosphere, oceans. Chemical evolution leading to origin of life (anaerobic bacteria).

4,600 mya. Origin of earth.

Figure 21.16   Summary of major events in the evolution of the earth and of life.

SUMMARY

1. The full story of life begins with the "big bang," a model of the origin of the universe.

a. By this model, all matter and all of space were once compressed in a fleeting state of enormous heat and density. Time began with the near-instantaneous distribution of all matter and energy throughout the known universe, which has been expanding ever since.

b. Nearly all of the helium and other light elements, the most abundant elements of the universe, were produced immediately after the big bang. The heavier elements originated during the formation, evolution, and death of stars.

c. Every element of the solar system, the planet earth, and life itself are products of the physical and chemical evolution of the universe and its stars.

2. Four billion years ago, the earth had a high-density core, a mantle of intermediate density, and a thin, extremely unstable crust of low-density rocks. Probably gaseous hydrogen, nitrogen, carbon monoxide, and carbon dioxide made up the first atmosphere. Free oxygen and water could not have accumulated under the prevailing conditions.

3. After the crust cooled, water accumulated and rains began. Runoff from rains carried mineral salts and other compounds from the crustal rocks to depressions in the crust and formed the early seas. Life could not have originated without this salty liquid water.

4. Many studies and experiments provide indirect evidence that life originated under conditions that presumably existed on the early earth.

a. Comparative analyses of the composition of cosmic clouds and of rocks from other planets and the earth's moon suggest that precursors of the complex molecules associated with life were available.

b. In laboratory tests that simulated primordial conditions, including the absence of free oxygen, the precursors spontaneously assembled into glucose, sugars, amino acids, and other organic compounds.

c. Known chemical principles as well as computer simulations indicate that metabolic pathways could have evolved as a result of chemical competition for the limited supplies of organic molecules that had accumulated in the seas.

d. Self-replicating systems of RNA, enzymes, and coenzymes have been synthesized in the laboratory. How DNA entered the picture is not yet understood.

e. Lipid and lipid-protein membranes that show properties of cell membranes form spontaneously under the prescribed conditions.

5. Life originated by 3.8 billion years ago. Since then, life's history has been influenced profoundly by major changes in the earth's crust, atmosphere, and oceans. Forces of change have included tectonic movements, asteroid impacts, and activities of organisms—such as oxygen-producing photosynthesizers and, currently, the human species.

6. Abrupt discontinuities in the fossil record mark the times of global mass extinctions. They serve as boundary markers for five great eras in the geologic time scale. Radiometric dating has allowed us to assign absolute dates to this time scale:

a. Archean: 4.6 to 2.5 billion years ago
b. Proterozoic: 2.5 billion to 550 million years ago
c. Paleozoic: 550 to 240 million years ago
d. Mesozoic: 240 to 65 million years ago
e. Cenozoic: 65 million years ago to the present

7. The first living cells were prokaryotic. Not long after they appeared, divergences began that led to three great lineages: the archaebacteria, the eubacteria, and the prokaryotic ancestors of all eukaryotes. Among the eubacteria were species that used a cyclic pathway of photosynthesis.

8. In the Proterozoic, the noncyclic pathway of photosynthesis had evolved in some lineages, and oxygen started accumulating in the atmosphere.

a. In time, the atmospheric concentration of free oxygen prevented further spontaneous formation of large organic molecules. The spontaneous origin of life was no longer possible on earth.

b. The abundance of atmospheric oxygen was a selective pressure that brought about the evolution of aerobic respiration. Aerobic respiration was a key innovation in the origin of the first eukaryotic cells.

c. Mitochondria and chloroplasts, two key eukaryotic organelles, may have evolved following endosymbiosis between aerobic bacteria and the anaerobic forerunners of eukaryotes.

d. The oxygen-rich atmosphere promoted the formation of a layer of ozone (O_3). In time, that shield against destructive ultraviolet radiation allowed some lineages to move out of the seas, into low wetlands.

9. Early in the Paleozoic, diverse organisms of all five lineages had become established in the seas. By the end of the era, the invasion of land was under way. From that time forward, there have been pulses of mass extinctions and adaptive radiations. Asteroid bombardments and other catastrophes triggered many of these events. So did tectonic movements that changed the distribution of oceans and land as well as the prevailing global and regional climates.

Review Questions

1. Describe the chemical and physical characteristics of the earth four billion years ago. How do we know what it may have been like? *324*

2. Describe the experimental evidence for the spontaneous origin of large organic molecules, the self-assembly of proteins, and the formation of organic membranes and spheres, under conditions similar to those of the early earth. *324–327*

3. How does continental drift occur, and in what ways does this process influence changes in biological communities? *328–329*

4. Summarize the theory of endosymbiosis. Cite evidence in favor of this theory. *331–332*

5. When did plants, insects, and vertebrates invade the land? *342–343*

6. The Atlantic Ocean is widening, and the Pacific Ocean and Indian Ocean are closing. Many millions of years from now, the continents will collide and form a second Pangea. Write a short essay on what conditions might be like on that future supercontinent and what types of organisms might survive there.

8. Which of the following statements is false?
   a. The first living cells were prokaryotes.
   b. The cyclic pathway of photosynthesis first appeared in early eubacterial species.
   c. Oxygen began accumulating in the atmosphere after the noncyclic pathway of photosynthesis evolved.
   d. During the Proterozoic, an increasing amount of atmospheric oxygen promoted further spontaneous generation of organic molecules.

9. Match the era with the events listed.
   ____ Archean
   ____ Proterozoic
   ____ Paleozoic
   ____ Mesozoic
   ____ Cenozoic

   a. major radiations of dinosaurs, origin of flowering plants and mammals
   b. formation of earth's crusts, oceans, early atmosphere, chemical evolution, origin of life
   c. major radiations of flowering plants, insects, birds, mammals, emergence of human forms
   d. oxygen present, origin of aerobic metabolism, protistans, algae, fungi, animals
   e. origin of amphibians, origin of reptiles, spread of early plants

Self-Quiz *(Answers in Appendix IV)*

1. Through study of the geologic record, we know that the evolution of life is linked to _____ .
   a. tectonic movements of the earth's crust
   b. bombardment of the earth by celestial objects
   c. profound shifts in land masses, shorelines, and oceans
   d. physical and chemical evolution of the earth
   e. all of the above

2. The geologic time scale begins with the _____ era.
   a. Paleozoic   d. Proterozoic
   b. Mesozoic   e. Cenozoic
   c. Archean

3. The first scientist to find indirect evidence that organic molecules could have been formed on the early earth was _____ .
   a. Darwin   c. Fox
   b. Miller   d. Margulis

4. One type of "proto-cell," the microsphere, was first produced in a laboratory by _____ .
   a. Darwin   c. Fox
   b. Miller   d. Margulis

5. An abundance of _____ was conspicuously absent from the earth's atmosphere four billion years ago.
   a. hydrogen   c. carbon monoxide
   b. nitrogen   d. free oxygen

6. Life originated by _____ .
   a. 4.6 billion years ago   c. 3.8 billion years ago
   b. 2.8 million years ago   d. 3.8 million years ago

7. Abrupt discontinuities in the fossil record indicate _____ .
   a. the death of all organisms when the atmospheric oxygen levels dropped
   b. the times of global mass extinctions
   c. boundary markers for five great eras of the geologic time scale
   d. both b and c

Selected Key Terms

archaebacterium *330*
Archean *329*
Cenozoic *329*
eubacterium *330*
eukaryotic cell *330*
geologic time scale *329*
mass extinction *343*
Mesozoic *329*

Paleozoic *329*
plate tectonic theory *328*
prokaryotic cell *330*
Proterozoic *329*
protistan *333*
radiation *343*
stromatolite *330*
theory of endosymbiosis *331*

Readings

Bambach, R., C. Scotese, and A. Ziegler. 1980. "Before Pangea: The Geographies of the Paleozoic World." *American Scientist* 68(1): 26–38.

Dobb, E. February 1992. "Hot Times in the Cretaceous." *Discover* 13: 11–13.

Dott, R., Jr., and R. Batten. 1988. *Evolution of the Earth*. Fourth edition. New York: McGraw-Hill.

Hartman, W., and Chris Impey. 1994. *Astronomy: The Cosmic Journey*. Fifth edition. Belmont, California: Wadsworth.

Horgan, J. February 1991. "Trends in Evolution: In the Beginning . . ." *Scientific American* 264(2): 116–125.

22 BACTERIA AND VIRUSES

The Unseen Multitudes

Did a friend ever mention that you are nearly 1/1,000 of a mile tall? Probably not. What would be the point of measuring people in units as big as miles? Yet we think this way, in reverse, when we measure **microorganisms**. These are mostly single-celled organisms too small to be seen without a microscope.

The bacteria in Figure 22.1 are a case in point. To measure them, you'd have to divide a meter into a thousand units (millimeters). Then you'd have to divide one millimeter into a thousand smaller units (micrometers). One millimeter would be as small as the dot of this "i." A thousand bacteria would fit side by side across the dot!

Of all organisms, bacteria generally are the smallest. To be sure, viruses are smaller. We measure them in nanometers (billionths of a meter). But viruses are not living things. We consider them here because they infect just about every kind of organism in the great spectrum of life.

Bacteria vastly outnumber all other organisms combined. Their reproductive potential is staggering. Under ideal conditions, some bacteria can divide about every twenty minutes. At that rate, a single bacterium could have nearly a billion descendants in ten hours! So why don't bacteria take over the world? Sooner or later, their activities typically ruin the conditions favoring their reproduction. Also, bacteria eat one another, other organisms attack them, and viruses and seasonal changes in living conditions help keep them in check.

Many kinds of bacteria are **pathogens** (infectious, disease-causing agents). They invade and multiply inside other organisms, and disease follows when their

Figure 22.1 How small are bacteria? Shown here, *Bacillus* cells on the tip of a pin. The cells in (**c**) are magnified 14,000 times. (**d**) How small are viruses? Shown here, bacteriophage particles, each about 225 nanometers tall, infecting a bacterial cell.

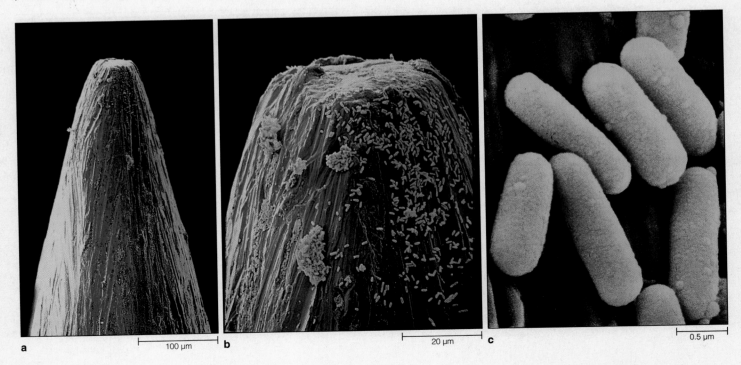

a | 100 µm | b | 20 µm | c | 0.5 µm

activities damage tissues and interfere with normal body functions. However, the pathogenic types should not give the whole kingdom a bad name.

Think back on the vast numbers of photosynthetic bacterial cells in the seas (page 117). They provide food and oxygen for entire communities, and they have major roles in the global cycling of carbon. Other bacteria decompose organic debris and so help cycle nutrients that sustain entire communities. Trace the lineages of *any* organism back far enough in time, and you discover bacterial ancestors.

And so, from *Escherichia coli* to elephants, clams, and coast redwoods, all organisms interconnect with bacteria—regardless of differences in size, numbers, and evolutionary distance.

(page 117)

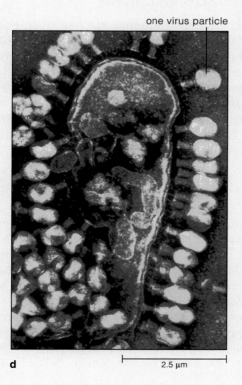

one virus particle

d

2.5 μm

KEY CONCEPTS

1. Bacteria are microscopically small, single-celled, prokaryotic organisms. They do not have a profusion of internal, membrane-bound organelles in the cytoplasm, as eukaryotic cells do. Yet bacteria show great metabolic diversity, and many make complex behavioral responses to the environment.

2. Bacteria reproduce by binary fission, a cell division mechanism that immediately follows DNA replication and that divides the parent cell into two genetically equivalent daughter cells.

3. Today there are two great bacterial lineages, the archaebacteria and eubacteria. Archaebacteria live in such inhospitable, anaerobic habitats that they may resemble the first living cells on earth. Most bacterial species are eubacteria, which inhabit nearly all existing environments.

4. From the human perspective, bacteria are good, bad, or dangerous. Basically, however, they simply are busy at surviving and reproducing like the rest of us.

5. A virus is a nonliving infectious particle that consists of nucleic acid and a protein coat (and sometimes an outer envelope). Viruses cannot be replicated without pirating the metabolic machinery of a specific type of host cell.

6. Nearly all viral multiplication cycles include five steps: attachment to a suitable host cell, penetration, viral DNA or RNA replication and protein synthesis, assembly of new viral particles, and release from the infected cell.

22.1 CHARACTERISTICS OF BACTERIA

Of all organisms, bacteria are the most abundant and far-flung. Thousands of species live in places ranging from deserts to hot springs, snow, the seafloor, and the ocean. Billions of bacterial cells may be present in a handful of rich soil. The ones in your gut and on your skin outnumber your body cells! (Your cells are larger, however, so you are only "a few percent bacterial" by weight.) And of all organisms, bacteria have the longest evolutionary histories. Two great lineages, **archaebacteria** and **eubacteria**, have endured from the time of life's origin to the present. A third lineage gave rise to eukaryotic cells (page 330). Table 22.1 and Figure 22.2 introduce the main features that help characterize these remarkable organisms.

Splendid Metabolic Diversity

All organisms must take in energy and carbon to meet their nutritional requirements. However, compared to other organisms, bacteria show the greatest diversity in their modes of securing these resources.

Like plants, some bacteria are **photoautotrophs**. These "self-feeders" can make their own biological molecules by using sunlight as an energy source and carbon dioxide as a carbon source for photosynthesis. Some species use a noncyclic pathway, based on light-trapping pigments, transport systems, and ATP synthases embedded in their plasma membrane. In this pathway, water molecules give up electrons and hydrogen that are used for the reactions, and oxygen is released as a by-product (page 112). Other photoautotrophic species use a cyclic pathway. They either cannot use oxygen or die in its presence. These anaerobes strip electrons and hydrogen from gaseous hydrogen (H_2), sulfur, hydrogen sulfide (H_2S), and other inorganic compounds in their environment.

Other bacteria are **photoheterotrophs**. As their name suggests, they are not self-feeders. They *can* use sunlight as an energy source for a cyclic pathway of photosynthesis—but they can't harness carbon dioxide from the surroundings. Instead, they get carbon from fatty acids, carbohydrates, and other organic compounds that other organisms produced.

Then there are the self-feeding **chemoautotrophs**. These self-feeders harness carbon dioxide as the main carbon source. They get the required electrons and hydrogen from a variety of inorganic substances, including gaseous hydrogen, sulfur, sulfur compounds, nitrogen compounds, and ferrous iron (Fe^{++}).

And then there are the **chemoheterotrophs**. These are parasites or saprobes, not self-feeders. The *parasitic* types live on or in a living host and draw glucose and other nutrients from it. The *saprobic* types obtain nutri-

Table 22.1   Characteristics of Bacterial Cells
1. Bacterial cells are prokaryotic (they have no membrane-bound nucleus or other organelles in the cytoplasm).
2. Bacterial cells have a single chromosome (a circular DNA molecule); many species also have plasmids (page 248).
3. Most bacteria have a cell wall composed of peptidoglycan.
4. Most bacteria reproduce by binary fission.
5. Collectively, bacteria show great diversity in their modes of metabolism.

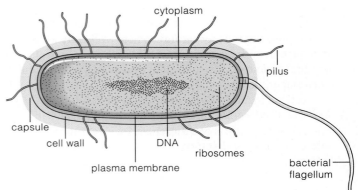

Figure 22.2   Generalized body plan of a bacterium.

ents from the organic products, wastes, or remains of other organisms.

Bacterial Sizes and Shapes

From the chapter introduction, you already have a sense of the microscopically small sizes of bacteria. Typically, the width or length of these cells falls between 1 and 10 micrometers.

Three basic shapes are common among bacteria. A spherical shape is a **coccus** (plural, cocci; from a word that really means berries). A rod shape is a **bacillus** (plural, bacilli, meaning small staffs). A bacterium with one or more twists in the cell body is a **spiral**:

coccus   bacillus   spiral

Don't let these simple categories fool you. Cocci may also be oval or partly flattened, and bacilli may be skinny (like straws) or tapered (like cigars). After they divide, daughter cells remain stuck together in various ways. For example, chains of daughter cells are called *strepto*cocci, and sheets of them are called *staphylo*cocci. Some spiral bacteria (vibrios) are curved, like a comma, and others (spirilla) are stiff, flagellated, and twisted helically, like a corkscrew. Other helically twisted

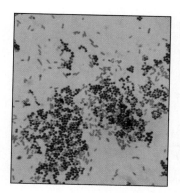

Figure 22.3 Example of Gram staining. Gram-positive *Staphylococcus aureus* remains purple when washed with organic solvents. Gram-negative *Escherichia coli* loses color easily. *E. coli* cells shown here appear red because they have been treated with a light-red dye (safranin) after being washed. Without this "counter-stain," they would be colorless.

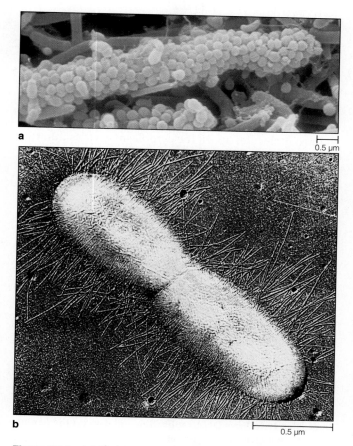

a   0.5 μm

b   0.5 μm

Figure 22.4 (**a**) Surface view of numerous bacteria that have become attached, by means of their sticky glycocalyx, to the surface of a human tooth—and doesn't this micrograph make you want to grab a toothbrush? (**b**) An abundance of the filamentous structures called pili, projecting from the surface of a cell (*Escherichia coli*) that is dividing in two.

forms, the spirochetes, have a flexible body with a sheathed filament attached.

Structural Features

Bacterial cells alone are **prokaryotic**, meaning they were around before the origin of the nucleus and of eukaryotic cells. (*Pro-*, before; *karyon*, nucleus.) Most bacteria are not internally subdivided into metabolic compartments. Reactions proceed in the cytoplasm or at the plasma membrane. (For example, proteins are synthesized at ribosomes distributed through the cytoplasm or attached to the plasma membrane.) This doesn't mean bacteria are somehow inferior to eukaryotic cells. Being tiny, fast reproducers, they do very well without great internal complexity.

Nearly all bacteria have a semirigid, permeable **cell wall** around the plasma membrane (Figure 22.2). The wall helps the cell maintain its shape and resist rupturing when internal fluid pressure increases. Eubacterial cell walls consist of **peptidoglycan**, a type of molecule in which peptides crosslink many polysaccharide strands to one another.

The precise structure and composition of the wall helps clinicians identify many bacterial species. In one staining reaction, the Gram stain, cells are exposed to a purple dye, then washed with alcohol, then exposed to a pink dye. Walls of *Gram-positive* bacteria retain the purple stain after the alcohol wash. Walls of *Gram-negative* bacteria lose color after the wash and become pink (Figure 22.3).

Surrounding the cell wall is a **glycocalyx**, a sticky mesh of polysaccharides, polypeptides, or both. When the mesh is highly organized and attached firmly to the wall, it is called a capsule. When less organized and loosely attached, it's called a slime layer. The glycocalyx helps bacterial cells attach to teeth, assorted membranes, rocks in streambeds, and other surfaces (Figure

22.4*a*). In some encapsulated types, it deters a host's infection-fighting cells.

Some bacteria have one or more motile structures called **bacterial flagella** (singular, flagellum). These don't have the same structure as eukaryotic flagella, and they don't operate the same way. They move the cell by rotating like a propeller. Many bacteria have **pili** (singular, pilus), short, filamentous proteins that project above the cell wall. Figure 22.4*b* shows an example. Some pili help cells adhere to surfaces. Others help them attach to one another as a prelude to conjugation, described next.

Bacteria are microscopic, prokaryotic cells having one bacterial chromosome and, often, a number of plasmids. The cells of nearly all species have a wall around the plasma membrane, and a capsule or slime layer around the cell wall.

22.2 BACTERIAL REPRODUCTION

Most bacterial cells reproduce by **binary fission**, a cell division mechanism. Each cell has only one bacterial chromosome, which is a circular, double-stranded DNA molecule with only a few proteins attached to it. Just after a parent cell replicates its DNA and starts moving the two molecules apart, the cytoplasm divides into two genetically equivalent daughter cells. As Figure 22.5 shows, the division requires suitable membrane growth and deposition of wall material at the midsection of the dividing cell.

In many species, daughter cells also inherit one or more plasmids. A **plasmid** is a small, self-replicating circle of extra DNA with only a few genes (page 248). Usually the genes confer a survival advantage, as when they specify an enzyme that allows a cell to synthesize or use some nutrient. Some plasmid genes confer resistance to antibiotics, substances that kill or inhibit the growth of bacteria. Others confer the means to conjugate. In **bacterial conjugation**, one bacterial cell transfers plasmid DNA to another cell, even of a different species (page 350 and Figure 22.6).

Bacteria use binary fission, a cell division mechanism that immediately follows DNA replication and divides the parent cell into two genetically equivalent daughter cells.

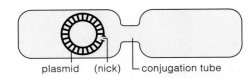

a Conjugation tube unites a donor and a recipient cell

b Replication starts on plasmid DNA in donor cell; displaced single DNA strand enters recipient cell

c Replication starts on transferred DNA strand

d Cells separate; plasmids circularize

Figure 22.6 Bacterial conjugation—transfer of a plasmid from a donor to a recipient cell. A long pilus brings the two cells into close contact so that a conjugation tube can form between them. For clarity, the bacterial chromosome is not shown and the plasmid in the diagrams is enormously enlarged (*compare* Figure 16.2).

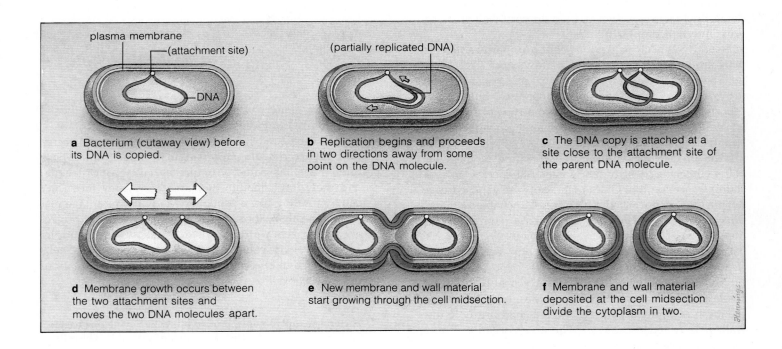

a Bacterium (cutaway view) before its DNA is copied.

b Replication begins and proceeds in two directions away from some point on the DNA molecule.

c The DNA copy is attached at a site close to the attachment site of the parent DNA molecule.

d Membrane growth occurs between the two attachment sites and moves the two DNA molecules apart.

e New membrane and wall material start growing through the cell midsection.

f Membrane and wall material deposited at the cell midsection divide the cytoplasm in two.

Figure 22.5 Bacterial reproduction by binary fission, a cell division mechanism.

22.3 BACTERIAL CLASSIFICATION

A few decades ago, reconstructing the phylogeny of bacteria seemed to be an impossible task. Except for stromatolites, early bacteria are not spectacularly represented in the fossil record. (Figure 21.9 shows some stromatolites—the fossilized, stacked mats of bacterial populations and sediments.) Most bacterial groups are not represented at all. Given their elusive evolutionary histories, the many thousands of known bacterial species traditionally have been grouped together on the basis of cell shape, modes of nutrition, metabolic patterns, staining attributes of their cell wall, and other traits. Table 22.2 lists representatives from some of the major groupings. Comparative biochemistry, including nucleic acid hybridization studies, is now providing insight into bacterial phylogeny. Such studies provide strong evidence that three great bacterial lineages diverged from a common ancestor not long after living cells originated. These led to the ancestors of eukaryotes, and to the archaebacteria and eubacteria.

Table 22.2 Some Major Groups of Bacteria

Group	Main Habitats	Characteristics	Representatives
Archaebacteria			
Methanogens	Anaerobic sediments of lakes, swamps; also animal gut	Chemosynthetic; methane producers; used in sewage treatment facilities	*Methanobacterium*
Halophiles	Brines (extremely salty water)	Heterotrophic; also have photosynthetic machinery of a unique sort	*Halobacterium*
Extreme thermophiles	Acidic soil, hot springs, hydrothermal vents on seafloor	Heterotrophic or chemosynthetic; use inorganic substances as electron donors	*Sulfolobus, Thermoplasma*
Photoautotrophic eubacteria			
Cyanobacteria, green sulfur bacteria, and purple sulfur bacteria	Mostly lakes, ponds; some marine, terrestrial	Photosynthetic; use sunlight energy and carbon dioxide (carbon source); cyanobacteria produce oxygen as by-product	*Anabaena, Nostoc, Chloroflexus, Rhodopseudomonas*
Photoheterotrophic eubacteria			
Purple nonsulfur and green nonsulfur bacteria	Anaerobic, organically rich muddy soils, and sediments of aquatic habitats	Use sunlight, organic compounds (electron donors); some purple nonsulfur may also grow chemotrophically	*Rhodospirillum, Chlorobium*
Chemoautotrophic eubacteria			
Nitrifying, sulfur-oxidizing, and iron-oxidizing bacteria	Soil; freshwater, marine habitats	Use carbon dioxide; electrons from inorganic compounds; some have roles in agriculture, cycling of nutrients in ecosystems	*Nitrosomonas, Nitrobacter, Thiobacillus*
Chemoheterotrophic eubacteria			
Spirochetes	Aquatic habitats; parasites of animals	Helically coiled, motile; free-living and parasitic species; some major pathogens	*Spirochaeta, Treponema*
Gram-negative, aerobic rods and cocci	Soil, aquatic habitats; parasites of animals, plants	Some major pathogens; some (e.g., *Rhizobium*) fix nitrogen	*Pseudomonas, Neisseria, Rhizobium, Agrobacterium*
Gram-negative, facultative anaerobic rods	Soil, plants, animal gut	Many are major pathogens; one (*Photobacterium*) is bioluminescent	*Salmonella, Proteus, Escherichia, Photobacterium*
Rickettsias and chlamydias	Host cells of animals	Intracellular parasites; many pathogens	*Rickettsia, Chlamydia*
Myxobacteria	Decaying plant, animal matter; bark of living trees	Gliding, rod-shaped; cells aggregate and migrate together	*Myxococcus*
Gram-positive cocci	Soil; skin and mucous membranes of animals	Some major pathogens	*Staphylococcus, Streptococcus*
Endospore-forming rods and cocci	Soil; animal gut	Some major pathogens	*Bacillus, Clostridium*
Gram-positive nonsporulating rods	Fermenting plant, animal material; gut, vaginal tract	Some important in dairy industry, others major contaminators of milk, cheese	*Lactobacillus, Listeria*
Actinomycetes	Soil; some aquatic habitats	Include anaerobes and strict aerobes; major producers of antibiotics	*Actinomyces, Streptomyces*

22.4 ARCHAEBACTERIA

There are three intriguing groups of archaebacteria—methanogens, halophiles, and extreme thermophiles. In many respects, their cell structure, metabolism, and nucleic acid sequences are unique. (For example, none has a cell wall of peptidoglycan.) They differ as much from other bacteria as they do from eukaryotes. They probably resemble the first living cells, which must have originated in extremely hot, acidic, and salty habitats. Archaebacteria live in such habitats—hence their name (*archae-* means beginning).

The **methanogens**, or "methane-makers," live in the muck of swamps, sewage, the animal gut, and other anaerobic habitats. They make ATP by converting carbon dioxide and hydrogen gases to methane (CH_4). Pungent stockyard fumes testify to their presence. So does the "marsh gas" that hangs over swamps and sewage treatment plants. As a group, methanogens produce about 2 billion tons of methane gas each year. As described on page 857, they affect atmospheric levels of carbon dioxide and the global carbon cycle.

Halophiles, or "salt-lovers," live in brackish ponds, salt lakes, near volcanic vents on the seafloor, and other high-salinity habitats (Figure 22.7). They can contaminate salted fish and animal hides as well as commercially produced sea salt.

Most halophiles are heterotrophs that form ATP by aerobic pathways. When oxygen levels decline, however, some strains produce ATP by photosynthesis. Patches of bacteriorhodopsin, a light-trapping pigment, form in the plasma membrane. Absorption of light energy triggers an increase in the hydrogen ion gradient across the membrane. When the ions flow down the gradient, through other membrane proteins, ATP forms (compare the discussion of chemiosmosis on page 114).

Extreme thermophiles (heat lovers) live in seemingly inhospitable places as hot springs, highly acidic soils, and sediments near hydrothermal vents. Vents are volcanic fissures in the ocean floor, where water temperature can exceed 250°C. They spew hydrogen sulfide, which thermophiles use as a source of electrons for ATP formation. Thermophiles inhabiting such places are the basis of remarkable food webs (page 885). They are cited as evidence that life could have originated deep in the early oceans. Waste heaps of coal mines are the only known habitat of *Thermoplasma*. Because this is a habitat of recent origin—coal mines have been around for only a few hundred years—*Thermoplasma* must have evolved in places we don't know about.

22.5 EUBACTERIA

The *eu-* in eubacteria implies "typical." Eubacteria are far more common than the three bacterial groups just described. Let's look at a few types, using modes of nutrition as a conceptual framework.

Photoautotrophic Eubacteria

Cyanobacteria (also called blue-green algae) are the most common photoautotrophic bacteria. Most species live in freshwater ponds, where they often grow as chains of cells, sheathed in mucus. The chains form dense, slimy mats near the surface of nutrient-enriched water (Figure 22.8). *Anabaena* and other types produce oxygen during photosynthesis. They also can convert nitrogen to ammonia, a nitrogen source for biosynthesis. When nitrogen-containing compounds are in short supply, some cells develop into **heterocysts**. These modified cells can make a nitrogen-fixing enzyme. Heterocysts share nitrogen compounds with the photosynthetic cells and get carbohydrates in return. Substances move freely through junctions that connect the cytoplasm of neighboring cells.

a

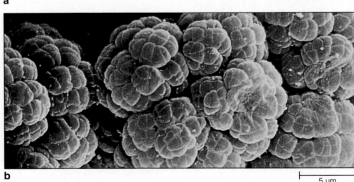

b

5 μm

Figure 22.7 Archaebacteria. (**a**) Pinkish, saline water in Great Salt Lake, Utah—a sign of colonies of halophilic bacteria and certain algae that contain pinkish to red-orange carotenoid pigments. (**b**) A colony of methanogenic cells (*Methanosarcina*).

Photoautotrophs that do not produce oxygen include green bacteria and purple bacteria. They cannot use water as a source of electrons. Green bacteria strip electrons from hydrogen sulfide or hydrogen gas. They may resemble the ancient anaerobic bacteria in which the cyclic pathway of photosynthesis first emerged.

Chemoautotrophic Eubacteria

Chemoautotrophs, recall, get their carbon from carbon dioxide and electrons from inorganic substances. Many eubacteria in this category affect the global cycling of nitrogen, sulfur, phosphorus, and other nutrients. Consider nitrogen, a building block for amino acids and proteins. Without it, there would be no life. Nitrifying bacteria, such as the one shown in Figure 22.9, strip electrons from ammonia. Plants can use the end product, nitrate, as a nitrogen source.

Chemoheterotrophic Eubacteria— Wonderfully Beneficial Types

Most bacteria are chemoheterotrophs, and many are beneficial. Most species are major decomposers. Their enzymes break down organic compounds, even pesticides in soil. Pseudomonads are an example (Figure 22.10). We also use species of *Lactobacillus* when manufacturing pickles, sauerkraut, buttermilk, and yogurt. We use actinomycetes as sources of antibiotics. *Escherichia coli*, a gut dweller, produces vitamin K and compounds useful in fat digestion, and it helps newborns digest milk. Its activities keep many food-borne pathogens from colonizing the gut. Sugarcane and corn plants benefit from a symbiotic association with a nitrogen-fixing spirochete, *Azospirillum*. Plants use some of the nitrogen and the spirochete uses some of the plant's carbohydrates. Beans and other legumes benefit from the nitrogen-fixing activities of *Rhizobium*, which dwells in their roots.

a

b

resting spore   heterocyst

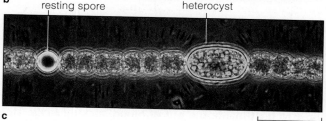

c

5 μm

Figure 22.8 *Cyanobacterium*—a premier photoautotroph. (**a**) A cyanobacterial population near the surface of a nutrient-enriched pond. (**b**,**c**) Resting spores form when conditions do not favor growth. A nitrogen-fixing heterocyst is shown in (**c**).

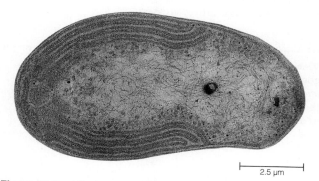

2.5 μm

Figure 22.9 *Nitrobacter*, a chemoautotroph. This nitrifying bacterium has a role in the global cycling of nitrogen.

Figure 22.10 One of the diverse chemoheterotrophs: a *Pseudomonas* cell, nearly completing binary fission.

1.2 μm

22.6 REGARDING THE PATHOGENIC EUBACTERIA

Also in the category of chemoheterotrophs are most of the pathogenic bacteria. We admire pseudomonads when they grow in soil, not when they grow on "our" carbon-containing materials, including antiseptics and bits of soap. Pseudomonad infections can be serious, because they transmit plasmids that carry antibiotic-resistant genes (see the *Focus* essay). *Listeria monocytogenes*, a relative of *Lactobacillus*, can contaminate even refrigerated foods. Some *E. coli* strains cause a form of diarrhea that is the leading cause of infant death in developing countries.

Or consider *Clostridium tetani* (Figure 22.11). It causes *tetanus*, a terrible disease described on page 569. As part of its life cycle, *C. tetani* forms an **endospore** around its DNA and a bit of cytoplasm. Endospores are structures that resist heat, drying, boiling, and radiation. New bacterial cells form after the endospore germinates. A relative, *C. botulinum*, can taint fermented grain as well as food in improperly sterilized or sealed cans and jars. Its toxins cause *botulism*, a form of poisoning that can lead to respiratory failure and death.

Also consider *Borrelia burgdorferi*, which taxis from host to host inside the gut of blood-sucking ticks. Tick bites transmit this spirochete from deer, field mice, and some other wild animals to humans, who develop *Lyme disease*. The first sign is a circular "bull's-eye" rash around the tick bite (Figure 22.12). Severe headaches, backaches, chills, and fatigue follow. Without prompt treatment, the condition worsens.

Viewed through the prism of human interests, bacteria are good, bad, or dangerous. Yet their lineages are the most ancient, their adaptations are breathtakingly diverse, and they simply are surviving and reproducing like the rest of us.

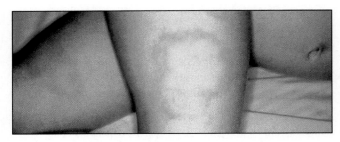

actual size: ●

Figure 22.12 An extreme reaction to an infection by *Borrelia burgdorferi*, a spirochete. The "bull's-eye" rash is a classic symptom of what may now be the most common tickborne disease in the United States: Lyme disease. Tick bites deliver the spirochete from one host to another. Statewide incidences in 1989 are mapped (*beige*, less than 10; *yellow*, between 15 and 90; *gold*, more than 90).

Focus on Health

Antibiotics

When your grandparents were children, bacterial agents of tuberculosis, pneumonia, and scarlet fever may have caused 25 percent of all deaths each year in the United States. Bacterial agents of dysentery, diphtheria, and whooping cough also were common killers. Bacterial infections during childbirth killed or maimed women by the thousands. From the 1940s onward, we started treating these and other diseases with antibiotics.

An **antibiotic** is a normal metabolic product of actinomycetes and other microorganisms, and it kills or inhibits the growth of other microbial species. For example, streptomycins block protein synthesis in their targets. Penicillins disrupt formation of covalent bonds that hold bacterial cell walls together. Penicillin derivatives cause the wall to weaken until it ruptures. The known antibiotics

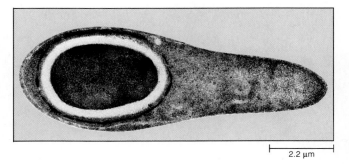

├─────────┤
2.2 μm

Figure 22.11 An endospore developing in *Clostridium tetani*. This chemoheterotroph is a dangerous pathogen.

22.7 REGARDING THE "SIMPLE" BACTERIA

Bacteria are small. Their insides are not elaborate. *But bacteria are not simple.* A brief look at their behavior will reinforce this point. Bacteria move toward regions with more nutrients. Aerobes move toward oxygen; anaerobes move away from it. Photosynthetic species move toward more intense light (or away from light if it is too intense). Many species tumble away from toxins.

Many bacterial behaviors involve membrane receptors that change shape when they absorb light or encounter chemical compounds. When a bacterium changes direction, its receptors are stimulated in a different way. This triggers a fleeting "memory," a changing biochemical condition that can be compared against that of the immediate past.

Magnetotactic bacteria contain a chain of magnetite particles that serves as a tiny compass (Figure 22.13). The compass helps them sense which way is north and also down. These bacteria swim toward the bottom of a body of water, where oxygen concentrations are lower and therefore more suitable for their growth.

In a jarring imitation of multicellularity, some species even show collective behavior. Millions of *Myxococcus xanthus* cells form "predatory" colonies. Their enzyme secretions digest "prey"—cyanobacteria and other microorganisms—that become stuck to the colony.

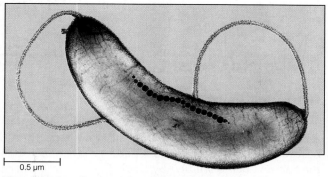

Figure 22.13 Transmission electron micrograph of a magnetotactic bacterium, showing a chain of magnetite particles that acts like a tiny compass.

Figure 22.14 *Chondromyces crocatus*, a myxobacterium. Cells of this species aggregate and form a fruiting body.

don't have comparable effects on protein coats of viruses. If you have a viral infection, antibiotics won't help.

Antibiotics must be carefully prescribed. Besides performing their intended function of counterattacking pathogens, some can disrupt the normal populations of bacteria in the intestines and of yeast cells in the vaginal canal. Such disruptions can lead to secondary infections.

Overprescribed antibiotics have lost their punch. Over time, they destroyed the most susceptible cells of target populations—and favored their replacement by more resistant ones. Antibiotic-resistant strains have made tuberculosis, typhoid, gonorrhea, "staph" infections, and other bacterial diseases more difficult to treat. In a few cases, "superbugs" that cause tuberculosis cannot be treated successfully.

The cells absorb the breakdown products. What's more, the cells migrate, change direction, and move as a single unit toward what may be food!

Many myxobacterial species also form **fruiting bodies**, which are a type of spore-bearing structure (Figure 22.14). Under appropriate conditions, some cells in the colony differentiate and produce a slime stalk, other cells form branches, and still others form clusters of single-celled spores. When the clusters burst open, the spores are dispersed; each may form a new colony. As you will see in the next chapter, certain eukaryotic organisms also form such structures.

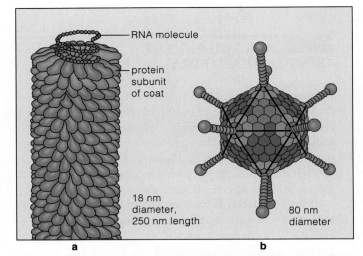

22.8 THE VIRUSES

Defining Characteristics

In ancient Rome, *virus* meant "poison" or "venomous secretion." In the late 1800s, this rather nasty word was bestowed on newly discovered pathogens, smaller than the bacteria being studied by Louis Pasteur and others. Many viruses deserve the name. They attack humans, cats, cattle, insects, crop plants, fungi, bacteria, even other viruses. You name it, and there probably are viruses that can infect it.

Today we define a **virus** as a noncellular infectious agent having two characteristics. First, each viral particle consists of a protein coat around a nucleic acid core—that is, its genetic material. Second, a virus cannot reproduce itself. It can be reproduced only after its genetic material and maybe a few enzymes enter a host cell and subvert the cell's biosynthetic machinery. Thus a virus is no more alive than a chromosome is alive.

The genetic material of a virus is either DNA or RNA. A viral coat consists of one or more types of protein subunits, organized into a rodlike or many-sided shape (Figure 22.15). It protects the genetic material during the journey from one host cell to the next. It also contains proteins that are capable of binding with specific receptors on host cells. Some coats are enclosed in an envelope of mostly membrane remnants from the previously infected cell. Protein-carbohydrate structures spike out from some of these envelopes. Coats of complex viruses have sheaths, tail fibers, and other accessory structures attached.

The immune system of vertebrates can detect certain viral proteins. The problem is, genes for many viral proteins mutate frequently, so a virus may elude the immune fighters. People who are vulnerable to lung infections get new "flu shots" each year because envelope spikes on influenza viruses keep changing.

Examples of Viruses

Each virus can multiply only in cells of particular species. Unless researchers culture living host cells, they cannot easily study viruses. This is why much of our knowledge of viruses comes from **bacteriophages**, which infect bacterial cells. Unlike the cells of humans and other complex, multicelled organisms, bacterial cells can be cultured easily and rapidly. This also is why bacteriophages and bacteria were used in the experiments to determine DNA function (page 210). They are still used as research tools in genetic engineering.

Table 22.3 lists the main groups of animal viruses, which cause herpes, influenza, warts, the common cold, some forms of cancer, and many other diseases. Figure 22.16 shows examples.

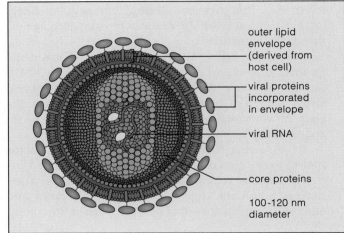

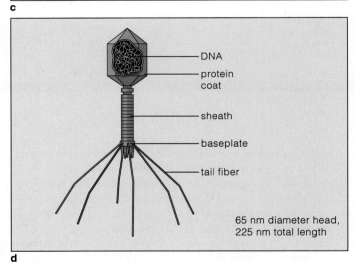

Figure 22.15 Body plans of viruses. (**a**) *Helical* viruses have a rod-shaped coat of protein subunits, coiled helically around the nucleic acid. For this portion of a tobacco mosaic virus, the upper subunits have been removed to reveal the RNA. (**b**) *Polyhedral* viruses, such as this adenovirus, have a many-sided coat. (**c**) *Enveloped* viruses, such as HIV, have an envelope around a helical or polyhedral coat. (**d**) *Complex* viruses, such as the T-even bacteriophages, have additional structures attached to the coat.

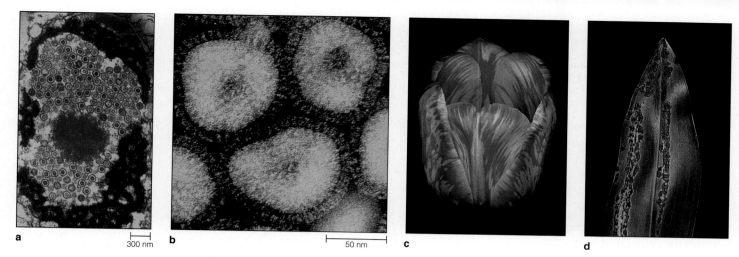

Figure 22.16 (**a**) Particles of a DNA virus that causes a herpes infection. This virus has a many-sided protein coat. (**b**) Particles of an RNA virus that causes influenza. Spikes project from the lipid envelope. Visible effects of two plant viruses: (**c**) Streaking of a tulip blossom. A harmless virus infected pigment-forming cells in the colorless parts. (**d**) An orchid leaf infected by a rhabdovirus.

Table 22.3 Classification of Animal Viruses	
DNA Viruses	**Some Diseases**
Adenoviruses	Respiratory infections
Hepatitis virus	Liver diseases
Herpesviruses:	
H. simplex type I	Oral herpes, cold sores
H. simplex type II	Genital herpes
Varicella-zoster	Chicken pox, shingles
Epstein-Barr	Infectious mononucleosis, implicated in some cancers
Papovaviruses	Benign and malignant warts
Parvoviruses	Roseola (fever, rash) in small children; aggravation of sickle-cell anemia
Poxviruses	Smallpox, cowpox
RNA Viruses	**Some Diseases**
Picornaviruses:	
Enteroviruses	Polio, hemorrhagic eye disease; hepatitis A (infectious hepatitis)
Rhinoviruses	Common cold
Togaviruses	Encephalitis, yellow fever, dengue fever
Paramyxoviruses	Measles, mumps
Rhabdoviruses	Rabies
Coronaviruses	Respiratory infections
Orthomyxoviruses	Influenza
Arenaviruses	Hemorrhagic fevers
Reoviruses	Respiratory, intestinal infections
Retroviruses:	
HTLV I, II	Associated with cancer (leukemia)
HIV	AIDS

Animal viruses range in size from parvoviruses (18 nanometers) to poxviruses (350 nanometers across). Their genetic material, double- or single-stranded DNA or RNA, is replicated in different ways. HIV, one of the RNA viruses, is the trigger for AIDS. As described in Chapter 40, it attacks certain white blood cells and so weakens the human immune system. Infections that otherwise might not kill the individual end up doing so. Most often, researchers attempting to find cures for AIDS and other diseases must use living animals for their experiments. For some viruses, they can use HeLa cells (page 145) and other immortal cell lines instead.

Plant viruses must breach plant cell walls to cause diseases. They typically hitch rides on the piercing or sucking devices of insects that feed on plant juices. RNA viruses infect tobacco plants (tobamovirus, or the tobacco mosaic virus), barley (hordeivirus), potatoes (potax-virus), and many other major crops. So do DNA viruses (caulimovirus infects cauliflower, and geminivirus infects maize). Figures 22.16*c* and *d* show the visible effects of other viral infections.

Some pathogens are even more stripped down than viruses. Prions, which are protein particles, cause rare, fatal degeneration of the nervous system in humans, sheep, and other animals. Viroids are strands or circles of RNA, smaller than any viral DNA or RNA molecule, and they have no protein coat. Viroids are plant pathogens that can destroy entire fields of citrus, potatoes, and other crop plants.

A virus is a nonliving infectious particle that consists of nucleic acid enclosed in a protein coat and sometimes an outer envelope. It cannot be replicated without pirating the metabolic machinery of a specific type of host cell.

VIRAL MULTIPLICATION CYCLES

Viruses multiply in a variety of ways. But nearly all multiplication cycles include five basic steps:

1. The virus attaches to a host cell. Any cell is a suitable "host" if a virus can chemically recognize and lock onto specific molecular groups at its surface.

2. The whole virus or its genetic material alone penetrates the cell's cytoplasm.

3. In an act of molecular piracy, the viral DNA or RNA directs the host cell into producing many copies of viral nucleic acids and proteins, including enzymes.

4. The viral nucleic acids and proteins are put together to form new virus particles.

5. Newly formed virus particles are released from the infected cell.

Two pathways are common among bacteriophages. In a **lytic pathway**, steps 1 through 4 proceed rapidly, with new particles released by lysis (Figure 22.17). "Lysis" means the cell's plasma membrane is damaged and its cytoplasm leaks out. Cell death is quick. In **lysogenic pathways**, an infection enters a latent period; the host cell isn't killed outright. Genetic recombination may take place during latency. A viral enzyme cuts the host chromosome and integrates viral genes into it. The recombinant DNA is replicated and passed on to all of the infected cell's descendants. Later, if viral genes move out of the chromosome, the multiplication cycle will resume.

Multiplication cycles differ among animal viruses. Figure 22.18 shows an example of an enveloped DNA animal virus. It penetrates a cell by endocytosis, becomes uncoated, and undergoes replication. Enzymes and other proteins are assembled from viral gene products. Virus particles are released by exocytosis.

By contrast, RNA viruses multiply only in the cytoplasm. Their RNA is used as a template either for mRNA or protein synthesis. Some animal viruses also enter latency. Herpesviruses that cause recurring cold sores in just about everybody are like this. So is HIV, a retrovirus. It carries along its own enzymes, which assemble DNA on viral RNA by the process of reverse transcription, as described on page 690.

Nearly all viral multiplication cycles include five steps: attachment to a suitable host cell, penetration, viral DNA or RNA replication and protein synthesis, assembly of new viral particles, and release from the infected cell.

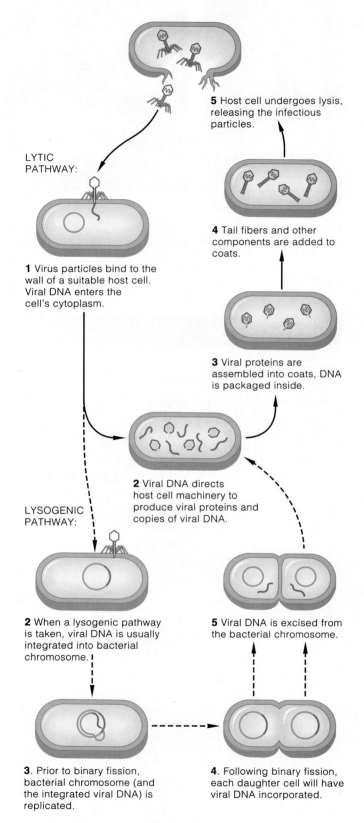

LYTIC PATHWAY:

5 Host cell undergoes lysis, releasing the infectious particles.

4 Tail fibers and other components are added to coats.

1 Virus particles bind to the wall of a suitable host cell. Viral DNA enters the cell's cytoplasm.

3 Viral proteins are assembled into coats, DNA is packaged inside.

2 Viral DNA directs host cell machinery to produce viral proteins and copies of viral DNA.

LYSOGENIC PATHWAY:

2 When a lysogenic pathway is taken, viral DNA is usually integrated into bacterial chromosome.

5 Viral DNA is excised from the bacterial chromosome.

3. Prior to binary fission, bacterial chromosome (and the integrated viral DNA) is replicated.

4. Following binary fission, each daughter cell will have viral DNA incorporated.

Figure 22.17 Multiplication cycle for some of the bacteriophages. New viral particles may be produced and released by a lytic pathway alone. For certain viruses, the lytic pathway may be expanded to include a lysogenic pathway.

Human Disease—Modes and Patterns of Infection

Humans are hosts for many viruses and other pathogens, including bacteria, fungi, protozoans, and parasitic worms. **Infection** means a pathogen has invaded the body and is multiplying in cells and tissues. Its outcome, **disease**, results when defenses cannot be mobilized fast enough to prevent the pathogen's activities from interfering with normal body functions. Pathogens reach hosts in several ways:

1. *Direct contact*, as by coming into contact with open sores caused by an infection. ("Contagious disease" comes from the Latin *contagio*, meaning touch or contact.) Gonorrhea, syphilis, and other sexually transmitted diseases are spread primarily by direct contact.

2. *Indirect contact*, as by touching doorknobs, food, diapers, hypodermic needles, or other objects that were previously in contact with an infected person. Food that is moist, not refrigerated, and not too acidic can be contaminated by various pathogens, including the ones responsible for amoebic dysentery and typhoid fever.

3. *Inhaling airborne pathogens*, which may be spread by coughing and sneezing.

4. *Transmission by biological vectors.* Mosquitoes, flies, fleas, ticks, and other arthropods transport pathogens from infected people or contaminated material to new hosts. Many pathogens use vectors as taxis; others also use them as hosts. They reproduce sexually in a *definitive* host, then must be transferred to an *intermediate* host, where they complete the infectious cycle. Agents of malaria, for example, use mosquitoes and humans as such hosts (page 369).

Some diseases occur in patterns. Whooping cough and other *sporadic* diseases break out irregularly and affect few people. Tuberculosis and other *endemic* diseases occur more or less continuously, but they do not spread far in large populations. Impetigo, a highly contagious bacterial infection, is like this; it is often confined to day-care centers. During an **epidemic**, a disease abruptly spreads through large portions of a population for a limited period. When influenza breaks out along the east coast of North America, this is an epidemic. When epidemics break out in several countries around the world in a given period, as is happening in the case of AIDS, this is a **pandemic**.

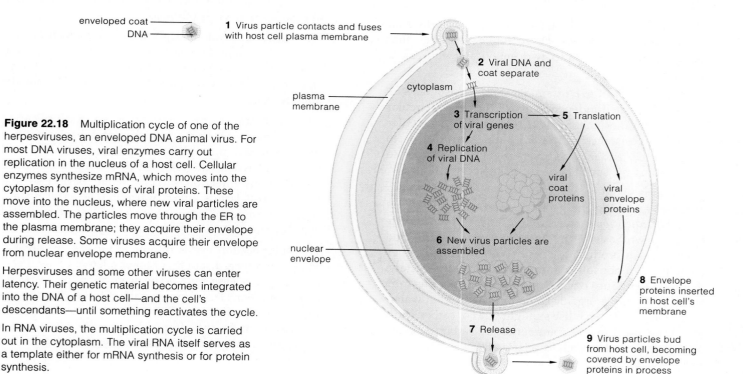

Figure 22.18 Multiplication cycle of one of the herpesviruses, an enveloped DNA animal virus. For most DNA viruses, viral enzymes carry out replication in the nucleus of a host cell. Cellular enzymes synthesize mRNA, which moves into the cytoplasm for synthesis of viral proteins. These move into the nucleus, where new viral particles are assembled. The particles move through the ER to the plasma membrane; they acquire their envelope during release. Some viruses acquire their envelope from nuclear envelope membrane.

Herpesviruses and some other viruses can enter latency. Their genetic material becomes integrated into the DNA of a host cell—and the cell's descendants—until something reactivates the cycle.

In RNA viruses, the multiplication cycle is carried out in the cytoplasm. The viral RNA itself serves as a template either for mRNA synthesis or for protein synthesis.

enveloped coat
DNA

1 Virus particle contacts and fuses with host cell plasma membrane

2 Viral DNA and coat separate

plasma membrane

cytoplasm

3 Transcription of viral genes

5 Translation

4 Replication of viral DNA

viral coat proteins

viral envelope proteins

nuclear envelope

6 New virus particles are assembled

8 Envelope proteins inserted in host cell's membrane

7 Release

9 Virus particles bud from host cell, becoming covered by envelope proteins in process

SUMMARY

1. Microorganisms are mostly single-celled species of bacteria and protistans that are too small to be seen without the aid of a microscope.

2. Soon after the origin of life, divergences led to three great prokaryotic lineages: archaebacteria, eubacteria, and the forerunners of eukaryotes.

   a. Archaebacteria (methanogens, halophiles, and extreme thermophiles) live in extreme environments, like those in which life probably originated.

   b. All other existing bacteria are eubacteria. These more common types differ from archaebacteria in wall structure and other features.

3. All bacteria are single-celled prokaryotes. Some have membrane infoldings and other structures in the cytoplasm, but none has a profusion of organelles.

   a. Three basic bacterial shapes are common: cocci (spheres), bacilli (rods), and spirals.

   b. Nearly all eubacteria have a cell wall composed of peptidoglycan. It protects the plasma membrane and helps it resist rupturing. Its composition and structure help clinicians identify particular bacterial species.

   c. A sticky mesh of polysaccharides (glycocalyx) surrounds the wall as a capsule or slime layer. It helps bacteria attach to substrates and sometimes to resist a host organism's infection-fighting mechanisms.

   d. Some species have bacterial flagella, unique motile structures that rotate like a propeller. Many species have pili, filamentous proteins that help cells adhere to a surface or facilitate conjugation.

4. Bacteria reproduce by binary fission, a cell division mechanism that follows DNA replication and that divides a parent cell into two genetically equivalent daughter cells.

5. Many species have plasmids, small circles of DNA that are replicated independently of the single, circular bacterial chromosome. Some genes on plasmids confer antibiotic resistance. Plasmids are transmitted to daughter cells and may be transferred by conjugation to cells of the same species or a different species.

6. Bacteria show great metabolic diversity, as in their modes of acquiring energy and carbon.

   a. *Photoautotrophs* use sunlight and carbon dioxide for photosynthesis. They include cyanobacteria and other oxygen-producing species of eubacteria. They include green nonsulfur and purple nonsulfur bacteria that do not produce oxygen; these obtain electrons from sulfur and other inorganic compounds, not from water.

   b. *Photoheterotrophs* use sunlight but not carbon dioxide. They use organic compounds as carbon sources. They include certain archaebacteria and eubacteria.

   c. *Chemoautotrophs* such as the nitrifying bacteria use carbon dioxide but not sunlight. They get energy by stripping electrons from a variety of inorganic substances, such as nitrogen compounds.

   d. *Chemoheterotrophs* are parasites (which draw carbon and energy from living hosts) or saprobes (which feed on the products, wastes, or remains of other organisms). Most bacteria are in this category. They include major decomposers and pathogens.

7. Bacteria make behavioral responses to stimuli, as when they move toward regions with more nutrients. The responses depend on activation of membrane receptors. Some bacteria have a magnetic compass in the cytoplasm that assists in directional behavior. Some species show collective behavior, as when they aggregate and move as a unit toward prey or form fruiting bodies by which spores are dispersed.

8. Viruses are nonliving, noncellular agents that infect particular species of nearly all organisms. They have two defining traits:

   a. Each virus particle consists of a nucleic acid core and a protein coat that sometimes is enclosed in a lipid envelope. Spikes of proteins and carbohydrates project from these envelopes. The coats of complex viruses have sheaths, tail fibers, and other accessory structures attached.

   b. A virus particle cannot reproduce itself. Its genetic material must enter a host cell and direct the cellular machinery to synthesize the materials necessary to produce new virus particles.

9. Nearly all viral multiplication cycles include five steps: attachment to a suitable host cell, penetration, viral DNA or RNA replication and protein synthesis, assembly of new viral particles, and release from the infected cell.

10. Two pathways are common in the multiplication cycle of bacteriophages, which infect bacteria. In a lytic pathway, multiplication proceeds rapidly and new viral particles are released by lysis. In a lysogenic pathway, the infection enters a latent period so that the host cell is not killed outright. The viral nucleic acid may undergo genetic recombination with a host cell chromosome.

11. Multiplication cycles of animal viruses are diverse. They may proceed rapidly or enter a latent phase. Penetration and release may be by way of endocytosis and exocytosis. For DNA viruses, part of the cycle proceeds in the nucleus of a host cell. For RNA viruses, it proceeds exclusively in the cytoplasm. The viral RNA serves as a template for mRNA and for protein synthesis.

Review Questions

1. Label the major structures on the generalized bacterial cell below. *348*

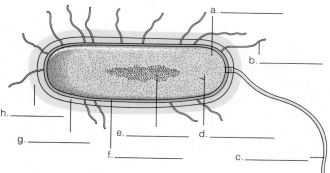

2. Describe the key characteristics of bacteria. What are some differences between archaebacteria and eubacteria? *348–355*

3. Name a few photoautotrophic, chemoautotrophic, and chemoheterotrophic eubacteria. Describe some that are likely to give humans the most trouble medically. *351–355*

4. What is a virus? Why is a virus considered to be no more alive than a chromosome? *356*

Self-Quiz *(Answers in Appendix IV)*

1. Nondividing bacteria have _____ chromosome(s) and may have extra circles of _____ called plasmids.
   - a. one; RNA
   - b. two; RNA
   - c. one; DNA
   - d. two; DNA

2. _____ live in extreme environments, much like those of the early earth.
   - a. Cyanobacteria
   - b. Eubacteria
   - c. Archaebacteria
   - d. Protozoans

3. Which of the following does not belong in the Archaebacteria?
   - a. Halophiles
   - b. Cyanobacteria
   - c. Thermophiles
   - d. Methanogens

4. Bacteria reproduce by _____ .
   - a. mitosis
   - b. meiosis
   - c. binary fission
   - d. longitudinal fission

5. Eubacterial cell walls are composed of _____ ; a sticky mesh of polysaccharides, the _____ , surrounds the wall as a capsule or slime layer.
   - a. peptidoglycan; plasma membrane
   - b. cellulose; glycocalyx
   - c. cellulose; plasma membrane
   - d. peptidoglycan; glycocalyx

6. Most bacteria are _____ and include major decomposers and pathogens.
   - a. photoautotrophs
   - b. photoheterotrophs
   - c. chemoautotrophs
   - d. chemoheterotrophs

7. Viruses are _____ .
   - a. the simplest living organisms
   - b. infectious particles
   - c. nonliving
   - d. both a and b
   - e. both b and c

8. Viruses have _____ .
   - a. DNA cores and carbohydrate coats
   - b. DNA or RNA cores; and plasma membranes
   - c. DNA-containing nuclei; and lipid envelopes
   - d. DNA or RNA cores; and protein coats

9. Which of the following is *not* characteristic of a viral lysogenic pathway?
   - a. bacteria are infected
   - b. the infection enters a latent phase
   - c. multiplication of virus particles proceeds rapidly
   - d. the host cell is not killed outright

10. Match each term appropriately.
   - _____ archaebacteria
   - _____ eubacteria
   - _____ viruses
   - _____ cyanobacteria
   - _____ plasmids

   - a. bacteria other than archaebacteria
   - b. nonliving infectious particle, nucleic acid core, protein coat
   - c. small circles of bacterial DNA
   - d. methanogens, halophiles, extreme thermophiles
   - e. photoautotrophs

Selected Key Terms

antibiotic *354*
archaebacteria *348*
bacillus *348*
bacterial conjugation *350*
bacterial flagella *349*
bacteriophage *356*
binary fission *350*
cell wall *349*
chemoautotroph *348*
chemoheterotroph *348*
coccus *348*
disease *359*
endospore *354*
epidemic *359*
eubacteria *348*
extreme thermophiles *352*
fruiting body *355*
glycocalyx *349*

halophiles *352*
heterocyst *352*
infection *359*
lysogenic pathway *358*
lytic pathway *358*
methanogens *352*
microorganism *346*
pandemic *359*
pathogen *346*
peptidoglycan *349*
photoautotroph *348*
photoheterotroph *348*
pili *349*
plasmid *350*
prokaryotic *349*
spiral *348*
virus *356*

Readings

Brock, T., and M. Madigan. 1988. *Biology of Microorganisms.* Fifth edition. Englewood Cliffs, New Jersey: Prentice-Hall.

Frankel-Conrat, H., P. Kimball, and J. Levy. 1988. *Virology.* Second edition. Englewood Cliffs, New Jersey: Prentice-Hall.

Frazier, W., and D. Westoff. 1988. *Food Microbiology.* Fourth edition. New York: McGraw-Hill. Good reference on the microbes that have major effects on our food supplies.

Stanier, R., et al. 1986. *The Microbial World.* Fifth edition. Englewood Cliffs, New Jersey: Prentice-Hall.

Woese, C. 1981. "Archaebacteria." *Scientific American* 244(6): 98–125.

23 PROTISTANS

Kingdom at the Crossroads

More than 2 billion years ago, in tidal flats and soils, in estuaries, streams, lakes, and lagoons, prokaryotic cells were inconspicuously changing the world. Ever since the time of life's origin, the earth's atmosphere had been free of oxygen, and anaerobic bacteria had reigned supreme. Now their realm was about to shrink, drastically. An oxygen-producing pathway of photosynthesis had been operating in vast populations of cells. Gaseous oxygen had been dissolving in the surrounding waters and escaping into the air—and its atmospheric concentration was beginning to approach its current level.

The oxygen-enriched atmosphere was a selection pressure of global dimensions. Anaerobic species that could not neutralize the highly reactive, potentially lethal gas were thereafter restricted to black muds, stagnant waters, and other anaerobic habitats. Oxygen tolerance developed in other species. And it must have been a short evolutionary step from having the capacity to neutralize oxygen to using it in metabolic reactions, for at least some aerobic species emerged in nearly all bacterial groups.

This was the start of rampant competition for resources. In the presence of so much oxygen, energy-rich organic compounds could no longer accumulate at the earth's surface through geochemical processes. Now the organic compounds formed by living cells became the premier source of carbon and energy. Through evolutionary experimentation, novel ways of acquiring and using those compounds developed. Diverse groups of bacteria engaged in new kinds of partnerships, in predatory and parasitic interactions. Within 500 million years, some experiments led to the first eukaryotic cells. Such cells were alike in having a nucleus and other organelles, some of which resulted from the coevolution of bacterial symbionts (page 331). Even so, the eukaryotic lineages soon branched in confounding directions.

Of all existing organisms, protistans are most like those early eukaryotes. Figure 23.1 will give you an idea of the range of diversity in this kingdom. Many are microscopic members of aquatic communities. The photosynthetic types range from single cells to giant seaweeds. Saprobic types resemble certain bacteria and fungi. Some predatory and parasitic types resemble animals.

As far as we can determine, ancestral forms very much like them were once poised at the evolutionary crossroads leading from prokaryotic lineages to the kingdoms of plants, fungi, and animals.

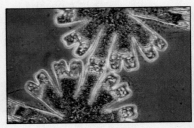

a

b

Figure 23.1 A sampling of the confounding diversity among protistans. (**a**) *Micrasterias*, a single-celled freshwater green alga, here dividing in two. (**b**) *Postelsia palmaeformis*, a multicelled brown alga. (**c**) *Physarum*, a plasmodial slime mold. This aggregation of cells is migrating along a rotting log. (**d**) Mealtime for *Didinium*, a ciliated protozoan with a big mouth. *Paramecium*, a different ciliated protozoan, is poised at the mouth (*left*) and swallowed (*right*).

c

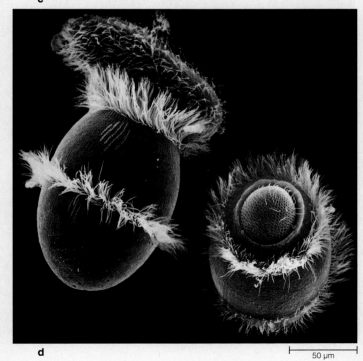

d

⊢————⊣
50 µm

1. The kingdom Protista includes all of the single-celled eukaryotes and some of the structurally simple multicelled forms. Many species live in aquatic habitats, where they are key components of phytoplankton and zooplankton.

2. Protistans are easily distinguishable from bacteria but are difficult to classify with respect to other eukaryotes. They also differ enormously from one another in their morphology and life-styles.

23.1 WHAT IS A PROTISTAN?

Through no fault of their own, protistans are looked upon as the rag-bag of classification schemes, a confusing bunch of eukaryotes that are assigned their own kingdom because they don't seem to fit anywhere else. Most protistans are single celled. Yet nearly every lineage includes multicelled forms, and some have close evolutionary ties to other kingdoms.

It's been said that it is easier to classify protistans by what they are *not*—as in, "They are not bacteria, fungi, plants, or animals." Remember, though, *we* impose the boundaries of kingdoms. *We make our cuts across continuous, unbroken lines of descent that extend from the origin of life to the present.* Some boundaries, such as the one between bacteria and protistans, are obvious; protistans are the simplest eukaryotes. Other boundaries are not as obvious, although they are now coming into focus through comparative biochemistry and other investigations.

Recall that protistans and other eukaryotes differ from bacteria in several respects. At the least, eukaryotic cells have a double-membraned nucleus, mitochondria, endoplasmic reticulum, and large ribosomes. They have microtubules, which have uses as cytoskeletal elements, as spindles for chromosome movements, and in the 9 + 2 core of flagella and cilia. These cells contain two or more chromosomes, which consist of DNA complexed with many histones and other proteins. Eukaryotic cells alone engage in mitosis and meiosis. Many types have chloroplasts and other plastids.

Later chapters will describe structural and functional traits that distinguish plants, fungi, and animals from protistans. By a process of elimination, "protistans" are organisms that do not show these traits and that are not bacteria. The ones called chytrids, water molds, slime molds, protozoans, and sporozoans are diverse heterotrophs. Euglenoids are photosynthetic, heterotrophic, or both. Nearly all chrysophytes, the dinoflagellates, and the red, brown, and green algae are evolutionarily committed to photosynthesis.

23.2 CHYTRIDS AND WATER MOLDS

Most of the **chytrids** (phylum Chytridiomycota) live in freshwater and marine habitats. Most of the 575 species are saprobic decomposers; some are parasites. Like fungi, all chytrids secrete enzymes that digest organic compounds of other organisms, such as marsh grasses, then they absorb the breakdown products.

Single-celled species produce flagellated asexual spores. The spores settle onto a host cell, germinate, then develop into globe-shaped cells with rhizoids, which are rootlike absorptive structures (Figure 23.2). At maturity, the globe-shaped cells become spore-producing structures. The cells of more complex species grow and develop into a mesh of absorptive filaments, called a **mycelium** (plural mycelia). Cell walls usually do not cut across the filaments, so there is an uninterrupted flow of cytoplasm that distributes enzymes and nutrients throughout the mycelium.

Chitin reinforces the cell wall of some chytrids. This, together with biochemical evidence, suggests an evolutionary link between chytrids and fungi.

The 580 species of **water molds** (phylum Oomycota) may be distantly related to red algae. Most produce an extensive mycelium, and some of the filaments develop into gamete-producing structures. At fertilization, a male and female gamete fuse to form a diploid zygote, which develops into a thick-walled resting spore. Following spore germination, a new mycelium develops. Water molds also produce flagellated, asexual spores.

Most water molds are key decomposers of aquatic habitats. Some parasitize aquatic animals and land plants (Figure 23.3). The cottony growths you may have seen on goldfish or tropical fish are mycelia of a parasitic type, *Saprolegnia*.

Some water molds have influenced human affairs. More than a century ago, Irish peasants cultivated potatoes as their main food source. Between 1845 and 1860, growing seasons were cool and damp, year after year. The cool conditions encouraged the rapid spread of *Phytophthora infestans*, a water mold that causes *late blight*—a rotting of potato and tomato plants. The mold produced abundant spores, spore dispersal through the watery film on plants went unimpeded, and destruction was rampant. During a fifteen-year period, a third of Ireland's population starved to death, died in the outbreak of typhoid fever that followed as a secondary effect, or fled to the United States and other countries.

Most chytrids and water molds are saprobic decomposers of aquatic habitats. Some are single cells. Others form a mesh of absorptive filaments (a mycelium) during the life cycle.

23.3 SLIME MOLDS

Call someone a slimy scum and you may get punched in the nose. Yet that is an apt description of the **slime molds**. During the life cycle of these heterotrophic protistans, free-living, amoebalike cells are produced. The cells crawl about on rotting plant parts, such as decaying leaves and bark. Like true amoebas, they are phagocytes; they engulf bacteria, spores, and organic compounds. When cells are starving, they aggregate into a slimy mass that may migrate to a new place. It moves through contractions of myosin molecules, present in secretions that sheath the mass. Later, the amoebalike cells develop into a few cell types that

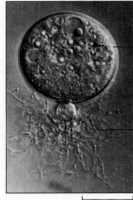

globe-shaped cell which, when mature, is a spore-producing structure

rhizoids, rootlike absorptive structures that grow through organic material

Figure 23.2 A common chytrid (*Chytridium confervaie*).

25 μm

a

b

Figure 23.3 Effects of two water molds. (**a**) A parasitic water mold (*Saprolegnia*) has destroyed tissues of this aquarium fish. (**b**) *Plasmopara viticola* causes downy mildew in grapes. At times it has threatened large vineyards in Europe and North America. Since the late 1800s, mixtures of copper sulfate, lime, and other chemicals have been applied to vines to control the disease.

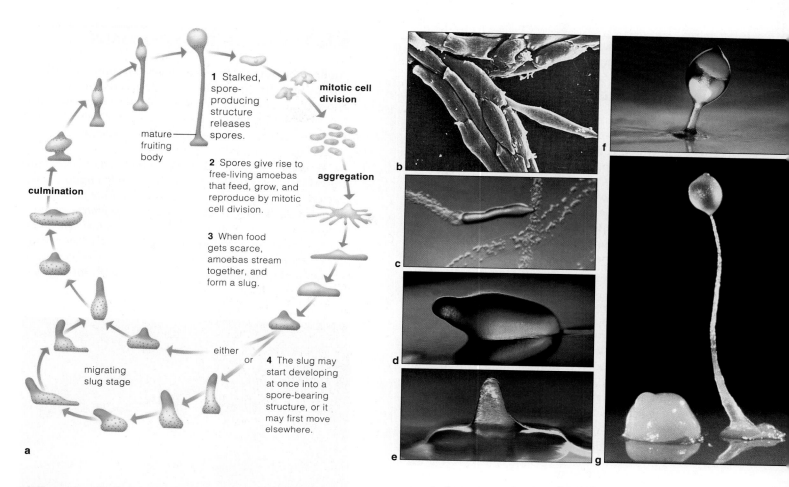

1 Stalked, spore-producing structure releases spores.

mature fruiting body

mitotic cell division

culmination

2 Spores give rise to free-living amoebas that feed, grow, and reproduce by mitotic cell division.

aggregation

3 When food gets scarce, amoebas stream together, and form a slug.

migrating slug stage

either or

4 The slug may start developing at once into a spore-bearing structure, or it may first move elsewhere.

a

b

c

d

e

f

g

Figure 23.4 *Dictyostelium discoideum*, a cellular slime mold. (**a**) The life cycle includes a spore-producing stage. Spores give rise to free-living amoebas, which grow and divide until food (soil bacteria) dwindles. Then, in response to a chemical signal (cyclic AMP) that they themselves secrete, amoebas stream toward one another. As amoebas aggregate, their plasma membranes become sticky, and they adhere to one another. A cellulose sheath forms around the aggregation, and it starts crawling, like a slug (**b–d**). Some of these slugs may consist of 100,000 or more amoebas.

As a slug migrates, the amoebas develop into prestalk, prespore, and anteriorlike cells. Prestalk cells (shaded *red* in **a**) form an anterior tissue. Prespore cells (*white*) and anteriorlike cells (*scattered brown dots*) form a posterior tissue. Prestalk cells secrete ammonia, in amounts that vary with temperature and light intensity. The slug moves most rapidly in response to intermediate amounts of ammonia (not too little, not too much), which correspond to warm, moist conditions (not too cold or hot, not soppy or dry).

Where such conditions exist at the surface of a substrate, prestalk cells and prespore cells differentiate and form a stalked, spore-bearing structure (**e–g**). Anteriorlike cells sort into two groups at each end of the posterior tissue. Possibly they function in elevating the nonmotile spores for dispersal from the top of the spore-bearing structure.

differentiate and form a stalked, spore-bearing reproductive structure. After a spore lands on a suitably warm, damp surface, it germinates and gives rise to an amoebalike cell. Sexual reproduction (by gametes) also is common among slime molds.

Figure 23.4 provides a close look at *Dictyostelium discoideum* at different stages of the life cycle. This is the best known of seventy species of *cellular* slime molds (phylum Acrasiomycota).

There are about 500 species of *plasmodial* slime molds (phylum Myxomycota), of obscure ancestry. When their amoebalike cells aggregate, the plasma membranes and cell walls break down. The cytoplasm flows as a unit, distributing nutrients and oxygen through the entire mass. This streaming mass (the plasmodium) may grow over several square meters. When food runs out, it migrates. You may have seen one crossing lawns or roads, even climbing trees. During dry seasons, the mass becomes encysted and often turns bright yellow and orange (Figure 23.1*c*).

During a slime mold life cycle, amoeboid cells aggregate to form a migrating mass. Cells in the mass differentiate, forming reproductive structures and spores or gametes.

23.4 CONCERNING THE ANIMAL-LIKE PROTISTANS

Protozoan Classification

Many chytrids, water molds, and slime molds are free-living predators or parasites some of the time and simple experiments in multicellularity at other times. About 65,000 other species of protistans are unequivocally single-celled predators and parasites. Collectively, they are called **protozoans** ("first animals") because they may resemble the single-celled, heterotrophic protistans that gave rise to animals. By one scheme, they are classified as the amoeboid, ciliated, and flagellated protozoans and the sporozoans.

Protozoan Life Cycles

Asexual reproduction dominates protozoan life cycles. Following increases in size, DNA replication, and multiplication of organelles, cells undergo fission or budding. Some species undergo multiple fission. More than two nuclei form, then each nucleus and a bit of cytoplasm around it separate to form a daughter cell. Sexual reproduction also occurs.

During the life cycle, many parasitic types form **cysts**, protective capsules that allow them to survive adverse conditions. Many species spread through host populations in encysted form.

Fewer than two dozen protozoans cause diseases in humans, but they receive our serious attention because there are no effective vaccines against them. In any given year, hundreds of millions of humans are affected by protozoan-caused diseases.

23.5 AMOEBOID PROTOZOANS

Among the **amoeboid protozoans** (phylum Sarcodina) are phagocytic amoebas, foraminiferans, heliozoans, and radiolarians. Adults move or capture prey by sending out pseudopods ("false feet"), temporary cytoplasmic extensions of the cell body. Most species feed on algae, bacteria, and other protozoans.

Amoebas are "naked," soft-bodied cells that live in freshwater, seawater, and soil. There they engulf algae, bacteria, and other protozoans. Amoebas are among the simplest protistans. They include *Amoeba proteus* of biology laboratory fame (Figure 23.5a). *Entamoeba histolytica* causes a severe intestinal disorder, *amoebic dysentery*. This parasite travels in cysts within feces, which may contaminate water and soil in regions with inadequate sewage treatment.

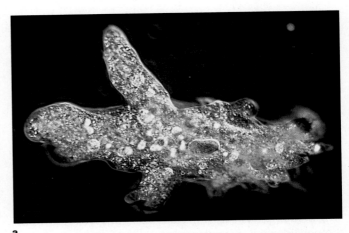
a

c

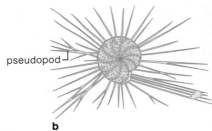
pseudopod
b

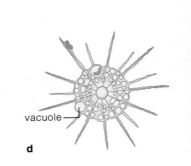
vacuole
d

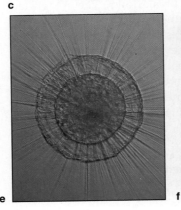
e

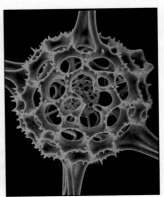
f

Figure 23.5 Amoeboid protozoans. (**a**) *Amoeba proteus*. The earliest eukaryotes may have been as soft-bodied as this. (**b,c**) Shells and body plan of foraminiferans. Needlelike spines support the pseudopods. (**d,e**) A living heliozoan. Its pseudopods have a supportive core of microtubules. Fluid-filled vacuoles make it neutrally bouyant so that it can remain suspended in the water. (**f**) A radiolarian shell.

Foraminiferans live mostly in the seas. Their hardened shells are often peppered with hundreds of thousands of tiny holes through which sticky, threadlike pseudopods extend. Figures 23.5b and c show some foraminiferan shells. Often the shells bear spines, which in some species are long enough that the shell can be seen with the naked eye.

The **heliozoans**, or "sun animals," have fine, needlelike pseudopods that radiate from the body like sun rays (Figure 23.5e). These largely freshwater protozoans are generally floaters or bottom-dwellers. Part of the cytoplasm forms an outer sphere around a core composed of denser cytoplasm and the bases of microtubular rods.

Among the structurally stunning protozoans are the **radiolarians**, which are found mostly in marine plankton. The cells basically resemble the heliozoans, but most also have a skeleton of silica (Figure 23.5f). Some radiolarian species form colonies in which many individual cells are cemented together. Accumulated shells of radiolarians and foraminiferans are key components of many ocean sediments, and are testimony to the abundance of these organisms in the past.

23.6 CILIATED PROTOZOANS

About 8,000 species of **ciliated protozoans** (phylum Ciliophora) have distinctive arrays of cilia, which are used as motile structures (Figure 23.1d and 23.6). Various organelles extrude mucus, toxins, and other materials at the cell surface.

The ciliates abound in freshwater and marine habitats, where they feed on bacteria, algae, and one another. *Paramecium* is a typical member of the group. Its rows of cilia beat in synchrony, sweeping water laden with bacteria and food particles into a gullet. The gullet is a cavity that leads into the cell body. There, food is enclosed in enzyme-filled vesicles and digested. Wastes move to an "anal pore" and are eliminated. Like the amoeboid protozoans, *Paramecium* has **contractile vacuoles**. Excess water that has entered the cytoplasm by osmosis collects in these organelles. A filled vacuole contracts, forcing water through a small pore to the outside (page 84).

When under attack or otherwise stressed, *Paramecium* discharges many sticky protein threads from organelles called trichocysts. This does not notably reduce stress or deter predators. Perhaps the threads function as anchors at feeding time.

Ciliates reproduce sexually as well as asexually. They have a large macronucleus and one or more haploid micronuclei. During **protozoan conjugation**, two cells exchange micronuclei, which fuse to form a diploid macronucleus in each cell. When the cells later divide by fission, the daughter cells are diploid, with genetic instructions from two parents.

In their movements and behavior, the hypotrichs are the most animal-like ciliates. They run around on leglike tufts of cilia. Some have a distinct "head" end equipped with modified, sensory cilia (Figure 23.6c). Hypotrichs prey on bacteria, but they also scavenge bits of plant and animal material.

Figure 23.6 Ciliated protozoans. (**a**) Body plan of *Paramecium* and (**b**) surface view. (**c**) From Bimini, the Bahamas, a hypotrich with long, stiff, leglike tufts of cilia.

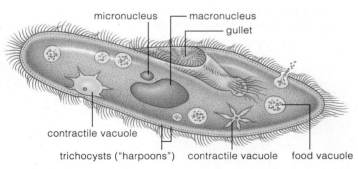

micronucleus   macronucleus

gullet

contractile vacuole

trichocysts ("harpoons")   contractile vacuole   food vacuole

a

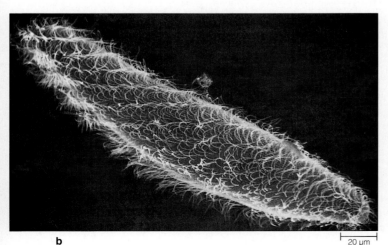

b   20 µm

Amoeboid and ciliated protozoans are single-celled predators and parasites, some of which have intriguing animal-like traits.

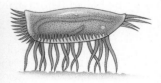

c

23.7 FLAGELLATED PROTOZOANS

The **flagellated protozoans** (phylum Mastigophora) are equipped with one to several flagella. Many species are internal parasites; others are free-living members of either freshwater or marine habitats.

The parasitic types include the trypanosomes. *Trypanosoma brucei* (Figure 23.7a) causes a dangerous disease called *African sleeping sickness*. Bites of the tsetse fly transmit the parasite from one host to another. Early symptoms include fever, headaches, rashes, and anemia. Later, the central nervous system becomes damaged. Without treatment, death follows. Another trypanosome (*Trypanosoma cruzi*) is prevalent in Mexico and South America, where it causes *Chagas disease*. A variety of bugs pick up the parasite when they feed on infected humans, armadillos, opossums, and other animals. The parasite multiplies in the insect gut, then the insect may excrete it onto the skin of a host animal. Scratched or abraded skin regions are open doors for infection. A terrible disease follows infection. The liver and spleen enlarge, the face and eyelids swell, then the brain and heart become severely damaged. There is no specific treatment or cure.

Flagellated protozoans also include trichomonads, of the sort shown in Figure 23.7b. *Trichomonas vaginalis*, a worldwide nuisance, is transferred to new human hosts during sexual intercourse. Without treatment, trichomonad infection damages the membranes in the urinary and reproductive tracts. This sexually transmitted disease and others are described on page 791.

By some estimates, about 10 percent of the human population in the United States is infected with another flagellated protozoan, *Giardia lamblia* (Figure 23.7c). Infection leads to mild intestinal disturbances, including diarrhea, but it has severe and sometimes fatal consequences in a few susceptible people. This flagellated protozoan also is common among wild animals and foraging cattle. It forms cysts that leave the body in feces. It infects a new host who drinks water or ingests food that has become contaminated with the feces. Even the water of remote mountain streams may contain the cysts and should be boiled before drinking.

Figure 23.7 A few flagellated protozoans. (**a**) *Trypanosoma brucei* causes African sleeping sickness. (**b**) *Giardia lamblia* causes intestinal disturbances. (**c**) *Trichomonas vaginalis* causes a sexually transmitted disease, trichomoniasis.

23.8 SPOROZOANS

Sporozoan is an informal designation for parasitic protistans that must live part of the time inside specific cells of host species. All of these intracellular parasites produce infective, motile stages called sporozoites. Many sporozoans also become encysted during some phase of the life cycle.

Trout, salmon, and other commercially important freshwater fishes are hosts for some sporozoans. The infective stage develops into an amoebalike form that may become quite massive.

The life cycle of *Plasmodium*, a sporozoan that causes the disease malaria, is described in the *Focus* essay. *Toxoplasma*, another sporozoan, completes the sexual phase of its life cycle in cats and asexual phases in humans as well as cattle, pigs, and other animals. It may enter new hosts by traveling along in infected meat that is either raw or undercooked. Sporozoite-containing cysts in the feces of infected cats are spread by houseflies, cockroaches, and other insects. The resulting disease, *toxoplasmosis*, is not prevalent in the population but is a major cause of birth defects. A woman who becomes infected during pregnancy may transmit the disease to her fetus, which may suffer brain damage or die. A pregnant woman should never empty litterboxes or otherwise clean up after any cat.

Certain parasitic species of flagellated protozoans and sporozoans cause dangerous human diseases.

a

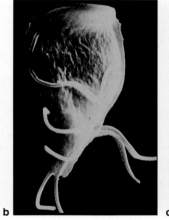

b

c

Malaria and the Night-Feeding Mosquitoes

Each year, about 150 million people come down with *malaria*, and 1 to 2 million die from it. At one time, this long-lasting disease was prevalent mostly in tropical and subtropical parts of Africa. However, the number of cases in North America and elsewhere is increasing dramatically, owing to globe-hopping travelers and unprecedented levels of immigration.

A sporozoan (*Plasmodium*) causes malaria. The mosquito *Anopheles* transmits its sporozoites to human (or bird) hosts. The sporozoites repeatedly engage in multiple fission, producing thousands of progeny that infect and multiply inside of red blood cells. The cells rupture and release the parasite's metabolic wastes, which cause fever and chills.

The *Plasmodium* life cycle (Figure *a*) is attuned to the host's body temperature, which normally rises and falls rhythmically every 24 hours. Gametocytes, the infective stage that is transmitted to new hosts, are ready for travel at night—when mosquitoes feed.

Travelers in countries with high rates of malaria are advised to use antimalarial drugs such as chloroquine. Drug resistance is common, however. And a vaccine has been difficult to develop. Vaccines induce the body to build up its resistance to a specific pathogen. Experimental vaccines for malaria have not been equally effective against all the different stages that develop during sporozoan life cycles. This is true of most parasites with complex life cycles.

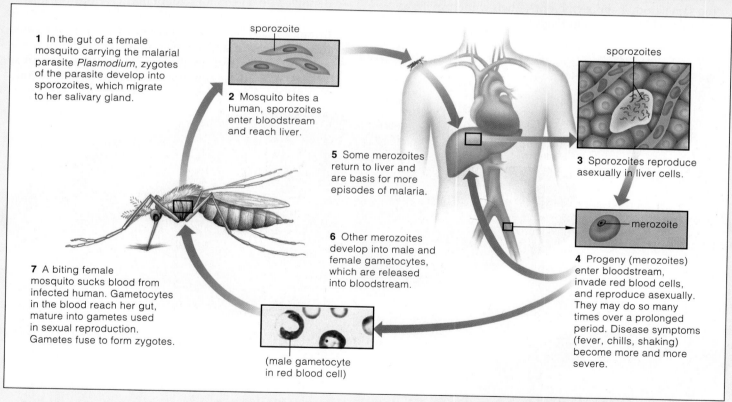

1 In the gut of a female mosquito carrying the malarial parasite *Plasmodium*, zygotes of the parasite develop into sporozoites, which migrate to her salivary gland.

sporozoite

2 Mosquito bites a human, sporozoites enter bloodstream and reach liver.

5 Some merozoites return to liver and are basis for more episodes of malaria.

6 Other merozoites develop into male and female gametocytes, which are released into bloodstream.

7 A biting female mosquito sucks blood from infected human. Gametocytes in the blood reach her gut, mature into gametes used in sexual reproduction. Gametes fuse to form zygotes.

(male gametocyte in red blood cell)

sporozoites

3 Sporozoites reproduce asexually in liver cells.

merozoite

4 Progeny (merozoites) enter bloodstream, invade red blood cells, and reproduce asexually. They may do so many times over a prolonged period. Disease symptoms (fever, chills, shaking) become more and more severe.

a *Plasmodium*, a sporozoan that causes malaria. Its life cycle requires a human host and an insect intermediate host.

23.9 EUGLENOIDS

The **euglenoids** (phylum Euglenophyta) are a classic example of evolutionary experimentation. About 1,000 species abound in freshwater or stagnant ponds and lakes. Like the flagellated protozoans, many are heterotrophs. Most, however, are photosynthetic.

Figure 23.8 shows one *Euglena* cell, which is 40,000 times larger and more complex than a bacterium. Notice the profusion of organelles, including the large chloroplasts and contractile vacuole. Beneath the plasma membrane are spiral strips of a translucent, protein-rich material. The strips help form a pellicle, a firm yet flexible layer. Light readily passes through the pellicle. An "eyespot" of carotenoid pigment granules partly shields a light-sensitive receptor. By moving a long flagellum, the cell keeps the receptor exposed to light and so keeps itself where light is most suitable for its activities.

Yet *Euglena* also can subsist heterotrophically, on dissolved organic compounds. Were its ancestors heterotrophs that acquired chloroplasts by way of endosymbiosis? Probably. The chloroplasts of green algae and plants may be stripped-down descendants of the same kind of endosymbionts. Their chlorophyll pigments are like the ones in *Euglena*.

23.10 CONCERNING "THE ALGAE"

Algae is a term that originally was used to define simple aquatic "plants." It no longer has formal meaning in classification schemes. The phyla once grouped under the term are now classified as follows:

Monera	Cyanobacteria	(blue-green algae)
Protista	Chrysophyta	(golden algae, diatoms, yellow-green algae)
	Pyrrhophyta	(dinoflagellates)
	Rhodophyta	(red algae)
	Phaeophyta	(brown algae)
	Chlorophyta	(green algae)

Many of the single-celled species are suspended in vast numbers in the ocean, seas, lakes, and smaller bodies of water. Collectively, the photosynthetic types are **phytoplankton**, the food producers that are the basis of nearly all food webs in aquatic habitats. (The word is derived from the Greek *planktos*, meaning to wander.) The food producers provide organic compounds *and* dissolved oxygen for **zooplankton**—mostly microscopic heterotrophs, drifting or weakly swimming through the water. These include bacteria, protozoans, tiny crustaceans, and animal larvae, which in turn are food for fishes and, in the oceans, most of the great whales. All of these communities are hurt by pollution.

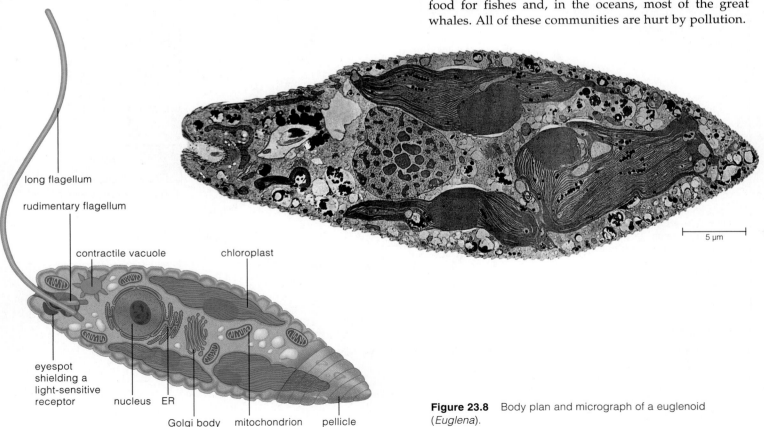

long flagellum

rudimentary flagellum

contractile vacuole

chloroplast

5 µm

eyespot shielding a light-sensitive receptor   nucleus   ER

Golgi body   mitochondrion   pellicle

Figure 23.8 Body plan and micrograph of a euglenoid (*Euglena*).

23.11 CHRYSOPHYTES AND DINOFLAGELLATES

Chrysophytes are mostly photosynthetic single cells with chlorophylls *a* and *c*. Some members have a cellulose wall, others have no wall at all, and still others have scales or mineral-impregnated shells.

Many of the 500 known species of **golden algae** have scales or skeletal elements of silica. Nearly all are photosynthetic. Fucoxanthin, a golden-brown carotenoid pigment, largely masks the chlorophylls. A few amoeboid types engulf bacteria. Golden algae are important in freshwater habitats and in "nanoplankton," marine communities composed of tremendous numbers of extremely tiny cells.

Similarly, the 5,600 existing species of **diatoms** are abundant in aquatic habitats. Although most are photosynthetic, many are temporary or permanent heterotrophs. Each diatom has an outer "shell"—two perforated, glasslike structures that overlap like a pillbox. Substances move to and from the plasma membrane through perforations in the shell.

We know of 35,000 species of extinct diatoms. For about 100 million years, their silica shells have been accumulating at the bottom of lakes and seas. Many sediments contain deposits of finely crumbled diatom shells, which we use in abrasives, filters, and insulating materials. More than 270,000 metric tons of this crumbly material are quarried annually near Lompoc, California.

Fucoxanthin is absent from the 600 species of **yellow-green algae**. One genus, *Vaucheria*, is common in aquatic habitats; it even grows as continuous filaments over wet soils (Figure 23.9*d*).

Of 1,200 species of **dinoflagellates**, most are members of marine phytoplankton. A few photosynthesizers live in freshwater, and a few heterotrophs live in the seas. Some have flagella that fit in grooves between stiff cellulose plates at the body surface (Figure 23.10*a*).

Dinoflagellates appear yellow-green, green, brown, blue, or red, depending on the photosynthetic pigments. Every so often, the red types undergo population explosions and color the seas red or brown. Because some forms produce a neurotoxin, the resulting **red tides** can have devastating effects (Figure 23.10*b*). Hundreds of thousands of fish that feed on plankton may be poisoned and wash up along the coasts. The neurotoxin does not affect clams, oysters, and other mollusks, but it builds up in their tissues. Humans who eat the tainted mollusks may die.

The single-celled photosynthetic protistans, including most euglenoids, chrysophytes, and dinoflagellates, are members of phytoplankton—the "pastures" of most aquatic habitats.

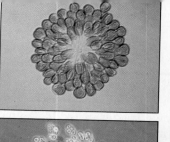
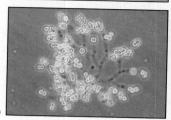

Figure 23.9 Chrysophytes. (**a**) *Synura*, a golden alga that grows as colonies in phytoplankton. It has a fishy odor. (**b**) *Mischococcus*, a yellow-green alga that forms colonies in phytoplankton. (**c**) Diatom shells. (**d**) *Vaucheria*, a yellow-green alga growing over red sandstone in Arizona.

Figure 23.10 (**a**) *Gymnodinium breve*, a dinoflagellate that causes red tides along Florida's coast. (**b**) Part of a fish kill resulting from a dinoflagellate "bloom."

23.12 RED ALGAE

Nearly all of the 4,000 known species of **red algae** live in the seas; fewer than 100 species are found in freshwater lakes, streams, and springs. Red algae are especially abundant in warm currents or tropical seas, often at surprising depths—more than 265 meters—when the water is clear. A few types are planktonic, and a few form colonies. Some colonial types contribute to the formation of coral reefs and coral banks (page 886). They have stonelike cell walls, hardened with calcium carbonate deposits.

The name of their phylum (Rhodophyta) is taken from the Greek *rhodon*, meaning rose, and *phyton*, meaning plant. Figure 23.11 shows two of the "red" species, but others appear green, purple, or greenish-black. They differ in the types and amounts of accessory pigments—mainly phycobilins—which tend to mask their chlorophyll *a*. Phycobilins are good at absorbing green and blue-green wavelengths, which can penetrate deep waters. (Chlorophylls are more efficient at absorbing red and blue wavelengths that may not reach as far below the water's surface.) The chloroplasts bear strong biochemical and structural resemblances to cyanobacteria. The resemblance suggests that these organelles had endosymbiotic origins.

Red algae have complex reproductive modes. For example, a typical life cycle includes haploid and diploid phases. In haploid individuals, structures near the tips of some filaments produce gametes. The male gametes are unflagellated. When released, currents carry them to egglike structures on the same haploid individual or a different one. There, after fertilization, a zygote develops into a diploid, spore-producing structure. Spores develop and are released into the water. After germinating, a spore grows by mitosis into a diploid, spore-producing individual that looks very much like the haploid ones. Some of its cells undergo meiosis, giving rise to *another* set of spores. After release and germination, a spore grows to become a new haploid individual.

Most species are multicelled, with a filamentous, often branching body plan. Unlike plants, the cells do not form tissues and organs. The cell walls incorporate mucous material, which gives red algae a flexible, slippery texture. **Agar** is made from extracts of wall material from a number of species. We use this inert, gelatinous substance as a moisture-preserving agent in cosmetics and bakery goods, as a setting agent for jellies and desserts, and as a culture medium. We also shape it into soft capsules (for packaging drugs and food supplements). We use **carrageenan**, extracted from the red alga *Euchema*, to stabilize paints, dairy products, and many other emulsions.

23.13 BROWN ALGAE

Walk along a rocky coast at low tide and you are likely to come across brown-tinged seaweeds. They are examples of **brown algae**. Nearly all of the 1,500 species inhabit temperate or cool marine waters throughout the world. Figure 23.12 shows examples. Depending on their photosynthetic pigments, brown algae appear olive-green, golden, or dark brown. Like the chrysophytes to which they may be related, they contain carotenoids, especially fucoxanthin (a xanthophyll), as well as chlorophylls *a* and *b*.

Brown algae range from microscopically small, filamentous *Ectocarpus* to giant kelps, 20 to 30 meters tall. The giant kelps, including *Macrocystis* and *Laminaria*, are the largest and most complex of all protistans. Dur-

a

b

Figure 23.11 Red algae. (**a**) *Bonnemaisonia hamifera*, showing the most common growth pattern among red algae: filamentous and branched. This alga reproduces asexually when the hooklike branchlets break off and grow into new plants after they become caught on branches of other algae. (**b**) From a tropical reef, a red alga showing sheetlike growth.

Figure 23.12   Brown algae. (**a**) Close-up of *Postelsia palmaeformis*, commonly called the sea palm. Figure 23.1*b* shows a population of sea palms growing in an intertidal zone, where they are alternately submerged and exposed to air.

(**b**) *Macrocystis*, a kelp that stays submerged even at low tide. (**c**) Diagram of *Macrocystis*, showing the structural organization that is characteristic of many of the large brown algae, which are the most structurally complex members of the protistan kingdom.

a

b

blade

stipe

bladder

holdfast

c

ing the life cycle, the formation of gametophytes (gamete-producing bodies) alternates with the formation of sporophytes (spore-producing bodies). The large, multicelled sporophytes have tissues that are differentiated into **holdfasts** (anchoring structures), **stipes** (stemlike parts), and **blades** (photosynthetic, leaf-shaped parts). Hollow, gas-filled bladders impart buoyancy to stipes and blades and help keep them upright. The stipes contain tubelike arrays of elongated cells. Like the food-conducting tissues of plants, the tubes rapidly distribute dissolved sugars and other products of photosynthesis through the individual.

Large populations of giant kelps are productive ecosystems. Think of them as underwater forests in which great numbers of diverse bacteria and protistans, as well as fishes and other animals, carry out their lives. Extensive masses of another brown alga, *Sargassum*,

serve as floating ecosystems in the vast Sargasso Sea, which lies between the Azores and the Bahamas.

If you enjoy ice cream, pudding, salad dressing, canned and frozen foods, jellybeans, or beer, if you use cough syrup, toothpaste, cosmetics, paper, or floor polish, thank the brown algae. The cell walls of some species contain **algin**, which is used as a thickening, emulsifying, and suspension agent. Especially in the Far East, kelps are harvested as sources of mineral salts and as a fertilizer for crops.

Red algae and brown algae are conspicuous photosynthetic members of aquatic habitats, especially the seas. Some species are microscopically small; others are among the largest of the multicelled protistans.

23.14 GREEN ALGAE

Of all protistans, **green algae** bear the greatest structural and biochemical resemblances to plants and may be their nearest relatives. For example, as is true of plants, their chlorophylls are the molecular versions designated *a* and *b*. Their chloroplasts contain starch grains. And the cell walls of some species contain cellulose, pectins, and other polysaccharides typical of plants.

With at least 7,000 species, green algae show more diversity than other algal groups. Figures 23.13 through 23.15 show examples. Most species live in freshwater. But you also can find green algae at the surface of open oceans, just below the surface of soil and marine sediments, on rocks, on tree bark as well as on other organisms, even on snow. Look closely and you find some living as symbionts with fungi, protozoans, and even a few marine animals. Look at the colonial form *Volvox*, a hollow whirling sphere of 500 to 60,000 flagellated cells.

Sink your feet in a white, powdery beach in the tropics; it is largely the work of countless green algal cells (*Halimeda*) that formed calcified cell walls, then went on to die and disintegrate. Green algae may even accompany astronauts on long journeys. They could grow in small spaces on light, carbon dioxide, and some minerals. Besides giving off oxygen, space-traveling algae would take up carbon dioxide exhaled by the aerobically respiring crew.

You won't see many species without the aid of a microscope. For example, Figure 23.1a shows one of thousands of microscopically small, mostly single-celled species of desmids. These freshwater algae are important food producers in nutrient-poor ponds and peat bogs. Figures 23.13a and b show larger members of the same class of algae. Unlike desmids, these two species are siphonous (no crosswalls separate their multinucleate cells). Another siphonous species, *Codium magnum*, is taller than you are.

a

b

c

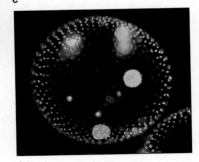

d

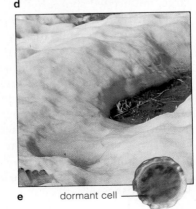

e   dormant cell ──

Figure 23.13   Green algae. (**a**) A marine species (*Codium*) that has a branched form. *Codium* from Puget Sound, Washington, was accidentally introduced into the mouth of the Connecticut River in 1956. There are no native species of *Codium* in that part of the Atlantic— and no predators that evolved with and are adapted to eating them. The introduced alga spread from the coasts of Maine to North Carolina in just over two decades.

(**b**) Also from a marine habitat, *Acetabularia*, fancifully called the mermaid's wineglass. Each individual in the cluster is a multinucleate cell mass with a rootlike structure, stalk, and cap in which gametes form. (**c**) Sea lettuce (*Ulva*), common to shallow seas throughout the world.

(**d**) *Volvox*, a colony of interdependent cells that bear resemblances to free-living, flagellated cells of the genus *Chlamydomonas*.

(**e**) One of the snow algae. "Red snow" above the summer timberline in Utah indicates the presence of dormant cells of *Chlamydomonas nivalis*. This green alga has an abundance of red accessory pigments that protect its chlorophyll.

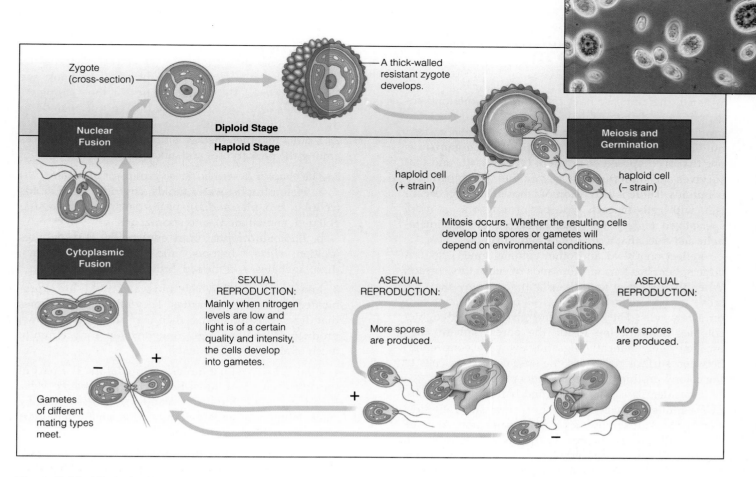

Zygote
(cross-section)

A thick-walled
resistant zygote
develops.

Nuclear Fusion

Diploid Stage

Haploid Stage

Meiosis and Germination

haploid cell
(+ strain)

haploid cell
(– strain)

Cytoplasmic Fusion

Mitosis occurs. Whether the resulting cells
develop into spores or gametes will
depend on environmental conditions.

SEXUAL
REPRODUCTION:
Mainly when nitrogen
levels are low and
light is of a certain
quality and intensity,
the cells develop
into gametes.

ASEXUAL
REPRODUCTION:

More spores
are produced.

ASEXUAL
REPRODUCTION:

More spores
are produced.

–

+

+

Gametes
of different
mating types
meet.

–

Figure 23.14 Life cycle of a single-celled species of *Chlamydomonas*, one of the most
common green algae of freshwater habitats. *Chlamydomonas* reproduces asexually most of
the time. It also reproduces sexually under certain environmental conditions.

Green algae employ diverse modes of reproduction.
Chlamydomonas, shown in Figure 23.14, provides a classic example. This freshwater alga is single celled, no more than 25 micrometers across. It can reproduce sexually. But most of the time it engages in asexual reproduction, with as many as sixteen daughter cells forming (by mitotic cell division) within the confines of the parent cell wall. The daughter cells may live at home for a while. But sooner or later they leave by secreting enzymes that digest what's left of their parent. Figure 23.15 shows how a filamentous green alga, *Spirogyra*, reproduces sexually.

Green algae show great diversity in size, morphology, lifestyles, and habitats. The structure and biochemistry of some groups indicate they may be evolutionarily linked with the plant kingdom.

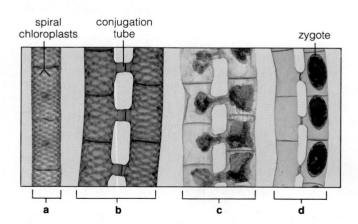

spiral
chloroplasts

conjugation
tube

zygote

a   b   c   d

Figure 23.15 One mode of sexual reproduction in *Spirogyra*, commonly called watersilk. (**a**) This green alga has ribbonlike, spiral chloroplasts. (**b**) A conjugation tube forms between cells of adjacent haploid filaments of different mating strains. (**c,d**) The cellular contents of one strain pass through the tubes into cells of the other strain, where zygotes form. The zygotes will develop thick walls. Later, as they germinate, they will undergo meiosis and give rise to new haploid filaments.

23.15 ON THE ROAD TO MULTICELLULARITY

So far in this unit, we have only hinted at the kinds of experiments that must have gone on among the first protistans, their prokaryotic ancestors, and their earliest multicellular descendants. A **multicellular organism** is one that is composed of different types of cells and that survives and reproduces by way of their combined contributions. There is a division of labor in such an organism, with cells of each type performing one or more specialized tasks even while engaging in basic metabolic activities that assure their own survival.

Reflect on *Volvox* and other colonial green algae. A large sphere has tens of thousands of cells at its surface. When the cells beat their flagella in synchrony, they can move the sphere forward. The task of reproduction falls on a few cells. They divide and give rise to daughter colonies that develop inside the parent sphere, then escape by enzymatically digesting away the connections between surface cells. In some species, certain cells of the colony produce eggs and others produce sperm.

Consider *Trichoplax*, a flattened ball of ciliated cells, half a millimeter across. It has no right side, left side, front, or back. It simply moves in any direction. And yet, as you will read in Chapter 26, *Trichoplax* is an animal. It may even bear close resemblances to one of the novel forms of ancient protozoans that gave rise to animals.

As a final example, green algae are eaten by *Placobranchus*, a marine mollusk. But the algal chloroplasts escape digestion. They accumulate in one of the mollusk's tissues—where they continue to function and provide their "host" with oxygen:

dorsal projections pushed back to show functional chloroplasts incorporated in tissues

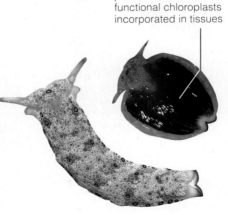

In short, the experimenting hasn't stopped.

SUMMARY

1. In this chapter and in other parts of the book, we have considered the characteristics of the "simplest" eukaryotes—the protistans—and have speculated on their evolutionary links with other kingdoms. Table 23.1 pulls together the key similarities and differences among the prokaryotes and eukaryotes.

2. The chapter described these groups of protistans:
   a. Heterotrophs: water molds, chytrids, slime molds (cellular and plasmodial), protozoans (amoeboid, flagellated, and ciliated), and sporozoans.
   b. Photoautotrophs: most euglenoids, chrysophytes (golden algae, diatoms, and yellow-green algae), dinoflagellates, and the red, brown, and green algae.

3. Like fungi, most chytrids and water molds are saprobic decomposers or parasites. They all secrete enzymes that digest organic matter, then they absorb breakdown products. Also like fungi, some develop a mycelium (a mesh of absorptive filaments).

4. Like animals, slime molds are predators. For part of the life cycle, they are phagocytic, amoebalike cells. Like fungi, their life cycle includes a spore-producing stage. Numerous single cells aggregate and differentiate into a spore-bearing structure.

5. Like animals, the 65,000 species of single-celled protozoans are predators and parasites. Fewer than two dozen cause diseases in humans, but these affect hundreds of millions of people each year.

Table 23.1 Comparison of Prokaryotes with Eukaryotes		
	Prokaryotes	Eukaryotes
Organisms represented:	Bacteria only	Protistans, fungi, plants, and animals
Ancestry:	Two major lineages (archaebacteria and eubacteria) that began more than 3.5 billion years ago	Equally ancient prokaryotic ancestors gave rise to the forerunners of eukaryotes, which emerged more than 1.2 billion years ago
Level of organization:	Single celled	Mostly multicelled, often with division of labor among differentiated cells and tissues
Typical cell size:	Small (1–10 micrometers)	Large (10–100 micrometers)
Cell wall:	Mostly distinctive sugars and peptides	Cellulose or chitin; none in animal cells
Membrane-bound organelles:	Very rarely	Typically profuse
Modes of metabolism:	Both anaerobic and aerobic	Aerobic modes predominate
Genetic material:	Single bacterial chromosome (and sometimes plasmids)	Complex chromosomes (DNA and many associated proteins) within a nucleus
Mode of cell division:	Binary fission, mostly	Mitosis, meiosis, or both

a. Amoeboid protozoans are amoebas and shelled foraminiferans, heliozoans, and radiolarians.

b. Ciliated protozoans, such as *Paramecium* and hypotrichs, use arrays of cilia as motile structures.

c. Flagellated protozoans are internal parasites or are free living in aquatic habitats. The trichomonads, trypanosomes, *Giardia*, and others cause human diseases that range from annoying to lethal.

d. Sporozoans are intracellular parasites that produce infective, motile stages and that may form cysts during the life cycle. *Plasmodium*, which causes malaria, is an example.

6. Most euglenoids and the chrysophytes and dinoflagellates are single-celled photosynthesizers. Many are members of phytoplankton.

7. Most red and brown algae are multicelled, photosynthetic members of aquatic habitats, especially the seas. Some are the largest protistans.

8. Green algae show great diversity in size, body plan, life-styles, and habitats. Plants may be descended from ancient green algae.

Review Questions

1. Generally describe the types of organisms classified in the kingdom Protista. In what habitats are they found? *363*

2. How do protistans differ from bacteria? *363*

3. Name the major categories of protistans. Think about where most of them live, then correlate some of their structural features with environmental conditions. *364–375*

Self-Quiz *(Answers in Appendix IV)*

1. Protistans and other eukaryotes differ from bacteria in which one of the following ways?
   a. Eukaryotes possess cytoplasm in their cells.
   b. Some protistans and other eukaryotes have cell walls.
   c. Protistans and other eukaryotes have DNA.
   d. Protistans and other eukaryotes engage in mitosis and meiosis.

2. Most chytrids and water molds are _____ decomposers of aquatic habitats.
   a. parasitic
   b. saprobic
   c. autotrophic
   d. chemosynthetic

3. Free-living, amoebalike cells crawl around on rotting plant parts as they engulf bacteria, spores, and organic compounds. This description best fits the _____ .
   a. water molds
   b. amoeboid protozoans
   c. sporozoans
   d. slime molds

4. During a _____ life cycle, amoeboid cells aggregate and form a migrating mass. Cells in the mass then differentiate, forming reproductive structures and spores or gametes.
   a. slime mold
   b. water mold
   c. protozoan
   d. chytrid

5. Amoebas, foraminiferans, radiolarians, and heliozoans are all classified as _____ .
   a. ciliated protozoans
   b. flagellated protozoans
   c. amoeboid protozoans
   d. sporozoans

6. Parasitic, flagellated protozoans of tropical regions known as trypanosomes are associated with which of the following diseases?
   a. African sleeping sickness and Chagas disease
   b. toxoplasmosis
   c. malaria
   d. amoebic dysentery

7. Euglenoids and chrysophytes are mostly _____ .
   a. photoautotrophic
   b. chemoautotrophic
   c. heterotrophic
   d. omnivorous

8. Single-celled photosynthetic protistans, including most euglenoids, chrysophytes, and dinoflagellates, are members of the _____ , the "pastures" of most aquatic habitats.
   a. zooplankton
   b. red algae
   c. brown algae
   d. phytoplankton

9. Algin is used in ice cream, pudding, salad dressing, jelly beans, beer, cough syrup, toothpaste, cosmetics, and other products. The source of algin is the _____ .
   a. green algae
   b. brown algae
   c. red algae
   d. dinoflagellates

Selected Key Terms

agar *372*
algin *373*
amoeba *366*
amoeboid protozoan *366*
blade *373*
brown algae *372*
carrageenan *372*
chytrid *364*
ciliated protozoan *367*
contractile vacuole *367*
cyst *366*
diatom *371*
dinoflagellate *371*
euglenoid *370*
flagellated protozoan *368*
foraminiferan *367*
golden algae *371*

green algae *374*
heliozoan *367*
holdfast *373*
multicellular organism *376*
mycelium *364*
phytoplankton *370*
protozoan *366*
protozoan conjugation *367*
radiolarian *367*
red algae *372*
red tide *371*
slime mold *364*
sporozoan *368*
stipe *373*
water mold *364*
yellow-green algae *371*
zooplankton *370*

Readings

Margulis, L. 1993. *Symbiosis in Cell Evolution*. Second edition. New York: Freeman. Paperback.

Margulis, L., and K. Schwartz. 1992. *Five Kingdoms*. Second edition. New York: Freeman. Paperback.

24 FUNGI

Dragon Run

When the first winter storm blasts through southeastern Virginia, Dragon Run—part swamp, part marsh, part old-growth woodland—pays tribute to the wind. Oaks, maples, gums, and beeches release dead leaves by the millions and these shower to earth, where they pile up as crisp, ankle-deep mounds. Branches snap off; sometimes trees crash down. Sheaves of dead grasses sink into marsh muds, and other plants buckle into the shallow, murky waters of the bottomlands. The storm kills uncounted numbers of insects, some birds, a few squirrels. By late December, Dragon Run is partially buried in organic debris.

Dig down through the debris and you will discover the accumulated litter of many past seasons—moist cushions of decayed leaves, bits of spiders and insects, mouldering branches, the carcasses of small mammals.

A new growing season begins with the warm rains of February. Buds develop on shrubs and trees, including a massive silver beech that has been budding each spring for 300 years. Woodland violets are the first to sprout in the thawing soil. By April, the resurgence of growth has imparted such a sharp, green freshness to Dragon Run that you might overlook the organisms on which the resurgence depends. On logs, under leaves, in the soil, fungi are commandeering resources and engaging in vegetative growth (Figure 24.1).

Fungi are nature's premier decomposers. Like other heterotrophs, they feast on organic compounds produced by other organisms. But few organisms besides fungi digest their dinner out on the table, so to speak. As fungi grow in or on organic matter, they secrete enzymes that digest it into bits that their individual cells can absorb. This "extracellular digestion" of organic matter liberates carbon and other nutrients *that also can be absorbed by plants*—the primary producers of Dragon Run and nearly all other ecosystems on earth.

Keep the global perspective in mind. Why? The metabolic activities of some fungi do cause diseases in humans, pets and farm animals, ornamental plants, and important crop plants. Some species are notorious spoilers of food supplies. Others have uses in the commercial production of substances ranging from antibiotics to excellent cheeses. We tend to assign "value" to fungi and other organisms in terms of their direct effect on our lives. There is nothing wrong with battling dangerous species and admiring beneficial ones—as long as we do not lose sight of the greater roles of fungi or any other kind of organism in nature.

a Sulfur shelf fungus (*Polyporus*) on a living tree

b Rubber cup fungus (*Sarcosoma*)

c Purple coral fungus (*Clavaria*)

Figure 24.1 Fungal species from southeastern Virginia. This small sampling merely hints at the rich diversity within the kingdom Fungi.

d Scarlet hood (*Hygrophorus*)

g Frost's bolete (*Boletus*)

e Yellow coral fungus (*Clavaria*)

h Trumpet chanterelle (*Craterellus*)

f Big laughing mushroom (*Gymnophilus*)

KEY CONCEPTS

1. Fungi are heterotrophs. Together with heterotrophic bacteria, they are decomposers of the biosphere. Saprobic types obtain nutrients from nonliving organic matter. Parasitic types obtain them from tissues of living hosts.

2. Fungi secrete enzymes that digest food outside their body, then fungal cells absorb breakdown products. Their metabolic activities also release carbon dioxide to the atmosphere and return many nutrients to the soil, where they become available to producer organisms.

3. Most fungi are multicelled. They form a mycelium, which is the food-absorbing part of the fungal body. A mycelium is a mesh of hyphae, which are elongated filaments that develop by repeated mitotic cell divisions.

4. Commonly, modified hyphae form a reproductive structure in or upon which spores develop. A "mushroom" is such a structure. Germinating spores grow and develop into a new mycelium.

5. Many fungi are symbionts. Some are part of lichens. Many are part of mycorrhizae; they are locked in mutually beneficial relationships with young roots of land plants. Fungal hyphae provide the plants with nutrients, and the plants provide the fungi with carbohydrates.

24.1 MAJOR GROUPS OF FUNGI

For many of us, our thoughts about fungi are limited to deciding between whole or sliced brown mushrooms in grocery stores. Yet those drab mushrooms are produced by a fungus that belongs to a huge group of diverse species. Figure 24.1 shows just a few of its relatives. These don't begin to do justice to the 80,000 fungal species we know about. And there may be at least a million more species we don't know about!

In this chapter we consider the three major groups of fungi. These are the zygomyctes (Zygomycota), sac fungi (Ascomycota), and club fungi (Basidiomycota). We also consider some of the "imperfect fungi," the members of which cannot be assigned to any of the recognized taxonomic groups.

We know of funguslike fossils 900 million years old. Together with simple plants, species resembling the zygomycetes started invading the land about 430 million years ago. Apparently this was a trigger for major radiations. About 100 million years later, all three of the major lineages were well established, with a number of diverse species.

24.2 | CHARACTERISTICS OF FUNGI

Nutritional Modes

Fungi are heterotrophs, meaning they require organic compounds synthesized by other organisms. Most are **saprobes**; they obtain nutrients from nonliving organic matter and so cause its decay. Others are **parasites**; they extract nutrients from tissues of a living host.

Fungal cells grow in or on organic matter. As they do, they secrete digestive enzymes, then absorb the breakdown products. This "extracellular digestion" benefits plants, which absorb some of the nutrients being released. They also release great quantities of carbon dioxide, which also benefits plants.

Together with heterotrophic bacteria, fungi are nature's decomposers. Without them, communities would become buried in their own garbage, nutrients would not be cycled, and life could not go on.

Fungal Body Plans

The vast majority of fungi are multicelled, land-dwelling species. During the life cycle, a mesh of branching filaments develops (Figure 24.2). The mesh, a **mycelium** (plural, mycelia), functions in food absorption. It rapidly grows over or into organic matter and has a good surface-to-volume ratio for absorption. Each filament in a mycelium is a **hypha** (plural, hyphae). Commonly, hyphae consist of tube-shaped cells with chitin-reinforced walls. The cytoplasm of these cells interconnects, so that nutrients flow unimpeded throughout the mycelium.

Reproductive Modes

Fungi reproduce asexually most often but, given the opportunity, also engage in sexual reproduction (Figure 24.3). They produce great numbers of nonmotile spores. Generally, **spores** are reproductive cells or multicelled structures, often walled, that germinate following dispersal from the parent body. Those of single-celled fungi simply form inside the parent cell. Most multicelled fungi produce asexual spores on or in reproductive structures called **sporangia** (singular, sporangium), which are made of interwoven hyphae. Sexual reproduction proceeds through the formation of gametes as well as spores. The gamete-producing structures are **gametangia** (singular, gametangium).

Fungi are spore-producing saprobes or parasites. Nearly all are multicelled. Together with heterotrophic bacteria, they are the decomposers of the biosphere.

24.3 | ZYGOMYCETES

There are about 765 species of **zygomycetes**. Their name refers to a thick-walled structure that forms when they undergo sexual reproduction.

Consider Figure 24.4, which shows the life cycle of the black bread mold, *Rhizopus stolonifer*. The sexual phase begins when hyphae of two different mating strains grow into each other and fuse. Two gametangia form between the hyphae, and several haploid nuclei are produced inside each. Later, their nuclei fuse, forming a zygote. A thick, protective wall forms around the zygote. This thick-walled structure is called a **zygosporangium** (plural, zygosporangia). Meiosis proceeds and spores are produced when this structure germinates. Each resulting spore can give rise to stalked structures that can produce many asexual spores, each of which

Figure 24.2 An example of the filaments (hyphae) of a mycelium, the food-absorbing portion of many fungi.

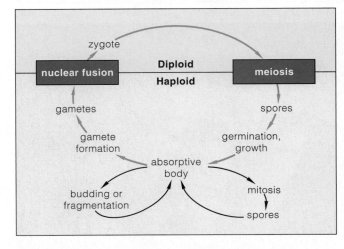

Figure 24.3 Generalized life cycle for many fungi. Asexual reproductive events (*black arrows*) are most common. Sexual reproduction (*blue arrows*) is less frequent.

can be the start of an extensive mycelium. The spores are very small, dry, and easily dispersed by winds.

R. stolonifer is a familiar saprobe with a bad reputation as a spoiler of baked goods. Most saprobic zygomycetes are at home in soil, decaying plant and animal matter, and stored food. *Pilobolus* prefers animal feces, from which it disperses its spores in a blastful way (Figure 24.5). Parasitic zygomycetes prefer living on houseflies and other insects.

Rhizopus is normally harmless to humans but, like other saprobic fungi, it is an opportunist. If its spores are inhaled or if they land on cuts in skin, they can cause diseases, especially in diabetics or individuals with weakened immune systems. Fungal infections are usually chronic (long lasting).

Zygosporangia (zygotes surrounded by a thick protective wall) are the key defining feature of the zygomycetes.

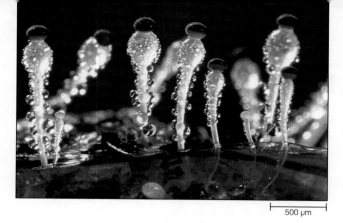

Figure 24.5 *Pilobolus*, a zygomycete. *Pilobolus* grows on animal feces and forms stalked, spore-bearing structures. Dark sacs at the end of each structure contain the spores. A stalk grows in such a direction that rays of sunlight converge at the base of its swollen portion, just below the dark sacs. Turgor pressure inside a vacuole in the swollen portion becomes so great that the spore sac can be blasted 2 meters away—a remarkable feat, considering the stalk is less than 10 millimeters tall!

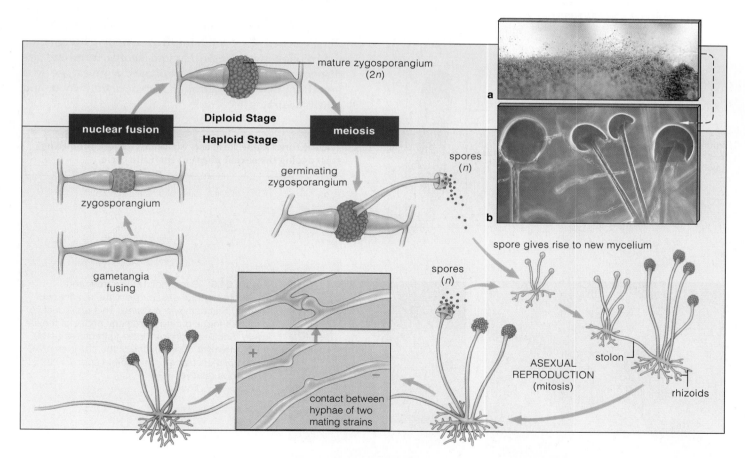

Figure 24.4 Life cycle of the black bread mold, *Rhizopus stolonifer*. As for other zygomycetes, the haploid phase dominates. Asexual reproduction is common, but different mating strains (+ and −) also reproduce sexually. Either way, haploid spores form and give rise to new mycelia. Chemical attraction between a + hypha and a − hypha causes them to fuse. Two gametangia form, each with several haploid nuclei inside. Later their nuclei fuse to form a zygote. The zygote develops a thick wall, thereby becoming a zygosporangium, and may remain dormant for several months. Meiosis occurs as the zygosporangium germinates, and new spores form.

24.4 SAC FUNGI

There are more than 30,000 known species of **sac fungi**. About 500 of these are single-celled yeasts. (Many other yeasts are classified with the club fungi.) The yeasts live in the nectar of flowers and on fruits and leaves. Bakers and vintners rely on the by-products of vast populations of busily fermenting yeasts (page 133). For example, the carbon dioxide by-product of *Saccharomyces cerevisiae* leavens bread. Its ethanol end product is central to the commercial production of wine, beer, and other alcoholic beverages. Many yeast strains with desirable properties have been developed through artificial selection and genetic engineering. Yeasts also are used in genetics research.

The vast majority of sac fungi are multicelled. They include most of the red, bluish-green, and brown molds that spoil stored food. One of them, the salmon-colored *Neurospora sitophila*, causes considerable damage when it invades bakeries or research laboratories. It produces so many easily dispersed spores, it is difficult to eradicate. One of its relatives, *N. crassa*, is an important organism in genetic research.

Multicelled sac fungi also include the edible truffles and morels, one of which is shown in Figure 24.6. Trained pigs and dogs have been used to snuffle out truffles, which grow underground as symbionts with the young roots of oak and hazelnut trees. Truffles are now cultivated commercially on the roots of inoculated seedlings in France. Even so, these fungi remain one of the most expensive luxury foods; they are priced by the gram, not by the pound.

As is true of other fungi, asexual reproduction is common among the sac fungi. Yeasts can reproduce sexually by budding. Multicelled species form specialized spores of a type called **conidia** (singular, conidium).

When sac fungi reproduce sexually, they form distinctive reproductive structures called **ascocarps**. In multicelled species, some hyphae become tightly interwoven into structures that resemble flasks, globes, and shallow cups of the sort shown in Figures 24.1*b* and 24.6. Spore-producing sacs called **asci** (singular, ascus) usually form on the inner surface of these structures. (Single-celled yeast cells simply fuse and so become the sac.) Following meiosis, haploid spores form and are dispersed. After a spore germinates, it gives rise to a mycelium that grows through soil, decaying wood, and other substrates.

The sac fungi alone form asci (distinctive spore-producing sacs) during the sexual phase of their life cycle.

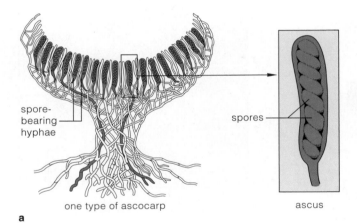

spore-bearing hyphae

spores

one type of ascocarp

ascus

a

b

c

Figure 24.6 Sac fungi. (**a**) Diagram and (**b**) photograph of ascocarps of the scarlet cup fungus, *Sarcoscypha coccinia*. This reproductive structure forms during the sexual phase of the life cycle. Saclike spore-producing structures (asci) form within the cup. (**c**) One of the choice morels (*Morchella esculenta*). This edible species has a poisonous relative.

Commentary

A Few Fungi We Would Rather Do Without

You know you are a serious student of biology when you view organisms objectively in terms of their place in nature, not in terms of their impact on humans generally and yourself in particular. As a student you can indeed respect saprobic fungi as vital decomposers and salute parasitic fungi that keep populations of destructive insects and weeds in check.

The true test is when you open the refrigerator for a dish of high-priced raspberries and discover that a fungus beat you to them or when another fungus starts feeding on warm, damp tissues between your toes. Who among us praises a fungus that makes skin reddened, scaly, and cracked? Which home gardener waxes poetic about black spot or powdery mildew on roses? Which farmers happily lose millions of dollars each year to rust or smut invaders of their crops?

Who would knowingly inhale the airborne spores of *Histoplasma capsulatum*, a pathogen that causes *histoplasmosis*? This disease passes for minor respiratory infections most of the time and for tuberculosis some of the time. When the fungal spores land on soil, they give rise to mycelia. When airborne spores are inhaled, they become decidedly yeastlike. At first, the fungus takes up residence in moist lung tissues. In rare cases, it ends up damaging most organs in the body. *H. capsulatum* is common worldwide. In the United States, it is most prevalent in many regions drained by the Mississippi and Ohio rivers, where it is especially partial to the nitrogen-rich droppings of birds and bats.

And who denies that fungi have influenced the course of human history? *Claviceps purpurea* is a fungal parasite of rye and other grains. We use its metabolic by-products (alkaloids) to treat migraine headaches and to shrink the uterus (to prevent hemorrhaging) after childbirth. But large amounts of the alkaloids are toxic. Eat a lot of bread made with contaminated rye flour and you end up with *ergotism*. Disease symptoms include hysteria, hallucinations, convulsions, vomiting, diarrhea, dehydration, and gangrenous limbs. Severe cases are fatal.

Ergotism epidemics were common in Europe in the Middle Ages, when rye was a major crop. Ergotism thwarted Peter the Great, the Russian czar who was obsessed with conquering ports along the Black Sea for his vast and nearly landlocked empire. Soldiers laying siege to the ports ate mostly rye bread and fed rye to their horses. The former went into convulsions and the latter into "blind staggers." Quite possibly, outbreaks of ergotism were an excuse to launch the Salem witch-hunts in colonial Massachusetts.

Some Pathogenic and Toxic Fungi

Zygomycetes

Rhizopus	Food spoilage

Sac fungi

Ophiostoma ulmi	Dutch elm disease
Cryphonectria parasitica	Chestnut blight
Venturia inaequalis	Apple scab (Figure *a*)
Claviceps purpurea	Ergot of rye, ergotism
Monilinia fructicola	Brown rot of stone fruits

Club fungi

Puccinia graminis	Black stem wheat rust
Ustilago maydis	Smut of corn
Amanita (some)	Severe mushroom poisoning

Imperfect fungi

Verticillium	Plant wilt
Microsporum, Trichophyton, Epidermophyton	Various species cause ringworms, including athlete's foot
Candida albicans	Infection of mucous membranes
Histoplasma capsulatum*	Histoplasmosis

*Now being reassigned to the sac fungi.

a telltale evidence of *V. inaequalis*

24.5 CLUB FUNGI

A Sampling of Spectacular Diversity

You probably are familiar with some of the 25,000 or so species of **club fungi**, the largest and most diverse group of the fungal kingdom. They include the grocery-store-variety mushrooms as well as the more exotic mushrooms, shelf fungi, and coral fungi shown in Figures 24.1 and 24.7. Other splendid types include bird's nest fungi and puffballs, as well as stinkhorns of the sort shown in Figure 1.6d.

Some of the saprobic types are major decomposers of plant debris. As you will see, others are symbionts that live in association with the young roots of forest trees. Still others, including the rust and smut fungi, cause serious plant diseases that can destroy entire fields of wheat, corn, and other major crops. Then again, cultivation of the common mushroom (*Agaricus brunnescens*) is a multimillion-dollar business.

Have you ever asked which organisms are the oldest and the largest? The club fungus *Armillaria bulbosa* is among them. The mycelium of one specimen, discovered in a northern Michigan forest, extends through at least 15 hectares. A hectare is 10,000 square meters. By one estimate, this individual weighs more than 10,000 kilograms and has been spreading through the forest soil for more than 1,500 years!

Figures 24.7c and d show two species of *Amanita*. Both contain toxins and intoxicants. *A. muscaria*, the fly agaric mushroom, causes hallucinations when ingested. It was used ritualistically in ancient societies in Central America, Russia, and India. The death cap mushroom (*A. phalloides*) can kill. Within eight to twenty-four hours of ingesting even as little as 5 milligrams of its toxin, a person starts vomiting and suffering diarrhea.

Figure 24.7 Club fungi. (**a**) Bird's nest fungus. (**b**) Light-red coral fungus (*Ramaria*). (**c**) Shelf fungus (*Polyporus*) growing on a rotting log. (**d**) From California, *Amanita ocreata*, which has caused fatalities when eaten. (**e**) The death cap mushroom (*A. phalloides*) is usually fatal when ingested. (**f**) The fly agaric mushroom (*A. muscaria*) causes hallucinations when eaten.

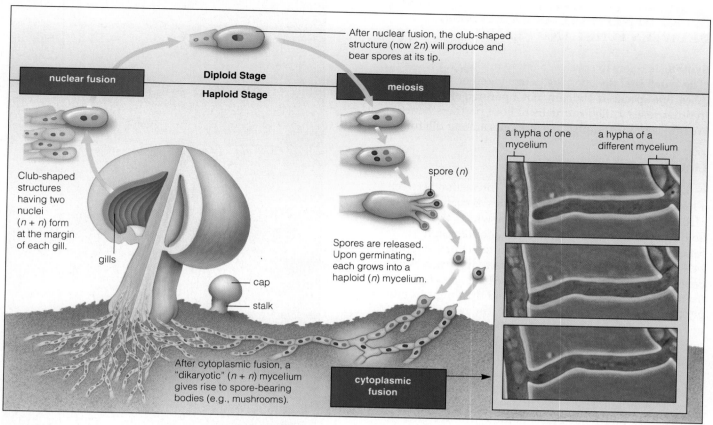

Figure 24.8 Generalized life cycle for many club fungi. When two compatible mating strains grow together, the cytoplasm (but not the nuclei) of two hyphal cells may fuse. Cell divisions produce a "dikaryotic" mycelium (its cells have two nuclei). When mushrooms develop, club-shaped structures form over the surface of the gills. Each structure contains two nuclei, which fuse to form a diploid zygote.

Later, kidney and liver cells start to degenerate; death can follow within a few days.

No general rule enables us to distinguish harmless from deadly mushrooms. *No one should eat mushrooms gathered in the wild unless they have been accurately identified as edible.* Said another way, there are old mushroom hunters and bold mushroom hunters, but no old, bold mushroom hunters.

Generalized Life Cycle

The spore-producing cells of club fungi are called **basidia** (singular, basidium). They usually are club-shaped, and they always bear the sexual spores on their outer surface (Figure 24.8). The club-shaped cells typically develop on a short-lived reproductive structure, the **basidiocarp**. Basidiocarps are merely the visible portion of the fungus; the living mycelia are buried in soil or decaying wood.

What most of us call a "mushroom" is a short-lived basidiocarp. About 10,000 species of club fungi produce them. Each consists of a stalk and a cap. Its spore-

producing cells are on the sides of gills, which are sheets of tissue in the cap.

When a spore dispersed from a mushroom lands on a suitable site, it germinates and gives rise to a haploid mycelium. When hyphae of two compatible mating strains grow next to each other, they may undergo cytoplasmic fusion (Figure 24.8). Nuclear fusion does not follow at once, so the hyphae form a **dikaryotic mycelium**. Each cell in this mycelium contains one nucleus of each mating type. After an extensive mycelium develops and when conditions are favorable, mushrooms form. At first, each spore-producing cell of a mushroom is dikaryotic, but then its two nuclei fuse to form a short-lived zygote. The zygote quickly undergoes meiosis and haploid spores are produced, then dispersed by air currents.

Club fungi, the fungal group with the greatest diversity, produce club-shaped, spore-bearing structures during the sexual phase of their life cycle.

24.6 BENEFICIAL ASSOCIATIONS BETWEEN FUNGI AND PLANTS

Symbiosis refers to species that live in close association. (The word literally means "living together.") In many cases, one species is a victim, not a partner, of a parasite. In other cases, called mutualism, both partners benefit. We find classic examples of symbionts among the fungi.

Lichens

A **lichen** is often called a mutualistic interaction between a fungus and a photosynthetic species. But it may well be a form of controlled parasitism, in which the fungus holds a cyanobacterium, green alga, or both captive. Sac fungi commonly enter into such interactions. Figures 24.9 and 24.10 are examples. Lichens form after a hypha penetrates a host cell and starts absorbing carbohydrates from it. If the cell survives, both it and the fungus multiply together into crusty formations. The fungus continues to extract nutrients from the cell's descendants. The photosynthesizer suffers in terms of its growth, although it might benefit a bit from the lichen's sheltering effect.

Sheltering is sometimes vital. Lichens colonize places that are too hostile for other organisms. They grow (slowly) on bare rocks in deserts and mountains, on tree bark and fence posts. Some types live close to the South Pole! They endure by suspending activities when it becomes extremely hot, cold, or bright. Then, their crusty part thickens and blocks sunlight, so photosynthesis stops. They become active only when moistened with raindrops, fog, or dew. Ancient lichens may have been the first invaders of land.

Lichens absorb minerals from rocks and nitrogen from the air. Their metabolic activities can slowly change the composition of their substrates. They may contribute to soil formation, thus setting the stage for colonization by different species. Lichens also are early warning signals that environmental conditions are deteriorating. They absorb but cannot rid themselves of toxins. When lichens die around cities, air pollution is getting bad. We know this from extensive studies in industrialized regions of England and in New York City.

Mycorrhizae

The small, young roots of nearly all vascular plants associate symbiotically with fungi in permanent, mutually beneficial ways. Such associations are called **mycorrhizae**, or "fungus-roots." Figure 24.11 shows an example. The fungus absorbs carbohydrates from the host plant, which absorbs mineral ions from the fungus. Collectively, fungal hyphae have an enormous surface area for taking up water and dissolved ions. The fungus takes up ions when they are abundant in soil and

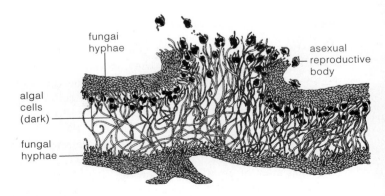

Figure 24.9 Lichens. (**a**) *Usnea*, old man's beard. (**b**) *Cladonia rangiferina*, sometimes called reindeer moss.

Figure 24.10 One type of lichen in cross section.

Where Have All the Fungi Gone?

Throughout European forests, wild mushrooms are getting smaller, and species are declining at an alarming rate. Mushroom gatherers aren't to blame; toxic as well as edible species may be on the road to extinction. Also, throughout Europe, driving cars, burning coal, prodding crop plants with nitrogen fertilizers, and other human practices are pumping ozone, nitrogen oxides, and sulfur oxides into the air. The decline of fungi is correlated with the rise in air pollution.

Forest trees and assorted mycorrhizal fungi are symbiotic partners. The fungi promote enhanced uptake of water and nutrients, and the trees provide the fungi with carbohydrates. Normally, as a tree ages, one fungal species gradually supplants another in predictable patterns. If fungi are indeed disappearing, trees will lose their support system and will become highly vulnerable to severe frost and drought. Entire forests may already be at risk. In Europe, collectors have been recording information about the wild mushroom populations since the 1900s. Similarly extensive records have not been kept in the United States, but comparable environmental conditions suggest that North American forests also are at risk.

releases them to the plant when ions are scarce. Many plants cannot grow as efficiently when mycorrhizae are not available to help them absorb phosphorus and other crucial ions (Figure 24.12).

The example in Figure 24.11 is an *exo*mycorrhiza, in which hyphae form a dense net around living cells in roots but do not penetrate them. Other hyphae form a velvety wrapping around the root, and the mycelium radiates outward from it. Exomycorrhizae are common in temperate regions, where beeches, oaks, willows, cottonwood, poplars, pines, and eucalyptus grow. The interaction makes the trees more resistant to adverse seasonal changes in temperature and rainfall. About 5,000 species of fungi enter these close associations. Most types are club fungi, including those truffles described earlier.

*Endo*mycorrhizae are much more common. They form on the roots of about 80 percent of all vascular plants. In this case, the fungal hyphae penetrate plant cells, as they do in lichens. Fewer than 200 species of zygomycetes serve as the fungal partner. Their hyphae branch extensively, forming tree-shaped absorptive structures within cells. They also extend for several centimeters into the surrounding soil.

As the *Focus* essay suggests, air pollution damages mycorrhizae, and this is affecting the world's forests. We return to this topic in Chapter 50.

Lichens and mycorrhizae are examples of mutualistic interactions between fungi and other organisms.

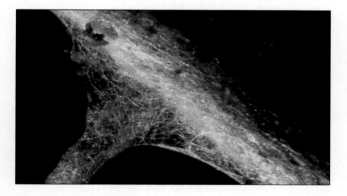

Figure 24.11 Mycorrhiza of a hemlock tree. The white threads are hyphal strands around a small, young root.

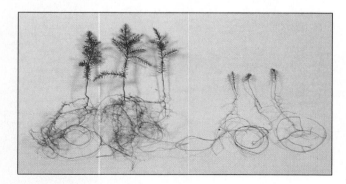

Figure 24.12 Effects of the presence or absence of mycorrhizae on plant growth. The juniper seedlings shown at the left are six months old. They were grown in sterilized, phosphorus-poor soil with a mycorrhizal fungus. The seedlings at the right were grown without the fungus.

24.7 "IMPERFECT" FUNGI

Some fungi are not easily classified. If a species has no sexual phase in its life cycle, it is called an "imperfect" fungus. If researchers later detect a sexual phase, the fungus is assigned to a recognized group—the sac or club fungi, most often.

Some imperfect fungi cause human diseases. *Candida albicans* is a notorious cause of vaginal infections (see Figure 24.13 and the *Commentary* on page 383). Other, predatory types ensnare tiny worms (Figure 24.13*b*). *Aspergillus* produces citric acid for candies and soft drinks, and it ferments soybeans for soy sauce. Certain *Penicillium* species "flavor" Camembert and Roquefort cheeses; and others produce penicillins, which we use as antibiotics.

SUMMARY

1. Fungi are heterotrophs. Many species are major decomposers of organic matter. Saprobic types feed on nonliving organic matter. Parasitic types obtain nutrients from living organisms. Mutualistic types are partners with other organisms in lichens and mycorrhizae.

2. The cells of all fungi secrete digestive enzymes that break down food into small molecules, which are absorbed across the plasma membrane.

3. Nearly all fungi are multicelled. Their food-absorbing part (mycelium) is a mesh of filaments (hyphae). Hyphae often interweave to form aboveground reproductive structures.

4. Fungi commonly reproduce asexually by spore formation or by budding from the parent body. When they reproduce sexually, distinctive spore-producing structures form.

5. The major groups of fungi are the zygomycetes, sac fungi, and club fungi. These groups are distinguished from one another largely on the basis of the reproductive structures that form during the sexual phase of the life cycle.

    a. Zygomycetes form a thick wall around the zygote, thus producing a zygosporangium.

    b. Sac fungi form saclike, spore-producing structures (asci), often in reproductive structures shaped like flasks, globes, and cups.

    c. Club fungi usually form club-shaped, spore-producing structures (basidia).

Figure 24.13 Imperfect fungi. (**a**) *Candida albicans*, cause of "yeast infections" of the vagina and mouth. In a more recent scheme, this fungal species has been assigned to ascomycetes. (**b**) Hyphae of *Arthrobotrys dactyloides*, a predatory fungus, form nooselike rings that swell rapidly with incoming water when stimulated. The "hole" in the noose shrinks and captures this worm. (**c**) Rows of conidia (asexual spores) of *Penicillium*.

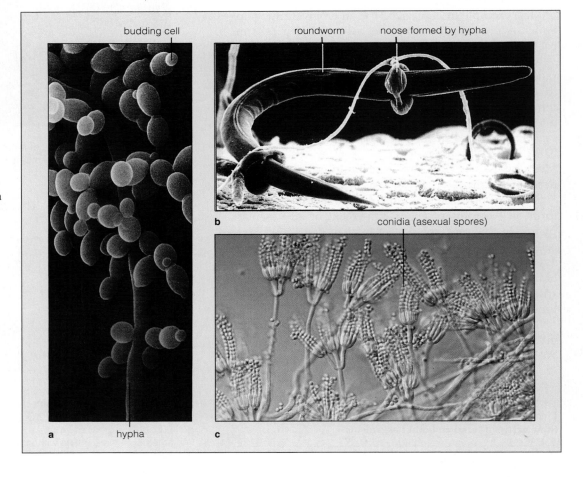

6. When a sexual phase cannot be detected or is absent from the life cycle of a fungal specimen, that specimen is assigned to an informal category called the "imperfect" fungi.

7. Fewer than 200 of the zygomycetes enter into mutualistic relationships with cyanobacteria or green algae. A lichen is an intimate association between such a fungus and its photosynthetic partner.

8. Many sac fungi and club fungi are symbiotic with young roots of shrubs and trees. In these associations, called mycorrhizae, fungal hyphae provide the plant with nutrients; the plant provides the fungus with carbohydrates.

Review Questions

1. Describe the fungal mode of nutrition. 379–380

2. How is a mycelium constructed and what is its function? 380

3. Describe one of the major groups of fungi. 379–385

4. Some fungi trap living nematodes as food. Do these fungi feed as parasites or saprobes? Why? 388

5. Define sporangium, gametangium, and basidiocarp. 380, 385

6. How does a mycorrhiza differ from a lichen? 386–387

Self-Quiz (Answers in Appendix IV)

1. The major groups of fungi are categorized mainly on the basis of their _____ .
   a. habitat
   b. color and size
   c. reproductive structures
   d. digestive enzymes

2. Fungi can reproduce asexually by _____ .
   a. forming spores
   b. budding
   c. fragmenting of parent bodies
   d. all of the above

3. New mycelia form following the germination of _____ .
   a. hyphae
   b. spores
   c. mycelia
   d. mushrooms

4. A "mushroom" is _____ .
   a. the food-absorbing part of a fungal body
   b. the part of the fungal body not constructed of hyphae
   c. a reproductive structure
   d. a nonessential part of the fungus

5. A mycorrhiza is a _____ .
   a. fungal disease of the foot
   b. fungus-plant relationship
   c. parasitic water mold
   d. fungus endemic to barnyards

6. A lichen is an intimate symbiotic association between a fungus and a _____ .
   a. mycorrhiza
   b. green alga
   c. parasitic fungus
   d. cyanobacteria
   e. both b and d

7. Parasitic fungi obtain nutrients from _____ .
   a. tissues of living host organisms
   b. nonliving organic matter
   c. only living plants
   d. only living animals
   e. none of the above

8. Saprobic fungi _____ .
   a. feed on nonliving organic matter
   b. feed on nutrients from living organisms
   c. are partners with other organisms
   d. are rarely multicelled

9. When the sexual phase is undiscovered or undetected from the life cycle of a fungus, that specimen is assigned to an informal category called _____ .
   a. zygomycetes
   b. ascomycetes
   c. imperfect fungi
   d. basidiomycetes

10. Match these terms appropriately.
   ____ zygomycetes
   ____ dikaryotic mycelium
   ____ hypha
   ____ basidiomycetes
   ____ mutualism
   ____ ascomycetes

   a. refers to an interaction between species in which positive benefits flow each way
   b. some yeasts and morels
   c. term for each filament in a mycelium
   d. black bread mold
   e. mushrooms, puffballs, shelf fungi, stinkhorns
   f. each cell contains two nuclei (one of each mating type)

Selected Key Terms

asci 382
ascocarp 382
basidia 385
basidiocarp 385
club fungi 384
conidia 382
dikaryotic mycelium 385
gametangia 380
hypha 380
lichen 386
mutualism 386
mycelium 380
mycorrhizae 386
parasite 380
sac fungi 382
saprobe 380
sporangia 380
spore 380
symbiosis 386
zygomycetes 380
zygosporangium 380

Readings

Bold, H., and J. LaClaire. 1987. *The Plant Kingdom*. Fifth edition. Englewood Cliffs, New Jersey: Prentice-Hall. Paperback.

Kendrick, B. 1991. *The Fifth Kingdom*. Second edition. Waterloo: Mycologue Publications.

Moore-Landecker, E. 1990. *Fundamentals of the Fungi*. Third edition. Englewood Cliffs, New Jersey: Prentice-Hall. Well-written introduction to the kingdom Fungi.

25 PLANTS

Pioneers in a New World

Seven hundred million years ago, no shorebirds stirred and noisily announced the dawn of a new day. No crabs clacked their tiny claws together and skittered off to burrows. The only sounds were the rhythmic muffled thuds of waves in the distance, at the outer limits of another low tide. Nearly 3 billion years before, life had its beginnings in the waters of the earth—and now, quietly, the invasion of the land was under way.

Astronomical numbers of photosynthetic cells had come and gone, and oxygen-producing ones had slowly changed the atmosphere. High above the earth, the sun's energy had converted much of the oxygen into a dense ozone layer. This became a shield against lethal doses of ultraviolet radiation—which had kept early organisms below the water's surface.

By one scenario, cyanobacteria were the first to adapt to intertidal zones, where the land dried out with each retreating tide. They were the first to move into shallow, freshwater streams meandering down to the coasts. Later, certain green algae and fungi made the same journey together. Every land plant around you is a descendant of green algae that lived near the water's edge or made it onto land. Fungi are still associated with nearly all of them.

We have some tantalizing fossils of the pioneers. We also are learning about them through comparative biochemistry and studies of existing species. Today, as in Precambrian times, cyanobacteria and green algae grow in mats in nearshore waters and on the banks of freshwater streams (Figure 25.1). When a volcanic eruption or some other event exposes rocks for the first time, cyanobacteria colonize them. Mutually beneficial interactions of green algae and fungi follow. The biochemical activities and accumulated remains of these

Figure 25.1 (**a**) Filaments of a green alga, massed in a shallow stream. More than 400 million years ago, green algae that may have been ancestral to plants lived in similar streams that meandered down to the shore of early continents. (**b**) A land-dwelling descendant of those ancestral forms—a ponderosa pine growing on a mountaintop above Yosemite Valley, California.

a

b

organisms slowly enrich the sediments around them with organic and inorganic substances. In this way they actually create soils in which mosses and other plants can take hold.

With this chapter, we turn to the plant kingdom. With few exceptions, its members are multicelled photoautotrophs. They produce their own organic compounds, using sunlight as their energy source and carbon dioxide as their carbon source. Plants have the metabolic machinery for the noncyclic as well as the cyclic pathway of photosynthesis. In other words, by splitting water molecules, they can obtain the huge numbers of electrons and hydrogen atoms required for their growth into multicellular forms as tall as giant redwoods, as vast as an aspen forest that is one continuous clone.

There are more than 275,000 known species of plants, and the vast majority reside on land. Be glad their ancient ancestors left the water. Without them, we humans and other land-dwelling animals never would have made it onto the evolutionary stage.

KEY CONCEPTS

1. Nearly all plants are multicelled photoautotrophs. Together with photosynthetic bacteria and protistans, they are the primary producers for nearly all communities.

2. Nearly all plants live on land, but their earliest ancestors were aquatic. Adaptations among different land-dwelling species include a waxy cuticle, root and shoot systems, and internal tissues for conducting water and solutes. They include special means of nourishing, protecting, and dispersing gametes and offspring.

3. Existing plants are bryophytes, seedless vascular plants, and seed-bearing vascular plants. The seed producers have been the most successful in radiating into drier environments.

25.1 CLASSIFICATION OF PLANTS

The plant kingdom includes 275,000 known species of photoautotrophs and a few heterotrophs. Most are **vascular plants**, with internal tissues that conduct and distribute water and solutes. These plants have roots, stems, and leaves, which are defined in part by the presence of vascular tissues. Fewer than 16,000 species are **bryophytes**, the "nonvascular" plants. Together with the photosynthetic bacteria and protistans, plants are the primary producers of organic compounds for nearly all communities.

This chapter surveys the major groups of existing plants. The liverworts, hornworts, and mosses are bryophytes. The whisk ferns, lycophytes, horsetails, and ferns are seedless vascular plants. The cycads, ginkgo, gnetophytes, and conifers are seed-bearing plants of a type called **gymnosperms**. A different type of seed-bearing plant, the **angiosperms**, also produces flowers. There are two classes of flowering plants, informally called the dicots and monocots.

Bryophytes, seedless vascular plants, and seed-bearing vascular plants constitute a kingdom of multicelled photoautotrophs, nearly all of which live on land.

25.2 EVOLUTIONARY TRENDS AMONG PLANTS

From the time their protistanlike ancestors arose until about 435 million years ago, plants evolved in the seas. Evolutionarily speaking, the pace picked up after that. Stalked species evolved along coasts and streams, in partnership with fungi. Within 60 million years, plants cloaked much of the land. Some long-term changes in structure and reproductive events help explain how the diversity came about.

Evolution of Roots, Stems, and Leaves

Underground structures started evolving among the first colonizers of land. In the lineages leading to vascular plants, these developed into **root systems**. Most root systems consist of underground, cylindrical absorptive structures that have a large surface area for rapidly taking up soil water and scarce mineral ions, and that often anchor the plant. Aboveground, other structures evolved into **shoot systems**. Among other things, shoot systems consist of stems and leaves, which function in the absorption of sunlight energy and carbon dioxide from the air.

The evolution of roots, stems, and leaves involved the development of pipelines through the plant. Such pipelines evolved as components of two vascular tissues—xylem and phloem. **Xylem** distributes water and dissolved ions through plant parts. **Phloem** distributes sugars and other photosynthetic products.

Extensive growth of stems and branches became possible when plants started producing **lignin**. This organic compound strengthens cell walls. Today, lignin-reinforced tissues structurally support plant parts that display the leaves and help increase the surface area that can intercept sunlight.

Life on land also depended on water conservation, which had not been a problem in most aquatic habitats. Stems and leaves became protected by a **cuticle**, a waxy surface covering that reduces water loss on hot, dry days. Numerous, tiny passageways called **stomata**

became the main route for absorbing carbon dioxide while controlling evaporative water loss. The next unit describes these tissue specializations.

From Haploid to Diploid Dominance

As early plants moved into higher, drier parts of the world, modifications accumulated in their life cycles. Think about the green algae. They spend most of their life producing and releasing gametes into the surrounding water. Actually, their gametes cannot get together *except* in liquid water. Figure 25.2*a* shows how you might diagram the life cycle of some green algae. The haploid (*n*) phase, which starts at meiosis and gamete formation, dominates the cycle. A diploid (2*n*) phase starts when gametes fuse at fertilization.

By contrast, the diploid phase dominates the life cycles of vascular plants (Figure 25.2*c*). Diploid dominance is an adaptation to land environments, most of which do not have unlimited supplies of water and nutrients. Long ago, through natural selection, complex sporophytes must have been favored. Recall that **sporophyte** means "spore-producing body." In plants, these form when a new zygote embarks on a course of mitotic cell divisions, which in time produce a large, multicelled body. A pine tree is an example. Such complex sporophytes have well-developed root systems. The young roots interact with mycorrhizal fungi, as described on page 386. This symbiotic relationship provides sporophytes with water and scarce nutrients, even in seasonally dry habitats.

For land plants, the haploid phase of the life cycle starts in reproductive parts of the sporophyte. There, haploid spores form by meiosis. Such spores divide by mitosis and give rise to gamete-producing structures called **gametophytes**. These multicelled, haploid game-

Figure 25.2 Comparison of life cycles for (**a**) some algae, (**b**) bryophytes, and (**c**) vascular plants. The diagrams highlight an evolutionary trend from haploid to diploid dominance during the colonization of land.

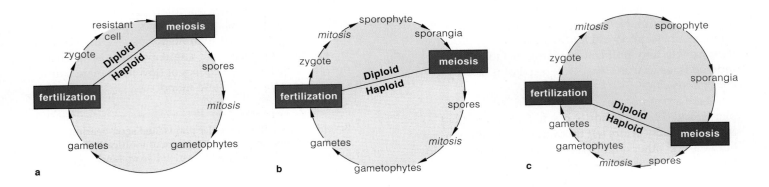

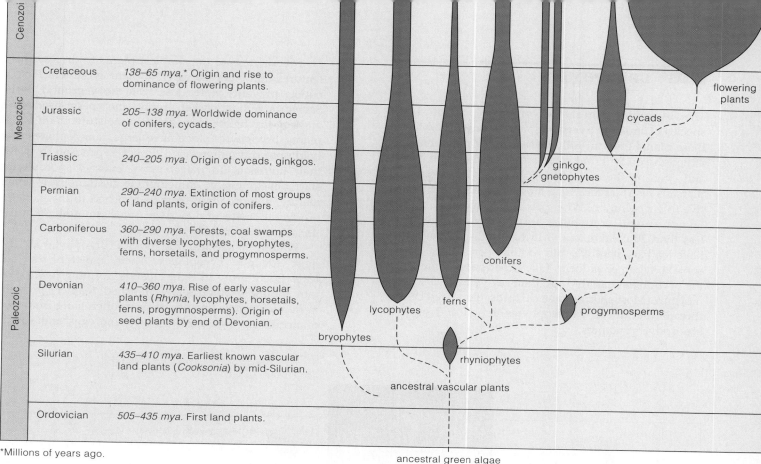

Cenozoic		
Mesozoic	Cretaceous	138–65 mya.* Origin and rise to dominance of flowering plants.
	Jurassic	205–138 mya. Worldwide dominance of conifers, cycads.
	Triassic	240–205 mya. Origin of cycads, ginkgos.
Paleozoic	Permian	290–240 mya. Extinction of most groups of land plants, origin of conifers.
	Carboniferous	360–290 mya. Forests, coal swamps with diverse lycophytes, bryophytes, ferns, horsetails, and progymnosperms.
	Devonian	410–360 mya. Rise of early vascular plants (Rhynia, lycophytes, horsetails, ferns, progymnosperms). Origin of seed plants by end of Devonian.
	Silurian	435–410 mya. Earliest known vascular land plants (Cooksonia) by mid-Silurian.
	Ordovician	505–435 mya. First land plants.

*Millions of years ago.

Figure 25.3 Milestones in plant evolution.

tophytes nourish and protect the forthcoming generation. Thus, unlike algae, plants retain spores and gametophytes until environmental conditions favor dispersal and fertilization.

Evolution of Pollen and Seeds

Gymnosperms and angiosperms have radiated into nearly all high and dry habitats. The evolution of pollen grains and seeds were factors in their successful radiations. Both kinds of seed-bearing plants produce not one but two kinds of spores. This condition is called heterospory, as opposed to homospory. One kind of spore develops into **pollen grains**, which become mature, sperm-bearing male gametophytes. The other kind of spore develops into female gametophytes, the plant parts where eggs form and become fertilized. Pollen grains hitch rides on air currents or insects, birds, and so on. Unlike algae, they don't require liquid water to meet up with the eggs.

Also in seed-bearing plants, a parent sporophyte holds onto its female gametophytes. It nourishes and protects fertilized eggs as they develop into young embryos. The nutritive tissues, protective tissues, and the embryo itself constitute a **seed**. It's probably no coincidence that the dominant seed plants arose during Permian times, when climatic fluctuations were extreme (page 335). Seeds are packages that endure hostile conditions.

Putting the key points of this overview together,

1. The evolution of land plants involved structural adaptations to dry conditions. In different lineages, these adaptations included waxy cuticles, stomata, vascular tissues, and lignin-reinforced tissues.

2. Sporophytes with well-developed systems of roots and shoots (stems and leaves) came to dominate the life cycle of complex land plants. Such plants had the means to nourish and protect their spores and gametophytes through unfavorable conditions.

3. Some plants started producing two types of spores instead of one. This led to the evolution of male gametes specialized for dispersal without liquid water. It led to the evolution of seeds—embryos packaged with protective and nutritive tissues.

Before turning to the spectrum of diversity among these plants, take a look at Figure 25.3. You can use it as a map for the branching evolutionary roads.

25.3 BRYOPHYTES

The bryophyte lineage encompasses about 16,000 species of **mosses**, **liverworts**, and **hornworts**. Most of these plants grow in fully or seasonally moist habitats, although you will find some mosses growing in deserts and even on the excruciatingly cold, windswept high plateaus of Antarctica. The mosses especially are sensitive to air pollution. Where air is bad, mosses are often few or absent. Bryophytes are small plants, generally less than 20 centimeters (8 inches) tall. Although they have leaflike, stemlike, and rootlike parts, these do not contain xylem or phloem. Like lichens and some algae, the bryophytes can dry out, then revive after absorbing moisture. Most species have **rhizoids**, elongated cells or threads that attach gametophytes to soil and serve as absorptive structures.

Bryophytes are the simplest plants to display three features that emerged early in plant evolution. *First*, a cuticle prevents water loss from aboveground parts. *Second*, a cellular jacket around the sperm-producing and egg-producing parts holds in moisture. *Third*, of all plants, bryophytes alone have large gametophytes that hold onto sporophytes and do not depend on them for their nutrition. To the contrary, embryo sporophytes start developing in gametophyte tissues. Even when mature, a sporophyte remains attached to the gamete-producing body and derives some amount of nutritional support from it.

With 9,500 species, true mosses (Bryophyta) are the most common bryophytes. Their gametophytes are leafy. Some grow in tight clusters, forming low, cushiony mounds. Others show branched, feathery growth; in humid regions, masses of them often hang from tree branches. As Figure 25.4 shows, the eggs and sperm

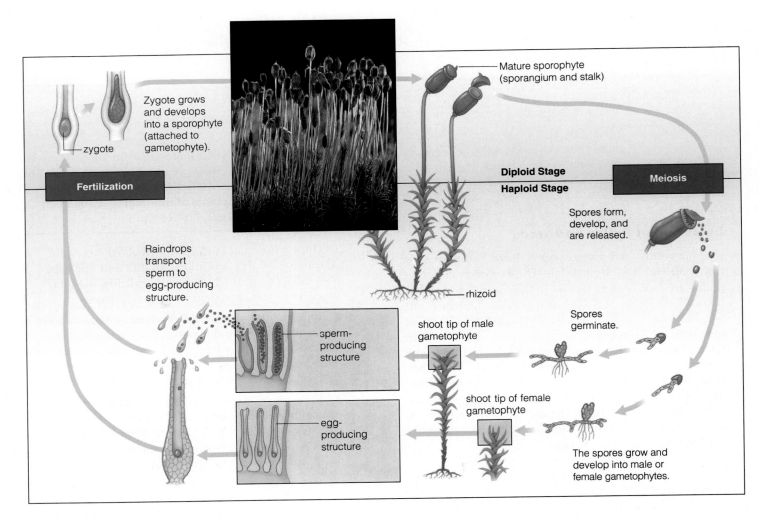

Figure 25.4 Life cycle of a moss (*Polytrichum*), a representative bryophyte. The sporophyte remains attached to and depends on the gametophyte for water and nutrients.

a

b

Figure 25.5 (**a**) From Alaska, a forest floor cloaked with a thick mat of peat—excessively moist, compressed organic matter that resists decomposition. (**b**) One of the peat mosses (*Sphagnum*). These gametophytes have several sporophytes attached to them. Along with grasses and other plants, such mosses have contributed to the formation of extensive peat bogs in cold and temperate regions. In Ireland and elsewhere, dried peat is used as a fuel source.

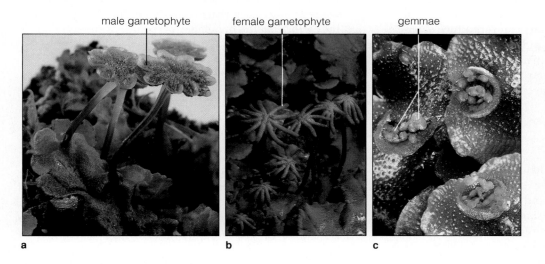

male gametophyte   female gametophyte   gemmae

a   b   c

Figure 25.6 *Marchantia*, a liverwort. Like other liverworts, it reproduces sexually. Unlike others, it produces (**a**) male and (**b**) female reproductive structures on separate plants. (**c**) *Marchantia* also can reproduce asexually by way of gemmae, multicelled vegetative bodies that develop in tiny cups on the plant body. Gemmae grow into new plants after splashing raindrops transport them to suitable sites.

develop at shoot tips, in **gametangia** (singular, gametangium, meaning "gamete vessel"). Sperm reach eggs by swimming through a film of water on the plant parts. After fertilization, zygotes give rise to sporophytes. Each sporophyte consists of a stalk and **sporangium** (plural, sporangia), a jacketed structure in which spores develop.

Figure 25.5 shows one of 350 species of peat mosses. These structurally simple mosses live in bogs, where turgor pressure keeps them erect. They soak up five times as much water as cotton does, owing to large, dead cells in their leaflike parts. Home gardeners use

peat moss to increase the water-holding capacity as well as the acidity of soils.

So as not to slight the liverworts, Figure 25.6 shows *Marchantia*, a type that has some interesting ways of reproducing.

Bryophytes are rather unspecialized, nonvascular plants that require liquid water for fertilization. Their sporophytes develop in gametophyte tissues and remain attached to them, and so continue to receive nutrients.

25.4 SEEDLESS VASCULAR PLANTS

By the late Silurian, some 420 million years ago, seedless vascular plants were established. Diverse plants of this lineage flourished for the next 60 million years, then most kinds became extinct.

Cooksonia, one of the rhyniophytes, was among their earliest ancestors (Figure 25.7a). Although leafless, members of this genus resembled many existing vascular plants in internal structure. Some were pin-sized, and others were a few centimeters tall. *Psilophyton*, one of the trimerophytes, was taller and more complex. By Devonian times, the progymnosperms evolved; they may have been ancestral to seed-bearing plants. Figures 20.2c and 21.11 show a few fossils of these extinct plants.

Existing seedless vascular plants are commonly called **whisk ferns**, **lycophytes**, **horsetails**, and **ferns**. Like their ancestors, they differ from bryophytes in some important respects. First, the sporophyte doesn't remain attached to the gametophyte. Second, it has complex vascular tissues. Third, it is the larger, longer lived phase of the life cycle.

Most seedless vascular plants live in wet, humid regions. Their gametophytes have no vascular tissues for water transport, and their flagellated sperm require ample water to reach the eggs. The few species in deserts and other extreme habitats can reproduce sexually only during brief, seasonal pulses of heavy rains or summer thaws. Thus the whisk ferns, lycophytes, horsetails, and ferns are the "amphibians" of the plant kingdom. *They have not fully escaped the aquatic habitats of their ancestors.*

Whisk Ferns

Whisk ferns (Psilophyta), which are not true ferns, resemble whisk brooms. These common ornamental plants grow in moist tropical and subtropical regions, including Hawaii, Florida, Louisiana, Texas, and Puerto Rico. One genus, *Psilotum*, is unique among vascular plants. Its sporophytes have no roots and leaves. Its members have branching photosynthetic stems with scalelike projections, not leaves (Figure 25.7b). The stems contain xylem and phloem. Belowground are **rhizomes**. These short, branching, mostly horizontal stems function in absorption. Mycorrhizal fungi assist the rhizomes in this task.

Lycophytes

About 350 million years ago, lycophytes (Lycophyta) included tree-sized members of those swamp forests that eventually became transformed into the world's coal deposits (see the *Focus* essay). Today there are 1,000

Focus on the Environment

Ancient Carbon Treasures

Between 360 and 280 million years ago, during the Carboniferous, swamp forests carpeted the wet lowlands of continents. Among the diverse species were the ancient ancestors of lycophytes, horsetails, ferns, and possibly the seed-bearing plants (Figure *a*). This was a period when sea levels rose and fell fifty times. When the seas moved out, swamp forests flourished. When the seas moved in, forest plants were submerged and became buried in sediments that protected them from decay. Gradually, the sediments compressed the saturated, undecayed remains into what we now call **peat**. As more sediments accumulated, increased heat and pressure made the peat even more compact. It became **coal** (Figure *b*).

Coal has a high percentage of carbon; it is energy-rich. It is one of our premier "fossil fuels." It took a fantastic amount of photosynthesis, burial, and compaction to form each major seam of coal in the earth. It has taken us only a few centuries to deplete much of the known coal deposits. Often you will hear about our annual "production rates" for coal or some other fossil fuel. But how much do we really produce each year? None. We simply *extract* it from the earth. Coal is a nonrenewable source of energy.

species. The most familiar are club mosses—tiny, inconspicuous members of communities in the Arctic, the tropics, and regions in between. Many types form mats on forest floors. Club mosses were once grouped in the same genus (*Lycopodium*), but as many as fifteen genera are now recognized.

Most club moss sporophytes have a branching rhizome that gives rise to vascularized roots and stems. All have tiny leaves. Sporophytes of some species also have nonphotosynthetic, cone-shaped clusters of leaves that contain spore sacs. Each cluster is a **strobilus** (plural, strobili). Figure 25.7c shows examples. Following dispersal, the spores germinate and develop into small, free-living gametophytes.

Members of one genus alone (*Selaginella*) are heterosporous. The two kinds of spores develop in the same cone-shaped cluster of leaves. In Texas, New Mexico, and Mexico, one species (*S. lepidophylla*) is commonly known as the resurrection plant.

a Reconstruction of a Carboniferous forest.

b Coal

Figure 25.7 (**a**) *Cooksonia*, the earliest known vascular plant, no more than a few centimeters tall. It probably grew in mud flats. Its upright, branching stems had a cuticle. Its spores were produced in sporangia at stem tips. Compare Figure 20.2c. (**b**) Sporophytes of a whisk fern (*Psilotum*), a seedless vascular plant. Pumpkin-shaped, spore-producing structures form at the ends of stubby branchlets. (**c**) Sporophytes of one of the lycophytes (*Lycopodium*).

a   b

c

Figure 25.8 (**a**) Vegetative shoots of *Equisetum*, the shape of which provides you with a clue to why someone thought "horsetails" would be a suitable name for these seedless vascular plants. (**b**) Nonphotosynthetic, fertile shoots of *Equisetum*. At the stem tips are strobili, the spore-bearing structures. (**c**) Closer look at a strobilus, showing the umbrella-shaped clusters of sporangia.

Horsetails

The seedless vascular plants called sphenophytes (Sphenophyta) barely squeaked through to the present. Gone are the tree-size calamites and other forms that flourished in ancient swamp forests. All that remain are fifteen species of a single genus (*Equisetum*). These are the horsetails. They may be the oldest of all existing plant genera; they have scarcely changed over the past 300 million years.

Horsetails grow in muds of streambanks and disturbed habitats, such as vacant lots, roadsides, and beds of railroad tracks. When their spores land on nutrient-rich mud, they grow into free-living, pinhead-sized gametophytes. Sporophytes usually have rhizomes, hollow photosynthetic stems, and scalelike leaves. In the stem tissue, clusters of xylem and phloem are arranged as a ring. Silica-reinforced ribs support stems and give them a gritty quality, like sandpaper. Pioneers of the American West used horsetails to scrub their cooking pots and pans.

Figure 25.8*a* shows the vegetative, photosynthetic stems of one horsetail species. Figure 25.8*b* shows the fertile stems, which have strobili at the tips. In the strobili, spores form along the edge of umbrella-shaped, spore-bearing branches (Figure 25.8*c*). Air currents disperse them.

Ferns

With 12,000 or so species, the ferns (Pterophyta) are the largest and most diverse group of plants. All but about 380 species are native to the tropics, but they are popular houseplants all over the world. Their size range is stunning. Some floating species are less than 1 centimeter across. Some tropical tree ferns are 25 meters (82 feet) tall. One climbing fern has a modified leaf stalk about 30 meters long.

Most species have underground, vascularized rhizomes that give rise to the leaves and roots. Exceptions include tropical tree ferns and epiphytes. ("Epiphyte" refers to any aerial plant that grows on tree trunks or branches.) Young fern leaves are coiled, in the shape of a "fiddlehead." The mature leaves (fronds) commonly are divided into leaflets.

You may have noticed rust-colored patches on the lower surface of many fern fronds. Each patch is a **sorus** (plural, sori), which is a cluster of sporangia. At dispersal time, the sporangia snap open, causing the spores to catapult through the air. Each germinating spore develops into a small gametophyte, such as the green, heart-shaped type shown in Figure 25.9.

We now leave the seedless vascular plants. Keep these summary points in mind:

Whisk ferns, lycophytes, horsetails, and ferns have sporophytes adapted to land, but they have not entirely escaped their aquatic ancestry.

The life cycle of these seedless vascular plants can be completed only when their flagellated sperm have ample water to reach the eggs.

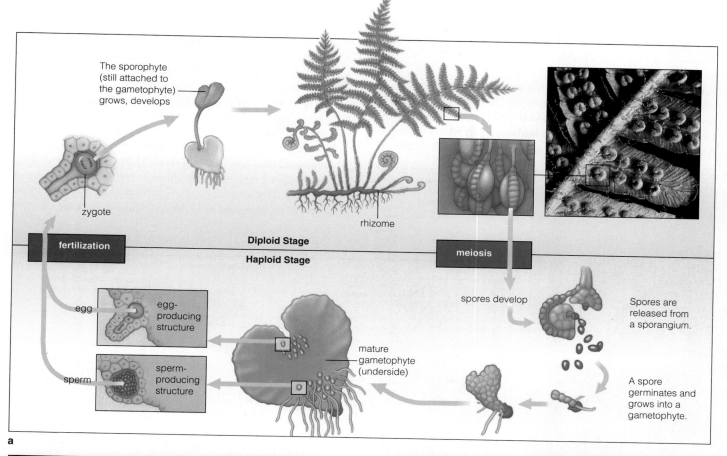

The sporophyte (still attached to the gametophyte) grows, develops

zygote

rhizome

fertilization

Diploid Stage

Haploid Stage

meiosis

spores develop

Spores are released from a sporangium.

egg

egg-producing structure

sperm

sperm-producing structure

mature gametophyte (underside)

A spore germinates and grows into a gametophyte.

a

b

c

Figure 25.9 (**a**) Life cycle of a fern. (**b**) Ferns in a moist habitat in Indiana. (**c**) Tree ferns in a temperate rain forest of Tasmania.

25.5 THE SEED-BEARING PLANTS

As indicated earlier, gymnosperms and angiosperms are seed-bearing plants that arose in Devonian times. In terms of diversity, numbers, and distribution, they have become the most successful plants, owing to an enormous advantage they have over seedless types. Air cur-rents or insects carry their sperm—packaged in pollen grains—to eggs. *Thus, seed-bearing plants have escaped dependency on free water for fertilization.* In addition, their sporophytes produce reproductive structures called ovules. An **ovule** contains the egg-producing female gametophyte, surrounded by nutritive tissue and a jacket of cell layers (Figure 25.10). As a fertilized egg develops, the outer layers form a coat. The mature, coat-enclosed ovule is the seed. *Thus, embryo sporophytes are protected by a seed coat during their dispersal, and they can tap stored nutrients at the critical time of germination, before their roots and shoots become fully functional.*

Figure 25.10 Life cycle of a gymnosperm (ponderosa pine).

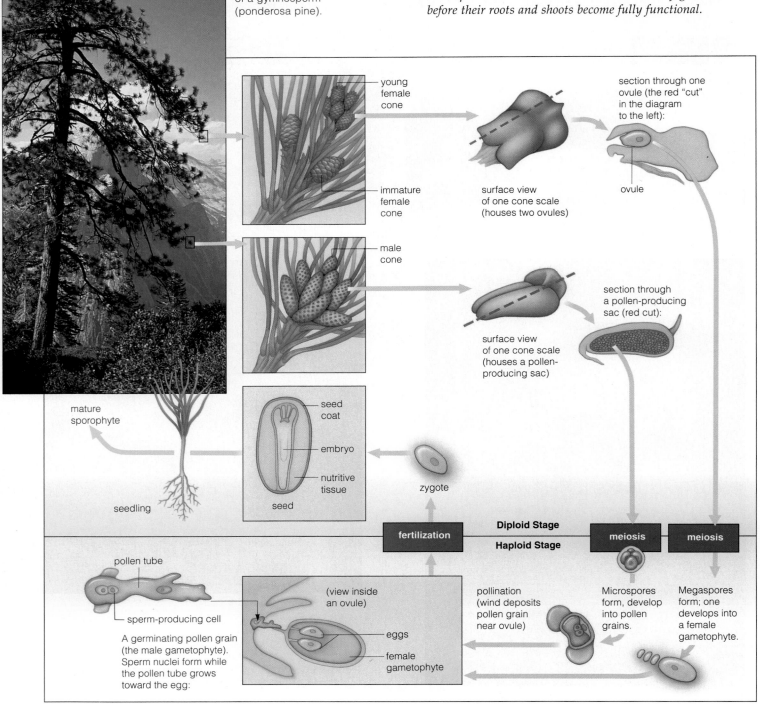

25.6 GYMNOSPERMS

"Gymnosperm" is a word derived from the Greek *gymnos* (meaning naked) and *sperma* (which is taken to mean seed). The name refers to the exposed location of their ovules and seeds. They are borne on surfaces of spore-producing reproductive structures. Conifers are familiar members of the lineage. Less well known are the cycads, ginkgos, and gnetophytes.

Conifers

Conifers (Coniferophyta) are "evergreen," woody trees and shrubs with needlelike or scalelike leaves. Most shed old leaves throughout the year but retain enough to distinguish them from "deciduous" species (which shed all their leaves in the fall). Conifers include diverse pines, spruces, firs, hemlocks, junipers, cypresses, and redwoods. Nearly all of their **cones** are clusters of modified leaves in which spore-producing structures develop. As the Figure 25.10 example suggests, conifers are heterosporous. Their **microspores** form in male cones and develop into pollen grains. Their **megaspores** form in the shelflike scales of female cones and develop into the female gametophytes.

Each spring, millions of pollen grains drift off male cones, and some land on ovules. The arrival of pollen on female reproductive parts is called **pollination**. Now pollen grains germinate. Each develops into a long, tubelike structure that grows down through the female tissues. This pollen tube transports sperm toward female gametophytes. In many species of conifers, fertilization does not occur until months to as much as a year after pollination.

Recall, from Chapter 21, that conifers radiated into many land environments during the Mesozoic. Their leisurely reproductive pace may have put them at a competitive disadvantage when flowering plants began their spectacular adaptive radiations.

Coniferous forests still cloak many regions, including the far north, high altitudes, and parts of the Southern Hemisphere. However, besides having to compete with flowering plants for resources, conifers are now vulnerable to global **deforestation**—the removal of all trees from vast tracts of land, as by clear cutting (Figure 25.11). Conifers are the premier source of lumber, furniture, paper, and many other products. We return to this topic in Chapter 50.

Conifers are woody, mostly evergreen shrubs and trees that produce pollen grains, bear seeds on cone scales, and are key sources of lumber, paper, and other manufactured products.

Figure 25.11 From North America, a small sampling of the rampant deforestation now under way throughout the world. (**a,b**) From the Nahmint Valley of British Columbia, an ancient forest before and after logging. From Alaska (**c**) and Washington (**d**), denuded mountains. (**e**) From America's heartland, logged-over acreage in Arkansas. From the eastern seaboard, a denuded piece of North Carolina (**f**). These are not isolated examples. In the early 1980s, 400 million board feet of timber were being cut annually in Washington's Olympic Peninsula. In Arkansas, approximately one-third of the Ouachita National Forest was clear-cut. Its once-diverse forest communities have been replaced by "tree farms" of a single species of pine. Throughout the world, large tracts that were deforested years ago still show no signs of recovery.

Lesser Known Gymnosperms

Figure 25.12 shows a few relatives of the conifers. The **cycads** (Cycadophyta) flourished with dinosaurs during the Mesozoic. About 100 species have survived to the present in tropical and subtropical regions. They superficially resemble palm trees (which are flowering plants). Cycads have massive cones that bear either pollen or ovules. Air currents or insects transfer pollen from "male" plants to developing seeds on "female" plants. In parts of Asia, people eat cycad seeds and a starchy flour made from cycad trunks, but only after they rinse out the plant's poisonous alkaloids.

Ginkgos (Ginkgophyta) were diverse during dinosaur times. The maidenhair tree, *Ginkgo biloba*, is the only surviving species. Several thousand years ago, ginkgo trees were planted in cultivated grounds around temples in China. Then the natural populations almost became extinct. This is puzzling, for ginkgos seem hardier than many other trees. Perhaps, as human populations grew, ginkgos were cut for firewood. Male ginkgo trees are now planted in cities. They have attractive, fan-shaped leaves and resistance to insects, disease, and air pollutants. Female trees are not favored; their fleshy-coated seeds produce an awful stench when stepped on.

The seventy species of **gnetophytes** (Gnetophyta) are divided into three genera: *Gnetum*, *Ephedra*, and *Welwitschia*. Thirty or so species of *Gnetum*, either trees or leathery leafed vines, grow in humid tropical regions. About thirty-five species of *Ephedra* grow in deserts and other arid regions around the world, including parts of California. Of all gymnosperms, *Welwitschia* is the most bizarre. This seed-producing plant grows in hot deserts of south and west Africa. The bulk of the plant is a deep-reaching taproot. The only exposed part, a woody disk-shaped stem, bears cone-shaped strobili and leaves. The plant never produces more than two strap-shaped leaves, which split lengthwise repeatedly as the plant ages.

Cycads, ginkgo, and gnetophytes bear seeds on exposed surfaces of spore-bearing structures, but they differ greatly from one another and from conifers in other features.

Figure 25.12 Lesser known gymnosperms. (**a**) Massive female cone of a cycad (*Encephalartos ferox*). (**b**) *Ephedra viridis*, growing in California. (**c**) Ginkgo trees. (**d**) Fleshy-coated ginkgo seeds. (**e**) A mature specimen of *Welwitschia mirabilis*. The inset shows the appearance of its female cones.

25.7 ANGIOSPERMS

Flowering Plant Diversity

Angiosperms, the flower-producing seed plants, have dominated the land for more than 100 million years. "Angiosperm" is derived from the Greek *angeion* (a vessel) and *sperma*. The "vessel" refers to tissues that surround and protect ovules and developing seeds. We know of about 235,000 existing species and are discovering new ones almost weekly in previously unexplored tropical forests. Untold numbers may soon be extinct, owing to deforestation.

Angiosperms are the most diverse of all plants. They live in mountains and deserts, wetlands, and seawater. They range in size from duckweeds (about a millimeter long) to *Eucalyptus* trees more than 100 meters tall. Most are free-living and photosynthetic, but there are exceptions (Figure 25.13). Some, including mistletoe, are parasites on other plants.

In terms of sheer diversity, numbers, and distribution, the angiosperms are the most successful plants. These seed-bearing plants alone produce flowers.

Figure 25.13 Sampling of flowering plant diversity. (**a**) Plants of alpine tundra. (**b**) Water lily (*Nymphea*). (**c**) Indian pipe (*Monotropa uniflora*), a nonphotosynthetic plant that obtains nutrients by way of mycorrhizal fungi that are partners with photosynthetic plants. (**d**) Dwarf mistletoe (*Arceuthobium*), a parasitic plant that limits the growth of forest trees in the western United States. Highly prized species: (**e**) *Theobroma cacao* fruits, source of cocoa beans; (**f**) an exotic orchid plant; and (**g**) sugarcane plants growing in Hawaii.

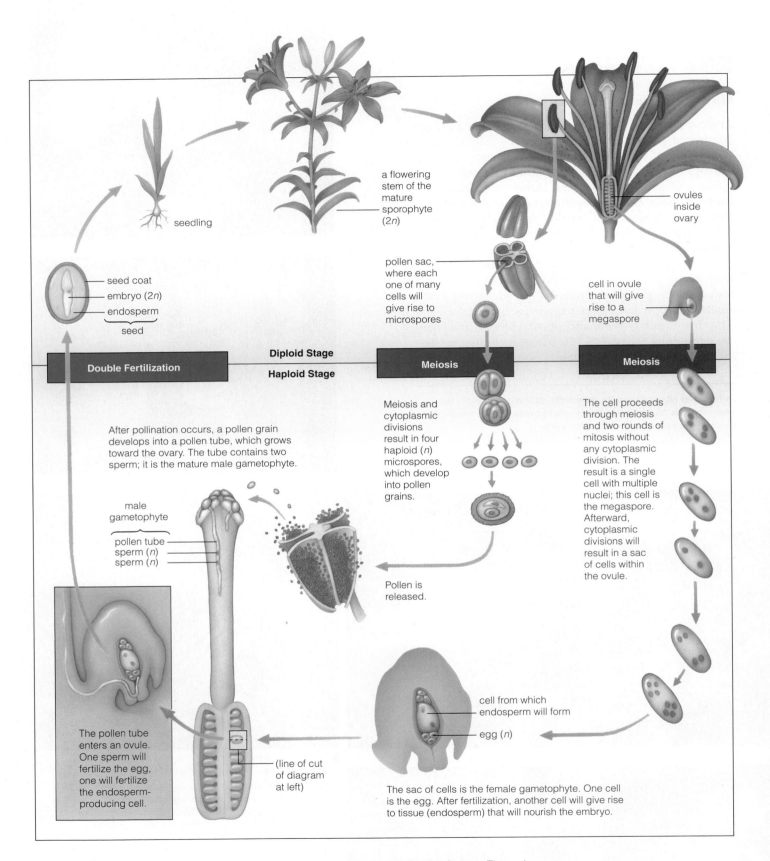

seedling

a flowering stem of the mature sporophyte (2n)

ovules inside ovary

seed coat
embryo (2n)
endosperm
seed

pollen sac, where each one of many cells will give rise to microspores

cell in ovule that will give rise to a megaspore

Double Fertilization | **Diploid Stage** | **Meiosis** | **Meiosis**
Haploid Stage

Meiosis and cytoplasmic divisions result in four haploid (n) microspores, which develop into pollen grains.

The cell proceeds through meiosis and two rounds of mitosis without any cytoplasmic division. The result is a single cell with multiple nuclei; this cell is the megaspore. Afterward, cytoplasmic divisions will result in a sac of cells within the ovule.

After pollination occurs, a pollen grain develops into a pollen tube, which grows toward the ovary. The tube contains two sperm; it is the mature male gametophyte.

male gametophyte

pollen tube
sperm (n)
sperm (n)

Pollen is released.

The pollen tube enters an ovule. One sperm will fertilize the egg, one will fertilize the endosperm-producing cell.

(line of cut of diagram at left)

cell from which endosperm will form

egg (n)

The sac of cells is the female gametophyte. One cell is the egg. After fertilization, another cell will give rise to tissue (endosperm) that will nourish the embryo.

Figure 25.14 Life cycle of a monocot (*Lilium*). "Double" fertilization is a distinctive feature. The male gametophyte delivers two sperm to an ovule. One sperm fertilizes the egg, and the other fertilizes a cell that gives rise to a tissue (endosperm) that will nourish the forthcoming embryo. Figure 31.5 provides a closer look at flowering plant life cycles, using a dicot as the example.

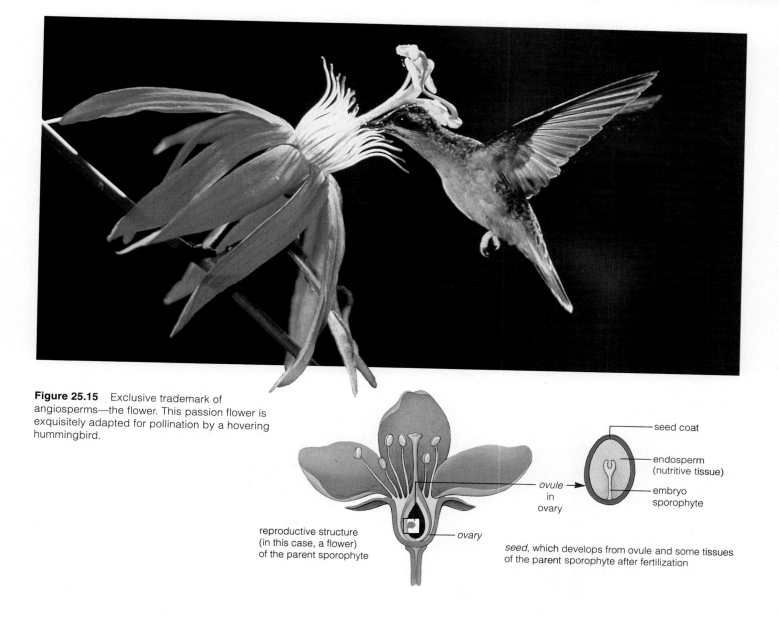

Figure 25.15 Exclusive trademark of angiosperms—the flower. This passion flower is exquisitely adapted for pollination by a hovering hummingbird.

reproductive structure
(in this case, a flower)
of the parent sporophyte

ovary

ovule
in
ovary

seed coat

endosperm
(nutritive tissue)

embryo
sporophyte

seed, which develops from ovule and some tissues
of the parent sporophyte after fertilization

Classes of Flowering Plants

There are two classes of flowering plants—the Dicotyledonae and Monocotyledonae (or, informally, dicots and monocots). Among the 200,000 species of dicots are most shrubs and trees, most nonwoody (herbaceous) plants, cacti, and water lilies. The 65,000 species of monocots include palms, lilies, and orchids. They also include grasses, such as wheat, corn, rice, rye, sugarcane, barley, and other major crop plants.

Key Aspects of the Life Cycles

Figure 25.14 shows a monocot life cycle. A typical dicot life cycle is described in detail in the next unit, which focuses on the structure and function of flowering plants. As with gymnosperms, a large diploid sporophyte dominates the cycle, it retains and nourishes gametophytes, and its sperm travel in pollen grains.

Unlike gymnosperms, flowering plants provide their seeds with a unique, nutrient-packed tissue (endosperm). They also package seeds in **fruits**, which aid in protection and dispersal. Figure 25.13*e* shows one of many diverse kinds of fruits. Generally, the flowers that develop during the life cycle have coevolved with insects, bats, birds, and other animals that assist in pollination (Figure 25.15). These **pollinators** go after nectar or pollen in flowers. As they do, they transfer pollen grains to female reproductive parts. The "recruitment" of animals probably was a key factor in the rise of angiosperms.

Flowering plants produce flowers, endosperm-rich seeds, and fruits. Most species have coevolved in mutually beneficial ways with animal pollinators.

SUMMARY

1. Nearly all members of the plant kingdom are multicelled, photoautotrophic species that live on land. They probably evolved from multicelled green algae more than 400 million years ago. Table 25.1 summarizes the major phyla (divisions) of existing plants.

2. As summarized in Table 25.2 and listed below, several trends in plant evolution can be identified by comparing the characteristics of different lineages:

    a. Structural adaptations to dry conditions, especially the development of the vascular tissues called xylem and phloem.

    b. A shift from haploid to diploid dominance. Complex sporophytes evolved that could hold onto, nourish, and protect spores and gametophytes.

    c. A shift from the production of one kind to two kinds of spores (homospory to heterospory). Among gymnosperms and angiosperms (flowering plants), this led to the evolution of pollen grains and seeds.

3. Existing nonvascular land plants (those without well-developed xylem and phloem) include the bryophytes—the mosses, liverworts, and hornworts. Existing seedless vascular land plants include the whisk ferns, lycophytes, horsetails, and ferns. In all cases, their flagellated sperm require ample water, so that they can swim to and fertilize the eggs.

4. Vascular land plants typically have a cuticle and stomata that help control water loss. They have sporangia (protective tissue layers around their spores). They also have ovules or similar structures with protective and nutritive tissue layers around their gametophytes. The embryo sporophyte begins its development *within* gametophyte tissues.

5. Gymnosperms and angiosperms are seed-bearing vascular plants. The evolution of pollen grains freed them from dependence on water for fertilization. Their seeds are efficient means of dispersing the new generation and helping it through hostile conditions. Pollen grains and seeds were key adaptations underlying the move into high, dry habitats.

6. The female gametophytes of gymnosperms and angiosperms are attached to and protected by the

---

Table 25.1 Comparison of Major Plant Groups*

Nonvascular land plants. Fertilization requires free water. Haploid dominance. Cuticle, stomata present in some.

Bryophytes	16,000 species. Moist, humid habitats.

Seedless vascular plants. Fertilization requires free water. Diploid dominance. Cuticle, stomata present.

Lycophytes	1,000 species with simple leaves. Mostly wet or shady habitats.
Horsetails	15 species of single genus. Swamps, disturbed habitats.
Ferns	12,000 species. Wet, humid habitats in mostly tropical, temperate regions.

Vascular plants with "naked seeds" (gymnosperms). Diploid dominance. Cuticle, stomata present.

Conifers	550 species, mostly evergreen, woody trees and shrubs having pollen- and seed-bearing cones. Widespread distribution.
Cycads	100 slow-growing species. Tropics, subtropics.
Ginkgo	1 species, a tree with fleshy-coated seeds.
Gnetophytes	70 species, limited distribution in deserts and tropics.

Vascular plants with flowers and protected seeds (angiosperms). Diploid dominance. Cuticle, stomata present.

Flowering plants:

Monocots	65,000 species. Floral parts often arranged in threes or multiples of three; one seed leaf; parallel leaf veins common.
Dicots	Nearly 200,000 species. Floral parts often arranged in fours, fives, or multiples of these; two seed leaves; net-veined leaves common.

*More than 275,000 known species total.

Table 25.2 Summary of Trends Among Plants

	Bryophytes	Ferns	Gymnosperms	Angiosperms
Nonvascular	———→ Vascular		——————→	
Haploid dominance	———→ Diploid dominance		——————→	
Spores of one type	——————————→ Spores of two types			——→
Motile gametes	——————————————→		Nonmotile gametes*	——→
Seedless	——————————————→ Seeds			——→

*Require pollination by wind, insects, etc.

sporophyte. With its root and shoot systems, the sporophyte is well adapted to conditions on dry land.

7. Angiosperms alone produce flowers. Their seeds contain a distinctive nutritive tissue (endosperm) and are often surrounded by fruits, which aid in dispersal. Most flowering plants have coevolved with animal pollinators, which enhance the efficiency of transferring pollen grains to female reproductive parts.

Review Questions

1. Describe the evolutionary trends among plants that figured in the invasion of land. *392–393*

2. What are some differences between bryophytes and the vascular plants? *392, 394–395, 396*

3. How does the life cycle of a gymnosperm (such as pine) differ from that of an angiosperm (such as a lily)? *400, 404*

Self-Quiz *(Answers in Appendix IV)*

1. Which of the following was *not* a major trend in the evolution of complex land plants?
   a. from nonvascular to vascular plant bodies
   b. from sporophyte to gametophyte dominance
   c. from one spore type to two specialized types
   d. away from reliance on free-standing water for fertilization
   e. among some lineages, reliance on pollinators, seed formation, and seed dispersal mechanisms

2. Which of the following statements is *not* true?
   a. Monocots and dicots are two classes of angiosperms.
   b. Bryophytes are nonvascular plants.
   c. Lycophytes, horsetails, ferns, gymnosperms, and angiosperms are vascular plants.
   d. Horsetails and gymnosperms are the simplest vascular plants.

3. Of all land plants, bryophytes alone have independent _____ and attached, dependent _____ .
   a. sporophytes; gametophytes
   b. gametophytes; sporophytes
   c. rhizoids; zygotes
   d. rhizoids; sporangia with stalks

4. Psilophytes, lycophytes, horsetails, and ferns are _____ .
   a. multicelled aquatic plants
   b. nonvascular seed plants
   c. seedless vascular plants
   d. seed-bearing vascular plants

5. Gymnosperms and angiosperms are _____ .
   a. multicelled aquatic plants
   b. nonvascular seed plants
   c. seedless vascular plants
   d. seed-bearing vascular plants

6. Which does *not* apply to both gymnosperms and angiosperms?
   a. vascular tissues   c. single spore type
   b. diploid dominance   d. all of the above

7. A seed is _____ .
   a. a female gametophyte   c. a mature pollen tube
   b. a mature ovule   d. an immature embryo

8. Which does *not* apply to both gymnosperms and angiosperms?
   a. produce pollen grains and seeds
   b. sporophyte well adapted to conditions on dry land
   c. female gametophytes are attached to and protected by the sporophyte
   d. produce flowers

9. Which plant's life cycle shows haploid dominance?
   a. conifer   d. lycophyte
   b. bryophyte   e. monocot
   c. horsetail

10. Match the terms appropriately.
    _____ gymnosperms   a. produces haploid gametes
    _____ sporophyte   b. control water loss
    _____ seedless vascular   c. "naked" seeds
        plants   d. protect, disperse embryo
    _____ ovary   sporophyte
    _____ bryophytes   e. produces haploid spores
    _____ gametophyte   f. nonvascular land plants
    _____ stomata   g. lycophytes
    _____ angiosperm seed   h. usually a fruit at maturity

Selected Key Terms

angiosperm *391*	microspore *401*
bryophyte *391*	moss *394*
coal *396*	ovule *400*
cone *401*	peat *396*
conifer *401*	phloem *392*
cuticle *392*	pollen grain *393*
cycad *402*	pollination *401*
deforestation *401*	pollinator *405*
fern *396*	rhizoid *394*
fruit *405*	rhizome *396*
gametangia *395*	root system *392*
gametophyte *392*	seed *393*
ginkgo *402*	shoot system *392*
gnetophyte *402*	sorus *398*
gymnosperm *391*	sporangium *395*
hornwort *394*	sporophyte *392*
horsetail *396*	stomata *392*
lignin *392*	strobilus *396*
liverwort *394*	vascular plant *391*
lycophyte *396*	whisk fern *396*
megaspore *401*	xylem *392*

Readings

Bold, H., and J. LaClaire. 1987. *The Plant Kingdom*. Fifth edition. Englewood Cliffs, New Jersey: Prentice-Hall. Paperback.

Gensel, P., and H. Andrews. 1987. "The Evolution of Early Land Plants." *American Scientist* 75: 478–489.

Raven, P., R. Evert, and S. Eichhorn. 1986. *Biology of Plants*. Fourth edition. New York: Worth. Lavishly illustrated.

26 ANIMALS: THE INVERTEBRATES

Animals of the Burgess Shale

It happened in Cambrian times, deep in a submerged basin between a massive reef and the coast of an early continent. Protected from ocean currents, great banks of sediments and mud had formed in the basin. About 500 feet below the water's surface, the waters were dimly lit, oxygenated, and clear. Tiny marine animals flourished in, on, and above muddy sediments piled against the steep flank of the reef.

Like castles built from wet sand along a seashore, their home was unstable. About 530 million years ago, part of the bank slumped suddenly (Figure 26.1). The underwater avalanche deposited sediments over an entire community, and these prevented scavengers from reaching and obliterating all traces of the dead animals. Muddy silt gradually rained down on the natural tomb. In time, through increased pressure and chemical changes, the sediments became compacted into finely stratified shale. And the soft parts of the flattened carcasses of animals became shimmering films of calcium aluminosilicate.

All the while, imperceptibly, the continents were on the move. By the dawn of the Cenozoic, a crustal plate under the Pacific Ocean was plowing under the western edge of the North American plate (page 341). The mountain ranges of western Canada were forming.

By the year 1909, the fossils were high up in the mountains that run along the eastern border of British Columbia. In that year a lucky fossil hunter, Charles Wolcott, tripped over a chunk of shale on a mountain path. The shale layers split apart easily, and Wolcott made the discovery of a lifetime. The Burgess Shale animals had come to light.

Wolcott and others soon recovered fossils of more than 120 species of animals. Some forms resembled existing species. Others were bizarre evolutionary experiments that apparently led nowhere. *Opabinia* was one of these. About as long as a tube of lipstick, *Opabinia* had five eyes perched on its head and a grasping organ that possibly was used in prey capture (Figure 26.1c). The smaller *Hallucigenia* sported seven pairs

Figure 26.1 (**a**) Reconstruction of a few marine organisms that ultimately became fossilized in the Burgess Shale. (**b**) Underwater slide off the coast of Baja California, reminiscent of the slumping that entombed these organisms. (**c**) Two Burgess Shale fossils: the five-eyed *Opabinia*, with a grasping organ on its head, and *Hallucigenia*, with multiple spines and soft protuberances.

a

b

of sharp spines on one side and seven soft organs on the other. For a long time, no one figured out which side was up.

The Burgess Shale animals are testimony to a great adaptive radiation, which may have been triggered by the breakup of a supercontinent just before the Cambrian era. Think of the adaptive zones that must have opened up in the waters along the vast new coastlines!

Yet where are signs of the *first* animals—the ones predating the Burgess Shale species? Here the fossil record is a closed book. One thing is clear, however. Animals representing every major animal phylum had evolved before the time when the Burgess Shale animals were buried. Some of these were on the evolutionary roads that led to modern sponges, cnidarians, flatworms, roundworms, mollusks, annelids, arthropods, echinoderms, and other invertebrates. Others were animals with muscles arrayed in a zigzag pattern and a stiffened rod running partway down their back. They were on or near the evolutionary road leading to vertebrates—to humans and all other animals with backbones. These are the roads traced in this chapter and the next.

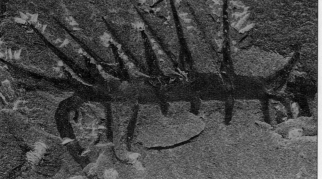

KEY CONCEPTS

1. Animals are multicelled, motile heterotrophs that pass through embryonic stages during their life cycle.

2. There are probably more than 2 million existing species of animals. Fewer than 50,000 species are vertebrates (animals with a backbone). The rest are invertebrates (animals with no backbone).

3. The invertebrates described in this chapter range from *Trichoplax* and the sponges to cnidarians, comb jellies, flatworms, roundworms, ribbon worms, rotifers, mollusks, annelids, arthropods, and echinoderms.

4. Several trends occurred in animal evolution. These are evident through comparisons of the body plans of existing animals, in conjunction with the fossil record.

5. The most revealing aspects of an animal's body plan are its type of symmetry, gut, and cavity (if any) between the gut and body wall; whether it has a distinct head end; and whether it is divided into a series of segments.

6. Sponges have no body symmetry or tissue organization. Cnidarians are radial animals with some organized tissues. Nearly all other invertebrates are bilateral animals with well-developed tissues, organs, and organ systems.

7. In terms of numbers and distribution, insects are the most successful invertebrates. These arthropods have a hardened exoskeleton, specialized segments, and jointed appendages. Many have efficient respiratory and nervous systems and a division of labor in the life cycle.

26.1 OVERVIEW OF THE ANIMAL KINGDOM

General Characteristics of Animals

Generally speaking, an **animal** is defined by the following characteristics:

---

1. Animals are multicelled. In most groups, cells form tissues that are arranged as organs and organ systems.

2. Animals are heterotrophs that must obtain carbon and energy by eating other organisms or absorbing nutrients from them.

3. Animals require oxygen (for aerobic respiration).

4. Animals reproduce sexually and, in many cases, asexually. Most are diploid organisms.

5. Animal life cycles include a period of embryonic development. In brief, cell divisions transform the zygote into a multicelled embryo. The embryo's cells become arranged as primary tissue layers: **ectoderm**, **endoderm**, and, in most species, **mesoderm**. These give rise to the adult's tissues and organs, as described on page 756.

6. Most animals are motile during at least part of the life cycle.

Diversity in Body Plans

The most familiar animals—mammals, birds, reptiles, amphibians, fishes—are **vertebrates**, the only animals with a "backbone." And yet, of probably more than 2 million species of animals, fewer than 50,000 are vertebrates! The **invertebrates** (animals whose ancestors evolved before backbones did) are lesser known, yet spectacularly diverse species.

Table 26.1 lists the groups of animals described in this chapter. Their shared characteristics arose early in time, before divergences from a common ancestor gave rise to separate lineages. Later, morphological differences accumulated among the lineages, and they were the foundation for today's bewildering diversity.

How can we get a conceptual handle on animals as different as flatworms, hummingbirds, platypuses, humans, and giraffes? We can compare similarities and differences with respect to five basic features. These are body symmetry, cephalization, type of gut, type of body cavity, and segmentation.

Body Symmetry and Cephalization Nearly all animals are radial *or* bilateral. Animals with **radial**

Table 26.1 Animal Phyla Described in This Chapter

Phylum	Some Representatives	Number of Known Species
Placozoa (*Trichoplax*)	Simplest animal; like a tiny plate, but with tissue layers	1
Porifera (poriferans)	Sponges	8,000
Cnidaria (cnidarians)	Hydrozoans, jellyfishes, corals, sea anemones	11,000
Ctenophora (comb jellies)	Radial animals with comblike structures of modified cilia	100
Platyhelminthes (flatworms)	Turbellarians, flukes, tapeworms	15,000
Nemertea (ribbon worms)	Proboscis-equipped worms closely related to flatworms	800
Nematoda (roundworms)	Pinworms, hookworms	20,000
Rotifera (rotifers)	Species with crown of cilia	1,800
Mollusca (mollusks)	Snails, slugs, clams, squids, octopuses	110,000
Annelida (segmented worms)	Leeches, earthworms, polychaetes	15,000
Arthropoda (arthropods)	Crustaceans, spiders, insects	1,000,000+
Echinodermata (echinoderms)	Sea stars, sea urchins	6,000
Chordata (chordates)	Invertebrate chordates: Tunicates, lancelets	2,100
	Vertebrates: Fishes	21,000
	Amphibians	3,900
	Reptiles	7,000
	Birds	8,600
	Mammals	4,500

symmetry have body parts arranged regularly around a central axis, like spokes of a bike wheel. Thus a cut down the center of a hydra (Figure 26.2a) divides it into equal halves; another cut at right angles to the first divides it into equal quarters. Radial animals are aquatic. With their body plan, they can respond to food suspended in the water all around them.

Animals with **bilateral symmetry** have right and left halves that are mirror images of each other. Most of these animals have an *anterior* end (head) and an opposite, *posterior* end. They have a *dorsal* surface (a back) and an opposite, *ventral* surface. Figure 26.2b shows this body plan, which evolved after forward-creeping animals emerged. Most often, their forward end was the first to encounter food and other stimuli. So we can

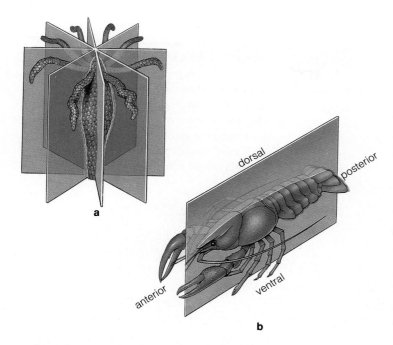

Figure 26.2 (**a**) Radial symmetry of a hydra. (**b**) Bilateral symmetry of a crayfish.

imagine there was strong selection for **cephalization**. By this evolutionary process, sensory structures and nerve cells became concentrated in a head. Bilateral body plans and cephalization evolved together. Their joint evolution resulted in *pairs* of muscles, sensory structures, nerves, and brain regions. We will return to this topic on page 576.

Type of Gut A **gut** is a region where food is digested and absorbed. Saclike guts have only one opening (a mouth) for taking in food and expelling residues. Other guts are part of tubelike systems with openings at both ends (mouth and anus). These "complete" digestive systems are more efficient. Their different regions have specialized functions, such as preparing, digesting, and storing food. The evolution of such systems helped pave the way for increases in body size and activity.

Body Cavities A body cavity separates the gut and body wall of most bilateral animals (Figure 26.3). One type of cavity, a **coelom**, has a distinctive lining called a peritoneum. The lining also covers organs in the coelom and helps hold them in place. You have a coelom. A sheetlike muscle divides it into an upper, *thoracic* cavity and a lower, *abdominal* cavity. Your thoracic cavity holds a heart and lungs; your abdominal cavity holds a stomach, intestines, and other organs.

Some types of worms do not have a body cavity. The region between their gut and body wall is packed with tissues. Other types have a "false coelom" (pseudocoel)—a body cavity but no peritoneum. A true coelom was a major step in the evolution of animals that were larger and more complex than worms. By cushioning and protecting organs, the coelom favored increases in their size and activity.

Segmentation Earthworms and other kinds of "segmented" animals consist of a series of body units that may or may not be similar to one another. Most earthworm segments have the same appearance on the outside. Insect segments are organized in three body regions (head, thorax, and abdomen), and they differ greatly from one another. Insects are fine examples of how diverse head parts, legs, wings, and other appendages evolved from less specialized segments.

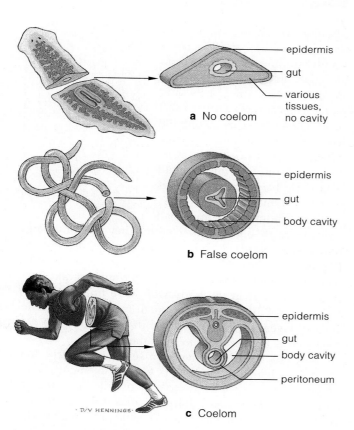

a No coelom

— epidermis
— gut
— various tissues, no cavity

b False coelom

— epidermis
— gut
— body cavity

c Coelom

— epidermis
— gut
— body cavity
— peritoneum

· D/V HENNINGS ·

Figure 26.3 Type of body cavity (if any) in animals.

The body plans of animals differ with respect to five features: body symmetry, cephalization, type of gut, type of body cavity, and segmentation.

26.2 AN EVOLUTIONARY ROAD MAP

Animals arose from protistanlike ancestors during the 200-million-year span before the Cambrian era. Most likely, the first ones were soft-bodied, so we may never find evidence of them. Yet by 560 million years ago, all major phyla were established.

There are more than thirty phyla of animals that range from placozoans and sponges to the chordates. Suppose we compare a number of these phyla in terms of body symmetry, cephalization, type of gut and body cavity, and segmentation. Doing so will reveal certain trends in animal evolution. As Figure 26.4 indicates, the trends did not show up in every line of descent. But their absence doesn't mean an animal is "primitive" or evolutionarily stunted. As you will see, even "simple" sponges are exquisitely adapted to their environment.

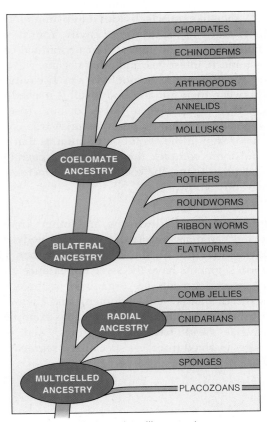

single-celled, protistanlike ancestors

Figure 26.4 Evolutionary relationships among major groups of animals. Take a moment to study this family tree. We will use it repeatedly as our road map through discussions of each group.

26.3 THE ONE PLACOZOAN

The first specimen of *Trichoplax adhaerens*, a soft-bodied marine animal no more than 3 millimeters across, was discovered gliding about in an aquarium tank. It is the only known **placozoan** (Placozoa, after the Greek *plax*, meaning plate; and *zoon*, meaning animal). *Trichoplax* is a plate-shaped, two-layered animal with no symmetry and no organs:

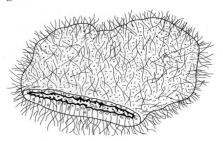

As this animal glides over food, glandular cells in its lower layer secrete digestive enzymes on it, then the cells absorb the breakdown products. Structurally and functionally, this is as simple as animals get.

26.4 SPONGES

Sponges (Porifera) are one of nature's success stories. Since Cambrian times, they have been abundant in the seas, especially in the waters off coasts and along tropical reefs. Of the 8,000 or so known species, about 100 live in freshwater habitats. Many animals, including a variety of worms, shrimps, small fishes, and barnacles, make their home in or on sponges.

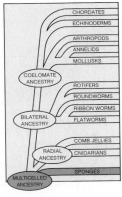

Some sponges are fingernail size; others are big enough to sit in. They are sprawling, flattened, compact, lobed, tubular, cuplike, or vaselike (Figure 26.5). Regardless of the shape, the sponge body has no symmetry or organs. Its outer surface and portions of cavities in the body wall have linings of flattened cells. But these are not much more complex than the cell layers of *Trichoplax*. And they certainly are not the same as the tissue linings of other animals.

Between the two linings is a semifluid matrix and glasslike elements (spicules), fibrous elements, and amoebalike cells. Spicules consist of silica or calcium carbonate. The pointed, sharp spicules function in support and protection. Fibers of a flexible protein give natural bath sponges a soft, squeezable texture.

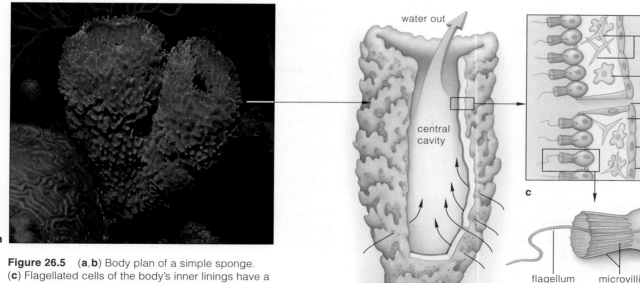

Figure 26.5 (**a**,**b**) Body plan of a simple sponge. (**c**) Flagellated cells of the body's inner linings have a collar of food-trapping microvilli. (**d**) A basket sponge of the Caribbean, releasing a cloud of sperm into the water. (**e**) A sprawling, red-orange sponge, one of many types that encrust underwater ledges in temperate seas.

water out

central cavity

water in

glasslike structural elements

amoeboid cell

pore

semifluid matrix

flattened surface cells

flagellum   microvilli   nucleus

Collar cell

Water flows into the sponge body through microscopic pores and chambers, then out through one or more openings called oscula (singular, osculum). The collective beating of thousands or millions of flagellated cells keeps water flowing this way. These **collar cells** help make up the sponge's inner linings. Besides flagella, they have "collars" of microvilli, which are absorptive structures (Figure 26.5c). Bacteria and other bits of food dissolved in the water become trapped in the collars. Some food gets transferred to the amoebalike cells for further breakdown, storage, and distribution.

Sponges reproduce sexually when collar cells give rise to eggs and to sperm, which are released into the water (Figure 26.5d). After being fertilized, the eggs are released. The young sponges pass through a microscopic, swimming larval stage. A **larva** (plural, larvae) is a sexually immature form that precedes the adult stage of many animal life cycles.

Some sponges reproduce asexually when small fragments break away from the parent body, then grow into new sponges. Most freshwater species also produce gemmules. These are clusters of sponge cells, some of which form a hard covering around others. The ones inside are protected from extreme cold or drying out. Under favorable conditions, gemmules germinate and establish a new colony of sponges.

Next to placozoans, sponges have the simplest body plan of all animals, yet they are among the most successful.

e

26.5 CNIDARIANS

All **cnidarians** (Cnidaria) are radial, tentacled animals, and most live in the seas. Of about 11,000 species, fewer than fifty live in freshwater habitats. Members of this phylum include scyphozoans (the jellyfishes), anthozoans (such as sea anemones and corals), and hydrozoans (such as *Hydra*). Figures 26.6 and 26.7 show examples. Of all the animals, cnidarians alone produce **nematocysts**. Their tentacles bristle with these thread-discharging capsules, which have roles in capturing prey and fending off predators. The toxin-laden threads of some species can give humans a painful sting (hence the phylum name Cnidaria, after the Greek word for "nettle").

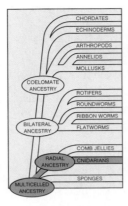

Cnidarian Body Plans

Two body forms are common among cnidarians. Both are organized around a saclike gut. As Figure 26.6a shows, one form is a **medusa** (plural, medusae). Some look like tentacle-fringed bells; others look like upside-down saucers. All medusas float in the water. Their mouth, centered under the bell, may have oral arms (extensions that assist in prey capture and in feeding). The other body form, a tubelike **polyp**, has a tentacle-fringed mouth at one end. The opposite end is usually attached to a substrate (Figure 26.6b).

Visualize the platelike *Trichoplax* draped over and digesting a mound of food, and you see that this simple animal is rather like a gut on the run. By comparison, the saclike cnidarian gut is a permanent food-processing chamber. It has a complex lining with glandular cells that secrete digestive enzymes. This sheetlike lining, a "gastrodermis," is one of the cnidarian's **epithelial tissues**. *All animals more complex than sponges have epithelial tissues.* Another lining, the "epidermis," covers the rest of the body's surfaces (Figure 26.7).

Threading through both epithelial tissues are interacting **nerve cells**. The ones in cnidarians form a "nerve net," a simple nervous system that coordinates responses to stimuli. Together with **sensory cells** and **contractile cells**, they control movements and shape changes. In some jellyfishes, structures composed of sensory cells detect changes in orientation.

Between the epidermis and the gastrodermis is a layer of secreted material, the mesoglea ("middle jelly"). Jellyfishes have enough mesoglea to impart buoyancy and serve as a firm yet deformable skeleton against which contractile cells act. Contractions narrow the bell, forcing a jet of water out from it, and the animal is propelled forward. Mesoglea is not abundant in most polyps, which typically use the water in their gut as a hydrostatic skeleton. A **hydrostatic skeleton** is a fluid-filled cavity or cell mass. When appropriate cells contract, its volume remains the same but is shunted about, and this changes the body's shape. Figure 26.7d shows how this feature can be put to use.

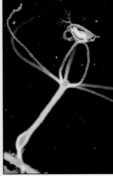

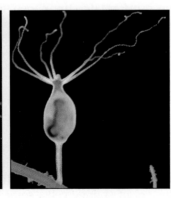

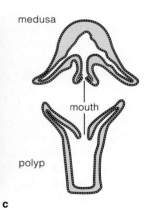

Figure 26.6 Representative cnidarians. (**a**) Medusa of the sea nettle (*Chrysophora*), one of the jellyfishes. Long, oral arms extend from the mouth, which is centered below the bell-shaped medusa. (**b**) A hydrozoan polyp (*Hydra*), firmly attached to a substrate. The polyp is shown capturing and then digesting a tiny aquatic animal. (**c**) This generalized diagram shows how a medusa and a polyp might look when sliced lengthwise, through their midsection.

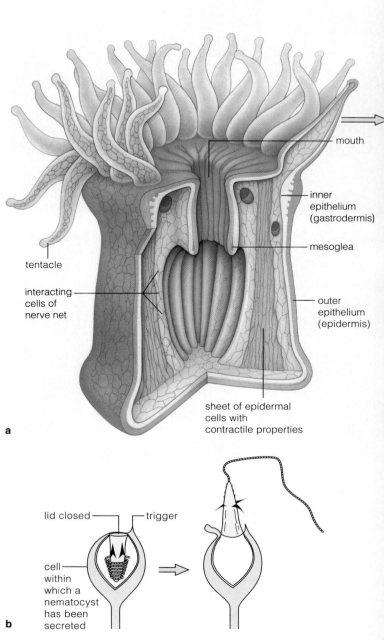

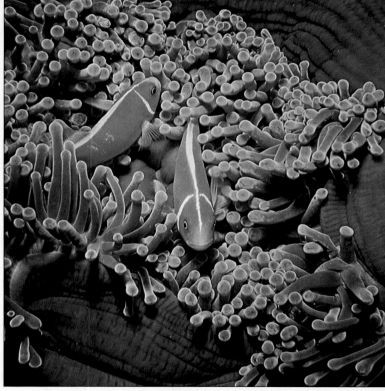

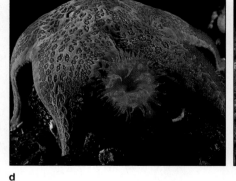

Figure 26.7 (**a**) Body plan of a sea anemone. The cutaway diagram shows aspects of its tissue organization.

(**b**) Nematocysts are embedded in the tentacles. These capsules contain an inverted tubular thread. The one shown has a bristlelike trigger. When prey touch the trigger, the capsule becomes more "leaky" to water. Water diffuses inward, pressure inside the capsule builds up, and the thread is forced to turn inside out. The thread's tip may penetrate prey and release a toxin.

(**c**) Tentacles fringing a sea anemone mouth. Sea anemones eat many fishes, but not clownfishes. Clownfishes swim away, capture food, and return to the tentacles, which protect them from predators. The sea anemone eats food scraps falling from fish mouths. This is a case of mutualism, a two-way flow of benefits between species.

(**d**) A sea anemone using its hydrostatic skeleton to escape from a sea star. The sea anemone closes its mouth, and the force generated by contractile cells in its epithelial tissues acts against water in the gut. The body changes shape, allowing the anemone to thrash about, generally in a direction away from the predator.

a

b

c   mouth

d

Figure 26.8 Corals. (**a**) A reef-building coral, with dinoflagellates in its tissues. In this mutually beneficial arrangement, the coral protects the dinoflagellates, which provide the coral with oxygen and recycle its mineral wastes. (**b**) A barrier reef, formed mainly of countless coral skeletons. Not all corals are reef builders. (**c**) A cup coral, the polyps of which look like tiny sea anemones. This one is *Tubestrea*. (**d**) *Telesto*, one of the tall, branching corals of deep or sheltered waters.

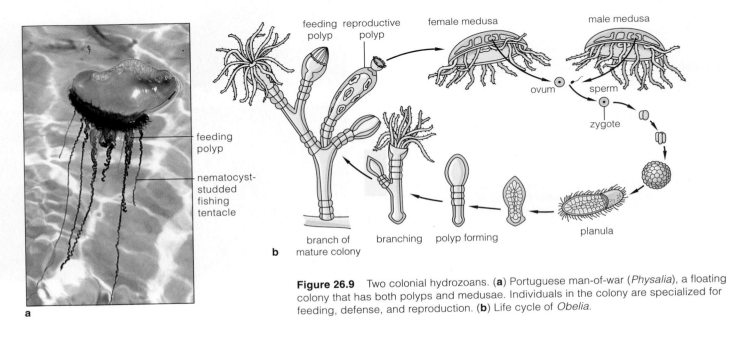

feeding polyp

nematocyst-studded fishing tentacle

a

feeding polyp   reproductive polyp   female medusa   male medusa

ovum   sperm

zygote

branch of mature colony   branching   polyp forming   planula

b

Figure 26.9 Two colonial hydrozoans. (**a**) Portuguese man-of-war (*Physalia*), a floating colony that has both polyps and medusae. Individuals in the colony are specialized for feeding, defense, and reproduction. (**b**) Life cycle of *Obelia*.

Some Colonial Cnidarians

Colonial anthozoans show interesting variations on the basic body plan. For example, most reef-building corals consist of interconnected polyps that secrete their own calcium-reinforced external skeletons. Uncounted numbers of coral skeletons serve as the main building material for reefs (Figure 26.8). Their most extravagant accomplishment, the Great Barrier Reef, parallels the eastern Australian coast for about 1,600 kilometers. The reef-forming corals have photosynthetic protistans in their tissues and must live in clear, warm water (at least 20°C).

Physalia, the hydrozoan shown in Figure 26.9a, has a remarkable colonial organization. You may have heard it called the Portuguese man-of-war. The colony consists of several kinds of polyps and medusae. They are specialized for different functions, such as feeding, defense, and reproduction.

Cnidarian Life Cycles

Many cnidarians have only a polyp stage or a medusa stage in the life cycle. *Physalia, Obelia,* and some other types have both, with the medusa being the sexual form (Figure 26.9b). Simple **gonads** (reproductive organs) are associated with the epidermis or gastrodermis. They just rupture and release the gametes. A zygote formed at fertilization nearly always develops into a swimming or creeping larva called a **planula**, which usually has ciliated epidermal cells. In time a mouth opens at one end, the larva is transformed into a polyp or medusa, and the cycle begins anew.

Cnidarians are tentacled, radial animals with a saclike gut, nerve net, and epithelial tissues. They are the only animals to produce nematocysts.

26.6 COMB JELLIES

The **comb jellies** belong to the phylum Ctenophora (a word that means "comb-bearing"). These are marine predators that show modified radial symmetry. You can slice them into two equal halves. Slice them into quarters, and two of the quarters will be mirror images of the other two.

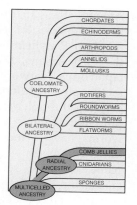

Their name refers to eight rows of comblike structures that are made of thick, fused cilia (Figure 26.10). All the combs in each row beat in waves and propel the animal forward, usually mouth first. Some species have a feeding net (two long, muscular tentacles with many branches that are equipped with sticky cells). Comb jellies don't produce nematocysts, but sometimes they opportunistically save and use the ones from jellyfish they have eaten. Others use sticky lips to capture prey, including other comb jellies or jellyfishes.

Evolutionarily, comb jellies are interesting because they have cells with multiple cilia. This trait evolved in many of the complex animals. Besides this, they are the simplest animals with embryonic tissues that resemble mesoderm. In complex animals, mesoderm gives rise to muscles and to organs of the circulatory, excretory, and reproductive systems.

Comb jellies have a modified radial body and comblike structures of modified cilia. They have an embryonic tissue that resembles the mesoderm of more complex animals.

---

row of combs   mouth

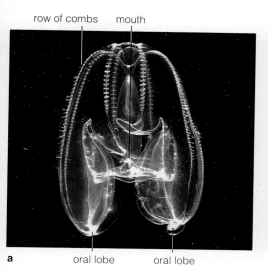

a   oral lobe   oral lobe

b

Figure 26.10 Two of the comb jellies. (**a**) *Mnemiopsis*, with its expanded, contractile mouth lobes. These waft prey into the mouth. (**b**) *Pleurobrachia*, with its two long, sticky tentacles. This one is about actual size.

26.7 FLATWORMS

Among the 15,000 or so known species of **flatworms** (phylum Platyhelminthes) are the turbellarians, flukes, and tapeworms. Flatworms have a saclike gut, with food usually entering through a pharynx (a muscular tube). Most flatworms have a more or less flattened body, hence the name. Unlike sponges or cnidarians, the flatworms are bilateral and cephalized. Three primary tissue layers develop in flatworm embryos. The midlayer, mesoderm, is the source of many tissues. Flatworms in fact are the simplest animals with **organs**, which consist of different tissues that are organized in specific proportions and patterns. They have **organ systems**, which consist of two or more organs interacting in a common task.

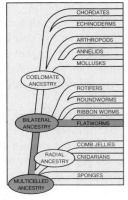

Turbellarians

Most turbellarians live in the seas; the planarians and a few others live in freshwater. The carnivores eat small animals or suck tissues from dead or wounded ones. Herbivorous types eat diatoms.

As is true of your urinary system, one planarian organ system regulates the composition and volume of body fluids. It has one or more protonephridia (singular, protonephridium). These branched tubes extend from pores at the body surface to many bulb-shaped flame cells in body tissues (Figure 26.11). Flame cells get their name from a tuft of cilia that "flickers" inside the bulbous, hollow interior. When excess water moves into a bulb, the flickering drives it through the tubes, to the outside. Some solutes may be reabsorbed from the tubes. Nitrogenous wastes leave with water and also are eliminated across the epidermis and gut lining.

Planarians commonly reproduce asexually by way of fission. A worm divides in half; then each half regenerates the missing part. However, most flatworms rely on sexual reproduction. Each worm is a **hermaphrodite**, with both female and male gonads, and two worms typically reproduce by the mutual transfer of sperm. The reproductive system includes a penis (a sperm-delivery structure) and glands that help produce a protective capsule around fertilized eggs.

Flukes

Flukes are parasites. A parasite, recall, lives on or in a living organism. It obtains nutrients from its host's tissues and may or may not end up killing it. Most adult flukes (trematodes) live in the vertebrate gut, liver, lungs, bladder, or blood vessels. Flukes have complex life cycles, with sexual and asexual phases and at least two kinds of hosts. They reach sexual maturity in a *definitive* host. Their larval stages develop or become encysted in an *intermediate* host.

a branching gut
pharynx (protruded)

b protonephridia

c brain
nerve cord

d ovary testis oviduct genital pore
penis

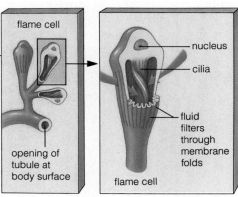

Figure 26.11 Organ systems in a planarian, a type of flatworm. (**a**) A pharynx opens to the gut. It projects to the outside, then retracts into a chamber between feedings. (**b**) Water-regulating system, (**c**) nervous system, and (**d**) reproductive system.

flame cell
opening of tubule at body surface
nucleus
cilia
fluid filters through membrane folds
flame cell

Tapeworms

Tapeworms (cestodes) are parasites of vertebrate intestines. Ancestral tapeworms probably had a gut, but they presumably lost it as they evolved in environments rich with predigested food. Existing tapeworms attach to the intestinal wall by a **scolex**, a structure with suckers, hooks, or both (Figure 26.12 and page 421). Just behind the scolex, **proglottids** (new tapeworm body units) form by budding. Proglottids are hermaphroditic; they mate and transfer sperm. Older proglottids (farthest from the scolex) store eggs; then they leave the body in feces and so carry the eggs to the outside. There, eggs may meet up with an intermediate host.

In some respects, the simplest turbellarians, larval flukes, and larval tapeworms resemble the planulas of cnidarians. The resemblance inspires speculation that bilateral animals evolved from planula-like ancestors, through increased cephalization and the development of tissues derived from mesoderm.

26.8 ROUNDWORMS

The **roundworms** (nematodes) live just about everywhere, from snowfields to deserts and hot springs. A cupful of rich soil has thousands of them; a dead earthworm or rotting fruit can hold many interesting scavenging types. Roundworms also parasitize living plants and animals. Thirty or so species, including pinworms and hookworms, parasitize humans.

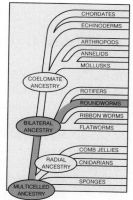

The roundworm body plan works well in many different habitats and has not changed much over time. All roundworms have a bilateral, cylindrical body, usually tapered at both ends. They have a **cuticle**, which in animals is a tough, flexible body covering. Roundworms are the simplest animals with a complete digestive system (Figure 26.13). Between the gut and body wall is a false coelom, often packed with reproductive organs. Fluid in the cavity circulates nutrients through the body.

We turn next to a *Focus* essay on some diseases caused by a few notorious flatworms and roundworms.

Flatworms and roundworms are among the simplest bilateral, cephalized animals with organ systems.

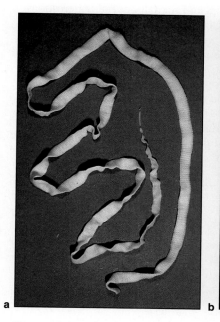

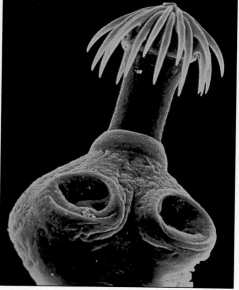

Figure 26.12 (**a**) A sheep tapeworm. (**b**) Hooks and suckers of a scolex, the part of a tapeworm that attaches to a primary host (in this case, a shorebird).

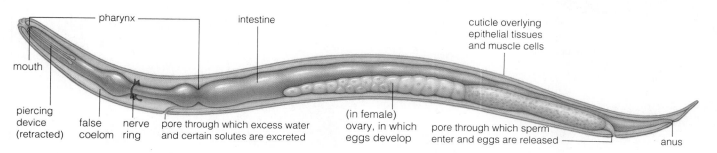

Figure 26.13 Body plan of a roundworm (*Paratylenchus*) that parasitizes the roots of certain plants.

A Rogues' Gallery of Parasitic Worms

To our enormous discomfort, many parasitic flatworms and roundworms call the human body home. In a given year, about 200 million people house blood flukes responsible for *schistosomiasis*. Figure *a* shows the life cycle of a Southeast Asian blood fluke (*Schistosoma japonicum*). The cycle requires a human definitive host, standing water through which larvae can swim, and an aquatic snail that serves as intermediate host. Flukes reproduce sexually and their eggs mature in the human body (1). Eggs leave the body in feces, then hatch into ciliated, swimming larvae (2). These burrow into a snail and multiply asexually (3). In time, many fork-tailed larvae develop (4). These leave the snail and swim until they encounter human skin (5). They bore inward and migrate to thin-walled intestinal veins, and the cycle begins anew. In infected humans, white blood cells that defend the body attack the masses of fluke eggs, and grainy masses form in tissues. In time, the liver, spleen, bladder, and kidneys deteriorate.

Tapeworms also parasitize humans. One species uses pigs as intermediate hosts; another uses cattle. Humans become infected when they eat insufficiently cooked pork or beef (Figure *b*). They become infected in other ways. For example, tiny crustaceans called copepods eat the larvae of one tapeworm species. The larvae escape digestion; then they mature in fishes that eat the copepods. Suppose

humans dine on raw, improperly pickled, or insufficiently cooked fish that were infected by the larvae. They can become hosts for the adult tapeworms.

Or consider the roundworms. One species that infects humans produces thin, serpentlike ridges at the surface of skin. For thousands of years, healers removed the "serpents" by winding them out slowly, painfully, around a stick. The symbol of the medical profession continues to be a serpent wound around a staff.

Then there are the types of roundworms called pinworms and hookworms. *Enterobius vermicularis*, a type of pinworm, parasitizes humans in temperate regions. It lives in the large intestine. At night, the centimeter-long female pinworms migrate to the anal region and lay eggs. Their presence causes itching, and scratchings made in response will transfer some eggs to other objects. Newly laid eggs contain embryos, but within a few hours they are juveniles and ready to hatch if another human inadvertently ingests them.

Hookworms are a serious problem in impoverished parts of the tropics and subtropics. Adult hookworms live in the small intestine. They feed on blood and other tissues after teeth or sharp ridges bordering their mouth cut into the intestinal wall. Adult females, about a centimeter long, can release a thousand eggs daily. These leave the body in

a Life cycle of a dangerous blood fluke, *Schistosoma japonicum*. The micrograph shows an adult male fluke.

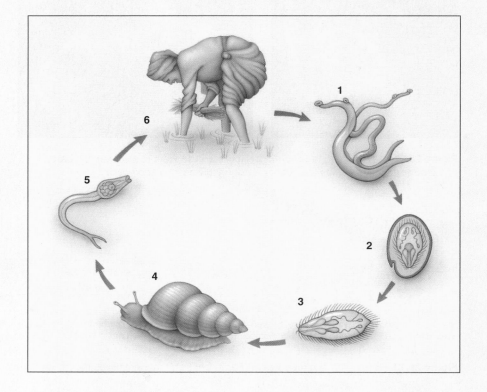

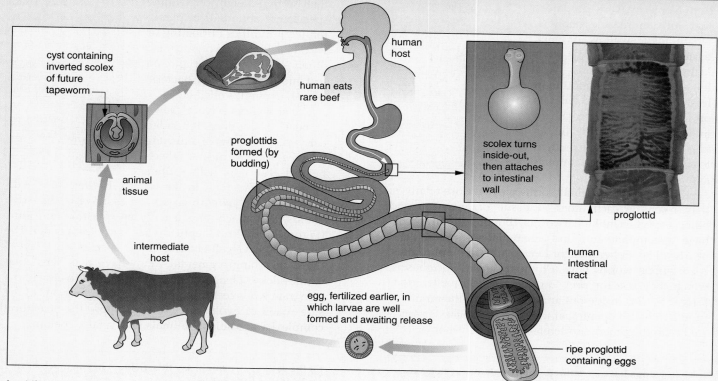

b Life cycle of a beef tapeworm, *Taenia saginata*.

cyst containing inverted scolex of future tapeworm

animal tissue

intermediate host

human eats rare beef

human host

proglottids formed (by budding)

scolex turns inside-out, then attaches to intestinal wall

proglottid

human intestinal tract

egg, fertilized earlier, in which larvae are well formed and awaiting release

ripe proglottid containing eggs

feces, then hatch into juveniles. A juvenile may penetrate the skin of a barefoot person. Inside a host, the parasite travels the bloodstream to the lungs, where it works its way into the air spaces. After moving up the windpipe, the parasite is swallowed. Soon it is in the small intestine, where it may mature and live for several years.

Another roundworm, *Trichinella spiralis*, causes painful, sometimes fatal symptoms. Adults live in the lining of the small intestine. Female worms release juveniles (Figure *c*), and these work their way into blood vessels and travel to muscles. There they become encysted (they produce a covering around themselves and enter a resting stage). Humans usually become infected by eating insufficiently cooked meat from pigs or certain game animals. The presence of encysted juveniles cannot easily be detected when fresh meat is examined, even in a slaughterhouse.

Figure *d* shows the results of prolonged, repeated infections by *Wuchereria bancrofti*, a roundworm. Adult worms live in the lymph nodes, where they can obstruct the flow of a fluid (lymph) that normally is returned to the bloodstream. The obstruction causes fluid to accumulate in legs and other body regions. These undergo grotesque enlargement, a condition called *elephantiasis*. A mosquito is the intermediate host. Female *Wuchereria* produce active young that travel the bloodstream at night. If a mosquito

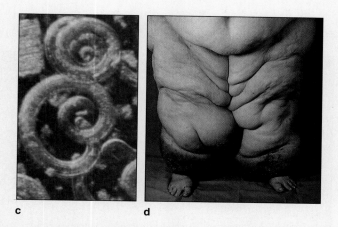

c Juveniles of a roundworm, *Trichinella spiralis*, living inside the muscle tissue of a host animal. (**d**) Legs of a woman who was infected by *Wuchereria bancrofti*, a roundworm that causes the disease elephantiasis.

sucks blood from an infected human, the juveniles may enter the insect's tissues. In time they move near the insect's sucking device, where they are ready to enter another human host when the mosquito draws blood again.

26.9 RIBBON WORMS

Nearly all **ribbon worms** (phylum Nemertea) are predators that lurk in marine habitats, where they swallow or suck tissue fluids from worms, mollusks, and crustaceans. Very few species have been found in freshwater or in damp habitats on land. The specimen shown in Figure 26.14 gives you an idea of why someone thought to name these bilateral animals after ribbons.

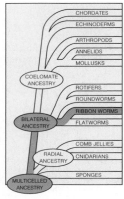

Ribbon worms and flatworms may be close relatives; they resemble each other in tissue organization and other respects. Unlike flatworms, though, ribbon worms have a complete gut and a circulatory system, and nearly all are either male or female, not hermaphrodites. Also, ribbon worms have a tubular device, a proboscis, which they use for prey capture. Muscle contractions force it to turn inside-out and project from the mouth or from a different opening at the head end. Glands associated with the proboscis secrete a paralytic venom.

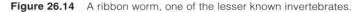

Ribbon worms have a complete gut, circulatory system, and other traits that are notable departures from flatworms.

26.10 ROTIFERS

Most rotifers (Rotifera) are less than a millimeter long. Some species are abundant in plankton (communities of floating or weakly swimming organisms); others crawl on algae and on wet mosses. They eat bacteria as well as single-celled algae.

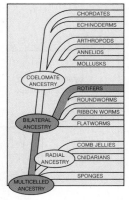

Most rotifers have a crown of cilia. Its motions reminded early microscopists of a turning wheel ("rotifer" means wheel-bearer). The cilia help rotifers swim and waft food to the mouth. Rotifers have salivary glands, a jawed pharynx, an esophagus, digestive glands, a stomach, and usually an intestine and anus (Figure 26.15). Protonephridia remove excess water. The head has nerve cell clusters. Some species have "eyes" (light-absorbing pigments). Many have two "toes" that exude sticky substances and allow rotifers to attach to a substrate while feeding. Rotifers are probably one of the side roads of evolution, but they show the structural complexity possible in animals with a false coelom.

Of the rotifers, we might say this: Seldom has so much been packed in so little space.

Figure 26.14 A ribbon worm, one of the lesser known invertebrates.

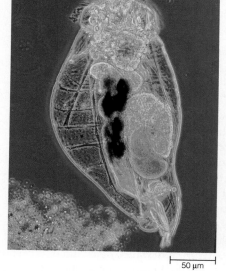

Figure 26.15 Lateral view of a rotifer, busily laying eggs. Males are unknown in many species; the females produce diploid eggs that develop into diploid females. Females of other species do the same, but they also produce haploid eggs that develop into haploid males. If a haploid egg happens to be fertilized by a male, it develops into a female. The male rotifers appear only occasionally, and are dwarfed and short-lived.

50 µm

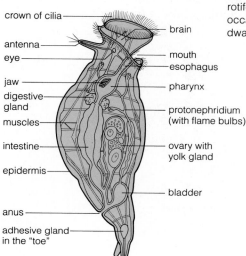

crown of cilia
antenna
eye
jaw
digestive gland
muscles
intestine
epidermis
anus
adhesive gland in the "toe"

brain
mouth
esophagus
pharynx
protonephridium (with flame bulbs)
ovary with yolk gland
bladder

26.11 A MAJOR DIVERGENCE

When we turn to the animals with coeloms, we find complexity on a spectacular scale. During Cambrian times, bilateral animals not much more complex than flatworms evolved. Not long afterward, they diverged into two distinct evolutionary lineages, the **protostomes** and **deuterostomes**. Mollusks, annelids, and arthropods represent the protostome lineage. Echinoderms and chordates represent the deuterostome lineage.

The two lineages differ partly in the way their embryos develop (Figure 26.16 and page 757). In deuterostomes, cell divisions divide a zygote as you might divide an apple, by cutting it into four wedges from top to bottom. Each wedge is cut in half at its midsection; then each piece is cut at *its* midsection. Cell divisions made parallel and perpendicular to an embryo's main body axis are called **radial cleavage**. In protostomes, the "wedges" are cut at oblique angles to the original axis, a pattern called **spiral cleavage**.

Besides this, the mouth and anus form differently, starting with openings at two indentations that appear in early embryos. In deuterostomes, the first opening becomes an anus, and the second becomes a mouth. In protostomes, the first opening becomes a mouth; an anus forms elsewhere. As a final example, a deuterostome coelom starts forming as outpouchings of the gut wall. A protostome coelom arises within solid masses of tissue at the sides of the gut.

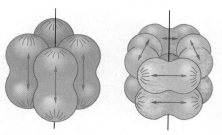

a Radial cleavage, a cell division pattern that transforms the zygotes of deuterostomes into the early embryo.

pouch destined to form mesoderm and enclose coelom
embryonic gut
mesoderm
coelom

c How the coelom forms in a deuterostome embryo. The diagrams show what you would see if you sliced an embryo through its midsection at two successive stages of development.

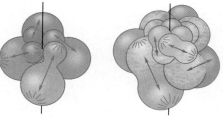

b Spiral cleavage, a cell division pattern that transforms the zygotes of protostomes into the early embryo.

solid mass of mesoderm
embryonic gut
mesoderm
coelom

d How the coelom forms in a protostome embryo.

Figure 26.16 Two characteristics of the early embryonic development of species that belong to the deuterostome and protostome branches of the animal family tree.

26.12 MOLLUSKS

With well over 100,000 known species, the fleshy, soft-bodied **mollusks** are the second largest invertebrate group. (Mollusca means "soft-bodied.") In this phylum we find the greatest array of marine invertebrates as well as successful occupants of land habitats. Among its better-known members are snails, clams, oysters, octopuses, and squids. These mollusks and others share the same bilateral ancestry, although the details are highly modified from one group to the next. Most have a head, foot, and visceral mass (mainly a gut and a heart, kidneys, and reproductive organs). Species with a well-developed head also have eyes and tentacles.

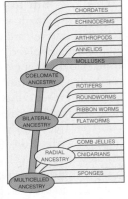

A distinctly molluscan feature is a **mantle**, a tissue fold that hangs, rather like a skirt, around some or all of the body (Figure 26.17). Another is a pair of gills, of a kind called ctenidia (singular, ctenidium). Most mollusks have these respiratory organs, each a series of thin-walled leaflets across which oxygen and carbon dioxide are exchanged. Most mollusks have a shell (or an evolutionary remnant of one) composed mainly of calcium carbonate. Many have a radula, a tonguelike organ with hard teeth that preshreds food.

Gastropods

Snails and slugs make up the largest molluscan group, the gastropods ("belly foots"). They are so named because their foot spreads out when the animal crawls about. Many of the snails have spirally coiled or cone-shaped shells (Figure 26.17a). Coiling is a way of compacting the organs into a mass that can be balanced above the rest of the body, much as you would balance a backpack full of books.

As most young gastropods grow, some body parts undergo a strange internal realignment. In a process called torsion, certain muscles contract and different body parts grow at different rates. The outcome? The posterior mantle cavity twists around to the right and then forward, until it is above the head:

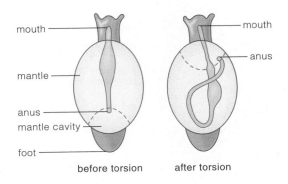

before torsion   after torsion

Figure 26.17 Features of the molluscan body plan, as represented by snails. (**a**) A land snail. (**b**) Body plan of an aquatic snail. (**c**) Snails and some other mollusks have a radula, which is protracted and retracted in a rhythmic way. Food is rasped off and drawn to the gut on the retraction stroke.

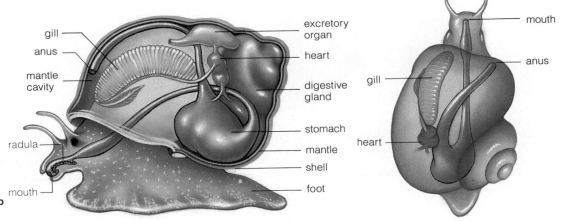

a

b

Figure 26.18 Two of the soft-bodied gastropods called nudibranchs.

(**a**) A sea hare (*Aplysia*), widely used in experimental studies of behavior and physiology. The two flaps over its dorsal surface are extensions of the foot. They undulate and help ventilate the underlying mantle cavity. Like many other gastropods, *Aplysia* is hermaphroditic. It may act as a male, a female, or both during encounters with others like itself. Neurobiologists are much taken with *Aplysia* and have tracked several of its nerve pathways. (**b**) These two sea slugs are "Mexican dancers," engaged in mating behavior.

The twisting produces a cavity into which the head can withdraw in times of danger. It also brings the gills, anus, and kidney openings above the head. This could create a sanitation problem, given that wastes are dumped near the respiratory structures and the mouth. In most species, the beating of cilia in this region creates currents that sweep wastes away.

Some of the gastropods, including the colorful nudibranchs (sea slugs), seem to have undergone "detorsion" during their evolution. To some extent, the larval body still twists. Then muscle contractions and asymmetrical growth make the body untwist. Also among nudibranchs, the mantle cavity largely disappeared. So did the ctenidia, although other outgrowths function as secondary gills in most species. Figure 26.18 shows the array of outgrowths on a few of the more striking nudibranchs.

Chitons

Chitons are slow-moving or sedentary marine mollusks. Most species use a radula to graze on algae, hydrozoans, and other low-growing organisms. A large, broad foot enables them to creep over and cling tenaciously to rocks and other hard surfaces. Chitons have a dorsal shell divided into eight plates (Figure 26.19). The divisions impart flexibility. When a chiton is disturbed or exposed by a receding tide, its foot muscles can pull down the body mass tightly. The mantle's edge, which more or less covers the plates, now functions like the rim of a suction cup, so the chiton is difficult to detach. If you manage to separate a pulled-down chiton from its substrate, it will roll up into a ball, protecting itself until it can safely unroll and become reattached elsewhere.

Figure 26.19 Chitons—one variation on the basic molluscan plan. Walk along a rocky shore and you will probably see chitons. The ones shown here live in the intertidal zone of Monterey Bay, California. Members of this class of mollusks show beautiful variations in the color and patterns of their elongated shells.

Bivalves

Clams, scallops, oysters, and mussels are well-known bivalves (animals with a "two-valved shell"). Humans have been eating one type or another since prehistoric times. The shell of many bivalves, including a few pearl-producing types, has an inner lining of iridescent mother-of-pearl. Some species are only 1 or 2 millimeters across. A few giant clams of the South Pacific are more than a meter across and weigh 225 kilograms (close to 500 pounds).

The bivalve head is not much to speak of, but the foot is usually large and specialized for burrowing. In nearly all bivalves, gills function in collecting food and in respiration. As cilia in the gills move water through the mantle cavity, mucus on the gills traps tiny bits of food (Figure 26.20). Tracts of cilia move the mucus and food to the palps, where final sorting takes place before acceptable bits are driven to the mouth.

Bivalves hunkered in mud or sand have a pair of siphons (extensions of the mantle edges, fused into tubes). Water is drawn into the mantle cavity through one siphon and leaves through the other, carrying wastes from the anus and kidneys. Siphons of the giant geoduck (pronounced "gooey duck") of the Pacific Northwest may be more than a meter long.

Scallops and a few other bivalves can swim by clapping the valves together and producing a localized jet of water that propels the animal for some distance (Figure 26.21).

Cephalopods

The smallest mollusks are snails and clams less than a millimeter long. Among the cephalopods we find species at the other extreme. The giant squid, which may measure 18 meters (about 60 feet) from tip to tip, is the largest invertebrate known. Also among the cephalopods are the swiftest invertebrates (jet-propelled squids) and the smartest ones (octopuses). Squids, octopuses, nautiluses, and cuttlefish—these are all fast-swimming predators of the oceans. Figures 26.22 and 26.23 show representative species.

Instead of a foot, cephalopods have tentacles, usually equipped with suction pads of the sort shown in Figure 26.23. They use the tentacles to capture prey, a radula to draw it into the mouth, and a beaklike pair of jaws to bite or crush it. Venomous secretions often speed the captive's death.

Cephalopods move rapidly by a kind of jet propulsion. They force a stream of water out of the mantle cavity through a funnel-shaped siphon. When muscles in the mantle relax, water is drawn into the mantle cavity. When they contract, a jet of water is squeezed out. If, at the same time, the free edge of the mantle is closed down on the animal's head and siphon, water shoots

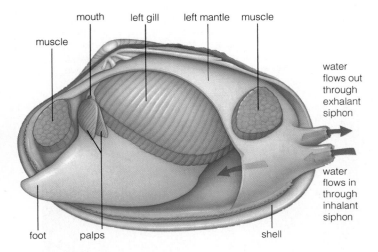

Figure 26.20 Body plan of a clam, with half of its shell (the left valve) removed for this diagram. The palps sort particles of food that become trapped in mucus on the gills. Cilia on the palps sweep suitable tidbits to the mouth.

Figure 26.21 A scallop moving away from a predator, a sea star. By clapping their valves together, scallops and a few other bivalves create a strong jet of water that can propel them for some distance.

out in a jet through the siphon. By manipulating the siphon, the animal partly controls the direction of its own movement.

Being highly active, cephalopods have great demands for oxygen, and they alone among the mollusks have a *closed* circulatory system (page 654). In such systems, blood is confined within the walls of one or more hearts and blood vessels. In cephalopods, blood is pumped from the main heart to the two gills. Each gill has an accessory, "booster" heart at its base that speeds the flow and therefore the uptake of oxygen and elimination of carbon dioxide.

Even when a cephalopod rests, it rhythmically draws water into its mantle cavity and then expels it. The water circulation enhances gas exchange and also helps to eliminate wastes that excretory organs and the anus discharge into the mantle cavity.

Cephalopods have a well-developed nervous system. Compared to other mollusks, they have the largest brains relative to body size. Giant nerve fibers connect the brain with muscles used in jet propulsion, making it possible for a cephalopod to respond quickly to food or danger. Cephalopod eyes bear some resemblance to yours, although they are formed in a different way. Finally, cephalopods can learn. Give an octopus a mild electric shock after showing it an object with a distinctive shape, for example, and it will thereafter avoid the object. In terms of memory and learning, the cephalopods are the world's most complex invertebrates.

Cephalopods have separate sexes. Male squids even rush at each other in mock battles that may get serious enough for a loser to lose an arm.

Mollusks make up the second largest animal phylum, next to arthropods, and have the greatest array of marine species.

All mollusks are soft-bodied, bilateral animals with a shared ancestry, but they vary enormously in size, life-styles, and details of their body plan.

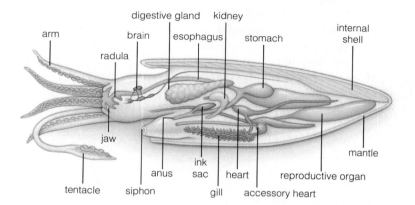

Figure 26.22 General body plan of a cuttlefish, one of the cephalopods. The tentacles are more slender than the arms and are specialized for prey capture. The photograph shows a female and a male cuttlefish (*Sepia*) head end to head end. They are engaged in mating behavior.

Figure 26.23 Another cephalopod—an octopus.

26.13 ANNELIDS

When heavy rains soak the soil, earthworms make their presence known by wriggling to the surface and so avoid drowning. At such times, you may suddenly notice these worms, which otherwise leave their burrows only at night. They are among the 15,000 or so species of **annelids** (Annelida). Also in this group are polychaetes (which are far more diverse but less familiar) and a few species

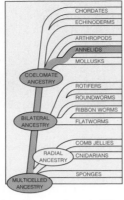

of leeches (some of which are the most notorious). The name of their phylum means "ringed forms." However, what appear to be rings actually are a series of body segments.

Annelids are bilateral worms with pronounced segmentation. Except for leeches, each side of their body has pairs or clusters of chitin-reinforced bristles on nearly every segment. Sometimes the bristles are called *setae* (a Latin term) or *chaetae* (a latinized Greek term), but these are merely formal names for bristles.

The bristles provide the traction required for crawling and burrowing through soil, and in swimming species they often are broadened into paddles. Although earthworms have only a few setae per body segment, the marine polychaete worms typically have many of them (*poly-* means many).

Earthworms: A Case Study of Annelid Adaptations

Earthworms, of the sort shown in Figure 26.24, are the usual textbook examples of annelids. They belong to a class of worms called oligochaetes, which has land-dwelling and aquatic members. The earthworms are scavengers on land. They burrow in moist soil and mud, where they feed on decomposing plant material and other organic matter. Every twenty-four hours, an earthworm can ingest its own weight in soil. Earthworms aerate soil, to the benefit of many plants. They also make nutrients available to other organisms by carrying subsoil to the surface.

As is true of most annelids, an earthworm's body segments have a cuticle, a layer of secreted material that wraps around the body's outer surface (Figure 26.25). The annelid cuticle is thin and flexible. It can bend easily, and it also permits the exchange of gases to and from the body surface. Inside the body, partitions separate the segments

from one another, so that there is a series of coelomic chambers. The gut extends continuously through all the chambers, from the earthworm's mouth to its anus.

As for other annelids, the coelomic chambers serve as a hydrostatic skeleton against which muscles can operate. As Figure 26.25*a* shows, circular muscles are located mostly within the body wall of each segment. Longitudinal muscles span several segments. When these contract and the circular muscles relax, segments shorten and fatten. The segments lengthen when the pattern reverses.

Muscle contractions are not enough to move an earthworm forward. Locomotion also involves the protraction and retraction of bristles on different segments. The body moves forward when its first few segments elongate. The segments just behind them plunge their bristles into the ground and hold their position. Then the first segments contract and plunge their bristles into the ground. Simultaneously, the segments behind them retract their bristles and are pulled forward. Alternating contractions and elongations proceed along the length of the body and move the worm forward.

A brain at the head end integrates sensory input for the whole worm (Figure 26.25*c*). Leading away from the brain is a double nerve cord that extends the length of the body. A **nerve cord** is a bundle of slender extensions of nerve cell bodies. In each segment, the cord broadens into a **ganglion** (plural, ganglia), a cluster of nerve cell bodies that controls local activity, as described in Chapters 34 and 35.

Figure 26.24 A well-known annelid, an earthworm.

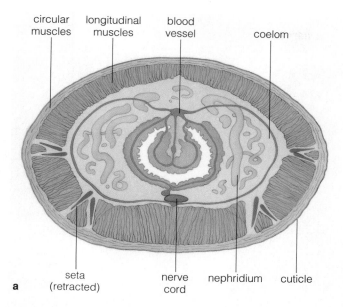

circular muscles · longitudinal muscles · blood vessel · coelom

seta (retracted) · nerve cord · nephridium · cuticle

a

Figure 26.25 Body plan of an earthworm, a representative annelid. (**a**) Section through the midbody. (**b**) Head end of the digestive system. (**c**) A portion of the nervous system. (**d**) Portion of the closed circulatory system, which is functionally linked with a system of nephridia. The nephridium in (**e**) is one of many functional units that maintain the volume and composition of body fluids.

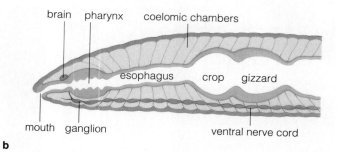

brain · pharynx · coelomic chambers

esophagus · crop · gizzard

mouth · ganglion · ventral nerve cord

b

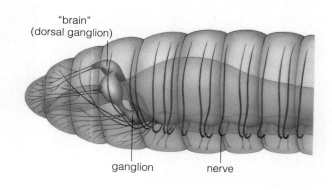

"brain" (dorsal ganglion)

ganglion · nerve

c

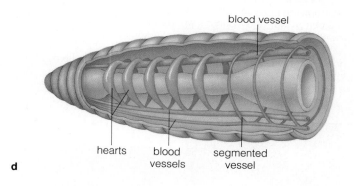

blood vessel

hearts · blood vessels · segmented vessel

d

Like most annelids, earthworms have a closed circulatory system (Figure 26.25*d*). Contractions of muscularized blood vessels keep blood circulating in one direction. Smaller blood vessels lead to and from the gut, nerve cord, and body wall.

A system of **nephridia** (singular, nephridium) regulates the volume and composition of body fluids. In many annelids, the units of this system have cells similar to flame cells, which may imply an evolutionary link between flatworms and annelids. More commonly, the beginning of each nephridium is a funnel-like structure that collects fluid from a coelomic chamber. The funnel leads into a tubular portion of the nephridium that carries the fluid to a pore at the body surface (Figure 26.25*e*). The funnel is located in one coelomic chamber, but its terminal pore is in the body wall of the next chamber in line.

From comparisons of the tissues and organ systems of annelids with those of other animals, it seems likely that annelids were an early offshoot of the protostome branch of the family tree. Their features remind us of developments that led to increased size and more complex internal organs.

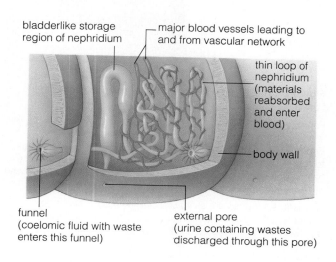

bladderlike storage region of nephridium · major blood vessels leading to and from vascular network

thin loop of nephridium (materials reabsorbed and enter blood)

body wall

funnel (coelomic fluid with waste enters this funnel) · external pore (urine containing wastes discharged through this pore)

e

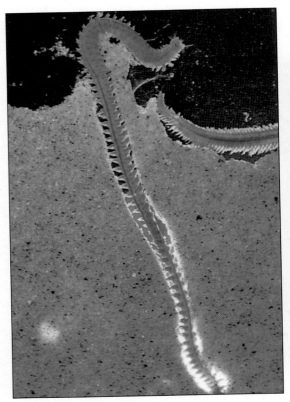

a

b

Figure 26.26 (a) A marine polychaete in a burrow of its own making. The profusion of bristles along the sides of its body grip the burrow wall and aid in the worm's movements. (b) A tube-dwelling polychaete with featherlike structures that are coated with mucus. After the mucus traps bacteria and other bits of food, the coordinated beating of cilia sweeps them to the nearby mouth.

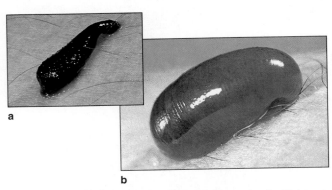

a

b

Figure 26.27 A leech before and after gorging itself with blood from a human host.

For at least 2,000 years, medical practitioners have used *Hirudo medicinalis*, a freshwater leech, as a blood-letting tool. Especially during the nineteenth century, doctors used leeches to "cure" problems ranging from nosebleeds to obesity. By one estimate, leeches were drawing more than 300,000 liters of blood annually from patients in France alone.

Nowadays, leeches are still used, but more selectively. For example, after surgeons reattach a severed ear, lip, or fingertip, leeches may be employed to draw off pooled blood. The body cannot do this on its own until the severed blood circulation routes are reestablished.

Polychaetes

The many-bristled polychaetes are the most diverse annelids. Most live in marine habitats. In fact, they are the most common animals along coasts, although you won't find them unless you turn over rocks or dig them up. Some excavate burrows in sand or soft mud; others build tubes from calcium carbonate secretions or from sand grains or bits of shells. Some are attached to substrates.

Sea nymphs and some other crawling polychaetes also can swim by undulating their long body. The swimmers typically have a specialized head end with notable sensory organs, sharp jaws and a muscular pharynx, and other structures. Their bristles project from well-developed parapods, a word taken to mean "closely resembling feet." Unlike your feet, parapods are fleshy-lobed appendages reinforced with chitin, not bone.

Many polychaetes dine on small invertebrates, some graze on algae, and others scavenge organic matter. The predatory types evert ("pop out") their jawed pharynx to capture prey. Herbivorous types grasp and tear algae. Most of the tube dwellers collect food suspended in the water with the help of ciliated, mucus-coated tentacles or featherlike structures (Figure 26.26).

Leeches

Leeches (Figure 26.27) are among the few annelids with no bristles; they inch about by using suckers at both ends of their muscular, segmented body. Different kinds

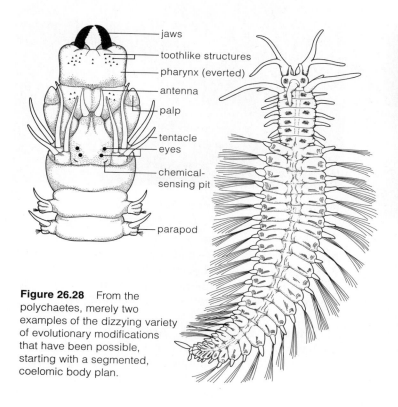

- jaws
- toothlike structures
- pharynx (everted)
- antenna
- palp
- tentacle
- eyes
- chemical-sensing pit
- parapod

Figure 26.28 From the polychaetes, merely two examples of the dizzying variety of evolutionary modifications that have been possible, starting with a segmented, coelomic body plan.

Figure 26.29 A velvety "walking worm" of the small phylum Onychophora, the members of which resemble both the annelids and the arthropods. These worms live mostly in humid forests, especially of the tropics. They hide by day under rotting logs and in leaf litter and hunt for prey at night. The largest is about 15 centimeters (nearly 6 inches) long. Depending on the species, there are between fourteen and forty-three pairs of stumpy, unjointed legs with claws at the tip (onychophoran means "claw bearer").

Like annelids, these worms have ciliated nephridia. Like arthropods, they molt and they have an open circulatory system, with blood entering spaces in body tissues and reentering the heart through pores. Moreover, they have tracheal tubes similar to those of millipedes, centipedes, and insects.

Such similarities do not mean that onychophorans are "missing links" between annelids and arthropods. They are probably one of many invertebrates that evolved from annelid ancestors and have survived to the present.

live in freshwater, the seas, and, in tropical regions, moist habitats on land. Most swallow small animals or kill them and suck out their juices.

The leeches that most people have heard about feed on vertebrate blood (Figure 26.27). Blood-sucking leeches have sharp jaws. They plunge a sucking apparatus into incisions they make in prey, then secrete a substance that prevents blood from coagulating while they eat. The leech gut has many side branches for storing food taken in during a big meal, this being handy for an animal that may have long waits between meals.

Regarding the Segmented, Coelomic Body Plan

When you look carefully at Figure 26.28, you can sense that a segmented, coelomic body plan must be terrific working material for evolutionary forces. When the body is divided this way, different segments can become specialized in the performance of different tasks.

Most of the segments in an annelid are much the same, but among some species we have evidence of evolutionary forays into specialization. Leeches have suckers at both ends. Polychaetes have specialized parapods, and many have truly spectacular elaborations at their head end. In sea nymphs and their relatives, differentiated segments at the posterior end of the body perform reproductive functions.

Long ago, among the ancestors of annelids and other protostomes, evolutionary experimentation must have been rampant. We find further evidence of this in the onychophorans, a small, distinct phylum that seems to have evolutionary links to annelids and some adaptations like those of arthropods (Figure 26.29).

1. Annelids include the earthworms, polychaetes, and leeches. All are elongated, bilateral worms having segmentation and a true coelom.

2. Annelids resemble flatworms in some features, including their water-regulating system. They resemble arthropods in certain aspects of their segmented, coelomic body plans.

3. Given their array of traits, the annelids were probably one of the first groups to evolve on the protostome branch of the animal family tree.

26.14 ARTHROPODS

Of all invertebrate phyla, the arthropods (Arthropoda) show the most diversity. We have discovered more than a million species, most of which are insects. New ones are being discovered weekly, especially in the tropics. It seems likely that ancient annelids (or animals very much like them) gave rise to arthropod lineages. The trilobites, one of the four major lineages, are now extinct (Figure 21.10d). The other three lineages are the chelicerates (such as spiders), crustaceans (such as crabs), and uniramians (the centipedes, millipedes, and insects).

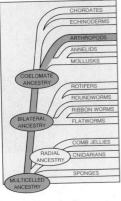

Figure 26.30 Molting of a centipede exoskeleton.

Adaptations of Insects and Other Arthropods

Arthropods are truly abundant, they engage in enormously different life-styles, and they have been evolutionary success stories in nearly all habitats. Six adaptations contributed to the success of the group in general and insects in particular:

1. A hardened exoskeleton
2. Specialized segments
3. Jointed appendages
4. Specialized respiratory structures
5. Efficient nervous system and sensory organs
6. Division of labor in the life cycle

Hardened Exoskeletons   The arthropod cuticle is actually an external skeleton, or **exoskeleton**. Although it may be hardened, its protein and chitin components make it light, flexible—and protective. Think of the calcium-stiffened exoskeleton of a lobster or crab. It is like armor plating. Probably exoskeletons evolved as defenses against predators. They took on added functions when certain arthropods first invaded the land. An exoskeleton restricts evaporative water loss, and it can support a body deprived of water's buoyancy. It does restrict increases in size. But arthropods grow in spurts, by **molting**. At different stages in the life cycle, they grow a new, soft exoskeleton under the old one, which they shed (Figure 26.30). Before this new one hardens, aquatic arthropods enlarge by swelling with water; land-dwelling types swell with air.

Specialized Segments   As arthropods evolved, body segments became fewer in number, grouped or fused in various ways—and more specialized in function. For example, in insects, different segments combined to form a head, thorax, and abdomen. In spiders, segments fused as a forebody and hindbody.

Jointed Appendages   Arthropods have highly specialized, jointed appendages. (Arthropod means "jointed foot.") For example, some paired appendages are used in feeding and detecting stimuli. Others are used in locomotion, feeding, sensing stimuli, transferring sperm to females, or spinning silk.

Respiratory Structures   Aquatic arthropods rely on gills for gas exchange. The evolution of tracheas helped bring about diversity among the land-dwelling arthropods, insects especially. Insect **tracheas** begin as pores on the body surface and branch into narrow tubes that deliver oxygen directly to body tissues (page 698). They help support such energy-consuming activities as insect flight.

Specialized Sensory Structures   Intricate eyes and other sensory organs contributed to arthropod success. Many species have a wide angle of vision and can process visual information from many directions.

Division of Labor   Many insects have a division of labor among different stages of their life cycle. They first develop as sexually immature larvae that molt and change as they grow. Then larvae enter a pupal stage, which involves massive reorganization and remodeling of body tissues. Growth and major transformation of a larva into the adult form is called **metamorphosis**.

Moths, butterflies, beetles, and flies are examples of metamorphosing insects (Figure 1.4). Their larval stages specialize in *feeding and growth*, whereas the adult is concerned mostly with *dispersal and reproduction*. This division of labor is adaptive to seasonal changes in the environment, such as variations in food sources.

With this overview in mind, let's now sample the diversity among the major groups of arthropods.

Figure 26.31 Chelicerates. (**a**) Horseshoe crab, one of the few living aquatic chelicerates. Beneath its hard, shieldlike cover are five pairs of legs, a defining feature of the group. (**b**) Wolf spider. Like most spiders, it plays a role in keeping insect populations in check and is harmless to humans. (**c**) Brown recluse. Its bite can be severe to fatal for humans. This North American spider lives under bark and rocks, and in and around buildings. A violin-shaped mark is present on its forebody. (**d**) A female black widow. Its bite can be painful and sometimes dangerous. (**e**) A book lung, a type of respiratory organ of many spiders and scorpions.

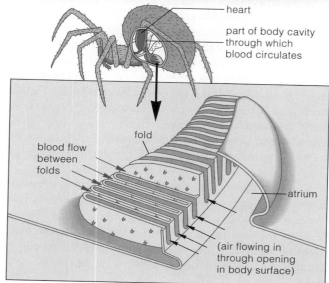

heart

part of body cavity through which blood circulates

fold

blood flow between folds

atrium

(air flowing in through opening in body surface)

Chelicerates

Chelicerates originated in shallow seas in the early Paleozoic. Except for a few mites, the only marine survivors are the horseshoe crabs (Figure 26.31a) and sea spiders. Of the familiar existing chelicerates—spiders, scorpions, ticks, and mites—we might say this: never have so many been loved by so few.

Like scorpions, spiders are predators, and many help control populations of insect pests. Ever so often, venomous types bite humans, and they have given the whole group a bad reputation. Ticks are blood-sucking parasites of vertebrates. Some transfer the bacteria that cause Rocky Mountain spotted fever, Lyme disease, and other diseases to humans (page 354). Most mites are free-living scavengers.

Spiders and scorpions pump digestive enzymes from their gut into prey, then suck up the liquefied remains. Their forebody has several eyes, four pairs of legs, a pair of chelicerae (which inflict wounds and discharge venom), and a pair of pedipalps (for grasping). Appendages on the spider hindbody spin out threads of silk for webs and egg cases. Most webs are netlike, but one spider spins a vertical thread with a ball of sticky material at the end. It uses one of its legs to swing the ball at passing insects!

Spiders and scorpions have pocketlike respiratory organs called book lungs (Figure 26.31e). The folded walls of the pockets resemble book pages, and they greatly increase the surface area available for gas exchange. Blood circulating between the moist folds picks up oxygen diffusing in from the surrounding air.

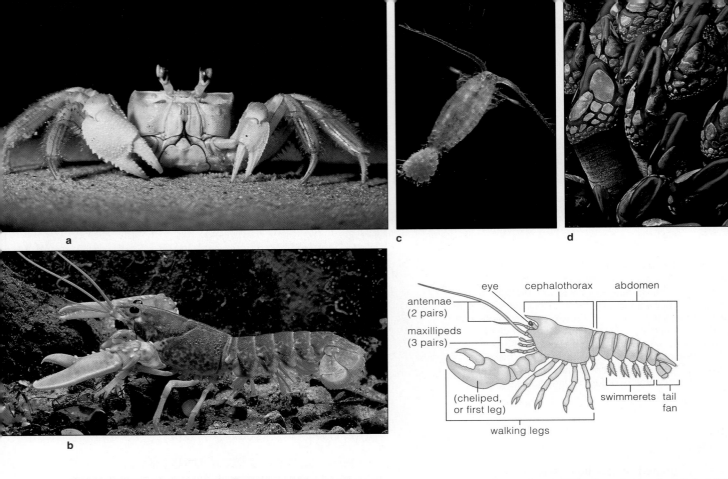

Figure 26.32   Representative crustaceans. (**a**) A crab and (**b**) a lobster, displaying their splendid appendages. (**c**) A marine copepod. (**d**) Stalked barnacles that grow in intertidal zones of the Pacific Ocean.

Diagram labels: antennae (2 pairs), maxillipeds (3 pairs), eye, cephalothorax, abdomen, (cheliped, or first leg), walking legs, swimmerets, tail fan

Crustaceans

Shrimps, crayfishes, lobsters, crabs, and pillbugs are well-known crustaceans. Most live in the seas; some live in freshwater or on land. All have roles in food webs, and humans harvest many edible types.

The simplest crustaceans have unspecialized body segments and may resemble their ancient ancestors. In most crustaceans, many segments have become highly modified. For example, crabs have strong claws that shred seaweed, collect organic debris, intimidate other animals, and sometimes dig burrows. Barnacles have feathery appendages that comb microscopic bits of food from the water.

Crustaceans commonly have sixteen to twenty segments; some have over sixty. The segmentation is not obvious in crabs, lobsters, and some other types because the head's dorsal cuticle extends backward, covering some or all segments with a shieldlike cover (carapace). The head has two pairs of antennae with mostly sensory functions, although in some species they are used in swimming. It has a pair of jawlike

appendages (mandibles) and a pair of food-handling appendages (maxillae). Figure 26.32*a* and *b* shows how these parts are arranged in crabs and lobsters.

With their ten pairs of legs projecting from the thorax, the crabs, lobsters, shrimps, crayfish, and their relatives are decapods (*deca*- means ten). Compared to other decapods, crabs are rather broad and flat, but their weight is distributed so evenly over their legs that they are agile walkers.

Figure 26.32*c* shows one of the 7,500 species of copepods, which are less than 2 millimeters long. A single eye in the middle of their head is a distinctive feature. These crustaceans are the most abundant aquatic animals. About 1,500 species are parasites of fishes and various invertebrates. The rest are weakly swimming members of plankton. Huge numbers of them feed on tiny photosynthetic protistans. By converting this food into their own tissues, they become food themselves for other organisms, and so on up through vital aquatic food webs.

Of all the arthropods, only barnacles have a strong, calcified "shell." It is actually a modified exoskeleton, and it protects them from predators, drying out, and the force of waves or currents. Some barnacles attach to rocks and other surfaces by an elongated stalk (Figure 26.32*d*). Others are stalkless forms that attach directly to rocks, wharf pilings, ship hulls, and similar surfaces. A few species prefer to attach themselves only to the skin of whales.

Insects and Their Relatives

Millipedes and Centipedes Closely related to insects are the millipedes and centipedes, which typically live under damp rocks, logs, or forest litter. As Figure 26.33 shows, both kinds of arthropods have a long, segmented body with many legs. Millipedes have a cylindrical or slightly flattened body. Different species range from about 2 millimeters to nearly 30 centimeters in length. All are slow-moving, nonaggressive scavengers of decaying plant material.

Centipedes have a streamlined, flattened body. All are fast-moving, aggressive carnivores, complete with fangs and venom glands. They prey on insects, earthworms, and snails, and some tropical species can subdue small lizards and toads. A few centipedes inflict painful bites.

Insects As a group, insects share the adaptations listed on page 432. Here we add a few details to the list. Insects have a head, a thorax, and an abdomen. The head has a pair of sensory antennae and paired mouthparts, specialized for biting, chewing, sucking, or puncturing (Figure 26.34). The thorax has three pairs of legs and usually two pairs of wings. Usually the only abdominal appendages are reproductive structures, such as egg-laying devices.

The insect gut is divided into foregut, midgut, and hindgut. Most digestion proceeds in the midgut; water is reabsorbed in the hindgut. Insects get rid of waste material through **Malpighian tubules**, small tubes that connect with the midgut. Nitrogen-containing wastes from protein breakdown diffuse from the blood into the tubules, where they are converted into harmless crystals of uric acid. The crystals enter the midgut and are eliminated with feces. This system allows land-dwelling insects to get rid of potentially toxic wastes without losing precious water.

a

b

Figure 26.33 (**a**) A mild-mannered millipede that scavenges on decaying plant parts. (**b**) An aggressive centipede of Southeast Asia that preys on small frogs and lizards.

Figure 26.34 From the insects, examples of specialized arthropod appendages. These headparts are used in feeding.

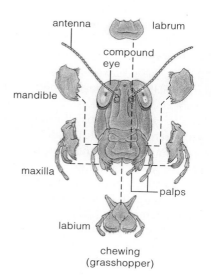

chewing
(grasshopper)

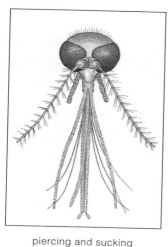

piercing and sucking
(mosquito)

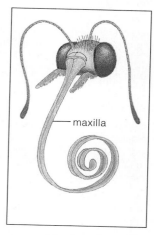

siphoning tube
(butterfly)

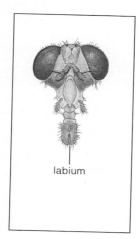

sponging
(housefly)

Figure 26.35   Representative insects.

(**a**) Mediterranean fruit fly (order Diptera). Its larvae destroy citrus fruit and other valuable crops.

(**b**) Flea (order Siphonaptera), with big strong legs for jumping onto and off animal hosts.

(**c**) A honeybee (order Hymenoptera) attracting its hive mates with a dance, as described in Chapter 52.

(**d**) Duck louse (order Mallophaga). It eats feather particles and bits of skin.

(**e**) Off with the old, on with the new. On the side of a tree trunk, a green cicada (order Homoptera) wriggling out of its outgrown cuticle during a molting cycle.

(**f**) A male praying mantid (order Mantodea) mating with a larger female. He may be eaten during or immediately after mating with her.

(**g**) European earwig (order Dermaptera), a common household pest.

(**h**) Stinkbugs (order Hemiptera), newly hatched.

(**i**) Ladybird beetles (order Coleoptera) swarming. These beetles are raised commercially and released in great numbers as biological controls of aphids and other pests. Also in this order, the scarab beetle (**j**). With more than 300,000 named species, Coleoptera is the largest order in the animal kingdom.

(**k**) Luna moth (order Lepidoptera), a flying insect of North America. Like most other moths and butterflies, its wings and body are covered with microscopic scales.

(**l**) Dragonfly (order Odonata). This swift aerialist captures and eats other insects in midflight.

Insect Diversity

More than 800,000 species of insects have already been catalogued. Figure 26.35 shows just a few representatives from the major orders of insects.

If we use sheer numbers and distribution as the criteria, which are the most successful insects? The most successful types are small in size and have a staggering reproductive capacity. Many might grow and reproduce in great numbers on a single plant that might be only an appetizer for another animal. By one estimate, if all the progeny of a single female fly were to survive and reproduce through six more generations, that fly would have more than 5 trillion descendants! Metamorphosis also is common in insect life cycles. This developmental

h

i

j

k

l

route allows the young and the adults to use different resources at different times.

Besides this, the most successful insect species are winged. In fact, they are the *only* invertebrates that have wings. They are able to move among food sources that are too widely scattered to be exploited by other animals. Their capacity for flight contributed greatly to their success on land.

The factors that contribute to their success also make insects our most aggressive competitors. They destroy vegetable crops, stored food, wool, paper, and timber. They draw blood from us and from our pets, and transmit pathogenic microorganisms as they do so. On the bright side, many insects pollinate flowering plants in general and crop plants in particular. And many "good" insects attack or parasitize the ones we would rather do without.

This completes our survey of the insects and other arthropods. Although our sampling necessarily has been limited, it has been enough to reinforce the following key point:

Arthropods, especially insects, owe their evolutionary success to hardened exoskeletons, specialized segments with jointed appendages, efficient respiratory and sensory structures, and often a division of labor in the life cycle.

26.15 ECHINODERMS

We turn now to a different lineage of coelomate animals, the deuterostomes. The major invertebrate members of this lineage include **echinoderms** (Echinodermata, meaning spiny skinned). Sea urchins, sea cucumbers, feather stars, brittle stars, and sea stars belong to this phylum (Figure 26.36). All contain calcium-stiffened spines, spicules, or plates in their body wall.

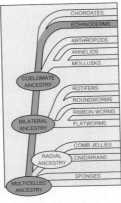

Notice the radially arranged arms of the sea star in Figure 26.36*e*. Oddly, adult echinoderms are radial with some bilateral features. Some even produce bilateral larvae during their life cycle. (Did bilateral invertebrates give rise to the ancestors of echinoderms? Maybe.)

Adult echinoderms have no brain. However, their decentralized nervous system does allow them to respond to stimuli coming from different directions. For instance, any arm of a sea star that senses food or danger can become the leader, directing the rest of the body to move in a suitable direction.

If you turn a sea star over, you will see **tube feet**. These are fluid-filled, muscular structures with suckerlike adhesive disks. Tube feet are used for walking, burrowing, clinging to a rock, or gripping a clam or snail about to become a meal. They are part of a **water vascular system** unique to echinoderms (Figure 26.37). In sea stars, the system includes a main canal in each arm. Short side canals extend from them and deliver water to the tube feet. Each tube foot has an ampulla, a fluid-filled, muscular structure something like the rubber bulb on a medicine dropper. When the ampulla contracts, it forces fluid into the foot, which thereby lengthens.

Tube feet change shape constantly as muscle action redistributes fluid through the water vascular system. Hundreds of tube feet may move at a time. After being released, each one swings forward, reattaches to the substrate, then swings backward and is released before swinging forward again.

Some sea stars swallow their prey whole. Many can push part of their stomach outside the mouth and around their prey, then start digesting the prey even before swallowing it. Sea stars get rid of coarse, undigested residues through the mouth. They do have a small anus, but this is of no help in getting rid of empty clam or snail shells.

With their curious assortment of traits, the echinoderms are a suitable point of departure for this chapter.

Figure 26.36 Echinoderms. (**a**) Sea urchin, which moves about on spines and tube feet. The spines also protect it from predators. If you step on a sea urchin and a spine breaks off under your skin, it can trigger inflammation. (**b**) Sea cucumber, with rows of tube feet along its body. (**c**) Feather star, with its finely branched food-gathering appendages. (**d**) Brittle stars. Their arms (rays) make rapid, snakelike movements. (**e**) Five-armed sea star, and a close view of its tube feet (**f**) and of the feeding apparatus on its ventral surface (**g**).

Even though we can identify broad trends in animal evolution, we should keep in mind that there are always confounding exceptions to the general rule.

Echinoderms, one of evolution's puzzling experiments, are mostly radial (but partly bilateral) animals with a coelom and with spines, spicules, or plates in the body wall.

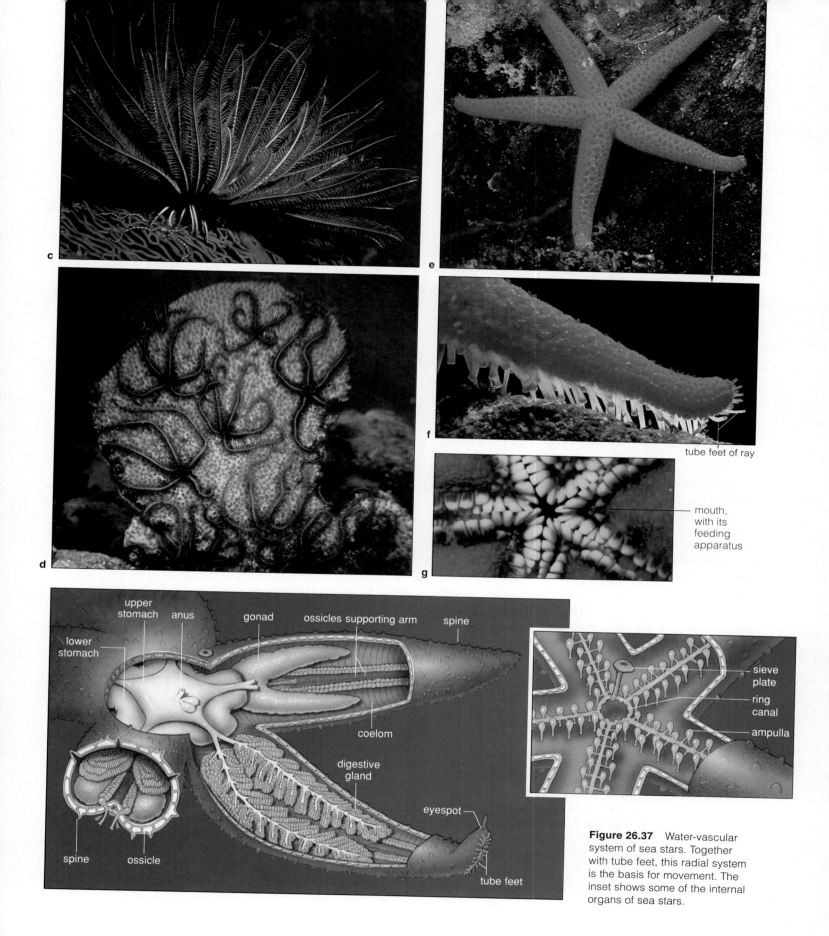

tube feet of ray

mouth, with its feeding apparatus

Figure 26.37 Water-vascular system of sea stars. Together with tube feet, this radial system is the basis for movement. The inset shows some of the internal organs of sea stars.

SUMMARY

1. Multicellular animals range from sponges to vertebrates. The members of each animal phylum share certain characteristics. By comparing phyla and integrating the information with fossil evidence, we can identify major trends that unfolded during the evolution of different groups.

2. The most revealing aspects of the animal body plan are its type of symmetry, gut, and cavity (if any) between the gut and body wall; whether it has a distinct head end; and whether it is divided into a series of segments (Figure 26.38).

3. Sponges, the simplest multicellular animals, do not have a gut or nervous system. Although they have several kinds of cells, these are not organized into tissues. Some cells trap food particles from the surrounding water. The beating of their flagella moves water through a system of pores, canals, and chambers.

4. Jellyfishes, sea anemones, and their relatives have radial symmetry, and their cells form tissues. The outer and inner tissue layers are separated by mesoglea, a secreted material. Cnidarians have contractile cells, sensory cells, and nerve cells, usually organized as a nerve net. They alone produce stinging capsules (nematocysts).

5. Most animals more complex than cnidarians are bilaterally symmetrical, and a type of tissue called mesoderm forms in their embryos. Mesoderm gives rise to muscle and other tissues that lie between the lining of the gut and the epidermis. The gut may be saclike, as it is in flatworms, but usually it is complete, with an anus as well as a mouth.

6. Most animals more complex than flatworms have a coelom or false coelom (cavities that separate the gut from the body wall). A true coelom originates within pouches or masses of mesoderm, and is lined by a peritoneum.

7. Most notably among the annelids and arthropods, the body is segmented (partitioned into a series of units). Arthropod segments tend to be highly specialized, and groups of successive segments form distinct body regions, such as the head, thorax, and abdomen of an insect. Arthropods also have a well-developed exoskeleton and jointed appendages.

8. Of the animals with a true coelom, the echinoderms are the only ones that are radially symmetrical and that lack a distinct head end. The nervous system is decentralized, there being no brain. This arrangement is favorable for radial animals, which can respond to food and danger coming from any direction.

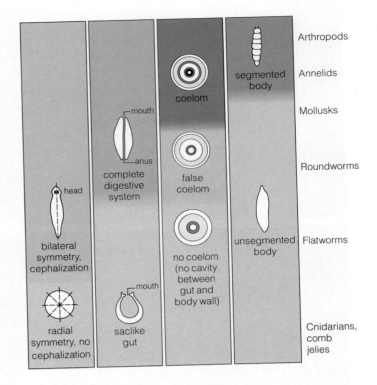

Figure 26.38 Summary of key trends in animal evolution, identified by comparing body plans of major phyla.

9. Table 26.2, on pages 442 and 443, summarizes the characteristics of the major groups of invertebrates described in this chapter. To provide perspective on what a small branch we humans occupy on the animal family tree, the phylum Chordata is included as the last entry for this table.

Review Questions

1. In what ways do invertebrates and vertebrates differ? *410*

2. Describe the features of one group of invertebrates. *428–431*

3. Name some of your paired body parts that evolved in your bilateral, cephalized ancestors. *410–411*

4. What is a coelom? Why was it important in the evolution of certain animal lineages? *411*

5. Name some animals with a saclike gut. Evolutionarily, what advantages does a complete gut afford? *414–418*

6. Choose an insect that thrives in your neighborhood and describe some of the adaptations that underlie its success. *432, 435–437*

1. Five body features that help us identify trends in animal evolution are _____ , _____ , _____ , _____ , and _____ .

2. Animals are _____ .
   a. autotrophs
   b. heterotrophs
   c. chemotrophs
   d. saprobes

3. Which is *not* characteristic of the animal kingdom?
   a. multicellularity; cells form tissues, organs
   b. exclusive reliance on sexual reproduction
   c. motility at some stage of the life cycle
   d. embryonic development during the life cycle

4. Jellyfishes, sea anemones, and their relatives have _____ symmetry, and their cells form _____ .
   a. radial; mesoderm
   b. bilateral; tissues
   c. radial; tissues
   d. bilateral; mesoderm

5. In sheer numbers and distribution, _____ are the most successful animals.
   a. arthropods
   b. sponges
   c. snails and clams
   d. sea stars
   e. vertebrates

6. Bilateral, segmented bodies and hardened exoskeletons occur among the _____ .
   a. arthropods
   b. sponges
   c. snails and clams
   d. sea stars
   e. vertebrates

7. Which phylum contains members that are notorious for causing serious diseases in humans?
   a. cnidarians
   b. flatworms
   c. segmented worms
   d. chordates

8. More complex animals have a _____ between the gut and body wall.
   a. pharynx
   b. planula
   c. coelom
   d. archenteron

9. Most animals that are more complex than cnidarians have _____ symmetry, and _____ forms in their embryos.
   a. radial; mesoderm
   b. bilateral; endoderm
   c. bilateral; mesoderm
   d. radial; endoderm

10. Match the terms with the appropriate groups.
    ____ sponges
    ____ cnidarians
    ____ flatworms
    ____ roundworms
    ____ rotifers
    ____ mollusks
    ____ annelids
    ____ arthropods
    ____ echinoderms
    ____ chordates

    a. spiny-skinned
    b. vertebrates and kin
    c. flukes and tapeworms
    d. no tissue organization
    e. cilia crowns; no males for some
    f. nematocysts, radial symmetry
    g. hookworms, elephantiasis
    h. jointed appendages
    i. "belly-foots" and kin
    j. segmented worms

Selected Key Terms

animal *410*
annelid *428*
bilateral symmetry *410*
cephalization *411*
cnidarian *414*
coelom *411*
collar cell *413*
comb jelly *417*
contractile cell *414*
cuticle *419*
deuterostome *423*
echinoderm *438*
ectoderm *410*
endoderm *410*
epithelial tissue *414*
exoskeleton *432*
flatworm *418*
ganglion *428*
gonad *417*
gut *411*
hermaphrodite *418*
hydrostatic skeleton *414*
invertebrate *410*
larva *413*
Malpighian tubule *435*
mantle *424*
medusa *414*
mesoderm *410*
metamorphosis *432*
mollusk *424*
molting *432*
nematocyst *414*
nephridium *429*
nerve cell *414*
nerve cord *428*
organ *418*
organ system *418*
placozoan *412*
planula *417*
polyp *414*
proglottid *419*
protostome *423*
radial cleavage *423*
radial symmetry *410*
ribbon worm *422*
roundworm *419*
scolex *419*
sensory cell *414*
spiral cleavage *423*
sponge *412*
trachea *432*
tube foot *438*
vertebrate *410*
water-vascular system *438*

Readings

Hickman, C. P., and L. S. Roberts. 1993. *Integrated Principles of Zoology*. Ninth edition. St. Louis: Mosby.

Kozloff, E. 1990. *Invertebrates*. Philadelphia: Saunders.

Pearse, V. and J., and Buchsbaum, M. and R. 1987. *Living Invertebrates*. Palo Alto, California: Blackwell.

Pough, F. H., J. Heiser, and W. McFarland. 1989. *Vertebrate Life*. Third edition. New York: Macmillan.

Table 26.2 Summary of Characteristics of the Major Animal Phyla

Phylum (and some representatives)	Typical Environment	Typical Life-Style of Adult Form	Nervous System	Support and Movement
Porifera (8,000)* sponges	Most marine, some freshwater	Attached filter feeders	No nervous system (only cell-to-cell communication)	Support by spicules, protein fibers, or both; contractile cells change openings at body surfaces
Cnidaria (11,000) hydras, sea anemones, jellyfishes, corals	Most marine, some freshwater	Attached, creeping, swimming, or floating carnivores	Nerve net, cell-to-cell transmission	Hydrostatic support (by fluid in gut), by secreted jellylike mesoglea or by skeletal elements; contractile fibers in epithelial cells
Ctenophora (100) comb jellies	Marine	Mostly planktonic; few attached or creeping	Nerve net	Support by mesoglea; muscles separate from epithelia; locomotion by cilia in eight rows of comblike structures; sometimes by muscular activity
Platyhelminthes (15,000) flatworms, flukes, tapeworms	Marine, freshwater, some terrestrial in moist places; many parasitic in or on other animals	Herbivores, carnivores, scavengers, parasites	Brain, nerve cords	Hydrostatic support (no secreted skeleton); well-developed muscle tissue
Nemertea (800) ribbon worms	Most marine; few freshwater, terrestrial	Mostly carnivores	Brain, nerve cords	Hydrostatic support; well-developed muscle tissue
Nematoda (20,000) roundworms	Marine, freshwater, terrestrial; many parasitic	Scavengers, carnivores parasites	Nerve ring, nerve cords	Hydrostatic support; (by false coelom); tough cuticle; longitudinal muscle in body wall
Rotifera (1,800) rotifers	Marine, freshwater, moisture on mosses	Mostly filter feeders, capturing bacteria, unicellular algae	Brain, nerve cords	Locomotion by cilia; muscles for shape changes
Mollusca (110,000) snails, slugs, clams, squids, octopuses	Marine, freshwater, terrestrial	Herbivores, carnivores, scavengers, detritus or filter feeders; mostly free-moving, some attached	Brain, nerve cords, major ganglia other than brain	Hydrostatic skeleton in most; well-developed musculature in foot, mantle, other structures
Annelida (15,000) earthworms, leeches, polychaetes	Marine, freshwater, terrestrial in moist places	Herbivores, carnivores, scavengers, detritus or filter feeders; mostly free-moving	Brain, double ventral nerve cord	Hydrostatic skeleton (using coelom); well-developed musculature in body wall
Arthropoda (1,000,000+) crustaceans, spiders, insects	Marine, freshwater, terrestrial	Herbivores, carnivores, scavengers, detritus or filter feeders, parasites; mostly free-moving	Brain, double ventral nerve cord	Exoskeleton (of cuticle); jointed appendages; muscles mostly in bundles
Echinodermata (6,000) sea stars, brittle stars, sea urchins, sea lilies, sea cucumbers	Strictly marine	Mostly carnivores, detritus feeders, few herbivores; most free-moving, some attached	Radially arranged nervous system	Endoskeleton (of spines, etc.); muscles for body movement, tube feet often used in locomotion
Chordata (47,100) tunicates, lancelets, jawless fishes, jawed fishes, amphibians, reptiles, birds, mammals	Marine, freshwater, terrestrial	Herbivores, carnivores, scavengers, filter feeders; generally free-moving (most tunicates attached as adults)	Well-developed brain, dorsal and tubular nerve cord in most	Notochord or a bony or cartilaginous endoskeleton; well-developed musculature in most

*Number in parentheses indicates approximate number of known species.

Digestive System	Respiratory System	Circulatory System	Mode of Reproduction
No gut; microscopic food particles secured by individual cells	None; respiration by individual cells	None	Sexual (certain cells become or produce gametes); some asexual budding; production of resistant bodies (gemmules)
Saclike gut (may be branched)	None; respiration by individual cells; gut may distribute oxygen	None, other than via gut	Sexual (usually separate sexes; gonads discharge gametes into gut or to exterior); asexual
Saclike, but branched	None; respiration by individual cells; gut may distribute oxygen	None, other than via gut	Sexual (usually hermaphroditic); gonads closely associated with gut
Saclike gut (may be branched)	None; gas exchange across body surface	None	Sexual (usually hermaphroditic, with complex reproductive system); some asexual
Complete gut	None; gas exchange across body surface	Closed system	Sexual (sexes usually separate; reproductive system simple)
Complete gut	None; gas exchange across body surface	False coelom	Sexual (sexes separate; reproductive system fairly complex)
Usually complete gut (sometimes saclike)	None; gas exchange across body surface	False coelom	Mostly parthenogenetic (eggs develop without fertilization); males appear only occasionally
Complete gut	Ctenidia, other gills; mantle can be modified as lung; gas exchange across body surface	Usually open (closed system in cephalopods)	Sexual (hermaphroditic or separate sexes; reproductive system usually complex)
Complete gut	Gas exchange across body surface; varied outgrowths of surface in many	Usually closed; coelom also may function in distribution	Sexual (hermaphroditic or sexes separate; reproductive system simple or complex); asexual also in many
Complete gut	Gills; tracheal tubes; book lungs; general body surface in some	Open system	Sexual (usually separate sexes; reproductive system fairly complex)
Usually complete gut (sometimes saclike)	Gas exchange across general body surface or surface of outgrowths of it (such as tube feet)	Coelom around viscera, also water-vascular coelom	Sexual (reproductive system simple; gonads usually discharge gametes directly to exterior); some asexual
Complete gut	Lungs in most vertebrates other than fishes; perforated pharynx; gills; gas exchange across body surface	Closed system in most (open system in most tunicates); lymphatic system in many vertebrates	Sexual (sexes usually separate, except in most tunicates); asexual in some tunicates

27 ANIMALS: THE VERTEBRATES

Making Do (Rather Well) With What You've Got

It has taken the platypus nearly two centuries to earn a little respect. In 1798, skeptical naturalists at the British Museum in London poked and probed a specimen, looking for signs that a prankster had stitched the bill of an oversized duck onto the pelt of a small furry mammal. No wonder. At first blush the platypus looks like one of nature's practical jokes (Figure 27.1*a*).

The platypus is a web-footed mammal, about half the size of a housecat. Like other mammals, it has mammary glands and hair. Yet, like birds and reptiles, it has a cloaca, a single external opening that functions in both reproduction and excretion. Like birds and most reptiles, it lays shelled eggs. Its young hatch from eggs early in their development, pink and helpless (Figure 27.1*b*). And about that fleshy platypus bill! It looks like it belongs to a duck. The broad, flat, furry tail seems borrowed from a beaver.

With its unusual array of traits, the platypus is a fine example of an animal that has been judged according to *our* ideas of what "an animal" is supposed to be. Those who assume that the platypus doesn't quite measure up are not using the real yardstick. *Its collection of traits happens to be exquisitely adapted for survival and reproduction under a particular set of environmental conditions.*

The platypus is a predator of streams and lagoons in remote parts of Australia and Tasmania. It dives into the water at night, forages skillfully for small edible mollusks and other prey, then hides during the day in underground burrows.

Its dense, blackish-brown fur functions in insulating the body. When a platypus is submerged all night long in cold water or sequestered all day in a cool burrow, the fur holds in heat and so helps maintain body temperature within tolerable limits. The broad, thick tail functions as a rudder, enhancing maneuverability through the water. It also functions as a storehouse for

Figure 27.1 One of evolution's success stories—the platypus, underwater (**a**) and in its burrow (**b**).

a

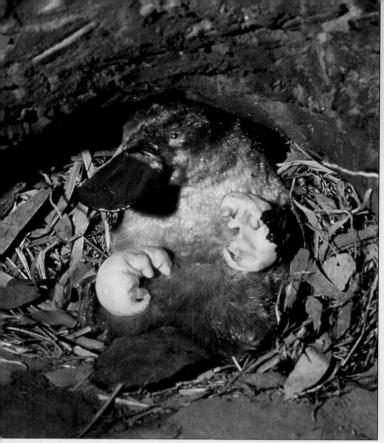

b

energy-rich fats. The webbed feet can function as paddles when they flare out. The strong, clawed hind feet
are great for digging.

When submerged in water, the platypus nostrils
close, a fleshy groove around the eyes and ears snaps
shut, and the bill takes over sensory functions. The
platypus has a sensory system for land and a separate
sensory system for underwater! More than 800,000
sensory receptors are built into the bill. Many detect
mechanical pressure—waves produced by swimming
prey. Others detect weak electric currents; even the tiny
electric field created by the flick of a shrimp tail will do
the trick. A platypus can zero in on a meal with awe-
some precision. It uses the flattened bill to scoop up
freshwater snails, mussels, insect larvae, worms, and
shrimps, then grinds them up on the horny pads of its
jaws.

This remarkable body plan started evolving about
100 million years ago, when ancestors of the platypus
coexisted with dinosaurs on Gondwana (page 336).
That supercontinent broke apart into huge fragments,
and the ancestral forms happened to be on the land
mass that eventually became Australia. The platypus
lineage has endured ever since. With that track record,
who are we to chuckle over the collection of platypus
traits? As is the case for all other animals, the traits
meet nature's most important test. They work.

KEY CONCEPTS

1. The chordate branch of the animal family tree includes
invertebrate and vertebrate species. All are bilateral
animals.

2. All chordates have a notochord (a supporting rod for
the body), dorsal nerve cord, pharynx, and gill slits in
the pharynx wall. These features appear in all chordate
embryos, and some or all persist in the adult forms of
different species. Vertebrate chordates additionally have a
cartilaginous or bony backbone and protective chamber
for the brain.

3. The existing invertebrate chordates include tunicates
and lancelets.

4. The existing vertebrates include jawless fishes,
cartilaginous fishes, bony fishes, amphibians, reptiles,
birds, and mammals. Another group, the placoderms,
became extinct early in vertebrate history.

5. Five major trends occurred during the evolution of
vertebrates, although not in every group. Structural
support and locomotor functions came to depend less on
the notochord and more on a vertebral column. Jaws
evolved. The nerve cord expanded into a spinal cord
and brain. Among aquatic ancestors of land-dwelling
vertebrates, gas exchange came to depend less on gills and
more on lungs and an efficient circulatory system. Also,
fleshy fins with skeletal supports evolved into legs,
which became modified further in specialized ways in
amphibians, reptiles, birds, and mammals.

27.1 THE CHORDATE HERITAGE

Characteristics of Chordates

The preceding chapter concluded with a look at the echinoderms, some of our distant relatives on the deuterostome branch of the animal family tree. We turn now to the vertebrates and their close relatives. All are bilateral animals, grouped together as **chordates** (phylum Chordata).

There are more than 47,000 species of "vertebrate chordates." These have a backbone of cartilage or bone, and they have skull bones that protect the brain. The 2,100 or so species of "invertebrate chordates" share certain features with the dominant members of this phylum, although a backbone isn't one of them.

Four features are evident in chordate embryos, and in many species these persist into adulthood. First, a **notochord**, a long rod of stiffened tissue (not cartilage or bone), helps support the body. Second, the nervous system is based on a tubular, dorsal **nerve cord** that forms parallel to the notochord and gut. The anterior end of the cord expands during development, forming the brain. Third, a muscular tube called a **pharynx** functions in feeding, respiration, or both. The chordate pharynx has slits in the wall. Fourth, a tail forms and extends past the anus.

Chordate Classification

Nearly all chordates belong to three subphyla: the Urochordata (such as tunicates), Cephalochordata (lancelets), and Vertebrata (vertebrates). As Figure 27.2 shows, there are eight classes of vertebrates:

Agnatha	*Jawless fishes*
Placodermi	*Jawed, armored fishes (extinct)*
Chondrichthyes	*Cartilaginous fishes*
Osteichthyes	*Bony fishes*
Amphibia	*Amphibians*
Reptilia	*Reptiles*
Aves	*Birds*
Mammalia	*Mammals*

Appendix I has an expanded classification scheme for vertebrates. Unit VI provides details of their body plans and functions. Here we become acquainted with major trends in their evolution. And we find clues to those trends among invertebrate chordates.

The embryos of chordates alone have this combination of features: a notochord, a tubular dorsal nerve cord, a pharynx with slits in the wall, and a tail extending past the anus.

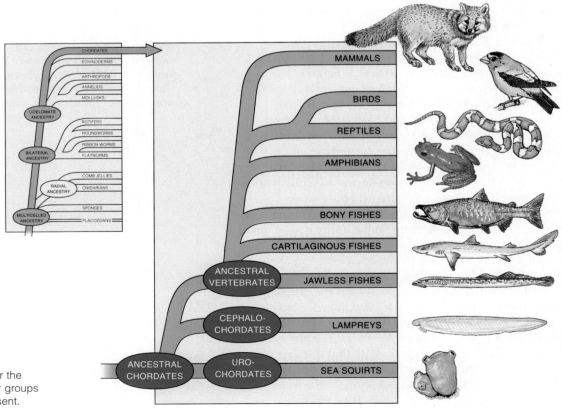

Figure 27.2 A family tree for the chordates, showing the major groups that have survived to the present.

INVERTEBRATE CHORDATES

Tunicates

The 2,000 species of existing urochordates are baglike animals, at least 2 centimeters long. They are often called **tunicates**, a name referring to the gelatinous or leathery "tunic" that adults secrete around themselves. The most common species are called sea squirts, because the adults tend to squirt out water through a siphon when something irritates them.

All tunicates live in marine habitats, ranging from the intertidal zone to surprising depths. Most adults remain attached to rocks, ship hulls, and other suitably hard substrates. Some live solitary lives; others are colonial. Figure 27.3 shows the body plan of one of the solitary types. It starts out as a bilateral, free-swimming larva that looks rather like a tadpole. (A larva, recall, is an immature stage between embryonic and adult stages of the life cycle.) The firm yet flexible notochord develops as a series of fluid-filled cells in the tail. It functions like a torsion bar. When muscles on one side or the other of the tail contract, the notochord bends. When the muscles relax, it springs back. The strong, side-to-side motion propels the animal forward. Most fishes use this kind of propulsive motion.

A sea squirt larva swims briefly, then attaches to a substrate and undergoes metamorphosis. The major tissue reorganization and remodeling transforms the larva into an adult. Its tail and notochord disappear; a tunic is secreted. The pharynx enlarges, and perforations in the pharynx wall become subdivided into many small slits. The tubular nerve cord regresses, leaving the adult with a much simplified nervous system. Figure 27.3 shows these events.

Water flows into and out of the adult body through two siphons. Incoming water enters a ciliated pharynx that is finely divided into openings called **gill slits**. As water flows through the sievelike pharynx, diatoms and other bits of food are strained from it. A heart and blood vessels service the thin-walled pharynx, which also serves as a respiratory organ.

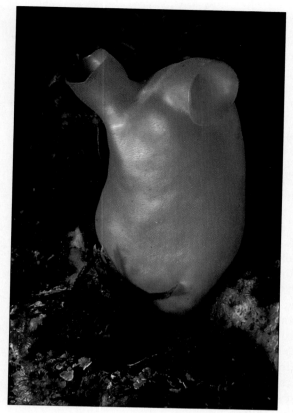

a Adult tunicate

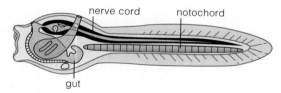

b A tunicate larva

Figure 27.3 (**a**) An adult tunicate. (**b**) The tadpolelike larva. It swims for a few minutes or days until it locates a suitable living site. (**c–e**) It attaches its head to a substrate. And now metamorphosis begins. The tail, notochord, and most of the nervous system are resorbed (recycled to form new tissues). The slits in the pharynx multiply. Organs become rotated until the openings through which water enters and leaves are directed away from the substrate.

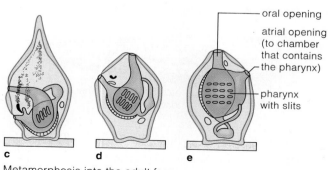

Metamorphosis into the adult form

Lancelets

Just offshore, on seafloors around the world, we find twenty-five or so species of cephalochordates. These translucent, fish-shaped animals are seldom more than 5 centimeters (about 2 inches) long. They spend much of the time buried almost up to their mouth in sand and sediments. Their common name, **lancelets**, refers to the sharp tapering of their body at both ends. As is true of tunicate larvae, the notochord and muscles of a lancelet can work together to produce swimming motions.

Figure 27.4 includes a diagram of the lancelet body plan. The four distinctly chordate features are present. Besides this, its muscles are arranged in a segmented pattern on both sides of the notochord. In addition, lancelets have a closed circulatory system (but no red blood cells). Slow waves of contraction drive blood forward through some blood vessels and backward through others. Blood from all parts of the body collects in a sac that resembles the vertebrate heart, but this does not function as a muscular pump. Although a complex brain is nowhere in sight, the head end of the dorsal nerve cord is expanded and pairs of nerves extend into each muscle segment. Lancelets do not have an internal respiratory system as you do. Carbon dioxide and oxygen simply diffuse across their comparatively thin skin.

Lancelets are more like tunicates in their feeding mechanisms. They are filter feeders. Cilia lining the mouth cavity create a current that draws water through the mouth. The water passes through a pharynx, where food becomes trapped in mucus. Trapped food is delivered to the rest of the gut, where digestion proceeds, nutrients are absorbed, and waste material is compacted for elimination.

Cilia are microscopically small structures, and the driving force of their collective beating cannot by itself provide a filter-feeding animal with sufficient food. In such animals, great numbers of cilia must work in conjunction with a large food-trapping surface area. As you can see from Figure 27.4, the lancelet pharynx is quite large, relative to the overall body length, and it is perforated by up to 200 ciliated, food-trapping gill slits.

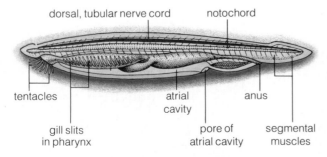

Figure 27.4 labels: dorsal, tubular nerve cord — notochord — tentacles — gill slits in pharynx — atrial cavity — pore of atrial cavity — anus — segmental muscles

Figure 27.4 Cutaway view and photograph of a lancelet, showing nerve cord and flexible notochord.

27.3 ORIGIN OF VERTEBRATES

Are lancelets "living fossils," the earliest members of the vertebrate lineage? You might think so, given their fishlike shapes, segmented muscles, circulatory system, pairs of nerves, and other features. However, even though lancelets might share some similarities with the early vertebrates, they are on a different evolutionary road. Several lines of evidence indicate that the similarities are an outcome of convergent evolution. By this process, recall, different body parts in separate lineages evolve in similar directions, because they are put to similar uses in similar kinds of environments.

Where, then, did the vertebrate road begin? We find tantalizing clues among the members of a rather obscure invertebrate phylum, the **hemichordates** (*hemi*- meaning half, as in "halfway to chordates"). Most people have never seen a hemichordate. The 100 or so species of these soft-bodied marine animals live on the seafloor. The acorn worm shown in Figure 27.5 is an example.

Evolutionarily, the hemichordates seem to be midway between the echinoderms and chordates. Like tunicates, they rely on a ciliated filter-feeding apparatus (Figure 27.6). They don't have a notochord, but they are decidedly like chordates in having a gill-slitted pharynx and a dorsal tubular nerve cord. Also, their larval stages resemble those of echinoderms and tunicates. Do the resemblances suggest a pathway of evolution? Possibly.

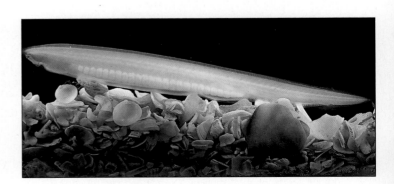

Suppose, in ancient echinoderms, mutations in regulatory genes brought about an increase in the rate at which sex organs developed and matured. If sex organs became functional early on, in the *larval* body, there would no longer be much advantage to metamorphosis, the wholesale reconstruction leading to the adult. Over evolutionary time, the original adult form simply could be dispensed with. This scenario is not so far-fetched. Among some existing tunicates—even among some amphibians—we find larvae with functional sex organs. These sexually precocious larvae can reproduce, generation after generation!

Ancestral tunicates, lancelets, and hemichordates were not on the evolutionary road leading to vertebrates, but they provide tantalizing clues to the animals that were.

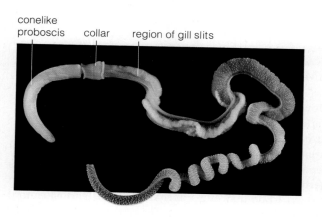

Figure 27.5 An acorn worm, of the phylum Hemichordata. Acorn worms often are found in mud and tidal flats, where they live in U-shaped burrows. One species was discovered near a hydrothermal vent, deep on the ocean floor. Many acorn worms are filter feeders. Food suspended in the water becomes trapped by sticky mucus on their proboscis and is carried to the mouth. The gill slits of acorn worms open from the pharynx and resemble the gill slits of chordates.

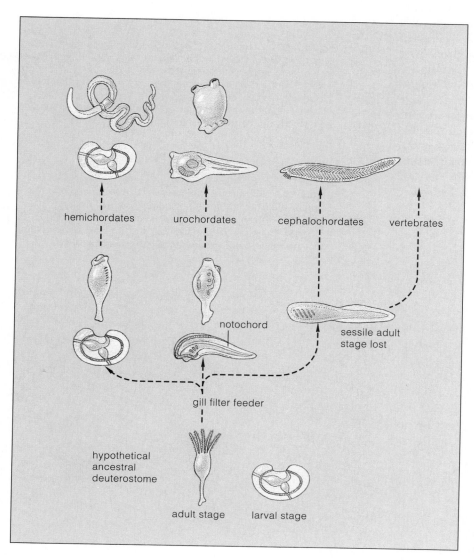

Figure 27.6 One hypothesis regarding the evolutionary origins of hemichordates and chordates.

27.4 EVOLUTIONARY TRENDS AMONG THE VERTEBRATES

How do we get from tadpole-shaped chordates to the vertebrates? Let's start with a broad evolutionary picture. One evolutionary trend involved a shift away from the notochord to reliance on a skeletal column of separate, hard segments. We call the skeletal elements **vertebrae** (singular, vertebra). The vertebral column proved to be a strong internal skeleton (endoskeleton) for muscles to work against. *The vertebral column was the foundation for fast-moving predators—some of which were ancestral to all other vertebrate animals.*

In a related trend, part of the nerve cord expanded and developed into a complex brain. The expansion began after **jaws** evolved. In early fishes, the first jaws arose through modification of the first in a series of structural elements that helped support the gill slits (Figure 27.7). Jaws led to new feeding possibilities—and to intensified competition among predators. Fishes able to recognize food or predators from a distance were favored. Over time, they developed better senses of smell and vision. The brain became better at processing information (Figure 27.8). *The trend toward complex sensory organs and nervous systems began in fishes and continued among land vertebrates.*

Another trend began when paired fins evolved. **Fins** are appendages that help propel, stabilize, and guide the body through water. Among some fishes, ventral fins became fleshy and equipped with skeletal supports—the forerunners of limbs. *Paired, fleshy fins were the starting point for the legs, arms, and wings seen among amphibians, reptiles, birds, and mammals.*

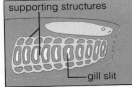

a Early jawless fish (an agnathan)

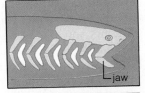

b Early jawed fish (a placoderm)

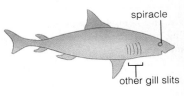
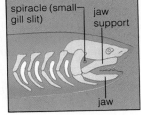

c Modern jawed fish (a shark)

Figure 27.7 Comparison of gill-supporting structures in jawless fishes and jawed fishes. In the placoderms and other early jawed vertebrates, cartilage supported the rim of the mouth. In modern jawed fishes, the gill slit between the jaws and an adjacent supporting element serves as a spiracle, an opening through which water is drawn. In (**a**), the gill supports are just under the skin. In (**b**) and (**c**), they are internal to the gill surface.

Figure 27.8 Evolutionary trend toward an expanded, more complex brain, as suggested by comparisons of existing vertebrates. These are dorsal views (looking down on the top of the head). Think about the head size of a frog and a horse, and you know these drawings are not to the same scale.

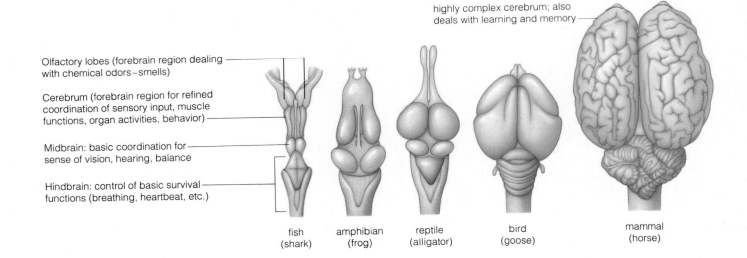

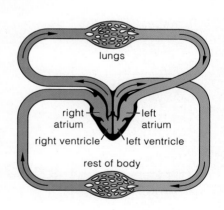

capillaries (network of fine blood vessels where gas exchange occurs)

a In fishes, a heart with two chambers (atrium and ventricle) pumps blood in one circuit. Blood picks up oxygen in gills, then delivers it to rest of body. Oxygen-poor blood flows back to the heart.

b In amphibians, the heart pumps blood through two partially separated circuits. Some blood is pumped to lungs. It picks up oxygen, returns to heart, and mixes with oxygen-poor blood still in heart. The heart pumps this partly oxygenated blood to rest of body.

c In some reptiles and all birds and mammals, a four-chambered heart pumps blood in two completely separate circuits. In one circuit, the heart's right ventricle pumps oxygen-poor blood from the body to lungs. Oxygen-rich blood travels from lungs back to heart. In the other circuit, the left ventricle pumps this blood to rest of body.

Figure 27.9 Circulatory systems of vertebrates. All vertebrates have a "closed" circulatory system, in which blood is confined within blood vessels and a heart. Some of these systems are more efficient than others at providing body tissues with oxygen.

richly endowed with blood vessels, and they provide a large surface area for gas exchange.

For example, five to seven pairs of a shark's gill slits contain gills, which extend from the pharynx to the outer body surface. When a shark opens its mouth and closes the external gill openings, its pharynx expands and oxygen-rich water flows in through the mouth. Now oxygen diffuses into the gills, and carbon dioxide diffuses out. When muscle action constricts the pharynx, water depleted of oxygen but loaded with carbon dioxide is forced out through the mouth.

As fishes became larger and more active, oxygen uptake and distribution improved. Gills became more efficient in aquatic lineages. But gills cannot work out of water. They stick together unless water flows through them and keeps them moist. In the fishes that were ancestral to land vertebrates, pouches developed on the gut wall. These pouches evolved into **lungs**—internally moistened sacs for gas exchange. In a related trend, modifications to the heart enhanced the pumping of oxygen and carbon dioxide through the body (Figure 27.9). *Ancestors of land vertebrates relied less on gills and more on lungs. More efficient circulatory systems accompanied the evolution of lungs.*

Another trend involved modifications to respiratory structures. Think of the gill slits of a lancelet. Except when this chordate buries itself in sand, oxygen dissolved in the surrounding water and carbon dioxide from metabolically active tissues simply diffuse across the body's surface, down their concentration gradients. In most vertebrate lineages, however, **gills** of one sort or another evolved. All of these respiratory structures have a moist, thin, intricately folded surface, they are

---

The emergence of a vertebral column, jaws, paired fins, and lungs were pivotal in the evolution of certain vertebrate groups.

27.5 FISHES

Submerged as they are in the waters of the earth, the fishes might not seem to be the dominant vertebrates. And yet, in sheer numbers, they surpass all other vertebrate groups. And fishes show far greater diversity; there are more than 21,000 existing species of bony fishes alone.

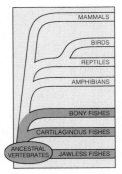

The body form and behavior of a fish tell us about the kinds of challenges it faces in the water. Being about 800 times denser than air, water resists rapid movements. Predatory fishes of the oceans are streamlined for pursuit, with a long, trim body that reduces friction. Their tail muscles are organized for propulsive force and forward motion. Bottom-dwelling fishes spend most of their lives hiding from predators or prey. We can deduce that their flattened body is easy to conceal, and that these fishes are sluggish, not the Ferraris of the deep.

Or consider a motionless trout, suspended in shallow water. It is a fine example of an adaptation to water's density. Like many fishes, a trout can maintain neutral buoyancy with a **swim bladder**. This is an adjustable flotation device that exchanges gases with the blood. When a trout gulps air at the water's surface, it is busily adjusting the gas volume in its swim bladder.

The First Vertebrates

As Figure 27.10 indicates, free-swimming vertebrates originated during the Cambrian period. They soon gave rise to two kinds of fishes—those without and those with jaws. One or the other kind probably gave rise to all vertebrate lineages that followed.

Ostracoderms were among the earliest jawless fishes (Agnatha). They were bottom-dwelling filter feeders. Water drawn into the mouth entered the pharynx, where food was strained out before the water left through gill openings between the head plates (Figure 27.11).

Figure 27.10 Evolutionary history of the fishes.

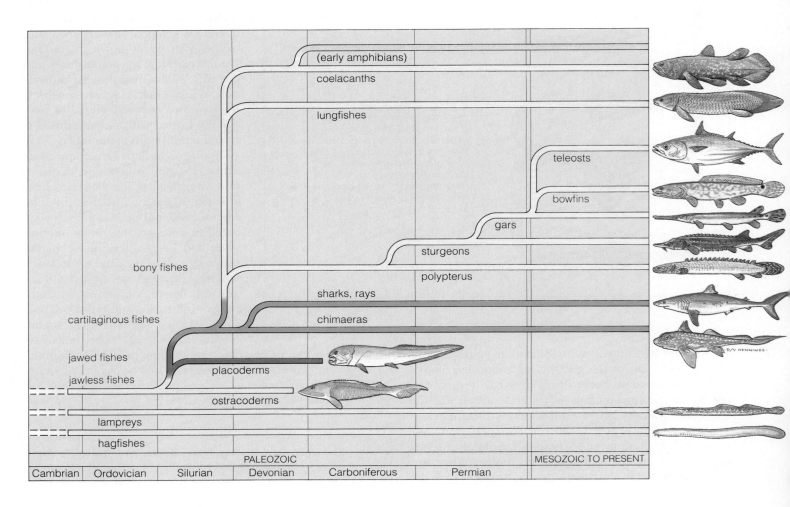

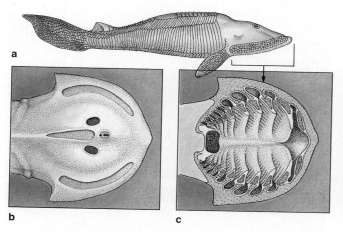
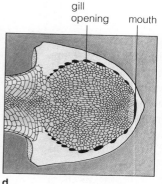

Figure 27.11 (**a**) Reconstruction of an ostracoderm and its skull. This one lived about 100 million years after the earliest known ostracoderms. The dorsal view of the skull in (**b**) shows the openings for three eyes (*blue*) and a single nostril (*red*). The ventral view (**c**) shows the small mouth and relatively large, food-straining pharynx; the flattened bones covering all but the gill openings and a slitlike mouth have been removed, but are shown in place in (**d**).

The ostracoderms did not have much of an internal skeleton; it probably consisted of a notochord and a protective covering for the newly enlarged brain and sensory organs. Armorlike plates covered the body surface. The plates were composed of bony tissue and dentin, a hard tissue that is still present in the teeth of living vertebrates. The armor afforded some protection against the giant pincers of sea scorpions, but it apparently was not much good against jaws. Ostracoderms disappeared when jawed fishes began their adaptive radiations.

Among the first fishes with jaws and paired fins were the **placoderms**. These fishes were bottom-dwelling scavengers and predators. Bony elements reinforced their notochord. The first in a series of gill-supporting structures had become enlarged and equipped with bony projections, something like teeth (Figure 27.7*b*). They functioned as jaws.

Before placoderms evolved, feeding strategies had been limited to filtering, sucking, or rasping bits of food material. Now, placoderms could bite and tear up large chunks of prey—and so get a splendid return on the energy they invested into securing food.

Placoderms with scalelike or bony-plated armor diversified during Silurian and Devonian times. Then, during the Carboniferous, they all became extinct. New kinds of predators—the cartilaginous and bony fishes—replaced them in the seas.

Existing Jawless Fishes

Descendants of some of the early jawless fishes made it to the present. We call them the **lampreys** and **hagfishes**. All of the seventy-five or so species have a cylindrical, eel-like body with no paired fins. They all have a skeleton—an internal structural framework—of cartilage.

Lampreys are specialized predators, almost parasites. Their suckerlike oral disk has horny, toothlike parts that rasp flesh from prey (Figure 27.12). Some

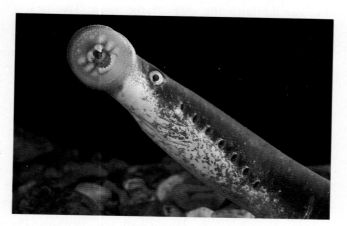

Figure 27.12 A lamprey, pressing its toothed oral disk to the glass wall of an aquarium.

types latch onto salmon, trout, and other commercially valuable fishes, then suck out juices and tissues. Just before the turn of the century, lampreys began invading the Great Lakes of North America. Wherever they have been introduced, populations of lake trout and other large fishes have collapsed. Today, in fisheries of the Great Lakes and other regions, the water is treated with a chemical that poisons lamprey larvae. But the battle goes on.

Visualize a large worm with feelers around the mouth and you have an idea of what the scavenging hagfishes look like. They are not the favorite of fishermen. They burrow into fish trapped by setlines or nets. The ones brought on deck secrete copious amounts of sticky, slimy mucus.

The first vertebrates were jawless and then jawed fishes. Among them were the ancestors of all existing vertebrate species.

Cartilaginous Fishes

Cartilaginous fishes (Chondrichthyes) include about 850 species of skates, sharks, and chimaeras. Most are specialized marine predators with a streamlined body. Figure 27.13 shows examples. All these fishes have pronounced fins, a skeleton of cartilage, and five to seven gill slits on both sides of the pharynx. Most species have a few to many rows of **scales**, small, bony plates at the body surface that often provide protection without weighing the fish down.

The skates and rays are mostly bottom dwellers with flattened teeth, suitable for crushing hard-shelled invertebrates. Both types have distinctive, enlarged fins that extend onto the side of the head. The largest species, the manta ray, measures up to 6 meters from fin tip to fin tip. Other rays have electric organs in the tail or fins that can deliver up to 200 volts of electricity— quite enough to stun prey. The stingray tail has a spine (a modified scale) with a venom gland at its base. Stingrays eat invertebrates, mostly, so the spine is probably used in defense against predators.

At 15 meters from head to tail, some sharks are among the largest living vertebrates. That is longer than two pickup trucks parked end to end. Sharks have formidable jaws, they can detect even traces of blood in water, and the relatively few shark attacks on humans have given the whole group a bad reputation. Yet most of the larger types feed on small invertebrates. The "man-eaters" obviously have not survived by eating humans alone. For many millions of years the large sharks have been eating large fishes and marine mammals, including seals, but not surfboards with legs dangling over the side. They use their sharp, triangular teeth to capture prey and rip off chunks of flesh. They continually shed and replace these teeth, which are modified scales.

The thirty or so species of chimaeras inhabit moderately deep waters, and most feed on hard-shelled mollusks. With their bulky body and long, slender tail, they do resemble a rat (hence their common name, ratfishes). A venom gland associated with a spine is present in front of the dorsal fin.

Bony Fishes

In terms of diversity and sheer numbers, the **bony fishes** (Osteichthyes) are the most successful of all vertebrates. Their ancestors arose during the Silurian, and before that period drew to a close, they had diverged into three lineages: ray-finned fishes, lobe-finned fishes, and lungfishes. Their descendants radiated into nearly every aquatic habitat. With at least 21,000 existing species, the bony fishes represent all but a paltry 4 percent of the modern fishes.

Figure 27.13 Cartilaginous fishes: (**a**) shark, (**b**) blue-spotted reef ray, and (**c**) chimaera (ratfish).

Figure 27.14 is a sampling of the stunning morphological diversity among the ray-finned fishes. The members of this group have paired fins supported by rays that originate from the dermis (one of the skin's layers). Most have notably maneuverable fins and light, flexible scales. Both adaptations contribute to the ability to make complex movements. A well-developed respiratory system rapidly delivers oxygen to metabolically active tissues.

Enormous variations exist on the basic body plan. Predators of the open ocean typically have torpedo shapes; strong, flexible bodies; and powerful tail fins that function in swift pursuit. Many reef dwellers have boxy shapes and fins adapted for maneuvering among narrow passageways of their habitats. The long, flexible bodies of eels are suitable for wriggling through mud

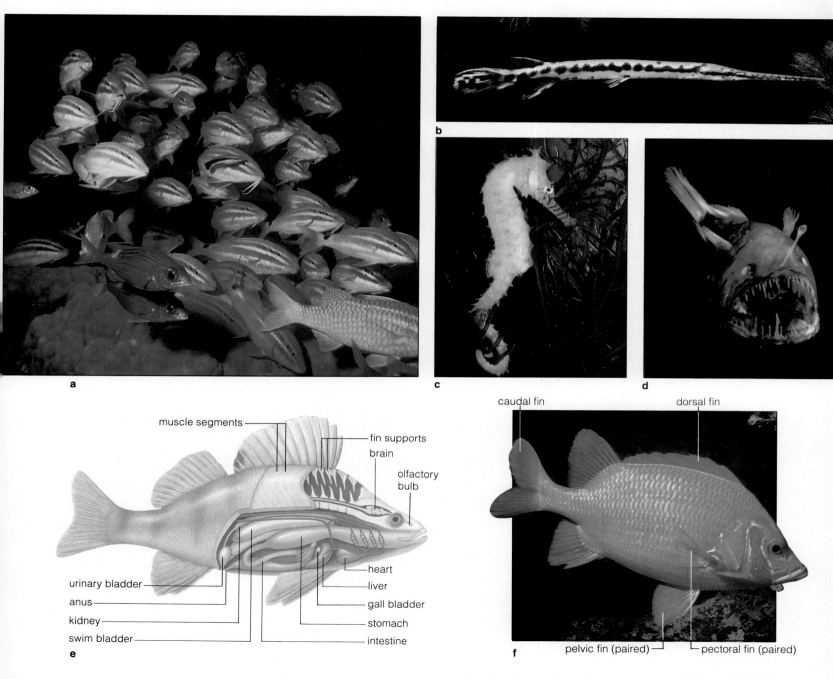

Figure 27.14 Bony fishes. (**a**) Goatfish showing schooling behavior, (**b**) a long-nose gar, (**c**) a seahorse, and (**d**) a deep-sea angler fish. Most ray-finned fishes have the same general body plan, represented here by a perch (**e**) and soldierfish (**f**).

and into nooks and crannies. Bottom dwellers have flattened bodies; sea horses and other types have bizarre body forms. All of these are suitable for concealment from prey and predators.

One group of ray-finned fishes still resembles the ancestral stock. It includes sturgeons and the paddle-fishes of the Mississippi River basin. Another group, the thin-scaled or scaleless teleosts, are the most abundant. Existing members of the thirty-five orders include

such diverse types as salmon, tuna, rockfish, catfish, minnows, moray eels, perch, flying fish, sculpins, scorpionfish, blennies, and pikes.

By contrast, there is only one existing species of lobe-finned fishes and three genera of lungfishes. As their name suggests, a **lobe-finned fish** is unique in having paired fins that incorporate fleshy extensions from the body. As you will see next, neither it nor the lungfishes have changed much from ancient forms.

Of all existing vertebrates, the bony fishes (ray-finned fishes especially) are the most spectacularly diverse in their morphology and behavior.

bony or cartilaginous structures in lobed fin undergoing modification

limb bones of early amphibian

Figure 27.15 (**a**) Coelacanth (*Latimeria*), a "living fossil" that resembles early lobe-finned fishes. (**b,c**) Proposed evolution of the skeletal elements inside the lobed fins of certain fishes into the limb bones of early amphibians.

About Those Lobed-Finned, Air-Gulping Fishes . . .

Figure 27.15 shows a coelacanth, the only existing species of lobe-finned fishes. Along with lungfishes, it is a relic of a pivotal time in evolution, when certain vertebrates first ventured onto land.

The ancestors of lobe-finned fishes arose during the Devonian, when sea levels rose and fell repeatedly and swamps that fringed the land were alternately flooded and drained. They must have used their lobed fins to pull themselves from disappearing pond to pond (Figure 21.12). But it was not only lobed fins that made pond-to-pond lurchings possible. These animals also had sac-shaped surfaces inside their body that supplemented respiration. The lobe-finned fishes had simple lungs.

An Australian lungfish provides additional clues to how the ancestral forms might have made it through stressful times. It lives in stagnant water but swims to the surface to gulp air. During the dry season, when streams shrink to mud, the lungfish encases itself in a mixture of mud and slime that protects it from drying out until the next rainy season.

27.6 AMPHIBIANS

Challenges of Life on Land

The lobe-finned fishes of the Devonian traveled across land simply as a way to reach more water. Yet their crucial travels favored the evolution of more efficient lungs and stronger fins. Among the evolving forms were the ancestors of amphibians. An **amphibian** is a vertebrate that is somewhere between fishes and reptiles in its body plan and reproductive mode. An amphibian has a mostly bony endoskeleton and four legs (or a four-legged ancestor).

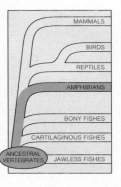

The ancestral amphibians began their forays onto land at the dawn of the Carboniferous. For them, the land was dangerous—and promising. Temperatures fluctuated more on land than in water, air didn't support the body as water did, and water was not always available. But air has far more oxygen. Lungs continued to evolve in ways that enhanced uptake of oxygen. Circulatory systems evolved and became better at distributing oxygen. Both modifications increased the energy base for more active life-styles.

The first amphibians also encountered novel forms of sensory information. Swampy forests abounded with tasty insects and other invertebrate prey. Animals with good vision, hearing, and balance—senses that are advantageous in land environments—were favored. Brain regions concerned with those senses expanded dramatically.

There are now three groups of amphibians—the salamanders, frogs and toads, and caecilians (Figure 27.16). None has escaped water entirely. Their thin skin dries out easily. To develop properly, the eggs of most species must be shed into water or laid in moist places. The larvae are still adapted to aquatic habitats. The species that spend most of their life in shallow water have gills and lungs, and they also use the skin as a respiratory surface. Among them we find somewhat flattened bodies, strong tails, and hindlimbs with webbed feet. The ones that spend most of their life on land use lungs, skin, and the lining of the pharynx to breathe.

Salamanders

Like fishes and the first amphibians, salamanders bend from side to side when they walk (Figure 27.16*a*). The first four-legged vertebrates probably walked this way, also. The adults of some species retain many larval features. Also, the larvae of some groups are sexually pre-

a   fish   salamander

Figure 27.16 Amphibians. (**a**) Forward movement of a salamander compared to that of a coelacanth. (**b**) Terrestrial stage in the life cycle of the red-spotted salamander. (**c**) A frog, splendidly jumping. (**d**) American toad. (**e**) A caecilian.

cocious; they can breed. This is true of the Mexican axolotl, for example. It retains the larval tail and external gills, and the development of its teeth and bones is arrested at an early stage.

Frogs and Toads

With more than 3,000 species, frogs and toads are the most successful amphibians. Their long hindlimbs and powerful muscles allow them to catapult into the air or move forcefully through water. Most often, their sticky-tipped, prey-capturing tongue flips out from the front of the mouth. An adult eats just about any animal it can catch; only its head size dictates the upper limit of prey size. One frog has such a large head, it is called a walking mouth. Skin glands of some species produce toxins, and poisonous types often have bright coloration that advertises their inedibility. The skin of the South African clawed frog (*Xenopus laevis*) is known to contain antibiotics that afford protection against the stew of microbes in its swampy habitat.

Frogs have a closed circulatory system with a three-chambered heart, in which oxygen-rich and oxygen-poor blood mix in the third chamber (Figure 27.9*b*). This may seem inefficient, compared with your four-chambered heart, but it keeps some oxygen circulating when these animals are underwater and cannot breathe. At that time, some gas exchange also takes place at the skin and pharynx.

Caecilians

The ancestors of caecilians lost their limbs and most of their scales, and they gave rise to decidedly worm-shaped amphibians (Figure 27.16*e*). Nearly all of the 150 or so species burrow through soft, moist soil in pursuit of insects and earthworms. A few live in shallow fresh-water habitats.

Regardless of how far they ventured onto land, amphibians have not fully escaped dependency on water.

27.7 REPTILES

The Rise of Reptiles

During the Late Carboniferous, insects began an adaptive radiation into lush habitats on land. At about the same time, the **reptiles** (Reptilia) evolved from amphibians. Like modern species, the amphibious forms probably were carnivores. The huge quantities and selections of edible insects represented a major, untapped food source.

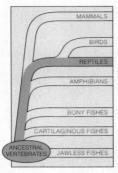

Compared to amphibians, early reptiles pursued prey with greater cunning and speed. Their jaw bones and muscles were better at applying sustained, crushing force. They had well-developed teeth, suitable for securing insects and fellow vertebrates. The limbs of nearly all types were better adapted to support the body on land. The circulatory system of those with a four-chambered heart was more efficient (Figure 27.9). Early reptiles relied fully on more efficient lungs, and they were the first vertebrates to suck in air rather than force it in by mouth muscles. Their cerebrum's thin surface layer, the **cerebral cortex**, was more highly developed (Figure 27.8). This brain region is responsible for the most complex integration of sensory information.

Reptiles were the first vertebrates to escape dependency on standing water. They did so through four adaptations that still distinguish reptiles from fishes and amphibians. *First*, they have tough, scaly skin that limits moisture loss. *Second*, they have some type of copulatory organ that permits internal fertilization. That is, sperm are deposited into a female's body; they do not have to swim through water to reach eggs. *Third*, reptilian kidneys are good at conserving water (the urine of many species is a concentrated paste, not liquid). *Fourth*, reptiles produce **amniote eggs**, in which the embryo develops to an advanced stage before being hatched or born into dry habitats. Such eggs have specialized membranes, and most (but not all) have a leathery or calcified shell (Figure 27.17). The membranes retain water and protect or metabolically support the embryo.

As Figure 27.18 indicates, the reptiles underwent a major adaptive radiation in the Mesozoic era. The types called dinosaurs as well as related forms emerged and ruled the land for the next 125 million years. Their domination finally ended when the Cretaceous drew to a close. The *Focus* essay on page 337 describes their dramatic story. Reptiles that survived to the present day include the turtles, lizards and snakes, tuataras, and crocodilians.

a

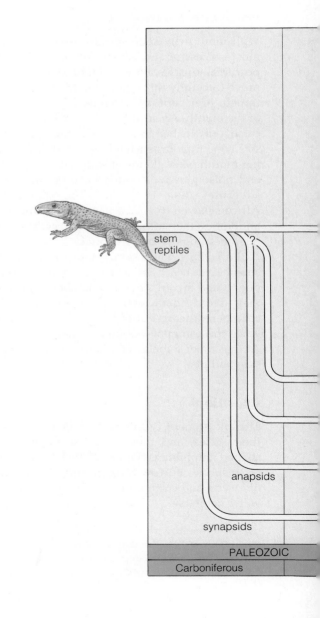

stem reptiles

anapsids

synapsids

PALEOZOIC

Carboniferous

liveborn snake in egg sac

b

Figure 27.17 (**a**) Eastern hognose snakes, emerging from leathery-shelled amniote eggs. Shelled or not, this type of egg contributed to the successful colonization of land by reptiles and, later, birds and mammals. (**b**) Not all snakes lay eggs. For example, offspring of this female copperhead were nourished by yolk reserves in an egg sac, then were born live.

Figure 27.18 (*Below*) Evolutionary history of the reptiles.

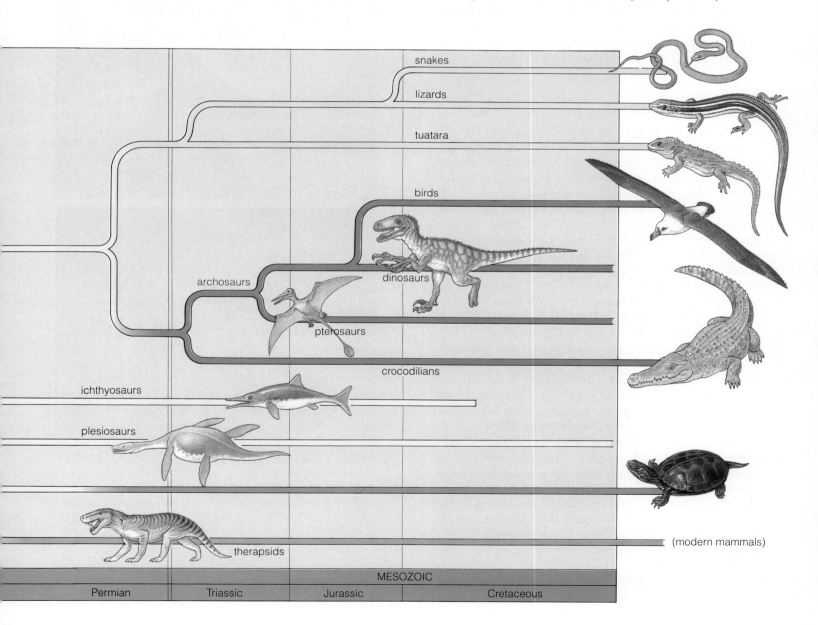

Turtles

The 250 existing species of turtles have inherited a body plan that works well; it has been around since Triassic times. Turtles live in a mobile home, a shell that actually is attached to the skeleton (Figure 27.19*a* and *b*). The shell consists of an inner, bony layer and an outer, horny layer of keratin. When threatened, most species can pull their head and limbs into their shell. Although turtles are toothless, they have tough, horny plates suitable for gripping and chewing food. They also have powerful jaws and sometimes a fierce disposition that helps keep predators at bay. For land-dwelling turtles, the shell also holds in moisture and body heat. Only among sea turtles and other notably mobile types is the shell reduced in size and strength.

All turtles lay eggs on land, then leave them. Predators eat most of the eggs, so few new turtles hatch. Prospects are especially bad for sea turtles. They are endangered species, on the brink of extinction (page 710). Humans kill them for their shell and meat.

Lizards and Snakes

Of all living reptiles, about 95 percent are lizards and snakes—distant relatives of dinosaurs. Most species are small, but the Komodo monitor lizard is large enough to hunt young water buffalo. The longest snake would stretch across 10 yards of a football field.

Most of the 3,750 known species of lizards are insect eaters of deserts and tropical forests. They include aggressive, adhesive-toed geckos that help keep walls and ceilings of homes in the tropics free of spiders and insects. They include iguanas; page 545 shows one of the more colorful types. Most lizards grab prey with small, peglike teeth. Chameleons catch them by accurately flicking at them with their tongue, which is longer than their body (Figure 47.12*a*).

Being small themselves, lizards are prey for many other animals. When grabbed by a predator, many lizards give up their tail. This wriggles for a bit and may be distracting enough to permit a getaway. Some lizards attempt to startle predators—and intimidate rivals—by flaring their throat fan (Figure 27.19*c*).

Short-legged, long-bodied lizards of the late Cretaceous gave rise to the elongated, limbless forms we call snakes. Some modern types of snakes retain bony remnants of the ancestral hindlimbs. Given their body plan, all of the 2,300 or so existing species are limited in the way they can move. Most move in S-shaped waves, much like salamanders. "Side-winding" vipers make surprisingly rapid J-shaped movements across loose sand and sediments.

Besides rattlesnakes, coral snakes are among the species that use venom to subdue prey (Figure 27.19*d*

Figure 27.19 A sampling of reptiles. (**a**) A marine turtle, with its shell streamlined for swimming. (**b**) A heavily shelled Galápagos tortoise. This land-dwelling reptile made a lasting impression on Charles Darwin. (**c**) A frilled lizard, flaring a ruff of neck skin in a defensive display.

On the facing page: (**d**) Coral snake, one of the most venomous types. (**e**) Tuatara, a "living fossil" that hasn't changed much since the age of dinosaurs. (**f**) Rattlesnake of the American Southwest. (**g**) An American alligator with offspring.

and *f*). Pythons and boas coil tightly around a prey animal and suffocate it. Snake jaws are highly movable—some species swallow animals wider than they are. Generally, snakes are not aggressive toward humans, but each year as many as 40,000 people die from bites by the venomous ones.

Tuataras

The two existing species of tuataras (*Sphenodon*) are restricted to small, windswept islands near New Zealand. Figure 27.19*e* shows their body plan, which hasn't changed much since the Mesozoic. Although they resemble lizards, their lineage is more ancient. Tuataras don't engage in sex until they are twenty years old. This works well because tuataras, like turtles, may live for sixty years or more.

Crocodilians

Among the modern crocodiles and alligators are the largest living reptiles. All live in or near water. The feared "man-eater" of southern Asia and the Nile crocodile weigh as much as 1,000 kilograms. They drag a mammal or bird into the water, tear it apart by violently

turning over and over, then gulp down the torn chunks. Alligators of the southern United States were once hunted for their belly skin. They were placed under protection after they neared extinction. Their now-increasing numbers put them at odds with some rural community developments.

All crocodilians have a slender snout, powerful jaws, and sharp teeth (Figure 27.19*g*). Even though they look like big lizards, they are evolutionarily closer to birds. For example, unlike most reptiles, which have conserved the amphibian circulatory system, crocodilians have a four-chambered heart.

Like other reptiles, crocodilians adjust their body temperature through behavioral and physiological mechanisms (page 746). They show complex social behavior, as when parents guard nests and assist hatchlings in their move out of the eggshell and into the water. An unguarded nest is vulnerable to egg-eating mammals, and hatchlings are appetizing to large fishes.

With their scaly skin, reliance on internal fertilization, water-conserving kidneys, and amniote eggs, the reptiles were the first vertebrates to escape dependency on standing water.

27.8 BIRDS

By definition, **birds** alone are animals with feathers (Figure 27.20). **Feathers** are lightweight structures used in flight, insulation, or both. Judging from fossils of *Archaeopteryx*, birds descended from two-legged reptiles that lived about 160 million years ago (page 268). They still resemble reptiles in many internal structures, their horny beaks and scaly legs, and their habit of laying eggs. One of Darwin's champions, Thomas Huxley, argued that birds are glorified reptiles. Today, many biologists do indeed classify birds as a branch of the reptilian lineage.

There are nearly 9,000 known species of birds. They show stunning variation in size, proportions, coloration, and capacity for flight (Figure 27.21). A small hummingbird barely tips the scales at 2.25 grams (0.08 ounce). The largest bird, the ostrich, weighs about 150 kilograms (330 pounds). As Figure 17.2 illustrates, ostriches cannot fly, but they are impressively long-legged sprinters. Warblers and other perching birds

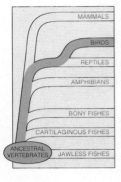

Figure 27.20 Feathers—the defining characteristic of birds. This male pheasant has flamboyant plumage, an outcome of sexual selection. It is a native of the Himalaya Mountains of India, and an endangered species. As is the case for many birds, its jewel-colored feathers end up adorning humans—in this case, on the caps of native tribespeople.

Figure 27.21 A few more characteristics of birds. (**a**) Of all living vertebrates, only birds and bats fly by flapping wings. (**b**) Many birds, including these Canada geese, show migratory behavior. These birds are at their wintering grounds in New Mexico. (**c**) Speckled eggs of a magpie. All birds lay hard-shelled eggs of the sort shown in the generalized diagram.

a

b

c

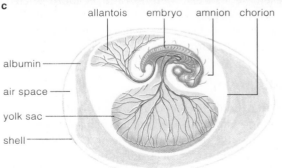

allantois   embryo   amnion   chorion

albumin

air space

yolk sac

shell

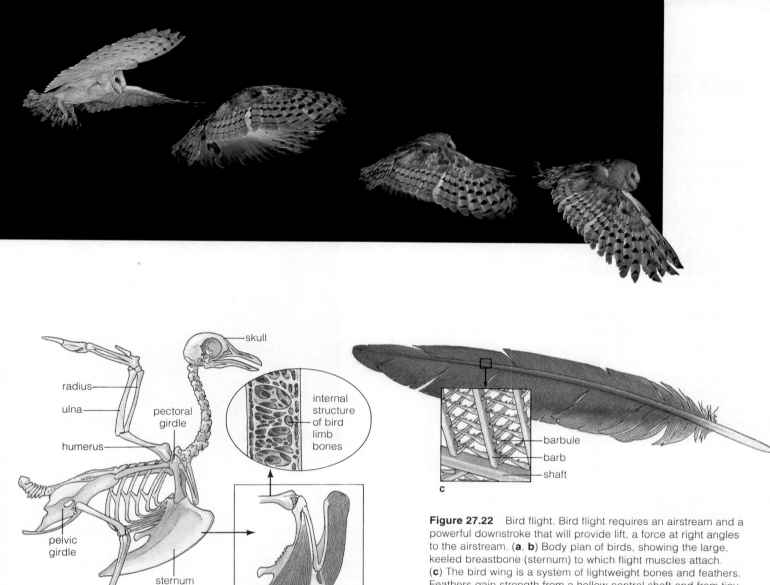

radius

ulna

humerus

skull

pectoral girdle

internal structure of bird limb bones

pelvic girdle

sternum

two main flight muscles attached to keel of sternum

a

b

barbule

barb

shaft

c

Figure 27.22 Bird flight. Bird flight requires an airstream and a powerful downstroke that will provide lift, a force at right angles to the airstream. (**a, b**) Body plan of birds, showing the large, keeled breastbone (sternum) to which flight muscles attach. (**c**) The bird wing is a system of lightweight bones and feathers. Feathers gain strength from a hollow central shaft and from tiny barbules interlocked in a latticelike array. With its long flight feathers, the bird wing serves as an airfoil. Usually, feathers are spread out on the downstroke; this increases the size of the surface pushing against air. On the upstroke, feathers fold somewhat, so the wing presents the least possible profile against the air.

especially differ markedly in feather coloration and in their territorial songs. Bird songs and other splendidly complex social behaviors are topics of later chapters.

Flight demands high metabolic rates, which require a good deal of oxygen. All birds have a large, strong, four-chambered heart that pumps oxygen-enriched blood to the lungs. They also have a unique respiratory system that greatly enhances oxygen uptake. This system is described on page 699. Flight also demands low weight and high power. The bird wing (a forelimb) consists of feathers, powerful muscles, and lightweight bones. Flight muscles attach to an enlarged breastbone

and the upper limb bones adjacent to it (Figure 27.22). When they contract, they produce the powerful downstroke for flight. Bird bones are strong and yet weigh very little because of air cavities in the bone tissue. The skeleton of a frigate bird, which has a 7-foot wingspan, weighs only 4 ounces. That's less than the feathers weigh!

Of all animals, birds alone have feathers, which they use in flight, in heat conservation, and in socially significant visual displays.

27.9 MAMMALS

We turn now to the **mammals**, the most recent of the existing classes of vertebrates. Ancestors of mammals evolved during the Carboniferous, from synapsid reptiles. As Figure 27.18 shows, those ancestral forms diverged from other lineages that led, ultimately, to the dinosaurs and to the modern reptiles and birds. Mammal-like reptiles called **therapsids** appeared on their evolutionary road. They gave rise to many small mammals that survived in the shrubbery during the late Mesozoic, when dinosaurs dominated the land. Mammal-eating dinosaurs probably kept them from moving into more diverse habitats.

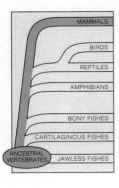

When the Cretaceous period abruptly crashed to a close, the last of the dinosaurs became extinct. Pages 336 through 341 cover this part of the evolutionary story. Afterward, a great mammalian radiation began that continued on into the modern era (Figure 19.11). It was then that the mammalian brain began to reveal its potential. Especially among primates, interconnected masses of information-encoding and information-processing cells expanded dramatically. The expansion was the foundation for our own capacity for memory, learning, and conscious thought.

With well over 4,500 existing species, mammals show great morphological diversity. They range in size from Kitti's hognosed bat, which weighs in at a mere 1.5 grams, to whales that exceed 100 tons. Like other vertebrates, all of these mammals have an internal skeleton. The skeleton's main axis is a bony column (backbone) that has a cordlike bundle of nerves threading through it. Perched above the backbone is a skull. Inside the skull is a complex brain that is functionally divided into three regions (hindbrain, midbrain, forebrain). Also inside are sensory organs concerned with sight, sound, balance, and smell.

Of all vertebrates, only female mammals feed the young with milk, a nutritious fluid produced by their mammary glands (Figure 27.23a). Hence the name of this vertebrate class, from the Latin *mamma*, meaning breast. The females and males care for the young for an extended period and serve as models for their behavior. Young mammals have an inborn capacity to learn and to repeat a set of behaviors that have survival value. Mammals in general show behavioral flexibility. That is, they can expand on the basics with novel forms of behavior.

Mammals typically have hair or thick skin that conserves heat (the trait was lost in most whales). And unlike reptiles, which generally swallow prey whole,

a

most mammals secure, cut, and sometimes chew food before they swallow it. As Figure 27.23c shows, reptiles and mammals differ in their **dentition** (the type, number, and size of teeth). Mammals have four types of upper and lower teeth that match up and work together to crush, grind, or cut food. Their incisors are like flat chisels or cones; they nip or cut food. Horses and other mammals that graze in open grasslands have pronounced incisors. Canines have piercing points. Meat-eating mammals use long, sharp canines to pierce prey. Premolars and molars (cheek teeth) are a platform for food. Their surface bumps (cusps) help crush, grind, and shear food. If a mammal has large, flat-surfaced cheek teeth, you can assume its ancestors evolved in places where fibrous plants were abundant food sources.

Teeth fossilize very well. As you will see in the chapter that follows, fragments of jaws and teeth from human ancestors give clues to their life-styles.

Mammals alone feed their young with milk from mammary glands. They have distinctive dentition and, typically, hair as well as an internal skeleton, a nerve cord, a three-part brain, and sensory organs inside the skull. Their young require an extended period of dependency and learning.

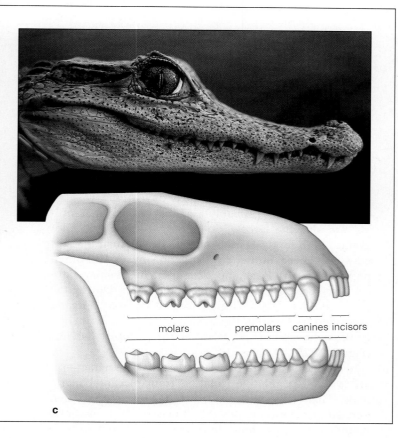

b

Figure 27.23 Distinctly mammalian traits. (**a**) A human baby, busily demonstrating the key defining feature. It derives nourishment from mammary glands. (**b**) An arctic fox, which survives murderously low winter temperatures with the help of hair. Its thick, insulative coat, which remains white during winter, also helps the fox "hide in the open" (camouflage itself) from prey by blending in with its snow-covered habitat. (**c**) Mammals descended from reptiles, which have peglike upper teeth that don't match up with peglike lower ones. Mammalian teeth do, as shown by the generalized diagram.

molars   premolars   canines incisors

c

We turn now to the three major groups of this vertebrate lineage: the egg-laying, pouched, and placental mammals.

Egg-Laying Mammals

Two major types of egg-laying mammals (subclass Prototheria) have survived to the present day. They are the platypus, described at the start of this chapter, and several species of spiny anteaters, which are burrowing animals of Australia and New Guinea (Figure 27.24). Both types lost most or all of their teeth during the course of evolution. The loss is correlated with their specialized diets. Whereas the platypus feeds mostly on small aquatic invertebrates, the spiny anteaters feed on termites and ants, which they capture with their long, sticky tongue.

Several rather archaic traits have persisted in both types of egg-laying mammals. Besides being egg layers, these animals do not show the same kind of skeletal developments of other mammals, and their metabolic rates are lower. Yet their young are suckled after hatching. Like other mammals, the platypus is covered with hair, and the spiny anteaters are covered with spines derived from hair. Despite their low metabolic rates, both mammals maintain a relatively constant body temperature that is well above that of their surroundings.

Figure 27.24 An egg-laying mammal of arid habitats of Australia: the short-nosed anteater (*Tachyglossus aculeatus*). The platypus in Figure 27.1 is another egg-laying mammal.

Pouched Mammals

There are about 260 species of pouched mammals, or marsupials (subclass Metatheria). Nearly all are native to Australia and nearby islands; a few live in North and South America. Marsupial young are born live but not quite "finished"—they are tiny, blind, and hairless. However, the newborns have an excellent sense of smell and strong forelimbs, which they use to locate and reach the mother's pouch, which is some distance away from the birth canal. They are suckled in that pouch, and there they complete their early development.

The kangaroos, probably the most familiar marsupials, are also the largest. Some species weigh 90 kilograms. Even larger species once lived in Australia and may have been hunted to extinction by early human populations. At least one marsupial has managed to coexist successfully with humans, however. During the past century, the Virginia opossum has greatly expanded its range in the United States. It now thrives in urban areas as well as in its native forest habitats (Figure 27.25).

For at least 50 million years, the Australian marsupials evolved in relative isolation from placental mammals. Their ancestors had crossed a narrow sea that opened up between two land masses, but placental mammals stayed behind. In the absence of competition, the marsupials radiated freely into adaptive zones in their own land mass, which became Australia.

Kangaroos, wallabies, and other plant-eating (herbivorous) mammals now occupy adaptive zones comparable to the ones filled by deer, antelope, and other herbivores on other continents. Siberian and North American forests have wolves; Australia has (or had)

Figure 27.26 Representative placental mammals, which occupy diverse aquatic and terrestrial habitats. (**a**) The manatee lives in the ocean and eats submerged seaweed. (**b**) Bats, the only flying mammals, dominate the night sky vacated by birds. (**c**) Camels traversing an extremely hot desert with ease. (**d**) Zebras, sometimes called walking hamburgers for predators of the African savanna. (**e**) Walruses swim in frigid waters and sunbathe on ice. (**f**) Raccoons climb trees.

the Tasmanian "wolf," a marsupial that is now probably extinct. Other marsupials glide like flying squirrels, and some climb like monkeys. In Australia, humans have introduced cattle, sheep, horses, and other placental mammals from other places. Many native marsupials are being threatened with displacement.

Placental Mammals

More than 4,500 placental mammals (subclass Eutheria) are known. One or more members of this group live in virtually every kind of aquatic and terrestrial environment (Figure 27.26).

Chapter 45 will describe the reproduction of humans, a placental mammal, in detail. For now, it is enough to know that a **placenta** is a spongy tissue in a pregnant female's uterus that develops functional connections with the embryo. This tissue, composed of maternal tissue and embryonic membranes, is the

Figure 27.25 A pouched mammal—a female opossum with her young.

c

d

e

means by which the embryo receives nutrients and oxygen and gets rid of metabolic wastes. Placental mammals grow faster in the uterus than marsupials do in a pouch. At birth, many are fully developed and can move about almost immediately. Even then, they all remain with the mother for some time.

Take a quick look at Appendix I, and you will see that placental mammals include a great variety of species—many familiar to us, and others obscure. In terms of diversity, the rodents are most successful. These include rats, mice, squirrels, and prairie dogs. Bats are next in terms of diversity. Other familiar placental mammals are the carnivorous dogs, bears, cats, walruses, and dolphins, as well as the herbivorous horses, camels, deer, and elephants. Exotic types, including manatees and anteaters, also hold membership in this group.

The physiology, behavior, and ecology of many of these mammals will occupy our attention in chapters to follow.

f

SUMMARY

1. Chordates are animals with four distinguishing characteristics. Nearly all chordate embryos, and commonly the adults, have a notochord, a dorsal hollow nerve cord, a pharynx with gill slits (or hints of these), and a tail that extends past the anus.

2. Invertebrate chordates include tunicates (such as sea squirts) and lancelets.

3. There are eight classes of vertebrates:
   a. Jawless fishes, such as lampreys (Agnatha).
   b. Jawed armored fishes (Placodermi); extinct.
   c. Sharks, rays, and other cartilaginous fishes (Chondrichthyes).
   d. Salmon, tuna, coelacanths, and other bony fishes (Osteichthyes).
   e. Frogs, toads, and other amphibians (Amphibia).
   f. Turtles, snakes, crocodiles, and other existing and extinct reptiles (Reptilia).
   g. Robins, eagles, ostriches, penguins, and other flying and flightless birds (Aves).
   h. Platypuses, kangaroos, opossums, camels, bats, humans, and other mammals (Mammalia).

4. We find clues to vertebrate origins in the body plan and embryonic development of tunicates as well as the wormlike hemichordates. In several respects, the earliest vertebrates may have resembled some of the larval forms and adults of these animals.

5. The first vertebrates were jawless fishes that arose during the Cambrian era, as represented by the ostracoderms. Their only living descendants are lampreys and hagfishes.

6. Jawed fishes also arose during the Cambrian. A dominant lineage, the placoderms, is now extinct. Other lineages gave rise to the cartilaginous and bony fishes.

7. The following trends occurred during the evolution of certain vertebrate lineages:
   a. A vertebral column supplanted the notochord as the structural element against which muscles act. The vertebral column foreshadowed the evolution of fast-moving, predatory animals.
   b. Jaws evolved from gill-supporting elements. They led to increased predator-prey competition. This in turn favored the evolution of more efficient nervous systems and sensory organs.
   c. In one lineage of bony fishes, paired fins evolved into fleshy lobes with internal structural elements—the forerunners of paired limbs.
   d. In certain fishes, lungs evolved and supplemented respiration by way of gills. Lungs proved adaptive in the invasion of land. In a related development, the circulatory system became more efficient at distributing oxygen.

8. With staggering numbers of individuals distributed among 21,000 or so known species, the bony fishes (especially the ray-finned species) are the most successful vertebrates. They exhibit spectacular variation in morphology and behavior.

9. Amphibians were the first vertebrates to invade the land, but they never fully escaped the water. Their skin drys out, and aquatic stages persist in the life cycle of all species.

10. Reptiles evolved from amphibious forms. They were the first vertebrates to escape dependency on standing water, owing to these adaptations:
    a. Tough, scaly skin that conserves body moisture.
    b. Copulatory organs that permit internal fertilization.
    c. Extremely efficient, water-conserving kidneys.
    d. Amniote eggs, often leathery or shelled, that protect and metabolically support the embryos.

11. Reptiles, and the birds and mammals that descended from certain reptilian lineages, have efficient circulatory and respiratory systems and a well-developed nervous system and sensory organs.

12. Of all vertebrates, birds alone have feathers, which they use in flight, heat conservation, and social displays.

13. Mammals alone have milk-producing mammary glands, and they have hair or thick skin that functions in insulation. They have distinctive dentition and a highly developed brain. Adults nurture their young through an extended period of dependency and learning.

Review Questions

1. List and describe the features that distinguish chordates from other animals. *446*

2. Name and describe the characteristics of an organism from each of the two groups of invertebrate chordates. *447–448*

3. List four evolutionary trends that occurred in certain vertebrate lineages. *450–451*

4. Which evolutionary modifications in fishes set the stage for the emergence of amphibians? *452–456*

5. List some of the identifying characteristics of reptiles, birds, and mammals. *458–467*

6. List the eight classes of vertebrates and cite examples of each. *446*

7. Which class of vertebrates was the first to invade the land? *456*

8. What specific evolutionary adaptations allowed reptiles to evolve from amphibians? *458*

1. The embryos and often the adults of _____ have a notochord, a tubular dorsal nerve cord, a pharynx with slits in the wall, and a tail extending past the anus.
   a. echinoderms
   b. tunicates and lampreys
   c. vertebrates
   d. both b and c
   e. all are correct

2. Gill slits function in _____ .
   a. respiration
   b. circulation
   c. food trapping
   d. water regulation
   e. both a and c

3. Existing aquatic vertebrates include _____ .
   a. jawed, armored fishes
   b. jawless fishes
   c. bony fishes
   d. both a and b
   e. both a and c
   f. both b and c

4. _____ appeared early on the evolutionary road leading to vertebrates.
   a. Tunicates and lancelets
   b. Hemichordates
   c. both a and b
   d. neither is correct

5. A shift from a reliance on _____ to reliance on _____ was pivotal in the evolution of all vertebrates.
   a. the notochord; a backbone
   b. filter feeding; jaws
   c. gills; lungs
   d. all are correct

6. The first vertebrates were _____ .
   a. bony fishes
   b. jawless fishes
   c. jawed fishes
   d. both a and b

7. The bony fishes include _____ .
   a. ray-finned fishes
   b. lobe-finned fishes
   c. lungfishes
   d. all are correct

8. Of all existing vertebrates, _____ are the most diverse.
   a. cartilaginous fishes
   b. bony fishes
   c. amphibians
   d. reptiles
   e. birds
   f. mammals

9. A four-chambered heart is characteristic of _____ .
   a. bony fishes
   b. amphibians
   c. birds
   d. mammals
   e. both b and c
   f. both c and d

10. The only amphibians to entirely escape dependency on aquatic habitats are _____ .
    a. salamanders
    b. desert toads
    c. caecilians
    d. none is correct

11. Adaptations that permitted reptiles to escape dependency on aquatic habitats were _____ .
    a. tough skin
    b. internal fertilization
    c. good kidneys
    d. amniote eggs
    e. both b and d
    f. all are correct

12. _____ have highly efficient circulatory and respiratory systems, and a complex nervous system and sensory organs.
    a. Reptiles
    b. Birds
    c. Mammals
    d. all are correct

13. Birds use feathers in _____ .
    a. flight
    b. heat conservation
    c. social functions
    d. all are correct

14. Various mammals _____ .
    a. hatch
    b. complete embryonic development in pouches
    c. complete embryonic development in the uterus
    d. both b and c
    e. all are correct

15. Match the organisms with the appropriate features.
    _____ jawless fishes
    _____ cartilaginous fishes
    _____ bony fishes
    _____ amphibians
    _____ reptiles
    _____ birds
    _____ mammals
    a. complex three-part brain, thick skin or hair
    b. respiration by skin and lungs
    c. include coelocanths
    d. include hagfishes
    e. include sharks and rays
    f. complex social behavior, feathers
    g. first with amniote eggs

Selected Key Terms

amniote egg 458
amphibian 456
bird 462
bony fish 454
cartilaginous fish 454
cerebral cortex 458
chordate 446
dentition 464
feather 462
fin 450
gill 451
gill slit 447
hagfish 453
hemichordate 448
jaw 450
lamprey 453

lancelet 448
lobe-finned fish 455
lung 451
mammal 464
nerve cord 446
notochord 446
ostracoderm 452
pharynx 446
placenta 466
placoderm 453
reptile 458
scale 454
swim bladder 452
therapsid 464
tunicate 447
vertebra 450

Readings

Carroll, R. L. 1988. *Vertebrate Paleontology and Evolution.* New York: Freeman.

Hoffman, E. 1990. "Paradox of the Platypus." *International Wildlife* 20(1):18–21.

Romer, A. S., and T. S. Parsons. 1986. *The Vertebrate Body.* Sixth edition. Philadelphia: Saunders.

Welty, J., and L. Baptista. 1988. *The Life of Birds.* Fourth edition. New York: Saunders.

28 HUMAN EVOLUTION: A CASE STUDY

The Cave at Lascaux and the Hands of Gargas

Half a century ago, on a warm autumn day, four boys out for a romp stumbled into a cave near Lascaux, a town in the Perigord region of France. What they discovered inside that intricately tunneled cave stunned the world.

Magnificent sketches, engravings, and paintings swept out across the cave walls (Figure 28.1). The red, yellow, purple, and brown pigments were as vivid as if they had been painted only recently. Yet radioisotope measurements revealed that they were 17,000 to 20,000 years old. The prehistoric artists worked deep within the cave, where sunlight could not fade the images and winds and water could not wear them away. By the light of crude oil lamps, they captured the graceful, dynamic lines of bison, stags, stallions, ibexes, lions, a rhinoceros, and a heifer, now known as the Great Black Cow.

Caves throughout southern France, northern Spain, and Africa hold treasures from even earlier times. About 25,000 years ago, for example, prehistoric peoples carefully committed more than 150 imprints and outlines of their hands to walls in the cave of Gargas, in the Pyrenees.

Who were the people who did this? From their fossilized remains, we know that they were anatomically like us. From the way they planned and executed their

Figure 28.1 Part of the human cultural heritage—prehistoric cave paintings, a unique outcome of a long history of biological evolution.

art, we sense a level of abstract thinking that is unique to humans.

The quality of "humanness" did not materialize out of thin air. The story of the human species began more the 60 million years ago, with the origin of primates in tropical forests. In turn, the primate story began more than 250 million years ago, with the origin of mammals. The story extends back to the origin of animals at some time before 750,000,000 years ago—and so on back in time, to the origin of the first living cells.

As we poke about the branches of our family tree, keep this greater evolutionary story in mind. *Our "uniquely" human traits emerged through modification of traits that had already evolved in ancestral forms.* At each branch in the family tree, certain mutations produced workable changes in traits, which proved useful in prevailing environments.

From this perspective, "ancient" cave paintings are the legacy of individuals who departed only yesterday, so to speak. The artists of Lascaux and Gargas are not remote from us. They *are* us.

KEY CONCEPTS

1. The primate branch of the mammalian lineage includes prosimians, tarsioids, and anthropoids, which include the monkeys, apes, humans, and ancestral humanlike forms. Apes and humans, which share common ancestry, are hominoids. Only humans and their humanlike ancestors are further classified as hominids.

2. Unlike other primates, many early hominids did not become restrictively specialized in a single habitat. They remained flexible and became adapted to a wide range of challenges in complex, unpredictable environments.

3. The capacity to make generalized responses that work in different environments is an outcome of several trends that occurred in certain primate lineages. They involved changes in bones, muscles, teeth, sensory systems, and the brain itself.

4. Ancestral, four-legged primates underwent skeletal modifications that led to an upright stance, which in turn freed forelimbs and hands for new functions. Modifications in handbones and muscles led to a capacity to hold, carry, use, and make objects.

5. Teeth became less specialized, and this allowed a greater variety of food sources to be tapped.

6. Reorganization of the skull and eye sockets led to increased reliance on daytime vision.

7. Brain tissue increased in volume and became more complex. The evolution of the brain was interlocked with cultural evolution.

In the preceding chapters, we likened the history of life to a great tree. Each branch of the tree represents a line of descent, and the branch points represent a divergence leading to new species. There are more than 47,000 existing species of fishes, amphibians, reptiles, birds, and mammals on the vertebrate branch of that tree. When we turn to the evolution of any one of those species—as we do here—it helps to keep a key point in mind. At each crossroad leading to a new species, complex traits were already in place and functioning—*and new traits emerged only through modification of traits that already were in place.* The evolution of the human species speaks eloquently of this characteristic of life.

PRIMATE CLASSIFICATION

The order **Primates** includes prosimians, tarsioids, and anthropoids. Figure 28.2*a* through *c* shows a few of these distinctive mammals, and Figure 28.2*d* shows their presumed evolutionary relationships.

Prosimians are the oldest lineage (*pro*, before; *simian*, ape). For millions of years, prosimians dominated the trees in North America, Europe, and Asia. That was before monkeys and apes evolved and almost displaced them entirely. Tarsioids, represented by tarsiers of southeastern Asia, are small primates with features that place them between prosimians and anthropoids.

Monkeys, apes, and humans are **anthropoids**. In structural details and biochemistry, apes are far more similar to humans than to monkeys. That is why apes and all species of the human lineage are classified together, as **hominoids**. A divergence from their common ancestor began many millions of years ago. All species on the separate evolutionary road leading to humans are classified as the **hominids**.

Figure 28.2 Representative primates. (**a**) Gibbons have limbs and a body adapted for swinging arm over arm through the trees. (**b**) Monkeys are quadrupedal (four-legged) climbers, leapers, and runners, as this spider monkey demonstrates. (**c**) Tarsiers are vertical clingers and leapers. (**d**) Evolutionary tree for the primates. *Green* boxes show the two highest groupings (the prosimians, and the tarsioids and anthropoids). *Tan* boxes show the major groups of living primates. A few representative members are named. Monkeys, apes, and humans are all anthropoids. The only hominids are humans and now-extinct species of their lineage.

a

b

c

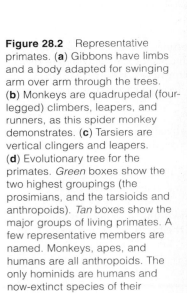

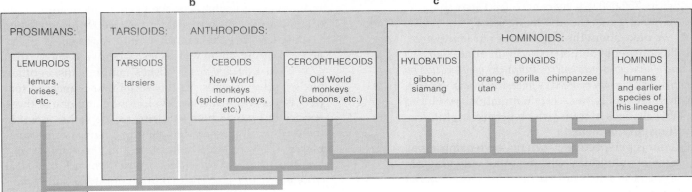

PROSIMIANS:	TARSIOIDS:	ANTHROPOIDS:		HOMINOIDS:		
LEMUROIDS	TARSIOIDS	CEBOIDS	CERCOPITHECOIDS	HYLOBATIDS	PONGIDS	HOMINIDS
lemurs, lorises, etc.	tarsiers	New World monkeys (spider monkeys, etc.)	Old World monkeys (baboons, etc.)	gibbon, siamang	orang- gorilla chimpanzee utan	humans and earlier species of this lineage

d

tree-dwelling, rodentlike primate of Paleocene

28.2 FROM PRIMATE TO HUMAN: KEY EVOLUTIONARY TRENDS

Most primates live in tropical or subtropical forests, woodlands, or **savannas** (open grasslands with a few stands of trees). Like their ancient ancestors, the vast majority are tree dwellers. Yet no one feature sets "the primates" apart from other mammals. Each primate lineage evolved in a distinct way and has its own defining traits. Five trends define the lineage we are concerned with here. They were set in motion when primates first started adapting to life in the trees, and they contributed to the emergence of modern humans.

---

1. Skeletal changes led to upright walking, which freed the hands for new functions.

2. Changes in bones and muscles led to refined hand movements.

3. There was less reliance on the sense of smell and more on daytime vision.

4. Changes led to fewer, less specialized teeth.

5. Brain elaboration and changes in the skull led to speech. These developments became interlocked with each other and with cultural evolution.

Upright Walking

Of all primates, only humans can stride freely on two legs for a long time. Their habitual two-legged gait is called **bipedalism**. By contrast, monkeys are adapted to life in the trees. Their skeleton permits rapid climbing, leaping, and running along branches. Their armbones and legbones are about the same length (Figure 28.2). This means monkeys can run palms-down. Try this yourself and see what happens.

Unlike monkeys, apes hang onto overhead branches and use their long arms to carry some body weight. The arms often support body weight when an ape is on the ground. Because of the way their shoulder blades are positioned, apes can swivel the arms freely above the head when the body is erect or semi-erect.

Compared with monkeys and apes, humans have a shorter, S-shaped, and somewhat flexible backbone. The position and shape of their backbone, shoulder blades, and pelvic girdle are the basis of bipedalism (Figure 28.3). These skeletal traits emerged not long after the divergence that led to the hominids.

Figure 28.3 Comparison of the skeletal organization and stance of a monkey, ape (gorilla), and human. Their modes of locomotion resulted from modifications of the basic mammalian plan. The quadrupedal monkeys climb and leap; apes climb and swing by forelimbs. Both modes are suited for life in the trees. Humans are two-legged striders. The drawings are not to the same scale.

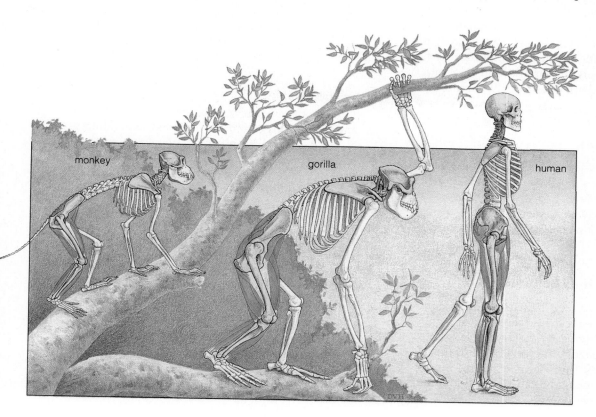

monkey gorilla human

Precision Grips and Power Grips

The first mammals spread their toes apart to help support the body as they walked or ran on four legs. Primates still spread their toes or fingers. Many also make cupping motions, as when monkeys bring food to the mouth. Two other hand movements developed in ancient tree-dwelling primates. Through alterations in handbones, fingers could be wrapped around objects (*prehensile* movements), and the thumb and tip of each finger could touch (*opposable* movements).

Hands began to be freed from load-bearing functions when early primates lived in the trees. Later, when hominids were evolving, refinements in hand movements led to the precision grip and power grip:

These hand positions gave early humans the capacity to make and use tools. They were a foundation for unique technologies and cultural development.

Enhanced Daytime Vision

Early primates had an eye on each side of the head. Later ones had forward-directed eyes, an arrangement that is better for sampling shapes and movements in three dimensions. Through additional modifications, the eyes were able to respond to variations in color and light intensity (dim to bright). These visual stimuli are typical of life in the trees.

Teeth for All Occasions

Monkeys have rectangular jaws and long canines (page 464). Humans have a bow-shaped jaw and smaller teeth of about the same length. Jaws and teeth became modified on the road from early primates to humans. There was a shift from eating insects, then fruit and leaves, and on to a mixed diet.

Better Brains, Bodacious Behavior

Living on tree branches favored shifts in reproductive and social behavior. Imagine the advantages of single births over litters, for example, or of clinging longer to the mother. In many lineages, parents started to invest more in fewer offspring. They formed strong bonds with their young, maternal care became intense, and the learning period grew longer (Figure 28.4).

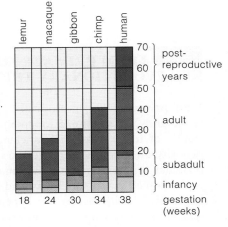

Figure 28.4 Trend toward longer life spans and longer periods of infant dependency among existing primates.

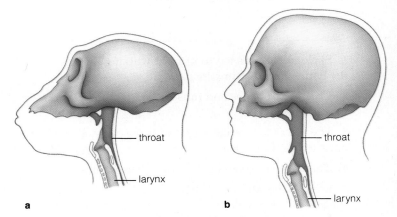

Figure 28.5 Structural basis of speech. (**a**) Like modern chimpanzees, early humans had a skull with a flattened base. Their larynx (a tube leading to the lungs) was not far below the skull, so the throat volume was small. (**b**) In modern humans, the skull's base angles down sharply as it develops. This moves the larynx down also. The result is an increase in the volume of the throat—the area in which sounds are produced.

And now brain regions concerned with encoding and processing information started expanding greatly. New behavior stimulated development of those regions—which in turn stimulated more new behavior. And so brain modifications and behavioral complexity became highly interlocked. The interlocking is most evident in the parallel evolution of the human brain and culture. **Culture** is the sum total of behavior patterns of a social group, passed between generations by learning and by symbolic behavior—especially language. The capacity for language arose among ancestral humans. And it arose through changes in the skull and expansion of parts of the brain (Figure 28.5).

These features emerged along the evolutionary road leading to humans: upright walking, refined hand movements, generalized dentition, refined vision, and the interlocked elaboration of brain regions and cultural behavior.

28.3 PRIMATE ORIGINS

Primates evolved from ancestral mammals more than 60 million years ago, during the Paleocene. Fossils indicate that the first ones resembled small rodents or tree shrews (Figures 28.6 and 28.7). Like tree shrews, they probably had huge appetites and foraged at night for insects, seeds, buds, and eggs beneath the trees of tropical forests. They had a long snout and a good sense of smell, suitable for detecting food or predators. They could claw their way up through the shrubbery, although not with much speed or grace.

Between 54 and 38 million years ago (the Eocene), some primates were staying in the trees. Fossils provide tangible evidence of increased brain size, a shorter snout, enhanced daytime vision, and refined grasping movements. How did these traits evolve?

Consider the trees. Trees offered food and safety from ground-dwelling predators. They also were a habitat of uncompromising selection. Imagine dappled sunlight, boughs swaying in the wind, colorful fruit tucked among the leaves, perhaps predatory birds. A long, odor-sensitive snout would not have been of much use up in the trees, where air currents disperse odors. But a brain that could assess movement, depth, shape, and color would have been a definite plus. So would a brain that worked fast when its owner was running, swinging, and leaping (especially!) from branch to branch. Distance, body weight, winds, and suitability of the destination had to be estimated, and any adjustments for miscalculations had to be quick.

By the dawn of the Oligocene, 35 million years ago, tree-dwelling anthropoids had evolved in the forests. The ancestors of monkeys, apes, and humans were among them. Some forms lived above swamps that were infested with predatory reptiles. Perhaps that is why they rarely ventured to the ground—and why it

Figure 28.7 A night-foraging tree shrew of Indonesia.

was imperative to think fast and grip strongly. Slip-ups were always possible; a surprising number of primates still fall out of the trees.

During the Miocene, which extended from 25 million to 5 million years ago, continents began to assume their current positions (Figure 21.16). Climates were becoming cooler and drier. An adaptive radiation of apelike forms—the first hominoids—took place, and by 13 million years ago, ape populations were scattered through Africa, Europe, and southern Asia. Among them were chimpanzee-sized **dryopiths** (Figure 28.6). Most Miocene apes became extinct around that time. However, fossils and biochemical studies point to three divergences that occurred between 10 million and 5 million years ago. Two branchings gave rise to gorillas and chimpanzees. The third gave rise to early hominids—including the ancestors of humans.

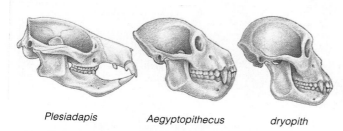

| Plesiadapis | Aegyptopithecus | dryopith |

Figure 28.6 Comparison of skull shape and teeth of some extinct primates. *Plesiadapis*, a Paleocene primate, had rodentlike teeth. *Aegyptopithecus*, an Oligocene anthropoid, probably predates the divergence leading to Old World monkeys and the apes. An adaptive radiation that began in the Miocene produced many dryopiths and other apelike forms. The drawings are not to the same scale. In size, *Plesiadapis* was like a tiny tree shrew. *Aegyptopithecus* was monkey-sized, and the dryopiths, chimpanzee-sized.

---

1. Small, rodentlike mammals started moving into arboreal habitats about 60 million years ago. The founders of the primate lineage were among them.

2. After 25 million years, during Miocene times, their descendants included the anthropoid ancestors of monkeys, apes, and the hominids.

3. A divergence led to the first hominids—the earliest ancestors of humans. It began between 10 and 5 million years ago, before the Miocene-Pliocene boundary.

28.4 THE FIRST HOMINIDS

Most of the earliest known hominids lived in the East African Rift Valley, where a 3,200-kilometer (2,000-mile) fracture in the earth's crust runs from the Red Sea down into Mozambique. For the past 20 million years, volcanoes have intermittently spewed, the valley floor has buckled upward and downward, and lakewaters have collected in the lowland basins.

At the Miocene-Pliocene boundary, the long-term shift to a cooler, drier climate triggered the breakup of the once-vast tropical forests, with their bounty of soft fruits and insects. Now there were scattered woodlands and open, grassy plains. There were pronounced seasonal variations in food availability. The African savanna had emerged.

The survivors in these challenging environments were able to locate and exploit new kinds of food and hide from new kinds of predators. This must have been a "bushy" time of evolution, with branchings and radiations into new adaptive zones. Why? Fossil hunters have found and accurately dated the fragmented remains of a variety of humanlike forms.

We don't have enough fossils to discern how all the forms were related (Figure 28.8). Even so, many had three features in common. *First*, they were upright walkers; their hands were freed for new functions. *Second*, modifications in their teeth, jawbones, and jaw muscles allowed them to vary their diet. *Third*, their brain must have been more elaborate than that of their forerunners. At the least, they had to be able to think ahead, to plan when and where to get foods when seasonal supplies ran out. All three features emerged through modifications of traits that can be observed among the other primates. *They were based on the primate heritage.*

We call the first known hominids **australopiths** (southern apes). They fall into two broad categories:

1. Gracile forms (slightly built), called *Australopithecus ramidus*, *A. afarensis*, and *A. Africanus.*

2. Robust forms (muscular, heavily built), including *A. boisei* and *A. robustus.*

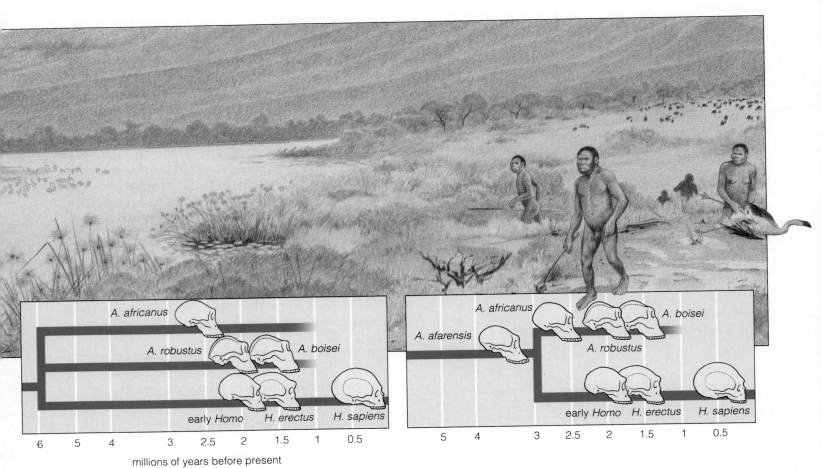

millions of years before present

Figure 28.8 Reconstruction of one of the environments in which the first humanlike forms (early hominids) evolved. Their evolutionary connections with one another and with later hominids are not understood. Several phylogenetic trees have been proposed, including the two shown here for australopiths and for species on the road to modern humans.

With an average volume of 400 cubic centimeters, the australopith brain was a long way from Einstein's. But these hominids were fine striders, like later ones. Legbone fragments, 4 million years old, have muscle attachment sites similar to yours. Figures 28.9 and 28.10 show the reconstructed form and skeleton of a female dubbed Lucy. Her thighbones angled inward, so her body weight was centered directly beneath the pelvis— a sure sign of bipedalism. Thighbones angle outward in apes, which have a waddling, four-legged gait. Most telling, the australopiths left footprints.

Also like later hominids, some australopiths had a bowed jaw. Different types ate different things. *A. robustus* had the strong jaw muscles and the grinding platform typical of plant eaters. *A. boisei*, with its huge, heavily cusped molars, may have been adapted to chewing dry seeds, nuts, and other tough plant material. The cheek teeth of *A. africanus* served as a grinding platform for plants—but the relatively large incisors were more typical of carnivores.

Australopiths, the earliest hominids, were small brained and bipedal, apelike in some ways and humanlike in others.

Australopiths endured for at least 3 million years in the African savanna. The ancestors of humans may have been among these transitional forms.

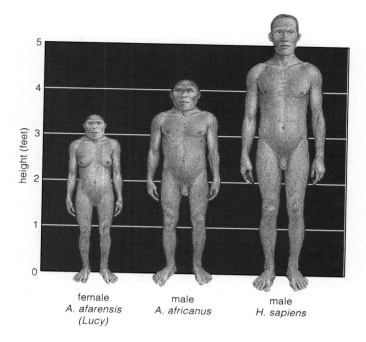

Figure 28.9 Size and stature of two australopiths, compared with a modern human. In 1994, an Ethiopian research team led by Tim White of the University of California at Berkeley found several fossils of a different australopith, *A. ramidus*. This form existed 4.4 million years ago, which puts it close to the presumed time of divergence from the last common ancestor of apes and humans. At that time, tropical forests were giving way to woodlands and grasslands. *A. ramidus* was chimplike in some respects. Yet its flattened face, smaller canines, and other features seem to place it on the evolutionary road that led to humans.

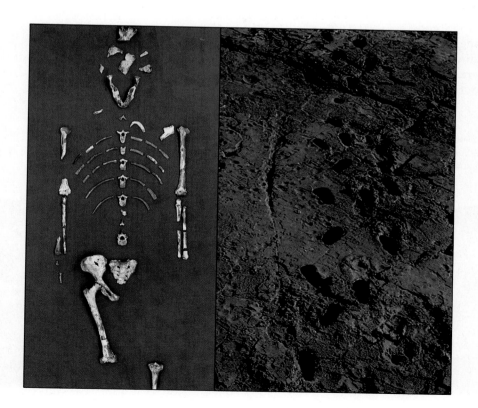

Following the path produces, at least for me, a kind of poignant time wrench. At one point, and you need not be an expert tracker to discern this, the traveler stops, pauses, turns to the left to glance at some possible threat or irregularity, then continues to the north. This motion, so intensely human, transcends time. Three million seven hundred thousand years ago, a remote ancestor—just as you or I—experienced a moment of doubt.

—Mary Leakey

Figure 28.10 Evidence of early hominids. The photograph at far left shows the remains of Lucy, one of the earliest known australopiths. The limb bone density indicates she had strong muscles. The adjacent photograph shows footprints made in soft, damp volcanic ash 3.7 million years ago at Laetoli, Tanzania, as discovered by Mary Leakey. The arch, big toe, and heel marks were made by bipedal hominids.

28.5 ON THE ROAD TO MODERN HUMANS

Early *Homo*

By 2.5 million years ago, some hominids were making stone tools. Maybe they were using sticks and other perishable items before then, much like modern apes do, but we have no way of knowing. We also don't know *which* species actually made stone tools. Maybe it was **early *Homo***. (This is only one of its names. In other taxonomic schemes the same form is called "late australopith" or *H. habilis*.) Early *Homo* may have been on or near the road leading to modern humans.

Some skull fragments of early *Homo* are dated at 2.4 million years. Compared to australopiths, this hominid had a smaller face, more generalized teeth, and a larger brain (Figure 28.11). It ate plants and may have scavenged the remains of small animals.

Early *Homo* may have started down a toolmaking road by picking up rocks to crack open animal bones and expose the soft, edible marrow. Perhaps it started using sharp-edged flakes, formed naturally as rocks tumbled through a river or down a hill. Flakes could scrape flesh from animal bones. And then some hominid started *shaping* stone implements. The earliest known "manufactured" tools were crudely chipped pebbles, first discovered by Mary Leakey at Olduvai Gorge (Figure 28.12). This African gorge cuts through a sequence of sedimentary deposits. The crudest tools are in the deepest layers. They may have been used in different tasks, such as digging the ground for roots and tubers, smashing bones for marrow, and poking insects out from tree bark. In more recent layers we find more complex tools. They signify that hominids were getting serious about exploiting a major food source—the abundant game of open grasslands.

Figure 28.11 Comparison of skull shapes of early hominids with that of an anatomically modern human (*Homo sapiens*). The drawings are not all to the same scale. White indicates reconstructed portions of missing bone tissue.

From *Homo erectus* to *H. sapiens*

Where modern human populations arose is a hotly debated topic. By one model, they evolved in Africa, from populations of an early human species called ***Homo erectus***, then migrated into Asia, Europe, and other regions starting about 900,000 years ago. By another model, *H. erectus* migrated out of Africa, then distinctive subpopulations of modern humans arose later, as an outcome of genetic divergence in different geographic regions (Figure 28.13). In 1994, fossil hunters in Southeast Asia and in the former Soviet republic of Georgia unearthed *H. erectus* specimens that are 1.8 million and 1.6 million years old. The finds lend support to the second model.

H. erectus individuals had a long, chinless face, a thick-walled skull, a heavy browridge, and other archaic traits (Figure 28.11). Like some modern humans, they also were tall, rather narrow hipped, and long legged. They were suitably built for travel. In cold regions they became stockier, with bones adapted to muscular stress. Life must have been challenging, and strenuous. More than once, vast glaciers advanced and retreated in northern Europe, Asia, and North America. Some glaciers were 2 miles high!

This was the time of cultural lift-off for the human species. From southern Africa to England, *H. erectus* populations were using the same kinds of hand axes and other tools designed to pound, scrape, shred, cut, and whittle. *H. erectus* also learned to control fire.

Fully modern humans, ***Homo sapiens***, evolved 300,000 to 200,000 years ago. Older populations of *H. erectus* coexisted with them, then disappeared gradually. Early *H. sapiens* had smaller teeth and jaws. Many individuals had a chin. They had thinner facial bones, a larger brain, and a rounder, higher skull. Certain aspects of their skeletal organization indicate they had the capacity for complex human language (Figure 28.5).

One group of early humans, the Neandertals, lived in southern France, central Europe, and the Near East about 130,000 years ago. Although large brained, they

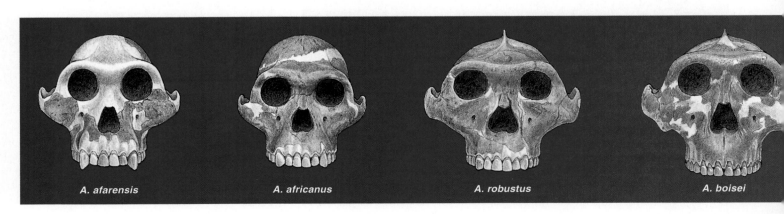

A. afarensis   A. africanus   A. robustus   A. boisei

Figure 28.12 A sampling of the 37,000+ stone tools from Olduvai Gorge. *First row, top to bottom:* The stone ball may herald a transition to aggressive tool use. It resembles Argentine bolas, which are strung on lengths of hide. Bolas are thrown at animals to entangle the legs and bring them down. Below the ball are three choppers. The last two are more advanced forms that have a joint as well as a sharp edge. *Second row:* A hand ax and a cleaver.

had heavy facial bones, often a large browridge, and a skull base flatter than that of *H. erectus.* Neandertals lived at the edge of forests, in caves, rock shelters, and open-air camps. Apparently they hunted or gathered food at their doorstep, so to speak. They did not learn how to exploit the abundant game herds. They disappeared 35,000 or 40,000 years ago, when anatomically modern humans arose. The more recent groups learned to store and share food among themselves. These newly evolved humans developed spectacularly rich cultures. They were the creators of the exquisite tools and artistic treasures at Lascaux and elsewhere.

From 40,000 years ago to the present, human evolution has been almost entirely cultural rather than biological, and so we leave the story. From the biological perspective, however, we can make these concluding remarks: Humans spread throughout the world by rapidly devising the cultural means to deal with a broad range of environments. Compared with their predecessors, modern humans developed rich and varied cultures, moving from "stone-age" technology to the age of "high tech." Yet hunters and gatherers persist in parts of the world, attesting to the great plasticity and depth of human adaptations.

Cultural evolution has outpaced the biological evolution of the only remaining human species. Humans everywhere rely on cultural innovation to adapt rapidly to a broad range of environmental challenges.

0.2%   0.1%   0% = genetic distance

NEW GUINEA, AUSTRALIA

PACIFIC ISLANDS

SOUTHEAST ASIA

ARCTIC, NORTHEAST ASIA

NORTH, SOUTH AMERICA

NORTHEAST ASIA

EUROPE, MIDDLE EAST

AFRICA

Figure 28.13 (*Above*) Proposed family tree for modern human populations that are native to different geographic regions. The branch points reflect divergences that began after the far-reaching migrations of *Homo erectus* populations. Data are based on biochemical and immunological comparisons (page 308). The first major divergence seems to have occurred after some *H. erectus* populations spread from Africa into Europe and others spread into Southeast Asia.

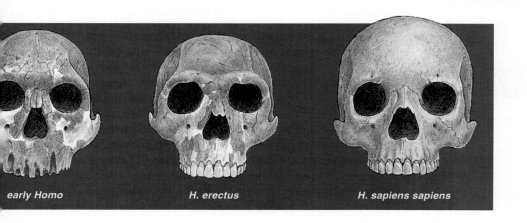

early Homo   H. erectus   H. sapiens sapiens

SUMMARY

1. Like other mammals, humans and other primates have an internal skeleton, a complex brain and sensory organs within a skull, mammary glands (in females), and distinctive teeth. Adults nourish, protect, and serve as behavioral models for the young.

2. Primates include prosimians (lemurs and related forms) and anthropoids (including tarsioids, monkeys, apes, and humans). Only apes and humans are hominoids. Only anatomically modern humans (*H. sapiens*) and others of their lineage (from *A. afarensis* to *H. erectus*) are further classified as hominids.

3. Unlike most primates, humans are not restrictively specialized. They remain flexible and adapt to a wide range of challenges in diverse environments. This capacity resulted from five evolutionary trends that occurred among certain primate lineages:

   a. From a four-legged gait to bipedalism.

   b. Increased manipulative skills owing to changes in the hands, which began to be freed from load-bearing functions among tree-dwelling primates.

   c. Less reliance on the sense of smell and more on enhanced daytime vision.

   d. From specialized to omnivorous eating habits.

   e. Increases in brain complexity and behavior.

4. The first primates—small, rodentlike mammals—evolved more than 60 million years ago. Anthropoids, including the ancestors of monkeys, apes, and humans, had evolved by 35 million years ago. The earliest known fossils of hominids (*A. ramidus*) date from approximately 4.4 million years ago.

5. About 2.4 million years ago, early *Homo* evolved. It may have been the first stone toolmaker. *Homo erectus*, the presumed ancestor of modern human populations, not only had evolved by 1.8 million years ago, some populations of this species had already moved out of Africa, into Asia and possibly Europe by then. Anatomically modern humans (*H. sapiens*) arose between 300,000 and 200,000 years ago. Starting about 40,000 years ago, cultural evolution has outstripped biological evolution of the human form.

Review Questions

1. What is the difference between "hominoid" and "hominid"? Are we hominoids, hominids, or both? 472

2. What are the general evolutionary trends that occurred among the primates as a group? What way of life apparently was the foundation for these trends? 473–474

3. What conditions seem to have been responsible for the great adaptive radiation of apelike forms during the Miocene? 475

Self-Quiz (Answers in Appendix IV)

1. Primates include _____ .
   a. lemurs   d. humans
   b. monkeys   e. all of the above
   c. apes

2. Adaptive flexibility is most characteristic of _____ .
   a. mammals   c. humans
   b. primates   d. all of the above

3. _____ are hominoids; _____ are hominids.
   a. Lemurs and monkeys; apes and humans
   b. Apes and humans; australopiths and humans
   c. Monkeys; apes and australopiths
   d. Monkeys, apes, and humans; apes only

4. Hominids showed plasticity. This means they _____ .
   a. could adapt to a wide range of demands in complex environments
   b. were adapted for a narrow range of demands in complex environments
   c. had flexible bones that cracked easily
   d. were limber enough to swing through the trees

5. The oldest known primates date from the _____ .
   a. Miocene   c. Oligocene
   b. Paleocene   d. Pliocene

6. The first known hominids were the _____ .
   a. *Homos*   c. cercopiths
   b. dryopiths   d. australopiths

7. The earliest known stone tools are _____ years old.
   a. 8 million   c. 4.2 million
   b. 6 million   d. 2.5 million

8. Match the primates with their descriptions.
   ____ prosimian   a. monkeys, apes, and humans
   ____ tarsioid   b. lemurs
   ____ anthropoid   c. humans and recent ancestors
   ____ hominoid   d. apes and humans
   ____ hominid   e. tarsiers

Selected Key Terms

anthropoid *472*   hominid *472*
australopith *476*   hominoid *472*
bipedalism *473*   *Homo erectus 478*
culture *474*   *Homo sapiens 478*
dryopith *475*   primate *472*
early *Homo 478*   savanna *473*

Readings

Stringer, C. December 1990. "The Emergence of Modern Humans." *Scientific American* 263:98–104.

Weiss, M., and A. Mann. 1990. *Human Biology and Behavior*. Fifth edition. New York: Harper Collins.

FACING PAGE: *A flowering plant (Prunus) busily doing what it does best: producing flowers for the fine art of reproduction.*

29 PLANT TISSUES

The Greening of the Volcano

In the spring of 1980, Mount Saint Helens in southwestern Washington exploded, and 540 million tons of ash blew skyward. Within minutes, the shock waves blew down or incinerated hundreds of thousands of mature trees near the volcano's northern flank. Rivers of ash and fire surged down the slopes faster than a hundred miles an hour. Twenty billion gallons of water, released when the intense heat melted snow and glacial ice, turned the nightmarish rivers into torrents of cement-like mud.

In one brief moment, nearly 100,000 acres of magnificent forests had been transformed into a scarred, barren sweep of land (Figure 29.1). And the aftermath of the volcano's eruption gave us a mind-numbing picture of what our world would be like without plants.

And yet, in less than a year, the seeds of fireweed, blackberry, and other flowering plants were sprouting around the grayed trunks of fallen trees around Mount Saint Helens. Within a decade, willows and alders were flourishing along the rivers, and low shrubs were growing across much of the land. Their presence began to provide the shade necessary for seedlings of slower growing species—the hemlocks and Douglas fir.

Figure 29.1*b* shows the view from a ridge overlooking Spirit Lake and Mount Saint Helens not long after the eruption. Figure 29.1*c* shows the same view nine years later. Within fifty years, new forest trees will be well established. Within a century, the forest will be as it once was.

With this chapter we open a unit that is dedicated to the most successful plants on earth—the flowering plants. No other kind of plant surpasses them in diversity or distribution. They have distinctive patterns of structural organization and splendid modes of reproduction, growth, and development. Their adaptations help them survive hostile conditions on land—even momentary takeovers by volcanoes.

Figure 29.1 (**a,b**) A stunning reminder of what our world would be like without plants—the devastation following the eruption of Mount Saint Helens in 1980. Nothing remained of the forest that had cloaked the volcano's flanks. (**c**) Less than a decade later, plants were making a comeback.

a

b

c

1. Flowering plants dominate the plant kingdom. They have aboveground shoots (which include stems, leaves, and flowers) and descending roots that typically grow downward and outward through soil.

2. Flowering plant tissues are organized into three main kinds of tissue systems. Ground tissue systems make up most of the plant body. Vascular tissue systems distribute water, minerals, and products of photosynthesis through roots, stems, and leaves. Dermal tissues cover and protect plant surfaces.

3. A simple plant tissue (parenchyma, sclerenchyma, or collenchyma) consists of one type of cell. Complex tissues have two or more types of cells. They include xylem and phloem (vascular tissues) and epidermis (a dermal tissue).

4. Plants grow by cell divisions and cell enlargements at meristems, which are localized regions of self-perpetuating embryonic cells. Their primary growth (lengthening of stems and roots) originates at apical meristems of stem and root tips. Their secondary growth (increases in diameter) originates at lateral meristems inside stems and roots.

5. Many plants show primary and secondary growth, year after year. Wood is the outcome of this secondary growth.

Earlier, in Chapter 25, we surveyed representatives of the 275,000 known species of plants. Even that fleeting run through plant diversity reveals why no one species can be used as a typical example of plant body plans. Even so, when we hear the word "plant," we mostly think of the well-known gymnosperms (such as pine trees) and angiosperms (which include roses, cactuses, corn plants, and apple trees).

Angiosperms, recall, are the only plants that produce flowers. And with more than 265,000 species, they dominate the plant kingdom. This chapter focuses mainly on the tissues and body plans of flowering plants. Chapter 30 explains how these plants take up water and nutrients, restrict water loss, and distribute organic substances through the plant body. Chapters 31 and 32 describe the key aspects of their growth, development, and reproduction.

29.1 OVERVIEW OF THE PLANT BODY

Shoots and Roots

Many flowering plants have a body plan like that shown in Figure 29.2. The aboveground parts, or **shoots**, consist of stems and their branchings, leaves, flowers, and other components. The stems are like structural beams for upright growth. An advantage of upright growth is that it favorably exposes photosynthetic cells in young stems and leaves to light. It also displays the flowers for pollinators.

The plant's descending parts are called **roots**. These specialized structures absorb water and dissolved nutrients. They typically penetrate downward and spread through the soil. Roots also anchor the plant's aboveground parts. Most roots store food as well, releasing it as required to root cells or for transport to aboveground parts.

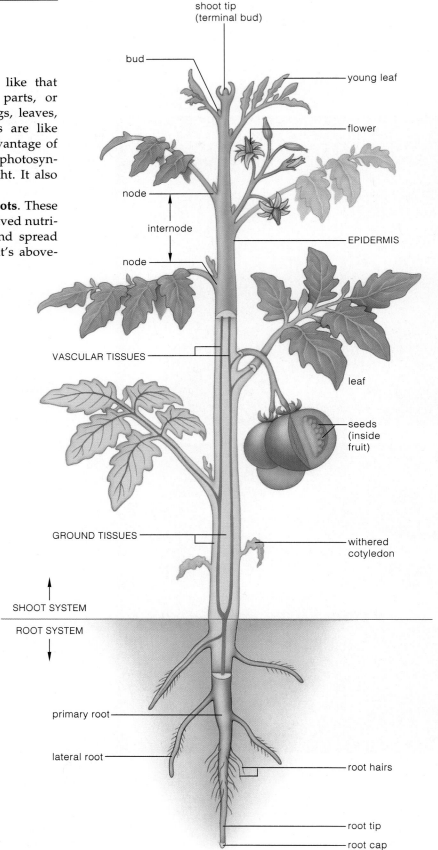

Figure 29.2 Body plan for one type of flowering plant, showing its shoot and root systems.

Vascular tissues (*purple*) conduct water, nutrients, and organic substances throughout the plant body. They thread through ground tissues, which make up most of the plant body.

Dermal tissues (in this case, epidermis) cover the surfaces of both the root system and shoot system.

Three Plant Tissue Systems

The stems, branches, leaves, and roots of flowering plants are similar in a key respect. Their main tissues are grouped into three systems. Their **ground tissue system** is the most extensive; its tissues make up the bulk of the plant body. The **vascular tissue system** contains two kinds of conducting tissues that distribute water and solutes through the plant body. The **dermal tissue system** covers and protects the plant's surfaces (Figures 29.2 and 29.3). Some tissues in these systems are simple, in that they contain one type of cell only. Parenchyma, collenchyma, and sclerenchyma fall in this category. Other tissues are complex, with highly organized arrays of two or more types of cells. Xylem, phloem, and epidermis are like this.

Meristems

Flowering plants grow at **meristems**: localized regions of self-perpetuating, embryonic cells. As they grow, stems and shoots lengthen and thicken. Increases in *length* originate at **apical meristems** located inside the dome-shaped tips of stems and roots. There, cell divisions and enlargements give rise to three primary meristems, which go on to produce the specialized tissue systems (Figure 29.4). In all, a stem or root lengthens through growth that originates at apical meristems and at meristematic tissues derived from them. This growth produces the *primary* tissues of the plant body.

Increases in diameter originate at lateral meristems, which form inside a stem or root (Figure 29.4). There are two kinds of lateral meristems: **vascular cambium** and **cork cambium**. Growth originating here produces *secondary* tissues of the plant body. Each spring, for example, primary growth lengthens a maple tree's stems, branches, and roots; and secondary growth thickens them. Some of the new cells that form during primary and secondary growth perpetuate the meristems. Others differentiate and become part of the tree's specialized tissues.

---

1. The roots and shoots of flowering plants consist of a *ground* tissue system (the bulk of the plant body), a *vascular* tissue system (which distributes water and solutes), and a *dermal* tissue system (which covers and protects plant surfaces).

2. *Primary* growth (lengthening of stems and roots) originates at apical meristems and at meristematic tissues that are derived from them.

3. *Secondary* growth (thickening of stems and roots) originates at lateral meristems, called vascular cambium and cork cambium.

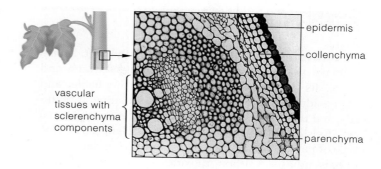

Figure 29.3 The locations of simple tissues and complex tissues in one kind of plant stem.

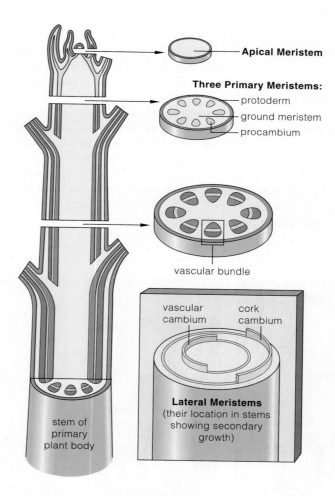

Figure 29.4 Approximate locations of primary meristems and lateral meristems in a plant showing both primary and secondary growth.

29.2 SIMPLE TISSUES

Parenchyma tissues make up most of the soft, moist, primary tissues of roots, stems, leaves, flowers, and fruits. Its cells typically are thin-walled, pliable, and many-sided (Figure 29.5a). The cells are alive at maturity, and they retain the capacity to divide. Parenchymal cell divisions often can heal wounded plant parts. One type of parenchyma, called mesophyll, is specialized for photosynthesis. Carbon dioxide and oxygen diffuse through abundant air spaces between its cells. Other types specialize in storage, secretion, and other tasks. Parenchyma also threads through the vascular tissue systems.

Collenchyma is a tissue that provides flexible support for primary tissues (Figure 29.5b). You will find patches or cylinders of this tissue near the surface of lengthening stems. Collenchyma also forms pliable ribs or strings in many leaf stalks. Collenchyma cells are mostly elongated, with unevenly thickened walls. Pectin in the walls imparts pliability to this tissue.

Sclerenchyma supports mature plant parts and often protects seeds. Generally, sclerenchyma cells have thick, lignin-impregnated walls (Figure 29.5c). **Lignin** strengthens and waterproofs cell walls. Without it, land plants would not have evolved to new heights, so to speak (page 392).

Sclerenchyma tissues consist of fibers or sclereids. *Fibers* are in vascular tissue systems of some stems and leaves. These long, tapered cells flex and twist without stretching. Fibers from flax plants are used to make rope, paper, cloth, and thread. *Sclereids* are shorter cells. Clusters of sclereids give pears and some other fruits a gritty texture (Figure 29.5d). They also form coconut shells, peach pits, and other seed coats.

29.3 COMPLEX TISSUES

Unlike ground tissues, each of which consists of one cell type only, the vascular and dermal tissues of plants are complex, with a variety of cell types.

Vascular Tissues

Two kinds of vascular tissues, called xylem and phloem, distribute substances through the plant. Often their conducting cells are bundled in a sheath composed of fibers and parenchyma cells.

Xylem conducts soil water and dissolved minerals, and it mechanically supports the plant. Figure 29.6a and b shows two examples of the conducting cells in xylem. The cells, called vessel members and tracheids, are not alive at maturity. The lignified walls of the dead cells connect with one another. They form the water-conducting pipelines, and they strengthen plant parts. Water flows between adjacent cells through pits in the side walls.

Phloem transports sugars and other solutes. Figure 29.6c shows an example of its main conducting cells, called sieve-tube members. These cells are alive at maturity. Adjacent sieve-tube members interconnect at perforations in their side walls and at open or perforated end walls.

In leaves, sieve-tube members become loaded with sugars from photosynthetic cells, often with the help of "companion" cells. Their functional companions are specialized, living parenchyma cells. The sugars travel throughout the plant and are unloaded from phloem in any region where cells are growing or storing food. The next chapter has more to say about the functions of xylem and phloem.

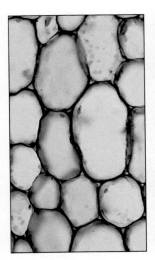

a Parenchyma

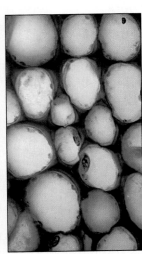

b Collenchyma

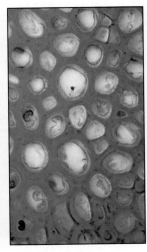

c Sclerenchyma

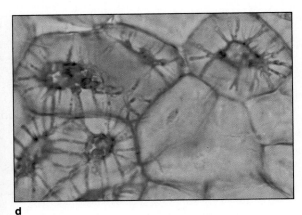

d

Figure 29.5 (**a–c**) From the stem of a sunflower plant (*Helianthus*), examples of ground tissues, in cross section. (**d**) From the flesh of a pear (*Pyrus*), some thick, lignin-impregnated stone cells, a type of sclereid.

Figure 29.6 (**a,b**) Examples of tracheids and vessel members, the main types of cells in xylem that conduct water and dissolved mineral salts. (**c**) One type of phloem cell that conducts sugars and other organic compounds.

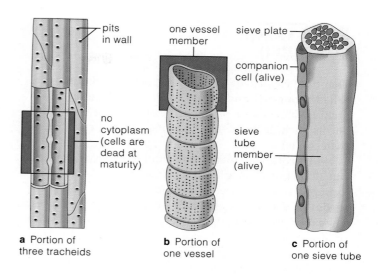

a Portion of three tracheids

b Portion of one vessel

c Portion of one sieve tube

Figure 29.7 (**a**) Light micrograph of a section through the upper surface of a kaffir lily leaf. The thick cuticle is composed of secretions from epidermal cells. (**b**) Scanning electron micrograph of a leaf surface, showing epidermal cells and stomata. Notice also the parenchyma tissue inside the leaf; it is composed of cells specialized for photosynthesis.

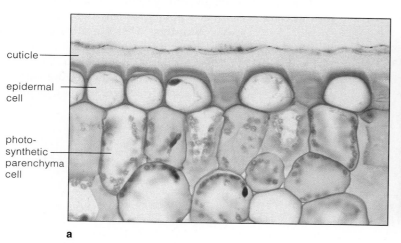

a

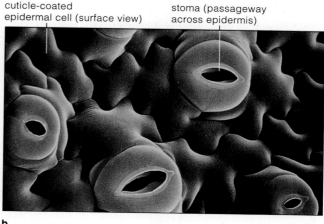

b

Dermal Tissues

All surfaces of primary plant parts are covered and protected by a dermal tissue system called **epidermis**. In most plants, epidermis is mainly a single layer of unspecialized cells. Waxes and cutin (a fatty substance) coat the outermost cell walls. The surface coating, called a **cuticle**, restricts water loss and resists attacks by some microorganisms (Figure 29.7).

Stem and leaf epidermis also contains specialized cells, including pairs of guard cells. Guard cells can change shape. When they do, a gap called a **stoma** (plural, stomata) opens up between them or closes. In the chapter to follow, you will see how the movement of water vapor, carbon dioxide, and oxygen across the epidermis is controlled at stomata.

As you will see later, **periderm** replaces the epidermis of stems and roots showing secondary growth. Most of this protective cover consists of walls of cork cells that are no longer alive. The walls are heavily impregnated with suberin, a fatty substance.

In short, these are the key points to remember about vascular plant tissues:

1. Most of the plant body (the ground tissue system) consists of simple tissues called parenchyma, collenchyma, and sclerenchyma. Each tissue is composed of one cell type only.

2. Tracheids and vessel members are the main cells of xylem, the vascular tissue that conducts water and dissolved mineral salts.

3. Sieve-tube members and companion cells are the main cells of phloem, the vascular tissue that conducts sugars and other organic compounds.

4. Of two dermal tissues, epidermis covers the surfaces of the primary plant body. Periderm replaces it on plant parts showing secondary growth.

29.4 MONOCOTS AND DICOTS COMPARED

The two classes of flowering plants commonly are called **monocots** and **dicots** (page 405). Grasses, lilies, orchids, irises, and palms are examples of monocots. Dicots include most familiar trees and shrubs, other than conifers.

Monocots and dicots are similar in structure and function, but they differ in some distinct ways. For example, monocot seeds have one cotyledon and dicot seeds have two. "Cotyledons" are leaflike structures. They form in seeds, as part of a plant embryo, and store or absorb food for it. They wither after the seed germinates and leaves start functioning. Figure 29.8 shows other differences between monocots and dicots.

The tissue organization of shoots and roots is described next. When looking at the accompanying photographs, keep in mind these terms, which identify the way tissue specimens are cut from plants:

transverse
(cross-section cut perpendicular to long axis of root or stem)

radial
(longitudinal section cut along the radius of root or stem)

tangential
(longitudinal section at right angles to root or stem)

29.5 SHOOT PRIMARY STRUCTURE

Picture yourself scanning a menu in a wood-framed restaurant, slouched comfortably in an old wicker chair. Grasscloth panels the walls. On the carved oak table, a woven basket holds bread, and a vase holds delicately stemmed flowers. Nearby, a waiter tosses a salad in a wooden bowl. Count the ways in which you are being pampered by stems and leaves (or what's left of them). Let's consider what stems and leaves of this sort look like before someone carves, weaves, snips, or chops them or mashes, moistens, and presses them flat.

Internal Structure of Stems

Recall that when a stem is lengthening, primary meristems give rise to its ground, vascular, and dermal tissues. Commonly, the primary xylem and phloem develop as **vascular bundles**. These multistranded, sheathed cords thread lengthwise through the ground tissue system. Inside each bundle, the phloem is often positioned near the side of the sheath facing the stem surface, and the xylem is positioned near the side facing the stem center.

Figure 29.8 Comparison of the main features that distinguish dicots from monocots.

a Typical dicot flower (*Hypericum*)

b Typical monocot flower (*Iris*)

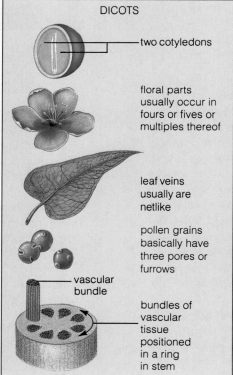

DICOTS

two cotyledons

floral parts usually occur in fours or fives or multiples thereof

leaf veins usually are netlike

pollen grains basically have three pores or furrows

vascular bundle

bundles of vascular tissue positioned in a ring in stem

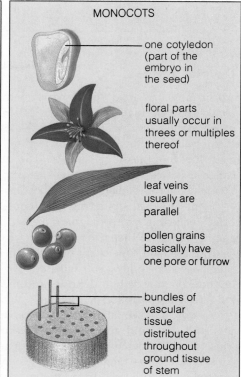

MONOCOTS

one cotyledon (part of the embryo in the seed)

floral parts usually occur in threes or multiples thereof

leaf veins usually are parallel

pollen grains basically have one pore or furrow

bundles of vascular tissue distributed throughout ground tissue of stem

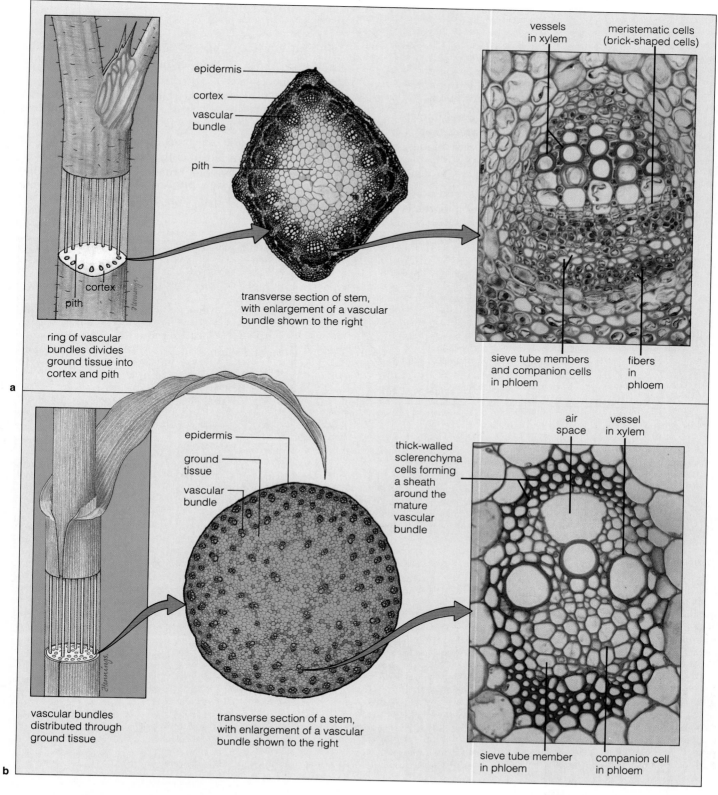

a

epidermis

cortex

vascular bundle

pith

cortex

pith

ring of vascular bundles divides ground tissue into cortex and pith

transverse section of stem, with enlargement of a vascular bundle shown to the right

vessels in xylem

meristematic cells (brick-shaped cells)

sieve tube members and companion cells in phloem

fibers in phloem

b

epidermis

ground tissue

vascular bundle

vascular bundles distributed through ground tissue

transverse section of a stem, with enlargement of a vascular bundle shown to the right

thick-walled sclerenchyma cells forming a sheath around the mature vascular bundle

air space

vessel in xylem

sieve tube member in phloem

companion cell in phloem

Examine the primary stems of flowering plants, and you see that two patterns predominate. In most dicot stems, vascular bundles have a ringlike arrangement. The ring divides the ground tissue into **cortex** (the part between the vascular bundles and the epidermis) and **pith** (the center of the root or stem, inside the ring of

Figure 29.9 Stem structure. (**a**) Alfalfa (*Medicago*), a dicot. (**b**) Corn (*Zea mays*), a monocot.

vascular bundles). In most monocots and even some dicots, vascular bundles are scattered through the ground tissue system. Figure 29.9*a* and *b* gives examples of the two patterns.

Characteristics of Leaves

All **leaves** are metabolic factories equipped with photosynthetic cells, but they vary enormously in size, shape, and texture. Duckweed leaves are less than a millimeter (0.04 inch) across; some water lily leaves are 2 meters (6.5 feet) wide. Among the leaf shapes are blades, spikes, needles, cups, feathers, and tubes. Many leaves are smooth-surfaced; others are sticky, hairy, or slimy. They often have hairs, scales, spikes, hooks, and other surface specializations; page 500 describes a particularly splendid example. Besides this, leaves run the gamut in terms of their colors, odors, and edibility (many are poisonous).

Figures 29.10 and 29.11 show just a few examples. The leaves of ryegrass, corn, and most other monocots are flat surfaced, like a knife blade. The base of the blade encircles and sheathes the stem. Generalizing about dicot leaves is difficult; they are quite diverse. However, many have one broad blade, attached to the stem by a stalk (petiole). Often the blade is lobed. Some "compound" leaves have stalked leaflets.

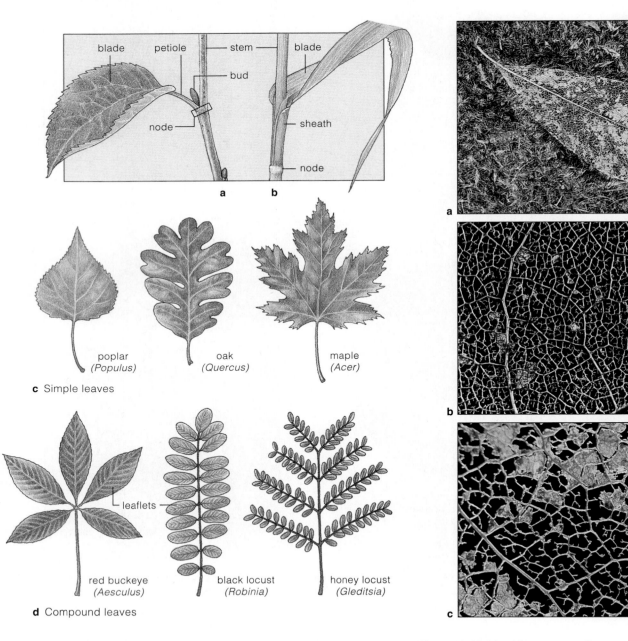

Figure 29.10 Common leaf forms among (**a**) dicots and (**b**) monocots. (**c,d**) Examples of simple and compound leaves.

Figure 29.11 Leaf from a magnolia tree, showing veins at increasing magnifications. Networks of minor veins thread through photosynthetic tissue.

As you probably know, the leaves of birches and other "deciduous" species drop away from the stem as winter approaches. Leaves of camellias and other "evergreen" species also drop, but they don't all drop at the same time (hence the name).

Leaf Internal Structure

The leaves just described have a large surface area exposed to sunlight and carbon dioxide. Their upper surface tissue is cuticle-covered epidermis (Figure 29.12). Next comes **mesophyll**, a parenchymal tissue with photosynthetic cells and an abundance of air spaces. The spaces enhance diffusion of carbon dioxide into cells and oxygen away from them.

Distinct vascular bundles called **veins** thread lacily through the interior of the leaf (Figure 29.11). They move water and nutrients to the cells and carry sugars and other substances away from them. Below the mesophyll is another cuticle-covered epidermis, the leaf's lower surface tissue. This often contains most of the stomatal passageways into and out of the leaf.

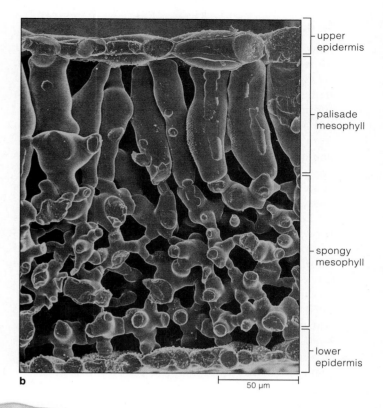

upper epidermis

palisade mesophyll

spongy mesophyll

lower epidermis

b

50 μm

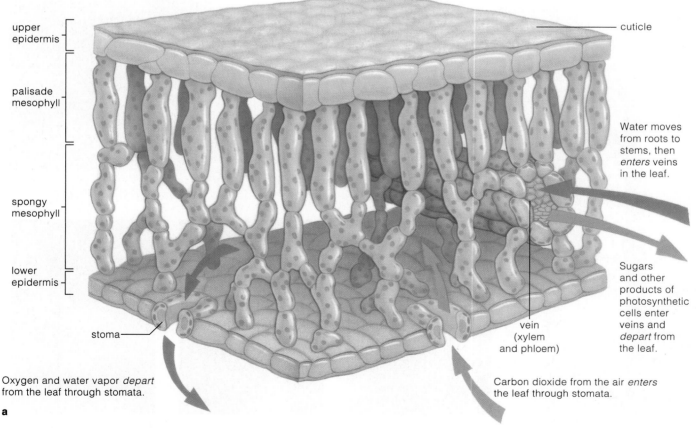

upper epidermis

palisade mesophyll

spongy mesophyll

lower epidermis

cuticle

Water moves from roots to stems, then *enters* veins in the leaf.

Sugars and other products of photosynthetic cells enter veins and *depart* from the leaf.

stoma

vein (xylem and phloem)

Oxygen and water vapor *depart* from the leaf through stomata.

Carbon dioxide from the air *enters* the leaf through stomata.

a

Figure 29.12 (**a**) Generalized diagram of leaf internal structure. (**b**) This micrograph shows the tissue layers of a leaf from the kidney bean plant (*Phaseolus*).

Origin of Leaves

How do leaves originate? The ones described so far form at the tip of a main stem or its youngest branches. As Figure 29.13 indicates, they start out as bulges on the flanks of the apical meristem, and each enlarges into a rudimentary leaf. The bulges are close together at first. But as primary growth continues, the leaves that form from them become spaced at intervals along the lengthening stem. Each stem region where one or more leaves are attached is a **node**. The stem region between two successive nodes is an "internode." Figure 29.2 shows an example.

Look at a twig of a walnut tree in winter, after all its leaves have dropped (Figure 29.14a). At the shoot tip is a *terminal* bud. A **bud** is an undeveloped shoot of mostly meristematic tissue, often covered and protected by scales (modified leaves). Besides this terminal bud, the stem has buds spaced at intervals along its length. Each *lateral* bud forms in the upper angle where a leaf was attached to the stem. Buds give rise to new stems and leaves, flowers, or both. Vascular bundles diverge from the main stem and develop in the new plant parts, also.

We conclude this section on leaves with a brief *Commentary* that invites reflection on some of the ways we have learned to use (and abuse) them.

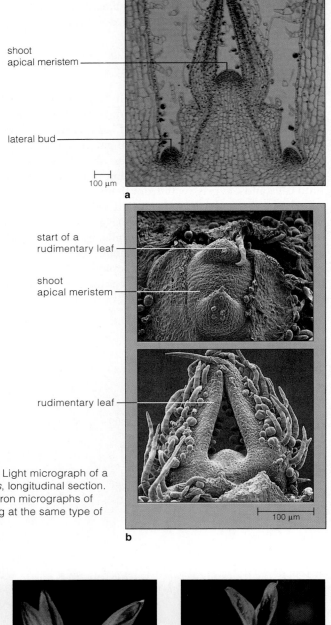

rudimentary leaf

shoot apical meristem

lateral bud

100 µm

a

start of a rudimentary leaf

shoot apical meristem

rudimentary leaf

100 µm

b

Figure 29.13 (**a**) Light micrograph of a shoot tip of *Coleus*, longitudinal section. (**b**) Scanning electron micrographs of new leaves forming at the same type of shoot tip.

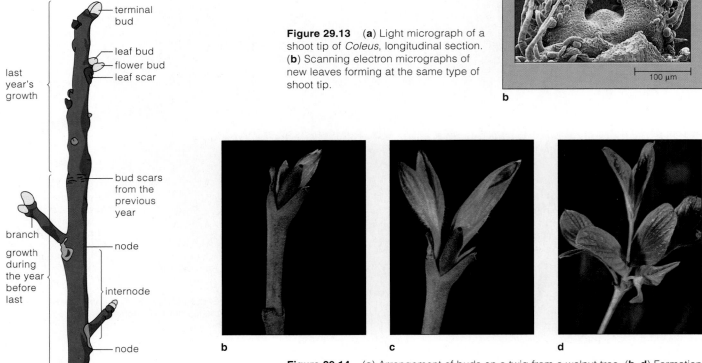

terminal bud

leaf bud
flower bud
leaf scar

last year's growth

bud scars from the previous year

branch

growth during the year before last

node

internode

node

a

b

c

d

Figure 29.14 (**a**) Arrangement of buds on a twig from a walnut tree. (**b–d**) Formation of leaves at a terminal bud of a dogwood tree.

Commentary

Uses and Abuses of Leaves

Imagine a leaf-deprived grocery cart. Out go the lettuces, parsley, spinach, and chard. No thyme, basil, oregano, and rosemary for your secret spaghetti sauce. No more mint, darjeeling, and other teas.

Beyond the table, oils from leaves of lavender scent perfumes and soaps. Oils of eucalyptus and camphor find their way into medicine chests. Leaves of nightshade plants (including *Atropa belladonna*) provide extracts to treat Parkinson's disease. Digitalis from leaves of foxgloves (*Digitalis purpurea*) helps stabilize heartbeat and blood circulation. Juices from the leaves of *Aloe vera* soothe sun-damaged skin. Alkaloids from rosy periwinkle leaves are used for patients with Hodgkin's disease.

Landscape architects and imaginative homeowners use leaves of various trees and shrubs much as an artist uses colors and textures. They plant shrubs next to buildings and help keep them from heating up in summer and losing heat in winter. We get twine and rope from century plant leaves (*Agave*), cords and textiles from leaf fibers of Manila hemp, hats from Panamanian palm fronds, and thatched roofs from leaves of palms and grasses.

Insecticides derived from Mexican cockroach plants kill cockroaches, fleas, lice, and flies. Extracts of neem tree leaves kill more than a hundred kinds of insects, mites, and nematodes without killing off natural predators of these common pests.

Through much of recorded history, people also have used leaves in nonbeneficial ways. Mayans introduced early European explorers to tobacco smoking. They cultivated tobacco plants, including *Nicotiana rusticum*. Mayan priests thought that the smoke rising from pipes carried their priestly thoughts to the gods. People continue to smoke, chew, or tuck into their mouth the leaves of tobacco plants—and so become candidates for the hundreds of thousands of annual deaths from lung, mouth, and throat cancers. Heavy smoking of *Cannabis sativa*, the source of marijuana and other mind-altering products, has been linked with low sperm counts. Cocaine, derived from coca leaves, is used medicinally and abused by increasing numbers of individuals— with devastating social and economic effects. The introduction to Chapter 35 and the *Focus* essay on page 588 provide more detail on this topic.

And what about that henbane (*Hyosyamus niger*) and belladonna? Their alkaloids have been tapped for murder as well as for medicinal purposes. Remember Hamlet's father? As he slept, he was sneakily dispatched by someone who poured a henbane solution into his ear. Remember Romeo's heartbroken, suicidal Juliet? She sipped a potion of deadly nightshade, then dropped dead.

Those self-proclaimed witches of the Middle Ages used henbane and nightshade in some of their suspect rituals. They used sticks to apply atropine-impregnated solutions to their body, and so induced a sensation of weightlessness. During their atropine-induced hallucinatory sprees, they assumed they were flying off to meet with demons. Hence those Halloween cartoons of witches flying about on broomsticks.

NICOTIANA RUSTICUM

DIGITALIS PURPUREA

CANNABIS SATIVA

ATROPA BELLADONNA

HYOSCYAMUS

29.6 ROOT PRIMARY STRUCTURE

Unless tree roots start to buckle a sidewalk or choke off a sewer line, most of us don't pay much attention to the elaborate root systems of flowering plants. For most species, roots extend to a depth of 2 to 5 meters. But in hot deserts, where water is scarce, one hardy mesquite shrub is known to have sent roots down 53.4 meters (175 feet) near a streambed. The shallow roots of some cacti may radiate out for 15 meters. Someone once measured the roots of a young rye plant that had been growing for four months in only 6 liters of soil. If the surface area of that root system were laid out as a single sheet, it would extend more than 600 square meters!

Taproot and Fibrous Root Systems

The first root to poke through the coat of a germinating seed is the **primary root**. In most dicot seedlings, it increases in diameter and grows downward. Later on, **lateral roots** start forming in internal tissues and erupt through the epidermis. The youngest of these lateral roots are closest to the root tip. A primary root and its lateral branchings are a **taproot system**. Oak trees, dandelions, and carrots are examples of plants that have a taproot system (Figure 29.15*a*).

What about monocots, including grasses such as the rye plant mentioned above? In general, they have short-lived primary roots. In their place, adventitious roots arise from the young plant's stem, then lateral roots branch from them. (*Adventitious* refers to a structure arising at an uncommon location, such as roots arising from stems or leaves.) These lateral roots, which are similar in diameter and length, form a **fibrous root system** (Figure 29.15*b*).

Internal Structure of Roots

Figure 29.16 shows the location of apical meristem in a root tip. Behind this region, cellular descendants of the apical meristem divide, enlarge, and become the specialized cells of the root's tissue systems. Figure 29.16 also shows the root cap, a dome-shaped cell mass at the root tip. Root apical meristem produces the cap and in turn is protected by it. As the root grows, it pushes the root cap forward against soil particles, so that some cells tear loose. The slippery remnants lubricate the cap as it moves through the soil.

A root's epidermis is the absorptive interface with soil. Some epidermal cells send out long extensions called **root hairs** (Figure 29.17*a*). Root hairs greatly increase the surface area available for taking up water and solutes. Gardeners learn never to yank a plant from the ground when transplanting it. Too much of the fragile absorptive surface would be torn off.

a   **b**

Figure 29.15 (**a**) Taproot system of a dandelion. (**b**) Fibrous root system of a grass plant.

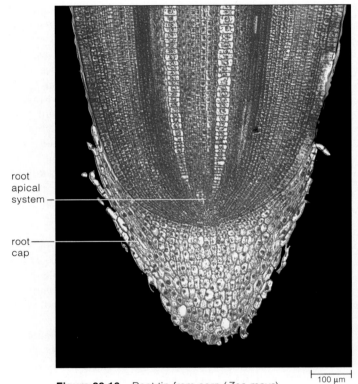

root apical system —

root — cap

⊢100 μm⊣

Figure 29.16 Root tip from corn (*Zea mays*).

Vascular tissues form a **vascular cylinder**, a central column inside the root. Ground tissues surrounding the cylinder are called the root cortex (Figures 29.17*b* and 29.18). An abundance of spaces in the cortex allows dissolved oxygen to reach living root cells. Like other cells in the plant, the cells require oxygen for aerobic respiration.

Water entering the root moves from cell to cell until it reaches the endodermis, the innermost part of the root cortex. Endodermis is a sheetlike layer, one cell

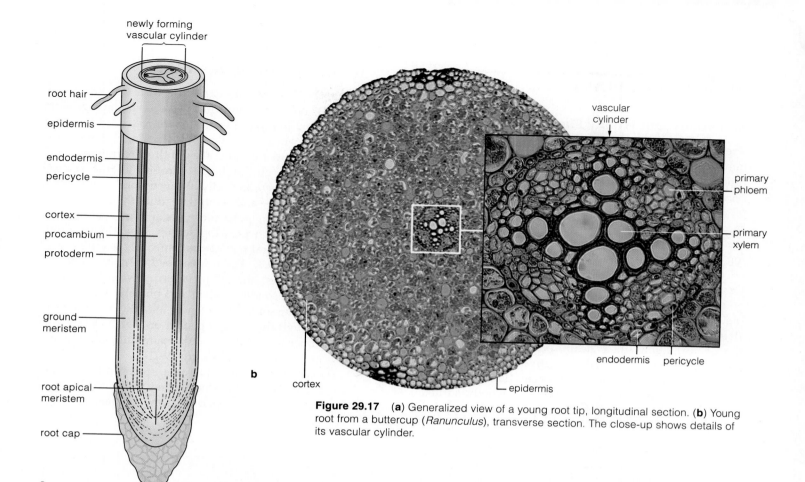

Figure 29.17 (a) Generalized view of a young root tip, longitudinal section. (b) Young root from a buttercup (*Ranunculus*), transverse section. The close-up shows details of its vascular cylinder.

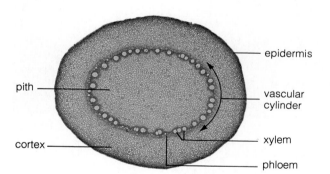

Figure 29.18 Root of a corn plant (*Z. mays*), transverse section. Notice how the vascular cylinder divides the ground tissue into cortex and pith.

thick, around the vascular cylinder. Abutting walls of its cells are waterproof, so they force incoming water to pass through the cytoplasm of endodermal cells. As described in Chapter 30, this arrangement helps control the movement of water and dissolved substances into the vascular cylinder.

Just inside the endodermis is the pericycle. This part of the vascular cylinder has one or more layers of cells that give rise to lateral roots, which erupt through the cortex and epidermis (Figure 29.19).

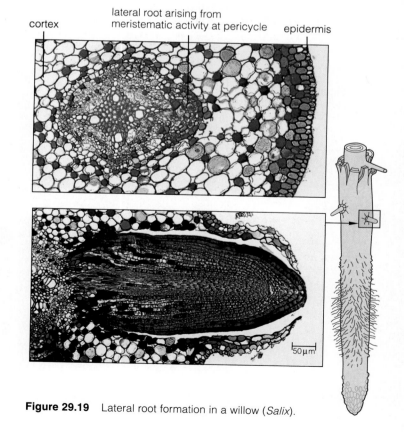

Figure 29.19 Lateral root formation in a willow (*Salix*).

29.7 WOODY PLANTS

Nonwoody and Woody Plants Compared

The life cycles of flowering plants extend from seed germination to seed formation, then eventual death. In that time, most monocots and some dicots show little or no secondary growth; they are nonwoody (herbaceous) plants. By contrast, many dicots (and all gymnosperms) show secondary growth during two or more growing seasons; they are woody plants. In this respect, flowering plants fall into three categories.

Annuals, such as corn and marigolds, complete their life cycle in one growing season and show little or no secondary growth. **Biennials**, such as carrots, complete their life cycle in two growing seasons. Roots, stems, and leaves form in the first season; seed and flower formation, then death occur in the second. **Perennials** show continued vegetative growth and seed formation year after year. Some species have secondary tissues. Examples are nonwoody cacti, woody shrubs (roses), vines (ivy, grape), and trees (elms, magnolias).

Growth at Lateral Meristems

Figure 29.20 shows the structure of a tree trunk—an older, woody stem with extensive secondary growth. Notice how living phloem cells are restricted to a thin zone just beneath the corky surface. Stripping off a band of phloem around a tree's entire circumference (an activity called girdling) cuts through all the vertical phloem. Photosynthetically derived food can't reach roots, which die—and so, in time, does the tree.

How do older stems and roots of plants get massive and woody? Each growing season, new tissues originate at their **lateral meristems**, especially vascular cambium (Figure 29.4). Fully developed vascular cambium is like a cylinder, one cell (or a few cells) thick. Its cells give rise to secondary xylem and phloem that conduct water and dissolved substances up, down, and sideways through the enlarging stem or root. Xylem forms on the inner face of the vascular cambium, and phloem forms on the outer face (Figure 29.21).

The layers of xylem increase season after season. They usually crush the thin-walled phloem cells from the preceding growth period. New phloem cells are

Figure 29.20 Structure of a woody stem with extensive secondary growth. *Heartwood*, the mature tree's core, has no living cells. *Sapwood* is the cylindrical zone of xylem between heartwood and vascular cambium. It contains some living parenchyma cells among nonliving conducting cells of xylem. Everything outside the vascular cambium is often called *bark*. Everything inside it is called *wood*.

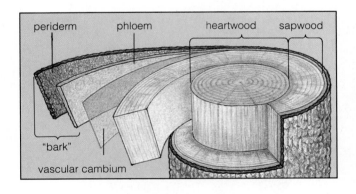

Figure 29.21 Composite drawing of the growth in a woody stem through successive seasons. Ongoing divisions displace the vascular cambium, moving its cells steadily outward as the enlarging core of xylem makes the stem or root thicker.

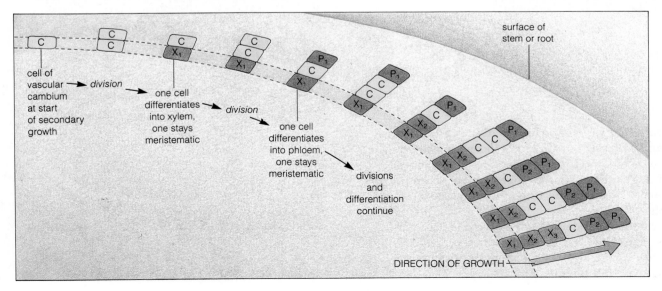

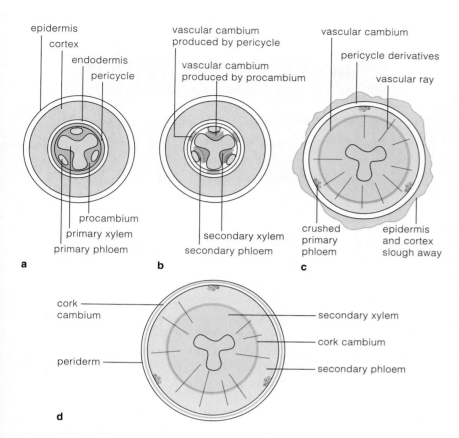

Figure 29.22 labels (a):
epidermis
cortex
endodermis
pericycle
procambium
primary xylem
primary phloem

(b):
vascular cambium produced by pericycle
vascular cambium produced by procambium
secondary xylem
secondary phloem

(c):
vascular cambium
pericycle derivatives
vascular ray
crushed primary phloem
epidermis and cortex slough away

(d):
cork cambium
periderm
secondary xylem
cork cambium
secondary phloem

Figure 29.22 Closer look at secondary growth in a dicot root. (**a**) Arrangement of tissues as primary growth ends. (**b,c**) A ring of vascular cambium forms and gives rise to secondary xylem and phloem. The root increases in diameter as cell divisions proceed parallel to the vascular cambium. The cortex ruptures as the tissue mass increases. (**d**) Periderm, which arises from cork cambium, replaces the epidermis.

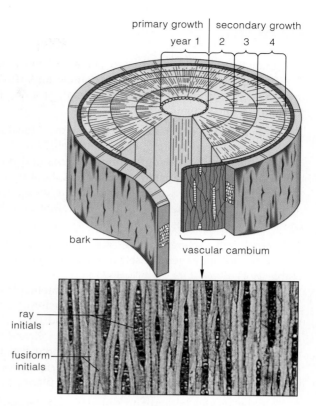

Figure 29.23 labels:
primary growth — year 1
secondary growth — 2 3 4
bark
vascular cambium
ray initials
fusiform initials

Figure 29.23 Another view of vascular cambium in an older stem showing secondary growth. This one has three annual rings, corresponding to secondary growth in years two through four. Some cells of the vascular cambium (the *fusiform* initials) produce the secondary xylem and phloem that conduct water and food vertically through the stem. Other cells (*ray* initials) produce parenchyma and other cells that serve as lateral channels for water and food, and as food storage centers.

added each year, outside the growing core of xylem. In time, the pressure exerted by the tissues accumulating inside a stem or root ruptures the cortex and outer phloem. Parts of the cortex split away, carrying epidermis with them. But cork cambium forms, and its cells produce periderm—a corky replacement for the lost epidermis (Figure 29.22). "Cork" is not the same as "bark," which refers to all living and nonliving tissues external to the vascular cambium (Figure 29.20).

Early and Late Wood

In regions having prolonged dry spells or cool winters, vascular cambium becomes inactive during parts of the year. The first xylem cells produced at the start of the growing season tend to have large diameters and thin walls. They form *early* wood. As the season progresses, cell diameters become smaller and their walls thicker. These cells form *late* wood.

In a full-diameter slice from an old tree trunk, early and late wood reflect light differently. You can identify them as alternating light and dark bands. The alternating bands represent **annual growth layers**, or tree rings (Figures 29.23 and 29.24).

Figure 29.24 labels:
vessels in late wood
vessels in early wood
toward stem's surface

Figure 29.24 Scanning electron micrograph of early and late wood in a block cut from red oak (*Quercus rubra*). Notice the different diameters of xylem's water-conducting vessels.

SUMMARY

1. Monocots and dicots are two classes of flowering plants (angiosperms). Their shoots (stems and leaves) and roots are composed mainly of dermal, ground, and vascular tissue systems.

2. Plant growth originates at meristems, localized regions of self-perpetuating, embryonic cells.

    a. Primary growth (lengthening of stems and roots) originates at apical meristems in root and shoot tips.

    b. Secondary growth (increases in diameter) originates at lateral meristems (vascular cambium and cork cambium) inside stems and roots.

3. Parenchyma, sclerenchyma, and collenchyma are simple tissues that each consist of only one type of cell (Table 29.1).

    a. Parenchyma cells are alive and metabolically active at maturity. Various types of parenchyma function in photosynthesis and other tasks, and they make up the bulk of ground tissue systems. Strands of parenchyma cells also thread through vascular tissues.

    b. Sclerenchyma mechanically supports growing plant parts. Collenchyma supports mature plant parts.

4. Vascular tissues (xylem and phloem) and dermal tissues (epidermis and periderm) are complex tissues, each consisting of two or more cell types (Table 29.1).

    a. Vascular tissues distribute water and dissolved substances through the plant. Typically, strands of conducting cells are bundled together inside sheaths of sclerenchyma and parenchyma. Many such vascular bundles thread through the ground tissues.

    b. Xylem's water-conducting cells are dead at maturity. Their lignified, pitted walls interconnect as pipelines for water and dissolved minerals.

    c. Phloem's conducting cells are alive at maturity. The cytoplasm of adjacent cells interconnects across perforated end walls and side walls. Dissolved sugars and other products of photosynthesis are loaded into these cells in leaves and unloaded in regions where cells are growing or storing food.

    d. Epidermis covers and protects all surfaces of primary plant parts. Periderm replaces epidermis on plants with secondary growth.

5. Stems favorably position photosynthetic cells of leaves to intercept light, and they display flowers. Their vascular bundles distribute substances to and from all plant parts. Most monocot stems have vascular bundles distributed throughout the ground tissue. Most dicot stems have a ring of bundles that divides the ground tissue into cortex and pith.

6. Leaves contain veins and mesophyll (photosynthetic parenchyma) between the upper and lower epidermis.

Table 29.1	Summary of Flowering Plant Tissues and Their Component Cells
Simple Tissues:	
Parenchyma	Parenchyma cells
Collenchyma	Collenchyma cells
Sclerenchyma	Fibers or sclereids
Complex Tissues:	
Xylem	Conducting cells (tracheids and vessel members); parenchyma cells; sclerenchyma cells
Phloem	Conducting cells (sieve-tube members); parenchyma cells; sclerenchyma cells
Epidermis	Mostly undifferentiated cells; also guard cells, other specialized cells
Periderm	Cork cells; cork cambium cells; parenchyma cells; sclerenchyma cells

Air spaces around the photosynthetic cells enhance gas exchange. Water vapor and gases move across the epidermis through numerous tiny openings (stomata).

7. Roots absorb water and dissolved minerals from the surroundings for distribution to aboveground parts. They anchor the plant. Most store food, and some help support the plant body.

Review Questions

1. Choose a flowering plant and list some functions of its roots and shoots. *484*

2. Name and define the basic functions of a flowering plant's three main tissue systems. *485*

3. Describe the differences between:
    a. apical and lateral meristems *485*
    b. parenchyma and sclerenchyma *486*
    c. xylem and phloem *486*
    d. epidermis and periderm *487*

4. Which of the following stem sections is typical of most monocots? of most dicots? Label the main tissue regions of each. *488*

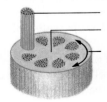

5. Label the component parts of this three-year-old tree section. Correctly label the annual growth layers. *497*

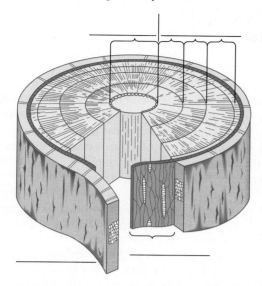

6. Identify early and late wood of this three-year-old oak stem. *497*

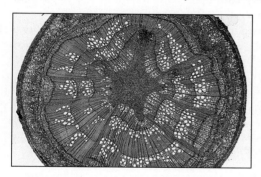

1. _____ and _____ are two classes of flowering plants.
   a. Angiosperms; gymnosperms   c. Shrubs; trees
   b. Monocots; dicots   d. Herbs; shrubs

2. Fleshy plant parts consist mostly of _____ cells.
   a. parenchyma   c. collenchyma
   b. sclerenchyma   d. epidermal

3. Xylem and phloem are _____ tissues.
   a. ground   c. dermal
   b. vascular   d. both b and c

4. _____ conducts water and minerals; _____ conducts food.
   a. Phloem; xylem   c. Xylem; phloem
   b. Cambium; phloem   d. Xylem; cambium

5. Roots and shoots lengthen through activity at _____ .
   a. apical meristems   c. vascular cambium
   b. lateral meristems   d. cork cambium

6. Older roots and stems thicken through activity at _____ .
   a. apical meristems   c. vascular cambium
   b. lateral meristems   d. both b and c

7. Buds give rise to _____ .
   a. leaves   c. stems
   b. flowers   d. all of the above

8. Mesophyll consists of _____ .
   a. waxes and cutin   d. photosynthetic
   b. lignified cell walls   parenchyma cells
   c. cork but not bark   e. sclerenchyma cells

9. Early wood forms from cells having _____ diameters and _____ walls.
   a. small; thick   c. large; thick
   b. small; thin   d. large; thin

10. Match the terms with the appropriate plant parts.
   ____ apical meristem   a. masses of xylem
   ____ lateral meristem   b. stems, roots lengthen
   ____ xylem, phloem   c. corky covering
   ____ periderm   d. older roots, stems thicken
   ____ vascular bundle   e. conduct water, food
   ____ wood   f. central column in roots

Selected Key Terms

angiosperm *483*
annual *496*
annual growth layer *497*
apical meristem *485*
biennial *496*
bud *492*
collenchyma *486*
cork cambium *485*
cortex *489*
cuticle *487*
dermal tissue system *485*
dicot *488*
epidermis *487*
fibrous root system *494*
ground tissue system *485*
lateral meristem *496*
lateral root *494*
leaf *490*
lignin *486*
meristem *485*
mesophyll *491*

monocot *488*
node *492*
parenchyma *486*
perennial *496*
periderm *487*
phloem *486*
pith *489*
primary root *494*
root *484*
root hair *494*
sclerenchyma *486*
shoot *484*
stoma (stomata) *487*
taproot system *494*
vascular bundle *488*
vascular cambium *485*
vascular cylinder *494*
vascular tissue system *485*
vein *491*
xylem *486*

Readings

Raven, P., R. Evert, and S. Eichhorn. 1992. *Biology of Plants*. Fifth edition. New York: Worth. Exquisite color micrographs and illustrations.

Rost, T. et al. 1984. *Botany: An Introduction to Plant Biology*. Second edition. New York: Wiley.

Simpson, B., and M. Conner-Ogorzaly. 1986. *Economic Botany*. New York: McGraw-Hill. Fascinating book on uses and abuses of plants.

Stehlin, D. October 1990. "Harvesting Drugs from Plants." *FDA Consumer* 24:20–23.

Stern, K. 1991. *Introductory Plant Biology*. Fifth edition. Dubuque, Iowa: Brown. Beautifully illustrated, accessible. Paperback.

30 PLANT NUTRITION AND TRANSPORT

Flies for Dinner

How often do we think that plants actually do anything impressive? Being mobile, intelligent, and emotional, we tend to be fascinated more with ourselves than with immobile, expressionless plants. Yet plants don't just stand around soaking up sunlight. Consider a flytrap that evolved long before humans even thought of building one.

The two-lobed, spine-fringed leaves of the Venus flytrap, a native of North and South Carolina, open and close much like a steel trap (Figure 30.1*a–c*). As is true of all organisms, this vascular plant cannot grow unless it takes up nitrogen and other essential minerals from its environment. Certain minerals happen to be scarce in the soggy soils where the Venus flytrap lives. But insects flying in from nearby areas are abundant.

Sticky, sugary substances ooze from glands on the surface of the flytrap's two-lobed leaf. The ooze is part bait, part snare. Insects enticed to land on the leaf brush against tiny hairlike structures projecting from its surface—and these are triggers for the trap. When an insect touches two hairs at the same time, or even when it touches the same hair twice in rapid succession, the leaf snaps shut. Then digestive juices pour out from leaf cells, pool around the insect—and dissolve its minerals. The plant, in short, makes its own mineral-rich water!

The Venus flytrap is one of several kinds of plants that are called carnivorous (although it takes a flying leap of the imagination to put their extracellular digestion and absorption in the same category as the chompings of lions, dogs, and similar meat eaters). In each case, the plant evolved in a habitat where particular minerals are hard to come by.

To expand on this point, think of the shallow waters of lakes and streams where the bladderworts grow.

Figure 30.1 Two "carnivorous" plants that turn the dinner table on animals. (**a–c**) A Venus flytrap (*Dionaea muscipula*) and (**d**) some of the insect-trapping bladders on the submerged stems of a bladderwort (*Utricularia*).

base of one of the hairlike projections—the triggers for the trap

surface of leaf

one of the secretory glands at the leaf surface

d

insect-trapping bladder

KEY CONCEPTS

1. Many aspects of plant structure and function are adaptive responses to low concentrations of water, minerals, and other environmental resources.

2. Leaves have passageways (stomata) across their epidermis. Plants lose water during the day, when stomata remain open and so allow carbon dioxide (used in photosynthesis) to move into the leaves. Most plants conserve water by closing the stomata at night.

3. Plants have mechanisms for water transport. The force of dry air causes evaporation from aboveground plant parts (transpiration). This pulls continuous columns of water molecules (which are hydrogen-bonded to one another) from roots to aboveground parts.

4. Sucrose and other organic compounds are distributed throughout the plant by translocation. In this energy-requiring process, organic compounds are loaded into conducting cells of phloem. They are unloaded at actively growing regions or storage regions of the plant body.

Minerals are present in the waters, but only in dilute concentrations. Bladderwort plants have hollow, pear-shaped bladders, each with an opening guarded by a bristle-fringed door (Figure 30.1*d*). When a small aquatic insect brushes against the bristles, the door springs open and water rushes into the bladder—carrying the insect with it. When the door snaps shut, cells in the bladder wall secrete digestive enzymes, as do mutualistic species of bacteria that have taken up residence in the traps.

With this chapter we turn to examples of plant physiology—to the adaptations by which plants function in their environment. As you already know, most plants are photoautotrophs. Using only sunlight, water, carbon dioxide, and some minerals, they nourish themselves. Yet plants, like people, don't have unlimited supplies of required resources. Of every million molecules of air, only 350 of these are carbon dioxide. Unlike the soggy home of Venus flytraps, most soils are frequently dry. Nowhere except in overfertilized gardens does soil water hold lavish amounts of dissolved minerals. As you will see in this chapter, *many aspects of plant structure and function are responses to low environmental concentrations of essential resources.*

30.1 UPTAKE OF WATER AND NUTRIENTS

Nutritional Requirements

It took you eighteen years or so to grow to your present height. A corn plant can grow more than that in three months! Of course, neither of you would be all that vigorous if you were deprived of certain elements.

Plant growth and functioning depend on sixteen **essential elements**. Three of these—oxygen, carbon, and hydrogen—are the main components of carbohydrates, lipids, proteins, and nucleic acids. Plants get them from water (H_2O), and from gaseous oxygen (O_2) and carbon dioxide (CO_2) in the air.

The other essential elements become available to plants as dissolved mineral ions. As Table 30.1 indicates, six of the mineral ions are **macronutrients**. Plants require them in concentrations that are great enough to be easily detectable in tissues. Seven are **micronutrients**. Only traces can be detected in plant tissues. Mineral ions have vital roles in photosynthesis, aerobic respiration, and other metabolic events. Taken as a whole, they also contribute to the solute concentration gradients that help move substances into and out of cells.

Availability of water and dissolved mineral ions profoundly affects root development, hence growth of the entire plant. Roots branch out through some parts of the surrounding soil; then they are replaced by roots that branch into different parts as conditions change. It is not that roots "explore" the soil for resources. Rather, areas that have greater concentrations of water and mineral ions stimulate outward growth.

Specialized Absorptive Structures

As you know from earlier chapters, certain bacteria and fungi help many plants take up nutrients, and they get something in return. This two-way flow of benefits between species is a type of symbiotic interaction called **mutualism**. Often the interaction is necessary for the survival and reproduction of the species, which have become locked in intimate dependency.

Consider how many crops suffer from nitrogen scarcity. There actually is plenty of gaseous nitrogen ($N\equiv N$) in air. But plants do not have the metabolic means to engage in **nitrogen fixation**. By this process, certain enzymes break apart the three covalent bonds in gaseous nitrogen and attach the nitrogen atoms to organic compounds. To get high yields of crop plants, farmers provide plants with nitrogen compounds in fertilizers or encourage the growth of nitrogen-fixing bacteria in soil. Such bacteria convert gaseous nitrogen to forms that they, and their plant neighbors, can use.

Table 30.1 Role of Mineral Elements in Plant Function

Macronutrient	Some Known Functions	Some Deficiency Symptoms	Micronutrient	Some Known Functions	Some Deficiency Symptoms
Nitrogen	Component of proteins, nucleic acids, coenzymes, chlorophylls	Stunted growth; light green older leaves; older leaves yellow and die (chlorosis)	Chlorine	Role in root, shoot growth; role in photolysis	Wilting; chlorosis; some leaves die
			Iron	Roles in chlorophyll synthesis, electron transport	Chlorosis; yellow and green striping in grasses
Potassium	Activation of enzymes, role in maintaining water-solute balance* and thus affecting osmosis	Reduced growth; curled, mottled, or spotted older leaves; burned leaf margins; weakened roots and stems	Boron	Roles in flowering, germination, fruiting, cell division, nitrogen metabolism	Terminal buds, lateral branches die; leaves thicken, curl, become brittle
Calcium	Roles in cementing cell walls, regulation of many cell functions	Leaves deformed; terminal buds die; reduced root growth	Manganese	Role in chlorophyll synthesis; coenzyme activity	Light green leaves with green major veins; leaves whiten and fall off
Magnesium	Component of chlorophylls; activation of enzymes	Chlorosis; drooped leaves	Zinc	Role in formation of auxin, chloroplasts, and starch; enzyme component	Chlorosis; mottled or bronzed leaves; abnormal roots
Phosphorus	Component of nucleic acids, phospholipids, ATP	Purplish veins in older leaves; fewer seeds and fruits; stunted growth	Copper	Component of several enzymes	Chlorosis; dead spots in leaves; stunted growth; terminal buds die
Sulfur	Component of most proteins, two vitamins	Light green or yellow leaves; reduced growth	Molybdenum	Component of enzyme used in nitrogen metabolism	Possible nitrogen deficiency; pale green, rolled or cupped leaves

*All mineral elements contribute to the water-solute balance, but potassium is notable because there is so much of it.

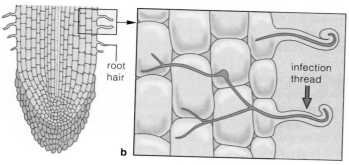

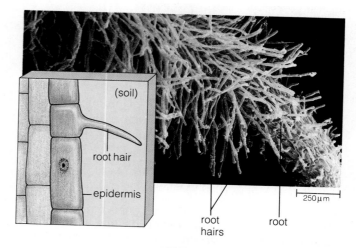

Figure 30.2 (**a**) Nutrient uptake at root nodules of legumes, which interact mutualistically with nitrogen-fixing bacteria (*Rhizobium* and *Bradyrhizobium*). (**b**) When bacterial cells infect slender extensions of root epidermal cells (root hairs), the cells are induced to form an "infection thread" of cellulose deposits. The bacteria use the thread like a highway to invade cells in the root cortex, some of which are tetraploid. (**c**) When infected, those cells and the bacteria within them divide repeatedly, forming a swollen mass that is destined to become a root nodule. The bacteria start fixing nitrogen when plant cell membranes surround them. The plant takes up some of the nitrogen, and the bacteria take up some compounds produced by the plant. (**d**) Soybean plants growing in nitrogen-poor soil. The plants on the right were inoculated with *Rhizobium* cells and developed root nodules.

String beans, peas, alfalfa, clover, and other legumes have an advantage in this respect. Nitrogen-fixing bacteria reside inside their roots, in localized swellings called **root nodules** (Figure 30.2*a*). The mutualistic bacteria withdraw some organic compounds from the plant tissues, but they provide the plants with some of their fixed nitrogen in return.

Or consider **mycorrhizae** (singular, mycorrhiza). As described on page 386, these are symbiotic interactions between young roots and a fungus (hence the name, meaning "fungus-root"). In some interactions, fungal filaments (hyphae) form a velvety covering around the roots. Collectively, hyphae have a large surface area for absorbing mineral ions from a large volume of soil. The fungus gets sugars and nitrogen-containing compounds from the roots. The roots get some scarce minerals that the fungus is better able to absorb.

Finally, remember that the plants themselves have numerous **root hairs** that greatly increase their surface area for absorbing water and nutrients. Root hairs are slender extensions of specialized epidermal cells (Figure 30.3 and page 494). A single root system might develop millions or billions of root hairs.

Root nodules, mycorrhizae, and root hairs help many plants absorb water and scarce mineral ions.

Figure 30.3 Scanning electron micrograph of root hairs on part of a small plant root.

30.2 CONTROL OF NUTRIENT UPTAKE

Roots do not absorb mineral ions willy-nilly. For example, certain salts as well as heavy metals would quickly destroy a plant if they were absorbed at the concentrations that exist in some soils.

Figure 30.4a shows a root's **vascular cylinder**—a central column of vascular tissue. A sheetlike layer of cells, the **endodermis**, surrounds the column. At this layer, a type of waxy band makes abutting walls of neighboring cells impenetrable to water and solutes. The band, a **Casparian strip**, stops water from trickling around the cells—it has to diffuse *through* their cytoplasm. Water can move into the vascular cylinder only by crossing the plasma membrane of endodermal cells, diffusing through the cytoplasm, then crossing the plasma membrane on the other side.

Plasma membranes, recall, permit the movement of some solutes but not others across the lipid bilayer. Depending on concentration gradients and other conditions, channel proteins and transport proteins help control the types of absorbed solutes that will become distributed through the plant (page 86).

Most flowering plants also have an **exodermis**, a layer of cells just inside the roots (Figure 30.4a). This layer has a Casparian strip that functions just like the one next to the root vascular cylinder.

Once nutrients enter a vascular cylinder, they are distributed to all living tissues in coordinated, interrelated ways that affect plant growth:

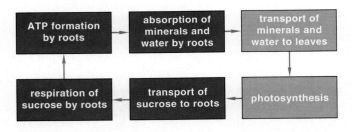

Thus living cells expend ATP energy and actively transport certain nutrients at membrane proteins. In photosynthetic cells, the required ATP is produced during photosynthesis and aerobic respiration. In other cells, ATP forms mostly by aerobic respiration.

Does all of this seem a bit abstract? The *Focus* essay reminds us of how the disappearance of growing, functioning roots can affect our lives.

Plants control nutrient uptake and distribution at a cell layer near the root surface, at another layer around the root vascular cylinder, then at the plasma membranes of living cells throughout the plant body.

Focus on the Environment

Putting Down Roots

All around you, trees, shrubs, and grasses are quietly going about the business of growing vast absorptive, anchoring networks of roots below the ground. They mine the good dirt—that is, topsoil, a complex mixture of mineral ions and the organic remains of organisms as well as sand, silt, clay, pebbles, and rocks. In deserts and some other places, good soil is only a few centimeters deep. In some grasslands, plants are lucky; their roots can spread through soil that is hundreds of meters deep.

Roots do more than absorb nutrients for individual plants. Collectively, they have profound influence over soil stability and nutrient storage in ecosystems. You will read about this in Chapter 48. For now, briefly consider the roots of trees in a young, undisturbed forest in New Hampshire. The calcium and other minerals they absorb become stored in plant tissues—instead of being leached from the soil during the spring rains. All plants were deliberately cleared from part of that forest. Before then, measurements showed that the annual loss of calcium from the

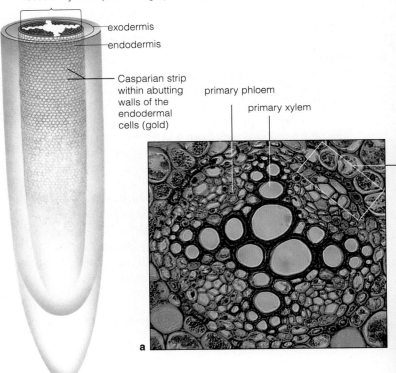

forest was 8 kilograms for each hectare (2.47 acres). After deforestation, the loss was six times higher.

Clear away the plants that cover the land, and wind will blow away soil and water will wash it away. On steep hillsides, even on fairly level farmland, the resulting erosion can be severe. Although soil erosion is a natural process, the vegetation cover in many natural ecosystems limits its effects.

Figure *a* shows the middle fork of the Salmon River in Idaho. The clear water at the right of the photograph comes from a wilderness area. The brown-colored water at the left is enriched with soil from a wilderness area that has been impacted by human activities, including cattle ranching. The dirty water is having bad effects on water quality and fish habitats.

Figure *b* shows gullies that have formed in an agricultural field. Gullies form by the erosive force of water. Their very existence channels the runoff from the land. As gullies grow deeper and wider, they promote increasingly rapid erosion. Soil nutrients are depleted, productivity declines, and fertilizers are brought in—all for the want of plants and their roots.

a

b

b Possible routes for the movement of soil water into a plant root. This diagram, a transverse section of the endodermis, corresponds to the boxed portion of the vascular cylinder shown in **a**.

c The only route for water and solutes into the vascular cylinder. Both are actively or passively transported in controlled ways across the plasma membrane of endodermal cells.

cortex

endodermis

inside vascular cylinder

Figure 30.4 Control over the uptake of water and dissolved nutrients in roots. (**a**) Roots of most flowering plants have an endodermis (a cell layer wrapped around the vascular cylinder) and an exodermis (a cell layer just beneath the epidermis). (**b**) The cells of both layers have a waxy Casparian strip embedded in their abutting walls. The strip keeps water from moving indiscriminately around the cells and into the vascular column. It makes water move through the cells (**c**). In this way, transport proteins that span the plasma membrane of those cells can selectively control the uptake of water and particular nutrients.

30.3 WATER TRANSPORT

Let's turn now to the means by which water—and the nutrients dissolved in it—moves from roots to stems, then into leaves. Of the water they absorb, plants use a fraction of it in growth and metabolism. But most of the water evaporates into air, by the mechanism described on page 26. When water evaporates from leaves as well as from stems and other plant parts, we call this process **transpiration**.

This brings up an interesting question. Assuming that water evaporates mainly from leaves, how does water get all the way to the top of leafy plants, including trees that are 100 meters tall?

The water moves through pipelines in the vascular tissue called **xylem**. Recall, from Chapter 29, that the water-conducting cells of xylem are called **tracheids** and **vessel members** (Figure 30.5). These cells are dead at maturity, and only their lignin-reinforced walls remain. Therefore, cells cannot be actively pulling the water "uphill."

Some time ago, the botanist Henry Dixon came up with a good explanation. According to his **cohesion-tension theory** of water transport, water in xylem is pulled up by the drying power of air, which creates continuous negative pressures (tensions) that extend downward from leaves to roots. Figure 30.6 illustrates this water transport mechanism. Think about the following key points of Dixon's explanation as you review this illustration:

---

1. The drying power of air causes transpiration, the evaporation of water from plant parts exposed to air.

2. Transpiration puts the water in xylem in a state of tension that extends from veins in leaves, down through the stems, to roots.

3. As long as water molecules continue to escape from the plant, the continuous tension in the xylem permits more molecules to be pulled up to replace them.

4. Unbroken, fluid columns of water show cohesion; they resist rupturing as they are pulled upward under tension. The collective strength of hydrogen bonds between water molecules, which are confined in the narrow, tubular xylem cells, imparts this cohesion.

5. Hydrogen bonds are enough to hold water molecules together in the xylem. But they are not strong enough to prevent the molecules from breaking away from each other during transpiration and then escaping from leaves.

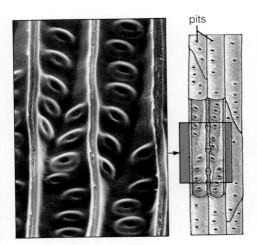

a Tracheids, with tapering, unperforated end walls. Small pits in the walls of neighboring tracheids match up. Although too small to permit rapid flow, they do confine air bubbles (which obstruct water transport) to individual tracheids.

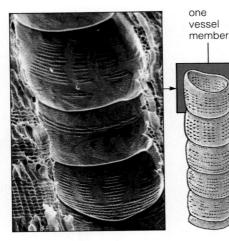

b Part of one type of vessel, with thick-walled cells (vessel members) that interconnect as a water-conducting tube.

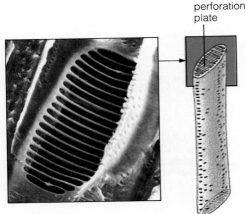

c A vessel member that has end wall perforations. These permit water—and air bubbles—to flow unimpeded through the vessel. This may be why natural selection has favored retention of tracheids along with vessels in the same plant.

Figure 30.5 Examples of tracheids and vessel members. These cells are dead at maturity. Their walls remain interconnected and form the water-conducting tubes of xylem. Tracheids probably evolved before vessel members, but both are present in nearly all flowering plants.

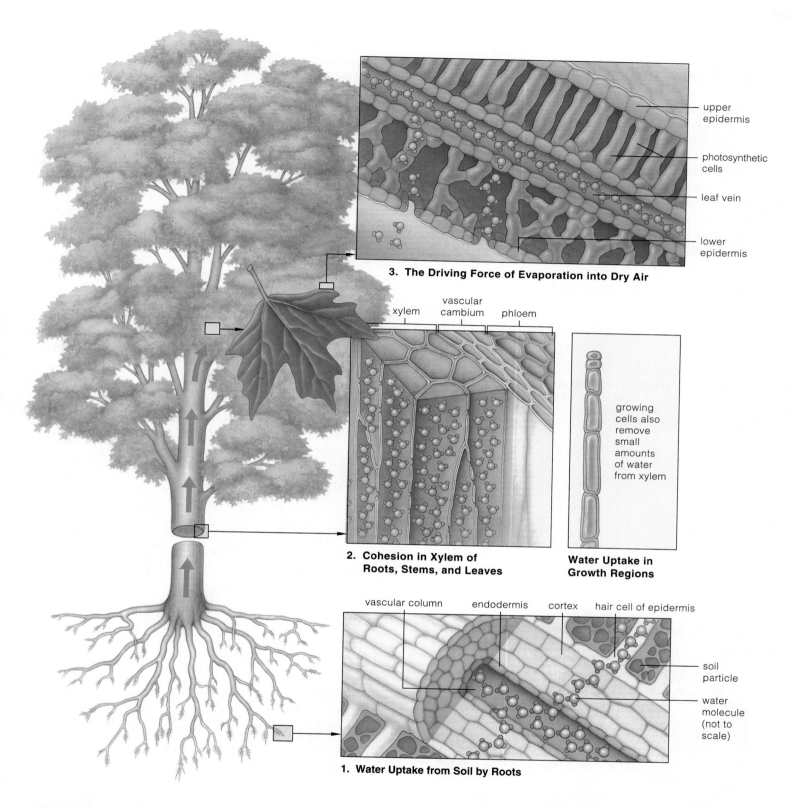

upper epidermis

photosynthetic cells

leaf vein

lower epidermis

3. The Driving Force of Evaporation into Dry Air

xylem vascular cambium phloem

growing cells also remove small amounts of water from xylem

2. Cohesion in Xylem of Roots, Stems, and Leaves

Water Uptake in Growth Regions

vascular column endodermis cortex hair cell of epidermis

soil particle

water molecule (not to scale)

1. Water Uptake from Soil by Roots

Figure 30.6 Cohesion-tension theory of water transport. Tensions in water within the xylem extend from leaf to root. The tensions result mostly from transpiration (the evaporation of water from plant parts). They allow columns of water molecules that are hydrogen-bonded to one another to be pulled upward. Hydrogen bonding gives water its cohesive properties (page 27).

30.4 CONTROL OF WATER LOSS

Of the water that moves into a leaf, 90 percent or more escapes (through transpiration). Of the remaining 10 percent, cells use only 2 percent during photosynthesis, membrane functions, and other activities. That tiny amount is vital. Plants wilt and water-dependent events are severely disrupted when transpiration exceeds water uptake at the plant's roots for very long. Yet land plants are not entirely at the mercy of the environment. They have a cuticle, and they have stomata (Figures 30.7 and 30.8).

The Water-Conserving Cuticle

Even mildly stressed plants would rapidly wilt and die were it not for their **cuticle**. This translucent, water-impermeable layer consists of secretions from epidermal cells. It coats their outer walls, which are exposed to the air. Figure 30.7 shows a fine example.

At the cuticle surface are deposits of waxes, which are water-insoluble lipids with long fatty-acid tails. The cuticle proper consists of waxes embedded in a matrix of **cutin**, an insoluble lipid polymer. Below this, cellulose threads through the matrix. Often a layer of pectins (a type of polysaccharide) helps bind the cuticle to the cell walls.

The cuticle does not bar the passage of light rays into photosynthetic parts of a plant. It does restrict water loss. It also restricts the *inward* diffusion of the carbon dioxide required for photosynthesis and the *outward* diffusion of oxygen by-products. (Recall, from page 118, that a buildup of oxygen in the air spaces inside a leaf has bad effects on the rate of photosynthesis.)

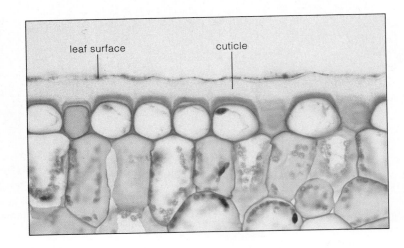

Figure 30.7 Example of a waxy cuticle on the upper epidermis of a leaf. Water, carbon dioxide, and oxygen cannot cross the cuticle proper. They enter or depart from the leaf mainly at stomata, of the sort shown in Figure 30.8.

leaf surface   cuticle

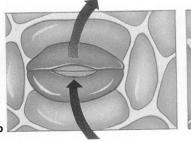

guard cell   nucleus   stomatal opening   guard cell

a

Figure 30.8 (**a**) Structure of a pair of guard cells from a beavertail cactus (*Opuntia*) stem. (**b**) Stomatal action. A stoma remains open or closed, depending on shape changes in the two guard cells that define the opening across the cuticle-covered epidermis.

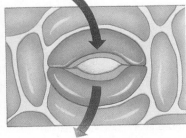

Stomatal closure. Water leaves the swollen guard cells, which collapse against each other and so close the gap (stoma) between them.

b

Stomatal opening. Water enters collapsed guard cells, which swell (under turgor pressure) and move apart, thus forming the stoma.

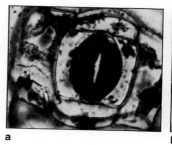
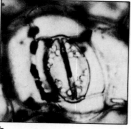

a   b

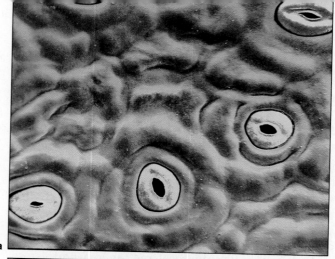

a

b

Figure 30.9 Experimental evidence for the accumulation of potassium in guard cells that are expanding with incoming water. Strips from the leaf epidermis of a dayflower (*Commelina communis*) were immersed in solutions containing dark-staining substances that bind preferentially with potassium ions. (**a**) In leaf samples having opened stomata, most of the potassium was concentrated in the guard cells. (**b**) In leaf samples having closed stomata, very little potassium was in guard cells; most of the potassium was present in normal epidermal cells.

Controlled Water Loss at Stomata

Given the requirement for water conservation, how do carbon dioxide and oxygen get past the cuticle-covered epidermis? They do so at tiny openings called **stomata** (singular, stoma). Water departs from the plant, and carbon dioxide moves in mainly at these openings. A pair of specialized cells, the **guard cells**, defines each opening (Figure 30.8). When the cells swell with water, the force of internal fluid pressure distorts their walls. As the cells bend under the pressure, a gap (the stoma) forms between them. When they lose water and internal fluid pressure drops, guard cells collapse against each other and so close the gap.

In most plants, stomata stay open during the day, when photosynthesis takes place. They lose water—but they gain carbon dioxide for the reactions. Stomata stay closed at night. Then, plants conserve water—and precious carbon dioxide accumulates inside the leaf as cells busily engage in aerobic respiration.

Whether a stoma opens or closes depends on how much water and carbon dioxide are in the two guard cells. Photosynthesis starts when the sun comes up. As the morning progresses, carbon dioxide levels drop in cells—including the guard cells. The decrease helps trigger active transport of potassium ions into the guard cells (Figure 30.9). So do blue wavelengths of light, which penetrate the atmosphere better as the sun arcs higher in the sky. Water follows potassium into the cells, by osmosis. The inward movement of water provides the fluid pressure to open the stoma.

Carbon dioxide levels rise when the sun goes down and photosynthesis stops. Potassium, then water, moves out of the guard cells—and the stoma closes.

Figure 30.10 Effects of air pollution on stomata. (**a**) This is a surface view of holly leaf stomata. (**b**) This is what the leaf looks like when a holly plant grows in industrialized regions. Gritty airborne pollutants clog its stomata. They also prevent much of the sun's rays from reaching photosynthetic cells inside the leaf.

CAM plants, which include most cacti, conserve water in a different way. They open stomata at night, when they fix carbon dioxide by a special metabolic C4 pathway (compare page 119). The cells of CAM plants use carbon dioxide in photosynthesis the following day, when their stomata close.

As this section makes clear, plant survival depends on stomata. Think about this when you are out and about on smog-shrouded days (Figure 30.10).

Transpiration and gas exchange take place mainly at stomata. These numerous small openings span the waxy cuticle, which covers all plant surfaces exposed to air.

By opening and closing stomata at different times, plants control water loss, carbon dioxide uptake, and oxygen disposal.

30.5 TRANSPORT OF ORGANIC SUBSTANCES

Storage and Transport Forms of Organic Compounds

What happens to sucrose and other organic products of photosynthesis? Leaf cells use some; the rest end up in roots, stems, buds, flowers, and fruits. Carbohydrates become stored as starch in most plant cells. Proteins and fats become stored in many seeds. Some fruits (avocados especially) also accumulate fats.

Starch molecules are too large for transport across the plasma membrane of cells that produce them. They also are too insoluble for transport to other plant parts. Fats (which are mostly insoluble) and protein reserves can't be transported away from storage sites. Cells convert storage forms of organic compounds to smaller solutes that are more readily transported. For instance, hydrolysis of starch liberates glucose units, which combine with fructose to form sucrose.

Experiments with aphids (a type of small insect) revealed the main carbohydrate in **phloem**, which is the tissue that transports dissolved organic compounds in vascular plants. Aphids feeding on the juices in the conducting tubes of phloem were exposed to high levels of carbon dioxide to anesthetize them. Then their bodies were severed from the mouthparts, which were left embedded in the tubes. For most plant species, sucrose turned out to be the most abundant carbohydrate in the fluid being forced out of the tubes.

Translocation

Translocation refers to the distribution of sucrose and other organic compounds through the plant body. This distribution apparently occurs under pressure in phloem. Like xylem, phloem has conducting tubes, fibers, and strands of parenchyma cells. Unlike xylem, the tubes consist of living cells positioned side by side and end to end. The cells are **sieve-tube members** (Figure 30.11). Dissolved organic compounds flow rapidly through large pores in their abutting end walls. **Companion cells**, specialized parenchyma cells adjacent to the sieve tubes, have supporting roles in translocation.

The pressure that forces sucrose and other organic compounds to flow through sieve tubes is high—often five times as high as in automobile tires. Aphids demonstrate this when they force a mouthpart into sieve tubes and feed on the dissolved sugars inside. Pressure can force the fluid through the aphid gut and out the other end as "honeydew" (Figure 30.12). Park your car under trees being attacked by aphids and it might get a spattering of sticky honeydew droplets, thanks to the high fluid pressure.

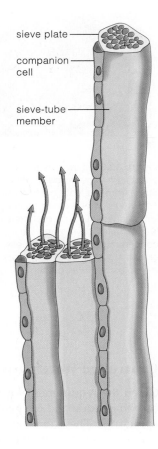

Figure 30.11 Example of cells that interconnect to form the conducting tubes (sieve tubes) of phloem. Companion cells expend energy to load sugars and other organic compounds into sieve-tube members.

sieve plate

companion cell

sieve-tube member

Figure 30.12 A honeydew droplet exuding from the tail end of an aphid feeding on sugars flowing under pressure through the phloem.

Pressure Flow Theory

Organic compounds are translocated along gradients of decreasing solute concentration and pressure. The flow's *source* is any region where organic compounds are being loaded into sieve tubes. Photosynthetic tissues in leaves are a common source. The flow ends at a *sink*—any region where organic compounds are being unloaded and used or stored. Young, growing flowers or developing fruits are common sink regions.

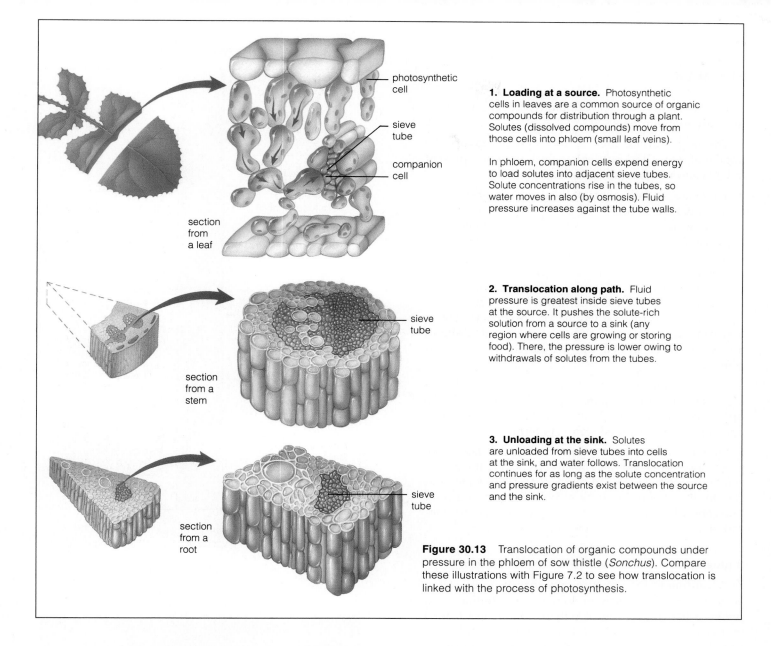

photosynthetic cell

sieve tube

companion cell

section from a leaf

1. Loading at a source. Photosynthetic cells in leaves are a common source of organic compounds for distribution through a plant. Solutes (dissolved compounds) move from those cells into phloem (small leaf veins).

In phloem, companion cells expend energy to load solutes into adjacent sieve tubes. Solute concentrations rise in the tubes, so water moves in also (by osmosis). Fluid pressure increases against the tube walls.

sieve tube

section from a stem

2. Translocation along path. Fluid pressure is greatest inside sieve tubes at the source. It pushes the solute-rich solution from a source to a sink (any region where cells are growing or storing food). There, the pressure is lower owing to withdrawals of solutes from the tubes.

3. Unloading at the sink. Solutes are unloaded from sieve tubes into cells at the sink, and water follows. Translocation continues for as long as the solute concentration and pressure gradients exist between the source and the sink.

sieve tube

section from a root

Figure 30.13 Translocation of organic compounds under pressure in the phloem of sow thistle (*Sonchus*). Compare these illustrations with Figure 7.2 to see how translocation is linked with the process of photosynthesis.

Why do organic compounds flow from a source to a sink? According to the **pressure flow theory**, pressure builds at the source end of a sieve-tube system and *pushes* the solute-rich solution toward a sink, where solutes are removed.

Experimental evidence for the pressure flow theory was obtained from studies of sow thistle (*Sonchus*). Figure 30.13 follows what happens after sucrose moves from photosynthetic cells into small veins in the leaves. There, companion cells expend energy to load sucrose into adjacent sieve-tube members. As the sucrose concentration increases in the tubes, water also moves in, by osmosis. More internal fluid pressure is exerted against the tube walls. As the pressure increases, it pushes the sucrose-laden fluid out of the leaf, into the stem, and on to the sink.

In short, keep these three central points in mind:

1. Plants store organic compounds in the form of starch, fats, and proteins. They convert the storage forms to sucrose and other small units that are soluble and easily translocated through the plant body.

2. The distribution of organic compounds to different plant regions depends on concentration and pressure gradients in the sieve-tube system of phloem.

3. These gradients exist as long as companion cells supply energy to load compounds into sieve tubes at sources (such as mature leaves) and as long as compounds are removed at sinks (such as roots).

SUMMARY

1. Plants require oxygen, carbon, hydrogen, and thirteen essential nutrients.

2. The large surface area of root systems favors the uptake of water and nutrients, which often are present in low concentrations in the surrounding soil.

3. These are the key points of the cohesion-tension theory of water transport:

   a. In water-conducting cells of xylem, continuous negative pressures (tensions) extend from leaves to roots. Transpiration (evaporation of water from leaves and other parts exposed to air) causes the tension.

   b. When water molecules escape from leaves, replacements are pulled under tension to the sites.

   c. The collective strength of hydrogen bonds between water molecules imparts cohesion that allows the water to be pulled upward as continuous fluid columns.

4. Water cannot cross a plant's waxy cuticle, which covers most aboveground parts. Plants transpire and take up carbon dioxide mostly at stomata. These are small openings across leaf (and stem) epidermis.

5. Stomata commonly are open during the day. Then, plants lose water but take in carbon dioxide for photosynthesis. Stomata close at night, conserving water and carbon dioxide produced by aerobically respiring cells. The controlled opening and closing of stomata balance carbon dioxide uptake with water conservation.

6. Translocation distributes organic compounds through plants; it occurs in sieve tubes of phloem.

7. According to the pressure flow theory, translocation is driven by differences in solute concentrations and pressure between a source (where solutes are loaded into sieve tubes) and a sink (where they are unloaded from them). The differences exist as long as companion cells expend energy to load solutes into the sieve tubes and as long as solutes are removed at the sink. Figure 30.14 summarizes this pressure-flow mechanism.

sieve tube of the phloem

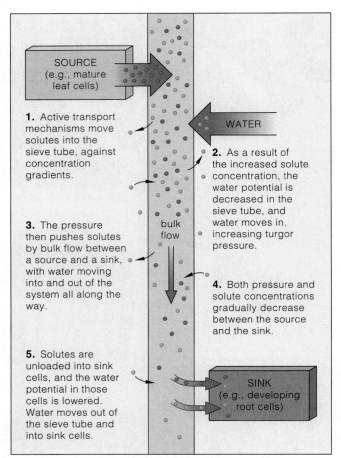

SOURCE (e.g., mature leaf cells)

WATER

1. Active transport mechanisms move solutes into the sieve tube, against concentration gradients.

2. As a result of the increased solute concentration, the water potential is decreased in the sieve tube, and water moves in, increasing turgor pressure.

3. The pressure then pushes solutes by bulk flow between a source and a sink, with water moving into and out of the system all along the way.

bulk flow

4. Both pressure and solute concentrations gradually decrease between the source and the sink.

5. Solutes are unloaded into sink cells, and the water potential in those cells is lowered. Water moves out of the sieve tube and into sink cells.

SINK (e.g., developing root cells)

Figure 30.14 Summary of the proposed mechanism of pressure flow in phloem, a vascular tissue of flowering plants.

Review Questions

1. List the three essential elements that are necessary for plant growth and functioning. Distinguish between macronutrients and micronutrients. *502*

2. Refer to Table 30.1. What are some signs that a plant suffers a deficiency in certain mineral elements? *502*

3. When moving a plant from one place to another, it helps to include some native soil around the roots. Explain why, given what you know about mycorrhizae and root hairs. *503*

4. What is the function of Casparian strips in the walls of root endodermal cells? *504–505*

5. Using botanist Henry Dixon's explanation, explain how water moves from the soil upward through leafy plants. In some cases, water moves to tree tops that are 100 meters tall. *506–507*

6. Most plants would die if their stomata remained open all the time or closed all the time. Explain why. *509*

7. Describe the pressure flow theory as a way of explaining translocation. *510–511*

1. Water can be pulled up through a plant due to the cumulative strength of _____ between water molecules.

2. Water leaves and carbon dioxide enters a plant through _____ (tiny openings across leaf epidermis).

3. In plants, mineral ions _____ .
   a. have roles in metabolism
   b. help establish gradients across membranes
   c. influence water movement into cells
   d. maintain cell shape and growth
   e. all of the above

4. The nutrition of some plants depends on a root-fungus association known as _____ .
   a. root nodules   c. root hairs
   b. mycorrhizae   d. root hyphae

5. The nutrition of some plants depends on a root-bacterium association known as _____ .
   a. root nodules   c. root hairs
   b. mycorrhizae   d. root hyphae

6. Water evaporation from plant parts is called _____ .
   a. translocation   c. transpiration
   b. expiration   d. tension

7. Upward water transport to the tops of tall trees is explained by the _____ .
   a. pressure flow theory
   b. differences in source and sink solute concentrations
   c. pumping force of xylem vessels
   d. cohesion-tension theory

8. Water entering roots is forced to diffuse through the cytoplasm of endodermal cells because of _____ in their walls.
   a. cutin strips   c. Casparian strips
   b. lignin strips   d. cellulose strips

9. During the day, most plants lose _____ and take up _____ .
   a. carbon dioxide; water   c. oxygen; water
   b. water; oxygen   d. water; carbon dioxide

10. At night, plants conserve _____ and _____ .
    a. carbon dioxide; oxygen   c. oxygen; water
    b. water; oxygen   d. water; carbon dioxide

11. In phloem, organic compounds flow through _____ .
    a. collenchyma cells   c. vessels
    b. sieve tubes   d. tracheids

12. Match the concepts of plant nutrition and transport.
    ____ stomata
    ____ sink
    ____ root system
    ____ hydrogen bonds
    ____ transpiration
    ____ translocation

    a. evaporation from plant parts
    b. balancing water loss with carbon dioxide requirements
    c. cohesion in water transport
    d. sugars unloaded from sieve tubes
    e. distribution of organic compounds through the plant
    f. response to scarce soil nutrients

CAM plant 509
Casparian strip 504
cohesion-tension theory 506
companion cell 510
cuticle 508
cutin 508
endodermis 504
essential element 502
exodermis 504
guard cell 509
macronutrient 502
micronutrient 502
mutualism 502
mycorrhiza 503

nitrogen fixation 502
phloem 510
pressure flow theory 511
root hair 503
root nodule 503
sieve-tube member 510
stoma (stomata) 509
symbiosis 502
tracheid 506
translocation 510
transpiration 506
vascular cylinder 504
vessel member 506
xylem 506

Epstein, E. 1973. "Roots." *Scientific American* 228(5):48–58.

Galston, A., P. Davies, and R. Satter. 1980. *The Life of a Green Plant.* Englewood Cliffs, New Jersey: Prentice-Hall.

Hewitt, E., and T. Smith. 1975. *Plant Mineral Nutrition.* New York: Wiley.

Hitz, W., and R. Giaquinta. 1987. "Sucrose Transport in Plants." *Bioessays* 6:217–221.

Salisbury, F., and C. Ross. 1991. *Plant Physiology.* Fourth edition. Belmont, California: Wadsworth.

31 PLANT REPRODUCTION

Chocolate From the Tree's Point of View

When tax collectors came around in ancient Mexico, they liked to be paid in seeds from the cacao tree (*Theobroma cacao*). Tax payments went straight to the king, Montezuma, and from him to the royal kitchen. Fermented, roasted, and ground to a paste, the seeds were the basis of chocolatl—a drink spiced with hot chili peppers, cinnamon, extracts from vanilla orchids, and a pinch of anise.

Montezuma sipped chocolatl from gold goblets after meals, during religious rites, and reportedly before visits to his harem. After each goblet had the privilege of touching the king's lips, it was ceremonially tossed into a lake outside the palace.

In 1519 Hernando Cortez dethroned Montezuma, drained the lake, and recovered the goblets. These he shipped to Spain, along with three chests filled with cacao seeds—and so began a worldwide fascination with chocolate.

T. cacao evolved in the undergrowth of tropical forests in Central America. Before Mayas and Aztecs domesticated the plant, wild animals ate its fruits and dispersed its seeds. Today it flourishes on vast plantations in Central America, the West Indies, and West Africa. Growers plant taller trees alongside the treasured cacao trees to provide filtered shade reminiscent of the native habitat (Figure 31.1*a*).

Figure 31.1 (**a**) A new generation of plants, completing their embryonic development inside fruits on the trunk of a parent cacao tree (*Theobroma cacao*), shown here in its native habitat. (**b**) Students, indirectly contributing to this plant's widespread distribution and reproductive success—which is far beyond what it could have accomplished on its own.

a

b

Like other angiosperms, *T. cacao* produces flowers. Unlike most other species, its flowers grow directly from buds on the tree trunk instead of on separate floral shoots. About six months after the flowers are pollinated and the eggs are fertilized, large, heavy fruits develop (Figure 31.1*a*). Each fruit, or "pod," contains as many as forty seeds—the cacao "beans." These are processed into cocoa butter and essences, which end up in chocolate products (Figure 31.1*b*).

Unknown interactions among the 1,000 or so compounds in chocolate exert compelling (some say addictive) effects on the human brain. Each year, for instance, the average American feels compelled to buy 8 to 10 pounds of chocolate, and the average Swiss citizen, a whopping 22 pounds. With that kind of demand, growers do their very best to keep cacao trees growing and reproducing.

And so, through interactions with humans, *T. cacao* is like orange trees, corn, and other domesticated crop plants that now reproduce in great numbers, in many different parts of the world. In this respect it is like flowering plants in the wild that interact with bees, birds, beetles, bats, and other animals in ways that promote reproductive success. And in evolutionary terms, reproductive success is what life is all about.

KEY CONCEPTS

1. Sexual reproduction of flowering plants requires spores and gametes, which develop in flowers. Commonly it depends on animals that help pollinate the flowers and disperse the seeds.

2. Male reproductive parts of flowers produce microspores. These develop into pollen grains, which give rise to sperm-bearing male gametophytes. Pollen grains are released, then carried by wind, water, or animals to the female reproductive parts of flowers.

3. Female reproductive parts of flowers produce megaspores. These give rise to egg-producing female gametophytes. The female gametophytes remain embedded within and are nourished by tissues of the parent plant.

4. After sperm fertilize the eggs, seeds develop. Each seed is an embryo sporophyte along with tissues that function in its nutrition, protection, and dispersal.

31.1 REPRODUCTIVE MODES

Although you probably don't think about this very often, flowering plants engage in sex. Like humans, they have splendid reproductive systems that produce, protect, and nourish sperm and eggs. Like human females, flowering plants house embryos during early development. Flowers serve as invitations to third parties—pollinators that help get sperm and egg together. Long before humans ever thought of it, flowering plants were using tantalizing colors and fragrances that improve the odds for sexual success.

Most plants also do something humans cannot do, at least not yet. They can reproduce asexually. In *sexual* reproduction, a fertilized egg has two sets of genetic instructions (from two gametes). *Asexual* reproduction proceeds by way of mitosis, so all offspring are genetically identical to the parent (they form a clone).

When we hear the word "plant," we often think of something like a cherry tree (Figure 31.2a). This is an example of a **sporophyte**, a vegetative body that grows, by mitotic cell divisions, from a fertilized egg. At some point in the life cycle, the sporophytes bear **flowers**, which are shoots specialized for reproduction. Flowers produce haploid spores, which develop into haploid bodies called **gametophytes**. Sperm form in male gametophytes, and eggs form in female gametophytes.

Sexual reproduction dominates most flowering plant life cycles, and it will be our focus here. But bear in mind, sporophytes also reproduce asexually—as when new plants grow along runners (aboveground stems) of strawberry plants or from underground stems of onions, lilies, and Bermuda grass. Many crop plants also are propagated asexually. For example, orchard pear trees are grown from cuttings or buds of a parent tree. Southern California's navel-orange industry is a huge clone from a single tree that grew in Riverside.

31.2 FLORAL STRUCTURE

Components of Flowers

Flowers grow at floral shoots. While they are growing, they differentiate into nonfertile parts (sepals and petals) and fertile parts (stamens and carpels). Figure 31.2b shows how these floral components are arranged. Directly or indirectly, all of these parts are attached to a receptacle, the modified base of a floral shoot.

Peel open a rosebud. Notice how its outermost whorl of parts, the leaflike sepals, enclose the petals, which in turn enclose the flower's fertile components (Figure 31.3). This is a common arrangement. The sepals are the flower's "calyx." The leaflike petals, which ring the male and female parts, are the flower's "corolla."

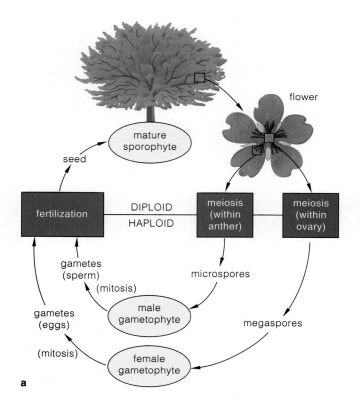

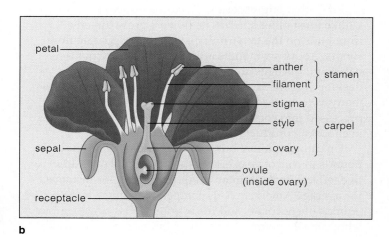

Figure 31.2 (**a**) Overview of flowering plant life cycles.

(**b**) Structure of a flower—a cherry blossom (*Prunus*). Like the flowers of many plants, it has a single carpel (a female reproductive part) and stamens (male reproductive parts). Flowers of other plants have two or more carpels, often united as a single structure. Single or fused carpels consist of an ovary and a stigma. Often the ovary extends upward as a slender column (style).

Figure 31.3 Location of a few floral parts in one of the most prized of all cultivated plants, the rose (*Rosa*).

corolla
(all petals
combined)

calyx
(all sepals
combined)

receptacle

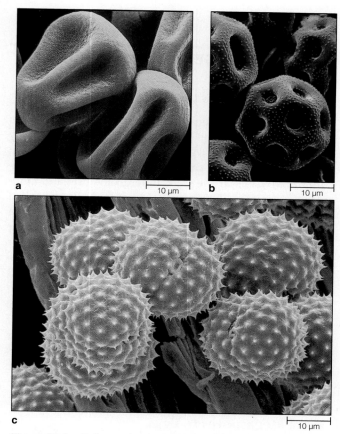

Figure 31.4 Pollen grains from (**a**) grass, (**b**) rose, and (**c**) ragweed plants. Pollen grains of most families of plants differ in size, wall sculpturing, and number of wall pores.

Like leaves, the sepals and petals have ground tissues, vascular tissues, and epidermis. What makes many flowers sweet smelling? Epidermal cells of their petals contain fragrant oils. What gives petals their color and shimmer? Various cells of the ground tissues contain pigments, such as carotenoids (yellow to red-orange) and anthocyanins (red to blue), and tiny, light-refracting crystals. What functions do the fragrances, colors, patterns, and arrangements of petals serve? As you will see, they attract pollinators.

Where Pollen and Eggs Develop

Just inside the flower's corolla are **stamens**, the male reproductive parts. Nearly all stamens consist of an anther and a single-veined stalk (filament). An anther is internally divided into pollen sacs, the chambers in which walled spores develop into haploid, sperm-producing bodies called **pollen grains** (Figure 31.4). Wind, water currents, or animals carry the pollen grains to the female reproductive parts.

Female reproductive parts are at the center of a flower. These parts are sometimes called pistils, but their more recent name is **carpels**.

Some flowers have a single carpel. Others have more, often fused together as a compound structure. The lower portion of single or fused carpels is the **ovary**, where eggs develop, fertilization takes place, and seeds mature. (The word "angiosperm" refers to the carpel; it is derived from the Greek *angeion*, meaning vessel, and *sperma*, meaning seed.) The carpel's upper portion, the stigma, is a sticky or hairy surface tissue that "captures" pollen grains and favors their germination. Commonly, a style (a slender, upward extension of the ovary wall) elevates the stigma.

Not all flowers have stamens and carpels. The species that have both male and female parts are called *perfect* flowers. Other species produce *imperfect* flowers, which contain male or female parts. In some species, such as oaks, the same plant bears male and female flowers. In willows and other species, they are on separate individual plants.

Flowers are specialized reproductive shoots in which sperm-producing pollen grains and eggs develop, fertilization takes place, and seeds mature.

A NEW GENERATION BEGINS

From Microspores to Pollen Grains

We turn now to pollen grain formation. While anthers grow, four masses of spore-producing cells form by mitotic cell divisions. Walls develop around them. Each anther now has four chambers, called pollen sacs (Figure 31.5a). Cells in the sacs undergo meiosis and cytoplasmic division. This results in haploid spores, called **microspores**, each of which becomes encased in a sculpted wall. The walled microspores divide once or twice by mitosis and become pollen grains, which enter a period of arrested growth. In time, they will be released from the anther. Their wall components will protect them from decomposer organisms.

As soon as they form, many types of pollen grains produce sperm nuclei, which are the male gametes of flowering plants. Other types don't do this until after they travel to a carpel and start growing toward its ovule. Thus, a pollen grain is a mature or immature male gametophyte, depending on the plant species.

From Megaspores to Eggs

Meanwhile, one or more masses of cells form on the inner wall of a flower's ovary. Each is the start of an ovule. An **ovule** is a structure that contains a female gametophyte and that may become a seed. As each cell mass grows, a tissue forms inside it, and one or two protective layers form around it. These layers are integuments. Inside the mass, a cell divides by meiosis, and four haploid spores form. Spores that form in flowering plant ovaries are called **megaspores**.

Commonly, all megaspores but one disintegrate. The one remaining undergoes mitosis three times without cytoplasmic division. At first it is a cell with eight nuclei (Figure 31.5b). Its cytoplasm divides after each nucleus migrates to a specific location. The result is a seven-celled embryo sac, the female gametophyte. One of the cells, the "endosperm mother cell," has two nuclei. It will help form **endosperm**, a nutritive tissue for the forthcoming embryo. Another cell is the egg.

From Pollination to Fertilization

Each spring, flowering plants release pollen. You are acutely aware of this reproductive event if you are one of the millions of people who suffer from hay fever. This is an allergic reaction to wall proteins of pollen grains released by ragweed and many other plants.

Pollination refers to the transfer of pollen grains to a receptive stigma. Air currents, water currents, insects, birds, or other pollinating agents make the transfer. The relationship between flowering plants and their pollina-

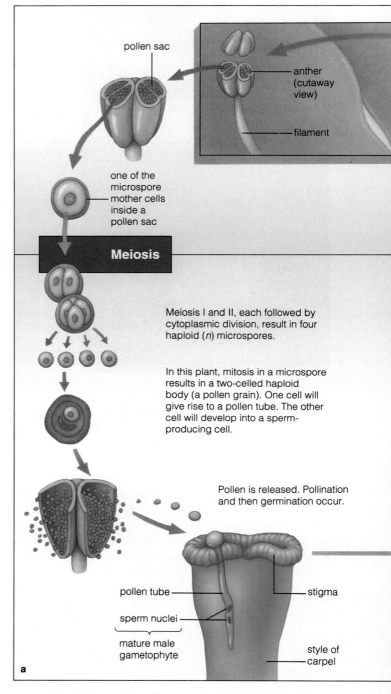

Figure 31.5 Life cycle of cherry (*Prunus*), one of the flowering plants classified as a dicot. (**a**) This part of the diagram shows how pollen grains (male gametophytes) develop and germinate. (**b**) This part shows what goes on inside the ovule in this ovary.

tors is one of the most intriguing of all evolutionary stories. It is the topic of the *Commentary* on page 520.

Once a pollen grain lands on a receptive stigma, it germinates. In this case, germination means that the pollen grain resumes growth and develops into a multicelled, tubular structure. This "pollen tube" starts bur-

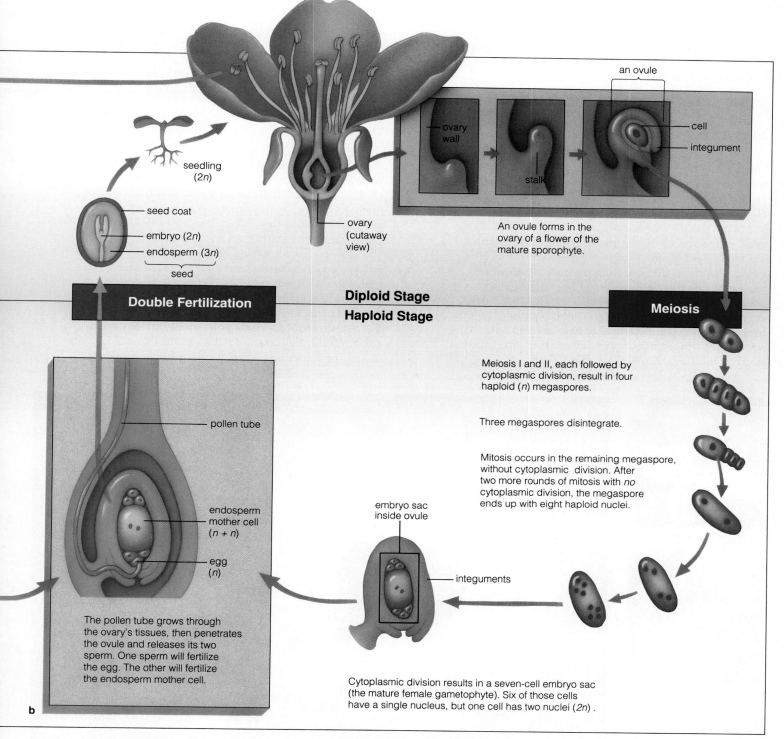

seedling
(2n)

seed coat
embryo (2n)
endosperm (3n)
seed

ovary
(cutaway
view)

an ovule

ovary
wall

cell
integument

stalk

An ovule forms in the
ovary of a flower of the
mature sporophyte.

Double Fertilization

Diploid Stage

Haploid Stage

Meiosis

pollen tube

endosperm
mother cell
(n + n)

egg
(n)

The pollen tube grows through
the ovary's tissues, then penetrates
the ovule and releases its two
sperm. One sperm will fertilize
the egg. The other will fertilize
the endosperm mother cell.

b

embryo sac
inside ovule

integuments

Meiosis I and II, each followed by
cytoplasmic division, result in four
haploid (n) megaspores.

Three megaspores disintegrate.

Mitosis occurs in the remaining megaspore,
without cytoplasmic division. After
two more rounds of mitosis with *no*
cytoplasmic division, the megaspore
ends up with eight haploid nuclei.

Cytoplasmic division results in a seven-cell embryo sac
(the mature female gametophyte). Six of those cells
have a single nucleus, but one cell has two nuclei (2n) .

rowing through tissues of the ovary, carrying sperm nuclei with it (Figure 31.5b). Chemical and molecular cues guide a pollen tube's growth through the tissues, toward the egg chamber and sexual destiny. When the tube reaches an ovule, it penetrates the embryo sac, its tip ruptures, and the two sperm are released.

"Fertilization" generally means the fusion of a sperm nucleus and an egg nucleus. **Double fertilization** occurs in flowering plants alone. In diploid species, one sperm nucleus fuses with that of the egg, producing a diploid (2n) zygote. Meanwhile, the other

sperm nucleus fuses with both nuclei of the endosperm mother cell, forming a cell with a triploid (3n) nucleus. That 3n cell gives rise to endosperm, a nutritive tissue.

In flowering plants, sperm cells form within pollen grains, the male gametophytes. Eggs develop inside ovules, which contain the female gametophytes.

After pollination and double fertilization, an embryo *and* nutritive tissue form inside the ovule, which becomes a seed.

Flowering Plants and Pollinators—A Coevolutionary Tale

Splendidly different flowering plants live almost everywhere, from icy tundra to deserts, ponds, even in the seas. Few organisms match their distribution and diversity. How did these plants manage it?

We can piece together a plausible answer, based on the fossil record and observations of existing species. About 435 million years ago, when plants began invading land, insects that feasted on decaying plant parts and spores were probably right behind them. Taller plants with cuticle-covered, lignified stems appeared, as did cuticle-covered insects adapted to land. The smorgasbord of lofty, tough-stemmed plants seemed to have favored natural selection of winged insects with an amazing assortment of sucking, piercing, and chewing mouthparts.

By 390 million years ago, the first seed-bearing plants were flourishing in humid coastal forests. The ancestors of existing gymnosperms and flowering plants were among them. Many species had separate conelike structures for pollen sacs and ovules. Pollen may have reached the ovules simply by drifting on air currents.

Pollen is rich in proteins. At some point, insects made the connection between "cone" and "food source." The plants lost some pollen to insects—but they gained a major reproductive advantage when pollen-dusted insects brushed up against ovules. The insects clambering over the cones weren't precision pollinators. But they made more accurate deliveries of pollen grains than air currents alone. The tastier the pollen, the more home deliveries,

(**a**) Shine ultraviolet light on a marsh marigold to reveal its bee-attracting pattern. (**b**) Broad landing platform of daisies. (**c**) The scotch broom landing platform corresponds to bee size and shape. A bee's weight forces its petals apart, releasing the stamens. These dust the bee with pollen. A bee grooms itself and packs pollen (the *orange mass*) inside "baskets" of leg hairs.

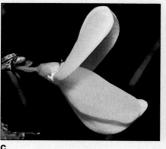

d Bahama woodstar, sipping nectar from a hibiscus blossom. Like other hummingbirds, it can forage for nectar in midflight. Its long, narrow bill coevolved with long, narrow floral tubes.

and the more seeds formed. *The greater the number of seeds, the greater the chance of reproductive success.*

What we are describing is **coevolution**. The word refers to two (or more) species jointly evolving as an outcome of close ecological interactions. When one species evolves, the change affects selection pressures operating between the two, so the other also evolves. In our coevolutionary tale, there was selection of more accurate pollen-delivering insects—and of plant structures attractive to them. Pollinators quick to recognize and locate particular plants had a competitive edge. So did plants with come-hither advertisements—fragrances and flowers.

Advertising also attracted pollen-eating beetles, which chewed on as well as pollinated ovules. Thus chewing behavior may have been another selective force in the evolution of flowers.

Today we can correlate many floral features with specific pollinators. Flowers pollinated by beetles and flies smell like decaying meat or moist dung. Their odors resemble the smells of decaying matter in forest litter, where beetles first evolved. By contrast, flowers with strong, sweet odors and bright yellow, blue, and purple parts commonly attract bees. Some pigments in the flowers that absorb ultraviolet light are distributed in ways that reflect it in specific patterns. Unlike us, bees can see ultraviolet light—and they find the patterns alluring (Figure *a*).

Clues to pollinators also exist in the structure of flowers. Consider a marsh marigold flower or daisy (Figure *a* or *b*). Its yellow center is a tight, flattened clump of tiny flowers (florets)—a landing platform for pollinators. As the insects bustle about, they transfer pollen from floret to floret, from daisy to daisy. Or consider scotch broom (Figure *c*). Its floral parts strategically position bee bodies so that they brush against pollen-laden anthers.

Beetles and honeybees don't respond to red colors, and they also can drown in the nectar of flowers with large, deep tubes. Birds do respond to flowers with red as well as yellow components. Red petals of the flower in Figure *d* form a tube for a tempting, sugar-rich fluid—nectar. Certain birds have a bill as long as the floral tube. This flower's anthers and stigma are located where birds brush against them. The birds visit nectar cups of many plants of this species and so promote cross-pollination. Bird-pollinated plants don't spend energy producing potent perfumes (birds have a poor sense of smell).

Butterflies forage by day. They are attracted to fragrant flowers with distinctive patterns and shapes, such as upright daisies with horizontal landing platforms. Some also are attracted to flowers with red and orange components. Most moths forage by night. They pollinate strong, sweet-smelling flowers with white or pale-colored petals, which are more visible in the dark. Butterflies and moths have long, narrow mouthparts, corresponding to narrow floral tubes or spurs. The uncoiled mouthpart of a Madagascar hawkmoth is 22 centimeters long—the same length as the floral tube of an orchid (*Angraecum sesquipedale*)! Like hummingbirds, hawkmoths don't use landing platforms; they hover near the floral tubes as they draw nectar from them.

Different insects exploit the cozy mutualisms between flowers and their pollinators. For example, crab spiders, ambush bugs, mantids, and other muggers may lurk in the flowers, waiting for meals. Or consider male and female blister beetles that reproduce on daisies (Figure *b*). When their larvae emerge, they become stuck in the fuzzy legs of foraging bees. The larvae are ferried to beehives, where they eat bee eggs and some pollen. The food fuels their development into adults, which return to daisies. Still other insects pierce the floret walls and steal nectar. This may explain the value of tight floret packing. Only a few florets are reachable by insects that take from the daisies but give nothing in return.

31.4 FROM ZYGOTE TO SEED

Formation of the Embryo Sporophyte

Following fertilization, the newly formed zygote embarks on a course of mitotic cell divisions that will lead to a mature embryo sporophyte. Take a look at Figure 31.6, which shows how an embryo forms in *Capsella*, a dicot. Early on, cell divisions produce a single row of cells. The lower cells develop into a stalklike structure that anchors the embryo and absorbs nutrients from the endosperm for it. The cells above it develop into the embryo proper. By the time the embryo reaches the stage shown in Figure 31.6*e*, its primary meristems have formed and two cotyledons have started to develop from two tissue lobes.

Cotyledons, or "seed leaves," develop as part of all flowering plant embryos. Dicot embryos have two; monocot embryos have one. In many plants, including peas and beans, large cotyledons absorb the endosperm and function in food storage. In wheat, corn, and other plants, thin cotyledons may produce enzymes for transferring stored food from the endosperm to the germinating seedling.

Seed and Fruit Formation

From the time a zygote forms until a mature embryo has developed, the parent plant transfers nutrients to tissues of the ovule. Food reserves accumulate in the expanding endosperm or cotyledons. Eventually, the physical connection between the ovule and the ovary wall separates. The ovule's integuments thicken and harden into a seed coat. The embryo and food reserves are now a self-contained package. The ovule has become a **seed**.

While seeds are forming, changes also occur in other parts of the flower. The ovary itself expands in size, and its tissues become modified. What we call a **fruit** is a mature ovary, with or without other floral structures that have become incorporated into it. Many fruits, including apples, are juicy and fleshy. Others, such as grains and nuts, are dry. Still others, including pineapples, are a multiple fruit, formed from clusters of many flowers. Different types of fruit are listed in Table 31.1. Figure 31.7 and the *Focus* essay that follows show examples.

A mature ovule, which encases an embryo sporophyte and food reserves inside a hardened coat, is a seed.

A mature ovary, with or without additional floral parts that have become incorporated into it, is a fruit.

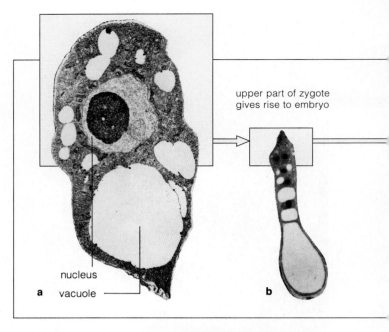

upper part of zygote gives rise to embryo

nucleus

a   vacuole

b

Figure 31.6   Stages in the development of shepherd's purse (*Capsella*), a dicot. The micrographs are not to the same scale. (**a**) The single-celled zygote. (**b–d**) The early embryo is identified by the *yellow* boxes. The row of cells below it transfers nutrients to the developing embryo from the parent plant. The embryo in (**e**) is well developed; the one in (**f**) is mature.

Table 31.1   Kinds of Fruits of Some Flowering Plants

Type	Characteristics	Some Examples
Simple (formed from single carpel, or two or more united carpels of one flower)	1. Fruit wall *dry; split* at maturity	Pea, magnolia, tulip, mustard
	2. Fruit wall *dry; intact* at maturity	Sunflower, wheat, rice, maple
	3. Fruit wall *fleshy,* sometimes with leathery skin	Grape, banana, lemon, cherry, orange
Aggregate (formed from many separate carpels of single flower)	*Aggregate* (cluster) of matured ovaries (fruits), all attached to receptacle (modified stem end)	Blackberry, raspberry
Multiple (formed from carpels of several associated flowers)	*Multiple* matured ovaries, massed together; may include accessory structures (such as receptacle, sepal, and petal bases)	Pineapple, fig, mulberry
Accessory (formed from one or more ovaries *plus* receptacle tissue that becomes fleshy)	1. *Simple:* a single ovary enclosed in receptacle tissue	Apple, pear
	2. *Aggregate:* swollen, fleshy receptacle with dry fruits on surface	Strawberry

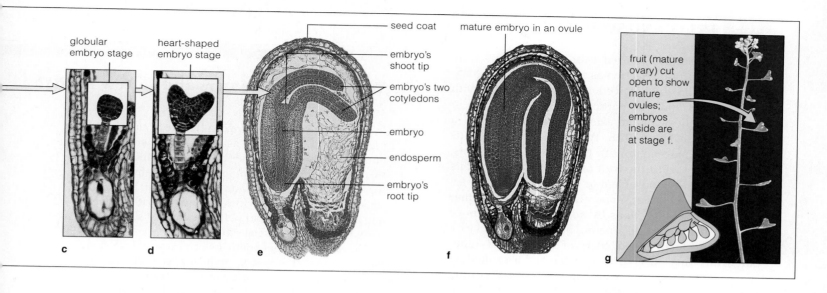

globular embryo stage

heart-shaped embryo stage

c

d

seed coat

embryo's shoot tip

embryo's two cotyledons

embryo

endosperm

embryo's root tip

e

mature embryo in an ovule

f

fruit (mature ovary) cut open to show mature ovules; embryos inside are at stage f.

g

a Apple blossoms

b Petals fallen away

c Enlarging ovaries

receptacle

ovary

e ovaries embedded in receptacle

remnants of petals, sepals

vascular bundles in outer portion of ovary

seeds

enlarged receptacle

d individual fruit

Figure 31.7 Fruits. (**a–c**) Fruit formation on an apple (*Malus*) tree. When petals drop, eggs usually are fertilized. Now the ovary and receptacle expand. Sepals and stamens remain on the immature fruit.

(**d**) Pineapple (*Ananas comosus*), a multiple fruit. Besides running into the New World, Christopher Columbus also ran into cultivated pineapples, viewed by native Americans as a symbol of hospitality. To Columbus, they resembled pine cones, hence the name. At the time, native populations were already using pineapple juice as a base for an alcoholic beverage and as an ingredient for arrow tip poison.

(**e**) Strawberry (*Fragaria*), a fragrant, wildly popular, easily bruised fruit. Genetic engineering may yield strains that retain the fragrance yet are firm enough to resist packing damage.

31.5 FRUIT AND SEED DISPERSAL

Long before humans started eating them, fruits were undergoing natural selection as seed-dispersing mechanisms. They have coevolved with air currents, with water currents, and with particular animals.

Consider the winglike extensions of the fruit of a maple (*Acer*) in Figure 31.8. When the fruit drops, its wings spin it sideways. With the help of air currents, the wings may whirl the fruit far enough away that its enclosed embryo won't have to compete with the parent plant for soil water, minerals, and sunlight. Other fruits taxi to new locations by adhering to feathers or fur with hooks, spines, hairs, and sticky surfaces. Embryos inside the seed coats of berries and other fleshy fruits must survive being eaten and assaulted by digestive enzymes in the animal gut. Besides digesting a fruit's flesh, the enzymes digest some of the seed coat. This will help the embryo break through its hard coat after seeds are expelled from the animal body and start to germinate. We return to this topic in the chapter to follow.

The structure of seeds and fruits enhances dispersal by wind or by particular kinds of animals.

Figure 31.8 Winged seeds of maple (*Acer*).

wing
seed (in carpel)

Focus on Science

Why So Many Flowers and So Few Fruits?

In the Sonoran Desert of Arizona (Figure *a*), the giant saguaro produces as many as a hundred flowers at the tips of its huge, spiny arms. The large, showy white flowers of this cactus (Figure *b*) bloom for only twenty-four hours. Insects visit the flowers by day, and bats visit by night (Figure *c*). The plant provides the animals with nectar—and the animals transport pollen grains that stick to their body from one cactus plant to another.

Flowers that do get pollinated, courtesy of insects or bats, now begin the task of producing seeds and fruits. Petals wilt, then wither as the many egg cells inside the ovary become fertilized. Each egg-containing ovule expands and matures into a seed. At the same time, the ovary develops into a fruit (Figures *d* and *e*). Weeks later, the plum-sized fruit splits open, exposing its bright red interior and dark seeds.

White-winged doves feast on the ripe fruits. When they fly away, they may carry hundreds of seeds in their gut. A few seeds may escape digestion in the gut, and later end up on the ground. There, each will have a tiny chance of growing into a giant saguaro.

One of the puzzling aspects of this annual event is the frequency with which saguaro flowers *fail* to give rise to fruit (see Figure *d*, for example). Why would a cactus invest energy in constructing a hundred flowers if only

a

c

b

thirty or so will set fruit? Compare one of these "inefficient" plants with a species that produces a hundred flowers, *all* of which develop into fruits. Which do you think
will leave more descendants?

Common sense might lead you to conclude, "The more
fruits, the better." Yet saguaros *and many other plants* commonly produce far more flowers than mature fruits. They
seem to be producing "excess" flowers and giving up
chances to make seeds and leave descendants. This is not a
reproductive adaptation that you would expect, based on
Darwinian evolutionary theory.

Well, suppose some plants never received enough
pollen to have all their egg cells fertilized. After all, unfertilized flowers cannot produce seeds and fruits. This
hypothesis has been experimentally tested for some plant
species. Researchers carefully brushed pollen on every
single flower. In many instances, the plants still failed to
produce fruit for every flower!

Alternatively, suppose the presumed "excess" flowers
are formed strictly to produce pollen for export to other
plants. Pollen grains are small. In energy terms, they are
inexpensive to produce, compared to large, calorie-rich,
seed-containing fruits. Thus, for a fairly small investment,
a plant might reap a large reward in offspring that carry its
genes. The idea that some plants actually set aside flowers
exclusively for pollen export has yet to be tested for
saguaros. Perhaps you might like to design and carry out
experiments that will help clear up the mystery of the
"excess" flowers of saguaros and similar plants.

d fruit failed to set, unlike fruit above it

e   maturing fruit

31.6 ASEXUAL REPRODUCTION OF FLOWERING PLANTS

Asexual Reproduction in Nature

Sexual reproduction dominates flowering plant life cycles. But most species also can reproduce asexually by various modes of **vegetative growth**. Table 31.2 lists some of these modes. In essence, new roots and shoots grow right out of extensions or fragments of a parent plant.

A "forest" of quaking aspen (*Populus tremuloides*) provides us with a stunning example of vegetative reproduction. The leaves of this flowering plant species tremble even in the slightest breeze, hence the name. Figure 31.9 provides a panoramic view of one individual. Its root system gave rise to many adventitious shoots, which became separate shoot systems. Barring rare mutations, individuals at the north end of this vast clone are genetically identical to individuals at the south end. Roots near the lake deliver water to shoot systems in much drier soil. Nutrients can travel in the opposite direction.

As long as environmental conditions favor growth and regeneration, such clones are as close as one can get to being immortal. No one knows how old aspen clones are. The oldest known clone, a ring of creosote bushes (*Larrea divaricata*) in the Mojave Desert, has been around for the past 11,700 years.

The possibilities are astounding. Raise strawberry plants and watch them send out runners (horizontal aboveground stems); then watch new roots and shoots develop at every other node. Eat an orange, and it may be from a tree that reproduced by **parthenogenesis** (the development of an embryo from an unfertilized egg). Parthenogenesis may be stimulated when pollen contacts a stigma even though a pollen tube has not grown down the style. Maybe hormones from the stigma or from pollen grains diffuse to the unfertilized egg and trigger embryo formation. The embryo becomes $2n$ by fusion of products of egg mitosis. A $2n$ cell outside the gametophyte also may be stimulated to develop into an embryo.

Induced Propagation

Most houseplants, woody ornamentals, and orchard trees are clones. They often are propagated from cuttings or fragments of shoot systems. Thus, with suitable encouragement, a severed African violet leaf forms a callus from which adventitious roots develop. A callus is meristematic tissue (a mass of undifferentiated cells that retain the potential for division). A twig or bud cut from one plant may be grafted onto different varieties

Figure 31.9 In the photograph to the right, a mere portion of Pando the Trembling Giant, so named by the researchers who studied its genetic makeup (Michael Grant and coworkers at the University of Colorado).

This 106-acre "forest" in Utah is actually a single asexually reproducing male organism, of a type called quaking aspen (*Populus tremuloides*). A root system connects its 47,000 genetically identical shoots (the trees). The clone weighs an estimated 13,000,000+ pounds.

Table 31.2	Asexual Reproductive Modes of Flowering Plants	
Mechanism	Representative	Characteristics
Vegetative reproduction on modified stems:		
1. Runner	Strawberry	New plants arise at nodes on an aboveground horizontal stem
2. Rhizome	Bermuda grass	New plants arise at nodes of underground horizontal stem
3. Corm	Gladiolus	New plant arises from axillary bud on short, thick, vertical underground stem
4. Tuber	Potato	New shoots arise from axillary buds on tubers (enlarged tips of slender underground rhizomes)
5. Bulb	Onion, lily	New bulb arises from axillary bud on short underground stem
Parthenogenesis	Orange, rose	Embryo develops without nuclear or cellular fusion (e.g., from unfertilized haploid egg; or develops adventitiously, from tissue surrounding embryo sac)
Vegetative propagation	Jade plant, African violet	New plant develops from tissue or organ (e.g., a leaf) that drops or is separated from plant
Tissue culture propagation	Carrot, corn, wheat, rice	New plant arises from cell in parent plant that is not irreversibly differentiated; laboratory technique only

of a closely related species. Thus French vintners graft prize grapevines onto disease-resistant root stock from America.

Frederick Steward and his coworkers pioneered in **tissue culture propagation**. They cultured bits of phloem from differentiated roots of carrot plants (*Daucus carota*) in rotating flasks. They used a liquid growth medium that contained sucrose, mineral ions, and vitamins. It also contained coconut milk, which Steward knew was rich in as-yet unidentified growth-inducing substances. As the flasks rotated, individual cells that were torn away from the tissue bits divided and formed multicelled clumps—which sometimes gave rise to roots (Figure 31.10a). These experiments were among the first to demonstrate that cells even of differentiated tissue contain the genetic instructions required to build new individuals.

Tissue culture propagations are now done with shoot tips or other parts of individual plants. The techniques are useful when an advantageous mutant arises. One such mutant might show resistance to a disease that is particularly crippling to wild-type plants of the same species. Tissue culture propagation can lead to hundreds and even thousands of identical plants from just one mutant specimen. This technique is already being used in efforts to improve major food crops, such as corn, wheat, rice, and soybeans. It also is being used to increase production of hybrid orchids, lilies, and other valued ornamental plants (Figure 31.10b).

a

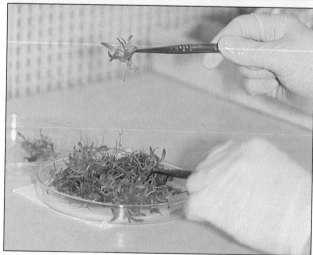
b

Besides reproducing sexually, flowering plants engage in asexual reproduction, as by vegetative growth.

Figure 31.10 (**a**) Cultured cells from a carrot plant. Roots and shoots of embryonic plants are already forming. (**b**) Young orchid plants, developed from cultured meristem and young leaf primordia. Orchids are one of the most prized cultivated plants. Before meristem cloning, they were difficult to hybridize. From the time seeds form, it can take seven long years or more until a new plant flowers. In the natural habitat, seeds normally will not germinate unless they interact with a specific fungus.

SUMMARY

1. Flowering plant life cycles proceed through a multicelled diploid stage (the sporophyte, or "spore-producing plant") and a multicelled haploid stage (the gametophyte, or "gamete-producing plant"). The sporophyte is a multicelled vegetative body with roots, stems, leaves, and, at some point, flowers.

2. A typical flower has sepals, petals, one or more stamens (male reproductive structures), and carpels (female reproductive structures), most or all attached to a receptacle (the modified end of a floral shoot).

 a. The leaflike sepals and petals are the outermost, nonfertile parts of a flower.

 b. The stalked anthers of stamens contain pollen sacs. In these chambers, cells divide by meiosis, and a wall develops around each resulting haploid cell. The walled cells are microspores. These develop into sperm-bearing pollen grains (male gametophytes).

 c. A carpel is a female reproductive part. A single carpel (or two or more carpels fused together) has an ovary. Here, eggs develop, fertilization takes place, and seeds mature. Above the ovary is a stigma, a sticky or hairy surface tissue that captures pollen grains and promotes their germination. Often the upper portion of the ovary is modified into a slender column (style).

3. One or more tissue masses, the forerunners of ovules, form on the inner ovarian wall. Prior to fertilization, an ovule consists of a female gametophyte with an egg cell, a surrounding tissue, and one or two protective layers (integuments). After fertilization, the ovule matures into a seed.

4. Female gametophytes commonly form as follows:

 a. Four haploid megaspores form after meiosis. All but one usually disintegrate.

 b. The remaining megaspore undergoes mitosis three times without cytoplasmic division. Its nuclei migrate to specified positions.

 c. Cytoplasmic division results in a female gametophyte—a seven-celled, eight-nuclei body. One cell is the egg. The cell with two nuclei (endosperm mother cell) will help form endosperm, a nutritive tissue for the forthcoming embryo.

5. Flowering plants have coevolved with air currents, animals, and other agents that help the transfer of pollen grains to a suitable stigma (pollination). So transferred, a pollen grain germinates. It becomes a pollen tube that grows down through tissues of the ovary, carrying the two sperm nuclei with it.

6. At double fertilization, one sperm nucleus fuses with one egg nucleus to form a diploid ($2n$) zygote. The other sperm nucleus and both nuclei of the endosperm mother cell fuse to form a cell that will give rise to triploid ($3n$) nutritive tissue in the forthcoming seed.

7. These events unfold after double fertilization:

 a. The endosperm forms, the ovule expands, the embryo sporophyte develops, and integuments harden and thicken. A fully matured ovule is a seed; its integuments have developed into the seed coat.

 b. Simultaneously with seed formation, the ovary develops into a fruit.

8. Fruits function to protect and disperse seeds, which germinate following dispersal from the parent plant. Fleshy fruits attract animals and other "dispersing" agents; lightweight fruits can be dispersed by winds. Some fruits have hooks and such that attach to animals.

9. Many flowering plants also reproduce asexually, as by vegetative growth. For example, new shoot systems may arise by mitotic divisions at nodes or buds along modified stems of the parent plant. New plants also may arise by vegetative propagation, from fragments or parts severed from a parent plant.

Review Questions

1. Label the floral parts in the following diagram. Explain floral function by relating some floral structures to events in the life cycle of flowering plants. *516–517*

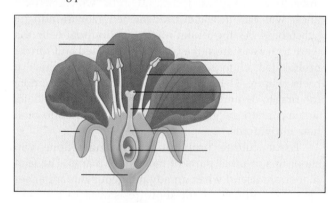

2. Distinguish between these terms:
 a. Megaspore and microspore *518*
 b. Pollination and fertilization *518–519*
 c. Pollen grain and pollen tube *518–519*
 d. Ovule and female gametophyte *516, 518–519*

3. Describe the steps in the formation of a seven-cell, eight-nucleate embryo sac (a female gametophyte). *518–519*

4. Observe several kinds of flowers growing in the area where you live. On the basis of what you have read about the likely coevolutionary links between flowering plants and their pollinators, can you perceive what kinds of pollination agents your floral neighbors might depend upon? *520–521*

5. Given how so much luscious, fleshy fruit surrounds their seeds, you might think that the peaches shown in the photograph below are in the same category as apples. Apples belong to the genus *Malus*. Peaches are in the same genus as cherries (*Prunus*). After reviewing Table 31.1 and Figure 31.7, identify the peach as being a simple, aggregate, multiple, or accessory fruit. In what key respect does it differ from apples? *522–523*

1. The _____ bears flowers, roots, stems, and leaves; it dominates the life cycle of flowering plants.
   a. sporophyte   b. gametophyte

2. The flowers of many species coevolved with insects, birds, and other agents that function as _____ .

3. Male gametophytes eventually produce _____ , and female gametophytes produce _____ .
   a. spores; eggs   c. eggs; sperm
   b. sperm; spores   d. sperm; eggs

4. Seeds are mature _____ and fruits are mature _____ .
   a. ovaries; ovules   c. ovules; ovaries
   b. ovules; stamens   d. stamens; ovaries

5. A _____ is a closed vessel that contains an ovary in which eggs develop, fertilization occurs, and seeds mature.
   a. pollen sac   c. receptacle
   b. carpel   d. sepal

6. The immediate products of meiosis within pollen sacs are haploid _____ .
   a. megaspores   c. microspores
   b. sperm   d. pollen grains

7. Following meiosis in carpels, _____ megaspores form.
   a. two   c. six
   b. four   d. eight

8. The seed coat forms from which structure(s)?
   a. integuments   c. endosperm
   b. ovary   d. residues of sepals

9. Cotyledons or "seed leaves" develop as part of all flowering plant _____ .
   a. seeds   c. fruits
   b. embryos   d. ovaries

10. Match the terms with their appropriate description.
   ____ double fertilization
   ____ ovule
   ____ mature female gametophyte
   ____ asexual reproduction
   ____ coevolution

   a. formation of zygote and first cell of endosperm
   b. outcome of two species interacting in close ecological fashion over geologic time
   c. contains a female gametophyte and has potential to be a seed
   d. an embryo sac, commonly with seven cells (one has two nuclei)
   e. mitotic cell division at bud or node produces a new plant

Selected Key Terms

carpel *517*
coevolution *521*
cotyledon *522*
double fertilization *519*
endosperm *518*
flower *516*
fruit *522*
gametophyte *516*
megaspore *518*
microspore *518*
ovary *517*
ovule *518*
parthenogenesis *526*
pollen grain *517*
pollination *518*
seed *522*
sporophyte *516*
stamen *517*
tissue culture propagation *527*
vegetative growth *526*

Readings

Grant, M. October 1993. "The Trembling Giant." *Discover.* 14(10):82–89.

Proctor, M., and P. Yeo. 1973. *The Pollination of Flowers.* New York: Taplinger. Beautifully illustrated introduction to pollination.

Raven, P., R. Evert, and S. Eichhorn. 1992. *Biology of Plants.* Fifth edition. New York: Worth. Outstanding illustrations.

Rost, T., et al. 1984. *Botany: A Brief Introduction to Plant Biology.* Second edition. New York: Wiley.

Salisbury, F., and C. Ross. 1991. *Plant Physiology.* Fourth edition. Belmont, California: Wadsworth.

32 PLANT GROWTH AND DEVELOPMENT

Foolish Seedlings and Gorgeous Grapes

A few years before the American stock market grew feverishly and then collapsed catastrophically in 1929, a researcher in Japan came across a substance that caused runaway growth and subsequent collapse of rice plants. Ewiti Kurosawa was studying what the Japanese called *bakane*—the "foolish seedling" effect on rice plants. Stems of rice seedlings that had become infected with *Gibberella fujikuroi*, a fungus, elongated twice the length of stems of uninfected plants. The long, spindly, weak stems eventually collapsed, and the plants died. Kurosawa discovered that applying extracts of the fungus to plants also could trigger the disease. Many years later, other researchers purified the disease-causing substance from fungal extracts. The substance was given the name gibberellin.

Gibberellin, as we now know, is one of the premier plant hormones. Researchers have isolated more than eighty different kinds from the seeds of flowering plants as well as from fungi. Possibly the same kind of substance is also present in even other plants.

Like all plant hormones, gibberellins are signaling molecules. They are produced by some cells and transported to different cells elsewhere in the multicelled body, where they trigger changes in metabolic activities. The particular changes that occur in gibberellins lead to stem lengthening (Figure 32.1a). In nature, these hormones help seeds and buds break dormancy and resume growth in spring. They might even help induce flowering.

Expose a radish plant to a suitable concentration of a gibberellin and it may grow as large as a beach ball. Do the same to a cabbage plant and it may grow 6 feet tall. Gibberellin applications can make celery stalks longer and crispier; they can prevent the skin of navel oranges from growing old too soon in orchards.

Walk past some large, plump seedless grapes in the produce bins of grocery stores and notice their market appeal (Figure 32.1b). The fruits of grape (*Vitus*) grow in bunches along stems. Gibberellin applications cause the stems to lengthen between internodes. With more space between one another, the individual grapes grow larger. Also, air circulates more freely between them—which makes it harder for disease-causing, grape-juice-loving fungi to become established.

At present, biologists do not know much about how plant hormones work. They do understand that at least five different types have major, predictable effects on flowering plants, beginning with the germination of

a

Figure 32.1 Hormonal effect on stem growth. (**a**) The California poppy plant shown above was left alone. Gibberellin, a plant hormone, was applied to the poppy plant to the right of it. (**b**) Seedless grapes, radiating market appeal. Gibberellin made their stems lengthen, which improved air circulation around the grapes and gave them more growing room. The grapes got larger and weighed more. This delights growers; grapes are sold by the pound.

each new seed. As with humans and other animals, plants grow and develop according to genetic information in their DNA. That information governs the synthesis of enzymes and other proteins necessary for metabolism and other aspects of cell function. How those enzymes function depends on hormonal action.

As you will see in this chapter, plant hormones operate through interactions with one another. They also operate in response to cues afforded by the rhythmic changing of the seasons, as when days become longer and warmer in spring, after the short, cold nights of winter. Those grapes, cabbage leaves, or celery stalks that find their way into your mouth are the culmination of a beautifully regulated program of plant growth and development.

KEY CONCEPTS

1. From the time a seed germinates, hormones influence growth and development. This is true of seed-bearing plants as well as for other organisms.

2. Hormones are signaling molecules between cells. That is, one cell type produces and secretes a particular kind of hormone, and its uptake by other cell types stimulates changes in their metabolic activities. These metabolic changes have predictable effects, as when they trigger cell divisions and other processes necessary for stem elongation.

3. The known plant hormones are auxins, gibberellins, cytokinins, abscisic acid, and ethylene. The existence of other kinds of hormones, including one that is tentatively named florigen, is suspected.

4. Plant hormones help bring about genetically programmed patterns of growth, including the extent and direction of growth of particular plant parts.

5. Plant hormones help adjust patterns of growth in response to environmental rhythms, including seasonal changes in daylength and temperature. They also help adjust those patterns in response to the environmental circumstances in which a plant finds itself—that is, the amount of sunlight or shade, moisture, and so on at a given site.

6. Commonly, two or more kinds of plant hormones must interact with one another to bring about specific effects on growth and development.

In the preceding chapter, we traced the events by which a zygote of a flowering plant forms and then develops into a mature embryo, housed inside a protective seed coat. At some point following its dispersal from the parent plant, that tiny embryo is transformed into a seedling, which in turn develops into a mature sporophyte. In due time, flowers, fruits, and then new seeds form. Depending on the species, old leaves drop away from the plant throughout the year or all at once, in autumn.

Let's begin this chapter with the kinds of heritable, internal mechanisms that govern these essential aspects of plant growth and development. Then we will look at environmental signals that set these mechanisms in motion.

32.1 PATTERNS OF GROWTH AND DEVELOPMENT

Seed Germination

Before or after seed dispersal, an embryo sporophyte's growth idles. Later on, if all goes well, it germinates. For seeds, **germination** is a resumption of growth after a time of arrested embryonic development. Genes *and* the environment govern this process.

Environmental factors that influence germination include soil temperature, moisture and oxygen levels, and the number of daylight hours (which varies seasonally). For example, mature seeds do not contain enough water for cell expansion or metabolism. In most places, ample water is available only on a seasonal basis, so germination must coincide with the return of spring rains. In a process called **imbibition**, water molecules move into the seed, being attracted mainly to the hydrophilic groups of stored proteins. As more water moves in, the seed swells and its coat ruptures.

Once the seed coat splits, more oxygen reaches the embryo, and aerobic respiration moves into high gear. Now cells rapidly grow and divide. Generally, the embryonic root cells are activated first. They divide, elongate, and give rise to a **primary root**, the first root of the seedling sporophyte. Germination is over when the primary root breaks through the seed coat.

Prescribed Growth Patterns

Growth involves cell divisions and enlargements. On the average, half of the resulting daughter cells do not increase in size, but they retain the capacity to divide. The other half enlarge, often by twenty times.

Water uptake drives cell enlargement. It increases internal fluid pressure and forces the cell's primary wall to expand. Imagine blowing up a balloon. If it's soft, the balloon inflates easily. Similarly, cells with a soft wall expand rapidly under pressure. However, a balloon wall gets thinner as it "grows." A cell wall does not. New polysaccharides are added to it, and more cytoplasm forms between the wall and the central vacuole:

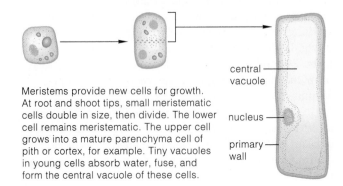

Meristems provide new cells for growth. At root and shoot tips, small meristematic cells double in size, then divide. The lower cell remains meristematic. The upper cell grows into a mature parenchyma cell of pith or cortex, for example. Tiny vacuoles in young cells absorb water, fuse, and form the central vacuole of these cells.

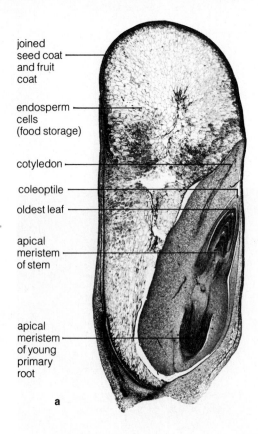

joined seed coat and fruit coat

endosperm cells (food storage)

cotyledon

coleoptile

oldest leaf

apical meristem of stem

apical meristem of young primary root

a

Figures 32.2 and 32.3 show the patterns of growth and development for a monocot and a dicot. Again, the basic patterns are heritable, dictated by the plant's genes. All cells in a new plant arise from the same cell (the zygote) and so carry the same genes. But unequal cytoplasmic divisions between daughter cells lead to differences in their metabolic equipment and output. Daughter cells start interacting in different ways, through selective gene expression (page 238). Genes governing, say, the synthesis of growth-stimulating hormones become activated in some cells and remain silent in others. Such events seal the developmental fate of various cell lineages. Their descendant cells divide in prescribed planes and expand in specific directions. The eventual outcome will be plant parts of specific shapes and functions.

Bear in mind, the prescribed patterns of growth can be adjusted in response to unusual environmental pressures. Suppose a seed germinates in a vacant lot and a heavy paper bag blows on top of it. The first shoot will bend and grow out from under the bag, toward sunlight. Enzymes, hormones, and other gene products in shoot cells carry out this growth response.

Plant growth involves cell divisions and cell enlargements.

Plant development requires cell differentiation, as brought about by selective gene expression.

Interactions among genes, hormones, and the environment govern how an individual plant grows and develops.

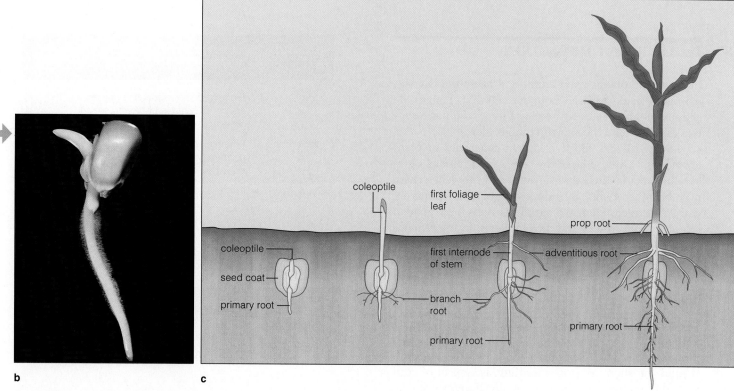

b

c

Figure 32.2 Growth and development of corn (*Zea mays*), a monocot. (**a**) An embryo sporophyte inside a corn grain, longitudinal section. (**b**) Growth resumes with germination of the corn grain, here oriented the same way as the one shown in **a**. (**c**) Following germination, the coleoptile protects young leaves when the new seedling grows through soil. Adventitious roots develop at the coleoptile's base.

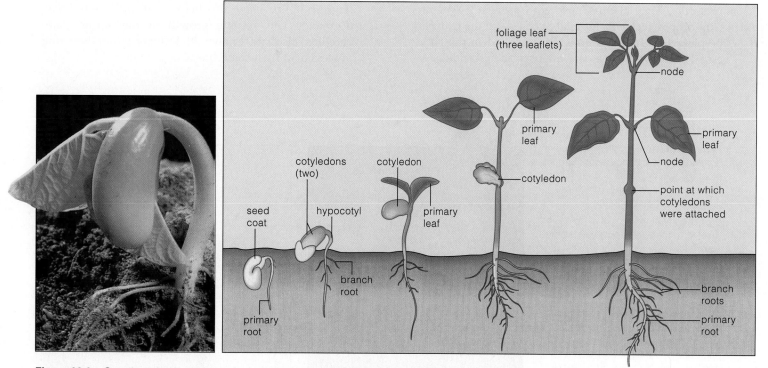

Figure 32.3 Growth and development of a common bean plant (*Phaseolus vulgaris*), a dicot. Below the cotyledon, a hypocotyl (a hook-shaped portion of the young stem) forces a channel through soil. Food-storing cotyledons are pulled up through the channel without being torn apart. At the surface, light causes the hook to straighten. The cotyledons serve in photosynthesis for several days; then they wither and fall off. In time, foliage leaves, each divided into three leaflets, take over the task. Flowers will develop in buds at the nodes.

In flowering plants, as in animals, a **hormone** is a signaling molecule released from one cell that changes the activity of target cells. Any cell is a target if it has receptors that can bind the signaling molecule.

Table 32.1 lists the five known categories of plant hormones. These hormones are called auxins, gibberellins, cytokinins, abscisic acid, and ethylene. There may be other types, including one that is tentatively named florigen. Here we simply survey their general roles in plant growth and development. In the remainder of this chapter, we will consider specific examples of their effects at different stages of the life cycle.

Auxins

The **auxins** stimulate stem lengthening, and they may influence responses to gravity and light. Auxins also promote coleoptile elongation. A **coleoptile**, a sheath around the shoot of a new seedling, is typical of corn and other members of the grass family. It helps keep the tender shoot from being torn apart during its upward growth through the soil (Figure 32.4).

Indoleacetic acid (IAA) is the most important naturally occurring auxin. Orchardists spray IAA and other auxins on fruit trees to thin out overcrowded seedlings in spring. They also use auxins to prevent premature fruit drop, which cuts farm labor costs—all the fruit can be picked at the same time. As described in the *Focus* essay, some synthetic auxins are used as herbicides.

Table 32.1	Main Plant Hormones and Some Known (or Suspected) Effects
Auxins	*Promote cell elongation in coleoptiles and stems; involved in phototropism and gravitropism*
Gibberellins	*Promote stem elongation; might help break dormancy of seeds and buds; stimulate breakdown of starch*
Cytokinins	*Promote cell division; promote leaf expansion and retard leaf aging*
Abscisic acid	*Promotes stomatal closure; promotes bud and seed dormancy*
Ethylene	*Promotes fruit ripening; promotes abscission of leaves, flowers, and fruits*
Florigen (?)	*Arbitrary designation for as-yet unidentified hormone (or hormones) thought to cause flowering*

Gibberellins

The examples used at the start of this chapter provide you with a good idea of the effects of **gibberellins**. Again, hormones of this category make stems lengthen (Figure 32.1a). They also help buds and seeds break dormancy and resume growth in the spring. Gibberellins also are known to influence the flowering process in at least some species.

Figure 32.4 Effects of auxin, a plant hormone. (**a**) A cutting from a gardenia plant (*left*), four weeks after an auxin was applied to its base. The other cutting (*right*) was untreated.

(**b**) Experiment showing that an auxin (IAA) in a coleoptile tip stimulates the cells below it to lengthen. (*1*) First, cut off the tip of an oat coleoptile. The stump of the cut coleoptile does not elongate much, compared to a normal oat coleoptile (*2*) used as a control. (*3*) Next, place a tiny block of agar under the cut-off tip and leave it for several hours. During that time, IAA diffuses into the agar. (*4*) Place the agar on another de-tipped coleoptile. You see that elongation proceeds about as rapidly as in an intact coleoptile growing alongside it (*5*).

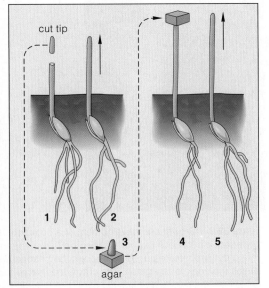

a   treated with auxin   untreated   b

About Those Herbicides

An **herbicide** is any compound which, at suitable concentration, kills plants. Many widely used types, including some of the synthetic auxins, kill some plants but not others. Consider 2,4-dichlorophenoxyacetic acid (2,4-D). Farmers use this synthetic auxin to control broadleaf weeds, which compete with crop plants for nutrients. Before the development of synthetic herbicides, farmers and gardeners controlled weeds with backbreaking hand labor or toxic chemicals.

After exhaustive testing, it appears that 2,4-D does not harm humans. This was not true of 2,4,5-T, a related compound. When mixed in equal proportions, the two compounds produce *Agent Orange*. The United States Armed Forces used this herbicide to defoliate thickly forested and nearly impenetrable war zones during the Vietnam conflict. Later, experiments with laboratory animals indicated that traces of dioxin, a contaminant of 2,4,5-T, may cause miscarriages, birth defects, leukemia, and liver disorders. The government banned the manufacture and use of 2,4,5-T in the United States. In 1994, results of comprehensive tests indicated that dioxin apparently is a carcinogenic substance.

Cytokinins

The **cytokinins** stimulate cell division (hence their name, which refers to cytokinesis). They are most abundant in root and shoot meristems and in the tissues of maturing fruits. Natural and synthetic versions are used to prolong the shelf life of cut flowers as well as lettuces, mushrooms, and other vegetables.

Abscisic Acid

The hormone called **abscisic acid** (ABA) inhibits cell growth, promotes bud dormancy, and keeps seeds from germinating prematurely. It also causes stomata to close and so helps conserve water when the plant is water stressed. Often ABA is applied to nursery stock about to be shipped. When plants are dormant, they are more resistant to damage.

Ethylene

Fruits, flowers, seeds, leaves, and roots contain ethylene. Under the influence of this hormone, green fruit ripens. Also under its influence, flowers, fruits, and leaves drop from plants at prescribed times of year.

Humans have been using ethylene to ripen picked fruit for centuries. Ancient Chinese knew that burning incense makes fruit ripen faster, although they didn't know the smoke contained ethylene. Food distributors now use ethylene to ripen green fruit after shipment. Fruit that is picked green doesn't bruise or deteriorate as fast. Oranges and other citrus fruits are exposed to ethylene to brighten the color of their rind before being displayed in the market.

Other Hormones and Hormone-Like Substances

We have evidence of other hormones besides the five types mentioned. Root and leaf cells produce their own hormones. An unidentified hormone (florigen) may trigger the flowering process.

An unknown hormone associated with shoot tips blocks growth of lateral buds, an effect called **apical dominance**. By pinching off shoot tips, gardeners prevent the hormone from diffusing down through the stems and exerting its effect. In the absence of this hormone, lateral buds are free to branch out, and gardeners get bushier plants.

Then there are the saponins. These are not plant hormones; they are toxic substances that may kill or repel herbivores or pathogenic microorganisms. Even so, their complex ring structures are similar to human sex hormones. Wild yams (*Dioscorea*) contain a modified saponin, and drug companies learned about its contraceptive properties from the native peoples of rural Mexico. For some time, birth control pills have been synthesized from saponin extracts, which have been less expensive than manufacturing synthetic versions from scratch.

Like animals, flowering plants produce hormones. These signaling molecules are released from one cell type and change the metabolic activities of target cells.

Auxins, gibberellins, cytokinins, abscisic acid, and ethylene are the known categories of plant hormones. The existence of others is suspected.

32.3 THE TROPISMS

Generally, young roots of land plants grow down through soil, and shoots grow upright through the air. But they also can adjust the direction of growth in response to environmental stimuli, as when a shoot turns toward sunlight. Through hormone-mediated shifts in the rates at which different cells grow and elongate, a root or shoot turns toward or away from the stimulus. Such responses to environmental stimuli are called tropisms (after the Greek *trope*, for turning).

Phototropism

Certain wavelengths of light drive photosynthesis. In their absence, photosynthesis stops, and so does plant growth. When a sun-loving plant is in shade, its stems start curving in the direction where the most light is available (Figure 32.5a). You can demonstrate this in no time at all by positioning a pot of bean sprouts next to a window where light is streaming in. When a plant adjusts its direction and rate of growth in response to light, this is a form of **phototropism**.

Charles Darwin was aware of phototropism. In the late 1800s, he noticed a coleoptile growing toward light striking one side of its tip. But it wasn't until the 1920s that Frits Went, a graduate student in Holland, made the connection between this response and a growth-promoting substance. He named this substance auxin (after the Greek *auxein*, meaning "to increase"). Went demonstrated that auxin moves down from a coleoptile's tip into cells less exposed to light. It stimulates them to elongate faster than cells on the illuminated side of the coleoptile. This brings about the bending toward light (Figure 32.5b–d).

Plants make the strongest phototropic response to light of blue wavelengths. **Flavoprotein**, a yellow pigment molecule that absorbs blue wavelengths, may be part of the phototropic bending mechanism.

Gravitropism

After seeds germinate, the first root always curves down and the coleoptile or stem always curves up. This is an example of **gravitropism**, a growth response to the earth's gravitational force.

Figure 32.6a shows what happens when you turn a potted seedling on its side in a dark room. The stem curves upward, even in the absence of light. How does it do this? Cells on the upper side of the horizontally positioned stem elongate more slowly—markedly so—and cells on the lower side elongate faster. The difference in the rate of elongation is enough to turn the stem in an upward direction. Does auxin have roles in this tropic response? Perhaps. If the stem cells are sensitive to auxin, the faster growing cells on the bottom of the sideways stem may have become *more* sensitive to auxin and those on top, *less* sensitive.

Auxin or some other growth-inhibiting hormone may also trigger the gravitropic response in roots. Turn a young root on its side and remove its root cap, and it will *not* curve downward. Put the cap back on, and the root will do so. Elongating cells don't stop growing when the root cap is removed; if anything, they grow faster. Suppose a growth inhibitor is present in root cap cells. Suppose it becomes *redistributed* in a root turned on its side. If gravity somehow causes the inhibitor to move out of the cap and accumulate in cells on the lower side of the root, those cells would not elongate as much as cells on the upper side—and the root would curve downward.

Generally, the gravity-sensing mechanism is based on **statoliths**. In plants, these are unbound starch grains within modified plastids. As Figure 32.6b shows, statoliths collect at the bottom of the plastid according to which way gravity is pulling on them. The plastids settle down through the cytoplasm until they rest on the lower part of the cell. Their redistribution may trigger the redistribution of auxin to bring about gravitropic responses.

a

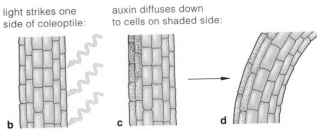

light strikes one side of coleoptile:

auxin diffuses down to cells on shaded side:

b   c   d

Figure 32.5 (**a**) Phototropism in seedlings. Frank Salisbury, a plant physiologist, grew these seedlings in darkness; then he allowed light to strike their right side for a few hours before photographing them. (**b**–**d**) Hormone-mediated differences in the rates of cell elongation bring about the bending toward light.

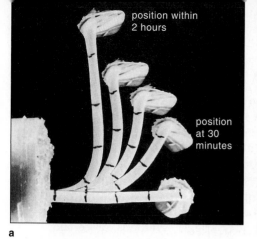
position within
2 hours

position
at 30
minutes

a

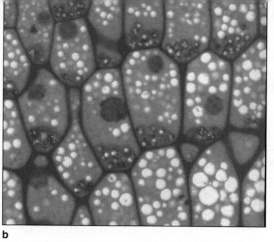

b

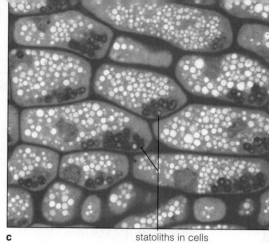

c

statoliths in cells
of the root cap

Figure 32.6 Gravitropism by young shoots and roots. (**a**) In one experiment, a newly emerged sunflower seedling was grown in the dark for five days. Then it was turned on its side and marked at 0.5-centimeter intervals. By some gravity-sensing mechanism, the young stem turned upright.

In a different experiment, a corn root was laid on its side. (**b**) This light micrograph reveals modified plastids (statoliths) in cells from the root's cap. It shows their orientation before the root was turned sideways. (They were elongating vertically.) (**c**) Within seconds after their reorientation, the plastids had settled to the bottom of the root cap cells. Statoliths may be part of a gravity-sensing mechanism that redistributes auxin through a root tip. A difference in auxin concentration may cause cells on the "top" of the sideways root to elongate faster than cells on the bottom. Different elongation rates could turn the root tip downward.

Thigmotropism

Plants shift the direction of growth when they contact solid objects. We call this **thigmotropism** (after the Greek *thigma*, meaning "touch"). Auxin and ethylene may take part in the response. **Vines**, stems too slender or soft to grow upright without support, make this contact response. So do **tendrils**, modified leaves or stems that wrap around objects and help support the plant (Figure 32.7). Suppose a tendril grows up against a stem of another plant. Cells on the contact side stop elongating, and within minutes the tendril starts curling around the stem, maybe several times. Then cells on both sides resume growth at the same rate.

Figure 32.7 Tendrils of a grapevine, twisting thigmotropically.

32.4 RESPONSES TO MECHANICAL STRESS

Prevailing strong winds, grazing animals, even farm machinery can inhibit plant growth. Trees growing near the snowline of windswept mountains show the effects of such mechanical stress. Compared to trees of the same species at lower elevations, they are stubby. Similarly, plant stems often are taller when grown indoors. Stems stop elongating when mechanically stressed. Shaking some plants daily for a brief period will inhibit growth of the whole plant (Figure 32.8).

Plants adjust their direction and rate of growth in response to environmental stimuli.

a   b   c

Figure 32.8 Effect of mechanical stress on tomato plants. (**a**) This plant, the control, grew in a greenhouse. (**b**) Each day for 28 days, this plant was mechanically shaken for 30 seconds at 280 revolutions per minute. (**c**) This one had two such shakings each day.

32.5 BIOLOGICAL CLOCKS AND THEIR EFFECTS

Like other organisms, plants have **biological clocks**. These internal time-measuring mechanisms have roles in adjusting daily activities. They also function in seasonal adjustments to the plant's patterns of growth, development, and reproduction.

Circadian Rhythms

Some biological activities recur in cycles of twenty-four hours or so. These are **circadian rhythms** (meaning "about a day"). For example, some plants position leaves horizontally each day, then fold them closer to stems at night (Figure 32.9). An internal time-keeping mechanism governs the response. Keep such a plant in constant light or darkness for a few days and it will continue to fold its leaves into the "sleep" position!

In plants, **phytochrome** is one alarm button for biological clocks. This blue-green pigment absorbs red or far-red wavelengths, with different results. It converts to active molecular form (Pfr) at sunrise, when red wavelengths dominate. It reverts to inactive form (Pr) at sunset, at night, even in shade, where far-red wavelengths of light dominate (Figure 32.10).

Phytochrome activation may stimulate plant cells to take up calcium ions (Ca^{++}), or it may induce certain plant organelles to release them. Either way, when free calcium ions combine with calcium-binding proteins, they may initiate at least some responses to light. For example, Ruth Satter, Richard Crain, and their coworkers at the University of Connecticut have experimental evidence that leaf movements governed by light and phytochrome are activated this way.

1:00 A.M.

6:00 A.M.

12:00 (noon)

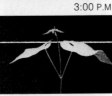
3:00 P.M.

10:00 P.M.

12:00 (midnight)

Figure 32.9 Rhythmic leaf movements. A bean plant was kept in constant darkness for 24 hours. The leaf movements continued independently of sunrise (6 A.M.) and sunset (6 P.M.). Leaves folded close to the stem may keep moonlight from activating phytochrome and interrupting the dark period that triggers flowering. Folding also may reduce heat loss from leaves exposed to cold night air.

Photoperiodism

There are more hours of daylight and warmer weather in summer than in winter. Wherever seasons change, biological clocks are reset, and plants adjust their growth, development, and reproductive activities. Any biological response to a change in the relative length of daylight and darkness in a cycle of twenty-four hours is an example of **photoperiodism**.

Phytochrome's active form, Pfr, may be involved in photoperiodism. It may trigger synthesis of certain enzymes in certain cells—and different enzymes are required for seed germination, stem lengthening and branching, leaf expansion, and formation of flowers, fruits, and seeds. The effects of Pfr deficiency show up when plants adapted to full sunlight are grown in darkness. The plants put fewer resources into leaf expansion or branching (Figure 32.11). We turn now to a few examples of Pfr's effects in nature.

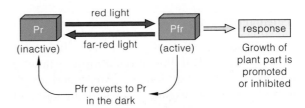

Figure 32.10 Interconversion of phytochrome from its active form (Pfr) to inactive form (Pr). This pigment molecule is part of a switching mechanism that promotes or inhibits growth of different plant parts.

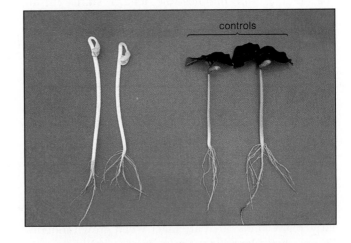

Figure 32.11 Effect of the absence of light on young bean plants. The two plants at the right, the control group, were grown in a greenhouse. The other two were grown in darkness for eight days. Dark-grown plants were yellow. They could form carotenoids but not chlorophylls in darkness. They had longer stems, smaller leaves, and smaller root systems. (Why do you suppose roots of dark-grown plants grow less, considering the roots of light-grown plants aren't exposed to light, either?)

32.6 THE FLOWERING PROCESS

As a flowering plant matures, it channels more energy and nutrients into producing flowers, seeds, and fruits. Most of these events are exquisitely predictable responses to changes in daylength through the year and to resulting changes in other environmental conditions. For example, Figure 32.12 correlates events in the life cycle with seasonal variations in daylength in temperate regions of North America.

Consider the flowering process. "Long-day" plants flower in spring, when daylength becomes longer than some critical value. "Short-day" plants flower in late summer or early autumn, when daylength becomes shorter than a critical value. "Day-neutral" plants flower when they are mature enough to do so.

Figure 32.13 shows what happens when spinach (a long-day plant) grows under short-day conditions and long-day conditions. A spinach plant will not flower and produce seeds unless it is exposed to fourteen hours of light every day for two weeks. These conditions do not exist in the tropics—which is why spinach is a poor choice for a seed farm there.

Consider how sensitively the cocklebur, a short-day plant, measures time. It flowers after a single night that's longer than eight and a half hours. But if artificial light interrupts that dark period for even a minute or two, cockleburs won't flower! Or consider why short-day poinsettias planted along California's interstate highways never did flower. Headlights of cars and trucks at night blocked the flowering response.

Probably at least one hormone, as yet unidentified, interacts with phytochrome to influence flowering and other growth responses. Evidence suggests that the hormone may be produced in leaves and transported to newly forming buds. Trim all but one leaf from a cocklebur plant, cover the remaining leaf with black paper for eight and a half hours, and the plant flowers. Cut off the leaf immediately after the dark period and the plant will not flower.

---

1. As with all organisms, flowering plants have internal time-keeping mechanisms—biological clocks.

2. Phytochrome, a blue-green pigment molecule, is part of a switching mechanism that responds to light of different wavelengths.

3. Pfr, the active form of phytochrome, may interact with one or more hormones to control which types of enzymes are being produced in particular cells.

4. The different enzymes are necessary to complete flowering and other growth responses that are influenced by sunlight and other environmental cues.

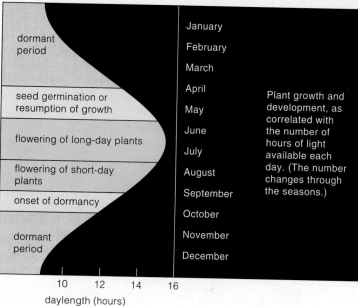

Plant growth and development, as correlated with the number of hours of light available each day. (The number changes through the seasons.)

dormant period	January
	February
	March
seed germination or resumption of growth	April
	May
flowering of long-day plants	June
	July
flowering of short-day plants	August
	September
onset of dormancy	October
	November
dormant period	December

daylength (hours): 10 12 14 16

Figure 32.12 Effects of daylength on flowering and other plant activities in temperate regions of North America. These regions show seasonal variation in rainfall and temperature.

flowers

Figure 32.13 Effect of daylength on spinach, a long-day plant. The plant on the left was grown under short-day conditions, and the one on the right, under long-day conditions.

32.7 VERNALIZATION

Bear in mind, in most parts of the world, temperatures as well as daylength also change through the seasons. Changing temperatures affect many plant responses, including flowering. For example, unless buds of some biennials and perennials are exposed to low winter temperatures, flowers do not form on stems in spring. Low-temperature stimulation of flowering is called **vernalization** (from the Latin *vernalis*, meaning "to make springlike"). Figure 32.14 shows how you can gather experimental evidence of this effect.

In 1915, the plant physiologist Gustav Gassner influenced the flowering of cereal plants by exposing their seeds to controlled temperatures. For example, he kept germinating seeds of winter rye (*Secale cereale*) at near-freezing temperatures. The seeds flowered the same summer even when they were planted in the late spring. Vernalization is now a common agricultural practice.

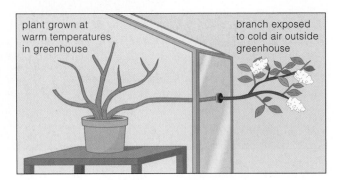

Figure 32.14 Localized effect of cold temperature on dormant buds of a lilac plant. For this experiment, a branch of the plant was positioned so that it protruded from a greenhouse during winter. The rest of the plant remained inside, at warm temperatures. Only the buds on the branch exposed to low outside temperatures resumed growth in spring.

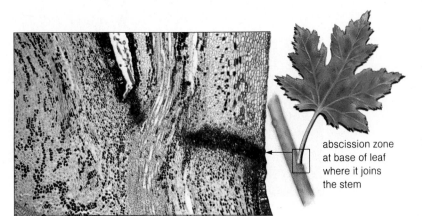

Figure 32.15 Abscission zone in a maple (*Acer*). This longitudinal section is through the base of a leaf's petiole.

32.8 SENESCENCE

Plants make major investments in reproduction. They actually withdraw nutrients from leaves, roots, and stems and distribute them to newly forming flowers, fruits, and seeds. As an outcome of the redistribution of nutrients, leaves usually wither and die. Among deciduous plants, which shed leaves at the end of each growing season, nutrients become transported to storage sites in twigs, stems, and roots before the leaves die and fall off.

The dropping of leaves or other parts from a plant is called **abscission** (after the Latin *abscissus*, meaning "to cut off"). The process proceeds at tissues in the base of leaves, flowers, fruits, or other plant parts. Figure 32.15 shows one of these abscission zones in a maple tree. Ethylene that forms in cells near the break points may trigger the process. Abscisic acid may cause cells near the break point to produce the required ethylene.

The sum total of processes leading to the death of plant parts or the whole plant is called **senescence**. The funneling of nutrients into reproductive parts may be a cue for senescence of leaves, stems, and roots. Stop the drain of nutrients by removing each newly emerging flower or seed pod, and a plant's leaves and stems will stay green and vigorous much longer (Figure 32.16). Gardeners routinely remove flower buds from many plants to maintain vegetative growth.

Figure 32.16 Experimental results showing that the removal of seed pods from a soybean plant delays its senescence.

32.9 DORMANCY

Entering Dormancy

As autumn approaches and days grow shorter, growth slows or stops in many perennial and biennial plants. It stops even if temperatures are still mild, the sky is bright, and water is plentiful. When a plant stops growing under conditions that seem (to us) quite suitable for growth, it has entered a state of **dormancy**. In this state, its metabolic activities idle. Ordinarily, its buds will not resume growth until there is a convergence of precise environmental cues in early spring.

Short days, long, cold nights, and dry, nitrogen-deficient soil are strong cues for dormancy. For example, when a short period of red light interrupts the long dark period for Douglas firs, the plants respond as if nights are shorter and days are longer. They continue to grow taller (Figure 32.17). In this experiment, conversion of Pr to Pfr by red light during the dark period prevented dormancy. In nature, buds may enter dormancy because less Pfr can form when daylength shortens in late summer.

The requirement for multiple dormancy cues has

Figure 32.18 A profusion of new leaves harvesting sunlight—a sure sign of spring and the resurgence of plant growth.

adaptive value. For example, if temperature were the only cue, plants might flower and seeds might germinate in warm autumn weather—only to be killed by winter frost. By contrast, through artificial selection, commercial growers have developed seeds that germinate promptly in greenhouses, at any time of year.

Breaking Dormancy

A dormancy-breaking process is at work between fall and spring. The temperature may become milder, and rains and nutrients become available again. With the return of favorable conditions, the cycle of life begins anew. Seeds germinate; buds resume growth and give rise to new leaves, then to flowers (Figure 32.18).

Depending on the species, breaking dormancy probably involves gibberellins and abscisic acid, and it requires exposure to low winter temperatures for specific periods. The actual temperature needed to break dormancy varies greatly among species. For example, Delicious apples grown in Utah require 1,230 hours near 43°F (6°C); apricots grown there require only 720 hours. Generally, trees growing in the southern United States require less cold exposure than those growing in northern states and Canada. If you live in Colorado and order a young peach tree from a Georgia nursery, the tree might start spring growth too soon and be killed by late frost or heavy snow.

Figure 32.17 Effect of the relative length of day and night on Douglas fir plant growth.

The plant at the left was exposed to twelve-hour light and twelve-hour darkness for a year; its buds became dormant because daylength was too short. The plant at the right was exposed to twenty-hour light and four-hour darkness; buds remained active and growth continued.

The middle plant was exposed to twelve-hour light, eleven-hour darkness, and one-hour light in the middle of the dark period. This light interruption of an otherwise long dark period also prevented bud dormancy. Such light causes Pfr formation at an especially sensitive time in the normal day-night cycle.

Daylength, temperature, moisture, and nutrient availability are environmental cues that stimulate or inhibit growth processes during the life cycles of flowering plants.

The Rise and Fall of a Giant

oak seed (acorn)

In this unit, you surveyed the spectrum of flowering plant tissues. You skimmed through the mechanisms and outcomes of plant growth, development, and reproduction. To pull all of this into a coherent picture, imagine walking near the central California coast, where the land rolls inward as steep, rounded hills. Those sandstone hills started forming 65 million years ago as a jagged new coastal range. Over time, rains and winds softened the stark contours of the inland hills and sent mineral-laden sediments into canyons. Grasses took hold, and their organic remains accumulated and enriched the soil. More than 10 million years ago, this is where the coast live oak (*Quercus agrifolia*) began to evolve.

As you walk, you come across a giant tree that is 300 years old. You stop briefly under trees more than 100 feet tall, with evergreen branches that spread even wider. It's early spring, so you see golden catkins—clusters of male flowers—among the leaves. A light wind is dispersing pollen from catkins to female flowers near the branch tips of the same tree or neighboring ones. Long after your walk, the sperm-bearing pollen tubes will still be growing toward the plant's ovaries. In time, newly formed zygotes will undergo the first of millions of cell divisions. Seed coats will form, and ovarian walls will become shells. By early fall, the giant trees will shed their seeds as hard-shelled acorns.

Three centuries ago, long before Gaspar de Portola sent landing parties ashore to found colonies throughout upper California, oaks were shedding the seeds of a new generation. By chance, one particular acorn escaped the attention of foraging squirrels and bluejays. The next spring, it germinated and embarked on a journey of continued growth.

Cells differentiated and developed into the cortex, epidermis, and a vascular cylinder through which water and ions would flow. Lateral roots emerged. As new roots grew longer, their absorptive surfaces increased. When the first shoot began its upward surge, vascular bundles began forming. In time they would form a continuous cylinder of secondary xylem and phloem.

The seedling's roots, stems, and leaves demanded more water and dissolved nutrients. Fungi formed velvety

coverings around the roots. And these fungus-roots—mycorrhizae—enhanced the uptake of water and mineral ions. Leaves started conserving precious water with the development of stomata. As the oak seedling grew, its parts became interconnected by vascular pipelines. Through xylem, water and minerals moved from roots to stems and leaves. Through phloem, sugars were shuttled to cells actively growing or storing food.

By chance, the seed had sprouted in a well-drained, sunlit basin at the foot of a canyon. Each winter, rainwater accumulated, and it kept the soil moist enough to encourage the plant's growth through spring and the dry summers. The sun's red wavelengths activated phytochrome, triggering hormonal events that encouraged stem branching and leaf expansion. Other hormone-mediated responses were made to the winds, the sun, the tug of gravity, the changing seasons.

And so the oak flourished every season. Century after century, roots snaked through a huge volume of the moist soil. Branches continued to spread beneath the sun. Leaves proliferated, and their photosynthetic cells put together food with sunlight energy, water, carbon dioxide, and the few simple minerals being mined from the soil.

In 1849, prospectors on their way to the gold fields rested in the shade of the oak's immense canopy. The great earthquake of 1906 scarcely disturbed the giant, anchored as it was by a root system extending 80 feet through the soil. The few small brush fires that swept through the canyon did no serious damage to the tree. Fungi that could have rotted its roots never took hold in the well-drained soil. Canyon birds and the tree's own protective chemicals kept insects in check.

Then, in the 1960s, the once-wild hills began to give way to suburban housing. A developer turned his tractors into the canyon but spared the giant oak. Death came later.

The new homeowners were ignorant of the ancient, delicate relationships between the giant tree and the land that sustained it. They graded the soil between the trunk and the drip line of the overhanging canopy. They mounded flower beds against the trunk and planted lawns beneath the branches, then kept the sprinklers busy. Overwatering in summer created standing water next to the great trunk. Before then, the giant had successfully resisted the oak root fungus (*Armillaria*). In the changed environment, the fungus took hold. With its roots rotting away, the oak began to suffer massive disruptions in the feedback relations among its roots, stems, and leaves. Eventually it had to be cut down. In their fifth winter, in their red brick fireplace, the homeowners began burning three centuries of firewood.

SUMMARY

1. Five types of plant hormones have been identified. Auxins and gibberellins promote stem elongation. Cytokinins stimulate cell division, promote leaf expansion, and retard leaf aging. Abscisic acid promotes bud and seed dormancy, and it limits water loss (by triggering stomatal closure). Ethylene promotes fruit ripening and abscission.

2. Plant hormones interact with one another to produce patterns of growth and development. They help adjust those patterns in response to environmental rhythms, including seasonal changes in daylength and temperature. They also adjust patterns to the amount of sunlight, shade, and other environmental circumstances.

3. Following dispersal from the parent, the seeds of flowering plants germinate: the embryo inside the seed absorbs water and resumes growth, and its primary root breaks through the seed coat.

4. Following germination, a plant increases in volume and mass. Tissues and organs of the seedling develop; later, flowers, fruits, and new seeds form, and then older leaves drop away from the plant. Plant hormones govern these developmental events.

5. Plants make tropic responses to environmental conditions. They turn or move through differences in the rate and direction of growth on two sides of an organ such as a stem or root. Phototropism and gravitropism are examples.

6. Plants have biological clocks, or internal time-measuring mechanisms that have a biochemical basis. They can "reset" the clocks and so make seasonal adjustments in their patterns of growth, development, and reproduction.

7. In photoperiodism, plants respond to a change in the relative length of daylight and darkness in a twenty-four-hour period. A switching mechanism involving phytochrome (a blue-green pigment) promotes or inhibits germination, stem elongation, leaf expansion, stem branching, and formation of flowers, fruits, and seeds.

8. Long-day plants flower in spring or summer, when daylength is long relative to night. Short-day plants flower when daylength is relatively short. Flowering of day-neutral plants is not regulated by light.

9. Senescence is the sum total of processes leading to the death of a plant or plant structure.

10. Dormancy is a state in which a perennial or biennial stops growing even though conditions appear to be suitable for continued growth. A decrease in Pfr levels may trigger dormancy.

1. List the five known types of plant hormones and describe the function of each. *534*

2. What is phytochrome, and what is its role in flowering or some other process? *538–539*

3. Define plant tropism and give a specific example. *536–537*

4. Distinguish among vernalization, senescence, and dormancy. *540–541*

5. "Solar tracking" refers to the observation that many plants are able to maintain the flat blades of their leaves at right angles to the sun throughout the day. This maximizes light harvest by leaves. Following study of this chapter, suggest the name of a molecule that might be involved in this tropism. *534, 536*

Self-Quiz *(Answers in Appendix IV)*

1. Seed germination is over when the _____ .
   a. embryo absorbs water
   b. embryo resumes growth
   c. primary root pokes out of seed coat
   d. cotyledons unfurl

2. Which of the following statements is false?
   a. Auxins and gibberellins promote stem elongation.
   b. Cytokinins promote cell division and leaf expansion but retard leaf aging.
   c. Abscisic acid promotes water loss and dormancy.
   d. Ethylene promotes fruit ripening and abscission.

3. Plant hormones _____ .
   a. interact with one another
   b. are influenced by environmental cues
   c. are active in plant embryos within seeds
   d. are active in adult plants
   e. all of the above are correct

4. Plant growth depends on _____ .
   a. cell division   c. hormones
   b. cell enlargement   d. all of the above are correct

5. Light of _____ wavelengths is the main stimulus for phototropism.
   a. red   c. green
   b. far-red   d. blue

6. Light of _____ wavelengths causes phytochrome to switch from inactive to active form; light of _____ wavelengths has the opposite effect.
   a. red; far-red   c. far-red; red
   b. red; blue   d. far-red; blue

7. The flowering process is a _____ response.
   a. phototropic   c. photoperiodic
   b. gravitropic   d. thigmotropic

8. Abscission occurs during _____ .
   a. seed germination   c. senescence
   b. flowering   d. dormancy

9. A decrease in _____ levels may trigger dormancy.
   a. Pr   c. Pfr
   b. gibberellin   d. abscisic acid

10. Match the plant reproduction and development terms.
    _____ gibberellin   a. promotes stem elongation
    _____ senescence   b. unequal growth following
    _____ phytochrome   contact with solid objects
    _____ phototropism   c. switching mechanism for
    _____ thigmotropism   photoperiodism
                           d. response to blue light, mainly
                           e. all processes leading to death
                              of plant or plant part

Selected Key Terms

abscisic acid *535*	herbicide *535*
abscission *540*	hormone *534*
apical dominance *535*	imbibition *532*
auxin *534*	photoperiodism *538*
biological clock *538*	phototropism *536*
circadian rhythm *538*	phytochrome *538*
coleoptile *534*	primary root *532*
cytokinin *535*	senescence *540*
dormancy *541*	statolith *536*
flavoprotein *536*	tendril *537*
germination *532*	thigmotropism *537*
gibberellin *534*	vernalization *540*
gravitropism *536*	vine *537*

Readings

Bowley, J. D., and M. Black. 1985. *Seeds: Physiology of Development and Germination.* New York: Plenum.

Nickell, L. 1982. *Plant Growth Regulators: Agricultural Uses.* New York: Springer-Verlag. Concise explanations of agricultural practices that include use of growth regulators.

Salisbury, F., and C. Ross. 1991. *Plant Physiology.* Fourth edition. Belmont, California: Wadsworth.

Villiers, T. 1975. *Dormancy and the Survival of Plants.* London: Edward Arnold.

FACING PAGE: *How many and what kinds of body parts does it take to function as a lizard in a tropical forest? Make a list of what comes to mind as you start reading Unit VI, then see how resplendent the list can become at the unit's end.*

Meerkats, Humans, It's All the Same

After a cold night in Africa's Kalahari Desert, animals small enough to fit in a coat pocket emerge stiffly from their burrows. These "meerkats" are a type of mongoose. They stand on their hind legs and face east, exposing their chilled bodies to the warm rays of the morning sun (Figure 33.1). Meerkats don't know it, but sunning behavior helps their enzymes. If their body temperature were to fall below or exceed a tolerable range, enzyme activity would drop sharply. With such a change in enzyme activity, metabolism would suffer.

Once meerkats warm up, they fan out from the burrows, looking for food. Into the meerkat gut go insects and an occasional lizard. These are pummeled, dissolved, and digested into glucose and other nutritious tidbits small enough to move across the gut wall,

into the bloodstream, and on to cells throughout the body. In cells, aerobic machinery cracks nutrients apart and so releases vital energy. A respiratory system and the bloodstream supply the machinery with oxygen and take away its carbon dioxide leftovers.

All of this activity changes the composition and volume of the "internal environment"—the bloodstream and fluids bathing the body's cells. Drastic changes in that fluid would kill cells, but a urinary system works to keep this from happening. Governing this system and all others are the body's command posts—a nervous system and an endocrine system. These work together and mobilize the body as a whole for everything from simple housekeeping tasks to heart-thumping flights from predators.

And so meerkats start us thinking about this unit's central topics: how an animal body is structurally put together (its *anatomy*) and how the body functions (its *physiology*). This chapter provides us with an overview of the animal tissues and organ systems that we will be considering. It also introduces the important concept of "homeostasis"—stability in the internal environment, brought about by the coordinated activity of the body's cells, tissues, organs, and organ systems.

Figure 33.1 In the Kalahari Desert, gray meerkats (*Suricata suricatta*) line up and face the warming rays of the morning sun, just as they do every morning. This simple behavior helps maintain internal body temperature. How animals function in their environment is the subject of this unit.

KEY CONCEPTS

1. The cells of most animals interact at three levels of organization—in *tissues*, many of which are combined in *organs*, which are components of *organ systems*.

2. Most animals are constructed of only four types of tissues: epithelial, connective, nervous, and muscle tissues.

3. Each cell engages in basic metabolic activities that assure its own survival. At the same time, cells of a tissue or organ perform activities that contribute to the survival of the animal as a whole.

4. The combined contributions of cells, tissues, organs, and organ systems help maintain a stable internal environment, which is required for individual cell survival. This concept helps us understand the functions of any organ system, regardless of its complexity.

33.1 ANIMAL STRUCTURE AND FUNCTION: AN OVERVIEW

Regardless of whether it is a flatworm or salmon, a meerkat or human being, each animal is structurally and physiologically adapted to perform these tasks:

1. Maintain internal operating conditions within a tolerable range even as external conditions change.

2. Acquire nutrients and other materials, distribute them through the body, and dispose of wastes.

3. Protect itself against injury or attack from viruses, bacteria, and other disease-causing agents.

4. Reproduce, and often help nourish and protect the new individuals during their early development.

Your body, and those of other complex animals, has only four basic types of tissues. These are epithelial, connective, muscle, and nervous tissues. A **tissue** is an interacting group of cells and intercellular substances that take part in one or more of the tasks listed above. An **organ**, such as a heart, consists of different tissues organized in specific proportions and patterns. An **organ system** consists of two or more organs that interact physically, chemically, or both in a common task, such as blood circulation. Cells, tissues, organs, and organ systems split up the work, so to speak, in ways that contribute to the survival of the animal as a whole. This is sometimes called a division of labor.

33.2 EPITHELIAL TISSUE

General Characteristics

An epithelial tissue is commonly called **epithelium** (plural, epithelia). This tissue has a free surface, which faces some kind of body fluid or the outside environment. *Simple* epithelium, with a single layer of cells, functions as a lining for body cavities, ducts, and tubes. *Stratified* epithelium, with two or more cell layers, typically functions in protection, as it does in skin. Figure 33.2 shows a few examples of this type of animal tissue.

All epithelial cells are close together, with little intervening material. As is true of nearly all animal tissues, specialized junctions provide structural and functional links between the cells, which absorb, synthesize, and secrete a variety of substances.

free surface
of epithelium

epithelial cells

basement membrane

connective tissue

a

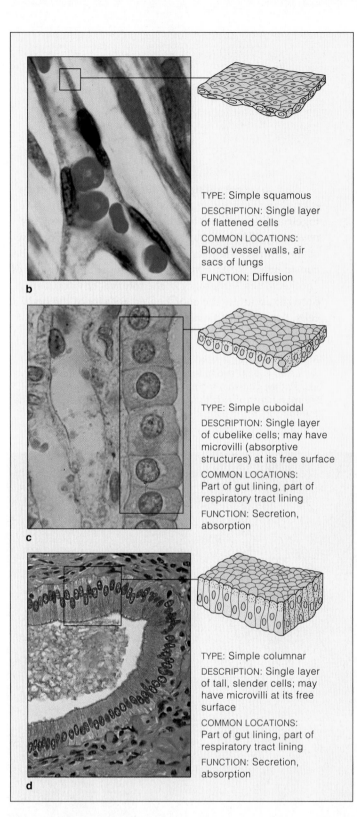

b

TYPE: Simple squamous

DESCRIPTION: Single layer of flattened cells

COMMON LOCATIONS: Blood vessel walls, air sacs of lungs

FUNCTION: Diffusion

c

TYPE: Simple cuboidal

DESCRIPTION: Single layer of cubelike cells; may have microvilli (absorptive structures) at its free surface

COMMON LOCATIONS: Part of gut lining, part of respiratory tract lining

FUNCTION: Secretion, absorption

d

TYPE: Simple columnar

DESCRIPTION: Single layer of tall, slender cells; may have microvilli at its free surface

COMMON LOCATIONS: Part of gut lining, part of respiratory tract lining

FUNCTION: Secretion, absorption

Figure 33.2 (**a**) Characteristics of epithelium. All epithelia have a free surface, and a basement membrane is interposed between the opposite surface and an underlying connective tissue. The diagram shows this arrangement in simple epithelium, which consists of a single layer of cells. The micrograph shows the upper portion of stratified epithelium, which has more than one layer of somewhat flattened cells.

(**b–d**) Examples of simple epithelium, showing the three basic cell shapes in this type of tissue.

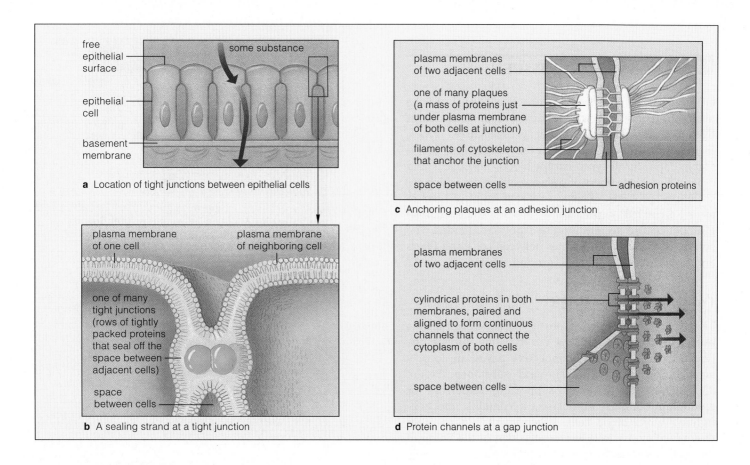

a Location of tight junctions between epithelial cells

free epithelial surface
epithelial cell
basement membrane
some substance

b A sealing strand at a tight junction

plasma membrane of one cell
plasma membrane of neighboring cell
one of many tight junctions (rows of tightly packed proteins that seal off the space between adjacent cells)
space between cells

c Anchoring plaques at an adhesion junction

plasma membranes of two adjacent cells
one of many plaques (a mass of proteins just under plasma membrane of both cells at junction)
filaments of cytoskeleton that anchor the junction
space between cells
adhesion proteins

d Protein channels at a gap junction

plasma membranes of two adjacent cells
cylindrical proteins in both membranes, paired and aligned to form continuous channels that connect the cytoplasm of both cells
space between cells

Cell-to-Cell Contacts

With few exceptions, the cells of epithelium as well as other animal tissues adhere strongly to one another by means of specialized attachment sites. These sites of cell-to-cell contact are especially profuse when substances must not leak from one of the body's compartments into another (Figure 33.3).

For example, if the potently acidic fluid in your stomach were to leak through its epithelial lining, it would start digesting your body's own proteins instead of the ones brought in with your meals. (Actually, this is an end result of a peptic ulcer, as described on page 719.) This epithelium and others have intricate cell-to-cell contacts that make up a leakproof barrier between cells near their free surface. As Figure 33.3 shows, epithelia also include junctions that serve as spot welds and as open channels between cells.

Glandular Epithelium

Glands are secretory cells or multicelled structures derived from epithelium and often connected to it. **Exocrine glands** secrete many products, such as mucus, saliva, earwax, oil, milk, and digestive enzymes. The products usually are released onto a free epithelial surface through ducts or tubes.

By contrast, **endocrine glands** have no ducts. Their products—hormones—are secreted directly into the

Figure 33.3 Examples of cell junctions.

(**a,b**) In some epithelia, protein strands form tight seals that ring each cell and seal it to its neighbors. The seals prevent substances from leaking across the free epithelial surface. The only way that substances can reach the tissues below is to pass through the epithelial cells—which have built-in mechanisms that can control their passage.

(**c**) Adhesion junctions are like spot welds that "cement" cells of epithelium (and all other tissues) together so that they function as a unit. They are abundant in the skin's surface layer and other tissues subjected to abrasion.

(**d**) Gap junctions promote diffusion of ions and small molecules from cell to cell. They are abundant in the heart, liver, and other organs in which cell activities must be rapidly coordinated.

fluid bathing the gland. Typically, the bloodstream picks up the hormones and distributes them to target cells elsewhere in the body.

Epithelia are sheetlike tissues with one free surface. Different types line the body's surface, cavities, ducts, and tubes.

Profuse cell-to-cell contacts bind epithelial cells closely together, with little intercellular material between them.

Glands are secretory cells or multicelled structures derived from epithelium and often connected to it.

33.3 CONNECTIVE TISSUE

Of all tissues in the body, connective tissues are the most abundant and widely distributed. They range from connective tissue proper to specialized types, including cartilage, bone, adipose tissue, and blood (Table 33.1). In all types except blood, fibroblasts and other cells secrete fibers of collagen or elastin, which are structural proteins. (Plastic surgeons use collagen to plump wrinkled skin, sunken acne scars, and even lips.) Fibroblasts also secrete modified polysaccharides. The secreted material accumulates between cells and fibers and so becomes the tissue's "ground substance."

Connective Tissue Proper

Tissues of this group have mostly the same components but in different proportions. As its name implies, the fibers and cells of **loose connective tissue** are loosely arranged in a semifluid ground substance (Figure 33.4a). Often this tissue is a supporting framework for epithelium. Besides fibroblasts, it contains infection-fighting white blood cells. When small cuts or other wounds allow bacteria to cross the skin or the lining of the gut, respiratory tract, or urinary tract, these cells mount an early counterattack.

Dense, irregular connective tissue consists of fibers, mostly collagenous, and a few fibroblasts. This tissue forms protective capsules around organs that are not stretched much. It also is present in the deeper portion of skin (Figure 33.4b).

Specialized Connective Tissue

With its parallel bundles of many collagen fibers, **dense, regular connective tissue** resists being torn apart. Rows of fibroblasts often are arranged between the bundles. This is true of tendons, which attach muscles to bones (Figure 33.4c) and of elastic ligaments, which attach bones to each other.

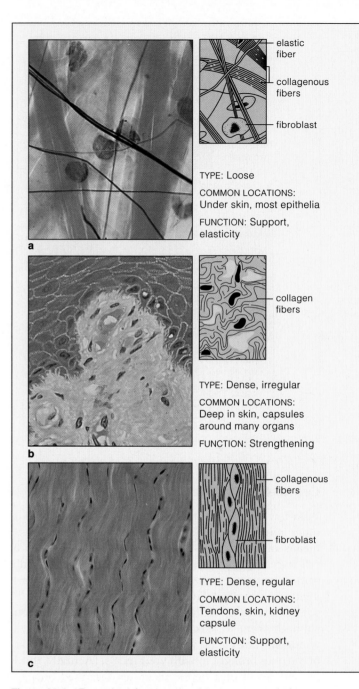

Figure 33.4 Examples of connective tissue proper and of specialized connective tissue.

Table 33.1 Types of Connective Tissue
Connective tissue proper:
Loose connective tissue
Dense, irregular connective tissue
Specialized connective tissue:
Dense, regular connective tissue (ligaments, tendons)
Cartilage
Bone
Adipose tissue
Blood

The intercellular material of **cartilage** is solid yet pliable, like a piece of solid rubber, and resists compression. The material is produced by cells that in time become imprisoned in small cavities within their own secretions (Figure 33.4d). Vertebrate embryos develop cartilage limbs. These are structural models for the bones that generally replace them. Cartilage also maintains the shape of the nose, outer ear, and other body parts. It cushions joints between adjacent bones of the vertebral column, limbs, hands, and elsewhere.

Bone (Figures 33.4e and 33.5) is the weight-bearing tissue of the vertebrate skeleton, which supports or pro-

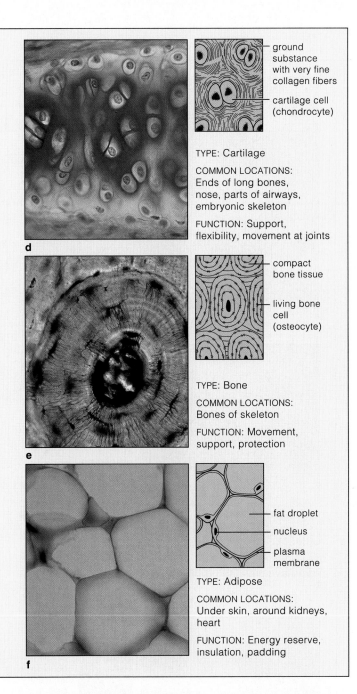

ground substance with very fine collagen fibers

cartilage cell (chondrocyte)

TYPE: Cartilage

COMMON LOCATIONS: Ends of long bones, nose, parts of airways, embryonic skeleton

FUNCTION: Support, flexibility, movement at joints

d

compact bone tissue

living bone cell (osteocyte)

TYPE: Bone

COMMON LOCATIONS: Bones of skeleton

FUNCTION: Movement, support, protection

e

fat droplet

nucleus

plasma membrane

TYPE: Adipose

COMMON LOCATIONS: Under skin, around kidneys, heart

FUNCTION: Energy reserve, insulation, padding

f

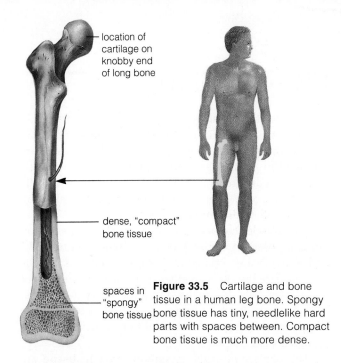

location of cartilage on knobby end of long bone

dense, "compact" bone tissue

spaces in "spongy" bone tissue

Figure 33.5 Cartilage and bone tissue in a human leg bone. Spongy bone tissue has tiny, needlelike hard parts with spaces between. Compact bone tissue is much more dense.

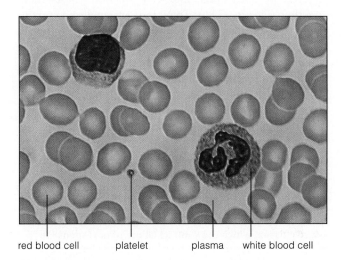

red blood cell   platelet   plasma   white blood cell

Figure 33.6 Some components of human blood, a vascular tissue. Its straw-colored, liquid matrix (plasma) is mostly water in which nutrients, diverse proteins, oxygen, carbon dioxide, ions, and other substances are dissolved.

tects softer tissues and organs. This connective tissue is hardened with minerals; its collagen fibers and ground substance are loaded with calcium salts. Limb bones, such as the ones in your legs, interact with attached muscles to bring about movements. Tissues in some bones also are production sites for blood cells.

Adipose tissue is so chockful of large fat cells, it no longer looks like a connective tissue (Figure 33.4f). The body's excess carbohydrates and proteins are converted to storage fats and tucked away in this tissue. Adipose tissue has a rich supply of blood, which serves as an immediately accessible "highway" for moving fats to and from the tissue's individual cells.

Blood is derived from connective tissue and serves transport functions. Circulating in its fluid medium

(plasma) are red blood cells, white blood cells, and platelets (Figure 33.6). Red blood cells deliver oxygen to all metabolically active tissues and carry carbon dioxide and other wastes away from them. Plasma also contains many dissolved proteins, ions, and other substances. Chapter 39 describes this complex tissue.

Diverse connective tissues bind together, support, strengthen, protect, and insulate other tissues in the body.

Most connective tissues consist of protein fibers and a variety of cells in a ground substance. One type (blood) is fluid. Another (adipose tissue) is a reservoir of stored energy.

33.4 MUSCLE TISSUE

In muscle tissue alone, cells contract (shorten) in response to stimulation, then lengthen and so return to their original state. Many long, cylindrical cells are arranged in parallel in these tissues. Their coordinated contraction and relaxation help move the body and its individual parts. There are three types, called skeletal, smooth, and cardiac muscle tissues.

The muscles connected to the bones of your skeleton are composed of **skeletal muscle tissue** (Figures 33.7 and 33.8a). Many striated (striped) skeletal muscle cells typically are bundled together. Then several bundles of the muscle cells are enclosed in a tough connective tissue sheath to form "a muscle," such as a biceps. The structure and function of skeletal muscle tissue are topics of the next chapter.

The contractile cells of **smooth muscle tissue** are tapered at both ends, and cell junctions hold them together (Figure 33.8b). They, too, are enclosed in connective tissue. The walls of blood vessels, the stomach, and other internal organs contain smooth muscle. We humans classify smooth muscle action as "involuntary," because we usually can't make it contract just by thinking about it (as we can do with skeletal muscle).

Cardiac muscle tissue is the heart's contractile tissue (Figure 33.8c). Cell junctions fuse together the plasma membranes of cardiac muscle cells. Some junctions at the fusion points allow the cells to contract as a unit. When one muscle cell receives a signal to contract, its neighbors are also stimulated into contracting.

Muscle tissue, which can contract (shorten) in response to stimulation, helps move the body and specific body parts.

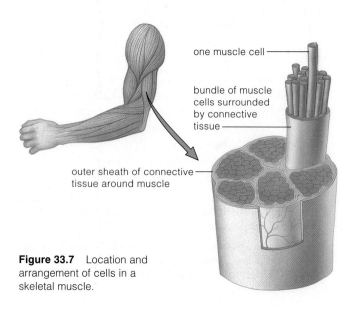

one muscle cell

bundle of muscle cells surrounded by connective tissue

outer sheath of connective tissue around muscle

Figure 33.7 Location and arrangement of cells in a skeletal muscle.

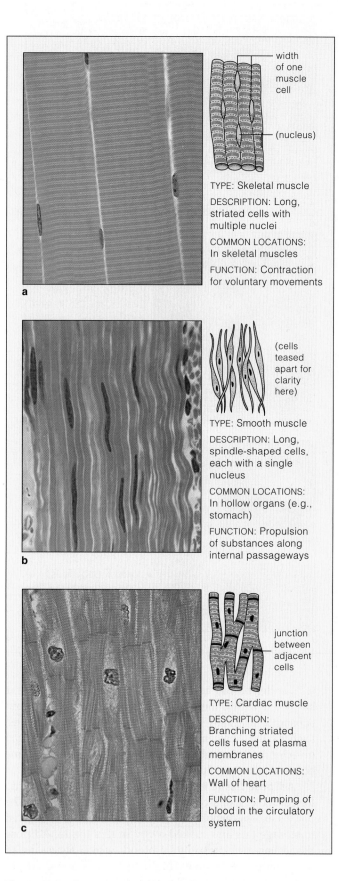

width of one muscle cell

(nucleus)

TYPE: Skeletal muscle
DESCRIPTION: Long, striated cells with multiple nuclei
COMMON LOCATIONS: In skeletal muscles
FUNCTION: Contraction for voluntary movements

a

(cells teased apart for clarity here)

TYPE: Smooth muscle
DESCRIPTION: Long, spindle-shaped cells, each with a single nucleus
COMMON LOCATIONS: In hollow organs (e.g., stomach)
FUNCTION: Propulsion of substances along internal passageways

b

junction between adjacent cells

TYPE: Cardiac muscle
DESCRIPTION: Branching striated cells fused at plasma membranes
COMMON LOCATIONS: Wall of heart
FUNCTION: Pumping of blood in the circulatory system

c

Figure 33.8 Examples of skeletal muscle, smooth muscle, and cardiac muscle tissues.

33.5 NERVOUS TISSUE

Of all tissues, **nervous tissue** exerts greatest control over the body's responsiveness to changing conditions. In this tissue, cells called **neurons** are organized as communication lines that extend through the body. Some neurons detect specific changes in environmental conditions. Others coordinate immediate and long-term responses to change. Neurons of the sort shown in Figure 33.9 relay signals to muscles and glands that carry out suitable responses. How these remarkable cells function is a topic of the next chapter.

Neurons are the basic units of communication in nervous tissue. Different kinds detect specific stimuli, integrate information, and issue or relay commands for response.

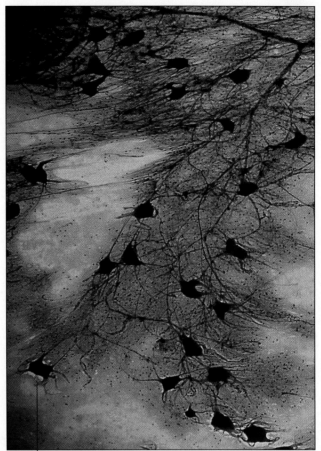

cell body of one of the motor neurons in this nervous tissue sample

Figure 33.9 A sampling of the millions of neurons that form communication lines within and between different regions of the human body. These motor neurons relay signals from the brain or spinal cord to muscles and glands. Collectively, neurons sense environmental change, integrate a great number and variety of signals about changes, and initiate suitable responses.

Focus on Science

Frontiers in Tissue Research

A tissue is more than a sum of its cells. As each new individual grows and develops, cells interact and become organized in specific ways. In time, cells of different tissues synthesize specific gene products that are vital for normal body functioning.

For many decades, researchers have attempted to construct artificial tissues in the laboratory. Today they can grow whole sheets of epidermis and use them to regenerate skin for patients with third-degree burns, bleeding ulcers, and other severely damaged tissues. A small section of epidermis from a patient's skin is cultured with nutrients and growth factors. The cells proliferate and form *laboratory-grown epidermis*. When an epidermal sheet is placed over a wound, it interacts biochemically and structurally with underlying cells to regenerate lost or damaged tissue.

On the horizon are *designer organs*—encapsulated groups of cells that will produce specific hormones, enzymes, and other substances. Such cells can be packaged within laboratory-grown epithelium derived from a patient, and so will not be rejected as foreign by his or her immune system.

Medical researchers are close to understanding how to make designer organs stick to suitable sites and become integral parts of normal body functioning. They hope to put together packages of cells that will produce specific life-saving substances in patients suffering from genetic disorders or chronic diseases. For example, imagine people with diabetes having a custom, insulin-secreting organ installed in the body. Their daily injections of insulin would be a thing of the past.

Tissue and Organ Formation

The tissues just described form organs and organ systems that are much the same in all vertebrates. To get a general sense of how this happens, start with a sperm and egg. Recall that these develop from germ cells, which are immature reproductive cells. (All other cells in the body are "somatic," after the Greek word for body.) After a zygote forms at fertilization, mitotic cell divisions produce an early animal embryo. In vertebrates, the cells soon become arranged as three primary tissues—ectoderm, mesoderm, and endoderm. These are embryonic forerunners of all tissues in the adult. **Ectoderm** gives rise to the skin's outer layer and to tissues of the nervous system. **Mesoderm** gives rise to the tissues of muscle, bone, and most of the circulatory, reproductive, and urinary systems. **Endoderm** gives rise to the gut's lining and organs derived from it.

Overview of the Major Organ Systems

Figure 33.10 provides an overview of the organ systems of a typical vertebrate, the adult human. Figure 33.11 illustrates some terms that are used when describing the positions of various organs, as well as the major body cavities in which they are located.

You might think we are stretching things a bit when we say that each one of those systems contributes to the survival of all living cells in the body. After all, what could the body's bones and muscles have to do with the life of a tiny cell? Yet interactions between the skeletal and muscular systems allow us to move about—toward sources of nutrients and water, for example. Parts of those systems assist blood circulation, as when contractions of leg muscles help move blood in veins back to the heart. The bloodstream transports nutrients and other substances to individual cells, and transports secreted products and wastes away from them.

Figure 33.10 Human organ systems.

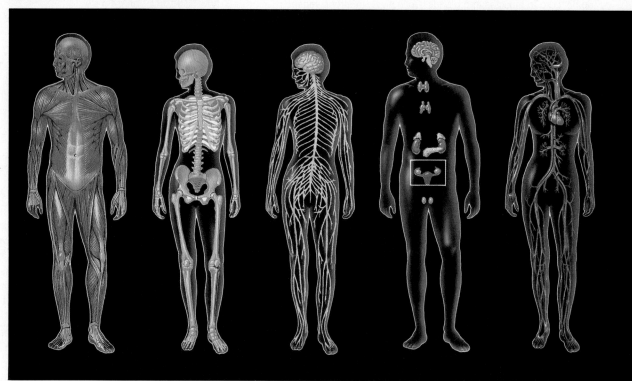

INTEGUMENTARY SYSTEM	MUSCULAR SYSTEM	SKELETAL SYSTEM	NERVOUS SYSTEM	ENDOCRINE SYSTEM	CIRCULATORY SYSTEM
Protection from injury, dehydration, and some microorganisms; control of body temperature; excretion of some wastes; reception of external stimuli	Movement of internal body parts; movement of whole body; maintenance of posture; heat production.	Support, protection of body parts; sites for muscle attachment, blood cell production, and calcium and phosphate storage.	Detection of external and internal stimuli; control and coordination of responses to stimuli; integration of activities of all organ systems.	Hormonal control of body functioning; works with nervous system in integrative tasks.	Rapid internal transport of many materials to and from cells; helps stabilize internal temperature and pH.

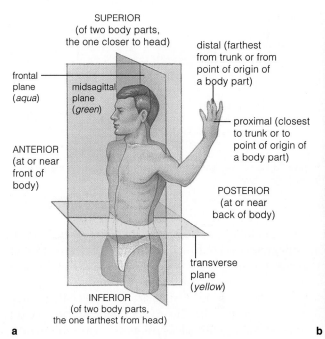

SUPERIOR
(of two body parts,
the one closer to head)

frontal
plane
(*aqua*)

midsagittal
plane
(*green*)

distal (farthest
from trunk or from
point of origin of
a body part)

proximal (closest
to trunk or to
point of origin of
a body part)

ANTERIOR
(at or near
front of
body)

POSTERIOR
(at or near
back of body)

transverse
plane
(*yellow*)

INFERIOR
(of two body parts,
the one farthest from head)

a

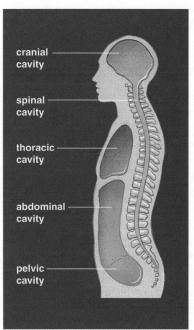

cranial
cavity

spinal
cavity

thoracic
cavity

abdominal
cavity

pelvic
cavity

b

Figure 33.11 (**a**) Directional terms and planes of symmetry for the human body. The midsagittal plane divides the body into right and left halves. The transverse plane divides it into anterior (front) and posterior (back) parts.

For humans, the main body axis is perpendicular to the earth. For rabbits and other animals that move with their main body axis parallel to the earth, *ventral* corresponds to anterior, and *dorsal* corresponds to posterior (compare Figure 26.2).

(**b**) Simplified diagram of the major cavities in the human body.

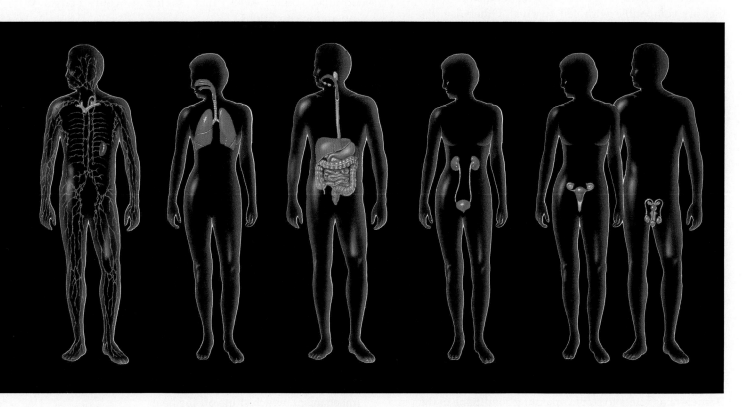

LYMPHATIC SYSTEM

Return of some tissue fluid to blood; roles in immunity (defense against specific invaders of the body).

RESPIRATORY SYSTEM

Delivery of oxygen to cells; removal of carbon dioxide wastes produced by cells; pH regulation.

DIGESTIVE SYSTEM

Ingestion of food, water; preparation of food molecules for absorption; elimination of food residues from the body.

URINARY SYSTEM

Maintenance of the volume and composition of extracellular fluid. Excretion of blood-borne wastes.

REPRODUCTIVE SYSTEM

Male: production and transfer of sperm to the female. Female: production of eggs; provision of a protected, nutritive environment for developing embryo and fetus. Both systems have hormonal influences on other organ systems.

Chapter 33 Tissues, Organ Systems, and Homeostasis   **555**

33.7 HOMEOSTASIS AND SYSTEMS CONTROL

The Internal Environment

To stay alive, your cells must be continually bathed in a fluid that supplies them with nutrients and carries away metabolic wastes. In this they are no different from an amoeba or any other free-living, single-celled organism. However, many *trillions* of cells crowd together in your body—and they all must draw nutrients from and dump wastes into the same 15 liters of fluid. That is less than 16 quarts.

The fluid *not* inside cells is called **extracellular fluid**. Much of it is *interstitial*, meaning it occupies spaces between cells and tissues. The rest is *plasma*, the fluid portion of blood. Interstitial fluid exchanges substances with blood and with the cells it bathes.

In functional terms, extracellular fluid is continuous with the fluid inside cells. That's why drastic changes in its composition and volume have drastic effects on cell activities. Its ion concentrations are especially important in this regard. They must be maintained at levels that are compatible with cell survival. Otherwise, the animal itself cannot survive.

It makes no difference whether an animal is simple or complex. *The component parts of any animal work together to maintain the stable fluid environment required by its living cells.* This concept is absolutely central to understanding the structure and function of animals, and its key points may be summarized this way:

---

1. Each cell of the animal body engages in basic metabolic activities that ensure its own survival.

2. Concurrently, the cells of a given tissue perform one or more activities that contribute to the survival of the whole organism.

3. The combined contributions of individual cells, organs, and organ systems help maintain the stable internal environment—that is, the extracellular fluid—required for individual cell survival.

Mechanisms of Homeostasis

Homeostasis refers to stable operating conditions in the internal environment. Three components, called sensory receptors, integrators, and effectors, interact to maintain this state. **Sensory receptors** are cells or cell parts that can detect a **stimulus**, which is a specific change in the environment. When someone kisses you, for example, there is a change in pressure on your lips. Receptors in the skin of your lips translate the stimulus

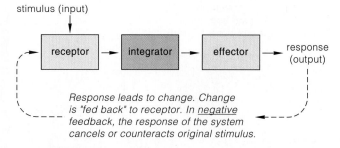

Figure 33.12 Components necessary for negative feedback at the organ level.

into a signal that can be sent to the brain. Your brain is an **integrator**, a control point where different bits of information are pulled together in the selection of a response. It can send signals to your muscles or glands (or both). Muscles and glands are **effectors**—they carry out the response. In this case, the response might include flushing with pleasure and kissing the person back. Of course, you cannot engage in a kiss indefinitely, for this would prevent you from eating and performing other tasks that maintain your body's operating conditions.

How does your brain reverse the physiological changes induced by the kiss? Receptors only provide it with information about how things *are* operating. The brain also receives information about how things *should be* operating—that is, information from "set points." When physical or chemical conditions deviate sharply from a set point, the brain functions to bring them back to an effective operating range. It does this by way of signals that cause specific muscles and glands to increase or decrease their activity.

Feedback mechanisms are among the controls that help keep physical and chemical aspects of the body within tolerable ranges. In a **negative feedback mechanism**, an activity alters a condition in the internal environment, and this triggers a response that reverses the altered condition (Figure 33.12). Think of a furnace with a thermostat. A thermostat senses the air temperature and "compares" it to a preset point on a thermometer built into the furnace control system. When the temperature falls below the preset point, the thermostat signals a switching mechanism that turns on the heating unit. When the air becomes heated enough to match the prescribed level, the thermostat signals the switching mechanism, which shuts off the heat.

Similarly, feedback mechanisms help keep the body temperature of meerkats, humans, huskies, and many other animals near 37°C (98.6°F), even during hot or cold weather. Imagine a husky running around on a hot summer day. Its body gets hot—and receptors trigger events that slow down the whole dog *and* its cells. The husky flops down and rests in the shade of a tree. Moisture

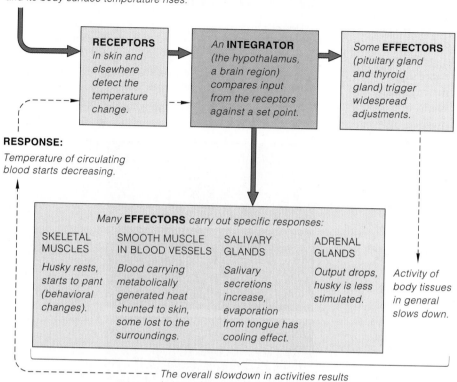

The husky is overactive on a hot, dry day
and its body surface temperature rises.

RECEPTORS
in skin and
elsewhere
detect the
temperature
change.

An **INTEGRATOR**
*(the hypothalamus,
a brain region)
compares input
from the receptors
against a set point.*

Some **EFFECTORS**
*(pituitary gland
and thyroid
gland) trigger
widespread
adjustments.*

RESPONSE:

*Temperature of circulating
blood starts decreasing.*

Many **EFFECTORS** *carry out specific responses:*

SKELETAL MUSCLES	SMOOTH MUSCLE IN BLOOD VESSELS	SALIVARY GLANDS	ADRENAL GLANDS
Husky rests, starts to pant (behavioral changes).	*Blood carrying metabolically generated heat shunted to skin, some lost to the surroundings.*	*Salivary secretions increase, evaporation from tongue has cooling effect.*	*Output drops, husky is less stimulated.*

*Activity of
body tissues
in general
slows down.*

*The overall slowdown in activities results
in less metabolically generated heat.*

Figure 33.13 Homeostatic controls over the internal temperature of a husky's body. Blue arrows indicate the main control pathways. The dashed line shows how a feedback loop is completed.

from its respiratory system evaporates from the tongue and carries some body heat with it (Figure 33.13). These and other control mechanisms counter overheating by curbing the body's heat-generating activities and by giving up excess heat to the surroundings.

Sometimes **positive feedback mechanisms** operate. These set in motion a chain of events that *intensify* a change from an original condition—and after a limited time, the intensification reverses the change. Positive feedback is associated with instability in a system. For example, during sexual intercourse, signals from a female's nervous system trigger intense physiological responses to her sexual partner. Those responses stimulate changes in her partner that further stimulate the female, and so on until an explosive, climax level is reached. As another example, during childbirth, a fetus exerts pressure on the wall of its mother's uterus. This stimulates production and secretion of oxytocin, a hormone. Oxytocin causes wall muscles to contract and exert pressure on the fetus, which exerts more pressure on the wall, and so on until the fetus is expelled.

What we have been describing here is a general pattern of monitoring and responding to a constant flow of

information about an animal's internal and external environments. During this activity, organ systems operate together in coordinated fashion. Throughout this unit, we will be asking the following questions about their operation:

1. What physical or chemical aspect of the internal environment are organ systems working to maintain as conditions change?

2. By what means are organ systems kept informed of the various changes?

3. By what means do they process the incoming information?

4. What mechanisms are set in motion in response?

As you will see in chapters to follow, the operation of all organ systems is under precise neural and endocrine control.

Homeostatic control mechanisms help maintain physical and chemical aspects of the body's internal environment within ranges that are most favorable for cell activities.

SUMMARY

1. A tissue is an aggregation of cells and intercellular substances that perform a common task. An organ is a structural unit with tissues combined in definite proportions and patterns that allow them to perform a common task. An organ system has two or more organs interacting chemically, physically, or both in ways that contribute to the survival of the body as a whole.

2. Epithelial tissues cover external body surfaces and line internal cavities and tubes. They have one free surface exposed to body fluids or to the environment.

3. A variety of connective tissues bind together, support, strengthen, protect, and insulate other tissues. Most consist of fibers of structural proteins (especially collagen), fibroblasts, and other cells in a ground substance.

   a. Loose connective tissue has a semifluid ground substance. It occurs under skin and most epithelia.

   b. Dense, irregular connective tissue has mostly collagen fibers and a few fibroblasts. It occurs in skin and forms protective capsules around many organs.

   c. Dense, regular connective tissue, with parallel bundles of collagen fibers, provides support and elasticity to tendons, skin, and other organs.

   d. Cartilage, with solid yet pliable intercellular material, has structural and cushioning roles. Bone, the weight-bearing tissue of vertebrate skeletons, works with skeletal muscle to bring about movement.

   e. Blood, a specialized connective tissue, is mostly plasma with cellular components and dissolved substances. Adipose tissue (mostly fat cells) is a reservoir of stored energy.

4. Muscle tissues contract (shorten), then return to the resting position. They help move the body or parts of it. The three types (skeletal muscle, smooth muscle, and cardiac muscle tissue) have long, cylindrical cells.

5. Nervous tissue detects and integrates information about internal and external conditions, and it governs the body's responses to change. Its neurons are the basic units of communication in nervous systems.

6. Tissues, organs, and organ systems work together to maintain the stable internal environment (the extracellular fluid) required for individual cell survival. At homeostasis, conditions in the internal environment are most favorable for cell activities.

7. Feedback controls help maintain internal operating conditions for the body's cells. For example, in negative feedback, a change in some condition triggers a response that reverses the change.

8. Homeostasis depends on receptors, integrators, and effectors. Receptors detect stimuli (specific changes in the environment). Integrating centers (such as a brain) process the information and direct muscles and glands (the body's effectors) to carry out responses.

Review Questions

1. Describe the general characteristics of epithelial tissue, the various types of epithelial tissues, and their specific characteristics and functions. *548–549*

2. List the major types of connective tissues; add the names and characteristics of their specific types. *550–551*

3. Identify and describe the following tissues. *548–552*

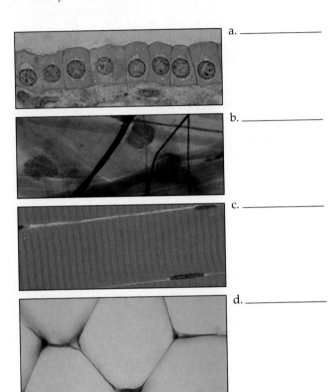

a. _____

b. _____

c. _____

d. _____

4. Identify this category of tissue and its characteristics. *552*

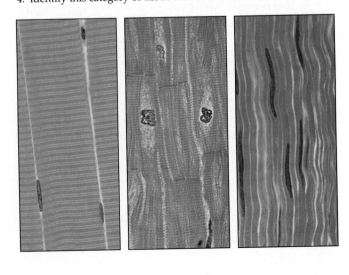

5. What type of cell serves as the basic unit of communication in nervous systems? *553*

6. Define an animal tissue, organ, and organ system. List and define the functions of the human body's major organ systems. *547, 554–555*

7. Define extracellular fluid and interstitial fluid. *556*

8. Define homeostasis. *547, 556–557*

9. Describe the major homeostatic mechanisms of the human body. *556–557*

Self-Quiz *(Answers in Appendix IV)*

1. _____ tissues have closely linked cells and one free surface.
   a. Muscle   c. Connective
   b. Nervous   d. Epithelial

2. Most _____ tissues have cells that secrete fibers of collagen and elastin.
   a. muscle   c. connective
   b. nervous   d. epithelial

3. _____ has a semifluid ground substance and occurs under skin and most epithelia.
   a. Dense, irregular connective tissue
   b. Loose connective tissue
   c. Dense, regular connective tissue
   d. Cartilage

4. _____ is a specialized connective tissue that is mostly plasma with cellular components and a variety of dissolved substances.
   a. Irregular connective tissue   c. Cartilage
   b. Blood   d. Bone

5. _____ tissues detect and coordinate information about change and control responses to those changes.
   a. Muscle   c. Connective
   b. Nervous   d. Epithelial

6. _____ tissues contract.
   a. Muscle   c. Connective
   b. Nervous   d. Epithelial

7. Cells of complex animals _____ .
   a. survive through their own metabolism
   b. contribute to the survival of the whole animal
   c. help maintain extracellular fluid
   d. all of the above

8. At _____ , physical and chemical aspects of the internal environment are being kept within tolerable ranges.
   a. positive feedback   c. homeostasis
   b. negative feedback   d. metastasis

9. When some changed condition in the internal environment triggers its own reversal back to normal, a _____ control mechanism has been operating.
   a. homeostatic   c. positive feedback
   b. negative feedback   d. both a and b

10. In negative feedback mechanisms, _____ .
   a. a detected change brings about a response that tends to return internal operating conditions to the original state
   b. a detected change suppresses internal operating conditions to levels below the set point
   c. a detected change raises internal operating conditions to levels above the set point
   d. fewer solutes are fed back to the affected cells

11. _____ detect specific environmental changes, an _____ pulls different bits of information together in the selection of a response, and _____ carry out the response.

12. Match the terms with their appropriate description.
   ____ epithelium   a. somewhat pliable, like rubber
   ____ cartilage   b. covers, lines body surfaces
   ____ homeostasis   c. stable internal environment
   ____ muscles and glands d. integrating center
   ____ positive feedback   e. the most common
   ____ negative feedback   homeostatic mechanism
   ____ brain   f. effectors
                         g. chain of events intensifies the original condition

Selected Key Terms

adipose tissue *551*
blood *551*
bone *550*
cardiac muscle tissue *552*
cartilage *550*
dense, irregular connective tissue *550*
dense, regular connective tissue *550*
ectoderm *554*
effector *556*
endocrine gland *549*
endoderm *554*
epithelium *548*
exocrine gland *549*
extracellular fluid *556*
homeostasis *556*

integrator *556*
loose connective tissue *550*
mesoderm *554*
negative feedback mechanism *556*
nervous tissue *553*
neuron *553*
organ *547*
organ system *547*
positive feedback mechanism *556*
sensory receptor *557*
skeletal muscle tissue *552*
smooth muscle tissue *552*
stimulus *556*
tissue *547*

Readings

Bloom, W., and D. W. Fawcett. 1986. *A Textbook of Histology*. Eleventh edition. Philadelphia: Saunders. Outstanding reference text.

Leeson, C. R., T. Leeson, and A. Paparo. 1985. *Textbook of Histology*. Philadelphia: Saunders.

Ross, M., and E. Reith. 1985. *Histology: A Text and Atlas*. New York: Harper & Row.

Vander, A., J. Sherman, and D. Luciano. 1990. "Homeostatic Mechanisms and Cellular Communication" in *Human Physiology*. Fifth edition. New York: McGraw-Hill.

34 INFORMATION FLOW AND THE NEURON

TORNADO!

It is spring in the American Midwest, and you are engrossed in photographing the wildflowers all around you in an expanse of shortgrass prairie. So intent are you on capturing all the different species on film that you fail to notice the rapidly darkening sky. By the time you finally look up, the sky is ominous. What's that rumbling you hear in the distance? It sounds like a freight train. Yet how can that be, when there is no train track, anywhere, in this part of the prairie? You turn to identify the source of the sound. And you see it—but you don't want to believe your eyes. A dark funnel cloud is advancing across the prairie and heading right for you! *TORNADO!* The terrifying image rivets your attention as nothing else has ever done (Figure 34.1).

You sense instantly that you cannot remain where you are and survive. Suddenly you remember that hiding in a low area is better than standing out in the open. Commands flash from your brain to your limbs: *GET MOVING OUT OF HERE!* With heart thumping, you start to run along a path, looking frantically for safety. You're in luck! Just ahead is a steep-banked creek. With a tremendous burst of speed you reach the creek in less than a minute, then scramble down the muddy bank. There you find a small ledge hanging over the water. You wedge yourself under it, hoping wildly to be inconspicuous, to be overlooked by a force of nature on the rampage.

Figure 34.1 A terrifying view across the Kansas prairie—a tornado about to touch down. Imagine yourself alone in the prairie when you first see the tornado, knowing you have but minutes to remove yourself from harm's way. By what means do you conceive of plans of action—and can they be put together and evaluated quickly enough? The answers begin with the functioning of neurons.

It takes a few minutes before you realize that the tornado has roared past the creek. You remain motionless, heart still thumping, fingers clutching mud. Finally you sit up and look around. Some distance away, you see a swath of twisted prairie grass that marks the tornado's path. Your heart no longer feels like it is slamming against your chest wall, although your legs feel like rubber when you stand up.

You can thank your nervous system for every perception, every memory, and nearly every action that helped you escape the tornado. You can thank the information that traveled swiftly, in suitable directions, among communication lines of interacting cells—the cells called neurons.

Were those lines silent until they received signals from the outside, much as telephone lines wait to carry calls from all over the country? Absolutely not. Even before you were born, your newly developing neurons became organized in vast gridworks, and they started chattering among themselves. All through your life, in moments of danger or reflection, excitement or sleep, their chatter has never ceased. Constant communication among neurons, such as the ones controlling the muscles concerned with breathing, keeps you alive. It keeps your body primed for rapid response to change—even to the totally unexpected appearance of a tornado.

This chapter begins with the structure and function of neurons. Then it starts us thinking about how neurons interact. In chapters to follow, we will consider the nervous system as well as the sensory structures and organs—including eyes—by which you detect tornadoes and other events that may have bearing on whether you survive from one day to the next.

KEY CONCEPTS

1. Neurons are the basic units of communication of nervous systems. Collectively, they detect and integrate information about external and internal conditions, then select or control muscles and glands in ways that produce suitable responses.

2. The inside of a neuron is negatively charged relative to the outside. When a neuron is stimulated, this polarity of charge across the membrane may briefly and abruptly reverse. Such reversals—action potentials—are the basis of messages sent through nervous systems.

3. Action potentials are self-propagating along a neuron. They cannot cross small gaps that exist *between* neurons. Chemical signals bridge the gaps and stimulate or inhibit the adjoining neuron, muscle cell, or gland cell.

4. Information flow through nervous systems depends on the moment-by-moment integration of excitatory and inhibitory signals that act upon the neurons of given pathways.

34.1 CELLS OF THE NERVOUS SYSTEM

All animals sense and respond to the environment. And if one has an elaborate life-style, you can safely bet it has a well-developed nervous system. In such systems, recall, **neurons** collectively monitor conditions and issue commands for responsive actions that benefit the body as a whole. Neurons are the communication cells of the brain, spinal cord, and nerves.

There are three classes of neurons. Each type of **sensory neuron** is adapted to respond to a specific stimulus (light, pressure, or another form of energy). They relay information about the stimulus to the spinal cord and brain. In the spinal cord and brain, **interneurons** receive sensory input, integrate it with other information, then influence the activity of other neurons. **Motor neurons** relay information away from the brain and spinal cord to muscles or glands. Muscles and glands, the body's effectors, carry out responses:

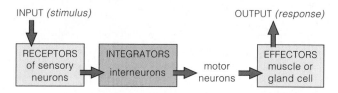

Neurons make up less than half the volume of vertebrate nervous systems. As you will see in the next chapter, the rest is mostly **neuroglia**—a variety of specialized cells that protect, structurally support, and functionally assist the neurons.

34.2 HOW THE NEURON RESPONDS TO STIMULATION

Functional Zones of a Neuron

Neurons have a nucleated cell body and cytoplasmic extensions, although these differ greatly in number and length (Figure 34.2). Typically, the cell body and the slender extensions called **dendrites** are *input* zones, where a neuron receives information. A slender, often longer extension called an **axon** is a *conducting* zone. Signals are rapidly propagated along axons. The signals themselves arise in a *trigger* zone. Except in sensory neurons, this is at the junction between the cell body and an axon. An axon's branched endings are *output* zones, where messages are sent to other cells.

A Neuron At Rest, Then Moved To Action

Different types of signals travel through a nervous system. Let's start with one that begins and ends on the same neuron, and is not transferred to another cell.

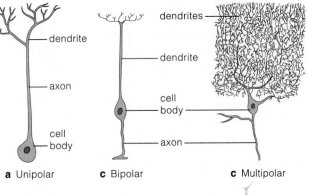

a Unipolar   **c** Bipolar   **c** Multipolar

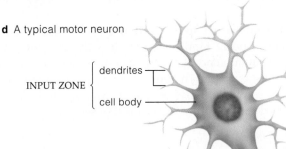

d A typical motor neuron

INPUT ZONE { dendrites
cell body

TRIGGER ZONE

Figure 34.2 (**a–c**) Categories of neurons, based on the number of cytoplasmic extensions of the cell body. *Unipolar* cells have a single, branched extension. One branch is the axon; others are dendritic. *Bipolar* cells have one dendrite and one axon. Many sensory neurons are like this. *Multipolar* cells, with one axon and many dendrites, predominate in vertebrate nervous systems. (**d**) Functional zones of a motor neuron, a multipolar cell.

axon

CONDUCTING ZONE

When a neuron is not being bothered, it works to maintain a voltage difference (a difference in electric charge) across the plasma membrane. It keeps the cytoplasmic fluid next to the membrane negatively charged, compared to the interstitial fluid outside. The amount of energy inherent in this steady voltage difference is the **resting membrane potential**. For many neurons, this amount is about −70 millivolts.

Suppose a weak signal reaches a patch of membrane in a neuron's input zone. The voltage difference across the patch changes only slightly, if at all. By contrast, a strong signal might trigger an **action potential**—an abrupt, short-lived reversal in the voltage difference across the plasma membrane. For a fraction of a second, the inside becomes positive with respect to the outside. The reversal triggers another action potential at the adjoining patch of membrane, this triggers another at the next patch, and so on away from the point of initiation.

In short, *stimulation of a neuron disturbs the distribution of electric charge across its plasma membrane.*

Restoring and Maintaining Readiness

In between action potentials, a neuron restores and maintains the voltage difference across each patch of membrane. It can do so because of two membrane properties. *First*, the membrane's lipid bilayer bars the passage of potassium ions (K^+), sodium ions (Na^+), and other charged substances. Thus the neuron can build up differences in ion concentrations across the membrane. *Second*, ions can flow from one side to the other in controllable ways—through the interior of proteins that span the bilayer (Figure 34.3).

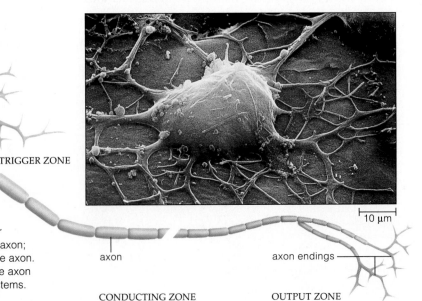

10 μm

axon endings

OUTPUT ZONE

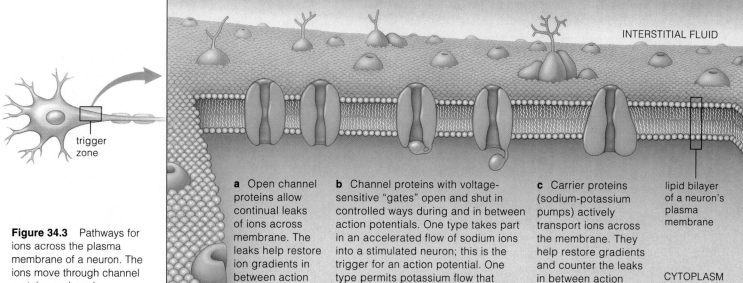

Figure 34.3 Pathways for ions across the plasma membrane of a neuron. The ions move through channel proteins and carrier proteins that span the membrane's lipid bilayer.

INTERSTITIAL FLUID

a Open channel proteins allow continual leaks of ions across membrane. The leaks help restore ion gradients in between action potentials.

b Channel proteins with voltage-sensitive "gates" open and shut in controlled ways during and in between action potentials. One type takes part in an accelerated flow of sodium ions into a stimulated neuron; this is the trigger for an action potential. One type permits potassium flow that helps end an action potential.

c Carrier proteins (sodium-potassium pumps) actively transport ions across the membrane. They help restore gradients and counter the leaks in between action potentials.

lipid bilayer of a neuron's plasma membrane

CYTOPLASM

Suppose a motor neuron has 15 sodium ions inside the membrane for every 150 outside. Suppose it has 150 potassium ions inside for every 5 on the outside. We can depict each ion's concentration gradient in this way (from the large to the small letter):

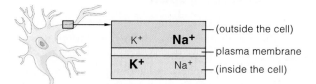

Such gradients dictate the direction in which sodium and potassium ions diffuse across the membrane. They can diffuse through the interior of channel proteins, of the sort described on page 86. Some channels never

shut, so ions leak (diffuse) through them all the time. Other channels are gated and open only when the neuron is being stimulated.

Suppose that a motor neuron is in between action potentials. At such times, its sodium channels are shut, so sodium cannot rush into the neuron. Some potassium is leaking out through a few open potassium channels, making the neuron a bit more negative inside—so some potassium is attracted back in. When the inward pull of electric charge balances the outward force of diffusion, there is no more net movement of potassium across the membrane. The concentration and electric gradients that now exist are the basis of the neuron's responsiveness to stimulation.

A resting neuron must expend energy to maintain these gradients. Why? A tiny fraction of the potassium that leaks out is not attracted back in. Also, a tiny fraction of sodium leaks in, through a few open channels (Figure 34.4). Unless the neuron counters the tiny leaks, the gradients would gradually disappear.

An active transport mechanism counters the leaks. Spanning the membrane are carrier proteins called **sodium-potassium pumps**. When they get an energy boost from ATP, they actively transport potassium in and sodium out at the same time.

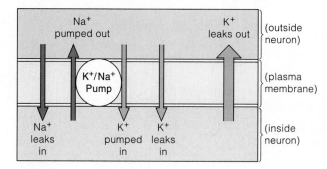

Figure 34.4 Pumping and leaking processes that maintain the distribution of sodium and potassium ions across the plasma membrane of a neuron at rest. Arrow widths indicate the magnitude of the movements. Notice how the total inward and outward movements for each kind of ion are balanced.

1. In an undisturbed neuron, a difference in electric charge across the plasma membrane has been established and is being maintained. An action potential is an abrupt, short-lived reversal in that difference.

2. Ion leaks and ion pumps keep the neuron in a state of readiness between action potentials.

34.3 ACTION POTENTIALS

Approaching Threshold

Stimulate a neuron at its input zone and you disturb the ion balance across the membrane, but not much. For instance, suppose you trip over a cat and put a bit of pressure on its skin. Sensory neurons have receptors (input zones) in tissues beneath the skin's epidermis. The pressure on the receptors deforms patches of their plasma membrane, and this allows some ions to flow across. The voltage difference across the membrane changes slightly, producing a graded, local signal.

Graded means that the signals at an input zone can vary in magnitude. They can be large or small; it depends on the intensity and duration of the stimulus.

Local means the signals don't spread far from the point of stimulation. It takes specialized types of ion channels to propagate a signal along the membrane, and input zones simply don't have them.

However, when a stimulus is intense or long-lasting, graded signals can spread out of the input zone and into an adjacent trigger zone. At this zone, a certain minimum amount of change in the voltage difference across the plasma membrane can trigger an action potential. That amount is the **threshold level**. Threshold can be reached at any membrane patch that has voltage-sensitive, gated channels for sodium ions.

As Figure 34.5 shows, a stimulus causes sodium ions to flow into the neuron. With the influx of these positively charged ions, the cytoplasmic side of the membrane becomes less negative. This causes more gates to open, more sodium to enter, and so on. The ever increasing inward flow of sodium is an example of positive feedback, whereby an event intensifies as a result of its own occurrence:

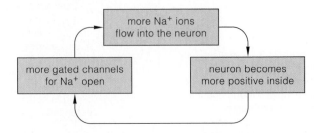

At threshold, the opening of more sodium gates no longer depends on the strength of the stimulus. Because the positive-feedback cycle is now under way, the inward-rushing sodium itself is enough to cause more sodium gates to open.

An All-or-Nothing Spike

Figure 34.6 is a recording of the voltage difference across the membrane before, during, and after an action potential. Notice how the membrane potential spikes once threshold is reached. All action potentials in a neuron spike to the same level above threshold as an *all-or-nothing* event. Once a positive-feedback cycle starts, nothing stops the full spiking. If threshold is not reached, however, the membrane disturbance will subside when the stimulus is removed.

Figure 34.5 Propagation of an action potential along the axon of a motor neuron.

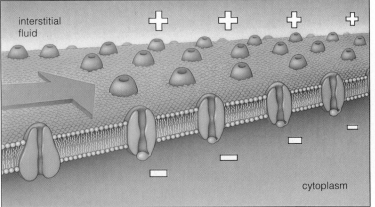

a Membrane at rest (inside negative with respect to the outside). An electrical disturbance (*red* arrow) spreads from an input zone to an adjacent trigger region of the membrane, which has many gated sodium channels.

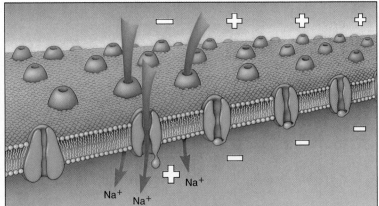

b A strong disturbance initiates an action potential. Sodium gates open, the sodium inflow decreases the negativity inside; this causes more gates to open, and so on, until threshold is reached and the voltage difference across the membrane reverses.

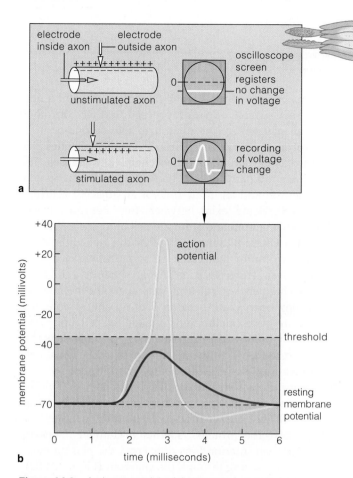

Figure 34.6 Action potentials. (a) When early researchers studied neural function, the squid *Loligo* provided them with evidence of action potential spiking. Its "giant" axons were large enough to slip electrodes inside. When such an axon was stimulated, electrodes positioned on the inside and outside detected voltage changes, which showed up as deflections in a beam of light across the screen of an oscilloscope.

(b) A typical waveform (*yellow* line) for an action potential. The *red* line represents a recording of a local signal that did not reach the threshold of an action potential; spiking did not occur.

Each spike lasts only for a millisecond or so. At the membrane site of the charge reversal, the gated sodium channels closed and shut off the sodium inflow. About halfway through the reversal, potassium channels opened, so many more potassium ions flowed out and restored the original voltage difference across the membrane. And sodium-potassium pumps restored the ion gradients.

Later, after the resting membrane potential has been restored, most potassium gates close and sodium gates are in their initial state, ready to be opened with the arrival of a suitable disturbance.

Propagation of Action Potentials

The membrane disturbances leading up to an action potential are self-propagating, and they don't diminish in magnitude. When they spread to an adjacent membrane patch, the opening of an equivalent number of gated channels is repeated. The disturbance causes gated channels to open in the next patch, and so on.

For a brief period after the disturbance, each membrane patch remains insensitive to stimulation because its sodium gates cannot open. That is why action potentials do not spread back into the trigger zone, but rather are self-propagating away from it.

The inside of a neuron at rest is more negative than the outside.

During an action potential, the inside is more positive than the outside at a disturbed patch of membrane.

Following an action potential, resting conditions are restored at the membrane patch.

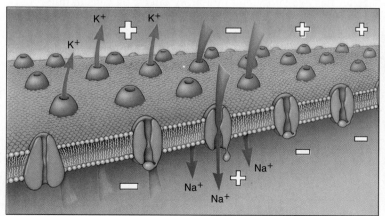

c The reversal causes sodium gates to shut and potassium gates to open (at *purple* arrows). Potassium follows its gradient (out of the neuron). Voltage is restored. The disturbance produced by the action potential triggers an action potential at the adjacent membrane site, and so on, away from the point of stimulation.

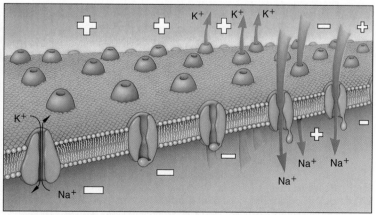

d The inside of the membrane becomes negative again following each action potential, but the sodium and potassium concentration gradients are not yet fully restored. Active transport at sodium-potassium pumps restores the gradients.

34.4 NODE-TO-NODE HOPPING ALONG SHEATHED AXONS

Action potentials arise rapidly along the axons of many sensory and motor neurons. As Figure 34.7 shows, the axons are wrapped in a **myelin sheath**. The sheath consists of the plasma membranes of neuroglial cells called Schwann cells. Each cell is separated from adjacent ones by a small, exposed gap (node) where the axon membrane is loaded with voltage-sensitive gated sodium channels. In a manner of speaking, the action potentials jump from node to node. (Hence the name *saltatory* conduction, after the Latin word meaning "to jump.") The sheathed regions between nodes hamper the movement of ions across the plasma membrane, so disturbances tend to flow along the plasma membrane until the next node in line—where ion flow can produce a new action potential. In the largest sheathed axons, action potentials propagate themselves at a remarkable 120 meters per second.

You may have heard of *multiple sclerosis*, a progressive deterioration of the myelin sheaths of neurons in the spinal cord. It may result from a viral infection combined with a genetic predisposition to the disease. Symptoms include muscle weakness, fatigue, numbness, and possibly slurred speech. Some patients become severely disabled.

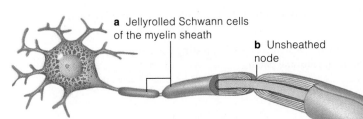

a Jellyrolled Schwann cells of the myelin sheath

b Unsheathed node

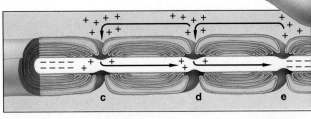

Figure 34.7 Propagation of an action potential along a motor neuron having a myelin sheath.

(a) The sheath is a series of Schwann cells, each wrapped like a jellyroll around the axon. Each "jellyroll" blocks ion movements across the membrane. But ions can cross it at nodes between Schwann cells **(b)**. The unsheathed nodes have dense arrays of gated sodium channels. **(c,d)** A disturbance caused by an action potential spreads down the axon. When it reaches a node, sodium gates open, sodium ions rush inward, and another action potential results. **(e)** This new disturbance spreads rapidly to the next node and triggers another action potential, and so on down the line.

34.5 CHEMICAL SYNAPSES

When action potentials reach the output zone of a neuron, they usually do not proceed farther. But their arrival may induce the neuron to release one or more **neurotransmitters**. These are signaling molecules that diffuse across junctions called **chemical synapses**. The junctions are narrow clefts between the output zone of one neuron and an input zone of an adjacent cell (Figure 34.8). Some exist between two neurons, others between a neuron and a muscle cell or gland cell.

At a chemical synapse, *one* of the two cells stores neurotransmitter molecules (in synaptic vesicles in its cytoplasm). Think of it as the *pre*synaptic cell. Here, gated channels for calcium ions span the membrane, and they open with the arrival of an action potential. There are more calcium ions outside the cell, and when they flow in (down their gradient), synaptic vesicles are induced to fuse with the plasma membrane. Then, neurotransmitter is released into the synaptic cleft.

Neurotransmitter molecules diffuse across the cleft. They bind with specific receptor proteins on the membrane of the *post*synaptic cell. Binding changes the shape of these proteins, so that a channel opens up through their interior. Ions now cross the plasma membrane by diffusing through the channels (Figure 34.8*d*).

How will the postsynaptic cell respond to a signal? The answer lies with the type of neurotransmitter and its concentration in the cleft, on the particular kinds of receptors and gated channels in the postsynaptic cell membrane, and on whether those channels are primed to open. Depending on such factors, a neurotransmitter may have an *excitatory* effect and help drive the postsynaptic cell's membrane toward the threshold of an action potential. Or it may have an *inhibitory* effect and drive the membrane away from the threshold.

Consider **acetylcholine** (ACh). This neurotransmitter has excitatory and inhibitory effects on cells in muscles, glands, the brain, and the spinal cord. Take another look at Figure 34.8, which highlights a chemical synapse between a motor neuron and a muscle cell. ACh released from the motor neuron diffuses across the cleft and binds to receptors on the muscle cell membrane. It has excitatory effects on this kind of cell. As described in Chapter 38, it triggers action potentials, which in turn initiate muscle contraction.

Neurotransmitters bridge the gap (synaptic cleft) between two neurons or between a neuron and a muscle cell or gland cell. These signaling molecules may have excitatory or inhibitory effects on different kinds of receiving cells.

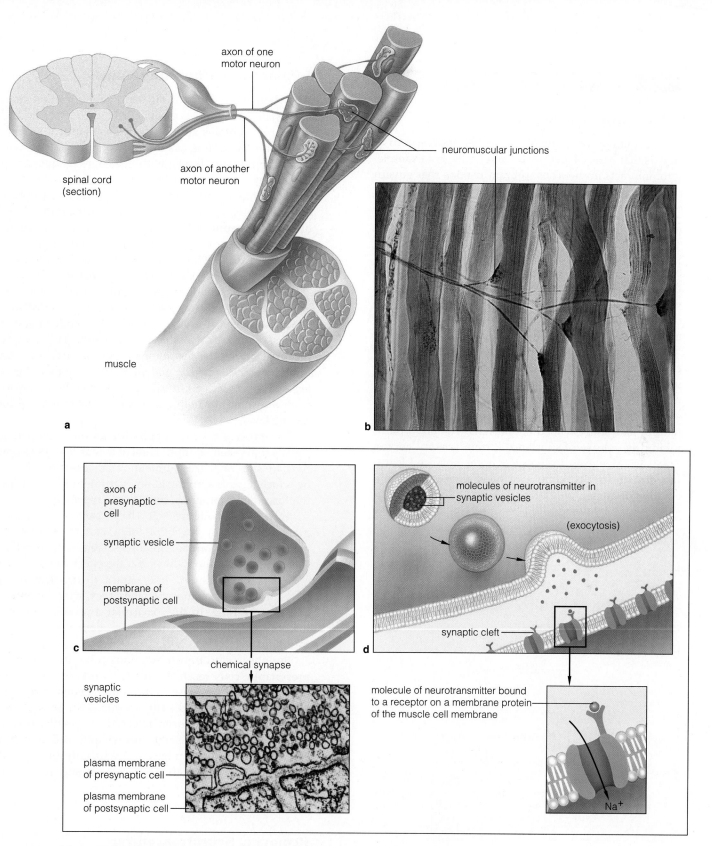

Figure 34.8 Chemical synapse between a neuron and another cell. (**a,b**) Drawing and micrograph of *neuromuscular* junctions. These are regions of chemical synapsing between a motor neuron and a muscle cell. The myelin sheath of the motor axon terminates before the junction. At the axon endings, the plasma membranes of the two cells are exposed to each other. (**c**) The *pre*synaptic cell is identifiable in photomicrographs by neurotransmitter molecules contained in cytoplasmic vesicles near the plasma membrane facing the junction. (**d**) Arrival of an action potential at the neuron's output zone (axon endings) triggers the release of neurotransmitter molecules. These molecules diffuse across the junction and bind to receptors on the adjoining cell.

34.6 SYNAPTIC INTEGRATION

A Smorgasbord of Signals

Acetylcholine is only one of a veritable smorgasbord of signals that neurons deliver to target cells. For example, serotonin acts on neurons in brain regions that govern sleeping, sensory perception, temperature control, and emotional states. Norepinephrine works in brain regions dealing with emotional states, dreaming, and arousal. Dopamine is the specialty of neurons in brain regions dealing with emotions. GABA (gamma aminobutyric acid) is the most common inhibitory signal in the brain. Valium and other antianxiety drugs may work by enhancing GABA's effects. Except for acetylcholine, all of these neurotransmitters and others are amino acids or are derived from them.

Neuromodulators, another group of signaling molecules, are the smorgasbord's garnishes. These act to magnify or reduce the effect of neurotransmitters on neighboring or distant neurons. Among them are substance P, which stimulates perception of pain, and the natural pain killers called endorphins. The endorphins inhibit the release of substance P from sensory nerves. They also may exert influence over memory and learning, sexual activity, control of body temperature, and emotional states.

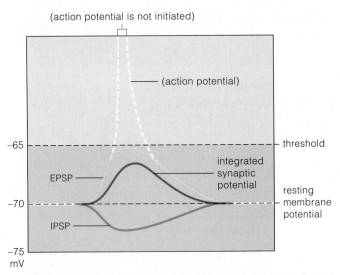

Figure 34.9 Synaptic integration. In this example, an excitatory synapse and an inhibitory synapse nearby are activated at the same time. The IPSP reduces the magnitude of the EPSP from what it could have been, pulling it away from threshold. The *red* line represents the integration of these two synaptic potentials. Threshold is not reached in this case; hence an action potential cannot be initiated.

Summation of Incoming Signals

Anywhere from 1,000 to 10,000 communication lines form synapses with a typical neuron in your brain. And your brain contains at least *100 billion* neurons! As long as you are alive, this astounding communications network continually hums with messages about doing what it takes to be a human.

At any given moment, a great number of excitatory and inhibitory signals are washing over the input zones of a postsynaptic cell. Some of the signals drive its membrane closer to threshold; others maintain the resting level or drive it away from threshold. *The signals are competing for control of the neuron's membrane.*

All of the synaptic signals are graded potentials. You may have come across the abbreviations **EPSP** and **IPSP**. They refer to an excitatory postsynaptic potential (EPSP), which has a *depolarizing* effect. This simply means it brings the membrane closer to threshold. An inhibitory postsynaptic potential (IPSP) either has a *hyperpolarizing* effect (it drives the membrane away from threshold) or helps maintain the membrane at its resting level.

In a process called **synaptic integration**, competing signals that reach an input zone of a neuron at the same time are summed. This summation process is the means by which signals arriving at any neuron in the body are reinforced or dampened, sent on or suppressed.

Take a look at Figure 34.9. The *yellow* line in this diagram shows how an EPSP of a certain magnitude would register on an oscilloscope screen *if it were acting alone*. The *purple* line shows what the effect would be for a single IPSP. The *red* line shows their effect on a postsynaptic cell membrane when they arrive together, at the same time. In this particular example, synaptic integration pulls the membrane potential away from threshold.

Suppose neurotransmitter molecules from several presynaptic cells reach the neuron's input zone at the same time. The response to the simultaneous molecular bombardment is called *spatial* summation. Or suppose neurotransmitter has been released repeatedly, over a short period, from a single presynaptic cell that has been whipped into a frenzy of excitability by a rapid series of action potentials. Then, the response is called a *temporal* summation.

Removing Neurotransmitter from the Synaptic Cleft

By now, it should be clear that synaptic integration depends on the prompt removal of neurotransmitter from the synaptic cleft. An accumulation of excitatory and inhibitory substances can have lethal effects, as the *Focus* essay makes clear.

Deadly Imbalances at Synapses

a

Clostridium botulinum, an anaerobic bacterium, lives in soil. It is an endospore former, and it can cause the disease *botulism*. When its endospores contaminate improperly stored, preserved, or canned food, they germinate and produce a dangerous toxin, which people can ingest and absorb. The toxin ends up binding to neurons that synapse with muscle cells and block their release of acetylcholine (ACh). Once this happens, muscles cannot contract. They undergo progressive, flaccid paralysis. Death may follow within ten days, usually from respiratory and cardiac failure. Recovery is possible when antitoxins are quickly administered. Extended use of an artificial respirator may also be required.

A related bacterium, *C. tetani*, lives in the gut of horses, cattle, and other grazing animals, even many people. Its endospores also survive in soil, especially manure-enriched types. They can persist for years if sunlight and oxygen don't reach them. They are exceptionally resistant to disinfectants, heat, even boiling water. If they enter the human body through a deep puncture or cut, they can germinate in dead (anaerobic) tissues. The bacteria themselves do not spread out from the dead tissue, and there is no inflammation. But they produce a toxin that can enter the blood or travel along nerves into the spinal cord and brain. There, the toxin acts on interneurons that help control motor neurons in the spinal cord. It stops them from releasing inhibitory neurotransmitters (GABA and glycine) and so frees the motor neurons from normal inhibitory control. The terrible symptoms of the disease *tetanus* are about to begin.

After four to ten days of overstimulation, an infected person experiences generalized muscle stiffness, then muscle spasms. Because muscles cannot be released from contraction, prolonged, spastic paralysis follows. Fists and jaws may stay clenched (the disease is also called lockjaw). The back may arch permanently. Muscle spasms may cause bones to break. When respiratory and cardiac muscles also become paralyzed, death nearly always follows.

Vaccines were not available for soldiers of early wars, when battlefields were littered with bodies of (and manure from) cavalry horses. Figure *a* is a painting of a young victim of a contaminated battle wound, as he lay dying in a military hospital.

Vaccines now prevent nearly all cases of tetanus in the United States. Since 1975, there are fewer than 100 cases a year, mostly among older women. (The vaccine was not available when they were children.) Many retired women dig in the garden and so are at greater risk of infection unless they are vaccinated.

In some developing countries, people cut the umbilical cord of newborns with knives that may be contaminated. If this were not bad enough, they daub mud on the raw stump of the cord. As many as 10 percent of all deaths within the first month of life are attributable to the neurotoxic effect of tetanus.

Normally, some amount of the neurotransmitter simply diffuses out of the cleft. Enzymes also degrade them in the cleft, as when acetylcholinesterase breaks down ACh. Besides this, membrane transport proteins actively pump neurotransmitter molecules back into presynaptic cells or into neighboring neuroglial cells.

Cocaine produces euphoria (intense pleasure) largely by blocking the neurotransporter uptake for dopamine. The dopamine that lingers in synaptic clefts just keeps on stimulating target cells—with disastrous effects, as you will read in the introduction to the next chapter.

Synaptic integration is the moment-by-moment combining of excitatory and inhibitory signals acting on a postsynaptic cell.

With this summation process, messages traveling through the nervous system can be reinforced or downplayed, sent onward or suppressed. The process is essential for normal body functioning.

34.7 PATHS OF INFORMATION FLOW

Blocks and Cables of Neurons

Through synaptic integration, signals arriving at any neuron in the body can be reinforced or dampened, sent on or suppressed. What determines the direction in which a given signal will travel? That depends on the organization of neurons in different body regions.

For example, your brain deals with many of its 100 billion neurons in a manner analogous to block parties. Hundreds or thousands of neurons are organized into regional blocks that receive excitatory and inhibitory signals, integrate them, and then send out signals in response. In some cases, neurons are organized as *divergent* circuits, as when they fan out from one block into another. In other cases, they form *convergent* circuits, with signals from many neurons sent on to just a few. In still other cases, neurons synapse back on themselves, repeating signals among themselves like gossip that just won't go away. Such neurons form *reverberating* circuits. Among them are the ones that make eye muscles twitch rhythmically as you sleep.

The axons of many other neurons are bundled into cables that provide long-distance communication between your brain and spinal cord and the rest of the body. The cables called **nerves** consist of the axons of sensory neurons, motor neurons, or both, bundled in parallel array in connective tissues (Figure 34.10). Within the brain and spinal cord, such bundled-together axons are called nerve "tracts."

Reflex Arcs

The chapter to follow provides splendid examples of the workings of your nervous system. In anticipation of that chapter, let's start thinking about its reflex arcs, one of the simplest paths of information flow.

The sensory and motor neurons of many nerves take part in reflexes, which are simple, stereotyped movements made in response to sensory stimulation. In the simplest **reflex arc**, sensory neurons directly synapse on motor neurons. The stretch reflex is an example; it works to contract a muscle when that muscle has been stretched.

Hold out a large bowl and keep it stationary when someone puts peaches into it. As the peaches add weight to the bowl and your hand starts to drop, a muscle in your arm (the biceps) is stretched. The stretching activates certain receptors in the muscle. The stretch-sensitive receptors are part of **muscle spindles**—sensory organs in which small, specialized cells are enclosed in a sheath that runs parallel with the muscle. The receptors are input zones of sensory neurons that synapse with motor neurons in the spinal cord (Figure 34.11). Axons of the motor neurons lead back to the stretched muscle, and action potentials that reach the axon endings trigger the release of ACh, which initiates contraction. Continued receptor activity excites the motor neurons further, allowing them to maintain your hand's position.

In the vast majority of reflexes, sensory neurons make connections with a number of interneurons, which then activate or suppress the motor neurons necessary for a coordinated response. An example is the withdrawal reflex, a rapid pulling away from an unpleasant or harmful stimulus. If you have ever accidentally touched a hot stove, you know this reflex action can be completed even before you are consciously aware that it has been going on.

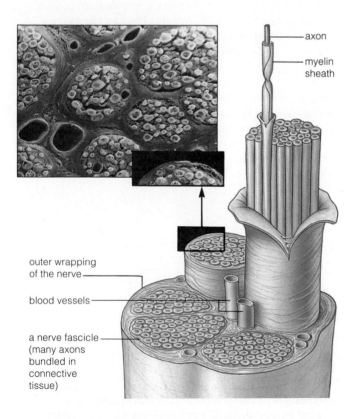

axon

myelin sheath

outer wrapping of the nerve

blood vessels

a nerve fascicle (many axons bundled in connective tissue)

Figure 34.10 Structure of a nerve. Axons in the nerve are bundled together inside wrappings of connective tissue.

Just as there is no such thing as one ant at a picnic, there is no such thing as one neuron acting alone.

In the human body, billions of neurons are organized into many information-processing blocks and long-distance communication cables.

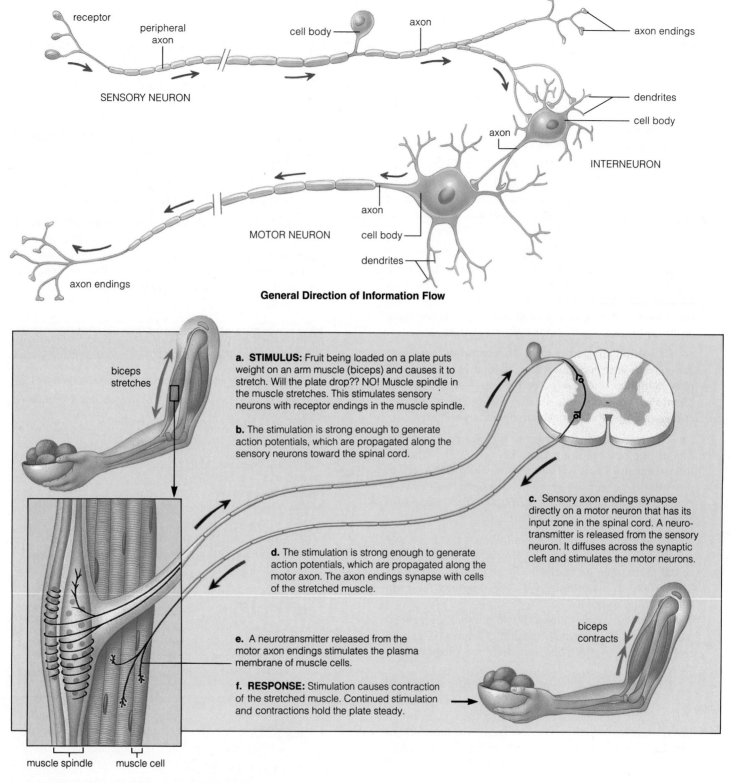

General Direction of Information Flow

a. STIMULUS: Fruit being loaded on a plate puts weight on an arm muscle (biceps) and causes it to stretch. Will the plate drop?? NO! Muscle spindle in the muscle stretches. This stimulates sensory neurons with receptor endings in the muscle spindle.

b. The stimulation is strong enough to generate action potentials, which are propagated along the sensory neurons toward the spinal cord.

c. Sensory axon endings synapse directly on a motor neuron that has its input zone in the spinal cord. A neurotransmitter is released from the sensory neuron. It diffuses across the synaptic cleft and stimulates the motor neurons.

d. The stimulation is strong enough to generate action potentials, which are propagated along the motor axon. The axon endings synapse with cells of the stretched muscle.

e. A neurotransmitter released from the motor axon endings stimulates the plasma membrane of muscle cells.

f. RESPONSE: Stimulation causes contraction of the stretched muscle. Continued stimulation and contractions hold the plate steady.

Figure 34.11 General direction of information flow through vertebrate nervous systems. As the *blue* arrows indicate, sensory nerves relay information *into* the spinal cord and brain, where their neurons synapse with interneurons. Signals are received and integrated by interneurons *within* the spinal cord and brain. Interneurons also relay information to motor neurons, which carry information *away* from the spinal cord and brain.

(**a–f**) Organization of nerves in a reflex arc dealing with muscle stretching. Inside a skeletal muscle, stretch-sensitive receptors of a sensory neuron are located in muscle spindles. Stretching disturbs them and generates action potentials that travel down to the axon endings. These synapse with a motor neuron that has axons leading back to the stretched muscle. The motor neuron can initiate muscle contraction.

SUMMARY

1. The nervous system senses, interprets, and issues commands for responses to specific aspects of the environment. In nearly all animals, its communication lines consist of sensory neurons, interneurons, and motor neurons (which activate muscle and gland cells).

2. A stimulus is a form of energy that the body detects by specific receptors of sensory neurons. Information about a stimulus travels as electrical and chemical signals through the nervous system.

3. Dendrites and the cell body are the input zones of a neuron. Arriving signals may spread to a trigger zone (the start of an axon, most often). They may give rise to an action potential, which travels to the neuron's output zone (axon endings). Action potentials are the basis of messages sent through the nervous system.

4. In an undisturbed neuron, concentration gradients for sodium and potassium ions hold steady across the plasma membrane. They are the basis of the resting membrane potential—a steady difference in electric charge (a voltage difference) across the membrane.

5. The cytoplasmic fluid next to the plasma membrane of a resting neuron is negatively charged with respect to interstitial fluid just outside the membrane. An action potential is an abrupt, short-lived reversal in the voltage difference across the membrane that occurs in response to adequate stimulation.

6. Action potentials arise as follows:

a. At the input zone of a resting neuron, stimulation of the membrane gives rise to local, graded potentials. These vary in magnitude and do not spread far from the stimulation point.

b. A disturbance caused by a number of graded potentials or by a rapid succession of them may spread to a neuron's trigger zone and drive the membrane to threshold.

c. In the trigger zone, an ever increasing number of gated sodium channels open until an action potential occurs. (The voltage difference across the resting membrane briefly and abruptly reverses.) The accelerated ion flow is self-propagating; it triggers an action potential in patch after patch of plasma membrane, down to the neuron's output zone.

7. Between action potentials, the gradients are *established* as some ions leak across the membrane, through channel proteins. The gradients are *restored and maintained* by sodium-potassium pumps.

8. Action potentials usually are confined to individual neurons. At a neuron's output zone, the arrival of action potentials triggers release of neurotransmitters. These signaling molecules diffuse across chemical synapses, where the neuron forms a junction with another neuron, a muscle cell, or a gland cell.

9. A neurotransmitter may excite that cell's membrane (drive it closer to threshold) or inhibit it (drive it away from threshold).

10. Integration is the moment-by-moment combining of all signals, excitatory and inhibitory, acting at all synapses on a neuron. It is the means by which information is played down or reinforced, suppressed or sent on to other neurons of the nervous system.

11. The human body has many billions of neurons, many of which are organized into information-processing blocks in the brain and as components of cablelike nerves that carry information to and from the spinal cord and brain, to all body regions.

Review Questions

1. Define sensory neuron, interneuron, and motor neuron. *561*

2. In the following diagram, label the terms dealing with chemical synapses. *567*

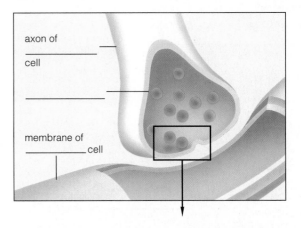

axon of _____

_____ cell

_____

membrane of _____ cell

_____

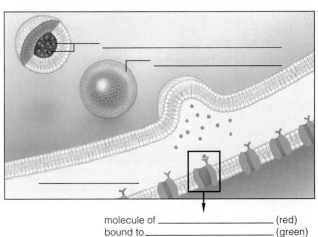

_____

molecule of _____ (red)
bound to _____ (green)

3. Label the input and output zones of this motor neuron. *562*

4. Two major concentration gradients exist across a neural membrane. What substances are involved, and how are the gradients maintained? *562–563*

5. Distinguish between an action potential and a graded potential. What is meant by "all-or-nothing" messages? *562, 564–565*

6. What is a synapse? Explain the difference between an excitatory and an inhibitory synapse. Define synaptic integration. *566–568*

9. Integration is the _____ .
   a. self-propagation of ion flow along a neuron
   b. trigger that releases neurotransmitters
   c. combining of all signals acting at all synapses on a neuron
   d. collection of all information about several stimuli

10. Match the terms with the suitable descriptions.
   _____ synaptic integration
   _____ neuromodulator
   _____ nerve
   _____ muscle spindle
   _____ graded, local potential
   _____ action potential
   _____ reverberating circuit

   a. block of neurons in brain that repeat signals among themselves
   b. cordlike bundle of axons to and from brain and spinal cord and the rest of the body
   c. spatial-temporal summation of signals arriving at a neuron
   d. magnifies or reduces effect of neurotransmitter
   e. occurs at neuron's trigger zone
   f. occurs at neuron's input zone
   g. stretch-sensitive receptor

Self-Quiz *(Answers in Appendix IV)*

1. An action potential is _____ across the plasma membrane of a neuron.
   a. a steady voltage difference
   b. an abrupt reversal in the voltage difference
   c. a short-lived reversal in the voltage difference
   d. both b and c

2. Action potentials occur when _____ .
   a. a neuron is adequately stimulated
   b. sodium gates open in an accelerating way
   c. sodium-potassium pumps kick into action
   d. both a and b

3. Compared to its surroundings, an undisturbed neuron carries a slight _____ charge inside the plasma membrane.
   a. positive   c. graded, local
   b. negative   d. both b and c

4. The resting membrane potential is maintained by _____ .
   a. ion leaks   c. neurotransmitters
   b. ion pumps   d. both a and b

5. Neurotransmitters diffuse across a _____ .
   a. chemical synapse   c. myelin sheath
   b. channel protein   d. both a and b

6. An action potential lasts only briefly because _____ at the membrane region where it occurred.
   a. gates for sodium open, gates for potassium close
   b. gates for sodium close, gates for potassium open
   c. sodium-potassium pumps restore gradients
   d. both b and c

7. _____ is an example of a neurotransmitter; _____ is an example of a neuromodulator.
   a. Serotonin; an endorphin   c. ACh; an endorphin
   b. Serotonin; GABA   d. both a and c

8. A nerve may consist of bundled-together axons of _____ .
   a. sensory neurons   c. sensory and motor neurons
   b. motor neurons   d. all of the above

Selected Key Terms

acetylcholine *566*
action potential *562*
axon *562*
chemical synapse *566*
dendrite *562*
excitatory postsynaptic potential (EPSP) *568*
inhibitory postsynaptic potential (IPSP) *568*
interneuron *561*
motor neuron *561*
muscle spindle *570*
myelin sheath *566*

nerve *570*
neuroglial cell *561*
neuromodulator *568*
neuron *561*
neurotransmitter *566*
reflex arc *570*
resting membrane potential *562*
sensory neuron *561*
sodium-potassium pump *563*
synaptic integration *568*
threshold level *564*

Readings

Berne, R., and M. Levy (editors). 1988. *Physiology*. Second edition. St. Louis: Mosby. Section II is an authoritative introduction to neural functioning.

Dunant, Y., and M. Israel. April 1985. "The Release of Acetylcholine." *Scientific American* 252(4):58–83. Experiments showing how this major neurotransmitter functions.

Lent, C., and M. Dickenson. June 1988. "The Neurobiology of Feeding Behavior in Leeches." *Scientific American* 258(6):98–103. Describes the relationship between serotonin (a transmitter substance) and feeding behavior in an invertebrate.

Nicholls, J., A. Martin, and B. Wallace. 1992. *From Neuron to Brain*. Third edition. Sunderland, Massachusetts: Sinauer.

35 INTEGRATION AND CONTROL: NERVOUS SYSTEMS

Why Crack the System?

Suppose your biology instructor asks you to volunteer for an experiment. A microchip will have to be implanted in your brain. It will make you feel really good. But it may mess up your health, lop ten years off your life, and destroy a good part of your brain. Your behavior will change for the worse, so you may have trouble completing school, getting or keeping a job, even having a normal family life.

The longer the chip is implanted, the less you will want to give it up. You won't get paid. You pay the experimenter—first at bargain rates, then a little more each week. The chip is illegal. If you get caught using it, you and the experimenter will go to jail.

Sometimes Jim Kalat, a professor at North Carolina State University, proposes this experiment (which of course is hypothetical). Hardly any students volunteer. Then he substitutes drug for "microchip" and dealer for "experimenter"—and an amazing number come forward! Like 30 million other Americans, the "volunteers" seem ready to engage in self-destructive uses of drugs that alter emotional and behavioral states.

The destruction shows up in unexpected places. Each year, for instance, about 300,000 newborns are already addicted to crack—thanks to their addicted mothers. *Crack* is a cheap form of cocaine. It causes relentless stimulation of brain regions that govern our

sense of "pleasure." It dampens normal urges to eat and sleep, and blood pressure rises. Elation and sexual desire intensify. In time, brain cells that produce the stimulatory chemicals can't keep up with the incessant demands. The chemical vacuum makes crack users frantic, then profoundly depressed. Only crack makes them "feel good" again.

Addicted newborns cannot know all of this, of course. They can only quiver with "the shakes." Their brain neurons are overstimulated, and this makes the newborns chronically irritable. Their body is small, abnormally so. As they were developing inside their mother's uterus, their body tissues simply were not provided with enough oxygen and nutrients—crack causes blood vessels to constrict. Paradoxically, even though crack babies are abnormally fussy, they do not respond to rocking and other kinds of normal stimulation. It may be a year or more before they recognize their mother. Without treatment they are likely to grow up as emotionally unstable children, prone to aggressive outbursts and stony silences.

Each of us possesses a body of great complexity. Its architecture, its functioning are legacies of millions of years of evolution. Its nervous system is unparalleled in the living world. One of its most astonishing products is language—the encoding of shared experiences of groups of individuals in time and space. Through the evolution of our nervous system, the sense of history was born, and the sense of destiny. Through this system we can ask how we have come to be what we are, and where we are headed from here. Perhaps the sorriest consequence of drug abuse is its implicit denial of this legacy—the denial of self when we cease to ask, and cease to care.

KEY CONCEPTS

1. Nervous systems provide a means of sensing specific information about external and internal conditions, integrating that information, and issuing commands for response from the body's effectors (muscles and glands).

2. The simplest nervous systems are the nerve nets of sea anemones and other cnidarians. The vertebrate nervous system shows pronounced cephalization and bilateral symmetry. It includes a brain, spinal cord, and many paired nerves. The brain is complex, with centers for receiving, integrating, processing, and responding to information.

3. The oldest parts of the vertebrate brain provide reflex control over breathing, blood circulation, and other essential functions. During the evolution of certain vertebrates, the brain expanded in complexity and its newer regions appropriated more and more control over the ancient reflex functions.

4. Nervous systems coevolved with sensory organs, such as eyes, and motor structures, such as legs and wings. The coevolution of nervous, sensory, and motor systems paved the way for increasingly intricate behavior.

Figure 35.1 Owners of an evolutionary treasure—a complex brain that is the foundation for our memory and reasoning, and our future.

35.1 INVERTEBRATE NERVOUS SYSTEMS

Life-Styles and Nervous Systems

Even if you stare at it for hours on end, a sea urchin will never dazzle you with spectacular bursts of speed or precision acrobatics. Because it seems only a bit livelier than the pincushion it resembles, you might suspect this animal is brainless, and in fact it is. However, sea urchins do have a nervous system, even if it is a decentralized one.

At this point in the book, you know that nearly all animals have some type of a **nervous system**, with communication lines of neurons (or simpler kinds of nerve cells) oriented relative to one another in signal-conducting and information-processing pathways. At the minimum, its component cells detect changing conditions outside and inside the body, then elicit suitable responses from muscle and gland cells.

To appreciate any animal's nervous system, you have to ask, *What does the animal do?* Many sea urchins spend most of their lives moving about slowly on little tube feet, scraping bits of algae off rocks with a marvelously toothy feeding apparatus (page 438 and Figure 35.2). Their nervous system includes three sets of nerves, which deal mainly with sensory functions, motor functions, and (you guessed it) getting food into the gut. The nerves interconnect with one another by a **nerve net**, a diffuse mesh of nerve cells intimately associated with epithelial tissue. Information flow through the nerve net is not highly directional. Signals travel diffusely, in all directions, from a point of stimulation. You might not find this type of nervous system impressive, but if you did little more than inch about eating algae, it would be quite sufficient.

Regarding the Nerve Net

Animals first evolved in the seas, and it is in the seas that we still find animals with the simplest nervous systems. They are sea anemones, jellyfishes, and other cnidarians (Chapter 26). These invertebrates show *radial* symmetry. Their body parts are arranged about a central axis, much like spokes of a bike wheel.

Like sea urchins, cnidarians have a nerve net (but no nerves). It extends through the body and is equally responsive to food or danger coming from any direction. Its nerve cells interact with sensory and contractile cells of the epithelium as part of **reflex pathways**. In such pathways, recall, sensory stimulation directly triggers simple, stereotyped movements (page 570).

For example, in jellyfishes, reflex pathways permit slow swimming movements, and they keep the body right-side up. In all cnidarians, a reflex pathway con-

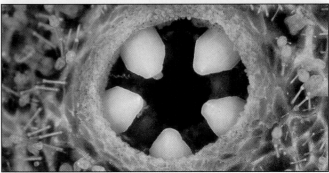

Figure 35.2 A spiny sea urchin, right-side up and flipped over to show its toothy, all-important mouth.

cerned with feeding behavior extends from sensory receptors present in the tentacles, along nerve cells, to contractile cells around the mouth. Figure 35.3 illustrates the cells involved in such a pathway.

On the Importance of Having a Head

Flatworms are the simplest animals with a bilateral nervous system (Figure 35.4). *Bilateral* symmetry means having equivalent parts on the left and right sides of the body's midsagittal plane. (Imagine sliding down a staircase banister that turns into a razor and you may never forget the location of the midsagittal plane.) Both sides have the same array of muscles that move the body forward. Both have the same array of nerves to control the muscles, and so on.

The ladderlike nervous system of flatworms includes two nerve cords (bundled-together axons of neurons). Some systems include **ganglia** (singular, ganglion). And some ganglia form a two-part brainlike structure at the head end. They coordinate signals from paired sensory organs, including two eyespots, and provide some control over the nerve cords.

Did bilateral nervous systems evolve from nerve nets? Maybe. Consider that some cnidarian life cycles include a self-feeding larval stage, called a **planula**.

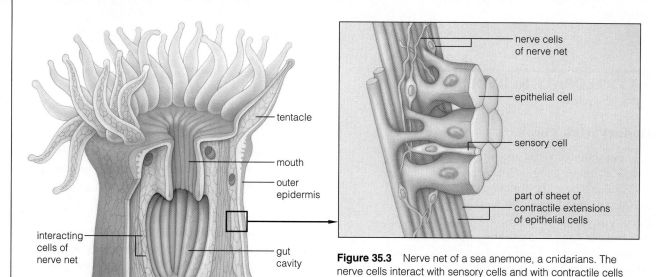

tentacle

mouth

outer epidermis

interacting cells of nerve net

gut cavity

nerve cells of nerve net

epithelial cell

sensory cell

part of sheet of contractile extensions of epithelial cells

Figure 35.3 Nerve net of a sea anemone, a cnidarians. The nerve cells interact with sensory cells and with contractile cells that are components of a sheetlike epithelium between the outer epidermis and a jellylike middle layer (mesoglea) of the body wall.

Figure 35.4 Bilateral nervous systems of a few invertebrates. The sketches are not to scale relative to one another.

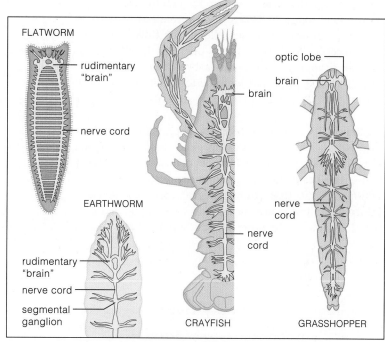

FLATWORM

rudimentary "brain"

nerve cord

EARTHWORM

rudimentary "brain"

nerve cord

segmental ganglion

brain

nerve cord

CRAYFISH

optic lobe

brain

nerve cord

GRASSHOPPER

Like flatworms, a planula has a flattened body and uses cilia to swim or crawl about:

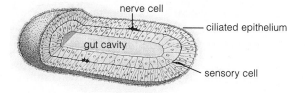

nerve cell

ciliated epithelium

gut cavity

sensory cell

Imagine a few ancient planulas crawling about on the seafloor in Cambrian times. By chance, mutations of regulatory genes blocked their transformation into adults but had no effect on maturation of reproductive organs. (As you read on page 456, this does happen in the larvae of certain salamanders and other animals.) Such planulas kept on crawling, they reproduced—and so they passed on the mutated genes responsible for their permanently forward mobility.

Planulalike animals that crawled far enough could put themselves in new, potentially dangerous or splendid situations. Having sensory cells located at the trailing end of the body wouldn't be helpful. But having them concentrated at the leading end would allow rapid, effective responses to stimuli. Possibly, natural selection favored planulas in which sensory cells became concentrated at the leading end. Cephalization (formation of a head region) and bilateral symmetry may have started in such a way.

As you will see next, *patterns of bilateral symmetry and cephalization have echoes in the paired nerves and mus-* *cles, paired sensory structures, and paired brain center of yourself and all other vertebrates.*

---

1. For any animal, the organization of its nervous system and the distribution of sensory organs are correlated with body symmetry.

2. Radial nervous systems are decentralized; they can respond to stimuli coming from any direction.

3. Bilateral nervous systems have centralized masses of nervous tissue, ranging from ganglia to brains, that connect with communication lines (nerves) that are arranged similarly on both sides of the body.

Evolutionary High Points

Hundreds of millions of years ago, the first fishlike vertebrates were evolving (page 450). A bony, segmented column was taking over the function of the notochord, a long rod of stiffened tissue that worked with muscles to bring about movements. Above the notochord, a hollow, tubular nerve cord was evolving. It was the forerunner of the vertebrate spinal cord and brain (Figure 35.5*a*). These developments were the foundation for new life-styles. In time, divergences from filter-feeding and scavenging ancestors led to swift, jawed predators of the seas.

At first, simple reflex pathways predominated. Sensory neurons that were tripped into action synapsed on motor neurons, which directly stimulated muscles to contract. No other neurons altered the flow of signals from reception of a stimulus to the response. However, in the changing world of fast-moving vertebrates, predators and prey that were better equipped to sense and respond to one another's presence had the competitive edge.

Especially among vertebrates that invaded the land, the senses of smell, hearing, and balance became keener. Their motor systems evolved in specialized ways. The brain itself became variably thickened with nervous tissue that could integrate the rich sensory information and issue commands for complex, coordinated responses. In time, the thickening brain tissues

became divided into three functionally specialized parts, called the forebrain, midbrain, and hindbrain.

The oldest parts of the vertebrate brain still deal with reflex coordination of breathing and other vital functions. Now, however, interneurons synapse on the sensory and motor neurons of ancient pathways. They form intricate synapses in the most recent additions to the brain. There, interneurons store and compare information about experiences. They come up with possibilities for brand-new responses. They give our own species the capacity to reason, remember, and learn.

A nerve cord persists in all vertebrate embryos. We call it the "neural tube." As an embryo grows and develops, the neural tube undergoes expansion and regional modification into the brain and spinal cord (Figure 35.5*b*). The spinal cord becomes enclosed within the vertebral column. Adjacent tissues in the embryo give rise to nerves that thread through all body regions and connect with the spinal cord and brain.

Functional Divisions of the Human Nervous System

Figure 35.6 gives you a sense of how intricate the communication lines have become in the human nervous system. At least 100 billion neurons interact in the brain alone.

Think of the system's component neurons as belonging to central or peripheral regions. Its interneurons are confined to the **central nervous system**, the spinal cord and brain. Its sensory and motor neurons belong to the **peripheral nervous system**, the commu-

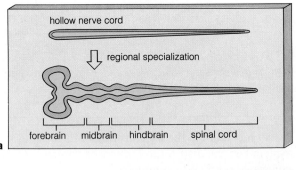

a

Figure 35.5 (**a**) Diagram, dorsal view, of the evolution of the vertebrate nerve cord into a spinal cord and specialized brain regions. (**b**) Some stages in the embryonic development of the human brain.

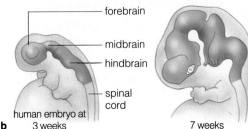

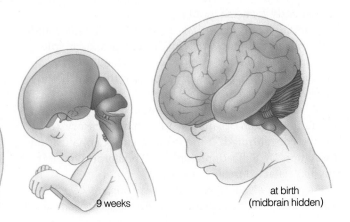

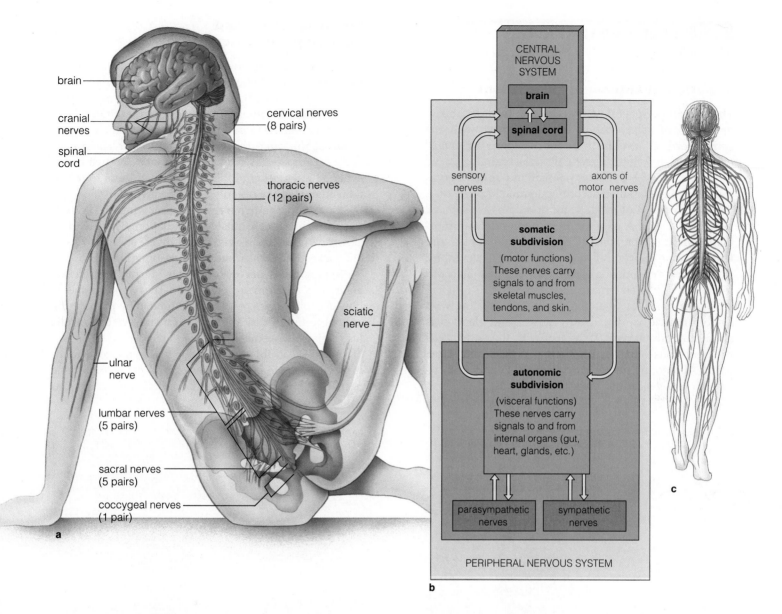

brain

cranial nerves

spinal cord

cervical nerves (8 pairs)

thoracic nerves (12 pairs)

sciatic nerve

ulnar nerve

lumbar nerves (5 pairs)

sacral nerves (5 pairs)

coccygeal nerves (1 pair)

a

CENTRAL NERVOUS SYSTEM

brain

spinal cord

sensory nerves

axons of motor nerves

somatic subdivision

(motor functions) These nerves carry signals to and from skeletal muscles, tendons, and skin.

autonomic subdivision

(visceral functions) These nerves carry signals to and from internal organs (gut, heart, glands, etc.)

parasympathetic nerves

sympathetic nerves

PERIPHERAL NERVOUS SYSTEM

b

c

nication lines that thread through the rest of the body and carry signals into and out of the central nervous system.

1. The coevolution of nervous, sensory, and motor systems was central to the development of more intricate life-styles among the vertebrates.

2. The vertebrate nervous system is so complex that the structure and functions of its central and peripheral regions are often described separately.

3. The central nervous system consists of the brain and spinal cord.

4. The peripheral nervous system consists of nerves that thread through the rest of the body and carry signals into and out of the central nervous system.

Figure 35.6 (**a**) Human nervous system, showing the brain, spinal cord, and some of the major peripheral nerves. The system also includes twelve pairs of cranial nerves that originate from the different regions of the brain stem. Similar arrangements exist in other vertebrates.

(**b**–**c**) Functional divisions of the human nervous system. The *central* nervous system is color-coded *blue*, the somatic nerves *green*, and the autonomic nerves *red*.

Sometimes the nerves carrying sensory input to the central nervous system are said to be *afferent* (a word meaning "to bring to"). The ones carrying motor output away from the central nervous system to muscles and glands are *efferent* ("to carry outward").

35.3 PERIPHERAL NERVOUS SYSTEM

Somatic and Autonomic Subdivisions

The human peripheral nervous system includes thirty-one pairs of *spinal* nerves, which connect with the spinal cord. It also includes twelve pairs of *cranial* nerves, which connect directly with the brain.

As Figure 35.6 indicates, cranial and spinal nerves can be further classified according to function. The ones that help move your head, trunk, and limbs are **somatic nerves**. The sensory axons of these nerves deliver information from receptors in the skin, skeletal muscles, and tendons into the central nervous system. Their motor axons deliver commands from the brain and spinal cord to the body's skeletal muscles.

By contrast, spinal and cranial nerves dealing with smooth muscle, cardiac (heart) muscle, and gland cells are **autonomic nerves**. They deal with the body's visceral portions—that is, its internal organs and structures. Autonomic nerves carry signals to and from its smooth muscle, cardiac muscle, and glands.

The Sympathetic and Parasympathetic Nerves

There are two categories of autonomic nerves. We call them parasympathetic and sympathetic. They normally work antagonistically, with the signals from one opposing those of the other. Both types carry excitatory and inhibitory signals to internal organs. Often their signals arrive at the same time at muscle cells or gland cells and so compete for control over them. In such cases, synaptic integration at the cellular level leads to minor adjustments in the organ's level of activity.

Parasympathetic nerves dominate when the body is not receiving much outside stimulation. They tend to slow down the body overall and divert energy to basic "housekeeping" tasks, such as digestion (Figure 35.7).

Sympathetic nerves dominate during heightened awareness, excitement, or danger. They tend to shelve housekeeping tasks. At the same time, they prepare the animal to fight or escape (when threatened) or to frolic (as in play and sexual behavior).

As you read this, sympathetic nerves are commanding your heart to beat a bit faster, and parasympathetic nerves are commanding it to beat a bit slower. Integra-

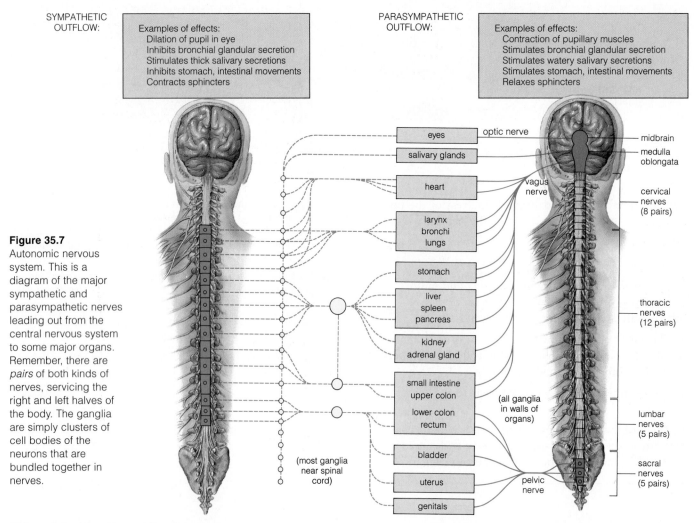

Figure 35.7
Autonomic nervous system. This is a diagram of the major sympathetic and parasympathetic nerves leading out from the central nervous system to some major organs. Remember, there are *pairs* of both kinds of nerves, servicing the right and left halves of the body. The ganglia are simply clusters of cell bodies of the neurons that are bundled together in nerves.

SYMPATHETIC OUTFLOW:

Examples of effects:
 Dilation of pupil in eye
 Inhibits bronchial glandular secretion
 Stimulates thick salivary secretions
 Inhibits stomach, intestinal movements
 Contracts sphincters

PARASYMPATHETIC OUTFLOW:

Examples of effects:
 Contraction of pupillary muscles
 Stimulates bronchial glandular secretion
 Stimulates watery salivary secretions
 Stimulates stomach, intestinal movements
 Relaxes sphincters

eyes
salivary glands
heart
larynx bronchi lungs
stomach
liver spleen pancreas
kidney adrenal gland
small intestine upper colon
lower colon rectum
bladder
uterus
genitals

optic nerve
vagus nerve
(all ganglia in walls of organs)
pelvic nerve

midbrain
medulla oblongata
cervical nerves (8 pairs)
thoracic nerves (12 pairs)
lumbar nerves (5 pairs)
sacral nerves (5 pairs)

(most ganglia near spinal cord)

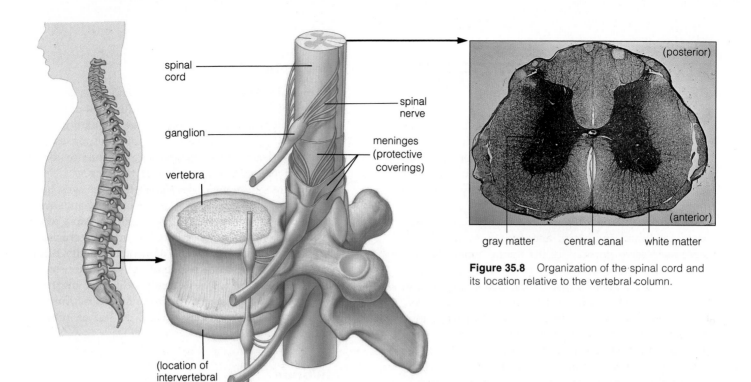

spinal cord

ganglion

vertebra

spinal nerve

meninges (protective coverings)

(location of intervertebral disk)

(posterior)

(anterior)

gray matter   central canal   white matter

Figure 35.8 Organization of the spinal cord and its location relative to the vertebral column.

tion of these opposing signals influences the heart rate. If something scares or excites you, parasympathetic input to your heart drops. Sympathetic nerves cause the release of epinephrine, a signaling molecule that makes your heart beat faster and makes you breathe faster and sweat. In this state of intense arousal, you are primed to fight (or play) hard or to get away fast. Hence the name **fight-flight response**.

When the stimulus for the fight-flight response stops, sympathetic activity may decrease abruptly and parasympathetic activity may rise suddenly. You might see this "rebound effect" after someone has been instantly mobilized to rush onto a highway to save a child from an oncoming car. The person may well faint as soon as the child has been swept out of danger.

The peripheral nervous system consists of the communication lines to and from the brain and spinal cord.

Its somatic nerves deal with skeletal muscle movements. Its autonomic nerves deal with the functions of internal organs, such as the heart and glands.

35.4 CENTRAL NERVOUS SYSTEM: THE SPINAL CORD

By definition, the **spinal cord** is a vital expressway for signals between the peripheral nervous system and the brain. It also is a center where some direct reflex connections are made between sensory and motor neurons.

The stretch reflex, illustrated in Figure 30.10*b*, arcs through the spinal cord in this manner.

As Figure 35.8 shows, the spinal cord threads through a canal formed by bones of the vertebral column. The bones and their attached ligaments protect the cord. So do the **meninges**, three tough, tubelike coverings around the spinal cord and brain. These coverings are not impervious to attack, however. *Meningitis*, an often-fatal disease, results when viruses or bacteria infect them. Disease symptoms start with severe headaches, fever, a stiff neck, nausea, and vomiting.

Signals travel rapidly up and down the spinal cord in bundles of sheathed axons. The glistening myelin sheaths of these axons distinguish the cord's "white matter" from its "gray matter" (Figure 35.8). The gray matter consists of dendrites, cell bodies of neurons, and neuroglial cells. It plays an important role in controlling the reflexes for limb movements (such as walking) and organ activity (such as bladder emptying).

Experiments with frogs give evidence of these pathways. Between the frog's spinal cord and brain are neurons that deal with straightening the legs after they have been bent. If these neurons are severed at the base of the brain, the legs become paralyzed—but only for about a minute. The so-called *extensor* reflex pathways in the spinal cord recover quickly and have the frog hopping around in no time. Recovery is minimal in humans and other primates—the vertebrates with the greatest cephalization.

The spinal cord is a vital expressway for signals between the peripheral nerves and the brain. Some of its interneurons also exert direct control over certain reflex pathways.

35.5 CENTRAL NERVOUS SYSTEM: THE VERTEBRATE BRAIN

The spinal cord merges with the **brain**, the body's master control panel. The brain receives, integrates, stores, and retrieves information. It also coordinates responses by intricately adjusting activities throughout the body. Like the spinal cord, the brain is protected by bones and membranes.

Look once more at Figures 27.8 and 35.5. The vertebrate forebrain, midbrain, and hindbrain form from three successive portions of the neural tube. The nervous tissue that evolved first in all three portions is called the **brain stem**. It is still recognizable in the adult brain, and it still contains many simple, basic reflex centers. Over evolutionary time, expanded layers of gray matter developed from the brain stem. We can correlate these more recent additions with increasing reliance on three major sensory organs: the nose, ears, and eyes. The forebrain and possibly parts of the cerebellum (a hindbrain region) contain the newest additions of gray matter. Figure 35.8 provides an overview of these interesting developments.

Hindbrain

The medulla oblongata, cerebellum, and pons are components of the hindbrain. The **medulla oblongata** has reflex centers for respiration, blood circulation, and other vital tasks. It coordinates motor responses and complex reflexes, such as coughing. It influences other brain regions that help you sleep or wake up.

The **cerebellum** integrates signals from the eyes, muscle spindles, and ears as well as motor commands from the forebrain. It helps control motor dexterity. More recent expansions of the human cerebellum may be crucial in language and some other forms of mental dexterity. Bands of axons extend from each side of the cerebellum to the **pons** (meaning bridge). Although lodged in the brain stem, the pons is a major traffic center for information passing between the cerebellum and the higher integrating centers of the forebrain.

Midbrain

The midbrain coordinates reflex responses to sights and sounds. The **tectum**, a more recent layering of gray matter, forms the midbrain's "roof." In fishes and amphib-

Divisions	Main Components	Some Functions
FOREBRAIN	Cerebrum	Two cerebral hemispheres. Centers for coordinating sensory and motor functions, for memory, and for abstract thought. Most complex coordinating center; intersensory association, memory circuits
	Olfactory lobes	Relaying of sensory input from the nose to olfactory structures of cerebrum
	Limbic system	Scattered brain centers. With hypothalamus, coordination of skeletal muscle and internal organ activity underlying emotional expression
	Thalamus	Major coordinating center for sensory signals; relay station for sensory impulses to cerebrum
	Hypothalamus	Neural-endocrine coordination of visceral activities (e.g., solute-water balance, temperature control, carbohydrate metabolism)
	Pituitary gland	"Master" endocrine gland (controlled by hypothalamus). Control of growth, metabolism, etc.
	Pineal gland	Control of some circadian rhythms; role in mammalian reproductive physiology
MIDBRAIN	Tectum	Largely reflex coordination of visual, tactile, auditory input; contains nerve tracts ascending to thalamus, descending from cerebrum
HINDBRAIN	Pons	"Bridge" of transverse nerve tracts from cerebrum to both sides of cerebellum. Also contains longitudinal tracts connecting forebrain and spinal cord
	Cerebellum	Coordination of motor activity underlying limb movements, maintaining posture, spatial orientation
	Medulla oblongata	Contains tracts extending between pons and spinal cord; reflex centers involved in respiration, cardiovascular function, gastric secretion, etc.

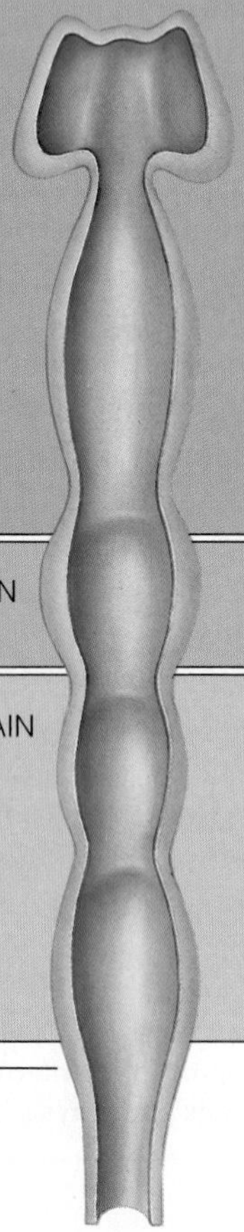

anterior end of spinal cord

Figure 35.9 Overview of the main subdivisions of the brain, correlated with a simplified diagram of the neural tube at an early stage of development. (Compare Figure 35.5.)

ians, the tectum is a center for coordinating nearly all sensory input and initiating motor responses. A frog surgically deprived of its highest brain center but with its tectum intact still does just about everything frogs do.

In most vertebrates—but not in mammals—the midbrain includes a pair of brain centers called optic lobes, which deal with sensory input from the eyes. In mammals, the eyes deserted the tectum, so to speak, and they formed major connections with higher integrating centers in the forebrain. The mammalian tectum has been reduced to a reflex center that swiftly relays sensory signals to those centers.

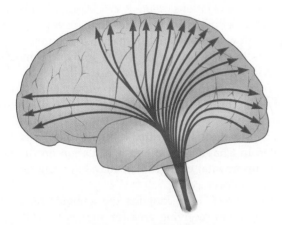

Figure 35.10 Communication pathways of the reticular formation, a diffuse, evolutionarily ancient network of neurons that extends from the anterior end of the spinal cord on up to the highest integrative centers of the cerebral cortex.

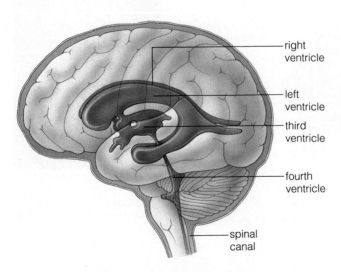

- right ventricle
- left ventricle
- third ventricle
- fourth ventricle
- spinal canal

Figure 35.11 Cerebrospinal fluid (*blue*) in the human brain. This extracellular fluid surrounds and cushions the spinal cord and the brain. It also fills four interconnected cavities (cerebral ventricles) within the brain and the spinal cord's central canal.

Forebrain

For much of vertebrate history, the sense of smell was most important for survival, and an olfactory lobe dominated the forebrain. It dealt mainly with odors emanating from predators, prey, and potential mates. The forebrain itself consisted of two rather small, paired outgrowths of the brain stem that integrated olfactory input and certain responses to it. In time those outgrowths began to expand spectacularly, especially during and after the vertebrate invasion of land. We now call them the two hemispheres of the **cerebrum**.

Another forebrain center, the **thalamus**, evolved as the coordinating center for sensory input and as a relay station for signals flowing into the cerebrum. As it does today, the **hypothalamus** (a subregion of the thalamus) monitored internal organs and influenced behavior related to organ activities, such as thirst, hunger, and sex. The last sections of this chapter provide a closer look at these interesting components of the forebrain.

The Reticular Formation

By now, you may be thinking that the brain is tidily subdivided into three main regions. This is not the case. An evolutionarily ancient mesh of interneurons extends from the uppermost part of the spinal cord, on through the brain stem, and up into the higher integrating centers of the cerebral cortex (Figure 35.10). This major network of interneurons is the **reticular formation**. It persists as a low-level pathway to motor centers of the medulla oblongata and spinal cord. It also activates centers in the cerebral cortex—and so helps govern activities in the nervous system as a whole.

Brain Cavities and Canals

The hollow neural tube persists in adults as a continuous system of fluid-filled cavities and canals. Inside this system is a clear extracellular fluid that bathes the brain as well as the spinal cord (Figure 35.11). This **cerebrospinal fluid** cushions the vital nervous tissue from sudden, jarring movements.

---

1. The brain stem, the oldest part of the vertebrate brain, provides reflex control over breathing, blood circulation, and other essential functions.

2. More recent layerings of gray matter in the hindbrain, midbrain, and forebrain have appropriated much of the brain stem's control over ancient reflex functions.

3. In all complex vertebrates, the highest integrating centers now reside in the forebrain—specifically, in the gray matter of the cerebral hemispheres.

A CLOSER LOOK AT THE HUMAN CEREBRUM

Organization of the Cerebral Hemispheres

Figure 35.12 shows a midsagittal section of the human brain. If your brain is average in size, it weighs about 1,300 grams (3 pounds) and contains about 100 billion neurons. The neurons that contribute most to your humanness reside in the cerebrum. The cerebrum, which resides within a chamber of hard skull bones, resembles the much-folded nut in a hard walnut shell. A deep fissure divides it into two halves, the left and right **cerebral hemispheres**.

Each cerebral hemisphere can function separately. It receives, processes, and coordinates responses to information mostly from the opposite side of the body. For example, sensory signals from your right eye travel into your left cerebral hemisphere. Signals traveling in both directions through the corpus callosum, a major transverse band of nerve tracts, coordinate the functioning of both hemispheres.

Most regions dealing with speech reside in the left cerebral hemisphere. Mathematics and analytical skills also have been localized to that part of the brain. Most regions dealing with nonverbal (abstract) skills, such as spatial abilities and music, have been localized to the right cerebral hemisphere.

The Cerebral Cortex

The **cerebral cortex** is a thin layer of gray matter—that is, the cell bodies of interneurons at or near the cerebrum's surface, above the information highways of axons. Figure 35.13 shows some of its centers for receiving, encoding, and processing information. These commonly are localized to four lobes of each cerebral hemisphere (called the occipital, temporal, parietal, and frontal lobes).

The rear portion, the *occipital* lobe, has centers for vision. Severe blows to it may cause loss of vision in some portion of the visual field (the outside world that the individual can actually see). The *temporal* lobe, near each temple, is a processing center for hearing and for complex associations related to vision. A blow here won't leave you blind, but it can impair your ability to recognize complex visual patterns (say, human faces). Centers in the temporal lobe also influence emotional behavior.

Compared to the odor-interpreting centers of the early vertebrates, the centers for vision and hearing have become far more important among humans and other primates. The primate fossil record, as highlighted in Chapter 28, invites speculation on the kinds of environmental changes that may have triggered the shift in importance.

The *parietal* lobe contains the somatosensory cortex—the main receiving area for signals from the skin

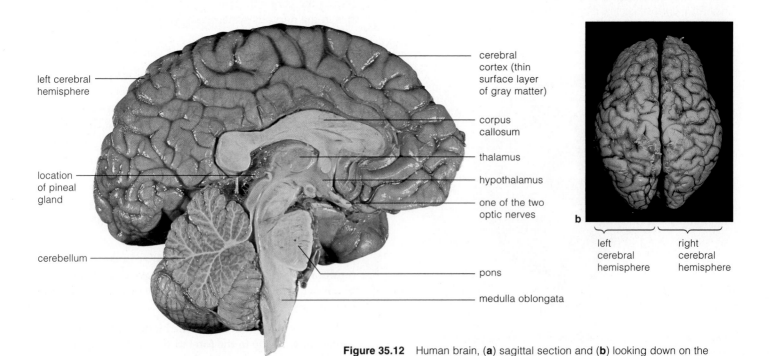

left cerebral hemisphere

location of pineal gland

cerebellum

a

(to spinal cord)

cerebral cortex (thin surface layer of gray matter)

corpus callosum

thalamus

hypothalamus

one of the two optic nerves

b

pons

medulla oblongata

left cerebral hemisphere

right cerebral hemisphere

Figure 35.12 Human brain, (**a**) sagittal section and (**b**) looking down on the longitudinal fissure between its two cerebral hemispheres. The corpus callosum, a major transverse nerve tract, connects the two hemispheres.

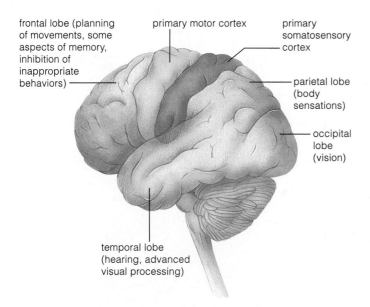

frontal lobe (planning of movements, some aspects of memory, inhibition of inappropriate behaviors)

primary motor cortex

primary somatosensory cortex

parietal lobe (body sensations)

occipital lobe (vision)

temporal lobe (hearing, advanced visual processing)

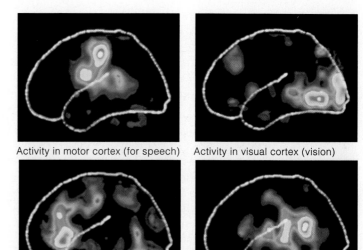

Activity in motor cortex (for speech)

Activity in visual cortex (vision)

Activity in auditory cortex (hearing)

Activity in association cortex

Figure 35.13 (**a**) Primary receiving and integrating centers for the human cerebral cortex. Primary cortical areas receive signals from receptors on the body's periphery. Association areas coordinate and process sensory input from different receptors. The PET scans (**b**) show which brain regions are active during the performance of specific tasks (speaking, hearing, generating, and observing words).

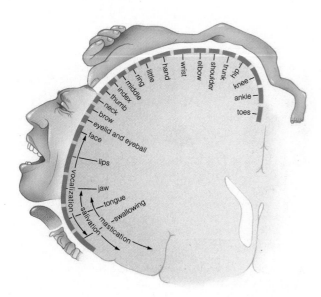

Figure 35.14 Body parts under the control of the primary motor cortex. This diagram is a slice through the motor cortex of the right hemisphere of someone facing you. It controls muscles on the body's left side. The distortions to the human body draped over it indicate which body parts receive the most precise control.

and joints. The *frontal* lobe includes the motor cortex, which coordinates instructions for motor responses. Thumb, finger, and tongue muscles get much of the brain's attention, which gives you an idea of how much control is required for hand movements and verbal expression (Figure 35.14). The prefrontal cortex, a region in front of the motor cortex, is crucial for planning and integrating information about movements, for the inhibition of unsuitable behaviors, and for some aspects of memory, as described shortly. PET scans suggest that this cortical region and part of the cerebellum (especially a structure called the dentate nucleus) may interact to govern motor abilities underlying language and some thought processes.

Many ingenious experiments have provided us with this information and a great deal more about the structure and functioning of the cerebral hemispheres. The *Focus* essay on the page that follows describes a classic example.

---

1. The cerebrum of humans and other complex vertebrates is divided into hemispheres. Two-way signals along a nerve tract (the corpus callosum) afford communication between both hemispheres.

2. The gray matter at and near the surface of the hemispheres, the cerebral cortex, contains the highest integrative centers of the nervous system.

3. The left hemisphere affords most of the control over speech, mathematics, and analytical skills. The right hemisphere affords most control over spatial abilities, music, and other abstract, nonverbal skills.

Sperry's Split-Brain Experiments

Roger Sperry and his coworkers demonstrated intriguing differences in perception between the two halves of the cerebrum of epileptics. Severe *epilepsy* is characterized by seizures, sometimes as often as every half hour. The seizures are analogous to an electrical storm in the brain. Sperry asked: Would *cutting* the corpus callosum of epileptics confine the electrical storm to one hemisphere, leaving at least the other hemisphere to function normally? Earlier studies of laboratory animals and of humans whose corpus callosum had been damaged suggested this might be so.

He performed the surgery, and the electrical storms did subside in frequency and intensity. Cutting the neural bridge ended what must have been positive feedback of ever intensifying electrical disturbances between the two hemispheres. The "split-brain" patients were able to lead what seemed, on the surface, entirely normal lives. But then Sperry devised some elegant experiments to test whether their conscious experience was indeed "normal." Given that the corpus callosum contains 200 million axons, surely *something* was different. Something was. "The surgery," he later reported, "left these people with two separate minds, that is, two spheres of consciousness. What is experienced in the right hemisphere seems to be entirely outside the realm of awareness of the left."

Sperry presented the two hemispheres of split-brain patients with two different portions of the same visual stimulus. It was known at the time that visual connections to and from one hemisphere are mainly concerned with the opposite half of the visual field (Figure *a*). Sperry projected words—say, COWBOY—onto a screen so that COW fell in the left half of the visual field, and BOY fell in the right (Figure *b*).

The subjects of this experiment reported *seeing* the word BOY. The left hemisphere, which controls language, perceived only the letters BOY. However, when asked to write the perceived word with the left hand—a hand that was deliberately blocked from a subject's view—the subject wrote COW. The right hemisphere "knew" the other half of the word (COW) and had directed the left hand's motor response. But it couldn't tell the left hemisphere what was going on because of the severed corpus callosum. The subject knew that a word was being written, but could not say what it was!

Thus Sperry showed that signals across the corpus callosum coordinate the functioning of the two cerebral hemispheres, each of which had responded to visual signals from the opposite side of the body.

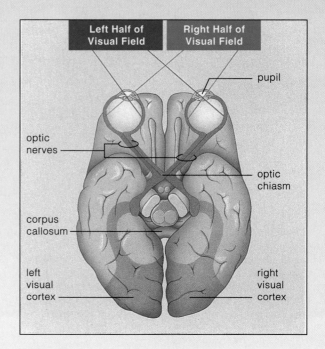

a The human eye gathers visual information at the retina, a layer of densely packed light receptors. Light from the *left* half of the visual field strikes receptors on the right side of both retinas. Parts of two optic nerves carry signals from the receptors to the right cerebral hemisphere. Light from the *right* half of the visual field strikes receptors on the left side of both retinas. Parts of the optic nerves carry signals from them to the left hemisphere.

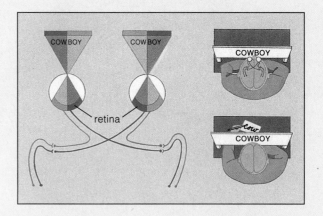

b Response of a split-brain patient to different portions of a visual stimulus.

35.7 MEMORY

Even before birth, the brain starts to accumulate **memory**—to store and retrieve information about the individual's unique experiences. Sensory stimulation triggers each memory. When signals about it arrive in the cerebral cortex, association begins. **Association** is the linkage of information into larger packages, much as chemical bonds link atoms into compounds. The packages are sent on to other brain regions for storage. Information is recovered in response to retrieval cues.

According to one prevailing theory, the brain stores information in stages. *Short-term* storage, a fleeting stage of neural excitation, lasts a few seconds to a few hours. It is limited to a few bits of information—numbers, words of a sentence, and so on. By contrast, in *long-term* storage, seemingly unlimited amounts of information become tucked away on a more or less permanent basis. This theory helps explain *retrograde amnesia*, which follows a severe head blow and loss of consciousness. Memory of what happened during a span of time that preceded the blow is lost. Yet memories of events before then often remain intact.

Information that seems most important to the individual moves most rapidly into long-term storage. When such information stimulates the brain, epinephrine secretions from sympathetic nerves pump the individual to a state of arousal. Up to a point, the blood level of epinephrine beefs up consolidation of a memory. (Among other things, it triggers conversion of glycogen into glucose, the brain's primary fuel.) Past a certain point, however, the blood level of epinephrine impairs memory storage. This is one reason why a person has trouble remembering details about a situation that brought on extreme panic.

Even if information is unused for decades, it still can be recalled—so memory must be encoded in a form resistant to degradation. Many experiments suggest that memory resides in neurons that have changed, chemically and physically, as an outcome of changes in synaptic connections. For example, laboratory mice were raised without visual stimulation. Later, micrographs revealed withered synapses in their visual cortex. Where the synapses fell into disuse, the functional link between neurons had been weakened or broken. Researchers also have evidence that intensely stimulated synapses form stronger connections, grow in size, or sprout buds or spines to form more connections.

The capacity to recover memories diminishes with *Alzheimer's disease*, which usually has its onset in later life. Affected people often can remember long-standing information, such as their Social Security number, but they have trouble remembering what just happened to them. In time they grow confused, depressed, and unable to complete a train of thought.

35.8 STATES OF CONSCIOUSNESS

Throughout the spectrum of consciousness, which includes sleeping and aroused states, neural chattering shows up as wavelike patterns in electroencephalograms (EEGs). EEGs are electrical recordings of the frequency and strength of membrane potentials at the brain surface (Figure 35.15). PET scans, of the sort shown in Figure 35.13, also provide information on the precise location of brain activity as it is proceeding. Part of the reticular formation controls the changing levels of consciousness. Inhibitory or excitatory chemical changes accompanying its flow of signals affects whether you stay awake or fall asleep. One of the reticular formation's "sleep centers" releases serotonin. This neurotransmitter inhibits other neurons that arouse the brain and maintain wakefulness. Thus, high serotonin levels are linked to drowsiness and sleep. Substances released from another brain center inhibit serotonin's effects and bring about wakefulness.

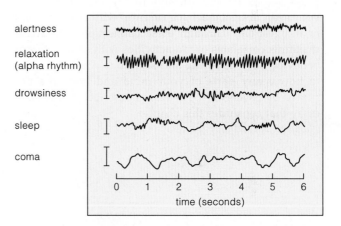

Figure 35.15 EEG patterns. Vertical bars indicate a range of 50 microvolts of electrical activity. The irregular horizontal graph lines indicate the electrical response with time.

The prominent wave pattern for someone who is relaxed, with eyes closed, is an *alpha rhythm*. The EEG waves are recorded in "trains" of one after the other, about ten per second. Alpha waves predominate during the state of meditation. With a transition to sleep, wave trains gradually become larger, slower, and more erratic. This *slow-wave sleep* pattern shows up about 80 percent of the total sleeping time for adults. It occurs when sensory input is low and the mind is more or less idling. Subjects awakened from slow-wave sleep usually report that they were not dreaming. Often they seemed to be mulling over recent, ordinary events.

Slow-wave sleep is punctuated by brief spells of *REM sleep*. *Rapid Eye Movements* accompany this pattern (the eyes jerk beneath closed lids), as do irregular breathing, faster heartbeat, and twitching fingers. Most people awakened from REM sleep report they were experiencing vivid dreams. The transition from sleep or deep relaxation into alert wakefulness is marked by a shift to low-amplitude, higher frequency wave trains. The transition, *EEG arousal*, occurs when conscious effort is made to focus on external stimuli or even on one's own thoughts.

35.9 THE BRAIN AND BEHAVIOR

You are an emotional animal. For this you can thank your cerebral cortex and **limbic system**, which consists of several brain regions (Figure 35.16). The limbic system is sometimes called our "emotional brain." It also plays significant roles in memory.

In evolutionary terms, the limbic system is distantly related to the olfactory lobes, and it still retains connections with the sense of smell. That is one reason why a sense of pleasure might wash over someone who recalls delicious odors emanating from grandmother's kitchen. It is why you may feel warm and fuzzy all over when your brain recalls the exact scent of cologne of a delightful person who wore it.

Besides monitoring your internal organs, the hypothalamus also is gatekeeper of the limbic system. Many connections from the cerebral cortex and from ancient brain centers pass through it. Through these connections, the reasoning possible in the cerebral cortex can dampen rage, hatred, and other "gut reactions." (Think of the signals passing through the hypothalamus when your heart and stomach are on fire—with either passion or indigestion.) Through these connections, the hypothalamus correlates organ activities with self-gratifying behavior, such as eating and sex. Thus you have a clue to how many parts of your mind and body are disrupted by psychoactive drugs, as described in the *Focus* essay that concludes this chapter.

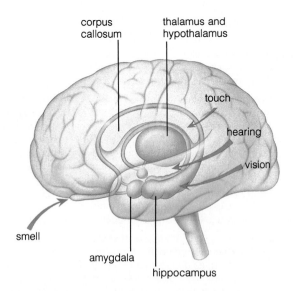

Figure 35.16 The limbic system, our "emotional brain." It includes brain regions called the thalamus, hypothalamus, amygdala, and hippocampus. The limbic system also plays key roles in memory.

Focus on Health

Drugs, The Brain, and Behavior

Broadly speaking, a drug is any substance introduced into the body to provoke a specific physiological response. Some drugs help a person cope with illness or stress. Others act on the brain, artificially fanning the pleasure we associate with sex and other self-gratifying behaviors.

Many drugs are habit-forming. Even if the body functions well without them, they continue to be used for the real or imagined relief they afford. Often the body develops tolerance of such drugs (it takes larger or more frequent doses to produce the same effect). Habituation and tolerance are signs of **drug addiction**—chemical dependence on a drug. *The drug has taken on an "essential" biochemical role in the body.* Abruptly deprive addicts of the drug, and they suffer physical pain and mental anguish. The entire body suffers major biochemical upheavals. Stimulants, depressants, hypnotics, narcotic analgesics, hallucinogens, and psychedelics have such effects.

Stimulants Caffeine, nicotine, amphetamines, cocaine, and other stimulants increase alertness and body activity. Then they cause depression.

Coffee, tea, chocolate, and many soft drinks contain caffeine, a widely used stimulant. Low doses stimulate the cerebral cortex first, leading to increased alertness and restlessness. Higher doses acting at the medulla oblongata disrupt motor coordination and mental coherence. Nicotine, a component of tobacco, has powerful effects throughout the nervous system. It mimics acetylcholine and can directly stimulate a variety of sensory receptors. Its short-term effects include water retention, irritability, increased heart rate and blood pressure, and gastric upsets.

Amphetamines (including "speed") resemble two neurotransmitters, dopamine and norepinephrine. In time, the brain produces less and less of these and depends more on artificial stimulation. Cocaine produces a rush of pleasure by *blocking* reabsorption of dopamine, norepinephrine, and other neurotransmitters. These accumulate in synaptic clefts and incessantly stimulate receptor cells for an extended period. Heart rate and blood pressure rise; sexual appetite increases. Then neurotransmitters diffuse away. Replacements can't be made fast enough to counter the loss. The sense of pleasure evaporates as the now-hypersensitive receptor cells demand stimulation. After prolonged, heavy use of cocaine, "pleasure" cannot be experienced. Addicts become anxious and depressed. They lose weight and cannot sleep properly. Their immune system weakens, and heart abnormalities set in.

Granular cocaine, which is inhaled (snorted), has been around for some time. Crack cocaine is burned, and the smoke is inhaled (Figure *a*). As suggested at the start of

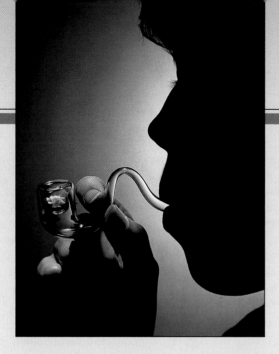

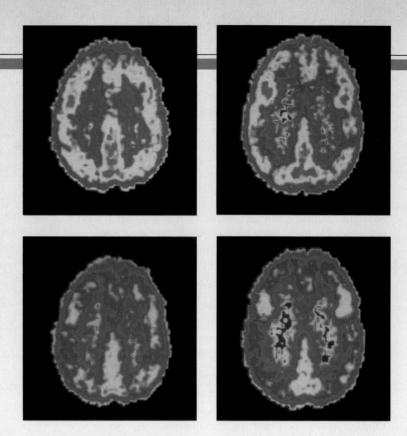

a Smoking crack. This form of cocaine reaches the brain in less than eight seconds. The top row of PET scans of horizontal sections of the brain reveal normal activity. The bottom row of the same sections show the effect of cocaine on brain activity. *Red* indicates greatest activity, followed by *yellow*, *green*, and *blue*.

this chapter, crack is extremely addictive. Its highs are higher, but the crashes are more devastating.

Depressants, Hypnotics These drugs lower activity in nerves and parts of the brain. Some act at synapses in the reticular formation and in the thalamus. Depending on the dosage, responses range from emotional relief, drowsiness, sleep, anesthesia, and coma to death. Low doses act more on inhibitory synapses, so the person feels excited or euphoric at first. Increased doses suppress excitatory synapses as well, leading to depression. Depressants and hypnotics have additive effects. One amplifies another, as when alcohol plus barbiturates increases the depression.

Alcohol (ethyl alcohol) acts directly on plasma membranes to alter cell function. Some people mistakenly think of it as a harmless stimulant (it produces an initial "high"). But alcohol is one of the most powerful psychoactive drugs and a major cause of death. Small doses even over the short term produce disorientation, uncoordinated motor functions, and diminished judgment. Long-term addiction can permanently damage the liver (cirrhosis).

Analgesics When stress causes physical or emotional pain, the brain produces natural analgesics (pain relievers). Endorphins and enkephalins are examples. They act on many parts of the nervous system, including brain centers that deal with emotions and pain. Narcotic analgesics, including codeine and heroin, sedate the body and relieve pain. They are extremely addictive. Deprivation following massive doses of heroin leads to fever, chills, hyperactivity and anxiety, violent vomiting, cramping, and diarrhea.

Psychedelics, Hallucinogens These drugs alter sensory perception by interfering with acetylcholine, norepinephrine, or serotonin activity. LSD (lysergic acid diethylamide) affects the action of serotonin, a neurotransmitter with roles in inducing sleep, in temperature regulation—and in sensory perception. Even in small doses, LSD warps perceptions. For example, some users "perceived" that they could fly and "flew" off buildings.

Marijuana, another hallucinogen, is made from crushed leaves, flowers, and stems of the plant *Cannabis*. In low doses marijuana is like a depressant. It slows down but does not impair motor activity; it relaxes the body and elicits mild euphoria. It also can produce disorientation, increased anxiety bordering on panic, delusions (including paranoia), and hallucinations.

Like alcohol, marijuana affects the performance of complex tasks, such as driving a car. In one study, pilots showed a marked deterioration in instrument-flying ability for more than two hours after smoking marijuana. Over time, marijuana smoking can suppress the immune system and impair mental functions.

SUMMARY

1. Nervous systems provide swift, specialized means of detecting and responding to stimuli. At their most basic operating level are reflexes: simple, stereotyped movements made directly in response to stimuli.

2. The simplest nervous systems are nerve nets, such as those of sea anemones and other radial animals. These meshworks of nerve cells form functional connections with contractile and sensory cells of the epithelium.

3. Bilateral, cephalized nervous systems exist among animals ranging from flatworms to humans. Such systems include nerve cells bodies concentrated as ganglia or a brain in a head region. Bundled-together axons form one or more longitudinal, cordlike nerves. Pairs of nerves also extend into the body's right and left halves.

4. Vertebrates have a *central* nervous system, consisting of the brain and spinal cord. Nerves of a *peripheral* nervous system functionally connect the central nervous system with all other body regions.

5. The peripheral nervous system has a somatic subdivision, dealing with skeletal muscles. It also has an autonomic subdivision, which deals with the heart, lungs, glands, and other internal organs.

   a. The autonomic system is composed of two categories of nerves (parasympathetic and sympathetic nerves), which often have opposing effects on the same organs.

   b. Parasympathetic nerves dominate when the body is not receiving much outside stimulation. They tend to divert energy to basic housekeeping activities, such as digestion.

   c. Sympathetic nerves increase overall body activities during times of heightened awareness or danger. They govern the fight-flight response.

6. Parts of the spinal cord direct reflex connections for limb movements and internal organ activity. Many nerve tracts in the spinal cord also carry signals between the brain and the peripheral nervous system.

7. During embryonic development, the brain develops through regional expansions of a neural tube. The anterior end of the tube becomes the brain stem, which is still discernible in the adult brain. Its most recent expansions of gray matter deal with storing, comparing, and using experiences to initiate novel, nonstereotyped action. These regions are the basis of memory, learning, and reasoning.

8. The brain of humans and other vertebrates has three main divisions (hindbrain, midbrain, and forebrain).

   a. The hindbrain includes the medulla oblongata, pons, and cerebellum and contains reflex centers for vital functions and muscle coordination.

   b. The midbrain functions in coordinating and relaying visual and auditory information.

   c. The forebrain includes the cerebrum, thalamus, hypothalamus, and limbic system. The thalamus relays sensory information and helps coordinate motor responses. The hypothalamus monitors internal organs and influences thirst, hunger, sexual activity, and other behaviors related to their functioning. It is gatekeeper to the limbic system, which has roles in learning, memory, and emotional behavior.

   d. The reticular formation, a diffuse network of interneurons, extends from the top of the spinal cord, through the brain stem, and on up to the cerebral cortex. It affords low-level control over motor activity and helps govern the nervous system as a whole.

9. The cerebral cortex, the outermost part of the cerebrum, contains the highest integrative centers. These centers receive signals from sensory receptors, process and integrate new information with memories of past events, and coordinate motor responses.

10. Memory apparently occurs in two stages: a short-term formative period and long-term storage, which depends on chemical or structural changes in the brain.

Review Questions

1. Generally describe the type of nervous systems found in radially symmetrical animals and in bilaterally symmetrical animals. *576–577*

2. What constitutes the central nervous system? The peripheral nervous system? *578–579*

3. Distinguish between the following:
   a. central and peripheral nervous system *578*
   b. cranial and spinal nerves *580*
   c. somatic and autonomic nerves *580*
   d. parasympathetic and sympathetic nerves *580*

4. Label the major parts of the human brain. *584*

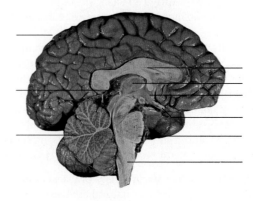

5. What is a psychoactive drug? Can you describe the effects of one such drug on the central nervous system? *588–589*

1. Sea anemones, hydras, and jellyfishes have simple nervous systems called _____ .

2. Structurally, the vertebrate nervous system shows pronounced _____ and _____ symmetry.

3. The oldest parts of the vertebrate brain provide _____ .
   a. reflex control of breathing, blood circulation, and other basic activities
   b. coordinating and relaying visual and auditory information
   c. storing, comparing, and using experiences to initiate novel, nonstereotyped action
   d. both a and c

4. The central nervous system includes _____ . The peripheral nervous system includes _____ .
   a. nerves and ganglia; brain and spinal cord
   b. brain and spinal cord; nerves and ganglia
   c. spinal cord; brain
   d. nerves and interneurons; brain and spinal cord

5. Overall, _____ nerves slow down the body and divert energy to digestion and other housekeeping tasks; _____ nerves slow down housekeeping tasks and increase overall activity during times of heightened awareness, excitement, or danger.
   a. autonomic; somatic
   b. sympathetic; parasympathetic
   c. parasympathetic; sympathetic
   d. peripheral; central

6. Parasympathetic and sympathetic nerves carry signals to and from the nervous system _____ to bring about minor adjustments in internal organ activity.
   a. continually
   b. in separate, alternating fashion
   c. only during a fight-flight response
   d. none of the above is correct

7. The _____ of the spinal cord consists of nerve tracts; the _____ consists of dendrites, neuron cell bodies, and neuroglial cells.
   a. gray matter; white matter
   b. white matter; gray matter

8. The hindbrain (medulla oblongata, cerebellum, and pons) contains _____ .
   a. part of the reticular formation
   b. major nerve tracts between brain centers
   c. reflex centers for limb movements, respiration, breathing, and other vital tasks
   d. all are correct

9. Extending from the uppermost part of the spinal cord and on through the brain stem is the _____ .
   a. reticular formation
   b. blood-brain barrier
   c. olfactory lobe
   d. tectum

10. The most highly developed part of the human brain (the forebrain includes the _____ .
    a. medulla oblongata, pons, and cerebellum
    b. cerebrum, thalamus, hypothalamus, and limbic system
    c. medulla oblongata, pons, and cerebral cortex
    d. cerebellum, medulla oblongata, pons, and limbic system
    e. hypothalamus, limbic system, pons, and cerebral cortex

11. Match the central nervous system region with some of its functions.
    ____ spinal cord
    ____ medulla oblongata
    ____ hypothalamus
    ____ limbic system
    ____ cerebral cortex

    a. receives sensory input, integrates it with stored information
    b. monitors internal organs and related behavior (e.g., hunger)
    c. with cerebral cortex, governs emotions
    d. reflex control of respiration, blood circulation, other basic activities
    e. expressway for signals between brain and peripheral nervous system

Selected Key Terms

association (in memory) *587*
autonomic nerve *580*
brain *582*
brain stem *582*
central nervous system *578*
cerebellum *582*
cerebral cortex *584*
cerebral hemisphere *584*
cerebrospinal fluid *583*
cerebrum *583*
drug addiction *588*
fight-flight response *581*
ganglia *576*
hypothalamus *583*
limbic system *588*
medulla oblongata *582*
memory *587*
meninge *581*
nerve net *576*
nervous system *576*
parasympathetic nerve *580*
peripheral nervous system *578*
planula *576*
pons *582*
reflex pathway *576*
reticular formation *583*
somatic nerve *580*
spinal cord *581*
sympathetic nerve *580*
tectum *582*
thalamus *583*

Readings

Bloom, F., and A. Lazerson. 1988. *Brain, Mind, and Behavior*. Second edition. New York: Freeman.

Churchland, P., and P. Churchland. January 1990. "Could a Machine Think?" *Scientific American* 262(1):32–37.

Fischbach, G. D. September 1992. "Mind and Brain." *Scientific American* 267(3):48–57. This issue is devoted to the development, function, and certain disorders of the brain.

Kalil, R. December 1989. "Synapse Formation in the Developing Brain." *Scientific American* 261(6):76–85.

Romer, A., and T. Parsons. 1986. *The Vertebrate Body*. Sixth edition. Philadelphia: Saunders. Insights into the evolution of vertebrate nervous systems.

Shepherd, G. 1988. *Neurobiology*. Second edition. New York: Oxford. Paperback.

36 SENSORY RECEPTION

Different Stokes for Different Folks

Even though you might be reluctant to pet a snake or scratch a bat behind the ears, you have to give it credit for being a vertebrate relative with some distinguishing sensory traits.

Consider the python, with thermoreceptors in rows of pits above and below its mouth (Figure 36.1a). Body heat (infrared energy) of prey—small, night-foraging mammals—activates these sensory receptors, which then signal the brain. The brain responds by assessing the signals and directing the aim of a strike with stunning accuracy. A motionless, edible frog does not have the same stimulatory effect, and the same snake will slither past it. Frog skin is cool and blends with background colors. The snake does not have receptors for detecting it or a neural program for responding to it.

Or consider bats, nearly all of which sleep by day and spread their webbed wings at dusk. Different kinds take to the air in search of nectar, fruit, frogs, or insects. Many sensory receptors in their eyes, nose, ears, mouth, and skin are not that different from yours. Others provide bats with a sense of hearing that you cannot begin to match. Even tiny-eyed, nearly blind species navigate and capture flying insects with precision in the dark! They are masters of **echolocation**. They emit calls, as the bat in Figure 36.1b is doing. When the sound waves of the calls bounce off insects, trees, and other objects, sensory receptors detect the echoes and send signals about them to the brain.

As an echolocating bat flies, it emits a steady stream of about ten clicking sounds per second—sounds you cannot hear. The clicks are intense "ultrasounds," above the range of sound waves that receptors in human ears can detect. When the bat hears a pattern of distant echoes from, say, an airborne mosquito or moth, it

Figure 36.1 Examples of diversity in sensory reception. (**a**) Python thermoreceptors in pits above and below the mouth detect body heat (infrared energy) of prey. (**b**) Some bats listen to echoes of their own high-frequency sounds. Their brain deciphers echoes bouncing back from objects in the environment as a "sound map." The map helps a flying bat capture mosquitoes and moths in midair, without even seeing them.

increases the rate of ultrasonic clicks to as many as 200 per second, faster than a machine gun fires bullets. In the few milliseconds of silence between clicks, receptors detect the echoes and send signals about them to the bat brain. The brain constructs a "map" of sounds that the bat uses in its maneuvers through the night world.

With this chapter, we turn to the means by which animals receive signals from the external and internal environments, and then decode the signals in ways that give rise to awareness of sounds, sights, odors, pain, and other sensations. Sensory neurons, nerve pathways, and brain regions are required for these tasks. Together, they represent the portions of the nervous system called sensory systems.

b

KEY CONCEPTS

1. Sensory systems are part of the nervous system. Each consists of specific types of sensory receptors, nerve pathways from those receptors to the brain, and brain regions that deal with sensory information.

2. A stimulus is a form of energy that activates a specific type of sensory receptor, which is a sensory neuron or a specialized cell adjacent to it. Photoreceptors detect light energy, thermoreceptors detect heat energy, and so on. Information about a stimulus travels in the form of electrical and chemical signals through the nervous system.

3. A sensation is a conscious awareness of change in external or internal conditions. It begins when sensory receptors detect a specific stimulus. The energy of the stimulus is converted to graded, local signals that can give rise to action potentials. Information about the stimulus becomes encoded in the number and frequency of action potentials sent to the brain along particular nerve pathways. Then specific brain regions translate the information into a sensation.

4. The somatic sensations include touch, pressure, temperature, pain, and muscle sense.

5. The special senses include taste, smell, hearing, and vision.

36.1 SENSORY SYSTEMS

Sensory systems are the front doors of the nervous system. They receive and notify the spinal cord and brain of specific changes outside and inside the body. Each consists of sensory receptors, nerve pathways leading from receptors to the brain, and brain regions where sensory information is processed and translated into sensation. **Sensation** means conscious awareness of a stimulus. It is not the same as perception (understanding what the sensation means). The reactions of a live lobster cooking in boiling water suggest that its brain is aware of a stimulus. But the lobster reacts the same way to *any* strong stimulus, as when you grip it firmly. The lobster probably does not "understand" its predicament or "feel" pain, as humans do.

Types of Sensory Receptors

Sensory receptors are the receptionists at the front door of the nervous system. There are five major types, based on the type of stimulus energy they detect:

1. **Chemoreceptors** detect chemical energy of specific substances dissolved in the fluid surrounding them.

2. **Mechanoreceptors** detect forms of mechanical energy (changes in pressure, position, or acceleration).

3. **Photoreceptors** detect visible and ultraviolet light.

4. **Thermoreceptors** detect infrared energy (heat).

5. **Nociceptors** (pain receptors) detect tissue damage.

Depending on the kinds and numbers of their sensory receptors, animals sample the environment in different ways and differ in their awareness of it. Unlike bees, you have no receptors for ultraviolet light and don't see many flowers the way they do (page 520). Unlike many bats, you and bees have no receptors for ultrasound. Unlike pythons, you, bees, and bats have no receptors to detect warm-blooded prey in the dark.

Sensory Pathways

Recall, from Chapter 34, that sensory axons carry signals from receptors to the brain. Before they can do so, the stimulus energy must be converted to action potentials, the basis of neural messages. Briefly, when a stimulus disturbs the plasma membrane of a receptor ending, some ions flow through protein channels across a local patch of the membrane. In other words, the disturbance has triggered a local, graded potential. Such signals don't spread far from the point of stimulation, and they vary in magnitude. However, when a stimulus is intense or repeated fast enough for summation of those local signals, action potentials may result. Action potentials propagate themselves from the receptors to the axon endings of sensory neurons (Figure 36.2). There, the release of neurotransmitter stimulates or inhibits the activity of interneurons (or motor neurons) adjacent to it. Action potentials may be generated in the interneurons, which may be part of nerve tracts leading to the brain.

The action potentials being propagated along sensory neurons are not like a wailing ambulance siren. *They do not vary in amplitude.* How, then, does the brain

Table 36.1	Receptors Associated With the Major Senses	
Category of Receptor	Examples	Stimulus
Chemoreceptors:		
Internal chemical senses	Carotid bodies in blood vessel wall	Substances (CO_2, etc.) dissolved in extracellular fluid
Taste	Taste receptors of tongue	Substances dissolved in saliva, etc.
Smell	Olfactory receptors of nose	Odors in air, water
Mechanoreceptors:		
Touch, pressure	Pacinian corpuscles in skin	Mechanical pressure against body surface
Stretch	Muscle spindle in skeletal muscle	Stretching
Auditory	Hair cells within ear	Vibrations (sound or ultrasound waves)
Balance	Hair cells within ear	Fluid movement
Photoreceptors:		
Visual	Rods, cones of eye	Wavelengths of light
Thermoreceptors	Cold or warm receptors in skin; thermoreceptors in hypothalamus	Presence of or change in radiant energy (heat)
Nociceptors	Free nerve endings in skin	Any stimulus that damages tissue and leads to sensation of pain

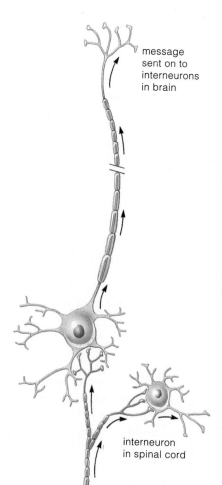

message sent on to interneurons in brain

interneuron in spinal cord

receptor endings of sensory neuron stimulated when foot lands on a tack

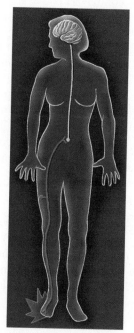

Figure 36.2 Example of a sensory nerve pathway leading from a sensory receptor to the brain. The sensory neuron is coded *red*; interneurons are coded *yellow*.

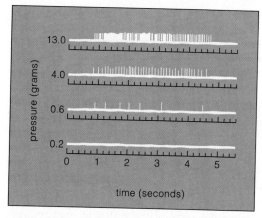

Figure 36.3 Action potentials recorded from a single pressure receptor of the human hand. The recordings correspond to variations in stimulus strength. A thin rod was pressed against the skin with the pressure indicated on the vertical axis of this diagram. Vertical bars above each thick horizontal line represent individual action potentials. Notice the increases in frequency, which correspond to increases in stimulus strength.

assess the nature of a given stimulus? The answer lies with *which* nerve pathways carry the incoming signals, the *frequency* of signals that are traveling along each axon of the pathway, and the *number* of axons that the stimulus recruited into action.

First, the genetically prescribed network of neurons in each animal's brain can interpret action potentials only in certain ways. That is why you "see stars" when your eye is poked, even in the dark. Photoreceptors in the eye were mechanically disturbed enough to trigger signals that traveled along an optic nerve to your brain. Your brain always interprets signals arriving from an optic nerve as "light."

Second, when a stimulus is stronger, receptors fire action potentials more frequently. The same receptor detects the sounds of a throaty whisper and a wild screech. The brain senses the difference through frequency variations in the signals that the receptor sends to it.

Third, strong stimulation recruits more receptors. Tap a spot of skin on your arm and you activate some receptors. Press hard on the same area and you activate

many more in a larger area. The increased disturbance sets off action potentials in many sensory axons at the same time. The brain interprets the combined activity as an increase in stimulus intensity (Figure 36.3).

Sometimes the frequency of action potentials decreases or stops even when a stimulus is being maintained at a constant strength. This is called **sensory adaptation**. After you have put on clothing, for example, you no longer are aware of its pressure against your skin. Some mechanoreceptors in your skin are a rapidly adapting type; they fire only at the onset of a stimulus. By contrast, pick up a warm muffin, and slowly adapting thermoreceptors in your skin will rapidly fire off action potentials. Then the firing frequency drops to a lower level, which is maintained for as long as you hold the muffin.

We turn now to some specific examples of the sensory receptors listed in Table 36.1. The ones that are present at more than one location in the body contribute to **somatic sensations**. Other receptors are restricted to special locations, such as the eyes or ears. They contribute to the **special senses**.

---

1. A sensory system has sensory receptors for specific stimuli, nerve pathways that conduct information from receptors to the brain, and brain regions where the information is processed.

2. The brain senses a stimulus based on which nerve pathways carry the incoming signals, the frequency of signals traveling along each axon of that pathway, and the number of axons that have been recruited.

36.2 SOMATIC SENSATIONS

When you become aware of touch, pressure, heat, cold, and pain, these are somatic sensations. So is the awareness of movements and positions in space. The sensations start with receptor endings in surface tissues, in skeletal muscles, and in walls of internal organs. Information from the receptors travels into the spinal cord and on up to the **somatosensory cortex**. Cells of this cortical region are laid out like a map corresponding to the body surface. The size of the different map regions corresponds to the functional importance of different body parts (page 584). In humans, much of the primary somatosensory cortex responds to receptors located in the fingers, thumbs, and lips, as Figure 36.4 indicates.

Touch and Pressure

Mechanoreceptors abound near the body's surface, although they are more abundant in some regions than others. Fingertips and the tip of the tongue have the greatest number, and so these regions show the greatest sensitivity to stimulation. The back of the hand and neck do not have nearly as many and are far less sensitive to stimulation.

Mechanoreceptors in the skin are the receiving end of sensory neurons. Some types have a surrounding capsule of epithelial or connective tissue. The free nerve endings and Pacinian corpuscles shown in Figure 36.5 are examples. These and other types respond easily to applied pressure that deforms their plasma membrane. Many fire off action potentials only when the stimulus is first encountered and when it ends. They make you aware of touch, even light tickles and vibrations. Many others fire off action potentials as long as the stimulus is present. They keep you aware of ongoing pressure.

Temperature

When temperatures near the body surface remain much the same, thermoreceptors in skin fire a steady barrage of signals to the brain. With increases in temperature, the firing frequency increases. Free nerve endings serve as heat receptors. Cold receptors have not been identified yet.

Pain

Perception of injury to some body region, or **pain**, begins with signals from nociceptors. Among these are free nerve endings, which detect any stimulus strong enough to damage the surrounding tissue. They respond to strong mechanical stimulation, intense heat or cold, and chemical irritation.

Responses to pain often depend on the brain's ability to identify the affected tissue. Get hit in the face with a snowball and you "feel" the contact on facial skin. For

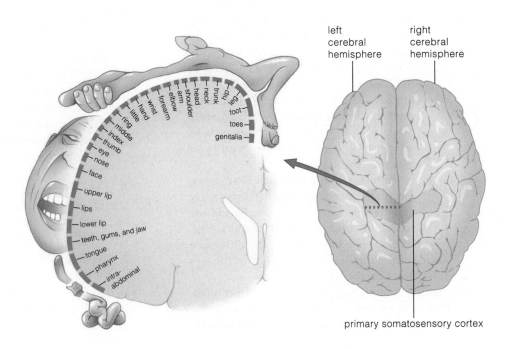

Figure 36.4 Body regions represented in the primary somatosensory cortex. This region is a strip a little more than an inch (2.5 centimeters) wide, running from the top of the head to just above the ear at the surface of each cerebral hemisphere.

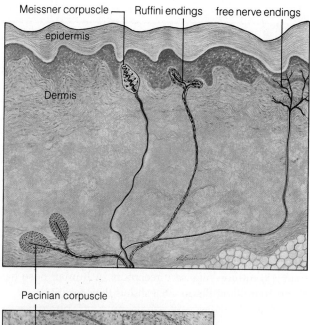

Meissner corpuscle   Ruffini endings   free nerve endings

epidermis

Dermis

Pacinian corpuscle

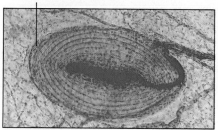

Figure 36.5 Tactile receptors in human skin. Signals from the types called free nerve endings lead to an awareness of changes in temperature, small amounts of pressure, and pain. Pacinian corpuscles detect vibration. Meissner corpuscles signal the onset and the end of sustained pressure. The Ruffini endings react continually to ongoing stimulation.

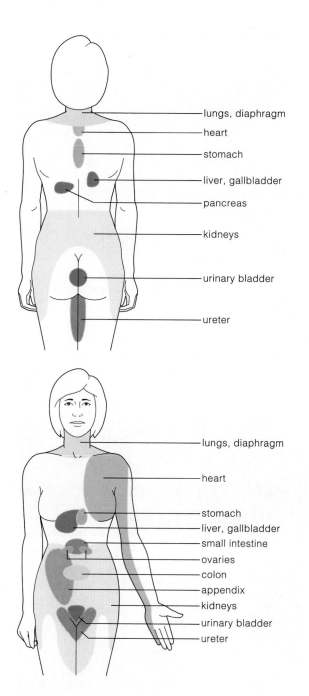

lungs, diaphragm
heart
stomach
liver, gallbladder
pancreas
kidneys
urinary bladder
ureter

lungs, diaphragm
heart
stomach
liver, gallbladder
small intestine
ovaries
colon
appendix
kidneys
urinary bladder
ureter

Figure 36.6 Referred pain. Instead of localizing the pain associated with certain disorders to the affected internal organ, it localizes the sensations to the skin areas indicated.

reasons not fully understood, the brain sometimes associates pain with a tissue located some distance away from an internal injury. This is "referred pain." For example, a heart attack can be felt as pain in skin above the heart and along the left shoulder and arm (Figure 36.6). In general, nerve pathways servicing both the injured tissue and a tissue to which pain is referred pass through the same segment of the spinal cord.

Muscle Sense

Sensing limb motions and the body's position in space requires mechanoreceptors in skeletal muscle, joints, tendons, ligaments, and skin. Among these are stretch receptors of the sensory organs called muscle spindles (Figure 30.10*b*). Such organs are embedded in skeletal muscle tissue and run parallel with the muscle cells. Their responses to stimulation depend on how much and how fast the muscle stretches.

Diverse mechanoreceptors detect touch, pressure, heat and cold, pain, limb motions, and changes in the body's position in space. Responses to signals from these receptors give rise to somatic sensations.

36.3 TASTE AND SMELL

We turn now to examples of the special senses, starting with the senses of taste and smell.

Different animals taste substances with their mouth, antennae, legs, tentacles, or fins. It all depends on where chemoreceptors called **taste receptors** are distributed. These detect differences in substances that become dissolved in fluid next to some body surface. The ones on animal tongues often occur in sensory organs called taste buds (Figure 36.7).

Animals smell substances with **olfactory receptors**. These provide clues or advance warning of predators, food, or anything else that gives off chemicals able to diffuse through water or air. A bloodhound nose has more than 200 million olfactory receptors. A human nose has about 10 million; Figure 36.8 shows their location.

Potential mates and rivals give off pheromones. **Pheromones**, a type of signaling molecule, function in social communication. As described in Chapter 52, these exocrine gland secretions from one animal can change the behavior of other, target animals of the same species. For now, simply consider the effect of a sex-attracting pheromone on the olfactory receptors of a male silk moth. Contact with merely one bombykol molecule per second sends action potentials to the moth brain. They can help a male locate a female in the dark, even if she is more than a kilometer upwind from him.

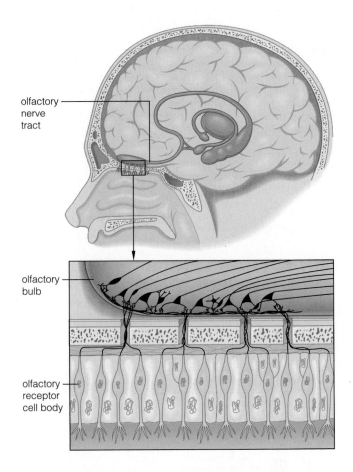

Figure 36.8 Sensory nerve pathway leading from olfactory receptors in the nose to primary receiving centers in the brain.

olfactory nerve tract

olfactory bulb

olfactory receptor cell body

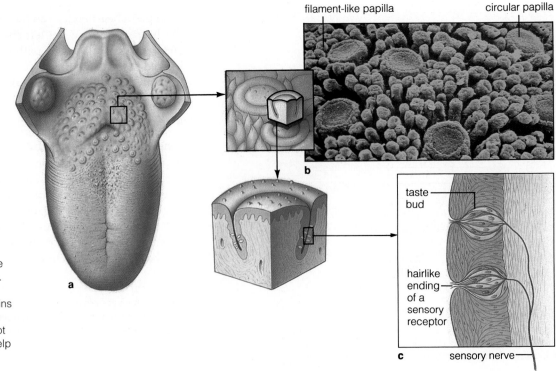

filament-like papilla

circular papilla

b

taste bud

hairlike ending of a sensory receptor

c

sensory nerve

Figure 36.7 Location of taste receptors inside taste buds in the human tongue. Circular papillae ring the epithelial tissue that contains numerous taste buds. The filament-like papillae do not contribute to taste (they help move food).

a

36.4 BALANCE

Almost all animals right themselves after their body is tilted or turned upside-down, so they must have a sense of their "natural" position. The baseline against which animals assess displacement from their natural position is called the *equilibrium* position. Then, the body is balanced in relation to gravity, velocity, acceleration, and other forces that influence position and movement. The same balancing act operates when wings and other body parts are on the move. The sense of balance depends on organs of equilibrium that incorporate **hair cells**. These mechanoreceptors fire off action potentials when suitably bent.

Amphibians, birds, and mammals rely partly on input from receptors in the eyes, skin, and joints for their sense of balance. They also rely on organs within a system of canals and fluid-filled sacs in each ear. As Figure 36.9 shows, **cristae** (singular, crista) are organs of *dynamic* equilibrium. They detect rotational, accelerated motions of the head. By contrast, **otolithic organs** are organs of *static* equilibrium; they detect linear movements in the head's position.

Motion sickness may result when monotonous linear, angular, or vertical motion overstimulates hair cells in the inner ear. Visual sensations often contribute to the sickness; fear and anxiety also play roles. Action potentials triggered by the sensory input reach a brain center that governs the vomiting reflex. Nausea and vomiting are the chief symptoms in individuals who get carsick, airsick, or seasick.

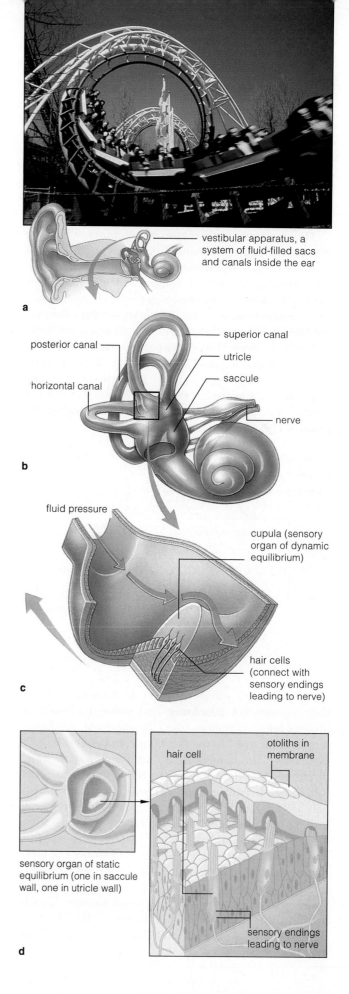

Figure 36.9 Organs of balance in the human ear.

(**a**) Cristae are organs of *dynamic* equilibrium. They respond to accelerated, rotational movements of the head, as when you ride a rollercoaster. (**b**) Each ear has a vestibular apparatus. Its three semicircular canals are oriented in three directions. (**c**) When your head rotates horizontally, vertically, or diagonally, fluid in the canal corresponding to that direction is displaced in the opposite direction. The fluid presses against a gelatinous mass (cupula). Mechanoreceptors called hair cells project into this mass, and when they bend under pressure, they send signals to the brain. The cupula and its hair cells are the key components of a crista.

(**d**) Otolithic organs of *static* equilibrium are activated when you start and stop moving in a straight line. Here again, hair cells project into a membrane. But embedded in this membrane are ear stones (crystals of calcium carbonate). The weighted membrane slides in response to linear acceleration. As it does, the hair cells bend and are activated.

36.5 HEARING

Many arthropods and nearly all vertebrates have *acoustical* organs and so can hear sounds. For example, humans and other land-dwelling mammals have a pair of distinctive **ears**. Small organs, flush with the body surface, serve comparable functions in insects and amphibians. Such organs are located on the front legs of crickets and near the hind legs of grasshoppers.

Hearing starts with **acoustical receptors**, which are vibration-sensitive mechanoreceptors. A vibration is a wavelike form of mechanical energy. For example, clapping produces waves of compressed air. Each time hands clap together, molecules are forced outward, so a low-pressure state is created in the region they vacated. The pressure variations can be depicted as a wave form, and the amplitude of its peaks corresponds to loudness. The *frequency* of a sound is the number of wave cycles per second. Each cycle extends from the start of one wave peak to the start of the next. The more cycles per second, the higher the frequency, and the higher the perceived *pitch* of the sound:

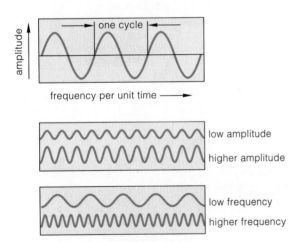

When sound waves arrive at an acoustical organ, they encounter a membrane and make it vibrate. In invertebrates, vibrations directly stimulate mechanoreceptors attached to the membrane. In vertebrates, the membrane vibrations cause a fluid inside the ear to be displaced. The fluid movement causes mechanoreceptors to bend. With enough deformation, action potentials are produced. They travel along an auditory nerve leading from the receptors to the brain.

The human ear has regions that receive, amplify, and sort out sound waves. Figure 36.10 shows the location of their specialized hair cells, the acoustical receptors that respond to disturbances created by sounds. Prolonged exposure to some intense sounds permanently damages the inner ears (Figure 36.11). The hair cells are not adapted to amplified music, jet planes, and other recent developments.

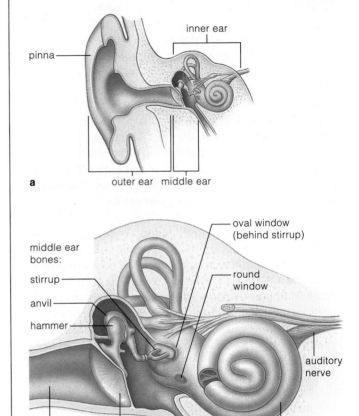

Figure 36.10 Sensory reception in the human ear. (**a**) Each ear is divided into three regions (outer, middle, and inner ear), which receive, amplify, and sort out acoustical stimuli. An ear's components are shown in (**b**). External flaps of the *outer* ear collect sound waves, which move into an auditory canal and then arrive at the eardrum (the tympanic membrane).

The introduction to this chapter described the sense of hearing that is associated with echolocation. Although we cannot hear them, the extremely high-frequency waves (ultrasound waves) produced by an echolocating bat are not weak. They have been measured at 100 decibels—which is in the same range as thunder or a freight train racing past.

Bear in mind, bats are not the only echolocating animals. Dolphins and whales also emit ultrasounds, which travel through their marine environments. By perceiving frequency variations in the echoes, all of these mammals also can pinpoint the distance and direction of movement of one another and of predators or prey.

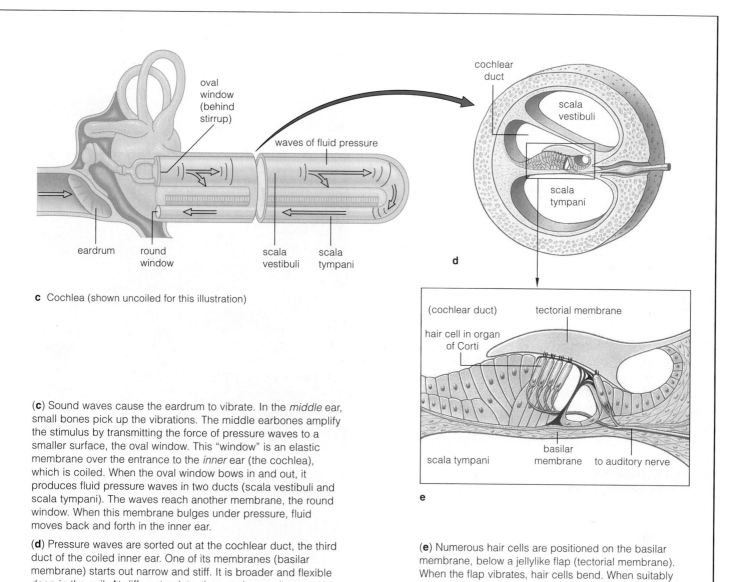

c Cochlea (shown uncoiled for this illustration)

(**c**) Sound waves cause the eardrum to vibrate. In the *middle* ear, small bones pick up the vibrations. The middle earbones amplify the stimulus by transmitting the force of pressure waves to a smaller surface, the oval window. This "window" is an elastic membrane over the entrance to the *inner* ear (the cochlea), which is coiled. When the oval window bows in and out, it produces fluid pressure waves in two ducts (scala vestibuli and scala tympani). The waves reach another membrane, the round window. When this membrane bulges under pressure, fluid moves back and forth in the inner ear.

(**d**) Pressure waves are sorted out at the cochlear duct, the third duct of the coiled inner ear. One of its membranes (basilar membrane) starts out narrow and stiff. It is broader and flexible deep in the coil. At different points, the membrane vibrates more strongly to sounds of different frequencies.

(**e**) Numerous hair cells are positioned on the basilar membrane, below a jellylike flap (tectorial membrane). When the flap vibrates, hair cells bend. When suitably bent, they fire off action potentials that travel through an auditory nerve to the brain.

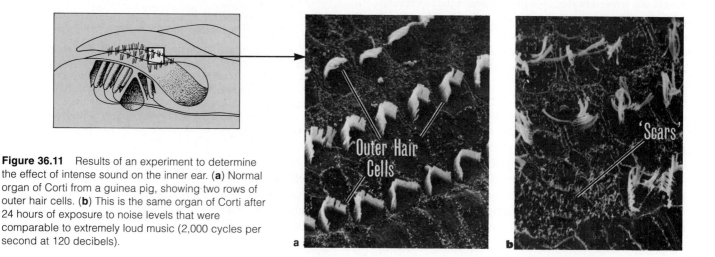

Figure 36.11 Results of an experiment to determine the effect of intense sound on the inner ear. (**a**) Normal organ of Corti from a guinea pig, showing two rows of outer hair cells. (**b**) This is the same organ of Corti after 24 hours of exposure to noise levels that were comparable to extremely loud music (2,000 cycles per second at 120 decibels).

36.6 VISION

All organisms, whether they see or not, are sensitive to light. Even a single-celled amoeba abruptly stops moving when you shine a light on it. What we call **vision**, however, requires (1) a complex system of photoreceptors and (2) complex brain centers that can receive and interpret the patterns of action potentials from different parts of the photoreceptor system. The sense of vision is an awareness of the position, shape, brightness, distance, and movement of visual stimulus.

Eyes are photoreceptor organs that contribute to image formation. As you will see, some types of eyes are better than others at doing this.

36.7 INVERTEBRATE EYES

Many invertebrates have **eyespots**, not eyes. Among these animals, eyespots are clusters of photosensitive cells that are arranged in a cuplike depression in the epidermis. The simplest animals with eyes are mollusks (Figures 36.12 and 36.13). Some molluscan eyes have a transparent lens with a transparent cover, a **cornea**, over it. They have a **retina**, a light-sensitive tissue with densely packed photoreceptors. Most of them incorporate an adjustable **lens**, a transparent cone or sphere that focuses incoming light onto a dense layer of photoreceptor cells behind it.

Just because an invertebrate has lens-equipped eyes, this doesn't necessarily mean it sees as you do. Often the lenses channel light either in front of or behind photoreceptors, the result being a very diffuse kind of stimulation. In such cases, the animal can detect a general change in light intensity, as when another animal passes overhead in the water. But it cannot discern the size or shape of objects.

Of all invertebrates, octopuses and other cephalopods have the most complex eyes and refined sense of vision. Each eye has a cornea, lens, retina, and an **iris**—a ring of contractile tissue that can be adjusted to admit more or less light through a **pupil**, an opening at the ring's center. Like you, they have "camera eyes." Light enters a chamber through a small opening, is focused at a lens, and then strikes a light-sensitive surface.

Insects and crustaceans (such as crabs) have **compound eyes**. These photoreceptor organs contain closely packed photosensitive units, of the sort shown in Figure 36.14a. Some compound eyes have many thousands of these units, which are called ommatidia (singular, ommatidium). According to the mosaic theory of image formation, each ommatidium samples only a small part of the overall visual field. An image is

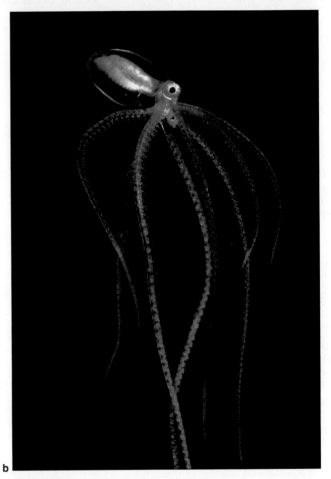

Figure 36.12 Examples of well-developed eyes among the invertebrates.

(a) This shelled mollusk, a conch, is peering into the waters of the Great Australian Barrier Reef.

(b) The octopus, one of the cephalopods, has a splendid visual sense. The pupils of its eyes constrict into narrow, rectangular slits. If you happen to bump into an octopus while snorkeling or diving, the large slits may flare open suddenly, one or both at a time, in response. The octopus uses its startling stare to secure the attention of a potential mate and possibly to warn away enemies.

built up according to signals about differences in light intensities across the field, with each unit contributing a separate bit to a visual mosaic (Figure 36.14b). The units must be very good at detecting movement; have you ever tried to sneak up on a fly?

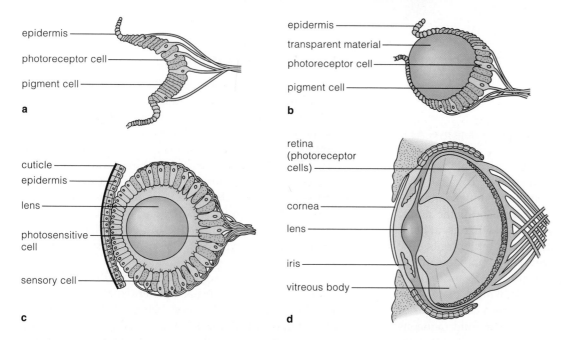

Figure 36.13 Structural organization of invertebrate eyespots and eyes, longitudinal section. There are many more photoreceptors than can be shown in these diagrams. (**a**) A limpet's eyespot, a shallow depression in the epidermis that incorporates light-sensitive receptors. (**b**) An abalone eye. The transparent material may serve as a lens. (**c**) Snail eye. (**d**) Octopus eye. Compare this with the structure of the human eye, shown in Figure 36.15.

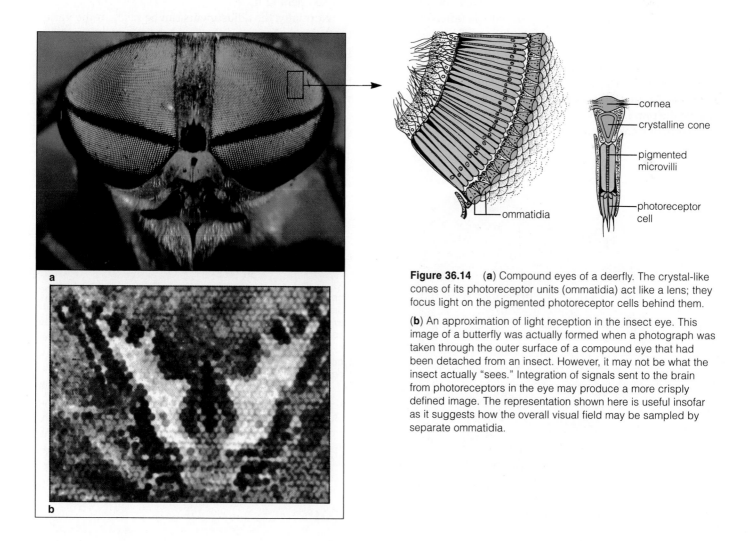

Figure 36.14 (**a**) Compound eyes of a deerfly. The crystal-like cones of its photoreceptor units (ommatidia) act like a lens; they focus light on the pigmented photoreceptor cells behind them.

(**b**) An approximation of light reception in the insect eye. This image of a butterfly was actually formed when a photograph was taken through the outer surface of a compound eye that had been detached from an insect. However, it may not be what the insect actually "sees." Integration of signals sent to the brain from photoreceptors in the eye may produce a more crisply defined image. The representation shown here is useful insofar as it suggests how the overall visual field may be sampled by separate ommatidia.

36.8 VERTEBRATE EYES

Eye Structure

Figure 36.15 shows the structure of the human eye. As is true of most vertebrate eyes, it has three layers, as listed in Table 36.2. The outer layer consists of a sclera and transparent cornea. The middle layer consists of a choroid and lens, as well as semifluid and jellylike substances. The key feature of the inner layer is the retina. As in invertebrates, the vertebrate retina is a light-sensitive tissue packed with photoreceptor cells. The retina is at the back of your eyeball and more toward the eyeball's roof in certain birds (Figures 36.15 and 36.16).

The sclera, the dense, fibrous "white" of the eye, protects most of the eyeball. The cornea covers the rest of it. In the eye's middle layer, a clear fluid (aqueous humor) bathes both sides of the lens. A jellylike substance (vitreous body) fills the chamber behind the lens. The choroid, a dark-pigmented tissue, absorbs the light that is not absorbed by photoreceptors (and so prevents it from scattering inside the eyeball).

Suspended beneath the transparent circle of the cornea is a doughnut-shaped, pigmented iris (after the Latin *irid*, referring to the rings of a rainbow). The dark "hole" in the center of the doughnut is the pupil, the entrance for light. When bright light hits the eye, circular muscles in the iris contract and so shrink the size of the pupil. In dim light, radial muscles contract and so enlarge the pupil. Behind the iris is the lens, with onion-like layers of transparent proteins.

Focusing Mechanisms

Because the cornea has a curved surface, light rays coming from the same source hit it at different angles, so their trajectories change. (As described in the *Focus* essay on page 54, rays of light bend at the boundaries between two different substances, and the bending sends them in new directions.) Owing to the changed angles of their trajectories, the rays converge at the back of the eyeball and stimulate the retina in a distinct pattern. That pattern of stimulation is upside-down and reversed left to right, relative to the original source of the light rays. Figure 36.17a is a simplified diagram of this pattern.

Light rays from sources at varying distances from the eye strike the cornea at different angles and will be focused at different distances behind it. Therefore, certain components of the eye must be adjusted, so that all of the incoming stimuli will be focused onto the retina. Such adjustment is known as **accommodation**. Without it, rays being transmitted from very distant objects would be erroneously focused in front of the retina, and rays being transmitted from very close objects would be focused behind it.

Normally, the adjustments can be made at the vertebrate lens. Among fishes and reptiles, eye muscles move the entire lens forward or back, like the focusing apparatus inside a camera. Increasing the distance between the lens and retina moves the focal point forward. Decreasing the distance moves it back.

By contrast, in both birds and mammals, the shape of the lens itself can be adjusted. A particular muscle (called the ciliary muscle) encircles the lens and is attached to it by fiberlike ligaments (Figure 36.17). Relaxation of this muscle makes the lens flatten and so moves the focal point farther back (Figure 36.17b). Contraction of the muscle makes the lens bulge and so moves the focal point forward (Figure 36.17c).

In some cases, the lens cannot be adjusted enough to make the focal point match up precisely with the retina. For example, in some people, the eyeball is not shaped quite right, and the position of the lens is too close or too far away from the retina. When this is the case, accommodation alone cannot bring about an exact match. As will be described shortly, in the *Focus* essay on page 606, eyeglasses can correct both problems, which are commonly called *nearsightedness* and *farsightedness*.

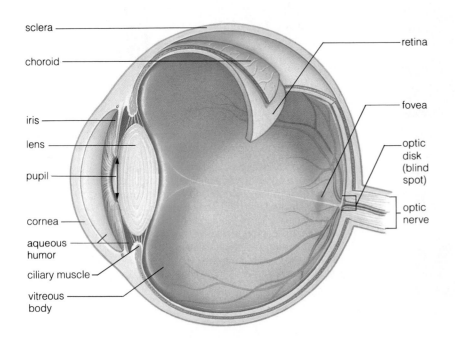

sclera · choroid · iris · lens · pupil · cornea · aqueous humor · ciliary muscle · vitreous body · retina · fovea · optic disk (blind spot) · optic nerve

Figure 36.15 Structure of the human eye.

Table 36.2 Vertebrate Eye Components

Eye Region	Functions
Outer Layer	
Sclera	Protect eyeball
Cornea	Focus light
Middle Layer	
Choroid:	
Pigmented tissue	Prevent light scattering
Iris	Control incoming light
Pupil	Let light enter
Lens	Finely focus light on photoreceptors
Aqueous humor	Transmit light, maintain pressure
Vitreous body	Transmit light, support lens, support eye
Inner Layer	
Retina	Absorb, convert light
Fovea	Increase visual acuity
Start of optic nerve	Transmit signals toward brain

Figure 36.16 A baby owl indirectly demonstrating that its photoreceptors are concentrated more on the roof of its eyeballs. It is advantageous for owls and other birds of prey to look down more than they look up when they are in flight, scanning the ground for a meal. When these birds are on the ground, they cannot see something overhead unless they turn their head almost upside-down.

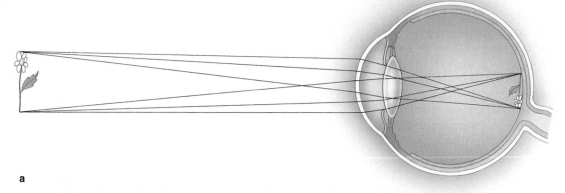

a

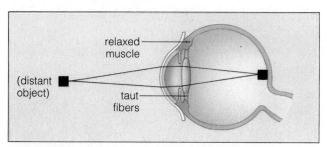

b Accommodation for distant objects (lens flattens)

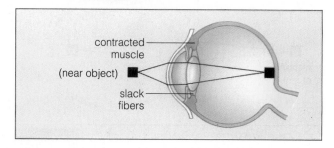

c Accommodation for near objects (lens bulges)

Figure 36.17 Pattern of stimulation of photoreceptor cells of the retina. The curved cornea changes the trajectories of incoming light rays in such a way that the pattern is upside-down and reversed left to right, compared to the object in the visual field that transmitted the rays.

(**a**) Focusing light on the retina through lens adjustments. A ciliary muscle that rings the lens attaches to it by fiberlike ligaments. (**b**) Distant objects are brought into focus when the muscle relaxes and makes the lens flatten, so the focal point moves farther back. (**c**) Close objects are brought into focus when this muscle contracts and makes the lens bulge, so the focal point moves closer.

Disorders of the Eye

Two-thirds of all the sensory receptors that your body requires are in your eyes. Those photoreceptors do more than detect light. They also allow you to see the world in a rainbow of colors. Your eyes are the single most important source of information about the outside world.

Injuries, disease, inherited abnormalities, and advancing age can disrupt functions of the eyes. The consequences range from relatively harmless conditions, such as near-sightedness, to total blindness. Each year, many millions of people must deal with such consequences.

Color Blindness   Consider a common heritable abnormality, *red-green color blindness*. It is an X-linked, recessive trait that shows up most often in males. The retina lacks some or all of the cone cells with pigments that normally respond to light of red or green wavelengths. Most of the time, color-blind persons merely have trouble distinguishing red from green in dim light. However, some cannot distinguish between the two even in bright light. The rare few who are totally color blind have only one of three kinds of pigments that selectively respond to red, green, or blue wavelengths. They see the world only in shades of gray.

Focusing Problems   Other heritable abnormalities arise from misshapen features of the eye that affect the focusing of light. *Astigmatism*, for example, results from corneas with an uneven curvature; they cannot bend incoming light rays to the same focal point.

Nearsightedness (myopia) commonly occurs when the horizontal axis of the eyeball is longer than the vertical axis. It also occurs when the ciliary muscle responsible for adjustments in the lens contracts too strongly. The outcome is that images of distant objects are focused in front of the retina instead of on it (Figure *a*). *Farsightedness* (hyperopia) is the opposite problem. The vertical axis of the eyeball is longer than the horizontal axis (or the lens is "lazy"), so close images are focused behind the retina (Figure *b*).

Even a normal lens loses some of its natural flexibility as a person grows older. That is why people over forty years old often start wearing eyeglasses.

Eye Diseases   The structure of the eye and its functions are vulnerable to infection and disease. Especially in the southeastern United States, for example, a fungal infection of the lungs (*histoplasmosis*) can lead to retinal damage. This complication can cause partial or total loss of vision. As another example, *Herpes simplex*, a virus that causes skin sores, also can infect the cornea and cause it to ulcerate.

Trachoma is a highly contagious disease that has blinded millions, mostly in North Africa and the Middle East. The culprit is a bacterium that also is responsible for the sexually transmitted disease chlamydia (page 782). The

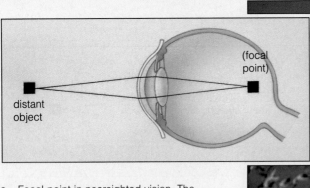

a   Focal point in nearsighted vision. The example shows flamingos in Tanzania, East Africa. In a nearsighted person, the birds in flight in the background would be out of focus.

eyeball and the lining of the eyelids (conjunctiva) become damaged. The damaged tissues are entry points for bacteria that can cause secondary infections. In time the cornea can become so scarred that blindness follows.

Age-Related Problems   *Cataracts*, a gradual clouding of the lens, is a problem associated with aging, although it also may arise through injury or diabetes. Possibly the condition arises when the transparent proteins making up the lens undergo structural changes. The clouding may skew the trajectory of incoming light rays. If the lens becomes totally opaque, light cannot enter the eye at all.

Glaucoma results when excess aqueous humor accumulates inside the eyeball. Blood vessels that service the retina collapse under the increased fluid pressure. Vision deteriorates as neurons of the retina and optic nerve die off. Although chronic glaucoma often is associated with advanced age, the problem actually starts in middle age. If detected early, the fluid pressure can be relieved by drugs or surgery before the damage becomes severe.

Eye Injuries   *Retinal detachment* is the eye injury we read about most often. It may follow a physical blow to the head or an illness that tears the retina. As the jellylike vitreous body oozes through the torn region, the retina becomes lifted from the underlying choroid. In time it may peel away entirely, leaving its blood supply behind. Early symptoms of the damage include blurred vision, flashes of light that occur in the absence of outside stimulation, and loss of peripheral vision. Without medical intervention, the person may become totally blind in the damaged eye.

New Technologies   Today a variety of tools are used to correct some eye disorders. In *corneal transplant surgery*, the defective cornea is removed, then an artificial cornea (made of clear plastic) or a natural cornea from a donor is stitched in place. Within a year, the patient is fitted with eyeglasses or contact lenses. Similarly, cataracts can be surgically corrected by removing the lens and replacing it with an artificial one, although the operation is not always successful.

Severely nearsighted people may opt for *radial keratotomy*, a still-controversial surgical procedure in which tiny, spokelike incisions are made around the edge of the cornea to flatten it more. When all goes well, the adjustment eliminates the need for corrective lenses. Sometimes, however, the result is overcorrected or undercorrected vision.

Retinal detachment may be treatable with *laser coagulation*, a painless technique in which a laser beam seals off leaky blood vessels and "spot welds" the retina to the underlying choroid.

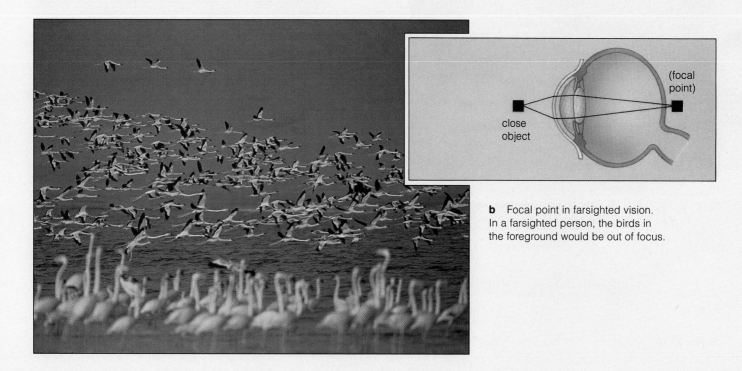

b   Focal point in farsighted vision. In a farsighted person, the birds in the foreground would be out of focus.

36.9 FROM SIGNALING TO VISUAL PERCEPTION

The pathway from the mammalian retina into the brain stands as one of the best examples of neuronal architecture. Along this sensory pathway, raw visual information is received, transmitted, and combined in ways that lead to awareness of light and shadows, of colors, of near and distant objects in the outside world.

Organization of the Retina

The flow of information begins as light reaches the retina, at the back of the eyeball. The retina's basement layer, a pigmented epithelium, covers the choroid. Resting on the epithelium are densely packed photoreceptors, called **rod cells** and **cone cells** (Figure 36.18). Rod cells detect very dim light. At night, they contribute to coarse perception of movement by detecting changes in light intensity across the field of vision. Cone cells detect bright light. They contribute to sharp daytime vision and color perception.

Sensory neurons are organized in distinct layers above the rods and cones. As Figure 36.19 shows, information flows linearly from these photoreceptors to the *bipolar* types of sensory neurons, then to the types called *ganglion* cells, the axons of which form the optic nerves to the visual cortex.

Before visual signals depart from the retina, they decrease dramatically in number, with input from 125 million photoreceptors converging on a mere 1 million ganglion cells. Also before leaving the retina, signals flow laterally among *horizontal* cells and *amacrine* cells. These cells act in concert to dampen or enhance the signals traveling to ganglion cells. Thus, *a great deal of synaptic integration and processing goes on even before visual information is sent to the brain.*

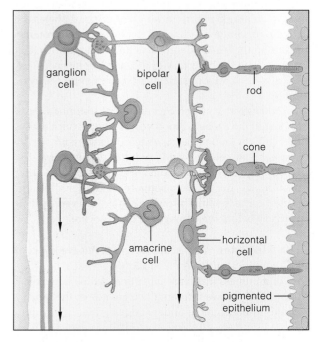

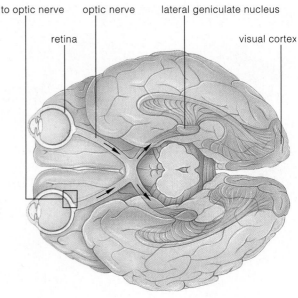

Figure 36.19 Sensory pathway from the retina to the brain.

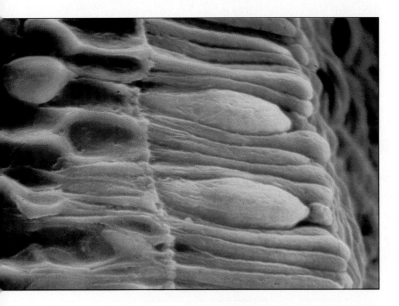

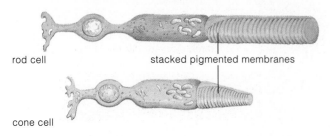

Figure 36.18 Mammalian photoreceptors: rods and cones.

Neuronal Responses to Light

Rod Cells A rod cell's outer segment consists of several hundred membranous disks, each peppered with 10^8 molecules of rhodopsin (a visual pigment). The membrane stacking and the sheer density of pigments increase the chance of photon interception. Signals resulting from even a few photons can lead to conscious awareness of incoming light.

Rhodopsin effectively absorbs wavelengths in the blue-to-green portion of the visible spectrum (Figure 7.3). Absorption changes the pigment's shape. This triggers a cascade of reactions that alter activity at ion channels and ion pumps across the membrane. Gated sodium channels now close, the voltage difference changes across the membrane, and a graded potential results. What does this signal do? It reduces the ongoing release of a neurotransmitter. This neurotransmitter otherwise suppresses activity of adjacent sensory neurons. Those neurons may now start sending signals about the visual stimulus.

Cone Cells Rhodopsin does not respond to bright light. Color and daytime vision depend on red, green, and blue cone cells, each with a different kind of visual pigment. Here again, photon absorption reduces the release of a neurotransmitter that inhibits signal formation in adjacent sensory neurons.

The fovea, a funnel-shaped depression near the center of the retina, consists of slender cone cells beneath much thinner layers of sensory neurons. This is the area of greatest visual acuity—the precise discrimination between adjacent points in space.

Ganglion Cells Visual perception begins when light strikes "receptive fields," restricted areas of the retinal surface that influence the activity of individual sensory neurons, including ganglion cells. The field for each ganglion cell is a tiny circle on the retina. In some cases, the cell responds best to a small spot of light, ringed by darkness, in the field's center. In other cases, motion, a rapid change in light intensity, or a spot of one color elicits the response. Figure 36.20 shows evidence of ganglion cells detecting the edge of a bar that crossed their receptive field. Such cells do not respond to diffuse, uniform illumination—which could produce a confusing array of signals to the brain.

Lateral Geniculate Nucleus Axons of the two optic nerves end in the lateral geniculate nucleus (Figure 36.19). Each half of this layered brain region has a distinctive bend to it (*geniculate* means "bent like a knee"). Within each layer, the positions of receptive fields correspond to those of the retina. Interneurons in each layer deal only with a certain aspect of a visual stimulus—form, movement, depth, color, texture, and

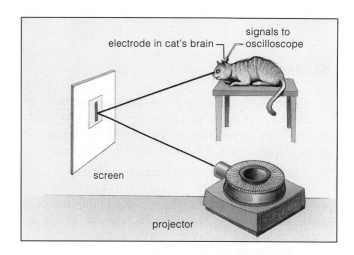

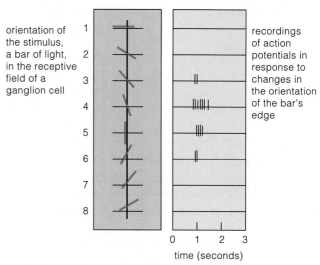

orientation of the stimulus, a bar of light, in the receptive field of a ganglion cell

recordings of action potentials in response to changes in the orientation of the bar's edge

time (seconds)

Figure 36.20 Example of experiments into the nature of receptive fields for visual stimuli. David Hubel and Torsten Wiesel implanted an electrode in the brain of an anesthetized cat. They positioned the cat in front of a small screen upon which different patterns of light were projected—in this case, a hard-edged bar. Light or shadow falling on a restricted portion of the screen excited or inhibited signals by a single neuron (a ganglion cell) in the visual pathway.

Tilting the bar at different angles produced changes in the neuron's activity. A vertical bar image produced the strongest signal (*numbered 5 in the sketch*). When the bar image tilted slightly, signals were less frequent. When the image tilted past a certain angle, signals stopped. In other experiments, one ganglion cell fired only when an image of a block was moved from left to right across the screen; another fired when the image was moved from right to left.

so on. After initial processing in the lateral geniculate nucleus, all signals travel rapidly, at the same time, to different portions of the visual cortex. There, final integration produces organized electrical activity—and the sensation of sight.

Each eye is an outpost of the brain, collecting and analyzing visual information, then sending it on for further processing in the brain through a highly organized sensory pathway.

SUMMARY

1. A stimulus is a specific form of energy that the body is able to detect by means of sensory receptors. A sensation is an awareness that stimulation has occurred. Perception is understanding what the sensation means.

2. Sensory receptors are endings of sensory neurons or specialized cells adjacent to them. They respond to specific stimuli, such as light and particular forms of mechanical energy. Animals can respond to specific events only if they have receptors that are sensitive to the energy of the stimulus.

    a. Chemoreceptors, such as taste receptors, detect chemical substances dissolved in the body fluids that are bathing them.

    b. Mechanoreceptors, such as free nerve endings, detect mechanical energy associated with changes in pressure, changes in position, or acceleration.

    c. Photoreceptors, such as rods and cones of the retina, detect light.

    d. Thermoreceptors detect the presence of or changes in radiant energy from heat sources.

3. At receptor endings, the stimulus triggers local, graded signals. When the stimulus is strong enough, summation of graded signals may produce action potentials. In vertebrates, the action potentials travel on particular nerve pathways from the receptors to parts of the cerebral cortex.

4. Variations in stimulus intensity are encoded in (a) the frequency of action potentials propagated along an information-carrying neuron and (b) the number of action potentials generated in a given tissue.

5. Somatic sensations include touch, pressure, temperature, pain, and muscle sense. The receptors associated with these sensations are not localized in a single organ or tissue. Stretch receptors, for example, occur in skeletal muscles throughout the body.

6. The special senses include taste, smell, hearing, balance, and vision. The receptors associated with these senses typically reside in sensory organs, such as eyes, or some other particular region.

1. Label the component parts of the human eye. *604*

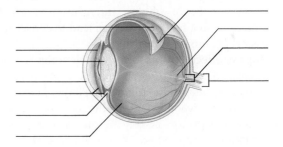

2. What is a stimulus? Receptor cells detect specific kinds of stimuli. When they do, what happens to the stimulus energy? *594–595*

3. Give some examples of chemoreceptors and mechanoreceptors. *594, 596–601*

4. What is sound? How are amplitude and frequency related to sound? Give some examples of animals that apparently perceive sounds. *600–601*

5. What is pain? Can you name one of the receptors associated with pain? *596–597*

6. How does vision differ from photoreception? What sensory apparatus does vision require? *602*

7. How does the vertebrate eye focus the light rays of an image? What is meant by *nearsighted* and *farsighted*? *604–606*

8. On the rim of a bell-shaped jellyfish are saclike structures that hold calcium particles next to a sensory cilium. When the bell tilts, the particles slide over the cilium. Do you think that these structures contribute to the sense of taste, smell, balance, hearing, or vision? *599*

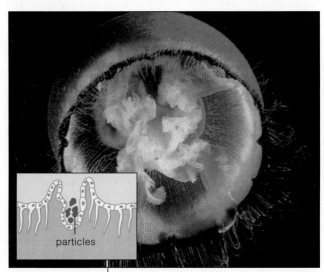

particles

section of the bell margin

1. A _____ is a specific form of energy that is capable of eliciting a response from a sensory receptor.

2. Conscious awareness of a stimulus is called a _____ .

3. _____ is understanding what particular sensations mean.

4. Each sensory system is composed of _____ .
   a. nerve pathways from specific receptors to the brain
   b. sensory receptors
   c. brain regions that deal with sensory information
   d. all of the above are components of sensory systems

5. _____ detect mechanical energy associated with changes in pressure, in position, or acceleration.
   a. Chemoreceptors   c. Photoreceptors
   b. Mechanoreceptors   d. Thermoreceptors

6. Detecting chemical substances present in the body fluids that bathe them is the function of _____ .
   a. thermoreceptors   c. mechanoreceptors
   b. photoreceptors   d. chemoreceptors

7. Name the special sense that is based on the following events: Membrane vibrations cause fluid movements, which bend mechanoreceptors and so trigger action potentials.
   a. taste   c. hearing
   b. smell   d. vision

8. The outer layer of the human eye includes the _____ .
   a. lens and choroid   c. retina
   b. sclera and cornea   d. both a and c

9. The middle layer of the human eye includes the _____ .
   a. lens and choroid   c. retina
   b. sclera and cornea   d. start of optic nerve

10. Match each term with the appropriate description.
    ____ somatic senses   a. produced by strong
    ____ stimulus   stimulation and summation
    ____ special senses   of graded signals
    ____ variations in   b. endings of sensory neurons
         stimulus intensity   or specialized cells next to them
    ____ action potential   c. taste, smell, hearing, balance,
    ____ sensory receptor   and vision
                               d. a specific form of energy that
                                  can elicit a response from a
                                  sensory receptor
                               e. encoded in frequency and
                                  number of action potentials
                               f. touch, pressure, temperature,
                                  pain, and muscle sense

Selected Key Terms

accommodation *604*	otolithic organ *599*
acoustical receptor *600*	pain *596*
chemoreceptor *594*	pheromone *598*
compound eye *602*	photoreceptor *594*
cone cell *608*	pupil *602*
cornea *602*	retina *602*
crista *599*	rod cell *608*
ear *600*	sensation *594*
echolocation *592*	sensory adaptation *595*
eye *602*	sensory system *594*
eyespot *602*	somatic sensation *595*
hair cell *599*	somatosensory cortex *596*
iris *602*	special sense *595*
lens *602*	taste receptor *598*
mechanoreceptor *594*	thermoreceptor *594*
nociceptor *594*	vision *602*
olfactory receptor *598*	

Readings

Kandel, E., and J. Schwartz. 1991. *Principles of Neural Science.* Third edition. New York: Elsevier. Advanced reading, but good coverage of sensory perception.

Nicholls, J., A. Martin, and B. Wallace. 1992. *From Neuron to Brain.* Third edition. Sunderland, Massachusetts: Sinauer.

Wright, Karen. April 1994. "The Sniff of Legend." *Discover* 15(4):60–67. Speculation on the existence of a sensory pathway activated by human pheromones.

Zeki, S. September 1992. "The Visual Image in Mind and Brain." *Scientific American* 267(3):68–76.

37 INTEGRATION AND CONTROL: ENDOCRINE SYSTEMS

Hormone Jamboree

In the 1960s, at her forest camp by the shore of Lake Tanganyika in Kenya, the primatologist Jane Goodall let it be known that bananas were available. One of the first chimpanzees attracted to the delicious food was a female—Flo, as she came to be called. Flo brought along her offspring, an infant female and a juvenile male, and exhibited commendable parental behavior toward them. Three years later, Flo's preoccupation with motherhood gave way to a preoccupation with sex. Now male chimpanzees followed Flo to the camp, and they stayed for more than the bananas.

Sex, Goodall discovered, is the premier force in the social life of chimpanzees. These primates don't mate for life as eagles do, or wolves. Before the rainy season begins, mature females that are entering a fertile cycle become sexually active. In response to changing con-

Figure 37.1 (**a**) Jane Goodall in Gombe National Park, near the shores of Lake Tanganyika, scouting for chimpanzees. (**b**) Flo and three of her offspring.

centrations of hormones in their blood, their external sexual organs become swollen and pink—a visual signal to males. The swellings are the flags of sexual jamborees, of great gatherings of highly stimulated chimps in which any males present may copulate in sequence with the same females.

The gathering of many flag-waving females draws together individuals that otherwise forage alone or in small family groups. It reestablishes bonds that unite their rather fluid community. As infants and juveniles play with one another and with adults, their aggressive and submissive jostlings help map out future dominance hierarchies. Consider Flo, a high-ranking member of the social hierarchy. Through her sexual attractiveness and direct solicitations, she built alliances with many male chimps. Through her high status and aggressive behavior, she helped her male offspring win confrontations with other young male chimps.

The hormone-induced swelling during a female's fertile period lasts somewhere between ten and sixteen days—yet she is fertile for only one to five days. Sex hormones induce swelling even after a female has become pregnant. Sexual selection has no doubt favored the prolonged swelling. Males groom a sexually attractive female more often, protect her, give her more food, and let her tag along to new foraging sites. The more that males accept a female, the higher she goes in the social hierarchy—and the more her individual offspring benefit.

Through their effects, hormones help orchestrate the growth, development, and reproductive cycles of nearly all animals, from invertebrate worms to chimpanzees to humans. They influence minute-by-minute and day-to-day metabolic functions. Through interplays with one another and with the nervous system, hormones have profound influence over the physical appearance, the well-being, and the behavior of individuals. And even the behavior of individuals affects whether they will survive, either on their own or in social groups.

This chapter focuses mainly on hormones—their sources, targets, and interactions as well as the mechanisms involved in their secretion. If the details start to seem remote, remember that this is the stuff of life. Hormones underwrote Flo's appearance, behavior, and rise through the chimpanzee social hierarchy—and just imagine what they have been doing for you.

KEY CONCEPTS

1. Hormones and other signaling molecules have roles in integrating the activities of many individual cells in ways that benefit the whole body. They operate by inducing changes in the activities of target cells.

2. Only cells with receptors for a specific hormone are its targets. Protein hormones alter enzyme activity, and steroid hormones induce gene activation and protein synthesis in target cells.

3. Some hormones help the body adjust to short-term changes in diet and in levels of activity. Other hormones help induce long-term adjustments in cell activities that bring about bodily growth, development, and reproduction.

4. Among vertebrates, the hypothalamus and pituitary gland interact in ways that coordinate the activities of many endocrine glands. Together, they exert control over many of the body's functions.

5. Neural signals, hormonal signals, chemical changes, and environmental cues are the triggers for hormonal secretion.

THE ENDOCRINE SYSTEM

Hormones and Other Signaling Molecules

Throughout their lives, cells respond to changing conditions by taking up and releasing chemical substances. In vertebrates, the responses of millions to many billions of cells must be integrated in ways that benefit the whole body.

Signaling molecules underlie the integration of cell activities. These molecules include hormones, neurotransmitters, local signaling molecules, and pheromones. All act on *target cells*, which are any cells that have receptors for a given type of signaling molecule and that may alter their activities in response to it. A target may or may not be adjacent to the cell that sends the signal.

By definition, **hormones** are secretions from endocrine glands, endocrine cells, and some neurons that the bloodstream distributes to nonadjacent target cells. They are this chapter's focus.

By contrast, *neurotransmitters*, released from neurons, act swiftly on abutting target cells. Chapter 34 describes their sources and their action. *Local signaling molecules*, released by many types of body cells, alter conditions within localized regions of tissues. Prostaglandins, for instance, target smooth muscle cells in bronchiole walls, which then constrict or dilate and so alter air flow in lungs. *Pheromones*, which are the secretions of certain exocrine glands, diffuse through water or air to targets outside the body. Pheromones act on cells of other animals of the same species and so help integrate social behavior (pages 598 and 922).

Discovery of Hormones and Their Sources

The word "hormone" dates back to the early 1900s. Then, W. Bayliss and E. Starling were trying to find out what triggers the secretion of pancreatic juices when food travels through the canine gut. As they knew already, acids mix with food in the stomach, and the pancreas secretes an alkaline solution after the acidic mixture moves into the small intestine. Was the nervous system or something else stimulating the pancreatic response?

To find the answer, Bayliss and Starling blocked nerves—but not blood vessels—leading to a laboratory animal's upper small intestine. Later, when acidic food entered the intestine, the pancreas was still able to respond to signals about this. More telling, extracts of cells from the intestinal lining—a glandular epithelium—also induced the response. Glandular cells had to be the source of the pancreas-stimulating substance.

Figure 37.2 On the facing page, an overview of some components of the human endocrine system and the primary effects of their major hormones. The system also includes endocrine cells of many organs, including the liver, kidneys, heart, and small intestine. The hypothalamus, a major component of the brain, also secretes some hormones.

The substance came to be called secretin. Proof of its existence and mode of action confirmed a centuries-old idea: *Internal secretions that are picked up by the bloodstream influence the activities of the body's organs.* Starling coined the word "hormone" for such internal glandular secretions (after *hormon*, meaning to set in motion).

Later, researchers identified other hormones and their sources. Figure 37.2 shows the locations of the following major sources of hormones in the human body, although these sources are typical of most vertebrates:

> Pituitary gland
> Adrenal glands (*two*)
> Pancreatic islets (*numerous cell clusters*)
> Thyroid gland
> Parathyroid glands (*in humans, four*)
> Pineal gland
> Thymus gland
> Gonads (*two*)
> Endocrine cells of the gut, liver, kidneys, placenta, and other organs

Collectively, these sources of hormones came to be viewed as the **endocrine system**. The name implies that there is a separate control system for the body, apart from the nervous system. (*Endon* means within; *krinein* means separate.) However, both biochemical studies and electron microscopy have since revealed that endocrine sources and the nervous system function in intricately connected ways, as you will see shortly.

Hormones are signaling molecules secreted by endocrine glands, endocrine cells, and some neurons. The bloodstream distributes hormones to nonadjacent target cells.

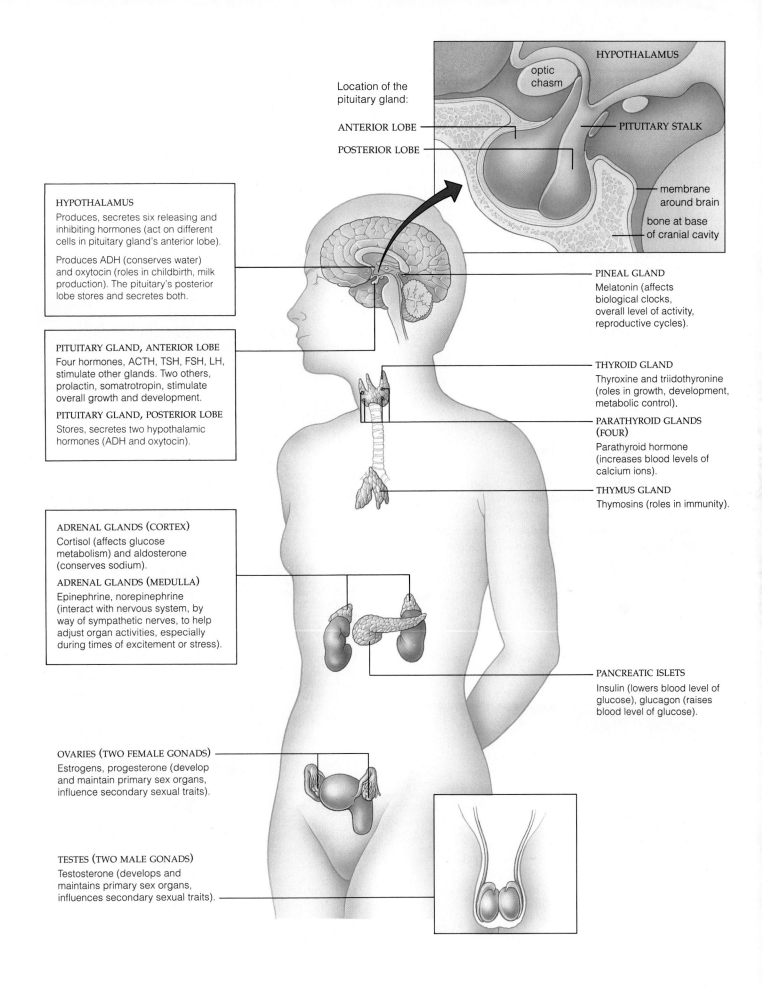

Location of the pituitary gland:

HYPOTHALAMUS

optic chasm

ANTERIOR LOBE

POSTERIOR LOBE

PITUITARY STALK

membrane around brain

bone at base of cranial cavity

HYPOTHALAMUS
Produces, secretes six releasing and inhibiting hormones (act on different cells in pituitary gland's anterior lobe).

Produces ADH (conserves water) and oxytocin (roles in childbirth, milk production). The pituitary's posterior lobe stores and secretes both.

PITUITARY GLAND, ANTERIOR LOBE
Four hormones, ACTH, TSH, FSH, LH, stimulate other glands. Two others, prolactin, somatrotropin, stimulate overall growth and development.

PITUITARY GLAND, POSTERIOR LOBE
Stores, secretes two hypothalamic hormones (ADH and oxytocin).

ADRENAL GLANDS (CORTEX)
Cortisol (affects glucose metabolism) and aldosterone (conserves sodium).

ADRENAL GLANDS (MEDULLA)
Epinephrine, norepinephrine (interact with nervous system, by way of sympathetic nerves, to help adjust organ activities, especially during times of excitement or stress).

OVARIES (TWO FEMALE GONADS)
Estrogens, progesterone (develop and maintain primary sex organs, influence secondary sexual traits).

TESTES (TWO MALE GONADS)
Testosterone (develops and maintains primary sex organs, influences secondary sexual traits).

PINEAL GLAND
Melatonin (affects biological clocks, overall level of activity, reproductive cycles).

THYROID GLAND
Thyroxine and triidothyronine (roles in growth, development, metabolic control).

PARATHYROID GLANDS (FOUR)
Parathyroid hormone (increases blood levels of calcium ions).

THYMUS GLAND
Thymosins (roles in immunity).

PANCREATIC ISLETS
Insulin (lowers blood level of glucose), glucagon (raises blood level of glucose).

Hormones and other signaling molecules have diverse effects. Some kinds induce their target cells to take up glucose or some other specific substance from the surroundings. Other kinds stimulate or inhibit their target cells in ways that lead to altered rates of protein synthesis, to modification of existing proteins or cytoplasmic structures, even to changes in the cell's shape.

Two factors exert major influence over the responses to hormonal signals. First, different hormones activate different cellular mechanisms. Second, not all cells can respond to a given signal. Many have receptors for cortisol, for instance, so this particular hormone has widespread effects. Only a few cell types have receptors for hormones that have highly directed effects.

Let's consider the effect of two types of hormone molecules. They are generally categorized as steroid and nonsteroid hormones. Table 37.1 lists just a few examples.

Steroid Hormone Action

Steroid hormones are synthesized from cholesterol. Being lipid-soluble, they diffuse directly across the lipid bilayer of a target cell's plasma membrane, as shown in Figure 37.3. Once inside the cytoplasm, a steroid hormone molecule usually moves into the nucleus, where it binds to some type of receptor. In some cases, the molecule binds to a receptor molecule that is located in the cytoplasm; then the hormone-receptor complex moves into the nucleus.

Either way, the molecular configuration of the complex allows it to interact with a specific region of the cell's DNA. Some complexes inhibit transcription of certain gene regions into mRNA. Other types stimulate transcription. In this case, enzymes and other proteins form by way of mRNA translation. These proteins carry out the cellular response to the hormonal signal.

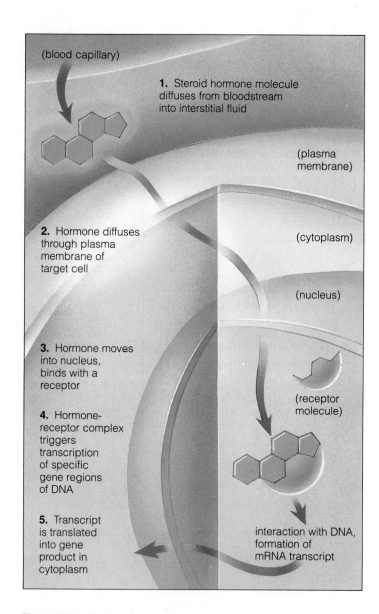

1. Steroid hormone molecule diffuses from bloodstream into interstitial fluid

(blood capillary)

(plasma membrane)

(cytoplasm)

(nucleus)

2. Hormone diffuses through plasma membrane of target cell

3. Hormone moves into nucleus, binds with a receptor

(receptor molecule)

4. Hormone-receptor complex triggers transcription of specific gene regions of DNA

5. Transcript is translated into gene product in cytoplasm

interaction with DNA, formation of mRNA transcript

Figure 37.3 Proposed mechanism of steroid hormone action on a target cell. This same type of mechanism is also thought to occur for thyroid hormones, with one qualification. Recent studies suggest that membrane proteins facilitate the movement of thyroid hormones across the plasma membrane.

Table 37.1	Two Main Categories of Hormones
Type of Hormone	Examples
Steroid	Estrogens, testosterone, aldosterone, cortisol
Nonsteroid:	
Amines	Norepinephrine, epinephrine
Peptides	ADH, oxytocin, TRH
Proteins	Insulin, somatotropin, prolactin
Glycoproteins	FSH, LH, TSH

Consider testosterone. The development of male sexual traits depends on this steroid hormone, and it proceeds normally when target cells have functional receptors for testosterone. In *testicular feminization syndrome*, the receptors are defective. Genetically, an affected individual is male (XY); he has functional testes that secrete testosterone. But target cells cannot respond to the hormone, so the secondary sexual traits that do develop are like those of females.

Nonsteroid Hormone Action

Nonsteroid hormones include certain amines, peptides, proteins, and glycoproteins. Like other water-soluble signaling molecules, nonsteroid hormones require the assistance of membrane proteins to exert their effect on target cells. For example, protein hormones bind to receptors at the plasma membrane. In some cases, the hormone-receptor complex moves into the cytoplasm by endocytosis (Figure 5.12); then further action takes place in the cell. In other cases, a hormone binds to receptors at the membrane surface. The hormone-receptor complex activates transport proteins or triggers the opening of channel proteins across the membrane. Either way, specific ions or other substances move in and their cytoplasmic concentrations change because of this. The changes influence specific cell activities.

Often, nonsteroid hormones activate **second messengers**. These are molecules within a cell that mediate the response to a hormone molecule that makes contact with a membrane protein of target cells.

Cyclic AMP (cyclic adenosine monophosphate) is an example. Suppose a protein hormone binds to a receptor on a target cell (Figure 37.4). Binding triggers activity at a membrane-bound enzyme system. Now an enzyme called adenyl cyclase speeds the conversion of ATP to cyclic AMP. The hormone-receptor complex activates many molecules of the enzyme, in a cascade of reactions. Each molecule speeds the conversion of many ATP molecules to cyclic AMP. Each cyclic AMP molecule switches on many enzyme molecules. These convert many substrate molecules into activated enzymes, and so on. Soon the number of molecules representing the final cellular response to the hormonal signal is enormous.

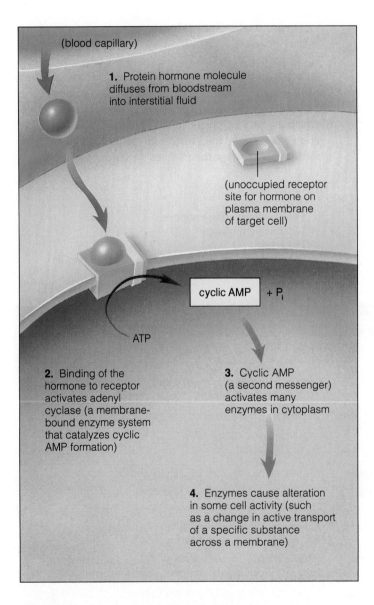

(blood capillary)

1. Protein hormone molecule diffuses from bloodstream into interstitial fluid

(unoccupied receptor site for hormone on plasma membrane of target cell)

cyclic AMP + P$_i$

ATP

2. Binding of the hormone to receptor activates adenyl cyclase (a membrane-bound enzyme system that catalyzes cyclic AMP formation)

3. Cyclic AMP (a second messenger) activates many enzymes in cytoplasm

4. Enzymes cause alteration in some cell activity (such as a change in active transport of a specific substance across a membrane)

Figure 37.4 Proposed mechanism of protein hormone action 1on a target cell. The response is mediated by a second messenger inside the cell—in this case, cyclic AMP. Other chemical messengers may be involved, depending on the particular hormone and its particular target cell.

1. *Steroid* hormones stimulate or inhibit protein synthesis by entering the nucleus of target cells and switching certain genes on or off.

2. *Protein* hormones enter cells either by receptor-mediated endocytosis or by activating membrane proteins. Often they bind to receptors and so activate second messengers in the cell, some of which trigger an amplified response to the hormone.

37.3 THE HYPOTHALAMUS AND THE PITUITARY GLAND

Deep in the brain is the **hypothalamus**. This brain region monitors internal organs and activities related to their functioning, such as eating and sexual behavior. It also secretes some hormones. Suspended from its base by a slender stalk is a pea-size lobed gland. The hypothalamus and this **pituitary gland** interact as a major neural-endocrine control center.

The *posterior* lobe of the pituitary stores and secretes two hormones that are synthesized in the hypothalamus. The *anterior* lobe produces and secretes its own hormones, most of which govern the release of hormones from other endocrine glands (Table 37.2). The pituitary of many vertebrates (but not humans) also has an intermediate lobe. In many cases, this lobe secretes a hormone that governs reversible changes in skin or fur color.

Posterior Lobe Secretions

Figure 37.5a shows the cell bodies of certain neurons in the hypothalamus. Their axons extend down into the posterior lobe, then terminate next to a capillary bed. The neurons produce antidiuretic hormone (ADH) and oxytocin, then store them in the axon endings. When either hormone is released, it diffuses through interstitial fluid and enters capillaries, then travels the bloodstream to its targets. ADH acts on cells of nephrons and collecting ducts in the kidneys. Kidneys filter blood and rid the body of excess water and salts in urine. ADH promotes water reabsorption when the body must conserve water. Oxytocin has reproductive roles. It triggers contractions in the uterus during labor and causes milk release when offspring are being nursed.

Anterior Lobe Secretions

Anterior Pituitary Hormones Other hypothalamic hormones enter a capillary bed in the pituitary stalk, which delivers them to a second capillary bed in the anterior lobe. There, the hormones leave the bloodstream and act on target cells. As Figure 37.6 shows, different cells of the anterior pituitary secrete six hormones that they themselves produce:

Corticotropin	ACTH
Thyrotropin	TSH
Follicle-stimulating hormone	FSH
Luteinizing hormone	LH
Prolactin	PRL
Somatotropin (or growth hormone)	STH (or GH)

Table 37.2	Hormones Released from the Mammalian Pituitary Gland			
Pituitary Lobe	Secretions	Designation	Main Targets	Primary Actions
Posterior Nervous tissue (extension of hypothalamus)	Antidiuretic hormone	ADH	Kidneys	Induces water conservation required in control of extracellular fluid volume (and, indirectly, solute concentrations)
	Oxytocin		Mammary glands	Induces milk movement into secretory ducts
			Uterus	Induces uterine contractions
Anterior Mostly glandular tissue	Corticotropin	ACTH	Adrenal cortex	Stimulates release of adrenal steroid hormones
	Thyrotropin	TSH	Thyroid gland	Stimulates release of thyroid hormones
	Gonadotropins: Follicle-stimulating hormone	FSH	Ovaries, testes	In females, stimulates egg formation; in males, helps stimulate sperm formation
	Luteinizing hormone	LH	Ovaries, testes	In females, stimulates ovulation, corpus luteum formation; in males, promotes testosterone secretion, sperm release
	Prolactin	PRL	Mammary glands	Stimulates and sustains milk production
	Somatotropin (also called growth hormone)	STH (GH)	Most cells	Promotes growth in young; induces protein synthesis, cell division; roles in glucose, protein metabolism in adults
Intermediate* Glandular tissue, mostly	Melanocyte-stimulating hormone	MSH	Pigmented cells in skin, other surface coverings	Induces color changes in response to external stimuli; affects behavior

*Present in most vertebrates (not humans). MSH is associated with the anterior lobe in humans.

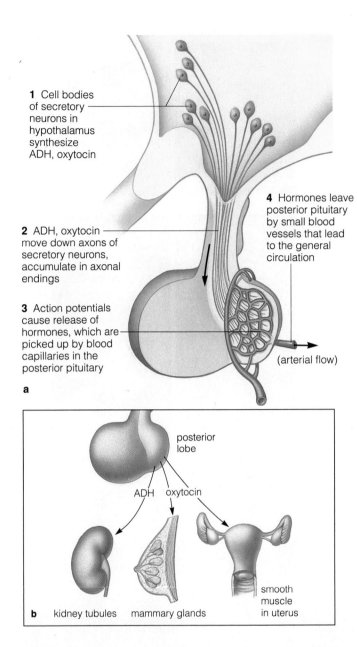

1 Cell bodies of secretory neurons in hypothalamus synthesize ADH, oxytocin

2 ADH, oxytocin move down axons of secretory neurons, accumulate in axonal endings

3 Action potentials cause release of hormones, which are picked up by blood capillaries in the posterior pituitary

4 Hormones leave posterior pituitary by small blood vessels that lead to the general circulation

(arterial flow)

a

posterior lobe

ADH   oxytocin

b   kidney tubules   mammary glands   smooth muscle in uterus

Figure 37.5 (**a**) Functional links between the hypothalamus and the posterior lobe of the pituitary. (**b**) Main targets of the posterior lobe secretions.

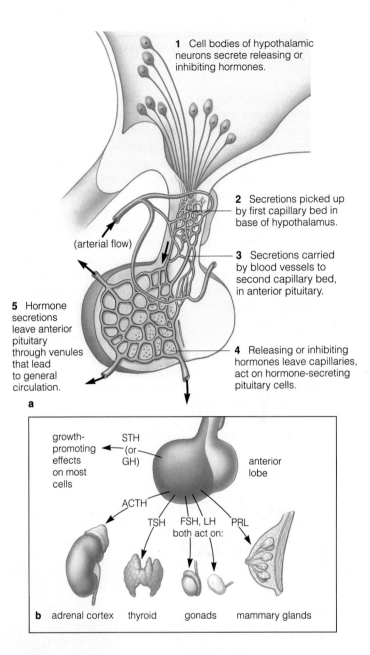

1 Cell bodies of hypothalamic neurons secrete releasing or inhibiting hormones.

2 Secretions picked up by first capillary bed in base of hypothalamus.

(arterial flow)

3 Secretions carried by blood vessels to second capillary bed, in anterior pituitary.

5 Hormone secretions leave anterior pituitary through venules that lead to general circulation.

4 Releasing or inhibiting hormones leave capillaries, act on hormone-secreting pituitary cells.

a

growth-promoting effects on most cells

STH (or GH)

anterior lobe

ACTH   TSH   FSH, LH both act on: PRL

b   adrenal cortex   thyroid   gonads   mammary glands

Figure 37.6 (**a**) Functional links between the hypothalamus and the anterior lobe of the pituitary. (**b**) Main targets of the anterior lobe secretions.

ACTH acts on adrenal glands, and TSH on the thyroid gland, as described shortly. FSH and LH have roles in reproduction, the central topic of Chapter 45.

Somatotropin affects body tissues in general (Table 37.2 and Figure 37.6). It works by stimulating protein synthesis and cell division and so has profound influence over growth, especially of cartilage and bone.

Prolactin also has general effects. But it is better known for its role in stimulating and sustaining milk production in mammary glands, after other hormones have primed the tissues. Prolactin also affects hormone production in ovaries.

About the Hypothalamic Triggers Most hypothalamic hormones acting in the anterior lobe of the pituitary are *releasers*; they stimulate secretion from their target cells. For example, the one called GnRH (short for gonadotropin-releasing hormone) brings about secretion of FSH and LH—both of which are classified as gonadotropins. Similarly, the one called TRH stimulates the secretion of thyrotropin.

Some hypothalamic hormones are *inhibitors* of secretion from their targets in the anterior pituitary. For instance, the one called somatostatin brings about a decrease in somatotropin and thyrotropin secretion.

37.4 EXAMPLES OF ABNORMAL PITUITARY OUTPUT

The body does not produce enormous numbers of hormone molecules. Roger Guilleman and Andrew Schally realized this when they isolated the first known releasing hormone. After four years of dissecting 500 tons of sheep brains, then 7 tons of hypothalamic tissue, they extracted a single milligram of TSH. Yet normal body function depends on those tiny amounts.

For example, when anterior pituitary cells produce too much somatotropin during childhood, *gigantism* results. Affected adults are proportionally similar to a normal person but much larger (Figure 37.7a). When not enough somatotropin is produced during childhood, *pituitary dwarfism* results. Affected adults are proportionally similar to a normal person but much smaller (Figure 37.7b).

What happens if somatotropin output becomes excessive during adulthood, when long bones no longer can lengthen? Then, *acromegaly* will be the result. Bone, cartilage, and other connective tissues in the hands, feet, and jaws will thicken abnormally. So will epithelia of the skin, nose, eyelids, lips, and tongue. Figure 37.8 shows an example.

age nine   sixteen

thirty-three   fifty-two

Figure 37.8 Acromegaly, which resulted from excessive production of STH during adulthood. Before this female reached maturity, she was symptom-free.

Figure 37.7 (**a**) Manute Bol, an NBA center, is 7 feet 6-3/4 inches tall owing to excessive STH production during childhood.

(**b**) Effect of somatotropin (STH) on overall body growth. The person at the center is affected by gigantism, which resulted from excessive STH production during childhood. The person at right displays pituitary dwarfism, which resulted from underproduction of STH during childhood. The person at the left is average in size.

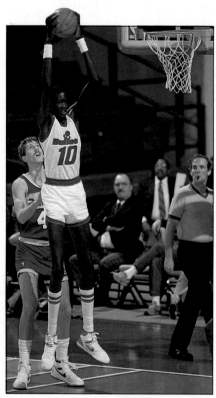

a

b

37.5 SELECTED EXAMPLES OF HORMONAL CONTROL

Table 37.3 lists hormones from endocrine glands other than the pituitary. The remainder of this chapter provides a few examples of the controls over their secretions. The examples will make more sense if you keep three points in mind. *First*, hormones often interact with one another. One or more hormones may oppose, add to, or prime target cells for another hormone's effects. *Second*, the hypothalamus and pituitary govern many responses through homeostatic feedback loops. One or both detect a change in a hormone's concentration in some region, then respond by inhibiting or stimulating its secretion. *Third*, responses vary, depending on the hormone's concentration at any given time and on the nature of receptors on the target cell.

Homeostatic feedback loops to the hypothalamus and the pituitary, variations in hormone levels, and the nature of receptors on target cells influence responses to hormones.

Table 37.3	Hormone Sources Other Than the Mammalian Hypothalamus and Pituitary		
Source	Secretion(s)	Main Targets	Primary Actions
Pancreatic islets	Insulin	Muscle, adipose tissue	Lowers blood sugar level
	Glucagon	Liver	Raises blood sugar level
	Somatostatin	Insulin-secreting cells	Influences carbohydrate metabolism
Adrenal cortex	Glucocorticoids (including cortisol)	Most cells	Promote protein breakdown and conversion to glucose
	Mineralocorticoids (including aldosterone)	Kidney	Promote sodium reabsorption; control salt–water balance
Adrenal medulla	Epinephrine (adrenalin)	Liver, muscle, adipose tissue	Raises blood level of sugar, fatty acids; increases heart rate, force of contraction
	Norepinephrine	Smooth muscle of blood vessels	Promotes constriction or dilation of blood vessel diameter
Thyroid	Triiodothyronine, thyroxine	Most cells	Regulate metabolism; have roles in growth, development
	Calcitonin	Bone	Lowers calcium levels in blood
Parathyroids	Parathyroid hormone	Bone, kidney	Elevates calcium levels in blood
Thymus	Thymosins, etc.	Lymphocytes	Have roles in immune responses
Gonads:			
Testes (in males)	Androgens (including testosterone)	General	Required in sperm formation, development of genitals, maintenance of sexual traits; influence growth, development
Ovaries (in females)	Estrogens	General	Required in egg maturation and release; prepare uterine lining for pregnancy; required in development of genitals, maintenance of sexual traits; influence growth, development
	Progesterone	Uterus, breasts	Prepares, maintains uterine lining for pregnancy; stimulates breast development
Pineal	Melatonin	Gonads (indirectly)	Influences daily biorhythms, seasonal sexual activity
Endocrine cells of stomach, gut	Gastrin, secretin, etc.	Stomach, pancreas, gallbladder	Stimulate activity of stomach, pancreas, liver, gallbladder
Liver	Somatomedins	Most cells	Stimulate cell growth and development
Kidneys	Erythropoietin*	Bone marrow	Stimulates red blood cell production
	Angiotensin*	Adrenal cortex, arterioles	Helps control blood pressure, aldosterone secretion
	Vitamin D$_3$*	Bone, gut	Enhances calcium resorption and uptake
Heart	Atrial natriuretic hormone	Kidney, blood vessels	Increases sodium excretion; lowers blood pressure

*These hormones are not produced in the kidneys but are formed when *enzymes* produced in kidneys activate specific substances in the blood.

37.6 PANCREATIC ISLETS

The pancreas is a gland with exocrine and endocrine functions. Its *exocrine* cells secrete digestive enzymes. About 2 million clusters of *endocrine* cells also are scattered through it. Each small cluster, a **pancreatic islet**, contains three types of hormone-secreting cells:

1. *Alpha* cells secrete the hormone glucagon. Between meals, cells throughout the body take up and use glucose from the blood. The glucose level in blood decreases. At such times, glucagon secretion causes glycogen (a storage polysaccharide) and amino acids to be converted to glucose in the liver. In such ways, *glucagon raises the glucose level.*

2. *Beta* cells secrete the hormone insulin. After meals, when the blood glucose level is high, insulin stimulates glucose uptake by liver, muscle, and adipose cells especially. It promotes synthesis of proteins and fats, and it inhibits protein conversion to glucose. Thus, *insulin lowers the glucose level.*

3. *Delta* cells secrete somatostatin, a hormone that helps control digestion. It also can block secretion of insulin and glucagon.

Figure 37.9 shows how pancreatic hormones interact to maintain blood glucose levels even though the times and amounts of food intake vary. Bear in mind, insulin is the only hormone that prods cells to take up and store glucose in forms that can be rapidly tapped when

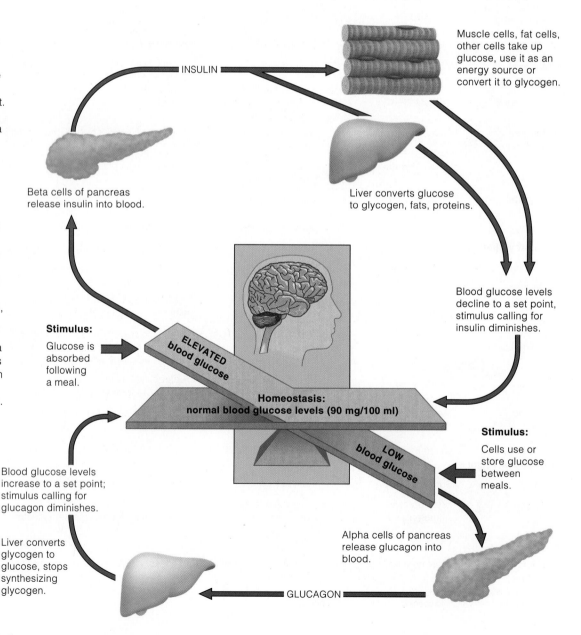

Figure 37.9 Some homeostatic controls over glucose metabolism.

Following a meal, glucose enters the bloodstream faster than cells can use it. The blood glucose level rises, and pancreatic beta cells are stimulated to secrete insulin. Insulin's targets (mainly liver, fat, and muscle cells) use glucose or store excess amounts as glycogen.

Between meals, the blood glucose level decreases. Pancreatic alpha cells are stimulated to secrete glucagon. This hormone's target cells convert glycogen back to glucose, which enters the blood. Also, the hypothalamus prods the adrenal medulla to secrete other hormones that slow down conversion of glucose to glycogen in liver, fat, and muscle cells.

INSULIN

Muscle cells, fat cells, other cells take up glucose, use it as an energy source or convert it to glycogen.

Beta cells of pancreas release insulin into blood.

Liver converts glucose to glycogen, fats, proteins.

Blood glucose levels decline to a set point, stimulus calling for insulin diminishes.

Stimulus:
Glucose is absorbed following a meal.

ELEVATED blood glucose

Homeostasis:
normal blood glucose levels (90 mg/100 ml)

LOW blood glucose

Stimulus:
Cells use or store glucose between meals.

Blood glucose levels increase to a set point; stimulus calling for glucagon diminishes.

Liver converts glycogen to glucose, stops synthesizing glycogen.

Alpha cells of pancreas release glucagon into blood.

GLUCAGON

required. Its central role in carbohydrate, protein, and fat metabolism becomes clear when we observe people who cannot produce enough insulin or who lack body cells that can respond to it.

Insulin deficiency can lead to *diabetes mellitus*, a disorder in which the blood glucose level rises and glucose accumulates in urine. Urination becomes excessive, so the body's water-solute balance is disrupted. Affected people become dehydrated and abnormally thirsty. Without a steady glucose supply, their cells start breaking down fats and proteins for energy. This leads to weight loss. Ketones, which are normal acidic products of fat breakdown, accumulate in the blood and urine. The accumulation promotes excessive water loss. The imbalance disrupts brain function. In extreme cases, death may follow.

In "type 1 diabetes," the body mistakenly mounts an autoimmune response against its own insulin-secreting beta cells. Certain lymphocytes identify the beta cells as "foreign" and destroy them. Genetic susceptibility and environmental triggers combine to produce the disorder, which is less common but more immediately dangerous than the other forms of diabetes. Symptoms usually appear during childhood and adolescence (the disorder also is called juvenile-onset diabetes). Type 1 diabetic patients survive with insulin injections.

In "type 2 diabetes," insulin levels are close to or above normal, but the target cells cannot respond to insulin. As affected persons grow older, their beta cells produce less and less insulin. Type 2 diabetes usually is manifested during middle age. It is less dramatically dangerous than the other type. Affected persons lead normal lives by controlling their diet and weight, and sometimes taking drugs to enhance insulin action or secretion.

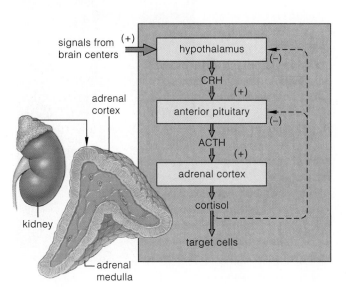

37.7 ADRENAL GLANDS

Adrenal Cortex

Humans have a pair of adrenal glands, one above each kidney (Figure 37.10). The **adrenal cortex** is the outer portion of each gland. Some of its cells secrete glucocorticoids and other hormones. Homeostatic feedback loops to the hypothalamus and pituitary govern their secretion. The glucocorticoids help maintain the blood level of glucose and help suppress inflammatory responses. Cortisol, for instance, blocks the uptake and use of glucose by muscle cells. It also stimulates liver cells to form glucose from amino acids.

In *hypoglycemia*, the glucose blood level falls below a set point, and a negative feedback mechanism kicks in. The hypothalamus detects the decrease and secretes CRH in response (Figure 37.10). This releasing hormone makes the anterior pituitary secrete corticotropin (ACTH). ACTH stimulates the adrenal cortex to secrete cortisol—which helps raise the level of glucose by preventing muscle cells from taking up more glucose from blood. These cells are major glucose users.

During severe stress, painful injury, or prolonged illness, the nervous system overrides feedback control of cortisol secretion. It initiates a *stress response* in which cortisol secretions help suppress inflammation. If unchecked, prolonged inflammation could damage tissues. Cortisol-like drugs, such as cortisone, counter asthma and other chronic inflammatory disorders.

Adrenal Medulla

Figure 37.10 also shows the **adrenal medulla**, the inner portion of the adrenal gland. Neurons located here release epinephrine and norepinephrine. Both are neurotransmitters in some contexts and hormones in others (compare page 568). Sympathetic nerves carry stimulatory and inhibitory signals to these neurons from the hypothalamus and other brain regions.

Epinephrine and norepinephrine help adjust blood circulation and carbohydrate metabolism when the body is excited or stressed. They increase heart rate, dilate arterioles in some regions and constrict them in others, and dilate airways to the lungs. Thus more blood volume is shunted to heart and muscle cells from other regions, and more oxygen flows to energy-demanding cells throughout the body. These are features of the *fight-flight response* (compare page 581).

Figure 37.10 Location of the adrenal glands. The diagram shows the negative feedback loop that governs cortisol secretion.

37.8 THE THYROID AND PARATHYROID GLANDS

Thyroid Functioning

The human **thyroid gland** is at the base of the neck in front of the trachea, or windpipe (Figure 37.11). Its main hormones, thyroxine and triiodothyronine, have widespread effects on most cells. They are critical for normal development of many tissues, and they control overall metabolic rates in humans and other warm-blooded animals. Their importance is brought into sharp focus by cases of abnormal thyroid secretion.

The synthesis of thyroid hormones requires iodine, which is obtained from the diet. In the absence of iodine, blood levels of these hormones decrease. The anterior pituitary responds to this by secreting a thyroid-stimulating hormone (TSH). Because the thyroid hormones cannot be synthesized, the feedback signal continues—and so does TSH secretion. Excess TSH overstimulates the thyroid gland and causes it to enlarge. The enlargement is a form of *goiter* (Figure 37.12*a*). Goiter caused by iodine deficiency is no longer common in countries where people use iodized salt.

Hypothyroidism results from insufficient concentrations of thyroid hormones. Hypothyroid adults often are overweight, sluggish, dry-skinned, intolerant of cold, and sometimes confused and depressed. Affected women often have menstrual disturbances. *Hyperthyroidism* results from excess concentrations of thyroid hormones. Hyperthyroid adults show an increased heart rate, elevated blood pressure, weight loss despite normal caloric intake, intolerance of heat, and profuse sweating. Typically they are nervous, agitated, and have trouble sleeping.

Parathyroid Functioning

Some endocrine glands do not respond directly to other hormones or nerves. They respond homeostatically to chemical change in the immediate surroundings. The **parathyroid glands** are like this. Four such glands are positioned next to the back of the human thyroid (Figure 37.11). They secrete parathyroid hormone (PTH) when the concentration of calcium ions in their surroundings decreases. Their action affects how much calcium is available for enzyme activation, muscle contraction, blood clotting, and many other tasks.

PTH stimulates bone cells to release calcium and phosphate ions, and stimulates the kidneys to conserve them. PTH also helps activate vitamin D. The activated form, a hormone, enhances calcium absorption from ingested food. In vitamin D deficiency, too little calcium and phosphorus is absorbed, so bones develop improperly. This ailment is called *rickets* (Figure 37.12*b*).

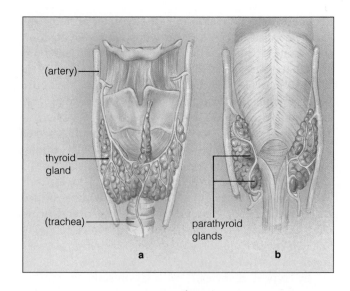

Figure 37.11 (**a**) Human thyroid gland, anterior view. (**b**) Four parathyroid glands, next to the back of the thyroid.

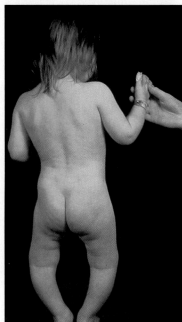

Figure 37.12 (**a**) A mild case of goiter, as displayed by Maria de Medici in the year 1625. A rounded neck was considered to be a sign of great beauty during the late Renaissance. It developed regularly in parts of the world where iodine supplies were insufficient for normal thyroid function.

(**b**) A child with rickets. Bowed legs are typical of the disorder.

37.9 PINEAL GLAND

Hormonal responses to the external environment influence growth, development, and reproduction in predictable ways. The **pineal gland**, a photosensitive organ in the brain, provides a good example of this (Figure 37.2). In the absence of light, the pineal gland secretes the hormone melatonin. Thus blood levels of melatonin vary from day to night, and they vary with the changing seasons. The variations influence the development of primary reproductive organs (gonads). They also influence reproductive cycles and reproductive behavior in a variety of species.

Consider the hamster. In winter, when nights are longest, melatonin levels are high and sexual activity is suppressed. In summer, when days are longer, melatonin levels are low—and hamster sex peaks. Or consider a male white-throated sparrow (Figure 37.13). Melatonin suppresses growth of the bird's gonads until spring, when days start to lengthen. Then, stepped-up gonadal activity leads to the production of hormones that influence singing behavior, in ways that are described on page 912. With his distinctive song, the sparrow defines his territory and may hold the interest of a mate.

Does melatonin also influence human behavior? Perhaps. It seems possible that decreased melatonin secretion triggers puberty, the age at which human reproductive organs and structures start to mature. In cases where disease has destroyed the pineal gland, puberty begins prematurely. The *Focus* essay mentions some other interesting possibilities.

Figure 37.13 A male white-throated sparrow, belting out a song that began, indirectly, with an environmentally induced decline in melatonin secretion.

Focus on the Environment

Rhythms and Blues

At sunset, when light is waning, a bump of tissue in the brain is stimulated to secrete a hormone. The bump is the pineal gland. The hormone, melatonin, acts on certain neurons that lower your body temperature and make you sleep. At sunrise, when melatonin secretion slows, body temperature increases and you wake up and become active.

An internal, biological clock governs the cycle of sleep and arousal. It seems to tick in synchrony with daylength. Think of night workers who try to sleep in the morning but end up staring groggily at sunbeams on the ceiling. Think of travelers from the United States to Paris who go through four days of "jet lag." Two hours past midnight they are sitting up in bed, wondering where the coffee and croissants are. Two hours past noon they are ready for bed. They will shift to a new routine when melatonin's signals start arriving at their target neurons on Paris time.

In winter, some people get depressed, go on carbohydrate binges, and have an overwhelming desire to sleep. Their "winter blues" may be symptoms of a biological clock out of sync with the season's shorter daylengths. Doses of melatonin make their seasonal symptoms worse. Exposure to intense light—which shuts down pineal activity—leads to dramatic improvements.

37.10 SOME FINAL EXAMPLES OF INTEGRATION AND CONTROL

Hormones of the Thymus

The lobed **thymus gland** is located behind the breastbone, between the two lungs (Figure 37.2). Its hormones, the thymosins, may induce the formation and early development of stem cells that give rise to certain white blood cells, the T lymphocytes. These are absolutely central players in the body's immune system (page 671). For example, after clinical tests, children with some immunodeficiency diseases who were provided with thymosins showed significant improvement.

Hormonal Control of the Gonads

Through homeostatic feedback loops, hormones of the **gonads**—the primary reproductive organs—govern reproductive function. Male gonads are called testes (singular, testis), and female gonads, ovaries. These produce gametes and secrete sex hormones—estrogens, progesterone, and androgens (including testosterone). Sex hormones also influence secondary sexual traits, as

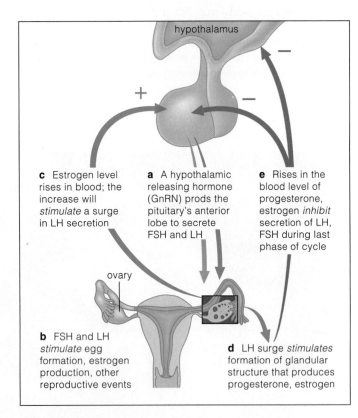

c Estrogen level rises in blood; the increase will *stimulate* a surge in LH secretion

a A hypothalamic releasing hormone (GnRN) prods the pituitary's anterior lobe to secrete FSH and LH

e Rises in the blood level of progesterone, estrogen *inhibit* secretion of LH, FSH during last phase of cycle

b FSH and LH *stimulate* egg formation, estrogen production, other reproductive events

d LH surge *stimulates* formation of glandular structure that produces progesterone, estrogen

hypothalamus

ovary

Figure 37.14 Feedback loops to the hypothalamus and pituitary gland from the ovaries during the menstrual cycle, a recurring reproductive event. Positive feedback triggers egg release from an ovary. Negative feedback after its release prevents release of another egg until the cycle is completed.

they did for Flo and other chimpanzees described at the start of this chapter. The example in Figure 37.14 gives you an idea of how sex hormones influence the menstrual cycle, a key topic of Chapter 45.

Effects of Some Local Signaling Molecules

Many cells detect changes in the surrounding chemical environment and alter their activity, often in ways that either counteract or amplify the change. The cells secrete **local signaling molecules**, the action of which is confined to the immediate vicinity of change. Most of the molecules are taken up rapidly, so few enter the general circulation. Prostaglandins and growth factors are examples.

A variety of prostaglandins are produced in tissues throughout the body. They are released continually, but the rate of synthesis often increases in response to local chemical changes. You will be reading about these molecules in the chapters on immunity and human reproductive function.

A variety of growth factors influence growth by regulating the rate at which certain cells divide. Epidermal growth factor (EGF), discovered by Stanley Cohen, influences the growth of many cell types. Nerve growth factor (NGF), discovered by Rita Levi-Montalcini, promotes the survival of neurons and influences the direction of their growth in an embryo. One experiment demonstrated that certain immature neurons survive indefinitely in tissue culture when NGF is present but die within a few days if it is not.

Comparative Look at a Few Invertebrate Hormones

From earlier chapters, you know that hormones are not exclusive to vertebrates. Bacteria, protistans, fungi, and plants also rely on hormones and other signaling molecules. So do invertebrates.

Consider the hormonal controls over **molting**, a periodic discarding and replacement of a hardened cuticle (page 432). Molting is part of the life cycle of all arthropods and certain crustaceans. It depends in part on the synthesis and secretion of some form of ecdysone, a steroid hormone. Special glands produce this hormone and release it to the circulatory system at molting time. Yet the production, release, and perhaps even the action of ecdysone are under the control of other hormones, released by hormone-secreting cells in the brain. Among crustaceans, these hormones suppress molting until certain times in the life cycle. Among insects, they have the opposite effect. They promote molting, in ways that are described in Figure 37.15, the concluding example for this chapter.

Figure 37.15 Hormonal controls at work during the life cycle of a silkworm moth, *Platysamia cecropia*.

A hatched larva eats until it completes five near-doublings in size, through cell division and enlargement. Secretions of chitin at the body surface form a cuticle that sets an upper limit on increases in mass. When the limit is reached, cell division idles and the larva molts (sheds its cuticle). Cell divisions resume, more chitin is secreted, and a larger cuticle results. The larva grows and molts repeatedly. Then chitin is deposited over the whole insect, legs and all; this is the pupal stage. The pupa lasts eight winter months, in a cocoon of its own making. In spring, drastic tissue changes transform the pupa into the adult.

Neural and endocrine controls dictate when molting and other events of the developmental program will proceed. Cells in the larval brain secrete a hormone that acts on paired prothoracic glands. The glands produce and secrete precursors of ecdysone, a hormone that activates gene expression and stimulates molting. (Compare page 242.) Hormones also act on two glands (the corpora allata) behind the brain. The glands secrete juvenile hormone (JH) which, at high levels, blocks the transformation into the adult.

High levels of both JH and ecdysone promote larval growth and development. High levels of ecdysone and low levels of JH arrest growth and promote cuticle formation. The amount of JH declines steadily and disappears by the fifth larval stage. The absence of JH triggers transformation into the adult.

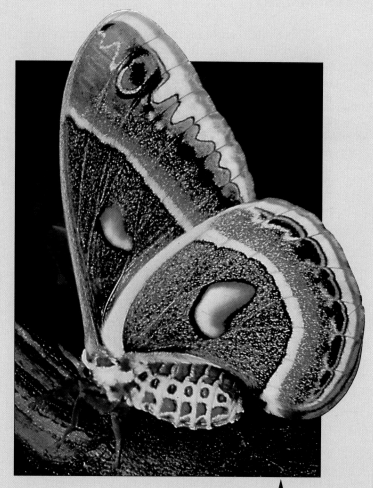

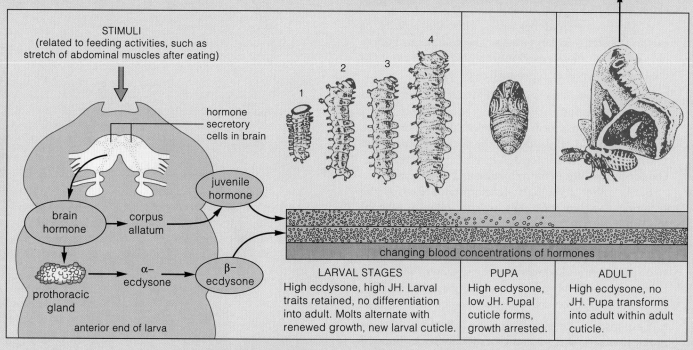

STIMULI
(related to feeding activities, such as stretch of abdominal muscles after eating)

hormone secretory cells in brain

juvenile hormone

brain hormone → corpus allatum

prothoracic gland

α–ecdysone → β–ecdysone

anterior end of larva

changing blood concentrations of hormones

LARVAL STAGES	PUPA	ADULT
High ecdysone, high JH. Larval traits retained, no differentiation into adult. Molts alternate with renewed growth, new larval cuticle.	High ecdysone, low JH. Pupal cuticle forms, growth arrested.	High ecdysone, no JH. Pupa transforms into adult within adult cuticle.

SUMMARY

1. This chapter concludes our survey of controls over the integration of activities in the multicelled animal body. Throughout the remainder of this unit, we will be looking at specific examples of these controls, so keep the following key points in mind:

2. Cells continually take up and release substances. In complex animals, myriad withdrawals and secretions must be integrated in ways that ensure cell survival through the whole body.

3. Integration requires the stimulatory or inhibitory effects of signaling molecules.

   a. Signaling molecules are chemical secretions by one cell that adjust the behavior of other, target cells.

   b. Any cell with receptors for the signal is the target. It may or may not be next to the signaling cell.

   c. Hormones, neurotransmitters, local signaling molecules, and pheromones are different kinds of signaling molecules.

4. Steroid hormones and nonsteroid hormones exert their effects on target cells by different mechanisms.

   a. Steroid (and thyroid) hormones have receptors inside their target cells. The hormone-receptor complex binds to the DNA. Binding triggers gene activation and protein synthesis.

   b. Nonsteroid hormones include certain amines, peptides, proteins, and glycoproteins. Their receptors are on the plasma membrane of target cells. Responses to them are often mediated by a second messenger, such as cyclic AMP, inside the cell.

   c. Most nonsteroid hormones alter the activity of existing proteins in target cells. The cellular responses help maintain the internal environment or contribute to the developmental or reproductive program.

5. The hypothalamus and pituitary gland interact to integrate many body activities.

   a. ADH and oxytocin, two hypothalamic hormones, are stored in and released from the posterior lobe of the pituitary. ADH influences extracellular fluid volume. Oxytocin has roles in reproduction and other events.

   b. Six other hypothalamic hormones control the secretions by different cells of the anterior lobe of the pituitary. They are called releasing and inhibiting hormones.

   c. Of the six hormones produced in the anterior lobe, two (prolactin and somatotropin, or growth hormone) have general effects on body tissues. Four (ACTH, TSH, FSH, and LH) act on specific endocrine glands.

6. Responses to hormones may be influenced by hormone interactions and homeostatic feedback loops to the hypothalamus and pituitary. They may be influenced by variations in hormone concentrations. They also may be influenced by the number and kind of receptors on a target cell.

7. Fast-acting hormones such as parathyroid hormone or insulin generally come into play when the extracellular concentration of a substance must be controlled homeostatically.

8. Slow-acting hormones such as somatotropin have more prolonged, gradual, and often irreversible effects, such as those on development.

Review Questions

1. Which secretions of the posterior and anterior lobes of the pituitary gland have the targets indicated? (*Fill in the blanks; see pages 618–619.*)

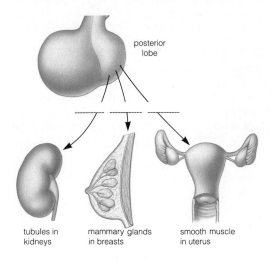

posterior lobe

tubules in kidneys — mammary glands in breasts — smooth muscle in uterus

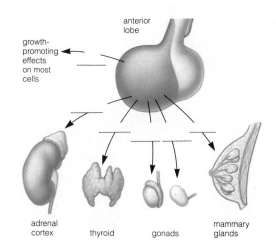

anterior lobe

growth-promoting effects on most cells

adrenal cortex — thyroid — gonads — mammary glands

2. Name the main endocrine glands and state where each is located in the human body. *615*

3. Distinguish among hormones, neurotransmitters, local signaling molecules, and pheromones. *614*

4. A hormone molecule binds to a receptor on a cell membrane. It doesn't enter the cell; rather, the binding activates a second messenger inside the cell that triggers an amplified response to the hormonal signal. Is the signaling molecule a sterioid or protein hormone? *616–617*

Self-Quiz *(Answers in Appendix IV)*

1. _____ are molecules released from a signaling cell that have effects on target cells.
   a. hormones   d. pheromones
   b. neurotransmitters   e. a and b
   c. local signaling molecules   f. all of the above

2. Hormones are products of _____ .
   a. endocrine glands and cells   d. a and b
   b. some neurons   e. a and c
   c. exocrine cells   f. a, b, and c

3. ADH and oxytocin are hypothalamic hormones secreted from the pituitary's _____ lobe.
   a. anterior   c. intermediate
   b. posterior   d. secondary

4. _____ and _____ have effects on body tissues in general.
   a. ACTH; FSH   c. LH; prolactin
   b. ACTH; TSH   d. Somatotropin; prolactin

5. Which do *not* stimulate hormone secretions?
   a. neural signals   d. environment cues
   b. local chemical changes e. all of the above can
   c. hormonal signals   stimulate hormone secretion

6. _____ lowers blood sugar levels; _____ raises it.
   a. Glucagon; insulin   c. Gastrin; insulin
   b. Insulin; glucagon   d. Gastrin; glucagon

7. The pituitary detects a rising hormone concentration in some region and inhibits the gland secreting the hormone. This is a _____ feedback loop.
   a. positive   c. long-term
   b. negative   d. b and c

8. Second messengers assist _____ .
   a. steroid hormones   c. releasing hormones
   b. protein hormones   d. both a and b

9. Match the hormone source with its closest description.
   ___ adrenal cortex   a. affected by daylength
   ___ adrenal medulla   b. cortisol source
   ___ thyroid gland   c. roles in immunity
   ___ parathyroids   d. raise blood calcium level
   ___ pancreatic islets   e. epinephrine source
   ___ pineal gland   f. insulin, glucagon
   ___ thymus gland   g. hormones require iodine

10. Match the endocrine control concepts.
    ___ oxytocin   a. work as neural-endocrine
    ___ ADH   control center
    ___ steroid hormone   b. influences extracellular
    ___ somatotropin (GH)   fluid volume
    ___ hypothalamus/   c. has general effects on growth
        pituitary   d. affects reproductive events
                           e. triggers protein synthesis

Selected Key Terms

adrenal cortex *623*   pancreatic islet *622*
adrenal medulla *623*   parathyroid gland *624*
endocrine system *614*   pineal gland *625*
gonad *626*   pituitary gland *618*
hormone *614*   second messenger *617*
hypothalamus *618*   steroid hormone *616*
local signaling molecule *626*   thymus gland *626*
molting *626*   thyroid gland *624*
nonsteroid hormone *617*

Readings

Goodall, J. 1986. *The Chimpanzees of Gombe*. Cambridge, Massachusetts: Belknap Press of Harvard University Press.

Hadley, M. 1992. *Endocrinology*. Third edition. Englewood Cliffs, New Jersey: Prentice-Hall.

Snyder, S. October 1985. "The Molecular Basis of Communication Between Cells." *Scientific American* 253(4):132–141.

Tepperman, J., and H. Tepperman. 1987. *Metabolic and Endocrine Physiology*. Fifth edition. Chicago: Year Book Medical Publishers. Paperback.

38 PROTECTION, SUPPORT, AND MOVEMENT

"All Right—GO!"

On that command, Susan Butcher's trained huskies leap forward for another race along Alaska's Iditarod Trail (Figure 38.1). For the next eleven days and nights, they will tow a 90-kilogram sled—that's 200 pounds—across 1,860 kilometers of snow, ice, and treacherous rivers between Anchorage and Nome. Three times in the past ten years, Butcher has mushed her team to victory.

Huskies have astounding strength and endurance, an outcome of artificial selection practices. Their leg bones are sturdy, yet lightweight. Their forelegs move freely, thanks to a rib cage that is deep but not too broad. Their hind legs have massive muscles. These are not the muscles of a sprinting greyhound or cheetah. They are the muscles of a load-pulling, long-distance runner. Huskies also have thick foot pads—cushions against sharp ice and frozen rock. And they have an insulating layer of thick, fine hairs next to the skin. Above this is a slightly oily layer of tougher, longer hairs that keep out biting winds and near-freezing moisture.

We humans don't even approach a husky's stamina and built-in protection against the elements. Long before each Iditarod race begins, Butcher must start a marathoner's regimen of diet and exercise to put her arm and leg muscles in peak condition for the extraor-

Figure 38.1 Susan Butcher and her Alaskan huskies, superbly illustrating their systems of support, movement, and protection along the Iditarod Trail.

dinary effort that lies ahead. Lacking the fur of mammals that are native to the Far North, she must acquire clothing that will insulate and protect her during a race without restricting her movements.

In this chapter, our examples will include human systems of protection, support, and movement. Human skin, which did not evolve in arctic environments, cannot withstand bitter cold. Muscles and bones in human legs are suitable for walking and striding, not for long-distance, load-pulling runs. From this perspective, it is mainly human ingenuity that keeps Butcher and others mushing down the Iditarod Trail—in the company of huskies supremely bred for it.

1. Nearly all animals have an integument (an outer covering, such as skin), muscles, and a skeleton.

2. Skin protects the body from abrasion, ultraviolet radiation, bacterial attack, and other environmental insults. It also contributes to overall body functioning, as when it helps control moisture loss.

3. Smooth muscle and cardiac muscle help move internal organs. Skeletal muscle helps move the body's limbs and other structural elements. When properly stimulated, cells of all three types of muscle tissue contract (shorten).

4. Skeletal systems interact with muscles to bring about movement. They protect and support soft organs and store minerals. Blood cells are produced in tissues of some bones.

Like most other animals, you have three systems to thank for your superficial features, shape, and movements. These systems are the topics of this chapter. Some simple diagrams of the human body will serve as the starting point for a tour of the structural organization and functions of these systems. Traveling from the outside in, we have an **integumentary system** (skin and its derivatives), a **muscle system**, and a **skeletal system**:

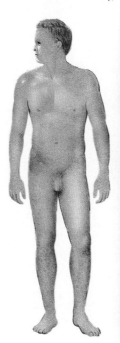
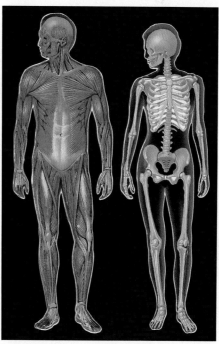

INTEGUMENTARY SYSTEM

Animals ranging from invertebrate worms to humans have an outer covering, or **integument** (after the Latin *integere*, meaning "to cover"). For most species, the integument is tough yet pliable, a barrier against a great variety of environmental insults. The integument of insects, crabs, and other arthropods is a hardened covering called a **cuticle**. Examples appear on pages 432 and 436. Arthropod cuticles consist of chitin, protein, and sometimes lipid secretions.

Vertebrate Skin and Its Derivatives

For vertebrates, the integument is skin and the structures derived from epidermal cells of its outer layers of tissue (Figures 38.2 and 38.3). Derivatives of skin range from scales, feathers, hair, beaks, hooves, horns, claws, nails, and quills to assorted glands and a lathering of slime. Variation exists within vertebrate groups as well as between them. Thus, as you read in Chapter 27, some species of fishes have hard scales, and others have bare skin coated with slimy mucus.

As Figure 38.2 shows, skin has two distinct regions—an outermost **epidermis** and an underlying **dermis**.

Below this, a tissue region called the hypodermis anchors the skin to underlying structures yet allows it to move a bit. Fat stored in the hypodermis insulates the body and cushions some body parts. Your own skin weighs about 4 kilograms (9 pounds). Stretched out, it would have a surface area of 15 to 20 square feet. For the most part, human skin is thin as a paper towel. It thickens only on the soles of the feet and in other regions that are subjected to pounding or abrasion.

Functions of Skin

No garment ever made comes close to skin's qualities. What besides skin holds its shape after repeated stretchings and washings, is waterproof, holds in moisture, kills many bacteria on contact, blocks harmful rays from the sun, repairs small cuts and burns on its own, and with a little care will last as long as you do?

Skin does more than protect against dehydration and various environmental insults. It helps stabilize the body's internal temperature. Its many small blood vessels are a reservoir for blood that can be shunted to metabolically active regions, such as muscles in legs that are running along the Iditarod Trail. Skin produces the vitamin D required for calcium metabolism. And signals from skin's sensory receptors help the brain assess what's going on in the outside world.

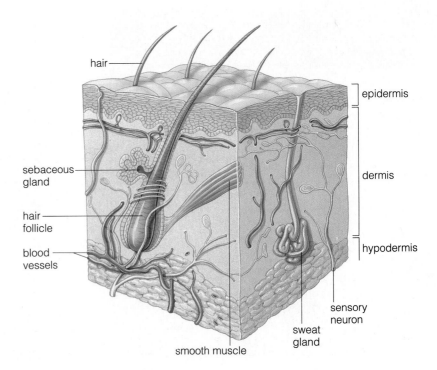

Figure 38.2 Structure of human skin. The uppermost portion consists of layers of epidermis. The lower portion is the dermis. The dermis itself rests on a subcutaneous layer (hypodermis) that attaches skin to underlying structures. The hairs and their follicles are specialized structures derived from skin.

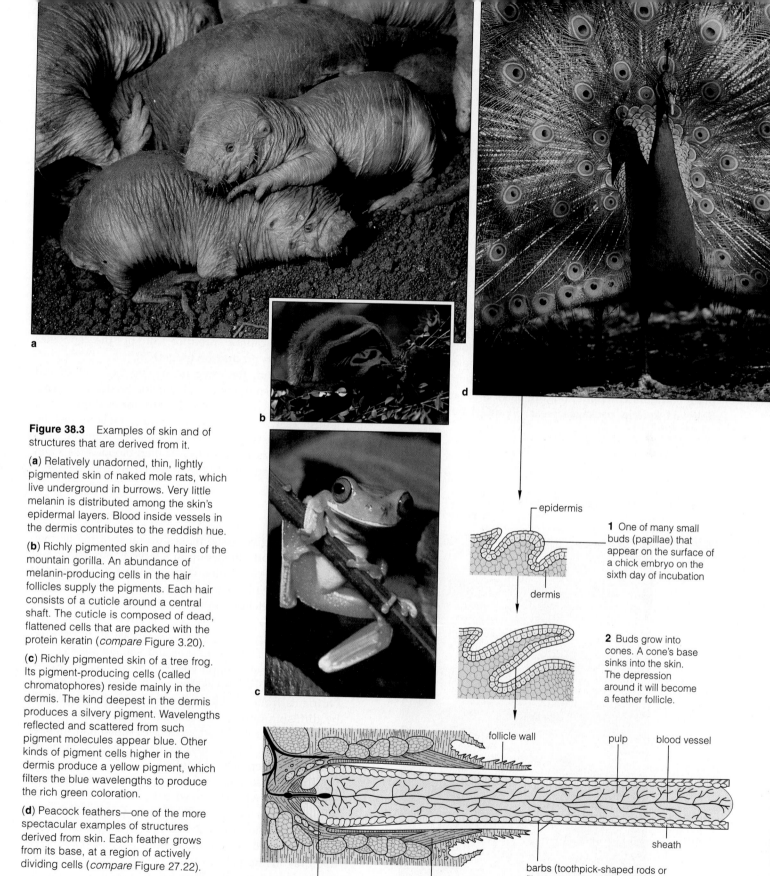

Figure 38.3 Examples of skin and of structures that are derived from it.

(**a**) Relatively unadorned, thin, lightly pigmented skin of naked mole rats, which live underground in burrows. Very little melanin is distributed among the skin's epidermal layers. Blood inside vessels in the dermis contributes to the reddish hue.

(**b**) Richly pigmented skin and hairs of the mountain gorilla. An abundance of melanin-producing cells in the hair follicles supply the pigments. Each hair consists of a cuticle around a central shaft. The cuticle is composed of dead, flattened cells that are packed with the protein keratin (*compare* Figure 3.20).

(**c**) Richly pigmented skin of a tree frog. Its pigment-producing cells (called chromatophores) reside mainly in the dermis. The kind deepest in the dermis produces a silvery pigment. Wavelengths reflected and scattered from such pigment molecules appear blue. Other kinds of pigment cells higher in the dermis produce a yellow pigment, which filters the blue wavelengths to produce the rich green coloration.

(**d**) Peacock feathers—one of the more spectacular examples of structures derived from skin. Each feather grows from its base, at a region of actively dividing cells (*compare* Figure 27.22).

epidermis

dermis

1 One of many small buds (papillae) that appear on the surface of a chick embryo on the sixth day of incubation

2 Buds grow into cones. A cone's base sinks into the skin. The depression around it will become a feather follicle.

follicle wall

pulp

blood vessel

sheath

barbs (toothpick-shaped rods or filaments that grow outwardly or diagonally in parallel fashion)

collar where barbs originate

feather muscle

3 A layer of thin, horny cells at the cone's surface differentiates into a sheath. An epidermal layer just beneath the sheath will give rise to the feather itself. The richly vascularized dermis beneath this layer becomes the pulp, which nourishes the growing feather but does not contribute to its structure.

38.2 A CLOSER LOOK AT THE STRUCTURE OF SKIN

Epidermis

Epidermis is mostly stratified epithelium. That is, like puff pastry, it consists of stacked sheets. An abundance of cell junctions knits the epithelial cells together, in the manner described on page 549. The cells themselves arise within the epidermis but are pushed toward its free surface, as rapid and ongoing mitotic divisions produce new cells beneath them.

Most cells of the epidermis are **keratinocytes**. Each is a tiny factory for manufacturing keratin, a tough, water-insoluble protein. Those cells start producing keratin when they are in mid-epidermal regions. By the time they reach the skin's free surface, they are dead and flattened. All that remain are fibers of keratin, packed inside plasma membranes. This is the composition of the outermost layer of skin—the tough, waterproof "stratum corneum" (Figure 38.4). Millions of the flattened keratin packages wear off daily, but rapid,

ongoing cell divisions in the epidermis continually push up replacements. The rapid divisions also contribute to skin's capacity to mend itself quickly after cuts or burns.

In the deepest epidermal layer, **melanocytes** produce the brownish-black pigment called melanin. These cells transfer the pigment to the keratin-producing cells. As keratin accumulates inside a cell, it forms a shield against ultraviolet radiation. As the *Focus* essay indicates, melanocyte activity has increased in suntanned skin.

Besides affording protection from harmful wavelengths, melanin also contributes to skin color. Although humans generally have the same number of melanocytes, they show variation in skin color owing to differences in the distribution and metabolic activity of these cells. Albinos, for example, lack melanin; their melanocytes cannot produce all of the enzymes required for its production (page 178).

Hemoglobin, the oxygen-carrying pigment of red blood cells, also influences skin color. So does carotene, a yellow-orange pigment. Pale skin, for example, has a pinkish cast. It does not have much melanin, so the presence of hemoglobin is not masked. Hemoglobin's red color shows through thin-walled blood vessels and through the epidermis, both of which are transparent.

Dermis

Dense connective tissue makes up most of the dermis, and it fends off damage from everyday stretching and other mechanical insults. There are limits to this protection. For example, the dermis tears when skin over the abdomen stretches too much during pregnancy. The tears heal, leaving white scars ("stretch marks"). As another example, with persistent abrasion, the epidermis separates from the dermis and you get a *blister*.

Blood vessels, lymph vessels, and the receptor endings of sensory nerves thread through the dermis. Nutrients from the bloodstream reach epidermal cells by diffusing through the dermal tissue. Sweat glands, oil glands, and the husklike cavities called hair follicles reside mostly in the dermis, even though they are derived from epidermal tissue.

The fluid secreted from **sweat glands** is 99 percent water, along with dissolved salts, traces of ammonia and other metabolic wastes, vitamin C, and other substances. You have about 2.5 million sweat glands, which are controlled by sympathetic nerves. One type abounds in the palms of the hands, soles of the feet, forehead, and armpits. They function mainly in temperature regulation (page 749). They also function in *cold sweats*, one of the responses you make when you are frightened, nervous, or merely embarrassed. The skin around the sex organs contains another type of sweat

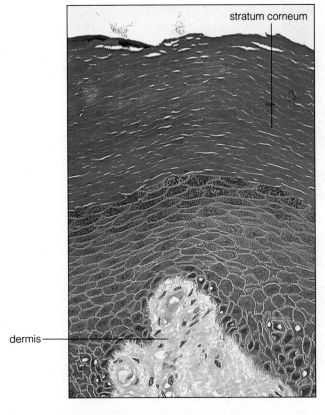

stratum corneum

dermis

Figure 38.4 Section through human skin, showing the uppermost layer of epidermis (stratum corneum), the deeper epidermal layers, and the underlying dermis.

Focus on Health

Sunlight and Skin

Are you someone who tans by rotating beneath the sun's rays or tanning lamps, like a chicken in an oven broiler? If so, remember what a broiler does to the chicken.

Exposure to the sun's ultraviolet light stimulates your melanin-producing cells. Continued exposure increases the amount of melanin in light skin, and so visibly darkens it (the sought-after tan). Tanning provides some protection against ultraviolet radiation. But prolonged exposure causes elastin fibers to clump together in the dermis, so skin loses its resiliency. In time it starts to look like old shoe leather.

Prolonged exposure to ultraviolet radiation also suppresses the immune system. For instance, infection-fighting phagocytes in the epidermis combat viruses and bacteria. Sunburns interfere with their functioning. This may be why sunburns can trigger small, painful blisters (*cold sores*) that announce the recurrence of a *Herpes simplex* infection. Nearly everyone harbors this virus. It remains hidden in the face, inside a ganglion (a cluster of neuron cell bodies). Stress factors, including sunburn, can activate the virus. When that happens, virus particles move down the neurons to their endings near the skin. There they infect epithelial cells and cause the painful skin eruptions.

Ultraviolet radiation also can activate proto-oncogenes that can trigger cancerous transformation of skin cells. *Epidermal skin cancers* start out as scaly, reddened bumps. They grow rapidly and can spread to adjacent lymph nodes unless they are surgically removed. *Basal cell carcinomas* start out as small, shiny bumps and slowly grow into ulcers with beaded margins (page 215). Their threat ceases if they are surgically removed in time.

gland. Here, glandular secretions increase during stress, pain, sexual foreplay, and the female reproductive cycle. Do they have functions similar to those of scent glands in other animals? No one knows.

Except on the palms of hands and the soles of feet, skin contains **oil glands** (also called sebaceous glands). Oil glands soften and lubricate both the hair and the skin, and they kill surface bacteria. The skin condition called *acne* is an inflammation resulting from bacterial infection of oil gland ducts.

The flexible structures called **hairs** consist mostly of keratinized cells. Each has a root embedded in skin and a shaft projecting above the skin surface. As living cells divide near the base of the root, older cells are pushed upward, then flatten and die. The outermost layer of the shaft consists of flattened cells that overlap one another like roof shingles (Figure 3.20). The most abused of these cells tend to frizz out near the end of the hair shaft; we call these "split ends."

The average scalp has about 100,000 hairs. However, genes, nutrition, and hormones influence hair growth and density. Protein deficiency causes hair to thin, for hair cannot grow without the amino acids required for keratin synthesis. Severe fever, emotional stress, and excessive vitamin A intake also cause hair thinning. Excessive hairiness (*hirsutism*) may result when the body produces abnormal amounts of testosterone. This hormone influences patterns of hair growth and other secondary sexual traits.

As we age, epidermal cells divide less often, and our skin becomes thinner and more susceptible to injury. Glandular secretions that had kept the skin soft and moistened start to dwindle. Collagen and elastin fibers in the dermis break down and become sparser, so the skin loses its elasticity and its wrinkles deepen. Excessive tanning, prolonged exposure to drying winds, and tobacco smoke accelerate the skin aging processes.

---

1. With its multiple layers of keratinized, melanin-shielded epidermal cells, skin helps the body conserve water, avoid damage by ultraviolet radiation, and resist mechanical stress.

2. Hairs, oil glands, sweat glands, and other structures associated with skin are derived from epidermal cells, but they are largely embedded in skin's underlying region, the dermis.

3. Blood and lymph vessels as well as receptor endings of sensory neurons also reside in the dermis.

38.3 INVERTEBRATE AND VERTEBRATE SKELETONS

Operating Principle for Skeletons

So far in this unit, we have looked at how the nervous system samples the external and internal environments with precision and keeps informed of change. Many responses to changes require movements—of either the whole body or some parts of it. Animals move by the activation, contraction, and relaxation of muscle cells. But muscle cells alone cannot produce movement. *Muscles require the presence of some medium or structural element against which the force of contraction can be applied.* An internal or external skeleton of the sort shown in Figures 38.5 through 38.8 fulfills this requirement.

Three kinds of skeletal systems predominate in the animal world:

hydrostatic skeleton	Internal body fluids receive the applied force of contraction and are redistributed within a limited space
exoskeleton	Rigid *external* body parts, such as shells or armor plates, receive the applied force of muscle contraction
endoskeleton	Rigid *internal* body parts, such as cartilage and bone, receive the applied force of contraction

a Relaxed position   **b** Feeding position

Figure 38.5 Outcomes of the force of muscle contraction applied against the hydrostatic skeleton of sea anemones. (**a**) Radial muscles (ringing the gut cavity) are relaxed, and longitudinal muscles (parallel with the body axis) are contracted. Anemones typically look like this at low tide, when currents cannot bring food morsels to them. (**b**) Radial muscles are contracted and longitudinal muscles are relaxed, so the body is stretched into an upright position. This is the feeding position of sea anemones.

Examples From the Invertebrates

Many soft-bodied invertebrates have hydrostatic skeletons, in which some type of fluid is confined to a limited space. Like a fully filled waterbed, the fluid resists compression, thereby serving as a medium against which muscles can work.

Consider the sea anemone, with its soft, vase-shaped body and saclike gut. The body wall contains longitudinal and radial muscles. Between meals, longitudinal muscles are contracted (shortened), the radial ones are relaxed (lengthened), and the animal looks short and squat (Figure 38.5*a*). It lengthens into its upright feeding position by contracting the radial muscles (which forces some fluid out of the gut cavity) and relaxing the longitudinal ones (Figure 38.5*b*).

Or consider the earthworm. This annelid, recall, has a series of coelomic compartments, each with its own muscles, nerves, and bristles. The body moves forward by contracting and relaxing its fluid-filled segments one after another (page 428). Besides this, by coordinating muscle contractions on one side or the other of different segments, the earthworm can thrash from side to side as well as move forward and back.

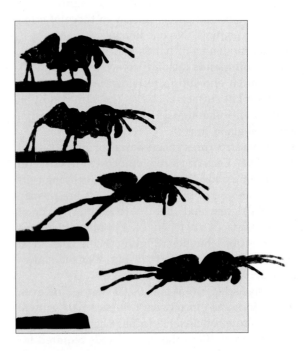

Figure 38.6 Leap of the jumping spider *Sitticus pubescens*, based on the hydraulic extension of its hind legs when blood surges into them under high pressure.

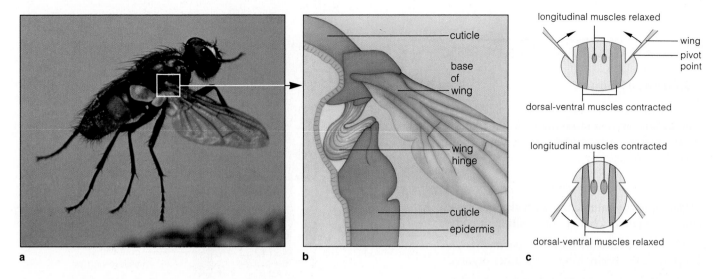

a

b

c

Figure 38.7 Housefly wing movement, an indirect outcome of antagonistic contraction and relaxation of different muscles. The contractile force is applied against the exoskeleton. The wings are attached to the exoskeleton near a hinge point. They "pop up" as dorsal-ventral muscles contract and so shorten the body segment. They swing down as longitudinal muscles contract.

Like other arthropods, jumping spiders have a hinged exoskeleton. They also use body fluids to transmit force when they leap at prey. Muscle contractions cause blood inside the body tissues to surge rapidly into the hind leg spines. It is a bit like giving a water-filled rubber glove a quick, hard squeeze, so that the glove's skinny fingers become rigidly erect. Figure 38.6 shows the outcome of this resourceful use of hydraulic pressure (*hydraulic* means fluid pressure inside tubes).

By itself, the hinged arthropod exoskeleton has advantages. Some of the hard components can be moved like levers by sets of muscles attached to them. Thus, small contractions can bring about large movements of wings or some other body parts. This is true of the cuticle of a winged insect. It extends over all the body segments and over gaps between segments (see Figure 38.7). At these gaps, the cuticle remains pliable. It acts like a hinge when muscles alternately raise and lower either the wing or the body parts to which the wings are attached.

Examples From the Vertebrates

Vertebrates have endoskeletons of one sort or another. For example, sharks have a skeleton of an opaque form of cartilage, hardened with calcium deposits (Figure 38.8). Some other fishes have a flexible skeleton of an elastic, translucent form of cartilage that almost looks like glass. For most vertebrates, however, the endoskeleton consists primarily of bone. Let's turn now to the functions and characteristics of bones. Afterward, we will consider how different types of bones are arranged in the human skeletal system.

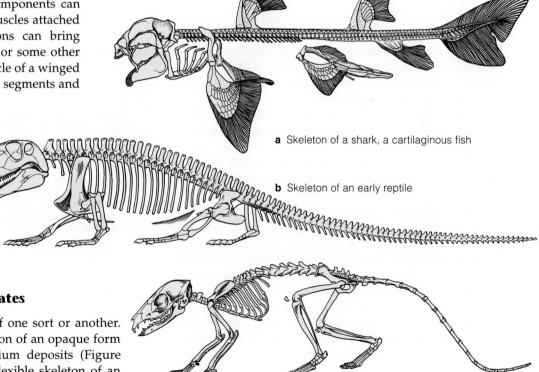

a Skeleton of a shark, a cartilaginous fish

b Skeleton of an early reptile

c Skeleton of a mammal

Figure 38.8 Comparison of the skeletons from (**a**) a shark, (**b**) a generalized early reptile, and (**c**) a generalized mammal. The components of the shark's skeleton are opaque cartilage hardened with calcium deposits.

CHARACTERISTICS OF BONE

Functions of Bone

Just as your skin is more than a baglike covering, so is your skeletal system more than a frame to hang muscles on. Its major parts, **bones**, are complex organs that serve these functions:

1. *Movement.* By interacting with skeletal muscles, bones maintain or change the position of body parts.

2. *Protection.* Bones are hard compartments that enclose and protect the brain, lungs, and other organs.

3. *Support.* Bones support and anchor muscles.

4. *Mineral storage.* Bone tissue is a bank for depositing and withdrawing mineral ions, and so helps maintain body fluids and support metabolic activities.

5. *Blood cell formation.* Some bones contain regions in which blood cells are produced.

Bone Structure

In size, human bones range from tiny earbones to club-like thighbones. In shape, bones are long, short (or cubelike), flat, and irregular. All contain epithelial and connective tissues as well as bone tissue. The calcium-hardened bone tissue consists of living cells and collagen fibers in a ground substance.

Figure 38.9 shows the structure of a thighbone. *Compact* bone tissue in its shaft and at its ends withstands mechanical shocks. The tissue's thin, dense layers form multiple cylinders around small, interconnected canals. Blood vessels and nerves in these "Haversian canals" service the living bone cells.

Within the bone ends and shaft, *spongy* bone tissue imparts strength without adding too much weight. With its abundant spaces, the tissue has a spongelike appearance, but its flattened parts are quite firm. **Red marrow** fills the spaces in some bones, including the breastbone. Red marrow is a major site of blood cell formation. Cavities in most mature bones contain **yellow marrow**. Although yellow marrow is mostly fat, it converts to red marrow and produces red blood cells when blood loss from the body is severe.

How Bones Develop As Figure 38.10 indicates, before a long bone develops in a growing embryo, a cartilage model for it is put together. Later, bone-forming cells (osteoblasts) secrete material inside the model's shaft and on its surface. A marrow cavity develops as cartilage breaks down inside the shaft.

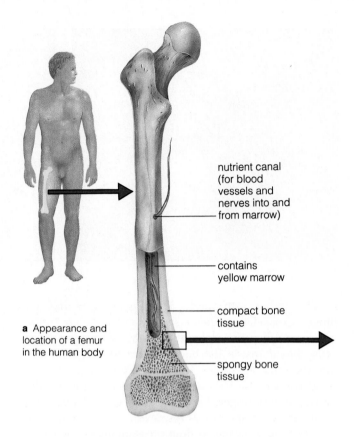

a Appearance and location of a femur in the human body

nutrient canal (for blood vessels and nerves into and from marrow)

contains yellow marrow

compact bone tissue

spongy bone tissue

Figure 38.9 Structure of a femur (thighbone), one of the long bones of mammals.

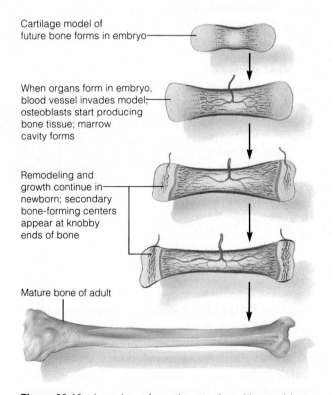

Cartilage model of future bone forms in embryo

When organs form in embryo, blood vessel invades model; osteoblasts start producing bone tissue; marrow cavity forms

Remodeling and growth continue in newborn; secondary bone-forming centers appear at knobby ends of bone

Mature bone of adult

Figure 38.10 Long bone formation, starting with osteoblast activity in a cartilage model (here, already formed in the embryo). Bone-forming cells are active first in the shaft region, then at the knobby ends. In time, the only cartilage left is at the ends.

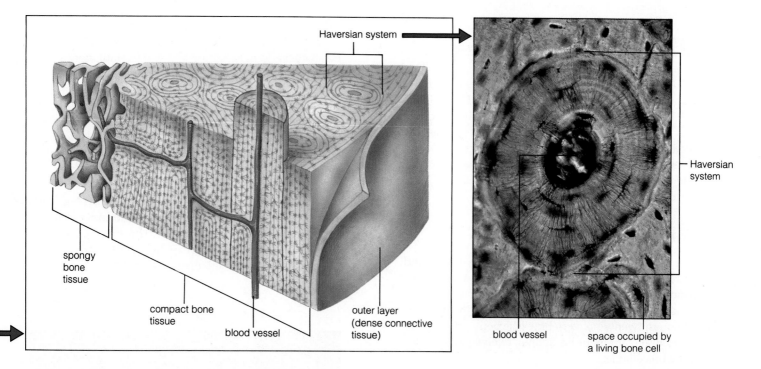

b Appearance of spongy bone tissue and compact bone tissue in a femur. The thin, dense layers of compact bone tissue are arranged as cylinders around interconnecting canals, which contain blood vessels and nerves. Each cylinder is a Haversian system.

c Micrograph of a Haversian system. The blood vessel services osteocytes, living bone cells within small spaces in bone tissue. Small tunnels connect neighboring spaces.

When bone-forming cells are finally surrounded by their own secretions, we call them **osteocytes** (living bone cells). Through their metabolic activities, these cells maintain mature bones.

Bone Tissue Turnover   Minerals are continually deposited in bone tissue and withdrawn from it. Exercising stimulates mineral deposition in adult bones and generally increases bone density. Stress or injury triggers withdrawals of mineral ions.

As children grow, their bones are remodeled. The remodeling proceeds through prescribed tissue turnovers. For instance, a thighbone is made thicker and stronger as bone cells deposit minerals at the surface of its shaft. At the same time, the bone is being made less heavy as other bone cells destroy bone tissue inside the shaft.

Bone turnover is coordinated to maintain calcium levels. Bone cells secrete enzymes that break down bone tissue, releasing calcium and other minerals required for metabolism into interstitial fluid. The bloodstream distributes the calcium through the body.

With increasing age, the backbone, hip bones, and other bones decrease in mass, especially in women. Figure 38.11 shows an example of the effect of the progressive deterioration on bone structure. This ailment is called *osteoporosis*. In time, the weakened backbone may collapse and curve abnormally. This lowers the rib cage, which causes problems for internal organs. Decreasing

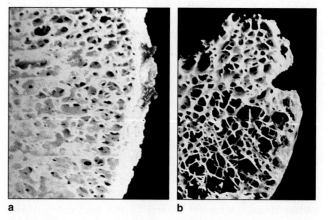

Figure 38.11   Osteoporosis. (**a**) In normal bone tissue, mineral deposits continually replace mineral withdrawals. (**b**) After the onset of osteoporosis, replacements cannot keep pace with withdrawals. In time the tissue erodes, and bones become hollow and brittle.

osteoblast activity, calcium loss, sex hormone deficiencies, excessive protein intake, and decreased physical activity may contribute to the disorder.

Bones are complex organs composed of living cells and their secretions. The collagen-rich extracellular matrix and the ground substance of bone tissue are richly mineralized.

38.5 HUMAN SKELETAL SYSTEM

Our distant four-legged ancestors started walking upright more than 4 million years ago. The shift to bipedal locomotion led to a pronounced S-shaped curve to the backbone of the human skeleton. This has not proved to be a perfectly workable modification of the four-legged plan. The older we get, the longer we have resisted gravity in a compromised way—and the more lower back pain we suffer.

Divisions of the Skeleton

Figure 38.12 lists a few of the 206 human bones and their functions. They belong to either the axial or the appendicular portion of the skeleton.

The *axial* portion of the human skeleton includes the skull bones, twenty-six vertebrae (the bony segments of the vertebral column, or backbone), twelve pairs of ribs, and the breastbone.

Of the twenty-nine skull bones, only the lower jawbone and three small earbones can move. The rest are tightly joined; they form protective chambers for the brain and other soft internal parts of the head.

The curved vertebral column extends from the base of the skull to the pelvic girdle, where it transmits the weight of the torso to the lower limbs. The delicate spinal cord threads through a series of bony protrusions at the rear of the column (compare Figure 35.8). Between the vertebrae are **intervertebral disks**, cartilage-containing shock absorbers and flex points that permit movement. Sometimes a severe or rapid shock will force a disk to slip out of place and possibly rupture. Such *herniated disks* may protrude into nerves or the spinal cord and cause severe pain.

The *appendicular* portion of the skeleton contains 126 bones. These include the pectoral girdles (at the shoulders), arms, hands, pelvic girdle (at the hips), legs, and feet. Slender collarbones and flat shoulder blades are rather flimsily arranged parts of the pectoral girdles. Fall on an outstretched arm and you might dislocate your shoulder or fracture a collarbone, which is the bone most frequently broken.

Skeletal Joints

Types of Joints   A skeletal **joint** is an area of contact or near-contact between bones. We recognize three kinds of joints, based on their bridging connective tissue. At fibrous joints, very short connecting fibers stitch bones together. At cartilaginous joints, cartilage holds them together. At synovial joints, **ligaments**—long straps of dense connective tissue—bridge the gap between bones.

Some fibrous joints hold teeth in place in their sockets. Others loosely connect the flat skull bones of a fetus. At childbirth, the loose connections allow the bones to slide over each other and so prevent skull fractures. The skull of a newborn still has fibrous joints and membranous areas that are known as "soft spots" (fontanels). During childhood, the fibrous tissue hardens completely, and the skull bones become fused into a single unit.

Cartilaginous joints occur between the vertebrae and between the breastbone and ribs. Cartilage fills spaces between bones and permits slight movement.

Synovial joints move freely. The knee joint is an example. Ligaments stabilize this joint, and a small bone (patella) in one ligament protects it:

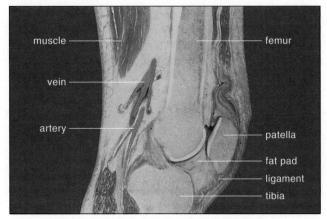

longitudinal section through the knee joint

Where one bone touches another, cartilage cushions both of them and absorbs shocks. A flexible capsule of dense connective tissue surrounds the region of contact. Cells of a membrane that lines the capsule's interior secrete a fluid that lubricates the joint.

Trouble at the Joints   As athletes know, many joints are vulnerable to stress. Consider the knee joint. It lets you swing, bend, and twist the long bones below it. When you run, it absorbs the force of your weight each time the foot hits the ground. But stretch or twist a knee joint too suddenly and too far, and you may *strain* it. Tear its ligaments or tendons and you *sprain* it. Move the wrong way and you may dislocate its bones. Also, blows to the knee during football and other collision sports frequently sever an entire ligament. Surgery must be done within ten days of the injury. Phagocytic cells in the joint's lubricating fluid clean up after everyday wear and tear. Present them with torn ligaments and they indiscriminately turn the tissue to mush.

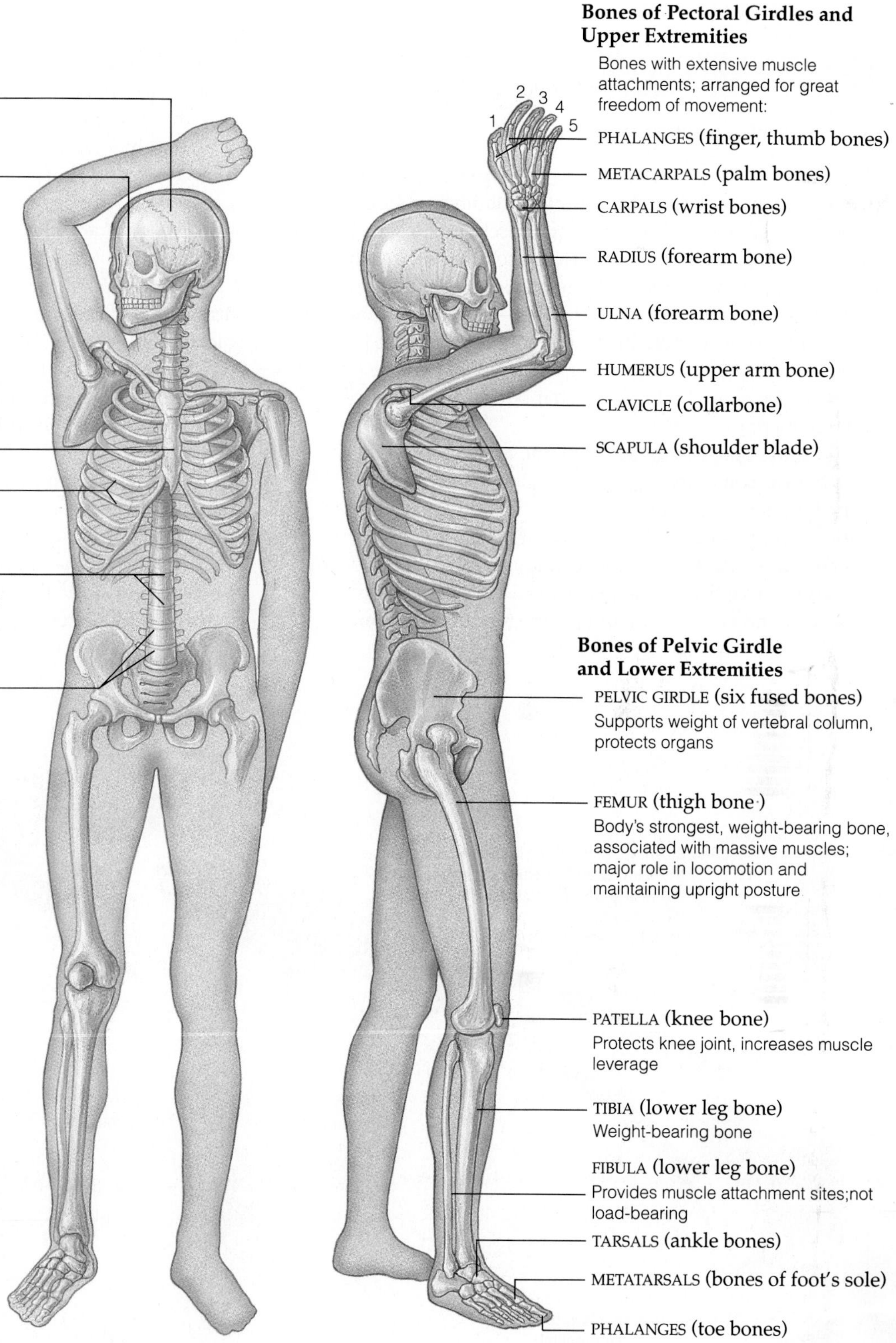

Bones of Skull

CRANIAL BONES —
Enclose, protect brain and sensory organs of head

FACIAL BONES —
Framework for facial region; support teeth

Bones of Rib Cage

Together with some vertebrae, these bones enclose and protect internal organs, assist breathing.

STERNUM (breastbone) —

RIBS (twelve pairs) —

Vertebral Column (backbone)

VERTEBRAE (thirty-three bones) —
Enclose, protect spinal cord; support skull, upper extremities; provide attachment for muscles

INTERVERTEBRAL DISKS —
Fibrous, cartilaginous structures between vertebrae; they absorb movement-related stress and lend flexibility to backbone

Bones of Pectoral Girdles and Upper Extremities

Bones with extensive muscle attachments; arranged for great freedom of movement:

PHALANGES (finger, thumb bones)

METACARPALS (palm bones)

CARPALS (wrist bones)

RADIUS (forearm bone)

ULNA (forearm bone)

HUMERUS (upper arm bone)

CLAVICLE (collarbone)

SCAPULA (shoulder blade)

Bones of Pelvic Girdle and Lower Extremities

PELVIC GIRDLE (six fused bones)
Supports weight of vertebral column, protects organs

FEMUR (thigh bone)
Body's strongest, weight-bearing bone, associated with massive muscles; major role in locomotion and maintaining upright posture

PATELLA (knee bone)
Protects knee joint, increases muscle leverage

TIBIA (lower leg bone)
Weight-bearing bone

FIBULA (lower leg bone)
Provides muscle attachment sites; not load-bearing

TARSALS (ankle bones)

METATARSALS (bones of foot's sole)

PHALANGES (toe bones)

Figure 38.12 The human skeletal system. Major bones of its axial portion are listed to the left. Major bones of its appendicular portion are listed to the right. Can you identify similar structures in Figure 38.8b and c?

Adding insult to injury, joints also are vulnerable to many types of inflammation or degeneration that are collectively called "arthritis." For example, in *osteo-arthritis*, the bone ends of the knees and other freely movable joints wear away as a person ages. The joints affected most often are in the fingers, knees, hips, and vertebral column.

As another example, in *rheumatoid arthritis*, synovial membranes become inflamed and thickened, cartilage degenerates, and bone is deposited in the joint. This degenerative disorder may be triggered by a bacterial or viral infection, but it also appears to have a genetic basis. It can begin at any age, but symptoms usually emerge before age fifty.

38.6 | MUSCLE SYSTEMS

How Muscles and Bones Interact

We turn now to skeletal muscles, the functional partners of bones. Each **skeletal muscle** contains bundles of hundreds to many thousands of muscle cells, which look like long, striped fibers. In muscle tissue, recall, these cells contract (shorten) in response to stimulation, then lengthen and so return to their original resting state. When you dance, breathe, scribble notes, or tilt your head skyward, contracting muscle cells are helping to move your body or to change the positions of some of its parts.

Skeletal muscles interact with one another as well as with bones. Some are arranged in pairs or groups that work together to promote the same movement. Others work in opposition; the action of one opposes or reverses the action of another (Figure 38.13a). Either way, connective tissue bundles the muscle cells in parallel and extends beyond them to form **tendons**. Tendons are cords or straps of dense connective tissue that attach muscle to bone. Most of these attachments are rather like the gearshift in a car. That is, they are a lever system, in which a rigid rod is attached to a fixed point but able to move about at it. The muscles connect with bones (rigid rods) near a joint (fixed point). When the muscles contract, they transmit force to the bones and so make them move.

Take a look at Figure 38.13b, then extend your right arm. Now place your left hand over the upper arm's biceps and slowly "bend your elbow," so that you can feel the biceps contract. Even if your biceps contracts only a bit, it can produce a large movement in the forearm bone connected to it. This is true of most leverlike arrangements.

Bear in mind, only skeletal muscle is the functional partner of bone. As mentioned on page 552, smooth muscle is present mostly in the walls of internal organs, such as the stomach and intestines. Cardiac muscle is present only in the heart wall, and its action is described in a later chapter.

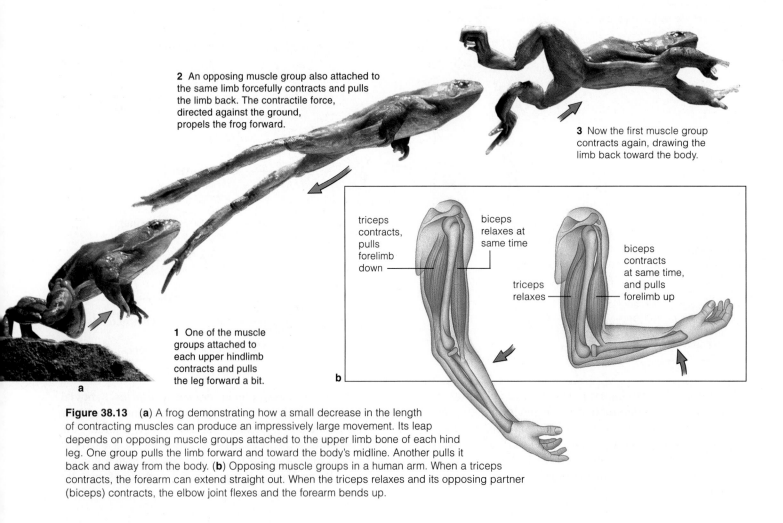

2 An opposing muscle group also attached to the same limb forcefully contracts and pulls the limb back. The contractile force, directed against the ground, propels the frog forward.

3 Now the first muscle group contracts again, drawing the limb back toward the body.

1 One of the muscle groups attached to each upper hindlimb contracts and pulls the leg forward a bit.

triceps contracts, pulls forelimb down

biceps relaxes at same time

triceps relaxes

biceps contracts at same time, and pulls forelimb up

a

b

Figure 38.13 (**a**) A frog demonstrating how a small decrease in the length of contracting muscles can produce an impressively large movement. Its leap depends on opposing muscle groups attached to the upper limb bone of each hind leg. One group pulls the limb forward and toward the body's midline. Another pulls it back and away from the body. (**b**) Opposing muscle groups in a human arm. When a triceps contracts, the forearm can extend straight out. When the triceps relaxes and its opposing partner (biceps) contracts, the elbow joint flexes and the forearm bends up.

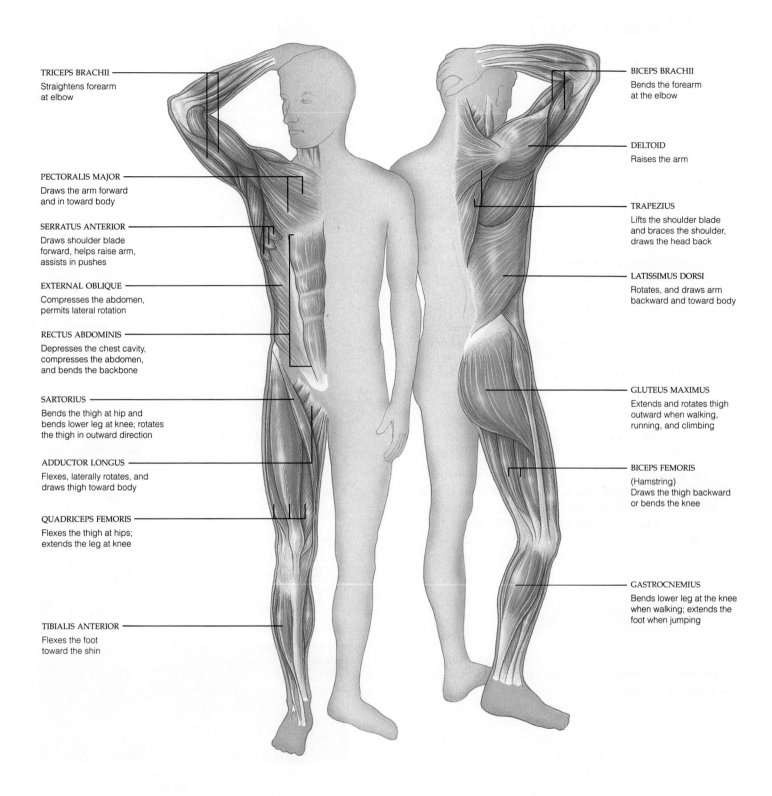

TRICEPS BRACHII
Straightens forearm
at elbow

PECTORALIS MAJOR
Draws the arm forward
and in toward body

SERRATUS ANTERIOR
Draws shoulder blade
forward, helps raise arm,
assists in pushes

EXTERNAL OBLIQUE
Compresses the abdomen,
permits lateral rotation

RECTUS ABDOMINIS
Depresses the chest cavity,
compresses the abdomen,
and bends the backbone

SARTORIUS
Bends the thigh at hip and
bends lower leg at knee; rotates
the thigh in outward direction

ADDUCTOR LONGUS
Flexes, laterally rotates, and
draws thigh toward body

QUADRICEPS FEMORIS
Flexes the thigh at hips;
extends the leg at knee

TIBIALIS ANTERIOR
Flexes the foot
toward the shin

BICEPS BRACHII
Bends the forearm
at the elbow

DELTOID
Raises the arm

TRAPEZIUS
Lifts the shoulder blade
and braces the shoulder,
draws the head back

LATISSIMUS DORSI
Rotates, and draws arm
backward and toward body

GLUTEUS MAXIMUS
Extends and rotates thigh
outward when walking,
running, and climbing

BICEPS FEMORIS
(Hamstring)
Draws the thigh backward
or bends the knee

GASTROCNEMIUS
Bends lower leg at the knee
when walking; extends the
foot when jumping

Human Muscle System

The human body is equipped with more than 600 skeletal muscles. Figure 38.14 shows a few and lists their functions. We turn next to the mechanisms underlying their contraction.

Figure 38.14 The human muscular system, showing some of the outermost skeletal muscles.

38.7 MUSCLE FUNCTION

Functional Organization of a Skeletal Muscle

Bones move—they are pulled in some direction—when the skeletal muscles that are attached to them shorten. When a skeletal muscle shortens, its component muscle cells are shortening. When a muscle cell shortens, many units of contraction within that cell are shortening. The basic units of contraction are called **sarcomeres**.

Take a look at Figure 38.15. It shows how the bundles of cells in a skeletal muscle run parallel with the muscle itself. It also shows that each muscle cell contains myofibrils, threadlike structures packed together in parallel array. Every one of the myofibrils is functionally divided into sarcomeres, which are arranged one after another along its length.

Look closely at the micrograph in Figure 38.15c, and you see that a sarcomere contains many filaments, side by side in parallel array. Some of the filaments are thin; others are thick. Each thin filament is actually two beaded strands, twisted together. The "beads" are ball-shaped molecules of the protein **actin**:

)— one actin molecule

} portion of one thin filament

Each thick filament consists of molecules of **myosin**, a protein with a head and a long tail. Myosin tails are packed together in parallel. The heads, which look like double-headed golf clubs, stick out to the sides:

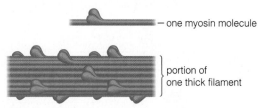

— one myosin molecule

} portion of one thick filament

The orderly arrays of actin and myosin filaments in the sarcomeres give skeletal muscle (and cardiac muscle) its striped appearance.

Think of the parallel orientation of all the myofibrils, muscle cells, and muscle bundles in a skeletal muscle. The orientation of a skeletal muscle's component parts focuses the force of contraction onto the bone in a particular direction.

A skeletal muscle shortens through the combined decreases in length of its sarcomeres, the basic units of contraction.

The parallel orientation of a muscle's component parts directs the force of contraction toward a bone that must be pulled in some direction.

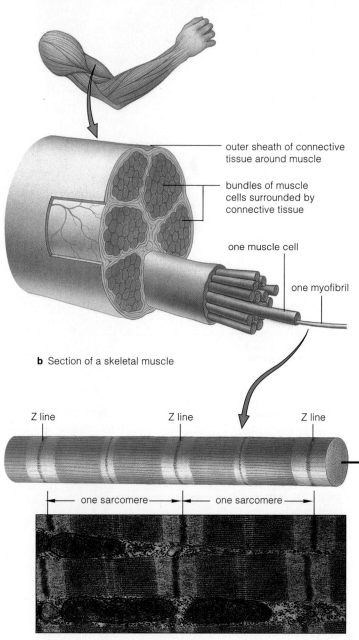

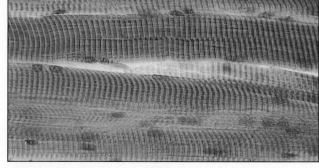

a Some of the fiberlike muscle cells in skeletal muscle

outer sheath of connective tissue around muscle

bundles of muscle cells surrounded by connective tissue

one muscle cell

one myofibril

b Section of a skeletal muscle

Z line   Z line   Z line

◄——— one sarcomere ———►◄——— one sarcomere ———►

c Two of the myofibrils inside a muscle cell. In each myofibril, the contractile units called sarcomeres are arranged one after the other. The oval-shaped organelles are mitochondria.

Figure 38.15 Components of a skeletal muscle. (**a**) Light micrograph of skeletal muscle cells. (**b**) Each muscle cell contains myofibrils. (**c**) Each myofibril is functionally divided into sarcomeres, the basic units of contraction. Dark "bands" (Z lines) define the two ends of each sarcomere.

Sliding-Filament Model of Muscle Contraction

How do sarcomeres shorten and so bring about contraction of a skeletal muscle? The answer lies with sliding and pulling interactions among the sarcomere's filaments. As Figure 38.16a shows, there are two sets of actin filaments, attached to opposite sides of the sarcomere and extending partway to its center. There is one set of myosin filaments. These don't extend all the way to the sides, but they partially overlap the actin filaments. During contraction, myosin filaments physically slide along and *pull* the two sets of actin filaments toward the center of the sarcomere, which thereby shortens. This is the key premise of the **sliding-filament model** of muscle contraction.

The interactions between myosin and actin filaments occur through **cross-bridge formation**. As indicated by Figure 38.16b, each cross-bridge is an attachment between a myosin "head" and a binding site on actin. When a muscle cell is stimulated, myosin heads are energized. They attach to an adjacent actin filament and

tilt in a short power stroke toward the sarcomere's center. An input of energy from ATP drives the power stroke. During the stroke, the heads pull the actin filament along with them. Then a new energy input makes the heads let go, attach to another region of the actin filament, tilt in another power stroke, and so on down the line. A single contraction of a sarcomere takes a whole series of power strokes.

Energy-driven interactions between myosin and actin filaments shorten the many sarcomeres of a muscle cell and collectively account for its contraction.

Figure 38.16 (**a**) Simplified picture of how actin and myosin filaments are arranged in a sarcomere. Interactions between the two kinds of filaments shorten (contract) the sarcomere. (**b**) Sliding-filament model of contraction in the sarcomeres of muscle cells. For simplicity, the action of only one myosin head is shown.

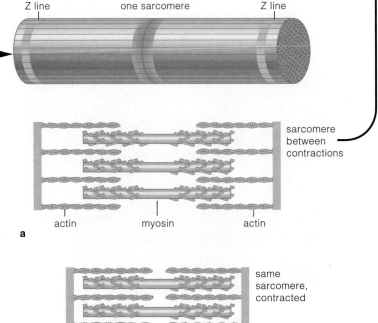

a

b

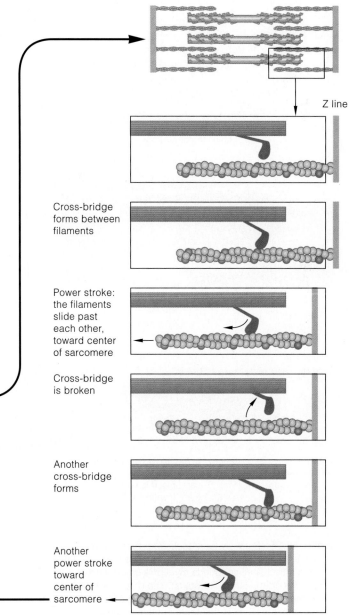

Cross-bridge forms between filaments

Power stroke: the filaments slide past each other, toward center of sarcomere

Cross-bridge is broken

Another cross-bridge forms

Another power stroke toward center of sarcomere

CONTROL OF CONTRACTION

The Control Pathway

Skeletal muscles contract to help move the body and its assorted parts at certain times, in certain ways—and they do so in response to commands from the nervous system. As described on pages 570 and 580, the nervous system communicates with muscles by way of motor neurons, which deliver signals that can stimulate or inhibit contraction of muscle cells.

Like neurons, muscle cells are "excitable." As is true of all cells, they show a difference in electric charge across their plasma membrane. (The cytoplasm just beneath the membrane is a bit more negative than the fluid outside the membrane.) But in excitable cells only, the difference in charge reverses suddenly and briefly in response to adequate stimulation. This sudden reversal in charge, an **action potential**, results from a flow of charged ions across the membrane. The commotion spreads without diminishing along the membrane, away from the point of stimulation.

Figure 38.17 shows what happens after commands from the nervous system stimulate action potentials in a muscle cell. Action potentials spread from the point of stimulation and rapidly reach small, tubular extensions of the plasma membrane. The small tubes connect with a system of membrane-bound chambers that thread lacily around the cell's myofibrils. That system—called the **sarcoplasmic reticulum**—takes up, stores, and releases calcium ions in controlled ways.

The Control Mechanism

When action potentials reach the sarcoplasmic reticulum, they trigger an outward flow of calcium ions from it. The released ions diffuse into the myofibrils and reach actin filaments. Before this, the muscle was resting; actin binding sites were blocked and myosin could not form cross-bridges with them. The arrival of calcium ions clears the binding sites—so contraction proceeds. Afterward, calcium ions are actively transported back into the membrane storage system.

What blocks the cross-bridge binding sites in a resting muscle? As Figure 38.18 shows, protein molecules (troponins and tropomyosins) are located in or near the surface grooves of actin filaments. At low calcium lev-

a Signals from the nervous system travel along spinal cord, down motor neuron.

section from spinal cord

motor neuron

b Endings of motor neuron terminate next to a muscle cell.

section from a skeletal muscle

part of one muscle cell

c Signals travel along muscle cell's plasma membrane to sarcoplasmic reticulum around cell's myofibrils.

Figure 38.17 Pathway for signals from the nervous system that stimulate contraction of skeletal muscle. The plasma membrane of each muscle cell surrounds myofibrils and connects with inward-threading tubes (T tubules). The membrane-bound tubes are close to the sarcoplasmic reticulum, a calcium-storing system that functions in the control of contraction.

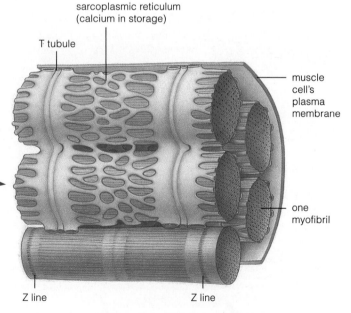

sarcoplasmic reticulum (calcium in storage)

T tubule

muscle cell's plasma membrane

one myofibril

Z line   Z line

d Signals trigger calcium release from sarcoplasmic reticulum lacing around the myofibrils. Calcium allows actin and myosin filaments inside the myofibrils to interact and bring about contraction.

els, the two kinds of proteins are joined tightly together. The arrangement puts tropomyosin slightly outside the groove—a position that blocks the cross-bridge binding site. When the calcium level rises, calcium binds to troponin and causes its shape to change. Thus changed, troponin releases its molecular grip on tropomyosin—which is free to move into the groove and so expose the binding site.

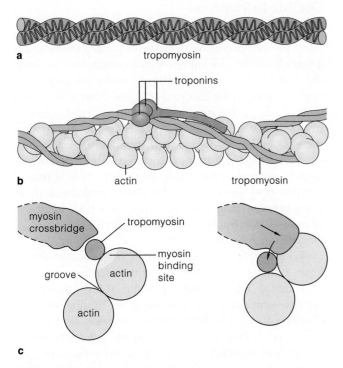

a tropomyosin

b troponins / actin / tropomyosin

c myosin crossbridge / tropomyosin / myosin binding site / groove / actin / actin

Figure 38.18 Arrangement of troponins, tropomyosin, and actin filaments in skeletal muscle cells. When calcium binds with a troponin, tropomyosin moves away from the actin and so exposes the cross-bridge binding sites.

Energy for Contraction

Even with high concentrations of calcium ions, muscles cannot contract without ATP. In a resting muscle cell, the supply of ATP is rather tiny. When called upon to contract, its demand for ATP skyrockets. At that time, the cell plucks phosphate from **creatine phosphate**, an organic compound, and attaches it to ADP. This ATP-forming reaction is simple, fast, and good for a few seconds' worth of contraction before the cell depletes its limited store of creatine phosphate. It also buys time for other, slower ATP-forming pathways to kick in (Figure 38.19).

During prolonged, moderate exercise, the oxygen-requiring reactions of aerobic respiration provide most of the ATP required for contraction. At first, the glucose required for the reactions comes from glycogen stores inside muscle cells. However, when contraction must be sustained, the cells rapidly switch to glucose and fatty acids delivered to them by the bloodstream. (Here you may wish to review page 134.)

Suppose the level of exercise is so intense that it exceeds the capacity of the body's respiratory and circulatory systems to deliver oxygen to muscles. Then, muscle cells produce more and more ATP by lactate fermentation. In this anaerobic pathway, recall, glucose is broken down to lactate. The ATP yield is small, and lactate builds up in the cell.

Commands from the nervous system initiate action potentials in muscle cells. These action potentials are the signals for cross-bridge formation—hence for contraction.

Although the nervous system governs skeletal muscle contraction, the availability of ATP in muscle cells affects whether and for how long contraction will proceed.

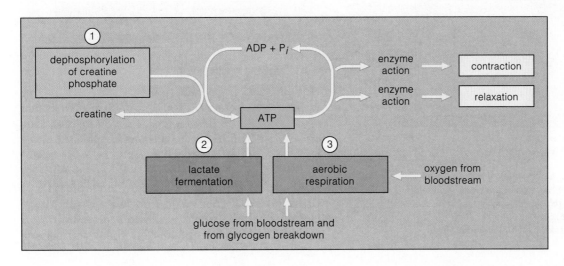

① dephosphorylation of creatine phosphate

creatine

ADP + P$_i$

ATP

② lactate fermentation

③ aerobic respiration

enzyme action → contraction

enzyme action → relaxation

oxygen from bloodstream

glucose from bloodstream and from glycogen breakdown

Figure 38.19 Three possible metabolic pathways by which ATP can form in muscles.

38.9 MUSCLE TENSION, STRENGTH, AND FATIGUE

Muscle Tension

Collectively, the cross-bridges that form during contraction exert **muscle tension**. This is a mechanical force that resists gravity, the weight of packages or dumbbells or other objects being lifted, and other forces that work against muscle tension. When muscle tension is greater than the forces opposing it, contracting muscle cells shorten. When opposing forces are stronger, muscle cells lengthen. People who engage in isometric exercises consciously maintain contraction of muscles at a constant length for a brief time.

Muscle Twitches and Tetanic Contraction

Arnold Schwarzenegger, with his bulging muscles, is Hollywood's icon of physical strength. His strength depends on how forcefully his muscles can contract. That, in turn, depends on (1) muscle size, (2) how many muscle cells are contracting, and (3) how rapidly the nervous system is stimulating them.

As indicated by Figure 38.17b, the endings of a motor neuron form junctions with more than one muscle cell. Each motor neuron and the muscle cells that receive its messages represent a **motor unit**.

Physiologists can record the responses of motor units to stimulation. They create "artificial" action potentials (electrical impulses), then record any changes in muscle contraction. With a single, brief stimulus, a muscle contracts briefly, then relaxes. This response is a **muscle twitch** (Figure 38.20a). A second stimulus applied before the response is over makes the muscle twitch again.

A **tetanus** is a large contraction in which a motor unit is stimulated repeatedly, so that twitches mechanically run together. (In a disease by the same name, toxins prevent muscle relaxation.) Figure 38.20d shows a recording of tetanic contraction.

Decline in Tension— Muscle Fatigue

Suppose continuous, high-frequency stimulation keeps a muscle cell in a state of tetanic contraction. In time, the result will be **muscle fatigue**—a decline in the tension resulting from its cross-bridges. After a few minutes of rest, a fatigued muscle will contract again in response to stimulation. But the extent of its recovery depends on how long and how frequently it was stimulated previously. Muscles associated with brief, intense exercise fatigue fast but recover rapidly. This is true of weightlifting. Muscles associated with prolonged, moderate exercise fatigue more slowly but take longer to recover—often by as much as 24 hours. This is true of long-distance hiking.

With regular exercise, muscles may become larger, have a higher metabolic capacity, and become more resistant to fatigue. Among other things, exercising improves circulation and increases the number of mitochondria in muscle cells. The *Focus* essay on the next page takes a look at what can happen when the desire for stronger muscles is carried to an extreme.

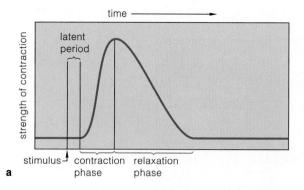

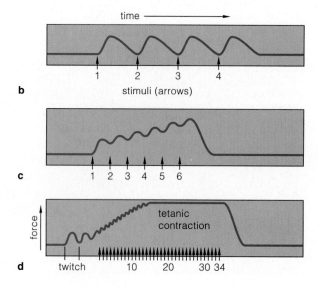

Figure 38.20 Recordings of twitches in artificially stimulated muscles. (**a**) A single twitch. (**b**) Two stimulations per second cause a series of twitches. (**c**) Six per second cause a summation of twitches. (**d**) About twenty per second cause tetanic contraction.

Muscle Mania

Some call it a will-to-win gone bonkers, a desire for muscles larger than life. Others say it's a modern necessity in athletic competition. Either way, many athletes illegally use performance-enhancing drugs, mainly stanazol and other anabolic steroids. Ten athletes were disqualified from the 1988 Olympics for using the banned drugs. Others dropped out rather than submit to stringent drug tests.

Each year in the United States alone, about 1 million athletes use anabolic steroids. Maybe 85 percent of all professional football players use or have used them. Even adolescent boys use them to gain a winning edge in wrestling, football, and weightlifting tournaments.

What Anabolic Steroids Are   Anabolic steroids are synthetic hormones, developed in the 1930s as therapeutic drugs that could mimic a sex hormone, testosterone. Secondary sexual traits, among other things, depend on testosterone. Under its influence, boys get a deeper voice; more hair on their facial, underarm, and pubic skin; more sweat gland secretions; and more massive muscles in the arms, legs, shoulders, and elsewhere. Testosterone also stimulates the more aggressive behavior often associated with maleness. Anabolic steroids can do these things as well—but at a physical and psychological price.

What Anabolic Steroids Do
Twenty or so varieties of anabolic steroids stimulate the synthesis of protein molecules, including muscle proteins. Supposedly, they induce rapid gains in muscle mass and strength during weight-training exercise programs. The claim is disputed (results of most studies are based on too few subjects). Even so, testimonials pour in from weightlifters, football players, and other athletes who specialize in events involving "brute power." Users commonly combine daily oral doses with a single hefty injection each month.

Yet athletes who use steroids suffer minor and major side effects. In men, acne, baldness, shrinking testes, and infertility are early signs of toxicity. These symptoms arise when high blood levels of anabolic steroids trigger a

sharp drop in normal testosterone production. The drugs also may cause early heart disease. Even brief or occasional use may damage kidneys or set the stage for cancer of the liver, testes, and prostate gland. In women, anabolic steroids deepen the voice and produce pronounced facial hair. Menstrual cycles become irregular. Breasts may shrink, and the clitoris may become grossly enlarged.

'Roid Rage   Not all steroid users have developed severe physical side effects. Far more common are mental difficulties, called 'roid rage or *body-builder's psychosis*. Some men become irritable and increasingly aggressive. Others become uncontrollably aggressive, delusional, and wildly manic. One steroid user deliberately drove his rapidly moving car into a tree.

Given the suspected dangers, why would people jeopardize themselves and their future? Possibly not everyone believes the drugs do enough damage to outweigh the "edge" they give in competition. What should a competitor do when "winning" accords athletes wealth and hero status but relegates others to the pile of also-rans? What would *you* do?

SUMMARY

1. Most animals have an integumentary system, which covers the body's surface. Skin (and the structures derived from it) is an example. Skin affords protection against abrasion, bacterial attack, ultraviolet radiation, and dehydration. It helps control internal temperature and serves as a blood reservoir. Its receptors detect specific stimuli in the external environment.

2. The outer portion of vertebrate skin is epidermis (stratified epithelium). It consists mainly of multiple layers of dead, keratinized, and melanin-shielded epithelial cells. Active cell divisions proceed in the inner portion, the dermis.

3. Hair, oil glands, sweat glands, and other derived structures associated with skin are embedded mainly in the dermis.

4. Animals move by way of the contraction of muscle cells. But muscle cells require some medium or structure against which the force of contraction can be applied. Three kinds of skeletons fulfill the requirement:

   a. In hydrostatic skeletons, internal body fluids accept the applied force of contraction and are redistributed within a confined space. Sea anemones and earthworms have such skeletons.

   b. In exoskeletons, rigid external body parts accept the force of contraction. Crab shells and insect cuticles are examples.

   c. In endoskeletons, rigid internal body parts receive the applied force of contraction. The bony human skeleton is an example.

5. Bones are structural elements of vertebrate skeletons. They consist of living cells, bone tissue, and connective tissues. Their collagen-rich extracellular matrix and ground substance are mineralized.

6. Bones function in movement (by interacting with skeletal muscles), protection and support of other body parts, mineral storage, and (in some cases) blood cell formation.

7. The human skeleton has axial and appendicular portions:

   a. The axial portion includes the bones of the skull, vertebral column, ribs, and breastbone. The vertebral column consists of bony segments (vertebrae) cushioned by intervertebral disks.

   b. The appendicular portion includes the limb bones, pelvic girdle, and pectoral girdle.

   c. Skeletal joints are areas of contact or near-contact between bones. They differ in the kind of connective tissue that bridges the gap between bones.

8. In response to action potentials (commands from the nervous system), cells of smooth, cardiac, and skeletal muscle tissues contract (shorten).

9. Each skeletal muscle cell contains many threadlike myofibrils, which contain actin and myosin filaments. The filaments are organized in orderly, parallel arrays in sarcomeres, the basic units of contraction.

   a. A skeletal muscle shortens through the combined decreases in the length of its sarcomeres.

   b. The parallel orientation of a skeletal muscle's component parts directs the force of contraction toward a bone that must be pulled in some direction.

   c. Thus skeletal muscles and bones work together like a system of levers, with rigid rods (bones) moving at fixed points (joints). Pairs and groups of muscles work together or in opposition to bring about movement of the body or positional changes in its parts.

10. Commands from the nervous system travel along motor neurons and initiate action potentials in muscle cells.

   a. Sarcomeres contract when the action potentials trigger the release of calcium ions from a membrane system (sarcoplasmic reticulum) that is arranged around the myofibrils.

   b. Calcium binds to actin filaments and induces changes at the binding sites that allow heads of adjacent myosin filaments to form cross-bridges.

   c. Each cross-bridge is a short-lived attachment between a myosin head and an actin binding site.

   d. Cross-bridges form during repeated, ATP-driven power strokes. The repeated strokes make actin filaments slide past myosin filaments and so shorten the sarcomere.

11. Muscle cells obtain the ATP required for contraction by three pathways: dephosphorylation of creatine phosphate, aerobic respiration, and lactate fermentation.

   a. The first pathway is simple, fast, and good for a few seconds of contraction.

   b. Aerobic respiration sustains prolonged, moderate exercise.

   c. Lactate fermentation is used when intense exercise exceeds the body's capacity to deliver oxygen to muscle cells.

12. Muscle tension refers to a mechanical force created by the cross-bridges formed during contraction. The force resists gravity, the weight of objects being lifted, and other opposing forces.

1. What are some of the functions of skin? List some derivatives of epidermis. *632–633*

2. What are some of the functions of bone tissue? *638*

3. Name the three types of muscle, then state the function of each and where they are located in the body. *642*

4. Look at Figures 38.15 and 38.16. Then, on your own, sketch and label the fine structure of a muscle, down to one of its individual myofibrils. Can you identify the basic unit of contraction in a myofibril? *644–645*

5. How do actin and myosin interact in a sarcomere to bring about muscle contraction? What role does ATP play? What role does calcium play? *644–647*

Self-Quiz *(Answers in Appendix IV)*

1. The _____ system protects the body from abrasion, ultraviolet radiation, bacterial attack, and other environmental stresses.

2. _____ and _____ systems work together to move the body and specific body parts.

3. The three types of muscle tissue are _____ , _____ , and _____ .

4. Which is *not* a function of skin?
   a. resist abrasion
   b. restrict dehydration
   c. produce movement
   d. act as blood reservoir

5. _____ are shock pads and flex points.
   a. Vertebrae
   b. Femurs
   c. Marrow cavities
   d. Intervertebral disks

6. Blood cells form in _____ .
   a. red marrow
   b. all bones
   c. many bones
   d. a and c

7. In skeletal muscle, the _____ is the basic unit of contraction.
   a. myofibril
   b. sarcomere
   c. muscle fiber
   d. myosin filament

8. Muscle contraction requires _____ .
   a. calcium ions
   b. ATP
   c. action potential arrival
   d. all of the above

9. _____ provides(s) ATP for muscle contraction.
   a. Aerobic respiration
   b. Lactate fermentation
   c. Creatine phosphate breakdown
   d. a and b only
   e. a and c only
   f. a, b, and c

10. Match the M words with their defining feature.
   ____ muscle
   ____ muscle twitch
   ____ muscle tension
   ____ melanin
   ____ myosin
   ____ marrow
   ____ metacarpals
   ____ myofibrils
   ____ muscle fatigue
   a. actin's partner
   b. all in the hands
   c. blood cell production
   d. tension decline
   e. pigment
   f. motor unit response
   g. force exerted by cross-bridges
   h. muscle cells bundled in connective tissue
   i. threadlike structures inside muscle cell

Selected Key Terms

actin *644*	motor unit *648*
action potential *646*	muscle fatigue *648*
bone *638*	muscle system *631*
creatine phosphate *647*	muscle tension *648*
cross-bridge formation *645*	muscle twitch *648*
cuticle *632*	myosin *644*
dermis *632*	oil gland *635*
endoskeleton *636*	osteocyte *639*
epidermis *632*	red marrow *638*
exoskeleton *636*	sarcomere *644*
hair *635*	sarcoplasmic reticulum *646*
hydrostatic skeleton *636*	skeletal muscle *642*
integument *632*	skeletal system *631*
integumentary system *631*	sliding-filament model *645*
intervertebral disk *640*	sweat gland *634*
joint *640*	tendon *642*
keratinocyte *634*	tetanus *648*
ligament *640*	yellow marrow *638*
melanocyte *634*	

Readings

Alexander, R. M. July–August 1984. "Walking and Running." *American Scientist* 72(4):348–354. The biomechanics of traveling on foot.

Brusca, R., and G. Brusca. 1990. *Invertebrates*. Sunderland, Massachusetts: Sinauer.

Eckert, R., D. Randall, and G. Augustine. 1988. *Animal Physiology: Mechanisms and Adaptations*. Third edition. New York: Freeman.

Huxley, H. E. December 1965. "The Mechanism of Muscular Contraction." *Scientific American* 213(6):18–27. Old article, great illustrations.

Spence, A., and E. Mason. 1992. *Human Anatomy and Physiology*. Fourth edition. New York: West.

Vander, A., J. Sherman, and D. Luciano. 1990. *Human Physiology*. Fifth edition. New York: McGraw-Hill.

Weeks, O. December 1989. "Vertebrate Skeletal Muscle: Power Source for Locomotion." *BioScience* 39(11):791–798.

39 CIRCULATION

Heartworks

For Augustus Waller, Jimmie the bulldog was no ordinary pooch (Figure 39.1). Connected to wires and soaked to his ankles in buckets of salty water, Jimmie was a four-footed window into the workings of the heart.

Feel the repeated thumpings of your heart at the chest wall. The same rhythms fascinated Waller and other physiologists of the nineteenth century. They wondered if each heartbeat might produce a characteristic pattern of electrical currents that could be recorded painlessly at the body surface. That is where Jimmie and the buckets came in.

Saltwater is an efficient conductor of electricity—so efficient that it carried faint signals from Jimmie's beating heart, through the skin of his legs, to a crude monitoring device. With this device, Waller made one of the world's first graphic recordings of the beating heart—an electrocardiogram, more commonly known as an ECG (Figure 39.1c).

A graph of your own heart's activity would look much the same. The pattern emerged a few weeks after you started growing as an embryo inside your mother. Patches of cardiac muscle started to contract. One patch took the

lead and has been the pacemaker ever since, and if all goes well, it will continue to do so until the day you die. That natural pacemaker sets, adjusts, and resets the

a JIMMIE   DR. WALLER

b   BUCKET

Figure 39.1 History in the making.
(**a**) Dr. Augustus Waller and Jimmie, beloved pet bulldog, sharing a quiet moment in Waller's study.

(**b**) Jimmie taking part in a painless experiment that yielded one of the world's first electrocardiograms (**c**).

rate at which blood is pumped from your heart, through a vast network of blood vessels, then back to the heart. When you rest, the rate is moderate, around 70 beats a minute. When you jog, your muscles demand more blood-borne oxygen and glucose. Then, your heart may start pounding 150 times a minute to deliver sufficient blood to them.

We have come a long way from Jimmie in our monitorings of the heart. Sensors now detect the faintest signals of an impending heart attack. Computers are used to analyze a patient's beating heart and build images of it on a video screen. Surgeons substitute battery-powered pacemakers for malfunctioning natural ones.

With this chapter, we turn to the circulatory system—the means by which substances are rapidly moved to and from the interstitial fluid that surrounds all of the living cells in complex animals. As you will see, the circulatory system is absolutely central to the body's ability to maintain operating conditions in the internal environment within a tolerable range—a state we call homeostasis.

KEY CONCEPTS

1. Cells survive by exchanging substances with their surroundings. In most animals, substances move rapidly to and from living cells by way of a closed circulatory system.

2. Blood, a fluid connective tissue that is commonly confined within a heart and blood vessels, is the transport medium of circulatory systems.

3. Human blood flows in two circuits. In the pulmonary circuit, the heart pumps oxygen-poor blood to the lungs, where it picks up oxygen; then blood flows back to the heart. In the systemic circuit, the heart pumps oxygen-enriched blood to all body regions. After giving up oxygen in those regions, blood flows back to the heart.

4. Besides oxygen, blood also transports carbon dioxide, plasma proteins, vitamins, hormones, lipids, and other solutes.

5. Arteries and veins are large-diameter transport tubes. Capillaries and, to some extent, venules are fine-diameter tubes for diffusion. Arterioles have adjustable diameters. They are control points for distributing different volumes of blood flow to different regions.

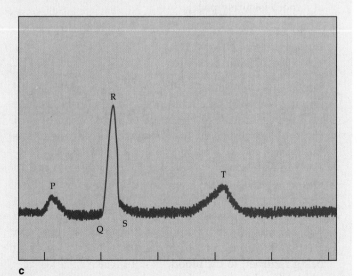

c

CIRCULATORY SYSTEMS—AN OVERVIEW

General Characteristics

Imagine that an earthquake has closed off the highways around your neighborhood. Grocery trucks can't enter and waste-disposal trucks can't leave, so food supplies dwindle and garbage piles up. Cells would face similar predicaments if your body's highways were disrupted. The highways are part of a **circulatory system**, which functions in the rapid internal transport of substances to and from cells.

A circulatory system helps maintain favorable neighborhood conditions, so to speak (Figure 39.2). This is important. Your body's differentiated cells, which perform specialized tasks, cannot fend for themselves.

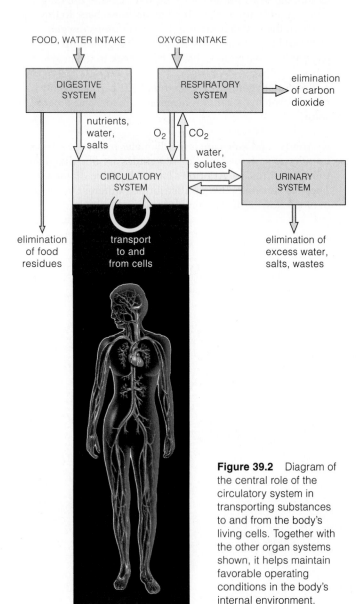

Figure 39.2 Diagram of the central role of the circulatory system in transporting substances to and from the body's living cells. Together with the other organ systems shown, it helps maintain favorable operating conditions in the body's internal environment.

Different types must interact in coordinated ways to maintain the composition, volume, and temperature of the tissue fluid surrounding them, the **interstitial fluid**. A circulating connective tissue—**blood**—interacts with that fluid, making continual deliveries and pickups that help keep conditions tolerable for cell activities. Together, blood and interstitial fluid are the body's internal environment.

Blood flows inside blood vessels—tubes that differ in wall thickness and diameter. A muscular pump, the **heart**, generates pressure that keeps blood flowing.

Like most animals, you have a *closed* circulatory system—blood flow is confined within the heart and blood vessels that have continuously connected walls. This is not true of the *open* circulatory systems of arthropods and most mollusks. In such systems, blood is pumped into tubes that open onto a space in the body's tissues. It mingles with tissue fluids, then moves into open-ended tubes leading back to the heart. Figure 39.3a is a diagram of this arrangement.

Think about the overall "design" of a closed system. The heart pumps incessantly, so the volume of flow through the entire system equals the volume returned to the heart. Yet the rate and volume have to be *adjusted* along the route. As Figure 39.3b suggests, blood flows rapidly through large-diameter vessels. Elsewhere in the system, it must flow slowly, so that there is time enough for substances to be exchanged with interstitial fluid. The required slowdown occurs at **capillary beds**, where the flow fans out through vast numbers of small-diameter blood vessels called **capillaries** (Figure 39.4). By dividing the blood flow, capillaries handle the same total volume of flow as the large-diameter vessels—but at a more leisurely pace.

Functional Links with the Lymphatic System

The heart's pumping action puts pressure on blood flowing through the circulatory system. Partly because of this pressure, small amounts of water and a few of the proteins dissolved in blood are forced out of the capillaries and become part of interstitial fluid. However, an elaborate network of drainage vessels picks up excess interstitial fluid and reclaimable solutes, then returns them to the circulatory system. This network is part of the **lymphatic system**. Later in the chapter, you will see how other parts of the lymphatic system help cleanse bacteria and other disease agents from fluid being returned to the blood.

A circulatory system, together with drainage vessels of the lymphatic system, transports substances to and from the interstitial fluid that bathes all living cells of the body.

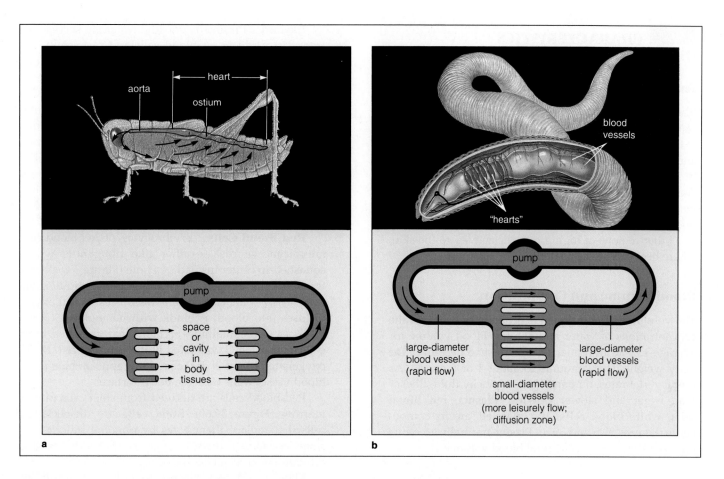

Figure 39.3 Examples of fluid flow in open and closed circulatory systems. (**a**) In the open system of a grasshopper, a "heart" pumps blood through a vessel (aorta), which dumps the blood into body tissues. Blood diffuses through the tissues, then through openings that lead back to the heart. (**b**) In an earthworm's closed system, blood vessels lead away from and back to several muscular "hearts" near its head end. Walls of the hearts and blood vessels interconnect.

Figure 39.4 Example of blood capillaries that are typical of a closed circulatory system. Notice the red blood cells moving single-file through these narrow-diameter blood vessels.

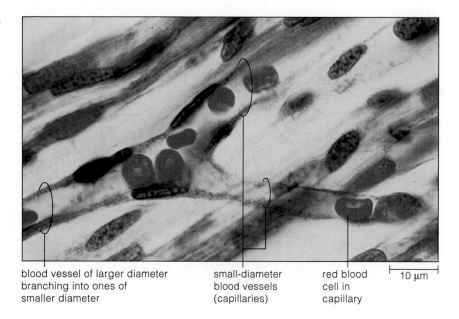

blood vessel of larger diameter branching into ones of smaller diameter

small-diameter blood vessels (capillaries)

red blood cell in capillary

10 μm

39.2 CHARACTERISTICS OF BLOOD

Functions of Blood

Blood, a connective tissue, has multiple functions. It transports oxygen, nutrients, and other solutes to cells. It carries away their secretions (including hormones) and metabolic wastes. Blood helps stabilize internal pH. It serves as a highway for phagocytic cells that scavenge tissue debris and fight infections. In birds and mammals, blood helps equalize body temperature. It does this by carrying excess heat from skeletal muscles and other regions of high metabolic activity to the skin, where heat can be dissipated.

Blood Volume and Composition

The volume of blood depends on body size and on the concentrations of water and solutes. Blood volume for average-size adult humans is about 6 to 8 percent of the body weight. That amounts to about 4 or 5 quarts. As for all vertebrates, human blood is a sticky fluid, thicker than water and slower flowing. **Plasma**, **red blood cells**, **white blood cells**, and **platelets** are its components (Figures 39.5 and 39.6). Plasma normally accounts for 50 to 60 percent of the total blood volume.

Figure 39.5 Components of blood. If a blood sample placed in a test tube is kept from clotting, it will separate into a layer of straw-colored liquid (the plasma) that floats over the red-colored cellular portion of blood.

Plasma Plasma is mostly water. It serves as a transport medium for blood cells and platelets. It also serves as a solvent for ions and molecules, including hundreds of different plasma proteins. Some plasma proteins transport lipids and fat-soluble vitamins. Others help clot blood or defend against disease agents. The movement of water between blood and interstitial fluid—hence the blood's fluid volume—is influenced by the concentrations of plasma proteins. Ions, glucose and other simple sugars, lipids, amino acids, vitamins, and hormones also are dissolved in plasma. So are oxygen, carbon dioxide, and nitrogen.

Red Blood Cells Erythrocytes, or red blood cells, are biconcave disks—rather like doughnuts with a squashed-in center instead of a hole (Figure 39.6a). They transport the oxygen used in aerobic respiration and carry away some carbon dioxide wastes. When oxygen diffuses into blood, it binds with hemoglobin (Figure 3.19), an iron-containing pigment that gives red blood cells their color. Oxygenated blood is bright red. Poorly oxygenated blood is darker red but appears blue inside blood vessel walls near the body surface.

Red blood cells are derived from stem cells in bone marrow (Figure 39.6b). **Stem cells** are unspecialized cells that replace themselves by ongoing mitotic divisions. Portions of their daughter cells also divide, then differentiate into specialized cells.

Mature red blood cells no longer have their nucleus, nor do they require it. They have enough hemoglobin, enzymes, and other proteins to function for about 120 days. Phagocytes continually engulf the oldest or already dead cells, but ongoing replacements keep the

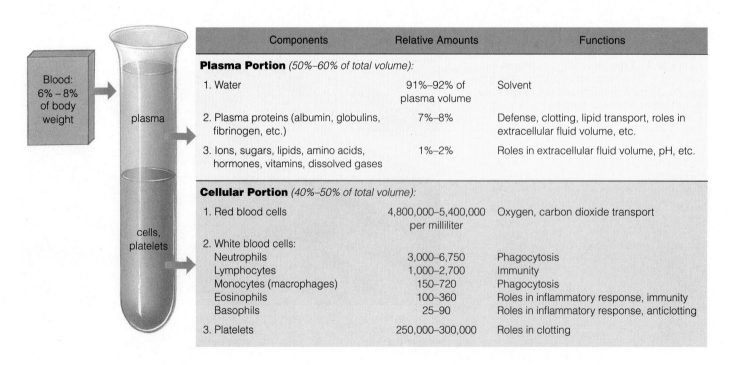

	Components	Relative Amounts	Functions
Plasma Portion *(50%–60% of total volume):*			
	1. Water	91%–92% of plasma volume	Solvent
	2. Plasma proteins (albumin, globulins, fibrinogen, etc.)	7%–8%	Defense, clotting, lipid transport, roles in extracellular fluid volume, etc.
	3. Ions, sugars, lipids, amino acids, hormones, vitamins, dissolved gases	1%–2%	Roles in extracellular fluid volume, pH, etc.
Cellular Portion *(40%–50% of total volume):*			
	1. Red blood cells	4,800,000–5,400,000 per milliliter	Oxygen, carbon dioxide transport
	2. White blood cells:		
	Neutrophils	3,000–6,750	Phagocytosis
	Lymphocytes	1,000–2,700	Immunity
	Monocytes (macrophages)	150–720	Phagocytosis
	Eosinophils	100–360	Roles in inflammatory response, immunity
	Basophils	25–90	Roles in inflammatory response, anticlotting
	3. Platelets	250,000–300,000	Roles in clotting

Blood: 6% – 8% of body weight

plasma

cells, platelets

cell count fairly stable. A **cell count** is the number of cells of a given type in a microliter of blood. The average number of red blood cells is 5.4 million in males and 4.8 million in females.

White Blood Cells Leukocytes, or white blood cells, function in daily housekeeping and defense. Some engulf old, damaged, or dead cells and anything that is chemically perceived as foreign to the body. Others target or destroy specific bacteria, viruses, and other agents of disease. All types arise from stem cells in bone marrow. As you will see, they take up stations in lymph nodes or go to work in inflamed tissues.

White blood cells differ in size, nuclear shape, and staining traits. There are five types: neutrophils, eosinophils, basophils, monocytes, and lymphocytes (Figure 39.6b). Their cell counts vary, depending on whether an individual is active, healthy, or under siege, as described in the next chapter. Neutrophils and monocytes are search-and-destroy cells. The monocytes follow chemical trails to inflamed tissues. There they develop into macrophages ("big eaters") that engulf invaders and debris. Two classes of lymphocytes, B cells and T cells, respond to specific invaders.

Platelets Some stem cells in bone marrow give rise to giant cells (megakaryocytes). These shed fragments of cytoplasm, which become enclosed in a bit of plasma membrane. The membrane-bound fragments are platelets. Each lasts five to nine days, but hundreds of thousands are always circulating in blood. Substances released from platelets initiate blood clotting.

Blood has roles in transport, in defense, in blood clotting, and in maintaining the volume, composition, and temperature of the internal environment.

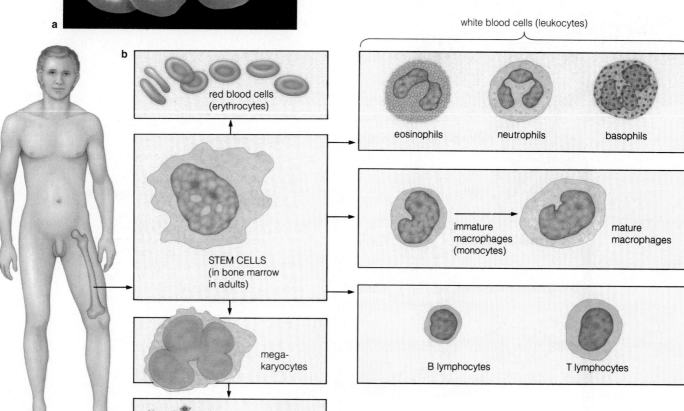

Figure 39.6 (**a**) Size and shape of red blood cells. (**b**) The cellular components of blood.

39.3 HUMAN CARDIOVASCULAR SYSTEM

General Organization

"Cardiovascular" comes from the Greek *kardia* (heart) and the Latin *vasculum* (vessel). In the human cardiovascular system, a heart pumps blood into large-diameter **arteries**. From there, blood flows into small, muscular **arterioles**, which branch into the even smaller diameter capillaries introduced earlier. Blood flows from capillaries into small **venules**, then into large-diameter **veins** that return blood to the heart.

Blood Circulation Routes

A muscular partition divides the human heart into two halves. The partition is the basis of two cardiovascular circuits. Each circuit has its own set of arteries, arterioles, capillaries, venules, and veins.

In the **pulmonary circuit**, the heart's right half pumps blood to the lungs. There, blood picks up oxygen and gives up carbon dioxide. From the lungs, the freshly oxygenated blood flows to the heart's left half (Figure 39.7*a*). In the **systemic circuit**, the heart's left half pumps the oxygenated blood to all regions where oxygen is used and carbon dioxide is produced. Then the oxygen-poor blood flows to the heart's right half (Figure 39.7*b*).

Figure 39.8 shows the main blood vessels of the two blood circulation routes. You might think that a given volume of blood passes through only one capillary bed during a trip away from and back to the heart. This is not always the case. For example, part of the systemic circuit threads around the intestines. There, blood flows through one capillary bed and picks up glucose and other absorbed substances. Then the blood moves on through another capillary bed, in the liver—an organ with a key role in nutrition. The second bed gives the liver time to process absorbed substances before blood is circulated further.

Figure 39.7 Diagram of the pulmonary and systemic circuits for blood flow through the human circulatory system.

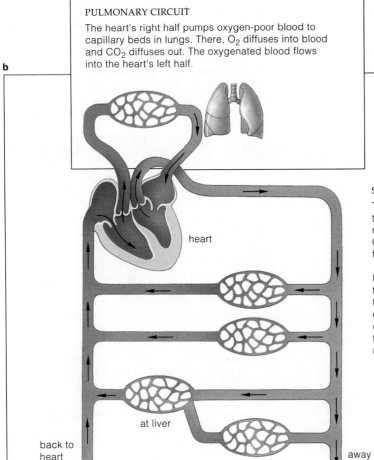

a

PULMONARY CIRCUIT

The heart's right half pumps oxygen-poor blood to capillary beds in lungs. There, O_2 diffuses into blood and CO_2 diffuses out. The oxygenated blood flows into the heart's left half.

b

heart

at liver

back to heart

at intestines

away from heart

SYSTEMIC CIRCUIT

The heart's left half pumps the oxygen-rich blood to capillary beds at organs and tissues in all body regions. There, O_2 diffuses out of blood and CO_2 diffuses in. Then the blood, now oxygen-poor, flows into the heart's right half.

In most cases, a given volume of blood passes through only one capillary bed when it makes the trip away from and back to the heart. There are exceptions, as when blood passes through a capillary bed at the intestines, then is shunted through a capillary bed at the liver before returning to the heart.

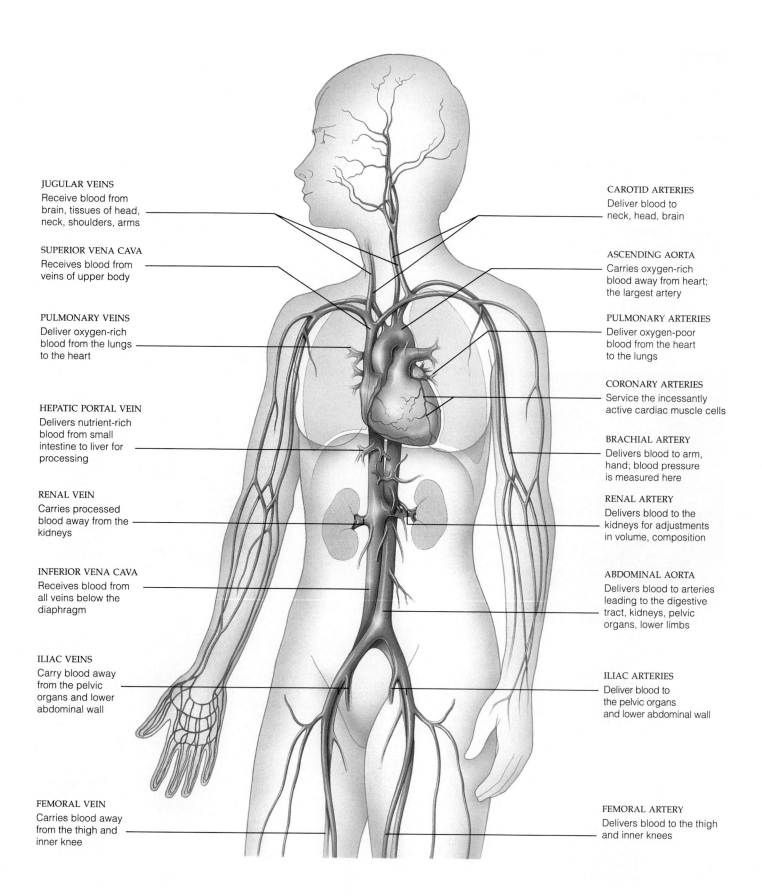

JUGULAR VEINS
Receive blood from brain, tissues of head, neck, shoulders, arms

SUPERIOR VENA CAVA
Receives blood from veins of upper body

PULMONARY VEINS
Deliver oxygen-rich blood from the lungs to the heart

HEPATIC PORTAL VEIN
Delivers nutrient-rich blood from small intestine to liver for processing

RENAL VEIN
Carries processed blood away from the kidneys

INFERIOR VENA CAVA
Receives blood from all veins below the diaphragm

ILIAC VEINS
Carry blood away from the pelvic organs and lower abdominal wall

FEMORAL VEIN
Carries blood away from the thigh and inner knee

CAROTID ARTERIES
Deliver blood to neck, head, brain

ASCENDING AORTA
Carries oxygen-rich blood away from heart; the largest artery

PULMONARY ARTERIES
Deliver oxygen-poor blood from the heart to the lungs

CORONARY ARTERIES
Service the incessantly active cardiac muscle cells

BRACHIAL ARTERY
Delivers blood to arm, hand; blood pressure is measured here

RENAL ARTERY
Delivers blood to the kidneys for adjustments in volume, composition

ABDOMINAL AORTA
Delivers blood to arteries leading to the digestive tract, kidneys, pelvic organs, lower limbs

ILIAC ARTERIES
Deliver blood to the pelvic organs and lower abdominal wall

FEMORAL ARTERY
Delivers blood to the thigh and inner knees

Figure 39.8 Human circulatory system, showing some of the major blood vessels. Arteries are color-coded *red*. Veins are color-coded *blue*.

39.4 THE HUMAN HEART

Heart Structure

During a seventy-year life span, the human heart beats some 2.5 billion times and rests only briefly between heartbeats. Its structure, shown in Figure 39.9, reflects its role as a durable pump.

The heart is mostly cardiac muscle tissue protected by a tough outer membrane. Connective tissue and endothelium line its inner chambers. (*Endothelium* is a layer of epithelial cells found only in the heart and blood vessels.) There are *two* chambers in each half of the heart—an **atrium** (plural, atria), located above a ventricle. Membrane flaps separate the two chambers and serve as a one-way valve between them. These flaps are called an AV valve (short for atrioventricular). Membrane flaps also are positioned between the ventricle and the artery leading away from it. These flaps are called a semilunar valve.

During a heartbeat, the AV valves and semilunar valves open and close in ways that help keep blood moving in one direction through the body.

The heart has its own "coronary circulation." Two coronary arteries lead into a capillary bed that services most of its cardiac muscle cells. They branch off the **aorta**, the major artery carrying oxygen-enriched blood away from the heart.

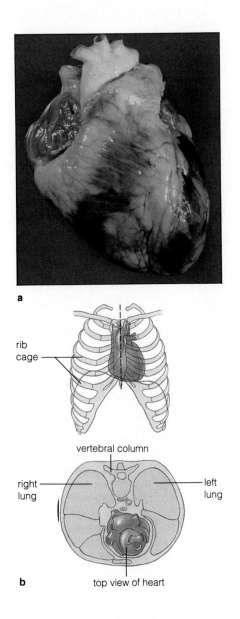

rib cage

vertebral column

right lung — left lung

b top view of heart

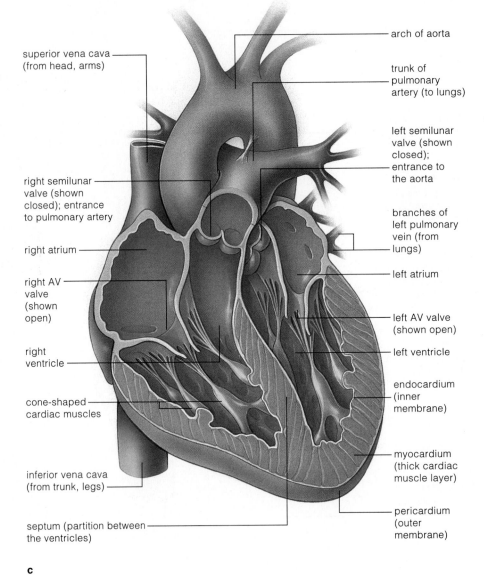

superior vena cava (from head, arms)

right semilunar valve (shown closed); entrance to pulmonary artery

right atrium

right AV valve (shown open)

right ventricle

cone-shaped cardiac muscles

inferior vena cava (from trunk, legs)

septum (partition between the ventricles)

arch of aorta

trunk of pulmonary artery (to lungs)

left semilunar valve (shown closed); entrance to the aorta

branches of left pulmonary vein (from lungs)

left atrium

left AV valve (shown open)

left ventricle

endocardium (inner membrane)

myocardium (thick cardiac muscle layer)

pericardium (outer membrane)

c

Figure 39.9 (**a**) The human heart and (**b**) its location in the thoracic cavity. (**c**) Cutaway view showing the heart's internal organization.

Cardiac Cycle

Each time the heart beats, its four chambers go through phases of contraction (systole) and relaxation (diastole). The sequence of contraction and relaxation is a **cardiac cycle**.

When relaxed atria are filling, fluid pressure increases inside and forces the AV valves open. Blood flows into the ventricles, which fill completely when the atria contract (Figure 39.10a). When the filled ventricles start to contract, fluid pressure inside them increases and forces the AV valves to shut. Their continued contraction causes pressure to rise sharply above that in blood vessels leading away from the heart. The pressure forces the semilunar valves open, and blood leaves the heart (Figure 39.10b). Now the ventricles relax, the semilunar valves close—and the already filling atria are ready to repeat the cycle.

Thus, during a cardiac cycle, atrial contraction simply helps fill the ventricles. *Contraction of the ventricles is the driving force for blood circulation.*

Mechanisms of Contraction

In skeletal muscle tissue, the ends of the fiberlike cells are attached to bones. In **cardiac muscle tissue**, the ends of cardiac muscle cells branch, then connect with one another. Communication junctions bridge the plasma membranes on the ends of abutting cells (Figure 39.11a). Signals for contraction spread across these junctions. With each heartbeat, the signals spread so rapidly that cardiac muscle cells contract together, almost as if they were a single unit.

Cardiac and skeletal muscle cells differ in another way. The nervous system makes skeletal muscle contract. But cardiac muscle contracts spontaneously—the nervous system can only *adjust* the rate and strength of its contraction. Even if all nerves leading to the heart are severed, the heart will keep on beating! How? Some self-excitatory cardiac muscle cells continue to produce and conduct signals for contraction.

Cell bodies of these pacemaking cells are clustered mainly in part of the wall of the right atrium (Figure 39.11b. This region is the SA node (short for sinoatrial). It generates waves of excitation, usually seventy or eighty times a minute. Each wave spreads over both atria, causes them to contract, then reaches the AV node. The wave spreads more slowly here. The delay allows the atria to finish contracting before a wave of excitation spreads over the ventricles.

Thus the SA node is the **cardiac pacemaker**. Its spontaneous, rhythmic signals are the basis for the normal rate of heartbeat.

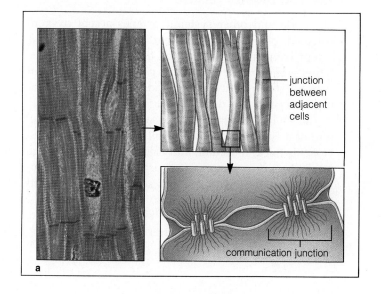

junction between adjacent cells

communication junction

a

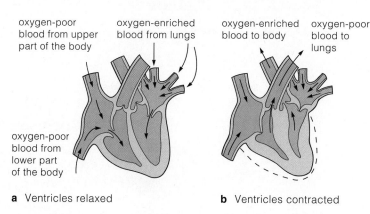

oxygen-poor blood from upper part of the body

oxygen-enriched blood from lungs

oxygen-enriched blood to body

oxygen-poor blood to lungs

oxygen-poor blood from lower part of the body

a Ventricles relaxed

b Ventricles contracted

Figure 39.10 Blood flow through the heart during part of a cardiac cycle. Blood and heart movements generate vibrations, producing a "lub-dup" sound that can be heard at the chest wall. At each "lub," AV valves are closing as ventricles contract. At each "dup," semilunar valves are closing as ventricles relax.

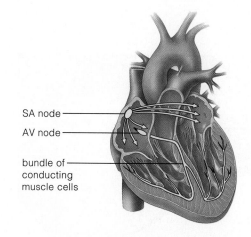

SA node

AV node

bundle of conducting muscle cells

b

Figure 39.11 (**a**) Communication junctions at the ends of abutting cardiac muscle cells. Signals travel rapidly across the junctions and cause cells to contract nearly in unison. (**b**) Location of cardiac muscle cells that conduct signals for contraction through the heart.

BLOOD PRESSURE IN THE CARDIOVASCULAR SYSTEM

Blood pressure is the fluid pressure generated by heart contractions. During the time it takes for a specified volume of blood to leave the heart and make the return trip, the pressure associated with it does not remain the same. For example, pressure normally is high in the aorta, then drops along the circuit away from and back to the heart. Figure 39.12 is a chart of the declining pressure.

What causes the drop? As blood passes through different kinds of blood vessels, energy in the form of pressure is lost as it overcomes resistance to the flow of blood. The structure of the blood vessels themselves, shown in Figure 39.13, is a factor in this loss of energy.

Arterial Blood Pressure

The aorta and other systemic arteries accept oxygen-enriched blood from the heart. Pulmonary arteries accept oxygen-poor blood for delivery to the lungs. Either way, the arteries serve as pressure reservoirs that "smooth out" pulsations in blood pressure. Such pulsations are generated during each cardiac cycle.

Take a look at Figure 39.12a, which shows the structure of a typical artery. The wall of this type of blood vessel is thick, muscular, and elastic. It is able to bulge somewhat under the pressure surge caused by ventricular contraction, then recoil and so force blood onward through the circuit. Arteries also have large diameters. They present little resistance to blood flow, so pressure does not drop much in the arterial portion of the systemic and pulmonary circuits.

Resistance at Arterioles

Arteries branch into vessels of smaller diameter, the arterioles. Track the blood flow along the systemic route and you will find the greatest pressure drop at arterioles, for these offer the greatest resistance to blood flow (Figure 39.12). By analogy, turn two open catsup bottles upside down. Both hold the same amount of catsup, but the neck of one is much narrower. Guess which bottle will be slowest to drain—that is, which will present more resistance to flow.

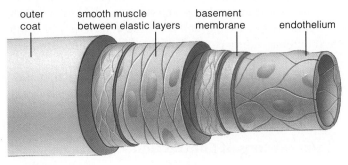

a ARTERY

outer coat · smooth muscle between elastic layers · basement membrane · endothelium

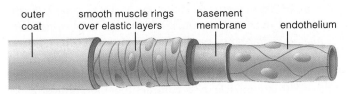

b ARTERIOLE

outer coat · smooth muscle rings over elastic layers · basement membrane · endothelium

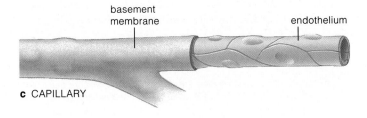

c CAPILLARY

basement membrane · endothelium

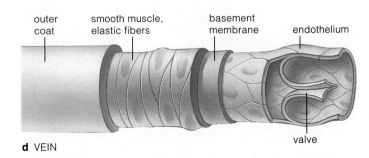

d VEIN

outer coat · smooth muscle, elastic fibers · basement membrane · endothelium · valve

Figure 39.13 Structure of blood vessels. The basement membrane around the endothelium is a noncellular layer, rich with proteins and polysaccharides.

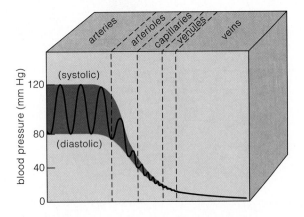

Figure 39.12 Blood pressure. This diagram plots measurements of the drop in fluid pressure for a given volume of blood moving through the systemic circuit.

With this slowdown, the blood flow can now be controlled in ways that divert greater or lesser portions of the total volume to different regions. Arteriole walls happen to contain rings of smooth muscle. As you will now see, in response to signals from the nervous system and endocrine systems, even to a change in local chemical conditions, smooth muscle cells contract or relax—and so increase or decrease the arteriole diameter.

Controls over Blood Pressure and Blood Distribution

Suppose you decide to measure your blood pressure every day while you are resting, in the manner shown in Figure 39.14. Assuming you are an adult in good health, the resting value remains fairly constant over a few weeks, even months—about 120/80 mm Hg, on the average. What mechanisms work to maintain the level of blood pressure over time and so ensure adequate blood flow to all regions of your body?

The resting level depends mainly on reflex centers in the medulla oblongata (page 582). These brain centers help coordinate the rate and strength of heartbeats with changes in the diameter of arterioles (and to some extent, of veins). They receive and integrate signals from sensory receptors that are located in cardiac muscle tissue and in certain arteries, such as the aorta and the **carotid arteries** in the neck.

When sensory receptors detect *increases* in the resting level of blood pressure, the brain centers command the heart to beat more slowly and contract less forcefully. They also command smooth muscle cells in the wall of arterioles to relax. This results in **vasodilation**. The word refers to an enlargement (dilation) of the blood vessel diameter. When a *decrease* in the resting level is detected, the heart is made to beat faster and contract more forcefully. Smooth muscle cells of arterioles are made to contract and so bring about a decrease in blood vessel diameter, or **vasoconstriction**.

Hormones help maintain blood pressure. Arterioles in specified regions have receptors for epinephrine, angiotensin, and other hormones. Epinephrine can trigger vasoconstriction or vasodilation. Angiotensin can trigger widespread vasoconstriction.

The nervous system and endocrine system also control the allocation of more or less blood to different body regions at different times. For example, after you have eaten a large meal, more blood is diverted to your digestive system. When your body is exposed to cold wind or snow for an extended time, blood is diverted away from the skin to deeper tissue regions, so that the metabolically generated heat that warmed the blood can be conserved.

Local controls also bring about diversions in blood flow. Think of what happens when you run. In skeletal

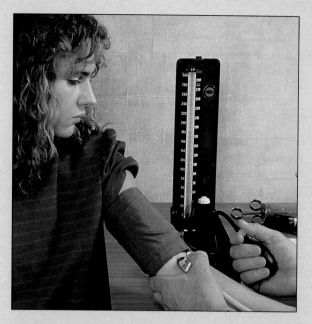

Figure 39.14 Measuring blood pressure. A hollow cuff attached to a pressure gauge is wrapped around the upper arm. Then it is inflated with air to a pressure above the highest pressure of the cardiac cycle (at systole, when the ventricles contract). Above this pressure, no sounds are heard through a stethoscope positioned above the artery (because no blood is flowing through it).

Air in the cuff is slowly released, so some blood flows into the artery. The turbulent flow causes soft tapping sounds, and when this first occurs, the value on the gauge is the systolic pressure—about 120 mm mercury (Hg) in young adults at rest. (This means the measured pressure would force mercury to move upward 120 millimeters in a narrow glass column.) More air is released from the cuff. Just after the sounds become dull and muffled, blood flows continuously. So the turbulence and tapping sounds stop. The silence corresponds to the diastolic pressure (at the end of a cardiac cycle, just before the heart pumps out blood). Generally the reading is about 80 mm Hg. In this example, the *pulse* pressure (the difference between the highest and lowest pressure readings) is 120 – 80, or 40 mm Hg.

muscle tissue, the oxygen level decreases and the levels of carbon dioxide, hydrogen ions, potassium ions, and other substances increase. The changes in local chemical conditions cause smooth muscle in arterioles to relax. With the resulting vasodilation, more blood flows past the active muscles, delivering more raw materials and carrying away cell products and wastes. Simultaneously, arteriole diameter is decreasing in tissues of your digestive tract and kidneys.

The nervous system, endocrine system, and changes in local chemical conditions control the level of blood pressure and the flow distribution to different regions.

The main controls operate on cardiac muscle and on smooth muscle in arterioles.

39.6 FROM CAPILLARY BEDS BACK TO THE HEART

Capillary Function

Capillary beds are *diffusion zones* for exchanges between blood and interstitial fluid. Of all blood vessels, capillaries have the thinnest wall—a single layer of flat endothelial cells, separated from one another by narrow clefts (Figure 39.12c). The capillary lumen is so small, blood cells must squeeze through it single file. Thus a single capillary presents high resistance to flow. Yet so many of these blood vessels are present in a capillary bed, their *combined* diameters are greater than the diameters of arterioles. Thus a capillary bed presents less total resistance to flow than the arterioles leading into it, and the total drop in blood pressure is not great in this region.

Capillaries thread through nearly every tissue, and at least one of them is as close as 0.001 centimeter to every living cell. Oxygen, carbon dioxide, and most other small solutes diffuse across the capillary wall. Certain proteins enter or leave by endocytosis or exocytosis. Certain ions probably cross the wall by passing through clefts between endothelial cells. So do white blood cells, as you will see in the chapter to follow. The clefts are wider in some capillary beds than in others. Such "leaky" beds serve important functions, as in inflammatory responses.

Bulk flow drives fluid across the capillary wall. (By this mechanism, fluid pressure forces molecules of water and solutes to move in the same direction.) Figure 39.15 describes the two opposing forces that bring about the fluid movements. Their action helps maintain the fluid balance between blood and interstitial fluid. The balance is important, for blood pressure is maintained only when there is adequate blood volume. Interstitial fluid can be tapped when blood volume drops rapidly, as during hemorrhage.

Venous Pressure

Capillaries merge into "little veins," or venules. In functional terms, the venules are a bit like capillaries. Some solutes diffuse across their wall, which is only a bit thicker than that of a capillary. Some control over capillary pressure also is exerted at these vessels.

Venules merge into veins (Figure 39.12d). Veins, the large-diameter transport tubes leading back to the heart, contain valves that prevent backflow. When gravity beckons, blood in veins tends to reverse direction—and pushes the valves shut.

Veins are blood volume reservoirs. Collectively, they contain 50 to 60 percent of the total blood volume. A vein wall is thin enough to bulge under pressure, more

Figure 39.15 Fluid movements in an idealized capillary bed. The movements play no significant role in diffusion. But they are important in maintaining the distribution of extracellular fluid between the bloodstream and interstitial fluid. The movements result from two opposing forces, called *filtration* and *absorption.*

At the arteriole end of a capillary, the difference between capillary blood pressure and interstitial fluid pressure leads to filtration. Because of the difference, some plasma (but very few plasma proteins) leaves the capillary. "Filtration" refers to this fluid movement *out* of the capillary.

"Absorption" refers to the movement of some interstitial fluid *into* the capillary. It results from a difference in the concentration of water between plasma and interstitial fluid. With its protein components, the plasma has a greater solute concentration, hence a lower concentration of water.

Absorption at the venule end of a capillary bed tends to balance filtration at the arteriole end. Normally, there is only a small *net* filtration of fluid, which the lymphatic system returns to the blood.

Edema results from an accumulation of excess fluid in interstitial spaces. This condition arises to some extent during exercise, when arterioles dilate in local tissue regions. Capillary pressure increases and triggers increased filtration. Edema also results from an obstructed vein or from heart failure. Again, capillary pressure and then filtration increase. Edema is extreme during *elephantiasis*, which is brought on by a roundworm infection and subsequent obstruction of lymphatic vessels (page 421).

so than an arterial wall. The wall also contains some smooth muscle. When blood must circulate faster (as during exercise), the smooth muscle contracts. The wall stiffens, the vein bulges less, and venous pressure rises and drives more blood to the heart.

When limbs move, skeletal muscles bulge against adjacent veins. Here again, venous pressure rises and drives blood back to the heart (Figure 39.16). Rapid breathing also increases venous pressure. When inhaled air pushes down on internal organs, it changes the pressure gradient between the heart and veins.

We turn next to a *Focus* essay on disorders of the heart and blood vessels we have just considered.

Capillary beds are diffusion zones for exchanges between blood and interstitial fluid. Venules overlap with capillaries in function.

Veins are highly distensible blood volume reservoirs and help adjust flow volume back to the heart.

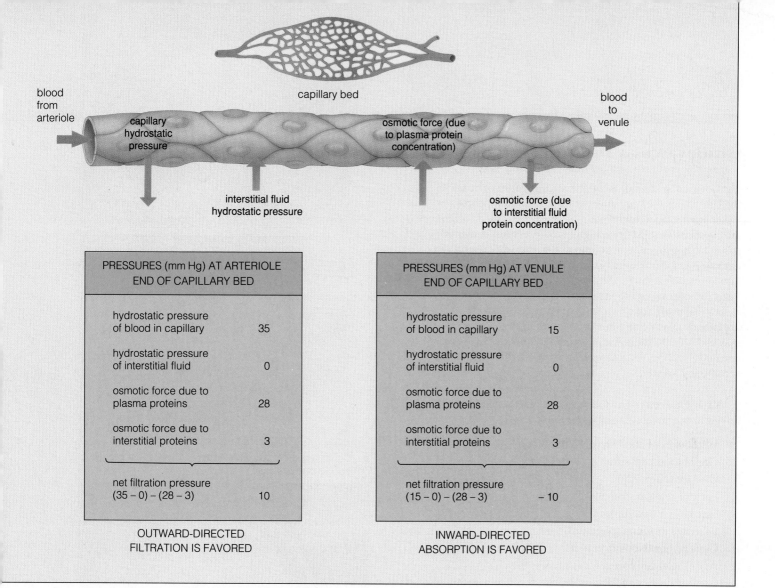

capillary bed

blood from arteriole

capillary hydrostatic pressure

osmotic force (due to plasma protein concentration)

blood to venule

interstitial fluid hydrostatic pressure

osmotic force (due to interstitial fluid protein concentration)

PRESSURES (mm Hg) AT ARTERIOLE END OF CAPILLARY BED	
hydrostatic pressure of blood in capillary	35
hydrostatic pressure of interstitial fluid	0
osmotic force due to plasma proteins	28
osmotic force due to interstitial proteins	3
net filtration pressure (35 − 0) − (28 − 3)	10

OUTWARD-DIRECTED
FILTRATION IS FAVORED

PRESSURES (mm Hg) AT VENULE END OF CAPILLARY BED	
hydrostatic pressure of blood in capillary	15
hydrostatic pressure of interstitial fluid	0
osmotic force due to plasma proteins	28
osmotic force due to interstitial proteins	3
net filtration pressure (15 − 0) − (28 − 3)	− 10

INWARD-DIRECTED
ABSORPTION IS FAVORED

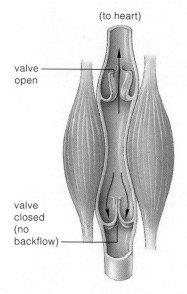

(to heart)

valve open

valve closed (no backflow)

a Skeletal muscle is contracted; assists blood flow to heart

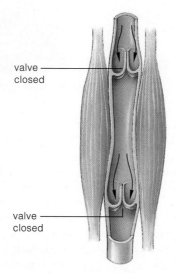

valve closed

valve closed

b Skeletal muscle is relaxed; valves shut and prevent backflow

Figure 39.16 How the bulging of contracting skeletal muscles can cause an increase in fluid pressure inside a vein. Notice the structure of the valve (membrane flaps) in the vein's lumen.

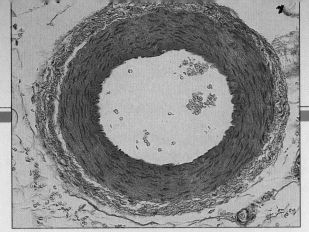

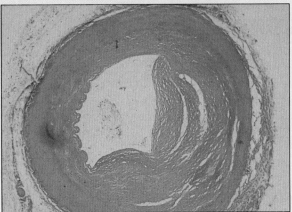

a Cross-section of a normal artery (*above*) and a partially obstructed one (*below*).

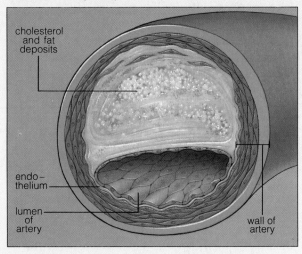

cholesterol and fat deposits

endo-thelium

lumen of artery

wall of artery

b Diagram of an atherosclerotic plaque.

Focus on Health

Cardiovascular Disorders

Cardiovascular disorders are the leading cause of death in the United States. They affect at least 40 million people and kill about a million each year. The most common disorders are *hypertension* (sustained high blood pressure) and *atherosclerosis* (progressive thickening of the arterial wall and narrowing of the arterial lumen). Both affect blood circulation and so cause most *heart attacks* (damage or death of heart muscle) and *strokes* (brain damage).

In most heart attacks, a "crushing" pain behind the breastbone lasts a half hour or more. Mild to severe pain may radiate into the left arm, shoulder, or neck. Sweating, nausea, vomiting, dizziness, or loss of consciousness may accompany an attack.

Risk Factors The following risk factors have been linked to cardiovascular disorders:

1. High level of cholesterol in the blood
2. High blood pressure
3. Obesity (page 728)
4. Lack of regular exercise
5. Smoking (page 706)
6. Diabetes mellitus (page 623)
7. Genetic predisposition to heart failure
8. Age (the older you get, the greater the risk)
9. Gender (until age fifty, males are at much greater risk than females)

Factors 1 through 5 are controllable by eating properly, exercising, and not smoking.

For example, the fatter you become, the more blood capillaries you get (they service the increased adipose tissue masses). So your heart has to work harder to pump blood through the increasingly divided vascular circuit. As another example, smoking can damage your heart. Nicotine in tobacco makes the adrenal glands secrete epinephrine, a hormone that constricts blood vessels and so triggers an accelerated heartbeat and a rise in blood pressure. Carbon monoxide in cigarette smoke outcompetes carbon dioxide for the binding sites on hemoglobin—so the heart has to pump harder to rid the body of carbon dioxide.

The next sections describe some of the tissue damage that results from cardiovascular disorders.

Hypertension Hypertension results from gradual increases in the resistance to blood flow through small arteries. In time, blood pressure remains elevated, even when a person is resting. Heredity may be a factor; the disorder tends to run in families. Diet is also a factor. For instance, high salt intake can raise blood pressure in sus-

ceptible people and increase the heart's workload. Eventually the heart may enlarge and fail to pump blood effectively. High blood pressure also may contribute to the "hardening" of arterial walls that hampers delivery of oxygen to the brain, heart, and other vital organs.

Hypertension is called the "silent killer" because affected persons may have no outward symptoms. Even when they know their blood pressure is high, some tend to resist medication, changes in diet, and regular exercise. Of 23 million hypertensive Americans, most don't undergo treatment. About 180,000 die each year.

Atherosclerosis In arteriosclerosis, arteries thicken and lose elasticity. In atherosclerosis, this condition worsens as cholesterol and other lipids build up in the wall of arteries and cause the lumen to narrow.

Recall, from page 32, that the liver produces enough cholesterol to satisfy the body's needs, and the diet adds to the cholesterol level. When circulating in blood, cholesterol is bound to proteins as *low-density lipoproteins* (LDLs). These bind to receptors on cells throughout the body. Cells take up LDLs and their cholesterol cargo for use in cell activities. Excess cholesterol is attached to proteins as *high-density lipoproteins* (HDLs) and transported back to the liver, where it can be metabolized.

For a variety of reasons, not enough LDL is removed from the blood in some people. As the blood level of LDL increases, so does the risk of atherosclerosis. With their bound cholesterol, the LDLs *infiltrate* arterial walls. Abnormal smooth muscle cells multiply and connective tissue components increase in the walls. Cholesterol accumulates in cells and extracellular spaces of the wall's endothelial lining. Calcium salts are deposited on top of the lipids, and a fibrous net forms over the mass. This *atherosclerotic plaque* sticks out into the arterial lumen (Figures *a,b*).

When platelets get caught on a plaque's rough edges, they secrete chemicals that initiate clot formation. Growth of the plaque and clot can narrow or block the artery. Blood flow to the tissues serviced by the artery may decrease to a trickle or stop entirely. A clot that stays in place is a *thrombus*. If it becomes dislodged and travels the bloodstream, it is an *embolus*.

Coronary arteries and their branches have narrow diameters. They are extremely vulnerable to clogging by a plaque or clot. When they narrow to one-quarter of their former diameter, the outcome ranges from *angina pectoris* (mild chest pains) to a heart attack.

Atherosclerosis involving coronary arteries can be diagnosed through stress electrocardiograms. These are recordings of the electrical activity of the cardiac cycle while a person is exercising on a treadmill. It also can be diagnosed by *angiography*. This procedure involves injections of a dye that is opaque to x-rays. Severe blockage may require surgery. In *coronary bypass surgery*, a section of an artery from the chest is stitched to the aorta and to the coronary artery below the narrowed or blocked region (Figure *c*). In *laser angioplasty*, laser beams vaporize the plaques. In *balloon angioplasty*, a small balloon is inflated within a blocked artery to break up plaques.

Arrhythmias ECGs reveal arrhythmias, irregular heart rhythms (Figure *d*). Not all arrhythmias are abnor-

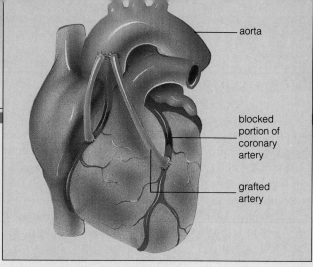

c Two coronary bypasses (*green*)

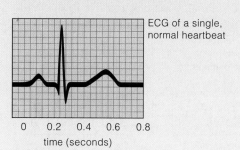

ECG of a single, normal heartbeat

time (seconds)

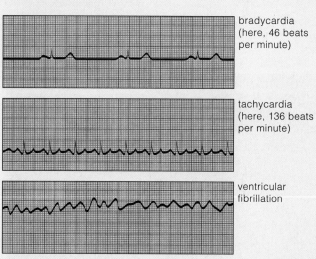

bradycardia (here, 46 beats per minute)

tachycardia (here, 136 beats per minute)

ventricular fibrillation

d Examples of ECG readings

mal. Endurance athletes may have a below-average resting cardiac rate (*bradycardia*). In an adaptation to ongoing strenuous exercise, their nervous system has adjusted the cardiac pacemaker's rate of contraction downward. Also, exercise or stress commonly causes 100+ heartbeats a minute (*tachycardia*). But a coronary occlusion or some other disorders may cause abnormal rhythms that can degenerate rapidly into a dangerous condition called *ventricular fibrillation*. In portions of the ventricles, cardiac muscle contracts haphazardly, and blood pumping suffers. Within seconds, the person loses consciousness, which may signify impending death. A strong electric shock to the chest may restore normal cardiac function.

39.7 HEMOSTASIS

Small blood vessels are vulnerable to ruptures, cuts, and other damage. **Hemostasis** repairs the damage and so prevents blood loss. This process includes blood vessel spasm, platelet plug formation, and blood coagulation. First, smooth muscle in a damaged wall contracts in an automatic response called a spasm. The blood vessel constricts, so blood flow through it temporarily stops. Second, platelets clump together as a temporary plug in the damaged wall. They also release substances that help prolong the spasm and attract more platelets. Third, blood coagulates (converts to a gel) and forms a clot (Figure 39.17). Finally, the clot retracts into a compact mass, and the breach in the wall is sealed.

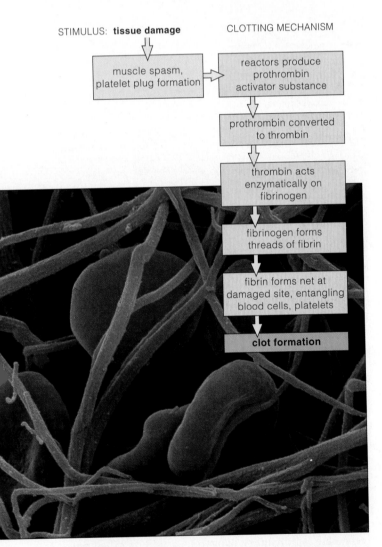

STIMULUS: **tissue damage**   CLOTTING MECHANISM

muscle spasm, platelet plug formation → reactors produce prothrombin activator substance

prothrombin converted to thrombin

thrombin acts enzymatically on fibrinogen

fibrinogen forms threads of fibrin

fibrin forms net at damaged site, entangling blood cells, platelets

clot formation

Figure 39.17 Plasma proteins involved in clot formation. Blood coagulates when damage exposes collagen fibers in blood vessel walls. This initiates a series of reactions that cause rod-shaped plasma proteins (fibrinogens) to stick together as long, insoluble threads. These stick to the exposed collagen, forming a net that traps blood cells and platelets. The entire mass is a clot.

39.8 BLOOD TYPING

Your body protects itself in another way, with "self" markers at the surface of its cells. These protein markers identify the cells as belonging to you. Your body also produces **antibodies**, a class of protein molecules that bind to a specific foreign marker and so target its bearer for destruction by the immune system. Bacteria, viruses, and anything else that isn't one of your own normal cells carry foreign markers.

During a *transfusion* (when blood from two people mixes), red blood cells bearing the "wrong" marker are recognized as foreign and destroyed, with serious consequences. The same thing will happen during pregnancy, if antibodies diffuse from the mother's circulatory system into that of her unborn child.

Based on an understanding of cell surface markers and antibodies, scientists have devised ways to analyze the forms of self markers on a person's red blood cells. Blood typing is based on these analyses.

ABO Blood Typing

Molecular variations in one kind of self marker on red blood cells are analyzed in **ABO blood typing**. The genetic basis of this variation was described earlier, on page 177. People with one form of the marker are said to have type A blood, and those with another form have type B blood. Many people have both forms of the marker on their red blood cells; they have type AB blood. Others have neither form of the marker; they have type O blood.

If you are type A, your antibodies ignore A markers but will act against B markers. If you are type B, your antibodies ignore B markers but will act against A markers. If you are type AB, your antibodies ignore both forms of the marker, so you can tolerate donations of type A, B, or AB blood. However, if you are type O, you have antibodies against both forms of the marker, so your options are limited to type O donations.

Figure 39.18 shows what happens when blood from incompatible donors and recipients intermingles. In a defensive response called **agglutination**, antibodies act against the foreign cells and cause them to clump. Unfortunately, these clumps can clog small blood vessels. They may lead to tissue damage and death.

Rh Blood Typing

Other kinds of surface markers on red blood cells also may cause agglutination responses. **Rh blood typing** is based on the presence or absence of an Rh marker (so named because it was first identified in blood samples of *Rh*esus monkeys). If you are type Rh$^+$, your blood cells bear this marker. If you are type Rh$^-$, they don't.

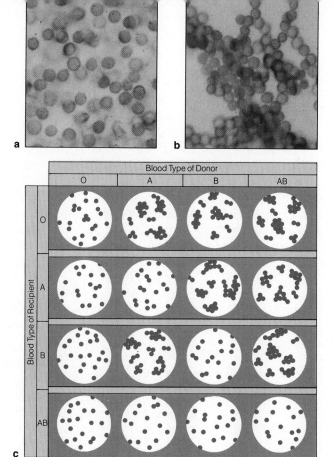

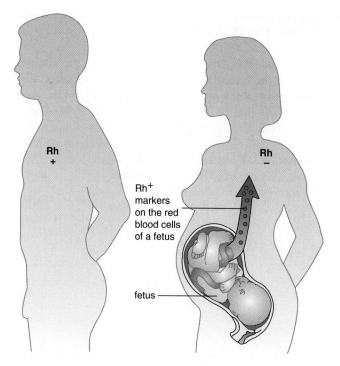

a A forthcoming child of an Rh⁻ woman and Rh⁺ man inherits the genetic instructions for the Rh⁺ marker. The placenta, a complex tissue that forms during pregnancy, allows the mother's blood to intermingle with that of the growing fetus. So fetal blood cells bearing the Rh⁺ markers enter her bloodstream.

Figure 39.18 Micrographs showing (**a**) an absence of agglutination in a mixture of two different but compatible blood types and (**b**) agglutination in a mixture of incompatible types. (**c**) Agglutination responses in blood types A, B, AB, and O when mixed with samples of the same and different types.

Ordinarily, people do not have antibodies against Rh markers. But a recipient of an Rh⁺ blood transfusion will produce antibodies against the marker, and these will continue circulating in the blood.

If an Rh⁻ woman becomes pregnant by an Rh⁺ man, there is a chance the fetus will be Rh⁺. During pregnancy or childbirth, some of its red blood cells may leak into the mother's bloodstream. If they do, her body will produce antibodies against Rh (Figure 39.19). If she becomes pregnant *again*, Rh antibodies will enter the bloodstream of this new fetus. If its blood is type Rh⁺, her antibodies will cause its red blood cells to swell, rupture, and release hemoglobin.

In *erythroblastosis fetalis*, an extreme case of this disorder, too many cells are destroyed and the fetus dies. If diagnosed before birth or delivered alive, it can survive by having its blood slowly replaced with transfusions that are free of Rh antibodies. Currently, a known Rh⁻ woman can be treated right after her first pregnancy with an anti-Rh gamma globulin (RhoGam) that will protect her next fetus. The drug will inactivate any Rh⁺ fetal blood cells circulating in the mother's bloodstream before she can become sensitized and begin producing potentially dangerous antibodies.

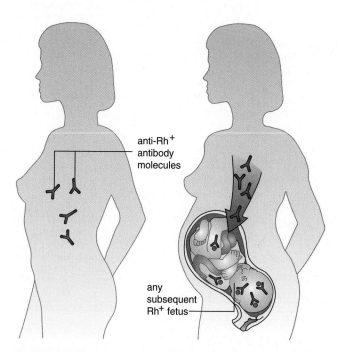

b The foreign markers in the mother's body stimulate production of antibody molecules. Suppose the woman becomes pregnant again. If the new fetus (or any other one) inherits instructions for the Rh⁺ marker, the anti-Rh⁺ antibodies will act against them.

Figure 39.19 Production of antibodies in response to Rh⁺ markers on the red blood cells of a developing fetus.

We conclude this chapter with a section on the manner in which the lymphatic system supplements blood circulation. But think of this section as a bridge to the next chapter, on immunity, for the lymphatic system also helps defend the body against injury and attack. As Figure 39.20 shows, the system consists of drainage vessels, lymphoid organs, and lymphoid tissues. Tissue fluid that has moved into the vessels is called **lymph**.

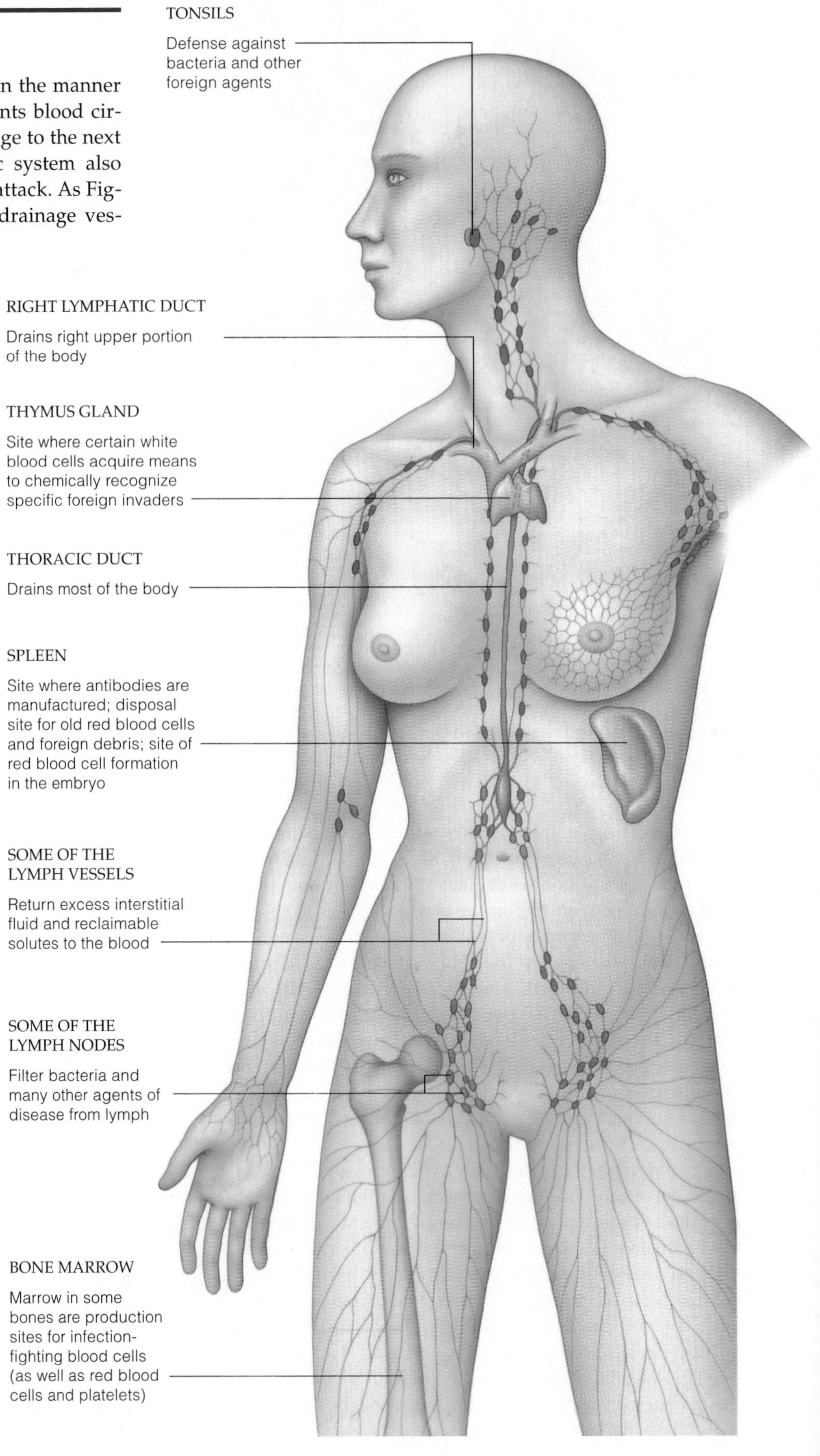

TONSILS

Defense against bacteria and other foreign agents

RIGHT LYMPHATIC DUCT

Drains right upper portion of the body

THYMUS GLAND

Site where certain white blood cells acquire means to chemically recognize specific foreign invaders

THORACIC DUCT

Drains most of the body

SPLEEN

Site where antibodies are manufactured; disposal site for old red blood cells and foreign debris; site of red blood cell formation in the embryo

SOME OF THE LYMPH VESSELS

Return excess interstitial fluid and reclaimable solutes to the blood

SOME OF THE LYMPH NODES

Filter bacteria and many other agents of disease from lymph

BONE MARROW

Marrow in some bones are production sites for infection-fighting blood cells (as well as red blood cells and platelets)

Figure 39.20 Components of the human lymphatic system and their functions.

The *green* dots show some of the major lymph nodes. Patches of lymphoid tissue in the small intestine and in the appendix also are part of the lymphatic system.

Lymph Vascular System

The vascular portion of the lymphatic system, which includes **lymph capillaries** and **lymph vessels**, has three functions. First, it takes up water and plasma proteins that have leaked out at capillary beds and returns them to the blood. Second, it transports fats absorbed from the small intestine, in the manner described on page 722. Third, it transports foreign particles and cellular debris to disposal centers (lymph nodes).

The lymph vascular system starts at capillary beds. Tissue fluid enters lymph capillaries. These have no pronounced entrance. Instead, their tips contain regions of overlapping endothelial cells (Figure 39.21*a*). Water and solutes move inward at these flaplike "valves."

Lymph capillaries merge with lymph vessels. These have a larger diameter, smooth muscle in their wall, and flaplike valves that prevent backflow. As Figure 39.20 shows, the lymph vessels eventually converge into collecting ducts, which drain into veins in the lower neck.

Lymphoid Organs and Tissues

The defense portion of the lymphatic system includes the **lymph nodes**, **spleen**, and **thymus**. It also includes the tonsils, adenoids, and patches of tissue in the small intestine and appendix.

The lymph nodes are located at intervals along lymph vessels (Figure 39.21*b*). Lymph trickles through at least one of these nodes before entering the bloodstream. Lymph nodes are sites where many white blood cells take up residence after they have been produced in bone marrow. During the course of an infection, they become battlegrounds where great armies of lymphocytes form and where foreign agents are destroyed.

The largest lymphoid organ, the spleen, is a filtering station for blood and a holding station for lymphocytes. The spleen has inner chambers filled with red and white "pulp." The red pulp is a large reservoir of red blood cells and macrophages. In human embryos, the spleen produces red blood cells.

In the thymus gland, lymphocytes multiply, differentiate, and mature into fighters of specific disease agents. The thymus produces hormones that influence these events. It is central to immunity, the focus of the chapter to follow.

The lymph vascular system returns tissue fluid to the blood, transports fats, and transports foreign material and debris to lymph nodes.

Lymph nodes and other lymphoid organs function in the body's systems of defense against infection and disease.

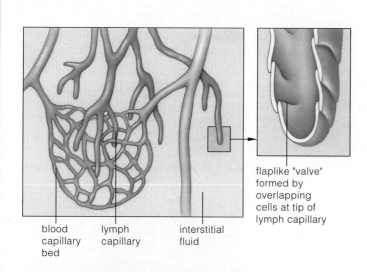

flaplike "valve" formed by overlapping cells at tip of lymph capillary

blood capillary bed   lymph capillary   interstitial fluid

a Lymph capillaries

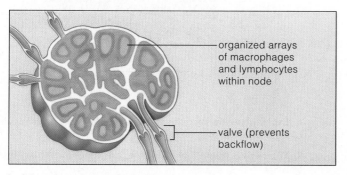

organized arrays of macrophages and lymphocytes within node

valve (prevents backflow)

b A lymph node, cross-section

Figure 39.21 (**a**) Some of the lymph capillaries at the start of the drainage network called the lymph vascular system.

(**b**) Cutaway diagram of a lymph node. Its inner chambers are packed with highly organized arrays of infection-fighting white blood cells.

SUMMARY

1. Humans and other vertebrates have a closed circulatory system. It consists of a heart (a muscular pump), blood vessels (arteries, arterioles, capillaries, venules, and veins), and blood. The system functions in rapid internal transport of substances to and from cells.

2. Blood, a fluid connective tissue, helps maintain favorable conditions for cells. It delivers oxygen and other substances to the interstitial fluid around cells. It also picks up cell products and wastes from that fluid.

3. Blood consists of plasma, red and white blood cells, and platelets.

   a. Plasma, the liquid component of blood, is a transport medium for blood cells and platelets. It is a solvent for plasma proteins, simple sugars, lipids, amino acids, mineral ions, vitamins, hormones, and several gases.

   b. Red blood cells transport oxygen from the lungs to the tissues of all body regions. They are packed with hemoglobin, an iron-containing pigment molecule that binds reversibly with oxygen. Red blood cells also transport some carbon dioxide from interstitial fluid to the lungs.

   c. Certain phagocytic white blood cells cleanse tissues of dead cells, cellular debris, and anything detected as not belonging to the body. Other white blood cells (lymphocytes) form armies that destroy specific bacteria, viruses, and other disease agents.

4. An internal partition divides the human heart into two halves, each with two chambers (an atrium and a ventricle). The partition separates the blood flow into two circuits, one pulmonary and the other systemic.

   a. In the pulmonary circuit, oxygen-poor blood in the heart's *right* half is pumped to capillary beds in the lungs. The blood picks up oxygen, then flows to the heart's left half.

   b. In the systemic circuit, oxygenated blood in the heart's *left* half is pumped to all body tissues. There, cells take up oxygen for aerobic respiration and give up carbon dioxide. The blood, now oxygen-poor, flows to the heart's right half.

5. Ventricular contraction drives blood through both circuits. Blood pressure is high in contracting ventricles, then drops in arteries, arterioles, capillaries, venules, and veins. It is lowest in relaxed atria.

   a. Arteries are an elastic pressure reservoir. They smooth out pressure changes resulting from heartbeats and so smooth out blood flow through capillaries.

   b. Arterioles are control points for distributing different volumes of blood to different regions.

   c. Capillary beds are diffusion zones where blood and interstitial fluid exchange substances.

   d. Venules overlap capillaries and veins somewhat in function.

   e. Veins are a blood volume reservoir that can be tapped to adjust the volume of flow back to the heart.

6. The lymphatic system has these functions:

   a. Its vascular portion takes up water and plasma proteins that seep out of blood capillaries, then returns them to the blood circulation. It transports absorbed fats and delivers agents of disease to disposal centers. Lymph capillaries and lymph vessels are components.

   b. Its lymphoid organs and tissues contain production centers for lymphocytes as well as battlegrounds.

Review Questions

1. Describe the cellular components of blood. Describe the plasma portion of blood. *656–657*

2. Define the functions of the circulatory system and the lymphatic system. *654, 658, 670–671*

3. Distinguish between:
   a. blood and interstitial fluid *654*
   b. systemic and pulmonary circuits *658*
   c. atrium and ventricle *660*

4. State the main functions of arteries, arterioles, capillaries, veins, venules, and lymph vessels. *658, 671*

5. Label the heart's components. *660*

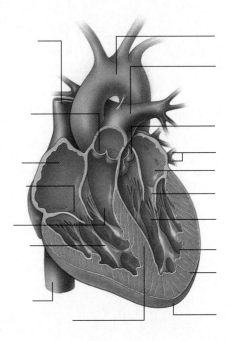

6. Explain how the medulla oblongata of the brain helps regulate blood flow to different body regions. *663*

7. What drives solutes out of and into capillaries in capillary beds? *664–665*

8. What forces work together in returning venous blood to the heart? *664*

1. Cells directly exchange substances with _____ .
   a. blood vessels   c. interstitial fluid
   b. lymph vessels   d. both a and b

2. Which are *not* components of blood?
   a. plasma
   b. blood cells and platelets
   c. gases and other dissolved substances
   d. all of the above are components of blood

3. The _____ produces red blood cells, which transport _____ and some _____ .
   a. liver; oxygen; mineral ions
   b. liver; oxygen; carbon dioxide
   c. bone marrow; oxygen; hormones
   d. bone marrow; oxygen; carbon dioxide

4. The _____ produces white blood cells, which function in _____ and _____ .
   a. liver; oxygen transport; defense
   b. lymph nodes; oxygen transport; pH stabilization
   c. bone marrow; housekeeping; defense
   d. bone marrow; pH stabilization; defense

5. In the pulmonary circuit, the heart's _____ half pumps blood to lungs, then _____ blood flows to the heart.
   a. right; oxygen-poor   c. right; oxygen-rich
   b. left; oxygen-poor   d. left; oxygen-rich

6. In the systemic circuit, the heart's _____ half pumps _____ blood to all body regions.
   a. right; oxygen-poor   c. right; oxygen-rich
   b. left; oxygen-poor   d. left; oxygen-rich

7. Blood pressure is high in _____ and lowest in _____ .
   a. arteries; veins   c. arteries; ventricles
   b. arteries; relaxed atria d. arterioles; veins

8. _____ contraction drives blood through the pulmonary circuit and the systemic circuit; blood pressure is highest in contracting _____ .
   a. Atrial; ventricles   c. Ventricular; arteries
   b. Atrial; atria   d. Ventricular; ventricles

9. Which of the following is *not* a function of the lymphatic system?
   a. Delivers disease agents to disposal centers
   b. Produces lymphocytes
   c. Delivers oxygen
   d. Returns water and plasma proteins to blood

10. Match the type of blood vessel with its major function.
    ____ arteries   a. diffusion
    ____ arterioles   b. control of blood volume distribution
    ____ capillaries
    ____ venules   c. transport, blood volume reservoirs
    ____ veins   d. overlap of capillary function
                       e. transport and pressure reservoirs

11. Match the circulation components with their descriptions.
    ____ capillary beds   a. two atria, two ventricles
    ____ lymph vascular   b. pressure reservoirs
         system   c. driving force for blood
    ____ heart chambers   d. zones of diffusion
    ____ veins   e. interstitial fluid
    ____ heart contractions   f. blood volume reservoirs
    ____ arteries

Selected Key Terms

ABO blood typing *668*
agglutination *668*
antibody *668*
aorta *660*
arteriole *658*
artery *658*
atrium *660*
blood *654*
blood pressure *662*
capillary *654*
capillary bed *654*
cardiac cycle *661*
cardiac muscle tissue *661*
cardiac pacemaker *661*
carotid artery *663*
cell count *657*
circulatory system *654*
heart *654*
hemostasis *668*
interstitial fluid *654*

lymph *670*
lymph capillary *671*
lymph node *671*
lymph vessel *671*
lymphatic system *654*
plasma *656*
platelet *656*
pulmonary circuit *658*
red blood cell *656*
Rh blood typing *668*
spleen *671*
stem cell *656*
systemic circuit *658*
thymus *671*
vasoconstriction *663*
vasodilation *663*
vein *658*
ventricle *660*
venule *658*
white blood cell *656*

Readings

Eisenberg, M. S., et al. May 1986. "Sudden Cardiac Death." *Scientific American* 254(5):37–43.

Golde, D. W., and J. C. Gasson. July 1988. "Hormones That Stimulate the Growth of Blood Cells." *Scientific American* 259(1):62–70.

Kapff, C. T., and J. H. Jandl. 1981. *Blood: Atlas and Sourcebook of Hematology.* Boston: Little, Brown. Beautiful micrographs of normal and abnormal blood and marrow cells.

Little, R., and W. Little. 1989. *Physiology of the Heart and Circulation.* Fourth edition. Chicago: Year Book Medical. Comprehensive coverage of cardiovascular physiology.

Raloff, J. September 1989. "Do You Know Your HDL?" *Science News* 136(11):171–173.

Robinson, T. F., et al. June 1986. "The Heart as a Suction Pump." *Scientific American* 254(6):84–91.

40 IMMUNITY

Russian Roulette, Immunological Style

Until about a century ago, smallpox epidemics swept repeatedly through the world's cities. Some outbreaks were so severe that only half of those who were stricken survived. The survivors had permanent scars on the face, neck, shoulders, and arms—but they seldom contracted the disease again. They were "immune" to smallpox.

No one knew what caused smallpox. But the idea of acquiring immunity was dreadfully appealing. In twelfth-century China, healthy people were gambling with deliberate infections. They sought out survivors of mild cases of smallpox (who were only mildly scarred), then removed crusts from the scars, ground them up, and inhaled the powder. By the seventeenth century, Mary Montagu, wife of the English ambassador to Turkey, was championing inoculation. She went so far as to inject bits of smallpox scabs into the veins of her children. Even the Prince of Wales did the same. Others soaked threads in fluid from the sores, then poked the threads into scratches on the body.

Individuals who survived these practices acquired immunity to smallpox—but many came down with raging infections. As if the odds were not dangerous enough, the crude inoculation procedures also invited the acquisition of syphilis, hepatitis, and other infectious diseases.

While this immunological version of Russian roulette was going on, Edward Jenner was growing up in the English countryside. At the time, it was known that people who contracted cowpox never got smallpox. (Cowpox is a mild disease that can be transmitted from cattle to humans.) No one thought much about this until 1796, when Jenner, by now a physician, injected material from a cowpox sore into the arm of an uninfected boy. Six weeks later, after the reaction subsided, Jenner injected fluid from smallpox sores into the boy (Figure 40.1a). He hypothesized that the earlier injection might provoke immunity to smallpox—and he was right. The boy remained free of smallpox.

The French mocked Jenner's procedure, calling it "vaccination" (which translates as "encowment"). Much later a French chemist, Louis Pasteur, devised similar procedures for other diseases. Pasteur also called his procedures vaccinations, and only then did the term become respectable.

By Pasteur's time, improved microscopes were revealing diverse bacteria, fungal spores, and other

a

previously invisible forms of life. As Pasteur discovered, microorganisms abound in ordinary air. Did some cause diseases? Probably. Could they settle into food or drink and cause it to spoil? Pasteur proved that they did. Boiling killed the microorganisms—Pasteur and others knew this. However, being a wine connoisseur, he also knew you cannot boil wine—or beer or milk, for that matter—and end up with the same beverage. He devised a way to heat food or beverages at a temperature low enough not to ruin them but high enough to kill most of the suspect microorganisms. We still depend on his partial sterilization methods, which were named *pasteurization* in his honor.

In the late 1870s, Robert Koch, a German physician, conducted experiments that linked a specific microorganism to a specific disease—namely, anthrax. Koch had injected blood from infected animals into uninfected ones. The recipients of the injections ended up with blood teeming with cells of a bacterium (*Bacillus anthracis*)—and they developed anthrax. Even more convincing, injections of bacterial cells that were cultured outside the body also caused the disease!

Thus, by the beginning of the twentieth century, the promise of understanding the basis of infectious disease and immunity loomed large. The battles against those diseases were about to begin in earnest. Those battles are the focus of this chapter.

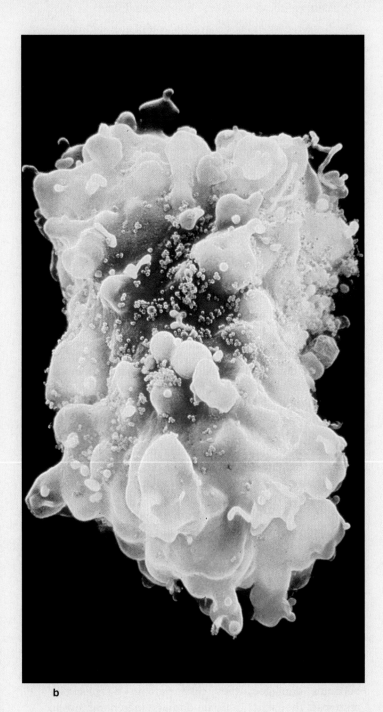

b

Figure 40.1 (**a**) Statue honoring Edward Jenner's development of an immunization procedure against smallpox, one of the most dreaded diseases in human history. (**b**) Micrograph of a white blood cell being attacked by the virus (*blue particles*) that causes AIDS. Immunologists are working to develop weapons against this modern-day scourge.

1. The vertebrate body has physical, chemical, and cellular defenses against invasion by viruses, bacteria, and other agents of disease.

2. During early stages of an invasion, white blood cells and plasma proteins take part in an inflammatory response. During this rapid, nonspecific counterattack, phagocytic white blood cells ingest invaders. Plasma proteins promote phagocytosis, and some also destroy invaders directly.

3. If the invasion persists, some white blood cells make immune responses. These cells can recognize distinct configurations on molecules that are abnormal or foreign to the body, such as those on bacteria and viruses. If the foreign or abnormal molecule triggers an immune response, it is called an antigen.

4. In one type of immune response, some of the white blood cells produce antibodies in huge amounts. Antibodies are molecules that bind to a specific antigen and tag it for destruction. In another type of immune response, executioner cells directly destroy body cells that have become abnormal, as by infection or by a tumor-producing process.

Throughout your life, amazingly diverse pathogens have been attacking you. **Pathogens** include viruses, bacteria, fungi, protozoans, and parasitic worms that cause disease. You and other vertebrates coevolved with most of them, so you need not lose sleep over this. Your body has barriers at its surface. When pathogens do breach the barriers, cells and chemical weapons destroy most of them. If all else fails, armies of elite white blood cells make highly focused counterattacks. Table 40.1 lists these three lines of defense.

Surface Barriers to Invasion

Most often, pathogens cannot get past skin or the linings of other body surfaces. Think of skin as a habitat of low moisture, low pH, and thick layers of dead cells. Normally harmless bacteria are adapted to life in this habitat. Few pathogens can compete with dense populations of the established types unless conditions change. Thus pathogenic fungi that lurk in gym locker rooms can thrive between toes that are repeatedly encased in warm, damp shoes. They can penetrate sodden, weakened tissues and give you *athlete's foot*.

The gut's mucous lining and its resident bacteria keep pathogens in check. So do *Lactobacillus* populations on the vagina's lining. Lactate, a by-product of their metabolism, helps maintain a low vaginal pH that most bacteria and fungi cannot tolerate.

Barriers in the tubular airways leading to your lungs stop airborne pathogens. As air rushes past the turns and branchings of the mucus-coated tube walls, it becomes turbulent and flings bacteria against the sticky sides. The mucus contains protective substances. One of these substances, **lysozyme**, is an enzyme that digests the cell wall of bacteria and so contributes to their death. In a final bit of housekeeping, broomlike cilia that line the airways sweep out the trapped and enzymatically whapped pathogens.

Lysozyme and other enzymes in tears, saliva, and fluid in the gut provide more protection. Tears give eyes a sterile washing. Urine, with its low pH and flushing action, helps keep pathogens from moving into the urinary tract. By its flushing action, diarrhea can rid the gut of irritating pathogens. Diarrhea should be controlled in young children to keep them from dehydrating, but taking steps to block its flushing action in adults can prolong infection.

Nonspecific and Specific Responses

All animals react to tissue damage. Even simple aquatic invertebrates contain phagocytic cells and antimicrobial substances, including lysozymes. The more complex the

---

Table 40.1   The Vertebrate Body's Three Lines of Defense Against Pathogens

Barriers at Body Surfaces (*nonspecific* targets):

1. Intact skin; mucous membranes at other body surfaces

2. Infection-fighting substances in tears, saliva, etc.

3. Normally harmless bacterial inhabitants of body surfaces that outcompete pathogenic visitors

4. Flushing effect of tears, urination, and diarrhea

Nonspecific Responses (*nonspecific* targets):

1. Inflammation

   a. Fast-acting white blood cells (neutrophils, eosinophils, basophils)
   b. Macrophages (also take part in immune responses)
   c. Complement proteins, blood-clotting proteins, other infection-fighting substances

2. Organs with phagocytic functions (e.g., lymph nodes)

Immune Responses (*specific* targets):

1. White blood cells (macrophages, T cells, B cells)

2. Communication signals (e.g., interleukins) and chemical weapons (e.g., antibodies, complement proteins)

---

animal body, however, the more complex are the systems that defend it. Consider this: After vertebrates invaded the land, selection favored greater efficiency in the circulation of body fluids (page 451). As circulatory and lymphatic systems evolved, so did the body's defenses. Great numbers of phagocytic cells could now travel rapidly, along with plasma proteins, to tissues under siege. The cells could also position themselves strategically along the lymph vascular highways—and so intercept pathogens trickling past.

Today, you and other mammals have sophisticated sets of plasma proteins. Some contribute to rapid clot formation following an injury, and others kill pathogens directly or target them for phagocytosis, as described next. Also, recognition of "wounded self" has expanded enormously to include recognition of subtle molecular differences among "nonself" pathogens.

Thus, internal defenses against a great variety of pathogens are in place even before an invasion occurs. The defenses include several kinds of white blood cells (leukocytes) as well as plasma proteins. When a tissue becomes damaged, they take part in a *nonspecific* response; they react to tissue damage in general, not to one pathogen or another. By contrast, a *specific* response is directed against only one kind of pathogen. Recognition of a molecular configuration on a specific pathogen's surface triggers their response, which runs its course whether tissues are damaged or not.

COMPLEMENT PROTEINS

A set of plasma proteins has roles in both nonspecific and specific defenses. Collectively, these proteins are called the **complement system**.

About twenty kinds of complement proteins circulate in blood in inactive form. If even a few molecules of one kind are activated, they trigger a huge cascade of reactions. They activate many molecules of another kind of complement protein. Each of these activates many molecules of another kind of protein at the next reaction step, and so on. Thus great numbers of molecules are deployed, with the following effects.

Some complement proteins join together to form pore complexes. These molecular structures have an interior channel (Figure 40.2). They become inserted into the plasma membrane of many pathogens, forming pores that induce lysis. **Lysis** refers to a gross structural disruption of a plasma membrane, and it leads to cell death. Pore complexes also become inserted into the wall of Gram-negative eubacteria. Such walls consist of a phospholipid-rich, outer membrane above a peptidoglycan layer. (Like the plasma membrane, the outer membrane selectively controls which substances cross it.) Lysozyme molecules diffuse through the pores and digest the peptidoglycan.

As you will read next, some activated complement proteins promote the nonspecific defense response called inflammation. Through their cascades of reactions, the proteins create concentration gradients that attract phagocytic white blood cells to an irritated or damaged tissue. The proteins also encourage the phagocytes to dine. Many invaders have numerous surface receptors to which complement proteins can bind. Through such binding, an invader ends up with a complement "coat." The coat happens to adhere to phagocytes, and this is rather like putting a basted turkey on a dinner table.

Figure 40.2 Formation of membrane attack complexes by complement proteins. (**a**) One reaction pathway starts when complement binds to bacterial surfaces; another operates during immune responses to specific invaders. Both produce membrane pore complexes that induce lysis in the target pathogen. They help defend the body against many bacteria, some parasitic protistans, and enveloped viruses. (**b**) Micrograph of a cell surface, showing pores formed by membrane attack complexes.

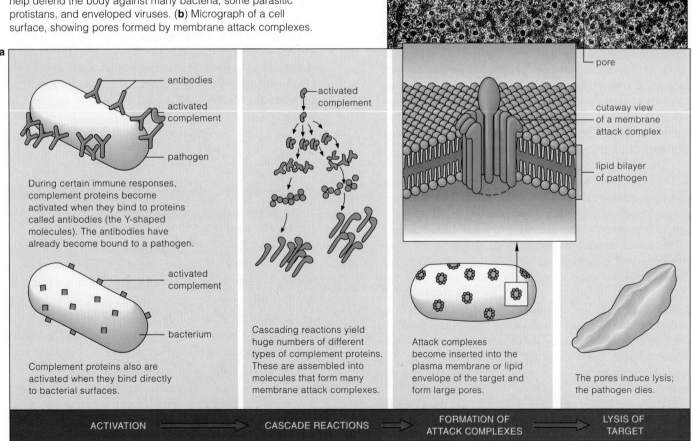

During certain immune responses, complement proteins become activated when they bind to proteins called antibodies (the Y-shaped molecules). The antibodies have already become bound to a pathogen.

Complement proteins also are activated when they bind directly to bacterial surfaces.

Cascading reactions yield huge numbers of different types of complement proteins. These are assembled into molecules that form many membrane attack complexes.

Attack complexes become inserted into the plasma membrane or lipid envelope of the target and form large pores.

The pores induce lysis; the pathogen dies.

ACTIVATION → CASCADE REACTIONS → FORMATION OF ATTACK COMPLEXES → LYSIS OF TARGET

40.3 INFLAMMATION

The Roles of Macrophages and Their Kin

Certain white blood cells take part in an initial response to tissue damage. White blood cells, recall, arise from stem cells in bone marrow. Many of these cells circulate in blood and lymph. A great many take up stations in the lymph nodes, spleen, liver, kidneys, lungs, and brain. (Here you may wish to refer to Figures 39.8 and 39.20.)

Three kinds of white blood cells are like SWAT teams—they act swiftly against danger in general but are not adapted for sustained battles. **Neutrophils**, the most abundant kind, phagocytize bacteria. They ingest, kill, and digest bacterial cells to simple molecular bits. **Eosinophils** secrete enzymes that punch holes in the surface of parasitic worms. **Basophils** secrete histamine and other substances that help keep inflammation going after it starts.

Although slower to act, **macrophages** are the "big eaters." Figure 40.3 shows one of them. A macrophage engulfs and digests just about any foreign agent. It also helps clean up damaged tissues. Immature macrophages circulating in blood are called monocytes.

The Inflammatory Response

An inflammatory response develops when a tissue's cells are damaged or killed, as happens during an infection. By a mechanism called **acute inflammation**, fast-acting phagocytes, complement proteins, and other plasma proteins escape from the bloodstream and enter interstitial fluid in the besieged tissue. Localized redness, swelling, heat, and pain are signs that acute inflammation is under way (Table 40.2).

During an inflammatory response, **histamine** and some other chemical substances that promote vasodilation are released from mast cells, which are a kind of tissue-dwelling basophil. Vasodilation, recall, refers to an enlargement of the diameter of small blood vessels. The vessels become engorged with blood, so the tissue reddens and becomes warmer. (Blood, remember, carries metabolic heat.) Vasodilation also causes cells that make up the blood vessel walls to pull apart slightly and become "leaky" to plasma. Some plasma, with its infection-fighting weapons, enters the tissue. Swelling and pain follow. Because voluntary movements aggravate the pain, the individual tends to avoid them. The avoidance of movement has the effect of promoting tissue repair.

Within hours of the first physiological responses to tissue damage, neutrophils are squeezing out through gaps in blood vessel walls. They swiftly go to work.

Table 40.2	Localized Signs of Inflammation and Their Causes
Redness	Vasodilation, increased blood flow to site
Warmth	Vasodilation, greater flow of blood carrying more metabolic heat to site
Swelling	Capillaries made more permeable, plasma and plasma proteins move into interstitial spaces
Pain	Increased fluid pressure and local chemical signals stimulate nocireceptors (pain receptors)

Monocytes arrive somewhat later, differentiate into macrophages, and engage in more sustained action (Figure 40.3).

While macrophages are busily engulfing pathogens, they secrete several substances. Among the secretions are **interleukins**, which are communication signals among white blood cells. One of these, interleukin-1, also signals a brain region that controls body temperature. Interleukin-1 induces an increase in the "set point" on the body's thermostat. A *fever* is a body temperature that has climbed to the higher set point. A fever of about 39°C (100°F) is not a bad thing. It promotes an increase in host defense activities. It also increases body temperature to a level that is "too hot" for the functioning of most pathogens.

Besides this, interleukin-1 induces drowsiness during a fever. Drowsiness reduces the body's demands for energy, so that more can be diverted to the tasks of defense and repair. Macrophages also take part in the cleanup and repair operations. So do blood-clotting proteins that help repair blood vessels, in the manner described on page 668.

1. An inflammatory response develops in a local tissue when cells are damaged or killed, as by infection.

2. Small blood vessels in the tissue dilate. As the cells making up their wall pull apart slightly, the blood vessels become leaky to plasma proteins and white blood cells.

3. Neutrophils, eosinophils, basophils, and then macrophages leave the bloodstream. They defend and help repair a besieged tissue.

4. Histamine, complement proteins, interleukins, blood-clotting proteins, and other chemicals take part in the inflammatory response or in the mopping-up and repair operations.

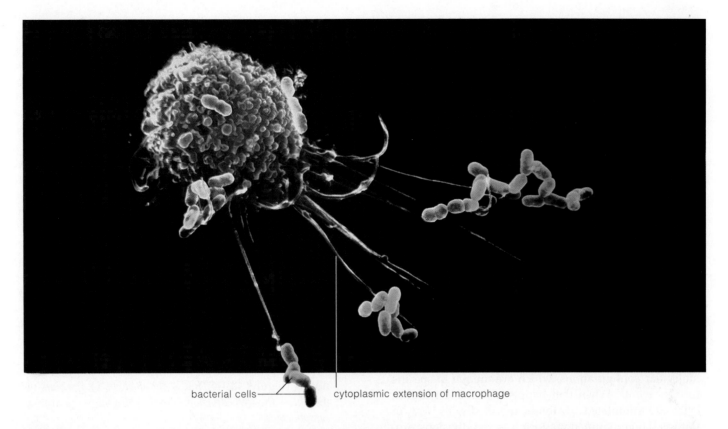

bacterial cells   cytoplasmic extension of macrophage

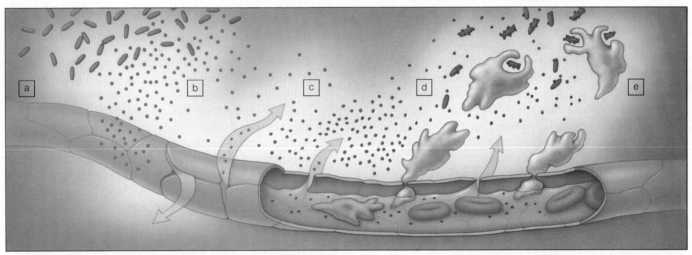

a Bacteria invade a tissue. They kill cells or release harmful metabolic by-products.

b The substances released by bacteria and by damaged or killed body cells accumulate in the tissue.

c The substances make the tissue's small blood vessels more permeable. Plasma fluid and various plasma proteins escape into the tissue.

d Some plasma proteins attack bacteria. Others create chemical gradients that facilitate migration of phagocytes to the tissue. Still others repair tissue damage (as by clotting mechanisms).

e Phagocytic white blood cells engulf bacteria.

Figure 40.3 (*Above*) Scanning electron micrograph of a macrophage, probing its surroundings with cytoplasmic extensions. This phagocytic cell engulfs bacteria that make contact with its surface.

(*Below*) Acute inflammation. In this example, a bacterial invasion induces chemical changes in a local tissue that trigger increased blood flow to the region. Small blood vessels become more permeable. Plasma fluid, certain plasma proteins, and phagocytic white blood cells leave the blood and enter the tissue. The invaders are killed and the tissue can be repaired.

THE IMMUNE SYSTEM

Defining Features

When physical barriers and inflammation are not enough to overwhelm an invader, an infection may become well established. Now armies of white blood cells called **B** and **T lymphocytes** form and join the battle. Lymphocytes are central to the body's third line of defense—the **immune system**. The immune system has two defining features. The first is called immunological *specificity*, whereby lymphocytes zero in on specific pathogens and eliminate them. The second feature is immunological *memory*, whereby some lymphocytes that form during a first-time confrontation are set aside for a future battle with the same pathogen.

The operating principle for the system is this: *Each kind of cell, virus, or substance bears unique molecular configurations that give it a unique identity.* An individual's lymphocytes recognize the body's own cells by unique configurations that serve as *self* markers. Lymphocytes normally ignore these. They also can recognize *nonself* molecular configurations, which are unique to specific foreign agents. When that happens, B and T lymphocytes are stimulated to divide repeatedly, by way of mitosis. Huge populations form. As the divisions proceed, subpopulations of the new cells become specialized to respond to the foreign agent in different ways. Some subpopulations are fully differentiated *effector* cells that engage and destroy the enemy. Others consist of *memory* cells that enter a resting phase. They will "remember" the specific agent and undertake a larger, more rapid response if it shows up again.

Any molecular configuration that triggers the formation of lymphocyte armies is called an **antigen**. Many antigens occur on proteins at the surface of pathogens or tumor cells. As you will see, lymphocytes produce receptor molecules that can bind to these configurations. This is how they "recognize" nonself.

In short, immunological specificity and memory involve three events: *recognition* of antigen, *repeated cell divisions* that form huge populations of lymphocytes, and *differentiation* into subpopulations of effector and memory cells with receptors that are specific for one kind of antigen.

Antigen-Presenting Cells—The Triggers for Immune Responses

The plasma membrane of every cell of an individual's body incorporates a variety of proteins. Among these membrane proteins are **MHC markers**, named after the genes coding for them. Some MHC markers are common to all of the body's cells. Others are unique to its macrophages and lymphocytes.

Suppose a splinter sinks into one of your fingers and bacteria enter the punctured tissue. They multiply, inflammation follows, and then phagocytes go to work. Some of the phagocytes are macrophages that engulf the foreign cells *and* do interesting things with the antigen. Their digestive enzymes cleave the antigen molecules into fragments, which become attached to MHC molecules. These **antigen-MHC complexes** are carted off in vesicles to the macrophage surface. And they become displayed at the cell surface.

Any cell that processes and displays antigen with a suitable MHC molecule is known as an **antigen-presenting cell**. When antigen fragments and a certain MHC marker are displayed together, lymphocytes take notice (Figure 40.4). *This is the antigen recognition that promotes the cell divisions leading to huge numbers of lymphocytes.*

Key Players in Immune Responses

The same kinds of white blood cells are called into action during every immune response. Figure 40.5 provides an overview of the interactions among these cells. In brief, recognition of antigen-MHC complexes activates **helper T cells**. These T lymphocytes produce and secrete substances that promote the formation of large populations of effector and memory cells.

Among the effectors are **cytotoxic T cells**. These T lymphocytes eliminate infected body cells and tumor cells by a lethal hit. Along with other "killers," they execute the *cell-mediated* immune responses. By contrast, **B cells** produce antigen-binding receptor molecules called **antibodies**. When a response is under way, effector B cells secrete staggering numbers of antibody molecules. Only B cells and their weapons carry out the *antibody-mediated* responses.

Figure 40.4 Molecular cues that stimulate lymphocytes to make immune responses.

MHC marker that designates self (only on body's own cells)

ignored

antigen (any foreign or abnormal molecular configuration lymphocytes recognize as nonself)

immune response

processed antigen bound with MHC marker

immune response

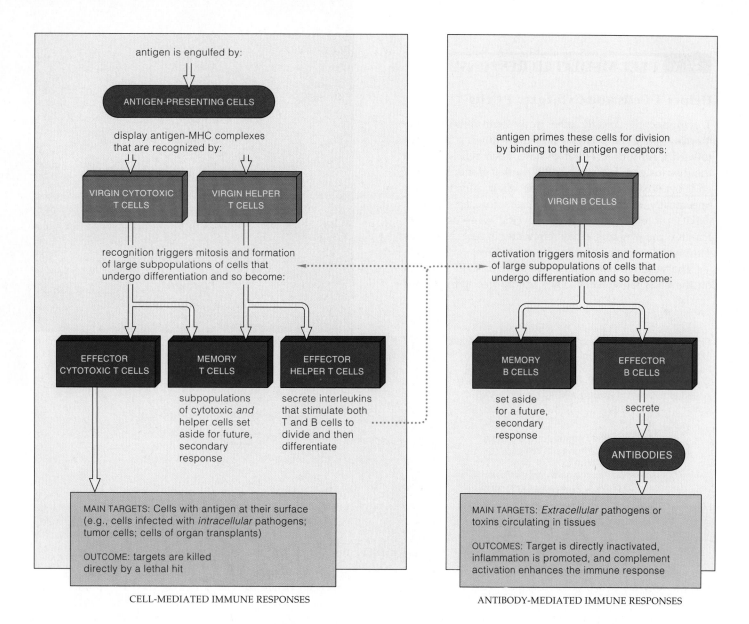

CELL-MEDIATED IMMUNE RESPONSES

ANTIBODY-MEDIATED IMMUNE RESPONSES

The figure contains the following labels and text:

antigen is engulfed by:

ANTIGEN-PRESENTING CELLS

display antigen-MHC complexes that are recognized by:

VIRGIN CYTOTOXIC T CELLS

VIRGIN HELPER T CELLS

recognition triggers mitosis and formation of large subpopulations of cells that undergo differentiation and so become:

EFFECTOR CYTOTOXIC T CELLS

MEMORY T CELLS

EFFECTOR HELPER T CELLS

subpopulations of cytotoxic *and* helper cells set aside for future, secondary response

secrete interleukins that stimulate both T and B cells to divide and then differentiate

MAIN TARGETS: Cells with antigen at their surface (e.g., cells infected with *intracellular* pathogens; tumor cells; cells of organ transplants)

OUTCOME: targets are killed directly by a lethal hit

antigen primes these cells for division by binding to their antigen receptors:

VIRGIN B CELLS

activation triggers mitosis and formation of large subpopulations of cells that undergo differentiation and so become:

MEMORY B CELLS

EFFECTOR B CELLS

set aside for a future, secondary response

secrete

ANTIBODIES

MAIN TARGETS: *Extracellular* pathogens or toxins circulating in tissues

OUTCOMES: Target is directly inactivated, inflammation is promoted, and complement activation enhances the immune response

Figure 40.5 Overview of the key interactions among B and T lymphocytes during an immune response. Most often, both types of white blood cells are activated when an antigen has been detected. An antigen is any large molecule that lymphocytes recognize as not being "self" (normal body molecules). A first-time encounter with antigen elicits a *primary* response. A subsequent encounter with the same type of antigen elicits a *secondary* immune response. This response is larger and more rapid. Memory cells that formed but were not used during the first battle can immediately engage in the second one.

Control of Immune Responses

Antigen provokes an immune response—and removal of antigen stops it. For instance, when the tide of battle turns, effector cells have destroyed most of the antigen-bearing agents in the body. With fewer antigen molecules around, stimulation of the response declines, then stops. Inhibitory signals from cells with suppressor functions also help shut down the response.

Let's turn now to the cell-mediated and antibody-mediated immune responses. Before we do, take a moment to fix these key concepts in your mind:

1. An antigen is a nonself molecular configuration that triggers an immune response.

2. An immune response involves *recognition* of antigen, *repeated mitotic cell divisions* that lead to the formation of huge populations of T and B lymphocytes, and *differentiation* into subpopulations of effector cells and memory cells, all of which are sensitized to that one kind of invader.

3. Helper T cells, cytotoxic T cells, and B cells are the key players in an immune response.

40.5 CELL-MEDIATED RESPONSES

Helper T Cells and Cytotoxic T Cells

T lymphocytes, recall, arise from stem cells in bone marrow (page 657). There, they start down a pathway toward a fully differentiated state. Each cell travels to the thymus gland (Figure 39.20). In this gland, helper T and cytotoxic T cells complete their development. Specifically, each acquires receptors for self markers (MHC proteins) and antigen-specific receptors. After leaving the thymus, they circulate in the blood as untouched, "virgin" T cells (Figure 40.6).

The virgin T cells ignore unadorned MHC markers on the body's own cells. They ignore free antigen. But

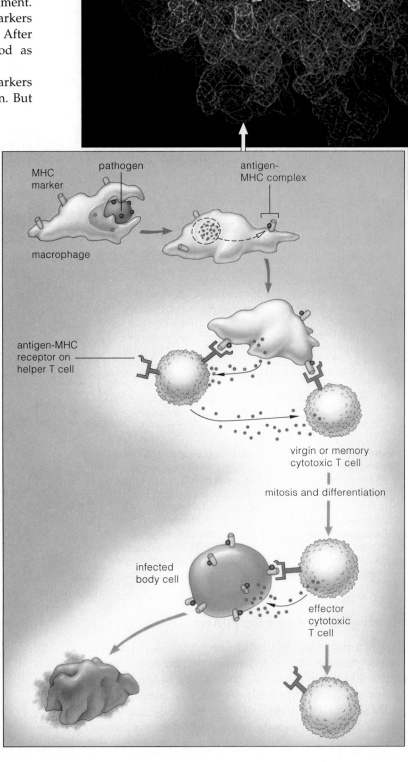

Figure 40.6 A cell-mediated immune response. In this example, the response begins after an antigen-MHC complex activates a cytotoxic T cell. The photograph shows a reconstruction of an antigen fragment (*pink*) bound to the cleft of a class 1 MHC protein.

a A macrophage engulfs and digests a pathogen, then cleaves its antigen into fragments that bind to MHC markers. The macrophage becomes an antigen-presenting cell; it displays processed antigen-MHC complexes at its surface.

b Receptors on T cells bind to the complexes. Binding stimulates the macrophage to secrete interleukin-1 (pink dots). This stimulates helper T cells to secrete other interleukins (blue dots). These stimulate virgin or memory cytotoxic T cells to divide and differentiate into large populations of effector and memory cells. Only the effector cytotoxic T cells have cell-killing abilities.

c An effector encounters a target: an infected body cell that has the processed antigens bound with MHC markers at its surface. And it delivers a lethal hit. It releases perforins and toxic substances (green dots) onto its target and so programs it for death.

d The effector disengages from the doomed cell and reconnoiters for new targets. Meanwhile, perforins make holes in the target's plasma membrane. Toxins move into the cell, disrupt organelles, and make the DNA disassemble. The infected cell dies.

Cancer and Immunotherapy

Carcinomas, sarcomas, leukemia—these chilling words refer to malignant tumors in skin, muscle, bone, and other tissues. Such tumors arise when viral attack, chemical change, or irradiation alters genes and cells turn cancerous (pages 234 and 244). The transformed cells divide again and again. Unless they are destroyed or surgically removed, they kill the individual.

Often a transformed cell bears abnormal proteins at its surface. Nonspecific and specific defense responses to transformed cells can make the tumor regress, but sometimes the responses are inadequate. Also, some tumors release many copies of the abnormal proteins. If these saturate antigen receptors on the defenders, the tumor escapes detection. If the tumor hides long enough and reaches a certain critical mass, it may overwhelm the immune system's capacity for an effective response.

Researchers are working to develop various procedures that will deliberately enhance immunological defenses against tumors as well as against certain pathogens. This prospect is called *immunotherapy*.

Suppose antibodies against tumor-specific antigens could be produced in quantity and injected into a cancer patient. Normal antibody-secreting B cells won't live long enough in culture to mass-produce pure antibody. Also, being end cells, they cannot reproduce. But Cesar Milstein and Georges Kohler showed how to make antibody "factories." They injected an antigen into a mouse. The mouse's B cells produced antibodies against it. Later, they extracted the antibody-producing mouse B cells and fused them with cells from B cell tumors. Some descendants of the hybrid cells divided nonstop, and they, too, produced the antibody. Clones of these proliferating hybrid cells are now being maintained indefinitely—and they make identical copies of antibodies in useful amounts. We call their products *monoclonal antibodies*.

A few cancer patients have been inoculated with monoclonal antibodies that are expected to home in on the malignant tumors. In some cases, cell-killing chemicals have been artificially attached to such antibodies. At this writing, the success of clinical trials is limited.

---

they recognize and bind with antigen-MHC complexes on antigen-presenting cells. The binding stimulates them to divide repeatedly and give rise to clones. (A *clone* is a population of genetically identical cells.) Their descendants differentiate into subpopulations of effector and memory cells—and every one of those descendants has receptors for the antigen.

The effector subpopulations of *helper T cells* secrete interleukins. These secretions fan the repeated cell divisions and differentiation. Effector subpopulations of *cytotoxic T cells* recognize antigen when it is combined with a particular MHC marker. The combination forms on body cells that have been infected by intracellular pathogens, including certain viruses, and on tumor cells (see the *Focus* essay). It serves as a "double signal" to destroy the cells that bear it.

Cytotoxic effectors destroy infected cells with a lethal hit. They secrete proteins called **perforins**. The perforin molecules form doughnut-shaped pores in a target's plasma membrane, similar to those shown in Figure 40.2. The cytotoxic effectors also secrete toxins that disrupt organelles and wreak havoc on the nuclear DNA of their target. Having made its hit, each cytotoxic effector quickly disengages and moves on to new targets (Figure 40.6d).

Cytotoxic T cells also are factors in the rejection of tissue and organ transplants. Parts of MHC markers on donor cells are different enough from the recipient's to be recognized as antigens—but other parts are similar enough to complete the double signal. MHC typing and matching donors to recipients minimize the risk.

Regarding the Natural Killer Cells

Other cytotoxic cells, including **natural killer cells** (NK cells), also arise from stem cells in bone marrow. They appear to be lymphocytes (but not T or B cells). Their arousal does not depend on actual encounters with antigen-MHC complexes. NK cells reconnoiter for tumor cells and infected cells, and kill them on contact. Apparently, their targets must be bearing odd molecular configurations at their surfaces.

Effector helper T cells secrete interleukins that promote the cell divisions and differentiations required for an immune response.

Effector cytotoxic T cells destroy infected cells, tumor cells, or foreign cells with a lethal hit.

40.6 ANTIBODY-MEDIATED RESPONSES

B Cells and Targets of Antibodies

Like T cells, the B cells also arise from stem cells in bone marrow and start down a pathway that will culminate in full differentiation. Along their pathway, however, B cells start synthesizing many copies of a single kind of antibody molecule.

All antibodies are proteins, but each kind has binding sites that match up to only a single antigen. As you will read later in the chapter, many of these molecules are more or less Y-shaped, with a tail as well as two arms that bear identical antigen receptors. After they are synthesized, antibody molecules move to the plasma membrane of a maturing B cell. There, the tail of each one becomes embedded in the lipid bilayer, and the two arms stick out above it. Bristling with antigen receptors (its bound antibodies), the cell enters the bloodstream as a virgin B cell.

Suppose the receptors encounter and lock onto antigen. When that happens, the B cell becomes primed for division. But it will *not* divide without the proper sig-

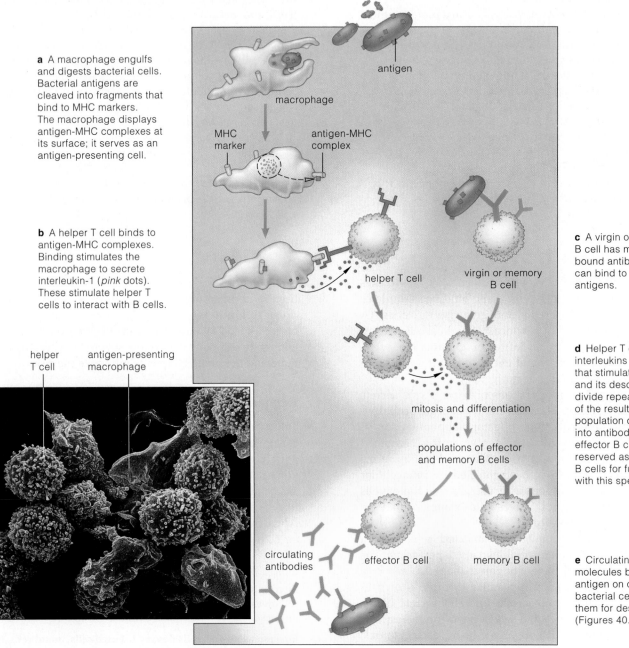

a A macrophage engulfs and digests bacterial cells. Bacterial antigens are cleaved into fragments that bind to MHC markers. The macrophage displays antigen-MHC complexes at its surface; it serves as an antigen-presenting cell.

b A helper T cell binds to antigen-MHC complexes. Binding stimulates the macrophage to secrete interleukin-1 (*pink* dots). These stimulate helper T cells to interact with B cells.

antigen

macrophage

MHC marker

antigen-MHC complex

helper T cell

virgin or memory B cell

c A virgin or memory B cell has membrane-bound antibodies that can bind to the bacterial antigens.

mitosis and differentiation

populations of effector and memory B cells

d Helper T cell secretes interleukins (*green* dots) that stimulate the B cell and its descendants to divide repeatedly. Part of the resulting B cell population differentiates into antibody-secreting effector B cells. Part is reserved as memory B cells for future battles with this specific invader.

helper T cell

antigen-presenting macrophage

circulating antibodies

effector B cell

memory B cell

e Circulating antibody molecules bind to antigen on other bacterial cells, tagging them for destruction (Figures 40.2a and 40.3).

Figure 40.7 Example of an antibody-mediated immune response to a bacterial invasion.

nals. As Figure 40.7 shows, the signal must come from a helper T cell that already has become activated by an antigen-presenting cell. In the presence of helper T cell secretions, the primed B cell and its descendants undergo repeated cell divisions. The resulting clonal B cell population differentiates into effector and memory B cells. These effectors (formerly called plasma cells) produce and secrete staggering numbers of antibody molecules. When freely circulating antibody binds antigen, it tags the invader for destruction, as by phagocytes and complement proteins (Figures 40.2 and 40.7e).

The main targets of antibody-mediated responses are extracellular pathogens and toxins, which are freely circulating in tissues or body fluids. Antibodies can't bind to pathogens or toxins hidden in a host cell.

The Immunoglobulins

B cells produce four classes of antibodies during each immune response. Collectively, the four classes are called **immunoglobulins**, or **Igs**. All have antigen-binding sites, but the molecules in each class also have other sites for specialized functions. The different classes are the protein products of gene shufflings, which occur during the cell divisions that give rise to subpopulations of effector and memory B cells.

IgM antibodies are the first to be secreted during immune responses. After binding antigen, they trigger the complement cascade. They also can bind invaders together in clumps, which are more handily eliminated by phagocytes.

IgG antibodies activate complement proteins and neutralize many toxins. The IgGs are long lasting. They are the only antibodies to cross the placenta during pregnancy and protect a fetus with the mother's acquired immunities. They also are secreted into the milk produced early on by mammary glands, then are absorbed into the suckling newborn's bloodstream.

IgA antibodies enter mucus-coated surfaces of the respiratory, digestive, and reproductive tracts and elsewhere. There they neutralize infectious agents. A mother's milk delivers them to the mucous lining of her newborn's gut.

IgE triggers inflammation when parasitic worms attack the body. As described later, it also figures in allergies. The tails of IgE antibodies bind to basophils and mast cells, and the antigen receptors face outward. IgE antibodies are like traps. When antigen springs the trap, the basophils and mast cells release substances that promote inflammation.

Antibodies that are secreted by B cells bind to antigens of extracellular pathogens or toxins and tag them for disposal, as by phagocytes and complement proteins.

The antigen-presenting cells and lymphocytes we have been describing interact in lymphoid organs that promote immune responses (Figure 40.8). Think of the location of tonsils and most other lymph nodules just beneath the mucous membranes of the respiratory, digestive, and reproductive systems. *The locations allow antigen-presenting cells and lymphocytes to intercept invaders just after they penetrate surface barriers.* Or think of antigen in tissue fluid that is entering the lymph vascular system. Because lymph vessels eventually drain into the expressways for blood transport, antigen could become distributed to every body region. *However, before antigen can reach the blood, it must trickle through lymph nodes—which are packed with defending cells.* Even in the few cases where antigen manages to enter the blood, defending cells in the spleen intercept it.

Cells are organized for maximum effect inside the lymphoid organs. Antigen-presenting cells make up the front line. There, antigen is engulfed and presented, and cell divisions produce populations of effector and memory lymphocytes. As lymph drains through an organ, it moves the effector activities to the back of the organ and beyond. All the while, virgin and memory cells circulate through the organ, reconnoitering at the front line.

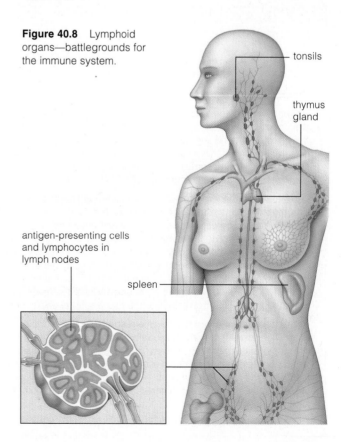

Figure 40.8 Lymphoid organs—battlegrounds for the immune system.

tonsils

thymus gland

antigen-presenting cells and lymphocytes in lymph nodes

spleen

40.8 SPECIFICITY AND MEMORY

How Lymphocytes Produce Antigen-Specific Receptors

Food, water, air, soil, people, pets—almost everything in your surroundings can house a mind-boggling variety of pathogens with unique antigens. How do your lymphocytes make the millions of different antigen-specific receptors required to detect these threats?

All T and B cells in an individual inherit the same genes. However, as the cells mature, a special recombination mechanism acts on the genes that specify antigen receptors (Figure 40.9). Different regions of these genes are shuffled at random into one of millions of possible combinations. In each T or B cell, the mechanism produces a nucleotide sequence that codes for one of millions of possible antigen receptors.

Consider antibodies, the receptors produced by B cells. Each arm of a Y-shaped antibody molecule consists of two polypeptide chains, as shown in Figure

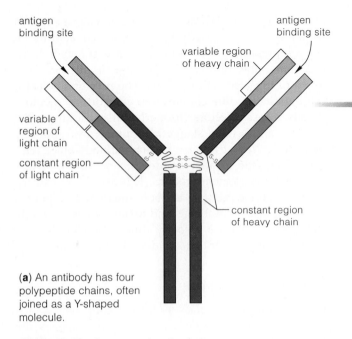

(a) An antibody has four polypeptide chains, often joined as a Y-shaped molecule.

Figure 40.10 Structure of antibodies.

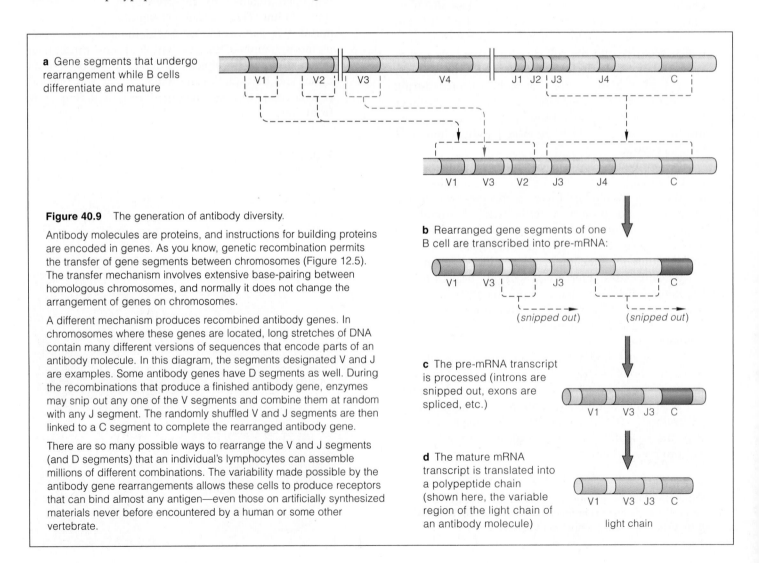

Figure 40.9 The generation of antibody diversity.

Antibody molecules are proteins, and instructions for building proteins are encoded in genes. As you know, genetic recombination permits the transfer of gene segments between chromosomes (Figure 12.5). The transfer mechanism involves extensive base-pairing between homologous chromosomes, and normally it does not change the arrangement of genes on chromosomes.

A different mechanism produces recombined antibody genes. In chromosomes where these genes are located, long stretches of DNA contain many different versions of sequences that encode parts of an antibody molecule. In this diagram, the segments designated V and J are examples. Some antibody genes have D segments as well. During the recombinations that produce a finished antibody gene, enzymes may snip out any one of the V segments and combine them at random with any J segment. The randomly shuffled V and J segments are then linked to a C segment to complete the rearranged antibody gene.

There are so many possible ways to rearrange the V and J segments (and D segments) that an individual's lymphocytes can assemble millions of different combinations. The variability made possible by the antibody gene rearrangements allows these cells to produce receptors that can bind almost any antigen—even those on artificially synthesized materials never before encountered by a human or some other vertebrate.

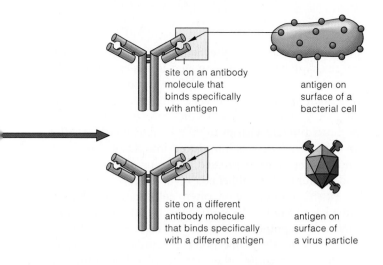

site on an antibody molecule that binds specifically with antigen

antigen on surface of a bacterial cell

site on a different antibody molecule that binds specifically with a different antigen

antigen on surface of a virus particle

(**b**) Some regions are almost the same in all antibodies. But in each kind of antibody, the molecular configuration in one region is unique; it is the binding site for one kind of antigen. Antigen fits into the site's grooves and onto its protrusions.

40.10a. In each maturing B cell, the randomly established gene sequence became translated into an amino acid sequence, which dictated the folding of the two chains into grooves and bumps having a particular charge distribution. This part of the antibody arm is an antigen-binding site. Only antigen having the appropriate shape, charge distribution, and so on will be able to bind with it (Figure 40.10b).

DNA recombinations help explain the **clonal selection theory**, first proposed by Macfarlane Burnet. All the clonal descendants of a virgin lymphocyte have the same randomly shuffled gene sequence. The sequence codes for the one receptor configuration that can bind to the specific antigen that "selected" the parent cell (Figure 40.11a).

Immunological Memory

The clonal selection theory explains how a person can have "immunological memory" of a first encounter with antigen. The term refers to the body's capacity to make a secondary immune response to any subsequent encounter with the same antigen. As Figure 40.11b shows, memory cells that form during a primary response do not engage in battle. They continue to circulate for years, sometimes for decades. Compared to the virgin cells that initiate a primary response, the patrolling battalions have far more cells, so antigen is intercepted much sooner (Figure 40.12). Effector cells form sooner, in greater numbers, so the infection is terminated before the host gets sick. Hence the advantage of immunity. Even greater numbers of memory T and antibody-producing B cells form during the secondary response (Figure 40.11b).

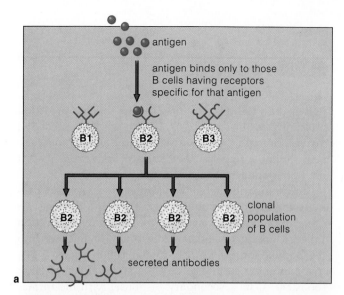

antigen

antigen binds only to those B cells having receptors specific for that antigen

B1   B2   B3

B2   B2   B2   B2

clonal population of B cells

secreted antibodies

a

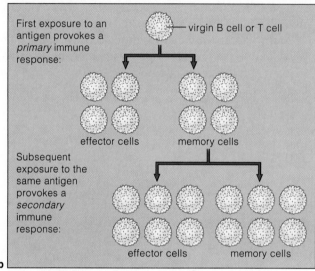

First exposure to an antigen provokes a *primary* immune response:

virgin B cell or T cell

effector cells   memory cells

Subsequent exposure to the same antigen provokes a *secondary* immune response:

effector cells   memory cells

b

Figure 40.11 (**a**) Clonal selection of a B cell that produced the specific antibody that can combine with a specific antigen. Only antigen-selected B cells (and T cells) are activated and give rise to a clonal population of immunologically identical cells.

(**b**) Immunological memory. Not all B and T cells are used in a primary immune response to an antigen. Many continue to circulate as memory cells, which become activated during a secondary immune response.

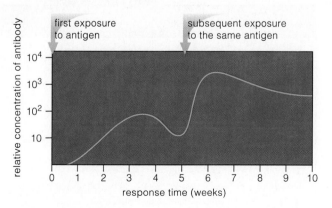

first exposure to antigen

subsequent exposure to the same antigen

relative concentration of antibody

10^4

10^3

10^2

10

0 1 2 3 4 5 6 7 8 9 10

response time (weeks)

Figure 40.12 Differences in magnitude and duration between a primary and a secondary immune response to the same antigen. (In this example, a secondary response starts at week 5.)

IMMUNIZATION

Jenner didn't know why his cowpox vaccine provided immunity against smallpox. Today we know that the viruses causing the two diseases are related, and they bear similar antigens at their surface. Let's express what goes on in modern terms.

Immunization refers to various processes that promote increased immunity against specific diseases. In *active* immunization, an antigen-containing preparation called a **vaccine** is either taken orally or injected into the body (Figure 40.13). A first injection elicits a primary immune response. Later, a subsequent injection (booster) elicits a secondary response, with the formation of more effector cells and memory cells that can provide long-lasting protection against the disease.

Many vaccines are manufactured from weakened or killed pathogens. For example, weakened polio virus particles are used for the Sabin polio vaccine. Other vaccines are based on inactivated forms of natural toxins, such as the bacterial toxin that causes tetanus. Still others are made with the help of harmless, genetically engineered viruses. These incorporate genes from three or more different viruses in their genetic material. After an engineered virus is introduced into a host, the incorporated genes are expressed, antigens are produced—and immunity is established.

Passive immunization often helps people who are already infected with pathogens, including the ones that cause diphtheria, tetanus, measles, and hepatitis B. A person receives injections of antibody molecules purified from some other source. The best source is another individual who already has produced a large amount of the required antibody molecules. The effects are not lasting, because the person's own B cells are not producing antibodies. However, injected antibody molecules may help counter the immediate attack.

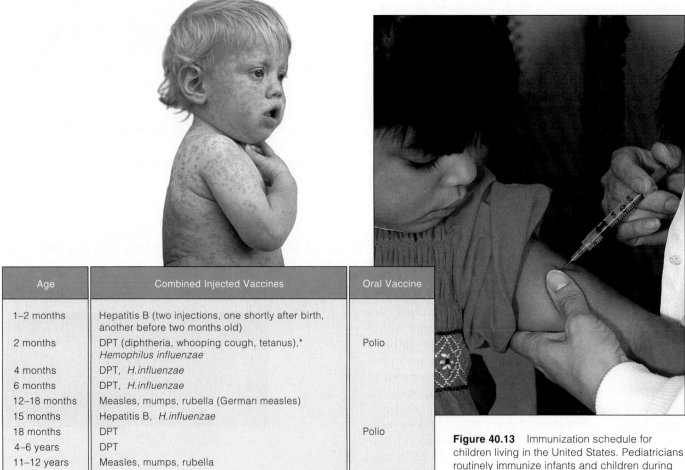

Age	Combined Injected Vaccines	Oral Vaccine
1–2 months	Hepatitis B (two injections, one shortly after birth, another before two months old)	
2 months	DPT (diphtheria, whooping cough, tetanus),* *Hemophilus influenzae*	Polio
4 months	DPT, *H.influenzae*	
6 months	DPT, *H.influenzae*	
12–18 months	Measles, mumps, rubella (German measles)	
15 months	Hepatitis B, *H.influenzae*	
18 months	DPT	Polio
4–6 years	DPT	
11–12 years	Measles, mumps, rubella	

* Also called tetramune.

Figure 40.13 Immunization schedule for children living in the United States. Pediatricians routinely immunize infants and children during office visits. Low-cost or free vaccinations also are available at many community clinics.

40.10 ABNORMAL OR DEFICIENT IMMUNE RESPONSES

Allergies

In many millions of people, normally harmless substances provoke immune responses. Such substances are known as **allergens**, and the response is called an **allergy**. Common allergens are pollen (Figure 40.14), a variety of drugs and foods, dust mites, fungal spores, insect venom, and cosmetics. Some responses to allergens start within minutes. Others are delayed. Either way, the allergens provoke inflammation.

Some people are genetically inclined to allergies. Also, infections, emotional stress, or changes in air temperature may trigger reactions that otherwise might not occur. Upon exposure to certain antigens, IgE antibodies are secreted, and these bind to mast cells. When IgE on mast cells bind to antigen, the cells are stimulated to secrete histamine, prostaglandins, and other substances that fan the inflammatory response. They also stimulate secretion of mucus and cause airways to constrict. In *asthma* and *hay fever*, congestion, sneezing, a drippy nose, and labored breathing are symptoms of the allergic response.

In some cases, inflammatory reactions proceed throughout the body and trigger a life-threatening condition called *anaphylactic shock*. For example, a person who is allergic to wasp or bee venom can die within minutes of a single sting. Air passages leading to the lungs constrict massively. Fluid escapes too rapidly from dilated, grossly permeable capillaries. Blood pressure plummets and may lead to circulatory collapse. Antihistamines (anti-inflammatory drugs) are often used to relieve the short-term symptoms of allergies. Desensitization programs may help over the long term. Here, skin tests may identify the offending allergens. Inflammatory responses to some of them can be blocked if the patient's body can be stimulated to make IgG instead of IgE. Over an extended period, larger and larger doses of specific allergens are administered. Each time, the body produces more circulating IgG molecules and IgG memory cells. IgG binds with the allergen and blocks its attachment to IgE—and thus blocks inflammation.

Autoimmune Disorders

In an **autoimmune response**, the powerful weapons of the immune system are unleashed against normal body cells. Consider *Grave's disorder*, in which the body overproduces thyroid hormones. These hormones influence the development of many tissues and overall metabolic rates, and delicate feedback mechanisms control their production (page 624). Affected persons produce antibodies that unfortunately bind to receptors on cells that

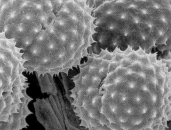
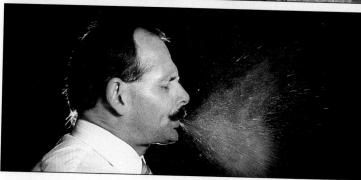

Figure 40.14 (*Above*) Common allergens—pollen grains—and a demonstration of one of their effects in sensitive people (*below*).

produce the hormones. The antibodies, which do not respond to feedback controls, stimulate the cells to produce hormones in excess. Typical disease symptoms include elevated metabolic rates, heart fibrillations, nervousness, excessive sweating, and weight loss.

Or consider *myasthenia gravis*. Antibodies are directed against acetylcholine receptors on skeletal muscle cells. Progressive muscular weakness follows.

Finally, consider *rheumatoid arthritis*, in which skeletal joints are chronically inflamed. Affected persons have a genetic (inherited) predisposition to the disorder. Macrophages, T cells, and B cells become activated by antigens associated with the joints. Immune responses are made against the body's own collagen and apparently against antibody that is bound to an unknown antigen. Complement activation and inflammation further damage joint tissues. So do skewed repair mechanisms. Over time, the joint becomes filled with synovial membrane cells and is eventually immobilized.

Deficient Immune Responses

When the body has inadequate numbers of lymphocytes, immune responses are not effective. This condition can be inherited or acquired. Either way, a person who makes deficient immune responses becomes highly vulnerable to infections that are not life threatening to the general population. This is what happens in *AIDS* (acquired immunodeficiency syndrome). AIDS is caused by the human immunodeficiency virus, or HIV. The *Focus* essay that follows describes how the virus replicates inside certain lymphocytes and destroys the capacity to fight infections. We return to this topic on page 796.

AIDS —The Immune System Compromised

AIDS is a constellation of disorders that follow infection by a type of virus designated HIV. The virus cripples the immune system, and the body becomes highly susceptible to usually harmless infections and some otherwise rare forms of cancer. At present there is no vaccine against the known forms of this virus (HIV-1 and HIV-2). There is no cure for those already infected.

About a million Americans are now infected with HIV. Worldwide, an estimated 13 million people have become HIV infected, and 2 million are already dead. By the turn of the century, the number of infected people may reach an estimated 40 million to 110 million.

How HIV Replicates   HIV infects macrophages, antigen-presenting cells, and helper T cells (also called CD4 lymphocytes). HIV is one of the retroviruses. It has a protein coat and an inner protein core that contains RNA and several copies of an enzyme, reverse transcriptase. Its outermost lipid envelope is a bit of plasma membrane that surrounded the virus particle when it departed from a previously infected cell. HIV proteins are incorporated in the envelope (compare Figure 22.18).

Once inside a host cell, the viral enzyme uses the RNA as a template for making DNA, which then is inserted into a host chromosome (Figure *a*). The inserted viral DNA may remain dormant in some cells for months to years. Or it may be activated. Then, transcription yields copies of

viral RNA that are translated into viral proteins. New virus particles are put together. They bud from the plasma membrane of the host cell and are released (Figures *b–d*). With each round of infection, more macrophages, more antigen-presenting cells, and more helper T cells are impaired or destroyed.

The host immune system produces antibodies in response to HIV antigenic proteins. (The antibodies are the basis of diagnostic tests to identify HIV infection.) However, antibodies do not eliminate the virus. Over the next decade or more, key lymphocytes are lost—and so, eventually, is the ability to mount immune responses.

At first an infected person might appear to be in good health, suffering no more than a bout of "the flu." Then he or she starts displaying symptoms that foreshadow AIDS. These include persistent weight loss, fever, fatigue, bed-drenching night sweats, and enlarged lymph nodes. In time, the diseases that follow certain opportunistic infections are signs of AIDS itself. Such diseases are rare in the population at large. Among these are widespread yeast infections and a form of pneumonia caused by *Pneumocystis carinii*. Spots that resemble bruises may appear, especially on the legs and feet. These are signs of Kaposi's sarcoma, a form of cancer that develops from endothelial cells of blood vessels. The immune system cannot control the infections or cancers, which end up killing the person with AIDS.

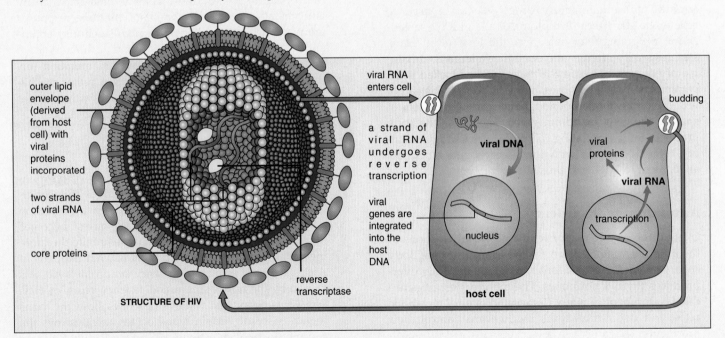

outer lipid envelope (derived from host cell) with viral proteins incorporated

two strands of viral RNA

core proteins

reverse transcriptase

STRUCTURE OF HIV

viral RNA enters cell

a strand of viral RNA undergoes reverse transcription

viral genes are integrated into the host DNA

viral DNA

nucleus

budding

viral proteins

viral RNA

transcription

host cell

a Life cycle of HIV, one of the retroviruses.

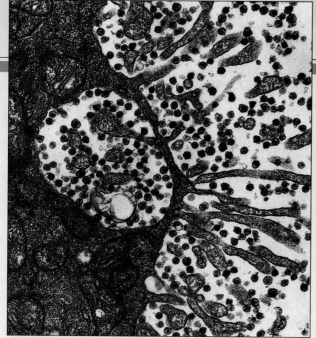

b

497 nm

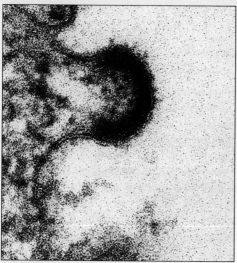

c

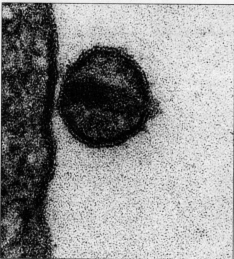

d

45 nm

b Transmission electron micrograph of HIV particles (*black specks*) escaping from an infected cell. (**c,d**) Closer views of a virus particle budding from the host cell's plasma membrane.

How HIV Is Transmitted   Like any human virus, HIV requires a medium that allows it to leave one host, survive in the environment into which it is released, then enter another host.

HIV is transmitted when body fluids of an infected person enter another person's tissues. Initially in the United States, transmission occurred most often between males during homosexual activities, especially anal intercourse. It has spread among intravenous drug abusers who share blood-contaminated syringes and needles. It also has spread among the heterosexual population, increasingly by vaginal intercourse. HIV also has traveled from infected mothers to offspring during pregnancy, birth, and breast-feeding. Contaminated blood supplies accounted for some cases before screening was implemented in 1985. Tissue transplants caused four infections in 1991. In several developing countries, health care providers have spread HIV by way of contaminated transfusions and reuse of unsterile needles and syringes.

The molecular structure of HIV is not stable outside the human body, so transmission must be quite direct. At this time, there is no evidence that HIV can be effectively transmitted by way of food, air, water, casual contact, or insect bites. The virus has been isolated from blood, semen, vaginal secretions, saliva, tears, breast milk, amniotic fluid, cerebrospinal fluid, and urine. It is likely to be present in other fluids, secretions, and excretions. However, only infected blood, semen, vaginal secretions, and breast milk contain the virus in concentrations that seem to be high enough for successful transmission.

Prospects for Treatment   Developing effective drugs or vaccines against HIV is a formidable challenge. HIV mutates rapidly, so it will be difficult to produce something that works against all of its mutated forms. In the meantime, drugs such as AZT (azidothymidine) and ddI (didanosine) are being used to slow down HIV replication and possibly to extend the life expectancy of men, women, and children.

Until effective vaccines and treatments are developed, checking the spread of HIV depends absolutely on persuading people to avoid or modify social behaviors that put them at risk. We return to this topic on page 796.

SUMMARY

1. Vertebrates are protected from many pathogens (infection-causing agents) by physical and chemical barriers at body surfaces. They also are protected by the nonspecific and specific responses of the white blood cells listed in Table 40.3.

   a. Nonspecific responses to tissue irritation or damage include inflammation and involve organs with phagocytic functions, such as the liver and spleen.

   b. Immune responses are made against specific pathogens, foreign cells, or abnormal body cells.

2. Intact skin and mucous membranes lining various body surfaces are physical barriers to infection. Glandular secretions (as in tears, saliva, and gastric fluid) are examples of chemical barriers. So are metabolic products of resident bacteria on body surfaces.

3. Tissue redness, warmth, swelling, and pain are signs of inflammation. The inflammatory response begins with changes in blood flow to a besieged tissue.

   a. Substances released by pathogens and by dead or damaged body cells make small blood vessels leaky. Circulating white blood cells enter the tissue and destroy invaders.

   b. Plasma proteins also enter the tissue. The complement proteins induce lysis of pathogens and attract phagocytes. Blood-clotting proteins help repair damaged blood vessels.

4. Immune responses have these characteristics:

   a. Each response is triggered by antigen, a unique molecular configuration that lymphocytes recognize as foreign (nonself). An example is a protein at the surface of a particular virus.

   b. Each immune response has specificity, meaning it is directed against one antigen alone.

   c. Each immune response has memory. A subsequent encounter with the same antigen will trigger a more rapid, secondary response, of greater magnitude.

   d. An immune response normally is not made against the body's own self-marker proteins.

5. Macrophages and other antigen-presenting cells process and display fragments of antigen with their own MHC markers. Lymphocytes have receptors that can bind to antigen-MHC complexes. Binding is the start signal for an immune response.

6. An immune response begins with recognition of antigen. It proceeds through repeated cell divisions that form clonal populations of lymphocytes, which differentiate into subpopulations of effector and memory cells. Interleukins and other signals among white blood cells drive the responses. Effector helper T and cytotoxic T cells, as well as effector B cells and antibodies, carry out the responses. Memory cells are set aside for secondary responses.

7. Effector cytotoxic T cells reconnoiter for any cell bearing antigen-MHC complexes. They directly destroy cells infected with intracellular pathogens, tumor cells, and cells of tissue or organ transplants.

8. Antibodies are Y-shaped protein molecules, each with binding sites for one kind of antigen. Only B cells produce them. When antibody binds to antigen, toxins may become neutralized, pathogens may be tagged for destruction, or the attachment of pathogens to body cells may be prevented.

9. In active immunization, vaccines provoke an immune response, with production of effector and memory cells. In passive immunization, injections of purified antibodies help patients through an infection.

10. An allergy is an immune response to a generally harmless substance. An autoimmune response is an attack by lymphocytes on the body's own cells. An immune deficiency is a weakened or nonexistent capacity to mount an immune response.

Table 40.3	Summary of Major White Blood Cells and Their Roles in Defense
Cell Type	Main Characteristics
Macrophage	Phagocyte; takes part in nonspecific defense responses; presents antigen to T cells; cleans up and helps repair tissue damage
Neutrophil	Fast-acting phagocyte; takes part in inflammation but not in sustained responses
Eosinophil	Secretes enzymes that attack parasitic worms
Basophil and mast cell	Secrete histamines, prostaglandins, other substances that act on small blood vessels, producing inflammation; also have roles in allergies
Lymphocytes:	(All take part in most immune responses; following antigen recognition, all form clonal populations of effector cells and memory cells.)
1. B cell	Effectors secrete four classes of antibodies (IgA, IgE, IgG, and IgM) that protect the host in specialized ways
2. Helper T cell	Effectors secrete interleukins that stimulate rapid divisions and differentiation of both B cells and T cells
3. Cytotoxic T cell	Effectors kill infected cells, tumor cells, and foreign cells by way of a lethal hit
Natural killer (NK) cell	Cytotoxic cell of undetermined affiliation (may be a kind of lymphocyte); kills infected cells and tumor cells by way of a lethal hit

Review Questions

1. While jogging barefoot along a seashore, your toes accidentally land on a jellyfish. Soon the bottoms of your toes are swollen, red, and warm to the touch. Describe the events that result in these signs of inflammation. *678–679*

2. Distinguish between:
   a. neutrophil and macrophage *678*
   b. cytotoxic T cell and natural killer cell *682–683*
   c. effector cell and memory cell *680*
   d. antigen and antibody *680, 686–687*

3. Describe antigen processing. *680–681*

4. HIV doesn't kill its host. Why are so many people dying of AIDS? *690–691*

5. Write a short essay on how the immune response contributes to homeostasis. (You may wish to refer to the discussion of homeostasis on pages *556–557*.)

6. Before each influenza season starts, you get an influenza vaccination. But this year you come down with "the flu" anyway. What do you suppose happened? (There are at least three explanations.) *688*

Self-Quiz *(Answers in Appendix IV)*

1. _____ are barriers to pathogens at body surfaces.
   a. Intact skin and mucous membranes
   b. Tears, saliva, and gastric fluid
   c. Resident bacteria
   d. Urine
   e. All of the above

2. Macrophages are derived from white blood cell precursors called _____ .
   a. lymphocytes   d. monocytes
   b. basophils   e. eosinophils
   c. neutrophils

3. Activated complement functions in defense by _____ .
   a. neutralizing toxins   c. promoting inflammation
   b. enhancing resident   d. forming holes in memory
   bacteria   lymphocyte membranes

4. _____ are large molecules that lymphocytes recognize as foreign and that elicit an immune response.
   a. Interleukins   d. Antigens
   b. Antibodies   e. Histamines
   c. Immunoglobulins

5. Immunoglobulins designated _____ increase antimicrobial activity in mucus.
   a. IgA   c. IgG   e. IgZ
   b. IgE   d. IgM

6. Antibody-mediated responses work best against _____ .
   a. intracellular pathogens   d. b and c
   b. extracellular pathogens   e. all of the above
   c. extracellular toxins

7. Antigens (nonself molecular markers) are _____ .
   a. nucleotides   c. steroids
   b. triglycerides   d. proteins

8. _____ would be a target of an effector cytotoxic T cell.
   a. Extracellular virus particles in blood
   b. Cervical tumor cells
   c. Parasitic flukes in the liver
   d. Bacterial cells in pus
   e. Pollen grains in nasal mucus

9. Development of a secondary immune response is based on populations of _____ .
   a. memory cells   d. effector cytotoxic T cells
   b. circulating antibodies   e. mast cells
   c. effector B cells

10. Match the immunity concepts.
    ____ inflammation   a. neutrophil
    ____ antibody secretion   b. effector B cell
    ____ a phagocyte   c. nonspecific response
    ____ immune memory   d. deliberately provoking
    ____ vaccination   memory cell production
    ____ allergy   e. basis of secondary
                                  immune response
                               f. nonprotective immune
                                  response

Selected Key Terms

acute inflammation *678*	histamine *678*
allergen *689*	immune system *680*
allergy *689*	immunization *688*
antibody *680*	immunoglobulin (Ig) *685*
antigen *680*	interleukin *678*
antigen-MHC complex *680*	lysis *677*
antigen-presenting cell *680*	lysozyme *676*
autoimmune response *689*	macrophage *678*
B cell *680*	MHC marker *680*
B lymphocyte *680*	natural killer cell *683*
basophil *678*	neutrophil *678*
clonal selection theory *687*	pathogen *676*
complement system *677*	perforin *683*
cytotoxic T cell *680*	T lymphocyte *680*
eosinophil *678*	vaccine *688*
helper T cell *680*	

Readings

Edelson, R., and J. Fink. June 1985. "The Immunologic Function of Skin." *Scientific American* 252(6):46–53.

Golub, E., and D. Green. 1991. *Immunology: A Synthesis.* Second edition. Sunderland, Massachusetts: Sinauer.

Kimball, J. 1990. *Introduction to Immunology.* Third edition. New York: Macmillan.

Tizard, I. 1992. *Immunology: An Introduction.* Third edition. Philadelphia: Saunders.

Tonegawa, S. October 1985. "The Molecules of the Immune System." *Scientific American* 253(4):122–131.

41 RESPIRATION

Conquering Chomolungma

To experienced climbers, Chomolungma may be the ultimate challenge (Figure 41.1). The summit of this Himalayan mountain, also known as Everest, is 9,700 meters (29,108 feet) above sea level. It is the highest place on earth.

Chomolungma's challenge is not merely iced-over vertical rock, driving winds, blinding blizzards, and heart-stopping avalanches. It is the extreme danger that oxygen-poor air poses to the brain.

Most of us live at low elevations. Of the air we breathe, one molecule in five is oxygen. In mountains higher than 3,300 meters (10,000 feet), the breathing game changes. The earth's gravitational pull is not as great and gas molecules spread out—so breathing the

way we do at sea level will not deliver enough oxygen to our lungs. The oxygen deficit can cause headaches, shortness of breath, heart palpitations, even loss of appetite, nausea, and vomiting. These are the telling symptoms of "altitude sickness."

At Chomolungma's base camp, climbers are 6,300 meters (19,000 feet) above sea level. More than half of the atmospheric oxygen is below them. At 7,000 meters (23,000 feet), oxygen and other gaseous molecules are extremely diffuse. The exceedingly low air pressure and scarcity of oxygen combine to make small blood vessels leaky. More plasma fluid trickles out through gaps between the endothelial cells that make up the blood vessel walls. In the brain and lungs, tissues swell with excess fluid. When the edema is not reversed, climbers become comatose and die.

During bad moments at high altitudes, stricken climbers inhale bottled oxygen. On Chomolungma's flanks, they have even been zipped inside airtight bags. Rescuers use a device that pumps in oxygen and removes carbon dioxide until "air" inside the bag more closely approximates the air at 6,000 to 9,000 feet.

Few of us will ever find ourselves near the peak of Chomolungma, pushing our reliance on oxygen to the limits. Here in the lowlands, disease, smoking, and other environmental insults push it in more ordinary ways. The risks can be just as great, as you will see by this chapter's end.

KEY CONCEPTS

1. Of all organisms, multicelled animals require the most energy. The energy comes mainly from aerobic metabolism, which requires oxygen and produces carbon dioxide wastes. In a process called respiration, animals move oxygen into their internal environment and give up carbon dioxide to the external environment.

2. Oxygen diffuses into the body as a result of a pressure gradient. The pressure of this gas is higher in air than it is in metabolically active tissues, where cells rapidly use oxygen. Carbon dioxide follows a gradient in the other direction. Its pressure is higher in tissues (where it is a by-product of metabolism) than it is in the air.

3. In most respiratory systems, oxygen and carbon dioxide diffuse across a respiratory surface, such as the thin, moist epithelium in the human lungs. Blood flowing through the body's circulatory system picks up oxygen and gives up carbon dioxide at this respiratory surface.

4. Respiratory systems differ in their adaptations for increasing gas exchange efficiency. They differ also in how they match air flow to blood flow.

Figure 41.1 A climber approaching the summit of Chomolungma, where oxygen is brutally scarce.

41.1 THE NATURE OF RESPIRATION

High in the mountains, in underground burrows, in the deep and shallow regions of the earth's water provinces, animals engage in aerobic respiration. This oxygen-requiring metabolic pathway yields energy and produces carbon dioxide wastes.

Most animals take in oxygen and rid themselves of carbon dioxide by a process called "respiration." Most have some type of **respiratory system** that functions in the exchange of gases. By interacting with other organ systems, their respiratory system helps maintain internal operating conditions for the body's living cells. Figure 41.2 is a diagram of the interactions among these systems.

Why Oxygen and Carbon Dioxide Move Into and Out of the Body

Partial Pressure Gradients Regardless of the kind of animal, respiration is based on the tendency of oxygen and carbon dioxide to diffuse down their concentration gradients—or, as we say for gases, their *pressure* gradients. When molecules of either gas are more concentrated in the surrounding air or water, more pressure is available to drive them into the body. When they are more concentrated inside, more pressure is available to drive them outside.

This is not to say that both gases exert the *same* pressure. Pump air into a flat tire near a beach in San Diego, and you are filling it with about 78 percent nitrogen, 21

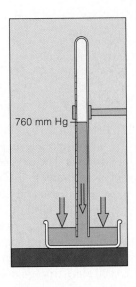

Figure 41.3 Atmospheric pressure as measured by a mercury barometer. At sea level, the level of mercury (Hg) in a glass column is about 760 millimeters (29.91 inches). At this level, the pressure exerted by the column of mercury equals atmospheric pressure outside the column.

760 mm Hg

percent oxygen, 0.04 percent carbon dioxide, and 0.96 percent of other gases. This is true of dry air anywhere at sea level. The numbers tell you that oxygen exerts only part of the total pressure on the tire wall, and its "partial pressure" is greater than that of carbon dioxide.

Said another way, atmospheric pressure at sea level is approximately 760 mm Hg, as measured by a mercury barometer. (Figure 41.3 shows a diagram of this device.) Thus the partial pressure of oxygen is (760 × 21/100), or about 160 mm Hg. The partial pressure of carbon dioxide is about 0.3 mm Hg.

Gases enter and leave the animal body by crossing a **respiratory surface**, such as a thin epithelial layer. The respiratory surface must be kept moist, for gases cannot diffuse across it unless they are dissolved in some fluid. What dictates the number of gas molecules moving across a respiratory surface in a given time? According to Fick's law, the more extensive the surface area and the larger the partial pressure gradient, the faster the diffusion rate will be.

Factors That Influence Gas Exchange

Surface-to-Volume Ratio All body plans of animals are adapted to promote favorable rates of inward diffusion of oxygen and outward diffusion of carbon dioxide. For example, animals without respiratory organs are tiny, tubelike, or flattened. Gases are exchanged directly across the body surface. And the surface-to-volume ratio, explained on page 53, has influenced their body plan.

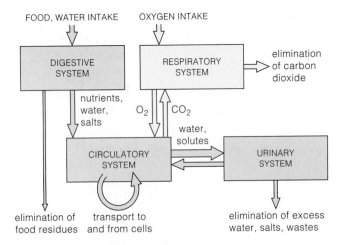

FOOD, WATER INTAKE   OXYGEN INTAKE

DIGESTIVE SYSTEM

RESPIRATORY SYSTEM

elimination of carbon dioxide

nutrients, water, salts

O_2 CO_2

water, solutes

CIRCULATORY SYSTEM

URINARY SYSTEM

elimination of food residues   transport to and from cells

elimination of excess water, salts, wastes

Figure 41.2 Interactions between the respiratory system and other organ systems in complex animals.

Figure 41.4 Respiratory adaptation of the flatworm, an aquatic animal. Unlike humans (**a**), flatworms (**b**) are small enough that they get along quite well without an oxygen-transporting circulatory system. Dissolved oxygen in their aquatic habitats can reach individual cells simply by diffusing across the flatworm's body surface. Unlike flatworms, humans do have a well-developed circulatory system, but it does them no good in aquatic habitats. They require great amounts of oxygen, and the oxygen that is dissolved in water cannot do the trick.

Imagine that the flatworm illustrated in Figure 41.4 has started to grow in all directions, like an inflating balloon. The worm's surface area does not increase at the same rate as its volume. Once its girth exceeds a single millimeter, the diffusion distance between the body surface and the internal cells is so great, the worm dies.

Ventilation For large-bodied, active animals, the requirement for gas exchange is great, and it cannot be maintained by diffusion alone. For example, a trout must move tissue flaps that are located above its respiratory organs (gills). As the flaps move back and forth, they stir the surrounding water. The stirring brings more dissolved oxygen closer to the body and moves more carbon dioxide away from it. Your own circulatory system rapidly delivers oxygen to cells and transports carbon dioxide to your lungs for disposal. Similarly, breathing ventilates your lungs.

Transport Pigments Rates of gas exchange get a boost with respiratory pigments, mainly **hemoglobin**, that help maintain steep pressure gradients across a respiratory surface.

For example, at the respiratory surfaces in human lungs, the oxygen concentration is high. There, each hemoglobin molecule binds loosely with as many as four oxygen molecules. When the circulatory system carries that same hemoglobin molecule past an oxygen-poor tissue, the oxygen is released. Thus, by transporting oxygen away from the respiratory surface, hemoglobin helps maintain the required pressure gradient that entices oxygen into the lungs.

---

1. By a process called respiration, animals take in oxygen for an energy-releasing pathway (aerobic respiration) and remove the pathway's carbon dioxide wastes. These gases enter and leave the body by diffusing across a moist respiratory surface.

2. Oxygen and other atmospheric gases tend to move down their pressure gradients. Each gas exerts only part of the total pressure on a system.

3. Gas exchange requires steep partial pressure gradients between the internal environment and the surroundings. The greater the area of the respiratory surface and the larger the partial pressure gradient, the faster diffusion will proceed.

41.2 MODES OF RESPIRATION

Integuments as Respiratory Surfaces

If an animal is not massive and if its life-style does not require high rates of metabolism, its demands for respiration are not great. Flatworms are like this. They rely on **integumentary exchange**, in which gases diffuse directly across the body covering (integument). Earthworms also rely on this mode of respiration. In addition, their circulatory system helps move gases to and from the body surface.

Animals in dry habitats have small, thick, or hardened integuments that are not well endowed with blood vessels. The integument conserves precious water but is not a good respiratory surface. This is true of the spiders, ticks, mites, and insects that rely on **tracheal respiration**. In such systems, small openings perforate the integument. Each is the start of a tube that branches inside the body (Figure 41.5). The terminal branchings have a fluid-filled tip, where gases diffuse directly into tissues. The tips abound in muscle and other tissues with high oxygen demands.

Gills

Gills are respiratory organs with a moist, thin, vascularized epidermis. Some insects, a few fish larvae, and a few amphibians have *external* gills that project into the surrounding water.

Adult fishes have a pair of *internal* gills. These are rows of slits or pockets at the back of the mouth that extend to the body surface (Figure 41.6). Water flows into the mouth and pharynx, then over filaments in the gills. Respiratory surfaces in a filament are richly endowed with blood vessels. When water flows past, it

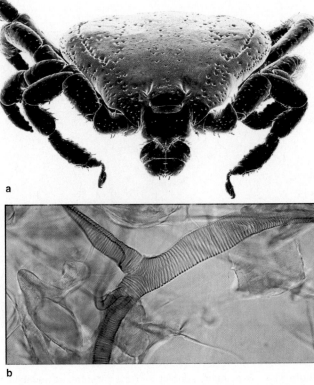

a

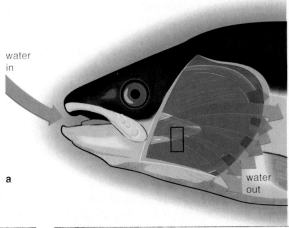

b

Figure 41.5 (**a**) Scanning electron micrograph of a tick. On its ventral surface, openings across the integument are the start of airways used in tracheal respiration. (**b**) A close-up view of an insect trachea. Rings of chitin reinforce this branching tube.

water in

water out

a

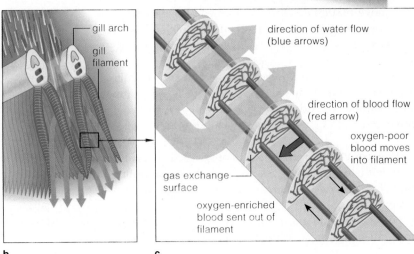

gill arch

gill filament

direction of water flow (blue arrows)

direction of blood flow (red arrow)

oxygen-poor blood moves into filament

gas exchange surface

oxygen-enriched blood sent out of filament

Figure 41.6 An example of the gills (respiratory organs) of adult fishes. (**a**) One of a pair of gills, each located under a bony lid (removed for this sketch).

(**b**) Each fish gill has filaments that contain richly vascularized respiratory surfaces. The filament tips are in contact, so that water flowing over them is directed past gas exchange surfaces before being exhaled. (**c**) A blood vessel carries oxygen-poor blood into each gill filament; another carries oxygenated blood away from it. Blood flowing from one vessel to the other runs counter to the direction of water flowing over the gas exchange surfaces. The arrangement favors the movement of oxygen (down its partial pressure gradient) into the blood.

b

c

first exchanges gases with a vessel carrying blood back to the body. This blood has less oxygen than the water does, so oxygen diffuses into it. Then the water flows over a vessel carrying blood from the body. The water has already given up oxygen, but it still has more than blood inside this vessel does—so more oxygen diffuses into the filament.

Movement of two fluids in opposing directions is called **countercurrent flow**. With this mechanism, fish extract about 80 to 90 percent of the dissolved oxygen flowing past. That is more than a fish would get from a one-way flow mechanism, at far less energy cost.

Lungs

Lungs are internal respiratory surfaces in the shape of a cavity or sac. They evolved in some fishes more than 450 million years ago (Figure 41.7). Lungfishes have gills *and* lungs that supplement respiration in oxygen-poor habitats. Amphibians have low metabolic rates and rely mostly on integumentary exchange, but they, too, have small lungs. Paired lungs are the dominant means of respiration in reptiles, birds, and mammals. In all lungs, air moves by bulk flow to the respiratory surface. Blood flow to and from the lungs enhances the diffusion of gases into and out of lung capillaries. In tissues, blood exchanges gases with interstitial fluid, which exchanges gases with cells.

We turn next to the human respiratory system. Its operating principles apply to most vertebrates. Birds are a notable exception. As Figure 41.8 shows, a unique system of air sacs helps ventilate bird lungs.

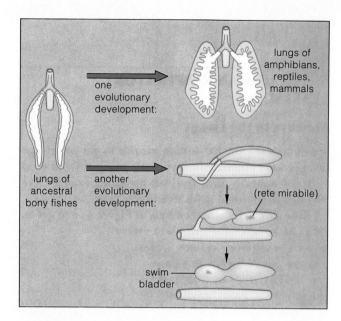

Figure 41.7 Evolution of vertebrate lungs and swim bladders. The esophagus (a tube leading to the stomach) is shaded *gold*; the respiratory tissues are shaded *pink*.

Lungs originated as pockets off the anterior part of the gut; they increased the surface area for gas exchange in oxygen-poor habitats. In some lineages, lung sacs became modified into swim bladders: buoyancy devices that help keep the fish from sinking. Adjusting gas volume in the bladders allows fishes to remain at different depths.

Trout and other less specialized fishes have a duct between the swim bladder and the esophagus; they replenish air in the bladder by surfacing and gulping air. Most bony fishes have no such duct; gases in the blood must diffuse into the swim bladder. Their swim bladder has a dense mesh of blood vessels (rete mirabile) in which arteries and veins run in opposite directions. Countercurrent flow through these vessels greatly increases gas concentrations in the bladder. Another region of the bladder allows reabsorption of gases by the body tissues.

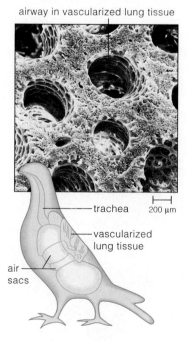

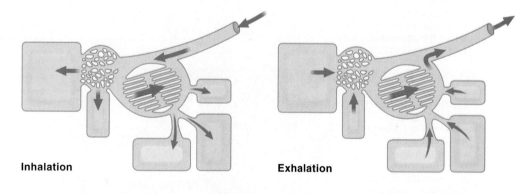

Inhalation   **Exhalation**

Figure 41.8 Respiratory system of birds. Typically, five air sacs are attached to each lung, which is small and inelastic. When the bird inhales, air is drawn into air sacs through tubes, open at both ends, that thread through the vascularized lung tissue. This tissue is the respiratory surface, where gases are exchanged. When the bird exhales, air is blown out of the sacs, through the small tubes, and out of the trachea. Thus, air is not merely drawn into the bird lungs. It is continuously drawn through them and across the respiratory surface. This unique ventilating system supports the high metabolic rates that birds require for flight and other energy-intensive activities.

41.3 HUMAN RESPIRATORY SYSTEM

Airways to the Lungs

It will take at least 300 million breaths to get you to age seventy-five. You may find yourself going without food for a few hours or days. But stop breathing even for five minutes and normal brain function is over.

Take a deep breath, then look at Figure 41.9 to get an idea of where the air will travel in your respiratory system. Unless you are out of breath and panting heavily, the air has just entered two nasal cavities, not your mouth. There it is warmed and picks up mois-

ture from mucus. A ciliated epithelium lining the nasal cavities filters dust and particles from it. So do those unseemly but functional nose hairs. Now the air is poised at the **pharynx**, or throat. This is the entrance to both the **larynx**, an airway, and the esophagus (the tube leading to the stomach). Right now the **epiglottis**, a flaplike structure at the start of the larynx, is pointing up. Air can move into your trachea, or windpipe. The trachea branches into two airways, one leading to each lung. Each airway is a **bronchus** (plural, bronchi). Its epithelial lining has an abundance of cilia and mucus-secreting cells (Figure 41.10). The lining serves as a barrier to infection. Bacteria and airborne particles stick in the mucus; then cilia sweep the debris-laden mucus up toward the mouth. Where it goes from there

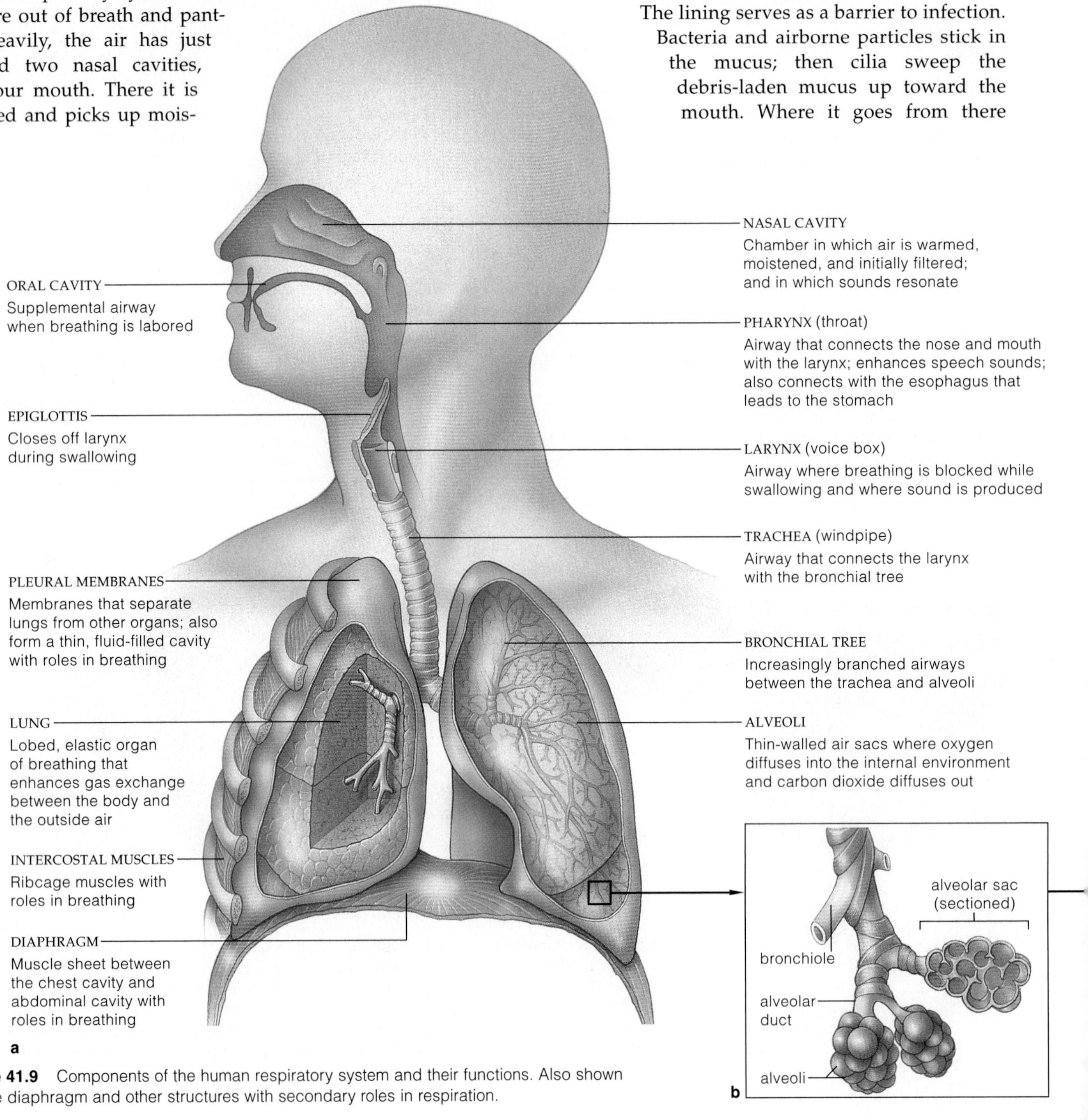

ORAL CAVITY
Supplemental airway when breathing is labored

EPIGLOTTIS
Closes off larynx during swallowing

PLEURAL MEMBRANES
Membranes that separate lungs from other organs; also form a thin, fluid-filled cavity with roles in breathing

LUNG
Lobed, elastic organ of breathing that enhances gas exchange between the body and the outside air

INTERCOSTAL MUSCLES
Ribcage muscles with roles in breathing

DIAPHRAGM
Muscle sheet between the chest cavity and abdominal cavity with roles in breathing

NASAL CAVITY
Chamber in which air is warmed, moistened, and initially filtered; and in which sounds resonate

PHARYNX (throat)
Airway that connects the nose and mouth with the larynx; enhances speech sounds; also connects with the esophagus that leads to the stomach

LARYNX (voice box)
Airway where breathing is blocked while swallowing and where sound is produced

TRACHEA (windpipe)
Airway that connects the larynx with the bronchial tree

BRONCHIAL TREE
Increasingly branched airways between the trachea and alveoli

ALVEOLI
Thin-walled air sacs where oxygen diffuses into the internal environment and carbon dioxide diffuses out

alveolar sac (sectioned)

bronchiole

alveolar duct

alveoli

a

b

Figure 41.9 Components of the human respiratory system and their functions. Also shown are the diaphragm and other structures with secondary roles in respiration.

really is up to you, but possibly the sidewalk is not a suitable destination.

When you swallow food, muscle contractions force the epiglottis down, to its closed position. This prevents food from going down the wrong tube and disrupting gas exchange. Such misdirected chunks of food can cause strangulation, as Figure 41.11 indicates.

Sites of Gas Exchange in the Lungs

Human lungs are elastic, cone-shaped organs of gas exchange inside the rib cage, to the left and right of the heart. They are positioned above the **diaphragm**, a muscular partition between the chest cavity and the abdominal cavity.

A thin membrane lines the outer surface of the lungs and the inner surface of the chest cavity wall. This is the pleural membrane. Visualize the lungs as two baseballs pushed into a partly inflated balloon. The balls take up so much space, they press the balloon's opposing sides together. Similarly, the pleural membrane is saclike, with the chest wall and the lungs pressing its opposing surfaces together. Only a thin film of lubricating fluid separates the two membrane surfaces, and it prevents friction between them. During *pleurisy*, a respiratory ailment, the pleural membrane becomes inflamed and swollen, friction follows, and breathing can be painful.

Inside each lung, air moves through finer and finer branchings of a "bronchial tree" (Figure 41.9a). These airways are **bronchioles**. Their endings, the *respiratory* bronchioles, have outpouchings from their walls. Each cup-shaped outpouching is an **alveolus** (plural, alveoli), and each lung has about 150 million of them. Most often, alveoli are clustered together as a larger pouch, an alveolar sac. The sacs are the major sites of gas exchange with lung capillaries (Figure 41.9b and c). Collectively, they represent a tremendous surface area for exchanging gases with blood. If your alveoli were stretched out as a single layer, they would cover the floor of a racquetball court!

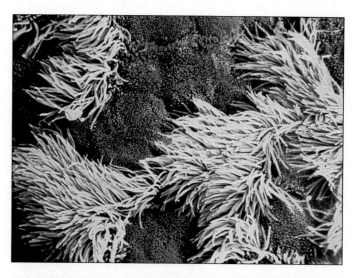

Figure 41.10 Color-enhanced scanning electron micrograph of cilia (*gold*) and mucus-secreting cells (*rust-colored*) in the respiratory tract.

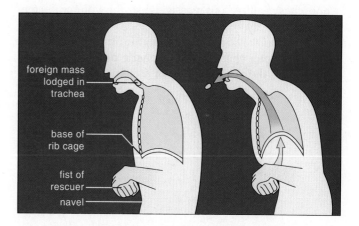

Figure 41.11 The Heimlich maneuver. Each year, several thousand people choke to death when food enters the trachea instead of the esophagus (compare Figure 42.4). Their air flow is blocked for as little as four or five minutes. The Heimlich maneuver, *an emergency procedure only*, often can dislodge the food. The idea is to elevate the diaphragm forcibly, causing a sharp decrease in the chest cavity volume and a sudden increase in alveolar pressure. The increased pressure forces air up the trachea and may be enough to dislodge the obstruction.

To perform the Heimlich maneuver, stand behind the victim, make a fist with one hand, then position the fist, thumb-side in, against the victim's abdomen. The fist must be slightly above the navel and well below the rib cage. Next, press the fist into the abdomen with a sudden upward thrust. Repeat the thrust several times if needed. The maneuver can be performed on someone who is standing, sitting, or lying down. Once the obstacle is dislodged, a physician must see the person at once, for an inexperienced rescuer can inadvertently cause internal injuries or crack a rib. It could be argued that the risk is worth taking, given that the alternative is death.

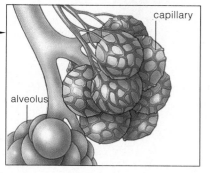

c Location of alveoli relative to the lung capillaries

41.4 VENTILATION

How Breathing Changes Air Pressure in the Lungs

When you breathe, you are ventilating your lungs. Air is *inhaled* (drawn into the airways), then *exhaled* (expelled from them). The chest cavity's volume increases and decreases in a rhythmic way. Each time you change the volume, you reverse pressure gradients between the lungs and air outside your body. Gases in the respiratory tract follow those gradients.

Figure 41.12 shows what happens when you start to inhale. The dome-shaped diaphragm contracts and flattens, and skeletal muscles lift the ribs upward and outward. Then, as the chest cavity expands, the rib cage moves away slightly from the lung surface. Pressure is lowered in the potential space between each lung and the pleural membrane. This pressure is transmitted to air spaces inside the alveoli, and it is lower than the atmospheric pressure at the mouth. Fresh air follows the pressure gradient. It flows down the airways, almost to the respiratory bronchioles, and causes the lungs to expand.

When you start to exhale, elastic tissue in the lungs and chest wall recoils passively, so the volume of the chest cavity decreases. The decrease compresses the air in the alveolar sacs. At that time, the air pressure in the sacs is greater than the atmospheric pressure, so now air follows the gradient—out of the lungs.

When oxygen demands increase, as during exercise, skeletal muscles of the rib cage and abdomen are stimulated to contract more often and more forcefully (page 623). When they do, they hasten the air flow.

Lung Volumes

When your body is resting and you are breathing normally, about 500 milliliters of air enter or leave the lungs with each breath. Although it takes a great leap of the imagination to compare the flow volume to the amount displaced by oceanic tides, this is nevertheless called the "tidal volume."

The maximum volume of air that can move out of your lungs after a single, maximal inhalation is called the "vital capacity." You rarely use more than half the total vital capacity, even when you take very deep breaths during strenuous exercise. To do so would exhaust the skeletal muscles with roles in respiration. Even at the end of the deepest exhalation, the lungs still would not be completely emptied of air. About 1,000 milliliters would remain (Figure 41.13).

How much of the 500 milliliters of inhaled air is actually available for gas exchange? About 150 milliliters remain in the air-conducting tubes between breaths. Thus only (500 – 150) or 350 milliliters of fresh air reach the alveoli with each inhalation. When you breathe, say, 10 times a minute, you are supplying your alveoli with (350 × 10) or 3,500 milliliters of fresh air per minute.

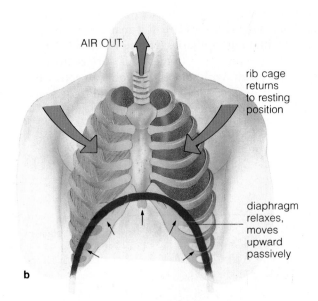

Figure 41.12 Changes in the size of the chest cavity during breathing. The *blue* line indicates the diaphragm's position when you (**a**) inhale and (**b**) exhale.

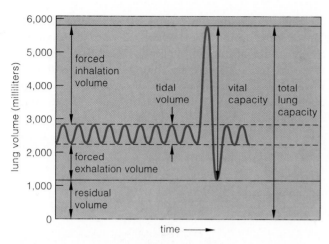

Figure 41.13 Capacities of the human lung. During normal breathing, a tidal volume of air enters and leaves the lungs. Forced inhalation can bring much larger quantities into the lungs, and forced exhalation can release some of the air normally kept in the lungs. A residual volume of gas remains trapped in partially filled alveoli despite the strongest exhalation.

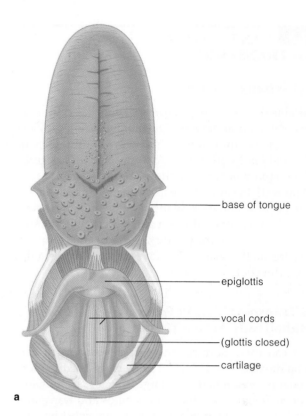

a

Breathing and Sound Production

Near the entrance to the larynx, part of a mucous membrane forms two pairs of horizontal folds. The lower pair of folds is called the **vocal cords** (Figure 41.14a). With each breath, air is forced through the **glottis**, the gap between the vocal cords. The air flow causes the cords to vibrate. By controlling the vibrations, we produce different kinds of sounds.

Within the folds are bands of elastic ligaments, connected to various cartilage tissues. When muscles of the larynx contract and relax, the ligaments tighten or slacken—and so change the extent to which the folds are stretched. Under commands from the nervous system, coordinated muscle action can narrow or widen the glottis. For example, by increasing muscle tension in the vocal cords, you decrease the gap between them and make high-pitched sounds or squeaks.

Sometimes an infection causes the mucous lining of the vocal cords to become irritated and inflamed. The tissues become swollen, and this interferes with the capacity of the vocal cords to vibrate. If hoarseness follows, this condition is called *laryngitis*.

You might assume that all mammals have vocal cords. Whales don't. However, they still produce sounds at a constriction between the back of the larynx and an airway to their lungs. Possibly the tissues in this region vibrate at low frequencies as air is forced past. Vibrations also may arise when muscles twitch at sonic frequencies. Such vibrations could travel through the skin and into the surrounding water.

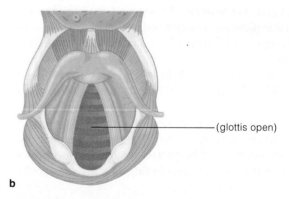

b

Figure 41.14 The paired vocal cords—where the sounds necessary for human speech originate. The glottis, the gap between the vocal cords, changes as a result of skeletal muscle action, which is under the control of the nervous system. The diagrams show what the glottis looks like when it is (**a**) closed and (**b**) opened.

41.5 GAS EXCHANGE AND TRANSPORT

Gas Exchange in Alveoli

Each alveolus is a single layer of epithelial cells, surrounded by a thin basement membrane. Lung capillaries also have thin walls, and only a thin film of interstitial fluid separates them from alveoli. Thus gases diffuse rapidly in both directions (Figure 41.15).

Figure 41.16 shows the partial pressure gradients for oxygen and carbon dioxide through the human respiratory system. Passive diffusion alone is enough to move oxygen across the respiratory surface and into the bloodstream. It is enough to move carbon dioxide in the reverse direction.

Gas Transport To and From Metabolically Active Tissues

Blood can carry only so much dissolved oxygen and carbon dioxide. Gas transport must be assisted to meet the energy requirements of large animals. In humans, the hemoglobin in red blood cells binds and transports both. This pigment increases oxygen transport by seventy times. It also increases carbon dioxide transport away from tissues by seventeen times.

Oxygen Transport   Inhaled air that reaches the alveoli has plenty of oxygen and not much carbon dioxide. The opposite is true of blood in the lung capillaries. Thus, in the lungs, oxygen diffuses into the plasma portion of blood and then into red blood cells, where it rapidly binds with hemoglobin. A hemoglobin molecule with oxygen bound to it is called **oxyhemoglobin**, or HbO_2. The amount of HbO_2 that forms depends on the partial pressure of oxygen. The higher the pressure, the more oxygen will be picked up. This will continue until all of the hemoglobin binding sites are saturated.

HbO_2 molecules hold onto oxygen rather weakly. They give it up in tissues where the partial pressure of oxygen is lower than in the lungs. They give it up even faster in tissues where the blood is warmer, the pH lower, and the partial pressure of carbon dioxide high. Such conditions exist in contracting muscle and other metabolically whipped-up tissues.

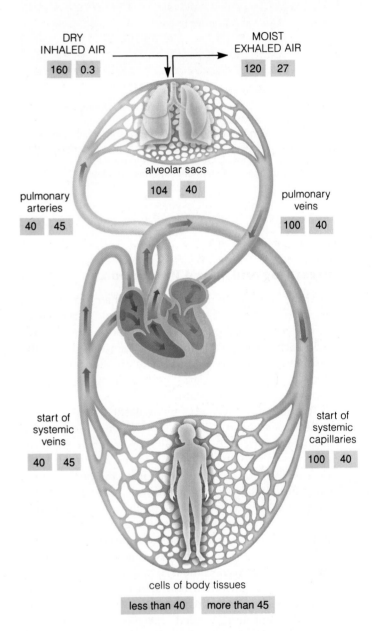

Figure 41.16 Partial pressure gradients for oxygen (*blue* boxes) and carbon dioxide (*pink* boxes) through the respiratory tract. This is the point to remember about the values shown: *Each gas moves from regions of higher to lower partial pressure.* That is why, for example, you become light-headed when you first visit places at high altitudes. The partial pressure of oxygen decreases with altitude, and your body does not function as well when the pressure gradient between the surrounding air and your lungs is lower than what you normally encounter.

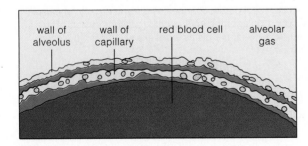

Figure 41.15 What a section through an alveolus and an adjacent lung capillary would look like. Compared to the red blood cell's diameter, the diffusion distance across the capillary wall, the interstitial fluid, and the alveolar wall is small.

Carbon Dioxide Transport Now think about a tissue where carbon dioxide's partial pressure is higher than in the blood flowing through the adjacent capillaries. Carbon dioxide diffuses into these blood capillaries, which transport it toward the lungs.

About 7 percent of the carbon dioxide remains dissolved in plasma. Another 23 percent or so binds with hemoglobin, forming carbaminohemoglobin ($HbCO_2$). But most of it—approximately 70 percent—is transported in the form of bicarbonate (HCO_3^-). The bicarbonate forms after carbon dioxide combines with water. The resulting carbonic acid dissociates (separates) into bicarbonate and hydrogen ions:

$$CO_2 + H_2O \rightleftharpoons \underset{\text{carbonic acid}}{H_2CO_3} \rightleftharpoons \underset{\text{bicarbonate}}{HCO_3^-} + H^+$$

The reactions just described do not amount to much in blood plasma, where only 1 of every 1,000 carbon dioxide molecules is converted. It's a different story in red blood cells. These cells contain **carbonic anhydrase**, an enzyme that increases the reaction rate by 250 times! In these cells, most of the carbon dioxide that is not bound to hemoglobin is converted to carbonic acid. The enzyme-mediated conversion makes the blood level of carbon dioxide drop swiftly. This helps maintain the gradient that keeps carbon dioxide diffusing from interstitial fluid into the bloodstream.

What happens to the bicarbonate ions that form? They tend to diffuse out of the red blood cells and into the blood plasma. What about the hydrogen ions? Hemoglobin acts as a buffer for them and keeps the blood from becoming too acidic. A buffer, recall, is a molecule that combines with or releases hydrogen ions in response to changes in cellular pH.

The reactions are reversed in the alveoli, where the partial pressure of carbon dioxide is lower than it is in neighboring blood capillaries. Water and carbon dioxide now form and diffuse into alveolar sacs. From there, carbon dioxide is exhaled from the body.

Controls Over Gas Exchange

Gas exchange is most efficient when the rate of air flow matches the rate of blood flow. The rates can be brought into balance by adjustments in local tissues as well as in tissues through the body as a whole.

Local Chemical Control Some local controls operate in the lungs, at alveoli. Suppose your heart is pounding and you don't breathe deeply enough. Blood flows too fast and the air too sluggishly for efficient disposal of carbon dioxide. An increase in the blood level of carbon dioxide affects smooth muscle in bronchiole walls. The wall diameter widens, enhancing the air flow.

Local controls also work on lung capillaries. When air flow is too great relative to blood flow, oxygen levels rise in some parts of the lungs. The increase affects smooth muscle in blood vessel walls. The wall diameter widens, so blood flow to the regions increases. If the volume of air flow is too small, the diameter shrinks, and blood flow decreases.

Neural Controls The nervous system governs oxygen and carbon dioxide levels in arterial blood for the entire body. Some sensory receptors notify the brain of rising carbon dioxide levels in blood. Others detect decreases in the partial pressure of oxygen dissolved in arterial blood. They include carotid bodies (at branches of the carotid arteries to the brain) and aortic bodies (in an arterial wall near the heart). The brain responds with signals to muscles in the diaphragm and chest wall. By bringing about changes in muscle activity, the nervous system adjusts rate and depth of breathing.

Where are the brain's **respiratory centers** located? One cell cluster in the reticular formation (page 583) coordinates signals dealing with inhalation. Another cluster coordinates signals dealing with exhalation. Other centers stimulate both of the cell clusters to fine-tune the resulting contractions of the diaphragm and rib cage muscles.

The controls over gas exchange are like a fine Swiss watch—intricately coordinated and smooth in their operation. The *Focus* essay on the page that follows describes a few respiratory ailments, some of which are serious enough to stop the watch from ticking.

---

1. Driven by its partial pressure gradient, oxygen diffuses from alveolar air spaces, through interstitial fluid, and into lung capillaries.

2. Carbon dioxide, driven by its partial pressure gradient, diffuses in the opposite direction.

3. Hemoglobin in red blood cells greatly enhances the oxygen-carrying capacity of blood.

4. Most carbon dioxide is transported in the blood in the form of bicarbonate. Nearly all of the bicarbonate forms through enzyme-mediated reactions in red blood cells.

5. Through neural controls and local controls (as in the lungs), the rates of air flow and blood flow can be adjusted in ways that enhance gas exchange.

When the Lungs Break Down

In large cities, in certain workplaces, even near a cigarette smoker, airborne particles and certain gases are present in abnormal amounts, and they put extra workloads on the respiratory system.

Bronchitis   The ciliated, mucous epithelium that lines your bronchioles helps protect you from respiratory infections. Toxins in cigarette smoke and other airborne pollutants irritate the lining and may lead to *bronchitis*. In this respiratory ailment, excess mucus is secreted when epithelial cells that line the airways become irritated. As mucus accumulates, so do bacteria and other particles stuck in it. Coughing brings up some of the gunk. If the irritation persists, so does the coughing. Initial attacks of bronchitis are treatable. When the aggravation is allowed to continue, bronchioles become inflamed. Bacteria, chemical agents, or both attack cells of the bronchiole walls. Ciliated cells are destroyed, and mucus-secreting cells multiply. Fibrous scar tissue forms; in time it may narrow or obstruct the airways.

Emphysema   When bronchitis persists, airways become clogged with mucus. Tissue-destroying bacterial enzymes attack the thin, stretchable walls of alveoli. As the walls break down, inelastic fibrous tissue surrounds them. Gas exchange now proceeds at fewer alveoli—which become enlarged. In time, the balance between air flow and blood flow becomes permanently compromised because the lungs remain distended and inelastic. Compare Figures *a* and *b* to get an idea of what goes on. Running, walking, even exhaling become difficult. These are symptoms of *emphysema*, a respiratory disease that affects about 1.3 million people in the United States alone.

A few people are genetically predisposed to emphysema. They do not have a functional gene for antitrypsin, an enzyme that can inhibit bacterial attack on the alveoli. Also, poor diet and chronic (persistent or recurring) colds and other respiratory infections make people susceptible to emphysema later in life. But smoking is the major cause of the disease. Emphysema can develop slowly, over twenty or thirty years. By the time it is detected, the damage to lung tissue cannot be repaired.

Effects of Cigarette Smoke   Every thirteen seconds, one of 50 million cigarette-puffing Americans dies of emphysema, chronic bronchitis, or heart disease. For every eight of them, one *nonsmoker* dies of ailments brought on by prolonged exposure to tobacco smoke in the surroundings. Children who breathe secondhand smoke can expect to suffer more allergies and lung ailments. Smoking is the

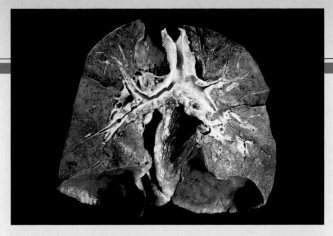

a   Normal appearance of tissues in the paired human lungs.

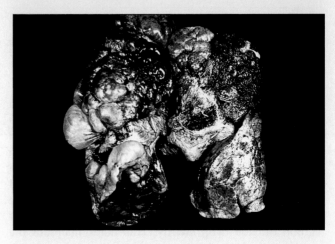

b   Lungs from someone who was affected by emphysema.

major cause of lung cancer, which is now the leading cause of death among women. Yet every day, 3,000 to 5,000 Americans light a cigarette for the first time. Children spend a billion dollars a year on cigarettes. The economy shrinks by 22 billion a year because of direct medical costs of treating smoke-induced respiratory disorders.

Consider how cigarette smoke affects the lungs. Noxious particles in the smoke from just one cigarette immobilize cilia in the bronchioles for several hours. The particles also stimulate mucus secretions that in time clog the airways. They can kill infection-fighting macrophages in the respiratory tract. What starts as "smoker's cough" can end in bronchitis and emphysema. Marijuana smoke also can cause extensive lung damage.

Or consider how cigarette smoke contributes to lung cancer. Inside the body, certain compounds in coal tar and cigarette smoke are converted to highly reactive intermediates. These are the carcinogens. They provoke the uncontrolled cell divisions in lung tissues. On the average, 90 of every 100 smokers who develop lung cancer will die from it. If you now smoke, or if you are thinking about starting or quitting, you may wish to consider the information presented in Figure *c*.

Risks Associated with Smoking: | **Reducing the Risks by Quitting:**

Shortened Life Expectancy: Nonsmokers live 8.3 years longer on average than those who smoke two packs daily from the midtwenties on

Cumulative risk reduction; after 10 to 15 years, life expectancy of ex-smokers approaches that of nonsmokers

Chronic Bronchitis, Emphysema: Smokers have 4–25 times more risk of dying from these diseases than do nonsmokers

Greater chance of improving lung function and slowing down rate of deterioration

Lung Cancer: Cigarette smoking is the major cause of lung cancer

After 10 to 15 years, risk approaches that of nonsmokers

Cancer of Mouth: 3–10 times greater risk among smokers

After 10 to 15 years, risk is reduced to that of nonsmokers

Cancer of Larynx: 2.9–17.7 times more frequent among smokers

After 10 years, risk is reduced to that of nonsmokers

Cancer of Esophagus: 2–9 times greater risk of dying from this

Risk proportional to amount smoked; quitting should reduce it

Cancer of Pancreas: 2–5 times greater risk of dying from this

Risk proportional to amount smoked; quitting should reduce it

Cancer of Bladder: 7–10 times greater risk for smokers

Risk decreases gradually over 7 years to that of nonsmokers

Coronary Heart Disease: Cigarette smoking is a major contributing factor

Risk drops sharply after a year; after 10 years, risk reduced to that of nonsmokers

Effects on Offspring: Women who smoke during pregnancy have more stillbirths, and weight of liveborns averages less (hence, babies are more vulnerable to disease, death)

When smoking stops before fourth month of pregnancy, risk of stillbirth and lower birthweight eliminated

Impaired Immune System Function: Increase in allergic responses, destruction of defensive cells (macrophages) in respiratory tract

Avoidable by not smoking

Bone Healing: Evidence suggests that surgically cut or broken bones require up to 30 percent longer to heal in smokers, possibly because smoking depletes the body of vitamin C and reduces the amount of oxygen reaching body tissues. Reduced vitamin C and reduced oxygen interfere with production of collagen fibers, a key component of bone. Research in this area is continuing.

Avoidable by not smoking

c From the American Cancer Society, a list of the risks incurred by smoking and the benefits of quitting. The photograph shows swirls of cigarette smoke poised at the entrance to the two bronchi that lead into the lungs.

41.6 RESPIRATION IN UNUSUAL ENVIRONMENTS

Breathing at High Altitudes

As you read in the introduction to this chapter, the partial pressure of oxygen decreases with increasing altitude. Unlike the occasional climbers of Chomolungma, humans, llamas, and other species accustomed to living at high altitudes have long-lasting adaptations to the thinner air.

For example, compared to the lungs of individuals who reside at lower elevations, the lungs of permanent residents of high mountains have many more air sacs and blood vessels, which developed while they were growing up. Besides this, the ventricles in their heart enlarged more, so that they pump larger volumes of blood. Llamas have an additional advantage. Compared to human hemoglobin, llama hemoglobin contains factors that give it a greater affinity for oxygen (Figure 41.17). It picks up oxygen more efficiently at the lower pressures characteristic of high altitudes.

Visitors who have not had time to adapt to the thinner air at high altitudes can suffer *hypoxia*, or cellular oxygen deficiency. Generally, at 2,400 meters (about 8,000 feet) above sea level, respiratory centers work to compensate for the oxygen deficiency by triggering *hyperventilation* (breathing much faster and more deeply than normal). At 3,300 meters (10,000 feet) or so, altitude sickness sets in.

Carbon Monoxide Poisoning

Cellular oxygen deficiency (hypoxia) is not restricted to high altitudes. It also arises when the partial pressure of oxygen in arterial blood falls as a result of *carbon monoxide poisoning*. Carbon monoxide, a colorless, odorless gas, is a component of automobile exhaust fumes and smoke from tobacco, coal, or wood burning. It binds to hemoglobin at least 200 times more strongly than oxygen does.

Even very small amounts of carbon monoxide can tie up half of the body's hemoglobin. It can dangerously impair the delivery of oxygen to tissues.

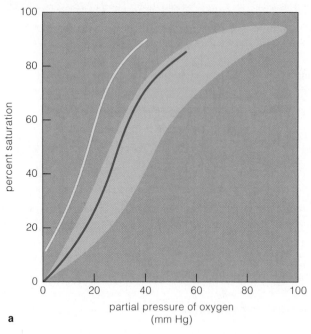

a

b

Figure 41.17 (**a**) Comparison of the binding and releasing capacity of human hemoglobin and llama hemoglobin. (**b**) Llamas high up in the Peruvian Andes.

Respiration in the Deep Ocean

As professional divers know, water pressure increases greatly with depth. When diving in deep water, they take along tanks of compressed air (air under pressure). They also make their ascents from the depths with utmost caution.

Why are divers so cautious? Because of the increased pressure at depths, more gaseous nitrogen (N_2) than usual has become dissolved in their body tissues. When the pressure decreases, N_2 moves back out of the tissues and into the bloodstream. If an ascent is too rapid, N_2 enters the blood faster than the lungs can dispose of it. When that happens, bubbles of nitrogen may form in the blood and tissues. Too many bubbles cause pain, especially at joints. Hence the common name, "the bends," for what is otherwise known as *decompression sickness*. If bubbles obstruct blood flow to the brain, deafness, impaired vision, and paralysis may result.

At depths greater than about 30 meters (98 feet), gaseous nitrogen poses still another threat. High partial pressures of this gas produce euphoria, even drunkenness. Divers have been known to offer the mouthpiece of their airtank to a fish.

Unlike humans, seals, whales, and other marine mammals have built-in respiratory adaptations that allow them to dive deeply with impunity. The dive of one sperm whale is known to have lasted 82 minutes. Another whale, tracked by sonar, reached a depth of 2,250 meters (Figure 41.18). Members of an even more ancient vertebrate lineage also exceed the diving capacity of humans, as described in the *Focus* essay on the page that follows.

Consider the respiratory system of a sperm whale, which is modified to collect and store oxygen. When a whale surfaces, it rapidly blows stale air from the lungs, through a tubelike epiglottis, and out through a blowhole at the body surface. Before the whale dives again, it breathes rapidly, so that its elastic, extensible lungs fill quickly with large volumes of air. During a dive, special valves close off its nostrils; and rings of muscle and of cartilage close off its bronchioles. The whale's respiratory surface is not that large. However, strategically arranged valves and plexuses (local networks of blood vessels) store and distribute the blood volume (and gases) in economical fashion. The heart rate slows, metabolism decreases, and so does the rate of oxygen use and carbon dioxide formation. Compared to land mammals, the whale's respiratory center (in the medulla oblongata) is less sensitive to carbon dioxide levels.

Last but not least, whale muscles store notably large amounts of **myoglobin**, an oxygen-binding protein. Their muscles can draw on their own large oxygen reservoirs during deep dives.

Figure 41.18 A sperm whale, supremely adapted for diving. This large aquatic mammal is more than 25 meters long, on the average. It can dive 2,250 meters below the ocean surface and stay there for well over an hour before coming up for oxygen.

The Leatherback Sea Turtle

Late at night, near the shoreline of a Caribbean island, biologist Molly Lutcavage and several colleagues move out on a turtle patrol. Finally a large, dark shape, with flippers flailing, emerges from the pale surf. A female Atlantic leatherback sea turtle (*Dermochelys coriacea*) is returning from the sea to nest in the sand.

Sea turtles are members of the reptilian lineage, which extends 300 million years back in time. Green turtles, ridleys, loggerheads, leatherbacks—all are endangered or threatened species. The race is on to gather information about their life histories and physiology that may help pull them back from the brink of extinction.

Leatherbacks are the largest and least understood. Adult females may weigh more than 400 kilograms (880 pounds). An adult male may weigh nearly twice that amount. Leatherbacks normally leave the water only to breed and lay eggs. The rest of the time they migrate across vast stretches of the open ocean. And leatherbacks do something no other reptile on earth can do. They can dive to 1,000 meters (3,000 feet) below sea level! Before biologists conducted tracking experiments in the mid-1980s, Weddell seals, fin whales, and some other marine mammals were the only nonhuman deep-sea divers known.

To reach such depths, a turtle would have to swim for nearly forty minutes without taking a breath! How does it dive so deep, for so long, and still have enough oxygen for aerobic metabolism?

Underwater staying power has nothing to do with lungs. (Even at depths between 80 and 160 meters, pressure exerted by the surrounding water causes air-filled lungs to collapse.) Consider that the skeletal muscle cells of marine mammals have large reservoirs of myoglobin, an oxygen-binding protein. Compared to other mammals, they also have more blood per unit of body weight, and their blood contains more red blood cells—hence more hemoglobin. Thus marine mammals can engage in aerobic respiration without having to come up for air. Could it be that the leatherback has similar adaptations for diving?

Imagine yourself observing Lutcavage's team. Their plan is to draw blood from a leatherback so that they can study the oxygen-binding capacity of its hemoglobin. Samples of respiratory gases dissolved in the blood may provide insight into how leatherbacks actually use oxygen. They know that a nesting female enters a trancelike state during the twenty or so minutes it takes her to deposit eggs in the sand. During that time, she will not resist being gently handled.

The team gets to work. They fit the head of a nesting female with a helmet that is equipped with an expandable plastic sleeve. The fit is snug but not restrictive, so the turtle breathes normally. With each exhalation, expired gases flow through a one-way valve into a collecting sac. At the same time, the researchers count the number of breaths required to completely fill the sac. (As a baseline, they also tallied the breathing frequency before placing the helmet over her head.) The gas samples are stored according to established procedures, then set aside for laboratory analysis.

Working quickly, a member of the team draws blood from a superficial vein in the turtle's neck, then packs the sample

Physiological Comparison of a Few Diving Reptiles and Mammals					
Characteristic	Leatherback Turtle	Green Turtle	Crocodile	Killer Whale	Weddell Seal
Maximum diving depth (meters)	Greater than 1,000	Less than 100	Less than 30	260	600
Red blood cell count (percent of total volume)	39	30	28	44	58
Hemoglobin level (grams/ deciliter of blood)	15.6	8.8	8.7	16.0	17–22
Myoglobin level (milligrams/ gram of muscle tissue)	4.9	—*	—*	—*	44.6
Oxygen-carrying capacity (volume percent)	21	7.5–11.9	12.4	23.7	31.6

*Not known.

in ice to preserve it. Back in the laboratory, the sample will be subjected to hemoglobin analysis and a red blood cell count. The oxygen level, carbon dioxide level, and pH will be measured.

Lutcavage and her coworkers gathered samples of blood and respiratory gases from several turtles. They also analyzed skeletal muscle tissue, taken earlier from a drowned turtle. The data gave insight into how leatherbacks manage their spectacularly deep dives.

Calculations of the oxygen uptake and total tidal volume during breathing revealed that a leatherback cannot inhale and hold enough air in the lungs to sustain aerobic metabolism during a prolonged dive. Other oxygen-supplying mechanisms had to be at work. Further analysis revealed that the myoglobin levels are so high in leatherbacks, oxygen is still available during a dive after air in their lungs gives out (or when their lungs collapse). Muscles could continue to bring about swimming movements.

Besides this, leatherbacks have an abundance of red blood cells and a type of hemoglobin with a high affinity for binding oxygen. A leatherback's blood may contain a remarkable 21 percent oxygen by volume. That amount is characteristic of a human, not a "plodding" turtle.

The leatherback's oxygen-carrying capacity is the highest ever recorded for a reptile—and it is close to the capacity of deep-diving mammals. Add to this the presence of a myoglobin-based oxygen reservoir, a streamlined body, and massive front flippers, and you have a reptile uniquely adapted for diving.

Studies of leatherbacks are only now beginning. Researchers are investigating tracking methods that may help them monitor a turtle's metabolism at sea and during dives. So far, it has proved extremely difficult and expensive to track individual turtles in the open ocean.

SUMMARY

1. Animal cells rely mainly on aerobic respiration, a metabolic pathway that provides enough energy for active life-styles. This pathway requires oxygen and produces carbon dioxide. The process by which the animal body as a whole acquires oxygen and disposes of carbon dioxide is called respiration.

2. Air is a mixture of oxygen, carbon dioxide, and other gases, each exerting a partial pressure. Each gas tends to move from areas of higher to lower partial pressure. Respiratory systems make use of this tendency.

3. In all respiratory systems, oxygen and carbon dioxide diffuse across a respiratory surface (a moist, thin layer of epithelium). In vertebrates, airways carry gases to and from one side of the respiratory surface, and blood vessels carry gases to and from the other side.

4. The airways of the human respiratory system are the nasal cavities, pharynx, larynx, trachea, bronchi, and bronchioles. Alveoli at the end of the terminal bronchioles are the main sites of gas exchange.

5. During inhalation, the chest cavity expands, lung pressure falls below atmospheric pressure, and air flows into the lungs. During normal exhalation, these events are reversed.

6. Driven by its partial pressure gradient, oxygen in the lungs diffuses from alveolar air spaces into the lung capillaries. Then it diffuses into red blood cells and binds weakly with hemoglobin. In tissues where cells are metabolically active, hemoglobin gives up oxygen, which diffuses out of the capillaries, across interstitial fluid, and into the cells.

7. Driven by its partial pressure gradient, carbon dioxide diffuses from cells, across interstitial fluid, and into the bloodstream. Most reacts with water to form bicarbonate, but the reactions are reversed in the lungs. There, carbon dioxide diffuses from lung capillaries into the air spaces of the alveoli, then is exhaled.

Review Questions

1. A few of your friends who have not taken a biology course ask you what insect lungs look like. What is your answer (assuming your instructor is listening)? *698*

2. Describe the features of the respiratory surface that are common to all respiratory systems. *696–697*

3. Distinguish between:
   a. aerobic respiration and respiration *696*
   b. pharynx and larynx *700*
   c. bronchiole and bronchus *700, 701*
   d. pleural sac and alveolar sac *701*

4. Explain why humans, including the female shown below, cannot survive on their own for very long in underwater environments. *696–697*

5. Label the components of the human respiratory system and the structures that enclose it: *700*

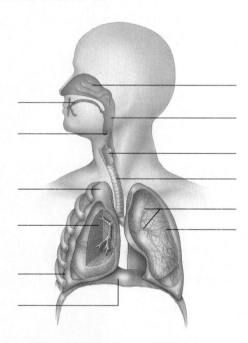

6. Some cigarette manufacturers have conducted public relations campaigns urging their customers to "smoke responsibly." What are some social and biological issues in this controversy? How do these issues apply to the nonsmoking spouse or children of a smoker? To nonsmoking patrons in a restaurant? To the unborn child of a pregnant smoker? In your opinion, what behavior would constitute "responsible smoking"? *706–707*

7. People occasionally poison themselves with carbon monoxide by building a charcoal fire in an enclosed area. Assuming help arrives in time, what would be the *most* effective treatment: placing the victim outdoors in fresh air or administering pure oxygen? Why? *704–705*

1. A partial pressure gradient of oxygen exists between _____ .
   a. air and lungs
   b. lungs and metabolically active tissues
   c. air at sea level and air at high altitudes
   d. all of the above

2. The _____ is an airway that connects the nose and mouth with the _____ .
   a. oral cavity; larynx   c. trachea; pharynx
   b. pharynx; trachea   d. pharynx; larynx

3. In humans, oxygen in the air must diffuse across _____ as it follows its partial pressure gradient into the internal environment.
   a. pleural sacs
   b. alveolar sacs
   c. a moist respiratory surface
   d. both b and c

4. Each human lung encloses a _____ .
   a. diaphragm   c. pleural sac
   b. bronchial tree   d. both b and c

5. Gas exchange occurs at the _____ .
   a. two bronchi   c. alveolar sacs
   b. pleural sacs   d. both b and c

6. The Heimlich maneuver may help people _____ .
   a. with altitude sickness   c. choking on food
   b. with the bends   d. quit smoking

7. Breathing _____ .
   a. ventilates the lungs
   b. draws air into airways
   c. expels air from airways
   d. causes reversals in pressure gradients
   e. all of the above

8. When you inhale, the diaphragm _____ .
   a. curves upward   c. relaxes
   b. flattens   d. remains stationary

9. After oxygen diffuses into lung capillaries, it diffuses into _____ and binds with _____ .
   a. interstitial fluid; red blood cells
   b. interstitial fluid; carbon dioxide
   c. red blood cells; hemoglobin
   d. red blood cells; carbon dioxide

10. Most carbon dioxide in blood is in the form of _____ .
    a. carbon dioxide   c. carbonic acid
    b. carbon monoxide   d. bicarbonate

11. Match the components with their descriptions.
    ____ trachea   a. airway leading into a lung
    ____ pharynx   b. throat
    ____ alveolus   c. fine bronchial tree branchings
    ____ hemoglobin   d. windpipe
    ____ bronchus   e. respiratory pigment
    ____ bronchiole   f. site of gas exchange

alveolus *701*
bronchiole *701*
bronchus *700*
carbonic anhydrase *705*
countercurrent flow *699*
diaphragm *701*
epiglottis *700*
gill *698*
glottis *703*
hemoglobin *697*
integumentary exchange *698*
larynx *700*
lung *699*
myoglobin *709*
oxyhemoglobin *704*
pharynx *700*
respiratory center *705*
respiratory surface *696*
respiratory system *696*
tracheal respiration *698*
vocal cord *703*

American Cancer Society. *Dangers of Smoking; Benefits of Quitting and Relative Risks of Reduced Exposure.* Revised edition. New York: American Cancer Society.

Mortality From Smoking in Developed Countries 1950–2000. A 1994 publication by scientists of Britain's Imperial Cancer Research Fund, the World Health Organization, and the American Cancer Research Fund. Research finds that worldwide, smoking now kills 3 million people every year. If current patterns do not change, then by the time today's young smokers reach middle age, 10 million may be dying annually because of their habit. That is one person every three seconds.

Vander, A., J. Sherman, and D. Luciano. 1990. *Human Physiology: The Mechanisms of Body Function.* Fifth edition. New York: McGraw-Hill. Clear introduction to the respiratory system.

West, J. 1989. *Respiratory Physiology: The Essentials.* Fourth edition. Baltimore: Williams & Wilkins. Excellent, brief introduction to respiratory functions. Paperback.

Young, J. Z. 1981. *The Life of Vertebrates.* Third edition. Oxford: Clarendon Press.

42 DIGESTION AND HUMAN NUTRITION

Sorry, Have To Eat and Run

Fall through winter, along mountain ridges from central Canada down into northern Mexico, pronghorn antelope (*Antilocapra americanus*) browse on wild sage. Come spring, herds of these medium-sized mammals move down to open grasslands and deserts, where they browse on new growth.

Coyotes, bobcats, and golden eagles prey on young antelope, so while antelope are browsing, they keep a constant eye out for danger. They can do this even while their head is bent low in the grasses, given how far back their eye sockets are positioned in the skull

(Figure 42.1a). And can they eat and run! With bursts of speed that can reach 95 kilometers per hour, antelope can leave predators in the dust.

The teeth of an antelope or any other mammal are clues to its life-style. Think of your own cheek teeth, each with a flattened crown that serves as a grinding platform. The crown of an antelope's molars—the cheek teeth—dwarfs yours (Figure 42.1b). You probably do not rub your mouth against the ground while eating. An antelope does, and abrasive bits of soil enter its mouth along with the tough plant material. Its teeth

Figure 42.1   (a) Pronghorn antelope (*Antilocapra americanus*) busily taking in nutrients.
(b) Comparison of human cheek teeth with those of plant-eating mammals, including antelope.

a

wear down more rapidly than yours do—and natural selection has favored more crown to wear down.

Antelopes are **ruminants**, a type of hoofed mammal with multiple stomach chambers in which cellulose—a tough, fibrous plant material—is slowly broken down. Cellulose breakdown starts in the first two of four stomach sacs. There, symbiotic bacteria produce cellulose-digesting enzymes to the antelope's benefit as well as their own. As enzyme action proceeds, the antelope regurgitates, rechews the contents of the first two stomach sacs, then swallows again. (This is what "chewing their cud" means.) Pummeling the plant material more than once exposes more surface area to the enzymes and gives them more time to act.

We humans, too, do amazing things with food. Somewhere in the world, one person might eat only raw whale blubber in a given day, another might eat only rice, and still another might eat only pizza and bananas, or couscous or snake meat, or drink dandelion wine. Through its metabolic magic, the human body converts these and a dizzying variety of other substances into usable energy and tissues of its own.

And with this interesting thought in mind, we start our tour of **nutrition**. The word encompasses processes by which food is ingested, digested, absorbed, and later converted to the body's own carbohydrates, lipids, proteins, and nucleic acids.

KEY CONCEPTS

1. Interactions among the digestive, circulatory, respiratory, and urinary systems supply the body's cells with raw materials, dispose of wastes, and maintain the volume and composition of extracellular fluid.

2. Most digestive systems have specialized parts for food transport, processing, and storage. Different parts mechanically break apart and chemically break down food, absorb breakdown products, and eliminate the unabsorbed residues.

3. To maintain an acceptable body weight and overall health, energy intake must balance energy output (by way of metabolic activity, physical exertion, and so on). Complex carbohydrates provide the most glucose, which typically is the body's main source of immediately usable energy.

4. Nutrition involves the intake of vitamins, minerals, and certain amino acids and fatty acids that the body itself cannot produce.

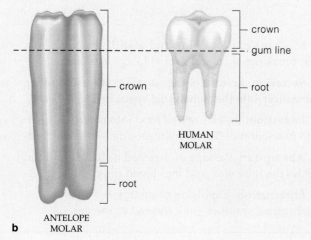

crown
gum line
root

HUMAN
MOLAR

crown

root

b ANTELOPE
MOLAR

DIGESTIVE SYSTEMS

General Features

A **digestive system** is a body cavity or tube in which food is reduced to particles, then to molecules small enough to be absorbed into the internal environment.

Some invertebrates have an **incomplete digestive system**—a saclike gut with one opening. Remember the flatworms? Their pharynx leads to a branched cavity,

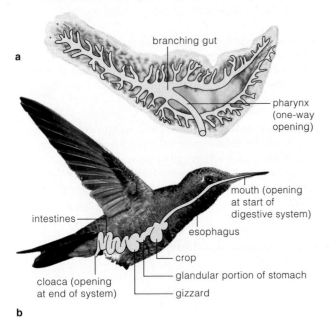

a

branching gut

pharynx (one-way opening)

mouth (opening at start of digestive system)

intestines

esophagus

crop

glandular portion of stomach

cloaca (opening at end of system)

gizzard

b

Figure 42.2 (**a**) Incomplete digestive system of a flatworm, with two-way traffic of food and undigested material through one opening. (**b**) Complete digestive system of a bird—basically a tube with regional specializations and an opening at each end.

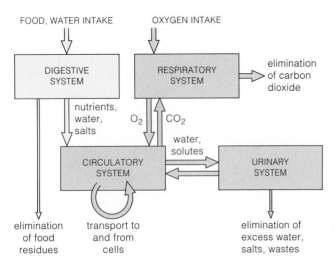

FOOD, WATER INTAKE   OXYGEN INTAKE

DIGESTIVE SYSTEM

RESPIRATORY SYSTEM

elimination of carbon dioxide

nutrients, water, salts

O_2 CO_2

water, solutes

CIRCULATORY SYSTEM

URINARY SYSTEM

transport to and from cells

elimination of food residues

elimination of excess water, salts, wastes

Figure 42.3 Links between the digestive, respiratory, circulatory, and urinary systems. These organ systems work together to supply cells with raw materials and eliminate wastes.

where food is partly digested and circulated to cells even as residues are sent back out through the pharynx (page 411 and Figure 42.2). The two-way traffic works for flatworms. But it would not be efficient enough for larger, more complex animals. Such animals have a **complete digestive system**—a tube with an opening at one end for food intake and an opening at the other end for eliminating unabsorbed residues. The tube itself is subdivided into specialized regions for the one-way transport, processing, and storage of food. Thus, part of a bird digestive system is modified into a crop, a food storage organ. Another part is modified into a gizzard, a muscular organ that grinds food into smaller bits.

The regional specializations correlate with feeding behavior. For example, the elaborate stomach of ruminants can accept a steady flow of plant material during lengthy feeding times, then slowly liberate nutrients during times of rest. Predators and scavengers have different specializations. When they get food, they gorge on it, then may not eat again for some time. Part of their digestive system stores food being gulped down too fast to be digested and absorbed. Other, accessory parts of the system help maintain an adequate distribution of nutrients between meals.

Bear in mind, a digestive system does not act alone. For most animals, the nutrients absorbed from it are distributed to cells by way of a circulatory system. A respiratory system helps the cells use nutrients by supplying them with oxygen (for aerobic respiration) and taking away carbon dioxide wastes. A urinary system counters variations in the kinds and amounts of absorbed nutrients. It helps maintain the composition and volume of the internal environment (Figure 42.3).

The Human Digestive System

Figure 42.4 shows the complete digestive system of an adult human and lists the functions of its component parts. As is true of similarly complex systems, its parts work in coordination and carry out these overall tasks:

---

1. **Mechanical processing and motility**. Movements that break up, mix, and propel food material.

2. **Secretion**. Release of digestive enzymes and other substances into the lumen (the space inside the tube).

3. **Digestion**. Breakdown of food into particles, then into nutrient molecules small enough to be absorbed.

4. **Absorption**. Passage of digested nutrients and fluid across the tube wall and into blood or lymph.

5. **Elimination**. Expulsion of undigested and unabsorbed residues from the end of the gut.

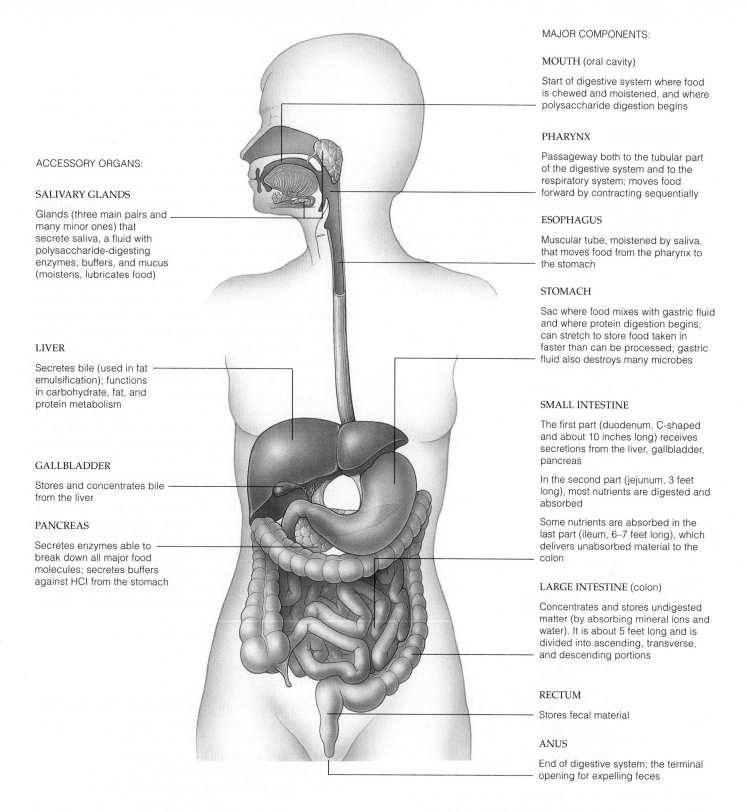

MAJOR COMPONENTS:

MOUTH (oral cavity)

Start of digestive system where food is chewed and moistened, and where polysaccharide digestion begins

PHARYNX

Passageway both to the tubular part of the digestive system and to the respiratory system; moves food forward by contracting sequentially

ESOPHAGUS

Muscular tube, moistened by saliva, that moves food from the pharynx to the stomach

STOMACH

Sac where food mixes with gastric fluid and where protein digestion begins; can stretch to store food taken in faster than can be processed; gastric fluid also destroys many microbes

SMALL INTESTINE

The first part (duodenum, C-shaped and about 10 inches long) receives secretions from the liver, gallbladder, pancreas

In the second part (jejunum, 3 feet long), most nutrients are digested and absorbed

Some nutrients are absorbed in the last part (ileum, 6–7 feet long), which delivers unabsorbed material to the colon

LARGE INTESTINE (colon)

Concentrates and stores undigested matter (by absorbing mineral ions and water). It is about 5 feet long and is divided into ascending, transverse, and descending portions

RECTUM

Stores fecal material

ANUS

End of digestive system; the terminal opening for expelling feces

ACCESSORY ORGANS:

SALIVARY GLANDS

Glands (three main pairs and many minor ones) that secrete saliva, a fluid with polysaccharide-digesting enzymes, buffers, and mucus (moistens, lubricates food)

LIVER

Secretes bile (used in fat emulsification); functions in carbohydrate, fat, and protein metabolism

GALLBLADDER

Stores and concentrates bile from the liver

PANCREAS

Secretes enzymes able to break down all major food molecules; secretes buffers against HCl from the stomach

Stretched out, the human digestive system would extend 6.5 to 9 meters (21 to 30 feet). Its main regions are the mouth, pharynx, esophagus, and gastrointestinal tract (the gut). Mucous epithelium lines the entire surface exposed to the lumen. The gut is subdivided into a stomach, small intestine, large intestine (colon), rectum, and anus. Accessory organs (salivary glands, gallbladder, liver, and pancreas) secrete substances into different parts of the system.

Figure 42.4 Major components of the human digestive system and their functions. Organs with accessory roles in digestion are also listed.

This is a complete digestive system—basically, a tube with openings at both ends and modified regions in between. Overall, material moves in one direction through it. From beginning to end, a thin, mucus-coated epithelium lines all surfaces that are exposed to the lumen (the space inside the tube). The mucus is thick enough to protect the tubular wall and moist enough to enhance diffusion across the lining.

42.2 FOOD PREPARATION AND STORAGE

Into the Mouth, Down the Tube

Food is chewed and polysaccharide breakdown begins in the **oral cavity**, or mouth. Inside is a **tongue**, an organ consisting of membrane-covered skeletal muscles that have roles in positioning food in the mouth, swallowing, and speech. Normally, the mouth of an adult human contains thirty-two teeth. As shown in Figure 42.5, each **tooth** has an enamel coat (hardened calcium deposits), dentin (a thick, bonelike layer), and an inner pulp (with nerves and blood vessels). It is an engineer-

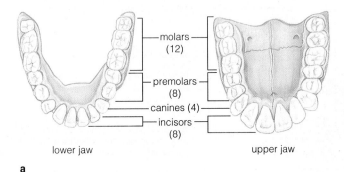

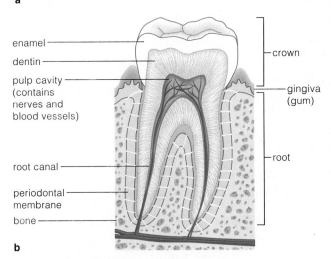

Figure 42.5 (**a**) Number and arrangement of human teeth. (**b**) Closer look at a molar. Its two main regions are called the crown and the root. Enamel caps the crown. It consists of calcium deposits and is the hardest substance in the body.

Normally harmless bacteria live on and between the teeth. Daily flossing, gentle brushing, and avoidance of too many sweets help keep their populations in check. In the absence of preventive measures, conditions favor bacterial infections. These may result in tooth decay (*caries*), inflamed gums (*gingivitis*), or both. Infections also can spread to the periodontal membrane, which anchors teeth to the jawbone. In *periodontal disease*, the infection slowly destroys the bone tissue around a tooth.

ing marvel, able to withstand years of chemical insults and mechanical stress. Recall that chisel-shaped incisors shear off chunks of food (page 464). Cone-shaped canines tear it. The premolars and molars, with their broad crowns and rounded cusps, are good at grinding and crushing food.

Chewing mixes food with fluid called saliva. **Saliva** contains an enzyme (salivary amylase), a buffer (bicarbonate, or HCO_3^-), mucins, and water. Salivary glands, located beneath and in back of the tongue, produce and secrete saliva through ducts that lead to the free surface of the mouth's lining. They secrete about 1,000 to 1,500 milliliters of saliva every day. Salivary amylase in this fluid breaks down starch. The bicarbonate helps maintain the mouth's pH when you eat acidic foods such as tomatoes. Mucins are modified proteins. They help form the mucus that binds food into a softened, lubricated ball, or bolus.

When you swallow, contractions push the tongue against the soft palate, the posterior roof of the mouth. When the soft palate moves upward, it blocks the passage of food and liquids into your nose. Contractions of tongue muscles also force softened balls of food down into your **pharynx**. This is the entrance for the tubular part of the digestive tract.

The pharynx also is the entrance to the trachea, the airway leading to the lungs. A flaplike valve, the epiglottis, closes off the airway and keeps you from breathing as food is moving into the esophagus. That is why you normally don't choke on food (page 701).

Contractions in the pharynx wall propel the food into the **esophagus**, the tubular route to the stomach. Contractions continue and propel food onward, past a sphincter and into the stomach. A **sphincter** is a circular array of muscles, the contractions of which can close off a passageway or an opening to the body surface.

The Stomach

The **stomach**, a muscular, stretchable sac, has three main functions. First, it stores and mixes food. Second, its secretions help dissolve and degrade food. Third, it helps control passage of food into the small intestine. Figure 42.6 compares the structure of the human stomach with the elaborate stomach of an antelope.

Stomach Acidity   Glandular epithelium lines the stomach wall that is exposed to the lumen. Each day, glandular cells in the lining secrete about two liters of hydrochloric acid (HCl), pepsinogens, mucus, and other substances. The substances make up the **gastric fluid**—that is, the stomach's own fluid. Stomach acidity increases when the HCl separates into H^+ and Cl^-, and this helps dissolve bits of food to form a liquid mixture called **chyme**. The acidity kills many microorganisms

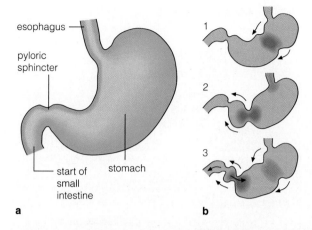

esophagus

pyloric sphincter

start of small intestine

stomach

a

1

2

3

b

Figure 42.6 (**a**) Human stomach, side view. (**b**) Peristaltic wave down the stomach, produced by alternating contractions and relaxation of muscles in the stomach wall.

(**c**) Multiple-chambered stomach of antelopes and other ruminants. The first chamber is a large pouch. The second chamber is smaller and has a honeycombed inner surface (this is what chefs call tripe). In both of these, food is mixed with fluid, kneaded, and exposed to fermentation activities of endosymbiotic bacteria and protozoans. (Some of the symbionts break down cellulose. Others synthesize organic compounds, fatty acids, and vitamins. The host animal also uses some molecules of these substances.) The kneaded food is regurgitated into the mouth, chewed further, then swallowed again. Now it enters the third stomach chamber, where it is pummeled once more before entering the final chamber, the true stomach.

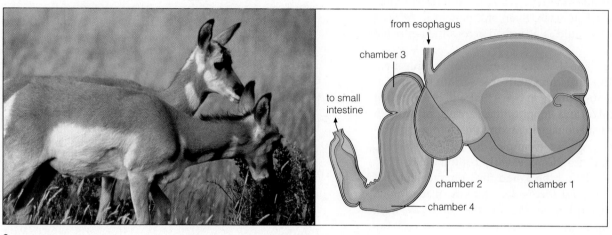

from esophagus

chamber 3

to small intestine

chamber 2

chamber 4

chamber 1

c

present in food. It may cause *heartburn* when gastric fluid from the stomach backs up into the esophagus.

In response to the sight, aroma, and taste of food, the brain signals the secretory cells to step up activity. Also, as food enters the stomach, it distends the wall and activates sensory receptors in the lining. This also leads to increased secretory activity.

Protein digestion starts in the stomach. The high acidity resulting from HCl secretions alters protein structure and exposes peptide bonds. It also converts pepsinogens to active enzymes (pepsins) that break the peptide bonds. Protein fragments accumulate and trigger secretion of gastrin, a hormone that stimulates the pepsinogen- and HCl-secreting cells even more.

A *peptic ulcer* is an eroded wall region in the stomach or small intestine. It results from insufficient secretion of mucus and buffers (or too much pepsin). Some people are genetically predisposed to the disorder. Chronic stress, smoking, and excessive intake of alcohol or aspirin may contribute to it. Nearly everyone with intestinal ulcers and 70 percent of those with stomach ulcers also have a bacterial infection. However, not everyone who harbors the infectious agent (*Helicobacter pylori*) develops peptic ulcers.

Stomach Emptying In the stomach, waves of contractions mix the chyme (Figure 42.6*b*). The waves build up force as they approach the pyloric sphincter at the start of the small intestine. The arrival of a strong contraction closes the sphincter, so most of the chyme is squeezed back. But a small amount moves into the small intestine. In time, most of the chyme moves out of the stomach; only alcohol and a few other substances are absorbed across the stomach lining.

Chyme's volume and composition affect how fast the stomach empties. For example, large meals activate more sensory receptors in the stomach wall, these call for more forceful contraction, and emptying speeds up. As another example, increased acidity (or fat content) in the small intestine stimulates hormone secretions that cause a slowdown in stomach emptying—so food is not moved along faster than it can be processed. Fear, depression, and other emotional upsets also trigger slowdowns.

42.3 DIGESTION IN THE SMALL INTESTINE

Table 42.1 lists the sites where the carbohydrates, lipids (including fats), proteins, and nucleic acids in chyme are digested by specific enzymes, then absorbed into the internal environment. As you can see, digestion is completed and most nutrients are absorbed in the small intestine.

The three regions of the small intestine are the duodenum, jejunum, and ileum (Figure 42.4). Ducts leading from the **pancreas**, **liver**, and **gallbladder** to the duodenum deliver secretions that assist in digestion. Each day, the small intestine receives approximately 9 liters of fluid from the stomach, liver, and pancreas. At least 95 percent of the fluid is absorbed across the epithelial lining.

Figures 42.7 and 42.8 are diagrams of the intestinal wall. Distinctive structures at its free surface function in digestion and absorption, in ways that will be described shortly. Circular and longitudinal arrays of smooth muscle in the wall engage in mixing the chyme and propelling it forward. For example, during *segmentation*, rings of circular muscle repeatedly contract and relax, creating an oscillating (back and forth) movement that constantly mixes and forces the contents of the lumen against the absorptive surface of the intestinal wall. During *peristalsis*, muscles produce waves of contraction that force the chyme onward. (This same type of movement also propels food forward in the esophagus.)

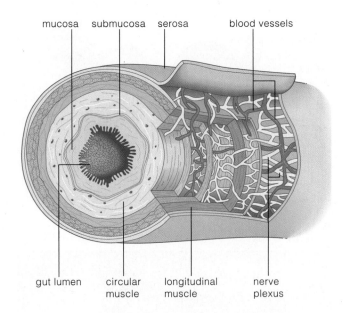

Figure 42.7 Wall structure of the small intestine. Facing the lumen is the mucosa (an epithelium and underlying layer of connective tissue). Surrounding this is the submucosa, a connective tissue layer with blood and lymph vessels and nerve plexuses (networks of neurons that provide local control). The wall contains two sublayers of smooth muscle, one circular and the other longitudinal in orientation. The serosa is an outer layer of connective tissue.

Table 42.1	Major Enzymes of Digestion			
Enzyme	Source	Where Active	Substrate	Main Breakdown Products*
Carbohydrate Digestion:				
Salivary amylase	Salivary glands	Mouth	Polysaccharides	Disaccharides
Pancreatic amylase	Pancreas	Small intestine	Polysaccharides	Disaccharides
Disaccharidases	Intestinal lining	Small intestine	Disaccharides	Monosaccharides (e.g., glucose)
Protein Digestion:				
Pepsins	Stomach lining	Stomach	Proteins	Protein fragments
Trypsin and chymotrypsin	Pancreas	Small intestine	Proteins	Protein fragments
Carboxypeptidase	Pancreas	Small intestine	Protein fragments	Amino acids
Aminopeptidase	Intestinal lining	Small intestine	Protein fragments	Amino acids
Fat Digestion:				
Lipase	Pancreas	Small intestine	Triglycerides	Free fatty acids, monoglycerides
Nucleic Acid Digestion:				
Pancreatic nucleases	Pancreas	Small intestine	DNA, RNA	Nucleotides
Intestinal nucleases	Intestinal lining	Small intestine	Nucleotides	Nucleotide bases, monosaccharides

*Yellow parts of table identify breakdown products that can be absorbed into the internal environment.

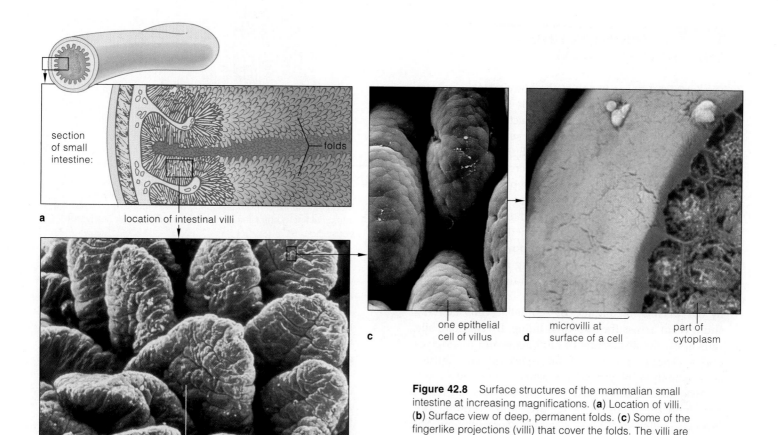

a section of small intestine:

folds

location of intestinal villi

b one villus

c one epithelial cell of villus

d microvilli at surface of a cell

part of cytoplasm

Figure 42.8 Surface structures of the mammalian small intestine at increasing magnifications. (**a**) Location of villi. (**b**) Surface view of deep, permanent folds. (**c**) Some of the fingerlike projections (villi) that cover the folds. The villi are so dense and numerous, they give the surface a velvety appearance. Individual epithelial cells are visible. (**d**) A dense crown of microvilli at the surface of a single cell.

Pancreatic Secretions

Pancreatic enzymes are essential for digestion. One type or another breaks down carbohydrates, proteins, nucleic acids, or fats. For example, like pepsin in the stomach, two of these enzymes (trypsin and chymotrypsin) digest protein molecules into peptide fragments. Another, carboxypeptidase, is one of the enzymes that cleave the fragments to free amino acids.

The pancreas also secretes bicarbonate, a buffer that helps neutralize HCl arriving from the stomach. You may have heard of two other pancreatic secretions, insulin and glucagon, but these have roles in nutrition, not digestion (page 622).

Bile and the Digestion of Fats

Fat digestion depends on more than enzyme action. It depends also on **bile**, which is secreted continually by the liver into the duodenum. Bile contains bile salts, bile pigments, cholesterol, and lecithin, which is a phospholipid. When food is not traveling through the gut, a sphincter closes off the main bile duct from the liver. At such times, bile backs up into the saclike gallbladder, where it is stored and concentrated.

By a process called **emulsification**, bile salts speed up fat digestion. Most fats in the human diet are triglycerides. Triglyceride molecules are insoluble in water, and they tend to aggregate into large fat globules in the chyme. When movements of the intestinal wall agitate chyme, fat globules break up into small droplets, which become coated with bile salts. Because bile salts carry negative charges, the coated droplets repel each other and stay separated. This suspension of fat droplets is the "emulsion."

Compared to fat globules, emulsion droplets give fat-digesting enzymes a much greater surface area to act upon. Thus triglycerides can be broken down much more rapidly to fatty acids and monoglycerides.

Digestion is completed and most nutrients are absorbed in the small intestine.

42.4 ABSORPTION FROM THE SMALL INTESTINE

The intestinal lining is one of the body's marvels. Take another look at Figure 42.8, which shows its dense folds. The folds are absorptive structures called **villi** (singular, villus). Villi increase the surface area available for interactions with chyme. Epithelial cells line these structures, and each cell has a crown of microvilli. Each **microvillus** is a threadlike projection of the plasma membrane. Collectively, they greatly increase the surface area available for absorption.

By the time nutrients are halfway through this elaborate tube, most have been broken down and digested. Some breakdown products, including monosaccharides and amino acids, are absorbed in straightforward fashion. So are water and mineral ions. Transport proteins in the plasma membrane of epithelial cells actively shunt them across the intestinal lining. By contrast, bile salts assist the absorption of fatty acids and monoglycerides. These products of fat digestion can diffuse across the lipid bilayer of the plasma membrane.

By a process called micelle formation, bile salts combine with the products of fat digestion, forming tiny droplets called **micelles**. Product molecules in the micelles continuously exchange places with product molecules that are dissolved in the chyme. When con-

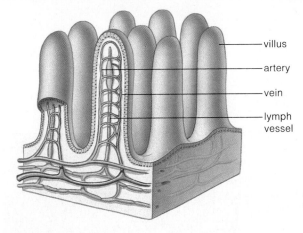

Figure 42.9 Blood and lymph vessels in intestinal villi. Monosaccharides and most amino acids that have crossed the intestinal lining enter the blood vessels. Fats enter the lymph vessels.

centration gradients favor it, molecules diffuse out of micelles, out of the chyme, and into epithelial cells. There, fatty acids and monoglycerides recombine as triglycerides. Then the triglycerides combine with proteins into particles (chylomicrons) that leave the cell by exocytosis and enter the internal environment.

Once absorbed into the internal environment, glucose and amino acids enter blood vessels. The triglycerides enter lymph vessels, which eventually drain into blood vessels (Figure 42.9).

Figure 42.10 summarizes the digestion and absorption processes that proceed in the small intestine.

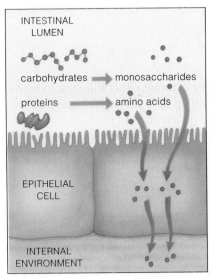

1 Digestion of carbohydrates to monosaccharides, and proteins to amino acids, completed by pancreatic enzymes.

2 Active transport of monosaccharides, amino acids across plasma membrane of cells, then out of cells, into internal environment.

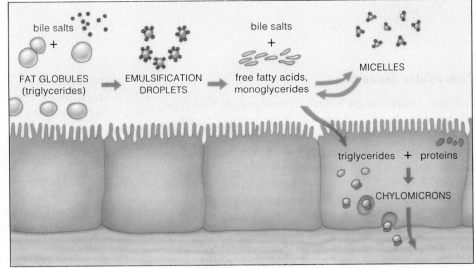

1 Emulsification. Wall movements break up fat globules into small droplets. Bile salts keep globules from re-forming. Pancreatic enzymes digest droplets to fatty acids and monoglycerides.

2 Formation of micelles (as bile salts combine with digested products and phospholipids). Products readily slip into and out of micelles

3 Concentration of monosaccharides, fatty acids in micelles enhances gradients that lead to their diffusion across lipid bilayer of cells' plasma membranes.

4 Reassembly of products into triglycerides inside cells. These join with proteins, then are expelled (by exocytosis) into the internal environment.

Figure 42.10 Summary of the processes of digestion and absorption in the small intestine.

42.5 FUNCTIONS OF THE LARGE INTESTINE

Material *not* absorbed in the small intestine moves into the large intestine, or **colon**. The colon concentrates and stores feces, a mixture of undigested and unabsorbed material, water, and bacteria. The mixture becomes concentrated as water moves across the colon's lining. Cells in the lining actively transport sodium ions out of the lumen. As ion concentrations in the lumen drop, the water concentration increases. Thus water also moves out of the lumen, by osmosis.

The colon starts out as a cup-shaped pouch known as the cecum. The **appendix** is a slender projection from that pouch, as the following sketch indicates:

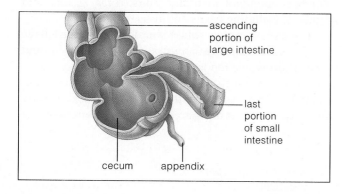

The appendix has no known digestive functions. It may have roles in defense against infections.

The colon ascends on the right side of the abdominal cavity, continues across to the other side, then descends and connects with a short tube, the rectum (Figure 42.4). Distention of the rectal wall triggers a reflex pathway that leads to expulsion of feces from the body. The nervous system controls the expulsion. It stimulates or inhibits contractions of a muscle sphincter at the anus, the terminal opening of the gut.

Bulk refers to a volume of fiber and other undigested material that cannot be decreased by absorption in the colon. As bulk increases, it puts more pressure on the colon wall. Without adequate bulk, it takes longer for fecal material to move through the colon, with irritating and perhaps carcinogenic effects. People of rural Africa and India rarely suffer colon cancer or appendicitis. (*Appendicitis* is an infection of the appendix. If the infected appendix ruptures, bacteria normally living in the colon can enter the abdominal cavity and cause severe problems.) Most people who live in those areas cannot afford to eat much more than whole grains, which are high in fiber. When they move to wealthier nations and leave behind their fiber-rich diet, they are *more* likely to suffer colon cancer and appendicitis.

42.6 CONTROLS OVER THE SYSTEM

Recall that homeostatic controls respond to changes in the internal environment. By contrast, controls over the digestive system act *before* food is absorbed into the internal environment. They respond to the volume and composition of material in the gut lumen. The nervous system, the nerve plexuses in the gut wall, and the endocrine system interact to exert control.

For example, after a meal, food distends the stomach wall. Signals from mechanoreceptors travel on short reflex pathways that are confined to nerve plexuses. They also may travel on long reflex pathways to the central nervous system. Signals along one or both types of pathways can lead to muscle contractions in the gut wall or secretion of enzymes and other substances into the gut lumen. The hypothalamus and other brain regions monitor such activities and coordinate them with events that are taking place elsewhere in the body (Figure 42.11).

Four gastrointestinal hormones are known. Gastrin is secreted by endocrine cells in the stomach's lining when amino acids and peptides are in the stomach. It mainly stimulates the secretion of acid into the stomach. Secretin, a peptide hormone, stimulates the pancreas to secrete bicarbonate (page 614). CCK (cholecystokinin) enhances the actions of secretin and stimulates gallbladder contractions. GIP (glucose insulinotropic peptide) is released in response to the presence of glucose and fat in the small intestine. It stimulates insulin secretion, hence glucose uptake by cells.

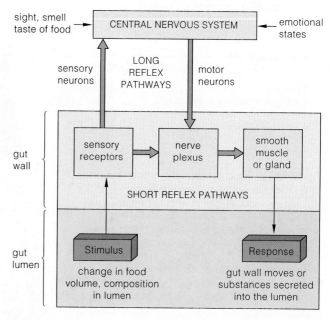

Figure 42.11 Local and long-distance reflex pathways called into action when food is in the digestive tract.

A few million years ago, our hominid ancestors dined mostly on fresh fruits and other fibrous plant material. From this nutritional beginning, humans in many parts of the world have become habituated to poor diets. Consider Americans. On the whole, they are among the world's best-fed people, yet they show a high incidence of digestive disorders. Along with affluence, bad eating habits are rampant. Americans generally skip meals, eat too much and too fast when they sit down at the table, and give the gut erratic workouts. Their diet tends to be loaded with cholesterol, sugar, and salt—and short on bulk. Remember, too little bulk means delayed expulsion of feces, with irritating and possibly harmful results. As epidemiologist Denis Burkitt put it, if you pass a small volume of feces, you have large hospitals.

New, Improved Food Pyramids

From elementary school onward, you probably have encountered "food pyramids," or charts of presumably well-balanced diets. Every so often, the charts are revised to reflect new findings of nutritional studies. Figure 42.12 shows the most recently favored chart.

Suppose you are an adult male of average size. According to Prerna Uppal of the Palo Alto Medical Foundation, your diet should include the following foods in the proportions indicated:

Complex carbohydrates:	58–60 percent
Proteins (less for females):	12–15 percent
Fats and other lipids:	20–25 percent

Carbohydrates

Starch and, to a lesser extent, glycogen should be the main carbohydrates in our diet. These complex carbohydrates are easily broken down to glucose units, which are the body's main energy source (Chapter 6). Starch is abundant in fleshy fruits, cereal grains, and legumes, including beans and peas.

The carbohydrates called simple sugars don't have the fiber of complex carbohydrates or the vitamins and minerals of whole foods. Each week, the average American eats as much as 2 pounds of refined sugar (sucrose). You may think this a far-fetched statement, but start looking at the ingredients listed on packages of

MILK, YOGURT, CHEESE GROUP

Choose 2–3 servings from these alternatives:

1 cup nonfat or lowfat milk
1 cup nonfat or lowfat yogurt
1-1/2 ounces lowfat, unprocessed cheese
2 ounces lowfat, processed cheese

FRUIT GROUP

Choose 2–4 servings from these alternatives:

1 medium-size fruit, such as apple, orange, mango
1 wedge cantaloupe or pineapple
1 cup fresh berries
3/4 cup unsweetened fruit juice
1/2 cup canned unsweetened fruit
1/4 cup dried fruit

ADDED FATS AND SIMPLE SUGARS

Restrict intake (other foods provide ample amounts of both)

Limit salad dressings, butter, cream, margarine, syrups, table sugar, soft drinks, candies, sweet desserts

LEGUME, NUT, POULTRY, FISH, MEAT GROUP

Choose 2–3 servings from these alternatives:

1-1/4 to 1-1/2 cups cooked legumes
2-1/2 to 3 ounces cooked poultry (no skin)
2-1/2 to 3 ounces cooked fish
2-1/2 to 3 ounces trimmed, cooked, lean red meat
1 egg (or two egg whites)
5 tablespoons peanut butter

VEGETABLE GROUP

Choose 3–5 servings from these alternatives:

1/2 cup chopped raw or cooked vegetables
1 cup raw, leafy vegetables

BREAD, CEREAL, RICE, PASTA GROUP

Choose 6–11 servings from these alternatives:

1 slice whole-grain bread
1/2 cup cooked rice (brown rice has more nutrients)
1/2 cup cooked pasta
1/2 cup cooked unsweetened cereal, such as oatmeal
1 ounce dry unsweetened cereal, such as shredded wheat

Figure 42.12 Food pyramid diagram, as revised in 1992. The daily portions of the different foods shown will help give you an idea of what constitutes a well-balanced diet. The alternatives listed are only a general guide to the kinds of foods in each group. Good health requires daily selections from all groups *except* the "red-hot" tip of the pyramid. (Added fats and simple sugars pile on the calories but provide few vitamins and minerals.)

Where there is a range of serving sizes, choosing the smallest one will help keep the total caloric intake to about 1,600 kilocalories. Choosing the largest will raise the total to about 2,800 kilocalories.

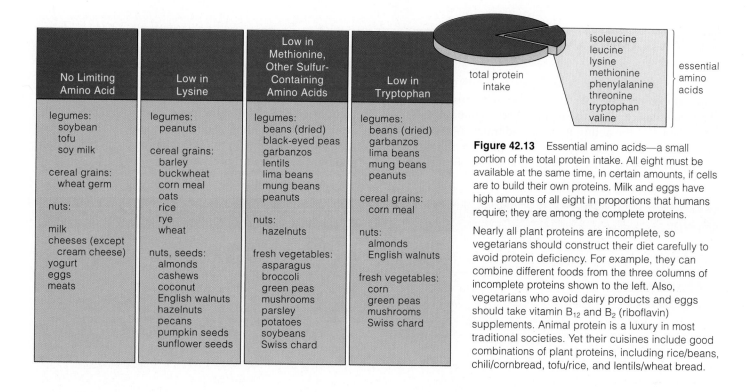

No Limiting Amino Acid	Low in Lysine	Low in Methionine, Other Sulfur-Containing Amino Acids	Low in Tryptophan

legumes:
  soybean
  tofu
  soy milk
 	
cereal grains:
  wheat germ
 	
nuts:
 	
milk
cheeses (except
  cream cheese)
yogurt
eggs
meats
	legumes:
	  peanuts
	
	cereal grains:
	  barley
	  buckwheat
	  corn meal
	  oats
	  rice
	  rye
	  wheat
	
	nuts, seeds:
	  almonds
	  cashews
	  coconut
	  English walnuts
	  hazelnuts
	  pecans
	  pumpkin seeds
	  sunflower seeds
		legumes:
		  beans (dried)
		  black-eyed peas
		  garbanzos
		  lentils
		  lima beans
		  mung beans
		  peanuts
		
		nuts:
		  hazelnuts
		
		fresh vegetables:
		  asparagus
		  broccoli
		  green peas
		  mushrooms
		  parsley
		  potatoes
		  soybeans
		  Swiss chard
			legumes:
			  beans (dried)
			  garbanzos
			  lima beans
			  mung beans
			  peanuts
			
			cereal grains:
			  corn meal
			
			nuts:
			  almonds
			  English walnuts
			
			fresh vegetables:
			  corn
			  green peas
			  mushrooms
			  Swiss chard

total protein intake

essential amino acids

isoleucine
leucine
lysine
methionine
phenylalanine
threonine
tryptophan
valine

Figure 42.13 Essential amino acids—a small portion of the total protein intake. All eight must be available at the same time, in certain amounts, if cells are to build their own proteins. Milk and eggs have high amounts of all eight in proportions that humans require; they are among the complete proteins.

Nearly all plant proteins are incomplete, so vegetarians should construct their diet carefully to avoid protein deficiency. For example, they can combine different foods from the three columns of incomplete proteins shown to the left. Also, vegetarians who avoid dairy products and eggs should take vitamin B_{12} and B_2 (riboflavin) supplements. Animal protein is a luxury in most traditional societies. Yet their cuisines include good combinations of plant proteins, including rice/beans, chili/cornbread, tofu/rice, and lentils/wheat bread.

cereal, frozen dinners, soft drinks, and other prepared foods. Many sugars are "hidden" as corn syrup, corn sweeteners, dextrose, and so on.

Lipids

The body can't function without fats and other lipids. For instance, the phospholipid lecithin is a required component of cell membranes. Fats serve as energy reserves, they cushion many organs, and they provide insulation beneath the skin. Dietary fats also help the body store fat-soluble vitamins. Yet the body can synthesize most of its own fats, including cholesterol, from protein and carbohydrates. The ones it cannot produce are called **essential fatty acids**, but whole foods provide plenty of them. Linoleic acid is an example. One teaspoon a day of corn oil, olive oil, or some other polyunsaturated fat in food provides enough of it.

Today, butter and other fats make up 40 percent of the average diet in the United States. Most of the medical community agrees the proportion should be less than 30 percent. Like other animal fats, butter is a saturated fat that tends to raise the level of cholesterol in the blood. Cholesterol is used in the synthesis of bile acids and steroid hormones, and it is a required component of our cell membranes. However, too much cholesterol may damage the circulatory system (page 667). Work is under way to develop edible but nondigestible oils, such as sucrose polyester, for those who have difficulty cutting back on the fats.

Proteins

The amino acid components of dietary proteins are required for the body's own protein-building programs. Of twenty common types, eight are **essential amino acids**. Our cells cannot synthesize them, so we must obtain them from food. The eight are methionine (or an equivalent called cysteine), isoleucine, leucine, lysine, phenylalanine (or tyrosine), threonine, tryptophan, and valine.

Most animal proteins are "complete," meaning their ratios of amino acids match human nutritional needs. Plant proteins are "incomplete," meaning they lack one or more of the essential amino acids. To get enough protein, vegetarians must eat certain combinations of different plants (Figure 42.13).

A measure called **net protein utilization** (NPU) is used to compare proteins from different sources. NPU values range from 100 (all essential amino acids present in ideal proportions) to 0 (one or more absent; when eaten alone, the protein is not complete).

Because enzymes and other proteins are vital for the body's structure and function, protein-deficient diets may have severe consequences. Protein deficiency is most damaging among the young, for the brain grows and develops rapidly early in life. Unless enough protein is taken in just before and just after birth, irreversible mental retardation occurs. Even mild protein starvation can retard growth and affect mental and physical performance.

42.8 VITAMINS AND MINERALS

Metabolic activity depends on small amounts of more than a dozen organic substances called **vitamins**. Most plants synthesize all of these substances. Animals generally have lost the ability to do so and must obtain vitamins from food. Human cells need at least thirteen different vitamins, each with specific metabolic roles.

Many metabolic activities require several vitamins, so the absence of one affects the functions of others. Metabolic activities also require certain inorganic substances called **minerals**. For example, most of your cells use calcium and magnesium in many reactions. All cells require iron in electron transport chains. Red blood cells can't function without iron in hemoglobin, the oxygen-carrying pigment in blood. Neurons can't function without sodium and potassium.

Table 42.2 Vitamins: Sources, Functions, and Effects of Deficiencies or Excesses*

Vitamin	Common Sources	Main Functions	Signs of Severe Long-Term Deficiency	Signs of Extreme Excess
Fat-Soluble Vitamins:				
A	Its precursor comes from beta-carotene in yellow fruits, yellow or green leafy vegetables; also in fortified milk, egg yolk, fish liver	Used in synthesis of visual pigments, bone, teeth; maintains epithelia	Dry, scaly skin; lowered resistance to infections; night blindness; permanent blindness	Malformed fetuses; hair loss; changes in skin; liver and bone damage; bone pain
D	D_3 formed in skin and in fish liver oils, egg yolk, fortified milk; converted to active form elsewhere	Promotes bone growth and mineralization; enhances calcium absorption	Bone deformities (rickets) in children; bone softening in adults	Retarded growth; kidney damage; calcium deposits in soft tissues
E	Whole grains, dark green vegetables, vegetable oils	Possibly inhibits effects of free radicals; helps maintain cell membranes; blocks breakdown of vitamins A and C in gut	Lysis of red blood cells; nerve damage	Muscle weakness, fatigue, headaches, nausea
K	Colon bacteria form most of it; also in green leafy vegetables, cabbage	Blood clotting; ATP formation via electron transport	Abnormal blood clotting; severe bleeding (hemorrhaging)	Anemia; liver damage and jaundice
Water-Soluble Vitamins:				
B_1 (thiamin)	Whole grains, green leafy vegetables, legumes, lean meats, eggs	Connective tissue formation; folate utilization; coenzyme action	Water retention in tissues; tingling sensations; heart changes; poor coordination	None reported from food; possible shock reaction from repeated injections
B_2 (riboflavin)	Whole grains, poultry, fish, egg white, milk	Coenzyme action	Skin lesions	None reported
Niacin	Green leafy vegetables, potatoes, peanuts, poultry, fish, pork, beef	Coenzyme action	Contributes to pellagra (damage to skin, gut, nervous system, etc.)	Skin flushing; possible liver damage
B_6	Spinach, tomatoes, potatoes, meats	Coenzyme in amino acid metabolism	Skin, muscle, and nerve damage; anemia	Impaired coordination; numbness in feet
Pantothenic acid	In many foods (meats, yeast, egg yolk especially)	Coenzyme in glucose metabolism, fatty acid and steroid synthesis	Fatigue, tingling in hands, headaches, nausea	None reported; may cause diarrhea occasionally
Folate (folic acid)	Dark green vegetables, whole grains, yeast, lean meats; colon bacteria produce some folate	Coenzyme in nucleic acid and amino acid metabolism	A type of anemia; inflamed tongue; diarrhea; impaired growth; mental disorders	Masks vitamin B_{12} deficiency
B_{12}	Poultry, fish, red meat, dairy foods (not butter)	Coenzyme in nucleic acid metabolism	A type of anemia; impaired nerve function	None reported
Biotin	Legumes, egg yolk; colon bacteria produce some	Coenzyme in fat, glycogen formation and in amino acid metabolism	Scaly skin (dermatitis), sore tongue, depression, anemia	None reported
C (ascorbic acid)	Fruits and vegetables, especially citrus, berries, cantaloupe, cabbage, broccoli, green pepper	Collagen synthesis; possibly inhibits effects of free radicals; structural role in bone, cartilage, and teeth; role in carbohydrate metabolism	Scurvy, poor wound healing, impaired immunity	Diarrhea, other digestive upsets; may alter results of some diagnostic tests

*The guidelines for appropriate daily intakes are being worked out by the Food and Drug Administration.

People who are in good health can get all the vitamins and minerals they need from a balanced diet of whole foods. Generally, specific vitamin and mineral supplements are necessary only for strict vegetarians, the elderly, and individuals suffering from a chronic illness or taking medication that affects the body's use of specific nutrients. For example, in one comprehensive study, vitamin K supplements helped older women retain calcium and so diminish bone loss (osteoporosis) by 30 percent. Preliminary research suggests two vitamins (C and E) and beta-carotene (the precursor for vitamin A) may inactivate free radicals—those rogue metabolic fragments that may contribute to aging and may impair immune function (page 92).

No one should take massive doses of any vitamin or mineral supplement except under medical supervision.

As Tables 42.2 and 42.3 indicate, excess amounts of many vitamins and minerals are harmful. For example, large doses of at least two vitamins (A and D) can actually damage the body. Like all fat-soluble vitamins, excess amounts can accumulate in tissues and interfere with normal metabolic function. Similarly, sodium is present in plant and animal tissues, and it is a component of table salt. Sodium has roles in the body's salt–water balance, muscle activity, and nerve function. Yet prolonged, excessive intake of sodium may contribute to high blood pressure.

Severe shortages or self-prescribed, massive excesses of vitamins and minerals can disturb the delicate balances in body function that promote health.

Table 42.3	Major Minerals: Sources, Functions, and Effects of Deficiencies or Excesses*			
Mineral	Common Sources	Main Functions	Signs of Severe Long-Term Deficiency	Signs of Extreme Excess
Calcium	Dairy products, dark green vegetables, dried legumes	Bone, tooth formation; blood clotting; neural and muscle action	Stunted growth; possibly diminished bone mass (osteoporosis)	Impaired absorption of other minerals; kidney stones in susceptible people
Chloride	Table salt (usually too much in diet)	HCl formation in stomach; contributes to body's acid-base balance; neural action	Muscle cramps; impaired growth; poor appetite	Contributes to high blood pressure in susceptible people
Copper	Nuts, legumes, seafood, drinking water	Used in synthesis of melanin, hemoglobin, and some transport chain components	Anemia, changes in bone and blood vessels	Nausea, liver damage
Fluorine	Fluoridated water, tea, seafood	Bone, tooth maintenance	Tooth decay	Digestive upsets; mottled teeth and deformed skeleton in chronic cases
Iodine	Marine fish, shellfish, iodized salt, dairy products	Thyroid hormone formation	Enlarged thyroid (goiter), with metabolic disorders	Goiter
Iron	Whole grains, green leafy vegetables, legumes, nuts, eggs, lean meat, molasses, dried fruit, shellfish	Formation of hemoglobin and cytochrome (transport chain component)	Iron-deficiency anemia, impaired immune function	Liver damage, shock, heart failure
Magnesium	Whole grains, legumes, nuts, dairy products	Coenzyme role in ATP-ADP cycle; roles in muscle, nerve function	Weak, sore muscles; impaired neural function	Impaired neural function
Phosphorus	Whole grains, poultry, red meat	Component of bone, teeth, nucleic acids, ATP, phospholipids	Muscular weakness; loss of minerals from bone	Impaired absorption of minerals into bone
Potassium	Diet provides ample amounts	Muscle and neural function; roles in protein synthesis and body's acid-base balance	Muscular weakness	Muscular weakness, paralysis, heart failure
Sodium	Table salt; diet provides ample to excessive amounts	Key role in body's acid-base balance; roles in muscle and neural function	Muscle cramps	High blood pressure in susceptible people
Sulfur	Proteins in diet	Component of body proteins	None reported	None likely
Zinc	Whole grains, legumes, nuts, meats, seafood	Component of digestive enzymes; roles in normal growth, wound healing, sperm formation, and taste and smell	Impaired growth, scaly skin, impaired immune function	Nausea, vomiting, diarrhea; impaired immune function and anemia

*The guidelines for appropriate daily intakes are being worked out by the Food and Drug Administration.

42.9 ENERGY AND BODY WEIGHT

The body grows and maintains itself when it receives energy and materials from foods of certain types, in certain amounts. Nutritionists measure energy in units called **kilocalories**. Each unit is 1,000 calories of heat energy (that is, the amount needed to raise the temperature of 1 kilogram of water by 1°C).

To maintain an acceptable weight and keep the body functioning normally, caloric intake must be balanced with energy output. The output varies from one person to another because of differences in physical activity, basic rate of metabolism, age, sex, hormone activity, and emotional state. Some of these factors are influenced by a person's social environment. Others have a genetic basis. For instance, researchers have studied many identical twins (with identical genes) who, owing to family problems, were separated at birth and raised apart, in different households. At adulthood, the body weights of the separated twins were still similar.

When caloric input balances the output, body weight doesn't change dramatically over long periods. As any dieter knows, the body behaves as if it has a set point for what that weight is going to be—and it works to counteract deviations from its set point.

How many kilocalories should you take in each day to maintain "acceptable" body weight? First, multiply the desired weight (in pounds) by 10 if you are not very active physically, by 15 if you are moderately active, and by 20 if you are quite active. Then, depending on your age, subtract the following amount from the value obtained from the first step:

Age:	25–34	*Subtract*:	0
	35–44		100
	45–54		200
	55–64		300
	Over 65		400

For example, if you want to weigh 120 pounds and are very active, 120 × 20 = 2,400 kilocalories. If you are thirty-five years old, you would take in (2,400 − 100) or 2,300 kilocalories a day. Such calculations provide a rough estimate of caloric intake. Other factors, such as height, must be considered also. An active person who is 5 feet, 2 inches tall doesn't need as much energy as an active person who weighs the same but is 6 feet tall.

By definition, **obesity** is an excess of fat in the body's adipose tissues, most often caused by imbalances between caloric intake and energy output. Yet what is too fat or too thin? What is an "ideal weight"? Insurance companies have developed many charts (Figure 42.14), mostly to identify overweight people who

Woman's Height	Size of Frame		
	Small	Medium	Large
4'10"	102–111	109–121	118–131
4'11"	103–113	111–123	120–134
5' 0"	104–115	113–126	122–137
5' 1"	106–118	115–129	125–140
5' 2"	108–121	118–132	128–143
5' 3"	111–124	121–135	131–147
5' 4"	114–127	124–138	134–151
5' 5"	117–130	127–141	137–155
5' 6"	120–133	130–144	140–159
5' 7"	123–136	133–147	143–163
5' 8"	126–139	136–150	146–167
5' 9"	129–142	139–153	149–170
5'10"	132–145	142–158	152–173
5'11"	135–148	145–159	155–176
6' 0"	138–151	148–162	158–179

Man's Height	Size of Frame		
	Small	Medium	Large
5' 2"	128–134	131–141	138–150
5' 3"	130–136	133–143	140–153
5' 4"	132–138	135–145	142–156
5' 5"	134–140	137–148	144–160
5' 6"	136–142	139–151	146–164
5' 7"	138–145	142–154	149–168
5' 8"	140–148	145–157	152–172
5' 9"	142–151	148–160	155–176
5'10"	144–154	151–163	158–180
5'11"	146–157	154–166	161–184
6' 0"	149–160	157–170	164–188
6' 1"	152–164	160–174	168–192
6' 2"	155–168	164–178	172–197
6' 3"	158–172	167–182	176–202
6' 4"	162–176	171–187	181–207

Figure 42.14 "Ideal" weights for adults according to one insurance company in 1983. Values shown are for people twenty-five to fifty-nine years old wearing shoes with 1-inch heels and 3 pounds of clothing (for women) or 5 pounds (for men). Extreme obesity puts severe strain on the circulatory system. The body produces many more blood capillaries to service the increased tissue masses. The heart becomes more stressed. It must pump harder to keep blood circulating.

might be insurance risks. Such charts factor in height. People who are 25 percent heavier than "ideal" are viewed as obese.

Some researchers who study causes of death suspect that the "ideal" may be ten to fifteen pounds heavier than the charts indicate. Some dietitians are convinced the chart values should be less. Whatever the ideal range may be, serious disorders do arise with extremes at either end of that range (see the *Focus* essay).

To maintain an acceptable body weight, energy input (caloric intake) must be balanced with energy output (as through exercise and loss of metabolically generated heat).

Focus on Health

When Does Dieting Become a Disorder?

By current standards, how much of our total tissue mass should be fat? The proportion of fat should be no more than 18 to 24 percent for a female who is less than thirty years old. For a male, it should be no more than 12 to 18 percent. Yet an estimated 34 million Americans are over-weight to the point of obesity (Figure *a*). Many sincerely attempt to shed the excess fat by dieting. Yet most eventually regain what they lost—and sometimes put on more.

Dieting and Exercise Why is it so difficult to lose weight permanently? The difficulty is partly a result of our evolutionary heritage. The fat-storing cells of adipose tissue are an adaptation for survival. Collectively, they represent an energy reserve that may carry an animal through times when food just isn't available. Remember this when you start to put on unwanted fat. Once a fat-storing cell is added to your body, food intake may affect how empty or full it gets—but apparently that cell is in your body to stay. Dieting *can* decrease the fat stores. But research now suggests that the body interprets dieting as "starvation."

Dieting triggers changes in metabolism. With these changes, food is used more conservatively—so that *fewer* calories are burned. Meanwhile, the dieter's appetite surges, and "starved" fat cells quickly refill when a diet ends. In *yo-yo dieting*, a person repeatedly gains and loses weight. This may alter cell metabolism in ways that make it *more* difficult to shed weight with each new round of dieting. Besides this, frequent changes in body weight may also increase the risk of heart disorders.

Some people opt to shed pounds by exercising, only to discover the going is slow indeed. Losing just a single pound of fat requires an energy expenditure of about 3,500 kilocalories. You can do this by, say, jogging for four hours or playing tennis for nearly eight hours straight. A more feasible way to keep off the fat is to combine a *moderate* reduction in caloric intake with increased physical activity. This approach might minimize the "starvation" response and increase the rate at which the body uses energy. Exercise also increases muscle mass. Even a resting muscle burns more calories than other types of tissues. Once you lose fat, the "starvation" response kicks in. To keep off unwanted weight, you have to continue to eat in moderation and

a

maintain a program of regular exercise.

Anorexia Nervosa
When dieting becomes obsessive, one result can be a potentially fatal eating disorder called *anorexia nervosa* (Figure *b*). The disorder is most common among women in their teens and early twenties.

People with anorexia nervosa have an abnormal perception of their body weight. Their fear of being fat and hungry becomes overwhelming. They start starving themselves and often overexercise. Emotional factors contribute to the disorder. Some anorexic people fear growing up in general and maturing sexually in particular. Others have irrational measures of what to expect of themselves and of what they might accomplish.

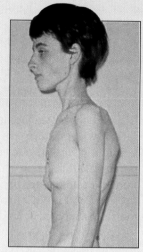
b

Bulimia At least 20 percent of college-age women now suffer to varying degrees from *bulimia* ("an oxlike appetite"). Outwardly they may look healthy, but their food intake is out of control. During an hour-long eating binge, a bulimic may take in more than 50,000 kilocalories—then vomit or use laxatives to purge the body. The binge-purge routine may occur once a month; it may occur several times a day. Repeated purgings can damage the gut. Repeated vomiting, which brings gastric fluid into the mouth, can erode teeth to stubs. Extreme bulimics can die of heart failure, stomach rupturing, or kidney failure.

Some women start the routine because it seems like a simple way to lose weight. Others have emotional problems. Often bulimics are well-educated and accomplished, but they strive for perfection and may have problems with family members who exert control over them. Eating may actually be unpleasant for bulimics, but the purging (which they themselves control) relieves them of anger and frustration.

Treatment for both anorexia nervosa and bulimia usually includes long-term psychotherapy and diet counseling. Severe cases may require hospitalization.

42.10 NUTRITION AND ORGANIC METABOLISM

Earlier, you considered a few conversion pathways by which carbohydrates, fats, and proteins can be broken apart (page 134). Figure 37.9 provided an example of mechanisms that govern organic metabolism—specifically, the disposition of glucose in the body. Here, Figure 42.15 rounds out the picture by showing all of the major routes by which organic compounds are shuffled and reshuffled in the body as a whole.

At any time, the body is breaking down most of its own carbohydrates, lipids, and proteins, then using the breakdown products as energy sources or building blocks. The nervous and endocrine systems work together to integrate this massive molecular turnover. A discussion of mechanisms controlling this turnover would be beyond the scope of this book. However, the *Focus* essay will give you a sense of their splendid intricacy.

For now, simply consider a few key points about organic metabolism. When you eat, your body builds up pools of organic compounds. Excess carbohydrates and other molecules provided by your diet are transformed mostly into fats, which are stored in adipose tissue. Some are converted to glycogen in the liver and in muscle tissue. When organic compounds are being absorbed and stored, most cells use glucose as the main energy source. There is no net breakdown of protein in muscle or other tissues during this period.

Between meals, the body taps into the fat stores. Fats are broken down to glycerol and fatty acids, which are released into blood. Glycerol is converted to glucose in the liver. Cells can take up the circulating fatty acids and use them for ATP production.

Bear in mind, the liver does more than store and interconvert organic compounds. For example, it helps maintain their concentrations in blood. It inactivates most hormone molecules, which are sent to the kidneys for excretion, in urine. It removes worn-out blood cells and inactivates many toxic compounds from blood. Ammonia (NH_3), produced during amino acid breakdown, also can be toxic in high concentrations. The liver converts it to urea, a less toxic waste product that leaves the body by way of the kidneys, in urine.

During a meal, glucose moves into cells, where it can be used for energy and where the excess can be stored.

Between meals, the body metabolizes fat molecules as the main energy source.

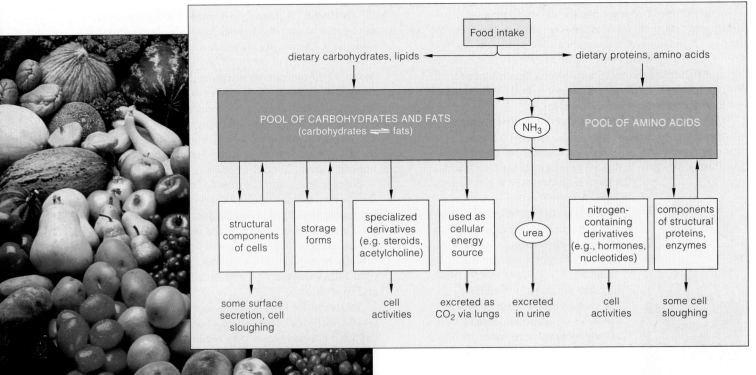

Figure 42.15 Summary of major pathways of organic metabolism. Carbohydrates, fats, and proteins are continually broken down and synthesized. Urea forms mainly in the liver. Compare Figure 8.12.

Focus on Health

Case Study: Feasting, Fasting, and Systems Integration

At any time you care to think about it, you depend absolutely on controls that are integrating organic metabolism with everything else going on in your life. Suppose, this morning, you are vacationing in the mountains and decide on impulse to follow a forested trail. You fail to notice a wooden trail marker with the intriguing name "Fat Man's Misery." Having eaten a large breakfast, you have assured your cells of ongoing nourishment; you need not forage constantly amongst the ferns as, say, a round-worm must do. Food partly digested in the stomach has already entered the small intestine. Right now, amino acids, simple sugars, and fatty acids are moving across the intestinal wall, then into the bloodstream.

Now glucose is surging into the bloodstream faster than your cells can use it, and its blood level rises a bit. No problem. Pancreatic beta cells secrete insulin, and the hormonal targets (liver, fat, and muscle cells) quickly begin using or storing glucose. Alpha cells don't secrete as much glucagon (which slows the liver's tapping of glycogen stores).

And so, as you walk along, excess glucose starts to leave the bloodstream. Even so, your brain cells still demand a lot of glucose. So do muscle cells, which are getting a strenuous workout. Now endocrine control shifts in the pancreas. With less glucose binding to them, beta cells secrete less insulin. With less glucose to inhibit them, alpha cells secrete more glucagon. In the liver, glucagon stimulates conversion of glycogen back to glucose—which enters the blood and travels to cells.

Yet the best-laid balance of internal conditions can go astray when external conditions change. The "miserable" part of the trail has begun! You scramble higher and higher on steep inclines. Suddenly you stop, surprised, in pain. You forgot to reckon with the lower oxygen pressure of mountain air, and your leg muscles cramped. Your body has already detected a deficiency of oxygen-carrying red blood cells at this altitude, but it will take days before more red blood cells are available. Meanwhile, your muscle cells are not getting enough oxygen. They are relying on a less efficient anaerobic pathway—lactate fermentation.

Again, controls work to restore conditions. Receptors detect the reduced oxygen pressure and an increase in H^+ concentrations in cerebrospinal fluid. Signals race toward respiratory centers in the medulla oblongata, then to your diaphragm and other muscles associated with breathing. You breathe faster now, and more deeply. Meanwhile, back in the liver, lactate is converted to glucose—which is returned to the blood.

On checking the sun's position, you see it is well past noon. And guess what: you didn't pack a lunch. When you start the long walk back, the drop in blood glucose trips more control mechanisms into action.

Under hypothalamic commands, the adrenal medulla secretes epinephrine and norepinephrine. The main targets are the liver, adipose tissue, and muscles. In the liver, glycogen synthesis stops. In body tissues generally, glucose uptake is blocked. In adipose tissue, fats undergo conversion. Now fatty acids become energy sources in the liver, muscles, and other tissues. For every fatty acid molecule used, several glucose molecules are held in reserve for the brain.

You do get back to the start of the trail by sundown. Your body had enough stored energy to sustain you for many more days, so the situation was never really desperate. Even after several days of fasting, your energy supplies would not have run out. Another command from the hypothalamus would have prodded your anterior pituitary into secreting ACTH. The ACTH would have signaled cells of the adrenal cortex to secrete glucocorticoid hormones, which have a potent effect on the synthesis of carbohydrates from proteins and on the further breakdown of fat. Slowly, in muscles and other tissues, your body's own proteins would have been dismantled. The liver would have used amino acids from these structural proteins to build new glucose molecules—and once more your brain would have been kept active.

As extreme as this last pathway might be, it would be a small price to pay for keeping your brain functional enough to figure out how to take in more nutrients and get you back to safety.

731

SUMMARY

1. *Nutrition* involves all the processes by which the body takes in, digests, absorbs, and uses food.

2. Mammals have a complete digestive system—basically a tube with two openings (mouth and anus) and regional specializations. Protective, mucus-coated epithelium lines all exposed surfaces of the tube and facilitates diffusion across the tube wall.

3. As summarized in Table 42.4, the human digestive system includes the mouth, pharynx, esophagus, stomach, small intestine, large intestine (colon), rectum, and anus. The salivary glands, liver, gallbladder, and pancreas have accessory roles in the system's functions.

4. These activities proceed in the digestive system:

   a. Mechanical processing and motility (movements that break up, mix, and propel food material through the system).

   b. Secretion (release of digestive enzymes and other substances from the pancreas, liver, and glandular epithelium).

   c. Digestion (breakdown of food into particles, then into nutrient molecules small enough to be absorbed).

   d. Absorption (passage of digested organic compounds, fluid, and ions into the internal environment).

   e. Elimination (expulsion of undigested and unabsorbed residues at the end of the system).

5. Starch digestion starts in the mouth and protein digestion starts in the stomach. Digestion is completed and most nutrients are absorbed in the small intestine. The pancreas secretes the main digestive enzymes. Bile from the liver assists in fat digestion.

6. During absorption, cells of the intestinal lining actively transport glucose and most amino acids out of the gut lumen. Fatty acids and monoglycerides diffuse across the lipid bilayer of these cells. In the cells' cytoplasm they are recombined as triglycerides, then are released, by exocytosis, into interstitial fluid.

7. The nervous system, endocrine system, and nerve plexuses in the gut wall interact to govern activities of the digestive system. Many controls operate in response to the volume and composition of food passing through the stomach and intestines. The controls trigger changes in muscle activity and in the rate at which hormones and enzymes are secreted.

8. To maintain a suitable body weight and overall health, caloric intake must balance energy output.

9. Many nutritionists recommend a daily food intake in certain proportions. For example, for an adult male of average body weight: 58–60 percent complex carbohydrates, 12–15 percent protein, and 20–25 percent fats and other lipids. A well-balanced diet of whole foods normally provides all required vitamins and minerals.

Table 42.4	Summary of the Digestive System
MOUTH (oral cavity)	Start of digestive system where food is chewed and moistened, and where polysaccharide digestion begins
PHARYNX	Passageway both to the tubular part of the digestive system and to the respiratory system; moves food forward by contracting sequentially
ESOPHAGUS	Muscular tube, moistened by saliva, that moves food from pharynx to stomach
STOMACH	Sac where food mixes with gastric fluid and where protein digestion begins; can stretch to store food taken in faster than can be processed; gastric fluid also destroys many microbes
SMALL INTESTINE	The first part (duodenum, C-shaped and about 10 inches long) receives secretions from the liver, gallbladder, pancreas
	In the second part (jejunum, 3 feet long), most nutrients are digested and absorbed
	Some nutrients are absorbed in the last part (ileum, 6–7 feet long), which delivers unabsorbed material to the colon
LARGE INTESTINE (colon)	Concentrates and stores undigested matter (by absorbing mineral ions and water). It is about 5 feet long and is divided into ascending, transverse, and descending portions
RECTUM	Stores fecal material

Accessory Organs:

SALIVARY GLANDS	Glands (three main pairs and many minor ones) that secrete saliva, a fluid with polysaccharide-digesting enzymes, buffers, and mucus (moistens, lubricates food)
PANCREAS	Secretes enzymes able to break down all major food molecules; secretes buffers against HCl from the stomach
LIVER	Secretes bile (used in fat emulsification); roles in carbohydrate, fat, and protein metabolism
GALLBLADDER	Stores and concentrates bile from the liver

Review Questions

1. What are the main functions of the stomach? the small intestine? the large intestine (colon)? *718–719, 720–722, 723*

2. A glass of whole milk contains lactose, proteins, butterfat (mostly triglycerides), vitamins, and minerals. Explain what happens to each type of component when it passes through your digestive tract. *720–721, 722*

3. Name the kinds of breakdown products that are small enough to be absorbed across the intestinal lining and into the internal environment. *720, 722*

4. Name the organs of the human digestive system, as well as glandular organs with accessory roles in its functioning: *717*

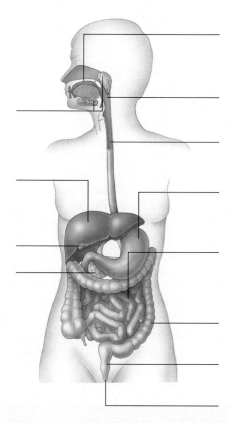

5. As a person ages, the number of body cells steadily decreases, and energy needs decline. If you were planning an older person's diet, what foods would you emphasize, and why? Which ones would you deemphasize? *724–728*

Self-Quiz *(Answers in Appendix IV)*

1. The _____ provides cells with tolerable conditions in the internal environment, raw materials, and waste disposal.
   a. digestive system
   b. circulatory system
   c. respiratory system
   d. urinary system
   e. interaction of all of the systems listed

2. Most digestive systems have regions for _____ food.
   a. transporting
   b. processing
   c. storing
   d. all of the above

3. Maintaining good health and normal body weight requires that _____ intake be balanced by _____ output.

4. Most of our caloric intake should come from _____ .
   a. complex carbohydrates
   b. simple carbohydrates
   c. proteins
   d. lipids

5. The human body cannot produce all of its own _____ .
   a. vitamins and minerals
   b. essential fatty acids
   c. essential amino acids
   d. a through c
   e. a and c

6. Secretions from the _____ do *not* assist in digestion and absorption.
   a. salivary glands
   b. thymus gland
   c. liver
   d. pancreas

7. Digestion is completed and most nutrients are absorbed in the _____ .
   a. mouth
   b. stomach
   c. small intestine
   d. colon

8. Glucose and most amino acids are absorbed across the gut lining _____ .
   a. by active transport
   b. by diffusion
   c. at lymph vessels
   d. as fat droplets

9. Bile has roles in _____ digestion and absorption.
   a. carbohydrate
   b. fat
   c. protein
   d. amino acid

10. Match the organ with its key digestive function(s).
   ____ gallbladder
   ____ stomach
   ____ colon
   ____ pancreas
   ____ salivary gland
   ____ small intestine
   ____ liver

   a. secrete bile and bicarbonate
   b. digest, absorb most nutrients
   c. store, mix, dissolve food; start protein breakdown
   d. store, concentrate bile
   e. concentrate undigested matter
   f. secrete substances that moisten food and start polysaccharide breakdown
   g. secrete digestive enzymes, bicarbonate

Selected Key Terms

appendix *723*
bile *721*
bulk *723*
chyme *718*
colon *723*
complete digestive system *716*
digestive system *716*
emulsification *721*
esophagus *718*
essential amino acid *725*
essential fatty acid *725*
gallbladder *720*
gastric fluid *718*
incomplete digestive system *716*
kilocalorie *728*
liver *720*
micelle *722*

microvillus *722*
mineral *726*
net protein utilization *725*
nutrition *715*
obesity *728*
oral cavity *718*
pancreas *720*
pharynx *718*
ruminant *715*
saliva *718*
sphincter *718*
stomach *718*
tongue *718*
tooth *718*
villi *722*
vitamin *726*

Readings

Campbell-Platt, G. May 1988. "The Food We Eat." *New Scientist* 19:1–4.

Cohen, L. 1987. "Diet and Cancer." *Scientific American* 257(5):42–68.

Wardlaw, G., P. Insel, and M. Seyler. 1992. *Contemporary Nutrition: Issues and Insights.* St. Louis: Mosby.

Weiss, P. September 1, 1990. "Fat and Fiction." *Science News* 138(9):138–139.

43 WATER-SOLUTE BALANCE AND TEMPERATURE CONTROL

Tale of the Desert Rat

About 375 million years ago, some animals left the seas for life on land. Their cells were exquisitely adapted to life in a salty fluid, yet the invasion succeeded anyway. Why? The animals brought salty fluid along with them, as an *internal* environment for their living cells. Still, it wasn't easy. On land they found intense sunlight, dry winds, water of dubious salt content, and sometimes no water at all. How could they keep from drying out? How could they replace the water and specific salts lost through everyday activities? How could they maintain the volume and composition of that precious internal environment and so prevent cellular anarchy?

Any of their existing descendants provides you with some answers. Think of a kangaroo rat in an isolated desert of New Mexico (Figure 43.1). After a brief rainy season, the sun bakes the sand for months. The only obvious water is imported, sloshing in canteens of an

occasional researcher or tourist. Yet with nary a sip of free water, this tiny mammal counters threats to its internal environment.

The kangaroo rat waits out the daytime heat in burrows, then forages in the cool of night for dry seeds and maybe a succulent. It is not sluggish about this. It hops rapidly and far, searching for seeds and fleeing from coyotes and snakes. All that hopping requires ATP energy and water. Seeds, chockful of energy-rich carbohydrates, supply both. Metabolic reactions that release energy from carbohydrates and other organic compounds also yield water. Each day, this "metabolic water" represents about 12 percent of your own total water intake. It represents a whopping 90 percent of the total for a kangaroo rat.

Inside its cool burrow, the kangaroo rat conserves and recycles water. As the animal breathes cool air into

Figure 43.1 A kangaroo rat (**a**), master of water conservation in the desert (**b**).

a

its warm lungs, water vapor condenses on the epithelial lining inside its nose—and some of it diffuses back into the body. Also, after a night of foraging, the animal empties its cheek pouches of seeds—which soak up water vapor that does escape from the nose. Eating the seeds recycles the water.

The kangaroo rat can't lose water by perspiring; it has no sweat glands. It can lose water by urinating—but specialized kidneys don't let it piddle away much. Kidneys filter the blood's water and dissolved salts (solutes). They adjust *how much* water and *which solutes* return to the blood or leave the body as urine.

Overall, the kangaroo rat and every other animal takes in enough water and solutes to replace its daily losses (Figure 43.1c). How the balancing acts are accomplished will be our initial focus in this chapter. Later on, we will take a look at some of the means by which mammals withstand hot, cold, and sometimes unpredictable changes in environmental temperatures on land.

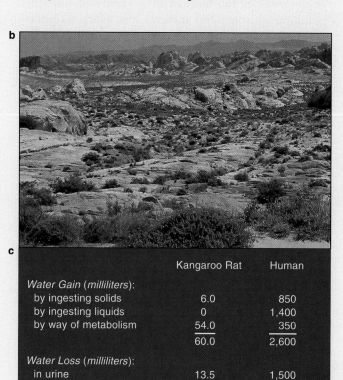

b

c

	Kangaroo Rat	Human
Water Gain (*milliliters*):		
by ingesting solids	6.0	850
by ingesting liquids	0	1,400
by way of metabolism	54.0	350
	60.0	2,600
Water Loss (*milliliters*):		
in urine	13.5	1,500
in feces	2.6	200
by evaporation	43.9	900
	60.0	2,600

KEY CONCEPTS

1. Animals continually gain and lose water and dissolved substances (solutes). They continually produce metabolic wastes. Even with all these inputs and outputs, the body's extracellular fluid does not change much in its overall volume and composition.

2. In humans and other mammals, a urinary system helps balance the intake and output of water and solutes. This system eliminates excess water and solutes as a fluid called urine. It also conserves water when necessary.

3. Urine forms in tubelike nephrons. These structures are packed inside a pair of kidneys, which are blood-filtering organs. Nephrons receive water and solutes from a set of blood capillaries. They return most of the filtrate to a second set of capillaries; the rest leaves the body as urine.

4. Body temperatures of animals depend on the balance between heat produced through metabolism, heat absorbed from the environment, and heat lost to the environment.

5. Body temperature is maintained within a favorable range through controls over metabolic activity and adaptations in structure, physiology, and behavior.

43.1 MAINTAINING THE EXTRACELLULAR FLUID

In most animals, recall, *interstitial fluid* fills the spaces between living cells and other components of the body's tissues. Another fluid, *blood*, circulates inside blood vessels. Taken together, interstitial fluid and blood are the body's **extracellular fluid**.

In humans, as in all mammals, a well-developed **urinary system** helps keep the volume and composition of extracellular fluid within tolerable ranges. It does this by balancing the daily losses of water and solutes with daily intakes. Other organ systems, especially those indicated in Figure 43.2, interact with the urinary system in the performance of this homeostatic task.

Water Gains and Losses

Ordinarily, and on a daily basis, humans and other mammals take in as much water as they give up. Figure 43.1c provides just two examples of this balancing act. How does the body *gain* water? It does so by two processes:

1. Absorption from solid food and liquids

2. Metabolism

A great deal of water is absorbed from solids and liquids that travel through the gut lumen. Among land-dwelling mammals, a thirst mechanism affects how much water enters the gut in the first place. When too much water is lost, the individual seeks out streams, water holes, cold drinks in the refrigerator, and so on. We will return to the thirst mechanism later in the chapter. Water also forms

as a by-product of many metabolic reactions, including carbohydrate breakdown.

How does the mammalian body *lose* water? It does so mainly by the following processes:

1. Urinary excretion

2. Evaporation from lungs and through skin

3. Sweating

4. Elimination (in feces)

Of these, **urinary excretion** affords the greatest control over water loss. This process eliminates excess water and excess or harmful solutes in the form of urine. Besides this, some water is always evaporating from the respiratory surfaces in lungs. Water also evaporates from sweaty skin. Normally, very little water that entered the gut is lost; most is absorbed, not eliminated in feces.

Solute Gains and Losses

The body's extracellular fluid *gains* solutes mainly as a result of four processes:

1. Absorption from solid food and liquids

2. Secretion

3. Respiration

4. Metabolism

Eating provides the body with a variety of nutrients (including glucose) and mineral ions (such as potassium and sodium ions). These are absorbed from the gut, as are drugs and food additives. Throughout the body, living cells secrete substances into interstitial fluid and the blood. The respiratory system puts oxygen into the blood, and aerobically respiring cells put carbon dioxide into it.

Extracellular fluid *loses* mineral ions and metabolic wastes in these ways:

1. Urinary excretion

2. Respiration

3. Sweating

In certain mammals, mineral ions are lost in sweat. Carbon dioxide, the most abundant metabolic waste, is exhaled from the lungs. Ammonia, urea, uric acid, and other major wastes leave in urine. The ammonia forms when amino groups are stripped from amino acids. It becomes toxic when it accumulates in the body. Urea forms in the liver when two ammonia molecules are combined with carbon dioxide. It is the main product of protein breakdown, and relatively harmless. Uric acid forms during nucleic acid breakdown. It tends to crystallize at high concentrations and may build up in skeletal joints. The urinary system also eliminates the phosphoric and sulfuric acids produced during protein breakdown.

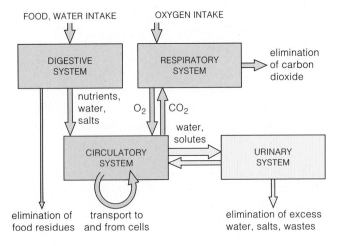

Figure 43.2 Organ systems that interact to maintain homeostasis, or favorable operating conditions in the body.

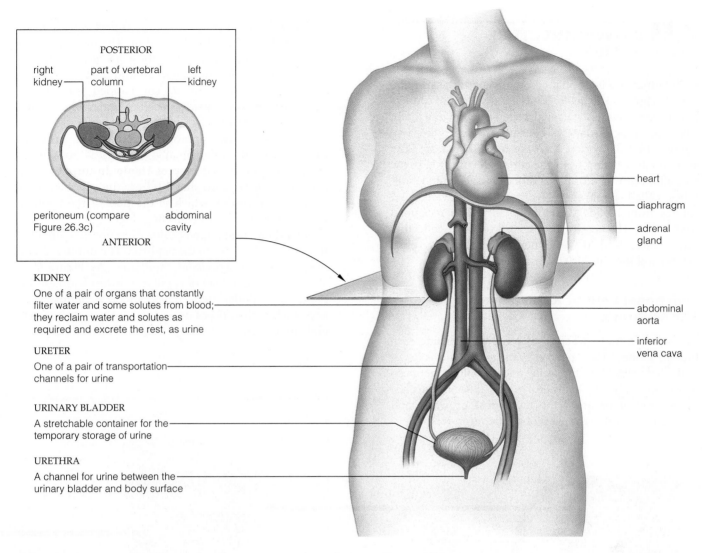

KIDNEY

One of a pair of organs that constantly filter water and some solutes from blood; they reclaim water and solutes as required and excrete the rest, as urine

URETER

One of a pair of transportation channels for urine

URINARY BLADDER

A stretchable container for the temporary storage of urine

URETHRA

A channel for urine between the urinary bladder and body surface

heart

diaphragm

adrenal gland

abdominal aorta

inferior vena cava

Figure 43.3 Components of the human urinary system and their functions. The two kidneys, two ureters, and urinary bladder are located outside the abdominal cavity's lining (the peritoneum).

Urinary System of Mammals

Different solid foods and fluids intermittently hit your gut. Afterward, differing amounts of absorbed water, nutrients, and other substances move into blood, then into interstitial fluid throughout your body. Any unwanted shifts in the volume and composition of this extracellular fluid are dealt with swiftly by a pair of organs called **kidneys**. In all mammals, kidneys are the central components of the urinary system.

Kidneys filter water, mineral ions, organic wastes, and other substances out of blood. They adjust the filtrate's composition, then return all but about 1 percent of it to the blood. This tiny portion of unreclaimed water and solutes is urine. By definition, **urine** is a fluid that rids the body of water and solutes that are in excess of the amounts required to maintain extracellular fluid.

Take a look at Figure 43.3, which shows the location of the kidneys in the human urinary system. Urine flows from each kidney into a **ureter**, a tubular channel. The ureters lead to the **urinary bladder**. Here urine is temporarily stored before moving out through the **urethra**, a muscular tube that opens at the body surface. All mammals have this arrangement—two kidneys, two ureters, a urinary bladder, and a urethra.

Urination (urine flow from the body) is a reflex response, although it can be consciously controlled to a great extent. As the urinary bladder fills, it stretches, and this activates mechanoreceptors in its wall. Their signals reach motor neurons, which send commands to smooth muscles in the bladder wall. Urethral sphincters open. At the same time, muscles in the wall contract, and so force urine through the urethra.

***

By adjusting blood's volume and composition, kidneys help maintain conditions in the extracellular fluid.

***

43.2 KIDNEY STRUCTURE AND FUNCTION

Each kidney is a bean-shaped organ, about the size of a fist. As Figure 43.4 shows, it has a tough coat of connective tissue. This is the renal capsule (from the Latin *renes*, meaning kidneys). Beneath the coat are two regions, first the cortex, then the medulla. Notice how several lobes extend from the cortex down through the medulla. Each lobe contains blood vessels and slender tubes called **nephrons**. Nephrons filter water and solutes from blood. Most of the filtrate is reclaimed from them. The rest (urine) enters tubelike collecting ducts. These lead to the kidney's central cavity (renal pelvis) and the entrance to a ureter.

Functional Units of Kidneys— The Nephrons

More than a million nephrons are packed inside each human kidney. The nephron wall is a single layer of epithelial cells, but the cells and junctions between them are not all the same. As you will see, the differences prevent or assist the movement of water and solutes across different parts of the nephron wall.

The beginning of each nephron is part of a **glomerulus**, which is a blood-filtering unit. The other part is a tiny set of blood vessels called the *glomerular* capillaries. The nephron wall forms a cup around them, as you can see in Figure 43.5c and *d*. The cupped wall region, called **Bowman's capsule**, receives water and solutes filtered from the blood.

The rest of the nephron is tubular. The **proximal tubule** (closest to Bowman's capsule) is continuous with a hairpin-shaped **loop of Henle**. In turn, the loop is continuous with the **distal tubule** (most distant from Bowman's capsule), which is the part of the nephron that delivers urine to a collecting duct.

Blood entering a glomerulus does not give up all of its water and solutes to the nephron. The unfiltered portion flows into *peritubular* capillaries. As Figure 43.5c shows, this second set of capillaries threads around the nephron's tubular parts. Here, blood reclaims water and solutes, then flows into veins and back to the general circulation.

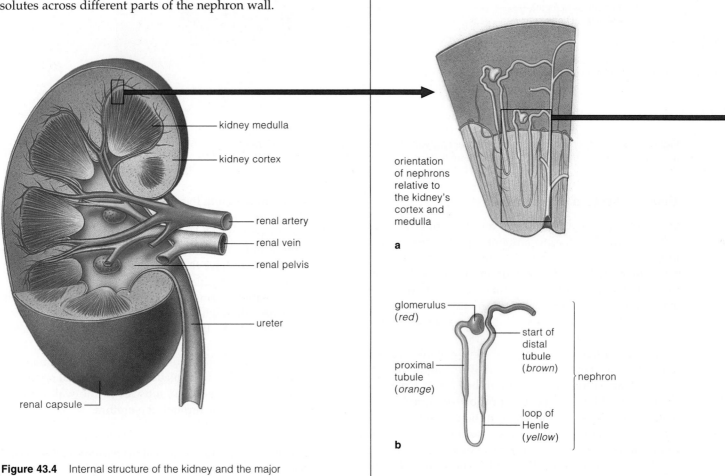

Figure 43.4 Internal structure of the kidney and the major blood vessels leading into and out of it.

Urine-Forming Processes—An Overview

Urine forms through three processes, called filtration, tubular reabsorption, and secretion.

Filtration starts and ends at the glomerulus. Blood pressure, generated by heart contractions, drives the process. It "filters" blood by forcing water and small solutes out of the glomerular capillaries, leaving blood cells as well as proteins and other large solutes behind.

Table 43.1	Average Daily Reabsorption Values for a Few Substances		
	Filtered	Excreted	Reabsorbed
Water	180 liters	1.8 liters	99%
Glucose	180 grams	None, normally	100%
Sodium ions	630 grams	3.2 grams	99.5%
Urea	54 grams	30 grams	44%

After this, the filtrate moves from the cupped part of the nephron into the proximal tubule.

Reabsorption proceeds along the nephron's tubular parts. Most of the filtrate's water and solutes move out of the nephron, then into adjacent blood capillaries. Table 43.1 lists some daily reabsorption values.

Secretion also proceeds at tubular wall regions but in the opposite direction. Substances move out of the capillaries and into cells of the wall. Then they move across the cells, which secrete them into the forming urine. Among other things, this highly controlled process rids blood of excess hydrogen ions (H^+) and potassium ions. It also prevents some metabolic wastes (such as uric acid) and foreign substances (such as penicillin and other drugs) from accumulating in blood.

In the kidneys, a concentrated or dilute urine forms through the processes of filtration, reabsorption, and secretion.

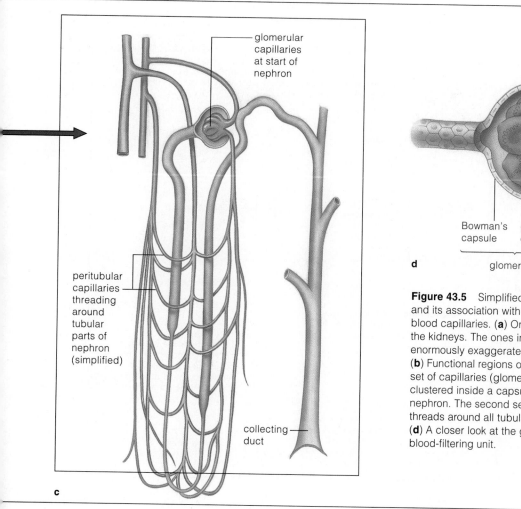

Figure 43.5 Simplified diagrams of a nephron and its association with two connected sets of blood capillaries. (**a**) Orientation of nephrons in the kidneys. The ones in this diagram are enormously exaggerated in size for clarity. (**b**) Functional regions of a nephron. (**c**) The first set of capillaries (glomerular capillaries) is clustered inside a capsule at the start of the nephron. The second set (peritubular capillaries) threads around all tubular parts of the nephron. (**d**) A closer look at the glomerulus, the nephron's blood-filtering unit.

glomerular capillaries at start of nephron

peritubular capillaries threading around tubular parts of nephron (simplified)

collecting duct

arteriole entering glomerulus

Bowman's capsule   glomerular capillaries   arteriole leaving

glomerulus

c

d

Factors Influencing Filtration

About 1.5 liters (1-1/2 quarts) of blood flow through an adult's kidneys every minute. From this enormous flow volume, 120 milliliters of water and small solutes are diverted every minute into the nephrons (Figure 43.6*a*). That's 180 liters of filtrate per day! Two mechanisms are responsible for the high rate of filtration.

First, compared to other capillaries, glomerular capillaries are 10–100 times more permeable to water and small solutes. *Second*, blood pressure inside them is high. Arterioles delivering blood to the glomerulus have a wider diameter and less resistance to flow than most of the other arterioles in the body. So the hydrostatic pressure resulting from heart contractions does not drop as much as it does elsewhere.

At any time, the flow volume to the kidneys affects the filtration rate. Neural, endocrine, and local controls keep that volume fairly constant even when blood pressure changes. For example, when you run a race or dance until dawn, the nervous system diverts an above-normal volume of blood away from the kidneys, toward your heart and skeletal muscles. The neural signals direct arterioles in different parts of the body to vasoconstrict or vasodilate in coordinated ways, so that less of the flow volume reaches the kidneys.

As another example, cells in the walls of arterioles leading to glomeruli respond to arterial blood pressure. When the pressure decreases, they secrete chemicals that induce vasodilation, so more blood flows in. When it rises, they constrict, so less blood flows in.

Finally, the filtration rate depends on how fast the kidney tubules are reabsorbing water. As you will see next, reabsorption is partly under hormonal control.

Reabsorption of Water and Sodium

Reabsorption Mechanism Kidneys precisely adjust how much water and sodium your body excretes or conserves. It makes no difference if you drink too much or too little water at lunch, or wolf down salty potato chips, or lose too much sodium by sweating.

Adjustments start promptly, when filtrate enters a nephron's proximal tubule. Cells in the tubule wall actively transport some sodium ions out of the filtrate into interstitial fluid. Other ions follow. Then water fol-

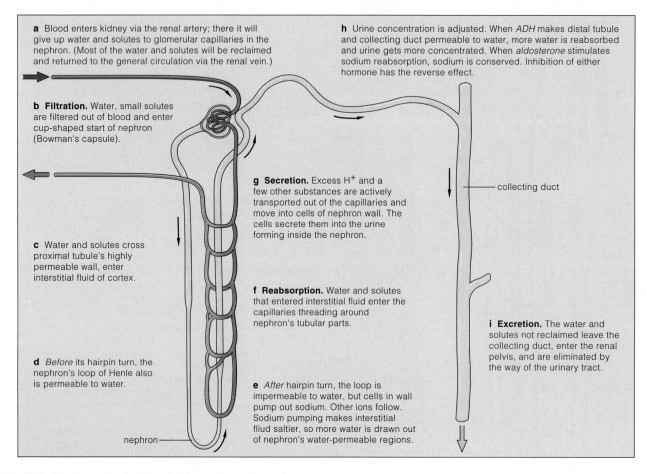

a Blood enters kidney via the renal artery; there it will give up water and solutes to glomerular capillaries in the nephron. (Most of the water and solutes will be reclaimed and returned to the general circulation via the renal vein.)

b **Filtration.** Water, small solutes are filtered out of blood and enter cup-shaped start of nephron (Bowman's capsule).

c Water and solutes cross proximal tubule's highly permeable wall, enter interstitial fluid of cortex.

d *Before* its hairpin turn, the nephron's loop of Henle also is permeable to water.

nephron

e *After* hairpin turn, the loop is impermeable to water, but cells in wall pump out sodium. Other ions follow. Sodium pumping makes interstitial fluid saltier, so more water is drawn out of nephron's water-permeable regions.

f **Reabsorption.** Water and solutes that entered interstitial fluid enter the capillaries threading around nephron's tubular parts.

g **Secretion.** Excess H$^+$ and a few other substances are actively transported out of the capillaries and move into cells of nephron wall. The cells secrete them into the urine forming inside the nephron.

h Urine concentration is adjusted. When *ADH* makes distal tubule and collecting duct permeable to water, more water is reabsorbed and urine gets more concentrated. When *aldosterone* stimulates sodium reabsorption, sodium is conserved. Inhibition of either hormone has the reverse effect.

collecting duct

i **Excretion.** The water and solutes not reclaimed leave the collecting duct, enter the renal pelvis, and are eliminated by the way of the urinary tract.

Figure 43.6 Processes involved in urine formation and excretion.

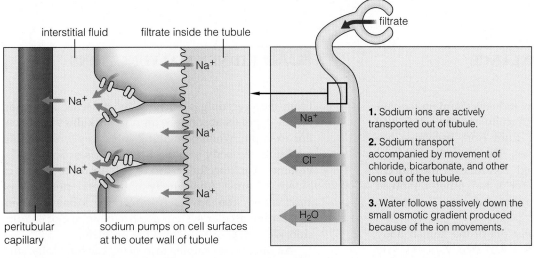

interstitial fluid   filtrate inside the tubule

Na⁺

Na⁺

Na⁺

Na⁺

Na⁺

peritubular capillary   sodium pumps on cell surfaces at the outer wall of tubule

filtrate

Na⁺

Cl⁻

H_2O

1. Sodium ions are actively transported out of tubule.

2. Sodium transport accompanied by movement of chloride, bicarbonate, and other ions out of the tubule.

3. Water follows passively down the small osmotic gradient produced because of the ion movements.

Figure 43.7 Reabsorption of solutes as a result of the active transport of sodium out of the proximal tubule.

lows the ions, by osmosis. This wall region is highly permeable to water, so most of the filtrate's water is reabsorbed here (Figure 43.6*c–h*).

Now the filtrate enters the loop of Henle, which plunges into the medulla. Where the loop makes its turn, the interstitial fluid is saltiest. Water moves out by osmosis before the turn. The fluid left behind gets saltier, until it matches the interstitial fluid. After the turn, no more water can cross the tubule wall—but *sodium* is pumped out, by active transport mechanisms (Figure 43.7). As the surroundings get saltier, more water is drawn out of fluid just entering the loop.

When fluid remaining in the nephron reaches the distal tubule in the kidney cortex, it is dilute. The stage is set for adjustments that lead to highly dilute or concentrated urine—or anywhere in between.

Hormone-Induced Adjustments Cells in the walls of distal tubules and collecting ducts have receptors for two hormones, ADH and aldosterone.

When water must be conserved, the hypothalamus triggers secretion of antidiuretic hormone, or **ADH**. This hormone makes the walls more permeable to water (Figure 43.6*h*). More water is reabsorbed, so the urine becomes more concentrated. When the body must lose excess water, ADH secretion is inhibited. Then, the walls remain impermeable to water, less water is reabsorbed, and the urine remains dilute.

Aldosterone enhances sodium reabsorption. The extracellular fluid volume decreases when too much sodium is lost. Sensory receptors in blood vessels and the heart detect the decrease and signal certain kidney cells (Figure 43.8). The cells secrete renin, an enzyme that lops off part of a plasma protein. Other reactions convert the protein to a hormone that acts on aldosterone-secreting cells of the adrenal cortex. This is the outer portion of a gland on each kidney (Figure 43.3). Aldosterone makes cells of the distal tubules and collecting ducts reabsorb sodium faster, so less sodium is excreted. Conversely, when the body has too much sodium, aldosterone secretion is inhibited. Less sodium is reabsorbed, and more is excreted.

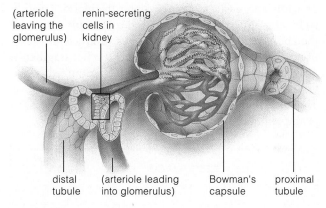

(arteriole leaving the glomerulus)   renin-secreting cells in kidney

distal tubule   (arteriole leading into glomerulus)   Bowman's capsule   proximal tubule

Figure 43.8 Location of renin-secreting cells that play a role in sodium reabsorption.

Finally, when solutes increase in extracellular fluid, the hypothalamus inhibits saliva production. The brain interprets the ensuing dryness in the mouth as "thirst." A cottony mouth, an early sign of dehydration, induces thirst behavior—you seek fluids.

1. Filtration rates in the kidneys depend on the high hydrostatic pressure at the start of nephrons and on the high permeability of glomerular capillaries. They depend on neural, endocrine, and local control of the flow volume directed to them at a given time.

2. A reabsorption mechanism and hormone-induced adjustments help maintain extracellular fluid. They form concentrated or dilute urine—and so conserve or rid the body of suitable amounts of water and solutes.

3. ADH conserves water by enhancing its reabsorption at distal tubules and collecting ducts. Excess water is excreted in urine when ADH secretion is inhibited.

4. Aldosterone conserves sodium by enhancing its reabsorption at distal tubules and collecting ducts. Less sodium is lost when aldosterone secretion is inhibited.

43.4 ACID-BASE BALANCE

So far, we've focused on how kidneys maintain the volume and composition of extracellular fluid. They also help keep the extracellular fluid from becoming too acidic or too basic (alkaline). This overall **acid-base balance** is maintained through control over the concentrations of hydrogen ions (H^+) and other dissolved ions. Buffer systems, respiration, and urinary excretion provide the control.

Normally, the extracellular pH of the human body must be maintained between 7.37 and 7.45. As you know, acids lower the pH and bases raise it. A variety of acidic and basic substances enter the blood. They do so by absorption from the gut and as a result of normal metabolism. Typically, cell activities produce an excess of acids, which dissociate into H^+ and other fragments. This lowers the pH. The effect is minimized when excess hydrogen ions react with buffer molecules. The **bicarbonate–carbon dioxide buffer system** is an example:

$$H^+ + \underset{\text{bicarbonate}}{HCO_3^-} \rightleftharpoons \underset{\text{carbonic acid}}{H_2CO_3} \rightleftharpoons H_2O + CO_2$$

In this case, the excess H^+ is neutralized, and the carbon dioxide that forms during the reactions is exhaled from lungs. Like other buffer systems, however, this one has a temporary effect; it does not eliminate acid. Only the urinary system eliminates excess H^+ and restores buffers.

The same reactions proceed in reverse in cells of the nephron's tubular walls. The HCO_3^- resulting from the reverse reactions moves into interstitial fluid, then into capillaries around the nephron. After this, it moves into the general circulation, where it buffers excess acid. The H^+ that formed in the cells is secreted into the nephron. There, it may combine with bicarbonate ions to form CO_2—which is returned to the blood and exhaled by the lungs. Or it may combine with phosphate ions or ammonia (NH_3), then leave the body in urine. In such ways, hydrogen ions are permanently removed from extracellular fluid.

---

1. Along with buffering systems and the respiratory system, kidneys help keep the extracellular fluid from becoming too acidic or too basic (alkaline).

2. Buffering systems can do no more than *neutralize* acids. Only the urinary system can *eliminate* excess hydrogen ions and restore the body's buffers.

43.5 KIDNEY MALFUNCTIONS

From the preceding sections, you can sense that good health depends absolutely on normal kidney function. For example, when something interferes with sodium excretion, the body's sodium content rises. So does the volume of extracellular fluid. This leads to a rise in blood pressure. Abnormally high blood pressure, or *hypertension*, can adversely affect the kidneys, brain, and cardiovascular system. Restricted intake of sodium chloride—table salt—lessens the problem.

As another example, uric acid, calcium salts, and other wastes can settle out of urine and collect in the renal pelvis as *kidney stones*. These hard deposits may lodge in the ureter (or urethra), then disrupt urine flow. They usually are passed naturally in urine. If not, they must be removed by medical or surgical procedures.

Whether by illness or as a result of accidents, about 13 million people in the United States alone suffer from kidney malfunctions that seriously disrupt the controls over the volume and composition of extracellular fluid. Among other things, toxic by-products of protein breakdown may accumulate in the bloodstream. Nausea, fatigue, and loss of memory are outcomes. In advanced cases, death may follow.

A **kidney dialysis machine** often can restore the proper solute balances. Like the kidney, this machine helps maintain extracellular fluid by selectively removing solutes from blood and adding solutes to it. "Dialysis" refers to an exchange of substances across a membrane that is positioned between solutions of differing compositions.

In *hemodialysis*, the machine is connected to an artery or a vein; then blood is pumped through tubes made of a material similar to sausage casing or cellophane. The tubes are submerged in a warm-water bath. The precise mix of salts, glucose, and other substances in the bath sets up the correct gradients with blood.

In *peritoneal dialysis*, fluid of the proper composition is put into the patient's abdominal cavity, left in place for a specific length of time, then drained out. The cavity's lining (the peritoneum) serves as the membrane for dialysis.

For hemodialysis to have optimum effect, it must be performed three times a week. Each time, the procedure takes about four hours to complete, for the patient's blood must circulate repeatedly in order to improve the solute concentrations in the body.

Bear in mind, hemodialysis is used as a temporary measure in reversible kidney disorders. In chronic cases, the procedure must be used for the rest of the patient's life or until a functional kidney is transplanted. With treatment and controlled diets, many patients resume fairly normal activity.

43.6 ON FISH, FROGS, AND KANGAROO RATS

Now that you have an idea of how your own body maintains water and solute levels, consider what goes on in some other vertebrates, including that kangaroo rat hopping about at the start of the chapter.

In freshwater, bony fishes and amphibians gain water and lose solutes (Figure 43.9a). They don't drink. Water moves in by osmosis, through thin gill membranes or, in adult amphibians, across skin. Excess water leaves as dilute urine, formed in kidneys. Solute losses are balanced by solutes gained from food and by the inward pumping of sodium by cells of the gills. Figure 43.10 describes the wonderful balancing act in a type of fish that makes its home in both freshwater *and* seawater.

Body fluids of herring, snapper, and other marine fishes are about three times less salty than seawater. Such fishes lose water by osmosis, drink replacements, and excrete ingested solutes against concentration gradients (Figure 43.9b). Fish kidneys do not have loops of Henle, so urine cannot ever become saltier than body fluids. Cells in the fish gills actively pump out most of the excess solutes in blood.

And about that kangaroo rat! Its nephrons have extremely long loops of Henle. The solute concentration in interstitial fluid around the loops becomes very high. The osmotic gradient between this fluid and the urine is so steep, nearly all the water that reaches the equally long collecting ducts is reabsorbed. Kangaroo rats give up only a tiny volume of urine, and it's three to five times more concentrated than that of humans.

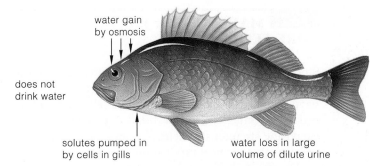

a Freshwater bony fish (body fluids much saltier than the surroundings).

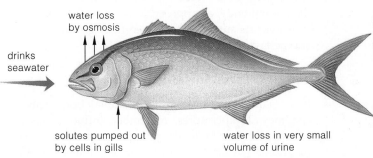

b Marine bony fish (body fluids less salty than the surroundings).

Figure 43.9 Water-solute balancing acts in fishes.

salmon avoiding grizzly while maintaining solute-water balance

Figure 43.10 Water-solute balance by salmon, a type of fish that can live in saltwater and in freshwater. Salmon hatch in streams and later move downstream to the seas, where they feed and mature. They return to their streams to spawn.

For most salmon, salt tolerance arises through changes in hormone concentrations. The changes seem to be triggered by increasing daylength in spring. Prolactin, a pituitary hormone, plays a role in sodium retention in freshwater habitats. When a freshwater fish has its pituitary gland removed, it dies from sodium loss—but that fish will live if prolactin is administered to it.

Cortisol, a steroid hormone secreted by the adrenal cortex, is crucial to the development of salt tolerance in salmon. Cortisol secretions correlate with increases in sodium excretion, in sodium-potassium pumping by cells in the salmon's gills, and in absorption of ions and water in the gut. In young salmon, cortisol secretion increases prior to the seaward movement—and so does salt tolerance.

43.7 MAINTAINING THE BODY'S CORE TEMPERATURE

Stop for a minute and think about the temperature around your body. Is the air too hot? Too cold? If you have been sitting for some time, your cells haven't been producing much metabolic heat. If you just finished exercising strenuously, metabolic rates have soared, and so has metabolic heat production.

Despite such differences, and assuming that you are in good health, your **core temperature** remains much the same. ("Core" refers to the internal temperature of the animal body, as opposed to temperatures of the tissues near its surface.) It does so through physiological and behavioral responses to change.

Temperatures Suitable for Life

In a manner of speaking, the animal body runs on enzymes. Temperatures, remember, influence enzyme activity. In most animals, enzymes commonly remain functional within the 0°–40°C range (Table 43.2). When temperatures rise above 41°C or so, this is an invitation to denaturation. Then, chemical interactions holding a molecule in its required three-dimensional shape are disrupted, and so is its function. Also, the rate of enzyme activity generally decreases by at least half when the temperature drops by ten degrees. Clearly, metabolism can be upset if the core temperature exceeds or falls below the proper range.

Heat Gains and Heat Losses

Enzyme-mediated reactions proceed simultaneously in the millions or billions of cells of a large-bodied animal. Heat is an inevitable by-product of all that metabolic activity. Even as you quietly read this book, you are producing roughly 1 kilocalorie per hour per kilogram of body weight. If that heat were to accumulate internally, your core temperature would steadily rise. However, a warm body tends to lose heat to a cooler environment. Its temperature will hold steady if the rate of heat loss strikes a balance with the rate of metabolic heat production.

Table 43.2	Temperatures Favorable for Metabolism, Compared with Environmental Temperatures
Temperatures generally favorable for metabolism:	0°C to 40°C (32°F to 104°F)
Air temperatures above land surfaces:	−70°C to +85°C (−94°F to +185°F)
Surface temperatures of open oceans:	−2°C to +30°C (+28.4°F to +86°F)

Generally speaking, the heat content of a complex animal depends on the balance between heat gains and losses, as summarized here:

change in body heat	=	heat produced	+	heat gained	−	heat lost

In this summary equation, the gains and losses refer to exchanges that take place at the outer surface of the body, which includes part of the respiratory tract.

Heat gains and heat losses depend on four processes: radiation, conduction, convection, and evaporation.

With **radiation**, the body gains heat through exposure to intense radiant energy (as from the sun or a heat lamp) or to any surface that is warmer than its own surface temperature. Indoors, radiation typically accounts for well over a third of your total heat loss.

With **conduction**, there is a direct transfer of heat between the body and a solid object in contact with it. Because heat moves down thermal gradients, animals gain heat by conduction when they rest on a warm substrate or squirm for any length of time across it (Figure 43.11a). Animals lose heat by resting on cold substrates, such as frozen deck chairs (Figure 43.11b).

Convection requires moving air or water to transfer heat. This process involves conduction (heat moves down the thermal gradient between the body surface and the air or water next to it). It also involves mass transfer, with currents carrying heat away from or toward the body. When skin temperature is higher than

Figure 43.11 (a) A sidewinder making its signature J-shaped track across hot desert sand at dusk. As near as you can tell, how might this rattlesnake be gaining or losing heat? (b) How might the intrepid tourist sitting on a deck chair of a ship off the coast of Antarctica be gaining or losing heat?

a

Figure 43.12 Evaporative water loss and sweating. Horses, humans, and some other mammals have sweat glands that move water and specific solutes through pores to the skin surface. A human of average size can produce 1 to 2 liters of sweat per hour, for hours on end. For every liter of sweat that evaporates, the body loses 600 kilocalories of heat energy. During intense exercise, this mechanism balances the high rates of heat production in skeletal muscle. In itself, sweat dripping from skin does not dissipate body heat by evaporation. When you exercise on hot, humid days, the evaporation rate cannot match the rate of sweat secretion; the air's high water content slows it down.

b

the air temperature, heat is lost by convection. Even when there is no breeze, heat is lost because any movement of the body creates its own convective current. Air becomes less dense as it is heated and rises from the body. Indoors, you typically lose as much heat by convection as you do by radiation.

Evaporation is the conversion of a liquid to a gaseous state. The conversion requires energy, which is supplied from the heat content of the liquid (page 26). Evaporation from the body surface has a cooling effect, because the water molecules that are escaping carry away some energy with them. For land animals, some evaporative water loss occurs at moist respiratory surfaces and across skin. Animals that sweat, pant, or lick their fur also lose water by this process (Figure 43.12). Evaporation rates depend on humidity and on the rate of air movement. If air next to the body is already saturated with water (that is, when the local relative humidity is 100 percent), water will not evaporate. If air next to the body is hot and dry, evaporation may be the only means of countering metabolic heat production and heat gains from radiation and convection.

---

1. The body's core temperature is maintained when there is a balance between heat gains and heat losses.
2. Heat moves to or away from the body by radiation, conduction, and convection. Evaporation carries heat away from the body.

43.8 CLASSIFICATION OF ANIMALS BASED ON TEMPERATURE

How much heat is exchanged between the animal body and its surroundings depends on the processes of radiation, conduction, convection, and evaporation. Even so, animals are not entirely at the mercy of these physical processes. At any given time, they can *adjust* the amount of heat lost or gained by behavioral and physiological means. As you will now see, different kinds of animals make the adjustments in different ways, with greater or less efficiency.

Ectotherms

Most animals have low metabolic rates. On top of that, they are poorly insulated. Such animals rapidly absorb and gain heat, especially when their body is a comparatively small one. Maintaining the core temperature depends primarily on heat gains from the environment, not from metabolism. Hence the classification of these animals as **ectotherms**, which means "heat from outside." Lizards, snakes, and other reptiles are good examples of ectotherms.

Ectotherms adjust behaviorally to changes in external temperatures. We call this **behavioral temperature regulation**. Lizards move about, putting themselves in places where they minimize heat or cold stress. To warm up, they move out of shade and keep reorienting their body to expose the maximum surface area to the sun's infrared radiation. They gain heat through con-

Figure 43.13 Meerkats keeping warm on a cold winter night in a zoo in Germany. In all complex animals, a variety of behavioral adjustments supplement built-in adaptations that help maintain the body's core temperature. Compare Figure 33.1.

duction by basking on rocks that absorbed heat earlier in the day. The splendid green iguana shown on page 545 is doing this. In such ways, the lizard body can warm up as fast as 1°C per minute.

Lizards lose heat just as rapidly when the sun goes down and temperatures drop. Then, metabolic activity decreases and the animal becomes almost immobilized. Before that happens, it usually crawls into crevices or under rocks. There, heat loss is not as great, and it is not as vulnerable to predators.

Endotherms

Most birds and mammals are **endotherms**, which means "heat from within." In endotherms, body temperature is controlled mainly by metabolic activity and by controls over heat conservation and dissipation. Endotherms typically can make complex behavioral adjustments that supplement the physiological controls (Figures 43.13 and 43.14).

Most endotherms have an active life-style, made possible by high metabolic rates. It is a costly adaptation. A foraging mouse uses up to thirty times more energy than a foraging lizard of the same weight does. Yet such energy outlays have advantages. They are the main reason why endotherms can be active under a wide range of temperatures. Cold nights or cold seasons don't stop them from flying, foraging, escaping from predators, or digging a burrow.

Endotherms have morphological adaptations that help them conserve or dissipate heat associated with their high metabolic rates. Fur, feathers, and layers of fat help reduce heat loss. Clothing can reduce heat loss or heat gain. Some of the mammals living in cold habitats have more massive bodies than closely related species in warmer regions do. Compared to the streamlined, thin-legged jackrabbits of the American Southwest, for example, you might think that the arctic hare is rather blocky. (In case you don't know what these animals look like, pages 462 and 749 show two representatives.) However, the arctic hare's more massive, rounder body has a greater volume of cells for generating heat and less surface area for losing it.

Like ectotherms, an endotherm also can adjust behaviorally to heat stress. During the day, the core temperatures of kangaroo rats and some other desert mammals of north temperate regions are much lower than temperatures of the air and the ground surface. Outside temperatures often are lower during the night as well as during winter. However, the soil well below the surface never heats up much, and it is here that most desert rodents and other mammals find refuge from the daytime heat. Typically they forage by night and spend the hottest part of the day in burrows or in the shade of bushes or rock outcroppings.

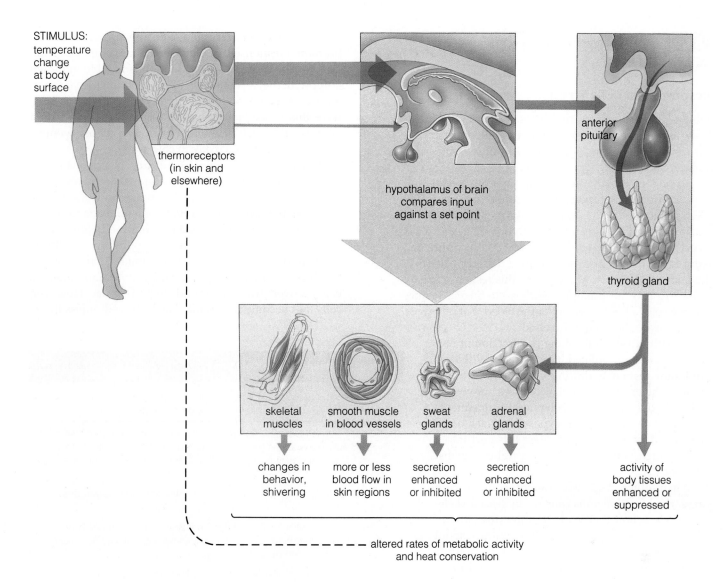

STIMULUS:
temperature
change
at body
surface

thermoreceptors
(in skin and
elsewhere)

hypothalamus of brain
compares input
against a set point

anterior
pituitary

thyroid gland

skeletal
muscles

smooth muscle
in blood vessels

sweat
glands

adrenal
glands

changes in
behavior,
shivering

more or less
blood flow in
skin regions

secretion
enhanced
or inhibited

secretion
enhanced
or inhibited

activity of
body tissues
enhanced or
suppressed

altered rates of metabolic activity
and heat conservation

Figure 43.14 Physiological and behavioral adjustments to changing environmental temperatures. This example shows homeostatic control mechanisms and behavioral responses for a typical mammal—a human.

Heterotherms

Certain birds and mammals fall in between the ectothermic and endothermic categories. Part of the time, these **heterotherms** allow the core temperature to fluctuate as ectotherms do; at other times, they control heat exchanges as endotherms do.

For example, given their small size, hummingbirds have very high metabolic rates. They devote much of the day to locating and sipping nectar as an energy source for metabolism. Because hummingbirds do not forage at night, they could rapidly run out of energy unless their metabolic rates decreased considerably. At night, however, they may enter a sleeplike state and become almost as cool as their surroundings.

Thermal Strategies Compared

In general, ectotherms are at an advantage in the warm, humid tropics. They do not have to expend much energy to maintain body temperature, and more energy can be devoted to other tasks, including reproduction. Indeed, reptiles far exceed mammals in terms of head counts and species diversity in the tropics. However, endotherms have the advantage in moderate to cold environments. They are notably more abundant in these places than ectotherms. With their high metabolic rates, polar bears and some other endotherms occupy even the polar regions, where you would never find a lizard.

Besides being morphologically adapted to their habitats, animals can make behavioral and physiological adjustments to environmental temperatures.

43.9 TEMPERATURE REGULATION IN MAMMALS

Responses to Cold Stress

Table 43.3 lists the major responses that birds and mammals make to cold stress. They are called peripheral vasoconstriction, the pilomotor response, shivering, and nonshivering heat production.

Among mammals, **peripheral vasoconstriction** follows a drop in temperature. Thermoreceptors at the body surface detect the decrease. Signals reach the hypothalamus, which commands smooth muscle in the skin's blood vessels to contact. The resulting vasoconstriction reduces the bloodstream's convective delivery of heat to the body's surface. How effective is this response to cold stress? To give an example, when your fingers or toes become cold, all but 1 percent of the blood that otherwise would flow to their skin is diverted.

Smooth muscle controlling the erection of hair or feathers is stimulated to contract. With this **pilomotor response**, fluffing of feathers or of fur creates a layer of still air that reduces convective and radiative heat loss. Behavioral adjustments that minimize the body's exposed surface areas further reduce heat loss, as when cats curl into a ball or when you hold both arms tightly against your body.

When other responses cannot counter cold stress, the hypothalamus calls for increased skeletal muscle action that leads to **shivering**. Rhythmic tremors begin as the muscles contract about ten to twenty times per second, and heat production throughout the body increases several times over. Shivering comes at a high energy cost and is not effective for very long.

Prolonged or severe cold exposure also can lead to a hormonal response that elevates the rate of metabolism. This **nonshivering heat production** is notable in brown adipose tissue, a connective tissue of animals that hibernate or become acclimatized to cold. Human infants have this tissue; adults have little unless they are cold adapted. It is well developed in Korean diving women, who spend six hours a day in cold water.

When defenses against cold are not adequate, the result is *hypothermia*, a condition in which the core temperature falls below normal. In humans, a drop of only a few degrees affects brain function and leads to confusion; further cooling can lead to coma and death (Figure 43.15). Some victims of extreme hypothermia, children particularly, survived prolonged immersion in water much colder than the core temperature. Many animals can recover from profound hypothermia. However, cells that become frozen may be destroyed unless thaw-

---

Table 43.3 Responses to Cold Stress

Core Temperature	Responses Made
36°–34°C (about 95°F)	Shivering response, increase in respiration. Increase in metabolic heat output. Constriction of peripheral blood vessels; blood routed to deeper regions. Dizziness and nausea set in.
33°–32°C (about 91°F)	Shivering response stops. Metabolic heat output drops.
31°–30°C (about 86°F)	Capacity for voluntary motion is lost. Eye and tendon reflexes inhibited. Consciousness is lost. Cardiac muscle action becomes irregular.
26°–24°C (about 77°F)	Ventricular fibrillation sets in (page 667). Death follows.

Figure 43.15 Two tragic episodes of hypothermia.

In 1912, the ocean liner *Titanic* set out from Europe on her maiden voyage across the cold Atlantic to America. In that same year, a huge chunk of the leading edge of a Greenland glacier broke off and began floating out to sea. Late at night on April 14, off the coast of Newfoundland, the iceberg and the *Titanic* made an ill-fated rendezvous. The *Titanic* was the largest ship afloat and was believed to be unsinkable. Survival drills had been neglected, and there were not enough lifeboats to hold even half the 2,200 passengers. The *Titanic* sank in about 2-1/2 hours.

Within two hours, rescue ships were on the scene, yet 1,517 bodies were recovered from a calm sea. All were wearing life jackets. None had drowned. Probably every one of those people died from hypothermia—from a drop in body temperature below tolerance levels.

Late in 1994, hypothermia also contributed to the deaths of more than 900 men, women, and children when a ferry sank rapidly in the cold, storm-tossed waters of the Baltic Sea.

ing is precisely controlled (this sometimes can be done in hospitals). Tissue destruction through localized freezing is called *frostbite*.

Responses to Heat Stress

The main responses to heat stress are peripheral vasodilation and evaporative heat loss (Table 43.4).

With **peripheral vasodilation**, hypothalamic signals cause blood vessels in skin to dilate. More blood flows from deeper body regions to skin, where the excess heat it carries is dissipated (Figure 43.16). With **evaporative heat loss**, the hypothalamus activates sweat glands. Human skin has 2-1/2 million or more sweat glands, and considerable heat is dissipated when the water they give up to the skin surface evaporates. With extreme sweating, as might occur in a marathon race, the body loses an important salt—sodium chloride —as well as water. Such losses may so alter the internal environment that the runner may collapse and faint.

What about mammals that sweat very little or not at all? Some make behavioral responses, such as licking fur, panting, or resting in shade. "Panting" refers to shallow, rapid breathing that increases evaporative water loss from the respiratory tract (compare Figure 33.14). Cooling takes place when the water evaporates from the nasal cavity, mouth, and tongue.

Sometimes peripheral blood flow and evaporative heat loss are not enough to counter heat stress, and *hyperthermia* results. This is a condition in which the core temperature increases above normal. For humans and other endotherms, an increase of only a few degrees above normal can be dangerous.

Fever

During a *fever*, the hypothalamus has reset the body's "thermostat" that dictates what the core temperature is supposed to be. The normal response mechanisms are brought into play, but they are carried out to maintain a higher temperature. At the onset of fever, heat loss decreases and heat production increases. At this time, the individual feels chilled. When a fever "breaks," peripheral vasodilation and sweating increase as the body attempts to restore the normal core temperature; then, the person feels warm.

Fever may be an important defense mechanism against infections, and perhaps against cancer. Following infection, macrophages and other cells secrete signaling molecules, including interleukin-1 and interferons. The secretions somehow influence the hypothalamus. We know that aspirin and other drugs that interfere with the synthesis of prostaglandins can block their influence. (Actually, prostaglandins injected

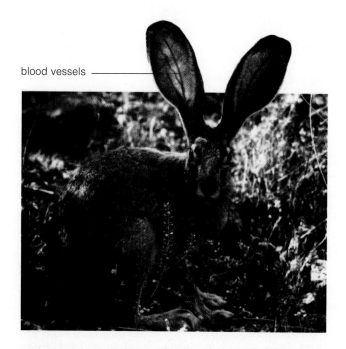

blood vessels

Figure 43.16 A jackrabbit (*Lepus californicus*) cooling off on a hot summer day in the mountains of Arizona. Notice the dilated blood vessels in its large ears. Both the large surface area of the ears and the extensive vascularization are useful for dissipating heat (by way of convection and radiation).

Table 43.4	Summary of Mammalian Responses to Cold Stress and Heat Stress	
Stimulus	Main Responses	Outcome
Cold stress	Widespread vasoconstriction in skin; adjusted behavior (e.g., minimizing surface parts exposed to cold)	Body heat is conserved
	Increased muscle action; shivering; nonshivering heat production	Heat production increases
Heat stress	Widespread vasodilation in skin; sweating; adjusted behavior; panting	Heat is dissipated from body
	Decreased muscle action	Heat production decreases

directly into the hypothalamus can induce fever.) The controlled increase in body temperature during a fever seems to enhance the body's immune response. Therefore, the widespread practice of administering aspirin and other drugs may interfere with beneficial effects of fever. But, without question, these drugs are necessary when the temperature increases that are associated with severe fevers approach dangerous levels.

SUMMARY

Control of Extracellular Fluid

1. Inside the animal body, the cellular environment consists of certain types and amounts of substances dissolved in water. This extracellular fluid fills tissue spaces and blood vessels. Its volume and composition are maintained only when the animal's daily intake and output of water and solutes are in balance. In mammals, the following processes maintain the balance:

   a. Water is gained by absorption (from the gut) and metabolism. It is lost by urinary excretion, evaporation from lungs and skin, sweating, and elimination of feces.

   b. Solutes are gained by absorption from the gut, secretion, respiration, and metabolism. They are lost by excretion, respiration, and sweating.

   c. Losses of water and solutes are controlled mainly by adjusting the volume and composition of urine.

2. The human urinary system consists of two kidneys, two ureters, a urinary bladder, and a urethra.

3. Blood is filtered and urine forms in many nephrons in the kidneys. Each nephron interacts closely with two sets of blood capillaries (glomerular and peritubular).

   a. A nephron has a cup-shaped beginning (Bowman's capsule), then three tubelike regions (proximal tubule, loop of Henle, and distal tubule, which empties into a collecting duct).

   b. Bowman's capsule surrounds a set of highly permeable capillaries. Together, they are a blood-filtering unit (glomerulus).

   c. Blood pressure forces water and small solutes out of the capillaries, into the cup. Most of the filtrate is reabsorbed by the tubules and returned to the blood. A portion is excreted as urine.

4. Urine forms in the nephron by three processes:

   a. Filtration of blood at the glomerulus, which puts water and small solutes into the nephron.

   b. Reabsorption. Water and solutes to be retained leave the nephron's tubular parts and enter capillaries that thread around them. A small volume of water and solutes remains in the nephron.

   c. Secretion. A few substances leave peritubular capillaries and enter the nephron, for disposal in urine.

5. Urine becomes more or less concentrated by the action of two hormones (ADH and aldosterone). These act on cells of distal tubules and collecting ducts:

   a. ADH conserves water by enhancing reabsorption across the nephron wall. Inhibition of ADH allows more water to be excreted.

   b. Aldosterone conserves sodium by enhancing reabsorption. Inhibition of aldosterone allows more sodium to be excreted.

6. Together with the respiratory system and other mechanisms, the kidneys also help maintain the body's overall acid-base balance.

Control of Body Temperature

1. Body temperature of animals is determined by the balance between metabolically produced heat and the heat absorbed from and lost to the environment.

2. Animals exchange heat with the environment by these processes:

   a. Radiation: the emission of energy in the form of infrared and other wavelengths that, following absorption at the surface of the animal body or some other object, is converted to heat energy.

   b. Conduction: the direct transfer of heat energy between two objects in direct contact with each other.

   c. Convection: heat transfer by air or water currents; involves conduction and mass transfer of heat-bearing currents away from or toward the animal body.

   d. Evaporation: the conversion of a liquid to the gaseous state; requires energy, supplied from the heat content of the liquid.

3. The body temperature of different animals depends on the rate of metabolic activity and anatomical, behavioral, and physiological adaptations.

   a. Ectotherms are animals whose body temperature is determined more by heat exchange with the environment than by metabolic heat.

   b. The body temperature of endotherms is determined largely by metabolic activity and by precise controls over heat produced and heat lost.

   c. Heterotherms allow their body temperature to fluctuate at some times, and at other times they control heat balance.

Review Questions

1. Label the regions of the nephron, and identify where filtration, reabsorption, and secretion occur: *738–739*

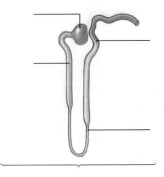

2. How does urine formation help maintain the body's internal environment? *736*

3. Define the components of the human urinary system. Then label the kidney's component parts: *737–738*

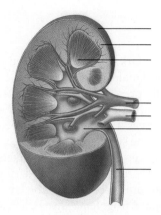

4. Which hormone influences water reabsorption and conservation? sodium reabsorption and conversion? *741*

5. Fatty tissue holds the kidneys in place. Extremely rapid weight loss may cause the tissue to shrink, so that the kidneys slip from their normal position. On rare occasions, the slippage may put a kink in one or both ureters and block urine flow. Speculate on what might happen to the kidneys if this happens. *742*

Self-Quiz *(Answers in Appendix IV)*

1. In mammals, water intake depends on _____ .
   a. absorption from gut   c. a thirst mechanism
   b. metabolism   d. all of the above

2. In mammals, water is lost by way of the _____ .
   a. skin   d. urinary system
   b. respiratory system   e. c and d
   c. digestive system   f. a through d

3. Water and small solutes return to blood during _____ .
   a. filtration   c. secretion
   b. reabsorption   d. both a and b

4. Water and small solutes leave blood during _____ .
   a. filtration   c. secretion
   b. reabsorption   d. both a and c

5. A few substances move out of the capillaries threading around tubular parts of the nephron. These substances are moved into the nephron during _____ .
   a. filtration   c. secretion
   b. reabsorption   d. both a and c

6. A nephron's reabsorption mechanism depends on _____ .
   a. osmosis across nephron wall
   b. active transport of sodium across nephron wall
   c. a steep solute concentration gradient starting at kidney cortex and descending into medulla
   d. all of the above

7. Water is conserved and urine becomes more concentrated by the action of _____ .
   a. ADH   c. aldosterone
   b. renin   d. both b and c

8. Sodium is conserved and urine becomes more dilute by the action of _____ .
   a. ADH   c. aldosterone
   b. renin   d. both b and c

9. Match the term with the appropriate description.
   ____ glomerulus   a. surrounded by saltiest fluid
   ____ distal tubule   b. extra-long loops of Henle
   ____ loop of Henle   c. involves buffer systems
   ____ acid-base balance   d. blood-filtering unit
   ____ kangaroo rat   e. ADH, aldosterone act here

10. Match the term with the appropriate description.
    ____ ectotherm   a. heat transfer by air/water currents
    ____ endotherm   b. body temperature fluctuates or is
    ____ evaporation   controlled
    ____ heterotherm   c. emission of wavelength energy
    ____ radiation   d. heat transfer between two objects in
    ____ conduction   contact
    ____ convection   e. metabolism dictates body temperature
                         f. environment dictates body
                            temperature
                         g. conversion of liquid to gas

Selected Key Terms

acid-base balance *742*
aldosterone *741*
antidiuretic hormone
   (ADH) *741*
behavioral temperature
   regulation *746*
bicarbonate–carbon dioxide
   buffer system *742*
Bowman's capsule *738*
conduction *744*
convection *744*
core temperature *744*
distal tubule *738*
ectotherm *746*
endotherm *746*
evaporation *745*
evaporative heat loss *749*
extracellular fluid *736*
filtration *739*
glomerulus *738*
heterotherm *747*

kidney *737*
kidney dialysis machine *742*
loop of Henle *738*
nephron *738*
nonshivering heat
   production *748*
peripheral
   vasoconstriction *748*
peripheral vasodilation *749*
pilomotor response *748*
proximal tubule *738*
radiation *744*
reabsorption *739*
secretion *739*
shivering *748*
ureter *737*
urethra *737*
urinary bladder *737*
urinary excretion *736*
urinary system *736*
urine *737*

Readings

Flieger, K. March 1990. "Kidney Disease: When Those Fabulous Filters Are Foiled." *FDA Consumer* 24:26–29.

Schmidt-Nielsen, K. 1990. *Animal Physiology*. Fourth edition. New York: Cambridge. Chapters 8 and 9 provide an excellent introduction to water-solute balances in animals.

Smith, H. 1961. *From Fish to Philosopher*. New York: Doubleday. Paperback.

Vander, A., J. Sherman, and D. Luciano. 1990. "The Kidneys and Regulation of Water and Inorganic Ions" in *Human Physiology*. Fifth edition. New York: McGraw-Hill, Chapter 15.

44 PRINCIPLES OF REPRODUCTION AND DEVELOPMENT

From Frog to Frog and Other Mysteries

With a quavering, low-pitched call that only a female of its kind could find seductive, a male frog proclaims the onset of warm spring rains, of ponds, of sex in the night. By August the summer sun will have parched the earth, and his pond dominion will be gone. But tonight is the hour of the frog!

Through the dark, a hormone-primed female moves toward the vocal male. They meet, then they dally in the behaviorally prescribed ways characteristic of their species. He clamps his forelegs about her swollen abdomen and gives it a prolonged squeeze (Figure 44.1a). Out into the surrounding water streams a ribbon of hundreds of eggs, and the male blankets them with a milky cloud of sperm. Soon afterward, tiny fertilized eggs—zygotes—are suspended in the water.

For the leopard frog, *Rana pipiens*, a drama now begins to unfold that has been reenacted each spring, with only minor variations, for many millions of years. Within a few hours after fertilization, each zygote divides into two cells, then four, then many more. In less than twenty hours, mitotic cell divisions have produced a ball of cells, no larger than the zygote. This is the early embryo.

a

b

developing embryo

c

Figure 44.1 Development of the leopard frog, *Rana pipiens*. (**a**) A male clasping a female in a behavior called amplexus. When the female releases her eggs into the water, the male releases his sperm over the eggs. (**b**) Frog embryos. (**c**) A larval form called a tadpole. (**d**) Transitional form between the tadpole and the young adult frog (**e**).

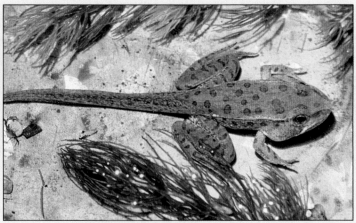

d

Soon the cells are signaling one another with different chemical substances. They continue to divide, but many of their descendants are now changing shape and migrating to prescribed positions within the embryo. Those cells, so recently developed from a single zygote, are becoming different from one another in both appearance and function! Gradually they form distinctive layers of embryonic tissues, then embryonic organs. Certain cells at the surface of the embryo interact with interior cells beneath them—and together they give rise to eyes. Within the embryo a tiny heart is forming. Very soon now, it will start an incessant, rhythmic beating.

In less than a week's time, that embryo has become a tadpole—a swimming, algae-eating larva (Figure 44.1c). Several months pass. Legs start to grow. The tail shortens, then disappears. The mouth develops jaws and now snaps shut on insects and worms. Eventually the transformations lead to an adult frog. With luck the frog will avoid predators, disease, and other threats in the months ahead. In time it may even call out quaveringly across a moonlit pond, and the cycle will begin again.

How does the single-celled zygote of a frog or any other complex animal become transformed into all the specialized cells and structures of the adult form? With this question we turn to one of life's greatest dramas—to the development of offspring in the image of their parents.

KEY CONCEPTS

1. Although many animals can reproduce asexually, most species rely mainly on sexual reproduction. Their specialized reproductive structures and often their behavior assist fertilization and, later, survival of the young.

2. Commonly, the life cycles of animals include six stages of embryonic development—from gamete formation, through fertilization, cleavage, gastrulation, and organ formation, and finally growth and tissue specialization.

3. In a developing embryo, the fate of each cell type depends partly on cleavage (when daughter cells inherit different regions of the fertilized egg's cytoplasm) and partly on cell interactions. These two events are the basis of cell differentiation and morphogenesis.

4. In cell differentiation, each type of cell selectively uses certain genes and synthesizes proteins not found in other types, and so becomes unique in structure and function.

5. In morphogenesis, tissues and organs change in size, shape, and proportion and become organized in prescribed patterns.

6. Each stage of embryonic development builds on tissues and structures that formed during the stage preceding it.

e

44.1 THE BEGINNING: REPRODUCTIVE MODES

In earlier chapters, you read about the cellular basis of **sexual reproduction**, in which offspring are produced by way of meiosis, gamete formation, and fertilization. You also read about **asexual reproduction**, in which offspring are produced by means other than gamete formation. Consider now a few structural, behavioral, and ecological aspects of these two different reproductive modes.

Think of a new sponge growing asexually from a fragment of parent sponge (page 413). Or think of flatworm fission. As certain flatworms glide through water, their body starts to constrict below the midsection. The part behind the constriction grips a substrate and initiates a tug-of-war with the part in front of it. After several hours it splits off, both parts go their separate ways—and both regenerate the missing part to become a whole worm. In such cases of asexual reproduction, offspring are genetically the same as parents, or nearly so. This absence of variation is useful when gene-encoded traits are highly adaptive to a limited and more-or-less consistent set of environmental conditions.

Most animals live in places where opportunities, resources, and danger vary in complicated ways. Such animals rely mainly on sexual reproduction, with offspring getting different mixes of alleles from male and female parents. The resulting variation in traits improves the odds that some offspring, at least, will be able to survive and reproduce.

Separation into sexes is not without cost. Sex means constructing special reproductive structures. It means engaging in behavior, such as courtship, that promotes fertilization. It means having control mechanisms to synchronize the timing of sperm and egg production, sexual readiness, even parental behavior.

Consider the question of *reproductive timing*. How do mature sperm in one individual become available exactly when eggs mature in a different individual? Timing depends on energy outlays for sensory structures and rather involved hormonal controls in both parents. Both parents must produce mature gametes in response to the same cues, such as changes in daylength that mark the onset of the best season for reproduction. Moose, for example, become sexually active in late summer or early fall. This assures that their offspring will be born the following spring, when the weather improves and food will be plentiful for many months.

Or consider the challenge of finding and recognizing a potential mate of the same species. Different species meet the challenge by investing energy in the production of chemical signals, structural signals such as feathers of certain colors and patterns, and sensory

a

b

receptors able to pick up the signals that are being sent. Besides this, the males often expend astonishing amounts of energy on elaborate courtship routines, as you will see in Chapter 52.

Assuring survival of offspring is also costly. Many invertebrates and bony fishes release eggs and motile sperm into the water. The chance of successful fertilization would not be good if adults produced only one sperm or one egg each season. Such species invest energy in producing gametes by the hundreds of thousands. As another example, nearly all land-dwelling animals depend on internal fertilization, the union of sperm and egg *inside* the body of the female parent. They invest energy on elaborate reproductive organs, such as a penis (by which a male deposits its sperm within the female) and a uterus (a chamber in the female where the embryo grows and develops).

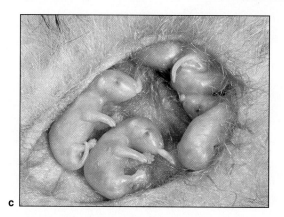

c

d

e

Figure 44.2 Examples of where invertebrate and vertebrate embryos develop, how they are nourished, and how (if at all) parents protect them. (**a**) Snails are *oviparous*; they produce eggs. Parents release the fertilized eggs, which develop on their own, unprotected. (**b**) Birds are oviparous. Their fertilized eggs, which have large yolk reserves, develop and hatch outside the mother's body. Parents expend considerable energy feeding and caring for the young.

(**c–e**) Most mammals are *vivaparous*; their young are born live (*viva-*, alive, *parous*, produce). The mother retains fertilized eggs, and her tissues nourish the embryo until the time of birth. The kangaroo, a marsupial, also nourishes embryos, but they are born in unfinished form. The young complete development in a pouch on the mother's ventral surface, where they are nourished from mammary glands.

Some fishes, lizards, and many snakes are *ovoviviparous*. Their fertilized eggs develop in the mother; then offspring are born live. Yolk reserves, not the mother's tissues, sustain such eggs. The copperhead shown in Figure 27.17*b* is an example. Her liveborn emerge in the relics of egg sacs.

Finally, energy is set aside for *nourishing some number of offspring*. Nearly all animal eggs contain **yolk**, a protein-rich, lipid-rich substance that can nourish the embryo. The eggs of some species have more yolk than others. Sea urchins produce small eggs without much yolk, release them in large numbers, and limit the biochemical investment in each one. A fertilized egg develops into a self-feeding, free-moving larval stage in less than a day. This reproductive strategy probably correlates with the fact that predators consume most of the eggs. Sea stars especially go into feeding frenzies when sea urchin eggs are released.

By contrast, mother birds lay eggs with a large yolk reserve. This nourishes the embryo through a longer period of development, inside an eggshell that forms after fertilization. Your own mother put great demands on her body to protect and nourish you through many months of early development, inside her body, from the time you were an egg with almost no yolk. You became attached to her uterus, and as you developed, physical exchanges with her tissues supported you during an extended pregnancy (Figure 44.2*e*).

As these examples suggest, animals show great diversity in reproduction and development. However, some patterns are widespread in the animal kingdom, and they will serve as a framework for our reading.

Separation into male and female sexes requires special reproductive structures, control mechanisms, and behaviors. The biological cost is offset by a selective advantage: variation in traits among the offspring.

44.2 STAGES OF DEVELOPMENT

You don't look like a frog. You didn't look like one when the two of you were embryos, either. **Embryos** are transitional forms on the road from a fertilized egg to the adult. Yet despite the differences, it is possible to identify certain patterns in the way that the embryos of nearly all species of animals develop.

Figure 44.3 provides a simple overview of the stages of animal development. In **gamete formation**, the first stage, eggs or sperm form within reproductive organs in the parent's body. The second stage, **fertilization**, begins when a sperm penetrates an egg. It ends when the egg nucleus and sperm nucleus fuse. Their fusion marks the formation of a zygote, the first cell of a new individual. Next comes **cleavage**—mitotic cell divisions that convert the zygote into a multicelled embryo. In many species of tiny animals, cleavage produces a blastula, a tiny ball of cells that is no wider in diameter than the zygote was.

As cleavage draws to a close, the pace of cell division slackens. The embryo enters **gastrulation**, which is a stage of major reorganization. Now cells become arranged into two or three primary tissues, or "germ layers," which will give rise to all of the tissues of the adult:

1. **Ectoderm**. This is the outermost primary tissue layer, and the one that forms first in all animal embryos. Ectoderm is the embryonic forerunner of the cell lineages that give rise to neural tissues and the outer layer of the integument.

2. **Endoderm**. This is the innermost primary tissue layer. It is the embryonic forerunner of the gut's inner lining and of the organs derived from it.

3. **Mesoderm**. This intermediate germ layer is the forerunner of muscle; of circulatory, reproductive, and excretory organs; of most of the skeleton; and of connective tissue layers of the gut and integument. Hundreds of millions of years ago, the origin of mesoderm was a pivotal step in the evolution of most large, complex animals (Chapter 26).

Next, the germ layers split into subpopulations of cells. This marks the onset of **organ formation**. By now, different sets of cells have become unique in structure and function, and their descendants are giving rise to different kinds of tissues and organs. During the final stage of development, called **growth and tissue specialization**, organs expand and take on specialized properties. This stage continues into adulthood.

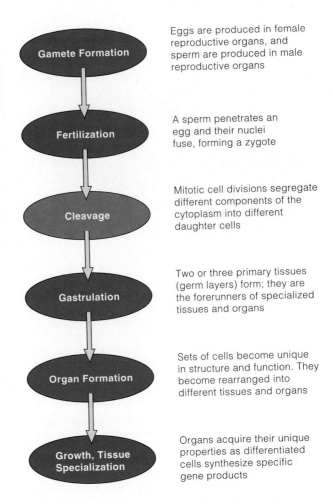

Figure 44.3 Overview of the stages of animal development.

Figure 44.4 shows examples of some of the stages of embryonic development. Take a moment to study these examples, for they are revealing of an important concept. By the end of each developmental stage, the embryo has become more complex than it was before. As you will see in sections to follow, the structures that develop during one stage serve as the foundation for the stage that follows it. Successful embryonic development absolutely depends on the formation of these structures according to normal patterns, in a prescribed sequence.

Each stage of embryonic development builds on structures that were formed during the stage preceding it. Development cannot proceed properly unless each stage is successfully completed before the next begins.

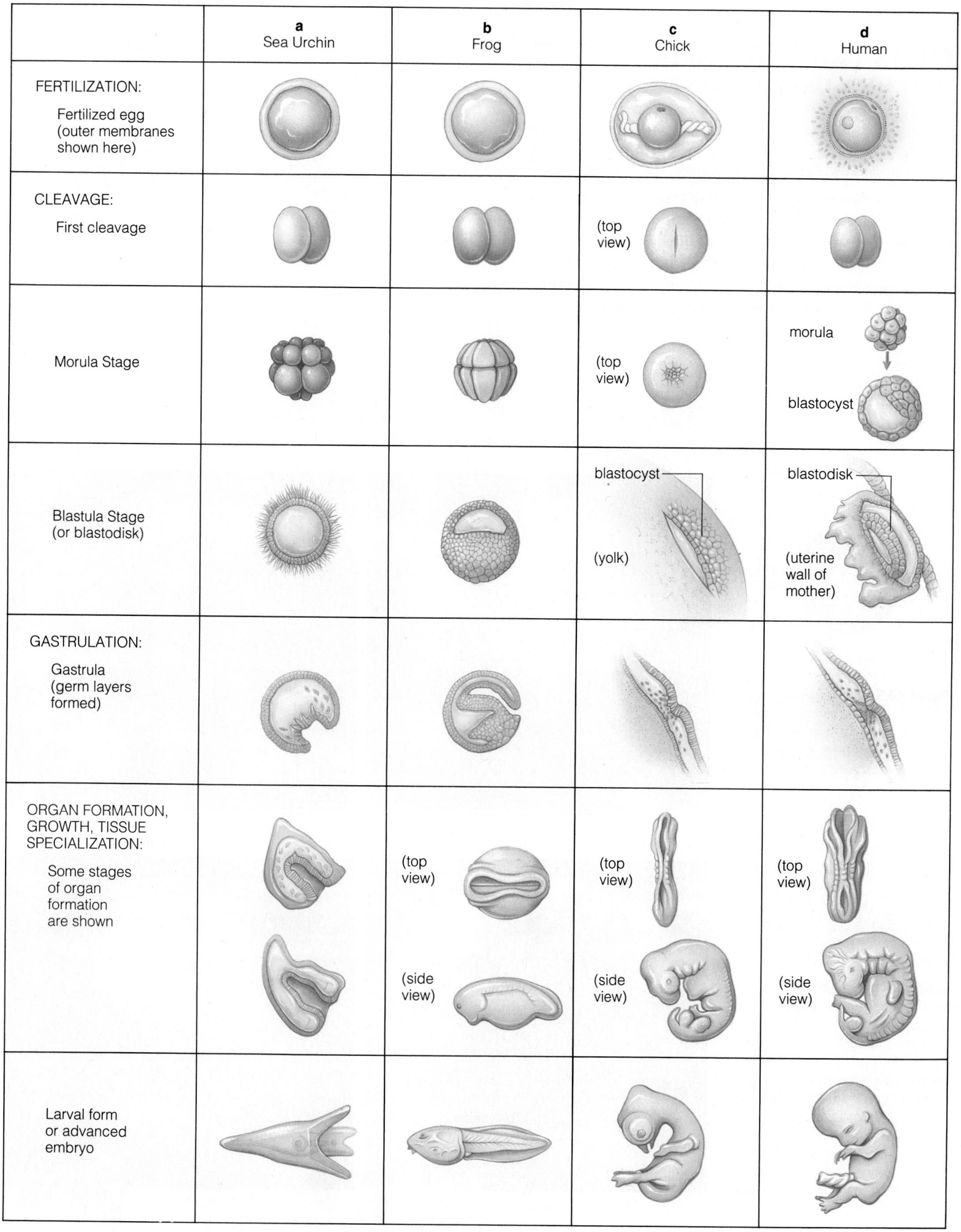

	a Sea Urchin	**b** Frog	**c** Chick	**d** Human
FERTILIZATION: Fertilized egg (outer membranes shown here)				
CLEAVAGE: First cleavage			(top view)	
Morula Stage			(top view)	morula blastocyst
Blastula Stage (or blastodisk)			blastocyst (yolk)	blastodisk (uterine wall of mother)
GASTRULATION: Gastrula (germ layers formed)				
ORGAN FORMATION, GROWTH, TISSUE SPECIALIZATION: Some stages of organ formation are shown		(top view) (side view)	(top view) (side view)	(top view) (side view)
Larval form or advanced embryo				

Figure 44.4 Comparison of embryonic development in four different animals. The developmental patterns are common to all four types. For clarity, membranes that normally surround the embryo are not shown from cleavage onward. Blastula and gastrula stages are shown in cross-section. The drawings are not to the same scale.

44.3 A VISUAL TOUR OF FROG AND CHICK DEVELOPMENT

In the sections that follow, you will be reading about the key mechanisms underlying the different stages of animal development. You also will be considering some of the experiments that provided crucial insights into those mechanisms. When reading through such details, it is sometimes easy to lose sight of the stunning magnitude of the transformations and of how rapidly some of the stages may proceed. The selection of photographs in Figures 44.5 and 44.6 may help you visualize what goes on.

Figure 44.6 Onset of organ formation in a chick embryo during the first seven days of development. The heart begins to beat between 30 and 36 hours. You may have observed such embryos at the yolk surface of raw, fertilized eggs.

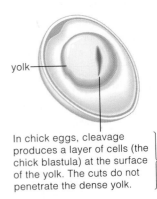

yolk

In chick eggs, cleavage produces a layer of cells (the chick blastula) at the surface of the yolk. The cuts do not penetrate the dense yolk.

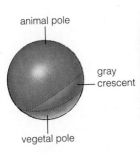

animal pole

gray crescent

vegetal pole

Figure 44.5 Micrographs and diagrams of the early embryonic development of a frog. For these micrographs, the jellylike layer surrounding the egg has been removed, with the exception of the micrograph in (**i**).

(**a**) Within about an hour after fertilization, the gray crescent establishes the body axis for the embryo, and gastrulation will begin here. (**b–f**) Cleavage leads to a blastula, a ball of cells in which a cavity (blastocoel) has appeared.

(**g,h**) Cells move about and become rearranged during gastrulation. Tissue layers form; a primitive gut cavity (archenteron) develops. (**i,j**) Neural developments now take place, and the fluid-filled body cavity in which vital organs will be suspended appears. Differentiation proceeds, moving the embryo on its way to becoming a functional larval form.

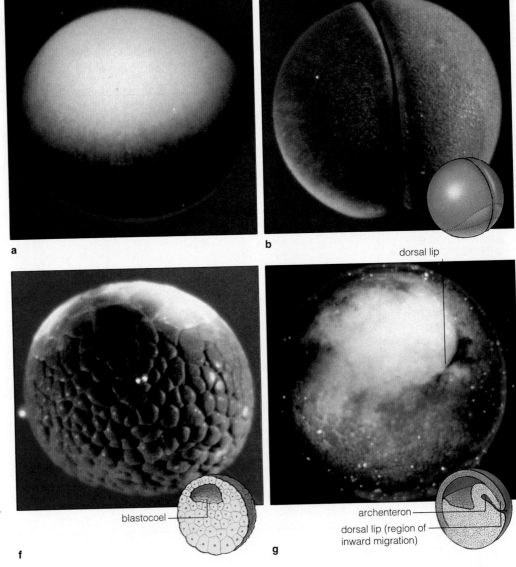

a

b
dorsal lip

f
blastocoel

g
archenteron
dorsal lip (region of inward migration)

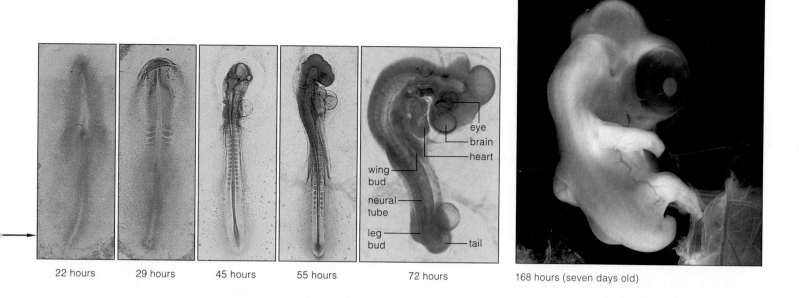

22 hours 29 hours 45 hours 55 hours 72 hours 168 hours (seven days old)

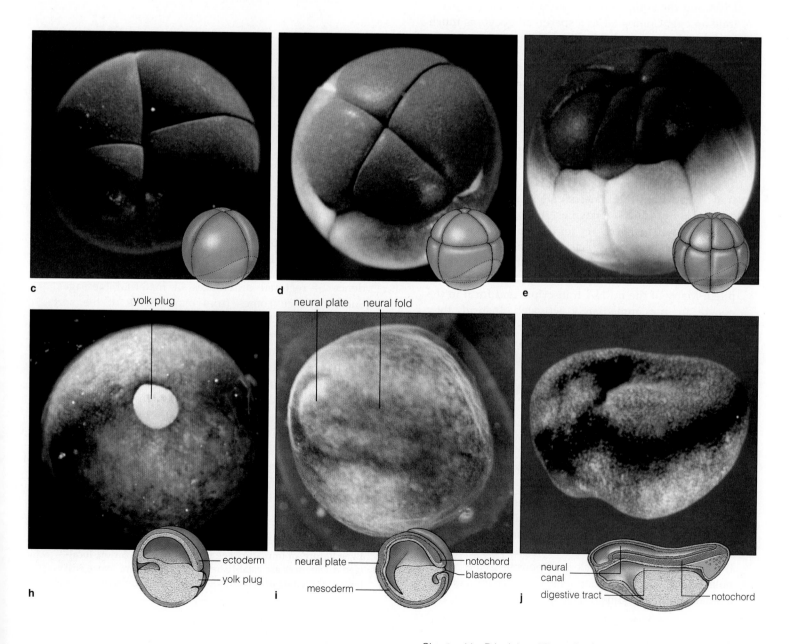

You probably don't have an arm attached to your nose or toenails growing from your navel. Your body parts generally are arranged in predictable patterns. For this you owe some thanks to the cytoplasm of an immature egg, or **oocyte**, that started to form decades ago in your mother's body.

Developmental Information in the Egg Cytoplasm

To a large extent, the development of a future embryo is mapped out in the cytoplasm of an oocyte, even before a sperm enters the picture.

A sperm, recall, consists of little more than paternal DNA and the equipment necessary to reach and penetrate an egg. Compared to a sperm, an oocyte is much larger and more complex. Page 139 and Figure 10.10 show examples.

While an oocyte is maturing, it increases in volume. And it stockpiles "maternal instructions" that will influence the embryo's shape and the arrangement of its body parts. The instructions are embodied in specific enzymes, mRNA transcripts, and other components— and they become distributed in different regions of the egg cytoplasm.

For example, many of the mRNA molecules being transcribed from the maternal DNA are translated at once. The products—enzymes, histones, and other proteins—will be necessary for the repeated rounds of DNA replications and cell divisions in the early embryo. Ribosomal subunits and other cytoplasmic components necessary for protein synthesis are stockpiled. Many of the mRNA transcripts that form in the oocyte are not immediately activated. Instead, they accumulate in different regions of the egg cytoplasm. They will become activated after fertilization, and only after early cell divisions have divided the egg cytoplasm.

Also present in the maturing oocyte are numerous microtubules and other cytoskeletal elements, oriented in specific directions. These elements will influence the first cell divisions of the embryo. Whenever a cell divides in two, it does so at a prescribed angle relative to the adjacent cells, based partly on the orientation of microtubules of the mitotic spindle.

Even the amount and distribution of yolk within the egg cytoplasm will influence cleavage. For example, the position of both the yolk and the nucleus in a frog oocyte imparts polarity to it. The *animal* pole is simply the one closest to the nucleus. Opposite is the *vegetal* pole, where yolk and other substances accumulate. All animal eggs show some degree of polarity—that is, two identifiable poles. As you will see, such polarity will influence the structural development of the embryo.

Reorganization of the Egg Cytoplasm at Fertilization

When a sperm penetrates an egg, it triggers structural reorganization in the egg cytoplasm. You can observe indirect signs of this reorganization in frog eggs at the time of fertilization.

Frog eggs contain dark pigment granules in their cortex, which includes the plasma membrane and the cytoplasm just beneath it. Pigment granules are concentrated near one pole of the egg, and yolk is concentrated near the other pole. At fertilization, a portion of the granule-containing cortex shifts away from the yolk. The shift exposes lighter-colored cytoplasm in a crescent-shaped gray area:

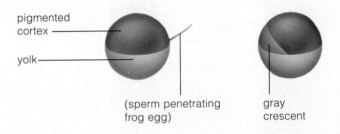

pigmented cortex

yolk

(sperm penetrating frog egg)

gray crescent

The fertilized egg now has a distinctive **gray crescent**, a region of intermediate pigmentation near its equator. Its formation establishes the body axis of the frog embryo.

In itself, the formation of a gray crescent is not evidence of regional differences in maternal messages. Such evidence comes from experiments of the sort shown in Figure 44.7. It also comes from observations of various embryos in which localized cytoplasmic differences are pronounced enough to be tracked easily during the course of development. Thus, if you were to continue tracking the development of fertilized frog eggs, you would find that the gray crescent is always the site where gastrulation normally begins. Similarly, embryos of the tunicate *Styela* have a distinctive region of cytoplasm—a *yellow* crescent (Figure 44.8). The yellow crescent always becomes divided among a small group of cells during normal cleavage. Those cells are the *only* ones in the embryo that give rise to muscles in the larva.

Even before an immature egg undergoes fertilization, localized differences are established in its cytoplasm. These will help seal the fate of cells in the forthcoming embryo.

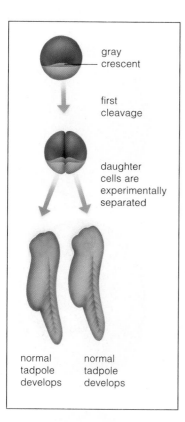

a Experiment 1

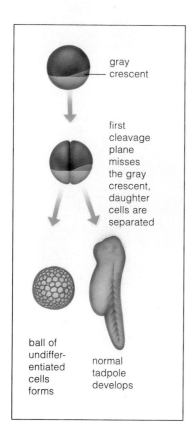

b Experiment 2

Figure 44.7 Examples of experiments that illustrate how the cytoplasm of a fertilized egg must have localized differences that help determine the fate of cells in a developing embryo. The cortex of frog eggs contains granules of dark pigment, which are concentrated near one pole of the egg. At fertilization, a portion of the granule-containing cortex shifts away from the yolk. This exposes lighter colored cytoplasm in a crescent-shaped gray area.

Normally, the first mitotic cell division after fertilization divides the gray crescent between two daughter cells.

(**a**) In one experiment, the first two daughter cells of the embryo were separated from each other. Each still gave rise to a complete tadpole.

(**b**) In another experiment, a fertilized egg was manipulated so that the cleavage plane of the first cell division missed the gray crescent entirely. Only one of the two daughter cells received the gray crescent. It alone developed into a normal tadpole. The daughter cell deprived of substances in the gray crescent gave rise to a ball of undifferentiated cells.

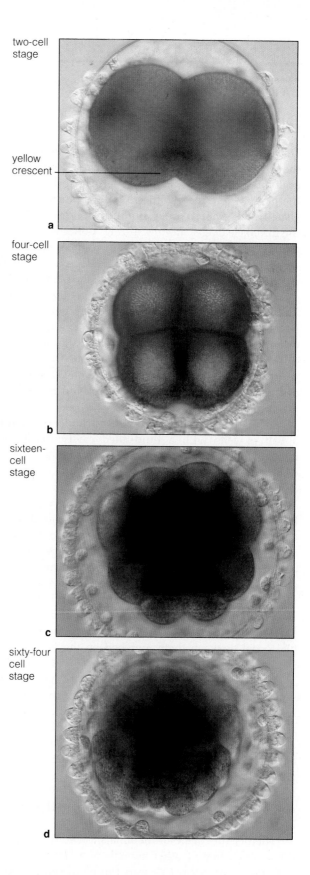

Figure 44.8 Visual signs of localized instructions about development in the egg cytoplasm of *Styela*. Fertilized eggs of these tunicates have a pigmented region called the yellow crescent. During cleavage, the region is allocated to a small group of cells that will give rise to muscles of the larva.

44.5 EMERGENCE OF THE EARLY EMBRYO

Cleavage

Once fertilization is over, mitotic cell divisions begin the zygote. Recall that a cleavage furrow forms during each mitotic cell division (page 150). The furrow defines the plane where the cytoplasm will be pinched in two. Usually an embryo does not increase in size while the initial cleavages are proceeding. Collectively, all the daughter cells occupy the same volume as did the zygote, although they differ in size, shape, and activity.

Simply by virtue of their location, the cells that form during cleavage receive different maternal instructions. This outcome, **cytoplasmic localization**, helps seal the developmental fate of each cell's descendants. For example, only the portion of cytoplasm allocated to one cell may contain molecules of an enzyme that can activate a gene coding for a certain hormone. Descendants of that cell alone will be the only ones in the body to produce the hormone.

Successive cleavages commonly produce a **blastula**, a ball of cells with a fluid-filled cavity (blastocoel) inside. However, the patterns of cleavage vary among different animal groups, often as a result of dramatic differences in the density and distribution of yolk.

For example, cleavage of a sea urchin egg, which has little yolk, produces a hollow, single-layered sphere of cells (Figures 44.4a and 44.9). In the eggs of frogs and other amphibians, the concentrated yolk impedes cleavage near the vegetal pole. A blastocoel forms near the animal pole (Figures 44.4b and 44.5). Reptile, bird, and most fish eggs have a large volume of yolk that restricts cleavage to a tiny, caplike region at the animal pole. Cleavage produces two flattened layers with a thin blastocoel between them, perched on the yolk surface (Figures 44.4c and 44.6).

The eggs of fruit flies (*Drosophila*) also have a large volume of yolk that influences cleavage in a distinct way. At first the nucleus of a fertilized egg divides repeatedly. However, the cytoplasm does not, so cleavage furrows do not form. Within hours, the fertilized egg has become a bag of nuclei, crowded together in the yolky cytoplasm (Figure 44.10). Then most of the nuclei migrate toward the egg surface. Cytoplasmic division now proceeds, and a layer of cells forms. This layer, a blastoderm, is the embryo.

The blastula stage of mammals, called a blastocyst, differs from that of other animals in a key respect. By this stage, *two* distinct regions can be discerned. Some cells form a hollow sphere that is not part of the embryo. A cluster of different cells attaches to the inner surface of the blastocyst wall. As described in Chapter 45, the embryo develops from this "inner cell mass."

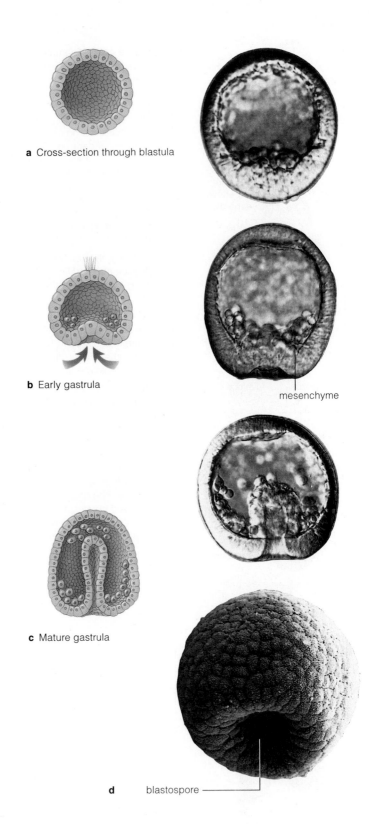

a Cross-section through blastula

b Early gastrula

mesenchyme

c Mature gastrula

d blastospore

Figure 44.9 (a) Blastula of a sea urchin (*Lytechinus*). The inward migration of surface cells during gastrulation is apparent in the cross-sections in (b) and (c). Some of the cells, designated mesenchyme, will ultimately develop into a third germ layer, the mesoderm. (d) This scanning electron micrograph shows a surface view of individual cells and the inward cell migrations of an early gastrula.

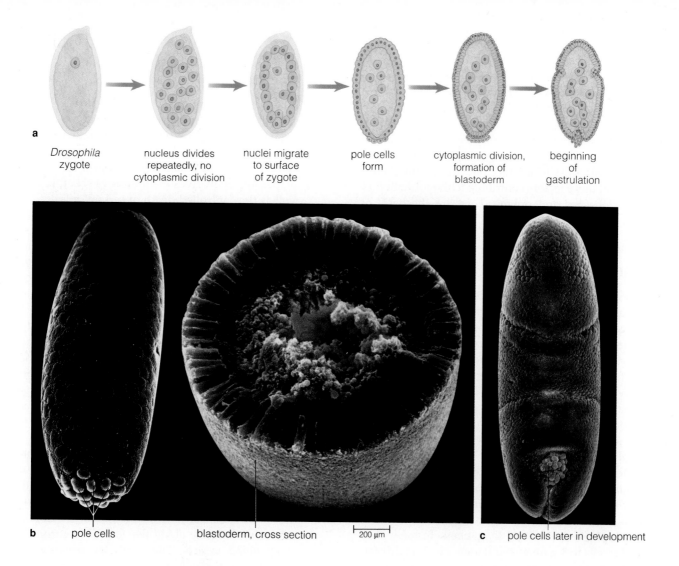

a

Drosophila zygote

nucleus divides repeatedly, no cytoplasmic division

nuclei migrate to surface of zygote

pole cells form

cytoplasmic division, formation of blastoderm

beginning of gastrulation

b pole cells   blastoderm, cross section   ⊢ 200 µm ⊣   **c** pole cells later in development

Figure 44.10 (**a**) Delayed cleavage in a *Drosophila* zygote. The nucleus divides repeatedly. Cleavage is postponed until the nuclei migrate toward the zygote's surface. A cell layer, the blastoderm, forms above the yolky cytoplasm. It is the early embryo.

(**b,c**) As the blastoderm forms, cytoplasmic components (polar granules) migrate and become isolated in pole cells that form at one end of the zygote. In certain mutant embryos, polar granules do not migrate properly. Later, reproductive cells never do form. The adult fly is sterile.

Gastrulation

As cleavage draws to a close, the pace of cell division slackens. Cells start to migrate and change positions relative to one another. Gastrulation, the stage of cell rearrangements, has begun. The embryo's size increases little, if at all. In sea urchins, surface cells migrate inward. Some form a lining of a cavity (archenteron) that will become the gut. In other species, the rearrangements establish a long axis with later roles in development. For example, in vertebrates, a neural tube—the forerunner of a brain and spinal cord—will form along this axis.

This last example underscores the significance of gastrulation. Nearly all animals have an internal region of cells, tissues, and organs that function in digestion and absorption of nutrients. They have a surface region with tissues that protect internal parts and sensory receptors that detect outside changes. In between these regions, most animals have numerous organs, such as those concerned with movement, support, and blood circulation. These regions develop from three germ layers—the endoderm, ectoderm, and mesoderm. *This three-layered organization is typical of most animals, and it arises through gastrulation.*

1. After fertilization, cleavage produces a number of daughter cells at prescribed locations in the embryo.

2. The particular cleavage planes that form dictate which maternal instructions each cell will inherit. They also influence its size and spatial position.

3. Cytoplasmic localization influences how each cell will interact with others during the gastrula stage—and so on through subsequent stages of development.

FORMATION OF SPECIALIZED TISSUES AND ORGANS

Maternal instructions guide the cleavages that form the early embryo. Then, as cleavage draws to a close, groups of cells start to interact physically and chemically. Both of these factors contribute to the formation of specialized tissues and organs. They are the basis of two processes that are crucial to development: cell differentiation and morphogenesis.

Cell Differentiation

All cells of an animal embryo descend from the same zygote, and they generally have the same number and kind of genes. From gastrulation onward, different cells and their descendants use a fraction of the genes in different ways. In **cell differentiation**, a cell selectively uses certain genes and synthesizes proteins that are not found in other cell types. Chapter 15 describes the molecular basis of this selective gene expression. Here it is enough to remember that the process results in proteins that are the foundation for distinct cell structures, products, and functions.

For example, when your eyes were developing, some cells started synthesizing quantities of crystallin, a family of proteins that become incorporated in transparent fibers of the lens. Only these cells could activate the required genes. As the long fibers formed, they forced the cells to lengthen and flatten out. These differentiated cells are the basis of the unique optical properties of your pair of lenses. Crystallin-producing cells are only one of 150 or so differentiated cell types now present in your body.

With few exceptions, cells become fully differentiated without any loss of genetic information. We know this from many experiments. For example, John Gurdon and his colleagues took unfertilized eggs from African clawed frogs (*Xenopus laevis*) and removed or inactivated the nucleus. They also isolated intestinal cells from *Xenopus* tadpoles and ruptured the plasma membrane, leaving the nucleus and most of the cytoplasm intact. Then they inserted the tadpole cell nucleus into the enucleated egg. In some cases, the transplanted nucleus, *which was from a differentiated intestinal cell*, directed the developmental program leading to a whole frog! Clearly the intestinal cell nucleus still had the same genes as the zygote nucleus.

Or consider what happens when a human embryo spontaneously splits during the first cleavage into two separate cells. The result is not two half-embryos but **identical twins**—two complete, normal individuals having the same genetic makeup. (By contrast, *nonidentical* twins form when two different eggs are fertilized at the same time by two different sperm.)

Morphogenesis

As an embryo continues to develop, tissues and organs change in size, shape, and proportion, and they become organized in patterns. This process, called **morphogenesis**, involves cell divisions and tissue growth. It involves cell migrations and changes in cell size and shape. And it involves the folding of sheetlike tissues as well as the controlled death of certain cells.

Cell Migrations During morphogenesis, cells and tissues migrate from one site to another. In active cell migration, cells move about by means of pseudopods. These transient extensions of the cell body are bulbous or fingerlike. Cells migrate over prescribed pathways, reach a prescribed destination, and establish contact with cells already there. For instance, the forerunners of nerve cells make the billions of connections that will allow the human nervous system to function.

How do cells "know" where to move? In part, they respond to chemical gradients. Specific substances released from target tissues may create the gradients. Figure 23.4 describes such chemotactic behavior for *Dictyostelium discoideum*, one of the slime molds.

Cells also move in response to adhesion proteins at the surface of other cells and in the extracellular matrix. In vertebrate embryos, pigment cells follow adhesive cues when they move along blood vessels but not along the axons of neurons. Schwann cells follow similar cues when they migrate along the axons of neurons but not along blood vessels. The synthesis, secretion, deposition, and removal of specific extracellular substances probably help coordinate the active migration of cells in a developing embryo.

How do migrating cells know when to stop moving? Adhesive cues must be involved. Migrating cells move to locations where adhesive interactions are strongest. Once they become arranged in a manner that maximizes adhesion, further migration is impeded.

Changes in Cell Size and Shape Another morphogenetic movement is the inward or outward folding of sheets of cells. Coordinated changes in cell shapes bring about the folding. Consider what happens after three germ layers form in the embryos of amphibians, reptiles, birds, and mammals. At an embryo's midline, ectodermal cells elongate and form a **neural plate**, the first indication that a region of ectoderm is on its way to becoming nervous tissue. Lengthening microtubules in the cytoplasm bring about the elongation. Next, cells near the middle become wedge-shaped when a ring of microfilaments constricts them at one end. Collectively, changes in shape cause the neural plate to fold over and meet at the midline, thus forming a **neural tube**. This tube gives rise to the brain and spinal cord (Figure 44.11).

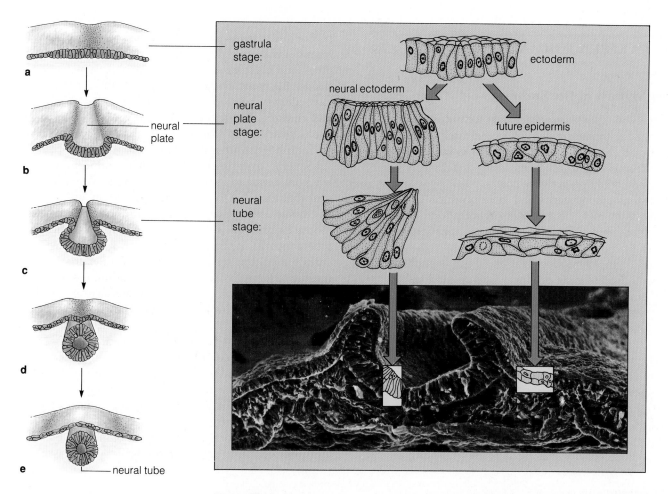

Figure 44.11 Example of morphogenesis—changes in cell shape during the formation of a neural tube, the forerunner of the brain and spinal cord. (**a**) When gastrulation ends, the ectoderm is a uniform sheet of cells. Now some cells elongate, forming a neural plate. Some constrict at one end and become wedge shaped. (**b–e**) Collectively, the changes make part of the sheet fold over the neural plate and form a tube. Other ectodermal cells flatten and become part of the epidermis. The scanning electron micrograph shows the neural tube and epidermis forming in a chick embryo.

Localized Growth Localized growth contributes to changes in the sizes, shapes, and proportions of body parts. How growth proceeds in some tissues more than others is not fully understood, but regulatory genes are surely involved. Recall that chimpanzees and humans have very few differences in their DNA. Yet they differ considerably in the proportions of body parts, including the skull, even though their fetuses develop at about the same rate, in parallel ways (page 304). Possibly mutated regulatory genes have brought about differences in the proportional changes in the skull.

Cell Death Elimination of tissues and cells that are used for only short periods in the embryo or adult is called **controlled cell death**. Such events are genetically programmed. For example, puppies are born with eyes sealed shut; their eyelids form as an unbroken layer of skin. Just after birth, cells stretching in a thin line across the middle of each eyelid die on cue. As the dead cells degenerate, a slit forms in the skin, and then the upper and lower lids part company.

In human embryos, hands and feet start out as paddle-shaped structures (Figure 9.11). Certain cells in the paddles die on cue, leaving separate toes and fingers. Between the time a death signal is sent and the actual time of death, protein synthesis declines dramatically in the doomed cells. Duck embryos also have paddlelike appendages, but cell death normally does not operate in them; that is why ducks have webbed feet instead of separated toes. In some mice and some humans, a gene mutation blocks cell death in the paddles, and the digits remain webbed.

Cell differentiation and morphogenesis are crucial aspects of development. Both processes depend on cytoplasmic localization during cleavage, followed by physical and chemical interactions among embryonic cells.

PATTERN FORMATION

Inducer Signals in the Embryo

Cell differentiation, *morphogenesis*—what technical names for the dazzling transformation of a nondescript clump of cells into an animal with skin and muscle, with head, trunk, and limbs—with all parts fitting together as a spatially organized, integrated whole!

Interactions among genes and gene products bring about this astounding feat of self-organization. They accomplish **pattern formation**—the specialization of tissues and their orderly positioning in space.

Through cytoplasmic localization, recall, signaling molecules and other gene products take up stations in different embryonic cells. Then, at prescribed times, some of the molecules switch on and suppress different genes in different groups of cells. *Their action allows the cell groups to sense where they are positioned in the body and which body parts they must construct.* The name for this selectively induced gene expression during development is **embryonic induction**.

In 1994, researchers identified **morphogens**, a class of signaling molecules that serve as inducers. When morphogens diffuse through embryonic tissues, they trigger the gene activity that leads to the sculpting of specialized tissues and body parts. Genes that specify morphogens were discovered through a mutation that gives fruit flies a bristly surface, much like a hedgehog's. (Hence the name, hedgehog genes.) At this writing, they have already been found in fruit flies, mice, zebra fish, and chickens. These organisms are widely separated in evolutionary time, which tells us that the genes coding for morphogens are highly conserved and vital.

Several experiments show that inducers diffuse from one tissue to the other. For example, when two tissues that give rise to pancreatic epithelium and connective tissue are surgically separated from each other in a mouse embryo, the future epithelium does not differentiate properly. When the two are grown together in a culture medium, separated only by a filter that is permeable to large molecules, the responding tissue differentiates as it should. Even in the absence of physical contact with the inducing tissue, the signaling molecules still reach it.

An experiment by Hans Spemann, a pioneer in developmental biology, provided a stunning example of embryonic induction. In all vertebrates, the eye's retina originates from the forebrain. Its lens, which focuses light onto the retina, originates from ectoderm (Figure 44.12). Spemann surgically removed retina-forming tissue from the forebrain region of a salamander embryo and inserted it into ectoderm on the belly. Afterward, a

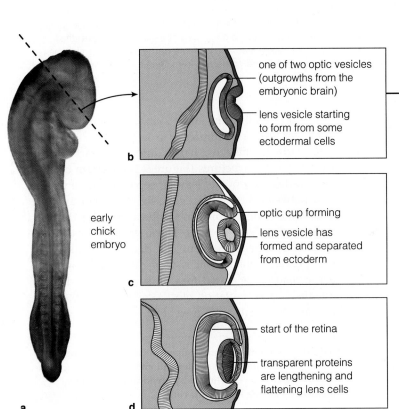

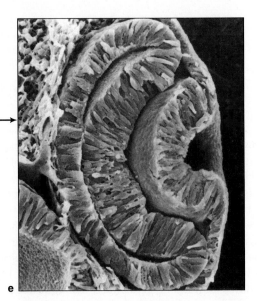

one of two optic vesicles (outgrowths from the embryonic brain)

lens vesicle starting to form from some ectodermal cells

b

early chick embryo

optic cup forming

lens vesicle has formed and separated from ectoderm

c

start of the retina

transparent proteins are lengthening and flattening lens cells

a   **d**   **e**

Figure 44.12 Eye formation in a chick embryo. (**a**) Each eye starts forming at a fluid-filled pouch (optic vesicle) beneath ectoderm of the head. (**b**) When ectoderm cells come into contact with cells of the optic vesicle, they are induced to elongate, fold inward, and form a lens vesicle. (**c**) At the same time, the contact induces cells of the optic vesicle to sink inward, forming an optic cup from which most of the eyeball will form. (**d**) The cup's inner layer will produce the retina. (**e**) Scanning electron micrograph of a forming lens in the right optic cup of the embryo's head.

lens formed on the belly! As Spemann correctly concluded, a molecular signal that emanates from retinal cells normally induces ectodermal cells to develop into a lens.

The positioning of bones in a developing chick wing is another example of embryonic induction. Each wing forms as a primordial bud, through cell divisions in mesoderm and in ectoderm. Surgically remove certain groups of ectodermal cells before the bud is fully grown, and further development here ceases:

normal wing

surgically altered wing

As new cells are added to the outwardly growing tip, they sense which bones already formed, then they become the next bones in line. Remove the ridge of ectoderm at the tip, and there will be no new mesodermal cells to form the remaining bones. Only the bones already formed at the time of the surgery differentiate; the wing skeleton will be incomplete.

Homeobox Genes

Edward Lewis was the first to perceive that certain genes control the developmental responses of groups of cells along the body's anterior-posterior axis. He also perceived that their sequence in chromosomes parallels the linear order of the cell groups. In animals as distant in evolutionary time as roundworms (*Caenorhabditis elegans*), fruit flies, and vertebrates, these **homeobox genes** and others control blocks of genes necessary for pattern formation. Their products interact with regulatory elements in the cellular DNA to activate or inhibit blocks of genes in similar ways.

Consider a *Drosophila* zygote. A "fate map" of its surface shows where each kind of differentiated cell in the forthcoming body segments originates (Figure 42.13*a*). During early development, before the embryonic cells show any sign of their fate, different homeobox genes are activated in successive stripes along the body's long axis. In one stripe, the protein products of those genes assign the cells to develop into components of the head. In another stripe, they assign cells to develop into a thorax, and so on.

Researchers experimenting with fruit flies found that mutated homeobox genes can transform one body segment into the likeness of another. (Hence the name; *homeo* means "likeness.") For example, mutations in the *antennapedia* gene lead to abnormal responses to inducer signals. One mutation activates the wrong set of genes in cells that are supposed to give rise to a pair

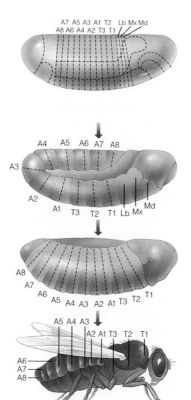

Figure 44.13 (**a**) Fate map of a *Drosophila* zygote. Dashed lines indicate the regions that will develop into different body segments, many with highly specialized appendages. The fate of each segment is sealed at the time of cytoplasmic localization. The drawings are not to scale.

(**b**) Experimental evidence of embryonic induction in *Drosophila*. During a larval stage of development, a cluster of cells that normally gives rise to antennae was surgically removed and exposed to abnormal tissue environments. Then it was reinserted into a different location in a larva undergoing metamorphosis. Instead of antennae, legs appeared on the head.

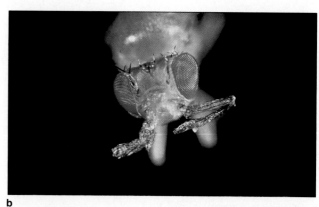

b

of antennae on the head—they give rise to a pair of legs instead (Figure 44.13*b*).

1. Signaling molecules, including morphogens, induce embryonic cells to differentiate into tissues of specific shapes and patterns.

2. Morphogens act on homeobox genes in different cells. These complexes activate or suppress blocks of genes that are required to form the body's head, thorax, and abdomen.

3. Homeobox genes are organized in similar ways in invertebrate and vertebrate DNA. And they are expressed in similar patterns.

44.8 FROM THE EMBRYO ONWARD

Post-Embryonic Development

Embryonic development is over when all of the organs required for feeding and other vital activities are formed and functioning. Now a young animal matures into an adult, the sexually mature form of the species.

Larvae, nymphs, and pupae are examples of immature, post-embryonic stages. Like human infants, some larvae are shaped like miniature adults. Unlike human children, they mature through episodes of growth and **molting** (page 432 and Figure 44.14a).

Other types of larvae, including those of tunicates and frogs, undergo reactivated growth, tissue reorganization, and remodeling of body parts. A resumption of growth and transformation into the adult form is called **metamorphosis**. As you can see from Figure 44.14b and c, the transformation is more drastic in some kinds than in others.

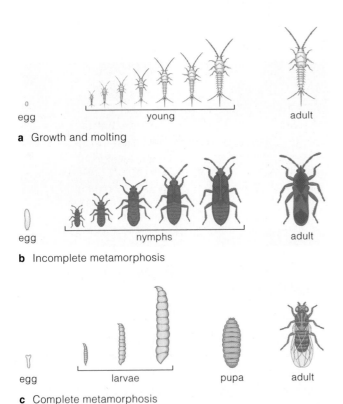

a Growth and molting

b Incomplete metamorphosis

c Complete metamorphosis

Figure 44.14 Some examples of post-embryonic development. (**a**) Silverfish are adults in miniature; they change little except in size and proportion as they mature into adults. (**b**) Bugs show *incomplete* metamorphosis, which involves gradual, partial change from the first immature form until the last molt. (**c**) Fruit flies show *complete* metamorphosis. Tissues of the immature forms are destroyed and replaced before emergence of the adult.

Aging and Death

Late in life, the body gradually deteriorates. Through processes collectively called **aging**, cells start to break down, and this leads to structural changes and gradual loss of body functions. The introduction to Chapter 6 provided an example of this. All animals with extensive cell differentiation undergo aging.

No one knows what causes aging, although research gives us interesting things to think about. More than two decades ago, Paul Moorhead and Leonard Hayflick cultured normal human embryonic cells. All of the cell lines proceeded to divide about fifty times; then the entire population died off. Hayflick took some cultured cells that were partway through the series of divisions and froze them for several years. Then he thawed the cells and placed them in a culture medium. The cells proceeded to complete the cycle of fifty doublings—whereupon they all died on schedule.

Such experiments suggest that normal cell types have a **limited division potential**. That is, mitosis may be genetically programmed to decline at a certain time in the life cycle. But does the change in mitosis cause aging or is it a result of the aging process? Think of neurons, which stop dividing after birth. They still deteriorate over time. Maybe age-related changes also affect dividing cells throughout the body.

It may be that cells gradually lose the capacity for DNA self-repair as a result of an accumulation of environmental insults. Over time, DNA mutations could thwart the production of enzymes and other proteins required for proper cell functioning. Consider how cells depend on smooth exchanges of materials between the cytoplasm and the extracellular fluid. Collagen is present in extracellular spaces throughout the body. If deteriorating DNA regions code for defective collagen molecules, it is conceivable that the movement of oxygen, nutrients, hormones, and so forth to and from cells could be hampered. Repercussions from one defective protein could ripple through the entire body.

Finally, consider what might happen if genes coding for membrane proteins deteriorate. What if the proteins serve as self markers? If they become altered, does the immune system then perceive the body's own cells as "foreign" and attack them? According to one theory, such autoimmune responses might increase over time. And so would the increased vulnerability to disease and stress associated with aging.

After embryonic development, animals show different patterns of maturation into the adult form.

All animals that show extensive cell differentiation undergo a process of aging and death.

Commentary

Death in the Open

By Lewis Thomas (Printed by permission from the author and the New England Journal of Medicine, January 11, 1973, 288:92–93)

Everything in the world dies, but we only know about it as a kind of abstraction. If you stand in a meadow, at the edge of a hillside, and look around carefully, almost everything you can catch sight of is in the process of dying, and most things will be dead long before you are. If it were not for the constant renewal and replacement going on before your eyes, the whole place would turn to stone and sand under your feet.

There are some creatures that do not seem to die at all; they simply vanish totally into their own progeny. Single cells do this. The cell becomes two, then four, and so on, and after a while the last trace is gone. It cannot be seen as death; barring mutation, the descendants are simply the first cell, living all over again. . . .

There are said to be a billion billion insects on the earth at any moment, most of them with very short life expectancies by our standards. Someone has estimated that there are 25 million assorted insects hanging in the air over every temperate square mile, in a column extending upward for thousands of feet, drifting through the layers of atmosphere like plankton. They are dying steadily, some by being eaten, some just dropping in their tracks, tons of them around the earth, disintegrating as they die, invisibly.

Who ever sees dead birds, in anything like the huge numbers stipulated by the certainty of the death of all birds? A dead bird is an incongruity, more startling than an unexpected live bird, sure evidence to the human mind that something has gone wrong. Birds do their dying off somewhere, behind things, under things, never on the wing.

Animals seem to have an instinct for performing death alone, hidden. Even the largest, most conspicuous ones find ways to conceal themselves in time. If an elephant missteps and dies in an open place, the herd will not leave him there; the others will pick him up and carry the body from place to place, finally putting it down in some inexplicably suitable location. When elephants encounter the skeleton of an elephant in the open, they methodically take up each of the bones and distribute them, in a ponderous ceremony, over neighboring acres.

It is a natural marvel. All of the life on earth dies, all of the time, in the same volume as the new life that dazzles us each morning, each spring. All we see of this is the odd stump, the fly struggling on the porch floor of the summer house in October, the fragment on the highway. I have lived all my life with an embarrassment of squirrels in my backyard, they are all over the place, all year long, and I have never seen, anywhere, a dead squirrel.

I suppose that is just as well. If the earth were otherwise, and all the dying were done in the open, with the dead there to be looked at, we would never have it out of our minds. We can forget about it much of the time, or think of it as an accident to be avoided, somehow. But it does make the process of dying seem more exceptional than it really is, and harder to engage in at the times when we must ourselves engage.

In our way, we conform as best we can to the rest of nature. The obituary pages tell us of the news that we are dying away, while birth announcements in finer print, off at the side of the page, inform us of our replacements, but we get no grasp from this of the enormity of the scale. There are 4 billion of us on the earth in this year, 1973, and all 4 billion must be dead, on a schedule, within this lifetime. The vast mortality, involving something over 50 million each year, takes place in relative secrecy. We can only really know of the deaths in our households, among our friends. These, detached in our minds from all the rest, we take to be unnatural events, anomalies, outrages. We speak of our own dead in low voices; struck down, we say, as though visible death can occur only for cause, by disease or violence, avoidably. We send off for flowers, grieve, make ceremonies, scatter bones, unaware of the rest of the 4 billion on the same schedule. All of that immense mass of flesh and bone and consciousness will disappear by absorption into the earth, without recognition by the transient survivors.

Less than half a century from now, our replacements will have more than doubled in numbers. It is hard to see how we can continue to keep the secret, with such multitudes doing the dying. We will have to give up the notion that death is a catastrophe, or detestable, or avoidable, or even strange. We will need to learn more about the cycling of life in the rest of the system, and about our connection in the process. Everything that comes alive seems to be in trade for everything that dies, cell for cell. There might be some comfort in the recognition of synchrony, in the information that we all go down together, in the best of company.

SUMMARY

1. Most animals reproduce sexually. Specialized reproductive structures and forms of behavior assist fertilization and nutritionally support the young.

2. Animal development commonly proceeds through six stages:

a. Gamete formation, during which the egg and sperm mature within reproductive organs of the parents. Localized molecular and structural differences are established in the egg cytoplasm that will help seal the fate of cells in the forthcoming embryo.

b. Fertilization, which begins when a sperm penetrates an egg and is completed when the sperm and egg nuclei fuse to form the zygote.

c. Cleavage, when the fertilized egg undergoes mitotic cell divisions that form the early multicelled embryo. Most genes are inactive during cleavage. The destiny of cell lineages is partly established by their position in the embryo and by cytoplasmic localization, the allocation of different components of the cytoplasm to different daughter cells.

d. Gastrulation, when the organizational framework of the whole animal is laid out. Endoderm, ectoderm, and usually mesoderm form. All the tissues of the adult body will develop from these germ layers.

e. Organogenesis, the onset of organ formation. The different organs start developing by a tightly orchestrated program of cell differentiation and morphogenesis.

f. Growth and tissue specialization, when organs enlarge overall and acquire their specialized chemical and physical properties. The maturation of tissues and organs continues into post-embryonic stages.

3. Embryonic development cannot proceed properly unless each stage is successfully completed before the next one begins.

4. In cell differentiation, a cell selectively uses certain genes and synthesizes proteins not found in other cell types. The outcome is subpopulations of specialized cells that exhibit structural, biochemical, and functional differences.

5. In morphogenesis, the size, shape, and proportion of tissues and organs change and become spatially organized in prescribed, orderly patterns. This pattern formation involves cell divisions, cell migrations, changes in cell shapes, localized growth, and controlled cell death.

6. Cell differentiation and morphogenesis depend on cytoplasmic localization, followed by chemical and physical interactions among embryonic cells.

a. In embryonic induction, morphogens and other signaling molecules induce embryonic cells to differentiate into tissues of specific shapes and patterns.

b. Morphogens act on homeobox gene complexes, which activate or suppress blocks of genes that are required to form specific structures and organs along the body's long axis.

7. Following embryonic development, some animals grow directly into the adult form. Other animals show metamorphosis, in which reactivated growth and tissue reorganization produce larvae or some other immature forms before emergence of the sexually mature adult.

8. All complex animals gradually show changes in structure and a decline in efficiency (aging). Aging is part of the life cycle of all animals having extensively specialized cell types.

Review Questions

1. Define and describe the key events during gamete formation, fertilization, cleavage, gastrulation, and organ formation. At which stages are the frog embryos shown in the following photographs? *756, 758–759*

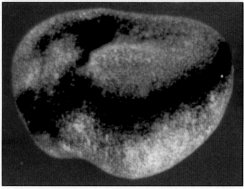

2. Define cell differentiation and morphogenesis. Describe the processes on which they are are based. *764, 765*

3. Describe some of the molecular mechanisms responsible for pattern formation. *766–767*

4. Experimentally, it is possible to divide an amphibian egg so that the gray crescent is wholly within one of the two cells formed. If the two cells are separated from each other, only the cell with the gray crescent will form an embryo with a long axis, notochord, nerve cord, and back musculature. The other cells form a shapeless mass of immature gut and blood cells. What do you think is the explanation of these outcomes? (*Figure 44.7; pages 760, 766*)

Self-Quiz *(Answers in Appendix IV)*

1. Sexual reproduction among animals _____ .
   a. is biologically costly   c. has evolutionary advantages
   b. is diverse in its details   d. all of the above

2. Development cannot proceed properly unless each stage is successfully completed before the next begins, starting with _____ .
   a. gamete formation   d. gastrulation
   b. fertilization   e. organ formation
   c. cleavage   f. growth, tissue specialization

3 During cleavage, daughter cells are allocated different components of the fertilized egg. This is called _____ .
   a. cytoplasmic localization   c. cell differentiation
   b. embryonic induction   d. morphogenesis

4. The primary tissue layers first appear in the _____ .
   a. egg cortex   c. gastrula
   b. blastula   d. primary organs

5. The astonishing internal complexity characteristic of most animals became possible following the evolution of _____ .
   a. ectoderm   c. myoderm
   b. mesoderm   d. endoderm

6. During development, the formation of subpopulations of different cell types is the outcome of _____ .
   a. selective gene expression   c. morphogenesis
   b. cell differentiation   d. a and b are correct

7. As an embryo develops, tissues and organs change in size, shape, and proportions. This process is called _____ .
   a. gastrulation   c. morphogenesis
   b. pattern formation   d. metamorphosis

8. The specialized tissues and body parts that form in an embryo become positioned in orderly ways relative to one another. This feat of self-organization is called _____ .
   a. gastrulation   c. morphogenesis
   b. pattern formation   d. metamorphosis

9. In organisms ranging from fruit flies to humans, the body's organization relative to its long axis depends on _____ .
   a. morphogens   c. embryonic induction
   b. homeobox genes   d. all of the above

10. A reactivation of growth and tissue reorganization following embryonic development is called _____ .
   a. gastrulation   c. morphogenesis
   b. pattern formation   d. metamorphosis

11. Match the development stage with its description.
   ____ cleavage
   ____ gametogenesis
   ____ organ formation
   ____ growth, tissue specialization
   ____ gastrulation
   ____ fertilization

   a. egg and sperm mature in parents
   b. sperm, egg nuclei fuse
   c. formation of primary tissue layers
   d. most genes are inactive but fate of cell lineages sealed partly by cytoplasmic localization
   e. organs, tissues acquire specialized properties
   f. starts when primary tissue layers split into subpopulations of cells

Selected Key Terms

aging *768*
asexual reproduction *754*
blastula *762*
cell differentiation *764*
cleavage *756*
controlled cell death *765*
cytoplasmic localization *762*
ectoderm *756*
embryo *756*
embyronic induction *766*
endoderm *756*
fertilization *756*
gamete formation *756*
gastrulation *756*
gray crescent *760*
growth and tissue specialization *756*

homeobox gene *767*
identical twin *764*
limited division potential *768*
mesoderm *756*
metamorphosis *768*
molting *768*
morphogen *766*
morphogenesis *764*
neural plate *764*
neural tube *765*
oocyte *760*
organ formation *756*
pattern formation *766*
sexual reproduction *754*
yolk *755*

Readings

Balinsky, B. 1981. *An Introduction to Embryology*. Fifth edition. Philadelphia: Saunders.

Browder, L., C. Erickson, and W. Jeffrey. 1991. *Developmental Biology*. Third edition. Philadelphia: Saunders.

Caldwell, M. November 1992. "How Does a Single Cell Become a Whole Body?" *Discover* 13(11): 86–93.

Carlson, B. 1988. *Patten's Foundations of Embryology*. Fifth edition. New York: McGraw-Hill.

Gilbert, S. 1994. *Developmental Biology*. Fourth edition. Sunderland, Massachusetts: Sinauer.

McGinnis, W., and M. Kuziora. February 1994. "The Molecular Architects of Body Design." *Scientific American* 270(2):58–66.

Raff, R., and T. Kaufman. 1983. *Embryos, Genes, and Evolution*. New York: Macmillan.

Saunders, J. 1982. *Developmental Biology: Patterns, Problems, Principles*. New York: Macmillan.

45 HUMAN REPRODUCTION AND DEVELOPMENT

The Journey Begins

At first nothing seems to be happening to the egg, so recently fertilized in the billowing folds of an oviduct (Figure 45.1a). Before the clock ticks off twenty-four hours, however, a spectacular journey is under way. That single cell starts *carving itself up*. Cleavage furrows cut the cytoplasm in one location, then another, and another. By the fifth day, a fluid-filled ball of thirty-two tiny cells is tumbling along the moist, soft lining of the uterus. It has a thin surface layer, like a Ping-Pong ball, and a small bunch of cells that are huddled against one side of the interior wall. This is the **blastocyst**—the forerunner of an adult body that will have many *trillions* of cells. It is smaller than the tip of a pin.

By week's end, in response to molecular cues, the blastocyst has anchored itself to the uterine lining—and has started burrowing into it. Soon, fingerlike projections grow out from the ball's surface and reach blood vessels in the mother's tissues. Connections are now established that will metabolically support the new individual through the months ahead.

Now, in a remarkable feat of self-organization, the inner mass of cells transforms itself into a three-layered, oval disk. Cells are interacting, changing shape, migrating to new locations. Gradually, through cell differentiation and morphogenesis, a pale, crescent-shaped embryo emerges. Day after day, the structural molding continues. Three weeks after fertilization, the tubelike forerunner of the nervous system starts forming. By the fourth week, buds of tissue are starting to grow and sculpt themselves into upper and lower limbs. A patch

a

of cells in a tubular, primordial heart starts beating (Figure 45.1b). From now on, through birth, adulthood, until the time of death, those cells will beat incessantly, as the cardiac pacemaker.

By now, the embryo displays bilateral symmetry, cephalization, a tail, and other features that are the hallmarks of vertebrates. But is it a human? A minnow? A duckling? The answer will not be readily apparent until eight weeks into the journey, when the embryo is recognizably a human in the making.

This was *your* beginning. Later, embryonic and fetal events filled in the details, rounded out contours, added flesh and fat and hair and nails to your peanut-sized body. That beginning, and all the subsequent events that unfolded inside your mother's body, is the story of this chapter.

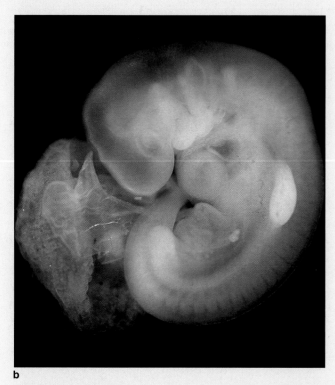

b

Figure 45.1 (a) Within an oviduct of the female reproductive system, the billowing mucosal folds on the road to the uterus. On that road, a sperm traveling from the opposite direction encountered an egg, and a remarkable developmental journey began.

(b) A human embryo, as it appears just four weeks after the moment of fertilization.

KEY CONCEPTS

1. The human reproductive system consists of a pair of primary reproductive organs (testes in males, ovaries in females), accessory glands, and ducts. Testes produce sperm; ovaries produce eggs. Both release sex hormones in response to signals from the hypothalamus and pituitary gland.

2. Human males continually produce sperm from puberty onward. The hormones testosterone, LH, and FSH control male reproductive functions.

3. Human females are fertile on a cyclic basis. Each month during their reproductive years, an egg is released from an ovary, and the lining of the uterus is prepared for pregnancy. The hormones estrogen, progesterone, FSH, and LH control this cyclic activity.

4. Human development proceeds through gamete formation, fertilization, cleavage, gastrulation, organ formation, and growth and tissue specialization.

The preceding chapter introduced some principles of animal reproduction and development. We turn now to a case study of how these principles apply to human reproduction and development.

For both men and women, the reproductive system consists of a pair of primary reproductive organs (gonads), accessory glands, and ducts. Male gonads are **testes** (singular, testis), and female gonads are **ovaries**. Testes produce sperm; ovaries produce eggs. Both secrete sex hormones that influence reproductive functions and the development of **secondary sexual traits**. Such traits are distinctly associated with maleness and femaleness, although they do not play a direct role in reproduction. Examples are the amount and distribution of body fat, hair, and skeletal muscle.

Gonads look the same in all early human embryos. After seven weeks of development, activation of genes on the sex chromosomes and hormone secretions trigger their development into testes *or* ovaries, in the manner described earlier on page 190. The gonads and accessory organs are already formed at birth, but they will not reach full size and become functional until twelve to sixteen years later.

MALE REPRODUCTIVE SYSTEM

Where Sperm Form

Figure 45.2 shows the organs of the male reproductive system and lists their functions. In an embryo that is genetically destined to become male, a pair of testes form on the abdominal cavity wall. Before birth, the testes descend into the scrotum, an outpouching of skin below the pelvic region. Figure 45.2 shows where the scrotum is positioned in an adult male.

Sperm develop properly when the temperature in the interior of the scrotum remains a few degrees cooler than the rest of the body's normal core temperature. A mechanism that controls the contractions of smooth muscles in the scrotum helps assure that the internal temperature does not stray far from 95°F. When it is cold just outside the body, contractions draw the pouch closer to the (warmer) body. When it is warm outside, the muscles relax and thereby lower the pouch.

Packed inside each testis are a great number of small tubes, each of which is highly coiled. These are called the **seminiferous tubules**. As you will read in Section 45.2, sperm formation begins in the seminiferous tubules.

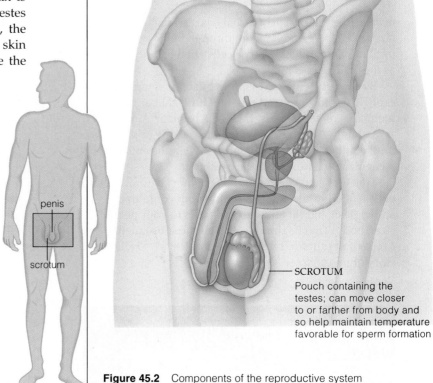

penis

scrotum

SCROTUM
Pouch containing the testes; can move closer to or farther from body and so help maintain temperature favorable for sperm formation

Figure 45.2 Components of the reproductive system of the human male and their functions.

Where Semen Forms

Mammalian sperm are not quite mature when they leave the testes. First they enter a pair of long, coiled ducts, the epididymides (singular, epididymis). Secretions from glandular cells in the wall of these ducts trigger the finishing touches on sperm. Until the time that sperm depart from the body, they become stored in the last stretch of each epididymis.

When a male becomes sexually aroused, muscle contractions in the walls of reproductive organs propel sperm into and through a pair of thick-walled tubes, the vas deferentia (singular, vas deferens). From there, contractions propel sperm through a pair of ejaculatory ducts, and then through the urethra. This last tube threads through the penis, the male sex organ, and opens at its tip. The urethra, recall, also functions in urine excretion.

During the trip to the urethra, glandular secretions become mixed with sperm. The result is **semen**, a thick fluid that is eventually expelled from the penis. A pair of seminal vesicles secrete the sugar fructose, which nourishes sperm cells. Seminal vesicles also secrete cer-

tain prostaglandins. These signaling molecules can induce muscle contractions. (It might be that the prostaglandins trigger contractions in the female reproductive tract and assist sperm movement through it.) Secretions from a prostate gland probably help buffer the acidic environment that sperm encounter in the vagina. Vaginal pH is about 3.5–4.0, but sperm motility improves at pH 6. A pair of bulbourethral glands secrete some mucus-rich fluid into the urethra when the male is sexually aroused. This fluid lubricates the penis and facilitates its penetration into the female's vagina during sexual engagement.

Cancer of the Testis

You may be surprised to learn that *testicular cancer* is a frequent cause of death among young men. About 5,000 cases are diagnosed in a given year in the United States alone. During its early stages, cancer of the testes is

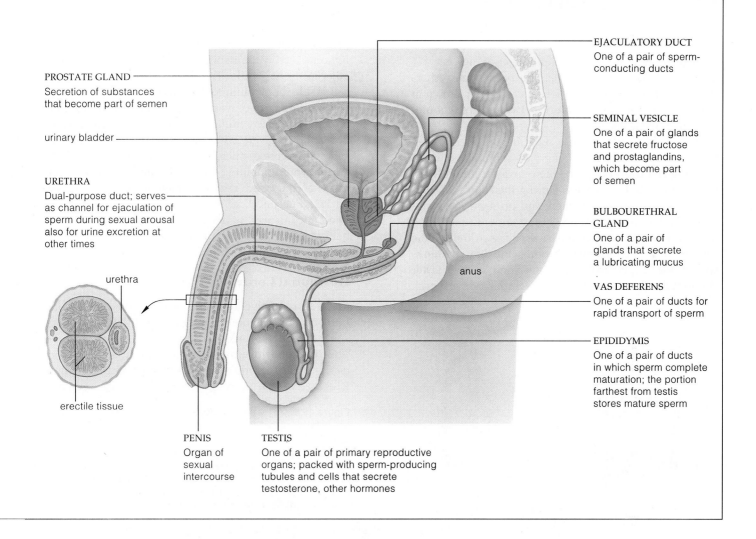

PROSTATE GLAND
Secretion of substances
that become part of semen

urinary bladder

URETHRA
Dual-purpose duct; serves
as channel for ejaculation of
sperm during sexual arousal
also for urine excretion at
other times

urethra

erectile tissue

PENIS
Organ of
sexual
intercourse

TESTIS
One of a pair of primary reproductive
organs; packed with sperm-producing
tubules and cells that secrete
testosterone, other hormones

EJACULATORY DUCT
One of a pair of sperm-
conducting ducts

SEMINAL VESICLE
One of a pair of glands
that secrete fructose
and prostaglandins,
which become part
of semen

BULBOURETHRAL
GLAND
One of a pair of
glands that secrete
a lubricating mucus

VAS DEFERENS
One of a pair of ducts for
rapid transport of sperm

EPIDIDYMIS
One of a pair of ducts
in which sperm complete
maturation; the portion
farthest from testis
stores mature sperm

anus

painless. If not detected in time, however, it can spread silently into lymph nodes in the abdomen, chest, neck, and, eventually, the lungs. Once the cancer has metastasized, it kills as many as 50 percent of those who have been stricken by it.

Once a month, from high school onward, men should examine each testis separately after a warm bath or shower, when muscles of the scrotum are relaxed. The testis should be rolled gently between the thumb and forefinger to check for any type of lump, enlargement, or hardening. Changes of that sort may or may not cause discomfort—but they must be reported to a physician, who can make a complete examination. Treatment of testicular cancer has one of the highest rates of success, when the cancer is caught before it can spread.

Table 45.1 is a summary list of the organs of the male reproductive system and the glands with accessory roles in its functioning.

Table 45.1	Organs and Accessory Glands of the Male Reproductive Tract
Organs:	
Testis (2)	Production of sperm, sex hormones
Epididymis (2)	Sperm maturation site and storage
Vas deferens (2)	Rapid transport of sperm
Ejaculatory duct (2)	Conduction of sperm
Penis	Organ of sexual intercourse
Accessory Glands:	
Seminal vesicle (2)	Secretions large part of semen
Prostate gland	Secretions part of semen
Bulbourethral gland (2)	Production of lubricating mucus

45.2 MALE REPRODUCTIVE FUNCTION

Sperm Formation

Each testis is only about 5 centimeters long, yet 125 meters of seminiferous tubules are packed into it! As many as 300 wedge-shaped lobes, of the sort shown in Figure 45.3a, partition its interior. Two or three tubules are coiled inside each lobe.

Just inside each tubule's wall are undifferentiated cells, the spermatogonia (singular, spermatogonium). Ongoing cell divisions force them away from the wall, toward the tubule's interior. During their forced departure, the cells are transformed into primary spermatocytes. As Figure 45.3c shows, they undergo meiosis I, the result being secondary spermatocytes. Although the cells are now haploid, each chromosome they contain is still in the duplicated state, consisting of two sister chromatids (Figure 10.9). Sister chromatids of each

chromosome separate during meiosis II. The resulting cells become haploid spermatids, which gradually develop into sperm—the male gametes. The entire process takes nine to ten weeks. All the while, developing cells receive nourishment and chemical signals from adjacent **Sertoli cells**. These are the only other type of cell inside the seminiferous tubules.

From puberty onward, males continuously produce sperm, with many millions in different stages of development on a given day. A mature sperm is a flagellated cell, with a core of microtubules in its tail (Figure 45.3d). An enzyme-containing cap called an acrosome covers most of the head region, which is mainly a DNA-packed nucleus. Enzymes of the cap help the sperm penetrate an egg at fertilization. In a midpiece just behind the head, arrays of mitochondria supply energy for the tail's whiplike movements.

Hormonal Controls

Male reproductive function depends on testosterone, LH, and FSH (Figure 45.4). **Leydig cells**, located in tissue between the lobes in testes, secrete **testosterone**. This hormone governs the growth, form, and functions of the male reproductive tract (Figure 12.3). It stimulates sexual and aggressive behavior. It also promotes

Figure 45.3 (a) Male reproductive tract, posterior view. Arrows show the route that sperm take before ejaculation from a sexually aroused male. (b) Micrograph of cells in seminiferous tubules. (c) Sperm formation, starting with a diploid germ cell. Mitotic cell divisions, then meiosis, produce a clone of haploid cells that differentiate into sperm. Leydig cells and Sertoli cells interact chemically to produce testosterone, which promotes sperm formation. (d) Structure of the mature human sperm.

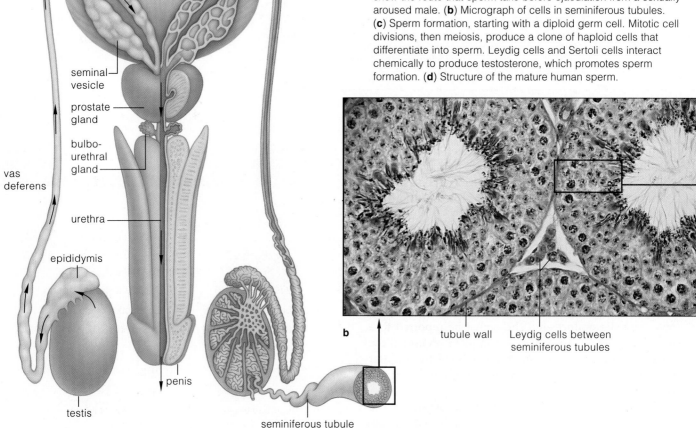

seminal vesicle

prostate gland

bulbo-urethral gland

vas deferens

urethra

epididymis

penis

testis

seminiferous tubule

a

b

tubule wall   Leydig cells between seminiferous tubules

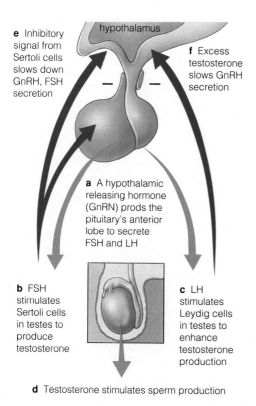

e Inhibitory signal from Sertoli cells slows down GnRH, FSH secretion

hypothalamus

f Excess testosterone slows GnRH secretion

a A hypothalamic releasing hormone (GnRN) prods the pituitary's anterior lobe to secrete FSH and LH

b FSH stimulates Sertoli cells in testes to produce testosterone

c LH stimulates Leydig cells in testes to enhance testosterone production

d Testosterone stimulates sperm production

Figure 45.4 Negative feedback loops to the hypothalamus and pituitary gland from the testes. Through these loops, excess testosterone production shuts off the mechanisms leading to its production. This helps maintain the testosterone level in amounts required for sperm formation.

development of secondary sexual traits, including facial hair growth and deepening of the voice at puberty.

LH and **FSH** are secreted by the anterior lobe of the pituitary gland. They were initially named for their effects in females. (As indicated in Chapter 37, the two abbreviations stand for luteinizing hormone and follicle-stimulating hormone.) However, their molecular structure is identical in males and females.

As shown in Figure 45.4, the hypothalamus governs interactions among testosterone, LH, and FSH and so controls sperm formation. It secretes GnRH when the blood level of testosterone decreases. This releasing hormone stimulates the pituitary to release LH and FSH, which the bloodstream distributes to the testes. There, LH stimulates the Leydig cells to secrete testosterone, which enters the tubes where sperm form. FSH enters the same tubes and diffuses into Sertoli cells. It improves testosterone uptake by those cells, in ways that enhance sperm production.

Over the long term, the testosterone level doesn't fluctuate much. Whenever it increases past a set point, feedback loops to the hypothalamus slow testosterone secretion. The hormone balancing act assures that sperm are available when a female becomes receptive.

Sperm formation depends on the hormones testosterone, LH, and FSH. Feedback loops from the testes to the hypothalamus and pituitary gland control their secretion.

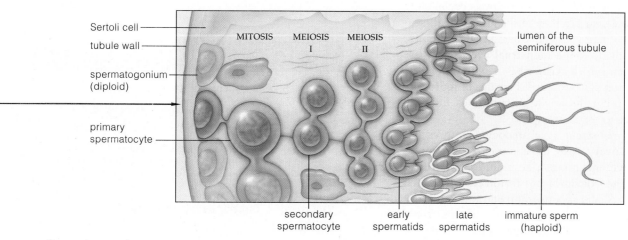

Sertoli cell

tubule wall

spermatogonium (diploid)

primary spermatocyte

MITOSIS

MEIOSIS I

MEIOSIS II

lumen of the seminiferous tubule

secondary spermatocyte

early spermatids

late spermatids

immature sperm (haploid)

c Spermatogenesis

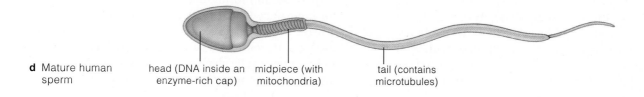

d Mature human sperm

head (DNA inside an enzyme-rich cap)

midpiece (with mitochondria)

tail (contains microtubules)

45.3 FEMALE REPRODUCTIVE SYSTEM

The Reproductive Organs

Figure 45.5 shows the components of the female reproductive system and lists their functions. Table 45.2 summarizes the major organs of this system.

Eggs are produced in the pair of ovaries. After an immature egg, or **oocyte**, is released from an ovary, it enters a neighboring **oviduct**. The pair of oviducts are channels to the **uterus**, a hollow, pear-shaped organ where the embryo grows and develops. The uterus has a thick layer of smooth muscle (myometrium). The inner lining of the uterus, the **endometrium**, consists of connective tissue, glands, and blood vessels. The narrowed part of the uterus is the cervix. A muscular tube, the vagina, extends from the cervix to the body surface. This tube receives sperm and functions as part of the birth canal.

At the body surface are external genitals (vulva) that include organs for sexual stimulation. Outermost are a pair of fat-padded skin folds, the labia majora. They enclose a smaller pair of skin folds (labia minora) that are highly vascularized but have no fatty tissue. The smaller folds partly enclose the clitoris, an organ sensitive to stimulation. At the body's surface, the urethra's opening is about midway between the clitoris and the vaginal opening.

Overview of the Menstrual Cycle

Most mammalian females follow an *estrous* cycle. Only at certain times of year are they fertile and in heat (sexually receptive to males). By contrast, female primates, including humans, follow a **menstrual cycle**. They are fertile intermittently, on a cyclic basis. The times of heat and fertility are not synchronized. Females of reproductive age can get pregnant only at certain times, but they may be receptive to sex at any time.

During each menstrual cycle, an oocyte matures and escapes from an ovary. Also, the endometrium becomes primed to receive and nourish an embryo, *if* fertilization occurs. If the oocyte is not fertilized, blood-rich fluid (about 4 to 6 tablespoons total) starts flowing out through the vaginal canal. Menstruation means "there is no embryo at this time," and it marks the first day of a new cycle. The uterine nest is being sloughed off and is about to be constructed once again.

The events just outlined proceed through three phases. First is the **follicular phase**, which includes menstruation, endometrial breakdown and rebuilding, and maturation of an oocyte. The next phase, **ovulation**, is restricted to the release of an oocyte from the ovary. During the **luteal phase**, an endocrine structure (the

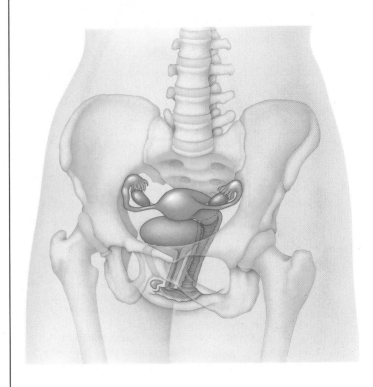

Figure 45.5 Components of the human female reproductive system and their functions.

Table 45.2	Female Reproductive Organs
Ovaries	Oocyte production, sex hormone production
Oviducts	Conduction of oocyte from ovary to uterus
Uterus	Chamber in which new individual develops
Cervix	Secretion of mucus that enhances sperm movement into uterus and (after fertilization) reduces the embryo's risk of bacterial infection
Vagina	Organ of sexual intercourse; birth canal

corpus luteum) forms, and the endometrium is primed for pregnancy (Table 45.3). All of these phases are governed by feedback loops to the hypothalamus and pituitary gland from the ovaries. As you will see, FSH and LH promote cyclic changes in ovaries. They also stimulate the ovaries to secrete **estrogens** and **progesterone**.

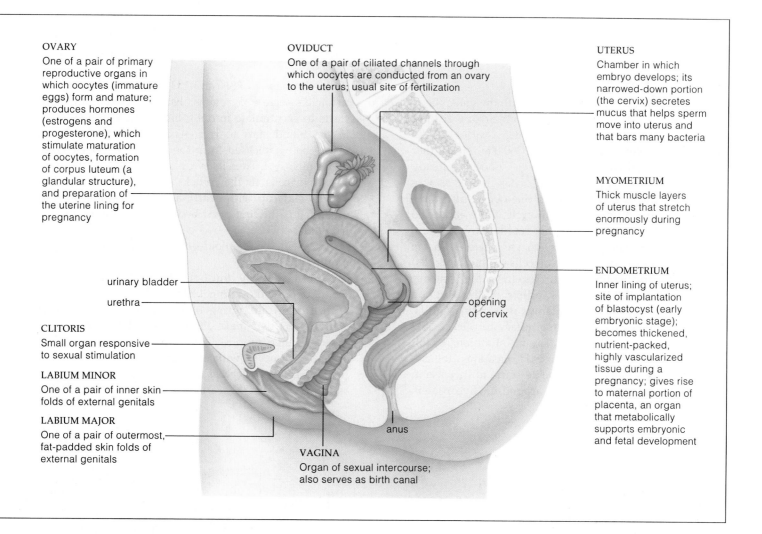

OVARY
One of a pair of primary reproductive organs in which oocytes (immature eggs) form and mature; produces hormones (estrogens and progesterone), which stimulate maturation of oocytes, formation of corpus luteum (a glandular structure), and preparation of the uterine lining for pregnancy

OVIDUCT
One of a pair of ciliated channels through which oocytes are conducted from an ovary to the uterus; usual site of fertilization

UTERUS
Chamber in which embryo develops; its narrowed-down portion (the cervix) secretes mucus that helps sperm move into uterus and that bars many bacteria

MYOMETRIUM
Thick muscle layers of uterus that stretch enormously during pregnancy

ENDOMETRIUM
Inner lining of uterus; site of implantation of blastocyst (early embryonic stage); becomes thickened, nutrient-packed, highly vascularized tissue during a pregnancy; gives rise to maternal portion of placenta, an organ that metabolically supports embryonic and fetal development

urinary bladder

urethra

opening of cervix

CLITORIS
Small organ responsive to sexual stimulation

LABIUM MINOR
One of a pair of inner skin folds of external genitals

LABIUM MAJOR
One of a pair of outermost, fat-padded skin folds of external genitals

anus

VAGINA
Organ of sexual intercourse; also serves as birth canal

Table 45.3	Events of the Menstrual Cycle	
Phase	Events	Days of Cycle*
Follicular phase	Menstruation; endometrium breaks down	1–5
	Follicle matures in ovary; endometrium rebuilds	6–13
Ovulation	Oocyte released from ovary	14
Luteal phase	Corpus luteum forms; endometrium thickens and develops	15–28

*Assuming a 28-day cycle.

These sex hormones promote the cyclic changes in the endometrium.

Estrogen acts on any endometrial tissue—and this tissue can spread and grow outside the uterus. This condition, called *endometriosis*, affects as many as 10 million American women annually. Its symptoms include sensations of pain during menstruation, sex, or urination. Also, endometrial scar tissue may form on the ovaries or oviducts and lead to infertility. The ailment may arise when some menstrual flow backs up through the oviducts and spills into the pelvic cavity. Or perhaps some embryonic cells were positioned in the wrong place before birth and are stimulated to grow at puberty, when sex hormones become active.

A human female's menstrual cycles start between ages ten and sixteen. Each cycle lasts about twenty-eight days, but this is simply the average. It runs longer for some women and shorter for others. The cycles continue until a woman is in her late forties or early fifties, when her egg supply is dwindling and hormone secretions slow down. This is the onset of menopause, the twilight of reproductive capacity.

During a menstrual cycle, an oocyte matures and is released from an ovary, and the endometrium is primed for pregnancy.

45.4 FEMALE REPRODUCTIVE FUNCTION

Cyclic Changes in the Ovary

Take a moment to review Figure 10.10, the generalized picture of meiosis in an oocyte. A normal baby girl has about 2 million primary oocytes in her ovaries. By the time she is seven years old, about 300,000 remain (her body has resorbed the rest). These oocytes already entered meiosis I, but then the division process was arrested. Meiosis will resume in one oocyte at a time, starting with the first menstrual cycle. Only about 400 or 500 will be released during her reproductive years.

Figure 45.6 shows a *primary* oocyte located near the surface of an ovary. A layer of cells (granulosa cells) surrounds and nourishes it. Together, the primary oocyte and the cell layer are a **follicle**. At the start of a menstrual cycle, the hypothalamus is secreting GnRH, which stimulates the anterior pituitary to release FSH and LH (Figure 45.7). These hormones stimulate the follicle to grow. Hence the name FSH, meaning "follicle-stimulating hormone." The oocyte starts to increase in size, and more cell layers form around it. Glycoprotein

deposits build up between the oocyte and the cell layers, widening the space between them. In time they form a noncellular coating, the **zona pellucida**, around the oocyte.

In response to FSH and LH, cells outside the zona pellucida secrete estrogens. Estrogen-containing fluid starts to accumulate in the follicle, and estrogen levels in the blood start to rise. About eight to ten hours before being released from the ovary, the oocyte completes meiosis I. Then its cytoplasm divides, so there are now two cells in the follicle. One cell, the *secondary* oocyte, ends up with nearly all of the cytoplasm. The other cell is the first polar body. Chromosomes are distributed between the two cells so that the secondary oocyte will be haploid—the chromosome number required of gametes.

About midway through the cycle, the pituitary gland detects the rising estrogen level and responds with a brief outpouring of LH. This triggers cellular contractions that make the fluid-filled follicle balloon outward, then rupture. The fluid escapes, carrying the secondary oocyte with it (Figure 45.6). *The midcycle surge of LH has triggered ovulation—the release of a secondary oocyte from the ovary.*

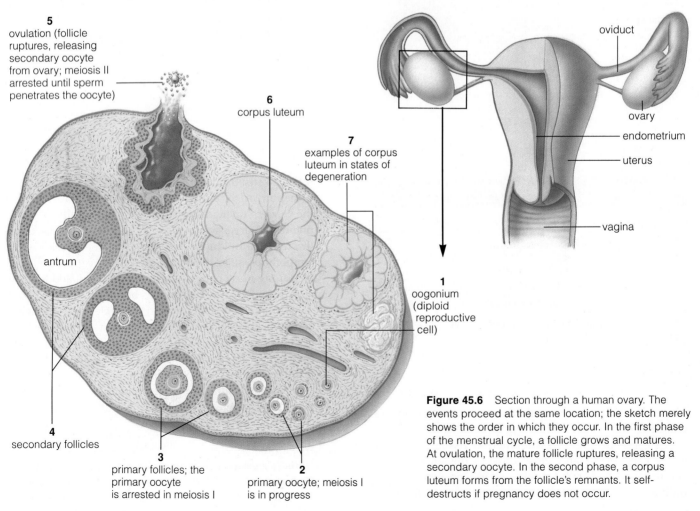

5 ovulation (follicle ruptures, releasing secondary oocyte from ovary; meiosis II arrested until sperm penetrates the oocyte)

6 corpus luteum

7 examples of corpus luteum in states of degeneration

1 oogonium (diploid reproductive cell)

antrum

4 secondary follicles

3 primary follicles; the primary oocyte is arrested in meiosis I

2 primary oocyte; meiosis I is in progress

oviduct

ovary

endometrium

uterus

vagina

Figure 45.6 Section through a human ovary. The events proceed at the same location; the sketch merely shows the order in which they occur. In the first phase of the menstrual cycle, a follicle grows and matures. At ovulation, the mature follicle ruptures, releasing a secondary oocyte. In the second phase, a corpus luteum forms from the follicle's remnants. It self-destructs if pregnancy does not occur.

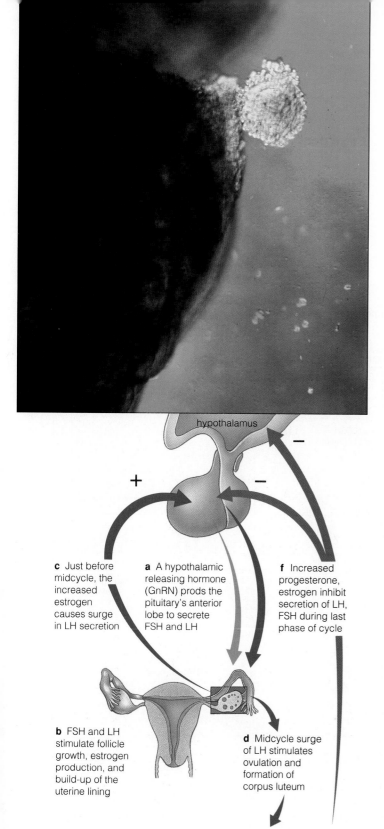

c Just before midcycle, the increased estrogen causes surge in LH secretion

a A hypothalamic releasing hormone (GnRN) prods the pituitary's anterior lobe to secrete FSH and LH

f Increased progesterone, estrogen inhibit secretion of LH, FSH during last phase of cycle

hypothalamus

b FSH and LH stimulate follicle growth, estrogen production, and build-up of the uterine lining

d Midcycle surge of LH stimulates ovulation and formation of corpus luteum

e Progesterone, estrogen from corpus luteum maintain uterine lining unless no pregnancy

Figure 45.7 Feedback loops to the hypothalamus and pituitary gland from the ovaries during a menstrual cycle. A positive feedback loop causes a surge in LH secretion that triggers ovulation. The micrograph shows a secondary oocyte escaping from the ovary at this time. Negative feedback loops after ovulation inhibit FSH and LH secretion to prevent another follicle from maturing until the cycle is completed.

Cyclic Changes in the Uterus

The estrogens released during the early phase of the menstrual cycle also help pave the way for pregnancy. They stimulate growth of the endometrium and its glands. Then, just before the midcycle LH surge, cells of the zona pellucida also start secreting progesterone as well as estrogens. The progesterone causes blood vessels to grow rapidly in the thickened endometrium. At ovulation, estrogens act on tissue around the cervical canal, a narrow opening to the uterus from the vagina. The cervix now secretes large amounts of a thin, clear mucus—an ideal medium for sperm travel.

After ovulation, another hormone becomes dominant. Granulosa cells left behind in the follicle differentiate into a yellowish glandular structure, the **corpus luteum** (meaning "yellow body"). Formation of the corpus luteum results from the midcycle surge of LH. Hence the name, luteinizing hormone.

The corpus luteum secretes progesterone and some estrogen. Progesterone prepares the reproductive tract for the arrival of a blastocyst. For example, it causes mucus in the cervix to become thick and sticky. The mucus may prevent normal bacterial inhabitants of the vagina from entering the uterus. Progesterone also maintains the endometrium during a pregnancy.

A corpus luteum persists for about twelve days. During that time, the hypothalamus signals for a decrease in FSH secretion, which prevents other follicles from developing. If a blastocyst does not arrive and burrow into the endometrium, the corpus luteum self-destructs during the last days of the menstrual cycle. It secretes prostaglandins, which apparently disrupt its own functioning.

After this, progesterone and estrogen levels fall rapidly, so the endometrium starts to break down. Deprived of oxygen and nutrients, its blood vessels constrict, and its tissues die. Blood escapes from the ruptured walls of weakened capillaries. The blood and sloughed endometrial tissues make up the menstrual flow, which continues for three to six days. Then the cycle begins anew, and rising estrogen levels stimulate the repair and growth of the endometrium.

By menopause, the egg supply of oocytes is dwindling, hormone secretions slow down, and in time menstrual cycles (and fertility) are over. The eggs that are still to be released late in a woman's life are at some risk of acquiring chromosome abnormalities when meiosis finally resumes. As described on page 202, a newborn with Down syndrome is one of the possible outcomes.

Coordinated secretions of estrogen, progesterone, LH, and FSH bring about changes in the ovary and uterus during the menstrual cycle. A midcycle surge of LH triggers ovulation.

45.5 | SUMMING UP—KEY EVENTS OF THE MENSTRUAL CYCLE

By now, you probably have come to the conclusion that the menstrual cycle is not a simple tune on a biological banjo. It is a full-blown hormonal symphony! Figure 45.8 may leave you with a better understanding of the integrated nature of the changing hormone levels during each menstrual cycle, as correlated with the changes that they bring about in the ovary and uterus.

Figure 45.8 Changes in the ovary and uterus, correlated with changing hormone levels during the menstrual cycle. *Green* arrows indicate which hormones dominate the cycle's first phase (when the follicle matures), then the second phase (when the corpus luteum forms). (**a,b**) FSH and LH cause changes in ovarian structure and function. (**c,d**) Estrogen and progesterone from the ovary cause changes in the endometrium.

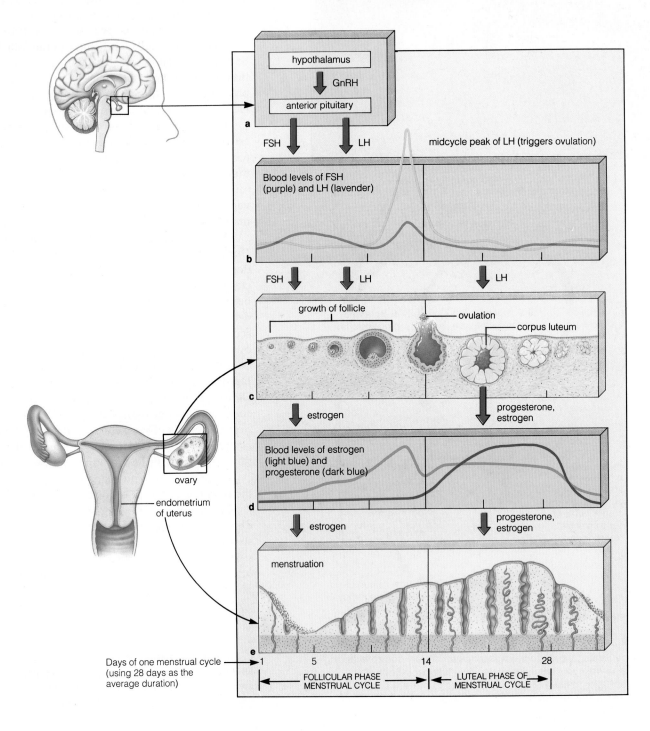

PREGNANCY HAPPENS

Sexual Intercourse

Suppose a secondary oocyte is on its way down an oviduct when a female and male are engaged in sexual intercourse (coitus). Within seconds of sexual arousal, the male's penis undergoes changes that assist its penetration into the vaginal canal. This organ contains cylinders of spongy tissue (Figure 45.2). One cylinder has a mushroom-shaped tip (glans penis) loaded with sensory receptors that are activated by friction. In sexually unaroused males, blood vessels leading into the cylinders are constricted and the penis is limp. In aroused males, blood flows into the cylinders faster than it flows out. As blood collects in the spongy tissue, the penis lengthens and stiffens.

During coitus, pelvic thrusts stimulate the penis as well as the female's vaginal walls and clitoral region. Mechanical stimulation causes rhythmic, involuntary contractions in the male reproductive tract. Sperm are moved rapidly out of each epididymis. The contents of seminal vesicles and the prostate gland are forced into the urethra. And semen is ejaculated into the vagina. (During ejaculation, a sphincter closes and prevents urine from being excreted from the bladder.)

The muscular contractions, ejaculation, and associated sensations of release, warmth, and relaxation are called "orgasm." During female orgasm, similar events occur, including intense vaginal awareness, involuntary uterine and vaginal contractions, and sensations of relaxation and warmth. Even if the female does not get that excited, she can still get pregnant.

Fertilization

Now sperm are in the vagina. A single ejaculation can put 150 million to 350 million there. If it is a few days before or after ovulation, or anytime in between, fertilization may be the outcome. Within thirty minutes after ejaculation, muscle contractions move sperm deeper into the female reproductive tract. Only a few hundred sperm actually reach the upper portion of the oviduct, where fertilization most often occurs.

The stunning micrograph on page 139 shows living sperm at the surface of a secondary oocyte. Upon contacting an oocyte, sperm release acrosomal enzymes that clear a path through the zona pellucida (Figure 45.9). Although many sperm might get this far, usually only one fuses with the oocyte. Only its nucleus and centrioles actually enter the oocyte's cytoplasm. The secondary oocyte and the first polar body are stimulated to complete meiosis II. There are now three polar bodies and a mature egg, or **ovum** (plural, ova). The sperm nucleus and egg nucleus fuse. Their chromosomes intermingle and so restore the diploid number for a brand new zygote.

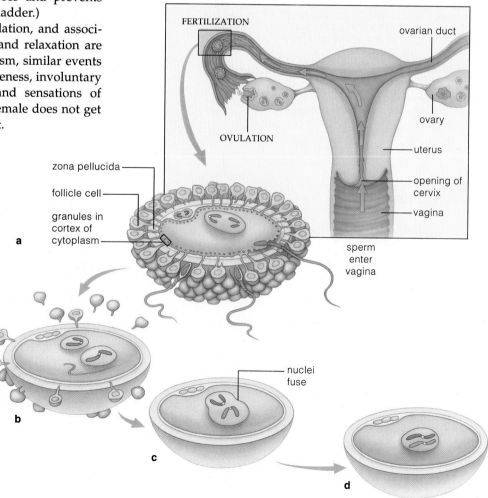

Figure 45.9 Fertilization.
(**a**) A number of sperm surround a secondary oocyte. Acrosomal enzymes clear a path through the zona pellucida. (**b**) When a sperm manages to penetrate the secondary oocyte, cortical granules in the egg cytoplasm release substances that make the zona pellucida impenetrable to other sperm. Penetration also stimulates the second meiotic division of the oocyte's nucleus. (**c**) The sperm tail degenerates. The sperm nucleus enlarges and fuses with the oocyte nucleus. (**d**) With that fusion, fertilization is over. The zygote has formed.

45.7 FORMATION OF THE EARLY EMBRYO

Pregnancy lasts an average of thirty-eight weeks. It takes about two weeks for the blastocyst to form. After this, an **embryonic period** of organ formation lasts from the third to the end of the eighth week. From the eighth week on, the embryo has distinctly human features and is called a fetus.

During the time extending from the eighth week until birth, the **fetal period**, organs grow in size and become specialized. We often call the first three months of pregnancy the first trimester. The second trimester extends from the start of the fourth month to the end of the sixth. The third trimester extends from the seventh month until birth. Starting with Figure 45.10, the next series of illustrations shows characteristic features of the new individual at progressive stages of development.

Early Cleavage and Implantation

Three or four days after fertilization, the zygote is moving through the oviduct, and cleavage has begun. By

Figure 45.10 From fertilization through implantation. Early cleavage produces the blastocyst, which develops into a disklike early embryo. Extraembryonic membranes (the amnion, chorion, and yolk sac) start forming.

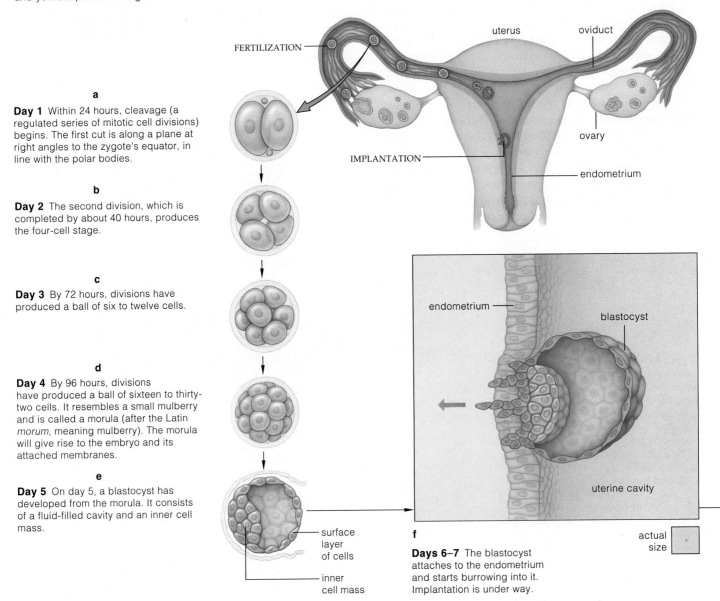

a

Day 1 Within 24 hours, cleavage (a regulated series of mitotic cell divisions) begins. The first cut is along a plane at right angles to the zygote's equator, in line with the polar bodies.

b

Day 2 The second division, which is completed by about 40 hours, produces the four-cell stage.

c

Day 3 By 72 hours, divisions have produced a ball of six to twelve cells.

d

Day 4 By 96 hours, divisions have produced a ball of sixteen to thirty-two cells. It resembles a small mulberry and is called a morula (after the Latin *morum*, meaning mulberry). The morula will give rise to the embryo and its attached membranes.

e

Day 5 On day 5, a blastocyst has developed from the morula. It consists of a fluid-filled cavity and an inner cell mass.

surface layer of cells

inner cell mass

uterus   oviduct

FERTILIZATION

ovary

IMPLANTATION

endometrium

endometrium

blastocyst

uterine cavity

actual size

f

Days 6–7 The blastocyst attaches to the endometrium and starts burrowing into it. Implantation is under way.

the time the cluster of dividing cells reaches the uterus, it is a solid ball, called the morula (Figure 45.10). A fluid-filled cavity forms inside the ball, which thereby becomes the blastocyst. A **human blastocyst** consists of a surface layer of cells and an interior cluster called the inner cell mass.

About five or six days after fertilization, **implantation** is under way. By this process, the blastocyst adheres to the uterine lining, some of its cells send out projections that invade the mother's tissues, and connections start forming that will metabolically support the developing embryo through the many months ahead. As the invasion proceeds, the inner cell mass is transformed into two cell layers having a flattened, somewhat circular shape. The embryo proper will develop from this **embryonic disk**.

Extraembryonic Membranes

As implantation progresses, some vital membranes start forming outside the embryo. First, a fluid-filled cavity opens up between the embryonic disk and the blastocyst's surface (Figure 45.10g). It becomes lined with a membrane, the **amnion**, that later will surround the embryo. The amniotic fluid will be a buoyant cradle where the embryo can grow, move freely, and be protected from sudden temperature shifts and mechanical impacts.

As implantation proceeds, migrating cells also form a lining on the wall of the blastocyst cavity. This lining becomes another extraembryonic membrane, the **yolk sac**. The yolk sac is part of the evolutionary heritage of land vertebrates (page 458). In most shelled eggs, it holds nutritive yolk. During human development, however, some of the yolk sac becomes a site of blood cell formation, and some of its cells give rise to germ cells, the forerunners of gametes.

Before the blastocyst is fully implanted, spaces open in maternal tissues and fill with blood seeping in from ruptured capillaries. In the blastocyst itself, another cavity opens up around the amnion and yolk sac. Now fingerlike projections start forming on this cavity's membranous lining, which is called the **chorion**. The chorion will become part of a spongy, blood-engorged tissue called the placenta.

A fourth extraembryonic membrane will form after a blastocyst is fully implanted. As you will see, it becomes an outpouching of the yolk sac, called the **allantois**. For reptiles, birds, and some mammals, the allantois will function in respiration and storage of metabolic wastes. In humans, it will function in early blood formation and in the formation of the urinary bladder.

One more point should be made here. Cells of the blastocyst secrete **HCG** (short for *Human Chorionic Gonadotropin*). This hormone stimulates the corpus luteum to keep on secreting estrogen and progesterone. Thus the blastocyst itself prevents menstrual flow and so works to avoid being sloughed off. By the start of the third week of pregnancy, HCG can be detected in the mother's blood or urine. At-home pregnancy tests are based on a "dip-stick" that changes color when HCG is present in urine.

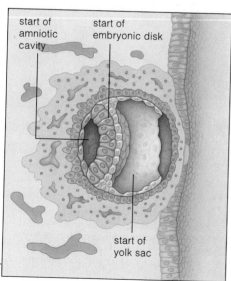

g
Days 10–11 The yolk sac, embryonic disk, and amniotic cavity start to form.

actual size

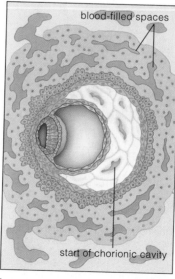

h
Day 12 Blood-filled spaces start to form in maternal tissue; chorionic cavity starts to form.

actual size

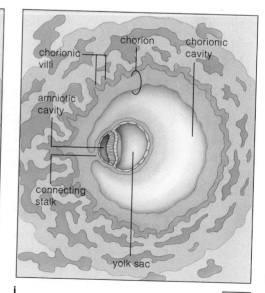

i
Day 14 A connecting stalk has formed between the embryonic disk and the chorion. Chorionic villi which will be features of the placenta, start to form.

actual size

45.8 EMERGENCE OF THE VERTEBRATE BODY PLAN

By the time a woman has missed her first menstrual period, gastrulation is under way. **Gastrulation**, recall, is the stage of development when cells become arranged into primary tissue layers, which later split into subpopulations of cells that give rise to all organs (page 756). As the third week of development begins, a two-layered embryonic disk, consisting of ectoderm and endoderm has already been produced. Two fluid-filled sacs—the amnion and chorion—surround the disk, except where a stalk connects the disk to the inner wall of the chorionic cavity.

Now, through cell divisions and cell migrations, a faint streak appears at the disk's midline (Figure 45.11). This "primitive streak" is the forerunner of a **neural tube**, which will give rise to the brain and spinal cord. Some of its cells also give rise to a notochord. The human notochord is simply a structural framework; the vertebral column will form around it.

These events establish the body's long axis and its forthcoming bilateral symmetry. In other words, *the embryonic disk is undergoing morphogenetic events that will provide the body with the basic form that is characteristic of all vertebrates.*

Early in the third week the allantois, a sausage-shaped outpouching, appears on the yolk sac. (The word is from the Greek *allas*, meaning "sausage".) It remains small in humans. Again, the allantois has roles in early blood formation and emergence of the urinary bladder.

Meanwhile, on the embryonic disk surface facing the yolk sac, the third primary tissue layer—mesoderm—has been forming. Toward the end of the third week, some mesoderm gives rise to **somites**—paired segments that will give rise to most bones and skeletal muscles of the head and trunk, as well as to the dermis overlying these regions. Pharyngeal arches start to form. Spaces open up in parts of the mesoderm, and in time these will coalesce to form the coelomic cavity.

During the third week after fertilization (a time when the woman has missed her first menstrual period), the basic vertebrate body plan emerges in the new individual.

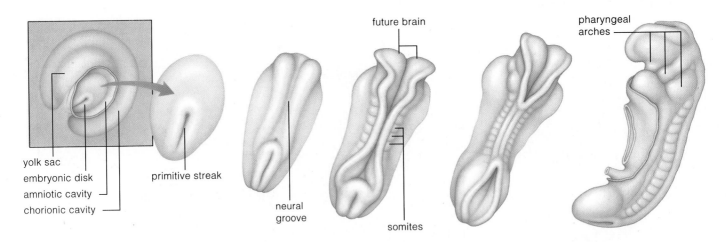

yolk sac
embryonic disk
amniotic cavity
chorionic cavity
primitive streak
neural groove
somites
future brain
pharyngeal arches

Day 15 A primitive streak appears along the axis of the embryonic disk. This thickened band of cells marks the onset of gastrulation.

Day 19–23 Cell migrations, tissue folding, and other morphogenetic events lead to the formation of a hollow neural tube and to somites (bumps of mesoderm). The neural tube gives rise to the brain and spinal cord. Somites give rise to most of the axial skeleton, skeletal muscles, and much of the dermis.

Day 24–25 By now, some cells have given rise to pharyngeal arches, which contribute to the face, neck, mouth, nasal cavities, larynx, and pharynx.

Figure 45.11 Hallmarks of the embryonic period of development—the appearance of a primitive streak—foreshadowing the brain and spinal cord—and the formation of somites and pharyngeal arches. These are dorsal views (of the embryo's back). Compare Figure 44.4.

45.9 ON THE IMPORTANCE OF THE PLACENTA

Even before the onset of the embryonic period, extra-embryonic membranes have been collaborating with the uterus to sustain the embryo's rapid growth. By the third week of development, fingerlike projections from the chorion have grown profusely into the maternal blood that has pooled in endometrial spaces. These projections, the chorionic villi, enhance the exchange of substances between the mother and the new individual. They are part of the **placenta**—a blood-engorged organ consisting of endometrial tissue and extraembryonic membranes. At full term, the placenta will make up one-fourth of the inner surface of the woman's uterus (Figure 45.12).

The placenta is the body's way of sustaining a new individual while allowing its blood vessels to develop apart from the mother's. Oxygen and nutrients diffuse out of the maternal blood vessels, across the placenta's blood-filled spaces, then into the embryonic blood vessels. (These vessels converge in the umbilical cord, the lifeline between the placenta and the new individual.) Carbon dioxide and other wastes diffuse in the opposite direction, and they are disposed of quickly by the mother's lungs and kidneys. After the third month of pregnancy, the placenta also secretes progesterone and estrogens that continue to maintain the uterine lining.

The placenta is a blood-engorged organ composed of endometrial and extraembryonic membranes. It provides for exchanges between the mother and the new individual while keeping their blood vessels separated.

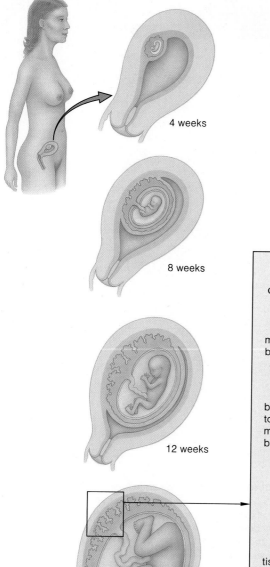

4 weeks

8 weeks

12 weeks

appearance of the placenta at full term

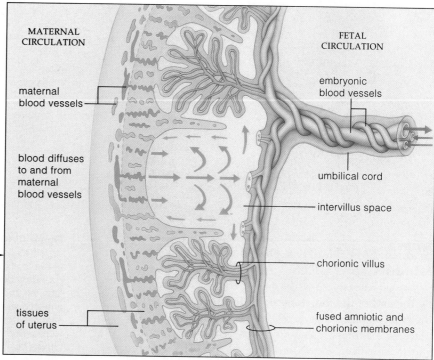

MATERNAL CIRCULATION

maternal blood vessels

blood diffuses to and from maternal blood vessels

tissues of uterus

FETAL CIRCULATION

embryonic blood vessels

umbilical cord

intervillus space

chorionic villus

fused amniotic and chorionic membranes

Figure 45.12 Relationship between fetal and maternal blood circulation in a full-term placenta. Blood vessels extend from the fetus, through the umbilical cord, and into chorionic villi. Maternal blood spurts into spaces between villi. Oxygen, carbon dioxide, and other small solutes diffuse across the placental membrane surface; there is no gross intermingling of the two bloodstreams.

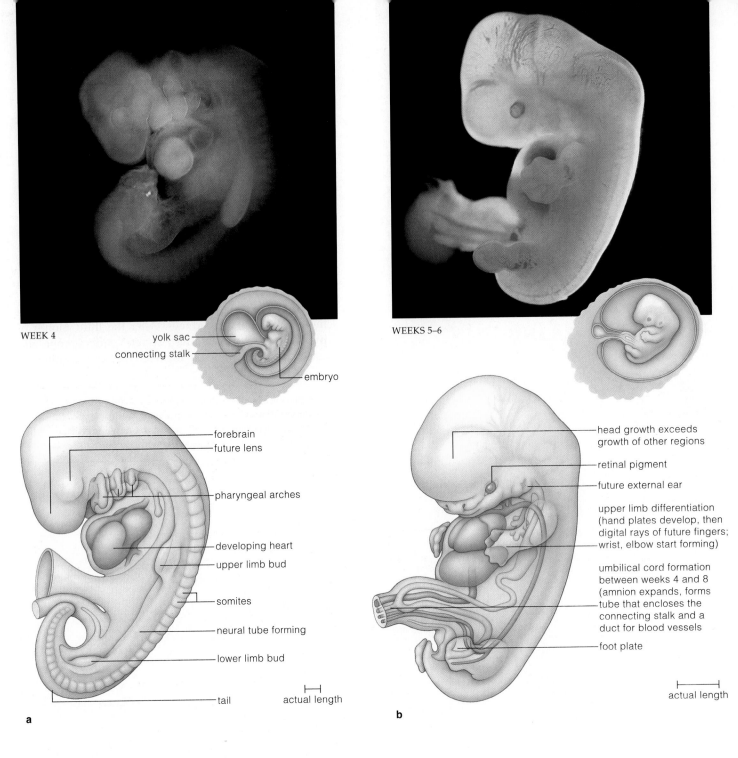

WEEK 4

yolk sac
connecting stalk
embryo

forebrain
future lens
pharyngeal arches
developing heart
upper limb bud
somites
neural tube forming
lower limb bud
tail
actual length

a

WEEKS 5–6

head growth exceeds
growth of other regions
retinal pigment
future external ear
upper limb differentiation
(hand plates develop, then
digital rays of future fingers;
wrist, elbow start forming)
umbilical cord formation
between weeks 4 and 8
(amnion expands, forms
tube that encloses the
connecting stalk and a
duct for blood vessels
foot plate
actual length

b

45.10 EMERGENCE OF DISTINCTLY HUMAN FEATURES

By the end of the fourth week of the embryonic period, the embryo has grown to 500 times its original size. The placenta has been sustaining this remarkable growth spurt. However, now the pace slows. The embryo embarks on a prescribed, intricate program of cell differentiation and morphogenesis. Limbs develop, and fingers and toes are sculpted from paddle-shaped structures. The circulatory system becomes more complex, and the umbilical cord forms. Much emphasis is placed

Figure 45.13 (**a**) Human embryo at four weeks. As is true of all vertebrates, it has a tail and pharyngeal arches. (**b**) The embryo at five to six weeks after fertilization. (**c**) An embryo poised at the boundary between the embryonic and fetal periods. It now has features that are distinctly human. It is floating in fluid within the amniotic sac. The chorion, which normally covers the amniotic sac, has been opened and pulled aside. (**d**) The fetus at sixteen weeks. During the fetal period, movements begin as soon as nerves establish functional connections with developing muscles. Legs kick, arms wave, fingers grasp, the mouth puckers. These reflex actions will be vital skills in the world outside the uterus.

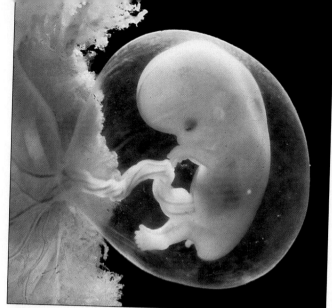

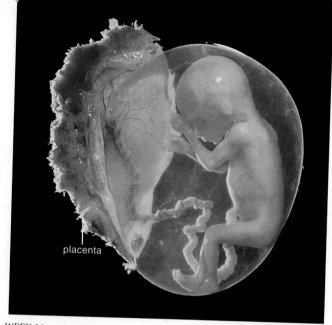

placenta

WEEK 8

final week of embryonic period; embryo looks distinctly human compared to other vertebrate embryos

upper and lower limbs well formed; fingers and then toes have separated

primordial tissues of all internal, external structures now developed

tail has become stubby

actual length

c

WEEK 16
Length: 16 centimeters
         (6.4 inches)
Weight: 200 grams
         (7 ounces)

during fetal period, length measurement extends from crown to heel (for embryos, it is the longest measurable dimension, as from crown to rump)

WEEK 29
Length: 27.5 centimeters
         (11 inches)
Weight: 1,300 grams
         (46 ounces)

WEEK 38 (full term)
Length: 50 centimeters
         (20 inches)
Weight: 3,400 grams
         (7.5 pounds)

d

on the all-important head; its growth surpasses that of any other body region (Figure 45.13).

The embryonic period is over as the eighth week draws to close. The embryo is no longer merely "a vertebrate." As you can see from Figure 45.13c, its features now clearly define it as a human individual—a human fetus.

During the second trimester, the fetus is moving facial muscles. It frowns; it squints. It busily practices the sucking reflex, shown in the *Focus* essay that follows. Now the mother easily senses movements of its arms and legs. When the fetus is five months old, its heart can be heard through a stethoscope on the mother's abdomen. Soft, fuzzy hair (the lanugo) covers its body. Its skin is wrinkled, reddish, and protected from abrasion by a thick, cheesy coating. In the sixth month, eyelids and eyelashes form. In the seventh month, the eyes open.

The fetus cannot survive if it is born prematurely before twenty-two weeks have passed. The situation also is grave for fetuses born before twenty-eight weeks, mainly because the lungs have not developed sufficiently. Even with the best medical care, they have difficulty breathing normally and maintaining a normal core temperature. By the ninth month, survival chances increase to about 95 percent.

Mother as Protector, Provider, Potential Threat

A woman who decides to become pregnant is committing a large part of her body's resources and functions to the development of a new individual. From the time of fertilization to birth, her future child will be at the mercy of her diet, health habits, and life-style (Figure *a*).

Some Nutritional Considerations   How does a pregnant woman best provide nutrients for the embryo, and then the fetus? Normally, the same balanced diet that is good for her provides her future child with all required carbohydrates, lipids, and proteins (page 724). Her own need for vitamins and minerals increases, but the placenta absorbs enough of these from the bloodstream, even at her own expense. If her diet is marginally poor, more nutritious, protein-rich food will do more good than vitamin pills and other food supplements.

A pregnant woman also must eat adequate amounts of food, so that she gains between 20 and 25 pounds, on the average. If she gains a great deal less than this, she is stacking the deck against her future child. Compared to newborns of normal weight, the significantly underweight ones go through more postdelivery complications. They also are more at risk of being mentally impaired in later life.

As birth approaches, the fetus makes greater nutritional demands of the mother, and her diet profoundly influences the remaining developmental events. As is true of most other fetal organs, the brain is especially vulnerable in the weeks just before and after birth, when it undergoes its greatest expansion. All of its neurons are formed: that is why poor nutrition now will have repercussions on intelligence and other brain functions later in life.

a   Sensitivity to teratogens during pregnancy. *Teratogens* are drugs and any other environmental factors that may induce deformities in the embryo or fetus. They usually have no effect before the onset of organ formation. After that, they can block or abnormally stimulate growth as well as the programmed remodeling and resorption of tissues.

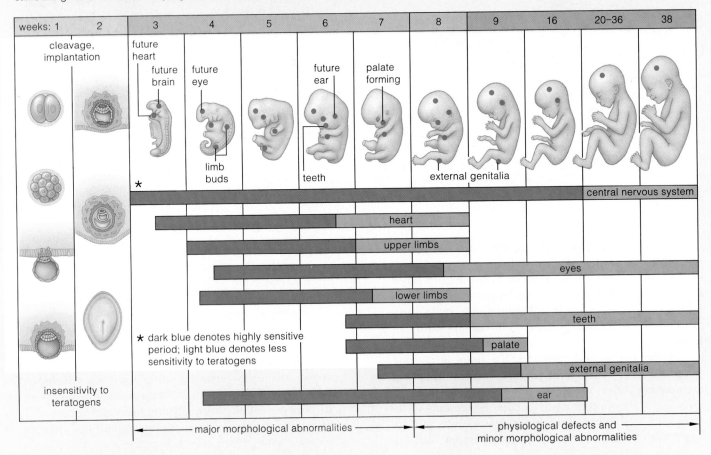

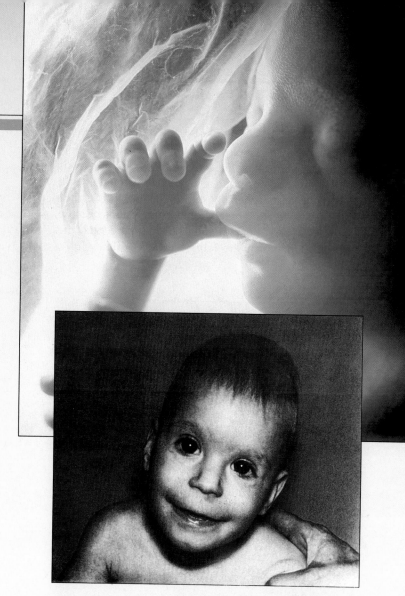

Risk of Infections

Antibodies circulating in a pregnant woman's bloodstream continually cross the placenta. They help protect a developing individual from all but the most severe bacterial infections. But certain viral diseases may be dangerous during the first six weeks after fertilization—the critical time of organ formation. Suppose a woman contracts *rubella* (German measles) during that period. There is a 50 percent chance some organs won't form properly. For example, if embryonic ears are forming, her child might be born deaf. (Vaccination *before* pregnancy can prevent rubella.) If the disease is contracted during or after the fourth month, it has no discernible effect.

Effects of Prescription Drugs

A pregnant woman should not take any drugs at all unless they are prescribed by a good physician. Consider what happened after *thalidomide* started being prescribed in Europe. Women who used this tranquilizer during the first trimester gave birth to infants with missing or severely deformed arms and legs. Thalidomide was withdrawn from the market, but other tranquilizers, sedatives, and barbiturates are still prescribed. They may cause similar, although less severe, damage. Even certain anti-acne drugs increase the risk of facial and cranial deformities. Tetracycline, a commonly prescribed antibiotic, yellows the teeth. Streptomycin causes hearing problems and may affect the nervous system.

Effects of Alcohol

As a fetus grows, its physiology becomes increasingly like the mother's. Alcohol passes freely across the placenta and has the same effect on the fetus as it has on her. *Fetal alcohol syndrome* (FAS) is a set of deformities that result from excessive alcohol intake during pregnancy. FAS is the third most common cause of mental retardation in the United States. The number of reported cases is increasing. Newborns have facial deformities, poor coordination and, sometimes, heart defects (Figure *c*). Between 60 and 70 percent of newborns of alcoholic women have FAS. Some researchers suspect that any alcohol at all may be dangerous for the fetus. Increasingly, physicians are urging near- or total abstinence during pregnancy.

Effects of Cocaine

A mother who uses cocaine, particularly crack, disrupts the nervous system of her future child as well as her own. Here you may wish to read again the introduction to Chapter 35.

Effects of Cigarette Smoke

Cigarette smoking impairs fetal growth and development. Newborns of women who smoked every day during pregnancy have a

b The fetus at eighteen weeks, (**c**) An infant affected by FAS. Symptoms include a small head, low and prominent ears, poorly developed cheekbones, and a long, smooth upper lip. The child can expect growth problems and abnormalities of the nervous system. About 1 in 750 newborns in the United States is affected by this disorder.

low birth weight. They do even if the woman's weight, nutritional status, and all other relevant variables are identical with those of pregnant nonsmokers. Smoking has other effects. In Great Britain, all infants born during a certain week were tracked for seven years. Besides being smaller, newborns of smokers had a 30 percent greater incidence of postdelivery deaths and a 50 percent greater incidence of heart abnormalities. At age seven, their average "reading age" was nearly half a year behind that of the children of nonsmokers.

In this last study, newborns of women who stopped smoking by the middle of the second trimester were indistinguishable from those born to nonsmokers. The mechanisms by which smoking affects the fetus are not known. But its demonstrated effects are further evidence that the placenta, marvelous structure that it is, cannot prevent all assaults on the fetus that the human mind can dream up.

The Process of Birth

Pregnancy draws to a close thirty-eight weeks after fertilization, give or take a few weeks. The birth process is called **labor** (and for good reason, as any mother will tell you). It starts when the uterus begins to contract. For the next two to eighteen hours, contractions become more intense and more frequent. Inside the cervix, the canal through which the fetus will pass dilates fully.

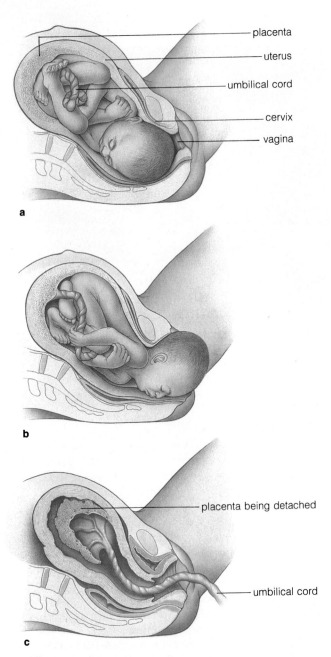

a

b

c

Figure 45.14 Expulsion of the fetus during the birth process. The placenta, fluid, and blood are expelled shortly afterward (this is the "afterbirth").

Just before birth, the amnion usually ruptures, and "water" (amniotic fluid) gushes from the vagina. The fetus is typically expelled within an hour after full dilation (Figure 45.14). Then contractions force fluid, blood, and the placenta from the body, typically within fifteen minutes after birth. Now the umbilical cord—the lifeline to the mother—is severed. The newborn embarks on a nurtured existence in the outside world.

In some cases, births are multiple. *Identical* twins can form at the two-cell stage of cleavage or later, when a split yields two embryos from one inner cell mass. They are genetically identical and look alike as they grow up. *Fraternal* twins are as genetically similar as brothers or sisters, no more. They arise when two secondary oocytes are released in the same menstrual cycle and are fertilized by different sperm. Fertility drugs also stimulate multiple ovulation.

Lactation

During **lactation**, hormone-primed glands produce milk. Estrogen and progesterone stimulated the growth of mammary glands and ducts in the mother's breasts during the pregnancy (Figure 45.15). For the first few days after birth, the glands produce a fluid rich in proteins and lactose. Then prolactin secreted by the pituitary stimulates milk production (page 619). When a newborn suckles, the pituitary also releases oxytocin. This hormone causes breast tissue to contract and force

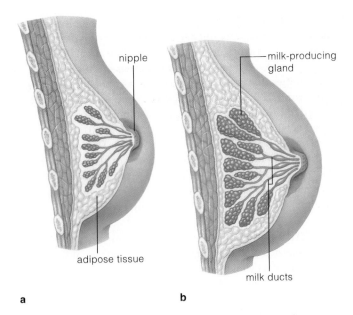

a   b

Figure 45.15 (a) Breast of a woman who is not pregnant. (b) Breast of a lactating woman. This cutaway view shows the mammary glands and ducts.

milk into the ducts. It also triggers uterine contractions that shrink the uterus back to normal size.

Each year in the United States, well over 100,000 women develop *breast cancer*. In some undetermined way, obesity, high cholesterol, and excessively high levels of estrogen and possibly other hormones play roles in the cancer's development. Chances for cure are excellent if the cancer is detected early and treated promptly. A woman should examine herself once a month, about a week after each menstrual period. She can contact her physician or the American Cancer Society (listed in the telephone directory) for a pamphlet on recommended examination procedures.

Postnatal Development and Aging

Following birth, the new individual follows a prescribed course of further growth and development that leads to the adult, the mature form of the species. Table 45.4 summarizes the *prenatal* (before birth) stages and the *postnatal* (after birth) stages. Figure 45.16 is an example of how the body undergoes proportional changes as the life cycle unfolds.

Postnatal growth is most rapid between ages thirteen and nineteen. Then, secondary sexual traits and sexual maturity emerge through the stepped-up secretions of sex hormones. It is not until adulthood that bones become fully mature. Generally, body tissues are maintained in peak condition during early adulthood. As the years pass, it becomes more and more difficult for the body to maintain and repair existing tissues, and the individual starts to age. Here you may wish to refer to pages 92 and 768, which describe some possible mechanisms that bring about aging.

Generally, a person who sticks to a low-fat diet and a low-impact, aerobic exercise program throughout adulthood is more likely to reduce some of the common effects of aging. Regular exercise enhances cardiovascular and respiratory functions. It also helps maintain higher-than-average bone density and so minimizes the bone loss seen in older adults.

Table 45.4   Stages of Human Development: A Summary

Prenatal Period:

1.	Zygote	Single cell resulting from fusion of sperm nucleus and egg nucleus at fertilization
2.	Morula	Solid ball of cells produced by cleavages
3.	Blastocyst	Ball of cells with surface layer and inner cell mass
4.	Embryo	All developmental stages from two weeks after fertilization until end of eighth week
5.	Fetus	All developmental stages from the ninth week until birth (about thirty-eight weeks after fertilization)

Postnatal Period:

6.	Newborn	Individual during the first two weeks after birth
7.	Infant	Individual from two weeks to about fifteen months after birth
8.	Child	Individual from infancy to about ten or twelve years
9.	Pubescent	Individual at puberty, when secondary sexual traits develop; girls between ten and fifteen years, boys between twelve and sixteen years
10.	Adolescent	Individual from puberty until about three or four years later; physical, mental, emotional maturation
11.	Adult	Early adulthood (between eighteen and twenty-five years); bone formation and growth completed. Changes proceed very slowly afterward
12.	Old age	Aging follows late in life

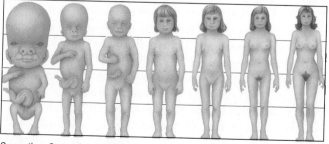

2 months   3 months   newborn   2   5   13   22 years

Figure 45.16   Observable changes in the proportions of the human body during prenatal and postnatal growth. The changes in the body's overall physical appearance are gradual but quite noticeable until the teenage years. For example, the head becomes proportionally smaller, compared to what it was during the embryonic period. The legs become longer and the trunk, shorter.

Some Ethical Considerations

The transformation of a zygote into an intricately detailed adult raises profound questions. *When does development begin?* As we have seen, key developmental events occur even before fertilization. *When does life begin?* During her lifetime, a human female can produce as many as 400 eggs, all of which are alive. During one ejaculation, a human male can release a quarter of a billion sperm, which also are alive. Even before sperm and egg merge by chance and establish the genetic makeup of a new individual, they are as much alive as any other form of life. It is scarcely tenable, then, to say "life begins" when they fuse. *Life began billions of years ago; and each gamete, each zygote, each mature individual is only a fleeting stage in the continuation of that beginning.*

This fact cannot diminish the meaning of conception. It is no small thing to entrust a new individual with the gift of life, wrapped in the unique evolutionary threads of our species and handed down through an immense sweep of time.

Yet how can we reconcile the marvel of individual birth with growing awareness of the astounding birth rate for our whole species? While this book is being written, an average of 10,700 newborns enter the world every single hour. By the time you go to bed tonight, there will be 257,000 more people on earth than there were last night at that hour. Within a week, the number will reach 1,800,000—about as many people as there are now in the entire state of Massachusetts. *Within one week.* Worldwide population growth has outstripped resources, and each year millions face the horrors of starvation. Living as we do on one of the most productive continents on earth, few of us can know what it means to give birth to a child, to give it the gift of life, and have no food to keep it alive.

And how can we reconcile the marvel of birth with the confusion surrounding unwanted pregnancies? Even highly developed countries have inadequate educational programs concerning fertility. And a great number of people are not inclined to exercise control. Each year in the United States alone, there are more than 100,000 "shotgun" marriages, about 200,000 unwed teenage mothers, and perhaps 1,500,000 abortions. Many parents encourage early boy-girl relationships, ignoring the risk of premarital intercourse and unplanned pregnancy. Advice is often condensed to a terse "Don't do it. But if you do it, be careful!"

The motivation to engage in sex has been evolving for more than 500 million years. A few centuries of moral and ecological reasoning that call for its suppression have not prevented unwanted pregnancies. And complex social factors have contributed to a population growth rate that is out of control.

How will we reconcile our biological past and the need for a stabilized cultural present? Whether and how fertility is to be controlled is one of the most volatile issues of our time. We will return to this issue in the next chapter, in the context of principles governing the growth and stability of populations. Here, we briefly consider some control options.

Birth Control Options

The most effective method of birth control is complete *abstinence*, no sexual intercourse whatsoever. It is unrealistic to expect many people to practice it.

A modified form of abstinence is the *rhythm method*. The idea is to avoid intercourse during the woman's fertile period, starting a few days before ovulation and ending a few days after. The woman identifies and tracks her fertile period. She keeps records of the length of her menstrual cycles, takes her temperature each morning when she wakes up, or both. (Body temperature rises by one-half to one degree just before the fertile period.) But ovulation may not be regular, and miscalculations are frequent. Also, sperm deposited in the vaginal canal a few days before ovulation may survive until ovulation. The method is inexpensive; it costs nothing after buying a thermometer. It doesn't require fittings and periodic checkups by a physician. But its practitioners do run a large risk of pregnancy, as Figure 45.17 indicates.

Withdrawal, or removing the penis from the vagina before ejaculation, dates back at least to biblical times. But withdrawal requires very strong willpower, and the method may fail anyway. Fluid released from the penis just before ejaculation may contain some sperm.

Douching, or rinsing out the vagina with a chemical right after intercourse, is next to useless. Sperm can move past the cervix and out of reach of the douche within ninety seconds after ejaculation.

Controlling fertility by surgical intervention is less chancy. In *vasectomy*, a tiny incision is made in a man's scrotum, and each vas deferens is severed and tied off. The simple operation can be performed in twenty minutes in a physician's office, with only a local anesthetic. After vasectomy, sperm cannot leave the testes and so will not be present in semen. So far there is no firm evidence that vasectomy disrupts hormonal interactions in males. There seems to be no difference in sexual activity. Vasectomies can be reversed. But half of those who submit to surgery later develop antibodies against sperm and may not be able to regain fertility.

For females, surgical intervention includes *tubal ligation*, in which the oviducts are cauterized or cut and tied off. Tubal ligation is usually performed in a hospi-

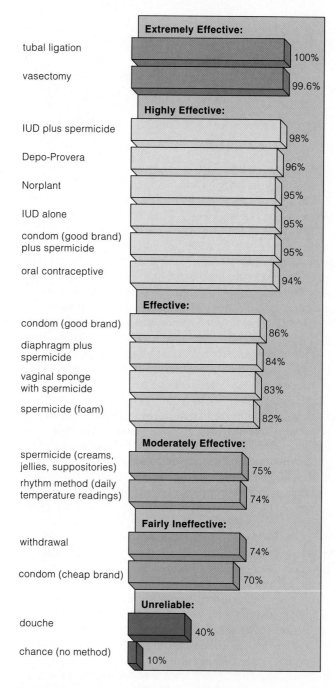

Figure 45.17 Comparison of the effectiveness of some methods of contraception in the United States. Percentages are based on the number of unplanned pregnancies per 100 couples who used only that method of birth control for a year. For example, "94% effectiveness" for oral contraceptives (the Pill) means that 6 of every 100 women will still become pregnant, on the average.

tal. A small number of women who have had the operation suffer recurring bouts of inflammation and pain in the pelvic region where surgery was performed. The operation sometimes can be reversed.

Other, less drastic methods of controlling fertility involve physical or chemical barriers to prevent sperm from entering the uterus and moving to the ovarian ducts. *Spermicidal foam* and *spermicidal jelly* are toxic to sperm. They are transferred from an applicator into the vagina just before intercourse. These products are not always reliable unless used with another device, such as a diaphragm or condom.

A *diaphragm* is a flexible, dome-shaped device, inserted into the vagina and positioned over the cervix before intercourse. A diaphragm is relatively effective when fitted initially by a doctor, used with foam or jelly before each sexual contact, inserted correctly each time, and left in place for a prescribed length of time.

Condoms, thin, tight-fitting sheaths worn over the penis during intercourse, are about 85-93 percent reliable. Only the kinds made of latex help protect against sexually transmitted diseases, as described in the *Focus* essay that follows. But condoms can tear and leak, at which time they become absolutely useless.

The *birth control pill* is an oral contraceptive of synthetic estrogens and progesterone-like hormones called progestins. Taken daily, except for the last five days of the menstrual cycle, it suppresses egg maturation and ovulation. Often it corrects erratic menstrual cycles and reduces cramping, but in some women it causes nausea, weight gain, tissue swelling, and headaches. More than 50 million women take "the Pill," making it the most-used fertility control method. It is 94 percent effective. Early formulations were linked with blood clots, high blood pressure, and possibly breast cancer. Newer, low-dose formulations may lessen the risk of endometrial and ovarian cancer; but a possible link between the Pill and breast cancer is still under investigation.

Progestin injections or implants also inhibit ovulation. A *Depo-Provera* injection works for 3 months and is 96 percent effective. *Norplant*, six rods implanted under the skin, works for 5 years and is 95 percent effective. Both may cause sporadic, heavy bleeding, and doctors have some trouble surgically removing the rods.

A pregnancy test does not register positive until after implantation. It has been argued that a woman is not pregnant until that time. From this perspective, *RU-486*, the "morning-after pill," intercepts pregnancy. It interferes with hormonal signals that govern events between ovulation and implantation. Three pills, taken within seventy-two hours after sexual intercourse, either block fertilization or prevent a blastocyst from burrowing into the uterine lining. RU-486 has side effects, mainly nausea, vomiting, and breast tenderness. These are temporary, and not all women experience them. But RU-486 disrupts complex hormonal interactions and should only be taken under medical supervision. As is true of oral contraceptives, it may trigger elevated blood pressure, blood clots, or breast cancer in some women. At this writing, RU-486 is available in Europe. In the United States, the Federal Drug Administration has just approved clinical trials.

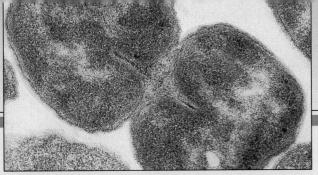

Focus on Health

Sexually Transmitted Diseases

Sexually transmitted diseases (STDs) have reached epidemic proportions in the United States. Urban poverty, prostitution, intravenous drug abuse, and sex-for-drugs are fanning the epidemic. STDs are rampant even on college campuses, where students who are not poor, not promiscuous, and well informed still think, "It can't happen to me." Ominously, antibiotic-resistant strains of bacteria are on the rise, and some viral diseases cannot be cured. At this writing, an estimated 57 million Americans have some form of sexually transmitted disease. Two-thirds of those infected are under age twenty-five; one-fourth are teenagers. Women and children are hardest hit.

The economics of this health problem are staggering. In 1993, the Centers for Disease Control informed us that the annual cost of treating the most prevalent STDs is as follows: *Herpes*, $759 million; gonorrhea, $1 billion; chlamydial infection, $2.4 billion; and pelvic inflammatory disease, $4.2 billion. This does not include the accelerating cost of treating AIDS patients. In many developing countries, AIDS alone threatens to overwhelm health care delivery systems and to unravel decades of economic progress. The social consequences are sobering. Mothers bestow a chlamydial infection on 1 of every 20 newborns in the United States. They bestow type II *Herpes* virus on 1 of every 10,000 newborns; one-half of the infected babies die, and one-fourth have severe neurological defects. Each year 1 million female Americans contract pelvic inflammatory disease.

AIDS   Someone may become infected by HIV, the human immunodeficiency virus, and not even know it. At first there may be no outward symptoms. But five to ten years later, a set of chronic disorders develops. Collectively, these are called *AIDS* (acquired immune deficiency syndrome). The virus has slowly crippled the immune system, in the manner described in Chapter 40. It has opened the door to "opportunistic" infection by the body's resident and normally harmless bacteria. These are positioned to be the first to take advantage of the lowered resistance. Dangerous pathogens also take their toll. In time, infections simply overwhelm the immune-compromised person.

Most commonly, HIV spreads by sexual intercourse (vaginal, anal, and oral) and by IV drug users. Most infections occur through the transfer of blood, semen, urine, or vaginal secretions between people. HIV enters the internal environment through cuts or abrasions on the penis, vagina, rectum, and maybe in membranes in the mouth.

At present there is no vaccine against HIV and no effective way to treat AIDS. If you get it, you die. *There is no cure.*

HIV wasn't identified until 1981, although it seems to have been present in some parts of Central Africa for at least several decades. In the 1970s and early 1980s, it

a   Bacterial agents of gonorrhea.

spread to the United States and other developed countries. Most of those initially infected were male homosexuals. Today, a significant portion of the heterosexual population is infected or at risk.

Worldwide, 10 million to 13 million may now be HIV infected. By early 1992, AIDS had become the second leading cause of death among men and the fifth leading cause among women between 25 and 44. Even so, in one poll of American high schools, two-thirds of the students said they don't use condoms. More than 40 percent of the teenagers polled reported having two or more sex partners.

Free or low-cost, confidential testing for exposure to HIV is available through public health facilities and many physicians' offices. People who suspect they are at risk should know there may be a time lag, from a few weeks to six months or more, until detectable antibodies form in response to infection. The presence of antibodies only indicates exposure to the virus; by itself, it doesn't mean AIDS will develop. Even so, anyone who tests positive for HIV should be considered capable of spreading the virus.

Public education programs are under way to stop the spread of HIV. Most advocate chastity or safe sex. Yet there is confusion about what "safe" means. High-quality latex condoms, used with a spermicide containing nonoxynol-9, may help prevent transmission, but carry a small risk of infection. Open-mouthed kissing with someone who tests positive for HIV should be avoided. Caressing carries no risk, *if* there are no lesions or cuts through which body fluids containing the virus can enter the body. Certain lesions are symptoms of other sexually transmitted diseases. They may increase susceptibility to HIV infection.

In sum, AIDS reached epidemic proportions mainly for three reasons. First, it took a while to discover that the virus is transmitted by semen, blood, and vaginal fluid and that *behavioral* controls can limit its spread. Second, it took time to develop ways to test symptom-free carriers, who can infect others. Third, many still don't understand that the medical, economic, and social consequences of AIDS affect everyone.

Genital Warts   More than sixty types of the human papillomaviruses (HPV) have been identified. A few cause benign, bumplike growths called *genital warts*. HPV infection of the genitals and anus has become the most prevalent STD in the United States. Type 16 HPV does not usually cause obvious warts but may be linked to precancerous sores and cancers of the cervix, vagina, vulva, penis, and anus. In one Seattle study, 22 percent of the female college students who were examined tested positive for the virus.

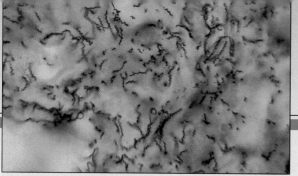

b Bacterial agents of syphilis.

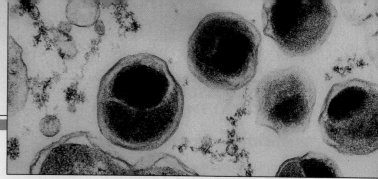

c Bacterial agents of chlamydia.

Gonorrhea During sexual intercourse, *Neisseria gonorrhoeae* can enter the body at mucous membranes of the urethra, cervix, and anal canal. This bacterium, shown in Figure *a*, causes *gonorrhea*. An infected female may only notice a slight vaginal discharge or a burning sensation while urinating. If the bacterium spreads to her oviducts, it may induce severe cramps, fever, vomiting, and scar tissue formation, which may cause sterility. Symptoms are more noticeable in males. Within a week after infection, the penis discharges yellow pus. Urinating is more frequent and may be painful.

This STD is rampant even though prompt treatment quickly cures it. Why? There are no troubling symptoms among women in early stages. Also, infection doesn't confer immunity to the bacterium, perhaps because there are sixteen or more different strains of it. Contrary to common belief, someone can get gonorrhea over and over again. Finally, oral contraceptives, which are widely used, encourage infection by altering vaginal pH. Resident bacterial populations decline, and *N. gonorrhoeae* can move in.

Syphilis *Syphilis*, a dangerous STD, is caused by a spirochete, *Treponema pallidum* (Figure *b*). Sex with an infected partner puts the motile, spiral bacterium onto the surface of genitals or into the cervix, vagina, or oral cavity. It can enter the body through tiny cuts in the epidermis. One to eight weeks later, new treponemes are twisting about in a flattened, painless *chancre* (local ulcer). This first chancre is a symptom of the primary stage of syphilis. By then, treponemes are in the blood. Treponemes can cross the placenta of an infected, pregnant woman. The result will be a miscarriage, stillbirth, or syphilitic newborn.

Usually the chancre heals, but treponemes are now multiplying in mucous membranes, joints, bones, and the eyes, spinal cord, and brain. More chancres and a skin rash develop in this highly infectious, secondary stage of syphilis. Then symptoms subside. In about 25 percent of the cases, the immune system cures the disease. Another 25 percent remain infected but symptom-free. In the rest, mild to major lesions and scars appear in the skin, liver, bones, aorta, and other internal organs. Few treponemes are present during this tertiary stage, but the immune system is hypersensitive to them. Chronic immune reactions can severely damage the brain and spinal cord and bring about general paralysis.

Probably because the symptoms are so alarming, more people seek early treatment for syphilis than for gonorrhea. Later stages require prolonged treatment.

Pelvic Inflammatory Disease *Pelvic inflammatory disease* (PID) is one of the most serious complications of gonorrhea, chlamydial infections, and other STDs. It also can arise when bacteria that normally inhabit the vagina ascend to the pelvic region. Most often, the uterus, oviducts, and ovaries are affected. There is bleeding and vaginal discharge. Pain in the lower abdomen may be so severe, infected women often think they are having an attack of acute appendicitis. The oviducts may become scarred, and so invite abnormal pregnancies as well as sterility.

Genital Herpes The type II *Herpes simplex* virus has already infected about 25 million people in the United States. It causes *genital herpes*. Infection requires direct contact with active *Herpes* viruses or sores that contain them. Mucous membranes of the mouth or genital area are highly susceptible to invasion. Symptoms often are mild or absent. Among infected women, small, painful blisters may appear on the vulva, cervix, urethra, or anal tissues. Among men, blisters form on the penis and anal tissues. Within three weeks, the virus enters latency, and the sores crust over and heal.

The virus is reactivated sporadically. Each time, it causes new, painful sores at or near the original site of infection. Sexual intercourse, menstruation, emotional stress, or other infections can trigger new *Herpes* infections. Acyclovir, an antiviral drug, decreases the healing time and often decreases the pain and viral shedding.

Chlamydial Infection *Chlamydia trachomatis* is a parasitic bacterium (Figure *c*). It must spend part of its life cycle inside cells of the genital and urinary tracts. It causes several diseases, including *NGU* (short for chlamydial nongonococcal urethritis). NGU is far more common than syphilis or gonorrhea. Often, *N. gonorrhoeae* and *C. trachomatis* are transmitted at the same time. Prompt penicillin treatment cures the gonorrhea—but not NGU, which requires tetracycline and sulfonamide treatment.

NGU leads to inflammation of the cervix and, in both sexes, the urethra. An infected person may experience a burning sensation while urinating. But sometimes there are no symptoms, treatment is not sought, and serious complications develop. In males, lymph vessels become blocked, so testes and the prostate become swollen and inflamed. In females, the infection may spread into the uterus and oviducts to cause pelvic inflammatory disease.

Symptom-free infected persons are unwittingly spreading destructive chlamydial infections through all ethnic groups, among poor and affluent alike.

797

45.13 TWO ISSUES OF CLINICAL AND ETHICAL IMPORTANCE

Termination of Pregnancy

Once implantation has occurred, the only way to terminate a pregnancy is *abortion*, the dislodging and removal of the embryo from the uterus. At one time, abortions were generally forbidden by law in the United States unless the pregnancy endangered the mother's life. The Supreme Court has ruled that the government does not have the right to forbid abortions during the early stages of pregnancy (typically up to five months). Before this ruling, there were dangerous, traumatic, and often fatal attempts to abort embryos, either by pregnant women themselves or by quacks.

Vacuum suctioning and other methods make abortion relatively rapid, painless, and free of complications when performed during the first trimester. Abortions in the second and third trimesters will probably remain extremely controversial unless the mother's life is threatened.

For both medical and humanitarian reasons, people in this country generally agree that the preferred route to birth control is not through abortion. It is through sexually responsible behavior that prevents unwanted pregnancy from happening in the first place.

In Vitro Fertilization

Control of fertility extends in the other direction—to help childless couples who are desperate to conceive a child. In the United States, about 15 percent of all couples cannot conceive because of sterility or infertility. For example, hormonal imbalances may prevent ovulation in females, or the sperm count in the male may be too low to assure fertilization.

With *in vitro fertilization*, external conception is possible, provided sperm and oocytes obtained from the couple are normal. A hormone is administered that prepares the ovaries for ovulation. A physician locates and removes the preovulatory oocyte with a suction device. Before this is done, sperm from the male are placed in a solution that simulates fluid in oviducts. When sperm and the suctioned oocyte meet, fertilization may result a few hours later. Twelve hours later, the zygote is transferred to a solution that supports development. Two to four days after that, the blastocyst is transferred to the female's uterus.

Each attempt costs about $8,000 and most are not successful. In 1994, investigators found that each "test-tube" baby costs the nation's health care system about $60,000 to $100,000, on average. Is the cost too great for society to bear? Is it a bargain for the childless couple? That depends on who is answering the questions.

SUMMARY

1. Most animals reproduce sexually. Separation into male and female sexes involves specialized reproductive structures and forms of behavior that help assure successful fertilization and that lend initial nutritional support to offspring.

2. Humans have a pair of primary reproductive organs (sperm-producing testes in males, egg-producing ovaries in females), accessory ducts, and glands. Testes and ovaries also produce hormones that influence reproductive functions and secondary traits.

3. The hormones testosterone, LH, and FSH control sperm formation. They are part of feedback loops among the hypothalamus, anterior pituitary, and testes.

4. The hormones estrogen, progesterone, FSH, and LH control egg maturation and release, as well as changes in the lining of the uterus (endometrium). They are part of feedback loops involving the hypothalamus, anterior pituitary, and ovaries.

5. The following events occur during a menstrual cycle:
   a. A follicle, which is an oocyte surrounded by a cell layer, matures in an ovary, and the endometrium starts to rebuild. (It breaks down at the end of each menstrual cycle when pregnancy does not occur.)
   b. A midcycle peak of LH triggers the release of a secondary oocyte from the ovary. This event is called ovulation.
   c. A corpus luteum forms from the remainder of the follicle. Its secretions prime the endometrium for fertilization. When fertilization occurs, the corpus luteum is maintained, and its secretions help maintain the endometrium.

6. As is true of animals generally, human embryonic development proceeds through six stages: gamete formation, fertilization, cleavage (formation of the blastula), gastrulation, organ formation, and growth and tissue specialization.

7. Embryos develop by way of cell differentiation and morphogenesis. Both processes depend on the segregation of cytoplasm during cleavage and on interactions among embryonic cells.

8. All tissues and organs in the newly developing individual arise from three primary tissue layers: the endoderm, ectoderm, and mesoderm (Table 45.5).

9. As is true of vertebrates generally, four kinds of extraembryonic membranes form during human embryonic development:
   a. Amnion: a fluid-filled sac; it surrounds and protects the embryo from mechanical shocks, abrupt temperature changes, and drying out.

Table 45.5	Summary of Body Parts Derived From the Three Primary Tissue Layers
Ectoderm	Central and peripheral nervous systems; sensory epithelia of the eyes, ears, and nose; epidermis and its derivatives (including hair and nails), mammary glands, pituitary gland, subcutaneous glands, tooth enamel, and adrenal medulla
Endoderm	Various epithelia, as in the gut, respiratory tract, urinary bladder and urethra, and parts of the inner ear; also portions of the tonsils, thyroid and parathyroid glands, thymus, liver, and pancreas
Mesoderm	Cartilage, bone, muscle, and various connective tissues; gives rise to cardiovascular system (including blood), lymphatic system, spleen, and adrenal cortex

Review Questions

1. Distinguish between:
   a. Sertoli cell and Leydig cell *776, 777*
   b. sperm and semen *774*
   c. primary oocyte and ovum *780, 783*
   d. follicle and corpus luteum *780, 781*
   e. ovulation and implantation *780, 785*

2. Which hormones influence male reproductive function? *776–777*

3. What is the menstrual cycle? Which four hormones influence this cycle? *778–779*

4. List four events that are triggered by the surge of LH at the mid-point of the menstrual cycle. *780–781*

5. What changes occur in the endometrium during the menstrual cycle? *779*

6. Label the components of the human male and female reproductive systems and state their functions. *774–775, 778–779*

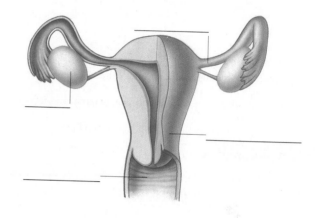

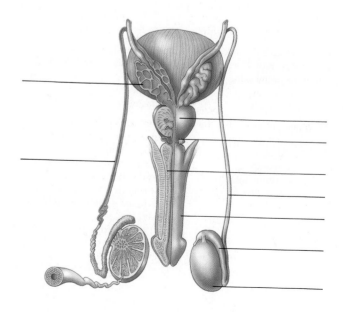

b. Yolk sac: it stores nutritive yolk in most shelled eggs; but in humans, part becomes a major site of blood formation and some of its cells give rise to germ cells. (Eventually the germ cells give rise to sperm and eggs.)

c. Chorion: it becomes a protective membrane around the embryo and the other membranes. It also becomes a major component of the placenta.

d. Allantois: in humans, this extraembryonic membrane functions in early blood formation and in development of the urinary bladder.

10. The placenta is a blood-engorged organ, composed of endometrium and extraembryonic membranes. It keeps the maternal and embryonic circulatory systems separate while allowing oxygen, nutrients, and wastes to diffuse between them.

11. The placental barrier provides some protection for the fetus, but it cannot protect it from the harmful effects of the mother's nutritional deficiencies, infections, intake of prescription drugs, illegal drugs, alcohol, and cigarette smoking.

12. At delivery, uterine contractions dilate the cervical canal and expel the fetus and afterbirth. Estrogen and progesterone stimulate growth in the mammary glands. After delivery, nursing stimulates the release of the hormones prolactin and oxytocin, which in turn stimulate milk production and release.

13. Control of human sexuality and fertility raises important ethical questions. These questions extend to physical, chemical, surgical, and behavioral interventions used to prevent or end pregnancies.

1. The _____ and the pituitary gland control secretion of sex hormones from the testes and the ovaries.

2. Sperm formation depends on _____ secretions.
   a. FSH   c. testosterone   e. both b and c
   b. LH   d. both a and b   f. all are correct

3. During a menstrual cycle, a midcycle surge of _____ triggers ovulation.
   a. estrogen   c. LH
   b. progesterone   d. FHS

4. During a menstrual cycle, _____ and _____ prime the uterus for pregnancy.
   a. FSH; LH   c. estrogens; progesterone
   b. FSH; testosterone   d. estrogens only

5. During implantation, a _____ burrows into the endometrium.
   a. zygote   c. blastocyst
   b. gastrula   d. morula

6. The _____ , a fluid-filled sac, surrounds and protects the embryo from mechanical shocks and keeps it from drying out.
   a. yolk sac   c. amnion
   b. allantois   d. chorion

7. At full-term, a placenta _____ .
   a. consists of endometrial tissue and extraembryonic membranes
   b. contains direct connections between maternal and fetal blood vessels
   c. keeps maternal and fetal blood vessels separated
   d. both a and b
   e. both a and c

8. The neural tube, somites, and other features of the basic vertebrate body plan are evident in the embryo during the _____ week following fertilization.
   a. second   c. fourth   e. eighth
   b. third   d. fifth   f. sixteenth

9. Distinctively human features emerge in the embryo during the _____ week following fertilization.
   a. second   c. fourth   e. eighth
   b. third   d. fifth   f. sixteenth

10. _____ is the forerunner of muscles, most of the skeleton, the circulatory, reproductive, and excretory organs, parts of the gut, and parts of the skin.
    a. Ectoderm   c. Endoderm
    b. Mesoderm   d. Plasmoderm

11. Match the term with the most suitable description.
    ____ seminiferous tubule
    ____ allantois
    ____ corpus luteum
    ____ somites
    ____ yolk sac

    a. glandular structure formed after ovulation
    b. most bones, skeletal muscles of head and trunk form from these segments
    c. in humans, helps form blood, urinary bladder
    d. blood cell formation site, source of germ cells
    e. sperm form here

allantois 785
amnion 785
blastocyst, human 785
chorion 785
corpus luteum 781
embryonic disk 785
embryonic period 784
endometrium 778
estrogen 778
fetal period 784
follicle 780
follicular phase 778
FSH (follicle-stimulating hormone) 777
gastrulation 786
HCG (human chorionic gonadotropin) 785
implantation 785
labor 792
lactation 792
Leydig cell 776

LH (luteinizing hormone) 777
luteal phase 778
menstrual cycle 778
neural tube 786
oocyte 778
ovary 773
oviduct 778
ovulation 778
ovum 783
placenta 787
progesterone 778
secondary sexual trait 773
semen 774
seminiferous tubule 774
Sertoli cell 776
somite 786
testis 773
testosterone 776
uterus 778
yolk sac 785
zona pellucida 780

Aral, S., and K. Holmes. February 1991. "Sexually Transmitted Diseases in the AIDS Era." *Scientific American* 2(264):62–69.

Caldwell, M. November 1992. "How Does a Single Cell Become a Whole Body?" *Discover* 13(11):86–93.

Cohen, J., and I. Stewart. April 1994. "Our Genes Aren't Us." *Discover* 15(4):78–84. Argument that DNA means nothing in the absence of a developmental context.

Gilbert, S. 1994. *Developmental Biology.* Fourth edition. Sunderland, Massachusetts: Sinauer.

Larsen, W. 1993. *Human Embryology.* New York: Churchill Livingston. Paperback. Strongly recommended for the serious student, but maybe not for the faint-hearted.

Moore, K. 1988. *The Developing Human: Clinically Oriented Embryology.* Fourth edition. Philadelphia: Saunders. Ditto.

Nilsson, L. et al. 1986. *A Child Is Born.* New York: Delacorte Press/Seymour Lawrence.

Zack, B. July 1981. "Abortion and the Limitations of Science." *Science* 213:291.

FACING PAGE: *Two organisms—a fox in the shadows cast by a snow-dusted spruce tree. What are the nature and consequences of their interactions with each other, with other organisms, and with their environment? By the end of this last unit, you possibly will see worlds within worlds in such photographs.*

46 POPULATION ECOLOGY

A Tale of Nightmare Numbers

Suppose this year the United States Congress passes legislation to control population growth. By law, there can be no more than three children per family. After the third child is born, the father must be sterilized, with or without his consent. *It would never happen here*, you may be thinking. Such an invasion of privacy would never be tolerated in our society. Besides, family size is not much of an issue in North America, where rates of food production and standards of hygiene and medical care are among the world's highest.

Elsewhere in the world, especially where living conditions are already marginal, many populations are growing at alarming rates. Consider India. Its population, which already surpasses that of North and South America combined, grows by about 2 percent annually.

Most people there do not have adequate food, shelter, or medical care. Each *week*, 100,000 enter the job market, with little hope for gainful employment. Each *day*, 100 acres of croplands that provide food for the population are removed permanently from agriculture. Why? Too many salts have built up in the intensively irrigated soil; India does not get enough rain to flush them out.

Birth control programs sponsored by the government of India have not worked well. Administering the programs has been difficult, for many people live in remote villages. Information must be conveyed by word of mouth because of widespread illiteracy. Many children die of disease and starvation, so villagers often resist limiting family size. Without a large family, they ask, who will survive and help a father tend fields?

Who will go to the cities and earn money to send back home? Who will care for parents when they are too old to work? How can a father otherwise know he will be survived by a son? By Hindu tradition, a son must conduct the last rites so the soul of his dead father will rest in peace.

In 1976, out of desperation, government officials subjected some men to compulsory vasectomies. Public outrage over the policy contributed to the eventual downfall of Indira Gandhi's government, and the law was rescinded. By the end of 1994, there were more than 5.6 billion people on earth. Of those, 812 million were living in India. More than 350 million live in abject poverty in rat-infested shantytowns, where they are forced to wash clothes, even dishes, in open sewers.

Is there a way out of such dilemmas? Should the wealthier, less densely populated nations that now use most of the world's resources learn to get by more efficiently, on less? Should they donate surplus food to less fortunate nations? Would donations help, or would they encourage dependency and further population growth? What would happen if the benefactor nations suffered severe droughts year after year and had trouble meeting their own resource requirements?

Whether we consider humans or any other kind of organism, *certain ecological principles govern the growth and sustainability of populations over time.* This chapter describes these principles, then shows how they apply to the past, present, and future growth of the human population.

Figure 46.1 Bathers crowding the banks of the Ganges River in India—a tiny sampling of the more than 5.6 billion humans on earth. In this chapter we turn to the principles governing the growth and sustainability of populations, including our own.

KEY CONCEPTS

1. Ecological principles govern the growth and sustainability of all populations, including our own.

2. A population may display an exponential growth pattern, in which it increases in size by ever larger amounts over increments of time. Alternatively, it may display a logistic growth pattern. By this pattern, a low-density population rapidly increases in numbers, then levels off in size as resource scarcity limits its further increase or triggers a decline in numbers. Finally, some populations display large fluctuations in numbers that are not easily explained.

3. All populations face limits to growth, for no environment can indefinitely sustain a continuously increasing number of individuals. Competition for resources, disease, predation, and other factors act as controls over population growth. The controls vary in their relative effects on populations of different species, and they vary over time.

With this unit, we turn to the ways in which organisms interact with one another and with the physical and chemical aspects of their environment. The study of these interactions, called **ecology**, is undertaken at the levels of populations, communities, ecosystems, and the biosphere. A **population**, recall, is a group of individuals of the same species occupying a given area. The place where the population (or individual) lives is its **habitat**. The populations of all species that occupy a habitat make up a **community**. Ecologists also use this term for evolutionarily related or ecologically similar groups of organisms in a habitat, such as a community of birds or of animals that feed on grasses. An **ecosystem** is a community and its environment. It has a *biotic* component (all of its living organisms) and *abiotic* (non-living) components, such as nutrients, temperature, and rainfall. The **biosphere** is the sum total of all places in which organisms live—from the lower atmosphere and the waters of the earth to the surface rocks, soils, and sediments of the earth's crust.

In Units V and VI, we considered ways in which individuals are physiologically adapted to their environments. We turn now to relationships that influence the size, structure, and distribution of the populations to which they belong. Communities, ecosystems, and the biosphere are the focus in chapters to follow.

46.1 CHARACTERISTICS OF POPULATIONS

Each population has certain characteristics, including size, density, and distribution. **Population size** is the number of individuals that make up the population's gene pool. **Population density** is the number of individuals in a given area or volume, such as the number of guppies in each liter of water in a stream. Density is not the same as **population distribution**, which refers to the general pattern in which its members are dispersed through their habitat.

A population also has a characteristic **age structure**. The term refers to the number of individuals in each of several to many age categories. For example, ecologists may divide a population into pre-reproductive, reproductive, and post-reproductive ages. Individuals in the first category have the potential to produce offspring when they mature. As you will see later in the chapter, together with the actually and potentially reproducing members in the second category, we count them as part of the population's **reproductive base**.

Regarding Population Density

The same area may be home to populations of a great number of different species, but these generally differ quite a bit in terms of their density. You probably know this if you have ever collected seashells, counted butterflies or birds in your backyard, or waged a running battle with snails in a garden. For every well-represented species, you probably noticed that several others were represented only rarely.

Later chapters show how variations in population density depend largely on ecological relationships among the species occupying the same area. When studying the effects of these relationships, it helps to define density in a given area at a given point in time. Such information serves as a baseline for tracking *changes* in density over time. For a sparsely represented population of trees, snakes, or some other organism, a simple "head count" will do. For dense populations, ecologists typically perform counts in small, randomly selected sampling areas, then use the data to estimate overall density.

Spatial Distribution

Most commonly, the individuals of a population clump together at specific sites. Sometimes, depending on the species, they are uniformly dispersed through the habitat. Only rarely are they randomly dispersed.

Clumped Dispersion For most populations, you will find clumps of individuals in different parts of the habitat. There are three reasons for this.

First, each species is adapted to a rather limited range of abiotic and biotic conditions—but such conditions usually exist in patches, not uniformly throughout the habitat. For example, certain pasture plants grow luxuriantly in small, scattered patches of soil. There, cowpats fell weeks or months before and enriched the soil with nitrogen. As other examples, some parts of a habitat might offer more shade, more water, or better hiding places for prey organisms and hunting possibilities for predators.

Figure 46.2 Dispersion patterns. (**a**) Clumping, typical of patchy habitats and of social species. (**b**) Nearly uniform dispersion, as among creosote bushes near Death Valley, California. (**c**) A rare example of random dispersion: one member of a population of free-ranging, predatory wolf spiders.

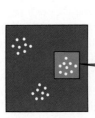

a Clumping

Second, many animals form social groups that offer distinct advantages for individual survival and reproduction (Figure 46.2a). Such groups might afford better possibilities for mating. Their individuals might engage in mutual defense against predators and assist one another in raising offspring. Also, with more pairs of eyes to detect danger, their individuals can forage more efficiently for food.

Third, many species simply do not have the means to disperse seeds, larvae, or other forms of the new generation over large distances. For example, even though sponge larvae can swim, they cannot swim that far from the parent sponge body. They settle down on substrates near the parent and one another.

Uniform Dispersion   Evenly spaced trees in an orchard are a good example of uniform dispersion—but you rarely see this distribution pattern in nature. Where the pattern does exist, it seems to be an outcome of competition among its individuals. Consider creosote (*Larrea*) plants that grow in dry scrub deserts of the American Southwest (Figure 46.2b). Large, mature plants deplete the soil around them of available water. Also, seed-eating ants and rodents concentrate their activities near the plants. It appears that seeds and seedlings often cannot survive near mature plants, so clumps do not form.

Random Dispersion   For a few populations, abiotic conditions do not vary across the habitat, and individual members are neither attracting nor repelling one another. Where such conditions exist, dispersion may be random. For example, wolf spiders are solitary hunters on the floor of many forests. The individuals of each generation may thus be randomly spaced (Figure 46.2c). However, random dispersion is the exception, not the rule. In population studies, this distribution pattern better serves as a theoretical baseline for measuring the degree of clumping or uniformity among members of a population.

Some Qualifiers

Bear in mind, our perception of population distribution usually depends on how wide a view we take. At the scale of a few square meters, acorns under an oak tree appear randomly spaced. At the scale of a forest of mixed hardwood trees, we perceive the same acorns as clustered near their parent trees.

Also, population distribution varies over time, often in response to environmental rhythms. Few environments yield abundant resources all year long, and many animals often move from one local habitat to another with the seasons. For example, in parts of deciduous tropical forests, many birds and mammals crowd into narrow "gallery forests" along watercourses during a dry season. There, trees are often evergreen and provide food and shelter. The animal density becomes much greater than it is during the wet season.

Each population has a characteristic size, density, distribution pattern, and age structure.

The scale of an investigation and the time of year it is undertaken may influence perceptions of population distribution.

b Nearly uniform dispersion

c Random dispersion

46.2 POPULATION SIZE AND EXPONENTIAL GROWTH

How the Number of Individuals Can Change

Over a specified time interval, any change in the size of a population depends on how many individuals enter and leave, one way or another. Population size increases by *births* and by *immigration*, whereby new individuals enter and take up residence. Population size decreases by *deaths* and by *emigration*, whereby individuals leave and take up residence elsewhere.

For this discussion, assume that immigration and emigration balance each other, so we can ignore their effects on population size. With this qualification in mind, let's view **zero population growth** as an interval during which the number of births and the number of deaths are in balance. For that interval, population size would be stabilized; the number of individuals would be neither increasing nor decreasing.

The Exponential Growth Pattern

We can measure births, deaths, and other variables that affect populations in terms of rates per individual—that is, as "per capita" rates. Think of 2,000 mice living in a cornfield. These small rodents reproduce rapidly. About twenty days after fertilization, females produce a litter of offspring, nurse them for about a month, then may get pregnant again. Suppose 1,000 mice are born during one month's time. The birth rate would be 1,000/2,000 = 0.5 per mouse per month. Suppose 200 of the 2,000 die during the same period, putting the death rate at 200/2,000 = 0.1 per mouse per month.

If we assume the birth rate and death rate remain constant, we can combine both into a single variable. This variable is the **net reproduction per individual per unit time**, or r for short. For the population in the cornfield, then, r is $0.5 - 0.1 = 0.4$ per mouse per month. This example gives us a way to represent population growth:

$$\begin{pmatrix} \text{population} \\ \text{growth per} \\ \text{unit time} \end{pmatrix} = \begin{pmatrix} \text{net population} \\ \text{growth rate} \\ \text{per individual} \\ \text{per unit time} \end{pmatrix} \times \begin{pmatrix} \text{number of} \\ \text{individuals} \end{pmatrix}$$

or, more simply, $G = rN$.

As the next month begins, 2,800 mice are scurrying about. The net increase of 800 furry fertile critters means the reproductive base has expanded. If r doesn't change, population size will expand also, giving us a net increase of $0.4 \times 2,800 = 1,120$ mice this month. Now the population stands at 3,920. Through the

months ahead, it just so happens that r remains constant, and a growth pattern emerges:

			Net Monthly Increase:		New Population Size:
$G = r \times$	3,920	=	1,568	=	5,488
$r \times$	5,488	=	2,195	=	7,683
$r \times$	7,683	=	3,073	=	10,756
$r \times$	10,756	=	4,302	=	15,058
$r \times$	15,058	=	6,023	=	21,081
$r \times$	21,081	=	8,432	=	29,513
$r \times$	29,513	=	11,805	=	41,318
$r \times$	41,318	=	16,527	=	57,845
$r \times$	57,845	=	23,138	=	80,983
$r \times$	80,983	=	32,393	=	113,376
$r \times$	113,376	=	45,350	=	158,726
$r \times$	158,726	=	63,490	=	222,216
$r \times$	222,216	=	88,887	=	311,103
$r \times$	311,103	=	124,441	=	435,544
$r \times$	435,544	=	174,218	=	609,762
$r \times$	609,762	=	243,905	=	853,667
$r \times$	853,667	=	341,467	=	1,195,134

In less than two years from the time we first started our counts, the number of mice running around in the cornfield increased from 2,000 to more than a million.

If we were to plot these monthly increases against time, we would end up with a graph line in the shape of a "J." When population growth over increments of time shows a constant rate of multiplication, this is a case of **exponential growth**:

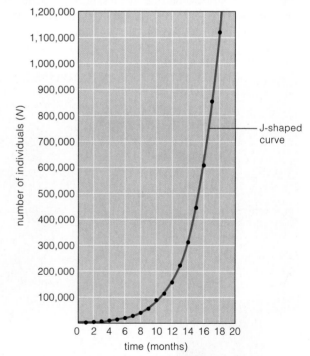

"Exponential" refers to a relationship in which one variable increases much faster than another in a specific

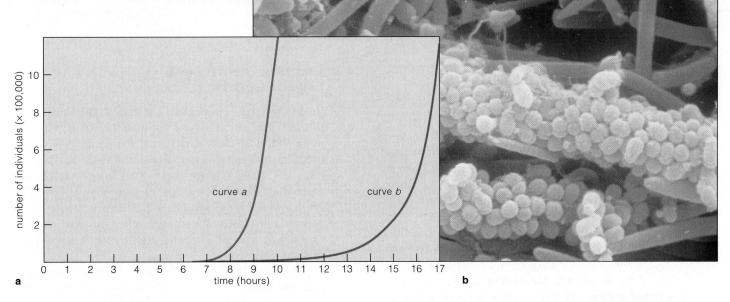

a

b

Figure 46.3 (**a**) Two curves that signify exponential growth. Growth curve *a* represents a population of bacterial cells that reproduced every half hour. Growth curve *b* represents a different population in which cells also divided every half hour, but 25 percent died between divisions. The deaths slowed the rate of increase but did not stop exponential growth. (**b**) Do growth curves seem far removed from your personal experience? Think of bacteria growing in, say, your mouth when they are presented with a buffet of nutrients in candy or some other sugar-laden food.

mathematical way. In this case, population growth is a variable that depends on the number of individuals making up the population's reproductive base in successive increments of time. *The larger the reproductive base, the greater will be the expansion in population size during each specified interval.*

Let's consider some other aspects of exponential growth by supplying a lone bacterium in a culture flask with all the nutrients that it requires for growth. Thirty minutes later, that one cell divides into two new cells. Thirty minutes later, these two divide, and so on every thirty minutes. Assuming no cells die between divisions, the population size will double during each interval—from 1 to 2, then 4, 8, 16, 32, 64, and so on. The length of time it takes for a population to double in size is its **doubling time**.

The larger the bacterial population becomes, the more cells there are to divide. After only 9-1/2 hours (nineteen doublings), the population exceeds 500,000. After 10 hours (twenty doublings), it exceeds 1 million. When we plot the doublings in size against time, we get the J-shaped curve characteristic of unrestricted, exponential growth (curve *a* in Figure 46.3).

What if some bacteria in the culture flask die? To examine the effect of deaths on the growth rate, let's start over with a single bacterium. This time, assume that 25 percent of the cells die during each thirty-minute interval. With this death rate, it takes almost two hours (not thirty minutes) to double the population size. So it takes thirty hours (not ten) to arrive at a million bacteria. *But only the time scale changes.* We still end up with a J-shaped curve that signifies exponential growth (curve *b* in Figure 46.3).

Regarding the Biotic Potential

Suppose a population has plenty of food, living space, and other resources for all of its members. Suppose the habitat is free of predators and pathogens. Under such ideal conditions, a population would show a maximum

rate of increase per individual. That rate is its **biotic potential**, and it differs greatly among species.

For many bacteria, the maximum rate of increase is 100 percent every thirty minutes or so. For humans and other large mammals, it is between 2 and 5 percent per year. The rate depends on the age at which each new generation starts reproducing, how many times each individual reproduces, and how many offspring are produced each time.

Bear in mind, a population that is not growing at its biotic potential may still grow exponentially. For example, each human female is biologically capable of bearing twenty or more children, but many do not reproduce at all. Even so, as you will see shortly, the human population has been growing exponentially since the mid-eighteenth century.

---

1. During exponential growth, the size of a population expands by ever increasing increments during successive time intervals as the reproductive base gets larger.

2. A plot of population size against time has a characteristic J-shaped curve if the population is growing exponentially.

3. As long as the per capita birth rate remains even slightly above the per capita death rate, a population will grow exponentially.

46.3 LIMITS ON THE GROWTH OF POPULATIONS

Limiting Factors

Most often, environmental circumstances prevent any population from fulfilling its biotic potential. That is why humans will not end up covering the planet, even with our capacity for exponential growth. That is why female sea stars, which can produce 2,500,000 eggs each year, do not fill up the oceans with sea stars.

In natural environments, complex interactions exist within and between populations of different species, so it is not easy to identify the factors working to limit population growth. To get a sense of what some of those factors might be, start again with a lone bacterium in a culture flask, where you can control the variables. First you enrich the culture medium with glucose and other nutrients required for bacterial growth. Then you allow bacterial cells to reproduce for many generations. At first the growth pattern appears to be exponential. Then growth slows, and population size remains rather stable. After the stable period, the population size starts to decline rapidly—and all the bacteria die.

What happened? As the population expanded by ever increasing amounts, the growing number of cells used more and more nutrients. When the nutrients dwindled, cell divisions decreased. When the supply was exhausted, the bacterial cells starved to death.

Any essential resource that is in short supply is a **limiting factor** on population growth. Food, micronutrients, refuge from predators, living space, and a pollution-free environment are examples. The number of such factors can be huge, and their effects can vary. Even so, one factor alone is often enough to put the brakes on a population's growth at any given time.

Suppose you kept on freshening the supply of all required nutrients for the growing bacterial culture. After an episode of exponential growth, the population would still crash. Like all organisms, bacteria produce metabolic wastes. The wastes produced by such a huge population of bacterial cells would be so great, they would drastically alter conditions in the culture. By its own activities, the population would end up polluting the surrounding environment— and put a stop to its exponential growth.

Figure 46.4 Idealized S-shaped curve characteristic of logistic growth. After a rapid growth phase (time *b* to *c*), growth slows and the curve flattens out as the carrying capacity is reached (to time *d*). Variations can occur in S-shaped growth curves, as when changed environmental conditions bring about a decrease in the carrying capacity (time *d* to *f*). This happened to the human population of Ireland before 1900, when a disease destroyed potato crops that were the mainstay of the diet.

Carrying Capacity and Logistic Growth

Visualize a small population, with members dispersed through the habitat. As the population increases in size, more and more individuals must share nutrients, living quarters, and other resources. As the share available to each one shrinks, fewer individuals may be born, and more may die by starvation or lack of nutrients. The population's rate of growth will decline until births are balanced or outnumbered by deaths. The final population size will largely depend on the sustainable supply of resources.

Biologists have a name for the maximum number of individuals of a population (or species) that can be sustained indefinitely by a given environment. They call it **carrying capacity**.

The effect of carrying capacity on a population is evident in the pattern of **logistic growth**. By this pattern, a population at low density starts growing slowly in size, then grows rapidly, and finally levels off in size once the carrying capacity is reached. We can represent logistic growth in this simplified way:

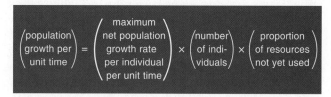

$$\begin{pmatrix} \text{population} \\ \text{growth per} \\ \text{unit time} \end{pmatrix} = \begin{pmatrix} \text{maximum} \\ \text{net population} \\ \text{growth rate} \\ \text{per individual} \\ \text{per unit time} \end{pmatrix} \times \begin{pmatrix} \text{number} \\ \text{of indi-} \\ \text{viduals} \end{pmatrix} \times \begin{pmatrix} \text{proportion} \\ \text{of resources} \\ \text{not yet used} \end{pmatrix}$$

or $G = r_{max} N [(K - N)/K]$. Here, K represents the carrying capacity. The term inside the parentheses is close to 1 when a population is small. It approaches zero when population size is close to carrying capacity.

A plot of logistic growth gives an S-shaped curve (Figure 46.4). Such curves are only an approximation of what goes on in nature. For example, sometimes a population grows so rapidly, it overshoots the carrying capacity. Then, the death rate skyrockets and the birth rate plummets, driving the number of individuals down to the carrying capacity—or lower (Figure 46.5).

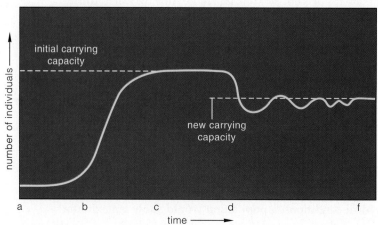

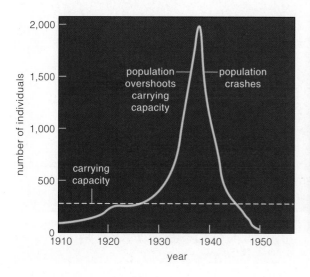

Figure 46.5 Carrying capacity and a reindeer herd. In 1910, four male and twenty-two female reindeer were introduced on one of Alaska's Pribilof Islands. In less than thirty years, the herd increased to 2,000. Its members had to compete for dwindling vegetation, and overgrazing destroyed most of it. In 1950, the herd plummeted to eight members. The growth pattern reflects how population size overshot the carrying capacity, then crashed.

Density-Dependent Controls

The logistic growth equation describes *density-dependent* population growth. When a population is not large, it may grow rapidly. When it finally bumps up against its carrying capacity, so to speak, it grows very slowly or even declines in size. However, carrying capacity is not the only factor that influences population size. Other natural controls come into play that put the number of individuals *below* the maximum sustainable level.

Consider that high density and overcrowding put individuals at greater risk of being killed by predators, colonized by parasites, or infected by microorganisms that cause contagious diseases. At such times, predators, parasites, and pathogens interact more intensely with the population and so bring about a decline in its density. Once population size declines, the density-dependent interactions become less common, and the population may grow once again.

The *bubonic plague* that swept through Europe in the fourteenth century is a classic example of density-dependent control. *Yersinia pestis*, the bacterium that causes the disease, normally lives in wild rodents. Hungry fleas transmit it to new hosts. *Y. pestis* multiplies in the flea gut and blocks digestion, the fleas attempt to feed more and more often, and so the disease spreads. The plague raced through cities where humans were crowded together, sanitary conditions were poor, and rats were abundant. By the plague's end, the populations in European cities had declined by 25 million.

Late in 1994, bubonic plague and a related disease, pneumonic plague, raced through rat-infested cities in India where garbage and animal carcasses had piled up for months in the streets. Thousands of residents fled in terror; some carried the diseases as far away as London. Global panic ensued before concerted efforts to burn the garbage, poison rats and fleas, and liberally dispense antibiotics averted a pandemic.

Density-Independent Controls

Some events cause more deaths or fewer births whether the members of a population are overcrowded or not. Freak summer snowstorms in the Rocky Mountains of Colorado provide us with an example of such density-independent control. They often destroy entire populations of butterflies. Similarly, heavy sprays of pesticides in your backyard may kill most insects, mice, cats, birds, and other animals regardless of how dense their populations are.

---

1. Resources in short supply put limits on population growth. Collectively, all of the limiting factors acting on a population dictate how many of its individuals can be sustained indefinitely in a given area.

2. *Carrying capacity* refers to the maximum number of individuals of a population that can be sustained indefinitely by the resources in a given environment.

3. A low-density population may increase slowly in size, go through a rapid growth phase, then level off in size once the carrying capacity is reached. This is a logistic growth pattern.

4. The size of most populations varies. Often the carrying capacity becomes lower or higher, owing to changes in resource availability over time. Density-dependent and density-independent controls bring about decreases in population size.

46.4 LIFE HISTORY PATTERNS

So far, we have been considering populations as if each were made up of identical individuals during a given interval. However, individuals of most species pass through different stages of development. And each stage proceeds in the context of interactions with other organisms and with the environment.

Let's look at a few examples of age-specific, life history patterns. They will help us appreciate the *Focus* essay that follows, on page 812. That essay will give you a glimpse into the kinds of environmental variables that influence life history patterns.

Life Tables

Each species has a characteristic life span, but few individuals reach the maximum age possible. Death looms more at some ages and less so at others. Also, individuals tend to reproduce or leave a population at certain ages, which vary from one species to the next.

Age-specific patterns in populations first intrigued life insurance and health insurance companies, but they also interest ecologists. Typically, researchers keep track of a **cohort**, a group of individuals from the time of birth until the last one dies. They also track the number of offspring born to individuals at each age interval. Computing the age-specific birth and death rates produces birth and death "schedules." A death schedule can be transformed to a "survivorship" schedule, which more cheerily lists the number of individuals that live long enough to reach age *x*.

The life table in Table 46.1 was constructed for a cohort of 996 phlox plants in Texas. The life table in Table 46.2 lists similar data for the human population of the United States, as reported in 1992.

To ecologists, dividing a population into age classes and assigning birth rates and mortality risks to each class have practical applications. Unlike a crude census (head count), such data can be the basis for informed policy decisions on conservation of endangered species, pest management, social planning for human populations, and other concerns. For example, birth and death schedules for the spotted owl figured in the court decisions that halted logging in old-growth forests, the owl's habitat.

Studies of cohorts reveal distinctive patterns of reproduction, death, and migration that are characteristic of the populations of each species.

Table 46.1   Life Table for a Cohort of Annual Plants (*Phlox drummondii*)

Age Interval (days)	Survivorship (number surviving at start of interval)	Number Dying During Interval	Death Rate per Individual During Interval	"Birth" Rate (number of seeds produced per individual) During Interval
0 – 63	996	328	0.329	0
63 – 124	668	373	0.558	0
124 – 184	295	105	0.356	0
184 – 215	190	14	0.074	0
215 – 264	176	4	0.023	0
264 – 278	172	5	0.029	0
278 – 292	167	8	0.048	0
292 – 306	159	5	0.031	0.33
306 – 320	154	7	0.045	3.13
320 – 334	147	42	0.286	5.42
334 – 348	105	83	0.790	9.26
348 – 362	22	22	1.000	4.31
362 –	0	0	0	0
		996		

Data from W. J. Leverich and D. A. Levin, *American Naturalist* 1979, 113:881–903.

Table 46.2   Life Table for the United States Human Population, 1989*

Age Interval (category for individuals between the two ages listed)	Survivorship (number alive at start of age interval, per 100,000 individuals)	Mortality (number dying during the age interval)	Life Expectancy (average lifetime remaining at start of age interval)	Reported Live Births for Total Population
0 – 1	100,000	896	75.3	
1 – 5	99,104	192	75.0	
5 – 10	98,912	117	71.1	
10 – 15	98,795	132	66.2	11,486
15 – 20	98,663	429	61.3	506,503
20 – 25	98,234	551	56.6	1,077,598
25 – 30	97,683	606	51.9	1,263,098
30 – 35	97,077	737	47.2	842,395
35 – 40	96,340	936	42.5	293,878
40 – 45	95,404	1,220	37.9	44,401
45 – 50	94,184	1,766	33.4	1,599
50 – 55	92,418	2,727	28.9	
55 – 60	89,691	4,334	24.7	
60 – 65	85,357	6,211	20.8	
65 – 70	79,146	8,477	17.2	
70 – 75	70,669	11,470	13.9	
75 – 80	59,199	14,598	10.9	
80 – 85	44,601	17,448	8.3	
85 +	27,153	27,153	6.2	Total: 4,040,958

*Compiled by Marion Hansen, based on data from U.S. Bureau of the Census, *Statistical Abstract of the United States*, 1992 (edition 112).

a

b

c

Figure 46.6 Three generalized types of survivorship curves. For Type I populations, there is high survivorship until some age, then high mortality. Type II populations show a fairly constant death rate at all ages. For Type III populations, survivorship is low early in life.

Survivorship Curves and Reproductive Patterns

Plots of the age-specific survival of a cohort in a given environment are called **survivorship curves**. Three types are common in nature.

Type I curves reflect high survivorship until fairly late in life, then a large increase in deaths. Such curves are typical of elephants and other large mammals that produce only one or a few large offspring at each reproduction and provide them with extended parental care (Figure 46.6*a*). Female elephants produce only four or five calves in a lifetime, and they devote several years of parental care to each one. Type I curves also are typical of human populations with access to good health care services. Historically, and where health care is poor today, infant deaths cause a sharp drop at the start of the curve. Following the drop, the curve then levels off from childhood to early adulthood.

Type II curves reflect a fairly constant death rate at all ages. They are typical of organisms that are just as likely to be killed or to die of disease at any age. This is true of lizards, small mammals, and some songbirds (Figure 46.6*b*).

Type III curves reflect a high death rate early in life. They are typical of animals that produce many small offspring, with little or no parental care. Figure 46.6*c* shows how the curve plummets for sea stars. Although sea stars produce mind-boggling numbers of tiny offspring, their young must feed and grow rapidly on their own, without support, protection, or guidance from parents. Most offspring are quickly eaten by corals and other animals. Plummeting survivorship curves are typical of many other marine invertebrates, most insects, and many fishes, plants, and fungi.

At one time, ecologists thought that selection processes favored either the early, rapid production of many small offspring or the late production of only a few large ones. These two patterns are now known to be extremes, at opposite ends of a range of possible life histories. Also, *both* patterns—and intermediate ones—may be evident in different populations of the same species, as the following *Focus* essay makes clear.

Survivorship curves reveal differences in age-specific survival among species. Such differences can exist even between different populations of the same species.

Natural Selection Among the Guppies of Trinidad

Several years ago, drenched with sweat and with fish nets in hand, two evolutionary biologists were busily engaged in field work in Trinidad, an island in the southern Caribbean Sea. They were after guppies—small, live-bearing fishes that darted about in shallow mountain streams (Figure *a*). John Endler and David Reznick were about to identify environmental variables that influence the life history patterns of the guppies.

The researchers could easily distinguish the male guppies from the females. The males are smaller. Their scales are brightly colored, in patterns that serve as visual signals during mating behavior. The males also engage in intricate courtship maneuvers. In contrast to the males, which stop growing once they reach sexual maturity, the females are drab colored. And they continue to grow larger as they reproduce.

In the mountains of Trinidad, guppies living in different streams—and even in different parts of the same streams—are subject to different dangers. In some streams, a small killifish preys heavily on immature guppies but does not have much success with the larger adults. In other streams, a larger pike-cichlid prefers mature, larger guppies and tends not to waste time hunting small ones (Figure *b*).

As Reznick and Endler predicted from their understanding of natural selection, predation has had profound influence over the life history patterns of guppies. In streams where the pike-cichlids reign supreme, the

researchers found that the individuals of guppy populations mature faster, compared to their counterparts in killifish streams. Besides this, those guppies are smaller in size at maturity. They reproduce at a younger age. They produce far more offspring and do so more often (compare Figures *c* through *e*).

Could the differences in guppy populations be a result of some other, unknown differences between the streams? The researchers decided to check out this possibility when they returned home.

They carefully packed up groups of guppies and shipped them back to the laboratory. There they raised guppies from the two kinds of streams for two generations. Both of the experimental populations were raised under identical conditions—that is, in the absence of predation.

Even when predators were absent, the same life history differences emerged between the two experimental populations. The experimental results provided evidence of a genetic (heritable) basis for the differences.

The researchers also investigated the role of predators in the evolution of differences in size among guppies. They raised guppies for many generations in the laboratory— some alone, some with killifish, and some with pike-cichlids.

As predicted, the guppy lineage that was subjected to predation over time by killifish became larger at maturity. The lineage that was raised with pike-cichlids showed a trend toward earlier maturity.

a Part of a shallow stream in the mountains of Trinidad.

b Pike-cichlid, a big-time danger if you are a guppy. Adults are much larger than shown here.

Finally, when they first visited Trinidad, Endler and Reznick had introduced guppies from a pike-cichlid stream to another stream that contained killifish but no pike-cichlids or guppies. Eleven years later, the guppies had evolved. As the researchers predicted, the guppies had become larger in size and slower at reproducing. And these characteristics are typical of natural guppy populations that live and die with killifish.

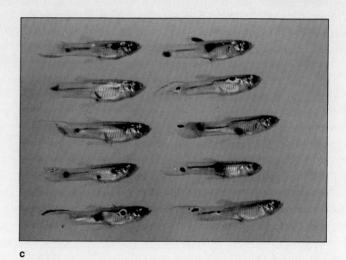

c

In places where predators are dangerous, guppies are smaller, more streamlined, and duller in color patterning. The guppies shown in (c) belong to this group. In places where predators eat guppies only occasionally, the guppies are larger, less streamlined, and more brightly colored. The guppies shown in (d) belong to this group. The differences between the groups are accompanied by pronounced differences in life history patterns.

(c,d) Samplings of the kind of guppies (*Poecilia reticulata*) that live in the mountain streams of Trinidad. They are approximately the size shown in the photographs. Guppies devote considerable time to courting and avoiding predators and must be adapted to do both. Depending on where they live, guppies might meet up with weak or dangerous predatory fishes.

(e) Experimental evidence of differences in body size, interval between broods, and embryo weights for guppies from streams with pike-cichlids (indicated by the *green* bars) or killifish (*red* bars). Pike-cichlids prey on larger guppies and killifish prey on smaller ones, selecting for the differences shown here.

d

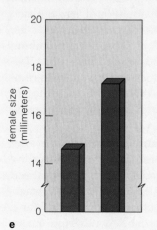 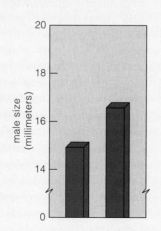 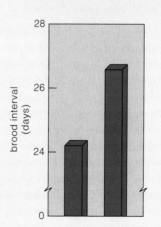 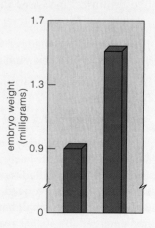

e

46.5 HUMAN POPULATION GROWTH

In 1994, the human population reached 5.6 billion. In that year alone, 89,600,000 individuals were added to it. These people are not dispersed uniformly through the environment. About 4.5 billion are clumped together on only 10 percent of the land's surface. About 3 billion crowd together within 300 miles of the seas.

Our staggering population growth continues while at least a billion people already on the planet are malnourished or starving, without clean drinking water and adequate shelter. Growth continues when 1.5 billion already go without the benefits of health care delivery and sewage treatment facilities.

Suppose it were possible, by monumental efforts, to double the food supply to keep pace with growth. We would do little more than maintain marginal living conditions for most, and annual deaths from starvation could still be 20 million to 40 million. Even this would come at great cost, for we are drastically modifying the very environment that must sustain us. Salted-out cropland, desertification, deforestation, pollution—these are some of the consequences you will read about in Chapter 50, and they do not bode well for our future.

For a while, it would be like the Red Queen's garden in Lewis Carroll's *Through the Looking Glass*, where one is forced to run as fast as one can to remain in the same place. But what happens when the human population doubles again? Can you brush this picture aside as being too far in the future to warrant your concern? It is no farther removed from you than the sons and daughters of the next generation.

How We Began Sidestepping Controls

How did we get into this predicament? For most of our history, our population grew slowly. But over the past two centuries, increases in growth rates have become astounding. There are three possible reasons for this:

1. We steadily developed the capacity to expand into new habitats and new climate zones.

2. We increased the carrying capacities in existing habitats.

3. We sidestepped several limiting factors.

The human population did all three of these things. Consider the first point. Early humans lived mostly in the open grasslands called savannas. They were vegetarians who added scavenged bits of meat to their diet. By 200,000 years ago, small bands of hunters and gatherers had emerged. By 40,000 years ago, hunter-gatherers had spread through much of the world. Most other species could not have expanded into such a broad range of habitats. Humans did so by using their complex brains. They applied learning and memory to problems such as how to build fires, assemble shelters, make clothing and tools, and plan community hunts. Learned experiences were not confined to individuals but spread quickly from one band to another because of language—the ability for cultural communication. *Thus, the human population expanded into diverse new environments in an extremely short time span, compared with the long-term geographic dispersal of other kinds of organisms.*

What about the second possibility? About 11,000 years ago, humans began to shift from the hunting and gathering way of life to agriculture. They stopped following game herds and moving about to harvest fruits and grains. They settled down and developed a more dependable basis for existence in more favorable settings. A milestone was the domestication of wild grasses, including species ancestral to modern wheats and rice. Seeds were harvested, stored, and planted in one place. Animals were domesticated and kept close to home for food and pulling plows. Water was diverted into hand-dug ditches to irrigate crops.

The emerging agricultural practices increased productivity. With larger, more dependable food supplies, population growth rates increased. As towns and cities developed, a social hierarchy emerged that provided a labor base for more intensive agriculture. Much later, food supplies increased again through the use of fertilizers and pesticides. Transportation for food distribution improved. *Even at its simplest, managing food supplies through agriculture increased the carrying capacity for the human population.*

What about the third possibility—sidestepping limiting factors? Consider what happened when medical practices and sanitary conditions improved. Until about 300 years ago, malnutrition, contagious diseases, and poor hygiene kept death rates high enough to more or less balance birth rates. Contagious diseases (density-dependent factors) swept through crowded settlements and cities. Without proper hygiene and sewage disposal, and plagued with disease-carrying fleas and rats, population growth increased slowly. Then plumbing and sewage treatment methods were developed. Vaccines, antitoxins, and antibiotics against many pathogens were developed. And the death rate dropped sharply. Births now exceeded deaths, and rapid population growth was under way.

In the mid-eighteenth century, people discovered how to harness the energy stored in fossil fuels, starting with coal. Within a few decades, large industrialized societies emerged in western Europe and North America. Efficient technologies developed after World War I. Cars, tractors, and other affordable goods were mass-produced in factories. Machines replaced many of the

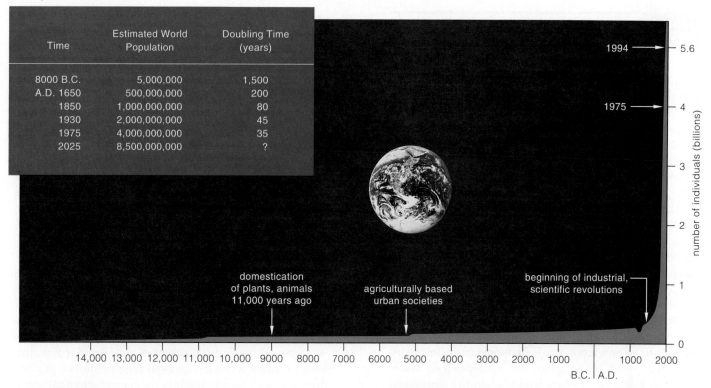

Time	Estimated World Population	Doubling Time (years)
8000 B.C.	5,000,000	1,500
A.D. 1650	500,000,000	200
1850	1,000,000,000	80
1930	2,000,000,000	45
1975	4,000,000,000	35
2025	8,500,000,000	?

domestication
of plants, animals
11,000 years ago

agriculturally based
urban societies

beginning of industrial,
scientific revolutions

1994 — 5.6
1975 — 4

number of individuals (billions)

14,000 13,000 12,000 11,000 10,000 9000 8000 7000 6000 5000 4000 3000 2000 1000   1000 2000
B.C. | A.D.

Figure 46.7 Growth curve for the human population. The diagram's vertical axis represents world population, in billions. (The slight dip between A.D. 1347 and 1351 is when 25 million people died in Europe as a result of bubonic plague.) Phenomenally, growth of the human population over the past two centuries has been even greater than exponential, owing to increases in the rate of multiplication. Agricultural revolutions, industrialization, and improvements in health care have sustained the increases. The *blue* box tells you how long it took for population size to double at different times in our history.

farmers needed to produce food, and fewer farmers could support a larger population.

Thus, *by bringing many disease agents under control and by tapping into concentrated, existing stores of energy, humans sidestepped certain factors that had previously limited their population growth.*

Present and Future Growth

Where have our farflung migrations and advances in agriculture, industrialization, and health care taken us? As you can see from Figure 46.7, it took *2 million* years for the human population to reach 1 billion. It took only 80 years to reach the second billion, 30 years to reach the third, 15 years to reach the fourth—*and only 12 years to reach the fifth billion!*

From what we know about the principles governing population growth, and unless technological breakthroughs once again increase the carrying capacity, we should expect to see a dramatic increase in death rates. *Although the stupendously accelerated growth of the human population continues, it cannot be sustained indefinitely.*

Besides having adverse effects on resource supplies, our skyrocketing numbers invite density-dependent controls. For example, the largest cholera epidemic of this century is sweeping through South Asia. Like the six previous epidemics, it began in India and may spread through Africa and the Middle East, and into Mediterranean countries before peaking. It may claim 5 million lives. Existing vaccines do not work against a current strain of *Vibrio cholerae*, the bacterial pathogen.

People get infected by drinking water or eating food contaminated by raw sewage. *V. cholerae* multiplies in the intestines and produces a toxin that causes massive fluid loss, by diarrhea. Two to seven days later, untreated people can die of extreme dehydration. For centuries, *V. cholerae* has thrived and mutated in sewage-rich rivers in India (Figure 46.1). In Calcutta's slums, millions are forced to bathe in stagnant ponds and polluted waterways. In 1992, cholera struck tens of thousands in that city alone.

At this writing, a new era of human migration has begun. By some estimates, 50 million people are on the move within and between countries, owing to economic hardship and civil strife. Will they be relocated peaceably? Where will they find sustainable supplies of food, clean water, and other basic resources?

By expansion into new habitats, cultural intervention, and technological innovation, the human population has temporarily skirted environmental resistance to growth. Its accelerated growth cannot be sustained indefinitely.

46.6 CONTROL THROUGH FAMILY PLANNING

Figure 46.8 shows the rates at which populations in different regions grew during 1994. The annual growth rate for the entire human population averaged out at 1.6 percent. If that rate is maintained, the human population may approach *8.5 billion* just thirty years from now (Figure 46.9). It is mind-numbing to think about the resources required to sustain 8.5 billion people. We will have to increase food production, supplies of drinkable water, energy reserves, and all the wood, steel, and other materials we use to meet everyone's basic needs—something we are not even doing now. Again, this gross manipulation of resources is likely to intensify pollution, which will adversely affect water supplies, the atmosphere, and productivity on land and in the seas.

Today there is growing realization that population growth, resource depletion, pollution, and the quality of life are all interconnected. As evidence of this, consider that most governments are attempting to lower the birth rates, as through **family planning programs**. The programs educate individuals about choosing the number of children they will have, and when. They vary in their details from country to country, but all provide information on methods of birth control, as described on page 794. When thoughtfully developed and carefully administered, family planning programs may indeed bring about a decline in birth rates.

To achieve zero population growth, the average "replacement rate" would be slightly higher than two children per couple, because some female children die before they reach reproductive age. The replacement rate is about 2.5 children per woman in less developed countries, and 2.1 in more developed countries.

Suppose family planning programs were successful beyond our wildest imagination, and each couple on the planet decided to have only two children. The human population would keep on growing for at least another sixty years! Why? An immense number of existing children will eventually be reproducing.

Take a look at Figure 46.10, which shows age structure diagrams for populations that are growing at different rates. In these diagrams, 15–44 is the average range for childbearing years. The diagram for Kenya and other rapidly growing populations has a broad base. The central portion of each diagram includes women and men of reproductive age. Besides this, the lower portion includes an even larger number of children who will be moving into the reproductive category during the next fifteen years. Said another way, *more than a third of the world population fall in this broad reproductive base*. This gives you an idea of the magnitude of the effort it will take to control population growth.

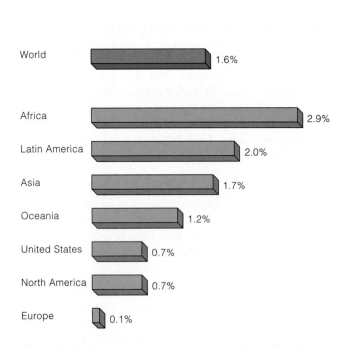

Figure 46.8 Average annual population growth rate in various groups of countries in 1994.

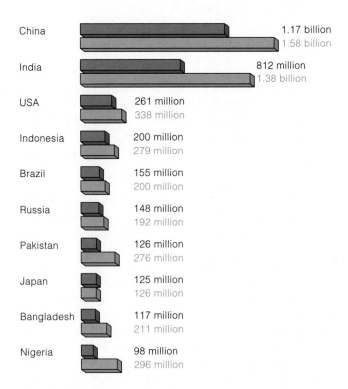

Figure 46.9 The ten most populous countries in 1994 (*red* bars). *Blue* bars indicate their population size projected for 2025.

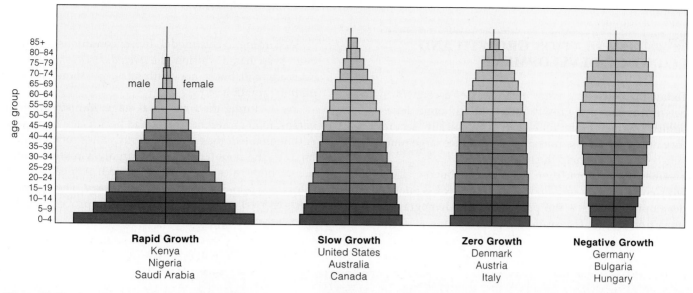

age group

85+
80–84
75–79
70–74
65–69
60–64
55–59
50–54
45–49
40–44
35–39
30–34
25–29
20–24
15–19
10–14
5–9
0–4

male   female

Rapid Growth
Kenya
Nigeria
Saudi Arabia

Slow Growth
United States
Australia
Canada

Zero Growth
Denmark
Austria
Italy

Negative Growth
Germany
Bulgaria
Hungary

Figure 46.10 Age structure diagrams for countries with rapid, slow, zero, and negative population growth rates. *Dark green* indicates the pre-reproductive years. *Purple* indicates reproductive years; *light blue*, the post-reproductive years. The portion of the population to the left of the vertical axis in each diagram represents males; the portion to the right represents females. The width of each bar indicates the proportion of males and females within each age group.

Figure 46.11 Alongside a city street in China, a billboard reminding couples to produce no more than one child. Couples who comply are rewarded economically and socially; those who do not suffer economic penalties.

One way to slow the birth rate is to bear children in the early thirties, rather than in the mid-teens or early twenties. Delayed reproduction slows the tendency toward growth and lowers the average number of children in each family. In China, for example, the government has established the most extensive family planning program in the world. Premarital sex is strongly discouraged, couples are strongly urged to postpone the age at which they marry, and they are encouraged to have one child only (Figure 46.11). Married couples have ready access to free contraceptives, abortion, and sterilization. Paramedics and mobile units ensure access in remote rural areas. Couples who take the pledge to have one child only are given extra food, better housing, free medical care, and salary bonuses. Their child will be granted free tuition and preferential treatment when he or she enters the job market. Those who break the pledge forego benefits and sometimes pay penalties.

Are these outrageous measures? Think about this: Family planning has been China's way of avoiding mass starvation. Between 1958 and 1962 alone, an estimated 30 million Chinese died because of famine. Even so, the population time bomb has not stopped ticking. China's population now numbers 1.17 billion—and 340 million of its young women are moving into the reproductive age category. By the year 2025, China is projected to have a population of 1.58 billion people.

Family planning programs on a global scale can help stabilize the size of the human population.

Even with zero population growth, the population will continue to grow for sixty years, for its reproductive base already consists of a staggering number of individuals.

46.7 POPULATION GROWTH AND ECONOMIC DEVELOPMENT

Today, there also is greater awareness of a connection between population growth rates and economic development. When individuals are economically secure, they appear to be less pressured to produce large numbers of children to help them survive.

Changes in population growth can be correlated with changes that unfold in four stages of economic development. This is the premise of the **demographic transition model** (Figure 46.12).

According to this model, living conditions are difficult, even harsh, during the *preindustrial* stage. Birth rates are high, but so are death rates, and therefore population growth is slow.

Next, during the *transitional* stage, industrialization begins, food production rises, and health care improves. Although death rates drop, the birth rates remain fairly high, so the population grows rapidly. Growth continues at high rates over a long period. Annual growth rates are 2.5 to 3 percent, on the average. Then growth starts to level off as living conditions improve and birth rates begin to decline.

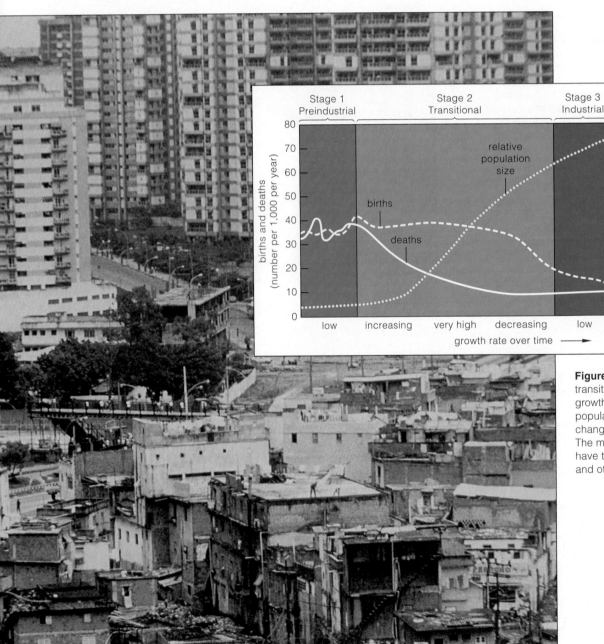

Figure 46.12 Demographic transition model of changes in the growth characteristics and size of populations, as correlated with changes in economic development. The model explains changes that have taken place in western Europe and other industrialized regions.

Focus on the Environment

Populations and Resource Consumption

This chapter began with a brief look at the appalling conditions confronting most of the people of India, with its whopping 16 percent of the human population. By comparison, the United States has merely 4.7 percent. Yet which country is the most "overpopulated"—not in terms of numbers, but rather in terms of resource consumption and environmental damage?

In his book *Living in the Environment*, G. Tyler Miller considers reports that the average American consumes fifty times as much as the average person in India. The United States produces 21 percent of all goods and ser- vices. It uses about 25 percent of the available processed minerals and nonrenewable energy resources. It also pro- duces at least 25 percent of the global pollution and trash. By contrast, India produces 1 percent of all goods and services. It uses about 3 percent of the available minerals and nonrenewable energy resources. And India produces about 3 percent of the global pollution and trash. Extrapo- lating from these numbers, Miller writes that it would take 12.9 billion impoverished people in India to match the present impact on the environment by 258 million average Americans. Think about it.

Population growth slows dramatically during the *industrial* stage, when industrialization is in full swing. The slowdown emerges mostly because people move from the countryside to cities—and urban couples tend to control family size. Many decide that raising more than a few children is costly and puts them at a disad- vantage in the accumulation of goods (which is a sticky, related issue, as the *Focus* essay suggests).

In the *postindustrial* stage, zero population growth is reached. The birth rate falls below the death rate, and population size slowly decreases.

Today, the United States, Canada, Australia, Japan, the Soviet Union, and most countries of western Europe are in the industrial stage. Their growth rate is slowly decreasing. In Germany, Bulgaria, Hungary, and some other countries, birth rates are lower than death rates, and the populations are getting smaller.

Mexico and other less-developed countries are in the transitional stage. They do not have enough skilled workers to complete the transition to a fully industrial economy. Fossil fuels and other resources that drive industrialization are being used up there as well as in the industrialized countries. Fuel costs may become pro- hibitive for countries at the bottom of the economic lad- der before they can enter the industrial stage. If population growth keeps outpacing economic growth, death rates will increase. Therefore, many countries may now be stuck in the transitional stage. And some may return to the harsh conditions of the preceding stage.

Enormous disparities in economic development are driving forces for immigration and emigration. Think of what happened in California between 1950 and 1993. Its population tripled and now stands at 32 million. Already the most highly populated state in the nation, California expects its population to increase by an addi- tional 20 million by the year 2020. Also, unlike the rest of the nation, its annual growth rate has been increas- ing, especially among legal and illegal newcomers from less-developed countries. That rate now stands at 2.5 percent.

For some time, California has been symbolic of the good life, a magnet drawing people from all over the world. Yet its economy is stagnating. Funding for social services is shrinking, schools are overcrowded, sewage treatment facilities are nearing capacity, and water shortages are severe and frequent. Salinization and air pollution are eroding the agricultural base, which is a key factor in California's economic health.

Recognizing the connection between population growth and economic well-being, many governments restrict immigration. Only the United States, Canada, Australia, and a few other countries allow large annual increases. (In 1991 alone, the United States admitted 1.8 million legal immigrants.) Elsewhere, poor living con- ditions, civil strife, and government policy promote emigration.

Differences in population growth among countries correlate with levels of economic development, hence with economic security (or lack of it) of individuals.

46.8 QUESTIONS CONCERNING ZERO POPULATION GROWTH

For us, as for all species, the biological implications of extremely rapid growth are truly staggering. Yet so are the social implications of what will happen when we achieve and maintain zero population growth or when the human population declines.

For instance, as you have seen, most individuals of an actively growing population fall in the younger age brackets. Under conditions that promote constant growth, the age distribution for the population as a whole guarantees the availability of a future work force. It takes a large work force to support older, nonproductive people with social security, low-cost housing, health care, and other social programs. However, if zero population growth is maintained over the long term, a larger proportion of the population will end up in the older age brackets.

Suppose this social condition is realized. Will the nonproductive members continue to be provided with goods and services if productive ones are asked to carry a greater and greater share of the economic burden? These are not abstract questions. Put them to yourself. How much are you willing to bear for the sake of your parents? Your grandparents? How much will your children be willing or able to bear for you?

We have arrived at a major turning point, not only in our biological evolution but in our cultural evolution as well. The decisions awaiting us are among the most difficult we will ever have to make, yet it is clear that they must be made, and soon.

All species face limits to growth. We may think we are different from the rest, for our unique ability to undergo cultural evolution has allowed us to postpone the action of most of the factors limiting growth. But the key word here is *postpone*. No amount of cultural intervention can hold back the ultimate check of limited resources and a damaged environment.

We have sidestepped a number of the smaller laws of nature. In doing so, we have become more vulnerable to those laws which cannot be repealed. Today there may be only two options available. Either we make a global effort to limit population growth in accordance with environmental carrying capacity, or we wait until the environment does it for us.

In the final analysis, no amount of cultural intervention can repeal the ultimate laws governing population growth, as imposed by the carrying capacity of the environment.

SUMMARY

1. A population's growth rate depends on the rates of birth, death, and immigration. Putting aside the effects of immigration and emigration, if the birth rate per individual exceeds the death rate per individual by a constant amount, a population will grow exponentially; it will show a constant rate of multiplication over time.

2. Carrying capacity is the maximum number of individuals in a population that can be sustained indefinitely by the resources that are available in a given environment.

3. Population size is determined by the availability of sustainable resources, predation, competition, and other factors that limit population growth. Limiting factors vary in their relative effects and they vary over time, so population size also changes over time.

4. Limiting factors such as competition for resources, disease, and predation are density dependent. Other factors, such as weather on a rampage, tend to increase the death rate or decrease the death rate more or less independently of population density.

5. Patterns of reproduction, death, and migration vary over the life span characteristic of a species. Environmental variables help shape its life history (age-specific) patterns.

6. Currently, human population growth varies from below zero in a few developed countries to more than 4 percent per year in some less-developed countries. In 1994 the annual growth rate for the entire human population was 1.6 percent.

7. Rapid growth of the human population during the past two centuries was possible largely because of a capacity to expand into new environments, and because of agricultural and technological developments that increased the carrying capacity. Ultimately, we must confront the reality of the carrying capacity and limits to our population growth.

Review Questions

1. Why do populations that are not restricted in some way tend to grow exponentially? *806–807*

2. Define carrying capacity, then describe its effect as evidenced by a logistic growth pattern. *808*

3. Give examples of limiting factors that come into play when a population of mammals (for example, rabbits or humans) reaches very high density. *808–809*

4. At present growth rates, how long will it take to add another billion individuals to the human population? *816*

5. How did human populations develop the means to expand steadily into new environments? How did they increase the carrying capacity of their habitats? How have they avoided some of the limiting factors on population growth? Is the avoidance an illusion? *814–815*

6. If a third of the world population is now below age fifteen, what effect will this age distribution have on the future growth rate of the human population? What sorts of humane recommendations would you make that would encourage individuals of this age group to limit the number of children they plan to have? What are some of the social, economic, and environmental factors that might keep them from following those recommendations? (*Reflect upon pages 802–803, 816–819.*)

7. Write a short essay about a population that shows one of the following age structures. Describe what might happen to younger and older age groups when individuals move into new categories. *816–817*

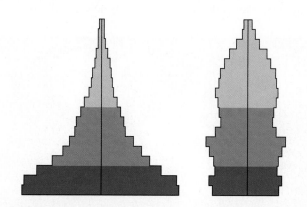

1. _____ is the study of how organisms interact with one another as well as with their physical and chemical environment.

2. A _____ is a group of individuals of the same species that occupy a certain area.

3. The rate at which a population grows or declines depends upon the rate of _____ .
   a. births   d. emigration
   b. deaths   e. all of the above
   c. immigration

4. Populations grow exponentially when _____ .
   a. birth rate exceeds death rate and neither changes
   b. death rate remains above birth rate
   c. immigration and emigration rates are equal
   d. emigration rates exceed immigration rates
   e. both a and c

5. The maximum number of individuals in a population that can be sustained indefinitely by the resources in a given environment is the _____ .
   a. biotic potential   c. environmental resistance
   b. carrying capacity   d. density control

6. Resource competition, disease, and predation are _____ controls on population growth rates.
   a. density-independent   c. age-specific
   b. population-sustaining   d. density-dependent

7. Which of the following factors does *not* affect sustainable population size?
   a. predation   d. pollution
   b. competition   e. all of the above can affect
   c. resources   population size

8. In 1994, the average annual growth rate for the human population was _____ percent.
   a. 0   c. 1.5   e. 2.7
   b. 1.0   d. 1.6   f. 4.0

9. If housecats that have not been neutered or spayed live up to their biotic potential, two can be the start of great numbers of kittens—12 the first year, 72 the second year, 429 the third, 2,574 the fourth, 15,416 the fifth, 92,332 the sixth, 553,019 the seventh, 3,312,280 the eighth, and 19,838,741 kittens in the ninth year. This is a case of _____ .
   a. logistic growth   c. irresponsible cat owners
   b. exponential growth   d. both b and c

10. Match the population ecology terms with the appropriate description.
   _____ carrying capacity   a. disease, predation
   _____ exponential growth   b. group of individuals of same
   _____ population growth   species occupying a given area
          rate   c. depends on rates of birth,
   _____ density-dependent   death, immigration, emigration
          controls   d. maximum number of
   _____ population   individuals that can be sustained
                                      indefinitely by a given
                                      environment's resources
                                  e. population growth over
                                      time plots out as J-shaped
                                      curve

age structure *804*   habitat *803*
biosphere *803*   limiting factor *808*
biotic potential *807*   logistic growth *808*
carrying capacity *808*   net reproduction per
cohort *810*   individual per unit time *806*
community *803*   population *803*
demographic transition   population density *804*
   model *818*   population distribution *804*
doubling time *807*   population size *804*
ecology *803*   reproductive base *804*
ecosystem *803*   survivorship curve *811*
exponential growth *806*   zero population growth *806*
family planning program *816*

Miller, G. T. 1994. *Living in the Environment*. Eighth edition. Belmont, California: Wadsworth. This author consistently pulls together information on human population growth into a coherent picture.

Polgar, S. 1972. "Population History and Population Policies from an Anthropological Perspective." *Current Anthropology* 13(2):203–241. Analyzes often-ignored cultural barriers to programs for population control.

47 COMMUNITY INTERACTIONS

No Pigeon Is an Island

Flying through the rain forests of New Guinea is an extraordinary pigeon with cobalt blue feathers and lacy plumes on its head (Figure 47.1). It is about as big as a turkey, and it flaps so slowly and noisily that its flight sounds like an idling truck. Like eight species of smaller pigeons living in the same forest, it perches on branches to eat fruit.

Why are there nine species of large and small fruit-eating pigeons living in the space of the same forest? Why hasn't competition for food left one species the winner? In fact, in that forest, each species lives, grows, and reproduces in a particular way, as defined by its relationships with other individuals and with the surroundings. Big pigeons perch on sturdy branches when they feed, and they eat big fruit. Smaller ones eat fruit hanging from branches too thin to support the weight of a turkey-size pigeon, and their bill is too small to open big fruit. Also, forest trees differ in the thickness of their fruit-bearing branches and fruit size, so they attract different pigeons. In such ways, the fruit supply is partitioned among the nine pigeon species.

And what about the forest's trees? Their pigeon-enticing fruits contain seeds with tough coats that resist digestion. While seeds pass through the pigeon gut, the pigeons are flying about and are likely to disperse them to new places. Dispersal improves the odds that seedlings won't compete directly with parent trees for sunlight, water, and nutrients. With their extensive roots and leafy crowns, the parent trees might well outcompete their offspring for resources.

Leaf-eating, fruit-munching, and bud-nipping insects interact with other organisms and the

surroundings in certain ways, as do nectar-drinking, tree-pollinating bats, birds, and insects. The same is true of diverse decomposers that busily extract energy from the remains and wastes of other organisms on the forest floor—and so cycle nutrients back to the trees.

Like humans, then, no pigeon is an island, isolated from the rest of the living world. The nine species of New Guinea pigeons eat fruit of different sizes, so they disperse seeds of different sorts. Dispersal influ-

ences where different trees grow and where decomposers flourish—and the tree distribution and the decomposition activities influence how the entire forest community is organized. Directly or indirectly, populations of each kind of organism are affected by interactions with neighboring populations. With this chapter, we turn to community interactions that influence populations over time and in the space of their environment.

Figure 47.1 The turkey-size Victoria crowned pigeon, one of nine pigeon species in the same tropical rain forest of New Guinea. Within this habitat, each species has its own niche.

KEY CONCEPTS

1. A habitat is the type of place where individuals of a species normally live. A community is an association of all the populations of all species that occupy the same habitat.

2. Each species has its own niche in the community, as defined by the full range of the physical and biological conditions under which its individuals live and reproduce.

3. The structure of a community begins with adaptive traits that allow each species to respond to the habitat's physical features, chemical features, and levels and patterns of resource availability over time.

4. Interactions among species influence the structure of a community. The interactions include mutually beneficial activities, competition, predation, and parasitism.

5. The geographic location and size of the habitat, the rates at which species arrive and disappear, and the history of physical disturbances influence the structure of a community.

6. The first species to occupy a habitat are replaced by others, which are replaced by others, and so on in orderly progression. This process, called primary succession, produces a climax community—a stable, self-perpetuating array of species that are in balance with one another and with the environment.

7. Different climax stages often exist in the same habitat, owing to local differences in soil and other environmental factors, recurring disturbances (such as fires), and chance events.

COMMUNITY CHARACTERISTICS

Factors That Shape Community Structure

Think of a damselfish darting above a coral reef, a maple tree on a Vermont hillside, or a mole burrowing under your lawn. The type of place where you will normally find a damselfish, maple, or mole is its **habitat**. The habitat of any organism is characterized by physical and chemical features, as well as by the array of other species living in the same place.

Directly or indirectly, populations of all species in a habitat associate with one another as a **community**. Five factors contribute to the structure of a community. First, interactions between climate and topography help dictate temperatures, rainfall, soil composition, and other physical conditions that characterize the habitat. Second, the kinds and amounts of food and other resources that are available through the year influence the species living there. Third, the individuals of each species have adaptive traits that allow them to survive the conditions of the habitat and to exploit specific resources. Fourth, the occupants of the habitat interact, as by competition, predation, and mutually beneficial activities. Fifth, community structure is influenced by the overall pattern and actual history of changing population sizes, the arrival and disappearances of species, and physical disturbances to the habitat.

How do these factors actually shape community structure? For one thing, they help dictate the number of species at different "feeding levels," starting with producers and continuing on up through assorted levels of consumers. In addition, they help dictate the overall number of species. (For example, the high solar radiation, temperatures, and humidity of tropical habitats favor growth of many kinds of plants, which in turn support many kinds of animals. Conditions in arctic habitats do not favor such great numbers of species.) Finally, these factors influence population size for each species.

The next chapters deal with energy flow through feeding levels and the geographic factors influencing community structure. Here we begin with interactions among species, using the niche concept as our guide.

The Niche

In any habitat, certain combinations of environmental conditions and resources allow individuals of a species to engage fully in the business of surviving and reproducing—as long as other species do not interfere with their required activities. For a predatory fish, the conditions might be water of suitable temperature and salinity, prey of acceptable size and nutritional value, and an absence of fishermen. The full range of environmental and biological conditions under which its members can live, grow, and reproduce is called the **niche** of that species. As you will see, conditions that define a niche are not static. They shift in large and small ways over time, creating a mosaic of changes to which members of the species must respond.

Types of Species Interactions

Even in a simple community, dozens to hundreds of species interact in diverse ways. In spite of this diversity, we can identify six categories of interactions that have different effects on population growth.

Often a **neutral interaction** links two species. For example, eagles and meadow grasses have no direct effect on each other, even though they are linked indirectly through relationships with other species. Eagles prey on populations of grass-eating rabbits and so help the grasses. In turn, the grasses feed rabbits and help provide eagles with fattened prey.

Commensalism is a rather lopsided interaction. It directly benefits one species but does not harm or help the other much, if at all. This is true of tree-roosting birds. The birds gain roosting sites. The trees get nothing in return but are not harmed by the interaction.

In **mutualism,** *benefits* flow both ways between two interacting species. Don't think of mutualism as cozy cooperativeness. The benefits are an outcome of two-way exploitation. By contrast, in **interspecific competition,** *disadvantages* flow both ways between two species. Finally, the extreme interactions called **predation** and **parasitism** directly benefit one species (the predator or parasite) and directly harm the other (the prey or host).

Commensalism, mutualism, and parasitism are all forms of **symbiosis**, a word that means "living together." Individuals of one species live with, in, or on individuals of another species for at least part of their life cycle.

---

1. A habitat is the type of place where individuals of a species normally live. A community is an association of all populations of all species in the same habitat.

2. Community structure is shaped by the habitat's physical and chemical features, resource availability over time, the adaptive traits of its various members, how those members interact, and the history of the habitat and all of its occupants.

3. The niche of a species is the full range of shifting environmental and biological conditions under which its individual members can survive and reproduce.

4. When two species interact, each may have neutral, positive, or negative effects on the other.

47.2 MUTUALISM

Mutualistic interactions, in which positive benefits flow both ways, abound in nature. The chapter introduction described how trees provide New Guinea pigeons with fruit. Such interactions exist between many kinds of plants and animals, which obtain food and help the plants by dispersing seeds to new germination sites. Most flowering plants and their pollinators are mutualists. Consider the yucca plant and the yucca moth. Each yucca plant species can be pollinated by only one yucca moth species. Yucca moth larvae can grow nowhere except in yucca plants; they eat only yucca seeds (Figure 47.2).

Some forms of mutualism are *obligatory*. That is, the individuals of one species cannot grow and reproduce unless they spend their entire life with individuals of other species, in intimate dependency. This is true of the mycorrhiza, an intermeshing of two kinds of absorptive structures (fungal hyphae and young roots). Hyphae of a fungal species penetrate the roots of certain plants and become virtual extensions of them. In fact, the plant cannot take up enough vital nutrients without its fungal symbiont (page 502). In turn, the fungus depends on sugar molecules produced by its photosynthetic partner. When photosynthesis stops, the fungus stops producing spores. As another example, think back on the presumed endosymbiotic origins of eukaryotes (page 331). If cells of different bacterial species had not evolved in such intimate, mutually beneficial ways, you and all other eukaryotic organisms would not even be around today.

In forms of mutualism, each of the participating species reaps benefits from the interaction.

Figure 47.2 A mutualistic interaction in the high desert of Colorado.

Each species of yucca plant (**a**) is pollinated exclusively by one species of yucca moth (**b**), an insect that cannot complete its embryonic development in any other plant.

The adult stage of the moth life cycle coincides with blossoming of yucca flowers. Using specialized mouthparts, a female moth gathers sticky pollen and rolls it into a ball. Then she flies to another flower. She pierces the wall of the flower's ovary, where seeds will develop, and lays eggs inside. As she crawls out of the flower, she pushes a ball of pollen onto a pollen-receiving surface.

The pollen grains germinate and grow down through the ovary tissues, carrying sperm to the flower's eggs. After fertilization, seeds develop.

Meanwhile, the moth eggs develop to the larval stage. When larvae emerge (**c**), they eat a few seeds, then gnaw their way out of the ovary. Uneaten seeds give rise to new yucca plants.

a

b

c

Categories of Competition

As you probably concluded from the preceding chapter, *intraspecific* competition can be fierce. The term refers to competition among individuals of the same population or species. By contrast, *interspecific* competition—that is, between populations of different species—usually is less intense. Why? The requirements of two species might be similar, but they never will be as close as they are for individuals of the same species.

Competitive interactions within or between species take two forms. In some cases, all individuals have equal access to a required resource, but some are better than others at exploiting it. This interaction tends to reduce the supply of a shared resource, unless the resource is abundant. (If you and a friend were sharing a 10-gallon milkshake, each drinking from your own straw, you wouldn't care how fast your friend drank.) In other cases, certain individuals control access to a resource. They partially or wholly prevent others from using it, regardless of its scarcity or abundance. (Even if you shared a 10-gallon milkshake, you would object if your friend pinched your straw.)

Competition abounds in nature. Corals poison and grow over other species of neighboring corals. The strangler fig tree uses other trees as a framework for upright growth, eventually surrounding and killing them. From the time wildflowers start blooming in the Rocky Mountains until late summer, a male broadtailed hummingbird aggressively chases other males and females of its species out of his blossom-rich feeding territory. Then, in August, male rufous hummingbirds migrate from their breeding grounds in the Pacific Northwest, and they cross through the high Rockies on their way to wintering grounds in Mexico. Rufous males are even more competitive. Resident broadtails are evicted from their territories all along the migratory hummingbird's route!

Competitive Exclusion

To a greater or lesser extent, any two species differ in their adaptations for securing food or avoiding enemies, so one usually competes more effectively for scarce resources. The two are less likely to coexist in the same habitat when they use resources in very similar ways. G. Gause demonstrated this by growing two species of *Paramecium* separately, then together (Figure 47.3). Both species exploited the same food (bacterial cells) and competed intensely for it. The test results suggested that two species that require identical resources cannot coexist indefinitely, a concept now known as **competitive exclusion**.

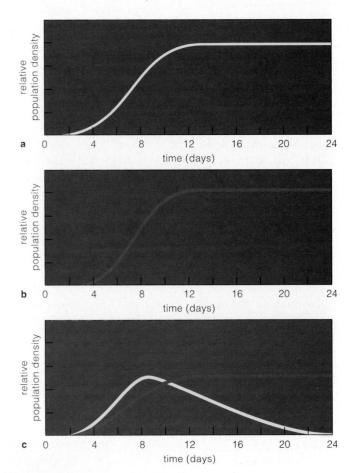

Figure 47.3 Competitive exclusion, as demonstrated by two bacterial species that compete for the same food resource. First, *Paramecium caudatum* (**a**) and *P. aurelia* (**b**) were grown apart, in separate culture tubes. They both established stable populations, as indicated by the S-shaped growth curves in (**a**) and (**b**). Next, populations of the two species were grown together in the same culture tube. *P. aurelia* (the *red* growth curve in **c**) drove the other species toward extinction (as indicated by the *yellow* growth curve in **c**).

From this experiment and others, it appears that two species cannot coexist indefinitely in the same habitat if they require identical resources. If their requirements do not overlap as much, one might influence the population growth rate of the other, but both may still coexist.

In other experiments, Gause used two other species of *Paramecium* that did not overlap as much in their requirements. When grown together, one species tended to feed on bacteria suspended in the liquid in the culture tube. The other tended to feed on yeast cells at the bottom of the tube. The rate of population growth was slower for both species, but not enough for one to exclude the other. They continued to coexist.

Field experiments also reveal effects of competitive interactions. For example, A. Tansley showed how competition influences the distribution of two species of bedstraw plants. One species normally grows in acidic soils and the other, in alkaline soils. Tansley transplanted one species to the different soil. It grew there—

as long as the other species was absent! Also, when the two were grown together in either soil, the survivor was the one that normally grew there. Thus, each species was excluding the other from the portion of the habitat in which it grows best.

In another field experiment, N. Hairston studied competitive coexistence between salamander species in the Great Smoky Mountains and Balsam Mountains (Figure 47.4). *Plethodon glutinosus* generally lives at lower elevations than its relative, *P. jordani,* but at some sites the ranges overlap. Hairston removed one or the other species from different test plots in the overlap areas. He also left some plots untouched, as controls. After five years, nothing had changed in the control plots; the two species were coexisting. But in plots that had been cleared of one species (*P. jordani*), the other species (*P. glutinosus*) increased in numbers. Plots cleared of *P. glutinosus* had proportionally more young *P. jordani* salamanders—evidence of a growing population. Thus, where the species coexist in nature, competitive interactions apparently suppress the growth rate of both populations.

Resource Partitioning

Think back to the chapter introduction, with its nine species of fruit-eating pigeons in the same forest in New Guinea. The different species require the same resource—fruit. Yet they overlap only slightly in the competition for this resource, because each specializes in fruits of a particular size. In this case, competing species coexist by sharing a required resource in different ways or at different times. This type of competitive situation is called **resource partitioning**.

As another example, consider how three species of annual plants partitioned resources in a plowed, abandoned field. As for other plants, the resources were sunlight, water, and dissolved mineral salts. Each species exploited different parts of the habitat. Drought-tolerant foxtail grasses have a shallow, fibrous root system that absorbs rainwater quickly. They grew where soil moisture varied from day to day. Mallow plants, with a taproot system, grew where deeper areas of soil were moist early in the growing season but drier later on. Smartweed, with a taproot system that branches in topsoil and in soil below the roots of the other species, grew where soil was continuously moist (Figure 47.5).

In short, the niches of two or more species living in the same habitat may overlap in several respects. The mere existence of overlap is not a sure indicator of competition among them. Birds overlap in their need for oxygen but don't compete for it. Tansley's bedstraws compete when experimentally grown together, but they don't overlap (and compete) in their use of the habitat. Hairston's salamanders compete yet coexist at sup-

a

b

Figure 47.4 Two coexisting salamander species: (**a**) *Plethodon glutinosus* complex and (**b**) *P. jordani.*

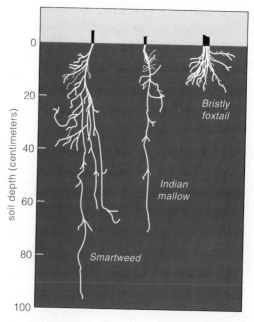

Figure 47.5 Resource partitioning by three annual plants in a plowed, abandoned field.

pressed population sizes. New Guinea pigeons coexist through resource partitioning.

1. In some competitive situations, all individuals have equal access to a required resource, but some are better than others at exploiting it. In other situations, some individuals control access to a resource.

2. The more two species in the same habitat differ in their use of resources, the more likely they can coexist.

3. Two competing species may also coexist by sharing a resource in different ways or at different times.

47.4 PREDATION

"Predator" Versus "Parasite"

Of all community interactions, predation is the most riveting of our attention, not to mention the prey's. But what, exactly, is a predator-prey interaction? Consider the leopard in Figure 47.6 as it closes in on a baboon, one of its preferred foods. A goat pulling up a thistle plant for breakfast, though less dramatic, is also a predator. Its prey is a living organism, killed for food. Can the same be said of a horse grazing on but not killing plants? What about a mosquito taking blood from your arm before it flies off? What about ticks or fleas taking blood for long periods before they get off one host and lay their eggs elsewhere? What about tapeworms, mistletoe, and other parasites that live in or on other organisms?

For simplicity, let's use two broad definitions for interactions between consumers and their victims. A **predator** is an animal that feeds on other living organisms (its prey) but does *not* take up residence on or in them. Prey may or may not die from the interaction. A **parasite** takes up residence in or on other living organisms (its hosts) and feeds on their tissues. It does this for a good part of its life cycle, and host organisms may or may not die as a result of the interaction.

Dynamics of Predator-Prey Interactions

Some predator and prey populations coexist at more or less steady population levels. Others undergo recurring cycles of abundance and population crashes, erratic population cycles, or prey extinction. Three

Figure 47.6 Idealized cycling of abundances of predator and prey. (The diagram's scale exaggerates predator density; predators usually are less common than prey at all points in the cycle.) The pattern arises through time lags in predator responses to changes in prey abundance. Starting at time *a*, prey density is low, so predators have more difficulty capturing food and their population is declining. In response to the predator decline, prey start increasing, but predators do not start increasing until reproduction gets under way (time *b*). Both populations grow until the increased number of predators causes the prey population to decline (time *c*). Predators continue to increase and take out more prey, but the lower prey density leads to starvation among predators and their growth rate slows (time *d*). At time *e*, a new cycle begins.

Figure 47.7 Predator-prey interactions between the Canadian lynx and snowshoe hare. Abundances of both populations, shown in the diagram, are based on counts of pelts that trappers sold to Hudson's Bay Company over a ninety-year period. The dashed line represents the abundance of lynx, and the solid line, the abundance of hares.

This figure is a good test of how willing you are to accept conclusions without questioning their scientific basis. (Remember the discussion of scientific methods in Chapter 1?) What other factors may have influenced the relative abundances of lynx and hare? Did weather vary greatly, with more rigorous winters imposing greater demand for food (required for the animal to keep warm) and higher death rates? Did competition between lynx and other predators (owls, goshawks, coyotes, foxes) complicate the lynx cycle? Did predators turn to alternate prey species during low points of the hare cycle? Did trapping increase with rising fur prices in Europe, and did it decrease as the pelt supply outstripped the demand?

factors influence the outcome of predator-prey interactions:

1. Carrying capacity of the prey population. Carrying capacity, recall, is the maximum number of individuals that can be maintained indefinitely by the resources available in a given environment.

2. Reproductive rates of the predator and prey (predators rarely reproduce as fast as prey).

3. Response of individual predators to increases in prey density (for example, they might eat more or move to areas where prey are more abundant).

When predation keeps a prey population from overshooting its carrying capacity, the two populations tend to coexist at fairly steady levels. This happens when predators reproduce promptly and eat more when more prey organisms are around to eat.

By contrast, population densities tend to fluctuate when predators do not reproduce as fast as the prey, when they can eat only so many prey organisms in a given time interval, and when the carrying capacity for the prey is high.

The graph lines in Figure 47.6 denote an idealized case of cyclic changes brought about by time lags in a predator's response to changes in prey abundance. In nature, we sometimes find such a correspondence between the rise and fall of predator and prey populations. Other factors also contribute to the cyclic changes.

For example, long-term studies of a snowshoe hare population show that the density of this population changes every nine to ten years. The cycle tends to be synchronized across much of Canada and Alaska. Records of pelts gathered by trappers and sold to the Hudson's Bay Company seem to suggest a similar rise and fall in Canadian lynx populations (Figure 47.7). However, lynx are not the only predators in the cycle. When hares are near their peak abundance, they also are feasted upon by great horned owls, goshawks, foxes, coyotes, and other predators. After the hare population declines, predator populations recover only when the density of hares begins to increase once again.

To see if food scarcity causes the declines in the hare population, researchers provided extra food when hares were at peak densities. The hare population declined anyway. Most likely, the quality and quantity of the hare's food supply affects hare population density. But most hares die of predation, not starvation. As food becomes scarce, hares are forced to take more risks to reach the remaining edible plants—and so expose themselves to increasing risk of predation.

When predation keeps a prey population from overshooting its carrying capacity, the predator and prey populations tend to coexist at relatively stable levels.

When there are time lags in a predator's response to changes in prey abundance, the populations tend to undergo cyclic or irregular fluctuations in density.

Predators and prey exert continual selection pressure on each other. Suppose a new, heritable defense arises in a prey population. Individual predators that cannot respond well to the defense won't eat. Thus a predator population also evolves to some extent, for the change affects selection pressures operating between the two. Of course, more efficient predators exert selective pressure that favors better defenses among individuals of the prey population, and so on in the arms race.

What we have been outlining is an example of **coevolution**—the joint evolution of two (or more) species interacting in close ecological fashion. In this section and the next, we take a look at some of the outcomes of their coevolution.

Camouflage

Predation pressure has favored the evolution of prey that can **camouflage** themselves—that is, hide in the open. Diverse adaptations in form, patterning, color, and behavior help prey organisms blend with their surroundings and escape detection. Figure 47.8 shows some classic examples, including a desert plant that looks like a small rock. This plant flowers only during a brief rainy season, when other plants and water are available for plant eaters.

Figure 47.8 Prey organisms demonstrating the fine art of camouflage. (**a**) What bird??? When a predator approaches its nest, the least bittern stretches its neck (colored much like the surrounding withered reeds), thrusts its beak upward, and sways gently like reeds in the wind. (**b**) An unappetizing bird dropping? No. The body coloration and positioning of this caterpillar help it hide in the open from bird predators. (**c**) Another insect, this time camouflaged as a chewed-up leaf. (**d**) Find the plants (*Lithops*) hidden from herbivores owing to their stonelike form, pattern, and coloring.

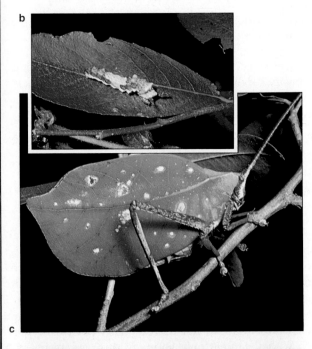

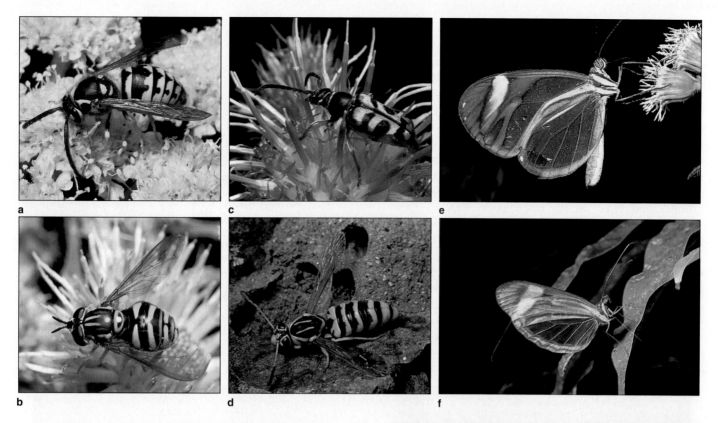

a   c   e

b   d   f

Figure 47.9 Mimicry. Many animals avoid predation by having a bad taste, obnoxious secretion, or painful bite or sting. Each young predator learns this the hard way, by unpleasant taste trials. If dangerous or unpalatable prey animals were not easy to recognize and remember, many would be lost while inexperienced predators were learning their lessons. Many avoid being eaten by displaying bright colors and bold markings. Many do not even bother hiding.

Many edible, unrelated species avoid predation by mimicking the behavior and appearance of the dangerous or unpalatable species. The aggressive yellow jacket shown in (**a**) is the probable model for similar-looking but edible flies (**b**), beetles (**c**), and wasps (**d**). The inedible butterfly in (**e**) is a model for the edible mimic *Dismorphia* (**f**).

Figure 47.10 Moment-of-truth defensive behavior. A cornered short-eared owl spreads its wings in a startling display.

Warning Coloration and Mimicry

Predation pressure has favored prey species that are bad-tasting, toxic, or able to inflict pain on attackers. Many toxic types have bold patterns and conspicuous colors—warning signals to predators. Inexperienced predators might attack a skunk, yellow-banded wasp, or orange-patterned monarch butterfly—once. They quickly learn to associate the colors and patterning with pain or digestive upsets. Repugnant species make no effort to conceal themselves. Sometimes they even deliberately flash their warning colors.

Often, weaponless prey species strikingly resemble unpalatable or dangerous species. Such resemblances are a type of **mimicry**. Figure 47.9 shows how closely the mimics can resemble repugnant species that serve as their models. Mimicry has many forms, however. For example, in *speed* mimicry, sluggish, easy-to-catch prey species closely resemble swift species that predators have given up trying to catch.

Moment-of-Truth Defenses

When cornered, owls and some other species startle or intimidate predators with display behavior (Figure 47.10). A baboon being run down by a leopard might do this also. By turning suddenly and displaying its formidable canines, the baboon may startle and confuse the predator long enough for one last chance at a getaway (Figure 47.6).

Other cornered animals release chemicals as warning odors, repellants, or poisons. Earwigs, skunks, and stink beetles produce awful odors. Several beetles take aim and let loose with noxious sprays. Similarly, many plants emit predator repellants. Tannins in the foliage and seeds of certain plants taste bitter and make the plant tissues hard to digest. Nibble on the delicate yellow petals of a buttercup (*Ranunculus*) and you will badly irritate the lining of your mouth.

47.6 ADAPTIVE RESPONSES TO PREY

In the coevolutionary arms race, predators counter prey defenses with their own marvelous adaptations, including clever avoidance of repellants, stealth, and camouflage. Remember the beetles that direct sprays of noxious chemicals at their attackers? Grasshopper mice grab such beetles, plunge the "sprayer" end into the ground, and feast on the unprotected head (Figure 47.11). Chameleons can remain motionless for so long, prey may not "see" them until zapped by the amazing chameleon tongue (Figure 47.12). And it is no accident that stealthy predators blend with their background. Think of polar bears camouflaged against snow, tigers against tall-stalked, golden grasses, and pastel insects against pastel flowers. And hope you never step on a dangerous scorpionfish, concealed on the seafloor.

Figure 47.11 Example of a behavior that counters an otherwise effective defense against predation. Some beetles spray noxious chemicals at their attackers, which works some, but not all, of the time. Grasshopper mice plunge the chemical-spraying tail end of their prey into the ground and feast on the head end.

Figure 47.12 A sampling of stealthy predators. (**a**) This chameleon remained motionless long enough to avoid detection. Before its prey could make a getaway, the chameleon uncoiled its long, sticky tongue with astonishing speed and accuracy. (**b**) Find the scorpionfish, a venomous predator with camouflaging fleshy flaps, multiple colors, and profuse spines. (**c**) This prey tidbit failed to detect the yellow crab spider, lurking motionless against yellow petals. (**d**) Do attempt to discern which parts are flowers and which parts are the praying mantid.

47.7 PARASITIC INTERACTIONS

Parasites

The flukes and tapeworms described in Chapter 26 are prime examples of parasites—symbionts that live on or in a host organism and use its tissues for nutrients. Sometimes these parasites and others indirectly cause death, as when a weakened host dies from secondary infections. In terms of reproductive success, this is not good for the parasite; hosts able to live longer spread more of its offspring. Thus parasites tend to coevolve with hosts in ways that have less-than-fatal effects. Generally, death results only if a parasite attacks a novel host (with no coevolved defenses against it) or when many parasitic individuals infect a host.

Not all parasites consume host tissues. Rather, they complete their life cycle by exploiting the social behavior of another species. Such animals are *social* parasites. Consider a North American brown-headed cowbird. It never builds a nest, incubates its eggs, or cares for its offspring. It removes an egg from the nest of another kind of bird and lays one as a "replacement." Some birds can't tell the difference, so they end up hatching the egg and raising a young cowbird. The large, aggressive cowbird often pushes the remaining rightful occupants out of the nest or gets most of the food.

Parasitoids

Somewhere between predators and parasites are the **parasitoids**, insect larvae that always kill what they eat. While parasitoids are growing up, they consume all the soft tissues of their hosts. This sounds quite horrendous, but fortunately their hosts are not humans but the larvae or pupae of other insect species.

Peter Price studied the evolutionary effects of a parasitoid wasp that lays its eggs on sawfly cocoons. The sawflies lay *their* eggs in trees. In due time, the fertilized eggs give rise to fly larvae, which drop to the forest floor. The larvae spin cocoons after they burrow into the leaf litter. Some burrow deeper than others. The first adult sawflies emerge from the cocoons that were the least deeply buried. Later in the season, more sawflies emerge from the more deeply buried ones (Figure 47.13).

It happens that parasitoid wasps tend to lay their eggs on the cocoons closest to the surface of the leaf litter. They exert strong selection pressure on the sawfly population, for the deep-burrowing sawfly individuals are more likely to escape detection. There are only so many cocoons near the surface, so the wasp able to locate cocoons deeper in the litter will be competitive in securing food for her larvae.

As Price points out, the host stays ahead in this coevolutionary contest. Each time the sawfly larvae

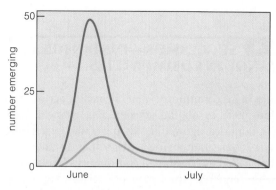

Figure 47.13 Effect of parasitoid wasp activity on sawfly emergence. The *brown* line represents emergence when attack by parasitoids was experimentally prevented; the *pink* line represents emergence after parasitoid attack. Notice also that most attacks occurred on the would-be early emergers, which did not burrow as deeply into the forest litter.

burrow deeper, the female wasps have to spend more time searching for them. Fewer wasp eggs are laid—so the wasp population is held in check.

Parasites as Biological Control Agents

As you might well conclude, parasites and parasitoids exert control over the population growth of other insect species. In fact, many kinds are commercially raised and selectively released, as an alternative to chemical pesticides. Less than 20 percent of these qualify as effective defenses against insect pests, however. As suggested by C. Huffaker and C. Kennett, the effective control agents have five attributes. Generally, they are well adapted to the host species and to their habitat. They are exceptionally good at searching for hosts. Their population growth rate is high relative to that of the host species, and their offspring are mobile enough for adequate dispersal. Finally, the lag time between responses to changes in the numbers of the host population is minimal.

Releasing more than one kind of biological control agent in a given area may trigger competition among them and eventually lessen their overall level of effectiveness. Besides, a shotgun approach to biological control is risky. There is a chance that the parasites may attack other, nontargeted species. In 1983, F. Howarth reported that populations of butterflies and moths native to the Hawaiian Islands were declining, partly because of the introduction of wasps that were supposed to serve as controls over something else. We will return to the effects of such species introductions later in the chapter.

Like predators and their prey victims, parasites and their hosts are locked into long-term, coevolutionary contests.

47.8 SUCCESSION—DIRECTIONAL CHANGE IN COMMUNITIES

Think of a community that stretches across a large region, such as all the species of a tropical forest in South America. How did this particular assemblage of species come into being? According to the classical model, called **succession**, a community develops in predictable sequence, from pioneer species to an end array of species that remains in equilibrium over an entire region. A **pioneer species** is an opportunistic colonizer of vacant or vacated habitats, and it is notable for its high dispersal rate and rapid growth. In time, more competitive species replace the pioneers, then are themselves replaced in orderly progression until the array of species stabilizes under prevailing conditions in the habitat. This most persistent array of species is known as the **climax community**.

Primary and Secondary Succession

A process of *primary* succession begins when pioneer species colonize a barren habitat. A new volcanic island is such a habitat. So is land, once buried under ice for thousands of years, that becomes exposed by a retreating glacier (Figure 47.14). The pioneer species are adapted to grow in exposed areas with intense sunlight, wide swings in temperature, and nutrient-deficient soil. Typically they are small plants with short life cycles. Each year they produce an abundance of small seeds, which are quickly dispersed.

Once established, the pioneers improve conditions for other species and commonly set the stage for their own replacement. For example, many pioneer plants are mutualists with nitrogen-fixing bacteria, so they have a competitive edge over other plants in nitrogen-poor habitats. In time, however, the accumulation of their wastes and remains adds volume to the soil and enriches it with nutrients that allow other species to take hold. Also, the pioneers form low-growing mats. These shelter the seeds of later species yet cannot shade out the seedlings. Later successional species crowd out the pioneers, whose seeds travel as fugitives on the wind or water—destined, perhaps, for a new but temporary habitat.

In *secondary* succession, a disturbed site in a community recovers and moves once again in a direction toward the climax state. This successional pattern is typical of ponds, shallow lakes, and abandoned fields. It is typical of patches of established forests when falling trees and other disturbances open up part of the continuous canopy of leaves. Sunlight reaches seeds and seedlings already on the forest floor and spurs their growth.

All models of succession hold that pioneer species are replaced, but they differ in how replacement

Figure 47.14 Primary succession in Alaska's Glacier Bay region (**a**), where changes in newly deglaciated regions have been carefully documented. Comparison of maps from 1794 onward shows that ice has been retreating at annual rates ranging from 8 meters at the glacier's sides to 600 meters at its tip over bays.

(**b**) When a glacier retreats, a constant flow of meltwater tends to leach the newly exposed soil of minerals, including nitrogen. Less than ten years ago, the soil here was still buried below ice. (**c,d**) The first invaders of these nutrient-poor sites are feathery seeds of mountain avens (*Dryas*), drifting on the winds. Mountain avens is a pioneer species that benefits from the nitrogen-fixing activities of mutualistic microbes. It grows and spreads rapidly over glacial till.

(**e**) Within twenty years, young alders take hold. These deciduous shrubs are symbiotic with nitrogen-fixing microbes. Young cottonwood and willows also become established (**f**). In time, alders form dense thickets (**g**). As the thickets mature, cottonwood and hemlock trees grow rapidly. So do a few evergreen spruce trees.

(**h**) By eighty years, spruce crowd out the mature alders. (**i**) In areas deglaciated for more than a century, dense forests of Sitka spruce and western hemlock dominate. By this time, nitrogen reserves are depleted, and much of the biomass is tied up in *peat*—excessively moist, compressed organic matter that resists decomposition and forms a thick mat on the forest floor.

proceeds. For example, by the model just described, colonizers *facilitate* their own replacement. By the *inhibition* model, the earliest colonizers compete against species that could replace them, and so the sequence of succession depends on who gets there first. Apparently no single model can explain the entire sequence. As summarized by the ecologist Charles Krebs, succession is a dynamic process, resulting from a balance between the colonizing ability of some species and the competitive capacity of others. And it may not always proceed from the simple to the complex.

The Climax-Pattern Model

Think of a climax community as a stable, self-perpetuating array of species that are in equilibrium with one another and with the environment. At one time, some ecologists thought that the same general type of community will always develop in a given region because of constraints imposed by climate. However, others

pointed out that stable communities *other* than "the climax community" often exist in a given area—as when persisting stretches of tallgrass prairie extend into the shortgrass prairie region of Indiana. By the **climax-pattern model**, a community is adapted to a total pattern of environmental factors—including climate, soil, topography, wind, species interactions, recurring disturbances such as fires, chance events, and so on. Thus, a continuum of climax stages of succession can exist. The stages are similar to one another, but they vary along environmental gradients.

A climax community is a stable, self-perpetuating array of species in equilibrium with one another and with the environment.

Similar climax stages can persist in the same area, along gradients dictated by environmental factors and by species interactions.

47.9 REGARDING COMMUNITY STABILITY

Community stability is an outcome of forces that have come into an uneasy balance. Resources are sustained, as long as populations do not flirt dangerously with the carrying capacity. Predators and prey coexist, as long as neither wins. Competitors have no sense of fair play. Even mutualists are stingy, as when plants produce as little nectar as necessary to attract pollinators, and pollinators take as much nectar as they can for the least effort. Even in the most stable climax communities, fires, overgrazing, and insect infestations can create local successional patches. Such disturbances may favor population growth for some species and hamper or eliminate others.

Besides this, climate or some other environmental variable may undergo long-term changes that destabilize the community. If the instability becomes great enough, a community may change in ways that will persist even if the disturbance ends or is reversed. If some of its member species are rare or are poor competitors, they may become extinct.

Cyclic, Nondirectional Change

Many changes recur in patches of the habitat, over and over again. Such small-scale changes are part of the internal dynamics of the community as a whole. Observe a disturbed patch, and you might conclude that great shifts in species composition are afoot. Observe the community on a larger scale, and you might discover that the overall composition includes all of the pioneer species that are colonizing the patch as well as the dominant climax species.

Consider how tropical forests develop through phases of colonization by pioneer species, increases in species diversity, maturity, then degeneration. Dancing sporadically through the successional pattern are heavy winds that cause treefalls. Where trees fall, gaps open in the forest canopy and more light reaches the understory. There, local conditions favor growth of previously suppressed small trees and the germination of pioneers or shade-intolerant species.

Or consider the groves of sequoia trees in the Sierra Nevada in California. Some trees of this type of climax community are giants, more than 4,000 years old. They persist partly because of recurring brush fires that sweep through parts of the forests. Sequoia seeds germinate only in the absence of smaller, shade-tolerant plant species. Too much litter on the forest floor inhibits their germination. Modest fires eliminate trees and shrubs that compete with the young sequoias but do not damage the sequoias themselves. Mature sequoias have exceptionally thick bark, which burns poorly and insulates the living phloem cells of these giant trees against modest heat damage.

At one time, fires were prevented in many sequoia groves in national and state parks—not just accidental fires from campsites and discarded cigarettes, but also natural fires touched off by lightning. Litter builds up when small fires are stopped, fire-susceptible species take hold, and dense underbrush forms. Sequoia seeds cannot germinate in underbrush—which is fuel for hotter fires that damage the giants. Controlled fires are now set to eliminate underbrush and promote conditions that are favorable for cyclic replacements.

So Many Communities, So Few All-Encompassing Models

The preceding examples might lead you to believe that all communities change and become stabilized in predictable ways. This is not always the case.

For example, sometimes ongoing moderate predation *promotes* diversity of prey species in a community. This is often true when community structure is dictated by a single, dominant species, or **keystone species**. Roger Paine identified one such species through removal experiments. Paine set up control plots in communities of rocky intertidal zones on the west coast of North America. Pounding surf, storms, and changing tides continually disturb these zones, and living space is scarce. The control plots included the sea star *Pisaster* and its invertebrate prey. This keystone species is very efficient at controlling the abundances of the resident mussels (*Mytilus*), limpets, chitons, and assorted barnacles.

Paine also removed the sea stars from experimental plots in the same intertidal zones. In the predator-free plots, mussels ended up monopolizing the available sites and crowding out seven of the other invertebrate species. Although mussels are the main prey of sea stars, they proved to be the strongest competitors in a community when sea stars were absent. In this case, predation (by sea stars) normally maintains the diversity of prey species by preventing competitive exclusion (by mussels). Remove sea stars, and the community shrinks from fifteen species to eight.

Jane Lubchenko studied a different community of the rocky intertidal zone in Massachusetts. Periwinkles (*Littorina littorea*), a relative of land snails, dominate this community. They graze on algae, and they increase *or* decrease the diversity of algal species in different settings. In tidepools, periwinkles feed on the dominant algal species (*Enteromorpha*), and so help many other, less competitive algal species survive. By contrast, certain red algae, including *Chondrus*, dominate rocks that are exposed only during low tide. Periwinkles leave

a Periwinkles in a tidepool

d *Enteromorpha*, a filamentous green alga

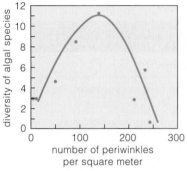

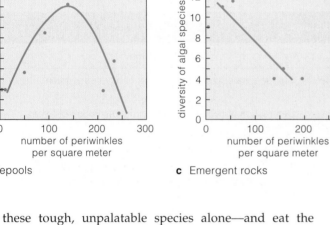

b Tidepools   **c** Emergent rocks

e *Chondrus*, another alga

these tough, unpalatable species alone—and eat the same competitively weak algal species that they pass up in tidepools. Thus, periwinkles increase algal diversity in tidepools and reduce it on recurringly emergent rocks (Figure 47.15).

To further complicate the picture, the rapid introduction of a species often tips the balance of community organization in truly unexpected ways. The section to follow provides fine examples of this.

---

1. Community structure is an outcome of a balance of forces, including predation and competition, that have been operating over time.

2. Recurring, small-scale changes are built into the overall workings of many communities.

3. Other changes, such as long-term shifts in climate or the rapid introduction of a successful new species, may permanently alter the community's configuration.

Figure 47.15 Example of the effect of competition and predation on community organization.

(**a**) Periwinkles (*Littorina littorea*) influence the number of algal species in certain marine habitats, depending on where they graze. (**b**) In tidepools, periwinkles graze on a competitively dominant alga (*Enteromorpha*) and so increase the number of less competitive algal species that might otherwise be overwhelmed. (**c**) On rocks exposed only at low tide, periwinkles do not eat the dominant algal species (including *Chondrus*); they eat other species and so cause a decrease in algal diversity. The photographs show *Enteromorpha* (**d**) and *Chondrus* (**e**) in their natural habitat.

47.10 EFFECTS OF DISPERSAL

Modes of Dispersal

One way or another, some residents of established communities undergo **geographic dispersal**: they move outside their home range and successfully take up residence elsewhere. Ecologists have identified three modes of dispersal. First, over several generations, a population may expand its home range by gradually diffusing into hospitable outlying regions. Second, during the course of a lifetime, individuals may be rapidly transported across great distances. This mode, called *jump* dispersal, commonly takes the individual across a region where it could not survive, as when a ship transports an insect from the mainland to Maui. Third, with such slowness as to be nearly imperceptible, a population may move out from its home range over geologic time, as by continental drift.

On the local scale, dispersal and colonization of vacant places may be quite rapid. In one experiment in the Bahamas, Amy Schoener set out plastic sponges in barren sands of Bimini Lagoon. How fast did aquatic species take up residence in the internal chambers of the artificial hotels? Schoener recorded occupancy by 220 species in less than 30 days.

Examples of Species Introductions

Through accidental or deliberate jump dispersal, about 4,500 nonindigenous species have become successfully established in the United States. These are just the ones we know about. Some, including soybeans, rice, wheat, corn, and potato plants, have been put to good use. Most of the others have disrupted community organization and agriculture (Table 47.1).

Consider the water hyacinth. In the 1880s, this blue-flowered aquatic plant from South America was displayed at an exposition in New Orleans. Flower fanciers took home clippings and set them out in ponds and streams. Unchecked by natural predators, the fast-

---

Table 47.1   Detrimental Effects of Species Introduced into the United States

Species Introduced	Origin	Mode of Introduction	Outcome
Water hyacinth	South America	Intentionally introduced (1884)	Clogged waterways; shading out of other vegetation
Dutch elm disease:			
Ophiostoma ulmi (the fungal pathogen)	Europe	Accidentally imported on infected elm timber (1930)	Destruction of millions of elms; great disruption of forest ecology
Bark beetle (carrier of the pathogen)		Accidentally imported on unbarked elm timber (1909)	
Chestnut blight fungus	Asia	Accidentally imported on nursery plants (1900)	Destruction of nearly all eastern American chestnuts; disruption of forest ecology
Argentine fire ant	Argentina	In coffee shipments from Brazil? (1891)	Crop damage; destruction of native ant communities; death of ground-nesting birds
Camphor scale insect	Japan	Accidentally imported on nursery stock (1920s)	Damage to nearly 200 species of plants in Louisiana, Texas, and Alabama
Japanese beetle	Japan	Accidentally imported on irises or azaleas (1911)	Defoliation of more than 250 plant species, including commercially important species such as citrus
Carp	Germany	Intentionally released (1887)	Displacement of native fish; uprooting of water plants with concurrent loss of waterfowl populations
Sea lamprey	North Atlantic Ocean	Through Erie Canal (1860s), then through Welland Canal (1921)	Destruction of lake trout and lake whitefish in Great Lakes
European starling	Europe	Released intentionally in New York City (1890)	Competition with native songbirds; crop damage; transmission of swine diseases; airport runway interference; extremely noisy and messy in large flocks
House sparrow	England	Released intentionally (1853)	Crop damage; displacement of native songbirds; transmission of some diseases
European wild boar	Russia	Intentionally imported (1912); escaped captivity	Destruction of habitat by rooting behavior; crop damage
Nutria (a large rodent)	Argentina	Intentionally imported (1940); escaped captivity	Alteration of marsh ecology; damage to earthen dams and levees; crop destruction

After David W. Ehrenfeld, *Biological Conservation*, 1970, Holt, Rinehart and Winston; and *Conserving Life on Earth*, 1972, Oxford University Press.

Focus on the Environment

Hello Lake Victoria, Goodbye Cichlids

Given the exponential growth of the human population, it seems imperative that we find better ways to manage our food supply. Such efforts are well intentioned, but they can have disastrous consequences when ecological principles are not taken into account.

Consider what happened several years ago, when someone thought it would be a great idea to introduce the Nile perch into Lake Victoria in East Africa. People had been using simple, traditional methods of fishing there for thousands of years, but now they were taking too many fish. Soon there would be too few fish to feed the local populations and no excess catches to sell for profit. But Lake Victoria is a very big lake, and the Nile perch is a very big fish (more than 2 meters long). This seemed an ideal combination to attract commercial fishermen from the outside, with their big, elaborate nets—right? Wrong.

Native fishermen had been harvesting native fishes called cichlids, which eat mostly detritus and aquatic plants. But the Nile perch eats other fish—including cichlids. Having had no prior evolutionary experience with the Nile perch, the cichlids that were native to the lake simply had no defenses against it.

And so the Nile perch ate its way through cichlid populations and destroyed the natural fishery. Dozens of cichlid species found nowhere else are now extinct. By wiping out its food source, the Nile perch undercut its own population growth. It ceased to be a potentially large, exploitable food source for people living around the lake.

As if that weren't enough, the Nile perch is an oily fish. Unlike cichlids, which can be sun-dried, it must be preserved by smoking—and smoking requires firewood. And so the people started cutting down more trees in the local forests, and trees are not rapidly renewable resources. To add insult to injury, people living near Lake Victoria never liked to eat Nile perch anyway; they preferred the flavor and texture of cichlids.

What is the lesson? A little knowledge and some simple experiments in a contained setting could have prevented the whole mess at Lake Victoria.

growing hyacinths spread through nutrient-rich waters, displaced many native species, then choked off rivers and canals. They have spread as far west as San Francisco. Similarly, kudzu, a well-behaved legume in Asia, now blankets trees, hills, even houses in the southern United States—although calling it the vine that swallowed the South is a bit of an exaggeration.

In another wholesale disaster, in 1859 a landowner in northern Australia released two dozen wild English rabbits. Good food and sport hunting, that was the idea. Perfect habitat with no natural predators—that was the reality. Six years later, the landowner had killed 20,000 rabbits and was besieged by 20,000 more. Rabbits displaced livestock, even kangaroos. Did the construction of a 2,000-mile-long fence protect western Australia? No. Rabbits made it to the other side before the fence was completed. In 1951, someone brought in the myxoma virus by way of mildly infected South American rabbits, its normal hosts. Lacking coevolved defenses, the English rabbits died in droves. As you might expect, however, natural selection has since favored the rise of populations of virus-resistant rabbits. This coevolutionary tale is still unfolding.

If you live in the northeastern United States, you know about imported gypsy moths, which escaped from a research facility near Boston in 1869. Their descendants have defoliated entire forests. You may know about zebra mussels, which probably entered the Great Lakes on the hull of a cargo ship. They attach to and block water intake pipes, and it may take 5 billion dollars to eradicate them. *Cryphonectria parasitica*, apparently introduced to North America nearly a century ago, killed nearly all American chestnut trees. And remember those African bees released in South America? Their Africanized, "killer bee" descendants, described on page 122, have already spread hundreds of miles beyond the border between Mexico and the United States. At this writing, they have killed several hundred people in Latin America. They have attacked 140 Americans, one of whom died from multiple stings.

Community organization can change irrevocably through the natural, accidental, or deliberate introduction of new species.

47.11 PATTERNS OF BIODIVERSITY

As you might well conclude from the preceding discussions, biodiversity differs greatly from one habitat to the next. Field investigations by ecologists also show that biodiversity differs greatly between geographic regions.

Mainland and Marine Patterns

The most striking pattern of biodiversity is related to distance from the equator. The number of coexisting species on land and in the seas is highest in the tropics and systematically declines toward the poles. Figure 47.16 shows examples. What factors underlie this pattern of biodiversity? Three are paramount.

First, tropical latitudes intercept more sunlight of consistently greater intensity, rainfall is high, and the growing season is long. Thus, in the tropics, resource availability tends to be higher and more reliable. Tropical trees provide new leaves, flowers, and fruit all year in wet tropical forests. Year in and year out, they support many diverse herbivores, nectar foragers, and fruit consumers. In temperate and arctic regions, such specializations would never evolve.

Second, species diversity may be self-reinforcing. The diversity of trees in tropical forests is far greater than in comparable forests at higher latitudes. When more plant species compete and coexist, more species of herbivores evolve, partly because no one herbivore can overcome the chemical defenses of all plants. Typically, more predators and parasites evolve in response to the diversity of prey and hosts. The same is true of diversity on tropical reefs.

Third, in terms of evolutionary history, the rates of speciation in the tropics have exceeded the rates of background extinction. At higher latitudes, biodiversity has been suppressed during episodes of mass extinction, as described in Chapter 21. However, bear in mind that millions of species in tropical forests may disappear in the next decade, for reasons that will be described in chapters to follow.

Island Patterns

Islands are splendid laboratories for the study of biodiversity. For example, in 1965, a volcanic eruption formed a new island southwest of Iceland. Within six months, bacteria, fungi, seeds, flies, and some seabirds were established on it. A vascular plant appeared after two years, and a moss two years after that (Figure 47.17). As soils improved, the number of plant species increased. All species were colonists from Iceland. None originated

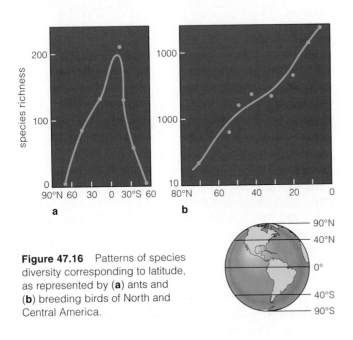

Figure 47.16 Patterns of species diversity corresponding to latitude, as represented by (**a**) ants and (**b**) breeding birds of North and Central America.

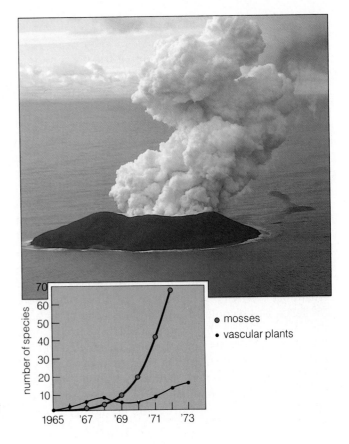

Figure 47.17 Surtsey, a volcanic island, at the time of its formation. Such islands are natural laboratories for ecologists. The chart shows the number of species of mosses and vascular plants recorded on the new island from 1965 to 1973.

a

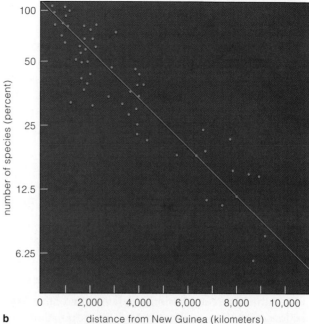

b

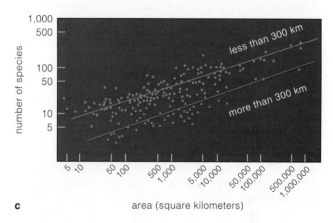

c

Figure 47.18 (**a**) One of the travel agents for jump dispersals. Seabirds that island-hop over long distances might have a few seeds stuck in their feathers, as shown here. If those seeds successfully germinate in a new island community, they will give rise to a population of new immigrants.

(**b**) The distance effect. Biodiversity on islands of a given size declines with increasing distance from the source of colonizing species. Each graph point denotes the number of species of land birds of lowland areas of South Pacific islands. The huge island of New Guinea is the source of colonists for these islands, each of which is at least 500 kilometers away. To correct for differences in island size (area), the actual number of bird species on each of the distant islands is expressed as a percentage of the number of bird species on an island of equivalent size close to New Guinea.

(**c**) The area effect. Among islands similarly distant from the source of colonizing species, the larger islands support larger numbers of species. The graph points denote the number of bird species on tropical and subtropical islands. The *solid green* circles and the upper line are for islands less than 300 kilometers from their source of colonists. The *orange* triangles and the lower line denote islands more than 300 kilometers from source areas. Thus this graph incorporates data on the distance effect as well as on the area effect.

on the island itself, which was named Surtsey. The number of new species will not increase indefinitely. Why is this so? Studies of community patterns on islands around the world provide us with two ideas.

First, islands far from a source of potential colonists receive few colonizing species; the few that do arrive are adapted for long-distance dispersal (Figure 47.18). This is called the **distance effect**. Second, larger islands tend to support more species than smaller islands at equivalent distances from source areas. This is the **area effect**. The larger islands tend to be physically more complex and often higher above sea level, and so offer more habitats that favor species diversity. Also, being bigger targets, large islands may "intercept" more colonists.

Most importantly extinctions suppress biodiversity on small islands. The smaller populations are more vulnerable to storms, volcanic eruptions, diseases, and random shifts in birth and death rates. On Surtsey or any other island, the ultimate number of species is a balance between the immigration rate for species new to the

island and the extinction rate for established species. Because small islands distant from a source of colonists have low immigration rates and high extinction rates, they support few species once this balance has been struck.

---

1. The number of coexisting species is highest in the tropics and systematically declines toward the poles.

2. A region's biodiversity depends on such factors as climate, topographical variation, possibilities for dispersal, and evolutionary history (including extinctions and the time available for speciation).

3. Disturbances, which prevent competitive exclusion, tend to favor increases in biodiversity.

4. Biodiversity on islands represents a balance between the immigration rates for new species and extinction rates for established species.

SUMMARY

1. A habitat is the type of place where individuals of a given species normally live. The populations of all species in a habitat, which directly or indirectly associate with one another, represent a community.

2. Each species has its own niche, as defined by the full range of conditions in its habitat and its relations with other organisms.

3. Two species that require the same limited resource are likely to compete, either by using up the resource as rapidly or efficiently as possible or by preventing the other from using it.

   a. Species are more likely to coexist when they differ in their use of resources.

   b. Two competing species also may coexist by using a shared resource in different ways or at different times.

4. Mutualism, competition, predation, and other interactions directly or indirectly link the populations in a community. Table 47.2 summarizes the effects of these interactions on the participating species.

5. Predators and prey tend to coexist at relatively stable levels over time when predation keeps the prey population from overshooting its carrying capacity.

6. Time lags in a predator's response to changes in prey density may lead to irregular or cyclic fluctuations in the abundance of both populations.

7. Predators and prey are locked in coevolutionary races. When a novel, heritable trait appears and gives prey an edge in the race, selection pressure operates on the predator population, which also may evolve in response. The same thing happens when a novel, heritable trait appears in the predator population.

8. Evolved prey defenses include threat displays, chemical weapons, mimicry, and camouflaging. Predators overcome the defenses through adaptive behavior (including stealth), camouflaging, and so on.

9. Parasites and hosts are similarly locked in coevolutionary races. Coevolution tends to favor resistant hosts and only moderately harmful parasites.

10. A climax community is a stable, self-perpetuating array of species in equilibrium with one another and with a particular environment.

11. Similar climax stages may persist in the same region. They may vary as a result of environmental gradients and species interactions.

12. Community structure is an outcome of predation, competition, and other forces that have been operating over time.

13. According to the classical model of succession, a community develops in predictable sequence, from pio-

Table 47.2 Summary of Types of Interactions Between Two Species

Type of Interaction	Direct Effect of Interaction*	
	Species 1	Species 2
Neutral	0	0
Commensalism **	+	0
Mutualism **	+	+
Interspecific competition	−	−
Predation	+	−
Parasitism **	+	−

*0 indicates no direct effect on population growth, + indicates positive effect, − indicates negative effect.

**A form of symbiosis; one of the participating species lives in close association with the other species.

neer species to an end array of species that remain in equilibrium with one another and with their environment over an entire region.

14. Recurring, small-scale changes are a part of the internal dynamics of communities. Other changes, such as long-term shifts in climate or species introductions, may permanently alter the community configuration.

15. Community structure is affected by dispersals of species from their home ranges. Some dispersals occur very slowly as a population gradually expands its home range. Jump dispersals rapidly put individuals into new, distant habitats. Dispersals also have occurred extremely slowly, over evolutionary time, as by continental drift.

16. The number of species in a given community depends on the size of a region, the colonization rate, disturbances, and extinction rates. It depends also on the level and pattern of resource availability.

17. Biodiversity is highest in the tropics and systematically declines toward polar regions.

Review Questions

1. What is the difference between the habitat and the niche of a species? Why do you suppose it might be difficult to define "the human habitat"? *824*

2. Describe competitive exclusion. How might two species that compete for the same resource coexist? Can you think of some possible examples besides the ones used in the chapter? For example, consider some of the animals and plants living in your own neighborhood. *826–827*

3. Define primary and secondary succession. *834*

4. What is a climax community, and how does the climax-pattern model help explain its structure? *834, 835*

5. Flesh flies (Figure *a*) have gray and black bodies, red eyes, and red tail ends. They are fast fliers, and bird predators soon give up trying to catch them. Many sluggish insects, including the weevil *Zygops rufitorquis*, resemble flesh flies (Figure *b*). This appears to be a case of _____ . *831*

a

b

1. A habitat _____ .
   a. has distinguishing physical and chemical features
   b. is where individuals of a species normally live
   c. is occupied by various species
   d. both a and b
   e. all are correct

2. Conditions that define a niche _____ .
   a. include the full range of conditions under which individuals of a species can live, grow, and reproduce
   b. are unvarying for a given species
   c. shift in large and small ways
   d. both a and b
   e. both a and c

3. The two-way flow of benefits in mutualistic interactions between species is an outcome of _____ .
   a. close cooperativeness
   b. two-way exploitation
   c. resource partitioning
   d. competitive coexistence

4. Two species in the same habitat can coexist when they _____ .
   a. differ in their use of resources
   b. share the same resource in different ways
   c. use the same resource at different times
   d. all are correct

5. A predator population and prey population _____ .
   a. always coexist at relatively stable levels
   b. may undergo cyclic or irregular changes in density
   c. cannot coexist indefinitely in the same habitat
   d. both b and c

6. All parasites _____ .
   a. tend to kill their hosts
   b. can kill novel hosts
   c. consume host tissues
   d. both a and c

7. In _____ , a disturbed site in a community recovers and moves again toward the climax state.
   a. the area effect
   b. the distance effect
   c. primary succession
   d. secondary succession

8. The number of coexisting species is _____ in the tropics and _____ toward the poles.
   a. highest; declines systematically
   b. lowest; increases systematically
   c. highest; declines sporadically
   d. lowest; increases sporadically

9. Match the terms with the most suitable descriptions.
   _____ jump dispersal
   _____ distance effect
   _____ pioneer species
   _____ climax community
   _____ keystone species

   a. opportunistic colonizer of vacant or newly vacated places
   b. dominates community structure
   c. rapid transport over great distance
   d. more biodiversity on large islands than small ones at same distance from source areas
   e. stable, self-perpetuating array of species

area effect *841*
camouflage *830*
climax community *834*
climax-pattern model *835*
coevolution *830*
commensalism *824*
community *824*
competitive exclusion *826*
distance effect *841*
geographic dispersal *838*
habitat *824*
interspecific competition *824*
keystone species *836*
mimicry *831*
mutualism *824*
neutral interaction *824*
niche *824*
parasite *828*
parasitism *824*
parasitoid *833*
pioneer species *834*
predation *824*
predator *828*
resource partitioning *827*
succession *834*
symbiosis *824*

Readings

Begon, M., J. Harper, and C. Townsend. 1990. *Ecology: Individuals, Populations, and Communities.* Second edition. Sunderland, Massachusetts: Sinauer.

Krebs, C. 1994. *Ecology.* Fourth edition. New York: Harper-Collins.

Moore, P. 1987. "What Makes a Forest Rich?" *Nature* 329:292.

Smith, R. 1992. *Elements of Ecology.* Third edition. New York: Harper-Collins.

Stiling, P. 1992. *Introductory Ecology.* Englewood Cliffs, New Jersey: Prentice-Hall.

48 ECOSYSTEMS

Crêpes for Breakfast, Pancake Ice for Dessert

Think of Antarctica, and you think of ice. Mile-thick slabs of the stuff hide all but a small fraction of a continent whipped by fierce winds and kept frozen by murderously low temperatures, on the order of −100°F. And yet, on patches of exposed rocky soil and on nearby islands, mosses and lichens grow. There, during the breeding season, penguins as well as seals form great noisy congregations, then reproduce and raise offspring. They cruise offshore or venture out in the open ocean in pursuit of food—krill, fishes, and squids.

The first explorers of Antarctica called it The Last Place on Earth. For all of its harshness, though, the eerie, unspoiled beauty fired the imaginations of those first travelers and others who followed them.

In 1961, thirty-eight nations signed a treaty to set aside Antarctica as a reserve for scientific research. This was the start of scientific outposts—and of the trashing of Antarctica. At first, researchers discarded a few oil drums and old tires. Then prefabricated villages went up. Onto the ice or into the water went garbage and used equipment. Just offshore, marine life became acquainted with sewage and chemical wastes.

Antarctica also became a destination of cruise ships. Each summer thousands of tourists, fortified by three sumptuous meals a day plus snacks, are ferried from ship to land across channels glistening with pancake ice (Figure 48.1). They trample vegetation and bob around penguins, cameras clicking. At the communal breeding ground at Cape Royds alone, the penguin population has declined by half.

Meanwhile, nations looking for new sources of food started licking their chops over Antarctica's krill. Word got out about potentially rich deposits of uranium, oil, and gold—and by 1988, treaty nations were poised to authorize digs and drillings. Of course, oil spills or krill harvesting could destroy the fragile ecosystem. For example, the tiny, shrimplike krill are all that Adélie penguins eat. They are a key food for baleen whales

a

and other marine animals that are, in turn, food for still others.

In 1991, the treaty nations thought about all of this and imposed a fifty-year ban on mineral exploration. Research stations started burying or incinerating wastes, treating sewage, and taking other measures to curb pollution. Tour operators promised to supervise tourists. Krill are not being harvested on a massive scale, yet.

Because Antarctica seems so remote from the rest of the world, it is easier to see how life is interconnected there. We can ask whether harm to one species or one habitat will lead to collapse of the whole, and be fairly sure of the answer. What about places not as sharply defined? Does interconnectedness prevail there, also? Are other places as vulnerable to disturbance or more resilient? These topics will now occupy our attention.

1. An ecosystem is an association of organisms and their physical environment, linked by a flow of energy and a cycling of materials. It is an open system, with inputs and outputs of both energy and nutrients.

2. Energy flows in one direction through an ecosystem. Most often the flow begins when photosynthetic autotrophs harness sunlight energy and convert it to forms that they and other organisms of the ecosystem can use. These "primary producer" organisms directly or indirectly nourish an array of consumers, decomposers, and detritivores, which interact as part of food webs.

3. Water and major nutrients move from the physical environment, through organisms, then back to the environment. Their movements, called biogeochemical cycles, are global in scale. Human activities are disrupting these natural cycles and so are endangering ecosystems.

b

Figure 48.1 (**a**) A boatload of tourists crossing a channel of "pancake ice" off the coast of Antarctica. (**b**) On the shore of an island near Antarctica, tourists get close to nature—maybe too close.

48.1 THE NATURE OF ECOSYSTEMS

Overview of the Participants

Diverse natural systems abound on the earth's surface. In climate, soil, vegetation, and animal life, deserts differ from hardwood forests, which differ from arctic tundra and prairies. In their physical properties and biodiversity, seas differ from reefs, which differ from lakes. *Yet despite the differences, such systems are alike in many aspects of their structure and function.*

With few exceptions, each system runs on energy from the sun. Plants and other photosynthesizers are the most common autotrophs (self-feeders). The photoautotrophs capture sunlight energy and convert it to forms they can use to build organic compounds from simple inorganic substances. By securing energy from the environment, autotrophs are **primary producers** for the entire system (Figure 48.2).

All other organisms in the system are heterotrophs, not self-feeders. They extract energy from compounds that were put together by primary producers. The **consumers** feed on tissues of other organisms. Consumers called *herbivores* eat plants, *carnivores* eat animals, *parasites* reside in or on living hosts and extract energy from them, and *omnivores* partake of a variety of edibles. Other heterotrophs, the **decomposers**, include fungi and bacteria that obtain energy by breaking down the remains or products of organisms. Still others, called **detritivores**, eat decomposing particles of organic matter. Crabs and earthworms are examples.

The autotrophs secure nutrients as well as energy for the entire system. During growth, they take up water and carbon dioxide (as sources of oxygen, carbon, and hydrogen) along with dissolved minerals, including nitrogen and phosphorus. Such materials are building blocks for carbohydrates, lipids, proteins, and nucleic acids. When decomposers and detritivores get their turn at consuming this organic matter, they break it down to small inorganic molecules. If the molecules are not washed away or otherwise removed from the system, autotrophs can use them again as nutrients.

What we have just described in broad outline is an **ecosystem**. An ecosystem is an array of organisms and their physical environment, all interacting through a flow of energy and a cycling of materials.

Ecosystems are open systems and so are not self-sustaining. They require an *energy input* (as from the sun) and often *nutrient inputs* (as from minerals carried by erosion into a lake). They also have energy and nutrient *outputs*. Energy cannot be recycled; in time, most of the energy originally fixed by autotrophs is lost to the environment, as metabolic heat. Nutrients are cycled, but some are still lost. This chapter deals with the inputs, internal transfers, and outputs of ecosystems.

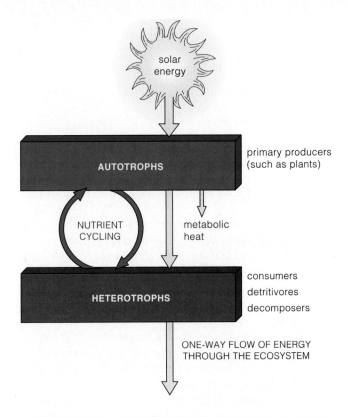

Figure 48.2 Model of an ecosystem, showing the one-way flow of energy through it and the cycling of nutrients between its autotrophic and heterotrophic organisms.

Table 48.1 Examples of Trophic Levels

Members of Trophic Levels	Energy Source	Representative Organisms
Primary producers:		
Photoautotrophs	Sunlight energy	Grasses, diatoms, cyanobacteria
Chemoautotrophs	Electrons from inorganic substances	Nitrifying bacteria
Primary consumers:		
Herbivores	Primary producers	Grasshoppers, deer, krill
Secondary consumers:		
Primary carnivores	Herbivores	Spiders, foxes, small squids
Tertiary consumers:		
Secondary carnivores	Primary carnivores	Emperor penguins

Structure of Ecosystems

Trophic Levels In all ecosystems, feeding relationships are structured in much the same way. Their members fit in a hierarchy of energy transfers, called **trophic levels** (from *troph*, meaning nourishment). "Who eats whom?" we can ask. When organism *B* eats organism *A*, energy gets transferred to *B* from *A*. At a given trophic level, all organisms are the same number of transfer steps away from the energy input into the system.

Primary producers (closest to the initial energy source) make up the first trophic level. Photoautotrophs, such as algae and cyanobacteria in a pond, are examples. Rotifers, snails, and other herbivores feeding directly on them are at the next trophic level. Birds and other primary carnivores that prey directly on the herbivores make up a third trophic level.

Many organisms, including decomposers, can extract energy from more than one source, so we cannot assign them to a single trophic level. They feed at several levels and must be partitioned among them or put in a separate group. Even so, the categories in Table 48.1 are a good starting point for understanding feeding relations.

Food Webs A straight-line sequence of who eats whom in an ecosystem is often called a **food chain**. You will have a hard time finding such simple, isolated sequences. Why? Most often, the same food resource is part of more than one chain. This is especially true of a food resource at low trophic levels. It is much more accurate to view food chains as cross-connecting with one another—as **food webs**. Figure 48.3 shows some of the participants in a food web in the seas that surround Antarctica.

A bad-luck story about a fisherman can clarify the difference between a food chain and food web. The fisherman nets fish that were feeding on algae near the ocean's surface. At lunchtime, he cooks some fish but later loses his footing and falls into the water, where sharks lurk. You might think this is a simple food chain: algae ⟶ fish ⟶ fisherman ⟶ shark. Yet the chain excludes alternate feeding relationships. Crustaceans also were grazing on algae, small squids and midsized fishes were feeding on crustaceans, and larger fishes were feeding on the smaller ones. Sharks may have been about to feed on the large and midsized fishes.

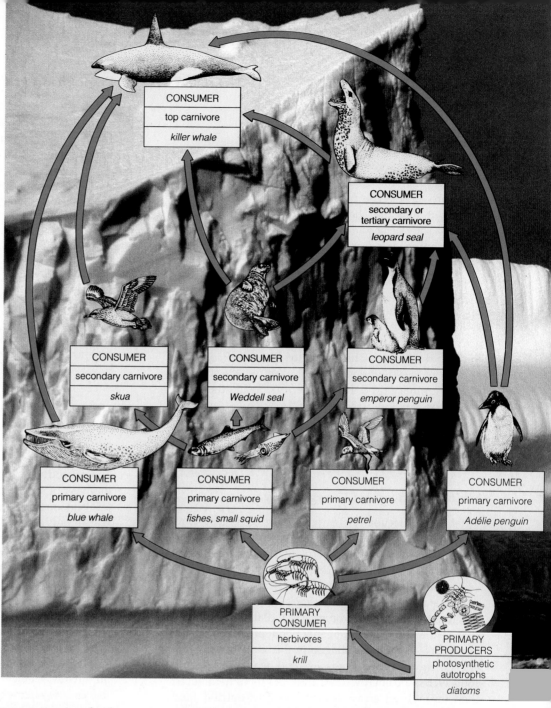

Figure 48.3 Food web in the waters of the Antarctic. Many more participants, including decomposers, are not shown.

The fisherman, who ate cooked fish and onions, shifted between carnivore and herbivore. He became even more omnivorous by drinking wine, derived from the fermenting activities of decomposers (yeasts).

An ecosystem consists of producers, consumers, decomposers, and detritivores and their physical environment, all interacting through energy flow and materials cycling.

A food web is a network of crossing, interlinked food chains involving primary producers, consumers, and decomposers.

48.2 ENERGY FLOW THROUGH ECOSYSTEMS

Primary Productivity

To get an idea of how energy flow is studied, consider a land ecosystem for which multicelled plants are the primary producers. The plants trap sunlight energy and convert it into the chemical energy of organic compounds. The rate at which the ecosystem's primary producers capture and store a given amount of energy in a given length of time is its **primary productivity**. How much energy actually gets stored depends on how many plants are present and on the balance between photosynthesis and aerobic respiration in the plants.

Other factors affect the amount of net primary production, its seasonal patterns, and its distribution through the habitat. This is true of ecosystems on land and in the seas (Figure 48.4a). For example, the size and form of primary producers are factors. So are the availability of minerals, the temperature range, and the amount of sunlight and rainfall during the growing season. The harsher the environment, the fewer shoots will be produced—and the lower the productivity.

Major Pathways of Energy Flow

What is the direction of energy flow through ecosystems on land? Only a small part of the energy from sunlight becomes fixed in plants. The plants themselves use as much as half of what they fix, and they lose metabolically generated heat. Other organisms tap into the energy stored in plant tissues, remains, or wastes. They, too, lose heat to the environment. All of these heat losses represent a one-way flow of energy out of the ecosystem.

Figure 48.4b and c diagrams this one-way flow of energy through two kinds of food webs. In **grazing food webs**, the energy flows from plants to herbivores, then through an array of carnivores. In **detrital food webs**, it flows mainly from plants through detritivores and decomposers. Usually the two kinds cross-connect, as when a small crab of a detrital food web wanders in front of a herring gull of a grazing food web.

The amount of energy moving through food webs differs from one ecosystem to the next and often varies with the seasons. In most cases, however, the greatest portion of net primary production passes through detrital food webs. You may doubt this; after all, when cattle graze heavily on pasture plants, about half the net primary production enters a grazing food web. But cattle don't use all the stored energy. Quantities of undigested plant parts and feces become available for decomposers and detritivores. Also consider marshes, where most of the stored energy is not used until plant parts die and become available for detrital food webs.

Ecological Pyramids

Often the trophic structure of an ecosystem is represented as an **ecological pyramid**, in which producers form a base for successive tiers of consumers above them. Some pyramids are based on biomass (the weight of all the members at each trophic level). Here is a pyramid of biomass (grams/square meter) for the aquatic ecosystem at Silver Springs, Florida:

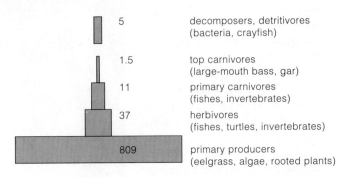

5	decomposers, detritivores (bacteria, crayfish)
1.5	top carnivores (large-mouth bass, gar)
11	primary carnivores (fishes, invertebrates)
37	herbivores (fishes, turtles, invertebrates)
809	primary producers (eelgrass, algae, rooted plants)

Some pyramids of biomass are "upside down," with the smallest tier on the bottom. Think of phytoplankton (including diatoms) and zooplankton (including rotifers) in a pond. A small biomass of rapidly growing and reproducing phytoplankton may support a greater biomass of larger zooplankton that grow more slowly and consume less energy per unit of weight.

An energy pyramid is a more useful way to depict an ecosystem's trophic structure. Such pyramids show the energy losses at each transfer to a different trophic level. They have a large energy base at the bottom and are always "right-side up." They give a more accurate picture of ever diminishing amounts of energy flowing through successive trophic levels of the ecosystem.

---

1. Energy flows into the food webs of ecosystems from an outside source—the sun, in most cases. Energy leaves ecosystems mainly by loss of metabolic heat, which each organism generates.

2. *Gross* primary productivity is the total rate of photosynthesis in an ecosystem in a specified interval. *Net* primary productivity is the rate of energy storage in plant tissues in excess of the rate of aerobic respiration by the plants themselves. Heterotrophic consumption affects the rate of energy storage.

3. Living tissues of photosynthesizers are the basis of grazing food webs. The remains of photosynthesizers and consumers are the basis of detrital food webs.

4. The amount of useful energy flowing through consumer trophic levels declines at each energy transfer. It declines as metabolically generated heat is lost and as food energy is shunted into organic wastes.

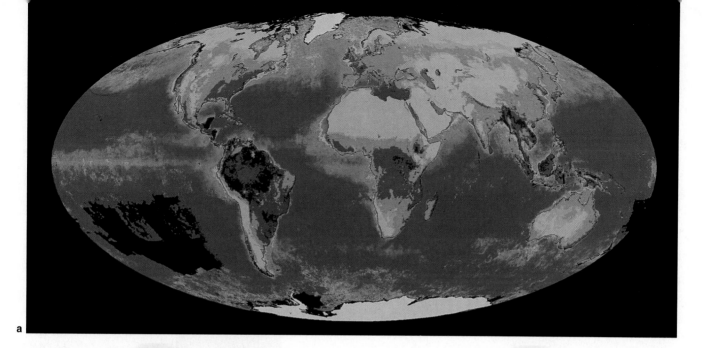

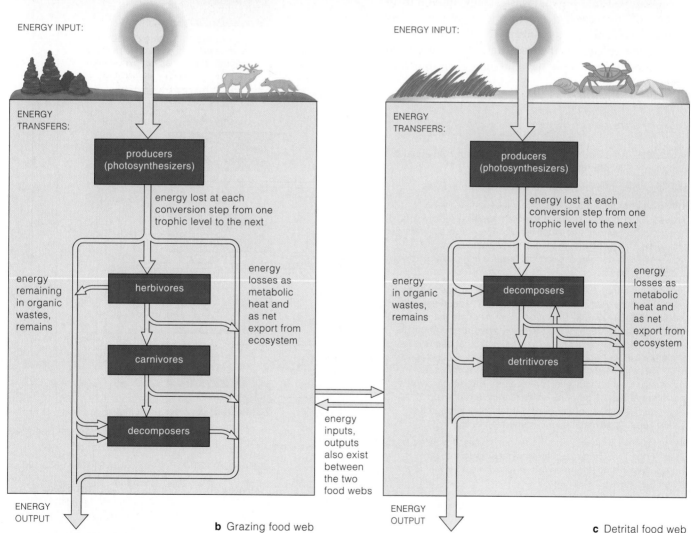

Figure 48.4 Energy bases and energy flow in ecosystems. (**a**) Summary of three years of satellite data on the earth's primary productivity. *Dark green* denotes rain forests and other highly productive regions. *Yellow* denotes deserts (low productivity). Productivity in oceans, ranging from high to low, is coded *red* down through *orange*, *yellow*, *green*, and *blue*. (**b,c**) The one-way flow of energy through two kinds of cross-connected food webs in ecosystems.

48.3 | CASE STUDY: ENERGY FLOW AT SILVER SPRINGS, FLORIDA

To gather data for an energy pyramid, researchers measure the energy that each type of individual in the ecosystem takes in, burns during metabolism, and stores in body tissues. They measure the energy remaining in its waste products. They also multiply the energy per individual by population size. Energy inputs and outputs are calculated so that energy flow can be expressed per unit of land (or water) per unit of time.

The pyramid in Figure 48.5a is based on a long-term study of a grazing food web in an aquatic ecosystem in Florida. Figure 48.5b shows some of the calculations used to construct it. Such studies tell us that, given the metabolic demands of organisms and the amount of energy shunted into organic wastes, only about 6 to 16 percent of the energy entering one trophic level becomes available to organisms at the next level. Because the efficiency of these energy transfers is so low, ecosystems generally have no more than four consumer trophic levels.

Figure 48.5 Measurements of energy flow through an aquatic ecosystem—Silver Springs, Florida.

(a) The pyramid of energy flow, as calculated for one year.

(b) Annual energy flow, in kilocalories/square meter/year. Producers in this small spring are mostly aquatic plants. Carnivores are insects and small fishes; top carnivores are larger fishes. The energy source (sunlight) is available all year. Detritivores and decomposers cycle organic compounds from other trophic levels.

The producers trap 1.2 percent of incoming solar energy, and only a little more than a third of this amount is fixed in new plant biomass (4,245 + 3,368). The producers use more than 63 percent of the fixed energy for their own metabolism. About 16 percent of it is transferred to herbivores, and most of this is used for metabolism or transferred to detritivores and decomposers. Only 11.4 percent of the energy transferred to herbivores reaches the next trophic level (carnivores), and the carnivores use all but about 5.5 percent, which is transferred to top carnivores. In time, all of the energy transferred through the system appears as metabolically generated heat.

This diagram is oversimplified, for no community is isolated from others. Organisms and materials constantly drop into the springs. Organisms and materials are slowly lost by way of a stream that leaves the springs.

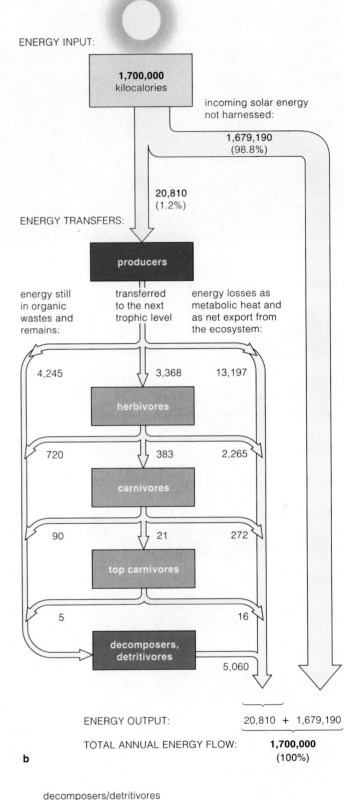

ENERGY INPUT:

1,700,000 kilocalories

incoming solar energy not harnessed: 1,679,190 (98.8%)

20,810 (1.2%)

ENERGY TRANSFERS:

producers

| energy still in organic wastes and remains: | transferred to the next trophic level | energy losses as metabolic heat and as net export from the ecosystem: |

herbivores — 4,245 / 3,368 / 13,197

carnivores — 720 / 383 / 2,265

top carnivores — 90 / 21 / 272

decomposers, detritivores — 5 / 16 / 5,060

ENERGY OUTPUT: 20,810 + 1,679,190

TOTAL ANNUAL ENERGY FLOW: **1,700,000** (100%)

b

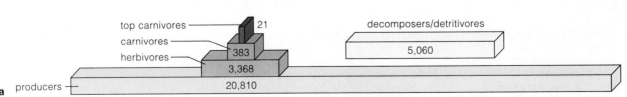

a

top carnivores — 21
carnivores — 383
herbivores — 3,368
producers — 20,810

decomposers/detritivores — 5,060

48.4 BIOGEOCHEMICAL CYCLES— AN OVERVIEW

Availability of *nutrients* as well as energy profoundly influences the structure of ecosystems. Photosynthetic producers require carbon, oxygen, and hydrogen, which they get from water and air. They require nitrogen, phosphorus, and other minerals (Table 30.1). Because mineral deficiencies lower primary productivity, they adversely affect the ecosystem at large.

Nutrients move in **biogeochemical cycles**. In such cycles, ions or molecules of a nutrient are transferred from the environment to organisms, then back to the environment—part of which serves as a reservoir for them. They generally move slowly through the reservoir, compared to their rapid exchange between organisms and the environment.

Figure 48.6 is a model of the relationship between geochemical cycles and most ecosystems. This model is based on four factors:

1. Elements used as nutrients usually become available to producer organisms as mineral ions, such as ammonium (NH_4^+).

2. Inputs from the physical environment and the cycling activities of decomposers and detritivores maintain an ecosystem's nutrient reserves.

3. The amount of a nutrient being cycled through most major ecosystems is greater than the amount entering or leaving in a given year.

4. Inputs to an ecosystem's nutrient reserves occur by rainfall or snowfall, metabolism (such as nitrogen fixation), and weathering of rocks. Outputs for land ecosystems include losses by runoff.

There are three types of biogeochemical cycles. In the *hydrologic* cycle (or water cycle), oxygen and hydrogen move in the form of water molecules. In *atmospheric* cycles, a large portion of a nutrient is in the form of an atmospheric gas. This is true of carbon and nitrogen, for example. In *sedimentary* cycles, the nutrient does not exist in gaseous form. Rather, it moves from land to the seafloor and only "returns" to land through geological uplifting, which may take millions of years. The earth's crust is the main storehouse for such nutrients, which include phosphorus.

In a biogeochemical cycle, ions or molecules of a nutrient move from the environment to organisms, then back to the environmental reservoir for them.

Nutrients generally move slowly through the environment but rapidly between organisms and the environment.

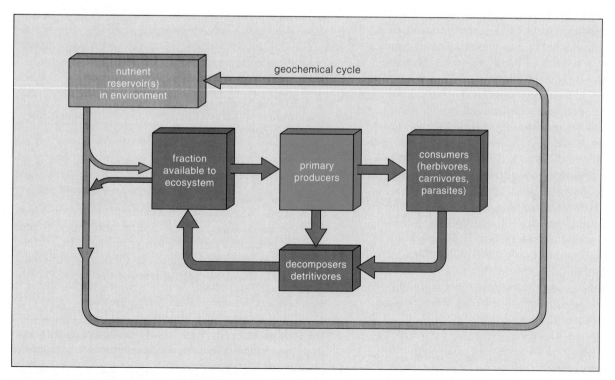

Figure 48.6 Generalized model of nutrient flow through a land ecosystem. The overall movement of nutrients from the physical environment, through organisms, and back to the environment constitutes a biogeochemical cycle.

HYDROLOGIC CYCLE

Driven by solar energy, the waters of the earth move slowly and on a vast scale from the ocean, into the atmosphere, to the land, and back to the ocean. Water evaporating into the atmosphere remains aloft as vapor, clouds, and ice crystals. It falls to earth as precipitation—rain and snow, mostly. Ocean currents and wind patterns play roles in this global **hydrologic cycle**, shown in Figure 48.7.

Airborne water molecules stay aloft for no more than ten days, on average. Water reaching the land as rain or snow stays there for about 10 to 120 days, depending on the season and the location, then evaporates or flows to the ocean. Water evaporates mainly from the world's ocean.

Water in itself is vital for ecosystems. It also is a transport medium that moves nutrients into and out of ecosystems. This became clear through studies in **watersheds**, where all precipitation in a specified region becomes funneled into a single stream or river. Watersheds may be any size. The Mississippi River watershed extends across roughly one-third of the United States. Watersheds at Hubbard Brook Valley in the White Mountains of New Hampshire average 14.6 hectares (36 acres). Most of the water entering a watershed seeps into soil or becomes surface runoff that moves into streams (Figure 48.8). Plants take up water and dissolved minerals from soil, then lose water by transpiration.

Measurements of water inputs and outputs at watersheds have practical application. Cities that depend on surface water supplies in watersheds can adjust water usage on the basis of seasonal variations.

Watershed studies also showed that plants greatly influence the movement of nutrients through the ecosystem phase of biogeochemical cycles. For example, you might think that water draining a watershed would rapidly leach calcium ions and other minerals. Yet in studies of young, undisturbed forests in Hubbard Brook watersheds, each hectare lost only about 8 kilograms of calcium, and rainfall and the weathering of rocks brought in calcium replacements. Tree roots were also "mining" the soil, so calcium was being stored in a growing biomass of tree tissues.

Nutrient outputs changed when experimental watersheds were stripped of vegetation. The soil was left undisturbed, and herbicides were applied for three years to prevent regrowth. Compared to undisturbed watersheds, *six times* as much calcium was lost by way of stream outflow (Figure 48.9c).

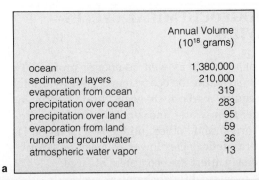

	Annual Volume (10^18 grams)
ocean	1,380,000
sedimentary layers	210,000
evaporation from ocean	319
precipitation over ocean	283
precipitation over land	95
evaporation from land	59
runoff and groundwater	36
atmospheric water vapor	13

a

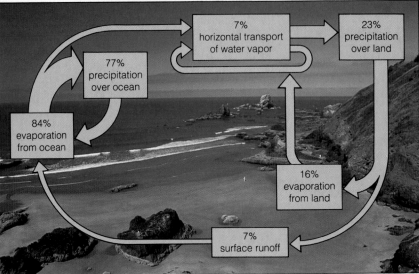

b

Figure 48.7 (a) Global water budget. Values reflect the annual movement of water into and out of the atmosphere.

(b) The hydrologic cycle. Values in boxes indicate total quantities of water present at a given time in the atmosphere, in the ocean, and on land. There is an annual net rate of transfer of 37.3 x 10^3 cubic kilometers of water from the atmosphere to the land. Balancing this is a comparable net loss from the ocean to the atmosphere and a net gain of that amount by the ocean (through runoff).

Because calcium and other nutrients move so slowly through geochemical cycles, deforestation may disrupt nutrient availability for an entire ecosystem. This is true of forests that cannot regenerate themselves over the short term. The coniferous forests of the Pacific Northwest are like this.

In the hydrologic cycle, water slowly moves on a global scale from the ocean reservoir, through the atmosphere, onto land, then back to the ocean.

By stabilizing the soil and absorbing dissolved minerals, plants minimize the loss of soil nutrients in runoff from land.

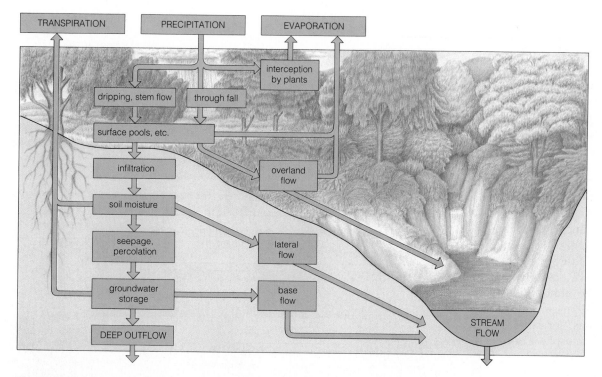

Figure 48.8 Movement of water through a watershed.

a

b

Figure 48.9 Studies of disturbances to a forest ecosystem.

(**a**) In experimental watersheds in the Hubbard Brook Valley, researchers studied the effects of deforestation. (**b**) All water that drains from an area being studied must flow over the V-notched concrete structure, where measurements are taken. This area was deforested, then herbicides were applied to prevent regrowth for three years.

(**c**) The *green* arrow indicates the time of deforestation. The concentrations of calcium ions and other mineral ions in the water passing through the catchment were compared against those in water passing through a control catchment in an undisturbed part of the same forest. As you can see, calcium losses were six times greater from the deforested region.

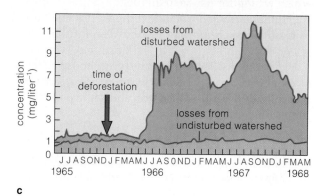

c

CARBON CYCLE

In one of the most important of all atmospheric cycles, carbon moves from reservoirs in the atmosphere and oceans, through organisms, then back to reservoirs. Figure 48.10 shows this global movement, which is called the **carbon cycle**.

Carbon enters the atmosphere via aerobic respiration, fossil fuel burning, and volcanic eruptions that release carbon from rocks deep in the earth's crust. Most of the carbon is dissolved in the ocean (Figure 48.10*b*). Soil, the atmosphere, and plant biomass represent other large "holding stations" for carbon. Most carbon in the atmosphere is in the form of carbon dioxide (CO_2).

Through carbon dioxide fixation, photosynthesizers lock up billions of metric tons of carbon atoms into organic compounds every year. However, the average length of time that a carbon atom is held in any given ecosystem varies greatly. In tropical forests, leaves decompose rapidly, so not much carbon is tied up in litter on the soil surface. In marshes, bogs, and other anaerobic areas, organic compounds cannot be broken down completely, so carbon slowly accumulates in compressed organic matter, such as peat.

In aquatic food webs, carbon becomes incorporated into shells and other hard parts. When shelled organisms die, they sink and become buried in sediments at different depths. Carbon in deep sediments may stay buried for millions of years, until tectonic movements bring it to the surface. More carbon is slowly converted to long-standing reserves of gas, petroleum, and coal, which we tap for use as fossil fuels.

Human activities, including the worldwide burning of fossil fuels, are putting more carbon into the atmosphere than can be cycled to the ocean reservoir and to holding stations. This adds to the greenhouse effect and so may help bring on global warming. The *Focus* essay that follows describes this effect and some possible outcomes of increases in it.

Fossil fuel burning and other human activities may be contributing to imbalances in the global carbon budget.

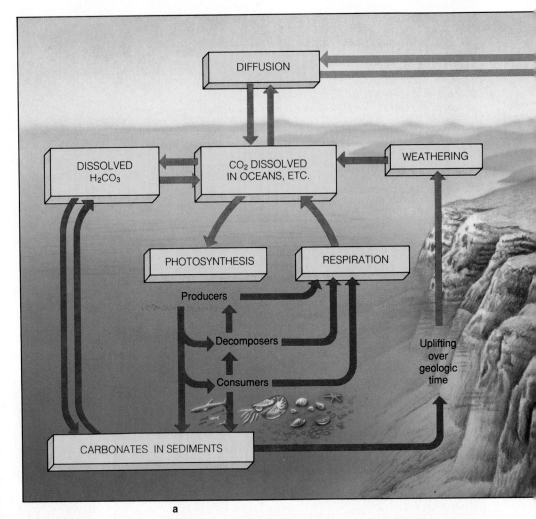

a

Figure 48.10 (**a**) Global carbon cycle. The portion of the diagram on this page illustrates the movement of carbon through marine ecosystems. The portion of the diagram on the facing page illustrates the movement of carbon through ecosystems on land.

(**b**) The present-day global carbon budget. The photograph shows a Los Angeles freeway system under its self-generated blanket of smog at twilight. The exhaust of vehicles as well as fossil fuel burning by industries and homes add carbon as well as other substances to the atmosphere.

Data in (**b**) from W. Schlesinger, *Biogeochemistry: An Analysis of Global Change* (1991). New York: Academic Press.

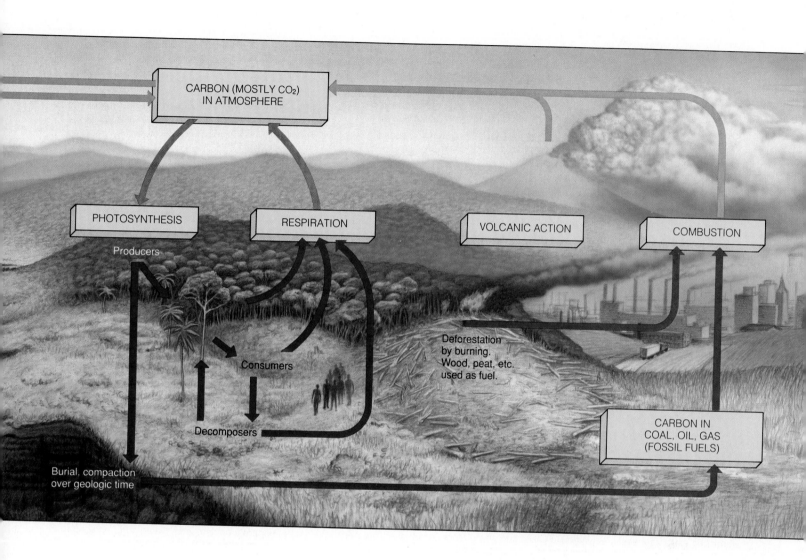

CARBON (MOSTLY CO₂)
IN ATMOSPHERE

PHOTOSYNTHESIS

Producers

RESPIRATION

VOLCANIC ACTION

COMBUSTION

Consumers

Decomposers

Deforestation
by burning.
Wood, peat, etc.
used as fuel.

Burial, compaction
over geologic time

CARBON IN
COAL, OIL, GAS
(FOSSIL FUELS)

Carbon reservoirs and holding stations:	Amount per Year (10¹⁵ grams)
Dissolved in ocean	38,000
Present in soil	1,500
Present in atmosphere	720
Plant biomass	560
Annual fluxes:	
From atmosphere to plants (carbon fixation)	120
From atmosphere to ocean	107
To atmosphere from ocean	105
To atmosphere from plants	60
To atmosphere from soil	60
To atmosphere from fossil fuel burning	5
To atmosphere from net destruction of plants	2
To ocean from runoff	0.4
Burial in ocean sediments	0.1

b

From Greenhouse Gases to a Warmer Planet?

Near the earth's surface, atmospheric concentrations of gaseous molecules play a profound role in shaping the average global temperature, which in turn has enormous effect on the global climate. Molecules of carbon dioxide, water, ozone, methane, nitrous oxide, and chlorofluorocarbons are the key players. Collectively, they act somewhat like a pane of glass in a greenhouse (hence their name, "greenhouse gases"). They let wavelengths of visible light reach the earth's surface, but they impede the escape of longer, infrared wavelengths—that is, heat—from the earth into space. They absorb infrared wavelengths, and much of the energy inherent in those wavelengths is reradiated back toward the earth (Figure *a*). In short, greenhouse gases cause heat to build up in the lower atmosphere, a warming action known as the **greenhouse effect**.

Without greenhouse gases, the earth would be cold and lifeless. But there can be too much of a good thing. Largely as a result of human activities, greenhouse gases are building up to higher levels in the atmosphere (Figure *b*). The increase may be contributing to an alarming trend toward global warming.

What is so alarming about a warmer planet? Suppose the temperature of the lower atmosphere were to rise by only 4°C (7°F). Sea levels could rise by about 2 feet, or 0.6

a The greenhouse effect

1. Sunlight penetrates the atmosphere and warms the earth's surface.

2. The earth's surface radiates heat (infrared wavelengths) to the atmosphere. Some heat escapes into space. Greenhouse gases and water vapor absorb some infrared wavelengths and reradiate a portion back toward the earth.

3. When greenhouse gases build up in the atmosphere, more heat is trapped near the earth's surface. Ocean surface temperatures rise, more water vapor enters the atmosphere, and the earth's surface temperature rises.

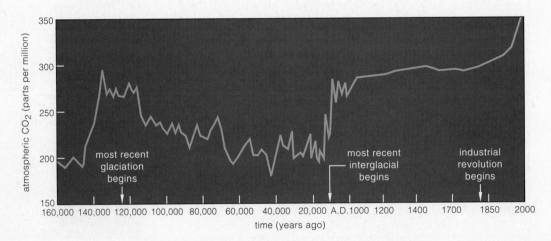

b Shifts in atmospheric concentrations of carbon dioxide, correlated with the most recent glaciation and interglacial period during the past 160,000 years.

meter. Why? The warming would increase ocean surface temperatures—and water expands when heated. Global warming also could make glaciers and the Antarctic ice sheet melt faster, so low coastal regions could flood.

Think of a long-term rise in sea level, combined with high tides and storm waves. Waterfronts of Vancouver, Boston, San Diego, Galveston, and other coastal cities would be submerged. So would agricultural lowlands and deltas in India, China, and Bangladesh, where much of the world's rice is grown. Huge tracts of Florida and Louisiana would face saltwater intrusions. Besides this, global warming could disturb regional patterns of precipitation and temperature. Crop yields would decline in currently productive regions, including parts of Canada and the United States, and increase in others.

In the late 1950s, researchers on a mountaintop in the Hawaiian Islands started measuring concentrations of different greenhouse gases, and their monitoring activities are still going on. They chose the remote site because it was free of local contamination and reflected average conditions for the Northern Hemisphere.

Consider what they found out about carbon dioxide alone. Atmospheric levels of carbon dioxide follow the annual cycle of plant growth in the Northern Hemisphere. They are lower in summer, when photosynthesis rates are highest. They are higher in winter, when aerobic respiration continues and photosynthesis slows. In Figure c (part 1), the peaks and troughs around the graph line represent the highs and lows. For the first time, scientists saw the integrated effects of the carbon balances for the land and water ecosystems of an entire hemisphere. The midline of the peaks and troughs in the cycle steadily increased. Many scientists take this as evidence of a buildup of carbon dioxide that may intensify the greenhouse effect over the next century.

The global burning of fossil fuels is probably contributing most to increasing carbon dioxide levels. Deforestation adds to it; carbon is released when wood burns. Today, vast tracts of forests throughout the world are being cleared and burned at a rapid rate (refer to Figure 50.12). More importantly, the plant biomass is plummeting— and this affects global absorption of carbon dioxide in photosynthesis.

Many scientists wonder whether atmospheric levels of greenhouse gases will continue to increase until the middle of the twenty-first century—and whether the global temperature will rise by several degrees. If this is indeed a trend already in motion, we will not be able to reverse it now by stopping fossil fuel burning and deforestation. So there is widespread agreement that we should begin preparing for the consequences. For example, we might step up genetic engineering studies to develop drought-resistant and salt-resistant plants. Such plants may prove crucial in regions of saltwater intrusions and climatic change.

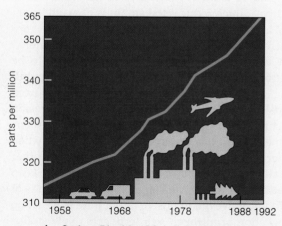

1. Carbon Dioxide (CO_2). Fossil fuel burning, factory emissions, car exhaust, and deforestation are contributing to the increased atmospheric concentration.

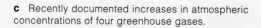

c Recently documented increases in atmospheric concentrations of four greenhouse gases.

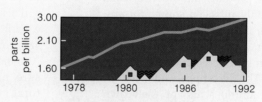

2. Chlorofluorocarbons (CFCs). These are used in plastic foams, air conditioners, refrigerators, and industrial solvents (page 897).

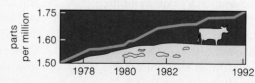

3. Methane (CH_4). This is produced by anaerobic bacteria in swamps, landfills, and termite activities. It is produced also by bacteria in the digestive tract of cattle and other ruminants.

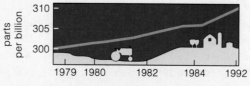

4. Nitrous Oxide (N_2O). This is a natural by-product of denitrifying bacteria. It also is released in great amounts from fertilizers and animal wastes, as in livestock feedlots.

48.7 NITROGEN CYCLE

Since the time of life's origin, the atmosphere and oceans have contained nitrogen. This component of all proteins and nucleic acids moves in an atmospheric cycle called the **nitrogen cycle**. Gaseous nitrogen (N_2) makes up about 80 percent of the atmosphere, the largest nitrogen reservoir. Triple covalent bonds hold its two atoms together ($N\equiv N$), and few organisms can break them. Only certain bacteria, volcanic action, and lightning convert N_2 into forms that enter food webs.

Of all nutrients required for plant growth, nitrogen often is the scarcest. Today, nearly all nitrogen in soils has been put there by nitrogen-fixing organisms. Ecosystems lose it through the activities of bacteria that "unfix" the fixed nitrogen. Land ecosystems lose more through leaching of soils, although this is the basis of nitrogen inputs to aquatic ecosystems such as streams, lakes, and the oceans (Figure 48.11).

Cycling Processes

Let's follow nitrogen atoms through the ecosystem part of the nitrogen cycle. They move through organisms by nitrogen fixation, assimilation and biosynthesis, decomposition, ammonification, and nitrification.

In **nitrogen fixation**, a few kinds of bacteria convert N_2 to ammonia (NH_3), which dissolves quickly in the cytoplasm to form ammonium (NH_4^+). Cyanobacteria, including *Anabaena* and *Nostoc*, are nitrogen fixers of aquatic ecosystems. *Rhizobium* and *Azotobacter* fix nitrogen in many land ecosystems. Collectively these organisms fix about 200 million metric tons of nitrogen each year! Plants assimilate and use this nitrogen in the biosynthesis of amino acids, proteins, and nucleic acids. Plant tissues serve as the only nitrogen source for animals, which feed directly or indirectly on plants.

Through **decomposition** and **ammonification**, bacteria and fungi break down nitrogen-containing wastes and remains of organisms. The decomposers use part of the released proteins and amino acids for their own metabolism. But most of the nitrogen is still in the decay products, in the form of ammonia or ammonium, which plants take up. Nitrifying bacteria also act on ammonia or ammonium. In **nitrification**, they strip these compounds of electrons, and nitrite (NO_2^-) is the result. Other bacteria use the nitrite in metabolism and produce nitrate (NO_3^-), which plants take up.

Remember, certain plants are better than others at securing nitrogen. For example, peas, beans, clover, and other legumes are mutualists with nitrogen-fixing bacteria. Page 502 describes their interaction. Also, most land plants are mutualists with fungi, forming mycorrhizae that enhance nitrogen uptake. Page 503 describes the formation and effects of these fungus roots.

Nitrogen Scarcity

You'd think that land plants can get enough nitrogen, given the cycling processes. However, the ammonium, nitrite, and nitrate that form during the cycle are highly vulnerable to leaching and runoff. With leaching, recall, soil water moves out of an area, which thereby loses the nutrients dissolved in it.

Also, some nitrogen is lost to the air by **denitrification**. Here, bacteria convert nitrate or nitrite to N_2 and a bit of nitrous oxide (N_2O). Ordinarily, most denitrifying bacteria rely on aerobic respiration. When soil is waterlogged and poorly aerated, they switch to anaerobic pathways and use nitrate, nitrite, or nitrous oxide as the final electron acceptor instead of oxygen. (Chapter 8 describes this type of metabolic pathway.) In these reactions, fixed nitrogen is converted to N_2, much of which escapes into the atmosphere.

Besides this, nitrogen fixation comes at high metabolic cost to plants that are mutualists with nitrogen fixers. In exchange for nitrogen, they give up sugars and other photosynthetic products that require large ATP and NADPH investments. Such plants do have the competitive edge in nitrogen-poor soil. In nitrogen-rich soil, however, species that do not have to pay the metabolic price often displace them.

Human Intervention in the Nitrogen Cycle

Humans are altering the cycling of nitrogen in natural ecosystems. Consider how air pollutants contribute to changes in soil. Vehicles, fossil fuel burning power plants, and nitrogenous fertilizers are sources of pollutants, including oxides of nitrogen. These contribute to soil acidity—which reduces the amounts of magnesium, calcium, and potassium that plants can take up. Figure 48.11 merely hints at the ion interactions in soil water.

And what about those nitrogen fertilizers? To be sure, nitrogen losses from soil are enormous in agricultural regions. With each harvest, nitrogen departs from the fields (in tissues of harvested plants). Soil erosion and leaching remove more. In Europe and North America, farmers traditionally have rotated crops, as when they alternate wheat with legumes. Along with other conservation practices, crop rotation has helped keep soils stable and productive, sometimes for thousands of years. Today, however, intensive agriculture is based on the application of nitrogen-rich fertilizers. New strains of crop plants are bred for an ability to take up fertilizers, and crop yields per hectare have doubled and even quadrupled over the past forty years. Whether pest control and soil management technologies can sustain high yields indefinitely remains uncertain, for reasons that will become apparent in Chapter 50.

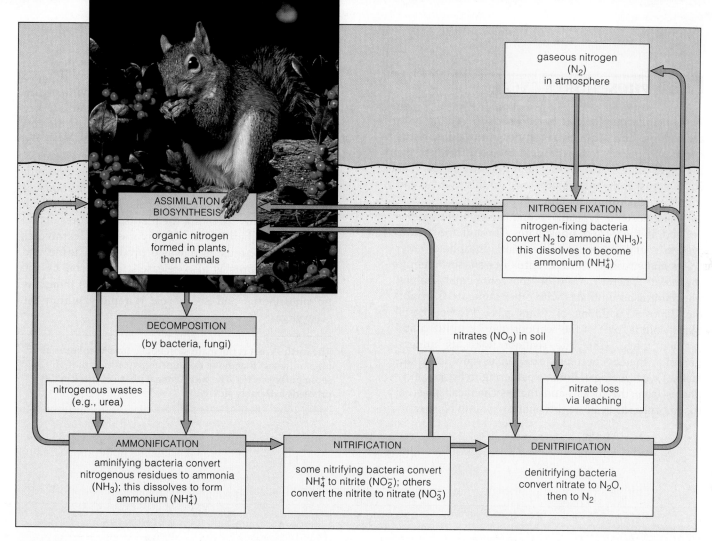

Figure 48.11 Above, the nitrogen cycle. The diagram to the right shows how nitrogen oxides and ammonia, together with ozone, sulfur oxides, and other air pollutants, are contributing to the decline of Central European forests. Spruce seedlings growing in acidified soils develop magnesium deficiencies, especially when ammonium levels are high. Tree needles yellow, then drop; photosynthesis suffers. Trees weakened by nutrient imbalances become susceptible to diseases.

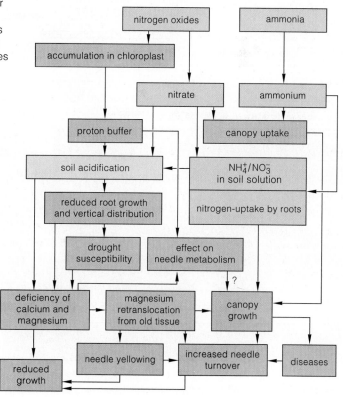

We can't get something for nothing. Fertilizer production requires huge amounts of energy from fossil fuels—not from free, unending sunlight. Few once believed that fossil fuel supplies might run out, so there was little concern about fertilizer costs. It still is common to pour more energy into soil (as in fertilizers) than we get out of it (in the form of food). As long as the human population continues to grow exponentially, farmers will be engaged in a race to grow as much food as they can for as many people as possible. Soil enrichment with nitrogen-containing fertilizers is part of the race, as it is now being run.

The cycling of nitrogen in natural ecosystems depends on the activity of nitrogen-fixing bacteria and on mycorrhizae. Human disruptions to that cycling may harm the ecosystems.

48.8 PHOSPHORUS CYCLE

We conclude our look at biogeochemical cycling with an example of a sedimentary cycle. In the **phosphorus cycle**, phosphorus moves from land to sediments in the seas and then back to the land (Figure 48.12). The earth's crust is the main storehouse for this mineral and for others, including calcium and potassium.

Phosphorus is typically present in rock formations on land, in the form of phosphates. Through the natural processes of weathering and erosion, phosphates enter rivers and streams, which eventually transport them to the ocean. There, mainly on the continental shelves, phosphorus accumulates with other minerals as insoluble deposits. Millions of years pass. Where crustal plates collide, part of the seafloor may be uplifted and drained (page 328). The seafloor, with its mineral deposits, thereby becomes exposed as new land surfaces. Over geologic time, weathering releases phosphates from the rocks—and the geochemical phase of the phosphorus cycle begins again.

The ecosystem phase of the cycle is far more rapid than the long-term geochemical phase. All living organisms require small amounts of phosphorus. It is a key component of ATP, NADPH, phospholipids, nucleic acids, and other organic compounds. Plants have the metabolic means to take up dissolved, ionized forms of phosphorus. Actually, they do this so rapidly and efficiently that they often reduce soil concentrations of phosphorus to extremely low levels. Herbivores obtain phosphorus only by eating the plants; carnivores obtain it by eating herbivores. Herbivores and carnivores excrete phosphorus as a waste product in urine and feces. Phosphorus is also released to the soil by the decomposition of organic matter. The plants then take up phosphorus and so recycle it rapidly within the ecosystem.

The earth's crust is the main storehouse for phosphorus and other minerals that move through ecosystems as part of sedimentary cycles. The geochemical phase of these cycles proceeds extremely slowly.

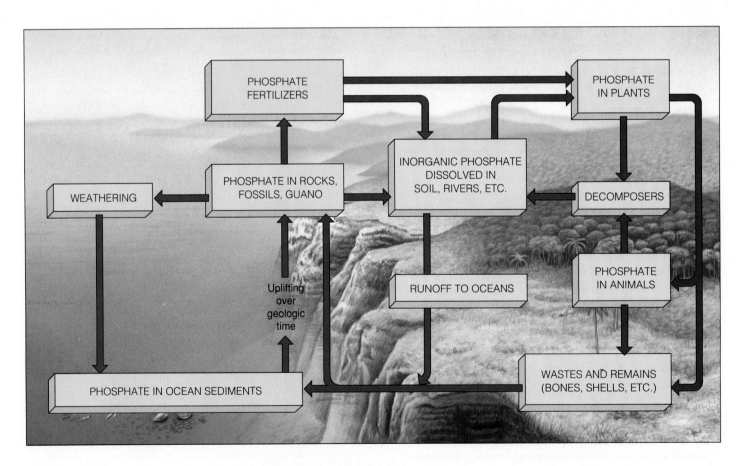

Figure 48.12 The phosphorus cycle. This is an example of a sedimentary cycle.

48.9 ECOSYSTEM MODELING

We now come full circle to the promise of the story that opened this chapter—that disturbances to one part of an ecosystem can have unexpected effects on other, seemingly unrelated parts. One approach to predicting unforeseen effects is through **ecosystem modeling**. By this method, researchers identify crucial bits of information about different ecosystem components. They use computer programs and models to combine the information, then use the resulting data to predict the outcome of the next disturbance. For example, an analysis of which species feed on others in the Antarctic food web shown in Figure 48.3 can be turned into a series of equations that describe how many of each population are consumed. Such equations can be used to predict, say, the effect of overharvesting whales or of greatly expanding the harvesting of krill.

As we attempt to deal with larger and more complex ecosystems, it becomes more difficult and expensive to run experiments in the field. The temptation is to run them instead on the computer. This is a valid exercise if the computer model adequately represents the system. The danger is that we may not have identified all of the key relationships in the ecosystem and incorporated them accurately into the model. The most important fact may be one that we do not yet know. The *Focus* essay reinforces this point.

Focus on the Environment

Transfer of Harmful Compounds Through Ecosystems

DDT, the first of the synthetic organic pesticides, was first used during World War II. In mosquito-infested regions of the tropical Pacific, people were vulnerable to a dangerous disease, malaria. DDT helped control the mosquitoes that were transmitting the sporozoan disease agents (*Plasmodium japonicum*). In war-ravaged cities of Europe, people were suffering from the crushing headaches, fevers, and rashes associated with typhus. DDT helped control the body lice that were transmitting *Rickettsia rickettsii*, the bacterial agent of this terrible disease. After the war, it seemed like a good idea to use DDT against insects that were agricultural or forest pests, transmitters of pathogens, or merely nuisances in homes and gardens.

DDT is a relatively stable hydrocarbon compound. It is nearly insoluble in water, so you might think that it would stay put and act only where applied. But winds can carry DDT in vapor form; water can transport fine particles of it. DDT also is highly soluble in fats, so it accumulates in the tissues of organisms. Thus, as we now know, DDT can show **biological magnification**. The term refers to an increase in concentration of a nondegradable (or slowly degradable) substance in organisms as it is passed along food chains. Most of the DDT from all the organisms that a consumer eats during its lifetime will become concentrated in its tissues. Besides this, many organisms have the means to partially metabolize DDT to DDE and other modified compounds with different but still disruptive effects. Both DDT and the modified compounds are toxic or physiologically disruptive to *many* aquatic and terrestrial animals.

After the war, DDT began to move through the global environment, infiltrate food webs, and affect organisms in ways that no one had predicted. In cities where DDT was sprayed to control Dutch elm disease, songbirds started dying. In streams flowing through forests where DDT was sprayed to control spruce budworms, salmon started dying. In croplands sprayed to control one kind of pest, new kinds of pests moved in. DDT was indiscriminately killing off the natural predators that had been keeping pest populations in check! It took no great leap of the imagination to make the connection. All of those organisms were dying at the same time and the same places as the DDT applications.

Then side effects of biological magnification started showing up in places far removed from the areas of DDT application—and much later in time. Most devastated were species at the end of food chains, including bald eagles, peregrine falcons, ospreys, and brown pelicans. One product of DDT breakdown interferes with physiological processes. As one consequence, birds produced eggs with brittle shells—and many of the chick embryos didn't make it to hatching time. Some species were at the brink of extinction.

Since the 1970s, DDT has been banned in the United States, except for restricted applications where public health is endangered. Many hard-hit species have partially recovered in numbers. Even today, however, some birds lay thin-shelled eggs. They pick up DDT at their winter ranges in Latin America. As recently as 1990, the California State Department of Health recommended that a fishery off the coast of Los Angeles be closed. DDT from industrial waste discharges that ended twenty years before is still contaminating that ecosystem.

SUMMARY

1. An ecosystem is an entire complex of producers, consumers, detritivores, and decomposers and their physical environment, all interacting through a flow of energy and a cycling of materials.

2. Ecosystems are open systems, with inputs and outputs of energy and nutrients.

a. With few exceptions, sunlight is the source of energy, and photoautotrophs are the primary producers. They convert the energy of sunlight to ATP and other forms that can be used to synthesize large organic compounds from simple inorganic substances.

b. Primary producers also assimilate much of the nutrients required by all other members of the system.

3. Directly or indirectly, primary producers nourish an array of heterotrophs in an ecosystem.

a. The heterotrophs include consumers: herbivores that feed on algae and plants, carnivores that feed on animals, and omnivores that have eclectic diets.

b. The heterotrophs also include decomposers (mainly certain fungi and bacteria that digest organic substances), and detritivores (such as crabs and earthworms, which feed on particles of dead or decomposing material).

4. Feeding relationships within ecosystems are structured as trophic levels: a hierarchy of energy transfers, sometimes referred to as "Who eats whom."

a. Primary producers make up the first trophic level, herbivores make up the next, carnivores, the next, and so on.

b. Decomposers, humans, and many other organisms get energy from more than one source and cannot be assigned to a single trophic level.

5. An isolated food chain (a straight-line sequence of who eats whom in an ecosystem) is rare in nature. Instead, food chains cross-connect with one another, forming food webs.

6. Ecosystems generally are most open for inputs and outputs of energy, water, and carbon.

a. The rate at which primary producers capture and store a given amount of energy in a given time interval is called the primary productivity. (The total rate is called the *gross* primary productivity. The rate of energy storage in plants in excess of the rate of aerobic metabolism by the plants is the *net* primary productivity.)

b. Most mineral nutrients are cycled within a natural ecosystem.

7. Energy fixed by photosynthesizers passes through grazing food webs and detrital food webs. Both types of food webs typically are interconnected in the same ecosystem (Figure 48.13).

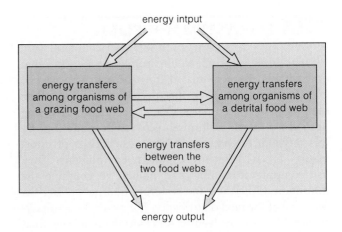

energy intput

energy transfers among organisms of a grazing food web

energy transfers among organisms of a detrital food web

energy transfers between the two food webs

energy output

Figure 48.13 Summary of the one-way flow of energy through two kinds of cross-connected food webs in ecosystems.

a. Both types of food webs lose energy (as heat) through metabolism of all organisms in the ecosystem.

b. In either type of food web, the amount of useful energy flowing through consumer levels declines at each energy transfer. It declines through the loss of metabolically generated heat and as food energy is shunted into organic wastes.

8. In biogeochemical cycles, substances move from the physical environment, to organisms, then back to the environment.

a. Water moves through the hydrologic cycle. Substances that exist primarily in gaseous phases move through atmospheric cycles. Phosphorus and other minerals move through sedimentary cycles.

b. Nutrients generally move slowly through the geochemical phase of these cycles but rapidly between organisms and the environment.

9. Ecosystems on land have predictable rates of nutrient losses that generally increase when the land is cleared or otherwise disturbed.

10. Fossil fuel burning and conversion of natural ecosystems to cropland or grazing land are contributing to increased atmospheric concentrations of carbon dioxide. The increase may be contributing to a global warming trend.

11. Nitrogen availability is often a limiting factor for the total net primary productivity of land ecosystems. Gaseous nitrogen is abundant in the atmosphere, but it must be converted to ammonia and to nitrates that primary producers can use. Some bacteria as well as volcanic action and lightning can cause the conversion.

1. Define ecosystem, and name the central roles that autotrophs play in all ecosystems. *846*

2. Define and give examples of trophic levels in ecosystems. *846–847*

3. Distinguish between food chain and food web. Can you imagine an extreme situation whereby you would be a participant in a food chain? *847*

4. If you were growing a vegetable garden, what variables might affect its net primary production? *848, 851*

5. Characterize grazing and detrital food webs. Indicate how energy leaves each one and how the two are interconnected. *848*

6. Describe the greenhouse effect. Make a list of agricultural products and manufactured goods that you depend on. Are any implicated in the amplification of the greenhouse effect? *856–857*

7. Describe the reservoirs and organisms involved in one of the biogeochemical cycles. *854-860*

8. Define nitrogen fixation, nitrification, ammonification, and denitrification. *858*

Self-Quiz *(Answers in Appendix IV)*

1. _____ is an ecosystem.
   a. A freshwater spring
   c. A city
   b. Antarctica
   d. All of the above

2. Ecosystems have _____ .
   a. energy inputs and outputs
   c. one trophic level
   b. nutrient cycling but not outputs
   d. a and b

3. The _____ is a sedimentary cycle.
   a. hydrologic cycle
   c. phosphorus cycle
   b. carbon cycle
   d. b and c

4. Trophic levels can be described as _____ .
   a. structured feeding relationships
   b. who eats whom in an ecosystem
   c. a hierarchy of energy transfers
   d. all of the above

5. A feeding relationship that proceeds from algae to a fish, then to a fisherman and then to a shark is _____ .
   a. a food chain
   b. a food web

6. Primary productivity is affected by _____ .
   a. photosynthesis and respiration by plants
   b. how many plants are neither eaten nor decomposed
   c. rainfall
   d. temperatures
   e. all of the above

7. Deforestation of watersheds _____ most nutrient outputs.
   a. lessens
   c. increases
   b. equalizes
   d. stabilizes

8. Match the ecosystem terms with the suitable description.
   _____ primary producers
   _____ consumers
   _____ decomposers
   _____ detritivores

   a. herbivores, carnivores, omnivores, parasites
   b. feed on partly decomposed organic particles
   c. break down remains or products of other organisms
   d. photoautotrophs

9. Match the ecosystem terms with the suitable description.
   _____ nitrogen availability
   _____ ecosystem components
   _____ phosphorus
   _____ ammonium
   _____ biogeochemical cycle

   a. movement of water or nutrients from the environment, to organisms, and back
   b. form of nitrogen plants can take up
   c. key limiting factor for net primary production in land ecosystems
   d. producers, consumers, detritivores decomposers
   e. moves through a sedimentary cycle

Selected Key Terms

ammonification *858*	food chain *847*
biogeochemical cycle *851*	food web *847*
biological magnification *861*	grazing food web *848*
carbon cycle *854*	greenhouse effect *856*
consumer *846*	hydrologic cycle *852*
decomposer *846*	nitrification *858*
decomposition *858*	nitrogen cycle *858*
denitrification *858*	nitrogen fixation *858*
detrital food web *848*	phosphorus cycle *860*
detritivore *846*	primary producer *846*
ecological pyramid *848*	primary productivity *848*
ecosystem *846*	trophic level *846*
ecosystem modeling *861*	watershed *852*

Readings

Begon, M., J. Harper, and C. Townsend. 1990. *Ecology: Individuals, Populations, and Communities*. Second edition. Sunderland, Massachusetts: Sinauer.

Botkin, D. B. 1990. *Discordant Harmonies: A New Ecology for the Twenty-First Century*. New York: Oxford University Press.

Krebs, C. 1994. *Ecology*. Fourth edition. New York: Harper-Collins.

Post, W. et al. 1990. "The Global Carbon Cycle." *American Scientist* 78:310–326.

49 THE BIOSPHERE

Does a Cactus Grow in Brooklyn?

Suppose you live in the American Southwest but find yourself touring the deserts of Africa (Figure 49.1). There you come across flowering plants with spines, tiny leaves, and columnlike, fleshy stems—just like some cactus plants back home. Or suppose you live in the coastal hills of California and decide to tour the Mediterranean coast, Africa's southern tip, or even central Chile. There you come across woody, many-branched plants—very much like the chaparral plants back home.

In both cases, the plants are separated by enormous geographic and evolutionary distances. Why are they so much alike? The question intrigues you, so you decide to compare their locations on a global map. As you quickly discover, American and African desert plants grow about the same distance from the equator. Chaparral plants and their distant look-alikes grow along western or southern coasts of continents between latitudes 30° and 40°. As Charles Darwin and other naturalists did before, you have stumbled onto one of many patterns in the world distribution of species.

In part, "accidents of history" put many species in particular places. Go back to Pangea's colossal breakup, more than 100 million years ago. When vast chunks of that supercontinent drifted apart, many species had no choice but to go along for the ride. Over evolutionary time, the species that were dispersed to isolated locations diverged in splendid ways from the parent populations. Among them were the ancestors of eucalyptus trees, wombats, and kangaroos on the chunk that became Australia.

But species also owe their distribution to topography, climate, and species interactions. With diligence, you might grow a cactus under artificial lights in a heated room in Brooklyn or some other New York City borough. Plant that cactus outside and it won't last one winter.

This last example reminds us that we humans tinker with the distribution of species. Not all of our tinkering is as harmless as growing a cactus in Brooklyn. Think of our predatory effects on the world's fisheries or the effects of our pesticide battles with insect competitors for food. Earlier chapters provided you with a general picture of predation, competition, and other species interactions. Consider now the physical forces shaping the biosphere itself. This will serve as a foundation for addressing the impact of the human species on the biosphere—the topic of the chapter to follow.

Figure 49.1 Morphological convergence in plants that are geographically and evolutionarily distant. (**a**) *Echinocereus*, of the cactus family, grows in deserts of the American Southwest. (**b**) *Euphorbia*, of the spurge family (Euphorbiaceae), grows in deserts of southwestern Africa. Although the plants seem related, their lineages are separate. Long ago, parts of plants in both lineages were put to comparable uses in similar environments, and they ended up resembling each other in form and function.

a

b

1. Besides being the main energy source for ecosystems, solar radiation influences their global distribution. It does so by providing heat energy that warms the atmosphere and drives the earth's weather systems.

2. Global air circulation patterns, ocean currents, and topographic features interact to produce regional variations in patterns of temperature and rainfall. These patterns influence the composition of soils and sediments, the growth and distribution of primary producers, and, through them, the distribution of ecosystems.

3. A biome is a large, regional unit of land that can be characterized by the climax vegetation of the ecosystems within its boundaries. Deserts and broadleaf forests are examples. Their distribution corresponds roughly with regional variations in climate, topography, and soil type.

4. The water provinces cover more than 70 percent of the earth's surface. They include oceans, inland seas, and bodies of standing freshwater and moving freshwater. Each freshwater and marine ecosystem has gradients in light availability, temperature, and dissolved gases. The gradients, which vary daily and seasonally, influence primary productivity and the composition of species.

The **biosphere** is the sum total of all places in which organisms live. It extends through the **hydrosphere**—the waters of the earth, including the ocean, polar ice caps, and other forms of liquid and frozen water. It extends into soils and sediments of the **lithosphere**—the earth's outer, rocky layer. It also extends into the lower **atmosphere**—the gases and airborne particles that envelop the earth. About 80 percent of the atmosphere's molecular components are distributed within 17 kilometers of the earth's surface.

Ecosystems of the biosphere range from continent-straddling forests to tiny pools. Climate profoundly influences all of these ecosystems except for a few at hydrothermal vents on the ocean floor. **Climate** refers to prevailing weather conditions, such as temperature, humidity, wind speed, cloud cover, and rainfall.

As you will see, many factors contribute to climate. The major ones are (1) variations in the amount of incoming solar radiation, (2) the earth's daily rotation and its path around the sun, (3) world distribution of continents and oceans, and (4) elevations of land masses. These factors interact to produce prevailing winds and ocean currents that influence global patterns of climate. Climate affects the development of soils and sediments. *Together, climate and the composition of soils and sediments affect the growth of primary producers—hence the distribution of entire ecosystems.*

49.1 AIR CIRCULATION PATTERNS AND REGIONAL CLIMATES

Each winter, Pacific Grove, California, is host to great gatherings of tourists and monarch butterflies (Figure 49.2). Similarly, caribou, Canadian geese, sea turtles, and whales undertake vast migrations, moving to and from overwintering grounds. Other kinds of animals stay put, having adaptations in form, physiology, and behavior that allow them to endure seasonal change. Throughout Canada and the United States, flowering plants form leaves, flower, bear fruit, and drop leaves. In the ocean, uncountable numbers of photosynthetic microorganisms show seasonal bursts of primary productivity. Clearly, organisms are exquisitely attuned to climatic conditions, which differ among regions and with the seasons. Those conditions begin with the incoming rays from the sun.

Of the total amount of solar radiation arriving at the outer atmosphere, only about half gets through to the earth's surface. Atmospheric molecules of ozone (O_3) and oxygen (O_2) absorb most of the wavelengths of ultraviolet radiation, which are lethal for most forms of life. Absorption is greatest in the **ozone layer**, a region between 17 and 25 kilometers above sea level where ozone is most concentrated (Figure 49.3). Clouds, dust, and water vapor absorb a portion of the other types of wavelengths or reflect them back into space.

Radiation that penetrates the atmosphere warms the earth's surface, which gives up heat by radiation or evaporation. Molecules of the lower atmosphere absorb some heat, then reradiate part of it toward the earth. The effect is a bit like heat retention in a greenhouse, which lets in the sun's rays while retaining heat that is being lost from the plants and soil inside (page 856). Why is this important? *Heat energy derived from the sun warms the atmosphere—and that energy drives the earth's weather systems.*

Incoming rays from the sun have different heating effects at different latitudes. The rays are more concentrated at the equator than at the poles—so air is heated more at the equator (Figure 49.4). The global pattern of air circulation begins when warm equatorial air rises and spreads northward and southward. The earth's rotation modifies the circulation into worldwide belts of prevailing east and west winds. Together, the differences in solar heating at different latitudes and the modified air circulation patterns define the world's major **temperature zones** (Figure 49.5).

Global air circulation patterns also cause differences in rainfall at different latitudes. Consider that warm air can hold more moisture than cool air. As air heats up at the equator, it picks up moisture and rises to cooler altitudes; then gives up moisture as rain. That rain supports luxuriant tropical forests. The now-drier air moves away from the equator, then it descends at latitudes of about 30°. As it descends, the air gets warmer

Figure 49.2 Monarch butterflies, migrants that gather in winter in trees of California coastal regions and central Mexico. Monarchs typically travel hundreds of kilometers south to these regions, which are cool and humid in winter; if they stayed in their breeding grounds, they would risk being killed by more severe conditions.

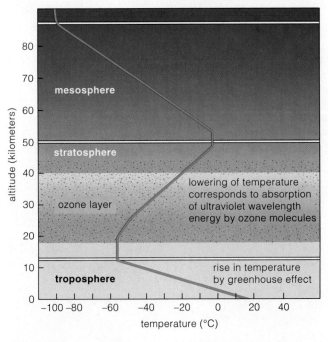

Figure 49.3 Earth's atmosphere. Most global air circulation occurs in the troposphere, where temperatures decrease rapidly with altitude. Ultraviolet wavelengths from the sun are absorbed in the upper atmosphere mainly at the ozone layer, between 17 and 25 kilometers above sea level.

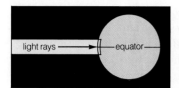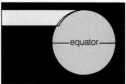

a

Figure 49.4 Global air circulation patterns, as brought about by three interrelated factors. *First*, the sun's rays are less spread out in equatorial regions than in polar regions (**a**). Warm equatorial air rises and spreads northward and southward to produce the initial pattern of air circulation (**b**).

Second, the nonuniform distribution of land masses creates variations in air pressure. Land absorbs and gives up heat faster than oceans do, so some parcels of air rise (or sink) faster than others. Air pressure decreases where warm air rises (and increases where it sinks). The differences give rise to winds, which disrupt the overall air movement from the equator to the poles.

Third, the earth's rotation and overall shape introduce easterly and westerly deflections in wind directions. Each time the ball-shaped earth makes a full rotation, its surface turns faster beneath air masses at the equator and slower beneath those at the poles. Thus, a rising air mass can't really move "straight north" or "straight south." It is deflected to the east or west. This is the source of prevailing east and west winds (**c**).

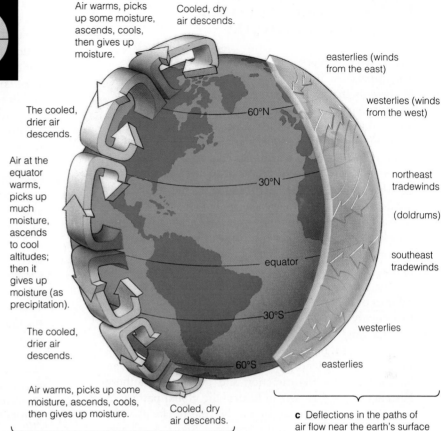

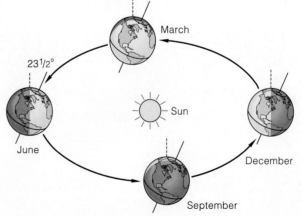

b Initial pattern of air circulation

and drier. Deserts tend to form at these latitudes. Air even farther from the equator picks up moisture, ascends to high altitudes, then creates another moist belt at latitudes of about 60°. Finally, air descends in polar regions, where low temperatures and almost nonexistent precipitation give rise to the cold, dry, polar deserts.

Finally, the amount of solar radiation reaching the earth's surface varies annually, owing to the earth's rotation around the sun (Figure 49.6). This leads to seasonal changes in daylength, wind directions, and temperature. Primary productivity alternately rises and falls on land and in the seas—and migrations shift many kinds of animals to new locations.

Latitudinal differences in the amount of solar radiation reaching the earth produce global air circulation patterns, which produce latitudinal belts of temperature and rainfall.

The earth's annual rotation around the sun introduces seasonal changes in winds, temperatures, and rainfall.

These factors influence the locations of different ecosystems.

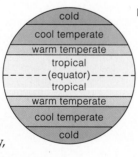

Figure 49.5 World temperature zones.

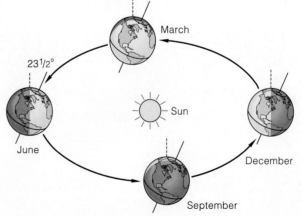

Figure 49.6 Annual variation in the amount of incoming solar radiation. The northern end of the earth's fixed axis tilts toward the sun in June and away from it in December. Thus the position of the equator relative to the boundary of illumination between day and night varies annually. Such variations in the intensity and duration of daylight lead to seasonal variations in temperature in the two hemispheres. Seasonal change becomes more pronounced with distance from the equator. It is greatest in the central regions of continents, where the moderating effects of oceans are minimal.

49.2 THE OCEAN'S EFFECT ON REGIONAL CLIMATES

Ocean water covers 71 percent of the earth's surface. The uppermost 10 percent of the world ocean circulates in currents that influence regional climates and distribute nutrients in marine ecosystems. Ultimately, the sun and wind friction drive the surface currents.

Latitudinal and seasonal variations in solar heating cause ocean water to warm and cool on a vast scale. A given volume of water increases when warmed and decreases when cooled. At the equator, the sea level is about 8 centimeters (3 inches) higher than it is at the poles. The sheer volume of water associated with this "slope" is enough to get the surface waters moving in response to gravity. They tend to move from the equator to the poles—and they warm air parcels above them during the journey. At midlatitudes, about *10 million billion* calories of heat energy are transferred from warm water to air every second!

The "tug" of mainly the trade winds and westerlies causes a rapid, mass flow of water. The earth's rotation, the positions of land masses, even the shapes of ocean basins influence the direction and properties of these currents. As Figure 49.7 indicates, the global pattern of circulation is clockwise in the Northern Hemisphere and counterclockwise in the Southern. Swift, deep, narrow currents of nutrient-poor waters run parallel with the east coasts of continents. For example, each second, the Gulf Stream moves 55 million cubic meters of warm water northward, along the eastern coast of North America. Slower, shallow, broad currents paralleling the west coasts of continents move cold water toward the equator. As you will see, they have roles in moving deep, nutrient-rich waters to the surface.

Why do cool, mild, foggy summers prevail along the Pacific Northwest coast? Track the currents shown in Figure 49.7. The California Current moves cold water near the coast as it flows toward the equator. Winds approaching the coast give up heat to the cold waters and so become cooler. Why is Washington, D.C., muggy in summer? The air above the Gulf Stream gains heat and moisture—which winds from the south and east carry to the city. Compared to Ontario in central Canada, why are the winters milder in London and Edinburgh even though they are at the same latitude? The North Atlantic Current picks up warm water from the Gulf Stream, flows past northwestern Europe—and gives up heat energy to winds blowing in from the east.

Surface ocean currents, in combination with global air circulation patterns, influence regional climates and help distribute nutrients in marine ecosystems.

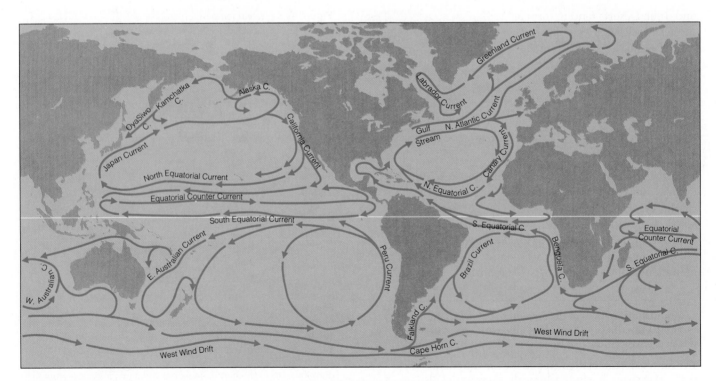

Figure 49.7 Surface currents of the world ocean. Warm water moves from the equator toward the poles. The direction of flow depends on winds, the earth's rotation, gravity, the shape of ocean basins, and the positions of land masses.

49.3 EFFECTS OF TOPOGRAPHY ON REGIONAL CLIMATES

Mountains, valleys, and other aspects of topography also influence regional climates. "Topography" means the physical features of a region, such as its elevation.

For example, more than 2 billion people in Africa and Asia depend on heavy summer rains for drinking water and irrigation. The very wet summers alternate with dry winters in seasonal patterns of air circulation called **monsoons** (after the Arabic *mausim*, meaning "season"). They exist in equatorial regions where trade winds converge and where intense heat causes air over the ocean to expand and rise. Enormous volumes of warm, moist, air move in over the land and produce a months-long deluge. In parts of North America, humid air from the Gulf of Mexico produces much smaller monsoons. Recurring sea breezes along coasts are "mini-monsoons." Here again, the heat capacities of land and water differ. In the morning, the temperature of water does not rise as fast as it does on land. As warmed air above the land rises, cooler marine air moves in. After sunset, the land loses heat faster, and land breezes flow in the reverse direction (Figure 49.8).

As another example, imagine a warm air mass picking up moisture off California's western coast. After moving inland, it reaches the Sierra Nevada, a mountain range that parallels the coast. As the air ascends to higher altitudes, it cools and loses moisture (Figure 49.9). Vegetation belts at different elevations correspond to the changes in air temperature and moisture. The western base of the range supports plants that can live in semiarid conditions. At higher elevations, moisture levels and cool temperatures support forests of deciduous and evergreen trees. The subalpine belt higher up supports only evergreen trees adapted to the rigors of a cold climate. Only low, nonwoody plants and dwarfed shrubs can grow above the subalpine belt.

After air flows over the mountain crests and starts to descend, it becomes warmer and can hold more moisture. The air now draws moisture out of plants and soil rather than giving it up as rain. Only plants that are adapted to arid or semiarid conditions grow in this **rain shadow**. The term refers to the reduction in rainfall on the leeward side of high mountains. (Leeward is the direction not facing a wind; windward is the direction from which the wind is blowing.) The high mountain ranges of Europe, the Himalayas of Asia, and the Andes of South America also create rain shadows.

Atmospheric circulation patterns, ocean currents, and landforms interact in ways that influence regional temperatures and moisture levels—which affect the distribution and dominant features of ecosystems.

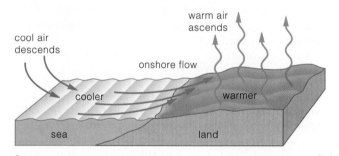

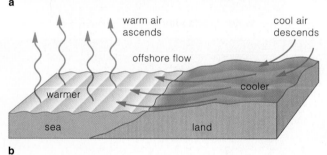

Figure 49.8 Coastal breezes in the afternoon (**a**) and at night (**b**).

Figure 49.9 Rain shadow effect, a reduction of rainfall on the side of high mountains facing away from the prevailing wind. Only plants adapted to arid or semiarid conditions grow in such places. *Blue* numbers are the average yearly precipitation (in centimeters) measured at different locations on both sides of the mountain range. *Black* numbers signify the elevation (in meters).

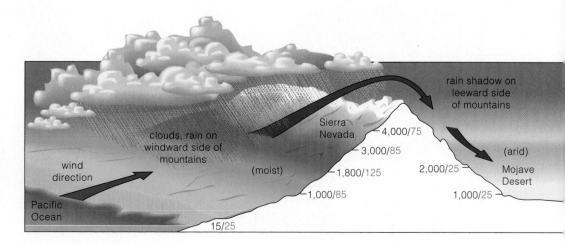

49.4 THE WORLD'S BIOMES

The circulation patterns in the atmosphere and at the ocean surface, in concert with topography, give rise to regional differences in temperature and moisture. This information helps explain why grasslands, deserts, forests, and tundra exist in some places but not others. It helps explain the global distribution of species, each of which is adapted to regional conditions. And it provides insight into why certain evolutionarily distant species look alike. Such species are examples of convergent evolution. Their ancestors faced similar environmental pressures and, by natural selection, became modified in similar ways (page 307). The ancestors of those cactus and spurge plants shown in Figure 49.1 evolved in hot, dry deserts where water is scarce. Fleshy plant stems with thick cuticles conserve precious water—and rows of sharp spines help prevent herbivores from chewing on juicy plant parts.

Long ago, W. Sclater and then Alfred Wallace attached the name **biogeographic realms** to six vast land areas, each with distinguishing plants and animals (Figure 49.10). The realms retain their identity partly because of climate—and partly because oceans, mountain ranges, and other major barriers tend to restrict gene flow and so keep their component species isolated. Each realm may be further divided into biomes (Figure 49.11). A **biome** is a large region of land characterized

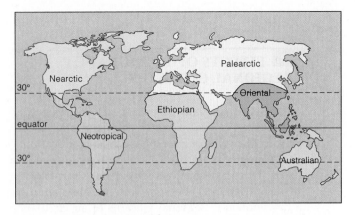

Figure 49.10 Major biogeographic realms.

mainly by the climax vegetation of ecosystems within its boundaries. Distinctive biomes prevail at certain latitudes and elevations. Thus, short plant species often prevail in dry regions, at high elevations, and at high latitudes. Biomes dominated by tall, leafy plant species prevail at tropical and temperate latitudes, and at low elevations with warm temperatures and high rainfall (Figure 49.12). As you will see, soils also affect the distribution of biomes.

The distribution and key features of biomes are an outcome of temperatures, soils, and moisture levels (which vary with latitude and elevation), and of evolutionary history.

arctic tundra   boreal coniferous forest   temperate deciduous forest

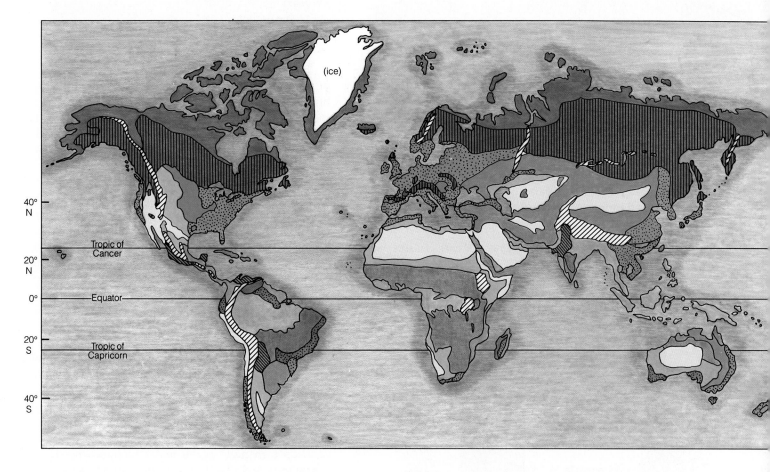

Figure 49.11 Distribution of the world's major biomes. The overall pattern corresponds roughly with distribution patterns for climate and soil type. Compare this map with the two-page photograph of the earth's surface that precedes Chapter 1.

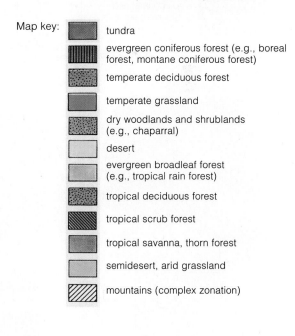

Map key:

- tundra
- evergreen coniferous forest (e.g., boreal forest, montane coniferous forest)
- temperate deciduous forest
- temperate grassland
- dry woodlands and shrublands (e.g., chaparral)
- desert
- evergreen broadleaf forest (e.g., tropical rain forest)
- tropical deciduous forest
- tropical scrub forest
- tropical savanna, thorn forest
- semidesert, arid grassland
- mountains (complex zonation)

Figure 49.12 Changes in plant form along environmental gradients for North America. Shown here, gradients in water availability and elevation that influence primary productivity. Mean annual temperature, soil drainage, and other factors also have effects.

49.5 SOILS OF MAJOR BIOMES

Soil is a mixture of rock, mineral ions, and organic matter in some state of physical and chemical breakdown. The rocks range from coarse-grained gravel to sand, silt, and fine-grained clay. Water, air, and diverse organisms occupy spaces among the particles and grains. The organic matter, which is in various stages of decomposition, is called humus.

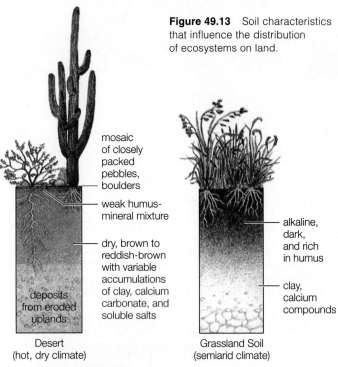

Figure 49.13 Soil characteristics that influence the distribution of ecosystems on land.

mosaic of closely packed pebbles, boulders

weak humus-mineral mixture

dry, brown to reddish-brown with variable accumulations of clay, calcium carbonate, and soluble salts

deposits from eroded uplands

Desert
(hot, dry climate)

alkaline, dark, and rich in humus

clay, calcium compounds

Grassland Soil
(semiarid climate)

Figure 49.13 shows some vertical structures of soils, or soil profiles. Topsoil, the upper region with the most humus, is most vulnerable to weathering. It may be less than a centimeter deep on steep slopes to more than a meter deep in grasslands. Water percolating down from the topsoil deposits clay, minerals, and some organic matter in the next layer. Beneath this are weathered, broken rocks from which true soil forms.

The growth of most plants suffers in poorly aerated, poorly draining soils. Loam topsoils have the best mix of sand, silt, and clay for agriculture. They have enough coarse particles to promote good drainage and enough fine particles to retain water-soluble mineral ions that serve as nutrients for plant growth. Gravelly or sandy soils promote rapid leaching that can deplete them of mineral ions. Clay soils with fine, closely packed particles are poorly aerated and do not drain well. Few plants can grow in waterlogged clay soils.

As farmers know, soil affects primary productivity. They grow most crops in cleared, former grasslands. They also clear tropical forests for agriculture (page 902). But these biomes have little topsoil above a poorly draining clay layer. Clearing exposes the topsoil, and heavy rains leach most of the nutrients.

Let's turn now to the major biomes—the deserts, shrublands, woodlands, grasslands, forests, and tundra. Bear in mind, no biome is uniform throughout. Within its borders, local climates, landforms, and other physical features favor patches of distinct communities.

Primary productivity depends on the soil profile and its proportions of sand, silt, clay, gravel, and humus.

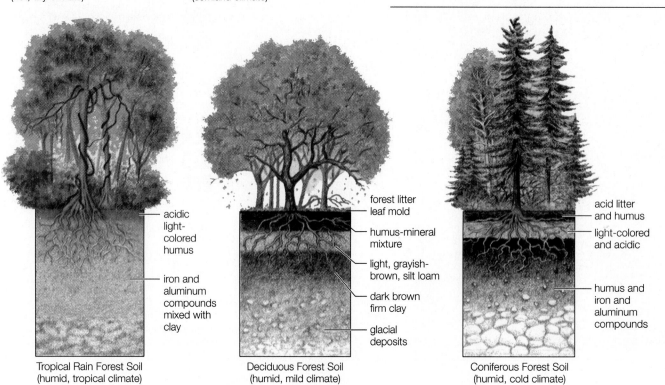

acidic light-colored humus

iron and aluminum compounds mixed with clay

Tropical Rain Forest Soil
(humid, tropical climate)

forest litter
leaf mold

humus-mineral mixture

light, grayish-brown, silt loam

dark brown firm clay

glacial deposits

Deciduous Forest Soil
(humid, mild climate)

acid litter and humus

light-colored and acidic

humus and iron and aluminum compounds

Coniferous Forest Soil
(humid, cold climate)

49.6 DESERTS

Deserts form in land regions where the potential for evaporation greatly exceeds rainfall. Such conditions prevail at latitudes of about 30° north and south. There we find the great deserts of the American Southwest, northern Chile, Australia, northern and southern Africa, and Arabia. Farther north are the vast Gobi of Asia, the Kyzyl-Kum east of the Caspian Sea, and the high deserts of eastern Oregon. The arid conditions in the northern deserts are largely an outcome of rain shadow effects of extensive mountain ranges.

Deserts do not have lush vegetation. Heavy, brief, infrequent pulses of rain swiftly erode the exposed topsoil. Humidity is so low, the sun's rays easily penetrate the air and quickly heat the ground surface. The surface radiates heat and cools quickly at night.

Although the arid or semiarid conditions do not favor large, leafy plants, deserts show plenty of diversity. A patch of Arizona desert may have deep-rooted evergreen shrubs, including creosote, and fleshy-stemmed, shallow-rooted cacti (Figure 49.14). It may have tall saguaro, short prickly pear, and ocotillo, which drops leaves more than once a year, then grows new ones after a rain. Desert annuals and perennials flower spectacularly but briefly after the rains. Mesquite, cottonwood, and other deep-rooted species may grow near streambeds with a permanent underground water supply.

Of all land surfaces, more than one-third is arid or semiarid, without enough rainfall to support crops. The crops growing in California's Imperial Valley and some other deserts require unrelenting soil management and extensive irrigation. Without proper drainage, such croplands become unproductive from waterlogging and salt buildup. Alarmingly, many parts of the world are undergoing **desertification**, the wholesale conversion of grasslands and other productive biomes to dry wastelands. This trend is described on page 901.

Figure 49.14 Warm desert near Tucson, Arizona. The plants include creosote bushes, multistemmed ocotillo, columnlike saguaro cacti, and prickly pear cacti with rounded pads.

49.7 DRY SHRUBLANDS AND WOODLANDS

Dry shrublands and **dry woodlands** prevail in western or southern coastal regions of continents between latitudes 30° and 40°. These and comparably semiarid regions get more rain than deserts, but not much more. Most rain falls during mild winters, and summers are long, hot, and dry. The dominant plants often have hard, tough, evergreen leaves.

Dry shrublands get less than 25 to 60 centimeters of rain per year. They have exotic local names, including fynbos (near the Mediterranean) and chaparral. California alone has 2.4 million hectares (6 million acres) of chaparral. In summer, lightning-sparked, wind-driven firestorms often sweep through these biomes (Figure 49.15). Many of the shrubs have highly flammable leaves, and their aboveground parts burn swiftly. Yet they are highly adapted to episodes of fire and quickly resprout from their root crowns. Trees do not fare as well in firestorms. The shrubs, which "feed" the fires, have the competitive edge.

Dry woodlands dominate when annual rainfall is about 40 to 100 centimeters. The dominant trees can be tall, but they do not form a dense, continuous canopy. Eucalyptus woodlands of southwestern Australia and oak woodlands of California and Oregon are like this.

49.8 GRASSLANDS AND SAVANNAS

Grasslands sweep across much of the interior of continents, in the zones between deserts and temperate forests. The main types, **shortgrass prairie** and **tallgrass prairie**, usually form where the land is flat or rolling. Warm temperatures prevail in summer, but winters are cold. The 25 to 100 centimeters of annual rainfall prevent deserts from forming in this zone but are not enough to support forests. The dominant animals are grazing and burrowing types. Grazing and periodic fires keep shrublands and forests from encroaching on the fringes of many grasslands.

Shortgrass prairie prevails when winds are strong, rainfall light and infrequent, and evaporation rapid (Figure 49.16a). The roots of drought-tolerant perennial plants extend profusely through the topsoil. In the 1930s, shortgrass prairie of the American Great Plains was overgrazed and plowed under for wheat, which needs more water than the region sometimes provides. Strong prevailing winds, prolonged drought conditions, and poor farming practices turned much of the prairie into a Dust Bowl. Steinbeck's *The Grapes of Wrath* and Michener's *Centennial*, two historical novels, speak eloquently of the consequences.

In North America's interior, tallgrass prairie once extended west from zones of temperate deciduous forest (Figure 49.16b). This richer, moister grassland boasted an abundance of legumes and composites, such as daisies. Most tallgrass prairie was converted to farmland, but large areas are being restored.

Between the tropical forests and deserts of Africa, South America, and Australia are broad belts of **savannas**. Rainfall averages 90 to 150 centimeters a year, and prolonged seasonal droughts are common. Where rainfall is low, fast-growing grasses dominate (Figure 49.16c). Patches of acacia and other shrubs grow in regions that get slightly more moisture. Where rainfall is higher, savannas grade into tropical woodlands with tall, coarse grasses, shrubs, and low trees. **Monsoon grasslands** exist in parts of southern Asia where heavy rains alternate with a dry season. The climate favors dense stands of tall, coarse grasses. These die back and often burn in the dry season.

Figure 49.15 (**a**) California chaparral. Dominant plants are multibranched, woody, and a few meters tall at most. Often they form a nearly impenetrable vegetation cover. The boxed inset shows a firestorm racing through a chaparral-choked canyon above Malibu. (**b**) Closer view of a chaparral plant community.

Figure 49.16 (**a**) Rolling shortgrass prairie to the east of the Rocky Mountains. Bison—60 million at one time—were the dominant large herbivore of North American grasslands. (**b**) Natural tallgrass prairie in eastern Kansas. (**c**) African savanna—warm grasslands with scattered stands of shrubs and trees. More large ungulates (hooved, plant-eating mammals) live here than anywhere else. They include migratory wildebeests (shown), giraffes, Cape buffalo, zebras, and impalas.

BROADLEAF FORESTS

In forest biomes, tall trees grow close together, forming a fairly continuous canopy over a broad region. In general, there are three types of trees. Which prevails in a region depends partly on distance from the equator. *Evergreen broadleafs* dominate between latitudes 20° north and south. *Deciduous broadleafs* are common at moist, temperate latitudes where winters are not severe. *Evergreen conifers* are common at high, colder latitudes and in mountains of temperate zones.

Evergreen Broadleaf Forests

These biomes sweep across tropical zones of Africa, the East Indies, southeast Asia, the Malay Archipelago, South America, and Central America. Annual rainfall can exceed 200 centimeters and is never less than 130 centimeters. Highly productive **tropical rain forests** exist where rainfall is regular and heavy, the annual mean temperature is 25°C, and the humidity is 80 percent or more (Figure 49.17). Evergreen trees produce new leaves and shed old ones through the year, so tropical rain forests produce more litter than other forests.

bromeliad

Figure 49.17 Tropical forests, with their spectacular biodiversity. Among the millions of known species are jaguars and *Rafflesia*, a leafless, foul-smelling plant with a fly-pollinated flower that can be 3 meters across. Bromeliads, orchids, and other plants grow on tree branches, obtaining minerals from organic remains of leaves, insects, and other litter that have dissolved in tiny pockets of water.

Decomposition and mineral cycling are rapid in the hot, humid climate, so soils are highly weathered, humus-poor, and poor reservoirs of nutrients. We take a look at these forests in the next chapter.

Deciduous Broadleaf Forests

Leaving tropical rain forests, we enter regions where temperatures stay mild but rainfall dwindles during part of the year. Here we find **tropical deciduous forests**, in which many trees drop some or all of their leaves during a pronounced dry season. The **monsoon forests** of India and southeastern Asia also have such trees. Farther north, in the temperate zone, rainfall is even lower, and temperatures become cold during the winter. Here we find **temperate deciduous forests** (Figure 49.18). Decomposition is not as rapid as in the humid tropics, and nutrients are conserved in accumulated litter on the forest floor.

Forests of ash, beech, birch, chestnut, elm, and deciduous oaks once stretched across northeastern North America, Europe, and eastern Asia. They declined drastically when farmers cleared the land. In North America, diseases spread by introduced species and destroyed nearly all chestnuts and many elms (Table 47.1). Now, maple and beech predominate in the Northeast. Farther west, oak-hickory forests prevail, then oak woodlands that grade into tallgrass prairie.

Spring

Summer

Figure 49.18 The changing character of a temperate deciduous forest in spring, summer, autumn, and winter. The one shown here is south of Nashville, Tennessee.

Winter

Autumn

49.10 CONIFEROUS FORESTS

Evergreen conifers are primary producers of boreal forests, montane coniferous forests, temperate rain forests, pine barrens, and many other forest biomes. Conifers are cone-bearing trees, and most have needle-shaped leaves adapted to arid conditions. The needles have a thick cuticle and recessed stomata that help the trees conserve water.

Boreal forests are expanses of coniferous trees that stretch across northern Europe, Asia, and North America. They also are known as the taiga (meaning "swamp forest"). Most of these biomes are in glaciated regions with cold lakes and streams (Figure 49.19a). It rains mostly in summer, and evaporation is low in the cool summer air. The cold, dry winters are more severe in eastern parts of these biomes than in the west, where oceanic winds moderate the climate. Spruce and balsam fir dominate North America's boreal forests. Pines, along with birches and aspens, take hold in burned or logged areas. Acidic bogs, dominated by peat mosses, shrubs, and stunted trees, develop in poorly drained areas. Boreal forests become much less dense to the north, where they grade into arctic tundra.

In the Northern Hemisphere, **montane coniferous forests** extend southward through the great mountain ranges. Spruce and fir dominate in the north and at higher elevations. They give way to fir and pines in the south and at lower elevations (Figure 49.19b).

Coniferous forests also grow in some temperate lowlands. For example, a **temperate rain forest** parallels the coast from Alaska into northern California. It includes some of the world's tallest trees—Sitka spruce to the north and redwoods to the south (Figure 1.6c). Heavy logging has destroyed much of this biome. In 1994, a major timber company surrendered its rights to log one of the largest intact temperate rain forests outside the tropics. This forest, with trees 800 years old, cloaks the Kitlope Valley of British Columbia.

Sandy, nutrient-poor soil of New Jersey's coastal plain supports **pine barrens**—scrub forests in which grasses and low shrubs grow beneath the open stands of pitch pine and oak trees. Pine barrens are adapted to recover quickly from lightning-triggered fires, which are frequent. Today, managed pine plantations have replaced most of the natural, fire-maintained pine forests that originally dominated the coastal plains of North Carolina, South Carolina, Georgia, and Florida.

a   b

Figure 49.19 (a) Example of a boreal forest. Spruce trees dominate this one. (b) The montane coniferous forest of Yosemite Valley in the Sierra Nevada of California.

49.11 TUNDRA

"Tundra" is derived from the Finnish *tuntura*, which refers to the treeless plain between the polar ice cap and belts of boreal forests in Europe, Asia, and North America. Much of this **arctic tundra** is flat, windswept, and wet (Figure 49.20*a*). Temperatures are cool in summer and below freezing in winter, so there is little evaporation. Little rain or snow falls. In summer's nearly continuous sunlight, short plants grow and then flower profusely, and seeds ripen fast.

Snow does not cloak arctic tundra all year long, but summers are too short to thaw much more than surface soil. Just beneath the surface is a perpetually frozen layer, the **permafrost**. It is more than 500 meters thick in some places. The soil above it cannot drain and remains waterlogged. Anaerobic conditions and low temperatures limit nutrient cycling. Organic matter decomposes so slowly, it accumulates in soggy masses of organic material. All but about 5 percent of the carbon in the arctic tundra is locked up in peat.

A similar type of biome prevails at high elevations in mountains throughout the world, although there is no permafrost beneath the soil. Figure 49.20*b* shows an example of this **alpine tundra**. The dominant plants often form cushions and mats that withstand the buffeting of strong winds. Winter temperatures are below-freezing, and even in summer, shaded patches of snow persist. The thin, fast-draining soil is nutrient-poor, and primary productivity is low.

b

a

Figure 49.20 (**a**) Ponding in the arctic tundra. Rain and snowmelt cannot percolate downward because of the permafrost. (**b**) Short, hardy plants typical of alpine tundra.

49.12 REGARDING THE WATER PROVINCES

The water provinces are far more extensive than the earth's biomes. They include the world's lakes, rivers, ponds, estuaries, and wetlands. They include rocky and sandy shores, coral reefs, parts of the open ocean, even hydrothermal vents on the ocean floor. "Typical" examples are difficult to find. Some ponds can be waded across; Lake Baikal in Siberia is more than 1.7 kilometers deep. Although all aquatic ecosystems have gradients in light penetration, temperature, and dissolved gases, these differ greatly from one to the next. All we can do here is sample the diversity.

a

b

Figure 49.21 (a) In the Canadian Rockies, a lake basin formed by glacial action during the last ice age. (b) In Oregon, Crater Lake, a collapsed volcanic cone now filled with water from rains and melting snow. Like other volcanoes of the Cascade Range, it started forming at the dawn of the Cenozoic, when crustal plates were undergoing major reorganization.

49.13 LAKE ECOSYSTEMS

The topography, climate, and geologic history of a lake dictate the kinds and numbers of its residents, how they are distributed, and how nutrients are cycled among them. All lakes form in land basins (Figure 49.21). In time, most are doomed to fill with sediments or drain when erosion deepens their outlets.

Lake Zonation

A **lake** is a body of freshwater with littoral, limnetic, and profundal zones. Its littoral is a shallow, usually well-lit zone extending all around the shore to the depth at which rooted aquatic plants stop growing (Figure 49.22). Diversity is greatest in this zone; it is the habitat of assorted plants, decomposers, and consumers, including snails and frogs. The limnetic is the open, sunlit water beyond the littoral, to a depth where photosynthesis is insignificant. Suspended in the waters of the limnetic are **plankton**, communities of mostly microscopic organisms. Cyanobacteria, diatoms, green algae, and other photoautotrophs make up the *phyto*plankton. Rotifers, copepods, and other heterotrophs make up the *zoo*plankton.

The profundal zone is all the open water below the depth at which wavelengths suitable for photosynthesis can penetrate. Detritus sinks through the profundal to bottom sediments, which contain communities of bacterial decomposers. Decomposition activities release nutrients into the water.

Seasonal Changes in Lakes

In temperate regions, which have warm summers and cold winters, lakes show seasonal changes in density and temperature from the surface to the bottom. A layer of ice forms over many of the lakes in midwinter. Water near the freezing point is the least dense and accumulates beneath the ice. Water at 4°C is the most dense. It accumulates in deeper layers, which are a bit warmer than the surface layer in midwinter.

In spring, daylength increases and the air is less cold. Ice on a lake melts, and the surface water slowly warms to 4°C so that temperatures become uniform throughout. Winds blowing across the lake surface now cause a **spring overturn**. In such overturns, strong vertical movements carry dissolved oxygen from the lake's surface layer to its depths, and nutrients released by decomposition are brought from the bottom sediments to the surface layer.

By midsummer, the surface layer is well above 4°C. Now the lake has a thermocline, a middle layer that changes abruptly in temperature and that prevents vertical mixing (Figure 49.23). Being warmer and less dense, the surface water floats on the thermocline.

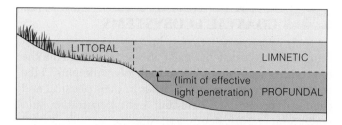

Figure 49.22 Lake zonation. The littoral includes all areas around the edge of the lake, from the shore to the depth where aquatic plants stop growing. The profundal includes areas below the depth of light penetration. Above the profundal are open, sunlit waters of the limnetic.

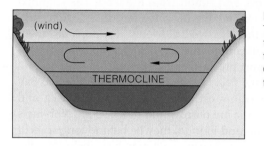

Figure 49.23 Thermal layering of the water in a temperate lake in Connecticut during the summer.

Figure 49.24 Experiment demonstrating eutrophication of a lake in Ontario, Canada. Researchers stretched a plastic curtain across a narrow channel between two basins of the same lake. They added phosphorus, carbon, and nitrogen to one basin (in the background) and carbon and nitrogen only to the other (in the foreground). Within two months, the phosphorus-enriched basin showed accelerated eutrophication, with a dense algal bloom turning the water green.

Decomposers may deplete deeper water of dissolved oxygen. In autumn, the upper layer cools, becomes denser, and sinks. The thermocline vanishes. During this **fall overturn**, water mixes vertically. Once again dissolved oxygen moves down and nutrients move up.

Primary productivity corresponds with the seasons. After a spring overturn, the longer daylengths and cycled nutrients support increased primary productivity. Then, phytoplankton and rooted aquatic plants quickly take up phosphorus, nitrogen, and other nutrients. As the growing season progresses, the thermocline cuts off vertical mixing. Nutrients tied up in the remains of microscopic producers and consumers sink to deeper waters. Nutrients dwindle in the upper waters, and primary productivity declines. By late summer, nutrient shortages are limiting photosynthesis.

After a fall overturn, the nutrient cycling drives another burst of primary productivity. However, with autumn's shorter daylight hours and declining temperatures, this burst does not last long. Primary productivity will not increase again until spring.

Trophic Nature of Lakes

Geologic processes create lakes, as when advancing glaciers carve basins in the earth. After the glaciers retreat, the exposed basins fill with water. In time, erosion, sedimentation, and other processes change a lake's dimensions. Soils of the basin and surrounding regions contribute to the type and amount of nutrients available to support the lake's primary producers.

Conditions ranging from oligotrophy to eutrophy result from interactions among soils, basin shape, and climate. *Oligotrophic* lakes are often deep and nutrient-poor, with low primary productivity. *Eutrophic* lakes are often shallow and nutrient-rich, with high primary productivity. As sediments accumulate, conditions may progress from oligotrophy to eutrophy, then on to a final successional stage with a filled-in basin.

Human activities can disrupt the geologic, climatic, and biological interactions that dictate the trophic condition of lakes. For example, dumping sewage into a lake or logging over the surrounding land can lead to **eutrophication**. The term refers to nutrient enrichment of a lake or some other body of water that typically reduces transparency and favors a phytoplankton-dominated community. The field experiment shown in Figure 49.24 nicely demonstrates this outcome.

As documented by Val Smith, humans caused eutrophication of Lake Washington in Seattle—and they also reversed it. From 1941 to 1963, large amounts of phosphorus-rich sewage were discharged into the lake. The discharge promoted large blooms of cyanobacteria, which formed thick mats over the lake each summer and made it useless for recreation. With phosphorus enrichment, nitrogen became the limiting resource, and cyanobacteria—which are superior competitors for nitrogen because of their ability to fix N_2—became dominant. From 1963 to 1968, sewage discharges were cut back, and finally stopped. By 1975, the lake had almost fully recovered, and lower density populations of diatoms and green algae became dominant.

49.14 STREAM ECOSYSTEMS

The flowing-water ecosystems called **streams** start out as freshwater springs or seeps. They grow and merge as they flow downslope, then often combine to form a river. Between the headwaters and the river's end, we find three kinds of habitats—riffles, pools, and runs. Riffles are shallow, turbulent stretches where water flows swiftly over a rough bottom of sand and rock. Figure 9.1 shows an example. Pools have deep water flowing slowly over a smooth, sandy, or muddy bottom. Runs are smooth-surfaced but fast-flowing stretches over bedrock or rock and sand.

A stream's average flow volume and temperature depend on rainfall, snowmelt, geography, altitude, and even the shade cast by plants. Its solute concentrations are influenced by the streambed's composition as well as by agricultural, industrial, and urban wastes.

Especially in forests, streams import most of the organic matter that supports food webs. Where trees cast shade, photosynthesis is limited. Most of the organic matter is litter that enters detrital food webs. Aquatic organisms continually take up and release nutrients as water flows downstream. (Nutrients move upstream only as components of migrating fishes and other animals.) Think of nutrients as spiraling between water and aquatic organisms as the stream flows on its one-way course, usually to the sea.

Ever since cities formed, streams have been sewers for industrial and municipal wastes. The wastes, as well as other pollutants from poorly managed farmlands, have left many streams choked with sediment and poisoned by chemicals. Streams are resilient, however, and show impressive recovery when pollution is controlled.

49.15 COASTAL ECOSYSTEMS

Remarkable ecosystems exist in estuaries and in the intertidal zones along the coasts of continents. Like freshwater ecosystems, they differ in their physical and chemical properties, including light penetration and water temperature, depth, and salinity.

Estuaries

An **estuary** is a partly enclosed coastal region where seawater mixes with nutrient-rich freshwater from rivers, streams, and runoff from the land. The confined conditions, slow mixing of water, and tidal action combine to trap the dissolved nutrients. The replenishment of nutrients in continually freshened water allows estuaries to support productive ecosystems.

Chesapeake Bay, Mobile Bay, and San Francisco Bay are broad, shallow estuaries. Estuaries in Alaska and British Columbia are narrow and deep; so are Norway's fjords. In Texas and Florida, they lie behind long spits of sand and mud. Estuarine salt marshes, including the one shown in Figure 49.25, are common along the New England coast.

Primary producers in estuaries include phytoplankton, salt-tolerant plants that withstand submergence at high tide, and algae that grow in mud and on plant surfaces. Much of the primary production enters detrital food webs in which bacterial and fungal decomposers are the first to feed. The detritus (and bacteria clinging to it) is food for nematodes, snails, crabs, and fish. Clams, oysters, barnacles, and other filter feeders focus on edible particles suspended in the slowly moving water. So many larval and juvenile stages of invertebrates and some fishes are found here that estuaries are often called marine nurseries. Migratory birds use estuaries as rest stops.

Many estuarine ecosystems are under assault from sewage, agricultural runoff, industrial and urban wastes, and upstream diversion of freshwater for human use. Normal conditions, including salinity levels, cannot be maintained without inflows of unpolluted freshwater.

Figure 49.25 Salt marsh of a New England estuary. The marsh grass *Spartina* is the major producer. Its microbe-enriched litter provides food for consumers in the creeks and adjacent sound.

SALT MARSH (estuary)

open ocean   sound   shallow bay   creek   tidal river

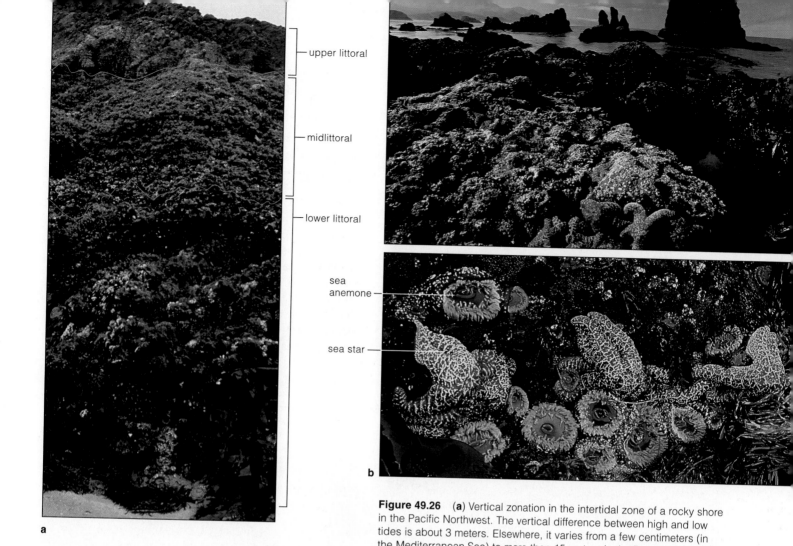

upper littoral

midlittoral

lower littoral

sea anemone

sea star

a

b

Figure 49.26 (**a**) Vertical zonation in the intertidal zone of a rocky shore in the Pacific Northwest. The vertical difference between high and low tides is about 3 meters. Elsewhere, it varies from a few centimeters (in the Mediterranean Sea) to more than 15 meters in the Bay of Fundy, next to Nova Scotia. (**b**) Typical tidepool residents in the Pacific Northwest.

Life Along the Coasts

Along rocky and sandy coastlines, we find ecosystems of the **intertidal zone**, which is not exactly renowned for creature comforts. The resident organisms are battered by waves, fiercely so during storms. They are alternately submerged and exposed by tides. The higher up they are on the shore, the more they might dry out, freeze in winter, or bake in summer, and the less food comes their way. The lower they are, the more they must compete in the limited space. At low tides, birds, rats, and raccoons move in to feed on them. High tides bring the predatory fishes.

Generalizing about life along the coasts is difficult, because waves and tides constantly resculpt a dizzying array of habitats. Vertical zonation is about the only feature that rocky and sandy shores have in common.

Rocky shores often have three zones (Figure 49.26*a*). The highest zone, the upper littoral, is submerged only during the highest tide of the lunar cycle. The upper littoral is sparsely populated. Every day, the middle zone—the midlittoral—is submerged during the highest regular tide, then exposed during the lowest. In the

tidepools characteristic of this zone we find red, brown, and green algae, small invertebrates such as hermit crabs and nudibranchs, and small fishes (Figure 49.26*b*). Diversity is greatest in the lower littoral, which is exposed only during the lowest tide of the lunar cycle. As is true of the other two zones, erosion prevents detritus from accumulating, so grazing food webs tend to predominate here.

Sandy and *muddy* shores are stretches of loose sediments, constantly rearranged by waves and currents. Few large plants grow in these unstable places, so you won't find many grazing food webs. Organic debris imported from offshore or from nearby landforms is the basis of detrital food webs. Vertical zonation is less obvious than along rocky shores. Along the coasts in temperate regions, blue crabs and sea cucumbers live below the low tide mark. Marine worms, crabs, and other invertebrates live between the high and low tide marks. At night, at the high tide mark, beach hoppers and ghost crabs leave their burrows and bound or lurch about the beach, seeking food.

49.16 THE OPEN OCEAN

Beyond the intertidal are two vast provinces of the open ocean. The **benthic province** includes all sediments and rocks of the ocean bottom (Figure 49.27). It starts with the continental shelf and extends down to deep-sea trenches. The **pelagic province** is the entire volume of ocean water. Its neritic zone consists of all the water above the continental shelves. Its oceanic zone is the water filling the ocean basins.

Walk along the ocean and you may be humbled by its vastness (Figure 49.28). You see an immense expanse stretching out to the distant horizon, with no mountains and plains and valleys to break it up visually into something less overwhelming. What you do not see are the submerged mountains and valleys and plains, just as varied as the ones configuring the dry continents.

In the upper surface waters of the ocean, photosynthesis proceeds on a stupendous scale, just as it does on land. Primary productivity varies seasonally, just as it does on land (page 117 and Figure 49.29). Its vast "pastures" of phytoplankton are the start of food webs that include zooplankton, copepods, shrimplike krill, whales, squids, and fishes. Organic remains and wastes from these marine communities sink through the water to become the basis of detrital food webs for most communities in the benthic province.

In 1977, researchers studying the ocean floor near the Galápagos Islands discovered a distinct ecosystem. In the Galápagos Rift, a volcanically active boundary between two of the earth's crustal plates, they found communities thriving near **hydrothermal vents**. At this boundary, near-freezing water seeps into fissures and becomes heated to very high temperatures. As pressure forces the heated water upward, minerals are leached

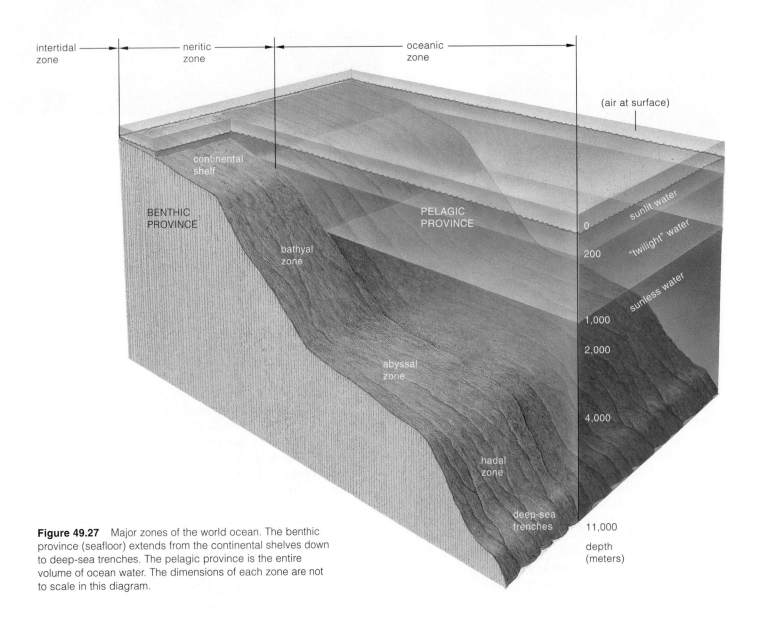

Figure 49.27 Major zones of the world ocean. The benthic province (seafloor) extends from the continental shelves down to deep-sea trenches. The pelagic province is the entire volume of ocean water. The dimensions of each zone are not to scale in this diagram.

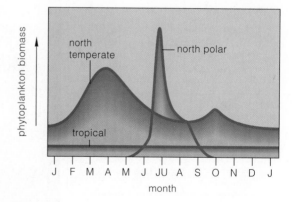

Figure 49.28 (**a**) Looking out from a rocky shore toward the open ocean. (**b**) A whale breaching (leaping out of the water) near a British Columbia coastline. From a hydrothermal vent ecosystem on the deep ocean floor, crustaceans (**c**) and tube worms (**d**).

from rocks before the water spews out through vents in the seafloor. The hydrothermal outpouring is enriched with zinc, iron, and copper sulfides as well as calcium and magnesium sulfates. These settle out, forming rich mineral deposits. Hydrogen sulfide in the deposits serves as an energy source for chemoautotrophic bacteria that are the starting point for hydrothermal vent communities. The bacteria are primary producers in a food web that includes tube worms, crustaceans, clams, and fishes.

Researchers have located other hydrothermal vent ecosystems in the South Pacific, near Easter Island; in the Gulf of California, about 150 miles south of the tip of Baja California, Mexico; and in the Atlantic. In 1990, a team of United States and Russian scientists located another in Lake Baikal, the world's deepest lake. This tectonically formed lake basin seems to be splitting apart (hence the hydrothermal vents) and may mark the beginning of a new world ocean.

Did life originate near such hospitable places on the seafloor? Certainly conditions at the surface of the early earth were about as inhospitable as you might imagine. At the least, the first living cells would have been protected from destructive forms of radiation that bombarded the earth before the oxygen-rich atmosphere formed. Were they slowly carted closer to the surface by the uplifting of the seafloor during episodes of crustal crunchings? These are some of the tantalizing, unanswered questions that some evolutionary detectives are now asking.

Figure 49.29 Seasonal variations in primary production in the ocean corresponding to latitude. The strongest peak in the graph line for north polar and temperate seas corresponds to phytoplankton blooms, brought about by a seasonal increase in daylength. The smaller peak corresponds to an increase in nutrient availability, something like the fall overturn in lakes. In most tropical seas, daylength and nutrient availability do not vary much, and neither does primary production (compare Figure 49.6).

CORAL REEFS AND BANKS

The wave-resistant formations called **coral reefs** are the accumulated remains of countless corals and other organisms. They include fringing reefs, barrier reefs, and atolls, which are shown in Figure 49.30a. Most coral reefs are located in clear, warm waters between latitudes 25° north and south. Long ago, corals began to grow and reproduce in nearshore waters. Their remains were a substrate for more corals to grow upon. Secretions of other organisms helped cement things together. Skeletons and residues accumulated, the reef grew, and tides and currents carved ledges and caverns.

Today, a reef's spine may have as many as 750 species of corals, along with a variety of red algae and other organisms. Figure 49.30b shows examples of warning colors, spines, tentacles, and stealthiness that are clues to the fierce competition for resources among species that are packed together in limited space. Chapters 26 and 27 include photographs of more reef organisms.

Farther north and south, **coral banks** have formed at the edges of continental shelves. Among these are the smooth, vertical banks in cold, deep waters near Japan, California, Norway, England, and New Zealand.

Figure 49.30 (a) The three types of coral reefs. *Fringing reefs* form next to the land's edge in regions of limited rainfall, as on the leeward side of tropical islands. *Barrier reefs* form around islands or parallel with the shore of a continent (page 416). A calm lagoon forms behind them. *Atolls* are ring-shaped coral reefs that enclose or almost enclose a shallow lagoon. (b) A sampling of tropical reef diversity.

b

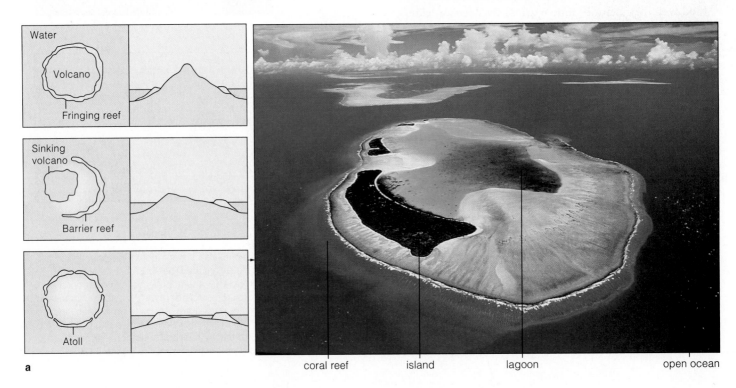

a

coral reef   island   lagoon   open ocean

lionfish

crab

daisy coral

chambered nautilus

sea anemone

pillar coral

moray eel

crown-of-thorns sea star

We return now to the unifying concept of this chapter—that interactions among the atmosphere, ocean, and land profoundly influence the world of life. We reinforce this concept with a story that begins with **upwelling**, an upward movement of deep, nutrient-rich water along the margins of continents.

In the Northern Hemisphere, winds from the north parallel the west coasts of continents, and they tug on the ocean surface. The wind friction causes the surface waters to begin moving. Under the influence of the earth's rotation, the moving water is deflected west, away from a coastline—and cold, deep water moves in vertically to replace it (Figure 49.31). Often the deep water is rich in nutrients.

The great and famous fogbanks of San Francisco are evidence that upwelling of cold water is proceeding along California's coast. The cold water cools the air above it, and contributes to the formation of fog.

In the Southern Hemisphere, the great anchoveta industry of Peru also is evidence of wind-induced upwelling. Along the Peruvian coast, prevailing winds from the south and southeast force surface water away from shore, and cold, deeper water brought to the continental shelf by the Humboldt current moves toward the surface. Here, upwelling pulls up tremendous amounts of nitrate and phosphate. Phytoplankton using the nutrients are the basis of one of the world's richest fisheries; the populations of anchoveta and other fishes are huge.

Every three to seven years, however, warm surface waters of the western equatorial Pacific move eastward. The massive displacement of warm water influences the prevailing winds, which accelerate the eastward flow.

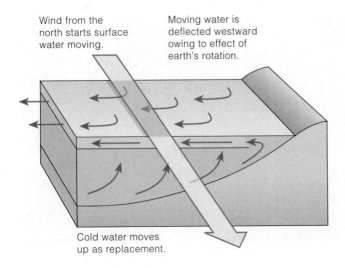

Wind from the north starts surface water moving.

Moving water is deflected westward owing to effect of earth's rotation.

Cold water moves up as replacement.

Focus on the Environment

El Niño and Seesaws in the World's Climates

The winter of 1982–1983 was one for the books. Record amounts of rain caused massive mudslides along the California coast. Month after month, storms caused major floods along the normally arid and semiarid coasts of Ecuador and Peru. In India, the life-sustaining monsoon rains hardly materialized. Droughts devastated natural ecosystems and farms in Australia and the Hawaiian Islands. An ongoing drought in Africa intensified, with consequent and appalling amplification of human starvation and death.

What caused the massive dislocations in the global patterns of rainfall? The answer lies with interactions among sea surface temperature, air circulation patterns, and drought-related conditions on land. Every three to seven years, the interactions produce a climatic event of global proportions. Meteorologists call it the *El Niño Southern Oscillation* (ENSO).

The "Southern Oscillation" part of the name refers to a seesaw in atmospheric pressure in the western equatorial Pacific. This area is the world's largest reservoir of warm water (Figure *a*). More warm, moisture-laden air rises here than anywhere else. Rainfall is heavy, and it releases much of the heat energy that drives the world's air circulation system.

The Southern Oscillation may be triggered by pulses of heat from the earth, possibly at clusters of a thousand or more active volcanoes recently discovered on the ocean floor. Whatever the cause, heat moves upward and warms the surface waters.

Normally, the warm reservoir and the heavy rainfall associated with it move westward. But now they move to the east (Figure *b*). This causes prevailing surface winds in the western equatorial Pacific to pick up speed. The stronger winds have a more pronounced effect on "dragging" the ocean surface waters to the east. Upper ocean currents are affected to the extent that the westward trans-

Figure 49.31 Upwelling as it frequently occurs in the Northern Hemisphere along the western coastlines of continents. Prevailing winds from the north start surface water moving. Owing to the effect of the earth's rotation, the moving water is deflected westward. Cold, deeper water moves up to replace it.

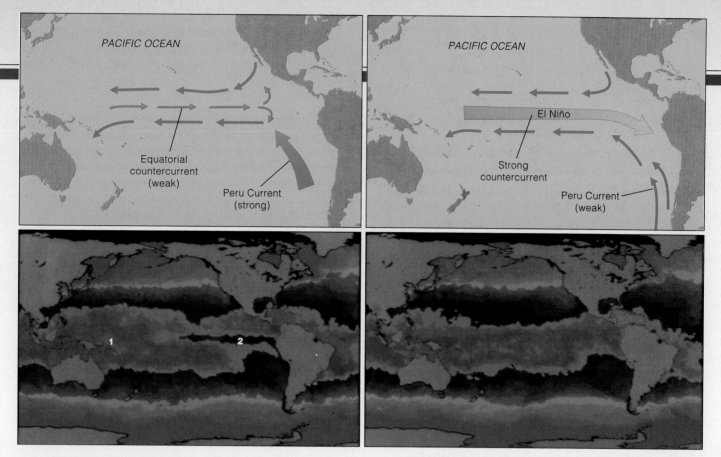

a Diagram and satellite image of the distribution of ocean surface temperatures in normal years. Normally, the warmest waters are found in the western equatorial Pacific (1 in the diagram), and a tongue of relatively cold surface water flows westward along the equator from South America (2).

b Diagram of the distribution of ocean surface temperatures when the Southern Oscillation develops. The trade winds weaken and then reverse, leading to a massive, eastward movement of warm water along the equator. Surface waters of the central and eastern Pacific become warmer. This satellite image shows the distribution of ocean surface temperatures that were associated with the 1982–1983 ENSO episode.

port of water slows down and the eastward transport increases. *More* warm water in the vast reservoir moves east—and so on in a feedback loop between the ocean and the atmosphere.

The reversal in the usual westward flow of air and water displaces the cold, deep Humboldt current—and so stops the upwelling of nutrients along the western coast of South America. Peruvian fishermen refer to the warm, nutrient-

poor current from the east as "El Niño." Usually it reaches their coast around Christmas (hence the name, meaning "the Christ child").

Numerical models are used to study these and other episodes of climatic change a few seasons in advance. More reliable forecasting from this ecosystem modeling should follow when more and better observations are made of the interrelated systems of the ocean, land, and atmosphere.

The flow is enough to influence the movement of water along the coastlines of Central and South America.

Water driven toward a coastline will be forced downward, then flow seaward along the continental shelf. Along the coast of Peru, "downwelling" is prolonged. It displaces the cooler waters of the Humboldt current—and so prevents upwelling. This phenomenon, which local fishermen named **El Niño**, has catastrophic

effects on productivity, on anchoveta-eating birds, and on the anchoveta industry. The *Focus* essay provides a closer look at the El Niño phenomenon.

Shifts in the circulation patterns of the ocean and atmosphere can have repercussions on the functioning of ecosystems that are global in scope.

SUMMARY

1. The biosphere encompasses the earth's waters, lower atmosphere, and uppermost portions of its crust in which organisms live. It contains diverse ecosystems that are influenced by the flow of energy and the movement of materials on a global scale.

2. Earth history, climate, topography, and species interactions shaped the world distribution of species.

3. Climate refers to prevailing weather conditions, including temperature, humidity, wind velocity, cloud cover, and rainfall. It is an outcome of differences in the amount of solar radiation reaching equatorial and polar regions, the earth's daily rotation and its annual path around the sun, the distribution of continents and oceans, and the elevation of land masses.

4. Interacting climatic factors produce prevailing winds and ocean currents, which influence global weather patterns. Weather affects soil composition, sedimentation, and water availability, which influence the growth and distribution of primary producers. Through these interactions, climate influences ecosystems.

5. Land masses are classified as six major biogeographic realms. Each realm has a characteristic array of plant and animal species, and each is more or less isolated by oceans, mountain ranges, or desert barriers.

6. Biomes are categories of ecosystems shaped by regional variations in climate, landforms, and soil composition. Plant species adapted to a certain set of conditions dominate each one. Deserts, dry shrublands and dry woodlands, grasslands and savannas, forests, and tundra are major types.

7. Water provinces cover more than 70 percent of the earth's surface. They include standing freshwater (such as lakes), running freshwater (such as streams), and oceans and seas. All of their aquatic ecosystems have gradients in light penetration, temperature, salinity, and dissolved gases. These vary daily and seasonally, and they influence primary productivity.

8. Estuaries, intertidal zones, rocky and sandy shores, tropical reefs, and scattered ecosystems of the open ocean are major marine ecosystems. Of these, photosynthetic activity is greatest in shallow coastal waters and in regions of upwelling along the margins of continents. Upwelling is an upward movement of deep, cooler water that carries nutrients to the surface.

9. The interrelatedness of ocean surface temperatures, the atmosphere, and the land is evident in studies of the El Niño Southern Oscillation. This recurring phenomenon is accompanied by abnormal drought conditions in many parts of the world.

Review Questions

1. Define climate, then list the major interacting factors that influence climate. *865*

2. List a few ways in which air currents, ocean currents, or both probably influence the biome in which you live. *866–869*

3. How does the composition of regional soils affect ecosystem distribution? *872*

4. What points favor the hypothesis that life may have originated near hydrothermal vents on the ocean floor? *890*

5. Spend some time outdoors observing the land or any freshwater or marine ecosystem near you. Write a short essay on some of its features.

6. This sandy shore along the Pacific Ocean is submerged at high tide. What kind of organisms would you expect to find living here? What kind of organisms compete for living space on the rocks? *883*

7. Observe or do some research on a lake near your home or a place where you vacation. Using the following table as a guide, would you judge it to be oligotrophic or eutrophic?

Oligotrophic	Eutrophic
Deep, steeply banked	Shallow with broad littoral
Large deep-water volume relative to surface-water volume	Small deep-water volume relative to surface-water volume
Highly transparent	Limited transparency
Water blue or green	Water green to yellow or brownish-green
Low nutrient content	High nutrient content
Oxygen abundant through all levels at all times	Oxygen depleted in deep water in summer
Not much phytoplankton; green algae and diatoms dominant	Abundant, massed phytoplankton; cyanobacteria dominant
Abundant aerobic decomposers in profundal	Anaerobic decomposers favored
Low biomass in profundal	High biomass in profundal

9. Match the terms with the suitable description.
_____ boreal forest
_____ permafrost
_____ chaparral
_____ desert
_____ deciduous broadleaf forest
_____ tropical rain forest

a. high productivity, poor mineral cycling
b. swamp forest
c. common at moist, temperate latitudes with mild winters
d. evaporation greatly exceeds infrequent rainfall
e. a type of dry shrubland
f. feature of arctic tundra

10. Match the terms with the suitable description.
_____ plankton
_____ upwelling
_____ eutrophication
_____ estuary
_____ benthic province

a. deep, nutrient-rich water moves up along coast
b. sediments, rocky formations of ocean bottom
c. seawater, freshwater slowly mix here
d. nutrient enrichment of body of water; reduced transparency, phytoplankton blooms
e. community of mostly microscopic floating or weakly swimming organisms

Self-Quiz (Answers in Appendix IV)

1. Solar radiation drives the distribution of weather systems and so influences the distribution of _____ .
   a. every single ecosystem
   b. ecosystems on land only
   c. all ecosystems except those at hydrothermal vents

2. The _____ is like a shield against harmful ultraviolet wavelengths from the sun.
   a. upper atmosphere   c. ozone layer
   b. lower atmosphere   d. greenhouse effect

3. Regional variations in patterns of rainfall and temperature depend on _____ .
   a. global air circulation   c. topography
   b. ocean currents   d. all of the above

4. A rain shadow is a reduction in rainfall on the _____ of a mountain range.
   a. windward side   c. highest elevation
   b. leeward side   d. lowest elevation

5. Biogeographic realms are _____ .
   a. land and water provinces   c. divided into biomes
   b. six major land provinces   d. b and c

6. Biome distribution corresponds roughly with regional variations in _____ .
   a. climate   c. topography
   b. soils   d. all of the above

7. _____ are highly adapted to episodes of fire.
   a. Dry shrublands   c. Pine barrens
   b. Grasslands   d. All of the above

8. During _____ , strong vertical movements transport dissolved oxygen in the surface layer of a body of water down to decomposers in its depths, and transport nutrients from the depths up to photosynthesizers.
   a. upwelling   c. fall overturns
   b. spring overturns d. b and d

Selected Key Terms

alpine tundra 879	lithosphere 865
arctic tundra 879	monsoon 869
atmosphere 865	monsoon forest 877
benthic province 884	monsoon grassland 874
biogeographic realm 870	montane coniferous forest 878
biome 870	ozone layer 866
biosphere 865	pelagic province 884
boreal forest 878	permafrost 879
climate 865	pine barren 878
coral bank 886	plankton 880
coral reef 886	rain shadow 869
desert 873	savanna 874
desertification 873	shortgrass prairie 874
dry shrubland 874	soil 872
dry woodland 874	spring overturn 880
El Niño 889	stream 882
estuary 882	tallgrass prairie 874
eutrophication 881	temperate deciduous forest 877
fall overturn 881	temperate rain forest 878
hydrosphere 865	temperature zone 866
hydrothermal vent 884	tropical deciduous forest 877
intertidal zone 883	tropical rain forest 876
lake 880	upwelling 888

Readings

Garrison, T. 1993. *Oceanography: An Invitation to Marine Science.* Belmont, California: Wadsworth.

Gibbons, B. September 1984. "Do We Treat Our Soil Like Dirt?" *National Geographic* 166(3):350–388.

Smith, R. 1989. *Ecology and Field Biology.* Fourth edition. New York: Harper & Row.

50 HUMAN IMPACT ON THE BIOSPHERE

Tropical Forests—Disappearing Biomes?

On the first night of an expedition into a tropical forest, a young graduate student carefully wrapped himself inside cotton mosquito netting, then fell asleep. He woke up before dawn with a sudden, gut-wrenching awareness that things were crawling all over him. During the hot, humid night, a platoon of large, voracious insects had devoured most of the drenched net, a feat that truly focused the student's attention on the exotic wonders awaiting him.

Figure 50.1 is a sweeping view of one tropical forest; the cover photograph of this book provides another. Such forests contain the greatest variety and numbers

of hungry insects and the world's biggest ones. They are home to the greatest variety of birds and to plants with the largest flowers. Slinking or bounding through their canopy and understory are splendidly varied monkeys, tapirs, and jaguars in South America and apes, okapi, and leopards in Africa. Specimens of trees from these forests may have made their way into your home or your dentist's waiting room—*Ficus benjamina*, rubber plants, and tree ferns, to name a few. In the forests, vines twist around tree trunks and grow toward sunlight. Orchids, mosses, lichens, and other plants grow on tree branches, absorbing minerals brought to

Figure 50.1   El Yunque rain forest, Puerto Rico.

them in rainwater. Entire communities of insects, spiders, and amphibians live, breed, and die in the small pools of water that collect in furled leaves.

Developing countries in Latin America, Africa, and Southeast Asia have growing populations and limited food, fuel, and lumber. Thanks to the invention of chainsaws, tractors, and logging trucks, they have been making concerted assaults on their forest ecosystems over the past four decades. Most of these forests, which evolved millions of years ago, may disappear within your lifetime.

Why does it matter? For purely ethical reasons, many individuals condemn the obliteration of a major chunk of the biosphere. For practical reasons, the destruction will affect your own life. Our food supply is based on a very limited number of species of crop plants and livestock. By developing new or hybrid crop plants, we can make the food supply less vulnerable. With genetic engineering and tissue culturing, we can tap tropical rain forests for genetic "resources." We can tap the forests to develop new antibiotics and vaccines.

Many tropical plants already give us alkaloids for treating cardiovascular disorders and cancer. Aspirin, the most widely used drug in the world, is based on a chemical "blueprint" of a compound extracted from tropical willow leaves. Coffee, bananas, cinnamon, cocoa, sweeteners, Brazil nuts, and many other spices and foods we take for granted originated in the tropics. So did many ornamental plants. So did the latex, gums, resins, dyes, waxes, and many oils used in ice cream, toothpaste, shampoo, condoms, cosmetics, perfumes, compact discs, tires, shoes, and other diverse products.

If aspirin or condoms don't catch your attention, think about this: wholesale destruction of tropical forests is changing the atmosphere. Burning trees happens to release carbon dioxide, carbon monoxide, hydrocarbons, nitric oxide, and nitrogen dioxide. Increasing atmospheric concentrations of these gases are contributing to acid rain, smog, and maybe global warming.

As an individual, you might choose to cherish, brood about, or ignore any aspect of the world of life. Whatever the choice, the bottom line is that you and all other organisms are in this together. Our lives interconnect, to degrees that we are only now starting to comprehend.

KEY CONCEPTS

1. The human population has been growing exponentially since the mid-eighteenth century. Today we have the population size, technology, and cultural inclination to use energy and modify the environment at astonishing rates.

2. Pollutants are substances with which ecosystems have had no prior evolutionary experience, in terms of kinds or amounts, so adaptive mechanisms are not in place that can deal with them. In a more restricted sense, pollutants are substances that have adverse effects on human health, activities, or survival.

3. Many pollutants, including those involved in smog formation, exert harmful effects on a local scale. The effects of other pollutants, including chlorofluorocarbons that attack the ozone layer, are global in scale.

4. Conversion of marginal lands for agriculture, rampant deforestation, and other practices required to sustain the growing human population are resulting in a decline in the quality and quantity of one of the most basic of all resources—fresh water. They also are leading to loss of soil fertility and desertification.

5. The world of life ultimately depends on energy from the sun. That energy drives the complex interactions among the atmosphere, ocean, and land. We may be disrupting those interactions, with serious consequences in the near future.

6. We as a species must come to terms with the principles of energy flow and resource utilization that govern all systems of life on earth.

50.1 HUMAN POPULATION GROWTH AND THE ENVIRONMENT

Of all the concepts introduced in the preceding chapter, the one that should be foremost in your mind is this: Interactions among the atmosphere, oceans, and land are the engines of the biosphere. Driven by energy from the sun, they create global circulation patterns and temperatures upon which life ultimately depends. With this chapter, we turn to a related concept of equal importance. Simply put, human population growth has been straining the global engines even while we do not fully comprehend how the engines work.

To gain perspective on what is happening, think of something we take for granted—the air around us. The composition of the present atmosphere is the outcome of geologic and metabolic events, including photosynthesis, that began billions of years ago. The first humans evolved about 2-1/2 million years ago. Like us, they breathed oxygen from an atmosphere of ancient origins. Like us, they were protected from ultraviolet radiation by an ozone shield in the stratosphere. Their population sizes were not much to speak of, and their interactions with the biosphere were trivial. About 10,000 years ago, however, agriculture began in earnest, and it laid the foundation for increases in population growth. With agriculture, and with the medical and industrial revolutions that followed, human population growth became exponential in a mere blip of evolutionary time (Figures 46.7 and 50.2).

Today, the burgeoning human population may be demanding more than the biosphere can sustain. As we take energy and resources from it, we put back monumental amounts of wastes. In the process, we threaten the stability of ecosystems on land, taint the hydrosphere, and change the composition of the atmosphere. Our carbon dioxide wastes alone may be modifying the global climate (page 854).

In a few developed countries of Europe, population growth has more or less stabilized. Rates of increase in the resource use per individual have slowed somewhat, but their resource utilization levels are already high. Population growth and demands for resources are increasing rapidly in Central America, South America, Asia, Africa, and elsewhere—even though millions in these developing countries are already starving to death and hundreds of millions more suffer from malnutrition and inadequate health care.

Many of the problems sketched out in this chapter are not going to disappear tomorrow. It will take decades, even centuries, to reverse some trends already in motion, and not everyone is ready to make the effort. A few enlightened individuals in Michigan or Alberta or New South Wales can make good attempts at resource conservation and at minimizing pollution—but scattered attempts will not be enough. Individuals of all nations will unite to reverse global trends only when they perceive that the dangers of not doing so outweigh the personal benefits of ignoring them.

Does this seem pessimistic? Think of the exhaust fumes released into the air each time you drive a car. Think of oil refineries, food-processing plants, and paper mills that supply you with goods—and also release chemical wastes into the nation's waterways. Think of Mexico and other developing countries that produce cheap food by using an unskilled labor force and toxic pesticides. Unregulated pesticide applications poison people who work the land, the land itself, and perhaps those who buy exported produce. Who changes behavior first? We have no answer to the question. We can suggest, however, that a strained biosphere can rapidly impose an answer upon us.

Figure 50.2 Tracking human population growth in one small part of the world only. *Red* denotes areas of dense settlement in and around the San Francisco Bay area and Sacramento since 1900, as compiled from historical data and satellite imaging.

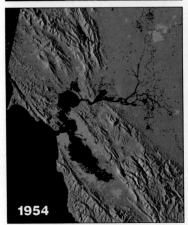

1954

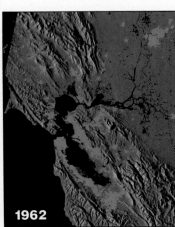

1962

1974

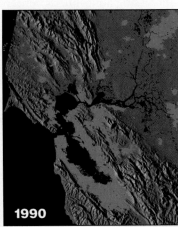

1990

50.2 LOCAL EFFECTS OF AIR POLLUTION

Pollutants are substances with which ecosystems have had no prior evolutionary experience, in terms of kinds or amounts, so adaptive mechanisms are not in place that can deal with them. From the human perspective, pollutants are substances that have adverse effects on our health, activities, or survival.

Consider the main classes of air pollutants, listed in Table 50.1. They include carbon dioxide, nitrogen oxides, sulfur oxides, and chlorofluorocarbons (CFCs). Also among them are photochemical oxidants formed when the sun's rays interact with certain chemicals. The United States alone releases 700,000 metric tons of such pollutants into the atmosphere, every day.

Air pollutants may be dispersed through the atmosphere or concentrated at their source. What happens in a given interval depends on local climate and topography. For example, during a **thermal inversion**, weather conditions trap a layer of cool, dense air under a layer of warm air (Figure 50.3). Pollutants in the trapped air cannot be dispersed by winds or move higher in the atmosphere, and they may accumulate to dangerous levels. By intensifying a phenomenon called smog, thermal inversions have contributed to some of the worst local air pollution disasters.

Two types of smog—gray or brown air—form in major cities. Where winters are cold and wet, **industrial smog** forms as a gray haze over industrialized cities that burn coal and other fossil fuels for heating, manufacturing, and generating electric power. The burning releases airborne pollutants, including dust, smoke, soot, ashes, asbestos, oil, bits of lead and other heavy metals, and sulfur oxides. Pollutants not dispersed by winds and rain may reach lethal concentrations. Industrial smog was the cause of London's 1952 air pollution

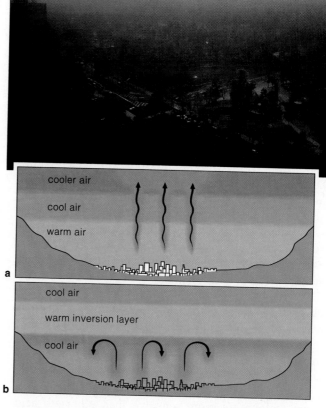

Figure 50.3 (**a**) A normal pattern of air circulation in smog-forming regions. (**b**) Airborne pollutants trapped under a thermal inversion layer. The photograph shows Mexico City on an otherwise bright, sunny morning, under its self-generated blanket of smog. Topography, staggering numbers of people and motor vehicles, and industrialization combine to make its air among the world's dirtiest. Breathing here is like smoking two packs of cigarettes a day.

disaster, in which 4,000 people died. New York, Pittsburgh, and Chicago, like London, were gray-air cities until coal burning was restricted. Today, most industrial smog forms in cities of China, India, and other developing countries, as well as in Hungary, Poland, and other coal-dependent countries of eastern Europe.

In warm climates, **photochemical smog** develops as a brown, smelly haze over large cities. Where the surrounding land forms a natural basin, as it does around Los Angeles and Mexico City, photochemical smog can be dangerous. The main culprit is nitric oxide from cars and other vehicles with internal combustion engines. When nitric oxide reacts with oxygen in the air, nitrogen dioxide forms. When exposed to sunlight, this compound reacts with hydrocarbons to form photochemical oxidants. (Most of the hydrocarbons come from spilled or partly burned gasoline.) The main oxidants in smog are ozone and **PANs** (short for peroxyacyl nitrates). PANs are similar to tear gas. Even traces can sting the eyes, irritate the lungs, and damage crops.

Table 50.1	Major Classes of Air Pollutants
Carbon oxides	Carbon monoxide (CO), carbon dioxide (CO_2)
Sulfur oxides	Sulfur dioxide (SO_2), sulfur trioxide (SO_3)
Nitrogen oxides	Nitric oxide (NO), nitrogen dioxide (NO_2), nitrous oxide (N_2O)
Volatile organic compounds	Methane (CH_4), benzene (C_6H_6), chlorofluorocarbons (CFCs)
Photochemical oxidants	Ozone (O_3), peroxyacyl nitrates (PANs), hydrogen peroxide (H_2O_2)
Suspended particles	Solid particles (dust, soot, asbestos, lead, etc.), liquid droplets (sulfuric acid, oils, dioxins, pesticides)

50.3 GLOBAL EFFECTS OF AIR POLLUTION

Acid Deposition

Oxides of sulfur and nitrogen are among the most harmful air pollutants. Coal-burning power plants, factories, and metal smelters emit most of the sulfur dioxides. Motor vehicles, power plants that burn fossil fuels, and nitrogen fertilizers produce nitrogen oxides. Depending on climatic conditions, particles of oxides may stay airborne for a while, then fall to earth as **dry acid deposition**. Most sulfur and nitrogen dioxides dissolve in atmospheric water to form a weak solution of sulfuric and nitric acids. Winds may distribute them over great distances before they fall to earth in rain and snow. This is wet acid deposition, or **acid rain**.

Normal rainwater has a pH of about 5. Acid rain can be 10 to 100 times more acidic—sometimes as much as lemon juice! The acids chemically attack marble buildings, metals, mortar, rubber, plastic, even nylon stockings. They also can disrupt the physiology of organisms and the chemistry of ecosystems.

Depending on the soil type and vegetation cover, some regions are more sensitive than others to acid deposition (Figure 50.4). Highly alkaline soil neutralizes acids before they enter the streams and lakes of watersheds. Water with a high carbonate content also neutralizes acids. However, in watersheds throughout much of northern Europe and southeastern Canada, and in scattered regions of the United States, thin soils overlie solid granite and do not buffer the acids much.

Rain in much of eastern North America is thirty to forty times more acidic than it was several decades ago. Crop production is suffering. Also, fish populations have vanished from more than 200 lakes in the Adirondack Mountains of New York. Some biologists predict that fish will disappear from 48,000 lakes in Ontario, Canada, within the next two decades.

Besides this, the forests in northern Europe are under attack by airborne acids and other pollutants from industrial regions of England and Germany. High in the Appalachian Mountains, such pollutants are damaging trees as well as mycorrhizae that support new growth (pages 387 and 859). Airborne acids also are becoming problems in heavily industrialized parts of Asia, Latin America, and Africa.

Researchers confirmed some time ago that power plants, factories, and vehicles are the main sources of acid pollutants, and that acid deposition does adversely affect the environment. Unfortunately, some of the responses to local air pollution problems that were recognized decades ago actually contributed to the current problems. For example, tall smokestacks were added to power plants and smelters. The idea was to dump the

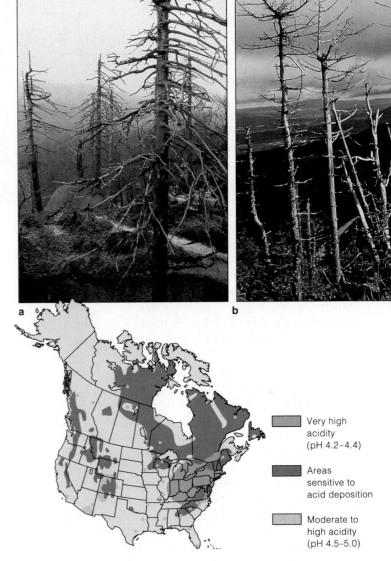

Very high acidity (pH 4.2–4.4)

Areas sensitive to acid deposition

Moderate to high acidity (pH 4.5–5.0)

Figure 50.4 Diagram of average acidities of precipitation and soil sensitivities to acid deposition, for regions of North America, in 1984. Prolonged exposure to multiple air pollutants is contributing to the rapid destruction of trees in Germany (**a**), New York's Whiteface Mountains (**b**), and elsewhere. Figure 48.11 shows merely a few of the destructive interactions. Weakened trees are more vulnerable to drought, disease, and insect attack.

smoke high in the atmosphere so winds could distribute it elsewhere—which winds readily did. The winds helped transform local air pollution into regional acid deposition. Today the world's tallest smokestack, in Sudbury, Ontario, by itself accounts for about 1 percent (by weight) of the annual worldwide emissions of sulfur dioxide.

Canada cannot be singled out in this issue. Canada presently receives more acid deposition from industrialized regions of the midwestern United States than it sends across its southern border. Most acidic pollutants in Finland, Norway, Sweden, the Netherlands, Austria, and Switzerland arise in industrialized parts of western and eastern Europe. *Prevailing winds do not stop at national boundaries; the problem is global.*

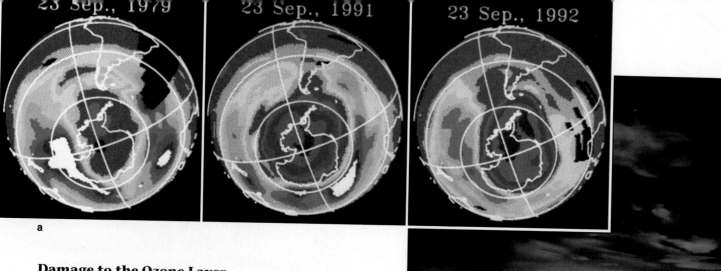

23 Sep., 1979 **23 Sep., 1991** **23 Sep., 1992**

a

b

Figure 50.5 (**a**) Expansion of the seasonal ozone hole over Antarctica, as recorded by special high-altitude planes. The lowest ozone values (the "hole") are indicated by *magenta* and *purple*. Droplets of sulfuric acid, released during the 1991 volcanic eruption of Mount Pinatubo, contributed to the thinning. (**b**) The photograph shows the ice clouds over Antarctica that have a role in the ozone thinning each spring.

Damage to the Ozone Layer

The ozone layer in the lower stratosphere is almost twice as high above sea level as the summit of Mount Everest, the highest place on earth. Each year, from September through mid-October, it thins down by as much as *half* above Antarctica. The pronounced seasonal thinning is called an **ozone hole**. In 1992, it spanned an area three times the size of the continental United States (Figure 50.5*a*). In 1993, the ozone level was the lowest ever recorded. Within sixty years, it may decrease by as much as 30 percent above heavily populated regions of North America, Europe, and Asia.

Why is the ozone reduction so alarming? It lets more ultraviolet radiation reach the earth's surface. A dramatic rise in skin cancers is one consequence (page 635). Cataracts, an eye disorder, are becoming more common. Besides this, ultraviolet radiation weakens the immune system and the ability to fight many infections. It harms photosynthesizers and so may alter the atmosphere. Collectively, phytoplankton alone take up huge amounts of carbon dioxide and release huge amounts of oxygen on an annual basis (page 855).

More than any other factor, **chlorofluorocarbons** (CFCs) are bringing about the ozone reduction. CFCs are odorless, invisible, and otherwise harmless compounds of chlorine, fluorine, and carbon. They have been widely used as propellants in aerosol spray cans, coolants in refrigerators and air conditioners, and in industrial solvents and plastic foams. CFCs enter the atmosphere slowly and resist breakdown. When a free CFC molecule absorbs ultraviolet light, it gives up a chlorine atom. Chlorine can react with ozone, forming an oxygen molecule and a chlorine monoxide molecule. When chlorine monoxide reacts with a free oxygen atom, another chlorine atom is released that can attack another ozone molecule. Each chlorine atom released in the reactions can convert as many as 10,000 molecules of ozone to oxygen!

Chlorine monoxide levels above Antarctica are 100 to 500 times higher than at mid-latitudes. Why? At high altitudes, ice clouds form there during the winter. Winds that rotate around the South Pole for most of the winter months keep the ice clouds isolated from other latitudes (Figure 50.5*b*). The same thing happens, on a lesser scale, in the Arctic. Ice provides a surface that promotes the breakdown of chlorine compounds. So chlorine is free to destroy ozone when the Antarctic air warms somewhat in the Antarctic spring. Hence the ozone hole.

By some estimates, nearly all of the CFCs released between 1955 and 1990 are still making their way up to the stratosphere. Even if we stop using CFCs today, it may be a century before ozone levels return to pre-1985 levels. Since 1978, CFCs in aerosol spray cans have been banned in the United States, Canada, and most Scandinavian countries but not in western Europe. Nonaerosol uses of CFCs continue throughout the world. In 1992, the bad news about the ozone thinning prodded more than half of the world's nations to agree to phase out CFC production except for a few essential uses. Such agreements are steps in the right direction, although some call them too little, too late. CFCs already in the air will be there for over a century, before natural processes neutralize them. You, your children, and your grandchildren will be living with their destructive effects. Are you part of the problem? Think of how you might take an active role in finding solutions instead.

50.4 WATER SCARCITY AND WATER POLLUTION

The world has a tremendous supply of water, but most is too salty for human consumption or for agriculture. Imagine that water filling a bathtub. The fresh, renewable portion would barely fill a teaspoon.

Consequences of Large-Scale Irrigation

As described on page 814, an expanding food-producing capability has helped fan the human population's exponential growth. How long can expansion continue? Today, we produce about one-third of our food supply on irrigated land (Figure 50.6). We must pipe water into agricultural fields from groundwater or from lakes, reservoirs, and other sources of surface waters. But irrigation often changes the land's suitability for agriculture. Irrigation water typically is loaded with mineral salts. In regions with poorly draining soils, evaporation may cause **salination**, or salt buildup, in the soil. Salination can stunt growth, decrease yields, and in time kill crop plants.

Besides this, irrigated lands that drain poorly become waterlogged. Water accumulating underground can gradually raise the **water table**, the upper limit at which the ground is fully saturated with water. When the water table is close to the ground's surface, soil becomes saturated with saline water that can damage plant roots. Salinity and waterlogging can be corrected with proper management of the water-soil system. The economic cost to do this is high.

We tap groundwater for many purposes, but irrigation is paramount. Consider American farmers who withdraw water from the vast Ogallala aquifer (Figure 50.7). On average, their annual overdraft—the amount that nature does not replenish—is nearly equal to the annual flow of the Colorado River! Today, water tables that were low to begin with are dropping rapidly, and stream and underground spring flows are drying up. Where will the water come from when the aquifer is depleted?

Why not remove salt from seawater? The supply of seawater is essentially unlimited, and desalination technologies already exist. However, it takes fuel energy to distill seawater or force it through membranes (a method called reverse osmosis). Desalination may be feasible in Saudi Arabia and a few other countries with limited population sizes, large energy reserves, and cash to spare. In some situations it may be the only alternative to running out of water, as Santa Barbara

Figure 50.6 Irrigation-dependent crops growing in the Sahara Desert, in Algeria.

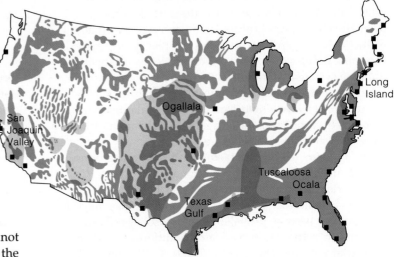

Figure 50.7 Underground aquifers (*blue* areas) that contain 95 percent of all freshwater in the United States. *Gold* shading indicates regions where aquifers are being depleted, mainly for agriculture. *Black* boxes indicate aquifers being contaminated by saltwater intrusion.

and some other California cities nearly did during a severe, prolonged drought. Even so, such efforts cannot solve the core problem. Desalination may never be cost-effective for large-scale agriculture, which accounts for almost two-thirds of the human population's annual use of freshwater.

Maintaining Water Quality

Water pollution compounds the problem of water scarcity. Human sewage, animal wastes, and toxic chemicals make water unfit to drink, even to swim in. Pollutants encourage contamination by pathogens. Agricultural runoff pollutes water with sediments, pesticides, and plant nutrients. Power plants and factories pollute water with chemicals, radioactive materials, and excess heat (thermal pollution).

Pollutants collect in lakes, rivers, and bays before reaching the oceans. Many cities throughout the world dump untreated sewage into coastal waters. Cities along rivers and harbors maintain shipping channels by dredging the polluted muck and barging it out to sea. They also barge out sewage sludge: coarse, settled solids that contain bacteria, viruses, and toxic metals. Sometimes black sludge washes ashore after storms.

In the United States, about 15,000 facilities treat liquid wastes from about 70 percent of the population and 87,000 industries. The remaining wastes are mostly from suburban and rural populations. They are treated in lagoons or septic tanks, or they are directly discharged—untreated—into waterways.

There are three levels of **wastewater treatment**. In *primary* treatment, screens and settling tanks remove sludge, which is dried, burned, dumped in landfills, or treated further. Chlorine often is used to kill pathogens, but it does not kill them all. Also, it can react with some industrial chemicals. Some of the resulting chlorinated organic compounds are carcinogens.

In *secondary* treatment, microbial populations are used to break down organic matter after primary treatment but before chlorination. Wastewater is sprayed and trickled through microbe-containing gravel beds, or it is aerated in tanks and seeded with microbes. Toxic solutes sometimes poison the microbial helpers. Then, the facilities are shut down until the microbial populations become reestablished.

Primary and secondary treatment remove most suspended solids and oxygen-demanding wastes. They do not remove all of the nitrogen, phosphorus, and toxic substances, including heavy metals, pesticides, and industrial chemicals. Usually the water is chlorinated before being released into the waterways.

Tertiary treatment adequately reduces pollution levels but is largely experimental and expensive. It is applied to only 5 percent of the nation's wastewater.

In short, most wastewater is not being treated adequately. A pattern is repeated thousands of times along our waterways. Water for drinking is removed upstream from a city, and wastes from industry and sewage treatment are discharged downstream. It takes no great leap of the imagination to see that pollution intensifies as rivers flow toward the oceans. In

Figure 50.8 An experimental wastewater treatment facility in Providence, Rhode Island, that uses solar heating and artificial lagoons to purify water.

Louisiana, waters drained from the central states flow toward the Gulf of Mexico. Its pollution levels are high enough to threaten public health. Water destined for drinking does get treated to remove pathogens, but this does not remove toxic wastes dumped by numerous factories upstream. *You may find it illuminating to investigate where your own city's supply of water comes from and where it has been.*

This rather bleak picture might be numbing to most of us, but not to John Todd. This biologist has constructed experimental wastewater treatment facilities in greenhouses, using sunlight and artificial lagoons (Figure 50.8). Treatment begins when sewage flows into rows of large water tanks in which water hyacinths, cattails, and other aquatic plants are growing. Decomposers in the tank break down wastes, which contain nutrients that promote growth of the plants. Heat, derived from incoming sunlight, speeds up the decomposition. From these tanks, water flows through an artificial marsh of sand, gravel, and bulrushes that filter out algae and organic wastes. Then it flows into aquarium tanks where zooplankton and snails consume microorganisms suspended in the water—and where zooplankton become food for crayfishes, tilapia, and other fishes that can be sold as bait. After ten days, the now-clear water flows into a second artificial marsh for final filtering and cleansing. When working properly, the solar-aquatic treatment system produces water fit to drink.

Such natural alternatives cannot solve the problems of large urban areas. But they are an attractive alternative for small towns and rural areas.

50.5 CHANGES IN THE LAND

What to Do With Solid Wastes?

In developing countries, many resources are scarce, so people discard few materials. In the United States, as in some other developed countries, a throwaway mentality prevails. People use something once, discard it, then buy another. Each year, millions of metric tons of solid wastes are dumped, burned, and buried. Paper products make up half the total volume, which also includes 50 billion nonreturnable cans and bottles.

It takes more than 500,000 trees to supply a Sunday newspaper to Americans. Every week. If everyone recycled even one out of ten newspapers, 25 million trees a year could be left standing. Also, recycling paper reduces by 95 percent the air pollutants released during paper manufacture. And recycling takes 30 to 50 percent less energy, compared to making new paper.

A throwaway mentality poses a unique problem in the world of life—what to do with the solid wastes. Unlike natural ecosystems, cities do not cycle materials. They bury them in landfills. Or they burn them in incinerators, which produce toxic air pollutants and ashes that must be disposed of safely. Besides this, all landfills eventually "leak" and pose a threat to groundwater supplies.

Moving away from a throwaway mentality to one based on recycling and reuse is feasible. It is affordable, and we have the technology to do so. Consumers can influence manufacturers by refusing to buy goods that are lavishly wrapped and boxed, packaged in indestructible containers, and designed for one-time use. Individuals can contact their local post office about reducing the flow of junk mail. Such unsolicited mail wastes an astounding amount of paper, time, and energy, and means higher postage rates for everyone.

Individuals also can ask local governments for well-designed, large-scale resource recovery centers. Such a plant operates in Saugus, Massachusetts.

Conversion of Marginal Lands for Agriculture

Agriculture is expanding onto marginally productive land. Nearly 21 percent of the earth's land surfaces are already used for cropland or for grazing. Another 28 percent is said to be potentially suitable for these practices, but productivity would be so low that conversion may not be worth the cost (Figure 50.9).

Asia and some other heavily populated regions have severe food shortages, yet more than 80 percent of their productive land is already cultivated. Valiant efforts have been made to improve crop production on existing land. Under the banner of the **green revolu-**

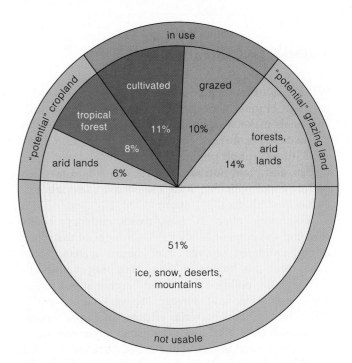

Figure 50.9 Classification of the earth's land. Theoretically, the world's cropland could be more than doubled by clearing tropical forests and irrigating arid lands. But converting this marginal land to cropland would destroy valuable forest resources, cause serious environmental problems, and possibly cost more than it is worth.

tion, research has been directed toward (1) improving crop plants for higher yields and (2) exporting modern agricultural practices and equipment to developing countries. Many developing countries rely on *subsistence* agriculture, which runs on energy inputs from sunlight and human labor. They also rely on *animal-assisted* agriculture, with energy inputs from oxen and other draft animals.

By contrast, *mechanized* agriculture requires massive inputs of fertilizers, pesticides, and ample irrigation to sustain high-yield crops. It requires fossil fuel energy to drive farm machines. Crop yields are four times as high. But the modern practices use up a hundred times more energy and mineral resources.

Many farmers in developing countries cannot afford to take advantage of the new, high-yield crop strains. Of necessity, costs of fertilizers and machinery are reflected in market food prices—which are too high for much of the country's own population. Farmers who make the investment end up depending on industrialized producers of fertilizers and machinery that often must be imported. Such importations can add to the foreign debt of developing countries.

Pressures for increased food production are greatest in parts of Central and South America, Asia, the Middle East, and Africa where human populations are rapidly expanding into marginal lands. The repercussions extend beyond national boundaries, as examples in the following sections will indicate.

50.6 DESERTIFICATION

Desertification refers to the conversion of large tracts of grasslands, rain-fed cropland, or irrigated cropland to a more desertlike state, with a 10 percent or greater drop in agricultural productivity. In the past fifty years, 9 million square kilometers worldwide have become desertified. People are converting at least 200,000 square kilometers annually. Prolonged drought can accelerate the process, as it did in the American Great Plains many years ago (Figure 50.10). Today, overgrazing on marginal lands is the main cause of large-scale desertification.

In Africa, for example, there are too many cattle in the wrong places. Cattle require more water than the wild herbivores that are native to the region. This means the cattle have to move back and forth between grazing areas and watering holes. As they do, they trample grasses and compact the soil surface (Figure 50.11). By contrast, gazelles and other native herbivores get most (if not all) of their water from plants. They also are better water conservers; they lose little in feces, compared to cattle.

In 1978 a biologist, David Holpcraft, formed a ranch composed of antelopes, zebras, giraffes, ostriches, and other native herbivores. He raised cattle as "control groups" in order to compare costs and meat yields on the same land. His initial results exceeded expectations. Native herds increased steadily and yielded meat. Range conditions improved rather than deteriorating. Vexing problems remained. African tribes have their own idea of what constitutes "good" meat, and some tribes view cattle as the symbols of wealth in their society.

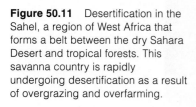

Figure 50.10 Dust storm approaching Prowers County, Colorado, in 1934. The Great Plains of the American Midwest are dry, windy, and subject to severe recurring droughts. Their conversion to agriculture started in the 1870s. Overgrazing left the ground bare through vast tracts of the natural grassland. In May 1934, a cloud of topsoil that blew off the land blanketed the entire eastern portion of the United States, giving the Great Plains a new name—the Dust Bowl. About 3.6 million hectares (9 million acres) of cropland were destroyed. Today, without massive irrigation and intensive conservation farming, desertlike conditions could prevail in this region.

Figure 50.11 Desertification in the Sahel, a region of West Africa that forms a belt between the dry Sahara Desert and tropical forests. This savanna country is rapidly undergoing desertification as a result of overgrazing and overfarming.

50.7 DEFORESTATION

The world's great forests profoundly influence the biosphere. Like giant sponges, forested watersheds absorb, hold, and gradually release water. By intervening in the downstream flow of water, forests help control soil erosion, flooding, and sediment buildup in rivers, lakes, and reservoirs.

Deforestation is the removal of all trees from large tracts of land for logging, agricultural, or grazing operations. The loss of vegetation cover exposes the soil, and this promotes leaching of nutrients and erosion, especially on steep slopes. Figures 50.12 and 50.13 provide close-up and panoramic views of deforestation in South America's Amazon basin. The photographs in Figure 25.11 give examples of the outcome of deforestation in North America.

In the tropics, clearing forests for agriculture leads to a long-term loss in productivity. The irony is that tropical forests are one of the worst places to grow crops and raise pasture animals. The high temperatures and heavy, frequent rainfall favor decomposition. In intact forests, organic remains and wastes decompose too fast for litter to build up. The forest trees and other plants absorb and assimilate nutrients as they become available.

Shifting cultivation (once called slash-and-burn agriculture) disrupts the forest ecosystem. People cut and burn trees, then till the ashes into the soil. The nutrient-rich ashes sustain crops for one to several seasons. Then, as a result of leaching, the cleared plots become infertile and are abandoned. Shifting cultivation on small, widely scattered plots may not damage forest ecosystems much, but fertility plummets when large areas are cleared and when plots are cleared again at shorter intervals.

Shifting rates of evaporation, transpiration, and runoff may even disrupt regional patterns of rainfall. Between 50 and 80 percent of the water vapor above tropical forests alone is released from the trees. Without

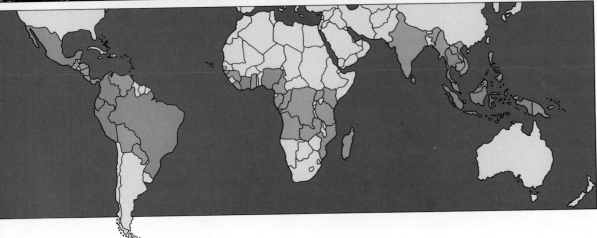

Figure 50.12 Countries permitting the largest destruction of tropical forests. *Red* shading denotes where 2,000 to 14,800 square kilometers are deforested annually. *Orange* denotes "moderate" deforestation (100 to 1,900 square kilometers).

Figure 50.13 The vast Amazon River Basin of South America in September 1988. Its features were completely obscured by smoke from fires that had been set to clear tropical forests, pasturelands, and croplands during the dry season.

Smoke extends to the Andes Mountains on the western horizon, about 650 miles (1,046 kilometers) away. The smoke cover was the largest that astronauts had ever before observed. It extended almost 175 million square kilometers (1,044,000 square miles).

The smoke plume near the center of the photograph alone covered an area comparable to the huge forest fire in Yellowstone National Park in that same year.

Massive deforestation is not confined to equatorial regions. For example, in the past century, 2 million acres of redwood forests along California's coast have been logged over. Most of the destruction of such temperate forests has been occurring since 1950, owing to the widespread use of chainsaws and tractors and the practice of exporting the logs to lumbermills overseas, where wages are low.

trees, annual precipitation declines. Rain rapidly runs off the bare soil. As the region gets hotter and drier, soil fertility and moisture decline even more. In time, sparse grassland or even desertlike conditions prevail instead of a rich tropical forest.

Widespread tropical forest destruction may have global repercussions. These forests absorb much of the sunlight reaching equatorial regions of the earth's surface. When the forests are cleared, the land becomes shinier, so to speak, and reflects more incoming energy back into space. Also, by their photosynthetic activity, the great numbers of trees in these vast forest biomes help sustain the global cycling of carbon and oxygen. When trees are harvested or burned, carbon stored in their biomass is released to the atmosphere in the form of carbon dioxide—and this may be amplifying the greenhouse effect.

Almost half the world's tropical forests have been destroyed for cropland, grazing land, timber, and fuelwood. Deforestation is greatest in Brazil, Indonesia, Colombia, and Mexico. If clearing and destruction continue at present rates, only Brazil and Zaire will have large tropical forests in the year 2010. By 2035, most of their forests will be gone.

A QUESTION OF ENERGY INPUTS

Paralleling the J-shaped curve of human population growth is a steep rise in total and per capita energy consumption. It is due to increased numbers of energy users and to extravagant consumption and waste. For example, in one of the most pleasant of all climates, a major university constructed seven- and eight-story buildings with narrow, sealed windows. The windows can't be opened to catch prevailing ocean breezes. The buildings and windows were not designed or aligned to use sunlight for passive solar heating and breezes for passive cooling. Massive energy-demanding cooling and heating systems were installed.

When you hear talk of abundant energy supplies, keep in mind that there is a huge difference between the *total* and the *net* amounts available. Net energy is that left over after subtracting the energy used to locate, extract, transport, store, and deliver energy to consumers. Some energy sources, such as direct solar energy, are renewable. Others, such as coal and petroleum, are not. Currently, 83 percent of the energy stores being tapped are in the second category (Figure 50.14).

Fossil Fuels

Forests that existed hundreds of millions of years ago gave us **fossil fuels** (page 397). Their carbon-containing remains became buried and compressed in sediments,

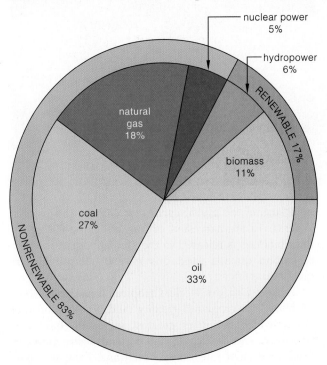

Figure 50.14 World consumption of nonrenewable and renewable energy sources in 1991.

then were transformed into coal, petroleum (oil), and natural gas. They are nonrenewable resources.

Even with strict conservation, known petroleum and natural gas reserves may be used up in the next century. As known reserves run out in accessible areas, we explore wilderness areas in Alaska and other fragile environments, such as continental shelves. Net energy declines as costs of extraction and transportation to and from remote areas increase. Environmental costs of extraction and transportation escalate. The long-term impact of the 11-million-gallon spill from the tanker *Valdez* off Alaska's coast is still not understood.

What about coal? In theory, world reserves can meet the energy needs of the human population for at least several centuries. But coal burning has been the main source of air pollution. Most coal reserves contain low-quality, high-sulfur material. Unless sulfur is removed before or after burning, sulfur dioxides are released into the air. They add to the global problem of acid deposition. Fossil fuel burning also releases carbon dioxide and adds to the greenhouse effect.

Extensive strip mining of coal reserves close to the earth's surface carries its own problems. It removes land from agriculture, grazing, and wildlife. Restoration is difficult and expensive in arid and semiarid lands, where much of the strip mining is proceeding.

Nuclear Energy

Nuclear Reactors As Hiroshima burned in 1945, the world recoiled in horror from the destructive force of nuclear energy. By the 1950s, nuclear energy was being viewed as an instrument of progress. Many energy-poor industrialized nations, France included, now depend heavily on nuclear power. Yet new nuclear plants have been delayed or cancelled in most other countries. Since 1970, in the United States alone, plans for 117 nuclear power plants were cancelled. Other plants were abandoned before completion. A few are being converted at great cost to fossil fuel burning.

Questions surround nuclear energy's costs, efficiency, safety record, and environmental impact. By 1990 in the United States, it cost slightly more to generate electricity by nuclear energy than by using coal. Energy from electricity now costs less. By the year 2000, solar energy with natural gas backup also should cost less.

What about safety? Less radioactivity and carbon dioxide normally escape from nuclear plants than from coal-burning plants of the same capacity and emit no sulfur dioxide. The danger lies with a potential for **meltdown**. As nuclear fuel breaks down, it releases considerable heat. Typically, water circulating over the fuel absorbs heat and produces steam that drives electricity-

Focus on the Environment

Incident at Chernobyl

On April 26, 1986, errors in judgment during a routine test led to runaway reactions at the Chernobyl power plant in the Ukraine. Explosions led to a total meltdown that lasted nearly ten days and continued right through the six-foot-thick steel and gravel barrier beneath the plant. Between 185 and 250 million curies of radioactive matter may have escaped. Inhaling as little as *one-millionth* of a curie of plutonium can cause cancer.

Thirty-one people died instantly; others died of radiation sickness during the following weeks. Inhabitants of entire villages were relocated; their former homes were bulldozed under.

After the accident, the number of people opposed to nuclear plants rose sharply, even in France. Figure *a* gives us an indication of why opposition increased. The radioactive fallout put an estimated 300 million to 400 million people at risk of leukemia and other radiation-induced disorders. Besides this, hundreds of millions of dollars were lost throughout eastern and western Europe when the fallout made crops and livestock unfit for consumption.

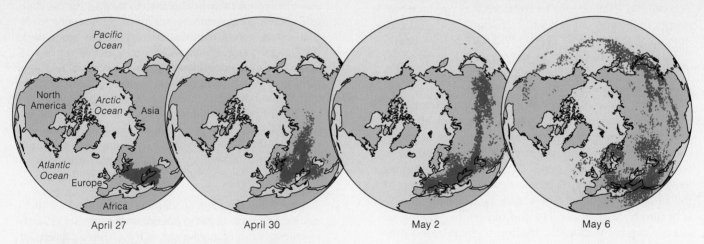

April 27   April 30   May 2   May 6

a Global distribution of radioactive fallout after explosions ripped through the Chernobyl nuclear plant on April 26, 1986.

generating turbines. Should a leak develop in the circulating water system, water levels might plummet around the fuel, which might then heat past its melting point. On the generator floor, melting fuel would instantly convert the remaining water to steam. Together with other reactions, the formation of steam could blow the system apart and release radioactive material. Also, an overheated reactor core could melt through its thick concrete containment slab and contaminate groundwater.

Is this scenario far-fetched? The *Focus* essay suggests not, but this is a controversial issue.

Nuclear Waste Disposal Unlike coal, nuclear fuel cannot be burned to harmless ashes. The fuel elements are spent after about three years, but they still contain uranium fuel and hundreds of new radioisotopes formed during the reactions. The wastes are extremely radioactive and dangerous. As they undergo radioactive decay, they become extremely hot. They are plunged at once into water-filled pools. The water cools them and keeps radioactive material from escaping. After being stored for several months, the remaining isotopes are still lethal. Some must be isolated for at least 10,000 years. If one kind of plutonium isotope (239^{Pu}) is not removed, the wastes must be kept isolated for a quarter of a million years!

After nearly fifty years of research, scientists still cannot agree on the best way to store high-level radioactive wastes. Even if they could do so, there is no politically acceptable solution. No one wants radioactive wastes anywhere near where they live.

Finally, following the Soviet Union's breakup, underpaid workers of a Russian nuclear power plant have been selling fuel elements on the black market. The buyers appear to be developing nations that want to produce nuclear weapons—and possibly put them in the hands of terrorist organizations.

50.9 ALTERNATIVE SOURCES OF ENERGY

Alternative energy sources for the human population are on the drawing boards. For example, in theory, **breeder reactors** may "breed" fuel by converting an abundant isotope of uranium into a fissionable isotope of plutonium. However, unlike a conventional nuclear reactor (which cannot explode like an atomic bomb), breeder reactors could produce a small nuclear explosion. Also, it may cost several times more to put them into operation. It may take 100 to 200 years to produce enough plutonium to fuel them.

As another example, **fusion power** might provide energy. The idea is to fuse hydrogen atoms to form helium atoms, with the release of considerable energy. The process is analogous to reactions that produce heat energy in the sun. The scientific, technological, and economic obstacles are great. Without major breakthroughs, fusion power is not expected to produce electricity on a commercial basis until the last half of the next century, if ever.

As a final example, consider that for more than a billion years, photosynthetic cells have stayed alive by harnessing sunlight energy and splitting water molecules for their building programs. Imagine a future in which we do the same thing. We have a virtually limitless supply of water; it covers nearly three-fourths of the earth's surface. We know that its molecules can be split into hydrogen gas (H_2) and oxygen by passing an electric current through it. Hydrogen gas released during this process can be used as an energy source.

Suppose we develop low-cost solar cells that can harness sunlight energy, then use that energy to generate the electricity required to produce hydrogen gas. **Solar-hydrogen energy** would assure the human population of an affordable, renewable energy source, and the cleansing of the environment could begin in earnest. Why? In the words of G. Tyler Miller, Jr., an environmental scientist, "If we can make the transition to an energy-efficient solar-hydrogen age, we can say goodbye to smog, oil spills, acid rain, and nuclear energy, and perhaps to global warming. The reason is simple. When hydrogen burns in air, it reacts with oxygen gas to produce water vapor—not a bad thing to have coming out of tailpipes, chimneys, and smokestacks. If clean solar-hydrogen technology comes on line soon enough, it could help heavily populated developing countries raise their living standards without severely disrupting the earth's life support systems."

Researchers in Germany, Japan, and elsewhere have been working intensively to develop the required technologies for hydrogen-fueled vehicles, factories, and homes. Prototype photovoltaic cells, which convert sun-

Focus on Bioethics

Biological Principles and the Human Imperative

Molecules, cells, tissues, organs, organ systems, multicelled organisms, populations, communities, ecosystems, the biosphere. These are architectural systems of life, assembled in increasingly complex ways over the past 3.5 billion years. We are latecomers to this immense biological building program. Yet, within the relatively short span of 10,000 years, we have been changing the character of the land, oceans, and atmosphere—even the genetic character of species.

It would be presumptuous to think we alone have had major effects on the world of life. As long ago as the Proterozoic era, photosynthetic organisms irrevocably changed the course of biological evolution by gradually enriching the atmosphere with oxygen. In the past as well as the present, competitive adaptations assured the rise of some groups, whose dominance assured the decline of others. Thus change is nothing new to this biological building program. What is new is the accelerated, potentially cataclysmic change accompanying the stupendous growth of the human population. We now have the population size, the technology, and the cultural inclination to use energy and modify the environment at frightening rates.

Where will rampant, accelerated change lead us? Will feedback controls begin to operate as they do, for example, when population growth exceeds the carrying capacity of the environment? In other words, will negative feedback controls come into play and keep things from getting too far out of hand?

Feedback control will not be enough, for it operates only when deviation already exists. Our population growth rates and patterns of resource consumption are founded on an illusion of unlimited resources and a forgiving environment. A prolonged, global shortage of food or the passing of a critical threshold for the global engines can come too fast to be corrected. At some point, deviations may have too great an impact to be reversed.

light energy into electricity, are in place in thousands of villages in southeast Asia, 30,000 homes around the world, and even drilling platforms in the Persian Gulf. Solar-hydrogen systems that will meet the heating, cooling, and other electrical needs of homes may be on the production line in Germany before the year 2000. Mercedes, BMW, and Mazda now have hydrogen-powered prototype vehicles on the road.

What about feedforward mechanisms, which might serve as early warning systems? For example, skin receptors sense a drop in outside air temperature. Each sends messages to the nervous system, which responds by triggering mechanisms that raise the core temperature before the body itself becomes dangerously chilled. If we could develop feedforward control mechanisms, maybe we could begin corrective measures before we alter the environment too significantly.

By themselves, feedforward controls won't work, for they start operating when change is already under way. Think of the DEW line—the Distant Early Warning system. It is like a sensory receptor, one that detects intercontinental ballistic missiles that may be launched against North America. By the time the system detects what it is designed to detect, it may be too late to stop widespread destruction.

It would be naive to assume we can ever reverse who we are at this point in evolutionary time, to de-evolve ourselves culturally and biologically into becoming less complex in the hope of averting disaster. Yet there is reason to believe we can avert disaster by using a third kind of control mechanism, one that is uniquely our own. We have the capacity to anticipate events before they happen. We are not locked into responding only after irreversible change has begun. We have the capacity to anticipate the future—it is the essence of our visions of utopia or of nightmarish hell. *We all have the capacity to adapt to a future that we can partly shape.* We can, for example, learn to live with less. Far from being a return to primitive simplicity, it would be one of the most complex and intelligent behaviors of which we are capable.

Having that capacity and using it are not the same thing. We have already put the world of life on dangerous ground because we have not yet mobilized ourselves as a species to work toward self-control.

Our survival depends on predicting possible futures. It depends on designing and constructing ecosystems that are in harmony with our definition of basic human values and with the biological models available to us. Human values can change; our expectations can and must be adapted to biological reality. *For the principles of energy flow and resource utilization, which govern the survival of all systems of life, do not change.*

It is our biological and cultural imperative that we come to terms with these principles, and ask ourselves what our long-term contribution will be to the world of life.

With serious commitment, we could expect solar cells to be supplying as much energy as nuclear power plants by the year 2010. By 2050, solar cells could be satisfying half the energy requirements of the United States, at lower cost and at much lower risk than the existing alternatives. However, cuts in the federal budget have resulted in a decline in research efforts and applications in the United States.

In sum, what are the best options? For now and the near future, we must reduce consumption by using more energy-efficient vehicles, architecture, heating and cooling systems, and appliances. For the long term, some argue for hydrogen fuel systems based on energy from the sun, wind, flowing water, heated steam being vented from the earth, or some other renewable resource. The *Focus* essay gives other arguments.

SUMMARY

1. Accompanying the exponential growth of the human population are increased energy demands and environmental pollution.

2. Pollutants are substances with which ecosystems have had no prior evolutionary experience (in terms of kinds and amounts) and so have no mechanisms for absorbing or cycling them. Many pollutants are an outcome of human activities, and they adversely affect the health, activities, or survival of human populations.

3. Industrial smog and photochemical smog are examples of local air pollution. Acid rain, thinning of the ozone layer, and possible enhancement of the greenhouse effect are examples of global air pollution.

4. Human population growth presently depends on the expansion of agriculture, made possible by large-scale irrigation and heavy applications of fertilizers and pesticides. Global freshwater supplies are limited, yet they are being polluted by agricultural runoff (which includes sediments as well as pesticides and fertilizers), industrial wastes, and human sewage.

5. Human populations are damaging land surfaces by the mind-boggling accumulation of solid wastes and by conversion of marginal lands for agriculture. Widespread desertification and the destruction of tropical and temperate forest biomes may be altering regional soils and patterns of rainfall.

6. Energy supplies in the form of fossil fuels are nonrenewable, dwindling, and environmentally costly to extract and use. Nuclear energy in itself is less polluting, but the costs and risks associated with fuel containment and with storing radioactive wastes are enormous. The challenge is to develop affordable alternatives based on renewable resources, such as solar-hydrogen energy.

Review Questions

1. What are CFCs? Describe how they have helped form the ozone hole above Antarctica. *895, 897*

2. Make a list of advantages you personally enjoy as member of an affluent, industrialized society. List some of the drawbacks. Do you believe the benefits outweigh the costs? This is not a trick question.

3. List six activities you pursue each day that harm the environment. List six ways in which you might reduce the harmful effects.

4. After reading this chapter, write your own caption about the flow of energy and materials into and out of the human ecosystem shown in the photograph below. Or write a caption about the effects of human activities on a natural ecosystem, as suggested by the photograph on page 909.

Self-Quiz *(Answers in Appendix IV)*

1. Since the mid-eighteenth century, human population growth has been _____ .
   a. leveling off   c. growing exponentially
   b. growing slowly   d. not much to speak of

2. Pollutants disrupt ecosystems because _____ .
   a. their component elements differ from those of natural substances
   b. only humans have uses for them
   c. there are no evolved, established mechanisms that can deal with them
   d. their only effect is on ecosystems but not humans

3. During a thermal inversion, weather conditions trap a layer of _____ air under a layer of _____ air.
   a. warm; cool
   c. warm; sooty
   b. cool; warm
   d. cool; sooty

4. _____ is (are) a case of local air pollution.
   a. Smog
   c. Ozone layer thinning
   b. Acid rain
   d. a and b

5. _____ is (are) a case of air pollution with global effects.
   a. Smog
   c. Ozone layer thinning
   b. Acid rain
   d. b and c

6. Two-thirds of the water used annually by the human population goes to _____ .
   a. urban centers
   c. treatment facilities
   b. agriculture
   d. a and c

7. The upper limit at which the ground is fully saturated with water is called _____ .
   a. groundwater
   c. the water table
   b. an aquifer
   d. the salination limit

8. Energy from fossil fuels is _____ ; their extraction and use come at _____ cost to the environment.
   a. renewable; low
   c. renewable; high
   b. nonrenewable; low
   d. nonrenewable; high

9. Nuclear energy normally pollutes _____ than fossil fuels; it poses _____ dangers than fossil fuels.
   a. less; lesser
   c. more; lesser
   b. more; greater
   d. less; greater

10. Match each term with its appropriate description.
    _____ desertification
    _____ deforestation
    _____ green revolution
    _____ pollutant
    _____ solar-hydrogen power

    a. substance with adverse effects on environment, humans, or both
    b. one of our best options
    c. soil loss, watershed loss, altered rainfall patterns follow
    d. attempt to improve crop production on existing land
    e. conversion of marginal land for agriculture, with at least 10 percent drop in productivity

Selected Key Terms

acid rain *896*
breeder reactor *906*
CFC (chlorofluorocarbon) *897*
deforestation *902*
desertification *901*
dry acid deposition *896*
fossil fuel *904*
fusion power *906*
green revolution *900*
industrial smog *895*
meltdown *904*

ozone hole *897*
PAN (peroxyacyl nitrate) *895*
photochemical smog *895*
pollutant *895*
salination *898*
shifting cultivation *902*
solar-hydrogen energy *906*
thermal inversion *895*
wastewater treatment *899*
water table *898*

Readings

Collins, M. 1990. *The Last Rain Forests.* New York: Oxford University Press.

Gruber, D. 1989. "Biological Monitoring and Our Water Resources." *Endeavour* 13(3):135–140.

Miller, G. T., Jr. 1994. *Living in the Environment.* Eighth edition. Belmont, California: Wadsworth. Chapter 17 is especially valuable reading. This author consistently puts information from numerous sources into a current, accessible survey of the present and future state of the environment.

Mohnen, V. August 1988. "The Challenge of Acid Rain." *Scientific American* 259(2):30–38.

Western, D., and M. Pearl. 1989. *Conservation for the Twenty-First Century.* New York: Oxford University Press.

Wilson, E. 1988. *Biodiversity.* Washington, D.C.: National Academy of Sciences.

51 AN EVOLUTIONARY VIEW OF BEHAVIOR

Deck the Nest With Sprigs of Green Stuff

About a century ago, the European starling hitched a boat ride to North America. Ever since, starlings have been multiplying and evicting great numbers of native birds from scarce nest sites in the cavities of trees.

Once a male and female starling commandeer a tree cavity, they build a nest of dead, dry grasses and twigs (Figure 51.1*a*). Theirs is no ordinary nest. They proceed to *decorate* the nest bowl with many green sprigs, freshly plucked.

Why do starlings do this? Do the sprigs of greenery camouflage the nest from predators? Not likely. The nests are already concealed, inside the tree cavities. Do the sprigs serve as insulative material that will help keep the forthcoming eggs warm? Actually, moist, green plant parts would promote heat loss, not heat conservation. Well, then, do the sprigs combat parasites? After all, mites parasitize birds and infest their

nest cavities (Figure 51.1*b*). Even a few tiny mites can rapidly produce thousands of descendants. In large numbers, the mites can suck enough blood from a nestling to weaken it and interfere with its growth and development.

Two biologists, Larry Clark and Russell Mason, decided to test the third hypothesis. As they knew, starlings do not weave just any green plant material into the nests. They choosily favor the leaves of certain plants, including wild carrot (Figure 51.1*c*). So Clark and Mason built a set of experimental nests, some with fresh-cut wild carrot leaves and some without. Then they removed the natural nests that pairs of starlings had constructed and had already started using. Half of the nesting pairs received replacement nests, decorated with sprigs of wild carrot. No sprigs decorated the replacement nests for the other pairs of starlings.

a

b

c

Figure 51.1*d* shows the results. The number of mites in the greenery-free nests was consistently greater than in nests with greenery. For example, at the end of one experiment, nests with no carrot sprigs teemed with an average of 750,000 mites—and the ones with sprigs contained a mere 8,000.

Shoots of wild carrot happen to contain a highly aromatic steroid compound. Almost certainly the compound repels herbivores and helps the plant survive. Coincidentally, the compound also prevents mites from becoming sexually mature—and therefore prevents a mite population explosion in nests festooned with wild carrot. Far from being a trivial behavior, "decorating" with a bit of aromatic greenery helps fumigate a nest and increases a starling's chance of producing healthy, surviving offspring.

And so starlings lead us into the world of behavioral research. As you will see in this chapter and the next, some behavioral studies focus on structural and functional mechanisms that enable individuals to behave as they do. Others focus on the adaptive value of some behavior to an individual's reproductive success.

KEY CONCEPTS

1. "Behavior" refers to the coordinated responses that an animal makes to stimuli. The responses are instinctive, learned, or a combination of both.

2. Through mechanisms underlying instinctive behavior, an animal can make complete responses to certain key stimuli the first time it encounters them. Through mechanisms underlying learned behavior, an animal can modify its behavior by acquiring information from certain experiences.

3. Forms of behavior have a heritable (genetic) basis, as when instructions encoded in certain genes govern the development of the nervous and endocrine systems. These organ systems allow an animal to detect, process, and issue commands for behavioral responses to stimuli.

4. Like other traits having a genetic basis, forms of behavior have evolved by way of natural selection—the measure of which is reproductive success. Thus sexual selection and other evolutionary processes have favored behavioral mechanisms that enhance the ability of individuals to pass on their genes to offspring.

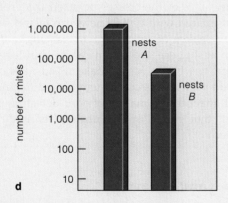

Figure 51.1 **(a)** A most excellent fumigator in nature—the European starling (*Sturnus vulgaris*). It combats infestations of mites **(b)** by decorating previously owned nests with fresh sprigs of wild carrot **(c)**.

(d) Results of one experiment. In this case, nests designated *A* were kept free of fresh green sprigs of wild carrot, which contains chemicals that prevent baby mites from reaching adulthood. Experimental nests designated *B* had fresh sprigs added every seven days. Head counts of mites (*Ornithonyssus sylviarum*) infesting the nests were made at the time the starling chicks left the nest, twenty-one days after this particular experiment was under way.

51.1 THE HERITABLE BASIS OF BEHAVIOR

Genes and Behavior

As you know, information encoded in genes directs the formation of the tissues and organs that make up the animal body, including its nervous system. A nervous system is the means by which an animal detects and processes specific information about conditions outside and inside its body. The system processes and integrates the information, then commands muscles and glands to make appropriate responses. Thus, through the operation of the nervous system, genes influence the observable, coordinated responses that animals make to stimuli. We call these responses **animal behavior**.

a

b

c

Figure 51.2 (**a**) Banana slug, food for (**b**) a grown-up garter snake of coastal California. (**c**) A newborn garter snake from a coastal population, tongue-flicking at a cotton swab drenched with banana slug fluids.

Some experiments conducted by Stevan Arnold serve as an elegant example of how genes contribute in an indirect yet major way to behavior. Arnold was interested in the basis of feeding preferences among different populations of a species of garter snake that lives in California. For snake populations living along the coast, the food of choice is the banana slug (Figure 51.2a). Populations of garter snakes that live farther inland prefer tadpoles and small fishes. Offer them a banana slug and they ignore it.

In one set of experiments, Arnold offered captive newborn garter snakes a chunk of slug as the first meal. Offspring of coastal snakes usually ate the chunk. They even flicked their tongue at cotton swabs that were drenched in essence of slug. (Snakes "smell" by tongue-flicking, which draws chemical odors into the mouth.) By contrast, offspring of the snakes living inland ignored the swabs, and only rarely did they eat the slug meat.

Here was a clear difference between captive baby snakes that had no prior, direct experience with slugs. It suggested that even before birth, garter snakes are genetically programmed to accept or reject slugs. (Certainly in Arnold's experiments, the snakes did not learn this feeding behavior as a result of "taste trials.")

To test this hypothesis, Arnold conducted some breeding experiments. He crossed coastal male snakes with inland female snakes. He also crossed coastal females with inland males. If the difference in food preferences between the snake populations has a genetic basis, then the "hybrid" offspring might make an *intermediate* response to slug chunks and odors.

The outcomes of the crosses matched the predicted results. Compared with typical newborn inland snakes, many baby snakes of mixed parentage tongue-flicked more often at slug-scented cotton swabs—but less often than typical newborn coastal snakes. Thus, the difference in food preferences could stem from differences in genes that affect the formation of odor-detecting mechanisms in the garter snake during its embryonic development.

Hormones and Behavior

The gene products called hormones contribute directly and indirectly to behavior. Imagine it is early spring in a Canadian forest. A male white-throated sparrow (Figure 51.3a) whistles a song that sounds rather like "Sam Peabody, Peabody, Peabody." He repeats the same song thousands of times, with such clarity and consistency that you wonder how he does it.

In a roundabout way, his singing starts with seasonal differences in the secretion of melatonin, a hormone product of the pineal gland (page 625). At high

Figure 51.3 Sound spectrograms of territorial songs. They are visual records of the pitch (frequency, as measured in kilohertz) of each note.

(**a**) A male white-throated sparrow belting out the Sam Peabody, Peabody, Peabody song. (**b**) A visual record of this territorial song.

(**c**) Zebra finch and (**d**) a visual record of the male's full song. (**e**) A sound spectrogram of a female zebra finch that had been experimentally converted to a singer. She was exposed to estrogen when she was a nestling, then given a testosterone implant when she reached adulthood.

concentration in the blood, melatonin suppresses the growth and functions of songbird gonads. This effect is reversed when sunlight acts on photoreceptors of cells in the pineal gland. Signals from these photoreceptors result in low melatonin concentrations.

There are fewer hours of daylight in winter than in spring, and the amount of sunlight is not enough to block melatonin secretion. When daylength increases in spring, the gonads are released from hormonal control. Now they grow in size and step up *their* secretion of estrogen and testosterone—which trigger a gender-related difference in singing behavior. How? While the embryos of most songbirds are developing, estrogen secretion controls the formation of brain regions that will govern muscles of a vocal organ. These regions represent a **sound system**—and their size and structure are notably different between males and females.

Among songbirds, females are XY and males, XX. And estrogen can be produced only in the absence of genes on the Y chromosome. Before a male bird is even hatched, a high estrogen level in his blood stimulates the development of a masculinized brain. Later, the

male's gonads enlarge. By the start of the breeding season, the gonads are increasing their testosterone secretion. Testosterone binds to receptors on cells in the sound system. The binding induces metabolic changes that eventually prepare the male to sing when suitably stimulated.

Experiments provide evidence of hormonal effects on singing behavior. For example, female zebra finches that received testosterone implants sang when they reached adulthood—but only if their brain had been masculinized through exposure to extra estrogen early in life (Figure 51.3e). In short, estrogen *organizes* the song system in developing embryos. Then testosterone *activates* the song system in adults.

By influencing the development of the nervous system, genes contribute in an indirect yet major way to behavior.

Hormones influence the organization and activation of mechanisms required for particular forms of behavior.

51.2 INSTINCTIVE AND LEARNED BEHAVIOR

Each animal depends on structural and physiological mechanisms that govern behavioral responses to environmental cues. Traditionally, such responses have been split into *instinctive* and *learned* behavior.

In **instinctive behavior**, components of the nervous system allow an animal to carry out complex, stereotyped responses to certain environmental cues, which are often simple. Consider the cuckoo, a social parasite. Adult females lay eggs in nests of other species. Young cuckoos eliminate the natural-born offspring, then get the undivided attention of their unsuspecting foster parents. Newly hatched cuckoos are blind, but when they contact an egg, they maneuver it onto their back, then push it out of the nest (Figure 51.4). This is an instinctive response, triggered by a well-defined, simple stimulus. Once set in motion, it is performed in its entirety. We call it a **fixed action pattern**.

With their tongue-flicking, body orientation, and strikes at prey, newborn garter snakes give us more examples of instinctive behavior. So do humans. For example, when human infants are only two or three weeks old, they smile instinctively when an adult's face comes close (Figure 51.5a). They will make the same smiling response to a simple stimulus—a flat, face-size mask with two dark spots where eyes would be on a real human face. A mask with one "eye" will not do the trick. As shown in Figure 51.5b, older infants also show instinctive behavior.

Yet each animal also processes and integrates information gained from specific experiences, then it uses the information to vary or change responses to stimuli. Such responses are called **learned behavior**.

Think of a young toad, with neural wiring that commands it to flip its sticky tongue instinctively at any dark object moving across its field of vision. In the toad world, such objects are usually edible insects. But suppose one object is a bumblebee that stings the toad's tongue. The experience leads the toad to perceive that "Bumblebee-size objects with black and yellow bands can sting." Or think of the time-dependent form of learning called **imprinting**. It requires exposure to key stimuli in the environment, usually during a sensitive period when animals are young. Figure 51.6 gives examples of imprinting and other forms of learning.

Don't fall into the trap of separating instinctive and learned behavior, as if one is determined by genes and the other by the environment. Behavior is an outcome of gene expression *and* of individual experiences with the environment. For example, all normal male white-crowned sparrows sing—but individuals in different habitats belt out variations ("dialects") of the species song. As Peter Marler discovered, the males acquire the full song by *hearing it* ten to fifty days after hatching. During this sensitive period, a learning mechanism is primed to receive a bit of information from the environment.

As Marler also discovered, the birds learn parts of the song by picking up acoustical cues from other males. He raised male nestlings to maturity in sound-proof chambers so they would not hear adult males singing. At maturity, the captives produced a song with none of the detailed structure of a typical adult's song. Marler also exposed isolated, captive males to recordings of white-crown sparrows and song sparrows. At maturity, the captive males sang only the white-crown song. They even mimicked the dialects. Evidently, territorial songs depend on learning some—but not all—acoustical cues.

In a different experiment, young, hand-reared white-crowns were allowed to interact with a "social tutor," as opposed to listening to taped songs. They tended to learn the song of the alien species. Their learning mechanisms must be primed to be influenced by social experience as well as by acoustical cues.

Figure 51.4 (a) A newly hatched cuckoo making a complex, innate behavioral response to an environmental cue—spherical objects in the nest. The European cuckoo lays eggs in the nests of other species. Even before a cuckoo hatchling opens its eyes, it responds to the shape of the host's eggs by shoving them out of the nest. (b) The foster parents keep feeding the usurper.

In instinctive behavior, animals make complex, stereotyped responses to specific, often simple environmental cues.

In learned behavior, responses may vary or change as a result of individual experiences in the environment.

Whether instinctive or learned, behavior is an outcome of genes *and* experience in the environment.

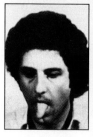

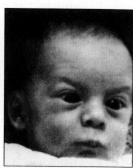

Figure 51.5 Examples of instinctive responses in humans. (a) Smiling in very young infants is a fixed action pattern that can be triggered automatically by certain simple stimuli. (b) Somewhat older infants also have the instinctive capacity to imitate facial expressions of adults.

a

b

a No one can tell these imprinted baby geese that Konrad Lorenz is not Mother Goose.

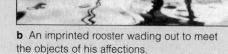

b An imprinted rooster wading out to meet the objects of his affections.

Figure 51.6 A few traditional categories of learned behavior.

Imprinting. For many animals, this is a time-dependent form of learning that requires exposure to key stimuli, usually early in development. (a) As an example, baby geese formed an attachment to the ethologist Konrad Lorenz—or to any moving object—if they were separated from their mother shortly after hatching. But they did this only after being exposed to a moving object during a short, sensitive period early in life. At that time only, they were primed to learn a valuable bit of information—the identity of the individual they would follow in the months ahead. In nature, that individual is normally the mother or father.

Imprinting has later consequences for the sexual behavior of various birds. For example, many birds direct sexual behavior toward members of whatever species they had been sexually imprinted upon in their youth (normally their own, because they encounter their mother soon after hatching). The rooster shown in (b) had been imprinted on a mallard duck early in life. He courted ducks instead of hens of his own species.

Classical conditioning. Ivan Pavlov's experiments with dogs are an example. Dogs salivate just before eating. Pavlov's dogs were conditioned to salivate even in the absence of food. They did so in response to the sound of a bell or a flash of light that was

initially associated with the presentation of food. In this case, the animals learned to associate an automatic, unconditioned response with a novel stimulus that does not normally trigger the response.

Operant conditioning. An animal learns to associate a voluntary activity with its consequences, as when a toad learns to avoid bad-tasting insects after voluntary attempts to eat them.

Habituation. Through experience, an animal learns not to respond to a situation if the response has no positive or negative consequences. Many of the birds living in cities learn not to flee from people who pose no real threat to them.

Spatial or latent learning. Through inspection of its environment, an animal acquires a mental map of a region, often by learning the position of local landmarks. Blue jays, for instance, can store information about the position of dozens, if not hundreds, of places where they have stashed food.

Insight learning. An animal abruptly solves a problem without trial-and-error attempts at the solution. Chimpanzees often exhibit insight learning in captivity when they suddenly solve a novel problem created for them by their captors. For example, some abruptly stacked together several boxes and used a stick to reach a bunch of bananas suspended out of their reach.

51.3 THE ADAPTIVE VALUE OF BEHAVIOR

We turn now to evolutionary mechanisms by which diverse forms of behavior come about. We begin by defining a few terms used in animal behavior studies:

---

1. **Reproductive success** refers to the survival and production of the offspring of an individual.

2. **Adaptive behavior** is a behavior that promotes the propagation of an individual's genes and tends to occur at increased frequency in successive generations.

3. **Selfish behavior** is a behavior by which an individual protects or increases its own chance of producing offspring, regardless of the consequences for the group to which it belongs.

4. **Altruistic behavior** is self-sacrificing behavior. The individual behaves in a way that helps others but decreases its chance to produce its own offspring.

5. **Natural selection** is a measure of the difference in survival and reproduction among individuals that differ from one another in their heritable traits.

When behavioral biologists speak of a "selfish" or an "altruistic" individual, they don't mean the individual is consciously aware of what it is doing or of the behavior's reproductive goal. A lion doesn't have to know that eating zebras is good for reproductive success. Its nervous system simply calls for hunting behavior when the lion is hungry and sees a zebra.

Whether selfish or altruistic, a behavior persists in a population because genes responsible for it are persisting also. Think of Norwegian lemmings, which disperse when population density skyrockets and food is scarce. As they move out of an overcrowded habitat, many die. Are they committing suicide to help their species? Or are they dying by way of starvation, predation, and accidental drowning? A wonderfully instructive cartoon by Gary Larson shows lemmings plunging into water, presumably in the act of suicide—but one has an inflated inner tube about its waist! If lemmings tended to be altruistic, only the "selfish" ones would reproduce—and genes underlying the altruistic behavior would disappear from the population. A more likely explanation is this: Overcrowded lemmings disperse to less crowded places, where individuals have a better chance to survive and reproduce.

In sum, when studying behavior, you usually will find it more profitable to look for evidence of natural (individual) selection than for something that benefits the species. A few examples of feeding and reproductive behavior will reinforce this point.

Selection Theory and Feeding Behavior

In northern forests, ravens scavenge carcasses of deer, elk, or moose, which are few and far between. When one of these large birds comes across a carcass—even in winter when food is scarce—it often calls loudly and attracts a crowd of similarly hungry ravens. The calling behavior seemed puzzling to Bernd Heinrich, for it seems to go against the caller's interests. Wouldn't a quiet raven get more food, increase its chance of surviving, and help it leave more descendants than an "unselfish" vocalizer? If the behavior is an outcome of natural selection, the cost of lost calories and nutrients must be offset by some reproductive benefit for the individual caller. But what benefit?

Maybe a lone bird picking at a carcass is vulnerable to predators that lie in wait for it. If so, other ravens called in could help keep an eye out for danger. But ravens are large and agile birds with very few known enemies. Heinrich stealthily watched ravens feed singly or in pairs at a carcass, and he never saw predators attack any of them.

Then he realized that territory has something to do with it. A **territory** is an area that one or more individuals defend against competitors. After hauling a cow carcass into a Maine forest, Heinrich saw that single or paired ravens don't always vocally advertise food. Maybe these silent birds were adults that had already staked out a large territory—which happened to include the spot where he put the carcass.

A pair of ravens would gain nothing by attracting others to their territory. But what if the forest had been *subdivided* into territories and was defended by powerful adults? A wandering young bird would be lucky to feed in an aggressive pair's territory. But recruiting a gang of other, nonterritorial ravens might overwhelm the resident pair's defensive behavior.

As it turns out, only the wandering ravens advertise carcasses. Basically, their calling behavior is selfish. It gives them a shot at otherwise off-limits food.

Selection Theory and Mating Behavior

Competition among members of one sex for access to mates is common. So is choosiness in selecting a mate. Both activities are forms of **sexual selection**. This microevolutionary process favors traits that give the individual a competitive edge in reproductive success.

Typically, male animals produce great numbers of tiny sperm, and the females produce considerably fewer but larger eggs. Generally, reproductive success for a male depends on how many eggs he fertilizes. For a female, success depends largely on how many eggs she produces or how many offspring she can care for.

a

c

b

Figure 51.7 (**a**) Sexual competition between male bison, which are fighting for access to a cluster of females. (**b**) Dancing display by a male sage grouse, performed at his own vigorously defended spot inside a compact mating area called a lek. Females (the smaller brown birds) at the lek observe the prancing males before choosing the one they will mate with. (**c**) A male hangingfly dangling a moth as a nuptial present for a future mate. Females of certain hangingfly species choose sexual partners on the basis of the size of prey that males offer to them.

The prime factor influencing her sexual preference usually is the quality of a mate, not the quantity of partners. With these points in mind, let's look at a few mating tactics in terms of sexual selection theory.

We find competition for access to clusters of sexually receptive females in many species. Thus competition for ready-made harems has favored combative male lions, sheep, elk, and bison (Figures 1.6*f* and 51.7*a*).

What if females don't cluster in defendable groups? For example, sage grouse females are dispersed through their habitat in the western United States. The males make no attempt to maintain a large territory. During the breeding season they congregate in a **lek**, a communal display ground. There, each male stakes out a few square meters as his territory. Females are attracted to the lek—not to feed or nest, but rather to observe the males. With tail feathers erect and neck pouches puffed, each male emits booming calls and stamps about like a big wind-up toy on his display ground (Figure 51.7*b*). Female sage grouse tend to select and mate with one male only. Then they go off to nest

by themselves, unassisted by their sexual partner. Many females often choose the same male, so most of the males never mate.

In some species, females select males that offer them superior material goods. Consider male hangingflies (*Harpobittacus apicalis*). They capture and kill a moth or some other insect, then release a chemical signal (a sex pheromone) that attracts females to the "nuptial gift" (Figure 51.7*c*). Females choose males that offer the larger, calorie-rich offerings. A female permits a male to mate with her—but only after she eats for about five minutes. Then she accepts and stores sperm in a reproductive organ—but only as long as the food holds out. Before twenty minutes are up, she can break off the mating at any point. If so, she will mate again and accept another male's sperm—and dilute the reproductive success of her first partner.

Thus, among hangingflies, sage grouse, and many other species, the females dictate the rules of male competition—and the males employ tactics that may help them fertilize as many eggs as possible.

SUMMARY

1. The observable, coordinated responses an animal makes to external and internal stimuli are called animal behavior.

2. Behavior originates with genes that provide instructions for the development of the nervous system and endocrine system, which interact to govern responses to stimuli.

3. An instinctive behavior is a complex, stereotyped response to a specific, often simple environmental stimulus. A learned behavior is a response that may vary or change as a result of the individual's experience in the environment.

   a. The mechanisms underlying instincts trigger complete responses to key stimuli the first time that an individual reacts to the stimuli. The mechanisms underlying learned behavior enable the individual to store information, which is used to modify the animal's response based on past experience.

   b. Both instinctive and learned behavior are outcomes of genetic and environmental inputs.

4. Behavior commonly is adaptive, and its heritable component makes it subject to evolution by natural selection.

5. Some behavioral researchers focus on the internal genetic and physiological mechanisms that control behavioral responses. Others deal with the evolution and adaptive value of behavioral abilities.

6. Researchers interested in the adaptive value of behavior, rather than behavioral mechanisms, usually use a research approach based on the theory of evolution by natural selection. This is a key premise of their approach: Traits that have evolved as a result of individual differences in reproductive success in the past should have reproductive benefits that exceed their reproductive costs (or disadvantages).

7. Several alternative hypotheses can usually be proposed on why benefits of a behavioral ability may exceed its costs. Discriminating among competing explanations requires using the hypotheses to generate predictions, which can then be tested by collecting new information and checking whether the actual results match the expected observations.

8. Behavioral biologists have tested many hypotheses on the possible adaptive value of various aspects of reproductive behavior. In developing their hypotheses, they recognize that members of the same species often create obstacles to reproductive success for each other, either through competition for access to mates or through selective mate choice.

Review Questions

1. What role does the environment play in the development of an instinct, such as the egg-ejecting behavior of baby cuckoos? *914*

2. Contrast altruistic behavior and selfish behavior. Why does either kind of behavior persist in a population? *916*

3. Give an example of feeding behavior or mating behavior that can be explained in terms of natural (individual) selection. *916–917*

4. A very large moth caterpillar resting on a vine in the tropics responds to being poked by partly letting go of the vine and puffing up part of its body:

Provide a hypothesis on the underlying mechanism that enables the caterpillar to behave this way. Provide a hypothesis on the possible adaptive value of the trait. How would you test both hypotheses?

Self-Quiz *(Answers in Appendix IV)*

1. "Starlings festoon their nests with sprigs of wild carrot and so interfere with the development of nest mites." This statement is _____ .
   a. an untested hypothesis
   b. a prediction
   c. a test of a hypothesis
   d. a proximate conclusion

2. Genes affect the behavior of individuals by _____ .
   a. influencing the development of nervous systems
   b. affecting the kinds of hormones in individuals
   c. governing the development of muscles and skeletons
   d. all of the above

3. Many kinds of female mammals live in herds. Which of these mating systems might you see in such herds?
   a. a lek mating system
   b. males competing for clusters of receptive females
   c. territorial defense of feeding sites by males
   d. males capturing food to lure females

4. Which of the following environmental factors do *not* influence the acquisition of song by a white-crown sparrow male reared in captivity?
   a. the experience of hearing the taped song of mature white-crown sparrow males
   b. the experience of hearing the taped song of mature song sparrow males
   c. the age of the bird when it is exposed to songs of its species
   d. the experience of interacting socially with a singing male of its own species

5. When evolutionary biologists claim that a particular human activity is "adaptive," they mean that the activity _____ .
   a. will promote the survival of our species
   b. increases the probability that an individual will behave altruistically
   c. increases the survival chances of the individual that behaves this way
   d. increases the frequency with which an individual will transmit genes to the next generation

6. The statement that lemmings, by dispersing at high population densities, reduce their population and bring it in line with available resources _____ .
   a. has been proven to be true
   b. is consistent with Darwinian evolutionary theory
   c. is based on the theory of evolution by group selection
   d. is supported by the finding that most animals behave altruistically during their lives

7. Match the animal behavior concepts.
   _____ fixed action pattern
   _____ altruism
   _____ basis of instinctive and learned behavior
   _____ animal behavior
   _____ imprinting

   a. time-dependent form of learning requiring exposure to key stimulus
   b. genes and environmental experience
   c. coordinated responses to external and internal stimuli
   d. instinctive response triggered by a well-defined, simple stimulus
   e. assisting another individual at one's own expense

Readings

Alcock, J. 1989. *Animal Behavior: An Evolutionary Approach.* Fourth edition. Sunderland, Massachusetts: Sinauer. A broad-ranging survey of the many topics that make up the study of behavior, with a strong emphasis on evolutionary theory.

Dawkins, R. 1989. *The Selfish Gene.* New York: Oxford University Press. An updated, annotated edition of an entertainingly written book. Dawkins explains why natural selection favors behavioral traits that are associated with individual success in propagating the individual's genes.

Krebs, J., and N. Davies (editors). 1991. *Behavioral Ecology: An Evolutionary Approach.* Cambridge, Massachusetts: Blackwell Scientific. A more advanced text consisting of chapters written by experts on many of the issues covered in this chapter.

Welty, C., and L. F. Baptista. 1988. *The Life of Birds.* Fourth edition. Philadelphia: Saunders. This edition contains a thorough review of many of the concepts discussed in this chapter as they relate to avian behavior.

Williams, G. C. 1966. *Adaptation and Natural Selection.* Princeton, New Jersey: Princeton University Press. The classic book that brought about a revolution in thinking about animal behavior in terms of individual selection.

52 ADAPTIVE VALUE OF SOCIAL BEHAVIOR

Consider the Termite

Imagine yourself in a eucalyptus forest in Queensland, Australia, chipping a few fragments from a narrow, brittle tube that runs along the trunk of a dead tree. As sunlight pours in through the breach in the tube wall, you notice some small, nearly white insects scurrying away from it—and a few brown ones rushing toward it. The brown insects have a swollen, eyeless, tapering head with a long, pointed "nose" (Figure 52.1). They must be there to fight off intruders. You tap the tunnel and thin jets of silvery goo shoot out of their nose!

The insects in that ruptured tunnel belong to a complex termite colony. The brown ones do indeed protect the colony from enemies—ants, mostly. They are soldier termites, and the white ones are workers. The workers build tunnels that lead to places where they can safely gather fibers of wood. They carry the fibers to an

underground nest, where termites cultivate an edible fungus. Fungal hyphae grow into the wood fibers and absorb nutrients from them. The termites eat some portions of the fungus. They also chew wood into small particles, which they swallow. Unlike most termite species, they do not produce enzymes that can digest the cellulose of wood. Microbial symbionts in their gut do this, and both the microorganisms and the termites absorb the released nutrients.

The soldiers and workers are both sterile; neither can reproduce. A single queen (with a huge, egg-producing abdomen) and one or more kings serve as "parents" for the entire society.

This glimpse into life as a termite raises two basic questions. *First*, by what means do members of a termite society or any other social unit cooperate and so benefit the group and themselves? For example, how do soldiers communicate with one another, locate a threat, and act together against it? Many social biologists attempt to identify communication signals among animals as well as the manner in which individuals respond to them.

Second, what is the adaptive value of social life? In the preceding chapter, you considered the premise that animals behave in ways that increase their individual chances of reproductive success. Yet in many animal societies, individuals appear to sacrifice their own chances by helping others reproduce! Among termites and some other extreme forms of sociality, helpful individuals are sterile. Addressing the costs and benefits of social life is a key focus of this chapter.

KEY CONCEPTS

1. Social interactions are based on diverse modes of communication that have developed over evolutionary time. A communication signal is an action that has net benefits for both the sender and the receiver of the signal.

2. Social living has costs and benefits, which can be measured in terms of an individual's ability to pass on its genes to offspring. Not every environment favors the evolution of social life. Under some circumstances, solitary individuals can leave more descendants than social ones would be able to do.

3. Sociality can evolve through the increased reproductive success of individuals. In other words, social animals need not exhibit altruism. Altruism means helping others but, in doing so, sacrificing personal reproductive success.

4. The evolution of altruism requires circumstances that enable individuals to propagate their genes indirectly by helping *relatives* reproduce successfully.

Figure 52.1 Soldier termites defending their social colony by guarding a break in a foraging tunnel. Members of this caste shoot out strands of glue from a pointed "nose." The glue entangles intruders, such as invading ants.

SIGNALING MECHANISMS

Functions of Communication Signals

Social behavior refers to cooperative, interdependent relationships among individuals of the same species. The relationships may be lifelong or as brief as a one-time sexual encounter. Regardless of the form it takes, social behavior requires **communication signals**. These are actions or cues sent by one individual that can change the behavior of another individual of the same species. Such interacting individuals are called **signalers** and **signal receivers**.

For example, when you ruptured the termite tunnel, pale workers banged their heads against the tunnel's ceiling and floor. The resulting vibrations alerted soldier termites and helped them locate the disturbance. Each soldier pointed its nose in the direction of danger, usually as announced by ant scents or a brush against an intruder. As a soldier shoots goo from its nose, chemical odors are released that attract more soldiers, which release more odorous chemicals.

Head banging and odorous goo are communication signals between termites. Is a scent from an ant also a communication signal for termites? Certainly the soldiers respond to it. However, even though the ant scent may have some evolved function, announcing the ant's presence to termites is not one of them. Termites use this scent as a *cue*, not as a signal. When a soldier termite detects the scent and kills the ant, the termite is an **illegitimate receiver** of a communication signal between individuals of a different species. The scent identifies the ant as a member of a particular colony and induces cooperative behavior from other *ants*.

Nature also has its share of **illegitimate signalers**. Certain assassin bugs hook dead and drained bodies of termite prey on their dorsal surface and so acquire the odor of their victims. They falsely signal themselves as members of a termite colony. The deception enables assassin bugs to hunt termite victims more easily.

A communication signal evolves or persists when it tends to increase the reproductive success of both the sender and the receiver. If it proves disadvantageous to either party, natural selection will favor the individuals that do not signal or do not respond.

Types of Communication Signals

Chemical, visual, acoustical, and tactile signals abound in the animal kingdom. Figures 52.2 and 52.3 show a few examples.

As you know, the most ancient sensory systems dealt with chemical odors. As they evolved, most animals began to use **pheromones**, chemical signals between individuals of the same species. *Signaling* pheromones may bring about an immediate behavioral response from a receiver. Some stimulate or suppress aggressive or defensive behavior. Termite alarm signals are like this. Other signaling pheromones are sex attractants. *Priming* pheromones, as discovered in rodents, elicit a generalized physiological response. For example, a chemical odor in the urine of male mice triggers and enhances estrus in female mice.

a

b

Figure 52.2 (**a**) Threat display by a male baboon. Exposing the canines, a visual signal of aggressive intent, can resolve conflict without fighting. (**b**) A male albatross in a courtship display. He spreads his wings and points his head to the sky in a visual signal to the female. Figure 19.4 shows more examples. Such displays precede copulation. They have visual, acoustical, and tactile components.

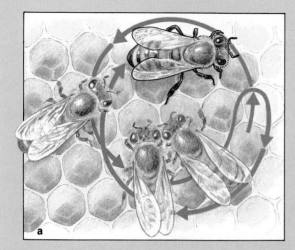

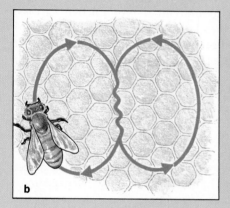

Figure 52.3 Dances of honeybees, which convey information through tactile signals. (**a**) Bees trained to visit feeding stations close to a hive perform a *round* dance on the honeycomb. Worker bees that maintain contact with the forager through the dance will search for food close to the hive. (**b**) Bees trained to visit feeding stations more than 100 meters from the hive perform a *waggle* dance. During the portion of the dance when the bee moves in a straight line, it waggles its abdomen. The slower the waggles, the more distant the food source.

(**c**) As discovered by Karl von Frisch, the orientation of the straight run varies, depending on the direction in which food is located. When he put a dish of honey on a direct line between the hive and the sun, foragers that located it returned to the hive and oriented their straight runs right up the honeycomb. When he put the honey at right angles to a line drawn between the hive and the sun, the foragers made their straight runs at 90 degrees to vertical. Thus, a honeybee "recruited" into searching for food could orient its flight *with respect to the sun and the hive*—and so waste less time and energy during the search. Waggle dancers also vary the speed of their dance in relation to the distance of the food source from the hive. When a site is 150 meters from a hive, the dance is much faster, with more waggles per straight run, compared with a dance concerning a food source that is 500 meters away.

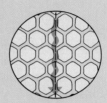
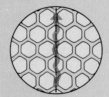
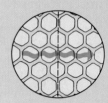

When bee moves straight up comb, recruits fly straight toward the sun. | When bee moves straight down comb, recruits fly to source directly away from the sun. | When bee moves to right of vertical, recruits fly at 90° angle to right of the sun.

Animals active during the day tend to use **visual signals**—observable actions or cues. A dominant male baboon "yawns" and exposes formidable canines when a rival for a receptive female confronts him (Figure 52.2*a*). His yawn, a *threat* display, may precede a physical attack. Suppose the rival backs down. The signaler benefits by retaining access to the female without a fight. The signal receiver also benefits by avoiding a serious beating, infection, even death.

Visual signals dominate *courtship* displays, which allow animals to assess the sexual overtures of potential partners. Lekking sage grouse and albatross provide examples (Figures 51.7*b* and 52.2*b*). Fireflies and some other nocturnal animals use bioluminescent flashes (page 103). A male firefly's light-generating organ emits a complex signal. A few seconds later, a receptive female of the same species may answer with a flash. The two flash back and forth until they meet. Females of some predatory fireflies are illegitimate signalers. When one sees a flash from a male firefly, she flashes back. If she lures him into attack range, he becomes a meal instead of a mate. This is an evolutionary cost of having an otherwise beneficial come-hither signal.

Acoustical signals are sounds with precise, species-specific information. Think of how a male white-throated sparrow's song attracts mates and secures territory. Or think of a tungara frog, sending acoustical signals at night to females and rival males. Its call is a "whine," followed by a "chuck." An illegitimate receiver, the fringe-lipped bat, may intercept the call. As the "chuck" attracts female frogs, this frog-eating bat may also be tracking it to its source.

With **tactile signals**, a signaler touches the receiver in socially significant ways. Such signals are vital in honeybee societies. After finding a rich source of pollen or nectar, a foraging honeybee returns to the hive and performs a complex dance. It moves in circles, jostling in the dark through a crowd of workers. Other bees may follow and maintain physical contact with the dancer. In this way, bees acquire information about the general location, distance, and direction of pollen or nectar (Figure 52.3).

A communication signal between individuals of the same species is an action that has a net beneficial effect on both the signaler and the signal receiver.

Natural selection tends to favor communication signals that promote reproductive success.

52.2 COSTS AND BENEFITS OF SOCIAL LIFE

Survey the animal kingdom and you find a range of social behaviors. Members of some species are mostly solitary or live in small family groups. Members of others, including termites and honeybees, live in huge groups of thousands of related individuals. Others, like modern humans, live in large social units composed of many unrelated individuals.

What is the basis for the diversity? To find answers, evolutionary biologists may take a **cost-benefit approach**. That is, they may measure the costs and benefits of social life in terms of the *reproductive success of the individual*, as measured by its contribution to the gene pool of the next generation.

Disadvantages to Sociality

In certain environments, the benefits of social life come at great cost. Consider the reduced odds of reproductive success for each individual in a large population. For example, each breeding pair of herring gulls in a huge nesting colony will cannibalize the eggs or young chicks of their neighbors in an instant if given the opportunity to do so. Besides this, many herring gulls crowded together in the same habitat will deplete the available food more rapidly than they would if they were dispersed. The overcrowded conditions also promote diseases.

The same is true of the huge colonies of cliff swallows, prairie dogs, or penguins (Figure 52.4). In such colonies, an individual and its offspring are more likely to be weakened by parasites, which are more readily transmitted from host to host in large groups. Our own history tells us that plagues spread like wildfire through the densely crowded cities of earlier times, before widespread reliance on plumbing, sewage treatment, and medical care (page 814).

Advantages of Sociality

Given the costs to individual reproductive success, why would the members of any species live close to others? As you will now see, *it may be that the benefits of social life are great enough to overcome the costs in certain environments.*

Cooperative Predator Avoidance   A group of animals acting cooperatively against a predator can reduce the net risk to any one individual. Simply having more pairs of eyes around helps individuals detect predators sooner. Individuals also may engage in a group counterattack or combine their defenses, as the musk oxen shown in Figure 52.5 are doing.

Figure 52.4   A colony of royal penguins on Macquarie Island, between New Zealand and Antarctica.

Figure 52.5 Social defensive behavior of musk oxen, which form a "ring of horns" against predators.

To gain insight into the benefits of sociality, the biologist Birgitta Sillen-Tullberg studied Australian sawfly caterpillars, which live together in clumps (Figure 52.6). When disturbed, the caterpillars collectively rear up, writhe about, and regurgitate partly digested food. Their food, eucalyptus leaves, contains chemical compounds that are toxic to most other animals—including predatory birds.

Sillen-Tullberg hypothesized that the caterpillars benefit from the coordinated repulsion of predators. She used her hypothesis to predict that insect-eating songbirds would more likely consume a solitary caterpillar, not a cluster of them. She tested her prediction by offering young, hand-reared Great Tits (*Parus major*) a chance to feed on caterpillars offered either one by one or in a group of twenty per offering. She did this for a standard number of presentations. Ten birds that were offered one sawfly at a time consumed an average of 5.6 sawflies. Ten birds presented with sawfly clumps ate 4.1, on average. In this experiment, individuals were, as expected, somewhat safer in a group than on their own.

The Selfish Herd Simply by their physical position in a group, some individuals form a living shield against predation on others in the group. They are part of a **selfish herd**, a simple society held together by reproductive self-interest, although not consciously so.

The selfish-herd hypothesis has been tested for male bluegill sunfishes, which build adjacent nests on lake bottoms. Each male uses its fins to hollow a depression in the muddy sediments. There his mate or mates deposit eggs.

Figure 52.6 Social defensive behavior of Australian sawfly caterpillars. These have smeared chemical secretions on one another in response to a disturbance.

If the colony of bluegill males is a selfish herd, we can predict competition for the "safe" sites—at the center of the colony. Compared to the periphery, eggs laid in nests at the center are less likely to be attacked by snails and largemouth bass. Competition of this sort does indeed exist. The largest, most powerful males tend to claim the central locations. Other, smaller males assemble around them and bear the brunt of predatory attacks. Even then, they are better off in the group than on their own, fending off a bass singlehandedly, so to speak.

Costs and benefits of social life may be evaluated in terms of reproductive success, as measured by the genes each individual contributes to the next generation.

52.3 SOCIAL LIFE AND SELF-SACRIFICE

Self-Sacrificing Behavior and Dominance Hierarchies

Members of a selfish herd make no personal sacrifice for the others. The personal benefits of living with others simply seem to outweigh the costs. By contrast, members of some social groups help other individuals survive and reproduce at personal cost. This is **self-sacrificing behavior**, and it may be more apparent than real. Individuals may *not* be giving up their chance at reproductive success entirely. We may hypothesize that making the sacrifice is really a cost of belonging to the group.

Does the hypothesis hold up when we examine self-sacrificing behavior in nature? Consider a baboon troop or wolf pack. In these social groups, individuals help others, but reproductive opportunity is unequal. Some members of a baboon troop give up safe sleeping places, choice bits of food, and even receptive females to others upon receiving a threat signal from another

troop member. In fact, a **dominance hierarchy** exists, in which some individuals have adopted a subordinate status to others (Figure 52.7). Jane Goodall provided us with details of one such hierarchy when she studied chimpanzee society (page 612).

Similarly, members of a wolf pack show helpful behavior, as when they fend off predators or share food. Nevertheless, only *one* male and *one* female wolf typically produce the pups (Figure 52.8). Other members of the pack do not breed, even though they sometimes bring back food to members that have stayed at the den, guarding the pups.

Why should subordinate, nonbreeding adults remain in a group and make sacrifices for their dominant peers? As for small bluegill sunfish, their benefits from group living may offset the sacrifices. Besides, challenging a stronger member could lead to life-shortening injury. It may not be possible to survive alone, outside of the group. A solitary baboon surely quickens the pulse of the first leopard that sees it.

Just as importantly, self-sacrificing behavior may give subordinates a chance to reproduce if they live long enough and if predation, weakness in old age, or some other event removes dominant peers in the hierarchy. Some subordinate wolves and baboons do indeed move up the social ladder when dominant members slip down a rung or fall off. Thus, acceptance of subordinate status may pay off in the long term for the patient individual.

Figure 52.7 Appeasement behavior between baboons. Notice the assured position of the dominant animal and the abject stare and groveling posture of the subordinate one, who is making little conciliatory smacking noises with its lips.

a   b

Figure 52.8 (**a**) A dominant male and female member of a wolf pack. Typically, the dominant male is the only pack member likely to breed successfully. (**b**) Subordinate members of the pack greet a dominant male.

Figure 52.9 Male and female Caspian terns, protecting their chicks. Parental care has costs as well as benefits.

Costs and Benefits of Parenting

In some species, parents take care of their offspring until the offspring have developed enough to survive on their own. Adult Caspian terns (Figure 52.9) incubate the eggs, then shelter the nestlings, feed them, and accompany and protect them after they start flying. Their parental behavior drains time and energy that might otherwise be allocated to improving their own chance of living to reproduce another time.

Yet parental behavior benefits the individual by improving the likelihood that the current generation of offspring will survive. The benefit of immediate reproductive success may outweigh the cost of reduced reproductive success at some later time.

Self-sacrificing behavior, in which individuals give up their chance to reproduce while helping others of their social group, may be more apparent than real.

Such individuals may be better off with the group than without it—and they may be presented with an opportunity to reproduce later on.

THE EVOLUTION OF ALTRUISM

With **altruistic behavior**, recall, an individual behaves in a self-sacrificing way that helps others but decreases its own chance of reproductive success.

For example, a termite soldier or honeybee worker is sterile and helps others at great personal cost. Given that such individuals cannot contribute their own genes to the next generation, how does the genetic basis for altruistic behavior persist over evolutionary time? By one theory, individuals can indirectly pass on their genes by helping *relatives* survive and reproduce. This is especially true of the sterile worker members of insect societies.

Consider that when a sexually reproducing parent helps offspring, it is not helping *exact* genetic copies of itself. Each offspring has only half of its genes. If other individuals of the social group have the same ancestors, they also have genes in common with the parents. Genetically, two siblings are as similar as a parent and one of its offspring. Nephews and nieces are like their uncle in about one-fourth of their genes.

According to William Hamilton's **theory of indirect selection**, genes associated with caring for relatives that are not direct descendants—that is, not one's own off-spring—are favored. Think of this type of altruistic behavior as an extension of parental behavior. For example, suppose an uncle helps a niece survive long enough to reproduce. He has effectively made a genetic contribution to the next generation, even though it is an indirect one. His contribution can be measured in the genes that he and his relative share. Altruism costs him; he may lose his own opportunities to reproduce. But if the cost is less than the benefit, the action will propagate the uncle's genes and favor the spread of his kind of altruism in the species. If an uncle saves two nieces, this is equivalent to saving his own daughter.

Thus, workers of insect societies indirectly promote their "self-sacrifice" genes through altruistic behavior directed toward relatives. Colonies of honeybees, ants, and termites are actually large families (Figure 52.10).

a

b

c

d

The family's worker force labors on behalf of siblings, some of which are future kings and queens. When a guard bee drives her stinger into a raccoon, she dies—but siblings perpetuate some of her genes.

Sterility and extreme self-sacrifice are rare among vertebrate societies. Among the known exceptions are naked mole-rats, the type of mammal described in the *Focus* essay on page 930.

By one theory of indirect selection, genes associated with behavior that helps nondescendant relatives are favored.

(**b**) Transfer of food from bee to bee, a helpful action in honeybee society. (**c**) Bee dance that recruits workers into searching for food, as described in Figure 52.3. (**d**) Guard bees. Worker females at the colony entrance, positioned to repel intruders.

(**e**) The queen is much larger than the workers, in part because her ovaries are fully developed (unlike those of her sterile daughters). (**f**) Stingless drones are produced at certain times of the year. They do not work for the colony but instead attempt to mate with queens of other hives.

(**g**) Worker bees forage, feed larvae, guard the colony, construct honeycomb, and clean and maintain the nest. Between 30,000 and 50,000 are present in a colony. They live about six weeks in the spring and summer, and they can survive about four months in an overwintering colony.

(**h**) Scent-fanning, another cooperative action by a worker. Fanned air passes over the bee's exposed scent gland. Pheromones released from the gland help other bees orient to the colony entrance. (**i**) Worker bees construct new honeycomb from wax secretions. Here, honey or pollen is stored or new generations are cared for from the egg stage to the emergence of the adult.

(**j**) Initial stages in the honeybee life cycle, revealing eggs and larvae of various ages. Larvae are fed by young worker bees. (**k**) Worker pupae. Cell caps have been removed, exposing pupae that will metamorphose into future workers. (**l**) Sequence of developmental stages of a worker bee, from the egg to a six-day-old larva (fourth from top), to a twenty-one-day-old pupa about to become an adult.

Figure 52.10 Life in a honeybee colony. (**a**) This queen bee, the only fertile female in the hive, is surrounded by her court of sterile worker daughters. Daughters feed her and relay her pheromones throughout the hive. The pheromones influences the activity of all members of the colony.

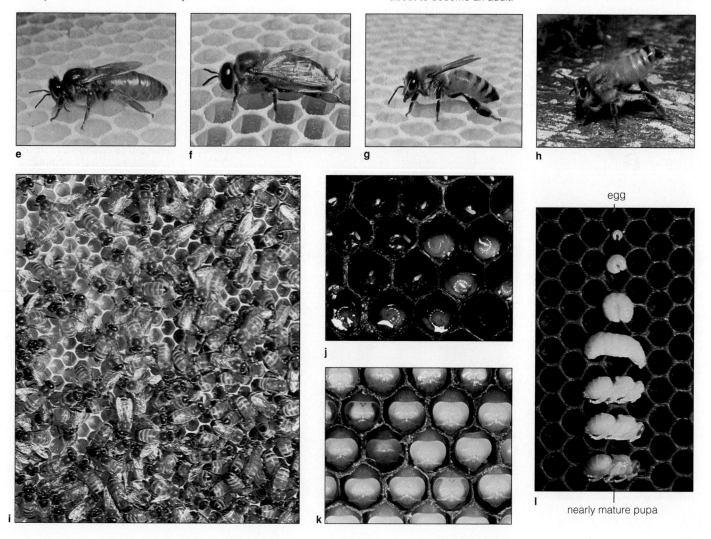

About Those Self-Sacrificing Naked Mole-Rats

Naked mole-rats look rather like bucktoothed sausages with wrinkled, pink skin from which a few hairs sprout (Figure *a*). Their behavior is as fascinating as their looks. Unlike any other known vertebrate, many naked mole-rat individuals spend their entire lives as nonbreeding helpers in their social group.

These highly social mammals live underground in certain arid regions of eastern Africa. They are always found in colonies of about 25 to 300 members. Each colony has only a single reproducing female that mates with one to three males. All other members of the colony care for the "queen" and "king" (or kings) and their offspring.

The nonreproducers dig extensive subterranean tunnels and special chambers that serve as living rooms or waste-disposal centers. They also locate and chop up large, underground tubers—the food for the colony. Workers deliver bits of root to the queen and her retinue of males and offspring. They also deliver food to some other helpers that spend time loafing about, shoulder to shoulder (and belly to back), with the reproductives. The "loafers" are usually larger than the diggers. They spring to action when a snake or some other enemy threatens the colony. At great personal risk, they collectively chase away or kill the predator.

How can such helpful behavior persist if helpful members fail to reproduce—and so fail to have offspring that

carry on the genetic basis for helping? It can if the helpers are related to the offspring. Applying the theory of indirect selection to naked mole-rats, let's formulate a hypothesis with a testable prediction:

1. If helping is genetically advantageous to some individuals in a naked mole-rat colony (*the hypothesis*), then it follows that the helpers will be related to the reproductive members of the colony that benefit from their altruism (*the prediction*).
2. Determine the genetic relationships among subordinate, sterile helpers and the dominant, reproducing queen and king or kings. (*This would be an appropriate test of the prediction.*)

Could the hypothesis be tested by marking each individual in a naked mole-rat colony and gradually establishing a family tree for the group? Probably not. It isn't likely that all the individuals could even be identified, given all the intricate, hidden tunnels.

Instead, H. Kern Reeve and his colleagues used *DNA fingerprinting*, a method of establishing degrees of genetic relatedness among individuals. The method uses restriction fragments, of the sort described on page 252. A visual record can be constructed of different sets of fragments from the DNA of different individuals. Identical twins have essentially identical DNA fingerprints (their DNA has the same base sequence). Individuals with the same father and mother have similar DNA—thus similar but not identical DNA fingerprints. On the average, the DNA fingerprints of genetically unrelated individuals should differ much more than those of siblings or other relatives.

DNA fingerprints were constructed for naked mole-rats. As it turns out, individuals from a given colony are *very* close relatives, and they are very different genetically from members of other colonies. These findings suggest that each naked mole-rat colony is highly inbred, a result of generations of brother-sister, mother-son, or father-daughter matings. Such lineages have extremely reduced genetic variability.

Therefore, an altruistic naked mole-rat is helping to perpetuate a very high proportion of the forms of genes (alleles) that it carries. The genotypes of helpers and the helped might be 90 percent identical!

52.5 AN EVOLUTIONARY VIEW OF HUMAN SOCIAL BEHAVIOR

If we can analyze the behavior of termites and naked mole-rats, is it possible to analyze human behavior also? Many people resist this idea. Apparently they believe that attempts to identify the adaptive value of a particular human trait is an attempt to define its moral or social advantage. But clearly there is a difference between trying to explain something in terms of its evolutionary history and trying to justify it. "Adaptive" does not mean "moral," it only means valuable in the transmission of an individual's genes.

Consider adoption, an example of an altruistic human behavior. If we wish to gain insight into this behavior, we can use an evolutionary approach to generate testable hypotheses about it. Whether adoption is moral or socially desirable is an entirely separate issue, about which biologists have no more to say than anybody else.

Start with the premise that all adaptations have costs and benefits. One cost is that certain adaptive behaviors might be *redirected* under rare or unusual circumstances. In many species, adults that have lost their offspring will, if presented with a substitute, adopt it. Thus cardinals have been known to feed goldfish. A whale tried to lift a log out of water, as if it were a distressed infant that needed help in reaching the water's surface. And emperor penguins fight to adopt an orphan (Figure 52.11).

As John Alcock suggests, such examples of redirected parental behavior allow us to formulate a hypothesis about adoptive behavior among humans. Specifically, "Husbands and wives who have lost an only child or who fail to produce children themselves should be especially prone to adopt strangers. Their frustrated desire to be parents could act as a proximate basis for adopting a substitute for genetic offspring."

Now consider this. For most of their evolutionary history, humans have lived in small groups or, later, small villages. They probably had few opportunities or showed little inclination to adopt strangers. They were likely to accept the child of a deceased relative.

How would such adoptive parents gain genetic representation in the next generation? According to theories of natural selection and indirect selection, individuals act in ways that promote their genetic self-interest. We might therefore predict that parents with some dependent, care-requiring children of their own are much less likely to adopt, compared to adults who have no children or already raised children and are living by themselves. We could test the prediction by securing a sufficient sample of data, ideally from a variety of existing human cultures, on the relationship between childlessness and adoption.

Figure 52.11 Two adult emperor penguins competing to adopt an orphan, which penguins accept as substitute offspring.

Indirect selection also favors adults who direct parenting behavior toward relatives and so indirectly perpetuate their shared genes. We might even predict that such adults are more likely to be related to the adopted child than would be expected by chance alone. Joan Silk did test this prediction. She found that in some traditional societies, children are adopted overwhelmingly by relatives.

Especially in large, industrialized societies, with their agencies and other means of assisted adoption, individuals become parents of nonrelated children. Such societies are recent novelties. Strong parenting mechanisms evolved in the past, and it may be that their redirection toward a nonrelative says more about our evolutionary history than it does about the transmission of one's genes.

The point is, evolutionary hypotheses about the adaptive value of behavior lend themselves to testing. And through such testing, we can gain additional understanding about the evolution of human behavior.

---

1. It is possible to test evolutionary hypotheses about the adaptive value of human behaviors.

2. There may be greater acceptance of such testing when more people understand that *adaptive* behavior and *socially desirable* behavior are separate issues. "Adaptive" means only that a given trait has proved beneficial in the transmission of an individual's genes.

SUMMARY

1. Social behavior is a cooperative interdependency among individuals of the same species.

2. Social behavior requires communication signals—actions or cues sent by one individual (the signaler) that can change the behavior of another individual of the same species (the signal receiver). A communication signal has a beneficial effect on both the signaler and the receiver.

3. Signaling animals employ many different sensory modes as channels of communication. Among these are chemical, visual, acoustical, and tactile signals.

    a. Pheromones are examples of chemical signals. The signaling types, such as termite alarm signals or sex attractants, can immediately change the behavior of a signal receiver. The priming pheromones elicit a generalized physiological response by the signal receiver.

    b. Visual signals are observable actions or cues. They are among the premier components of courtship displays and threat displays.

    c. Acoustical signals are sounds with precise, species-specific information, such as the mating calls of frogs.

    d. Tactile signals are forms of physical contact between a signaler and a receiver. Interactions among honeybees during a round dance or waggle dance in the hive are examples.

4. Natural selection tends to favor communication signals that promote reproductive success.

5. The costs and benefits of social life may be discerned in terms of individual reproductive success, as measured by the individual's genetic contribution to the next generation.

6. Sociality carries high risks of contagious disease and competition for limited resources. Benefits outweigh the costs of social life under some circumstances, as when predation pressure is severe.

7. Self-sacrificing behavior, in which individuals give up chances to reproduce while helping others of their social group, may be more apparent than real.

    a. In most societies, individuals do not sacrifice reproductive chances to help others in their group.

    b. Helpful acts often may be the result of dominant individuals forcing subordinates to relinquish food or some other resource.

    c. When subordinates give in to dominant members of their group, they may receive compensatory benefits of group living, such as safety from predators.

    d. In some species, subordinates may eventually reproduce if they live long enough and if dominant ones slip in the hierarchy or die off.

8. By one theory of indirect selection, genes associated with self-sacrificing behavior that helps nondescendant relatives are favored.

9. Indirect selection can lead to extreme altruism. This is evident among workers of some species of social insects and naked mole-rats. Such workers normally do not reproduce. Instead, they help their reproducing siblings survive and thereby pass on "by proxy" the genes underlying the development of altruism.

Review Questions

1. A hyena places scent marks on vegetation in its territory by releasing specific chemicals from certain glands. What evidence would you need to demonstrate that this action is an evolved communication signal? *922*

2. Explain how a threat display can be considered an example of cooperation. *923*

3. Why don't the members of a "selfish herd" live apart if each member is "trying to take advantage" of the others? *925*

4. How can a self-sacrificing individual of a social group still pass on more of its genes than a noncooperative individual? *926*

5. Is parental behavior always adaptive? *927*

6. How can an individual propagate its genes even if it is sterile? *928–929*

Self-Quiz *(Answers in Appendix IV)*

1. A communication signal affects the behavior of _____ .
    a. legitimate receivers
    b. illegitimate receivers
    c. illegitimate signalers
    d. both a and b

2. Social behavior evolves because _____ .
    a. social species are more advanced evolutionarily than solitary ones
    b. under some ecological conditions, the costs of social life are less than its benefits to the group
    c. under some ecological conditions, the costs of social life are less than its benefits to the individual
    d. predator pressures always favor social living

3. Which is an evolutionary cost of a loud acoustical signal?
    a. the energy expended by the caller
    b. exploitation of the call by an illegitimate receiver
    c. the time spent calling that cannot be spent feeding
    d. all of the above

4. The behavior of subordinate animals in groups with dominance hierarchies puzzles Darwinian biologists because _____ .
    a. subordinates use valuable resources that the more adaptive dominants should receive if the group is to persist
    b. dominants often physically attack subordinates
    c. subordinates are usually younger or weaker individuals
    d. subordinates often do not even attempt to reproduce when dominants are present

5. The theory of evolution by indirect selection _____ .
   a. focuses on evolution for group benefit
   b. explains that altruism is always adaptive
   c. examines the consequences of the effects of differences among individuals on the reproductive success of relatives
   d. shows how families with many descendants are superior to self-sacrificing families

6. Altruism is (in terms of evolutionary biology) helpful _____ .
   a. behavior that cannot evolve
   b. behavior that can spread through a species only if it raises the altruist's reproductive success
   c. behavior that reduces an individual's reproductive success
   d. behavior that always helps spread the altruist's genes

7. The discovery that members in large colonies of cliff swallows are more likely to have nests infested with nest parasites provides evidence that _____ .
   a. sociality has disadvantages as well as possible benefits
   b. cliff swallows will never nest solitarily
   c. the costs of sociality are greater than its benefits
   d. social life forms are more advanced evolutionarily than solitary ones

8. The finding that predators eat more sawflies when the prey are offered to them experimentally one by one rather than in groups of ten is _____ .
   a. a hypothesis   c. a test of a prediction
   b. a prediction   d. a proven hypothesis

9. The degree of genetic similarity between an uncle and his nephew is _____ .
   a. the same as between a parent and his offspring
   b. greater than between two full siblings
   c. dependent on how many other nephews the uncle has
   d. less than that between a mother and her daughter

10. You discover a new species of insect in which groups consist of sterile males and females that cooperate to help rear the young of four pairs of reproducing males and females. You might use DNA fingerprinting to test which of the following predictions based on the theory of indirect selection:
   a. The sterile males will not be closely related to the sterile females.
   b. The sterile individuals will be more closely related to each other than to the reproducing males and females.
   c. The reproducing individuals will be cousins.
   d. All the members of one group will be much more closely related genetically to each other than to members of any other group.

Selected Key Terms

acoustical signal *923*	self-sacrificing behavior *926*
altruistic behavior *928*	selfish herd *925*
communication signal *922*	signal receiver *922*
cost-benefit approach *924*	signaler *922*
dominance hierarchy *926*	social behavior *922*
illegitimate receiver *922*	tactile signal *923*
illegitimate signaler *922*	theory of indirect selection *928*
pheromone *922*	visual signal *923*

Readings

Alcock, J. 1989. *Animal Behavior*. Fourth edition. Sunderland, Massachusetts: Sinauer.

Daly, M., and M. Wilson. 1983. *Sex, Evolution, and Behavior*. Second edition. Boston: Willard Grant Press.

Dawkins, R. 1989. *The Selfish Gene*. New York: Oxford University Press.

Frisch, K. von. 1961. *The Dancing Bees*. New York: Harcourt Brace Jovanovich. A classic on the natural history and behavior of honeybees.

Maho, Y. 1977. "Emperor Penguin: A Strategy to Live and Breed in the Cold." *American Scientist* 65(6) 680–693.

Trivers, R. 1985. *Social Evolution*. Menlo Park, California: Benjamin/Cummings. A complete and readable account of all the topics that make up an evolutionary approach to social behavior.

Wilson, E. O. 1975. *Sociobiology: The New Synthesis*. Cambridge, Massachusetts: Harvard University Press. The pivotal book in presenting a new approach to the study of social behavior.

Wittenberger, J. F. 1981. *Animal Social Behavior*. Boston: Willard Grant Press. A book that covers much the same ground as *Social Evolution* by Robert Trivers but at a somewhat more advanced level. The book does an excellent job of laying out competing hypotheses on aspects of social behavior and evaluating the evidence for them.

APPENDIX I
A Brief Classification Scheme

The following classification scheme is a composite of several that are used in microbiology, botany, and zoology. The major groupings are more or less agreed upon. There is not always agreement on what to call a given grouping or where it fits in the overall hierarchy. There are several reasons for this.

First, the fossil record varies in its quality and completeness. Therefore, the relationship of one group to others is sometimes open to interpretation. Comparative studies at the molecular level are firming up the picture, but this work is still under way.

Second, since the time of Linnaeus, classification schemes have been based on perceived morphological similarities and differences among organisms. Although some original interpretations are now open to question, we are so used to thinking about organisms in certain ways that reclassification proceeds slowly. Traditionally, for example, birds and reptiles are separate classes (Reptilia and Aves). Yet there are compelling arguments for grouping lizards and snakes as one class, and crocodilians, dinosaurs, and birds as another.

Finally, microbiologists, mycologists, botanists, zoologists, and other researchers have inherited a wealth of literature, based on classification schemes peculiar to their fields. Most see no good reason to give up established terminology and so disrupt access to the past. Until very recently, botanists were using *division*, and zoologists, *phylum* for groupings that are equivalent in the hierarchy. Opinions are polarized with respect to an entire kingdom (the Protista), certain members of which could just as easily be called single-celled plants, fungi, or animals. Indeed, the term protozoan is a holdover from earlier schemes that ranked the amoebas and some other single cells as simple animals.

Given the problems, why do we bother imposing artificial frameworks on the history of life? We do this for the same reason that a writer might decide to break up the history of civilization into several volumes, many chapters, and many more paragraphs. Both efforts are attempts to impart structure to what might otherwise be an overwhelming body of information.

Bear in mind, we include this classification scheme mainly for your reference purposes. It is by no means complete. Numerous existing and extinct organisms of the so-called lesser phyla are not represented. Our strategy is to focus mainly on organisms mentioned in the text. A few examples of organisms also are listed under the entries.

SUPERKINGDOM PROKARYOTA. Prokaryotes. Single-celled organisms with DNA concentrated in a cytoplasmic region, not in a membrane-bound nucleus.

KINGDOM MONERA. Bacteria, either single cells or simple associations of cells. Both autotrophs and heterotrophs (refer to Table 22.2). *Bergey's Manual of Systematic Bacteriology,* the authoritative reference in the field, calls this "a time of taxonomic transition." It groups bacteria mainly on the basis of form, physiology, and behavior, not on phylogeny. The scheme presented here does reflect the growing evidence of evolutionary relationships for at least some bacterial groups.

SUBKINGDOM ARCHAEBACTERIA. Methanogens, halophiles, thermophiles. Strict anaerobes, distinct from other bacteria in cell wall, membrane lipids, ribosomes, and RNA sequences. *Methanobacterium, Halobacterium, Sulfolobus.*

SUBKINGDOM EUBACTERIA. Gram-negative and gram-positive forms. Peptidoglycan in cell wall. Photosynthetic autotrophs, chemosynthetic autotrophs, and heterotrophs.

PHYLUM GRACILICUTES. Typical Gram-negative, thin wall. Autotrophs (photosynthetic and chemosynthetic) and heterotrophs. *Anabaena* and other cyanobacteria. *Escherichia, Pseudomonas, Neisseria, Myxococcus.*
PHYLUM FIRMICUTES. Typical Gram-positive, thick wall. Heterotrophs. *Bacillus, Staphylococcus, Streptococcus, Clostridium, Actinomycetes.*
PHYLUM TENERICUTES. Gram-negative, wall absent. Heterotrophs (saprobes, pathogens). *Mycoplasma.*

SUPERKINGDOM EUKARYOTA. Eukaryotes (single-celled and multicelled organisms. Cells typically have a nucleus (enclosing the DNA) and other membrane-bound organelles.

KINGDOM PROTISTA. Mostly single-celled eukaryotes, some colonial forms. Diverse autotrophs and heterotrophs. Many lineages apparently related evolutionarily to certain plants, fungi, and possibly animals.

PHYLUM CHYTRIDIOMYCOTA. Chytrids. Heterotrophs; saprobic decomposers, parasites. *Chytridium.*
PHYLUM OOMYCOTA. Water molds. Heterotrophs. Decomposers, some parasites. *Saprolegnia, Phytophthora, Plasmopara.*
PHYLUM ACRASIOMYCOTA. Cellular slime molds. Heterotrophs with amoeboid and spore-bearing stages. *Dictyostelium.*
PHYLUM MYXOMYCOTA. Plasmodial slime molds. Heterotrophs with amoeboid and spore-bearing stages. *Physarum.*
PHYLUM SARCODINA. Amoeboid protozoans. Heterotrophs. Soft-bodied; some shelled. Amoebas, foraminiferans, radiolarians, heliozoans. *Amoeba, Entomoeba.*
PHYLUM CILIOPHORA. Ciliated protozoans. Heterotrophs. Distinctive arrays of cilia, used as motile structures. *Paramecium,* hypotrichs.
PHYLUM MASTIGOPHORA. Flagellated protozoans. Some free-living, many internal parasites; all with one to several flagella. *Trypanosoma, Trichomonas, Giardia.*

SPOROZOANS. Parasitic protozoans, many intracellular. "Sporozoans" is the common name with no formal taxonomic status. *Plasmodium, Toxoplasma*.

PHYLUM EUGLENOPHYTA. Euglenoids. Mostly heterotrophs, some photosynthetic types. Flagellated. *Euglena*.

PHYLUM CHRYSOPHYTA. Golden algae, yellow-green algae, diatoms. Photosynthetic. Some flagellated, others not. *Mischococcus, Synura, Vaucheria*.

PHYLUM PYRRHOPHYTA. Dinoflagellates. Photosynthetic, mostly, but some heterotrophs. *Gymnodinium breve*.

PHYLUM RHODOPHYTA. Red algae. Photosynthetic, nearly all marine, some freshwater. *Porphyra. Bonnemaisonia, Euchema*.

PHYLUM PHAEOPHYTA. Brown algae. Photosynthetic, nearly all temperate or marine waters. *Macrocystis, Fucus, Sargassum, Ectocarpus, Postelsia*.

PHYLUM CHLOROPHYTA. Green algae. Photosynthetic. Most freshwater, some marine or terrestrial. *Chlamydomonas, Spirogyra, Ulva, Volvox, Codium, Halimeda*.

KINGDOM FUNGI. Mostly multicelled eukaryotes. Heterotrophs (mostly saprobes, some parasites). Major decomposers of nearly all communities. Reliance on extracellular digestion of organic matter and absorption of nutrients by individual cells.

PHYLUM ZYGOMYCOTA. Zygomycetes. All produce nonmotile spores. Bread molds, related forms. *Rhizopus, Philobolus*.

PHYLUM ASCOMYCOTA. Ascomycetes. Sac fungi. Most yeasts and molds; morels, truffles. *Saccharomycetes, Morchella Neurospora, Sarcoscypha. Claviceps, Ophiostoma*.

PHYLUM BASIDIOMYCOTA. Basidiomycetes. Club fungi. Mushrooms, shelf fungi, stinkhorns. *Agaricus, Amanita, Puccinia, Ustilago*.

IMPERFECT FUNGI. Sexual spores absent or undetected. The group has no formal taxonomic status. If better understood, a given species might be grouped with sac fungi or club fungi. *Verticillium, Candida, Microsporum, Histoplasma*.

LICHENS. Mutualistic interactions between a fungus and a cyanobacterium, green alga, or both. *Usnea, Cladonia*.

KINGDOM PLANTAE. Nearly all multicelled eukaryotes. Photosynthetic autotrophs, except for a few parasitic types.

PHYLUM RHYNIOPHYTA. Earliest known vascular plants; extinct. *Cooksonia, Rhynia*.

PHYLUM TRIMEROPHYTA. Trimerophytes. *Psilophyton*.

PHYLUM PROGYMNOSPERMOPHYTA. Progymnosperms. Ancestral to early seed-bearing plants; extinct. *Archaeopteris*.

PHYLUM CHAROPHYTA. Stoneworts.

PHYLUM BRYOPHYTA. Liverworts, hornworts, mosses. *Marchantia, Polytrichum, Sphagnum*.

PHYLUM PSILOPHYTA. Whisk ferns. *Psilotum*.

PHYLUM LYCOPHYTA. Lycophytes, club mosses. *Lycopodium, Selaginella*.

PHYLUM SPHENOPHYTA. Horsetails. *Equisetum*.

PHYLUM PTEROPHYTA. Ferns.

PHYLUM PTERIDOSPERMOPHYTA. Seed ferns. Fernlike gymnosperms; extinct.

PHYLUM CYCADOPHYTA. Cycads. *Zamia*.

PHYLUM GINKGOPHYTA. Ginkgo. *Ginkgo*.

PHYLUM GNETOPHYTA. Gnetophytes. *Ephedra, Welwitchia*.

PHYLUM CONIFEROPHYTA. Conifers.
  Family Pinaceae. Pines, firs, spruces, hemlock, larches, Douglas firs, true cedars. *Pinus*.
  Family Cupressaceae. Junipers, cypresses. *Juniperus*.
  Family Taxodiaceae. Bald cypress, redwoods, Sierra bigtree, dawn redwood. *Sequoia*.
  Family Taxaceae. Yews.

PHYLUM ANTHOPHYTA. Flowering plants.
  Class Dicotyledonae. Dicotyledons (dicots). Some families of several different orders are listed:
    Family Nymphaeaceae. Water lilies.
    Family Papaveraceae. Poppies.
    Family Brassicaceae. Mustards, cabbages, radishes.
    Family Malvaceae. Mallows, cotton, okra, hibiscus.
    Family Solanaceae. Potatoes, eggplant, petunias.
    Family Salicaceae. Willows, poplars.
    Family Rosaceae. Roses, apples, almonds, strawberries.
    Family Fabaceae. Peas, beans, lupines, mesquite.
    Family Cactaceae. Cacti.
    Family Euphorbiaceae. Spurges, poinsettia.
    Family Cucurbitaceae. Gourds, melons, cucumbers, squashes.
    Family Apiaceae. Parsleys, carrots, poison hemlock.
    Family Aceraceae. Maples.
    Family Asteraceae. Composites. Chrysanthemums, sunflowers, lettuces, dandelions.
  Class Monocotyledonae. Monocotyledons (monocots). Some families of several different orders are listed:
    Family Liliaceae. Lilies, hyacinths, tulips, onions, garlic.
    Family Iridaceae. Irises, gladioli, crocuses.
    Family Orchidaceae. Orchids.
    Family Arecaceae. Date palms, coconut palms.
    Family Cyperaceae. Sedges.
    Family Poaceae. Grasses, bamboos, corn, wheat, sugarcane.
    Family Bromeliaceae. Bromeliads, pineapples, Spanish moss.

KINGDOM ANIMALIA. Multicelled eukaryotes. Heterotrophs (herbivores, carnivores, omnivores, parasites, detritivores).

PHYLUM PLACOZOA. Small, organless marine animal. *Trichoplax*.

PHYLUM MESOZOA. Ciliated, wormlike parasites, about the same level of complexity as *Trichoplax*.

PHYLUM PORIFERA. Sponges.

PHYLUM CNIDARIA.
  Class Hydrozoa. Hydrozoans. *Hydra, Obelia, Physalia*.
  Class Scyphozoa. Jellyfishes. *Aurelia*.
  Class Anthozoa. Sea anemones, corals. *Telesto*.

PHYLUM CTENOPHORA. Comb jellies. *Pleurobrachia*.

PHYLUM PLATYHELMINTHES. Flatworms.
  Class Turbellaria. Triclads (planarians), polyclads. *Dugesia*.
  Class Trematoda. Flukes. *Schistosoma*.
  Class Cestoda. Tapeworms. *Taenia*.

PHYLUM NEMERTEA. Ribbon worms.

PHYLUM NEMATODA. Roundworms. *Ascaris, Trichinella*.

PHYLUM ROTIFERA. Rotifers.

PHYLUM MOLLUSCA. Mollusks.
  Class Polyplacophora. Chitons.
  Class Gastropoda. Snails (periwinkles, whelks, limpets, abalones, cowries, conches, nudibranchs, tree snails, garden snails), sea slugs, land slugs.
  Class Bivalvia. Clams, mussels, scallops, cockles, oysters, shipworms.
  Class Cephalopoda. Squids, octopuses, cuttlefish, nautiluses. *Loligo*.

PHYLUM BRYOZOA. Bryozoans (moss animals).

PHYLUM BRACHIOPODA. Lampshells.

PHYLUM ANNELIDA. Segmented worms.
  Class Polychaeta. Mostly marine worms.
  Class Oligochaeta. Mostly freshwater and terrestrial worms, but many marine. *Lumbricus* (earthworms).
  Class Hirudinea. Leeches.

PHYLUM TARDIGRADA. Water bears.

PHYLUM ONYCHOPHORA. Onychophorans. *Peripatus*.

PHYLUM ARTHROPODA.
  Subphylum Trilobita. Trilobites; extinct.
  Subphylum Chelicerata. Chelicerates. Horseshoe crabs, spiders, scorpions, ticks, mites.
  Subphylum Crustacea. Shrimps, crayfishes, lobsters, crabs, barnacles, copepods, isopods (sowbugs).
  Subphylum Uniramia.
    Superclass Myriapoda. Centipedes, millipedes.
    Superclass Insecta.
      Order Ephemeroptera. Mayflies.
      Order Odonata. Dragonflies, damselflies.

Order Orthoptera. Grasshoppers, crickets, katydids.
Order Dermaptera. Earwigs.
Order Blattodea. Cockroaches.
Order Mantodea. Mantids.
Order Isoptera. Termites.
Order Mallophaga. Biting lice.
Order Anoplura. Sucking lice.
Order Homoptera. Cicadas, aphids, leafhoppers, spittlebugs.
Order Hemiptera. Bugs.
Order Coleoptera. Beetles.
Order Diptera. Flies.
Order Mecoptera. Scorpion flies. *Harpobittacus*.
Order Siphonaptera. Fleas.
Order Lepidoptera. Butterflies, moths.
Order Hymenoptera. Wasps, bees, ants.

PHYLUM ECHINODERMATA. Echinoderms.
Class Asteroidea. Sea stars.
Class Ophiuroidea. Brittle stars.
Class Echinoidea. Sea urchins, heart urchins, sand dollars.
Class Holothuroidea. Sea cucumbers.
Class Crinoidea. Feather stars, sea lilies.
Class Concentricycloidea. Sea daisies.

PHYLUM HEMICHORDATA. Acorn worms.

PHYLUM CHORDATA. Chordates.
Subphylum Urochordata. Tunicates, related forms.
Subphylum Cephalochordata. Lancelets.
Subphylum Vertebrata. Vertebrates.
Class Agnatha. Jawless vertebrates (lampreys, hagfishes).
Class Placodermi. Jawed, heavily armored fishes; extinct.
Class Chondrichthyes. Cartilaginous fishes (sharks, rays, skates, chimaeras).
Class Osteichthyes. Bony fishes.
Subclass Dipnoi. Lungfishes.
Subclass Crossopterygii. Coelacanths, related forms.
Subclass Actinopterygii. Ray-finned fishes.
Order Acipenseriformes. Sturgeons, paddlefishes.
Order Salmoniformes. Salmon, trout.
Order Atheriniformes. Killifishes, guppies.
Order Gasterosteiformes. Seahorses.
Order Perciformes. Perches, wrasses, barracudas, tunas, freshwater bass, mackerels.
Order Lophiiformes. Angler fishes.
Class Amphibia. Mostly tetrapods; embryo enclosed in amnion.
Order Caudata. Salamanders.
Order Anura. Frogs, toads.
Order Apoda. Apodans (caecilians).
Class Reptilia. Skin with scales, embryo enclosed in amnion.
Subclass Anapsida. Turtles, tortoises.
Subclass Lepidosaura. *Sphenodon*, lizards, snakes.
Subclass Archosaura. Dinosaurs (extinct), crocodiles, alligators.
Class Aves. Birds. (In more recent schemes, dinosaurs, crocodilians, and birds are grouped in the same category.)

Order Struthioniformes. Ostriches.
Order Sphenisciformes. Penguins.
Order Procellariiformes. Albatrosses, petrels.
Order Ciconiiformes. Herons, bitterns, storks, flamingoes.
Order Anseriformes. Swans, geese, ducks.
Order Falconiformes. Eagles, hawks, vultures, falcons.
Order Galliformes. Ptarmigan, turkeys, domestic fowl.
Order Columbiformes. Pigeons, doves.
Order Strigiformes. Owls.
Order Apodiformes. Swifts, hummingbirds.
Order Passeriformes. Sparrows, jays, finches, crows, robins, starlings, wrens.
Class Mammalia. Skin with hair; young nourished by milk-secreting glands of adult.
Subclass Prototheria. Egg-laying mammals (duckbilled platypus, spiny anteaters).
Subclass Metatheria. Pouched mammals or marsupials (opossums, kangaroos, wombats).
Subclass Eutheria. Placental mammals.
Order Insectivora. Tree shrews, moles, hedgehogs.
Order Scandentia. Insectivorous tree shrews.
Order Chiroptera. Bats.
Order Primates.
Suborder Strepsirhini (prosimians). Lemurs, lorises.
Suborder Haplorhini (tarsioids and anthropoids).
Infraorder Tarsiiformes. Tarsiers.
Infraorder Platyrrhini (New World monkeys).
Family Cebidae. Spider monkeys, howler monkeys, capuchin.
Infraorder Catarrhini (Old World monkeys and hominoids).
Superfamily Cercopithecoidea. Baboons, macaques, langurs.
Superfamily Hominoidea. Apes and humans.
Family Hylobatidae. Gibbons.
Family Pongidae. Chimpanzees, gorillas, orangutans.
Family Hominidae. Humans and most recent ancestors of humans.
Order Carnivora. Carnivores.
Suborder Feloidea. Cats, civets, mongooses, hyenas.
Suborder Canoidea. Dogs, weasels, skunks, otters, raccoons, pandas, bears.
Order Proboscidea. Elephants; mammoths (extinct).
Order Sirenia. Sea cows (manatees, dugongs).
Order Perissodactyla. Odd-toed ungulates (horses, tapirs, rhinos).
Order Artiodactyla. Even-toed ungulates (camels, deer, bison, sheep, goats, antelopes, giraffes).
Order Edentata. Anteaters, tree sloths, armadillos.
Order Tubulidentata. African aardvarks.
Order Cetacea. Whales, porpoises.
Order Rodentia. Most gnawing animals (squirrels, rats, mice, guinea pigs, porcupines).

APPENDIX II
Units of Measure

Metric-English Conversions

Length

English		Metric
inch	=	2.54 centimeters
foot	=	0.30 meter
yard	=	0.91 meter
mile (5,280 feet)	=	1.61 kilometer

To convert	multiply by	to obtain
inches	2.54	centimeters
feet	30.00	centimeters
centimeters	0.39	inches
millimeters	0.039	inches

Weight

English		Metric
grain	=	64.80 milligrams
ounce	=	28.35 grams
pound	=	453.60 grams
ton (short) (2,000 pounds)	=	0.91 metric ton

To convert	multiply by	to obtain
ounces	28.3	grams
pounds	453.6	grams
pounds	0.45	kilograms
grams	0.035	ounces
kilograms	2.2	pounds

Volume

English		Metric
cubic inch	=	16.39 cubic centimeters
cubic foot	=	0.03 cubic meter
cubic yard	=	0.765 cubic meters
ounce	=	0.03 liter
pint	=	0.47 liter
quart	=	0.95 liter
gallon	=	3.79 liters

To convert	multiply by	to obtain
fluid ounces	30.00	milliliters
quart	0.95	liters
milliliters	0.03	fluid ounces
liters	1.06	quarts

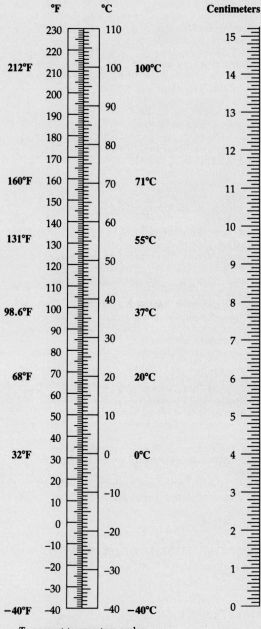

To convert temperature scales:

Fahrenheit to Celsius: $°C = 5/9 \, (°F - 32)$

Celsius to Fahrenheit: $°F = 9/5 \, (°C) + 32$

APPENDIX III
Answers to Genetics Problems

Chapter 11

1. a. *AB*
   b. *AB* and *aB*
   c. *Ab* and *ab*
   d. *AB*, *aB*, *Ab*, and *ab*

2. a. All the offspring will be *AaBB*.
   b. 25% *AABB*; 25% *AaBB*; 25% *AABb*; 25% *AaBb*
   c. 25% *AaBb*; 25% *Aabb*; 25% *aaBb*; 25% *aabb*
   d.

1/16	*AABB*	(6.25%)
1/8	*AaBB*	(12.5%)
1/16	*aaBB*	(6.25%)
1/8	*AABb*	(12.5%)
1/4	*AaBb*	(25%)
1/8	*aaBb*	(12.5%)
1/16	*AAbb*	(6.25%)
1/8	*Aabb*	(12.5%)
1/16	*aabb*	(6.25%)

3. Yellow is recessive. Because the first-generation plants must be heterozygous and have a green phenotype, green must be dominant over the recessive yellow.

4. a. The mother must be heterozygous for both genes; the father is homozygous recessive for both genes. The first child is homozygous recessive for both genes.
   b. The probability that the second child will not be a tongue roller and will have detached earlobes is 1/4 (25%).

5. a. *ABC*
   b. *ABc* and *aBc*
   c. *ABC*, *aBC*, *ABc*, and *aBc*
   d. *ABC*, *aBC*, *AbC*, *abC*, *ABc*, *aBc*, *Abc*, and *abc*

6. The F_1 plants must all be heterozygous for both genes. When these plants self-pollinate, 1/4 (or 25%) of the F_2 plants will be heterozygous for both genes.

7. The most direct way to do this would be to allow a true-breeding yellow-furred mouse to mate with a true-breeding brown-furred mouse. (Strains that breed true for yellow fur and for brown fur could be obtained through repeated inbreeding—that is, mating of relatives, such a male and a female of the same litter. This should yield homozygous yellow and homozygous brown mice.)

   All the F_1 offspring of a cross between the true-breeding individuals should be heterozygous. If their phenotype is either yellow or brown, then dominance is simple or complete; the phenotype reflects the dominant allele. If it is intermediate between yellow and brown, this points to incomplete dominance. A yellow and brown mixed phenotype points to codominance.

8. a. The mother must be heterozygous ($I^A i$). The male with type B blood could have fathered the child if he, too, were heterozygous ($I^B i$).
   b. If the male were heterozygous, then he *could* be the father. However, any other type B heterozygous male also could be the father, so one cannot say that this particular individual absolutely must be. Actually, any male who could contribute an O allele (*i*) could have fathered the child. This would include males with type O blood (*ii*) or heterozygotes with type A blood ($I^A i$).

9. a. F_1 genotypes (and phenotypes): 100% *BbCc (all brown)*. The F_2 phenotypes: 9/16 brown, 3/16 tan, 4/16 albino. The F_2 genotypes:

   1/16 *BBCC* + 2/16 *BBCc* + 2/16 *BbCC* + 4/16 *BbCc*
   1/16 *bbCC* + 2/16 *bbCc*
   1/16 *BBcc* + 2/16 *Bbcc* + 1/16 *bbcc*

   b. Backcross phenotypes: 1/4 brown, 1/4 tan, 2/4 albino. The backcross genotypes:

   1/4 *BbCc* (1/4 brown)
   1/4 *bbCc* (1/4 tan)
   1/4 *Bbcc* and 1/4 *bbcc* (1/2 albino)

10. Fred should mate his black guinea pig with a white one. This would be a testcross. The white one has the genotype *ww*, and the black one may be homozygous dominant (*WW*) or heterozygous (*Ww*). If any offspring from the cross are white, then his black guinea pig is *Ww*. If all offspring are black after enough matings to produce a significant number of offspring, there is a good probability that Fred's guinea pig is *WW*. (If ten of the offspring are all black and none is white, then there is about a 94 percent chance that his guinea pig is *WW*. The greater the number of offspring, the greater the degree of confidence in the conclusion.)

11. a. $R^1 R^1 \times R^1 R^2$ yields 1/2 red flowers, 1/2 pink flowers.
    b. $R^1 R^1 \times R^2 R^2$ yields all pink.
    c. $R^1 R^2 \times R^1 R^2$ yields 1/4 red, 1/2 pink, and 1/4 white.
    d. $R^1 R^2 \times R^2 R^2$ yields 1/2 pink and 1/2 white.

12. Both parents are heterozygotes ($Hb^A Hb^S$).
    a. 1/4
    b. 1/4
    c. 1/2

13. a. Both parents must be heterozygous (*Aa*). They may produce albino (*aa*) and unaffected (*AA* or *Aa*) children.

b. Both parents and all their children must be *aa*.

c. The albino male must be *aa*. Because they have albino children, the woman must be *Aa*. (If she were *AA*, they would have no albino children.) The two albinos must be *aa*, and the two unaffected children must be *Aa* because they must inherit an *a* allele from their father).

14. 9/16 walnut (pea-rose); 3/16 rose; 3/16 pea; 1/16 single.

15. a. All offspring of the mating will have an intermediate amount of pigmentation corresponding to genotype *AaBb*.

b. All possible genotypes could occur among offspring of this mating, in these expected proportions:

1/16 *AABB*
1/8 *AABb*
1/16 *AAbb*
1/8 *AaBB*
1/4 *AaBb*
1/8 *Aabb*
1/16 *aaBB*
1/8 *aaBb*
1/16 *aabb*

Chapter 12

1. a. Males inherit their X chromosome from their mother.

b. With respect to an X-linked gene, a male can produce two kinds of gametes. One kind will contain a Y chromosome and will not have the gene. The other kind will contain an X chromosome—hence the X-linked gene.

c. All gametes of a female who is homozygous for an X-linked gene will have an X chromosome (hence the gene).

d. A female heterozygous for an X-linked gene will produce two kinds of gametes. One kind will have an X chromosome with the dominant allele. The other kind will have an X chromosome with the recessive allele.

2. a. Because this gene is only carried on Y chromosomes, we would not expect females to have hairy pinnae (females normally do not have Y chromosomes).

b. Because sons always inherit a Y chromosome from their father and daughters never do, a male having hairy pinnae will always transmit the gene for the trait to his sons and never to his daughters.

3. A 0% crossover frequency means that 50% of the gametes will be *AB* and 50% will be *ab*.

4. The F_1 females must be heterozygous for both genes. The 42 red-eyed, vestigial-winged and the 30 purple-eyed, long-winged offspring arose from recombinant gametes. They must have produced 600 gametes (there were 600 offspring). Because 42 + 30 of these were recombinant, the percentage of recombinant gametes is 72/600, or 12%, which implies that 12 map units separate the two genes.

5. The rare vestigial-winged flies may have arisen as a result of the x-rays, which may have induced a deletion of the dominant allele from one of the chromosomes. Alternatively, the radiation may have induced a mutation in the dominant allele.

6. If a translocation attached most of chromosome 21 to the end of a normal chromosome 14, then the individual would develop Down syndrome. The cells of that individual would contain the attached chromosome 21 as well as two normal chromosomes 21. The chromosome number would still be 46.

7. Using *c* as the symbol for color blindness and *C* for normal color vision, the cross can be diagrammed as follows:

C(Y) female × *Cc* male.

In mugwumps, a son receives one sex-linked allele from each of his parents, but a daughter inherits her unpaired sex-linked allele from her father. In this cross, half of the sons will be *CC* and half *Cc*, but none will be color blind. Of the daughters, half will be *C*(Y) and half *c*(Y). There is a 50% chance that a daughter will be color blind. Note: unlike humans, in this sex chromosome designation is not the same as it is in humans; but it is correct not only for mugwumps, all birds, moths, butterflies, and some other organisms, the Y chromosome is associated with females, not males.

8. To produce a daughter who will be affected by childhood muscular dystrophy, the prospective father as well as the mother must carry the allele. Few, if any, male carriers who survive to adulthood are capable of having children.

APPENDIX IV
Answers to Self-Quizzes

Chapter 1
1. cell
2. Metabolism
3. Homeostasis
4. adaptive
5. mutations
6. d
7. d
8. d
9. c

Chapter 2
1. electrons
2. d
3. electrons
4. Isotopes
5. c
6. b
7. d
8. b
9. b
10. d
11. c, e, a, b, d

Chapter 3
1. carbon
2. a
3. complex carbohydrates, lipids, proteins, nucleic acids
4. c
5. c
6. d
7. b
8. c, e, b, d, a

Chapter 4
1. a
2. c
3. c
4. c
5. d
6. d
7. b
8. d
9. h, g, a, b, d, e, i, f, c

Chapter 5
1. c
2. a
3. c
4. a
5. a
6. a
7. centrifuge
8. d
9. d
10. b
11. f, d, e, a, b, c

Chapter 6
1. metabolism
2. thermodynamics
3. c
4. a
5. c
6. d
7. d
8. c
9. c, d, a, b

Chapter 7
1. carbon
2. carbon dioxide, sunlight
3. d
4. c
5. b
6. e (b and d)
7. c
8. c
9. c, d, e, b, a

Chapter 8
1. ATP
2. pyruvate
3. d
4. c
5. d
6. c
7. b
8. b
9. b, c, a, d

Chapter 9
1. b
2. a
3. b
4. b
5. a
6. c
7. a
8. d
9. b
10. d, b, c, a

Chapter 10
1. d
2. d
3. a
4. b
5. d
6. c
7. c
8. d
9. d, a, c, b

Chapter 11
1. a
2. b
3. a
4. c
5. b
6. a
7. d
8. b, d, a, c

Chapter 12
1. c
2. e
3. e
4. c
5. e
6. c
7. d
8. d
9. d
10. c, e, d, b, a

Chapter 13
1. c
2. d
3. d
4. c
5. a
6. a
7. d
8. d, e, b, c, a

Chapter 14
1. three
2. e
3. b
4. c
5. c
6. a
7. c
8. b
9. a
10. c
11. c
12. a
13. e, c, d, b, a

Chapter 15
1. d
2. d
3. b
4. a
5. a
6. d
7. d
8. e
9. c
10. b
11. d
12. a, f, d, c, e, b

Chapter 16
1. Plasmids
2. c
3. d
4. a
5. b
6. a
7. b
8. d
9. b
10. d, e, a, c, b

Chapter 17
1. c
2. e
3. e
4. c, d, e, b, a

Chapter 18
1. populations
2. c
3. e
4. a
5. e
6. d
7. b
8. c
9. c
10. c, d, a, b

Chapter 19
1. e
2. a
3. d
4. d
5. e
6. c
7. a
8. d
9. d
10. b
11. d, a, b, e, c

Chapter 20
1. c
2. e
3. a
4. b
5. a
6. b
7. c
8. d
9. e, b, c, f, a, d

Chapter 21
1. e
2. c
3. b
4. c
5. d
6. c
7. d
8. d
9. b, d, e, a, c

Chapter 22
1. c
2. c
3. b
4. c
5. d
6. d
7. e
8. d
9. c
10. d, a, b, e, c

Chapter 23
1. d
2. b
3. d
4. a
5. c
6. a
7. a
8. d
9. b

Chapter 24
1. c
2. d
3. b
4. c
5. b
6. e
7. a
8. a
9. c
10. d, f, c, e, a, b

Chapter 25
1. b
2. d
3. b
4. c
5. d
6. c
7. b
8. d
9. b
10. c, e, g, h, f, a, b, d

Chapter 26
1. body symmetry, cephalization, type of gut, type of body cavity, segmentation
2. b
3. b
4. c
5. a
6. a
7. b
8. c
9. c
10. d, f, c, g, e, i, j, h, a, b

Chapter 27
1. d
2. e
3. f
4. d
5. a
6. b
7. d
8. b
9. e
10. d
11. f
12. d
13. d
14. e
15. d, e, c, b, g, f, a

Chapter 28
1. e
2. c
3. b
4. a
5. b
6. d
7. d
8. b, e, a, d, c

Chapter 29
1. b
2. a
3. b
4. c
5. a
6. d
7. d
8. d
9. d
10. b, d, e, c, f, a

Chapter 30
1. hydrogen bonds (cohesion)
2. stomata
3. e
4. b
5. a
6. c
7. d
8. c
9. d
10. d
11. b
12. b, d, f, c, a, e

Chapter 31
1. a
2. pollinators
3. d
4. c
5. b
6. c
7. b
8. a
9. b
10. a, c, d, e, b

Chapter 32
1. c
2. c
3. e
4. d
5. d
6. a
7. c
8. c
9. c
10. a, e, c, d, b

Chapter 33
1. d
2. c
3. b
4. b
5. b
6. a
7. d
8. c
9. b
10. a
11. Receptors, integrator, effectors
12. b, a, c, f, g, e, d

Chapter 34
1. d
2. d
3. b
4. d
5. a
6. d
7. d
8. d
9. c
10. c, d, b, g, f, e, a

Chapter 35
1. nerve nets
2. cephalization, bilateral symmetry
3. a
4. b
5. c
6. a
7. b
8. d
9. a
10. b
11. e, d, b, c, a

Chapter 36
1. stimulus
2. sensation
3. Perception
4. d
5. b
6. d
7. c
8. b
9. a
10. f, d, c, e, a, b

Chapter 37
1. f
2. d
3. b
4. d
5. e
6. b
7. b
8. b
9. b, e, g, d, f, a, c
10. d, b, e, c, a

Chapter 38
1. integumentary
2. Skeletal, muscular
3. smooth, cardiac, skeletal
4. c
5. d
6. d
7. b
8. d
9. f
10. h, f, g, e, a, c, b, i, d

Chapter 39
1. c
2. d
3. d
4. c
5. c
6. d
7. b
8. d
9. c
10. e, b, a, d, c
11. d, e, a, f, c, b

Chapter 40
1. e
2. d
3. c
4. d
5. a
6. d
7. d
8. b
9. a
10. c, b, a, e, d, f

Chapter 41
1. d
2. d
3. d
4. b
5. c
6. c
7. e
8. b
9. c
10. d
11. d, b, f, e, a, c

Chapter 42
1. e
2. d
3. energy; energy
4. a
5. d
6. b
7. c
8. a
9. b
10. d, c, e, g, f, b, a

Chapter 43
1. d
2. f
3. b
4. d
5. c
6. d
7. a
8. d
9. d, e, a, c, b
10. f, e, g, b, c, d, a

Chapter 44
1. d
2. a
3. a
4. c
5. b
6. d
7. c
8. b
9. d
10. d

Chapter 45
1. hypothalamus
2. f
3. c
4. c
5. c
6. c
7. e
8. b
9. e
10. b
11. e, c, a, b, d

Chapter 46
1. Ecology
2. population
3. e
4. e
5. b
6. d
7. e
8. d
9. d
10. d, e, c, a, b

Chapter 47
1. e
2. e
3. b
4. d
5. b
6. b
7. d
8. a
9. c, d, a, e, b

Chapter 48
1. d
2. a
3. c
4. d
5. a
6. e
7. c
8. d, a, c, b
9. c, d, e, b, a

Chapter 49
1. c
2. c
3. d
4. b
5. d
6. d
7. d
8. d
9. b, f, e, d, c, a
10. e, a, d, c, b

Chapter 50
1. c
2. c
3. b
4. a
5. d
6. b
7. c
8. d
9. d
10. e, c, d, a, b

Chapter 51
1. a
2. d
3. b
4. b
5. d
6. c
7. d, e, b, c, a

Chapter 52
1. d
2. c
3. d
4. d
5. c
6. c
7. a
8. c
9. d
10. d

APPENDIX V
A Closer Look at Some Major Metabolic Pathways

ENERGY-
REQUIRING
STEPS OF
GLYCOLYSIS

(two ATP
invested)

ENERGY-
RELEASING
STEPS OF
GLYCOLYSIS

(four ATP
produced)

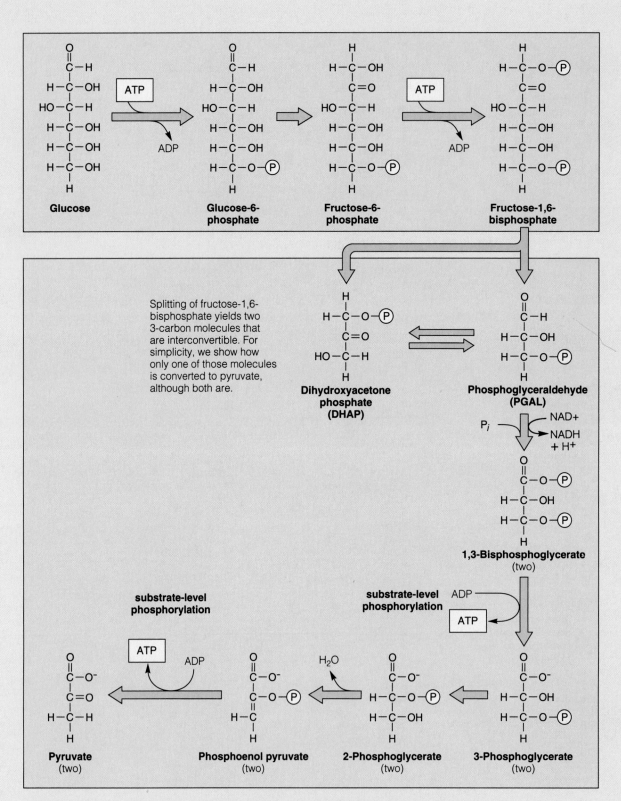

Figure a Glycolysis, ending with two 3-carbon pyruvate molecules for each 6-carbon glucose entering the reactions. The *net* energy yield is two ATP molecules (two invested, four produced).

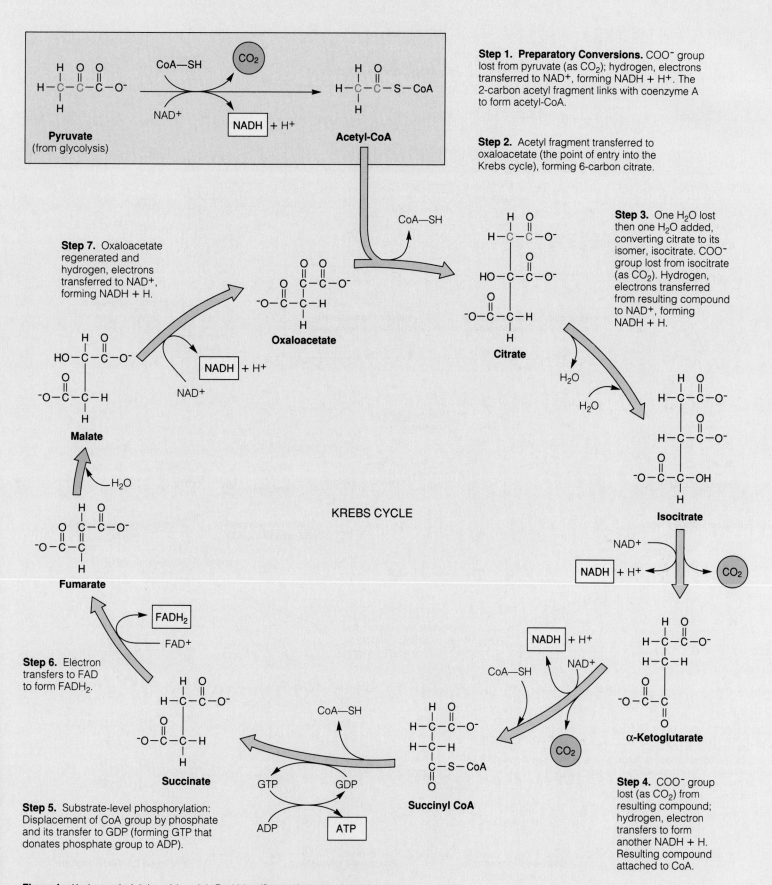

Step 1. Preparatory Conversions. COO^- group lost from pyruvate (as CO_2); hydrogen, electrons transferred to NAD^+, forming $NADH + H^+$. The 2-carbon acetyl fragment links with coenzyme A to form acetyl-CoA.

Step 2. Acetyl fragment transferred to oxaloacetate (the point of entry into the Krebs cycle), forming 6-carbon citrate.

Step 3. One H_2O lost then one H_2O added, converting citrate to its isomer, isocitrate. COO^- group lost from isocitrate (as CO_2). Hydrogen, electrons transferred from resulting compound to NAD^+, forming $NADH + H$.

Step 7. Oxaloacetate regenerated and hydrogen, electrons transferred to NAD^+, forming $NADH + H$.

Step 6. Electron transfers to FAD to form $FADH_2$.

Step 5. Substrate-level phosphorylation: Displacement of CoA group by phosphate and its transfer to GDP (forming GTP that donates phosphate group to ADP).

Step 4. COO^- group lost (as CO_2) from resulting compound; hydrogen, electron transfers to form another $NADH + H$. Resulting compound attached to CoA.

Pyruvate (from glycolysis)

Acetyl-CoA

Oxaloacetate

Citrate

Malate

Isocitrate

KREBS CYCLE

Fumarate

α-Ketoglutarate

Succinate

Succinyl CoA

Figure b Krebs cycle (citric acid cycle). Red identifies carbon entering the cycle by way of acetyl-CoA.

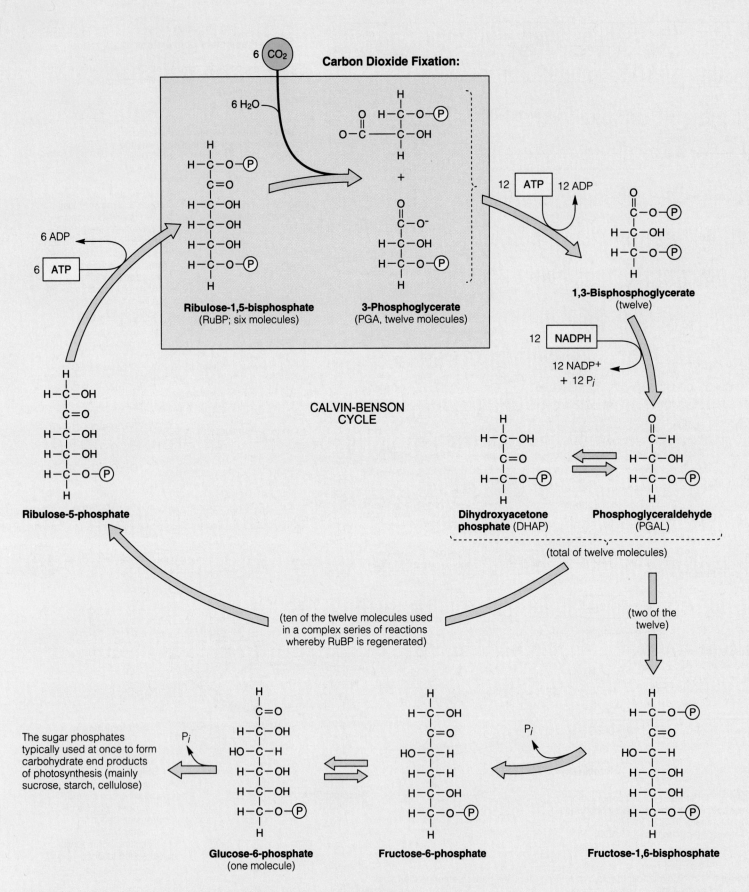

Figure c Calvin-Benson cycle of the light-independent reactions of photosynthesis.

APPENDIX VI
Periodic Table of the Elements

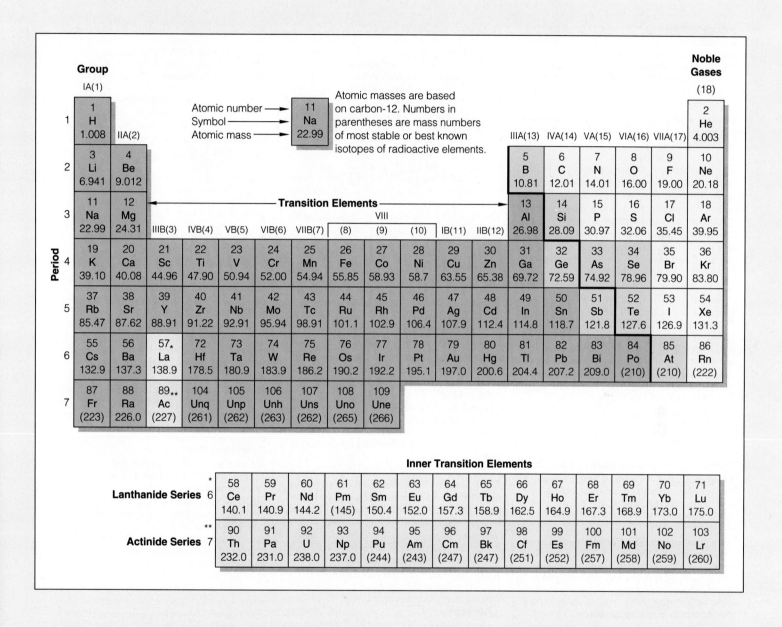

Group

IA(1)

Atomic number ——► 11
Symbol ——► Na
Atomic mass ——► 22.99

Atomic masses are based on carbon-12. Numbers in parentheses are mass numbers of most stable or best known isotopes of radioactive elements.

Noble Gases

(18)

Period	IA(1)	IIA(2)	IIIB(3)	IVB(4)	VB(5)	VIB(6)	VIIB(7)	(8)	(9)	(10)	IB(11)	IIB(12)	IIIA(13)	IVA(14)	VA(15)	VIA(16)	VIIA(17)	(18)

VIII

Transition Elements

Period 1
1 H 1.008 — 2 He 4.003

Period 2
3 Li 6.941 — 4 Be 9.012 — 5 B 10.81 — 6 C 12.01 — 7 N 14.01 — 8 O 16.00 — 9 F 19.00 — 10 Ne 20.18

Period 3
11 Na 22.99 — 12 Mg 24.31 — 13 Al 26.98 — 14 Si 28.09 — 15 P 30.97 — 16 S 32.06 — 17 Cl 35.45 — 18 Ar 39.95

Period 4
19 K 39.10 — 20 Ca 40.08 — 21 Sc 44.96 — 22 Ti 47.90 — 23 V 50.94 — 24 Cr 52.00 — 25 Mn 54.94 — 26 Fe 55.85 — 27 Co 58.93 — 28 Ni 58.7 — 29 Cu 63.55 — 30 Zn 65.38 — 31 Ga 69.72 — 32 Ge 72.59 — 33 As 74.92 — 34 Se 78.96 — 35 Br 79.90 — 36 Kr 83.80

Period 5
37 Rb 85.47 — 38 Sr 87.62 — 39 Y 88.91 — 40 Zr 91.22 — 41 Nb 92.91 — 42 Mo 95.94 — 43 Tc 98.91 — 44 Ru 101.1 — 45 Rh 102.9 — 46 Pd 106.4 — 47 Ag 107.9 — 48 Cd 112.4 — 49 In 114.8 — 50 Sn 118.7 — 51 Sb 121.8 — 52 Te 127.6 — 53 I 126.9 — 54 Xe 131.3

Period 6
55 Cs 132.9 — 56 Ba 137.3 — 57* La 138.9 — 72 Hf 178.5 — 73 Ta 180.9 — 74 W 183.9 — 75 Re 186.2 — 76 Os 190.2 — 77 Ir 192.2 — 78 Pt 195.1 — 79 Au 197.0 — 80 Hg 200.6 — 81 Tl 204.4 — 82 Pb 207.2 — 83 Bi 209.0 — 84 Po (210) — 85 At (210) — 86 Rn (222)

Period 7
87 Fr (223) — 88 Ra 226.0 — 89** Ac (227) — 104 Unq (261) — 105 Unp (262) — 106 Unh (263) — 107 Uns (262) — 108 Uno (265) — 109 Une (266)

Inner Transition Elements

Lanthanide Series 6*
58 Ce 140.1 — 59 Pr 140.9 — 60 Nd 144.2 — 61 Pm (145) — 62 Sm 150.4 — 63 Eu 152.0 — 64 Gd 157.3 — 65 Tb 158.9 — 66 Dy 162.5 — 67 Ho 164.9 — 68 Er 167.3 — 69 Tm 168.9 — 70 Yb 173.0 — 71 Lu 175.0

Actinide Series 7**
90 Th 232.0 — 91 Pa 231.0 — 92 U 238.0 — 93 Np 237.0 — 94 Pu (244) — 95 Am (243) — 96 Cm (247) — 97 Bk (247) — 98 Cf (251) — 99 Es (252) — 100 Fm (257) — 101 Md (258) — 102 No (259) — 103 Lr (260)

A GLOSSARY OF BIOLOGICAL TERMS

ABO blood typing Method of characterizing an individual's blood according to whether one or both of two protein markers, A and B, are present at the surface of red blood cells. The O signifies that neither marker is present.

abortion Spontaneous or induced expulsion of the embryo or fetus from the uterus.

abscisic acid (ab-SISS-ik) Plant hormone that promotes stomatal closure, bud dormancy, and seed dormancy.

abscission (ab-SIH-zhun) [L. *abscindere*, to cut off] The dropping of leaves, flowers, fruits, or other plant parts due to hormonal action.

absorption Of complex animals, the movement of nutrients, fluid, and ions across the gut lining and into the internal environment.

accessory pigment A light-trapping pigment that contributes to photosynthesis by extending the range of usable wavelengths beyond those absorbed by the chlorophylls.

acid [L. *acidus*, sour] A substance that releases hydrogen ions (H^+) in water.

acid rain The falling to earth of snow or rain that contains sulfur and nitrogen oxides. Also called wet acid deposition (as opposed to dry acid deposition, or the falling to earth of airborne particles of sulfur and nitrogen oxides).

acoelomate (ay-SEE-luh-mate) Type of animal that has no fluid-filled cavity between the gut and body wall.

acoustical signal Sounds that are used as a communication signal.

actin (AK-tin) A globular contractile protein. In muscle cells, actin interacts with another protein, myosin, to bring about contraction.

action potential An abrupt, brief reversal in the steady voltage difference across the plasma membrane (that is, the resting membrane potential) of a neuron and some other cells.

activation energy The minimum amount of collision energy required to bring reactant molecules to an activated condition (the transition state) at which a reaction will proceed spontaneously. Enzymes enhance reaction rates by lowering the activation energy (they put substrates on a precise collision course).

active site A crevice on the surface of an enzyme molecule where a specific reaction is catalyzed.

active transport The pumping of one or more specific solutes through a transport protein that spans the lipid bilayer of a cell membrane. Most often, the solute is transported against its concentration gradient. The protein is activated by an energy boost, as from ATP.

adaptation [L. *adaptare*, to fit] In evolutionary biology, the process of becoming adapted (or more adapted) to a given set of environmental conditions. Of sensory neurons, a decrease in the frequency of action potentials (or their cessation) even when a stimulus is maintained at constant strength.

adaptive behavior A behavior that promotes the propagation of an individual's genes and that tends to increase in frequency in a population over time.

adaptive radiation A burst of speciation events, with lineages branching away from one another as they partition the existing environment or invade new ones.

adaptive trait Any aspect of form, function, or behavior that helps an organism survive and reproduce under a given set of environmental conditions.

adaptive zone A way of life, such as "catching insects in the air at night." A lineage must have physical, ecological, and evolutionary access to an adaptive zone to become a successful occupant of it.

adenine (AH-de-neen) A purine; a nitrogen-containing base found in nucleotides.

adenosine diphosphate (ah-DEN-uh-seen die-FOSS-fate) ADP, a molecule involved in cellular energy transfers; typically formed by hydrolysis of ATP.

adenosine phosphates Any of several relatively small molecules, some of which function as chemical messengers within and between cells, and others that function as energy carriers.

adenosine triphosphate *See* ATP.

ADH Antidiuretic hormone. Produced by the hypothalamus and released by the posterior pituitary, it stimulates reabsorption in the kidneys and so reduces urine volume.

adipose tissue A type of connective tissue having an abundance of fat-storing cells and blood vessels for transporting fats.

ADP Adenosine diphosphate. A nucleotide coenzyme that accepts unbound phosphate or a phosphate group to become ATP.

ADP/ATP cycle In cells, a mechanism of ATP renewal. When ATP donates a phosphate group to other molecules (and so energizes them), it reverts to ADP, then forms again by phosphorylation of ADP.

adrenal cortex (ah-DREE-nul) Outer portion of the adrenal gland; its hormones have roles in metabolism, inflammation, maintaining extracellular fluid volume, and other functions.

adrenal medulla Inner region of the adrenal gland; its hormones help control blood circulation and carbohydrate metabolism.

aerobic respiration (air-OH-bik) [Gk. *aer*, air, + *bios*, life] The main energy-releasing metabolic pathway of ATP formation, in which oxygen is the final acceptor of electrons stripped from glucose or some other organic compound. The pathway proceeds from glycolysis through the Krebs cycle and electron transport phosphorylation. A typical net yield is 36 ATP for each glucose molecule.

age structure Of a population, the number of individuals in each of several or many age categories.

agglutination (ah-glue-tin-AY-shun) Clumping together of foreign cells that have invaded the body (as pathogens or in tissue grafts or transplants). Clumping is induced by cross-linking between antibody molecules that have already latched onto antigen at the surface of the foreign cells.

aging A range of processes, including the breakdown of cell structure and function, by which the body gradually deteriorates. All organisms showing extensive cell differentiation undergo aging.

AIDS Acquired immunodeficiency syndrome. A set of chronic disorders following infection by the human immunodeficiency virus (HIV), which destroys key cells of the immune system.

alcoholic fermentation Anaerobic pathway of ATP formation in which pyruvate from glycolysis is broken down to acetaldehyde, which accepts electrons from NADH to become ethanol, and NAD^+ is regenerated. Its net yield is two ATP.

aldosterone (al-DOSS-tuh-rohn) Hormone secreted by the adrenal cortex that helps regulate sodium reabsorption.

allantois (ah-LAN-twahz) [Gk. *allas*, sausage] Of vertebrates, one of four extraembryonic membranes that form during embryonic development. It functions in respiration and storage of metabolic wastes in reptiles, birds, and some mammals. In humans, it functions in early blood formation and development of the urinary bladder.

allele (uh-LEEL) For a given location on a chromosome, one of two or more slightly different molecular forms of a gene that code for different versions of the same trait.

allele frequency Of a given gene locus, the relative abundances of each kind of allele carried by the individuals of a population.

allergy An immune response made against a normally harmless substance.

allopatric speciation [Gk. *allos*, different, + *patria*, native land] Speciation that follows geographic isolation of populations of the same species.

allosteric control (AL-oh-STARE-ik) Control over a metabolic reaction or pathway that operates through the binding of a specific substance at a control site on a specific enzyme.

alpine tundra A type of biome that exists at high elevations in mountains throughout the world.

altruistic behavior (al-true-ISS-tik) Self-sacrificing behavior; the individual behaves in a way that helps others but decreases its own chances of reproductive success.

alveolar sac (al-VEE-uh-lar) Any of the pouch-like clusters of alveoli in the lungs; the major sites of gas exchange.

alveolus (ahl-VEE-uh-lus), plural **alveoli** [L. *alveus*, small cavity] Any of the many cup-shaped, thin-walled outpouchings of respiratory bronchioles. A site where oxygen diffuses from air in the lungs to the blood, and carbon dioxide diffuses from blood to the lungs.

amino acid (uh-MEE-no) A small organic molecule having a hydrogen atom, an amino group, an acid group, and an R group covalently bonded to a central carbon atom. The subunit of polypeptide chains, which represent the primary structure of proteins.

ammonification (uh-moan-ih-fih-KAY-shun) Together with decomposition, a process by which certain bacteria and fungi break down nitrogen-containing wastes and remains of other organisms.

amnion (AM-nee-on) Of land vertebrates, one of four extraembryonic membranes. It becomes a fluid-filled sac in which the embryo (and fetus) can grow, move freely, and be protected from sudden temperature shifts and impacts.

amniote egg A type of egg, often with a leathery or calcified shell, that contains extraembryonic membranes, including the amnion. An adaptation that figured in the vertebrate invasion of land.

amphibian A type of vertebrate somewhere between fishes and reptiles in body plan and reproductive mode; salamanders, frogs and toads, and caecilians are the existing groups.

anaerobic pathway (an-uh-ROW-bik) [Gk. *an*, without, + *aer*, air] Metabolic pathway in which a substance other than oxygen serves as the final acceptor of electrons that have been stripped from substrates.

analogous structures Body parts, once different in separate lineages, that were put to comparable uses in similar environments and that came to resemble one another in form and function. They are evidence of morphological convergence.

anaphase (AN-uh-faze) The stage at which microtubules of a spindle apparatus separate sister chromatids of each chromosome and move them to opposite spindle poles. During anaphase I of *meiosis*, the two members of each pair of homologous chromosomes separate. During anaphase II, sister chromatids of each chromosome separate.

aneuploidy (AN-yoo-ploy-dee) A change in the chromosome number following inheritance of one extra or one less chromosome.

angiosperm (AN-gee-oh-spurm) [Gk. *angeion*, vessel, and *spermia*, seed] A flowering plant.

animal A heterotroph that eats or absorbs nutrients from other organisms; is multi-celled, usually with tissues arranged in organs and organ systems; is usually motile during at least part of the life cycle; and goes through a period of embryonic development.

annelid The type of invertebrate classified as a segmented worm; an oligochaete (such as an earthworm), leech, or polychaete.

annual plant A flowering plant that completes its life cycle in one growing season.

anther [Gk. *anthos*, flower] In flowering plants, the pollen-bearing part of the male reproductive structure (stamen).

antibiotic A normal metabolic product of certain microorganisms that kills or inhibits the growth of other microorganisms.

antibody [Gk. *anti*, against] Any of a variety of Y-shaped receptor molecules with binding sites for specific antigens. Only B cells produce antibodies, then position them at their surface or secrete them.

anticodon In a tRNA molecule, a sequence of three nucleotide bases that can pair with an mRNA codon.

antigen (AN-tih-jen) [Gk. *anti*, against, + *genos*, race, kind] Any molecular configuration that is recognized as foreign to the body and that triggers an immune response. Most antibodies are protein molecules at the surface of infectious agents or tumor cells.

antigen-presenting cell A macrophage or some other white blood cell that engulfs and digests antigen, then displays it with certain MHC molecules at its surface. Recognition of antigen-MHC complexes by T and B cells triggers an immune response.

aorta (ay-OR-tah) [Gk. *airein*, to lift, heave] Main artery of systemic circulation; carries oxygenated blood away from the heart to all body regions except the lungs.

apical dominance The inhibitory influence of a terminal bud on the growth of lateral buds.

apical meristem (AY-pih-kul MARE-ih-stem) [L. *apex*, top, + Gk. *meristos*, divisible] Of most plants, a mass of self-perpetuating cells responsible for primary growth (elongation) at root and shoot tips.

appendicular skeleton (ap-en-DIK-yoo-lahr) In vertebrates, bones of the limbs, hips, and shoulders.

appendix A slender projection from the cup-shaped pouch at the start of the colon.

archaebacteria One of three great prokaryotic lineages that arose early in the history of life; now represented by methanogens, halophiles, and thermophiles.

arctic tundra A type of biome that lies between the polar ice cap and boreal forests of North America, Europe, and Asia.

arteriole (ar-TEER-ee-ole) Any of the blood vessels between arteries and capillaries. They are control points where the volume of blood delivered to different body regions can be adjusted.

artery Any of the large-diameter blood vessels that conduct oxygen-poor blood to the lungs and oxygen-enriched blood to all body

tissues. Their thick, muscular wall allows them to smooth out pulsations in blood pressure caused by heart contractions.

arthropod An invertebrate having a hardened exoskeleton, specialized segments, and jointed appendages. Spiders, crabs, and insects are examples.

asexual reproduction Mode of reproduction by which offspring arise from a single parent, and inherit the genes of that parent only.

atmosphere A region of gases, airborne particles, and water vapor enveloping the earth; 80 percent of its mass is distributed within seventeen miles of the earth's surface.

atmospheric cycle A biogeochemical cycle in which the atmosphere is the largest reservoir of an element. The carbon and nitrogen cycles are examples.

atom The smallest unit of matter that is unique to a particular element.

atomic number The number of protons in the nucleus of each atom of an element; it differs for each element.

ATP Adenosine triphosphate (ah-DEN-uh-seen try-FOSS-fate). A nucleotide composed of adenine, ribose, and three phosphate groups. As the main energy carrier in cells, it directly or indirectly delivers energy to or picks up energy from nearly all metabolic pathways.

atrium (AYE-tree-um) Of the human heart, one of two chambers that receive blood. *Compare* ventricle.

australopith (OHSS-trah-low-pith) [L. *australis*, southern, + Gk. *pithekos*, ape] Any of the earliest known species of hominids, that is, the first species of the evolutionary branch leading to humans.

autoimmune response Misdirected immune response in which lymphocytes mount an attack against normal body cells.

autonomic nervous system (auto-NOM-ik) Those nerves leading from the central nervous system to the smooth muscle, cardiac muscle, and glands of internal organs and structures, that is, to the visceral portion of the body.

autosomal dominant inheritance Condition arising from the presence of a dominant allele on an autosome (not a sex chromosome). The allele is always expressed to some extent, even in heterozygotes.

autosomal recessive inheritance Condition arising from a recessive allele on an autosome (not a sex chromosome). Only recessive homozygotes show the resulting phenotype.

autosome Any of the chromosomes that are of the same number and kind in both males and females of the species.

autotroph (AH-toe-trofe) [Gk. *autos*, self, + *trophos*, feeder] An organism able to build its own large organic molecules by using carbon dioxide and energy from the physical environment. Photosynthetic autotrophs use sunlight energy; chemosynthetic autotrophs extract energy from chemical reactions involving inorganic substances. *Compare* heterotroph.

auxin (AWK-sin) Any of a class of growth-regulating hormones in plants; auxins promote stem elongation as one effect.

axial skeleton (AX-ee-uhl) In vertebrates, the skull, backbone, ribs, and breastbone (sternum).

axon Of a neuron, a long, cylindrical extension from the cell body, with finely branched endings. Action potentials move rapidly, without alteration, along an axon; their arrival at axon endings may trigger the release of neurotransmitter molecules that influence an adjacent cell.

B lymphocyte, or **B cell** The only white blood cell that produces antibodies, then positions them at the cell surface or secretes them as weapons in immune responses.

bacterial conjugation The transfer of plasmid DNA from one bacterial cell to another.

bacterial flagellum Of many bacterial cells, a whiplike motile structure that does not contain a core of microtubules.

bacteriophage (bak-TEER-ee-oh-fahj) [Gk. *baktērion*, small staff, rod, + *phagein*, to eat] Category of viruses that infect bacterial cells.

balanced polymorphism The maintenance of two or more forms of a trait in fairly stable proportions over generations.

Barr body In the cells of female mammals, a condensed X chromosome that was inactivated during early embryonic development.

basal body A centriole which, after having given rise to the microtubules of a flagellum or cilium, remains attached to its base in the cytoplasm.

base A substance that releases OH⁻ in water.

base pair A pair of hydrogen-bonded nucleotide bases in two strands of nucleic acids. In a DNA double helix, adenine pairs with thymine, and guanine with cytosine. When an mRNA strand forms on a DNA strand during transcription, uracil (U) pairs with the DNA's adenine.

base sequence The particular order in which one nucleotide base follows the next in a strand of DNA or RNA. The order differs to some extent for each kind of organism.

basophil Fast-acting white blood cells that secrete histamine and other substances during inflammation.

behavior, animal A response to external and internal stimuli, following integration of sensory, neural, endocrine, and effector components. Because behavior has a genetic basis, it is subject to natural selection and commonly can be modified through experience.

benthic province All of the sediments and rocky formations of the ocean bottom; begins with the continental shelf and extends down through deep-sea trenches.

biennial A flowering plant that lives through two growing seasons.

bilateral symmetry Body plan in which the left and right halves of the animal are mirror-images of each other.

binary fission Of bacteria, a mode of asexual reproduction in which the parent cell replicates its single chromosome, then divides into two genetically identical daughter cells.

biogeochemical cycle The movement of an element such as carbon or nitrogen from the environment to organisms, then back to the environment.

biogeographic realm [Gk. *bios*, life, + *geographein*, to describe the surface of the earth] Any of six major land regions, each having distinguishing types of plants and animals and generally retaining its identity because of climate and geographic barriers to gene flow.

biological clocks Internal time-measuring mechanisms that have roles in adjusting an organism's daily activities, seasonal activities, or both in response to environmental cues.

biological magnification The increasing concentration of a nondegradable or slowly degradable substance in body tissues as it is passed along food chains.

biological systematics Branch of biology that assesses patterns of diversity based on information from taxonomy, phylogenetic reconstruction, and classification.

bioluminescence A flashing of light that emanates from an organism when excited electrons of luciferins, or highly fluorescent substances, return to a lower energy level.

biomass The combined weight of all the organisms at a particular trophic (feeding) level in an ecosystem.

biome A broad, vegetational subdivision of a biogeographic realm shaped by climate, topography, and composition of regional soils.

biosphere [Gk. *bios*, life, + *sphaira*, globe] All regions of the earth's waters, crust, and atmosphere in which organisms live.

biosynthetic pathway A metabolic pathway in which small molecules are assembled into lipids, proteins, and other large organic molecules.

biotic potential Of a population, the maximum rate of increase per individual under ideal conditions.

bipedalism A habitual standing and walking on two feet, as by ostriches and humans.

bird A type of vertebrate, the only one having feathers, with strong resemblances and evolutionary connections to reptiles.

blastocyst (BLASS-tuh-sist) [Gk. *blastos*, sprout, + *kystis*, pouch] In mammalian development, a blastula stage consisting of a hollow ball of surface cells and an inner cell mass.

blastula (BLASS-chew-lah) An embryonic stage consisting of a ball of cells produced by cleavage.

blood A fluid connective tissue composed of water, solutes, and formed elements (blood cells and platelets); it carries substances to and from cells and helps maintain an internal environment that is favorable for cell activities.

blood pressure Fluid pressure, generated by heart contractions, that keeps blood circulating.

blood-brain barrier Set of mechanisms that helps control which blood-borne substances reach neurons in the brain.

bone The mineral-hardened connective tissue of bones.

bones In vertebrate skeletons, organs that function in movement and locomotion, protection of other organs, mineral storage, and (in some bones) blood cell production.

bottleneck An extreme case of genetic drift. A catastrophic decline in population size leads to a random shift in the allele frequencies among survivors. Because the population must rebuild from so few individuals, severely limited genetic variation may be an outcome.

Bowman's capsule The cup-shaped portion of a nephron that receives water and solutes filtered from blood.

brain Of most nervous systems, the most complex integrating center; it receives, processes, and integrates sensory input and issues coordinated commands for response.

brainstem The vertebrate midbrain, pons, and medulla oblongata, the core of which contains the reticular formation that helps govern activity of the nervous system as a whole.

bronchiole Of most vertebrates, a component of the finely branched bronchial tree inside each lung.

bronchus, plural **bronchi** (BRONG-CUSS, BRONG-kee) [Gk. *bronchos*, windpipe] Tubelike branchings of the trachea that lead into the lungs of most vertebrates.

brown alga A type of aquatic plant, found in nearly all marine habitats, that has an abundance of xanthophyll pigments.

bryophyte A nonvascular land plant that requires free water to complete its life cycle.

bud An undeveloped shoot of mostly meristematic tissue; often covered and protected by scales (modified leaves).

buffer A substance that can combine with hydrogen ions, release them, or both. Buffers help stabilize the pH of blood and other fluids.

bulk Of human digestion, a volume of fiber and other undigested material that absorption processes in the colon cannot decrease.

bulk flow In response to a pressure gradient, a movement of more than one kind of molecule in the same direction in the same medium (as in blood, sap, or air).

C4 pathway Of many plants, a pathway of photosynthesis in which carbon dioxide is fixed twice, in two different cell types. Carbon dioxide accumulates in the leaf and helps counter photorespiration. The first compound formed is the 4-carbon oxaloacetate.

Calvin-Benson cycle Cyclic reactions that are the "synthesis" part of the light-independent reactions of photosynthesis. In land plants, RuBP, or some other compound to which carbon has been affixed, undergoes rearrangements that lead to formation of a sugar phosphate and to regeneration of the RuBP. The cycle runs on ATP and NADPH from light-dependent reactions.

CAM plant A plant that conserves water by opening stomata only at night, when it fixes carbon dioxide by way of a C4 pathway.

cambium (KAM-bee-um), plural **cambia** In vascular plants, one of two types of meristems that are responsible for secondary growth (increases in stem and root diameter). Vascular cambium gives rise to secondary xylem and phloem; cork cambium gives rise to periderm.

camouflage An outcome of an organism's form, patterning, color, or behavior that helps it blend with its surroundings and escape detection.

cancer A type of malignant tumor, the cells of which show profound abnormalities in the plasma membrane and cytoplasm, abnormal growth and division, and weakened capacity for adhesion within the parent tissue (leading to metastasis). Unless eradicated, cancer is lethal.

canine A pointed tooth that functions in piercing food.

capillary [L. *capillus*, hair] A thin-walled blood vessel that functions in the exchange of gases and other substances between blood and interstitial fluid.

capillary bed A diffusion zone, consisting of numerous capillaries, where substances are exchanged between blood and interstitial fluid.

carbohydrate [L. *carbo*, charcoal, + *hydro*, water] A simple sugar or large molecule composed of sugar units. All cells use carbohydrates as structural materials, energy stores, and transportable forms of energy. The three classes of carbohydrates include: monosaccharides, oligosaccharides, or polysaccharides.

carbon cycle A biogeochemical cycle in which carbon moves from its largest reservoir in the atmosphere, through oceans and organisms, then back to the atmosphere.

carbon dioxide fixation First step of the light-independent reactions of photosynthesis. Carbon (from carbon dioxide) becomes affixed to a carbon compound (such as RuBP) that can enter the Calvin-Benson cycle.

carcinogen (kar-SIN-uh-jen) An environmental agent or substance, such as ultraviolet radiation, that can trigger cancer.

cardiac cycle (KAR-dee-ak) [Gk. *kardia*, heart, + *kyklos*, circle] The sequence of muscle contraction and relaxation constituting one heartbeat.

cardiac pacemaker Sinoatrial (SA) node; the basis of the normal rate of heartbeat. The self-excitatory cardiac muscle cells that spontaneously generate rhythmic waves of excitation over the heart chambers.

cardiovascular system Of most animals, an organ system that is composed of blood, one or more hearts, and blood vessels and that functions in the rapid transport of substances to and from cells.

carnivore [L. *caro*, *carnis*, flesh, + *vovare*, to devour] An animal that eats other animals; a type of heterotroph.

carotenoids (kare-OTT-en-oyds) Light-sensitive, accessory pigments that transfer absorbed energy to chlorophylls. They absorb violet and blue wavelengths but transmit red, orange, and yellow.

carpel (KAR-pul) The female reproductive part of a flower; sometimes called a pistil. The lower portion of a single carpel (or of a structure composed of two or more carpels) is an ovary, where eggs develop and are fertilized and where seeds mature. The upper portion has a stigma (a pollen-capturing surface tissue) and often a style (a slender extension of the ovary wall).

carrier protein Type of transport protein that binds specific substances and changes shape in ways that shunt the substances across a plasma membrane. Some carrier proteins function passively, others require an energy input.

carrying capacity The maximum number of individuals in a population (or species) that can be sustained indefinitely by a given environment.

cartilage A type of connective tissue with solid yet pliable intercellular material that resists compression.

Casparian strip In the exodermis and endodermis of roots, a waxy band that acts as an impermeable barrier between the walls of abutting cells.

cDNA Any DNA molecule copied from a mature mRNA transcript by way of reverse transcription.

cell [L. *cella*, small room] The smallest living unit; an organized unit that can survive and reproduce on its own, given DNA instructions and suitable environmental conditions, including appropriate sources of energy and raw materials.

cell count The number of cells of a given type in a microliter of blood.

cell cycle Events during which a cell increases in mass, roughly doubles its number of cytoplasmic components, duplicates its DNA, then undergoes nuclear and cytoplasmic division. It extends from the time a new cell is produced until it completes its own division.

cell differentiation The developmental process in which different cell types activate and suppress a fraction of their genes in different ways and so become specialized in composition, structure, and function. Regulatory proteins, enzymes, hormonal signals, and control sites built into DNA interact to bring about this selective gene expression.

cell junction Of multicelled organisms, a point of contact that physically links two cells or that provides functional links between their cytoplasm.

cell plate Of a plant cell undergoing cytoplasmic division, a disklike structure that forms from remnants of the spindle; it becomes a crosswall that partitions the cytoplasm.

cell theory A theory in biology, the key points of which are that (1) all organisms are composed of one or more cells, (2) the cell is the smallest unit that still retains a capacity for independent life, and (3) all cells arise from preexisting cells.

cell wall A rigid or semirigid wall outside the plasma membrane that supports a cell and imparts shape to it; a cellular feature of plants, fungi, protistans, and most bacteria.

central nervous system The brain and spinal cord of vertebrates.

central vacuole Of mature, living plant cells, a fluid-filled organelle that stores amino acids, sugars, ions, and toxic wastes. Its enlargement during growth causes the increases in surface area that improve nutrient uptake.

centriole (SEN-tree-ohl) A cylinder of triplet microtubules that gives rise to the microtubules of cilia and flagella.

centromere (SEN-troh-meer) [Gk. *kentron*, center, + *meros*, a part] A small, constricted region of a chromosome having attachment sites for microtubules that help move the chromosome during nuclear division.

cephalization (sef-ah-lah-ZAY-shun) [Gk. *kephalikos*, head] During the evolution of bilateral animals, the concentration of sensory structures and nerve cells in the head.

cerebellum (ser-ah-BELL-um) [L. diminutive of *cerebrum*, brain] Hindbrain region with reflex centers for maintaining posture and refining limb movements.

cerebral cortex Thin surface layer of the cerebral hemispheres. Some regions of the cortex receive sensory input, others integrate information and coordinate appropriate motor responses.

cerebrospinal fluid Clear extracellular fluid that surrounds and cushions the brain and spinal cord.

cerebrum (suh-REE-bruhm) Part of the vertebrate forebrain that originally integrated olfactory input and selected motor responses to it. In mammals, it evolved into the most complex integrating center.

channel protein Type of transport protein that serves as a pore through which ions or other water-soluble substances move across the plasma membrane. Some channels remain open, while others are gated and open and close in controlled ways.

chemical bond A union between the electron structures of two or more atoms or ions.

chemical synapse (SIN-aps) [Gk. *synapsis*, union] A small gap, the synaptic cleft, that separates two neurons (or a neuron and a muscle cell or gland cell) and that is bridged by neurotransmitter molecules released from the presynaptic neuron.

chemiosmotic theory (kim-ee-OZ-MOT-ik) Theory that an electrochemical gradient across a cell membrane drives ATP formation. Metabolic reactions cause hydrogen ions (H^+) to accumulate in a compartment formed by the membrane. The combined force of the resulting concentration and electric gradients propels hydrogen ions down the gradient, through channel proteins. Through enzyme action at these proteins, ADP and inorganic phosphate combine to form ATP.

chemoreceptor (KEE-moe-ree-sep-tur) Sensory receptor that detects chemical energy (ions or molecules) dissolved in the surrounding fluid.

chemosynthetic autotroph (KEE-moe-sin-THET-ik) One of a few kinds of bacteria able to synthesize its own organic molecules using carbon dioxide as the carbon source and certain inorganic substances (such as sulfur) as the energy source.

chlorofluorocarbon (KLORE-oh-FLOOR-oh-carbun), or **CFC** One of a variety of odorless, invisible compounds of chlorine, fluorine, and carbon, widely used in commercial products, that are contributing to the destruction of the ozone layer above the earth's surface.

chlorophylls (KLOR-uh-fills) [Gk. *chloros*, green, + *phyllon*, leaf] Light-sensitive pigment molecules that absorb violet-to-blue and red wavelengths but that transmit green. Certain chlorophylls donate the electrons required for photosynthesis.

chloroplast (KLOR-uh-plast) An organelle that specializes in photosynthesis in plants and certain protistans.

chordate An animal having a notochord, a dorsal hollow nerve cord, a pharynx, and gill slits in the pharynx wall for at least part of its life cycle.

chorion (CORE-ee-on) Of placental mammals, one of four extraembryonic membranes; it becomes a major component of the placenta. Absorptive structures (villi) that develop at its surface are crucial for the transfer of substances between the embryo and mother.

chromatid Of a duplicated eukaryotic chromosome, one of two DNA molecules and its associated proteins. One chromatid remains attached to its "sister" chromatid at the centromere until they are separated from each other during a nuclear division; then each is a separate chromosome.

chromosome (CROW-moe-some) [Gk. *chroma*, color, + *soma*, body] Of eukaryotes, a DNA molecule with many associated proteins. A bacterial chromosome does not have a comparable profusion of proteins associated with the DNA.

chromosome number Of eukaryotic species, the number of each type of chromosome in all cells except dividing germ cells or gametes.

chytrid A type of single-celled fungus of muddy and aquatic habitats.

ciliated protozoan One of four major groups of protozoans.

cilium (SILL-ee-um), plural **cilia** [L. *cilium*, eyelid] Of eukaryotic cells, a short, hairlike projection that contains a regular array of microtubules. Cilia serve as motile structures, help create currents of fluids, or are part of sensory structures. They typically are more profuse than flagella.

circadian rhythm (ser-KAYD-ee-un) [L. *circa*, about, + *dies*, day] Of many organisms, a cycle of physiological events that is completed every 24 hours or so, even when environmental conditions remain constant.

circulatory system Of multicelled animals, an organ system consisting of a muscular pump (heart, most often), blood vessels, and blood; the system transports materials to and from cells and often helps stabilize body temperature and pH.

cladistics An approach to biological systematics in which organisms are grouped according to similarities that are derived from a common ancestor.

cladogram Branching diagram that represents patterns of relative relationships between organisms based on discrete morphological, physiological, and behavioral traits that vary among taxa being studied.

classification system A way of organizing and retrieving information about species.

cleavage Stage of animal development when mitotic cell divisions convert a zygote to a ball of cells, the blastula.

cleavage furrow Of an animal cell undergoing cytoplasmic division, a shallow, ringlike depression that forms at the cell surface as contractile microfilaments pull the plasma membrane inward. It defines where the cytoplasm will be cut in two.

climate Prevailing weather conditions for an ecosystem, including temperature, humidity, wind speed, cloud cover, and rainfall.

climax community Following primary or secondary succession, the array of species that remains more or less steady under prevailing conditions.

clonal selection theory Theory that lymphocytes activated by a specific antigen rapidly multiply and differentiate into huge subpopulations of cells, all having the parent cell's specificity against that antigen.

cloned DNA Multiple, identical copies of DNA fragments that have been inserted into plasmids or some other cloning vector.

club fungus One of a highly diverse group of multicelled fungi, the reproductive structures of which have microscopic, club-shaped cells that produce and bear spores.

cnidarian A radially symmetrical invertebrate, usually marine, that has tissues (not organs), and nematocysts. Two body forms (medusae and polyps) are common. Jellyfishes, corals, and sea anemones are examples.

coal A nonrenewable source of energy that formed more than 280 million years ago from submerged, undecayed plant remains that were buried in sediments, compressed, then compacted further by heat and pressure.

codominance Condition in which a pair of nonidentical alleles are both expressed even though they specify two different phenotypes.

codon One of a series of base triplets in an mRNA molecule, most of which code for a sequence of amino acids of a specific polypeptide chain. (Of sixty-four codons, sixty-one specify different amino acids and three of these also serve as start signals for translation; one other serves only as a stop signal for translation.)

coelum (SEE-lum) [Gk. *koilos*, hollow] Of many animals, a type of body cavity located between the gut and body wall and having a distinctive lining (peritoneum).

coenzyme A type of nucleotide that transfers hydrogen atoms and electrons from one reaction site to another. NAD$^+$ is an example.

coevolution The joint evolution of two or more closely interacting species; when one species evolves, the change affects selection pressures operating between the two species, so the other also evolves.

cofactor A metal ion or coenzyme; it helps catalyze a reaction or serves briefly as an agent that transfers electrons, atoms, or functional groups from one substrate to another.

cohesion Condition in which molecular bonds resist rupturing when under tension.

cohesion theory of water transport Theory that water moves up through vascular plants due to hydrogen bonding among water molecules confined inside the xylem pipelines. The collective cohesive strength of those bonds allows water to be pulled up as columns in response to transpiration (evaporation from leaves).

collenchyma One of the simple tissues of flowering plants; lends flexible support to primary tissues, such as those of lengthening stems.

colon (CO-lun) The large intestine.

commensalism [L. *com*, together, + *mensa*, table] Two-species interaction in which one species benefits significantly while the other is neither helped nor harmed to any notable extent.

communication signal Of social animals, an action or cue sent by one member of a species (the signaler) that can change the behavior of another member (the signal receiver).

community The populations of all species occupying a habitat; also applied to groups of organisms with similar life-styles in a habitat (such as the bird community).

companion cell A specialized parenchyma cell that helps load dissolved organic compounds into the conducting cells of the phloem.

comparative morphology [Gk. *morph*, form] Anatomical comparisons of major lineages.

competitive exclusion Theory that populations of two species competing for a limited resource cannot coexist indefinitely in the same habitat; the population better adapted to exploit the resource will enjoy a competitive (hence reproductive) edge and will eventually exclude the other population from the habitat.

complement system A set of about twenty proteins circulating in blood plasma with roles in nonspecific defenses and in immune responses. Some induce lysis of pathogens, others promote inflammation, and others stimulate phagocytes to engulf pathogens.

compound A substance in which the relative proportions of two or more elements never vary. Organic compounds have a backbone of carbon atoms arranged as a chain or ring structure. The simpler, inorganic compounds do not have comparable backbones.

concentration gradient A difference in the number of molecules (or ions) of a substance between two adjacent regions, as in a volume of fluid.

condensation reaction Enzyme-mediated reaction leading to the covalent linkage of small molecules and, often, the formation of water as a by-product.

cone cell In the vertebrate eye, a type of photoreceptor that responds to intense light and contributes to sharp daytime vision and color perception.

conifer A type of plant belonging to the dominant group of gymnosperms; mostly evergreen, woody trees and shrubs with pollen- and seed-bearing cones.

connective tissue proper A category of animal tissues, all having mostly the same components but in different proportions. These tissues contain fibroblasts and other cells, the secretions of which form fibers (of collagen and elastin) and a ground substance (of modified polysaccharides).

consumers [L. *consumere*, to take completely] Of ecosystems, heterotrophic organisms that obtain energy and raw materials by feeding on the tissues of other organisms. Herbivores, carnivores, omnivores, and parasites are examples.

continuous variation A more or less continuous range of small differences in a given trait among all the individuals of a population.

contractile vacuole (kun-TRAK-till VAK-you-ohl) [L. *contractus*, to draw together] In some protistans, a membranous chamber that takes up excess water in the cell body, then contracts, expelling the water outside the cell through a pore.

control group In a scientific experiment, a group used to evaluate possible side effects of a test involving an experimental group. Ideally, the control group should differ from the experimental group only with respect to the variable being studied.

convergence, morphological *See* morphological convergence.

cork cambium A type of lateral meristem that produces a tough, corky replacement for epidermis on parts of woody plants showing extensive secondary growth.

corpus callosum (CORE-pus ka-LOW-sum) A band of axons (200 million in humans) that functionally link two cerebral hemispheres.

corpus luteum (CORE-pus LOO-tee-um) A glandular structure; it develops from cells of a ruptured ovarian follicle and secretes progesterone and some estrogen, both of which maintain the lining of the uterus (endometrium).

cortex [L. *cortex*, bark] In general, a rindlike layer; the kidney cortex is an example. In vascular plants, ground tissue that makes up most of the primary plant body, supports plant parts, and stores food.

cotyledon A seed leaf, which develops as part of a plant embryo; cotyledons provide nourishment for the germinating seedling.

courtship display Social behavior by which individuals assess and respond to sexual overtures of potential partners.

covalent bond (koe-VAY-lunt) [L. *con*, together, + *valere*, to be strong] A sharing of one or more electrons between atoms or groups of atoms. When electrons are shared equally, the bond is nonpolar. When electrons are shared unequally, the bond is polar—slightly positive at one end and slightly negative at the other.

cross-bridge formation Of muscle cells, the interaction between actin and myosin filaments that is the basis of contraction.

crossing over During prophase I of meiosis, an interaction between a pair of homologous chromosomes. Their nonsister chromatids break at the same place along their length and exchange corresponding segments at the break points. Crossing over breaks up old combinations of alleles and puts new ones together in chromosomes.

culture The sum total of behavior patterns of a social group, passed between generations by learning and by symbolic behavior, especially language.

cuticle (KEW-tih-kull) A body covering. Of land plants, a covering of waxes and lipid-rich cutin deposited on the outer surface of epidermal cell walls. Of annelids, a thin, flexible surface coat. Of arthropods, a hardened yet lightweight covering with protein and chitin components that functions as an external skeleton.

cycad A type of gymnosperm of the tropics and subtropics; slow growing, with massive, cone-shaped structures that bear ovules or pollen.

cyclic AMP (SIK-lik) Cyclic adenosine monophosphate. A nucleotide that has roles in intercellular communication, as when it serves as a second messenger (a cytoplasmic mediator of a cell's response to signaling molecules).

cyclic pathway of ATP formation Photosynthetic pathway in which excited electrons move from a photosystem to an electron transport system, and back to the photosystem. The electron flow contributes to the formation of ATP from ADP and inorganic phosphate.

cyst Of some microorganisms, a walled, resting structure that forms during the life cycle.

cytochrome (SIGH-toe-krome) [Gk. *kytos*, hollow vessel, + *chrōma*, color] Iron-containing protein molecule; a component of electron transport systems used in photosynthesis and aerobic respiration.

cytokinesis (SIGH-toe-kih-NEE-sis) [Gk. *kinesis*, motion] Cytoplasmic division; the splitting of a parental cell into daughter cells.

cytokinin (SIGH-tow-KY-nin) Any of the class of plant hormones that stimulate cell division, promote leaf expansion, and retard leaf aging.

cytomembrane system [Gk. *kytos*, hollow vessel] Organelles, functioning as a system to modify, package, and distribute newly formed proteins and lipids. Endoplasmic reticulum, Golgi bodies, lysosomes, and a variety of vesicles are its components.

cytoplasm (SIGH-toe-plaz-um) [Gk. *plassein*, to mold] All cellular parts, particles, and semifluid substances enclosed by the plasma membrane except for the region of DNA (which in eukaryotes, is the nucleus).

cytosine (SIGH-toe-seen) A pyrimidine; one of the nitrogen-containing bases in nucleotides.

cytoskeleton Of eukaryotic cells, an internal "skeleton." Its microtubules and other components structurally support the cell, organize and move its internal components. The cytoskeleton also helps free-living cells move through their environment.

cytotoxic T cell A T lymphocyte that eliminates infected body cells or tumor cells with a single hit of toxins and perforins.

decomposers [L. *de-*, down, away, + *companere*, to put together] Of ecosystems, heterotrophs that obtain energy by chemically breaking down the remains, products, or wastes of other organisms. Their activities help cycle nutrients back to producers. Certain fungi and bacteria are examples.

deforestation The removal of all trees from a large tract of land, such as the Amazon Basin or the Pacific Northwest.

degradative pathway A metabolic pathway by which molecules are broken down in stepwise reactions that lead to products of lower energy.

deletion A change in a chromosome's structure after one of its regions is lost as a result of irradiation, viral attack, chemical action, or some other factor.

demographic transition model Model of human population growth in which changes in the growth pattern correspond to different stages of economic development. These are a preindustrial stage, when birth and death rates are both high, a transitional stage, an industrial stage, and a postindustrial stage, when the death rate exceeds the birth rate.

denaturation (deh-NAY-chur-AY-shun) Of any molecule, the loss of three-dimensional shape following disruption of hydrogen bonds and other weak bonds.

dendrite (DEN-drite) [Gk. *dendron*, tree] A short, slender extension from the cell body of a neuron.

denitrification (DEE-nite-rih-fih-KAY-shun) The conversion of nitrate or nitrite by certain bacteria to gaseous nitrogen (N_2) and a small amount of nitrous oxide (N_2O).

density-dependent controls Factors such as predation, parasitism, disease, and competition for resources, which limit population growth by reducing the birth rate, increasing the rates of death and dispersal, or all of these.

density-independent controls Factors such as storms or floods that increase a population's death rate more or less independently of its density.

dentition (den-TIH-shun) The type, size, and number of an animal's teeth.

dermal tissue system Of vascular plants, the tissues that cover and protect the plant surfaces.

dermis The layer of skin underlying the epidermis, consisting mostly of dense connective tissue.

desert A type of biome that exists where the potential for evaporation greatly exceeds rainfall and vegetation cover is limited.

desertification (dez-urt-ih-fih-KAY-shun) The conversion of grasslands, rain-fed cropland, or irrigated cropland to desertlike conditions, with a drop in agricultural productivity of 10 percent or more.

detrital food web Of most ecosystems, the flow of energy mainly from plants through detritivores and decomposers.

detritivores (dih-TRY-tih-vorez) [L. *detritus*; after *deterere*, to wear down] Of ecosystems, heterotrophs that consume dead or decomposing particles of organic matter. Earthworms, crabs, and nematodes are examples.

deuterostome (DUE-ter-oh-stome) [Gk. *deuteros*, second, + *stoma*, mouth] Any of the bilateral animals, including echinoderms and chordates, in which the first indentation in the early embryo develops into the anus.

diaphragm (DIE-uh-fram) [Gk. *diaphragma*, to partition] Muscular partition between the thoracic and abdominal cavities, the contraction and relaxation of which contribute to breathing. Also, a contraceptive device used temporarily to prevent sperm from entering the uterus during sexual intercourse.

dicot (DIE-kot) [Gk. *di*, two, + *kotyledon*, cup-shaped vessel] Short for dicotyledon; class of flowering plants characterized generally by seeds having embryos with two cotyledons (seed leaves), net-veined leaves, and floral parts arranged in fours, fives, or multiples of these.

differentiation *See* cell differentiation.

diffusion Net movement of like molecules (or ions) down their concentration gradient. In the absence of other forces, molecular motion and random collisions cause their net outward movement from one region into a neighboring region where they are less concentrated (because collisions are more frequent where the molecules are most crowded together).

digestive system An internal tube or cavity from which ingested food is absorbed into the internal environment; often divided into regions specialized for food transport, processing, and storage.

dihybrid cross An experimental cross in which offspring inherit two gene pairs, each consisting of two nonidentical alleles.

dinoflagellate A photosynthetic or heterotrophic protistan, often flagellated, that is a component of plankton.

diploid number (DIP-loyd) For many sexually reproducing species, the chromosome number of somatic cells and of germ cells prior to meiosis. Such cells have two chromosomes of each type (that is, pairs of homologous chromosomes). *Compare* haploid number.

directional selection Of a population, a shift in allele frequencies in a steady, consistent direction in response to a new environment or to a directional change in the old one. The outcome is that forms of traits at one end of the range of phenotypic variation become more common than the intermediate forms.

disaccharide (die-SAK-uh-ride) [Gk. *di*, two, + *sakcharon*, sugar] A type of simple carbohydrate, of the class called oligosaccharides; two monosaccharides covalently bonded.

disruptive selection Of a population, a shift in allele frequencies to forms of traits at both ends of a range of phenotypic variation and away from intermediate forms.

distal tubule The tubular section of a nephron most distant from the glomerulus; a major site of water and sodium reabsorption.

divergence Accumulation of differences in allele frequencies between populations that have become reproductively isolated from one another.

divergence, morphological *See* morphological divergence.

diversity, organismic Sum total of variations in form, function, and behavior that have accumulated in different lineages. Those variations generally are adaptive to prevailing conditions or were adaptive to conditions that existed in the past.

DNA Deoxyribonucleic acid (dee-OX-ee-RYE-bow-new-CLAY-ik). For all cells (and many viruses), the molecule of inheritance. A category of nucleic acids, each usually consisting of two nucleotide strands twisted together helically and held together by hydrogen bonds. The nucleotide sequence encodes the instructions for assembling proteins, and, ultimately, new individuals of a particular species.

DNA-DNA hybridization *See* nucleic acid hybridization.

DNA fingerprint Of each individual, a unique array of RFLPs, resulting from the DNA sequences inherited (in a Mendelian pattern) from each parent.

DNA library A collection of DNA fragments produced by restriction enzymes and incorporated into plasmids.

DNA ligase (LYE-gaze) Enzyme that seals together the new base-pairings during DNA replication; also used by recombinant DNA technologists to seal base-pairings between DNA fragments and cut plasmid DNA.

DNA polymerase (poe-LIM-uh-raze) Enzyme that assembles a new strand on a parent DNA strand during replication; also takes part in DNA repair.

DNA probe A short DNA sequence that has been assembled from radioactively labeled nucleotides and that can base-pair with part of a gene under investigation.

DNA repair Following an alteration in the base sequence of a DNA strand, a process that restores the original sequence, as carried out by DNA polymerases, DNA ligases, and other enzymes.

DNA replication Of cells, the process by which the hereditary material is duplicated for distribution to daughter nuclei. An example is the duplication of eukaryotic chromosomes during interphase, prior to mitosis.

dominance hierarchy Form of social organization in which some members of the group have adopted a subordinate status to others.

dominant allele In a diploid cell, an allele that masks the expression of its partner on the homologous chromosome.

dormancy [L. *dormire*, to sleep] Of plants, the temporary, hormone-mediated cessation of growth under conditions that might appear to be quite suitable for growth.

double fertilization Of flowering plants only, the fusion of one sperm nucleus with the egg nucleus (to produce a zygote), *and* fusion of a second sperm nucleus with the two nuclei of the endosperm mother cell, which gives rise to triploid (3n) nutritive tissue.

doubling time The length of time it takes for a population to double in size.

drug addiction Chemical dependence on a drug, following habituation and tolerance of

it; the drug takes on an "essential" biochemical role in the body.

dry shrubland A type of biome that exists where annual rainfall is less than 25 to 60 centimeters and where short, woody, multi-branched shrubs predominate; chaparral is an example.

dry woodland A type of biome that exists when annual rainfall is about 40 to 100 centimeters; there may be tall trees, but these do not form a dense canopy.

dryopith A type of hominoid, one of the first to appear during the Miocene about the time of the divergences that led to gorillas, chimpanzees, and humans.

duplication A change in a chromosome's structure resulting in the repeated appearance of the same gene sequence.

early *Homo* A type of early hominid that may have been the maker of stone tools that date from about 2.5 million years ago.

echinoderm A type of invertebrate that has calcified spines, needles, or plates on the body wall. It is radially symmetrical but with some bilateral features. Sea stars and sea urchins are examples.

ecology [Gk. *oikos*, home, + *logos*, reason] Study of the interactions of organisms with one another and with their physical and chemical environment.

ecosystem [Gk. *oikos*, home] An array of organisms and their physical environment, all of which interact through a flow of energy and a cycling of materials.

ecosystem modeling Analytical method of predicting unforeseen effects of disturbances to an ecosystem, based on computer programs and models.

ectoderm [Gk. *ecto*, outside, + *derma*, skin] Of animal embryos, the outermost primary tissue layer (germ layer) that gives rise to the outer layer of the integument and to tissues of the nervous system.

effector Of homeostatic systems, a muscle (or gland) that responds to signals from an integrator (such as the brain) by producing movement (or chemical change) that helps adjust the body to changing conditions.

effector cell Of the differentiated subpopulations of lymphocytes that form during an immune response, the type of cell that engages and destroys the antigen-bearing agent that triggered the response.

egg A type of mature female gamete; also called an ovum.

El Niño A recurring, massive displacement to the east of warm surface waters of the western equatorial Pacific, which in turn displaces the cooler waters of the Humboldt Current off the coast of Peru.

electron Negatively charged unit of matter, with both particulate and wavelike properties, that occupies one of the orbitals around the atomic nucleus. Atoms can gain, lose, or share electrons with other atoms.

electron transport phosphorylation (FOSS-for-ih-LAY-shun) Final stage of aerobic respiration, in which ATP forms after hydrogen ions and electrons (from the Krebs cycle) are sent through a transport system that gives up the electrons to oxygen.

electron transport system An organized array of enzymes and cofactors, bound in a cell membrane, that accept and donate electrons in sequence. When such systems operate, hydrogen ions (H^+) flow across the membrane, and the flow drives ATP formation and other reactions.

element Any substance that cannot be decomposed into substances with different properties.

embryo (EM-bree-oh) [Gk. *en*, in, + probably *bryein*, to swell] Of animals generally, the stage formed by way of cleavage, gastrulation, and other early developmental events. Of seed plants, the young sporophyte, from the first cell divisions after fertilization until germination.

embryo sac The female gametophyte of flowering plants.

emulsification Of chyme in the small intestine, a suspension of droplets of fat coated with bile salts.

end product A substance present at the end of a metabolic pathway.

endangered species A species poised at the brink of extinction, owing to the extremely small size and severely limited genetic diversity of its remaining populations.

endergonic reaction (en-dur-GONE-ik) Chemical reaction showing a net gain in energy.

endocrine gland Ductless gland that secretes hormones into interstitial fluid, after which they are distributed by way of the bloodstream.

endocrine system System of cells, tissues, and organs that is functionally linked to the nervous system and that exerts control by way of its hormones and other chemical secretions.

endocytosis (EN-doe-sigh-TOE-sis) Movement of a substance into cells; the substance becomes enclosed by a patch of plasma membrane that sinks into the cytoplasm, then forms a vesicle around it. Phagocytic cells also engulf pathogens or prey in this manner.

endoderm [Gk. *endon*, within, + *derma*, skin] Of animal embryos, the inner primary tissue layer, or germ layer, that gives rise to the inner lining of the gut and organs derived from it.

endodermis A sheetlike wrapping of single cells around the vascular cylinder of a root; it functions in controlling the uptake of water and dissolved nutrients. An impermeable barrier (Casparian strip) prevents water from passing between the walls of abutting endodermal cells.

endometrium (EN-doh-MEET-ree-um) [Gk. *metrios*, of the womb] Inner lining of the uterus, consisting of connective tissues, glands, and blood vessels.

endoplasmic reticulum or **ER** (EN-doe-PLAZ-mik reh-TIK-yoo-lum) An organelle that begins at the nucleus and curves through the cytoplasm. In rough ER (which has many ribosomes on its cytoplasmic side), many new polypeptide chains acquire specialized side chains. In many cells, smooth ER (with no attached ribosomes) is the main site of lipid synthesis.

endoskeleton [Gk. *endon*, within, + *sklēros*, hard, stiff] In chordates, the internal framework of bone, cartilage, or both. Together with skeletal muscle, supports and protects other body parts, helps maintain posture, and moves the body.

endosperm (EN-doe-sperm) Nutritive tissue that surrounds and serves as food for a flowering plant embryo and, later, for the germinating seedling.

endospore Of certain bacteria, a resistant body that forms around DNA and some cytoplasm; it germinates and gives rise to new bacterial cells when conditions become favorable.

endosymbiosis A permanent, mutually beneficial interdependency between two species, one of which resides permanently inside the other's body.

energy The capacity to do work.

energy carrier A molecule that delivers energy from one metabolic reaction site to another. ATP is the most widely travelled of these; it readily donates energy to nearly all metabolic reactions.

energy flow pyramid A pyramid-shaped representation of an ecosystem's trophic structure, illustrating the energy losses at each transfer to a different trophic level.

entropy (EN-trow-pee) A measure of the degree of disorder in a system (how much energy has become so disorganized and dispersed, usually as heat, that it is no longer readily available to do work).

enzyme (EN-zime) One of a class of proteins that greatly speed up (catalyze) reactions between specific substances, usually at their functional groups. The substances that each type of enzyme acts upon are called its substrates.

eosinophil Fast-acting, phagocytic white blood cell that takes part in inflammation but not in immune responses.

epidermis The outermost tissue layer of a multicelled plant or animal.

epiglottis A flaplike structure at the start of the larynx, the position of which directs the movement of air into the trachea or food into the esophagus.

epistasis (eh-PISS-tih-sis) A type of gene interaction, whereby two alleles of a gene influence the expression of alleles of a different gene.

epithelium (EP-ih-THEE-lee-um) An animal tissue consisting of one or more layers of adhering cells that covers the body's external surfaces and lines its internal cavities and tubes. Epithelium has one free surface; the opposite surface rests on a basement membrane between it and an underlying connective tissue. Epidermis or skin is an example.

equilibrium, dynamic [Gk. *aequus*, equal, + *libra*, balance] The point at which a chemical reaction runs forward as fast as it runs in reverse; thus the concentrations of reactant molecules and product molecules show no net change.

erythrocyte (eh-RITH-row-site) [Gk. *erythros*, red, + *kytos*, vessel] Red blood cell.

esophagus (ee-SOF-uh-gus) Tubular portion of a digestive system that receives swallowed food and leads to the stomach.

essential amino acid Any of eight amino acids that certain animals cannot synthesize for themselves and must obtain from food.

essential fatty acid Any of the fatty acids that certain animals cannot synthesize for themselves and must obtain from food.

estrogen (ESS-trow-jen) A sex hormone that helps oocytes mature, induces changes in the uterine lining during the menstrual cycle and pregnancy, and maintains secondary sexual traits; also influences bodily growth and development.

estrus (ESS-truss) [Gk. *oistrus*, frenzy] For mammals generally, the cyclic period of a female's sexual receptivity to the male.

estuary (EST-you-ehr-ee) A partly enclosed coastal region where seawater mixes with freshwater from rivers, streams, and runoff from the surrounding land.

ethylene (ETH-il-een) Plant hormone that stimulates fruit ripening and triggers abscission.

eubacteria The subkingdom of all bacterial species except the archaebacteria; one of the three great prokaryotic lineages that arose early in the history of life.

euglenoid A type of flagellated protistan, most of which are photosynthesizers in stagnant or freshwater ponds.

eukaryotic cell (yoo-CARRY-oh-tic) [Gk. *eu*, good, + *karyon*, kernel] A type of cell that has a "true nucleus" and other distinguishing membrane-bound organelles. *Compare* prokaryotic cell.

eutrophication Nutrient enrichment of a body of water, such as a lake, that typically results in reduced transparency and a phytoplankton-dominated community.

evaporation [L. *e-*, out, + *vapor*, steam] Conversion of a substance from the liquid to the gaseous state; some or all of its molecules leave in the form of vapor.

evolution, biological [L. *evolutio*, act of unrolling] Change within a line of descent over time. A population is evolving when some forms of a trait are becoming more or less common, relative to the other kinds of traits. The shifts are evidence of changes in the relative abundances of alleles for that trait, as brought about by mutation, natural selection, genetic drift, and gene flow.

evolutionary tree A treelike diagram in which branches represent separate lines of descent from a common ancestor.

excitatory postsynaptic potential or **EPSP** One of two competing signals at an input zone of a neuron; a graded potential that brings the neuron's plasma membrane closer to threshold.

excretion Any of several processes by which excess water, excess or harmful solutes, or waste materials leave the body by way of the urinary system or certain glands.

exergonic reaction (EX-ur-GONE-ik) A chemical reaction that shows a net loss in energy.

exocrine gland (EK-suh-krin) [Gk. *es*, out of, + *krinein*, to separate] Glandular structure that secretes products, usually through ducts or tubes, to a free epithelial surface.

exocytosis (EK-so-sigh-TOE-sis) Movement of a substance out of a cell by means of a transport vesicle, the membrane of which fuses with the plasma membrane, so that the vesicle's contents are released outside.

exodermis Layer of cells just inside the root epidermis of most flowering plants; helps control the uptake of water and solutes.

exon Of eukaryotic cells, any of the nucleotide sequences of a pre-mRNA molecule that are spliced together to form the mature mRNA transcript and are ultimately translated into protein.

exoskeleton [Gk. *exo*, out, + *sklēros*, hard, stiff] An external skeleton, as in arthropods.

experiment A test in which some phenomenon in the natural world is manipulated in controlled ways to gain insight into its function, structure, operation, or behavior.

exploitation competition Interaction in which both species have equal access to a required resource but differ in how fast or efficiently they exploit it.

exponential growth (EX-po-NEN-shul) Of populations, a pattern of growth based on a constant rate of multiplication over increments of time. One variable (population size) increases much faster than another variable (the number of individuals in the reproductive base) in a specific mathematical way. The larger the reproductive base, the greater the expansion in population size.

extinction, background A steady rate of species turnover that characterizes lineages through most of their histories.

extinction, mass An abrupt increase in the rate at which major taxa disappear, with several taxa being affected simultaneously.

extracellular fluid In animals generally, all the fluid not inside cells; includes plasma (the liquid portion of blood) and interstitial fluid (which occupies the spaces between cells and tissues).

extracellular matrix A material, largely secreted, that helps hold many animal tissues together in certain shapes; it consists of fibrous proteins and other components in a ground substance.

FAD Flavin adenine dinucleotide, a nucleotide coenzyme. When delivering electrons and unbound protons (H^+) from one reaction to another, it is abbreviated $FADH_2$.

fall overturn The vertical mixing of a body of water in autumn. Its upper layer cools, increases in density, and sinks; dissolved oxygen moves down and nutrients from bottom sediments are brought to the surface.

family pedigree A chart of genetic relationships of the individuals in a family through successive generations.

fat A lipid with a glycerol head and one, two, or three fatty acid tails. The tails of saturated fats have only single bonds between carbon atoms and hydrogen atoms attached to all other bonding sites. Tails of unsaturated fats additionally have one or more double bonds between certain carbon atoms.

fatty acid A long, flexible hydrocarbon chain with a —COOH group at one end.

feedback inhibition Of cells, a control mechanism by which the production (or secretion) of a substance triggers a change in some activity that in turn shuts down further production of the substance.

fermentation [L. *fermentum*, yeast] A type of anaerobic pathway of ATP formation, it starts with glycolysis, ends when electrons are transferred back to one of the breakdown products or intermediates, and regenerates the NAD^+ required for the reaction. Its net yield is two ATP per glucose molecule degraded.

fern One of the seedless vascular plants, mostly of wet, humid habitats; requires free water to complete its life cycle.

fertilization [L. *fertilis*, to carry, to bear] Fusion of a sperm nucleus with the nucleus of an egg, which thereupon becomes a zygote.

fever A body temperature higher than a set point that is preestablished in the brain region governing temperature.

fibrous root system Of most monocots, all the lateral branchings of adventitious roots, which arose earlier from the young stem.

filtration Of urine formation, the process by which blood pressure forces water and solutes out of glomerular capillaries and into the cupped portion of a nephron wall (Bowman's capsule).

fin Of fishes generally, an appendage that helps propel, stabilize, and guide the body through water.

first law of thermodynamics [Gk. *therme*, heat, + *dynamikos*, powerful] Law stating that the total amount of energy in the universe remains constant. Energy cannot be created and existing energy cannot be destroyed. It can only be converted from one form to another.

fish An aquatic animal of the most ancient vertebrate lineage; jawless fishes (such as lampreys and hagfishes), cartilaginous fishes (such as sharks), and bony fishes (such as coelacanths and salmon) are the three existing groups.

fixed action pattern An instinctive response that is triggered by a well-defined, simple stimulus and that is performed in its entirety once it has begun.

flagellated protozoan A member of one of four major groups of protozoans, many of which cause serious diseases.

flagellum (fluh-JELL-um), plural **flagella** [L. whip] Tail-like motile structure of many free-living eukaryotic cells; it has a distinctive 9 + 2 array of microtubules.

flatworm A type of invertebrate having bilateral symmetry, a flattened body, and a saclike gut; a turbellarian, fluke, or tapeworm.

flower The reproductive structure that distinguishes angiosperms from other seed plants and often attracts pollinators.

fluid mosaic model Model of membrane structure in which proteins are embedded in a lipid bilayer or attached to one of its surfaces. The lipid molecules give the membrane its basic structure, impermeability to water-soluble molecules, and (through packing variations and movements) fluidity. Proteins carry out most membrane functions, such as transport, enzyme action, and reception of signals or substances.

follicle (FOLL-ih-kul) In a mammalian ovary, a primary oocyte (immature egg) together with the surrounding layer of cells.

food chain A straight-line sequence of who eats whom in an ecosystem.

food web A network of cross-connecting, interlinked food chains, encompassing primary producers and an array of consumers, detritivores, and decomposers.

forebrain Brain region that includes the cerebrum and cerebral cortex, the olfactory lobes, and the hypothalamus.

forest A type of biome where tall trees grow together closely enough to form a fairly continuous canopy over a broad region.

fossil Recognizable evidence of an organism that lived in the distant past. Most fossils are skeletons, shells, leaves, seeds, and tracks that were buried in rock layers before they decomposed.

fossil fuel Coal, petroleum, or natural gas; a nonrenewable source of energy formed in sediments by the compression of carbon-containing plant remains over hundreds of millions of years.

founder effect An extreme case of genetic drift. By chance, a few individuals that leave a population and establish a new one carry fewer (or more) alleles for certain traits. Increased variation between the two populations is one outcome. Limited genetic variability in the new population is another.

free radical A highly reactive, unbound molecular fragment with the wrong number of electrons.

fruit [L. after *frui*, to enjoy] Of flowering plants, the expanded and ripened ovary of one or more carpels, sometimes with accessory structures incorporated.

FSH Follicle-stimulating hormone. Produced and secreted by the anterior lobe of the pituitary gland, this hormone has roles in the reproductive functions of both males and females.

functional group An atom or group of atoms that is covalently bonded to the carbon backbone of an organic compound and that influences its behavior.

Fungi The kingdom of fungi.

fungus A eukaryotic heterotroph that uses extracellular digestion and absorption; it secretes enzymes able to break down an external food source into molecules small enough to be absorbed by its cells. Saprobic types feed on nonliving organic matter; parasitic types feed on living organisms. Fungi as a group are major decomposers.

gall bladder Organ of the digestive system that stores bile secreted from the liver.

gamete (GAM-eet) [Gk. *gametēs*, husband, and *gametē*, wife] A haploid cell that functions in sexual reproduction. Sperm and eggs are examples.

gamete formation Generally, the formation of gametes by way of meiosis. Of animals, the first stage of development, in which sperm or eggs form and mature within reproductive tissues of parents.

gametophyte (gam-EET-oh-fite) [Gk. *phyton*, plant] The haploid, multicelled, gamete-producing phase in the life cycle of most plants.

ganglion (GANG-lee-un), plural **ganglia** [Gk. *ganglion*, a swelling] A distinct clustering of cell bodies of neurons in regions other than the brain or spinal cord.

gastrulation (gas-tru-LAY-shun) Of animals, the stage of embryonic development in which cells become arranged into two or three primary tissue layers (germ layers); in humans, the layers are an inner endoderm, an intermediate mesoderm, and a surface ectoderm.

gene [short for German *pangan*, after Gk. *pan*, all + *genes*, to be born] A unit of information about a heritable trait that is passed on from parents to offspring. Each gene has a specific location on a chromosome.

gene flow A microevolutionary process; a physical movement of alleles out of a population as individuals leave (emigrate) or enter (immigrate), the outcome being changes in allele frequencies.

gene frequency More precisely, allele frequency: the relative abundances of all the different alleles for a trait that are carried by the individuals of a population.

gene locus A given gene's particular location on a chromosome.

gene mutation [L. *mutatus*, a change] Change in DNA due to the deletion, addition, or substitution of one to several bases in the nucleotide sequence.

gene pair In diploid cells, the two alleles at a given locus on a pair of homologous chromosomes.

gene pool Sum total of all genotypes in a population. More accurately, allele pool.

gene therapy Generally, the transfer of one or more normal genes into the body cells of an organism in order to correct a genetic defect.

genetic code [After L. *genesis*, to be born] The correspondence between nucleotide triplets in DNA (then in mRNA) and specific sequences of amino acids in the resulting polypeptide chains; the basic language of protein synthesis.

genetic disorder An inherited condition that results in mild to severe medical problems.

genetic drift A microevolutionary process; a change in allele frequencies over the generations due to chance events alone.

genetic engineering Altering the information content of DNA through use of recombinant DNA technology.

genetic equilibrium Hypothetical state of a population in which allele frequencies for a trait remain stable through the generations; a reference point for measuring rates of evolutionary change.

genetic recombination Presence of a new combination of alleles in a DNA molecule compared to the parental genotype; the result of processes such as crossing over at meiosis, chromosome rearrangements, gene mutation, and recombinant DNA technology.

genome All the DNA in a haploid number of chromosomes of a given species.

genotype (JEEN-oh-type) Genetic constitution of an individual. Can mean a single gene pair or the sum total of the individual's genes. *Compare* phenotype.

genus, plural **genera** (JEEN-US, JEN-er-ah) [L. *genus*, race, origin] A taxon into which all species exhibiting certain phenotypic similarities and evolutionary relationship are grouped.

geologic time scale A time scale for earth history, the subdivisions of which have been refined by radioisotope dating work.

germ cell Of animals, one of a cell lineage set aside for sexual reproduction; germ cells give rise to gametes. *Compare* somatic cell.

germ layer Of animal embryos, one of two or three primary tissue layers that form during gastrulation and that gives rise to certain tissues of the adult body. *Compare* ectoderm; endoderm; mesoderm.

germination (jur-min-AY-shun) Generally, the resumption of growth following a rest stage, of seed plants, the time at which an embryo sporophyte breaks through its seed coat and resumes growth.

gibberellin (JIB-er-ELL-un) Any of a class of plant hormones that promote stem elongation.

gill A respiratory organ, typically with a moist, thin vascularized layer of epidermis that functions in gas exchange.

ginkgo A type of gymnosperm with fan-shaped leaves and fleshy coated seeds; now represented by a single species of deciduous trees.

gland A secretory cell or multicelled structure derived from epithelium and often connected to it.

glomerular capillaries The set of blood capillaries inside Bowman's capsule of the nephron.

glomerulus (glow-MARE-you-luss) [L. *glomus*, ball] The first portion of the nephron, where water and solutes are filtered from blood.

glucagon (GLUE-kuh-gone) Hormone that stimulates conversion of glycogen and amino acids to glucose; secreted by alpha cells of the pancreas when the flow of glucose decreases.

glyceride (GLISS-er-eyed) One of the molecules, commonly called fats and oils, that has one, two, or three fatty acid tails attached to a glycerol backbone. They are the body's most abundant lipids and its richest source of energy.

glycerol (GLISS-er-oh) [Gk. *glykys*, sweet, + L. *oleum*, oil] A three-carbon molecule with three hydroxyl groups attached; together with fatty acids, a component of fats and oils.

glycogen (GLY-kuh-jen) In animals, a storage polysaccharide that is a main food reserve; can be readily broken down into glucose subunits.

glycolysis (gly-CALL-ih-sis) [Gk. *glykys*, sweet, + *lysis*, loosening or breaking apart] Initial reactions of both aerobic and anaerobic pathways by which glucose (or some other organic compound) is partially broken down to pyruvate, with a net yield of two ATP. Gly-

colysis proceeds in the cytoplasm of all cells, and oxygen has no role in it.

gnetophyte A type of gymnosperm limited to deserts and tropics.

Golgi body (GOHL-gee) Organelle in which newly synthesized polypeptide chains as well as lipids are modified and packaged in vesicles for export or for transport to specific locations within the cytoplasm.

gonad (GO-nad) Primary reproductive organ in which gametes are produced.

graded potential Of neurons, a local signal that slightly changes the voltage difference across a small patch of the plasma membrane. Such signals vary in magnitude, depending on the stimulus. With prolonged or intense stimulation, they may spread to a trigger zone of the membrane and initiate an action potential.

granum, plural **grana** Within many chloroplasts, any of the stacks of flattened, membranous compartments with chlorophyll and other light-trapping pigments and reaction sites for ATP formation.

grassland A type of biome with flat or rolling land, 25 to 100 centimeters of annual rainfall, warm summers, and often grazing and periodic fires that regenerate the dominant species.

gravitropism (GRAV-ih-TROPE-izm) [L. *gravis*, heavy, + Gk. *trepein*, to turn] The tendency of a plant to grow directionally in response to the earth's gravitational force.

gray matter Of vertebrates, the dendrites, neuron cell bodies, and neuroglial cells of the spinal cord and cerebral cortex.

grazing food web Of most ecosystems, the flow of energy from plants to herbivores, then through an array of carnivores.

green alga One of a group or division of aquatic plants with an abundance of chlorophylls a and b; early members of its lineage may have given rise to the bryophytes and vascular plants.

green revolution In developing countries, the use of improved crop varieties, modern agricultural practices (including massive inputs of fertilizers and pesticides), and equipment to increase crop yields.

greenhouse effect Warming of the lower atmosphere due to the presence of greenhouse gases—carbon dioxide, methane, nitrous oxide, ozone, water vapor, and chlorofluorocarbons.

ground meristem (MARE-ih-stem) [Gk. *meristos*, divisible] Of vascular plants, a primary meristem that produces the ground tissue system, hence the bulk of the plant body.

ground substance Of certain animal tissues, the intercellular material made up of cell secretions and other noncellular components.

ground tissue system Tissues that make up the bulk of the vascular plant body; parenchyma is the most common of these.

guanine A nitrogen-containing base; present in one of the four nucleotide building blocks of DNA and RNA.

guard cell Either of two adjacent cells having roles in the movement of gases and water vapor across leaf or stem epidermis. An opening (stoma) forms when both cells swell with water and move apart; it closes when they lose water and collapse against each other.

gut A body region where food is digested and absorbed; of complete digestive systems, the gastrointestinal tract (the portions from the stomach onward).

gymnosperm (JIM-noe-sperm) [Gk. *gymnos*, naked, + *sperma*, seed] A plant that bears seeds at exposed surfaces of reproductive structures, such as cone scales. Pine trees are examples.

habitat [L. *habitare*, to live in] The type of place where an organism normally lives, characterized by physical features, chemical features, and the presence of certain other species.

hair cell Type of mechanoreceptor that may give rise to action potentials when bent or tilted.

halophile A type of archaebacterium that lives in extremely salty habitats.

haploid number (HAP-loyd) The chromosome number of a gamete which, as an outcome of meiosis, is only half that of the parent germ cell (it has only one of each pair of homologous chromosomes). *Compare* diploid number.

HCG Human chorionic gonadotropin. A hormone that helps maintain the lining of the uterus during the menstrual cycle and during the first trimester of pregnancy.

heart Muscular pump that keeps blood circulating through the animal body.

helper T cell One of the T lymphocytes; when activated, it produces and secretes interleukins that promote formation of huge populations of effector and memory cells for immune responses.

hemoglobin (HEEM-oh-glow-bin) [Gk. *haima*, blood, + L. *globus*, ball] Iron-containing, oxygen-transporting protein that gives red blood cells their color.

hemostasis (HEE-mow-STAY-sis) [Gk. *haima*, blood, + *stasis*, standing] Stopping of blood loss from a damaged blood vessel through coagulation, blood vessel spasm, platelet plug formation, and other mechanisms.

herbivore [L. *herba*, grass, + *vovare*, to devour] Plant-eating animal.

heterocyst (HET-er-oh-sist) Of some filamentous cyanobacteria, a type of thick-walled, nitrogen-fixing cell that forms when nitrogen is scarce.

heterotroph (HET-er-oh-trofe) [Gk. *heteros*, other, + *trophos*, feeder] Organism that cannot synthesize its own organic compounds and must obtain nourishment by feeding on autotrophs, each other, or organic wastes. Animals, fungi, many protistans, and most bacteria are heterotrophs. *Compare* autotroph.

heterozygous condition (HET-er-oh-ZYE-guss) [Gk. *zygoun*, join together] For a given trait, having nonidentical alleles at a particular locus on a pair of homologous chromosomes.

hindbrain One of the three divisions of the vertebrate brain; the medulla oblongata, cerebellum, and pons; includes reflex centers for respiration, blood circulation, and other basic functions; also coordinates motor responses and many complex reflexes.

histone Any of a class of proteins that are intimately associated with DNA and that are largely responsible for its structural (and possibly functional) organization in eukaryotic chromosomes.

homeostasis (HOE-me-oh-STAY-sis) [Gk. *homo*, same, + *stasis*, standing] Of multicelled organisms, a physiological state in which the physical and chemical conditions of the internal environment are being maintained within tolerable ranges.

homeostatic feedback loop An interaction in which an organ (or structure) stimulates or inhibits the output of another organ, then shuts down or increases this activity when it detects that the output has exceeded or fallen below a set point.

hominid [L. *homo*, man] All species on the evolutionary branch leading to modern humans. *Homo sapiens* is the only living representative.

hominoid Apes, humans, and their recent ancestors.

Homo erectus A hominid lineage that emerged between 1.5 million and 300,000 years ago and that may include the direct ancestors of modern humans.

Homo sapiens The hominid lineage of modern humans that emerged between 300,000 and 200,000 years ago.

homologous chromosome (huh-MOLL-uh-gus) [Gk. *homologia*, correspondence] Of sexually reproducing species, one of a pair of chromosomes that resemble each other in size, shape, and the genes they carry, and that line up with each other at meiosis I. The X and Y chromosomes differ in these respects but still function as homologues.

homologous structures The same body parts, modified in different ways, in different lines of descent from a common ancestor.

homozygous condition (HOE-moe-ZYE-guss) Having two identical alleles at a given locus (on a pair of homologous chromosomes).

homozygous dominant condition Having two dominant alleles at a given locus (on a pair of homologous chromosomes).

homozygous recessive condition Having two recessive alleles at a given gene locus (on a pair of homologous chromosomes).

hormone [Gk. *hormon*, to stir up, set in motion] Any of the signaling molecules secreted from endocrine glands, endocrine cells, and some neurons that the bloodstream distributes to nonadjacent target cells (any cell having receptors for that hormone).

horsetail One of the seedless vascular plants, which require free water to complete the life cycle; only one genus has survived to the present.

human genome project A basic research project in which researchers throughout the world are working together to sequence the estimated 3 billion nucleotides present in the DNA of human chromosomes.

hydrogen bond Type of chemical bond in which an atom of a molecule interacts weakly with a neighboring atom that is already taking part in a polar covalent bond.

hydrogen ion A free (unbound) proton; a hydrogen atom that has lost its electron and so bears a positive charge (H^+).

hydrologic cycle A biogeochemical cycle, driven by solar energy, in which water moves slowly through the atmosphere, on or through surface layers of land masses, to the ocean, and back again.

hydrolysis (high-DRAWL-ih-sis) [L. *hydro*, water, + Gk. *lysis*, loosening or breaking apart] Enzyme-mediated reaction in which covalent bonds break, splitting a molecule into two or more parts, and H^+ and OH^- (derived from a water molecule) become attached to the exposed bonding sites.

hydrophilic substance [Gk. *philos*, loving] A polar substance that is attracted to the polar water molecule and so dissolves easily in water. Sugars are examples.

hydrophobic substance [Gk. *phobos*, dreading] A nonpolar substance that is repelled by the polar water molecule and so does not readily dissolve in water. Oil is an example.

hydrosphere All liquid or frozen water on or near the earth's surface.

hydrothermal vent ecosystem A type of ecosystem that exists near fissures in the ocean floor and is based on chemosynthetic bacteria that use hydrogen sulfide as the energy source.

hypha (HIGH-fuh), plural **hyphae** [Gk. *hyphe*, web] Of fungi, a generally tube-shaped filament with chitin-reinforced walls and, often, reinforcing cross-walls; component of the mycelium.

hypodermis A subcutaneous layer having stored fat that helps insulate the body; although not part of skin, it anchors skin while allowing it some freedom of movement.

hypothalamus [Gk. *hypo*, under, + *thalamos*, inner chamber or possibly *tholos*, rotunda] Of vertebrate forebrains, a brain center that monitors visceral activities (such as salt-water balance, temperature control, and reproduction) and that influences related forms of behavior (as in hunger, thirst, and sex).

hypothesis A possible explanation of a specific phenomenon.

immune response A series of events by which B and T lymphocytes recognize a specific antigen, undergo repeated cell divisions that form huge lymphocyte populations, and differentiate into subpopulations of effector and memory cells. Effector cells engage and destroy antigen-bearing agents. Memory cells enter a resting phase and are activated during subsequent encounters with the same antigen.

immunization Various processes, including vaccination, that promote increased immunity against specific diseases.

immunoglobulins (Ig) Four classes of antibodies, each with binding sites for antigen and binding sites used in specialized tasks. Examples are IgM antibodies (first to be secreted during immune responses) and IgG antibodies (which activate complement proteins and neutralize many toxins).

implantation A process by which a blastocyst adheres to the endometrium and begins to establish connections by which the mother and embryo will exchange substances during pregnancy.

imprinting Category of learning in which an animal that has been exposed to specific key stimuli early in its behavioral development forms an association with the object.

incisor A tooth, shaped like a flat chisel or cone, used in nipping or cutting food.

incomplete dominance Of heterozygotes, the appearance of a version of a trait that is somewhere between the homozygous dominant and recessive conditions.

independent assortment Mendelian principle that each gene pair tends to assort into gametes independently of other gene pairs located on nonhomologous chromosomes.

indirect selection A theory in evolutionary biology that self-sacrificing individuals can indirectly pass on their genes by helping relatives survive and reproduce.

induced-fit model Model of enzyme action whereby a bound substrate induces changes in the shape of the enzyme's active site, resulting in a more precise molecular fit between the enzyme and its substrate.

industrial smog A type of gray-air smog that develops in industrialized regions when winters are cold and wet.

inflammation, acute In response to tissue damage or irritation, fast-acting phagocytes and plasma proteins, including complement proteins, leave the bloodstream, then defend and help repair the tissue. Proceeds during both nonspecific and specific (immune) defense responses.

inheritance The transmission, from parents to offspring, of structural and functional patterns that have a genetic basis and are characteristic of each species.

inhibiting hormone A signaling molecule produced and secreted by the hypothalamus that controls secretions by the anterior lobe of the pituitary gland.

inhibitor A substance that can bind with an enzyme and interfere with its functioning.

inhibitory postsynaptic potential, or **IPSP** Of neurons, one of two competing types of graded potentials at an input zone; tends to drive the resting membrane potential away from threshold.

instinctive behavior A complex, stereotyped response to a particular environmental cue that often is quite simple.

insulin Hormone that lowers the glucose level in blood; it is secreted from beta cells of the pancreas and stimulates cells to take up glucose; also promotes protein and fat synthesis and inhibits protein conversion to glucose.

integration, neural [L. *integrare*, to coordinate] Moment-by-moment summation of all excitatory and inhibitory synapses acting on a neuron; occurs at each level of synapsing in a nervous system.

integrator Of homeostatic systems, a control point where different bits of information are pulled together in the selection of a response. The brain is an example.

integument Of animals, a protective body covering such as skin. Of flowering plants, a protective layer around the developing ovule; when the ovule becomes a seed, its integument(s) harden and thicken into a seed coat.

integumentary exchange (in-teg-you-MEN-tuh-ree) Of some animals, a mode of respiration in which oxygen and carbon dioxide diffuse across a thin, vascularized layer of moist epidermis at the body surface.

interference competition Interaction in which one species may limit another species' access to some resource regardless of whether the resource is abundant or scarce.

interleukin One of a variety of communication signals, secreted by macrophages and by helper T cells, that drive immune responses.

intermediate compound A compound that forms between the start and the end of a metabolic pathway.

intermediate filament A cytoskeletal component that consists of different proteins in different types of animal cells.

interneuron Any of the neurons in the vertebrate brain and spinal cord that integrate information arriving from sensory neurons and that influence other neurons in turn.

internode In vascular plants, the stem region between two successive nodes.

interphase Of cell cycles, the time interval between nuclear divisions in which a cell increases its mass, roughly doubles the number of its cytoplasmic components, and finally duplicates its chromosomes (replicates its DNA). The interval is different for different species.

interspecific competition Two-species interaction in which both species can be harmed due to overlapping niches.

interstitial fluid (IN-ter-STISH-ul) [L. *interstitus*, to stand in the middle of something] In multicelled animals, that portion of the extracellular fluid occupying spaces between cells and tissues.

intertidal zone Generally, the area on a rocky or sandy shoreline that is above the low water mark and below the high water mark; organisms inhabiting it are alternately submerged, then exposed, by tides.

intervertebral disk One of a number of disk-shaped structures containing cartilage that serve as shock absorbers and flex points between bony segments of the vertebral column.

intraspecific competition Interaction among individuals of the same species that are competing for the same resources.

intron A noncoding portion of a newly formed mRNA molecule.

inversion A change in a chromosome's structure after a segment separated from it was then inserted at the same place, but in reverse. The reversal alters the position and order of the chromosome's genes.

invertebrate Animal without a backbone.

ion, negatively charged (EYE-on) An atom or a compound that has gained one or more electrons, and hence has acquired an overall negative charge.

ion, positively charged An atom or a compound that has lost one or more electrons, and hence has acquired an overall positive charge.

ionic bond An association between ions of opposite charge.

isotonic condition Equality in the relative concentrations of solutes in two fluids; for two fluids separated by a cell membrane, there is no net osmotic (water) movement across the membrane.

isotope (EYE-so-tope) For a given element, an atom with the same number of protons as the other atoms but with a different number of neutrons.

J-shaped curve A curve, obtained when population size is plotted against time, that is characteristic of unrestricted, exponential growth.

joint An area of contact or near-contact between bones.

karyotype (CARRY-oh-type) Of eukaryotic individuals (or species), the number of metaphase chromosomes in somatic cells and their defining characteristics.

keratin A tough, water-insoluble protein manufactured by most epidermal cells.

keratinization (care-AT-in-iz-AY-shun) Process by which keratin-producing epidermal cells of skin die and collect at the skin surface as keratinized "bags" that form a barrier against dehydration, bacteria, and many toxic substances.

kidney In vertebrates, one of a pair of organs that filter mineral ions, organic wastes, and other substances from the blood, and help regulate the volume and solute concentrations of extracellular fluid.

kilocalorie 1,000 calories of heat energy, or the amount of energy needed to raise the temperature of 1 kilogram of water by 1°C; the unit of measure for the caloric value of foods.

kinetochore A specialized group of proteins and DNA at the centromere of a chromosome that serves as an attachment point for several spindle microtubules during mitosis or meiosis. Each chromatid of a duplicated chromosome has its own kinetochore.

Krebs cycle Together with a few conversion steps that precede it, the stage of aerobic respiration in which pyruvate is completely broken down to carbon dioxide and water. Coenzymes accept the unbound protons (H^+) and electrons stripped from intermediates during the reactions and deliver them to the next stage.

lactate fermentation Anaerobic pathway of ATP formation in which pyruvate from glycolysis is converted to the three-carbon compound lactate, and NAD^+ (a coenzyme used in the reactions) is regenerated. Its net yield is two ATP.

lactation The production of milk by hormone-primed mammary glands.

lake A body of fresh water having littoral, limnetic, and profundal zones.

lancelet An invertebrate chordate having a body that tapers sharply at both ends, segmented muscles, and a full-length notochord.

large intestine The colon; a region of the gut that receives unabsorbed food residues from the small intestine and concentrates and stores feces until they are expelled from the body.

larva, plural **larvae** Of animals, a sexually immature, free-living stage between the embryo and the adult.

larynx (LARE-inks) A tubular airway that leads to the lungs. In humans, contains vocal cords, where sound waves used in speech are produced.

lateral meristem Of vascular plants, a type of meristem responsible for secondary growth; either vascular cambium or cork cambium.

lateral root Of taproot systems, a lateral branching from the first, primary root.

leaching The movement of soil water, with dissolved nutrients, out of a specified area.

leaf For most vascular plants, a structure having chlorophyll-containing tissue that is the major region of photosynthesis.

learned behavior The use of information gained from specific experiences to vary or change a response to stimuli.

lethal mutation A gene mutation that alters one or more traits in such a way that the individual inevitably dies.

LH Leutinizing hormone. Secreted by the anterior lobe of the pituitary gland, this hormone has roles in the reproductive functions of both males and females.

lichen (LY-kun) A symbiotic association between a fungus and a captive photosynthetic partner such as a green alga.

life cycle A recurring, genetically programmed frame of events in which individuals grow, develop, maintain themselves, and reproduce.

life table A tabulation of age-specific patterns of birth and death for a population.

ligament A strap of dense connective tissue that bridges a joint.

light-dependent reactions First stage of photosynthesis in which the energy of sunlight is trapped and converted to the chemical energy of ATP alone (by the cyclic pathway) or ATP and NADPH (by the noncyclic pathway).

light-independent reactions Second stage of photosynthesis, in which sugar phosphates form with the help of the ATP (and NADPH, in land plants) that were produced during the first stage. The sugar phosphates are used in other reactions by which starch, cellulose, and other end products of photosynthesis are assembled.

lignification Of mature land plants, a process by which lignin is deposited in secondary cell walls. The deposits impart strength and rigidity by anchoring cellulose strands in the walls, stabilize and protect other wall components, and form a waterproof barrier around the cellulose. Probably a key factor in the evolution of vascular plants.

lignin A substance that strengthens and waterproofs cell walls in certain tissues of vascular plants.

limbic system Brain regions that, along with the cerebral cortex, collectively govern emotions.

limiting factor Any essential resource that is in short supply and so limits population growth.

lineage (LIN-ee-age) A line of descent.

linkage The tendency of genes located on the same chromosome to end up in the same gamete. For any two of those genes, the probability that crossing over will disrupt the linkage is proportional to the distance separating them.

lipid A greasy or oily compound of mostly carbon and hydrogen that shows little tendency to dissolve in water, but that dissolves in nonpolar solvents (such as ether). Cells use lipids as energy stores and structural materials, especially in membranes.

lipid bilayer The structural basis of cell membranes, consisting of two layers of mostly phospholipid molecules. Hydrophilic heads force all fatty acid tails of the lipids to become sandwiched between the hydrophilic heads.

liver Glandular organ with roles in storing and interconverting carbohydrates, lipids, and proteins absorbed from the gut, maintaining blood; disposing of nitrogen-containing wastes; and other tasks.

local signaling molecules Secretions from cells in many different tissues that alter chemical conditions in the immediate vicinity where they are secreted, then are swiftly degraded.

locus (LOW-cuss) The specific location of a particular gene on a chromosome.

logistic population growth (low-JIS-tik) Pattern of population growth in which a low-density population slowly increases in size, goes through a rapid growth phase, then levels off once the carrying capacity is reached.

loop of Henle The hairpin-shaped, tubular region of a nephron that functions in reabsorption of water and solutes.

lung An internal respiratory surface in the shape of a cavity or sac.

lycophyte A type of seedless vascular plant of mostly wet or shade habitats; requires free water to complete its life cycle.

lymph (LIMF) [L. *lympha*, water] Tissue fluid that has moved into the vessels of the lymphatic system.

lymph capillary A small-diameter vessel of the lymph vascular system that has no pronounced entrance; tissue fluid moves inward by passing between overlapping endothelial cells at the vessel's tip.

lymph node A lymphoid organ that serves as a battleground of the immune system; each lymph node is packed with organized arrays of macrophages and lymphocytes that cleanse lymph of pathogens before it reaches the blood.

lymph vascular system [L. *lympha*, water, + *vasculum*, a small vessel] The vessels of the lymphatic system, which take up and transport excess tissue fluid and reclaimable solutes as well as fats absorbed from the digestive tract.

lymphatic system An organ system that supplements the circulatory system. Its vessels take up fluid and solutes from interstitial fluid and deliver them to the bloodstream; its lymphoid organs have roles in immunity.

lymphocyte Any of various white blood cells that take part in nonspecific and specific (immune) defense responses.

lymphoid organs The lymph nodes, spleen, thymus, tonsils, adenoids, and other organs with roles in immunity.

lysis [Gk. *lysis*, a loosening] Gross structural disruption of a plasma membrane that leads to cell death.

lysosome (LYE-so-sohm) The main organelle of digestion, with enzymes that can break down polysaccharides, proteins, nucleic acids, and some lipids.

lysozyme An infection-fighting enzyme that digests bacterial cell walls. Present in mucous membranes that line the body's surfaces.

lytic pathway During a viral infection, viral DNA or RNA quickly directs the host cell to produce the components necessary to produce new virus particles, which are released by lysis.

macroevolution The large-scale patterns, trends, and rates of change among groups of species.

macrophage One of the phagocytic white blood cells. It engulfs anything detected as foreign. Some also become antigen-presenting cells that serve as the trigger for immune responses by T and B lymphocytes. *Compare* antigen-presenting cell.

mammal A type of vertebrate; the only animal having offspring that are nourished by milk produced by mammary glands of females.

mass extinction An abrupt rise in extinction rates above the background level; a catastrophic, global event in which major taxa are wiped out simultaneously.

mass number The total number of protons and neutrons in an atom's nucleus. The relative masses of atoms are also called atomic weights.

maternal chromosome One of the chromosomes bearing the alleles that are inherited from a female parent.

mechanoreceptor Sensory cell or cell part that detects mechanical energy associated with changes in pressure, position, or acceleration.

medulla oblongata Part of the vertebrate brainstem with reflex centers for respiration, blood circulation, and other vital functions.

medusa (meh-DOO-sah) [Gk. *Medousa*, one of three sisters in Greek mythology having snake-entwined hair; this image probably evoked by the tentacles and oral arms extending from the medusa] Free-swimming, bell-shaped stage in cnidarian life cycles.

megaspore Of gymnosperms and flowering plants, a haploid spore that forms in the ovary; one of its cellular descendants develops into an egg.

meiosis (my-OH-sis) [Gk. *meioun*, to diminish] Two-stage nuclear division process in which the chromosome number of a germ cell is reduced by half, to the haploid number. (Each daughter nucleus ends up with one of each type of chromosome.) Meiosis is the basis of gamete formation and (in plants) of spore formation. *Compare* mitosis.

meltdown Events which, if unchecked, can blow apart a nuclear power plant, with a release of radioactive material into the environment.

membrane excitability A membrane property of any cell that can produce action potentials in response to appropriate stimulation.

memory The storage and retrieval of information about previous experiences; underlies the capacity for learning.

memory cell One of the subpopulations of cells that form during an immune response and that enters a resting phase, from which it is released during a secondary immune response.

memory lymphocyte Any of the various B or T lymphocytes of the immune system that are formed in response to invasion by a foreign agent and that circulate for some period, available to mount a rapid attack if the same type of invader reappears.

Mendel's theory of independent assortment Stated in modern terms, during meiosis, the gene pairs of homologous chromosomes tend to be stored independently of how gene pairs on other chromosomes are sorted for forthcoming gametes. The theory does not take into account the effects of gene linkage and crossing over.

Mendel's theory of segregation Stated in modern terms, diploid cells have two of each kind of gene (on pairs of homologous chromosomes), and the two segregate during meiosis so that they end up in different gametes.

menopause (MEN-uh-pozz) [L. *mensis*, month, + *pausa*, stop] End of the period of a human female's reproductive potential.

menstrual cycle The cyclic release of oocytes and priming of the endometrium (lining of the uterus) to receive a fertilized egg; the complete cycle averages about 28 days in female humans.

menstruation Periodic sloughing of the blood-enriched lining of the uterus when pregnancy does not occur.

mesoderm (MEH-so-derm) [Gk. *mesos*, middle, + *derm*, skin] In most animal embryos, a primary tissue layer (germ layer) between ectoderm and endoderm. Gives rise to muscle; organs of circulation, reproduction, and excretion; most of the internal skeleton (when present); and connective tissue layers of the gut and body covering.

mesophyll Of vascular plants, a type of parenchyma tissue with photosynthetic cells and an abundance of air spaces.

messenger RNA A linear sequence of ribonucleotides transcribed from DNA and translated into a polypeptide chain; the only type of RNA that carries protein-building instructions.

metabolic pathway One of many orderly sequences of enzyme-mediated reactions by which cells normally maintain, increase, or decrease the concentrations of substances.

Different pathways are linear or circular, and often they interconnect.

metabolism (meh-TAB-oh-lizm) [Gk. *meta*, change] All controlled, enzyme-mediated chemical reactions by which cells acquire and use energy. Through these reactions, cells synthesize, store, break apart, and eliminate substances in ways that contribute to growth, survival, and reproduction.

metamorphosis (met-uh-MOR-foe-sis) [Gk. *meta*, change, + *morphe*, form] Transformation of a larva into an adult form.

metaphase Of mitosis or meiosis II, the stage when each duplicated chromosome has become positioned at the midpoint of the microtubular spindle, with its two sister chromatids attached to microtubules from opposite spindle poles. Of meiosis I, the stage when all pairs of homologous chromosomes are positioned at the spindle's midpoint, with the two members of each pair attached to opposite spindle poles.

metazoan Any multicelled animal.

methanogen A type of archaebacterium that lives in oxygen-free habitats and that produces methane gas as a metabolic by-product.

MHC marker Any of a variety of proteins that are self-markers. Some occur on all body cells of an individual; others are unique to the macrophages and lymphocytes.

micelle formation Formation of a small droplet that consists of bile salts and products of fat digestion (fatty acids and monoglycerides) and that assists in their absorption from the small intestine.

microevolution Changes in allele frequencies brought about by mutation, genetic drift, gene flow, and natural selection.

microfilament [Gk. *mikros*, small, + L. *filum*, thread] In animal cells, one of a variety of cytoskeletal components. Actin and myosin filaments are examples.

microorganism An organism, usually single-celled, that is too small to be observed without the aid of a microscope.

microspore Of gymnosperms and flowering plants, a haploid spore, encased in a sculpted wall, that develops into a pollen grain.

microtubular spindle Of eukaryotic cells, a bipolar structure composed of organized arrays of microtubules that forms during nuclear division and that moves the chromosomes.

microtubule Hollow cylinder of mainly tubulin subunits; a cytoskeletal element with roles in cell shape, motion, and growth and in the structure of cilia and flagella.

microtubule organizing center, or **MTOC** Small mass of proteins and other substances in the cytoplasm; the number, type, and location of MTOCs determine the organization and orientation of microtubules.

microvillus (MY-crow-VILL-us) [L. *villus*, shaggy hair] A slender, cylindrical extension of the animal cell surface that functions in absorption or secretion.

midbrain Of vertebrates, a brain region that evolved as a coordination center for reflex responses to visual and auditory input; together with the pons and medulla oblongata,

part of the brainstem, which includes the reticular formation.

migration Of certain animals, a cyclic movement between two distant regions at times of year corresponding to seasonal change.

mimicry (MIM-ik-ree) Situation in which one species (the mimic) bears deceptive resemblance in color, form, and/or behavior to another species (the model) that enjoys some survival advantage.

mineral An inorganic substance required for the normal functioning of body cells.

mitochondrion (MY-toe-KON-dree-on), plural **mitochondria** Organelle that specializes in ATP formation; it is the site of the second and third stages of aerobic respiration, an oxygen-requiring pathway.

mitosis (my-TOE-sis) [Gk. *mitos*, thread] Type of nuclear division that maintains the parental chromosome number for daughter cells. It is the basis of bodily growth and, in many eukaryotic species, asexual reproduction.

molar One of the cheek teeth; a tooth with a platform having cusps (surface bumps) that help crush, grind, and shear food.

molecular clock With respect to the presumed regular accumulation of neutral mutations in highly conserved genes, a way of calculating the time of origin of one species or lineage relative to others.

molecule A unit of matter in which chemical bonding holds together two or more atoms of the same or different elements.

mollusk A type of invertebrate having a tissue fold (mantle) draped around a soft, fleshy body; snails, clams, and squids are examples.

molting The shedding of hair, feathers, horns, epidermis, or a shell (or some other exoskeleton) in a process of growth or periodic renewal.

Monera The kingdom of bacteria.

moneran A bacterium; a single-celled prokaryote.

monocot (MON-oh-kot) Short for monocotyledon; a flowering plant in which seeds have only one cotyledon, whose floral parts generally occur in threes (or multiples of three), and whose leaves typically are parallel-veined. *Compare* dicot.

monohybrid cross [Gk. *monos*, alone] An experimental cross in which offspring inherit a pair of nonidentical alleles for a single trait being studied, so that they are heterozygous.

monophyletic group A set of independently evolving lineages that share a common evolutionary heritage.

monosaccharide (MON-oh-SAK-ah-ride) [Gk. *monos*, alone, single, + *sakharon*, sugar] The simplest carbohydrate, with only one sugar unit. Glucose is an example.

monosomy Abnormal condition in which one chromosome of diploid cells has no homologue.

morphogenesis (MORE-foe-JEN-ih-sis) [Gk. *morphe*, form, + *genesis*, origin] Processes by which differentiated cells in an embryo become organized into tissues and organs, under genetic controls and environmental influences.

morphological convergence A macroevolutionary pattern of change in which separate lineages adopt similar lifestyles, put comparable body parts to similar uses, and in time resemble one another in structure and function. Analogous structures are evidence of this pattern.

morphological divergence A macroevolutionary pattern of change from a common ancestral form. Homologous structures are evidence of this pattern.

motor neuron A type of neuron; it delivers signals from the brain and spinal cord that can stimulate or inhibit the body's effectors (muscles, glands, or both).

mouth An oral cavity; in human digestion, the site where polysaccharide breakdown begins.

multicelled organism An organism that has differentiated cells arranged into tissues, organs, and often organ systems.

multiple allele system Three or more different molecular forms of the same gene (alleles) that exist in a population.

muscle fatigue A decline in tension of a muscle that has been kept in a state of tetanic contraction as a result of continuous, high-frequency stimulation.

muscle tension A mechanical force, exerted by a contracting muscle, that resists opposing forces such as gravity and the weight of objects being lifted.

muscle tissue Tissue having cells able to contract in response to stimulation, then passively lengthen and so return to their resting stage.

mutagen (MEW-tuh-jen) An environmental agent that can permanently modify the structure of a DNA molecule. Certain viruses and ultraviolet radiation are examples.

mutation [L. *mutatus*, a change, + *-ion*, result or a process or an act] A heritable change in the DNA. Generally, mutations are the source of all the different molecular versions of genes (alleles) and, ultimately, of life's diversity. *See also* lethal mutation; neutral mutation.

mutualism [L. *mutuus*, reciprocal] An interaction between two species that benefits both.

mycelium (my-SEE-lee-um), plural **mycelia** [Gk. *mykes*, fungus, mushroom, + *helos*, callus] A mesh of tiny, branching filaments (hyphae) that is the food-absorbing part of a multicelled fungus.

mycorrhiza (MY-coe-RISE-uh) "Fungus-root;" a symbiotic arrangement between fungal hyphae and the young roots of many vascular plants. The fungus obtains carbohydrates from the plant and in turn releases dissolved mineral ions to the plant roots.

myelin sheath Of many sensory and motor neurons, an axonal sheath that affects how fast action potentials travel; formed from the plasma membranes of Schwann cells that are wrapped repeatedly around the axon and are separated from each other by a small node.

myofibril (MY-oh-FY-brill) One of many thread-like structures inside a muscle cell; each is functionally divided into sarcomeres, the basic units of contraction.

myosin (MY-uh-sin) A type of protein with a head and long tail. In muscle cells, it interacts with actin, another protein, to bring about contraction.

NAD$^+$ Nicotinamide adenine dinucleotide; a nucleotide coenzyme. When carrying electrons and unbound protons (H$^+$) between reaction sites, it is abbreviated NADH.

NADP$^+$ Nicotinamide adenine dinucleotide phosphate; a phosphorylated nucleotide coenzyme. When carrying electrons and unbound protons (H$^+$) between reaction sites, it is abbreviated NADPH$_2$.

nasal cavity Of a respiratory system, the region where air is warmed, moistened, and filtered of airborne particles and dust.

natural selection A microevolutionary process; a difference in survival and reproduction among members of a population that vary in one or more traits.

negative feedback mechanism A homeostatic feedback mechanism in which an activity changes some condition in the internal environment and so triggers a response that reverses the changed condition.

nematocyst (NEM-ad-uh-sist) [Gk. *nema*, thread, + *kystis*, pouch] Of cnidarians only, a stinging capsule that assists in prey capture and possibly protection.

nephridium (neh-FRID-ee-um), plural **nephridia** Of earthworms and some other invertebrates, a system of regulating water and solute levels.

nephron (NEFF-ron) [Gk. *nephros*, kidney] Of the vertebrate kidney, a slender tubule in which water and solutes filtered from blood are selectively reabsorbed and in which urine forms.

nerve Cordlike communication line of the peripheral nervous system, composed of axons of sensory neurons, motor neurons, or both packed within connective tissue. In the brain and spinal cord, similar cord-like bundles are called nerve pathways or tracts.

nerve cord Of many animals, a cordlike communication line consisting of axons of neurons.

nerve impulse *See* action potential.

nerve net Cnidarian nervous system.

nervous system System of neurons oriented relative to one another in precise message-conducting and information-processing pathways.

nervous tissue A type of connective tissue composed of neurons.

net energy Of energy resources available to the human population, the amount of energy that is left over after subtracting the energy used to locate, extract, transport, store, and deliver energy to consumers.

net population growth rate per individual (r) Of population growth equations, a single variable in which birth and death rates, which are assumed to remain constant, are combined.

neuroglial cell (NUR-oh-GLEE-uhl) Of vertebrates, one of the cells that provide structural and metabolic support for neurons and that collectively represent about half the volume of the nervous system.

neuromodulator Type of signaling molecule that influences the effects of transmitter substances by enhancing or reducing membrane responses in target neurons.

neuromuscular junction Chemical synapses between axon terminals of a motor neuron and a muscle cell.

neuron A nerve cell; the basic unit of communication in nervous systems. Neurons collectively sense environmental change, integrate sensory inputs, then activate muscles or glands that initiate or carry out responses.

neurotransmitter Any of the class of signaling molecules that are secreted from neurons, act on immediately adjacent cells, and are then rapidly degraded or recycled.

neutral mutation A gene mutation that has neither harmful nor helpful effects on the individual's ability to survive and reproduce.

neutron Unit of matter, one or more of which occupies the atomic nucleus, that has mass but no electric charge.

neutrophil Fast-acting, phagocytic white blood cell that takes part in inflammatory responses against bacteria.

niche (NITCH) [L. *nidas*, nest] Of a species, the full range of physical and biological conditions under which its members can live and reproduce.

nitrification (nye-trih-fih-KAY-shun) A chemosynthetic process in which certain bacteria strip electrons from ammonia or ammonium present in soil. The end product, nitrite (NO$_2^-$), is broken down to nitrate (NO$_3^-$) by different bacteria.

nitrogen cycle Biogeochemical cycle in which the atmosphere is the largest reservoir of nitrogen.

nitrogen fixation Process by which a few kinds of bacteria convert gaseous nitrogen (N$_2$) to ammonia. This dissolves rapidly in their cytoplasm to form ammonium, which can be used in biosynthetic pathways.

NK cell Natural killer cell, possibly of the lymphocyte lineage, that reconnoiters and kills tumor cells and infected body cells.

nociceptor A receptor, such as a free nerve ending, that detects any stimulus causing tissue damage.

node In vascular plants, a point on a stem where one or more leaves are attached.

noncyclic pathway of ATP formation (non-SIK-lik) [L. *non*, not, + Gk. *kylos*, circle] Photosynthetic pathway in which excited electrons derived from water molecules flow through two photosystems and two transport chains, and ATP and NADPH form.

nondisjunction Failure of one or more chromosomes to separate properly during mitosis or meiosis.

nonsteroid hormone A type of water-soluble hormone, such as a protein hormone, that cannot cross the lipid bilayer of a target cell. These hormones enter the cell by receptor-mediated endocytosis, or they bind to receptors that activate membrane proteins or second messengers within the cell.

notochord (KNOW-toe-kord) Of chordates, a rod of stiffened tissue (not cartilage or bone) that serves as a supporting structure for the body.

nuclear envelope A double membrane (two lipid bilayers and associated proteins) that is the outermost portion of a cell nucleus.

nucleic acid (new-CLAY-ik) A long, single- or double-stranded chain of four different kinds of nucleotides joined one after the other at their phosphate groups. They differ in which nucleotide base follows the next in sequence. DNA and RNA are examples.

nucleic acid hybridization The base-pairing of nucleotide sequences from different sources, as used in genetics, genetic engineering, and studies of evolutionary relationship based on similarities and differences in the DNA or RNA of different species.

nucleoid Of bacteria, a region in which DNA is physically organized apart from other cytoplasmic components.

nucleolus (new-KLEE-oh-lus) [L. *nucleolus*, a little kernel] Within the nucleus of a nondividing cell, a site where the protein and RNA subunits of ribosomes are assembled.

nucleosome (NEW-klee-oh-sohm) Of eukaryotic chromosomes, one of many organizational units, each consisting of a small stretch of DNA looped twice around a "spool" of histone molecules, which another histone molecule stabilizes.

nucleotide (NEW-klee-oh-tide) A small organic compound having a five-carbon sugar (deoxyribose), nitrogen-containing base, and phosphate group. Nucleotides are the structural units of adenosine phosphates, nucleotide coenzymes, and nucleic acids.

nucleotide coenzyme A protein that transports hydrogen atoms (free protons) and electrons from one reaction site to another in cells.

nucleus (NEW-klee-us) [L. *nucleus*, a kernel] Of atoms, the central core of one or more positively charged protons and (in all but hydrogen) electrically neutral neutrons. In eukaryotic cells, a membranous organelle that physically isolates and organizes the DNA, out of the way of cytoplasmic machinery.

nutrition All those processes by which food is selectively ingested, digested, absorbed, and later converted to the body's own organic compounds.

obesity An excess of fat in the body's adipose tissues, caused by imbalances between caloric intake and energy output.

oligosaccharide A carbohydrate consisting of a short chain of two or more covalently bonded sugar units. One subclass, disaccharides, has two sugar units. *Compare* monosaccharide; polysaccharide.

omnivore [L. *omnis*, all, + *vovare*, to devour] An organism able to obtain energy from more than one source rather than being limited to one trophic level.

oncogene (ON-coe-jeen) Any gene having the potential to induce cancerous transformations in a cell.

oocyte An immature egg.

oogenesis (oo-oh-JEN-uh-sis) Formation of a female gamete, from a germ cell to a mature haploid ovum (egg).

operator A short base sequence between a promoter and the start of a gene; interacts with regulatory proteins.

operon Of transcription, a promoter-operator sequence that services more than a single gene. The lactose operon of *E. coli* is an example.

orbitals Volumes of space around the nucleus of an atom in which electrons are likely to be at any instant.

organ A structure of definite form and function that is composed of more than one tissue.

organ formation Stage of development in which primary tissue layers (germ layers) split into subpopulations of cells, and different lines of cells become unique in structure and function; foundation for growth and tissue specialization, when organs acquire specialized chemical and physical properties.

organ system Two or more organs that interact chemically, physically, or both in performing a common task.

organelle Of cells, an internal, membrane-bounded sac or compartment that has a specific, specialized metabolic function.

organic compound In biology, a compound assembled in cells and having a carbon backbone, often with carbon atoms arranged as a chain or ring structure.

osmosis (oss-MOE-sis) [Gk. *osmos*, act of pushing] Of cells, the tendency of water to move through channel proteins that span a membrane in response to a concentration gradient, fluid pressure, or both. Hydrogen bonds among water molecules prevent water *itself* from becoming more or less concentrated; but a gradient may exist when the water on either side of the membrane has more substances dissolved in it.

ovary (OH-vuh-ree) In female animals, the primary reproductive organ in which eggs form. In seed-bearing plants, the portion of the carpel where eggs develop, fertilization takes place, and seeds mature. A mature ovary (and sometimes other plant parts) is a fruit.

oviduct (OH-vih-dukt) Duct through which eggs travel from the ovary to the uterus. Formerly called Fallopian tube.

ovulation (AHV-you-LAY-shun) During each turn of the menstrual cycle, the release of a secondary oocyte (immature egg) from an ovary.

ovule (OHV-youl) [L. *ovum*, egg] Before fertilization in gymnosperms and angiosperms, a female gametophyte with egg cell, a surrounding tissue, and one or two protective layers (integuments). After fertilization, an ovule matures into a seed (an embryo sporophyte and food reserves encased in a hardened coat).

ovum (OH-vum) A mature female gamete (egg).

oxidation-reduction reaction An electron transfer from one atom or molecule to another. Often hydrogen is transferred along with the electron or electrons.

ozone hole A pronounced seasonal thinning of the ozone layer in the lower stratosphere above Antarctica.

pancreas (PAN-cree-us) Gland that secretes enzymes and bicarbonate into the small intestine during digestion, and that also secretes the hormones insulin and glucagon.

pancreatic islets Any of the two million clusters of endocrine cells in the pancreas, including alpha cells, beta cells, and delta cells.

parasite [Gk. *para*, alongside, + *sitos*, food] An organism that obtains nutrients directly from the tissues of a living host, which it lives on or in and may or may not kill.

parasitism A two-species interaction in which one species directly harms another that serves as its host.

parasitoid An insect larva that grows and develops inside a host organism (usually another insect), eventually consuming the soft tissues and killing it.

parasympathetic nerve Of the autonomic nervous system, any of the nerves carrying signals that tend to slow the body down overall and divert energy to basic tasks; also work continually in opposition with sympathetic nerves to bring about minor adjustments in internal organs.

parathyroid glands (PARE-uh-THY-royd) In vertebrates, endocrine glands embedded in the thyroid gland that secrete parathyroid hormone, which helps restore blood calcium levels.

parenchyma Most abundant of the simple tissues in flowering plant roots, stems, leaves, and other parts. Its cells function in photosynthesis, storage, secretion, and other tasks.

parthenogenesis Development of an embryo from an unfertilized egg.

passive immunity Temporary immunity conferred by deliberately introducing antibodies into the body.

passive transport Diffusion of a solute through a channel or carrier protein that spans the lipid bilayer of a cell membrane. Its passage does not require an energy input; the protein passively allows the solute to follow its concentration gradient.

paternal chromosome One of the chromosomes bearing alleles that are inherited from a male parent.

pathogen (PATH-oh-jen) [Gk. *pathos*, suffering, + *-genēs*, origin]. An infectious, disease-causing agent, such as a virus or bacterium.

pattern formation Of animals, mechanisms responsible for specialization and positioning of tissues during embryonic development.

PCR Polymerase chain reaction. A method used by recombinant DNA technologists to amplify the quantity of specific fragments of DNA.

peat An accumulation of saturated, undecayed remains of plants that have been compressed by sediments.

pedigree A chart of genetic connections among individuals, as constructed according to standardized methods.

pelagic province The entire volume of ocean water; subdivided into neritic zone (relatively shallow waters overlying continental shelves) and oceanic zone (water over ocean basins).

penis A male organ that deposits sperm into a female reproductive tract.

perennial [L. *per-*, throughout, + *annus*, year] A flowering plant that lives for three or more growing seasons.

perforin A type of protein, produced and secreted by cytotoxic cells, that destroys antigen-bearing targets.

pericycle (PARE-ih-sigh-kul) [Gk. *peri-*, around, + *kyklos*, circle] Of a root vascular cylinder, one or more layers just inside the endodermis that gives rise to lateral roots and contributes to secondary growth.

periderm Of vascular plants showing secondary growth, a protective covering that replaces epidermis.

peripheral nervous system (per-IF-ur-uhl) [Gk. *peripherein*, to carry around] Of vertebrates, the nerves leading into and out from the spinal cord and brain and the ganglia along those communication lines.

peristalsis (pare-ih-STAL-sis) A rhythmic contraction of muscles that moves food forward through the animal gut.

peritoneum A lining of the coelom that also covers and helps maintain the position of internal organs.

peritubular capillaries The set of blood capillaries that threads around the tubular parts of a nephron; they function in reabsorption of water and solutes back into the body and in secretion of hydrogen ions and some other substances in the forming urine.

permafrost A permanently frozen, water-impenetrable layer beneath the soil surface in arctic tundra.

peroxisome Enzyme-filled vesicle in which fatty acids and amino acids are digested first into hydrogen peroxide (which is toxic), then to harmless products.

PGA Phosphoglycerate (FOSS-foe-GLISS-er-ate). A key intermediate in glycolysis and in the Calvin-Benson cycle.

PGAL Phosphoglyceraldehyde. A key intermediate in glycolysis and in the Calvin-Benson cycle.

pH scale A scale used to measure the concentration of free hydrogen ions in blood, water, and other solutions; pH 0 is the most acidic, 14 the most basic, and 7, neutral.

phagocyte (FAG-uh-sight) [Gk. *phagein*, to eat, + *kytos*, hollow vessel] A macrophage or certain other white blood cells that engulf and destroy foreign agents.

phagocytosis (FAG-uh-sigh-TOE-sis) [Gk. *phagein*, to eat, + *kytos*, hollow vessel] Engulfment of foreign cells or substances by amoebas and some white blood cells by means of endocytosis.

pharynx (FARE-inks) A muscular tube by which food enters the gut; in land vertebrates, the dual entrance for the tubular part of the digestive tract and windpipe (trachea).

phenotype (FEE-no-type) [Gk. *phainein*, to show, + *typos*, image] Observable trait or traits of an individual; arises from interactions between genes, and between genes and the environment.

pheromone (FARE-oh-moan) [Gk. *phero*, to carry, + *-mone*, as in hormone] A type of signaling molecule secreted by exocrine glands that serves as a communication signal between individuals of the same species. Signaling pheromones elicit an immediate behavioral response. Priming pheromones elicit a generalized physiological response.

phloem (FLOW-um) Of vascular plants, a tissue with living cells that interconnect and form the tubes through which sugars and other solutes are conducted.

phospholipid A type of lipid that is the main structural component of cell membranes. Each has a hydrophobic tail (of two fatty acids) and a hydrophilic head that incorporates glycerol and a phosphate group.

phosphorus cycle Movement of phosphorus from rock or soil through organisms, then back to soil.

phosphorylation (FOSS-for-ih-LAY-shun) The attachment of unbound (inorganic) phosphate to a molecule; also the transfer of a phosphate group from one molecule to another, as when ATP phosphorylates glucose.

photochemical smog A brown-air smog that develops over large cities when the surrounding land forms a natural basin.

photolysis (foe-TALL-ih-sis) [Gk. *photos*, light, + *-lysis*, breaking apart] A reaction sequence of the noncyclic pathway of photosynthesis, triggered by photon energy, in which water is split into oxygen, hydrogen, and electrons.

photoperiodism A biological response to a change in the relative length of daylight and darkness.

photoreceptor Light-sensitive sensory cell.

photosynthesis The trapping of sunlight energy and its conversion to chemical energy (ATP, NADPH, or both), followed by synthesis of sugar phosphates that become converted to sucrose, cellulose, starch, and other end products. It is the main biosynthetic pathway by which energy and carbon enter the web of life.

photosynthetic autotroph An organism able to synthesize all organic molecules it requires using carbon dioxide as the carbon source and sunlight as the energy source. All plants, some protistans, and a few bacteria are photosynthetic autotrophs.

photosystem One of the clusters of light-trapping pigments embedded in photosynthetic membranes. Photosystem I operates during the cyclic pathway; photosystem II operates during both the cyclic and noncyclic pathways.

phototropism [Gk. *photos*, light, + *trope*, turning, direction] Adjustment in the direction and rate of plant growth in response to light.

photovoltaic cell A device that converts sunlight energy into electricity.

phycobilins (FIE-koe-BY-lins) A class of light-sensitive, accessory pigments that transfer absorbed energy to chlorophylls. They are abundant in red algae and cyanobacteria.

phylogeny Evolutionary relationships among species, starting with most ancestral forms and including the branches leading to their descendants.

phytochrome Light-sensitive pigment molecule, the activation and inactivation of which triggers plant hormone activities governing leaf expansion, stem branching, stem length and often seed germination and flowering.

phytoplankton (FIE-toe-PLANK-tun) [Gk. *phyton*, plant, + *planktos*, wandering] A freshwater or marine community of floating or weakly swimming photosynthetic autotrophs, such as cyanobacteria, diatoms, and green algae.

pigment A light-absorbing molecule.

pineal gland (py-NEEL) A light-sensitive endocrine gland that secretes melatonin, a hormone that influences reproductive cycles and the development of reproductive organs.

pioneer species Typically small plants with short life cycles that are adapted to growing in exposed, often windy areas with intense sunlight, wide swings in air temperature, and soils deficient in nitrogen and other nutrients. By improving conditions in areas they colonize, pioneers invite their own replacement by other species.

pituitary gland Of endocrine systems, a gland that interacts with the hypothalamus to coordinate and control many physiological functions, including the activity of many other endocrine glands. Its posterior lobe stores and secretes hypothalamic hormones; the anterior lobe produces and secretes its own hormones.

placenta (play-SEN-tuh) Of the uterus, an organ composed of maternal tissues and extraembryonic membranes (the chorion especially); it delivers nutrients to the fetus and accepts wastes from it, yet allows the fetal circulatory system to develop separately from the mother's.

plankton [Gk. *planktos*, wandering] Any community of floating or weakly swimming organisms, mostly microscopic, living in freshwater and saltwater environments. *See* phytoplankton; zooplankton.

plant The type of eukaryotic organism, usually multicelled, that is a photosynthetic autotroph—it uses sunlight energy to drive the synthesis of all its required organic compounds from carbon dioxide, water, and mineral ions. Only a few nonphotosynthetic plants obtain nutrients by parasitism and other means.

Plantae The kingdom of plants.

plasma (PLAZ-muh) Liquid component of blood; consists of water, various proteins, ions, sugars, dissolved gases, and other substances.

plasma cell Of immune systems, any of the anitbody-secreting daughter cells of a rapidly dividing population of B cells.

plasma membrane Of cells, the outermost membrane. Its lipid bilayer structure and proteins carry out most functions, including transport across the membrane and reception of extracellular signals.

plasmid Of many bacteria, a small, circular molecule of extra DNA that carries only a few genes and replicates independently of the bacterial chromosome.

plasmodesma (PLAZ-moe-DEZ-muh) Of multi-celled plants, a junction between linked walls of adjacent cells through which nutrients and other substances flow.

plasticity Of the human species, the ability to remain flexible and adapt to a wide range of environments.

plate tectonics Arrangement of the earth's outer layer (lithosphere) in slablike plates, all in motion and floating on a hot, plastic layer of the underlying mantle.

platelet (PLAYT-let) Any of the cell fragments in blood that release substances necessary for clot formation.

pleiotropy (PLEE-oh-troe-pee) [Gk. *pleon*, more, + *trope*, direction] A type of gene interaction in which a single gene exerts multiple effects on seemingly unrelated aspects of an individual's phenotype.

polar body Any of three cells that form during the meiotic cell division of an oocyte; the division also forms the mature egg, or ovum.

pollen grain [L. *pollen*, fine dust] Depending on the species, the immature or mature, sperm-bearing male gametophyte of gymnosperms and flowering plants.

pollen sac In anthers of flowers, any of the chambers in which pollen grains develop.

pollen tube A tube formed after a pollen grain germinates; grows down through carpel tissues and carries sperm to the ovule.

pollination Of flowering plants, the arrival of a pollen grain on the landing platform (stigma) of a carpel.

pollutant Any substance with which an ecosystem has had no prior evolutionary experience, in terms of kinds or amounts, and that can accumulate to disruptive or harmful levels. Can be naturally occurring or synthetic.

polymer (POH-lih-mur) [Gk. *polus*, many, + *meris*, part] A molecule composed of three to millions of small subunits that may or may not be identical.

polymerase chain reaction or **PCR** DNA amplification method; DNA containing a gene of interest is split into single strands, which enzymes (polymerases) copy; the enzymes also act on the accumulating copies, multiplying the gene sequence by the millions.

polymorphism (poly-MORE-fizz-um) [Gk. *polus*, many, + *morphe*, form] Of a population, the persistence through the generations of two or more forms of a trait, at a frequency greater than can be maintained by new mutations alone.

polyp (POH-lip) Vase-shaped, sedentary stage of cnidarian life cycles.

polypeptide chain Three or more amino acids joined by peptide bonds.

polyploidy (POL-ee-PLOYD-ee) A change in the chromosome number following inheritance of three or more of each type of chromosome.

polysaccharide [Gk. *polus*, many, + *sakcharon*, sugar] A straight or branched chain of hundreds of thousands of covalently linked sugar units, of the same or different kinds. The most common polysaccharides are cellulose, starch, and glycogen.

polysome Of protein synthesis, several ribosomes all translating the same messenger RNA molecule, one after the other.

population A group of individuals of the same species occupying a given area.

population density The number of individuals of a population that are living in a specified area or volume.

population distribution The general pattern of dispersion of individuals of a population throughout their habitat.

population size The number of individuals that make up the gene pool of a population.

positive feedback mechanism Homeostatic mechanism by which a chain of events is set in motion that intensifies a change from an original condition; after a limited time, the intensification reverses the change.

post-translational controls Of eukaryotes, controls that govern modification of newly formed polypeptide chains into functional enzymes and other proteins.

predation A two-species interaction in which one species (the predator) directly harms the other (its prey).

predator [L. *prehendere*, to grasp, seize] An organism that feeds on and may or may not kill other living organisms (its prey); unlike parasites, predators do not live on or in their prey.

prediction A claim about what you can expect to observe in nature if a theory or hypothesis is correct.

premolar One of the cheek teeth; a tooth having a platform with cusps (surface bumps) that can crush, grind, and shear food.

pressure flow theory Of vascular plants, a theory that organic compounds move through phloem because of gradients in solute concentrations and pressure between source regions (such as photosynthetically active leaves) and sink regions (such as growing plant parts).

primary growth Plant growth originating at root tips and shoot tips.

primary immune response Actions by white blood cells and their products elicited by a first-time encounter with an antigen; includes both antibody-mediated and cell-mediated responses.

primary productivity, gross Of ecosystems, the rate at which the producer organisms capture and store a given amount of energy during a specified interval.

primary productivity, net Of ecosystems, the rate of energy storage in the tissues of producers in excess of their rate of aerobic respiration.

primate The mammalian lineage that includes prosimians, tarsioids, and anthropoids (monkeys, apes, and humans).

probability With respect to any chance event, the most likely number of times it will turn out a certain way, divided by the total number of all possible outcomes.

procambium (pro-KAM-bee-um) Of vascular plants, a primary meristem that gives rise to the primary vascular tissues.

producers, primary Of ecosystems, the organisms that secure energy from the physical environment, as by photosynthesis or chemosynthesis.

progesterone (pro-JESS-tuh-rown) Female sex hormone secreted by the ovaries.

prokaryotic cell (pro-CARRY-oh-tic) [L. *pro*, before, + Gk. *karyon*, kernel] A bacterium; a single-celled organism that has no nucleus or any of the other membrane-bound organelles characteristic of eukaryotic cells.

promoter Of transcription, a base sequence that signals the start of a gene; the site where RNA polymerase initially binds.

prophase Of mitosis, the stage when each duplicated chromosome starts to condense, microtubules form a spindle apparatus, and the nuclear envelope starts to break up.

prophase I Of meiosis, the stage at which the microtubular spindle starts to form, the nuclear envelope starts to break up, and each duplicated chromosome also condenses and pairs with its homologous partner. At this time, their sister chromatids typically undergo crossing over and genetic recombination.

prophase II Of meiosis, a brief stage after interkinesis during which each chromosome still consists of two chromatids.

protein Large organic compound composed of one or more chains of amino acids held together by peptide bonds. Proteins have unique sequences of different kinds of amino acids in their polypeptide chains; such sequences are the basis of a protein's three-dimensional structure and chemical behavior.

Protista The kingdom of protistans.

protistan (pro-TISS-tun) [Gk. *protistos*, primal, very first] Single-celled eukaryote.

proto-oncogene A gene sequence similar to an oncogene but that codes for a protein required in normal cell function; may trigger cancer, generally when specific mutations alter its structure or function.

proton Positively charged particle, one or more of which is present in the atomic nucleus.

protostome (PRO-toe-stome) [Gk. *proto*, first, + *stoma*, mouth] A bilateral animal in which the first indentation in the early embryo develops into the mouth. Includes mollusks, annelids, and anthropods.

protozoan A type of protistan, some predatory and others parasitic; so named because they may resemble the single-celled heterotrophs that presumably gave rise to animals.

proximal tubule Of a nephron, the tubular region that receives water and solutes filtered from the blood.

pulmonary circuit Blood circulation route leading to and from the lungs.

Punnett-square method A way to predict the possible outcome of a mating or an experimental cross in simple diagrammatic form.

purine Nucleotide base having a double ring structure. Adenine and guanine are examples.

pyrimidine (phi-RIM-ih-deen) Nucleotide base having a single ring structure. Cytosine and thymine are examples.

pyruvate (PIE-roo-vate) A compound with a backbone of three carbon atoms. Two pyruvate molecules are the end products of glycolysis.

r Designates net population growth rate; the birth and death rates are assumed to remain constant and so are combined into this one variable for population growth equations.

radial symmetry Body plan having four or more roughly equivalent parts arranged around a central axis.

radioisotope An unstable atom that has dissimilar numbers of protons and neutrons and that spontaneously decays (emits electrons and energy) to a new, stable atom that is not radioactive.

rain shadow A reduction in rainfall on the leeward side of high mountains, resulting in arid or semiarid conditions.

reabsorption Of urine formation, the diffusion or active transport of water and usable solutes out of a nephron and into capillaries leading back to the general circulation; regulated by ADH and aldosterone.

receptor Of cells, a molecule at the surface of the plasma membrane or in the cytoplasm that binds molecules present in the extracellular environment. The binding triggers changes in cellular activities. Of nervous systems, a sensory cell or cell part that may be activated by a specific stimulus.

receptor protein Protein that binds a signaling molecule such as a hormone, then triggers alterations in cell behavior or metabolism.

recessive allele [L. *recedere*, to recede] In heterozygotes, an allele whose expression is fully or partially masked by expression of its partner; fully expressed only in the homozygous recessive condition.

recognition protein Protein at cell surface recognized by cells of like type; helps guide the ordering of cells into tissues during development and functions in cell-to-cell interactions.

recombinant technology Procedures by which DNA (genes) from different species may be isolated, cut, spliced together, and the new recombinant molecules multiplied in quantity in a population of rapidly dividing cells such as bacteria.

red blood cell Erythrocyte; an oxygen-transporting cell in blood.

red marrow A substance in the spongy tissue of many bones that serves as a major site of blood cell formation.

reflex [L. *reflectere*, to bend back] A simple, stereotyped movement elicited directly by sensory stimulation.

reflex pathway [L. *reflectere*, to bend back] Type of neural pathway in which signals from sensory neurons directly stimulate or inhibit motor neurons, without intervention by interneurons.

refractory period Of neurons, the period following an action potential at a given patch of membrane when sodium gates are shut and potassium gates are open, so that the patch is insensitive to stimulation.

regulatory protein A protein that enhances or suppresses the rate at which a gene is transcribed.

releasing hormone A hypothalamic signaling molecule that stimulates or slows down secretion by target cells in the anterior lobe of the pituitary gland.

repressor protein Regulatory protein that provides negative control of gene activity by preventing RNA polymerase from binding to DNA.

reproduction In biology, processes by which a new generation of cells or multicelled individuals is produced. Sexual reproduction requires meiosis, formation of gametes, and fertilization. Asexual reproduction refers to the production of new individuals by any mode that does not involve gametes.

reproduction, sexual Mode of reproduction that begins with meiosis, proceeds through gamete formation, and ends at fertilization.

reproductive isolating mechanism Any aspect of structure, functioning, or behavior that restricts gene flow between two populations.

reproductive isolation An absence of gene flow between populations.

reproductive success The survival and production of the offspring of an individual.

reptile A type of carnivorous vertebrate; its ancestors were the first vertebrates to escape dependency of standing water, largely by means of internal fertilization and amniote eggs. They include turtles, crocodiles, lizards and snakes, and tuataras.

resource partitioning A community pattern in which similar species generally share the same kind of resource in different ways, in different areas, or at different times.

respiration [L. *respirare*, to breathe] In most animals, the overall exchange of oxygen from the environment for carbon dioxide wastes from cells by way of circulating blood. *Compare* aerobic respiration.

respiratory surface The surface, such as a thin epithelial layer, that gases diffuse across to enter and leave the animal body.

respiratory system An organ system that functions in respiration.

resting membrane potential Of neurons and other excitable cells that are not being stimulated, the steady voltage difference across the plasma membrane.

restriction enzymes Class of bacterial enzymes that cut apart foreign DNA injected into them, as by viruses; also used in recombinant DNA technology.

reticular formation Of the vertebrate brainstem, a major network of interneurons that helps govern activity of the whole nervous system.

reverse transcriptase Viral enzyme required for reverse transcription of mRNA into DNA; used in recombinant DNA technology.

reverse transcription Assembly of DNA on a single-stranded mRNA molecule by viral enzymes.

RFLPs Restriction fragment length polymorphisms. Of DNA samples from different individuals, slight but unique differences in the banding pattern of fragments of the DNA that have been cut with restriction enzymes.

Rh blood typing A method of characterizing red blood cells on the basis of a protein that serves as a self-marker at their surface; Rh^+ signifies its presence and Rh^-, its absence.

rhizoid Rootlike absorptive structure of some fungi and nonvascular plants.

ribosomal RNA (rRNA) Type of RNA molecule that combines with proteins to form ribosomes, on which the polypeptide chains of proteins are assembled.

ribosome In all cells, the structure at which amino acids are strung together in specified sequence to form the polypeptide chains of proteins. An intact ribosome consists of two subunits, each composed of ribosomal RNA and protein molecules.

RNA Ribonucleic acid. A category of single-stranded nucleic acids that function in processes by which genetic instructions are used to build proteins.

rod cell A vertebrate photoreceptor sensitive to very dim light and that contributes to coarse perception of movement.

root hair Of vascular plants, an extension of a specialized root epidermal cell; root hairs collectively enhance the surface area available for absorbing water and solutes.

root nodule A localized swelling on the roots of certain legumes and other plants that contain symbiotic, nitrogen-fixing bacteria.

roots Descending parts of a vascular plant that absorb water and nutrients, anchor aboveground parts, and usually store food.

rotifer An invertebrate common in food webs of lakes and ponds.

roundworm A type of parasitic or scavenging invertebrate with a bilateral, cylindrical, cuticle-covered body, usually tapered at both ends. Some cause diseases in humans.

RuBP Ribulose biphosphate. A compound with a backbone of five carbon atoms that is required for carbon fixation in the Calvin-Benson cycle of photosynthesis.

S-shaped curve A curve, obtained when population size is plotted against time, that is characteristic of logistic growth.

sac fungus A type of fungus, usually multicelled, with spores that develop inside the cells of reproductive structures shaped like globes, flasks, or dishes. Yeasts (single-celled) are also in this group.

salination A salt buildup in soil as a result of evaporation, poor drainage, and often the importation of mineral salts in irrigation water.

salivary gland Any of the glands that secrete saliva, a fluid that initially mixes with food in the mouth and starts the breakdown of starch.

salt An ionic compound formed when an acid reacts with a base.

saltatory conduction In myelinated neurons, rapid, node-to-node hopping of action potentials.

saprobe Heterotroph that obtains its nutrients from nonliving organic matter. Most fungi are saprobes.

sarcomere (SAR-koe-meer) Of vertebrate muscles, the basic unit of contraction; a region of myosin and actin filaments organized in parallel between two Z lines of a myofibril inside a muscle cell.

sarcoplasmic reticulum (sar-koe-PLAZ-mik reh-TIK-you-lum) In muscle cells, a membrane system that takes up, stores, and releases the calcium ions required for cross-bridge formation in sarcomeres, hence for contraction.

Schwann cells Specialized neuroglial cells that grow around neuron axons, forming a myelin sheath.

sclerenchyma One of the simple tissues of flowering plants, generally with cells having thick, lignin-impregnated walls. It supports mature plant parts and often protects seeds.

sea squirt An invertebrate chordate with a leathery or jellylike tunic and bilateral, free-swimming larvae that look like tadpoles. Ancient sea squirts may have resembled animals that gave rise to vertebrates.

second law of thermodynamics Law stating that the spontaneous direction of energy flow is from organized (high-quality) to less organized (low-quality) forms. With each conversion, some energy is randomly dispersed in a form, usually heat, that is not as readily available to do work.

second messenger A molecule inside a cell that mediates and generally triggers amplified response to a hormone.

secondary immune response Rapid, prolonged response by white blood cells, memory cells especially, to a previously encountered antigen.

secondary sexual trait A trait that is associated with maleness or femaleness but that does not play a direct role in reproduction.

secretion Generally, the release of a substance for use by the organism producing it. (Not the same as *excretion*, the expulsion of excess or waste material.) Of kidneys, a regulated stage in urine formation, in which ions and other substances move from capillaries into nephrons.

sedimentary cycle A biogeochemical cycle without a gaseous phase; the element moves from land to the seafloor, then returns only through long-term geological uplifting.

seed Of gymnosperms and flowering plants, a fully mature ovule (contains the plant embryo), with its integuments forming the seed coat.

segmentation Of earthworms and many other animals, a series of body units that may be externally similar to or quite different from one another.

segregation, Mendelian principle of [L. *se-*, apart, + *grex*, herd] The principle that diploid organisms inherit a pair of genes for each trait (on a pair of homologous chromosomes) and that the two genes segregate during meiosis and end up in separate gametes.

selective gene expression Of multicelled organisms, activation or suppression of a fraction of the genes in unique ways in different cells, leading to pronounced differences in structure and function among different cell lineages.

selfish behavior A behavior by which an individual protects or increases its own chance of producing offspring, regardless of the consequences of the group to which it belongs.

selfish herd A simple society held together by reproductive self-interest.

semen (SEE-mun) [L. *serere*, to sow] Sperm-bearing fluid expelled from a penis during male orgasm.

semiconservative replication [Gk. *hēmi*, half, + L. *conservare*, to keep] Reproduction of a DNA molecule when a complementary strand forms on each of the unzipping strands of an existing DNA double helix, the outcome being two "half-old, half-new" molecules.

senescence (sen-ESS-cents) [L. *senescere*, to grow old] Sum total of processes leading to the natural death of an organism or some of its parts.

sensation The conscious awareness of a stimulus.

sensory neuron Any of the nerve cells that act as sensory receptors, detecting specific stimuli (such as light energy) and relaying signals to the brain and spinal cord.

sensory system The "front door" of a nervous system; that portion of a nervous system that receives and sends on signals of specific changes in the external and internal environments.

sessile animal Animal that remains attached to a substrate during some stage (often the adult) of its life cycle.

sex chromosome Of most animals and some plants, a chromosome whose presence determines a new individual's gender. *Compare* autosomes.

sexual dimorphism Phenotypic differences between males and females of a species.

sexual reproduction Production of offspring from the union of gametes from two parents, by way of meiosis, gamete formation, and fertilization.

sexual selection A microevolutionary process; natural selection favoring a trait that gives the individual a competitive edge in reproductive success.

shifting cultivation The cutting and burning of trees, followed by tilling of ashes into the soil; once called slash-and-burn agriculture.

shoots The above-ground parts of vascular plants.

sieve tube member Of flowering plants, a cellular component of the interconnecting conducting tubes in phloem.

sink region In a vascular plant, any region using or stockpiling organic compounds for growth and development.

sister chromatids Of a duplicated chromosome, two DNA molecules (and associated proteins) that remain attached at their centromere only during nuclear division. Each ends up in a separate daughter nucleus.

skeletal muscle In vertebrates, an organ that contains hundreds to many thousands of muscle cells, arranged in bundles that are surrounded by connective tissue. The connective tissue extends beyond the muscle (as tendons that attach it to bone).

sliding filament model Model of muscle contraction, in which myosin filaments physically slide along and pull two sets of actin filaments toward the center of the sarcomere, which shortens. The sliding requires ATP energy and cross-bridge formation between the actin and myosin.

slime mold A type of heterotrophic protistan with a life cycle that includes free-living cells that at some point congregate and differentiate into spore-bearing structures.

small intestine Of vertebrates, the portion of the digestive system where digestion is completed and most nutrients absorbed.

smog, industrial Gray-colored air pollution that predominates in industrialized cities with cold, wet winters.

smog, photochemical Form of brown, smelly air pollution occurring in large cities with warm climates.

social behavior Cooperative, interdependent relationships among animals of the same species.

social parasite Animal that depends on the social behavior of another species to gain food, care for young, or some other factor to complete its life cycle.

sodium-potassium pump A transport protein spanning the lipid bilayer of the plasma membrane. When activated by ATP, its shape changes and it selectively transports sodium ions out of the cell and potassium ions in.

solute (SOL-yoot) [L. *solvere*, to loosen] Any substance dissolved in a solution. In water, this means spheres of hydration surround the charged parts of individual ions or molecules and keep them dispersed.

solvent Fluid in which one or more substances is dissolved.

somatic cell (so-MAT-ik) [Gk. *sōmā*, body] Of animals, any cell that is not a germ cell (which gives rise to gametes).

somatic nervous system Those nerves leading from the central nervous system to skeletal muscles.

sound system Of birds, the brain regions that govern muscles of the vocal organ.

source region Of vascular plants, any of the sites of photosynthesis.

speciation (spee-cee-AY-shun) The evolutionary process by which species originate. One speciation route starts with divergence of two reproductively isolated populations of a species. They become separate species when accumulated differences in allele frequencies prevent them from interbreeding successfully under natural conditions. Speciation also may be instantaneous (by way of polyploidy, especially among self-fertilizing plants).

species (SPEE-sheez) [L. *species*, a kind] Of sexually reproducing organisms, a unit consisting of one or more populations of individuals that can interbreed under natural conditions to produce fertile offspring that are reproductively isolated from other such units.

sperm [Gk. *sperma*, seed] A type of mature male gamete.

spermatogenesis (sperm-AT-oh-JEN-ih-sis) Formation of a mature sperm from a germ cell.

sphere of hydration Through positive or negative interactions, a clustering of water molecules around the individual molecules of a substance placed in water. *Compare* solute.

sphincter (SFINK-tur) Ring of muscle between regions of a tubelike system (as between the stomach and small intestine).

spinal cord Of central nervous systems, the portion threading through a canal inside the vertebral column and providing direct reflex connections between sensory and motor neurons as well as communication lines to and from the brain.

spindle apparatus A type of bipolar structure that forms during mitosis or meiosis and that moves the chromosomes. It consists of two sets of microtubules that extend from the opposite poles and that overlap at the spindle's equator.

spleen One of the lymphoid organs; it is a filtering station for blood, a reservoir of red blood cells, and a reservoir of macrophages.

sponge An invertebrate having a body with no symmetry and no organs; a framework of glassy needles and other structures imparts shape to it. Distinctive for its food-gathering, flagellated collar cells.

sporangium (spore-AN-gee-um), plural **sporangia** [Gk. *spora*, seed] The protective tissue layer that surrounds haploid spores in a sporophyte.

spore Of land plants, a type of resistant cell, often walled, that forms between the time of meiosis and fertilization. It germinates and develops into a gametophyte, the actual gamete-producing body. Of most fungi, a walled, resistant cell or multicelled structure, produced by mitosis or meiosis, that can germinate and give rise to a new mycelium.

sporophyte [Gk. *phyton*, plant] Of plant life cycles, a vegetative body that grows (by mitosis) from a zygote and that produces the spore-bearing structures.

sporozoan One of four categories of protozoans; a parasite that produces sporelike infectious agents called sporozoites. Some cause serious diseases in humans.

spring overturn Of certain lakes, the movement of dissolved oxygen from the surface layer to the depths and movement of nutrients from bottom sediments to the surface.

stabilizing selection Of a population, a persistence over time of the alleles responsible for the most common phenotypes.

stamen (STAY-mun) Of flowering plants, a male reproductive structure; commonly consists of pollen-bearing structures (anthers) on single stalks (filaments).

start codon Of protein synthesis, a base triplet in a strand of mRNA that serves as the start signal for mRNA translation.

stem cell Of animals, one of the unspecialized cells that replace themselves by ongoing mitotic divisions; portions of their daughter cells also divide and differentiate into specialized cells.

steroid (STAIR-oid) A lipid with a backbone of four carbon rings and with no fatty acid tails. Steroids differ in their functional groups. Different types have roles in metabolism, intercellular communication, and (in animals) cell membranes.

steroid hormone A type of lipid-soluble hormone, synthesized from cholesterol, that diffuses directly across the lipid bilayer of a target cell's plasma membrane and that binds with a receptor inside that cell.

stigma Of many flowering plants, the sticky or hairy surface tissue on the upper portion of the ovary that captures pollen grains and favors their germination.

stimulus [L. *stimulus*, goad] A specific change in the environment, such as a variation in light, heat, or mechanical pressure, that the body can detect through sensory receptors.

stoma (STOW-muh), plural **stomata** [Gk. *stoma*, mouth] A controllable gap between two guard cells in stems and leaves; any of the small passageways across the epidermis through which carbon dioxide moves into the plant and water vapor moves out.

stomach A muscular, stretchable sac that receives ingested food; of vertebrates, an organ between the esophagus and intestine in which considerable protein digestion occurs.

stop codon Of protein synthesis, a base triplet in a strand of mRNA that serves as the stop signal for translation, so that no more amino acids are added to the polypeptide chain.

stream A flowing-water ecosystem that starts out as a freshwater spring or seep.

stroma [Gk. *strōma*, bed] Of chloroplasts, the semifluid interior between the thylakoid membrane system and the two outer membranes; the zone where sucrose, starch, and other end products of photosynthesis are assembled.

stromatolite Of shallow seas, layered structures formed from sediments and large mats of the slowly accumulated remains of photosynthetic populations.

substrate A reactant or precursor molecule for a metabolic reaction; a specific molecule or molecules that an enzyme can chemically recognize, bind briefly to itself, and modify in a specific way.

substrate-level phosphorylation The direct, enzyme-mediated transfer of a phosphate group from the substrate of a reaction to another molecule. An example is the transfer of phosphate from an intermediate of glycolysis to ADP, forming ATP.

succession, primary (suk-SESH-un) [L. *succedere*, to follow after] Orderly changes from the time pioneer species colonize a barren habitat through replacements by various species until the climax community, when the composition of species remains steady under prevailing conditions.

succession, secondary Orderly changes in a community or patch of habitat toward the climax state after having been disturbed, as by fire.

surface-to-volume ratio A mathematical relationship in which volume increases with the cube of the diameter, but surface area increases only with the square. Of growing cells, the volume of cytoplasm increases more rapidly than the surface area of the plasma membrane that must service the cytoplasm. Because of this constraint, cells generally remain small or elongated, or have elaborate membrane foldings.

survivorship curve A plot of the age-specific survival of a group of individuals in a given environment, from the time of their birth until the last one dies.

symbiosis (sim-by-OH-sis) [Gk. *sym*, together, + *bios*, life, mode of life] A state in which two or more species live together in close association (one lives with, in, or on the other). Commensalism, mutualism, and parasitism are examples of symbiotic interactions.

sympathetic nerve Of the autonomic nervous system, any of the nerves generally concerned with increasing overall body activities during times of heightened awareness, excitement, or danger; also work continually in opposition with parasympathetic nerves to bring about minor adjustments in internal organs.

sympatric speciation [Gk. *sym*, together, + *patria*, native land] Speciation that follows after ecological, behavioral, or genetic barriers arise within the boundaries of a single population. This can happen instantaneously, as when polyploidy arises in a type of flowering plant that can self-fertilize or reproduce asexually.

synaptic integration (sin-AP-tik) The moment-by-moment combining of excitatory and inhibitory signals arriving at a trigger zone of a neuron.

systematics Branch of biology that deals with patterns of diversity among organisms in an evolutionary context; its three approaches include taxonomy, phylogenetic reconstruction, and classification.

systemic circuit (sis-TEM-ik) Circulation route in which oxygen-enriched blood flows from the lungs to the left half of the heart, through the rest of the body (where it gives up oxygen and takes on carbon dioxide), then back to the right side of the heart.

T lymphocyte A white blood cell with roles in immune responses.

tactile signal A physical touching that carries social significance.

taproot system A primary root and its lateral branchings.

target cell Of hormones and other signaling molecules, any cell having receptors to which they can bind.

taxonomy (tax-ON-uh-mee) Approach in biological systematics that involves identifying organisms and assigning names to them.

telophase (TEE-low-faze) Of mitosis, the final stage when chromosomes decondense into threadlike structures and two daughter nuclei form. Of meiosis I, the stage when one of each pair of homologous chromosomes has arrived at one or the other end of the spindle pole. At telophase II, chromosomes decondense and four daughter nuclei form.

telophase II Of meiosis, final stage when four daughter nuclei form.

temperate pathway A viral infection that enters a latent period; the host is not killed outright.

tendon A cord or strap of dense connective tissue that attaches muscle to bones.

territory An area that one or more individuals defend against competitors.

test An attempt to produce actual observations that match predicted or expected observations.

testcross Experimental cross to reveal whether an organism is homozygous dominant or heterozygous for a trait. The organism showing dominance is crossed to an individual known to be homozygous recessive for the same trait.

testis, plural **testes** Male gonad; primary reproductive organ in which male gametes and sex hormones are produced.

testosterone (tess-TOSS-tuh-rown) In male mammals, a major sex hormone that helps control male reproductive functions.

tetanus Of muscles, a large contraction in which repeated stimulation of a motor unit causes muscle twitches to mechanically run together. In a disease by the same name, toxins prevent muscle relaxation.

theory A testable explanation of a broad range of related phenomena. In modern science, only explanations that have been extensively tested and can be relied upon with a very high degree of confidence are accorded the status of theory.

thermal inversion Situation in which a layer of dense, cool air becomes trapped beneath a layer of warm air; can cause air pollutants to accumulate to dangerous levels close to the ground.

thermophile A type of archaebacterium that lives in hot springs, highly acidic soils, and near hydrothermal vents.

thermoreceptor Sensory cell that can detect radiant energy associated with temperature.

thigmotropism (thig-MOTE-ruh-pizm) [Gk. *thigm*, touch] Of vascular plants, growth orientation in response to physical contact with a solid object, as when a vine curls around a fencepost.

threshold Of neurons and other excitable cells, a certain minimum amount by which the voltage difference across the plasma membrane must change to produce an action potential.

thylakoid membrane system Of chloroplasts, an internal membrane system commonly folded into flattened channels and disks (*grana*) and containing light-absorbing pigments and enzymes used in the formation of ATP, NADPH, or both during photosynthesis.

thymine Nitrogen-containing base in some nucleotides.

thymus gland A lymphoid organ with endocrine functions; lymphocytes of the immune system multiply, differentiate, and mature in its tissues, and its hormone secretions affect their functions.

thyroid gland Of endocrine systems, a gland that produces hormones that affect overall metabolic rates, growth, and development.

tissue Of multicelled organisms, a group of cells and intercellular substances that function together in one or more specialized tasks.

tonicity The relative concentrations of solutes in two fluids, such as inside and outside a cell. When solute concentrations are isotonic (equal in both fluids), water shows no net osmotic movement in either direction. When one fluid is hypotonic (has less solutes than the other), the other is hypertonic (has more solutes) and is the direction in which water tends to move.

tooth Of the mouth of various animals, one of the hardened appendages used to secure or mechanically pummel food; sometimes used in defense.

tracer A radioisotope used to label a substance so that its pathway or destination in a cell, organism, ecosystem, or some other system can be tracked, as by scintillation counters that detect its emissions.

trachea (TRAY-kee-uh), plural **tracheae** An air-conducting tube that functions in respiration; of land vertebrates, the windpipe, which carries air between the larynx and bronchi.

tracheal respiration Of insects, spiders, and some other animals, a respiratory system consisting of finely branching tracheae that extend from openings in the integument and that dead-end in body tissues.

tracheid (TRAY-kid) Of flowering plants, one of two types of cells in xylem that conduct water and dissolved minerals.

transcript-processing controls Of eukaryotic cells, controls that govern modification of new mRNA molecules into mature transcripts before shipment from the nucleus.

transcription [L. *trans*, across, + *scribere*, to write] Of protein synthesis, the assembly of an RNA strand on one of the two strands of a DNA double helix; the base sequence of the resulting transcript is complementary to the DNA region on which it was assembled.

transcriptional controls Of eukaryotic cells, controls influencing when and to what degree a particular gene will be transcribed.

transfer RNA (tRNA) Of protein synthesis, any of the type of RNA molecules that bind and deliver specific amino acids to ribosomes *and* pair with mRNA code words for those amino acids.

translation Of protein synthesis, the conversion of the coded sequence of information in mRNA into a particular sequence of amino acids to form a polypeptide chain; depends on interactions of rRNA, tRNA, and mRNA.

translational controls Of eukaryotic cells, controls governing the rates at which mRNA transcripts that reach the cytoplasm will be translated into polypeptide chains at ribosomes.

translocation Of cells, a change in a chromosome's structure following the insertion of part of a nonhomologous chromosome into it. Of vascular plants, conduction of organic compounds through the plant body by way of the phloem.

transpiration Evaporative water loss from stems and leaves.

transport control Of eukaryotic cells, controls governing when mature mRNA transcripts are shipped from the nucleus into the cytoplasm.

transposable element DNA element that can spontaneously "jump" to new locations in the same DNA molecule or a different one. Such elements often inactivate the genes into which they become inserted and give rise to observable changes in phenotype.

trisomy (TRY-so-mee) Of diploid cells, the abnormal presence of three of one type of chromosome.

trophic level (TROE-fik) [Gk. *trophos*, feeder] All the organisms in an ecosystem that are the same number of transfer steps away from the energy input into the system.

tropical rain forest A type of biome where rainfall is regular and heavy, the annual mean temperature is 25°C, and humidity is 80 percent or more; characterized by great biodiversity.

tropism (TROE-prizm) Of vascular plants, a growth response to an environmental factor, such as growth toward light.

true-breeding Of sexually reproducing organisms, a lineage in which the offspring of successive generations are just like the parents in one or more traits.

tumor A tissue mass composed of cells that are dividing at an abnormally high rate.

turgor pressure (TUR-gore) [L. *turgere*, to swell] Internal pressure applied to a cell wall when water moves by osmosis into the cell.

uniformitarianism The theory that existing geologic features are an outcome of a long history of gradual changes, interrupted now and then by huge earthquakes and other catastrophic events.

upwelling An upward movement of deep, nutrient-rich water along coasts to replace surface waters that winds move away from shore.

uracil (YUR-uh-sill) Nitrogen-containing base found in RNA molecules; can base-pair with adenine.

ureter A tubular channel for urine flow between the kidney and urinary bladder.

urethra A tubular channel for urine flow between the urinary bladder and an opening at the body surface.

urinary bladder A distensible sac in which urine is temporarily stored before being excreted.

urinary excretion A mechanism by which excess water and solutes are removed by way of a urinary system.

urinary system An organ system that adjusts the volume and composition of blood, and so helps maintain extracellular fluid.

urine Fluid formed by filtration, reabsorption, and secretion in kidneys; consists of wastes, excess water, and solutes.

uterus (YOU-tur-us) [L. *uterus*, womb] Chamber in which the developing embryo is contained and nurtured during pregnancy.

vaccine An antigen-containing preparation, swallowed or injected, that increases immunity to certain diseases. It induces formation of huge armies of effector and memory B and T cell populations.

vagina Part of a female reproductive system that receives sperm, forms part of the birth canal, and channels menstrual flow to the exterior.

variable Of a scientific experiment, the only factor that is not exactly the same in the experimental group as it is in the control group.

vascular bundle Of vascular plants, the arrangement of primary xylem and phloem into multistranded, sheathed cords that thread lengthwise through the ground tissue system.

vascular cambium Of vascular plants, a lateral meristem that increases stem or root diameter.

vascular cylinder Of plant roots, the arrangement of vascular tissues as a central cylinder.

vascular plant Plant having tissues that transport water and solutes through well-developed roots, stems, and leaves.

vascular tissue system Xylem and phloem; the conducting tissues that distribute water and solutes through the body of vascular plants.

vein Of the circulatory system, any of the large-diameter vessels that lead back to the heart; of leaves, one of the vascular bundles that thread through photosynthetic tissues.

ventricle (VEN-tri-kuhl) Of the vertebrate heart, one of two chambers from which blood is pumped out. *Compare* atrium.

venule A small blood vessel that accepts blood from capillaries and delivers it to a vein; also overlaps capillaries somewhat in function.

vernalization Of flowering plants, stimulation of flowering by exposure to low temperatures.

vertebra, plural **vertebrae** Of vertebrate animals, one of a series of hard bones arranged with intervertebral disks into a backbone.

vertebrate Animal having a backbone of bony segments, the vertebrae.

vesicle (VESS-ih-kul) [L. *vesicula*, little bladder] Within the cytoplasm of cells, one of a variety of small membrane-bound sacs that function in the transport, storage, or digestion of substances or in some other activity.

vessel member One of the cells of xylem, dead at maturity, the walls of which form the water-conducting pipelines.

villus (VIL-us), plural **villi** Any of several types of absorptive structures projecting from the free surface of an epithelium.

viroid An infectious nucleic acid that has no protein coat; a tiny rod or circle of single-stranded RNA.

virus A noncellular infectious agent, consisting of DNA or RNA and a protein coat; can replicate only after its genetic material enters a host cell and subverts its metabolic machinery.

vision Precise light focusing onto a layer of photoreceptive cells that is dense enough to sample details concerning a given light stimulus, followed by image formation in the brain.

visual signal An observable action or cue that functions as a communication signal.

vitamin Any of more than a dozen organic substances that animals require in small amounts for normal cell metabolism but generally cannot synthesize for themselves.

vocal cord One of the thickened, muscular folds of the larynx that help produce sound waves for speech.

water mold A type of saprobic or parasitic fungus that lives in fresh water or moist soil.

water potential The sum of two opposing forces (osmosis and turgor pressure) that can cause the directional movement of water into or out of a walled cell.

water table The upper limit at which the ground in a specified region is fully saturated with water.

watershed Any specified region in which all precipitation drains into a single stream or river.

wax A type of lipid with long-chain fatty acid tails that help form protective, lubricating, or water-repellent coatings.

white blood cell Leukocyte; of vertebrates, any of the macrophages, eosinophils, neutrophils, and other cells which, together with their products, comprise the immune system.

white matter Of spinal cords, major nerve tracts so named because of the glistening myelin sheaths of their axons.

wild-type allele Of a population, the allele that occurs normally or with greatest frequency at a given gene locus.

wing Of birds, a forelimb of feathers, powerful muscles, and lightweight bones that functions in flight. Of insects, a structure that develops as a lateral fold of the exoskeleton and functions in flight.

X chromosome Of humans, a sex chromosome with genes that cause an embryo to develop into a female, provided that it inherits a pair of these.

X-linked gene Any gene on an X chromosome.

X-linked recessive inheritance Recessive condition in which the responsible, mutated gene occurs on the X chromosome.

xylem (ZYE-lum) [Gk. *xylon*, wood] Of vascular plants, a tissue that transports water and solutes through the plant body.

Y chromosome Of humans, a sex chromosome with genes that cause the embryo that inherited it to develop into a male.

Y-linked gene Any gene on a Y chromosome.

yellow marrow A fatty tissue in the cavities of most mature bones that produces red blood cells when blood loss from the body is severe.

yolk sac Of land vertebrates, one of four extraembryonic membranes. In most shelled eggs, it holds nutritive yolk. In humans, part becomes a site of blood cell formation and some of its cells give rise to the forerunners of gametes.

zero population growth A population for which the number of births is balanced by the number of deaths over a specified period, assuming immigration and emigration also are balanced.

zooplankton A freshwater or marine community of floating or weakly swimming heterotrophs, mostly microscopic, such as rotifers and copepods.

zygospore-forming fungus A type of fungus for which a thick spore wall forms around the zygote; this resting spore germinates and gives rise to stalked, spore-bearing structures.

zygote (ZYE-goat) The first cell of a new individual, formed by the fusion of a sperm nucleus with the nucleus of an egg (fertilization).

CREDITS AND ACKNOWLEDGMENTS

Front Matter

Page i Thomas D. Mangelsen/Images of Nature / **Pages vi–vii** Stock Imagery / **Pages viii–ix** James M. Bell/Photo Researchers / **Pages x–xi** S. Stammers/ SPL/Photo Researchers / **Pages xii–xiii** © 1990 Arthur M. Greene / **Pages xiv–xv** Bonnie Rausch/ Photo Researchers / **Pages xvi–xvii** Lennart Nilsson from *A Child Is Born,* © 1966, 1977 Dell Publishing Company, Inc. / **Pages xviii–xix** Thomas D. Mangelsen/Images of Nature / **Pages xx–xxi** Jim Doran

Page 1 Tom Van Sant/The GeoSphere Project, Santa Monica, CA

Chapter 1

1.1 Frank Kaczmarek / **Page 3** Art by American Composition & Graphics, Inc. / **1.3 (b)** Walt Anderson/Visuals Unlimited; (c) Gregory Dimijian/Photo Researchers; (d) Alan Weaving/Ardea, London / **1.4** Jack deConingh / **1.5** J. A. Bishop and L. M. Cook / **1.6 (a)** Tony Brain/SPL/Photo Researchers; (b) M. Abbey/Visuals Unlimited; (c) Dennis Brokaw; (d), (e) Edward S. Ross; (f) Pat & Tom Leeson/Photo Researchers / **1.7** Levi Publishing Company / **Page 13** Photograph Jack deConingh; art by Raychel Ciemma / **Page 15** James M. Bell/Photo Researchers

Chapter 2

2.1 Martin Rogers/FPG / **2.2** Jack Carey / **Page 20 (a) (left)** Kingsley R. Stern; (right) Chip Clark / **Page 21 (c) (above)** Dr. Harry T. Chugani, M.D., UCLA School of Medicine; (below) Hank Morgan/Rainbow / **2.8** Art by Palay/Beaubois / **2.9** Photograph Richard Riley/FPG / **2.10** Art by Palay/Beaubois; photograph Colin Monteath, Hedgehog House, New Zealand / **2.11** H. Eisenbeiss/Frank Lane Picture Agency / **2.12** Art by Raychel Ciemma / **2.14** Michael Grecco / Picture Group

Chapter 3

3.1 Lewis L. Lainey / **3.5** Peter Steyn/Ardea, London / **3.6** Micrograph Biophoto Associates/Science Source/Photo Researchers / **3.9 (a)** David Scharf/ Peter Arnold, Inc.; (b) Robert C. Simpson/Nature Stock / **3.10, 3.11** Art by Precision Graphics / **3.12** Clem Haagner/Ardea, London / **3.13** Art by Precision Graphics / **3.14 (a)** Kenneth Lorenzen; (b) Larry Lefever/Grant Heilman / **3.15, 3.16** Art by Precision Graphics / **3.19** Art by Palay/Beaubois / **3.20** CNRI/SPL/Photo Researchers; art by Robert Demarest / **3.21, 3.22** Art by Precision Graphics / **3.23** A. Lesk/SPL/Photo Researchers

Chapter 4

4.1 (a) (left) National Library of Medicine; (right) Armed Forces Institute of Pathology; (b) The Bettmann Archive; (d) The Francis A. Countway Library of Medicine / **4.2** Art by Precision Graphics / **Page 55 (d–g)** Jeremy Pickett-Heaps, School of Botany, University of Melbourne / **4.5 (a)** Micrograph G. Cohen-Bazire; art by Palay/Beaubois; (b) K. G. Murti/Visuals Unlimited / **4.6 (a)** Gary Gaard and Arthur Kelman; (b) R. Calentine/Visuals Unlimited / **4.7** Art by Raychel Ciemma / **4.8** Micrograph M. C. Ledbetter, Brookhaven National Laboratory; art by Raychel Ciemma / **4.9** Art by Raychel Ciemma / **4.10** Micrograph G. L. Decker; art by Raychel Ciemma / **4.11** Micrograph Stephen L. Wolfe; art by Raychel Ciemma / **4.12 (a) (left)** Don W. Fawcett/Visuals Unlimited; (right) A. C. Faberge, *Cell and Tissue Research,* 151:403–415, 1974 / **4.13 (b)** Art by Raychel Ciemma / **4.14 (a), (b)** Micrographs Don W. Fawcett/Visuals Unlimited; (below right) art by Robert Demarest after a model by J. Kephart; micrograph Gary W. Grimes / **4.15 (left)** Art by Robert Demarest after a model by J. Kephart; micrograph Gary W. Grimes / **4.16** Gary W. Grimes / **4.17** Micrograph Keith R. Porter / **4.18** Micrograph L. K. Shumway; (below) art by Palay/ Beaubois / **4.19 (a)** Andrew S. Bajer; (b) J. Victor Small and Gottfried

Rinnerthaler; (c) art by Precision Graphics / **4.20 (a)** C. J. Brokaw; (b) Sidney L. Tamm / **4.21** Art by Precision Graphics after Stephen L. Wolfe, *Molecular and Cellular Biology,* Wadsworth, 1993 / **4.22 (a)** H. A. Core, W. A. Coté, and A. C. Day, *Wood Structure and Identification,* Second Editon, Syracuse University Press, 1979; (b) sketch by D. & V. Hennings after P. Raven et al., *Biology of Plants,* Third Edition, Worth Publishers, 1981; (c) P. A. Roelofsen

Chapter 5

5.1 (a) Runk/Schoenberger/Grant Heilman; (b) Inigo Everson/Bruce Coleman Ltd. / **5.3** Art by Raychel Ciemma / **Page 81** Art by Palay/Beaubois; micrograph P. Pinto da Silva and D. Branton, *Journal of Cell Biology,* 45:598, by copyright permission of The Rockefeller University Press / **5.4** Art by Palay/Beaubois / **5.5** Micrograph M. Sheetz, R. Painter, and S. Singer, *Journal of Cell Biology,* 70:193, by copyright permission of The Rockefeller University Press / **5.6** Art by Precision Graphics after Stephen L. Wolfe, *Molecular and Cellular Biology,* Wadsworth, 1993 / **Page 84 (a)** Photographs Frank B. Salisbury; (b) Frieder Sauer/Bruce Coleman Ltd. / **5.7** Art by Leonard Morgan; (above) after Alberts et al., *Molecular Biology of the Cell,* Second edition, Garland Publishing Co., 1989 / **5.10** Art by Leonard Morgan / **5.12** M. M. Perry and A. B. Gilbert / **5.13** Art by Palay/Beaubois

Chapter 6

6.1 Gary Head / **6.2** Evan Cerasoli / **6.3 (above)** NASA; (below) Manfred Kage/Peter Arnold, Inc. / **6.4** Art by Palay/Beaubois / **6.8** Thomas A. Steitz / **6.9** Art by Palay/Beaubois / **6.11** Photograph Douglas Faulkner/Sally Faulkner Collection / **6.14** Art by Raychel Ciemma and American Composition & Graphics, Inc. after B. Alberts et al., *Molecular Biology of the Cell,* Garland Publishing Co., 1983 / **Page 103 (a),** (b) © Oxford Scientific Films/Animals Animals; (c) © Raymond Mendez/Animals Animals; (d) Keith V. Wood

Chapter 7

7.1 David R. Frazier/Photo Researchers / **7.2 (a)** Hans Reinhard/Bruce Coleman Ltd.; (e), (f) micrographs David Fisher; art by Raychel Ciemma and American Composition & Graphics, Inc. / **7.3 (a)** Barker-Blakenship/FPG; (b) art by Precision Graphics after Govindjee / **7.4** Photograph Carolina Biological Supply Company; art by Raychel Ciemma / **7.5** Larry West/FPG / **7.6** Art by Illustrious, Inc. / **7.7** E. R. Degginger / **7.8** Art by Raychel Ciemma and Precision Graphics / **7.9** Art by Raychel Ciemma / **7.10** Art by Raychel Ciemma and Precision Graphics / **Page 117** NASA / **7.11** Art by Raychel Ciemma / **7.12** Martin Grosnick/Ardea, London

Chapter 8

8.1 Stephen Dalton/Photo Researchers / **8.3 (left) (above)** Dennis Brokaw; (below) Janeart/Image Bank; (right) Paolo Fioratti; art by Precision Graphics / **8.4 (right)** Art by Palay/Beaubois / **8.5 (a)** Micrograph Keith R. Porter; (b) art by L. Calver; (c) art by Raychel Ciemma / **8.11 (b)** Adrian Warren/Ardea, London; (c) David M. Phillips/Visuals Unlimited / **8.12** Photograph Gary Head / **Page 137** R. Llewellyn/Superstock, Inc. / **Page 139** © Lennart Nilsson

Chapter 9

9.1 (left and right center) Chris Huss; (right top and bottom) Tony Dawson / **9.2** C. J. Harrison et al., *Cytogenetics and Cell Genetics* 35:21–27, copyright 1983 S. Karger A.G., Basel / **9.3** CNRI/SPL/Photo Researchers / **9.5** Andrew S. Bajer, University of Oregon / **9.6** Micrographs Ed Reschke; art by Raychel Ciemma / **9.9** B. A. Palevitz and E. H. Newcomb, University of Wisconsin/BPS/Tom Stack & Associates / **9.10** Micrographs H. Beams and R. G. Kessel,

American Scientist, 64:279–290, 1976 / **9.11 (a–c), (e)** Lennart Nilsson from *A Child Is Born* © 1966, 1967 Dell Publishing Company, Inc.; (d) Lennart Nilsson from *Behold Man,* © 1974 by Albert Bonniers Forlag and Little, Brown and Company, Boston

Chapter 10

10.1 (a) Jane Burton/Bruce Coleman Ltd.; (b) Dan Kline/Visuals Unlimited / **10.2** Courtesy of Kirk Douglas/The Bryna Company / **10.3** Art by Precision Graphics / **10.4** CNRI/SPL/Photo Researchers / **10.5** Art by Raychel Ciemma / **10.6** Art by Raychel Ciemma and American Composition & Graphics, Inc. / **10.7** Art by Raychel Ciemma / **10.10** Micrograph David M. Phillips/Visuals Unlimited / **10.11** Art by Raychel Ciemma

Chapter 11

11.1 (a) Frank Trapper/Sygma; (b) Focus on Sports; (c) Fabian/Sygma; (d) Moravian Museum, Brno / **11.2** Photograph Jean M. Labat/Ardea, London; art by Jennifer Wardrip / **11.5, 11.7** Art by Hans & Cassady, Inc. / **11.8** Art by Raychel Ciemma / **11.10** Photographs William E. Ferguson / **11.11** Photograph Bill Longcore/Photo Researchers / **11.12 (a),** (b) Michael Stuckey/Comstock Inc.; (c) Russ Kinne/Comstock Inc. / **11.13** David Hosking / **11.14** Tedd Somes / **11.15 (top to bottom)** Frank Cezus; Frank Cezus; Michael Keller; Ted Beaudin; Stan Sholik/all FPG / **11.16 (a)** Dan Fairbanks, Brigham Young University / **11.17** After John G. Torrey, *Development in Flowering Plants,* by permission of Macmillan Publishing Company, copyright © 1967 by John G. Torrey / **11.18** Photograph Jane Burton/Bruce Coleman Ltd.; art by D. & V. Hennings / **Page 184** Evan Cerasoli

Chapter 12

12.1 Eddie Adams/AP Photo / **Page 189 (a)** Photograph Omikron/Photo Researchers / **12.3 (a)** From Lennart Nilsson, *A Child Is Born,* © 1966, 1977 Dell Publishing Company, Inc. (b) Redrawn by Robert Demarest by permission from page 126 of Michael Cummings, *Human Heredity: Principles and Issues,* Third Edition. Copyright © 1994 by West Publishing Company. All rights reserved.; (c) art by Robert Demarest after Patten, Carlson, and others / **12.4** Photographs Carolina Biological Supply Company / **12.7** Photograph Dr. Victor A. McKusick / **12.11** After Victor A. McKusick, *Human Genetics,* Second edition, copyright 1969. Reprinted by permission of Prentice-Hall, Inc., Englewood Cliffs, NJ; photograph The Bettmann Archive / **12.12 (a, b)** Courtesy of G. H. Valentine; (c) C. J. Harrison / **12.13** Art by Raychel Ciemma / **12.14 (a)** Cytogenetics Laboratory, University of California, San Francisco; (b) after Collman and Stoller, *American Journal of Public Health,* 52, 1962 / **12.15 (left)** Used by permission of Carole Iafrate; (center and right) Courtesy of Peninsula Association for Retarded Children and Adults, San Mateo Special Olympics, Burlingame, CA / **Page 205 (a)** Art by Palay/Beaubois; (b) From Fran Heyl Associates, © Jacques Cohen, Computer enhanced by © Pix*Elation / **Page 207 (above)** Bonnie Kamin/Stuart Kenter Associates; (below) Carolina Biological Supply Company

Chapter 13

13.1 Photograph A. C. Barrington Brown © 1968 J. D. Watson; model A. Lesk/SPL/Photo Researchers / **13.3 (b)** Micrograph Lee D. Simon/Science Source/Photo Researchers / **13.5** Micrograph Biophoto Associates/SPL/Photo Researchers / **13.6, 13.8** Art by Precision Graphics / **13.9** Ken Greer/ Visuals Unlimited / **13.10 (a) (left)** C. J. Harrison et al., *Cytogenetics and Cell Genetics* 35:21–27, copyright 1983 S. Karger A.G., Basel; (right) U. K. Laemmli from *Cell,* 12:817–828, copyright 1977 by MIT/Cell Press; (b) B. Hamkalo; (c) O. L. Miller, Jr. and Steve L. McKnight; art by Nadine Sokol

Chapter 14

14.1 (left) Dennis Hallinan/FPG; (right) Kevin Magee/Tom Stack & Associates / **Page 220** (a) Gary Head; (b) from Stephen L. Wolfe, *Molecular and Cellular Biology*, Wadsworth, 1993 / **14.3** Art by Hans & Cassady, Inc. / **14.7** (a) Courtesy of Thomas A. Steitz from *Science*, 246:1135–1142, December 1, 1989 / **14.9** Art by Raychel Ciemma and American Composition & Graphics, Inc. / **14.10** Micrograph Dr. John E. Heuser, Washington University School of Medicine, St. Louis, MO; art by Palay/Beaubois / **14.13** Peter Starlinger / **14.14** Art by Raychel Ciemma

Chapter 15

15.1 Lennart Nilsson © Boehringer Ingelheim International GmbH / **15.2** Art by Palay/Beaubois and Hans & Cassady / **15.3** Art by Raychel Ciemma / **15.4** (a) M. Roth and J. Gall / **15.5** Art by Palay/Beaubois / **15.6** (a), (b) Stuart Kenter Associates / **15.7** Jack Carey / **15.8** W. Beerman / **15.9** Frank B. Salisbury / **15.10** Art by Precision Graphics

Chapter 16

16.1 Secchi-Lecague/Roussel-UCLAF/CNRI/SPL/Photo Researchers; (inset) Ted Thai/Time Magazine / **16.2** Dr. Huntington Potter and Dr. David Dressler / **16.5** After Stephen L. Wolfe, *Molecular and Cellular Biology*, Wadsworth, 1993 / **Page 252** (a) Damon Biotech, Inc. / **16.9** Michael Maloney/San Francisco Chronicle / **16.10** (a) W. Merrill; (b) Keith V. Wood / **16.11** (a), (b) Monsanto Company; (c), (d) Calgene, Inc. / **16.12** R. Brinster and R. E. Hammer, School of Veterinary Medicine, University of Pennsylvania / **Page 259** S. Stammers/SPL/Photo Researchers

Chapter 17

17.1 (above) Art by Leonard Morgan after R. H. Dott, Jr. and R. L. Batten, *Evolution of the Earth*, Third edition, McGraw-Hill, 1981. Reproduced by permission of McGraw-Hill, Inc.; (below) Werner Stoy/Bruce Coleman Ltd. / **17.2** (a) Dave Watts/A. N. T. Photo Library; (b) Kenneth W. Fink/Photo Researchers; (c) Jen & Des Bartlett/Bruce Coleman Ltd. / **17.4** (a) (left) Courtesy George P. Darwin, Darwin Museum, Down House; (right) Heather Angel; (b) Christopher Ralling; (c) D. Barrett/Planet Earth Pictures; (d) Heather Angel; (e) photograph Dieter & Mary Plage/Survival Anglia; art by Leonard Morgan / **17.5** (left) Lee Kuhn/FPG; (right) Field Museum of Natural History, Chicago, and the artist, Charles R. Knight (Neg. No. CK21T) / **17.6** (a) Heather Angel; (b) David Cavagnaro; (c) George W. Cox; (d) Alan Root/Bruce Coleman Ltd. / **17.7** Down House and The Royal College of Surgeons of England / **17.8** John H. Ostrom, Yale University

Chapter 18

18.1 Elliott Erwitt/Magnum Photos, Inc. / **18.2** Alan Solem / **18.3** (a) William E. Ferguson; (b) Eric Crichton/Bruce Coleman Ltd. / **18.5** Art by Precision Graphics / **18.6** (above) David Neal Parks; (below) W. Carter Johnson / **Page 277** Kjell B. Sandved/Visuals Unlimited / **18.8** J. A. Bishop and L. M. Cook / **18.10** (a) Thomas N. Taylor; (b) Edward S. Ross; (c) After M. Karns and L. Penrose, *Annals of Eugenics*, 15:206–233, 1951 / **18.12** (a), (b) Thomas Bates Smith / **18.13** C. W. Fowler / **18.14** Bruce Beehler / **Page 283** D. Avon/Ardea, London

Chapter 19

19.1 (a) Adapted from R. K. Selander and D. W. Kaufman, *Evolution*, Vol. 29, No. 3, December 31, 1975; (b) photograph Gary Head; snail courtesy of Larry Reed / **19.2** Tim Davis/Photo Researchers / **19.3** Jen & Des Bartlett/Bruce Coleman Ltd. / **19.4** G. Ziesler/ZEFA / **19.5** After V. Grant, *Organismic Evolution*, W. H. Freeman and Co., 1977 / **19.6** After A. D. Bradshaw and T. McNeilly, *Evolution and Pollution*, Edward Arnold, London, 1981 / **Page 291** (right) Diagram after F. Ayala and J. Valentine, *Evolving*, Benjamin-Cummings, 1979 / **Page 293** (a) Nancy Sefton/Photo Researchers; (b) Patrice Geisel/Visuals Unlimited; (right) Tom Van Sant/The Geosphere Project, Santa Monica, CA / **19.7** (left) Art by Precision Graphics based on U. Schliewen et al., *Nature*, 368:629–632, 14 April 1994. Used by permission of Macmillan Magazines, Ltd.; (right) Tom Van Sant/The Geosphere Project, Santa Monica, CA / **19.8** After W. Jensen and F. B. Salisbury, *Botany: An Ecological Approach*, Wadsworth, 1972 / **19.9** Photograph Robert C. Simpson/Nature Stock / **19.11** After P. Dodson, *Evolution: Process and Product*, Third edition, Prindle, Weber & Schmidt

Chapter 20

20.1 (left) Vatican Museums; (right) Martin Dohrn/SPL/Photo Researchers / **20.2** (a) Donald Baird, Princeton Museum of Natural History; (b) A. Feduccia, *The Age of Birds*, Harvard University Press, 1980; (c) H. P. Banks; (d) Jonathan Blair/Woodfin Camp & Associates / **20.3** David Noble/FPG / **20.4** (b) From T. Storer et al., *General Zoology*, Sixth edition, McGraw-Hill, 1979. Reproduced by permission of McGraw-Hill, Inc. / **20.5** Art by Victor Royer / **20.6** (a), (b) Gary Head; (c–e) art by Precision Graphics after E. Guerrant, *Evolution*, 36:699–712 / **20.7** (a–f) Art by Joel Ito; (g) art by Raychel Ciemma / **20.8** (top) Douglas P. Wilson/Eric & David Hosking; (center) Superstock, Inc.; (bottom) E. R. Degginger / **20.10** (top) Kjell B. Sandved/Visuals Unlimited; (center) Jeffrey Sylvester/FPG; (bottom) Thomas D. Mangelsen/Images of Nature / **20.11** Photograph Kevin Schafer/Tom Stack & Associates / **20.13** Edward S. Ross / **20.14** Art by Raychel Ciemma after C. T. Regan and E. Trewavas, 1932 / **20.16** (a) Suzanne L. Collins and Joseph T. Collins; (b), (c) from *The Amphibians and Reptiles of Missouri* by Tom R. Johnson. Copyright © 1987 by the Conservation Commission of the State of Missouri. Reprinted by permission. / **20.17** Photographs (clockwise from top left) Peter Scoones/Seaphot Limited: Planet Earth Pictures; C. B. & D. W. Frith/Bruce Coleman Ltd.; © John Gurche 1989; Roger K. Burnard; Jane Burton/Bruce Coleman Ltd.; W. A. Banaszewski / Visuals Unlimited; Hans Reinhard/ZEFA / **Page 316** Art by D. & V. Hennings / **Page 317** Art by John & Judy Waller / **Page 319** (above) (left) Larry Lefever/Grant Heilman Inc.; (right) Bruce Coleman Ltd.; (below) (left) R. I. M. Campbell/Bruce Coleman Ltd.; (right) Runk/Schoenberger/Grant Heilman Inc. / **Page 321** © 1990 Arthur M. Greene

Chapter 21

21.1 (a) © David Malin, Anglo-Australian Observatory; (b) NASA/Space Telescope Science Institute / **21.2** Painting by William K. Hartmann / **21.3** (a) Painting by Chesley Bonestell / **21.4** Art by Precision Graphics / **21.5** (a) Sidney W. Fox; (b) W. Hargreaves and D. Deamer / **21.7** Art by Leonard Morgan / **21.8** Maps by Lloyd K. Townsend after Krohn and Sündermann, 1983 / **21.9** (a) Stanley W. Awramik; (b) M. R. Walter / **Page 331** Art by Raychel Ciemma / **Pages 332–333** Micrograph Robert K. Trench; art by Precision Graphics / **21.10** (a), (b) Neville Pledge/South Australian Museum; (c), (d) Chip Clark / **21.11** Patricia G. Gensel / **21.12** From *Evolution of Life*, Linda Gamlin and Gail Vines (Eds.), Oxford University Press, 1987; art by D. & V. Hennings; photograph Rod Salm/Planet Earth Pictures / **Page 335** Maps by Lloyd K. Townsend after Krohn and Sündermann / **21.13** (above) Painting by Megan Rohn courtesy of David Dilcher; (below) art by Precision Graphics / **Page 337** (a) NASA; (b) NASA Galileo Imaging Team / **Pages 338–339** (c) (left) Art by Raychel Ciemma; (right) © John Gurche 1989 / **21.14** (a), (b) Field Museum of Natural History, Chicago, and the artist Charles R. Knight (Neg. No. CK46T and CK8T) / **21.15** Data from J. J. Sepkoski, Jr., *Paleobiology*, 7(1):36–53 and J. J. Sepkoski, Jr. and M. L. Hulver in Valentine, ed., *Phanerozoic Diversity Patterns: Profiles in Macroevolution*, Princeton University Press, 1985 / **Page 341** Map by Lloyd K. Townsend after Ziegler, Scotese, and Barrett, 1983 / **21.16** (left) Maps by Lloyd K. Townsend after A. M. Ziegler, C. R. Scotese, and S. F. Barrett, "Mesozoic and Cenozoic Paleogeographic Maps" and J. Krohn and J. Sündermann, "Paleotides Before the Permian" in F. Brosche and J. Sündermann (Eds.), *Tidal Friction and the Earth's Rotation II*, Springer-Verlag, 1983

Chapter 22

22.1 (a–c) Tony Brain and David Parker/SPL/Photo Researchers; (d) Lee D. Simon/Photo Researchers / **22.2** Art by L. Calver / **22.3** L. J. LeBeau, University of Illinois Hospital/BPS / **22.4** (a) Stanley Flegler/Visuals Unlimited; (b) CNRI/SPL/Photo Researchers / **22.5** Art by D. & V. Hennings / **22.7** (a) © 1994 Barrie Rokeach; (b) R. Robinson/Visuals Unlimited / **22.8** (a) John D. Cunningham/Visuals Unlimited; (b) Tony Brain/SPL/Photo Researchers; (c) P. W. Johnson and J. McN. Sieburth, University of Rhode Island/BPS / **22.9** Stanley W. Watson, *International Journal of Systematic Bacteriology*, 21:254–270, 1971 / **22.10** J. J. Cardamone, Jr./BPS / **22.11** T. J. Beveridge, University of Guelph/BPS / **22.12** (above) Centers for Disease Control; (below) Edward S. Ross / **22.13** Richard Blakemore / **22.14** Hans Reichenbach, Gesellschaft für Biotechnologische Forschung, Braunschweig, Germany / **22.15** Art by Nadine Sokol / **22.16** (a) George Musil/Visuals Unlimited; (b) K. G. Murti/Visuals Unlimited; (c), (d) Kenneth M. Corbett / **22.17, 22.18** Art by Palay/Beaubois and Precision Graphics

Chapter 23

23.1 (a) Ronald W. Hoham, Dept. of Biology, Colgate University; (b) Steven C. Wilson/Entheos; (c) Edward S. Ross; (d) Gary W. Grimes and Steven L'Hernault / **23.2** M. S. Fuller, *Zoosporic Fungi in Teaching and Research*, M. S. Fuller and A. Jaworski (eds.), 1987, Southeastern Publishing Company, Athens, GA / **23.3** (a) Heather Angel; (b) W. Merrill / **23.4** (a) Art by Leonard Morgan; (b) M. Claviez, G. Gerish, and R. Guggenheim; (c) London Scientific Films; (d–f) Carolina Biological Supply Company; (g) Photograph courtesy Robert R. Kay from R. R. Kay et al. *Development*, 1989 Supplement, pp. 81–90, © The Company of Biologists Ltd. 1989 / **23.5** (a) M. Abbey/Visuals Unlimited; (b, d) From V. & J. Pearse and M. & R. Buchsbaum, *Living Invertebrates*, The Boxwood Press, 1987. Used by permission. (c) John Clegg/Ardea, London; (e) T. E. Adams/Visuals Unlimited; (f) Manfred Kage/Bruce Coleman Ltd. / **23.6** (a) Art by Raychel Ciemma redrawn from V. & J. Pearse and M. & R. Buchsbaum, *Living Invertebrates*, The Boxwood Press, 1987. Used by permission. (b) Gary W. Grimes and Steven L'Hernault; (c) photograph by Ralph Buchsbaum and sketch from V. & J. Pearse and M. & R. Buchsbaum, *Living Invertebrates*, The Boxwood Press, 1987. Used by permission. / **23.7** (a) John D. Cunningham/Visuals Unlimited; (b) Jerome Paulin/Visuals Unlimited; (c) David M. Phillips/Visuals Unlimited / **Page 369** Art by Leonard Morgan; micrograph Steven L'Hernault / **23.8** (a) P. L. Walne and J. H. Arnott, *Planta*, 77:325–354, 1967; (b) T. E. Adams/Visuals Unlimited; art by Palay/Beaubois / **23.9** (a), (b), (d) Ronald W. Hoham, Dept. of Biology, Colgate University; (c) Jan Hinsch/SPL/Photo Researchers / **23.10** (a) Florida Department of Environmental Protection, Florida Marine Research Institute, St. Petersburg; (b) C. C. Lockwood / **23.11** (a) D. P. Wilson/Eric & David Hosking; (b) Douglas Faulkner/Sally Faulkner Collection / **23.12** (a) J. R. Waaland, University of Washington/BPS; (b) Dennis Brokaw; (c) from Tom Garrison, *Oceanography: An Invitation to Marine Science*, Wadsworth, 1993 / **23.13** (a), (c) Hervé Chaumeton/Agence Nature; (b) Alex Kerstitch/Tom Stack & Associates; (d) Manfred Kage/Peter Arnold, Inc.; (e) Ronald W. Hoham, Dept. of Biology, Colgate University / **23.14** Photograph D. J. Patterson/Seaphot Limited: Planet Earth Pictures; art by Raychel Ciemma / **23.15** Carolina Biological Supply Company / **Page 376** Richard W. Greene

Chapter 24

24.1 Robert C. Simpson/Nature Stock / **24.2** Micrograph G. T. Cole, University of Texas, Austin/BPS / **24.4** Photographs (above) John D. Cunningham/Visuals Unlimited; (below) David M. Phillips/Visuals Unlimited; art by Raychel Ciemma / **24.5** John Hodgin / **24.6** (a) After T. Rost et al., *Botany*, Wiley, 1979; (b), (c) Robert C. Simpson/Nature Stock / **Page 383** Eric Crichton/Bruce Coleman Ltd. / **24.7** (a) Victor Duran; (b), (c) Robert C. Simpson/Nature Stock; (d), (e) Thomas J. Duffy; (f) Jane Burton/Bruce Coleman Ltd. / **24.8** Photographs Martyn Ainsworth from A. D. M. Rayner, *New Scientist*, November 19, 1988; art by Raychel Ciemma / **24.9** (a) Mark Mattock/Planet Earth Pictures; (b) Edward

S. Ross / **24.10** After Raven, Evert, and Eichhorn, *Biology of Plants*, Fourth edition, Worth Publishers, New York, 1986 / **24.11** © 1990 Gary Braasch / **24.12** F. B. Reeves / **24.13** (a) G. T. Cole, University of Texas, Austin/BPS; (b) N. Allin and G. L. Barron; (c) G. L. Barron, University of Guelph

Chapter 25

25.1 (a) Pat & Tom Leeson/Photo Researchers; (b) Edward S. Ross / **25.4** Photograph Jane Burton/Bruce Coleman Ltd.; art by Raychel Ciemma / **25.5** (a) Roger K. Burnard; (b) John D. Cunningham/Visuals Unlimited / **25.6** (a), (b) Kingsley R. Stern; (c) John D. Cunningham/Visuals Unlimited / **Page 397** (a) Field Museum of Natural History, Chicago; (Neg. #7500C); (b) Brian Parker/Tom Stack & Associates / **25.7** (b) Kingsley R. Stern; (c) Edward S. Ross / **25.8** (a) Edward S. Ross; (b) W. H. Hodge; (c) Kratz/ZEFA / **25.9** (a) Art by Raychel Ciemma; photograph A. & E. Bomford/Ardea, London; (b) Lee Casebere; (c) Jean Paul Ferrero/Ardea, London / **25.10** Photograph Edward S. Ross; art by Raychel Ciemma / **25.11** (a), (b) © 1989, 1991 Clinton Webb; (c) © 1994 Robert Glenn Ketchum; (d) © 1993 Trygve Steen; (e) David Hiser, Photographers/Aspen, Inc.; (f) Terry Livingstone / **25.12** (a) Ed Reschke; (b) Edward S. Ross; (c) John H. Gerard; (d) Kingsley R. Stern; (e) Edward S. Ross; (inset) F. J. Odendaal, Duke University/BPS / **25.13** (a) Hans Reinhard/Bruce Coleman Ltd.; (b) Heather Angel; (c) Peter F. Zika/Visuals Unlimited; (d) L. Mellichamp/Visuals Unlimited; (e), (f) Edward S. Ross; (g) Dick Davis/Photo Researchers / **25.14** Art by Raychel Ciemma / **25.15** Photograph M. P. L. Fogden/Ardea, London; art by Jennifer Wardrip

Chapter 26

26.1 (a) Courtesy of Department of Library Services, American Museum of Natural History (Neg. # K10273); (b) Jim Stewart/Scripps Institution of Oceanography; (c) Chip Clark / **26.2, 26.3** Art by D. & V. Hennings / **Page 412** (above) Laszlo Meszoly in L. Margulis, *Early Life*, Jones and Bartlett Publishers, / Inc., Boston, © 1982 / **26.5** (a) David C. Haas/Tom Stack & Associates; (b) art by Raychel Ciemma; (d) Marty Snyderman/Planet Earth Pictures; (e) Bruce Hall / **26.6** (a) Frieder Sauer/Bruce Coleman Ltd.; (b) Kim Taylor/Bruce Coleman Ltd. / **26.7** (a) Art by Raychel Ciemma; photograph Francois Gohier/Photo Researchers; (c) Douglas Faulkner/Sally Faulkner Collection; (d) F. S. Westmorland/Tom Stack & Associates / **26.8** (a) Christian Della-Corte; (b) Douglas Faulkner/Photo Researchers; (c) Bill Wood/Seaphot Limited: Planet Earth Pictures; (d) Walter Deas/Seaphot Limited: Planet Earth Pictures / **26.9** (a) Andrew Mounter/Seaphot Limited: Planet Earth Pictures; (b) art by Precision Graphics after T. Storer et al., *General Zoology*, Sixth edition, © 1979 McGraw-Hill / **26.10** (a) Kathie Atkinson/Oxford Scientific Films; (b) Larry Madin/Planet Earth Pictures / **26.11** Photograph Kim Taylor/Bruce Coleman Ltd.; art by Raychel Ciemma / **26.12** (a) Robert & Linda Mitchell; (b) Cath Ellis, University of Hull/SPL/Photo Researchers / **26.13** Art by Raychel Ciemma / **Page 420** (a) Photograph Robert L. Calentine; art by Raychel Ciemma / **Page 421** (b) Art by Nadine Sokol; photograph Carolina Biological Supply Company; (c) Lorus J. and Margery Milne; (d) Dianora Niccolini / **26.14** Kjell B. Sandved / **26.15** Photograph J. Solliday/BPS; art by Raychel Ciemma / **26.16** Art by Palay/Beaubois / **26.17** (a) Gary Head; (c) Kjell B. Sandved / **26.18** (a) Hervé Chaumeton/Agence Nature; (b) Alex Kerstitch / **26.19** Jeff Foott/Tom Stack & Associates / **26.20** Art by Raychel Ciemma / **26.21** Hervé Chaumeton/Agence Nature / **26.22** Photograph Douglas Faulkner/Sally Faulkner Collection; art by Raychel Ciemma / **26.23** J. Grossauer/ZEFA / **26.24** © Cabisco/Visuals Unlimited / **26.25** Art by Raychel Ciemma / **26.26** (a) Hervé Chaumeton/Agence Nature; (b) Jon Kenfield/Bruce Coleman Ltd. / **26.27** J. A. L. Cooke/Oxford Scientific Films / **26.28** From Eugene N. Kozloff, *Invertebrates,* copyright © 1990 by Saunders College Publishing. Reproduced by permission of the publisher. / **26.29** C. B. & D. W. Frith/Bruce Coleman Ltd. / **26.30** Jane Burton/Bruce Coleman Ltd. / **26.31** (a) (above) Angelo Giampiccolo/FPG; (below) Jane Burton/Bruce Coleman Ltd. (b) P. J. Bryant, University of California, Irvine/BPS; (c) John H. Gerard; (d) Ken Lucas/Seaphot Limited: Planet Earth Pictures / **26.32** (a) Franz Lanting/Bruce Coleman Ltd.; (b) Hervé Chaumeton; (c) Agence Nature; (d) Fred Bavendam/Peter Arnold, Inc. / **26.33** (a) Z. Leszczynski/Animals Animals; (b) Steve Martin/Tom Stack & Associates / **26.34** Art by D. & V. Hennings / **26.35** (a) David Maitland/Seaphot Limited: Planet Earth Pictures; (b), (d) Edward S. Ross; (c) Kenneth Lorenzen; (e) Robert & Linda Mitchell; (f) Ralph A. Reinhold/FPG; (g–j), (l) Edward S. Ross; (k) C. P. Hickman, Jr. / **26.36** (a) John Mason/Ardea, London; (b) Kjell B. Sandved; (c) Chris Huss/The Wildlife Collection; (d) Ian Took/Biofotos; (e), (f) Hervé Chaumeton/Agence Nature; (g) Jane Burton/Bruce Coleman Ltd. / **26.37** Art by L. Calver

Chapter 27

27.1 (a) Tom McHugh/Photo Researchers; (b) Jean Phillipe Varin/Jacana/Photo Researchers / **27.2** Art by D. & V. Hennings and Precision Graphics / **27.3** (a) Photograph Rick M. Harbo; (b) photograph Peter Parks/Oxford Scientific Films/Animals Animals; (b–e) sketches after *Living Invertebrates*, V. & J. Pearse and M. & R. Buchsbaum, The Boxwood Press, 1987. Used by permission. / **27.4** Photograph Hervé Chaumeton/Agence Nature; art by Laszlo Meszoly and D. & V. Hennings / **27.5** C. R. Wyttenbach, University of Kansas/BPS / **27.6** Art by D. & V. Hennings / **27.7** After C. P. Hickman, Jr. and L. S. Roberts, *Integrated Principles of Zoology*, Seventh edition, St. Louis: Times Mirror/Mosby College Publishing, 1984; art by Palay/Beaubois / **27.8** Art by Raychel Ciemma / **27.9** Art by Precision Graphics / **27.10** Art by D. & V. Hennings / **27.11** After A. S. Romer and T. S. Parsons, *The Vertebrate Body*, Sixth edition, Saunders College Publishing, © 1986 CBS College Publishing; art by Laszlo Meszoly and D. & V. Hennings / **27.12** Heather Angel / **27.13** (a) Erwin Christian/ZEFA; (b) Allan Power/Bruce Coleman Ltd.; (c) Tom McHugh/Photo Researchers / **27.14** (a) Douglas Faulkner/Sally Faulkner Collection; (b) Patrice Ceisel/© 1986 John G. Shedd Aquarium; (c) Robert & Linda Mitchell; (d) William H. Amos; (e) art by Raychel Ciemma; (f) Bill Wood/Bruce Coleman Ltd. / **27.15** (a) Peter Scoones/Seaphot Limited: Planet Earth Pictures; (b), (c) art by Laszlo Meszoly and D. & V. Hennings / **27.16** (a) After A. S. Romer and T. S. Parsons, *The Vertebrate Body*, Sixth edition, Saunders College Publishing, © 1986 CBS College Publishing; art by Leonard Morgan; (b) Jerry W. Nagel; (c) Stephen Dalton/Photo Researchers; (d) John Serraro/Visuals Unlimited; (e) Juan M. Renjifo/Animals Animals / **27.17** (a) Zig Leszczynski/Animals Animals; (b) Leonard Lee Rue III / **27.18** Art by D. & V. Hennings / **27.19** (a) Peter Scoones/Seaphot Limited: Planet Earth Pictures; (b) D. Kaleth/Image Bank; (c) Andrew Dennis/A. N. T. Photo Library; (d) W. J. Weber/Visuals Unlimited; (e) Heather Angel; (f) Stephen Dalton/Photo Researchers; art by Raychel Ciemma; (g) W. A. Banaszewski/Visuals Unlimited / **27.20** Rajesh Bedi / **27.21** (a) Gerard Lacz/A. N. T. Photo Library; (b) Thomas D. Mangelsen/Images of Nature; (c) J. L. G. Grande/Bruce Coleman Ltd. / **27.22** (c) Art by D. & V. Hennings / **27.23** (a) Sandy Roessler/FPG; (b) Christopher Crowley; (c) Kevin Schafer/Tom Stack & Associates; art by Raychel Ciemma after M. Weiss and A. Mann, *Human Biology and Behavior*, Fifth edition, HarperCollins, 1990 / **27.24** D. & V. Blagden/A.N.T. Photo Library / **27.25** Jack Dermid / **27.26** (a) Douglas Faulkner/Photo Researchers; (b) J. Scott Altenbach, University of New Mexico; (c) Clem Haagner/Ardea, London; (d) Roger K. Burnard; (e), (f) Leonard Lee Rue III/FPG

Chapter 28

28.1 (left) FPG; (right) Douglas Mazonowicz/Gallery of Prehistoric Art / **28.2** (a) Bruce Coleman Ltd.; (b) Tom McHugh/Photo Researchers; (c) Larry Burrows/Aspect Picture Library / **28.3** Art by D. & V. Hennings / **28.7** © Time Inc. 1965/Larry Burrows Collection / **28.8, 28.9** Art by D. & V. Hennings / **28.10** (left) Dr. Donald Johanson, Institute of Human Origins; (right) Louise M. Robbins / **28.11** Art by D. & V. Hennings / **28.12** Photographs by John Reader copyright 1981 / **Page 325** Bonnie Rauch/Photo Researchers

Chapter 29

29.1 (a) Roger Werth; (b) © 1980 Gary Braasch; (c) © 1989 Gary Braasch / **29.2** Art by Raychel Ciemma / **29.3** Micrograph James D. Mauseth, *Plant Anatomy*, Benjamin-Cummings, 1988 / **29.4** Art by D. & V. Hennings / **29.5** (a–c) Biophoto Associates; (d) Kingsley R. Stern / **29.7** (a) George S. Ellmore; (b) Jeremy Burgess/SPL/Photo Researchers / **29.8** (a), (b) Edward S. Ross; art by D. & V. Hennings / **29.9** Art by D. & V. Hennings; (a) (center) Ray F. Evert; (right) James W. Perry; (b) (center) Carolina Biological Supply Company; (right) James W. Perry / **29.10** Art by D. & V. Hennings / **29.11** Heather Angel / **29.12** (a) Art by Raychel Ciemma; (b) C. E. Jeffree et al., *Planta*, 172(1):20–37, 1987. Reprinted by permission of C. E. Jeffree and Springer-Verlag / **29.13** (a) Robert & Linda Mitchell; (b) Roland R. Dute / **29.14** (b–d) E. R. Degginger / **Page 493** Redrawn by Precision Graphics from Beryl B. Simpson and Molly C. Ogorzaly, *Economic Botany: Plants in Our World*, © 1986 McGraw-Hill Inc. Used by permission of McGraw-Hill, Inc. / **29.15** John E. Hodgin / **29.16** Ripon Microslides, Inc. / **29.17** (a) Art by Hans & Cassady, Inc.; (b) micrographs Chuck Brown / **29.18** Carolina Biological Supply Company / **29.19** Ripon Microslides; sketch after T. Rost et al., *Botany: A Brief Introduction to Plant Biology*, Second edition, © 1984, John Wiley & Sons / **29.23** Micrograph Jerry D. Davis / **29.24** H. A. Core, W. A. Coté, and A. C. Day, *Wood Structure and Identification*, Second edition, Syracuse University Press, 1979 / **Page 499** Ripon Microslides

Chapter 30

30.1 (a), (c) Robert & Linda Mitchell; (b) John N. A. Lott, *Scanning Electron Microscope Study of Green Plants*, St. Louis: C. V. Mosby Company, 1976; (d) Robert C. Simpson/Nature Stock / **30.2** (a) Adrian P. Davies/Bruce Coleman Ltd.; (c) Mark E. Dudley and Sharon R. Long; (d) NifTAL Project, University of Hawaii, Maui; art by Jennifer Wardrip / **30.3** Micrograph Jean Paul Revel / **30.4** (a) Chuck Brown; art by Leonard Morgan / **Page 505** (a) James T. Brock; (b) U. S. Department of Agriculture / **30.5** Micrographs H. A. Core, W. A. Coté, and A. C. Day, *Wood Structure and Identification*, Second edition, Syracuse University Press, 1979 / **30.6** Art by Raychel Ciemma / **30.7** George S. Ellmore / **30.8** (a) W. Thomson, *American Journal of Botany*, 57(3):316, 1970; (b) art by Palay/Beaubois / **30.9** T. A. Mansfield / **30.10** Jeremy Burgess/SPL/Photo Researchers / **30.11** Art by Hans & Cassady, Inc. / **30.12** Martin Zimmermann, *Science*, 133:73–79, © AAAS 1961 / **30.13** Art by Palay/Beaubois

Chapter 31

31.1 (a) Thomas D. W. Friedmann/Photo Researchers; (b) Gary Head / **31.3** John Shaw/Bruce Coleman Ltd. / **31.4** (a), (b) David M. Phillips/Visuals Unlimited; (c) David Scharf/Peter Arnold, Inc. / **31.5** Art by Raychel Ciemma / **Page 520** (a) Thomas Eisner, Cornell Univerity; (b) Raymond Mendez/Animals Animals; (c) Edward S. Ross / **Page 521** (d) Robert A. Tyrrell / **31.6** (a), (b) Patricia Schulz; (c), (d) Ray F. Evert; (e), (f) Ripon Microslides, Inc.; (g) Kingsley R. Stern / **31.7** (a–c) Janet Jones; (d) B. J. Miller, Fairfax, VA/BPS; (e) Richard H. Gross / **31.8** R. Carr/Bruce Coleman Ltd. / **Page 524** (a) John Alcock / **Page 525** (b), (d), (e) John Alcock; (c) Merlin D. Tuttle, Bat Conservation International / **31.9** Russell Kaye/ © 1993 The Walt Disney Co. Reprinted with permission of Discover Magazine / **31.10** (a) Runk/Schoenberger/Grant Heilman, Inc.; (b) Kingsley R. Stern

Chapter 32

32.1 (a) R. Lyons/Visuals Unlimited; (b) Michael A. Keller/FPG / **32.2** (a) B. Bracegirdle and P. Miles, *An Atlas of Plant Structure*, Heinemann Educational Books, 1977; (b) Carolina Biological Supply Company; art by Hans & Cassady, Inc. / **32.3** Photograph Hervé Chaumeton/Agence Nature; art by

Hans & Cassady, Inc. / **32.4** (a) Kingsley R. Stern / **32.5** (a) Frank B. Salisbury / **32.6** (a) John Digby and Richard Firn; (b), (c) micrographs courtesy of Randy Moore from "How Roots Respond to Gravity," M. L. Evans, R. Moore, and K. Hasenstein, *Scientific American*, December 1986 / **32.7** Gary Head / **32.8** Cary Mitchell / **32.9, 32.11** Frank B. Salisbury / **32.12** Photograph Diana Starr / **32.13** Jan Zeevart / **32.15** N. R. Lersten / **32.16** Larry D. Nooden / **32.17** R. J. Downs / **32.18** Diana Starr / **Page 542** (above) Edward S. Ross; (below) Gary Head / **Page 544** Grant Heilman Inc. / **Page 545** © Kevin Schafer

Chapter 33

33.1 David Macdonald / **33.2** (a) Focus on Sports; (inset) Manfred Kage/Bruce Coleman Ltd.; art by Palay/Beaubois; photographs (b) Lennart Nilsson from *Behold Man*, © 1974 by Albert Bonniers Forlag and Little, Brown and Company, Boston; (c) Manfred Kage/Bruce Coleman Ltd.; (d) Ed Reschke/Peter Arnold, Inc. / **33.3** Art by Palay/Beaubois / **33.4** Photographs (a–c), (e), (f) Ed Reschke; (d) Fred Hossler/Visuals Unlimited / **33.5** (left) Art by Joel Ito; (right) art by L. Calver / **33.6** Photograph Ed Reschke / **33.7** Art by Robert Demarest / **33.8** Photographs Ed Reschke / **33.9** Lennart Nilsson from *Behold Man*, © 1974 Albert Bonniers Forlag and Little, Brown and Company, Boston / **33.10** Art by L. Calver / **33.11** Art by Palay/Beaubois / **33.13** Photograph Fred Bruemmer

Chapter 34

34.1 Adrian Warren/Ardea / **34.2** Manfred Kage/Peter Arnold, Inc. / **34.3, 34.5** Art by Raychel Ciemma / **34.8** (a) Art by Kevin Somerville; (b) Ed Reschke; (c) micrograph J. E. Heuser and T. S. Reese; (c), (d) art by Leonard Morgan / **Page 569** Painting by Sir Charles Bell, 1809, courtesy of Royal College of Surgeons, Edinburgh / **34.10** Micrograph from *Tissues and Organs: A Text-Atlas of Scanning Electron Microscopy*, by R. G. Kessel and R. H. Kardon. Copyright © 1979 by W. H. Freeman and Company. Reprinted with permission. Art by Robert Demarest / **34.11** (a) Art by Kevin Somerville; (b) art by Robert Demarest

Chapter 35

35.1 Comstock, Inc. / **35.2** Kjell B. Sandved / **35.3** Art by Raychel Ciemma / **35.5, 35.6** Art by Kevin Somerville / **35.8** Micrograph Manfred Kage/Peter Arnold, Inc.; art by Robert Demarest / **35.11** Art by Kevin Somerville / **35.12** (left) C. Yokochi and J. Rohen, *Photographic Anatomy of the Human Body*, Second edition, Igaku-Shoin Ltd., 1979; (right) Colin Chumbley/Science Source/Photo Researchers / **35.13** (left) From James W. Kalat, *Introduction to Psychology*, Third edition, Brooks/Cole Publishing Company, 1993; (right) Marcus Raichle, Washington University School of Medicine / **35.14** Art by Palay/Beaubois after Penfield and Rasmussen, *The Cerebral Cortex of Man*, copyright © 1950 Macmillan Publishing Company, Inc. Renewed 1978 by Theodore Rasmussen / **35.16** Art by Robert Demarest / **Page 589** (left) Ogden Gigli/Photo Researchers; (right) From Edythe D. London et al., *Archives of General Psychiatry*, 47:567–574 (1990)

Chapter 36

36.1 (a) Eric A. Newman; (b) Merlin D. Tuttle, Bat Conservation International / **36.2** Art by Kevin Somerville / **36.3** From Hensel and Bowman, *Journal of Physiology*, 23:564–568, 1960 / **36.4** Art by Palay/Beaubois after Penfield and Rasmussen, *The Cerebral Cortex of Man*, copyright © 1950 Macmillan Publishing Company, Inc. Renewed 1978 by Theodore Rasmussen / **36.5** Art by Ron Ervin; micrograph Ed Reschke / **36.6** Art by Raychel Ciemma / **36.7** Art by Robert Demarest; micrograph Omikron/SPL/Photo Researchers / **36.9** Photograph Edward W. Bower/© 1991 TIB/West; (a–c) art by Kevin Somerville / **36.10** Art by Robert Demarest / **36.11** (a), (b) Robert E. Preston, courtesy Joseph E. Hawkins, Kresge Hearing Research Institute, University of Michigan Medical School / **36.12** (a) Keith Gillett/Tom Stack & Associates; (b) Chris Newbert / **36.13** After M. Gardiner, *The Biology of Vertebrates*, McGraw-Hill, 1972 / **36.14** (a) Photograph E. R. Degginger; sketch after M. Gardiner, *The Biology of Vertebrates*, McGraw-Hill, 1972; (b) G. A. Mazohkin-Porshnykov (1958). Reprinted with permission from *Insect Vision*, © 1969 Plenum Press / **36.15** Art by Robert Demarest / **36.16** Chase Swift / **36.17** Art by Kevin Somerville / **Pages 606–607** Photographs Gerry Ellis/The Wildlife Collection / **36.18** Micrograph Lennart Nilsson © Boehringer Ingelheim International GmbH / **36.19** (a) Art by Nadine Sokol; (b) art by Palay/Beaubois / **36.20** Art by Palay/Beaubois after S. Kuffler and J. Nicholls, *From Neuron to Brain*, Sinauer, 1977 / **Page 610** Photograph Douglas Faulkner/Sally Faulkner Collection

Chapter 37

37.1 Hugo van Lawick / **37.2** Art by Kevin Somerville / **37.5, 37.6** Art by Robert Demarest / **37.7** (a) Mitchell Layton; (b) Syndication International (1986) Ltd. / **37.8** Photographs courtesy of Dr. William H. Daughaday, Washington University School of Medicine, from A. I. Mendelhoff and D. E. Smith, eds., *American Journal of Medicine*, 20:133 (1956) / **37.9** Art by Leonard Morgan / **37.11** Art by Joel Ito / **37.12** (a) The Bettmann Archive; (b) Biophoto Associates/SPL/Photo Researchers / **37.13** John S. Dunning/Ardea, London / **Page 625** Evan Cerasoli / **37.15** Sketches after Willier, Weiss, and Hamburger, *Analysis of Development*, Philadelphia, W. B. Saunders Co., 1955; photograph Roger K. Burnard

Chapter 38

38.1 Jeff Schultz/AlaskaStock Images / **Page 631** Art by L. Calver / **38.2** Art by Robert Demarest / **38.3** (a) Gregory Dimijian/Photo Researchers; (b) Tom McHugh; (c) M. P. L. Fogden/Bruce Coleman Ltd.; (d) Chaumeton-Lanceau/Agence Nature; sketches from *The Life of Birds*, Fourth edition, by Luis Baptista and Joel Carl Welty, copyright © 1988 by Saunders College Publishing. Reproduced by permission of the publisher. / **38.4** Ed Reschke / **Page 635** Michael Keller/FPG / **38.5** Linda Pitkin/Planet Earth Pictures / **38.6** D. A. Parry, *Journal of Experimental Biology*, 36:654, 1959 / **38.7** (a) Stephen Dalton/Photo Researchers; (b) art by Raychel Ciemma / **38.8** Art by D. & V. Hennings / **38.9** (a) (left) Art by L. Calver; (a) (right), (b) art by Joel Ito; (c) Ed Reschke / **38.10** Art by K. Kasnot / **38.11** National Osteoporosis Foundation / **Page 640** Photograph C. Yokochi and J. Rohen, *Photographic Anatomy of the Human Body*, Second edition, Igaku-Shoin Ltd., 1979 / **38.12** Art by Kevin Somerville / **38.13** (a) N.H.P.A./A. N. T. Photo Library; (b) art by Robert Demarest / **38.14** Art by Kevin Somerville / **38.15** (a) John D. Cunningham; (b) Art by Robert Demarest; (c) D. W. Fawcett, *The Cell*, Philadelphia: W. B. Saunders Co., 1966 / **38.16** Art by Nadine Sokol / **38.17** Art by Robert Demarest / **38.18** After Stephen L. Wolfe, *Molecular and Cellular Biology*, Wadsworth, 1993 / **Page 649** Photograph Michael Neveux

Chapter 39

39.1 (a) photograph courtesy of The New York Academy of Medicine Library; (b) from A. D. Waller, *Physiology, The Servant of Medicine*, Hitchcock Lectures, University of London Press, 1910 / **39.3** (b) (below) After M. Labarbera and S. Vogel, *American Scientist*, 70:54–60, 1982 / **39.4** Lennart Nilsson from *Behold Man*, © 1974 by Albert Bonniers Forlag and Little, Brown and Company, Boston / **39.5** Art by Palay/Beaubois / **39.6** (a) CNRI/SPL/Photo Researchers; (b) Art by Raychel Ciemma / **39.7, 39.8** Art by Kevin Somerville / **39.9** (a) C. Yokochi and J. Rohen, *Photographic Anatomy of the Human Body*, Second edition, Igaku-Shoin Ltd, 1979; (b), (c) art by Kevin Somerville / **39.11** (a) Ed Reschke / **39.13** Art by Robert Demarest based on A. Spence, *Basic Human Anatomy*, Benjamin-Cummings, 1982 / **39.14** Sheila Terry/SPL/Photo Researchers / **39.15** Art by Raychel Ciemma / **39.16** Art by Kevin Somerville / **Page 666** (a) (above) Ed Reschke; (below) F. Sloop and W. Ober/Visuals Unlimited / **39.17** Lennart Nilsson © Boehringer Ingelheim International GmbH / **39.18** (a), (b) Lester V. Bergman & Associates, Inc.; (c) after F. Ayala and J. Kiger, *Modern Genetics*, © 1980 Benjamin-Cummings / **39.19** Art by Nadine Sokol after Gerard J. Tortora and Nicholas P. Anagnostakos, *Principles of Anatomy and Physiology*, Sixth edition, Copyright © 1990 by Biological Sciences Textbooks, Inc., A & P Textbooks, Inc. and Elia-Sparta, Inc. Reprinted by permission of HarperCollins Publishers / **39.20, 39.21** Art by Raychel Ciemma

Chapter 40

40.1 (a) The Granger Collection, New York; (b) Lennart Nilsson © Boehringer Ingelheim International GmbH / **40.2** (a) Art by Nadine Sokol; (b) micrograph Robert R. Dourmashkin, courtesy of Clinical Research Centre, Harrow, England / **40.3** (above) Lennart Nilsson © Boehringer Ingelheim International GmbH; (below) art by Raychel Ciemma / **40.4** Art by Raychel Ciemma / **40.6** Photograph courtesy Don C. Wiley, Harvard University; art by Raychel Ciemma / **40.7** Art by Raychel Ciemma; micrograph Morton H. Nielsen and Ole Werdlin, University of Copenhagen / **40.8** Art by Raychel Ciemma / **40.10** Art by Hans & Cassady, Inc. / **40.11** Art by Palay/Beaubois after B. Alberts et al., *Molecular Biology of the Cell*, Garland Publishing Company, 1983 / **40.13** (left) Lowell Georgia/Science Source/Photo Researchers; (right) Matt Meadows/Peter Arnold, Inc. / **40.14** (a) David M. Phillips/Visuals Unlimited; (b) David Scharf/Peter Arnold, Inc.; (c) Kent Wood/Photo Researchers / **Page 690** (a) Art by Nadine Sokol / **Page 691** (b–d) Micrographs Z. Salahuddin, National Institutes of Health

Chapter 41

41.1 Galen Rowell/Peter Arnold, Inc. / **41.4** (a) Steve Lissau/Rainbow; (b) Peter Parks/Oxford Scientific Films / **41.5** (a) David Scharf/Peter Arnold, Inc.; (b) Ed Reschke / **41.6** Art by Nadine Sokol / **41.7** After C. P. Hickman et al., *Integrated Principles of Zoology*, Sixth edition, St. Louis: C. V. Mosby Co., 1979 / **41.8** Micrograph H. R. Duncker, Justus-Liebig University, Giessen, Germany; art by Palay/Beaubois adapted from H. Scharnke, *Z. vergl. Physiol.*, 25:548–583 (1938) in *Form and Function in Birds*, Vol. 4, A. King and J. McLelland, Eds., Academic Press, 1989 / **41.9** Art by Kevin Somerville / **41.10** CNRI/SPL/Photo Researchers / **41.12** Art by K. Kasnot / **41.13** From L. G. Mitchell, J. A. Mutchmor, and W. D. Dolphin, *Zoology*, copyright © 1988 by The Benjamin-Cummings Publishing Company. Reprinted by permission. / **41.14** After A. Spence and E. Mason, *Human Anatomy and Physiology*, Fourth edition, 1992, West Publishing Company / **41.16** Art by Leonard Morgan / **Page 706** (a), (b) O. Auerbach/Visuals Unlimited / **Page 707** (c) Lennart Nilsson from *Behold Man*, © 1974 by Albert Bonniers Forlag and Little, Brown and Company, Boston / **41.17** Giorgio Gualco/Bruce Coleman Ltd. / **41.18** From Tom Garrison, *Oceanography: An Invitation to Marine Science*, Wadsworth, 1993 / **Page 711** Christian Zuber/ Bruce Coleman Ltd.

Chapter 42

42.1 (a) D. Robert Franz/Planet Earth Pictures; (b) art by D. & V. Hennings adapted from *Mammology*, Third edition, by Terry Vaughan, copyright © 1986 by Saunders College Publishing. Used by permission of the publisher. / **42.2** (a) Kim Taylor/Bruce Coleman Ltd.; (b) Wardene Weisser/Ardea, London; art by Precision Graphics / **42.4** Art by Kevin Somerville / **42.6** Art by Nadine Sokol; (b) after A. Vander et al., *Human Physiology: Mechanisms of Body Function*, Fifth edition, McGraw-Hill, 1990. Used by permission. (c) photograph D. Robert Franz/Planet Earth Pictures; sketch after A. Romer and T. Parsons, *The Vertebrate Body*, Sixth edition, copyright © 1986 by Saunders College Publishing, reprinted by permission of the publisher / **42.7** From page 763 of *Human Anatomy and Physiology*, Fourth edition, by A. Spence and E. Mason. Copyright © 1992 by West Publishing Company. All rights reserved. / **42.8** (a) Art by Victor Royer; (b), (d) Lennart Nilsson © Boehringer Ingelheim International GmbH; (c) Biophoto Associates/SPL/Photo Researchers / **42.9** Art by Robert Demarest / **42.10** Art by Raychel Ciemma / **Page 729** (left) Photograph Steven Jones/FPG;

(right) CNRI/Phototake / **42.15** Modified after A. Vander et al. *Human Physiology*, Fourth edition, McGraw-Hill, 1985; photograph Ralph Pleasant/FPG / **Page 731** Photograph courtesy of David Steinberg

Chapter 43

43.1 (a) Claude Steelman/Tom Stack & Associates; (b) David Noble/FPG / **43.3–43.5** Art by Robert Demarest / **43.6** Art by Precision Graphics / **43.8** Art by Joel Ito / **43.9** From Tom Garrison, *Oceanography: An Invitation to Marine Science*, Wadsworth, 1993 / **43.10** Thomas D. Mangelsen/Images of Nature / **43.11** (a) Bob McKeever/Tom Stack & Associates; (b) Colin Monteath, Hedgehog House, New Zealand / **43.12** David Jennings/Image Works / **43.13** Everett C. Johnson / **43.14** Art by Kevin Somerville / **43.15** The Bettmann Archive / **43.16** Terry Vaughan

Chapter 44

44.1 (a) Hans Pfletschinger; (b–e) John H. Gerard / **44.2** (a) Frieder Sauer/Bruce Coleman Ltd.; (b) Wisniewski/ZEFA; (c) Carolina Biological Supply Company; (d) Fred McKinney/FPG; (e) Evan Cerasoli / **44.4** Art by Palay/Beaubois adapted from R. G. Ham and M. J. Veomett, *Mechanisms of Development*, St. Louis, C. V. Mosby Co., 1980 / **44.5** Carolina Biological Supply Company / **44.6** (left) Photographs Carolina Biological Supply Company; (far right) Peter Parks/Oxford Scientific Films/Animals Animals / **44.8** J. R. Whittaker / **44.9** Micrographs J. B. Morrill; art by Raychel Ciemma after V. E. Foe and B. M. Alberts, *Journal of Cell Science*, 61:32, © The Company of Biologists 1983 / **44.10** Micrographs F. R. Turner; art by Raychel Ciemma / **44.11** (right) Sketches after B. Burnside, *Developmental Biology*, 26:416–441, 1971; micrograph K. W. Tosney / **44.12** (a) Peter Parks/Oxford Scientific Films/Animals Animals; (b–d) art by Hans & Cassady, Inc. adapted from W. Freeman and Brian Bracegirdle, *An Atlas of Embryology*, Third edition, Heinemann Educational Books, 1978; (e) S. R. Hilfer and J. W. Yang, *The Anatomical Record*, 197:423–433. 1980 / **44.13** (a) Art by Palay/Beaubois after Robert F. Weaver and Philip W. Hedrick, *Genetics*. Copyright © 1989 Wm. C. Brown Publishers. (b) Carolina Biological Supply Company / **44.14** From Georges Pasteur, "Jean Henri Fabre," *Scientific American*, July 1994. Copyright © 1994 by Scientific American, Inc. All rights reserved. / **Page 769** Dennis Green/Bruce Coleman Ltd.

Chapter 45

45.1 Lennart Nilsson from *A Child Is Born*, © 1966, 1977 Dell Publishing Company, Inc. / **45.2** Art by Raychel Ciemma / **45.3** Art by Raychel Ciemma; (b) micrograph Ed Reschke / **45.5** Art by Raychel Ciemma / **45.6** Art by Robert Demarest / **45.7** Photograph Lennart Nilsson from *A Child Is Born*, © 1966, 1977 Dell Publishing Company. Inc. / **45.8** Art by Robert Demarest / **45.9–45.12** Art by Raychel Ciemma / **45.13** Photographs Lennart Nilsson, *A Child Is Born*, © 1966, 1977 Dell Publishing Company, Inc.; art by Raychel Ciemma / **Page 790** (a) Art by Raychel Ciemma modified from Keith L. Moore, *The Developing Human: Clinically Oriented Embryology*, Fourth edition, Philadelphia: W. B. Saunders Co., 1988 / **Page 791** (b) From Lennart Nilsson, *A Child Is Born*, © 1966, 1977 Dell Publishing Company, Inc.; (c) James W. Hanson, M.D. / **45.14** Art by Robert Demarest / **45.15** Art by Raychel Ciemma / **45.16** Art by Raychel Ciemma adapted from L. B. Arey, *Developmental Anatomy*, Philadelphia, W. B. Saunders Co., 1965 / **Page 796** (a) CNRI/SPL/Photo Researchers / **Page 797** (b) John D. Cunningham/Visuals Unlimited; (c) David M. Phillips/Visuals Unlimited / **Page 801** Alan and Sandy Carey

Chapter 46

46.1 Antoinette Jongen/FPG / **46.2** (a) Fran Allan/Animals Animals; (b) E. R. Degginger; (c) P. J. Bryant/BPS / **46.3** (b) Stanley Flegler/Visuals Unlimited / **46.5** Photograph E. Vetter/ZEFA / **Page 810** Photograph Eric Crichton/Bruce Coleman

Ltd. / **46.6** (a) Jonathan Scott/Planet Earth Pictures; (b) Wisniewski/ZEFA; (c) Fred Bavendam/Peter Arnold, Inc. / **Page 812–813** Photographs John A. Endler / **46.7** Photograph NASA / **46.8, 46.9** Data from Population Reference Bureau and World Bank after G. T. Miller, Jr., *Living in the Environment*, Ninth edition, Wadsworth, 1995 / **46.10** Data from Population Reference Bureau after G. T. Miller, Jr., *Living in the Environment*, Eighth edition, Wadsworth, 1993 / **46.11–46.12** Photographs United Nations

Chapter 47

47.1 (left) Donna Hutchins; (right) Edward S. Ross / **47.2** (a), (c) Harlo H. Hadow; (b) Bob and Miriam Francis/Tom Stack & Associates / **47.3** After G. Gause, 1934 / **47.4** Stephen G. Tilley / **47.5** After N. Weland and F. Bazazz, *Ecology*, 56:681–188, © 1975 Ecological Society of America / **47.6** Photograph John Dominis, Life Magazine, © Time Inc. / **47.7** Photograph Ed Cesar/Photo Researchers / **47.8** (a) James H. Carmichael; (b), (c) Edward S. Ross; (d) W. M. Laetsch / **47.9** Edward S. Ross / **47.10** Roger T. Peterson/Photo Researchers / **47.11** Thomas Eisner, Cornell University / **47.12** (a) Kim Taylor/Bruce Coleman Ltd.; (b) Douglas Faulkner/Sally Faulkner Collection; (c), (d) Edward S. Ross / **47.13** Data from P. Price and H. Tripp, *Canadian Entymology*, 104:1003–1016, 1972 / **47.14** (a–f), (i) Roger K. Burnard; (g), (h) E. R. Degginger / **47.15** (a), (d) Jane Burton/Bruce Coleman Ltd.; (b), (c) Jane Lubchenco, *American Naturalist*, 112:23–29, © 1978 by The University of Chicago Press; (e) Heather Angel / **Page 839** R. Slavin/FPG / **47.16** (a) After M. Kusenov, *Evolution*, 11:298–299, 1957; (b) after T. Dobzhansky, *American Scientist*, 38:209–221, 1950 / **47.17** (above) Dr. Harold Simon/Tom Stack & Associates; (below) after S. Fridriksson, *Evolution of Life on a Volcanic Island*, Butterworth: London, 1975 / **47.18** (a) David Cavagnaro; (b), (c) After J. M. Diamond, *Proceedings of the National Academy of Sciences*, 69:3199–3201, 1972 / **Page 843** (a), (b) Edward S. Ross

Chapter 48

48.1 (a), (b) Wolfgang Kaehler / **48.2** Art by Precision Graphics / **48.3** Photograph Sharon R. Chester / **48.4** (a) Gene C. Feldman and Compton J. Tucker/NASA, Goddard Space Flight Center / **48.7** (b) Photograph © 1991 Gary Braasch / **48.8** Art by Raychel Ciemma / **48.9** (a) Photograph by Gene E. Likens from G. E. Likens and F. H. Bormann, *Proceedings First International Congress of Ecology*, pp. 330–335, September 1974, Centre Agric. Publ. Doc. Wagenigen, The Hague, The Netherlands; (b) photograph by Gene E. Likens from G. E. Likens et al., *Ecology Monograph*, 40(1):23–47, 1970; (c) after G. E. Likens and F. H. Bormann, "An Experimental Approach to New England Landscapes" in A. D. Hasler (ed.), *Coupling of Land and Water Systems*, Chapman & Hall, 1975 / **48.10** (b) Photograph John Lawler/FPG / **Page 856–857** Art by Precision Graphics; (b) after W. Dansgaard et al., *Nature*, 364:218–220, 15 July 1993; D. Raymond et al., *Science*, 259:926–933, February 1993; W. Post, *American Scientist*, 78:310–326, July–August 1990 / **48.11** (above) Photograph William J. Weber/Visuals Unlimited; (right) after E. Schulze, *Science*, 244:776–783, 1989

Chapter 49

49.1 (a) (left) Edward S. Ross; (right) David Noble/FPG; (b) (left) Edward S. Ross; (right) Richard Coomber/Planet Earth Pictures / **49.2** Edward S. Ross / **49.4** (b) Art by L. Calver / **49.7, 49.8** From Tom Garrison, *Essentials of Oceanography*, Wadsworth, 1995 / **49.9** Art by Raychel Ciemma / **49.11** Art by Victor Royer / **49.12** Art by Raychel Ciemma / **49.13** Art by D. & V. Hennings after Whittaker; Bland; and Tilman / **49.14** Harlo H. Hadow / **49.15** (above) John D. Cunningham/Visuals Unlimited; (inset) AP/Wide World Photo; (below) Jack Wilburn/Animals Animals / **49.16** (a) Kenneth W. Fink/Ardea, London; (b) Ray Wagner/Save the Tall Grass Prairie, Inc.; (c) Jonathan Scott/Planet Earth Pictures / **49.17** (left) © 1991 Gary Braasch; (right) Thase Daniel; (below) (left) Adolf Schmidecker/FPG; (right) Edward S. Ross /

49.18 Thomas E. Hemmerly / **49.19** (a) Dennis Brokaw; (b) Jack Carey / **49.20** (a) Fred Bruemmer; (b) Doug Sokell/Visuals Unlimited / **49.21** (a) D. W. MacManiman; (b) Jack Carey / **49.23** Modified after Edward S. Deevy, Jr., *Scientific American*, October 1951 / **49.24** D. W. Schindler, *Science*, 184:897–899 / **49.25** E. R. Degginger; art by D. & V. Hennings / **49.26** (a) Courtesy of J. L. Sumich, *Biology of Marine Life*, Fifth edition, William C. Brown, 1992; (b) (above) © 1991 Gary Braasch; (below) Phil Degginger / **49.27** Art by Lloyd K. Townsend / **49.28** (a) Dennis Brokaw; (b) McCutcheon/ZEFA; (c) Robert Hessler; Woods Hole Institution of Oceanography; (d) J. Frederick Grassle, Woods Hole Institution of Oceanography / **49.30** (a) Photograph Douglas Faulkner/Sally Faulkner Collection; (b) (top) Jim Doran; (chambered nautilus) Alex Kerstitch; all other photographs Douglas Faulkner/Sally Faulkner Collection / **49.31** Adapted from Tom Garrison, *Essentials of Oceanography*, Wadsworth, 1995 / **Page 889** Illustrations from Tom Garrison, *Oceanography: An Invitation to Marine Science*, Wadsworth, 1993; photographs R. Legeckis/NOAA / **Page 890** Photograph © 1991 Gary Braasch

Chapter 50

50.1 Gerry Ellis/The Wildlife Collection / **50.2** U.S. Geological Survey / **50.3** Photograph United Nations / **50.4** (a) Heather Angel; (b) USDA Forest Service; art after G. T. Miller, Jr., *Environmental Science: An Introduction*, Wadsworth, 1986, and the Environmental Protection Agency / **50.5** (a) NASA; (b) National Science Foundation / **50.6** Dr. Charles Henneghien/Bruce Coleman Ltd. / **50.7** From Water Resources Council / **50.8** Ocean Arks International / **50.9** Data from G. T. Miller, Jr. / **50.10** USDA Soil Conservation Service/Thomas G. Meier / **50.11** Agency for International Development / **50.12** (above) R. Bieregaard/Photo Researchers; (below) after G. T. Miller, Jr. *Living in the Environment*, Eighth edition, Wadsworth, 1993 / **50.13** NASA / **50.14** Data from G. T. Miller, Jr. / **Page 905** (a) Art by Precision Graphics after Marvin H. Dickerson, "ARAC: Modeling an Ill Wind" in *Energy and Technology Review*, August 1987. Used by permission of University of California, Lawrence Livermore National Laboratory and U. S. Department of Energy / **Page 907** J. McLoughlin/FPG / **Page 908** © 1983 Billy Grimes / **Page 909** David Steinberg

Chapter 51

51.1 (a) Robert Maier/Animals Animals; (b) Jack Clark/Comstock, Inc; (c) John Bova/Photo Researchers; (d) from L. Clark, *Parasitology Today*, Vol. 6, No. 11, 1990, Elsevier Trends Journals, Cambridge, U.K. / **51.2** (a) Eugene Kozloff; (b), (c) Stevan Arnold / **51.3** (a) John S. Dunning/Ardea, London; (b) sonogram J. Bruce Falls and Tom Dickson, University of Toronto; (c) Hans Reinhard/Bruce Coleman Ltd.; (d), (e) from G. Pohl-Apel and R. Sussinka, *Journal for Ornithologie*, 123:211–214 / **51.4** (a) Eric Hosking; (b) Stephen Dalton/Photo Researchers / **51.5** (a) Evan Cerasoli; (b) From A. N. Meltzoff and M. K. Moore, "Imitation of Facial and Manual Gestures by Human Neonate," *Science*, 198:75–78. Copyright 1977 by the AAAS. / **51.6** (a) Nina Leen in *Animal Behavior*, Life Nature Library; (b) F. Schutz / **51.7** (a) Michael Francis/The Wildlife Collection; (b) Ray Richardson/Animals Animals; (c) John Alcock / **Page 918** Lincoln P. Brower

Chapter 52

52.1 John Alcock / **52.2** (a) Edward S. Ross; (b) E. Mickleburgh/Ardea, London / **52.3** Art by D. & V. Hennings / **52.4** A. E. Zuckerman/Tom Stack & Associates / **52.5** Fred Bruemmer / **52.6** John Alcock / **52.7** Timothy Ransom / **52.8** Patricia Caulfield / **52.9** Frank Lane Agency/Bruce Coleman Inc. / **52.10** Kenneth Lorenzen / **Page 930** Gregory D. Dimijian/Photo Researchers / **52.11** Yvon Le Maho

INDEX

Italic numerals refer to illustrations.

A

Abdomen
  animal, 411
  human, *555*
  insect, 435
Abdominal aorta, *659*
Abdominal cavity, 411, *555*
Abiotic component, of
  ecosystems, 803
Abiotic formation, organic
  compounds, 324–325, *325, 327*
ABO blood typing, *177,* 668
Abortion
  and genetic disorders, 205
  induced, 205, 795
  methods, 798
  and population control, 817
  timing, 798
Abscisic acid, *534,* 535, 540–541
Abscission, 540
Absorption
  in capillary bed, *664–665*
  by mycorrhizae, 503
  in small intestine, 722, *722*
  solute gains by, 736
  structures in plants, 502–503
Absorption spectrum,
  photosynthetic pigments, *110*
Abstention, sexual, 794
Abyssal zone, ocean, *884*
Acacia tree, 874
Accessory organ, digestive, *717*
Accommodation, visual, 604, *605*
Accutane, 790
Acetabularia, 374
Acetylase, *237*
Acetylcholine, 569
  excitatory effects in muscle, 566
Acetylcholinesterase, 569
Acetyl-CoA, *128–129, 135*
Achondroplasia, *196,* 197
Acid
  defined, 28
  salts, 29
Acid-base balance, 742
Acid deposition, 896, *896,* 904
Acidity
  effect on flower color, *273*
  gastric fluid, 29
  stomach, 718–719
Acidosis
  defined, 29
  lung disease-related, 29
Acid rain, 337, 893, 896, *896*
  pH, 28
  sulfur dioxides in, *29*
Acid stomach, 29
Acne, 635
Acoelomate, *410*
Acorn, 277, *542*
Acorn worm, *449*
Acoustical receptor, 600
Acoustical signal, 923
Acquired immunodeficiency
  syndrome, *see* AIDS
Acromegaly, 620, *620*
Acrosome, 776, *777, 783, 783*
Actin
  arrangement in sarcomeres,
    645, *645*
  cross-bridge formation, 645
  microfilaments, 70
  role in contraction, 70, *646*
  structure, 644
  subunits in microfilaments, 69, *69*
Actinomycete, *351, 354*

Action potential
  all-or-nothing spike, 565
  defined, 562
  in muscle, 646
  at nodes, *566*
  propagation, *564,* 565, *566,*
    594–595, *595*
  recording, *565*
  threshold level, 564
  waveform, *565*
Activation energy, defined, 99
Activator protein, 237, 242
Active immunization, 688
Active transport
  ATP input, 87
  calcium pump, 87
  defined, 85
  mechanisms, 87, *87*
  occurrence, 85
  overview, *85*
  by sodium-potassium pump, 87,
    563
Acute inflammation, 678
Acyclovir, 796
Adaptation, sensory, 595
Adaptive behavior, *313,* 501,
  916–917, 931
Adaptive radiation, *296*
  and adaptive zones, 297, 334
  animals, 334, 408–409
  conifers, 401
  defined, 297
  dinosaurs, 336–338
  flowering plants, 401
  hominid, 471
  hominoid, 471, 475
  insects, 336, *340–341,* 458
  jawed fish, 453
  mammals, *297,* 336, *340–341,*
    464
  plants, 334–335, *340–341,* 501
  pouched mammals, 466
  reptiles, 336, 456–458
Adaptive response, to environmental
  conditions, *313,* 501
Adaptive trait
  defined, 7
  and mass extinction, 297
  in moths, 7, *7*
  and promotion of survival, 278
  selection, 267
Adaptive zone, 297, 334, 337, 476
Addiction, drug, 589
Addictive behavior, 575–576
Adductor longus, *643*
Adélie penguin, 844
Adenine, *211,* 212, *212,* 225
Adenosine
  diphosphate, *104*
  monophosphate, 246
  phosphate, *48*
  triphosphate, *see* ATP
    (adenosine triphosphate)
Adenosine deaminase, 246
Adenovirus, *356–357*
Adhesion junction, 72, *549*
Adhesion protein, *79,* 80
Adhesive, mussel derivatives,
  218–219
Adhesive cue, in cell migration,
  764
Adipose tissue
  components, *550,* 551
  functions, 551
  obesity, 728
  triglycerides, 40, 134
Adolescence (human), defined, *793*
Adoptive behavior, 931, *931*
ADP (adenosine diphosphate),
  structure, *104*

Adrenal cortex, hormone
  secretions, *621*
Adrenal glands
  cortex, 618, *618,* 623, *623*
  as effector, 557
  fight-flight response, 623
  hormone secretion, 614, *616*
  innervation, *580*
  role in metabolism, 731
Adrenal medulla, hormone
  secretion, *621–622,* 623, *623*
Adrenocorticotropic hormone, *618,*
  731
Adult (developmental stage), 768
  793
Adventitious root, 494, *533*
Aegyptopithecus, 475
Aerobic respiration, 331, 504, 509
  ATP transfer in, 5
  ATP yield from, *131*
  by bacteria, *111*
  carbon cycle, 854, *854–855*
  defined, 124
  and early atmosphere, 136
  energy yield, *125*
  evolution, 113
  links with photosynthesis, *106,*
    136
  mitochondrial, 124, *124*
  and muscle contraction, 647, *647*
  net energy yield, *131*
  origin, 136
  overview, 124, *125*
  and oxygen, *111,* 125, *127–129,*
    330
  reactions, *123–127, 131*
  stages, 123–126, *123–127,* 130,
    131
  summary equation for, 124
Aerosol spray, and ozone layer, 897
Afferent nerve, *579,* 580
Africanized bee, *122,* 122–123, 839
African sleeping sickness, 368
African violet, *526*
Afterbirth, 792, *792*
Agar, 372
Agaricus brunnescens, 379, 384
Agave, 493
Agent Orange, 365
Age spots, 93
Age structure
  defined, 804
  and population growth,
    816–817, *817*
Agglutination response, 668
Aggressive behavior
  anabolic steroid-related, 649
  testosterone effects, 776
Aging
  and bone turnover, 639
  and cancer resistance, 683
  and cardiovascular disorders,
    666
  and death, 768
  defined, 768, 793, *793*
  exercise effects, 793
  free radical effects, 92–93
  glandular secretion changes, 635
  and glaucoma, 607
  and posture, 640
  skin changes in, 635
  sun tanning effects, 635
  and tissue repair, 793
  visual system changes, 607
Agnatha, 452
Agriculture
  animal-assisted, 900
  converting marginal lands for,
    900
  and deforestation, 902, *902*

  and desertification, 901, *901*
  and genetic engineering, 247,
    254–255
  global cropland, *900*
  green revolution, 900
  large-scale irrigation effects,
    898, *898*
  mechanized, 900
  overgrazing, 901, *901*
  and population growth, 814, 898
  slash-and-burn, 902, *903*
  and soil salination, 898
  and strip mining, 904
  subsistence, 900
  in tropical forests, 902
Agrobacterium
  A. tumefaciens, 255
  classification, *351*
Agrostis tenuis, 291
AIDS (acquired immunodeficiency
  syndrome)
  deficient immune response in,
    689
  economics of, 796
  infection statistics, 796
  pandemic nature, 359
  spread, 796
  testing for, 796
  transmission, 691, 796
  treatment, 691, 796
Air circulation
  global, *866,* 866–867, *867*
  in smog-forming regions, *895*
Air pollutants
  and acid deposition, 896
  classes, *895*
  deforestation-related, 893, *903*
  and ozone layer, 897
  from paper manufacturing, 900
  and smog, 895
Air pollution
  deforestation-related, 893, *903*
  and directional selection, 279
  effects on mycorrhizae, 387,
    387
  effects on stomata, *509*
  and lichen death, 279, 386
  local effects, 895–896
  scale of, 893
Aix sponsa, 288
Albatross, *290,* 922, *923*
Albinism
  consequences, *196*
  defined, 178
  phenotypic treatment, 204
  rattlesnake, *179*
Albumin
  blood, 656, *656*
  egg, 46
Alcock, J., 931
Alcohol (ethyl alcohol)
  addiction, 588
  classification, 35
  effects on central nervous
    system, 588
  effects on digestive system, 718
  fetal alcohol syndrome, 791
  functional groups, 35
  and pregnancy, 790, 791
Alcoholic fermentation, 133, *133*
Aldehyde group
  locations, 35
  structural formula, 35
Alder, *834–835*
Aldosterone
  actions, *621,* 741
  effects on sodium reabsorption,
    741
  secretion, *616,* 741
  sources, *621*

Aldosterone *continued*
  targets, *621*
  and water–solute balance, 741
Alfalfa, stem structure, *489*
Algae
  blue-green, 370
  brown, 372–373, *406*, 883
  classification, 370
  golden, 370–371
  green, *111, 374–375, 375,* 376, *406,* 880–881, 883
  photosynthetic, 370
  red, 372, *372, 406,* 836–837, *837,* 883
  species, *406*
  trophic level, 847
  yellow-green, 370–371
Algal bloom, *881*
Algin, 373
Alkaloid
  fungal parasite, 383
  plant, 402
Allantois, 785
Allele
  autosomal dominant, 197
  autosomal recessive, 196–197
  codominance, 176
  combination by fertilization, 273
  combinations at fertilization, 162, 164
  crossing over effects, 161
  defined, 156, 171, 272
  dominant, 168, 171
  and heredity, 273
  of homologous chromosome, 188
  incomplete dominance, 176
  multiple allele systems, *177*
  pair, *171*
  recessive, 168, 171, *192*
  survival value, 283
  wild-type, defined, *192*
  X-linked, 198
Allele frequency
  defined, 274
  and directional selection, 278–279
  divergence, 286–287
  and evolution, 261
  and gene flow, 277
  and genetic drift, 276
  at genetic equilibrium, 274–275
  and geographic isolation, 292
  Hardy–Weinberg principle, 274–275
  mutation effects, 275
  and stabilizing selection, 280
Allele loss
  endangered species, 276–277
  and genetic drift, *176*
Allergen, 689, *689*
Allergy, 689
Alligator, *450*
Allopatric speciation, 292–293
All-or-nothing event, 564
Allosteric enzyme, 100–101
Aloe vera, 493
Alpha-1 antitrypsin, 256
Alpha cell, 622, *622,* 731
Alpha rhythm, electroencephalographic, *587*
Alpine tundra, *403, 871,* 879
Alternative splicing, *239*
Altitude sickness, 695, 708
Altruistic behavior, 916, 921, 927–931
Aluminosilicates, 325
Alveolar sac, 701, *701*
Alveolus
  gas exchange in, 704–705
  pulmonary, 701, *701*
Alzheimer's disease, 202, 587
Amacrine cell, *608*
Amanita, 383
  A. muscaria, 384
  A. ocreata, 384

A. phalloides, 384, *384*
Amazon River Basin, *903*
Amino acid
  abiotic synthesis, 324–325
  blood, *656*
  defined, 42
  essential, 725, *725*
  functional groups in, *35*
  and polypeptide chain assembly, 226
  polypeptide chains, 42
  R group, 42, *42,* 44
  and self-replicating systems, 326–327
  sequence, 42, *43*
  structural formulas for, *42*
  substitutions in sickle-cell anemia, 221
Amino acid sequence
  cytochrome *c, 308–309*
  and evolution, 326
  polypeptide chains, 221
Amino group
  buffering action, 35
  locations, *35*
  structural formula, *35*
Aminopeptidase, *720*
Ammonia
  early Earth, 324–325
  in nitrogen cycle, 858, *859*
  in urine, 736, 742
Ammonification, 858, *859*
Ammonifying bacterium, *859*
Ammonite, fossilized, *301*
Ammonium, 858, *859*
Amnesia, retrograde, 587
Amniocentesis, 204, *205*
Amnion, 205, *458, 784,* 785, 792, *792*
Amniote egg, reptile, 458
Amniotic fluid, 205, *205, 788,* 792
Amniotic sac, *788*
Amoeba
  aggregation, *365*
  defined, 366
  habitat, 366
  parasitic, 366
Amoeba proteus, 366, *366*
Amoebic dysentery, 359, 366
Amoeboid motion, defined, 70
AMP (adenosine monophosphate), formation, *104*
Amphetamine, effects on central nervous system, 588
Amphibian (Amphibia)
  brain size, *450*
  characteristics, 456–457
  classification, 446, 456
  defined, 456
  egg, 762
  evolution, 450, 456–457
  germ layer, 765
  groups, 456
  skin, 456
  water–solute balance, 743, *743*
Amplexus, *752*
Amplification, DNA, *238,* 249–250, *249–250*
Ampulla, 438, *438–439*
Amygdala, *585*
Amyloplast, 73
Amylose, structure, 39
Anabaena, 351
Anabaena, 352, 858
Anabolic steroid, 41, 649
Anaerobic electron transport, 133
Anaerobic pathway, 330
Anagenesis, 296
Anal canal, 715
Analgesic, narcotic, 589
Analogous structure, 306–307
Anal pore, 367
Ananas comosus, 523
Anaphase
  mitotic, *147,* 149

sister chromatid separation, 149, *149*
Anaphylactic shock, 689
Anatomy
  comparative, 262–263
  defined, 547
Anchoveta industry, 888–889
Androgen
  actions, *616, 621*
  secretion, 626
  sources, *621*
  targets, *621*
Anemia, sickle-cell, 220–221
Angina pectoris, 667
Angiography, 667, *667*
Angioplasty, 667
Angiosperm, *see also* Flowering plant; Plant (Plantae)
  characteristics, 403
  classification, 391, *406*
  diversity, 403, *403*
  evolution, 336, 393, *393,* 483
  extinction, 403
  parasitic, 403
  radiations, 393, *393*
Angiotensin
  actions, *621*
  sources, *621*
  targets, *621*
Angler fish, 312
  sexual dimorphism, *312*
Angracecum sesquipedale, 521
Anhidrotic ectodermal dysplasia, 241
Animal (Animalia)
  acoelomate, *410*
  adaptive radiations, 408–409
  body cavities, 411, *555*
  body plans, 410–411
  body symmetry, 410–411
  body temperature, 744, 746
  Cambrian, *334*
  cephalization, 411, 576–577
  circulatory system, 654
  classification, 9
  classification, temperature-based, 746–747
  coelomate, *410*
  as consumer, 5
  defining characteristics, 410, *442–443*
  DNA viruses, *357*
  egg, 756
  evaporative water loss, 745, *745*
  evolution, 136, 333–334, 408, *412,* 444–446, 578, 734–735
  five-kingdom classification, 318, *318*
  gamete formation, 162
  genetic engineering, 256, *256, 256–257*
  gut, 411
  life cycle, *162,* 410
  organ systems, *554–555*
  polyploidy in, 295
  pseudocoelomate, *410*
  representative, *9*
  reproductive isolation mechanisms, 286–287
  ruminant, 715
  segmentation, *577*
  spermatogenesis, generalized, *163*
  with sweat glands, 744, *744*
  without sweat glands, 744, *744, 749*
  world distribution, 262
Animal-assisted agriculture, 900
Animal behavior, defined, 912
Animal cell
  ATP formation lactate fermentation, 132
  cleavage, 150, *151*
  components, 60, *60–61, 73*
  cytoplasmic division, *151*

diploid, *146–147*
  division mechanisms, 150, *151*
  endocytosis, 88
  intercellular material, 72
  organelles, *60, 65*
  pinocytosis, 88
  vs. plant cell, 60
  RNA viruses, 357
  transport proteins, 86
Animal development
  aging and death, 768
  and cell differentiation, 764
  characteristics, 423
  cleavage, 423, 756
  comparative embryology, 304
  deuterostome, 423
  early embryo, 762–763
  and fertilization, 754–756
  gametogenesis, 756
  gastrulation, 756
  information in egg cytoplasm, 760
  metamorphosis, 768
  and morphogenesis, 764
  organogenesis, *757,* 764–765
  overview, 754–756, *756*
  post-embryonic, 768
  protostome, 423
  stages, *757*
Animal pole, 760
Animal reproduction
  asexual, 754
  chick, *759*
  cytoplasmic information, 760
  cytoplasmic reorganization, 760
  developmental stages, 756, *756–757*
  embryo formation, 762–763
  energy outlay for, 754–755
  frog, *758*
  modes, 754
  organogenesis, 764–765
  pattern formation, 766–767
  sexual, 754
  strategy problems, 754–755
  tissue formation, 764–765
Annelid (Annelida)
  adaptations, 428
  body plan, *410*
  circulatory system, *443*
  digestive system, *443*
  evolution, *412,* 429
  groups, 428
  movement, *442*
  nervous system, *442*
  representative, *410*
  reproduction, *443*
  respiratory system, *443*
Annual growth layers, tree rings, 497
Annual plant, defined, 496
Anopheles, 369
Anorexia nervosa, 729
Ant
  altruistic behavior, 928–929
  species diversity, 840
  species introduction, *838*
  vs. termite, 920, 922
Antacid, 29
Antagonistic muscular action, 642, 646
Antarctica
  disruption, *844,* 844–845, *845*
  food web in, *847*
  krill, 844–845, *847,* 884
  ozone hole over, 897, *897*
Antelope, *714,* 714–715, *719*
Antenna
  crustacean, 434
  rotifer, *423*
antennapedia gene, 767
Anterior (anatomy), 410, *411, 555*
Anteriorlike cell, *365*
Anther, *404, 516,* 518, *519*
Anthozoan, 414, *415,* 417

Anthrax, 674
Anthropod, evolution, *412*
Anthropoid, 471–472, *472*, 475
Antianxiety drug, 568
Antibiotic
  defined, 354
  effectiveness, 355
  functions, 354
  from fungus, 378
  overprescription, 355
Antibody
  agglutination response, 668
  defined, 680
  diversity, 686, *686*
  formation, 680
  functions, *48*, 668
  heavy chain, *686*
  immunoglobulins, 685
  light chain, *686*
  monoclonal, 683
  and pregnancy, 668–669
  during pregnancy, 790
  production by B lymphocytes,
    238, 684–685
  recombined genes, *686*
  structure, *686*
Antibody-mediated immune
    response, *681*, 684–685, *685*
Anticodon, tRNA, 225
Antidiuretic hormone, *see also*
    Aldosterone
  actions, *618*
  inhibition effects, 741
  secretion, 741
  targets, *618*
  and water–solute balance, 741
Antigen
  antibody binding, *686*, 686–687
  defined, 680
Antigen–MHC complex, 680, 682,
    683
Antigen-presenting cell, 680, 685
Antilocapra americanus, *714*,
    714–715
Antitrypsin, 706
Antocyanin, *111*
Anus, 717
  deuterostome, 423
  gastropod, *424*
  mollusk, *424*
  protostome, 423
  rotifer, *423*
  sea star, *438–439*
  snail, *424*
  squid, *427*
Aorta, 660, *660*
Ape
  characteristics, 334–335
  classification, *472*
  evolution, 336–337
  skeleton, *473*
Apex, of heart, *660*
Aphid, *155*, 255, 510, *510*
Apical dominance, 535
Apical meristem, 485, *485*, 492,
    492, 494, *533*
Aplysia, *425*, 425
Apolipoproteins
  Apo E2, 33
  Apo E3, 33
  Apo E4, 33
  genes for, 33
  lipoprotein formation, 33
Appeasement behavior, 926,
    926–927
Appendage
  arthropod, *435*
  jointed, 432, *435*
Appendicitis, 723
Appendicular skeleton, 640
Appendix, *670*, 722, 723
Apple, 522–523, *523*
Apple blossom, *523*
Apple scab, *383*
Aquatic ecosystem, *see also* Lake

bacteria, *351*
  coastal, 882–883
  diversity, 880
  estuary, 882, *882*
  food web, 854
  lake, 880–881
  open ocean, *884*, 884–885
  stream, 882
Aqueous body (humor), *604–605*,
    607
Aqueous humor
  buildup in glaucoma, 607
  and retinal detachment, 606
Aquifer, *898*
Arceuthobium, *403*
Archaeanthus linnenbergeri, *336*
Archaebacteria, 330, *351*,
    352–353, 357
Archaeopteris, *335*
Archaeopteryx, 268, *268*, 462
Archean, 323, 329, *329*, 330
Archenteron, *758*, 762, 763
Archosaur, *456*
Arctic fox, *465*
Arctic tundra, *870*, 879, *879*
Area effect, species diversity, 841,
    841
Arenavirus, *357*
Areola, *792*
Argentine fire ant, *838*
Arid grassland, *see* Grassland,
    arid
Aristotle, 262
Armadillo, 266
Armillaria, 384, 543
Armor plate, 334
Arnold, S., 912
Arrhythmia, 667
Arteriole
  basement membrane, *662*
  cardiovascular, 658
  diameter changes, 662–663
  elastic layers in, *662*
  endothelium, *662*
Arteriosclerosis, 667
Artery
  atherosclerotic plaque, 667
  basement membrane, *662*
  blood pressure in, 664–665
  coronary, *660*
  defined, 658
  elastic layers in, *662*
  endothelium, *662*
  hardening, 666
  major, diagram, *659*
  pulmonary, *660*
Arthritis
  defined, 641
  osteoarthritis, 641
  rheumatoid, 641, 689
Arthrobotrys dactyloides, *388*
Arthropod (Arthropoda), *see also*
    Insect; Spider
  adaptations, 432
  characteristics, 432
  circulatory system, *443*, 654
  digestive system, *443*
  as disease vectors, 359
  evolution, *412*
  major groups, 432
  movement, *442*
  nervous system, *442*
  representative, *410*
  reproduction, *443*
  respiratory system, *443*
  sensory organs, 432
Artificial selection, 10, 247, 267,
    270–271, 630
Asbestos
  and air pollution, 895, *895*
  carcinogenicity, 244
Ascending aorta, *659*
Ascocarp, 382, *382*
Ascorbic acid, *726*
Ascus (asci), formation, *382*

Asexual reproduction, *see also*
    Reproduction
  clones, 156
  defined, 754
  eukaryote, *142*
  flatworm, 754
  flowering plants, 516, *526*,
    526–527, *527*
  fungus, 380, *381*
  induced propagation, 526
  in nature, 526
  offspring, 156
  planarians, 418
  and polyploidy, 295
  prokaryote, *142*
  sac fungus, 382
  sponge, *413*, 754
Ash tree, *111*, 877
Aspen tree, *111*
Aspirin, 719
Aspartame, 204
Aspergillus, 388
Assassin bug, 922
Association, defined, 587
Association cortex, *585*, 586
Asteroid
  Galileo photograph of, *337*
  Manicougan crater, *337*
  and mass extinction, 337–338
Asthma
  allergic response, 689
  drugs for, 623
Astigmatism, 606
Atherosclerosis, 33, 666–667
Atherosclerotic plaque
  appearance, *33*, 666
  formation, 33, 667
Athlete, steroid use, 41
Athlete's foot, *383*, 676
Atmosphere
  carbon dioxide concentrations,
    352, 357, 854, *856*, 857, 903
  and climate, 866–867, *867*,
    888–889
  composition, 136, 894
  defined, 865
  early, 324, 330
  greenhouse gases in, *856*,
    856–857, *857*
  and hydrologic cycle, 852, *852*
  nitrogen concentration, 858
  and nitrogen cycle, 858
  pollution, 893, 895–897,
    895–897, 903
  pressure, 696, *696*
  structure, 865, *866*
  thermal inversion, 895, *895*
Atmospheric cycle
  carbon, 854–855, *854–855*
  defined, 851
  nitrogen, 858–859, *859*
Atoll, 886, *886*
Atom
  carbon, 34–35, *34–35*
  in chemical bonds, 22
  components, 19
  defined, 19
  electronegative, 25
  ionization, 24
  isotopes, 19
  negatively charged, 24
  positively charged, 24
  structure, 19, 22, *23*
Atomic number
  common elements, *19*
  defined, 19
Atomic weight, defined, 19
ATP (adenosine triphosphate)
  in active transport, 87, *87*
  energy transfer by, 4–5
  formation, 104, 504
  functions, 46, *48*, 104
  in muscle contraction, 647
  phosphorylation by, 104
  in plant function, 504

role in muscle contraction,
    645–646
  structure, 104, *104*
  yields of Africanized honeybee,
    123
ATP/ADP cycle
  defined, 104
  energy transactions in, 104
ATP formation
  by aerobic respiration, 124–131,
    127
  by alcoholic fermentation, 133,
    133
  by anaerobic electron transport,
    124, 133
  by bacteria, 352
  chemiosmotic theory, *114*, 130
  in chloroplast, 68, 114
  and concentration gradients,
    114
  cyclic pathway, 112
  and electric gradients, *114*
  by electron transport
    phosphorylation, 126,
    130–131
  in electron transport
    phosphorylation, *131*
  and exercise, 648
  by fermentation, 124, 132–133
  by glycolysis, 126
  by lactate fermentation, 132,
    132
  mitochondrial, 67, *67, 73*
  in muscle cells, 647, *647*
  during muscle contraction,
    645–646
  in muscles, 647
  noncyclic pathway, 112,
    113–114
  during photosynthesis, 68
  by substrate-level
    phosphorylation, 126
  yield from glucose metabolism,
    131
ATP synthase, *128*, 130, *130*
Atrial natriuretic peptide
  actions, *621*
  sources, *621*
  targets, *621*
Atrioventricular valve (AV valve),
    660, 660–661
Atrium
  contraction, 661
  heart, 660, *660*
  SA (sinoatrial) node, 661
Atropa belladonna, *493*
Atropine, 493
Auditory cortex, *585*
Auditory nerve, *600–601*
Aurelia, *415*
Australian biogeographic realm,
    870
Australian sawfly caterpillar, 925,
    925
Australopith, *476*, 476–477, *477*
Australopithecus
  A. afarensis, 476, *477*
  A. africanus, 476, *477*
  A. boisei, 476–477
  A. ramidus, 476, *477*
  A. robustus, 476–477
Autoimmune disorder, 623, 689
Autoimmune response
  and aging, 768
  in type 1 diabetes, 623
Autonomic nervous system, *579*
  innervation, *580*
  motor axons, 580–581
  parasympathetic nerves, *580*
  parasympathetic outflow, *580*
  somatic system, 580
  sympathetic nerves, *580*
  sympathetic outflow, *580*
Autosomal dominant inheritance,
    196, 196–197

Autosomal recessive inheritance, *196*, 197
Autosome, defined, 189
Autotroph
  chemosynthetic, *351, 846*, 858, 885
  defined, 107
  photosynthetic, 845–846, *846–847*
Auxin, 534, *534*, 536–537
  synthetic, 365
AV valve (atrioventricular valve), *660*, 660–661
Axial skeleton, 640
Axolotl, 457
Axon
  bundles, 570
  giant, *565*
  interneuron, *571*
  motor, *579*
  motor neuron, *562*
  myelinated, *570–571*
  myelin sheath, 566, *566*
  nerve tracts, 570
  sensory, *579*
  sensory neuron, *571*
  structure, *571*
Azidothymidine (AZT), 691
Azospirillum, 353
Azotobacter, 858
AZT (azidothymidine), 691

B

Baboon
  appeasement behavior, 926, *926*
  vs. leopard, *828*, 831, 926
  threat display, *922*, 923
Bacillus
  appearance in micrographs, *346*
  B. anthracis, *674*
  rod-like shape, 348
Backbone, *641*
  carbon, 34–35
  mammalian, 464
  phospholipid, 41
Background extinction, 297
Bacterial infection
  antibiotic-resistant, 355
  eye diseases caused by, 606
  pseudomonad, 354
  spirochete, *354*
  teeth, *718*
Bacteriophage
  appearance in micrographs, *211*
  components, *211*
  defined, 210, 356
  infectious cycle, *211*
  lysogenic pathway, 358, *358*
  lytic pathway, *358*
  multiplication cycle, *358*
  mutations, *228*
  radiolabeling, *211*
  as research tools, *211*, 356
  T4, *211, 357*
Bacteriorhodopsin, 352
Bacterium, *see also* Moneran; Prokaryote
  aerobe, 355
  ammonifying, 858, *859*
  anaerobe, *351*, 355
  anaerobic, 569
  anaerobic electron transport, 124, 133
  in animal gut (enteric), *351*
  antibiotic resistant, 103, 355, 796
  appearance in micrographs, 355
  archaebacteria, 348, *351*
  autotrophic, *351*
  binary fission, 350, *350*
  biotic potential, 807
  body plan, *348*
  capsule, *348*

cell wall, 56, *211, 348*, 349, *351*
characteristics, *8, 348*
chemical barriers, *676*
chemoautotrophic, 348, *351*, 885
chemoheterotrophic, 348, *351*
chromosome, *248, 348*
classification, 351, *351*
collective behavior, 355
conjugation, 350, *350*
connectedness with other organisms, 347
cytoplasm, 56–57, *348*, 349
as decomposer, 5, 353
denitrifying, 858, *859*
disease-causing, 56
diversity, 57
DNA, 57, *348*
endospore-forming, *351*
eubacteria, 348, *351*, 352–353, *353*
evolution, 362
flagella, 56, *56, 348*, 349
fruiting body, 355
genetic engineering, 254, *254*
gene transfers using, 103
Gram-negative, *348, 351*
Gram-positive, *348, 351*
habitats, *351*, 352
halophiles, *351*, 352
harmful, 676
heterotrophic, *351*
ice-minus, 254, *254*
iron-oxidizing, *351*
lactate-producing, 133
lineages, 348
magnetotactic, 355
major groups, *351*
metabolism, 348, 352–355
methanogens, *351*
mutualistic relationship with plants, 502–503
nitrifying, *351*, 353, *353*, 858, *859*
nitrogen-fixing, 502, 858, *859*, 881
parasitic, 348, *351*, 362
pathogenic, 346, *351*, 354, *354*
photoautotrophic, 348, *351*
photoheterotrophic, 348
photosynthetic, *351*, 355
physical barriers, *676*
pili, *348–349*
plasma membrane, 56, *348*, 349
restriction enzymes, 249
ribosome, 57, 226
saprobic, 348, *351*, 362
shapes, 348–349
simple, 355
size, 56, 348
skin, UV radiation effects, *635*
sulfate-reducing, 133
sulfur-oxidizing, *351*
superbugs, 355
thermoacidophile, *351*
thermophilic, 352
toxins, 354
transfer of antibiotic-resistant genes, 354
transformation, 210, *210*
transport proteins, 86
Balance, mechanoreceptors, 596
Balanced polymorphism, defined, 282–283
Balanced polymorphism, defined, 282–283
Baleen whale, 844
Balloon angioplasty, 667
Balsam fir, 878
Baltimore oriole, 295
Banana slug, 912, *912*
Barbiturate, 790
Barbule, *463*
Bark, 496, 497
Bark beetle, *838*
Barnacle, 434, *434*, 836, 882
Barometer, *696*
Barr, M., 241

Barr body, 241, *241*
Barrier
  chemical (bacterial), *676*
  to gene flow, 289–292
  integumentary, *676*
  physical (bacterial), *676*
  reproductive isolating mechanisms, 289–292
Barrier reef, 886, *886*
Barrier to gene flow, 289–291
Basal cell carcinoma, *215, 234*, 635
Base
  defined, 28
  nitrogen-containing, *211*, 221
Basement membrane, *548*, 550
  arteriole, *662*
  artery, *662*
  capillary, *662*
  vein, *662*
Base pairing, DNA
  characterization, 213, *213*
  and diversity of life, 229
  errors in, 214
  nucleic acid hybridization, 252–253, *253*
  in replication, 222
  substitution, 228, *228*
  transcription, 222
Base sequence, DNA, 212–213, *214*, 221
  enhancers, 242
Basidiocarp, 385, *385*
Basidium, 385
Basilar membrane, cochlear duct, 600–601
Basket sponge, *413*
Basophil, *656–657*, 678
Bat, *466*, 524, *524, 592*, 592–593, 923
  fossil, *302*
  and morphological divergence, 306
  wing, 263, 306
Bateson, W., 179
Bathyal zone, ocean, *884*
Bayliss, W., 614
B cell, *see* B lymphocyte
Beach hopper, 883
Beadle, G., 220
Beak, bird, 267, 280–281
Bean plant, *533*, 538
Bear, *309*, 310, 464, 832
Bear lineage, *309*
Bedstraw plant, 826–827
Bee, *521*
  Africanized, *122*, 122–123, 839
  honeybee, 923, *923*, 928–929, *928–929*
  pollinators, 521
Beech tree, 877
Beeswax, 41, *41*
Beetle, *437*
  bombardier, 831–832, *832*
  Japanese, *838*
  ladybird, 255
  as pollinator, 521
Behavior, *see also* Social behavior
  adaptive, 916–917, 931
  adoptive, 931, *931*
  altruistic, 916, 921, 927–931
  amplexus, *752*
  animal, defined, 912
  anti-predator, 831, *831*, 925
  appeasement, 926, *926–927*
  bacteria, 355
  camouflaging, 830, *830*
  chewing, and floral structures, 521
  cooperative defense, 924–925, *925*
  courtship, *see* Courtship behavior
  defensive, 460
  defined, 911
  dominance hierarchies, 926, *927*

and ecology, 803
feeding, *847*, 912, 916
fight-flight, 623
genetic, *912*, 912–913, *913*
grazing, 837
herding, 925, *925*
hormonal effects on, *618*, 623–624, 912–913
human, 471, 474, 478–479
imprinting, 914, *915*
instinctive, 911, 914, *914–915*
learned, 911, 914, *915*
learning, 474
mimicry, 831, *831*
motor, *582*
and natural selection, 916–917
parental, *755, 927, 927*, 931, *931*
reproductive, 287, 754–756
as reproductive isolating mechanism, 286
safe sex, 797
selfish, 916, 925
self-sacrificing, 916, 926–931
self-subordinating, 926, *926–927*
sexual, 287, 612–613, 625, 776, 794–797
social, 340, 463, 612–613
territorial, 916
and thermal regulation, 746, *747*, 747–749
warning, 831, *831*
Behavioral isolation, *289*, 290, *290*, 313
Behavioral temperature regulation, 746
Behavioral trait, 272
Belief systems, scientific, 12–13
Belladonna, 493
Bell-shaped curve, for trait variation, 180, *181*
Bends (decompression sickness), 709
Benthic province, 884, *884*
Benzene, 895
Beta-amyloid protein, 202
Beta-carotene, 727
Beta cell, 622–623, *622*, 731
Bicarbonate
  buffering action, 29
  formation, 705
  functions, 721
  pancreatic secretion, 721, 723
  in saliva, 718
Bicarbonate–carbon dioxide buffer system, 742
Biceps
  bone interaction, 642
  contraction, *646*
  stretch reflex, *571*
  tricep interaction, *642*
Biceps brachii, *643*
Biceps femoris, *643*
Biennial plant, 496, 541
Bighorn sheep, *9*
Bilateral symmetry, 410, *411–412*, 418, 576–577, 786
Bile, 721
Bile acid, 725
Bile salts, 41, 721
Binary fission, *350, 376*
Binge-purge disorder, 729
Biochemistry, comparative, 308–309, 390
Biodiversity, *see* Diversity of life; Speciation; Species diversity
Biogeochemical cycle
  atmospheric, 851
  carbon, 854–855, *854–855*
  defined, 845, 851
  hydrologic, 851, *852*, 852–853
  model, 851, *851*
  nitrogen, 858–859, *859*
  phosphorus, 860, *860*
  sedimentary, 851

Biogeographic realms, 870, *870*
Biogeography, defined, 262
Biological clock, 6, 538, 625
Biological controls, 279
Biological diversity, *see* Diversity of life; Speciation; Species diversity
Biological imperative for human beings, 906–907
Biological magnification, 861
Biological molecule, *see also* Carbohydrate; Lipid; Nucleic acids; Protein
　characteristic of life, 36
　covalent bonding, 24–25
　hydrogen bonding, 25, *26*
　ionic bonding, 24
　oxygen effects on, 324, 330
Biological perspective
　on human gene therapy, 246–247, 257
　on life in the biosphere, 123
Biological science, emergence, 260, 262
Biological species concept, 288
Biological vector, 359
Bioluminescence, 103, *103*, 255, *351*, 923
Biome, *see also* Water provinces
　defined, 865, 870
　environmental gradients, *870–871*
　global distribution, *871*
　soils, 872, *872*
　types, 870–879
Biosphere, *see also* Biome; Climate; Earth; Water provinces
　defined, 4, 803, 865
　energy flow through, *5*
　human impact on, 892–908
　unity of life throughout, 123
Biosynthesis, energy for, 326
Biotic component, of ecosystems, 803
Biotic potential, 807
Biotin, *726*
Bipedalism, 338, 473, 476–477
Bipolar cell, *562, 608*
Birch tree, *111, 877*
Bird (Aves)
　body plan, 463, *463*
　bones, 463, *463*
　brain size, *450*
　camouflage by, *830*
　classification, 446, 462
　courtship behavior, *917, 922,* 923
　and DDT, 861
　development, *755, 765*
　egg, 756, 762
　egg shells, 861
　endotherms, 746
　and estuaries, 882
　evolution, *450, 459*
　evolutionary link with reptiles, 462
　feathers, 462–463, *463*
　flight requirements, *462–463,* 463
　germ layer, 765
　heart, 463
　heterotherms, 747
　imprinting, 914, *915*
　and morphological divergence, 306
　nest infestations, 910–911, *910–911*
　oviparous, *755*
　patterns of species diversity, *840–841*
　pollinators, *520–521,* 521
　reproduction, 755
　respiratory system, 699, *699*
　response to cold stress, 748
　singing by, 912–913, *913*
　social parasitism, 833, 914, *914*
　song, 463

survivorship curves, 811, *811*
　wing, 463, *463*
Bird of Paradise, *283*
Birth canal, *778*
Birth control
　in China, 817, *817*
　in India, 802–803
　methods, 794–795
Birth control pill, 535
Birth defect
　dioxin-related, 365
　toxoplasmosis-related, 368
Birth rate
　and economic development, 818–819
　and family planning, 816–817
　lowering, 816–819
　and population growth, 806–807
Birth weight, human, 280, *280*
Bison, *917*
Biston betularia, 278–279, *279*
Bittern, *830*
Black bear, 310
Black-bellied seedcracker, 280
Black bread mold, 220, *381, 383*
Black spot, 383
Black stem wheat rust, *383*
Black widow spider, *433*
Bladder
　cancer, 707
　rotifer, *423*
　urinary, *580*
Bladderwort, *500,* 500–501
Blade (sporophyte), *373*
Blastocoel, *758, 762, 762*
Blastocyst, *793*
　human, *784–785,* 785
　mammalian, 762
　structure, *784–785,* 785
Blastoderm, 762, *763*
Blastodisk, *757*
Blastopore, *757*
Blastula, *758–759, 762, 762*
Blindness
　bacterial disease-related, 606
　red-green, *196*
　viral infection-related, 606–607
Blind staggers, ergotism-related, 383
Blister, *635*
Blood
　agglutination response, 668
　carbon dioxide transport, 704
　cell count, 657
　cellular portion, 656, *656*
　clotting, *192*
　coagulation, 668
　components, 551, 656, *656*
　contaminated, 691
　distribution, 663
　filtration, *664–665,* 740
　functions, 551, 656, *656*
　glucose levels, 86, *622,* 731
　hemostasis, 668
　in intestinal villi, *722*
　leatherback turtle, *710,* 710–711
　low-density lipoprotein levels, 667
　oxygen level, sickle-cell anemia effects, 176
　oxygen transport, 704
　pH, 28
　plasma, 556
　plasma component, 656, *656*
　self markers, *177*
　sugar levels, 6, 39
　transfusion, 668
　volume in body, 656
Blood circulation, *see also* Blood flow; Circulatory system
　cardiac, 660–661, *661*
　in closed systems, *655*
　fetal, 787, *787*
　flow controls, 663
　flow rate, 656

in open systems, *655*
　placental, 787, *787*
　pulmonary, 658, *658*
　skin, 748–749
　systemic, 658, *658*
Blood clotting
　fibrinogen role, 656
　and hemostasis, 668
　plasma proteins in, *668*
　platelet role, 657
　thrombin effects, 98
Blood filtration, renal, 739, *739*
Blood flow, *see also* Blood circulation; Circulatory system
　control of, 706–707
　renal, *739–740*
Blood fluke, 420, *420*
Blood pressure
　arterial, 662–664, 666
　in arterioles, 664
　in capillary bed, 664
　controls over, 663
　diastolic, 660, *662, 664–665*
　in hypertension, 666
　in hyperthyroidism, 624
　kidney malfunction effects, 742
　measurement, *662–665*
　and obesity, 666
　risk factor in cardiovascular disease, 666
　and smoking, 666
　systolic, 660, *662*
　vascular system, 662–663, *662–663,* 664
　venous, 664
Blood typing, 668–669
　ABO typing, 668
　Rh typing, 660, 668–669
Blood vessel
　earthworm, 429
　in intestinal villi, *720*
　lung, 706–707
　major, in circulatory system, *659*
　renal, *738*
　vasoconstriction, 663
　vasodilation, 663
Bloom, of dinoflagellates, *371*
Blue crab, 883
Blue-footed booby, *264*
Bluegill sunfish, *925–926*
Blue-green algae, 370
Blue-headed wrasse, *293*
Blue whale, *847*
B lymphocyte
　activated, 235
　antibody production, *238,* 684–685
　antigen-specific receptor production, 686–687
　Burkitt's lymphoma, 235
　clonal selection, 687, *687*
　functions, 680
　immune functions, 657
　interaction with T lymphocyte, *681*
　memory, 685, 687, *687*
　origin, 684
　virgin, *681*
Boa constrictor, 461
Boar, wild, *838*
Body-builder's psychosis, 649
Body plan
　bacterial, *57,* 348
　bilateral symmetry, 410–411
　bird, 463, *463*
　bony fishes, 454–455
　cavities, 411
　cephalization, 411
　clam, *426*
　coevolution with cephalization, 411
　cuttlefish, *427*
　earthworm, 428, *429*
　echinoderm, 438, *438–439*
　emergence in vertebrates, 786

Escherichia coli, 56
　euglenoid, *370*
　flattened, evolution, 334
　flatworm, 418
　fungal, 380, *380*
　insect, 435, *435*
　invertebrate, evolutionary trends, *440*
　lancelet, *448*
　molluscan, *424*
　Paramecium, 367
　platypus lineage, 443
　prokaryotic, 56
　radial symmetry, 410
　red algae, 372
　sea anemone, *415*
　segmentation, 411
　snail, aquatic, *424*
　spider, 433
　sponge, *413*
　surface-to-volume constraints, 53
　surface-to-volume ratio, 696
　symmetry, 410–411
　tunicate, *447*
　turtle, 460
　vertebrate, 786
　and vestigial bones, 263
　virus, *356*
Body weight
　and energy needs, 728–729
　ideal, *728,* 728–729
　percent fat, 729
Bog
　acidic, 878
　formation, *395*
Bol, M., 620
Bolus, 718
Bombardier beetle, 831–832, *832*
Bombykol, 598
Bond, chemical
　defined, 22
　electron transport, 102
　strained, 98
Bone marrow, *685*
　cell cycle in, 144
　functions, 670
　immune function, *685*
　transplants, 247
Bone (osseous tissue)
　bird, 463
　calcium, 624
　classification, 638
　compact, *551,* 638, *638*
　defined, 550–551
　development, 639
　disorders, 624
　functions, 551, 638
　ground substance, 638
　Haversian system, *638*
　hormonal effects on, 624
　intercellular material, 72
　lamellae, *638*
　long, *638,* 638–639
　loss, 727
　mature, 72
　pectoral girdle, *641*
　pelvic girdle, *641*
　red marrow, 638
　rib cage, *641*
　and skeletal structure, 641
　skull, *641*
　spongy, *551,* 638, *638*
　structure, 638, *638*
　turnover, 639
　vertebral column, *641*
　vestigial, and body plans, 263
　Volkmann's canal, *638*
　yellow marrow, 638
Bonnemalisonia hamifera, 372
Bony fish, 426, 454, *454,* 743, 754, 756
Book lung, 433
Boreal forest, *870–871,* 878, *878*
Boron, in plant function, *502*
Borrelia burgdorferi, 351, 354

Bothrops asper, 310, *310*
Bottleneck, 276–277, 296
Botulism, 354, 569
Bowman's capsule, 738, *739*
Brachial artery, *659*
Bradycardia, 667
Bradyrhizobium, 503
Brahma chicken, 179, *179*
Brain
   auditory centers, 582
   canals, 583
   cavities, 583
   development, 764
   and drug addiction, 588
   earthworm, *429*
   electroencephalograms, *587*
   embryonic, *578*
   emotional, 588, *588*
   evolution, *450,* 450–451, 464,
     578, 582
   flatworm, *418*
   forebrain, *582,* 583
   functions, 582
   and gas exchange, 707
   hemispheres, 582
   hindbrain, 582, *582*
   human, 582–585
   information processing, 556–557
   as integrator, 556
   invertebrate, *577*
   left, 584
   mammalian, 464
   and memory, 582–583
   midbrain, *580, 582,* 582–583
   PET scanning, *21*
   posterior view, *579*
   respiratory centers, 705
   right, 584
   rotifer, *423*
   sagittal section, *584*
   subdivisions, *582, 582*
   vertebrate, *582,* 585
   vertebrate, evolution, 578
   visual centers, 582
   visual perception in, *609*
Brain size
   early *Homo,* 474, 478
   hominids, 471
   Homo erectus, 478
   Neandertal, 478–479
   during primate evolution, 471,
     474, 478–479
Brain stem, 582
Bread mold, 220, *381, 383*
Breast
   anabolic steroid effects, 649
   lactation, 792, *792*
Breast cancer, 793
Breathing, *see also* Respiratory
   system
   air pressure changes during, 702
   exhalation, 702
   inhalation, 702
   neural control, *582*
   and sound production, 703
Breeder reactor, 906
Brenner, S., 256
Brinster, R., 256
Bristle
   earthworm, 428, *429*
   polychaete, 430, *430*
Bristly foxtail, 827, *827*
Brittle star, 438, *442–443*
Broadleaf forest, *see* Deciduous
   forest, temperate; Deciduous
   forest, tropical; Tropical
   rain forest
Broca's area, *585*
Bromeliad, *876*
Bronchial tree, 701
Bronchioles
   defined, 701
   functions, 701
Bronchitis, 706–707
Bronchus, defined, 700

Brown, R., 50
Brown algae (Phaeophyta), *362,*
   370, 372–373, *373, 406,* 883
Brown bear, 310
Brown recluse spider, *433*
Brown rot, *383*
Bryophyte, 394, *394*
   characteristics, 394
   classification, 391
   comparison with other plant
     groups, *406*
   evolution, *393,* 394
   species diversity, 391
Bubonic plague, 809
Bud
   arrangement, *492*
   defined, 492
   dormancy, 541, *541*
   lateral, 492
   terminal, *484, 492*
Budding, fungal, *388*
Buffer
   bicarbonate-carbon dioxide
     system, 742
   defined, 29
   hemoglobin, 705
   and pH, 29
   systems, 29
Buffon, G.L. de, 263
Bufo americanus, 310
Bulb, *526*
Bulbourethral glands
   functions, 774, *774–775*
   location, *774–775*
   secretions, 774
Bulimia, 729
Bulk
   dietary, 724
   intestinal, 723
Bulk flow, defined, 83
Bullock's oriole, 295, *295*
Bull's eye rash, *354*
Bundle sheath cell, *118,* 119
Burgess Shale, *334, 408,* 408–409
Burkett, D., 724
Burkitt's lymphoma, 235
Burnet, M., 687
Burning
   forests, 857, 893, *902,* 903
   fossil fuel, 854, 857, 895
Bursa, 640
Butcher, S., *630,* 630–631
Butter, 725
Buttercup, *182, 495,* 831
Butterfly, 280, *312,* 429, 435, 508,
   831, 866, 866
Byssus, 218–219

C

Cabbage, 530
Cacao, *514,* 514–515
Cactus, *524,* 524–525, 873, *873*
Caecilian, 457, *457*
Caffeine, effects on central
   nervous system, 588
Calcitonin
   actions, *621*
   targets, *621*
Calcium
   atomic number, *19*
   and bone structure, 639
   deficiency, *727*
   excess, *727*
   in human nutrition, *727*
   leaching, 852, *853*
   mass number, *19*
   in muscle function, *646–647*
   and parathyroid hormone, 624
   in plant function, *502*
   resorption by bone, 639
   role in muscle contraction, 646
   sources, *727*
   symbol for, *19*
   in urine, *739*

Calcium aluminosilicate, 408
Calcium carbonate, 423
Calcium pump system, 87
Calico cat, 241, *241*
California coast redwood, 8
Callus, 526
Caloric intake
   calculation, 728
   reducing through diet, 729
Calvin–Benson cycle, 115,
   115–116, 118
Calyx, 516
Cambium (Cambia)
   cork, 485, *485,* 497
   vascular, 485, 489, 496,
     496–497, 507
Cambrian, 330, 334, *334,* 408, 423
Camel, *467*
Camouflage, 830, *830*
Camphor scale insect, *838*
CAM plant, 119
Camptodactyly, 180–181,
   180–181, 196
Canadian geese, 462–463
Canadian lynx, 829, *829*
Cancer
   amoeboid motion of cells, 70
   bladder, 707
   breast, 793
   carcinogen effects, 244
   cell divisions, 234, *234*
   colon, 244, 723
   deaths due to, 235
   defined, 682
   DNA transformations, 244
   drug treatments, 683
   epidermal skin, *635*
   and genetic engineering, 256
   and immune system, 683
   immunotherapy, 683
   laryngeal, 707
   lung, 707
   metastasis, 234
   oncogenes, 235
   oral, 707
   pancreatic, 707
   proto-oncogene expression, 244
   skin, *215,* 635, 897
   testicular, 774–775
   treatment, 683
   tumors, 234–235
   uncontrolled cell cycling, 144
Candida infection, *383, 388*
Cannabis sativa, 493, 589
CAP activator protein, 237
Capillary
   basement membrane, *662*
   cardiac, 658
   endothelium, *662*
   glomerular, 738, *739*
   lymph, 671
   peritubular, 738, *739*
Capillary bed
   absorption in, *664–665*
   blood flow through, 654
   diffusion in, 664, *664–665*
   filtration in, *664–665*
   fluid movements in, *664–665*
   function, 664
   lymph vessels near, *655*
Capsella, 522, *522*
Capsid
   defined, 356
   forms, 356, *356*
Carbaminohemoglobin, formation,
   705
Carbohydrate
   classes, 37
   complex, 724
   defined, *48*
   dietary sources, 724–725
   digestion, *720*
   functions, 48
   monosaccharide, *48*
   polysaccharide, *48*

   simple sugars, 724
   storage in plants, 510
Carbohydrate metabolism
   and aerobic pathway, 134–135,
     135
   enzymes in, *720*
   and glycolysis, *127, 131*
   and Krebs cycle, 128ff
   and nutrition, *730*
Carbon
   abundance in human body, 34
   in arctic tundra, 879
   global fluxes in, *855*
   global reservoirs in, *855*
Carbon atom
   atomic number, *19*
   bonding behavior, 34–35,
     34–35
   double covalent bond, 35
   electron distribution, *23*
   isotopes, 19
   mass number, *19*
   numbering, *37*
   single covalent bond, 34
   in sugar, *37*
   symbol for, *19*
Carbon compound, *see also*
   Biological molecule; Organic
   compound
   backbone, 34–35, *35*
Carbon cycle, 352, 536, 854–855,
   854–855
Carbon dioxide
   acquisition by plants, 487, 493
   arterial blood levels, 705
   atmospheric concentration, *856,*
     857, *857,* 903
   classification, *895*
   conversion to carbonic acid,
     705
   diffusion, 115, *115,* 705
   and evolution, 336
   exchange in human body,
     696–697
   fixation, 854
   and greenhouse effect, 854,
     856–857, 894, 903–904
   hemoglobin transport, 705
   and methanogens, 352
   partial pressure in respiratory
     tract, 704, *704*
   and photorespiration, *118*
   and photosynthesis, 115–119,
     118, 487
   plasma, 705
   pressure gradient, 696
   stomatal control, 487
   tissue transport, 705
Carbon dioxide fixation, 115, *115*
   in CAM plants, 119
   in C3 plants, 118–119
   in C4 plants, 118–119
Carbonic acid, formation, 29, 705
Carbonic anhydrase, 705
Carboniferous, 334–335, *393,* 458,
   459, 464
Carbon-14 isotope
   half-life, *20*
   in plant research, 21
Carbon monoxide, *895*
   in cigarette smoke, 666
   poisoning, 708
Carbon oxide air pollutants, *895*
Carboxyl group, *35*
   locations, *35*
   structural formula, *35*
Carboxypeptidase, *720,* 721
Carcinogen, defined, 244
Carcinoma
   basal cell, *215,* 234, 635
   defined, 683
Cardiac cycle, 661
Cardiac muscle
   autonomic control, 580–581
   contraction, 642

Cardiac muscle *continued*
striation, 644
tissue, 552, *552*
Cardiac pacemaker, 21, 653
Cardiovascular disorders
anabolic steroid-related, 649
arrhythmias, 667
atherosclerosis, 33, 666–667
atherosclerotic plaque, 33, *33, 666, 667*
hypertension, 666–667, 742
risk factors, 666
Cardiovascular system
blood circulation, 658, *659*
blood pressure, arterial, 662–663
blood pressure, venous, 664
cardiac cycle, 661
disorders, *see* Cardiovascular disorders
heart structure, 660, *660*
human, *658(f)*
Caries, dental, *718*
Carnegiea gigantea, 119
Carnivore, 846, *847–850*, 860
Carnivorous plant, *500*
Carotene, 634
Carotenoid pigment, 68, *110*, 352, 371–372, *538*
Carotid artery, *659*, 663
Carp, *838*
Carpal bones, *641*
Carpel, *170*, 516, *516*, 517
Carrageenan, 372
Carrier protein, 80, *85, see also* Transport protein
types, 86
Carroll, L., 814
Carrot, *526–527*, 910–911, *910–911*
Carrying capacity, 808, 814, 829
Cartilage
characteristics, 550
components, 72
fish, 452
formation, 550
functions, 550, 640
joint, 640
Cartilaginous joint, 640
Casparian strip, 504, *504*
Caspian tern, 927, *927*
Catalase, functions, 92
Cataracts, 607, 897
Catastrophism, theory, 264
Caterpillar, *830, 918, 925, 925*
Catfish, *335*
Catkin, 542
Cattle farming, and desertification, 901, *901*
Cave paintings, 470–471, *470–471*
Cavity, body, 411
CD4 lymphocyte, 690
Cecum, *427*, 723
Celery, 530
Cell, *see also* Animal cell; Plant cell; specific cell type
aging and death, 768
bone-forming, 638–639
components, *73*
components, overview, 60
cultures, 4
cytoskeleton, *73*
death, 765
defined, 4, 51
diagram, *60*
endocrine, 723
eukaryotic, 52, 58, 330, 363
lipid bilayer, *see* Lipid bilayer
membranes, *see* Membrane, cell
migration, 764
organelles, 60, *60*
origin, 324
pH, 28–29
phagocytic, 88, *88*
photosynthetic, *95*
prokaryotic, 52, 330
protistan, 363

proto-cell, 327
shape, 53
sickled, 176
size, 53
starch usage, 36, 68
structure-function summary, *73*
surface-to-volume ratio, 53, *53*
survival, 51
temperature stabilization, 26
Cell count, defined, 657
Cell culture, HeLa cells, 145
Cell cycle
DNA replication during, 144
duration, 144
G1 phase, *144*
G2 phase, *144*
interphase, 144, *144*
limited division potential, 768
myc gene effects, 235
S (synthesis) phase, *144*
uncontrolled, 144
Cell differentiation
embryonic, 764
fetal, 788–789
gene control, 238, *238–239*
in immune response, 680
plant, *519*
Cell division
bacterial binary fission, *350*
and cancer, 234
in cell cultures, 145
in deuterostomes, 423
eukaryotic, *142, 376*
growth factor effects, 626
in immune response, 680
limited division potential, 768
mechanisms, 142, *142*
meiotic, 164, *164*
mitotic, 142–150, *142–150, 164, 164, 762*
mitotic spindle effects, 760
in plant growth, 532
prokaryotic, *142, 376*
in protosomes, 423
Cell fate, embryonic, *761*
Cell junction
adhesion, 72, *549*
animal, 72
formation, 80
gap, 72, *549*
plant, 72
tight, 72, *549*
Cell-mediated immune response, *681, 682, 682*
Cell plate formation, 150
Cell theory, 51
Cellular slime mold, 365
Cellulose
fibers, 38
functions, *48*
properties, 38
structure, *38*
Cell wall
appearance in micrographs, *58–59*
bacterial, 56, *211, 348*, 349
cellulose, 38, *38*
chrysophyte, 371
chytrid, 364
functions, *58–59, 73*
fungal, 39, *39*
green algae, 374
occurrence in nature, *73*
plant, 38, *38, 58–59*, 72, 83, 357, 532
plasmolysis, 84
primary, 72
prokaryote *vs.* eukaryote, *376*
red algae, 372
secondary, 72
viral attack on, 357
viral enzyme effects, *211*
Cenozoic, 297, *297*, 329, *329, 340, 341, 393*, 408
Centipede, *432*, 435, *435*

Central nervous system
and autonomic system, *579–580*
brain, 582–583
cerebral cortex, 584–585
cerebral hemispheres, 584
components, 578
psychoactive drugs affecting, 589
and somatic system, *579–580*
spinal cord, 580, *580*
Central vacuole
appearance in micrographs, *58–59*
functions, *58–59, 73*
occurrence in nature, *73*
Centrifuge, 80
Centriole
functions, *73*, 148
location, *60*
in microtubule organizing centers, 71
Centromere
defined, 142–143
location, 142–143, *144*
Centrosome
defined, 71
location, 71
Cephalization, 411, 418–419, 576–577
Cephalochordate, 446, 448
Cephalopod, 426–427
Cerebellum, 582, *582, 584*
Cerebral cortex, 582, *582, 585*
association centers, 585
and emotional states, 585
functional regions, *585*
premotor area, 585
reptile, 450, 458
supplementary motor area, 585
and visual perception, *609*
Cerebral hemisphere, 582, 584, *584*
and corpus callosum, 586
Cerebrospinal fluid, 583, *583*
Cerebrum, 582, 583, *584*
Certainty, relative, 11
Cervical nerve, *579–580*
Cervix
during birth process, 792
location, 784
reproductive functions, *778, 781*
sexually transmitted diseases, 797
CFCs, *see* Chlorofluorocarbons (CFCs)
CFTR protein, 256
Chagas disease, 368
Chain elongation (protein synthesis), *226–227*
Chain of Being, 262–264, 310, 318
Chain termination (protein synthesis), *227*
Chalmydia, 606
Chambered nautilus, *887*
Chameleon, 460, 832, *832*
Channel protein, 79, 80, *85*, 86
Chaparral, *871, 874, 874*
Chargaff, E., 212
Chase, M., *211*
Cheek, dimpling, *169*
Cheese, as protein source, 725
Cheetah, 277
Chelicerate, 432–433, *433*
Chemical equilibrium, 96–97, *97*
Chemical evolution
and living cells, 327
self-replicating systems, 327, *327*
spontaneous assembly of porphyrin, 326, *326*
Chemical messenger
defined, 46
masked, *239*
Chemical signal, 922
Chemical synapse, 567
Chemiosmotic theory, *130*, 352
Chemoautotroph

bacterial, 348, *351*
defined, 107
energy acquisition, 119
eubacterial, *351*, 353
in hydrothermal vent communities, 885
in nitrogen cycle, 858
as primary producer, *846*, 885
Chemoheterotroph
bacterial, 348, *351*
eubacterial, *351*, 353, *353*
Chemoreceptor
smell, 598, *598*
stimuli for, *594*
taste, 598, *598*
Chemosynthesis
by chemoautotrophs, 119
defined, 119
Chemosynthetic autotroph, *351, see* Chemoautotroph
Chemotaxis, 764
Chemotherapy, 682
Chemotroph, *see* Chemoautotroph
Chernobyl nuclear power plant, 905, *905*
Cherry, *41, 516, 519*
Chestnut blight, *383*, 838
Chestnut tree, 877
Chiasmata, *160*
Chicken
comb shape, 179, *179*
egg, *757*
embryo, *757, 765*
Chickenpox, *357*
Chihuahua, *270*
Childbirth
fetal skull joint motions, 640
oxytocin secretion during, *616*
positive feedback mechanisms, 557
process, 792, *792*
protein intake and, 725
Chimaera, *452, 454, 454*
Chimpanzee
comparative morphology, 304
evolution, 475
learning by, *915*
skull, *305, 474*
Chin
dimpling, *169*
fissure, *156*
Chitin
annelid, 428, *697*
chytrid, 364
functions, 39
parapod, 430
silkworm moth, *627*
structure, 39
Chiton, 836
Chlamydia, 351
C. trachomatis, 797, *797*
Chlamydial infection, 797
Chlamydomonas
C. nivalis, 374
life cycle, *375*
Chloride, in urine, *739*
Chlorination, wastewater, 899
Chlorine
atomic number, 19
deficiency, *727*
electron distribution, *23*
excess, *727*
functions, *727*
in human nutrition, *727*
mass number, 19
in plant function, *502*
sources, *727*
symbol for, 19
Chlorine monoxide, and ozone destruction, 897
Chlorofluorocarbons (CFCs)
atmospheric concentration, *857*
classification, *895*
and greenhouse effect, 856–857
and ozone layer, 893, 897

Chlorophyll
  in chloroplasts, 68
  light absorption, 110, 372
  phytochrome deficiency effects,
    538
  synthesis, 242
Chlorophyll *a*, 110, *110*, 371–372,
  374
Chlorophyll *b*, 110, *110*, 372, 374
Chlorophyll *c*, 371
Chlorophyll P680, 112
Chlorophyll P700, 112
Chloroplast
  appearance in micrographs,
    58–59, 68, 108
  ATP formation in, 68, 114
  chemiosmotic theory of ATP
    formation, *114*
  defined, 68
  electron transport system, 102
  euglenoid, 370, *370*
  evolution, 68
  functions, *58–59, 68, 73*
  green color, *110–111*
  membranes, 109
  membrane system, 68
  occurrence in nature, *73*
  photosynthesis in, 109
  pigments, 68
  structure, 68
  thylakoid membrane, *111*
  thylakoid membrane system,
    109
Chloroquine, 369
Chocolate, 514–515, 588
Chocolatl, 514–515
Cholecystokinin, 723
Cholera, 815
Cholesterol
  bile, 721
  biosynthesis, 725
  and cardiovascular disorders,
    666–667
  deposits, arterial, 666–667
  excess, 725
  familial hypercholesterolemia, 197
  functions, 32–33, 41, *48*
  hepatic synthesis, 32
  plaque formation and, 667
  precursor role, 41
  structure, 78, *78*
  transport, 33
Choline, *726*
Chomolungma, 694–695
Chondrichthyes, *452*
Chondromyces crocatus, 355
Chondrus, 837
Chordate (Chordata)
  characteristics, 446
  circulatory system, *443*
  classification, 446
  digestive system, *443*
  evolution, *412, 449*
  groups, 446
  invertebrate, *410*, 446–448
  lancelets, 448
  movement, *442*
  nervous system, *442*
  representative, *410*
  reproduction, *443*
  respiratory system, *443*
  tunicates, 447
  vertebrate, 446, 448–451
Chorion, *784, 785, 787, 787–788*
Chorionic villi, 787, *787*
Chorionic villi sampling, 204
Choroid, 604, *604–605*
Chromatids
  appearance in micrographs, *143*
  formation, 142
  nonsister, 160, *160*
  sister, 142, 149, *149*, 157, *159*
Chromatin
  defined, 63, 142
  functions, 142

Chromatophore, *633*
Chromoplast
  defined, 68
  functions, *73*
  pigments in, 68
Chromosome
  alignment at metaphase I, 161
  alignment during metaphase,
    147
  anaphase, *147*
  appearance changes in, 63
  appearance in micrographs,
    143
  autosome, 189
  bacterial, *248, 348*
  Barr body, 241, *241*
  centromere location, 143
  chromatin, 142
  coiling, 216
  condensation, 240–241
  condensed, 63, 142, 146
  configurations, 142
  crossing over, 160, *160*
  defined, 63, 142
  deletions, 200
  differences between species,
    313
  DNA organization in, 216, *216,
    238–239*
  duplicated, 63, 142–143, *143,
    157*
  functional domains, 216, *216*
  gene linkage mapping, 194
  genetic recombination, 194
  histone-DNA interactions, 216
  homologous, *142*, 143–144,
    157, *157*
  kinetochores, 148, *148*
  lampbrush, 240, *240*
  loops, *216*
  maternal, 161
  metaphase, 147, *147*
  paternal, 161
  polytene, 242, *242*
  prokaryote *vs.* eukaryote, *376*
  prophase, 146
  sex, 189
  sister chromatids, 142
  staining, 188–189
  structure, *189*
  telophase, *147*
  threadlike, 142
  transcription in, *238–239*
  unduplicated, 142
Chromosome abnormalities, *see
  also* Chromosome number;
  Genetic disorder
  deletions, 200
  duplications, 200
  and genetic counseling,
    204–205
  and genetic screening, 204
  and instantaneous speciation,
    295
  inversions, 200
  numerical disorders, 201–202
  prenatal diagnosis, 205
  structural disorders, 200
  translocations, 200
  types, *195*
Chromosome number, *see also*
  Chromosome abnormalities;
  Genetic disorder
  aneuploidy, 201
  changes during sexual
    reproduction, *166*, 201, *201*
  defined, 143
  diploid, 143, *167*
  at fertilization, 162
  genetic disorders, *196*
  and genetic variation, 273
  nondisjunction effects, 201
  polyploidy, 201
  tetraploidy, 201
  trisomic, 201

Chrysaora, 414
Chylomicron, 722
Chyme, 718–719
Chymotrypsin, *720*, 721
Chytrid, 364, *364*
Chytridium confervaie, 364
Cicada, 287, *436*
Cichlid, 294, 839
Cigarette smoke
  effects on lungs, 706
  and heart damage, 666
  risks related to, 706–707
Cigarette smoking
  effects during pregnancy, 791
  and heart damage, 666
  and pregnancy, 790
  risks related to, 706–707
Cilium (Cilia)
  airway, 676, 701, *701*
  bending, 70, *71*
  bilvalve, 426
  functions, *73*
  gastropod, 425
  human respiratory tract, *701*
  lancelet, 448
  occurrence in nature, *73*
  protozoan, 367
  rotifer, 422, *423*
  structure, 70, *71*
Circadian rhythm, 538
Circulation, air, *see* Air circulation
Circulatory system, *see also* Blood
  circulation; Blood flow
  amphibian, *442*, 456–457, *457*
  annelids, *443*
  arthropods, *443*
  avian, *442*
  bird, 463
  cardiovascular, 658–659,
    658–659, 664
  cephalopod, 427
  chordates, *443*
  closed, 427, *429*, 448, 451, *451*,
    654, *655*
  cnidarian, *443*
  coral, *443*
  crocodile, 461
  ctenophoran, *443*
  earthworm, *429, 655*
  echinoderms, *443*
  fish, *457*
  flatworm, *443*
  flow rate in, 654
  fluke, *443*
  frog, *457*
  functions, *554*
  human, *554*
  insect, *443*
  jelly fish, *443*
  lancelet, 448
  links with other systems, 654,
    654
  major blood vessels, diagram,
    659
  mammalian, *442, 457*
  mollusks, *443*
  nematode, *443*
  obesity effects, *728*
  open, *431*, 654, *655*
  polychaete, *443*
  pulmonary circuit, 658
  reptile, *451*, 458
  reptilian, *442*
  ribbon worm, 422, *443*
  rotifer, *443*
  roundworm, *443*
  sea anemone, *443*
  spider, *443*
  sponges, *443*
  systemic, *658*
  tapeworm, *443*
  tunicate, *442, 447*
  vertebrate, 451, *451*
Cirrhosis, alcohol addiction-
  related, 589

Citrate (ionized form of citric acid),
  129
Cladistics
  cladogram construction, 316–317
  cladograms, 316–317
  defined, 314
  derived traits, 316
  ingroups, 316
  monophyletic lineages, 316
  outgroups, 314
  time line, 317
Cladogenesis, pattern of
  speciation, 296
Cladogram
  construction, 316–317
  defined, 314
  interpretation, 317
Cladonia rangiferina, 386
Cladophora, 111
Clam, *410, 426, 426, 442–443,
  882, 885*
Clark, L., *910–911, 910–911*
Classical conditioning, *915*
Classification
  amphibians, 446, 456
  animals, 9
  Aristotle's descriptions, 262
  bacteria, 351, *351*
  binomial system, 310
  birds, 446
  and Chain of Being, 310
  chordates, 446
  contemporary systems, 318
  fishes, 446
  five-kingdom, 8–9, 318, *318*
  fungi, 9
  and information retrieval, 314,
    327
  land, *900*
  Linnean scheme, 310
  mammals, 331–332, 446
  monerans, 8
  phylogenetic, 314, 327–328
  plants, 9
  primates, 472, 475
  protistans, 9, 363
  and retrieval, 314, 327–328
  as retrieval system, 314,
    327–328
  and systematics, 314–318
  vertebrates, 331–332
  viruses, 356
  white blood cells, 680
Class (taxon), 8, 311
Clathrin, *89*
Clavaria, 378
Claviceps purpurea, 383, *383*
Clavicle, *641*
Clay, templates for protein
  synthesis, 325–326
Cleavage
  animal cell, *151*
  blastula formation, 762
  and cell fate, 762
  comparative (animal cell), *757*
  cytoplasmic localization, 762
  defined, 36, 150, 756
  human embryo, *784*, 784–785
  planes, 762
  radial, 423, *423*
  spiral, 423, *423*
  spontaneous splitting at, 764
Cleavage furrow, 150, 762
Cleft lip, 204
Cliff swallow, 924
Climate, *see also* Rainfall
  air circulation effects, 866–867,
    867, 888–889
  defined, 865
  factors affecting, 865
  global oscillations, 888–889
  global patterns, 865–867
  ocean effects, 868
  and primary productivity, 865
  topography effects, 869, *869*

Climax community
    climax-pattern model, 835
    defined, 823, 834
Clitoris
    anabolic steroid effects, 649
    functions, 778, *778*
    location, 778
    structure, 778
Cloaca, 444
Clonal selection theory, 687, *687*
Clone
    from asexual reproduction, 156,
        164, *526,* 526–527
    defined, 164
    T cell, 682–683
Cloning, plant, 526, *526,* 527, *527*
Clostridium, 351
    C. botulinum, 354, 569
    C. tetani, 354, *354,* 569
Clotting, blood, *see* Blood clotting
Clotting factor VIII, 198
Clownfish, *415*
Club fungus (Basidiomycota), *383,*
    384–385, *385*
Club moss, 396
Clumping, populations, *804,*
    804–805
Cnidarian (Cnidaria)
    body form, 414
    body plan, *412*
    circulatory system, *443*
    colonial, *416,* 417
    digestive system, *443*
    epithelia, 414
    evolution, *412*
    gut, 414
    hydrostatic skeleton, 414
    life cycle, 416
    movement, *442*
    nematocysts, *415*
    nerve net, 414
    nervous system, *442,* 576–577,
        577
    representative, *410, 414–415*
    reproduction, *443*
    respiratory system, *443*
Coal
    burning, 854, *855,* 857, 895,
        904
    depletion, 396
    formation, 335, 396, 854, *855,*
        904
    strip mining, 904
Coal mine habitat, 352
Coastal breezes, *869*
Coastal ecosystem, 882–883
Coast live oak, 542–543
Coat color
    determinants, 178, *178, 618*
    and gene transcription, 241,
        241
    melanin effects, 178, *178*
    melanocyte-stimulating
        hormone effects, *618*
Coated pit, 88, *89*
Cobalt-60 isotope, 21
Cocaine, 493
    addiction, 575
    effects during pregnancy, 791
    effects on central nervous
        system, 588–589
    granular, 588
    mode of action, 569
Coca plant, 493
Coccus (cocci), 348, *351*
Coccygeal nerve, *579*
Cochlea, *600*
Cochlear duct, *601*
Cocklebur, 539
Cockroach plant, 493
Cocoa bean, *403*
Codeine, 589
Codfish, *450*
Codium, 374, *374*
Codominance (genetics), 176

Codon
    anticodon interactions, 226
    defined, 224
Coelacanth, *452,* 456, *456*
Coelom
    characteristics, 411
    deuterostome *vs.* protostome,
        423
    earthworm, *429*
    echinoderm, *438–439*
    false, 411, 422
    species diversity, 423, *423*
Coelomate, *410*
Coelomic cavity, human, 786
Coelomic chamber, earthworm,
    428–429, *429*
Coelomoduct, *424*
Coenzyme, *see also specific*
        coenzyme
    defined, 46, 101
    and electron transfer, *125*
    electron transfer, 46
    functions, 48, 101
    in glycolysis, 126, *127*
    nucleotide, 46, *46,* 48
Coenzyme, nucleotide
    and electron transfer, *128–129*
    and self-replicating systems,
        327
Coevolution, *see also* Evolution,
        biological
    bacterial symbionts, 362
    bilateral body plans and
        cephalization, 411
    defined, 521
    flowering plants and pollinators,
        520–521, 524, 825
    parasitoid and host, 833
    predator and prey, 830–832
    yucca moth and yucca plant,
        825
Cofactor, defined, 98
Coffee, 588
Cohesion theory of water
    transport, 27, 506, *507*
Cohort, 810
Cortus, 783
Cold
    chronic, 706
    common, *357*
    rhinovirus-related, *357*
    thermoreceptors, 596
Cold sore, *357,* 358, *635*
Cold stress, 748
Cold sweats, 634
Coleoptera, *436*
Coleoptile, *532,* 534, 536
Coleus, 492
Collagen
    aging effects, 768
    bone, 638–639
    in bone and cartilage, 48
    in connective tissue, 550
    dermal, aging effects, 635
    functions, 44, 48
    genetic engineering, 256
    structure, 44
Collar cell, sponge, 413
Collecting duct, renal, *740,* 741
Collenchyma, 486, *486*
Colon
    ascending, 723
    constipation, 723
    descending, 723
    functions, 723
    location, 723
    movement through, 723
Colon cancer, 244, 723
Color
    hair, 178, *178*
    skin, 632, *633,* 634
Coloration
    bird, 463
    feather, 12–13
    moth, 7, *7*

    plant, 110, *111*
    warning, 831, *831*
Color blindness
    red-green, *196,* 606
    total, 606
    X-linked, 144
Color perception, *196,* 606,
    608–609
Comb jelly, *412,* 417, *417,* 442–443
Comb shape, in poultry, 179, *179*
Commensalism, 824, *824*
Common cold, *357*
Commotion, defined, 94
Communication junction, cardiac
    muscle cell, 661, *661*
Communication signal, *see also*
        Signaling molecule
    chemical, 922
    defined, 921–922
    functions, 922
    illegitimate, 922
    tactile, 923, *923*
    types, 922–923
    visual, *922,* 923
Community, *see also* Human
        population; Population
    climax, 823, 834–835
    climax-pattern model, 835
    cyclic replacement in, *834–835,*
        836
    defined, 4, 803, 823–824
    dispersal, 838–839
    feeding levels, 824
    interactions, 824–842
    patterns of diversity, 840–841,
        841
    and resource partitioning, 827
    role of disturbance in, 836
    species introductions into, *838,*
        838–839
    stability, 836–837
    structure, 824
    succession, 834–835, *834–835*
Companion cell, 489, 510, *510*
Comparative anatomy, 262–263
Comparative biochemistry, 308–309
Comparative embryology, 304–305
Comparative morphology, 304–307
    analogous structures, 306–307
    homologous structures, 306
    reconstructions, 304
    stages of development, 304
Competition, *see also* Parasite;
        Predation; Prey defense
    categories, 824, 826
    competitive coexistence, 827
    competitive exclusion, 826–827,
        827
    predation effects on, 836–837,
        839
    resource partitioning, 827, *827*
Complementary DNA, 253, *253,*
    256
Complement system
    activation, 677, *677*
    lysis induction by, 677, *677*
    pore complexes, 677, *677*
    reaction cascade, 677
Compound
    defined, 19
    organic, *see* Organic
        compound
Compound eye, 602, *602*
Compound microscope, *50,* 54
Computer simulation, natural
    selection, 327
Concentration, defined, 82
Concentration gradient
    and chemiosmotic theory, *130*
    defined, 82
    diffusion, 82, 563
    in thylakoid membrane, *114*
    tonicity effects, 83
    water, 83
Conch, *602*

Condensation reaction
    chromosomal, 240–241
    enzyme action in, 36
    fatty acids into triglyceride, *40*
    mechanisms, 36, *36*
    monosaccharides into
        disaccharide, *37*
    water formation, 36
Conditioning, *see also* Learned
        behavior
    classical, *915*
Condom, 795, *795*
Conduction
    heat transfer by, 744
    saltatory, 566
Cone
    conifer, 401
    cycad, 402, *402*
Cone photoreceptor, 606, 608,
    608, 609
Conidium, 382, *388*
Conifer (Coniferophyta)
    classification, 401
    comparison with other plant
        groups, *406*
    evergreen, 876, 878
    evolution, *393*
    life cycle, *400*
    radiations, *393,* 401
    structure, 401, 878
    uses, 401
    vulnerability to deforestation, 401
Coniferous forest
    biome, *870,* 878
    global distribution, *871, 876,* 878
    soil, *872*
Conjugation
    bacterial, 350, *350*
    protozoan, 367
Conjunctivitis, 607
Connective tissue
    adipose, 551
    appearance, *550–551*
    blood, *see* Blood
    cartilage, 550
    dense irregular, 549, *549*
    dense regular, 550, *550*
    ground substance, 550
    ligaments, 640
    loose, 550, *550*
    osseous, 550–551
    types, 550
Conservation of mass, law, 22
Constellation Orion, *322*
Constipation, fiber deficiency-
    related, 723
Consumer, 5
    examples, *846–848*
    functions, 846, *846*
    trophic levels, *846,* 846–847,
        847–848
Consumption, human population,
    819, 894, 904, *904*
Continental drift, 328, 336, 341
Continental shelf, *884*
Continuous variation in traits,
    180–181, *180–181*
Contraception, 817
Contraceptive methods, 794–795
Contractile cell, cnidarian, 414
Contractile protein, *239*
Contractile vacuole, 84, *84,* 367,
    370, *370*
Contraction, muscle
    acetylcholine effects, 566, 570
    actin filaments, 70
    ATP demands, 645–646
    ATP requirements, 647
    autonomic control, 580–581
    and calcium, *646*
    calcium role, 646
    cardiac, 661, *661*
    and childbirth, 557
    control, 646–647
    energy for, 648

Contraction, muscle *continued*
exercise effects, 648
hormonal effects on, 557
mechanisms, 70
myosin–actin interactions during, 645, *645*
myosin filaments, 70
sliding-filament model, 645, *645*
somatic control, 580
stomach, 718–719
stretch reflex, 570, *571*
tension, 646
tetanus effects, 569
Control group, experimental, 12
Controlled cell death, 765
Convection, heat dissipation by, 744–745, *749*
Convergence, 307, 317
Convergent circuit, neuronal, 570
Convergent evolution, *307,* 448, *864–865,* 870
Cooksonia, 302, 335, 396, 397
Cooperative predator avoidance, 924–925, *925*
Cooperative regulation, of transcription, 236
Copepod, 434, *434,* 880, 884
Copernicus, N., 13
Copper
deficiency, *727*
excess, *727*
functions, *727*
in human nutrition, *727*
in plant function, *502*
sources, *727*
Copperhead, *755*
Coprolite, 302
Coral, *410,* 414, *415–416, 442–443,* 826, *887*
Coral bank, 886
Coral fungus, *384*
Coral reef, 886, *886–887*
Coral snake, 460–461, *461*
Core temperature, defined, 744
Cork, 481–482, *487, 496*
Cork cambium, *485, 497*
Corm, *526*
Corn, *see also Zea mays*
classification, *311*
C4 pathway, *118*
leaf, 476
as protein source, *725*
reproduction, *526*
root, *494–495*
root cap, *537*
root tip, *537*
seed, *538*
smut, *383*
species introductions, 838
stem structure, *489*
Cornea, 602, *602, 604–605*
artificial, 606
astigmatism, 607
diseases, 606–607
farsightedness, 606
nearsightedness, 606
transplantations, 606
Corneal transplant surgery, 607
Corn sweetener, 725
Corn syrup, 725
Corolla, 516
Coronary artery, *659–660, 666–667*
Coronary artery disease, 707
Coronary bypass surgery, 667
Coronary circulation, 660, *660*
Coronavirus, 357
Corpus callosum, *582,* 586
Corpus luteum, 778, *780,* 781, *782*
Cortex
root, 494, *494, 495, 503, 507*
stem, *489*
Cortez, H., 514
Cortez rainbow wrasse, *293*
Corticotropin (ACTH), *616,* 618, *618, 623*

Corticotropin-releasing hormone, effects on anterior pituitary, *618*
Corticotropin-stimulating hormone, *618*
Cortisol
actions, 616, *616, 621,* 623
classification, *616*
and salt tolerance in salmon, *743*
secretion, *616,* 621, *623*
Cortisone, 623
Cotton plant, 255, *255*
Cottonwood, *834–835,* 873
Cottony mold, 378
Cotyledon (seed leaf), *484,* 522, *532*
Coughing, brain areas affecting, 582
Counseling, diet, 729
Countercurrent flow, in gills, 699, *699*
Courtship behavior
and behavioral isolation, *290*
as competition, 916–917, *917*
hormone effects, 612–613
and imprinting, *915*
Rana pipiens, 752
and sex separation, 754
and sexual selection, *283*
signals, *922,* 922–923
Covalent bond
carbon, 34–35, *34–35*
in cellulose, 38
in condensation reaction, *36*
defined, 24
double, 24
energy content, 94
in formulae, 24
in hydrolysis, *36*
nonpolar, 24
between nucleotides within DNA, *46*
polar, 25
in starch, 38
triple, 24
Cowbird, 833
Cowpox, 674, 688
C3 plant, 119
C4 plant, *118,* 119, 494
Crab, *410,* 432, 434, *434,* 882–883
Crack cocaine, 575, 588–589, *589*
Cranial bones, *641*
Cranial cavity, human, *555*
Cranial nerve, 580, *580*
Crassulacean acid metabolism (CAM) plant, 119
Crater
asteroid-related, *337,* 337–338
lakes, sympatric speciation in, 294, *294*
Crater Lake (Oregon), 880
Crayfish, 411
Creatine, in urine, *739*
Creatine phosphate, dephosphorylation, 647, *647*
Creosote, 526, 873, *873*
Cretaceous, 336, *336, 338, 393,* 458, *459,* 464
Cretaceous–Tertiary boundary, 338
Crete, destruction by earthquake, 260
Cri-du-chat syndrome, *196,* 200
Crista, 67, *67,* 599
Crocodile, *459, 461,* 710
Crop, food
and acid deposition, 896
fertilizer, kelp-derived, 373
and genetic engineering, 247, 254–255
and global warming, 857
high-yield, 900
improving yields, 255, 858
and insecticide resistance, 279
irrigation, 898, *898*
irrigation-dependent, 898

and nitrogen scarcity, 858
RNA viruses affecting, 357
viroid effects, 357
Cropduster, *17*
Cropland
converting marginal lands to, 900
desertification, 901, *901*
global, *900*
Cross-bridge formation, 645–646
Cross-connecting food webs, 847–848
Crossing over
defined, 160
effects, *160*
homologous chromosomes, 188
and linkage groups, 193–194
during prophase I, *160,* 164
and variation in populations, 273
Crown, tooth, *718*
Crown gall tumor, 255, *255*
Crown-of-thorns sea star, 887
Crustacean, *see also* Arthropod (Arthropoda)
body plan, 434
characteristics, summary, *442–443*
classification, 432
eye, 602, *602*
in food web, 885
molting, 626
representative, *434*
species diversity, 434
Cryphonectria parasitica, 383, 839
Crystallin-producing cell, 764
Ctenidium, 424, *424,* 425
Ctenophora, *410*
circulatory system, *443*
digestive system, *443*
movement, *442*
nervous system, *442*
reproduction, *443*
respiratory system, *443*
Cuckoo, 914, *914*
Cuspid, 718
Cuticle
arthropod, 432
cherry, *41*
chitin-reinforced, *39*
earthworm, 428, *429*
insect, *637*
plant, 392, 394, 487, 508, *508*
roundworm, 419
silkworm moth, *627*
tick, *39*
Cutin, *48,* 487, 508
Cuttlefish, 426, *427*
Cuvier, G., 264
CVS (chorionic villi sampling), 205
Cyanobacterium, *351,* 352–353, *353,* 390, 847, 858, 880–881
Cycad (Cycadophyta), 335
characteristics, 402, *402*
comparison with other plant groups, *406*
evolution, *393*
Cyclic AMP
CAP–cAMP complex, 237
and lactose operon, 237
nonsteroid hormone effects, 617
as second messenger, 617
Cyclic pathway, photosynthesis, 330
Cyclic replacement, community, *834–835,* 836
Cyst, *421*
Entamoeba histolytica, 366
Giardia lamblia, 368
sporozoite, 368
Cysteine, 725
Cystic fibrosis
CFTR protein, 256
RFLP applications, 252
Cytochrome
distribution, 101

and electron transport systems in cells, 326, 331
functions, 101
Cytochrome *b,* 294
Cytochrome *c,* 308
human, *308–309*
structure, *308–309*
wheat, *308–309*
yeast, *308–309*
Cytokinesis, 142, 150, *150–151*
Cytokinin, *534,* 535, 541
Cytomembrane system, *64*
Cytoplasm
bacterial, 56–57, *348*
connections between cells, 72
cytokinesis, 142, 150, *150–151,* 155
defined, 52
egg, 760, *761*
plasmolysis, 84
Cytoplasmic localization, 762, 766
Cytoplasmic streaming, 70
Cytosine, *211, 212, 212,* 225
Cytoskeleton
appearance in micrographs, *69*
assembly, 69
changes during cell cycle, 69
components, 69, *69*
egg, 760
functions, 60, 69, *73*
intermediate filaments, 69
occurrence in nature, *73*
organization, 760
plasma membrane, *79*
structure, 69, *69*
Cytotoxic T lymphocyte, 680, 683

D

2,4-D (2,4-dichlorophenoxyacetic acid), 365
Daisy, *520*
Daisy coral, 887
Dance, honeybee, 923, *923, 928–929*
Dandelion, *494*
Darwin, C., 10, 261, *264,* 536, 864
development of theory, 266–268
early studies, 264–265
plant growth studies, 536
voyage of H.M.S. *Beagle,* 264–265
Darwin's finches, 267
Dating methods
iridium, 338
radioisotope, 301, 470
da Vinci, L., 300
Daylength
and dormancy, 541
and flowering, 539, *539*
Day-neutral plant, 539
Daytime vision, and human evolution, 473–474
ddI (didanosine, dideoxyinosine), 691
DDT, effects on ecosystems, 861
Death
and aging, 768
at cellular level, 768
commentary on, 769
Death cap mushroom, *384*
Death rate, and population growth, 806–807, 819
Deciduous forest, temperate
biome, 870, 877, *877*
global distribution, *871,* 876–877
soil, *872*
Deciduous forest, tropical
biome, *871, 877, 877*
global distribution, *871,* 877
Deciduous plant, 491
Decomposer
bacterial, 353
defined, 5, 846

Decomposer *continued*
  in ecological pyramid, *848*
  function, 846, *846*, 858
  fungus, 378, 380
  saprobic, 364
  trophic level, 847, *848–850,*
    860
  water molds, 364
Decomposition, 858, 902
Decompression sickness, 709
Deerfly, *603*
Defensive behavior, 460
Definitive host, *359*, 418, 420
Deforestation
  and air pollution, 857, 893, *903*
  conifers, 401
  defined, 902
  global, 401, *401*, 902
  from gypsy moths, 839
  in North America, *903*
  nutrient loss from, 852, *853*
  and rainfall, 902–903
  redwood forest, *903*
  tropical forest, 893, *902,*
    902–903, 903
Delbruck, M., 210
Delphinium
  D. decorum, 304–305, *305*
  D. nudicaule, 305, *305*
Delta cell, hormone secretion, 622
Deltoid, *643*
Demographic transition model,
  818, 818–819
Denaturation
  and body temperature, 744
  defined, 46
  nucleic acid, 253
  protein, 46
  reversibility, 46
Dendrite
  interneuron, *571*
  sensory neuron, *571*
Denitrification, 858, *859*
Denitrifying bacterium, 858, *859*
Density-dependent controls, 809
Density-independent controls, 809
Dentin, 453
Dentition, 473–474, 476
  mammalian, 464–465, *465*
Depolarizing effect, of EPSPs, 568
Depo-Provera, 795
Depressant drug, 589
Depression
  and biological clock, 625
  and daylength, 625
  drug-related, 588–589
Derived trait, 316
Dermis
  component of skin, *632, 634*
  plant, 485, 487
Dermochelys coriacea, 710,
  710–711
Desalination, water, 898
Desert
  biome, 873, *873*
  climate, 873
  and global air circulation,
    866–867
  global distribution, *871,* 873
  plants, 873, *873*
  soil, *872,* 873
Desertification, 873, 901, *901*
Desmid, *374*
Detachment, retinal, 606
Detorsion, gastropod, 425
Detrital food web, 848, *849,*
  882–883
Detritivore, 846, *848–850*
Detritus, 880, 882–883
Deuterostome, *412,* 423
Development, *see* Animal
  development; Embryonic
  development; Fetal
  development; Human

development; Plant
  development
Devonian, 334, *335, 393,* 400
Devonian-Carboniferous boundary,
  335
DEW line (Distant Early Warning
  system), 907
Dextrose, 725
Diabetes
  juvenile-onset, 623
  mellitus, 623, 666
  type 1, 623
  type 2, 623
Dialysis
  hemodialysis, 742
  kidney, 742
  peritoneal, 742
Diaphragm
  functions, *700–701*
  location, *700,* 701
Diaphragm (contraceptive), 794,
  795
Diarrhea
  Escherichia coli-related, 354
  protozoan infection-related, 368
Diastole, 660, *662*
Diastolic pressure, *664–665*
Diatom, 370–371, *371,* 848,
  880–881
Dicot (Dicotyledonae)
  A. tumefaciens-infected, 255
  comparison with other plant
    groups, *406*
  development patterns, 532, *533*
  embryo, 522
  function, 488
  growth patterns, 532, *533*
  herbaceous, 481
  vs. monocot, *406*
  primary root, 494
  representative, 405, *406*
  root, *497*
  stem, *489*
  structure, *406,* 488
  taproot system, 494
  vascular bundle, *488*
  woody, 496–497
Dictyostelium discoideum, 365, 764
Didanosine (dideoxyinosine, ddl),
  691
Didinium, 363
Diet
  balanced, 727
  bulk in, 723–724
  and cholesterol blood level, 33
  cholesterol content, 33
  and dieting, 729
  and emphysema, 706
  and exercise, 729
  fiber-deficient, 723
  fiber-poor, 723
  fiber-rich, 723
  and genetic disorders, 204
  and pregnancy, 790
  protein-deficient, 725
  salt intake, 666
  vegetarian, 725, *725*
  well-balanced, 724
Diffusion
  across lipid bilayer, *85,* 504
  in capillary beds, 654, 664,
    664–665
  carbon dioxide, 115, *115,* 704
  defined, 82
  direction, 82
  and dynamic equilibrium, 82
  electric gradient effects, 82
  facilitated, 86
  gas exchange via, 696
  in human respiratory system,
    704–705
  neuronal, 563–564, *563–564*
  oxygen, 704
  passive, 86
  placental, 787

rate of, 82
simple, *85*
surface-to-volume ratio, 696
temperature effects, 82
Digestive system, *see also* Gut
  accessory organs, 717, *717*
  amphibian, *442*
  annelids, *443*
  arthropods, *443*
  avian, *442*
  chordates, *443*
  cnidarian, *443*
  colon, 723
  complete, 716, *716–717*
  control, 718–719
  coral, *443*
  ctenophoran, *443*
  deuterostomes, 423
  disorders, 724
  earthworm, *429*
  echinoderm, 438, *443*
  endocrine controls, 723
  and endocrine function,
    718–719, 720
  extracellular (fungi), 380
  fat digestion, 134, *720,* 721
  flatworm, *443,* 716, *716*
  fluke, *443*
  functions, *555,* 716
  fungi, 380
  hormonal control, 718–719
  human, *555,* 716–723, *732*
  incomplete, 716, *716*
  innervation, *580*
  insect, 435, *443*
  jelly fish, *443*
  large intestine, 723
  links with other systems, *642,*
    654, 716
  major components, *717*
  mammalian, *442*
  mollusks, *443*
  nematode, *443*
  nervous system controls, 723
  polychaete, *443*
  processes, *720*
  protosomes, 423
  reptilian, *442*
  ribbon worm, *443*
  rotifer, *423, 443*
  roundworm, 420, *443*
  sea anemone, *443*
  sea star, *438–439*
  small intestine, 722, *722*
  spider, *443*
  sponges, *443*
  tapeworm, *443*
  tunicate, *442*
Digitalis, 493
Digitalis purpurea, 493
Dihybrid cross
  garden pea, 174–175, *175*
  independent assortment,
    174–175
  outcome prediction, 174–175,
    175
Dikaryotic mycelium, 385, *385*
Dilation, cervical, 792
Dimorphism, sexual, 282–283
Dimpling
  cheek, *169*
  chin, *169*
Dinoflagellate, 370, *416*
Dinosaur, 336–339, *339, 458, 459*
Diondea muscipula, 500
Dioscorea, 535
Dioxin, *895*
  carcinogenicity, 365
  exposure effects, 365
Diphtheria, 688
Diploid cell
  chromosome number in, 143
  defined, 143
Directional selection, *278,*
  278–279, *281,* 296, 304

Disaccharidase, *720*
Disaccharide, defined, 37
Disease
  biological vectors, 359
  contagious, 359, 814, 924
  defined, 359
  endemic, 359
  epidemic, 359
  foodborne, 359
  genetic, 196
  pandemic, 359
  patterns, 359
  and population growth, 814
  sexually transmitted, *see*
    Sexually transmitted disease
  and sociality, 924
  sporadic, 359
  transmission modes, 359
  viral, *357,* 359
Disease resistance
  and genetic engineering, 254
  plant, 254, 273
Disk, intervertebral, 640, *641*
Dismorphia, 831
Disorder, eating, 729
Dispersal, community, 838–839
Dispersion
  clumped, *804,* 804–805
  of populations, 804–805,
    804–805
  random, 805, *805*
  uniform, 805, *805*
Disruptive selection, 280–281, *281*
Dissolved, defined, 27
Distal tubule
  hormone receptors, 741
  location, 738, *739*
Distance effect, species diversity,
  841, *841*
Distant Early Warning system
  (DEW line), 907
Disturbance
  community, 861
  modeling, 861
Divergence
  defined, 289
  example, *291*
  genetic, 289, 475
  and molecular clock, 308
  morphological, 306–307
  and speciation, 289
Divergent circuit, neuronal, 570
Diver's euphoria, 709
Diversity of life, *see also*
    Speciation; Species diversity
  biological, 297
  during Cenozoic, 341
  and DNA base pairing, 297
  and evolution, 10
  five-kingdom classification, 8–9
  and genetic recombination, 247
  and geologic record, 260
  molecular basis, 229
  patterns, 314
  perspective on, 136–137, 229
  plant, during Mesozoic, *336*
  in reproduction and
    development, 755, *755*
  seafloor spreading effects, 303
  sensory systems, *592*
  sources, 229, 247
  and species relatedness, 301
  and unity, 8, 208
Division of labor
  during arthropod life cycle, 432
  in multicellular organisms, 376
Division (taxon), 8, 293
Dixon, H., 506
DNA (deoxyribonucleic acid)
  amplification, 249–250,
    249–250, 250
  bacterial, 56–57
  banding patterns, 252
  base pairing, *213*
  base sequence, 213, *214,* 221

DNA continued
carcinogen effects, 244
cellular, 244
cloned, 249
complementary, 253, 253, 256
components, 212
covalent bonds between
nucleotides, 46
defined, 3
discovery, 209
double helix, 142, 213–214
duplication before meiosis I, 157
duplication by cell division, 142
duplication during mitotic cell
division, 146
eukaryotic, 216
eukaryotic cells, 52
evolution, 10, 229
fingerprint, 252
functional groups, 35
functions, 48, 73, 210–211
gel electrophoresis, 250–251
helical coiling, 213
helical coiling model, 47, 47
heritable changes in, 10
histone-DNA spools, 216
and histones, 238–239
hybridization, 249
hydrogen bonding, 25
hydrogen bonds between
nucleotides, 46
and inheritance, 7
injections, 255
loops, 216, 216
mitochondrial, 67, 294
molecules in animal cells, 62
mutagens, 228
mutations, 7
novel sequences, 229
nuclear, 62
nucleosomes, 216
nucleotide arrangement in, 47,
213, 213
nucleotide subunits, 212
organization in chromosomes,
216, 216, 238–239
and organization of life, 136
physical dimensions, 213
polymerase chain reaction, 250,
250
probes, 256
prokaryotic cells, 52
proteins bound to, 216
rearrangements, 238
recombination, 249, 249
regulatory protein binding, 236
replication fork, 214–215
self-replicating systems,
326–327, 333
sequencing, 251, 251
in sperm nucleus, 776
structure, 212–213, 212–214
thymine dimers, 215
transcription, see Transcription
transformations, carcinogen-
related, 244
viral, 211, 244, 247, 356, 356
Watson–Crick model, 208, 208,
213
DNA-DNA hybridization, 308, 309
DNA fingerprinting, 930
DNA library, 249, 249
DNA ligase, 214, 215, 249, 249
DNA polymerase, 214, 215, 250
DNA repair
aging effects, 768
base-pair substitutions, 228, 228
enzymes for, 214, 228, 228
thymine dimers, 215
DNA replication
bacterial, 350
base pairing, 222
continuous, 214–215
discontinuous, 214–215
enzymes for, 214–215, 214, 228

errors in, 228
nucleotide strand assembly, 214
origin and direction, 214–215
semiconservative, 214, 214
and transcription, 222
DNA virus
animal, 357, 357
diseases caused by, 357
herpes-related, 357
infections, 357
replication, 358, 359
types, 357
Doldrums, 867
Dolphin, 600
Dominance, apical, 535
Dominance hierarchy, 926, 927
Dominance relations, 176
Dominant allele, 168, 171
Dopamine, 568–569
effects on anterior pituitary, 618
Dormancy, 541, 541
Dorsal (anatomy), 410–411, 411,
555
Double fertilization, 509, 519, 519
Double helix, DNA, 142, 213,
213–214
Douching, as birth control method,
794, 795
Douglas fir, 541, 541
Dove, 10, 524
Down syndrome, 202, 202
consequences, 196
prenatal diagnosis, 204
risk factors, 781
Downwelling, 888–889
Downy mildew, 364, 383
Dragonfly, 437
Dragon run, 378
Drinking water, 899
Drone, honeybee, 929
Drosophila melanogaster
breeding experiments, 192
chromosomes, 192
cleavage, 762
delayed cleavage, 763
egg, 762
eye color, 192, 193
fate map, 767
linked genes, 192, 193–194
mutants, 192
pattern development, 767
photograph, 192
Drowsiness, during fever, 678
Drug
analgesic, 589
antimalarial, 369
cortisol-like, 623
depressant, 589
hypnotic, 589
IV, and AIDS, 796
and pregnancy, 790
psychoactive, 589
psychedelic, 589
Drug addiction
crack cocaine, 575
defined, 588
Drug therapy, targeted, 683
Dry acid deposition, 896
Dryas, 834
Dryopith, 475, 475
Dry shrubland
biome, 874
global distribution, 871, 874
Dry woodland
biome, 874
global distribution, 871, 874
Duchenne muscular dystrophy,
196, 198
Duck, 915
Duckbilled dinosaur, 337
Duckbilled platypus, 755
Duck louse, 436
Dung, cycling, 5, 5
Dung beetle, 5
Duodenum, 719, 720

Duplication, chromosomal, 200
Dust storm (Dust Bowl), 901
Dutch elm disease, 383, 838
Dwarfism
achondroplasia, 196, 197
pituitary, 620, 620
Dye, for karyotype preparation, 189
Dynamic equilibrium, 82, 599, 599

E

Ear
components, 600–601
hair cells, 600–601
human, 600, 600–601
inner, 600–601, 600–601
intense sound effects, 601
organs of balance, 599, 599
outer, 600–601, 600–601
structure, 601
Earlobe
attached, 168, 168–169
detached, 168, 168–169
Early Homo, 478
Early wood, 497, 497
Earth, see also Biosphere
age, 324
atmosphere, 324, 330
evolution, 342–343
during formation, 325
land classification, 900
rotation around sun, 866, 867
species diversity on, 840, 840
tectonic plates, 328, 328, 334,
336
uniformitarianism, 265
water coverage, 865
Earthquake, 260–261, 328, 336–338
Earth's crust
Cenozoic era, 408
elements in, 18
faults in, 260, 328
formation, 324
geologic change and, 266,
336–338
uplifting, 328, 341
Earthworm, see also Annelid
(Annelida)
adaptations, 428
body plan, 428, 429
circulatory system, 443, 655
classification, 410
digestive system, 443
habitat, 428
movement, 442
nervous system, 442, 577
reproduction, 443
respiratory system, 443
segmentation, 411, 428
setae, 428
skeletal system, 636
species, 410
Earwax, 48
Earwig, 436, 831
Easterlies, 867
Eating disorders, and diet, 729
Ecdysone
effects on DNA transcription, 242
functions, 242
and molting, 626, 627
ECG, see Electrocardiogram
Echinocereus, 864
Echinoderm (Echinodermata), 438
body plan, 410
circulatory system, 443
digestive system, 443
evolution, 412
locomotion, 438–439
movement, 442
nervous system, 442
representative, 410
reproduction, 443
respiratory system, 443
water vascular system, 438, 439

Echolocation, 592–593, 600
Ecological isolation, 289, 290
Ecological pyramid, 848, 848, 850
Ecology
defined, 803
population, 802–820
Ecosystem, see also
Biogeochemical cycle;
Community; Lake; Trophic
levels
coastal, 882–883
components, 803
coral bank, 886
coral reef, 886, 886–887
DDT effects, 861
defined, 4, 803, 845–847
energy flow through, 846, 848,
848–850, 849–850, 862
hydrothermal vent, 884–885, 885
inputs and outputs, 846, 846
modeling, 861, 889
model of nutrient flow, 851
nature, 846–847
open ocean, 884, 884–885, 885
primary productivity in, 848
stream, 882
Ecosystem analysis
at Hubbard Brook Valley, 852,
853
at Silver Springs, 850
Ectocarpus, 372
Ectoderm, 410, 756, 765, 765
Ectotherm, 746
Edema, 664–665
altitude-related, 695
Ediacara Hills, fossils, 334
EEG, see Electroencephalogram
Effector, in homeostatic control,
556, 557
Effector cell, 680, 681, 685
Efferent nerve, 579, 580
Egg, see also Oocyte
Africanized honeybee, 122
amniote, 458
amphibian, 238, 762
animal, 163
animal pole, 760
bird, 462, 756, 757, 762
blood fluke, 420, 420
cleavage, 757
cytoplasm, 760
cytoskeleton, 760
defined, 162
developmental information in, 760
duckbilled platypus, 755
Florida panther, 277
flowering plant, 502, 504, 516
fluke, 420, 420
formation, 163
frog, 752
human, 757
insect, 762
mature, 784
pinworm, 420
polarity, 760
as protein source, 725
reptile, 762
rotifer, 423
sea urchin, 762
shelled, 459
snail, 755
snake, 755
tapeworm, 419, 421
turtle, 460
vegetal pole, 760
Wuchereria bancrofti, 421
yolk, 755, 755, 756
Ejaculatory duct, 774, 774–775
EKG, see Electrocardiogram
Elastin, 635
in connective tissue, 550
dermal, aging effects, 635
Electric gradient
and chemiosmotic theory, 130
defined, 82

Electric gradient *continued*
  effects on diffusion, 82
  in thylakoid membrane, *114*
Electrocardiogram (ECG), 652,
  652, 667
Electroencephalogram (EEG)
  defined, 587
  EEG arousal pattern, *587*
  normal patterns, *587*
  sleep patterns, *587*
Electromagnetic spectrum, *110*
Electron
  in atomic structure, 19, *19, 23*
  in carbon bonds, 34
  charge, 19
  in chemical bonds, 22
  energy level, 22, 112
  excited, 102–103
  orbitals, 22, *23*
  shuttle, *131*
  vacancies, 22, *23*
Electron microscope
  high-voltage, 55
  scanning, 55, *55*
  transmission, 54–55, *55*
Electron microscopy
  freeze-fracturing for, *80,* 81
  sample preparation for, 80–81,
  80–81
Electron transfer
  in chemical bonding, 24–25
  by coenzymes, 46
  by proteins, *79*
Electron transport
  anaerobic, 133
  in cells, 326, 330–331
  in glycolysis, 126
  in photosynthesis, 112
  steps in, 102, *102*
  system of, 102
  and water formation, 130
Electron transport phosphorylation
  in aerobic respiration, 130,
  130–131
  summary of, 124, *125, 135*
Electrophoresis, 220–221, 293
Element
  in chemical formulae, *22*
  chemical symbols for, 18, *19*
  defined, 18
  essential for human nutrition, *727*
  essential for plant nutrition,
  500–505, *502*
  half-life, 20
  isotopes, 19
  naturally occurring, 18
  and origin of life, 322–323
  radioactive, 20
  trace, 18
Elephant, 811, *811*
Elephantiasis, 421, *421, 664*
Elephant seal, 279, *283*
Elk, *917*
Elm tree, 877
El Niño Southern Oscillation
  (ENSO), 888–889, *889*
Elodea, 111, *113*
Elongation, polypeptide chains in
  translation, 226
Embryo, animal
  body axis, 760
  chicken, *765*
  chordate, 446
  development, 755–756,
  772–773, *773, 791*
  and egg polarity, 760
  erythrocyte formation, 671
  eye formation, *766*
  fishlike structures in, *304*
  growth rates, 304
  inducer signals, 766–767
  inner cell mass, 762
  neural tube, 578
  organogenesis in, *759*

teratogen sensitivity, *790*
  vertebrate, comparison, *304*
Embryo, plant, 518–519, 522, 524
Embryogenesis
  chick embryo, *759*
  early cleavage, 784–785
  frog embryo, *758*
Embryology, comparative, 304–305
Embryonic development
  axis for, *758*
  chick, *757*
  comb jelly tissue, 417
  comparative, 304–305
  deuterostome *vs.* protostome,
  423, *423*
  frog, *757*
  human, *757*
  sea urchin, *757*
  similarities among vertebrates,
  304
Embryonic disk, *784,* 785–786
Embryonic induction
  defined, 766
  in *Drosophila,* 767
  experiment with, *766*
Embryo sac, *519*
Emotional state
  brain areas affecting, *582*
  drug effects, 588–589
  light effects, 625
Emperor penguin, *847, 931*
Emphysema, 256, 706
Emu, *262–263*
Emulsification, 721, *722*
Emulsion, defined, 721
Enamel, tooth crown, *718*
Encephalartos ferox, 402
Encephalitis, *357*
Endangered species, allele loss,
  276–277
Endemic disease, defined, 359
Endergonic reaction, 96, *96*
Endler, J., 288, 812–813
Endocrine gland, 549, 622
Endocrine system
  and blood flow, 663
  control of digestive system,
  718–719, 723
  defined, 614
  and digestion, 723
  functions, *554*
  human, *554*
  links with other systems, 614
Endocytosis
  animal viruses, 358
  defined, 88
  at plasma membrane, *85*
  receptor-mediated, 88, *89*
Endoderm, 410, 756, *762*
Endodermis, 494, 504, *504–505,*
  507
Endometriosis, 779
Endometrium
  estrogen effects, 779
  during fertilization and
  implantation, 784
  functions, 778, *778*
  location, *778*
  in menstrual cycle, *782*
  structure, 778
Endomycorrhizae, 387
Endoplasmic reticulum
  appearance in micrographs, *65*
  functions, 60, *64, 73*
  occurrence in nature, 73
  structure, 65
Endorphins, 568, 589
Endoskeleton
  bird, 463
  evolution, 450
  frog, 457
  shark, 637
  vertebrate, 636–637
Endosperm, 402, *404,* 504–505,
  509, 518, *518–519, 532*

Endospore, *351*
  Clostridium botulinum, 569
  Clostridium tetani, 354, *354,* 569
  defined, 354
Endosymbiosis, 331–333
Endothelium
  arteriole, *662*
  artery, *662*
  capillary, *662*
  vein, *662*
Endotherm, 746
End product
  defined, 98
  in metabolic reactions, 96–97
Energy, *see also* Fossil fuel
  ATP transfer, 4–5
  in covalent bonds, 94
  defined, 3, 93
  directional flow, 94–95
  entropy, 95
  first law of thermodynamics, 94
  heat, 94–95
  high quality forms, 95
  hill, 99, *99*
  light, in phototropism, 536
  low-quality, 95
  and metabolism, 4
  metabolism in muscle, 647
  net, 904
  net gain reactions, 96, *96*
  net loss reactions, 96, *96*
  nonrenewable, 396
  nuclear, 904–906
  second law of thermodynamics,
  95
  solar-hydrogen, 906
  sources, *904,* 904–907
Energy flow (transfer)
  through biosphere, 133,
  136–137
  through detrital food web, *849*
  through ecosystems, *846,*
  848–850, *848–850, 862*
  and electron activity, 19
  through grazing food web, *849*
  and interdependency among
  organisms, 5
  and organization of life, 136–137
  pyramid, *848, 850*
Enerobius vermicularis, 420
Englemann, T., *111*
English bulldog, 271
Enhancer, defined, 242
Enkephalin, 589
ENSO (El Niño Southern
  Oscillation), 888–889, *889*
Entamoeba histolytica, 366
Enteropmorpha, 837
Enterovirus, *357*
Entropy, defined, 95
Envelope, of viral capsid, 356, *356*
Environment
  adaptation to, 279, 292–293, 471
  cellular, 556
  effect on population size, 266,
  501
  effects on phenotype, 182, *182*
  and gene mutations, 275,
  292–293
  human impact on, 892–909
  internal, of animals, 547,
  556–557, *557,* 744
  internal, of fish, 743, *743*
  and phenotypic variation, 273,
  273
  plant adaptations to, 488,
  536–538
Enzyme
  acrosomal, 783, *783*
  active site, 98
  allosteric, 100–101
  bacterial, *248,* 355
  and body temperature, 744
  bone cell secretions, 639
  chemical evolution, 325

coenzymes, 101
  common features, 98
  control over activity, 136
  defined, 36, 98
  digestive, *720*
  DNA repair, 214, *215*
  DNA replication, *215*
  energy hill diagram, *99*
  feedback inhibition, 100–101,
  101
  fish muscle cells, 292–293
  functions, 36, *48*
  fungal, 378
  hydrolytic, 37
  induced-fit model, 98, *99*
  lactose-degrading, 236, *237*
  metal ion cofactors, 101
  pancreatic, *720,* 721
  against pathogens, 676
  pH effects, 100
  protective, 676
  reactions mediated by, 36
  replication, 250
  restriction, 249, *249, 252*
  ribosomal surface, 57
  spider, 433
  starch-degrading, 718
  stomach, 718
  substrate interactions, 98–99
  synthesis, control of, 100–101,
  101
  temperature effects, 100
  viral, *211, 358*
Eocene, *340, 475*
Eosinophil, *656–657*
Ephedra, 402, *402*
Epidemic
  AIDS, 796
  defined, 359
  ergotism, 383
  influenza, 359
Epidermal growth factor, 626
Epidermal skin cancer, 635
Epidermis, animal
  artificial, 553
  cnidarian, 414
  component of skin, *632,* 634
  laboratory-grown, 553
  melanin-producing cells, *635*
  rotifer, *423*
Epidermis, plant
  defined, 487
  examples, *484, 489, 491,*
  507–508
  fragrant oils in, 517
  vs. periderm, 487
  root, 495, *503*
Epidermophyton, 383
Epididymis (epididymides)
  functions, 775
  location, *774–775*
  role in semen formation, 774,
  774–775
  structure, 774
Epiglottis
  functions, 700
  location, *700*
Epilepsy, 586
Epinephrine, 731
  actions, 616, *621, 623*
  and memory storage, 587
  secretion, *616,* 623
  sources, *621*
  targets, *621*
Epistasis, *78–179,* 178–179
Epithelium
  basement membrane, *548,* 550
  cell cycle in, 144
  cell shapes, *548–549*
  cell-to-cell contacts, 549, *549*
  characteristics, 548, *548,*
  548–549
  cnidarian, 414
  designer organs from, 553
  gap junctions, *549*

Epithelium *continued*
  glandular, 549
  invertebrate, *577*
  pigmented, *608*
  simple, 548–549
  stratified, 548–549
  tight junctions, *549*
EPSP, *see* Excitatory postsynaptic
  potential
Epstein-Barr virus, *357*
Equation, chemical, *22*
Equilibrium
  chemical, 96–97, *97*
  dynamic, 82, *509*, 599
  static, 599, *599*
Equilibrium position, 599
Equisetum, 280, *280*, *398*
Ergotism, 383
Erosion, 504–505
Erwin, D., 310
Erythroblastosis fetalis, 669
Erythrocyte, *see* Red blood cell
Erythropoietin
  actions, *621*
  sources, *621*
  targets, *621*
Escherichia coli
  appearance in micrographs, *349*
  bacteriophage infections, 210,
    211, 212
  body plan, *56*
  classification, *351*
  DNA, 57
  functions in gut, 353
  gene activity in, *237*
  infections caused by, 354
  lactose breakdown, 236, *237*
  lactose operon, *237*
  meat contamination by, *56*
  pili, *348–349*
  plasmids, *248*
  staining characteristics, *349*
  synthetic insulin gene, 254
Esophagus
  function in digestion, *717*
  human, 718
  squid, *427*
Essential amino acid, *725*
Essential fatty acid, 725
Estrogen
  actions, *616*, *621*, 779
  and bird sound system, 913, *913*
  menstrual cycle secretion, 781
  and ovarian function, *782*
  and the pill, 795
  placental secretion, 787
  precursors, 41
  secretion, *616*, 626, 778
  sources, *621*
  targets, *621*, 781
  and uterine function, *782*
Estrous cycle, 778
Estuary, ecosystem, 882
Ethanol, formation, *133*
Ethics
  fertility control, 794
  in vitro fertilization, 798
  pregnancy termination, 798
  recombinant DNA technology,
    257
Ethiopian biogeographic realm, *870*
Ethylene, *534*, 535, 537, 540
Ethyl group
  locations, *34–35*
  structural formula, *34–35*
Eubacterium (eubacteria)
  evolution, 330
  heterocysts, 352
  pathogenic, 354, *354*
  photoautotrophic, *351*, 352–353
Eucalyptus oil, 477
Eucalyptus tree, 403, 874, 920
Euchema, 372
Eugenic engineering, 257
Euglena (Euglenophyta), 370, *370*

Eukaryote
  asexual reproduction, *142*
  cell components, *73*
  cell division, 142, *142*, *376*
  cell features, *60–61*
  cell wall, *376*
  defined, 58
  DNA, *216*
  evolution, 331–333, 362
  metabolism, *376*
  organelles, 58–60
  vs. prokaryote, *376*
  ribosome, 226
Euphorbia, *864*
European starling, *838*, 910–911
European wild boar, *838*
Eutheria, 466–467
Eutrophication, 881
Eutrophic lake, 881, *881*, *891*
Evaporation
  behavioral adjustments, 746,
    747
  from body surface, 745
  defined, 745
  in hydrologic cycle, 852, *852*
  physiological adjustments, 746,
    747
  water, 26
  from a watershed, *853*
  and water–solute balance, 745
Evaporative heat loss, 749
Evergreen broadleaf forest, *see*
  Tropical rain forest
Evergreen coniferous forest, *see*
  Coniferous forest
Evergreen plant, 491
Evergreen tree, 401
Evolution
  and altruism, 927–931
  and behavior, 916–917,
    920–932
  populations, 272–273
  and self-sacrifice, 926–931
Evolution, biological, *see also*
  Coevolution; Macroevolution;
  Microevolution; Natural
  selection
  adaptive radiations, *340–341*
  aerobic respiration, 113
  amino acid sequences and,
    308–309
  amphibians, *450*
  angiosperms, 336, *340–341*
  animals, 333–334, 336, *340–341*,
    408–409, *412*, 734–735
  annelids, 429
  bacteria, 353
  bacterial symbionts, 362
  and balanced polymorphism,
    282–283
  and base pairing, 229
  and bilateral symmetry, 576–577
  birds, *450*
  bottlenecks in, 276–277
  cacao tree, 515
  and cephalization, 419, 576–577
  and changing atmosphere, 136
  chemical, *326*
  chloroplasts, 68, 332
  chromosome structural changes
    during, 200
  and coelom formation, 411
  comb jellies, 417, *417*
  convergent, 448, *864–865*, 870
  correlated with earth's history,
    390
  defined, 10, 261, 271
  deuterostomes, 423
  through directional selection,
    278–279, 304
  disruptive selection, 280
  DNA, 229
  through DNA mutation, 10
  and Earth's history, 260–263,
    265, 328, *342–343*

  fishes, 335, *340–341*, 450, 452
  flatworms, 418
  fossil evidence, 268
  founder effect, 276
  fungi, 333
  gastropods, 425
  through gene flow, 277
  and gene mutation, 449
  through gene mutation, 275, 765
  through genetic drift, 276
  gradualistic model, 296
  Hardy–Weinberg principle, 274,
    274
  hemichordates, 448, *449*
  insecticide resistance effects,
    279
  insects, 335–336, *340–341*,
    436–437
  Lamarck's theory, 264
  limbic system, 588
  lobe-finned fishes, 335, *456*
  lung, 456, 699
  mammals, 336, *340–341*, 450,
    464
  mass extinctions, 334–338,
    340–341, 341
  metabolic functions, 136–137
  milestones, summary, *342–343*
  mitochondria, 67
  through mutation, 229
  through natural selection,
    278–283
  nautiloids, 334
  nervous system, vertebrate, 578
  oxygen usage in metabolic
    reactions, 362
  perspective on, 136–137
  photosynthetic pathway, 136
  placoderm, 453
  plants, 336, *340–341*, 520
  and plate tectonics, 328, 336
  pollen grain, 393
  protistans, 362–363
  protostomes, 423
  punctuational model, 296
  rate, 274–275
  and recombinant DNA
    technology, 257
  reptiles, 335, *340–341*, 450,
    458–459, *459*
  root systems, 392
  and seafloor spreading, 328
  seeds, 393
  segmented body plans, 431
  self-replicating systems,
    326–327
  and sexual dimorphism,
    282–283
  through sexual selection,
    282–283
  shoot systems, 392
  skull, 765
  snakes, 460
  stabilizing selection, 280
  through stabilizing selection,
    280, 304
  systematics, 314
  theories, 264
  time scale for, 266, 301
  trends in, *440*
  tribolites, 334
  and variation in populations,
    269–271
  vertebrate, 444–446, 448–451,
    578
Evolution, cultural, 330–331, 335,
  339–341
Evolutionary tree, *see also* Tree of
  descent
  adaptive radiation of mammals,
    297
  diagrams, 296, *296–297*
  eukaryotes, *333–334*
  five-kingdom classification, 318
  primates, *472*

Excitable cell, 646
Excitatory postsynaptic potential,
  568, *568*
Exclusion, competitive, 826–827
Excretion
  extracellular fluid loss by, 736
  urinary, 736
Exercise
  and aging, 793
  ATP formation, 648
  benefits to muscular system, 648
  and bone tissue turnover, 639
  brief, intense, 647
  and cardiovascular disorders,
    666
  and dieting, 729
  effect on venous pressure, 664
  evaporative water loss during,
    745, *745*
  glycogen uptake during, 39
  metabolic effects, 729
  muscle contraction during, 647
  muscle fatigue, 648
  prolonged, moderate, 647–648
Exergonic reaction, 96, *96*
Exhalation, 702
Exocrine gland, 549, 622
Exocytosis
  defined, 88
  at plasma membrane, *85*
Exodermis, root, 494
Exomycorrhizae, 387
Exon, defined, 223
Exoskeleton, 636–637
  arthropod, 432
  barnacle, 434, *434*
  centipede, 432
Experiments
  auxin effects on plant growth,
    534
  bacterial transformation, 210,
    210
  bacteriophages, 210, *211*
  bird nest fumigation, 910–911,
    910–911
  cell aging and death, 768
  cell development and
    differentiation, 764
  cell fate in developing embryos,
    761
  circadian rhythms in plants, 538
  competitive coexistence, 827
  competitive exclusion, 826–827
  daylength effects on plant
    growth, *541*
  dihybrid crosses, 174–175, *175*
  DNA functions, 211, *211*
  dormancy, *541*
  ecosystem modeling, 861, 889
  embryonic induction, 766, *766*
  feather coloration effects, 12–13
  food scarcity and population
    decline, 829
  gene localization on
    chromosomes, 193
  genetic contribution to behavior,
    912, 912–913, *913*
  genetic drift, *286*, 286–287
  genetic engineering, 247,
    254–256, *254–256*
  global warming, 857
  gravitropism, 536, *537*
  heart beat recording, 652
  Hubbard Brook Valley
    watersheds, 852, *853*
  intense sound effects on inner
    ear, *601*
  life history patterns, 810–813
  mark-release-recapture, 279,
    279
  mathematical models, *806*, *808*
  mechanical stress, 537, *537*
  Mendelian, 170
  monohybrid crosses, 172–173,
    172–173

Experiments *continued*
  native herd ranching, 901
  nature's, 247, 376, 431
  nerve growth factor effects on cultured neurons, 626
  neural function, 586
  nuclear transplant, 764
  nutrient enrichment effects on a lake, 881, *881*
  parasitoid wasp effects on sawfly cocoons, 833, *833*
  perception, conscious, 586
  photoperiodism, *538*
  photosynthesis, *111*
  phototropism, 536, *536*
  phytochrome control of gene transcription, 243
  pollen export by flowers, 524–525
  predation, 836–837
  processing visual information, *609*
  radioactive labeling, *211*
  reciprocal crosses, *192*
  red blood cell membrane structure, 80–81, *80–81*
  resource partitioning, 827
  role in scientific inquiry, 12
  senescence, *540*
  song learning, 912–913, *913, 914*
  split-brain, 586
  stem elongation, 536, *537*
  stem growth, *534*
  tactile signaling of bees, *923*
  vernalization, *540*
  wilting, 84
  X-linked traits, *192*
Exponential growth, *806,* 806–807, *807*
External oblique muscle, *643*
Extinction, 297
  as agent of evolution, 329, 338, 341
  angiosperms, 403
  background, 297
  Cretaceous–Tertiary boundary, 338
  DDT-related, 861
  Devonian, 335
  dinosaurs, 337–338
  Jurassic, 337
  mass, 334–337, 341, *342–343*
  mushrooms, 387
  Ordovician, 334
  ostracoderms, 453
  Permian, 335
  placoderms, 453
  Triassic–Jurassic boundary, 336
  tropical, 840
Extracellular fluid
  and acid–base balance, 742
  and cell survival, 556
  changes in, 736
  composition, 556
  defined, 556, 736
  factors affecting, 736
  gains and losses, 736
  human body, 556
  interstitial, 556
  ion concentration, 556, 736
  pH level, 742
  stability, 556
  volume, 556, *618,* 742
  water gains and losses, 736
Extracellular matrix
  bone, 638–639
  components, 550
  ground substance, 550
Extraembryonic membrane, *784*
Eye
  components, 602
  compound, 602, *602*
  diseases, 606
  disorders, 897

  focusing mechanisms, 604, *605*
  formation, embryonic, 764, 766–767, *766–767*
  human, 604, *604–605*
  injuries to, 606
  innervation, *580*
  insect, 602, *602*
  invertebrate, *602,* 602–603, *603*
  mammal, *604*
  mollusk, 424, *424,* 602, *602*
  rotifer, 422
  vertebrate, 604, *604,* 604–609, *605*
Eye color
  continuous variation in, 180, *180*
  Drosophila, 192, 193
  melanin role, 180
Eyespot, 370, *370,* 438–439, 602, *602–603*

F

Facial bones, *641*
FAD (flavin adenine dinucleotide), 46, 101, *128,* 129
FADH2 (flavin adenine dinucleotide, reduced), 101, *125,* 128–129, 130, *131*
Fail-safe gene, 254
Fallout (radioactive), from Chernobyl, *905*
Fall overturn, lake, 880–881
Familial hypercholesterolemia, 197
Family pedigree, 195, *195, 199,* 205
Family planning, 816–817
Family (taxon), 8, 311
Family tree, *see also* Tree of descent
  for chordates, *446*
  and evolutionary relationships, 316
Farsightedness, 604, 606
Fascicle, nerve, *570*
Fat
  as alternative energy source, 134, *135*
  animal, 725
  in bloodstream, 666–667
  body composition, 729
  deposits, arterial, 666–667
  dietary, 725
  and dieting, 729
  digestion, 134, *720,* 721
  digestion products, 722, *722*
  energy yield from, 134
  excess, in obesity, 728
  functional groups, *35*
  functions, *48*
  hypodermal, 632
  metabolism, 134, *135,* 720
  neutral, *see* Triglyceride
  nondigestible, 725
  saturated, 725
  storage in animals, 546
Fate map
  defined, 767
  Drosophila zygote, *767*
Fatigue, muscular, 648
Fat-soluble vitamin, *726*
Fatty acid
  condensation into triglyceride, *40*
  conversion to energy, 134
  essential, 725
  saturated, 40
  structure, 40, *40,* 134
  unsaturated, 40
Faulty enamel trait, *196,* 198
Feather
  bird, 462–463, *463*
  coloration, 12–13
  peacock, *633*
Feather star, 438

Feces
  concentration mechanisms, 723
  diet effects, 724
  expulsion, 723
  fossilized, 302
  hookworm transmission in, 420
  Pilobolus growth on, 381, *381*
  tapeworm transmission in, 419
Feedback
  and accelerated change, 906–907
  and body temperature, 556–557, *557,* 748–749
  in epilepsy, 586
  and homeostasis, 620, 623
  negative, *556,* 556–557, *623,* 906
  positive, 557
Feedback loop
  homeostatic, 621, 623
  inhibition of metabolic pathways, 100–101, *101*
  during menstrual cycle, *781*
  testosterone production, *777*
Feedforward control, 906–907
Feeding behavior
  food chains, 847
  genetic contribution, 912
  mammalian, 464
  placoderm, 453
  snake, 460, 912
Feeding levels, *see* Trophic levels
Feeding net, comb jellies, 417
Femoral artery, 659
Femoral vein, *659*
Femur, 640
  location, *641*
  structure, *638*
Fermentation pathway
  alcoholic, 133, *133*
  energy yield from, 132
  lactate, 132, *132*
  organisms using, 132, 382
Fern, *393,* 396, 398, *399, 406*
Ferrous iron, 101
Fertility control
  abortion, 798
  contraceptive methods, 794–795
  in vitro fertilization, 798
Fertilization
  chromosome number at, 162
  combining alleles, 273
  comparative (animal cell), *757*
  defined, 156, 519
  double, 519, *519*
  egg cytoplasm changes, 760
  egg cytoplasmic activity, 760
  external, 290, 754–755
  and gene flow, 289
  human, *784*
  internal, 456–457, 754
  internal (reptile), 458
  in vitro, 205, 798
  isolating mechanisms, postzygotic, *289,* 291
  isolating mechanisms, prezygotic, *289,* 290–291
  sea urchin, 290
Fertilization, plant, *see also* Hybridization
  asexual reproduction, 526–527
  double, *404*
  and genetic engineering, 254–255
  isolating mechanisms, postzygotic, *289,* 291
  isolating mechanisms, prezygotic, *289,* 290–291
  pollination, *see* Pollination
  reproductive modes, 516
Fertilizer
  kelp-derived, 373
  production cost, 859
  reliance on, 858–859
Fetal alcohol syndrome, 791, *791*
Fetal development

  birth process, 792, *792*
  human hand, *151*
  stages, *793*
Fetus
  blood circulation, 787, *787*
  defined, *793*
  development, 772–773, *773*
  premature, 789
  Rh+ markers, 669
  teratogen sensitivity, 790
Fever, *357,* 749
  defined, 678
  in inflammatory response, 678
Fever blister, *357*
Fiber
  dietary, 723
  insoluble, 723
  role in intestinal health, 723
  sclerenchyma cells, *489*
  soluble, 723
Fibrillation, ventricular, 667
Fibrinogen
  in blood plasma, 656
  functions, 656, *656*
Fibroblast, cytoskeleton, *69*
Fibroma, *357*
Fibrous joint, 640
Fibrous root system, 494, *494*
Fibula, *641*
Fick's law, 696
Fight-flight response, 581, 623–624
Fighting, reproductive benefits, 283
Fig tree, strangler, 826
Filament, of stamen, *516*
Filtration
  blood, 740
  in capillary bed, *664–665*
  urine formation by, 739–740, *740*
Fin
  evolution, 450
  fleshy, 450
  paired, 450, 453
  ventral, 450
Finch
  African, beak size, 280–281
  beak size and shape, 267
  Darwin's, 267
  directional selection, *281*
  disruptive selection, 280–281
  species, *267*
  zebra, 913, *913*
Fingerprint, DNA, 252
Fingerprinting, genetic, 930
Fire, adaptation to, 836
Fire ant, *838*
Firefly, 255, 923
First law of thermodynamics, 94
Fir tree, 878
Fish
  and acid deposition, 896
  bony, 454, *454,* 743, 754, 756
  cartilaginous, 454, *454,* 637, *637*
  classification, 446
  comparative embryology, 304, *304*
  and DDT, 861
  egg, 762
  evolution, *450, 452,* 452–453
  gills, 698, *698,* 743
  internal environment, 743
  jawed, *450, 452*
  jawless, *450,* 452–453
  lobe-finned, 335, *335,* 454–456, *456*
  lungfish, 454–456
  as protein source, *725*
  ray-finned, 454–455
  respiratory system, 698, *698*
  salt tolerance, 743
  salt-water balance, 743
  sexual dimorphism, 312
  sporozoan parasites, 368

Fish *continued*
summary of characteristics, *442–443*
survivorship curves, 811, *811*
water–solute balance, 743, *743*
Fission
binary, *350, 376,* 754
prokaryotic, *142*
Fissure of Rolando, *584*
Five-kingdom classification, 318, *318*
Five-toed limb, 306
Fixed action pattern, 914
Flagellum
bacterial, 56, *56,* 349
bending, 70, *71*
functions, *73*
occurrence in nature, *73*
sperm, *70,* 776
structure, 70, *71*
Flame cell, 418, *418–419*
Flatworm (Platyhelminthes)
asexual reproduction, 154, 754
body plan, 418
characteristics, *577,* 754
circulatory system, *443*
classification, 410
digestive system, *443,* 716, *716*
evolution, *412,* 576–577
infections in humans, 420
major groups, 418
movement, *442*
nervous system, *442,* 576–577, *577*
reproduction, 418, *443,* 754
respiratory adaptations, *697*
respiratory system, *443*
Flavoprotein, 536
Flea, 359, *436*
Flemming, W., 187
Flesh fly, 843, *843*
Flight, insect, *637*
Flight, requirements for, 463, *463*
Florida panther, and inbreeding, 277
Florigen, *534,* 535–536
Flow
bulk, 83
in circulatory system, *655*
countercurrent, in gills, 699, *699*
lipids through cytomembrane system, *64*
proteins through cytomembrane system, *64*
Flower
adaptations, *405*
chromoplast, 68
color, *273*
components, *516,* 516–517, *517*
development, and evolution, 304–305
egg development in, 517
excess, 524–525
vs. fruit, 524
imperfect, 517
monocot, *404*
nectar, 520–521
perfect, 517
pollen formation, 517
pollination, *516,* 520–521
structure, *516,* 516–517, 521
Flowering plant, *see also* Angiosperm; Plant (Plantae)
accessory, *522*
aggregate, *522*
Agrobacterium-infected, 255
asexual reproduction, 526–527, *526–527*
blue, 838–839
body plan, *484,* 484–485
characteristics, *406*
classification, 391
coevolution with pollinators, 402, 506–508, 520–521, 825
collenchymal tissue, 486

day-neutral, 539
dermal tissue, 487, *487*
dicot, 405
diversity, 403, *403*
dormancy, 541, *541*
double fertilization, *404*
embryo sporophyte, 522
endodermis, 490
evolution, *393,* 482, 864
exodermis, 490
florigen effects, 535
flowering process, *291,* 539
fruit, 522–523, *522–523*
gibberellin effects, 534
hormones, 534
life cycle, *404, 516*
long-day, 539
meristem, 485, *485*
monocot, 405
monocot *vs.* dicot, 474
multiple, *522*
mutualistic relationships, 502–503
parenchymal tissue, 486
pollination, 520–521
pressure flow in, 510–511
root nodules, 503
root structure, 494–495
root system, 484, *484*
sclerenchymal tissue, 486
seed formation, 522–523
shoot structure, 488–492
shoot system, 484, *484*
short-day, 539
simple, *522*
species introductions, 838–839
sperm, 516, *516*
vascular tissues, 486, *486*
vernalization response, 540
water uptake, 502–505, *504*
woody, 496–497
Fluid
body, as source of viral transmission, 691
flow in circulatory systems, *655*
interstitial, 736
Fluida, 264
Fluke, *410,* 418, 420, *420, 442–443,* 833
Fluorine
deficiency, *727*
excess, *727*
functions, *727*
sources, *727*
Fly, 521
flesh, 843, *843*
hanging fly, 917, *917*
Fly agaric mushroom, *384*
Folic acid, *726*
Follicle-stimulating hormone
actions, 618, 780
feedback loops affecting, *781*
and female reproductive function, 780
and male reproductive function, 776–777
secretion, *616,* 777, 780
source, *615,* 618-619
targets, *618,* 619, 780
Fontanel, 640
Food chain, 847, *see also* Food web
Food pyramid, 724, *724*
Food web, 352
Antarctic, *847*
aquatic, 854
coastal, 883
crustaceans in, 434
and DDT, 861
defined, 847
detrital, 848, *849,* 882–883
estaurine, 882
grazing, 848, *849–850,* 883
in hydrothermal vents, 885
intertidal zone, 883

Foot, clam, *426*
Footprint, early hominid, 477, *477*
Foraminiferan, 366, 367
Forced exhalation volume, *703*
Forced inhalation volume, *703*
Forebrain, *582*
Forelimb
evolution, 306–307
morphological convergence in, 307, *307*
morphological divergence in, 306, *306*
Forest
and acid deposition, 896, *896*
biomes, 876–878
boreal, 878, *878*
burning, 836, 857, 893, *902, 903*
conifer, 401
defoliation with 2,4,5-T, 365
deforestation, *see* Deforestation
evergreen coniferous, 878, *878*
function, 902
monsoon, 877
montane coniferous, 878, *878*
sequoia groves, 836
soil, *872*
temperate deciduous, *877*
Forest tree, symbiosis with mycorrhizal fungi, 387
Formaldehyde, chemical evolution into porphyrin, *326*
Formula, chemical, *22*
Fossil
ammonite, *301*
Archaeopteris, 335
Archaeopteryx, 268
Burgess Shale, *408,* 408–409
Cambrian, *334*
cells, *330*
defined, 302
derivation of term, 302
Devonian, *335*
discussed, 302–303
examples, *302*
fungus, 379
hominid, 477–478
living, *456, 461*
Lucy, *605, 477*
plant, 390
platypus lineage, 445
radioactive dating, 301, 470
requirements, 303
sequences, 263
spores, 330
teeth, 476–477
tooth, 464
trace, 302
Fossil fuel
burning, 854, *855,* 857, 895
depletion, 396
formation, 854, *855,* 904
Fossil record
analysis, 314
australopith, 476–477
bacteria, 352
and catastrophism, 264
completeness, 303
and Darwin's theory, 268
environmental change, 292
and evolutionary thought, 260–261, 263, 268
genetic histories, 288
geologic time scale, 303
interpretations, 300, 303, 314
macroevolution, 301
missing links in, 268
morphological gaps in, 304
and stratification, 303
Founder effect, 276, 296, 304
Founder species, 294
Fovea, *605,* 609
Fox, S., 327
Foxglove, 493
Foxtail, 827, *827*

Fragaria, 523
Fragile X syndrome, *196,* 200
Frameshift mutation, 228, *228*
Franklin, R., 212
Fraternal twins, 792
Free nerve endings, 596, *597*
Free radicals
and age spots, 93
effects, 92–93
formation, 92
gene mutation caused by, 228
vitamin effects, 727
Freeze-etching, 81, *81*
Freeze-fracturing, 81, *81*
Freezing, engineered resistance, 254, *254*
Frequency, sound, 600
Fresh water, 893, 898, *898,* 899
Frigate bird, 463
Frilled lizard, *460*
Fringe-lipped bat, 923
Fringing reef, 886, *886*
Frisch, Karl von, *923*
Frog
brain, *450*
circulatory system, 457
courtship behavior, 923
egg, 752, 760
embryonic development, 752–753
endoskeleton, 457
evolution, *450*
heart, 457
reproduction, 752–753
skin, 457
tadpole, *752*
Frontal lobe, *584,* 585
Frostbite, 749
Frost tolerance, genetically engineered, 254
Fructose
functions, *48*
ring forms, *37*
straight-chain form, *37*
Fructose-1,6-bisphosphate, *127*
Fructose-6-phosphate, *127*
Fruit
adaptations, 524–525
characteristics, *522*
coevolutionary mechanisms, 524
defined, 522
from flowering plants, *522*
formation, 522–523, *522–523*
ripening, 535
saquaro cactus, 524–525, *524–525*
Fruit fly, *436*
Fruiting body, bacterial, 355
Fucoxanthin, 371–372
Fuel
alternative sources, 906–907
fossil, *see* Fossil fuel
nuclear, 904–906
Fumarate, *129*
Fumigation, natural, 910–911, *910–911*
Functional groups, of organic compounds, 34–35, *34–35*
Fungal infection
duration, 381
eye disease caused by, 606
transmission mode, 381
Fungus (Fungi), *see also* Club fungus; Sac fungus; *specific fungus;* Zygomycete
ammonifying, 858
antibiotics from, 378
beneficial associations with plants, 386–387
body plan, 380
cell components, *73*
cell wall, 39, *39*
classfication, 9
as decomposer, 5, 378, 380
diseases caused by, 381

Fungus continued
  enzymes, 378
  evolution, 318, 333, 364
  five-kingdom classification, 318, 318
  harmful, 676
  hyphae, 386–387, 387, 825
  imperfect, 379, 383, 388, 388
  lichens, 386, 386
  major groups, 379
  mycorrhizae, 386–387, 387, 403
  nutrition, 380
  parasitic, 380–381
  pathogens, 383
  predatory, 388
  representative, 9
  reproduction, 380
  saprobic, 380–381
  spores, 388
  stinkhorn, 9
  survivorship curves, 811, 811
  and termites, 920–921
  value, 378
  yeasts, 133, 133, 382
Fungus roots, 386–387, 387
Fur color, see Coat color

G

Galactose, in lactose metabolism, 196
Galactosemia, 196, 196–197, 204
Galactose-1-phosphate, 197
Galactosidase, 237
Galápagos Islands, 265, 267, 460
Galápagos Rift, 884–885
Galileo Galilei, 13, 50
Gallbladder, 717
Gametangium, 380, 381, 395
Gamete
  defined, 156
  formation in animals, 162
  formation in flowers, 518
  formation in plants, 162
  haploid, 157, 162
  hybrid infertility, 289
  isolation, 289
  mature, 1162
  structures producing, 156
Gametic isolation, 289, 290
Gametic mortality, 289
Gametocyte, 369
Gametogenesis
  animal, 756
  plant, 392
Gametophyte, 162
  female, 518
  flowering plant, 516, 516
  formation, 516
  functions, 393
  immature, 518
  in life cycle, 392, 392–393
  mature, 518
Gamma aminobutyric acid (GABA), 568
Gamma globulin
  anti-Rh, 669
  functions, 656
Gamma ray, DNA damage caused by, 244
Gandwanaland, 443, 445
Ganglion
  defined, 428, 580
  earthworm, 428
  invertebrate, 576–577, 577
  mollusk, 424
  receptive field, 609
  retinal, 608, 608, 609, 609
Gap junction, 72, 549
Gargas cave, 470–471
Garrod, A., 220
Garter snake, 912, 912
Gas, natural, 854, 904
Gas exchange

alveolar, 704
  chemical controls, 705
  by diffusion, 696
  integumentary, 696, 698
  neural controls, 705
  respiration, 696–699
  transport pigments, 697
  by ventilation, 697
Gasoline, and smog, 895
Gassner, G., 540
Gastric fluid, 718–719, 729
  acidity, 29
Gastrin
  actions, 621
  secretion, 723
  sources, 621
  targets, 621
Gastrocnemius, 643
Gastrodermis, cnidarian, 414
Gastroenteritis, 357
Gastrointestinal tract
  components, 717
  human, 580
  and nervous system, 580
  structure, 717
Gastropod, 424–425, 425
Gastrula, 757, 762
Gastrulation, 760, 763, 765, 786
  comparative, 757
  defined, 756
Gause, G., 826
Geese, 463, 915
Gel electrophoresis, 220–221, 250-251, 252, 293
Gemmule, 413
Gender
  and cardiovascular disorders, 666
  determination, 190
Gene
  adenosine deaminase, 246–247
  allele, 156
  amplification, 238
  antibiotic-resistance, 252
  antibody, 686
  antitrypsin, 706
  Apo, 33
  and behavior, 912–913, 912–913
  chromosomal locus, 171
  defined, 156, 171, 188
  distance between, 193
  of endangered species, 277
  evolutionary relationships, 308
  exons, 223
  fail-safe, 254
  functions, 220
  glucose breakdown, 237
  hedgehog, 766
  hemoglobin, 220
  hemophilia A, 198
  hok, 254
  homeobox, 767
  housekeeping, 238
  human, possible combinations, 272–273
  ice-forming, 254
  information organization in, 169
  insulin, synthetic, 254
  interactions, 182
  introns, 223
  jumping, 228
  lethal, 229
  linkage groups, 193
  linkage mapping, 194
  linked, 192
  localization, 193
  locus, 171
  luciferase, 103
  male-determining, 190
  morphogens, 766
  mutation, 156, 187
  mutation rate, 229
  myc, 235
  nucleic acid hybridization, 252–253, 253

oncogene, 244
  operator, 236
  operon, 236, 237
  pairs, 171
  pleiotropy, 176
  promoter, 236
  and protein synthesis, 220
  proto-oncogene, 244
  regulatory, 304
  sex-linked, 193
  somatotropin, 256, 256
  spotting, 241
  for traits, 168–169
  transcription, 222–223, 222–223
  transposable elements, 228
  tyrosinase, 178, 182
  viral protein, 356
  X-linked, 192
  Y-linked, 192
Gene amplification, 249, 250, 250
Gene control, eukaryotic
  cell differentiation, 238
  chemical modifications, 238
  DNA rearrangements, 238–239
  gene amplification, 238–239
  levels, 238–239
  negative control systems, 236
  positive control systems, 236
  post-transcriptional, 238, 238–239
  post-translational, 239
  transcriptional, 238–239
  and transcript processing, 238–239
  translational, 238–239
Gene control, prokaryotic
  lactose operon, 236–237, 237
  mechanisms, 237
  negative control, 237
  negative control systems, 236
  positive control systems, 236
Gene expression
  developmental stage specificity, 240
  environmental effects, 273
  factors affecting, 236
  and genetic engineering, 256
  hormonal signals, 242
  molecular signals, 242–243
  and mutation, 275
  negative control, 236
  in plants, environmental effects, 182, 182
  in plants, phytochrome effects, 243, 243
  polydactyly, 195
  positive control, 236–237
  proto-oncogenes, 244
Gene flow
  barriers to, 289–295, 291
  disruptive selection effects, 281–282
  effect on genetic equilibrium, 274
  between hominids, 341
  physical barriers, 292
  between populations, 289
  postzygotic isolating mechanisms, 289, 291
  prezygotic isolating mechanisms, 289, 290–291
  process, 277
  stabilizing selection effects, 280
Gene mutation
  and aging, 768
  antennapedia gene, 767
  beneficial, 228–229, 275
  defined, 228
  Drosophila, 192
  embryonic, 304
  and evolution, 229
  fish enzymes, 292–293
  frameshift, 228, 228
  frequency, 228
  and genetic variation, 275
  harmful, 228–229, 275

hedgehog, 766
  heritability, 228
  lethal, 275
  mutagens, 228
  neutral, 228–229, 275, 292
  in proto-oncogenes, 244
  sickle-cell anemia, 220–221
  sickle-cell related, 282
  spontaneous, 228
  and variation in populations, 275
  and variation in traits, 182
Gene pair
  epistasis, 78–179, 178–179
  on homologous chromosomes, 172–173
  independent assortment, 174–175
  interactions, 178–179, 178–179
Gene pool, 272–273, 278, 288
Genera (taxon), 310, 322
Gene shuffling
  at metaphase I, 161
  at prophase I, 160–161
  and trait variations, 161
Gene therapy, 247, 257
Genetic abnormality
  defined, 196
  vs. genetic disorder, 196
Genetic code
  components, 224, 224
  defined, 224
Genetic counseling, 204–205
Genetic disorder, see also Chromosome abnormalities
  achondroplasia, 196, 197
  albinism, 178, 179
  autosomal dominant, 196
  autosomal recessive, 196
  camptodactyly, 180
  chromosome number-related, 196
  chromosome structure-related, 196
  defined, 196
  designer organs for, 553
  Down syndrome, 202
  early detection, 204
  familial hypercholesterolemia, 197
  gene therapy for, 247, 257
  vs. genetic abnormality, 196
  genetic counseling, 204
  and genetic disease, 196
  genetic screening, 204
  HbS/HbA heterozygotes, 282
  human, 196
  Huntington's disorder, 197
  Klinefelter syndrome, 203
  and life, 199
  nondisjunction-related, 203, 203
  phenotypic treatments, 204
  prenatal diagnosis, 204–205
  progeria, 186–187, 197
  RFLP applications, 252
  sickle-cell anemia, 176, 220–221
  Turner syndrome, 203
  xeroderma pigmentosum, 215
  X-linked recessive, 196
  XXY condition, 203
Genetic divergence, 289, 291
Genetic drift
  among snails, 286–287
  effect on genetic equilibrium, 274
  example, 276
  and extreme genetic uniformity, 277
  gene flow effects, 277
  process, 276
  and speciation, 289
  stabilizing selection effects, 280
Genetic engineering
  animals, 256, 256–257
  bacteria, 254

Genetic engineering *continued*
  benefits and risks, 254, 256
  ethical/social issues, 257
  mussel genes, 218
  and pest control, 279
  and pesticide reduction, 279
  plants, 254–255, *254–255*
  risks, 254
Genetic equilibrium, 274, *274–275*
Genetic fingerprinting, 930
Genetic recombination
  and crossing over, 194
  in nature, 247
Genetics, terminology, 171
Genetic screening, 204
Genetic variation
  and chromosome number, 273
  and chromosome structure, 273
  and crossing over, 273
  defined, 272
  and fertilization between
    different gametes, 273
  and gene mutation, 273
  and independent assortment of
    chromosomes, 273
Gene transfer
  in animals, 256, *256*
  in bacteria, 254, *254*
  luciferase gene, 103
  in plants, *254*, 254–255, *255*
Genital herpes, 797
Genital warts, 796
Genome, human genome project,
  252, 256
Genotype, defined, 171
Genus (taxon), 8, 310–311
Geoduck, 426
Geographic dispersal, 838–839
Geographic isolation, 290, 293
Geography, biological, 262
Geologic time scale, 303, 329,
  329
Geology, and evolution, 262–263,
  265, 328, 336, 341
German measles, *357*, 688, 791
Germ cell
  chromosomal combinations
    possible, 161
  chromosome number, 157
  cytoplasmic division, 155
  defined, 142
  meiosis, 155
Germination, 524, 530, 538, 836
  abscisic acid effects, 535
  and day length, 539
  pollen grain, 518
  seed, 532, *533*
  spore, 162
Germ layer, 418, 756, 763, 765
Gey, G., 145
Gey, M., 145
Ghost crab, 883
Giant oak, 542–543
Giant panda, *309*
Giant sequoia, 836
Giant squid, 426
Giardia lamblia, 368, *368*
Gibberella fujikuroi, 530
Gibberellin, 530, 534, *534*, 541
Gibbon, *472*
Gibbons, R., *71*
Giemsa dye, 189
Gigantism, 620, *620*
Gilford progeria syndrome, *196*
Gill
  bilvalve, 426
  clam, *426*
  club fungus, *385*
  evolution, 451
  external, 698
  fish, 698, 743
  gastropod, 425
  internal, 698
  mollusk, 424
  ostracoderm, *453*

snail, *424*
squid, *427*
water–solute balance, 743, *743*
Gill slits, 447, *449*
Gingivitis, *718*
Ginkgo biloba, 402
Ginkgo (Ginkgophyta), 335, *393*,
  402, *402, 406*
Glacial lake, *880*
Glacier Bay, *834–835*
Gland
  defined, 549
  endocrine, 549
  exocrine, 549
  innervation, *580*
Glans penis, 783
Glaucoma
  causes, 607
  chronic, 607
  treatments, 607
Global air circulation, *866,
  866–867, 867*
Global warming, 117, 854, *856,
  856–857, 893*
Globulin
  in blood plasma, 656
  functions, 656, *656*
Glomerular capillary, 738, *739*
Glomerular capsule, *739*
Glomerulus, 738, *739*
Glottis, 703, *703*
Glucagon, *616*
  actions, *621*
  secretion, *616*, 622
  sources, *621*
  targets, *621*
Glucocorticoids
  actions, *621*, 623
  sources, *621*
  targets, *621*
Glucose
  anaerobic breakdown, 132
  blood levels, 86, *622*, 623, 731
  breakdown reactions, 94
  carbon backbone, 126
  energy yield, *126, 131*
  functions, *48*
  homeostatic controls over, *622*
  metabolism, 86, *125*, 126 ff,
    127–129, 134, 135, 622, 730
  passive transport, 86
  reabsorption, *740*
  reabsorption values, *739*
  ring forms, *37*
  straight-chain form, *37*
Glucose insulinotropic peptide,
  718–719, 723
Glucose-1-phosphate, 197
Glucose-6-phosphate, *127*
  conversions, 134
  functions, 37
Glue, termite, 920, *920–921*
Glutamate, 220
Gluteus maximus, *643*
Glyceride
  functions, *48*
  structure, *48*
Glycerol, functions, 37
Glycocalyx, 349, *349*
Glycogen
  functions, 39
  and glucose metabolism, 546
  storage in animals, 546
  structure, 39
Glycolipid, in lipid bilayer, 78
Glycolysis
  ATP yield, 126, *127*
  defined, 126
  end products, *127*
  energy-releasing step, 126, *126*
  intermediates, *125*, 197
  net energy yield, 126, *127*
  reactions, *127*
  substrate-level phosphorylation,
    126, *126*

Glycoproteins
  adhesion, 80
  distribution, 46
  formation, 46
Glyoxysome
  distribution, 65
  functions, 65
Glyptodont, 266
Gnetophyte (Gnetophyta), *393*,
  402, *402, 406*
Gnetum, 402
Goiter, 624
Golden algae, 370–371
Golgi body
  appearance in micrographs, *66*
  budding, 66
  functions, 60, *64, 73*
  occurrence in nature, *73*
Gombe National Park, 613
Gonad
  cnidarian, 417
  as endocrine target, *618*
  female, 626, 773
  flatworm, 418
  hormone secretion, 614, *616*, 626
  male, 626, 773
  maturation, 773
  melatonin effects, 625
  mollusk, *424*
  sea star, *438–439*
  secretions, 773
  songbird, 912–913
  squid, *427*
Gonadotropin-releasing hormone,
  618, 619, 777, 780
Gonadotropins, 618, *618, 782*
Gondwana, 334, 445
Gonorrhea
  antibiotic-resistant strains, 355
  infection cycle, 797
  transmission modes, 359, 797
Goodall, J., *612*, 613, 926
Goose barnacle, 76, *76*
Gorilla, *316–317, 473*, 475, *633*
Gosling, *915*
G1 phase, cell cycle, 144, *144*
G2 phase, cell cycle, 144, *144*
Gradient
  defined, 82
  electric, 82
  pressure, 82
Gradualistic model, evolutionary
  change, 296
Gradual model, speciation, 296
Gram-negative eubacteria, 349, *351*
Gram-positive eubacteria, 349, *351*
Gram staining, 349, *349*
Grand Canyon, *303*
Grant, M., *526*
Granulocyte, *657*
Granulosa cell, 780
Granum, chloroplast, 68
Grape, *364, 383*, 530, *531*
Grass, *494*
Grasshopper
  circulatory system, *655*
  nervous system, *577*
  segmentation, *577*
Grasshopper mouse, 832, *832*
Grassland, arid
  desertification, 901, *901*
  global distribution, *871*
  soil, *872*
Grassland, temperate
  biome, 874, *875*
  desertification, 901, *901*
  global distribution, *871*, 874
  monsoon, 874
  plants, 874, *875*
Grave's disorder, 689
Graves, J., 292–293
Gravitropism, 536, *537*
Gravity-sensing mechanisms, in
  plants, 536, *537*
Gray crescent, 760

Gray matter
  cerebral cortex, 582
  spinal cord, 581
Gray treefrog, 313, *313*
Grazing
  and grasslands, 901, *901*
  by periwinkles, 837
Grazing food web, 848, *849–850*,
  883
Great Barrier Reef, 417
Great Dane, *270*
Great Deluge, *300*, 301
Great Salt Lake, *352*
Great Tit (songbird), 925
Green algae (Chlorophyta)
  classification, 370
  desmids, 374, *374*
  in ecosystems, 880–881, 883
  evolution, 390
  filaments, 390, *390*
  life cycle, 392
  and multicellularity, 376
  representative, *362, 374–375*
  reproduction, 374–375
  siphonous, 374
  wavelengths for photosynthesis,
    111
Green bacteria, *351*, 353
Greenhouse effect
  and carbon dioxide, 854,
    856–857, 894, 903–904
  operation, *856, 856–857, 866*
Greenhouse gases, *856, 856–857,
  857*
Green revolution, 900
Griffith, F., 210
Gross primary productivity, 848
Ground meristem, *485*
Ground substance, bone, 638
Ground tissue, plant, *484,
  485–487, 486, 489, 489*
Groundwater contamination, 898,
  898
Grouse, 917, *917*, 923
Growth
  control, *582*
  embryonic, comparative, *757*
  growth factor effects, 626
  localized, 765
Growth factors
  and cell division, 626
  signaling by, 626
Growth hormone
  actions, *616*
  effects on anterior pituitary, *618*
  role in development, *618*
  source, *618*
  targets, *618*
Growth hormone-releasing
  hormone, effects on anterior
  pituitary, *618*
Growth patterns
  exponential, *806, 806–807, 807*
  logistic, 808, *808*
  plant, 536–540
  of populations, 806–809
Growth rate, population, 816, *816*
GTP (guanosine 5'-triphosphate),
  129
Guanine, *211, 212, 212*, 225
Guanosine 5'-triphosphate (GTP),
  129
Guard, honeybee, *928*
Guard cell, 487, 508–509
Gull, herring, 924
Gum, gingivitis, *718*
Guppy, *812*, 812–813, *813*
Gurdon, J., 764
Gut, *see also* Digestive system
  animal, 411, *412*
  chordate, *447*
  cnidarian, 414
  flatworm, 418, *418*
  formation, 763
  insect, 435

Gut continued
  leech, 431
  motility, 718
  mucous epithelium, 717
  mucus lining, 676
  parasites, 419
  snail, 424
Gut reaction, 588
Gycolysis
  overview, 124
  summary, 126
Gymnodinium breve, 371
Gymnophilus, 39
Gymnosperm
  classification, 391, *406*
  conifer, *393*, 401
  cycads, 402, *402*
  ginkgos, *393*, 402, *402*
  gnetophytes, 402, *402*
  life cycle, *400*
  radiations, 393, *393*
  representative, *400–402*
  woody, 496
Gypsy moth, 839

H

H. M. S. *Beagle*, 265
Habitat, defined, 803, 823–824
Habituation, 915
Hadal zone, ocean, *884*
Haemanthus, 145
Hagfish, *452*, 453
Hair
  aging effects, 768
  epidermal, 635
  follicle, *632*
  nasal cavity, *700*
  protein deficiency effects, 635
  root, 494, 503, *503*
  scalp, 635
  shaft, *635*
  split ends, 635
  structure, *45*
  thinning, 635
Hair cell, 599–600, *601*
Hair color
  determinants, 178, *178*
  in mammals, 178, *178*
  melanin effects, 178
Hairston, N., 827
Half-life, of radioactive elements, 20
Halimeda, 374
Hallucigenia, 408, *408*
Hallucinogen, *384*, 589
Halobacterium, *351*, 352
Halophile, 352, *352*
Hamilton, W., 928
Hand
  evolution, 471, 474
  fetal development, *151*
  movements, 473–474
Hanging fly, 917, *917*
Hard coral, *415*
Hardy–Weinberg formula, 274, *274–275*
Hare, 829, *829*
Harpobittacus apicalis, 917, *917*
Haversian canal, 638, *638*
Haversian system, *638*
Hawking, S., 199
Hawkmoth, 521
Hay fever, 689
Hayflick, L., 768
HCG, *see* Human chorionic gonadotropin
Head
  bivalve, 426
  insect, 435
  sperm, 776, *777*
Head banging, termite, 922
Hearing
  ear structure, *601*
  echolocation, 600

mechanism, 601, *601*
  receptors for, 600, *600*
Hearing loss, *601*
Heart
  amphibian, 457
  arteries, *660*
  atrial contraction, 661
  atrium, 660
  AV valve, *660*, 660–661
  bird, 463
  contraction, 661
  crocodile, 461
  earthworm, *429*
  glucose energy yield, *131*
  innervation, *580*
  mollusk, *424*
  pacemaker cell, 661
  SA node, 661
  squid, *427*
  structure, *658*, *660*
  veins, *660*
  ventricle, 660
  ventricular contraction, 661
Heart attack, 653, 666–667
Heartbeat
  fetal, 789
  pacemaker for, 661
  rate, 661
  recording, 652–653
Heartburn, 719
Heart disease, *see* Cardiovascular disorders
Heart rate, hypothyroidism effects, 624
Heartwood, *496*
Heat, thermoreceptors, *594*, 596
Heat balance, body, 744, 746
Heat loss, evaporative, 749
Heat stress, 749
Hedgehog gene, 766
Heimlich maneuver, *701*
Heinrich, B., 916
HeLa cells
  cultures, 145
  history, 145
Helianthus, 486
Heliozoan, *366*, 367
Helium, 322
  atomic structure, 22
  electron distribution, 23
  structure, 19
Helix aspersa, 286, 286–287
Helper T lymphocyte, 680, *681*, 683, 690
Hemichordate, 448, *449*
Hemisphere, brain, 582
Hemlock, *387*, 834–835
Hemodialysis, 742
Hemoglobin
  alpha chain, 221
  beta chain, 221
  buffering capacity, 705
  carbaminohemoglobin formation, 705
  carbon dioxide transport, 705
  functions, 48
  genes coding for, 220–221
  HbA, 220–221
  HbS, 220–221
  HbS/HbA heterozygotes, 282
  and homeostasis, 656
  human *vs.* llama, *708*
  marine mammals and reptiles, *710*, 710–711
  oxygen transport, 697, 704
  oxyhemoglobin formation, 704
  polypeptide chains, 44, *45*
  quaternary structure, 44, *45*
  sickle-cell, 220–221
  and skin color, 634
Hemophilia, *196*, 199
Hemophilia A, *196*, 198
Hemorrhagic fever, *357*
Hemostasis, 668
Henslow, J., 264, 266

Hepatic portal vein, *659*
Hepatitis A, 688
Hepatitis B, 688
Herbaceous plant, 496
Herbicide
  beneficial, 365
  defined, 365
  deleterious, 365
Herbivore, 846, *847–850*, 860
Herd
  farming native, 901
  selfish, 925, *925*
Heredity, *see also* Inheritance
  theory of A. Weismann, 187
Heritable trait, 7
Hermaphroditism, 419
  Aplysia, 425
  flatworms, 418
Hermit crab, 883
Herniated (slipped) disk, 640
Heroin, 589
Herpes viruses
  DNA virus causing, *357*
  Epstein-Barr, *357*
  Herpes simplex, 606, *635*, 797
  Herpes simplex type I, *357*
  Herpes simplex type II, *357*, 797
  multiplication cycle, *359*
  Varicella-zoster, *357*
Hershey, A., 210, *211*
Heterocyst, 352, *353*
Heterospory, 393
Heterotherm, 747
Heterotrophic organism, 107, *351*, 846, *846*
Heterozygote, defined, 282
Heterozygous, defined, 171
Heterozygous condition, 171
Hexokinase, substrate interaction, 99
Hickory tree, 877
High blood pressure, 624, 666
High-density lipoprotein, 33, 667
High-voltage electron microscope, 55
Hindbrain, *582*
Hippocampus, *585*
Hippocrates, 262
Hirudo medicinalis, *430*
Histamine, 678
Histone, DNA interactions, 216, *238–239*
Histoplasma capsulatum, 383, *383*
Histoplasmosis, 383, *383*, 606
HIV, *see* Human immunodeficiency virus (HIV)
Hognose snake, *458*
hok gene, 254
Holdfast (sporophyte), 373
Holpcraft, D., 901
Homarus americanus, 310
Homeobox gene, 767
Homeostasis
  components, *556*, 556–557, *557*
  defined, 6, 547, 556, 653
  extracellular fluid, 736
  feedback loops, 621, 626
  glucose metabolism, *622*
  hemoglobin role, 656
  and hormone secretion, 557
  and internal environment, 556
  negative feedback, *556*, 556–557
  overview, 556–557
  positive feedback, 557
  role of nutrition in, 731
Homeostatic mechanisms
  and body temperature, *747–748*, 748–749
  and calcium balance, 624
Hominid
  adaptive radiation, 342
  classification, 471–472

early, 476–477, *476–477*
  fossils, *477*
  skull types, *478–479*
  tools, 478
  transitional forms, 476–477
Hominoid, 471–472
Homo erectus, 478–479, *479*
Homo habilis, 478
Homologous chromosomes
  alleles, 172–173, *172–173*, 188
  crossing over, 160, *160*, 188, 193
  defined, *142*
  gene pairs on, 172–173
  human, 143, *143*
  in human germ cell, 157, *157*
  and linkage groups, 193
  in meiosis, *142*
  during meiosis I and II, 157, *157*
  in mitosis, *142*
  pair, *171*, 188
Homologous structure, 301, 306, 317, 324
Homo sapiens, *477*
  classification, *311*
  evolution, 478–479
  skull, compared, *478*
Homo sapiens neanderthalensis, 478–479
Homospory, 393
Homozygote
  defined, 171
  environmental pressures, 282
  malaria infections, 282
Homozygous dominant condition, 171
Homozygous recessive condition, 171
Honeybee
  Africanized, *122*, 122–123
  altruistic behavior, 928–929
  colony life, *928–929*
  dance of, *436*
  pollinator, 305
  tactile signals, 923, *923*
Honeycomb, 41, *41*
Honeydew, 510, *510*
Hooke, R., 50, *50*
Hookworm infection, 410, 420
Hordei virus, 357
Horizontal cell, 608, *608*
Hormone, animal
  amoebic, 365
  and behavior, 618, 624–625, 912–913, *913*
  blood, 656
  and blood pressure, 663
  defined, 614
  derivation of term for, 614
  in digestion, 718–719
  discovery, 614
  enhancers, 242
  and female reproductive function, 779–780
  gastrointestinal, 718–719, 723
  homeostatic control, 556–557, 621, 623
  hypothalamic, 619
  invertebrate, 626
  and male reproductive function, 776–777
  menstrual cycle, 778–779
  nonsteroid, *616*, 616–617
  and ovarian function, 780
  pituitary secretions, *618*, 618–619
  protein, *616*, 617
  responses to, 620
  signaling mechanisms, 614, 616–617
  signals, 242
  sources, *621*
  steroid, 41, 616–617, 725
  targets, 614, *618*, 621
  thyroid, 689

Hormone, plant
  abscisic acid, 535
  apical dominance, 535
  auxin, 534, *534*
  cytokinins, 535
  effects on stem growth, 530,
    530–531
  ethylene effects, 535
  gibberellins, *534,* 534–535
  in growth and development,
    534–537
  growth-inhibiting, 536
  major, *534*
  in root cap, 536
  types, 536
Hornwort, 394
Horse, 289, *340, 450, 745*
Horseshoe crab, 433
Horsetail, 280, *280,* 396, 398, *398,*
    406
Housefly, *311, 435*
House sparrow, *838*
Howarth, F., 833
HTLV I & II (Human T-cell leukemia
    I & II), 357
Hubbard Brook Valley watershed,
    852, *853*
Hubel, D., *609*
Huffaker, C., 833
Human
  adaptive capacity, 907
  agent of gene flow, 286
  altruistic behavior, 931
  biological imperative for, 906–907
  birth weight, 282
  cladistics, *316–317*
  classification, *311, 472*
  control mechanisms for,
    906–907
  cytochrome *c,* 308
  early social behavior, 470–471,
    474, 478–479
  evolution, 470–479
  impact on other species, 341
  integumentary system, 620–635
  migration, 815
  nervous system, 578–585
  populations, tree of descent, *479*
  regulatory gene mutations, 304
  reproductive system, 773–789
  skeletal system, 640–641
  skeleton, 263, *473*
  skin, 620–635
  skull, *305, 474*
  smiling in infants, 914, *915*
  sociality, 924
  survivorship curves, 811, *811*
  traits, 471
Human body
  balance senses, 599–600,
    599–600
  blood volume in, 656
  changes during prenatal and
    postnatal growth, *793*
  circulatory system, *554*
  core temperature, 744–745
  digestive system, *555*
  directional terms for, *555*
  elements in, *18*
  endocrine system, *554*
  homeostasis mechanisms,
    556–557
  integumentary system, *554*
  internal environment, 6, 556
  internal temperature,
    homeostatic controls,
    556–557, *557*
  lymphatic system, *555, 655*
  major nerves, *579*
  muscular system, *554*
  negative feedback
    mechanisms, *556,* 556–557
  nervous system, *554*
  organ systems, overview,
    554–555

  planes of symmetry, *555*
  positional senses, 597
  positive feedback mechanisms,
    557
  reproductive system, *555*
  respiratory system, *555*
  skeletal system, *554*
  surface barriers against
    pathogens, 676, *676*
  surface-to-volume constraints, 53
  urinary system, *555*
Human chorionic gonadotropin,
    785
Human chromosome
  abnormalities, 195
  sex, 190
  sex-determining region (SRY),
    190
Human development
  aging, 768
  alcohol effects on, 791
  beginnings, 794–795
  birth control methods, 794–795
  blastocyst, *784*
  and bone formation, 638–639
  cleavage, *784*
  critical periods (embryonic),
    790, 791
  early cleavage, 784–785
  egg, 756
  embryonic, *761, 773,* 790,
    790–791
  embryonic disk, *784*
  extraembryonic membranes,
    785
  fertilization, *784*
  fetal, *773,* 788–789, *788–789*
  and fetal infections, 790
  gonadal, 190
  hormonal effects on, *618*
  implantation, *784,* 784–785
  infant dependency, *474*
  and nutrition, 790
  ovaries, 190–191, *191*
  postnatal, 793
  pregnancy, risks associated
    with, *790*
  sex determination, 190, *190*
  smoking effects on, 791
  stages, summary, 793, *793*
  tail, embryonic, 263
  testes, 190–191, *191*
  X chromosome inactivation,
    240–241
Human genome project, 252, 256
Human immunodeficiency virus
    (HIV)
  capsid, *356*
  carriers, 796
  classification, *357*
  fluids present in, 691
  forms, 690
  lipid envelope, *356*
  micrograph, *675*
  mode of action, 357, 690
  replication, 690, 796
  research, 357
  structure, *356,* 691
  transmission modes, 691, 796
Human population, *see also*
    Community; Population
  distribution, 814, *816*
  and food production, 898
  growth, *see* Population growth
  impact on biosphere, 892–909
  resource consumption, 819,
    894, 904, *904*
  total (1994), 814
  various countries, *816*
Human reproduction
  birth process, 792
  early cleavage and
    implantation, 784–785
  female reproductive system,
    778–782, *778–782*

  fertilization, 783, *783–784*
  gastrulation, 786–787
  lactation, 792–793
  male reproductive system,
    774–777, *776*
  oogenesis, 782
  postnatal period, summary, *793*
  prenatal period, summary, *793*
  sexual intercourse, 783
  uterine function, 782
Human T-cell leukemia I & II (HTLV
    1 & II), 357
Humboldt current, 888, *889*
Humerus, *463, 641*
Hummingbird, 305, *405,* 462, *521,*
    746–747, 826
Humus, 872
Huntington's disorder, *196, 197*
Hutchinson–Gilford progeria
    syndrome, 186–187
Hutton, J., 265
Huxley, T., 462
Hybiscus, *521*
Hybrid
  defined, 289
  inviability, *289,* 291
  offspring, *171, 289,* 291
Hybridization, *see also* Fertilization
  dihybrid crosses, 174–175, *175*
  DNA, *249*
  DNA-DNA, 308
  with *Drosophila melanogaster,*
    193
  interspecific, barriers to,
    289–291
  monohybrid crosses, 172–173,
    172–173
  nucleic acid, 252–253, *253,* 351
  testcrosses, 173
  wheat, 295
Hybrid zone, 295, *295*
Hydra (hydrozoan), 410, *410,* 411,
    414, *414,* 442–443
Hydrangea macrophylla, 273
Hydrocarbon
  and air pollution, 895
  characteristics, 35
  structure, 35
Hydrochloric acid, 718–719
Hydrogen
  in abiotic synthesis, 324–325
  atomic number, *19*
  atomic structure, 22, *23*
  electron distribution, *23*
  mass number, *19*
  structure, 19
  symbol for, *19*
Hydrogen bond
  below 0°C, 26–27
  and cohesion theory, 493
  defined, 25
  in DNA, 25
  in polypeptide chains, 44, *44*
  between two nucleotide strands
    in DNA, *46*
  in water, 25–26, *26*
Hydrogen ion
  in aerobic respiration, *128*
  in ATP formation, *128*
  in digestion, 718–719
  effects on acid–base balance,
    742
  in extracellular fluid, 742
  neutralization, 742
  in photosynthesis, 114, *114*
  and pH value, 28
  shuttle, *131*
Hydrogen peroxide
  catalase effects, 92
  conversion to water and
    oxygen, 65
  photochemical oxidant, 895
  superoxide dismutase effects, 92
Hydrogen power, 906–907
Hydrogen sulfide, 352, 885

  Hydrologic cycle, 851, *852,*
    852–853
  Hydrolysis
    enzyme action in, 36
    mechanisms, 36, *36*
  Hydrophilic interaction, 26
  Hydrophobic interaction, 26, 78
  Hydrosphere, 865
  Hydrostatic skeleton, 414, 428,
    636, 636–637
  Hydrothermal vent, 884–885, *885*
  Hydroxide ion (OH-), 28–29, 126,
    127
  Hydroxyl group
    in alcohols, 35
    locations, *35*
    structural formula, *35*
  Hydrozoan, 414, *414, 416*
  Hyla
    H. chrysoscelis, 313, *313*
    H. versicolor, 313, *313*
  Hypercholesterolemia, 197
  Hyperopia, 606
  Hyperpolarizing effect, of IPSPs,
    568
  Hypertension, 666, 742
  Hyperthermia, 749
  Hyperthyroidism, 624
  Hypertonic solution, 83
  Hyperventilation, 708
  Hypha, fungal, 380, *380,* 382, *382,*
    385, 386–387, 388, 825
  Hypnotic drug, 589
  Hypodermis, 632, *632*
  Hypoglycemia, 623
  Hypothalamus, *585*
    and body temperature, 748–749
    functional links with pituitary,
    619
    functions, 583, *782*
    homeostatic feedback loop, 621
    hormone secretion, *616,* 619,
    777
    inhibiting hormones, *618*
    as integrator, *557*
    and limbic function, 588
    location, *582, 584*
    pleasure center, 589
    releasing hormones, *618*
    and sperm formation, 777
    and thirst, 741
  Hypothermia, 748, *749*
  Hypothesis
    alternative, 12
    evolutionary, 931
    role in phylogeny, 314
    scientific, 11
    testing, 12
  Hypothyroidism, 624
  Hypotonicity, 80, 744
  Hypotonic solution, 83
  Hypotrich, 367, *367*
  Hypoxia
    and carbon monoxide
    poisoning, 708
    defined, 708
    and hyperventilation, 708

  ## I

  Ice, hydrogen bonding in, 26–27
  Ice age
    Cretaceous, 338
    Ordovician, 334
  Ice cloud, and ozone destruction,
    897, *897*
  Ice-minus bacteria, 254, *254*
  Ichthyosaur, *302, 307, 459*
  Icterus, 295, *295*
  Identical twins, 273, 764, 792
  Iditarod, 631–632
  IgA, 685
  IgE, 689
  IgG, 685, 689

IgG, 685, 689
IgM, 685
Iguana, 460, *545*
Ileum, 720
Iliac artery, *659*
Iliac vein, *659*
Imbibition (water molecules), 532
Immune response
    antibodies, 685
    antibody-mediated, *681, 684,*
        684–685
    B cells, 685
    cell-mediated, *681, 682, 682*
    control, 681
    cytotoxic T cells, 683
    deficient, 689
    helper T cells, 683
    immunoglobulins, 685
    inflammation during, *681*
    macrophage function, 680
    nonspecific, 676, *676*
    overview, 680, *681*
    primary, *681, 682–683,* 688
    secondary, *681,* 688
    specific, 676, *676*
    targets, *681*
Immune system
    amphetamine abuse effects,
        588
    antibodies, *681*
    antibody formation, 668
    cells, *681*
    defining features, 680
    gamma globulin role, 656
    marijuana effects, 589
    overview, *681*
    smoking-related impairments,
        707
    thymosin effects, 626
    thymosin role in, 626
    UV radiation effects, *635*
Immunity
    leukocyte role, 657
    smallpox, 674, *674*
Immunization
    active, 688
    defined, 688
    passive, 688
    schedule for children, 688
Immunoglobulins
    classes, 685
    in mother's milk, 685
Immunotherapy, 683
Impact cratering, 337, *337*
Imperfect fungus, 379, *383,* 388,
    388
Impetigo, 359
Implantation (human), *784,*
    784–785
Imprinting, 914, *915*
Inactivation, X chromosome,
    240–241
Inbreeding, and genetic uniformity,
    277
Incisor, 718
    canine, *465*
    mammalian, 464
Incomplete dominance, 176
Independent assortment, at
    metaphase I, *174*
Independent assortment, Mendel's
    theory, 174–175, *174–175*
Indian mallow, 827, *827*
Indian pipe (*Monotropa uniflora*),
    403
Indirect selection, 928–931
Indoleacetic acid, 534
Induced-fit model, of
    enzyme–substrate
    interactions, 98, *99*
Inducer molecule, 236
Induction, embryonic, 766
Industrial pollution, 28, *29,* 279,
    895, 899, *see also* Air
    pollution; Water pollution

Infant
    defined, *793*
    dependency, and life span, *474*
    smiling response, 914, *915*
Infection
    fever effects, 749
    lymph node role in, 671
    mucous membrane, *383*
    opportunistic, 796
    sexually transmitted, *see*
        Sexually transmitted disease
Infectious mononucleosis, *357*
Inferior (anatomy), *555*
Inferior vena cava, *659*
Inflammation
    acute, 678
    allergen-related, 689
    chronic, 689
    stress response, 623
Inflammatory response, 678, *681*
Influenza
    epidemics, 359
    RNA virus causing, *357*
Influenza virus, 246, *357*
Information flow
    and memory, 587
    through nerve nets, 576
    nervous system, 82, *571*
    reflex arcs, 570, *571*
Information retrieval, and
    classification, 311
Ingram, V., 220–221
Ingroup, 314
Inhalation, 702
Inheritance
    autosomal dominant, *196,* 197
    autosomal recessive, *196,*
        196–197
    blending theory, 169
    Lamarckian theory, 264
    Mendel's insight, 170–171
    patterns, *192,* 195
    and reproduction, 7
    terminology, 171
    X-linked, *192*
    X-linked dominant, *196,* 198
    X-linked recessive, *196,* 198
Inhibitory postsynaptic potential,
    568, *568*
Initiation
    of protein synthesis, *226,* 227
    of RNA translation, 226
Injury, nociceptors, 596
Ink sac, squid, *427*
Inner cell mass, 784
Inner ear, 600–601
Insect, *see also* Arthropod
    (Arthropoda); *specific type*
    adaptations, 432, 435
    adaptive radiation, 335–336
    altruistic behavior, 928–929
    beneficial, 437
    body plan, 435, *435*
    body segmentation, 410
    circulatory system, *443*
    coevolution with plants,
        520–521
    colony life, 920, 920–921,
        928–929
    communication signals, 923, *923*
    compound eye, 602
    cooperative predator
        avoidance, 925
    cuticle, *637*
    developmental stages, 6, *6*
    digestive system, 435, *443*
    diversity, 436–437
    eggs, 762
    evolution, *340–341*
    evolutionary success, 436–437
    exoskeleton, *637*
    flight, *637*
    harmful, 437
    insecticide-resistant, 279
    larvae, *929*

light reception, *603*
metamorphosis, 432, 436
movement, *442*
nervous system, *442*
pollinators, 388, *520,* 520–521,
    521, 524, *524*
pupae, *929*
representative, *436–437*
reproduction, *443*
respiratory system, *443*
segmentation, 411
sterility, 921, 928–930
survivorship curves, 811, *811*
trachea, 432
tracheal respiration, 698
wing, *637*
winged species, 437
Insect control, biological, 279
Insecticide, 493
Insecticide resistance, and
    directional selection, 279
Insight learning, 915
Instinctive behavior, 911, 914,
    914–915
Insulin
    actions, 616, 621
    amino acid sequence, 42, *43*
    deficiency, 623
    functions, 6, *48*
    polypeptide chains, 42
    receptors, 6
    secretion, 616, 622, 723
    sources, *621*
    synthetic gene, 254
    targets, *621–622*
Integration, drug action on, 589
Integrator, in homeostatic control,
    556, *557*
Integument, seed, 518, *519*
Integumentary exchange, 696, 698
Integumentary system, *see also*
    Skin
    dermal tissue, 632, 634–635
    epidermal tissue, 632, 634
    functions, *554,* 632
    human, *554, 634*
    hypodermis, 632, *632*
    overview, 631
Intercostal muscles
    functions, *700*
    location, *700*
Intercostal nerve, *579*
Interferon, 682, 749
Interleukin
    defined, 678
    in inflammatory response, 678
    secretion, 681, 683
Interleukin-1, 678, 749
Intermediate filament, 69
Intermediate lobe, pituitary, *618*
Intermediate (metabolic pathway),
    98
Internal environment
    cold stress effects, 748, *749*
    defined, 556
    and evolution, 734–735
    heat balance, 744, 746
    heat stress effects, 749, *749*
    homeostatic control, 556–557
    solute balance, 744
    water balance, 744
Interneuron
    axons, *571*
    defined, 561
    dendrites, *571*
    distribution, 578
    as integrator, 561, 578
    lateral geniculate nucleus, 609
    in reflex arcs, 570, *571*
    tetanus effects, 569
Internode, *484, 533*
Interphase
    defined, 144
    duration, 144, *144*
    mitotic, 144, *144*

nucleus, male *vs.* female, *241*
Interspecific competition, 824,
    824, 826
Interstitial fluid, 556, 654, 736
    at capillary bed, *664–665*
    composition (vertebrate), 556
    defined, 556
Intertidal zone, 836–837, *837,* 883,
    883–884, 890
Intervertebral disk, 640, *641*
Intestinal nuclease, *720*
Intestine, large, *see also* Colon
    appendix, 723
    bulk effects, 723
    cecum, 723
    functions, *717,* 723
    innervation, *580*
    role in digestion, 723
Intestine, small
    absorption processes, 722, *722*
    digestive processes, *717,* 722,
        722
    innervation, *580*
    lymphoid tissue, *670*
    peristalsis, 718, *719,* 720
    segmentation, 720
    surface structures, *721*
    wall structure, 720, *720*
Intracellular fluid, defined, 736
Intraspecific competition, 826
Intron, defined, 223
Inversion, chromosomal, 200
Inversion layer, 895, *895*
Invertebrate
    cephalization, 576–577
    estaurine, 882
    evolution, *440*
    eye, 602–604
    fertilization, 756
    nervous system, 576–577, *577*
    photoreceptors, 602
In vitro fertilization, 205
Iodine
    atomic number, *19*
    blood levels, 624
    deficiency, 624, *727*
    functions, *727*
    mass number, *19*
    sources, *727*
    symbol for, *19*
Iodine-123 isotope, thyroid
    scanning with, 21
Ion
    bonding, 24
    concentration gradient, *see*
        Concentration gradient
    concentration in extracellular
        fluid, 556
    cushioning effects, 30, *30*
    defined, 24
    diffusion, 82
Ionic bond, defined, 24
Ionization, water, 28
Ionizing radiation, mutagenicity,
    228
IPSP, *see* Inhibitory postsynaptic
    potential
Iridium, and dating, 338
Iris, eye, 604, *604–605*
Iron
    atomic number, *19*
    deficiency, *727*
    excess, *727*
    functions, *727*
    in human nutrition, *727*
    mass number, *19*
    in plant function, *502*
    sources, *727*
    symbol for, *19*
Irradiation, for cancer treatment,
    682
Irrigation, 898, *898*
Island, species diversity, *840,*
    840–841, *841*
Isocitrate, *129*

Isolating mechanisms, reproductive
    categories, *289*
    and genetic divergence, 289–291, 336
Isoleucine, 725
Isotope
    bacteriophage labeling with, *211*
    defined, 19
    radioactive, *see* Radioisotope
Isthmus of Panama, speciation, 292–293
Itano, H., 220
IUD, *795, 797*

J

Jackrabbit, 746, *749*
Jade plant, *526*
Jaguar, *876*
Japanese beetle, *838*
Jaw
    evolution, 450
    rotifer, *423*
    snake, 460, *460*
    vertebrate, *577*
Jawed fish, *450, 452*
Jawless fish, 452–453
Jejunum, 720
Jellyfish, *410,* 414, *414–415, 442–443*
Jenner, E., 674, *675,* 688
Jet lag, 625
Jet propulsion, of cephalopods, 426
Jogging, kilocalories expended by, 729
Joint
    arthritis, 641
    cartilaginous, 640
    fibrous, 640
    sprains, 640
    strains, 640
    synovial, 640
    vertebrate, 640
J-shaped curve, exponential growth, *806, 806–807*
Jugular vein, *659*
Jump dispersal, 838–839
Jumping gene, 228, *229*
Jumping spider, *636,* 637
Juniper seedling, *387*
Jurassic, 336–337, *393, 459*
Juvenile hormone, *627*
Juvenile-onset diabetes, 623

K

Kangaroo, 466, *755*
Kangaroo rat, 734, *734,* 743
Kaposi's sarcoma, 690
Karyotype
    defined, 189
    diagram preparation, 188–189, *189*
    human male, *189*
    trisomy 21, *202*
Kelp, 372–373
Kennett, C., 833
Kentucky bluegrass, 118
Keratin, *48,* 634
    functions, *48*
    synthesis, 45
Keratinization, 635
Keratinocyte, 634
Keratotomy, radial, 606
α-Ketoglutarate, *129*
Ketone, accumulation, 623
Ketone group, *35*
    locations, *35*
    structural formula, *35*
Kettlewell, H. B., 279
Keystone species, 836
Kidney
    aging effects, 768
    anabolic steroid effects, 649
    blood vessels in, *738*
    controls over, *580, 618*
    dialysis, 742
    disorders, 649, 742
    functions, 737, *737,* 738
    gastropod, 425
    glucose energy yield, *131*
    innervation, *580*
    location, *737*
    malfunctions, 742
    mollusk, *424*
    nephron structure, 738, *738–739*
    squid, *427*
    structure, 738, *738–739*
    transplantation, 742
Kidney bean, *725*
Kidney stone, 742
Killer whale, *710,* 847
Killifish, 812–813
Kilocalorie, defined, 95, 728
Kinetochore, 148, *148*
Kingdom (taxon), 8, 311, 318
Kittyboo beetle, 103, *103*
Klinefelter syndrome, *196,* 203, *203*
Knee cap (patella), 640
Knee joint, 640, *640*
Kock, R., 674
Kohler, G., 683
Koshland, D., 98
Kreb, H., 128
Krebs cycle
    aerobic pathway, *125,* 128
    in aerobic respiration, 128, 130
    ATP yield, 129–130
    defined, 128
    functions, 128–129
    overview, 124, *125*
Krill, 844–845, *847,* 884
K–T (Cretaceous–Tertiary) boundary, 338
Kudzu, 839
Kurosawa, E., 530

L

Labia majora, 778, *778*
Labia minora, 778, *778*
Labrador retriever, 178, *178*
Lacks, H., 144
Lactate, 676
Lactate fermentation, 132, *132,* 647, *647,* 731
Lactation, 792, *792,* 793
Lactobacillus, 132, *351,* 353, 676
Lactoferrin, gene for, 256
Lactose
    blood levels, 197
    metabolism, 196–197, 237
    operon, 236, *237,* 254
Lactose operon, *237*
Lacuna, *638*
Ladybird beetle, 255, *437*
Lake
    crater, *880*
    ecosystem, 880–881
    eutrophic, 881, *881, 891*
    fall overturn, 880–881
    formation, 881
    glacial, *880*
    oligotrophic, 881, *881, 891*
    photosynthesis, 880–881
    pollution, 881
    seasonal changes in, 880–881
    spring overturn, 880
    thermal layering, 880–881, *881*
    trophic nature, 881, *881*
    zonation, 880, *881*
Lake Baikal, 880, 885
Lake Barombi Mbo, Cameroon, *294*
Lake Tanganyika, 613
Lake Victoria, 839
Lake Washington, 881
Lamarck, J.-B., 264
Lamarckian theory of inheritance, 264
Laminaria, 372
Lamprey, 316–317, 453, *453, 838*
Land, *see* Cropland; Grassland; Shrubland; Woodland
Land snail, *424*
Language
    brain areas affecting, 582, 584, *585*
    and culture, 474
    of epileptics, 586
    hemispheric dominance in, 586
    and nervous system, 575
    structural basis, *474,* 478
Lanugo, 789
Large intestine, *see* Intestine, large
Larkspur, comparative embryology, 304–305
Larrea divaricata, 526
Larson, G., 916
Larva, 768
    defined, 413, 447
    ecdysone effects, 242
    echinoderm, 438
    fluke, 418, 420
    honeybee, *929*
    insect, *238*
    lamprey, 452
    metamorphosis, 432
    parasitoid, 831, 833
    salamander, 456–457
    sawfly, 833, *833*
    sea slug, 425
    sea star, *757*
    sexually precocious, 449
    silkworm moth, *627*
    sponge, 413
    tapeworm, 419
    tunicate, 447
    yucca moth, *825*
Laryngitis, 703
Larynx, 707
    innervation, *580*
    location, 701
Lascaux cave, 470–471, 479
Laser angioplasty, 667
Laser coagulation, 607
Late blight, potato, 364
Latent learning, *915*
Lateral geniculate nucleus, 609
Lateral meristem, 485, *485,* 496
Lateral root, *484,* 494–495, *495*
Late wood, 497, *497*
Latimeria, 456
Latissimus dorsi, *643*
Latitude, and species diversity, *840*
Laurentia, 330, 334
Lava flow, 260–261, 328, 336–337
Law of conservation of mass, 22
Laws of thermodynamics, 94–95
Laxative, 729
Leaching, 852, *853,* 858
Leaf
    arrangement, *492*
    central vacuole, *111*
    characteristics, 490–493
    chlorophyll, *111*
    compound, *490*
    epidermis, *487*
    formation, 492
    forms, monocots and dicots, *490, 492*
    mesophyll, *118*
    oils, 493
    photosynthesis in, 490
    pigments, *110*
    simple, *490*
    structure, internal, *487, 491, 491*
    surface, electron micrograph, *487*
    uses, 493
    veins, *118, 490*
    young, *484, 492*
Leak, ion, 562–563, *563*
Leakey, M., 477–478
Learned behavior, 911, 914, *915*
Learning
    evolution, 474
    mammalian, extended period, 474, *474*
Least bittern, *830*
Leatherback sea turtle, 710–711, *710–711*
Leave
    evolution, 392
    stomata, 392
Lecithin, 721, 725
Leech, *410,* 429–430, *430–431,* 442–443
Leeward direction, 869
Legume, 503, *503*
Lek (display ground), 917, *917,* 923
Lemming, dispersal, 916
Lens
    clouding, 606
    embryonic, 766–767, *766–767*
    eye, 602, 604, *604–605, 766, 766–767*
    lazy, 606
Leopard, *828, 926*
Leopard frog, 752–753, *752–753*
Leopard seal, *847*
Lepus californicus, 749
Leucine, 725
Leucine aminopeptidase, *286*
Leukemia, defined, 683
Levels of biological organization
    Aristotle's observations, 262
    and Chain of Being, 261–262
    and history of life, 136–137
Levi-Montalcini, R., 626
Lewis, E., 767
Leydig cell, 776, *776*
Lichen, 279, 386, *386, 388*
Life
    characteristics, 3–7
    defined, 3
    defining characteristics, 326
    diversity, *see* Diversity of life
    energy flow into, 94–95, *95*
    and genetic disorders, 199
    metabolic control, 136–137
    molecular basis, 136–137
    molecules of, *see* Biological molecule
    organization, 136–137
    origin and evolution, 136–137, 322–343, *342–343*
    perspective on, 136–137
    shared characteristics, 9
Life cycle
    angiosperm, *404*
    animal, *162,* 410
    arthropod, 432
    beef tapeworm, *421*
    blood fluke, *420*
    brown algae, 373
    cell divisions during, *163*
    cherry tree, *519*
    Chlamydomonas, 375
    Clostridium tetani, 354
    club fungus, 385, *385*
    cnidarian, 417, 576–577
    conifer, *400*
    defined, 142
    Dichyostelium discoideum, 365, *365*
    fern, *399*
    flowering plant, *404,* 516, *519*
    fungus, 380
    fungus without sexual phase, 388
    green algae, 392
    gymnosperm, *400*
    human, 794
    human immunodeficiency virus, *691*

Life cycle *continued*
  insect, 436–437
  Lilium, 404
  metamorphosis in, 436–437
  monocot, *404*
  moss, *394*
  obelia, 416
  perspective on, 769
  pioneer species, 834
  plant, *162, 392*
  Plasmodium, 368, *369*
  protozoan, 366
  red algae, 372
  Rhizopus stolonifer, 380, *381*
  salamander, *457*
  Schistosoma japonicum, 420
  silkworm moth, *627*
  slime mold, 365, *365*
  Taena saginata, 421
  tapeworm, 419
  tunicate, 447
  vascular plant, 392
  yucca moth, *825*
Life expectancy, 707
Life history patterns, 810–813
Life span, and infant dependency, *474*
Life table, 810, *810*
Ligament
  defined, 640
  functions, 640
  joint, 640
  torn, 640
Light
  and emotional state, 625
  and flowering process, 539
  micrographs, *55*
  microscopes, 54, *54*
  and photoperiodism, 538, *538, 539*
  and photoreception, 604, *605*
  and phototropism, 536, *536*
  spectrum, *110*
  wavelengths, 54, *110–111*
  and winter blues, 625
Lignification, 486
Lignin, 486, *486*
  formation, 392
  functions, 392
Lilium, 404
Limb, forerunners, 450
Limbic system, *582, 584–585,* 588, *588*
Limited division potential, defined, 768
Limnetic zone, lake, 880, *881*
Limpet, *155, 603,* 836
Lindow, S., 254
Lineage
  adaptive radiations, 297
  analogous structures, 306–307
  cell, selective gene control, 244
  changes, and evolution, 244
  cladograms, 316–317
  comparative morphology, 304
  defined, 296, 320
  eukaryotic, *333–334*
  evolutionary relationships, 314
  evolutionary tree of life, *333*
  extinctions, 297
  homologous structures, 306
  mutations in, 275
  speciation patterns, 287, 296
  true-breeding, *171*
Linkage groups
  and crossing over, 193–194
  defined, 193
Linkage mapping, defined, 194
Linnaeus, 310
Linné, C. von (Linnaeus), 310
Linoleic acid, 725
Linolenic acid, structural formula, *40*
Linophryne, 312
Lion, 917

Lipase, *720*
Lipid, *see also* Phospholipid;
    Triglyceride; Wax
  in blood, *656*
  defined, *48*
  envelope, viral, 356, *356*
  essential fatty acid, 725
  with fatty acids, 40, 48, *48*
  flow through cytomembrane
    system, *64*
  functions, 40, *48*
  hydrophilic head, *78*
  hydrophobic tail, *78*
  kinked, 79
  metabolism, *730*
  with no fatty acids, 41, 48
  in plaque formation, 666–667
  self-assembly, 324, 327, *327*
  synthesis, 65
  without fatty acids, *48*
Lipid bilayer
  components, 41, 52, 78
  diffusion across, 85, 504
  formation, 78
  functions, 52, 63, 136, 504, 562
  ion transport, 562–563
  and life, 78
  mitochondrial, 67, *67*
  nuclear envelope, 63, *63*
  and nutrient uptake by plants, 504
  origin, 327
  phospholipids, 41, 78, 136
  plasma membrane, 80–81, *80–81*
  proteins in, 52
  structure, 52
Lipoprotein
  components, 46
  formation, 33, 46
  high-density, 33
  low-density, 33, 667
  very-low-density, 33
Liquid waste, 899
Listeria, 351
Listeria monocytogenes, 354
Lithops, 830
Lithosphere, 865
Littoral zone
  coast, *883*
  lake, 880, *881*
Littorina littorea, 837
Liver
  activities dependent upon, *730*
  bile secretion, 721
  cell cross-section, *60*
  cirrhosis, 589
  disorders, dioxin-related, 365
  functions, 731
  glucose energy yield, *131*
  glycogen stores, 134
  innervation, *580*
  role in digestion, *717*
  role in metabolism, 731
  urea formation, *730*
Liverwort, 394, *395*
Lizard, *459,* 460, *460, 762*
Llama, *708*
Lobe-finned fishes, 335, *335, 452,
    455–456, 456*
Lobster, 310, *410, 432,* 434, *434*
Local signaling molecule, 626
Lockjaw, 569
Locomotion
  amphibian, 456–457, *457*
  annelids, *442*
  arthropods, *442*
  bivalve, 426
  chordates, *442*
  cnidarian, *442*
  coral, *442*
  ctenophoran, *442*
  earthworm, 428, *636*
  echinoderms, *442*
  flatworm, *442*

  fluke, *442*
  insect, *442*
  jelly fish, *442*
  mollusks, *442*
  nematode, *442*
  polychaete, *442*
  ribbon worm, *442*
  rotifer, *442*
  roundworm, *442*
  sea anemone, *442*
  snake, 460
  spider, *442*
  sponges, *442*
  squid, 426
  tapeworm, *442*
  water-vascular system (sea
    star), *438–439*
Logistic growth, 808, *808*
Loligo, 565
Long-day plant, 539
Longitudinal muscle, 636, *636*
Long-term memory, 587
Looped domain, chromosomal, 216, *216*
Loop of Henle, 738, *739, 741,* 744
Lorenz, K., *915*
Loudness, and inner ear, *601*
Low-density lipoprotein, 667
  defined, 33
  and plaque formation, 33
Lower littoral zone, rocky coast, 883, *883*
LSD (lysergic acid diethylamide), 589
Lubchenko, J., 836–837
Luciferase, 103, 255
Lucy (australopith), 338, *477*
Lumbar nerve, *579–580*
Luna moth, *437*
Lung, *see also* Respiratory system
  air pressure changes in, 702
  airways, 700, *700*
  amphibian, 456
  blood vessels, 706–707
  cancer, 706–707
  capacities, 702–703, *703*
  dioxin-related disorders, 365
  disorders, 365, 706–707
  evolution, 451, 456, 699
  functions, *700*
  innervation, *580*
  location, *700*
  tidal volume, 702
  vital capacity, 702
Lungfish, *316–317,* 454–456
Luria, S., 210
Luteinizing hormone
  actions, *618,* 780
  feedback loops affecting, *781*
  and female reproductive
    function, 780
  and male reproductive function, 776–777
  secretion, *616,* 777, 780
  sources, *618*
  targets, *618, 619,* 780
Lycophyte, *393,* 396, *406*
Lyell, C., 266
Lyme disease, 354, *354,* 433
Lymph
  defined, 670
  flow through nodes, 671, 685
Lymph capillary, *655,* 671
Lymph node, *655, 685*
  functions, *670,* 671
  locations, *670,* 671
Lymphatic duct, 655
Lymphatic system
  components, *655,* 670
  defined, 670
  diagram, *659*
  functions, *555,* 654
  human, *555, 655, 659*
  and immunity, 670
  links with circulatory system, 654

  lymph, 670
  lymph capillaries, 671
  lymph vascular system, 671
Lymphoblast
  leukocyte formation from, 246
  use in gene therapy, 247
Lymphocyte
  antibody-mediated immune
    response, *681*
  antigen-specific receptor
    production, 686–687
  B cell, *see* B lymphocyte
  cell-mediated immune
    response, *681*
  clonal selection theory, 687, *687*
  component of blood, *656*
  cytotoxic, 682
  functions, *656,* 657
  in lymphatic system, *655*
  in lymph nodes, 671
  memory cell, *681*
  plasma cell, *681*
  splenic, 671
  T cell, *see* T lymphocyte
  in thymus gland, 671
Lymphoid organs
  locations, *670, 685*
  lymphocyte actions, 671, *685*
Lymph vascular system
  components, *670,* 671
  functions, 671
Lymph vessel
  functions, *670,* 671
  in intestinal villi, *722*
  locations, *655, 720*
  structure, 671
Lynx, 829, *829*
Lyon, M., 241
Lyonization, 241
Lysergic acid diethylamide (LSD), 589
Lysine, 725, *725*
Lysis
  complement-induced, 677, *677*
  defined, 677, *677*
  virus-induced, 358
Lysogenic pathway, *211, 357,* 358, *358*
Lysosome
  components, 66
  defined, 66
  enzymes in, 66, 88
  functions, 60, 64, 73
  occurrence in nature, 73
Lysozyme, 676
Lystrosaurus, 337
Lytechinus, 762
Lytic pathway, of bacteriophage
    replication, 358, *358*

M

Macrocystis, 372, *373*
Macroevolution, *see also*
    Evolution; Evolution,
    biological
  cladistics, 314–317, *316–317*
  comparative embryology, 304–305
  defined, 301
  fossil record, 302–303
  morphological divergence, 306–307
  patterns, summary, *298*
  reconstructions, 314–315
  species identification, 312–313
  summary of processes and
    patterns, *298*
  time scale for, 301
Macronutrient, 502
Macrophage
  as antigen-presenting cell, 680
  appearance in micrographs, *679*
  in blood, *656*

Macrophage *continued*
  formation, 657
  functions, 171, 657, 677–678
  immature, *657*
  immune functions, 680
  mature, *657*
Magnesium
  atomic number, *19*
  deficiency, *727*
  electron distribution, *23*
  excess, *727*
  in human nutrition, *727*
  mass number, *19*
  in plant function, *502*
  sources, *727*
  symbol for, *19*
Magnesium ion, 29
Magnetite particle, 355
Magnetotactic bacterium, 355
Magnification, biological, 861
Magpie, *462*
Maidenhair tree, 402
Mainland, species diversity on, 840, *840*
Major histocompatibility complex (MHC) marker, 680, 682–683, *683*
Malaria, 282, 359, 368, *369*, 861
Malate, *129*
Mallard, *915*
Mallow plant, 827, *827*
Malpighian tubule, 435
Malt brewery, 37
Malthus, T., 266
Maltose, 37
Malus, 523
Mammal (Mammalia)
  adaptive radiations, 464
  brain, *584*
  brain size, *450*
  cell types in, 764
  classification, 446
  classification, temperature-based, 746–747
  comparative embryology, 304
  defined, 464
  dentition, 464–465, *465*
  distinctive traits, *464–465*
  diversity, 464
  ear, 600, 607
  egg-laying, 465, *465*, 755
  endotherms, 746
  estrous, 778
  evolution, *450*, 464
  eye, *604*
  feeding behavior, 464
  germ layer, 765
  heterotherms, 747
  marine, *710*
  placental, *466*, 466–467, *755*
  pouched, 466, *466*, 755
  response to cold stress, 748, *748*
  skeleton, 464
  survivorship curves, 811, *811*
  teeth, 464, *464–465*
  temperature regulation, 748–749
  urinary system, 737, *737*, 738
Mammary gland, 444, 464, *618*, 755, 792, *792*
Mandible, crustacean, 434
Manganese, in plant function, *502*
Manicougan crater, 337
Mantle
  cephalopod, 426
  chiton, 425
  clam, *426*
  gastropod, 425
  mollusk, 424, *424*
  squid, *427*
Mantle cavity
  bilvalve, 426
  cephalopod, 426–427
  gastropod, 425
  mollusk, *424*
Maple tree, *111*, 524, *524*, *540*, 877

Marchantia, 395
March marigold, *520*
Margulis, L., 332
Marijuana, 589
Marine turtle, 460, *460*
Marker
  MHC, 680, 682–683
  nonself, 680
  self, 680
Marler, P., 914
Marrow, bone, 638
Marsh, salt, 882, *882*
Marsh gas, 352
Marsupial, 466, *466*, *755*
Mason, R., 910–911, *910–911*
Mass extinction, 334–338, 341, *342–343*
Mass number
  common elements, *19*
  defined, 19
Mast cell, 678
Mating behavior, 916–917, *917*, *922*, 922–923
  cuttlefish, *427*
Maxillae, crustacean, 434
Mayr, E., 288
McClintock, B., 228
Measles, *357*, 688
Meat, as protein source, *725*
Mechanical isolation, 289, 290, *290*
Mechanical stress, in plants, 537
Mechanized agriculture, 900
Mechanoreceptor, 596–597, 600, *600*
  skin, *594*
  stimuli for, *594*
Medicago sativa, 489
Mediterranean fruit fly, *436*
Medulla, renal, 738
Medulla oblongata, 582, *582*, 584
  location, *580*
Medusa, cnidarian, 414, *414*, 416
Meerkat, 546–547, *546–547*, 556, *746*
Megakaryocyte, 657, *657*
Megaspore, *404*, 518, *518–519*
  conifer, 401
Meiosis
  and chromosome number, *142–159*
  comparison with mitosis, 164, *164*
  and crossing over, 193
  defined, 156
  divisions, 157
  germ cells, 142
  and independent assortment, 174–175
  nondisjunction effects, 201, *201*
  and oogenesis, *163*
  overview, 156–157, *158–159*
  plant, 518, *518–519*
  sites, 142
Meiosis I
  anaphase I, *158*
  events, 157, *158*
  metaphase I, *158*, 161
  prophase I, *158*
  telophase I, *158*
Meiosis II
  anaphase II, *159*
  events, 157, *159*
  metaphase II, *159*
  prophase II, *159*
  telophase II, *159*
Meiospore, *516*
Meissner corpuscle, 589, *597*
Melanin
  and albinism, 634
  and eye color, 180
  and hair color, 178
  metabolic pathway, 178
  production, 634
  in skin, *635*
  synthesis, *241*

Melanocyte, 634
Melanocyte-stimulating hormone, *618*
Melatonin, 912–913
  functions, 625
  secretion, *616*, 625
Meltdown, nuclear, 904–905
Membrane, cell, *see also* Plasma membrane
  calcium pump, 87
  and cell evolution, 136
  chloroplast, 109
  coated pits, 88, *89*
  diffusion, 82
  extraembryonic, 785
  fluid mosaic model, *79*, 79–80
  interaction with water, 136
  and life, 77
  lipid bilayer, *see* Lipid bilayer
  lipid cycling, *89*
  nonself markers, 80
  passive transport across, 86, *86*
  phospholipids, 78
  protein coat model, 80
  protein cycling, *89*
  selective permeability, 82
  self markers, 80, *177*
  shuttle, *131*
  sodium-potassium pump, 87, *743*
  sterols in, 41
  structure, *79*
  tonicity effects, 83
Membrane, tectorial, *600–601*
Membrane potential, brain surface, 587, *587*
Membrane transport
  active, 85
  passive, 85
  by pumps, 490
Memory
  Alzheimer's disease effects, 587
  brain areas affecting, *582*
  defined, 587
  epinephrine effects, 587
  immunological, 680, *681*, 687
  limbic system role, 588, *588*
  long-term, 587
  short-term, 587
Mendel, G., 170, 172–174
Mendelian theory of independent assortment, 174–175, *174–175*
Mendelian theory of segregation, 172–173, *172–173*, *192*
Meninges, *580*, 581
Meningitis, *357*, 581
Menopause, defined, 779
Menstrual cycle
  anabolic steroid effects, 649
  duration, 779
  follicular phase, 778, *782*
  hormone secretions during, *626*, 778–779, 781
  luteal phase, 778, *782*
  ovarian function, 780
  ovulation, 778, 780
  phases, 780, *780*
  summary of events in, *779*, 782, *782*
  uterine function, 781
Menstruation, 778
Mental retardation
  chromosome structure-related, *196*, 200, *200*
  phenylpyruvic acid-related, 204
  protein deficiency-related, 725
  thyroid-related, 624
Meristem
  apical, 485, *485*, 492, *492*, 532
  defined, 485
  derivatives, 485
  ground, *485*
  lateral, 485, *485*, 496
  and plant growth, 485
Mesenchyme, *762*

Mesoderm, 756, *762*
  in animal evolution, 418
  formation, 410
Mesoglea, cnidarian, 414
Mesophyll, *118*, 119, 486, 491, *491*, *507*
Mesozoic, *297*, 329, *329*, 336, *336*, 337, *393*, 458, *459*
Mesquite, 873
Messenger RNA (mRNA)
  alternative splicing, *239*
  assembly, 222
  base triplets, 224
  cDNA formation, 253, *253*
  codons, 224
  exons, 223
  formation, 221
  hybrid DNA–mRNA molecules, 253, *253*
  introns, 223
  nucleotide cap, 222, *223*
  oocyte, 760
  poly-A-tail, 222, *239*
  pre-mRNA, 222–223, *223*, *239*
  translation, *see* Translation
  tRNA interaction, 225
Metabolic activity
  goals, 5
  PET scans for, *21*
  requirements, 5
Metabolic pathway
  biosynthetic, 98, 136–137
  blocked, 220
  cyclic, 98
  defined, 98
  degradative, 98, 102
  electron transfers in, 102
  evolution, 136–137
  feedback inhibition, 100–101, *101*
  linear, 98
  melanin, 178
  participants in, 98
Metabolic reaction
  endergonic, 96, *96*
  equilibrium states, 96–97, *97*
  exergonic, 96, *96*
  reversible, 96–97, *97*
Metabolism
  aging effects, 92–93, 768
  bacterial, 348, *348*, 352–355
  carbohydrate, *730*
  control, 136
  DDT, 861
  defined, 4, 93
  dieting effects, 729
  effects on extracellular fluids, 736
  eubacterial, 352–353
  exercise effects, 729
  heat gain through, 744
  lipid, *730*
  mineral needs, 726
  in muscle cells, 647, *647*
  organic, major pathways, 730, *730*
  origins, 326
  prokaryote *vs.* eukaryote, *376*
  protein, *730*
  solute gains by, 736
  synthetic pathway, 326
  temperatures favorable for, *744*
  vitamin needs, 726
  waste products, and population growth, 808
Metacarpal bones, *641*
Metal ion cofactors, 101
Metal shadowing, *81*
Metamorphosis
  defined, 432
  insect, 432, 436–437
  seq squirt, 447
  and vertebrate evolution, 449
Metaphase
  karyotype preparation, 188
  mitotic, 147, *147*

Metaphase I
  gene shufflings, 161
  independent assortment at, *174*
  random alignment of
    homologous chromosomes,
    161
Metastasis, defined, 234
Metatarsal bones, *641*
Metatheria, 466
Methane
  atmospheric concentration, *857*
  classification, *895*
  early Earth, 324–325
  and greenhouse effect, 856–857
  methanogens, 352
Methanogen, *351*, 352
Methionine, 725, *725*
Methods, scientific, *see* Scientific
  method
Methyl group
  locations, *34–35*
  structural formula, *34–35*
Methylxanthin, 515
MHC marker, 680, 682–683
Micelle
  defined, 722
  formation, 720, *722*
Michaelangelo, *300*
Micrasterias, 362
Microbial symbiont, 921
Microbody, *64*
Microevolution, *see also* Evolution,
  biological
  balanced polymorphism,
    282–283
  defined, 274
  and natural selection, 270–283
  sexual dimorphism, 282–283
  summary of processes and
    patterns, *284, 298*
Microfilament
  actin, 70
  cytoskeletal, 69, *69*
  myosin, 70
  sliding mechanisms, 70
Micrograph, defined, 54
Micrometer, defined, 346
Micronutrient, 502
Microorganism, defined, 346
Micropyle, *519*
Microscope
  compound, *50*
  compound light, 54
  light, 54, *54*
  phase-contrast, 54
  resolution, 54
  scanning electron, 55
  transmission electron, 54–55
Microscopy
  chromosome staining, *189*
  Gram staining, 349, *349*
  units of measurement in, *53*
Microspore, 401, *404*, 518, *518–519*
Microsporum, 383
Microtubule
  central, 70, *71*
  chromosome positioning by, 161
  cytoskeletal, 69, *69*
  doublets, *71*
  formation, 71
  formation during mitosis, 148
  functions, 147
  9 + 2 arrays, 70, *71*
  sliding mechanisms, 70
Microtubule organizing center, 71,
  148
Microvillus
  intestinal, 722
  photoreceptor, *602*
  sponge, 413
Midbrain, *582*
Middle ear, *600–601*, 607
Midlittoral zone, rocky coast, 883,
  883
Midsagittal plane, *555*

Miescher, J. F., 209
Migraine headache, 383
Migration
  animal, 867
  cell, during morphogenesis, 764
  human, 815
Migratory behavior, Canadian
  geese, *462*
Milk, as protein source, *725*
Milk of magnesia, 29
Miller, G. T., 906
Miller, S., 325, *325*
Millipede, 432, 435, *435*
Milstein, C., 683
Mimicry
  butterflies, 312, *312*
  Dismorphia, 831
  forms of, 831
  as prey defense, 831
Mineral
  absorption, 500–505, *503*
  deficiency in plants, *502*
  defined, 726
  excess, 727
  from hydrothermal vents, 885
  intake during pregnancy, 791
  ion gains and losses, 736
  leaching, 852, *853*
  and plant nutrition, 500–505,
    502, 509
  recommended daily allowance,
    727
  required for cell functioning, *727*
  in tropical rain forest, *876*
Mineralocorticoids
  actions, *621*
  sources, *621*
  targets, *621*
Miocene, 475
Miocene-Pliocene boundary, 476
Miscarriage, 365, 535
Mischococcus, 371
Missing links, 268
Mississippi River watershed, 852
Mistletoe, 403, *403*
Mite, 432–433, 910–911, *910–911*
Mitochondrion
  Africanized honeybee, 123
  appearance in micrographs,
    58–59, 128
  ATP formation in, 67, *67,*
    127–128, *128, 131*
  ATP supplier, 331
  chemiosmotic theory, *130*
  crista, 67
  diagram, *60*
  DNA, 67
  double-membrane system, 67
  euglenoid, 370, *370*
  evolution, 67
  functional zones, *128*
  functions, *58–59*, 60, *73*
  membrane shuttle, *131*
  membrane system, 126–127,
    128, 131
  muscle cells, *646*
  occurrence in nature, *73*
  plant cell, *58–59*
  self-replicating, 332
Mitochondrion electron transport
  system, 102
Mitosis
  in African blood lily, *145*
  anaphase, *147*
  appearance in micrographs, *145*
  and asexual reproduction, 142,
    164
  and chromosome number,
    142–139
  at cleavage, 756
  comparison with meiosis, 164,
    164
  metaphase, 146, *147*
  nondisjunction effects, 201
  overview, 146, *146–147*

plant, *519*
  prometaphase, 148
  prophase, 146, *146*
  telophase, *147*
Mnemiopsis, 417
Molar, *465*, 718
Mold, bread, 220
Mole, defined, *22*
Molecular clock, 308
Molecule, *see also* Biological
  molecule
  defined, 19
  signaling, 242–243, 530, 534,
    614, 766
Mole rat, *633*, 929–930, *930*
Mollusk (Mollusca)
  bivalves, 426
  body plan, 424, *424*
  cephalopods, 426–427
  characteristics, 424
  chitons, 425
  circulatory system, *443*, 654
  ctenidia, 424, *424*
  digestive system, *443*
  evolution, *412*
  gastropods, 424–425
  mantle, 424, *424*
  movement, 442
  nervous system, *442*
  photoreceptor, *602*
  reproduction, *443*
  respiratory system, *443*
  vision, 602, *602*
Molting
  arthropod exoskeleton, 432
  hormonal controls, 626
  and maturation, 768
Molybdenum, in plant function,
  502
Moment-of-truth defense, *831,*
  831–832, *832*
Monarch butterfly, 866, *866*
Moneran (Monera), *see also*
  Bacterium; Prokaryote
  cell components, *73*
  evolutionary tree for, *318, 333*
  five-kingdom classification, 318
  representative, *8*
Monilinia fructicoli, 383
Monkey, 308, 334, 472, *472–473*
Monoclonal antibody, 683
Monocot (Monocotyledonae)
  adventitious roots, 478
  characteristics, *406*
  coevolution with pollinators, 405
  comparison with other plant
    groups, *406*
  cotyledon, *488*
  development patterns, 532, *533*
  embryo, 522
  fruits, 405
  function, 488
  growth patterns, 532, *533*
  herbaceous, 481
  life cycle, *404*, 405, *405*
  primary root, 478
  representative, 405, *406*
  stem, *488–489*
  structure, 488
  vascular bundle, *488*
Monocyte, *656*, 657, *657*, 678
Monohybrid crosses, 172–173,
  172–173
Monomer, defined, 36
Mononucleosis, infectious, *357*
Monophyletic lineage, 316
Monosaccharide
  carbon backbone, 37
  defined, 37
  functions, 48
Monosomy, 201
Monotropa uniflori, 403
Monsoon, 869
Monsoon forest, 877
Monsoon grassland, 874

Montane coniferous forest, *871,*
  878, 878
Moorhead, P., 768
Moray eel, *887*
Morchella esculanta, 382
Morel fungus, *382*
Morgan, T. H., *192,* 193
Morning-after pill, 795
Morphogenesis
  cell migrations, 764
  and cell shape, 764
  and cell size, 764
  fetal (human), 788–789
  and localized growth, 765
Morphogens, 766
Morphological convergence, 307,
  307, 317
Morphological divergence, 306
Morphological gap, 304
Morphological trait, 272
Morphology, comparative, 304–307
Morula, *793*
Mosquito, 282, 359, 421, *435*, 861
Moss, 394, *840*
  club, 396
  peat, 396
Moth, 598
  coloration, 7, *7*
  gypsy, 839
  lung, *437*
  mutations, 7
  peppered (*B. betularia*),
    278–279, *279*
  silkworm, *5–6*, 598, *627*
  yucca, 825, *825*
Mother-of-pearl, 426
Mother's milk, immunoglobulins in,
  685
Motility, sperm, 774
Motor cortex, *585*
Motor neuron, *554,* 561
  axons, 562
  cell body, 554
  distribution, 578
  functions, *553,* 561
  and muscle contraction, 646
  reflex arc, 570, *571*
  stretch reflex, *571*
Moufet, T., 262, 318
Mountain aven, *834*
Mount St. Helens, 482, *482–483*
Mouse
  gene transfer in, 256, *256*
  Grasshopper, 832, *832*
  priming pheromone, 922
Mouth
  cancer, 707
  clam, *426*
  deuterostome, 423
  functions, *717*, 718
  gastropod, *424*
  human, 718
  protosome, 423
  rotifer, *423*
  yeast infections, *388*
mRNA, *see* Messenger RNA
  (mRNA)
Mucin, 718
Mucous
  fungal infections, *383*
  membrane, *383*
Mucus, 718
  human respiratory tract, 700,
    701
  lamprey, 452
  membrane, 676
  polychaete, 430, *430*
Muddy shore, coast, 883
Mule, 248, 291
Multicellularity
  defined, 376
  division of labor in, 376
  origin, 376
Multiple allele system, *177*
Multiple sclerosis, 566

Multiplication cycles, viral, 358–359, *359–359*
Multipolar cell, *562*
Mumps, *357*, 688
Muscle
    anabolic steroid effects, 649
    antagonistic, 642
    cardiac, 552, *552*
    cardiac tissue, 661
    cardiac vs skeletal, 661
    contraction, *see* Contraction, muscle
    defined, 552
    Duchenne muscular dystrophy, *196*, 198
    earthworm, 428
    excitable cells, 646
    exercise effects, 648
    fatigue, 648, 731
    flight, 463, *463*
    functions, 552, *552*
    glycogen stores, 134
    and heat production, *748*
    lactate buildup in, 647
    lengthening, 646
    longitudinal, 636, *636*
    membrane system, *646*
    radial, 636, *636*
    response to cold stress, 748
    shivering, 748, *748*
    skeletal, 552, *552*
    smooth, 552, *552*
    strength, 648
    synergistic, 642
    twitch, 648, *648*
Muscle spindle, 570, *571*
Muscular system
    anabolic steroid effects, 649
    arm (human), *642*
    functions, *554*
    human, *554*, 643, *643*
    major muscles, *643*
    skeletal system interactions, 642
Musculocutaneous nerve, 580
Mushroom, *see also* Fungus
    air pollution effects, 387
    cell wall, 39, *39*
    common, 384
    extinction, 387
    extracellular digestion, 378
    harmful, identification, 385
    poisoning, 383
    toxins, 384
Musk ox, social defense, 924, *925*
Mussel, 218–219, 426, 836, 839
Mutagen, *see also* Teratogen
    cancer-inducing, 244
    defined, 228
Mutation
    beneficial, 7, 228–229, 275
    defined, 156, 275
    and diversity, 229
    Drosophila, *192*
    and environment, 275
    and evolution, 10, 229, 275, 449
    gene, 156, 187, *192*, 199, 229
    gene flow effects, 277
    harmful, 7, 228–229, 275
    lethal, 275
    metabolic pathway, 220
    and molecular clock, 308
    neutral, 228–229, 275, 292, 308
    novel gene creation by, 275
    nutritional, 220
    random, 275
    rate, 229
    rate, culturing effects, 254
    regulatory genes, 304
    stabilizing selection effects, 280
Mutualism
    coral–dinoflagellates, *416*
    defined, 386, 824, *824*
    flower–pollinator, 520–521
    legumes–bacteria, 502–503

lichens, 386
    mycorrhizae, 386–387
    pigeon–fruit tree, 822, 825
    sea anemone–clownfish, *415*
    yucca moth–yucca plant, 825, *825*
Myasthenia gravis, 689
Mycelium
    chytrid, 364
    dikaryotic, 385, *385*
    fungal, 380, *380*, 384
myc gene, 235
Mycobacterium tuberculosis, 103
Mycorrhiza, 386–387, *387*, 392, *403*, 490, 543, 825, 859, 896
Myelin sheath
    action potential propagation, *566*
    composition, 566
    in multiple sclerosis, 566
Myofibril, 644, *644, 646*
Myoglobin, 709, *710*, 711
Myometrium
    functions, 778, *778*
    location, *778*
Myopia, 606
Myosin
    actin-binding, 645–646
    arrangement in sarcomeres, 645, *645*
    cross-bridge formation, 645
    high-energy form, 645
    microfilaments, 70
    role in contraction, 70
    structure, 644
    subunits in microfilaments, 69, *69*
Mytilus, 836
Myxobacterium, 355
Myxococcus, *351*, 355
Myxomycota, 365
Myzoma virus, 839

N

NADH (nicotinamide adenine dinucleotide, reduced), 101, *125*, 126, *127–128*, 130, *131*, 133
NAD⁺ (nicotinamide adenine dinucleotide, oxidized), 46, *48*, 101, 126, *128*, 129, 133
NADPH (nicotinamide adenine dinucleotide phosphate, reduced), 101, 108, 112, 115, *115*
NADP⁺ (nicotinamide adenine dinucleotide phosphate, oxidized), 108, 112
    functions, *48*, 101
Nahmint Valley (British Columbia), *401*
Naked mole-rat, 929–930, *930*
Nanometer, defined, 346
Nanoplankton marine community, 371
Narcotic analgesic, 589
Nasal cavity, *598*
    functions, *700*
    location, *700*
Native herd farming, 901
Natural gas
    burning, 854, *855*, 857, 895, 904
    formation, 854, *855*, 904
Natural killer cell, 683
Natural phenomena, interpretation, 260, 264
Natural selection, *see also* Evolution, biological
    allele frequency changes, 274–275
    and balanced polymorphism, 282–283
    basis, 261
    computer simulation, 327
    Darwin's view, 10, 266–267

defined, 10, 267, 278, 916, 931
    directional selection, 278–279
    disruptive selection, 280–281
    effect on genetic equilibrium, 274
    evidence, 278–283
    evolution by, 278–283
    gene flow effects, 277
    indirect, 928–931
    and life history patterns, 810–813
    measure, 911
    methods, 278–283
    and morphological convergence, 307
    and mutation, 229
    sexual, 916–917, *917*
    and sexual dimorphism, 282–283
    stabilization selection, 280
    summary, 278
    transitional forms, 268
    and variation in populations, 266–267
    Wallace's theory, 268
Nature
    levels of organization, 4–5
    scientific approach to, 11
Nautiloid, 334
Nautilus, 334, 426
Neandertals, 478–479
Nearctic biogeographic realm, *870*
Nearsightedness, 604, 606
Nebula, Orion, *323*
Nectar, *521*, 524
Neem tree, 493
Negative feedback, 556, *556–557*
    in cortisol secretion, 623, *623*
    during menstrual cycle, 781
    testosterone production, 777
Neisseria gonorrhoeae, *351*, 797, *797*
Nematocyst, cnidarian, 414, *414–415*
Nematode (Nematoda), *see also* Roundworm
    circulatory system, *443*
    digestive system, *443*
    in estuarine ecosystems, 882
    evolution, *412*
    movement, *442*
    nervous system, *442*
    as prey, 388
    representative, *410*
    reproduction, *443*
    respiratory system, *443*
Nembutal, 589
Neon, electron distribution, *23*
Neotropical biogeographic realm, *870*
Nephridium
    earthworm, *429*
    onychophoran, *431*
Nephridium, earthworm, 429
Nephron
    functional regions, *739*
    structure, 738, *739*
Neritic zone, ocean, 884, *884*
Nerve
    afferent, *579*, 580
    autonomic, *579–580*
    cnidarian, 414
    components, *570*
    cranial, 580
    defined, 570
    efferent, *579*, 580
    fascicle, 570
    fluida, Lamarckian, 264
    free nerve endings, 596, *597*
    invertebrate, 576–577, *577*
    major, in human body, *579*
    mollusk, 424
    musculocutaneous, 580
    myelinated axons in, *570*
    parasympathetic, 580, *580*
    in reflex arc, *571*
    rotifer, 422

sensory, *579*, 580
    somatic, 580
    spinal, 580
    structure, *570*
    sympathetic, 580, *580*
    tract, 570
Nerve cord, *see also* Neural tube
    chordate, 446–447, *447*
    defined, 428
    earthworm, 428, *429*
    evolution in vertebrates, 450
    flatworm, 576
    lancelet, *448*
    mammalian, 464
    persistence in vertebrate embryos, 578
Nerve growth factor, 626
Nerve net
    cnidarian, 414
    defined, 414
    evolution, 576
    information flow through, 576
    sea anemone, *577*
Nerve pathway
    of olfactory receptors, *598*
    sensory, 594–595, *595*
    stretch reflex, *571*
    to visual association cortex, *608*
Nerve plexus, *720*, 723
Nervous system
    amphibian, *442*
    annelid, *442*
    arthropod, 432, *442*
    autonomic, *579–580*
    avian, *442*
    bilateral, 576–577, *577*
    and blood flow, 663
    and carbon dioxide levels, 707
    cell types in, 561
    central, 581–586, *581–586*
    cephalopod, 427
    chordate, *442*
    cnidarian, *442*
    components, *579*
    coral, *442*
    ctenophoran, *442*
    defined, 912
    and digestion, 723
    earthworm, 428, *429*
    echinoderm, 428, 438, *442*
    evolution, 578
    flatworm, *418, 442*
    fluke, *442*
    functional divisions, *579*
    functions, *554*
    human, *554*, 578–585
    Huntington's disorder, *196*, 197
    information flow, 82, *571*
    insect, *442*
    invertebrate, 576–577, *577*
    jelly fish, *442*
    major nerves, *579*
    mammalian, *442*
    molluscan, *442*
    nematode, *442*
    neuroglia, 561
    and oxygen levels, 707
    polychaete, *442*
    reflex pathways, *579*, 723
    reptilian, *442*
    ribbon worm, *442*
    rotifer, *442*
    roundworm, *442*
    sea anemone, *442*
    and segmentation, *577*
    somatic, *579–580*
    spider, *442*
    sponge, *442*
    stress response, 623
    tapeworm, *442*
    tunicate, *442*, 447
    vertebrate, 578–585
Net energy, 904
Net primary productivity, 848
Net protein utilization, 725, *725*

Net reproduction per individual per unit time, 806
Neural plate, 764, *765*
Neural tube, *see also* Nerve cord
  cerebrospinal fluid in, 583
  defined, 578
  formation, 764, *765*
  primitive streak, 786
Neuroglial cell, 561
Neuromodulator
  defined, 568
  functions, 568
Neuromuscular junction, *567*
Neuron
  action potential, 562
  bipolar cell, *562*
  categories, *562*
  convergent circuits, 570
  defined, 553
  divergent circuits, 570
  functional zones, 562
  functions, 553, *553*, 726
  ganglia, *580*
  interneuron, 561
  interneuron communication, 560–561
  motor, 552, *552*, 561
  in multiple sclerosis, 566
  multipolar cell, *562*
  neurotransmitter release, 566, *567*
  organization, 570
  resting membrane potential, 562
  reverberating circuits, 570
  sensory, 561, 609
  synaptic clefts, 566, *567*
  unipolar cell, *562*
Neurospora crassa, 220, 382
Neurospora sitophila, 382
Neurotoxin, dinoflagellate-related, 371
Neurotransmitter
  actions, 614
  defined, 566
  excitatory effects, 566, 568
  inhibitory effects, 566, 568
  neuromodulation, 568
  release, 566
  removal from synaptic cleft, 568
  spatial summation response, 568
  temporal summation response, 568
  toxins affecting, 569
  visual, 609
Neutral interactions, 824, *824*
Neutral mutations, 275, 292, 308
  rate, 342
Neutron
  in atomic structure, 19, *19*
  mass number, 19
Neutrophil, *656*, 657, *657*, 678
Newborn
  AIDS-infected, 796
  crack cocaine-addicted, 575
  defined, *793*
  of ex-smoker, 791
  phenylketonuria screening, 97
Newt, *240*
New World monkey, *472*
NGU (chlamydial nongonococcal urethritis), 797
Niacin, *726*
Niche
  defined, 824
  overlapping, 827
Nicolson, G., 79
Nicotiana rusticum, 493
Nicotine effects
  on the central nervous system, 589
  on the peripheral nervous system, 589
Nightshade, 493
Nile perch, 839

Nitrate, 858, *859*
Nitric oxide, and air pollution, 895, *895*
Nitrification, 858, *859*
Nitrifying bacterium, *351*, 353, *353*, 858, *859*
Nitrite, 858, *859*
Nitrobacter, *332*, *353*
Nitrogen
  atmospheric concentration, 858
  atomic number, *19*
  bacterial conversion, 352
  bubbles, 709
  decompression sickness, 709
  deficiency, and dormancy, 541
  electron distribution, *23*
  global cycling, 353
  losses from soil, 858
  mass number, *19*
  in plant function, *502*
  in polypeptide chains, 42
  symbol for, *19*
Nitrogen cycle, *351*, 858–859, *859*
Nitrogen dioxide, and air pollution, 895, *895*, 896
Nitrogen fixation, 502–503, 858, *859*, 881
Nitrogen-fixing bacterium, 502–503, *503*, 858, *859*, 881
Nitrogen oxide air pollutants, 895, *895*, 896
Nitrosomas, *351*
Nitrous oxide
  atmospheric concentration, *857*
  classification, *895*
  and greenhouse effect, 865–867
Nociceptor, *594*, 596
Node, stem, *484*, *492*, *533*
Node of Ranvier, *566*
Nodule, root, 503
Nomarski process, 55, *55*
Nomenclature, scientific, *310*
Nondisjunction
  defined, 201
  mechanisms, 201
  sex chromosomes, 203, *203*
Nonpolar covalent bond, 24
Nonrenewable energy sources, 904, *904*
Nonself marker, 680
Nonshivering heat production, 748
Nonsister chromatid, 160, *160*
Nonsteroid hormone, 617
Nonvascular plant
  bryophytes, 394–395, *395*
  species diversity, 391, 394, *406*
Norepinephrine
  actions, 568, *616*, *621*, 623
  secretion, *616*, 623
  sources, *621*
  targets, *621*, 731
Northern elephant seal, *283*
Northern fur seal, *283*
Nose, olfactory receptors, 598
Nostoc, *57*, *351*, 858
Notochord, *577*, 578
  chordate, 446
  lancelet, 448
  ostracoderm, 453
  placoderm, 453
  tunicate, 447, *447*
Notophthalmus viridiscens, *240*
Nuclear energy
  breeder reactor, 906
  fission power, 904–905
  fusion power, 906
Nuclear envelope
  appearance in micrographs, *58–60*, *62–63*
  diagram, *60*
  during mitotic metaphase, 147
  during mitotic prophase, 146
  plant cell, *58–59*
  pore complexes in, 63, *63*
  during prometaphase, 148

during prophase I, *160*
  structure, *63*
Nuclear medicine, radioisotope use in, 21
Nuclear power plant
  Chernobyl accident, 905, *905*
  meltdown danger, 904–905
  opposition to, 905
Nuclear waste disposal, 905
Nucleic acid hybridization, *249*, 252–253, *253*
Nucleic acids
  digestion, *720*
  functions, 48
  metabolism, *720*
  nucleotide arrangement in, 47
  viral, 356
Nuclein, defined, 209
Nucleoid, defined, 57
Nucleolus
  appearance in micrographs, *58–60*
  diagram, *60*
  functions, 62
  location, 62
  occurrence in nature, *73*
  plant cell, *58–59*
Nucleoplasm
  defined, 63
  diagram of, *60*
Nucleosome
  changes during transcription, *238–240*
  components, 240, *240*
  defined, 216
Nucleotide
  anticodons, 225
  arrangement in nucleic acids, 47
  base pairing in DNA, 213, *213*
  base sequence in DNA, 212–213, *214*
  components, 46
  defined, *48*
  of DNA, *211*
  functions, 46, *48*
  in mRNA, 222–224
  radioactively labeled, 252
  sequencing of DNA fragments, 251, *251*, 252
  strand assembly, *214*
  structure, 46, *46*
  types, 46
Nucleotide coenzymes, functions, 46
Nucleus, atomic, 19
Nucleus, cell
  appearance in micrographs, *58–59*, 62
  components, 62–63, *63*
  diagram, *60*
  euglenoid, 370, *370*
  eukaryote *vs.* prokaryote, *73*
  functions, *58–59*, 60, 62, *73*
  at interphase, male *vs.* female, *241*
  location, 52
  oocyte, 783, *783*
  plant cell, *58–59*, *72–73*, *519*, *532*
  sperm, 783
  structure, 62
  true, 59
Nudibranch, 425, *425*, 883
Nutria, *838*
Nutrient availability
  and dormancy, 541
  and senescence, 540
Nutrient flow, *see* Biogeochemical cycle; Energy flow (transfer)
Nutrition
  carbohydrate requirements, 724–725
  eubacteria, 352–353
  food pyramid chart, 724, *724*
  fungus, 380

and homeostasis, 731
  lipid requirements, 725
  and metabolism, *730*, 731
  mineral requirements, 726–727
  plant, 500–512, *503*, *505*
  during pregnancy, 790
  protein requirements, 725
  vitamin requirements, 726–727
Nutritional mutation, 220
Nymphea, *403*

O

Oak root fungus, 543
Oak tree, 277, *497*, 542–543, 874, 877–878
Obelia, 414
Obesity
  and cardiovascular disease, 666
  defined, 728
  effects on circulatory system, *728*
Occipital lobe, 584
Oceanic zone, 884, *884*
Ocean (sea)
  and climate, 868
  color, phytoplankton-related, *371*
  Pacific, 888–889, *889*
  pollution, 899
  and seafloor spreading, 328
  species diversity in, 840
  surface currents, 868, *868*, 888–889, *889*
  surface temperature, 888–889, *889*
  tides, 883, *883*
  upwelling, 888, *888*
  zones, 884, *884*
Ocotillo, 873, *873*
Octopus, 154, *410*, 426, *427*, 442–443, 603
Offspring
  hybrid, *171*
  nourishment, 755
  survival, 754–755
  true-breeding lineage, *171*
Ogallala aquifer, 898
Oil
  burning, 854, *855*, 857, 895, 904
  formation, 336, 854, *855*, 904
  functional groups, *35*
  functions, *48*
  from leafy plants, 493
  nondigestible, 725
  spills, 904
Oil glands, 635
Okazaki, R., *215*
Old man's beard (lichen), *386*
Olduvai Gorge, 478, *479*
Old World monkey, *472*
Olfaction, 598
Olfactory area, *585*
Olfactory bulb, *582*
Olfactory lobe, *582*, 583, 584
Olfactory nerve, *582*, 598
Olfactory receptor, 598, *598*
Oligocene, 475
Oligosaccharide
  defined, 37
  functions, 37, *48*
  side chain, 37
Oligotrophic lake, 881, *881*, 891
Olympic Peninsula (Washington), *401*
Ommatidium, 602, *602*
Omnivore, 846
Oncogene, 256
  alterations, 235
  defined, 244
  proto-oncogenes, 244
Onychophoran, 431, *431*
Oocyte, *see also* Egg
  cytoplasm, developmental information, 760

Oocyte *continued*
cytoplasm, reorganization at fertilization, 760, *761*
defined, 162
developmental information in, 760
follicle, 780, *780*
gray crescent, 760
human, *784*
maturation, 760
mature, 783
meiosis in, 780
in menstrual cycle, 778
ovulation, 780
penetration by sperm, 783, *783–784*
primary, *163*, 780
secondary, 162, *163*, 780
zona pellucida, 780
Oogenesis
defined, 162
generalized picture, *163*
Oogonium, *163*
Opabinia, 408, *408*
Operant conditioning, 915
Operator sequence, 236
Operon
defined, 236, *237*
lactose, 236, *237*
Ophiostoma ulmi, *383*, 838
Opium, 589
Opossum, 466, *466*
Opportunistic infection, 796
Optic chiasm, *584*, 586
Optic cup, 766
Optic lobe, *582*, 583
Optic nerve, 580, *586*, *605*, *608*, 609
Optic vesicle, 766
Opuntia, *119*
Oral arm, cnidarian, 414
Oral cavity, *700*, 718
Orange, 525, *526*
Orbital, electron, 22, *23*
Orcein dye, 189
Orchid, *403*, 521, 876
Order (taxon), 8, 311
Ordovician, 334, *393*
Organ
acoustical, 600
balance (ear), *599*, 599–600, *600*
designer, 553
otolithic, 599, *599*
Organelle
animal, 60, *60–61*
defined, 58, 60
endosymbiosis, 331
evolution, 331, 361–362
functions, 58, 60
plant, 58, *58–59*
types, *58–59*
Organic compound
cell usage, 36
chlorinated, 899
defined, 34
and eukaryotic evolution, 362
families, *48*
functional groups, 34–35, *34–35*
shape, 34–35
size, 36
spontaneous assembly, 324–327, *325*, *327*
stability, 34
synthesis, 36
transport by plants, 510–511, *511*
Organism
interdependency, 5, 347
response to change, 6
single-cell, 4
Organism, multicellular
defined, 4
division of labor in, 376
internal environment, 6
pH controls, 29

Organ of Corti, *600–601*
Organogenesis
chick embryo, *759*
comparative (animal cell), *757*
defined, 756
embryonic period, 784
fetal period, 784
frog embryo, *758*
Organ system, *see also specific system*
components, 547
defined, 418
flatworm, 418
human, *554–555*
interactions, 555
and internal environment, 556
overview, *554–555*
planarian, *418*
Organ transplant, 683
Orgasm
defined, 783
and pregnancy, 783
Oriental biogeographic realm, *870*
Origin, DNA replication, *214–215*
Orion, *322–323*
Ornithonyssus sylviarum, *910–911*
Orohippus, *340*
Oscilloscope, *565*
Osculum, 413
Osmosis
in capillary bed, *664–665*
defined, 83
effects on plant cells, 84
in *Paramecium*, 84
and plasma proteins, 656
Ossicle, *438–439*
Osteichthyes, *452*, 454
Osteoarthritis, 641
Osteoblast, 638
Osteocyte, *638*, 639
Osteoporosis, 639, *639*, 727
Ostracoderm, *452*, 452–453, *453*
Ostrich, 262–263, 462
Otolithic organ, 599, *599*
Ouachita National Forest (Arkansas), *401*
Outer ear, 600, 601
Outgroup, 314, 316
Oval window, *600–601*
Ovarian duct, *772*
Ovary
changes during menstrual cycle, *782*
cyclic changes in, 780
development, 190–191, *191*
endometrial scar tissue, 779
flatworm, *418*
functions, 778, *778*
human, *784*
location, *616*, 778
plant, *516*, 517, 519, 522, *523*
rotifer, *423*
secretions, *616*, 773, 778
structure, 780
Overgrazing, 901, *901*
Oviduct, *418*
functions, 778, *778*
location, 778
Oviparity, 755
Ovoviviparity, 755
Ovulation, 778, 780, 783
and birth control, 794
Ovule
components, 400
conifer, *400*
defined, 400, 518
developing, *519*
flowering plant, *404*, *516*, 518, *519*, 520
Lilium, *404*
mature, 522
Ovum
defined, 162
at fertilization, 783
Owl, *605*, 831, *831*

Ox, musk, 924, *925*
Oxaloacetate, 118–119, 128, *129*
Oxidation-reduction reaction, defined, 102
Oxygen
and aerobic respiration, *111*, *125*, 127–129, 136, 330
arterial blood levels, 705
atmospheric, accumulation, 136
atmospheric, early Earth, 308, 324, 330
atomic number, *19*
blood, sickle-cell anemia effects, 176
by-product of photosynthesis, 113, 136
deficit effects, 695
effect on biological molecules, 324, 330
as electron acceptor, *125*, *131*, 136
electron distribution, *23*
exchange in human body, 696–697
free radicals, 92–93
mass number, *19*
partial pressure in respiratory tract, 704, *704*
and photorespiration, *118*
in plant function, 494
pressure gradient, 696
and selection, 330
transport by hemoglobin, 704
utilization by leatherback turtle, 710–711
Oxyhemoglobin, formation, 704
Oxytocin, 557, 792
actions, *616*
secretion during childbirth, *616*
targets, *618*
Oyster, 426, 882
Ozone
and air pollution, 895, *895*
and greenhouse effect, 856–857
and ultraviolet radiation, 866
Ozone hole, 897, *897*
Ozone layer
damage to, 893, 897, *897*
defined, 866, *866*

P

Pacemaker, 21, 653
Pacemaking cell, 661
Pacific Ocean, 888–889, *889*
Pacinian corpuscles, 589, *594*, *597*
Pain
nociceptors, 596
referred, 597, *597*
Paine, R., 836
Palearctic biogeographic realm, *870*
Paleocene, 475
Paleozoic, 295, 329, *329*, 334–335, *393*
Palisade mesophyll, *491*
Palmiter, R., 256
Palp, 426, *426*
Panama, Isthmus, and speciation, 292–293
Pancake ice, *844*
Pancreas
bat, *128*
cancer, 707
endocrine cells, 622
exocrine cells, 622
function in digestion, 717
Pancreatic amylase, 720
Pancreatic islets, 614, *616*, 622–623, *622–623*, 625, *774–775*
Pancreatic nuclease, *720*

Panda
giant, *309*
red, *309*
Pandemic disease
AIDS, 359
defined, 359
Pangea, 335–336, 341, 864
PANs (peroxyacyl nitrates), and air pollution, 895, *895*
Panting, 748
Paper recycling, 900
Papilio dardanus, 312, *312*
Papilla, tongue, 598
Papovavirus, *357*
Paradisaea raggiana, *283*
Paramecium, *70*, 84, *84*, 826, *826*
body plan, 367
habitat, 367
Paramyxovirus, *357*
Parapatric speciation, 292, 295, *295*
Parapod, 430–431
Parasite
biological control with, 833
copepod, 434
defined, 418, *428*, 828, 846
flatworm, 420
fluke, 418, 420
hookworm, 420
intestinal, 419
mite, 910–911, *910–911*
pinworm, 420
vs. predator, 828
roundworm, 420–421
social, 833, 914, *914*
and sociality, 924
tapeworm, 414–421
tick, 433
Parasitic bacteria, 348, *351*
Parasitic fungus, 380–381
Parasitism, defined, 824
Parasitoid, 833, *833*
Parasympathetic nerve
functions, 580, *580*
locations, *580*
Parathyroid glands, 614, *616*, 621, 624, *624*
Parathyroid hormone
actions, *616*, 621, 624
secretion, *616*, 624
sources, *621*
targets, *621*
Parenchyma, 486, *486*, *496*
Parental behavior
costs and benefits, 927, *927*
redirected, 931, *931*
Parietal lobe, 584, *584*
Parthenogenesis, 526, *526*
Partial pressure gradient
carbon dioxide, 704, *704*
defined, 696
oxygen, 704, *704*
Parus major, 925
Parvovirus, *357*
Passion flower, *405*
Passive immunization, 688
Passive transport, 85, *85*
Pasteur, L., 674
Patella, 640, *641*
Pathogen
airborne, 359, 676
antibiotic effects, 354–355
chemical barriers, *676*
defined, 346, 676
fungal, *383*
lysis, 677, *677*
nonspecific responses, 678
physical barriers, *676*
transmission modes, 359
Pattern formation
embryonic, 766–767
homeobox genes, 767
Pauling, L., 220
Pavlov, I., 915
PCR, *see* Polymerase chain reaction (PCR)

Pea, garden (Pisum sativum), 489, 537
  dihybrid crosses, 174–175, 175
  floral structure, 170
  monohybrid crosses, 172, 172, 173, 173
  testcrosses, 173
Peacock, 633
Pear, 525
Peat, 834–835, 854, 879
  composition, 395
  formation, 396
  as fuel source, 395, 396
Peat moss, 878
Pectin, 508
  deposits on cell walls, 72
Pectoral girdle, 641
  bird, 463
Pectoralis major, 643
Pedigree, 195, 195, 199, 205
Pelagic province, ocean, 884, 884
Pelvic cavity, human, 555
Pelvic girdle, 641
  bird, 462–463
  human, 263, 473
  snake, 263, 263
Pelvic inflammatory disease, 797
Pelvic nerve, 580
Pelvis, renal, 738
Penguin, 40, 41, 306–307
  adoptive behavior, 931, 931
  colony life, 845, 924, 924
  trophic level, 844, 847
Penicillin, 388
  mode of action, 354
Penicillium, 388
Penis
  flatworm, 418
  human, 774, 774–775
Peppered moth, 278–279, 279
PEP (phosphoenol pyruvate), 126
Pepsin, 718, 720
Pepsinogen, 718
Peptic ulcer, 549, 719
Peptide bond
  defined, 42
  formation, 42
  formation during protein synthesis, 42, 43
Peptidoglycan, bacterial, 349
Percy, E., 640
Perennial plant, 496, 541
Perforation plate, 486, 487
Perforin, 683
Pericycle, 494, 495
Periderm, 487, 496, 497
Periodontal disease, 718
Peripheral nervous system, 578, 589
  autonomic system, 580–581
  cranial nerves, 580, 580
  diagram, 579–580
  parasympathetic nerves, 580, 580
  somatic system, 580–581
  spinal nerves, 580, 580
  sympathetic nerves, 580, 580
Peripheral vasoconstriction, 748
Peripheral vasodilation, 749
Peristalsis, 718, 719, 720
Peritoneal dialysis, 742
Peritoneum, 411
Peritonitis, 723
Peritubular capillary, 738, 739
Periwinkle, 836–837, 837
Permafrost, 879, 879
Permeability, selective, 82
Permease, 237
Permian, 334–335, 393, 459
Peroxisomes
  functions, 65
  metabolic roles, 65
Peroxyacyl nitrates (PANs), and air pollution, 895, 895
Perspiration, evaporation, 26

Peru current, 888, 889
Pest
  biological control, 279
  resurgence, 279
Pesticide
  beneficial, 17
  biological magnification, 861
  classification, 895
  and directional selection, 279
  harmful, 17
  resistance, 17, 279
  use reduction methods, 279
Petal, 516–517, 517, 523
Petrel, 847
Petroleum, see Oil
PET scans, see Positron emission tomography
pH
  buffering mechanisms, 29
  control mechanisms, 29
  extracellular, 742
  vaginal, 676, 774
Phagocyte, 654, 656, 676, 678, 679, 680
Phagocytosis, 88
Phalanges, 641
Pharyngeal arch, 786, 786
Pharynx
  chordate, 446, 447
  defined, 701
  earthworm, 429
  flatworm, 418
  functions, 700, 717, 718
  human, 700, 700, 718
  lancelet, 448
  rotifer, 423
Phase-contrast microscope, 54
Phaseolus, 491
Phenotype, 272–273, 276, 280–281
  defined, 171
  environmental effects, 182
Phenotypic variation, 278, 280
Phenylalanine, 204, 725
  phenylketonuria, 97
Phenylketone, formation, 97
Phenylketonuria, 97, 196, 204
Phenylpyruvic acid, 204
Pheromone, 598, 614
  honeybee, 929
  types, 922
pH (H+ level)
  effects on enzyme action, 100, 100
  effects on hemoglobin–oxygen binding, 704
  internal, 654
Philobolus, 381
Phloem
  companion cell, 489, 510, 510
  defined, 486
  example, 487, 494, 496, 507
  functions, 392, 486
  pressure flow in, 510–511, 510–511
  and source-to-sink pattern, 510, 512
  structure, 486
  translocation in, 501
Phosphate
  in phosphorus cycle, 860, 860
  in urine, 736, 742
Phosphate group
  locations, 35
  structural formula, 35
Phosphatidylcholine, structure, 78, 78
Phosphoenol pyruvate, 126
Phosphoglyceraldehyde (PGAL), 115, 116, 126, 127, 134
2-Phosphoglycerate, 127
3-Phosphoglycerate, 127
Phosphoglycerate (PGA), 115, 116
Phospholipid
  bilayers, see Lipid bilayer

  cell membrane, 78
  functions, 48
  hydrophobic interactions, 78
  structure, 41, 48, 78
Phosphoric acid, urinary elimination, 736
Phosphorus
  atomic number, 19
  deficiency, 727
  electron distribution, 23
  and eutrophication, 881
  excess, 727
  functions, 727
  global cycling, 353
  in human nutrition, 727
  mass number, 19
  in plant function, 502
  radioactive isotope, 211
  sources, 727
  symbol for, 19
Phosphorus cycle, 851, 860, 860
Phosphorylation
  defined, 104
  electron transport, 126, 130
  substrate-level, 126, 127
Phosphosglycerate, 118
Photoautotroph
  behavior, 355
  characteristics, 351
  energy sources, 107, 501
  eubacterial, 348, 351, 352–353
  habitats, 351
  heterocyst formation, 352
  as primary producer, 845–846, 846–847
  types, 107, 351
Photoautotrophism, 501
Photochemical oxidants, 895, 895
Photochemical smog, 854–855, 895
Photoheterotroph
  characteristics, 351
  eubacterial, 348, 351
  habitats, 351
  types, 351
Photolysis
  defined, 112
  mechanisms, 114
Photon
  defined, 110
  photosystem use, 112
  wavelength, 110
Photoperiodism
  in animals, 912–913
  in plants, 538–539
Photoreceptor
  cone, 608
  invertebrate, 602
  mollusk, 602
  ommatidia, 602, 602
  rod, 608
  stimuli for, 594
  and visual perception, 602
Photorespiration, 118, 118
Photosynthesis, 326
  algal, 370
  ATP formation, 68
  bacterial, 352–353
  brown algae, 372–373
  chrysophytes, 371
  and Cladophora experiments, 111
  in C3 plants, 118
  in C4 plants, 118
  14C research, 21
  effective wavelengths for, 110
  end products, 109, 115, 115–116
  end products, 117
  energy conversions, 104
  energy transfer during, 4–5
  evolution, 353
  intermediates, 115, 116, 117
  in lakes, 880–881
  light-dependent reactions, 108, 112–114

  light-independent reactions, 108, 115–116
  links with aerobic respiration, 106, 136
  links with respiration, 131
  macronutrients in, 502
  micronutrients in, 502
  ocean populations, 117, 347
  in oceans, 884
  in parenchyma cells, 486
  pigments, 352
  in plants, 486, 509
  reactants, 116
  role in ecosystems, 136, 845–846, 854
  site, 109
  and stomatal action, 118, 508
  summary, 109
  and ultraviolet radiation, 897
Photosynthetic bacteria, see Photoautotroph; Photoheterotroph
Photosystem I, 107, 112
Photosystem II, 107, 112
Phototroph, see Photoautotroph
Phototropism, 536, 536
Photovoltaic cells, 906–907
pH scale, defined, 28
Phycobilin, 372
Phycocyanin, 110
Phycoerythrin, 110
Phylogenetic reconstruction, 314–316
Phylogenetic tree, 476
Phylogeny
  cladistics, 314–316
  defined, 311
  and homologous structures, 317
  hypotheses in, 314
  and phenetics, 316
Phylum (taxon), 311
Physalia, 415–416, 417
Physarum, 362
Physiological trait, 272
Physiology, defined, 547
Phytochrome, 539
  active form, 243, 538
  control of gene transcription, 243, 243
  deficiency effects, 538
  inactive form, 243, 538
  interconversion of forms, 538
  and photoperiodism, 538, 543
Phytophthora infestans, 364
Phytoplankton, 370, 370, 371, 371, 848, 881, 884, 885, 888, 897
Phytosterol, 78
Pigeon, 10
  in New Guinea, 822–823, 823, 827
Pigment, animal
  amphibian egg, 757, 760
  bile, 721
  carotene, 634
  and coat color, 178, 178, 633
  and hair color, 178, 178
  hemoglobin, 634
  respiratory, 697
  skin, 618
  visual, 609
Pigment, plant
  anthocyanin, 517
  bacteriorhodopsin, 352
  carotenoid, 68, 110, 110–111, 352, 517, 538
  chlorophyll, 68, 110, 370, 538
  chlorophylls, 110, 110
  chloroplast, 68
  chromoplast, 68
  defined, 110
  flavoprotein, 536
  functions, 68, 73
  green, 111
  halobacterium, 352
  photosynthetic, 110, 110, 352

Pigment, plant *continued*
  phycobilins, 110, *110*
  phytochrome, 243, *243*, 538, *538*, 539, 543
  xanthophyll, *111*
Pike-cichlid, *812*, 812–813
Pill (birth control), 535, 795, *795*
Pilobolus, 381
Pilomotor response, 748
Pilus (pilli), *348*, 349, *349*
Pineal gland, *582*, *584*, 912–913
  functions, 625
  hormone secretion, 614, *616*, 625
  secretions, 625
Pineapple plant, *523*
Pine barrens, 878
Pine tree, *162*, 392, *392*, 878
Pinkeye, 606
Pinna (pinnae), *600*
Pinocytosis, 88
Pinworm infection, *410*, 420
Pioneer species, 834
Pistil, *see* Carpel
Pit, in vessel members, 486, *487*
Pitch, sound, 600
Pith, stem, *489*
Pituitary dwarfism, 620, *620*
Pituitary gland
  disorders, 620
  as effector, *557*
  functional links with hypothalamus, *619*
  functions, *582*
  homeostatic feedback loop, 621, 623
  hormone secretion, 614, *616*, *618*, 777
  intermediate lobe, *618*
  location, *584*, *616*
  stress response, 623
Pituitary gland, anterior
  location, *616*
  secretions, *616*, 618–619, *619*, 731, 777, *782*
Pituitary gland, posterior
  location, *616*
  secretions, *616*, 618, *619*
PKU, 204
Placazoa, *410*
Placenta, 762
  blood circulation, 787, *787*
  chorionic villi development, *787*
  composition, 466
  defined, 466
  diffusion along, 787
  functions, 787
Placental mammal, 466–467, *467*, 755
Placoderm, *452*, 453
Placozoan, 412
Plague, 809
Plakobranchus, 376
Planarian, 418, *418*
Plankton, 848, 880–881
Plant cell
  abscisic acid effects, 535
  amyloplasts, 68
  vs. animal cell, 60
  appearance in micrographs, *59*
  cell junctions, 72
  cell plate formation, 150
  cell wall, 72
  cellulose, 38
  central vacuole, *58–59*
  chloroplast, *58–59*, 68, *68*
  chromoplast, 68
  components, *73*
  cytoplasmic division, *150*
  cytoplasmic streaming, 70
  cytoskeleton, 69, *69*
  division mechanisms, 150, *150*
  enlargement, 532
  fertilization, *see* Fertilization, plant

glyoxysomes, 65
mesophyll, 119
mitochondrion, *58–59*
mitosis, *145*
nuclear envelope, *58–59*
nucleolus, *58–59*
nucleus, *58–59*
organelles, 58, *58–59*
osmosis effects, 83–84
plasma membrane, *58–59*
polyploidy, 201
ribosome, *58–59*
starch in, 38, 58, 68
transport proteins, 86
turgor pressure, 84
wall, *58–59*
water movement in, 83–84
Plant development
  asexual reproduction, 516, 526
  bean plant, *533*
  comparative embryology, 304–305
  corn, *533*
  corn plant, *533*
  embryonic, 518–519, 522–523, *522–523*, 533
  endosperm formation, 519, *519*
  and environment, 536, 541
  evolutionary, 276, 304–305, 336, *340–341*
  fertilization, 518–519, *519*, *see* Fertilization, plant
  fruit formation, 522–523
  hormonal effects, 530
  and meristems, 485
  nutritional requirements, 487–488
  oak as case study, 542–543
  patterns, 532–533, *532–533*
  pollination, 518–519, *519*
  prescribed growth patterns, 532
  primary growth, 473
  seasonal adjustments in, 538
  secondary growth, 473, 481–483
  seed formation, 522–523
  sexual reproduction, 516–524
  woody growth, 473
Plant growth
  apical dominance effect, 535
  hormonal effects, 534–535
  mycorrhizae effects, 386–387, *387*
  patterns, 532–533, *532–533*
  senescence, 540
  vernalization response, 540
Plant (Plantae), *see also* Flowering plant; *specific division*
  adaptations, 393
  amphibian-like, 396
  biennial, 481, 541
  in biogeochemical cycle, 852
  biological clock, 538
  biomass, 857
  bird-pollinated, 521
  body, overview, 484–485
  C3, 118
  C4, 118, *118*, 509
  CAM, 119, 509
  carbon-fixing adaptations, 118–119
  carnivorous, 486–487
  circadian rhythms, 538
  classification, 9, 391
  cloning, 254
  coloration, 110, *111*
  desert, 873, *873*
  disease resistance, 254
  diversity, 391
  and environment, 487–488, 536–538
  estuarine, 882
  evolution, 276, 304–305, 336, *340–341*, 394
  evolutionary tree for, *318*
  five-kingdom classification, 318

form, along environmental gradients, *870–871*
founder effect, 276
fungus interactions, 386–387
gamete formation, 162
genetic engineering, 254–255, *254–255*
gene transcription controls, 243, *243*
grassland, 874, *875*
gravity-sensing mechanisms, 536, *537*
herbaceous, 496
hormones, 530, 534–536, 543
hybridization, 254
life cycle, *162*, 392
meristem, 485, *485*
mineral uptake by, 500–505, 509
mutualistic relationship with bacteria, 489
nonvascular, 394
nutritional requirements, 500–512, *502*
perennial, 541–542
photorespiration, *118*
pollination, *290*, 825
polyploidy, 295
as producer, 5, 378
regeneration from cultures, 254–255
representative, *8*
root nodules, 503
root system, 484, 494–495, *504–505*
savanna, 874, *875*
seed, evolution, 393, *393*
seedbearing, 520
senescence, 540–541
shoot system, 484
shrubland, 874
stabilizing selection in, 280
structure and function, 482–497
survivorship curves, 811, *811*
symbiosis with fungi, 386–387
tissues, 484–487, 489, 492, 494–497
true-breeding, 170
viroid effects, 357
viruses, 357
water absorption, 502
water loss, 508–509
water uptake, 502–507, *504–505*, 510
wilting, 84
woodland, 874
woody, 496–497
world distribution, 262
worm-resistant, 255
Planula, *415*, 576–577
  cnidarian, 417
Plaque, atherosclerotic, 667, *667*
Plasma
  blood, 551, 556, 656, *656*
  defined, 556
Plasma cell, *681*
Plasma membrane, *see also* Membrane, cell
  adhesion proteins, 80
  appearance in micrographs, 60, *79*
  bacterial, 56
  carrier proteins, 80
  channel proteins, *79*, 80
  cytoskeleton, *79*
  defined, 52
  electron transfer proteins, *79*
  endocytic processes, 85, 88, *88–89*
  fluid behavior, 78
  fluid mosaic model, *79*, *79*, 81
  functions, *58–60*, *73*
  infoldings, 331, *332*
  ion pathways across, 562–563, *562–563*
  and life, 77

lipid bilayer, *see* Lipid bilayer
lysis, 358
muscle cell, *646*
neuron stimulation effects, 562
nutrient movement across, 504, *505*
occurrence in nature, *73*
plant cell, *58–59*, 504, *505*
polysaccharide groups, *79*
protein coat model, 80–81
receptor proteins, *79*, 80
recognition proteins, *79*, 80
structure, 80
transport across, 77, 82, 85, 88, *88–89*
transport proteins, *79*, 80
Plasma protein, 656, *656*, *668*, 676
Plasmid
  bacterial conjugation, 350, *350*
  defined, 350
  DNA transfers, 331
  integration into bacterial chromosome, 248
  recombinant, 249, 252–253, *253*
  and recombinant DNA technology, *249*
  Ti (tumor-inducing), 255, *255*
Plasmodesma, defined, 72
Plasmodial slime mold, *363*, 365
Plasmodium japonicum, 365, 368, *369*, 861
Plasmolysis, defined, 84
Plasmopara viticola, *364*
Plasticity, in human adaptations, 479
Plastid, functions, *73*
Platelet, *657*
  formation, 657
  functions, 656, *668*
Plate tectonics, *328*, 336
  seafloor spreading, 328
Platyhelminthes
  circulatory system, *443*
  digestive system, *443*
  movement, *442*
  nervous system, *442*
  representative, *410*
  reproduction, *443*
  respiratory system, *443*
Platypus, 442–443, *444*, 444–445, 465
Platysamia cecropia, 5, 598, *627*
Pleiotropy, 176
Pleistocene, 340, *340*
Plesiadapis, *475*
Plethodon
  P. glutinosus, 827, *827*
  P. jordani, 827, *827*
Pleural membrane, 701–702
Pleurisy, 701
Pleurobrachia, *410*, 417
Plutonium-238 isotope, 21
Plutonium-239 isotope, 905
Plutonium isotope, fissionable, 905
Pneumocystis carinii, 690
Pneumonia, *351*
Pneumonic plague, 809
Poecilia reticulata, 812–813, *812–813*
Poinsetta, 539
Poisoning, carbon monoxide, 708
Poisons, *see also* Toxins
  mushroom, 383
Polar bear, 310, 832
Polar body, *163*, 780, 783, *784*
Polar covalent bond, 25
Polarity
  animal egg, 760
  water molecule, 26, *26*
Pole cell, formation, *763*
Polio, *357*, 688
Pollen grain
  allergenicity, 689, *689*
  appearance in micrographs, *517*
  conifer, 401

Pollen grain *continued*
  development, 517–518
  evolution, 393
  export by saquaros, 524–525
  and fertilization, 518–519,
    518–519
  and flowering plant coevolution,
    520–521
  germination, 518–519
  gymnosperm, *400*
  in Mendel's experiments, *170*
  monocot, *404*
  transfer by pollinators, 405
  transport, 517–519
  variations, *517*
Pollen sac, 518, 520
Pollen tube, *404*, 518–519
Pollination
  characterization, *518*, 518–519,
    519
  coast live oak, 542
  coevolution of plant and
    pollinators, 520–521
  conifers, 401
  defined, 401
  and fruit production, 524, *524*
  gymnosperm, 401
  monocot, *404*
Pollinator
  animals, 524, *524*
  coevolution with plants, 405,
    520–521
  and flowering plant evolution,
    305
  limitations, *290*
Pollutant
  air, *see* Air pollutants; *specific
    pollutant*
  water, *see* Industrial pollution;
    Water pollution
Pollution, *see also* Air pollution;
    Water pollution
  global, 894
  industrial, 28, *29*, 279, 895, 899
Polychaete, *410*, 428, 430, *430*,
    442–443
Polydactyly, *195*
Polymer
  defined, 36
  subunits, 36
Polymerase, *48*, 214, *215*, 222,
    223, 228, 250
Polymerase chain reaction (PCR),
    250, *250*
Polymorphism, balanced, 282–283
Polyp, cnidarian, 414, *414*
Polypeptide chain
  abnormal hemoglobin HbS, 221
  amino acid sequence, 220
  gene coding for, 220
  post-translational modifications,
    239
Polypeptide chains
  bacterial, 57
  defined, 42
  fibrous proteins, 44
  formation, 42
  globular proteins, 44
  hemoglobin, 44, *45*
  hydrogen bonds, *44*
  insulin, *43*
  keratin, *45*
  protein, 44, 57
  in rough endoplasmic reticulum,
    65
Polyploidy
  in animals, 295
  through nondisjunction, 201, *201*
  in plants, 201, 295
  speciation via, 294–295, *295*
Polyporus, 378, 384
Polysaccharide
  bacterial cell wall, 56
  defined, 38
  functions, *48*

plasma membrane, *79*
  sources, *48*
  structure, *48*
  uses, *38*
Polysome, formation, 227
Polytene chromosome, 242, *242*
Polytrichum, 394
Ponderosa pine, *391, 400*
Pons, 582, *582, 584*
Pool
  in stream, 882
  tidepool, 836–837, *837*, 883, *883*
Population, *see also* Community;
    Human population
  biotic potential, 807
  carrying capacity, 808
  characteristics, 804–805
  continuous variation, 180, *180*
  defined, 4, 272, 803
  density, 804, 829
  dispersion patterns, 804–805,
    804–805
  distribution, 804–805
  evolution, 10, 272–273, 812–813
  impact on resources, 266
  life history patterns, 810–811
  life tables, *810*
  size, 804
  survivorship curves, 811, *811*
  variations in, 10
Population growth
  and age structure, 816–817, *817*
  and agriculture, 814, 898
  annual, 816, *816*
  in California, 819
  controlling, 816–820
  demographic transition model,
    818, 818–819
  density-dependent controls, 809
  density-independent controls,
    809
  and disease, 814
  doubling time, 807
  and economic development,
    818, 818–819
  exponential, *806*, 806–807, *807*
  and family planning, 816–817
  human, 814–820, *815*
  in India, 802–803
  industrial stage, 819
  limiting factors, 803, 808–809,
    814–815
  logistic, *808*
  overshoot, *809*
  parasitism effects, 833
  patterns, 803
  postindustrial stage, 819
  predation effects, 829, *829*
  preindustrial stage, 818
  transitional stage, 818–819
  zero, *806*, 816, 820
Population size, environmental
    limits, 266
Populus tremuloides, 526, 527
Pore complex, nuclear envelope,
    63, *63*
Porifera
  characteristics, *442–443*
  representative, *410*
Porphyrin, possible synthesis from
    formaldehyde, *326*
Porpoise, *306–307*
Portuguese man-of-war, *416*, 417
Positive feedback loop, 557, *781*
Positron emission tomography, 21,
    21, 585, *585, 589*
Postelsia palmaeformis, 362, 373
Posterior (anatomy), 410, *411, 555*
Postnatal development, stages,
    793, *793*
Postsynaptic cell, 566, *567*
Post-translational control, of gene
    expression, *238–239*
Posture
  and aging, 640

human body, *582*
Postzygotic isolation, *289*, 291
Potassium
  atomic number, *19*
  deficiency, *727*
  excess, *727*
  in human nutrition, *727*
  mass number, *19*
  and nephron function, 739
  neuronal diffusion, 563–564
  in plant function, *502*
  sodium-potassium pump, 563,
    563
  sources, *727*
  and stomatal action, 509, *509*
  symbol for, *19*
Potassium-40 isotope, *20*
Potato, 838
Potato late blight, 364
Potax virus, 357
Potentilla glandulosa, 320
Pouched mammal, 466, *466*
Poultry, comb shape, 179, *179*
Powdery mildew, 383
Pox virus, *357*
Prairie
  shortgrass, 874, *875*
  tallgrass, 874, *875*, 877
Prairie dog, 924
Praying mantid, *436, 832*
Precipitation, *see also* Rainfall
  in a watershed, *853*
Predation
  defined, 824, *824*
  effects on competition, 836–837
  human, 864
  through illegitimate signaling, 922
  and population growth, 829, *829*
Predator
  cooperative avoidance,
    924–925, *925*
  defined, 828
  vs. parasite, 828
  stealthy, 832, *832*
Predator–prey interaction
  ant–termite, 920, 922
  bat–frog, 923
  camouflage in, 830, *830*
  fungus–nematode, *388*
  leopard–baboon, *828*, 831
  lynx–hare, 829, *829*
  mouse–beetle, 832, *832*
  population dynamics, *828*,
    828–829, *829*
  sea anemone–sea star, *415*
  sea star–scallop, *426*
  spider–fly, *832*
Prediction, scientific, 11
Prefrontal cortex, 585
Pregnancy
  alcohol effects, 790–791
  among primates, 756
  and antibodies, 668–669
  cocaine effects on, 791
  critical periods, *790*
  effects of prescription drugs,
    790
  embryonic development,
    772–773, *772–773*
  embryonic period, 784, 789
  fertility control, 794–795
  fetal alcohol syndrome, 791,
    791
  fetal development, 772–773,
    772–773
  fetal period, 784, 789
  first week, *784*
  and genetic counseling, 204–205
  and genetic screening, 204
  HIV transmission during, 691
  human chorionic gonadotropin
    detection, 785
  and intercourse, 783
  mammary gland growth during,
    792, *792*

mineral intake during, 791
  and nutrition, 790
  prenatal diagnosis of genetic
    disorders, 205
  and prescription drugs, 791
  Rh blood typing, 668–669
  Rh− women, 669, *669*
  risks of infection during, 790–791
  sensitivity to teratogens during,
    790
  smoking effects, 707, 791
  termination, 798
  tests for, 795
  toxoplasmosis transmission
    during, 368
  trimesters, 784, 789
  unwanted, 794–795
Preimplantation diagnosis, 205
Premotor cortex, *585*
Prenatal diagnosis
  defined, 204
  DNA probes for, 256
  RFLP applications, 252
  trisomy 202, 204–205
  types, 205–206
Prespore cell, *365*
Pressure
  atmospheric, 696, *696*
  mechanoreceptors, 596
Pressure flow theory, 510–511, *512*
Pressure gradient
  defined, 82
  respiratory, 696
Prestalk cell, *365*
Presynaptic cell, 566, *567*
Prey defense
  adaptive responses to, 832, *832*
  camouflage, 830, *830*
  chemical, 831
  and coevolution, 830
  cooperative, 924–925, *925*
  mimicry, 312, *312*, 831, *831*
  moment-of-truth, *831*, 831–832,
    832, 918
  warning coloration, 831, *831*
Prezygotic isolation, *289*, 290–291
Pribilof Islands, 809
Price, P., 833
Prickly pear cactus, *119*, 873, *873*
Primary consumer, 846
Primary immune response, *681*,
    682–683, 688
Primary oocyte, *163*, 780
Primary producer
  estuarine, 882
  examples, *846*, 848
  function, 845–846, 848
  trophic level, *846*, 847, *848*
Primary productivity
  and climate, 865, 867
  defined, 848
  global patterns, 849
  in lakes, 881
  in oceans, 884, *885*
Primary root, *484*, 532, *533*
Primary spermatocyte, 162, *163*
Primary structure, proteins, 42–44
Primary succession, 823, 834,
    834–835
Primate (Primates)
  characteristics, 472, *472*
  classification, 472
  compared, 473–474
  origin and evolution, 475
  skull comparisons, *475*
Primer (DNA synthesis), *215*, 250
Priming pheromone, 922
Primitive streak, 786, *786*
Principles of Geology (C. Lyell), 266
Prion
  defined, 357
  effects in animals, 357
Probability (genetics)
  defined, 173
  monohybrid crosses, 173

Probability (genetics) *continued*
  Punnett-square method, 173, *173*
Probosis, ribbon worm, 422
Procambium, *485, 494*
Processes, neuronal, *571*
Prochlorobacteria, *351*
Prochloron, 351
Producer, *see also* Primary
      producer
  examples, *848, 850*
  functions, *849–850*
  plant, 378
Product, in chemical equations, *22*
Productivity, primary, *see* Primary
      productivity
Profundal zone, lake, 880, *881*
Progeria, 186–187, *196, 197*
Progesterone
  actions, *616, 621*
  menstrual cycle secretion, 781
  placental secretion, 787
  secretion, *616,* 778
  sources, *621*
  targets, *621,* 781
  and uterine function, *782*
Proglottid, 419, *421*
Progymnosperm, *393,* 396
Prokaryote, *see also* Bacterium;
      Moneran
  binary fission, *376*
  body plan, 56
  cell components, 52, *73*
  cell division, *142*
  cell wall, *376*
  defined, 8, 56
  vs. eukaryote, 376, *376*
  metabolism, *376*
Prokaryotic cell, 330
Prolactin, *743*
  actions, *616, 618,* 619
  secretion effects, 242
  and sodium retention in salmon,
      743
  targets, *618,* 619
Prometaphase, defined, 148
Promoter, 222, 236, *237*
Pronghorn antelope, *714,*
      714–715
Propagation
  action potentials, 565, 594–595,
      595
  asexual (flowering plant), 516
  tissue culture, *526, 527*
  vegetative, *526,* 526–527
Prophase
  chromatid changes during, 148
  early, *146*
  late, *146*
  mitotic, 146, 362, *601*
Prophase I
  events, 160, *160,* 161
  gene shufflings, 160–161
Propulsion
  cephalopod, 426
  fish, 447
  lancelet, 448
  tunicate, 447
Prosimian, 471–472, *472*
Prostaglandin
  and body temperature, 749
  sources, 614, *615*
  synthesis, 626
Prostate gland
  disorders, anabolic steroid-
      related, 649
  functions, 774, *774–775*
  location, *774–775*
  secretions, 774
Protein
  activator, 237, 242
  adhesion, *79,* 80
  as alternative energy source,
      134, *135*
  amino acid components, 725
  amino acid sequence, 42

animal, 725
backbone formation, 35
bound to DNA, 216
carrier, 80
channel, *79,* 80, *563*
complement, 677, *677*
complete, 725, *725*
defined, *48*
denaturation, 46
dietary, 725
digestion, 134, 718, *720*
and essential amino acids, *725*
fibrous, 44, *48*
flow through cytomembrane
      system, *64*
functional groups, *35*
functions, 42, *48*
gel electrophoresis, 298–299
globular, 44, *48*
histone, *238–239*
incomplete, 725, *725*
in lipid bilayer, 52
metabolism, 134, *135, 720, 730*
mutations in, 136
plant, 725
plasma, 656, *656,* 676
polypeptide chains, 44, 57
primary structure, 42–43
quaternary structure, 44
radiolabeling, *211*
receptor, *79,* 80
recognition, *79,* 80
recommended minimum daily
      requirements, *725*
regulatory, 236, 244
repressor, 236
secondary structure, 44
sources, 725, *725*
storage in plants, 504, 510
surface regions, 30
tertiary structure, 44
three-dimensional structure, 35,
      44
transport, *79,* 80, *563*
types, 42
utilization, human, *725*
Protein deficiency
  effects on hair, 635
  essential amino acids, 725
Protein hormone, *616,* 617
Protein synthesis
  bacterial, 57
  chain elongation, 226, *226–227*
  chain termination, 226, *227*
  control, 136
  and DNA transcription, *238–239*
  and DNA translation, *238–239*
  initiation, 226, *226*
  mutation effects, 228, *228*
  requirements, 221
  self-assembly, 324–325
  self-assembly under abiotic
      conditions, *327, 327*
  stages, *227*
  templates for, 325
  transcription, 222–223, *222–223*
  translation, 226–227, *226–227*
Proterozoic, 323, 329, *329,* 330
Protgesterone, secretion, 626
Protistan (Protista)
  cell components, *73, 363*
  classification, 9, 318, *333,* 363
  evolutionary links, 362–363
  evolutionary tree for, *318, 333*
  parasitic, 368
  representative, *8*
Proto-cell, 327
Protoderm, *494*
Proton, *see also* Hydrogen ion
  atomic number, 19
  in atomic structure, 19, *19*
  charge, 19
  in glycolysis, 126, *127*
Protonephridium, 418, *418,* 422,
      423

Proto-oncogene
  activation by UV radiation, 635
  defined, 244
  expression, 244
Protoplast
  cultures, 256
  genetic engineering, 256
Protostome, *412,* 423
Protozoan
  amoeboid, *366,* 366–367
  ciliated, *363,* 367, *367*
  conjugation, 367
  flagellated, 368, *368*
  infections, 359, 368
  life cycle, 366
Proximal tubule
  blood flow through, *739*
  location, 738
  sodium transport, 740, *741*
Prunus, 516
Pseudocoelom, *423*
Pseudocoelomate, *410*
Pseudomonads, 353
Pseudomonas
  binary fission, *353*
  P. marginalis, 57
  P. syringae, 354
Pseudopod, 70, 88, *366,* 764
Psilophyton, 335, 396
Psilotum, 396, *397*
Psychedelic drug, 589
Psychoactive drug, 575, 589
Psychosis, body-builder's, 649
Psychotherapy, for eating
      disorders, 729
Pterosaur, *306, 459*
Puberty, 625, *793*
Puccinia graminis, 383
Puffball, 378, 384
Pulmonary artery, *659–660*
Pulmonary circulation, 658, *658*
Pulmonary vein, *659–660*
Pumpkin, *18*
Punctuational model, of
      evolutionary change, 296
Punctuation model of speciation,
      296
Punnett, R., 179
Punnett-square method, 173, *173*
Pupa, *929*
Pupil, *580,* 604–605
  defined, 602
  eye, 604
Purple bacteria, *351,* 353
Purple coat fungus, *378*
Pyloric sphincter, *719*
Pyracantha shrub, 133
Pyramid, ecological, 848, *848, 850*
Pyrenestes ostrinus, 280
Pyrophorus noctilucus, 103
Pyrus, 486
Pyruvate, destinations, 128, *135*
Python, *263,* 461, *592*

Q

Quadriceps femoris, *643*
Quaking aspen, 526, *527*
Quaternary structure, proteins, 44
Queen bee, *928–929*
Quercus
  Q. agrifolia, 542
  Q. alba, 8
  Q. rubra, 497

R

Rabbit
  Himalayan, *182*
  introduction into Australia, 839
  population dynamics, 829, *829*
Rabies, *357*
Racoon, *309, 467*
Radial cleavage, 423, *423*

Radial keratotomy, 607
Radial muscle, 636, *636*
Radial symmetry, 410, *411–412,*
      414, 417, 438, *438*
Radiation
  heat gains and losses through,
      744, *749*
  solar, 865, *866,* 866–867, *867*
Radioactive dating
  radioisotopes used in, 20, *21*
  reliability, 20
Radioactive fallout
  Chernobyl, *905*
  from meltdowns, 904
Radioactive waste
  decay, 905
  disposal, 905
Radioisotope dating, 301, 470
Radioisotopes, *see also specific
      isotope*
  bacteriophage labeling, *211*
  decay, 19
  defined, 19
  half-life, 20
  medical applications, 21
  in nuclear medicine, 21
  for radioactive dating, 20, *21*
  as tracers, 21
Radiolarian, *366,* 367
Radish, 530
Radium-226 isotope, 21
Radius, *463, 641*
Radula
  cephalopod, 426
  chiton, 425
  mollusk, 424, *424*
  squid, *427*
Rafflesia, 876
Rage, anabolic steroid-related, 649
Rain, pH, 28
Rainfall
  acid, 893, 896, *896*
  boreal forest, 878
  and deforestation, 902–903
  desert, 873
  dry shrubland, 874
  dry woodland, 874
  and global air circulation,
      866–867
  global oscillations in, 888–889
  grasslands, 874
  savanna, 874
  tropical deciduous forest, 877
  tropical rain forest, 876
  tundra, 879
Rain forest
  temperate, *see* Coniferous forest
  tropical, *see* Tropical rain forest
Rain shadow, 869, *869,* 873
Rainwater, acidity, 896, *896*
Ramaria, 384
Rana pipiens 752–753, *752–753*
Ranunculus, 182, 495, 831
Rash
  bull's eye, 354
  roseola, 357
Ratfish, *452, 454, 454*
Rattlesnake, 460, *461*
Raven, feeding behavior, 916
Ray-finned fishes, *452,* 454–456
Ray (fish), *452, 454, 454*
Reabsorption
  sodium, *739–741,* 740–741
  urine formation by, 739, *740,*
      740–741
  water, 740–741
Reactant
  in chemical equations, *22*
  in metabolic reactions, 96–97
Reactor
  breeder, 906
  nuclear, 904–905
Readsorption
  glucose, *739*
  sodium, *739,* 740–741

Readsorption *continued*
  urea, *739*
Rebound effect, 581
Receptive field, visual, 609
Receptor, cell
  for antigen, 682
  defined, 6
  for transmitter substance, *567*
  types, 6
Receptor, sensory
  adaptation, 595
  chemoreceptors, 596
  cone (photoreceptor), *608*
  defined, 6
  ear (human), *600*
  functions, 556
  hair cell, *600*
  hearing, *600*
  in homeostatic control, 556, *557*
  mechanoreceptors, 596
  nociceptors, 596
  olfactory, 598, *598*
  pathways, 594–595
  pharynx, 718
  photoreceptors, 602, *602, 608*
  platypus, 445
  recruitment, 595
  rod (photoreceptor), *608*
  skin, 556, 632
  tactile, *597*
  taste, 598, *598*
  thermoreceptors, *488,* 596,
    748–749
  types, 594, *594, 654*
Receptor protein, *79*
Recessive allele, 168, 171, *192*
Reciprocal cross, *192*
Recognition, immunological, 680
Recognition protein, *79,* 80
Recombinant DNA technology, 247
  applications, 247
  benefits and risks, 254, 256
  defined, 248
  ethical issues, 257
  vs. natural selection, 248
  social issues, 257
Recombination, by restriction
    enzymes, 249, *249*
Recommended daily allowance
  minerals, *727*
  vitamins, *726*
Rectum, *580, 717,* 723
Rectus abdominis, *643*
Recycling, 900
Red algae, 370, 372, *372,*
    836–837, *837,* 883
Red blood cell (erythrocyte)
  appearance in micrographs, *81,*
    657
  carbonic anhydrase in, 704
  cell count, 657
  components, 81
  freeze-fractured and etched, 81,
    81
  functions, 551, *656*
  in hypertonic solution, *83*
  in hypotonic solution, 80, *83*
  iron requirements, 726
  in isotonic solution, *83*
  in leatherback turtle, *710*
  life span, 81
  Plasmodium infection, 369
  shape, *656–657*
  shriveling, *83*
  size, 53, *657*
  splenic, 671
  swelling, 80, *83*
Red-green blindness, *196,* 606
Red marrow, 638
Red oak, *497*
Red tide, 371, *371*
Redwood tree, California coast, 8
Reef-building coral, *416,* 417
Reef organisms, evolution, 334
Reeve, H. K., 930

Referred pain, 597, *597*
Reflex
  stretch, 570, *571,* 581
  withdrawal, 570
Reflex arc, 570, *571*
Reflex center, *582*
Reflex pathway
  digestive tract, *723*
  extensor, 581
  interneurons, 581
  invertebrate, 576
  nervous system, *579*
  vertebrate, 578
Regeneration, 768
Regulatory genes, 304
Regulatory protein, 191, 236
Reindeer, *809*
Reindeer moss, *386, 388*
Relatives, caring for, 928
REM (rapid eye movement) sleep,
    587
Renal artery, 659, *738*
Renal capsule, *738*
Renal pelvis, *738*
Renal vein, *659*
Renewable energy sources, *904*
Renin
  actions, 741
  role in sodium readsorption, *741*
Replacement rate, for zero
    population growth, 816
Replication, viral, *see* Viral
    replication
Replication enzyme, *214–215*
Replication fork, *214–215*
Repressor protein, 236
Reproduction, *see also* Asexual
    reproduction; Human
    reproduction; Sexual
    reproduction
  asexual, 142
  basic rule, 142
  brown algae, 373
  cnidarian, 417
  continuum, 6
  defined, 6
  defining characteristic of life, 326
  defining function of species,
    288
  fungus, 380
  green algae, 375
  hermaphroditic, 419
  and inheritance, 7
  and life cycle, 142
  red algae, 372
  reproductive isolating
    mechanisms, 288–291, 336
Reproductive behavior, 754–756,
    916–917, *917,* 922, 922–923
Reproductive rate, *see* Birth rate
Reproductive success, 911, 916,
    927
Reproductive system
  female, 778–780, *779–782, 782,*
    784
  flatworm, 418, *418*
  functions, 555
  hermaphroditic, 418–419
  male, 774–775, *774–775,*
    774–777, *776*
Reproductive timing, 754
Reptile (Reptilia)
  adaptations, 458
  brain size, *450*
  circulatory system, *451,* 458
  classification, 446
  ectothermy, 746
  egg, 762
  evolution, *450,* 458–459, *459*
  evolutionary link with birds, 462
  existing groups, 457–460
  mammal-like, 464
  marine, physiology, *710,* 710–711
  skeleton, *637*
  skin, 458

Research tools
  bacteria, 210, 356
  bacteriophages, *211,* 356
  barometer, *696*
  cell cultures, 254–255
  centrifuge, 80
  cladograms, 316–317
  comparative morphology,
    306–307
  compound light microscope, 54
  compound microscope, *50*
  computer simulations, 861
  dating, 338, 470
  DNA probes, 252
  Drosophila, 192
  electrodes, *565, 609*
  electrophoresis, 220–221, 293
  freeze-etching, 81, *81*
  freeze-fracturing, 81, *81*
  gel electrophoresis, 250–252
  in gene therapy, 247
  gravity maps, 338
  iridium levels in soil, 338
  karyotype diagrams, 188–189
  kitchen blender, *211*
  light microscopes, 54, *54*
  metal shadowing, *81*
  micrometer, *53*
  monoclonal antibodies, 683
  nanometer, *53*
  Neurospora crassa, 220, 382
  nucleic acid hybridization,
    252–253, *253*
  oscilloscope, *565*
  phase-contrast microscope, 54
  polymerase chain reaction, 250,
    250
  positron emission tomography,
    21, *21,* 585, *589*
  radioactive labeling, *211,* 252
  radioisotopes, 20, 470
  RFLP analysis, 252
  scanning electron microscope,
    55, *55*
  scintillation counters, 21
  transmission electron
    microscope, 54–55, *55*
  ultraviolet light, *520*
  viruses, 210–211, *211*
  Xenopus laevis, 764
  X-linked genes, *192*
  x-ray diffraction, 212–213
Residual volume, lung, *703*
Resource availability, and
    population limits, 808–809
Resource consumption, human
    population, 819, 894, 904, *904*
Resource partitioning, 827, *827*
Resource recovery center, 900
Resources, and populations,
    266–267, 278, 501
Respiration
  arthropod, 432
  control, *582*
  evaporative loss in, 744, *744*
  at high altitudes, 708
  in marine mammals and
    reptiles, *710,* 710–711
  modes, 697–698
  tracheal, 698, *698*
  underwater, 709
  waste products, 705, 736
Respiratory bronchiole, 701
Respiratory centers, brain, 705
Respiratory surface, defined, 696
Respiratory system, *see also* Lung
  annelids, *443*
  arthropods, *443*
  bird, 463, 697
  chordates, *443*
  and cigarette smoke, 706–707
  cnidarian, *443*
  components, *700*
  coral, *443*
  ctenophoran, *443*

  disorders, 706–707
  echinoderms, *443*
  fish, 698
  flatworm, *443*
  fluke, *443*
  functions, 555, 696
  human, 555, 700–705, *701*
  insect, *443*
  jelly fish, *443*
  lancelet, 448
  links with other systems, *654,*
    696, *696*
  mollusks, *443*
  nematode, *443*
  polychaete, *443*
  ribbon worm, *443*
  rotifer, *443*
  roundworm, *443*
  sea anemone, *443*
  sperm whale, 709, *709*
  spider, *443*
  sponges, *443*
  tapeworm, *443*
  water loss by, 744
Restriction enzyme, 249, 252
Restriction fragment, 249, *249,*
    251, *251,* 252
Restriction fragment length
    polymorphism
  applications, 252
  characterization, 252
  DNA fingerprints, 252
Reticular formation, 583, *583,* 587
Retina
  defined, 602
  detachment, 607
  embryonic, 766–767, *766–767*
  functions, *605*
  human, *604, 608*
  light receptors, *586*
  organization, 608, *608*
  vertebrate, 604, *605, 608*
Retrograde amnesia, 587
Retrovirus, *357*
  HIV actions, 690
  oncogene identification in, 244
Reverberating circuit, neuronal, 570
Reverse transcriptase, 253, *253,*
    690, *691*
Reverse transcription, 253, *253,* 358
Reversible reactions, 96–97, *97*
Reznick, D., 812–813
R group, amino acid, 42, *42,* 44
Rhabdovirus, *357*
Rh blood typing, 668–669, *669*
Rhea, *262–263*
Rheumatoid arthritis, 641, 689
Rhinovirus, *357*
Rhizobium, 351, 353, 503, 858
Rhizoid, *381,* 394
Rhizome, 396, 398, *399,* 525, *526*
Rhizopus stolonifer, 380, *381, 383*
Rh markers, antibodies against, 282
Rhodopsin, 609
RhoGam, 669
Rhyniophyte, *393*
Rhythm method, of birth control,
    794, *795*
Ribbon worm, *421,* 422, *422*
Rib cage, *641*
Riboflavin, *725*
Ribosomal RNA (rRNA)
  formation, 221
  ribosome subunit assembly, 225
Ribosome
  appearance in micrographs,
    58–59
  bacterial, 57, 226
  components, 57
  eukaryotic, 226
  functions, 60, 63, 226
  model, *225*
  plant cell, *58–59*
  polysome formation, 227, *227*
  structure, 226

Ribosome *continued*
    subunits, 225, *225*
Ribs, *641*
Ribulose bisphosphate (RuBP),
    115, *115–116*
Rice plant, 838
    aerobic respiration, 5
    photosynthesis, 4
Rickets, 624, *624*
Rickettsia rickettsii, 351, 861
Riffle (in stream), 882
Rifting, geologic, *328*
Right lymphatic duct, *670*
Ringworm, *383*
Risk factors
    AIDS, 796
    cardiovascular disorders, 666
    Down syndrome, 781
    eating disorders, 729
    obesity, 728
RNA polymerase, 222, 223, 237
RNA (ribonucleic acid)
    assembly, 221–222
    covalent bonds between
        nucleotides, *46*
    functional groups, *35*
    functions, 47, *48*, 73
    hybrid mRNA–DNA molecules,
        253, *253*
    messenger (mRNA), 221
    nucleotide arrangement in, 47
    nucleotide structure, *222*
    ribosomal (rRNA), 221, 225
    and self-replicating systems,
        326–327
    transfer (tRNA), 221
    translation, 226–227, *226–227*
    viral, 356, *356*
RNA virus, *691*
    animal, 357, *357*
    diseases caused by, *357*
    HIV-related, *357*
    infections, 357
    influenza-related, *357*
    multiplication cycle, *359*
    plant, 357
    replication, 358, *359*
    types, *357*
Robin, 133
Rock dove, *10*
Rocky Mountain spotted fever, 433
Rocky shore, coast, 883, *883*
Rod (bacterial), *351*
Rod photoreceptor, 608, *608*, 609
Roid rage, 649
Root
    adventitious, 494, *533*
    cap, *484*, 494, 537, *537*, 542
    Casparian strip, 490
    cortex, *494*, 494, 495, *503*
    endodermis, *494*
    epidermis, *494*, 495, *503*
    evolution, 392
    flowering plant, 484, *484*,
        494–495
    functions, 484
    hairs, *484*, 494, *494*, 503, *503*
    internal structure, 494–495
    lateral, *484*, 494
    meristems, *494*, 494–495
    nodules, 503, *503*
    overview, 504–505
    pericycle, *494*
    periderm, 487
    primary, *484*, 494, 532, *532*
    procambium, *494*
    protoderm, *494*
    secondary structure, *497*
    and soil erosion, 504–505
    structure, *484*, 494, *494*
    system, 494–495
    taproot, 494
    tip, 494–495, 537, *537*
    tooth, *718*
    vascular cylinder, 494–495, *495*

Rose, *526*
Roseola, *357*
Rotifer, *410*, *412*, 422, 847, 880
    circulatory system, *443*
    digestive system, *443*
    movement, *442*
    nervous system, *442*
    reproduction, *443*
    respiratory system, *443*
Rough endoplasmic reticulum, 65,
    65
Round dance (honeybee), 923,
    923, *928–929*
Round-Up herbicide, 365
Round window, cochlear, *600*, 601
Roundworm, *410*, 420–421, *see
    also* Nematode
    body plan, 419
    cuticle, 419
    evolution, *412*
    infections in humans, 420–421,
        664
    representative, *421*
    species, 419
Royal penguin, *924*
rRNA, *see* Ribosomal RNA (rRNA)
RU-486, 795
Rubber cup fungus, *378*
Rubella, *357*, 688, 791
Rubidium-87 isotope, *20*
Ruminant animal, 715, *719*
Run (in stream), 882
Runner, 525, *526*
Runner's knee, 640
Rust, fungal, *383*

S

Saber-tooth cat, *340*
Saccharomyces, 133
Sac fungus, 382, *382*, 383, *383*
Sacral nerve, *579–580*
Sacred disease, 262
Safe sex, 797
Sage, *290*
Sagittaria sagittifolia, 313, *313*
Saguaro cactus, *524*, 524–525,
    873, *873*
Sahel, desertification, *901*
Salamander, 456–457, *457*
    species coexistence, 827, *827*
Salination, soil, 898
Salisbury, F., *536*
Saliva, *717*, 718
    hypothalamic inhibition, 741
    insect larvae, 242
Salivary amylase, 718, *720*
Salivary gland
    as effector, 557
    functions, *717*, 718
    human, 718
    innervation, *580*
    location, *717*
Salix, 495
Salmon, *743*
Salmonella, *351*
Salt, *see also* Sodium
    dietary, 666
    dissolution, 29
    formation, 29
    iodized, 624
    table, 24, *24*
    tolerance, by fish, *743*
Saltatory conduction, 566
Salt marsh, 882, *882*
Saltwater, electrical conduction,
    652
Salvia
    S. apiana, *290*
    S. mellifera, *290*
Sandy shore, coast, 883
Sanger, F., *43*, 251
Saponin
    extracts, 535

synthetic, 535
Saprobic bacterium, 348, *351*,
    362
Saprobic fungus, 380–381
Saprolegnia, 364, *364*
Sapwood, *496*
Saquaro cactus, *119*
Sarcoma, defined, 683
Sarcomere
    contraction mechanisms, 645,
        645
    thick filaments, 644, *644*
    thin filaments, 644, *644*
Sarcoplasmic reticulum, 65, 646,
    646
Sarcoscypha coccinia, *382*
Sargasso Sea, 373
Sargassum, 373
Sartorius, *643*
SA (sinoatrial) node, 661
Saturated fat, 725
Saturated fatty acid, 40, *40*
Savanna
    African, *5*
    biome, 874, *875*
    desertification, *901*
    global distribution, *871*, 874
    and human evolution, 476
Sawfly
    caterpillar, 925, *925*
    and wasp, 833, *833*
Scala tympani, *600–601*
Scala vestibuli, *600*
Scale, fish, 454
Scallop, 426
Scalp, hair, 635
Scanning electron micrograph,
    487, *491–492*, *497*, *503*
Scanning electron microscope, 55,
    55
Scapula, *641*
Scarlet cup fungus, *382*
Scenedesmus, 55, *55*
Scent-fanning, honeybee, *929*
Schistosoma japonicum, *420*
Schistosomiasis, 420
Schleiden, M., 50
Schwann, T., 50
Schwann cell, *566*, 578, 764
Schwarzenegger, A., 648
Sciatic nerve, *579*
SCIDs (severe combined immune
    deficiency), 246, *246–247*
Scientific method
    hypotheses, 11
    limits, 12–13
    predictions, 11
    and subjectivity, 11
    theories, 11
Scientific names, *vs.* common
    names, *310*
Scintillation counter, 21
Sclater, W., 870
Sclera, 604, *604–605*
Sclereid, 486
Sclerenchyma, 486, *486*, 489
Scolex, 419, *419*, 421
Scorpian, 433, *433*
Scorpionfish, 832, *832*
Scotch broom, *521*
Scrotum
    functions, *774–775*
    location, *774–775*
Scrub forest
    pine barrens, 878
    tropical, *871*
Scrubjay, 542
    role in oak population, *277*
Scyphozoan, 414, *415*
Sea anemone, *410*, 414, *415*,
    442–443, *636*, 885, *887*
Sea cucumber, *410*, 438, *438*,
    442–443, 883
Sea hare, 425, *425*

Seal, 41, 844, *847*
    northern elephant, *283*
    northern fur, *283*
Sea lamprey, *838*
Sea nymph, 430–431
Sea palm, *373*
Sea salt, 352
Sea slug, 425
Seasonal variation in climate, *867*
Sea spider, 433
Sea squirt, 443, 445
Sea star, *410*, *415*, 426, 438,
    438–439, 756
    on coral reef, *887*
    in intertidal zone, 836, *883*
    survivorship, 808, 811
Sea turtle, 710–711, *710–711*
Sea urchin, 290, *410*, 438,
    442–443, 576, *576*, 756, *757*,
    762, *762*, 763
Seawater, salt concentration, *76*
Sebaceous gland, *632*
Secondary consumer, *846*
Secondary immune response, *681*,
    688
Secondary oocyte, *163*
Secondary sexual trait, defined,
    773
Secondary spermatocyte, *163*
Secondary structure, proteins, 44
Secondary succession, 834
Second messenger, 617
Secretin
    actions, *621*
    sources, *621*
    targets, *621*
Secretion
    digestive system, 718
    hormone, 557
    hydrochloric acid, 718–719
    solute gains by, 736
    urine formation by, 739–740,
        740
Sedative, 790
Sedimentary beds, 263, 303, 478
Sedimentary cycle
    defined, 851
    phosphorus, 860, *860*
Sedimentary rocks, 296, *303*
Seed
    coat, *404*, *519*, *532*
    dispersal, 522–523
    evolution, 393
    formation, 518, 522–523
    germination, 524, 530, 532, *533*,
        538–539, 542
    integument, *519*
    monocot, 405
    oak, 543
    Pine tree, 392
    plant, defined, 393
    sequoia, 836
Seedless grape, *531*
Seedling
    bean, *538*
    corn, *538*
    gravitropism, *537*
    growth responses, 536–538
    oak, 542
Segmentation
    annelid, 428
    arthropod, 432
    in body plans, 411, *577*
    crustacean, 434
    earthworm, 411
    and evolution, 431
    insect, 411
    and specialization, 431
Segregation, Mendel's theory,
    172–173, *172–173*
Seismic activity, 260–261,
    336–338, 482
Seizure, epileptic, 586
Selaginella lepidophylla, 396
Selander, R., 286

Selection, *see also* Natural selection
  artificial, 10, 247, 267, 271, 630
  directional, *278*, 278–279, 296, 304
  disruptive, 280–281
  and divergence, 289
  indirect, 928–931
  and mutation, 229
  and oxygen, 330
  sexual, 282–283, 916–917, *917*
  stabilizing, 280, *280*, 281, 304
Selective breeding, 271
Selfish behavior, 916, 925
Selfish herd, 925
Self marker, 668, 680
Self-replicating system
  laboratory experiments, 327
  origins of, 326–327
Self-sacrificing behavior, 916, 926–931
Self-subordinating behavior, 926, *926–927*
Semen, formation, 774
Semiconservative nature, of DNA replication, 214, *214*
Semiconservative replication, *214*
Semidesert
  global distribution, *871*
  soil, *872*
Semilunar valve, *660*, 661
Seminal vesicle
  functions, 774, *774–775*
  location, *774–775*
Seminiferous tubule, 774, 776, *776*
Senescence
  defined, 540
  delay in, 540
Sensation
  defined, 594
  somatic, 596–597
Sensory neuron
  action potential propagation, 594–595, *595*
  adaptations, 561
  axons, *571*
  cnidarian, 414
  distribution, 578
  functions, 561
  reflex arc, 570, *571*
  retinal, 609
  stretch reflex, *571*
Sensory receptor, *see* Receptor, sensory
Sensory system
  arthropod, 432
  components, 593
  functions, 594
  mammalian, 464
  pathways, 594–595, *608*
  platypus, 445
  receptors in, 594–595, *595*
Sepal, *523*
Sepia, 427
Septum, heart, *660*
Sequoia grove, 836
Serosa, *720*
Serotonin, 568, *587*, 589
Serratur anterior, *643*
Sertoli cell, 776, *776–777*
Sessile, 447
Setae, earthworm, 428, *429*
Severe combined immune deficiency syndrome (SCIDs), *246*, 246–247
Sewage dumping, 881–882, 899
Sex chromosome, *see also* X chromosome; Y chromosome
  defined, 189
  differences in, 144
  Drosophila, 192
  homologous, 144
  human, 190
  Klinefelter syndrome, 203
  Turner syndrome, 203
  XXY condition, 203

Sex determination, 190, *190*
Sex hormone
  effects, 613, 778–779
  homeostatic feedback loop, 626
  in menstrual cycle, *626*
  production, 191, 626
Sex-linked gene, 193
Sexual behavior, *see also* Courtship behavior
  and imprinting, *915*
Sexual characteristics, *see also* Trait, sexual
  development, 190
Sexual dimorphism, 282–283, *288*, *312*, 320, 340
Sexual intercourse, 783
Sexually transmitted disease, *see also specific disease*
  patterns of infection, 359
  types of, 796–797
Sexual reproduction, *see also* Human reproduction; Reproduction
  amphibian, *442*
  annelid, *443*
  arthropod, *443*
  asexual episodes in, 154
  avian, *442*
  chordates, *443*
  chromosome number changes during, *166*
  club fungus, 385
  cnidarian, 417, *443*
  coral, *443*
  ctenophoran, *443*
  defined, 754
  echinoderm, *443*
  flatworm, 418, *443*
  flowering plant, 516–524
  fluke, *443*
  frog, 752–753
  fungus, 380
  green algae, 375, *375*
  hallmarks, 155
  hormone effects, 625
  insect, *443*
  jelly fish, *443*
  mammalian, *442*
  mollusk, *443*
  nematode, *443*
  pineal gland effects, 625
  polychaete, *443*
  and polyploidy, 294–295
  reproductive isolating mechanisms, 286–287
  reptilian, *442*
  ribbon worm, *443*
  roundworm, *443*
  sac fungus, 382
  sea anemone, *443*
  spider, *443*
  sponge, 413, *443*
  tapeworm, *443*
  tunicate, *442*
  and variation in traits, 156
Sexual selection, 282–283, *283*, 916–917, *917*
Shale, 408
Shark, *307*, 452, 454, *454*, *637*
Sheep, bighorn, *9*
Sheep tapeworm, *419*
Shelf fungus, *384*
Shell
  amniote egg, 458, *459*
  barnacle, 434, *434*
  bivalve, 426
  chiton, 425
  clam, *426*
  cuttlefish, *427*
  electron orbital, 22, *23*
  mollusk, 424
  snail, 424, *424*
  squid, *427*
  turtle, 460
Shepard's purse, 522, *522–523*

Shifting cultivation, 902
Shigella, 351
Shingles, *357*
Shivering, 748
Shock, anaphylactic, 689
Shoot system
  evolution, 392
  flowering plant, *484*, 488–489
  functions, 484
  woody plant, *496*, 496–497
Shore, rocky, 883, *883*
Short-day plant, 539
Shortgrass prairie, 874, *875*
Short-term memory, 587
Shrew, tree, *475*
Shrubland, dry
  biome, 874
  global distribution, *871*, 874
Siberian huskie, 630–631
Sickle-cell anemia, 220–221
  and balanced polymorphism, 282
  defined, 176
  effects blood oxygen, 176
  genetic basis, *196*, 220, 282
  RFLP applications, 252
  and stabilizing selection, 282
  symptoms, *177*
  treatment for, 204
Sidewinder, *744*
Side-winding viper, 460
Sieve plate, *510*
Sieve tube, 496, *510*
Sieve tube member, 486, *489*, 510
Signal, *see also* Communication signal
  cardiac, 661, *661*
  hormonal, 242
  molecular, in gene expression, *238*, 242–243
Signaling molecule
  actions, 626
  during embryonic induction, 766
  hormonal, 534
  local, 614, 626
  mechanism of action, 616–617
  morphogens, 766
  pheromones, 922, *929*
  plant hormone, 530
Silica shell, diatom, 371
Silk, J., 931
Silkworm moth, *5*, 598, *627*
Sillen-Tullberg, B., 925
Silt, 300
Silurian, 334, *393*
Singer, S. J., 79
Sink region, plant, 510, *512*
Sinoatrial node (SA node), 661
Siphon
  bivalve, 426, *426*
  cephalopod, *426*
  squid, *427*
Siphonous algae, 374
Sister chromatids
  defined, 142
  formation, 142
  during meiosis II, 157, *159*
  separation at anaphase, 149, *149*
Sistine Chapel, painting of Great Deluge, *300*
Sitka spruce, *834–835*, 878
Sitticus pubescens, 636
Size, human body, during prenatal and postnatal growth, *793*
Skate, *452*, 454
Skeletal muscle
  actin component, 644–645, *644–645*, 645, *645*
  appearance in micrographs, 644
  bone–muscle interactions, 642
  vs. cardiac muscle, 642
  components, 644–645
  contraction, 648

  contraction mechanisms, 647
  as effector, 557
  functional organization, 644–645
  innervation, 580
  major muscles, *643*
  myofibrils, 644, *644*
  myosin component, 644–645, *644–645*, 645, *645*
  sarcomeres, 644, *644*
  sarcoplasmic reticulum, *646*
  shivering response, 748
  stretch reflex, *579*
  striated, 644
  tissue, 552, *552*
Skeletal-muscular system
  evolution, 576–577
  interactions, *646*
Skeletal system
  appendicular, *641*
  axial, *641*
  comparative (ape, human, monkey), *473*
  functions, *554*
  human, *554*
  invertebrate, 636, *636*, 637
  major bones, *641*
  muscular system interactions, 642
  overview, *631*
  ribs, 93
  sternum, *641*
  types, 636
  vertebrate, 637, *637*
Skeleton
  appendicular, 640
  axial, 640
  hydrostatic (cnidarian), 414
  intervertebral disks, 640
  invertebrate, *577*
  joints, 640
  mammal, *637*
  reptile, *637*
  shark, *637*
  structure, 640
Skin, *see also* Integumentary system
  age spots, 93
  and aging, 635, *635*, 768
  amphibian, 457
  anhydrotic ectodermal dysplasia, 241
  as barrier, 676, *676*
  body temperature regulation, 744–745, 748–749
  cancers, *215*, 635, *635*, 897
  color, 634
  derivatives, 632, *633*
  dermis, 634
  disorder, pituitary-related, 620
  epidermis, 634
  evaporative water loss, 736
  free radical effects, 92–93
  frog, 457
  functions, 556, 632
  and heat balance, 744, 748, *748*
  mammalian, 464
  mechanoreceptors, 594
  oil glands, 635
  pigment, 618, *633*
  receptors, 556
  reptile, 458
  sensory receptors, *557*, 632
  structure, 632, *632*
  sunlight and, *635*
  sweat glands, 634
  tactile receptors, 596–597, *597*
  ultraviolet radiation effects, *215*, 635
  warts, 234
  weight (human), 632
Skua, *847*
Skull
  extinct primates, *475*
  fetal, 640

hominid and modern human compared, *478*
Homo erectus, 478
human, 640
human *vs.* chimp, *305, 474*
mammalian, 464
ostracoderm, *453*
Skunk, 831
Slash-and-burn agriculture, 902, *903*
Sled dog, 630–631
Sleep
  EEG patterns in, *587*
  eye muscle twitch during, 570
  REM (rapid eye movement), *587*
  slow-wave, *587*
  transition to wakefulness, *587*
Sleep center, 587
Sliding-filament mechanism, muscle contraction, 645, *645*
Slime mold
  cell movement in, 764
  cellular, 365
  life cycle, *365*
  nucleosome, *238–239*
  plasmodial, *363*, 365
Slime stalk, bacterial, 355
Slipper limpet, *155*
Slow-wave sleep, *587*
Slug, *410*, 424, *442–443*
Small intestine, *see* Intestine, small
Smallpox, *357*, 674
Smartweed, 827, *827*
Smell
  brain areas affecting, 583
  chemoreceptors, 598
Smiling in human infants, 914, *915*
Smilodon, *340*
Smith, T., 280
Smog, *854–855*, 893, 895
Smoke
  carbon monoxide in, 666
  cigarette, 666, 706
  marijuana, 706
  secondhand, 706
Smoker's cough, 706
Smoking
  and bronchitis, 706
  and cardiovascular disorders, 666
  effects, 589, *790*
  and emphysema, 706
  marijuana, 589
  and peptic ulcers, 719
Smooth endoplasmic reticulum, 65, *65*
Smooth muscle
  arterial, *662*
  arteriole, *662*
  as effector, *557*
  functions, 642
  intestinal, 720
  lymph vessels, 671
  peristalsis, 718, *719*, 720
  segmentation, 720
  tissue, 552, *552*
  venous, *662*
Smut, *383*
Snail
  body plan, 424, *424*
  characteristics, summary, *442–443*
  classification, *410*
  eye structure, 602, *602*
  in food web, 847, 882
  genetic variation, *286*
  oviparous nature, *755*
  shell patterns, *272*
  speciation, 286–287
Snake
  egg, *755*
  evolution, *459*, 460
  feeding behavior, 460
  genetic food preferences, 912,

912
  movement, 460
  pelvic girdle, 263, *263*
  reproduction, *755*
  species, 460
  tongue-flicking, 912, 914
  venomous, 460
Snapdragon, 176, *176*
Snowshoe hare, 829, *829*
Social behavior
  adaptive value, 920–932
  adoptive, 931, *931*
  altruistic, 916, 921, 927–931
  cooperative defense, 924–925, *925*
  courtship, *915*, 916–917, *917*, 922, 922–923
  crocodile, 461
  defined, 922
  herding, 925, *925*
  parental, 927, *927*, 931, *931*
  range, 924
  self-sacrificing, 916, 926–931
  self-subordinating, 926, *926–927*
  signaling mechanisms, 922–923
  sociality, 924–925
Social parasite, 833, 914, *914*
Sodium, *see also* Salt
  atomic number, *19*
  atomic structure, 22, *23*
  deficiency, *727*
  electron distribution, *23*
  excessive intake, 727
  in human nutrition, *727*, 737
  mass number, *19*
  metabolic need for, 727
  neuronal diffusion, 563–564
  reabsorption, 739, *740–741*
  symbol for, *19*
  in urine, *739*
Sodium chloride, 24, *24*, 727
Sodium-potassium pump, 87, 563, *563*, 743
Sodium–potassium pump, *743*
Soft drinks, 588
Soft spots (fontanel), 640
Soil
  and acid deposition, 896, *896*
  acidity, 858
  fertilizers, 858–859
  leaching, *853*, 858
  salination, 898
  structure, 872, *872*
Solar energy, 904, 906–907
Solar-hydrogen energy, 906
Solar radiation, 865, *866*, 866–867, *867*
Solar tracking, 544
Soldier, termite, *920*, 920–921, 928
Solid waste, 900
Solute
  concentration gradients, 82–83
  defined, 27
  regulation, 736
  in soil, 84
  tonicity, 83, *83*
  transport proteins, 86
Solution
  acidic, 28
  alkaline, 28
  hypertonic, 83
  hypotonic, 83
  pH value, 28
Somatic cell, defined, 142
Somatic nervous system
  movements controlled by, 580
  nerves in, *579–580*, 580
  neurons, 580
  somatic sensations, 596–597
Somatomedins
  actions, *621*
  sources, *621*
  targets, *621*

Somatosensory cortex, *585*, 596, *596*
Somatostatin
  actions, *621*
  secretion, 622
  sources, *621*
  targets, *621*
Somatotropin
  effects, *618*, 620, *620*
  and genetic engineering, 256, *256*
  secretion, 242, *618*, 618–619
  targets, *618*, 619
Somite, embryonic, 786, *786*
Sonchus, 108, 511, *511*
Song, bird, 912–913, *913*
Sorus, 398, *399*
Sound
  frequency, 600
  pitch, 600
  production, human, 703, *703*
Sound system, avian, 913
Source region, plant, 510, *512*
Southern Oscillation (El Niño), 888–889, *889*
Sow thistle, 108, 511, *511*
Soybean, *503*, 539, *725*, 838
Sparrow
  singing behavior, *625*, 912–913, *913*, 914
  species introduction, 838
Spartina, 882
Spatial learning, *915*
Spatial summation, 568
Speciation, *see also* Diversity of life; Species diversity
  allopatric, 292–294
  anagenesis, 296
  cladogenesis, 296
  and continental drift, 328, 336, 341
  defined, 287, 291
  divergence, 289, *296*, 336
  gradualist model, 296
  instantaneous, 295
  island, 840–841
  and Isthmus of Panama, 292–293
  mainland and marine, 840
  models, 292–296
  parapatric, 292, 295
  patterns, 296–297
  and polyploidy, 294–295
  punctuation model, 296
  rapid, 296, 336
  reproductive isolation mechanisms, 289–291
  sympatric, 292, *294*, 294–295
Species, *see also* Community
  analogous structures, 306–307
  biotic potential, 807
  catastrophic changes in, 264
  in Chain of Being, 262
  chromosome differences, 313
  classification, 310–311
  and creation, 264
  defined, 288
  distribution, 864
  diversity, 8, *342–343*, 363, *363*, 379
  endangered, allele loss, 276–277
  estimated durations, *297*
  extinction, 840, 861
  and fossil record, 261
  higher taxa, 310
  hominoid and hominid, 472
  homologous structures, 306
  identification, 310–311
  imperfections, 263
  interactions, 824–842
  introductions, *838*, 838–839
  keystone, 836
  and lineages, 320
  morphological gaps between, 304

mutualism, 386
  pioneer, 834
  related, *262–263*
  symbiosis, 386
  in taxonomic nomenclature, 8, 311
  variation within populations, 320
Species differences, in DNA base sequence, 212–213
Species diversity, *see also* Diversity of life; Speciation
  factors affecting, 297
  intertidal zone, 883, *883*
  island, *840*, 840–841, *841*
  mainland, 840, *840*
  marine, 840
  and predation, 836–837, *837*
  tropical rain forest, *892*, 892–893
  tropical reef, 886, *886*
Specificity, immunological, 680, 686–687
Speech
  brain areas affecting, 584
  split-brain experiments, 586
Speed (amphetamine), 589
Speed mimicry, 831
Spemann, H., 766–767
Sperm
  acrosome, 776, *777, 783, 783*
  appearance in micrographs, *70, 139*
  fertilization events, 760
  flagellum, *70*, 776
  formation, 162–163, 774, 776, 779–780, *779–780*
  gene insertion into, 256
  hormonal controls, 776–777, *777*
  motility, 774
  penetration of egg, *784*
  plant, *516*
  and reproductive strategy, 754
  structure, *776–777*
Spermatid
  formation, 162
  haploid, 776
Spermatocyte
  primary, 162, *163*, 776
  secondary, *163*, 776
Spermatogenesis
  animal, generalized, *163*
  defined, 162
Spermatogonium, *163*, 776
Spermicidal foam, 795, *795*
Spermicidal jelly, 795, *795*
Sperm whale, 709, *709*
Sperry, R., 586
Sphagnum, *395*
Sphenodon, 461
Sphenophyte, 398, *398*
Spheres of hydration, 27
Sphincter, 580
  defined, 718
  stomach, 718, *719*
Sphygmomanometer, *663*
Spicule, 412, 438
Spider, *410*, 432–433, *433*, *442–443*
  jumping, *636*, 637
  tracheal respiration, 698
  yellow crab, 832
Spike, neuron, 565, *565*
Spinach, 539, *539*
Spinal cavity, human, *555*
Spinal cord
  development, 764, *765*
  evolution, 578
  extensor reflex pathway, 581
  functions, 581
  gray matter, 581
  location, *581*
  organization, *581*
  white matter, 581
Spinal nerve, 580, *580*
Spindle, muscle, *571*

Spindle apparatus
  bipolar, 148
  and cell division, 760
  chromosome alignment on, 147
  formation, 148
  functions, 146–147
  during metaphase, 147
  during prophase, 146
Spine, sea star, 438, *438–439*
Spiny anteater, 465, *465*
Spiral bacteria, 348, *351*
Spiral cleavage, 423
Spirilla, 348
Spirochete, 349, *351, 353, 354*
Spirogyra, 375, *375*
Spleen, functions, *670, 671, 685*
Split ends, 635
Sponge (Porifera)
  body plan, *413*
  circulatory system, *443*
  classification, *410*
  digestive system, *443*
  evolutionary relationships, *412*
  movement, *442*
  nervous system, *442*
  reproduction, *443*, 754
  respiratory system, *443*
  structure, 412–413
Spongy mesophyll, *491*
Spontaneous assembly of organic
    compounds, 323–327
Sporangium, 380, *392, 394, 398*
Spore
  bacterial, 355
  chytrid, 364
  defined, 162
  Dichyostelium discoideum, 365
  evolution, 308
  formation, 162
  fungal, 380, 385, *388*
  pollen grain, 393
  red algae, 372
  types, 392–393
  water mold, 364
Sporophyte, *162*, 373, *392*
  bryophyte, 394
  club moss, 396
  defined, 392
  embryo, 522
  flowering plant, 516, *516, 519*
  in life cycle, 394, *394, 516*
  lycophyte, 396, *397*
  seed coat, 400
  sphenophyte, 398
  vascular plant, 392
  whisk fern, *397*
Sporozoan, 368
Sporozoite, 369
Spotting gene, *241*
Sprain, joint, 640
Spring overturn, lake, 880
Spruce tree, *834–835*, 878
Squid, *410*, 426–427, *427,
    442–443*, 884
SRY gene region, 191
S-shaped curve, logistic growth,
    808
Stabilizing selection, 280, *280*,
    281–282, 304
Staining, in microscopy, *348*
Stalk, *519*
Stamen, 170, *170*, 516, *516*, 517,
    523
Staph infection, antibiotic-resistant
    strains, 355
Staphylococcus, 348, *351*
Staphylococcus aureus, 349
Starch
  formation, 510
  functions, *48*
  in plant tissue, 37, 58, 68
  structure, *48*
Starling, *838*, 910–911, *910–911*
Starling, E., 614
Stars, formation, 322–323

Starvation
  and dieting, 729
  population growth effects, 814
Static equilibrium, 599, *599*
Statolith, 536, *537*
Stearic acid, structural formula, *40*
Steinbeck, J., 874
Stem
  cortex, *489*
  dicot, *489*
  elongation, 536, *537*, 538–539
  internal structure, 488–489, *489*
  monocot, *489*
  pith, *489*
  primary structure, 489, *489*
  secondary structure, 496
  simple and complex tissues,
    485
  vascular bundles, *485*, 488, *489*
  woody, *496*
Stem cell
  bone marrow, 671
  in gene therapy, 247
  location, 657
  T cells from, 682
  use in gene therapy, 247
Sterility, 291, 295
  in insects, 921, 928–930
Sterilization, 817
Sternum, *463*, 641
Steroid hormone
  actions, *616*, 617
  anabolic, 649
  athletes' use of, 41
  biosynthesis, 725
  functions, *48*
  mechanism of action, *616*,
    616–617
  receptor binding, 616
  structure, *48*
  synthesis, 616
Sterol
  in lipid bilayer, 78
  structure, 41
  types, 41
Steward, F., 254, 527
Sticky end, of restriction fragment,
    249, *249*
Stigma, *516*
Stillbirth, achondroplasia-related,
    197
Stimulant (drug), 588
Stimulus
  chemoreceptor, *594*
  defined, 556
  mechanoreceptor, *594*
  nociceptor, *594*
  photoreceptor, *594*
  thermoreceptor, *594*
  visual, split-brain experiments,
    586
Stingray, 454
Stink beetle, 831
Stinkbug, *437*
Stinkhorn fungus, *9*, 384
Stipe (sporophyte), 373
Stolon, *381*
Stoma, *491*, 501, 508–509
  defined, 487
  functions, 392, 487
  guard cells, 487, *508*
  location, 392
Stomach
  acid, 29
  acidity, 718–719
  emptying, 719
  endocrine cell lining, 723
  enzymes active in, 718–719
  functions, *717*, 718
  gastric fluid, 718
  human, 718–719, *719*
  hydrogen ion effects, 718
  innervation, *580*
  lining, 718–719
  protein digestion, 718–719

ruminant, *719*
  sea star, *438–439*
Stone tool, 478–479, *479*
Strain, joint, 640
Strangler fig tree, 826
Stratification, 262, 303
Stratum corneum, 634, *634*
Strawberry plant, 254, *254, 523,
    526*
Stream
  ecosystem, 882
  pollution, 882, 899
Strength, muscular, 648
Streptococcus
  infections, *351*
  shape, 348
Streptomyces, 351
Streptomycin, 354, 791
Stress, and skin disorders, 635
Stress electrocardiogram, 667
Stress response, 623
Stretching, mechanoreceptors, 596
Stretch marks, 634
Stretch reflex, *571*, 581
Striation
  cardiac muscle, 644
  skeletal muscle, 644
Strip mining, coal, 904
Strobilus, 396, *397, 398, 398*
Stroke, 666
Stroma, 68, 109, *109*, 115, *115*
Stromatolite, 330, *330*
Sturnus vulgaris, 910–911
Styela, 760
Style (of carpel), *516*, 517
Suberin, 487
Submucosa, intestinal, *720*
Suborder (taxon), 311
Subphylum (taxon), 311
Subsistence agriculture, 900
Substance P, 568
Substrate
  activation energy, 99
  defined, 98
  energy hill, 99, *99*
  enzyme–substrate interactions,
    98–99
  transition state, 98–99, *99*
Substrate-level phosphorylation,
    126, *127*
Succession
  defined, 834
  primary, 823, 834, *834–835*
  secondary, 834
Succinate, *129*
Succinyl-CoA, *129*
Succulent plant, *see* CAM plants
Sucrose
  components, 37
  formation, 37
  functions, *48*
  transport by plants, 501
Sucrose, functions, *48*
Sucrose polyester, 725
Sugar
  blood levels, 6, 39
  functional groups in, *35*
  phosphate group formation,
    115, *115–116*
  refined, 725
  simple, 37, 724
  table, 37
  yeast effects on, 133
Sugarcane, 37, *403*
Sulfhydryl group
  locations, *35*
  structural formula, *35*
Sulfolobus, 351
Sulfur
  atomic number, *19*
  deficiency, *727*
  electron distribution, *23*
  excess, *727*
  functions, *727*
  global cycling, 353

in human nutrition, *727*
  mass number, *19*
  in plant function, *502*
  radioactive isotope, *211*
  sources, *727*
  symbol for, *19*
Sulfur dioxide
  in acid rain, *29*
  and air pollution, *895*, 896, 904
Sulfuric acid, 736, *895*
Sulfur oxides, 895, *895, 896*
Sulfur shelf fungus, *378*
Sulfur trioxide, *895*
Sumac tree, *111*
Summation
  spatial, 568
  temporal, 568
Summers, K., 71
Sun, early, representation, *324*
Sunbird, *520*
Sunburn, *635*
Sunfish, 925–926
Sunflower
  seed, *537*
  stem, *486*
Sunlight
  effects on DNA, *215*
  effects on skin, 635
  energy from, 95
  as molecular signal, 242–243
Superfamily (taxon), 311
Superior (anatomy), *555*
Superior vena cava, *659*
Superoxide dismutase, functions,
    92
Surface-to-volume ratio, 53, *53*, 696
Surgery
  corneal transplantation, 606
  for skin cancers, *635*
Surtsey, *840*, 840–841
Survivorship curves, 811, *811*
Suspended particle air pollutants,
    895
Swallow, cliff, 924
Swamp forest, 878
Sweat glands, 241, *632, 634, 734,
    744, 745, 749*
Sweating
  control center, 736
  extreme, 749
  water loss by, 736
Swim bladder, *699*
Symbiont
  bacterial, 362
  microbial, 921
Symbiosis
  defined, 386, 502, 825
  examples, 825
  forest trees and mycorrhizal
    fungi, 387
Symmetry
  animal body, 410, *411*
  bilateral, 410, *411–412*,
    576–577, 786
  planes in human body, *555*
  radial, 414, 417
Sympathetic nerve
  functions, 580, *580*
  location, *580*
Sympatric speciation, 292, 294,
    294–295
Synapse, chemical
  defined, 566
  excitatory, 568, *568*
  inhibitory, 568, *568*
  intense stimulation effects, 587
  at neuromuscular junctions, *567*
  psychoactive drugs affecting,
    588–589
Synaptic cleft, 566, *567*, 568–569
Synaptic integration, 568, *568*, 608
Synaptic vesicle, 566, *567*
Synergistic muscular action, 642
Synovial joint, 640
Synthetic pathway, metabolic, 326

Synura, 371
Syphilis, 359, 797
Systematics
    cladistics, 314–315
    cladograms, *316–317*
    and classification, 314
    defined, 314
    phylogenetic reconstruction, 314, 316
    relative relationships, 314, *315*
Systemic circulation, 658, *658*
Systole, 660, *662*

T

2,4,5-T (Agent Orange), 365
T4 bacteriophage, *211, 357*
Table salt, 24, *24,* 727
Tachycardia, 667
Tachyglossus aculeatus, 465
Tactile signal, 923, *923*
Tadpole, *752,* 764
Taena saginata, 421
Taiga, 878
Tail, sperm, *777*
Tallgrass prairie, 874, *875,* 877
Tangelo, 247–248
Tanning, skin, 635, *635*
Tanning lamp, effects on DNA, *215*
Tansley, A., 826–827
Tapeworm, *410,* 419–420, *421, 442–443,* 833
Taproot system, 494, *494*
Target cell, for hormones, 614, *616–617*
Targeted drug therapy, *683*
Tarsal bones, *641*
Tarsier, *472*
Tarsioid, 471–472, *472*
Tasmanian wolf, 466
Taste bud, *598*
Taste receptor, 598, *598*
Tatum, E., 220
Taxa, shared traits, 317
Taxon, higher, 310
Taxonomy
    defined, 312
    and systematics, 312–314
T cell, *see* T lymphocyte
Tea, 588
Tectorial membrane, *600–601*
Tectum, 582, *582, 584*
Teeth
    australopith, 476–477
    early *Homo,* 339
    faulty enamel trait, *196,* 198
    Homo sapiens, 341
    human, *717, 718, 718*
    and life-style, 474, 476
    mammalian, 464–465, *465*
    number and arrangement, *718*
    primate, 471, 473–474, 476
Teleost, 455
Telesto, 416
Telophase
    mitotic, *147,* 149
    nuclei formation during, 149
Temperate deciduous forest, *see* Deciduous forest, temperate
Temperate grassland, *see* Grassland, temperate
Temperate rain forest, *see* Coniferous forest
Temperate temperature zone, *867*
Temperature, animal body
    behavioral regulation, 746, *747,* 748–749
    blood effects, 654
    cellular, 26
    core, 744
    effects on diffusion, 82
    effects on enzyme action, 100, *100*

effects on hemoglobin-oxygen binding, 704
and enzymatic activity, 744
homeostatic controls, 556–557, *557*
maintenance, 744, 746
mammalian, 465
metabolic regulation, 744
nonshivering heat production, 748
physiological adjustments, 746, *747*
platypus, 444
range suitable for metabolism, 744
regulation, 425, 748–749
response to cold stress, 748
response to heat stress, 749, *749*
scrotum, 774
skin effects, 632
suitable for metabolism, *744*
sweat gland role, 634
and sweat glands, 634, *749*
thermoreceptors for, 596
and thyroid function, 624
Temperature, environmental
    behavioral adjustments to, 746, *747*
    and dormancy, 541
    and flowering, 539–540
    and greenhouse effect, 856–857
    physiological adjustments to, 746, *747*
    and vernalization, 540
    water, defined, 26
Temperature zone, 866, *867*
Template, for protein synthesis, 325–326
Temporal isolation, *289,* 290
Temporal lobe, 584, *584*
Temporal summation, 568
Tendon, 642
Tendril, 537, *537*
Tennis, kilocalories expended by, 729
Tentacle
    cephalopod, 426, *426*
    mollusk, 424, *424*
    squid, *427*
Teratogen, *see also* Mutagen
    defined, *790*
    sensitivity during pregnancy, *790*
Teratornis, 340
Terminal bud, 476
Termination, RNA translation, 226
Termite
    altruistic behavior, 920, *920,* 928–929
    vs. ant, 920, 922
    colony life, 920–921
Tern, 927, *927*
Territory, defined, 916
Tertiary consumer, *846*
Tertiary structure, proteins, 44
Testability, of hypotheses, 12
Testcrosses, 173, *173*
Testes
    anabolic steroid effects, 649
    cancer, 774–775
    development, 190–191, *191*
    flatworm, *418*
    functions, *774–775,* 775
    location, *616,* 774–775
    secretions, *616,* 776
Testicular feminization syndrome, *196,* 617
Testing
    AIDS, 797
    pregnancy, 795
Testosterone
    actions, *616, 621,* 776
    and bird sound systems, 913, *913*

feedback loops affecting, *777*
fluctuations, 777
and male reproductive function, 776–777
negative feedback loops, *777*
precursors, 41
receptors, 617
role in development, 191, 617, 649
secretion, *616,* 626
and sexual traits, 649
sources, *621*
targets, *621*
Test-tube baby, 205, 798
Tetanus, 354, 569, 648, 688
Tetracycline, 791
Tetraploidy, 201
Thalamus, *582,* 583, *584–585*
Thalassoma
    T. bifasciatum, 293
    T. lucasanum, 293
Thalidomide, 791
Theater of Insects (T. Moufet), 262
Theobroma cacao, 403, 514, 514–515
Theobromine, 515
Theory
    scientific, 11
    and subjectivity, 12
Thera (island), 260–261, *260–261*
Therapsid, 464
Thermal inversion, 895, *895*
Thermal layering, lake, 880–881, *881*
Thermoacidophile, *351*
Thermocline, 880–881, *881*
Thermodynamics
    first law, 94
    second law, 95
Thermophile, 352
Thermoplasma, 351, 352
Thermoreceptor, *594,* 596, 748–749
Thigmotropism, 537
Thirst behavior, 741
Thomas, L., 769
Thoracic cavity, 411, *555*
Thoracic duct, *670*
Thoracic nerve, *579–580*
Thorax, insect, 435
Thorium-232 isotope, *20*
Thorn forest, global distribution, *871*
Threonine, 725
Threshold level, 564
Thrombin, actions, 98
Thrombus, 667
Throw-away mentality, 900
Thylakoid membrane system
    concentration gradients, *114*
    electric gradients, *114*
    photosystems, 112
    structure, 109
Thymine, *211, 212, 212*
Thymine dimer, *215*
Thymosin, 626
    actions, *616, 621*
    secretion, *616*
    sources, *621*
    targets, *621*
Thymus gland, 682, *685*
    functions, *670,* 671
    hormone secretion, 614, *616,* 626
    location, *615,* 626
Thyroid gland
    and body temperature, 624
    disorders, 623–624
    as effector, *557*
    endocrine functions, 624
    as endocrine target, *618*
    hormone secretion, 614, *616, 621,* 624
    location, 624, *624*
    scanning with 123
    iodine, 21
    structure, *624*

Thyroid hormone, production in Grave's disorder, 689
Thyroid-stimulating hormone, *618,* 624
Thyrotropin, *618*
Thyrotropin-releasing hormone, *618*
Thyroxine, 623–624
    actions, *616, 621*
    secretion, *616,* 624
    sources, *621*
    targets, *621*
Tibia, 640
Tibialis anterior, *643*
Tick, *39,* 359, 432–433, *698*
Tick bite, 354, *354*
Tidal volume, lung, 702, *703*
Tidepool, 836–837, *837,* 883, *883,* 890
Tight junction, 72, *549*
Timothy grass, *59*
Ti plasmid (*Agrobacterium tumefaciens*), 255
Tissue, animal
    artificial, 553
    carbon dioxide content, 705
    connective, 550–551, *550–551*
    defined, 547
    desmosomes, *550*
    epithelial, 548–549, *548–549*
    fat percentage, 729
    gap junctions, *549*
    and germ layers, 763
    grafts, 682
    laboratory-grown, 553
    lyonization, 241
    morphogenesis, 764
    mosaic effect, X inactivation-related, 241
    muscle, 552, *552*
    nervous, 553, *553*
    oxygen diffusion, 704
    placental, 762
    regeneration, 768
    tight junction, *549*
Tissue, plant
    dermal, 485, 487
    ground, 485–487
    meristems, 485
    primary growth, 492
    secondary growth, 496–497
    specimens, 488
    vascular, 484–487
    woody, 496–497
Tissue culture propagation, *526,* 527
Tissue transplants, HIV transmission during, 691
Titanic, 748
Ti (tumor-inducing) plasmid, 255
T lymphocyte, 247
    antigen-specific receptor production, 686–687
    cytotoxic, 680, 682–683
    effector, 683
    functions, 680
    helper, 680, *681,* 682–683
    immune functions, 657
    interaction with B lymphocyte, *681*
    killer, *681, 683*
    memory, 687, *687*
    in SCIDs patients, 247
    suppressor, 682
    virgin, *681,* 682
Toad, 310, 457
Tobacco mosaic virus, *356,* 357, *357*
Tobacco plant, 255, *255,* 493
    hemoglobin-producing, 271
Tobacco smoke, DNA damage caused by, 228, 244
Tobacco smoking
    benefits of quitting, 707
    effects, 589
    effects during pregnancy, 791

Tobacco smoking *continued*
   and skin aging, 635
Tobamovirus, 357
Tobin, E., 243
Todd, J., 899, *899*
Toe, rotifer, 422, *423*
Togavirus, 357
Tomato plant, *537*
Tongue, taste receptors, 598
Tonicity, defined, 83, *83*
Tonsils, *670*, 685, *685*
Tooth decay, *718*
Topography, and climate, 869, *869, 895*
Topsoil, 872, *872*
Tornado, 560–561
Torsion, gastropod, *424*, 424–425, *425*
Total lung capacity, *703*
Touch, mechanoreceptors, 596
Toxins, *see also* Poisons
   bacterial, 354
   cigarette smoke, 706
   ciliated protozoan, 367
   Clostridium botulinum-related, 569
   Clostridium tetani-related, 569
   frog, 457
   fungal, 384
   mushroom, 384
   sea anemone, *415*
   in tetanus, 648
Toxoplasma, 368
Toxoplasmosis, 368
Tracer, radioisotopic, 21
Trachea
   functions, *700*
   human, 718
   insect, 432, *698*
   location, 700, *700*
Tracheal respiration, 698
Tracheid
   example, *487, 506*
   function, 506
Trachoma, 606
Tradewinds, *867*, 888–889
Trait
   adaptive, 7, *7*, 278
   behavioral, 272
   continuous variation, 180–181, *180–181*
   derived, 316–317
   gene pairs for, 171
   genes for, 168–169
   heritable, 7, 266–267, 275, 288, *291*
   homologous, 317
   human, 471, 473
   morphological, 272
   and morphological convergence, 326
   observable, 169
   phenotype for, 171
   physiological, 272
   sexual, hormonal effects, 626
   sexual, secondary, 773
   variation in, 7, 10, 156, 161–162, 164, 180
   X-linked, 144
Transcription
   bacterial, 236
   base pairing, 222
   controls over, *237–239*
   cooperative regulation, 236
   vs. DNA replication, 222
   in lampbrush chromosomes, 240, *240*
   mRNA modifications, 222–223
   negative controls, 236
   nucleosome change during, *240*
   in oocytes, 760
   phytochrome effects, 243, *243*
   in plants, 243, *243*
   positive controls, 236–237
   promoter region, 222

rapid shifts in, 236
rates, 238, *238*
RNA assembly, 222
site, 221
Transcript processing, 226, *226, 239*
Transfer RNA (tRNA)
   anticodon, 225
   formation, 221
   model, computer-generated, *225*
   mRNA interaction, 225
   structure, 225, *225*
Transfusion
   blood, 668
   and blood type, *177*
   HIV transmission during, 691
Translation
   controls over, *238–239*
   elongation, 226
   initiation, 226
   overview, 221
   polysome formation, 227
   stages, 226, *226–227*
   termination, 226
Translocation
   chromosomal, 200
   plant solute transport by, 501, 510, *511*
Transmission electron microscope, 54–55, *55*
Transpiration
   plant, 501, 506
   in a watershed, *853*
Transplantation
   bone marrow, 247
   cornea, 606
   cytotoxic T cell role, 683
   and immune response, 683
   kidney, 742
   organ, 683
   tissue, 683, 691
Transport pigments, 697
Transport protein, *79*, 80, *563, see also* Carrier protein
   in active transport, 87, *87*
   difference in, 86
   functions, 86
   in passive transport, 86, *86*
   selectivity, 86
Transport systems, electron, 326, 330–331
Transport vesicle, *64*
Transposable element, 228, *229*
Trapezius, *643*
Tree
   influence on human evolution, 475
   structure of trunk, *496*
   symbiosis with mycorrhizal fungi, 387
Tree farm, *401*
Tree fern, *399*
Tree frog, 313, *313, 633*
Tree of descent, *see also* Evolutionary tree
   for animals, *412, 423*
   branches, 471
   categories of relationships in, 316
   for chordates, *446*
   for eukaryotes, *333–334*
   and evolutionary systematics, 316
   for giant panda, 309
   modern human populations, *479*
   and phylogenetic hypothesis, 316
   for prokaryotes, *333–334*
   for vertebrates, *446*
Tree rings, 497
Tree shrew, *475*
Trematode, 418
Treponema pallidum, 351, 797, *797*

Triassic, 336–337, *393, 459*
Triassic-Jurassic boundary, 336
Tribolite, 334, *334*, 432
Triceps
   biceps interaction, *642*
   contraction, *646*
   location, *643*
Triceratops, 337
Trichinella spiralis, 420–421, *421*
Trichocyst, 367
Trichomonad, 8, 368, *368*
Trichomonas vaginalis, 368, *368*
Trichomoniasis, 148, 368
Trichophyton, 383
Trichoplax, 376
   evolution, *412*
   T. adhaerens, 408, *410*, 412
Triglycerides
   in adipose tissue, 40
   as alternative energy source, 134
   breakdown, 721
   chylomicron formation, 722
   defined, 40
   emulsification, 721
   energy value, 40
   functions, 40, *41*
Triiodothyronine
   actions, *616, 621*, 624
   secretion, *616*, 624
   sources, *621*
   targets, *621*
Tripe, *719*
Trisomy, 201
Trisomy 21, 202, *202*
Triticum
   T. aestivum, 295
   T. monococcum, 295
   T. tauschii, 295
   T. turgidum, 295
tRNA, *see* Transfer RNA (tRNA)
Trophic levels
   Antarctic food web, *847*
   as community structure, 824, *846*, 846–847
   ecological pyramid, *848, 850*
   savanna, *875*
Trophoblast, *784*
Tropical deciduous forest
   biome, *871, 877, 877*
   global distribution, *871*, 877
Tropical rain forest
   agriculture in, 902
   biome, *871, 876*, 876–877, *892*
   decomposition in, 902
   destruction, 893, *902*, 902–903, *903*
   global distribution, *871*, 876
   resources, 893
   soil, *872*
   species diversity, *876, 892*, 892–893
Tropical reef, 886, *886*
Tropical savanna, *see* Savanna
Tropical scrub forest, global distribution, *871*
Tropical temperature zone, 840, *867*
Tropism, defined, 536
Tropomyosin, 646, *646*, 647
Troponin, 646, *646–647*
Troponin T, *239*
Trout, *316–317*
True-breeding
   lineage, *171*
   plant, 170
Truffle, 382
Truth, in science, 11
Trypanosoma
   T. brucei, 368, *368*
   T. cruzi, 368
Trypanosome, 368, *368*
Trypsin, *720*, 721
Tryptophan
   feedback inhibition, 100–101, *239*

requirements, 725
   sources, *725*
T tubule, *646*
Tuatara, *459*, 461, *461*
Tubal ligation, 794–795
Tubastrea, 416
Tube feet, echinoderm, 438
Tube nucleus, *519*
Tuber, *526*
Tuberculosis
   antibiotic-resistant strains, 355
   patterns, 359
Tube worm, 885, *885*
Tubule, nephron, *739*
Tubulin
   formation, 148
   subunits in microtubules, 69, *69*
Tumor
   benign, 234
   defined, 234
   HeLa cells, 144
   malignant, 234–235
   viruses, *357*
Tundra
   biome, *870–871*, 879, *879*
   global distribution, *871*
Tungara frog, 923
Tungsten particles, 255
Tunicate, 447
Turbellarian, *410*, 418
Turgor pressure, 84
Turner syndrome, *196, 203, 203*
Turtle, 308, 316–317, *459–460*, 710–711, *711*
Twins
   fraternal, 792
   identical, 764, 792
Twitch, muscle, 648, *648*
Tympanic membrane, *600–601*
Typhoid, antibiotic-resistant strains, 355
Typhoid fever, 359
Typhus, 861
Typing, blood, 668–669
Tyrannosaurus rex, 313
Tyrosinase, gene for, 178, *182*
Tyrosine, 204, 725

U

Uintatherium, 340
Ulna, 641
Ulnar nerve, *579*
Ultrasaur, 337
Ultrasound, animal, 592–593, 600
Ultraviolet light
   and pollination, 521
   wavelengths, *110*
Ultraviolet radiation
   effects on DNA, *215*, 244
   effects on skin, 635
   and melanin, 634
   mutagenicity, 228
   ozone depletion-related, 897
Umbilical cord, 792, *792*
   formation, 787
   neurotoxic effects of tetanus, 569
Ungulate, *875*
Uniformitarianism, 266
Unipolar cell, *562*
Uniramian, 432
Unity of life
   at biochemical level, 123, 136–137
   DNA as source, 229
   in ecological interdependency, 136–137
   energetic basis, 136–137
   molecular basis, 136–137, 208, 229
   perspective on, 136–137
   role of death in, 769
Unsaturated fatty acid, 40, *40*

Upper littoral zone, rocky coast, 883, *883*
Upright walking, 473, 476
Upwelling, 888, *888*
Uracil, 222, 225
Uranium fuel, breeding, 906
Uranium-235 isotope, *20*
Uranium-238 isotope, *20*
Urea
    formation, 134, *730*, 736
    reabsorption, *740*
    reabsorption values, *739*
    role in organic metabolism, *730*
    in urine, *739*
Ureter, 738
    functions, 737, *737*
    location, *737*
Urethra, 738
    functions, 737, *737*, 774, *774–775*
    location, *737*, *774–775*
Uric acid
    excretion, 739
    human, 736
    insect, 435
Urinary bladder, 738
    functions, 737, *737*
    location, *737*
Urinary system
    components, *737*, 738, *738–739*, 739–740
    functions, *555*, 738
    human, *555*
    links with other system, *654*
    links with other systems, *654*, *736*
    mammalian, 738
Urination, 737
Urine
    composition, 739
    excretion, 736
    functions, 738
    role in organic metabolism, *730*
    secretion, 740–741, *741*
    waste products in, 736
Urine formation
    filtration processes, 739
    in Kangaroo rats, 743
    reabsorption processes, 739
    secretion processes, 739–740
Urochordate (Urochordata), 446
Ursus americanus, 310
Ursus arctos, 310
Ursus maritimus, 310
Usnea, 386
Ustilago maydis, 383
Uterus
    changes during menstrual cycle, *782*
    at childbirth, *618*
    cyclic changes in, 781
    functions, 778, *778*, 782
    hormonal effects on, *782*
    innervation, *580*
    location, *778*
    structure, *778*
Utricularia, 500

V

Vacancy, in electron orbitals, 22, *23*
Vaccination
    first, 674
    and pregnancy, 790
    smallpox, 674
Vaccine
    defined, 688
    genetically engineered, 688
    malaria, 369
    manufacture, 688
    polio, 688
    schedule for children, 688
    tetanus, 569

Vacuole, contractile, 84, *84*
Vacuum suctioning, 798
Vagina, *784*
    functions, 778, *778*
    pH, 676, 774
Vaginal infection
    Candida albicans-related, 388
    yeast, *388*
Vagus nerve, *580*
Valdez oil spill, 904
Valine, 220, 725
Valium, 568, 589
Valve, venous, 664, *665*
Vanilla orchid, classification, *311*
Van Leeuwenhoek, A., 50
Variable, experimental, 12
Variation
    environmental influences, 273
    in shell patterns of snails, *272*
Variation in populations
    and balanced polymorphism, 282–283
    behavioral traits, 272
    bell-shaped curve for, 180
    and directional selection, 278–279
    and disruptive selection, 280–281
    divergence effects, 289
    and evolution, 266–267, 272
    and gene flow, 277
    and gene mutations, 275
    genetic basis, 272–273
    and genetic drift, 276
    and genetic variation, 272–273
    morphological traits, 272
    and natural selection, 266–268
    phenotypic, 272, 312
    physiological traits, 272
    reproductive isolating mechanisms, 289–291
    of same species, 312
    and sexual reproduction, 754–755
    and stabilizing selection, 280
Varicella-zoster virus, *357*
Vascular bundle
    defined, 488–489
    development, 492
    dicot, *488–489*
    monocot, *488–489*
Vascular cambium, *485*, 489, 496, *496–497*, 507
Vascular cylinder (column), 494–495, *495*, 504, *504*, 507
Vascular plant
    ancestral, *393*
    characteristics, *406*
    common names for, 396
    defined, 485
    defining characteristics, 391
    dermal tissues, 485
    with flowers, *406*
    ground tissues, 485–487
    life cycle, 392, *392*
    with naked seeds, *406*
    phloem, 486
    root system, 484
    seedbearing, 400–405, *400–405*
    seedless, 396–399, *396–399*
    shoot system, 484
    species, *406*, 840
    tissue, *484*, 485–487
    xylem, 486
Vas deferens
    functions, 774–775, *775*
    location, 774, *774–775*
Vasectomy, 795
Vasoconstriction, 663, 748
Vasodilation, 663, 749
Vasopressin, *618*
Vaucheria, 371, *371*
Vector, biological, 359
Vegetal pole, 760, 762

Vegetarian diet, 725, *725*
Vegetative propagation, *526*, 526–527
Vein
    basement membrane, *662*
    in cardiovascular system, 658, *659*
    defined, 658
    endothelium, *662*
    fluid pressure within, 664, *665*
    function, 664
    leaf, *118*, *490*, 491, *491*, *507*
    pulmonary, *660*
    valves in, 664, *665*
Velociraptor, 337
Vena cava, *660*
Venom
    cephalopod, 426
    platypus, 443, 445
    ribbon worm, 422
    snake, *460*, 460–461
    spider, 433
    stingray, 454
Venous pressure (blood), 664
Venter, J., 256
Ventilation, 697, 702, 708
Ventral (anatomy), 410, *411*, *555*
Ventricle, heart, 660, *660*, 661
Ventricular contraction, 661
Ventricular fibrillation, 667
Venturia inaequalis, 383
Venule, 658, 664, *664–665*
Venus flytrap, 219, *500–501*
Vernalization, 540
Vertebral column, 450, 578, *641*
Vertebrate
    analogous structures, 306–307
    circulatory system, 451
    classes, 446
    classification, 314–317
    comparative morphology, 304–305
    defense against pathogens, 676
    developmental stages, 304
    embryo, *582*
    evolution, 445ff, 446, 448–451, 578
    eye, 604, *605*
    fin, 450
    integuments, 620–635
    jaw, 450
    joints, 640
    nervous system, 578–585
    vertebral column, 450
    visual system, 604–609
Verticillium, 383
Very-low-density lipoprotein, 33
Vesicle
    in cytomembrane system, 64
    endocytic, 64, 88, *89*
    exocytic, 64, 88, *89*
    functions, 60
Vespula arenaria, 831
Vessel, lymph, 671
Vessel member, xylem, *487*, 489, 506, *506*
Vibora (Bothrops asper), 310
Vibrio, 348
Vibrio cholerae, 815
Victoria crowned pigeon, 822–823, *823*
Villus, intestinal, *721*, 722
Vine, plant, 537
Viral infection
    defined, 259
    DNA viruses, 357, *357*
    eye diseases caused by, 606–607
    during pregnancy, *790*, 791
    RNA viruses, 357, *357*
Viral replication
    HIV, 690
    lysogenic pathway, 358, *358*
    lytic pathway, 358, *358*

multiplication cycles, 358–359, *359–359*
    stages, 356, 358
Virchow, R., 50
Virgin T lymphocyte, *681*
Viroid
    defined, 357
    effects in plants, 357
Virus, *see also specific virus*
    animal, classification, *357*
    animal, multiplication, 258
    bacteriophage, 211
    body plan, 356
    characteristics, 211, 347
    classification, *357*
    complex, *356*
    defined, 356
    DNA, 247, 356, *356*
    envelope, 347, 356, *356*
    enzymes, 358
    gene transfer, 247
    helical, *356*
    latency, 358, *359*
    lytic pathway, *211*
    plant, 357
    polyhedral, *356*
    replication, 211, 356
    RNA, 356, *356–357*
    size, 346
    skin, UV radiation effects, *635*
    structure, *211*
    transmission modes, 359
    viral coat, 356
Visceral mass, mollusk, 424, *424*
Vision
    invertebrate, 602–603
    neural basis, 586
    and photosensitivity, *110*
    primate, 471, 473–474
    requirements, 602
    vertebrate, 604, *605*
Visual cortex, 584, *585–586*
Visual perception, *609*
Visual signal, *922*, 923
Visual system
    accommodations, 604, *605*
    eye, *see* Eye
Vital capacity, lung, 702, *703*
Vitamin
    in blood, *656*
    defined, 726
    deleterious, 727
    excess, 727
    fat-soluble, 725, *726*
    intake during pregnancy, 791
    supplements, 727, 790
    water-soluble, *726*
Vitamin A, *726*, 727
Vitamin B1, *726*
Vitamin B2, *725–726*
Vitamin B6, *726*
Vitamin B12, *725–726*
Vitamin C, 37, *726*, 727
Vitamin D, *726*, 727
    deficiency, 624, *624*
    formation, 41
    functions, 41
    production, 632
Vitamin D3
    actions, *621*
    sources, *621*
    targets, *621*
Vitamin E, *726*
Vitamin K, *726*, 727
Vitreous body (humor), *604*
Viviparity, *755*
Vocal cords, 701, 703, *703*
Volatile organic compound air pollutants, *895*
Volcanoes, *328*, 336, 338, 341, *482*
Voltage difference, across neuronal membranes, *475–566*, 562–563
Volvox, 95, 374, *374*, 376

Vomiting, bulimic, 729
Vulva, 778

W

Waggle dance (honeybee), 923, *923, 928–929*
Walking worm, *431*
Wallabie, 466
Wallace, A., 261, 268, *268*, 870
Waller, A., *652*
Walrus, *467*
Warts
    genital, 796
    skin, 234
    viruses causing, *357*
Wasp, *831, 833, 833*
Waste
    metabolic, 808
    radioactive, 905
    solid, 900
Wastewater treatment, 899, *899*
Water
    in abiotic synthesis, 324–325
    absorption by root hairs, 494
    absorption by roots, 495
    aquifers, *898*
    atmospheric, 856
    bulk flow, 83
    cohesive properties, 27
    concentration gradient, 83
    conservation by human body, 740–741
    conservation by plants, 508–509
    cushioning effects, 30, *30*
    density, 880
    desalination, 898
    diffusion, 82–83, *83*
    evaporation, 26
    extracellular, regulation, 736
    formation in aerobic respiration, *128–129*, 130, 136–137
    fresh water, 893, 898, *898*
    and greenhouse effect, 856–857
    hydrologic cycle, 852
    intercellular movements, 83
    ionization, 28
    loss by urinary excretion, 734, 736, 743
    metabolic, 734
    molecular polarity, 25–26, *26*
    molecule formation, 36
    and origin of life, 324
    polarity, 136
    properties, 136
    rainwater, 896, *896*
    solvent properties, 27
    spheres of hydration, 27, *27*
    temperature-stabilizing effects, 26
    transport, cohesion theory, 506, *507*
    transport by plants, 501, 506–507
    uptake by plants, 502–507, *504–505*
Water gain, processes affecting, 736
Water hyacinth, *838, 838–839*
Water lily, *403*
Water loss
    insensible, 736
    processes affecting, 736
    by urinary excretion, 736
Water mold, 364, *383*
Water pollution
    effects on ocean communities, 370
    in estuarine ecosystems, 882
    lake eutrophication effects, 881
    large-scale irrigation effects, 898

and water quality maintenance, 899
    and water scarcity, 898–899
Water potential, defined, 84
Water provinces, 865, 880–889
Water purification, 899, *899*
Water quality, 893, 899
Watershed
    and acid deposition, 896
    Hubbard Brook Valley, 852, *853*
    water movement through, 852, *853*
Watersilk, 375, *375*
Water-soluble vitamin, *726*
Water–solute balance
    in amphibians, 743, *743*
    in fish, 743, *743*
    in flatworms, 418
    gains and losses, 734, *734*, 736
    hormonal adjustments, *740*, 741
    and protonephridium, 418
    reabsorption mechanisms, 740–741
Water strider, *27*
Water table, 898
Water treatment, 899, *899*
Water-vascular system, of echinoderms, 438, *439*
Wavelength, light
    absorption by plant pigments, *110–111*
    absorption by red algae, *372*
    measure of, 54
Wax
    functional groups, 35
    functions, *48*, 72
    secretion, 41
    structure, 41, *48*
Web, spider, 433
Weddell seal, *710, 847*
Weevil, *843, 843*
Weismann, A., 187
Welwitschia, 402
W. mirabilis, 402
Went, F., 536
Westerlies, *867*
Wet acid deposition, 896
Whale, 600, *709*, 844, *847, 885*
Wheat, *383*, 838
Whisk fern, 396
White, T., 477
White blood cell (leukocyte)
    appearance in micrographs, *234*
    B cells, *see* B lymphocyte
    classes, *657*
    classification, 680
    formation, 246
    functions, 246–247, *656*, 657
    in lymph nodes, 671
    origins, 657
    phagocytic, *66*
    recognition of nonself proteins, 80, 676
    types, 657, *657*, 680
White-crowned sparrow, singing by, 914
White matter
    cerebral cortex, *582*
    spinal cord, 581
White oak, 8
White-throated sparrow, singing by, 912–913, *913*, 923
Whittaker, R., 318
    five-kingdom system of classification, *318*
Whooping cough, 359, 688
Wiesel, T., *609*
Wild-type allele, defined, *192*
Wilkins, M., 212
Willow tree, 255, *255*
Wilting, of plants, 84
Wind
    coastal breezes, *869*
    tradewinds, *867, 888–889*

Windward direction, 869
Wing
    bird, 463, *463*
    butterfly, 312
    evolution, 305
    insect, *637*
    morphological divergence, 306
Wing, bird, 463, *463*
Withdrawal, as birth control method, 794, *795*
Withdrawal reflex, 570
Wolcott, C. D., 408
Wolf, 926, *927*
Wolf spider, *433*
Women, bone tissue turnover, 639
Wood, 496–497
Wood duck, *288*
Woodland, dry
    biome, 874
    global distribution, *871*, 874
Woody plant, 496–497
Worker
    honeybee, 923, *923, 928–929, 928–929*
    termite, *920*, 920–921
Worm resistance, 255
Wrasse, species, *293*
Wuchereria bancrofti, 421, 421
Wyandotte chicken, 179, *179*

X

Xanthophyll, *110*
X chromosome, *see also* Sex chromosome
    genes on, 190
    as homologue of Y chromosome, 188
    inactivation, *238*, 240–241
    lyonization effects, 241, *241*
    during prophase I, *160*
    and sex-linked traits, 190
Xenopus laevis, 457, 764, see also Frog
Xeroderma pigmentosum, 215
X-linked gene, *192*, 193
X-linked recessive inheritance, *196*
X-ray diffraction, DNA, 212–213
X rays
    DNA damage caused by, 244
    in electromagnetic spectrum, *110*
XXY condition, 203, *203*
Xylem
    defined, 486
    example, *487, 494, 506–507*
    functions, 392, 486, 506
    structure, 486
    water tensions in, 492–493
    woody plants, *496*
XYY condition, *196*

Y

Yam, wild, 535
Yawn, as visual signal, 923
Y chromosome, *see also* Sex chromosome
    genes on, 190
    as homologue of X chromosome, 188
    male-determining gene, 190
    during prophase I, 160
    and sex determination, 190, *190*
    sex-determining region (SRY), 191
    and sex-linked traits, 190–191
    SRY gene, 191
Yeast, *see also* Sac fungus
    alcoholic fermentation, 133, *133*
    manufacturing use, 133
    trophic level, 847
Yeast infection, *388*
    oral, *388*

in PWAs, 690
    vaginal, *388*
Yellow crab spider, *832*
Yellow crescent, tunicate, 760, *761*
Yellow fever, *357*
Yellow-green algae, 370–371
Yellowjacket, *831*
Yellow marrow, 638
Yersinia pestis, 809
Yield symbol, in chemical equations, *22*
Y-linked gene, *192*, 193
Yolk, 762
    amount, 755, 760
    chick, *757, 759*
    defined, 755
    distribution, 755, 760, 762
    functions, 755
Yolk sac, *784*, 785
Yucca moth, 825, *825*
Yucca plant, 825, *825*

Z

Zea mays, see also Corn
    classification, *311*
    C4 pathway, *118*
    developmental stages, *533*
    linkage groups, 193
    muations, *229*
    root structure, *495*
    stem structure, *489*
Zebra, *289*
Zebra finch, 913, *913*
Zebra mussel, 839
Zebroid, 289, *289*
Zero evolution, 275
Zero population growth, 806, 816, 820
Zinc
    deficiency, *727*
    functions, *727*
    in human nutrition, 727
    in plant function, *502*
    sources, *727*
Z line, *646*
Zona pellucida, 780, *784*
Zonation
    intertidal zone, 883, *883*
    lake, 880, *881*
Zooplankton, 370, 848, 880, 884
Zygomycete
    food spoilage caused by, 381, *383*
    life cycle, *381*
    parasitic, 381
    reproduction, 380–381, *381*
    saprobic, 381
    species diversity, 380
Zygops rufitorquis, 843, 843
Zygosporangium, 380, *381*
Zygospore, *381*
Zygote
    defined, *793*
    formation, 162
Zygotic mortality, *289*, 291

APPLICATIONS INDEX

ABO blood typing, *177, 184*, 668
Abortion, 205, 794, 795, 798, 817, *798*
Achondroplasia, *196, 197*
Acid deposition, 896, *896*, 904
Acid rain, 28, *29*, 337, 893, 896, *896*
Acid stomach, 29
Acidosis, 29
Acne, 635
Acromegaly, 620, *620*
Active immunization, 688
Acute inflammation, 678
ADA deficiency, 246-247
Addiction, drug, 589
Addictive behavior, 575-576
Adoptive behavior, *931, 931*
Aerosols, and ozone layer, 897
Africanized bees, *122, 122*-123, 839
African sleeping sickness, 368
Afterbirth, 792, *792*
Age spots, *92, 93*
Aggressive behavior
    anabolic steroid-related, 649
    testosterone effects, 776
Aging
    and bone turnover, 639
    and cancer resistance, 683
    and cardiovascular disorders, 666
    effects of exercise on, 793
    effects of free radicals on, 92-93
    and glaucoma, 607
    and posture, 640
    skin changes in, 635
Agriculture
    and deforestation, 902, *902*
    and desertification, 901, *901*
    and genetic engineering, 247,
        254-255
    green revolution, 900
    irrigation effects, 898, *898*
    overgrazing, 901, *901*
    and population growth, 814, 898
    shifting cultivation (slash-and-
        burn), 902, *903*
    and strip mining, 904
AIDS (acquired immunodeficiency
    syndrome)
    deficient immune response in, 689
    economic cost of, 796
    infection statistics, 796
    modes of transmission, 691, 796
    pandemic nature, 359
    testing for, 796
    treatment for, 691, 796
    viral agent of, 674, *675*, 690-691, *691*
Air pollutants
    and acid deposition, 896
    classes of, *895*
    and ozone layer, 897
    from paper manufacturing, 900
    and smog, 895
Albinism, 178, 204
Alcohol (ethyl alcohol)
    addiction, 588
    effects on central nervous system,
        588
    effects on digestive system, 718
    fetal alcohol syndrome, 791
    and pregnancy, *790, 791*
Allergy, 689
Aloe vera, 493
Altitude sickness (hypoxia), 695, 708
Alzheimer's disease, 202, 587
Amniocentesis, 204, *205*
Amoebic dysentery, 359
Amphetamine, 588
Anaphylactic shock, 689
Anemia, sickle-cell, 220-221
Angina pectoris, 667
Angiography, *667, 667*

Angioplasty, 667
Anhidrotic ectodermal dysplasia, 241
Anorexia nervosa, 729
Antarctica
    disruption of, *844, 844*-845, *845*
    ozone hole over, *897, 897*
Anthrax, 674
Anti-acne drugs, and pregnancy, 791
Antianxiety drug, 568
Antibiotic
    effectiveness, 354, 355
    sources, 353, *354*, 378, 893
    overprescription of, 355
Antibody
    agglutination response, 668
    monoclonal, 683
    and pregnancy, 668-669, 790
Antibody-mediated immune
    response, *681*, 684-685, *685*
Appendicitis, 723
Apple scab, 383
Arrhythmia, 667
Arteriosclerosis, 666
Arthritis, 641, 689
Asbestos
    and air pollution, 895, *895*
    carcinogenicity, 244
Aspartame, 203
Aspirin, 719, 893
Asthma, 689
Atherosclerosis, 33, 666-667
Atherosclerotic plaque, 33, *33*, 667
Athlete's foot, *383*, 676
Athletes, steroid use by, 41
Atmosphere
    greenhouse gases, *856*, 856-857,
        857
    pollution, 893, 895-897, *895, 897, 903*
    and thermal inversion, *895, 895*
Autoimmune disorder, 623, 689
Autoimmune response, 623, 768
Azidothymidine (AZT), 691

Bacteria
    and antibiotics, 354-355
    disease-causing, 354, *354*, 606
    immune responses to, 674, 682-685
    in mouth, *349*
    natural defenses against, 676, 700
    and STDs, 796-797
Balance, sense of, *599, 599*
Balloon angioplasty, 667
Barbiturate, 790
Behavior, human, 588, 931
Belladonna, 493
Bends (decompression sickness), 709
Beta-carotene, 727
Biological clock, 6, 538, 625
Biological controls, 279
Biological magnification, 861
Biological perspective
    on fertility control, 794
    on gene therapy, 246-247, 257
    in vitro fertilization, 798
    on life in the biosphere, 123
    pregnancy termination, 798
    recombinant DNA technology, 257
Bioluminescence, as research tool,
    103, 103
Birth control methods, 794-795
Birth control pill, 535
Birth rate, human
    and economic development,
        818-819
    and family planning, 816-817
    and population growth, 806-807
Birth weight, human, 280, *280*
Blindness, *196*, 606-607
Blister, 635
Blood clotting, 98, 656, 657, 668, *668*

Blood fluke infection, 420, *421*
Body-builder's psychosis, 649
Body weight
    and energy needs, 728-729
    ideal, 728, *728*, 728-729
Bone marrow transplants, 247
Botulism, 354, 569
Bradycardia, 667
Breast, and anabolic steroids, 649
Breast cancer, 793
Breeder reactors, 906
Brown rot, *383*
Bubonic plague, 809, *815*
Bulimia, 729
Bulk, in diet, 723-724

Caffeine, and nervous system, 588
Caloric intake, and diet, 728, 729
Camptodactyly, 180
Cancer
    breast, 793
    colon, 244, 723
    deaths due to, 235
    DNA transformations, 244
    and genetic engineering, 256
    laryngeal, 707
    lung, 707
    metastasis, 234
    oral, 707
    pancreatic, 707
    and proto-oncogene expression, 244
    skin, *215*, 635, *635*, 897
    testicular, 774-775
    prospects for treatment, 683
Candida infection, *383*, 388
Carbon monoxide, *895*
    in cigarette smoke, 666
    poisoning, 708
Cardiovascular disorders
    anabolic steroid-related, 649
    arrhythmias, 667
    atherosclerosis, 33, 666-667,
        666-667
    hypertension, 666-667, 742
    risk factors, 666
Cervix
    during birth process, 792
    and sexually transmitted diseases,
        797
Chancre, of syphilis, 797
Chemotherapy, 682
Chernobyl, 905, *905*
Chestnut blight, 383, *838*
Childbirth
    process, 557, 616, 640, 792, *792*
    protein intake and, 725
Chin
    dimpling, *169*
    fissure, *156*
Chlamydial infection, 797
Chlorofluorocarbons (CFCs), *895*
    and greenhouse effect, 856-857
    and ozone layer, 893, 897
Cholesterol
    and cardiovascular disorders,
        666-667
    deposits, arterial, 666-667
    excess, 725
    familial hypercholesterolemia, 197
Cholera, 815
Chorionic villi sampling (CVS), 204
Chromosome abnormalities
    and genetic counseling, 204-205
    and genetic screening, 204
    prenatal diagnosis, 205
Cigarette smoke
    effects on lungs, 706
    and heart damage, 666
    risks related to, 706-707

Cigarette smoking
    and bone healing, 707
    and heart damage, 666, 706-707
    and lungs, 706-707
    and pregnancy, 790
    risks associated with, 706-707
Cirrhosis, and alcohol addiction, 589
Clear-cutting, 902
Cleft lip, 204
Cloning, DNA, 250
Clotting, blood, 668
Clotting factor VIII, 198
Coal (fossil fuel), 396, *397*, 904
Cocaine, 493
    effects of addiction, 575, 588-589
    effects during pregnancy, 791
    granular, 588
    mode of action, 569
Codeine, 589
Coffee, effect on nervous system, 588
Coitus, 783
Cold sore, *357, 358, 635*
Cold stress, 748
Cold sweats, 634
Collagen
    aging effects, 768
    and genetic engineering, 256
Colon cancer, 244, 723
Color blindness
    red-green, *196*, 606
    total, 606
    X-linked, 144
Color perception, *196*, 606, 608-609
Common cold, *357*
Condom, 795, *795*
Conjunctivitis, 607
Constipation, 723
Continuous variation
    in eye color, 180, *180*
    in height, 180, *181*
Contraceptive methods, 794-795
Cornea
    artificial, 606
    and astigmatism, 607
    and farsightedness, 606
    and nearsightedness, 606
Corneal transplant surgery, 607
Coronary artery disease, 707
Coronary bypass surgery, 667
Counseling, diet, 729
Crack cocaine, 575, 588-589, *589*
Cri-du-chat syndrome, *196*, 200
Crop, food
    and acid deposition, 896
    and genetic engineering, 247,
        254-255
    and global warming, 857
    improving yields, 255, 858
    and insecticide resistance, 279
    irrigation-dependent, *898*
    viruses and viroids affecting, 357
Cropland, *900*
    converting marginal lands to, 900
    desertification, 901, *901*
CVS (chorionic villi sampling), 205
Cystic fibrosis
    CFTR protein, 256
    RFLP applications, 252

2,4-D (2,4-dichlorophenoxyacetic
    acid), 365
ddI (didanosine, dideoxyinosine), 691
Deforestation
    and air pollution, 857, 893, *903*
    in Amazon basin, *903*
    global, 401, *401, 902*
    of temperate forests, *903*
    of tropical forests, 893, *902*,
        902-903, *903*, also see *Frontispiece*
Depressant, drug, 589

APPLICATIONS INDEX
(Continued)

Depression
  and daylength, 625
  drug-related, 588–589
Desertification, 901, *901*
Designer organs, 553
Diabetes
  juvenile-onset, 623
  mellitus, 623, 666
Dialysis, kidney
  hemodialysis, 742
  peritoneal, 742
Diaphragm (contraceptive), 794, *795*
Diarrhea
  as body's defense response, 676
  from *Escherichia coli* infection, 354
  from protozoan infection, 368
Diet
  balanced, 724, 727
  and cholesterol blood level, 33
  deficiencies, 723, 725
  and dieting, 729
  and emphysema, 706
  and exercise, 729
  and genetic disorders, 204
  and pregnancy, 790
  and salt intake, 666
  vegetarian, 725, *725*
Dieting, 729
Digestive disorders, 724
Digitalis, 493
Diphtheria, 688
Disease resistance
  and genetic engineering, 254
  of crop plants, 254, 273
DNA fingerprinting, 252, 930
DNA virus diseases, 357, *357*
Doldrums, *867*
Dopamine, 568–569, *618*
Douching, 794, *795*
Down syndrome, 202, *202*
  consequences, *196*
  prenatal diagnosis, 204
  risk factors, 781, 794, *795*
Drugs
  addiction to, 575, 586
  analgesic, 589
  antimalarial, 369
  cortisol-like, 623
  depressant, 589
  hypnotic, 589
  intravenous, and STDs, 796
  and pregnancy, 790
  psychoactive, 589
  psychedelic, 589
Drug therapy, targeted, *683*
Dry acid deposition, 896
Dust Bowl, *901*
Dutch elm disease, *383*, *838*
Dwarfism
  achondroplasia, *196*, 197
  pituitary, 620, *620*

Earlobe attachment trait, 168, *168*
Eating disorders, 729
ECG (electrocardiogram), 587, 667
Edema, *664–665*, 695
Electroencephalogram, 587, *587*, 667
Elephantiasis, 421, *421*
El Niño Southern Oscillation (ENSO), 888–889, *889*
Embolus, 667
Emphysema, 256, 706
Encephalitis, *357*
Endometriosis, 779
Endorphins, 568, 589
Entamoeba histolytica infection, 366
Epidermal skin cancer, 635
Epilepsy, 586

Epstein-Barr virus, *357*
Ergotism, 383
Erythroblastosis fetalis, 669
Escherichia coli
  diseases caused by, 354
  food contamination by, *56*
  and synthetic insulin gene, 254
Estrogen, and the Pill, 795
Estuary pollution, 882
Eugenic engineering, 257
Eutrophication, 881
Exercise
  and aging, 793
  benefits to muscular system, 648
  and cardiovascular disorders, 666
  and dieting, 729
  evaporative water loss in, 745, *745*
  metabolic effects of, 729
  and muscle fatigue, 648
Eye
  color, 180, *180*
  disorders, 606–607
  shape, 168, 604

Familial hypercholesterolemia, 197
Family planning, 816–817
Fat
  dietary, 725
  and dieting, 729
Faulty enamel trait, *196*, 198
Feedforward control, 906–907
Fertility, human
  and abortion, 798
  and contraception, 794–795
Fertilization, human, 783, 798
Fertilizer, kelp-derived, 373
Fetal alcohol syndrome (FAS), 791, *791*
Fever, *357*, 678, 749
Fiber, and intestinal health, 723
Folic acid, in diet, *726*
Food web, and DDT, 861
Fossil fuels
  burning, 854, *855*, 857, 895
  depletion, 396
  formation, 396-397, 854, *855*, 904
Free radicals, and aging, 92–93
Fungus
  antibiotics from, 378
  diseases caused by, 381
Fusion power, 906

Galactosemia, *196*, 196–197, 204
Gene amplification, *249*, 250, *250*
Gene mutation
  and aging, 768
  beneficial, 228–229, 275
  in embryo, 304
  harmful, 228–229, 275
  lethal, 275
  mutagens causing, 228
  and proto-oncogenes, 244
Gene therapy, 247, 257
Genetic counseling, 204–205
Genetic disorder
  autosomal dominant, *196*
  autosomal recessive, *196*
  designer organs for, 553
  gene therapy for, 247, 257
  and genetic counseling, 204
  and genetic disease, 196
  phenotypic treatment of, 204
  and prenatal diagnosis, 204–205
  and RFLP applications, 252
  screening for, 204
  X-linked recessive, *196*
Genetic engineering
  of animals, 256, *256*, 256–257
  of bacteria, 254
  benefits and risks, 254, 256
  ethical and social issues, 257
  and pest control, 279
  and pesticide reduction, 279

Genetic fingerprinting, 930
Genital warts, 796
German measles, *357*, 688, 791
Giardia lamblia infection, 368, *368*
Gigantism, 620, *620*
Gilford progeria syndrome, *196*
Global warming, 117, 854, *856*, 856–857, 893
Goiter, 624
Gonorrhea, 355, 359, 797
Grave's disorder, 689
Greenhouse effect, 854, 856–857, 866, 894, 903–904
Green revolution, 900
Groundwater contamination, 898, *898*

Hair
  aging effects, 768
  protein deficiency effects, 635
  split ends, 635
  thinning, 635
Hair color, 178, *178*
Hallucinogen, *384*, 589
Hay fever, 689
HCG (human chorionic gonadotropin), 785
HDLs, 33, 667
Hearing
  loss, *601*
  sense of, 600, *600–601*
Heimlich maneuver, 701
HeLa cells, 145
Hemodialysis, 742
Hemorrhagic fever, 357
Herniated (slipped) disk, 640
Heroin, 589
Herpesviruses
  Epstein-Barr, *357*
  Herpes simplex, 357, 606, *635*, 797
  Varicella-zoster, *357*
High-density lipoprotein, 33, 667
Hirsutism (excessive hairiness), 635
Histoplasmosis, 383, *383*, 606
Hookworm infection, *410*, 420
Hormone
  and behavior, *618*, 624–625, 912–913
  and female reproduction, 779–780
  and male reproduction, 776–777
  effects on menstrual cycle, 778–779
  nonsteroid, *616*, 616–617
  steroid, 41, 616–617, 725
HTLV I and II (human T-cell leukemia I and II), *357*
Human behavior
  and brain, 588
  evolutionary theory applied to, 931
  sexual, 783, 794, 796
Human chorionic gonadotropin (HCG), 785
Human embryonic development
  events during, 772-773, 784-789, *793*
  risks during pregnancy, 791
Human genome project, 252, 256
Human immunodeficiency virus (HIV)
  carriers of, 796
  mode of action, 357, 690
  modes of transmission, 691, 796
Human reproduction, 772–783
Huntington's disorder, *196*, 197
Hypertension, 666, 742
Hyperthermia, 749
Hyperthyroidism, 624
Hyperventilation, 708
Hypnotic drug, 589
Hypothermia, 748, *749*
Hypothyroidism, 624
Hypoxia, 708, 806, 816, 820

Ice-minus bacteria, 254, *254*
Immune response
  deficient, 689
  inflammation during, *681*

Immunization
  active, 688
  passive, 688
  schedule for children, 688
Immunotherapy, 683
Impetigo, 359
Industrial pollution, 28, 279, 895, 899
Infection
  fever effects, 749
  lymph node role in, 671
  mucous membrane, *383*
  opportunistic, 796
Infectious mononucleosis, *357*
Influenza, 246, *357*, 359
Insect control, biological, 279
Insecticide resistance, and directional selection, 279
Interferon, 682, 749
Interleukin, 678, *681*, 683
Inversion layer, 895, *895*
In vitro fertilization, 205
Irrigation, large-scale, 898, *898*
IUD, *795*, 797

Jet lag, 625
Juvenile-onset diabetes, 623
Kaposi's sarcoma, 690
Kidney
  and aging, 768
  and anabolic steroids, 649
  dialysis, 742
  disorders, 649, 742
  transplantation, 742
Knee cap (patella), 640
Lactation, 792, *792*, 793
Lake
  eutrophic, 881, *881*, *891*
  pollution, 881
Language
  brain areas affecting, 582, 584, *585*
  of epileptics, 586
  hemispheric dominance in, 586
  structural basis of, *474*, 478
Laryngitis, 703
Laser angioplasty, 667
Late blight, potato, 364
Laxative, 729
Leaching, of soil nutrients, 852, *853*, 858
LDLs, 33, 667
Life history patterns, human, 810–813
Liver
  and cirrhosis, 589
  and dioxin-related disorders, 365
Lockjaw, 569
Low-density lipoproteins, 33, 667
Lung cancer, 706–707
Lyme disease, 354, *354*, 433
LSD, 589

Malaria, 282, 359, 368, *369*, 861
Marijuana, 589
Measles, *357*, 688
Meltdown, nuclear, 904–905
Meningitis, *357*, 581
Menopause, 779
Mental retardation
  and chromosome abnormalities, *196*, 200, *200*
  and PKU, 204
  and protein deficiency, 725
  thyroid-related, 624
Metastasis, 234
Mineral
  deficiencies, in plants, 502
  excess, in human diet, 727
  intake during pregnancy, 791
  loss through deforestation, *876*
Miscarriage, 535
Monoclonal antibody, 683
Mononucleosis, infectious, *357*
Morning-after pill, 795
Mount St. Helens, 482, *482–483*